Viral Infections of Humans

Epidemiology and Control

FOURTH EDITION

Viral Infections of Humans

Epidemiology and Control

FOURTH EDITION
completely revised and expanded

Edited by

Alfred S. Evans

Late of Yale University
School of Medicine
New Haven, Connecticut

and

Richard A. Kaslow

University of Alabama at Birmingham
School of Public Health and School of Medicine
Birmingham, Alabama

PLENUM MEDICAL BOOK COMPANY

New York and London

Library of Congress Cataloging-in-Publication Data

Viral infections of humans : epidemiology and control / edited by
   Alfred S. Evans and Richard A. Kaslow. -- 4th ed., completely rev.
 and expanded.
       p. cm.
   Includes bibliographical references and index.
   ISBN 0-306-44855-6 (hardbound). -- ISBN 0-306-44856-4 (pbk.)
   1. Virus diseases--Epidemiology. 2. Virus diseases--Prevention.
 I. Evans, Alfred S., 1917-   . II. Kaslow, Richard A.
   [DNLM: 1. Virus Diseases--epidemiology. 2. Virus Diseases-
 -prevention & control.   WC 500 V8155 1997]
   RA644.V55V57 1997
   614.5'75--dc21
 DNLM/DLC
 for Library of Congress   97-17146
                                                              CIP

ISBN 0-306-44855-6 (Hardbound)
ISBN 0-306-44856-4 (Paperback)

© 1997 Plenum Publishing Corporation
233 Spring Street, New York, N. Y. 10013

Plenum Medical Book Company is an imprint of Plenum Publishing Corporation

http://www.plenum.com

10 9 8 7 6 5 4 3 2 1

Printed in the United States of America

Alfred S. Evans
1917–1996

Contributors

Miriam J. Alter, Hepatitis Branch, Division of Viral and Rickettsial Diseases, National Center for Infectious Diseases, Centers for Disease Control and Prevention, Atlanta, Georgia 30333

David M. Asher, Laboratory of Method Development, Center for Biologics Evaluation and Research, Food and Drug Administration, Rockville, Maryland 20852-1148; and Laboratory of Central Nervous System Studies, National Institute of Neurological Disorders and Stroke, National Institutes of Health, Bethesda, Maryland 20892

Robert L. Atmar, Acute Viral Respiratory Disease Unit, Departments of Medicine and Microbiology and Immunology, Baylor College of Medicine, Houston, Texas 77030

Abram S. Benenson, Graduate School of Public Health, San Diego State University, San Diego, California 92182

Francis L. Black, Department of Epidemiology and Public Health, Yale University School of Medicine, New Haven, Connecticut 06519

William A. Blattner, Institute of Human Virology, University of Maryland, Medical Biotechnology Center, Baltimore, Maryland 21201-1192

Kevin E. Brown, Hematology Branch, National Heart, Lung, and Blood Institute, Bethesda, Maryland 20892-1652

Robert D. Burk, Departments of Pediatrics, Microbiology, and Immunology and Obstetrics and Gynecology, Albert Einstein College of Medicine of Yeshiva University, Bronx, New York 10461

Robert B. Couch, Influenza Research Center, Department of Microbiology and Immunology, Baylor College of Medicine, Houston, Texas 77030

Floyd W. Denny, Department of Pediatrics, School of Medicine, University of North Carolina, Chapel Hill, North Carolina 27514

Janet A. Englund, Acute Viral Respiratory Disease Unit, Departments of Microbiology and Immunology and Pediatrics, Baylor College of Medicine, Houston, Texas 77030

Alfred S. Evans,[†] Department of Epidemiology and Public Health, Yale University School of Medicine, New Haven, Connecticut 06510

Hjordis M. Foy, Department of Epidemiology, School of Public Health and Community Medicine, University of Washington, Seattle, Washington 98195

Anne A. Gershon, Department of Pediatrics, Columbia University College of Physicians and Surgeons, New York, New York 10032

Clarence J. Gibbs, Jr., Laboratory of Method Development, Center for Biologics Evaluation and Research, Food and Drug Administration, Rockville, Maryland 20852-1148; and Laboratory of Central Nervous System Studies, National Institute of Neurological Disorders and Stroke, National Institutes of Health, Bethesda, Maryland 20892

[†]*Deceased.*

W. Paul Glezen, Influenza Research Center, Department of Microbiology and Immunology, Baylor College of Medicine, Houston, Texas 77030

Eli Gold, Department of Pediatrics, Harborview Medical Center, University of Washington, Seattle, Washington 98104

Jack M. Gwaltney, Jr., Department of Internal Medicine, University of Virginia School of Medicine, Charlottesville, Virginia 22908

Stephen C. Hadler, Epidemiology and Surveillance Division, National Immunization Program, Centers for Disease Control and Prevention, Atlanta, Georgia 30333

Cathleen A. Hanlon, New York State Department of Health, Zoonoses Program, Albany, New York 12237

Sandra J. Holmes, Center for Pediatric Research, Eastern Virginia Medical School, Children's Hospital of The King's Daughters, Norfolk, Virginia 23510-1001

Peter B. Jahrling, USAMRID, Fort Detrick, Frederick, Maryland 21702

Daniel M. Jorgensen, Division of Infectious Diseases, Epidemiology and Immunology, Department of Pediatrics, Emory University School of Medicine, Atlanta, Georgia 30303

Albert Z. Kapikian, Epidemiology Section, Laboratory of Infectious Diseases, National Institute of Allergy and Infectious Diseases, National Institutes of Health, Bethesda, Maryland 20892

Richard A. Kaslow, Department of Epidemiology, School of Public Health, University of Alabama at Birmingham, Birmingham, Alabama 35294-0008

James W. LeDuc, National Centers for Infectious Diseases, Centers for Disease Control and Prevention, Atlanta, Georgia 30333

Paul H. Levine, Viral Epidemiology Branch, National Cancer Institute, Bethesda, Maryland 20892

Harold S. Margolis, Hepatitis Branch, Division of Viral and Rickettsial Diseases, National Center for Infectious Diseases, Centers for Disease Control and Prevention, Atlanta, Georgia 30333

Kenneth McIntosh, Division of Infectious Diseases, The Children's Hospital, and Department of Pediatrics, Harvard Medical School, Boston, Massachusetts 02115

James M. Meegan, Division of Microbiology and Infectious Diseases, National Institute of Allergy and Infectious Diseases, National Institutes of Health, Bethesda, Maryland 20892

Joseph L. Melnick, Division of Molecular Virology, Baylor College of Medicine, Houston, Texas 77030

Arnold S. Monto, Departments of Epidemiology and International Health, School of Public Health, University of Michigan, Ann Arbor, Michigan 48109

Nancy E. Mueller, Department of Epidemiology, Harvard School of Public Health, Boston, Massachusetts 02115

André J. Nahmias, Division of Infectious Diseases, Epidemiology and Immunology, Department of Pediatrics, Emory University School of Medicine, Atlanta, Georgia 30303

George A. Nankervis, Department of Pediatrics, Northeastern Ohio University College of Medicine at Children's Hospital Medical Center, Akron, Ohio 44308

James C. Niederman, Department of Epidemiology and Public Health, Yale University School of Medicine, New Haven, Connecticut 06510

Thomas R. O'Brien, Viral Epidemiology Branch, Epidemiology and Biostatistics Program, National Cancer Institute, Bethesda, Maryland 20892

Walter A. Orenstein, Centers for Disease Control and Prevention, The National Immunization Program, Atlanta, Georgia 30340

Charles E. Rupprecht, Centers for Disease Control and Prevention, Viral and Rickettsial Zoonoses Branch, Atlanta, Georgia 30333

Mark H. Schiffman, Epidemiology and Biostatistics Program, National Cancer Institute, Bethesda, Maryland 20892

Robert E. Shope, Department of Pathology, The University of Texas Medical Branch at Galveston, Galveston, Texas 77555

Lawrence R. Stanberry, Division of Infectious Diseases, Department of Pediatrics, University of Cincinnati College of Medicine, and Children's Hospital Medical Center, Cincinnati, Ohio 45229

G. de-Thé, The Pasteur Institute, 75015 Paris, France

Thomas H. Weller, Department of Tropical Public Health, Harvard School of Public Health, Boston, Massachusetts 02115; *present address*: 56 Winding River Road, Needham, Massachusetts 02192

Neal S. Young, Hematology Branch, National Heart, Lung, and Blood Institute, Bethesda, Maryland 20892-1652

Preface to the Fourth Edition

I prepared this preface to the fourth edition of *Viral Infections of Humans* alone and with mixed emotions. It was immensely gratifying when Al Evans invited me to join him in producing the new edition. After following at a distance his exemplary career as a scholar and gentleman in every sense of those words, I was filled with anticipation at the privilege of observing close at hand the thoughtful dedication he had previously brought to the editorial process. However, our work together was soon overshadowed by the condition that would slowly take him from our task and our midst. As we both increasingly realized that his declining health might prevent him from seeing this final product, the initial gratification that I had felt gave way to a sense of awesome responsibility for perpetuating his masterful blend of science and literacy.

From the earliest days in the planning of this revision, Al and I recognized the profound changes taking place both in the knowledge of viral infections and in the way that knowledge was being gathered. With all of biomedical science moving in the rush of revolution, every discipline generating information relevant to this text—from virology, immunology, and pharmacology to epidemiology and neuropsychology—has been swept along in the torrent, propelled principally by two synergizing technologic forces: molecular biology and cybernetics. Although the text does not make frequent direct reference to these two forces, every chapter reflects their technical application through the use of DNA amplification methods and electronic data manipulation. For virtually every virus mentioned, our insight into the identification, classification, structural organization, and functional interaction with the host has lately been altered somehow by investigators using essential tools of that technology—the polymerase chain reaction and the computer. Equally exciting is the increasingly swift translation of knowledge generated by that technology into measurable impact on large numbers of people in the real world.

This edition has expanded to encompass those advances. The introductory part contains a new chapter devoted to the profusion of laboratory methods relevant to epidemiologic inquiry, and another heavily revised chapter documents the transformation of epidemiologic research and disease surveillance also being wrought by those tools of revolution. The remainder of text is also much enlarged, with both additional material in a number of chapters and entirely new chapters added to cover the proliferation of recently discovered viruses, such as herpesviruses, variants of several different classes of vector-borne agents, those responsible for acute and chronic hepatitis, and those ever more tightly linked to specific cancers, to name a few. Completion of the chapter on hemorrhagic fevers was delayed to permit more timely coverage of recent investigations into the outbreaks of infections with Ebola and related filoviruses.

This fourth edition was finished without the man for whom this text was inspiration and joy itself. The contributing authors and I hope that it has reached successfully toward the high standards of scholarship he set in the first three editions. We also hope others will help us strive to preserve the Evans legacy of wisdom about viral infections.

Richard A. Kaslow

Birmingham, Alabama

Contents

Chapter 2   **Laboratory Methods for the Diagnosis of Viral Diseases**

Robert L. Atmar and Janet A. Englund

Chapter 3   **Surveillance and Seroepidemiology**

Richard A. Kaslow and Alfred S. Evans

II. ACUTE VIRAL INFECTIONS

Chapter 4

Adenoviruses

Hjordis M. Foy

Chapter 5

African Hemorrhagic Fevers Caused by Marburg and Ebola Viruses

Robert E. Shope and James M. Meegan

Chapter 6

Arboviruses

Robert E. Shope and James M. Meegan

Chapter 7   **Arenaviruses**

Peter B. Jahrling

Chapter 8

Coronaviruses

Arnold S. Monto

Chapter 9

Cytomegalovirus

Anne A. Gershon, Eli Gold, and George A. Nankervis

Chapter 10

Epstein–Barr Virus

James C. Niederman and Alfred S. Evans

Chapter 11

Viral Gastroenteritis

Albert Z. Kapikian

Chapter 16   **Influenza Viruses**

W. Paul Glezen and Robert B. Couch

Chapter 19

Parainfluenza Viruses

W. Paul Glezen and Floyd W. Denny

Chapter 20

Parvovirus B19 Infection

Kevin E. Brown and Neal S. Young

Chapter 23

Respiratory Syncytial Virus

Kenneth McIntosh

Chapter 24

Retroviruses—Human Immunodeficiency Virus

William A. Blattner, Thomas R. O'Brien, and Nancy E. Mueller

Chapter 27

Rubella

Sandra J. Holmes and Walter A. Orenstein

Chapter 28

Smallpox: The End of the Story?

Abram S. Benenson

Chapter 29

Varicella–Herpes Zoster Virus

Thomas H. Weller

III. VIRAL INFECTIONS AND MALIGNANT DISEASES

Chapter 30   **Epstein–Barr Virus and Malignant Lymphomas**

Alfred S. Evans and Nancy E. Mueller

Chapter 33 | **Human Papillomaviruses**

Mark H. Schiffman and Robert D. Burk

IV. VIRAL INFECTIONS AND CHRONIC DISEASES

Chapter 34

Chronic Neurological Diseases Caused by Slow Infections

David M. Asher and Clarence J. Gibbs, Jr.

PART I

Concepts and Methods

CHAPTER 1

Epidemiologic Concepts and Methods

Richard A. Kaslow and Alfred S. Evans[†]

1. Introduction

The epidemiology of infectious diseases is concerned with the circumstances under which both infection and disease occur in a population and the factors that influence their frequency, spread, and distribution. This concept distinguishes between infection and disease because the factors that govern their occurrence may be different and because infection without disease is common with many viruses. Infection indicates the introduction and multiplication of a biological agent within a host, leading to an interaction often manifest as an immune response. It is determined largely by factors that govern exposure to the agent and by the susceptibility of the host. Disease represents the host response to infection when it is severe enough to evoke a recognizable pattern of clinical symptoms. The factors that influence the occurrence and severity of this response vary with the particular viruses involved and their portal of entry, but the most important determinants for many common infections lie within the host itself. Of these, the age at the time of infection, genetic background, and immune status of the host are the most crucial.

This first chapter deals in a general way with concepts, methods, and control techniques that are explored in detail in individual chapters concerned with specific viruses or groups of viruses. For fuller presentations of the epidemiologic principles, see references 120, 169, 186, and 213 and texts from Suggested Reading by Friedman, Hennekens, Last (1992), Lillienfeld, and Rothman, and for widely accepted definitions, see Last (1988).

2. Definitions and Methods

Incidence is the number of new events (instances of infection or cases of disease) occurring in some time interval. Generally, the *incidence rate* is the number of new events divided by the number of people at risk. The incidence rate may be expressed more specifically as a number of events per unit of population per unit of time or as the number of events in a fixed total population during a fixed total time period. The latter is considered a cumulative incidence but is often called an "attack rate" in an epidemic setting, where the total time period under consideration is established by the circumstances. In the public health environment, the numerator in the incidence rate of the disease in question is often based on reported *clinical cases* as recognized by physicians and the denominator represents the population under surveillance, commonly the total population of the geographic area encompassed by the reporting system. Public health agencies generally tabulate disease statistics in the form of annual rates. In more focused studies of viral infection, the numerator may signify infection (with or without disease) as determined by viral excretion and/or appearance of antibody during a brief defined time interval. In these studies the denominator may include those who are both exposed and known to be susceptible (i.e., lack antibody).

[†]*Deceased.*

Richard A. Kaslow • Department of Epidemiology, School of Public Health, University of Alabama at Birmingham, Birmingham, Alabama 35294-0008. **Alfred S. Evans** • Department of Epidemiology and Public Health, Yale University School of Medicine, New Haven, Connecticut 06510.

Prevalence is the number of cases existing at one time. The *prevalence rate* is the number of such cases divided by the population at risk. The time period involved may be 1 year or other fixed period (period prevalence) or a given instant of time (point prevalence). The term *period prevalence* involves both the number of new cases (incidence) and the duration of illness (number of old cases persisting from the previous reporting period). It is used most commonly for chronic diseases.

In serological surveys, *prevalence* represents the presence of an antibody, antigen, chemical marker, or other component in blood samples from a given population at the time of the collection. The *prevalence rate* is the number of sera with that component divided by the number of persons whose blood was tested. For viral infections, the prevalence of antibody represents the cumulative infection rate over recent and past years depending on the persistence of the antibody. For neutralizing or other long-lasting antibody, it reflects the lifetime or cumulative experience with that agent. If the antibody measured is present only transiently, as is often true of immunoglobulin M (IgM) antibody, then prevalence indicates infection acquired within a recent period.

Descriptive epidemiology deals with the characteristics of the agent, the environment, and the host and with the distribution of the resultant disease in terms of place, season, and secular trends. It is concerned with what the late John R. Paul[226] called "the seed, the soil and the climate." The delineation of these attributes of infection and disease in a population is the "meat" of epidemiology. Highly sensitive and specific molecular methods are increasingly being employed to define the agent and the host response (see Chapter 2 and reference 261). Public health professionals are adopting increasingly elaborate computer-based systems to capture morbidity and mortality data for descriptive and enumerative purposes.

Analytical epidemiology is concerned with planned epidemiologic investigations designed to weigh various risk factors or to evaluate a hypothesis of causation. Two methods of analytical study are commonly employed: the prospective or cohort and the retrospective or case-control study. These are discussed in detail in a recent book on methods in observational epidemiology dealing with both infectious and chronic disease epidemiology.[169] Laboratory diagnostic methods applicable in most clinical and epidemiologic settings are presented in Chapter 2.

The prospective method is a means of measuring incidence in a population or a cohort observed over time. Incidence studies permit the direct assessment of the risk of infection or disease, or both, in a defined population group over time in terms of age, sex, socioeconomic level, and other factors. Both the numerator and the denominator are known. In practice, incidence rates of clinical disease are often calculated retrospectively by using data on cases and populations that have been filed away; in virology, total infection rates, with or without clinical illness, can be determined by carrying out virus isolations or serological tests, or both, on materials that have been frozen away and for which data on the population sampled are available. Since such studies are not "prospective" in terms of the observer, calling them "cohort," "longitudinal," or "incidence" studies is more appropriate in a semantic sense. In addition to directly measuring risk, this type of investigation avoids the potential pitfalls of selecting controls, because the occurrence of infection or disease is recorded more comprehensively in persons with different characteristics. Large prospective and generally multicenter follow-up studies of homosexual men,[141,164,168,171,276] hemophiliacs,[126] intravenous drug users,[270] high-risk heterosexuals,[230] and others[9] have been invaluable in establishing risk factors for acquisition, changing incidence, clinical manifestations, determinants of progression, and many other aspects of human immunodeficiency virus (HIV) infection. The disadvantages of incidence studies are that they are expensive because an entire population must be kept under observation and appropriate specimens collected; the lower the incidence of the disease, the larger the denominator requiring observation, and the higher the expense. They are sometimes laborious to conduct and may require much technical help.

Retrospective or case-control studies compare the relative frequencies of certain suspected etiologic factors in patients with and subjects without a certain disease. An example is the relationship of smoking to the occurrence of lung cancer. When both the disease and the characteristic are already present at the time of observation, the data obtained represent prevalence rather than incidence rates. The absolute risk of the disease in persons with different characteristics cannot be measured because no denominators are available. Only the relative prevalence of the disease in persons having the characteristic can be calculated unless specific effort is made to identify incident cases. The measure best used to quantify the relationship under study is the "odds ratio," which has been shown to approximate the "relative risk" of a (rare) disease associated with exposure to the risk factor. The selection and identification of appropriate controls in retrospective studies often pose difficulties because unrecognized biases may be present. In virology, an example of the case-control method would be the evaluation of the etiologic role of a given virus in a certain disease by comparison of the frequency of viral excretion and/or antibody rises in patients having this disease with their frequency in those not having the disease. In evaluating this relationship, it must be remembered that infection without clinical disease is common in viral infections and might be occurring in the control group. Another recent example is comparison of the frequency of elevated viral antibody titers in the sera of patients with certain malignant or chronic diseases with those of age- and sex-matched controls as a clue to causation. Examples of this are the relationship of higher levels of antibody to Epstein–Barr virus (EBV) in cases of Burkitt's lymphoma and nasopharyngeal cancer as compared to controls, or of measles antibody titers in cases of subacute sclerosing panencephalitis and multiple sclerosis in relation to controls. In general, retrospective or case-control analyses are cheaper, are more quickly performed, and require smaller numbers than incidence studies but measure relative rather than absolute risk.

Case-control studies have been used creatively for a variety of purposes and with impressive success.[68,148,151] For diseases of low incidence like cancer, a case-control study constructed or "nested" within a larger cohort can achieve much of the value of a cohort study with greater efficiency. With this design a large diverse cohort of healthy persons is identified, say through a hospital record system or a serum bank. The outcome of the study, in the form of cases of disease, are ascertained through records or some independent source (e.g., a registry). Information on possible risk factors (e.g., serum antibody measurements) is gathered on the cases and on a comparable sample of members of the cohort who remain unaffected. Insofar as possible, measurements of antibody or other risk factors are performed by persons unaware of the outcome in the individual being evaluated or according to tightly standardized protocols. Three diverse examples of case-control studies include the original nationwide effort to identify the risk factors involved in transmission of HIV infection among homosexual men prior to the discovery of the virus itself,[151] the confirmation of the early link between aspirin use and Reye's syndrome,[148] and the nested case-control studies conducted on 240,000 persons whose sera were stored before the development of certain cancers thought to be associated with EBV infection but were tested only after those cancers occurred (e.g., non-Hodgkin's lymphoma[209] and nasopharyngeal carcinoma[61]).

Traditionally, the existence of a possible causal association between a factor and a disease is usually recognized in a clinical setting, and its statistical significance is determined by comparison with controls using the case-control or retrospective method. If the results suggest an important association, incidence and other studies are then undertaken to evaluate or confirm the observation. Thus, the risk of smoking in lung cancer and that of rubella infection in congenital abnormalities were discovered by clinical and case-control methods and confirmed by incidence and cohort analyses. Retrospective case-control investigations such as those on the relationship between certain blood groups and influenza[116] have yielded results that could not be confirmed when tested using incidence data. Similarly, the strength of the early associations of herpes simplex virus type 2 (HSV-2) with cervical cancer diminished when two prospective studies showed no difference between women with and without antibodies to HSV-2 in the occurrence of subsequent malignancy.[3,271] It is now widely accepted that most cases of cervical carcinoma are due to human papillomavirus infection (see Chapter 33).

Experimental epidemiology utilizes epidemiologic models and is the most elegant and sophisticated approach because all the variables should be subject to control. Unfortunately, animal models may be difficult or impossible to establish in the laboratory, and even if they are established, there is sometimes the question of the applicability of the results to the human host. Theoretically, the ideal experiment would employ volunteers. In the past, human subjects have participated in studies of yellow fever, malaria, hepatitis, infectious mononucleosis, acute respiratory infections, measles, rubella, and even syphilis. Such investigations involved important technical, medical, ethical, and moral issues. On the technical level, there is the question of the susceptibility of the volunteer to the disease under study; i.e., volunteer adults may already be immune as a consequence of childhood infection. Second, the host response to many infections may result in disease in only a small percentage of those exposed or even of those infected, thus requiring a large volunteer group. Medically, there is concern for the seriousness of the disease produced and for the possibility, however remote, of permanent disability or even death. Finally, the moral and ethical right to use human subjects in any medical experimentation is under debate. In today's climate, experimental studies in volunteers are subject to very strict control, and work being supported by government, foundation, or institutional funds must be scrupulously reviewed by a committee of professional along with lay and religious representatives. This peer group is required to weigh the benefits of the experiment against the risks involved and to ensure that the experimental subjects are fully aware of all possible consequences before signing a statement of "informed consent."

Seroepidemiology is a term applied to the systematic

testing of blood specimens from a defined sample of a healthy population for the presence or level of various components. These include antigens, antibodies, proteins, biochemical and genetic markers, and other biological characteristics (see Chapter 3). The same epidemiologic principles would apply to studies of other biological substances.

3. Epidemics

An *epidemic* or outbreak of disease is said to exist when the number of cases is in excess of the expected number for that population based on past experience. This determination obviously requires a knowledge of the number of both current and past cases. The definition of "excess" is an arbitrary one and depends on the concentration of cases in any given place, time period, or population group. The occurrence of a large number of cases, compressed in time, as when a new influenza strain is introduced, is readily identified as an "epidemic." Indeed, for influenza, a more sophisticated index has been set up by the national Centers for Disease Control in the United States by which an expected threshold of deaths from influenza and pneumonia in 121 cities has been established based on a 5-year average.[13] When this threshold is exceeded, an influenza outbreak is said to exist. In contrast, even a few cases of encephalitis or a single case of poliomyelitis in a summer may constitute an "outbreak" in areas where no cases previously existed. When several continents are involved, a disease is said to be "pandemic." The current global distribution of the acquired immune deficiency (AIDS) represents such a pandemic.

Chronic diseases pose more difficult problems in definition because their scale of occurrence must be viewed over years rather than months or weeks. In such a perspective, we do have current "epidemics" of chronic illnesses such as coronary artery disease, lung cancer, and intravenous drug abuse. The use of cocaine, especially in its free-base form or "crack," is posing an epidemic threat in the United States. The key words are "an unusual increase in the expected numbers of cases," irrespective of whether the time period involved is short or long.

Three essential requirements for an outbreak of viral disease are the presence of an infected host or contaminated reservoir, an adequate number of susceptibles, and an effective method of contact and transmission between them. If the agent is not endemic within the community, then the introduction of an infected person, animal, insect, or other vector of transmission is needed to initiate an outbreak. This is particularly important in a remote island

or isolated population group, where a virus disappears after no more persons remain susceptible, if persistent viral excretion does not occur to permit infection of newborns. Rubella, for example, disappeared from Barbados for 10 years despite an accumulation in the number of susceptibles to a level representing about 60% of the population and despite the existence of a large tourist trade.[109] In an isolated Indian tribe in Brazil, antibodies to respiratory-transmitted viruses including measles, influenza, and parainfluenza were essentially absent from the entire tribe.[29] The introduction of more susceptibles or of more infected persons may tip this balance. However, antibodies to viruses characterized by persistent or recurrent viral excretion, such as herpesviruses and adenoviruses, have been present in every population thus far tested, no matter how remote or isolated.[29]

The cumulative proportion of persons immune to a given disease within a community has been termed the *herd immunity* level, but the concept has limited applicability because of the variables involved. These include the probability of contact between a source of infection and the susceptible person, the portal of entry accessible, the contagiousness of the agent, and the degree of individual host immunity. If the prevalence of antibody of sufficient titer is high among persons in a given community, then the occurrence of an outbreak has been regarded as highly unlikely. For highly communicable infections such as rubella or measles, the level of herd immunity must be of the order of 95% or higher to be effective. For example, in an open college community, a preexisting level of immunity to rubella of 75% failed to prevent an outbreak of this disease.[115] Indeed, the rubella infection rate of 64% among those completely susceptible (i.e., without detectable antibody) was even higher than the 45% infection rate in the same community for a new influenza strain to which the entire population was susceptible.[115] A rubella outbreak has occurred among military recruits in the presence of a 95% level of immunity: 100% of the susceptibles were infected.[146] And even a documented 98% level of immunity did not prevent an outbreak in college students.[140] The spread of infection is apparently so efficient under these circumstances of close and prolonged contact that a high level of herd immunity does not deter its progress. Another possibility is that reinfection of partially immune persons results in pharyngeal excretion and further spread of virus to susceptible persons. Other principles are also worth emphasizing: (1) the concept of herd immunity is even less valid where several strains of virus exist and cross-protection is not complete; (2) even the identical strain of virus that does not naturally confer complete immunity (e.g., certain herpesviruses) may reinfect; (3) reactivation of latent infection may produce disease, especially in immunocompromised hosts; and (4) the potential for virus to coexist in the blood in the presence of antibody, as in HIV or hepatitis C virus infection, indicates that protection from reinfection or reactivation is incomplete at best.

For smallpox, the induction of herd immunity by vaccination has resulted in the complete global eradication of the disease through the efforts of the World Health Organization (WHO). The last case occurred in Somalia on October 26, 1977.[278] No new natural cases have been reported for more than 19 years since then, although laboratory infections have occurred and remain a hazard to laboratory personnel. Intentional maintenance of prototype virus in the laboratory has been controversial (see Chapter 28). Continued vigilance is clearly needed for unsuspected laboratory sources, biological warfare, and animal reservoirs of smallpox-related viruses.

Mathematical models have long been sought to fit the epidemic spread and incidence patterns of certain infectious diseases or as a basis for immunization programs.[2,5,6,12,66,88] For diseases in which most infections are clinically expressed, the immunity is good, the means of transmission is limited to one or two routes, the mixture of susceptibles and immunes is homogeneous and equally distributed and where the age at the time at which infection occurs is figured in the calculations, then such mathematical models may be useful in planning control measures. They require the input of a good mathematician and good epidemiologist, both of whom understand the dynamic interplay of these various factors. Even then, the model must be based on a particular population group with consideration for such factors as socioeconomic status, population mixing, vaccination programs, and behavioral characteristics. The model must then be tested over time in that population group against the actual number of cases reported in a good surveillance system. A reasonably accurate prediction of actual events has been achieved in a model developed for rubella and measles vaccination programs by Anderson and May.[5] Models for measles are discussed further in Chapter 17. Data from cohort studies have validated models of transmission of influenza in communities and secondary attack rates within households.[133] In anticipation of wide use of varicella vaccine, an extensive theoretical assessment of routine immunization has suggested the need for additional data on residual susceptibility and infectiousness and on the booster effect on immunized individuals reinfected with wild-type virus.[136] But in other situations, where there are many inapparent infections or the disease results from reactivated rather than primary infection or in which the agent is

intermittently excreted in the infected host or intermittently present in some environment or arthropod vector, then the events leading to infection and disease are so complex and variable that a mathematical model is difficult to construct. Considerable effort has been invested in development of models for sexual and other modes of transmission of HIV infection in various settings,[33,39,217] for estimating the period of latency between infection and onset of AIDS,[10,11] and for describing the trajectory of the epidemic.[122] However, for the reasons mentioned above, these models have proved difficult to evaluate using actual data (see also Chapter 24). The limitations for such models and recommendations for their improvement based on better data has been well reviewed by Singer[247] using malaria as the example.

4. Investigation of an Epidemic

The investigation of an epidemic is addressed only briefly here. More extensive description of the methods employed can be found elsewhere.[43,86,128] Epidemic investigations have similar but not identical objectives. Table 1 summarizes the sequence of steps usually taken, but they do not represent the appropriate order of execution. It may not be possible to establish a definitive diagnosis early, so a rather specific, simple working definition should be established using key epidemiologic and clinical features as a case-finding device. This definition can be expanded and made more sensitive later, when laboratory studies are possible. Control measures should be instituted as soon as the means of spread is reasonably established. Not surprisingly many epidemics of viral infection result from person-to-person spread, partic-

---

Table 1. Epidemic Investigation

1. *Define the problem.* Diagnosis? Is it an epidemic?
2. *Appraise existing data*
   Time: date (and hour) of onset; make epidemic curve
   Place: spot map of cases; home, work, and recreational places; special meetings
   Person: age, sex, occupation, ethnic groups
   Incidence rates: infection, cases, deaths
   Possible means of transmission
   Seek common denominator and unusual exceptions
3. *Formulate hypothesis.* Source of infection, method of spread, possible control
4. *Test the hypothesis.* Search for added cases; evaluation; laboratory investigation
5. *Conclusion and practical application.* Long-term surveillance and prevention

ularly by the respiratory route, in open communities or in relatively closed populations like health or extended care facilities. Vector-borne spread also accounts for a significant portion of epidemic viral infection throughout the world (see Chapters 6, 7, and 12). The role of the murine reservoir in the recent emergence of hantavirus in the southwest and possibly elsewhere in the United States (see Chapter 12 and reference 55) emphasizes the transmission potential of continuous indirect and incidental contact with animals carrying human pathogens. Common source outbreaks of viral infections from water, food, milk, or environmental sources are not nearly as common as with bacterial infections. However, they do occur. Some examples include spread of adenoviruses by eye tonometers in eye clinics or via swimming pools, of hepatitis A by public water supplies or by seafood, of hepatitis B by virus-contaminated yellow fever vaccines, or of enteroviruses by fecally contaminated foodstuffs or milk.

The pandemic of AIDS involving HIV constitutes one of the most significant threats to human health ever faced. In the United States alone, 500,000 clinical cases of AIDS were recognized between 1981 and the end of 1995, and the majority have died. About 80,000 cases are projected to have occurred in 1995 in the United States alone. Broadly recruited cohorts of infected homosexual men, hemophiliacs, and other young adults in developed countries have demonstrated consistent annual rates of occurrence of AIDS of about 4–5%[141,212]; the median intervals between onset of AIDS and death have been prolonged somewhat by specific antiviral and supportive care but under favorable circumstances remain about 2 years. The prevalence of HIV infection in the United States appears to have remained constant at approximately 1 million for several years and similar proportions have been observed in other parts of the developed world. New infections are occurring in different population subgroups than those initially affected and the potential for wider spread remains serious. Worldwide, only the crudest estimates are available, but they suggest a prevalence of about 14 million or more as of the end of 1994.[282] The WHO has projected upward of 40 million infections by the year 2000, and pessimism extends beyond the direct epidemiologic repercussions to the broader impact on whole societies.

Historically, the closest recorded epidemiologic precedent for such a pervasive assault by an agent transmitted primarily through sexual contact was the 16th-century pandemic of presumed syphilis. Obviously, at that time the void in medical and epidemiologic knowledge left the affected populations helpless. For HIV infection, in contrast, during the past decade biomedical scientists around the world have mounted the most intense investigative and control effort ever directed at any disease. Chapter 24, devoted to HIV infection, reviews the knowledge gained about the infection and the disease itself through this unprecedented mobilization of resources. Although efforts to control HIV infection have resulted in important successes, e.g., nearly complete interruption of transmission from transfused blood products and substantially decreased acquisition among older men having sex with other men in the United States, the unique plasticity of the virus has tempered optimism that imminent vaccine development, pharmacotherapeutic progress, or behavioral enlightenment will soon diminish the force of morbidity among the population groups most severely affected.

The collective investigative response to epidemic HIV infection, despite its inadequacies, has produced invaluable progress in public health and infectious disease epidemiology. Although Chapter 24 provides additional detail, and a more complete picture can be found in the voluminous literature of HIV/AIDS, the following developments are among the important ones: (1) innovative approaches to surveillance (e.g., systems for obtaining unbiased seroprevalence figures for women of reproductive age by screening newborn infants); (2) improved analytical tools for empirical projections of cases, for estimating sexual transmission efficiency, for clinical prognostication, etc.; (3) application of programs that prevent transmission of other infections as well as HIV (e.g., universal precautions, blood donor screening, wide distribution of condoms and other strategies for control of sexually transmitted diseases, injection needle exchange systems, and assistance with substance abuse); (4) improved formulation of practical screening diagnostic reagents and vehicles (e.g., filter paper serological methods); (5) revolutionary advances in fundamental knowledge of and research techniques in virology, immunology, and genetics, as well major progress in the relevant social and behavioral sciences; (6) novel approaches to development, testing, and distribution of therapeutic agents; (7) deeper understanding of human behavior; and (8) on the whole, increasingly enlightened medical, legal, political, and cultural responses to a public health crisis.

5. The Agent

This section is concerned primarily with those general properties of viruses that are important to an understanding of their epidemiology and not with their basic clinical chemistry, morphology, genetics, or multiplication. Advances in molecular virology are clearly yielding critical insights into how viruses infect and produce disease. Most of the chapters of this book provide information on developments pertinent to their topic, but a basic virology text, such as Field's *Virology*,[119] or any of several other textbooks of infectious diseases, microbiology, and virology are excellent supplemental sources (see Suggested Reading).

The chief characteristics of viruses that are of importance in the production of infection in man are (1) factors that promote efficient transmission within the environment; (2) the ability to enter one or more portals in man; (3) the capacity for attachment to, entry into, and multiplication within a wide variety of host cells; (4) the excretion of infectious particles into the environment; and (5) a means of developing alternative mechanisms of survival in the face of antibody, cell-mediated immunity, chemotherapeutic agents, interferon, or other hostile elements. Survival of the virus might be achieved through mutation, recombination, basic properties of resistance, or the availability of alternative biochemical pathways.

Intensive studies of the various parts of the viral genome responsible for particular functions, the dynamics of infection at the cellular level, and the specificity and complexity of the immune response in the susceptible host are now being carried out for many viral infections. For example, it is now clear that a minor change in a single gene of a particular virus, such as the reovirus, or even in a single nucleotide or amino acid, such as in rabies virus, may have a profound effect on the pathogenicity of the agent and the pattern of clinical disease that develops (see texts by Notkins, Thomas, and others listed in Suggested Reading). Similarly, subtle changes in the immune system can alter the host's response to the same virus.[165,166]

The spread of viruses depends on (1) the stability of the virus within the physical environment required for its transmission, including resistance to high or low temperatures, desiccation, or ultraviolet; (2) the amount of virus expelled into the proper vehicle of transmission; (3) the virulence and infectivity of the agent; and (4) the availability of the proper vector or medium for its spread.

After entry through an appropriate portal, the virus must escape from ciliary activities, macrophages, and other primary defense mechanisms during its sojourn to the target cell, find appropriate receptors on the cell surface for its attachment, and be able to penetrate and multiply within the cell. The steps then include initiation of transcription of messenger ribonucleic acid (mRNA), translation of early proteins, replication of viral nucleic acids, transcription of mRNA, translation of late proteins, assembly of virions, and then viral release.[118] These aspects fall into the province of basic virology and are not discussed in detail here. What is important in pathogenesis is the efficiency of spread from cell to cell, either by direct involvement of contiguous cells or by transport via body fluids to other susceptible cells; the number of cells infected; and the consequences of viral multiplication on the cell itself and on the organism as a whole. The long-term survival of a virus in human populations depends on its ability to establish a chronic infection without cell death or on an effective method of viral release into the environment in a manner ensuring its transport to a susceptible host, or on a highly adaptive system for biological adversity. The prime examples of adaptability among animal viruses are influenza A and HIV. Without its propensity for antigenic variation, influenza virus would probably behave like measles or rubella viruses and be dependent for survival on the temporal accumulation of new susceptibles.

6. The Environment

The external environment exerts its influences on the agent itself, on the manner of its spread, and on the nature of the host response to infection. Although viruses survive or die within defined ranges of certain physical factors such as temperature and humidity, there is much variability from one viral group to another. A simple environmental factor such as cold may have different effects on the survival of different viruses and on their ability to multiply within cells. Although environmental characteristics play an important role in the survival of a virus, they are probably of much greater significance in their influence on the routes of transmission and on the behavior patterns of the host.

For infections that require an insect vector, such as the arboviruses, the environment exerts an obvious role in restricting the occurrence of infection and disease to those areas that have the proper temperature, humidity, vegetation, amplifying animal hosts, and other features necessary for the insect involved. For viruses potentially transmitted by water, such as hepatitis A virus and Norwalk agent, a warm environment attended by poor sanitation and fecal contamination clearly enhances the degree of exposure and the efficiency of transmission.

Perhaps the most crucial effect of climate on common viral diseases is exerted on the social behavior of the host. In tropical settings and in the summer season in temperate climates, the opportunity for transmission of gastrointestinal diseases is increased through contact with water, as in swimming in and drinking from the polluted areas. Warm weather also brings closer contact with insect vectors of arboviruses and with dogs and other animal sources of rabies. In the cooler seasons, people collect indoors, promoting transmission of airborne and droplet-borne infections. This spread is amplified by the assembly and dispersal of students coinciding with the periodic openings and closings of educational institutions. In addition, the environment within most houses and buildings tends to be hot and dry, which impairs the protective mechanisms of human mucous surfaces and may permit easier entry and attachment of certain respiratory viruses. Cultural as well as physical environment can contribute to the spread of infection, as exemplified by the patterns of spread of HIV infection among gay bathhouse patrons, injectable substance abusers in "shooting" galleries, and women selling sexual favors in order to purchase crack-cocaine.[218]

Just as winter clearly brings with it an increase in viral respiratory illnesses, heavy rains and the monsoon similarly influence these same diseases in tropical settings. Indeed, the incidence of common upper respiratory diseases in college students was as high in the warm climate at the University of the Philippines as in the intemperate winters at the University of Wisconsin.[106,107] Viruses that cause respiratory infections in children have also been found to be active in all climates around the world.[63] Community studies in India,[221] Trinidad,[28] and Panama[203] have indicated a high morbidity from influenza and other respiratory diseases in tropical settings. As in temperate climates, factors that tend to aggregate people inside, such as heavy rainfall or schooling, also coincide with the highest incidence of respiratory-transmitted infections in the tropics.[107,203]

7. The Host

The factors that influence infection involve primarily exposure to the infectious agent and the susceptibility of the host. The opportunity for a susceptible host to come in contact with a source of infection depends on the means of transmission. Respiratory-transmitted agents are usually general in their exposure; those transmitted by gastrointestinal routes are related to exposure to food or water and to the hygienic and socioeconomic level of the host; those that depend on arthropod-borne transmission may involve persons in special settings or special occupational exposures. Others, such as sexually transmitted agents, require specific behavioral acts of the host; still others require specialized exposures such as transfusions, rabid animals, or specialized environments. The factors that influence exposure are therefore mostly extrinsic to the host, but not all fully exposed persons will develop infec-

tion, as manifested by the appearance of antibody and the isolation and/or demonstration of the causative agent. The agent factors affecting the outcome of its encounter with the host include the dose, infectivity, and virulence of the virus and the number of surviving infectious particles that enter an appropriate portal and find viral receptors on susceptible cells. Host factors include the vigor of the primary defense system, such as cilia, mucus, and non-specific viral inhibitors, the genetic susceptibility to the virus, and the presence or absence of antibody and cell-mediated immunity.

Those factors that determine whether clinical illness will develop in a person already infected depend in part on the dosage, virulence, and portal of entry of the agent, but more important, they depend on certain intrinsic properties of the host.[102] Some of these characteristics are listed in Table 2. Age at the time of infection is a critical host factor and influences whether clinical illness develops following infection with such agents as Epstein–Barr virus, hepatitis viruses, and poliomyelitis viruses. In general, the probability that clinical illness will develop increases as the age at the time of infection increases; in a similar fashion, the severity of the clinical response also increases with age at the time of illness. The nature of the immune response to a virus can be either beneficial to the host in limiting the infection or detrimental if the clinical disease is caused by certain immunopathological consequences of infection such as immune complexes or auto-immune mechanisms. The vigor of the humoral and cell-mediated immune responses may also determine when a virus becomes persistent or is eradicated from the body.

The severity of the clinical response to viral infections is greatly enhanced when the immune system is compromised as a result of an inherited or acquired immunodeficiency, by immunosuppressive drug therapy as in renal transplant patients, or by infection and destruction of

**Table 2. Factors that Influence
the Clinical Host Response**

1. Dosage, virulence, and portal of entry of the agent
2. Age at the time of infection
3. Preexisting level of immunity
4. Nature and vigor of the immune response
5. Genetic factors controlling the immune response, the presence of receptor sites, and cell-to-cell spread
6. Nutritional status of the host
7. Preexisting disease
8. Personal habits: smoking, alcohol, exercise, drugs
9. Dual infection or superinfection with other agents
10. Psychological factors (e.g., motivation, emotional crises, attitudes toward illness)

key lymphocytes involved in cell-mediated immunity such as the elimination of susceptible CD4$^+$ lymphocytes by HIV. Other host attributes also affect the occurrence or severity of certain infections: smoking increases the severity of acute respiratory infections, as does the presence of maternal antibody in respiratory syncytial viral infections. Alcohol appears to increase the risk of chronicity for certain hepatitis viruses, and exercise predisposes to the development of paralytic poliomyelitis in the exercised limb. Psychosocial factors manifested by increased motivation toward a career, overachieving fathers, and poor academic performance have been shown to increase the frequency of clinical infectious mononucleosis as well as its severity in cadets at the West Point Military Academy infected with Epstein–Barr virus.[102,163] The ability to identify which persons are susceptible at the time of exposure and those who are infected, the frequency of clinical disease among the infected, and the availability of psychosocial data at the start of the study permitted the delineation of factors that would have been obscured if just exposure and disease had been considered. Alteration of the immune system with reactivation of EBV may be a risk factor for later development of lymphoma.[113,170]

Nutrition and genetic susceptibility are probably of importance in tipping the scale toward clinical illness, but few studies have been done to measure this. The determinants of clinical expression among the infected have been termed "clinical illness promotion factors" or "cofactors" and have been reviewed.[102] Knowledge about the role and mechanism of these factors varies greatly from one infectious agent to another. There has been particularly intense interest in the "cofactors" for expression or rate of progression of HIV infection. As with nearly all infections, age at the time of acquisition of HIV has consistently been associated with the pace of immunologic deterioration.[238] There is also growing evidence that certain alleles or variants of genes in the highly polymorphic human leukocyte antigen (HLA) region of the genome are associated with the rate of progression— some with an accelerated and others with a retarded course.[165,166,174] Our knowledge of the actual cellular and molecular basis for these events is meager, but new virological and immunologic techniques are making rapid advances in our understanding both in humans and in animal models[166,201,202,244] (see also the text by Thomas listed in Suggested Reading).

8. Routes of Transmission

The major routes of transmission of viral infections are listed in Table 3. Many viruses have several alternate routes, thus enhancing the chance of infection. The sequence of events in transmission involves release of the virus from the cell, exit from the body, transport through the environment in a viable form, and appropriate entry into a susceptible host.

Some viruses are released from cells at the end of the cycle of multiplication. Others do not complete this cycle (incomplete viruses), and some do not effect efficient escape (cell-bound viruses). Many viruses are released from cells by budding, acquiring a lipoprotein coat or envelope as they go through the cell membrane; these include herpesviruses, togaviruses, myxoviruses, paramyxoviruses, and coronaviruses. Nonenveloped viruses not released by budding are the adenoviruses, parvoviruses, poxviruses, picornaviruses, and reoviruses. Some of these latter are released by cell lysis. Once released, viruses find their way to new hosts via one or more portals such as the respiratory tract (influenza), skin [varicella-zoster virus (VZV) and smallpox virus], blood [HIV, human T-cell leukemia virus types I and II (HTLV-I and HTLV-II), hepatitis B virus (HBV), hepatitis C virus (HCV), and arboviruses], gastrointestinal tract (enteroviruses), genital tract (HIV, HTLV-I, HSV-2), and placenta [HIV, rubella, and cytomegalovirus (CMV)]. A more detailed presentation of these major routes of spread is now given.

8.1. Respiratory

The respiratory route is probably the most important method of spread for most common viral diseases of man and is the least subject to effective environmental control. For influenza virus, the degree of transmissibility varies from one strain to another and seems to be independent of other attributes of the virus. Schulman[243] compared the features of a strain with high transmissibility (Jap 305) and one with low transmissibility (Ao/NWS) in an experimental mouse model. The virus titer in the lung was similar for both strains, but the virus content in the bronchial secretion was low for the Ao/NWS strain compared to the Jap 305 strain. This higher degree of release into the respiratory portal of exit resulted in detectable virus in the air surrounding mice infected by the Jap 305 but not these infected by the Ao/NWS strain. Once an aerosol was created, the stability of both strains was similar. Protein analysis also revealed differences in the neuraminidase of the two strains; this component is associated with dissociation of viruses from the cell and thus perhaps with its transmissibility. However, high transmissibility did not go along with transfer of the gene for neuraminidase, so it was concluded that other factors were also involved in the efficacy of spread.

Table 3. Transmission of Viral Infections

Routes of exit	Routes of transmission	Example[a]	Factors	Routes of entry[b]
Respiratory	Bite	Rabies	Animal	Skin
	Saliva	EBV	Kissing	Mouth
			Prechewed food, infants	
		HBV	Dental work	
		?HIV	Sexual	
	Aerosol	Influenza, measles	Cough, sneeze	Respiratory
	Oropharynx to hands, surfaces	HSV, RSV, rhinovirus	Fomites	Oropharynx
Gastrointestinal	Stool to hands	Enteroviruses	Poor hygiene	Oropharynx
	Stool to water, milk food	HAV, rhinoviruses	Seafood, water, etc.	Mouth
		HAV, HEV		
	Thermometer	HAV	Nurses	Rectal
Skin	Air	Pox viruses	Vesicles	Respiratory
	Skin to skin	Mulluscum contagiosum warts	Abrasions	Abraded skin
Blood	Mosquitoes	Arboviruses	Extrinsic	Skin
			Incubation period	
	Ticks	Group B togaviruses	Transovarial transmission	Skin
	Transfusions of blood and its products	HIV, HBV, HCV, HTLV-I/II, CMV, EBV	Carrier in plasma or lymphs	Skin
	Needles for injection	HIV, HBV, HDV	Drug addicts, tatooing	Skin
Urine	Rarely transmitted	CMV, measles, mumps, rubella	Unknown	Unknown
Genital	Cervix	HSV, CMV, HBV, HIV, HPV, rubella	Sexual, perinatal	Genital
	Semen	CMV, HBV, HIV	Heterosexual	Genital
			Homosexual	Rectal
Placental	Vertical to fetus	CMV, HBV, HIV, rubella	Infection in pregnancy	Blood
Eye	Tonometer	Adenovirus	Glaucoma test	Eye
	Corneal transplant	Rabies, Creutzfeldt–Jakob disease	Surgery	
Breast	Breast feeding	CMV, HIV, HTLV-I	Maternal viremia	Mouth

[a]CMV, cytomegalovirus; EBV, Epstein–Barr virus; HAV, hepatitis A virus; HBV, hepatitis B virus; HCV, hepatitis C virus; HDV, delta hepatitis virus; HEV, hepatitis E virus; HIV, human immunodeficiency virus; HPV, human papilloma virus; HTLV-I/II, human T-lymphotropic virus, type I; RSV, respiratory syncytial virus.
[b]Transmission does not always follow standard routes (see Section 8.1).

Other aspects that affect the transmission of respiratory viruses are the intensity and method of propulsion of discharges from the mouth and nose, the size of the aerosol droplets created, and the resistance of the airborne virus to desiccation. Much of the early work on the transmission of respiratory viruses was done by Knight[177] and his group. At one extreme is the direct transmission of infection via personal contact such as kissing, touching of contaminated objects (hands, handkerchiefs, soft drink bottles), and direct impingement of large droplets produced by coughing or sneezing. This last method is regarded as a form of personal contact because of the short range of the heavy droplets formed. Sneezing and coughing also create aerosols varying in size from about 1 to more than 20 μm that permit transmission of infection at a distance. The dispersion of an aerosol depends on wind currents and on particle size. In still air, a spherical particle of unit density of 100-μm diameter requires 10 sec to fall the height of the average room (3 m), 40-μm particles require 1 min, 20-μm particles 4 min, and 10-μm particles

17 min. This means that particles under 10 μm have a relatively long circulation time in the ordinary room. Particles 6 μm or more in diameter are usually trapped in the nose, whereas those 0.6–6.0 μm in diameter are deposited on sites along the upper and lower respiratory tract.

Hygroscopic particles of 1.5-μm diameter discharged in large numbers by coughing or sneezing lose moisture and shrink in ambient air but regain their original dimensions from the saturated air in the respiratory tract. The site of disposition of an aerosol containing virus particles does not necessarily represent the level in the respiratory tree where the greatest number of susceptible cells exist for that agent. Quantitative studies have indicated that with four different respiratory viruses, the number of viral particles necessary to produce infection in the respiratory tract is relatively small. With adenoviruses, for example, it is on the order of seven virions. The lower infective dose required for nasal implantation of rhinoviruses and coxsackievirus indicates that this route, perhaps by personal contact, leads to their effective transmission.[177] The high concentrations of rhinovirus particles on fingers, hands, and hard surfaces as opposed to the lower concentrations found in aerosols suggest that infection via hands may be an important route of spread. This is supported by the frequent inadvertent contact of hands with the nose or eyes.[138] Although the importance of this mechanism is uncertain, frequent hand washing may help control the spread of the common cold, as may certain chemicals impregnated into cleaning tissues. Hands and other fomites are also important in the transmission of respiratory syncytial virus.[134]

The size and number of viral particles in sneezes and coughs have varied from study to study depending on the methodology employed. In one study, 1,940,000 particles were present in sneezes and 90,765 in coughing, a ratio of 2.14:1.[124] Despite the high level of particles, the recovery of coxsackievirus A21 itself was more frequent from coughs than from sneezes.[177] Many questions on the mechanics of transmission of respiratory viruses remain unanswered, and any generalizations are premature, but the methodology to answer some of these is becoming available.

Aerosolization of certain viral agents may occur from suction devices and from catheters in intensive care units and from blood products in dialysis units. These include not only respiratory and intestinal agents but also agents such as hepatitis viruses that circulate cell-free in the blood and cell-associated viruses such as CMV, EBV, HIV, and HTLV-1.

Finally, hantavirus and other arenaviruses may be spread by aerosols created from soil containing the rodent urine in which those viruses are excreted. The recent outbreak of hantavirus pulmonary syndrome in the southwestern United States and sporadic cases in the East[83] have been attributed to environmental conditions favoring such spread (see Chapter 12).

8.2. Gastrointestinal

Transmission by the oral–fecal route is probably the second most frequent means of spread of common viral infections, and the gastrointestinal tract is the second great portal of entry of infection. Viruses can directly infect susceptible cells of the oropharynx, but to induce intestinal infection, virus-containing material must be swallowed, successfully resist the hydrochloric acid in the stomach and the bile acids in the duodenum, and progress to susceptible cells in the intestine. These cells may be the epithelial cells in the intestinal mucosa or in the intestinal lymphatics, as with adenoviruses. Viruses with envelopes do not normally survive exposure to these acids, salts, and

enzymes in the gut. The major enteric viruses are rotaviruses, Norwalk agent, poliomyelitis, echo, coxsackie, and hepatitis A and E viruses. It is known that under conditions of close and prolonged contact, hepatitis A, B, and E viruses may also be transmitted in this way. Multiplication and excretion in the intestinal tract also occur with adenoviruses and reoviruses, but this route of transmission is not usually of epidemiologic importance. The rhinoviruses are acid labile and do not survive passage through the stomach. Unlike the respiratory viruses, the enteroviruses rarely produce evidence of local disease as a consequence of their multiplication in cells lining that area. Thus, diarrhea, vomiting, and abdominal pain are highly unusual features of infection with these agents. Instead, their major target organs and the site of major symptomatology are at a distance: hepatitis viruses in the liver and enteroviruses in the central nervous system and skin. The HIV virus may be introduced into the rectum during passive anal intercourse and enter the blood through abrasions in the mucous membrane.

Viruses excreted via the gastrointestinal tract must successfully infect other susceptible persons via the oral–intestinal route through fecally contaminated hands, food, water, milk, flies, thermometers, or other vehicles. Viruses spread via these routes are subject to much greater environmental control than are agents transmitted by the respiratory route. Thus, good personal hygiene, especially washing of hands after defecation, proper cleanliness and cooking of food, pasteurization of milk, good waste disposal, and purification of drinking water supplies are effective preventive measures. Hepatitis A virus (HAV) and hepatitis E virus (HEV) are stable viruses in water and, when present in sufficient dosage, may not be inactivated by ordinary levels of chlorine. Furthermore, HAV, at least, can persist in oysters and clams over long periods. This is especially hazardous because these foods are so often eaten without having been cooked. Hepatitis viruses and the enteroviruses also flourish in certain institutional settings (mental hospitals, institutions for retarded children, some prisons) and in countries where personal hygiene is lacking or difficult to practice or where poor environmental control is present. Since some enteroviruses may also multiply in the respiratory tract and be transmitted by the respiratory route, this alternate pathway is of epidemiologic importance even in the face of good personal and environmental hygiene.

8.3. Skin

Skin is the third important area for the entry and exit of viral infections. Although penetration of the intact skin is an unlikely mechanism of infection, the introduction of virus particles via a bite, as with rabies, or via a mosquito, as with the arboviruses, or via a needle or blood transfusion, as with all types of hepatitis viruses and HIV makes this route a very important one. CMV, EBV, and HTLV-I may also be transmitted through blood transfusions. The abraded skin may serve as the entry point of human papovavirus, which causes warts, of hepatitis B virus, and probably for the agent of kuru.

The skin serves as a portal of exit only for those viruses that produce skin vesicles or pox lesions that release infectious particles on rupture. These include herpes simplex, smallpox, varicella-zoster, and vaccinia viruses. The viruses of certain maculopapular exanthems may also be present in the skin, as in rubella, but this does not seem to be an important avenue of escape, since vesicles are not formed and skin involvement occurs late in the disease, at a time when the virus may be bound by antibody; indeed, the antigen–antibody complex may be responsible for the rash itself.

8.4. Genital

The genital tract serves as a portal of infection for both heterosexual and homosexual partners during sexual activity and is a source of infection as the fetus passes down the birth canal. Herpes simplex, types 1 and 2, CMV, HBV, HIV, and rubella virus are present in cervical secretions and can infect infants during delivery or shortly thereafter; CMV, HBV, and HIV are present in the semen and can be transmitted during either heterosexual or homosexual intercourse.[32,152,160,184,259,277] Long-term asymptomatic cervical or semen carrier states exist and make recognition and control difficult. Receptive anal intercourse is an important method of spread of CMV and HIV infections.

The presence of other genital infections in either partner has been shown in repeated studies to predispose to acquisition of HIV infection. Much but not all of the epidemiologic data indicate that the ulcerative lesions of syphilis, chancroid, and HSV-2 infection mechanically facilitate penetration of the epithelial barriers of the genital tract by HIV.[145,167,230] However, suggestions of predisposition by nonulcerative infections like gonorrhea, chlamydia, or even HBV[263] are consistent with an alternative to enhanced epithelial penetration, namely, recruitment and activation of macrophages and other cells responsive to the breach in the immunologic barrier. Although there is now compelling evidence that cervical infection with certain types of human papillomavirus (HPV), principally 16 and 18, causes a substantial proportion of cervical carcinoma, the inability to propagate the virus or measure

a serological response has forced reliance on more indirect documentation of genital transmission of these two, or of the more common wart-associated types (see Chapter 33).

8.5. Intrauterine or Transplacental

Viruses may infect the fetus either by direct contact via the birth canal as discussed in Section 8.4 or by hematogenous spread via the placenta to the fetus within the uterus. Viruses that produce intrauterine infection include CMV, hepatitis B, herpes simplex, rubella, varicella viruses, and HIV; CMV and rubella viruses are the most common congenital infections in that order, with congenital rubella decreasing with increasing vaccination. Infection of the fetus may result in no symptoms, in abortion and stillbirth, in developmental abnormalities, in persistent postnatal infection, and in some later manifestations (see also Section 12.2.6). Acquisition of EBV or HBV in the early postnatal period is associated with persistence of infection and substantially increased risk of subsequent African Burkitt's lymphoma and hepatocellular carcinoma, respectively.[112,208] At least one textbook[236] provides a thorough and authoritative approach to perinatal infection.

8.6. Urinary

Although viruses such as CMV and measles are excreted in the urine, this portal of exit has not been established as being of epidemiologic or clinical importance. Considering the wide variety of viruses that can multiply in human kidney tissue cultures *in vitro*, it is surprising that renal infections in man from these viruses are virtually nonexistent or at least are nonrecognized. It seems possible that viruses may play a role in immune complex nephritis in man as they do in experimental animal models, but to date this has not been clearly demonstrated, nor has it been reflected in abnormally high viral antibody levels in such patients.[274] Adenovirus types 11 and 21 have been implicated as the cause of hemorrhagic cystitis in children (see Chapter 4).

8.7. Personal Contact

Direct transfer of infected discharges from the respiratory or gastrointestinal tract to a susceptible person is often included under "transmission by personal contact." Many viruses regarded as "respiratory or airborne" in spread may in fact be more direct in their transmission mechanism, as has been previously mentioned for the rhinoviruses[132,138] and respiratory syncytial virus.[134] At the other extreme, casual person-to-person spread of HIV has not been shown in serological follow-up of families with an index case of AIDS. Although hepatitis C is most often transmitted parenterally, there are some suggestions that it may also be carried in biological fluids exchanged through intimate personal exposure that is neither typically parenteral nor typically sexual.[143]

8.8. Water and Food

Outbreaks of viral hepatitis have occurred from sewage-contaminated water, as in the huge outbreaks in New Delhi, India, in 1955[199] and subsequently, often due to HEV, and in less dramatic attacks by this virus in other locations.[36,48] These infections may also be acquired from seafood obtained from fecally contaminated waters, as shown in outbreaks associated with oysters in the United States and in Sweden[196] and with clams in New Jersey.[80] Milk and water have also served as vehicles of transmission of hepatitis, Norwalk agent, and poliomyelitis viruses. Summer outbreaks of adenovirus type 3 infections have occurred in association with swimming pools.[21]

8.9. Arthropod-Borne

Mosquitos, flies, ticks, and other insects may transmit viral infections. One kind of transmission is a passive type, simply involving survival of the virus in or on the insect that has picked it up from skin lesions or the blood. This type requires neither incubation time in the insect vector nor any specificity for either the arthropod host or the virus. Poliomyelitis and possibly hepatitis viruses may be carried in this way. On the other hand, some viruses require multiplication in the insect vector. In this instance, virus acquired from the blood of the human or animal host during viremia requires a period of multiplication within the arthropod vector before it is infectious, and there is a high degree of vector–virus–host specificity. Examples of this include the transmission of yellow fever virus by *Aedes aegypti* mosquitos and of the seasonally epidemic St. Louis, California, and equine encephalitis viruses.[52] Chapter 6 describes transmission by arthropods in more detail, including a possible explanation for overwinter survival by transovarian or sexual transmission within mosquito populations.

8.10. Nosocomial Transmission

The unique populations and physical circumstances found in hospitals lead to transmission of infections by several of the foregoing routes. Viruses are estimated to

cause at least 5% of nosocomial infections,[268] but few systematic studies have been done because of the technical requirements, cost, and time involved. More than two dozen viruses have been documented as being nosocomially transmitted,[82,268] but the relative importance of only a few of these has been evaluated in broad surveys of hospital-acquired infection or even in special populations. Several groups of viruses have been implicated. These include respiratory viruses (especially CMV, HSV-1, HSV-6, and VZV), hepatitis viruses (including HAV, HBV, and HCV), enteric viruses (mainly rotavirus),[284] the viruses of several exanthemata (rubella and measles), and picornaviruses. Recent concern has centered on nosocomial potential risk of transmission of HIV between patients and personnel. Present clinical and serological evidence suggests that HIV infection is acquired in only about 0.3% of health care workers following a single percutaneous (usually needle stick) injury and more rarely by any other exposure route.[137,195,260] Prompt initiation of postexposure zidovudine prophylaxis has unfortunately not proved uniformly effective.[123] The Centers for Disease Control (CDC) has led the effort to collect information data about those risks.[195,260] Fortunately, the much-publicized cluster of AIDS cases in patients of a single dentist has remained a puzzling exception to the extremely unlikely occurrence of provider-to-patient transmission of the virus.[223] A few instances of transmission of a slow virus (Creutzfeldt–Jakob) and of rabies virus have been documented under special circumstances such as corneal transplants. In the appropriate tropical setting, Lassa fever, Ebola virus, and Marburg virus have been transmitted as nosocomial infections. Lassa fever, in particular, has infected patients and staff, especially in the obstetrical wards via infected placentas.

One of the most carefully studied viruses is respiratory syncytial virus (RSV), which appears to be the major nosocomial agent on some pediatrics wards, where it can infect both patients and staff; outbreaks among adults and elderly patients in health care facilities have also been reported.[134,135] Most studies of nosocomial RSV infection have centered on certain wards or population groups, occasionally in adults,[130,256] and more often in the pediatric population of a neonatal unit, the nursery, and the intensive care wards, where susceptibility, crowding, and often an immature immune system make these high-risk groups. In these settings, infections, diseases, and outbreaks of CMV, HSV, VZV, enteroviruses, myxoviruses, parainfluenza viruses, and especially RSV have occurred.[275] Among neonates, RSV infections can be severe and atypical, with a high mortality. In one prospective study, 45% of exposed infants hospitalized for a week or longer acquired RSV infection, and it involved 42% of ward personnel.[134,135] Parainfluenza and rhinovirus infections may also be widely spread in these settings and may involve transmission by personal contact as well as by fomites, thus emphasizing the need for good hand-washing techniques. Rubella has spread to other infants and to susceptible nursery staff; thus, vaccine protection of female staff of child-bearing age is of importance.

A second setting is the hemodialysis unit and laboratory, where staff are exposed by aerosol or blood to HBV, HCV, and probably to HIV. A third setting is on any crowded ward or intensive care unit where both susceptible staff and patients are at risk for influenza and other respiratory agents. For this reason, routine influenza immunization should be carried out yearly for all hospital personnel, not only for their own protection but also to prevent spread to patients. Fourth, patients receiving multiple transfusions or transplanted organs may be infected with the parenterally transmissible hepatitis viruses, CMV, HIV, and sometimes EBV. Immunosuppressed patients are especially subject to both primary and reactivated infections, among which the herpesviruses are the most common.

9. Pathogenesis

Since each chapter on specific viruses deals with the subject of pathogenesis, this discussion is limited to a general consideration of infections involving certain local or systemic features. Brief reference is made in Section 11 and elsewhere to some of the pathophysiological events following interaction between microbe and host (e.g., cellular trafficking, cell–cell communication, cytokine and receptor interactions, and intracellular peptide processing). Good general presentations can be found in books by Fields, Mandell, Mims, Notkins, Paul, Thomas, and others (see Suggested Reading).

9.1. Respiratory

Infectious particles may be implanted directly on nasal surfaces from contaminated hands or from large droplets or may reach the lower respiratory passage from aerosols. Since man continually samples the environmental air about 20 times a minute in breathing, it is no wonder that exposure to and infection with respiratory viruses are common indeed. Furthermore, only a small number of infectious particles need to be implanted in appropriate areas to induce infection. This is on the order of three particles for influenza A by aerosol, six for Coxsackievirus A21 by intranasal implantation, and seven for adeno-

virus 4 by aerosol.[177] In general, aerosol particles 3 μm in size reach the alveolus, and those 6 μm or greater are retained in the upper respiratory tract. The mucociliary epithelium transports particles up from the lung or down from the nasal mucosa.[201] To reach susceptible cells, viruses must pass through the mucus film and make physical contact with the cell receptors. The mucus contains mucopolysaccharide and other inhibitors, such as specific immunoglobulin A (IgA) antibody in previously exposed persons. Influenza virus is assisted in its spread by its own neuraminidase, which hydrolyzes the polysaccharides of the inhibitors; the virus attaches to cell receptors by means of surface hemagglutinin spikes. In the alveolus, small aerosol particles are ingested by macrophages, and some viruses are digested and degraded by these cells; other viruses are even capable of multiplication within macrophages themselves.

Most respiratory viruses produce illness through the direct consequences of local multiplication, although there is increasing evidence that cell-mediated cytolytic events play a role. Necrosis and lysis occur with desquamation of the respiratory epithelium. Constitutional symptoms may then result from breakdown products of dying cells that are absorbed into the bloodstream; fever is produced by the liberation of cytokines resulting from viral action on leukocytes. This sequence of events may be modified or altered by interferon production in infected cells, by the appearance or preexistence of secretory or local antibody, by the presence of preexisting or produced humoral antibody, or by prior priming of cell-mediated pathways. If humoral antibody is present in the absence of local antibody, then a more severe reaction may occur, possibly through antigen–antibody deposition on the cell membrane. The mechanism of this is not clear, but the phenomenon has been observed in infants with passively acquired maternal respiratory syncytial antibody who subsequently develop an infection with this virus. It has also been seen following parenteral administration of an inactivated vaccine that produces humoral antibody but little or no local antibody, such as experimental respiratory syncytial and early measles vaccines when followed by natural or purposeful exposure to live virus.[64]

The multiplication and effect of respiratory viruses such as influenza virus, parainfluenza virus, rhinoviruses, and respiratory syncytial virus are generally limited to the respiratory tract. Influenza virus has been detected in the blood only rarely[81] but has been isolated from the spleen, lymph nodes, tonsils, liver, kidney, and heart in fatal cases of Asian influenza pneumonia.[222] There are sporadic reports of the presence of influenza virus on peripheral leukocytes and in other organs, such as skin and muscle.

Systematic spread of this type appears to be unusual and associated with overwhelming viral infection.[266] More examples may come to light with more widespread use of immunosuppressive drugs. Adenoviruses and the enteroviruses multiply both in the respiratory tract and in the gut; viremia and secondary multiplication in the central nervous system are common in the latter group. Among the enteroviruses, however, only coxsackievirus A21 acts primarily as a respiratory virus, and its importance is limited mainly to military recruits. Enterovirus 70 causes acute hemorrhagic conjunctivitis, and the virus is present in the conjunctiva and throat (see Chapter 21).

9.2. Gastrointestinal

Hepatitis A and E viruses, enteroviruses, adenoviruses, reoviruses, rotaviruses, calciviruses, and astroviruses multiply within the gut. Many of the same barriers that prevent cell attachment and penetration may exist there as in the respiratory tract, including local IgA antibody. Local, humoral, and cell-mediated immunity follows natural viral infections of the intestinal tract and is the basis for immunity following oral administration of live vaccines such as poliomyelitis and adenoviruses 4 and 7. Unlike the case with respiratory viruses, local multiplication of many enteric viruses does not produce local symptoms; these occur only after implantation has occurred in secondary sites of multiplication such as the liver for hepatitis virus and the central nervous system for enteroviral infections. Other viruses found primarily in the gastrointestinal tract do produce disease that is largely confined to that organ system. They appear to account for a relatively minor proportion of the total burden of diarrheal illness.

9.3. Systemic Infections

Systemic infections involve viremia with or without additional spread along other routes. Spread via the bloodstream is the major route by which many viruses locate in secondary habitats, where their principal effects are produced. Some viruses become closely associated with lymphocytes in the bloodstream during the viremic phase and may persist there for years; these include CMV, EBV, human retroviruses, measles, and poxviruses. Some produce a chronic proliferative infection of B lymphocytes (EBV), and some infect T lymphocytes (HTLV-I, HHV-6); others cause destruction of CD4$^+$ lymphocytes (HIV and sometimes HTLV-I). Some circulate, at least intermittently, in the form of free virus in plasma (arboviruses, enteroviruses, hepatitis viruses, HIV) or as immune complexes. Some have a special affinity for red blood cells (Colorado tick fever and Rift Valley fever viruses). Viremia may be maintained by continual or intermittent seeding from the liver, spleen, bone marrow, and other organs. The persistence of CMV, EBV, human retroviruses (especially HIV), and hepatitis viruses in the blood for years poses a hazard in their transmission via blood or blood products. Most of these occur when the viruses circulate in the presence of antibody. Persistent antigemia may result in other consequences. Immune complexes may form, deposit, fix complement, and cause local tissue injury, especially in small blood vessels as in HBV and periarteritis nodosa; HBV antigemia may also result in hepatocellular carcinoma with or without an intervening cirrhosis. Prospective studies in Taiwan have shown an enormously increased risk of hepatocellular cancer in those with antigemia as compared with those without (see Chapter 32 and references 18 and 19).

9.4. The Exanthem

The list of viral infections associated with an eruption of the visible surfaces of the body is long.[65] Our understanding of the pathogenesis of systemic viral infections associated with a rash such as the pox group, measles, and rubella has been enhanced by the fine studies of Fenner with mousepox.[118] In each such exanthem, there is an incubation period of 10–12 days before symptoms of illness appear. After multiplication of the virus at the site of implantation and in the regional lymph nodes, a primary viremia occurs within the first few days, resulting in seeding of organs such as the liver and spleen. A secondary viremia then follows, with focal involvement of the skin and mucous membranes, the appearance of a rash, and the onset of symptoms. In mousepox, a primary lesion then develops at the site of inoculation. Although the destruction of cells involved in viral multiplication and the release of pyrogens from leukocytes or other cytokines may be responsible for symptoms such as fever, the appearance of antibody at this time suggests that antigen–antibody complexes may play an important role in the pathogenesis of the rash. The viruses of smallpox, HSV-1 and -2, and VZV are present in the skin vesicles of each of these diseases.

9.5. Infections of the Central Nervous System

Comprehensive reviews of the pathogenesis of viral infections of the central nervous system (CNS)[127,154–156,159] emphasize that one or more routes of infection may be involved and that the pathways differ with the particular viruses, the host, and the portal of entry. Reviews of the pathogenesis of infection with representative arthropod-borne viruses provide further perspective.[129,228] In man, the hematogenous routes to the CNS from the portal of entry and from primary multiplication sites in the gut, respiratory tract, parotid, or lymph nodes are clearly of importance in enteroviral infections, mumps, lymphocytic choriomeningitis, primary herpes simplex infections, HIV infections and fetal infections with rubella virus and CMV. Secondary multiplication sites in the liver, spleen, muscle, or vascular tissue may augment or maintain the viremia; the brown fat has also received attention in this regard for a variety of viruses. Several mechanisms have been suggested as to how viruses enter the brain from the bloodstream. This may be a passive process, or the viruses may actually grow their way through the choroid plexus. Viral multiplication at this site or leakage into the cerebrospinal fluid (CSF) following growth in the meningeal cells may explain the presence of echovirus and coxsackievirus in the spinal fluid during CNS infections; the presence of viral-specific IgM antibody in the spinal fluid usually indicates active viral multiplication in the CNS.[154,155]

The blood–brain barrier is represented morphologically by the cerebral capillaries, whose endothelial cells lack fenestrations, are joined by tight junctions, and are surrounded by dense basement membranes.[155] This barrier inhibits viral invasion of the CNS and may deter viral clearance. The blood–brain barrier also isolates the CNS from systemic immune responses in the absence of disease, and in normal persons the immunoglobulins present in the CSF are derived from the blood and are dependent on the size of the immunoglobulin molecule: IgG and IgA are present at about 0.2 to 0.4% of the plasma levels, and IgM at a lower level.[155] During an inflammatory process, there is a change in the blood–brain barrier allowing transudation of serum proteins, including immunoglobulins. Once plasma cells are recruited to the CNS, synthesis of immunoglobulins occurs in the CNS, but of limited heterogeneity. T lymphocytes are also recruited that are sensitized to the invading virus and release lymphokines that attract macrophages; these constitute the majority of cells in the inflammatory response.

Neural spread along nerves can occur in rabies, poliomyelitis, and B virus infections of man. In rabies, it appears to be the predominant, if not the sole, method of spread to the CNS, whereas it seems to be relatively unimportant in poliomyelitis. The axons, lymphatics, and tissue spaces between nerve fibers represent three possible conduits for spread along the neural route. Transmission via the tissue spaces plus direct infection and involvement

of endoneural cells seem the most likely mechanisms. Spread along the olfactory pathway has also been experimentally demonstrated for poliomyelitis, herpes simplex, and certain arthropod-borne viruses. The role of this route in natural infections is uncertain. As with respiratory viruses, those that infect the CNS have different cell preferences: poliomyelitis has a predilection for anterior horn cells of the spinal cord and the motor cortex of the brain, and arboviruses have a predilection for cells of the encephalon. Herpes simplex appears to have more catholic tastes and multiplies in a wide variety of cell types. As is also true of respiratory cells, the existence of specific cell receptors for individual viruses may play a crucial role in susceptibility.

At different stages of HIV infection, different sites within the CNS are also damaged through various pathophysiological mechanisms initiated by the virus. The pathogenesis of AIDS dementia, a distinctive form of encephalopathy, and other CNS manifestations have been reviewed in recent clinical neurology texts and articles.[23,158] HTLV-I also produces distinctive forms of CNS disease, most notably spastic paraparesis (see Chapter 25 and reference 189).

9.6. Persistent Viral Infections

The pathogenetic mechanisms discussed thus far have dealt with infections in which an acute illness results, usually after a relatively short incubation period (except for rabies), and in which recovery ensues. The virus disappears and is often eliminated from the body. Another pathogenetic mechanism under increasing study is one in which the virus persists for months or years and may result in delayed host responses. Some of these persistent viruses are also capable of evoking an acute response; these include the herpesviruses, rubella virus, the adenoviruses, measles and other paramyxoviruses, HBV, and HIV. Other persistent viruses such as papovaviruses and polyoma viruses rarely produce any acute illness. Acquisition early in life, as with EBV, HBV, and HIV, or in other states of immunodeficiency may also occur silently and lead to persistence.

In persistent viral infections of the CNS, prolonged synthesis of specific IgG immunoglobulins may be found, as in subacute sclerosing panencephalitis or rubella panencephalitis. Included in this group of potentially persistent viral agents are the herpesviruses, adenoviruses, papovaviruses, paramyxoviruses, rhabdoviruses, retroviruses, coronaviruses, arenaviruses, togaviruses, and picornaviruses, all of which have been shown capable of long-term neural infections.[154]

Still other agents called "slow viruses" produce chronic degenerative disease years after exposure. This group includes kuru and Creutzfeldt–Jakob disease of man; Gerstmann–Sträusler–Scheinker syndrome; fatal familial insomnia in humans; scrapie disease of sheep; transmissible mink encephalopathy; chronic wasting disease of mink, deer, and elk; and bovine spongiform encephalopathy and exotic ungulate encephalopathy. Increasing evidence indicates that these diseases are actually caused by a new form of infectious agent called proteinaceous infectious particles, or "prions," rather than viruses.[231] These transmissible particles are apparently composed largely of an abnormal isoform of the prion protein, which is encoded on a chromosomal gene and may thereby be responsible for neurodegenerative diseases that are both transmissible and genetic in origin (see Chapter 34 and references 157 and 231).

Factors that favor persistence of certain viruses have been summarized by Mims[201]: (1) persistent viruses tend to have low or no pathogenicity for the cells they infect, in contrast to viruses with severe, destructive effects, which induce acute disease terminated by death or by recovery and the elimination of the virus; (2) there may be an ineffective antibody response possibly because of tolerance, autoimmunosuppression, production of nonneutralizing or blocking antibodies, not enough antigen on the surface of the infected (target) cell to induce adequate antibody formation, or spread of the virus directly from cell to cell where antibody does not reach it; (3) there may be an ineffective cell-mediated immune response for reasons similar to those involved in the poor antibody response [tolerance, autoimmunosuppression, blocking antibodies, too little antigen expressed on surface to infected cell, failure of immune cells to reach infected (target) cells]; (4) there may be a defective interferon response, such as in lymphocytic choriomeningitis in mice; other viruses may be relatively insensitive to interferon action even though it may be produced; (5) certain persistent viral infections induce neither an immune nor an interferon response; these include the "slow virus" infections such as kuru and Creutzfeldt–Jakob disease; and (6) lymphocytes and macrophages are often infected in persistent viral infections, such as with adenoviruses, EBV, CMV, and measles virus, thus altering the host's immune response. Several of the lentiviruses, including HIV, have recently become the most prominent examples of this last phenomenon. Interferon produced by infected macrophages may have no protective effect on other macrophages, although there is normal activity on normal cell types; certain virus–antibody complexes still remain infectious after phagocytosis by macrophages; infected macrophages may be less active in releasing the same virus from the blood, thus favoring persistent viremia.

Such persistent and latent viral infections may reactivate, producing the acute disease again, or may result in a chronic viral infection manifested by immune complex disease, degenerative diseases of the CNS, or certain malignancies. These infections will acquire greater visibility and importance as immunosuppressive drugs are used more widely in medical therapy and in organ transplant recipients. They now include the consequences of AIDS in which acute, chronic, and malignant manifestations may appear as a result of the reactivation of many types of microorganisms, including viruses, some of which are nonpathogenic in the normal host. Certain genetic disorders involving the immune system can also result in persistent and/or reactivated viral infections or in aberrant responses to them such as the X-linked lymphoproliferative syndrome.[254]

10. Incubation Period

The period from the time of exposure to the appearance of the first symptoms is called the *incubation period*. Viruses that do not require distant spread but are able to produce disease through multiplication at the site of implantation, such as the respiratory tract, have short incubation periods of the order of 2–5 days. Those that require hematogenous spread and involvement of distant target organs such as the skin or CNS have incubation periods of 2–3 weeks.

With HIV infection, a primary clinical response may occur within the first 2 months or so after infection in a substantial proportion of newly infected persons.[72,121,258] The syndrome resembles mononucleosis with fever, malaise, lymphadenopathy, rash, headache, neck and muscle ache, and other features. The majority of infected individuals then enter a period of clinical but usually not biological latency (see Chapter 24). During that period the $CD4^+$ lymphocytes are destroyed at highly variable rates. Accordingly, clinical latency continues until significant enough immunosuppression permits new opportunistic pathogens to superinfect or long-latent agents to reactivate and produce clinical disease.

Viruses such as rabies, dependent on spread along nerves, have very long and variable incubation periods ranging from 8 days to a year or more. The variation in incubation periods in different diseases is indicated in Fig. 1. In some diseases, early symptoms or even a rash may accompany the period of initial invasion or viremia. This has been seen in poliomyelitis, dengue, hepatitis, infec-

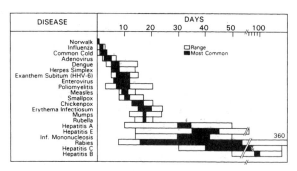

Figure 1. Incubation periods in viral diseases. Based mostly on data from Benenson.[22]

tious mononucleosis, and the acute syndrome of primary HIV infection. In such instances, the apparent incubation period to the appearance of these early features is much shorter than the usually accepted period; more often, this early phase is not clinically recognized or occurs before the patient visits the physician.

Knowledge of the incubation period has many practical uses. Epidemiologically, it helps define the period of infectiousness: a patient is not usually infectious until close to the time of the appearance of clinical symptoms. In epidemics, knowledge of the mean, minimum, and maximum incubation periods can be used to identify the probable time of exposure to the index case or other source of infection. The duration of infectivity depends on the persistence of the virus and its exit into the environment. Clinically, the duration of the incubation period helps to identify the likelihood of viral exanthem after a known exposure or to differentiate hepatitis A from hepatitis B infections. Prophylactically, it determines the feasibility of prevention of the clinical illness by immune serum as in hepatitis A, varicella-zoster infections, rubella, and rabies, as well as the potential success of rabies vaccination.

In addition to the viruses that produce acute infections, there are delayed effects of certain common viruses in which the "incubation period" represents a true or apparent interval of "latency" lasting several to many years, during which there is little if any viral replication. Examples include the relationship of measles virus to subacute sclerosing panencephalitis, in which infection in infancy may be associated with involvement of the CNS some 5–10 years later.[38] Certain papovaviruses cause widespread inapparent infections in childhood. Rarely, reactivation occurs later in life in the form of progressive multifocal leukoencephalopathy. This is seen in patients with Hodgkin's disease in association with depression of

cell-mediated immunity and more recently in AIDS patients[60,144] (see Chapter 30). In kuru, the period from exposure by ingestion of infected brain or other tissues or by absorption via abraded skin at a cannibalistic feast ranged up to 27 years or more.[175]

For HIV and perhaps other retroviral infections the agent is often not truly latent for long and this quiescent period is quite variable. Immunologic deterioration in HIV infection begins relatively early in the course and proceeds at different rates among individuals. The average time from initial HIV infection to that clinical event may be shorter or longer depending not only on the degree of immunosuppression but also on the other cofactors required for any specific clinical AIDS outcome. For example, Kaposi's sarcoma tends to occur at a somewhat earlier stage of immunosuppression than cerebral atrophy with dementia or lymphoma for reasons presumably related to the unknown determinants of these conditions. Besides differences in the properties of the virus itself, such as the capacity to penetrate cells or induce syncytium formation *in vitro*, the route of transmission and host factors determine the rate of progression of HIV infection. The average AIDS-free interval is shorter for infants and children, longest among the youngest adults, and then gradually shorter with increasing age.[238] In addition, there is increasing evidence for an immunogenetic predisposition. As with HIV infection, the latency period for expression of HTLV-I varies depending on the clinical outcome and other factors. For example, in adult T-cell leukemia/lymphoma, the interval from presumed infection via breast milk to the appearance of tumor differs from the interval between presumed sexual transmission and onset of tropical spastic paraparesis or myelopathy. The concept of a latency period is also applicable to long-delayed virus-induced cancer as seen with HBV and hepatocellular carcinoma and EBV-induced nasopharyngeal carcinoma (see Chapters 31 and 32).

11. Immune Response

The human immune system has elaborate mechanisms for defending against viral infections and their pathogenic consequences. The system consists of specialized cells producing molecules of many different classes, some of which demonstrate an extraordinary degree of variation. The distinctive cell types are distinguished by their combination of surface structures that determine how they interact with contiguous cells, by their secretion of specific cytokines and other substances that modulate functions of noncontiguous cells, by the antibodies and other proteins they generate to bind foreign antigenic material, and by other even less well-understood processes. Thus, paradoxically, the remarkably individualized molecular profiles confer remarkably broad population responsiveness. Both the success and the failure of the immune system at controlling viral infection derives from the exquisite specificity involve in these myriad interactions. Several excellent books have effectively organized the rapidly expanding knowledge of the immune system (see Golub, Paul, and Roitt, Suggested Reading); a particularly clear and thorough work is Paul's *Fundamental Immunology*. However, at the present pace of discovery in certain areas, the currency of textbooks may be particularly foreshortened.

The immune system comprises a few main classes of cells and a much larger variety of cell subsets. Lymphocytes provide direction for the main activities of the system and, for the most part, govern the specific nature of the immune response. They originate or develop in the bone marrow (B lymphocytes) or thymus (T lymphocytes). Other cells include those of the circulating monocyte or tissue macrophage line, dendritic and Langerhans' cells, natural killer cells, mast cells, and basophils. These cells originate, develop, and migrate to or reside in many organs, but thymus, bone marrow, lymph nodes, spleen, and mucosal surface clusters are the main locations.

11.1. B Lymphocytes and Humoral Immunity

The B lymphocytes are responsible for humoral immunity, i.e., production of antibodies in the form of immunoglobulin (Ig) that remains on the cell surface or circulates in the blood. The sequence of events leading to a mature, Ig-producing cell and the highly specific antibody structure have been relatively well characterized. Depending on the nature of the antigen, B cells are activated by binding to circulating antigen that is free or attached to a particle like a virus. The binding and interaction lead to further secretion of cytokines that regulate the B cell synthetic and transport machinery. Most B cells then either continue to secrete antibody or differentiate into memory cells in the environment of lymphoid organs. Other B cells may have distinctive functions.

The genetically programmed capacity to produce almost limitless variation in antibodies through simple structural rearrangements accounts for the host's great versatility in its response to foreign antigens. The sequence of events in the synthesis of antibodies has been studied in exhaustive detail (chapters in the texts by Paul and Roitt provide useful reviews; see Suggested Reading). Briefly, as B cells mature, they produce heavy and light chains. Two heavy chains and two light chains of either the κ or λ form combine to form the Ig molecule. Immunoglobulin chains are composed of several distinct regions encoded by genes capable of rearranging by an orderly process of translocation and deletion. Five different major Ig isotypes—M, D, G, E, and A—are produced by different B lymphocytes/plasma cells. Switching is induced by various cytokines, which may act in a type-specific manner, e.g., interferon gamma may stimulate switching from IgM to production of IgG of a particular subclass. The antigen-reactive ends of the two heavy and light chains together form the Fab fragment, where the specificity for individual antigens resides. The Fc fragment, consisting of the ends of the constant regions of the two heavy chains, binds to complement and other cell surface receptors, thereby bringing antibody-coated microorganisms to the surface of phagocytic cells.

Immunoglobulins of the IgM class appear in response to the primary infection, are of relatively short duration, over 3–4 months, and are commonly taken as a marker of a recent infection. They may reappear in lower titer in some reactivated viral infections, especially in the herpes group, and may sometimes persist over longer periods as in congenital infections or in the brain in measles infections early in life that may lead to subacute sclerosing panencephalitis. They remain confined to the vascular system except where they are produced locally as in infections of the CNS. Molecule for molecule, they have five times the number of antigen-reactive sites as IgG; the IgM molecule also has five times the number of Fc sites and therefore five times greater capacity to activate complement.

The IgM antibodies are the first to appear in response to initial infection, and thus may play a determining role in the course of the initial infection; they are then followed by appearance of IgG antibodies. Virus-specific IgG is the major circulating isotype and usually persists at some level for life. The IgG molecules cross the placenta, conferring temporary immunity on the newborn.

The presence of these antibodies in the blood constitutes a major deterrent to the spread of viruses to distant sites. They constitute the major basis for vaccines that induce humoral immunity and prevent the development of clinical illnesses dependent on viremic spread, such as poliomyelitis, hepatitis, and the viral exanthems. Infections characterized by viremia also produce the most marked and longest-lasting humoral antibody response.

In some but not all viral infections, the neutralizing (and presumably protective) capacity of these antibodies may be a function of their antigen-binding affinity, although the data supporting this concept are inconclusive.

Passive transfer of convalescent serum or immune globulins may also produce protection against those infections that are dependent on a viremia to reach target organs for their clinical expression. The amount of protection reflects the antibody level of the donors of the blood from which the serum or globulin is derived, and passive antibody does not protect against multiplication of the virus at the site of initial implantation in the respiratory or gastrointestinal tract. Lately there has been interest in the phenomenon of antibody-dependent enhancement of viral infection, best typified by the experience with dengue, in which reinfection precipitates a more severe illness in previously exposed (seropositive) individuals than is usually seen among those not previously infected.

11.2. Local Immunity (Mucosal Secretory IgA System)

Although ultimately under the same complex regulatory control of different cell types and cytokines, B cells secreting IgA are most immediately responsible for the phenomenon of local or mucosal immunity. Local immunity to virus is conferred by virus-specific secretory IgA found primarily on mucosal surfaces and in breast milk. They are critical in resistance to infection in the respiratory, intestinal, and urinogenital tracts. Their production follows natural infection or administration of live vaccines by the natural portal of entry. Less effective production occurs with live vaccines given parenterally, and usually poor responses occur with killed vaccines given parenterally; similarly, administration of passive antibody or immunoglobulin does not provide local immunity. This leaves these unprotected mucosal surfaces susceptible to primary infections with viral agents entering these portals. Although the individual may be protected against clinical illness by humoral antibody, epidemiologically the multiplication and excretion of the organism provides a continued method of spread in the community. Thus, protection against clinical disease but not against infection for the individual or his or her contacts may result from vaccines that fail to induce a satisfactory secretory IgA antibody response. In the submucosal tissues, antigens combine with specific antibody, form immune complexes, enter the blood, and are then filtered out and excreted in bile. The IgA molecule lacks the secretory piece and, after responding to local antigens, enters the blood via lymphatics to give increased serum IgA levels. The IgA-producing cells (B lymphoblasts) may also be carried via lymph and vascular routes to areas other than the local site such as the salivary glands, lung, mammary glands, and intestine or even to other sites of the same organ, i.e., other parts of the intestinal tract. There, they may be active in preventing local infection in these new locations.

11.3. Complement

A set of enzymes, receptors, and other proteins form the complement system. These proteins interact with each other in a well-defined pattern to initiate or mediate a number of specific functions at the cell membrane (e.g., attachment, lysis).

11.4. T Lymphocytes and Cell-Mediated Immunity

These cells perform key regulatory functions of the immune system. Through their T-cell receptors, the CD4 and CD8 molecules that distinguish major subsets of T lymphocytes and other surface proteins that function in concert with the T-cell receptor, these cells participate in the increasingly well-understood critical events of the immune response: interaction with antigen-presenting cells, activation, secretion of an array of cytokines, proliferation of other specialized response cells, and cytolytic destruction of target cells expressing foreign antigens. An important advance in our understanding of these events has been the rapid elucidation of the structure and function of the major histocompatibility complex (MHC), the T-cell receptor, and a number of the other cell surface components (e.g., CD3, CD4, LFA) (Fig. 2).

All nucleated mammalian cells carry genetically determined MHC molecules: HLA in man (Fig. 2). One set of cells, the macrophages and other antigen-presenting cells, use these molecules as carriers for viral and other peptides. When viruses enter these cells by direct attachment and fusion with the cell membrane or by receptor-mediated endocytosis, their antigenic proteins are digested internally. These constituents are degraded into peptide fragments within a protected intracellular compartment and then attached to and transported by HLA and related antigens. These carriers and other components assembled in a carrier protein complex enable antigen-presenting cells to manifest virus peptide fragments on their surface, displaying them to T lymphocytes [either to CD4$^+$ (helper) cells in the context of class II HLA molecules or, more likely for synthesized viral protein, to CD8$^+$ (cytotoxic) cells in the context of class I HLA molecules]. The highly variable genetically determined structure of the HLA and other surface markers, as with the Ig molecules secreted by B cells, equip human and other mammalian hosts for the great variety and specificity of reactions to foreign material they encounter. As a result of positive and negative selection in the milieu of

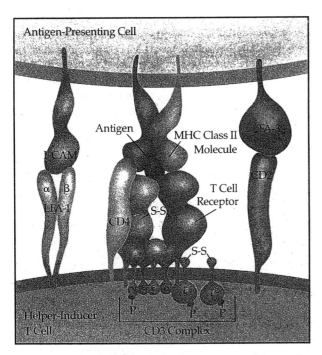

Figure 2. Principal cell surface molecular components of the complex involved in antigen presentation and host recognition.

the lymphoid tissue, the repertoire of the antigen-presenting cell and T-cell markers responsible for that great variety of reactions is limited; cells that tend to react to the human host's own peptides are eliminated, while cells that recognize foreign peptides presented by their own HLA markers are preserved. Work done in transgenic mice and other ingenious experimental systems has both supported that model of selection and begun to reveal the complexity of the events involved.

T-helper (CD4$^+$) cells appear to function by activating B cells, monocytes–macrophages, and other T-helper cells; by binding directly to those cells (e.g., through T-cell ligand–B-cell receptor interaction); or by secreting interleukins–cytokines that stimulate cell proliferation or regulate production of substances such as other cytokines, Ig, or still other effectors (e.g., tumor necrosis factor). A number of lymphokines and colony-stimulating growth factors are produced by different T-cell subsets and by other cells of the immune system. Some act on cells in the immediate vicinity and others have more remote effects. Cytotoxic T lymphocytes (usually CD8$^+$) lyse other cells that contain foreign protein, particularly viral protein synthesized endogenously. When the cytotoxic T lymphocyte response is confined to the relatively few infected cells, their destruction is clearly beneficial to the host; however,

more extensive damage may be harmful to the host in the short run. In addition to direct destruction of infected cells, lymphocytes may kill adjacent uninfected cells or retard the spread of virus by interrupting contiguous cellular connections.

Natural killer cells resemble lymphocytes but have some distinctive properties, such as expression of a specific receptor for the Fc portion of IgG. Under the right conditions, they kill virus-infected or neoplastic cells by receptor binding and secretion of interferon gamma (IFN-γ), especially when induced to do so by tumor necrosis factor and other cytokines produced by macrophages, even in the absence of specific antigen stimulation.

Macrophages and related monocytes perform the critical function of processing and delivering antigen for recognition by the lymphocyte. The macrophage is most effective at destroying virus when the particle is incorporated in the cell. There its antigenic peptides can be bound to MHC molecules intracellularly and then presented to T-cell receptors of T lymphocytes, whereupon IFN-γ secretion by the T cell in turn activates the macrophage's toxic oxygen and enzymatic pathways.

Granulocytes are involved in a wide range of functions related to the inflammatory response. The polymorphonuclear leukocytes ingest and kill extracellular bacteria but have a less well-defined role in protection against viral infection.

11.5. Viral Immunopathogenesis

Although each pathogenic virus expresses itself in a characteristic if not distinctive manner immunologically and clinically, there are some general principles that underlie these observed patterns of expression of virus–host interaction. The following examples have been drawn from accompanying chapters as well as the Thomas text and other references.[182,242,288] Some viruses replicate rapidly and destroy host cells efficiently; others continue to grow and spread slowly, causing only mild or intermittent damage in the host; still others may become dormant for months or years, after which disease may occur in the form of neoplasia or degenerative phenomena or not at all. During the primary response, in whatever form the clinical expression takes, the immunologic events involve recognition of viral antigen by the host. These events are mediated by immunoactive cells—antigen-presenting cells (e.g., macrophages) and T cells that are mobilized within 5–8 days after infection—and by antibodies generated during the 2–3 weeks following infection by B cells that show increasing specificity for core, capsid, or envelope epitopes.

Viruses probably enter cells by attachment to the target cell membrane followed by either direct penetration or incorporation into vesicles by receptor-mediated endocytosis. The mode of cell entry may, in turn, determine subsequent events because extracellular proteins entering by receptor-mediated endocytosis tend to activate the MHC class II and CD4$^+$ cell response pathway, whereas the viral proteins synthesized within the cell tend to activate the class I–CD8$^+$ cell response.

One common response pattern occurs with herpesviruses,[242] influenza,[198] HIV,[272] and others. By directly infecting antigen-presenting cells or delivering their antigens to these cells, such viruses provoke a cascade of immunochemical events often culminating in a strong cytopathic effect and prompt clinical consequences. This brief, highly effective, and well-characterized response is usually mediated by CD8$^+$ cytotoxic T lymphocytes, although CD4$^+$ cytotoxic T lymphocytes are also important in herpesviruses,[242] rotavirus,[220] influenza A,[35] and other infections; noncytolytic CD8$^+$ cells may also play a role in HIV immunopathogenesis.[192] The different cell subsets may recognize different antigenic components of the virus. In this milieu, virus may also be killed directly by IFN-γ and other cytokines secreted by the lymphocytes, suggesting that in the control of viral infection, the so-called Th1 pattern of cytokine production by subpopulations of T-helper cells is dominant over the Th2 pattern, contrasting with the response to helminthic infection.[227] Other forms of cytotoxicity have been demonstrated in experimental systems. However, what roles, if any, natural killer cells or antibody-dependent cell-mediated cytotoxicity play in the protection of the human host from serious damage remains uncertain. Nor is it yet known whether macrophages or their soluble products are as directly involved as cytotoxic T lymphocytes in containment and elimination of viruses.

Many viral antigens stimulate antibodies, but only certain antibodies to certain antigens are protective. In HSV infections, for example, the mere presence of neutralizing antibody in the circulation does not protect against the development of typical lesions[290] and antibody levels do not change significantly when lesions occur. Primary as well as secondary infections may elicit both T-cell-dependent and -independent neutralizing antibodies that are involved in protection of the host. Evolution from the initial IgM to a predominantly IgG response is also largely T cell dependent. In the circulation these antibodies may not prevent reinfection, but rather control the systemic spread of the infection and suppress cytopathic effects. Alternatively, containment may take place predominantly by specialized antibodies (e.g., IgA) produced locally.

Another response pattern is seen in infections with rabies and other neurotropic viruses, which penetrate immunologically inert cells and may elicit little or no cytopathic immune response for a relatively long interval following infection. In these situations the host may fail to control or eliminate the infection, either because virus sequesters itself in privileged sites where it can replicate freely until its destruction of infected cells finally exposes immune mechanisms to antigenic material or because it remains largely or completely inactive for years. In either case the host remains unable to recognize antigenic products on cell surfaces in the usual manner. In these situations the precise immunopathologic consequences (e.g., transformation, immune complex disease, autoimmune attack) probably depends heavily on the completeness with which the virus has usurped control of cell function or on the balance between antigen production and elimination. Still another mechanism by which the host response may paradoxically exacerbate the pathogenic process is antibody-dependent enhancement.[205]

12. Patterns of Host Response

The host responses to viral infections vary along a biological gradient in terms of both the severity and the nature of the clinical syndrome produced.

Although the emphasis on the biological gradient presented here is on the entire host, it is clear that different qualitative and quantitative responses may occur at the cellular level. Biochemical changes in the molecular composition of the virus, even a change in a single nucleotide, may alter its effect on susceptible cells, and genetic alterations in the host cell affecting the presence or absence of specific receptors for viral attachment and entry and probably the internal assembly and release of the viral particle may affect the nature and gradient of the cellular consequencs of viral infection.

12.1. The Biological Gradient

The host response to a virus may range from a completely inapparent infection without any clinical signs or symptoms to one of great clinical severity, even death. The ratio of these inapparent (or subclinical) to apparent (or clinical) responses varies from one virus to another; representative examples are shown in Table 4. At one end of the spectrum are certain infections that are almost completely asymptomatic or unrecognizable in their pattern until some special event provokes a clinical response. The response may appear long after the initial infection and arise from viral persistence or reactivation or both.

Table 4. Subclinical/Clinical Ratio in Selected Viral Infections (Inapparent/Apparent Ratio)

Virus	Clinical feature	Age at infection	Estimated subclinical/ clinical ratio	Percentage of infection with clinical features
Poliomyelitis	Paralysis	Child	±1000:1	0.1–1
Epstein–Barr	Heterophil-positive infectious mononucleosis	1–5	>100:1	1
		6–15	10–100:1	1–10
		16–25	2–3:1	35–50
Hepatitis A	Jaundice	<5	20:1	5
		5–9	11:1	10
		10–15	7:1	14
		Adult	2–3:1	35–50
Rubella	Rash	5–20	2:1	50
Influenza	Fever, cough	Young adult	1.5:1	60
Measles	Rash, fever	5–20	1:99	99+
Rabies	CNS symptoms	Any age	0:100	100

The BK and JC strains of papovavirus fall into this category: no known clinical disease has been associated with the initial infection, which is a common one in normal school children and adults, as reflected by high prevalence and rates of acquisition of antibody to the virus.[144,245] In addition to the inapparent or trivial infections occurring sporadically in normal individuals, primary or more often reactivated infection in immunocompromised patients (e.g., with AIDS, Hodgkin's disease, or renal transplantation) may develop a progressive multifocal leukoencephalopathy. The virus can be isolated from the brains of such persons, and high antibody titers may be present if the person survives long enough.[119,144]

A second group of viral infections are those that are predominantly mild or asymptomatic when exposure and infection occur in early childhood but that frequently result in symptomatic and sometimes severe clinical disease when infection is delayed until late childhood and young adult life. Examples of this are viral hepatitis, poliomyelitis, and EBV infections.

At the other end of the spectrum are infections caused by measles, rabies, and Lassa fever viruses, in which clinically recognized illness usually accompanies the infection. Indeed, in rabies infection of man, death is almost inevitable after characteristic symptoms develop.

The subclinical–clinical ratio for HIV infection defies the more straightforward categorization possible for many other viral infections. A syndrome resembling mononucleosis with fever, fatigue, headache, lymph node swelling, joint and muscle aching, rash, sore throat, and other features occurs frequently in newly infected individuals. However, the proportion who are reported to have experienced this syndrome is higher or lower depending on the method of ascertainment (e.g., presentation at an sexually transmitted disease clinic, follow-up of cohort of

initially uninfected homosexual men) and the clinical definition.[121,168,258] It is not yet clear whether the occurrence of this prodromal illness heralds earlier onset of opportunistic disease.[168]

The vast majority of HIV-infected individuals traverse a highly variable period—years more often than months—free of any serious illness. However, laboratory evidence usually indicates that immunologic deterioration is continuing at some rate even when the infection is clinically silent. With up to 16 years of follow-up available on some men in carefully followed cohorts, it now appears that most persons with uncontrolled HIV infection will eventually lose enough T-helper lymphocytes to permit emergence of life-threatening opportunistic diseases.[40,126,212,238] Age,[238] mode of transmission,[9,126,141] and HLA genes[166] are among the factors that appear to determine how rapidly immunodeficiency progresses and clinical illness supervenes. It has been estimated that, on the average, the median time to the first opportunistic condition formally defined as AIDS is about 10 years (i.e., it occurs at a rate of about 5% per year) in homosexual men infected around age 30.[141] The deterioration proceeds considerably faster in adults infected after the age of 40[9] and in infants and children with perinatally acquired infection (upward of 10–15% per year) but somewhat slower in younger adults. Specific alleles encoded in the HLA gene complex appear to accelerate or retard the process of immune destruction by several years.[165,166,174] Information derived from assessing genetic profiles of rapid and slow progressors may prove valuable for elucidating the determinants of the natural history, developing specific control measures, and predicting response to them.

In Africa and other less-developed areas, inadequate diagnostic facilities leading to substantial underascertainment and underreporting of serious HIV-induced illness in the face of high prevalence of infection probably accounts for a relatively high subclinical–clinical ratio. However, differences in the proportion of clinical cases by race and geography could also be due in part to differing distributions of the major genetic determinants or other host characteristics like nutritional status. Accumulating experience with the protease inhibitors and the other major antiretroviral agents, despite viral resistance and side effects among recipients, along with increased understanding of the insidious pathogenetic process, has generated optimism that interventions available now or on the immediate horizon can significantly improve the course of the infection, change the very low subclinical–clinical ratio, and alter the ultimately fatal outcome.

This biological gradient of host response is often pictured as an iceberg in which clinically apparent illness— i.e., above the water line—represents only a small proportion of the response pattern and the larger amount represents unrecognized and inapparent infections; a similar analogy may exist at the cellular level. Figure 3 portrays these concepts. The cellular responses shown might better be considered as differences in the nature rather than the severity of the response.

12.2. Clinical Syndromes: Frequency and Manifestations

The nature and severity of the host response vary widely in viral infections, even with the same virus. These various clinical responses may reflect variations in the strain of virus, even minor biochemical differences, different organ tropisms of the virus, different portals of entry, different ages at the time of infection, variations in the immune response, and differences in the genetic control of the characteristics of the agent and of the immune response of the host to it. The clinician faced with the diagnosis and management of a patient presenting with clinical syndromes involving various organ systems, or with a rash, may have great difficulty in making an etiologic diagnosis based on clinical features alone, even in distinguishing between viral and bacterial infections. This is because these target organs have only a limited number of ways to respond to infection, and any one of several viruses or other causative agents may trigger the same general response pattern. These causes will also vary with age, season, year, and geographic setting. The results of specific viral isolations and of serological tests may come too late to be useful during the acute illness, although advances in rapid, direct identification of many viral agents and demonstration of virus-specific IgM antibody are rapidly changing this situation. Such tests are often available in special laboratories.

Early knowledge of the viral etiology of a syndrome may avoid misuse of bacterial antibiotic therapy and for some viral infections may allow selection of an appropriate antiviral compound. Prevention of infection in exposed and susceptible contacts may also be possible. Often, however, the physician must rely on epidemiologic and clinical features and simple laboratory tests in making a tentative etiologic diagnosis. This diagnostic reasoning is based on the known frequency of a given causative

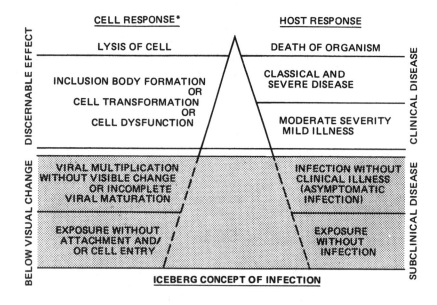

Figure 3. "Iceberg" concept of infectious diseases at level of the cell and at level of the host. Within any cell population, varying patterns of cell response also occur. *, hypothetical.

agent in that year, season, age group, or special setting and its epidemic behavior. The following sections present some of the etiologic agents involved in common clinical syndromes. However, the frequency distributions are generalizations that are unlikely to apply to such special settings as nursing homes and day-care facilities.

12.2.1. Common Respiratory Tract Infections. A great many viruses and viral groups can evoke respiratory symptoms and diseases, as can bacteria, rickettsiae, and certain fungi. The viral causes vary from season to season, from year to year, from place to place, between and within countries, and especially from one age group to another.[176] The etiology differs between infancy and childhood. Few studies have evaluated both viral and bacterial causes in the same population group at the same time and setting. Although studies of viral etiology have included minor and major illnesses within the family, community, and hospital, those of bacterial origin have focused mainly on the more severe and hospitalized cases. Lung aspirates have given different results from those of sputum or of oral/pharyngeal washings.[173] Thus, the generalizations made in this discussion must be accepted with caution. In general, however, the great majority of respiratory illnesses in infants, children, and young adults are caused by viruses with the exception of *M. pneumoniae* in older children and young adults. In the more severe and hospitalized cases, and in persons over 50 years of age, bacterial infections play the predominant role.

A number of investigators have tried to sort out the predominantly viral etiology of clinical syndromes of acute respiratory diseases in different age groups and population settings.[22,74,92,93,125,191,210]

In infants under 2 years old, respiratory syncytial virus (RSV) is the most important respiratory pathogen, producing bronchitis and bronchiolitis as well as pneumonia, croup, otitis media, and febrile upper respiratory disease. Parainfluenza virus type 3 is second to RSV as a cause of pneumonia and bronchiolitis in infants less than 6 months of age. Both viruses can reinfect and cause upper respiratory illnesses in older children and adults. Parainfluenza type 1 is the most important cause of croup (laryngotracheobronchitis) in children; type 2 resembles type 1 in clinical manifestations but less commonly causes serious illness. Parainfluenza 4 infections are encountered infrequently. Influenza and adenoviruses also cause bronchiolitis and other acute respiratory diseases in children and young adults.

Etiologic "pie" diagrams for four common respiratory syndromes of young adults are depicted in Fig. 4. A fair percentage of the causes remain unidentified. In unimmunized military recruits, adenoviruses types 4, 7, and 21 have been important causes of pneumonia and upper respiratory infections. Orally administered type-specific vaccines have been effective in preventing adenovirus infections in these high-risk populations. *Mycoplasma pneumoniae* is probably the most important cause of acute lower respiratory infections in older children and young adults. Influenza is of importance in all age groups, but the mortality is most associated with infections in infancy and in the aged: this can be caused by primary viral pneumonia, concomitant bacterial infection, or secondary bacterial infection. The predominance of viral infections in infancy and children explains the failure of antibiotic therapy for most respiratory diseases in these age groups. Newer antiviral compounds have shown clinical promise. Ribavirin appears effective in severe RSV infections, as does amantadine for the prophylaxis of influenza A infections in contained population groups such as nursing homes. Because of doubts about their efficacy and importance and concern about their adverse effects, however, these agents have not been widely embraced by clinicians.

12.2.2. Common Infections of the Central Nervous System. Multiple agents are also involved in the causation of clinical syndromes of the CNS as manifested by encephalitis and aseptic meningitis.[154] In the 10 years prior to 1993, from 900 to 1500 cases of encephalitis were reported annually to the CDC[52]; these cases were distributed throughout the United States. Cases of indeterminate etiology have generally accounted for about four fifths of those reported. Among the cases of known cause, herpes simplex has led the list, in part because of the vigor with which this etiology was sought because of its high mortality and because it is the only form of viral encephalitis that responds to specific antiviral therapy. However, the diagnosis requires a brain biopsy, so only more severe and hospitalized cases are likely to be identified as having herpesvirus. Enteroviruses have accounted for a small fraction, and the arboviruses for another small number, often caused by California virus.[257] The last major arbovirus outbreak occurred in 1976 and was caused by St. Louis encephalitis, but cases occur at lower frequencies every summer with great regularity.[52] New causes of arbovirus encephalitis such as Snowshoe hare and Jamestown Canyon viruses in North America continue to be recognized. For that reason, monitoring of the etiologic pattern should continue. Measles encephalitis is disappearing in the United States because of intensive measles vaccination programs.

The syndrome of aseptic meningitis showed no major change in incidence or etiologic pattern over a period of 6 years, 1986–91, although the large number of patients with symptomatic acquisition of HIV infection may pre-

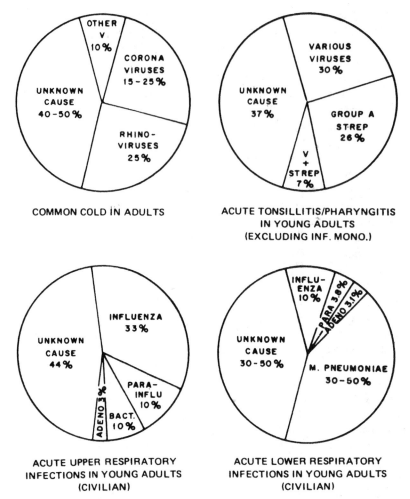

Figure 4. The causes of acute respiratory syndromes in young adults.

sent with this clinical picture. Yearly fluctuations reflected the activity of the enteroviruses and usually occurred in August or September. In 1991, 14,526 cases of aseptic meningitis were reported to the CDC. No etiologic breakdown was given for these cases, but past experience suggests that about 80% were of unknown or indeterminate cause, some 15% were probably caused by enteroviruses, and a small percent by mumps. The etiology of most encephalitis is still not documented because of the need in most cases for isolating the virus from the stool, identifying it, and then testing for an antibody rise to that specific agent. In the 1950s and 1960s, certain state laboratories carried out these procedures more intensively, and an etiologic agent was identified in about 65% of the cases.[187] Less intensive investigation in the 1970s resulted in identifying only about 20%. In tropical countries

arboviruses are more common, including a resurgence of yellow fever, and cause many infections involving the CNS.

12.2.3. Common Exanthems. Acute viral syndromes involving the skin are represented by the exanthems of childhood (measles, rubella, varicella, and erythema infectiosum or fifth disease), by various strains of coxsackieviruses and echoviruses,[204] by certain adenoviruses (such as type 7), occasionally during EBV mononucleosis (often brought on by a reaction to ampicillin), by human herpesvirus type 6 (HHV-6), the most well-established cause of roseola infantum (exanthem subitum),[283] and by human parvovirus B19, the cause of erythema infectiosum (see Table 5 and reference 65). Earlier speculations about EBV, other herpesviruses,[194] and about a retrovirus as possible causes of Kawasaki

Table 5. Viral Causes of Common Exanthems[a]

Type of rash	Examples
Macular/papular	CMV, HBV, HIV, HHV-6
	Measles, atypical measles (vaccine)
	Rubella
	Echovirus
	Enterovirus 71
	Coxsackievirus
	Adenovirus
	Parvovirus B19 (erythema infectiosum)
Vesicular	Varicella-zoster virus
	Smallpox
	Eczema herpeticum
	Eczema vaccinatum
	Coxsackievirus esp. A16
Petechial or purpuric	Coxsackievirus esp. A9
	Echovirus, esp. 9
	EBV
	Atypical measles (vaccine)
Erythema multiforme	Coxsackievirus A
	Echovirus
	Adenovirus
Other	Coxsackievirus A
	Echovirus

[a]Adapted from Cherry.[65]

disease (mucocutaneous lymph node syndrome) have not been confirmed. A relationship to parvovirus B19 has been suggested,[286] but so has involvement of a bacterial exotoxin,[188] which is attractive at the mechanistic level because the rash often mimics that of toxic shock syndrome, scarlet fever, and other conditions of bacterial origin. First recognized in Japan, outbreaks and cases are seen in the United States and around the world.[14,20] An unusual feature of the disease is cardiac involvement, with aneurysms of the coronary artery in 17–31% of the cases; the overall case mortality is 0.5–2.8%.

12.2.4. Hepatitis. At least five types of viral hepatitis are currently recognized (see Chapter 13, this volume; related chapters in text by Belshe; and references 22, 150, and 183). These include (1) hepatitis A virus (HAV); (2) hepatitis B virus (HBV); (3) hepatitis C virus (HCV), the non-A, non-B hepatitis agent resembling HBV in transmission and association with hepatocellular carcinoma; (4) the delta virus (HDV); and (5) hepatitis E virus (HEV), which clinically and epidemiologically resembles hepatitis A. Diagnostic tests are available for HAV, HBV, HCV, and HDV, and should lead to improved specificity in reporting of cases. Four categories of hepatitis are reportable to CDC: HAV, HBV, non-A, non-B by exclusion (since 1982), and hepatitis type not specified. In

the United States in recent years, HAV and HBV have each accounted for about 40% of all reported cases, and the other forms for the remaining 20% (see Chapter 13 and reference 52).

The recently identified agent of the major portion of non-A, non-B hepatitis (i.e., HCV) has been found worldwide. In the United States it accounts for the largest proportion of posttransfusion hepatitis cases. However, the absolute number of cases has declined dramatically since the elimination of commercial sources of blood products, the introduction of surrogate hepatocellular enzyme testing, and more recently the availability of direct antibody testing.[1,4] It also causes a substantial proportion of the sporadic community-acquired cases. Progression to chronic inflammation and cirrhosis is common, even in the absence of symptoms, and HCV appears to be responsible for a proportion of the virus-induced hepatocellular carcinoma (see Chapters 13 and 32).

Delta virus is an unusual RNA virus, dependent on the presence of HBV for its pathogenic expression if not for its multiplication. It was first described in Europe but is now recognized in a number of countries where it has been associated with rapidly progressive and severe liver disease. It has also been recognized in Africa and South America and in special risk groups such as hemophiliacs, injection drug users, and occasionally male homosexuals. Outbreaks have occurred in as dissimilar populations as an Indian community in Venezuela and drug users in Worcester, Massachusetts. Hepatitis E is a recently identified cause of disease transmitted by the oral–fecal route. Large waterborne outbreaks have occurred in east, south, and west Asia and, more recently, Central America.[48]

In 1991, there were 48,523 cases of viral hepatitis reported in the United States, corresponding to an annual rate of 19.5 cases/100,000 population. Of these, 50.2% were reported as HAV (9.8/100,000), 37.1% as HBV (7.2/100,000), 7.3% as non-A, non-B (1.4/100,000), and 2.6% as type unspecified (0.51/100,000). Routine specific diagnostic testing for HCV has not been in widely used long enough to establish patterns of occurrence reliably, but sentinel surveillance based on relatively thorough testing suggests that in the past 5 years the proportion of acute hepatitis cases due to HCV has been more like 1 in 5 and corresponding annual incidence about 5–8/100,000 per year (see Chapter 13).

12.2.5. Gastroenteritis. Rapid advances in our knowledge of the causes of acute viral gastroenteritis have occurred over the past few years, and these are presented in Chapter 11. The importance of rotaviruses as the most important cause of acute gastroenteritis in infants and children under 2 years of age worldwide has been firmly established through application of a variety of methods to identify the virus in the stool, including immune electron microscopy, the enzyme-linked immunosorbent assay (ELISA), and serological techniques (see Chapter 2 and references 26, 27, 30, 37, 181, 284, 285). In one of the early and seminal studies of 378 children with acute gastroenteritis in Melbourne, Australia[76] rotaviruses (then called duoviruses) were found in the stools of 52% of the cases as contrasted to their absence in the stools of 116 control children. Subsequent studies have amply confirmed these observations. For example, of 1537 children admitted with diarrhea to the Childrens Hospital National Medical Center in Washington, DC, from 1974 to 1982, rotaviruses were detected in the stools of 34.5%. The contribution of this virus to acute gastroenteritis in infants and children has varied some in different countries and different studies: in Canada 11.0%,[131] Japan 89%,[181] Venezuela 41.3%,[262] and the United States 89%.[161] It is important to remember that the substantial geographic and temporal variation in viral gastroenteritis attributable to specific causes reflects not only the great variability in methods used to detect enteric viruses but also the marked local seasonality in their appearance. Three-year monitoring of rotavirus prevalence in the United States, for example, showed prominent cold-weather surges, with peaking somewhat earlier in the western than the eastern part of the country.[149] Although rotaviruses (Norwalk-like agents or

caliciviruses) and astroviruses affect children more than adults (see Chapter 11 and references 75, 76, 87, 162, 180, 190, and 267), these agents clearly cause outbreaks in older persons as well.[180,190] Recent comprehensive studies of gastroenteritis in travelers discount the importance of viruses as major causes.[7,31]

Rotavirus is also common in developing countries, and the mortality is higher because of the lack of treatment centers where fluid replacement or oral salts are available. In Bangladesh, rotaviruses were implicated in 46% of 6352 patients seeking treatment at the Matlab Treatment Center.[30]

12.2.6. Perinatal Infections. Infections of the infant may be acquired from the mother *in utero* via placental transfer or during passage through the birth canal or from other individuals postpartum via nosocomial and other similar close contact. Estimates of their occurrence may vary according to location, personal hygienic and sexual activities, obstetric practices, utilization of vaccines, and other factors. When *in utero* infections are acquired early enough in embryogenesis, they may result in congenital anomalies and other sequelae. Infection later in pregnancy may lead to such adverse outcomes as intrauterine growth retardation and prematurity. Infection at or soon after birth may lead to persistent infection.[236] Tables 6 and 7 catalogue the major clinical consequences of these infections.

Table 6. Effects of Transplacental Fetal Infection[a]

Organism or disease	Effect of infection on the fetus and newborn infant[b]				
	Prematurity	Intrauterine growth retardation and low birth weight	Developmental anomalies	Congenital disease	Persistent postnatal infection
Viruses					
Rubella	−	+	+	+	+
Cytomegalovirus	+	+	+	+	+
Herpes simplex	+	−	−	+	+
Varicella-zoster	−	(+)	+	+	+
Mumps	−	−	−	(+)	−
Rubeola	+	−	−	+	−
Vaccinia	−	−	−	+	−
Smallpox	+	−	−	+	−
Coxsackieviruses B	−	−	(+)	+	−
Echoviruses	−	−	−	−	−
Polioviruses	−	−	−	+	−
Influenza	−	−	−	−	−
Hepatitis B	+	−	−	+	+
Human immunodeficiency virus	(+)	(+)	(+)	+	+
Lymphocytic choriomeningitis virus	−	−	−	+	−
Parvovirus	−	−	−	(+)	−

[a]Modified from Remington and Klein.[236]
[b]+, Evidence for effect; −, no evidence for effect; (+), association of effect with infection has been suggested and is under consideration.

Table 7. Clinical Manifestations of Perinatal Viral Infection Acquired *in Utero* or at Delivery[a]

Clinical sign	Virus[b]			
	Rubella virus	Cytomegalovirus	Herpes simplex virus	Enteroviruses
Hepatosplenomegaly	+	+	+	+
Jaundice	+	+	+	+
Adenopathy	+	−	−	+
Pneumonitis	+	+	+	+
Lesions of skin or mucous membranes				
Petechiae or purpura	+	+	+	+
Vesicles	−	+	+ +	−
Maculopapular exanthems	−	−	+	+
Lesions of nervous system				
Meningoencephalitis	+	+	+	+
Microcephaly	−	+ +	+	−
Hydrocephalus	+	+	+	−
Intracranial calcifications	−	+ +	−	−
Paralysis	−	−	−	+ +
Hearing deficits	+	+	−	−
Lesions of heart				
Myocarditis	+	−	+	+ +
Congenital defects	+ +	−	−	−
Bone lesions	+ +	−	−	−
Eye lesions				
Glaucoma	+ +	−	−	−
Chorioretinitis or retinopathy	+ +	+	+	−
Cataracts	+ +	−	+	−
Optic atrophy	−	+	−	−
Microphthalmia	+	−	−	−
Uveitis	−	−	−	−
Conjunctivitis or keratoconjunctivitis	−	−	+ +	+

[a]Modified from Remington and Klein.[236]

[b]−, Either not present or rare in infected infants; +, occurs in infants with infection; + +, has special diagnostic significance for this infection.

Historically, rubella infection was responsible for the numerically and clinically most important adverse outcomes of pregnancy. Among other abnormalities, it can result in abortion, stillbirth, and such anomalies as cataracts, deafness, cardiovascular anomalies, and psychomotor retardation (see Chapter 27). Congenital disease occurs in 15–20% of infants born of mothers infected during the first trimester; later manifestations may increase the total to 30–45%. Wherever they have been instituted, aggressive, widespread vaccination programs have greatly diminished the importance of congenital rubella.

Currently, the most common serious congenital disease due to viral infection is caused by CMV. Primary CMV infection has been documented in over 2% of middle-income and nearly 7% of low-income pregnant women.[249] In addition, previously infected women experience reactivation or recurrence of infection during pregnancy. The proportion of neonates who acquire infection from their mothers, estimated in a variety of locations around the world, ranges from 0.24 to 2.2%.[250] Although these infections are usually benign, the virus can produce anomalies such as microencephalopathy, chorioretinitis, deafness, and mental retardation in a small proportion of those infected. In the United States and England, CMV is transmitted in about 1 in 100 live births and about 1 in 1000 will show a congenital defect. In 1990, the CDC established formal surveillance for congenital CMV disease. During the first 2 years of its existence, 100 cases were reported. The most commonly noted manifestations was a petechial rash and thrombocytopenia, occurring in about half of the cases and often accompanied by enlarged liver or spleen or intracranial calcifications.[78]

Herpes simplex virus infections occur quite variably among pregnant women in different geographic, ethnic, and socioeconomic subpopulations. Frequencies of 1 per 1500 live births have been noted.[255] Infection is almost always complicated by mucocutaneous (skin, eye, and mouth) lesions (75%), encephalitis (57%), pneumonia (18%), or disseminated infection with combinations of the three (30%). Prompt recognition is important because antiviral therapy is often effective in reducing morbidity.

Primary or recrudescent infection with VZV is far less common during pregnancy, perhaps 5–7 per 10,000 pregnancies with only rare serious clinical consequences.

Congenital and intrapartum HBV infection transmitted by maternal carriers is considerably more common in Asia and Africa than in Europe and the Western hemisphere.[287] In the United States, antigen carriage has been seen in 1–3 per 1000 pregnancies, and 5–8% of the infants born of those mothers become hepatitis B surface antigen carriers during the first 6 months of life. High transmission rates (40–75%) in infants exposed to infected mothers appear to parallel differences in the levels of carriage, perhaps in turn a function of immune tolerance due to early exposure in the mother. Clinically, young adults infected in infancy are far more likely than those not infected until later to manifest not only persistent antigenemia but cirrhosis and, after an even longer latency period, hepatocellular carcinoma. Success of childhood immunization in reducing the rates of those complications later in adulthood will require years to realize and evaluate (see Chapters 13 and 32).

Retroviruses and in particular HIV are transmitted *in utero* and transplacentally to about 25–30% of the offspring of infected mothers. The mother's stage of immunodeficiency appears to be one factor in determining the likelihood of transmission. Adverse outcomes of pregnancy similar to those of perinatal herpesvirus and certain other infections occur with fetal and neonatal HIV infection as well. There are also distinctive features (e.g., lymphocytic interstitial pneumonitis) potentially attributable to the unique attack on the immune system. A major clinical trial of prenatal treatment with zidovudine was recently terminated early because of the striking (and subsequently well-confirmed) reduction in transmission frequency (8% infection of infants of treated mothers compared with 21% in infants of untreated).[71]

In a recent prospective study of 156 parvovirus B19-infected mothers, 88% delivered normal babies.[232] The investigators estimated the overall fetal risk to be 9%, although higher in the second trimester, and the transplacental transmission rate to be 33%. Adverse outcomes, particularly fetal loss and hydrops (a syndrome associated with destruction of red blood cells *in utero*), follow a proportion of the parvovirus B19 infections.[172,232]

The frequency of congenital and perinatal infection with mumps and measles viruses are declining in parallel with congenital rubella infection as simultaneous vaccination for all three is offered and accepted more widely.

12.2.7. Immunosuppressed and Surgical Patients. Reactivation of viral infections, especially herpes, is common in immunosuppressed, transplanted, transfused, or HIV-infected patients. There are many inapparent infections, but acute illness with a wide range of severity develops in late stages of HIV infection. Retinitis, pneumonia, and disseminated infection with CMV are especially common; both typical and atypical clinical forms of recrudescent VZV infection are also seen. Armstrong *et al.*[8] found the infection rates in 26 prospectively followed renal transplant recipients to be: CMV, 43%; HSV, 28%; EBV, 32%. With the exception of three primary CMV infections, all others represented reactivation. Clinically, five patients developed herpetic-type sores, three of whom showed HSV antibody rises; five had fever of unknown origin with rises in CMV antibody titer. Hematologically, seven patients showed atypical lymphocytosis associated with serological evidence of CMV in six. Of 13 episodes of rejection, five occurred in patients with CMV antibody rises. Fever and lymphocytosis caused by CMV also occur after cardiac surgery, and the mononucleosis syndrome occurs in up to one-third of patients after heart surgery with an extracorporeal pump. The source of CMV in transplant and surgery patients is unclear, and it might be exogenous in origin, be introduced with blood, result from reactivation in the blood of the recipient, or be present in the transplanted organ. Both severe graft-versus-host disease and pneumonitis have recently been described in conjunction with high levels of HIV-6 DNA in the lungs of bone marrow transplant patients.[69] Immunodeficiency also enhances the severity of induced infections in persons receiving live polio, measles, rubella, smallpox, or yellow fever vaccines. Measles, mumps, echoviruses and other enteroviruses, papovaviruses, and RSV have also been observed with greater severity of frequency in immunocompromised hosts.[89,284] In addition to acute illnesses resulting from reactivation of certain viruses, malignancies—especially lymphomas associated with EBV infection—occur in recent transplant recipients, patients receiving immunosuppressive agents like cyclosporin, those with X-linked lymphoproliferative syndrome, and persons with advanced HIV infection (see Chapters 24 and 30 and references 234 and 235).

12.2.8. Sexually Transmitted Diseases. Both heterosexual and homosexual intercourse effectively transmit a number of viruses: HBV, HIV, HTLV-I, HPV and herpesviruses, particularly CMV and HSV-1 and -2. Only diseases associated with the first two are currently reportable on a nationwide basis. The common clinical syndromes associated with those two are described in Section 12.2.4 of this chapter and Chapters 13 and 24. However, for clinical consequences in the genital tract itself, HPV and HSV-2 are more significant.

Human papillomavirus types (variants) 6 and 11 as well as other HPV types are responsible for genital warts (condyloma acuminata) on the surface of the genital area. This condition has apparently increased during the last two decades to a prevalence of 4–13% in young, sexually active men and women seen recently in sexually transmitted disease clinics (see Chapter 33). There has been a concomitant recent increase in transmission of infection through more casual sexual behavior with types 16 and 18 and others specifically associated with higher risk of cervical carcinoma.

In aggregate at least 10% and probably closer to 50% of young, sexually active women in the United States carry one or more types of HPV at any given time.[15] However, because estimates of frequency are highly method dependent, the precise figures are difficult to interpret. Most women who carry HPV will experience no symptoms. The specific HPV types occur at very low frequency (<3%) individually, but in aggregate could reach a prevalence of 10%. The proportion of women with cervical cytological abnormalities—the major concern in HPV infection—is considerably lower, and invasive cancer is still much less common.

Genital herpes caused by HSV infection produces characteristic painful, tender blisters on the epithelial surfaces of the genital organs and adjacent areas, including the anus and rectum. Up to 5.4% of men and 10% of women attending sexually transmitted disease clinics may be carrying HSV,[73] much of which may be persistent or recurrent infection. Symptoms are more frequent and prominent with primary than with recurrent infection. Although symptomatic genital HSV infection is not systematically reported to public health authorities, lately it has been seen more commonly than syphilis or gonorrhea in college students. During pregnancy, where there is greater vigilance because of the consequences to the neonate, about 1% of women are estimated to experience symptomatic genital infection at some time. Seroprevalence is lower in adult and adolescent males than females,[153] but this does not appear to be reflected in a proportionally lower frequency of symptoms.

12.2.9. Urinary Tract Syndromes. With diseases of the urinary tract, despite the occasional presence of viruses in urine, evidence of viral causation has not been firmly established in humans except for hemorrhagic cystitis, which is most characteristically caused by adenovirus 11. The role of immune complex formation of viruses and antibody in the causation of human glomerulonephritis is unknown, although there is ample precedent in animal models; except for elevated antibody titers to rubella virus in the nephritis of systemic lupus erythematosus, no other leads were found in a serological study of 106 cases of immune complex glomerulonephritis of unknown cause employing 13 different viral antigens.[274] It is likely that improved techniques of identifying viruses and immune complexes will lead to the discovery of a role for products of viruses and other foreign agents in both acute and chronic nephritis.

12.2.10. Febrile Illness with or without Hemorrhage. Epidemic febrile illnesses are far less common in the United States than most of the clinical entities addressed in the foregoing sections. However, the cluster of cases of dengue fever in U.S. military personnel stationed in Haiti (characterized primarily by combinations of high fever, headache, myalgia or arthralgia, and rash)[56] and the outbreak of Bolivian hemorrhagic fever (most frequently fever, chills, conjunctivitis, myalgia, arthralgia, back pain, and at times hemorrhagic sequelae)[57] serve as reminders that febrile illness may signify emergence of vector-borne viral infection in endemic but ordinarily quiescent places or through importation from outside.

13. Proof of Causation

The classic concepts of causation in infectious diseases are those elaborated by Jakob Henle (1809–1885) in 1840 and by his student Robert Koch (1843–1910) in 1884 and 1890, as well as by Edwin Klebs who carried out studies on tuberculosis similar to those of Koch.[16,179] These are termed the Henle–Koch postulates. The basic criteria (Table 8, column 1) included the consistent presence of the parasite in the disease in question under circumstances that can account for the pathological changes and clinical course, the absence of the parasite in other diseases as a fortuitous or nonpathogenic parasite, and the experimental reproduction of the disease by the organism after having been grown repeatedly in pure culture. The inability of many clear-cut causes of certain diseases to fulfill these criteria was recognized by Koch himself and other limitations were later recognized.[97] He recognized that whereas the bacteria of anthrax, tuberculosis, tetanus, and many animal diseases fulfilled the proof, those of many other diseases did not. These latter included typhoid fever, diphtheria, leprosy, relapsing fever, and Asiatic cholera. He felt particularly strongly about cholera because he himself had discovered the causative organism. For these diseases, he felt that fulfillment of only the first two criteria was needed and that experimental reproduction of the disease was not essential to proof of causation. Rivers[237] reviewed the Koch postulates in terms of viral infections in his presidential address to the American Immunological Society in 1937 and found them lacking. Included in his objections were (1) the idea that a disease is necessarily caused by only one agent, citing the work of Shope[246] with swine influenza, in which both a virus and a bacteria are required; (2) the necessity of demonstrating the presence of viruses in *every* case of the disease produced by it; and (3) the fact that the existence of virus carriers must be recognized. He set forth two conditions for establishing the specific relationship of a virus to a disease (Table 8, column 2): (1) a specific virus must be present with a degree of regularity in association with the disease, and (2) the virus must occur in the sick individual not as an incidental or accidental finding but as a cause of the disease. In support of the latter, he stressed the importance of the experimental reproduction of the disease in susceptible experimental hosts with the inclusion of suitable controls to eliminate the fortuitous presence of other viral agents either in the patient or in the experimental host. The absence of antibody to a virus in the patient's serum at the onset of illness and its appearance during recovery were recognized as an important but not absolute link in causation; Rivers was cautious in this statement because of the possible presence of passenger viruses to which antibody appeared but that were not of etiologic significance. He also noted that recovery from viral infection sometimes takes place without the development of

Table 8. Postulates of Causation

Bacteria[a] Henle (1840); Koch (1890)	Viruses[b] Rivers (1937)	Viruses[c] Immunologic proof (1973)
1. Parasite occurs in every case of the disease in question and under circumstances that can account for the pathological changes and clinical course of the disease.	1. A specific virus must be found associated with a disease with a degree of regularity.	1. Virus-specific antibody is regularly absent prior to illness.
2. Occurs in no other disease as fortuitous and nonpathogenic parasite.	2. Virus occurs in the sick individual not as incidental or accidental finding but as cause of the disease.	2. Antibody regularly appears during illness, including: a. Transient viral-specific IgM antibody b. Persistent IgG antibody c. Local antibody (IgA) at site of primary multiplication.
3. After being fully isolated from the body and repeatedly grown in pure culture, can induce the disease anew.	3. Transmissible infection is produced with a degree of regularity in susceptible experimental hosts by means of inoculation of material, free from ordinary microbes or rickettsiae, obtained from patients with the disease, and proper control and immunological studies demonstrate that the virus was neither fortuitously present in the patient nor accidentally picked up in the experimental animals.	3. Antibody production is accompanied by presence of viruses in appropriate tissues.
Only 1 and 2 were regarded as essential by Koch.		4. Absence of IgG antibody indicates susceptibility to the disease. 5. Presence of IgG antibody indicates immunity to the disease. 6. No other virus or antibody is similarly associated. 7. Production of the antibody (immunization) prevents the disease.

[a]Koch[178] (see Rivers[237]).
[b]Rivers.[237]
[c]Derived from Rivers[237] and Evans.[96,98]

antibodies and that occasionally an individual already possessing antibodies against a virus succumbs to a disease caused by it (i.e., reinfection or reactivation).

The "virologists' dilemma" was further discussed in 1957 by Huebner,[147] who revised the Koch and Rivers postulates into the following criteria: (1) the virus must be "real entity," i.e., well established on animal or tissue culture passage in the laboratory; (2) the virus must originate in human tissues and be repeatedly present therein and not in the experimental animals, cells, or the media used to grow it; (3) the agent should be characterized early to permit differentiation from other agents, including immunologic comparisons; (4) the virus should have a constant association with the clinical entity in question; (5) the clinical syndrome should be experimentally reproducible in volunteers inoculated with the agent in a "double-blind" study; (6) carefully conceived epidemiologic cross-sectional and longitudinal studies are indispensable in establishing the role of highly prevalent viruses in human diseases; (7) the disease should be prevented by a specific vaccine. He also added an eighth consideration—financial support—which is so needed to carry out the virological and epidemiologic analyses required in establishing proof of causation.

The problem of establishing causality for viral infections has been exemplified by the relationship of EBV to infectious mononucleosis. In the beginning, no method of virus isolation existed, no susceptible laboratory animal was known, and EBV antibody was already present at the time the patient with infectious mononucleosis was first seen by the physician. The proof of causation had to rest on prospective serological investigations that fulfilled certain immunologic criteria.[94,139,219,241,264] The most important of these were the regular absence of antibody prior to disease, its regular appearance during illness, and the relationship of antibody to susceptibility and immunity[114,241] (see Table 8, column 3). Advances in viral technology later permitted the identification of the presence and persistence of EBV in the pharynx of patients having acute infectious mononucleosis. Human and monkey transmission experiments with EBV have resulted in the reproduction of some but not all of the features of the disease (see Chapter 10). The web of causation is now tight that EBV causes all heterophil-antibody-positive infectious mononucleosis and most heterophil-negative cases.[95] To date, a vaccine to prevent the disease has not been developed, but phase I trials have begun with the promise of real progress.[104,206]

Similar seroepidemiologic techniques were needed in the early studies of the significance of hepatitis B surface antigen (HBsAg) because of the difficulty of isolating the HBV in the laboratory and the lack of a convenient experimental model (see Chapter 13 and reference 34).

Historically, claims for causation of disease have often been premature. What is therefore all the more remarkable and ironic a counterpoint to the typically hasty causal judgment is the resistance among a tiny group of scientists to conclusive proof that HIV is the etiologic agent of AIDS. A handful of eminent and experienced investigators[84,85,211] have offered their own fragmented alternative explanation of the epidemiologic data. In this instance, it has been the contrarians rather than the advocates who have selectively ignored or misinterpreted significant portions of the compelling evidence. They have adamantly insisted that the incomplete clinical expression or delayed occurrence of AIDS in many infected persons disproves the unequivocal causal role of HIV, as if every instance of an infection must demonstrate an obligatory, uniform natural history. These skeptics miss the crucial distinction of retrovirus as a necessary agent but retrovirus alone as an insufficient cause of the great variability in the nature and timing of the late clinical manifestations we designate as AIDS. Along the course of this distraction, appropriate semantic clarifications and systematic refutations of the fallacious reasoning were offered.[103] The fitful resurgences of these arguments fortunately seem to have been discounted by thoughtful scholars and policymakers.[67] However, it is difficult to know what impact such media attention has on a public already increasingly dubious about the reliability of both the scientific enterprise and government.

Brief concern over the possibility of alternative infectious agents as causes of AIDS was raised by anecdotal scientific reports of HIV-seronegative individuals who developed unexplained immunodeficiency as measured by low CD4 cell numbers, with or without clinical correlates. However, characteristics of infection due to a transmissible agent were not observed in those reported cases, and concerns about a significant new public health threat abated soon after strong counterarguments were published.[117,248]

The most difficult and challenging problems of causation are arising in establishing the possible relationship between certain viruses and various malignant and chronic diseases. A summary of the difficulties in proving causation and suggested guidelines for causal inferences is available,[103,105,112] and criteria for establishing the viral etiology of cancer at the molecular level are discussed in Zur Hausen's *Introduction to Viruses and Cancer*.[289] In the former category is the relationship between EBV and Burkitt's lymphoma, nasopharyngeal carcinoma,[92,95] and to lesser extent Hodgkin's disease[108,110,111]; of HBV to hepatocellular cancer[18,19]; of certain HPV types to cervical cancer; and of human T-cell leukemia/lymphoma virus (HTLV-1) to adult T-cell leukemia (see Chapters 30–33). In Burkitt's lymphoma it has been clearly shown that high EBV-viral capsid antigen IgG antibody elevations precede the development of the tumor such that a twofold titer elevation above normal constitutes a 30-fold risk for the tumor compared with children with normal levels.[77] The virus has also been consistently demonstrated in tumor tissue,[224] and a malignant tumor has been reproduced in nonhuman primates, as discussed in Chapter 31.

High EBV antibody levels have also been shown to precede the diagnosis of Hodgkin's disease in a pilot study of two cases,[108] and a large prospective study of 44 cases and matched controls has confirmed the increased risk of Hodgkin's disease in the presence of elevated EBV antibody titers.[207] However, any role of the virus in tumor causation is probably an indirect one, since the virus or its genomes have rarely been found in tumor tissue (see Chapter 30). Prospective studies of the relationship of HBV to hepatocellular cancer in Taiwan have clearly established the presence of HBsAg many years prior to the tumor, with at least a 100-fold increased risk of the cancer in those with antigenemia over those without.[18,19] This virus–tumor causal association is discussed in Chapter 32 and represents the strongest current proof that a virus can cause human cancer. This will be firmly established if the ongoing trials of HBV vaccine in infants can prevent the development of the tumor in young adult life; however, this will take many years to determine.

The persistence and/or reactivation of viruses under circumstances of impaired cell-mediated immunity have been postulated as a possible common mechanism for various chronic or delayed conditions (see Section 11 and references 91 and 265). Such an impairment could arise when the viral infection occurs very early in infancy or during pregnancy; it might also result from the presence of a concomitant infection (malaria) that depresses the immune response, from the use of immunosuppressive drugs, from genetic defects in the ability of cytotoxic T lymphocytes to recognize or respond to certain viruses, from serum inhibitors of cellular immunity, or from disease-induced immunosuppression (Hodgkin's disease, HIV infection).

In the field of chronic diseases, the importance of slow or unconventional viruses in causing kuru, Creutzfeldt–Jakob disease, and fatal infections of the nervous system has been well established, as have the causal relationship of measles virus to subacute sclerosing panencephalitis[70] and of papovavirus to progressive multifocal leukoencephalopathy (see Chapter 34). The origins of a number of other chronic diseases for which viral etiologies have been suspected remain obscure. They include multiple sclerosis, insulin-dependent diabetes mellitus and certain other autoimmune endocrine disorders, and rheumatoid arthritis, systemic lupus erythematosus, and other muscle and connective tissue diseases. There are many examples of the pitfalls of facile attribution of causality; for example, initial association of high serum antibody titers to EBV and certain other viruses with both sarcoidosis[41] and systemic lupus[229,239] was followed by recognition that polyclonal B-cell activation rather than a specific viral cause probably accounted for these findings. There have also been as yet unconvincing attempts to link EBV to such conditions as rheumatoid arthritis.

Elegant molecular techniques have now been brought to bear on the etiologic mystery of Kaposi's sarcoma seen in a variety of immunosuppressed states, particularly HIV infection, as well as in ostensibly immunologically normal hosts. The early results,[62] were promptly replicated and extended in various other settings, provide rather strong evidence that a herpesvirus identical (or nearly so) to the previously described animal agent, herpesvirus saimiri, causes several of the various forms of Kaposi's sarcoma. Even as their confirmation is sought, these observations have further supported the search for relationships between herpesvirus infection and cancer as well as other chronic conditions.

Current evidence thus suggests that certain cancers and certain chronic diseases of man are caused by the persistence and/or reactivation of common, ubiquitous viruses in an immunologically compromised host. Those viruses with a capacity for latency such as the herpes, papova, measles, rubella, and adenoviruses appear to be the most likely candidates for the causation of these conditions. Present and future work to determine the elements of causation include (1) large-scale multipurpose prospective studies of populations, seeking evidence of viral persistence, high viral antibody levels, and/or impaired lymphocyte response to viral agents as a possible prelude to malignancy and chronic disease, and then the appearance of the disease itself as more definitive proof of causation; (2) the demonstration of the virus or viral genome in afflicted tissues but not in normal tissues; (3) the occurrence of reproduction of the condition in man and/or experimental hosts, or both, under natural or induced viral infection. It must be stressed that cancer or a chronic disease will not always result even under propitious circumstances. The host response will probably fall along a biological gradient from very mild to severe.

It also seems likely that any given malignant or chronic condition may be produced by more than one cause or group of causes. The current evidence on viruses, cancer, and their relationship to chronic neurological diseases is discussed in later chapters of this book. The developments in our concepts of causation and the limitations of the Henle–Koch postulates have been reviewed.[96–98] A unified set of guidelines has been proposed for both infectious and noninfectious diseases.[96] However, existing postulates concentrate on the relationship between a suspected cause and the resulting clinical illness. Yet most viral infections result in many subclinical or inapparent infections for every one that is clinically manifest. Subclinical pathological expression is also common in bacterial infections as well as in many chronic diseases such as coronary heart disease, diabetes, and some malignancies.

Once the pathogenic process has been initiated, some additional factor or factors may be needed to result in clinical illness. These clinical illness promotion factors[102] or cofactors were discussed earlier (see Section 7). In infectious diseases, these factors are incompletely understood and vary from one disease to another. For some, the age at the time of infection is an important determinant (poliomyelitis, viral hepatitis, infectious mononucleosis); for others, genetic susceptibility to the infection and/or the disease among those infected probably plays an important role, perhaps operating through the immune system, as in the X-linked lymphoproliferative syndrome associated with EBV[233] and in the deterioration of the immune system during HIV infection.[166] Psychosocial factors have also been presumed to be important in the development of infectious mononucleosis among those infected with EBV.[163] Focus on the means of preventing the emergence of clinical illness among those infected is of special relevance to a virus like HIV because approximately one million persons in the United States and at least 10–15 times that number worldwide are currently infected. The search for clues to pathogenesis among genetic and other cofactors remains intense because only very recently has any intervention under evaluation promises to prevent those people from proceeding inexorably toward fatal disease.

A recent book has summarized the chronological development of concepts of causation in acute and chronic, immunologic, epidemic, malignant, and occupational diseases.[105]

14. Control and Prevention

The basic strategy for controlling a viral disease is to break a link in the chain of causation. Interruption of a single known essential link may effectively control a disease even if knowledge of other links, or of the etiology itself, is incomplete. Despite this, very little has been accomplished in most viral diseases by environmental changes except for the arboviruses, in which the appropriate insect vector can be controlled. Improved water supplies, proper sewage disposal, and improved personal hygiene could potentially decrease the incidence of poliomyelitis and other enterovirus and hepatitis A infections, but in general the results have been disappointing because so many pathways of infection exist. Furthermore, improved sanitation may delay the age of exposure to later childhood and young adult life, when infections are more often clinically apparent and more severe.

Perhaps the most significant major challenge is to control HIV infection and its consequences with an alluring but early hope of antiviral chemoprophylaxis and with no immediate prospects for an effective vaccine. Efforts must be directed at prevention of infection through culturally sensitive education coupled with programs to facilitate difficult changes in behavior. The key objectives must be: circumscribed sexual activity and more frequent and effective use of condoms; assurance of virus-free medical injections and infusions through appropriate safeguards in health care facilities; well-conceived and well-directed programs for blood collection, needle exchange, and drug withdrawal; and prenatal screening, counseling, and intervention for prospective mothers at high risk.

14.1. Immunization

The difficulty in the environmental control of viral infections spread by close personal contact, by the respiratory route, or even by oral–intestinal spread has directed the main thrust of prevention to immunization of the host. The requirements of a good vaccine are listed in Table 9. The overall objective is to create the same degree and duration of protection as with natural infection but without the accompanying clinical illness. Both live and killed

---

Table 9. Objectives of Immunization

1. Produce a good humoral, cellular, and local immune response similar to natural infection.
2. Produce protection against clinical disease and reinfection.
3. Give protection over several years, preferably a lifetime.
4. Result in minimal immediate side reactions or mild disease and with no delayed effects such as late reactivation, CNS involvement, or cancer.
5. Can be administered simply in a form and according to a schedule acceptable to the public.
6. Cost and benefit of administration should clearly outweigh the cost and risk of natural disease and the adverse consequences of immunization.

Table 10. Comparison of Live and Killed Vaccines[a]

	Live	Killed
Immune response		
Humoral antibody (IgG)	+++	+++
Local antibody (IgA)	+++	+
Cell-mediated immunity	+++	+
Duration of response	Long	Shorter
Epidemiologic response		
Prevents reinfection by natural route	+++	++
Stops spread of "wild" virus to others	++	+
Some vaccine viruses (polio) spread to others	+++	0
Creates herd immunity if enough persons are vaccinated	+++	0
Characteristics of the vaccine		
Usually heat stable	0	++
Vaccine virus may mutate or increase in virulence	+	+
Antigenic site limited or lost during preparation (e.g., formalin treatment)	0	+
Contraindicated in immunosuppressed persons	+++	0
Side reactions: systemic (viremia)	+	0
local	0	++
Number of doses for successful take	1[b]	2–3

[a]The table is a simplification and may not apply to all vaccines. Some live vaccines (polio) are relatively heat stable. The induction of local immunity is often dependent on the antigenic dose of killed vaccine; some induce sufficient immunity to lower reinfection rates and decrease spread of wild virus. Our knowledge of the presence and degree of cell-mediated immunity is inadequate for many vaccines.
[b]Certain killed vaccines (e.g., Japanese B) may provide protection of unvaccinated susceptible members of a well-vaccinated population.
[c]Several doses of polio vaccine are given to insure a take against the three types on at least one of these.

vaccines have been used. A comparison of live and killed vaccines is given in Table 10. In general, live viral vaccines are more desirable and induce a longer and broader immune response, especially if given by a natural route. Some of the problems include successful attenuation without reversion to virulence, avoidance of viral persistence and the risk of reactivation, and the elimination of possible oncogenicity. These are major hurdles for vaccines against herpesviruses, and it is difficult to measure some of these attributes in the laboratory. There are efforts to produce live vaccines with temperature-sensitive mutants for respiratory syncytial and influenza viruses that would multiply only in the colder temperature of the upper respiratory host but not in the lung where clinical disease might result. Tables 11 and 12 summarize information on the use of viral immunoprophylaxis in immunocompetent individuals as of 1994,[44,47,50,58] but recommendations are updated frequently.[47,54] For example, recommendations for use of the recently licensed, effective, attenuated, killed vaccine against hepatitis A will likely continue to be refined for specific subpopulations (e.g., day-care attendees and travelers from the United States to highly endemic areas). Table 11 covers standard recommendations for adults, and Table 12 addresses normal infants and children. Recommendations for immunization of immunocompromised patients have also been summarized.[49] In general, live virus vaccines are *not* recommended for persons with HIV infection, except for measles–mumps–rubella (MMR) vaccine; it should be given to those asymptomatic HIV-infected persons who would ordinarily receive it in the absence of HIV infection. Oral polio vaccine (OPV) should *not* be used in household or close nursing contacts. The WHO provides information on vaccination requirements for international travel.[280,281]

The most successful efforts toward viral vaccine development have used an attenuated live virus as the antigen (adenovirus, measles, mumps, poliovirus, rubella, smallpox, and VZV). Administration by the natural portal of entry to produce local immunity has also been important (poliovirus, adenovirus). Inactivated viral vaccines such as influenza vaccine have met with limited success, but highly purified and concentrated preparations and other newer constructs are giving more promising results. Killed poliovaccine has been successfully employed as the sole method of vaccination in several countries such as Sweden, Finland, and The Netherlands in the past, but some problems arising for religious reasons led to an outbreak in immunized persons that spread to Canada and the United States. In Finland, waning immunity was apparently the reason for an outbreak in which oral vaccine was added to the program. On the other hand, a newer killed vaccine with enhanced potency (eIPV) yielded high seroconversion rates after one or two injections and was field tested in Senegal in combination with diphtheria–pertussis–tetanus (DPT) in two injections 6 months apart.[251] It has been useful in areas where the response to OPV has been poor, in highly endemic areas where mass oral programs are difficult, in immunocompromised persons, or in susceptibles exposed to OPV. Both may be useful in some areas.

Passive immunization with an immunoglobulin (Ig) preparation is a short-term expedient useful in prevention primarily when it can be administered soon after exposure and when it contains a sufficiently high titer of specific antibody against the agent. In some instances, preparations are derived from persons known to be convalescent from the disease, from persons hyperimmunized against it, or by selecting only donors shown to have high antibody titers. Passive immunization in adults is generally limited to well-defined exposures in immunocompromised patients who are susceptible to HAV, HBV, VZV, CMV, vaccinia (unlikely now that immunization against the smallpox has been discontinued but potentially of renewed importance if the use of vaccinia virus as a carrier

Table 11. Immunobiologics and Schedules for Adults (≥18 years of age), United States[a,b]

Immunobiologic generic name	Primary schedule and booster(s)	Indications	Major precautions and contraindications[c]	Special considerations
Live virus vaccines				
Measles vaccine, live	One dose subcutaneously (SC); second dose at least 1 month later, at entry into college or post-high school education, beginning medical facility employment, or before traveling. Susceptible travelers should receive one dose.	All adults born after 1956 without documentation of live vaccine on or after first birthday, physician-diagnosed measles, or laboratory evidence of immunity; persons born before 1957 are generally considered immune.	Pregnancy; immunocompromised persons[d]; history of anaphylactic reactions following egg ingestion or receipt of neomycin.	Measles–mumps–rubella (MMR) is the vaccine of choice if recipients are likely to be susceptible to rubella and/or mumps as well as to measles. Persons vaccinated between 1963 and 1967 with a killed measles vaccine alone, killed vaccine followed by live vaccine, or with a vaccine of unknown type should be revaccinated with live measles virus vaccine.
Mumps vaccine, live	One dose SC; no booster.	All adults believed to be susceptible can be vaccinated. Adults born before 1957 can be considered immune.	Pregnancy; immunocompromised persons[d]; history of anaphylactic reaction following egg ingestion.	MMR is the vaccine of choice if recipients are likely to be susceptible to measles and rubella as well as to mumps.
Rubella vaccine, live	One dose SC; no booster.	Indicated for adults, both male and female, lacking documentation of live vaccine on or after first birthday or laboratory evidence of immunity, particularly young adults who work or congregate in places such as hospitals, colleges, and military, as well as susceptible travelers.	Pregnancy, immunocompromised persons[d]; history of anaphylactic reaction following receipt of neomycin.	Women pregnant when vaccinated or who become pregnant within 3 months of vaccination should be counseled on the theoretical risks to the fetus. The risk of rubella vaccine-associated malformations in these women is so small as to be negligible. MMR is the vaccine of choice if recipients are likely to be susceptible to measles or mumps as well as to rubella.
Smallpox vaccine (vaccinia virus)	THERE ARE NO INDICATIONS FOR THE USE OF SMALLPOX VACCINE IN THE GENERAL CIVILIAN POPULATION.			Laboratory workers working with orthopox viruses or healthcare workers involved in clinical trials of vaccinia-recombinant vaccines.
Yellow fever attenuated virus, live (17D strain)	One dose SC 10 days to 10 years before travel; booster every 10 years.	Selected persons traveling or living in areas where yellow fever infection exists.	Although specific information is not available concerning adverse effects on the developing fetus, it is prudent on theoretical grounds to avoid vaccinating a pregnant woman unless she must travel where the risk of yellow fever is high. Immunocompromised persons[d]; history of hypersensitivity to egg ingestion.	Some countries require a valid International Certificate of Vaccination showing receipt of vaccine. If the only reason to vaccinate a pregnant woman is an international requirement, efforts should be made to obtain a waiver letter.

(continued)

Table 11. (Continued)

Immunobiologic generic name	Primary schedule and booster(s)	Indications	Major precautions and contraindications[c]	Special considerations
Live virus and inactivated virus vaccines				
Polio vaccines: Enhanced potency inactivated poliovirus vaccine (eIPV) Oral poliovirus vaccine, live (OPV)	eIPV preferred for primary vaccination; two doses SC 4 weeks apart; a third dose 6–12 months after second; for adults with a completed primary series and for whom a booster is indicated, either OPV or eIPV can be administered. If immediate protection is needed, OPV is recommended.	Persons traveling to areas where wild poliovirus is epidemic or endemic. Certain healthcare personnel.	Although there is no convincing evidence documenting adverse effects of either OPV or eIPV on the pregnant woman or developing fetus, it is prudent on theoretical grounds to avoid vaccinating pregnant women. However, if immediate protection against poliomyelitis is needed, OPV is recommended. OPV should not be given to immunocompromised individuals or to persons with known or possibly immunocompromised family members.[d] eIPV is recommended in such situations.	Although a protective immune response to eIPV in the immunocompromised person cannot be assured, the vaccine is safe, and some protection may result from its administration.
Inactivated virus vaccines				
Hepatitis B (HB) inactivated virus vaccine	Two doses IM 4 weeks apart; third dose 5 months after second; booster doses not necessary within 7 years of primary series. Alternate schedule for one vaccine: three doses IM 4 weeks apart; fourth dose 10 months after the third.	Adults at increased risk of occupational, environmental, social, or family exposure.	Data are not available on the safety of the vaccine for the developing fetus. Because the vaccine contains only noninfectious HBsAg particles, the risk should be negligible. Pregnancy should *not* be considered a vaccine contraindication if the woman is otherwise eligible.	The vaccine produces neither therapeutic nor adverse effects on HBV-infected persons. Prevaccination serological screening for susceptibility before vaccination may or may not be cost effective depending on costs of vaccination and testing and on the prevalence of immune persons in the group.
Influenza vaccine (inactivated whole-virus and split-virus) vaccine	Annual vaccination with current vaccine. Either whole- or split-virus vaccine may be used.	Adults with high-risk conditions, residents of nursing homes or other chronic-care facilities, medical care personnel, or healthy persons ≥65 years.	History of anaphylactic hypersensitivity to egg ingestion.	No evidence exists of maternal or fetal risk when vaccine is administered in pregnancy because of an underlying high-risk condition in a pregnant woman. However, it is reasonable to wait until the second or third trimester, if possible, before vaccination.

Human diploid cell rabies vaccine (HDCV) inactivated, whole-virion; rabies vaccine, adsorbed (RVA)	Veterinarians, animal handlers, certain laboratory workers, and persons living in or visiting countries for >1 month where rabies is a constant threat.	Preexposure prophylaxis: two doses 1 week apart; third dose 3 weeks after second. If exposure continues, booster doses every 2 years, or an antibody titer determined and a booster dose administered if titer is inadequate (<5). Postexposure prophylaxis: All postexposure treatment should begin with soap and water. 1. Persons who have (a) previously received postexposure prophylaxis with HDCV, (b) received recommended IM preexposure series of HDCV, (c) recommended ID preexposure series of HDCV in the United States, or (d) have a previously documented rabies antibody titer considered adequate; two doses of HDCV, 1.0 ml IM, one each on days 0 and 3. 2. Persons not previously immunized as above: HRIG 20 IU/kg body weight, half infiltrated at bite site if possible; remainder IM; and five doses of HDCV, 1.0 mL IM one each on days 0, 3, 7, 14, 28.	If there is substantial risk of exposure to rabies, preexposure vaccination may be indicated during pregnancy. Corticosteroids and immunosuppressive agents can interfere with the development of active immunity; history of anaphylactic or type III hypersensitivity reaction to previous dose of HDCV. / Complete preexposure prophylaxis does not eliminate the need for additional therapy with rabies vaccine after a rabies exposure. The decision for postexposure use of HDCV depends on the species of biting animal, the circumstances of biting incident, and the type of exposure (e.g., bite, saliva contamination of wound). The type of and schedule for postexposure prophylaxis depends on the person's previous rabies vaccination status, or the result of a previous or current serological test for rabies antibody. For postexposure prophylaxis, HDCV should always be administered IM, *not* ID.
Immune globulins			
Cytomegalovirus immune globulin (intravenous)	As prophylaxis for bone marrow and kidney transplant recipients	Bone marrow transplant recipients: 1.0 g/kg weekly; kidney transplant recipients: 150 mg/kg initially, then 50–100 mg/kg every 2 weeks.	Prophylaxis must be continued for 3–4 months to be effective.
Immune globulin (IG)	Nonimmune persons traveling to developing countries.	Hepatitis A prophylaxis: *Preexposure:* one IM dose of 0.02 ml/kg for anticipated risk of 2–3 months; IM dose of 0.06 ml/kg for anticipated risk of 5 months; repeat appropriate dose at above intervals if exposure continues.	For travelers, IG is not an alternative to continued careful selection of foods and water. Frequent travelers should be tested for hepatitis A antibody. IG is not indicated for persons with antibody to hepatitis A.

(continued)

Table 11. (*Continued*)

Immunobiologic generic name	Primary schedule and booster(s)	Indications	Major precautions and contraindications[c]	Special considerations
	Postexposure: one IM dose of 0.02 ml/kg administered within 2 weeks of exposure.	Household and sexual contacts of persons with hepatitis A; staff, attendees, and parents of diapered attendees in day-care center outbreaks		
	Measles prophylaxis: 0.25 ml/kg IM (maximum 15 ml) administered within 6 days after exposure.	Exposed susceptible contacts of measles cases.	IG should *not* be used to control measles.	IG administered within 6 days after exposure can prevent or modify measles. Recipients of IG for measles prophylaxis should receive live measles.
Hepatitis B immune globulin (HBIG)	0.06 ml/kg IM as soon as possible after exposure (with HB vaccine started at a different site); a second dose of HBIG should be administered 1 month later (percutaneous/mucous membrane exposure) or 3 months later (sexual exposure) if the HB vaccine series has not been started.	Following percutaneous or mucous membrane exposure to blood known to be HBsAg positive (within 7 days); following sexual exposure to a person with acute HBV or an HBV carrier (within 14 days).		
Rabies immune globulin, human (HRIG)	20IU/kg, up to half infiltrated around wound; remainder IM.	Part of management of rabies exposure in persons lacking a history of recommended pre-exposure or postexposure prophylaxis with HDCV.		
Vaccinia immune globulin	0.6 ml/kg in divided doses over 24–36 hr; may be repeated every 2–3 days until no new lesions appear.	Treatment of eczema vaccinatum, vaccinia necrosum, and ocular vaccinia.		Of no benefit for postvaccination encephalitis.
Varicella-zoster immune globulin (VZIG)	Persons <50 kg: 125 U/10 kg IM; persons >50 kg: 625 U[e].	Immunocompromised patients known or likely to be susceptible with close and prolonged exposure to a household contact case or to an infectious hospital staff member or hospital roommate.		Although preferable to administer with the first dose of vaccine, can be administered up to the eighth day after the first dose of vaccine.

[a]Adapted from CDC.[(47)]

[b]Several vaccines and toxoids are in "Investigational New Drug" (IND) status and available only through the U.S. Army Research Institute for Infectious Diseases. These are: (a) eastern equine encephalitis vaccine (EEE), (b) western equine encephalitis vaccine (WEE), (c) Venezuelan equine encephalitis vaccine (VEE).

[c]When any vaccine or toxoid is indicated during pregnancy, waiting until the second or the third trimester, when possible, is a reasonable precaution that minimizes concern about teratogenicity.

[d]Persons immunocompromised because of immune deficiency diseases, HIV infection (who should primarily not receive OPV and yellow fever vaccines), leukemia, lymphoma, or generalized malignancy or immunosuppressed as a result of therapy with corticosteroids, alkylating drugs, antimetabolites, or radiation.

[e]Some persons have recommended 125 U/10 kg regardless of total body weight.

Table 12. Immuniobiologics and Schedules for Children, United States[a,b]

Vaccine	Birth	2 Months	4 Months	6 Months	12[c] Months	15 Months	18 Months	4–6 Years	11–12 Years	14–16 Years
Hepatitis B[d]	HB-1		HB-2		HB-3					
Diphtheria, tetanus, pertussis[e]		DTP	DTP	DTP	DTP or DTaP at ≥15 months			DTP or DTaP	Td	
H. influenzae type b[f]		Hib	Hib	Hib	Hib					
Poliovirus[g]		OPV	OPV	OPV				OPV		
Measles, mumps, rubella[h]					MMR			MMR or MMR		

[a]Modified from Advisory Committee on Immunization Practices, American Academy of Pediatrics, and American Academy of Family Physicians, 1995.

[b]Recommended vaccines are listed under the routinely recommended ages. Shaded bars indicate range of acceptable ages for vaccination.

[c]Vaccines recommended in the second year of life (12–15 months of age) may be given at either one or two visits.

[d]Infants born to hepatitis B surface antigen (HBsAg)-negative mothers should receive the second dose of hepatitis B vaccine between 1 and 4 months of age, provided at least 1 month has elapsed since receipt of the first dose. The third dose is recommended between 6 and 18 months of age. Infants born to HBsAg-positive mothers should receive immunoprophylaxis for hepatitis B with 0.5 ml hepatitis B immune globulin (HBIG) within 12 hr of birth, and 0.5 ml vaccine at a separate site. In these infants, the second dose of vaccine is recommended at 1 month of age and the third dose at 6 months of age. All pregnant women should be screened for HBsAg during an early prenatal visit.

[e]The fourth dose of diphtheria and tetanus toxoids and pertussis vaccine (DTP) may be administered as early as 1 month of age, provided at least 6 months have elapsed since the third dose of DTP. Combined DTP-Hib products may be used when these two vaccines are administered simultaneously. Diphtheria and tetanus toxoids and acellular pertussis vaccine (DTaP) is licensed for use for the fourth and/or fifth dose of DTP in children ages ≥15 months and may be preferred for these doses in children in this age group.

[f]Three H. influenzae type b conjugate vaccines are available for use in infants: (1) oligosaccharide conjugate Hib vaccine (HbOC); (2) polyribosylribitol phosphate-tetanus toxoid conjugate (PRP-T); and (3) Haemophilus conjugate vaccine (meningococcal protein conjugate) (PRP-OMPM). Children who have received PRP-OMP at 2 and 4 months of age do not require a dose at 6 months of age. After the primary infant Hib conjugate vaccine series is completed, any licenses Hib conjugate vaccine may be used as a booster dose at age 12–15 months.

[g]Recommendations for substituting IPV for the first and second dose of OPV will soon be adopted.

[h]The second dose of measles–mumps–rubella vaccine should be administered EITHER at 4–6 years OR at 11–12 years of age.

for other antigens increases), and rabies virus. In rabies this approach is important early in severe exposures, as it may limit local multiplication and subsequent spread of the virus to the CNS.

The rapidly expanding knowledge of molecular virology, of DNA technology, and of the function and cloning of various parts of the genome of viruses, of their insertion into carrier vehicles, and of the concept of preparing idiotypic vaccines directed against the receptor for the virus on the host cell have led to a plethora of new experimental vaccines that offer great hope for the future. To facilitate the further development of these methods, in 1982, the National Institute of Allergy and Infectious Diseases (NIAID), National Institutes of Health, in cooperation with the Institute of Medicine in the United States first composed a list of infections for which development or improvement of vaccines was considered high priority[214]; that list has been modified periodically. The vaccines assigned the highest priorities for use in the United States and the developing world and their status in 1994 are shown in Tables 13 and 14.

Rapid and encouraging progress is being made in the understanding of the biology of viruses and in applying novel strategies for constructing, enhancing, and delivering vaccines made from those agents. Progress on individual vaccines are addressed in individual chapters on the

Table 13. Progress in Vaccine Priorities for the United States

Disease	Vaccine status (1994)
Hepatitis B virus (HBV)	Two rDNA vaccines licensed
Respiratory syncytial virus (RSV)	Attenuated and rDNA-derived vaccines under study
Haemophilus influenzae type b (Hib)	Conjugate vaccines licensed; combination vaccine with DTP licensed
Influenza	Attenuated and rDNA-derived vaccine candidates
Varicella-zoster virus (VZV)	Application for licensing pending.
Group B streptococcus	Glycoconjugate vaccine candidates
Parainfluenza	Attenuated and rDNA-derived vaccine candidates

Table 14. Progress in International Vaccine Priorities

Disease	Vaccine status (1994)
Streptococcus pneumoniae	Multivalent conjugate vaccines in clinical trials
Malaria	New approaches under investigation
Rotavirus	Vaccine candidates in clinical trial
Typhoid fever	Ty21a and Vi vaccines licensed
Haemophilus influenzae type b (Hib)	Conjugate vaccines licensed; combination vaccine with DTP licensed
Hepatitis B virus (HBV)	Two rDNA vaccines licensed
Shigella	Basic research
Group A streptococcus	Basic research

agents. More generally, intensive efforts are underway on: (1) safe but immune-enhancing viral "vectors" (produced by inserting critical genetic information from the target virus into a carrier virus such as vaccinia, poliovirus, adenovirus, or herpesvirus); (2) conventional adjuvants as well as new lipid and surface-active substances, including combinations with HIV glycoprotein; (3) specific, chemically defined epitopes that can be linked synthetically or incorporated separately into a virus vector in order to stimulate both cellular and humoral immune response; (4) improved immunogenicity through microencapsulation for slow release of antigen at certain sites, formation of detergent-based carrier particles, and creation of yeast–virus antigen combinations; and (5) immune stimulation by addition of cytokines like interleukin-2 either separately or in combination with viral antigen. NIAID regularly updates its summary of recent progress in vaccine development (usually referred to as the Jordan Report[215]).

Despite these exciting new developments, it must be remembered that each new vaccine must be evaluated in carefully conducted field trials to prove that its efficacy, safety, cost, ease of administration, thermostability, and freedom from long- and short-term reactions or vaccine complications are better than those of existing vaccines. This will not be an easy task and in some instances may be an impossible one. More sobering is the reality that even in nations as affluent as the United States there are still formidable obstacles to achieving acceptable levels of routine immunization of the population. Especially intense efforts are being made in the Untied States by the Childhood Immunization Initiative to reach the point where 90% of children under 2 years of age have received the recommended series of vaccinations and by the Vaccines for Children program to reduce the financial barriers to achieving those levels.[59] Similarly, our greatest prob-

lem on a worldwide basis today is not the lack of efficacy of most available vaccines but in delivering them in a viable state and at an appropriate age to susceptible children before natural infection occurs.

14.1.1. Immunization in Developing Countries

The effective utilization of current vaccines, especially against childhood diseases, in tropical and developing countries presents many biological, economic, logistic, and political problems. Some of these are listed in Table 15. The need to initiate immunization in the very short period before natural infection occurs in these settings, the need for maintenance of the viability of live vaccines through an effective cold chain, the difficulty in transportation of vaccine to remote areas or during the rainy season or finding adequate health personnel to administer it on arrival, and the poor socioeconomic and educational levels in many settings are but a few of the difficulties. Despite these problems, in 1982 the WHO initiated the Expanded Programme on Immunization (EPI)[279] directed at achieving immunization of all children in the developing world against six targeted diseases by 1990 as part of their development of primary care programs. The

Table 15. Immunization Problems in Developing Countries

1. Inadequate surveillance of infectious diseases.
2. Inadequate diagnostic facilities.
3. Inadequate and unreliable transportation, maintenance problems in the tropical environment, and the difficulties of movement in the rainy season.
4. Inadequate health personnel for surveillance, diagnosis, and the delivery of vaccines.
5. Inadequate funds for immunization programs.
6. Remote and dispersed populations in many areas.
7. Problems in record keeping because of illiteracy rate.
8. Problems in communication.
9. Problems in maintaining the cold chain for vaccines and lack of sufficiently heat-stable preparations.
10. Poor antibody response to some vaccines such as OPV because of poor nutrition, poor immune response, presence of inhibitors(?), interference by other agents, loss of antigenicity in tropical areas, inadequte dose because of faulty equipment and other unknown reasons.
11. Early age of infection, requiring immunization in the first year of life, perhaps earlier, even at the time of birth.
12. Difficulty in getting people back for the follow-up doses after the first one.
13. Poor integration of immunization programs into other health activities.
14. Lack of political will and support.
15. Higher priority given to other health or economic programs.

diseases are measles, poliomyelitis, diphtheria, pertussis, tetanus, and tuberculosis in children. It also includes immunization of mothers to protect against neonatal tetanus. Although the target date was overly optimistic, important progress toward that goal has been made. The status of the program is frequently reviewed in the *WHO Weekly Epidemiological Record*.[280,281]

Recently, the WHO and the United Nations Children's fund declared the goal of immunization of 80% of the 1990 birth cohort had been met according to the original EPI prescription, but the ability to sustain that level of success in future years is a serious concern. In 1992, the US government launched a scientific program called the Children's Vaccine Initiative with the major objective of coordinating public and private efforts to develop new combinations of childhood vaccines that would substantially reduce the 500 million contacts currently necessary to provide complete protection to the world birth cohort of 125 million.

14.1.2. Eradication versus Control. The successful global eradication program against smallpox through the efforts of WHO has been a singular achievement in preventive medicine and has raised the hope that other diseases might be similarly controlled.[278] The term has been characterized as follows: "Eradication of an infection implies that the infection has disappeared from all countries of the world because transmission of the causative organism has ceased in an irreversible manner."[253] It involves the control of the clinical disease with its attendant morbidity, disability, and mortality, the control of the infection itself, and the control of the presence of the causative organism in the environment.[101] True eradication is achieved only when there is no risk of infection or disease in the absence of vaccination or any other control measure in the entire world. The disappearance of transmission in a given area is termed "elimination" but would not exclude the importation of infection from outside.

The biological features favoring the possibility of eradication or a high degree of control are listed in Table 16. Under the auspices of the Carter Center of Emory University, between 1989 and 1992, an International Task Force for Disease Eradication met to consider the feasibility of eradicating each of more than 90 diseases. They recently issued recommendations summarized in CDC publications.[45,51] The characteristics identified by the Task Force for assessing those candidates for global eradication are summarized in Table 16. Accordingly, 29 infectious diseases were examined in depth and classified into (1) those targeted for early eradication, (2) those with current potential for elimination of some major aspect,

Table 16. Factors Favoring Eradication of Communicable Diseases[a]

Infection and disease limited to human host and transmitted person to person (no animal or insect reservoir).
Characteristic clincal disease, usually serious, and easily diagnosed.
Few or no subclinical cases.
No long-term carrier states.
Only one causative agent or serotype.
Short period of infectivity pre- and postdisease.
Immunity following disease or immunization is:
    Of long duration.
    Not subject to reinfection or reactivation.
    Decreases or eliminates excretion of organism.
    Evidence of vaccine immunity detectable.
Disease has seasonality (permitting vaccine strategies).
Characteristics of vaccine needed:
    Simulates natural infection.
    Stable: resists physical and genetic change.
Eradication would be cost effective.

[a]From Evans.[100,101]

(3) those with some potential in the foreseeable future, and (4) those for which there is no hope of eradication in the foreseeable future. Table 17 includes the eight viral diseases considered.

Great progress has been made in several developed countries in the "elimination" or near elimination of measles. In the United States, language, ethnic, and socioeconomic barriers have recently led to major resurgences of measles and rubella, especially in the inner-city areas. Elimination of measles in the United States suffered a serious setback during the 1989–91 period when the 10–42% immunization-series completion frequencies in children at 12–15 months of age resulted in 55,622 reported cases, more than four times the number reported during the previous 3-year period.[50,52] Intensive public health efforts to improve series completion frequencies were supported by the National Vaccine Advisory Committee and other interested organizations. This renewed emphasis on vaccine delivery following the resurgence restored momentum toward the goal of elimination. Other pockets of susceptibility remain either because the immunization began too late in life or the young adults were not included and were not old enough to have had natural infection or because of refusal or religious or other grounds. In developing countries, however, there are major obstacles, even in addition to those listed in Table 15. Among the most difficult problems are that many cases of measles occur under the age of 1 and that maternal immunity is of much shorter duration, perhaps because the mother was also infected in infancy and the immunity has waned, or possibly there is more rapid loss of antibody.

**Table 17. Diseases Considered as Candidates for Global Eradication
by the International Task Force for Disease Eradication**[a]

Disease	Current annual toll worldwide	Chief obstacles to eradication	Conclusion
Diseases targeted for eradication			
Poliomyelitis	100,000 cases of paralytic disease; 10,000 deaths	No insurmountable technical obstacles; increased national/international commitment needed	Eradicable
Mumps	Unknown	Lack of data on impact in developing countries; difficult diagnosis	Potentially eradicable
Rubella	Unknown	Lack of data on impact in developing countries; difficult diagnosis	Potentially eradicable
Diseases/conditions of which some aspect could be eliminated			
Hepatitis B	250,000 deaths	Carrier state, infections *in utero* not preventable, need routine infant vaccination	Not now eradicable, but could eliminate transmission over several decades
Rabies	52,000 deaths	No effective way to deliver vaccine to wild animals that carry the disease	Could eliminate urban rabies
Diseases that are not eradicable now			
Measles	Almost 1 million deaths, mostly among children	Lack of suitably effective vaccine for young infants; cost; public misconception of seriousness	Not now eradicable·
Rotaviral enteritis	80 million cases; 870,000 deaths	Inadequate vaccine	Not now eradicable
Yellow fever	>10,000 deaths	Sylvatic reservoir; heat-labile vaccine	Not now eradicable
Diseases that are not eradicable			
Varicella-zoster	3 million cases in United States alone	Latency of virus; inadequate vaccine	Not eradicable

Whatever the reason, measles immunization is needed in that rather brief "window" of time between the loss of maternal antibody and exposure to natural measles infection, and this period may vary in different countries, even in the same setting and possibly from individual to individual. If vaccine is given too early, a poor antibody response may occur in the infant, and booster doses may be relatively ineffective or the immunity short-lived. The use of an intranasal vaccine may circumvent this issue, but there are technical problems in its proper administration; more potent injectable vaccines may overcome low levels of maternal antibody (see Chapter 17 for a fuller discussion).

The campaign against poliomyelitis has actually achieved remarkable success, in a number of countries, particularly in the Western hemisphere: eight of nine cases in 1991 occurred in Columbia, and the last paralytic case was seen in August 1991 in a 2-year-old boy.[225] In the United States itself, no natural case of paralytic polio has occurred in the last decade,[252] and only a handful of vaccine-related cases occur annually.[46] In the absence of new cases due to natural virus, in September 1994 a special commission certified the Americas polio-free (Fig. 5). It is too early to determine whether American populations will remain sufficiently protected to prevent significant reintroduction from the Eastern hemisphere. Elsewhere, the presence of many subclinical cases, three serotypes, and relatively long persistence of the virus in the intestine pose challenges to its control. Still the results of the concerted immunization initiative augur well for eradication of polio-induced paralysis if not wild poliovirus itself. However, given enough personnel, a massive 1- or 2-day countrywide immunization program can be mounted in some areas as it was in Brazil,[240] and remarkable control can be achieved in a short time. Such an effort simulates a large vaccine-induced epidemic, since the virus spreads to contacts. Whether such a program would overcome the poor seroconversion rate to oral vaccine found in many African countries has not been established, nor can every country afford to place primary emphasis on the control of one disease at the expense of other vaccination and primary care programs. The new inactivated vaccine in one or two doses in settings where the response to oral vaccine has been poor or where it has not been logistically possible to administer three of four doses of oral vaccine should be considered, at least as the first encounter with a vaccine. This can be accompanied or followed by oral vaccine. Enhanced inactivated polio vaccine (eIPV) has not routinely been recommended in the United States except for adults exposed to oral vaccine

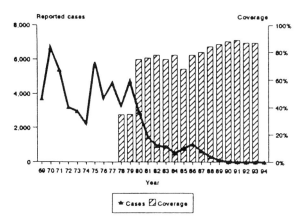

Figure 5. Three doses of oral poliovirus vaccine coverage and paralytic poliomyelitis incidence in children under 1 year of age in the Americas, 1969–1994.[225]

or immunocompromised patients and for those expecting to travel to places where transmission is taking place. Nor has eIPV been incorporated into the EPI regimen for the developing world.[280,281] These issues are discussed in Chapter 21. Recommendations will change soon.

Rubella is another possible candidate because it is limited to the human host and has one serotype, but about half the infections are subclinical, and disease is not of high economic impact; protection of congenital rubella in the newborn is the major objective. The number of cases reported to the CDC reached a nadir of 225 in 1988, whereupon cases in unvaccinated preschool and inadequately protected college-age individuals followed the pattern seen with measles (see Section 14.1.2) by multiplying to 1401 in 1991—a number comparable to those seen in the early 1980s.[52] With an intensive campaign to improve coverage with MMR, the numbers resumed their decline to fewer than 200. Immunization both in childhood and young adulthood may be necessary to give long enough protection.

14.1.3. Strategies for Vaccine Delivery. It seems unlikely that elimination of measles or poliomyelitis can be fully achieved in the developing world, but control of the clinical disease and its associated mortality seems a worthy and attainable objective over time. Vaccines will only be effective if they are administered to the persons who need them. The major reasons for ineffectiveness of immunization programs in the United States, as addressed in a recent review,[142] include failure to deliver potent vaccines properly to target populations and inadequate stimulation of the immune response due to compromising vaccine or host factors. Various strategies have evolved in developed and in developing countries to achieve this end.

They must be adjusted to the social, economic, cultural, religious, climatic, and logistic setting in which they are used. In the United States the various states, operating under the guidance and encouragement of the CDC, have initiated the requirement for completion of vaccines against childhood diseases as a criterion for entry into the school system. The proof required and the vigor of the enforcement of the regulations have been the major determinants of whether success has been achieved within the existing guidelines for preventable diseases.

As noted earlier[225] success against polio was recently highlighted with the official declaration of its eradication from the Western hemisphere, and reintroduction of measles and rubella into areas where they had been eliminated was effectively countered with aggressive targeting of vulnerable populations. In developing countries a variety of vaccine strategies have been tried in an effort to control the six diseases targeted by the World Health Organization. These include (1) integration of vaccines into the primary care program with special emphasis where necessary; (2) obtaining political endorsement for the vaccine program both at the national level and at the smallest administrative unit where vaccine is to be given; (3) seeking help from volunteers and from the community; (4) creating special vaccine days, a "national vaccine day," or "pulse vaccine days" such as used in some parts of India when intensive periodic programs are carried out in different communities; (5) using mobile teams to go from village to village; (6) vaccinating from household to household; (7) delivery to concentrated population groups and setting up satellite vaccine stations in remote areas; (8) providing extra services such as oral rehydration salts to mothers bringing infants for immunization; (9) vaccinating children when they are brought in to clinics or hospitals for medical care, because the risks of reactions to the vaccine are usually less than those of leaving the child unvaccinated; (10) establishing free hospital vaccination clinics for such children as well as for tetanus immunization of pregnant mothers; (11) creating a health registry indicating what children need to be vaccinated, their current vaccine status, when the next shot is due, and the child's height and weight; and (12) setting up means to preserve the viability of vaccine by preserving and monitoring the "cold chain." The anticipated efforts of the Children's Vaccine Initiative (see Section 14.1.1) to develop combinations of vaccine that will simplify delivery raise hopes of wider and more long-lasting levels of protection. In the meantime, documentation of successes and failures of immunization programs throughout the world continues to be an important mission of the local and regional components of WHO. The application of surveillance techniques to immunization program evaluation is addressed in Chapter 3.

14.2. Chemoprophylaxis and Therapy

Rapid progress is being made in the development of antiviral drugs effective against various points in the replicative cycle of viruses, a process that includes adsorption, penetration, uncoating, transcription, penetration, uncoating, transcription and translation of regulatory proteins, genome replication, transcription and translation of structural proteins, virion assembly, and maturation and release.[17,79,216,273] Pharmacological mechanisms, pharmacokinetic properties, and details of toxicity of various antiviral agents are beyond this summary (see texts in Suggested Reading and references 17 and 269). Increased knowledge of this replicative cycle has led to the production of many compounds in various stages of development and testing. The steps include *in vitro* experiments, animal testing, human clinical trials, and, finally, if all goes well, licensure. Rapid diagnostic techniques now permit early use of the drugs shown to be effective against a specific virus in a given clinical setting. The limitations to the development of antiviral agents is that *in vitro* and animal models do not always predict their effectiveness in actual human use, that different viruses, even strains of the same virus, may respond differently, that no drug is truly virucidal, and that resistance to the drug may emerge. Toxicity may present a problem because of the difficulty of drugs in distinguishing sufficiently between certain host cell functions and viral replication. Longer-term toxic effects such as oncogenicity and teratogenicity are also of concern. In contrast to bacterial antibiotics, few available antiviral agents are broad spectrum in their activity, but some are useful in both prophylaxis and therapy against a single virus. An obvious major challenge is to develop preparations that would suppress actively replicating HIV or in some other way prevent the clinical consequences of the retroviral infection.

14.2.1. Amantadine and Rimantadine.
Amantadine hydrochloride is effective in prophylaxis, and to a lesser extent in therapy, against most influenza A strains but not against influenza B strains. Rimantadine, a related drug, was licensed in the United States in 1993 after more extensive use in Europe, including Great Britain and Russia. It is thought to be more effective in both treating and preventing influenza A infection in adults but only in treatment of children, and it is less toxic than amantadine. The mechanism of action of these drugs involves interference with uncoating of the virus after it penetrates the cell and possibly with virus maturation and assembly.

When used in prophylaxis, these drugs have been at least 50% and up to 90% effective in preventing infection with influenza A and at least 60% effective against the development of clinical illness. This dissociation between prevention of infection and disease is actually beneficial to the extent that infection without disease confers immunity.[17,269] Treatment of symptoms accelerates recovery somewhat. For greatest therapeutic efficacy these agents must be given within 48 hr of onset of illness. There is growing concern about the emergence of resistance during the 2- to 4-week courses of therapy. Reactions to amantadine occur in 5–10% of recipients and consist of mild CNS symptoms such as anxiety, insomnia, and difficulty in concentrating. Rimantadine also produces insomnia, nausea, and dizziness. These side effects may limit the use of both compounds in elderly patients, especially those in nursing homes, where they would optimally be combined with influenza vaccination.

14.2.2. Ribavirin.
Ribavirin is a purine nucleoside analogue that has shown a wide spectrum of activity *in vitro* against both RNA and DNA viruses. Its precise mechanism of action is not clear. Its use as an aerosol has been approved by the U.S. Food and Drug Administration only for carefully selected cases of severe RSV infections of infants and young children. Placebo-controlled trials have shown significantly greater improvement in the severity of illness, in arterial oxygen saturation, and in shorter duration of virus shedding.[17] No significant toxicity has been noted during therapy. Careful respiratory monitoring must be maintained throughout treatment. A recent trial in infants requiring assisted ventilation emphasized the value of ribavirin relative to the earlier concerns that precipitation of the drug in the respiratory equipment might interfere with safe and effective ventilation of the patient. Deterioration of respiratory function has been associated with ribavirin use in infants and to some extent in adults with chronic obstructive lung disease or asthma. Ribavirin is not indicated in milder respiratory infections caused by RSV, in which the course runs less than the 3 to 7 days required for complete treatment with ribavirin. Results of therapy in influenza A and B infections have been conflicting. Animal and human experimental studies have suggested its usefulness in hepatitis A, measles, HSV infections, and, most notably, Lassa fever.[17] The drug has shown efficacy in the treatment of Korean hemorrhagic fever, a syndrome due to infection with a bunyavirus; however, during the 1993 outbreak of hantavirus pulmonary syndrome in the southwestern United States, too few cases of this disease occurred to provide a clear indication of efficacy. Because it crosses the blood–brain barrier well, it may prove useful for treating bunyavirus

(e.g., LaCrosse) encephalitis. Its efficacy as a single agent against HIV has been disappointing, but in combination with other effective drugs it may have some value. Reversible hematologic toxicity has been noted, and the regimen for RSV is rather expensive.

14.2.3. Vidarabine. Vidarabine is active against the human herpesviruses. It inhibits nucleic acid synthesis through one or more mechanisms. Clinically, it has several uses in herpetic infections. In placebo-controlled trials in proved cases of herpes simplex encephalitis, vidarabine reduced the mortality from 70 to 28% at 1 month and to 40% at the end of 6 months.[17] About half of the survivors had relatively normal function at the end of a year. In newborn infants with CNS or disseminated infections with HSV, the mortality dropped from 74 to 38%, but only 29% of the survivors were normal at the end of 1 year. The drug has also been used successfully in topical application for acute keratoconjunctivitis and stomatitis. Vidarabine is approved for use intravenously against VZV infection manifesting as zoster (shingles) in immunocompromised patients; its effectiveness has been demonstrated to benefit cutaneous and visceral manifestations, by reduction in new lesion formation, and by reduction in the duration of viral shedding as well as by shortening of episodes of postherpetic neuralgia.[273] Adverse effects have been relatively minor. More important limitations to its use have been the need for a brain biopsy to establish the diagnosis of encephalitis due to HSV and the large volume of infusion fluid required to administer the drug because of its low solubility. Results of trials of vidarabine in chronic hepatitis B infection have been more discouraging.

14.2.4. Antiherpesvirus Drugs. By early 1995, there were three agents licensed for use exclusively against HHV, and several other related compounds under active clinical investigation. Acyclovir is the most effective antiviral available. It is a potent and specific inhibitor of certain herpesviruses in which a virus-encoded thymidine kinase, present in infected tissues, phosphorylates the drug to its active form, acyclovir monophosphate. The drug has a high therapeutic index against HSV-1, HSV-2, and to a lesser extent VZV, all of which produce deoxypyrimidine kinase; it has more limited effect in attenuating infection with CMV, which does not produce this enzyme. Epstein–Barr virus is more sensitive than CMV to the acyclovir (although the former does not produce its own deoxypyrimidine), perhaps through the drug's action on EBV DNA polymerase. Acyclovir is available for intravenous, oral, and topical use. Intravenous administration has proved of marked benefit in primary genital infections and in mucocutaneous HSV infections in im-

munosuppressed patients. Topical therapy has been less effective in diminishing the duration of illness. Virus shedding, healing time, new lesion formation, and the duration of symptoms are reduced under treatment.[17,273] However, virus shedding and new lesions may develop after discontinuation of the drug. Intravenous acyclovir is effective against VZV infection and may suppress CMV in transplantation patients. Oral azcyclovir therapy is also effective in primary orofacial or genital HSV infections, especially in immunosuppressed patients, but of questionable value in reducing subsequent recurrences. It is not well absorbed through the gastrointestinal mucosa. Early patient-initiated therapy for recurrences may shorten the episode by about 30%. Selection of resistant variants in the presence of acyclovir is a significant threat to the sustained usefulness of the drug. Long-term prophylaxis is highly effective at suppressing symptoms even though it does not prevent shedding of virus.[17]

Ganciclovir is a nucleoside analogue with activity against the herpesviruses but with considerably greater efficacy than acyclovir against CMV. In the context of immunosuppression, particularly with HIV infection, it has been very effective in treating retinitis but less so in pneumonia due to CMV. Foscarnet is a pyrophosphate analogue that is likewise quite effective against CMV retinitis. It has been used against acyclovir-resistant HSV as well. Ongoing trials of these newer agents should clarify their therapeutic value, particularly in HIV-infected patients. Other drugs related to acyclovir (e.g., famciclovir and valacyclovir) are promising in their equivalent or superior efficacy and more favorable pharmacokinetics following oral administration.

14.2.5. Cytokines. As a class, compounds known as interferons have long and frequently been used against various viral infections, although it is not clear whether they are acting directly against the viruses as well as indirectly through immunomodulatory mechanisms. Some but not all patients with chronic hepatitis due to HBV infection respond to IFN-α, and early trials in HCV infection have been encouraging despite relapses in a proportion of those treated. Topical IFN preparations have proved widely effective against genital condylomata (warts); however, it is unknown whether these compounds reduce the risk of the more serious sequelae of HPV infection. Much work is needed to elaborate the indications and adverse effects of the IFNs, and various cytokines and other immunomodulatory agents are at even earlier stages of development for therapeutic use.

14.2.6. Antiretroviral Drugs. Other nucleoside analogues have been intensively studied as retroviral reverse transcriptase inhibitors. They are the most potent agents for the treatment of HIV infection currently available. Zidovudine (also known as azidothymidine or AZT), the most thoroughly examined, is available for oral and intravenous use. In many studies it has retarded progression of the infection, delayed onset of AIDS, reduced mortality, or diminished symptoms. For various reasons, however, including emergence of viral resistance, the efficacy is short-lived. Data from some studies suggest that the duration of its effectiveness is similar whether treatment is begun early or late in the course of infection. However, a recent randomized placebo-controlled study of the capacity of zidovudine administered early in pregnancy to prevent congenital infection was terminated abruptly when it became apparent that among the treated women the proportions of infants of infected mothers to whom virus was being transmitted had dropped from the more than 20% expected to 8%.[71] In earlier trials using higher doses, hematologic toxicity, especially a macrocytic anemia, forced interruption or termination of therapy, but in subsequent studies lower doses have been effective and better tolerated. Insufficient data are available to assess whether the antiviral benefit of zidovudine for postexposure prophylaxis, for example, following nosocomial needle-puncture injury or aerosol–mucous membrane contact, will outweigh the side effects.

Didanosine (DDI) and zalcitabine (DDC) are two other nucleoside analogue HIV reverse transcriptase inhibitors that are available for use alone or principally in combination. Both are as active as zidovudine against intracellular HIV but have longer half-lives and are effective against zidovudine-resistant isolates. Peripheral neuropathy has been a frequent complication for both, and pancreatitis has limited the usefulness of didanosine. Other nucleoside and non-nucleoside inhibitors have shown encouraging effects in trials. However, the greatest promise comes from the protease inhibitors, which, in combination with the other drugs, can substantially suppress viral replication for weeks to months, raising hopes of a more fundamental change in the outlook for optimally managed patients.

15. References

1. AACH, R. D., SZMUNESS, W., MOSELY, J. W., HOLLINGER, F. B., KAHN, R. A., STEVENS, C. E., EDWARDS, V. M., AND WELCH, J., Serum alanine aminotransferase of donors in relation to the risk of non-A, non-B hepatitis in recipients, *N. Engl. J. Med.* **304**:989–994 (1981).

2. ABBEY, H., An examination of the Reed–Frost theory of epidemics, *Hum. Biol.* **24**:201–233 (1952).

3. ADAM, E., KAUFMAN, R. H., ADLER-STORHZ, J. H., MELNICK, J. L., AND DREESMAN, G. R., A prospective study of association of herpes simplex virus and papilloma virus infection with cervical neoplasia in women exposed to diethylstilbesterol *in utero*, *Int. J. Cancer* **35**:19–26 (1985).

4. ALTER, H. J., Posttransfusion hepatitis: Clinical features, risk and donor testing, in: *Infection, Immunity and Blood Transfusion* (R. Y. DODD, AND L. F. BARKER, EDS.), pp. 47–61, Liss, New York, 1984.

5. ANDERSON, R. M., AND MAY, R. M., Vaccination against rubella and measles: Quantitative investigations of different policies, *J. Hyg. (Camb.)* **90**:259–325 (1983).

6. ANDERSON, R. M., GUPTA, S., AND MAY, R. M., Potential of community-wide chemotherapy or immunotherapy to control the spread of HIV-1, *Nature* **350**:356–359 (1991).

7. ARDUINO, R. C., AND DUPONT, H. L., Travellers' diarrhoea, *Baillieres Clin. Gastroenterol.* **7**:365–385 (1993).

8. ARMSTRONG, J. A., EVANS, A. S., RAO, N., AND HO, M., Viral infections in renal transplant recipients, *Infect. Immun.* **14**:970–975 (1976).

9. ASHTON, L. J., LEARMONT, J., LUO, K., WYLIE, B., AND STEWART, G., HIV infections in recipients of blood products from donors with known duration of infection, *Lancet* **344**:718–720 (1994).

10. BACCHETTI, P., Estimating the incubation period of AIDS by comparing population infection and diagnosis patterns, *J. Am. Stat. Assoc.* **85**:1002–1008 (1990).

11. BACHETTI, P., AND MOSS, A. R., Incubation period of AIDS in San Francisco, *Nature* **338**:251–253 (1989).

12. BAILEY, N. T. J., *The Mathematical Theory of Infectious Diseases*, 2nd ed., Macmillan, New York, 1975.

13. BARON, R. C., DICKER, R. C., BUSSELL, K. E., AND HERNDON, J. L., Assessing trends in mortality in 121 U.S. cities, 1970–79, from all causes and from pneumonia and influenza, *Public Health Rep.* **103**:120–128 (1988).

14. BARRON, K. S., Kawasaki disease: Epidemiology, late prognosis, and therapy, *Rheum. Dis. Clin. North Am.* **17**:907–919 (1991).

15. BAUER, H. M., TING, Y., GREER, C. E., CHAMBERS, J. C., TASHIRO, C. J., CHIMERA, J., REINGOLD, A., AND MANOS, M. M., Genital human papilloma virus infection in female university students as determined by a PCR-based method, *J.A.M.A.* **265**:472–477 (1991).

16. BAUMGARTNER, I., AND KLEBS, E., A centennial note, *N. Engl. J. Med.* **213**:60–63 (1935).

17. BEAN, B., Antiviral therapy: Current concepts and practices, *Clin. Microbiol.* **5**:146–182 (1992).

18. BEASLEY, P., HWANG, P.-Y., LIN, C.-Y., AND CHIEN, C.-S., Hepatocellular carcinoma and hepatitis B virus. A prospective study of 22707 men in Taiwan, *Lancet* **2**:1129–1132 (1981).

19. BEASLEY, R. P., AND HWANG, L.-Y., Overview of the epidemiology of hepatocellular carcinoma, in: *Viral Hepatitis and Liver Disease* (F. B. HOLLINGER, S. M. LEMON, AND H. S. MARGOLIS, EDS.), pp. 532–535, Williams and Wilkins, Baltimore, 1991.

20. BELL, D. M., BRINK, E. W., NITZKIN, J. L., HALL, C. B., WULFF, H., AND BERKOWITZ, I. D., Kawasaki syndrome: Description of two outbreaks in the United States, *N. Engl. J. Med.* **304**:1568–1575 (1981).

21. BELL, J. A., ROWE, W. P., ENGLER, J. L., PARROT, R. H., AND HUEBNER, R. J., Pharyngeal conjunctival fever: Epidemiological studies of a recent recognized disease entity, *J.A.M.A.* **175**:1083–1092 (1955).

22. BENENSON, A. S., (ED.), *Control of Communicable Diseases in*

Man, 16th ed., American Public Health Association, Washington, 1995.

23. BERGER, J. R., AND LEVY, J., The human immunodeficiency virus type 1—the virus and its role in neurologic disease, *Semin. Neurol.* **12**:1–9 (1992).

24. BERKELMAN, R., BRYAN, R. T., OSTERHOLM, M. T., LEDUC, J. W., AND HUGHES, J. M., Infectious diseases surveillance: A crumbling infrastructure, *Science* **264**:368–370 (1994).

25. BERMAN, S., AND MCINTOSH, K., Selective primary health care. Strategies for control of disease in the developing world. XXI. Acute respiratory infections, *Rev. Infect. Dis.* **7**:674–691 (1985).

26. BISHOP, R. F., DAVIDSON, G. P., HOLMES, I. H., AND RUCK, B. J., Virus particles in epithelial cells of duodenal mucosa from children with acute nonbacterial gastroenteritis, *Lancet* **2**:1281–1293 (1973).

27. BISHOP, R. F., DAVIDSON, G. P., HOLMES, I. H., AND RUCK, B. J., Detection of a new virus by electron microscopy of fecal extracts of children with acute gastroenteritis, *Lancet* **1**:149–151 (1974).

28. BISNO, A. L., BARRATT, N. P., SEVANSTON, W. H., AND SPENSE, L. P., An outbreak of acute respiratory disease in Trinidad associated with para-influenza virus, *Am. J. Epidemiol.* **91**:68–77 (1970).

29. BLACK, F. L., HIERHOLZER, W. J., PINEIRO, F. DEP., EVANS, A. S., WOODHALL, J. P., OPTON, E. M., EMMONS, J. E., WEST, B. S., EDSALL, G., DOWNS, W. G., AND WALLACE, G. D., Evidence for persistence of infections agents in isolated human populations, *Am. J. Epidemiol.* **100**:230–250 (1974).

30. BLACK, R. E., MERSON, M. H., RAHMAN, A. S. M., YUNUS, M., ALIM, A. R. M., HUQ, I., YOLKEN, R. H., AND CURLIN, G. T., A two-year study of bacterial, viral and parasitic agents associated with diarrhea in rural Bangladesh, *J. Infect. Dis.* **142**:660–664 (1980).

31. BLACK, R. E., Epidemiology of travelers' diarrhea and relative importance of various pathogens, *Rev. Infect. Dis.* **12**(Suppl. 1)**:**S73–S79 (1990).

32. BLATTNER, W. A., BIGGAR, R. J., WEISS, S. H., MELBYE, M., AND GOEDERT, J. J., Epidemiology of human T-lymphotropic virus type III and the risk of the acquired immunodeficiency syndrome, *Ann. Inter. Med.* **103**:665–670 (1985).

33. BLOWER, S. M., ANDERSON, R. M., AND WALLACE, P., Loglinear models, sexual behavior and HIV: Epidemiological implications of heterosexual transmission, *J. Acquir. Immune Defic. Syndr.* **3**:763–772 (1990).

34. BLUMBERG, B. S., ALTER, H. J., AND VISNICK, S., A "new" antigen in leukemia sera, *J.A.M.A.* **191**:541–546 (1965).

35. BOURGAULT, I., GOMEZ, A., GOMRAD, E., PICARD, F., AND LEVY, J. P., A virus-specific CD4+ cell-mediated cytolytic activity revealed by CD8+ cell elimination regularly develops in uncloned human antiviral cell lines, *J. Immunol.* **142**:252–256 (1989).

36. BRADLEY, D. W., Hepatitis E: Epidemiology, aetiology and molecular biology, *Rev. Med. Virol.* **2**:19–28 (1992).

37. BRANDT, C. D., KIM, H. W., RODRIGUEZ, W. J., ARROBIO, J. O., JEFFRIES, B. C., STALLINGS, E. P., LEWIS, C., MILES, A. J., CHANOCK, R. M., KAPIKIAN, A. Z., AND PARROTT, R. H., Pediatric viral gastroenteritis during eight years of study, *J. Clin. Microbiol.* **18**:71–78 (1983).

37a. BRANCON, B. M., Early intervention for persons infected with human immunodeficiency virus, *Clin. Infect. Dis.* **20**(Suppl.):S3–S22 (1995).

38. BRODY, J. A., AND DETELS, R., Subacute sclerosing panencephalitis: A zoonosis following aberrant measles, *Lancet* **2**:500–501 (1970).

39. BROOKMAYER, R., AND GAIL, M. H., *AIDS Epidemiology: A Quantitative Approach*, Oxford University Press, New York, 1994.

40. BUCHBINDER, S. P., KATZ, M. H., HESSOL, N. A., O'MALLEY, P. M., AND HOLMBERG, S. D., Long-term HIV-1 infection without immunologic progression, *AIDS* **8**:1123–1128 (1994).

41. BYRNE, E. B., EVANS, A. S., FONTS, D. W., AND ISRAEL, H. L., A seroepidemiological study of Epstein–Barr virus and other viral antigens in sarcoidosis, *Am. J. Epidemiol.* **97**:335–363 (1973).

42. CARPENTER, C. B., AND DAVID, J., Immunology—histocompatibility antigens and immune response genes, in: *Scientific American Medicine* (E. RUBENSTEIN AND D. D. LEDERMAN, EDS.), pp. 1–10, Scientific American, New York, 1991.

43. CENTERS FOR DISEASE CONTROL AND PREVENTION, Guidelines for investigating clusters of health events, *Morbid. Mortal. Week. Rep.* **39**(No. RR-11):1–23 (1990).

44. CENTERS FOR DISEASE CONTROL AND PREVENTION, Immunization Practices Advisory Committee, Update on Adult Immunization: Recommendations of the Immunization Practices Advisory Committee (ACIP), *Morbid. Mortal. Week. Rep.* **40**(No. RR-12):1–94 (1991).

45. CENTERS FOR DISEASE CONTROL AND PREVENTION, Update: International task force for disease eradication, 1990 and 1991, *Morbid. Mortal. Week. Rep.* **40**:40–42 (1992).

46. CENTERS FOR DISEASE CONTROL AND PREVENTION, Summary of notifiable diseases, United States, 1991, *Morbid. Mortal. Week. Rep.* **40**:1–64 (1992).

47. CENTERS FOR DISEASE CONTROL AND PREVENTION, Standards for pediatric immunization practices, *Morbid. Mortal. Week. Rep.* **42**(No. RR-5):1–13 (1991).

48. CENTERS FOR DISEASE CONTROL AND PREVENTION, Hepatitis E among U.S. travelers, *Morbid. Mortal. Week. Rep.* **42**:1–4 (1993).

49. CENTERS FOR DISEASE CONTROL AND PREVENTION, Recommendations of the Advisory Committee on Immunization Practices (ACIP): Use of vaccines and immune globulins in persons with altered immunocompetence, *Morbid. Mortal. Week. Rep.* **42**(No. RR-16):1–18 (1993).

50. CENTERS FOR DISEASE CONTROL AND PREVENTION, Standards for pediatric immunization practices. Recommended by the National Vaccine Advisory Committee, *Morbid. Mortal. Week. Rep.* **42**(No. RR-5):1–13 (1993).

51. CENTERS FOR DISEASE CONTROL AND PREVENTION, Recommendations of the International Task Force for Disease Eradication, 1990 and 1991, *Morbid. Mortal. Week. Rep.* **42**(No. RR-16) (1993).

52. CENTERS FOR DISEASE CONTROL AND PREVENTION, Summary of notifiable diseases, United States, 1993, *Morbid. Mortal. Week. Rep.* **42**:1–73 (1994).

53. CENTERS FOR DISEASE CONTROL AND PREVENTION, Addressing emerging infectious disease threats: A prevention strategy for the United States. Atlanta, Georgia: U.S. Department of Health and Human Services, Public Health Service, 1994.

54. CENTERS FOR DISEASE CONTROL AND PREVENTION, General recommendations on immunization, *Morbid. Mortal. Week. Rep.* **43**:1–38 (1994).

55. CENTERS FOR DISEASE CONTROL AND PREVENTION, Hantavirus pulmonary syndrome—northeastern United States, 1994, *Morbid. Mortal. Week. Rep.* **43**:548–556 (1994).

56. CENTERS FOR DISEASE CONTROL AND PREVENTION, Dengue fever among U.S. military personnel—Haiti, September–November 1994, *Morbid. Mortal. Week. Rep.* **43**:845–848 (1994).

57. CENTERS FOR DISEASE CONTROL AND PREVENTION, Bolivian hemorrhagic fever—El Beni department, Bolivia, 1994, *Morbid. Mortal. Week. Rep.* **43:**943–946 (1994).

58. CENTERS FOR DISEASE CONTROL AND PREVENTION, Recommended childhood immunization schedule—United States, January 1995, *Morbid. Mortal. Week. Rep.* **43:**959–960 (1995).

59. CENTERS FOR DISEASE CONTROL AND PREVENTION, Physician vaccination referral practices and vaccines for children—New York, 1994, *Morbid. Mortal. Week. Rep.* **44:**3–6 (1995).

60. CHAISSON, R. E., AND GRIFFIN, D. E., Progressive multifocal leukoencephalopathy in AIDS, *J.A.M.A.* **264:**79–82 (1990).

61. CHAN, C. K., MUELLER, N., EVANS, A., HARRIS, N. L., COMSTOCK, G. W., JELLUM, E., MAGNUS, K., ORENTREICH, N., POLK, B. F., AND VOGELMAN, J., Epstein–Barr virus antibody patterns preceding the diagnosis of nasopharyngeal carcinoma, *Cancer Causes Control* **2:**125–131 (1991).

62. CHANG, Y., CESARMAN, E., PESSIN, M. S., LEE, F., CULPEPPER, J., KNOWLES, D. M., AND MOORE, P. S., Identification of herpesvirus-like DNA sequences in AIDS-associated Kaposi's sarcoma, *Science* **266:**1865–1869 (1994).

63. CHANOCK, R., CHAMBON, L., CHANG, W., GERREIRA, F. G., GHARPURE, P., GRANT, L., HATEL, J., INMAN, I., KALRA, S., LIM, K., MADALENGOITIA, J., SPENSE, L., TENG, P., AND FERREIRA, W., WHO respiratory survey in children: A serological study, *Bull. WHO* **37:**363–369 (1967).

64. CHANOCK, R. M., PARROTT, R. H., KAPIKIAN, A. Z., KIM, H. W., AND BRANDT, C. D., Possible role of immunological factors in pathogenesis of RS virus lower respiratory tract disease, *Perspect. Virol.* **6:**125–135 (1968).

65. CHERRY, J. D., Contemporary infectious exanthems, *Clin. Infect. Dis.* **16:**199–207 (1993).

66. CHOI, K., AND THACKER, S. B., Improved accuracy and specificity of forecasting deaths attributed to pneumonia and influenza, *J. Infect. Dis.* **144:**606–608 (1981).

67. COHEN, J., Special news report: The Duesberg phenomenon, *Science* **266:**1642–1649 (1994).

68. COMSTOCK, G. W., Evaluating vaccination effectiveness and vaccine efficacy by means of case-control studies, *Epidemiol. Rev.* **16:**77–89 (1994).

69. CONE, R. W., HACKMAN, R. C., HUANG, M-L. W., BOWDEN, R. A., MEYERS, J. D., METCALF, M., ZEH, J., ASHLEY, R., AND COREY, L., Human herpesvirus 6 in lung tissue from patients with pneumonitis after bone marrow transplantation, *N. Engl. J. Med.* **329:**156–161 (1993).

70. CONNOLLY, J. H., ALLEN, I. V., HURWITZ, L. J., AND MILLAR, J. H. D., Measles-virus antibody and antigen in subacute sclerosing panencephalitis, *Lancet* **1:**542–544 (1967).

71. CONNOR, E. M., SPERLING, R. S., GELBER, R., KISELEV, P., SCOTT, G., O'SULLIVAN, M. J., VANDYKE, R., BEY, M., SHEARER, W., JACOBSON, R. L., JIMENEZ, E., O'NEILL, E., BAZIN, B., DELFRAISSY, J.-F., CULNANE, M., COOMBS, R., ELKINS, M., MOYE, J., STRATTON, P., AND BALSLEY, J., Reduction of maternal–infant transmission of human immunodeficiency virus type 1 with zidovudine treatment, *N. Engl. J. Med.* **331:**1173–1180 (1994).

72. COOPER, D. A., GOLD, J., MACLEAN, P., DONOVAN, B., FINLAYSON, R., BARNES, T. G., MICHELMORE, H. M., BROOKE, P., AND PENNY, R., FOR THE SYDNEY AIDS STUDY GROUP, Acute AIDS retrovirus infection. Definition of a clinical illness associated with seroconversion, *Lancet* **1:**537–540 (1985).

73. COREY, L., Genital herpes, in: *Sexually Transmitted Disease* (K. K. HOLMES, P-A. MARDH, P. F. SPARLING, AND P. J. WIESNER, EDS.), pp. 391–414, McGraw-Hill, New York, 1990.

74. CORRIEL, L. L., Clinical syndromes in children caused by respiratory infection, *Med. Clin. North Am.* **51:**819–830 (1967).

75. CUBITT, W. D., The candidate caliciviruses, *CIBA Found. Symp.* **128:**126–143 (1987).

76. DAVIDSON, G. P., BISHOP, R. F., TOWNLEY, R. R. W., HOLMES, I. H., AND RUCK, B. J., Importance of a new virus in acute sporadic enteritis in children, *Lancet* **1:**242–245 (1975).

77. DE THE, G., GESER, A., DAY, N. E., TUBEI, P. M., WILLIAMS, E. H., BIER, D. P., SMITH, D. G., DEAN, A. G., BORNHAMM, G. W., FEORINO, P., AND HENLE, W., Epidemiological evidence for causal relationship between Epstein–Barr virus and Burkitt's lymphoma from Ugandan prospective study, *Nature* **274:**756–761 (1978).

78. DOBBINS, J. G., AND STEWART, J. A., Surveillance of congenital cytomegalovirus disease, 1990–1991, *Morbid. Mortal. Week. Rep.* **41** (1992).

79. DOLIN, R., Antiviral chemotherapy and chemoprophylaxis, *Science* **227:**1296–1303 (1985).

80. DOUGHERTY, W. J., AND ALTMAN, R., Viral hepatitis in New Jersey 1960–1961, *Am. J. Med.* **32:**704–716 (1962).

81. DOUGLAS, R. G., JR., Influenza in man, in: *The Influenza Viruses and Influenza* (E. D KILBOURNE, ED.), pp. 395–447, Academic Press, New York, 1975.

82. DOUGLAS, R. G., BETTS, R. F., HRUSKA, J. F. AND HALL, C. B., Epidemiology of nosocomial viral infections, in: *Seminars in Infectious Disease*, Vol. II (L. WEINSTEIN AND B. H. FIELDS, EDS.), pp. 98–144, Stratton Intercontinental Medical Book Corp., New York, 1979.

83. DUCHIN, J. S., KOSTER, F. T., PETERS, C. J., SIMPSON, G. L., TEMPEST, B., ZAKI, S. R., KSIAZEK, T. T., ROLLIN, P. E., NICHOLS, S., UMLAND, E. T., MOOLENAAN, R. L., REEF, S. E., NOLTE, K. B., GALLERHER, M. M., BUTLER, J. C., BREIMAN, R. F., AND THE HANTAVIRUS STUDY GROUP, Hantavirus pulmonary syndrome: A clinical description of 17 patients with a newly recognized disease, *N. Engl. J. Med.* **330:**949–955 (1994).

84. DUESBERG, P., AIDS epidemiology. Inconsistencies with human immunodeficiency virus and with infectious disease, *Proc. Natl. Acad. Sci. USA* **88:**1575–1579 (1991).

85. DUESBERG, P., HIV is not the cause of AIDS, *Science* **241:**514–516 (1988).

86. DWYER, D. M., STRICKLER, H., GOODMAN, R. A., AND ARMENIAN, H. K., Use of case-control studies in outbreak investigations, *Epidemiol. Rev.* **16:**109–123 (1994).

87. EDITORIAL, Rotaviruses of man and animals, *Lancet* **1:**257–259 (1975).

88. ELVEBACK, L. R., ACKERMAN, E., YOUNG, G., AND FOX, J. P., A stochastic model for competition between viral agents in the presence of interference. 1. Live virus vaccine in randomly mixing population, model III, *Am. J. Epidemiol.* **87:**373–384 (1968).

89. ENGLUND, J. A., SULLIVAN, C. A., JORDAN, M. C., DEHNER, L. P., VERCELLOTTI, G. M., AND BALFOUR, H. H., JR., Respiratory syncytial virus infection in immunocompromised adults, *Ann. Intern. Med.* **109:**203–208 (1988).

90. EVANS, A. S., Clinical syndromes in adults caused by respiratory infection, *Med. Clin. North Am.* **5:**803–818 (1967).

91. EVANS, A. S., The spectrum of infections with Epstein–Barr virus: A hypothesis, *J. Infect. Dis.* **124:**330–337 (1971).

92. EVANS, A. S., Clinical syndromes associated with EB virus infection, *Adv. Intern. Med.* **18:**77–93 (1972).

93. EVANS, A. S., Diagnosis and prevention of common respiratory infection, *Hosp. Pract.* **10:**31–41 (1974).

94. EVANS, A. S., New discoveries in infectious mononucleosis, *Mod. Med.* **42**:18–24 (1974).

95. EVANS, A. S., EB virus, infectious mononucleosis and cancer: The closing of the web, *Yale J. Biol. Med.* **47**:113–122 (1974).

96. EVANS, A. S., Causation and disease: The Heenle–Koch postulates revisited, *Yale J. Biol. Med.* **49**:175–195 (1976).

97. EVANS, A. S., Limitations of Koch's postulates, *Lancet* **2**:1277–1278 (1977).

98. EVANS, A. S., Causation and disease: A chronological journey, *Am. J. Epidemiol.* **108**:249–258 (1978).

99. EVANS, A. S., AIDS: The alternative view, *Lancet* **339**:1547 (1992).

100. EVANS, A. S., Criteria for control of infectious diseases using poliomyelitis as an example, *Prog. Med. Virol.* **29**:141–165 (1984).

101. EVANS, A. S., The eradication of communicable diseases: Myth or reality? *Am. J. Epidemiol.* **122**:199–207 (1985).

102. EVANS, A. S., The clinical illness promotion factor. A third ingredient, *Yale J. Biol. Med.* **55**:193–199 (1985).

103. EVANS, A. S., Does HIV cause AIDS? An historical perspective, *J. Acquir. Immune Dis. Syndr.* **2**:107–113 (1989).

104. EVANS, A. S., Epstein–Barr vaccine: Use in infectious mononucleosis, in: *Epstein–Barr Virus and Associated Diseases* (T. TURSZ, J. S. PAGANO, D. V. ABLASHI, G. DE THE, G. LENOIR, AND G. PEARSON, EDS.), pp. 593–598, Collogue/INSERM, John Libbey, Eurotext, London, 1993.

105. EVANS, A. S., *Causation and Disease. A Chronological Journey*, Plenum Press, New York, 1993.

106. EVANS, A. S., AND CAMPOS, L. E., Acute respiratory disease in students at the University of the Philippines, *Bull. WHO* **45**:103–112 (1971).

107. EVANS, A. S., CAMPOS, L. E., D'ALLESSIO, D. A., AND DICK, E. C., Acute respiratory disease in University of the Philippines and University of Wisconsin students: A comparative study, *Bull. WHO* **36**:397–407 (1967).

108. EVANS, A. S., AND COMSTOCK, G., Presence of elevated antibody titres to Epstein–Barr virus before Hodgkin's disease, *Lancet* **1**:1182–1186 (1981).

109. EVANS, A. S., COX, F., NANKERVIS, G., OPTON, E., SHOPE, R., WELLS, A. V., AND WEST, B., A health and seroepidemiological survey of a community in Barbados, *Int. J. Epidemiol.* **3**:267–275 (1974).

110. EVANS, A. S., AND GUTENSOHN, N., A population-based case-control study of EBV and other viral antibodies among persons with Hodgkin's disease and their siblings, *Int. J. Cancer* **34**:149–157 (1984).

111. EVANS, A. S., KIRCHHOFF, L. V., PANUTTI, C. S., CARVALHO, R. P. S., AND MCCLELLAND, K. A., Case-control study of Hodgkin's disease in Brazil. II. Seroepidemiologic studies in cases and family members, *Am. J. Epidemiol.* **122**:609–618 (1980).

112. EVANS, A. S., AND MUELLER, N., Viruses and cancer. Causal associations, *Ann. Epidemiol.* **1**:71–92 (1989).

113. EVANS, A. S., AND MUELLER, N. E., The past is prologue: Use of serum banks in cancer research, *Cancer Res.* **52**:5557s–5560s (1992).

114. EVANS, A. S., NIEDERMAN, J. C., AND MCCOLLUM, R. W., Sero-epidemiologic studies of infectious mononucleosis with EB virus, *N. Engl. J. Med.* **279**:1123–1127 (1968).

115. EVANS, A. S., NIEDERMAN, J. C., AND SAWYER, R. N., Prospective studies of a group of Yale University freshmen. II. Occurrence of acute respiratory infections and rubella, *J. Infect. Dis.* **123**:271–278 (1971).

116. EVANS, A. S., SHEPARD, K. A., AND RICHARDS, V. A., ABO blood groups and viral diseases, *Yale J. Biol. Med.* **45**:81–92 (1972).

117. FAUCI, A. S., CD4$^+$ T lymphocytopenia without HIV infection—no lights, no camera, just facts, *N. Engl. J. Med.* **328**:429–431 (1993).

118. FENNER, F. J., AND WHITE, D. O., *Medical Virology*, 3rd ed., Academic Press, Orlando, FL, 1990.

119. FIELDS, B. N., KNIPE, D. M., CHANOCK, R. M., MELNICK, J. L., ROIZMAN, B., AND SHOPE, R. E. (EDS.), *Field's Virology*, 2nd ed., Raven Press, New York, 1990.

119a. FINKELSTEIN, D. M., AND SCHOENFELD, D. A. (EDS.), *AIDS Clinical Trials*, John Wiley, New York, 1995.

120. FOX, J. P., AND HALL, C. B., *Viruses in Families: Surveillance of Families as a Key to Epidemiology of Virus Infections*, PSG Publishing, Littleton, MA, 1980.

121. FOX, R., ELDRED, L. J., FUCHS, E. J., KASLOW, R. A., VISSCHER, B. R., HO, M., PHAIR, J. P., AND POLK, B. F., Clinical manifestations of acute infection with human immunodeficiency virus in a cohort of gay men, *AIDS* **1**:35–38 (1987).

122. GAIL, M. H., AND BROOKMEYER, R., Methods for projecting the course of acquired immunodeficiency syndrome epidemic, *J. Natl. Cancer Inst.* **80**:900–911 (1988).

123. GERBERDING, J., Is antiretroviral treatment after percutaneous HIV exposure justified? *Ann. Intern. Med.* **188**:979–980 (1993).

124. GERONE, P. J., COUCH, R. B., KEEFER, G. V., DOUGLAS, R. G., DERRENBACHER, E. B., AND KNIGHT, V., Assessment of experimental and natural viral aerosols, *Bacteriol. Rev.* **30**:576–584, 584–588 (1966).

125. GLEZEN, W., AND DENNY, F., Epidemiology of acute lower respiratory disease in children, *N. Engl. J. Med.* **288**:498–505 (1973).

126. GOEDERT, J. J., KESSLER, C. M., ALEDORT, L. M., BIGGAR, R. J., ANDES, W. A., WHITE, G. C., DRUMMOND, J. E., VAIDYA, K., MANN, D. L., EYSTER, M. E., RAGNI, M., LEDERMAN, M. M., COHEN, A. R., BRAY, G. L., ROSENBERG, P. S., FRIEDMAN, R. M., HILGARTNER, M. W., BLATTNER, W. A., KRONER, B., AND GAIL, M. H., A prospective study of human immunodeficiency virus type 1 infection and the development of AIDS in subjects with hemophilia, *N. Engl. J. Med.* **321**:1141–1148 (1989).

127. GONZALEZ-SCARANO, F., AND TYLER, K. L., Molecular pathogenesis of neurotropic viral infections, *Ann. Neurol.* **22**:565–574 (1987).

128. GREGG, M. B., The principles of epidemic investigation, in: *Oxford Textbook of Public Health*, 2nd ed. (W. W. HOLLAND, R. DETELS, AND G. E. KNOX, EDS.), pp. 284–297, Oxford University Press, Oxford, 1991.

129. GRIFFIN, D. E., Alphavirus pathogenesis and immunity, in: *The Togaviridae and Flaviviridae* (S. SCHLESINGER AND M. J. SCHLESINGER, EDS.), pp. 209–249, Plenum Press, New York, 1986.

130. GUIDRY, G. G., BLACK-PAYNE, C. A., PAYNE, D. K., JAMISON, R. M., GEORGE, R. B., AND BOCCHINI, J. A., JR., Respiratory syncytial virus infection among intubated adults in a university medical intensive care unit, *Chest* **100**:1377–1384 (1991).

131. GURWITH, M. J., AND WILLIAMS, T. W., Gastroenteritis in children: A two-year review in Manitoba. I. Etiology, *J. Infect. Dis.* **136**:239–247 (1977).

132. GWALTNEY, J. M., JR., AND HENDLEY, J. O., Rhinovirus transmission: One if by air, two if by hand, *Am. J. Epidemiol.* **107**:357–361 (1978).

133. HABER, M., LONGINI, I. M., JR., AND HALLORAN, M. E., Estimation of vaccine efficacy in outbreaks of acute infectious diseases, *Stat. Med.* **10**:1773–1784 (1991).

134. HALL, C. B., DOUGLAS, R. G., JR., GEINAN, J. M., AND MESSNER,

M. K., Nosocomial respiratory syncytial infections, *N. Engl. J. Med.* **293:**1343–1346 (1975).

135. HALL, W. J., HALL, C. B., AND SPEERS, D. M., Respiratory syncytial virus infections in adults, *Ann. Intern. Med.* **88:**203–205 (1978).

136. HALLORAN, M. E., COCHI, S. L., LIEU, T. A., WHARTON, M., AND FEHRS, L., Theoretical epidemiologic and morbidity effects of routine varicella immunization of preschool children in the United States, *Am. J. Epidemiol.* **140:**81–104 (1994).

137. HENDERSON, D. K., FAHEY, B. J., WILLY, M., SCHMITT, J. M., CAREY, K., KOZIOL, D. E., LANE, H. C., FEDIO, J., AND SAAH, A. J., Risk for occupational transmission of human immunodeficiency virus type 1 (HIV-1) associated with clinical exposures: A prospective evaluation, *Ann. Intern. Med.* **113:**740–746 (1990).

138. HENDLEY, J. O., WENZEL, R. P., AND GWALTNEY, J. M., JR., Transmission of rhinovirus colds by self-induction, *N. Engl. J. Med.* **288:**1361–1364 (1974).

139. HENLE, G., HENLE, W., AND DIEHL, V., Relation of Burkitt's tumor-associated herpes-type virus to infectious mononucleosis, *Proc. Natl. Acad. Sci. USA* **59:**94–101 (1968).

140. HERSH, B. S., MARKOWITZ, L. E., HOFFMAN, R. B., HOFF, D. R., DORAN, M. J., FLEISHMAN, J. C., PREBLUD, S. R., AND ORENSTEIN, W. A., A measles outbreak at a college with a prematriculation immunization requirement, *Am. J. Public Health* **81:**360–364 (1991).

141. HESSOL, N. A., KOBLIN, B. A., VAN GRIENSVEN, G. J. P., BACHETTI, P., LIU, J. Y., STEVENS, C. E., COUTINHO, R. A., BUCHBINDER, S. P., AND KATZ, M. H., Progression of human immunodeficiency virus type 1 (HIV-1) infection among homosexual men in hepatitis B vaccine trial cohorts in Amsterdam, New York City, and San Francisco, 1978–1991, *Am. J. Epidemiol.* **139:**1077–1087 (1994).

142. HINMAN, A. R., ORENSTEIN, W. A., AND MORTIMER, E. A., When, where, and how do immunizations fail? *Ann. Epidemiol.* **2:**805–812 (1992).

143. HO, M.-S., YANG, C.-S., CHEN, P.-J., AND MAU, Y.-C., Intrafamilial transmission of hepatitis C virus, *J. Clin. Microbiol.* **32:**2824–2826 (1994).

144. HOGAN, T. F., PADGETT, B. L., AND WALKER, D. L., Human polyomaviruses, in: *Textbook of Human Virology* (R. B. BELSHE, ED.), pp. 970–1000, Mosby-Yearbook, St. Louis, 1991.

145. HOLMBERG, S. D., STEWART, J. A., AND GERBER, A. R., Prior herpes simplex virus type 2 infection as a risk factor for HIV infection, *J.A.M.A.* **259:**1048–1050 (1988).

146. HORSTMANN, D. M., LIEBHABER, H., LEEOUVIER, G. L., ROSENBERG, D. A., AND HALSTEAD, S. B., Rubella: Reinfection of vaccinated and naturally immune persons exposed in an epidemic, *N. Engl. J. Med.* **283:**771–778 (1970).

147. HUEBNER, R. J., The virologist's dilemma, *Ann. N. Y. Acad. Sci.* **67:**430–442 (1957).

148. HURWITZ, E. S., BARRETT, M. J., BREGMAN, D., GUNN, W. J., SCHONBERGER, L. B., FAIRWEATHER, W. R., DRAGE, J. S., LaMONTAGNE, J. R., KASLOW, R. A., BURLINGTON, D. B., QUNNAN, G. V., PARKER, R. A., PHILLIPS, K., PINSKY, P., DAYTON, D., AND DOWDLE, W. R., Public Health Service study of Reye's syndrome and medications: Report of the pilot phase, *N. Engl. J. Med.* **313:**849–857 (1985).

149. ING, D., GLASS, R. I., LEBARON, C. W., AND LEW, J. F., Laboratory-based surveillance for rotavirus, United States, January 1989–May 1991, *Morbid. Mortal. Week. Rep.* **41**(SS-3)**:**47–56 (1992).

150. IWARSON, S., The main five types of viral hepatitis: An alphabetical update, *Scand. J. Infect. Dis.* **24:**129–135 (1992).

151. JAFFE, H. W., CHOI, K., THOMAS, P., HAVERKOS, H. N., AUERBACH, D. M., GUINAN, M. E., ROGERS, M. F., SPIRA, T. J., DARROW, W. W., KRAMER, M. A., FRIEDMAN-KIEN, A. E., LAUBENSTEIN, L. J., MARMOR, M., SAFAI, B., DRITZ, S. K., CRISPI, S. J., FANNIN, S. L., ORKWIS, J. P., KELTER, A., RUSHING, W. R., THACKER, S. B., AND CURRAN, J., National case-control study of Kaposi's sarcoma and *Pneumocystic carinii* pneumonia in homosexual men. Part 1: Epidemiologic results, *Ann. Intern Med.* **99:**145–151 (1983).

152. JENISON, S. A., LEMON, S. M., BAKER, L. N., AND NEWBOLD, J. E., Quantitative analysis of hepatitis B virus DNA in saliva and semen of chronically infected homosexual men, *J. Infect. Dis.* **156:**299–307 (1987).

153. JOHNSON, R. E., NAHMIAS, A. J., MAGDER, L. S., LEE, F. K., BROOKS, C. A., AND SNOWDEN, C. B., A seroepidemiologic survey of the prevalence of herpes simplex virus type 2 infection in the United States, *N. Engl. J. Med.* **321:**7–12 (1989).

154. JOHNSON, T. R., *Viral Infections of the Nervous System*, Raven Press, New York, 1982.

155. JOHNSON, R. T., Persistent viral infections and demyelinating disease, in: *Viruses and Demyelinating Diseases* (C. A. MIMS, M. L. CUZNER, AND R. E. KELLY, EDS.), pp. 7–19, Academic Press, New York, 1983.

156. JOHNSON, R. T., AND MIMS, C. A., Pathogenesis of viral infections of the nervous system, *N. Engl. J. Med.* **278:**23–30, 84–92 (1968).

157. JOHNSON, R. T., Prion diseases, *N. Engl. J. Med.* **326:**486–487 (1992).

158. JOHNSON, R. T., Questions and prospects related to HIV-1 and the brain, *Res. Publ. Assoc. Res. Nerv. Ment. Dis.* **72:**311–323 (1994).

159. JOHNSON, R. T., The pathogenesis of acute viral encephalitis and postinfectious encephalomyelitis, *J. Infect. Dis.* **155:**359–364 (1987).

160. JORDAN, M. C., ROUSSEAU, W. E., NOBLE, G. R., STEWART, J. A., AND CHIN, T. D. Y., Association of cervical cytomegaloviruses with venereal disease, *N. Engl. J. Med.* **288:**932–934 (1973).

161. KAPIKIAN, A. Z., KIM, H. W., WYATT, R. G., CLINE, W. L., ARROBIO, J. O., BRANDT, C. D., RODRIGUEZ, W. J., SACK, D. A., CHANOCK, R. M., AND PARROTT, R. H., Human reovirus-like agent as the major pathogen associated with "winter gastroenteritis" in hospitalized infants and young children, *N. Engl. J. Med.* **294:**965–972 (1976).

162. KAPIKIAN, A. Z., WYATT, R. G., DOLIN, R., THORNHILL, T. S., KALICA, A. R., AND CHANOCK, R. M., Visualization by immune electron microscopy of a 27-nm particle associated with acute infectious non-bacterial gastroenteritis, *J. Virol.* **10:**1075–1081 (1972).

163. KASL, S. V., EVANS, A. S., AND NIEDERMAN, J. C., Psychosocial risk factors in the development of infectious mononucleosis, *Psychosom. Med.* **41:**445–446 (1979).

164. KASLOW, R. A., OSTROW, D. G., DETELS, R., PHAIR, J. P., POLK, B. F., AND RINALDO, C. R., JR., The Multicenter AIDS Cohort Study: Rationale, organization, and selected characteristics of the participants, *Am. J. Epidemiol.* **126:**310–318 (1987).

165. KASLOW, R. A., DUQUESNOY, R., VANRADEN, M., KINGSLEY, L., MARRARI, M., FRIEDMAN, H., SU, S., SAAH, A. J., DETELS, R., PHAIR, J., AND RINALDO, C., A1, Cw7, B8, DR3 HLA combination associated with rapid decline of T-helper lymphocytes in HIV-1 infection, *Lancet* **335:**927–930 (1990).

166. KASLOW, R. A., CARRINGTON, M., APPLE, R., PARK, L., MUNOZ, A., SAAH, A. J., GOEDERT, J. J., WINKLER, C., O'BRIEN, S. J., RINALDO, C., DETELS, R., BLATTNER, W., PHAIR, J., ERLICH, H., AND MANN, D. L., Influence of combinations of human major histocompatibility complex genes on the course of HIV-1 infection, *Nature Med.* **2**:405–411 (1996).

167. KEET, I. P. M., LEE, F. K., VANGRIENSVEN, G. J. P., LANGE, J. M. A., NAHMIAS, A., AND COUTINHO, R. A., Herpes simplex virus type 2 and other genital ulcerative infections as a risk factor for HIV-1 acquisition, *Genitourin. Med.* **66**:330–333 (1990).

168. KEET, I. P. M., KRIJNEN, P., KOOT, M. LANGE, J. M. A., MIEDEMA, F., GOUDSMIT, J., AND COUTINHO, R. A., Predictors of rapid progression to AIDS in HIV-1 seroconverters, *AIDS* **7**:51–57 (1993).

169. KELSEY, J. L., THOMPSON, W. D., AND EVANS, A. S., *Methods in Observational Epidemiology*, Oxford University Press, New York, 1986.

170. KIECOLT-GLASER, J. K., DURA, J. R., SPEICHER, C. E., TRASK, O. J., AND GLASER, R., Spousal caregivers of dementia victims: Longitudinal changes in immunity and health, *Psychosom. Med.* **53**:345–362 (1991).

171. KINGSLEY, L. A., ZHOU, S. Y. J., BACELLAR, H., RINALDO, C. R., JR., CHMIEL, J., DETELS, R., SAAH, A., VANRADEN, M., HO, M., AND MUNOZ, A., Temporal trends in human immunodeficiency virus type 1 seroconversion 1984–1989. A report from the Multicenter AIDS Cohort Study (MACS), *Am. J. Epidemiol.* **134**(4): 331–339 (1991).

172. KINNEY, J. S., ANDERSON, L. J., FARRAR, J., STRIKAS, R. A., KUMAR, M. L., KLIEGMAN, R. M., SEVER, J. L., HURWITZ, E. S., AND SIKES, R. K., Risk of adverse outcomes of pregnancy after human parvovirus B19 infection, *J. Infect. Dis.* **157**:663–667 (1988).

173. KLEIN, J. O., Diagnostic lung puncture in the pneumonias of infants and children, *Pediatrics* **44**:486–492 (1969).

174. KLEIN, M. R., KEET, I. P. M., D'AMARO, J., BENDE, R. J., HEKMAN, A., MESMAN, B., KOOT, M., DEWAAL, L. P., COUTINHO, R., AND MIEDEMA, F., Associations between HLA frequencies and pathogenic features of human immunodeficiency virus type 1 infection of seroconverters from the Amsterdam cohort of homosexual men, *J. Infect. Dis.* **169**:1244–1249 (1994).

175. KLITZMAN, R. L., ALPERS, M. P., AND GAJDUSEK, D. C., The natural incubation period of kuru and episodes of transmission in three clusters of patients, *Neuroepidemiology* **3**:3–20 (1984).

176. KLOENE, W., BANG, F. B., CHAKRABORTY, S. M., COOPER, M. R., KULEMANN, H., OTA, M., AND SHAH, K. V., A two-year respiratory virus survey in four villages in West Bengal, India, *Am. J. Epidemiol.* **92**:307–320 (1970).

177. KNIGHT, V., (ED.), *Viral and Mycoplasma Infections of the Respiratory Tract*, Lea & Febiger, Philadelphia, 1973.

178. KOCH, R., Uber bacteriologische Forschung, *Int. Med. Cong. Berl.* **1**:35 (1891).

179. KOCH, R., *Uber die Aetiologie der Tuberculose, Verhandlungen des Kongresses fur Inner Medicin. Erst Kongress*, Verlag von J. F. Bergmann, Weisbaden, 1882.

180. KOHN, M. A., FARLEY, T. A., ANDO, T., CURTIS, M., WILSON, S. A., JIN, Q., MONROE, S. S., BARON, R. C., McFARLAND, L. M., AND GLASS, R. I., A large outbreak of Norwalk virus gastroenteritis associated with eating raw oysters: Implications for monitoring safe oyster beds, *J.A.M.A.* **273**:466–471 (1995).

181. KONNO, T., SUZUKI, H., IMAI, A., AND IHIDA, N., Reovirus-like agent in acute epidemic gastroenteritis in Japanese infants: Fecal shedding and serologic response, *J. Infect. Dis.* **135**:259–266 (1977).

182. KOSZINOWSKI, U. H., REDDEHASE, M. J., AND JONJIC, S., The role of CD4 and CD8 T cells in viral infections, *Curr. Opin. Immunol.* **3**:471–475 (1991).

183. KRUGMAN, S., Viral hepatitis: A, B, C, D and E-infection, *Pediatr. Rev.* **13**:203–212 (1992).

184. LANG, D. J., AND KUMMER, J. F., Demonstration of cytomegalovirus in semen, *N. Engl. J. Med.* **287**:756–758 (1972).

185. LAST, J. M., International Epidemiological Association, *A Dictionary of Epidemiology*, Oxford University Press, New York, 1988.

186. LAST, J. M., AND WALLACE, R. B., *Maxcy-Rosenau-Last Public Health and Preventive Medicine*, 13th ed., Appleton and Lange, Norwalk, C. T., 1992.

187. LENNETTE, E. H., MAGOFFIN, R. L., AND KNAUF, E. G., Viral central nervous system disease: An etiologic study conducted at the Los Angeles General Hospital, *J.A.M.A.* **179**:687–695 (1962).

188. LEUNG, D. Y. M., MEISSNER, H. C., FULTON, D. R., MURRAY, D. L., KOTZIN, B. L., AND SCHLIEVERT, P. M., Toxic shock syndrome toxin-secreting *Staphylococcus aureus* in Kawasaki syndrome, *Lancet* **342**:1385–1388 (1983).

189. LEVY, R. M., AND BERGER, J. R., HIV and HTLV infections of the nervous system, in: *Infectious Diseases of the Central Nervous System, Contemporary Neurology Series* (L. KERN AND J. B. TYLER, EDS.), F. A. Davis, Philadelphia, 1993.

190. LEWIS, D. C., LIGHTFOOT, N. F., CUBITT, D., AND WILSON, S. A., Outbreaks of astrovirus type 1 and rotavirus gastroenteritis in a geriatric in-patient population, *J. Hosp. Infect.* **14**:9–14 (1989).

191. MACASAET, F. F., KIDD, P. A., BOLANO, C. R., AND WENNER, H. A., The etiology of acute respiratory infections. II. The role of viruses and bacteria, *J. Pediatr.* **72**:829–839 (1968).

192. MACKEWICZ, C. E., YANG, L. C., LIFSON, J. D., AND LEVY, J., Noncytolytic CD8T-cell and HIV responses in primary HIV infection, *Lancet* **344**:1671–1673 (1994).

193. MANDELL, G. L., DOUGLAS, R. G., JR., AND BENNETT, J. E. (EDS.), *Principles and Practices of Infectious Diseases*, 4th ed., John Wiley & Sons, New York, 1994.

194. MARCHETTE, N. J., MELISH, M. E., HICKS, R., KIHARA, S., SAM, E., AND CHING, D., Epstein–Barr virus and other herpesvirus infections in Kawasaki syndrome, *J. Infect. Dis.* **161**:680–684 (1990).

195. MARCUS, R., AND THE CDC COOPERATIVE NEEDLESTICK SURVEILLANCE GROUP, Surveillance of health care workers exposed to blood from patients infected with the human immunodeficiency virus, *N. Engl. J. Med.* **318**:1118–1123 (1988).

196. MASON, J. O., AND MCLEAN, W. R., Infectious hepatitis traced to consumption of raw oysters: An epidemiologic study, *Am. J. Hyg.* **75**:90–111 (1962).

197. MAY, R., AND ANDERSON, R. M., Transmission dynamics of HIV infection, *Nature* **326**:137–142 (1987).

198. McMICHAEL, A. J., TING, A., ZWEERINK, H. J., AND ASKONAS, B. A., HLA restriction of cell-mediated lysis of influenza virus-infected cells, *Nature* **270**:524–526 (1977).

199. MELNICK, J. L., A water-borne urban epidemic of hepatitis, in: *Hepatitis Frontiers* (F. W. HARTMAN, ED.), pp. 211–225, Churchill, London, 1957.

200. MERMIN, J. H., HOLODNIY, M., KATZENSTEIN, D. A., AND MERRIGAN, T. C., Detection of human immunodeficiency virus DNA and RNA in semen by the polymerase chain reaction, *J. Infect. Dis.* **164**:769–772 (1991).

201. MIMS, C. A., *The Pathogenesis of Infectious Diseases*, 3rd ed., Academic Press, London, 1987.

202. MIMS, C. A., CURNER, M. L., AND KELLY, R. E. (EDS.), *Viruses and Demyelinating Diseases*, Academic Press, New York, 1983.

203. MONTO, A. S., AND JOHNSON, K. M., A community study of respiratory infections in the tropics. I. Description of the community and observation on the activity of certain respiratory agents, *Am. J. Epidemiol.* **86:**78–92 (1967).

204. MORENS, D. M., Boston exanthem agent: Echovirus 16, *Am. J. Dis. Child.* **131:**1306 (1977).

205. MORENS, D. M., Antibody-dependent enhancement of infection and the pathogenesis of viral disease, *Clin. Infect. Dis.* **19:**500–512 (1994).

206. MORGAN, A. J., Recent progress in EBV vaccine development, in: *Epstein–Barr Virus and Associated Diseases* (T. TURSZ, J. S. PAGANO, D. V. ABLASHIET, G. DE THE, G. LENOIR, AND G. PEARSON, EDS.), pp. 593–598, Collogue/INSERM, John Libbey, Eurotext, London, 1993.

207. MUELLER, N., Epidemiologic studies assessing the role of the Epstein–Barr virus in Hodgkin's disease, *Yale J. Biol. Med.* **60:**321–327 (1987).

208. MUELLER, N., EVANS, A. S., AND LONDON, W. T., Viruses, in: *Cancer Epidemiology and Prevention*, 2nd ed. (D. SCHOTTENFELD AND J. F. FRAUMENI, JR., EDS.), Oxford University Press, New York, 1996.

209. MUELLER, N., MOHAR, A., EVANS, A., HARRIS, N. L., AND THE MEMBERS OF THE EBV–NHL COLLABORATION, Epstein–Barr virus antibody patterns preceding the diagnosis of non-Hodgkin's lymphoma, *Int. J. Cancer* **49:**387–393 (1991).

210. MUFSON, M. A., CHANG, V., GILL, V., WOOD, S. C., ROMANSKY, M. J., AND CHANOCK, R. M., The role of viruses, mycoplasmas and bacteria in acute pneumonia in civilian adults, *Am. J. Epidemiol.* **86:**526–544 (1967).

211. MULLIS, K. B., JOHNSON, P. E., AND THOMAS, C. A., JR., Dissenting on AIDS, *San Diego Union-Tribune*, Section G, May 15, 1994.

212. MUNOZ, A., KIRBY, A. J., HE, Y. D., MARGOLICK, J. B., VISSCHER, B. R., RINALDO, C. R., KASLOW, R. A., AND PHAIR, J. P., Long-term survivors with HIV-1 infection: Incubation period and longitudinal patterns of CD4+ lymphocytes, *J. Acquir. Immune Dis. Syndr.* **8:**496–505 (1995).

213. NATHANSON, N., Epidemiology in: *Virology* (B. FIELDS, D. M. KNIPE, R. M. CHANOCK, M. S. HIRSCH, J. L. MELNICK, T. P. MONATH, AND B. ROIZMAN, EDS.), pp. 267–291, Raven Press, New York, 1990.

214. NATIONAL INSTITUTE OF ALLERGY AND INFECTIOUS DISEASES, Program for accelerated development of new vaccines, 1985 Progress Report, NIH, Bethesda, 1985.

215. NATIONAL INSTITUTE OF ALLERGY AND INFECTIOUS DISEASES, The Jordan Report, 1994.

216. NICHOLSON, K. G., Antiviral therapy. Respiratory infections, genital herpes, and hepatitis keratitis, *Lancet* **2:**617–621 (1984).

217. NICOLOSI, A., *HIV Epidemiology: Models and Methods*, Raven Press, New York, 1994.

218. NICOLOSI, A., LEITE, M. L. C., MUSICCO, M., MOLINARI, S., AND LAZZARIN, A., Parenteral and sexual transmission of human immunodeficiency virus in intravenous drug users: A study of seroconversion, *Am. J. Epidemiol.* **135:**225–233 (1992).

219. NIEDERMAN, J. C., MCCOLLUM, R. W., HENLE, G., AND HENLE, W., Infectious mononucleosis: Clinical manifestations in relation to EB virus antibodies, *J.A.M.A.* **203:**205–209 (1968).

220. OFFIT, P. A., AND DUDZIK, K. I., Rotavirus-specific cytotoxic T lymphocytes passively protect against gastroenteritis in suckling mice, *J. Virol.* **64:**6325–6328 (1990).

221. OLSON, L. C., LEXOMBOON, U., SITHISARN, P., AND NOYES, H. E., The etiology of respiratory tract infections in a tropical country, *Am. J. Epidemiol.* **97:**34–43 (1973).

222. OSEASOHN, R., ADELSON, L., AND KAJI, M., Clinical pathological study of 33 fatal cases of Asian influenza, *N. Engl. J. Med.* **260:**509–518 (1959).

223. OU, C.-Y., CIESIELSKI, C. A., MYERS, G., BANDEA, C. I., LUO, C.-C., KORBER, B. T. M., MULLINS, J. I., SCHOCHETMAN, G., BERKELMAN, R. L., ECONOMOU, A. N., WITTE, J. J., FURMAN, L. J., SATTEN, G. A., MACINNES, K. A., CURRAN, J. W., AND JAFFE, H. W., Molecular epidemiology of HIV transmission in a dental practice, *Science* **256:**1165–1170 (1992).

224. PAGANO, J. S., HUANG, C. H., AND LEVINE, P., Absence of Epstein–Barr viral DNA in American Burkitt's lymphoma, *N. Engl. J. Med.* **389:**1195–1199 (1973).

225. PAN AMERICAN HEALTH ORGANIZATION, The certification of wild poliovirus eradication from the Western hemisphere, *Epidemiol. Bull.* **15:**1–3 (1994).

226. PAUL, J. R., *Clinical Epidemiology*, 2nd ed., University of Chicago Press, Chicago, 1966.

227. PEARCE, E. J., CASPAR, P., GRZYCH, J.-M., LEWIS, F. A., AND SHER, A., Downregulation of Th1 cytokine production accompanies induction of Th2 responses by a parasitic helminth, *Schistosoma mansoni*, *J. Exp. Med.* **173:**159–166 (1991).

228. PETERS, C. J., AND ANDERSON, G. W., JR., Pathogenesis of Rift Valley fever, *Contrib. Epidemiol. Biostat.* **3:**21–41 (1981).

229. PHILIPS, P. E., AND CHRISTIAN, C. L., Myxovirus antibody increases in human connective tissue disease, *Science* **168:**982–984 (1970).

230. PLUMMER, F. A., SIMONSEN, J. N., CAMERON, D. W., NDINYA-ACHOLA, J. O., KREISS, J. K., GAKINYA, M. N., WAIYAKI, P., CHEANG, M., PIOT, P., RONALD, A. R., AND NGUGI, E. N., Cofactors in male–female transmission of human immunodeficiency virus type 1, *J. Infect. Dis.* **163:**233–239 (1991).

231. PRUSINGER, S. B., Molecular biology of prion diseases, *Science* **252:**1515–1522 (1991).

232. PUBLIC HEALTH LABORATORY SERVICE WORKING PARTY ON FIFTH DISEASE, Prospective study of human parvovirus (B19) infection in pregnancy, *Br. Med. J.* **300:**1166–1170 (1990).

233. PURTILO, D. T., HUTT, L., BHAWAN, J. P. S., ALLEGRO, S., AND ROSEN, F. S., Immunodeficiency to the Epstein–Barr virus in the X-linked recessive lymphoproliferative syndrome, *Clin. Immunol. Monogr.* **9:**147–156 (1978).

234. PURTILO, D. T., AND LAI, P. K., Clinical and immunopathological manifestations and detection of Epstein–Barr virus infection in immune deficient patients, in: *Medical Virology VI* (L. M. DE LA MAZA AND E. M. PETERSON, EDS.), pp. 121–167, Elsevier Publications, Amsterdam, 1987.

235. PURTILO, D. T., STROBACH, R. S., OKANO, M., AND DAVIS, J. R., Epstein–Barr virus-associated lymphoproliferative disorders, *Lab. Invest.* **67:**5–23 (1992).

236. REMINGTON, J. S., AND KLEIN, J. O. (EDS.), *Infectious Diseases of the Fetus and Newborn Infant*, 3rd ed., W. B. Saunders, Philadelphia, 1990.

237. RIVERS, T., Viruses and Koch's postulates, *J. Bacteriol.* **33:**1–12 (1937).

238. ROSENBERG, P. S., GOEDERT, J. J., AND BIGGAR, R. J., Effect of age at seroconversion on the natural AIDS incubation distribution, *AIDS* **8:**803–810 (1994).

239. ROTHFIELD, N. F., EVANS, A. S., AND NIEDERMAN, J. C., Clinical and laboratory aspects of raised virus antibody titers in systemic lupus erythematosus, *Ann. Rheum. Dis.* **32:**238–246 (1973).

240. SABIN, A. B., AND SILVA, E., Residual paralytic poliomyelitis in a tropical region of Brazil, 1969–1977: Prevalent surveys in different age groups as indicators of changing incidence, *Am. J. Epidemiol.* **117:**193–200 (1983).

241. SAWYER, R. N., EVANS, A. S., NIEDERMAN, J. C., AND MCCOLLUM, R. W., Prospective studies of a group of Yale University freshmen. I. Occurrence of infectious mononucleosis, *J. Infect. Dis.* **123:**263–270 (1971).

242. SCHMID, D. S., AND ROUSE, B. T., The role of T cell immunity in control of herpes simplex virus, *Curr. Top. Microbiol. Immunol.* **179:**57–74 (1992).

243. SCHULMAN, J., Transmissibility as a separate genetic attribute of influenza viruses, in: *Aerobiology* (I. H. SILVER, ED.), pp. 248–259, Academic Press, New York, 1970.

244. SHARPE, A. H., AND FIELDS, B. N., Pathogenesis of viral infections; basic concepts derived from a reovirus model, *N. Engl. J. Med.* **312:**486–496 (1985).

245. SHAH, K. V., DANIEL, R. W., AND WARZAWSKI, R. M., High prevalence of antibodies to BK virus, and SV40-related papova virus, in residents of Maryland, *J. Infect. Dis.* **128:**784–787 (1973).

246. SHOPE, R. E., Swine influenza. I. Experimental transmission and pathology, *J. Exp. Med.* **54:**349–359 (1931).

247. SINGER, B., Mathematic models of infectious diseases: Seeking new tools for planning and evaluating child survival control programs, *Popul. Dev. Rev.* **10**(Suppl.):347–365 (1984).

248. SMITH, D. K., NEAL, J. J., HOLMBERG, S. D., AND THE CENTERS FOR DISEASE CONTROL IDIOPATHIC CD4+ T-LYMPHOCYTOPENIA TASK FORCE, Unexplained opportunistic infections and CD4+ T-lymphocytopenia without HIV infection, *N. Engl. J. Med.* **328:**373–379 (1983).

248a. SPOONER, K. M., LANE, H. C., AND MASUR, H., Guide to major clinical trials of antiretroviral therapy administered to patients infected with human immunodeficiency virus, *Clin. Infect. Dis.* **23:**15–27 (1996).

249. STAGNO, S., CLOUD, G., PASS, R. F., BRITT, W. J., AND ALFORD, C. A., Factors associated with primary cytomegalovirus infection during pregnancy, *J. Med. Virol.* **13:**347–353 (1984).

250. STAGNO, S., PASS, R. F., DWORSKY, M. E., AND ALFORD, C. A., Maternal cytomegalovirus infection and perinatal transmission, *Clin. Obstet. Gynecol.* **25:**563–576 (1982).

251. STOECKEL, P., SCHLUMBERGER, M., PARENT, G., MAIRE, B., VAN WENZEL, A., VAN STEENIS, G., EVANS, A., AND SALK, D., Use of killed poliovirus vaccine in a routine immunization program in West Africa, *Rev. Infect. Dis.* **6**(Suppl. 2):S463–466 (1984).

252. STREBEL, P. M., SUTTER, R. W., COCHI, S. L., BIELLIK, R. J., KEW, O. M., PALLANSCH, M. A., ORENSTEIN, W. A., AND HINMAN, A. R., Epidemiology of poliomyelitis in the United States: One decade after the last reported case of indigenous wild virus-associated disease, *Clin. Infect. Dis.* **14:**568–579 (1992).

253. STUART-HARRIS, C., Can infectious disease be eliminated? A report on the international conference on the eradication of infectious disease, Foreword, *Rev. Infect. Dis.* **4:**913–914 (1982).

254. SULLIVAN, J. L., AND WODA, B. A., X-linked lymphoproliferative syndrome, *Immunodefic. Rev.* **1:**325–347 (1989).

255. SULLIVAN-BOLYAI, J., HULL, H. F., WILSON, C., AND COREY, L., Neonatal herpes simplex virus infection in King County, Washington: Increasing incidence and epidemiologic correlates, *J.A.M.A.* **250:**3059–3062 (1983).

256. TAKIMOTO, C. H., CRAM, D. L., AND ROOT, R. K., Respiratory syncytial virus infections on an adult medical ward, *Arch. Intern. Med.* **151:**706–708 (1991).

257. THOMPSON, W. H., AND EVANS, A. S., California virus studies in Wisconsin, *Am. J. Epidemiol.* **81:**230–234 (1965).

258. TINDALL, B., AND COOPER, D. A., Primary HIV infection: Host responses and intervention strategies, *AIDS* **5:**1–14 (1991).

259. TINDALL, B., EVANS, L., CUNNINGHAM, P., MCQUEEN, P., HURREN, L., VASAK, E., MOONEY, J., AND COOPER, D. A., Identification of HIV-1 in semen following primary HIV-1 infection, *AIDS* **6:**949–952 (1992).

260. TOKARS, J. I., MARCUS, R., CULVER, D. H., SCHABLE, C. A., MCKIBBEN, P. S., BANDEA, C. I., AND BELL, D. M., Surveillance of HIV infection and zidovudine use among health care workers after occupational exposure to HIV-infected blood, *Ann. Intern. Med.* **118:**913–919 (1993).

261. TOMPKINS, L. S., The use of molecular methods in infectious diseases, *N. Engl. J. Med.* **327:**1290–1297 (1992).

262. TORRES, B. V., ILIA, R. M., AND ESPARZA, J., Epidemiological aspects of rotavirus infection in hospitalized Venezuelan children with gastroenteritis, *Am. J. Trop. Med. Hyg.* **27:**567–572 (1978).

263. TWU, S. J., DETELS, R., NELSON, K., VISSCHER, B. R., KASLOW, R., PALENICEK, J., AND PHAIR, J., Relationship of hepatitis B virus infection to human immunodeficiency virus type 1 infection, *J. Infect. Dis.* **167:**299–304 (1993).

264. UNIVERSITY HEALTH PHYSICIANS AND PHLS LABORATORIES, A joint investigation: Infectious mononucleosis and its relationship to EB virus antibody, *Br. Med. J.* **4:**643–646 (1971).

265. UNTERMOHLEN, V., AND ZABRISKIE, J. F., Suppressed cellular immunity to measles antigen in multiple-sclerosis patients, *Lancet* **2:**1147–1148 (1973).

266. URQUHARDT, G. E. D., AND STOTT, E. J., Rhinoviraemia, *Br. Med. J.* **4:**28–30 (1970).

267. UTAGAWA, E. T., NISHIZAWA, S., SEKINE, S., HAYASHI, Y., ISHIHARA, Y., OISHI, I., IWASAKI, A., YAMASHITA, I., MIYAMURA, K., YAMAZAKI, S., INOUYE, S., AND GLASS, R. I., Astrovirus as a cause of gastroenteritis in Japan, *J. Clin. Microbiol.* **32:**1841–1845 (1994).

268. VALENTI, W. M., Selected viruses of nosocomial importance, in: *Hospital Infections*, 3rd ed. (J. V. BENNETT, S. P. SANFORD, AND P. S. BRACHMAN, EDS.), pp. 789–821, Little, Brown, Boston, 1992.

269. VAN DER SIJS, I. H., AND WILTINK, E. H., Antiviral drugs: Present status and future prospects, *Int. J. Biochem.* **26:**621–630 (1994).

270. VLAHOV, D., ANTHONY, J. D., MUNOZ, A., MARGOLICK, J., NELSON, K. E., CELENTANO, D. D., SOLOMON, L., AND POLK, B. F., The ALIVE study, a longitudinal study of HIV-1 infection in intravenous drug users, *NIDA Res. Monogr.* **109:**75–100 (1991).

271. VONKA, V., KANKA, J., JELINEK, J., SUBRT, I., SUCHANEK, A., HAVRANKOVA, A., VACHAL, M., HIRSCH, I., DOMORAZKOVA, E., ZAVADOVA, H., RICHTEROVA, V., NAPRSTKOVA, J., DVORAKOVA, V., AND SVOBODA, B., Prospective study on the relationship between cervical neoplasia and herpes simplex, type-2 virus. I. Epidemiological characteristics, *Int. J. Cancer* **33:**49–60 (1984).

272. WALKER, C. M., MOODY, D. J., STITES, D. P., AND LEVY, J. A., CD8+ lymphocytes can control HIV infection *in vitro* by suppressing virus replication, *Science* **234:**1563–1566 (1986).

273. WHITLEY, R. J., AND GNANN, J. W., JR., Acyclovir: A decade later, *N. Engl. J. Med.* **327:**782–789 (1992).

274. WILSON, C. B., DIXON, F. J., EVANS, A. S., AND GLASOCK, R. J., Anti-viral antibody responses in patients with renal disease, *Clin. Immunol. Immunopathol.* **2:**121–132 (1973).

275. WILSON, C. W., STEVENSON, D. K., AND ARVIN, A. M., A concurrent epidemic of respiratory syncytial virus and echovirus 7 infections in an intensive care nursery, *Pediatr. Infect. Dis. J.* **8:**24–29 (1989).

276. WINKELSTEIN, W., JR., WILEY, J. A., PADIAN, N. S., SAMUEL, M., SHIBOSKI, S., ASCHER, M., AND LEVY, J. A., The San Francisco Men's Health Study: Continued decline in HIV seroconversion rates among homosexual/bisexual men, *Am. J. Public Health* **78:**1472–1474 (1988).

277. WOFSY, C., COHEN, J., HANER, L. B., PADRA, N., MICHAELS, B., EVANS, L., AND LEVY, J. A., Isolation of AIDS-associated retrovirus from genital secretions of women and antibodies to the virus, *Lancet* **1:**527–528 (1986).

278. WORLD HEALTH ORGANIZATION, Smallpox surveillance, *Week. Epidemiol. Rec.* **54:**1–6 (1979).

279. WORLD HEALTH ORGANIZATION, Expanded Programme on Immunization, Document EB 69/25, March 25, 1982.

280. WORLD HEALTH ORGANIZATION, Expanded programme on immunization, global advisory group—Part I, *Week. Epidemiol. Rec.* **68:**1–6 (1993).

281. WORLD HEALTH ORGANIZATION, Expanded programme on immunization, global advisory group—Part II, *Week. Epidemiol. Rec.* **68:**11–15 (1993).

282. WORLD HEALTH ORGANIZATION, The current global situation of the HIV/AIDS pandemic, *Week. Epidemiol. Rec.* **70:**7–8 (1995).

283. YAMANISHI, K., OKUNO, T., SHIRAKI, K., TAKAHASHI, M., KONDO, T., ASANO, Y., AND KURATA, T., Identification of a human herpesvirus-6 as causal agent for exanthem subitum, *Lancet* **1:**1065–1067 (1988).

284. YOLKEN, R. H., BISHOP, C. A., TOWNSEND, T. R., BOLYARD, E. A., BARTLETT, J., SANTOS, G. W., AND SARAL, R., Infectious gastroenteritis in bone marrow-transplant recipients, *N. Engl. J. Med.* **306:**1009–1012 (1982).

285. YOLKEN, R. H., WYATT, R. G., ZISAIS, G., BRANDT, C. D., RODRIGUEZ, W. J., KIM, H. W., PARROTT, R. H., URITTA, J. J., MATA, L., GREENBERG, H. B., KAPIKIAN, A. Z., AND CHANOCK, R. M., Epidemiology of human rotavirus types 1 and 2 as studied by enzyme-linked immunosorbent assay, *N. Engl. J. Med.* **299:** 1156–1161 (1978).

286. YOTO, Y., KUDOH, T., HASEYAMA, K., SUZUKI, N., CHIBA, S., AND MATSUNAGA, Y., Human parvovirus B19 infection in Kawasaki disease, *Lancet* **344:**58–59 (1994).

287. ZELDIS, J. B., AND CRUMPACKER, C. S., Hepatitis, in: *Infectious Diseases of the Fetus and Newborn Infant*, 3rd ed. (J. S. REMINGTON AND J. O. KLEIN, EDS.), pp. 574–600, W. B. Saunders, Philadelphia, 1990.

288. ZINKERNAGEL, R. M., Immunity to viruses, in: *Fundamental Immunology*, 3rd ed., (W. PAUL, ED.), pp. 1211–1250, Raven Press, New York, 1993.

289. ZUR HAUSEN, H., Papilloma virus/host cell interaction in the pathogenesis of anorectal cancer, in: *Origins of Human Cancer: A Comprehensive Review* (J. BRUGGE, T. CURRAN, E. HADLOW, AND F. MCCORMICK, EDS.), pp. 685–705, Cold Spring Harbor Laboratory Press, Cold Spring Harbor, NY, 1991.

290. ZWEERINK, H. J., AND STANTON, L. W., Immune response to HSV infections: Virus specific antibodies in sera from patients with recurrent facial infections, *Infect. Immun.* **31:**624–630 (1981).

16. Suggested Reading

BELSHE, R. B. (ED.), *Textbook of Human Virology*, 2nd ed., Mosby Year Book Medical, St. Louis, 1991.

BENENSON, A. S. (ED.), *Control of Communicable Diseases in Man*, 15th ed., American Public Health Association, Washington, DC, 1990.

CHRISTIE, A. B., *Infectious Diseases: Epidemiology and Clinical Practice*, 3rd ed., E. & S. Livingston, Edinburgh, 1987.

DAVIS, B. D., DULBECCO, R., EISEN, H. N., AND GINSBERG, H. S. (EDS.), *Microbiology*, 4th ed., Lippincott, Philadelphia, 1990.

EVANS, A. S., AND BRACHMAN, P. S. (EDS.), *Bacterial Infections of Humans. Epidemiology and Control*, Plenum Press, New York, 1991.

FENNER, F. J., AND WHITE, D. O., *Medical Virology*, 3rd ed., Academic Press, Orlando, FL, 1990.

FIELDS, B. N., KNIPE, D. M., CHANOCK, R. M., MELNICK, J. L., ROIZMAN, B., AND SHOPE, R. E. (EDS.), *Field's Virology*, 2nd ed., Raven Press, New York, 1990.

FRAENKEL-CONRAT, H., AND KIMBALL, P. C., *Virology*, 2nd ed., Prentice-Hall, Englewood Cliffs, NJ, 1988.

FRIEDMAN, G. D., *Primer of Epidemiology*, 3rd ed., McGraw-Hill, New York, 1988.

GOLUB, E. S., *Immunology, a Synthesis*, 2nd ed., Sinauer Associates, Sunderland, MA, 1991.

HENNEKENS, C. H., BURING, J. E., AND MAYRENT, S. L., *Epidemiology in Medicine*, Little, Brown, Boston, 1987.

HOLLAND, W. W., DETELS, R., AND KNOX, G. E. (EDS.), *Oxford Textbook of Public Health*, 2nd ed., Oxford University Press, Oxford, 1991.

KELSEY, J. L., THOMPSON, W. D., AND EVANS, A. S., *Methods in Observational Epidemiology*, Oxford University Press, New York, 1986.

LAST, J. M., International Epidemiological Association, *A Dictionary of Epidemiology*, Oxford University Press, New York, 1988.

LAST, J. M., AND WALLACE, R. B., *Maxcy-Rosenau-Last Public Health and Prevention Medicine*, 13th ed., Appleton and Lange, Norwalk, CT, 1992.

LILIENFELD, A. M., AND LILIENFELD, D. E., *Foundations of Epidemiology*, 2nd ed., Oxford University Press, New York, 1982.

MANDELL, G. L., DOUGLAS, R. G., JR., AND BENNETT, J. E. (EDS.), *Principles and Practice of Infectious Diseases*, 4th ed., John Wiley & Sons, New York, 1994.

MIMS, C. A., *The Pathogenesis of Infectious Diseases*, 3rd ed., Academic Press, London, 1987.

NOTKINS, A., *Concepts in Viral Pathogenesis III*, Springer-Verlag, New York, 1989.

NOTKINS, A. L., AND OLDSTONE, B. A. (EDS.), *Concepts in Viral Pathogenesis*, Springer-Verlag, New York, 1984.

PAUL, W. (ED.), *Fundamental Immunology*, 3rd ed., Raven Press, New York, 1993.

REMINGTON, J. S., AND KLEIN, J. O. (EDS.), *Infectious Diseases of the Fetus and Newborn Infant*, 3rd ed., W. B. Saunders, Philadelphia, 1990.

ROITT, I., *Essential Immunology*, 8th ed., Blackwell Scientific Publications, Oxford, 1994.

ROTHMAN, K., *Modern Epidemiology*, Little, Brown, Boston, 1986.

THOMAS, D. B., *Viruses and the Cellular Immune Response*, Marcel Dekker, London, 1993.

VOLK, W. A., *Essentials of Medical Microbiology*, 4th ed., Lippincott, Philadelphia, 1991.

WEBSTER, R. G., AND GRANOFF, A., *Encyclopedia of Virology*, Academic Press, London, 1994.

ZUCKERMAN, A. J., BANATVALA, J. E., AND PATTISON, J. R. (EDS.), *Principles and Practices of Clinical Virology*, 3rd ed., John Wiley, Chichester, England, 1994.

CHAPTER 2

Laboratory Methods for the Diagnosis of Viral Diseases

Robert L. Atmar and Janet A. Englund

1. Introduction

The ability of the virology laboratory to document infection with viral agents has increased dramatically over the past decade. Standard virological techniques, such as the propagation of viruses in animals and determination of antibody using complement fixation, have been important historically in the study of the epidemiology, pathology, and assessment of clinical disease associated with viral infection. Improvements in cell culture techniques and antibody assays now permit the study of viruses and viral disease in many laboratories, not only specialized research laboratories. Viral cultivation techniques utilizing cytokines and other stimulating substances and genetically engineered cell lines or animal strains have enhanced the ability to isolate viruses that were previously unknown or only suspected on clinical grounds. Viruses such as human immunodeficiency virus (HIV), human herpesviruses types 6 and 7, and parvovirus have been better recognized and studied because of these innovations in viral cultivation.

Advances in molecular biotechnology and detection of nucleic acid sequences and gene products have revolutionized the ability to detect viral agents and to document infection more conclusively. The ability to trace the spread of viruses by "fingerprinting" or to sequence viral genomes directly has resulted in important new epidemiologic observations. Commercially available supplies of monoclonal antibodies, cytokines, growth factors, and other reagents, as well as the widespread clinical use of specific antiviral therapy, have created new opportunities and demands in viral laboratories to provide rapid, sensitive, and specific diagnostic tests. The development of simple, rapid, and often relatively inexpensive antigen detection test kits has revolutionized both clinical care and laboratory practice. An understanding of various detection methods is increasingly important in the design and interpretation of epidemiologic studies. The vast array of laboratory tests now permits enhanced detection of viral antigens, although the clarification of the classic issue of "causation" of disease remains blurred.

The significance of detection or lack of detection of a virus or viral antigen remains difficult to interpret. Isolation of a virus from a normally sterile site, such as tissue, cerebrospinal fluid (CSF), or blood, is generally highly significant and usually establishes the etiology of the infection. The isolation of certain viruses, such as influenza or respiratory syncytial virus (RSV) in respiratory specimens, also is diagnostic because an asymptomatic carrier state has not been shown to exist. However, prolonged and generally asymptomatic excretion or shedding of other viruses makes the determination of the effect of a particular virus on the disease process very difficult. Viruses such as cytomegalovirus (CMV), adenoviruses, and enteroviruses are shed in disease states but also may be shed asymptomatically for prolonged periods of apparent good health. An additional complicating factor is the differentiation of primary infection from reactivation of disease, a problem common to the study of infections with viruses such as herpes simplex (HSV) or CMV, especially

Robert L. Atmar • Acute Viral Respiratory Disease Unit, Departments of Medicine and Microbiology and Immunology, Baylor College of Medicine, Houston, Texas 77030. **Janet A. Englund** • Acute Viral Respiratory Disease Unit, Departments of Microbiology and Immunology and Pediatrics, Baylor College of Medicine, Houston, Texas 77030.

in immunocompromised hosts such as transplant recipients or patients infected with HIV. Interpretation of laboratory results in such situations requires a thorough understanding of the pathogenesis and epidemiology of the virus.

Failure to detect a virus or viral antigen does not necessarily mean that the virus was not present previously or did not cause disease. Although failure to detect a virus may be a result of inappropriate or inadequate specimen collection and handling, it may also be a function of the time course of the disease, the age or antibody status of the patient, and the technical resources available to detect or cultivate the virus. The investigator today has many options to diagnose the presence or past presence of a viral infection, but careful epidemiologic and laboratory studies are still required to ultimately link the viral agent to a specific disease process.

2. Diagnosis of Viral Diseases

Epidemiologic studies continue to rely on basic precepts of the causation of infectious diseases established at the turn of the century and on careful correlation of laboratory findings and clinical findings now well established in the field of virology (see Chapter 1). The determination of a viral etiology for a disease process usually requires evidence obtained from laboratory tests to definitively confirm the current or prior presence of the viral agent. However, a probable causative agent often can be suggested on clinical or epidemiologic grounds without laboratory confirmation. There are four circumstances in which a probable causative agent is suggested on clinical or epidemiologic grounds. First, some viral infections have distinctive clinical features that enable the recognition of disease if symptoms occur in the right geographic area, season, and/or age group. Such infections include varicella and herpes zoster, herpes simplex infection, measles, mumps, paralytic poliomyelitis, rabies, rubella, and viral hepatitis. Second, if an epidemic has occurred in which an etiologic agent has been isolated, then most clinical syndromes of the same type are usually caused by the epidemic virus; universal documentation of the virus in all cases is probably unnecessary. Examples include outbreaks of influenza, arbovirus infections, enteroviral exanthems, epidemic pleurodynia, and pharyngeal–conjunctival fever. Third, unique or special epidemiologic circumstances may indicate the probable diagnosis. Examples include bronchiolitis in an infant during the winter months being suggestive of RSV infection, or croup in a toddler representing parainfluenza virus infection. Finally, the site of organ involvement may be helpful in establishing the diagnosis; i.e., 80–90% of common respiratory infections are viral in origin, and nonpurulent infections of the central nervous system (CNS) in the normal host are likely to be viral, with the most likely candidates being enteroviruses, HSV, mumps virus, and arboviruses.

There are some common, but not pathognomonic, features of viral diseases: viral infections are usually nonpurulent and associated with mononuclear rather than polymorphonuclear infiltrates; the onset is more likely to be insidious than with a bacterial infection; prodromal symptoms occur frequently; and retrobulbar headache is common. In the clinical laboratory, the presence of a normal or low white count is suggestive of viral infection, although typhoid, tuberculosis, brucellosis, malaria, histoplasmosis, and overwhelming bacterial infections also produce leukopenia. The presence of absolute lymphocytosis and atypical lymphocytes also suggests a viral infection. Lymphocytosis of 50% or more and atypical lymphocytosis of 20% or more may be seen following infection by Epstein–Barr virus (EBV), CMV, and HIV, but also following infection with *Bordetella pertussis* and *Toxoplasma gondii*; drugs such as *p*-aminosalicylate (PAS), phenytoin, and tetrachloroethylene also may evoke lymphocytosis. Less intense lymphocyte responses are seen in a variety of viral infections such as rubella, hepatitis A, adenovirus, mumps, herpes simplex, and varicella infections. They may occasionally occur in tuberculosis, histoplasmosis, and other nonviral infections. Reversal of the typical CD4 : CD8 lymphocyte ratio or low absolute CD4+ cell counts may indicate infection with HIV, but similar findings are seen in patients with congenital immunodeficiencies and following viral infections, particularly with viruses from the herpes family, such as varicella or CMV.

The diagnostic procedures used for evaluation of specific viral infections are presented in individual chapters of this book. However, certain common aspects of specimen collection, laboratory evaluation, and interpretation of laboratory data are widely used and important for the design and implementation of epidemiologic studies. These general issues are addressed below.

3. Interpretation of Laboratory Tests

The classic criteria for the diagnosis of a viral infection require virus isolation or a fourfold or greater rise in antibody titer to the specific virus between the acute and convalescent sera. For some viral infections, isolation of the virus first and a subsequent test for a serological rise against that isolate are required. This is true of virus groups in which there are too many antigenically distinct strains to carry out a battery of serological tests, such as the echovirus, coxsackievirus, and rhinovirus groups. For viruses that have common group antigens, such as the adenoviruses and influenza A viruses, a single assay, particularly the complement-fixation test, may indicate a recent infection caused by one of the members of that viral group.

Difficulties arise in interpretation of serological tests when only a single convalescent serum sample is obtained and a high antibody titer is found, or if high titers are present in both acute and convalescent sera without a fourfold rise. These results can reflect either current infection or persistently high antibody levels from a previous infection. Significance may be attached to these findings if the disease is a rare one in which the presence of antibody is unique, if the test reflects a short-lasting antibody, or if specific immunoglobulin M (IgM) antibody can be demonstrated. A rapid drop in antibody titer in a subsequent specimen is also suggestive of a recent infection. Sequential testing of other family members also may be useful, since they may be in different stages of apparent or inapparent infection with the same virus. In an epidemic setting, comparison of the geometric mean antibody titer of sera collected early in illness from one group of patients with that from another group of patients convalescing from the same illness may permit rapid identification of the outbreak.

At times, a virus may be isolated or an antibody rise demonstrated that is not, in fact, causally related to the illness. Sometimes dual infections with two viruses, or with a virus and a bacterium, occur, and the interpretation of the role of an individual viral pathogen in the disease process may be very difficult. On other occasions, no virus is isolated or the serological rise is not sufficient to demonstrate whether a specific virus is the real cause of the illness. A list of common causes for false-positive and false-negative results is given in Table 1.

4. Specimen Collection

The appropriate collection of specimens is of the utmost importance for the successful identification of viruses in clinical samples. The source of the specimen, the timing of collection in relation to onset of symptoms, the rapidity and method of delivery to the laboratory, and the clinical and epidemiologic data provided to the laboratory all are important variables that relate to the likelihood

**Table 1. Viral Diagnosis:
Some Causes of False-Positive and False-Negative Tests**

False-positive
  Virus isolation
    1. Persistent or reactivated virus from prior and unrelated infection.
    2. A viral contaminant is present in the tissue culture or other isolation system.
    3. Nonspecific cytopathic effects occur because of toxicity of specimen or presence of bacteria or other causes and are mistaken for a virus.
    4. Two microbial agents are present, and the one isolated is not the cause of the disease.
  Serological rise
    1. Cross-reacting antigens.
    2. Nonspecific inhibitors.
    3. Double infection with only one agent producing the illness.
    4. Rise to vaccination rather than natural infection.
False-negative
  Virus isolation
    1. Viral specimen taken too late or too early in illness.
    2. Wrong site of multiplication sampled, e.g., throat rather than rectal swab.
    3. Improper transport or storage of specimen, e.g., not kept frozen.
    4. Wrong laboratory animal or tissue culture system selected for isolation.
    5. Toxicity of specimen kills the tissue culture, obscuring the presence of virus.
  Serological rise
    1. Specimens not taken at proper time, i.e., too late in illness or too close together to show antibody rise.
    2. Poor antibody response–low antigenicity of the virus or removal of antibody by immune-complex formation.
    3. Wrong virus or wrong virus strain used in the test.
    4. Nonspecific inhibitor obscures true antibody rise.
    5. Wrong test used for the timing of the serum specimens.

of successful isolation and/or identification of a viral pathogen.

4.1. Source

The clinical syndrome caused by a virus and its pathogenesis of infection determine the specimen(s) that is most appropriate for virus identification. Viruses that primarily cause disease at a mucosal surface or cause vesicular skin lesions generally can be identified in specimens taken from those sites. However, viruses causing generalized or congenital disease or causing symptoms in an internal organ (e.g., central nervous system) often can be identified in specimens taken from multiple different sites. Viruses that cause respiratory tract disease such as influenza viruses, RSV, and rhinoviruses are most frequently identified in samples of respiratory secretions;

viruses that cause gastroenteritis, such as rotaviruses, caliciviruses, and astroviruses, are identified in fecal specimens; and viruses that cause generalized or congenital diseases, such as measles, CMV, and mumps, are identified from respiratory secretions, urine, and blood.

Specimens may be obtained from an infected site by several methods. For example, the upper respiratory tract may be sampled by nasopharyngeal aspirate, nasal wash, nasopharyngeal swab, throat wash, pharyngeal swab, or by a combination of these methods. The selection of a sampling method may be influenced by the virus(es) targeted for identification and the identification method.[153] RSV is more likely to be identified in a nasopharyngeal specimen than from a pharyngeal swab, while adenoviruses are more likely to be identified from a pharyngeal swab.[110] A combination of sampling methods, such as nasopharyngeal aspirate or wash plus a pharyngeal swab, provides the greatest likelihood of detecting the broadest range of the more common respiratory viruses by culture. When swabs are used for sample collection, calcium alginate swabs should be avoided because they may decrease the recovery of HSV and may also be responsible for decreased rates of recovery of RSV from the nasopharynx.[2,28,53]

Specimens for viral diagnosis can be obtained from many sites. The lower respiratory tract is sampled using sputum specimens or bronchoalveolar lavage fluid, with the latter being particularly useful in immunocompromised hosts.[72] The gastrointestinal tract is generally sampled using fecal samples or rectal swabs. Fecal specimens have yielded a higher rate of recovery of enteroviruses than have rectal swabs, but rectal swabs are easier to obtain.[114] Cerebrospinal fluid can be useful in the diagnosis of a viral etiology for aseptic meningitis, and urine and blood specimens have been used to identify a number of systemic viral infections. Viruses also can be identified from samples taken from skin lesions and conjunctivitis. Tissue biopsy specimens may be essential for the diagnosis of certain diseases, such as CMV organ disease in the immunocompromised host or herpes encephalitis. Table 2 lists clinical specimens from which viruses may be identified.

Table 2. Sites from Which Selected Viruses May Be Identified[a]

Virus Group	Blood	CSF	Feces	Respiratory tract	Skin[b]	Tissue	Urine
Adenovirus			+	+	+	+	+
Arbovirus	+					+	
Coronavirus				+			
Cytomegalovirus	+	+	+	+		+	+
Enterovirus		+	+	+	+	+	+
Epstein–Barr virus	+			+			
Hepatitis A virus			+				
Hepatitis B virus	+						
Hepatitis C virus	+					+	
Herpes simplex virus		+		+	+	+	
Influenza virus				+		+	
Measles virus	+		+	+	+	+	+
Mumps virus	+	+		+		+	+
Norwalk virus			+				
Papillomavirus						+	
Parainfluenza virus				+		+	
Parvovirus	+					+	
Rabies virus		+			+	+	
Respiratory syncytial virus				+		+	
Retrovirus	+	+		+	+	+	
Rhinovirus				+		+	
Rotavirus			+			+	
Rubella virus	+	+		+		+	+
Varicella-zoster virus	+	+		+	+	+	

[a]Does not include sites from which identification is rare.
[b]Includes conjunctiva.

4.2. Timing of Collection

Specimens to be used for virus identification should be obtained early in the course of the illness. For many viral infections, viral shedding begins before the onset of symptoms, peaks during the illness, and disappears around the time that symptoms resolve. There are notable exceptions; enteroviruses and adenoviruses may be shed in the feces for weeks to months, and congenitally acquired CMV is shed in the urine for prolonged periods. Some factors that influence the likelihood of successful virus identification include the type of virus, the site from which the sample was obtained, the test being used, the age of the patient being sampled (e.g., younger children shed influenza viruses longer than adults), and the immune competence of the patient (e.g., immunocompromised hosts shed HSV from genital lesions longer than do immunocompetent adults).[26]

A serum sample should be collected early in the course of illness for potential use in identification of a viral infection. For some viral infections, the identification of IgM antibody or the presence of high titers of antibody is sufficient to confirm a virus infection. A second, or convalescent, serum sample should be obtained 2–4 weeks later to look for a rise in virus-specific antibody titer.

4.3. Transport to the Laboratory

4.3.1. Transport Conditions.
Once a clinical specimen has been collected, it should be transported to the virus laboratory as soon as is possible. A higher diagnostic yield can often be obtained when a specimen is processed expeditiously because virus viability decreases with time. Several steps may be taken to enhance the likelihood of preserving viruses during the transportation of a specimen to the laboratory.

1. Place the specimen on wet ice or in a refrigerator (at 4°C). A decline in infectivity for most viruses has been demonstrated to be temperature dependent so that cooling of the specimen improves virus stability.[68]
2. Use a good viral transport medium. A suitable transport medium generally contains protein, antibiotics to prevent bacterial growth, and a buffer to control the pH.[87] Many different viral transport media have been described and used successfully, but several studies have shown that charcoal-containing media may reduce virus recovery.[79,87,145] Some clinical specimens, such as CSF, urine, and feces, can be transported without the use of a viral transport medium.[87]
3. Avoid freezing the specimen unless the specimen will not be examined within 2–5 days of collection.[21,145,152] Baxter *et al.*[10] showed no loss of influenza virus or parainfluenza virus infec-

tivity after storage of specimens for up to 5 days at 4°C. The freeze–thaw cycle is harmful to some viruses, particularly enveloped viruses. For example, a single freeze–thaw of a laboratory strain of HSV held at −20°C for 1–3 days resulted in decreases in virus titer of 100-fold or more, while little loss in infectivity was seen in most specimens held at 4°C for the same period of time.[21] If specimens must be frozen, storage at ≤−70°C is usually advised, since warmer temperatures (e.g., −20°C) may lead to the rapid loss of infectivity.[87,189] Quick freezing a specimen in a dry ice–ethanol bath may enhance recovery of some labile viruses such as RSV.

4.3.2. Transportation Regulations.
Regulations governing the shipping of specimens have been published by the U.S. Public Health Service [*Fed. Register* 45(141): July 21, 1980] and can also be found in the Department of Transportation and Interstate Quarantine regulations (49 CFR, Section 173.386.388, and 42 CFR, Section 72.25, Etiologic Agents). Copies of these regulations may be obtained from the Biohazards Control Officer, Centers for Disease Control, Atlanta, GA. The regulations stipulate that the specimen should be wrapped in an absorbent material that can absorb all of the contents in case of a spill. This material must be placed in a watertight container that is secured in a second container using shock-absorbent material or tape.[63]

4.4. Storage of Specimens

Virus recovery from stored specimens can be optimized by using appropriate storage temperatures and cryoprotectants. As noted above, −70°C is a better storage temperature for many viruses than is −20°C, although the latter is adequate for the recovery of enteroviruses from stool specimens.[145] Cryoprotectants used to enhance virus recovery after freezing of a specimen have included dimethylsulfoxide (DMSO), serum, skim milk, other proteins, sucrose, glycerol, and sorbitol.[87]

4.5. Clinical Data

Clinical information may be useful in helping one choose the types of diagnostic assays that should be performed on a clinical specimen. The time of year and age of the patient are examples of epidemiologic information that will influence the likelihood of identifying certain viral infections. For example, rotavirus infections occur seasonally and are more common in young children, while Norwalk and Norwalk-like virus infections often occur in outbreaks and are more common in older individuals. Enteroviruses are the most common cause of viral meningitis and tend to occur seasonally in epidemics, whereas

HSV type 1 is the most common cause of sporadic viral meningoencephalitis. Knowledge of the pertinent epidemiologic information will permit use of appropriate tissue culture lines, enzyme immunoassays (EIA), immunofluorescence assays, and other diagnostic methods to identify a potential viral pathogen in the clinical specimen.

5. Laboratory Methods for Viral Detection

The detection of viruses or viral components is the foundation of diagnostic virology. Although the detection of antibodies to specific viral proteins remains an important element of viral epidemiology, the ability to isolate and/or characterize viral pathogens initially is critically important. The performance of any diagnostic test in a reproducible, sensitive, and specific manner is crucial in the study of viral diseases. Combinations of various techniques, including centrifugation-enhanced tissue culture, antibody–antigen detection, and detection of viral nucleic acid, can be used to supplement classic tissue culture

methods. A brief general description of commonly used laboratory techniques is outlined below (see also Table 3). Advantages of using a particular method or combination of methods to optimize results must be considered in designing any study, but the choice of method(s) remains highly dependent on the specific goal of a study, the viral agent, the patient population, and the available resources (Table 4).

5.1. Direct Visualization

5.1.1. Electron Microscopy. Viral diagnosis by direct or indirect visualization using electron microscopy relies on the identification of viruses based on typical morphological characteristics. The use of electron microscopy to differentiate one virus from another began in the late 1940s with the comparison of very different viral agents: the large smallpox virus and the small, round varicella-zoster virus.[119,176] Eventually, negative staining techniques were used to enhance viral resolution and reveal viral ultrastructure.[14] Further refinements using

Table 3. Laboratory Methods Used to Detect Selected Viruses

Virus	Diagnostic method[a]			
	Direct visualization	Culture	Ag detection (EIA, IF, RIA)	Nucleic acid detection
Adenoviruses	++	+++	+	+
Arboviruses[b]		++		(+)
Coronaviruses		+	+	+
Cytomegalovirus	+++	+++	+++	+
Enteroviruses[b]		+++	+	+
EBV[c]		+	++	+
Hepatitis B and C			+++	+
HSV	+++	+++	+++	+
HHV-6,7[c]		+++	++	+
Influenza		+++	++	+
Measles		++	+	+
Mumps		+		
Norwalk/Rotavirus	++	(Rota+)	+++	+
Papillomaviruses	++ ++			
Parainfluenza		+++	++	+
Parvovirus		+	+	+
Rabies[b]			++	+
RSV	+++	+++	+++	+
Retroviruses[c]		++	+++	++
Rhinoviruses		+++	+	+
Rubella		++		
Varicella-zoster virus	+++	+++	++	+

[a]+ Method mainly used in research setting; ++ method available in specialized reference laboratories; +++ method readily available and frequently used in many virology laboratories.
[b]Suckling mice are often necessary or desirable to use for isolation of these viruses.
[c]Human lymphocytes may be required to isolate the virus.

Table 4. Comparison of Laboratory Methods Used in the Diagnosis of Viral Infections

Method	Sensitivity[a]	Specificity[a]	Cost	Time to Dx	Availability
Electron microscopy	+	+	Moderate	<1 day	Reference lab
Direct cytology	+	+	Least	<1 day	Clinical lab
Tissue culture	++/+++	+++	Expensive	>3 days	Clinical lab
Animal culture	++	+	Expensive	>3 days	Reference lab
Immunofluorescence	++	++	Moderate	<1 day	Clinical lab
Immunocytology	++	++	Moderate	<1 day	Clinical lab
Radioimmunoassay	++	++	Moderate	1–3 days	Reference lab
Enzyme immunoassay	++	++	Least	<1 day	Clinical lab
Dot blot hybridization	++	++	Expensive	1–3 days	Reference lab
Southern hybridization	++	++	Expensive	1–3 days	Reference lab
In situ hybridization	++	++	Expensive	1–3 days	Reference lab
Polymerase chain reaction	+++	++/+++	Moderate/expensive	<1 day	Reference lab
Branched DNA amplification	++	++	Moderate/expensive	<1 day	Reference lab

[a]+, Relatively low; ++, moderate; +++, high.

immunologic tagging, ultracentrifugation, and other methods of enhancement have been utilized for the study of a number of different viruses.

Major advantages to the use of electron microscopy include the ability to visualize a virus quickly without the need for replication to take place. Using typical methodology for negative staining, for example, a clinical specimen can be prepared within minutes and a virus identified rapidly thereafter.[36] Furthermore, viruses can be detected without a preconceived idea of the agent, a factor frequently required with viral probes or cell culture. Because identification, at least to the family level, is based on morphology alone, there is no need to ensure specimen collection appropriate to maintain viral or host cell viability, and inactivated viruses or those difficult to culture, such as hepatitis B virus, Norwalk-like viruses, and rotaviruses, may be promptly identified. Advances in the design of various types of electron microscopes, including transmission electron microscopes, scanning electron microscopes, and high-voltage microscopes, now permit higher resolution and better discrimination of viral particles. Immune electron microscopic techniques, relying on the reactions between viruses and virus-specific antibodies coupled with electron-dense material, have resulted in the specific identification of smaller numbers of viral particles. Computer-enhanced imaging permits the resolution of virus particles at very high levels and, coupled with specific immunologic reagents, offers the advantage of high-resolution discrimination of various viral particles and their structural components.[130]

Limitations of diagnosis by electron microscopy include the need to have a relatively high concentration of

viral particles, with a titer of approximately $10^6–10^7$ particles/ml generally required.[110] This problem can be partially overcome by the use of physical measures to enhance viral concentrations such as ultracentrifugation, agar gel diffusion, cell culture amplification, and immune enhancement. Access to an electron microscope and the expertise of an experienced microscopist are essential to the successful use of any of these techniques. The cost and processing of multiple specimens is appreciable, and the processed specimens utilized for microscopy are not generally suitable for further evaluation using other methods. Finally, when a virus particle is identified by electron microscopy from a patient specimen, the determination of the virus beyond the family classification is generally not possible. This lack of specific identification is problematic, for example, in differentiating disease due to various paramyxoviruses.

Electron microscopy is widely utilized for the diagnosis of viruses associated with gastroenteritis, including rotavirus, enteric adenoviruses, and Norwalk virus, and remains a commonly used method for the identification of viruses that are generally not cultivated in the laboratory, such as small round viruses (SRVs) including Norwalk virus, other caliciviruses, and astroviruses. Another common use of electron microscopy is the rapid detection of virus particles within biopsy specimens obtained from patients with suspected viral encephalitis or pneumonia. Generally, glutaraldehyde fixation of the selected sample at the time the sample is obtained may adequately preserve tissue and viral architecture. Because thin-section specimens can be prepared from formalin-fixed and even paraffin-embedded tissue such as those routinely prepared in many pathology laboratories, the use of electron microscopic techniques can even be performed retrospectively on selected tissues, revealing such pathogens as herpesviruses, the human polyoma virus JC associated with progressive multifocal leukoencephalopathy, and the measleslike virus associated with subacute sclerosing panencephalitis.[36]

5.1.2. Cytology and Histopathology. Cellular changes induced by viruses have long been appreciated to be a reliable indicator of infection with certain viruses or virus families. Virus-infected cells obtained from skin, eyes, or mucous membranes, for example, may contain pathognomonic features that permit rapid diagnosis of a viral agent; changes in other tissues may be useful in suggesting viral infection or supporting the diagnosis of a viral etiology. Cytological detection can be extremely useful in the clinical setting because, in the hands of experienced laboratory personnel, this method offers prompt and clinically useful information. Histopathologic diagnosis, commonly based on small biopsy samples, also may prove diagnostic of a viral infection. Common samples include those obtained by direct scraping or aspiration of vesicles, exfoliative scraping of the genital tract or mucous membranes, direct smears of respiratory or other secretions, and cytocentrifugation of specimens obtained from endoscopy, urine, or stool. More invasive techniques such as biopsy of the respiratory or gastrointestinal tract can also provide useful information. Specific viral agents can be identified by the relatively viral-specific intranuclear inclusions seen in specimens obtained during active infection with CMV, HSV, and adenovirus; intracytoplasmic inclusions may be seen in multinucleated cells during measles and RSV infection and in uninucleate cells in parainfluenza virus infection.

The availability of semi-invasive procedures to obtain clinical specimens, such as bronchoalveolar lavage or endoscopy, has led to more demands for the rapid detection of viral infections from these specimens. Histological or cytological techniques may be combined with antigen or nucleic acid detection to obtain a diagnosis rapidly, with identification potentially at the level of viral subtype or strain. The high sensitivity of some nucleic acid or antigen detection techniques must be taken into account in the interpretation of results, particularly with latent viruses such as CMV and EBV, because these viruses may or may not be responsible for the disease process in the patient. In some settings (e.g., CMV pneumonitis), histopathologic examination combined with antigen detection may correlate better with disease than viral culture or viral nucleic acid detection because of increased specificity.[39,44,137]

5.2. Propagation of Viruses

Tissue culture remains the "gold standard" for much of clinical virology. Viruses are routinely detected by virus isolation in many diagnostic virology laboratories around the world. Isolation of a virus by culture confirms the presence of a complete infectious unit capable of further multiplication. Positive results may be obtained with as little as a single infectious virion. More than one virus can be detected in a single specimen, permitting the detection of unexpected viruses. The advantages of culture compared with the detection of viral antigens, genes, or gene products are numerous, but some viruses such as EBV, hepatitis B and C, rotaviruses, Norwalk virus, enteric adenoviruses, papillomaviruses, and parvoviruses cannot be or are not easily isolated in cell culture. Major limitations of viral isolation include the dependence on biological systems that may vary in quality and availability, the time, expense, and expertise required to isolate the virus, and the inability to propagate some viruses *in vitro*.

5.2.1. Isolation of Viruses in Animals. The transmission of viral disease by inoculation of a clinical specimen from one host to another has been the classical method used to document disease causation and pathology for centuries. Detection of viral agents in modern times began with the inoculation of filtered specimens into monkeys, ferrets, and mice in the late 19th and early 20th centuries; these animal models remain useful today. The direct inoculation of specimens from human to human, begun in the smallpox era in the 1700s (see Chapter 28) and used throughout the past two centuries to document the pathogenesis of many viruses, including yellow fever, EBV, varicella-zoster virus (VZV), and many respiratory viruses, is practiced today only under well-controlled clinical conditions.

The development of the embryonated hen's egg model for the isolation of influenza virus by Francis[52] in 1937 represented a breakthrough in viral propagation because the method was reproducible, practical, and sensitive. With this method, inoculation of clinical specimens or viral strains into the allantoic cavity, the amniotic cavity, or the chorioallantoic membrane is utilized for viral propagation, with virus replication recognized by the development of pocks of varying sizes on the chorioallantoic membrane, by the detection of hemagglutinins in the allantoic or amniotic fluid, or death of the embryo.[145] Embryonated eggs are still used today, although the advent of primary and continuous cell culture lines has supplanted this model in most clinical laboratories (see Chapter 16). It should be recognized, however, that several viruses or strains of the same virus, or a virus and a bacterium, may be involved in the same outbreak. An example is the occurrence of several strains of influenza A and influenza B in the same epidemic.

Intracerebral inoculation of suckling mice, a method introduced to document infections with enterovirus in 1948,[31] continues to be an important method in the detection of many arboviruses, rabies, and certain group A coxsackieviruses. Although specific cell culture systems are as sensitive for the isolation of some arboviruses and more readily adapted to routine virology laboratories, the suckling mouse model is used today because of certain advantages that are unlikely to change (Chapter 6). These advantages include the sensitive detection of a wide range of viral species from a limited amount of patient tissue or serum and the fact that many outbreaks occur in remote areas where tissue culture lines are less readily available.

The use of mammals and primates for the study of viral disease continues to be important in research but is generally not useful in epidemiologic studies. For example, the study of HIV in chimpanzees and other primates has been important in the study of the immunology of this infection because of the similarity of primates to the human host. However, limited availability of most primates, concern over the use of primates in research, and the expense involved in such research limit this practice substantially.

5.2.2. Isolation of Viruses by Cell Culture. Cell cultures were introduced during the late 1920s and expanded in the 1940s to meet the challenge of studying encephalitis, culminating with the important work on poliovirus by Enders *et al.*[41] in 1949. Cell cultures can be divided into three general types: primary cultures, semicontinuous cell cultures, and continuous cell lines. Primary cultures, or those derived directly from animal tissue, such as monkey kidneys or guinea pig embryos, were used for the routine isolation of viruses in the past but are used less commonly today because of the time and expense involved in preparation, as well as a general lack of availability. Disadvantages of primary cultures are that they can only be subcultured for a further one or two passages and they may be contaminated with adventitious viruses that are difficult to recognize and may be dangerous to the laboratory worker. The main role of primary cultures currently includes the use of human monocytes for the isolation of human viral pathogens such as EBV and HIV.

For routine virology laboratory studies, cell cultures are either semicontinuous (diploid), with the capacity for 20–50 passages, or continuous (heteroploid), with the capacity for indefinite passaging. Advantages of continuous or semicontinuous cell lines include ease and cost of preparation, general availability, and the ability to detect viruses or contaminants endogenous to the cell line prior to its use. The semicontinuous cell strains most commonly used in virology laboratories are those obtained from human foreskin, human embryonic kidney, and human embryonic lung fibroblasts (e.g., MRC-5 and WI-38). These cell lines are important for the recovery of HSV, CMV, VZV, adenovirus, and enterovirus. Continuous cell lines are often derived from an epithelial malignancy source [e.g., laryngeal (HEp-2 cells), lung (A549 cells), and cervical (HeLa) carcinoma] and vary significantly in their susceptibility to different viruses. The sensitivity of some continuous cell lines such as HEp-2 and Vero cell lines to various viruses such as RSV and arboviruses may change after serial passages so that careful quality control is required. The typical laboratory must use several different cell types in order to isolate different, clinically significant viruses. For example, cytomegalovirus can only be cultivated in diploid cultures, while influenza viruses are usually isolated in a heteroploid cell line such as the Madin-Darby canine kidney (MDCK) cell line.

5.2.2a. Cytopathic Effect. The presence of viral replication in cell cultures is traditionally detected in three ways: (1) by the observation of typical changes in the cultured cells (cytopathic effect), (2) by hemadsorption, and (3) by interference. The most common method used to identify virus isolates in the laboratory requires evaluation of the inoculated cell culture under light microscopy. Many viruses can be readily identified by the characteristic changes they produce in susceptible cell cultures, generally referred to as "cytopathic effect" (CPE).

When dealing with certain fastidious viruses or specimens with low viral inoculum, "blind passage" of a sample of the inoculated culture into a fresh cell culture may be advantageous for detecting a virus. For certain cell-associated viruses such as CMV or VZV, trypsinization and subsequent passage of the putative infected cell culture enhances viral recovery.[167] Subculture after freeze–thawing has been beneficial for the detection of adenovirus.[96]

5.2.2b. Hemadsorption. A variety of viruses, including influenza, parainfluenza, measles, arboviruses, and some picornaviruses, can agglutinate the erythrocytes of certain animal species (e.g., guinea pigs, goats, or chickens). This property allows the adherence of red blood cells to virus-infected cells, or hemadsorption. The presence of hemadsorbing viruses can be detected prior to or, in some cases, in the absence of detectable CPE. In the laboratory, a suspension of fresh RBCs (frequently guinea pig RBCs) is added at varying time points following inoculation of the clinical specimen into cell culture, and adherence of the RBC to the cell membrane is evaluated, generally using light microscopy. When hemadsorption is found in a cultured specimen, the supernatant fluid may be subcultured onto fresh tissue or tested using specific antigen detection methods to determine virus identity (see Section 6.2).

5.2.2c. Interference. Some viruses (e.g., rubella) do not cause easily detectable CPE and do not possess the property of hemagglutination, but they can be detected by their ability to interfere with the growth of a cytopathic virus inoculated into the same culture. This method of detection is called interference. In the typical assay for rubella virus, a standard dose of a challenge virus (commonly, echovirus 11) is inoculated onto a control tube of African green monkey kidney cells as well as a tube previously inoculated with a clinical specimen. When the challenge virus is inhibited in the culture tube containing the clinical specimen, an "interfering virus" is present.[96] Further analysis using antigen detection or serotyping is required for definitive results.

5.2.2d. Explant or Cocultivation Methods. Tissue specimens can be cultivated in the laboratory in the presence of established cell cultures (cocultivation) or as an explant culture. The tissue is minced, digested with trypsin, and inoculated directly onto cell culture monolayers or suspended in tissue culture media (as an explant). Cocultivation methods have been used for the detection of measleslike viruses in the brains of patients with subacute sclerosing panencephalitis.[76] This technique also is utilized for the demonstration of latent virus, but is not routinely available in diagnostic laboratories.

5.2.2e. Isolation of Viruses by Organ Culture. Culture of certain fastidious viruses using pieces of organs or tissue in which the normal histological architecture is preserved has been called "organ culture." Such culture methods have been useful for the detection of certain coronavirus (OC43) and rotavirus strains that were previously noncultivatable and for the detection of latent HSV in sensory ganglia.[106,159,172,187] With this method, pieces of tissue such as tracheal rings, embryonic intestine, or ganglia are placed intact into cell culture medium and the clinical specimen inoculated directly onto the tissue. Detection of viral replication is accomplished using a variety of techniques, including antigen detection, electron microscopy, serological tests, loss of ciliary function, or changes in histopathology. This technique is rarely used, however, because it is labor-intensive, time-consuming, and tissues are not readily available to most laboratories.

5.2.2f. Isolation of Viruses Using Physical and Chemical Refinements. Methods that enhance virus isolation or speed the detection of an agent are highly desirable for the virology laboratory. A variety of physical and chemical techniques may increase the speed and sensitivity of a viral diagnosis.[78] Roller culture methods, first introduced by Gey[58] in 1933, are known to increase rates of propagation of eukaryotic cells. This method also has been shown to increase both cellular RNA and viral RNA synthesis.[78] Low-speed rolling, high-speed rolling, rocking, and orbital motion have all been found to be advantageous in certain culture conditions.

Centrifugation of a clinical specimen or viral isolate onto a cell culture monolayer can enhance viral detection. The improved detection is probably not due to enhanced penetration of the virus into the cell but to stimulation of cell proliferation and activation of gene expression.[77,78] Methods utilizing centrifugation of clinical samples, such as described by Smith and co-workers,[44,46,59] are used routinely in many clinical virology laboratories, frequently in conjunction with viral antigen detection to provide rapid diagnosis. The use of short-term centrifugation of 10 min to 2 hr at 500 to 15,000 $\times$ *g* has been reported to increase detection of a variety of viruses, including HSV, VZV, CMV, rotavirus, RSV, HIV, polyomavirus, and measles. Culture vessels commonly used in this method are either shell vials (cylindrical, flat-bottomed vials that may contain a removable round coverslip at the bottom on which tissue culture cells are attached), or multiwell plates containing either 24, 48, or 96 wells. The centrifugation of viruses onto tissue culture monolayers has been shown to enhance infection by 3- to 10,000-fold, to reduce the time to viral detection, to be essential for isolating certain viruses (e.g., rotavirus), and to increase plaque or foci formation by 2- to 100-fold.[78]

Chemical measures to enhance viral replication include the use of DMSO and sodium butyrate.[146,164,165] The importance of these agents, however, in diagnostic settings requires further evaluation. The use of other agents, such as hormones, dexamethasone, and calcium may enhance cell growth, but these agents are not routinely utilized in the diagnostic laboratory. The addition of trypsin to culture medium is helpful in the isolation of certain influenza viruses, as well as rotaviruses, caliciviruses, and astroviruses, because the proteolytic cleavage of certain viral glycoproteins plays a role in cell fusion and enhances the initiation and spread of viral replication.[78]

5.3. Detection of Viral Antigens

Methods for the detection of virus-specific antigens have allowed rapid identification of a wide variety of viruses. Specific monoclonal antibodies conjugated to biochemical markers may provide high levels of sensitivity and specificity. The key to the success of the assays outlined below is the use of reliable virus-specific antibody. Molecular biological techniques that permit the production of relatively large quantities of avid monoclonal antibodies have facilitated viral antigen detection.

5.3.1. Immunofluorescence Techniques. Viral-specific antibodies conjugated to a fluorescein-labeled moiety have been used to identify viral pathogens since the late 1950s.[99] Immunofluorescence (IF) assays are widely used for the rapid detection of viruses in clinical samples and for definitive identification of a virus in tissue culture that may allow viral antigens to be sought before or after CPE is evident. In the direct IF test, specific antibody labeled with a fluorescent dye such as fluorescein isothiocyanate or, less commonly, rhodamine isothiocyanate, is allowed to react for a short time with cells obtained from a clinical specimen or from an inoculated cell culture. After allowing time for an antigen–antibody reaction to occur, the slides are washed and examined microscopically for direct visualization of the fluorescence of the infected cells in the specimen. In the indirect IF test, two different antisera are used: an unlabeled virus-specific antibody capable of binding to a specific viral antigen is used first and is followed by a fluorescein-labeled, species-specific antibody directed against the species in which the first antibody was raised. If a reaction occurs between the first antiserum and the clinical specimen, the second antibody will bind to the antigen–antibody complex and fluorescence of the virus-infected cells can be detected.

The two IF methods each have advantages and disadvantages. The indirect IF is usually more sensitive because several fluorescein-conjugated molecules potentially are able to bind to each virus-specific antibody molecule attached to the viral antigen, resulting in amplification of the fluorescence. The direct IF test may offer enhanced specificity due to lower background fluorescence. The indirect IF method requires more reagents and more time to perform. Whether monoclonal or polyclonal antisera are optimal for use in either test method is still debated. The use of monoclonal antibodies generally provides the lowest background, but may be limited by the high specificity of these reactions. This problem can usually be overcome by using a pool of monoclonal antibodies.

The use of IF for the direct detection of viral antigens in clinical samples and the confirmation of viral growth in cell cultures has increased with the widespread commercial availability of relatively inexpensive antibodies specific for many of the herpesviruses and respiratory viruses. The IF method has the advantage of allowing rapid

viral diagnosis in properly obtained specimens.[39,47,48,113] When working with large numbers of clinical specimens, the time required for sample collection, processing, and interpretation of the stained slide becomes substantial. The enthusiasm for this technique in clinical specimens has varied due to the time and degree of technical competence required to read such samples, the availability of other, less labor-intensive antigen detection methods, and the frequency of false-negative results obtained because of the dependence of the assay on having a high degree of viral antigen expression in the clinical sample. Nevertheless, the appropriate use of this test can result in reliable and sensitive rapid diagnosis from clinical samples. The use of IF for the detection of RSV in pediatric patients by an experienced laboratory can detect up to 90–95% of the samples positive by culture,[67,93] although many laboratories report rates of 70–80%.[37,163]

The combination of IF techniques with cell culture has increased the sensitivity of cell culture while providing a positive result in a shorter time period. With the use of centrifugation or other methods of enhancement of viral replication and pools of varying antibodies, cell cultures can be incubated between 1 and 3 days and then stained for a variety of virus antigens using indirect or direct IF methods. For some viruses, such as CMV or VZV, specific antibodies directed toward early or nonstructural antigens permit the rapid diagnosis within 48 hr, well before CPE would be visualized under routine cell culture conditions.[59,152] Disadvantages of IF techniques include the need for fluorescent microscopes, difficulty in the interpretation of clinical specimens that have a high level of nonspecific fluorescence, and the fact that prepared slides are not generally stable over periods longer than 1 month.[113]

Genetic engineering of cell culture systems to allow detection of a reporter gene after virus infection has allowed more rapid detection of viruses in clinical samples.[74,125,183] The *Escherichia coli* lac Z gene has been placed behind an early HSV promoter (ICP6 promoter from the UL39 gene of HSV-1) in a baby hamster kidney cell line. If infected by HSV, the cells produce β galactosidase and will turn blue when stained for β galactosidase activity.[74,183] A similar system for Sindbis virus using luciferase as the reporter (under the control of a Sindbis virus subgenomic promoter) also has been described.[125]

5.3.2. Immunocytochemical Staining. Immunocytochemical staining is a sensitive and specific method for detecting viral antigens with labeled antibodies. This technique, pioneered by Coons[24] in 1942, has been used to study the structure and function of a variety of viral proteins and continues to be utilized in both the research and clinical laboratories. It has been used both for detection of viral infection of a monolayer prior to the appearance of cytopathic effect and in rapid screening assays for drug resistance.[57,78] This method utilizes reagents similar to those used in the IF assay except that the fluorescent marker is replaced by an enzyme. When enzyme-specific substrates are provided, a colored precipitate forms at the site of reaction. Typical enzymes used to detect viral antigens include alkaline phosphatase and horseradish peroxidase. A major drawback of alkaline phosphatase-based reagents is their lack of stability; a major drawback of peroxidase as a marker is the fact that this enzyme is endogenous to some mammalian tissue, thus requiring either elimination of the endogenous enzyme or use of a nonmammalian enzyme, glucose oxidase.[55]

Advantages of immunoenzymatic staining compared to IF staining include the virtual permanence of stained preparations and the ability to view slides using an ordinary light microscope. Both direct and indirect staining with immunoperoxidase and other enzymes have been utilized to detect many viruses. Refinements have been developed that allow even greater sensitivity than that seen with indirect staining without the need to conjugate enzyme to an antibody. For example, a modification of these techniques has been a four-layer sandwich technique involving (1) virus-specific antibody raised in species X, (2) an excess amount of a second antibody raised against the species X antibody, (3) a complex of peroxidase and antiperoxidase antibody (raised in species X), and (4) reducing substrate for peroxidase.[158] The second antibody acts as a bridge, binding to both the virus and the antiperoxidase antibody. Similar unlabeled assay methods have been described for alkaline phosphatase–antialkaline phosphatase and glucose oxidase–antiglucose oxidase.[22,25] Increasing commercial availability of peroxidase kits for the diagnosis of viral diseases indicates that immunocytochemical methods will remain an option for the rapid, specific, and nonisotopic detection of viruses.

5.3.3. Radioimmunoassay. Radioimmunoassay (RIA) techniques have proven valuable for the detection of many compounds in laboratories and clinical medicine. Initially, the technique was developed for the determination of endogenous human plasma insulin levels.[188] The first important use of RIA in diagnostic virology was for the detection of hepatitis B surface antigen.[75] The original RIA described by Yalow and Berson[188] was a competitive binding assay in which the competition between an unlabeled antigen and a radiolabeled antigen reacting with a limited amount of antibody over a short period of time was monitored. Variations in RIA methods have been

developed, with the most common being the direct and indirect solid-phase RIA. In the direct solid-phase RIA, antigen or antibody are captured on a solid support and detected by radiolabeled (usually ^{125}I) antibody or antigen, respectively. The amount of signal increases proportionally to the amount of antigen or antibody present in the sample. In indirect assays, the capture of antigen or antibody to the solid phase prevents the binding of labeled antibody or antigen, respectively, so that the amount of signal detected is inversely proportional to the amount of antigen or antibody present. RIA methods currently are utilized mainly for the detection of antigens and antibodies of viral hepatitis.[116] The use of RIA for the detection of various hepatitis markers has demonstrated the assay's high degree of sensitivity. For the most part, RIA methods have been replaced by EIA for routine diagnostic purposes due to the complexity of the assay, the use of radioisotopes, lack of standardized commercially available reagents, and high equipment costs.

5.3.4. EIA. Enzyme immunoassays, or EIAs, have gained widespread acceptance in virology laboratories for the detection of a variety of viral antigens and antibodies. The assays used in this method rely on antibodies directed against a specific virus or viral antigen that are adsorbed or directly linked to polystyrene wells in microtiter plates, plastic beads, or membrane-bound material. When viral antigen is present in a specimen, it binds to the immobilized antibody and a second "detecting" antibody conjugated to an enzyme such as horseradish peroxidase or alkaline phosphatase then attaches to the antigen, forming a three-layer "sandwich" consisting of the immobilized antibody, the antigen, and the detecting antibody with enzyme attached. A substrate specific for the enzyme is added and a color reaction occurs that can be monitored by spectrophotometry or by direct visualization. The test is quite simple to run, requiring only standardization of reagents and techniques such as dilution, incubation, and washing. The principles involved in EIA are similar to those involved in immunofluorescence and RIA, but the EIA test has the distinct advantages of being simple to perform, utilizing reagents that have long shelf lives, are inexpensive, and do not require sophisticated technical evaluation to determine results. Advantages of the EIA technique also include sensitivity (less than 1 ng/ml), specificity, rapidity, safety, automation potential, and low cost, particularly when many specimens require evaluation.

Variations in the methodology for EIA testing include the materials used, the procedures for incubation and detection, and the interpretation of results. Many different test kits are commercially available and in widespread clinical use for the detection of common viral pathogens such as RSV, HSV, VZV, adenovirus, HIV, and rotavirus; EIA tests have been devised and reported for nearly all virus groups and continue to be used widely for clinical and research purposes.[123]

5.4. Detection of Viral Nucleic Acid

The detection of viral nucleic acid is another strategy for the identification of viruses in clinical samples. Use of these methodologies may add to and increase the diagnostic yield of antigen detection and tissue culture methods. The detection of viral nucleic acid may provide information that cannot otherwise be obtained.[121] In some instances, the virus cannot be isolated in tissue culture due to virus inactivation during collection, transport, or storage; contamination with other microbes, particularly other viruses; the lack of an available *in vitro* culture system; or slow growth of the virus. In other cases, no antigen detection assay is available or antigenic variation precludes the development of a broadly reactive antigen detection assay. Nucleic acid detection is also useful if evidence of a chronic carrier state is being sought, or if it is desirable to identify cells containing viral nucleic acids within a tissue sample.

5.4.1. Hybridization Assays.
Nucleic hybridization assays rely on the detection of a signal generated after the interaction of a labeled probe with the target nucleic acid. Hybridization occurs when the sequence of the nucleic probe is sufficiently similar to that of the target nucleic acid that a duplex is formed and held together by hydrogen bonds from nucleotide pairing. The specificity of the reaction is determined primarily by the sequence of the probe used, but is also affected by the temperature at which hybridization and subsequent washes take place, the salt concentration used in the hybridization mix and the washes, and the concentration of formamide used in the hybridization mix. These latter factors affect the stability of the duplex, with increased temperature and formamide concentration destabilizing less perfectly matched duplexes (increasing the stringency of the hybridization conditions) and increased salt concentrations stabilizing less well-matched duplexes (decreasing the stringency). More detailed description of hybridization conditions and procedures have been published.[69,142]

The target nucleic acid may be DNA or RNA and may be single- or double-stranded. All target nucleic acids must undergo a denaturing step before undergoing hybridization. This step ensures adequate binding with the probe because even single-stranded nucleic acids are in part double-stranded as a result of self-annealing. Heat and/or strong alkali are sufficient for denaturation when the target nucleic acid is DNA, but DMSO, methyl mercuric hydroxide, glyoxal, or formaldehyde are required for RNA.[3]

The probe also may be single-stranded or double-stranded RNA or DNA. Double-stranded probes may be made from vectors (plasmids, bacteriophages, cosmids) into which a virus-specific sequence has been cloned. The virus-specific sequence can be separated from the vector by digestion with a restriction endonuclease and gel purification before being labeled. Single-stranded bacteriophage vectors related to M13 also have been used but are limited by the size of the DNA that can be cloned into it.[156] Single-stranded RNA probes can be made by *in vitro* transcription using one of several DNA-dependent RNA polymerases.[4,17,142] Single-stranded DNA or RNA oligonucleotides, 20–40 bases in length, can now be made or obtained easily and at only minimal expense. In general, single-stranded probes are preferable to double-stranded probes because annealing of the probe to itself is less likely to occur. Single-stranded RNA probes offer the additional advantage of increased stability of RNA–RNA and RNA–DNA duplexes and the removal of vector sequences that may cause nonspecific hybridization if bacterial DNA is present in the sample.[112,135]

The probe must be labeled in order to detect its hybridization to the target nucleic acid. A number of different strategies for probe detection have been developed. Early assays relied on the use of radioisotopes (^{32}P, ^{35}S, ^{3}H), which are still among the most sensitive labels available. Nonisotopic methods have improved such that they approach the level of detection obtained with radioisotopes without the inherent hazards. The probe is linked to biotin, digoxigenin, or some other hapten. The biotin can then interact with avidin or streptavidin, or the hapten with a hapten-specific antibody, to which an enzyme (like alkaline phosphatase or β-galactosidase) is attached. A substrate is added and the test is interpreted based on the intensity of the resulting colorimetric or chemiluminescent reaction.[81]

Several hybridization strategies have been used for virus detection. The most common are dot (or slot) blot, Southern and Northern blot, and *in situ* hybridization. Solution hybridization methods also are being developed.

5.4.1a. Dot Blot Hybridization. In standard dot and slot blot hybridization assays, the target nucleic acid is immobilized on a solid substrate. Nitrocellulose, nylon, and chemically activated papers are the most common filters used for this purpose.[3] Although nucleic acid must be denatured to bind to the filter, the optimal conditions for binding the nucleic acid differ for the different types of filter. In sandwich hybridization assays, a virus-specific probe is attached to nitrocellulose to capture viral nucleic acids and a second probe (nonoverlapping with the first) is used to detect the captured nucleic acid.[177] This method has the advantage of high specificity, but is 10- to 100-fold less sensitive than standard methods.[145]

In some situations, dot blot hybridization assays can be performed on crude clinical specimens.[101] More frequently, partial purification of the nucleic acid is required, usually with phenol and chloroform followed by precipitation in ethanol or isopropanol. If known amounts of a virus standard are included in the test, the amount of target nucleic acid in an unknown sample can be semiquantitated using a densitometer or a liquid scintillation counter.[3]

The greatest success in the application of dot blot hybridization assays for viral diagnosis has been for DNA viruses such as papillomavirus, cytomegalovirus, HSV, parvovirus, and hepatitis B virus.[30,88,132,155,180] The use of RNase inhibitors allowed the development of assays for several RNA viruses, including rotavirus, hepatitis A virus, enterovirus, RSV, and coronavirus.[51,80,85,118,135,136,169,175] The sensitivity of these assays generally has been similar to that of antigen detection[112] enzyme-linked immunosorbent assays (ELISAs). Dot blot hybridization assays are now frequently used as an adjunct for detection of target nucleic acids after they have undergone an amplification procedure (see Section 5.4.2a).

5.4.1b. Southern and Northern Blot Hybridization. These hybridization procedures are very similar to dot blot hybridization except that fragments of the target nucleic acid are separated by size using gel electrophoresis before being transferred to a filter. First described by Southern[154] for the detection of DNA fragments, the method has since carried his name. The detection of RNA is called Northern blot hybridization. Such assays increase the specificity of a nucleic acid hybridization assay because they allow the recognition of fragments of specific size. However, the procedures are time consuming, often taking several days, and require a greater level of skill than the dot blot assay.

5.4.1c. In Situ Hybridization. *In situ* hybridization assays allow the detection of viral DNA or RNA within cells while preserving the histological detail of a sample. Specimens to be used should be frozen or fixed as soon as possible after collection so that degradation of the nucleic acid is minimized. If a specimen has been fixed and embedded in paraffin, it must be deparaffinized and rehydrated prior to hybridization. Target nucleic acid can be detected using a photographic emulsion to cover the slide for radiolabeled (^{3}H or ^{35}S) probes or by developing an enzyme-catalyzed, colorimetric reaction for noniso-

topically labeled probes (biotin, digoxigenin). These methods can be technically difficult, but have yielded valuable information about the pathogenesis of many different viral infections, including enteroviral myocarditis, papillomavirus-associated genital neoplasia, HSV and VZV latency, and EBV replication.[13,29,89,150,160] Modifications of *in situ* hybridization assays such as *in situ* transcription and *in situ* polymerase chain reaction (PCR) have been described in which new labeled nucleic acid is made using the virus-specific nucleic acid in the tissue as a template.[18,122,168]

5.4.2. Nucleic Acid Amplification Techniques.
Many nucleic acid hybridization assays have a sensitivity of 10^5–10^6 molecules, and even the most sensitive have a limit of detection of 10^3–10^4 molecules.[122,128] This level of detection is inadequate for many clinical situations. Thus, strategies to amplify the target (viral) nucleic acid have been developed. After amplification of the viral nucleic acid, many of these assays then rely on nucleic acid hybridization assays for detection of the virus-specific products. PCR is the most developed of the nucleic acid techniques, but the ligation chain reaction (LCR), transcript amplification system, the Qbeta replicase system, and signal amplification with compound probes have also been used successfully.[9,128,186]

5.4.2a. Polymerase Chain Reaction. PCR has revolutionized the detection of small amounts of nucleic acid. First described in 1985, PCR became much less cumbersome with the use of a thermostable DNA polymerase, Taq (from *Thermus aquaticus*) polymerase, and an automated thermocycler apparatus.[140,141] The use of a reverse transcription step to make a complementary DNA from RNA has allowed the detection of small quantities of RNA.[6,54] New innovations continue to broaden the potential applications for this method for the detection and characterization of viral nucleic acid.

PCR consists of three basic steps that, when repeated, result in exponential amplification of the target DNA. The first step is denaturation of the target, double-stranded DNA, and generally is carried out at 92–95°C. The second step is primer annealing; the reaction mix is cooled to 50–70°C to allow virus-specific oligonucleotide primers to anneal to the target nucleic acid. The third step is primer extension and usually occurs at 70–75°C. Every ten times these three steps are repeated, the amplified DNA can theoretically increase 1000-fold (2^{10}-fold). In practice, PCR is less efficient, but a million- to a billion-fold amplification of target DNA can be achieved.[65]

For standard reactions, two primers complementary to regions on opposite strands of the target DNA are used. The regions complementary to the primers are usually

200–500 base pairs apart. Hybridization of the primers to the denatured DNA is favored over that of the opposite strand because the primers are present in tremendous molar excess of the target DNA. The primers give PCR its specificity, and this specificity can be increased by using higher annealing temperatures in the reaction. The optimal annealing temperature for maximum sensitivity and specificity is approximately 5°C below the temperature (Tm) at which the primers are 50% dissociated from the target DNA.

PCR, as originally described, will amplify only DNA. Various modifications have been made that allow the detection of RNA. The first modification described was the synthesis of a complementary DNA (cDNA) from RNA using reverse transcriptase.[54,91] Using this methodology, less than one infectious dose of virus can be detected.[6] Another thermostable DNA polymerase that also has reverse transcriptase activity, the Tth polymerase, also has been described for RNA amplification.[117]

PCR products, or amplicons, are detected by visualization of silver- or ethidium bromide-stained gels after electrophoresis or by detection of a hybridization signal (e.g., dot blot, Southern blot, or liquid hybridization). Because nonspecifically amplified DNA may occasionally be the same length as the desired amplicon, the use of an additional test of specificity, such as by probe hybridization of restriction enzyme digestion of the amplicon, may be of value.[65] The hybridization assay may use a DNA probe in a standard dot blot or Southern hybridization assay, or it may use an RNA probe with biotinylation of one of the PCR primers so that detection of RNA–DNA hybrids can be accomplished with a monoclonal antibody in an enzyme immunoassay format.[27,90,174] If a restriction enzyme is used in a confirmation assay, one should be chosen that will cut the virus-specific amplicons into two fragments that can be detected by gel electrophoresis.

The extreme sensitivity of the PCR reaction has led to the recognition of problems of false-positives due to carryover contamination. Because the method can detect less than ten copies of DNA and generates millions to billions of copies of amplicons, the generation of aerosols containing amplified products can result in contamination of samples that originally did not contain viral nucleic acid. Several precautions have been used to overcome this problem. Physical separation of pre- and post-PCR areas, the use of positive displacement pipettes, the use of disposable gloves, and premixing and aliquotting reagents have been advocated to minimize false-positives.[95] UV irradiation of the PCR reaction mix after amplification and the use of dUTP in place of dTTP, allowing the degradation of contaminating, uracil-containing DNA with the enzyme uracil-N-glycosylase, are other methods that have been used to decrease false-positive reactions.[100,143] Use of the Tth polymerase and uracil-N-glycolase enzyme in a single-step assay may get around problems of carryover when RNA viruses are amplified by PCR.[191]

5.4.2b. Other Amplification Methods. The LCR utilizes many of the same principles as PCR. DNA is denatured at a high temperature, primers hybridize to the target DNA, and a heat-stable enzyme acts on the primers, resulting in an exponential increase in the LCR product. However, the LCR product is the result of the ligation of the primers, with the product length being the sum of the length of the ligated primers. Four primers must be used for exponential amplification to occur; the use of only two primers results in linear amplification.[9,186]

The transcript amplification system can detect RNA or DNA. A primer containing a promoter sequence for a DNA-dependent, RNA polymerase followed by a target-specific sequence is used for the synthesis of a cDNA for each target strand of RNA or DNA. After second-strand synthesis using a second target-specific primer, multiple copies of RNA per target strand are generated, using the appropriate RNA polymerase. The target nucleic acid may be amplified more than a million-fold in a short period of time.[66,94] More work needs to be done to determine the general applicability of this method, but it has been applied to the diagnosis of HIV infection.[92]

The Qbeta replicase system is a probe amplification system. Qbeta replicase is an RNA polymerase that will replicate RNA that has the secondary structure of the genomic RNA of the bacteriophage Qbeta. A small, target-specific insert can be made into the genomic RNA that allows its hybridization to a specific target. After a wash step, the Qbeta replicase will amplify the probe up to a billionfold. This method has the advantage of speed, high sensitivity, and the ability to quantify the amount of target nucleic acid; however, problems with background still exist.[128]

Branched DNA (bDNA) amplification is another amplification system in which the signal is amplified instead of the target nucleic acid. This is accomplished through a series of hybridization steps. Viral nucleic acid is captured by hybridization with an oligonucleotide anchored to a solid substrate. A second virus-specific probe containing a sequence complementary to a bDNA amplifier sequence then is hybridized to the viral nucleic acid. The bDNA amplifier has multiple chains branching in a treelike fashion and on each of the branches are multiple sites for hybridization with an enzyme-labeled probe. The enzyme reacts with a substrate to allow colorimetric or chemiluminescent detection. Assays for HCV, HIV, CMV, and hepatitis B virus have been described.[173] The assay has been used to monitor serum viral RNA levels after the administration of antiviral chemotherapy in HCV and HIV infections.[32,35]

6. Laboratory Methods for Virus Characterization

Further characterization of a virus obtained from a clinical specimen is frequently desirable once an agent has been isolated. This can be done in a variety of ways, depending on what is known about the virus and what additional information is being sought. For example, if a previously unrecognized virus is recovered, characterization of its physicochemical as well as biological, antigenic, and genomic properties would be useful.

6.1. Physicochemical Methods

Virus classification originally depended on the determination of physicochemical properties. When viruses were first discovered at the turn of the century, filterability was the only physical property that could be measured.[115] Since that time, additional physicochemical properties used for virus characterization include virion size and shape, virus stability at different pHs and temperatures, and virus stability in the presence of cations (magnesium and manganese), solvents, vital dyes, detergents, and radiation. Though virion size and shape are still used to initially classify viruses, other physicochemical properties are rarely used to characterize human viruses. An exception is the classification of human picornaviruses as rhinoviruses or enteroviruses, with rhinoviruses being distinguished from enteroviruses by their lability at low pH.

6.2. Immunologic Methods

The presence of a virus may be detected by changes in tissue culture, but further steps can be taken to identify the virus. This is important because more than one type of virus can produce similar changes (e.g., hemadsorption for influenza and parainfluenza viruses and similar CPE for CMV and adenovirus) and nonviral agents can produce viruslike CPE (*Trichomonas vaginalis* and herpes simplex, *Clostridium difficile* toxin and enteric viruses).[145] Various immunologic methods can be used for this purpose because of the general availability of immune reagents for most human viruses. Immunofluorescence, radioimmunoassay, and enzyme immunoassay formats may be used in a fashion similar to that described for the virus detection in clinical specimens (Section 5.3). Other methods for virus

identification and characterization include virus neutralization assays, hemagglutination-inhibition assays, and epitope-blocking enzyme immunoassays using monoclonal antibodies.

6.2.1. Neutralization Assays. Virus neutralization assays detect the loss of virus infectivity that results from the interaction of virus with specific antibody. Unknown viruses may be identified using virus-specific antisera, and antibody to a specific virus present in a serum sample can be detected or quantitated (see Section 7.1). The loss of infectivity can be measured in a number of ways, depending on the biological systems capable of supporting virus growth, the types of viruses being sought, and the capabilities of the laboratory performing the studies. The principal biological systems used for neutralization assays are tissue culture, embryonated chicken eggs, and adult and suckling mice.[8] Cell culture systems are used most commonly because they support the growth of a large number of viruses, are widely available, are easier to work with than the other two systems, and lack an immune system (that may influence test results). Embryonated hen's eggs and mice are used as for primary isolation (Section 5.2.1). Neutralization cannot be measured for some viruses (e.g., Norwalk-like viruses) because their infectivity cannot be measured in currently available culture systems.

Pools of virus-specific antisera have been used to decrease the number of neutralization assays needed to serotype enteroviruses.[98] Each serum pool contains antisera to a discrete number of enteroviruses, and antiserum to a given enterovirus is present in one to three pools. Thus, the pattern of neutralization obtained from the use of only eight intersecting serum pools allows the identification of 42 different enteroviruses.[108] Methods for the production of intersecting serum pools have been published.[109]

6.2.2. Hemagglutination and Hemagglutination-Inhibition Assays. The ability to agglutinate erythrocytes, a property shared by many viruses, can be used for the identification of some of these viruses. The differential hemagglutination of rat, human group O, and monkey erythrocytes by different adenovirus serotypes allows their separation into groups so that fewer type-specific sera need to be used in neutralization or hemagglutination-inhibition assays.[73] Type-specific antisera can be used to prevent hemagglutination (hemagglutination-inhibition) and permit the identification of influenza A and B viruses, parainfluenza viruses, and adenoviruses.

6.2.3. Agar Gel Immunodiffusion. Agar gel immunodiffusion, or agar gel precipitation, has been used for the characterization of a variety of viral antigens using standard, or reference, antisera. A thin layer of agarose is made in a plate or on a slide, and small wells are cut into the agarose. The unknown antigen and known antiserum are placed in separate wells, and the proteins in the wells diffuse through the agarose. If the antiserum reacts with the virus antigen, a precipitation band appears. Though less sensitive than other methods, this method offers high specificity and is simple to perform. It has been used for the identification of orthopoxviruses, typing of influenza viruses, and subtyping of hepatitis B viruses.[38,49,120] It also has been used to characterize unknown sera with known virus antigens.[19]

6.2.4. Antigenic Characterization. The antigenic differences or similarities between vaccine and wild-type strains or among virus strains that have been isolated from different geographic locations or at different times may be examined in a number of ways. The availability of monoclonal antibodies permits the examination of these relationships and may detect differences or similarities that cannot be detected by polyclonal antisera.[182,190] These assays examine the ability of a given monoclonal antibody to interact with a particular virus strain and are performed using the same formats used for polyclonal antisera: RIA or EIA, neutralization (if antibody is neutralizing), immunoprecipitation, hemagglutination-inhibition (if the virus has hemagglutination activity), and so forth.

Monoclonal antibodies also have been used to map antigenic sites on virus proteins. When a virus is grown in the presence of a monoclonal antibody that normally neutralizes it, the only progeny virus will be escape mutants, or viruses that are no longer neutralized by the antibody. Frequently, escape mutants arise after substitution of a single nucleotide, resulting in a single amino acid change, and the location of the change can be determined by sequencing the virus gene(s) encoding the viral protein(s) important in neutralization (e.g., rotavirus).[102,166] A less precise map of antigenic sites can be obtained through the use of a panel of monoclonal antibodies by determining whether an individual monoclonal antibody competes with other monoclones for an antigenic site and whether the antibody has activity against the escape mutants raised by a different monoclonal antibody.[161]

6.3. Genomic Characterization

The identification and characterization of viruses has been enhanced greatly by advances in molecular biology. PCR has had the greatest impact of all the newer technologies. Advances in the characterization of the viral genome have been made possible or greatly simplified by

this technology. Evaluation of the total genomic nucleic acid of some viruses by restriction enzyme digestion, hybridization studies, or polyacrylamide electrophoresis still may provide useful information, but similar evaluations can often be made using amplified portions of the viral genome. This section will review some approaches used for characterization of the viral genome in epidemiologic studies.

6.3.1. Restriction Enzyme Digestion. Restriction enzymes are bacterial endonucleases that cut double-stranded DNA at defined sites based on the nucleotide sequence. The recognized nucleotide sequence may be 4, 6, or more bases in length, and there is usually an axis of symmetry to the recognized sequence (i.e., the two opposite strands have the same nucleotide sequence for the recognized site). For example, the restriction enzyme BamHI recognizes the following DNA sequence:

$$5'\text{-GGATCC-}3'$$
$$| \ | \ | \ | \ | \ |$$
$$3'\text{-CCTAGG-}5'$$

After digestion, the DNA fragments are separated by gel electrophoresis and the size and pattern of the fragments are determined. Restriction enzymes have been applied to total genomic DNA in epidemiologic studies of adenovirus outbreaks to identify the epidemic strain.[15,129,178] They also have been applied to amplicons, the specific products resulting from PCR amplification, to document the specificity of a reaction or to examine the relatedness of virus strains.[6,65,162]

6.3.2. Hybridization Assays. The principles of hybridization are elaborated in Section 5.4.1. In addition, application for viral nucleic acid detection and as a test of amplicon specificity, hybridization studies also can be performed to determine the homology between different viruses. For this method, a portion of the genome of one virus is labeled and used to probe the genome of other virus strains. A greater homology between two viruses is shown when a more intense signal is obtained from a labeled probe after hybridization. This methodology has been used, for example, to examine the homologies of several different enterovirus strains and to differentiate wild-type from vaccine strains of poliovirus.[7,33,170]

6.3.3. PCR Assays. PCR technology can be used in a number of ways to identify and characterize viral nucleic acids. Nested PCR is the performance of two sequential PCR reactions, with the second reaction using two primers located between the two used in the first reaction. Nested PCR serves to increase the sensitivity of the PCR assay, and it also has been used as another test of specificity of the PCR assay (e.g., hepatitis C).[64,65] Although sensitivity is increased by the nested PCR assay, the potential for carryover contamination also is increased and may limit the usefulness of this methodology.

Multiplex PCR is used to detect several viruses or virus types in a single PCR assay. It utilizes several different primers or primer sets in the reaction mix, and a virus type is identified by the size of the amplicon detected after amplification. Such an assay has been used to identify different VP7 and VP4 serotypes of rotaviruses and for the identification of different serotypes of hantaviruses.[56,61,131] Unfortunately, the use of several different primer sets may adversely affect the efficiency of the nucleic acid amplification and decrease the overall sensitivity of a PCR assay.

Quantitative PCR measures the amount of a specific nucleic acid that is present in a sample; in viral studies such information might be used to correlate the amount of virus present with the degree of clinical illness or to measure the clinical response to an antiviral agent. Quantative PCR can be accomplished by several approaches.[23] The simplest is a dilution series until an endpoint is reached; this method only provides semiquantitative data.[86] Another approach is the generation of a standard curve using known quantities of a cloned DNA or *in vitro* transcribed RNA followed by the comparison of the amount of amplicon produced in an unknown sample to that in the standard curve (by densitometry or by incorporation of radioactivity into the amplicon).[147] This method is still subject to errors produced by variability of tube-to-tube or sample-to-sample variation. The ideal method for nucleic acid quantitation utilizes an internal standard in each tube that is amplified by the same primers as those used for the target nucleic acid. Amplification of the internal standard produces an amplicon that has a unique restriction site not present or measurably different sized than that produced by the target.[103,111] Serial dilutions of the unknown sample are amplified in the presence of a known amount of the internal standard, and the dilution at which the same amount of amplicons are produced for both target and internal standard is used to calculate the amount of target present in the sample. Methods for the production of internal standards have been described.[103,111]

A single primer set can sometimes be designed to allow the differentiation of virus strains based on amplicon size. This has been done for the two major genera of human picornaviruses.[124] Rhinoviruses and most enteroviruses can be separated into their respective genera using a primer set derived from conserved regions of the 5' end of their genome. Although this method does not amplify echovirus-22-like strains, it has compared favorably to the more traditional acid lability assay for differentiating rhinoviruses from enteroviruses isolated from the respiratory tract.[5,124]

6.3.4. Sequence Analysis. PCR technology has greatly simplified the sequencing of portions of the viral genome. The viral nucleic acid can be amplified and then cloned and sequenced or the sequence can be determined by directly sequencing the amplicons. This technology has been useful in many epidemiologic studies, including the investigation of HIV transmission in a dental practice and the identification of a hantavirus as the cause of a previously unrecognized respiratory illness with a high mortality rate.[20,126]

6.3.5. Other Methods of Viral Detection. Other methods of viral detection are being developed and refined for clinical use. The use of molecular viral detection methods to detect virus in stored paraffin or frozen sections is a useful tool for retrospective studies evaluating causal association such as between viruses and cancer. An example of such a method is the EBER probe for EBV. This test is useful on paraffin-fixed and frozen tissues and detects the EBV genome in about 50% of Reed-Sternberg cells in Hodgkin's disease.[127]

6.4. Antiviral Susceptibility

The development of clinical resistance to antiviral agents has been documented in a variety of viruses, including influenza, HSV, VZV, CMV, and HIV.[11,42,43,84,138,139] As the use of antiviral therapy becomes more widespread, the importance of antiviral resistance for the host and for the community increases.[42] Laboratory documentation of antiviral susceptibility may be useful for both clinical and epidemiologic reasons. The emergence of antiviral resistance is important since resistant forms of virus may be transmitted from person to person, as has been demonstrated with zidovudine-resistant strains of HIV.[45] Such documentation is also valuable when evaluating the effects of antiviral therapy on the virus and the host, as well as individuals subsequently infected.

A viral isolate or clinical specimen from an infected patient is required prior to determination of antiviral susceptibility. Frequently used methods of assessing viral replication in the presence of varying concentrations of antiviral drugs include plaque reduction, DNA or RNA hybridization, immunofluorescence, and dye uptake. Other methods include the detection by PCR of a genetic mutation responsible for resistance to an antiviral agent, such as that seen in the reverse transcriptase gene of HIV associated with resistance to zidovudine, the thymidine kinase or DNA polymerase genes associated with resistance to acyclovir, or the M2 gene in influenza A strains associated with resistance to amantadine and rimantadine.[12,71,134,138] Sensitivity testing systems for viruses are not standardized, and results may depend on multiple factors, including the assay system, cell line used, viral inoculum, and laboratory.[34] Results from different laboratories, in general, are not comparable. Inconsistent results may be seen with systems relying on either culture or nucleic acid detection because of differences in the numbers of passages of virus prior to analysis, concentrations of virus or drugs used, incubation times, and statistical analysis (i.e., 50% vs. 90% endpoint).

7. Serological Diagnosis

The detection of newly developed, virus-specific antibody or the detection of an increase in titer of preexisting antibody is important in viral diagnosis and is one of the most commonly used methods in epidemiologic studies of viruses. Most primary infections or reinfections result in the production of specific antibodies. In addition, viruses such as EBV, HIV, rubella, hepatitis A and B viruses, and arboviruses are difficult to detect directly and the serological diagnosis may be the only practical means of identifying the particular agent.

The detection of specific IgM antibody may be used to suggest a recent infection in a single serum specimen. Detection of specific IgM antibody in the neonate is useful to diagnose congenital infections, because maternal IgM antibody does not cross the placenta. IgM antibody also is useful to detect acute disease in a variety of other clinical situations, including infection with CMV, rubella, hepatitis A and B, and EBV. Limitations to the use of IgM detection include (1) IgM-specific antibody is not restricted to primary infection and may be seen with reactivated disease, particularly with herpesviruses such as HSV or varicella-zoster; (2) false-positive responses may occur in the presence of rheumatoid factor or false-negative results from competition by IgG antibody for binding sites on the antigen; and (3) heterotypic reactivation of IgM may be found with some infections (such as CMV or EBV). For example, removal of Coombs antibody from sera is necessary for the EBV–VCA–IgM test; otherwise, false-positive results may arise. Methods useful for the detection of viral-specific IgM will be described below, but, in general, diagnosis using a single IgM sample needs to be carefully controlled to exclude the detection of IgG or other interfering substances.

Many different serological techniques have been used in the diagnosis of viral infections (Table 5). Factors involved in the selection of a specific antibody assay include specificity, sensitivity, speed, technical complexity, cost, and availability of reagents (Table 6). All antibody assays rely on the proper collection and storage of sera and, ideally, the comparison of acute and convalescent specimens collected at an interval of at least 2 weeks. The development of newer techniques, such as EIA, for antibody determination is replacing some of the older methods, such as complement fixation, but an understanding of the available methods is important prior to choosing a laboratory test to evaluate a specific question.

Problems specific to serodiagnosis of a viral infection include the broad cross-reactivity among some virus groups, such as the coxsackie A viruses that cross-react with antibodies to coxsackie B and echoviruses. Another serious limitation of this approach to diagnosis is the failure of some individuals, particularly young children or immunocompromised patients, to mount a detectable antibody response to a specific infection. However, serological methods remain extremely important in epidemiologic studies because results are not dependent on obtaining a specimen at the peak of illness, tests can be performed retrospectively for a variety of agents simultaneously, and large-scale studies can be conducted in a timely and cost-effective manner.

7.1. Neutralization

Serum specimens may be assayed for neutralizing antibody against a given virus by testing serial dilutions of the serum against a standard dose of the virus. The antibody titer is expressed as the highest serum dilution that neutralizes the test dose of virus. As a bioassay, neutralization assays are highly specific and quite sensitive. For many viral agents, the neutralizing antibody level is directly correlated with immunity, an important clinical and epidemiologic endpoint. Disadvantages of the assay include the time required to obtain a result and the relatively high cost, due to the labor intensity and requirement for cell culture and titered viral stocks. Neutralization assays may be carried out in a variety of systems and the endpoint measured by a number of different procedures. Different neutralization systems include plaque reduction neutralization, sometimes using complement enhancement, where the number of virus plaques in control wells are compared with the number seen in cultures inoculated with the virus–serum mixture; microneutralization, an assay performed in microtiter plates requiring small amounts of sera; and colorimetric assays. Colorimetric assays rely on markers indicating metabolic inhibition of the virus in cell cultures or on antigen–antibody reactions

Table 5. Serological Diagnosis of Selected Viruses

Virus	Neutralization	Complement fixation	Hemagglutination-inhibition	Immunoassay (EIA, IF)
Adenoviruses	+	+ +		+
Arboviruses	+	+ +	+	+
Coronaviruses	+		+	+
Cytomegalovirus	+	+ +[b]		+ +
Enteroviruses	+			
EBV[c]				+ +
Hepatitis B and C				+ +
HSV	+	+ +[b]		+ +
Influenza	+	+ +[b]	+ +	+
Measles	+	+ +[b]	+	+ +
Mumps	+			+ +
Norwalk/Rotavirus				+ +
Parainfluenza	+	+ +[b]	+ +	+
Parvovirus				+
Rabies				+
RSV	+	+ +[b]		+ +
Retroviruses	+			+ +[d]
Rhinoviruses	+			+
Rubella			+	+ +
VZV		+ [b]		+ +

[a]+ Method used in research setting; + + Method in use and readily available in virology laboratories.
[b]Complement fixation method may lack sensitivity for these viruses.
[c]The absorbed heterophile test is commonly used for infectious mononucleosis, with the EBV-VCA-IgM needed if that test is negative.
[d]Western blot commonly used as confirmatory test.

with antibody tagged or reacted with enzyme-linked antibodies. Colorimetric assays are generally analyzed by measurement of optical density and have the advantage of being less time-consuming and costly to set up and analyze than other assays. Colorimetric assays often are more sensitive than assays relying on inhibition of CPE.[144]

The type of assay used to assess antibody is very important, particularly in the evaluation of susceptibility to vaccine-preventable or epidemic viruses such as measles or rubella, where low levels of antibody may be predictive of protection. Direct comparison of various laboratory methods used for the detection of virus-specific antibody may be important in designing studies or evaluating study results. For example, analysis of CMV antibody using neutralization by plaque reduction has shown poor correlation with CMV antibody using EIA.[50] Different neutralization methods also may give differing results, as has been shown in the analysis of antibody to RSV, where microneutralization assays appear to be more indicative of biological protection than either direct or competitive ELISA methods or complement-enhanced plaque reduction.[149]

Table 6. Comparison of Serological Methods Used to Detect Viral Antibodies

Method	Sensitivity[a]	Specificity	Cost	Time to Dx	Availability
Neutralization	+ + +	+ +	Expensive	>1 week	Research
Complement fixation	+	+/+ +	Inexpensive	<1 day	Widely available
Hemagglutination inhibition	+ +	+ +	Inexpensive	<1 day	Research/reference labs
Enzyme immunoassay	+ + +	+ +	Inexpensive	<1 day	Widely available
Immunofluorescence	+ +/+ + +	+ +	Moderate	<1 day	Research/reference labs
Radioimmunoassay	+ + +	+ +	Expensive	1–3 day	Research
Western blot	+ +/+ + +	+ + +	Expensive	<1–3 day	Research/reference labs

[a]+, Relatively low; + +, moderate; + + +, high.

In general, measurement of neutralizing antibody is the most specific method that reflects immunity, although other tests for some viruses may be surrogate markers for this. However, neutralization tests are rather expensive since a demonstration of inhibition of viral replication in cell culture, embryonated egg, or laboratory animal (such as the suckling mouse) is required.

7.2. Complement Fixation

The complement fixation (CF) test is a relatively simple technique that may be used successfully with a large variety of viral antigens. First developed in 1909 by Wasserman and co-workers[157,181] for the detection of syphilis antibodies, CF has been adapted to test for antibodies to many bacterial and fungal pathogens of animals and man. The CF test relies on competition between two antigen–antibody systems for a fixed amount of complement, with the result ultimately demonstrated by the lysis of erythrocytes. The serum is heated at 50°C for 30 min to inactivate any complement that may be present. Antigen and a known amount of complement are added to dilutions of serum. The complexes formed between the initial antigen and antibody bind the available free complement, thus preventing further reaction of the complement in the second step of the assay. In the second step, a hemolytic indicator system using red blood cells (RBC), which have been reacted with hemolysin to sensitize them to complement, is used to detect the free complement. The RBCs are reacted with hemolysin, or antibody to the RBC, and added to the assay. Lysis of the RBCs occurs if free complement is present. Thus, the presence of hemolysis at the conclusion of the assay is indicative of the absence of specific antibody, whereas the formation of clumped RBC (often referred to as a "RBC button") indicates a positive test reaction. Antibody titers can be calculated using standard endpoint determinations. Antibodies detected by CF are primarily of the IgG class and develop during the convalescent stages of illness. The greatest application of the CF test lies in the demonstration of a rise in antibody in convalescent compared with acute sera. The CF test has been widely used for the serodiagnosis of many human viral pathogens because of its broad reactivity and effectiveness in detecting changes in antibody titers and it is often the standard against which new methods are compared. The CF test continues to maintain its usefulness because reagents are relatively inexpensive and readily available and the test is rapid, reliable, and relatively easy to perform. The CF method may be the only method available for detecting antibody to less prevalent viruses or those with many serotypes (e.g., arboviruses, coxsackieviruses) (see Table 5).[83] Another advantage of this assay is that many antigens can be easily tested in the same sera samples simply by changing the antigen but keeping all other reagents and conditions the same. The CF test is also adaptable to microtiter and automated methods.

Despite the widespread use of the CF assay over time and in many epidemiologic studies, the assay has some unique problems: (1) the test relies on a cascade of interactions involving multiple biological reagents that must be carefully monitored; (2) it is relatively insensitive because high concentrations of antigen are required to produce CF complexes and the assay is unable to detect small changes in antibody concentrations or low levels of antibody that may be predictive of protection in other assay systems (such as VZV or measles[60,104]; (3) sera containing antibodies to host cell components or anticomplementary sera will not give a valid result. Newer laboratory methods, such as EIA methods, are now available commercially and are replacing CF tests for many viral pathogens because of their ability to discriminate between IgG and IgM antibody, increased sensitivity, and enhanced specificity at a similar cost.

7.3. Hemagglutination Inhibition

The hemagglutination inhibition test (HAI) is based on the ability of some viruses to attach to receptors on certain species of erythrocytes and cause hemagglutination (Section 6.2.2). While HAI may be used to identify an unknown virus with virus-specific antisera, it also is useful for the detection of virus-specific antibodies in the serum. HAI may be used to evaluate antibody titers of influenza viruses, parainfluenza viruses, adenoviruses, rubella, arboviruses, and some strains of picornaviruses. In this assay, serial dilutions of sera are allowed to react with a defined amount (4 HA units) of viral hemagglutinin. Subsequently, agglutinable RBCs are added and the ability of the virus to agglutinate the RBCs is measured. Properly treated sera containing antibody specific to the virus will prevent agglutination of the RBCs, resulting in formation of RBC buttons in the test wells; sera lacking specific antibody will result in RBC agglutination. Nonspecific viral inhibitors can give rise to false-positive results in some systems, requiring that the sera be properly prepared prior to use. The specificity of the assay varies somewhat with the particular virus, with influenza and parainfluenza virus systems being more specific than the arbovirus system.[107] For example, the HAI test can identify specific strains of influenza viruses, whereas it identifies only the group-specific antigens of the arboviruses (with neutralization tests required for strain identification). Advantages of the HAI assay are its simplicity, the low cost for reagents and equipment, and speed of the assay. Disadvantages of this assay include the fact that the system only works with those viruses that cause hemagglutination, and that nonviral-specific serum components may also inhibit hemagglutination, thereby invalidating the test results.

The immune adherence hemagglutination method is another method well suited to the clinical laboratory. After an initial reaction between viral antigens and specific antibodies is allowed to occur, complement is added and binds to the antigen–antibody complex, if present. Human erythrocytes then are added and reaction to the antigen–antibody–complement complex with the C3b receptor causes hemagglutination. This method, commonly used as a microtiter procedure, has a well-defined endpoint and is rapid, inexpensive, and more sensitive than CF.[97] Disadvantages of this method include the inability to differentiate between different immunoglobulin subclasses and its difficulties with viruses that themselves agglutinate RBCs.

The direct agglutination of sheep or horse RBCs by serum is a diagnostic test for the heterophile antibody of infectious mononucleosis. Removal of an inhibitory (Forssman) antibody by adsorption of the sera with guinea pig kidney extracts is required before testing. The hemolysis of beef cells by sera from patients with acute infectious mononucleosis due to EBV is another diagnostic test that does not require adsorption but is less sensitive (see Chapter 10).

7.4. Immunoassay Technique

7.4.1. Enzyme Immunoassay. Enzyme-based immunoassays (EIA), sometimes called enzyme-linked immunosorbant assays or ELISA, are widely used for many purposes, including the detection of antigen-specific antibody. This methodology has replaced other methods in many laboratories in part due to the commercial availability of test materials and complete test kits. The most commonly used immunoassays employ a four-layer approach: antigen is bound directly to a surface, the unknown serum sample is then added, followed by an enzyme-conjugated antihuman IgG or IgM, and an indicator system used to determine the amount of reaction between the enzyme-linked antibody and the antigen–serum reaction. As discussed in Section 5.3.4, this method has gained widespread use due to its sensitivity, specificity, safety, simplicity, low cost, and ability to be automated. The system lends itself to automation because of readily available microtiter diagnostic systems and because multiple tests for different antigens may be run by varying only

the initial antigen in the system. Clinical specimens from various sources besides serum or plasma, such as respiratory secretions, cerebrospinal fluid, and breast milk, also may be tested in this system. The EIA test may also be read visually and so is useful in developing countries unable to afford the photometric reader used in developed countries to accurately quantitate the antibody content.

Difficulties inherent in immunoassay techniques include those associated with obtaining and standardizing purified, sensitive, and specific IgG and IgM reagents. Specific antibody subclasses can be purified directly from serum samples using techniques such as column chromatography, sucrose gradient ultracentrifugation, ion-exchange chromatography, or adsorption with material such as staphylococcal protein that causes insoluble matrices to form between the reagent and specific IgG subclasses.[83] However, such methods require significant time, sample volume, and expense. Some EIA techniques use adsorption with anti-human IgG or aggregated human γ-globulin to decrease nonspecific IgM activity or false-negative IgG activity.[83] In general, results from different laboratories using different reagents are not directly comparable. Specific analysis for subclasses, particularly IgM, requires careful standardization and quality control to assure reliability and specificity of the assay. Attention to such detail is critically important in assays with life-threatening implications, such as the EIA assays currently used in blood banks to detect the presence of antibody to HIV or hepatitis B and C.[1,82] Evaluation and standardization of tests, including commercially available kits, and comparison among different products prior to use in research and clinical settings remains an important part of EIA.

Epitope-blocking assays using monoclonal antibodies have been used to examine serological responses to a number of viruses.[16,62,70,105,148,179] These assays measure the ability of a test serum to block the binding of a monoclonal antibody to a virus antigen. They have been particularly useful in determining serotype-specific immune responses to multivalent vaccines and in determining which of a number of cross-reactive virus strains is responsible for a natural infection in a given host.

Other variations of the immunoassay include isoelectric focusing and affinity immunoblotting. These techniques are useful because of enhanced sensitivity, particularly for diagnosis in the congenitally infected infant or for the differentiation of passively acquired antibody from endogenous antibody. Isoelectric focusing relies on the separation of serum antibody in thin-layer gels, with subsequent antibody detection by a reaction with antigen-coated membranes.[151] Clonal-specific antibodies can be detected by this method, which may discriminate, for example, between unique maternal and fetal antibodies.

Radioimmunoassay techniques for the sensitive detection of antibody to viral antigens, pioneered in the 1970s for the serodiagnosis of hepatitis B, are used by research laboratories but otherwise are not widely utilized.[75] Immunoassays that do not require radioisotopes and require less technical expertise, such as the EIA and IF tests, are more commonly available.

7.4.2. Immunohistochemical.

The most commonly used immunohistochemical technique to detect antibody is the immunofluorescence technique (IF). Whereas direct techniques are used commonly to detect antigen in infected tissue or cells, indirect immunohistochemical techniques are used to detect antibody in sera or other bodily fluids. Variations on the indirect IF test, such as amplification immunoassay systems utilizing various sandwich techniques (double indirect IF or anticomplement IF) or chemical amplification systems utilizing biotin–avidin complexes are used in research settings.[83] Indirect IF methods are available in many laboratories for a variety of assays, although more automated techniques such as EIA are replacing IF techniques in many clinical settings.

In the indirect IF test, tissue or cells containing viral antigen are fixed on a glass slide, a serum dilution is added, and a fluorochrome-conjugated antibody indicator system is used to detect the resulting reaction. The conjugated detector antibody can be varied to specifically measure the presence of antibody subclasses, such as IgG, IgM, and IgA. IF techniques provide sensitive methods to detect antibody that can be related to immunity to viruses, such as VZV, to detect congenital infections in newborns based on IgM-specific antibody, and to detect epitope-specific antibody.[83,184] In particular, the anticomplement-amplified indirect IF technique, utilizing a four-layer reaction including antigen, patient serum, complement, and fluorochrome-conjugated anti-C3 antibody, has been useful for the detection of the nuclear antigen of EBV, or EBNA.[83] The inclusion of standard positive and negative sera is needed in each test and independent reading by two observers is recommended to minimize errors in interpretations.

Advantages of indirect immunohistochemical methods include (1) the ability to use the system to detect antibody to many diverse viruses by varying only the initial step in the reaction; (2) the higher sensitivity and specificity compared with CF; (3) the simplicity and relative speed of an individual test; and (4) the reproducibility of the test by experienced personnel. Disadvantages of the test include the technical complexity, the lack of automation, the requirement for specialized cells and reagents, and the need for special equipment, such as a fluorescence microscope and darkroom.

7.4.3. Western Blot.

Western blot (WB) is another widely used method for the detection of antibody to specific viral antigens. This technique relies on the incubation of patient serum with partially purified whole virus from which the viral proteins have been separated by electrophoresis in a polyacrylamide gel and transferred onto nitrocellulose paper.[171] The assay is based on the same principal as EIA but has the advantage of identifying antibodies specific for several antigens of the same virus simultaneously. Quantitation of the specific reactions can be determined by a densitometer. This assay is more sensitive than routine EIA by as much as 100-fold in the HIV system.[82] Difficulties of WB include the expense and time required for the test, the technical requirements for performing and interpreting the test, and the problems encountered with preparing reagents such as purified virus and radiolabeled antihuman IgG. The time interval from virus acquisition and seroconversion by WB may vary for different patients, as well as different viruses, indicating that diagnosis of infection by WB analysis, while extremely sensitive, is not always definitive.[185] Furthermore, analysis by WB will not differentiate maternal from fetal infection in many cases. Despite these problems, the WB assay is widely used today in laboratories around the world as the "definitive" confirmatory test for HIV, and many different commercially available kits and reagents are available. Current techniques do not require the use of radioisotopes or prolonged incubations.

8. Safety Considerations

Laboratory safety should be a concern for all who work in a virology laboratory. Technicians work with both potentially hazardous reagents and clinical material. Exposure to infectious agents in clinical specimens, tissue culture harvests, and stock reagents; to chemicals such as radioisotopes, solvents, and various organic chemicals; and to equipment such as needles, knives, glass, and autoclaves is common. Constant awareness of routine laboratory procedure, use of "universal precautions" with potentially infective materials, and strict attention to updating and maintaining laboratory standards are mandatory for preventing accidents and minimizing the risks of working in the laboratory. Active or passive immunoprophylaxis also may be indicated for individuals working with certain viruses or under certain situations (e.g., rabies, hepatitis A and B, vaccinia). A serum sample should be obtained from each new employee who is working with infectious materials as a baseline for identifying a laboratory infection. For some highly infectious agents, periodic testing is advised to ensure that safety precautions are adequate. Guidelines for the design and implementation of laboratory safety programs have been published.[40,133]

9. Unresolved Problems

Many problems still must be resolved in the area of viral diagnosis. Some viruses cannot be cultivated in tissue culture systems, making their identification by isolation problematic. This also may make the generation of diagnostic reagents more difficult, although cloning and expression of the viral genome has helped circumvent this (e.g., Norwalk virus and hepatitis C virus). Another problem to be solved is the identification of the best manner in which newer molecular methods for viral diagnosis should be utilized. While molecular methodologies offer the potential for great sensitivity and specificity, they often require specialized equipment for their performance. Ultimately, the ideal diagnostic test will be rapid, easy to perform, inexpensive, sensitive, and specific, and will not require the use of specialized equipment. However, even if such a test is developed, the significance of the identified virus to the observed disease process will remain in the hands of the epidemiologist, virologist, and clinician.

10. References

1. AACH, R. D., STEVENS, C. E., HOLLINGER, F. B., MOSLEY, F. B., PETERSON, J. W., TAYLOR, D. A., JOHNSON, R. C., BARBOSA, R. G., AND NEMO, G. J., Hepatitis C virus infection in post-transfusion hepatitis: Analysis with first- and second-generation assays, *N. Engl. J. Med.* **325**:1325–1336 (1991).

2. AHLUWALIA, G., EMBREE, J., MCNICOL, P., LAW, B., AND HAMMOND, G. W., Comparison of nasopharyngeal aspirate and nasopharyngeal swab specimens for respiratory syncytial virus diagnosis by cell culture, indirect immunofluorescence assay, and enzyme-linked immunosorbent assay, *J. Clin. Microbiol.* **25**:763–767 (1987).

3. ANDERSON, M. L. M., AND YOUNG, B. D., Quantitative filter hybridisation, in: *Nucleic Acid Hybridisation: A Practical Approach* (B. D. HAMES AND S. J. HIGGINS, EDS.), pp. 73–111, IRL Press, Washington, DC, 1985.

4. ARRAND, J. E., Preparation of nucleic acid probes, in: *Nucleic Acid Hybridization: A Practical Approach* (B. D. HAMES AND S. J. HIGGINS, EDS.), pp. 17–45, IRL Press, Washington, DC, 1985.

5. ATMAR, R. L., AND GEORGHIOU, P. R., Classification of respiratory tract picornavirus isolates as enteroviruses or rhinoviruses using reverse transcription–polymerase chain reaction, *J. Clin. Microbiol.* **31**:2544–2546 (1993).

6. ATMAR, R. L., METCALF, T. G., NEILL, F. H., AND ESTER, M. K., Detection of enteric viruses in oysters using the polymerase chain reaction, *Appl. Environ. Microbiol.* **59**:631–635 (1993).

7. AUVINEN, P., STANWAY, G., AND HYYPIA, T., Genetic diversity of enterovirus subgroups, *Arch. Virol.* **104**:175–186 (1989).

8. BALLEW, H. C., Neutralization, in: *Clinical Virology Manual* (S. SPECTER AND G. LANCZ, EDS.), pp. 229–241, Elsevier, New York, 1992.

9. BARANY, F., Genetic disease detection and DNA amplification using cloned thermostable ligase, *Proc. Natl. Acad. Sci. USA* **88**:189–193 (1991).

10. BAXTER, B. D., COUCH, R. B., GREENBERG, S. B., AND KASEL, J. A., Maintenance of viability and comparison of identification methods for influenza and other respiratory viruses of humans, *J. Clin Microbiol.* **6**:19–22 (1977).

11. BELSHE, R. B., BURK, B., NEWMAN, F., CERRUTI, R. L., AND SIM, I. S., Resistance of influenza A virus to amantadine and rimantadine: Results of one decade of surveillance, *J. Infect. Dis.* **159**:430–435 (1989).

12. BELSHE, R. B., SMITH, M. H., HALL, C. B., BETTS, R., AND HAY, A. J., Genetic basis of resistance to rimantadine emerging during treatment of influenza virus infection, *J. Virol.* **62**:1508–1512 (1988).

13. BOWLES, N. E., RICHARDSON, P. J., OLSEN, E. G. J., AND ARCHARD, L. C., Detection of coxsackie-B-virus-specific RNA sequences in myocardial biopsy samples from patients with myocarditis and dilated cardiomyopathy, *Lancet* **1**:1120–1123 (1986).

14. BRENNER, S., AND HORNE, R. W., A negative staining method for high resolution electron microscopy of viruses, *Biochim. Biophys. Acta* **34**:103–110 (1959).

15. BUITENWERF, J., LOUWERENS, J. J., AND DE JONG, J. C., A simple and rapid method for typing adenoviruses 40 and 41 without cultivation, *J. Virol. Methods* **10**:39–44 (1985).

16. BURKE, D. S., NISALAK, A., AND GENTRY, M. K., Detection of flavivirus antibodies in human serum by epitope-blocking immunoassay, *J. Med. Virol.* **23**:165–173 (1987).

17. BUTLER, E. T., AND CHAMBERLIN, M. J., Bacteriophage SP6-specific RNA polymerase, *J. Biol. Chem.* **10**:5772–5778 (1982).

18. CARSTENS, J. M., TRACY, S., CHAPMAN, N. M., AND GAUNTT, C. J., Detection of enteroviruses in cell cultures by using *in situ* transcription, *J. Clin. Microbiol.* **30**:25–35 (1992).

19. CASALS, J., BUCKLEY, S. M., AND CEDENO, R., Antigenic properties of the arenaviruses, *Bull. WHO* **52**:421–427 (1975).

20. CENTERS FOR DISEASE CONTROL, Update: Outbreak of Hantavirus infection—southwestern United States, 1993, *Morbid. Mortal. Week. Rep.* **42**:477–479 (1993).

21. CHERNESKY, M. A., RAY, C. G., AND SMITH, T. F., Laboratory diagnosis of viral infections, in: *Cumulative Techniques and Procedures in Clinical Microbiology* (W. L. DREW, ED.), pp. 1–17, American Society for Microbiology, Washington, DC, 1982.

22. CLARK, C. A., DOWNS, E. C., AND PRIMUS, F. J., An unlabeled antibody method using glucose oxidase–antiglucose oxidase complexes (GAG): A sensitive alternative to immunoperoxidase for the detection of tissue antigens, *J. Histochem. Cytochem.* **30**:27–34 (1982).

23. CLEMENTI, M., MENZO, S., BAGNARELLI, P., MANZIN, A., VALENZA, A., AND VARALDO, P. E., Quantitative PCR and RT-PCR in virology, *PCR Methods Applic.* **2**:191–196 (1993).

24. COONS, A. H., CREECH, H. J., JONES, R. N., AND BERLINER, E., The demonstration of pneumococcal antigen in tissues by the use of fluorescent antibody, *J. Immunol.* **45:**159–170 (1942).

25. CORDELL, J. L., FALINI, B., ERBER, W. N., GHOSH, A. K., ABDUL-AZIZ, Z., MACDONALD, S., PULFORD, K. A. F., STEIN, H., AND MASON, D. Y., Immunoenzymatic labeling of monoclonal antibodies using immune complexes of alkaline phosphatase and monoclonal anti-alkaline phosphatase (APAAP complexes), *J. Histochem. Cytochem.* **32:**219–229 (1984).

26. COREY, L., ADAMS, H. G., BROWN, Z. A., AND HOLMES, K. K., Genital herpes simplex virus infections: Clinical manifestations, course, and complications, *Ann. Intern. Med.* **98:**958–972 (1983).

27. COUTLEE, F., YANG, B., BOBO, L., MAYUR, K., YOLKEN, R. H., AND VISCIDI, R., Enzyme immunoassay for detection of hybrids between PCR-amplified HIV-1 DNA and a RNA probe: PCR–EIA, *AIDS Res. Hum. Retroviruses* **6:**775–784 (1990).

28. CRANE, L. R., GUTTERMAN, P. A., CHAPEL, T., AND LERNER, A. M., Incubation of swab materials with herpes simplex virus, *J. Infect. Dis.* **141:**531 (1990).

29. CROEN, K. D., OSTROVE, J. M., DRAGOVIC, L. J., AND STRAUS, S. E., Patterns of gene expression and sites of latency in human nerve ganglia are different for varicella-zoster and herpes simplex viruses, *Proc. Natl. Acad. Sci. USA* **85:**9773–9777 (1988).

30. CUNNINGHAM, D. A., PATTISON, J. R., AND CRAIG, R. K., Detection of parvovirus DNA in human serum using biotinylated RNA hybridization probes, *J. Virol. Methods* **19:**279–288 (1988).

31. DALLDORF, G., SICKLES, G. M., PLOGER, H., AND GIFFORD, R., A virus recovered from the feces of "poliomyelitis" patients pathogenic for suckling mice, *J. Exp. Med.* **89:**567–582 (1949).

32. DAVIS, G. L., LAU, J. Y., URDEA, M. S., NEUWALD, P. D., WILBER, J. C., LINDSAY, K., PERRILLO, R. P., AND ALBRECHT, J., Quantitative detection of hepatitis C virus RNA with a solid-phase signal amplification method: Definition of optimal conditions for specimen collection and clinical application in interferon-treated patients, *Hepatology* **19:**1337–1341 (1994).

33. DE, L., NOTTAY, B., YANG, C.-F., HOLLOWAY, B. P., PALLANSCH, M., AND KEW, O., Identification of vaccine-related polioviruses by hybridization with specific RNA probes, *J. Clin. Microbiol.* **33:**562–571 (1995).

34. DECKER, C., ELLIS, M. N., McLAREN, C., HUNTER, G., ROGERS, J., AND BARRY, D. W., Virus resistance in clinical practice, *J. Antimicrob. Chemother.* **12:**137–152 (1983).

35. DEWAR, R. L., HIGHBARGER, H. C., SARMIENTO, M. D., TODD, J. A., VASUDEVACHARI, M. B., DAVEY, R. T., JR., KOVACS, J. A., SALZMAN, N. P., LANE, H. C., AND URDEA, M. S., Application of branched DNA signal amplification to monitor human immunodeficiency virus type 1 burden in human plasma, *J. Infect. Dis.* **170:**1172–1179 (1994).

36. DOANE, F. W., Electron microscopy and immunoelectron microscopy, in: *Clinical Virology Manual* (S. SPECTER AND G. LANCZ, EDS.), pp. 89–109, Elsevier, New York, 1992.

37. DOMINGUEZ, E. A., TABER, L. H., AND COUCH, R. B., Comparison of rapid diagnostic techniques for respiratory syncytial and influenza A virus respiratory infections in young children, *J. Clin. Microbiol.* **31:**2286–2290 (1993).

38. DOWDLE, W. R., GALPHIN, J. C., COLEMAN, M. T., AND SCHILD, G. C., A simple double immunodiffusion test for typing influenza viruses, *Bull. WHO* **51:**213–218 (1974).

39. EMANUEL, D., PEPPARD, J., STOVER, D., GOLD, J., ARMSTRONG, D., AND HAMMERLING, U., Rapid immunodiagnosis of cytomegalovirus pneumonia by bronchoalveolar lavage using human and murine monoclonal antibodies, *Ann. Intern. Med.* **104:**476–481 (1986).

40. EMMONS, R. W., AND HAGENS, S., Laboratory safety, in: *Diagnostic Procedures for Viral, Rickettsial and Chlamydial Infections* (N. J. SCHMIDT AND R. W. EMMONS, EDS.), pp. 37–49, American Public Health Association, Washington, DC, 1989.

41. ENDERS, J. F., WELLER, T. H., AND ROBBINS, F. C., Cultivation of the Lansing strain of poliomyelitis virus in cultures of various human embryonic tissues, *Science* **109:** 85–87 (1949).

42. ENGLUND, J. A., ZIMMERMAN, M. E., SWIERKOSZ, E. M., GOODMAN, J. L., SCHOLL, D. R., AND BALFOUR, H. H., Herpes simplex virus resistant to acyclovir: A study in a tertiary care center, *Ann. Intern. Med.* **112:**416–422 (1990).

43. ERICE, A., CHOU, S., BIRON, K. K., STANAT, S. C., BALFOUR, H. H., AND JORDAN, J. C., Progressive disease due to ganciclovir-resistant cytomegalovirus in immunocompromised patients, *N. Engl. J. Med.* **320:**289–293 (1989).

44. ERICE, A., HERTZ, M. I., SNYDER, L. S., ENGLUND, J., EDELMAN, C. K., AND BALFOUR, H. H., Evaluation of centrifugation cultures of bronchoalveolar lavage fluid for the diagnosis of cytomegalovirus pneumonitis, *Virology* **10:**205–212 (1988).

45. ERICE, A., MAYERS, D. L., STRIKE, D. G., SANNERUD, K. J., McCUTCHEON, F. E., HENRY, K., AND BALFOUR, H. H., JR., Brief report: Primary infection with zidovudine-resistant human immunodeficiency virus type 1, *N. Engl. J. Med.* **328:**1163–1165 (1993).

46. ESPY, M. J., HIERHOLZER, J. C., AND SMITH, T. F., The effect of centrifugation on the rapid detection of adenovirus in shell vials, *Am. J. Clin. Pathol.* **88:**358–360 (1987).

47. ESPY, M. J., SMITH, T. F., HARMON, M. W., AND KENDAL, A. P., Rapid detection of influenza virus by shell vial assay with monoclonal antibodies, *J. Clin. Microbiol.* **24:**677–679 (1986).

48. EVANS, A. S., AND OLSON, B., Rapid diagnostic methods for influenza virus in clinical specimens: A comparative study, *Yale J. Biol. Med.* **55:**391–403 (1982).

49. FEINSTONE, S. M., BARKER, L. F., AND PURCELL, R. H., Hepatitis A and B, in: *Diagnostic Procedures for Viral, Rickettsial and Chlamydial Infections* (E. H. LENNETTE AND N. J. SCHMIDT, EDS.), pp. 879–925, American Public Health Association, Washington, DC, 1979.

50. FILIPOVICH, A. H., PELTIER, M. H., BECHTEL, M. K., DIRKSEN, C. L., STRAUSS, S. A., AND ENGLUND, J. A., Circulating cytomegalovirus (CMV) neutralizing activity in bone marrow transplant recipients: Comparison of passive immunity in a randomized study of four intravenous IgG products administered to CMV-seronegative patients, *Blood* **80:**2656–2660 (1992).

51. FLORES, J., BOEGGEMAN, E., PURCELL, R. H., SERENO, M., PEREZ, I., WHITE, L., WYATT, R. G., CHANOCK, R. M., AND KAPIKIAN, A. Z., A dot hybridisation assay for detection of rotavirus, *Lancet* **1:**555–559 (1983).

52. FRANCIS, T., AND MAGILL, T. P., Direct isolation of influenza virus, *Proc. Soc. Exp. Biol. Med.* **36:**134–135 (1937).

53. FRAYHA, H., CASTRICIANO, S., MAHONY, J., AND CHERNESKY, M., Nasopharyngeal swabs and nasopharyngeal aspirates equally effective for the diagnosis of viral respiratory disease in hospitalized children, *J. Clin. Microbiol.* **27:**1387–1389 (1989).

54. GAMA, R. E., HORSNELL, P. R., HUGHES, P. J., NORTH, C., BRUCE, C. B., AL-NAKIB, W., AND STANWAY, G., Amplification of rhinovirus specific nucleic acids from clinical samples using the polymerase chain reaction, *J. Med. Virol.* **28:**73–77 (1989).

55. GAY, H., CLARK, W. R., AND DOCHERTY, J. J., Detection of herpes simplex virus infection using glucose oxidase–antiglucose oxi-

dase immunoenzymatic stain, *J. Histochem. Cytochem.* **32:**447–451 (1984).

56. GENTSCH, J. R., GLASS, R. I., WOODS, P., GOUVEA, V., GORZIGLIA, M., FLORES, J., DAS, B. K., AND BHAN, M. K., Identification of group A rotavirus gene 4 types by polymerase chain reaction, *J. Clin. Microbiol.* **30:**1365–1373 (1992).

57. GERNA, G., SARASINI, A., PERCIVALLE, E., ZAVATTONI, M., BALDANTI, F., AND REVELLO, M. G. Rapid screening for resistance to ganciclovir and foscarnet of primary isolates of human cytomegalovirus from culture-positive blood samples, *J. Clin. Microbiol.* **33:**738–741 (1995).

58. GEY, G. O., An improved technic for massive tissue cultures, *Am. J. Cancer* **17:**752–756 (1933).

59. GLEAVES, C. A., AND MEYERS, J. D., Comparison of MRC-5 and HFF cells for the identification of cytomegalovirus in centrifugation culture, *Diagn. Microbiol. Infect. Dis.* **6:**179–182 (1987).

60. GOLD, E., AND GODEK, G., Complement fixation studies with a varicella zoster antigen, *J. Immunol.* **95:**692–695 (1965).

61. GOUVEA, V., GLASS, R. I., WOODS, P., TANIGUCHI, K., CLARK, H. F., FORRESTER, B., AND FANG, Z.-Y., Polymerase chain reaction amplification and typing of rotavirus nucleic acid from stool specimens, *J. Clin. Microbiol.* **28:**276–282 (1990).

62. GREEN, K. Y., TANIGUCHI, K., MACKOW, E. R., AND KAPIKIAN, A. Z., Homotypic and heterotypic epitope-specific antibody responses in adult and infant rotavirus vaccinees: Implications for vaccine development, *J. Infect. Dis.* **161:**667–679 (1990).

63. GREENBERG, S. B., AND KRILOV, L. R., Laboratory diagnosis of viral respiratory disease, in: *Cumulative Techniques and Procedures in Clinical Microbiology* (W. L. DREW AND S. J. RUBIN, EDS.), pp. 1–21, American Society of Microbiology, Washington, DC, 1986.

64. GRETCH, D. R., WILSON, J. J., CARITHERS, R. L., JR., DE LA ROSA, C., HAN, J. H., AND COREY, L., Detection of hepatitis C virus RNA: Comparison of one-stage polymerase chain reaction (PCR) with nested-set PCR, *J. Clin. Microbiol.* **31:**289–291 (1993).

65. GUATELLI, J. C., GINGERAS, T. R., AND RICHMAN, D. D., Nucleic acid amplification *in vitro*: Detection of sequences with low copy numbers and application to diagnosis of human immunodeficiency virus type 1 infection, *Clin. Microbiol. Rev.* **2:**217–226 (1989).

66. GUATELLI, J. C., WHITFIELD, K. M., KWOH, D. Y., BARRINGER, K. J., RICHMAN, D. D., AND GINGERAS, T. R., Isothermal, *in vitro* amplification of nucleic acids by a multienzyme reaction modeled after retroviral replication, *Proc. Natl. Acad. Sci. USA* **87:**1874–1878 (1990).

67. HALSTEAD, D. C., TODD, S., AND FRITCH, G., Evaluation of five methods for respiratory syncytial virus detection, *J. Clin. Microbiol.* **28:**1021–1025 (1990).

68. HAMBLING, M. H., Survival of the respiratory syncytial virus during storage under various conditions, *Br. J. Exp. Pathol.* **45:**647–655 (1964).

69. HAMES, B. D., AND HIGGINS, S. J., *Nucleic Acid Hybridization*, IRL Press, Washington, DC, 1985.

70. HAWKES, R. A., ROEHRIG, J. T., BOUGHTON, C. R., NAIM, H. M., ORWELL, R., AND ANDERSON-STUART, P., Defined epitope blocking with Murray Valley encephalitis virus and monoclonal antibodies: Laboratory and field studies, *J. Med. Virol.* **32:**31–38 (1990).

71. HAYDEN, F. G., BELSHE, R. B., CLOVER, R. D., HAY, A. J., OAKES, M. G., AND SOO, W., Emergence and apparent transmission of rimantadine-resistant influenza A virus in families, *N. Engl. J. Med.* **321:**1696–1702 (1989).

72. HERTZ, M. I., ENGLUND, J. A., SNOVER, D., BITTERMAN, P. B., AND McGLAVE, P. B., Respiratory syncytial virus-induced acute lung injury in adult patients with bone marrow transplants: A clinical approach and review of the literature, *Medicine* **68:**269–281 (1989).

73. HIERHOLZER, J. C., Further subgrouping of the human adenoviruses by differential hemagglutination, *J. Infect. Dis.* **128:**541–550 (1973).

74. HODINKA, R. L., AND STETSER, R. L., Evaluation of a commercial enzyme-linked virus inducible system™ for the rapid detection of herpes simplex virus (HSV) in clinical samples, *Tenth Annual Clinical Virology Symposium and Annual Meeting, Pan American Group for Rapid Viral Diagnosis*, abstract T22 from meeting, Clearwater Beach, Florida (1994).

75. HOLLINGER, F. B., VORNDAM, V., AND DREESMAN, G. R., Assay of Australia antigen and antibody employing double-antibody and solid-phase radioimmunoassay techniques and comparison with the passive hemagglutination methods, *J. Immunol.* **107:**1099–1114 (1971).

76. HORTA-BARBOSA, L., FUCCILLO, D. A., SEVER, J. L., AND ZEMAN, W., Subacute sclerosing panencephalitis: Isolation of measles virus from a brain biopsy, *Nature* **221:**974 (1969).

77. HUDSON, J. B., MISRA, V., AND MOSMANN, T. R., Cytomegalovirus infectivity: Analysis of the phenomenon of centrifugal enhancement of infectivity, *Virology* **72:**235–243 (1976).

78. HUGHES, J. H., Physical and chemical methods for enhancing rapid detection of viruses and other agents, *Clin. Microbiol. Rev.* **6:**150–175 (1993).

79. HUNTOON, C. J., HOUSE, R. F., AND SMITH, T. F., Recovery of viruses from three transport media incorporated into culturettes, *Arch. Pathol. Lab. Med.* **105:**436–437 (1981).

80. HYYPIA, T., STALHANDSKE, P., VAINIONPAA, R., AND PETTERSSON, U., Detection of enteroviruses by spot hybridization, *J. Clin. Microbiol.* **19:**436–438 (1984).

81. JABLONSKI, E. G., Detection systems for hybridization reactions, in: *DNA Probes for Infectious Diseases* (F. C. TENOVER, ED.), pp. 15–30, CRC Press, Boca Raton, FL, 1989.

82. JACKSON, J. B., AND BALFOUR, H. H., Practical diagnostic testing for human immunodeficiency virus, *Clin. Microbiol. Rev.* **1:**124–138 (1988).

83. JAMES, K., Immunoserology of infectious diseases, *Clin. Microbiol. Rev.* **3:**132–152 (1990).

84. JAPOUR, A. J., MAYERS, D. L., JOHNSON, V. A., KURITZKES, D. R., BECKETT, L. A., ARDUINO, J. M., LANE, J., BLACK, R. J., REICHELDERFER, P. S., D'AQUILA, R. T., CRUMPACKER, C. S., THE RV-43 STUDY GROUP, AND THE AIDS CLINICAL TRIALS GROUP VIROLOGY COMMITTEE RESISTANCE WORKING GROUP, Standardized peripheral blood mononuclear cell culture assay for determination of drug susceptibilities of clinical human immunodeficiency virus type 1 isolates, *Antimicrob. Agents Chemother.* **37:**1095–1101 (1993).

85. JIANG, X., ESTES, M. K., AND METCALF, T. G., Detection of hepatitis A virus by hybridization with single-stranded RNA probes, *Appl. Environ. Microbiol.* **53:**2487–2495 (1987).

86. JIANG, X., WANG, J., GRAHAM, D. Y., AND ESTES, M. K., Detection of Norwalk virus in stool by polymerase chain reaction, *J. Clin. Microbiol.* **30:**2529–2534 (1992).

87. JOHNSON, F. B., Transport of viral specimens, *Clin. Microbiol. Rev.* **3:**120–131 (1990).

88. KAM, W., RALL, L. B., SMUCKLER, E. A., SCHMID, R., AND RUTTER, W. J., Hepatitis B viral DNA in liver and serum of asymptomatic carriers, *Proc. Natl. Acad. Sci. USA* **79**:7522–7526 (1982).

89. KANDOLF, R., AMEIS, D., KIRSCHNER, P., CANU, A., AND HOF-SCHNEIDER, P. H., *In situ* detection of enteroviral genomes in myocardial cells by nucleic acid hybridization: An approach to the diagnosis of viral heart disease, *Proc. Natl. Acad. Sci. USA* **84**:6272–6276 (1987).

90. KARRON, R. A., FROEHLICH, J. L., BOBO, L., BELSHE, R. B., AND YOLKEN, R. H., Rapid detection of parainfluenza virus type 3 RNA in respiratory specimens: Use of reverse transcription-PCR-enzyme immunoassay, *J. Clin. Microbiol.* **32**:484–488 (1994).

91. KAWASAKI, E. S., Amplification of RNA, in: *PCR Protocols: A Guide to Methods and Applications* (M. A. INNIS, D. H. GEL-FAND, J. J. SNINSKY AND T. J. WHITE, EDS.), pp. 21–27, Academic Press, New York, 1990.

92. KIEVITS, T., VAN GEMEN, B., VAN STRIJP, D., SCHUKKINK, R., DIRCKS, M., ADRIAANSE, H., MALEK, L., SOOKNANAN, R., AND LENS, P., NASBA isothermal enzymatic *in vitro* nucleic acid amplification optimized for the diagnosis of HIV-1 infection, *J. Virol. Methods* **35**:273–286 (1991).

93. KIM, H. W., WYATT, R. G., FERNIE, B. F., BRANDT, C. D., ARROBIO, J. O., JEFFRIES, B. C., AND PARROTT, R. H., Respiratory syncytial virus detection by immunofluorescence in nasal secretions with monoclonal antibodies against selected surface and internal proteins, *J. Clin. Microbiol.* **18**:1399–1404 (1983).

94. KWOH, D. Y., DAVIS, G. R., WHITFIELD, K. M., CHAPPELLE, H. L., DiMICHELE, L. J., AND GINGERAS, T. R., Transcription-based amplification system and detection of amplified human immuno-deficiency virus type 1 with a bead-based sandwich hybridization format, *Proc. Natl. Acad. Sci. USA* **86**:1173–1177 (1989).

95. KWOK, S., AND HIGUCHI, R., Avoiding false positives with PCR, *Nature* **339**:237–238 (1989).

96. LANDRY, M. L., AND HSIUNG, G. D., Primary isolation of viruses, in: *Clinical Virology Manual* (S. SPECTER AND G. LANCZ, EDS.), pp. 43–69, Elsevier, New York, 1992.

97. LENNETTE, E. T., AND LENNETTE, D. A., Immune adherence hem-agglutination, in: *Clinical Virology Manual* (S. SPECTER AND G. LANCZ, EDS.), pp. 251–261, Elsevier, New York, 1992.

98. LIM, K. A., AND BENYESH-MELNICK, M., Typing of viruses by combinations of antiserum pools. Application to typing of entero-viruses (coxsackie and echo), *J. Immunol.* **84**:309–317 (1959).

99. LIU, C., Rapid diagnosis of human influenza infection from nasal smears by means of fluorescein-labeled antibody, *Proc. Soc. Exp. Biol. Med.* **92**:883–887 (1956).

100. LONGO, M. C., BERNINGER, M. S., AND HARTLEY, J. L., Use of uracil DNA glycosylase to control carry-over contamination in polymerase chain reactions, *Gene* **93**:125–128 (1990).

101. LORINCZ, A. T., Nucleic acid hybridization, in: *Clinical Virology Manual* (S. SPECTER AND G. LANCZ, EDS.), pp. 285–298, Elsevier Science Publishing Company, Amsterdam, 1992.

102. MACKOW, R. E., SHAW, R. D., MATSUI, S. M., VO, P. T., DANG, M. N., AND GREENBERG, B. H., Characterization of the rhesus rotavirus VP3 gene: Location of amino acids involved in homologous and heterologous rotavirus neutralization and identification of a putative fusion region, *Proc. Natl. Acad. Sci. USA* **85**:645–649 (1988).

103. MARTINO, T. A., SOLE, M. J., PENN, L. Z., LIEW, C. C., AND LIU, P., Quantitation of enteroviral RNA by competitive polymerase chain reaction, *J. Clin. Microbiol.* **31**:2634–2640 (1993).

104. MATSON, D. O., BYINGTON, C., CANFIELD, M., ALBRECHT, P., AND FEIGIN, R. D., Investigation of a measles outbreak in a fully vaccinated school population including serum studies before and after revaccination, *Pediatr. Infect. Dis. J.* **12**:292–299 (1993).

105. MATSON, D. O., O'RYAN, M. L., PICKERING, L. K., CHIBA, S., NAKATA, S., RAJ, P., AND ESTER, M. K., Characterization of serum antibody responses to natural rotavirus infections in children by VP7-specific epitope-blocking assays, *J. Clin. Microbiol.* **30**:1056–1061 (1992).

106. McINTOSH, K., DEES, J. H., BECKER, W. B., KAPIKIAN, A. Z., AND CHANOCK, R. M., Recovery in tracheal organ cultures of novel viruses from patients with respiratory disease, *Proc. Natl. Acad. Sci.* **57**:933–940 (1967).

107. McLAREN, L. C., Hemagglutination inhibition and hemadsorption, in: *Clinical Virology Manual* (S. SPECTER AND G. LANCZ, EDS.), pp. 243–249, Elsevier, New York, 1992.

108. MELNICK, J. L., RENNICK, V., HAMPIL, B., SCHMIDT, N. J., AND HO, H. H., Lyophilized combination pools of enterovirus equine antisera: Preparation and test procedures for the identifcation of field strains of 42 enteroviruses, *Bull WHO* **48**:263–268 (1973).

109. MELNICK, J. L., WENNER, H. A., AND PHILLIPS, C. A., Entero-viruses, in: *Diagnostic Procedures for Viral, Rickettsial, and Chlamydial Infections* (E. H. LENNETTE AND N. J. SCHMIDT, EDS.), pp. 471–534, American Public Health Association, Washington, DC, 1979.

110. MENEGUS, M. A., AND DOUGLAS, R. G., JR., Viruses, rickettsiae, chlamydiae, and mycoplasmas, in: *Principles and Practice of Infectious Diseases* (G. L. MANDELL, R. G. DOUGLAS, JR. AND J. E. BENNETT, EDS.), pp. 193–205, Churchill Livingstone, New York, 1990.

111. MENZO, S., BAGNARELLI, P., GIACCA, M., MANZIN, A., VAR-ALDO, P. E., AND CLEMENTI, M., Absolute quantitation of viremia in human immunodeficiency virus infection by competitive reverse transcription and polymerase chain reaction, *J. Clin. Microbiol.* **30**:1752–1757 (1992).

112. METCALF, T. G., JIANG, X., ESTES, M. K., AND MELNICK, J. L., Nucleic acid probes and molecular hybridization for detection of viruses in environmental samples, *Prog. Med. Virol.* **35**:186–214 (1988).

113. MINNICH, L. L., AND RAY, C. G., Immunofluorescence, in: *Clinical Virology Manual* (S. SPECTER AND G. LANCZ, EDS.), pp. 117–128, Elsevier, New York, 1992.

114. MINTZ, L., AND DREW, W. L., Relation of culture site to the recovery of nonpolio enteroviruses, *Am. J. Clin. Pathol.* **74**:324–326 (1990).

115. MURPHY, F. A., AND KINGSBURY, D. W., Virus taxonomy, in: *Fields Virology* (B. N. FIELDS AND D. M. KNIPE, EDS.), pp. 9–36, Raven Press, New York, 1990.

116. MUSHAHWAR, I. K., AND BRAWNER, T. A., Radioimmunoassay, in: *Clinical Virology Manual* (S. SPECTER AND G. LANCZ, EDS.), pp. 129–151, Elsevier, New York, 1992.

117. MYERS, T. W., AND GELFAND, D. H., Reverse transcription DNA amplification by a *Thermus thermophilus* DNA polymerase, *Biochemistry* **30**:7661–7666 (1991).

118. MYINT, S., SIDDELL, S., AND TYRRELL, D., Detection of human coronavirus 229E in nasal washings using RNA:RNA hybridisation, *J. Med. Virol.* **29**:70–73 (1989).

119. NAGLER, F. P. O., AND RAKE, G., The use of the electron microscope in diagnosis of variola, vaccinia and varicella, *Bacteriology* **55**:45–51 (1947).

120. NAKANO, J. H., Poxviruses, in: *Diagnostic Procedures for Viral,*

Rickettsial and Chlamydial Infections (E. H. LENNETTE AND N. J. SCHMIDT, EDS.), pp. 257–308, American Public Health Association, Washington, DC, 1979.

121.   NORVAL, M., AND BINGHAM, R. W., Advances in the use of nucleic acid probes in diagnosis of viral diseases of man, *Arch. Virol.* **97:**151–165 (1987).

122.   NUOVO, G. J., MACCONNELL, P., FORDE, A., AND DELVENNE, P., Detection of human papillomavirus DNA in formalin-fixed tissues by *in situ* hybridization after amplification by polymerase chain reaction, *Am. J. Pathol.* **139:**847–854 (1991).

123.   O'BEIRNE, A. J., AND SEVER, J. L., Enzyme immunoassay, in: *Clinical Virology Manual* (S. SPECTER AND G. LANCZ, EDS.), pp. 153–188, Elsevier, New York, 1992.

124.   OLIVE, D. M., AL-MUFTI, S., AL-MULLA, W., KHAN, M. A., PASCA, A., STANWAY, G., AND AL-NAKIB, W., Detection and differentiation of picornaviruses in clinical samples following genomic amplification, *J. Gen. Virol.* **71:**2141–2147 (1990).

125.   OLIVO, P. D., FROLOV, I., AND SCHLESINGER, S., A cell line that expresses a reporter gene in response to infection by Sindbis virus: A prototype for detection of positive strand RNA viruses, *Virology* **198:**381–384 (1994).

126.   OU, C. Y., CIESIELSKI, C. A., MYERS, G., BANDEN, C. I., LUO, C. C., KORBER, B. T., MULLINS, J. I., SCHOCHETMAN, G., BERKELMAN, R. L., ECONOMOU, A. N., WITTE, J. J., FURMAN, L. J., SATTEN, G. A., MACINNES, K. A., CURRAN, J. W., JAFFE, H. W., LABORATORY INVESTIGATION GROUP, AND EPIDEMIOLOGIC INVESTIGATION GROUP, Molecular epidemiology of HIV transmission in a dental practice, *Science* **256:**1165–1171 (1992).

127.   PALLESON, G., HAMILTON-DUTOIT, S. J., AND SANDVEJ, K., EBV-related lymphomas and Hodgkin's disease, in: *Vth International Symposium on Epstein–Barr Virus and Associated Diseases, Annecy, France*, John Libby Eurotext, London, 1992.

128.   PERSING, D. H., AND LANDRY, M. L., *In vitro* amplification techniques for the detection of nucleic acids: New tools for the diagnostic laboratory, *Yale J. Biol. Med.* **62:**159–171 (1989).

129.   PIEDRA, P. A., KASEL, J. A., NORTON, H. J., GARCIA-PRATS, J. A., RAYFORD, Y., ESTES, M. K., HULL, R., AND BAKER, C. J., Description of an adenovirus type 8 outbreak in hospitalized neonates born prematurely, *Pediatr. Infect. Dis. J.* **11:**460–465 (1992).

130.   PRASAD, B. V. V., BURNS, J. W., MARIETTA, E., ESTES, M. K., AND CHIU, W., Localization of VP4 neutralization sites in rotavirus by three-dimensional cryo-electron microscopy, *Nature* **343:**476–479 (1990).

131.   PUTHAVATHANA, P., LEE, H. W., AND KANG, C. Y., Typing of hantaviruses from five continents by polymerase chain reaction, *Virus Res.* **26:**1–14 (1992).

132.   REDFIELD, D. C., RICHMAN, D. D., ALBANIL, S., OXMAN, M. N., AND WAHL, G. M., Detection of herpes simplex virus in clinical specimens by DNA hybridizations, *Diagn. Microbiol. Infect. Dis.* **1:**117–128 (1983).

133.   RICHARDSON, J. H., AND BARKLEY, W. E., *Biosafety in Microbiological and Biomedical Laboratories*, Department of Health and Human Services, Public Health Service, Centers for Disease Control and National Institutes of Health, US Government Printing Office, HHS Publication No. (CDC) 84-8395, Bethesda, MD, 1984.

134.   RICHMAN, D. D., GUATELLI, J. C., GRIMES, J., TSIATIS, A., AND GINGERAS, T., Detection of mutations associated with zidovudine resistance in human immunodeficiency virus by use of the polymerase chain reaction, *J. Infect. Dis.* **164:**1075–1081 (1991).

135.   ROTBART, H. A., Nucleic acid detection systems for enteroviruses, *Clin. Microbiol. Rev.* **4:**156–168 (1991).

136.   ROTBART, H. A., LEVIN, M. J., MURPHY, N. L., AND ABZUG, M. J., RNA target loss during solid phase hybridization of body fluids—a quantative study, *Mol. Cell. Probes* **1:**347–358 (1987).

137.   RUUTU, P., RUUTU, T., VOLIN, L., TUKIAINEN, P., UKKONEN, P., AND HOVI, T., Cytomegalovirus is frequently isolated in bronchoalveolar lavage fluid of bone marrow transplant recipients without pneumonia, *Ann. Intern. Med.* **112:**913–916 (1990).

138.   SACKS, S. L., WANKLIN, R. J., REECE, D. E., HICKS, K. A., TYLER, K. L., AND COEN, D. M., Progressive esophagitis from acyclovir-resistant herpes simplex, *Ann. Intern. Med.* **111:**893–899 (1989).

139.   SAFRIN, S., BERGER, T. R., GILSON, I., WOLFE, P. R., WOFSY, C. B., MILLS, J., AND BIRON, K. K., Foscarnet therapy in five patients with AIDS and acyclovir-resistant varicella-zoster virus infection, *Ann. Intern. Med.* **115:**19–21 (1991).

140.   SAIKI, R. K., GELFAND, D. H., STOFFEL, S., SCHARF, S. J., HIGUCHI, R., HORN, G. T., MULLIS, K. B., AND ERLICH, H. A., Primer-directed enzymatic amplification of DNA with a thermostable DNA polymerase, *Science* **239:**487–491 (1988).

141.   SAIKI, R. K., SCHARF, S., FALOONA, F., MULLIS, K. B., HORN, G. T., ERLICH, H. A., AND ARNHEIM, N., Enzymatic amplification of beta-globin genomic sequences and restriction site analysis for diagnosis of sickle cell anemia, *Science* **230:**1350–1354 (1985).

142.   SAMBROOK, J., FRITSCH, E. F., AND MANIATIS, T., *Molecular Cloning: A Laboratory Manual*, Cold Spring Harbor Laboratory Press, Cold Spring Harbor, NY, 1989.

143.   SARKAR, G., AND SOMMER, S. S., Shedding light on PCR contamination, *Nature* **343:**27 (1990).

144.   SCHMIDT, N. J., Cell culture procedures for diagnostic virology, in: *Diagnostic Procedures for Viral, Rickettsial, and Chlamydial Infections* (N. J. SCHMIDT AND R. W. EMMONS, EDS.), pp. 51–100, American Public Health Association, Washington, DC, 1989.

145.   SCHMIDT, N. J., AND EMMONS, R. W., General principles of laboratory diagnostic methods for viral, rickettsial and chlamydial infections, in: *Diagnostic Procedures for Viral, Rickettsial and Chlamydial Infections* (N. J. SCHMIDT AND R. W. EMMONS, EDS.), pp. 1–35, American Public Health Association, Washington, DC, 1989.

146.   SCHOLTISSEK, C., AND MUELLER, K., Effect of dimethylsulfoxide (DMSO) on virus replication and maturation, *Arch. Virol.* **100:**27–35 (1988).

147.   SESHAMMA, T., BAGASRA, O., TRONO, D., BALTIMORE, D., AND POMERANTZ, R. J., Blocked early-stage latency in the peripheral blood cells of certain individuals infected with human immunodeficiency virus type 1, *Proc. Natl. Acad. Sci. USA* **89:**10663–10667 (1992).

148.   SHAW, R. D., FONG, K. J., LOSONSKY, G. A., LEVINE, M. M., MALDONADO, Y., YOLKEN, R., FLORES, J., KAPIKIAN, A. Z., VO, P. T., AND GREENBERG, H. B., Epitope-specific immune responses to rotavirus vaccination, *Gastroenterology* **93:**941–950 (1987).

149.   SIBER, G. R., LESZCZYNSKI, J., PENA-CRUZ, V., FERREN-GARDNER, C., ANDERSON, R., HEMMING, V. G., WALSH, R. E., BURNS, J., McINTOSH, K., GONIN, R., AND ANDERSON, L. J., Protective activity of a human respiratory syncytial virus immune globulin prepared from donors screened by microneutralization assay, *J. Infect. Dis.* **165:**456–463 (1992).

150.   SIXBEY, J. W., NEDRUD, J. G., RAAB-TRAUB, N., HANES, R. A., AND PAGANO, J. S., Epstein–Barr virus replication in oropharyngeal epithelial cells, *N. Engl. J. Med.* **310:**1225–1230 (1984).

151. Slade, H. B., Pica, R. V., and Pahwa, S. G., Detection of HIV-specific antibodies in infancy by isoelectric focusing and affinity immunoblotting, *J. Infect. Dis.* **160**:126–130 (1989).

152. Smith, T. F., Rapid methods for the diagnosis of viral infections, *Lab. Med.* **18**:16–20 (1987).

153. Smith, T. F., Specimen requirements: Selection, collection, transport, and processing, in: *Clinical Virology Manual* (S. Specter and G. Lancz, eds.), pp. 19–41, Elsevier, New York, 1992.

154. Southern, E. M., Detection of specific sequences among DNA fragments separated by gel electrophoresis, *J. Mol. Biol.* **98**:503–517 (1975).

155. Spector, S., Rua, J. A., Spector, D. H., and McMillan, R., Detection of human cytomegalovirus in clinical specimens by DNA–DNA hybridization, *J. Infect. Dis.* **150**:121–126 (1984).

156. Spector, S. A., Nucleic acid probes, in: *Diagnostic Procedures for Viral, Rickettsial and Chlamydial Infections* (N. J. Schmidt and R. W. Emmons, eds.), pp. 203–218, American Public Health Association, Washington, DC, 1989.

157. Stark, L. M., and Lewis, A. L., Complement fixation test, in: *Clinical Virology Manual* (S. Specter and G. Lancz, eds.), pp. 203–228, Elsevier, New York, 1992.

158. Sternberger, L. A., Hardy, J., Cuculis, J. J., and Meyer, H. C., The unlabeled antibody enzyme method of immunochemistry. Preparation and properties of soluble antigen–antibody complex (horseradish peroxidase–antihorseradish peroxidase) and its use in identification of spirochetes, *J. Histochem. Cytochem.* **18**:315–333 (1969).

159. Stevens, J. G., and Cook, M. L., Latent herpes simplex virus in spinal ganglia of mice, *Science* **173**:843–845 (1971).

160. Stoler, M. H., and Broker, T. R., In situ hybridization detection of human papillomavirus DNAs and messenger RNAs in genital condylomas and a cervical carcinoma, *Hum. Pathol.* **17**:1250–1258 (1986).

161. Stone, M. R., and Nowinski, R. C., Topological mapping of murine leukemia virus proteins by competition-binding assays with monoclonal antibodies, *Virology* **100**:370–381 (1980).

162. Sullender, W. M., Sun, L., and Anderson, L. J., Analysis of respiratory syncytial virus genetic variability with amplified cDNAs, *J. Clin. Microbiol.* **31**:1224–1231 (1993).

163. Swierkosz, E. M., Flanders, R., Melvin, L., Miller, J. D., and Kline, M. W., Evaluation of the Abbott testpack RSV enzyme immunoassay for detection of respiratory syncytial virus in nasopharyngeal swab specimens, *J. Clin. Microbiol.* **27**:1151–1154 (1989).

164. Tanaka, J., Kamiya, S., Ogura, T., Sato, H., Ogura, H., and Hatano, M., Effect of dimethyl sulfoxide on interaction of human cytomegalovirus with host cell: Conversion of a nonproductive state of cell to a productive state for virus replication, *Virology* **146**:165–176 (1985).

165. Tanaka, J., Sadanari, H., Sato, H., and Fukuda, S., Sodium butyrate-inducible replication of human cytomegalovirus in a human epithelial cell line, *Virology* **185**:271–280 (1991).

166. Taniguchi, K., Hoshino, Y., Nishikawa, K., Green, K. Y., Maloy, W. L., Morita, Y., Urasawa, S. Z., Kapikian, A. Z., Chanock, R. M., and Gorziglia, M., Cross-reactive and serotype-specific neutralization epitopes on VP7 of human rotavirus: Nucleotide sequence analysis of antigenic mutants selected with monoclonal antibodies, *J. Virol.* **62**:1870–1874 (1988).

167. Taylor-Robinson, D., Chickenpox and herpes zoster. III. Tissue culture studies, *Br. J. Exp. Pathol.* **40**:521–532 (1959).

168. Tecott, L. H., Barchas, J. D., and Eberwine, J. H., In situ transcription: Specific synthesis of complementary DNA in fixed tissue sections, *Science* **240**:1661–1664 (1988).

169. Ticehurst, J. R., Feinstone, S. M., Chestnut, T., Tassopoulos, N. C., Popper, H., and Purcell, R. H., Detection of hepatitis A virus by extraction of viral RNA and molecular hybridization, *J. Clin. Microbiol.* **25**:1822–1829 (1987).

170. Tracy, S., Comparison of genomic homologies in the coxsackievirus B group by use of cDNA:RNA dot-blot hybridization, *J. Clin. Microbiol.* **21**:371–374 (1985).

171. Tsang, V. C. W., Peralta, J. M., and Simons, A. R., Enzyme-linked immunoelectrotransfer blot techniques (EITB) for studying the specificities of antigens and antibodies separated by gel electrophoresis, *Methods Enzymol.* **92**:377–391 (1983).

172. Tyrrell, D. A. J., and Bynoe, M. L., Cultivation of a novel type of common-cold virus in organ cultures, *Br. Med. J.* **1**:1467–1470 (1965).

173. Urdea, M. S., Synthesis and characterization of branched DNA (bDNA) for the direct and quantitative detection of CMV, HBV, HCV, and HIV, *Clin. Chem.* **39**:725–726 (1993).

174. Valentine-Thon, E., Evaluation of sharp signal system for enzymatic detection of amplified hepatitis B virus DNA, *J. Clin. Microbiol.* **33**:477–480 (1995).

175. Van Dyke, R. B., and Murphy-Corb, M., Detection of respiratory syncytial virus in nasopharyngeal secretions by DNA–RNA hybridization, *J. Clin. Microbiol.* **27**:1739–1743 (1989).

176. VanRooyen, C. E., and Scott, G. D., Smallpox diagnosis with special reference to electron microscopy, *Can J. Public Health* **39**:467–477 (1948).

177. Virtanen, M., Palva, A., Laaksonen, M., Halonen, P., Soderlund, H., and Ranki, M., Novel test for rapid viral diagnosis: Detection of adenovirus in nasopharyngeal mucus aspirates by means of nucleic-acid sandwich hybridisation, *Lancet* **1**:381–383 (1983).

178. Wadell, G., and De Jong, J. C., Restriction endonucleases in identification of a genome type of adenovirus 19 associated with keratoconjunctivitis, *Infect. Immun.* **27**:292–296 (1980).

179. Wang, M.-L., Skehel, J. J., and Wiley, D. C., Comparative analyses of the specificities of anti-influenza hemagglutinin antibodies in human sera, *J. Virol.* **57**:124–128 (1986).

180. Warford, A. L., and Levy, R. A., Use of commercial DNA probes, *Clin. Lab. Sci.* **2**:105–108 (1989).

181. Wasserman, A., Ueber die serodiagnostik der syphilis und ihne praktische bedeutung fur die medizin, *Verh. Kong. Inn. Med. (Wiesbaden)* **25**:181–191 (1908).

182. Webster, R. G., Kendal, A. P., and Gerhard, W., Analysis of antigenic drift in recently isolated influenza A (H1N1) viruses using monoclonal antibody preparations, *Virology* **96**:258–264 (1979).

183. Williams, B., Evaluation of the enzyme linked virus inducible system (elvis™) for rapid detection of herpes simplex virus, *Tenth Annual Clinical Virology Symposium and Annual Meeting, Pan American Group for Rapid Viral Diagnosis*, abstract T21 from meeting, Clearwater Beach, Florida (1994).

184. Williams, V., Gershon, A., and Brunell, P. A., Serologic response to varicella-zoster membrane antigens measured by indirect immunofluorescence, *J. Infect. Dis.* **130**:669–672 (1974).

185. Wolinsky, V., Rinaldo, C. R., Kwok, S., Snisky, J. J., Gupta, P., Imagawa, D., Farzadegan, H., Jacobson, L. P., Grovit, K. S., Lee, M. H., Chmiel, J. S., Ginzburg, H., Kaslow, R. A., and Phair, J. P., Human immunodeficiency virus type 1 (HIV-1) infection—a median of 18 months before a diagnostic western

blot. Evidence from a cohort of homosexual men, *Ann. Intern. Med.* **111:**961–972 (1989).

186. WU, D. Y., AND WALLACE, R. B., The ligation amplification reaction (LAR)—amplication of specific DNA sequences using sequential rounds of template-dependent ligation, *Genomics* **4:** 560–569 (1989).

187. WYATT, R. D., KAPIKIAN, A. Z., THORNHILL, T. S., SERENO, M. M., KIM, H. W., AND CHANOCK, R. M., *In vitro* cultivation in human fetal intestinal organ culture of a reovirus-like agent associated with nonbacterial gastroenteritis in infants and children, *J. Infect. Dis.* **130:**523–528 (1974).

188. YALOW, R. S., AND BERSON, S. A., Immunoassay of endogenous plasma insulin in man, *J. Clin. Invest.* **39:**1157–1175 (1960).

189. YEAGER, A. S., MORRIS, J. E., AND PROBER, C. G., Storage and transport of cultures for herpes simplex virus, type 2, *Am. J. Clin. Pathol.* **72:**977–979 (1979).

190. YEWDELL, J. W., AND GERHARD, W., Antigenic characterization of viruses by monoclonal antibodies, *Annu. Rev. Microbiol.* **35:**185–206 (1981).

191. YOUNG, K. K. Y., ARCHER, J. J., YOKOSUKA, O., OMATA, M., AND RESNICK, R. M., Detection of hepatitis C virus RNA by a combined reverse transcription PCR assay: Comparison with nested amplification and antibody testing, *J. Clin. Microbiol.* **33:**654–657 (1995).

CHAPTER 3

Surveillance and Seroepidemiology

Richard A. Kaslow and Alfred S. Evans†

1. Introduction

Surveillance has been described as the systematic collection of data pertaining to the occurrence of specific diseases, the analysis and interpretation of these data, and the dissemination of consolidated and processed information to contributors to the program and other interested persons.

The establishment of a surveillance system must have a purpose, usually the establishment of a method of control for the disease or diseases under surveillance or the evaluation of existing methods. The principles were well set forth initially by Langmuir[103] and subsequently other officials of the U.S. Public Health Service, Centers for Disease Control and Prevention (CDC),[122,151] and by Raška[131,132] for the World Health Organization (WHO). They were first a major focus of discussion of the 21st World Health Assembly in 1968.[160]

The techniques and uses of surveillance have been presented by Brachman[14] in the companion volume to this one for bacterial diseases. They have also been applied to occupational, chronic and malignant diseases, to evaluation of health services, and to evaluate progress in the control of infectious diseases.[44,70,83,130,150] The techniques of surveillance have become a part of national and international programs of disease control. In *Healthy Peo-*

†Deceased.

Richard A. Kaslow • Department of Epidemiology, School of Public Health, University of Alabama at Birmingham, Birmingham, Alabama 35294-0008. **Alfred S. Evans** • Department of Epidemiology and Public Health, Yale University School of Medicine, New Haven, Connecticut 06510.

ple, the principal document setting the health agenda for the nation for the next decade, the U.S. Public Health Service established goals for improvement in systems for collecting data pertinent to the assessment of maintenance of public health and prevention goals.[84]

The current and future need for improvement in surveillance techniques in developed countries and the need for their increasing establishment in developing countries have been given high priority by a group of epidemiologic experts,[54] by a series of reports from the Institute of Medicine,[96,116] and by others.[58,132,150,161] In parallel with the growing appreciation of the value of surveillance, the mechanics of recording and communicating surveillance data have also been reexamined in light of the new options for electronic transmission. The United States and France both have more than a decade of experience with electronic reporting of communicable diseases.[9,79,153] Early problems with standardization of definitions and reporting methods among official sources and with timeliness of primary reports have not proved major obstacles to electronic reporting.

This chapter discusses the background and elements of traditional surveillance, the concept and uses of serological and molecular epidemiology, and their application to the control of infectious diseases.

2. Surveillance

The traditional methods of reporting and surveillance are based on the occurrence of a case of clinical disease or of a death from clinical disease. They form the basis of public health control and immunization programs throughout the world, although they are increasingly being supplemented by techniques that rely on modern communications systems and laboratory methods.

2.1. Historical Background

The use of mortality and morbidity data as a basis for public health action goes back centuries. Brief highlights are offered here; a recent history and review[39] expands on some of this material. The occurrence of the "Black Death" or pneumonic plague in Europe in about 1348 resulted in the appointment of three guardians of public health by the Venetian Republic to exclude ships with affected persons aboard. The detention of travelers from plague-infected areas for 40 days in Marseilles (1377) and in Venice (1403) led to our current concept of quarantine.

The term *surveillance* has been employed for years in the restrictive sense of follow-up of persons who have had contact with plague or of infectious syphilis patients to determine whether disease developed within the limits of the incubation period. The dictionary defines the word in terms of police surveillance as meaning to "watch or guard over a person, especially a suspected person, a prisoner, or the like."[123] In public health practice, the suspect is the disease.

The principles of surveillance were first exemplified by William Farr, Superintendent of the Statistical Department of the General Registry in London, in a series of classic letters on the causes of death in England appearing from 1839 to 1870 and through a collection of papers on "Vital Statistics" published in 1885. The WHO Influenza Centers for recognition of influenza outbreaks and new viral strains were established in 1948 prior to the introduction and general use of the term. Formal development of the concept of surveillance is of more recent origin and was in response to national needs for disease surveillance or to major new epidemic problems. These needs involved the requirement for a nationally centralized clearinghouse of essential information in order to define the magnitude of the problem, to inform the appropriate authorities on whom responsibility fell for public health control measures, and as a means of evaluating the effectiveness of such measures. Use of the term in the United States began in 1949 with the development of a modified program at the CDC called "Surveillance and Appraisal of Malaria." In 1951, the concept was applied to the residual smallpox cases in the United States.

Surveillance really became an established concept and public health practice on April 28, 1955, when the Surgeon General directed the establishment of a "National Poliomyelitis Surveillance Program" in response to paralytic polio cases following the use of Salk vaccine (the "Cutter incident"). This program was set up at the CDC. The technique became an effective tool in following trends in the disease, in measuring the effectiveness of polio immunization programs, and in detecting suspected vaccine-associated cases.

On July 5, 1957, the Asian influenza surveillance program was initiated and consisted of bimonthly reports from the CDC to keep everyone informed of the progress of the outbreak, including the public press. It served as an essential system tying together the massive national program to control the pandemic. Influenza surveillance has continued at the CDC in conjunction with the WHO ever since and has provided critical data on the occurrence of influenza outbreaks throughout the world.

The surveillance of hepatitis similarly followed an epidemic in 1961 in which shellfish from contaminated waters were identified as the source of an outbreak. *Salmonella* surveillance was initiated in 1962 following 18 hospital outbreaks. Many other diseases were added to this list over time, and now the CDC publishes special surveillance reports on about 20 categories of infectious disease. In Europe, Dr. Kǎrél Raška was an enthusiastic supporter of the surveillance concept who initiated both traditional and serological surveillance in his own country, Czechoslovakia, and promoted the principles as Director, Division of Communicable Disease, WHO. A special unit called "Epidemiological Surveillance of Communicable Diseases" was established in the WHO by Dr. A. M.-M. Payne to coordinate and extend this program. Currently, WHO closely monitors and regularly reports in its bulletin on several diseases under international sanitary regulation, e.g., cholera, plague, and yellow fever. Other diseases are kept under surveillance with varying degrees of attention, and the organization actively supports regional, national, and international control programs of which surveillance is an essential component. In anticipation of European unification, the different approaches to surveillance in the individual countries has raised issues of comparability and coordination.[40]

The advent of AIDS has forced many countries into a serious surveillance mode for the first time or into reconsidering the effectiveness of previous approaches. In the United States and most other developed countries, the most intensive and elaborate machinery ever constructed for such purposes has been directed at the surveillance of human immunodeficiency virus (HIV) infection and acquired immunodeficiency syndrome (AIDS).[18,22,109,121,143] In Europe, joint efforts have been constructive,[45,46] offering a possible model for international coordination of such activities.[108] Unfortunately, in many less-developed countries, despite encouragement and some assistance from WHO, political, social, and economic constraints have kept health care officials from responding with anything like the full array of measures needed to monitor the evolution or maintain control of the epidemic in their severely affected populations.

While a set of legal and ethical principles of surveillance had been evolving gradually prior to the early 1980s, the pressing need for accurate information about HIV infection has led to a broad array of new surveillance practices that have continued to test the balance between individual and community interests and responsibilities.

2.2. Types of Surveillance

Surveillance can be classified according to the method of reporting, the criteria for initiating the report, the location or population studied, and the purpose of the system. The most common method of reporting is the passive one in which the reporter, usually a physician, transmits a weekly summary of all the cases of reportable diseases he or she has seen in the previous week. Preformatted reporting instruments are usually provided with a list of reportable conditions, as defined by that state (in the United States), and these are sent in at periodic intervals to the public health office designated. The reporter gives data on the number of cases seen in the interval prescribed, or sometimes the absence of cases. Reporting can be accomplished in print, by telephone, even using toll-free numbers or automatic recording devices available at all hours. Motivation for this passive system is usually the legal requirement, usually unenforceable, an interest in public health, or provision for adequate feedback of data of clinical or therapeutic interest to the physician. Time and lack of interest greatly limit such a system to a small percentage of most reportable diseases, but as long as the system and requirements remain unchanged, the changes in incidence may reflect meaningful patterns of disease.

Active surveillance is usually employed for special diseases, but some states use it for all reportable diseases. This system requires the health department to contact the reporters at regular intervals and request specific data on cases of specified diseases. It permits further data collection on epidemiologic features of certain diseases. Sometimes individual physicians are designated on a geographic or specialty basis to report the occurrence of certain diseases either on a passive or active basis. Public health officials have addressed the issue of case definition as part of the continuing effort to increase the effectiveness of surveillance.[156] The criterion for reporting is usually the occurrence of a clinical case of sufficient severity that medical care has been sought. This may be supported by laboratory data, but for diseases of unknown cause, at least in the beginning, standard definitions based on available clinical and epidemiologic data may be established, as was true for Legionnaire's disease, toxic shock syndrome, AIDS, and hantavirus pulmonary syndrome and still is for Reye's syndrome and Kawasaki syndrome. For a few diseases some special clinical feature of the disease may be used to estimate prevalence data, such as leg lameness for poliomyelitis or scarring for smallpox, or incidence data as with sudden severe acute lower respiratory illness for hantavirus infection. Sometimes surveillance is based on laboratory data, as with the isolation of the agent or results of a serological test. These represent special types of study, although routine laboratory diagnoses in public or even hospital laboratories may be used to supplement or compare with case reporting. Although most surveillance systems cover political jurisdictions (e.g., national, state, and county), certain population groups or locations may represent high risks and require special reporting methods as a basis for control. Examples are military populations, hospitals, day-care centers, and homes for the elderly. Most surveillance systems incorporate a list of reportable diseases, but the focus of the reporting may be on specific disease entities, such as congenital defects as a reflection of the impact of rubella and cytomegalovirus infections.[42]

2.3. Elements of Surveillance

The CDC has described the object of surveillance as "health data essential to the planning, implementation, and evaluation of public health practice, closely integrated with the timely dissemination of these data to those who need to know...."[21] As applied to communicable diseases, surveillance has been defined as "the exercise of continuous scrutiny of, and watchfulness over, the distribution and spread of infections and factors related thereto, of sufficient accuracy and completeness to be pertinent to effective control."[160] A wide variety of sources of data on disease occurrence and on the characteristics of the populations at risk contribute to surveillance. These sources vary from country to country depending on the stage of development and sophistication of the public health services, the quality and extent of laboratory facilities, the available funds, and the characteristics of the indigenous diseases. The major features have been summarized by the WHO[160] in ten "elements of surveillance" (Table 1); these are addressed briefly in the following subsections. Earlier reference was made to the recent survey of 17 European nations' surveillance activities,

Table 1. Elements of Surveillance

1. Mortality registration
2. Morbidity reporting
3. Epidemic reporting
4. Laboratory investigations
5. Individual case investigations
6. Epidemic field investigations
7. Surveys
8. Animal reservoir and vector distribution studies
9. Biologics and drug utilization
10. Knowledge of the population and environment

with an emphasis on the variety of approaches taken and on potential barriers to a standardized system even for the European community.[40]

2.3.1. Mortality Registration. Mortality registration is the oldest form of disease reporting and has the advantage of being legally required and of a high order of completeness in most countries. Wherever a physician or other health practitioner is in attendance, there is a reasonable expectation that most infectious diseases of sufficient severity to cause death would exhibit distinctive enough clinical characteristics to permit diagnosis. An autopsy may also contribute to the accuracy of identification of the disease process. On the other hand, some deaths such as those from coronary artery disease may be sudden and unattended by a physician; multiple causes of death may be involved, and the one of most public health significance may be lost in the order of causation recorded on the death certificate. There is often a long delay in the tabulation and publication of mortality data; autopsy information may not be added to amend information on the original death certificate. In general, mortality data reflect incidence only when there is some relatively constant ratio between deaths and cases. Except for AIDS, rabies, Lassa fever, and certain hemorrhagic fevers, most viral diseases are not fatal, so that mortality data have limited usefulness as a barometer of disease occurrence. However, the occurrence of an excess of deaths from influenza and pneumonia above the expected level has been a sensitive index of influenza. It was first used by William Farr in 1847 and has been consistently reported in the United States since 1918. In recent years, weekly data on deaths from influenza and pneumonia in 121 cities have been compared to the average number of deaths in the same week of the five previous years.[33] Current deaths that exceed this "epidemic threshold" can usually be attributed to epidemic influenza. Heat exhaustion and smog may also increase deaths above this level, but such events are temporally and geographically circumscribed and can

be readily identified from environmental data. Unfortunately, the time required for reporting and analyzing these mortality data results in a delay of a month or so in recognition of an outbreak; this delay is longer during holiday periods. For certain fatal conditions, death records may also contribute to the early assessment of the completeness of reporting or of the changes in the pattern of occurrence over time.[34] In viral diseases of longer latency, like AIDS and cervical cancer, mortality obviously measures more remote and selected consequences of the original infection.

2.3.2. Morbidity Reporting. The reporting of cases of specified communicable diseases is legally required in most countries, and as many as 40 conditions may be involved. A simple and effective reporting system is the backbone of surveillance for most health departments. The advantages are that (1) such reports are usually made by physicians who are best qualified to identify the diseases; (2) laboratory confirmation may be available; and (3) there is usually an organized system of regional or national tabulation and reporting. The disadvantages are (1) the absence of many viral diseases from the required list; (2) the notorious underreporting of diseases despite legal requirements, primarily because of physicians' lack of motivation, secretarial help, or time; (3) the uncertainty of diagnosis (especially without laboratory confirmation)—a major issue for many viral infections; and (4) the variability of reporting efficiency from one time period to another, being in general highest during epidemics. Lack of rapid, reliable, inexpensive diagnostic techniques has represented a discouraging obstacle to accurate reporting of viral diseases. There have been occasional efforts to examine the completeness and efficiency of existing surveillance systems. A careful analysis of the efficacy of reporting of 570 cases of notifiable communicable diseases from 11 hospitals in Washington, DC, revealed an overall reporting rate of 35%.[106] For individual diseases, the rates were as follows: viral hepatitis, 11%; *Haemophilus influenza* meningitis, 32%; meningococcal meningitis, 50%; shigellosis, 62%; and tuberculosis, 11%. As part of its recent effort to improve techniques, public health workers have explored the use of hospital records[32] to validate direct reporting of AIDS cases by local health authorities.

2.3.3. Epidemic Reporting. The recognition and identification of epidemic viral diseases are commonly more accurate than individual reports because public health officials and laboratory facilities are usually involved. This is true of outbreaks of yellow fever, influenza, rubella, hepatitis, viral exanthems, and certain arbovirus infections. Unfortunately, this may not always be the case, particularly if the viral agent produces primarily mild or inapparent infections or if the outbreak occurs in areas with poor medical care or inadequate public health and laboratory facilities. Unrecognized epidemics of dengue involving thousands of people have taken place under such circumstances. However, local outbreaks of diseases that have a high mortality or that involve tourists or other persons from outside the country are now being recognized through better surveillance, even in more remote areas. Examples are Lassa, Ebola, and Marburg fevers and certain animal pox viruses, such as camel, gerbil, and monkeypox, that may occasionally involve humans.

Historically, outbreaks of diseases such as poliomyelitis, hemorrhagic fevers, and influenza were recognized by local health authorities but not reported to the WHO because of fear of the economic impact of publicity on tourist or export trade. Soon after the advent of AIDS, numerous countries followed an all too predictable pattern of early widespread suppression of information exchange out of fear or denial, followed by more enlightened official acknowledgment of the reality as HIV infection spread across political borders and social strata. This pattern of response undoubtedly delayed early mobilization of international efforts to control the pandemic.

2.3.4. Laboratory Investigations. Laboratory identification of the causative agent is an almost absolute requirement for the etiologic diagnosis of individual cases and for most epidemics of viral diseases; the exceptions are poliomyelitis and certain viral exanthems in which the clinical features are characteristic enough to permit diagnosis. Sophisticated laboratory facilities and experienced personnel are therefore needed for the isolation and/or serological identification of the majority of viral infections. These may exist in the national or regional public health laboratories, in specialized virus diagnostic institutes, or in university settings. For a number of years WHO supported a broad network of regional, national, and international reference laboratories with expertise in specific viral infections to assist with influenza and other viral respiratory infections, arboviruses, and enteroviruses, as well as a group of about 15 other collaborating diagnostic laboratories around the world. Unfortunately, declining resources have forced WHO to withdraw funding from those facilities, only some of which have continued to provide reference services. The need for trained personnel, special equipment, standardized antigens and antisera, protection against laboratory infections, good water, reliable refrigeration, and proper sterilizing equipment make viral laboratories expensive and difficult to maintain without this type of support.

2.3.5. Individual Case Investigations. The occurrence of a disease of public health importance in an area previously free of the disease or where control measures have been established demands rapid and intensive investigation. This has included viral infections such as smallpox, yellow fever, certain types of viral encephalitis, the hemorrhagic fevers, rabies (either in humans or in a new animal species), and paralytic poliomyelitis. Of special importance is the follow-up of persons returning to their own country from areas where these infections are known to occur. This was done in the United States for smallpox through an alert card issued to incoming travelers from foreign countries; the card requires notification and investigation of any illness developing within a defined period after arrival. Since the discontinuance of the requirement for routine smallpox vaccination for returning American travelers and the confirmed global eradication of the natural disease as of October 1977, surveillance and accurate diagnosis of persons developing poxlike illness on return from potential geographic foci have ceased. Political upheaval, enemy action, or lack of funds or interest may disrupt the present high order of surveillance, immunization, and control in these "high-risk" areas, the value of monitoring the diseases mentioned above and others must be reassessed continually.

2.3.6. Epidemic Field Investigations. When there is an increase in the number of cases or deaths from a viral disease of public health significance, an epidemic team must be deployed to make further study. The team should include an epidemiologist and a virologist, with appropriate equipment for the collection and transportation of specimens. Rapid diagnostic techniques, such as fluorescent antibody or enzyme-linked immunosorbent assay (ELISA) tests on throat or skin specimens or on infected insects, may permit direct identification of the causative agent in the field. These teams are usually composed of experts from regional or national health services that operate in support of local health officials. In the United States, the Epidemic Intelligence Service and other components of the CDC in Atlanta, Georgia, fulfill this role on the request of state health departments and occasionally governments of other countries. On a worldwide basis, the WHO in Geneva, its regional branches, or WHO-designated laboratories may be able to render assistance. More routine outbreaks are handled by state or municipal health departments, often assisted by laboratory personnel. There is increasing need for better integration and cooperation between epidemiologists and public health laboratory personnel in this task.[48] The establishment of an epidemiologic investigation unit in affiliation with each public health laboratory should be promoted not only for epidemic analysis but also for effective implementation of immunization programs as well as day-to-day surveillance work involving both epidemiologic and laboratory data. A high degree of integration of epidemiologic and laboratory functions has been achieved in the United States not only at the CDC but also in many state health departments. Years ago in the United Kingdom,[35] and more recently in other European and a few developed countries, public health officers began to make significant efforts to coordinate field and laboratory activities. Not surprisingly, progress in less developed countries has been slower.

2.3.7. Surveys. Many types of surveys of infectious disease have been used in public health work. Well-standardized information can often be collected widely and rapidly, but often at greater cost than would be incurred using other methods. Clinical or epidemiologic markers have identified certain indicator conditions: splenomegaly or positive blood smears for malaria, scars for smallpox vaccination, and positive skin tests for tuberculosis. Immunization histories, personal interviews, or clinic records may also be used to assess the vaccination status of the population.

The Expanded Programme on Immunization (EPI) of the World Health Organization has developed an excellent and standardized system of cluster household sampling of children to assess immunization coverage.[85,86,161] The elements of the program are (1) identification of the geographic area(s) of interest, (2) identification of the age group(s) of interest, (3) random selection of 30 sites (termed "clusters") from within each geographic area for which individual results are desired, (4) random selection of a starting point ("household") within each site, and (5) selection of seven individuals of the appropriate age from within each of the 30 sites. Selection begins in the starting household until a total of seven individuals is obtained. All individuals of the appropriate age living in the last household falling into the sample are included, even if this means including eight or ten individuals within the cluster rather than the required minimum of seven. It was intended that the sample estimates have 95% confidence limits within 10% of the true population mean. The statistical validity was later established by computer simulation,[86] and refinement continues.[15] The sampling method has been of great use in developing countries where training programs and booklets have permitted adoption of the technique on a wide scale. Despite its high degree of effectiveness, proposed refinements in the method may improve its performance.[15] The cluster survey design has also been applied more generally to measurements of morbidity.[60,104]

In addition, WHO has pioneered in surveys of schools and households as a means of discovering the high prevalence of poliomyelitis in young children and the need for starting immunization near birth and prior to infection.[55,85,86] For many viral diseases, major reliance must be placed on antibody surveys or, in the case of hepatitis and HIV infection, tests on blood donors or blood available as residual from routine perinatal genetic screening programs (see Section 3.3.1 and references 91 and 120). The use and application of serological surveys to these ends are discussed in detail in Section 3.

2.3.8. Animal-Reservoir and Vector-Distribution Studies. Surveillance of human diseases acquired from animals or of diseases in which the vector is arthropod-borne requires the collection of data on the zoonoses and the presence of appropriate vectors in the area. Examples for which such data may be useful are yellow fever and other arthropod-borne diseases, especially dengue and the hemorrhagic fevers, rabies, and monkeypox. The emergence of a group of unusual but frequently lethal diseases such as Lassa fever, the Argentinian and Bolivian hemorrhagic fevers, and more recently the hantavirus pulmonary syndrome involves study of the rodents that are suspected as the reservoirs. The surveillance of these infections requires a special investigation team and/or close cooperation among existing epidemiologic, veterinary, and entomological services.

2.3.9. Biologics and Drug Utilization. In the United States, interest in tracking the use of vaccines has recently grown to the point where CDC now publishes data monthly.[27] The establishment of an effective system of determining the scale and utilization of viral vaccines and immune globulins may not only provide a lead to the immunization status of an area but might also give supplementary information to permit the recognition of an outbreak or of other special problems. The cluster sampling techniques developed by the EPI have been extended to this use.[59] It is also worth a reminder that it was a clear increase in the number of requests for a drug used to treat *Pneumocystis carinii* pneumonia that led to the detection of the earliest cases of AIDS.

2.3.10. Knowledge of the Population and Environment. The denominator used in determining incidence and prevalence rates is the population at risk to the disease in the area from which the cases are reported. Necessary information includes age, sex, ethnic, economic, and other demographic data in order to interpret disease trends. Other background information often needed relates to sanitary conditions, food and water supplies, housing, insects, nutrition, and cultural habits. The accessibility, utilization, and quality of medical care must be known in order to evaluate the potential efficacy of case reporting, mortality data, and other indices of the health of the population.

2.4. Other Surveillance Methods

Additional sources of data may be utilized in supplementing routine surveillance techniques or in evaluating special disease situations. Some of these are listed in Table 2.

2.4.1. Hospital and Medical Care. The existence of national health plans in many countries and the extension of prepaid health insurance schemes in other areas make computerized accounting necessary. This provides the opportunity for including morbidity and mortality information in the data system. Large-scale health plans such as the Kaiser–Permanente Plan in California, the Cooperative Group Health Insurance Plan in Seattle, and the Health Insurance Plan (HIP) of greater New York City have been utilized for these purposes, and the various health alliances formed in response to health care reform may increase the opportunities further. Centralized data-processing centers for hospitals such as the Professional Activities Services (PAS) operating out of Ann Arbor, Michigan, provide another opportunity; this system, which covers some several hundred hospitals and a substantial, albeit selected, proportion of persons hospitalized in the United States, is based on a hospital discharge form that incorporates much useful information on diagnosis, surgical procedures, complications, length of stay, laboratory data, and other factors. Formal systems for collecting similar data are increasingly being developed by state health agencies, about 20 of which make hospital discharge data available for research as well as routine public health purposes.[115] Other techniques capitalizing on computerized records have been explored.[154] Surveillance of the number of patients with acute respiratory infections in emergency rooms, outpatient clinics, and pediatric clinics in large community hospitals[134] combined with prospective virological surveillance of such patients[76,77] can provide sensitive and specific indicators of an influenza outbreak. Other sentinel systems for surveillance of both community-acquired and nosocomial infections have been effective for monitoring trends in certain viral infections such as influenza and HIV.

2.4.2. Panels of Cooperating Physicians. In some areas, data networks have been established by groups of cooperating physicians to record morbidity data and analyses of their medical care programs. For example, the *National Disease and Therapeutic Index*[117] is a product of one such network that provides analyses of the frequency of different diagnoses and the patterns of drug prescription in outpatients in over 1500 private physicians' offices in selected parts of the United States.

2.4.3. Public Health Laboratory Reports. The state and municipal public health laboratories provide a wide range of diagnostic facilities for communicable diseases. Complemented and assisted by the fine laboratories of the CDC, these represent the predominant diagnostic services in viral infections in the country. They are supplemented by virus research and diagnostic laboratories in universities and a few large hospitals. These usually operate in close communication with the public health laboratories of the area. The consolidation and utilization of information from these various sources represent the best ongoing method for surveillance of viral infections. This is especially true for viral diseases, since many are not reportable and because accurate diagnoses are often dependent on laboratory identification of the viral agent.[135] The use of sera sent to such laboratories for multipurpose viral testing is discussed in Section 3.5.3.

2.4.4. Absenteeism from Work or School. A sensitive barometer of any major epidemic in children is an increase in school absentee rates; in adults, it is a jump in absenteeism in industry. Since an absence of any duration in schools may be investigated by the school or public health nurse and any significant loss of time from work may require a physician's certificate of illness, additional data on the nature and duration of the condition may be obtained. The active cooperation of a few geographically representative schools and key industries may provide public health officials with valuable leads on epidemic illness.

2.4.5. Telephone and Household Surveys. The CDC has conducted telephone surveys to verify the presence and the extent of an epidemic such as influenza or to assess recent utilization of a specific vaccine. Research on the occurrence of minor illnesses, such as acute respiratory infections in Tecumseh, Michigan, have used telephone interviews as the basis of data collection.[113] The limitations of this method for disease surveillance are obvious: (1) the household must have a phone; (2) someone must be at home to answer it; (3) the person answering

---

Table 2. Other Sources of Surveillance Data

1. Hospital and medical care statistics—clinical, administrative
2. Panels of cooperating physicians
3. Public health laboratory reports
4. Absenteeism from work or school
5. Telephone and household surveys
6. Newspaper and newsbroadcasting reports

must be aware of illnesses in other members of the family; (4) the disease involved must be characteristic enough to permit recognition by a layman; (5) the presence of illness can be sought over only short periods because of imperfect memory; and (6) the person answering must be willing to cooperate in the survey.

Household surveys by skilled interviewers form the basis of the U.S. National Health Interview Survey, in which periodic carefully selected random samples consisting of many thousands of families are the source of data. An extensive series of morbidity and health analyses and much useful surveillance information have resulted from these repeated surveys and from the physical and laboratory examinations performed on a sizable subset of these individuals.[110,141] They are not helpful in providing immediate surveillance of common infections but reveal general patterns and long-term trends of importance.

2.4.6. Newspaper and Newsbroadcasting Reports.   The news media often report outbreaks of disease before they have been announced by the slower process of most health-reporting mechanisms. Furthermore, there may be epidemics of nonreportable diseases picked up by an active news surveillance system that may be missed or never officially reported to public health authorities. A systematic clipping and recording service from local news services may thus provide important leads to an alert epidemiologic surveillance program. The earliest reports of influenza outbreaks in Hong Kong, or of Lassa fever in Africa, or of an outbreak of hemorrhagic fever in South America may be found on the pages of newspapers with extensive coverage such as *The New York Times*. Indeed, in recent years electronic and print journalists have been particularly attentive to occurrences of new or emerging infections in remote as well as domestic locations.

2.5. Predictive Surveillance and Mathematical Models

The ability to predict the occurrence of infection or of an outbreak is theoretically possible if there are adequate epidemiologic data on hand. This has been tried with varying success for influenza epidemics; here the problem is confounded by our sparse knowledge of the factors governing the emergence of new antigenic variants of influenza virus. For infections in which quite precise requirements exist, such as an insect vector, intermediate hosts, animal reservoirs, or special terrain, the *potential* existence of the infection in certain geographic areas can be presumed, and its *probable absence* in other areas can be predicted. This approach has special interest to the armed forces because the introduction of suscep-

tible military units into areas in which few health data are available poses hazardous conditions for the men and for the success of the mission. In this connection, three infections were analyzed in detail by Baker,[5] employing computer analysis of existing epidemiologic and ecological information: schistosomiasis, malaria, and leptospirosis. Attempts have been made to predict patterns of transmission of arbovirus infections via insect vectors based on well-defined biological and ecological dynamics of the arthropod populations. Serological surveys can also provide valuable data on the potential danger of certain infections; these may be especially useful in areas in which health reporting is poor, or in the case of a virus that produces largely subclinical infections. For example, the presence in persons of all ages and especially young children of antibody to viruses such as poliomyelitis, hepatitis B virus (HBV), Epstein–Barr virus (EBV), or dengue indicates the continuing activity of that agent in that environment, irrespective of the apparent absence of clinical illness or of reported cases caused by these viruses. If soldiers, tourists, Peace Corps volunteers, or other visitors who lack antibody to the agent intermix with the local population and/or are exposed to the local environment and population, infection may result. Such infection is often accompanied by a higher risk of clinical disease than in the indigenous population because infection of older children and adults results in more severe host response than infection very early in life.

Mathematical models have long been used to predict the occurrence and spread of epidemics. Now with the use of computers, they are being developed to assess the impact of some intervention programs such as immunization or the effect of an environmental change such as a major dam or new highway (the Trans-Amazon Highway, for instance) on disease patterns, or the consequences of an aging population or large population movement on public health. These models are often developed before the program is started to assess its impact and often before all the epidemiologic data are known or available. A model for the global spread of influenza,[136] originally developed by investigators in the Soviet Union, has been reassessed in the United States.[105] Soon after the introduction of measles vaccine, Millar reported a model for measles control in West Africa that led to successful changes in vaccine strategy.[112] Models for control of measles, rubella, and other immunizable diseases in other countries have been prepared.[1,90,165] Longini,[105] Haber,[81] Halloran,[82] and others have devoted considerable attention to refining the measurement of vaccine efficacy under different circumstances. A model for the occurrence and control of diarrheal disease has been published.[47] An example in making a predictive model before all the facts are known was one comparing the impact of various programs for the control of AIDS.[20] In the decade that followed, nearly every aspect of exposure to and transmission, diagnosis, course, outcome, and cost of HIV infection has been intensively scrutinized with descriptive and predictive models.[16,118] Very few of these have been or can be subjected to rigorous empirical verification, and it is generally difficult to determine how the results of many of them actually altered those aspects of medical care or public health policy they were intended to influence.

2.6. Evaluation of Surveillance Systems

The general advantages and disadvantages of the existing sources of surveillance data listed in Tables 1 and 2 are summarized in Section 2.3. Declich and Carter[39] highlight five characteristics on which surveillance systems can be evaluated (Table 3). No system provides both high sensitivity and high specificity, and in few, except mortality data, does reporting approximate completeness. A combination of methods is required to obtain some picture of disease patterns and their dynamic change, and even here any change in the criteria for a case, the code used to classify it, the means of obtaining the data, or the method of analysis may result in serious errors of interpretation.

Public health authorities in the United States, Canada, and Europe have recently been more active in evaluating the objectives and methods of infectious disease surveillance.[19,23,26] A previously cited effort to assess the electronic reporting system that had been in place for several years demonstrated differences in the timeli-

**Table 3. Considerations
in Evaluation
of a Surveillance System**[a]

1. Importance
2. Objectives and components
3. Usefulness
4. Cost
5. Quality, dependent on:
   a. Simplicity
   b. Flexibility
   c. Acceptability
   d. Sensitivity
   e. Predictive value
   f. Representativeness
   g. Timeliness

[a]Adapted from Declich and Carter.[39]

ness of reporting by state and by disease.[9] Attempts to optimize use of combinations of surveillance sources and techniques have led to more critical assessment of incidence estimates arrived at by alternative approaches.[92,95,107] Increasing attention has been paid to refinement of capture–recapture methods of ascertainment as they might be applied in specific situations,[92] including surveillance of infectious diseases.[107] Formal statistical analysis of reporting compared with capture–recapture techniques, for example, suggested that over- or under-ascertainment is a function of the degree of independence of the reporting sources. One implication would appear to be that for situations common to reporting of infectious diseases, capture–recapture methods may be useful in correcting for underestimation. Other new methods of analyzing and interpreting surveillance data are also being considered.[147]

Few quantitative data are available to compare the disease prevalence or incidence rates based on different methods of surveillance of the same population group. Comparisons of case reporting with the actual number of hospitals[106] or of laboratory-identified data or data from clinics such as those for sexually transmissible diseases have all emphasized the underreporting of the current reporting system in the United States. It has been estimated that only 1% of cases of salmonellosis, 10% of cases of measles (before institution of the current active surveillance program), and 15–20% of cases of viral hepatitis are reported.[14,150] During the first decade of the AIDS epidemic in the United States, the completeness and accuracy of the elaborate surveillance mechanisms have been subjected to more thorough evaluation than any previous effort.[18,32,121] The shortcomings have been typical of surveillance programs for many diseases. But the visibility of these shortcomings has been magnified by the extraordinary difficulty of monitoring an infection that is so long silent, so deadly, yet so highly stigmatized as a result of its sexual and illicit parenteral modes of transmission.

Approaches to ascertainment of minor illnesses have, for obvious reasons, not received such careful attention. One study of the efficacy of alternative methods for ascertaining the occurrence of minor illnesses was carried out in a graduate student housing unit at the University of Wisconsin and compared (1) a weekly postcard sent in by a volunteer mother reporter from each of the 55 apartment units who visited the other families in the unit one day each week to collect data on a standard card, (2) a weekly telephone survey of successive 10% samples from the telephone book of the housing unit, (3) a calendar sheet the size of a dollar bill, filled out daily by the mother and picked up weekly, such as employed in the Cleveland

Family Study,[41] and (4) school absenteeism as reported by the school nurse, who then confirmed the reason on a personal visit to the family; this last technique was taken as the "gold standard." Even more than for diseases with significant morbidity, the success of any active surveillance system for a minor infectious disease will be heavily dependent on the purpose, setting, target subjects, timing, nature and complexity, feedback, and other incentives of the inquiry.

2.7. Applications of Surveillance

2.7.1. Basis of Public Health Control Programs. The continual monitoring of the patterns of incidence of communicable diseases and of key chronic, malignant, and occupational diseases provides important data bases on which decisions on immunization and other control programs can be made, including identification of those at highest risk. Factors such as the age, population group, behavioral and cultural characteristics, socioeconomic level, occupation, geographic location, seasonal distribution, and other factors affecting the disease must be considered in identifying the target group on the one hand and the health resources, cost effectiveness, political will, and logistics of delivering the program on the other hand. Unfortunately, with declining resources for prevention complicating general complacency about public health infrastructure,[8] the routine collection of surveillance data has not been accorded a high priority by governments at any level. Authorities are often forced to react tardily and inefficiently to each new natural disaster, potential or real epidemic, or unanticipated failure in a disease control program.[43]

2.7.2. Evaluation of Control Programs. Although the basic surveillance system should reflect the general effectiveness of control programs, special surveillance techniques may be desirable for monitoring new diseases or new control programs or for focusing on a particular public health problem. Criteria suggested for the control of poliomyelitis[54] and other infectious diseases[55] have been based on methods that would ascertain (1) the occurrence of the clinical disease (employing routine surveillance plus added methods such as household surveys, as used in the WHO Expanded Programme on Immunization,[69] or lameness surveys for poliomyelitis, (2) the occurrence of infection with or without clinical illness (as measured by laboratory tests, as through serological surveillance of the antibody level of a sample of the target population or measurement of the response to immunization), and (3) the occurrence of the organism in the environment (poliovirus in sewage, arboviruses in mosquitos, hepatitis viruses in water or blood products, etc.). Successful control at each of the three levels is necessary for the eradication or elimination of an infection (see also reference 56 and Chapter 1). The advantages and disadvantages of the methods employed have been reviewed.[55]

Examples of a large-scale control program for which the importance of systematic evaluation has been recognized include the EPI[15,37,69] and the Childhood Immunization Initiative (CII).[29] For the former, the standard data collection methods have included: (1) routine reporting of immunization, adverse events, and disease; (2) sentinel sites for disease surveillance; (3) community health facilities; (4) cluster sample surveys of immunization coverage and disease; and (5) outbreak reporting and investigation. WHO has developed program guidelines for improving surveillance and control of EPI target diseases among others,[162] and there has been close attention to successes of this and related programs.[28,149] In 1993, the United States government mounted a particularly intensive effort (the CII) to control six vaccine-preventable diseases for which current vaccination coverage was considered inadequate. This program has drawn on local and national surveillance data to monitor the progress toward achieving the goal of 90% coverage by 1996 for all but hepatitis B vaccination. In early 1994, CDC began publishing monthly summaries of data on reported cases of these illnesses.

Efforts to monitor and control specific diseases have begun to draw on technological innovations in fields as diverse as biotechnology and computer mapping. Applications of the latter include the use of geographic information systems techniques to study changes in the ecology of vector populations that may help predict changes in transmission patterns.

2.7.3. Surveillance in Developing Countries. The establishment of simple, reliable, and inexpensive means of recognizing disease incidence and prevalence has been recognized as a high priority for the control of the major diseases in developing countries and for the proper use of scarce medical and economic resources. The U.S. National Academy of Sciences has termed this application of surveillance "rapid epidemiologic assessment."[116] The key elements of the program are (1) the measuring instrument, (2) the sampling process, and (3) data analysis. The measuring instrument should reveal the basic demographic characteristics of the population, mortality data derived from death records, hospitals, morgues, newspaper reports, and church records, and morbidity data from health department records, hospitals, clinics, physician records, and sample surveys. Sampling techniques might use cluster or household samples and "minisurveys" by telephone of households, workers, students, etc. Exploration of standardized, modular methods is suggested; this might include a "century sample" consisting of 100 subjects from upper, middle, and lower socioeconomic groups. If necessary, the data analysis should not require more sophisticated instruments than a hand calculator and simple graphs.

Clearly, surveillance methods are of high importance in developing and least-developed countries, but the methods will vary from country to country. Where the health infrastructure is inadequate, surveillance based on hospital and clinic cases, small household surveys, and serological sampling from either of these sources are suggested approaches.

2.7.4. Recognition of Emerging Infectious Diseases.

Early dreams that identification and rational administration of antimicrobial substances coupled with manipulation of immune system (vaccination) could be expected to contain or even eliminate most infectious diseases have been replaced by a more sober, realistic view. We are finally acknowledging fully that incursions by pathogenic microorganisms into individual immunocompromised hosts and into entire human populations are inevitable. More acutely aware than ever before that these agents will continue to test human insight and innovation, we must monitor the microbial world continuously, scrutinize closely the unusual events that perturb the horizon, and respond swiftly to any ominous development. Of course, in recent years the infectious agent representing the most profound threat to have materialized "out of nowhere" is HIV. In less than a decade it has managed to spread globally, causing a devastating pandemic, and firmly embed itself for the forseeable future in populations where it had previously been present rarely if at all. Other previously unencountered or unrecognized infections such as *Legionella* sp. (Legionnaire's disease), *Borrelia burgdorferi* (Lyme disease), and hepatitis E virus have also securely established themselves. Moreover, old enemies have reappeared in new forms and new places [*Escherichia coli* 0157:H7 (hemolytic–uremic syndrome), hantavirus (adult respiratory distress syndrome), feral rabies, dengue, multidrug-resistant *Mycobacterium tuberculosis*, and the dreaded Ebola virus]. Taken together, the rapid emergence of new and the resurgence of old human pathogens have raised serious doubts about the current level of vigilance, that not only our global but even our rather highly developed domestic sentinel systems are inade-

quate. That concern, voiced first in reports by the Institute of Medicine advisory groups[116] and amplified in the latest sequel[96] has culminated in strategic proposals by the CDC[25] for addressing the dangers of emerging and reemerging infections. The CDC publication provides a coherent perspective on the problem, setting four broad goals and corresponding specific objectives (Table 4). Not surprisingly, the foremost goal features systematic comprehensive surveillance, the tactical pursuit of which is elaborated more fully in the document. Many of the recommendations therein further emphasize the importance of sustained vigilance in the early defense against "the restless tide … of the microbial world."[101]

2.7.5. Surveillance in Research Studies.

The methods thus far discussed contribute directly to official public health agencies. In research, specialized surveillance systems have been established for various viruses or groups of viruses. The Virus Watch Programs established by Fox and his associates[36,73] in New York City and then in Seattle, Washington, are excellent examples of this method, which involves systematic sampling of a population of families for enteric and respiratory viruses, antibody testing, and analyses of coincident illness patterns. An earlier effort of special surveillance of this type was the extensive analyses of common illnesses in a group of Cleveland families carried out by Dingle *et al.*[41] Special population groups, such as Tecumseh, Michigan, used for analyses of chronic disease have also been utilized for studies of viral infections, especially acute respiratory disease.[113] In children, the massive long-term study of viral infections in Junior Village carried out by Bell *et al.*[7] at the National Institutes of Health yielded sequential data on the behavior of common viruses in such a closed setting. Extensive studies of agents causing gastroenteritis in populations such as those selected for field sites in Dacca, Pakistan, and in Huascar, Peru,[10,11] and elsewhere have employed community surveillance techniques. Surveys of pregnant women have yielded valuable information about women at risk of transmitting congenital infection such cytomegalovirus (CMV).[72]

Recent population-based research on HIV has been conducted in large cohorts of usually self-selected individuals in different high-risk categories. These studies have used regular serological screening of seronegative participants to estimate ongoing risk and identify newly infected cohort members for inclusion in various research protocols (see Section 3.5.2). Surveillance of large populations for susceptibility to various infections and for other characteristics may also be performed in conjunction with estimates of vaccine efficacy or durability of protection.

Table 4. Summary of Goals and Objectives[a]

Goal I: Surveillance
Detect, promptly investigate, and monitor emerging pathogens, the diseases they cause, and the factors influencing their emergence.
Objectives:
  A. Expand and coordinate surveillance systems for the early detection, tracking, and evaluation of emerging infections in the United States.
  B. Develop more effective international surveillance networks for the anticipation, recognition, control, and prevention of emerging infectious diseases.
  C. Improve surveillance and rapid laboratory identification to ensure early detection of antimicrobial resistance.
  D. Strengthen and integrate programs to monitor and prevent emerging infections associated with food/water, new technology, and environmental sources.
  E. Strengthen and integrate programs to monitor, control, and prevent emerging vector-borne and zoonotic diseases.

Goal II: Applied Research
Integrate laboratory science and epidemiology to optimize public health practice.
Objectives:
  A. Expand epidemiologic and prevention effectiveness research.
  B. Improve laboratory and epidemiologic techniques for the rapid identification of new pathogens and syndromes.
  C. Ensure timely development, appropriate use, and availability of diagnostic tests and reagents.
  D. Augment rapid response capabilities for vaccine delivery and expand evaluation of vaccine efficacy and the cost effectiveness of vaccination programs.

Goal III: Prevention and Control
Enhance communication of public health information about emerging diseases and ensure prompt implementation of prevention strategies.
Objectives:
  A. Use diverse communication methods for wider and more effective delivery of critical public health messages.
  B. Establish the mechanisms and partnerships needed to ensure the rapid and effective development and implementation of prevention measures.

Goal IV: Infrastructure
Strengthen local, state, and federal public health infrastructures to support surveillance and implement prevention and control programs.
Objectives:
  A. Ensure the ready availability of the professional expertise and support personnel needed to better understand, monitor, and control emerging infections.
  B. Make available state-of-the-art physical resources (laboratory space, training facilities, equipment) needed to safely and effectively support the preceding goals and objectives.

[a]From CDC.[25]

2.8. Publications on Surveillance

The dissemination of data derived from surveillance programs is an essential requirement for public health action. Table 5 lists the common publications reporting current and long-term trends in infectious diseases. The most current and useful are the *WHO Weekly Epidemiological Record* and the *Morbidity and Mortality Weekly Report* from the U.S. Public Health Service. The latter is now available electronically to anyone with access to the Internet.[31]

Weekly or monthly surveillance reports are published in many other countries—Canada, the Caribbean area (CAREC), Kuwait (MER), Israel, and the European community nations, to name a few. The regular and special reports from the CDC are expanding to include addi-

Table 5. Publications Dealing with Surveillance Data

1. Worldwide
   WHO Weekly Epidemiological Record
   WHO Epidemiological and Vital Statistics Report (monthly)
   WHO Annual, Vol. II: *Infection Diseases; Cases, Deaths, Vaccinations*
2. Europe
   Bulletins published weekly, monthly, or quarterly by offical organizations of indivdual countries[a]
3. North America
   PAHO—*Weekly Epidemiological Report*
   PAHO—*Quarterly Information Bulletin*, Pan American Zoonoses Center
   PAHO—*Annual Report of the Director*
   CAREC—*Caribbean Epidemiology Center Surveillance Report* (monthly)
4. United States (USPHS)
   Morbidity and Mortality Weekly Report, CDC, U.S. Department of Health and Human Services
   Annual Supplement, Reported Incidence of Notifiable Diseases in the United States, CDC, U.S. Department of Health and Human Services
   Periodic Surveillance Reports (recent issues have included the following viral diseases) from CDC, U.S. Department of Health and Human Services:
       Congenital cytomegalovirus
       Hepatitis
       Measles
       Neurological viral diseases (aseptic meningitis, encephalitis, enteroviruses, poliomyelitis)
       Rabies
       Rubella
5. Other
   State health departments and private organizations (e.g., NDTI[117]) publish summaries on selected conditions at irregular intervals

[a]See Table V in Desenclos.[40]

tional infectious diseases of special interest [AIDS, influenza, CMV infection, human T-cell lymphotropic virus type I (HTLV-I)] as well as other selected congenital, chronic, and occupational diseases, and high-risk behavioral problems. The *WHO Technical Report Series*[159,161] provides up-to-date information on various viral infections as well as other diseases, prepared by expert committees.

3. Seroepidemiology

3.1. Introduction

The application of laboratory methods to the study of various components of blood of epidemiologic interest has been termed "seroepidemiology." In view of the variety of fluids and cells now harvested for investigative purposes and the profusion of tests now applied to them, the term is admittedly restrictive. However, other broader terms lately employed (e.g., "biomarkers", "molecular epidemiology") have their own connotations, and no euphonic alternative to seroepidemiology has been coined to convey both the appropriate diversity of the biological materials to which the techniques are applied and the epidemiologic context in which the testing takes place. For the purposes of illustrating applications to viral infections, and because serological testing is still the method most widely applied for surveillance purposes, this section draws heavily on examples from the realm of seroepidemiology. Methods of collection, storage, and testing of many types of biological specimens and some of their epidemiologic applications have been addressed in Chapter 2 and such other sources as the text by Schulte and Perera (see Suggested Reading).

Seroepidemiology can be defined as the systematic collection and testing of blood samples from a target population, or a representative sample thereof, to identify current and past experiences with infectious diseases by means of antibody and antigen tests and by measurement of other indices of immunity. Additional purposes are to seek biochemical markers for various chronic diseases, to measure certain nutritional components, and to characterize the genetic aspects of red cells, leukocytes, and serum proteins. In infectious diseases, serological epidemiology contributes to two broad and overlapping areas:

1. Serological surveillance to provide supplementary data as the basis for immunization and public health planning programs.
2. In research, as an epidemiologic tool to investigate the risk and occurrence of infectious diseases and to study the behavior of old and newly recognized microbiological agents in different population groups.

Serological surveys may be carried out to determine the distribution of a single agent such as poliomyelitis but are more commonly "multipurpose" in nature. This section considers the history, methods, and uses of seroepidemiology with particular reference to its application as an important adjunct to traditional methods of surveillance of infectious diseases.

Just as epidemiology is concerned with the occurrence and distribution of clinical cases in different populations, so serological epidemiology, as noted above, is concerned with the occurrence and distribution of various components of the blood that indicate past or current infection, that are biochemical markers for certain chronic diseases, or that reveal the genetic attributes of various population groups. The epidemiologic characteristics are detected in the laboratory rather than at the bedside. The name "serological surveys" has been used interchangeably with "immunologic surveys." No satisfactory expression has been found to refer to the whole spectrum of components (red cells, white cells, plasma, and serum) that may be examined in population surveys.

The use of laboratory techniques to identify the initiation of the pathological process and to identify individuals in whom the process results in inapparent rather than apparent clinical manifestations has been termed "subclinical epidemiology."[57] Laboratory tools are essential to the full understanding of the pathogenesis of infectious and chronic diseases and of their biological spectrum.

3.2. Historical Background

The introduction of serological tests for the diagnosis of disease provided the basis for later serological surveys. As early as 1916, the Wassermann test was applied routinely to patients attending a prenatal clinic at Johns Hopkins Hospital by Williams[158] but this was more of a case-finding procedure than an attempt to delineate disease patterns. In 1930, the development of a neutralization test for poliomyelitis led Aycock and Kramer[4] to use the procedure to define the immunity pattern of a given population; this is a landmark in the history of serum surveys. In 1932, Soper *et al.*[145] mapped out the occurrence of yellow fever in Brazil by antibody surveys under the auspices of the Rockefeller Foundation, and this technique has been widely used subsequently in studying arbovirus infections. Antibody surveys for influenza also date back to the mid-1930s. The discovery of swine influenza virus by Shope[142] in 1931 and of human influenza virus by Smith *et al.*[144] in 1933 was rapidly followed by population studies to measure antibody to these viruses in persons of different age groups.[2,17,75]

The Yale Poliomyelitis Study Unit under Dr. John R. Paul employed serological survey techniques as long ago as 1935,[129] and his analysis with Riordan of the poliomyelitis pattern in Alaskan Eskimos is a classic study.[128] He became one of the foremost users and promoters of the concept of serological epidemiology, and through his work and writing,[125,130] the utilization of this technique in public health practice and research studies has become a reality. The World Health Organization also took note of this development in 1960 and established three WHO Serum Reference Banks to practice and promote seroepidemiology in New Haven, Connecticut; Prague, Czechoslovakia; and Johannesburg, South Africa. An additional bank was established in 1971 in Tokyo, Japan. The activities and principles of these banks have been reviewed in two *WHO Technical Reports*,[159,161] in a book,[130] and in several other publications.[124,126] Although WHO no longer formally supports these banks, the rationale for proper collection, cataloging, and storage of specimens for use of both primary and collaborating investigators has gained wide application in public health and academic research institutions. In many such places recognition of the value of such biological resources has led to the creation of sizable repositories of serum and, increasingly, tissue or other cellular material containing nucleic acid suitable for molecular and genetic analysis.

3.3. Methodology

3.3.1. Sources of Biological Material.

A list of several sources of material for survey analysis is given in Table 6. By far the most important method is the collection of blood specimens and of health data from a carefully selected sample from the target population at risk. To achieve the highest yield from this type of study, serological surveys should be multipurpose in nature and can include measurement of antibodies and other markers of various prevalent infections. As an example of a large multipurpose survey, sera was collected in household surveys in rural areas by the WHO for the evaluation of the effectiveness of penicillin in mass eradication programs for yaws.[80] In the United States, the series of Health and Nutrition Examination Surveys (HANES) has produced a nationwide population-based health profile for more than two decades. As one of its many components, it has provided valuable seroprevalence data on a variety of viral infections [e.g., measles, poliomyelitis, herpes simplex virus (HSV), hepatitis A virus (HAV), HBV].[110,141]

Procurement of blood from a subsample of the households in the cluster sample used by the Expanded Programme on Immunization of WHO to measure immunization coverage is another excellent possibility in confirming the antibody status of those interviewed.[85] This sample is based on the random selection of 210 children in 30 clusters of seven children each. This number of sera is within the capabilities of a good laboratory to test for the presence of immunity to diphtheria–pertussis–tetanus (DPT), polio, and measles, five of the six vaccines included in the program (the sixth is BCG), and the results would reflect both naturally acquired and vaccine-acquired antibody or antitoxin levels.

Because of the cost of collecting sera from a properly selected sample of a population, other sources of sera have been utilized. These have included blood specimens collected for other purposes, especially for routine tests during physical examinations for the armed forces or industry or during an outpatient visit or admission to a hospital and from neonates screened for specific heritable disorders. Sera sent to a public health laboratory for serological tests for syphilis, viral diagnosis, or other diagnostic tests have also been employed. These collections of sera may not be representative of the age, sex, and geographic distribution of the entire population; the nature of the biases introduced must be recognized and evaluated. However, they are economical to obtain and sometimes may reveal important information on the presence or absence of a certain virus in the community or on the occurrence of a recent outbreak. For most multipurpose surveys, a representative sample carefully selected from the community at risk is important. A broad representation of all younger age groups is essential if the sera are to be used in evaluating the immunization needs of the population or in measuring the impact of a vaccination program. As a rough guide for

---

**Table 6. Sources of Human Biological
Materials for Surveillance**

1. Planned serum surveys from target populations
2. Entrance and periodic examinations of different groups:
    a. Military
    b. Industry
    c. Health clinics
3. Blood donors in Red Cross and similar programs (e.g., transplantation donors)
4. Public health laboratories:
    a. Serological tests for syphilis (e.g., premarital)
    b. Other immunologic and diagnostic tests
5. Hospitals:
    a. Entry tests for blood chemistries or syphilis
    b. Diagnostic tests for infectious diseases
    c. Blood banks and transplantation programs
    d. Prenatal clinics
    e. Employee health services

---

multipurpose surveys, the reports of the WHO have suggested a sample of 300–600 persons divided into 25 sera per age group (e.g., single-year groups under 5 years, 5-year groups up to 19 years, and broader groups thereafter).[159,161]

The blood must be collected and separated under sterile conditions. Aliquots of 0.5 ml each have been used by the WHO and CDC serum banks and are very useful for microtiter tests; several replicates of the entire collection may be prepared at the time of aliquoting so they can be shipped to other laboratories for testing. Sera are usually stored at −20°C, often in a commercial warehouse. Temperatures of −70°C are best but are more expensive to maintain. Lymphocytes can also be separated from anticoagulated blood, frozen at low temperatures in fetal calf serum and dimethylsulfoxide (DMSO), and later thawed for examination of stable cell surface markers and other cell-associated products. Cellular material, even in extremely low concentration and a nonviable state, may be quite suitable for amplification and identification of fragments of nucleic acid. These techniques offer powerful tools for detecting genes characteristic of specific infectious agents and other biological material of interest.

3.3.2. Laboratory Tests. The general principles and techniques of laboratory testing for viral infections are presented comprehensively in Chapter 2. The antibody tests most suitable to serological surveys of specific viruses are detailed in each chapter of this book. The criteria for a satisfactory test include simplicity, sensitivity, specificity, reliability, ability to detect long-lasting antibody, minimal interference from nonspecific inhibitors, the availability of satisfactory reagents, and the safety of the test for the laboratory technician.[49,161] The microtiter procedure developed by Takatsy in 1950 in Hungary and popularized in this country by Sever[140] in 1962 has become the standard method in serological survey laboratories. It is adaptable to a wide variety of antibody determinations, it requires a minimal amount of serum (usually 0.1 ml) and other ingredients, and large numbers of sera can be efficiently tested. Several automated methods of dilution and of adding various reagents have been introduced to speed the testing even more.[161]

The development of simple tests such as the ELISA for antibody measurement in microtiter plates has provided a sensitive method for identification of antibody in serum samples from survey populations and can indicate both past and current infections.[164] The use of monoclonal antibodies in this and other antibody tests permits highly specific identification of individual strains of the virus, a special advantage in determining whether a new strain has been introduced in a community or if reinfection or reactivation has occurred in the individual. Commercial kits for many antibodies are now available, and new formulations are continually being devised for a variety of clinical, public health, and research applications. Another test of importance is the radioimmunoassay assay (RIA), which often has high specificity but is more expensive and cumbersome to perform. These same tests can be used for antigen identification in sera, as for hepatitis B and other viruses, as well as other body fluids, excretions, and tissues.

3.4. Advantages and Limitations

The traditional methods of surveillance are based on cases of clinical disease reported by physicians or identified by some survey technique. The sequence usually involved includes the requirements listed in Table 7. In underdeveloped and developing countries, many of these requirements for surveillance may be missing or inadequate, and even in highly developed countries the reporting of communicable diseases is less than satisfactory and involves much variability.[9,26,40,106] The use of serological surveys is an important means of supplementing morbidity information, and their advantages are listed in Table 8. Because many viral infections may be clinically mild or inapparent, may require laboratory confirmation for accurate diagnosis of even overt cases, and may not be on the list of reportable diseases, the serological survey technique is an important tool: it reveals *total* burden of infection (apparent and inapparent), both currently and in the past. Selection of tests that reflect antibody of long duration permits measurement of the cumulative experience of the population tested with the disease in question; selection of a test based on short-lived antibody such as immunoglobulin M (IgM) allows identification of a recent epidemic or infection. Testing of two sera spaced in time permits measurement of the incidence of infection during the interval period.

The disadvantages of seroepidemiology are the cost

Table 7. Requirements for Surveillance Based on Clinical Cases

1. Occurrence of clinical illness
2. Sufficient severity to seek medical care
3. Availability of medical care
4. Capability of physicians to diagnose illness
5. Laboratory support of diagnosis
6. Reporting of disease to health department
7. Collection and analysis of data by health department

Table 8. Advantages of Seroepidemiologic Cohort Studies

1. Can identify who is susceptible and who is immune at the start of the study
2. Can yield incidence data on both infection and disease
3. Can relate reinfection rates to the level of preexisting antibody
4. Allow calculation of apparent to inapparent infection ratios
5. Permit recognition of the spectrum and elucidation of natural history of host response to infection
6. Can evaluate the absolute risk of various cofactors related to infection and disease separately
7. Can measure the secondary attack rate for infection and disease among household or other contacts of clinical cases included in the original cohort

and effort involved in the selection and bleeding of the target population, the collection and analysis of data, and the need for and cost of laboratory facilities equipped to carry out the tests. There must also be a satisfactory means of measuring antibody for the particular virus to be studied, and the method of carrying it out must be simple enough to allow performance on a large-scale basis. Because aliquots of sera from a collection can be shipped long distances in the frozen state to a number of specialized reference laboratories for testing, the work can be divided among participating laboratories.

As mentioned earlier, the discovery of human retroviruses and their relationship to AIDS and to adult T-cell leukemia has reawakened interest in the principles of seroepidemiology.[64] Many existing serum collections have been reactivated to help map out the global distribution of antibody to these viruses, and new prospective studies have been initiated in high-risk populations to identify the incidence of HIV infection and disease. These include a number of cohorts of homosexual men, injection drug users, hemophiliacs, transfusion recipients, and prostitutes (see Section 3.5.2). Investigators in many of these cohorts have regularly made banked serum and occasionally cells available to other scientists for collaborative or independent research.

3.5. Uses of Serological and Molecular Techniques

The uses discussed below encompass both public health and research applications, and there is some overlapping in the various categories. The utilization of biological markers in the surveillance of disease (serological surveillance) is clear in many of the "uses" described.

3.5.1. Prevalence. The presence in the serum of one or more antibodies to specific infectious agents at the time of collection is called *antibody prevalence* (or commonly, "seroprevalence"). The antibody prevalence is a proportion composed of the number of persons whose sera contain a particular antibody in the lowest dilution tested divided by the total number of persons examined. Unlike "case prevalence," which indicates the existence of disease at the time of the survey, antibody prevalence reflects the cumulative experience, past and present, with an infectious agent. The prevalence is a function both of prior and current infection and of the duration of the antibody tested.

Many antibodies such as the neutralization antibody for poliomyelitis or yellow fever virus; antibodies to the HIV core, polymerase, and envelope; and the hemagglutination-inhibition antibody for influenza, parainfluenza, rubella, measles, or arboviruses last for years, perhaps a lifetime. Thus, the cumulative experience of a population can be measured, and infection acquired in childhood can be detected in persons of middle or perhaps even old age. Some drop-off in antibody titer (sometimes below the lowest detectable levels) may occur in older age groups after a childhood infection. Similarly, the viral capsid antigen (VCA)-IgG antibody to EBV measured by the indirect immunofluorescence test has been found to be of long duration; even complement-fixing antibodies to CMV, herpesviruses, or dengue virus have been found to persist for years following infection.

Measurement of IgG and IgM antibody by tests such as the ELISA or immunofluorescent antibody tests provides a simple way in a single serum sample of reflecting current and past infection by use of tests for IgG and recent infections by tests for IgM antibody. Although IgM antibody usually denotes a primary infection, certain viruses such as herpesviruses may induce IgM on reactivation. It should also be reemphasized that unlike prevalence data for clinical infectious disease, serological prevalence data reflect total infection rates, representing both clinical and subclinical (or asymptomatic) infections.

Multipurpose antibody surveys have been carried out in a number of countries under WHO auspices as an extension of evaluation surveys for penicillin campaigns to eradicate yaws, as noted above. They have been largely in rural areas of Nigeria, Toga, Afghanistan, the Philippines, Samoa, Thailand, and Yugoslavia.[80] Unfortunately, the results of most of these surveys have not been published. Examples of multipurpose surveys include those of military recruits in the United States,[98,127,148] Brazil[59,71] Colombia,[60] and Argentina.[61] The successive HANES surveys described earlier (Section 3.3.1) have provided large numbers of serum specimens for estimating cumulative exposure to various infections in representative samples of the United States population. There have been sporadic health and serological surveys of Barbados[63,67] and of St. Lucia.[62] In the Barbados study, a 10% household sample was randomly selected from a middle- and lower-socioeconomic-level community of 10,000 persons in Bridgetown. The results indicate the type of information that can be derived from this type of study. Of 100 sera from children under age 10 tested, 30% lacked protective levels of antitoxin against both diphtheria and tetanus, indicating the need for intensifying the immunization program against these diseases. The prevalence of protective levels against tetanus is a good indicator of the level of public health practice, since this antitoxin is acquired almost exclusively by immunization procedures and not through natural infection. The age distribution of antibodies to various viruses may provide useful information on the behavior of these infections in the community and of the need for immunization programs. Antibodies to EBV were acquired very early in life, reaching a plateau of about 95% positive by age 5. Antibodies to CMV were present in 60% by age 5, rose to 78% by age 15, and reached a plateau of about 85% by age 30 (Fig. 1A). This means that clinical illness caused by these viruses would be rare because they usually cause mild and inapparent infection when acquired by young age groups. In contrast, rubella antibody, although present in 41.4% of the females, was essentially absent from children under 11 years old (Fig. 1A). This indicates that there had been no rubella infection for the previous 10 years and that a female population was entering childbearing age without any protection against rubella. On this basis, an active rubella immunization program was initiated in girls of 12 and under. In subsequent years through 1978, a few sporadic cases have been reported yearly, but no epidemic occurred. A similar age pattern was seen for all three types of dengue antibody: it was absent in persons under age 25, but antibody prevalence rose rapidly after this to reach levels of 50–60% (Fig. 1B). This suggested that dengue virus had not been introduced in the previous 25 years or that mosquito and other control measures had been effective, or both. However, in 1978, Barbados experienced a small outbreak of dengue as part of the Caribbean-wide outbreak that had severely affected Puerto Rico in late 1977. The information obtained on the patterns of susceptibility and immunity to these viral infections could not have been obtained by ordinary surveillance methods based on the reporting of clinical cases.

Initial tests for poliomyelitis antibody employing conventional microtiter neutralization procedures indicated that 27, 42, and 53.8% of those tested at 1:5 or 1:8 serum dilution lacked antibody to poliomyelitis types 1, 2,

and 3, respectively.[63] Subsequent tests on 304 sera using a 1:2 serum dilution and longer serum–virus incubation periods indicated that only 13.1% lacked type 1 antibody, 6.5% type 2 antibody, and 14.3% type 3 antibody.[67] This emphasizes the need for sensitive methods for detecting low levels of antibody. Two mass poliomyelitis programs were carried out after the 1972 survey, one in 1974–1975 and one in 1977–1978. There have been no reported cases of poliomyelitis since 1972.

The 1972 Barbados serum collection was tested for human retroviruses after the agents were discovered. The HTLV-1 antibody was found in 4.25% overall, rising in age from a 2.7% prevalence under age 10 to 9.0% in the 61–70 age group.[133] Females had a higher prevalence rate (5.8%) than males (2.3%). The adult pattern raised the possibility of sexual transmission, and this was strengthened by the finding of a prevalence rate of 14.1% in persons with a positive VDRL test for syphilis as compared with 3.5% in those VDRL negative. There was evidence of household clustering and of vertical transmission.

3.5.2. Incidence. Prospective serological and clinical cohort studies provide powerful tools to identify who is susceptible to a specific virus in a population, who becomes infected, and, in concert with traditional surveillance, who develops clinical illness. Tables 8 and 9 summarize the advantages of this seroepidemiologic approach. The techniques employed in prospective epidemiologic studies have been reviewed (see Chapter 1). Cohort studies may be retrospective or prospective according to the time of exposure. In retrospective studies the time of exposure occurred in the past, and the exposed and unexposed groups are identified and followed by records for the development of disease or other events of interest in the present. In prospective cohort studies, the exposure occurs in the present or after the observation period begins, and the exposed and unexposed cohorts are identified and followed forward with ongoing identification of the illness or illnesses under study. The size of the cohort needed depends on the percentage of susceptible (antibody-free) persons at the start of the study, the infection rate expected over the time period between samplings, and the percentage developing clinical illness or experiencing a defined event (if that is included in the follow-up).

There are also several methods of follow-up, testing, and analysis depending on the design of the study and the feasibility of testing the entire cohort initially and at intervals during the follow-up period, as shown in Table 10. One way is to test the entire cohort at the start and end of the study and at times of illnesses in between. Alternatively the sera from all bleedings can be kept frozen

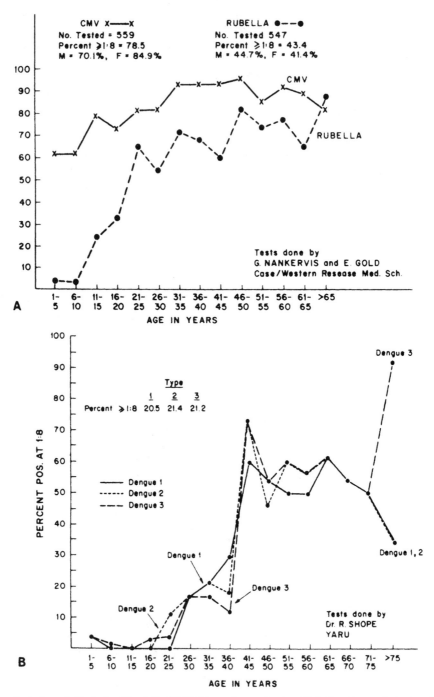

Figure 1. Results of serological tests in Barbados, West Indies.[63,67] CF, complement fixation; HI, indirect hemagglutination. (A) Age distribution of antibody to cytomegalo- and rubella (HI) viruses. (B) Survey of dengue (CF) titer in 336 persons.

Table 9. Strategies for Seroepidemiologic Cohort Studies of Viral Infections

1. The entire cohort is divided by laboratory tests into those previously infected (who either remain infected but seropositive where agent persists or possess protective antibody where agent is eliminated) and those susceptible to infection (who lack antibody)
2. The focus may be on:
   a. The infected—for studies of pathogenesis, late natural history, reactivation and recurrence, durability of immunity, morbidity, mortality, and treatment and secondary prophylaxis
   b. The susceptible—those who are exposed or not exposed to the agent for studies of exposure other rates, transmission efficiency, primary prophylaxis, incidence of infection, acute infection, clinical–subclinical ratio, timing of appearance of antibody, early natural history, morbidity, mortality, disability, and other aspects
3. Storage of serum permits delayed determination of susceptibility, patterns of immunity, and infection rates until follow-up is complete, until events of interest have occurred, or until more advanced technology is available

until the end of the study and tested all at once, avoiding laboratory variations that occur among tests done at different times.[64] This general approach is useful for establishing a causal relationship between a new agent and a disease. If those lacking antibody at the start of the study seroconvert at the time of illness and those with existing antibody do not develop the disease, then a causal association exists. The method would also reveal whether reinfection with or without illness occurred in those with low levels of preexisting antibody. For exploring relationships to acute self-limited conditions, only those susceptible (lacking antibody) to the presumed causal agent need be identified and followed. For diseases of long latency or chronic expression, such as the sequelae of retroviral infections, much can be learned from intensive investigation tailored to the intermediate manifestations of interest in persons carrying the agent.

A third technique is usually necessary in studies of rare diseases among a large cohort all of whom have had blood drawn at the start of the study, such as the relationships of various viruses to the development of a malignancy.[64] In this relationship it is usually the level of antibody and not just its presence that is a risk marker. With such a large population it is usually impossible to test for the presence and level of antibody in all sera taken at the start of the study. The initial sera are thus usually stored frozen. When cases of the disease occur, such as a malignancy, such persons have blood drawn, as do healthy age- and sex-matched controls from members of the original cohort. Then initial and final sera from cases and controls are mixed, coded, and tested at the same time to determine whether the earlier presence of antibody or of antigen for hepatitis B in the original serum is a predictor for the subsequent occurrence of the disease or the tumor. This is a case-control study nested within a prospective cohort. For example, in order to determine the role of high antibody levels in the development of African Burkitt's lymphoma to EBV, de-Thé and colleagues[38] bled 42,000 children in the West Nile district of Africa and followed them for the development of the tumor. Fourteen cases occurred over the next 5 years. Sera from these cases were then tested with age- and sex-matched controls from the original cohort for several EBV antibodies. It was found that persons with EBV IgG antibodies to the viral capsid antigen in the initial sera that were twofold or more higher than the controls had a 30-fold increased risk of subsequently developing the tumor. In another prospective study based on a follow-up of 240,000 sera stored in four serum banks, 43 cases of Hodgkin's disease were identified and matched with one or two controls.[114] Significantly higher levels of EBV antibodies were found in the sera of 25% of persons who subsequently developed the disease than were found in the controls who did not. This type of study requires very careful laboratory control

Table 10. Methods of Testing and Analysis in Prospective Seroepidemiologic Studies

Type of method	Method	Examples
Total cohort	1. Bleed and test entire cohort at beginning and end of study and during illness	EBV and IM at West Point Military Academy[67] HIV and AIDS in many settings[89,97,99]
	2. Determine incidence of infection and disease in between in relation to presence or absence of the antibody or antigen in first serum sample	HBV and HCC[6] HIV and AIDS
Susceptible cohort	Test all at start; just follow susceptibles	EBV and IM at Yale[138]
Case-control or case-cohort	1. Freeze initial and final sera until cases are identified	EBV and Burkitt's lymphoma[38]
	2. Select matched or a random subset of controls from initial cohort who remained healthy	HIV and AIDS[16]
	3. Mix case and control sera and test as unknowns	

including testing all sera as coded unknowns, inclusion of standard positive and negative reference sera in each run, and preferably two independent readers of the test. Further examples are found in chapters dealing with malignancies.

The appearance of antibody to a virus in the second of two sequentially collected specimens indicates infection with that agent somewhere between the two times of collection. A fourfold or greater rise in antibody titer over a preexisting level usually indicates *reinfection* with that agent. If surveillance of clinical illness can also be maintained between serum collections, then the ratio of apparent (i.e., clinical) to inapparent (i.e., subclinical) infections can be ascertained. An example of this is a study in which a group of Yale college students were followed clinically and serologically during their freshman year for acute respiratory infections, rubella, and infectious mononucleosis [66,138] (Table 7). During the year, epidemics of influenza and rubella occurred. The serological infection rate for influenza was 43.6% in susceptible students, with 59% of infected persons exhibiting clinical illness. For rubella, 25% lacked antibody on entry; of these, 61.4% became infected, of whom 39% had a clinical illness with rash. No clinical rubella cases occurred in persons with titers 1:64 or higher. With EBV an infection rate of 13.1% occurred among the 49% lacking antibody; this was clinically manifested as definite infectious mononucleosis in 74% of those infected.[138]

Other prospective serological studies have been made of respiratory infections, rubella, EBV, CMV, and other viral infections in a variety of settings. Such studies have been especially useful in defining the risk in pregnant women of CMV, rubella, and HIV infections and the risk in the infected fetus of subsequent congenital malformations or disease. They have been used to portray the infection pattern in military recruits during training[52] or in college students during various years.[66,138] The level of antibody in the first serum sample that protects against reinfection or clinical illness or both over the period until the next serum specimen is drawn has been used to evaluate the quality and duration of both natural and vaccine-induced antibody to influenza and rubella.[52,66,74,93,94]

Many prospective serological and clinical studies were and still are being used to study HIV infections and the development of AIDS and related clinical syndromes in high-risk populations such as homosexuals, drug users, hemophiliacs, and their contacts. Some of the earliest studies were based on cohort methods using sera originally collected for evaluation of HBV vaccine, and others started afresh with a new cohort.[97] The findings in the New York[146] and San Francisco cohorts[24] were as follows: (1) the prevalence level of HIV antibody was 6.6% in New York at entry into the study in 1978, and rose to 10.6% by 1984; in San Francisco the entering prevalence was 4.5% in 1978, but rose dramatically to 73.1% by 1985; (2) the rate of viral infection (seroconversion) among those lacking antibody on entry was 5.5–10.6% per year in New York and 11.2% in San Francisco. Among homosexual and bisexual men entering the Multicenter AIDS Cohort Study (MACS) in Baltimore, Chicago, Los Angeles, and Pittsburgh in late 1984 and early 1985,[97] seroprevalence ranged from 23 to 49%. In a pattern similar to those in other cohorts, the incidence rate of new infection in the MACS, documented at 3–5% per year in early 1985, declined rapidly during the next 3 years to about 1%.[99] In Africa and Thailand, measurements of seroprevalence and incidence in prostitutes and other high-risk groups have repeatedly demonstrated the alarming increases that have taken place over a very few years. Transfusion-associated HIV infection systematically documented through serosurveillance of hemophiliacs and other recipients[3,100] soon after known parenteral exposure provided an important resource for comparative research on the natural history. Taken together, these and a number of other cohort studies have yielded critical data for the empirically based projections about future burden of infection and disease.

3.5.3. Diagnostic Serology.   Sera sent to large hospitals or public health laboratories for various tests can be frozen and stored for later antibody testing against other antigens. The specimens must be adequate in amount and free of bacterial contamination. Specimens sent for viral antibody tests usually fulfill these criteria and are accompanied by minimal demographic and clinical data concerning the patient. There are many uses for this type of collection. All sera coming from patients with central nervous system, gastrointestinal, or respiratory infections or the exanthems can be tested at the time of receipt, or later, against a battery of viral antigens in order to reveal the profile of agents likely to have caused the syndromes. The later discovery of a new agent involved in one of these conditions permits a retrospective assessment of its importance using sera previously collected and stored. An example of this was the evaluation of the importance of California virus in the causation of infections of the central nervous system by testing of all sera received in a state public health laboratory for this syndrome. In Wisconsin, 5.7% of 351 sera received in the state laboratory over the period 1961–1964 revealed evidence of this infection[152]; in Minnesota, 4.1% of 1617 retrospectively tested sera contained this antibody. A second and related application is the determination of the clinical spectrum associated with a newly discovered virus; this is accomplished by testing stored sera from patients with a variety of clinical syndromes and looking for evidence of infection with the new agent.

A third application for sera stored over time is the measurement of *secular trends* or antigenic shifts in viruses over time. This is especially useful in relation to influenza viruses. A fourth use, employing freshly received sera for VDRL or other tests, is the search for influenza antibody patterns that may reveal the beginning of an outbreak or a change in the antigen composition of currently circulating strains; this was used by Widelock *et al.*[157] at the public health laboratories of New York City. Comparison of the geometric mean antibody titer to influenza sera from persons in the acute phase of an unidentified respiratory illness with the titer in others convalescing from a similar illness may permit early identification of an outbreak without waiting for serial samples from the same persons.

Finally, investigators in South Australia made efficient use of samples taken in conjunction with blood donation. They searched for evidence of Ross River virus activity in different locations during the arbovirus season by measuring IgA-, IgG-, and IgM-specific antibody in samples from Red Cross blood donors. Differences in antibody prevalence by region indicated prior activity and helped identify endemic areas but revealed that acute infection had not occurred in that year at frequencies high enough to be detected.[155]

3.5.4. Evaluation of Immunization Programs.

The effectiveness of an immunization program is traditionally judged on the basis of clinical cases or epidemic behavior. A program is regarded as effective when cases decrease or epidemics do not occur. This information may be biased by the possibility that the decrease in clinical cases is related to poor reporting or that insufficient time has elapsed for another epidemic to have occurred. Currently, our knowledge of the utilization of vaccines depends on such sources as sales records of manufacturers, public clinic and physicians' data, direct interviews, and school entry surveys.[30] Because of the inadequacy of these traditional surveillance techniques in determining the need for and the effectiveness of a given vaccine, serological surveys could play an even larger role in evaluation of immunization programs. Although the necessity for a venipuncture reduces the ease and acceptability of this method, public health professionals and physicians can help surmount such barriers by conveying the importance of seroepidemiology for such purposes (Table 11). Newer techniques that obviate the requirement for blood sampling may further encourage the application of biological tools to the evaluation of effectiveness.

The uses of serological epidemiology in immunization programs are summarized in Table 11. Much of this

Table 11. Uses of Seroepidemiology in Immunization Programs

1. Cross-sectional surveys to determine the need for immunization programs in:
   a. Different age groups
      - 6–20 months—measure duration of maternal antibody
      - School/college entry—identify omissions, failures, loss of protective antibody
   b. Different geographic areas
   c. Different socioeconomic groups
   d. High-risk occupational groups
2. Follow-up measurements in immunized persons to determine:
   a. Proportion developing local, humoral, cell-mediated immune responses
   b. Quality and extent of response
   c. Nature and degree of interaction between vaccine components
   d. Duration of response
   e. Level of protection against disease and asymptomatic infection
   f. Degree of spread of live vaccine strains to exposed and susceptible contacts
3. Periodic serological surveillance to identify groups who are not receiving vaccines or who have inadequate responses

information could be obtained in no other way. Immunoserological surveillance already has been of particular importance for the new epidemiologic settings created by substituting vaccine immunity for natural immunity as for poliomyelitis, measles, rubella and to a lesser extent mumps and influenza.[53] With vaccines against HBV now universally recommended for infants in the United States and a vaccine against varicella-zoster virus recently licensed, the patterns of susceptibility to and the distributions of these infections can be expected to change substantially in the near future. Patterns of susceptibility and immunity to all of these viruses may now vary from place to place, from age group to age group, and in various socioeconomic settings, depending on the immunization program instituted by the health department and the activities of physicians and clinics rather than on the inherent epidemiologic characteristics of the natural infection and disease. The methods of immunization practice, the frequency of repeated immunization programs, and the quality and duration of vaccine immunity now and will increasingly constitute the major determinants of the behavior of these diseases.

Over the years, serological surveys in American cities, such as Syracuse,[102] Cleveland,[78] and Houston,[111] and also in undeveloped areas have uncovered serious deficiencies in the antibody patterns for viral diseases for which vaccines are available. The U.S. military has used this approach to similar advantage. A national serosurvey of 1547 Army recruits entering in 1989 at ages 15 to 24 years documented 15–21% seronegativity for measles, mumps, and rubella; 7% for varicella-zoster virus; and 2.3, 0.6, and 14.6% for poliovirus types 1, 2, and 3, respectively.[98] Likewise, a survey of 281 Air Force entrants demonstrated comparable overall seronegativity for measles and mumps but two- to eightfold differences between black and white recruits in the proportions lacking antibodies.[148] Serosurveillance has also been used effectively in the developing world for such purposes as a general assessment of success in the WHO Expanded Programme on Immunization[68,139] and a specific comparison of HIV-infected and uninfected Zairian infants for responses to polio vaccine.[137] The importance of surveillance programs and serological surveys to evaluate immunization programs has long been recognized.[53,93]

3.5.5. Agents in Search of Disease. The application of new immunologic and microbiological techniques has sometimes led to the discovery of a new agent, an antigen, or a new antibody before anything is known of the disease, if any, with which it is associated. The use of seroepidemiologic techniques to identify possible associations is one important research application of the method. The discovery of EBV, hepatitis B antigen, and human herpesviruses 6 and 7 are examples. The sequence in developing an association between a serological response and illness is outlined in Table 12, but the general principles apply to other biological indicators (e.g., cytotoxic T-lymphocyte response) of exposure to infectious agents. In the case of EBV, a fluorescent antibody method developed by Henle and Henle[87] in 1966 provided the key tool. The virus had been found in cultured cells derived from Burkitt's lymphoma, but it could not be isolated in the usual tissue culture systems. By application of serological techniques, the presence of antibody to the virus was found commonly not only in sera from Burkitt's lymphoma patients but also in sera from healthy African

and American children. The fortuitous development of infectious mononucleosis in the Henles' technician provided the clue to a disease association[88] which was confirmed by prospective studies carried out at Yale University.[65,119,138] The sequence of events is presented in detail in Chapter 10 and has been reviewed.[50] Through this approach, EBV was firmly established as the sole cause of heterophil-positive infectious mononucleosis.[51]

A second example of the use of seroepidemiologic techniques is the discovery by Blumberg *et al.*[12] in 1965 of a particular antigen in the serum of an Australian aborigine; it was uncovered in the course of genetic studies of β-lipoprotein. Since the agent from which the antigen was derived could not be isolated or cultivated in the laboratory, serological surveys using immunodiffusion tests were carried out to detect its presence in the sera of different population groups and different disease entities. The results provided the sole initial evidence that this "Australia antigen" was associated causally with hepatitis B or "long-incubation hepatitis."[13] Recently identified herpesviruses, HHV-6 and HHV-7, were not immediately recognized in association with previously defined disease entities, although the former has been unequivocally shown to be the major, if not the only, causal agent of exanthem subitum (see Chapter 15 and reference 163).

3.5.6. Biomarkers beyond Antigens and Antibodies. Revolutionary advances in immunology and molecular biology have opened the possibilities for intensive analysis of humoral and cellular material—what some are calling broadly "molecular epidemiology"—analysis that will clarify the role of viral infection in the pathogenesis of autoimmune, degenerative, neoplastic, and even "genetic" diseases and will facilitate targeted drug and vaccine development (see Chapters 1 and 2 and Section 3). Although large-scale application of such techniques to routine surveillance activities are still limited by cost and logistical complexity, as suggested earlier, what began as seroepidemiology is now flowering into the application of a broad array of powerful biotechnological strategies for the study of host, agent, and their interactions.

Table 12. Protocol for Serological Approach to Agents in Search of Disease

1. Develop serological test for mass screening
2. Test representative population groups
3. Test paired sera from candidate infectious diseases
4. If associated syndrome is identified:
   a. Test for frequency of association with the agent
   b. Exclude presence of agent in other infections
   c. Determine frequency in close contacts
5. Carry out prospective study of susceptible persons at high risk of infection for:
   a. Incidence of infection
   b. Clinical/subclinical ratio
   c. Determination of clinical spectrum
   d. Pattern of incidence related to antibody level

---

4. References

1. ANDERSON, R. M., AND MAY, R. M., Vaccination against rubella and measles: Quantitative investigations of different policies, *J. Hyg. (Camb.)* **90:**259–325 (1983).
2. ANDREWS, C. H., LAIDLAW, P. P., AND SMITH, W., Influenza: Observations on the recovery of virus from man on the antibody content of human sera, *Br. J. Pathol.* **16:**566–582 (1935).
3. ASHTON, L. J., LEARMONT, J., LUO, K., WYLIE, B., AND STEWART,

G., HIV infections in recipients of blood products from donors with known duration of infection, *Lancet* **344:**718–720 (1994).

4. AYCOCK, W. L., AND KRAMER, S. D., Immunity to poliomyelitis in normal individuals in urban and rural communities as indicated by the neutralization test, *J. Prev. Med.* **4:**189–200 (1930).

5. BAKER, H., Infectious disease information and preparedness, presented at meeting of the Society of Medical Consultants to the Armed Forces, Washington, DC, 1973.

6. BEASLEY, R. P., AND HWANG, L-Y., Overview of the epidemiology of hepatocellular carcinoma, in: *Viral Hepatitis and Liver Disease* (F. B. HOLLINGER, S. M. LEMON, AND H. S. MARGOLIS, EDS.), pp. 532–535, Williams and Wilkins, Baltimore, 1991.

7. BELL, J. A., HUEBNER, R. J., ROSEN, L., ROWE, W. P., COLE, R. M., MASTROT, F. M., FLOYD, T. M., CHANOCK, R. M., AND SHOEDOFF, R. A., Illness and microbial experiences of nursery children at Junior Village, *Am. J. Hyg.* **74:**267–292 (1961).

8. BERKELMAN, R. L., BRYAN, R. T., OSTERHOLM, M. T., LeDuc, J. W., AND HUGHES, J. M., Infectious disease surveillance: A crumbling infrastructure, *Science* **264:**368–370 (1994).

9. BIRKHEAD, G., CHORBA, T. L., ROOT, S., KLAUCKE, D. N., AND GIBBS, N. J., Timeliness of national reporting of communicable diseases: The experience of the National Electronic Telecommunications System for Surveillance, *Am. J. Public Health* **81:**1313–1315 (1991).

10. BLACK, R. E., BROWN, K. H., BECKER, S., ABDUL ALIM, A. R. M., AND HUQ, I., Longitudinal studies of infectious diseases and physical growth of children in rural Bangladeshi, *Am. J. Epidemiol.* **115:**315–324 (1982).

11. BLACK R. E., LOPEZ DE ROMANA, G., BROWN, K. H., BRAVO, N., BALAZAR, O. G., AND KANASHIRO, H. C., Incidence and etiology of infantile diarrhea and major routes of transmission in Huascar, Peru, *Am. J. Epidemiol.* **129:**785–799 (1989).

12. BLUMBERG, B. S., ALTER, H. J., AND VISNICH, S., A "new" antigen in leukemia sera, *J.A.M.A.* **191:**541–546 (1965).

13. BLUMBERG B. S., GERSTEY, B. J. S., HUNGERFORD, D. A., LONDON, W. T., AND SUTNIK, A. I., A serum antigen (Australia antigen) in Down's syndrome, leukemia, and hepatitis, *Ann. Intern. Med.* **66:**924–931 (1967).

14. BRACHMAN, P. S., Surveillance, in: *Bacterial Infections of Humans Epidemiology and Control*, 3rd ed. (P. S. BRACHMAN AND A. S. EVANS, EDS.). Plenum Press, New York, in press.

15. BROGAN, D., FLAGG E. W., DEMING, M., AND WALDMANN, R., Increasing the accuracy of the Expanded Immunization Programme's cluster survey design, *Ann. Epidemiol.* **4:**302–311 (1994).

16. BROOKMEYER, R., AND GAIL M. H. (EDS.), *AIDS Epidemiology: A Quantitative Approach*, Oxford University Press, New York, 1994.

17. BROWN, H. W., The occurrence of neutralizing antibodies for human influenza virus in the sera of persons with various histories of influenza, *Am. J. Hyg.* **24:**361–380 (1936).

18. BUEHLER, J. W., BERKELMAN, R. L., AND STEHR-GREEN, J. K., The completeness of AIDS surveillance, *J. Acquir. Immune Dis. Syndr.* **5:**257–264 (1992).

19. CANADA (HEALTH AND WELFARE CANADA), Establishing goals, techniques and priorities for rational communicable disease surveillance, *Can. Dis. Week. Rep.* **17:**79–84 (1991).

20. CENTERS FOR DISEASE CONTROL AND PREVENTION, Changing patterns of acquired immunodeficiency disease in hemophilia patients—United States, *Morbid. Mortal. Week. Rep.* **34:**241–243 (1985).

21. CENTERS FOR DISEASE CONTROL AND PREVENTION, *Comprehensive Plan for Epidemiologic Surveillance*, Atlanta, 1986

22. CENTERS FOR DISEASE CONTROL AND PREVENTION, Revision of the CDC surveillance case definition for acquired immunodeficiency syndrome, *Morbid. Mortal. Week. Rep.* **36**(1S)**:**3S–15S (1987).

23. CENTERS FOR DISEASE CONTROL AND PREVENTION, Guidelines for evaluating surveillance systems, *Morbid. Mortal. Week. Rep.* **37** (1988).

24. CENTERS FOR DISEASE CONTROL AND PREVENTION, Update: Acquired immunodeficiency syndrome in the San Francisco cohort study, 1978–1985. *Morb. Mortal. Week. Rep.* **34:**573–575 (1985).

25. CENTERS FOR DISEASE CONTROL AND PREVENTION, *Addressing Emerging Infectious Disease Threats: A Prevention Strategy for the United States*, US DHHS, Public Health Service, Atlanta, GA, 1994.

26. CENTERS FOR DISEASE CONTROL AND PREVENTION, Update: Changes in notifiable disease surveillance data—United States, 1992–1993, *Morbid. Mortal. Week. Rep.* **42:**824–836 (1993).

27. CENTERS FOR DISEASE CONTROL AND PREVENTION, Monthly immunization table, *Morbid. Mortal. Week. Rep.* **43:**226 (1994).

28. CENTERS FOR DISEASE CONTROL AND PREVENTION, Progress toward poliomyelitis eradication—Egypt, 1993, *Morbid. Mortal. Week. Rep.* **43:**223–226 (1994).

29. CENTERS FOR DISEASE CONTROL AND PREVENTION, Reported vaccine-preventable diseases—United States, 1993, and the Childhood Immunization Initiative, *Morbid. Mortal. Week. Rep.* **43:**57–60 (1994).

30. CENTERS FOR DISEASE CONTROL AND PREVENTION, Vaccination coverage of 2-year-old children—United States, 1992–93, *Morbid. Mortal. Week. Rep.* **43:**282–284 (1994).

31. CENTERS FOR DISEASE CONTROL AND PREVENTION, Availability of electronical MMWR on Internet, *Morbid. Mortal. Week. Rep.* **44:**48–50 (1995).

32. CHAMBERLAND, M. E., ALLEN, J. R., MONROE, J. M., GARCIA, N., MORGAN, C., REISS, R., STEPHENS, H., WALKER, J., AND FRIEDMAN, S. M., Acquired immunodeficiency syndrome in New York City: Evaluation of an active surveillance system. *J.A.M.A.* **254:**383–387 (1985).

33. CHOI, K., AND THACKER, S. B., An evaluation of influenza mortality surveillance, 1962–79. I. Time series forecasts of expected pneumonia and influenza deaths, *Am. J. Epidemiol.* **113:**215–226 (1981).

34. CHU, S. Y., BUEHLER, J. W., AND BERKELMAN, R. L., Impact of the human immunodeficiency virus epidemic on mortality in women of reproductive age, United States, *J.A.M.A.* **264:**225–229 (1990).

35. COOK, G. T., AND PAYNE, A. M., Epidemiological control of infectious diseases, *Br. Med. Bull.* **7:**185–187 (1951).

36. COONEY, M. K., HALL, C. E., AND FOX, J. P., The Seattle Virus Watch Program. I. Infection and illness experience of Virus Watch Families during a community-wide epidemic of echovirus 30 aseptic meningitis. *Am. J. Public Health* **60:**1456–1465 (1970).

37. CUTTS, F. T., WALDMAN, R. J., AND ZOFFMAN, H. M. D., Surveillance for the Expanded Programme on Immunization, *Bull. WHO* **71:**633–639 (1993).

38. DE-THÉ, G., GESSER, A., DAY, N., INKEL, P. M., WILLIAMS, E. H., BERI, D. P., SMITH, P. G., DEAN, A. G., BORNKAMM, G. W., FEORINO, P., AND HENLE, W., Epidemiological evidence for casual relationship between Epstein–Barr virus and Burkitt's lym-

phoma from Ugandan prospective study, *Nature* **274**:756–766 (1978).

39. DECLICH, S., AND CARTER, A. O., Public health surveillance: Historical origins, methods and evaluation, *Bull. WHO* **72**:285–304 (1994)

40. DESENCLOS, J-C., BIJKERK, H., AND HUISMAN, J., Variations in national infectious diseases surveillance in Europe, *Lancet* **341**:1003–1006 (1993).

41. DINGLE, J. H., BADGER, G. F., AND JORDAN, W. S., JR., *Illness in the Home: A Study of 25,000 Illnesses in Group of Cleveland Families*, Press of Western Reserve University, Cleveland, 1964.

42. DOBBINS, J. G., AND STEWART, J. A., Surveillance of congenital cytomegalovirus disease, 1990–91, *Morbid. Mortal. Week. Rep.* **41** (1992).

43. DOLK, H., AND LECHAT, M. F., Health surveillance in Europe: Lessons from EUROCAT and Chernobyl, *Int. J. Epidemiol.* **22**:363–368 (1993).

44. DOWDLE, W. R., Surveillance and control of infectious diseases: Progress toward the 1990 objectives, *Public Health Rep.* **98**:210–221 (1985).

45. DOWNS, A. M., ANCELLE-PARK, R., AND BRUNET, J. B., Surveillance of AIDS in the European Community: Recent trends and predictions to 1991, *AIDS* **4**:1117–1124 (1990).

46. DOWNS, A. M., ANCELLE-PARK, R., COSTAGLIOLA, D., RIGAUT, J. P., AND BRUNET, J. B., Transfusion associated AIDS cases in Europe: Estimation of the incubation period distribution and prediction of future cases, *J. Acquir. Immune Dis. Syndr.* **4**:805–813 (1991).

47. ESREY, S. S., FEACHEM, R. G., AND HUGHES, J. M., Intervention for the control of diarrheal diseases among young children: Improving water supplies and excreta disposal facilities, *Bull. WHO* **63**:757–772 (1985).

48. EVANS, A. S., Epidemiology and the public health laboratory, *Am. J. Public Health* **57**:1041–1052 (1967).

49. EVANS, A. S., Serological techniques, in: *Serological Epidemiology* (J. R. PAUL AND C. WHITE, EDS.), pp. 42–54, Academic Press, New York, 1973.

50. EVANS, A. S., The history of infectious mononucleosis, *Am. J. Med. Sci.* **267**:189–195 (1974).

51. EVANS, A. S., New discoveries in infectious mononucleosis, *Mod. Med.* **42**:18–24 (1974).

52. EVANS, A. S., Serologic studies of acute respiratory infections in military personnel, *Yale J. Biol. Med.* **48**:201–209 (1975).

53. EVANS, A. S., The need for serologic evaluation of immunization programs, *Am. J. Epidemiol.* **112**:725–731 (1980).

54. EVANS, A. S., Criteria for assessing accomplishment of poliomyelitis control, *J. Infect. Dis.* **6**(Suppl 2):S571–S576 (1984).

55. EVANS, A. S., Criteria for control of infectious diseases with poliomyelitis as an example, *Prog. Med. Virol.* **29**:141–165 (1984).

56. EVANS, A. S., The eradication of communicable disease: Myth or reality? *Am. J. Epidemiol.* **122**:199–207 (1985).

57. EVANS, A. S., Subclinical epidemiology. The first Harry A. Feldman Memorial Lecture, *Am. J. Epidemiol.* **125**:545–555 (1987).

58. EVANS, A. S., AND BRACHMAN, P. S., Emerging issues in infectious disease epidemiology. *J. Chron. Dis.* **39**:1105–1124 (1986).

59. EVANS, A. S., CARVALHO, R. P. S., FROST, P., JAMRA, M., AND POZZI, D. H. B., Epstein–Barr virus infection in Brazil. II. Hodgkin's disease, *J. Natl. Cancer Inst.* **61**:19–26 (1978).

60. EVANS, A. S., CASALS, J., OPTON, E. M., BORMAN, E. K., LEVINE, L., AND CUADRADO, R., A nationwide survey of Colombian military recruits, 1966. 1. Description of sample and antibody patterns with arboviruses, polioviruses, respiratory viruses, tetanus, and treponematosis, *Am. J. Epidemiol.* **90**:292–303 (1969).

61. EVANS, A. S., CASALS, J., OPTON, E. M., BORMAN, E. K., LEVINE, L., AND CUADRADO, R., A nationwide survey of Argentine military recruits, 1965–1966. I. Description of sample and antibody patterns with arboviruses, polioviruses, respiratory viruses, tetanus, and treponematosis, *Am. J. Epidemiol.* **93**:111–121 (1971).

62. EVANS, A. S., COOK, J. A., KAPIKIAN, A. Z., NANKERVIS, G., SMITH, A. L., AND WEST, B., Serological survey of St. Lucia, *Int. J. Epidemiol.* **8**:327–332 (1979).

63. EVANS, A. S., COX, F., NANKERVIS, G., OPTON, E., SHOPE, R., WELLS, A. V., AND WEST B., A health and seroepidemiological survey of a community in Barbados, *Int. J. Epidemiol.* **3**:167–175 (1974).

64. EVANS, A. S., AND MUELLER, N. E., The past is prologue: Use of serum banks in cancer research, *Cancer Res.* **52**(Suppl.):5557s–5560s (1992).

65. EVANS, A. S., NIEDERMAN, J. C., AND MCCOLLUM, R. W., Seroepidemiologic studies of infectious mononucleosis with EB virus, *N. Engl. J. Med.* **279**:1121–1127 (1968).

66. EVANS, A. S., NIEDERMAN, J. C., AND SAWYER, R. N., Prospective studies of a group of Yale University freshman. II. Occurence of acute respiratory infections and rubella, *J. Infect. Dis.* **123**:271–278 (1971).

67. EVANS, A. S., WELLS, A. V., RAMSEY, F., DRABKIN, P., AND PLAMER, K., Poliomyelitis, rubella, and dengue antibody survey in Barbados: A follow-up study, *Inst. J. Epidemiol.* **8**:235–241 (1979).

68. EXPANDED PROGRAMME ON IMMUNIZATION, Diphtheria and measles control, *World Health Stat. Quart.* **38**:65–75 (1985).

69. EXPANDED PROGRAMME ON IMMUNIZATION, *Training for Mid-Level Managers: The EPI Coverage Survey*, WHO/EPI/MLM/91.10.

70. EYLENBOSCH, W. J., AND NOAH, N. D. (EDS.), *Surveillance in Health and Disease*, Oxford University Press, Oxford, 1988.

71. FLOREY, C. DUV., CUADRADO, R. R., HENDERSON, J. R., AND DE GOES, P., A nationwide survey of Brazilian military recruits, 1964: Method and sampling results, *Am. J. Epidemiol.* **86**:314–318 (1967).

72. FOWLER, K. B., STAGNO, S., AND PASS, R. F., Maternal age and congenital cytomegalovirus infection: Screening of two diverse newborn populations 1980–90, *J. Infect. Dis.* **168**:552–556 (1993).

73. FOX, J., ELVEBACK, L. R., SPIGLAND, I., FROTHINGHAM, T. E., STEVENS, D. A., AND HUGER, M., The Virus Watch Program: A continuing surveillance of viral infections in metropolitan New York families. 1. Overall plan, methods of collecting and handling information and a summary report of specimens collected and illnesses observed, *Am. J. Epidemiol.* **83**:389–412 (1966).

74. FOY, A. M., COONEY, M. K., MCMAHAN, R., BOR, R., AND GRAYSTON, J. T., Single-dose monovalent A$_2$/Hong Kong influenza vaccine—Efficacy 14 months after immunization, *J.A.M.A.* **217**:1067–1071 (1971).

75. FRANCIS, T. F., AND MAGILL, T. P., The incidence of neutralizing antibody for human influenza virus in the serum of human individuals of different ages, *J. Exp. Med.* **63**:655–668 (1936).

76. GLEZEN, W. P., AND COUCH, R. B., Interpandemic influenza in the Houston area, 1974–1976, *N. Engl. J. Med.* **298**:587–592 (1978).

77. GLEZEN, W. P., FALCAO, O., CATE, T. R., AND MINTZ, A. A., Nosocomial influenza in a general hospital for indigent patients, *Can. J. Infect. Control* **6**:65–67 (1991).

78. GOLD, E., FEVRIER, A., HATCH, M. H., HERMANN, K. L., JONES, W. L., KRUGMAN, R. D., AND PARKMAN, P. D., Immune status of

urban children determined by antibody measurement, *N. Engl. J. Med.* **289:**231–234 (1973).

79. GRAITCER, P. L., AND BURTON, A. H. The epidemiology surveillance project: A computer-based system for disease surveillance, *Am. J. Prev. Med.* **3:**123–127 (1987)

80. GUTHE, T., RIDET, J., VORST, F., D'COSTA, J., AND GRAB, B., Methods for the surveillance of endemic treponematosis and seroimmunological investigations of "disappearing" disease, *Bull. WHO* **46:**1–14 (1972).

81. HABER, M., LONGINI, I. M., JR., AND HALLORAN, M. E., Estimation of vaccine efficacy in outbreaks of acute infectious diseases, *Stat. Med.* **10:**1773–1784 (1991).

82. HALLORAN, M. E., HABER, M., LONGINI, I. M., JR., AND STRUCHINER, C. J., Direct and indirect effects in vaccine efficacy and effectiveness, *Am. J. Epidemiol.* **133:**323–331 (1991).

83. HALPERIN, W., BAKER, E. L., JR., AND MONSON, R. R., *Public Health Surveillance*, Van Nostrand Reinhold, New York, 1992.

84. *Healthy People 2000, National Health Promotion and Disease Prevention Objectives*, Public Health Service, Department of Health and Human Services, Washington, DC, 1990.

85. HENDERSON, R. H., The Expanded Programme of the World Health Organization, *Rev. Infect. Dis.* **6**(Suppl. 2):S475–S479 (1984).

86. HENDERSON, R. H., AND SUNDARESAN, T., Cluster sampling to assess immunization coverage: A review of experience with a simplified sampling method, *Bull. WHO* **60:**253–260 (1982).

87. HENLE, G., AND HENLE, W., Immunofluorescence in cells derived from Burkitt's lymphoma, *J. Bacteriol.* **91:**1248–1256 (1966).

88. HENLE, G., HENLE, W., AND DIEHL, V., Relationship of Burkitt's tumor-associated herpes-type virus to infectious mononucleosis, *Proc. Natl. Acad. Sci. USA* **59:**94–101 (1968).

89. HESSOL, N. A., KOBLIN, B. A., VAN GRIENSVEN, G. J. P., BACHETTI, P., LIU, J. Y., STEVENS, C. E., COUTINHO, R. A., BUCHBINDER, S. P., AND KATZ, M. H., Progression of human immunodeficiency virus type 1 (HIV-1) infection among homosexual men in hepatitis B vaccine trial cohorts in Amsterdam, New York City, and San Francisco, 1978–1991, *Am. J. Epidemiol.* **139:**1077–1087 (1994).

90. HETHCOTE, H. W., Measles and rubella in the United States, *Am. J. Epidemiol.* **117:**2–13 (1983).

91. HOFF, R., BERARDI, V. P., WEIBLEN, B. J., MAHONEY-TROUT, L., MITCHELL, M. L., AND GRADY, G. F., Seroprevalence of human immunodeficiency virus among childbearing women: Estimation by testing samples of blood from newborns, *N. Engl. J. Med.* **318:**525–530 (1988).

92. HOOK, E. B., AND REGAL, R. R., The value of capture–recapture methods even for apparent exhaustive surveys, *Am. J. Epidemiol.* **135:**1060–1067 (1992).

93. HORSTMANN, D. M., Need for monitoring vaccinated populations for immunity levels, *Prog. Med. Virol.* **16:**215–240 (1973).

94. HORSTMANN, D. M., LIEBHABER, H., LEBOUVIER, G. L., ROSENBERG, D. A., AND HALSTEAD, S. G., Rubella: Reinfection of vaccinated and naturally immune persons exposed in an epidemic, *N. Engl. J. Med.* **283:**771–778 (1970).

95. HUBERT, B., AND DESENCLOS, J. C., Evaluation the exhaustiveness and representativeness of a surveillance system using the capture–recapture methods, *Rev. Epidemiol. Sante Publique* **41:**241–249 (1993).

96. INSTITUTE OF MEDICINE, *Emerging Infections: Microbial Threats to Health in the United States*, National Academy Press, Washington, DC, 1992.

97. KASLOW, R. A., OSTROW, D. G., DETELS, R., PHAIR, J. P., POLK,

B. F., AND RINALDO, C. R., JR., The Multicenter AIDS Cohort Study: Rationale, organization, and selected characteristics of the participants, *Am. J. Epidemiol.* **126:**310–318 (1987).

98. KELLEY, P. W., PETRUCELLI, B. P., STEHR-GREEN, P., ERICKSON, R. L., AND MASON, C. J., The susceptibility of young adult Americans to vaccine preventable diseases. National serosurvey of U.S. Army recruits, *J.A.M.A.* **266:**2724–2729 (1991).

99. KINGSLEY, L. A., ZHOU, S. Y. J., BACELLAR, H., RINALDO, C. R., JR., CHMIEL, J., DETELS, R., SAAH, A., VANRADEN, M., HO, M., AND MUNOZ, A. Temporal trends in human immunodeficiency virus type 1 seroconversion 1984–1989. A report from the Multicenter AIDS Cohort Study (MACS), *Am. J. Epidemiol.* **134:**331–339 (1991).

100. KOPEC-SCHADER, E., TINDALL, B., LEARMONT, J., WYLIE, B., AND KALDOR, J. M., Development of AIDS in people with transfusion-acquired HIV infection, *AIDS* **7:**1009–1013 (1993).

101. KRAUSE, R. M., The restless tide: The persistent challenge of the microbial world, National Foundation for Infectious Diseases, Washington, DC, 1981.

102. LAMB, G. A., AND FELDMAN, H. A., Rubella vaccine response and other viral antibodies in Syracuse children, *Am. J. Dis. Child.* **122:**117–121 (1971).

103. LANGMUIR, A. D., The surveillance of communicable diseases of national importance, *N. Engl. J. Med.* **268:**182–192 (1963).

104. LEMESHOW, S., AND ROBINSON, D., Surveys to measure program coverage and impact: A review of the methodology used by the Expanded Programme on Immunization, *World Health Stat. Q.* **63:**225–267 (1988).

105. LONGINI, I. M., JR., FINE, P. E. M., AND THACKER, S. B., Predicting the global spread of new infectious agents, *Am. J. Epidemiol.* **123:**383–391 (1986).

106. MARIER, R., The reporting of communicable diseases, *Am. J. Epidemiol.* **105:**587–590 (1977).

107. MASTRO, T., KITAYAPORN, D., WENIGER, B. G., VANICHSENI, S., LAOSUNTHORN, V., UNEKLABH, T., UNEKLABH, C., CHOOPANYA, K., AND LIMPAKANNJANARAT, K., Estimating the number of HIV-infected injection drug users in Bangkok: A capture–recapture method, *Am. J. Public Health* **84:**1094–1099 (1994).

108. MAYNARD, A. M., FREDERICKSON, D., GARATTINI, S., MAKELA, H., MATTHEIS, R., AND PAPADIMITRIOU, M., Evaluation of the fourth Medical and Health Research Programme (1987–1991) (EUR 13001), Commission of the European Communities, Luxembourg, 1990.

109. MCDONALD, A. M., GERTIG, D. M., CROFTS, N., AND KALDOR, J. M., for the National HIV Surveillance Committee, A national surveillance system for newly acquired HIV infection in Australia, *Am. J. Public Health* **84:**1923–1928 (1994).

110. MCQUILLAN, G. M., TOWNSEND, T. R., FIELDS, H. A., CARROLL, M., LEAHY, M., AND POLK, B. F., Seroepidemiology of hepatitis B virus infection in the United States, *Am. J. Med.* **87**(Suppl. 3A):5S–8S (1989).

111. MELNICK, J. L., BURKHARDT, M., TABER, L. T., AND ERCKMAN, P. N., Developing gap in immunity to poliomyelitis in an urban area, *J.A.M.A.* **209:**1181–1185 (1969).

112. MILLAR, J. D., Theoretical and practical problems in measles control, *CDC Smallpox Eradication Program Report* **4:**165–176 (1970).

113. MONTO, A. S., AND ULLMAN, B. M., Acute respiratory illnesses in an American community, *J.A.M.A.* **227:**164–169 (1974).

114. MUELLER, N., Epidemiologic studies assessing the role of Epstein–Barr virus in Hodgkin's disease, *Yale J. Biol. Med.* **60:**321–327 (1987).

115. NAHDO (NATIONAL ASSOCIATION OF HEALTH DATA ORGANIZATIONS). Report to Congress: The Feasibility of Linking Research-Related Data Bases to Federal Medical Administrative Data Bases. Rockville, MD, Agency for Health Care Policy and Research, Public Health Service, AHCPR, Public Health Service, AHCPR Pub. No. 91–0003, April 1991.

116. NATIONAL ACADEMY OF SCIENCES, Rapid epidemiological assessment and evaluation. Conclusions from March 15, 1982 meeting. Memorandum dated March 28, 1982, National Academy of Sciences, Washington, DC.

117. *National Disease and Therapeutic Index*, Lea Associates, Ambler, PA, 1969.

118. NICOLOSI, A. (ED.), *HIV Epidemiology: Models and Methods*, Raven Press, New York, 1994.

119. NIEDERMAN, J. C., McCOLLUM, R. W., HENLE, G., AND HENLE, W., Infectious mononucleosis in relation to EB virus antibodies, *J.A.M.A.* **203:**205–209 (1968).

120. NOVICK, L. F., BERNS, D., STRICOF, R., STEVENS, R., PASS, K., AND WETHERS, J., HIV seroprevalence in newborns in New York State, *J.A.M.A.* **261:**1745–1750 (1989).

121. ONORATO, I. M., GWINN, M., DONDERO, T. J., JR., Applications of data from the CDC Family of Surveys, *Public Health Rep.* **109:**204–211 (1994).

122. ORENSTEIN, W. A., AND BERNIER, R. H., Surveillance in the control of vaccine-preventable diseases, in: *Public Health Surveillance*, (W. HALPERIN, E. L. BAKER, JR., AND R. R. MONSON, EDS.), Van Nostrand Reinhold, New York, 1992.

123. *Oxford Universal Dictionary*, Clarendon Press, Oxford, 1955.

124. PAUL, J. R., The story to be learned from blood samples: Its value to the epidemiologist, *J.A.M.A.* **175:**601–605 (1961).

125. PAUL, J. R., Aims, purposes and method of the World Health Organization Serum Banks, *Yale J. Biol. Med.* **36:**2–4 (1963).

126. PAUL, J. R., *Clinical Epidemiology*, rev. ed. University of Chicago Press, Chicago, 1966.

127. PAUL, J. R., NIEDERMAN, J. C., PEARSON, R. J. C., AND FLOREY, C. DU V., A nationwide serum survey of United States military recruits. 1. General considerations, *Am. J. Hyg.* **80:**286–292 (1964).

128. PAUL, J. R., AND RIORDAN, J. T., Observations on serological epidemiology. Antibodies to the Lansing strain of poliomyelitis virus in sera from Alaskan Eskimos, *Am. J. Hyg.* **52:**202–212 (1950).

129. PAUL, J. R., AND TRASK, J. E., Neutralization test in poliomyelitis, comparative results with four strains of the virus, *J. Exp. Med.* **61:**447–464 (1935).

130. PAUL, J. R., AND WHITE, C. (EDS.), *Serological Epidemiology*, Academic Press, New York, 1973.

131. RAŠKA, K., National and international surveillance of communicable diseases, *WHO Chron.* **20:**313–321 (1966).

132. RAŠKA, K., Epidemiological surveillance in the control of infectious diseases, *Rev. Infect. Dis.* **6:**1112–1117 (1983).

133. REIDEL, D. A., EVANS, A. S., SAXINGER, C., AND BLATTNER, W. A., A historical study of human T lymphotropic virus type I transmission in Barbados, *J. Infect. Dis.* **159:**603–609 (1989).

134. RUBIN, R. J., AND GREGG, M. B., A national influenza surveillance system: Methods and results, 1972–1974, *Am. J. Epidemiol.* **100:**516–517 (1974).

135. RUSHWORTH, R. L., BELL, S. M., RUBIN, G. L., HUNTER, R. M., AND FERSON, M. J., Improving surveillance of infectious diseases in New South Wales, *Med. J. Aust.* **154:**828–831 (1991).

136. RYACHEV, L. A., AND LONGINI, I. M., JR., A mathematical model for the global spread of influenza, *Math. Biosci.* **75:**3–22 (1985).

137. RYDER, R. W., OXTOBY, M. J., MVULA, M., BATTWR, V., BAENDE, E., NSA, W., DAVACHI, F., HASSIG, S., ONORATO, I., DEFOREST, A., KASHAMUKA, M., AND HAYWARD, W. L., Safety and immunogenicity of bacille Calmette–Guérin, diphtheria–tetanus–pertussis, and oral polio vaccines in newborn children in Zaire infected with human immunodeficiency virus type 1, *J. Pediatr.* **122:**697–702 (1993).

138. SAWYER, R. N., EVANS, A. S., NIEDERMAN, J. C., AND McCOLLUM, R. W., Prospective studies of a group of Yale University freshman. 1. Occurrence of infectious mononucleosis, *J. Infect. Dis.* **123:**263–270 (1971).

139. SEJDA, J., Control of measles in Czechoslovakia (CSSR), *Rev. Infect. Dis.* **5:**564–567 (1983).

140. SEVER, J. L., Applications of a microtechnique to viral serologic investigations, *J. Immunol.* **88:**320–329 (1962).

141. SHAPIRO, C. N., COLEMAN, P. J., McQUILLAN, G. M., ALTER, M. J., AND MARGOLIS, H. S., Epidemiology of hepatitis A: Seroepidemiology and risk groups in the USA, *Vaccine* **10**(Suppl. 1):S59–S62 (1992).

142. SHOPE, R. E., Swine influenza. 1. Experimental transmission and pathology, *J. Exp. Med.* **54:**349–359 (1931).

143. SMITH, E., RIX, B. A., MELBYE, M., Mandatory anonymous HIV surveillance in Denmark: The first results of a new system, *Am. J. Public Health* **84:**1929–1932 (1994).

144. SMITH, W., ANDREWS, C. H., AND LAIDLAW, P. O., A virus obtained from influenza patients, *Lancet* **2:**66–68 (1933).

145. SOPER, F. L., PENNA, H., CARDOSA, E., SERAFIM, J., JR., FROSBISHER, M., JR., AND PINHIERO, J., Yellow fever without *Aedes aegypti*: Study of a rural epidemic in the Valle do Chanaan, Espirito Santo, Brazil, *Am. J. Hyg.* **18:**555–587 (1932).

146. STEVENS, C. E., TAYLOR, P. E., AND ZANG, E. A., Human T-cell lymphotropic virus infection in a cohort of homosexual men in New York City, *J.A.M.A.* **255:**2167–2172 (1986).

147. STROUP, D. F., WHARTON, M., KAFADAR, K., AND DEAN, A. G., An evaluation of a method for detecting aberrations in public health surveillance data, *Am. J. Epidemiol.* **137:**373–380 (1993).

148. STRUEWING, J. P., HYAMS, K. C., TUELLER, J. E., AND GRAY, G. C., The risk of measles, mumps, and varicella among young adults: A serosurvey of U.S. Navy and Marine recruits, *Am. J. Public Health* **83:**1717–1720 (1993).

149. TASK FORCE FOR CHILD SURVIVAL AND DEVELOPMENT, Status Report on Polio Eradication, *Child Survival World Dev.* **10:**15 (1994).

150. THACKER, S. B., CHOI, K., AND BRACHMAN, P. S., The surveillance of infectious disease, *J.A.M.A.* **249:**1181–1185 (1983).

151. THACKER, S. B., AND BERKELMAN, R. L., Public health surveillance in the United States, *Epidemiol. Rev.* **10:**164–190 (1988).

152. THOMPSON, W. H., AND EVANS, A. S., California virus encephalitis studies in Wisconsin, *Am. J. Epidemiol.* **81:**230–244 (1965).

153. VALLERON, A.-J., BOUVET, E., GARNERIN, P., MENARES, J., HEARD, I., LETRAIT, S., AND LEFAUCHEUX, J., A computer network for surveillance of communicable diseases: The French experiment, *Am. J. Public Health* **76:**1289–1292 (1986).

154. WATKINS, M., LAPHAM, S., AND HOY, W., Use of a medical center's computerized health care database for notifiable disease surveillance, *Am. J. Pub. Health* **81:**637–639 (1991).

155. WEINSTEIN, P., WORSWICK, D., MACINTYRE, A., AND CAMERON, S., Human sentinels for arbovirus surveillance and regional risk classification in South Australia, *Med. J. Aust.* **160:**494–499 (1994).

156. WHARTON, M., CHORBA, T. L., VOGT, R. L., MORSE, D. L., AND

BUEHLER, J. W., Case definitions for public health surveillance, *Morbid. Mortal. Week. Rep.* **39**(RR13) (1990).

157. WIDELOCK, D., KLEIN, S., PEIZER, L. R., AND SIMONOVIC, O., Laboratory analyses of 1957–1958 influenza outbreak (A/Japan) in New York City. I. Preliminary report on seroepidemiologic investigation and variant A/Japan isolate, *J.A.M.A.* **167**:541–543 (1958).

158. WILLIAMS, J. W., The value of the Wassermann reaction in obstetrics, based upon the study of 4547 consecutive cases, *Johns Hopkins Hosp. Bull.* **31**:335–342 (1920).

159. WORLD HEALTH ORGANIZATION, Immunological and hematological surveys, *WHO Tech. Rep. Ser.* **181**:1–95 (1959).

160. WORLD HEALTH ORGANIZATION, *Proceedings of the Twenty-first World Health Assembly*, WHO, Geneva, 1968.

161. WORLD HEALTH ORGANIZATION, Multipurpose serological surveys and WHO Serum Reference Banks, *WHO Tech. Rep. Ser.* **454**:1–95 (1970).

162. WORLD HEALTH ORGANIZATION, Improving routine systems for surveillance of infectious diseases including EPI target diseases: Guidelines for national programme managers, unpublished document WHO/EPI/TRAM/93.1, 1993.

163. YAMANISHI, K., OKUNO, T., SHIRAKI, K., TAKAHASHI, M., KONDO, T., ASANO, Y., AND KURATA, T., Identification of a human herpesvirus-6 as a casual agent for exanthem subitum, *Lancet* **1**:1065–1067 (1988).

164. YOLKEN, R. H., Enzyme immunoassays for the detection of infectious agents in body fluids: Current limitations and future prospects, *Rev. Infect. Dis.* **4**:35–67 (1982).

165. YORKE, J. A., NATHANSON, N., PIANIGIANI, G., AND MARTIN, J., Seasonality and requirements for perpetuation and eradication of viruses in populations, *Am. J. Epidemiol.* **109**:103–123 (1979).

5. Suggested Reading

CENTERS FOR DISEASE CONTROL AND PREVENTION, *Addressing Emerging Infectious Disease Threats: A Prevention Strategy for the United States*, U.S. DHHS, Public Health Service, Atlanta, GA, 1994.

CUTTS, F. T., WALDMAN, R. J., AND ZOFFMAN, H. M. D., Surveillance for the Expanded Programme on Immunization, *Bull. WHO* **71**:633–639 (1993).

DECLICH, S., AND CARTER, A. O., Public health surveillance: Historical origins, methods and evaluation, *Bull. WHO* **72**:285–304 (1994).

DESENCLOS, J.-C., BIJKERK, H., AND HUISMAN, J., Variations in national infectious diseases surveillance in Europe, *Lancet* **341**:1003–1006 (1993).

EYLENBOSCH, W. J., AND NOAH, N. D. (EDS.), *Surveillance in Health and Disease*, Oxford University Press, Oxford, 1988.

HALPERIN, W., BAKER, E. L., JR., AND MONSON, R. R., *Public Health*, Van Nostrand Reinhold, New York, 1992.

INSTITUTE OF MEDICINE, *Emerging Infections: Microbial Threats to Health in the United States*, National Academy Press, Washington, DC, 1992.

SCHULTE, P. A., AND PERERA, F. P. (EDS.), *Molecular Epidemiology: Principles and Practice*, Academic Press, New York, 1993.

THACKER, S. B., AND BERKELMAN, R. L., Public health surveillance in the United States, *Epidemiol. Rev.* **10**:164–190 (1988).

PART II

Acute Viral Infections

CHAPTER 4

Adenoviruses

Hjordis M. Foy

1. Introduction

Adenoviruses (Ads) derived their name from the fact that they were first isolated from adenoid tissues (tonsils) and have a certain affinity for lymph glands, where they may remain latent for years. There are six subgenera with 47 species (previously serotypes) (Table 1). Symptoms of infections and epidemiologic patterns vary between subgenera and even for different species. Adenoviruses invade primarily the respiratory and gastrointestinal tracts and the conjunctiva. They may cause a variety of clinical manifestations ranging from pharyngitis to bronchitis, croup, pneumonia, and diarrhea. Some species invade the urinary tract or central nervous system or cause systemic infections. Adenovirus infections are widely distributed and common. Most infections occur in childhood.

A recognized syndrome seen especially among military recruits is febrile acute respiratory disease (ARD). Other disease syndromes caused by certain specific species of adenovirus are pharyngoconjunctival fever (PCF) and epidemic keratoconjunctivitis (EKC). In the acquired immunodeficiency syndrome (AIDS), systemic infection with adenoviruses, some of them newly discovered, is common.[95a] Several other disease syndromes that usually occur in children (hemorrhagic cystitis, pertussislike disease, skin rashes, gastroenteritis, intussusception) have been associated with adenoviruses.

The viruses are unusually resistant to inactivation and degradation, allowing purification and molecular investigation. As much detailed structure and biochemistry of adenovirus are known as for any microorganism.[81,131,166] Although many problems remain concerning production

and use of adenovirus vaccines, both inactivated and live vaccines have been developed and found effective. Subunit and genetically engineered vaccines are possible.

2. Historical Background

In 1953, a few years after cell culture became practical for isolation of viruses, Rowe et al.[136] described an agent that caused spontaneous degeneration of tissue culture originating from surgically removed human tonsils and adenoids. In 1954, Hilleman and Werner[78] reported isolation of similar agents from military personnel ill with febrile respiratory disease. Epidemics of this had been a serious problem in recruit training, and the Commission on Acute Respiratory Diseases of the U.S. Armed Forces Epidemiological Board initiated a series of studies revealing that the newly discovered agents were the cause of a large proportion of febrile acute respiratory disease among military recruit populations.[25,37] Several names were used for the new viruses but in 1956 the term *adenoviruses* was adopted.[43]

Although the laboratory techniques for diagnosing adenovirus infections were not available until the 1950s, human adenoviruses probably have been present for a long time. Most illnesses caused by adenoviruses cannot be diagnosed by clinical observation alone. However, epidemics of acute respiratory diseases were recognized as causing disruption in military-recruit training as far back as the Civil War. During World War II, the Commission on Acute Respiratory Diseases defined the entity termed "ARD" through epidemiologic and human volunteer investigations.[37] This syndrome was an entity often distinguishable from other definable acute respiratory tract diseases, had an incubation period of 5–6 days, and was caused by a filterable agent. The finding that this syndrome was caused by adenovirus suggests that this

Hjordis M. Foy • Department of Epidemiology, School of Public Health and Community Medicine, University of Washington, Seattle, Washington 98195.

Table 1. Properties of Human Adenovirus Subgenera A to F[a]

Subgenus	Species	DNA			Hemagglutination pattern[c]	Oncogenicity in newborn hamsters	Most common source of isolations and/or disease syndromes
		Homology (%)[b]	G + C (%)	Number of SmaI fragments			
A	12,18,31	48–69 (8–20)	48	4–5	IV	High (tumors in most animals in 4 months)	Stool specimens, diarrhea
B	3,7,11,14,16, 21,34,35	89–94 (9–20)	51	8–10	I	Weak (tumors in few animals in 4–18 months)	URI,[d] LRI,[e] ARD, cystitis (for types 11 and 21)
C	1,2,5,6	99–100 (10–16)	58	10–12	III	Nil	URI, LRI, endemic viruses
D	8–10,13,15,17, 19,20,22– 30,32,33,36– 39, 42–47	94–99 (4–17)	58	14–18	II	Nil	EKC, conjunctivitis, asymptomatic rectal carriage
E	4	(4–23)	58	16–19	III	Nil	ARD, PCF, pneumonia
F	40,41	62	Not done	9–12	IV	Nil	Gastroenteritis

[a]Adapted with permission from Wadell.[166]
[b]Percent homology within the subgenus. Figures in parenthesis: homology with members of other subgenera.
[c]I, Complete agglutination of monkey erythrocytes; II, complete agglutination of rat erythrocytes; III, partial agglutination of rat erythrocytes (fewer receptors); IV, agglutination of rat erythrocytes discernible only after addition of heterotypic antisera.
[d]URI, upper respiratory infections.
[e]LRI, lower respiratory infections.

agent was responsible for the similar disease at times of previous military mobilizations.

There is also evidence that other adenovirus infections were present long before the 1950s. The disease syndrome now classified as EKC was clinically described by German workers at the end of the 19th century.[80,86] Several epidemics of conjunctivitis with fever, pharyngitis, and systemic symptoms centering around swimming baths were reported by German workers in the 1920s,[126] a syndrome that fits the description of PCF (see Section 5.2.3). Adenoviruses, particularly subgenera B and D, are heterogeneous and subject to evolution, and new species and variants of established species appear from time to time.[74–76,89,92,105,106,122]

3. Methodology Involved in Epidemiologic Analysis

3.1. Sources of Data

Because the disease syndromes caused by adenoviruses are not easily distinguished purely on clinical criteria in the absence of epidemics, general epidemiologic studies require virus detection and serological evidence. Adenovirus disease is not reportable, and there are no national or other general-population mortality and mor-

bidity data. Many epidemiologic data come from selected representative population samples under appropriate surveillance. Important studies have included ARD in military populations (especially recruits),[5,25,37,38,51,104,112,162] children's institutions and day-care facilities,[10,22,42,71,110,150,164] groups of families,[27,52–54,66,79,90,133,168,173] and even an entire city.[114] Slow spread of Ad 21 has been observed in an isolated arctic station.[144] The World Health Organization (WHO) has collected data on isolation of adenoviruses on a global scale since the 1960s.[6,140] Such virus isolates are almost all disease associated, in contrast to those originating from studies in basically healthy families.

Adenovirus infections, especially with species Ad 4 and Ad 7, were first extensively studied in the military. Not only were illnesses diagnosed with the appropriate laboratory tests, but also serological specimens collected at the beginning and end of training provided data on susceptibility and infection rates (whether symptomatic or asymptomatic). The epidemiologic pattern of the disease in the four military services, including the species involved under various training conditions and at different geographic locations, including international differences, has been studied.[5,25,37,47,48,51,155,157,162]

The epidemiologic pattern of adenovirus infection in civilian populations is unlike that of the military.[60] Adenovirus disease was uncommon among college studies.[46]

On the other hand, longitudinal studies in orphanages and children's homes revealed high rates of infection with several low-numbered serotypes in infancy.[10,79,164] In such settings, certain viruses (Ads 1, 2, and 5) were often found to be endemic, whereas Ads 3 and 7 occurred in epidemics. Serological studies suggest that Ads belonging to subgenus A are also common in children.[166]

When asymptomatic children were studied as controls of clinic or hospitalized patients, adenoviruses were frequently isolated. It was quickly recognized that isolation of the virus from a sick patient did not establish a causal relationship. Much of our understanding of the epidemiology of adenovirus infection in children comes from continuous observations of panels of families conducted in Cleveland,[90] Kansas,[173] New Orleans,[66] New York,[52] Seattle,[27,53] and also India.[79,98] In the first studies, respiratory and serological specimens were utilized. The broadest approach in family studies is exemplified by the Virus Watch studies, conducted in New York and Seattle, in which biweekly collections of both fecal and respiratory specimens for virus isolation were included in the routine observations. The latter type of studies uncovered a large number of asymptomatic infections, particularly with the lower-numbered serotypes, and demonstrated the high frequency of recrudescent shedding of virus.

3.2. Interpretation of Laboratory Tests

Virus isolation with serotyping and antibody tests has been used to provide epidemiologic information on adenovirus infection.[81] Isolation provides the most conclusive evidence of infection and offers the opportunity to determine the species. Recovery of the virus is not difficult for the lower-numbered species in the acute phase of illness in the majority of cases with upper respiratory tract and eye infections. Both swab specimens and nasal washings are suitable for isolation from the respiratory tract. Most adenovirus species are isolated easily in cell cultures of human origin, especially HEK, Hep-2, and HeLa cells.[27,56] Monkey kidney cells are less suitable. Centrifugation into shell vials may lead to more rapid detection.[44] Enteric adenoviruses (EAds), previously recognized by electron microscopy only, can now be isolated in a HEK cell line transformed by adenovirus type 5 (229 cells) and in Chang conjunctival cells.[18,174,178] Stool suspensions give better yield than rectal swabs. The optimal temperature for virus isolation is 37°C.[56,61,175] Typical cytopathic effect (CPE)—rounding of cells and clumping in grapelike clusters—may take 1–2 weeks, and sometimes cultures must be observed as long as 4 weeks before

CPE is seen. Blind passage is recommended to recover the higher-numbered serotypes (above 7) and when the virus concentration in the specimen is low.[45,61] EAds can be detected more rapidly by latex, enzyme-linked immunosorbent assays (ELISA), monoclonal antibody tests, and polymerase chain reaction (PCR).[3,59]

Use of group-specific (antihexon) antigen identifies the isolate as a member of the adenovirus group.[81] Serotyping (speciation) has traditionally been performed in a neutralization test using species-specific rabbit antisera. DNA restriction analysis has been used for genome typing and to determine subgenus and sometimes species.[2,49,166,167] The latter type of analysis covers the whole genome, whereas serotyping reflects only that part of the genome that codes for hexons and fibers.[2] With the newer methods, species that appear intermediate between the established serotypes and even subgenera have been discovered.[2,36,73,82] Immunofluorescent antibody tests will detect group antigen and may be useful as rapid diagnostic tests, but they are not specific enough for epidemiologic studies. Newer methods for detection include ELISA, radio and enzyme immunoassays, and DNA probes, including dot–blot hybridization.[18,83,104,138,165] These methods also are rapid but do not allow speciation. Electron microscopy (EM), including immune electron microscopy (IEM), has been useful for viruses not easily grown in cell culture, such as the EAds.[18,68,174,178] Monoclonal antibodies can be used for both detection and typing.[34]

Adenoviruses are often present for a prolonged period in the gastrointestinal tract and may be recovered from fecal or rectal swab specimens after virus has disappeared from the upper respiratory tract. Fecal excretion, especially for subgenus C, may continue intermittently for months or even years after acquisition,[52,79] and the diagnostic significance of recovery from this site in illness is often in doubt. Isolation from the respiratory tract has a greater probability of association with illness (carriage rate can be as high as 7% in healthy children).[17,52,53] The virus usually disappears from the eye and pharynx as acute symptoms abate, but recrudescent infection may also occur in the throat. A significant rise in antibody titer indicates current infection or possibly reactivation.

The commonly applied serological tests in adenovirus epidemiology are the complement-fixation (CF), neutralization (N), and hemagglutination-inhibition (HI) antibody tests and ELISA tests. The CF test measures group-specific antibody. It is the easiest to perform and the most useful in diagnosis of acute infections, but it is not species specific. The production of CF antibodies in infants may be poor, and the test can be erroneously negative. In older persons, CF antibodies from previous infections can cause difficulty in demonstrating a new infection. Neutralization and HI tests are more laborious to carry out, but they are species specific and the antibody is long-lasting. ELISA methodology is generally easier to carry out and is species specific.[160]

The HI and N tests have been useful for serological epidemiology. A variety of such studies have been conducted, comparing age of acquisition of antibody by species in different population groups including international comparisons.[154] Restriction endonuclease analysis is being used for discrimination between isolates and as an epidemiologic tool.[89,92,123,166,167] Caution must be employed in comparing different studies, both of isolation and of antibody titers, because of variation in laboratory techniques. Isolation of adenovirus or significant antibody measurements by themselves may only indicate reactivation of the virus.[52,55] Molecular epidemiologic methods using restriction endonuclease analysis can be used to investigate relationships between different isolates.[89a]

4. Characteristics of the Virus

Human adenoviruses belong to the genus *Mastadenovirus*, family *Adenoviridae*. They are double-stranded DNA viruses lacking an envelope.[81,131] The capsid contains 252 capsomeres and shows icosahedral symmetry. The capsomeres consist of 240 hexons and 12 pentons with a projecting fiber on each penton. Hexons are group-specific antigens common to all mastadenoviruses and induce primarily group-specific CF antibodies; the fibers are largely responsible for type-specific antibodies. The pentons have mixed functions and are especially active in hemagglutination.[131] The size of the virus is 60–90 nm. Forty-seven immunologically distinct species are recognized to cause human infection. The last recognized species, Ad 42–47, have been isolated from fecal material of AIDS patients.[74,75] Mutation and recombination results in appearance of a new species.

Adenoviruses are recognized by a common group antigen, consisting primarily of hexons, in a CF test. Adenoviruses have been classified according to hemagglutination of rhesus and rat erythrocytes into groups I through IV and according to oncogenicity into groups A through E. However, hemagglutination and oncogenicity in hamsters reflect only a part of the genome. Perhaps more promising is grouping according to nucleic acid sequences or DNA genome homologies, which may be more relevant for pathogenesis.[2,63,166] Intermediate strains and genetic recombination have drawn attention in the 1980s.[2,63,73,82,123] Hybridization of viruses appears to

occur *in vivo*, and the recently discovered serotypes 34 and 35 may be recombinant with type 7.[36] Table 1 shows the classification of the 41 species into subgenera A through F on the basis of properties of DNA, hemagglutination pattern, and oncogenicity.[166]

Previously only subgroup A was considered oncogenic in animals, but the discovery that Ad 9 induces rat mammary tumors, which have many characteristics similar to human breast cancer, is a cause for new concern regarding dormant adenovirus infection.[85] Interaction, promotion, or suppression with other oncogenic viruses is also possible. Viral hybridization and cell transformation with adenovirus have been observed.[131] Analysis of human cancers for DNA sequences of human adenovirus have so far been negative.[176] Because of the latency of the infections, late sequelae are at least a theoretical possibility.

By use of restriction endonucleases, it is now clear that many variants of the same species occur, which may explain why a species in one locality causes symptoms slightly different from the same species in another geographic area, a circumstance that previously seemed puzzling. For instance, adenovirus type 7 has been divided into seven distinct genome types, with different distributions around the globe.[167] It is also plausible that new viruses arise by recombination, and in the case of immunosuppressed patients, reactivation of dormant viruses may occur in this process.[36,72,75,82,105,106,122,171]

Adenoviruses can also hybridize with simian virus 40 (SV40), a simian papovavirus known for its oncogenic potential.[131]

Adenoviruses are unusually stable to physical and chemical agents and adverse pH conditions, resulting in prolonged survival outside the host cells and great potential for spread. They are ether resistant but are destroyed by heat at 56°C for 30 min. Type 4 is especially heat resistant. The viruses endure a pH range of 5.0–9.0 and temperatures ranging between 4 and 36°C. They survive freezing with minimal loss of infectivity.

Adenovirus-associated viruses (AAVs) belong to parvoviruses (dependoviruses). They are small (18–28 nm) defective viruses, contain single-stranded DNA, and are found in association with adenoviruses, which act as "helpers" in replication. The AAVs were originally discovered as contaminants of adenovirus cultures by special staining techniques in electron microscopy. Now their presence is more easily indicated by serological techniques. These viruses do not cause CPE. They are resistant to heating, ether, and chloroform and can be stored at 1–4°C for many months. They are antigenically distinct from the adenoviruses. At least four immunotypes are known. So far, no disease syndrome has been associated with these viruses. Replication of AAVs inhibits adenovirus replication and oncogenicity. They may in fact have a protective function for the host. They can be useful as epidemiologic markers, since in an outbreak of PCF in Seattle, almost all those infected with adenovirus type 3 also had evidence of infection with AAV types 3.[139]

5. Descriptive Epidemiology

The epidemiologic characteristics and probably also tissue tropism vary among the subgenera.[166] Certain characteristics are common to them all; they are transmitted by direct contact, by the fecal–oral route, or sometimes by water. Adenoviruses 1, 2, 5, and 6 (subgenus C) are usually acquired early in childhood and are endemic; most of the other viruses occur in epidemics or sporadically. A large proportion of infections are asymptomatic, particularly for subgenera A and D, whereas those belonging to B and E tend to cause symptomatic respiratory disease. Enteric Ads 40 and 41, subgenus F, cause gastroenteritis, particularly in children.

The following synopsis outlines the major epidemiologic features of adenovirus infections. More detailed epidemiologic characteristics are presented under the various clinical syndromes.

5.1. Synopsis of Descriptive Epidemiology

5.1.1. Incidence and Prevalence. The incidence of infection is highest in children under age 2 for the common adenoviruses (Ads 1, 2, 5, and 6). A carrier state (prevalence) may result from such infant infection. For details see Section 5.2.

5.1.2. Epidemic Behavior. Adenoviruses 1, 2, 5, and 6 are endemic in most areas of the world; Ads 4, 7, 14, and 21, which cause ARD (see Section 5.2.2), occur mostly in epidemics; Ad 3 is sometimes endemic and sometimes epidemic (see Section 5.2.3). Adenovirus 8 is endemic in the Far East but occurs only in epidemics (and is usually iatrogenic) in Western countries (see Section 5.2.4).

5.1.3. Geographic Distribution. Most types have been recovered from all areas of the world where they have been sought. Some of the higher-numbered types (8 and above) have been reported more frequently in underdeveloped countries such as Saudi Arabia and India and in Africa.

5.1.4. Temporal Distribution. The incidence of adenovirus-associated respiratory disease is higher in the late winter, spring, and early summer than in the remaining seasons of the year. This holds true for endemic childhood diseases as well as for military-recruit epidemics (see Section 5.2.2). Type 3 epidemics (PCF), associated with swimming, have been described most frequently in the summer.

5.1.5. Age. Most children have been exposed to several types of the endemic viruses by the time they enter school (see Fig. 1). Adenovirus 4, 7, 8, 14, and 21 infections may occur later in life.

5.1.6. Sex. The WHO reports more illness-associated isolations from males than from females.[140] Otherwise, there is no significant difference between the sexes except for ARD, a disease of military personnel, to which primarily males are exposed.

5.1.7. Occurrence in Different Settings. The family constitutes the most important unit for transmission of the endemic Ads (1, 2, 5, and 6); however, rates of adenovirus infections (all types) are higher in children's institutions and day-care centers than in families. The incidence is higher in lower socioeconomic groups. Adenoviruses 4 and 7 (and to some extent Ads 14, 21, and 11) cause ARD in recruits. Adenovirus 3 and 7 epidemics have been associated with swimming (see Section 5.2.3).

5.1.8. Occupation. Acute respiratory disease caused by Ad 4 and Ad 7 (and to some extent Ads 14, 21, and 11) occurs primarily in military recruits. Epidemics of EKC, caused by Ads 8, 19, and 37, have been observed in shipyards and in the offices of ophthalmologists and their patients.

5.1.9. Other Factors. Persons with deficient cell-mediated immunity are at higher risk of severe adenovirus infection. This has been observed in congenital immunodeficiency, among immunosuppressed transplant patients,[72,111,143,148,171] with acquired immunodeficiency syndrome (AIDS),[36,72,75,82] and following measles.[21,141,145,169]

5.2. Epidemiologic and Clinical Aspects of Specific Syndromes

5.2.1. Endemic Adenovirus Infections. Most children become infected with some of the common adenoviruses early in life. Adenoviruses 1, 2, and 5 are endemic in those parts of the world where studies have been conducted. By the age of 1 year, 80% of children in the New Orleans area had acquired antibodies to these viruses, whereas the proportion with such antibodies was lower in New York and Seattle[27,52,66] and still slightly lower in Stockholm, Sweden.[150] In studies of adenovirus isolation in childhood, the most frequently occurring were

Ads 1 and 2, followed by types 5, 3, and 6, in that order. Neutralizing antibody studies in children and adults in four areas of the world are summarized in Fig. 1. The highest prevalence of adenovirus antibody was observed among crowded populations in developing countries.

Overall, only about 50% of childhood adenovirus infections result in disease.[17,52] However, this proportion approaches two thirds when infection is associated with pharyngeal location of the virus. A lower percentage of symptomatic infection has been reported in some studies for Ad 2 than for any other adenoviruses.[10,12,19] The most significant contribution of adenoviruses to illness is in childhood, particularly under age 5, when about 5% of all acute respiratory illnesses can be associated with these viruses. The symptoms are usually nonspecific, described as stuffy nose and cough. Older children acquiring adenovirus infection often have symptoms of pharyngitis, which may mimic and be clinically indistinguishable from streptococcal pharyngitis.[71] The virus has also been implicated in otitis media.[42,137]

Studies of ill children in several hospitals and clinics suggest that 2–7% of all lower-respiratory-tract illnesses in young children seeking medical care can be attributed to adenoviruses.[19,55] Such illnesses are rare under the age of 6 months, when the child is protected by maternal antibody. The disease syndromes attributed to adeno-

viruses include pharyngitis, bronchitis, bronchiolitis, croup, and pneumonia. Adenovirus pneumonia in children has been associated particularly with Ad 7, and occasionally the disease is fatal.[6,15,21,23,151,172] It sometimes occurs as a complication of measles, especially in developing countries.[87,141,169] Also, according to data collected by the WHO on viral disease leading to death, Ad 7 is the most pathogenic.[6] In the Netherlands, Ad 21 was frequently isolated from respiratory patients in the 1960s. When AD 21 reappeared in epidemic form in 1985, the endonuclease cleavage pattern was different, suggesting evolution of the virus.[161] Adenovirus infections occur all year round, but the incidence of disease is higher in late winter, spring, and early summer.[17,140] When exposure is equal, both sexes are usually equally affected.

Public health laboratories in England reported a prolonged epidemic of Ad 7 virus moving slowly through the country from 1971 to 1974; children manifested sore throats, conjunctivitis, and abdominal pain.[152] Localized outbreaks of adenovirus pneumonia (Ad 7 and Ad 21) have been reported in native Canadian,[172] New Zealand,[101] and northern Finnish infants and children.[145] It is not known whether this severe form of infection in these mostly nonwhite population groups is related to racial differences or to low socioeconomic status with crowding in a cold climate. There are no reports suggesting that American black children are more severely affected by adenovirus infection than white children. Sequelae in the form of bronchiectasis and radiolucent lung have been frequently reported in these special groups.[101,172]

The mode of transmission in early life is thought to be primarily fecal–oral.[52] A child born into a family in which other members harbor the virus in the intestinal tract will eventually become an excretor, but it may take several months before intrafamilial transmission occurs.[52,173] In family spread, the duration of shedding of the introducer seemed important: if shedding was brief, 42% of susceptibles and 6% of immunes were infected; if persistent, the rates were 65 and 21%, respectively.[52] Figure 2 shows examples of typical fecal—oral spread of type 2 in families with newborn infants. Transmission shown in Fig. 2 contrasts with a family transmission of PCF in Fig. 3, where all became infected with type 3 shortly after the onsets of the index cases, and all but the father were symptomatic. An infected child may first excrete the virus from the respiratory tract, but the virus usually disappears from this location and may instead be found intermittently in fecal specimens for extended time periods.[52,79] Many apparently purely enteric infections occur that are usually asymptomatic. Intermittent excretion for up to 906 days has been reported.[52] The fact that

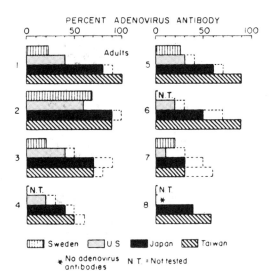

Figure 1. Comparison of adenovirus-neutralizing antibodies (types 1–8) in populations of four countries. The bar graphs indicate the percentage of antibody in children. If adults had a higher percentage of antibody, it is shown by a dashed-line extension. Reproduced with permission from Tai and Grayston[155] and the *Proceedings of the Society for Experimental Biology and Medicine* with data added from Sterner.[150]

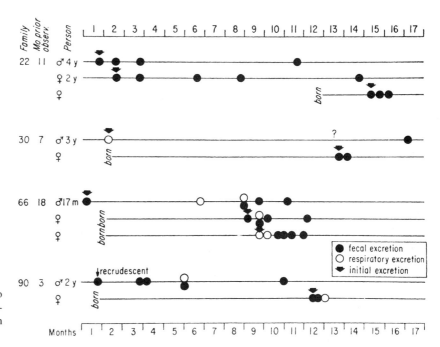

Figure 2. Spread of type 2 adenovirus to infants born into families with siblings excreting the virus. Reproduced with permission from Fox *et al.*[52]

the adenoviruses can be found in more than 50% of surgically removed tonsils suggests that the infection may lie dormant for many years.[45,136] The long-term effect of chronic adenovirus infections is unknown.

Children brought up in institutions acquire adenoviruses sooner than children living at home.[42,125] Epidemics of infection with Ad 5 and Ad 7 have been described in orphanages and day-care facilities.[10,150,164] Whereas studies in free-living families have rarely revealed the presence of other than Ads 1, 2, 3, 5, and 6, the studies in Junior Village,[10] an institution for homeless children in Washington, DC, revealed spread of higher-numbered species, and studies among African children showed significant fecal excretion of Ads 8, 12, and 16.[162] The higher-numbered species (from Ad 9) are usually recovered only from fecal specimens,[10] and with the exceptions of Ads 11, 12, 19, 22, 31, 34, 35, and 37, they have generally not been associated with respiratory, eye, or other illnesses. In 1975, an epidemic of Ad 31 was observed in New York families under continuous observation; the virus may have caused some mild respiratory disease.[168]

5.2.2. Acute Respiratory Disease of Military Recruits.
Soon after adenoviruses were first isolated from cases of ARD among military personnel, it became clear that these viruses were the primary cause of morbidity among recruits, especially among North American and northern European forces.[25,37] Morbidity rates were reported as high as 6–17/100 per week. Epidemics usually peaked at about 3–6 weeks after onset of training, although sometimes the epidemic occurrence was delayed. The seasons for highest rates of adenovirus infections were winter and spring, independent of geographic locations and climatic conditions of training posts within the United States.[38] In one study, it was shown that recruits from the southern United States were less susceptible than those from the North, probably reflecting a more intensive exposure to adenoviruses during childhood and subsequent immunity for those brought up in the South.[112] In contrast to recruits, seasoned troops have a low incidence of adenovirus infections, suggesting that lasting immunity is acquired early in military life[37] (Fig. 4).

The spectrum of clinical manifestations caused by adenoviruses in military recruits spans mild respiratory disease, febrile pharyngitis (often including adenitis), and pneumonia. Typical ARD is a febrile respiratory disease with symptoms of sore throat and cough, sometimes coryza, headache, and chest pain. Malaise is characteristic, and the illness lasts for approximately 10 days. White blood cell counts are normal or slightly elevated. It is estimated that 10% of recruits reporting to sick bay with ARD have pneumonitis on X-ray examination. Chest films characteristically have feathery or mottled infiltrates. Deaths from adenovirus pneumonia in the military are rare but do occur.[39]

Epidemics of ARD are usually caused by Ad 4 and

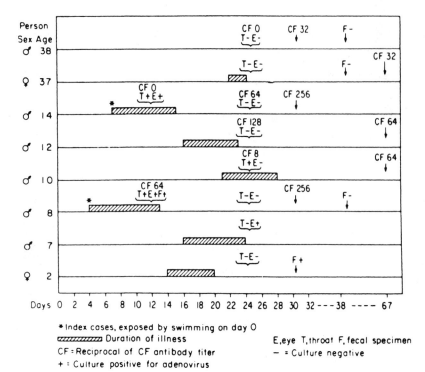

*Index cases, exposed by swimming on day 0
Duration of illness
CF = Reciprocal of CF antibody titer
+ = Culture positive for adenovirus

E,eye T,throat F,fecal specimen
− = Culture negative

Figure 3. Sequence of illnesses in a family in which the two index cases, boys of 8 and 14 years, contracted PCF through exposure in a swimming pool. The first secondary family case occurred 10 days after onset of the first index case; within 17 days, all family members became infected. Only the father was asymptomatic; the mother's illness lasted only 2 days and was mild. Isolates from throat and eye were obtained only during acute illness (or at the latest 1 day after), whereas fecal excretion lasted longer. Isolates from the eyes were obtained only from those with conjunctivitis. Not shown in the illustration is the fact that all members developed antibodies to AAV 3. Reproduced (modified) from Foy *et al.*[54]

less frequently by Ad 7. Occasionally, Ads 3, 11, 14, and 21 have caused epidemic ARD.[38,112,134,162]

The epidemiology of adenovirus infection among military populations has been studied extensively in the United States. Epidemics have been sharper and occurred sooner in Navy base camps, where extensive mingling of new and old recruits took place, than in Marine training camps, where new recruits were segregated in separate living quarters. The situation is similar in most European countries from which reports are available.[108,162] The incidence in the British Army and Navy was less than in the Air Force, reflecting different management of recruit populations, with less contact between previously and recently inducted recruits.

Incidence has been reported to be low in Argentina[47] and Colombia[48] and among Chinese recruits on Taiwan.[155] The types of adenoviruses prevalent in military forces of various countries have varied. Adenovirus 4 was rarely observed among South American forces. Adenoviruses 4 and 3 were epidemic in two different garrisons in Finland,[108] whereas Ad 7 was not. In the Netherlands, Ad 21 became epidemic in 1960–1961 among military recruits.[162] On Taiwan, Ad 5 was found in one outbreak.[155]

Despite the epidemic occurrence of adenoviruses in military populations, little spread has been noticed from the military bases to the civilian populations, including military dependents. An exception was the Netherlands, where epidemics of Ad 21 were observed among the military and children in civilian communities simultaneously.[162] Adenovirus 4 has been uncommonly associated with disease in civilians.

The sharp contrast between the epidemiology of adenovirus infections among civilian and military groups remains incompletely understood. Studies in adult civilian populations, including college students, have shown that adenoviruses can be associated with only a very small portion (0.5–3%) of respiratory disease.[46,60] In recruit training, a number of factors known to encourage epidemics exist. Persons are brought together from different geographic areas and backgrounds and subjected to crowding and stress. This situation is known to encourage spread in the military of many other infections often classified as childhood diseases, such as mumps, rubella, chickenpox, and meningococcal disease. Methods of processing recruits may potentiate an epidemic; thus, it has been found that when new recruits are allowed to mix with those already present and presumably infected, epidemics readily occur.[5] In the Marine Corps, when each training unit was kept separate, the overall incidence was lower.

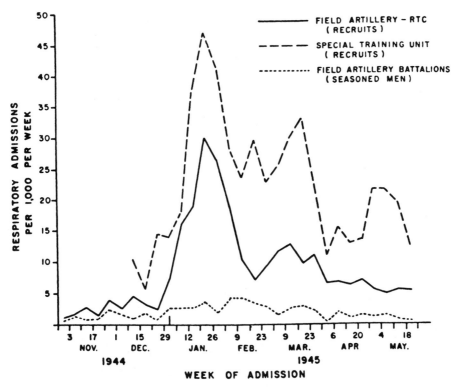

Figure 4. Respiratory admission rates for two recruit groups and one seasoned army group at Fort Bragg, North Carolina, November 1944 to May 1945. Reproduced with permission from Dingle and Langmuir.[37]

The prevalence of neutralizing antibodies to the adenovirus types causing epidemics has generally been low in the incoming recruits.

5.2.3. Pharyngoconjunctival Fever. Pharyngoconjunctival fever (PCF) is a syndrome characterized by pharyngitis, conjunctivitis, and spiking fever. Either one or both eyes may be involved, and only rarely is the cornea affected. Diarrhea, coryza, and occasionally otitis may be present. The tonsils may show an exudate, and lymphadenopathy is often observed.

In the 1920s, epidemics of a febrile disease with conjunctivitis centering around swimming activities were first described from Germany[126] and later from the United States.[7] The clinical epidemiologic characteristics of these epidemics are highly suggestive of adenovirus infections, and the epidemics were thought to be caused by inadequate chlorination. However, the only nonbacterial agent known to cause conjunctivitis at that time was trachoma-inclusion conjunctivitis agent (*Chlamydia trachomatis*), and it was believed that these epidemics were caused by such agents, since inclusions were seen in a few patients. After the 1950s, when adequate diagnostic tools for isolation of both inclusion con-

junctivitis agent and adenovirus were developed, all swimming-pool-centered epidemics have been traced to adenovirus infections.[31,54,91,127] Thus, the name "swimming pool conjunctivitis" for *C. trachomatis* inclusion conjunctivitis infections is probably a misnomer, and the term should apply to adenovirus infections.

The association of PCF with Ad 3 was first described by Bell *et al.*[11] in 1955. The syndrome is caused most frequently by Ad 3 or Ad 7 but has also been associated with other species, such as Ads 1, 4, and 14. The disease may be seen sporadically and cause epidemics in families and other closed population groups[11,54,150] but is best known as an epidemic disease centering around summer camps and especially swimming pools and small lakes.[54,91,96,127] It usually affects children and young adults. The fact that the youngest age groups (under age 4), who waded rather than swam, had the lowest attack rate in the original outbreak described by Parrott *et al.*[127] suggests that direct contact with the water, possibly allowing introduction of the virus into the eye or upper respiratory tract, is the important mode of spread. However, once the virus is introduced into a family, secondary spread to other family members frequently occurs. This secondary spread may be by direct contact or possibly droplet transmission. In a study conducted in Seattle, it was found that parents secondarily infected from children usually had milder symptoms, often only conjunctivitis.[54] This suggests that they possessed some immunity, most likely humoral, but that the conjunctiva lacked sufficient defense mechanism.

The incubation period is estimated at 6–9 days. In a common-source epidemic originating from a swimming pool open only for 1 day, the incubation period was estimated at an average of 6 days.[91] An incubation period of 6 days was also observed by Bell *et al.*[11] when a patient exposed his physician by coughing in his face. On the other hand, the incubation period observed in experimental inoculation of volunteers has been as short as 2 days.[13] This shortening of the incubation period may be dose related.

In eye clinics in Great Britain, the predominant adenoviruses isolated from conjunctivitis cases were Ads 7, 3, 4, and 8.[67] In similar clinics in Japan, Taiwan, and South Korea, on the other hand, Ad 8 predominated, followed by Ads 3 and 19 in the same time period.[84]

The virus has been isolated from the water by a sewage outlet at a lake and from a swimming pool[56] where outbreaks of PCF have occurred.[31] Several outbreaks have clearly been associated with unsatisfactory chlorination of swimming-pool water. In the laboratory, Ad 3 appears to be as susceptible to chlorination as *Escherichia coli* bacteria.

5.2.4. Epidemic Keratoconjunctivitis. Epidemic keratoconjunctivitis is a disease entity that was first described by German workers in the latter part of the 19th century.[80,86] During World War II, when shipbuilding flourished, it was observed in industrial outbreaks first on Hawaii, then on the United States West Coast, and finally on the East Coast.[86] Transmission probably took place in the medical facilities, where the workers sought treatment for foreign bodies and chemical irritation of the eye.

It was during this period that the disease was named "epidemic keratoconjunctivitis" because of its clinical characteristics, but it was often popularly called "shipyard eye." The illness characteristically has an incubation period of 8–10 days and starts with conjunctivitis that may be follicular, followed by edema of the eyelids, pain, photophobia, and lacrimation. After a couple of days, superficial erosions of the cornea may develop, followed in a full-blown case by deeper subepithelial corneal infiltrates with characteristic round shape, located in the center of the corna. These may interfere with vision and cause lasting visual impairment. Preauricular lymph gland swelling is common, and occasionally cervical and sub-maxillary lymph glands are involved. Frequently, only one eye is infected. Constitutional symptoms may occur among children but are usually mild.

Adenovirus 8 was the almost exclusive cause of the typical disease in Western countries until 1973, when Ad 19 and then Ad 37 appeared almost simultaneously in several European countries and North America[32,33,35,50,65,95,163] and also were reported from Japan. Mixed outbreaks and intermediate strains were found.[73] Possibly this change in predominant strains indicates an antigenic drift, similar to that observed in influenza. Rarely have other Ads such as Ads 4, 7, 10, 11, 14, and 15 been implicated in EKC.[33,62] The virus can be isolated during the acute disease from the eye, occasionally from the throat, and for a longer period from the feces. Adenoviruses 19 and 37 were also isolated from the genital tract.[35,69,70] Since EKC has sometimes appeared sporadically and often affected young adults, sexual transmission has to be considered a possibility.

Outbreaks of EKC have been reported centering around the offices of ophthalmologists.[30,33,50,86,88,95,110,147,153] Since the virus is unusually hardy, ether and alcohol treatment of ophthalmological instruments (in particular, tonometers) is insufficient as a means of sterilizing equipment, and heat sterilization (240°C or 465°F) is necessary. Rigorous infection-control measures are needed to prevent nosocomial spread, including gloving rather than washing of hands.[88,170] Spread may also occur directly by fingers during manipulation of the lids or by use of eye solutions and ointments. Patients acquiring the infection by a visit to the ophthalmologist's office only occasionally transmit the infection to family members. Concurrent outbreaks in the community and in day-care centers occurred in central Australia.[110] Direct inoculation into the eye appears necessary to cause disease.

In areas of Japan and Taiwan, yearly epidemics of EKC have been observed in the late summer and fall.[62] The spread presumably occurred by direct contact between children and between family members. In camps for recently arrived Vietnamese refugees, adenovirus keratoconjunctivitis has been rampant, spreading also to the indigenous Americans.[180] Seemingly, crowded and less hygienic living conditions promote the spread of the virus. It is also noteworthy that many of the higher-numbered serological adenovirus types were first isolated from eyes in Saudi Arabia, where trachoma is highly endemic, suggesting environmental and hygienic requirements for spread of both of these eye diseases.[14] Serological studies in the late 1950s revealed that 40–60% of Japanese and Taiwanese school children had acquired neutralizing antibody to type 8, whereas virtually no American children had antibodies to adenovirus type 8[154] (see Fig. 1).

5.2.5. Gastroenteritis. Many of the adenoviruses may cause diarrhea, such as Ad 31 and those belonging to subgroup B and C. Fastidious EAds Ad 40 and 41, subgenus F, cause a watery, dehydrating diarrhea primarily in infants.[18,22,34,99,133,159,160,174,178] In infantile diarrhea, this infection may be second in importance only to rotaviruses; however, unlike rotaviruses, these viruses are usually not seasonal, although outbreaks, including nosocomial epidemics, may occur.

EAds were first visualized by electron microscopy. They were initially isolated in transformed HEK cell line 293 Chang's conjunctival cells and detected by enzyme immunoassay.[14,18,97,120,131,135] Latex agglutination, ELISA, and monoclonal antibodies can be used for rapid diagnosis, and PCR is a reliable diagnostic tool.[3,59,76,81]

The duration of diarrhea is longer than for rotaviruses. The incubation period has been estimated to be 2–10 days. A shift of the dominating type of EAd from Ad 40 to type Ad 41 was observed in the Netherlands and possibly Great Britain, between 1981 and 1989.[34] The EAds have also been implicated in necrotizing enterocolitis and associated with gastrointestinal surgery.[177] Adenovirus infections have also been associated with appendicitis.[100]

5.2.6. More Unusual Syndromes Associated with Adenovirus Infections. *Acute hemorrhagic cystitis* is a rare self-limited disease primarily of childhood, characterized by polyuria, dysuria, and hematuria. Isolation of Ad 11 from this disease by use of special techniques was first accomplished in Sendai, Japan, in 1968.[124] The patients showed significant increases in antibody titers by N and CF but not by HI test. Subsequently, isolation of Ad 11 and also Ad 21 from the urine of children with this disease was reported from Chicago.[117] Urethritis may on occasion be caused by Ad 19 and Ad 37, and genital transmission is postulated.[69,70]

In *immunocompromised patients*, adenovirus infection may cause severe complications, and the viruses have been recovered from urine sporadically.[36,72,74,75,82,95a,143,148,179] In bone marrow transplants, Ads 1, 2, 5, and 31 commonly complicate recovery, and the infection is often fatal.[72] The infection may appear due to either reactivation of dormant virus or nosocomial spread. In liver transplant patients, Ads 1, 2, and 5 dominate, and Ad 5 is reported to cause hepatitis; in renal transplant patients, Ads 7, 33, 34, and 35 occur, often causing acute hemorrhagic cystitis.[72] In AIDS patients a multitude of adenoviruses have been isolated, including the recently recognized types 43–47. It has been postulated that these new viruses appearing in immunosuppressed patients are recombinants, seemingly with Ad 7.[36]

Encephalomeningitis is seen as a severe complication, usually to respiratory tract infections with adenoviruses, both in children and adults.[94] Adenoviruses belonging to subgenera B and C are usually implicated.[140]

Intussusception, an acute illness of infancy, is characterized by "telescoping" of one part of the intestine into the next distal portion. The disease usually requires prompt surgical intervention because of risk of necrosis as the blood supply of the involved bowel segment becomes obstructed. Lymph-tissue enlargement from adenovirus infection may initiate such telescoping, and reports of a high isolation rate of adenoviruses from such patients suggest an etiologic role for adenoviruses.[16,24,57,135] However, not all patients with intussusception have evidence of adenovirus infection, and the etiology of this syndrome may be complex.

Pertussislike syndrome in association with various serological types of adenovirus infection has been described.[26,97] Patients selected to serve as controls have generally had less evidence of virus infection. On the other hand, in an in-depth study of pertussis, *Bordetella pertussis* was isolated from 45 of 65 patients, and adenovirus was shed in eight cases. In convalescence, adenovirus was isolated in only 1 of 33 cases. The course was no worse in patients shedding adenovirus. It has been postulated that pertussis merely reactivates adenovirus infections.[9,121]

Neonatal adenovirus infection characterized by systemic symptoms and progressive pneumonia occurs primarily in premature infants and is usually fatal.[1]

Skin rashes have been associated with adenovirus infection, but in most such incidences, the rash could be attributed to preceding or concurrent measles or rubella.[21,64,119,141,145]

Reye's syndrome has occasionally been associated with adenovirus infection.[41]

6. Mechanisms and Route of Transmission

Adenovirus infections in man are spread from person to person by various routes. Many lower animals have their own adenoviruses, and there is no evidence that these are human in origin or that animal adenoviruses infect humans. There is no known spread by vectors. In childhood and in family transmission, the fecal–oral route undoubtedly plays the major role. This route is the most important for transmission of Ads 1, 2, 5, and 6. Respiratory transmission of these and other types is possible at all ages in association with ARD. The respiratory route is of prime importance when ARD becomes epidemic in military-recruit camps. Pharyngeal carriage rates as high as 18% have been reported among healthy recruits.[51] In volunteer studies, inhalation of small doses of adenoviruses in aerosols usually resulted in infection accompanied by febrile ARD, sometimes with pneumonia.[28] In contrast, nasopharyngeal administration was much less effective in producing disease[13]; inoculation of the conjunctiva was more successful. The eye is an important portal of entry for virus types Ad 3 and Ad 7, particularly when PCF is transmitted in swimming pools or lakes. Epidemic keratoconjunctivitis (types 8, 19, and 37) is often a iatrogenic disease spread by ophthalmological instruments, ointments, or fingers in physicians' offices. Adenoviruses occasionally cause nosocomial infections in hospitals.[11,21,50,88,151,156,163] Adenoviruses may be recovered from sewage, and environmental contamination may be important, particularly in developing countries. Ad 19 and Ad 37 can be isolated from the genital tract, and may be transmitted sexually.[69,70]

7. Pathogenesis and Immunity

The incubation period in volunteers challenged artificially can be as short as 2 days, which may be a dose-related response.[11] Under the natural challenge, the incubation period for respiratory syndromes averages 6–9 days with a mean of 6 days. The incubation for EKC is usually longer, varying between 3 and 22 days with a mean of 9 days. The incubation period for EAds is estimated to be 2–10 days.

Most disease caused by adenoviruses is acute and self-limited. Species-specific neutralizing antibody has been associated with prevention of symptomatic reinfection. Although adenovirus disease is short-lasting, infection may be prolonged, and asymptomatic infections are common. Viral shedding in the gastrointestinal tract may recur for years. Adenoviruses can be isolated from at least 50% of surgically removed tonsils,[45] occasionally from the kidney,[148] and also from lymphocytes,[4] suggesting that infection may remain latent for a very long time, possibly for life. In AIDS patients who have systemic infections, reactivation may occur, and recombination of virus species in this syndrome has been postulated.[36,75,82,105]

The virus may invade the bloodstream in the early stages of the disease, and viremia has been associated with a maculopapular skin rash.[15,21,64,145] However, these rashes may have been caused by measles and rubella, and the adenovirus merely reactivated. Adenovirus has been isolated from cerebrospinal fluid in patients with meningoencephalitis.[94]

In the rare fatal illness, the virus has been recovered from most body organs.[15] In such cases, extensive pathology is found in the lungs, with microscopic necrosis of tracheal and bronchial epithelium characterized as necrotizing bronchiolitis. In bronchial epithelial cells, acidophilic intranuclear inclusions have been described, as well as basophilic masses of cells surrounded by characteristic clear halos. The latter may represent aggregation of the larger amount of viral material.[15,119,141] Rosette formation, a mononuclear cell infiltrate, and focal necrosis of mucous glands appear to be characteristic. Typical intranuclear virus particles have been observed in alveolar lining and bronchiolar cells by electron microscopy.[119] In infants who recover from adenovirus pneumonia, severe sequelae may follow, including radiolucent lung syndrome, bronchiectasis, and persistent lobar collapse.[101,172]

Neurotropism of the virus, especially Ad 7, has been suggested by isolation from the central nervous system.[15,145]

Rates of adenovirus pneumonia in children are lower under the age of 6 months than in older infants, suggesting that maternally derived humoral antibodies are protective and that humoral antibody plays a major part in the defense mechanism. Studies of volunteers have shown that even artificial challenge with virus in persons already having neutralizing antibody for that type rarely results in symptomatic infection. In an outbreak in a boys' boarding school with Ad 3, previous infection provided 88% protection against reinfection.[128] Intracellular location probably protects the virus from the effect of humoral antibody and permits persistent latent infection. The severe adenovirus infections observed in immunosuppressed patients, including those with AIDS, indicate that cell-mediated immunity (CD4 cells) also is important.[143,148,179]

8. Patterns of Host Response

The host response to adenovirus infection is dependent on the route of transmission, the primary site of viral localization, and on Ad type and dose. Disease syndromes (ARD, PCF, EKC) occurring in special situations have been described above. In persons with normal immune systems, the presence of humoral antibodies correlates with at least partial immunity. Persons with poor IgA antibody response to infection have been shown to have increased severity of symptoms.[109] Immunosuppressed and immunodeficient patients, including those with AIDS, are at risk of disseminated infection.[111,143,148,179] Immune complexes have been found in severe infection, especially in patients with renal manifestations.[113] Postponement and spacing of the several routine immunizations given to military recruits have been associated with reduction of ARD, pointing out an interesting type of host stress immunologically associated with adenovirus disease.[112,130]

Primary localization of adenoviruses, subgenera B, C, and E, in the respiratory tract or conjunctiva results in a higher rate of symptomatic infection than gastrointestinal localization.

The relationship of species to disease severity is difficult to separate from host and environmental factors. Adenoviruses 8, 19, and 37 cause more severe eye disease than other types; Ads 4, 7, and 21 have been associated with severe respiratory disease.

Why some infections cause pneumonia and even death is not well understood. Host age of 6 months to 2 years is clearly associated with adenoviral pneumonia, as is general host deprivation, including lower socioeconomic status. A poor nutritional state and measles both lead to suppressed cell-mediated immunity and may explain why severe adenovirus pneumonia occurs in certain situations.[1,87,141] Adenovirus 7 and to some extent Ad 21 have been particularly associated with fatal infant pneumonia.[15,21,23,101]

Reactivation of adenovirus infection may possibly be evoked by other respiratory agents.[9,23,58,121] It has been reported that during mumps infection, adenovirus excretion reappeared.[52] In studies of pneumonia etiology, an unexpectedly high rate of serological titer rises to adenoviruses was found together with antibody rises to influenza, parainfluenza, and *Mycoplasma pneumoniae*, the latter agents being implicated as etiologic agents of the pneumonia by isolation or epidemiologic evidence.[55,77]

9. Control and Prevention

Of the various approaches to the prevention or control of adenovirus infections, bolstering the host's resistance through immunization has shown the most promise. In certain situations, environmental control has been effective.[102,107] Intensive hand washing may not remove the virus load completely, as observed in outbreaks of EKC, and rigorous control measures are needed in the ophthalmology clinics.[88,170] Control or treatment with antiviral compounds, interferon, or interferon inducers has so far shown no practical value. Ribavirin has been used intravenously in a case of hemorrhagic cystitis, seemingly with success.[118]

Because of the major disruption and economic impact of ARD epidemics in military recruits, most of the efforts in the development of human adenoviral vaccines have been directed at this population. The first widely used vaccines were inactivated adenoviruses grown in monkey kidney tissue culture.[20,129] They were shown to be effective against adenoviral infection with Ad 3, Ad 4, and Ad 7. However, potency varied among lots of vaccine because of poor growth in the culture system. Efforts to meet minimal standards for military use were often unsuccessful. A major production difficulty was contamination of the cell culture with SV40. This virus is recognized as oncogenic in animals, and in addition, it was found that adenoviruses were able to incorporate part of the SV40 genome (hybridization) when the viruses grew in the same culture.

The extensive molecular studies of adenoviruses have shown the possibility of developing a highly purified subunit vaccine.[93] However, production problems for such a vaccine have not been solved.

Major vaccine efforts have been made with live attenuated virus given orally.[20,40,129,134,146,158] This approach has been based on the theory that by bypassing the respiratory tract and introducing a live virus into the gastrointestinal tract, where the virus was known to multiply, respiratory disease could be avoided while the subject acquired solid immunity. For this purpose, live adenoviruses grown in human cells were placed in enteric-coated gelatin capsules. Following experimental studies, this approach to immunization has been shown to be highly effective in military-recruit populations. However, when Ad 4 military epidemics were controlled by live oral immunization, Ad 7 epidemics occurred. Bivalent immunization has successfully controlled both viruses.[157] The fear that another type of adenovirus, for example, Ad 21, would take over and fill the "ecological niche" has not yet been substantiated.[134]

The experimental studies with these vaccines have shown a number of interesting facts about adenoviral spread.[29,129,158] Neither Ad 4 nor Ad 7 spread from adults infected by enteric capsules to susceptible adults housed together for a prolonged period.[29,129] On the other hand, when enteric live Ad 4 vaccine was introduced to one partner of each of 39 married couples and a placebo was given to the other, viral isolates were obtained from 70% of the placebo recipients, suggesting that intimate physical contact facilitated transmission of the virus.[149] No serious symptoms were encountered in the placebo group. In another study, conjunctivitis occurred in two of six volunteers infected in the gastrointestinal tract, probably by fecal–conjunctival contamination.[29] When children in a household were immunized by this route, transmission of the virus (Ad 4) to other members of the household was demonstrated, including occasional illness.[116]

There have been no efforts to protect children or

other civilian populations from adenovirus infections by immunization with oral vaccine. Successful immunization by the gastrointestinal route has been demonstrated with Ad 1, Ad 2, and Ad 5 in adult volunteers.[142] Since none of the viruses in these vaccines has been demonstrated to be significantly attenuated, there is the risk that administration of live vaccines to infants may cause spread of adenoviral disease. The practical problems of spread of infection and the hypothetical concerns that remain concerning oncogenicity would make an effort to immunize infants and children with live adenoviruses a dubious undertaking.

Environmental control is effective in certain situations. Thus, the evidence suggests that adequate chlorination of swimming pools prevents the spread of PCF. The spread of keratoconjunctivitis through contaminated ophthalmological instruments, ointments, and solutions can be prevented by heat sterilization and appropriate hygienic measures. Adenoviruses may be recovered from surfaces for several weeks.[120]

Since adenovirus infections in military-recruit populations appear to be associated with crowding and airborne transmission, various environmental control attempts have been tried. Modification of the sleeping arrangements in barracks has met with, at best, limited success.[103] Attempts at dust suppression have been more successful,[107] whereas air purification with ultraviolet light and germicidal sprays has failed.[102,103] More extensive changes in the patterns of recruit housing and training can modify ARD occurrence but may be impractical in time of mobilization.[5]

Adenoviruses are being used as vectors in the development of new vaccines for other virus infections, such as hepatitis B and HIV[115] and for gene transfer.[8]

10. Unresolved Problems

The adenoviruses are subject to continuous evolution. New types and variants need to be characterized and their epidemiologic pattern elucidated.

Great strides have been made during the last decade in describing and understanding the molecular biology of adenoviruses, yet no effective antiviral agent has been found. Although several types of vaccines have been developed, none is ideal, and the possibility of using a purified subunit of the virus or a genetically controlled virus free of oncogenic genes for vaccine is an attractive prospect.

The role of concomitant or subsequent viral or bacterial infection in precipitating more severe adenoviral disease is not understood.

The long-term effect of latent adenovirus infection or reactivation later in life is open to speculation. What is the mechanism for latency: chronic low-grade infection, lysogenicity, or incorporation of the virus genome in the host cell? How long does latency last? Oncogenicity has been shown only when human adenoviruses have been introduced into animal models. Although the search for oncogenic potential in humans by investigation for T (tumor) antigens and DNA sequences so far has been negative, an oncogenic effect in humans cannot be totally ruled out. The recent finding that injection of Ad 9 may cause mammary tumors in rats, which respond to hormonal treatment similarly to humans in the development of breast cancer, merits further search for a viral etiology of breast cancer. Transactivation between viruses may lead to oncogenic suppression or promotion and deserves further study. A search for viral DNA sequences of subgenera A and B has been carried out in gastrointestinal cancers, but since adenovirus is dormant in lymphatic tissue, further investigation must be conducted in other malignancies, using newer techniques such as PCR, before the possibility of oncogenicity can be rejected. Adenovirus 19 and Ad 37 have been isolated from the genital tract, sometimes in association with adenovirus keratoconjunctivitis. Their role in the genital tract needs to be elucidated, including possible transmission to the newborn. Adenoviruses cause hepatitis in dogs, and occasional case reports of recovery of adenovirus, especially Ad 5 from humans with hepatitis, may warrant additional investigation of hepatitis of unknown cause.

11. References

1. ABZUB, M. J., AND LEVIN, M. J., Neonatal adenovirus infection: Four patients and review of the literature, *Pediatrics* **87:**890–896 (1991).
2. ADRIAN, T., WIGAND, R., AND HIERHOLZER, J. C., Immunological and biochemical characterization of human adenoviruses from subgenus B. II. DNA restriction analysis. *Arch. Virol.* **84:**79–89 (1985).
3. ALLARD, A., GIRONES, R., JUTO, P., AND WADELL, G., Polymerase chain reaction for detection of adenoviruses in stool samples, *J. Clin. Microbiol.* **28:**2659–2667 (1990).
4. ANDIMAN, W. A., AND MILLER, G., Persistent infection with adenovirus types 5 and 6 in lymphoid cells from humans and woolly monkeys, *J. Infect. Dis.* **145:** 83–88 (1982).
5. ARLANDER, T. R., PIERCE, W. E., EDWARDS, E. A., PECKINPAUGH, R. O., AND MILLER, L. F., IV, An epidemiologic study of respiratory illness patterns in Navy and Marine Corps recruits, *Am. J. Public Health* **55:**67–80 (1965).
6. ASSAD, F., AND BORECKA, I., Nine-year study of WHO virus reports on fatal viral infections, *Bull. WHO* **55:**445–453 (1977).
7. BAHN, C., Swimming bath conjunctivitis, *New Orleans Med. Sci. J.* **79:**586–590 (1927).

8. BAJOCCHI, G., FELDMAN, S. H., CRYSTAL, R. G., AND MASTRANGELI, A., Direct *in vivo* gene transfer to ependymal cells in the central nervous system using recombinant adenovirus vectors, *Nature Genet.* **3:**229–234 (1993).

9. BARAFF, L. J., WILKINS, J., AND WEHRLE, P. F., The role of antibiotics, immunizations, and adenoviruses in pertussis, *Pediatrics* **61:**224–230 (1978).

10. BELL, J. A., HUEBNER, R. J., ROSEN, L., ROWE, W. P., COLE, R. M., MASTROTA, F. M., FLOYD, T. M., CHANOCK, R. M., AND SHVEDOFF, R. A., Illness and microbial experiences of nursery children at Junior Village, *Am. J. Hyg.* **74:**267–292 (1961).

11. BELL, J. A., ROWE, W. P., ENGLER, J. I., PARROTT, R. H., AND HUEBNER, R. J., Pharyngoconjunctival fever: Epidemiological studies of a recently recognized disease entity, *J.A.M.A.* **175:**1083–1092 (1955).

12. BELL, J. A., ROWE, W. P., AND ROSEN, L., II. Adenoviruses, *Am. J. Public Health* **52:**902–907 (1962).

13. BELL, J. A., WARD, T. G., HUEBNER, R. J., ROWE, W. P., SUSKIND, R. G., AND PAFFENBARGER, R. S., JR., Studies of adenoviruses (APC) in volunteers, *Am. J. Public Health* **46:**1130–1146 (1956).

14. BELL, S. D., JR., ROTA, T. R., AND McCOMB, D. E., Adenoviruses isolated in Saudi Arabia. III. Six new serotypes, *Am. J. Trop. Med.* **9:**523–526 (1960).

15. BENYESH-MELNICK, M., AND ROSENBERG, H. S., The isolation of adenovirus type 7 from a fatal case of pneumonia and disseminated disease, *J. Pediatr.* **64:**83–87 (1964).

16. BHISITKUL, D. M., TODD, K. M., AND LISTERNICK, R., Adenovirus infection and childhood intussusception, *Am. J. Dis. Child.* **146:** 1331–1333 (1992).

17. BRANDT, C. D., KIM, H. W., JEFFRIES, B. C., PYLES, G., CHRISTMAS, E. E., REID, J. L., CHANOCK, R. M., AND PARROTT, R. H., Infections in 18,000 infants and children in a controlled study of respiratory tract disease. II. Variation in adenovirus infections by year and season, *Am. J. Epidemiol.* **95:**218–227 (1972).

18. BRANDT, C. D., KIM, H. W., RODRIGUEZ, W. J., THOMAS, L., YOLKEN, R. H., ARROBIO, J. O., KAPIKIAN, A. Z., PARROTT, R. H., AND CHANOCK, R. M., Comparison of direct electron microscopy, immune electron microscopy, and rotavirus enzyme-linked immunosorbent assay for detection of gastroenteritis viruses in children, *J. Clin. Microbiol.* **13:**976–981 (1981).

19. BRANDT, C. D., KIM, H. W., VARGOSKO, A. J., JEFFRIES, B. C., ARROBIO, J. O., RINDGE, B., PARROTT, R. H., AND CHANOCK, R. M., Infections in 18,000 infants and children in a controlled study of respiratory tract disease. I. Adenovirus pathogenicity in relation to serologic type and illness syndrome, *Am. J. Epidemiol.* **90:**484–500 (1969).

20. CHANOCK, R. M., LUDWIG, W., HUEBNER, R. J., CATE, T. R., AND CHU, L. W., Immunization by selective infection with type 4 adenovirus grown in human diploid tissue culture. I. Safety and lack of oncogenicity and tests for potency in volunteers, *J.A.M.A.* **195:**151–158 (1966).

21. CHANY, C., LEPINE, P., LELONG, M., VINH, L. T., SATGE, P., AND VIRAT, J., Severe and fatal pneumonia in infants and young children associated with adenovirus infections, *Am. J. Hyg.* **67:**367–378 (1958).

22. CHIBA, S., NAKATA, S., NAKAMURA, I., TANIGUCHI, K., URASAWA, S., FUJINAGA, K., AND NAKAO, T., Outbreak of infantile gastroenteritis due to type 40 adenovirus, *Lancet* **2:**954–957 (1983).

23. CHIN-HSIEN, T., Adenovirus pneumonia epidemic among Peking infants and pre-school children in 1958, *Chin. Med. J.* **80:**331–339 (1960).

24. CLARKE, E. J., PHILLIPS, I. A., AND ALEXANDER, E. R., Adenovirus infection in intussusception in children in Taiwan, *J.A.M.A.* **208:**1671–1674 (1969).

25. COMMISSION ON ACUTE RESPIRATORY DISEASES, Experimental transmission of minor respiratory illness to human volunteers by filter-passing agents. I. Demonstration of two types of illness characterized by long and short incubation periods and different clinical features, *J. Clin. Invest.* **26:**957–973 (1947).

26. CONNOR, J. D., Evidence for an etiologic role of adenoviral infection in pertussis syndrome, *N. Engl. J. Med.* **283:**390–394 (1970).

27. COONEY, M. K., HALL, C. E., AND FOX, J. P., The Seattle Virus Watch. III. Evaluation of isolation methods and summary of infections detected by virus isolations, *Am. J. Epidemiol.* **96:**286–305 (1972).

28. COUCH, R. B., CATE, T. R., FLEET, W. F., GERONE, P. J., AND KNIGHT, V., Aerosol-induced adenoviral illness resembling the naturally occurring illness in military recruits, *Am. Rev. Respir. Dis.* **93:**529–535 (1966).

29. COUCH, R. B., CHANOCK, R. M., CATE, T. R., LANG, D. J., KNIGHT, V., AND HUEBNER, R. J. Immunization with types 4 and 7 adenovirus by selective infection of the intestinal tract, *Am. Rev. Respir. Dis.* **88:**394–403 (1963).

30. D'ANGELO, L. J., HIERHOLZER, J. C., HOLMAN, R. C., AND SMITH, J. D., Epidemic keratoconjunctivitis caused by adenovirus type 8: Epidemiologic and laboratory aspects of a large outbreak, *Am. J. Epidemiol.* **113:**44–49 (1981).

31. D'ANGELO, L. J., HIERHOLZER, J. C., KEENLYSIDE, R. A., ANDERSON, L. J., AND MARTONE, W. J., Pharyngoconjunctival fever caused by adenovirus type 4: Report of a swimming pool-related outbreak with recovery of virus from pool water, *J. Infect. Dis.* **140:**42–47 (1979).

32. DAROUGAR, S., GREY, R. H. B., THAKER, U., AND McSWIGGAN, D. A., Clinical and epidemiological features of adenovirus keratoconjunctivitis in London, *Br. J. Ophthalmol.* **67:**1–7 (1983).

33. DAROUGAR, S., WALPITA, P., THAKER, U., VISWALINGAM, N., GARDNER, L., AND McSWIGGAN, D. A., Adenovirus serotypes isolated from ocular infections in London, *Br. J. Ophthalmol.* **67:**111–114 (1983).

34. DE JONG, J. C., BIJLSMA, K., WERMENBOL, A. G., VERWEIJ-UIJTERWAAL, M. W., VAN DER AVOORT, H. G. A. M., WOOD, D. J., BAILEY, A. S., AND OSTERHAUS, A. D. M. E., Detection, typing, and subtyping of enteric adenoviruses 40 and 41 from fecal samples and observation of changing incidences of infections with these types and subtypes, *J. Clin. Microbiol.* **31:**1562–1569 (1993).

35. DE JONG, J. C., WIGAND, R., WADELL, G., KELLER, D., MUZERIE, C. J., WERMENBOL, A. G., AND SCHAAP, G. J. P., Adenovirus 37: Identification and characterization of a medically important new adenovirus type of subgroup D, *J. Med. Virol.* **7:**105–118 (1981).

36. DE JONG, P. J., VALDERRAMA, G., SPIGLAND, I., AND HORWITZ, M. S., Adenovirus isolates from urine of patients with acquired immunodeficiency syndrome, *Lancet* **1:**1293–1296 (1983).

37. DINGLE, J., AND LANGMUIR, A. D., Epidemiology of acute respiratory disease in military recruits, *Am. Rev. Respir. Dis.* **97:**1–65 (1968).

38. DUDDING, B. A., TOP, F. H., JR., WINTER, P. E., BUESCHER, E. L., LAMSON, T. H., AND LEIBOVITZ, A., Acute respiratory disease in military trainees: The adenovirus surveillance program, 1966–1971, *Am. J. Epidemiol.* **97:**187–198 (1973).

39. DUDDING, B. A., WAGNER, S. C., ZELLER, J. A., GMELICH, J. T., FRENCH, G. R., AND TOP, F. H., JR., Fatal pneumonia associated

with adenovirus type 7 in three military trainees, *N. Engl. J. Med.* **286:**1289–1292 (1972).

40. EDMONDSON, W. P., PURCELL, R. H., GUNDERFINGER, B. F. LOVE, J. W. P., LUDWING, W., AND CHANOCK, R. M., Immunization by selective infection with type 4 adenovirus grown in human diploid tissue culture. II. Specific protective effect against epidemic disease, *J.A.M.A.* **195:**159–165 (1966).

41. EDWARDS, K. M., BENNETT, S. R., GARNER, W. L., BRATTON, D. L., GLICK, A. D., GREENE, H. L., AND WRIGHT, P. F., Reye's syndrome associated with adenovirus infections in infants, *Am. J. Dis. Child.* **139:**343–346 (1985).

42. EDWARDS, K. M., THOMPSON, J., PAOLINI, J., AND WRIGHT, P. F., Adenovirus infections in young children, *Pediatrics* **76:**420–424 (1985).

43. ENDERS, J. F., BELL, J. A., DINGLE, J. H., FRANCIS, T., JR., HILLEMAN, M. R., HUEBNER, R. J., AND PAYNE, A. M. M., "Adenoviruses": Group name proposed for new respiratory-tract viruses, *Science* **124:**119–120 (1956).

44. ESPY, M. J., HIERHOLZER, J. C., AND SMITH, T. F., The effect of centrifugation on the rapid detection of adenovirus in shell vials, *Am. J. Clin. Pathol.* **88:**358–360 (1987).

45. EVANS, A. S., Latent adenovirus infections of the human respiratory tract, *Am. J. Hyg.* **67:**256–266 (1958).

46. EVANS, A. S., Clinical syndromes in adults caused by respiratory infections, *Med. Clin. North. Am.* **51:**803–818 (1967).

47. EVANS, A. S., CENABRE, L., WANAT, J., RICHARDS, V., NEIDERMAN, J. C., AND ACTIS, A., Acute respiratory infections in different ecologic settings. I. Argentine military recruits, *Am. Rev. Respir. Dis.* **108:**1311–1319 (1973).

48. EVANS, A. S., JEFFREY, C., AND NIEDERMAN, J. C., The risk of acute respiratory infections in two groups of young adults in Colombia, South America: A prospective seroepidemiologic study, *Am. J. Epidemiol.* **93:**463–471 (1971).

49. FIFE, K. H., ASHLEY, R., SHIELDS, A. F., CURFMAN, M. F., SALTER, D., MEYERS, J. D., AND COREY, L., Comparison of neutralization and DNA restriction enzyme methods for typing clinical isolates of human adenovirus, *J. Clin. Microbiol.* **22:**95–100 (1985).

50. FORD, E., NELSON, K. E., AND WARREN, D., Epidemiology of epidemic keratoconjunctivitis, *Epidemiol. Rev.* **9:**244–261 (1987).

51. FORSYTH, B. R., BLOOME, H. H., JOHNSON, K. M., AND CHANOCK, R. M., Etiology of primary atypical pneumonia in a military population, *J.A.M.A.* **191:**364–369 (1965).

52. FOX, J. P., BRANDT, C. D., WASSERMANN, F. E., HALL, C. E., SPIGLAND, I., KOGON, A., AND ELVEBACK, L. R., The Virus Watch Program: A continuing surveillance of viral infections in metropolitan New York families. VI. Observations of adenovirus infections: Virus excretion patterns, antibody response, efficiency of surveillance, patterns of infection and relation to illness, *Am. J. Epidemiol.* **89:**25–50 (1969).

53. FOX, J. P., HALL, C. E., AND COONEY, M. K., The Seattle Virus Watch. VII. Observations of adenovirus infections, *Am. J. Epidemiol.* **105:**362–386 (1977).

54. FOY, H. M., COONEY, M. K., AND HATLEN, J. B., Adenovirus type 3 epidemic associated with intermittent chlorination of a swimming pool, *Arch. Environ. Health* **17:**795–802 (1968).

55. FOY, H. M., COONEY, M. K., MCMAHAN, R., AND GRAYSTON, J. T., Viral and mycoplasmal pneumonia in a prepaid medical care group during an eight-year period, *Am. J. Epidemiol.* **97:**93–102 (1973).

56. FUCHS, N., AND WIGAND, R., Virus isolation and titration at 33°C and 37°C, *Med. Microbiol. Immunol.* **161:**123–126 (1975).

57. GARDNER, P. S., KNOX, E. G., COURT, S. D. M., AND GREEN, C. A., Virus infection and intussusception in childhood, *Br. Med. J.* **2:**697–700 (1962).

58. GLEZEN, W. P., WULFF, H., LAMB, G. A., RAY, C. G., CHIN, T. D. Y., AND WENNER, H. A., Patterns of virus infections in families with acute respiratory illnesses, *Am. J. Epidemiol.* **86:**350–361 (1966).

59. GRANDIEN, M., PETTERSSON, C.-A., SVENSSON, L., AND UHNOO, I., Latex agglutination test for adenovirus diagnosis in diarrheal disease, *J. Med. Virol.* **23:**311–316 (1987).

60. GRAYSTON, J. T., LASHOF, J. C., LOOSLI, C. G., AND JOHNSTON, P. B., Adenoviruses. III. Their etiological role in acute respiratory disease in civilian adults, *J. Infect. Dis.* **103:**93–101 (1958).

61. GRAYSTON, J. T., LOOSLI, C. G., SMITH, M., MCCARTHY, M. A., AND JOHNSTON, P. B., Adenoviruses. I. The effect of total incubation time in HeLa cell cultures on the isolation rate, *J. Infect. Dis.* **103:**75–85 (1958).

62. GRAYSTON, J. T., YANG, Y. F., JOHNSTON, P. B., AND KO, L. S., Epidemic keratoconjunctivitis on Taiwan: Etiological and clinical studies, *Am. J. Trop. Med.* **13:**492–498 (1964).

63. GREEN, M., MACKEY, J. K., WOLD, W. S. M., AND RIGDEN, P., Thirty-one human adenovirus serotypes (Ad1–Ad31) form five groups (A–E) based upon DNA genome homologies, *Virology* **93:**481–492 (1979).

64. GUTEKUNST, R. R., AND HEGGIE, A. D., Viremia and viruria in adenovirus infections: Detection in patients with rubella or rubelliform illness, *N. Engl. J. Med.* **264:**374–378 (1961).

65. GUYER, B., O'DAY, D. M., HIERHOLZER, J. C., AND SCHAFFNER, W., Epidemic keratoconjunctivitis: A community outbreak of mixed adenovirus type 8 and type 19 infection, *J. Infect. Dis.* **132:**142–150 (1975).

66. HALL, C. E., BRANDT, C. D., FROTHINGHAM, T. E., SPIGLAND, I., COONEY, M. K., AND FOX, J. P., The Virus Watch Program: A continuing surveillance of viral infections in metropolitan New York families. IX. A comparison of infections with several respiratory pathogens in New York and New Orleans families, *Am. J. Epidemiol.* **94:**367–385 (1971).

67. HARDING, S. P., MALLINSON, H., SMITH, J. L. S., AND CLEARKIN, L. G., Adult follicular conjunctivitis and neonatal ophthalmia in a Liverpool eye hospital, 1980–1984, *Eye* **1:**512–521 (1987).

68. HARMON, M. W., AND PAWLIK, K. M., Enzyme immunoassay for direct detection of influenza type A and adenovirus antigens in clinical specimens, *J. Clin. Microbiol.* **15:**5–11 (1982).

69. HARNETT, G. B., AND NEWNHAM, W. A., Isolation of adenovirus type 19 from the male and female genital tracts, *Br. J. Vener. Dis.* **57:**55–57 (1981).

70. HARNETT, G. B., PHILLIPS, P. A., AND GOLLOW, M. M., Association of genital adenovirus infection with urethritis in men, *Med. J. Aust.* **141:**337–338 (1984).

71. HARRIS, D. J., WULFF, H., RAY, C. G., POLAND, J. D., CHIN, T. D. Y., AND WENNER, H. A., Viruses and disease. III. An outbreak of adenovirus type 7A in a children's home, *Am. J. Epidemiol.* **93:** 399–402 (1971).

72. HIERHOLZER, J. C., Adenoviruses in the immunocompromised host, *Clin. Microbiol. Rev.* **5:**262–274 (1992).

73. HIERHOLZER, J. C., AND RODRIGUEZ, F. H., JR., Antigenically intermediate human adenovirus strain associated with conjunctivitis, *J. Clin. Microbiol.* **13:**395–397 (1981).

74. HIERHOLZER, J. C., STONE, Y. O., AND BRODERSON, J. R., Anti-

genic relationships among the 47 human adenoviruses determined in reference horse antisera, *Arch. Virol.* **121:**179–197 (1991).

75. HIERHOLZER, J. C., WIGAND, R., ANDERSON, L. J., ADRIAN, T., AND GOLD, J. W. M., Adenoviruses from patients with AIDS: A plethora of serotypes and a description of five new serotypes of subgenus D (types 43–47), *J. Infect. Dis.* **158:**804–813 (1988).

76. HIERHOLZER, J. C., WIGAND, R., AND DE JONG, J. C., Evaluation of human adenoviruses 38, 39, 40, and 41 as new serotypes, *Intervirology* **29:**1–10 (1988).

77. HILLEMAN, M. R., HAMPARIAN, V. V., KETLER, A., REILLY, C. M., MCCLELLAND, L., CORNFIELD, D., AND STOKES, J., JR., Acute respiratory illness among children and adults: Field study of contemporary importance of several viruses and appraisal of the literature, *J.A.M.A.* **180:**445–453 (1962).

78. HILLEMAN, M. R., AND WERNER, J. H., Recovery of new agents from patients with acute respiratory illness, *Proc. Soc. Exp. Biol. Med.* **85:**183–188 (1954).

79. HILLIS, W. D., COOPER, M. R., AND BANG, F. B., Adenovirus infections in West Bengal: I. Persistence of viruses in infants and young children, *Indian J. Med. Res.* **61:**980–988 (1973).

80. HOGAN, M. J., AND CRAWFORD, J. W., Epidemic keratoconjunctivitis (superficial punctate keratitis, keratitis subepithelialis, keratitis maculosa, keratitis nummularis), *Am. J. Ophthalmol.* **25:**1059–1078 (1942).

81. HORWITZ, M. S., Adenoviruses, in: *Virology*, 2nd ed. (B. N. FIELDS *ET AL.*, EDS.), pp. 1723–1740, Raven Press, New York, 1990.

82. HORWITZ, M. S., VALDERRAMA, G., KORN, R., AND SPIGLAND, I., Adenovirus isolates from the urines of AIDS patients: Characterization of group B recombinants, in: *Acquired Immune Deficiency Syndrome: UCLA Symposia on Molecular and Cellular Biology*, New Series, Vol. 16 (M. S. GOTTLIEB AND J. E. GROOPMAN, EDS.), pp. 187–207, Alan R. Liss, New York, 1984.

83. HYPPIÄ, T., Detection of adenovirus in nasopharyngeal specimens by radioactive and nonradioactive DNA probes, *J. Clin. Microbiol.* **21:**730–733 (1985).

84. ISHII, K., NAKAZONO, N., FUJINAGA, K., FUJII, S., KATO, M., OHTSUKA, H., OAKI, K., CHEN, C. W., LIN, C. C., SHEU, M. M., LIN, K. H., OUM, B. S., LEE, S. H., CHUN, C. H., YOSHII, T., AND YAMAZAKI, S., Comparative studies on aetiology and epidemiology of viral conjunctivitis in three countries of East Asia—Japan, Taiwan and South Korea, *Int. J. Epidemiol.* **16:**98–103 (1987).

85. JAVIER, R., RASKA, K., JR., MACDONALD, G. J., AND SHENK, T., Human adenovirus type 9-induced rat mammary tumors, *J. Virol.* **65:**3192–3202 (1991).

86. JAWETZ, E., The story of shipyard eye, *Br. Med. J.* **1:**873–878 (1959).

87. JEN, K. F., TAI, Y., LIN, Y. C., AND WANG, H. Y., The role of adenovirus in the etiology of infantile pneumonia and pneumonia complicating measles, *Chin. Med. J.* **81:**141–148 (1962).

88. JERNIGAN, J. A., LOWRY, B. S., HAYDEN, F. G., KYGER, S. A., CONWAY, B. P., GRÖSCHEL, D. H. M., AND FARR, B. M., Adenovirus type 8 epidemic keratoconjunctivitis in an eye clinic: Risk factors and control, *J. Infect. Dis.* **167:**1307–1313 (1993).

89. JOHANSSON, M. E., BRUNDIN, M., ADAMSON, L., GRILLNER, L., LANDQVIST, M., THÖRNER, Å., AND ZWEYGBERG WIRGART, B., Characterization of two genome types of adenovirus type 31 isolated in Stockholm during 1987, *J. Med. Virol.* **28:**63–68 (1989).

89a. JOHANSSON, M. E., ANDERSSON, M. A., AND THORNER, P. A.,

Adenoviruses isolated in the Stockholm area during 1987–1992: Restriction endonuclease analysis and molecular epidemiology, *Arch. Virol.* **137:**101–115 (1994).

90. JORDAN, W. S., JR., The frequency of infection with adenoviruses in a family study population, *Ann. N.Y. Acad. Sci.* **67:**273–278 (1957).

91. KAJI, M., KIMURA, M., KAMIYA, S., TATEWAKI, E., TAKAHASHI, T., NAKAJIMA, O., KOGA, T., ISHIDA, S., AND MAJIMA, Y., An epidemic of pharyngoconjunctival fever among school children in an elementary school in Fukuoka Prefecture, *Kyushu J. Med. Sci.* **12:**1–8 (1961).

92. KAJON, A. E., AND WADELL, G., Characterization of adenovirus genome type 7h: Analysis of its relationship to other members of serotype 7, *Intervirology* **33:**86–90 (1992).

93. KASEL, J. A., ALFORD, R. H., LEHRICH, J. R., BANKS, P. A., HUBER, M., AND KNIGHT, V., Adenovirus soluble antigens for human immunization, *Am. Rev. Respir. Dis.* **94:**170–174 (1966).

94. KELSEY, D. S., Adenovirus meningoencephalitis, *Pediatrics* **61:**291–293 (1978).

95. KEMP, M. C., HIERHOLZER, J. C., CABRADILLA, C. P., AND OBIJESKI, J. F., The changing etiology of epidemic keratoconjunctivitis: Antigenic and restriction enzyme analyses of adenovirus types 19 and 37 isolated over a 10-year period, *J. Infect. Dis.* **148:**24–33 (1983).

95a. KHOO, S., BAILEY, A. S., DE-JONG, J. C., AND MANDAL, B. K., Adenovirus infections in human immunodeficiency virus-positive patients: Clinical features and molecular epidemiology, *J. Infect. Dis.* **172:**629–637 (1995).

96. KJELLEN, L., ZETTERBERG, B., AND SVEDMYR, A., An epidemic among Swedish children caused by adenovirus type 3, *Acta Paediatr. Scand.* **46:**561–568 (1957).

97. KLENK, E. L., GWALTNEY, J. M., AND BASS, J. W., Bacteriologically proved pertussis and adenovirus infection, *Am. J. Dis. Child.* **124:**203–207 (1972).

98. KLOENE, W., BANG, F. B., CHAKRABORTY, S. M., COOPER, M. R., KULEMANN, H., OTA, M., AND SHAH, K. V., A two-year respiratory virus survey in four villages in West Bengal, India, *Am. J. Epidemiol.* **92:**307–320 (1970).

99. KOTLOFF, K. L., LOSONSKY, G. A., MORRIS, J. G., WASSERMAN, S. S., SINGH-NAZ, N., AND LEVINE, M. M., Enteric adenovirus infection and childhood diarrhea: An epidemiologic study in three clinical settings, *Pediatrics* **84:**219–225 (1989).

100. KULCSÁR, G., VUTSKITS, ZS., NÁSZ, I., DÁN, P., AND LÉB, J., Viruses isolated from appendicitis cases in childhood, *Zentralbl. Bakteriol. I. Abt. Orig.* **215:**506–510 (1970).

101. LANG, W. R., HOWDEN, C. W., LAWS, J., AND BURTON, J. F., Bronchopneumonia with serious sequelae in children with evidence of adenovirus type 21 infection, *Br. Med. J.* **1:**73–79 (1969).

102. LANGMUIR, A. D., JARRETT, E. T., AND HOLLAENDER, A., Studies of the control of acute respiratory disease among Navy recruits, *Am. J. Hyg.* **48:**240–251 (1948).

103. LEHANE, D. E., NEWBERG, N. R., AND BEAM, W. E., Environmental modifications for controlling acute respiratory disease, *Am. J. Epidemiol.* **99:**139–144 (1974).

104. LEHTOMÄKI, K., JULKUNEN, I., SANDELIN, K., SALONEN, J., VIRTANEN, M., RANKI, M., AND HOVI, T., Rapid diagnosis of respiratory adenovirus infections in young adult men, *J. Clin. Microbiol.* **24:**108–111 (1986).

105. LI, Q. G., HAMBRAEUS, J., AND WADELL, G., Genetic relationship between thirteen genome types of adenovirus 11, 34 and 35 with different tropisms, *Intervirology* **32:**338–350 (1991).

106. LI, Q. G., AND WADELL, G., Comparison of 17 genome types of

adenovirus type 3 identified among strains recovered from six continents, *J. Clin. Microbiol.* **26:**1009–1015 (1988).

107. LOOSLI, C. G., LEMON, H. M., ROBERTSON, O. H., AND HAMBURGER, M., Transmission and control of respiratory disease in Army barracks, *J. Infect. Dis.* **90:**153–164 (1952).

108. MANTYJARVI, R., Adenovirus infections in servicemen in Finland, *Ann. Med. Exp. Fenn.* **44:**1–43 (1966).

109. McCORMICK, D. P., WENZEL, R. P., DAVIES, J. A., AND BEAM, W. E., Nasal secretion protein responses in patients with wild-type adenovirus disease, *Infect. Immun.* **6:**282–288 (1972).

110. McMINN, P. C., STEWART, J., AND BURRELL, C. J., A community outbreak of epidemic keratoconjunctivitis in central Australia due to adenovirus type 8, *J. Infect. Dis.* **164:**1113–1118 (1991).

111. MICHAELS, M. G., GREEN, M., WALD, E. R., AND STARZL, T. E., Adenovirus infection in pediatric liver transplant recipients, *J. Infect. Dis.* **165:**170–174 (1992).

112. MILLER, L. F., RYTEL, M., PIERCE, W. E., AND ROSENBAUM, M. J., Epidemiology of nonbacterial pneumonia among Naval recruits, *J.A.M.A.* **185:**92–99 (1963).

113. MISTCHENKO, A. S., LENZI, H. L., THOMPSON, F. M., MOTA, E. M., VIDAURRETA, S., NAVARI, C., AND GRINSTEIN, S., Participation of immune complexes in adenovirus infection, *Acta Paediatr.* **81:**983–988 (1992).

114. MONTO, A. S., NAPIER, J. A., AND METZNER, H. L., The Tecumseh study of respiratory illnesses. I. Plan of study and observations on syndromes of acute respiratory disease, *Am. J. Epidemiol.* **94:**269–279 (1971).

115. MORIN, J. E., LUBECK, M. D., BARTON, J. E., CONLEY, A. J., DAVIS, A. R., AND HUNG, P. P., Recombinant adenovirus induces antibody response to hepatitis B virus surface antigen in hamsters, *Proc. Natl. Acad. Sci. USA* **84:**4626–4630 (1987).

116. MUELLER, R. E., MULDOON, R. L., AND JACKSON, G. C., Communicability of enteric live adenovirus type 4 vaccine in families, *J. Infect. Dis.* **119:**60–66 (1969).

117. MUFSON, M. A., BELSHE, R. B., HORRIGAN, T. J., AND ZOLLAR, L. M., Cause of acute hemorrhagic cystitis in children, *Am. J. Dis. Child.* **126:**605–609 (1973).

118. MURPHY, G. F., WOOD, D. P., JR., McROBERTS, J. W., AND HENSLEE-DOWNEY, P. J., Adenovirus-associated hemorrhagic cystitis treated with intravenous ribavirin, *J. Urol.* **149:**565–566 (1993).

119. NAHMIAS, A. J., GRIFFITH, D., AND SNITZER, J., Fatal pneumonia associated with adenovirus type 7, *Am. J. Dis. Child.* **114:**36–41 (1967).

120. NAUHEIM, R. C., ROMANOWSKI, E. G., ARAULLO-CRUZ, T., KOWALSKI, R. P., TURGEON, P. W., STOPAK, S. S., AND GORDON, Y. J., Prolonged recoverability of desiccated adenovirus type 19 from various surfaces, *Ophthalmology* **97:**1450–1453 (1990).

121. NELSON, K. E., GAVITT, F., BATT, M. D., KALLICK, C. A., REDDI, K. T., AND LEVIN, S., The role of adenoviruses in the pertussis syndrome, *J. Pediatr.* **86:**335–341 (1975).

122. NIEL, C., MORAES, M. T. B., MISTCHENKO, A. S., LEITE, J. P. G., AND GOMES, S. A., Restriction site mapping of four genome types of adenovirus types 3 and 7 isolated in South America, *J. Med. Virol.* **33:**123–127 (1991).

123. NODA, M., MIYAMOTO, Y., IKEDA, Y., MATSUISHI, T., AND OGINO, T., Intermediate human adenovirus type 22/H10,19,37 as a new etiological agent of conjunctivitis, *J. Clin. Microbiol.* **29:**1286–1289 (1991).

124. NUMAZAKI, Y., KUMASAKA, T., YANO, N., YAMANAKA, M., MIYAZAWA, T., TAKAI, S., AND ISHIDA, N., Further study of acute hemorrhagic cystitis due to adenovirus type 11, *N. Engl. J. Med.* **289:**344–347 (1973).

125. PACINI, D. L., COLLIER, A. M., AND HENDERSON, F. W., Adenovirus infections and respiratory illnesses in children in group day care, *J. Infect. Dis.* **156:**920–927 (1987).

126. PADERSTEIN, R., Was ist Schwimmbad-Konjunktivitis? *Klin. Monatsbl. Augenheilkd.* **72:**634–642 (1925).

127. PARROTT, R. H., ROWE, W. P., HUEBNER, R. J., BERNTON, H. W., AND McCULLOUGH, N. B., Outbreak of febrile pharyngitis and conjunctivitis associated with type 3 adenoidal–pharyngeal–conjunctival virus infection, *N. Engl. J. Med.* **251:**1087–1090 (1954).

128. PAYNE, S. B., GRILLI, E. A., SMITH, A. J., AND HOSKINS, T. W., Investigation of an outbreak of adenovirus type 3 infection in a boys' boarding school, *J. Hyg. Camb.* **93:**277–283 (1984).

129. PIERCE, W. E., ROSENBAUM, M. J., EDWARDS, E. A., PECKINPAUGH, R. O., AND JACKSON, G. G., Live and inactivated adenovirus vaccines for the prevention of acute respiratory illness in Naval recruits, *Am. J. Epidemiol.* **87:**237–246 (1968).

130. PIERCE, W. E., STILLE, W. T., AND MILLER, L. F., A preliminary report on effects of routine military inoculations on respiratory illness, *Proc. Soc. Exp. Biol. Med.* **114:**369–372 (1963).

131. PHILIPSON, L., PETERSON, U., AND LINDBERG, U., Molecular biology of adenoviruses, *Virol. Monogr.* **14:**1–115 (1975).

132. RICHMOND, S. J., WOOD, D. J., AND BAILEY, A. S., Recent respiratory and enteric adenovirus infection in children in the Manchester area, *J. R. Soc. Med.* **81:**15–18 (1988).

133. RODRIGUEZ, W. J., KIM, H. W., BRANDT, C. D., SCHWARTZ, R. H., GARDNER, M. K., JEFFRIES, B., PARROTT, R. H., KASLOW, R. A., SMITH, J. I., AND TAKIFF, H., Fecal adenoviruses from a longitudinal study of families in metropolitan Washington, DC: Laboratory, clinical, and epidemiologic observations, *J. Pediatr.* **107:**514–520 (1985).

134. ROSE, H. M., LAMSON, T. H., AND BUESCHER, E. L., Adenoviral infection in military recruits: Emergence of type 7 and type 21 infections in recruits immunized with type 4 oral vaccine, *Arch. Environ. Health* **21:**356–361 (1970).

135. ROSS, J. G., POTTER, C. W., AND ZACHARY, R. B., Adenovirus infection in association with intussusception in infancy, *Lancet* **2:**221–223 (1962).

136. ROWE, W. P., HUEBNER, R. J., GILMORE, L. K., PARROTT, R. H., AND WARD, T. G., Isolation of a cytopathogenic agent from human adenoids undergoing spontaneous degeneration in tissue culture, *Proc. Soc. Exp. Biol. Med.* **84:**570–573 (1953).

137. RUUSKANEN, O., AROLA, M., PUTTO-LAURILA, A., MERTSOLA, J., MEURMAN, O., VILJANEN, M. K., AND HALONEN, P., Acute otitis media and respiratory virus infections, *Pediatr. Infect. Dis. J.* **8:**94–99 (1989).

138. SARKKINEN, H., RUUSKANEN, O., MEURMAN, O., PUHAKKA, H., VIROLAINEN, E., AND ESKOLA, J., Identification of respiratory virus antigens in middle ear fluids of children with acute otitis media, *J. Infect. Dis.* **151:**444–448 (1985).

139. SCHMIDT, O. W., COONEY, M. K., AND FOY, H. M., Adenovirus-associated virus in adenovirus type 3 conjunctivitis, *Infect. Immun.* **11:**1362–1370 (1975).

140. SCHMITZ, H., WIGAND, R., AND HEINRICH, W., Worldwide epidemiology of human adenovirus infections, *Am. J. Epidemiol.* **117:**455–466 (1983).

141. SCHONLAND, M., STRONG, M. L., AND WESLEY, A., Fatal adenovirus pneumonia: Clinical and pathological features, *S. Afr. Med. J.* **50:**1748–1751 (1976).

142. SCHWARTZ, A. R., TOGO, Y., AND HORNICK, R. B., Clinical evaluation of live, oral types 1, 2, and 5 adenovirus vaccines, *Am. Rev. Respir. Dis.* **109**:233–238 (1974).

143. SHIELDS, A. F., HACKMAN, R. C., FIFE, K. H., COREY, L., AND MEYERS, J. D., Adenovirus infections in patients undergoing bone-marrow transplantation, *N. Engl. J. Med.* **312**:529–533 (1985).

144. SHULT, P. A., POLYAK, F., DICK, E. C., WARSHAUER, D. M., KING, L. A., AND MANDEL, A. D., Adenovirus 21 infection in an isolated antarctic station: Transmission of the virus and susceptibility of the population, *Am. J. Epidemiol.* **133**:599–607 (1991).

145. SIMILÄ, S., YLIKORKALA, O., AND WASZ-HOCKERT, O., Type 7 adenovirus pneumonia, *J. Pediatr.* **79**:605–611 (1971).

146. SMITH, T. J., BUESCHER, E. L., TOP, F. H., JR., ALTEMEIER, W. A., AND MCCOWN, J. M., Experimental respiratory infection with type 4 adenovirus vaccine in volunteers: Clinical and immunological responses, *J. Infect. Dis.* **122**:239–248 (1970).

147. SPRAGUE, J. B., HIERHOLZER, J. C., CURRIER, R. W. II, HATTWICH, M. A. W., AND SMITH, M. D., Epidemic keratoconjunctivitis: A severe industrial outbreak due to adenovirus type 8, *N. Engl. J. Med.* **289**:1341–1346 (1973).

148. STALDER, H., HIERHOLZER, J. C., AND OXMAN, M. N., New human adenovirus (candidate adenovirus type 35) causing fatal disseminated infection in a renal transplant recipient, *J. Clin. Microbiol.* **6**:257–265 (1977).

149. STANLEY, E. D., AND JACKSON, G. G., Spread of enteric live adenovirus type 4 vaccine in married couples, *J. Infect. Dis.* **119**:51–59 (1969).

150. STERNER, G., Adenovirus infection in childhood: An epidemiological and clinical survey among Swedish children, *Acta Paediatr. Scand. [Suppl.]* **142**:1–30 (1962).

151. STRAUBE, R. C., THOMPSON, M. A., VAN DYKE, R. B., WADELL, G., CONNOR, J. D., WINGARD, D., AND SPECTOR, S. A., Adenovirus type 7b in a children's hospital, *J. Infect. Dis.* **147**:814–819 (1983).

152. SUTTON, R. N. P., PULLEN, H. J. M., BLACKLEDGE, P., BROWN, E. H., SINCLAIR, L., AND SWIFT, P. N., Adenovirus type 7: 1971–74, *Lancet* **2**:987–991 (1976).

153. SVARTZ-MALMBERG, G., AND GERMANIS, M., Adenovirus type 8-associated keratoconjunctivitis: Hospital infections and secondary spread in Stockholm, 1967, *Scand J. Infect. Dis.* **1**:161–168 (1969).

154. TAI, F. H., AND GRAYSTON, J. T., Adenovirus neutralizing antibodies in persons on Taiwan, *Proc. Soc. Exp. Biol. Med.* **109**:881–884 (1962).

155. TAI, F. H., GRAYSTON, J. T., JOHNSON, P. B., AND WOOLDRIDGE, R. L., Adenovirus infections in Chinese Army recruits on Taiwan, *J. Infect. Dis.* **107**:160–164 (1960).

156. TAKEUCHI, R., NOMURA, Y., KOJIMA, M., UCHIO, E., KOBAYASHI, N., AND MATUMOTO, M., A nosocomial outbreak of epidemic keratoconjunctivitis due to adenovirus type 37, *Microbiol. Immunol.* **34**:749–754 (1990).

157. TOP, F. H., JR., Control of adenovirus acute respiratory disease in US Army trainees, *Yale J. Biol. Med.* **48**:185–195 (1975).

158. TOP, F. H., JR., GROSSMAN, R. A., BARTELLONI, P. J., SEGAL, H. E., DUDDING, B. A., RUSSELL, P. K., AND BUESCHER, E. L., Immunization with live types 7 and 4 adenovirus vaccines. I. Safety, infectivity, and potency of adenovirus type 7 vaccine in humans, *J. Infect. Dis.* **124**:148–154 (1971).

159. UHNOO, I., SVENSSON, L., AND WADELL, G., Enteric adenoviruses, *Baillieres Clin. Gastroenterol.* **4**:627–642 (1990).

160. UHNOO, I., WADELL, G., SVENSSON, L., AND JOHANSSON, M. E., Importance of enteric adenoviruses 40 and 41 in acute gastroenteritis in infants and young children, *J. Clin. Microbiol.* **20**:365–372 (1984).

161. VAN DER AVOORT, H. G. A. M., ADRIAN, T., WIGAND, R., WERMENBOL, A. G., ZOMERDIJK, T. P. L., AND DE JONG, J. C., Molecular epidemiology of adenovirus type 21 in The Netherlands and the Federal Republic of Germany from 1960 to 1985, *J. Clin. Microbiol.* **24**:1084–1088 (1986).

162. VANDER VEEN, J., The role of adenoviruses in respiratory disease, *Am. Rev. Respir. Dis.* **88**:167–180 (1963).

163. VASTINE, D. W., WEST, C. E., YAMASHIROYA, H., SMITH, R., SAXTAN, D. D., GIESER, D. I., AND MUFSON, M. A., Simultaneous nosocomial and community outbreak of epidemic keratoconjunctivitis with types 8 and 19 adenovirus, *Trans. Am. Acad. Ophthalmol. Otolaryngol.* **81**:OP826–OP840 (1976).

164. VIHMA, L., Surveillance of acute viral respiratory disease in children, *Acta Paediatr. Scand. [Suppl.]* **192**:8–52 (1969).

165. VIRTANEN, M., PALVA, A., LAAKSONEN, M., HALONEN, P., SODERLUND, H., AND RANKI, M., Novel test for rapid viral diagnosis: Detection of adenovirus in nasopharyngeal mucus aspirates by means of nucleic-acid sandwich hybridisation, *Lancet* **1**:381–383 (1983).

166. WADELL, G., Molecular epidemiology of human adenoviruses, *Curr. Top. Microbiol. Immunol.* **110**:191–220 (1984).

167. WADELL, G., COONEY, M. K., DA COSTA LINHARES, A., DE SILVA, L., KENNETT, M. L., KONO, R., REN, G.-F., LINDMAN, K., NASCIMENTO, J. P., SCHOUB, B. D., AND SMITH, C. D., Molecular epidemiology of adenoviruses: Global distribution of adenovirus 7 genome types, *J. Clin. Microbiol.* **21**:403–408 (1985).

168. WANG, S. S., AND FELDMAN, H. A., Pharyngeal isolations of adenovirus 31 from a family population, *Am. J. Epidemiol.* **104**:272–277 (1976).

169. WARNER, J. O., AND MARSHALL, W. C., Crippling lung disease after measles and adenovirus infection, *Br. J. Dis. Chest* **70**:89–94 (1976).

170. WARREN, D., NELSON, K. E., FARRAR, J. A., HURWITZ, E., HIERHOLZER, J., FORD, E., AND ANDERSON, L. J., A large outbreak of epidemic keratoconjunctivitis: Problems in controlling nosocomial spread, *J. Infect. Dis.* **160**:938–943 (1989).

171. WEBB, D. H., SHIELDS, A. F., AND FIFE, K. H., Genomic variation of adenovirus type 5 isolates recovered from bone marrow transplant recipients, *J. Clin. Microbiol.* **25**:305–308 (1987).

172. WENMAN, W. M., PAGTAKHAN, R. D., REED, M. H., CHERNICK, V., AND ALBRITTON, W., Adenovirus bronchiolitis in Manitoba. Epidemiologic, clinical, and radiologic features, *Chest* **81**:605–609 (1982).

173. WENNER, H. A., BERAN, G. W., WESTON, J., AND CHIN, T. D. Y., with collaboration of ANDERSON, N. W., AND GOLDSMITH, R., The epidemiology of acute respiratory illness. I. Observations on adenovirus infections prevailing in a group of families, *J. Infect. Dis.* **101**:275–286 (1957).

174. WIGAND, R., BAUMEISTER, H. G., MAASS, G., KÜHN, J., AND HAMMER, H. J., Isolation and identification of enteric adenoviruses, *J. Med. Virol.* **11**:233–240 (1983).

175. WIGAND, R., AND SCHULZ, R., Laboratoriumspraxis be Adenoviren. II. Empfindlichkeit verschiedener Zellkulturen bei Endpunkttitration, *Zentralbl. Bakteriol. Parasitenkd. Infektionskr. Abt. 1 Orig. Reihe A* **231**:31–41 (1975).

176. WOLD, W. S. M., MACKEY, J. K., RIGDEN, P., AND GREEN, M., Analysis of human cancer DNAs for DNA sequences of human adenovirus serotypes 3, 7, 11, 14, 16, and 21 in group B, *Cancer Res.* **39**:3479–3484 (1979).

177. YOLKEN, R. H., AND FRANKLIN, C. C., Gastrointestinal adenovirus: An important cause of morbidity in patients with necrotizing enterocolitis and gastrointestinal surgery, *Pediatr. Infect. Dis.* **4:**42–47 (1985).

178. YOLKEN, R. H., LAWRENCE, F., LEISTER, F., TAKIFF, H. E., AND STRAUSS, S. E., Gastroenteritis associated with enteric type adenovirus in hospitalized infants, *J. Pediatr.* **101:**21–26 (1982).

179. ZAHRADNIK, J. M., SPENCER, M. J., AND PORTER, D. D., Adenovirus infection in the immunocompromised patient, *Am. J. Med.* **68:**725–732 (1980).

180. ZWEIGHAFT, R. M., HIERHOLZER, J. C., AND BRYAN, J. A., Epidemic keratoconjunctivitis at a Vietnamese refugee camp in Florida, *Am. J. Epidemiol.* **106:**399–407 (1977).

12. Suggested Reading

GINSBERG, H. S. (ED.), *The Adenoviruses*, Plenum Press, New York, 1984.

GINSBERG, H. S., Adenoviruses, in: *Microbiology*, 4th ed. (B. D. DAVIS *ET AL.*, EDS.), pp. 915–928, J. B. Lippincott, Philadelphia, 1990.

GWALTNEY, J. M., JR., AND HENDLEY, J. O., Acute respiratory infectious, in: *Communicable and Infectious Diseases* (P. F. WEHRLE, AND F. H. TOP, SR., EDS.), pp. 79–100, C. V. Mosby, St. Louis, 1981.

HIERHOLZER, J. C., Adenoviruses in the immunocompromised host, *Clin. Microbiol. Rev.* **5:**262–274 (1992).

HORWITZ, M. S., Adenoviridae and their replication, in: *Virology*, 2nd ed. (N. F. FIELD, ED.), pp. 1679–1719, Raven Press, New York, 1990.

HORWITZ, M. S., Adenoviruses, in: *Virology*, 2nd ed. (N. F. FIELD, ED.), pp. 1723–1740, Raven Press, New York, 1990.

SCHMITZ, H., WIGAND, R., AND HEINRICH, W., Worldwide epidemiology of human adenovirus infections, *Am. J. Epidemiol.* **117:**455–466 (1983).

WADELL, G., Molecular epidemiology of human adenoviruses, *Curr. Top. Microbiol. Immunol.* **110:**191–220 (1984).

African Hemorrhagic Fevers Caused by Marburg and Ebola Viruses

Robert E. Shope and James M. Meegan

1. Introduction

Marburg and Ebola viruses are morphologically and genetically similar, immunologically distinct rod-shaped agents in the family Filoviridae. They produce acute hemorrhagic fever in man.[41a] Although other viruses cause a rather similar disease in Africa and differential diagnosis of a sporadic case cannot be made on clinical grounds, the syndrome associated with infection by these two agents is sufficiently unique and unvarying to distinguish it from yellow fever, Lassa fever, and other infections whenever a cluster of cases occurs. For this reason, the term African hemorrhagic fever (AFHF), rather than Marburg or Ebola disease, is used here to refer to clinical infection caused by either virus. Explosive emergence, high mortality, nosocomial secondary transmission, and ecological mystery have combined to draw worldwide attention to these infections.

This chapter is a revised version of Chapter 4, Karl M. Johnson, *Viral Infections of Humans*, 3rd edition, Plenum Press, 1989.

Robert E. Shope • Department of Pathology, The University of Texas Medical Branch at Galveston, Galveston, Texas 77555. **James M. Meegan** • Division of Microbiology and Infectious Diseases, National Institute of Allergy and Infectious Diseases, National Institutes of Health, Bethesda, Maryland 20892.

2. Historical Background

These viruses have the briefest of histories. Information to be presented is derived largely from five rather dramatic epidemics plus sporadic cases. Marburg virus was first isolated during an epidemic in laboratory workers processing kidney cells from African monkeys in 1967. Cases occurred in Marburg and Frankfort am Main, Germany, and Belgrade, Yugoslavia.[46,71,72] A second focal outbreak took place in South Africa in 1975.[26] Sporadic cases have since been recognized in Kenya. In 1980, two cases were reported with one surviving[70]; in 1982, one nonfatal case was diagnosed near the site where the index case in 1975 was presumably exposed[81]; and in 1987, one fatality occurred.[41] Ebola virus was discovered nearly concurrently in Zaire and Sudan[7,39,54] in association with epidemics comprising more than 500 cases during 1976.[19,20]

A smaller Ebola outbreak of 34 patients (22 deaths; case fatality rate 64%) occurred in Sudan in 1979.[1b,14,47a] There were no outbreaks until 1989, when a shipment of monkeys infected with Ebola virus was imported from the Philippines by a commercial biological supply company in the United States.[1a,17,20b,20c,36b,75c] This outbreak was focused in Reston, Virginia, and hundreds of monkeys died or were sacrificed. The Ebola strain causing this outbreak was termed Ebola-Reston, and since seven animal handlers showed signs of infection but without clinical disease, there is a possibility that this Philippine monkey strain of Ebola is less virulent for man.[30,47b,75b] After this outbreak, there were only sporadic cases of Ebola reported in the first half of the decade. A small, quickly contained monkey outbreak occurred in 1992 in Italy in a shipment of Philippine monkeys.[30] A single case of Ebola was seen in a Swiss ethnologist who did an autopsy on a chimpanzee that died as a part of a chimpanzee outbreak in the Tai forest area of Ivory Coast.[44a,47c]

Then in early May, 1995, health authorities in Zaire noticed a cluster of hemorrhagic disease cases at a hospital in Kikwit (550 km east-southeast of Kinshasa). The vast majority of cases resulted from close contact with patients, and aerosol transmission did not appear to play a major role in virus transmission. Thus, the outbreak was controlled by limiting contact transmission, and the last case occurred in late June, 1995. In all, 315 cases were detected, and 77% (244) of these died.[50a,52a,75d]

In early 1996, an isolated outbreak occurred in Gabon with 37 cases and 21 deaths (case fatality rate of 56.8%) (unpublished reports, WHO). At the same time a limited monkey outbreak occurred in a commercial colony in Texas after importation of Philippine monkeys.[20a]

Because of the severe disease produced by Marburg and Ebola viruses and the high potential hazard incurred during laboratory manipulation of them, progress in understanding both viral biology and epidemiology has been limited. Few laboratories in the world possess the safety facilities necessary for making specific diagnosis of infection, much less the resources required for intensive research.

3. Methodology Used in Epidemiologic Analysis

3.1. Sources of Morbidity and Mortality Data

No country where these viruses are known or presumed to be endemic has established a formal requirement for reporting cases of viral hemorrhagic disease other than yellow fever. Nevertheless, all known cases of such disease caused by Marburg and Ebola viruses have been reported to the World Health Organization (WHO), and this body has urged that the clinical syndrome be added to the list of internationally notifiable communicable diseases.[79] It is likely that sporadic human illness caused by these and other viruses that produce hemorrhagic fever in Africa has been, and continues to be, unrecognizable because of the lack of specific diagnostic capability on much of that continent. Furthermore, despite the high mortality observed during outbreaks, which suggests that such events are likely to be recognized, serological sur-

veys disclose that human infection with these agents may be more common and geographically widespread than was previously believed.[4]

3.2. Laboratory Diagnosis

3.2.1. Recovery of Virus. Marburg and Ebola viruses can be isolated from acutely ill patients.[11,20,61,84] In the few cases examined, viremia was present until death or for an interval of at least 1 week during acute illness. In Kikwit, virus could be recovered from postmortem samples, and epidemiologically it appeared that many patients contracted the disease during ritual washing of bodies in preparation for burial.[50a,52a,75d] Viremia and antigenemia during the acute phase is very high, and antigen-capture ELISA has been very useful for diagnosis. In Kikwit, virus was recovered from blood-contaminated syringes that had sat at ambiant temperature for over 5 days. Biopsies and "skin snips" at postmortem have yielded virus, and those fixed and immunohistochemically stained have revealed large amounts of antigen in many organs.[52a,75d]

3.2.2. Measurement of Virus-Specific Antibodies. Persons who survive acute infection develop low-titered complement-fixing (CF) antibodies as well as antibodies detectable by indirect immunofluorescence (IIF) and enzyme immunoassay (EIA).[4,9,40,61,68] Antibodies appear sooner, reach higher levels, and persist for much longer when measured by the IIF and EIA methods. The IIF and EIA procedures also can be used to detect specific antibodies of the immunoglobulin M (IgM) class.[84] These rarely persist at high levels for longer than 3 months, and thus can serve to make valid retrospective diagnosis where specimens are not available during the acute stage of infection.

Despite much effort, neutralizing antibodies are hard to measure for Marburg and Ebola viruses.[80,82] There have been concerns about nonspecific positive results with the IIF. The enzyme-linked immunosorbent assay, Western blot, and blocking assay offer confirmatory evidence of the specificity of IIF.[4]

3.2.3. Interpretation. When appropriate specimens are obtained, Marburg and Ebola viruses are recovered from blood of nearly all acute patients tested. Similarly, antigens of these agents were readily detected in acute and postmortem cases.[22,39,43] Every virus-confirmed survivor of either infection has developed specific antibodies. Thus, the tools for recognition of infection, at least in its severe clinical form, are highly sensitive and reliable.

3.3. Surveys

Although IIF antibodies in titers of at least 1 : 32 uniformly appear in sera of patients who survive AFHF, the interpretation of data derived from population surveys using this method is somewhat less certain. This situation derives from the fact that a few Ebola-positive sera with titers of 1 : 4 to 1 : 64 were found in indigenous populations on a continent not thought to harbor this virus.[76,78] We now know, however, that Ebola-related viruses are found in the Philippines,[30,47b] and perhaps their distribution is even wider outside of Africa.[4,36a,38]

Serological surveys using the IIF in the Central African Republic,[28a,38] Ethiopia,[73] Madagascar,[51] Cameroon,[53] Nigeria,[74] Guinea,[5] Kenya,[37] and Liberia[78] indicate wide distribution of Ebola-reactive antibodies and, in lower prevalence, Marburg-reactive antibodies. The specificity of these reactions requires confirmation by the EIA.

3.4. Clinical Diagnosis of Acute Infection

The clinical features of AFHF are detailed in Section 8. The occurrence of acute fever, severe headache, malaise, myalga, hemorrhagic signs, and high mortality was used to estimate incidence of disease caused by Ebola virus in both Zaire and Sudan.[19,20] Residents of the affected regions repeatedly reported that this disease pattern in epidemic form was an event not common in their villages.

Clinical diagnosis of AFHF in epidemic situations is probably quite reliable, except possibly in infants less than 1 year of age, for whom no confirmatory laboratory data are available. This judgment is based on the high rate of laboratory confirmation of suspected cases and the absence or paucity of specific antibodies among persons suffering no or mild illness after direct contact with cases or even residence in the same area.

4. Characteristics of Marburg and Ebola Viruses

4.1. Morphology and Morphogenesis

Marburg and Ebola viruses are pleomorphic, usually seen as large rods, about 80–90 nm in diameter but varying in length from 600 nm to several micrometers.[1,22,39,56] Brushlike spikes protrude from an outer virionic membrane. Branching forms are commonly seen, as are twisted rods and bulbous protrusions at ends of particles. Infectivity is associated with Ebola filamentous forms of 970 nm and with forms of 790 nm for Marburg virus. These viruses contain RNA.[42,57,65] They have essential lipids on their surface membranes,[6,45] and contain an internal helical nucleocapsid core. Morphogenesis occurs in the cytoplasm of infected cells where inclusions composed of a matrix containing nucleocapsids are formed, and virus maturation and release occurs by budding through the host cell plasma membrane.[50,55]

There are 7 genes generally organized in a fashion similar to viruses of the family Paromyxoviridae.[59,59b] Replication occurs in a manner similar to that seen for rhabdoviruses and paramyxoviruses.[25,47d] Seven viral proteins have been well studied. The heavily glycosylated surface glycoprotein makes up the virion surface spikes.[27a,59b] A secreted glycoprotein seen during virus growth in cells is a unique feature.

4.2. Physical Properties

Marburg and Ebola viruses are only moderately thermolabile, but complete inactivation requires heating to 60°C for 1 hr. They are stable indefinitely at −70°C and persist well at 4°C or room temperature for several days.[6] Infectivity is preserved by lyophilization but destroyed after variable intervals of exposure to ultraviolet or γ irradiation[15a,45] (L. Elliott, M. Dudley, and K. M. Johnson, CDC, unpublished data, 1979). Virions have a buoyant density of about 1.14 g in potassium tartrate gradients.[41]

4.3. Chemical Properties

These viruses are inactivated by brief exposure to a variety of chemicals including phenol, peracetic acid, sodium hypochlorite, methyl alcohol, ether, and sodium deoxycholate.[6]

4.4. Biological Properties

Marburg and Ebola viruses infect a wide range of cultured cells from mammals but do not replicate in cells of birds, amphibians, reptiles, or arthropods so far tested.[35,77] Both viruses grow well in vero, MA-104, and SW-13, all of which have proved useful for virus isolation. Although a ragged cytopathic effect has been observed in some continuous cell lines, this property is not useful for assay of infectivity because IIF reveals infection well beyond the dilutions at which morphological cell changes are observed.[65,66,80] These viruses also infect most commonly used laboratory animals. Monkeys are highly sensitive and usually succumb to infection after an illness somewhat resembling that seen in man.[10,29,63] Serial passages are usually required to induce death in hamsters and guinea pigs, which otherwise serve as convenient hosts for preparation of immune reagents.[7,60,80,85] Marburg virus is not pathogenic for mice,[44] but a Zaire strain of Ebola virus induced lethal infection in suckling mice,[54] and some strains have become lethal for adult mice upon passage. Research is progressing toward correlating biological properties with nucleotide sequence data.[12a,16a,41,47a,57a,59,59a,59b]

4.5. Serological Relationships

Marburg virus and specific antisera to this agent were tested against reagents for all previously known viruses causing hemorrhagic fever and a large number of arthropod-transmitted viruses by CF, hemagglutination inhibition, or neutralization methods. No immunologic relationships were demonstrated.[13] Somewhat surprisingly, in view of their morphological similarity, Ebola and Marburg viruses were found not to share antigens when examined by IIF and CF techniques.[39,80]

5. Descriptive Epidemiology

5.1. Prevalence and Incidence

Current information based upon older serological assays is inadequate to indicate the prevalence and incidence of Marburg and Ebola virus infections in the general population in endemic areas. Their occurrence has thus far been recognized only through outbreaks of clinically typical cases that have occurred in localized geographic areas.

5.2. Epidemic Behavior and Contagiousness

5.2.1. Marburg Virus.

In August and September 1967, 30 cases of acute hemorrhagic fever occurred in Marburg and Frankfurt am Main, Germany, and Belgrade, Yugoslavia. There were seven deaths, and five of the illnesses resulted from secondary infection of persons in contact with patients. Each outbreak occurred among personnel of laboratories engaged in processing tissues from vervet monkeys (*Cercopithecus aethiops*) for production of poliovirus vaccine, and all the monkeys had come from a single source in Uganda during the month prior to the outbreak.[31,44,68] Use of gloves and gowns for handling animals and tissues, suspension of further processing of tissues from the animals, and destruction of remaining

monkeys brought the outbreak to an abrupt halt. From careful histories of exposure, an incubation of 3–7 days was determined for primary infection and 5–8 days for secondary cases.[32]

During January 1975, a chain of three Marburg infections occurred in Johannesburg, South Africa.[26] The index case was a young man who had made a hitchhiking trip from Johannesburg to Victoria Falls in Rhodesia and return. Exposure to the virus was deemed to have occurred in Rhodesia. A female traveling companion fell ill within 2 days of the death of the first patient but survived, and a nurse who attended this patient later acquired the disease. Incubation periods were similar to those seen in Europe, as were the clinical features of these infections.

5.2.2. Ebola Virus.

Severe outbreaks of acute hemorrhagic fever caused by Ebola virus took place in southwestern Sudan from July through November 1976 and in northwestern Zaire from September to November of that year.[19,20] Both regions are within 5° of the equator, and each outbreak was centered on a rural town with a large hospital: Maridi, Sudan, and Yambuku, Zaire. There were 284 clinically recognized cases in Sudan and 318 in Zaire. Repeated use of needles and unsterile surgical instruments was considered a major route of transmission during both outbreaks. Mortality rates were calculated as 53 and 88%, respectively. Attack rates were not estimated in Sudan, but most illnesses occurred among adults and many were "hospital" associated. A similar pattern was noted in Zaire, where 13 of 17 members of a hospital staff became ill and 11 died. Cases occurred in 55 of some 250 villages in the epidemic area of Zaire, and 56% of these were in females. Few cases occurred in children less than 10 years of age, and attack rates in adults ranged from 10 to 14 per 1000. The incubation period of disease was about 1 week in both Zaire and Sudan.

Another Ebola virus outbreak occurred during August–September 1979 in the area of Nzara, near Maridi, Sudan. Of 34 confirmed cases, 22 were fatal.[1b,14] As in 1976, transmission of virus occurred after close contact with a sporadic index case at a local hospital, and further dissemination of the virus was observed among family members caring for patients in homes.

The next Ebola outbreak occurred in 1989, when a shipment of infected monkeys was imported into the U.S. from the Philippines.[1a,17,20b,20c,36b,75c] The outbreak was focused at a biological supply house in Reston, Virginia, and hundreds of monkeys died or were sacrificed before U.S. Army scientists from the high containment facilities at Ft. Detrick, Maryland, controlled the outbreak and decontaminated the facility. However, shipments of monkeys from the same source in the Philippines were used to repopulate the facility, and they too were infected. Thus, the second wave of the outbreak lasted into 1990 before the facility was depopulated, decontaminated, and closed. A few monkeys from that shipment were sent to other states, but quick sacrifice of animals prevented additional outbreaks from occurring.[1a,17,36b,75a] The virus strain was called Ebola-Reston, and since seven animal handlers showed signs of infection but without clinical disease, there is a possibility that this Philippine monkey strain of Ebola is less virulent for man.[25b,30,47b] Interestingly, in this outbreak, as opposed to all previous outbreaks, airborne/droplet transmission appeared to be a major route of virus transmission.[17,36b,36d,75c] The incident caused worldwide publicity and spawned a series of best-selling books highlighting the emerging threat of Ebola virus.

There were only sporadic cases of Ebola reported in the first half of the 1990 decade. A small, quickly contained monkey outbreak occurred in 1992 in Italy in a shipment of Philippine monkeys.[30] In 1994, a single human case of Ebola was seen in a Swiss ethnologist who did an necropsy on a chimpanzee that died as a part of a chimpanzee outbreak (40 cases with 12 deaths) in the Tai forest area of Ivory Coast.[44a,47c]

Then in May, 1995, health authorities in Zaire noticed an unusual cluster of hemorrhagic disease cases at a hospital in Kikwit (a town of 150,000 located in Bandundu Province, 550 km east-southeast of the Zaire capital of Kinshasa). At first, the disease disproportionately struck hospital doctors, nurses, and Italian missionary nuns working at the hospital. As more and more cases occurred in the hospital staff and in people who had close contact with patients, the clinical symptoms of the disease, the transmission through patient contact, and the high fatality rate of 75% suggested Ebola. Fresh samples of blood and biopsy tissue were collected and transported to the U.S. Centers for Disease Control (CDC), which is one of a chain of collaborative reference laboratories established by the World Health Organization (WHO). Applying newly developed ELISA and PCR diagnostic assays, Ebola virus was detected at high levels in almost all the initial blood and tissue samples.[43,52a,75d]

In Kikwit, under intensive international media coverage, the Zaire Health authorities and the WHO coordinated multinational relief efforts. An international team of medical and scientific experts was organized, and the team's initial efforts focused on stopping the chain of virus transmission. The primary goal was to reestablish good hospital nursing procedures employing clean single-use needles, gloves, and other protective clothing. Simultaneously, community public health programs were estab-

lished to seek out patients (and potentially contaminated contacts), limit travel of those possibly infected, and safely bury the dead. As in other human outbreaks of Ebola, the vast majority of cases resulted from close contact with patients, and aerosol transmission did not appear to play a major role in virus transmission. Thus, the outbreak was controlled by limiting contact transmission, and the last case occurred in late June 1995. In all, 315 cases were detected, and 77% (244) of these died.[50a,52a,75d] Seventy-five of the cases were in medical staff. Although cases occurred in at least seven towns, the quarantine measures appeared to limit spread and all cases occurred in the general Kikwit area with only one case reported from Kinshasa.

In early 1996, an isolated outbreak occurred in Gabon with 37 cases and 21 deaths (case fatality rate of 56.8%) (unpublished reports, WHO). At the same time a limited monkey outbreak occurred in a commercial colony in Texas after importation of Philippine monkeys.[20a]

5.3. Other Epidemiologic Features

Since available information is limited to outbreaks of Marburg and Ebola viruses, it is not possible to delineate the age, sex, geographic distribution, and other features of these infections at the present time except as given above.

6. Mechanism and Route of Transmission

6.1. Spread of Virus

6.1.1. Marburg Virus. The original Marburg outbreaks in Europe were the direct result of human contact with infected green monkeys (*C. aethiops*) that originated from a single dealer in Uganda. All these animals had been captured in the district near Lake Kyoga. Twenty-five primary infections occurred among personnel of three laboratories where polio vaccines were in production. It was noteworthy that 20 of 29 persons having contact with the blood or organs of live monkeys became infected, whereas only 4 of 13 persons exposed exclusively to cultured kidney cells from these monkeys became ill.[33] Animal caretakers suffered no infections, and this fact, together with the observation that many monkeys apparently survived until sacrifice as part of a thorough "clean-up" of premises, strongly suggests that infectious aerosols were not important in virus transmission. A total of seven secondary human infections occurred in Europe and South Africa. Only one of these seems a possible aerosol transmission; the others resulted from continued close contact with patients and their body fluids. One notable event was sexual transmission from husband to wife some 3 months after clinical convalescence.[47] No instance of subclinical infection by Marburg virus has been recorded.

6.1.2. Ebola Virus. Transmission of Ebola virus in Sudan and Zaire was either by close contact in the course of patient care or by virus-contaminated syringe and needle. Each outbreak originated in one or a very few apparently sporadic cases, and amplification was the result of substandard hospital practice. In Zaire in 1976, none of 85 patients infected by accidental parenteral inoculation survived.[20] Secondary attack rates calculated for several generations of cases ranged from 3 to 14%. When close family relatives were considered, however, these rates often exceeded 20%, and hospital personnel attending patients suffered infection and illness at even higher rates.[12,19] The excess morbidity among adult females in Zaire in 1976 was largely related to infections accidently acquired parenterally at a prenatal clinic of the hospital where the outbreak was centered. In all human outbreaks, the institution of patient isolation and basic personnel precautions resulted in rapid termination of epidemics in both countries. Thus, available data point to the conclusion that aerosols were not an important vehicle of virus transmission during human outbreaks.

6.2. Reservoir

The true origin and the natural cycle of maintenance for Marburg and Ebola viruses remain unsolved. For Marburg virus, the experimental work with green monkeys disclosed that all animals that received even tiny amounts of virus experienced fatal infection within 12 days. Furthermore, neither virus nor specific antibodies were detected subsequent to the original outbreak among monkeys captured in the area of Uganda where the implicated group originated.[67] Despite exhaustive investigation, specimen collection, and testing, no trace of Marburg virus was uncovered in southern Africa.

For Ebola virus, the story remains equally elusive. Monkeys are highly sensitive to lethal infection, and more than 200 sera from several primate species in Zaire collected after the 1976 outbreak were negative for specific antibodies (K. M. Johnson, G. Van der Groen, L. Elliott, and B. Robbins, CDC, unpublished data, 1979). Exhaustive studies have been undertaken in the Kikwit area and in areas where the index case was believed to have worked and traveled. To date, tests show no evidence of a reservoir species. But sporadic cases of Ebola virus infection were documented in the Zaire River basin in 1977, and retrospectively (autopsy accident with illness, survival, and persistent antibody) in 1972.[34] Furthermore, possibly

specific Ebola antibodies have been found in 5–8% of persons in several localities in Zaire and Cameroon using the IFA assay. These data imply that the virus resides in the tropical rain forest of Central Africa and that mild or no disease may be a common event where secondary human transmission does not occur.

7. Pathogenesis and Immunity

7.1. In Guinea Pigs and Hamsters

Inoculation of unpassaged Marburg virus produces febrile infection in guinea pigs but no overt signs of disease in this species or the Syrian hamster. Specific antiviral antibodies appear within 2–3 weeks. After serial passage of the agent in guinea pigs and monkeys, however, a uniformly fatal disease is induced in adult guinea pigs and suckling hamsters. The pathogenesis of infection is generally similar in both species.[29,62,64,85] Animals become viremic within 2–4 days and die 5–8 days after inoculation. There is widespread necrosis without inflammatory reaction in the lymphoid elements of nodes and spleen and in the liver. Interstitial pneumonia is common, and there is evidence of diffuse intravascular clotting as well as hyperplasia of fixed macrophages often containing partially destroyed erythrocytes. In addition, hamsters display vascular changes in the central nervous system typical of viral encephalitis.

7.2. In Nonhuman Primates

Rhesus, vervet, and squirrel monkeys are highly susceptible to parenteral and/or intranasal infection by Marburg and Ebola viruses. The outcome of infection is always fatal, even with unpassaged virus, and the disease is very similar to that noted in man. Infected monkeys develop high fever within 2–5 days and generally die after 6–9 days.[10,29,63] Animals become anorectic and lethargic 1–2 days before death, and some develop a maculopapular skin rash during this time. Death is preceded by a drop in temperature to subnormal levels, strongly suggesting terminal shock. Mild leukopenia and thrombocytopenia are frequently observed, and gross impairment of blood coagulation is sometimes noted prior to death. Virus is present in blood of all monkeys from 1 to 4 days after infection until death. Titers in excess of 10^6 infectious units/ml were found in Ebola virus infection.[10] Large amounts of virus are present in most organs examined, but data on brain content have not been obtained. From 10^3 to 10^6 infectious doses of Marburg virus can be recovered from saliva and urine of infected monkeys.[63] Antibodies to the viruses are not present prior to death.

Pathological changes associated with infection consist of (1) necrosis of lymphoid elements with reticulo-endothelial hypertrophy in nodes and spleen; (2) focal, often severe necrosis of hepatocytes with little inflammatory response but with development of large eosinophilic inclusions reminiscent of Councilman bodies of yellow fever; (3) variable degrees of interstitial pneumonia; and (4) widespread microintravascular coagulation with extravasation of erythrocytes.[2,52] Immunofluorescent and ultrastructural examination of Marburg-infected animals reveals large amounts of viral antigen in liver, spleen, and lungs, with accumulation of virions principally in liver, only moderately in spleen, and not at all in the lung, where no evidence for virus replication is found.[48]

7.3. In Man

The clinical features of AFHF are described in Section 8. Pathologically, the disease caused in humans by both Marburg and Ebola viruses is remarkably similar to that documented for monkeys. High persistent viremia is typical, and large amounts of virus are present in many infected viscera at autopsy.[82] Histological features are also similar.[3,18,27,49] In addition, renal tubular necrosis was observed in several patients, but this may have been nonspecifically related to terminal shock. Marburg patients also display a diffuse glial-nodule type of encephalitis together with a mononuclear vasculitis reminiscent of a pattern produced by certain arthropod-borne viruses.[27] Small amounts of Marburg virus were present in the brain of a single case that was tested.[83] Pathological evidence for disseminated intravascular coagulation was regarded as definitive.[21,26,57b]

8. Patterns of Host Response

8.1. Clinical Features

Marburg and Ebola infections are marked by the appearance of headache, progressive fever, sore throat, myalgia, and diarrhea.[20,46,69,72a] Conjunctivitis is sometimes present, as is a papular exanthem of the palate. By the fourth or fifth day of evolution, there is chemical evidence of hepatitis, and over 50% of patients develop a centripetal maculopapular rash that rarely lasts more than 3 days. Some patients experience symptoms suggesting acute pancreatitis, and this has been documented by chemical tests in Marburg infection.[26] Most patients who survive fail to develop a hemorrhagic diathesis, although severe weight loss, asthenia, and psychological depression are common features of a convalescence requiring several weeks. Melena, hematemesis, and bleeding from other sites generally begin on the fifth or sixth day of fatal disease, and such patients rarely survive beyond 9 days of evolution. There is ample evidence for disseminated intravascular coagulation. Terminal shock is an unvarying finding. Patients infected with Ebola virus in Sudan in 1976 had a high incidence of dry, nonproductive cough, not seen in any other outbreak.[19]

8.2. Diagnosis

In fatal cases, electron microscopy can be expected to reveal virus particles in liver. Surviving patients generally develop specific IgM and IgG antibodies about 7–10 days after onset of symptoms.[43,79a,84] Complement-fixation antibodies generally appear about 1 week later. To date, measuring the low levels of virus-neutralizing antibodies has been of limited use for diagnosis.

Viremia and antigenemia during the acute phase is very high, and virions in blood can be visualized by the electron microscope.[9,43] In fact, some acute serum samples contain antigen at titers greater than 1:256 when titered in a rapid antigen-capture ELISA. Biopsies and "skin snips" at postmortem have yielded virus, and those fixed and immunohistochemically stained have revealed large amounts of antigen in many organs.[52a,75d] Although not yet applied in a epidemiological fashion, postmortem "skin snips" fixed in formalin might provide a safely transportable diagnostic specimen. Polymerase chain reaction assays proved reliable during the Kikwit outbreak and yielded information useful in studying the molecular epidemiology of the outbreak strain.[52a,75d]

9. Prevention and Control

9.1. General Concepts

In the absence of significant knowledge concerning the ecology of Marburg and Ebola infections, there is little that can be said at this time regarding control of infection. Monkeys imported for biomedical research should be quarantined and handled by personnel using universal precautions. The diagnosis of AFHF should be considered in monkeys dying in quarantine and in animal caretakers who develop fever >38.5°C for >2 days.[15] Basic hygienic practices in hospitals can be relied on to prevent further major epidemics. New guidelines on patient management have recently been published.[25a] Both viruses multiply to high titers in cell cultures, and it is possible to develop experimental vaccines for protection of key laboratory and clinical personnel.[36] Without data on incidence of these diseases in Africa, vaccination of human populations does not seem to warrant a high priority at present. Antiviral drug development should be considered.

9.2. Management and Disposition of Patients

Patients with AFHF should be placed in strict isolation. Patient management guidelines have recently been revised.[25a] Where air-circulating systems are employed, they should be unidirectional, nonrecirculating, and away from corridors. All materials leaving the patient's room must be disinfected chemically or, preferably, by steam under pressure. Medical personnel must use gloves, gowns, eye protection, and masks, and where possible, primary isolation systems for either the patient (bed isolator) or the medical team (positive-pressure hood or suit) should be utilized. Clinical pathological procedures should be carried out under strict isolation, and if maximum primary containment of potential aerosols is feasible, it should be employed.

Clinical management of patients is largely supportive. Fluid and electrolyte balance is an important consideration, and nasogastric suction with acid neutralization is indicated in patients with evidence of acute pancreatitis. Careful anticoagulation of two patients with Marburg virus infections with intravenous heparin prior to the potential onset of bleeding was associated with clinical recovery.[26] Whether or not there was a cause–effect relationship is not clear. The only known survivor of a confirmed clinical Ebola virus infection acquired by parenteral injection, a laboratory worker accidentally infected by a needle puncture, was treated with both plasma containing Ebola antibodies and large doses of interferon beginning on the third day of illness. The virus titer in this patient's blood fell from $10^{4.5}$ infectious units/ml to barely detectable levels within 24 hr after administration of plasma.[23] Subsequent experimental studies in monkeys showed that interferon was unable to prevent death following infection.[8] Thus, it may be that passive antibodies administered before the onset of bleeding are of specific value in treatment of AFHF.[40] Interestingly, this patient developed the typical rash at the expected time and also had virus in seminal fluid for 2 months after clinical recovery. In Kikwit, it appeared that patients given immune plasma improved, and many survived. However, only a few patients were treated, and most of these were already past their acute phase.

9.3. General Strategy for Isolation and Management of Suspect "Exotic" Infections

In view of the medical drama so far associated with "exotic" diseases, the residual uncertainty regarding their potential for transmission by aerosol, and the problems of differentiation from other hazardous viral infections such as Lassa fever, it is not at all clear what the optimum course of action should be whenever a patient presents for medical care after recent travel in rural Africa. The probability of such an event is quite low. Diagnostic services are available at present in Atlanta, Georgia; Frederick, Maryland; Porton Down, England; Antwerp, Belgium; Johannesburg, South Africa; and Novosibirsk, Russia. Containment facilities for patient care and clinical pathology also exist at or near all these centers, primarily as protection for laboratory workers.[16,24,28,75] Their utilization otherwise depends on availability of safe air transport from point of patient intake. Alternatively, patients might be isolated at the hospital where they are initially seen and portable, contained, clinical pathological instrumentation flown to the hospital.[58] Resolution of these issues can be expected to vary in different countries depending on geography, political and economic structure, and continuing expenditures of time and money on feasibility study.[86] As always, energy devoted to such problems, and the configurations reached, will depend on experience and the outcome of individual, and unusual, events.

10. Unresolved Problems: Detection of Virus Reservoirs

We still have little knowledge of the vertebrate reservoirs of Marburg and Ebola viruses and of the mode of transmission to reservoir hosts and to people.[36c,64a] Safe, inexpensive, sensitive, and specific tests are now available to detect antibodies, antigens, and nucleic acids of these viruses. Their use will undoubtedly lead to a more extensive understanding of these diseases. Ideally, modern methods will lead to the production and use of diagnostic tests without the requirement to handle infectious virus. There is strong circumstantial evidence that monkeys are somehow involved in the transmission cycle, and the possibility remains that nonhuman primates may be the reservoir. Field studies are needed in Africa and elsewhere to confirm the role of monkeys.

11. References

1. ALMEIDA, J. D., WATERSON, A. P., AND SIMPSON, D. I. H., Morphology and morphogenesis of the Marburg agent, in: *Marburg Virus Disease* (G. A. MARTINI AND R. SIEGERT, EDS.), pp. 84–97, Springer-Verlag, New York, 1971.

1a. ANDERSON, G. C., US shuts down monkey trade [news]. *Nature* **344:**369 (1990).

1b. BARON, R. C., MCCORMICK, J. B., AND ZUBEIR, O. A., Ebola virus disease in southern Sudan: Hospital dissemination and intra-familial spread. *Bull. World Health Organ.* **61**:997–1003 (1983).

2. BASKERVILLE, A., BOWEN, E. T. W., PLATT, G. S., MCARDELL, L. B., AND SIMPSON, D. I. H., The pathology of experimental Ebola virus infection in monkeys, *J. Pathol.* **125**:131–138 (1978).

3. BECHTELSHEIMER, H., JACOB, H., AND SOLCHER, H., The neuropathology of an infectious disease transmitted by African green monkeys (*Cercopithecus aethiops*), *Ger. Med. Monthly* **141**:10–12 (1969).

4. BECKER, S., FELDMANN, H., WILL, C., AND SLENCZKA, W., Evidence for occurrence of filovirus antibodies in humans and imported monkeys: Do subclinical filovirus infections occur worldwide? *Med. Microbiol. Immunol.* **181**:43–55 (1992).

5. BOIRO, I., LOMONOSSOV, N. N., SOTSINSKI, V. A., CONSTANTINOV, O. K., TKACHENKO, E. A., INAPOGUI, A. P., AND BALDE, C., Clinico-epidemiologic and laboratory research on hemorrhagic fevers in Guinea, *Bull. Soc. Pathol. Exot. Filiales* **80**:607–612 (1987).

6. BOWEN, E. T. W., SIMPSON, D. I. H., BRIGHT, W. F., ZLOTNIK, I., AND HOWARD, D. M. R., Vervet monkey disease: Studies on some physical and chemical properties of the causative agent, *Br. J. Exp. Pathol.* **50**:400–407 (1969).

7. BOWEN, E. T. W., PLATT, G. S., LLOYD, G., BASKERVILLE, A., HARRIS, W. J., AND VELLA, E. E., Viral haemorrhagic fever in southern Sudan and northern Zaire, *Lancet* **1**:571–573 (1977).

8. BOWEN, E. T. W., BASKERVILLE, A., CANTELL, K., MANN, G. F., SIMPSON, D. I. H., AND ZUCKERMAN, A. J., The effect of interferon on experimental Ebola virus infection in rhesus monkeys, in: *Ebola Virus Haemorrhagic Fever* (S. R. PATTYN, ED.), pp. 245–252, Elsevier/North-Holland, Amsterdam, 1978.

9. BOWEN, E. T. W., LLOYD, G., PLATT, G., MCARDELL, L. B., WEBB, P. A., AND SIMPSON, D. I. H., Virological studies on a case of Ebola virus infection in man and in monkeys, in: *Ebola Virus Haemorrhagic Fever* (S. R. PATTYN, ED.), pp. 95–100, Elsevier/North-Holland, Amsterdam, 1978.

10. BOWEN, E. T. W., PLATT, G. S., SIMPSON, D. I. H., MCARDELL, L. B., AND RAYMOND, R. T., Ebola haemorrhagic fever: Experimental infection of monkeys, *Trans. R. Soc. Trop. Med. Hyg.* **72**:188–191 (1978).

11. BOWEN, E. T. W., PLATT, G. S., LLOYD, G., MCARDELL, L., SIMPSON, D. I. H., SMITH, D. H., FRANCIS, D. P., HIGHTON, R. B., CORNET, M., DRAPER, C. C., ELTAHIR, B., MAYOM DENG, I., LOLIK, P., AND DUKU, O., Viral haemorrhagic fever in the Sudan, 1976: Human virological and serological studies, in: *Ebola Virus Haemorrhagic Fever* (S. R. PATTYN, ED.), pp. 143–151, Elsevier/North-Holland, Amsterdam, 1978.

12. BREMAN, J. G., PIOT, P., JOHNSON, K. M., WHITE, M. K., MBUYI, M., SUREAU, P., HEYMAN, D. L., VAN NIEUWENHOVE, S., MCCORMICK, J. B., RUPPOL, J. P., KINTOKI, V., ISAACSON, M., VAN DER GROEN, G., WEBB, P. A., AND NGUETE, K., The epidemiology of Ebola haemorrhagic fever in Zaire, 1976, in: *Ebola Virus Haemorrhagic Fever* (S. R. PATTYN, ED.), pp. 103–124, Elsevier-North-Holland, Amsterdam, 1978.

12a. BUKREYEV, A. A., VOLCHKOV, V. E., BLINOV, V. M., DRYGA, S. A., AND NETESOV, S. V., The complete nucleotide sequence of the Popp (1967) strain of Marburg virus: A comparison with the Musoke (1980) strain, *Arch. Virol.* **140**:1589–1600 (1995).

13. CASALS, J., Absence of serological relationship between Marburg virus and some arboviruses, in: *Marburg Virus Disease* (G. A. MARTINI AND R. SIEGERT, EDS.), pp. 98–104, Springer-Verlag, New York, 1971.

14. CENTERS FOR DISEASE CONTROL, Ebola hemorrhagic fever—southern Sudan, *Morbid. Mortal. Week. Rep.* **28**:557–559 (1979).

15. CENTERS FOR DISEASE CONTROL, Update: Ebola-related filovirus infection in nonhuman primates and interim guidelines for handling nonhuman primates during transit and quarantine, *Morbid. Mortal. Week. Rep.* **39**:22–30 (1990).

15a. CHEPURNOV, A. A., CHUEV IUP, P'IANKOV, O. V., AND EFIMOVA, I. V., [The effect of some physical and chemical factors on inactivation of the Ebola virus]. *Vopr. Virusol.* **40**:74–76 (1995).

16. CLAUSEN, L., BOTHWELL, T. H., ISAACSON, M., KOORNHOF, H. J., GEAR, J. H., MCMURDO, J., PAYN, E. M., MILLER, G. B., AND SHER, R., Isolation and handling of patients with dangerous infectious disease, *S. Afr. Med. J.* **53**:238–242 (1978).

16a. COX, N. J., MCCORMICK, J. B., JOHNSON, K. M., AND KILEY, M. P., Evidence for two subtypes of Ebola virus based on oligonucleotide mapping of RNA, *J. Infect. Dis.* **147**:272–275 (1983).

17. DALGARD, D. W., HARDY, R. J., PEARSON, S. L., PUCAK, G. J., QUANDER, R. V., ZACK, P. M., PETERS, C. J., AND JAHRLING, P. B., Combined simian hemorrhagic fever and Ebola virus infection in cynomolgus monkeys, *Lab. Anim. Sci.* **42**:152–157 (1992).

18. DIETRICH, M., SCHUMACHER, H. H., PETERS, D., AND KNOBLOCH, J., Human pathology of Ebola (Maridi) virus infection in the Sudan, in: *Ebola Virus Haemorrhagic Fever* (S. R. PATTYN, ED.), pp. 37–42, Elsevier-North-Holland, Amsterdam, 1978.

19. Ebola haemorrhagic fever in Sudan, 1976, *Bull. WHO* **56**:247–270 (1978).

20. Ebola haemorrhagic fever in Zaire, 1976, *Bull. WHO* **56**:271–293 (1978).

20a. Ebola-Reston virus infection among quarantined nonhuman primates—Texas, 1996. *Morb. Mortal. Wkly. Rep.* **45**:314–316 (1996).

20b. Ebola virus, *Wkly. Epidemiol. Rec.* **65**:45–47 (1990).

20c. Ebola virus infection in imported primates—Virginia, 1989, *Morb. Mortal. Wkly. Rep.* **38**:831–832, 837–838 (1989).

21. EGBRLNG, R., SLENCZKA, W., AND BALTZER, G., Clinical manifestations and mechanism of the haemorrhagic diathesis in Marburg virus disease, in: *Marburg Virus Disease* (G. A. MARTINI AND R. SIEGERT, EDS.), pp. 41–49, Springer-Verlag, New York, 1971.

22. ELLIS, D. S., SIMPSON, D. I. H., FRANCIS, D. P., KNOBLOCH, J., BOWEN, E. T. W., LOLIK, P., AND MAYOM DENG, I., Ultrastructure of Ebola virus particles in human liver, *J. Clin. Pathol.* **31**:201–208 (1978).

23. EMOND, R. T. D., EVANS, B., BOWEN, E. T. W., AND LLOYD, G., A case of Ebola virus infection, *Br. Med. J.* **2**:541–544 (1977).

24. EMOND, R. T. D., SMITH, H., AND WELSBY, P. D., Assessment of patients with suspected viral haemorrhagic fever, *Br. Med. J.* **1**:966–967 (1978).

25. FELDMAN, H., MUHLBERGER, E., RANDOLF, A., WILL, C., KILEY, M. P., SANCHEZ, A., AND KLENK, H. D., Marburg virus, a filovirus: Messenger RNAs, gene order, and regulatory elements of the replication cycle, *Virus Res.* **24**:1–19 (1992).

25a. From the Centers for Disease Control and Prevention. Update: Management of patients with suspected viral hemorrhagic fever—United States, *J.A.M.A.* **274**:374–375 (1995).

26. GEAR, J. S. S., CASSEL, G. A., GEAR, A. J., TRAPPER, B., CLAUSEN, L., MEYERS, A. M., KEW, M. C., BOTHWELL, T. H., SHER, R., MILLER, G. B., AND SCHNEIDER, J., Outbreak of Marburg virus disease in Johannesburg, *Br. Med. J.* **4**:489–493 (1975).

27. GEDIGIC, P., BECHTELSHEIMER, H., AND KORB, G., The morbid anatomy of Marburg-virus-disease, *Ger. Med. Monthly* **14**:68–77 (1969).

27a. GEYER, H., WILL, C., FELDMANN, H., KLENK, H. D., AND GEYER, R., Carbohydrate structure of Marburg virus glycoprotein, *Glycobiology* **2**:299–312 (1992).

28. GOMPERTS, E. D., ISAACSON, M. KOORNHOF, H. J., METZ, J., GEAR, J. H., SCHOUB, B. D., MCINROSFF, B., AND PROZESKY, O. W., Handling of highly infectious material in a clinical pathology laboratory and in a viral diagnostic unit, *S. Afr. Med. J.* **53**:243–248 (1978).

28a. GONZALEZ, J. P., JOSSE, R., JOHNSON, E. D., MERLIN, M., GEORGES, A. J., ABANDJA, J., DANYOD, M., DELAPORTE, E., DUPONT, A., AND GHOGOMU, A., Antibody prevalence against haemorrhagic fever viruses in randomized representative Central African populations, *Res. Virol.* **140**:319–331 (1989).

29. HASS, K., AND MAASS, G., Experimental infection of monkeys with the Marburg virus, in: *Marburg Virus Disease* (G. A. MARTINI AND R. SIEGERT, EDS.), pp. 136–143, Springer-Verlag, New York, 1971.

30. HAYES, C. G., BURANS, J. P., KSIAZEK, T. G., DEL ROSARIO, R. A., MIRANDA, M. E., MANALOTO, C. R., BARRIENTOS, A. B., ROBLES, C. G., DAYRIT, M. M., AND PETERS, C. J., Outbreak of fatal illness among captive macaques in the Philippines caused by an Ebola-related filovirus, *Am. J. Trop. Med. Hyg.* **46**:664–671 (1992).

31. HENNESSEN, W., A hemorrhagic disease transmitted from monkeys to man, *Natl. Cancer Inst. Monogr.* **29**:161–171 (1968).

32. HENNESSEN, W., BONIN, O., AND MAULER, R., Zur Epidemiologie der Erkrankung von Menschen durch Affen, *Dtsch. Med. Wochenschr.* **93**:582–587 (1968).

33. HENNESSEN, W., Epidemiology of "Marburg virus" disease, in: *Marburg Virus Disease* (G. A. MARTINI AND R. SIEGERT, EDS.), pp. 161–165, Springer-Verlag, New York, 1971.

34. HEYMANN, D. L., WEISFELD, J. S., WEBB, P. A., JOHNSON, K. M., CAIRNS, T., AND BERQUIST, H., Ebola hemorrhagic fever: Tandala, Zaire, 1977–1978, *J. Infect. Dis.* **142**:372–376 (1980).

35. HOFMANN, H., AND KUNZ, C. H., Cultivation of the Marburg virus (*Rhabdovirus simiae*) in cell cultures, in: *Marburg Virus Disease* (G. A. MARTINI AND R. SIEGERT, EDS.), pp. 112–116, Springer-Verlag, New York, 1971.

36. IGNATEV, G. M., STRELTSOVA, M. A., AGAFONOV, A. P., ZHUKOVA, N. A., KASHENTSEVA, E. A., AND VOROBEVA, M. S., The immunity indices of animals immunized with the inactivated Marburg virus after infection with homologous virus, *Vopr. Virusol.* **39**:13–17 (1994).

36a. IVANOFF, B., DUQUESNOY, P., LANGUILLAT, G., SALUZZO, J. F., GEORGES, A., GONZALEZ, J. P., AND MCCORMICK, J., Haemorrhagic fever in Gabon. I. Incidence of Lassa, Ebola and Marburg viruses in Haut-Ogooue, *Trans. R. Soc. Trop. Med. Hyg.* **76**:719–720 (1982).

36b. JAHRLING, P. B., GEISBERT, T. W., DALGARD, D. W., JOHNSON, E. D., KSIAZEK, T. G., HALL, W. C., AND PETERS, C. J., Preliminary report: Isolation of Ebola virus from monkeys imported to USA, *Lancet* **335**:502–505 (1990).

36c. JANSSENS, P. G., AND PATTYN, S. R., [Epidemic hemorrhagic fevers]. *Verh. K. Acad. Geneeskd. Belg.* **45**:31–200 (1983).

36d. JOHNSON, E., JAAX, N., WHITE, J., AND JAHRLING, P., Lethal experimental infections of rhesus monkeys by aerosolized Ebola virus, *Int. J. Exp. Pathol.* **76**:227–236 (1995).

37. JOHNSON, B. K., WAMBUI, C., OCHENG, D., GICHOGO, A., OOGO, S,., LIBONDO, D., GITAU, L. G., TUKEI, P. M., AND JOHNSON, E. D., Seasonal variation in antibodies against Ebola virus in Kenyan fever patients, *Lancet* **1**:1160 (1986).

38. JOHNSON, E. D., GONZALEZ, J. P., AND GEORGES, A., Filovirus activity among selected ethnic groups inhabiting the tropical forest of equatorial Africa, *Trans. R. Soc. Trop. Med. Hyg.* **87**:536–538 (1993).

39. JOHNSON, K. M., WEBB, P. A., LANGE, J. V., AND MURPHY, F. A., Isolation and partial characterization of a new virus causing acute haemorrhagic fever in Zaire, *Lancet* **1**:569–571 (1977).

40. JOHNSON, K. M., WEBB, P. A., AND HEYMANN, D. L., Evaluation of the plasmapheresis program in Zaire, in: *Ebola Virus Haemorrhagic Fever* (S. R. PATTYN, ED.), pp. 219–222, Elsevier/North-Holland, New York, 1978.

41. KILEY, M. P., COX, N. J., ELLIOTT, L. H., SANCHEZ, A., DEFRIES, R., BUCHMEIER, M. J., RICHMAN, D. D., AND MCCORMICK, J. B., Physicochemical properties of Marburg virus: Evidence of three distinct virus strains and their relationship to Ebola virus, *J. Gen. Virol.* **69**:1957–1967 (1988).

41a. KILEY, M. P., BOWEN, E. T., EDDY, G. A., ISAACSON, M., JOHNSON, K. M., MCCORMICK, J. B., MURPHY, F. A., PATTYN, S. R., PETERS, D., PROZESKY, O. W., REGNERY, R. L., SIMPSON, D. I., SLENCZKA, W., SUREAU, P., VAN DER GROEN, G., WEBB, P. A., AND WULFF, H., Filoviridae: A taxonomic home for Marburg and Ebola viruses? *Intervirology* **18**:24–32 (1982).

42. KISSLING, R. E., ROBINSON, R. Q., MURPHY, F. A., AND WHITFIELD, S. G., Agent of disease contracted from green monkeys, *Science* **160**:888–890 (1968).

43. KSIAZEK, T. G., ROLLIN, P. E., JAHRLING, P. B., JOHNSON, E., DALGARD, D. W., AND PETERS, C. J., Enzyme immunosorbent assay for Ebola virus antigens in tissues of infected primates, *J. Clin. Microbiol.* **30**:947–950 (1992).

44. KUNZ, C., HOFMANN, H., KOVAC, W., AND STOCKINGER, L., Biologische und morphologische Charackteristica des Virus de Deutschland aufgetretenen "Hamorrhagischen Fiebers," *Wi. Kim. Wochenschr.* **80**:161–166 (1968).

44a. LEGUENNO, B., FORMENTRY, P., WYERS, M., GOUNON, P., WALKER, F., AND BOESCH, C., Isolation and partial characterisation of a new strain of Ebola virus [see comments], *Lancet* **345**:1271–1274 (1995).

45. MALHERBE, H., AND STRICKLAND-CHOLMLEY, M., Studies on the Marburg virus, in: *Marburg Virus Disease* (G. A. MARTINI AND R. SIEGERT, EDS.), pp. 188–194, Springer-Verlag, New York, 1971.

46. MARTINI, G. A., KNAUFF, H. G., SCHMIDT, H. A., MAYER, G., AND BALTZER, G., A hitherto unknown infectious disease contracted from monkeys, *Ger. Med. Monthly* **13**:457–470 (1968).

47. MARTINI, G. A., AND SCHMIDT, H., Spermat Ubertragung des Marburg virus, *Klin. Wockenschr.* **46**:391–393 (1968).

47a. MCCORMICK, J. B., BAUER, S. P., ELLIOTT, L. H., WEBB, P. A., AND JOHNSON, K. M., Biologic differences between strains of Ebola virus from Zaire and Sudan, *J. Infect. Dis.* **147**:264–267 (1983).

47b. MIRANDA, M. E., WHITE, M. E., DAYRIT, M. M., HAYES, C. G., KSIAZEK, T. G., AND BURANS, J. P., Seroepidemiological study of filovirus related to Ebola in the Philippines [letter], *Lancet* **337**:425–426 (1991).

47c. MORELL, V., Chimpanzee outbreak heats up search for Ebola origin [news], *Science* **268**:974–975 (1995).

47d. MUHLBERGER, E., SANCHEZ, A., RANDOLF, A., WILL, C., KILEY, M. P., KLENK, H. D., AND FELDMANN, H., The nucleotide sequence of the L gene of Marburg virus, a filovirus: Homologies with paramyxoviruses and rhabdoviruses, *Virology* **187**:534–547 (1992).

48. MURPHY, F. A., SIMPSON, D. I. H., WHITFIELD, S. G., ZLOTNIK, I., AND CARTER, G. B., Marburg virus infection in monkeys: Ultrastructural studies, *Lab. Invest.* **24**:279–291 (1971).

49. MURPHY, F. A., Pathology of Ebola virus infection, in: *Ebola Virus Haemorrhagic Fever* (S. R. PATTYN, ED.), pp. 43–59, Elsevier/North-Holland, Amsterdam, 1978.

50. MURPHY, F. A., VAN DER GROEN, G., WHITFIELD, S. G., AND LANCE, J. V., Ebola and Marburg virus morphology and taxonomy, in: *Ebola Virus Haemorrhagic Fever* (S. R. PATTYN, ED.), pp. 61–82, Elsevier/North-Holland, Amsterdam, 1978.

50a. MUYEMBE, T., AND KIPASA, M., Ebola haemorrhagic fever in Kikwit, Zaire. International Scientific and Technical Committee and WHO Collaborating Centre for Haemorrhagic Fevers [letter], *Lancet* **345:**1448 (1995).

51. MATHIOT, C. C., FONTENILLE, D., GEORGES, A. J., AND COULANGES, P., Antibodies to haemorrhagic fever viruses in Madagascar populations, *Trans. R. Soc. Trop. Med. Hyg.* **83:**407–409 (1989).

52. OEHLERT, W., The morphological picture in livers, spleens, and lymph nodes of monkeys and guinea pigs after infection with the "vervet agent," in: *Marburg Virus Disease* (G. A. MARTINI AND R. SIEGERT, EDS.), pp. 144–156, Springer-Verlag, New York, 1971.

52a. Outbreak of Ebola viral hemorrhagic fever—Zaire, *Morb. Mortal. Wkly. Rep.* **44:**381–382 (1995).

53. PAIX, M. A., POVEDA, J. D., MALVY, D., BAILLY, C., MERLIN, M., AND FLEURY, H. J., Serological study of the virus responsible for hemorrhagic fever in an urban population of Cameroon, *Bull. Soc. Pathol. Exot. Filiales* **81:**679–682 (1988).

54. PATTYN, S., VAN DER GROEN, G., JACOB, W., PIOT, D., AND COURTEILLE, G., Isolation of Marburg-like virus from a case of haemorrhagic fever in Zaire, *Lancet* **1:**573–574 (1977).

55. PETERS, D., AND MULLER, G., Elektronenmikroskopische Erkennung und Charakterisierung des Marburger Erregers, *Dtsch. Aerztebl.* **65:**1831–1834 (1968).

56. PETERS, D., MULLER, G., AND SLENCZKA, W., Morphology, development and classification of the Marburg virus, in: *Marburg Virus Disease* (G. A. MARTINI AND R. SIEGERT, EDS.), pp. 68–83, Springer-Verlag, New York, 1971.

57. REGNERY, R. L., JOHNSON, K. M., AND KILEY, M. P., The nucleic acid of Ebola virus, *J. Virol.* **36:**465–469 (1980).

57a. RICHMAN, D. D., CLEVELAND, P. H., MCCORMICK, J. B., AND JOHNSON, K. M., Antigenic analysis of strains of Ebola virus: Identification of two Ebola virus serotypes, *J. Infect. Dis.* **147:**268–271 (1983).

57b. RIPPEY, J. J., SCHEPERS, N. J., AND GEAR, J. H., The pathology of Marburg virus disease, *S. Afr. Med. J.* **66:**50–54 (1984).

58. RUTTER, D. A., Safety cabinet for use in laboratory studies on hazardous infectious disease, *Br. Med. J.* **2:**24 (1971).

59. SANCHEZ, A., KILEY, M. P., HOLLOWAY, B. P., AND AUPERIN, D. D., Sequence analysis of the Ebola virus genome: Organization, genetic elements, and comparison with the genome of Marburg virus, *Virus Res.* **29:**215–240 (1993).

59a. SANCHEZ, A., KILEY, M. P., KLENK, H. D., AND FELDMANN, H., Sequence analysis of the Marburg virus nucleoprotein gene: Comparison to Ebola virus and other non-segmented negative-strand RNA viruses, *J. Gen. Virol.* **73:**347–357 (1992).

59b. SANCHEZ, A., TRAPPIER, S. G., MAHY, B. W., PETERS, C. J., AND NICHOL, S. T., The virion glycoproteins of ebola viruses are encoded in two reading frames and are expressed through transcriptional editing, *Proc. Natl. Acad. Sci. USA* **93:**3602–3607 (1996).

60. SIEGERT, R., SHU, H. L., SLENCZKA, W., PETERS, D., AND MULLER, G., The etiology of a hitherto unknown infectious disease transmitted from monkeys to man, *Ger. Med. Monthly* **13:**1–3 (1968).

61. SIEGERT, R., SHU, H. L., AND SLENCZKA, W., Isolierung und Identifierung des "Marburg virus," *Dtsch. Med. Wochenschr.* **93:**604–612 (1968).

62. SIMPSON, D. I. H., ZLOTNIK, I., AND RUTTER, D. A., Vervet monkey disease: Experimental infection of guinea-pigs and monkeys with the causative agent, *Br. J. Exp. Pathol.* **49:**458–464 (1968).

63. SIMPSON, D. I. H., Marburg virus disease: Experimental infection of monkeys, *Lab. Anim. Handb.* **4:**149–154 (1969).

64. SIMPSON, D. I. H., Vervet monkey disease: Transmission to the hamster, *Br. J. Exp. Pathol.* **50:**389–392 (1969).

64a. SIMPSON, D. I., The filovirus enigma [comment]. *Lancet* **345:** 1252–1253 (1995).

65. SLENCZKA, W., Growth of Marburg virus in Vero cells, *Lab. Anim. Handb.* **4:**143–147 (1969).

66. SLENCZKA, W., AND WOLFF, G., Biological properties of the Marburg virus, in: *Marburg Virus Disease* (G. A. MARTINI AND R. SIEGERT, EDS.), pp. 105–108, Springer-Verlag, New York, 1971.

67. SLENCZKA, W., WOLFF, G., AND SIEGERT, R., A critical study of monkey sera for the presence of antibody against the Marburg virus, *Am. J. Epidemiol.* **93:**496–505 (1971).

68. SMITH, C. E. G., SIMPSON, D. J. H., BOWEN, E. T. W., AND ZLOTNIK, I., Fatal human disease from vervet monkeys, *Lancet* **2:**1119–1121 (1967).

69. SMITH, D. H., FRANCIS, D. P., AND SIMPSON, D. I. H., African haemorrhagic fever in the southern Sudan: The clinical features, in: *Ebola Virus Haemorrhagic Fever* (S. R. PATTYN, ED.), pp. 21–26, Elsevier/North-Holland, Amsterdam, 1978.

70. SMITH, D. H., JOHNSON, B. K., ISAACSON, M. *ET AL.*, Marburg virus disease in Kenya, *Lancet* **1:**816–820 (1982).

71. STILLE, W., BOHLE, E., HELM, E., VAN REY, W., AND SIEDE, W., Uber eine durch *Cercopithecus aethiops* ubertragene infektiose Krankheit ("Grune-meirkatzen-krankheit," "Green monkey disease"), *Dtsch. Med. Wochenschur.* **93:**572–582 (1968).

72. STOJKOVIC, L. J., BURDIOSKI, M., GLUGIC, A., AND STEFANOVIC, Z., Two cases of *Cercopithecus*-monkeys-associated haemorrhagic fever, in: *Marburg Virus Disease* (G. A. MARTINI AND R. SIEGERT, EDS.), pp. 24–33, Springer-Verlag, New York, 1971.

72a. SUREAU, P. H., Firsthand clinical observations of hemorrhagic manifestations in Ebola hemorrhagic fever in Zaire, *Rev. Infect. Dis.* **11 Suppl 4:**S790–S793 (1989).

73. TIGNOR, G. H., CASALS, J., AND SHOPE, R. E., The yellow fever epidemic in Ethiopia, 1961–1962: Retrospective serological evidence for concomitant Ebola or Ebola-like virus infection, *Trans. R. Soc. Trop. Med. Hyg.* **87:**162 (1993).

74. TOMORI, O., FABIYI, A., SORUNGBE, A., SMITH, A., AND MCCORMICK, J. B., Viral hemorrhagic fever antibodies in Nigerian populations, *Am. J. Trop. Med. Hyg.* **38:**407–410 (1988).

75. TREXLER, P. C., EMOND, R. T. D., AND EVANS, R., Negative pressure isolator for patients with dangerous infections, *Br. Med. J.* **2:**559–561 (1977).

75a. Update: Ebola-related filovirus infection in nonhuman primates and interim guidelines for handling nonhuman primates during transit and quarantine, *Morb. Mortal. Wkly. Rep.* **39:**22–24, 29–30 (1990).

75b. Update: Evidence of filovirus infection in an animal caretaker in a research/service facility, *Morb. Mortal. Wkly. Rep.* **39:**296–297 (1990).

75c. Update: Filovirus infection associated with contact with nonhuman primates or their tissues, *Morb. Mortal. Wkly. Rep.* **39:**404–405 (1990).

75d. Update: Outbreak of Ebola viral hemorrhagic fever—Zaire, 1995, *Morb. Mortal. Wkly. Rep.* **44:**468–469, 475 (1995).

76. VAN DER GROEN, G., JOHNSON, K. M., WEBB, P. A., WULFF, H. T., AND LANGE, J. V., Results of Ebola antibody surveys in various

population groups in: *Ebola Virus Haemorrhagic Fever* (S. R. PATTYN, ED.), pp. 203–205, Elsevier/North-Holland, Amsterdam, 1978.

77. VAN DER GROEN, G., WEBB, P. A., JOHNSON, K. M., LANGE, J. V., LINDSAY, H., AND ELLIOTT, L., Growth of Lassa and Ebola viruses in different cell lines, in: *Ebola Virus Haemorrhagic Fever* (S. R. PATTYN, ED.), pp. 255–260, Elsevier/North-Holland, Amsterdam, 1978.

78. VAN DER WAALS, F. W., POMEROY, K. L., GOUDSMIT, J., ASHER, D. M., AND GAJDUSEK, D. C., Hemorrhagic fever virus infections in an isolated rainforest area of central Liberia. Limitations of the indirect immunofluorescence slide test for antibody screening in Africa, *Trop. Geogr. Med.* **38:**209–214 (1986).

79. Viral haemorrhagic fever, *WHO Weekly Epidemiol. Rep.* **52:**185–192 (1977).

79a. VLADYKO, A. S., CHEPURNOV, A. A., MAR'IANKOVA, R. F., BYSTROVA, S. I., EGORICHEVA, I. N., KUZ'MIN, V. A., AND LUKASHEVICH, I. S., [A comparison of the methods of fluorescent antibodies and solid-phase immunoenzyme analysis in the detection of Marburg virus antibodies]. *Vopr. Virusol.* **36:**326–328 (1991).

80. WEBB, P. A., JOHNSON, K. M., WULFF, H. T., AND LANGE, J. V., Some observations on the properties of Ebola virus, in: *Ebola Virus Haemorrhagic Fever* (S. R. PATTYN, ED.), pp. 91–94, Elsevier/North-Holland, Amsterdam, 1978.

81. WORLD HEALTH ORGANIZATION, Viral haemorrhagic fever surveillance, *Weekly Epidemiol. Rec.* **57:**359 (1982).

82. WULFF, H., AND CONRAD, L., Marburg virus disease, in: *Comparative Diagnosis of Viral Diseases* (E. KURSTAK AND C. KURSTAK, EDS.), pp. 3–33, Academic Press, New York, 1977.

83. WULFF, H., SLENCZKA, W., AND GEAR, J. H. S., Early detection of antigen and estimation of virus yield in specimens from patients with Marburg virus disease, *Bull. WHO* **56:**633–639 (1978).

84. WULFF, H., AND JOHNSON, K. M., Immunoglobulin M and G responses measured by immunofluorescence in patients with Lassa and Marburg virus infections, *Bull. WHO* **57:**631–635 (1979).

85. ZLOTNIK, I., AND SIMPSON, D. I. H., The pathology of experimental vervet monkey disease in hamsters, *Br. J. Exp. Pathol.* **50:**393–399 (1969).

86. ZWEIGHAFT, R. M., FRASER, D. W., HATTWICK, M. A. W., WINKLER, W. G., JORDAN, W. C., ALTER, M., WOLFE, M., WULFF, H., AND JOHNSON, K. M., Lassa fever: Response to an imported case, *N. Engl. J. Med.* **297:**803–807 (1977).

12. Suggested Reading

MARTINI, G. A., AND SIEGERT, R. (EDS.), *Marburg Virus Disease,* Springer-Verlag, New York, 1971.

PATTYN, S. R. (ED.), *Ebola Virus Haemorrhagic Fever,* Elsevier/North-Holland, Amsterdam, 1978.

MURPHY, F. A., KILEY, M. P., AND FISHER-HOCH, S. P., Marburg and Ebola viruses, in: *Virology,* 2nd ed. (B. N. FIELDS AND D. M. KNIPE, EDS.), pp. 933–942, Raven Press, New York, 1990.

CHAPTER 6

Arboviruses

Robert E. Shope and James M. Meegan

1. Introduction

Arthropod-borne viruses (Arboviruses) are an ecologically defined set of viruses that have in common replication in both arthropods and vertebrate hosts and transmission between vertebrate animals by the bite of mosquitoes, ticks, sandflies, and midges.[184] The vast majority of arboviruses belong to one of five families. These are *Togaviridae, Flaviviridae, Bunyaviridae, Reoviridae,* and *Rhabdoviridae.* Information on their isolation, morphology, sensitivity to inactivation by chemicals, arthropod vectors, vertebrate hosts, laboratory propagation, serological reactions, geographic distribution, clinical manifestations, and epidemiology is found in the *International Catalogue of Arthropod-Borne Viruses,* compiled by the American Committee on Arboviruses.[16,102] This exhaustive reference source has been used freely in preparing the text that follows.

The biological transmission of arboviruses is characterized by multiplication of the virus in the arthropod, a process that takes several days (usually 12 days for yellow fever virus, for instance). This period from ingestion of an infected blood meal until the virus replicates, reaches the salivary gland, and can be transmitted is the extrinsic incubation period. After being fed on by an infected arthropod, the vertebrate host is infected and becomes viremic. This part of the cycle takes from 2 days to more than a week and constitutes the intrinsic incubation period. Biological transmission should be distinguished from mechanical transmission in which the virus contaminates the mouth parts of the arthropod and can be transmitted immediately to a new vertebrate host.

Arboviruses are usually maintained in a reservoir cycle that consists of both arthropod and vertebrate host. Both are needed to maintain the virus in nature.[137] A subset of arboviruses, including members of the families *Bunyaviridae,*[193,194] *Flaviviridae,*[2,75,155,156,180] and *Togaviridae,*[77] are transmitted vertically through the egg of the arthropod. In these cases of transovarial transmission, the arthropod alone may be the reservoir of the virus and may maintain the virus in the absence of a vertebrate animal.

In the case of LaCrosse virus, venereal transmission from infected male mosquito to a female mosquito was demonstrated under laboratory conditions.[186] This is potentially an important mechanism in nature for maintaining transmission of this and other viruses.[168]

2. Historical Background

Yellow fever was a major plague of Caucasians exploring and settling the African and American continents. The virus was transmitted by *Aedes aegypti* mosquitoes in the major tropical cities during the 1700s and 1800s. Transmission often occurred on sailing ships carrying both the mosquito and the virus from Africa to America. Moreover, during the summer months, ships carried the mosquito into the major ports of North America, resulting in classic epidemics in New York in 1668, Boston in 1691, and Charleston in 1699, as well as later in the cities of the Gulf of Mexico and the Mississippi River. For over 200 years, yellow fever was the major infectious disease killer of the Americas. As late as 1905, there were over 1000 deaths in the port cities of the southern United States.[63,172,181,196]

The U.S. Army Yellow Fever Commission, under direction of Walter Reed and with assistance of Carlos

Robert E. Shope • Department of Pathology, The University of Texas Medical Branch at Galveston, Galveston, Texas 77555. **James M. Meegan** • Division of Microbiology and Infectious Diseases, National Institute of Allergy and Infectious Diseases, National Institutes of Health, Bethesda, Maryland 20892.

Finlay of Cuba, showed in 1900–1901 that yellow fever was caused by a virus found in the blood; that it was transmitted by a mosquito; and that there was a 12-day extrinsic incubation. With this information, General William C. Gorgas engineered the elimination of the mosquito initially from Havana and then later from the environs of the Panama Canal construction site. Yellow fever disappeared.[63,73,98,116,162,172,176,196] This was a dramatic demonstration that Reed and his co-workers were correct.

In 1932, Soper and colleagues showed that there was a jungle cycle of yellow fever in the absence of *Ae. aegypti*. This significant observation meant that yellow fever could not be stopped just by controlling the mosquitoes in the cities. Theiler and Smith[183] succeeded in cultivating the *Asibi* strain of yellow fever virus, first in monkeys and then in embryonated eggs, and attenuated it by passage. In 1937, they announced their discovery of an attenuated vaccine—the 17D strain.[181] This vaccine is used throughout the world today to prevent yellow fever.

By 1930, there were only three arthropod-borne virus human diseases known.[98,167] These were yellow fever, sandfly (pappataci) fever, and dengue fever. During the 1930s, the mouse was adopted as a laboratory host for viruses. Eastern equine, Western equine, and Venezuelan equine encephalomyelitis viruses were isolated from horses and humans.[29,189] Japanese encephalitis and tick-borne encephalitis viruses were also recovered from fatal human cases.[63,73]

By the 1950s, scientists of The Rockefeller Foundation, the U.S. Army and Navy, the U.S. Public Health Service, several U.S. universities, and many foreign governments recognized that arboviruses could be recovered in nature from apparently healthy arthropods and wild vertebrate animals. This search in the natural cycles of arboviruses yielded a bonanza. By 1985, over 500 different arboviruses had been discovered, about 100 of them causing human disease[16,31,32,102] (see Tables 1 and 2). Clinical syndromes included fever, arthritis, rash, hemorrhagic fever, retinitis, and encephalitis. In addition, serious arthropod-borne diseases of domestic animals included Rift Valley fever, African horse sickness, blue tongue of sheep, equine encephalitis, Nairobi sheep disease, Japanese encephalitis, and African swine fever.

Dengue fever has been a problem in the tropics since the 1600s.[154] It was well known for its severe muscle and joint pains, rash, and lengthy convalescence associated with mental depression. In 1953, a newly recognized form of dengue was noted in children in the Philippines. What started as mild dengue fever became more severe on the second or third day of illness, and the child developed hemorrhagic signs and shock. About 10% of these young-

sters died. The new disease was called dengue hemorrhagic fever and dengue shock syndrome (DHF/DSS).[93,127] Dengue is caused by four different serotypes of virus and any of the four can be associated with DHF/DSS. Through the pioneering work of Halstead,[93] we now understand that a major risk factor for DHF/DSS is sequential infection with different serotypes of dengue virus.[127,154] A child who recovered from infection with one serotype and was later infected with a different serotype can become severely ill.

Because of a combination of population growth, urbanization, and increased breeding of *Ae. aegypti* mosquitoes in Asian and American tropical cities, dengue is rapidly emerging as a major killer of children in the tropics. The disease is spreading in Asia and in the tropical Americas. Throughout the 1980s, the World Health Organization reported 2,512,123 cases and 42,751 deaths, and the numbers are increasing each year.[54,154]

As the numbers of recognized arboviruses increased rapidly during the 1950s and 1960s, the development of diagnostic techniques kept pace.[184] Early methods included the vaccination-challenge test in laboratory mice, the neutralization (N) test, the complement-fixation (CF) test, and the hemagglutination-inhibition (HI) test.[3,92,158,159,173,174] The HI test was especially useful because it was group reactive, permitting the separation of arboviruses into serological groups.[14,30–32,56] Initially, Casals[32] described groups A, B, and C; as new agents were discovered, there were so many new groups that letters were abandoned and the group was named after the prototype isolate. The *International Catalogue of Arthropod-Borne Viruses* lists over 500 viruses in at least 46 serogroups.[102]

Serological techniques evolved to include the immunofluorescence technique and the enzyme-linked immunosorbent assay (ELISA).[4,68] Today, the ELISA is widely used and has the advantage of being ideal to detect serum and spinal fluid IgM, which rises in titer early in the illness and is transient.[27,125] Thus, a rapid and early diagnosis can be made on a single serum specimen. The polymerase chain reaction (PCR) to detect nucleic acid is a very recent addition to the diagnostic armamentarium and will be a sensitive tool for rapid and early diagnosis.[29,48,49]

Unfortunately, the control of arbovirus infections has not kept pace with the discovery and spread of disease. Antiviral drugs for arboviruses are not commercially available. Except for yellow fever, tick-borne encephalitis, and Japanese encephalitis, there are no widely used vaccines available for humans.[94,167,175] Sandflies and mosquitoes were once controlled to a degree with DDT and environmental sanitation; however, resistance to pesticides is unrelenting and new chemical or biological

Table 1. Descriptive Epidemiology of Arboviruses Important as Causative Agents of Disease in Human Beings in the 48 Contiguous United States

Virus[a]	Arbovirus grouping	Incidence and prevalence data	Geographic distribution	Temporal distribution of cases	Age	Sex	Race	At-risk group	Occurrence in different settings
EEE	Togaviridae	Encephalitis rare; immunity rates low	Eastern seaboard from Massachusetts to Florida and Louisiana	Late summer and fall; sporadic cases or restricted sharp epidemics	Any age	—	—	—	Rural and suburban
WEE	Togaviridae	Encephalitis uncommon; immunity rates may be high locally	Virus widespread; human disease restricted to West and Southwest	Summer and fall; broadly endemic and occasional epidemic peaks	Cases more severe in young	—	—	—	Rural and suburban
SLE	Flaviviridae	Encephalitis uncommon; immunity rates may be high locally	Widespread in southern, central, and western states	Summer and fall; broadly endemic and occasional epidemic peaks	Cases more severe in older people	—	—	—	Rural and suburban
CE	Bunyamwera supergroup, California group bunyaviridae	Encephalitis uncommon; immunity rates low; infections sporadic	North central states; virus occurrence in some eastern and western states, but cases rare or absent	Summer; broadly endemic; sporadic cases	Any age, LAC in children	—	—	Sylvan settings	Forested regions; campsites
CTF	Coltivirus	Immunity rates low; infections sporadic	Rocky Mountain states	Spring through fall; broadly endemic; sporadic cases	Adults more likely to be exposed	Males more likely to be exposed	—	Hunters, hikers, fishermen, outdoor people	Mountain foothills and medium altitude slopes

[a]EEE, eastern equine encephalitis; WEE, western equine encephalitis; SLE, St. Louis encephalitis; CE, California encephalitis; LAC, LaCrosse; CTF, Colorado tick fever.

Table 2. Important Arbovirus Infections of Human Beings in South America, Europe, Asia, Africa, and Australia

Disease	Geographic region(s)	Vector(s)	Vertebrate host(s)	Features of disease in human beings			Control measures
				Disease pattern	Description of disease	Diagnosis	
Yellow fever, urban	New World and African cities (seaports usually)	*Aedes aegypti*	Man	Epidemic	Acute onset, high fever, prostration, later jaundice, proteinuria; fatalities common, although ratio of inapparent/apparent infections is high	Virus isolation, CF, HI, N, ELISA tests	Vaccination with 17D vaccine; *Aedes aegypti* control
Jungle	New World and African tropics	Mosquitoes: *Haemagogus* and aedines	Forest primates	Endemic	As above; cases occur sporadically in people exposed in forested regions in Africa and New World	Virus isolation, CF, HI, N, ELISA tests	Vaccination with 17D vaccine; mosquito control not practicable
Dengue(s)	New World and Old World tropics and subtropics	*Aedes aegypti* and other aedines	Man; possibly a "jungle" cycle in primates	Endemic and epidemic	Acute onset with rash in many cases and joint pains; simulates an influenzalike syndrome	Virus isolation, CF, HI, N, ELISA tests	Mosquito control and protection against mosquito bites; no vaccines yet
Dengue hemorrhagic fever	Southeast Asia	*Aedes aegypti*	Man	Endemic and epidemic	Serious illness with hemorrhagic complications, shock syndrome, and high mortality—almost exclusively in children, and following a second infection with a different dengue virus	CF, HI, N, ELISA tests, animals or cell-culture system	Mosquito control
Japanese encephalitis	Orient, Korea to India, and East Indies	*Culex tritaeniorhynchus* and other culicines	Wild birds; pigs can serve as amplifying hosts	Endemic and epidemic	Infection usually mild, but encephalitic complications can be serious in young and in elderly; very important disease in the Orient	CF, HI, N, ELISA tests	Mosquito control; vaccination with an inactivated vaccine

Disease	Geographic distribution	Vector	Reservoir	Epidemiology	Clinical features	Diagnosis	Prevention and control
Murray Valley encephalitis	Australia	*Culex annulirostris* and other mosquitos	Birds	Epidemic, sporadic, over wide areas	Infection usually mild, but encephalitis may occur, with greatest probability in children and high fatality rates in the young	CF, HI, N, ELISA tests	Mosquito control measures and protection against mosquito bite
Chikungunya	Africa, and Asia, tropics and subtropics	*Aedes aegypti*	Possibly primates	Epidemic	Acute onset, often with rash; rarely with hemorrhagic manifestations; joint aching and swelling are prominent features	CF, HI, N, ELISA tests and virus isolation	Mosquito control
Tick-borne encephalitis	Russia and northern Europe	Ticks—*Ixodes ricinus* and others	Small wild mammals and birds	Endemic in forested regions	Acute onset, violent headache, fever, nausea, vomiting, hyperesthesia, photophobia, drowsiness, delirium, coma may follow; mortality rate about 20%	Virus isolation CF, HI, N, ELISA tests	Tick control where possible; an inactivated vaccine has been tried
Kyasanur Forest disease	India (Mysore State)	Ticks—mainly *Haemaphysalis*	Monkeys; possibly also small mammals	Endemic and epidemic	Sudden onset, fever, headache, severe myalgia; there may be a disphasic course with second phase after an afebrile period of 7–15 days; mortality rates under 5%	Virus isolation, CF, HI, N, ELISA tests	Protection against tick bite
Crimean–Congo hemorrhagic fever	Southern former USSR; Bulgaria, Central and South Africa, Pakistan, Iraq	Ticks—*Hyalomma marginatum*	Probably small mammals	Endemic	Sudden onset, chills, fever, headache, nausea, vomiting, hemorrhagic manifestations common; mortality rate 5–10%	Virus isolation, CF test	Protection against tick bite
Venezuelan equine encephalitis	Central and South America and southern United States	Mosquitoes of several species	Horses; possibly small mammals	Probably endemic; sharply epidemic	Fever, encephalitic signs, usually mild, fatalities rare	Virus isolation, CF, HI, N, ELISA tests	Mosquito control and protection against mosquito bites; attenuated vaccine exists for equines

methods for control of arthropods have not kept up with resistance; nor does personal protection or environmental sanitation offer immediate promise for control.[48]

3. Methodology Involved in Epidemiologic Analysis

3.1. Sources of Mortality Data

Mortality data are collected systematically but passively by national governments for dengue, yellow fever, Rift Valley fever, the encephalidites, and other diseases such as epidemic polyarthritis. Data are published in the *Morbidity and Mortality Weekly Report* of the U.S. Centers for Disease Control and Prevention, in the *Weekly Epidemiological Report* of the Pan American Health Organization, and the *Weekly Epidemiological Report* of the World Health Organization.[38–54,131,199–202] The mortality data are grossly underreported, but may serve as a comparative data base, since underreporting may be uniform throughout much of the world. The reporting of yellow fever mortality in Africa, for instance, was found to be about 10% of the true figure.[63,167,189]

The information flow to the World Health Organization is sometimes facilitated by informal networks of scientists and interested citizens. Nevertheless, the organization is constrained from action until official reports are received. This constraint often means a delay in control of a disease of regional or world importance.

3.2. Sources of Morbidity Data

The same sources supply morbidity data as supply mortality data. In the United States, of the arbovirus diseases, only encephalitis is reportable. There has been a trend that the numbers of nonspecific encephalitic cases reported increases in the years of arbovirus epidemics. This may be because physicians request more diagnostic tests in these years, and that laboratories therefore diagnosed a higher percentage of cases. If so, this implies underdiagnosis in other years.[126]

Monath[123] documented the time elapsed between onset of an epidemic of St. Louis encephalitis and its recognition. That time varied between 2 and 8 weeks. In the Corpus Christi, Texas, epidemic of 1966, nearly 707 of the cases had occurred before recognition. In some countries the reporting of arbovirus cases is delayed or suppressed for political reasons or, in profitable resort areas, because of a conflicting ailment, acute remunerative forgetfulness.

3.3. Serological Surveys

Arboviruses are distributed focally throughout the world.[167] The distribution of any given arbovirus is ecologically limited by the range of the vector and vertebrate host. Serological surveys are ideally suited for arboviruses to determine the point prevalence in different geographical areas. Often distribution of the antibody will give clues to the ecological conditions necessary for maintenance of the virus. Surveys of different age groups will show if the virus is more prevalent as the population ages, typical of an endemically transmitted agent. Another prevalence pattern in which antibody is present only in persons born before a certain year may indicate an epidemic in that year. Alternatively, a constant percentage of antibody in each age group may indicate a recent introduction of the virus causing a virgin soil epidemic.

Rift Valley fever illustrates some of these concepts. In 1977, a Rift Valley fever epidemic was diagnosed in the delta region of Egypt.[105,120] A serosurvey of sera collected prior to the epidemic and stored frozen in a serum bank indicated the near absence of antibody prior to 1977.[121] It was therefore concluded that Rift Valley fever virus was introduced to the Nile Delta in 1977. In another example, Rift Valley fever antibody was found in an endemic pattern (see above) in residents of the Senegal River Valley where the Diama Dam was under construction. Transmission was apparently at a low enough level so that cases were not observed. In 1986–1987, when the dam was completed and the land inundated, providing new breeding conditions for mosquitoes, an epidemic of Rift Valley fever was observed.[100,132] In this example, the serosurvey results were used to predict a possible future epidemic.

Broadly based serological surveys of large populations can provide extensive information about virus distribution in different geographic areas, rural versus urban populations, different age and sex groups, and different occupational types. An excellent example is the project of the WHO Serum Bank at Yale University,[97] involving recruits from the United States and Argentina,[70] Brazil,[129] Colombia,[69] and on general populations in Barbados[72] and St. Lucia.[71]

Arbovirus serosurveys have limitations. Cross-reactions occur among viruses of the same serogroup. This is especially true of the HI test and ELISA with flaviviruses. The neutralization test is more specific and should be used where feasible. Surveys with HI or ELISA must be interpreted with caution unless one is certain that only one flavivirus is extant in the region, or unless the results have been confirmed by neutralization test with a portion of the negative and positive sera.

The serosurvey usually will not indicate when the infection responsible for the antibody occurred. If the antibody is suspected to be of recent origin, the IgM antibody-capture ELISA is useful for detection of infections originating within the prior 6 months.

A serosurvey is not a reliable indicator of natural infection in populations vaccinated for tick-borne encephalitis, yellow fever, or Japanese encephalitis. On the other hand, the survey is an excellent tool for determination of vaccination coverage.

3.4. Laboratory Methods

The laboratory is an all-important resource in the study of the epidemiology of arbovirus diseases. Diagnosis can rarely be made with certainty by clinical examination. Isolation of virus from arthropods, wild and domestic animals, and people is essential to determine the natural history of infection with these agents.

Positive serological findings on survey sera are greatly bolstered by virus isolations from the vector, vertebrate, and human being. Virus isolation procedures for serum specimens require inoculation of a small amount of serum into laboratory animals or cell cultures or both. The usual inoculum for intracerebral inoculation of a 1-day-old mouse is 0.02 ml, and for the adult mouse 0.03 ml. Mice are observed daily, or more than once a day, for evidence of illness. A subpassage may be made to attempt to enhance the virulence of the virus, and when one is assured that a virus has been isolated, a stock pool of virus is established and hyperimmune mouse serum or ascitic fluid is prepared if needed. In fatal cases, 40% suspension of tissue—brain, liver, lung, and spleen (purified by centrifugation)—may be inoculated as for serum.

Cell culture systems (vertebrate or insect cell cultures) can be used for virus isolation. For certain viruses and certain cell cultures, the cell culture systems, employing usually a 0.1-ml inoculum, are as sensitive as laboratory animals. Arbovirus outbreaks often occur far from established cell culture laboratories so that inoculation of more readily available laboratory animals is still more widely done than inoculation of cell culture systems. This situation is changing. It must be remembered, however, that intracerebral inoculation of infant mice will serve to isolate a much wider total range of arboviruses than will any single cell culture system. This factual position is not likely to change. Conversely, in the investigation of a specific virus, a cell culture system that has been predeter-mined to be suited to the virus can be easier, cheaper, and as reliable as or more reliable than techniques that require laboratory animals.

In the study of material derived from patients, it is highly desirable to have a pair of serum specimens to work with. The first should be taken early in the course of illness and can serve as material both for virus isolation and for the determination of baseline serum antibody levels before the patient has developed antibodies to the infecting virus. A second serum should be obtained at least 3 weeks after onset of illness, or as late as several months after onset. A seroconversion demonstrable between the early and later specimens is strong evidence of a recent virus infection. Demonstration of IgM in ELISA permits a presumptive diagnosis with a single serum.[27,125]

Details of the techniques of CF, HI, and virus neutralization relating specifically to arboviruses are available in current manuals.[37,56,167] Fluorescent antibody (FA) techniques, antigen ELISA, often coupled with monoclonal antibody, and PCR are widely used for antibody, antigen, and RNA detection. Such advances are greatly extending the scope of field and laboratory investigations, making possible the processing of large numbers of arthropods, often as individual insects rather than as "pools" of hundreds.[11,60,166]

4. Biological Characteristics of the Virus that Affect the Epidemiologic Pattern

Arboviruses by definition share one common feature, the requirement of propagating in some intact arthropod. Transmission by the arthropod is a necessary postlude. For some years, it was considered that all arboviruses were RNA viruses, although any explanation as to why only RNA viruses (and indeed only certain RNA viruses) would multiply in the arthropod is lacking. This position was challenged by the demonstration[1,134] that African swine fever virus, which has been shown to multiply in and be transmitted through the bite of certain ticks,[133] is a DNA virus.

The property of sensitivity to deoxycholic acid (DCA) and to other lipid solvents and detergents has been useful as a screening procedure for candidate arboviruses. However, various nonarboviruses, such as influenza virus and smallpox and ectromelia viruses, are also inactivated by such reagents. Theiler[181] and others also noted that several viruses, unquestionably able to multiply in arthropods and to be transmitted by arthropods, were partially resistant to the action of DCA and the lipid solvents. From this beginning, Borden, Murphy, Shope, and Harrison have proposed the group of orbiviruses,[20,128] a subset of reoviridae, which includes several important viruses such as African horsesickness, blue-tongue, and epidemic hemorrhagic disease of deer. The orbiviruses share morphological and physicochemical features, but serological relationships between or among them may be marginal or lacking.

Another challenging circumstance relates to the specificity of various arboviruses for particular arthropod species. Many workers have attempted to infect arthropods with many of the arboviruses. Flaviviruses can be conveniently divided into two sections: mosquito-transmitted and tick-transmitted. Attempts to infect ticks with certain mosquito-transmitted flaviviruses such as yellow fever virus have consistently failed. Attempts to infect mosquitos with the flavivirus of tick-borne encephalitis have likewise failed. However, Whitman and Aitken[197] have shown that an argasid tick can be infected by feeding on a viremic host and can transmit West Nile virus, a mosquito-transmitted flavivirus. Also, L'vov *et al.*[112] have shown that Tyuleniy virus, a tick-transmitted flavivirus, can multiply in and can be transmitted by a mosquito. Attempts to infect mosquitoes with *Phlebotomus*-transmitted viruses have yielded conflicting results. *Phlebotomus* flies are now being reared and manipulated successfully in the laboratory, and infection of them with mosquito-transmitted viruses has been tried. Attempts with Venezuelan equine, eastern equine, and yellow fever viruses were negative. Phlebotomines were readily infected with known *Phlebotomus*-transmitted viruses.[58,178]

When Grace succeeded in cultivating insect cells in culture systems and in 1966 reported establishment of a line of mosquito (*Aedes aegypti*) cells in continuous culture,[170] the susceptibility of such cell lines to arboviruses was tested. It was early established that the cells were susceptible to infection with mosquito-transmitted viruses. Singh, working with clones of mosquito cells, showed that tick-borne flaviviruses did not multiply in mosquito-cell culture.[171] Buckley[24] extended these observations and showed that even *Phlebotomus*-transmitted viruses did not multiply in mosquito cell cultures. Further observations on tick-transmitted viruses indicated that only Colorado tick fever virus, a tick-transmitted virus, gave evidence of growth in mosquito-cell cultures. Libiková and Buckley [108] later reported limited growth of Kemerovo virus, a tick-transmitted orbivirus, in a mosquito cell line. Řeháček and Pesek[142] have noted limited growth of eastern equine encephalitis virus in tick tissue explants. If efforts to cultivate tick tissues, *Phlebotomus* tissues, and tissues of other arthropods succeed, the problem of specificity can be explored from new directions.

A prerequisite for the survival of an arbovirus in nature is a period of viremia in a vertebrate host at a level sufficient to infect the arthropod vector. Even with a very virulent virus and a very receptive vector, a virus titer in excess of 100,000 infectious doses of virus per milliliter of blood is usually necessary. Such high levels of viremia are presumed to occur in the natural host system and often are attained in the more commonly used laboratory animals such as mice, hamster, guinea pigs, chicks, and monkeys. For many of the more esoteric arboviruses, however, the natural hosts or natural vectors, or both, are either unavailable or not known, so that the postulate of a level of viremia in the vertebrate adequate to infect a susceptible vector remains undemonstrated.

5. Epidemiology

The epidemiology of arbovirus infections in man is influenced by three major determinants: (1) the behavior of the arthropod vector, including the ecological setting in which its breeding occurs, its pattern and range of mobility, its biting habits and species preferences for feeding, its longevity, and the factors affecting the entry, multiplication, and excretion of virus within the arthropod host; (2) the frequency, nature, and duration of exposure of human beings to the infected arthropod vectors, as influenced by the presence, level, and specificity of humoral antibody and by use in the population of insecticides, insect repellents, and protective clothing; and (3) the requirements for the presence of a necessary and/or amplifying vertebrate host for the virus, such as horses, birds, or rodents, and of the availability of man as a diversion in the arthropod–vertebrate cycle.

The great variability in these three determinants does not permit broad generalization on the epidemiology of over 500 arboviruses. Instead, this section, Tables 1 and 3, and the sections on transmission, pathogenesis, host response, and control focus on the most common and important arbovirus infections of man in the United States and on certain common features. Subsequent sections present these in more detail and deal briefly in descriptive and tabular form with arboviruses important outside the United States.

5.1. Incidence and Prevalence

No general statement can be made about incidence, which varies greatly from area to area depending on the presence of an appropriate vector and the presence or absence of an outbreak. The occurrence of a fresh outbreak in a region with susceptible and exposed humans may result in very high infection rates. In epidemics of

Table 3. Mechanisms and Route of Transmission, Pathogenesis, and Immunity Features of Arboviruses Important as Causative Agents of Disease in Human Beings in the 48 Contiguous United States

Virus[a]	Transmission mechanism	Vector(s)	Animal reservoirs	Pathogenesis in man	Immunity	Patterns of host response		
						Key clinical features	Biological spectrum	Ratio of inapparent to apparent infections
EEE	Mosquito bite	*Culiseta melanura* and various *Aedes*	Wild birds; penned pheasants; equines affected but not reservoirs	Encephalitis, frequently fatal; nerve cell damage often focal	Long-lasting, probably lifetime	Abrupt onset, high fever, signs of encephalitis, often becoming severe or fatal in 3–5 days	Severe cases mostly in small children; sequelae common; adults who recover usually have few or no sequelae	Few mild or clinically inapparent infections; immunity rates in populations low
WEE	Mosquito bite	*Culex tarsalis* in the West; *Culiseta melanura* in the East; other mosquitoes	Wild birds; equines affected but not reservoirs	Encephalitis with mortality rates 2–18%; nerve cell damage often focal	Long-lasting, probably lifetime	Fever, drowsiness, and other signs of encephalitis, remission is sudden, with recovery in 5–10 days	Children have sequelae, more severe in younger children; adults rarely have sequelae	Many mild or inapparent infections occur (ratio 58:1 in children, 1150:1 in adults)
SLE	Mosquito bite	*Culex pipiens* and *C. quinquefasciatus*, *C. nigripalpus*, and other culicines	Wild birds	Encephalitis, with diffuse nerve cell damage	Long-lasting, probably lifetime	Brief febrile illness with severe encephalitis in small proportion of cases	Cases of greater severity more frequently in elderly patients; sequelae rare	Ratio high, 64:1; attack rate of clinically apparent infections 280/100,000
LAC	Mosquito bite	*Aedes triseriatus* and other *Aedes* and *Culex*; transovarial transmission demonstrated	Small mammals	Mild encephalitis	Probably long-lasting	Brief febrile illness and sometimes mild encephalitis	Encephalitis of mild to moderate severity in children and adults	Probably high
CTF	Tick bite	*Dermacentor andersoni*	Small mammals	Febrile illness of several days' duration; leukopenia	Probably long-lasting	Diphasic fever, leukopenia; CNS or hemorrhagic complications occasionally in young children	Affects all ages; cases in children more severe	Mild and inapparent infections occur; frequency undetermined

[a]For definitions of abbreviations, see Table 1, footnote a. LAC, LaCrosse.

St. Louis encephalitis, for example, a high proportion of susceptibles in a region where an epidemic is in progress may be infected, even though comparatively few clinical cases may be detected.[15,22,38]

The presence of antibody to most arboviruses reflects the cumulative and lifelong prevalence of infection. A classic arbovirus study was the mapping of the worldwide distribution of yellow fever by Sawyer *et al.*[162] in 1937. The virus neutralization test was performed using adult white mice as test animals. Yellow fever virus was found to be more widely distributed in South America and Africa than had been earlier suspected and was absent from Europe, Asia, and Australia. Much has been learned about group relationships of arboviruses since the 1937 report, and many serological cross relationships have been demonstrated, particularly among the flaviviruses, of which group yellow fever virus is a member. Such cross-reactivity has been demonstrated by CF and HI tests. Theiler and Downs[184] show, in Tables 24, 25, and 26 of their book, development of cross-immune reactions in proven primary and secondary cases of yellow fever. *Secondary* means an infection in a person who had had prior experience with another flavivirus. However, the serological response to the current incitant agent is usually more pronounced than the reaction to other members of the serogroup, enabling specific diagnosis (if the current agent is included in the test). The whole area of cross-immune responses is regarded with something akin to awe. It would be a courageous worker today who would attempt a project such as the 1937 yellow fever sero-survey, with just a single virus and a single test. The 1937 study relied on a relatively insensitive testing procedure, not very responsive to the nuances of cross-reacting antibodies, and the mapping of yellow fever that resulted has required but little alteration.

Neutralization tests measure durable antibody that persists for years. High rates of antibody prevalence could result from continued, widespread virus activity and/or from a large outbreak with a high attack rate. For example, yellow fever neutralization tests performed on sera collected from residents of Trinidad, West Indies, in 1953 revealed no immunes under the age of 15 years and therefore no apparent virus activity later than 1938. This situation changed dramatically with the reappearance of yellow fever on the island in epidemic form in 1954.[64]

Antibody patterns to California group virus infection in Wisconsin[187] revealed varying prevalence rates depending on the frequency and duration of exposure: 48% of Native American forest workers, 34% of wildlife conservation workers, 23% of veterinarians, and 11% of short-term summer workers in forestry camps. Interpretation is difficult, since it is known that those antibodies included not only one LaCrosse strain but also several nonpathogenic members of the California-virus complex.

5.2. Epidemic Behavior

Outbreaks of arbovirus infections in the United States involving human beings occur periodically and unpredictably, as evident from the experience from 1955 through 1985. Only 178 cases of eastern equine encephalitis (EEE) were reported during this time, of which 12 cases and 6 deaths were in 1968; during the same period, 982 cases of western equine encephalitis (WEE) were reported, of which 172 occurred in 1965, with 4 deaths, and 133 in 1975, with 6 deaths. Outbreaks of St. Louis encephalitis (SLE) were first recognized in St. Louis in 1932; other outbreaks occurred in 1966, in 1974 in Memphis, Tennessee, and the largest in 1975, mostly in Mississippi; with 1815 cases and 142 deaths. Venezuelan equine encephalitis (VEE) produced its first outbreak in the United States in Texas in 1971, with no epidemic activity in the United States since that date. Since recognition and reporting of California serotype viruses began in 1963, hundreds of cases have been reported each year with no epidemic peaks.

5.3. Geographic Distribution

Arbovirus infections are worldwide in distribution and may occur whenever the appropriate mosquito or other arthropod vectors abound in proximity to man and a suitable amplifying host. Table 1 includes the geographic distribution of arboviruses in the United States, and Table 2 that in South America, Europe, Asia, Africa, and Australia.

5.4. Temporal Distribution

In the United States, mosquito-borne arbovirus infections produce human infections primarily in late summer and fall; Colorado tick fever (CTF) has involved hunters, fishermen, and campers from spring through fall (see Table 1).

5.5. Age and Sex

Infections with arboviruses can occur at any age. The age distribution depends on the degree of exposure to the particular transmitting arthropod relating to age, sex, and occupational, vocational, and recreational habits of the individual or group of individuals. Adult males are also more commonly infected with CTF virus, since they dominate the hunter and fisherman population who become exposed to the ticks in the high-altitude forested areas of the Rocky Mountains.

Once humans have been exposed and infected, the severity of the host response may also be influenced by age: WEE tends to produce the most severe clinical infections in young persons, and SLE in older persons. Serious LaCrosse virus infections primarily involve children, especially boys who climb in trees, because the mosquito vector *Aedes triseriatus* breeds in small accumulations of water sometimes found where tree branches join the main trunk of a tree—an ideal spot for a child to sit or to build a treehouse. Adult male forest workers also get exposed and have high prevalence rates of antibody but usually show no serious illness.

5.6. Other Factors

Nutritional and genetic factors are not known to influence directly the epidemiology of arbovirus infections. Socioeconomic factors may play a part insofar as they relate to life patterns in seedy suburbs, barrios, or rural regions, where insect-vector populations may flourish under conditions of poor sanitation. It may emerge that cell-mediated immunity under genetic control plays a part in the nature and severity of the host response, the persistence of virus, and the formation of immune complexes, but few data are on hand to support these conjectures.

6. Mechanism and Route of Transmission

By definition, arboviruses must be transmitted by arthropod vectors within which multiplication of the virus is a necessary requirement. This biological transmission may be supplemented by mechanical transmission in which the virus is passively carried externally on the vector or even passively excreted. The primary vectors are mosquitoes, sandflies (*Phlebotomus* and *Culicoides*), and ticks. Of far less importance are horseflies, mites, and blackflies. There is often a high-level virus–vector specificity.

The duration of the necessary period of virus multiplication within the arthropod host before it becomes infectious varies from virus to virus and vector to vector and is also directly temperature dependent. For most viruses under average summer temperature conditions, the extrinsic incubation period falls in the 7- to 14-day range.

Once infected, vectors may remain infected and able to transmit for many weeks or months.

Transmission of virus transovarially in arthropods, often referred to as "vertical transmission," has been demonstrated (specific examples: tick-borne encephalitis in ticks, LaCrosse strain of California encephalitis in mosquitoes, and Sicilian sandfly fever virus in phlebotomine sandflies). Venereal transmission of LaCrosse and SLE viruses in vector mosquitoes has been described.[164,168] These mechanisms may be important for survival of some arboviruses in nature, permitting overwintering or survival over a protracted dry spell. Birds are important reservoir vertebrates for the viruses of EEE, WEE, and SLE, and small mammals for CE and CTF viruses.

Table 3 summarizes some features of arbovirus infections important in the United States.

7. Pathogenesis and Immunity

Arbovirus infections are transmitted by the bite of the appropriate vector, so that the skin represents the sole portal of entry. With penetration of the skin by the biting mouth parts of the arthropod, the virus is deposited directly into lymph or the bloodstream in addition to forming a local pool. With early wide dissemination throughout the host, multiplication follows in as yet inadequately identified target cells and tissues. Viremia in the host then provides the seedbed for infection of succeeding cohorts of biting arthropods. The incubation period is usually 5–15 days.

The site of multiplication of most arboviruses remains undetermined but is presumed to be in the vascular epithelium and the reticuloendothelial cells on the lymph nodes, liver, spleen, and elsewhere. Liberation of virus from these organs constitutes the "systemic phase of viremia," resulting after 4–7 days in fever, chills, and aching. A number of arbovirus infections have two phases—this early phase and then a second phase with or without a few days of freedom from symptoms. The second phase may be attended by encephalitis, joint involvement, rash (sometimes hemorrhagic), and involvement of liver and kidneys. In most arbovirus infections, only the first phase occurs, and the disease is mild and "nonspecific." In other instances, the early phase may be missed, and only the severe manifestations occur. The early phase is accompanied by leukopenia, and the second phase often by leukocytosis. Tissue injury may be the direct effect of viral multiplication in susceptible cells, as is the case with liver involvement in yellow fever. The role of immunologic injury is not clear, although antigen–antibody complexes are suspected of playing an important part in the pathogenesis of the dengue shock syndrome.[93,127,153]

Humoral antibodies regularly appear early in the course of arbovirus infection and constitute the major basis of immunity. Such immunity may be lifelong. No infection with yellow fever virus has been recorded in an individual who either had antibodies from an earlier infection or had a history of yellow fever vaccination with development of postvaccination antibody. The presence of antibodies in the blood at the time of exposure to an infected arthropod vector provides a primary deterrent to reinfection with the homologous virus. Later infection may occur with a virus strain related to but not identical with the original infecting virus. This has been reported for dengue (which has four "types"). Such an event has been demonstrated to produce an exaggerated antibody response, and such a response with the resulting antigen–antibody complexes is suspected of precipitating the host responses seen clinically as dengue hemorrhagic fever or dengue shock syndrome.[93,127] It has been shown that some cases of hemorrhagic fever and shock do occur after a primary dengue infection.[153]

The long persistence of CF antibody to dengue in the absence of reexposure and of reinfection raises the possibility of persistence of the virus, perhaps in association with circulating macrophages. Marchette et al.[114] have shown that dengue virus multiplies in lymphocytes from individuals previously sensitized (i.e., infected) to this virus but not in lymphocytes from those not so infected.

The role of cell-mediated immunity in arbovirus infections has been very little studied. It is possible that it may be important in controlling virus persistence and in determining the immunopathological lesions suspected in certain manifestations of arbovirus infection.

8. Patterns of Host Response

8.1. Clinical Features

Inapparent and subclinical human host responses predominate in most arbovirus infections. Clinical illness is frequently the exception rather than the rule. This varies from virus to virus. For example, infection with WEE, SLE, and CE viruses results principally in mild and inapparent infections, whereas in infection with EEE virus, the host response is likely to be clinically apparent and often severe (Table 4); the reasons for these differences are not known.

Table 4. Patterns of Host Response to Arbovirus Infections in Man

Response	Examples[a]
Asymptomatic infection	WEE, SLE, CE
Mild febrile illness	WEE, SLE, CE, yellow fever
Influenzalike illness with aching and joint pains	Dengue, chikungunya
Encephalitis, mild	CE, WEE, SLE
Encephalitis, severe	SLE, EEE, WEE, CE, tick-borne encephalitis
Jaundice, proteinuria	Yellow fever
Rash, sometimes with hemorrhagic manifestations	Dengue
Shock syndrome	Dengue (following secondary infection with a different dengue serotype)

[a]Certain viruses have been selected for this list particularly to illustrate the range of symptoms that may be seen in populations infected with a single virus.

8.2. Diagnosis

Cases of arbovirus infections in the United States are not likely to be diagnosed unless there is a high degree of clinical suspicion. Outbreaks of encephalitis in horses in summer, caused by EEE, WEE, or VEE viruses, serve to focus attention on febrile illness in humans associated with symptoms or signs indicating involvement of the CNS. Cases with such symptoms and signs occurring in children in the Midwestern states such as Wisconsin, Indiana, Minnesota, and Ohio in late summer should arouse suspicion of LAC virus involvement; SLE virus epidemic sweeps have few indicators in the natural scene. Birds, although widely infected, are not clinically ill, and the primary culicine mosquitos involved in transmission are diffusely spread throughout the United States, especially in areas with inadequate disposal of waste water such as slum areas and the outskirts of towns. The risk may thus be generalized and diffuse and the prediction or pinpointing of an actual outbreak impossible.

Recognition of the arbovirus infection acquired by the traveler outside the United States also depends on the alertness of the examining physician. Rapid jet transport now permits exposed overseas travelers to reach home and fall sick even within the short incubation period of such infections. The physician must maintain a high degree of suspicion when seeing CNS infections or influenzalike illnesses occurring in travelers recently returned from areas endemic for arboviruses. Since specific diagnosis depends on the laboratory, if the specific question is not asked, the laboratory is not likely to carry out the tests needed for specific diagnosis.

It should be emphasized that arbovirus infections constitute only a small fraction of the encephalitis cases seen in the United States. For example, of 938 cases of infectious encephalitis reported to the Centers for Disease Control (CDC) in 1977, only 10.7% were associated with arboviruses; 3.3% were mumps, 5.9% exanthem viruses, 7.1% other viruses, and an impressive 72.9% were of indeterminate etiology (Table 5).

The laboratory diagnosis depends on the isolation of the virus from the blood and/or a fourfold antibody rise in titer, or presence of specific IgM, in a serological test on sera taken during the acute and convalescent phases of illness. Often, the suspicion of an arbovirus infection in individual cases arises too late for virus isolation or for demonstration of a rise in antibody titer. Under these circumstances, the presence of a high antibody titer in a single serum may be significant if the infection is an uncommon one in that region, and particularly if antibody surveys reveal a low antibody prevalence or if prior surveys have demonstrated the absence of antibody in that community. The appropriate procedure in suspected cases is to (1) notify the health department and seek background epidemiologic and clinical data and (2) send acute and convalescent serum samples to the nearest public health laboratory (usually a state laboratory), with a request for antibody tests for arboviruses and other encephalitis-

producing viruses. Some state laboratories may not provide this testing, so a request for transshipment of sera to the CDC might be included. Usually, the specimens should not be shipped directly to the CDC.

9. Control and Prevention

Major control methods include (1) control of the arthropod vector, which may be by elimination of breeding sites or modification of them by application of insecticidal substances or by direct attack on the adult arthropods through residual insecticide treatment of adult resting places; (2) avoidance of exposure to vector bites by screening of houses, by use of protective clothing, and by application of insect repellent sprays or creams when outside in high-risk areas; and (3) immunization, a procedure widely used only for yellow fever and Japanese encephalitis in endemic areas. Specific control measures are discussed in appropriate sections below and are listed in Table 2 for arboviruses of importance outside the United States.

Control of vectors through biological approaches ranging from introduction of competing species, of parasites, including protozoa, helminths, bacteria, and viruses, or of genes deleterious to the vector population, or influ-

Table 5. Causes of Infectious Encephalitis in the United States in 1977, 1978, 1983, and 1984[a]

Causes	1977 Cases	1977 Percentage of total	1978 Cases	1978 Percentage of total	1983 cases	1984 cases
Arboviruses						
Western equine	41	4.4	3	0.2	7	2
St. Louis	23	2.5	26	1.8	19	33
Eastern equine	1	0.1	5	0.3	14	5
California group	35	3.7	109	7.5	62	74
Totals	100	10.7	143	9.0	102	114
Childhood infections						
Mumps	31	3.3	26	1.8		
Measles	17	1.8	13	0.9		
Chickenpox	39	4.1	40	2.7		
Rubella	0	0.0	1	0.1		
Totals	87	9.3	80	5.5		
Other						
Enteroviruses	22	2.3	40	2.8		
Herpes simplex	32	3.4	77	5.3		
Other known	13	1.4	23	1.6		
Totals	67	7.1	140	9.7		
Etiology undetermined	684	72.9	1071	74.3		
Totals	938	100.0	1441	100.0		

[a]From *Morbidity and Mortality in the United States, Annual Summary 1978*, and Centers for Disease Control, *Morbidity and Mortality Weekly Reports* 1978, 1983, and 1984.[23]

encing the vector behavior or capacity to be infected with a pathogen is receiving much attention.[186]

10. Characteristics of Selected Arboviruses

10.1. Arboviruses of Importance in the United States

Of the arboviruses found in North America, five are of primary importance in human infections: eastern equine encephalitis (EEE), western equine encephalitis (WEE), St. Louis encephalitis (SLE), California encephalitis (CE), as subtypes LaCrosse (LAC) and Jamestown Canyon (JC), and Colorado tick fever (CTF) viruses. Powassan (POW), vesicular stomatitis (Indiana) (VSI), and Venezuelan equine encephalitis (VEE) are less important as causes of human disease. In addition to these endemic viruses, clinicians also will encounter arbovirus infec-

tions, particularly dengue, in travelers, returning expatriates, and military personnel. Brief summaries of main features for these viruses are presented below, in Tables 1 and 3, and in Figs. 1 and 2.

10.1.1. Eastern Equine Encephalitis.   Clinical EEE in human beings has been sporadic and focal even under epidemic circumstances. EEE occurs predominantly in eastern and Gulf Coast states, and in northeastern inland states in the United States and Canada (Fig. 1).[43,44,49,136] It also circulates in focal areas of the Caribbean and in Central and South America. Serosurveys of populations, even when focused on population groups estimated to be at greater risk, reveal generally low rates of antibody prevalence. However, the high case fatality associated with this severe infection (30–70%), and the potential of changing disease distribution following ongoing environmental change, demands continuing epidemiologic vigilance. Outbreaks in equine populations in the endemic areas are kept in check through vaccination using

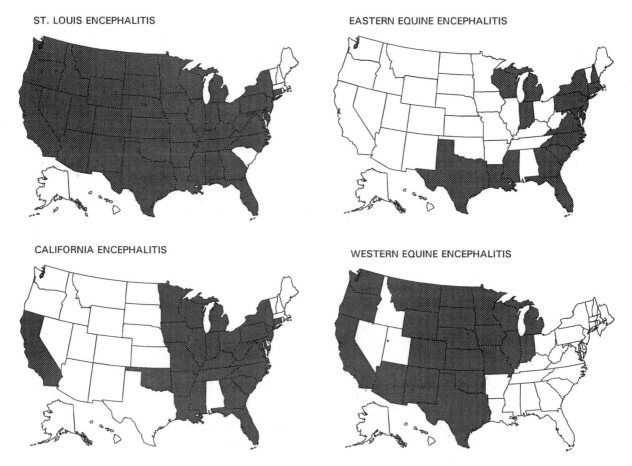

ST. LOUIS ENCEPHALITIS

EASTERN EQUINE ENCEPHALITIS

CALIFORNIA ENCEPHALITIS

WESTERN EQUINE ENCEPHALITIS

Figure 1. States reporting human arbovirus infections, 1964–1994.

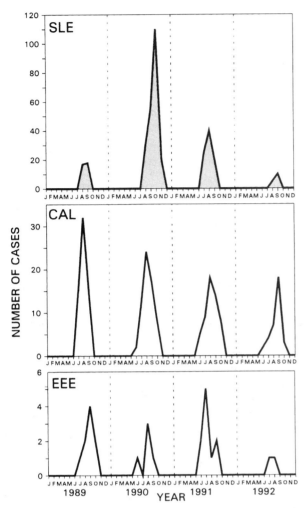

Figure 2. Monthly distribution of human cases of St. Louis, California, and eastern equine encephalitis, 1989–1992.

a formalin-inactivated EEE vaccine. Immunization of humans has not been considered feasible except for laboratory workers engaged in specific EEE virus manipulations.[94]

EEE is maintained in eastern North America in a cycle among wild birds and mosquitoes; although EEE virus activity can be documented in birds in the endemic areas, generally less than 10 human cases are reported in the United States each year.[43,44,49] The dominant epidemiologic feature that protects human populations from EEE is the localization and highly ornithophylic feeding habits of the principal enzootic vector, the mosquito *Culiseta melanura*. This mosquito breeds and remains localized to certain types of deep swamp habitats, in areas generally remote from humans, and feeds mainly on small passerine wild birds.[198] Other mosquitoes that feed on wild birds and larger vertebrates may transmit the virus from wild birds to horses and man, who are dead-end hosts. Passerine birds are the primary amplifying hosts, and the yearly cycling (Fig. 2) of EEE virus is often first detected in pheasant flocks. The peak of occurrence of epidemics and human cases is late summer and early fall (Fig. 2). The first heavy frosts usually terminate transmission. In the southern and midwestern states, *C. melanura* is not found, and the vectors that transmit virus to the wild bird populations include *Aedes* and other culicine species.[122,166] At least one analysis of rainfall patterns shows a general association of human cases and excess rainfall.[107] There has been speculation that landscape and faunal changes have caused EEE to emerge in the northeast after the 1930s.[104,136]

There now is considerable concern that recent environmental changes might alter the normal pattern of EEE human and equine disease and result in the emergence of increased human EEE cases. In some areas, new susceptible hosts, such as the emu, have been introduced.[8,190] Moreover, a new human-biting vector, *Aedes albopictus*, has been introduced into the United States and spread to many states.[87,122] EEE has been isolated from this new host and the new vector,[122] and experimental laboratory transmission studies demonstrate that *Ae. albopictus* is a good vector of EEE virus.[165]

The incubation period in human beings is short, usually 4 to 10 days.[136] The ratio of inapparent or mild infections to severe infections is low (2–4 : 1). Abrupt onset of severe fever, nausea, myalgias, and intense headache are followed by encephalitis in 1–2 days. Infants and children often present with convulsions. The encephalitis rapidly progresses to a coma, particularly in small children. Between 30 to 70% of clinical cases are fatal. Pathological features are those of a diffuse encephalitis with evidence of scattered neuronal destruction. Serious neurological sequelae occur most frequently in young children, and in this population, sequelae may persist in 50% of the survivors. Intensive supportive care is essential, since there is no specific treatment for EEE. Immunity is probably long-lasting, with no reinfections being described. Experimental, inactivated vaccines are available, but are only used to protect laboratory workers.[152,199]

After initial clinical suspicion, diagnosis is established by virus isolation or, more commonly, by demonstration of serological conversion to EEE virus, or the finding of high-titer IgM in serum or cerebrospinal fluid (CSF). Since the background prevalence of EEE in endemic areas is very low, a single serum demonstrating high-titer antibodies is highly suggestive of infection.

Evidence of an outbreak of disease involving birds and mosquitoes, and more particularly equines in a region, should alert clinicians.

10.1.2.   Western Equine Encephalitis.   Compared to EEE, WEE infection of humans (Table 4) occurs at a higher ratio of inapparent to apparent (diagnosed) infections, estimated as approximately 50:1 in children and 1000:1 in adults.[140,199] Clinical infections range from mild to severe meningoencephalitis. The case fatality rate is in the range of 3%.[189]

Like EEE, WEE virus is maintained in a wild bird–mosquito cycle. The virus is endemic in the western United States (Fig. 1), Canada, Mexico, and focal areas of South America. The major virus activity in the western and southwestern United States is closely tied to the presence and behavior of *Culex tarsalis* populations.[138,140,198] Since this mosquito also feeds readily on larger vertebrates, it requires no other mosquito species to generate an epidemic. The numbers of *C. tarsalis* mosquitoes in a region can be augmented by irrigation practices, and the major foci of virus activity involving human beings in the West and Southwest relate directly to major areas of crop irrigation. Proper water management procedures can greatly reduce *C. tarsalis* production, as can the coordinated application of larvicidal measures. Those who spend large amounts of time outdoors are at higher risk, with cases in males outnumbering cases in females by almost 2:1. As is the case for EEE, WEE cases appear from June to October, and generally peak in August to September. In addition to the main vector mosquito, several other mosquito species can be infected and transmit virus. Recent data suggest that one means of overwintering of the virus is by transovarial transmission in *Aedes* species.[77] There have been recent studies showing that recombination might play an important role in the evolution of these alphaviruses.[91] Nucleotide sequence analysis has shown that WEE virus appears to have evolved from recombination of the New World EEE virus and the Old World Sindbis virus. There also are a number of apparently distinct viruses (Highlands J, Aura, and Fort Morgan viruses) that are closely related to WEE and considered to be in the same antigenic complex.[102]

In equines, immunization with an inactivated WEE vaccine is widely used, and an attenuated vaccine strain has been shown to be effective.[19] Immunity conferred by natural infection is long-lasting. Although the highlands J subtype of WEE is found on the eastern seaboard, infections in human beings east of the Mississippi river are uncommon (Fig. 1). This may be because of the ornithophilic behavior of the vector (*Culiseta melanura*), as described above for EEE.

In view of the high ratio of inapparent or mild cases to severe cases, the low figures for human mortality, and the millions of people potentially at risk, vaccination has not been considered necessary in human populations. An experimental inactivated vaccine is available for laboratory workers.[152]

The clinical presentation is generally mild, and after a short incubation period of a few days, nonspecific symptoms of fever, headache, myalgias, nausea, and malaise may be followed by drowsiness and more pronounced signs of encephalitis.[138,140] Although a fatal outcome is rare, sequelae after encephalitis occur in 10% of the cases, particularly in infants and small children. Pathological findings are those of a diffuse encephalitis and focal neuronal destruction.

Specific diagnosis is based on virus isolation, and more commonly on serological grounds, with a fourfold or greater rise in antibodies between early and convalescent serum specimens, or high titers of IgM in serum or CSF, generally being considered an adequate criterion. Eastern equine encephalitis and VEE are the only other antigenically related alphaviruses in the United States, and the possibilities of confusion resulting from serological cross-reactivity are minimal.

10.1.3.   St. Louis Encephalitis.   The first recognized outbreak of SLE in St. Louis in 1932[111] provided epidemiologic data that have been added to but not changed in most essential features, up to the present time.[123] The disease is the leading cause of epidemic (as opposed to endemic) viral encephalitis in the United States (Fig. 1).[15,22,38,115,126]

The SLE virus is widely prevalent in the United States, with 42 states reporting cases since 1964 (Fig. 1).[123] It has caused disease in southern Canada, Mexico, and parts of Central and South America.[15,123,167,184] Although this is a flavivirus, and as such contrasts with the alphaviruses EEE and WEE, it shares many general ecological and clinical features with EEE and WEE. It overlaps with much of the distribution of WEE and EEE, and also circulates in a wild bird–mosquito cycle. Cases occur in the late summer months (Fig. 2).

The virus is transmitted by a number of culicine mosquitoes, including *Culex pipiens* and *C. quinquefasciatus* in the East and urban areas of the West, *C. tarsalis* in rural areas of the western states, and *C. nigripalpus* in Florida and southern areas.[15,38,39,41,42,123,184] In rural areas of the West and Southwest, it is closely tied to *C. tarsalis* and irrigation practices, much as is the case with WEE. There is a diffuse endemicity over vast rural areas, with low rates of seropositives in humans, but with occasional epidemics of hundreds to more than a thousand cases.

These periodic epidemics often occur on the outskirts of towns and cities. In such urban areas of the West, and in the East and Southeast, mosquitoes breed in great numbers under circumstances where there is inadequate disposal of waste water. Although many communities are at risk annually, and one or more annual outbreaks can be expected, the site(s) of such outbreak(s) cannot be predicted. As with EEE and WEE, the monitoring of sentinel flocks of birds for infection has proved useful as an early warning system.[145] Depending on the climate of the region, maintenance of the virus can occur by year-round horizontal transmission from bird to mosquito,[146] over-winter survival of infected mosquitoes,[9] or venereal or transovarian infection.[75,168]

St. Louis encephalitis affects all age groups, but only 1–2% of infections lead to clinically apparent infections.[15,22] Exposure to mosquitoes is a risk factor. Most cases start with a benign course, with fever for a few days, severe headache, nausea, myalgias, and complete recovery. Advanced age is the most significant risk for developing both symptomatic disease, and more severe encephalitis after infection.[115] More than 75% of explicit clinical cases progress to aseptic meningitis or encephalitis. Remission may be dramatic, and except in very young infants, recovery is usually complete. In 10–25% of these young infants, sequelae can include mental deterioration, personality changes, muscle weakness, and paralysis. Overall, the case fatality rate is about 6%, but the disease is generally milder in children, and the fatality rate for those under 5 years old is 1%. Short-lived sequelae of nervousness, headache, and memory impairment occur uncommonly in older children and adults.[38,126]

Diagnosis is usually made by serological means.[189] A leukopenia or mild leukocytosis and slight increase in spinal fluid pressure, cells, and protein lead to the suspicion of encephalitic involvement, but confirmation must come from virus isolation (very rarely accomplished) or demonstration of serological conversion or high-titer IgM in serum or CSF.[125] In geographic regions where other flaviviruses occur, serological tests may show confounding cross-reactivity.

There is no commercially available vaccine for SLE.[175] Historically, control in rural or an urban epidemics has been attempted by emergency application of mosquito-control measures. Long-term control for urban and suburban localities depends on good sanitation with respect to drainage and adequate disposal of waste water. In most rural areas, control measures, aside from general protection against mosquito bites, are not feasible. In more specific rural areas under irrigation, much can be accomplished through water management and directed application of insecticides to keep mosquito populations low.

10.1.4. California Encephalitis. California encephalitis is a general term for infections with distinct but related viruses of the California serogroup in the *Bunyaviridae* family.[81,82,95,101] Several California encephalitis-related viruses are recognized, but only LaCrosse, snowshoe hare, Jamestown Canyon, California, or trivittatus viruses have been implicated as disease-causing agents in the United States and Canada.[6,85,106] Of these agents, the most important in the United States is LaCrosse (LAC) virus. It annually causes 50–150 cases of "summer encephalitis" in the northern midwest (Fig. 2).[41–44,49] Over the last three decades, an increased awareness of the disease has led to recognition of LAC encephalitis in over 25 states, predominantly in the eastern half of the United States (Fig. 1). Most of these cases are in children less than 15 years of age. Reports of human disease caused by other California group viruses are infrequent. However, serological surveys indicate that there might be substantial underreporting. In Europe and central Asia, related viruses have been implicated in febrile and encephalitic disease.[26]

For California (CAL) group viruses in general, the pattern of disease occurrence and of human infection is one relating to exposure in sylvan settings.[139,187] The viruses are maintained in cycles involving small mammals and *Aedes* and *Culiseta* woodland mosquito species.[124] Cases are seen in workers and hikers who enter wooded areas, or frequently in children who live and play near woods or forests. Year-to-year virus survival in northern climates likely involves maintenance in the mosquito by transovarial and/or venereal transmission mechanisms.[164,186,193] Overwintering of infected mosquitoes may also play a role in more temperate climates.[194]

The best-studied CAL viruses are LAC and Jamestown Canyon (JC) viruses. LaCrosse encephalitis virus annually causes 50–150 cases of encephalitis in midwestern children. Its amplifying hosts are eastern chipmunks and squirrels.[5,59,60] The main vector, *Aedes triseriatus*, is a treehole-breeding mosquito of wide distribution, common in hardwood forest habitats and suburban woodlands. The vector also breeds in water present in old tires and in other man-made containers.[106,164]

Jamestown Canyon virus causes febrile illnesses and low numbers of adult meningitis cases in the Midwest and South.[86,88,89] It is amplified in white-tailed deer. The range of JC and LAC viruses overlap, but with different amplifying hosts and vectors; they exist sympatrically. As for the other viruses in the group, occasional febrile illnesses, aseptic meningitis, or encephalitis have been reported as the result of infection with snowshoe hare (predominantly in Canada), California, and trivittatus viruses.[6,106]

Most infections with LAC appear to be subclinical. When clinical illness occurs, it starts with often indefinite or mild symptoms of fever, anorexia, headache, and nausea.[90,101,188] With mild to severe encephalitis, ataxia and confusion are often first seen, and convulsions occur in about half of the cases. Fatality rates are less than 0.5%. One report of autopsy findings describes diffuse encephalitic changes of neuronal degeneration and patchy inflammatory response in the cortex and in the basal ganglia. Recovery after mild encephalitis is usually complete, but severe cases show residual effects on psychometric performance.

Supportive treatment is effective. No specific antiviral treatment is approved for use. Immunity rates are high only in special population groups with extensive woodland exposure.[124] Immunity is probably longlasting and reinfection with the same CAL group virus probably does not occur. Protective measures are limited to protection of individuals from the bite of mosquitoes[187] through the application of insect repellents. In specific local situations, such as extensively used campgrounds, area disinsectization can be considered and removal of trash and man-made containers is important. Children in some endemic areas are taught to play in open areas.

Diagnosis, following initial clinical suspicion, may be by virus isolation, but is much more commonly based on detection of seroconverion of IgG or detection of hightiter IgM in appropriate serum or CSF specimens.[27,65] There is no difficulty in narrowing diagnosis to the California subset of viruses, but it is more difficult to identify the type of CAL virus unless the virus itself has been isolated, either from the case (the least likely but most desirable situation) or from mosquitoes captured in the vicinity of the infection.

Unfortunately, in 1985, the Asian "tiger" mosquito, *Ae. albopictus*, was introduced in old tires imported from Japan to Houston. It has now spread to infest over 22 states, mostly in the South and Midwest.[87] This mosquito is an aggressive, opportunistic feeder with a wide host range that includes man. It adapts well to forest or urban settings, and potentially it can vector many different arboviruses.[11,60,122,165] In some areas, it already has replaced local mosquito species. As mentioned above for the other endemic arboviruses, there is concern that this new vector will transmit endemic viruses and change the current pattern of arbovirus disease.[59,60,87]

Aedes albopictus can be infected by a number of endemic arboviruses, including both LAC or JC. Both viruses are in the bunyavirus group, and as such have a tripartite RNA genome that can reassort and recombine in a fashion analogous to the segmented genome of influenza virus.[12,13,55,78,82,191] This evolutionary advantage of being able to exchange genome pieces has undoubtedly led to the large number of related bunyaviruses. The bunyaviruses include classic arthropod-borne viruses as well as rodent-borne viruses such as hantaviruses. Since *Ae. albopictus* has now invaded LAC and JC endemic areas, and will feed on either small or larger mammals, there is an increased opportunity that a dual infection of a mosquito will occur and a resulting increased risk that reassorted viruses could emerge. The ability to predict the phenotypic characteristics of a virus that might emerge from a reassortment is not high. It is possible that a new reassorted virus with a mixture of existing properties still might represent a new public health problem.

10.1.5.   Colorado Tick Fever.   Colorado tick fever (CTF) is a tick-transmitted coltivirus localized to mountainous regions of the western United States, Canada, and small parts of Mexico (a related virus exists in Asia).[22,23,61,67,118,119] Overall, antibody prevalence in the general population is low, and despite wide virus distribution in the endemic regions, human infections are usually sporadic and casual. CTF is reported often in hikers and those exposed to the outdoors (Table 1); one survey of 178 sera from sheepherders revealed 32% seropositives.[79] Fortunately, it rarely causes serious or fatal illness. The disease is difficult to diagnose on clinical grounds alone, so there is undoubtedly much underreporting. Infection occurs in hikers, foresters, or vacationers venturing into high plains or wooded mountainous habitats generally between 4,000 and 10,000 feet elevation. These areas are populated by various small mammals that are natural hosts for CTF.[67] The small mammals transmit the virus to the main vector, the wood tick *Dermacentor andersoni*.[23,61] The natural cycle between tick and small mammal occurs in open habitats of pine and shrubs on dry rocky surfaces.

Importantly, in diagnosed cases there is invariably a history of tick bite or exposure to ticks. Following a tick bite, the incubation period is 3–6 days, with sudden onset of fever, headache, acute retro-orbital pain, and severe muscle pains.[61,67] A painful skin rash is often reported, but no respiratory symptoms occur. In about half the cases, there may be a brief remission of 3–5 days followed by a second brief period of fever. The virus infects erythrocyte precursors and circulates in mature erythrocytes. A pronounced lymphopenia occurs during the febrile phase in 60–70% of the cases, and a modest thrombocytopenia is common. Serious complications such as encephalitis, myocarditis, severe bleeding, and death have been reported, but are rare and almost exclusively limited to children. Recovery can be protracted but uncomplicated, and immunity following recovery is presumably lifelong.[67]

Because of the association of virus with erythrocytes, diagnosis is commonly made by isolation of the virus from the blood.[68,79] The intracellular viremia is relatively long, with virus having been isolated from erythrocytes of patients as early as 1 day and as late as 120 days after onset of symptoms.[99] Serological techniques are useful where virus isolation facilities are lacking.[28] Perhaps because of the long viremia, antibodies appear relatively late, generally 1–2 weeks after onset of fever. Because of the relatively long viremia, there have been cases of CTF infections after blood transfusions.

An experimental, inactivated vaccine has been described,[185] but a commercial product is not available. Protection against CTF virus infection is largely a matter of protection against tick bite (repellents, protective clothing, avoidance of tick-infected regions). There is no specific therapy. Infected individuals should not donate blood until well after complete recovery.

10.2. Other Arboviruses in the United States that Affect Humans

Several other arboviruses capable of causing illness are increasing in importance worldwide, but thus far are considered of only minor importance as causative agents of human disease in the United States.[29,189] These diseases, particularly dengue, are seen in travelers, returning expatriates, and military personnel. However, the emergence of the acquired immunodeficiency syndrome (AIDS) epidemic, the appearance of hantavirus in the Southwest, and the emergence of Lyme disease in the Northeast have demonstrated the vulnerability of the United States to emerging diseases. Factors influencing the pattern of emergence and distribution of infectious diseases in general include those associated with the microbial agent itself, the agent's hosts and vectors, and the environment in which agent and host interact. The natural life cycle of many arboviruses is multifaceted and, in addition to the organism, may include one or several reservoir or amplifying hosts and an arthropod vector. A natural or manmade change affecting the interaction of any of these fundamental elements might lead to the emergence or reemergence of these now uncommon diseases.[48]

10.2.1. Venezuelan Equine Encephalitis. Venezuelan equine encephalitis is an alphavirus responsible for large outbreaks of encephalitis in horses, with associated human disease, in South and Central America and the West Indies.[66,109,130] The first reported outbreak in the United States occurred in 1971 in the Southwest, coincident with an extensive outbreak in Mexico. There were over 100 human cases and 1500 horse cases, but no epidemic activity has been detected in the United States since 1971. Human cases are diagnosed as encephalitis of mild or moderate severity, but fatal human infections have been described.[29] In endemic regions, thousands of human cases have been reported during some outbreak years.[50,53,66] There are nucleic acid sequence data suggesting that at least some of the outbreaks during 1969–1972 might have resulted from distribution of incompletely inactivated equine vaccine.[103,130] A subtype of VEE, Everglades virus, has been reported as cause of sporadic CNS disease in Florida.[192]

Enzootic and epizootic subtypes of VEE virus and epizootic and enzootic cycles have been described, but definitive reservoirs for each subtype of virus and the natural cycles of maintenance of the viruses are still under study.[18,205] In the enzootic cycle, virus is mosquito-transmitted, and natural reservoirs appear to be small mammals rather than birds, as has been described above for the other alphviruses WEE and EEE.[18,109,192] The epizootic cycle is less clear, and large and small wild and domestic mammals, birds, and multiple mosquito vectors all have been implicated as having a role in virus circulation.

In 1995, a large epizootic of VEE occurred in Venezuela and Colombia.[50,53] It was the largest epidemic since 1962–1971, with estimates of 13,000 to 45,000 human cases and widespread equine deaths. Molecular evidence suggested the emergence of a new subtype of epizootic VEE.[151]

Clinically, VEE is similar to WEE and EEE, and starts abruptly with a fever, severe headache, nausea, and myalgias. Less than 5% of cases progress to encephalitis, and most of these are children.[50,53,66,130] An estimated 20% of those patients with CNS involvement die. Treatment is supportive. Inactivated and attenuated vaccines are available for laboratory workers.[94] Diagnosis can be established by virus isolation, often from throat swabs, or by demonstration of seroconversion of IgG antibodies or high-titer IgM antibodies.[57]

10.2.2. Powassan. Powassan is a tick-transmitted virus of the tick-borne encephalitis subgroup of flaviviruses. It is related to tick-borne encephalitis virus, an important pathogen in Russia and northern Europe, and to Kyasanur Forest disease virus, a human pathogen in India.[199] The original virus isolation was from the brain of a child who died in Ontario in 1959.[117] Since that time, there have been only scattered cases reported in humans in North America.[(29,80,189)] Powassan virus has been found to be widely distributed in small mammals in Canada and the northern states of the United States. It is transmitted by hard ticks (*Ixodidae*). A factor that limits the invasiveness of the virus for man is undoubtedly the reluctance of the vector ticks, *Ixodes cookei* and *I. marxi*, to attach to and to feed on human beings. Diagnosis can be accomplished by serological means, being cautious to control for cross-reactions from other flavivirus infections (particularly SLE in the United States).

10.2.3. Vesicular Stomatitis Virus.

Vesicular stomatitis virus (VSV) is a rhabdovirus that causes disease in livestock that results in vesicles and ulceration on mucosal surfaces. Epizootics of VSV, both New Jersey and Indiana serotypes, have occurred repeatedly in the United States, and phlebotomine flies and *Culicoides* midges are suspected as vectors.[(58)] Clinically recognized infections in human beings are rare, but several infections were reported during the 1982 epizootic in Colorado.[(74,143)]

10.3. Arboviruses outside the United States that Affect Humans

Probably less than 20% of the more than 500 arboviruses are documented as a cause of recognizable disease in human beings. However, several of these are important human pathogens. Table 2 gives a brief summary of the most important exotic (for the United States) arboviruses, plus a selected group of viruses of lesser but still considerable importance. The epidemiologist or physician in the United States will not encounter any of these infections as endemic diseases. However, occasionally acutely ill patients may be evacuated from a foreign location for hospitalization in the United States, and with increasingly mobile populations, there exists the possibility that any of these diseases could enter the country with immigrants or returning travelers or military personnel. With the common feature of incubation periods of several days, it is possible that infected individuals, in the incubation period and asymptomatic, can enter the country and proceed to fall ill within the next few days. In the United States, imported dengue is an ongoing occurrence, and over the last decade, hundreds of confirmed cases have been reported from 32 states.[(45,51,54)]

In general terms, the infections listed in Table 2 are not protracted febrile illnesses but are acute. Frequently, physicians with a patient with a history of protracted fever or bouts of fever over a span of weeks request examination for possible arbovirus infection. The possibility of an arbovirus causing such an illness is limited. However,

with the recent development of sensitive IgM assays, it can be important to examine a specimen taken several weeks after an acute illness has resolved, even though there may exist no specimen from the acute phase of illness.

The recognition of an imported arbovirus illness is complicated by the fact that in the early stages, differential diagnosis is almost impossible. Various causes of febrile illness must be reviewed: malaria, trypanosomiasis, typhoid fever, typhus fever, leptospirosis, hepatitis, gastrointestinal infections with diarrhea and vomiting, and the grabbag of influenzalike illnesses. As outlined above, several of the arboviruses occur in the United States and in North, Central, and South America and some of the West Indian islands. The possibility of arriving at a correct diagnosis is dependent on determining an accurate travel history and of obtaining appropriate laboratory studies. A compounding diagnostic problem is that few laboratories are capable of performing the required diagnostic tests.

10.3.1. Dengue and Dengue Hemorrhagic Fever.

Dengue (DEN) viruses are the most widespread arthropod-borne viruses. They are members of the *Flaviviridae* family, which includes more than 70 related but distinct viruses, most of which are mosquito-borne. Other major pathogens in this family include yellow fever and Japanese encephalitis viruses. In 1994, DEN was found on most continents, and almost one half of all United Nations member-states were threatened by DEN. Epidemics continue to emerge, and this virus now causes severe infections in areas where periodic epidemics did not occur previously. The disease will continue to spread as newly urbanized areas become infested with mosquito vectors. In those areas in which DEN is endemic, more than 1.5 billion people—including about 600 million children—are at risk. Each year, it is estimated that from 35 million to 60 million people are infected with DEN, and that 2000 to 5000 children die from this viral infection. These figures most likely underestimate the scope of this problem.[(51,154,163,199,200)]

There are four closely related, but serologically distinct, DEN viruses (serotypes 1 through 4).[(168)] Since there is only transient, weak cross-protection among the four serotypes, a population could experience a DEN-1 epidemic in 1 year, followed by a DEN-2 (or others) epidemic during later years. Primary infection with any serotype often causes a debilitating, but nonfatal illness. To date, antiviral drug chemotherapy has not been successful; consequently, most currently used forms of therapy are mainly supportive in nature.

Some infected patients experience a more severe and often fatal form of the disease, called dengue shock syndrome or dengue hemorrhagic fever (DHF).[93,127] Unlike other infectious diseases, the presence of antibodies after recovery from one type of DEN infection predisposes some individuals to contract the more severe form of disease (DHF) when infected by a second, different DEN virus serotype. Thus, among those infected during a previous DEN epidemic, such "immune enhancement" would result in some children developing fatal DHF during a subsequent outbreak of the disease caused by another serotype. Although all age groups are susceptible to DEN fever, DHF is most common in children, and usually occurs in about 1–10% of all hospitalized cases. Some studies suggest that not all severe DEN illnesses occur via the immune-enhancement mechanism; but instead, DHF can occur after primary infection with some strains of DEN. However, this appears to be a rare event.[10,153]

DEN viruses are prevalent throughout the tropics, where the urban dwelling mosquito, *Ae. aegypti*, is a major vector. In parts of Asia, DEN may be associated also with *A. albopictus*.[154,157] Although the virus typically circulates in endemic cycles, it periodically causes acute, widespread epidemics, in which 30% or more of the population may be infected. An example of the explosive nature of these epidemics occurred in 1987, when 174,285 cases of DEN were reported to occur in Thailand; most involved children, less than 15 years of age, and 1007 deaths were reported. In that year, aside from diarrhea and fever, DEN was the third leading cause of illness in children (154,381 cases) and the leading cause of childhood death (925 cases) (extracted from Thai Ministry of Health published tables).

DHF has become more prevalent in Southeast Asia as new waves of epidemics caused by different DEN serotypes occur; this appears to be happening in the Western hemisphere as well. In the Americas, the first cases of severe DHF occurred in 1981. They were associated with a DEN-2 epidemic in Cuba that followed the DEN-1 epidemic of 1977. During the 1981 outbreak in Cuba, 116,151 hospitalized cases of DEN fever were reported and 10,312 cases were classified as severe DHF; 158 deaths occurred.[54] More recently, South America has experienced major outbreaks of DEN, with cases of fatal DHF being reported in Brazil, Venezuela, and Peru. In 1994 and 1995, Central and South America experienced unprecedented numbers of cases, often in newly infested regions.[52]

During the last few years, DEN has been spreading into areas previously considered to be endemic but not usually associated with major outbreaks of the disease. The westward expansion of DEN in Asia was first documented in the late 1980s by the appearance of epidemics of DEN, with DHF, in India and Sri Lanka. Africa and the Middle East also were considered to be areas with a low incidence of DEN; however, DEN has emerged in these areas in the early 1990s, as evidenced by the widespread occurrence of DEN infections in United States military personnel stationed in Somalia and Haiti, as well as by preliminary reports of DEN in Saudi Arabia [47,149] (personal communication, A. Zaki).

The control of DEN will be possible only after an efficient vaccine has been developed, since attempts to eradicate mosquito vectors have not been successful in developing countries. Clearly, the phenomenon of immune enhancement may be a major problem in developing an effective vaccine for DEN. It suggests that one may have to prepare a multivalent vaccine against all four serotypes of the DEN virus in order to avoid inducing the DHF associated with subsequent natural infections caused by other serotypes.

Diagnosis of DEN and DEN hemorrhagic fever and shock, on clinical grounds, may be reinforced by virus isolation or by serological findings in the CF, HI, N, or ELISA test, buttressed by use of monoclonal antibodies to specific dengue types. Diagnosis of specific DEN type can be done only in a few specialized laboratories.[54,199]

Control of DEN is basically control of the vector mosquito, *A. aegypti*, and avoidance of being bitten by mosquitoes. Rudnick,[157] working in Malaya, has evidence that forest monkeys are involved in a "jungle dengue" cycle with forest *Aedes* species serving as vectors. A similar jungle cycle occurs in West Africa, although there is no evidence that forest viruses are responsible for epidemics.[154,199] The possibility of a jungle DEN cycle postpones hopes of eventual eradication of the disease.

10.3.2. Yellow Fever. Yellow fever (YF) virus is a member of the *Flaviviridae* family of viruses.[199,201,202] This family contains over 70 related but distinct viruses, of which most are arthropod-borne. Other major pathogens in this classification include DEN and Japanese encephalitis viruses. Many synonyms have been used for YF, including fievre jaune, yellow jack, typhus amaril, typhus icteroide, black vomit, febris flava, typhus amaril, vomito prieto, vomit negro, Gelbfieber, and *Flavivirus hominis* fever.

The origin of YF is speculated to have been western Africa, although the disease was first distinguished from other tropical febrile diseases by Spanish explorers in the Yucatan peninsula in 1648. During the 18th and 19th centuries, it was one of the great plagues of the world,[184,201,202] occurring along the eastern U.S. seacoast, in Central America and South America, and in Africa throughout the tropical area. Major epidemics occurred in many seaports of the United States, and the 1905 outbreak in New Orleans was particularly severe. The last U.S. indigenous case occurred in 1911, and the last imported cases in 1924.[63,184] It never became established in Europe above the range of the vector, *Ae. aegypti* mosquitoes. Yellow fever has never been reported from Asia and Australia, despite the endemic presence of *Ae. aegypti* in tropical Asia.[162,167,201,202]

In 1900, a U.S. Army Yellow Fever Commission under Major Walter Reed was assigned to Cuba, where its studies demonstrated that a filterable agent present in the blood of acute-phase patients could be transmitted by *Ae. aegypti* mosquitoes.[63,202] The elucidation of the epidemiology of YF was principally achieved in Africa and South America under the auspices of the Rockefeller Foundation.[63] Theiler's attenuated live virus vaccine introduced in 1937 forms the basis for the present-day product.[181]

Extensive studies on the maintenance of YF virus have shown that the virus exists in two cycles: an urban cycle involving human beings and *Ae. aegypti* mosquitoes, and a sylvatic or jungle YF cycle involving forest primates, principally monkeys, and forest canopy mosquitoes, with human infections tangential to the transmission cycle.[201,202] In the Western hemisphere, urban YF was solely transmitted by *Ae. aegypti* mosquitoes.

In 1901, eradication efforts directed toward *Ae. aegypti* mosquitoes were launched under the direction of Dr. William Gorgas in Havana. These eradication efforts, with concomitant reduction of YF, were extended throughout Central and South America in the early 1900s. The chain of urban *Ae. aegypti*-transmitted YF was successfully broken by the eradication program. The last endemic focus of *Ae. aegypti*-transmitted urban YF was in northeastern Brazil in 1934.[201] The eradication of the vector, and the concomitant reduction in urban YF cases in the Americas, historically represents one of the most successful public health campaigns against infectious diseases.

However, jungle YF persists in the Western hemisphere and is transmitted chiefly among monkeys, marmosets, and possibly other forest-dwelling animals, commonly causing fatal infections. The vectors are mosquitoes of the forest canopy, chiefly of *Haemagogus* spp., and to a lesser extent, *A. leucocelaenus*, *Sabethes chloropterus*, and possibly *Ae. fulvus* in Brazil.[202] For the last few decades in South America, the vast majority of YF cases were in males over 15 years of age whose occupations increase their exposure to YF-infected mosquitoes in endemic forest and jungle areas.[203,204] Up to 500 unvaccinated forest workers were infected in peak years. The majority of cases occur in the first 3 months of the year in South America.[131,203]

Unfortunately, *Ae. aegypti* has now reinfested most of South and Central America, and occupies habitats just adjacent to the areas where endemic YF transmission occurs. A major threat is that this species could transmit YF in an urban cycle.

In contrast to the endemic–jungle circulation of YF in the Western hemisphere in the late 20th century, YF in Africa periodically explodes out of its endemic state to infect large numbers during major epidemics. During the decade 1980–1990, YF reemerged as a major health problem in Africa, and, as mentioned above, threatens to reemerge in South America.

In Africa, two control strategies have been attempted during the last 40 years. The first was routine immunization programs, and the second was emergency control programs after the start of an outbreak. A routine, mandatory YF immunization program was begun in the early 1940s in French West Africa, and the recurring pattern of epidemics in West Africa was interrupted in those immunizing countries. This strategy was abandoned in 1960, and the program switched to a postoutbreak, fire-fighting type of emergency immunization and control strategy. Since then, there has been a series of epidemics of varying severity.[202]

The period 1986–1990 represented an extraordinarily active period of YF. The worldwide total of 17,728 cases and 4,710 deaths [case fatality rate (CFR) = 26.6%] represents the greatest amount of YF activity reported to the WHO for any 5-year period since reporting began in 1948.[203,204] However, in Africa, numerous studies have shown that only a small percentage of African YF cases are reported.[202] Ironically, 1988 marked the 50th anniversary of the development of the live attenuated YF vaccine. There now has been a safe, efficacious, and carefully defined YF vaccine since 1937–1938.[148,181]

Due to the sylvatic cycle of jungle YF, worldwide eradication is not considered possible. Despite numerous studies, the question of maintenance of YF in nature remains somewhat obscure. Although a continual vertebrate–vector maintenance cycle is possible in some environments, in other areas overwintering and maintenance probably occur via other mechanisms. The laboratory and field studies that confirmed that YF virus can be transovarially transmitted in many of its mosquito vectors suggest that this mechanism might play an important role in nature.

Yellow fever patients have a characteristic but nondiagnostic febrile disease with fever, headache, backache, nausea, variable epistaxis, and a lack of correlation between pulse and body temperature (see reviews in refs. 184, 201, 202). The clinical course is 2 to 4 days, followed by uneventful recovery. Many tropical diseases, including a variety of arboviral infections, malaria, and relapsing tick fever, may present similar clinical syndromes, making a clinical diagnosis of suspected YF difficult unless seen during a recognized epidemic. In a serious complication of the disease, the development of icterus occurs following a remission of the general manifestations. It develops as a yellowish tinge of the sclera, a very important diagnostic sign in black people, and only rarely becomes marked. Hemorrhagic signs and vomiting of blood characterize this more serious form of disease. They are more common preceding death, which usually occurs within 9 days of onset. Mortality rates vary widely, and during epidemics reach from 20 to 80% of the cases. Icteric YF must be differentiated from infectious and serum hepatitis, leptospirosis, and poisoning. Because of these uncertain clinical criteria, laboratory diagnostic tests must be used. Isolation and identification of YF virus in blood samples or necropsy specimens or demonstration of specific antibody titer rises constitute definitive diagnosis. The virus generally grows well in standard cell cultures or suckling mice. Rapid diagnostic assays have shown promise for demonstration of YF-specific antigen, nucleic acid, or IgG and IgM antibodies. Cross-reactivity with other flaviviruses has been a historical serological problem. Demonstration of pathognomonic hepatic lesions in necropsy specimens is used when applicable, but needle biopsies of the liver have proved hazardous and the procedure is controversial. No specific therapeutic regimen is generally available for YF, and treatment is chiefly supportive. Prevention is based both on protection from exposure and on vaccination. The 17D YF vaccine was one of the earliest viral vaccines to be developed, and it has proved to be one of the safest and most efficacious live attenuated vaccines.[184,203,204]

Yellow fever in urban or jungle form is a continuing threat. The two forms are nonetheless the same virus and the same disease, distinguished on epidemiologic grounds. Human beings can be protected by immunization with 17D YF vaccine (not advised for infants under 1 year of age). A list of the centers in the United States authorized to give the vaccination can be obtained from the Public Health Service.[202–204]

10.3.3. Japanese Encephalitis.

Japanese encephalitis (JE), a mosquito-transmitted flavivirus (group B arbovirus), is closely related to St. Louis encephalitis (SLE) in the New World, to West Nile virus in Africa and the Middle East, and to Murray Valley encephalitis virus in Australia–New Guinea.[31–33,158,159,174] Not only are the viruses close serological relatives, but also the ecological pattern of disease is similar, with mosquito vectors (*Culex* species usually), wild-bird vertebrate reservoirs, and patterns of involvement of human beings with large numbers of inapparent or mild infections for each severe case seen.[184,199] JE affects populations in rice-growing, suburban, and rural regions of the Orient, and has been responsible for large outbreaks of encephalitis in Korea, Japan, China, Taiwan, and India and a more diffuse scattering of cases in Malaysia, Southeast Asia, and Indonesia. The outbreaks usually occur in the summer months and affect all age groups but peak mortality is in younger age groups and the elderly.[33,63,98,199]

Diagnosis is on clinical grounds, reinforced by laboratory findings in serology. There is no specific treatment. Control efforts include immunization and mosquito control. However, mosquito control can be very difficult in extensive rice-growing regions. In the last three decades, extensive vaccination campaigns have been initiated, in Japan in particular, using an inactivated vaccine prepared from infected mouse brains. These vaccination programs, concentrating on children, have been accompanied by a marked reduction in the incidence of serious disease.[152,175,184]

10.3.4. Chikungunya and O'Nyong Nyong.

Chikungunya (CHIK), an alphavirus, has been responsible for large outbreaks of DEN-like illness in wide areas of Africa and Asia.[30,199] It is spread by *Ae. aegypti*. It occurs in the same regions of Asia as does DEN and has been linked to a small proportion of the cases of hemorrhagic fever syndrome. Large, severe, but short-lived outbreaks are common. Often clinical features can help differentiate CHIK infections from DEN, in that with CHIK the pains are restricted more to the joints, the febrile period is shorter and not diphasic, and many patients experience persistent residual joint pains following the acute episode. The virus is readily isolated from acute serum samples and serological methods are well-established.[167,184]

A closely related virus, O'nyong nyong (ONN), associated with epidemics of disease in human beings in East Africa, is spread by anopheline mosquitoes. Wild primates are suspected of being reservoirs for both viruses. Diagnosis is often on clinical grounds, but it is difficult to differentiate from CHIK without the laboratory studies.[73,167]

No specific treatment exists for CHIK and ONN infections, and control is limited to vector control plus avoidance of being bitten by mosquitoes.

10.3.5. Murray Valley Encephalitis.

Murray Valley encephalitis (MVE) is a close serological relative of SLE, JE, and West Nile viruses, with a similar epidemiologic pattern and clinical syndrome.[62,76] It is localized to the Australian continent and islands to the north. The first human epidemic of MVE was described in 1952, and since that time case reports have been very sporadic. However, in 1974, there was widespread occurrence of cases in humans again in Australia.[62]

10.3.6. Tick-Borne Encephalitis. Tick-borne encephalitis (TBE) is a flavivirus responsible for a tick-transmitted disease of the more northerly Eurasian regions. The disease has been known under several names, including Russian spring–summer encephalitis and Far Eastern encephalitis. Clinically, an acute febrile illness is often followed by encephalitis with high mortality rates and has been seen in most of the northern European countries, Russia, and Siberia.[61–63] The endemic virus cycle involves small mammals (and to a lesser extent birds) in forested regions and the vector ticks, *Ixodes ricinus* and *I. persulcatus*. These ticks will feed readily on human beings. Most cases are reported in forest and construction workers in newly opened regions, woodsmen, trappers, and farmers.[199] Antibodies to TBE have been found in human populations in hilly or mountainous regions of Italy, Greece, and Turkey.[63] The closely related virus that causes Kyasanur Forest disease (KFD) in Mysore State, India, is also tick-transmitted and has a vertebrate host cycle in two monkey species of the region as well as in small mammals.[199] Despite the common phenomenon of tick transport on birds migrating between Europe and Africa, none of the TBE subset of viruses has been found in countries south of the equator. Transovarial transmission of TBE and KFD viruses has been demonstrated in vector ticks.

No specific therapy exists for TBE or KFD infections. Prevention is through control of tick populations, protection against tick bite (repellents, protective clothing), and avoidance of tick habitats. An inactivated vaccine is widely used in central Europe to protect at-risk groups. Diagnosis is made on clinical grounds, reinforced by virus isolation and, more commonly, by serological tests.[167,184,199]

10.3.7. Crimean Hemorrhagic Fever and Congo Virus. Crimean hemorrhagic fever (CHF) was first recognized in southern Russia as a clinical entity in 1944. The causative virus was not finally isolated until the mid-1960s.[62,200] In the meantime, workers in the Congo and Uganda had isolated several virus strains from febrile human beings, to which the name "Congo virus" was applied. Casals[35] showed in 1969 that the two viruses were indistinguishable. They are now referred to as C-CHF.

In southern Russia and Bulgaria, several hundred cases of CHF are seen annually, with a mortality rate as high as 15–20%. The cases present as febrile illness, with pronounced hemorrhagic manifestations in severe cases.[200] Virus isolation and seroconversions (CF particularly) es-

tablish the diagnosis. Cases are seen in rural regions and are almost invariably associated with bite of ixodid ticks. Tick control and avoidance measures are important to prevent infection.[62] An immune-globulin preparation and an experimental vaccine have been made and tested in a limited number of cases.[167,200]

The situation in Africa is still not well established. After the initial finding of several human cases in the 1950s, cases have been sporadic. Small but severe hospital-based outbreaks have occurred in South Africa, with high fatality rates. The known range of the virus is large, with numerous isolations from ticks, cattle, goats, and a hedgehog, and from *Culicoides* (biting gnats) in Nigeria and from ticks in Senegal.

Diagnosis is often by clinical criteria buttressed by virus isolation and by serological tests. Nosocomial infections are common in health care workers.

10.3.8. Rift Valley Fever Virus. Rift Valley fever (RVF) virus, a phlebovirus transmitted by mosquitoes, has long been recognized as causing serious illness in domestic ruminants and also in man in countries south of the Sahara. Until the 1970s, this disease was thought to be limited to southern Africa and a cause of only an uncomplicated febrile illness in humans. In 1977, it was responsible for a major outbreak of disease in animals and man in Egypt.[105,120,121] During this outbreak, an undetermined but probably small proportion of infected humans developed clinical complications after the acute febrile disease.[121] It is estimated that less than 1% of total cases develop either hemorrhagic fever, encephalitis, or an ocular disease.[169] In the 1977–1978 Egyptian epizootic, over 500 patients died of the hemorrhagic complication, although this was possibly underestimated.[121] In 1987, concurrent with the opening of a new dam project, RVF virus caused an outbreak in the Senegal river basin. Animals and humans were involved, and 224 fatalities were reported.[100,132] In 1993, RFV again appeared in Egypt, but the majority of human cases were either febrile illnesses or ocular disease.[46]

11. Unresolved Problems

The emergence of the AIDS epidemic and the appearance of hantavirus in the southwestern United States have demonstrated the vulnerability of the United States to emerging viruses. Two studies by the Institute of Medicine dealing with emerging diseases warned that the threat posed by disease-causing microbes may be expected to continue and intensify in coming years.

Factors influencing the pattern of emergence and

distribution of infectious diseases in general include those associated with the microbial agent itself, the agent's hosts and vectors, and the environment in which agent and host interact. However, for many infectious agents, the specific factors contributing to emergence are poorly understood. Nonetheless, knowledge of these principles is essential in planning strategies to prevent, treat, and control these diseases.

The natural life cycle of many arboviruses is multifaceted and, in addition to the virus, may include one or several reservoir or amplifying hosts and often an arthropod vector. A change affecting the interaction of these fundamental elements might lead to the emergence or reemergence of a viral disease. Natural or man-made changes to the environment typically impact on virus vectors or hosts. There have been several accounts of endemic viruses emerging and spreading as the result of such changes as: (1) development of dams and water projects resulting in altered water distribution patterns, (2) deforestation and changing land use associated with the development of new communities, and (3) the introduction of new virus-amplifying hosts or the expansion of new vectors.

Furthermore, commerce has often brought new viruses, vectors, or hosts into an area. A classic example is the spread of yellow fever in the Western hemisphere in the 19th and early 20th centuries. Impending expansion of worldwide commercial trade may facilitate the emergence of new viruses or increase the spread of previously known viruses to a more receptive environment.

In some instances, viruses might emerge as the result of selection of new genetic strains and variants with increased infectiousness, virulence, or transmissibility. This has been well-established as a cause for the emergence of new influenza outbreaks and, in an analogous fashion, probably contributes to the emergence of arboviruses, particularly the bunyaviruses.

Major impediments to meeting these emerging virus threats are the recent reduction of researchers addressing arthropod-borne and zoonotic viruses, and the formidable research problems posed by the need for input from multiple disciplines. To control emerging viruses in general, and arthropod-borne viruses in particular, there is a need for expanded: (1) basic and applied research that will help formulate coordinated strategies for anticipating, detecting, controlling, and preventing emergence or reemergence of viral diseases; and (2) basic and applied research on the virus, the infective process, and the host response to infection, which will be useful in development of vaccines and antiviral drugs.

Additional unresolved problems are discussed from several points of view, relating to the viruses, the vectors, the vertebrate hosts, and transmission cycles involving virus, vector, and host. The disease in the vertebrate host, which includes the host response to the pathogen, merits independent consideration. Problems relating to the epidemiology of each specific disease require a synthesis of many specific items. Finally, effective control exercised at the level of the virus, the vector, or the vertebrate requires thorough understanding of the epidemiologic background. Specific examples will help to illustrate problems.

11.1. The Viruses

Much progress has been made in recent years in cataloging the several hundred described arboviruses and determining the biochemical, growth, and morphological characteristics in intact vertebrates, in invertebrates, and in cell culture systems of vertebrate and invertebrate cells. However, further work is needed to understand the evolution and emergence of epidemiologically relevent strains. In particular, new research is needed on the nucleic acid homology that may exist among the numerous members of a given arbovirus grouping and on the mechanisms of recombination, reassortment, and selection of new virus strains with important virulence properties.

11.2. The Vectors

The factors that determine specific virus–vector associations are still being elucidated. Historically, mosquito-transmitted viruses were not thought to multiply in ticks and vice versa. However, in the large *Bunyaviridae* family there are viruses that are vectored by mosquitoes, ticks, sandflies, other arthropods, and rodents. Expanded research is needed before the principles governing virus–vector interactions are carefully delineated.

There is a continuing need for taxonomic refinements with respect to arthropods, such as the need for more information on both Old World and New World mosquitoes of the genera *Culex* and *Aedes* and on *Phlebotomus* and *Culicoides* sandflies. This need is generated by the increasing realization of their involvement with a large number of arboviruses. The same remarks are pertinent for the tick vectors. The need is equally great for more information on the biology, feeding preferences, longevity, flight range, and distribution of each arthropod species involved. The genetic constitution of each vector species is basic to an understanding of what constitutes a vector, both physiologically and behaviorally, and will become increasingly important as control of vectors through genetic manipulation is considered.[116]

Unfortunately, in 1985, the Asian "tiger" mosquito, *Ae. albopictus*, was introduced in old tires imported from Japan to Houston, Texas. It has now spread to infest over 22 states, mostly in the South and Midwest. This mosquito is an aggressive, opportunistic feeder with a wide host range that includes man. It adapts well to forest or urban settings and it can vector many different arboviruses. In some areas, it already has replaced local mosquito species, and there is a concern it will transmit endemic viruses. Because of its potential as a new vector of endemic or emerging arboviruses in the United States, research on it should receive high priority.

11.3. The Vertebrate Hosts

For most of the arboviruses, the primary vertebrate host, i.e., the host that serves as the basic unit for propagation of the virus, is not man. For many of the arboviruses, the vertebrate hosts are not yet determined or are recognized on the most tenuous of evidence. Identification of the host(s) is a primary need. Following this, an emphasis should be placed on elucidating a biological profile of the hosts, including the full range of biological and ecological considerations, as well as the degree of host susceptibility to the virus.

11.4. Transmission Cycles Involving Virus, Vector, and Vertebrate

The problem of virus persistence in nature is a particularly baffling one. For example, there are many theories but few facts to explain how a given virus manages to overwinter or survive past a long dry season, when vectors may practically disappear and vertebrate populations decline (or go into hibernation).

Current theories hypothesize that the virus persists in vector populations that overwinter with some members harboring virus, or the virus is in vertebrate populations that overwinter and some infected individuals respond to a reactivation stimulus. In other cases, existing data point to an important role for transovarial transmission, permitting passage of virus to generation after generation of a vector without the need for an intercalated vertebrate host. Certain of the tick-transmitted viruses apparently utilize mechanisms of long persistence in ticks plus transovarial transmission of virus to exist in an endemic form in defined geographic areas. An excellent review of this topic is presented by Reeves,[137] who discusses the epidemiologic problems of overwintering of arboviruses in northern countries and possible transport via infected vectors on migrating birds, with his discussion extending to

Old World as well as New World viruses. A related paper by Lord and Calisher[110] discusses the transport of arboviruses in infected migrating birds along the Atlantic coast flyway of the United States. Further research in these areas is vitally needed to determine any point in the natural maintenance of the virus where intervention might lead to control or eradication of the endemic disease.

Studies of transmission cycles are tied closely with simulation of cycles by models with carefully defined parameters. Such models may permit computer manipulation and simulation of field conditions by varying the values applied to defined parameters, following which epidemic curves can be generated. Further work on models is needed, with the hope of eventual prediction of emergence and spread of disease.

11.5. Disease in the Vertebrate Host

Studies of the human response to arboviral infections are difficult, since the epidemics that provide numbers of cases for study usually occur unpredictably in time and often far from modern facilities required for detailed clinical investigation. Classically, studies of infection in the vertebrate host (including man) have been part of research programs in the fields of pathology. However, animal models are available for only a limited number of the arboviruses. The development of these models will be crucial for the development modern vaccines and antiviral drugs.

The dengue hemorrhagic fever–shock phenomena awakened an interest in the immunological response of the host as a contributing factor to illness. However, only a limited number of research programs are currently studying dengue. Because of its emergence as a worldwide health problem, more research resources should be focused on dengue and dengue hemorrhagic fever.

Recognition of disease in the vertebrate host, as well as worldwide surveillance of disease, requires further development of simple diagnostic techniques and standardization of these assays using accepted reference reagents.

11.6. Control

Virus vaccines are a highly cost-effective means of disease control; yet for the arboviruses, there exists only an attenuated yellow fever (17D) vaccine in use on an international scale. The recent WHO recommendation that this vaccine be included in the Expanded Program of Immunization (EPI) in YF-endemic countries was a major step toward control of YF in Africa.[203,204] The threat of emergence of YF in South America has prompted some to call for consideration of inclusion of YF vaccine in selected EPI programs in South American countries.

In the case of JE, an inactivated vaccine has proved highly successful and is in wide-scale use in parts of Asia. An inactivated vaccine for tick-borne encephalitis also has been successfully used for decades in central Europe. However, inactivated vaccines are often costly and need recurring reimmunizations to booster immunity. Advanced testing of the new attenuated JE vaccine, now in partial use in China, should receive high priority. Research, safety testing, and standardized production studies needed to include such a vaccine in international immunization programs should be strongly supported.

Progress in research on DEN has been slowed, mainly because these viruses grow poorly in cell culture and there is no acceptable animal model for DEN infection or DHF. However, the genome sequences of both DEN and JE viruses have been determined; they appear to be similar to that described for the related YF virus. The genomes of all three viruses now have been sequenced, and full-length cDNAs have been demonstrated to be infectious. After defining the molecular structure of these viruses, researchers are now examining the possibility of expressing relevant antigens in suitable vector systems. Various viral proteins have been expressed in vaccinia, yeast, baculoviruses, and chimeric virus systems. The expressed proteins are immunogenic under some, but not all, experimental conditions. Further studies are needed to expand upon these most promising preliminary findings. With the development of infectious clones of YF, JE, as well as DEN-2 and DEN-4 viruses, there is much interest in accelerating the development of infectious clone technology, and such work undoubtedly will contribute significantly to the development of a live attenuated vaccine. The WHO-supported DEN vaccine development studies in Thailand have followed a more classic approach. DEN viruses are passaged several times in cell culture and many of the resulting live attenuated vaccine candidates now have been tested in humans. The results of these small preliminary clinical trials have been encouraging and they need to be expanded.

Control at the vector level involves continuing work on methodology for control of arthropods. Development of resistance to various insecticides has impaired many control programs, and recent regulatory actions limiting the use of insecticides have further intensified the need for exploration of alterative methods for vector control.

Limited studies utilizing biological mosquito control (such as *Bacillus thuringiensis*) to control selected arbovirus vectors have been encouraging and may lead to widespread control strategies. Other approaches that need further research include those that introduce genes into a mosquito population relating to: (1) increased insecticide susceptibility, (2) reduced capacity to support virus multiplication and/or to transmit virus, (3) alternatives in host feeding preferences, (4) reduction in numbers through mutations leading to reduction in reproductive success (sterile males, conditional lethal mutants), and (5) subversions of host feeding habits.

Control procedures at the level of the vertebrate reservoir have not always been well studied in the arbovirus field. Certainly, rodent control has long been successfully applied for plague control and more recently for control of the South American hemorrhagic fevers and hantavirus in the United States.

12. References

1. ADIDINGER, H. K., STONE, S. S., HESS, W. R., AND BACHRACH, H. L., Extraction of infectious deoxynucleic acid from African swine fever virus, *Virology* **30:**750–752 (1966).
2. AITKEN, T. H. G., TESH, R. B., BEATY, B., AND ROSEN, L., Transovarial transmission of yellow fever virus by mosquitoes *Aedes aegypti*, *Am. J. Trop. Med. Hyg.* **28:**119–121 (1979).
3. ARDOIN, P., CLARKE, D. H., AND HANNOUN, C., The preparation of arbovirus hemagglutinins by sonication and trypsin treatment, *Am. J. Trop. Med. Hyg.* **18:**592–598 (1969).
4. ARTSOB, H., SPENCE, L. P., AND TH'NG, C., Enzyme-linked immunosorbent assay typing of California serogroup viruses isolated in Canada, *J. Clin. Microbiol.* **20:**276–280 (1984).
5. ARTSOB, H., SPENCE, L., TH'NG, C., LAMPOTANG, V., JOHNSTON, D., MACINNES, C., MATEJKA, F., VOIGT, D., AND WATT, I., Arbovirus infections in several Ontario mammals, 1975–1980, *Can. J. Vet. Res.* **50:**42–46 (1986).
6. ARTSOB, H., Distribution of California serogroup viruses and virus infections in Canada, *Prog. Clin. Biol. Res.* **123:**277–290 (1983).
7. ASMAN, S., Cytogenetic observations in *Culex tarsalis*: Mitosis and meiosis, *J. Med. Entomol.* **11:**375–382 (1974).
8. AYERS, J. R., LESTER, T. L., AND ANGULO, A. B., An epizootic attributable to western equine encephalitis virus infection in emus in Texas, *J. Am. Vet. Med. Assoc.* **205:**600–601 (1994).
9. BAILEY, C. L., FARAN, M. E., GARGAN, T. P., AND HAYES, D. E., Winter survival of blood-fed and nonblood-fed *Culex pipiens* L, *Am. J. Trop. Med. Hyg.* **31:**1054–1061 (1982).
10. BARNES, W. J. S., AND ROSEN, L., Fatal hemorrhagic disease and shock associated with primary dengue infection on a Pacific island, *Am. J. Trop. Med. Hyg.* **23:**495–506 (1974).
11. BEAMAN, J. R., AND TURELL, M. J., Transmission of Venezuelan equine encephalomyelitis virus by strains of *Aedes albopictus* (Diptera: Culicidae) collected in North and South America, *J. Med. Entomol.* **28:**161–164 (1991).
12. BEATY, B. J., AND BISHOP, D. H., Bunyavirus–vector interactions, *Virus Res.* **10:**289–301 (1988).
13. BEATY, B. J., SUNDIN, D. R., CHANDLER, L. J., AND BISHOP, D. H. L., Evolution of bunyaviruses by genome reassortment in dually infected mosquitoes, *Science* **230:**548–550 (1985).

14. BEATY, B. J., SHOPE, R. E., AND CLARKE, D. H., Salt dependent hemagglutination with *Bunyaviridae* antigens, *J. Clin. Microbiol.* **5:**548–550 (1977).

15. BELL, R. L., CHRISTENSEN, B., HOLGUIN, A., AND SMITH, O., St. Louis encephalitis: A comparison of two epidemics in Harris county, Texas, *Am. J. Public Health* **71:**168–170 (1981).

16. BERGE, T. (ED.), AND AMERICAN COMMITTEE ON ARTHROPOD-BORNE VIRUSES, International Catalogue of Arboviruses Including Certain Other Viruses of Vertebrates, USPHS, Washington, 1975.

17. BIANCHI, T. I., AVILES, G., MONATH, T. P., AND SABATTINI, M. S., Western equine encephalomyelitis: Virulence markers and their epidemiologic significance, *Am. J. Trop. Med. Hyg.* **49:**322–328 (1993).

18. BIGLER, W. J., VENTURA, A. K., LEWIS, A. L., WELLINGS, F. M., AND EHRENKRANTZ, N. J., Venezuelan equine encephalomyelitis in Florida: Endemic virus circulation in native rodent populations of the Everglades Hammocks, *Am. J. Trop. Med. Hyg.* **23:**513–521 (1974).

19. BINN, L. N., SPONSELLER, M. L., WOODING, W. L., McCONNLL, S. J., SPERTZEL, R. O., AND YAEGER, R. H., Efficacy of an attenuated western equine encephalitis vaccine in equine animals, *Am. J. Vet. Res.* **27:**1599–1604 (1966).

20. BORDEN, E. C., SHOPE, R. E., AND MURPHY, F. A., Physicochemical and morphological relationships of some arthropod-borne viruses to bluetongue virus—a new taxonomic group: Physicochemical and serological studies, *J. Gen. Virol.* **13:**261–271 (1971).

21. BOWEN, G. S., McLEAN, R. G., SCHRINER, R. B., FRANCY, D. B., POKORNY, K. S., TRIMBLE, J. M., BOLIN, R. A., BARNES, A. M., CALISHER, C. H., AND MUTH, D. J., The ecology of Colorado tick fever in Rocky Mountain National Park in 1974. II. Infection in small mammals, *Am. J. Trop. Med. Hyg.* **30:**490–496 (1981).

22. BRODY, J. A., BURNS, K. F., BROWNING, G., AND SCHATTNER, J. D., Apparent and inapparent attack rates for St. Louis encephalitis in a selected population, *N. Engl. J. Med.* **261:**644–646 (1959).

23. BROWN, T. P., ROBERTS, W., AND PAGE, R. K., Acute hemorrhagic enterocolitis in ratites: Isolation of eastern equine encephalomyelitis virus and reproduction of the disease in ostriches and turkey poults, *Avian Dis.* **37:**602–605 (1993).

24. BUCKLEY, S. M., Susceptibility of the *Aedes albopictus* and *Aedes aegypti* cell lines to infection with arboviruses, *Proc. Soc. Exp. Biol. Med.* **131:**625–630 (1969).

25. BURGDORFER, W., AND EKLUND, C., Studies on the ecology of Colorado tick fever virus in western Montana, *Am. J. Hyg.* **69:**127–137 (1959).

26. BUTENKO, A. M., VLADIMIRTSEVA, E. A., LVOV, S. D., CALISHER, C. H., AND KARABATSOS, N., California serogroup viruses from mosquitoes collected in the USSR, *Am. J. Trop. Med. Hyg.* **45:**366–370 (1991).

27. CALISHER, C. H. PRETZMAN, C. I., MUTH, D. J., PARSONS, M. A., AND PETERSON, E. D., Serodiagnosis of La Crosse virus infections in humans by detection of immunoglobulin M class antibodies, *J. Clin. Microbiol.* **23:**667–671 (1986).

28. CALISHER, C. H., POLAND, J. D., CALISHER, S. B., AND WARMOTH, L. A., Diagnosis of Colorado tick fever virus infection by enzyme immunoassays for immunoglobulin M and G antibodies, *J. Clin. Microbiol.* **22:**84–88 (1985).

29. CALISHER, C. H., Medically important arboviruses of the United States and Canada, *Clin. Microbiol. Rev.* **7:**89–116 (1994).

30. CAREY, D. E., Chikungunya and dengue: A case of mistaken identity, *J. Hist. Med. Allied Sci.* **26:**243–262 (1971).

31. CASALS, J., Immunological relationship among central nervous system viruses, *J. Exp. Med.* **79:**341–359 (1944).

32. CASALS, J., The arthropod-borne group of animal viruses, *Trans N.Y. Acad. Sci.* **19:**219–235 (1957).

33. CASALS, J., Viral encephalitis, in: *Viral Encephalitis: A Symposium* (W. S. FIELDS, AND R. J. BLATTNER, EDS.), pp. 5–21, Charles C. Thomas, Springfield, IL, 1958.

34. CASALS, J., Antigenic variants of eastern equine encephalitis virus, *J. Exp. Med.* **119:**547–565 (1964).

35. CASALS, J., Antigenic similarity between the virus causing Crimean hemorrhagic fever and Congo virus, *Proc. Soc. Exp. Biol. Med.* **131:**233–236 (1969).

36. CASALS, J., AND BROWN, E. V., Hemagglutination with arthropod-borne viruses, *J. Exp. Med.* **99:**429–449 (1954).

37. CAUSEY, O. R., AND THEILER, M., Virus antibody survey of sera of residents of the Amazon Valley in Brazil, *Am. J. Trop. Med. Hyg.* **7:**36–41 (1958).

38. CENTERS FOR DISEASE CONTROL AND PREVENTION, Epidemiologic notes and reports: St. Louis encephalitis—Tennessee, *Morbid. Mortal. Week. Rep.* **23:**294, 299 (1974).

39. CENTERS FOR DISEASE CONTROL AND PREVENTION, Epidemiologic notes and reports: St. Louis encephalitis—Texas, Louisiana, *Morbid. Mortal. Week. Rep.* **29:**415–416 (1980).

40. CENTERS FOR DISEASE CONTROL AND PREVENTION, Epidemiologic notes and reports: Dengue—Texas, *Morbid. Mortal. Week. Rep.* **29:**451, 531–532 (1980).

41. CENTERS FOR DISEASE CONTROL, Arboviral infections of the central nervous system—United States, 1984, *Morbid. Mortal. Week. Rep.* **34:**283–294 (1985).

42. CENTERS FOR DISEASE CONTROL, Arboviral infections of the central nervous system—United States, 1985, *Morbid. Mortal. Week. Rep.* **35:**341–344 (1986).

43. CENTERS FOR DISEASE CONTROL AND PREVENTION, Arboviral diseases—United States, 1992, *J.A.M.A* **270**(3)**:**308 (1993).

44. CENTERS FOR DISEASE CONTROL AND PREVENTION, Arboviral disease—United States, 1993, *J.A.M.A* **272**(4)**:**262–264 (1994).

45. CENTERS FOR DISEASE CONTROL AND PREVENTION, Imported dengue—United States, 1992, *Morbid. Mortal. Week. Rep.* **43:**97–99 (1994).

46. CENTERS FOR DISEASE CONTROL AND PREVENTION, Rift Valley Fever—Egypt, 1993, *Morbid. Mortal. Week. Rep.* **43:**693–700 (1994).

47. CENTERS FOR DISEASE CONTROL AND PREVENTION, Dengue fever among U.S. military personnel—Haiti, September–November, 1994, *Morbid. Mortal. Week. Rep.* **43:**845–849 (1994).

48. CENTERS FOR DISEASE CONTROL AND PREVENTION, Addressing emerging infectious disease threats: A prevention strategy for the United States, *Morbid. Mortal. Week. Rep.* **43:**RR-5 (1994).

49. CENTERS FOR DISEASE CONTROL AND PREVENTION, Arboviral disease—United States, 1994, *J.A.M.A* **274**(14)**:**1110–1112 (1995).

50. CENTERS FOR DISEASE CONTROL AND PREVENTION, Venezuelan equine encephalitis—Colombia, 1995, *Morbid. Mortal. Week. Rep.* **44:**721–724 (1995).

51. CENTERS FOR DISEASE CONTROL AND PREVENTION, Imported dengue—United States, 1993–1994, *Morbid. Mortal. Week. Rep.* **44:**353–356 (1995).

52. CENTERS FOR DISEASE CONTROL AND PREVENTION, Dengue type-3 infection—Nicaragua and Panama, October–November 1994, *Morbid. Mortal. Week. Rep.* **44:**721–724 (1995).

53. CENTERS FOR DISEASE CONTROL AND PREVENTION, Update: Venezuelan equine encephalitis—Colombia, 1995, *Morbid. Mortal. Week. Rep.* **44:**775–777 (1995).

54. CENTERS FOR DISEASE CONTROL AND PREVENTION, Dengue surveillance—United States, 1986–1992, *Morbid. Mortal. Week. Rep.* **43**(no. SS-2):7–19 (1995).

55. CHANDLER, L. J., BEATY, B. J., BALDRIDGE, G. D., BISHOP, D. H., AND HEWLETT, M. J., Heterologous reassortment of bunyaviruses in *Aedes triseriatus* mosquitoes and transovarial and oral transmission of newly evolved genotypes, *J. Gen. Virol.* **71**:1045–1050 (1990).

56. CLARKE, D. H., AND CASALS, J., Techniques for hemagglutination and hemagglutination-inhibition with arthropod-borne viruses, *Am. J. Trop. Med. Hyg.* **7**:561–573 (1958).

57. COATES, D. M., MAKH, S. R., JONES, N., AND LLOYD, G., Assessment of assays for the serodiagnosis of Venezuelan equine encephalitis, *J. Infect.* **25**:279–289 (1992).

58. COMER, J. A., AND TESH, R. B., Phlebotomine sand flies as vectors of vesiculoviruses: A review, *Parasitologia* **33**(Suppl.):143–150 (1991).

59. CULLY, J. F., JR., HEARD, P. B., WESSON, D. M., AND CRAIG, G. B., JR., Antibodies to La Crosse virus in eastern chipmunks in Indiana near an *Aedes albopictus* population, *J. Am. Mosq. Control. Assoc.* **7**:651–653 (1991).

60. CULLY, J. F., JR., STREIT, T. G., AND HEARD, P. B., Transmission of La Crosse virus by four strains of *Aedes albopictus* to and from the eastern chipmunk (*Tamias striatus*), *J. Am. Mosq. Control. Assoc.* **8**:237–240 (1992).

61. DOAN-WIGGINS, L. Tick-borne diseases, *Emerg. Med. Clin. North Am.* **9**:303–325 (1991).

62. DOHERTY, R. L., CARLEY, J. G., FILIPPICH, C., BARROW, G. J., AND WILSON, R. K., Epidemiological studies of arthropod-borne viruses in: 29th Annual Report of Queensland Institute of Medical Research, p. 3, Queensland Institute of Medical Research, Brisbane, 1974.

63. DOWNS, W. G., The Rockefeller Foundation Virus Program: 1951–1971 with update to 1981, *Annu. Rev. Med.* **33**:1–29 (1982).

64. DOWNS, W. G., AITKEN, T. H. G., AND ANDERSON, C. R., Activities of the Trinidad Regional Virus Laboratory in 1953 and 1954 with special reference to the yellow fever outbreak in Trinidad, B.W.I., *Am. J. Trop. Med. Hyg.* **4**:837–843 (1955).

65. DYKERS, T. I., BROWN, K. L., GUNDERSEN, C. B., AND BEATY, B. J., Rapid diagnosis of LaCrosse encephalitis: Detection of specific immunoglobulin M in cerebrospinal fluid, *J. Clin. Microbiol.* **22**:740–744 (1985).

66. EHRENKRANTZ, N. J., AND VENTURA, A. K., Venezuelan equine encephalitis virus infection in man, *Annu. Rev. Med.* **25**:9–14 (1974).

67. EMMONS, R. W. Ecology of Colorado tick fever, *Annu. Rev. Microbiol.* **42**:49–64 (1988).

68. EMMONS, R. W., DONDERO, D. V., DEVLIN, V., AND LENNERRE, E. H., Serologic diagnosis of Colorado tick fever: A comparison of complement-fixation, immuno-fluorescence, and plaque reduction methods, *Am. J. Trop. Med. Hyg.* **18**:796–802 (1969).

69. EVANS, A. S., CASALS, J., OPTON, E. M., BORMAN, E. K., AND CUADRADO, R. R., A nationwide serum survey of Colombian military recruits, 1966: Description of sample and antibody patterns with arboviruses, polioviruses, respiratory viruses, tetanus and treponomatosis, *Am. J. Epidemiol.* **90**:292–303 (1969).

70. EVANS, A. S., CASALS, J., OPTON, E. M., BORMAN, E. K., AND CUADRADO, R. R., A nationwide serum survey of Argentinian military recruits, 1965–1966. 1. Description of samples and antibody patterns with arboviruses, polioviruses, respiratory viruses, tetanus and treponematosis, *Am. J. Epidemiol.* **93**:111–121 (1971).

71. EVANS, A. S., COOK, J. A., KAPIKIAN, A. Z., NANKERVIS, G.,

SMITH, A. L., AND WEST, B., A serological survey of St. Lucia, *Int. J. Epidemiol.* **8**:327–332 (1979).

72. EVANS, A. S., COX, F., NANKERVIS, G., OPTON, E. M., SHOPE, R. E., WELLS, A. V., AND WEST, B., A health and seroepidemiological survey of a community in Barbados, *Int. J. Epidemiol.* **3**:167–175 (1974).

73. FIELDS, B. N., KNIPE, D. M., AND HOWLEY, P. M. (EDS.-IN-CHIEF), *Virology*, 3rd ed., Lippincott-Raven Publishers, New York, 1996.

74. FIELDS, B. N., AND HAWKINS, K., Human infection with the virus of vesicular stomatitis during an epizootic, *N. Engl. J. Med.* **277**:989–994 (1967).

75. FRANCY, D. B., RUSH, W. A., MONTOYA, M., INGLISH, D. S., AND BOLIN, R. A., Transovarial transmission of St. Louis encephalitis virus by *Culex pipiens* complex mosquitoes, *Am. J. Trop. Med. Hyg.* **30**:699–705 (1981).

76. FRENCH, E. L., Murray Valley encephalitis, *Med. J. Aust.* **39**:100–103 (1952).

77. FULHORST, C. F., HARDY, J. L., ELDRIDGE, B. F., PRESSER, S. B., AND REEVES, W. C., Natural vertical transmission of western equine encephalomyelitis virus in mosquitoes, *Science* **263**:676–678 (1994).

78. GENTSCH, J., WYNNE, L. R., CLEWLEY, J. P., SHOPE, R. E., AND BISHOP, D. H. L., Formation of recombinants between snowshoe hare and LaCrosse bunyaviruses, *J. Virol.* **24**:893–902 (1977).

79. GERLOFF, R. K., AND EKLUND, C. M., A tissue culture neutralization test for Colorado tick fever antibody and the use of the test for serologic surveys, *J. Infect. Dis.* **104**:174–183 (1959).

80. GOLDFIELD, M., AUSTIN, S. M., BLACK, H. C., TAYLOR, B. F., AND ALTMAN, R., A nonfatal human case of Powassan virus encephalitis, *Am. J. Trop. Med. Hyg.* **22**:78–81 (1973).

81. GONZALEZ-SCARANO, F., ENDRES, M. J., AND NATHANSON, N., *Bunyaviridae*: Pathogenesis, *Curr. Top. Microbiol. Immunol.* **169**:217–249 (1991).

82. GONZALEZ-SCARANO, F., JACOBY, D., GRIOT, C., AND NATHANSON, N., Genetics, infectivity and virulence of California serogroup viruses, *Virus Res.* **24**:123–135 (1992).

83. GORMAN, B. M., TAYLOR, I., WALKER, T. I., AND YOUNG, P. R., The isolation of recombinants between selected orbiviruses, *J. Gen. Virol.* **4**:333–342 (1978).

84. GRACE, T. D. C., Establishment of a line of mosquito (*Aedes aegypti* L.) cells grown *in vitro*, *Nature* **211**:366–367 (1966).

85. GRIMSTAD, P. R., BARRETT, C. L., HUMPHREY, R. L., AND SINSKO, M. J., Serologic evidence for widespread infection with La Crosse and St. Louis encephalitis viruses in the Indiana human population, *Am. J. Epidemiol.* **119**:913–930 (1984).

86. GRIMSTAD, P. C., CALISHER, C. H., HARROFF, R. N., AND WENTWORTH, B. B., Jamestown Canyon (California serogroup) is the etiologic agent of widespread infection in Michigan humans, *Am. J. Trop. Med. Hyg.* **35**:376–386 (1986).

87. GRIMSTAD, P. R., KOBAYASHI, J. F., ZHANG, M. B., AND CRAIG, G. B., JR., Recently introduced *Aedes albopictus* in the United States: Potential vector of La Crosse virus (*Bunyaviridae*: California serogroup), *J. Am. Mosq. Control. Assoc.* **5**:422–427 (1989).

88. GRIMSTAD, P. R., SHABINO, C. L., CALISHER, C. H., AND WALDMAN, R. J., A case of encephalitis in a human associated with a serologic rise to Jamestown Canyon virus, *Am. J. Trop. Med. Hyg.* **31**:1238–1244 (1982).

89. GRIMSTAD, P. R., SCHMITT, S. M., AND WILLIAMS, D. G., Prevalence of neutralizing antibody to Jamestown Canyon virus (California group) in populations of elk and moose in northern Michigan and Ontario, Canada, *J. Wildl. Dis.* **22**:453–458 (1986).

90. GUNDERSEN, C. B., AND BROWN, K. L., Clinical aspects of La Crosse encephalitis: Preliminary report, *Prog. Clin. Biol. Res.* **123:**169–177 (1983).

91. HAHN, C. S., LUSTIG, S., STRAUSS, E. G., AND STRAUSS, J. H., Western equine encephalitis virus is a recombinant virus, *Proc. Natl. Acad. Sci. USA* **85:**5997–6001 (1988).

92. HALLAUER, C., Uber den Virusnachweis mit dem Hirst-Test, *Z. Parol. Bakteriol.* **9:**553–554 (1946).

93. HALSTEAD, S. B., Observations relating to pathogenesis of dengue hemorrhagic fever, VI. Hypotheses and discussion, *Yale J. Biol. Med.* **42:**350–362 (1970).

94. HAMMON, W. M., Present and future of killed and live arbovirus vaccines, in: *First International Conference on Vaccines against Viral and Rickettsial Infections of Man*, pp. 252–259, Pan American Health Organization, Washington, DC, 1967.

95. HAMMON, W. M., AND REEVES, W. C., California encephalitis virus, a newly described agent, *Calif. Med.* **77:**303–309 (1952).

96. HAMMON, W. M., RUDNICK, A., AND SATHER, G. E., Viruses associated with epidemic hemorrhagic fever in the Philippines and Thailand, *Science* **13:**1102–1103 (1960).

97. HENDERSON, J. R., LIGHTFOOT, P. R., AND LYONS, R. W., A nationwide serum survey of United States military recruits, 1962, *Am. J. Hyg.* **80:**308–313 (1964).

98. HORSFALL, F. L., AND TAMM, I. (EDS.), *Viral and Rickettsial Infections of Man*, 4th ed., J. B. Lippincott, Philadelphia, 1965.

99. HUGHES, L. E., CASPER, E. A., AND CLIFFORD, C. M., Persistence of Colorado tick fever virus in red blood cells, *Am. J. Trop. Med. Hyg.* **23:**530–532 (1974).

100. JOUAN, A., PHILIPPE, B., RIOU, O., COULIBALY, I., LEGUENNO, B., MEEGAN, J., MONDO, M., AND DIGOUTTE, J., Les formes cliniques benignes de la fievre de la Vallee du Rift pendant l'epidemie de Mauritanie. [Mild clinical forms of Rift Valley fever during the epidemic in Mauritania], *Bull. Soc. Pathol. Exot. Filiales.* **82**(5)**:**620 (1989).

101. KAPPUS, K. D., MONATH, T. P., KAMINSKI, R. M., AND CALISHER, C. H., Reported encephalitis associated with California serogroup virus infections in the United States, 1963–1981, *Prog. Clin. Biol. Res.* **123:**31–41 (1983).

102. KARABATSOS, N. (ED.), *International Catalogue of Arboviruses 1985: Including Certain Other Viruses of Vertebrates*, American Society of Tropical Medicine and Hygiene, Subcommittee on Information Exchange, San Antonio, 1985.

103. KINNEY, R. M., TSUCHIYA, K. R., SNEIDER, J. M., AND TRENT, D. W., Molecular evidence for the origin of the widespread Venezuelan equine encephalitis epizootic of 1969 to 1972, *J. Gen. Virol.* **73:**3301–3305 (1992).

104. KOMAR, N., AND SPIELMAN, A., Emergence of eastern encephalitis in Massachusetts, *Ann. N. Y. Acad. Sci.* **740:**157–168 (1994).

105. LAUGHLIN, L. W., MEEGAN, J. M., STRAUSBAUGH, L. J., MORENS, D. M., AND WATTEN, R. H., Epidemic Rift Valley fever in Egypt: Observations of the spectrum of human illness, *Trans. Roy. Soc. Trop. Med. Hyg.* **73:**630 (1979)

106. LEDUC, J. W., Epidemiology and ecology of the California serogroup viruses, *Am. J. Trop. Med. Hyg.* **37:**60S–68S (1987).

107. LETSON, G. W., BAILEY, R. E., PEARSON, J., AND TSAI, T. F., Eastern equine encephalitis (EEE): A description of the 1989 outbreak, recent epidemiologic trends, and the association of rainfall with EEE occurrence, *Am. J. Trop. Med. Hyg.* **49:**677–685, 1993.

108. LIBIKOVÁ, H., AND BUCKLEY, S. M., Studies with Kemerovo virus in Singh's aedes cell lines, *Acta. Virol.* **15:**393–403 (1971).

109. LORD, R. D., History and geographic distribution of Venezuelan equine encephalitis, *Bull. Pan Am. Health Org.* **8:**100–110 (1974).

110. LORD, R. D., AND CALISHER, C. H., Further evidence of southward transport of arboviruses by migratory birds, *Am. J. Epidemiol.* **92:**73–78 (1970).

111. LUMSDEN, L. L., St. Louis encephalitis in 1933: Observations on epidemiological features, *Public Health Rep.* **73:**340–353 (1958).

112. L'VOV, D. K., TIMOPHEEVA, A. A., CHERVONSKI, V. L., GROMASHEVSKI, V. L., KLISENKO, G. A., GOSTINSCHIKOVA, G. V., AND KOSTYRKO, I. N., Tyuleniy virus: A new group B arbovirus isolated from *Ixodes* (*Ceratiyodes*) *putus* Pick.-Camb. 1878 collected on Tyuleniy Island, Sea of Okhotsk, *Am. J. Trop. Med. Hyg.* **20:**456–460 (1971).

113. MACDONALD, G., *The Epidemiology and Control of Malaria*, Oxford University Press, London, 1957.

114. MARCHETTE, N. J., HALSTEAD, S. B., AND CHOW, J. S., Replication of dengue viruses in cultures of peripheral blood leukocytes from dengue-immune rhesus monkeys, *J. Infect. Dis.* **133:**274–282 (1976).

115. MARFIN, A. A., BLEED, D. M., LOFGREN, J. P., OLIN, A. C., SAVAGE, H. M., SMITH, G. C., MOORE, P. S., KARABATSOS, N., AND TSAI, T. F., Epidemiologic aspects of a St. Louis encephalitis epidemic in Jefferson County, Arkansas, 1991, *Am. J. Trop. Med. Hyg.* **49:**30–37 (1993).

116. MATTINGLY, P. F., Ecological aspects of the evolution of mosquito-borne virus diseases, *Trans. R. Soc. Trop. Med. Hyg.* **54:**97–112 (1960).

117. MCLEAN, D. M., AND DONOHUE, W. L., Powassan virus: Isolation of virus from a fatal case of encephalitis, *Can. Med Assoc. J.* **80:**708–711 (1959).

118. MCLEAN, R. G., FRANCY, D. B., BOWEN, G. S., BAILEY, R. E., CALISHER, C. H., AND BARNES, A. M., The ecology of Colorado tick fever in rocky Mountain National Park in 1974. I. Objectives, study design, and summary of principal findings, *Am. J. Trop. Med. Hyg.* **30:**483–489 (1981).

119. MCLEAN, R. G., SHRINER, R. B., POKORNY, K. S., AND BOWEN, G. S., The ecology of Colorado tick fever in Rocky Mountain National Park in 1974. III. Habitats supporting the virus, *Am. J. Trop. Med. Hyg.* **40:**86–93 (1989).

120. MEEGAN, J. M., HOOGSTRAAL, H., AND MOUSSA, M. I., An epizootic of Rift Valley fever in Egypt in 1977, *Vet. Rec.* **105:**124 (1979).

121. MEEGAN, J. M., The Rift Valley fever epizootic in Egypt 1977–1978. 1. Description of the epizootic and virological studies, *Trans. Roy. Soc. Trop. Med. Hyg.* **73:**618 (1979).

122. MITCHELL, C. J., NIEBYLSKI, M. L., SMITH, G. C., KARABATSOS, N., MARTIN, D., MUTEBI, J. P., CRAIG, G. B., JR., AND MAHLER, M. J., Isolation of eastern equine encephalitis virus from *Aedes albopictus* in Florida, *Science* **257:**526–527 (1992).

123. MONATH, T. P., *St. Louis Encephalitis* (T. P. MONATH, ED.), American Public Health Association, Washington, DC, 1980.

124. MONATH, T. P., NUCKOLLS, J. G., BERALL, J., BAUER, H., CHAPPELL, W. A., AND COLEMAN, P. H., Studies on California encephalitis in Minnesota, *Am. J. Epidemiol.* **92:**40–50 (1970).

125. MONATH, T. P., NYSTROM, R. R., BAILEY, R. E., CALISHER, C. H., AND MUTH, D. J., Immunoglobulin M antibody capture enzyme-linked immunosorbent assay for diagnosis of St. Louis encephalitis, *J. Clin. Microbiol.* **20:**784–790 (1984).

126. MONATH, T. P., AND TSAI, T. F., St. Louis encephalitis: Lessons from the last decade, *Am. J. Trop. Med. Hyg.* **37:**40S–59S (1987).

127. MORENS, D. M., Antibody dependent enhancement of infection

and the pathogenesis of viral disease, *Clin. Infect. Dis.* **19:**500–512 (1994).

128. MURPHY, F. A., BORDEN, E. C., SHOPE, R. E., AND HARRISON, A., Physicochemical and morphological relationships of some arthropod-borne viruses to bluetongue virus—a new taxonomic group: Electron microscopic studies, *J. Gen. Virol.* **13:**273–288 (1971).

129. NIEDERMAN, J. C., HENDERSON, J. R., OPTON, E. M., BLACK, F. L., AND SKORNOVA, K. A., A nationwide serum survey of Brazilian military recruits, 1964. II. Antibody patterns with arboviruses, polioviruses, measles and mumps, *Am. J. Epidemiol.* **86:**319–329 (1967).

130. OSORIO, J. E., AND YUILL, T. M., Venezuelan equine encephalitis, in: *Handbook of Zoonoses, Section B: Viral* (G. W. BERAN, ED.), pp. 33–46, CRC Press, Boca Raton, FL, 1994.

131. PAN AMERICAN HEALTH ORGANIZATION, Yellow fever epidemic in Bolivia, *PAHO Epidemiol. Bull.* **6:**13 (1985).

132. PHILIPPE, B., JOUAN, A., RIOU, O., COULIBALY, I., LEGUENNO, B., MEEGAN, J., MONDO, M., AND DIGOUTTE, J., Les formes hemorragiques de la fievre de la Vallee du Rift en Mauritanie. [Hemorrhagic forms of Rift Valley fever in Mauritania], *Bull. Soc. Pathol. Exot. Filiales.* **82**(5)**:**611 (1989).

133. PLOWRIGHT, W., PERRY, C. T., PIERCE, M. A., AND PARKER, J., Experimental infection of the argasid tick *Ornithodoros moubata porcinus*, with African swine fever virus, *Arch. Ges. Virusforsch.* **31:**33–50 (1979).

134. PLOWRIGHT, W., BROWN, F., AND PARKER, J., Evidence for the type of nucleic acid in African swine fever virus, *Arch. Ges. Virusforsch.* **29:**289–304 (1966).

135. PRINCE, A. M., METSELAAR, D., KAFUKO, G. W., MUKWAYA, L. G., LING, C. M., AND OVERBY, L. R., Hepatitis B antigen in wild-caught mosquitoes in Africa, *Lancet* **2:**247 (1972).

136. PRZELOMSKI, M. M., O'ROURKE, E., GRADY, G. F., BERARDI, V. P., AND MARKLEY, H. G., Eastern equine encephalitis in Massachusetts: A report of 16 cases, 1970–1984, *Neurology* **38:**736–739 (1988).

137. REEVES, W. C., Overwintering of arboviruses, *Prog. Med. Virol.* **17:**193–220 (1974).

138. REEVES, W. C., The discovery decade of arbovirus research in western North America, 1940–1949, *Am. J. Trop. Med. Hyg.* **37:**94S–100S (1987).

139. REEVES, W. C., EMMONS, R. W., AND HARDY, J. L., Historical perspectives on California encephalitis virus in California, *Prog. Clin. Biol. Res.* **123:**19–29 (1983).

140. REEVES, W. C., AND HAMMON, W. M., Epidemiology of the arthropod-borne viral encephalitides in Kern County, California, 1943–1952, *Univ. Calif. Publ. Public Health* **4:**1–257 (1962).

141. REEVES, W. C., HARDY, J. L., REISEN, W. K., AND MILBY, M. M., Potential effect of global warming on mosquito-borne arboviruses, *J. Med. Entomol.* **31:**323–332 (1994).

142. ŘEHÁČEK, J., AND PESEK, I., Propagation of eastern equine encephalomyelitis virus in surviving tick tissues, *Acta Virol.* **4:**241–254 (1960).

143. REIF, J. S., WEBB, P. A., MONATH, T. P., EMERSON, J. K., POLAND, J. D., KEMP, G. E., AND CHOLAS, G., Epizootic vesicular stomatitis in Colorado, 1982: Infection in occupational risk groups, *Am. J. Trop. Med. Hyg.* **36:**177–182 (1987).

144. REISEN, W. K., HARDY, J. L., REEVES, W. C., PRESSER, S. B., MILBY, M. M., AND MEYER, R. P., Persistence of mosquito-borne viruses in Kern County, California, 1983–1988, *Am. J. Trop. Med. Hyg.* **43:**419–437 (1990).

145. REISEN, W. K., MEYER, R. P., MILBY, M. M., PRESSER, S. B., EMMONS, R. W., HARDY, J. L., AND REEVES, W. C., Ecological observations on the 1989 outbreak of St. Louis encephalitis virus in the southern San Joaquin Valley of California, *J. Med. Entomol.* **29:**472–482 (1992).

146. REISEN, W. K., MILBY, M. M., PRESSER, S. B., AND HARDY, J. L., Ecology of mosquitoes and St. Louis encephalitis virus in the Los Angeles Basin of California, 1987–1990, *J. Med. Entomol.* **29:**582–598 (1992).

147. REISEN, W. K., HARDY, J. L., PRESSER, S. B., MILBY, M. M., MEYER, R. P., DURSO, S. L., WARGO, M. J., AND GORDON, E., Mosquito and arbovirus ecology in southeastern California, 1986–1990, *J. Med. Entomol.* **29:**512–524 (1992).

148. RICE, C. M., LONCHES, E. M., EDDY, C. R., SHIN, S. J., SHEETS, R. L., AND STRAUSS, J. H., Nucleotide sequence of yellow fever virus: Implications for flavivirus gene expression, *Science* **229:**726–733 (1985).

149. RICHARDS, A., MALONE, J., SHERIS, S., WEDDLE, J., ROSSI, C., KSIAZEK, T., LEDUC, J., DASCH, G., AND HYAMS, K., Arbovirus and rickettsial infections among combat troops during Operation Desert Shield/Desert Storm, *J. Infect. Dis.* **168**(4)**:**1080–1081 (1993).

150. RICHARDSON, J., SYLVESTER, E. S., REEVES, W. C., AND HARDY, J. L., Evidence of two inapparent nonoccluded viral infections of *Culex tarsalis*, *J. Invertebr. Pathol.* **23:**213–224 (1974).

151. RICO-HESSE, R., WEAVER, S., DESIGER, J., MEDINA, G., AND SALAS, R., Emergence of a new epidemic/epizootic Venezuelan equine encephalitis virus in South America, *Proc. Natl. Acad. Sci. USA* **92:**5278–5281 (1995).

152. ROEHRIG, J. T., Immunogens of encephalitis viruses, *Vet. Microbiol.* **37:**273–284 (1993).

153. ROSEN, L., The emperor's new clothes revisited, or reflections on the pathogenesis of dengue hemorrhagic fever, *Am. J. Trop. Med. Hyg.* **25:**337–343 (1977).

154. ROSEN, L., Dengue—an overview, in: *Viral Diseases in Southeast Asia and the Western Pacific* (J. S. MACKENZIE, ED.), pp. 484–493, Academic Press, New York, 1982.

155. ROSEN, L., Further observations on the mechanism of vertical transmission of flaviviruses by *Aedes* mosquitoes, *Am. J. Trop. Med. Hyg.* **39:**123–126 (1988).

156. ROSEN, L., TESH, R. B., LIEN, J. C., AND CROSS, J. H., Transovarial transmission of Japanese encephalitis virus by mosquitoes, *Science* **199:**909–911 (1978).

157. RUDNICK, A., Studies of the ecology of dengue in Malaysia: A preliminary report, *J. Med. Entomol.* **2:**203–208 (1965).

158. SABIN, A. B., Antigenic relationship of dengue and yellow fever viruses with those of West Nile and Japanese B encephalitis, *Fed. Proc.* **8:**410 (1949).

159. SABIN, A. B., AND BUESCHER, E. L., Unique physicochemical properties of Japanese B virus hemagglutinin, *Proc. Soc. Exp. Biol. Med.* **74:**222–230 (1950).

160. SASLOW, A., *A Survey of Encephalitis of Unknown Etiology in Connecticut June–September 1967*, M.P.H. thesis, Yale School of Medicine, New Haven, CT, 1968.

161. SAVAGE, H. M., SMITH, G. C., MOORE, C. G., MITCHELL, C. J., TOWNSEND, M., AND MARFIN, A. A., Entomologic investigations of an epidemic of St. Louis encephalitis in Pine Bluff, Arkansas, 1991, *Am. J. Trop. Med. Hyg.* **49:**38–45 (1993).

162. SAWYER, W. A., BAUER, J. H., AND WHITMAN, L., Distribution of yellow fever immunity in North America, Central America, West Indies, Europe, Asia and Australia, with special reference to

specificity of the protection test, *Am. Trop. Med. Hyg.* **17:**137–161 (1937).

163. SCHATZMAN, H. G., Epidemic dengue 1 in Brazil, in: *Dengue Surveillance Summary*, p. 1, San Juan Laboratories, Dengue Branch, Division of Vector-borne Viral Diseases, Center for Infectious Diseases, Centers for Disease Control, Atlanta, 1986.

164. SCHOPEN, S., LABUDA, M., AND BEATY, B., Vertical and venereal transmission of California group viruses by *Aedes triseriatus* and *Culiseta inornata* mosquitoes, *Acta Virol.* **35:**373–382 (1991).

165. SCOTT, T. W., LORENZ, L. H., AND WEAVER, S. C., Susceptibility of *Aedes albopictus* to infection with eastern equine encephalomyelitis virus, *J. Am. Mosq. Control. Assoc.* **6:**274–278 (1990).

166. SCOTT, T. W., AND WEAVER, S. C., Eastern equine encephalomyelitis virus: Epidemiology and evolution of mosquito transmission, *Adv. Virus Res.* **37:**277–328 (1989).

167. SHOPE, R. E., AND SATHER, G. E., Arboviruses, in: *Diagnostic Procedures in Virus and Rickettsial Infections*, 5th ed. (E. H. LENNETTE AND N. J. SCHMIDT, EDS.), pp. 767–814, American Public Health Association, New York, 1979.

168. SHROYER, D. A., Venereal transmission of St. Louis encephalitis virus by *Culex quinquefasciatus* males (Diptera: Culicidae), *J. Med. Entomol.* **27:**334–337 (1990).

169. SIAM, A. L., MEEGAN, J. M., AND GHARBAWI, K. F., Rift Valley fever ocular manifestations: Observations during the 1977 epidemic in Egypt, *Brit. J. Opthal.* **64:**366 (1980).

170. SINGH, K. R. P., Cell cultures derived from larvae of *Aedes albopictus* (Skuse) and *Aedes aegypti* (L.), *Curr. Sci.* **36:**506–508 (1967).

171. SINGH, K. R. P., AND PAUL, S. D., Multiplication of arboviruses in cell lines from *Aedes albopictus* and *Aedes aegypti*, *Curr. Sci.* **37:**65–67 (1968).

172. SMITH, H. H., Controlling yellow fever, in: *Yellow Fever* (G. K. STRODE, ED.), pp. 529–625, McGraw-Hill, New York, 1951.

173. SMITHBURN, K. C., Antigenic relationships among certain arthropod-borne viruses as revealed by neutralization tests, *J. Immunol.* **72:**376–388 (1954).

174. SMITHBURN, K. C., Differentiation of the West Nile virus from the viruses of St. Louis and Japanese B encphalitis, *J. Immunol.* **44:**25–31 (1942).

175. STEPHENSON, J. R., Flavivirus vaccines, *Vaccine* **6:**471–480 (1988).

176. STOKED, A., BAUER, J. H., AND HUDSON, N. B., Transmission of yellow fever to Macacus rhesus: A preliminary note, *J.A.M.A.* **90:**2253–2254 (1928).

177. SURTEES, G., SIMPSON, D. I. H., BOWEN, E. T. W., AND GRANINGLER, W. E., Rice field development and arbovirus epidemiology, Kano Plain, Kenya, *Trans. R. Soc. Trop. Med. Hyg.* **64:**511–518 (1970).

178. TESH, R. B., The genus Phlebovirus and its vectors, *Ann. Med. Entomol.* **33:**169–181 (1988).

179. TESH, R. B., PERALTA, P. H., AND JOHNSON, K. M., Ecologic studies of vesicular stomatitis virus. I. Prevalence of infection among animals and humans living in an area of endemic VSV activity, *Am. J. Epidemiol.* **90:**255–261 (1969).

180. TESH, R. B., ROSEN, L., AND ATKEN, T. H. G., Studies of transovarial transmission of yellow fever and Japanese encephalitis viruses in *Aedes* mosquitoes and their implications for the epidemiology of dengue, *Pan. Am. Health Org. Sci. Publ.* **375:**179–182 (1979).

181. THEILER, M., The development of vaccines against yellow fever, in: *Les Prix Nobel, 1951*, pp. 174–182, Konigl. Boktryckeriet P. A. Norstedt, Stockholm, 1952.

182. THEILER, M., Action of sodium deoxycholate on arthropod-borne viruses, *Proc. Soc. Exp. Biol. Med.* **96:**380–382 (1957).

183. THEILER M., AND SMITH, H. H., Effect of prolonged cultivation *in vitro* upon pathogenicity of yellow fever virus, *J. Exp. Med.* **65:**767–786 (1937).

184. THEILER, M., AND DOWNS, W. G., *The Arthropod-Borne Viruses of Vertebrates*, Yale University Press, New Haven, 1973.

185. THOMAS, L. A., PHILIP, R. N., PATZER, E., AND CASPER, E., Long duration of neutralizing antibody response after immunization of man with formalinized Colorado tick fever vaccine, *Am. J. Trop. Med. Hyg.* **16:**60–62 (1967).

186. THOMPSON, W. H., AND BEATY, B. J., Venereal transmission of LaCrosse (California encephalitis) arbovirus in *Aedes triserialus* mosquitoes, *Science* **196:**530–531 (1977).

187. THOMPSON, W. H., AND EVANS, A. S., California encephalitis studies in Wisconsin, *Am. J. Epidemioi.* **81:**230–244 (1965).

188. THOMPSON, W. H., KALFAYAN, B., AND ANSLOW, R. O., Isolation of California encephalitis group virus from a fatal human illness, *Am. J. Epidemiol.* **81:**245–253 (1965).

189. TSAI, T. F. Arboviral infections in the United States, *Infect. Dis. Clin. North Am.* **5:**73–102 (1991).

190. TULLY, T. N., JR., SHANE, S. M., POSTON, R. P., ENGLAND, J. J., VICE, C. C., CHO, D. Y., AND PANIGRAHY, B., Eastern equine encephalitis in a flock of emus (*Dromaius novaehollandiae*), *Avian. Dis.* **36:**808–812 (1992).

191. URQUIDI, V., AND BISHOP, D. H., Non-random reassortment between the tripartite RNA genomes of La Crosse and snowshoe hare viruses, *J. Gen. Virol.* **73:**2255–2265 (1992).

192. VENTURA, A. K., BUFF, E. E., AND EHRENKRANTZ, N. J., Human Venezuelan equine encephalitis virus infection in Florida, *Am. J. Trop. Med. Hyg.* **23:**507–512 (1974).

193. WATTS, D. M., PANTUWATANA, S., DEFOLIART, G. R., YUILL, T. M., AND THOMPSON, W. H., Transovarial transmission of LaCrosse virus (California encephalitis group) in the mosquito *Aedes triseriatus*, *Science* **182:**1140–1141 (1973).

194. WATTS, D. M., THOMPSON, W. H., YUILL, T. M., DEFOLIART, G. R., AND HANSON, R. P., Overwintering of LaCrosse virus in *Aedes triseriatus*, *Am. J. Trop. Med. Hyg.* **23:**694–700 (1974).

195. WELLINGS, F. M., SATHER, G. E., AND HAMMON, W. M., Immunoelectrophoretic studies of the California encephalitis virus group, *J. Immunol.* **107:**252–259 (1971).

196. WHITMAN, L., Arthropod vectors of yellow fever, in: *Yellow Fever* (G. K. STRODE, ED.), pp. 229–298, McGraw-Hill, New York, 1951.

197. WHITMAN, L., AND AITKEN, T. H. G., Potentiality of *Ornithodoros moubota* (Murray) (Acarina: Argasidae) as a reservoir–vector of West Nile virus, *Ann. Trop. Med. Parasitol.* **54:**192–204 (1960).

198. WILLIAMS, J. E., YOUNG, O. P., AND WATTS, D. M., Relationship of density of *Culiseta melanura* mosquitoes to infection of wild birds with eastern and western equine encephalitis viruses, *J. Med. Entomol.* **11:**352–354 (1974).

199. WORLD HEALTH ORGANIZATION, Arthropod-borne and rodent-borne viral diseases, *WHO Tech. Rep. Ser.* **719:**1–116 (1985).

200. WORLD HEALTH ORGANIZATION, Viral hemorrhagic fevers, *WHO Tech. Rep. Ser.* **721:**1–126 (1985).

201. WORLD HEALTH ORGANIZATION, Expert committee on yellow fever, 3rd Rep., *WHO Tech. Rep. Ser.* **479:**1056 (1971).

202. WORLD HEALTH ORGANIZATION, *Prevention and Control of Yellow Fever in Africa*, WHO, Geneva, 1986.

203. WORLD HEALTH ORGANIZATION, *Weekly Epidem. Rec.* **28:**213 (1990).

204. WORLD HEALTH ORGANIZATION, *Weekly Epidem. Rec.* **67**:245 (1992).
205. YOUNG, N. A., AND JOHNSON, K. M., Antigenic variants of Venezuelan equine encephalitis virus: Their geographic distribution and epidemiologic significance, *Am. J. Epidemiol.* **89**:286–307 (1969).

13. Suggested Reading

FIELDS, B. (ED.), *Virology*, Raven Press, New York, 1985.
FIELDS, B. N., KNIPE, D. M., AND HOWLEY, P. M. (EDS.-IN-CHIEF), *Virology*, 3rd ed., Lippincott-Raven, New York, 1996.

KARABATSOS, N., *International Catalogue of Arboviruses Including Certain Other Viruses of Vertebrates*, 3rd ed., American Society of Tropical Medicine and Hygiene, San Antonio, 1985.
SHOPE, R. E., AND SATHER, G., Arboviruses, in: *Diagnostic Procedures for Viral and Rickettsial Infections*, 5th ed. (E. H. LENNETTE AND N. J. SCHMIDT, EDS.), pp. 767–814, American Public Health Association, New York, 1979.
THEILER, M., AND DOWNS, W. G., *The Arthropod-Borne Viruses of Vertebrates*, Yale University Press, New Haven, 1973.
TSAI, T. F., Arboviral infections in the United States. *Infect. Dis. Clin. North Am.* **5**:73–102 (1991).
California serogroup viruses. Proceedings of an international symposium, Cleveland, Ohio, November 12 and 13, 1982. *Prog. Clin. Biol. Res.* **123**:1–399 (1983).

CHAPTER 7

Arenaviruses

Peter B. Jahrling

1. Introduction

Arenavirus is the designation for members of the family Arenaviridae.[61] Virions are round to ovoid pleomorphic particles usually 90–300 nm in diameter. They have an electron-dense membrane with spikes or projections and a variable number of dense inclusions that convey a sense that virions are sprinkled with sand (*arenosus*). Arenaviruses have a negative-stranded, bipartite RNA genome and are antigenically related. Lymphocytic choriomeningitis (LCM) virus, the first recognized member, is considered the prototype arenavirus. These viruses are important public health problems in geographically restricted regions of tropical Africa and South America. While exotic to North America, these viruses have the potential to be introduced via these and other international crossroads by travelers returning from the tropical foci where the viruses are endemic.

2. Historical Background

Lymphocytic choriomeningitis virus was discovered in 1933 and shortly thereafter was associated with benign aseptic meningitis in man.[129] Junin virus was recovered from patients suffering from Argentinian hemorrhagic fever (AHF) in 1958.[119] It was subsequently shown to share antigens with Tacaribe virus, not pathogenic for man, which had previously been recovered from bats in Trinidad.[39] Machupo virus, the cause of Bolivian hemorrhagic fever (BHF), was initially isolated in 1936.[74] Within the next 2 years, five additional arenaviruses antigenically related to Tacaribe virus were identified: Amapari in Brazil,[124] Latino in Bolivia, Parana in Para-

guay,[162] Pichinde in Colombia,[152] and Tamiami in the United States.[25] These agents do not cause clinically significant human disease, but can infect laboratory workers.[28] Lassa virus was recovered in 1969 from patients suffering severe acute disease in Nigeria.[24] Since that time, three other Lassa-related viruses, Mopeia,[173] Mobala,[54] and Ippy[149] (which may be identical with Mobala), have been identified from Africa, and three other agents were reported from South America, Flexal[125] and Sabia[34] from Brazil, and Guanarito[138] from Venezuela. Thus, the Arenaviridae currently comprise 17 named viruses, six of which cause acute disease in man. In addition, several unnamed arenaviruses with unknown human disease potential have been isolated from rodents in Brazil and Argentina more recently (P. B. Jahrling, unpublished observations).

Early studies of antigenic relationships among arenaviruses were based on broadly reactive antibody binding assays, historically the complement-fixation test[27] and more recently the indirect fluorescent antibody test (IFAT)[170] and now enzyme-linked immunosorbent assay (ELISA).[117] Finer discriminations among arenavirus strains have been based on neutralization tests, although recently characterized monoclonal antibodies (MAbs) have also been used for fine discriminations in IFAT or ELISA formats.[20,21] These serological studies revealed that viruses from the New World were closely related, resulting in the designation of a Tacaribe (or New World) antigenic group.[104] In 1969, LCM virus was recognized to share similar morphology and morphogenesis with Tacaribe group agents,[113] and an antigenic link was detected soon afterward.[133] Lassa virus was found to share the typical morphology,[146] and subsequent work firmly demonstrated that Old World arenaviruses, LCM, Lassa, and relatives, form an antigenic complex complementary to that of the New World arenaviruses.[24,172] Biochemical characterizations of prototype arenavirus strains have val-

Peter B. Jahrling • USAMRID, Fort Detrick, Frederick, Maryland 21702.

idated the taxonomic relationships established by serological tests; fine mapping of peptide sequences of prototype arenavirus strains has identified at least one antigen site that is conserved among all arenavirus strains tested.[163] Even more recently, phylogenetic grouping on the basis of nucleotide sequences (by primer extension of the viral S-RNA segments) has confirmed the validity of the serologically based taxonomy.

3. Methodology

3.1. Mortality

The case fatality proportions for AHF (5–16%) and BHF (5–25%) reflect total morbidity and infection due to those agents, because clinical diagnosis is reasonably accurate and has been confirmed by laboratory testing and population surveys reveal that little subclinical infection occurs. Deaths from LCM, however, are rare, and clinical diagnosis of this syndrome is difficult. Lassa fever case fatality was reported as 20–60% in early nosocomial outbreaks, but later laboratory-documented clinical studies revealed that fewer than 20% of hospitalized patients died.[99] Likewise, early reports of fatality for Venezuelan hemorrhagic fever (Guanarito) virus infection were based on 9 of 15 cases (60%), but during the subsequent 4-year period, the proportion fell to 26 of 105 (24%).[156] For Sabia virus, only two cases have been reported, one of them fatal.[34]

3.2. Morbidity

The epidemiology of AHF and BHF is reasonably accurate because these diseases are geographically circumscribed, seasonal, not easily confused with other acute diseases, and corroborated by specific diagnostic methods. For example, clinical suspicion of AHF is confirmed in at least 70% of instances by laboratory testing.[92,105]

In contrast, sporadic cases of LCM are rarely recognized because the clinical syndrome is shared by many other viruses and laboratory testing for LCM infection is not widely practiced. Similarly, Lassa fever is easily confused with other infectious diseases in Africa. There are few laboratory facilities available on the continent, and the malignant connotations of this disease provide, if anything, a disincentive for official reporting of Lassa fever. Even so, in hospitals near endemic foci in Sierra Leone, mortality among hospitalized patients is still estimated at 15–20%, and 30% of all medical deaths are attributed to Lassa fever. However, serological surveys in catchment areas suggest that many subclinical cases occur, resulting in a substantially lower overall mortality, perhaps less than 1%.[76,102]

3.3. Serological Surveys

The neutralization (N) test was employed wherever possible because of its high specificity and the long persistence of such antibodies. It was the basis for epidemiologic studies of AHF and BHF and, to a lesser degree, of LCM. Complement fixation (CF)[27] was historically the method of choice for arenavirus serology, particularly for early work with LCM, but today it has been supplanted by the indirect immunofluorescence test IFAT,[170,172] and the trend is now toward ELISA. The CF test was both technically arduous and relatively insensitive; this method suffers from the distinct limitation that CF antibodies may not persist for more than 1–3 years after infection.

Most studies, especially those of Lassa virus infection, have been done with the IFAT. This method was chosen because N antibodies to this virus are very difficult to detect and because the IFAT method is much less expensive and safer than N, for which live virus is required. Antibodies measured by the IFAT are usually the first to appear, often becoming detectable within the first few days of hospitalization for Lassa virus,[171] and somewhat later for Junin and Machupo viruses.[65] Presence of specific immunoglobulin M (IgM) antibodies or a rising immunofluorescent antibody (IFA) titer is a presumptive diagnosis of acute infection. IgM antibodies measured by IFAT decline to undetectable titers within several months, while IgG antibodies persist at least several years. A popular approach to surveys for antibodies to various viruses causing hemorrhagic fever in Africa is the use of "multispots" in which cells infected with different agents are mixed. Positive reactions are identified by subsequent resort to univalent fluorescent antibody (FA) tests.[67]

ELISA procedures for Lassa-specific IgG and IgM have been developed[117] and successfully employed on field-collected human sera.[63] In combination with the Lassa antigen-capture ELISA, virtually all Lassa fever patients can be specifically diagnosed within hours of hospital admission. As for the antigen-capture ELISA, success of the antibody ELISA is critically dependent on highly avid, purified capture antibodies or globulins.

3.4. Laboratory Diagnosis

In most instances, a specific diagnosis of these illnesses requires detection of virus or viral antigens in blood or other body fluids or virus-specific antibodies that can be temporally related to illness.

Virus isolation is rarely employed to diagnose BHF because virus is only sporadically present in readily accessible clinical specimens.[68] Blood yields Junin virus through the first week of most clinical AHF infections,[19] and LCM virus can be recovered from cerebrospinal fluid (CSF) in many cases of LCM exhibiting acute neurological disease. Large amounts of Lassa virus are often present in blood of patients with Lassa fever through at least 2 weeks of illness.[70] Arenaviruses may also be recovered occasionally from throat washings and urine of patients.[109] Suckling mice or hamsters and several types of continuous cell cultures, particularly Vero cells from African green monkeys, are used to isolate and identify arenaviruses. Immunologic diagnosis of acute arenavirus infection was classically done by use of the CF test employing acute and convalescent serum samples. This method has been superseded by the FA technique because of its simplicity, the fact that antibodies are detected earlier in the course of illness than by CF in some arenavirus infections, and the fact that single-sample IgM antibody diagnosis can be made early in clinical disease.[56,85,122,170,171] Enzyme immunoassay (ELISA) is rapidly being adapted to arenavirus diagnosis.[53,63,78] Unlike IFAT, ELISA can be measured objectively and has the added advantage of capability for early detection of viral antigens, if they are present in sufficient concentration.[62,63,117] Likewise, primer pairs for polymerase chain reaction techniques have been developed to detect viral nucleic acids in clinical materials, especially for Lassa[89,153] and Junin[18,86] viruses. Comparisons with conventional isolation and antigen capture sensitivity and specificity are needed.

4. The Viruses

4.1. Biochemical and Physical Properties

Arenaviruses have a simple complement of structural proteins.[22] There is a nonglycosylated protein of 63–72 kDa, which functions as a nucleoprotein (NP) adherent to viral RNA, as well as two major glycoproteins (G1 and G2) of about 65 and 38 kDa respectively. G1 and G2 are cleaved from the precursor protein GPC identified by radioimmune precipitation of cytosol from infected cells. These are involved in formation of virion membrane and surface spikes and contain antigens responsible for virus-neutralizing antibodies.[15]

Four RNA species with distinct oligonucleotide fingerprints can be isolated from intact virions. Two are virus-specific. The S-RNA (22S) codes for NP and GPC, while the L-RNS (31S) codes for a 200-K protein believed to be an RNA-dependent RNA polymerase.[22] The remaining two RNA species, 28S and 18S, are isolated in varying proportions and are ribosomal. The coding strategy for arenaviruses is unique and has been termed *ambisense*, since the 3' half of the S-RNA codes for NP in the viral complementary sequence and is separated by an intragenic hairpin structure from the 5' region, which codes for GPC in the viral-sense sequence.[16] Through this mechanism, GPC and NP gene expression is independently regulated. Viral RNA must be translated before GPC can be expressed, and this regulation may be fundamental to the maintenance of persistent infections in chronically infected hosts and cells in culture.

Arenaviruses are readily inactivated by lipid solvents (ethyl ether, chloroform, sodium deoxycholate), acid media (pH less than 5), short wavelength ultraviolet light, common disinfectants (hypochlorite, phenolics, quaternary ammonium compounds), and common fixatives (formaldehyde, glutaraldehyde). In addition, inactivation of viral infectivity with preservation of serological reactivity can be achieved by treatment with β-propiolactone,[157] psoralen in conjunction with long wavelength ultraviolet light,[58] or γ irradiation.[41] Viral infectivity is stable in serum samples maintained at room temperature for several hours, and declines slowly over several days. Arenaviruses are readily inactivated by heating to 56°C for 30 min.

4.2. Morphology and Morphogenesis

The similarities in morphology and morphogenesis are so marked and distinctive that they were the basis for first associating the viruses in the present taxon.[113,132,146]

Thin-section electron microscopy of Vero cells infected with all arenaviruses shows them to be indistinguishable from each other. Coinciding with the highest infectious titers of the inoculated cultures is the occurrence of a large number of virions. The virions are round, oval, or pleomorphic and range from 60 to 280 nm in diameter (average 110–130 nm). They have membranous envelopes with club-shaped surface projections or spikes approximately 10 nm long. A unique and prominent feature of arenavirus virions internally is the presence of electron-dense particles. These particles usually number from 2 to 10, are connected by fine filaments, and are 20–25 nm in diameter; they are identical to host cell ribosomes by biochemical and oligonucleotide analysis. No symmetry has been discerned with any of these viruses.[114–116]

The particles mature by budding from plasma membranes. Vero cells infected with each of the viruses contain distinctive intracytoplasmic inclusion bodies consisting of a smooth matrix in which dense granules are embedded similar to those seen in the virions and indistinguishable from the host cell ribosomes. These inclusion bodies seem to match in size and location the cytoplasmic inclusions observed under light microscopy in cells infected with the virus.[114,116] Using immunohistochemistry and monoclonal antibodies, these intracytoplasmic inclusions are demonstrably immunoreactive with antinucleocapsid, but not G1 or G2 antibodies.

Negative-contrast electron microscopy of unfixed infectious virus particles generally confirmed results of cell-section analysis. Pleomorphism (90–350 nm) was noted, and the surface spikes were shown to be club-shaped with hollow central axes.[115] No resolution of internal structure or definite surface symmetry of particles was achieved.

Electron-microscopic studies of whole animals infected with arenaviruses[116] have revealed the presence of particles similar to those described above in a number of tissues of *Calomys callosus* infected as newborns with Machupo and Latino viruses. No such particles have been seen in hamsters infected with Junin virus, and only few in the salivary gland tissue of mice infected with Tacaribe virus. In general, only occasional virus particles have been observed in the brain tissue of mice infected with LCM, Tacaribe, Lassa, or Tamiami virus, although parallel studies indicate the presence of specific antigen.

The particles associated with arenavirus infection of cells in culture have been shown by labeling procedures to contain specific antigen material, at least with LCM virus; whether all size particles are equally infectious cannot be decided by electron microscopy alone.[97] Estimates of infectious-size particles by centrifugation or filtration have given sizes between 37 and 60 nm for LCM virus[80] and by filtration have given sizes between 70 and 140 nm for Lassa virus.[24]

4.3. Antigenic Properties

Early studies with LCM virus demonstrated the existence of CF antigen distinct and separable from the infective particle by centrifugation; it was designated *soluble antigen*. The nature and properties of this antigen have been the subject of studies[57] that confirm that virion and CF antigen are distinct entities. The latter, on inoculation into experimental animals, induces formation of antibodies that react *in vitro* with the CF antigen but will not neutralize the virion; furthermore, repeated inoculations of CF antigen fail to induce any protection against subsequent challenge of guinea pigs. These studies, as well as more recent information with other arenaviruses, indicate that there are at least two distinct antigenic molecules in the virion. The ribonucleoprotein core appears to be the main CF determinant and is responsible for the group antigenic relationship; the envelop or surface proteins, including the spikes, are associated with virus neutralization and are highly type-specific.[22] No hemagglutinins have been found for the arenaviruses.

Antigenic relationships among arenaviruses, established originally by the CF test, were later confirmed by the IFAT, which has gained general acceptance. Table 1 is a composite table incorporating results from several sources,[138,150] and it is an attempt to illustrate the relative positions of the viruses in the taxon. Tacaribe, Junin, Machupo, Guanarito, Amapari, Parana, and Latino viruses are very closely related by IFAT with mouse hyperimmune sera; Pichinde and Tamiami viruses are not closely related to the others or to each other. The LCM and Lassa viruses are very distantly related to the other agents; only when the highest-titered antisera are used can cross-reactions be observed. From these and corroborating molecular phylogenetic data, the arenaviruses can be divided into New World and Old World subgroups, the latter comprising LCM, Lassa, Ippy, Mopeia, and Mobala viruses.

In contrast to the FA and CF tests, the results of N tests, many of them done with samples of the same sera that showed marked crossing by CF and IFAT, are very specific. In comprehensive serum dilution, plaque reduction tests[25,68] in which sera had homologous titers in the range from 1:32 to 1:2048 (generally 1:128–512), no cross-neutralizations have been noted, even between viruses that are very close by CF and IFAT, such as Machupo, Tacaribe, and Junin. The same marked specificity has been observed when constant serum was used with varying dilutions of virus (Table 2). Little cross-neutralization among the arenavirus pathogens was noted, with the exception of a one-way cross between Machupo and Junin (P. B. Jahrling, unpublished observations).

4.4. Biological Properties

The natural hosts and reservoirs of the arenaviruses that cause human disease are discussed in the corresponding sections. The remaining viruses have been isolated in nature from the following animals: Tacaribe from *Artibeus* bats, Amapari from *Oryzomys* and *Neacomys* rats, Pichinde from *Oryzomys* and *Thomasomys*, Parana from *Oryzomys*, Latino from *Calomys*, Mopeia from *Mas-*

Table 1. Results of Indirect Fluorescent Antibody Tests Among 12 Arenaviruses

Viral antigen	Antibody[a]											
	GNT	JUN	MAC	TCR	AMA	LAT	PAR	TAM	FLE	PIC	LCM	LAS
Guanarito	>1280[b]	640	>1280	640	640	640	1280	>1280	>1280	320	80	<10
Junin	>1280	320	640	80	1280	640	1280	640	640	640	80	<10
Machupo	>1280	>1280	>1280	640	640	160	>1280	>1280	320	80	80	<10
Tacaribe	>1280	1280	640	80	>1280	320	640	320	640	640	80	<10
Amapari	>1280	80	320	320	640	320	640	320	160	640	10	<10
Latino	>1280	320	320	40	80	>1280	640	160	320	40	10	
Parana	>1280	40	160	40	320	320	>1280	320	640	>1280	160	
Tamiami	>1280	40	80	80	40	160	>1280	>1280	320	320	40	20
Flexal	>1280	320	160	640	640	1280	>1280	1280	>1280	>1280	80	<10
Pichinde	640	40	80	<10	80	160	1280	320	320	>1280	10	
LCM	40	<10	10	20	20	40	40	20	20	20	>1280	80
LAS	<10	<10	<10	<10	<10	<10	<10	<10	<10	>10	40	640

[a]GNT, Guanarito; JUN, Junin; MAC, Machupo; TCR, Tacaribe; AMA, Amapri; LAT, Latino; PAR, Parana; TAM, Tamiami; FLE, Flexal; PIC, Pichinde; LCM, lymphocytic choriomeningitis; LAS, Lassa.
[b]Reciprocal of highest positive antibody dilution. Antigens were infected Vero cells.

tomys, Mobala from *Praomys*, Tamiami from *Sigmodon*, and Flexal from *Oryzomys*.* Attempts to isolate these viruses from other natural hosts, including arthropods, have been reported, largely with negative results: Pichinde virus has been isolated from ectoparasites taken from viremic hosts, Amapari has been isolated from mites (Gamasidae), and Tacaribe was reported to have been isolated from a mixed mosquito pool.

Among experimental hosts, 1- to 4-day-old mice develop fatal illness following intracerebral inoculation of most, but not all, arenaviruses. Latino virus does not infect mice, and Parana inoculation results in illness but no death; LCM virus strains, in general, are lethal when inoculated into young adult mice but not when inoculated into newborn mice.

Newborn hamsters are lethally infected by Junin, Latino, Machupo, Parana, and Pichinde viruses; guinea pigs are susceptible to lethal infections with LCM, Lassa, Machupo, Junin, and Guanarito viruses.

Vero cells (a continuous cell line derived from African green monkey kidneys) are readily infected with all arenaviruses. Typically, the majority of cells are produc-

*The genera *Akodon, Calomys, Neacomys, Oryzomys, Sigmodon*, and *Thomasomys* comprise mouselike and ratlike rodents, in the tribe Hesperomyinae, subfamily Cricetinae, family Cricetidae, and the entire tribe is New World only. The genera *Mus, Rattus, Mastomys*, and *Praomys* are rats and mice in the family Muridae and are found only in the Old World (except for the established New World immigrants in *Mus* and *Rattus*). To the untrained observer, many of the small mouselike or ratlike rodents of the Old and New World are indistinguishable from one another, but habitats, habits, and life histories may vary greatly.

tively infected within 4 days of inoculation, and peak infectivity titers, which may approach 10^6 to 10^8/ml, are attained within 4–7 days. Most of these viruses also produce plaques on Vero cells maintained under agarose. When inoculated Vero cells are maintained under fluid medium, most of the cells are productively infected but fail to develop any obvious cytopathic effect. Thus, virus replication is most readily detected by examination of the cells for viral antigen by immunohistochemical techniques,[35,172] or by back-titration of the supernatant fluids on cell monolayers maintained under agarose. While Vero cells are the most generally useful, other cells that permit the replication of arenaviruses and filoviruses include LLCMK-2, BHK-21, and diploid cells such as FRhL-2 and MRC-5. A special property of arenaviruses that cause disease of man, repeatedly described with LCM, Machupo, and Lassa, is their capacity to induce persistent

Table 2. Results of Cross-Neutralization among Arenavirus Pathogens

Specificity of antibody	LNI of antiserum against virus[a]				
	GNT	JUN	MAC	LCM	LAS[b]
Guanarito	2.6	0.2	0.0	0.0	0.0
Junin	0.2	2.7	0.2	0.0	0.3
Machupo	0.1	2.5	>3.9	0.0	0.1
LCM	0.0	0.1	0.1	3.7	0.1
LAS	0.0	0.0	.0.0	0.0	3.2

[a]LNI, $\log_{10}$ neutralization index.
[b]See Table 1 for abbreviations.

infection in their natural hosts with chronic viremia and viruria; the epidemiologic implications of this fact are evident. In cell culture, arenaviruses frequently establish carrier states, where the cells show little or no cytopathic effect (CPE) and are refractory to superinfection with other arenaviruses but not to unrelated viruses. This characteristic is thought to result from both the ambisense replication strategy where virion-specified cell surface antigens (G1 and G2) decline but NP antigen persists intracellularly,[16] coupled with defective interfering (DI) particles. The modulating role of DI particles has practical significance in attempting virus isolation in cell culture. It is important to test dilutions of infectious specimens as well as more concentrated material, since interfering particles present in the lower dilutions may totally inhibit virus replication and antigen expression.

5. Pathogenesis and Immunity

An LCM virus infection of the adult mouse is the classic example of virus-induced immunopathological disease.[118] Intracerebral inoculation results in fulminant choriomeningitis and death without direct damage to neurons. This effect is mediated by cytotoxic T lymphocytes, is determined by class I histocompatibility antigens on such cells, and can be suppressed by prior thymectomy, anti-T-lymphocyte antibodies, cyclophosphamide, cyclosporin A, or radiation.[23,81,118,140,159,174] Recognition and target cell lysis by virus-specific cytotoxic T lymphocytes are controlled by the S genome of LCM virus.[131] In the neonatal mouse, LCM virus infection generally leads to chronic infection without acute disease. Originally thought to represent immune tolerance with absent response to viral antigens, such chronic infection, which is marked by viremia and viruria of long duration, is now known to be characterized by the formation of circulating antigen–antibody complexes, which over time may accumulate in, and cause functional damage to, renal glomeruli.[23] A rather similar pattern has been observed for Machupo virus in its natural rodent host *C. callosus*; no antiviral antibodies have been found in this host–parasite system, however, and microcytic anemia rather than glomerulonephritis is the long-term pathological consequence of chronic disseminated virus infection.[73]

Given the present state of knowledge, no definite statements can be made concerning the mechanisms in humans responsible for control of infection or production of disease caused by arenaviruses. Control of virus infection might be mediated by cytotoxic T lymphocytes, activation of macrophages and nonspecific killer cells, humoral antibodies, interferon, production of DI virus particles, or combinations of these defense systems. Humoral antibodies appear to play a small role because they generally appear after acute illness, especially for the Old World pathogens, LCM and Lassa. Arenaviruses are generally insensitive to the actions of α or β interferon; indeed, interferon is thought to have an exacerbating effect on the clinical manifestations of disease.[83] Defective virus interference is a routine event for arenaviruses in culture, yet a role for DI particles has not been demonstrated in human infection. Similarly, the specific roles of macrophages and various lymphocytes remain to be elucidated, although multiple lines of investigation suggest that the cell-mediated immune system plays a pivotal role in recovery from arenavirus infection.[44]

In contrast with the mouse model for LCM, the pathogenesis of human arenaviral disease is generally attributable to direct or indirect viral damage rather than to an immunopathological process. The "second-wave" CNS disease seen in human LCM may represent an exception, and it is of interest that immunosuppressed cancer patients given LCM virus in an attempt to kill tumor cells sustained high viremia of several days' duration without subsequent CNS disease.[59] The late neurological syndrome associated with patients treated with AHF-immune serum[43,95] is thought by some to have an immunopathological component.

In both AHF and BHF, virus replicates in lymphoid tissues, producing variable viremia, inclusion bodies in Kupffer cells of the liver, and diffuse capillary endothelial swelling without any significant inflammatory response.[30,52,68] There is little or no evidence of direct parenchymal damage, minimal functional damage to liver, kidney, or heart, and no detectable virus in the brain. Scattered hemorrhages are seen in mucosal (gastrointestinal tract) and serosal surfaces, and there is clearly a capillary leak syndrome in which plasma protein escapes at a more rapid rate than erythrocytes, leading to hemoconcentration and functional hypovolemic shock. Blood loss is not usually significant. Disseminated intravascular coagulation, although reported in a few cases of AHF,[5] is probably not a major factor in pathogenesis.

Lassa fever differs from AHF and BHF in important respects. Although capillary leak and hypovolemic shock are important components of disease pathogenesis, patients with Lassa fever generally have much greater viremia of longer duration. There is direct damage to hepatocytes, and aspartate aminotransferase (AST) levels are correlated with outcome of human disease.[99,101,166] In addition, Lassa virus infection may cause permanent eighth nerve deafness, and some patients experience meningitis with virus recoverable from the CSF. The "split" humoral antibody response, in which antibodies to NP but not G1 and G2 formed during acute illness, together with the observed excess mortality noted in patients having more than $10^{3.5}$ ID_{50} of virus in blood, suggests that Lassa virus somehow exerts direct yet selective depressive effects on cells of the immune system.[70]

Whatever the eventual elucidation of arenaviral pathogenesis in man, it should be noted that with rare exceptions (CNS), recovery from acute infection, if achieved, is complete. Thus, understanding of this complex equation is of more than academic interest.

The humoral immune response to arenavirus infection in man displays two basically different patterns. In New World AHF, BHF, and Venezuelan hemorrhagic fever, no antibodies generally appear until 3–4 weeks after onset of disease. Antibodies to NP and G antigens thereafter develop nearly in synchrony and are of good concentration. In contrast, NP antibodies detectable by CF or FA are found within a week of clinical disease in Old World LCM and Lassa, rise to high titer, then slowly decline over many months. The G antibodies (neutralizing), in contrast, require 2–6 months to form and, in the case of Lassa fever, never achieve significant levels in most instances. Indeed, studies in animals have shown that arenaviruses may be divided into those that are readily neutralized and those that are not. The human pathogens, save Lassa, are well neutralized, as are Tacaribe and Amapari viruses. All the others either express G antigens poorly or are not inactivated by binding of "neutralizing" antibodies.

6. Lymphocytic Choriomeningitis

The LCM virus was first isolated in 1933 in the course of investigations on the etiology of an epidemic of encephalitis in St. Louis, Missouri[9]; the virus may have been present in the CNS tissue of a patient who died of that illness or, more likely, derived from monkeys inoculated during the study. An etiologic association between the virus and a disease of man, acute aseptic meningitis, was established[130,143] by isolation of the agent and demonstration of development of antibodies. Although at an early period it was assumed that LCM virus was the exclusive etiologic agent of acute aseptic meningitis, or Wallgren's disease, it soon became apparent that the virus caused only a small proportion of the cases. Traub[154] reported that a colony of laboratory albino mice was chronically infected with a virus subsequently identified as LCM; this finding was the beginning of a new concept, persistent tolerant virus infections, that has considerable

epidemiologic implications with respect to LCM virus and other arenavirus infections of man.

6.1. Descriptive Epidemiology

6.1.1. Incidence and Prevalence. Determination of infection or illness caused by LCM virus requires a laboratory-confirmed specific diagnosis; in general, this is not attempted, since the required laboratories are not always available. Efforts to obtain a specific diagnosis usually require special circumstances, such as a large number of clinically suspect cases appearing simultaneously[8] or the continuing interest of groups of investigators.[4,17,108]

Soon after the discovery of the virus and its association with cases of aseptic meningitis, it became apparent that clinical infection of man by LCM virus was a rare event; later surveys supported the view.

One of the most extensive surveys to determine the prevalence of clinical LCM virus infection in man was conducted in U.S. military personnel and dependents over an 18-year period, from 1943 to 1960.[4,108] Examination of nearly 1600 CNS illnesses revealed that only 8% were specifically diagnosed as LCM infections; on average, seven cases a year occurred during the entire period. No estimate can be made of undiagnosed cases or, if they existed, of subclinical infections. A study in the United States[167] showed that 5% of about 1200 sera from residents of various areas had neutralizing antibodies; it is conceivable that a certain degree of nonspecific neutralization of virus may have occurred in that study,[80] so that the results may not be specific. Recent studies in inner-city Baltimore, Maryland, have suggested that human seroprevalence in a U.S. urban center is still approximately 5%.[31]

Investigations in West Germany since 1960 by Ackerman et al.,[2] Scheidt et al.,[141] and Blumenthal et al.[17] indicate the extent of the distribution of LCM virus in that country and the close association between the incidence of infection of man and the presence of virus in the mouse. Early observations by these investigators had shown the rarity of the disease in a number of large hospitals in the country; furthermore, antibody surveys with sera from selected individuals revealed only about 1% of positives.[1] In a subsequent survey[17] done after the distribution of the virus in mice had been investigated, sera from about 2000 persons from rural districts were tested for neutralizing antibodies; 68 of these sera, 3.4%, were positive. On the basis of this survey, Ackerman[1] estimates that as many as 1000 new infections per year may occur in a population of about 6 million persons in rural German areas. Because only a few clinical cases of LCM disease are reported annually, the inference is that most LCM virus infections go undiagnosed or are subclinical.

6.1.2.   Geographic Distribution.   The virus of LCM may well have worldwide distribution, being present in all parts of the world where the house mouse is found. Well-documented proof of the virus's presence has been given for European countries and North and South America; its presence has also been reported in Asia, less convincingly in Africa, but not in Australia.[80]

6.1.3.   Age, Sex, and Occupation.   Since LCM is not usually reported, the effect of a number of variables on its spread and prevalence is difficult to appraise. A sero-epidemiologic report from West Germany[17] indicated that the distribution of antibodies was not influenced by sex or occupation, whether farm work or professional or office work; in that survey, few positives were found among persons under 20 years of age. On the other hand, no influence of age was seen in hamster-related outbreaks.[3,8] It is possible that mouse-associated infections are more common in rural populations or in lower socioeconomic urban groups and that hamster-associated cases are found principally in urban centers. Further, a seasonal fluctuation of cases has been suggested in man, more in winter than in summer. Perhaps this is associated with migratory habits of the house mouse[80] and possibly with closer contact with mice in the cold months in the temperate zone.

Special attention should be given to LCM as an occupational disease in laboratory personnel, either in persons who work with the virus or in those who work with other problems but who use animals—mice, hamsters, possibly monkeys—that may be infected, as well as cells derived from these animals.[40,64] Reported laboratory accidents may well represent only a fraction of all the occurrences; in the period between 1952 and 1966, 45 laboratory infections with five deaths were documented.[28] More recently, seven human LCM infections were documented in association with nude mice inoculated with hamster tumor cells, determined retrospectively to harbor latent LCM virus.[40]

6.2.   Mechanism and Route of Transmission

6.2.1.   Spread of Virus.   The only lifelong carrier of LCM virus is the mouse, from which man becomes infected. The hamster may develop a transient carrier state in the course of which it can also infect man. Man-to-man transmission seems unlikely. The mechanism of transmission from mouse to man cannot be stated with certainty. It appears that either the airborne route, through household dust contaminated with mouse urine and other excretions and secretions, or the contamination of food and drink by mouse excretions is the most likely source of human disease. The portal of entry in these instances would be the upper respiratory tract or, possibly, the upper digestive tract; the possibility of transmission through skin abrasions has also been considered.

6.2.2.   Reservoir.   In nature, the virus has been isolated from various animal hosts in addition to man, who is most likely a dead end. Chief among these, for epidemiologic implications including maintenance of the virus in nature, is the house mouse (*Mus musculus*). From the first demonstration of LCM virus in house mice trapped in the homes of two persons suffering from nonbacterial meningitis,[10] the abundance of isolations has left no doubt about the close association in nature between the virus and this rodent. Furthermore, it has been shown that experimental mouse colonies can be chronically infected.[155] Studies on the nature of the infection of laboratory mice by LCM virus extending over a period of 30 years have clearly shown that the mouse infected *in utero* or within a few hours after birth develops persistent infection; mice thus infected circulate virus in their blood and maintain an active and relatively normal health condition for a period of time representing a good fraction of a normal mouse's life span. The epidemiologically important feature of the persistent infection is that wild mice so infected shed virus continuously for the duration of their lives by way of urine, feces, and nasal and oral secretions. The virus thus excreted will contaminate households, including food, drink, dust, and fomites; from these, and by ways as yet undetermined, man becomes infected. In addition, new generations of mice become infected at birth or *in utero*, thus maintaining the carrier status of the mouse populations; mice so infected may become the source from which other species—hamsters, guinea pigs, monkeys—are infected, and they, in turn, may infect man.

The studies of Ackerman *et al.*[2] and Blumenthal *et al.*[17] are particularly illustrative of the association between infection of mice and infection of man. Of 1795 mice trapped between 1960 and 1962 in 44 of 376 areas in West Germany, 65 were LCM carriers, as shown by virus isolation; nearly all positive trapping areas were in northern and northwestern Germany, none in southern Germany. Serological surveys done at about the same time in which 1371 persons from rural districts were tested by neutralization test showed that of 511 sera from persons in north and northwestern Germany, in or near the places where LCM virus had been isolated from mice, 9.1% had antibodies. The second set of sera from 811 persons was from south Germany, where no LCM virus had been isolated from mice; only five (0.6%) were positive.

In recent years, the Syrian hamster (*Mesocricetus aurarus*) has emerged as an important source of human infection and illness caused by LCM virus if not as a true reservoir. Small outbreaks had occurred in the past involving persons participating in biomedical research work in which hamsters were used.[11,84] Between 1968 and 1971, 47 LCM infections were described in West Germany. These are specifically diagnosed by antibody detection as caused by LCM virus in persons who shortly before their illness had been in contact with pet hamsters; 45 of these infections were clinical, mainly influenzalike or aseptic meningitis, and two had no clinical manifestations.[3] Subsequent investigations in West Germany[45] on commercial breeding colonies showed that of 598 animals examined, representing 11 different breeders, LCM virus was isolated from members of six different colonies. According to the authors of that study, it is estimated that close to 1 million hamsters are sold annually as pets in West Germany; it is obvious that the importance of this animal as a source of infection of man cannot be overlooked.

Two outbreaks of LCM in man, associated with hamsters, have been observed in the United States. Early in 1973, an episode occurred in a laboratory where hamsters were used for cancer work.[8] The investigation of the outbreak revealed several interesting points. In all, 21 persons became ill with a severe influenzalike illness, of which 14 cases had occurred before LCM infection was suspected. In addition to the 21 clinical cases, confirmed by FA antibodies, where were 17 persons who had antibodies but no illness; in other words, inapparent infections occurred. The association with hamsters was clearly seen in that 75% of 20 persons admitting to having touched the hamsters were seropositive, whereas only 17% of 61 persons who had no contact other than entering the premises were positive; the latter may have been instances of airborne infections. The LCM virus was isolated from 11 of 24 hamsters tested. The animals appear to have been infected from virus present in the tumor line with which they had been inoculated. The recent report of a laboratory-centered outbreak related to nude mice, inoculated with latently infected hamster tumor cells, is a variation on this theme.[40,64]

In another extended episode, 93 human cases in seven states were diagnosed and specifically confirmed between December 1973 and April 1974.[42] The association with pet hamsters was established in every instance; in some instances, two or three cases occurred in a family. The episodes in the United States and Germany point to the importance of hamsters as a source of human infection with LCM virus; although this animal is not a true lifelong carrier, it can circulate and excrete virus for periods of 2 or 3 months after its infection.

Another recent reminder of the importance of maintaining suspicion for LCM virus came from recent outbreaks of a fatal viral disease among tamarins and marmosets. Ten different U.S. zoos suffered 12 separate epizootics of "Callitrichid hepatitis" between 1980 and 1990 before the causative virus was finally recognized as LCM virus.[148] The source of this virus was shown in some cases to be the suckling mice fed to the callitrichids; in other cases, epidemiologic data suggested a role for feral mice coinhabiting the zoo settings. Serosurveys of workers in the affected zoos may determine the degree of transmission and the spectrum of disease in man.

6.3. Patterns of Host Response

6.3.1. Clinical Aspects. Infection of many by LCM virus presents different clinical forms, and there may be inapparent infections. In one survey of 165 personnel potentially exposed to LCM virus, 32 were seropositive with no history of clinical disease. An additional 13 had experienced an influenzalike syndrome without clear-cut CNS involvement, plus two patients diagnosed to have aseptic meningitis. However, among the less seriously ill patients headache, photophobia, and subtle mental difficulties were common. Three major clinical forms seem to prevail: aseptic meningitis, influenzalike or non-nervous-system type, and meningoencephalomyelitic type.[80] The influenzalike and meningeal are the most frequent types. The incubation period is believed to be from 6 to 13 days. In the influenzalike (or grippal) type, there are fever, malaise, muscular pains, coryza, and bronchitis; in the meningeal type, which is the most common, there is a "grippe"-like beginning followed by defervescence and a second clinical episode with definite signs and symptoms of meningitis, with stiff neck, headache, and nausea, which may remain mild and of short duration or can be pronounced, last for 2 weeks or longer, and lead to considerable prostration.

The great majority of specifically diagnosed clinical infections follow a benign course; only a few fatal cases have been reported, either following CNS involvement[4] or after systemic generalized illness with hemorrhagic manifestations.[145] Chronic sequelae, although rare, have been reported, including paralyses, headaches, and personality changes; it appears that the documentation of most such cases is ambiguous.[80] The possibility that prenatal infection with LCM virus may result in hydrocephalus and chorioretinitis has been reported.[144]

6.3.2. Diagnosis. Clinical and routine laboratory analyses are only indicative of aseptic meningitis: the CSF is under increased pressure, with slightly increased protein, normal or slightly reduced sugar, and a moderate number of mononuclear cells, from 150 to 400/mm³.

The virus in man can be isolated from blood, CSF, and, in fatal cases, brain tissue. The best sources for isolation are blood during the febrile period and CSF during the period of meningeal manifestations.

The animal of choice for LCM virus isolation is the laboratory albino mouse, 3–5 weeks old; following intracerebral inoculation, the incubation period and signs of illness are nearly pathognomonic. Care must be taken to use mice from a colony known to be free from the virus. Inoculation of cells in cultures (particularly Vero) is also common; however, a problem with autointerference (thought to be due to DI particles) may inhibit viral replication. Dilution of the sample frequently circumvents this problem, permitting detection of the viral replication by measuring plaques or viral antigens by CF, IFAT, or antigen-capture ELISA.

The techniques employed for detection of antibody development are the CF, IFAT, and ELISA tests; IFAT and ELISA antibodies appear several days earlier than CF antibody. The presence of specific IgM antibodies detected by IFAT is indicative of recent infection, since IFAT-IgM titers persist for less than 3 months. The presence of specific IFAT-IgM in the CSF of LCM patients constitutes a definitive diagnosis. Neutralizing antibodies appear much later after onset; therefore, the N test is not helpful for an early diagnosis; it is most profitably used in serum surveys.[17]

6.4. Treatment and Prevention

There is no specific treatment advocated; in view of the definite association with mice, it appears that rodent control may minimize the risk. Pet hamsters may be a source of infection, particularly in children. Monitoring of hamster colonies for presence of virus or antibodies or both would be indicated. Laboratories using mice or hamsters or cells derived from them are advised to maintain a rigid surveillance and occupational health program.[28]

7. Argentinian Hemorrhagic Fever

A disease resembling AHF and with the same geographic location seems to have been first recognized in 1943.[126] Arribalzaga[12] gave the first detailed account of the disease and considered it a new nosological entity; his description included extremely accurate clinical and epidemiologic observations. The causal agent, Junin virus, was isolated in 1958.[95,119] Annual outbreaks of the disease have occurred since 1958, and the endemic zone has been progressively increasing in area.

7.1. Descriptive Epidemiology

Collection of data concerning clinically diagnosed cases of the disease and laboratory efforts to confirm the clinical diagnosis appear to be done efficiently through local, provincial, and national public health centers in Argentina.[37,126,168]

7.1.1. Prevalence in Man.
Argentinian hemorrhagic fever is predominantly a rural disease that affects adult males with agricultural occupations, particularly harvesting of maize; 80% of nearly 1000 cases analyzed in 1973 were of males, and 63% of the total number were in the age group between 20 and 49 years.[92]

7.1.2. Geographic and Seasonal Distribution.
The endemoepidemic zone was first recognized in the northwest of Buenos Aires province; by 1958, its area was estimated at 16,000 km². Since that year, the zone has spread west and north to include additional localities in Buenos Aires as well as sections of two adjacent provinces, Córdoba and Santa Fe; the affected area was estimated in 1970 to be 80,000 km² and to include a population of 800,000 persons.[136] The total number of cases reported from 1958 to 1984 was about 20,000, with annual fluctuations of between 100 and 3500 cases; the mortality rate for laboratory-confirmed cases studied at Pergamino has been from 10 to 20%.[92]

The disease is sharply seasonal, with the outbreaks beginning late in summer (February), reaching a peak in autumn (May), and ending early in winter. The seasonal distribution coincides with the intensification of agricultural labors, particularly harvest of maize, and with an influx of transient, farm workers; at the same time, there is an increase in the population of wild rodents, which are considered the principal reservoir of the virus. In recent years, maize harvesting in the affected region has become mechanized, leading to extremely high infection rates among a numerically diminished work force that operates combines and grain trucks. Disease patterns have also been modified downward by the advent of crop rotation between maize and soybeans. Cultivars of the latter have been found to support much lower populations of the reservoir rodents. Although AHF is overwhelmingly a rural disease, cases have been observed in an urban setting in the near absence of the main rodent reservoir (*Calomys*) of the virus[94,136]; however, the simultaneous existence of LCM and Junin virus in an area may create diagnostic problems.[7,93]

7.2. Mechanism and Route of Transmission

7.2.1. Spread to Man.
Chronic infection of rodents with associated viruria is the basic mechanism of

transmission of the virus to man; there is no evidence implicating arthropod transmission.[73,136] The mode of transmission from wild infected rodents to man has not been definitely established. It may be airborne, from dust contaminated by the excretions or secretions from rodents, or by the oral route through ingestion of food and drink equally contaminated. Since the disease has been transmitted to human volunteers by injection,[127] it may be possible that the disease is also acquired through skin abrasions in the course of farm work while materials contaminated with rodent excreta are being handled. Although the virus has been isolated from throat swabs and urine from patients, contact transmission between individuals is exceptional.[134]

7.2.2. Reservoir. The possible connection between wild rodents and AHF was first stated by Arribalzaga.[12] Accumulated observations beginning in 1958 support the close association between disease and rodents in the endemic areas. The main reservoir is two species of Cricetidae, *Calomys laucha* and *C. musculinus*, which are present in farm fields and along hedgerows, the latter species predominating in Córdoba province[136]; Junin virus has also been isolated from *Akodon azarae* and, rarely, *Mus musculus*.[37] Field and laboratory investigations show that Junin virus causes a chronic tolerant infection in *C. musculinus*, with persistent viremia and viruria and no development of antibodies.[136] Most likely, the virus is maintained in nature by infection of the rodents at birth. Although the virus has been isolated from mites,[120,127] it has not been established that they play a role in transmission between rodents or to man.

7.3. Patterns of Host Response

7.3.1. Clinical Features. The disease presents a syndrome that includes manifestations of renal, cardiovascular, and hematiologic involvement; pronounced neurological manifestations are also described. The disease lasts from 7 to 14 days and terminates either with complete recovery with no sequelae or with death. After an incubation period estimated at from 7 to 16 days,[135] there is an insidious and gradual onset with chills, asthenia, malaise, headache, retro-ocular pain, muscular pains often pronounced in the costovertebral angle, anorexia, nausea, and vomiting. The most prevalent signs at the outset are fever with temperatures up to 102–104°F, conjunctival injection, enanthem, exanthem on face, neck, and upper thorax, a few petechiae, particularly in the axilla, polyadenopathy, and muscular tenderness at the thigh. Three to 5 days after onset, the signs and symptoms become more pronounced in the severe cases, with dry

tongue, dehydration, oliguria, hypotension, relative bradycardia, and, in the worst cases, hemorrhages from the gums and nasal cavities, hematemesis, hematuria, and melena; oliguria may develop into anuria. In the severe cases, there are psychosensorial and motor alterations. Death is caused by hypotension and hypovolemic shock resulting from plasma leakage, not whole blood loss. In nonfatal cases, the fever diminishes by lysis, and there is marked diuresis and rapid improvement within days; however, convalescence is prolonged. Case fatality has been as high as 20%; usually, it is between 3 and 15% in different outbreaks. Clinically inapparent infections appear to be very rare.[68,106,136]

7.3.2. Simultaneous Occurrence of AHF and LCM. Investigations in areas of the AHF endemic zone to determine the source of virus in an urban setting[96] led to the finding of antibodies against LCM virus in mice (*M. musculus*); at about the same time, a strain of LCM virus was isolated from that species.[137] A reexamination of antibodies in acute and convalescent sera from nearly 3000 cases of AHF, using LCM and Junin antigens, revealed that in a substantial number of instances, AHF occurred in persons who showed evidence of previous infection with LCM virus.[92] Furthermore, there were a few cases previously diagnosed as AHF in which the serological diagnosis was changed to LCM.[92]

Additional evidence of the activity of LCM virus in the endemic AHF area is given by the simultaneous admission to a hospital of two agricultural workers with clinical diagnosis of AHF but with specific serological conversions to Junin virus in one and to LCM virus in the other.[93]

Since only between 60 and 70% of patients clinically diagnosed as having AHF are generally serologically confirmed, it had previously been suspected that other agents were active in the endemic area in epidemic times; serological evidence of infection with group B arboviruses has been reported[105] in a number of persons diagnosed clinically as AHF cases, and St. Louis encephalitis virus was isolated from one.[107] Evidence for the coexistence of Junin and LCM viruses was noted more recently during a field trial of AHF vaccine in Argentina in 1988–1989; 2.3% of 7227 volunteers had antibody specific for LCM.[7] This suggests the possibility that new arenaviral strains might arise in this rural setting, through recombination or reassortment of the segmented genomes in dually infected animal hosts. The feasibility of generating reassortants among arenavirus pathogens has been demonstrated experimentally.[88]

7.3.3. Diagnosis. Clinical diagnosis of AHF has been confirmed either serologically or by virus isolation in from 60 to 70% of reported cases.[106] In a thorough study involving 2249 reported cases over the period 1965–1972 in the city of Pergamino, the diagnosis was confirmed in 62% and was doubtful in 11% of cases.[92]

During the epidemic season of AHF, other diseases of viral etiology that at the early phase present clinical manifestations similar to those of AHF appear to occur in the same areas.[105,142] A study of signs and symptoms in a number of patients clinically diagnosed as having AHF, in whom laboratory confirmation was sought, showed that during the first week of illness a combination of asthenia, dizziness, petechiae in the axillary region or anterior chest wall, and conjunctival congestion was present in 71% of confirmed cases but in only 3.5% of nonconfirmed ones. When leukopenia, thrombocytopenia, and casts in the urine are also found, the diagnostic accuracy is further increased.[142]

Specific diagnosis is based on isolation of virus or demonstration of serological conversion. The virus is isolated from the blood during the acute period, probably from the third to the tenth day after onset, and, in fatal cases, from liver, spleen, kidney, and clotted blood. The materials are intracerebrally inoculated into newborn mice 1–3 days old or into guinea pigs by the peripheral or intracerebral route. Junin virus is also frequently isolated in Vero cells in culture, with viral replication assessed by plaque formation or antigen accumulations measured by CF, FA, or antigen-capture ELISA. The isolation frequency of Junin had been reportedly enhanced by cocultivation of peripheral blood leukocytes, isolated by hypaque–ficol separation, with Vero cells.[6] Various formats for reverse transcription–polymerase chain reaction (RT-PRC) have been devised to detect Junin RNA in clinical materials.[18,86] Some of these are reputed to be far more sensitive than conventional isolation procedures, especially in the presence of antibody; however, none of these procedures has been benchmarked yet against conventional techniques for Junin diagnosis.

The IFAT has been the serological test of choice for demonstration of antibody development; more recently, ELISA has been increasingly used,[53] although detection of IgM has not been satisfactory. AHF shares with other diseases caused by arenaviruses the characteristic that CF antibodies develop relatively late after onset. An early serum sample and a second one taken at least 30 days after onset offer the best possibility for detection of serological conversion. The presence of specific IgM antibodies detected by IFAT is indicative of recent infection, since IFAT-IgM titers persist for less than 3 months. As for all the arenavirus pathogens, a rising IFAT-IgM or -IgG titer to Junin constitutes a strong presumptive diagnosis. Since IFAT-IgM titers, as well as CF titers, do not persist long, a

decreasing titer suggests a recent infection that occurred perhaps several months previously. Investigations spanning several decades[7,92] have shown the simultaneous development of CF antibodies against Junin and LCM virus antigens in patients in the AHF endemic zone clinically diagnosed as cases of AHF, including some from whose acute-phase blood Junin virus was isolated. This fact creates serious difficulties in reaching a definite diagnosis not heretofore encountered with other arenavirus infections of man; as a result, epidemiologic evaluation of data may be faulty.

7.4. Treatment, Control, and Prevention

Plasma from patients convalescent from AHF has been shown to reduce mortality of acute disease dramatically. In a placebo-controlled trial, AHF fatality was reduced from 16% to 1% in patients who received Junin virus antibodies prior to the ninth day of illness.[43,95] Plasma therapy has now become standard practice in Argentina. Presently, convalescent plasma therapy is being further refined by optimizing the dose of immune plasma administered on the basis of neutralizing antibody titer and patient body weight. Despite the unequivocal success of immune plasma therapy for acute AHF, 11% of treated patients develop a late neurological syndrome (LNS) not observed in patients receiving normal plasma. Patients experiencing LNS return 4–6 weeks after acute onset with fever, headache, tremor, and ataxia, which usually all clear within several days. Forty such cases have been located; most were mild and transient, but one death was reported. The pathogenesis of LNS is being investigated, but it is speculated that immune plasma protects against the lethal effects of systemic disease, but not against possible CNS involvement in patients spared an acute fatal outcome. Administration of passive antibody to AHF patients suppresses the immune response, which makes subsequent recruitment of plasma donors with adequate neutralizing antibody titers more difficult. It is hoped that development of human MAbs[139] with neutralizing activity might offer a long-term solution.

The antiviral drug ribavirin is believed to be beneficial in treatment of AHF, as it has proven to be for Lassa fever.[100] Ribavirin is a relatively nontoxic nucleoside analogue with broad-spectrum antiviral activity.[147] The regimen successfully used for AHF calls for a loading dose of 30 mg/kg (intravenous), then 16 mg/kg every 6 hr for 4 days, followed by 8 mg/kg every 8 hr for 6 days.

An effective vaccine for AHF was recently developed and will be available soon for general use in Argentina. A live attenuated Junin strain (Candid #1)[121] has passed safety and immunogenicity tests in U.S. volunteers; its efficacy was proven in a double-blind trial in 15,000 agricultural workers at risk to natural infection in Argentina. Subsequently, over 100,000 persons were immunized with Junin vaccine in Argentina.[168]

Ecological control to reduce the number of rodents and exposure of man to them appears difficult to implement. The circumstances under which the bulk of the exposed population live and work at the time of harvest will not be likely to change in the near future.

8. Bolivian Hemorrhagic Fever

Bolivian hemorrhagic fever was first recognized in 1959 in two rural areas in the northeastern part of Bolivia, Department of Beni. In late 1962 or early 1963, cases began to appear in a nearby town, San Joaquin, developing into a large outbreak that continued until the middle of 1964; nearly 700 persons were ill, with a mortality of 18%. Several subsequent outbreaks have occurred in the Department of Beni.[68,73,90] An outbreak that occurred in 1971 in Cochabamba, Bolivia,[73] differed ecologically from previous ones and represented an extension of the virus to a new area.

8.1. Descriptive Epidemiology

8.1.1. Incidence in Man. Before 1962, the small outbreaks affected mainly adult males, and most cases occurred from April to September, which is the time of highest agricultural activity. From 1963, when the disease appeared in towns and villages and larger numbers of persons sickened, the pattern changed. Adult males still had a somewhat higher rate of morbidity, but persons were affected with little relationship to sex, age, and occupation; it was soon apparent that the disease was "house-associated," with the lower socioeconomic groups experiencing the highest incidence of disease. Although a seasonal pattern is evident, with the highest incidence from February to September, cases occur in each month.[68,90]

8.1.2. Geographic Distribution. The main epidemic centers, San Joaquin and Orobayaya, are located in the Department of Beni, in the northeastern section of Bolivia; these centers are on an immense flat plain east of the Andes. The prevailing vegetation type is that of a grassland broken with "islands" of forest and numerous tree-lined rivers and streams.[79] The human settlements where cases have occurred in the past are on slightly elevated sites that generally escape flooding during the heavy rains; the houses are on the edge of the forest,

overlooking the grass-covered marshlands. These villages and settlements are heavily infested with *Calomys callosus*, a mouselike rodent that, although pastoral, readily invades and lives in houses in a manner similar to that of the house mouse, *Mus musculus*.[79]

8.2. Mechanism and Route of Transmission

8.2.1. Spread to Man. Transmission from rodent to man is probably by contamination of food, water, or air with infected rodent urine or by inoculation through skin abrasions.[73] Human-to-human transmission can occur in rare cases of close contact,[38] but it is not considered important in the spread of the natural infection. There may be, however, circumstances that promote such transmission, as shown in a small outbreak in Cochabamba, Bolivia, which is outside the habitat of the rodent *Calomys callosus*. From an index case acquired in the endemic area, five secondary cases developed, including family and medical personnel. All cases were fatal except one.[73]

8.2.2. Reservoir. The distribution of cases in a town in the form of clusters definitely associated with certain houses, the absence of evidence of human-to-human transmission, and the equally negative evidence of an arthropod vector led to the inference that a reservoir species might be involved that lived in or near households; soon, the association of disease with a rodent, *Calomys callosus*, was firmly established. This rodent has been trapped in all households where cases have occurred; houses located in sites that did not favor the presence of this rodent were spared. Finally, the dramatic termination of the epidemic in San Joaquin in June 1964, 2 weeks after continuous trapping of *C. callosus* had been implemented, left little doubt about the association.[79,90] Fifty percent of *C. callosus* caught wild at the time of that epidemic were infected with the virus.[75] Experimental studies with colonized *Calomys* show that on inoculation of Machupo virus, the rodent develops a tolerant infection with persistent viremia and viruria and no development of antibodies.[75] *Calomys* is easily infected by oral and nasal routes and also by contact with infected cagemates; about 50% develop viremia and viruria for life.[72]

All efforts to isolate the virus from arthropods caught in the epidemic area have been negative.[79]

8.3. Patterns of Host Response

8.3.1. Clinical Features. The disease is clinically so similar to AHF that a joint description is often given.[66,68] The incubation period is estimated to be from 7 to 14 days. The onset is insidious, and the fever, which

has been carefully monitored in numerous etiologically confirmed cases, reaches a temperature between 102 and 105°F with little diurnal variation, remaining at that level for at least 5 days. About 30% of the patients present hemorrhagic manifestations consisting of petechiae on the upper part of the trunk and oral mucous membranes and, on occasion, bleeding from gums, nose, stomach, intestines, and uterus; blood loss, however, is not a threat to life.[68] Nearly half the patients exhibit a fine intention tremor of tongue and hands beginning 4 or 6 days from onset; about one fourth of these may develop a frank and extensive neurological disorder. Somnolence and coma are hardly ever seen except in very young children. The acute disease can last for 2–3 weeks; convalescence is long, with complaints of severe generalized weakness and manifestations of autonomic dysfunction. Probably because of the continuously elevated temperature, loss of hair and transverse grooving of the nails are common. The mortality has varied depending on outbreaks, ranging from 5 to 30%; clinically inapparent infection is very rare.

8.3.2. Diagnosis.   The clinical diagnosis by an experienced physician in the endemic area and in moderately severe or severe cases is fairly accurate; because of the toxic condition of the patient, BHF may resemble typhus or typhoid fever. Certain clinical laboratory data help the clinician: leukopenia, thrombocytopenia, and increased hematocrit. The last indicates a bad prognosis.[68]

A specific diagnosis is based on virus isolation or development of antibodies. The most successful animal for virus isolation has been the newborn hamster inoculated by the intracerebral or intraperitoneal route. Recovery of Machupo virus from acutely ill patients is, however, quite difficult; only one in five samples, mostly sera, from serologically confirmed cases yielded virus, most frequently between the 7th and 12th day after onset. Virus is rarely isolated from urine or throat swabs. In fatal cases, virus is easily and generally isolated from spleen and lymph nodes.[68] The most convenient and efficient means for establishing a specific diagnosis are the FA and CF tests. ELISA assays are feasible, using crude cell lysates as antigen,[78] although they have not been used extensively in the field. Although these tests are group-specific rather than type-specific, no diagnostic problems have arisen with Machupo virus because of its sharply localized geographic location. As with other diseases in this group of viruses, FA and CF antibodies are relatively late in appearing; although they have been found on the 14th day after onset, it is advisable to test a sample of serum between 40 and 60 days after onset.[90,91] Neutralization tests can distinguish Machupo from Junin infections; for Machupo, NP antibodies evolve simultaneously with FA and ELISA, and coincide with initiation of recovery.

8.4. Treatment, Prevention, and Control

Administration of convalescent plasma has been advocated and used; despite some impressions of favorable clinical responses, its efficacy has not been established or denied because of insufficient observation. Ribavirin has never been tested for efficacy in treatment of BHF, although experiments in primates and experience with AHF predict that ribavirin will work for human BHF as well.

No vaccine has been developed for Machupo. However, nucleic acid homologies for Machupo and Junin[32] and experimental cross-protection experiments in primates suggest that AHF vaccine (Candid #1) will protect against Machupo virus challenge (P. B. Jahrling, unpublished observation). Serious consideration is being given to immunizing workers at high risk of exposure to Machupo, such as laboratory and medical care providers and rodent trappers in the endemic zone.

Rodent control by continued trapping has been the most effective single means for preventing human infection and for terminating an epidemic. A routine surveillance program instituted in the principal towns and villages of the Department of Beni in 1972 has reduced the annual occurrence of BHF since that time to fewer than ten recognized cases annually.[103] This program is based on the fact that more than 75% of chronically viremic–viruric *C. callosus* have grossly enlarged spleens.[73] *Calomys* are trapped twice each year in a given locality; if animals with large spleens are encountered, a community-wide campaign is instituted until captures are reduced to a very low rate, usually 2–4 weeks of effort. This remarkable system is one of the least expensive effective disease control operations extant.

9. Venezuelan Hemorrhagic Fever

In September 1989, two cases of hemorrhagic fever were reported in males living in the rural municipality of Guanarito, Portuguesa State, Venezuela. Diagnosis was complicated by an ongoing epidemic of dengue hemorrhagic fever due to dengue type 2. Eventually, a new arenavirus was identified as the causative agent. The disease was named Venezuelan hemorrhagic fever (VHF) and the agent designated Gaunarito.[138]

9.1. Descriptive Epidemiology

During 3 years of epidemiologic surveillance, from September 1989 to December 1991, 88 cases of VHF were documented, with a fatality rate of 34%. Since then, only sporadic cases have emerged, with 105 cases and 26 deaths reported by the end of 1992. The age group affected ranged from 6 to 54 years old, with the highest rate in males over 15 years. old. Most VHF cases have been reported from the municipality of Guanarito and neighboring Barinas State. These areas are predominantly rural, with small farms used for agriculture and cattle production; thus, most homes are near fields. Field studies have shown that the cotton rat, *Sigmodon alstoni*, is the major host and natural reservoir of Guanarito virus.[151] In addition, the cane rat, *Zygdontomys breviocauda*, may serve as a transiently infected host with importance in maintenance of the viral cycle.

9.2. Mechanism and Route of Transmission

As for the other arenaviruses, transmission from rodents to man is thought to relate primarily to contact with food, water, or air contaminated with rodent urine; inoculation through skin abrasions is also possible. Even within the VHF endemic zone, the prevalence of Guanarito virus infection is very low, estimated to be between 0.1 and 3%,[151] although the percentage of infected people who develop VHF is thought to be high. From rodents trapped in the endemic zone, 10 of 49 *S. alstoni* and 12 of 106 *Z. brevicauda* yielded Guanarito virus; seven other species were negative. Of 195 people living near one of the rodent-trapping sites and tested for antibody, five (2.6%) had Guanarito virus antibodies. All were adults, and two had been diagnosed previously with overt VHF. Preliminary epidemiologic information suggests that person-to-person transmission is uncommon, based on low antibody rates among family contacts of cases and hospital personnel caring for infected patients. More extensive epidemiologic studies are in progress.

9.3. Patterns of Host Response

9.3.1. Clinical Features. The clinical syndrome for VHF is very similar to the descriptions for AHF and BHF.[156] The onset of disease is usually insidious, following an incubation period of 6 to 14 days. The fever is usually progressive, accompanied by severe myalgia, arthralgia, odynophagia, retro-orbital pain, severe headache, dizziness, and photophobia; petechiae occur occasionally. Various other manifestations of bleeding diathesis occur frequently, as with AHF and BHF. Pathological features in fatal cases of VHF include generalized vasocongestion with multiple hemorrhages in the gastrointestinal mucosa, uterus, other organs, and subcutaneous tissue.

9.3.2. Diagnosis. The differential diagnosis includes dengue, yellow fever, hepatitis, leptospirosis, and hantaviral renal syndrome. Guanarito virus has been isolated routinely in Vero cells, although infant mice are susceptible when inoculated intracerebrally. Viral antigens can be identified in cell culture by FA or antigen-capture ELISA. IHC has been used to demonstrate Guanarito viral antigens in tissues of experimentally infected primates as well as human cases. PCR methods are being developed that may preclude much of the need for biological amplification in the future. Serological diagnosis has depended on IFAT and ELISA. Both IgM and IgG appear 12 to 30 days after onset, correlating with clinical improvement. Neutralizing antibodies measured by plaque reduction assays appear by 30 days and persist indefinitely. Neutralizing antibodies clearly discriminate Guanarito virus from Junin and Machupo, unlike the IFAT and ELISA assays, which are group-reactive.

9.4. Treatment, Prevention, and Control

As with BHF, laboratory data suggest that ribavirin would be effective in treatment of acute VHF; clinical trials have been proposed, but no efficacy data are yet available. No vaccine effort has been initiated, although the Junin vaccine was tested for possible efficacy against Guanarito infection in experimentally infected primates, without success (P. B. Jahrling, unpublished observation).

Exposure of the human population to *Sigmodon* and *Zygodontomys* spp. reflects recent land use changes in the endemic area. Deforestation and large-scale agricultural activities have provided more favorable habitat for these granivorous rodents; this trend, coupled with human migration into the area, increases the opportunity for human contact with the rodent reservoirs.[151] This has become a repeated theme in the recent explosion of interest in the topic of emerging viruses.[60]

10. Lassa Fever

Lassa fever was first observed in 1969 in a missionary nurse stationed at a locality in northeastern Nigeria; following her admission to a hospital in Jos, Nigeria, two contact cases developed in nurses at that hospital.[47] Because of the circumstances surrounding this outbreak and the fact that two of the three persons affected died, the disease acquired from the outset a reputation for severity that subsequent events amply justified. In addition to the initial episode, four more outbreaks occurred between 1970 and 1974 in Nigeria, Liberia, and Sierra Leone[26,50,110]; furthermore, laboratory accidents have occurred in the United States.[13,82]

10.1. Descriptive Epidemiology

10.1.1. Geographic Distribution.
The disease has been observed in several localities in Nigeria since 1969 (Lassa, Jos, Onitsha, Zaria), in Liberia since 1972, and in Sierra Leone since 1970–1972. The number of cases seen or retrospectively diagnosed in the outbreaks has varied from 3 to about 60. Recent investigations indicate that the disease at present occurs in an endemo-epidemic form in the eastern section of Sierra Leone and in Liberia[46,48,102] and that it may have been seen as far back as 1956.[98] Other countries with serological evidence of Lassa fever include Guinea, Ivory Coast, Ghana, Senegal, Upper Volta, the Gambia, and Mali. Related viruses, possibly not pathogenic for man, have been isolated in Mozambique, Zimbabwe, and the Central African Republic.

10.1.2. Epidemic Types: Season, Age, and Sex Distribution.
Two patterns of disease have been observed. The first type, hospital-associated, develops as a result of exposure and spread from a hospitalized index case to other patients, visitors, and medical staff. The index case usually acquired the virus in the nearby community; between 10 and 20 days after admission to the hospital, a cluster of secondary cases develops.[26,47,110]

The second pattern, much more common, occurs in the community at large. Patients acquire their infection at home or in other community surroundings rather than by exposure or contact in the hospital with another patient. However, there is also the possibility of nosocomial transmission, particularly to the hospital staff.[50]

Tertiary cases in hospitals have been recorded, but with a few exceptions—notably by transmission to medical staff—they have been milder. No evidence has been reported of further propagation of the disease.[26]

Mortality during nosocomial outbreaks has been 20 to 66%, but it was about 16% during a 3-year prospective study of febrile adults admitted to two hospitals in Sierra Leone.[99] Furthermore, 30% of all medical deaths in these hospitals are attributed to Lassa fever. However, population-based studies in Sierra Leone and Liberia demonstrated that only 1 in 10–20 Lassa virus infections resulted in hospitalization, making the mortality-to-infection ratio no more than 0.02.[102]

Although endemic Lassa fever is observed throughout the year in Sierra Leone and Liberia, more cases are registered in the dry season months from January to May, the period during which most of the nosocomial outbreaks also have occurred.

Endemic Lassa fever occurs in persons of all ages and both sexes, although available data are much better for adults than children. Lack of census information precludes comparison of age-adjusted infection and clinical attack rates, but the steady rise in age-specific antibody prevalence in Sierra Leone villages with stable populations suggests that attack rates are lowest in small children and in the aged, who are largely immune.[102]

Evidence regarding risk of infection to hospital personnel differs. In Sierra Leone, the prevalence of Lassa virus antibodies among staff and residents of hospital town sites was similar (about 30%), but in Liberia, hospital staff had excess antibody prevalence in comparison to village residents.[49,102]

10.2. Mechanism and Route of Transmission

10.2.1. Spread of Virus.
The transmission in a hospital setting is undoubtedly from person to person by either the contact or the airborne route, including direct contact, droplet spread, or sharing of drink, food, clinical instruments, objects, and utensils. The same mode of transmission may be at work in contact infections acquired in the home.

Endemic Lassa virus transmission in Sierra Leone was correlated with the presence of *Mastomys* rodents, although there was not always a high correlation between simultaneous presence of virus-infected rodents and either human antibody prevalence or occurrence of acute disease. Clustering of infection in households was noted.[49] An attempt to reduce transmission within a household by vigorous trapping of rodents in houses of hospitalized "index" cases was not highly successful.[76] From these observations it is not possible to determine the relative contribution of rodent–human and human–human transmission to Lassa virus infection in the community.

The mode of transmission to man from the natural source or reservoir, a rodent, is still unknown. It may include direct contact with the rodent, its urine, and oral secretions, eating of uncooked rodent flesh, or contact with food and drink contaminated by the rodent; it could also be airborne.

Lassa fever is particularly severe among pregnant women, for whom mortality rates are somewhat higher, especially during the third trimester. Lassa virus has been isolated from milk, and there is a suggestion that mothers can infect their newborn infants.

The possibility of an arthropod vector appears extremely remote to explain transmission to man. Whether ectoparasites can transmit the infection between rodents is not known. Penetration through a cut while performing an autopsy appears to have been the mode of infection of a physician,[164] and infection through a cut on a finger may have occurred in a nurse.[47]

Most medical and nursing personnel probably acquire infection by direct contact with blood because virus concentration has been found to be very high in severely ill patients, and the frequency and concentration of virus in throat washings is much less.[70] Cough is not a common feature of acute illness.

10.2.2. Reservoir. During field investigations of Lassa fever in Sierra Leone in 1972, tissues were collected from 325 rodents and bats and tested for presence of virus. Materials from 10 of 46 *Mastomys natalensis* yielded Lassa virus, and no other species was infected.[111] This high prevalence of active infection was reminiscent of other rodent-borne arenaviruses. Subsequent field work in Nigeria, Liberia, and Sierra Leone, coupled with the experimental demonstration of persistent infection in *M. natalensis*, leaves little doubt that this African rodent is the primary, if not the exclusive, reservoir–vector of Lassa virus.[161,169] Interestingly, laboratory mice (*Mus musculus*) have been shown to support persistent infection,[24] and FA antibodies, but not virus, have been detected in wild *Mus* captured in houses of "infected" villages in Sierra Leone.

Since the discovery of Lassa virus, further study of African *Mastomys* indicates that there are at least three, perhaps four, distinct species. Two of these occur in West Africa[123]; both are infected with Lassa virus, but one is much more restricted to the human environment and is exclusively present in villages of the forested Eastern Province of Sierra Leone where Lassa fever is concentrated. Two other species of *Mastomys* are present in southern Africa, and at least one of these is the reservoir of Mopeia, a Lassa-related arenavirus.[55,71] Thus, as with rodent arenaviruses of South America, there is now a strong suggestion that *Mastomys* rodents and arenaviruses have coevolved.

10.3. Patterns of Host Response

10.3.1. Clinical Features. Lassa fever is a disease with generalized organ involvement manifested in severe cases by pharyngitis, pneumonitis, myositis, myocarditis, encephalopathy, nephropathy, and hemorrhagic diathesis. The overall spectrum of infection of man is not yet fully known; therefore, it is not possible to estimate the risk of severe illness following infection with the virus. All earlier reports stressed the severity of the disease; however, there is evidence of milder forms and inapparent infections.[102]

The incubation period is ordinarily between 6 and 14 days. The disease has an insidious onset; in variable order malaise, asthenia, lassitude, headache, sore throat, muscular aches, abdominal pains, loss of appetite, nausea, vomiting, and diarrhea appear. Fever appears early and ranges from 38 to 41°C for an interval of up to 3 weeks. About 70% of patients experience pharyngitis, and in more than half of these cases purulent exudates are observed in the tonsillar fauces. Conjunctivitis and proteinuria are other common features in hospitalized patients. Most patients have chemical evidence of acute hepatitis, and quantitative elevation of serum AST is directly related to fatal outcome of disease, although frank icterus is rarely present. In severely ill patients there is usually clinical and laboratory evidence for a capillary leak syndrome: scattered petechiae, minor gastrointestinal hemorrhage, increasing hematocrit, significant proteinuria, pleural effusion, ascites, edema of the neck and face, and finally hypovolemic circulatory collapse. Abortion at any stage of gestation occurs in about two thirds of pregnant women, and Lassa virus has been found in high concentration in placental and fetal tissues.[160] In children, the disease course is similar to that in adults, although the presenting complaint is usually cough and vomiting.[128] Among Lassa virus-infected infants, a condition described as "swollen baby syndrome" characterized by anasarca, abdominal distention, and bleeding has been described.[112] Patients who survive may exhibit the following complications in 5% or less of infections: unilateral or bilateral permanent eighth nerve deafness, meningoencephalitis with virus present in CSF,[70] and pericarditis, which is seen almost exclusively in males. The overall mortality in patients who contract Lassa fever in the endemic areas of West Africa is about 15%.

10.3.2. Diagnosis. The clinical differential diagnosis of Lassa fever in West Africa is complicated by the protean and variable manifestations of this viral illness and the endemic or sporadic occurrence of many other acute infectious diseases. Lassa fever may be confused with or accompanied by malaria. Typhoid fever is a major consideration, and other possibilities include leptospirosis, relapsing fever, rickettsiosis, bacterial sepsis, amebic liver disease, and viral diseases including yellow fever, Rift Valley fever, Congo–Crimean, Marburg, and Ebola fevers.

The pattern of nonspecific clinical findings exhibited by Lassa fever patients is of significant diagnostic value to the clinician in West Africa. Pharyngitis, conjunctivitis, proteinuria, chest or abdominal pain, vomiting, and facial edema all occur more frequently in Lassa fever than in hospitalized febrile non-Lassa patients. Nearly two thirds of Lassa fever patients had at least four of these abnormalities, and the odds of confirmed disease were 8 : 1 in such patients compared with those exhibiting fewer than three such findings.[99]

Specific laboratory tests are required to diagnose Lassa fever, particularly those cases initially presenting without hemorrhage, severe hepatitis, or purulent pharyngitis. Virus is readily isolated from blood of patients for up to 2 weeks after onset of fever, from pleural or abdominal effusions and from many tissues at autopsy, but less frequently from throat washings and urine.[48,70,109] Vero cell cultures are inoculated with clinical samples and examined daily by FA technique, yielding identifiable virus in 2–7 days, depending on virus content in a given specimen. Because Lassa virus is infectious by aerosol, such work should be limited to laboratories having special high-containment facilities for protection of personnel. Antigen-capture ELISAs for detection and presumptive identification of arenavirus antigens in viremic sera have also been developed for Lassa.[117] In an extensive series of field samples from Liberia, using conventional virus isolation of IFAT as comparative test, virtually all Lassa infections were diagnosed by a combination of antigen and IgM ELISA tests that could be completed within 6 hr of sample receipt. The threshold sensitivity of the test is approximately 2.1 $\log_{10}$ pfu/ml, so it is sufficiently sensitive to detect most acute-phase Lassa viremias, plus virus concentrations expected in throat wash and urine specimens as well. The antigen and IgM detection procedures do not require replication of virus and are reliable using inactivated samples. Thus, they are well suited for application in endemic areas where facilities for virus isolation are not available. Diagnostic tools based on PCR have been evaluated in a small series of Lassa fever patient sera from west Africa.[153] The sensitivity by PCR relative to conventional viral isolation was 0.82, specificity 0.68. PCR was positive longer in the disease course than isolation, perhaps due to its greater sensitivity in the presence of the early antibodies associated with acute Lassa fever.

10.4. Treatment and Disposition of Patients

The antiviral drug ribavirin has been demonstrated to be effective in treatment of Lassa fever, particularly if administered prior to the seventh day of illness. Ribavirin is a relatively nontoxic nucleoside analogue with broad-spectrum antiviral activity.[147] Among patients presenting with AST concentrations of >150 IU, mortality was reduced from about 50 to 5% by intravenous administration of 1 g ribavirin every 6 hr for 4 days and 0.5 g every 8 hr for a further 6 days.[100] It was found, in addition, that either intravenous or oral ribavirin (250 mg every 6 hr for 10 days) reduced mortality in patients with less than 150 IU of AST compared to previously studied untreated controls. The principal side effect of therapy was reversible anemia, which in no instance was sufficiently severe to cause interruption of treatment. Oral ribavirin has been proposed for prophylaxis of high-risk patient contacts (500 mg/kg by mouth every 6 hr for 7 days).[29] However, animal modeling data suggest that longer periods of treatment (perhaps 14 days) are warranted to prevent late onset of disease. Ribavirin thus represents the first dramatically successful treatment for a serious systemic viral disease in otherwise normal persons.

Because certain Lassa fever patients (particularly those most severely ill, who have high concentrations of virus in body fluids) pose a health hazard to persons in close contact with them, isolation of febrile patients suspected of having the disease is mandatory. Continuing observation in endemic areas now suggests that use of gloves and gowns and attention to maintenance of "enteric precautions" effectively prevent nosocomial spread of infection in virtually all instances.[29,102]

10.5. Prevention and Control

Prevention at the individual level is based on strict sanitary precautions; no vaccine is available. Since *Mastomys natalensis* is thus far the only known reservoir, measures that minimize contact of this rodent with man and his habitat will be helpful. However, it is unrealistic at this time to base too much hope on the effectiveness of the control of this rodent in the endemic areas. As mentioned above, special precautions must be taken in hospitals to prevent spread to the medical and nursing staff, especially those on obstetrical services with infected mothers.

No effective vaccine is available. For Lassa virus, a naturally attenuated strain (Mopeia virus, from Mozambique) protects rhesus monkeys against Lassa virus challenge,[77] but field studies are required to establish the extent and nature of natural human infection with this virus before it seriously can be considered a candidate for human vaccine development. Formidable questions related to safety, persistence, and mechanism of cross-protection remain. Alternative approaches, including the use of vaccinia virus vectors bearing the Lassa virus GPC or N-genes are being actively investigated,[14] although none has shown good protection in primate models of Lassa infection to date.

11. Unresolved Questions

11.1. Vaccines

With the exception of LCM virus, which appears to have worldwide distribution, the other arenavirus diseases of man have been found to be restricted to definite geographic areas that, although showing a tendency to increase, are still easily identifiable. In the case of Lassa virus, although the areas are multiple and possibly more extensive than now known, they are confined to one continent. These geographic considerations, added to the fact that some of the more exposed groups—maize harvesters for AHF and hospital personnel and diamond miners for Lassa fever—are well defined, make these diseases an excellent target for preventive vaccination. Indeed, successful demonstration of the live attenuated Junin vaccine strain efficacy in Argentina[168] was facilitated by the existence of a large risk group concentrated in a geographically circumscribed area. Use of this vaccine to protect against Machupo has been proposed, based on the cross-protection observed in laboratory primates. However, the demographics are far less favorable for demonstrating efficacy, due to the sporadic occurrence of BHF and the generally lower incidence in the endemic area. Attempts to develop a Lassa vaccine were discussed above. It is unlikely that the naturally attenuated strains will be used as vaccines because of lingering concerns about reversion to virulence, persistence, and reassortment of genomes among arenavirus strains. The nature of the protective immune response for Lassa is also unclear. While neutralizing antibody is widely regarded as protective, immunologic resistance to challenge is also believed to depend on cell-mediated immune mechanisms. Attempts to elicit a protective cell-mediated immune response through recombinant vaccinia techniques have been largely unsuccessful.[14] Also, there is now increasing concern about the use of any live viral vaccine, not only vaccinia, particularly in Africa where human immunodeficiency virus (HIV) prevalence may be very high. Clearly, a totally different approach will be required before a safe and efficacious vaccine will be available for use against Lassa in Africa.

11.2. Pathogenesis and Immunopathology

Although the pathogenesis of LCM in mice definitely involves an immunopathological component, there is no direct evidence for this in man. In experimentally infected mice, intrathecal levels of B-cell-stimulating factor 2 (BSF-2) and γ interferon have been correlated with the development of meningitis, and there is evidence that intrathecal BSF-2 also rises in the CSF of patients with acute LCM-associated meningitis. This suggests that one aspect of the CNS pathology of meningitis in humans is invasion by lymphocytes and plasma cells and their attendant cytokines.[51]

Disseminated intravascular coagulation has been reported in some fatal cases of AHF. Viremia in man occurs, the duration of which is not well established but presumably not long, and antibodies develop late, having seldom been reported before the end of the third week; although one may entertain the possibility that antigen–antibody complexes are formed, factual proof is not available.

Up to half of AHF and BHF patients have neurological symptoms. The pathology of CNS disease is obscure, but there is no evidence for direct viral infection. A late neurological syndrome has also been described, consisting mainly of cerebellar signs,[43,95] particularly for patients treated with AHF-immune plasma. Despite the unequivocal success of immune plasma therapy for acute AHF, 11% of treated patients develop a late neurological syndrome not observed in patients receiving normal plasma. Patients experiencing late neurological syndrome return, 4–6 weeks after acute onset, with fever, headache, tremor, and ataxia which usually all clear within several days. Forty such cases have been located; most were mild and transient, but one death was reported. The pathogenesis of late neurological syndrome is being investigated, but it is speculated that immune plasma protects against the lethal effects of systemic disease but not against possible CNS involvement in patients spared an acute fatal outcome. Alternatively, a more complex immunopathological mechanism may be involved.

There is evidence that antigen and antibody coexist in the blood in Lassa fever[63,70]; formation and deposition of complexes may result in immunologically mediated lesions by activation of complement and release of products that increase capillary permeability, which is one of the outstanding characteristics of the disease. In two separate studies it was concluded that hepatic lesions appeared to be the result of direct cell damage caused by the virus, with no cellular infiltration or other evidence of immunopathology.[87,160,165] A single case report describes an interesting complex of pericarditis and cardiac tamponade with pleural effusions and ascites 6 months after acute Lassa infection. Effusion fluids were devoid of recoverable virus, yet they contained high titers of Lassa-specific IgG and numerous lymphocytes, suggesting an immune-mediated mechanism. Studies on the physiopathology of arenavirus diseases of man are urgently needed in order to improve treatment. These studies should include immunologic mediators and cytokines, as well as circulating inhibitors such as that which inhibits both platelet function and superoxide generation in polymorphonuclear leukocytes in Lassa-infected patients.[36] The role of cytokines in the pathogenesis of arenaviruses is a largely neglected field; investigation, especially outside the experimental laboratory, is made more difficult by the extreme biohazard associated with these agents.

11.3. Geographic Distribution

Knowledge of the distribution of the arenaviruses pathogenic for man is mainly based on recognized and specifically diagnosed disease; this knowledge fails to supply information concerning the total actual prevalence of infections including mild forms of disease and inapparent infections. Furthermore, atypical or mild cases in localities not now considered to be within the endemic areas may go unrecognized, particularly if they appear only sporadically.

Seroepidemiologic surveys of LCM have been done in the past in Germany and the United States, but not within the past 15 or 20 years. There is a modest amount of recent information available regarding AHF, collected during the field trials of Junin vaccine in Argentina (WER). However, for BHF, current intelligence is much more sparse. Few reports concerning Lassa fever are forthcoming, in part due to civil unrest in the endemic region. Biopolitics aside, much of this information gap derives from the unavailability of practical and sensitive tests by which to safely assess the immune status of at-risk populations in the field. Neutralizing antibody tests carry the inherent risk of requiring manipulation of infectious virus; even under optimal conditions, determination of neutralizing antibodies by plaque reduction tests is difficult, as it is for LCM. While easier to perform, neutralizing antibody assays for AHF and BHF have not been applied to any remarkable extent in recent serosurveys. Yet, neutralizing antibodies may well be the most reliable indicator of the resistance of a population to reinfection. Among the group-specific antibody tests for arenaviruses, the IFAT gained practical acceptance over the CF more than a decade ago. Now, however, ELISA-based assays are widely regarded to provide a more objective method than fluorescence. IgM-capture ELISA, as well as antigen-capture ELISA, are rapidly becoming routine.

PCR-based assays will add another dimension to the capability of field laboratories to diagnose acute disease in almost "real-time." Proper tailoring of primers should permit design of tests with the proper degree of specificity. The recent emergence of two new arenaviral pathogens, Guanarito and Sabia, serves as a reminder that broadly reactive, grouping reagents are still required to augment the newly evolving tools of PCR and capture ELISAs based on extremely specific MAbs and gene sequences. An investment in rapid diagnosis should result in more timely intervention with effective treatment regimens, and through implementation of appropriate public health measures may reduce dissemination of these highly virulent viral pathogens.

12. References

1. ACKERMAN, R., Epidemiologic aspects of lymphocytic choriomeningitis in man, in: *Lymphocytic Choriomeningitis Virus and Other Arenaviruses* (F. LEHMANN-GRUBE, ED.), pp. 233–237, Springer-Verlag, New York, 1973.
2. ACKERMAN, R., BLOEHORN, H., KUPPER, B., WINKENS, I., AND SCHEID, W., Uber die Verbreitung des Virus der lymphocitaren Choriomeningitis under den Mausen in West-deutschland. I. Untersuchungen uberwiegend an Hausmausen (*Mus musculus*), *Zentralbl. Bacteriol. Parasitenkd. Infektionskr. Hyg. Abt. I* **194:**407–430 (1964).
3. ACKERMAN, R., STILLE, W., BLUMENTHAL, W., HELM, E. B., KELLER, K., AND BALDUS, O., Syrische Gold-hamster als ubertrager von lymphocytaren Choriomeningitis, *Dtsch. Med. Wochenschr.* **45:**1725–1731 (1972).
4. ADAIR, C. V., GAULD, R. L., AND SMADEL, J. E., Aseptic meningitis, a disease of diverse etiology: Clinical and etiologic studies on 854 cases, *Ann. Intern. Med.* **36:**675–704 (1953).
5. AGREST, A., AVALOS, J. C. S., ARCE, M., AND SLEPOY, A., Fiebre hemorragica Argentina y coagulopatia por consumo, *Medicina* **29:**194–201 (1969).
6. AMBROSIO, A. M., ENRIA, D. A., AND MAIZTEGUI, J. I., Junin virus isolation from lympho-mononuclear cells of patients with Argentine hemorrhagic fever, *Intervirology* **25:**97–102 (1986).
7. AMBROSIO, A. M., FEUILLADE, M. R., AND GAMBOA, G. S., AND MAIZTEGUI, J. I., Prevalence of lymphocytic choriomeningitis virus infection in a human population in Argentina, *Am. J. Trop. Med. Hyg.* **50:**381–386 (1994).
8. ANONYMOUS, Laboratory epidemic traced to hamsters, *J.A.M.A.* **228:**815–816 (1974).
9. ARMSTRONG, C., AND LILLIE, R. D., Experimental lymphocytic choriomeningitis of monkeys and mice produced by a virus encountered in studies of the 1933 St. Louis encephalitis epidemic, *Public Health Rep.* **49:**1019–1027 (1934).
10. ARMSTRONG, C., AND SWEET, L. K., Lymphocytic choriomeningitis: Report of two cases, with recovery of the virus from gray mice (*Mus musculus*) trapped in the two infected households, *Public Health Rep.* **54:**673–684 (1939).
11. ARMSTRONG, D., FORTNER, J. G., ROWE, W. P., AND PARKER, J. C., Meningitis due to lymphocytic choriomeningitis virus endemic in a hamster colony, *J.A.M.A.* **209:**265–267 (1969).
12. ARRIBALZAGA, R. A., Una nueva enfermedad epidemica a germen desconocido, hipertermica, nefrotoxica, leucopenica y enantematica, *Dia Med.* **27:**1204–1210 (1955).
13. ATKINS, J. L., FREEMAN, S., SCHRACK, D. W., JR., DOWNS, W. G., AND CORONA, R. C., Lassa virus infection, *Morbid. Mortal. Week. Rep.* **19:**123 (1970).
14. AUPERIN, D., Construction and evaluation of recombinant virus vaccines for Lassa fever, in: *The Arenaviridae* (M. S. SALVATO, ED.), pp. 259–280, Plenum Press, New York (1993).
15. BORROW, P., AND OLDSTONE, M. B. O., Characterization of lymphocytic choriomeningitis virus-binding protein(s): A candidate cellular receptor for the virus, *J. Virol.* **66:**7270–7281 (1992).
16. BISHOP, D. H. L., AND AUPERIN, D. D., Arenavirus gene structure and organization, in: *Current Topics in Microbiology and Immunology, Arenaviruses, Genes, Proteins, and Expression* (M. B. A. OLDSTONE, ED.), pp. 5–18, Springer-Verlag, New York (1987).
17. BLUMENTHAL, W., KESSLER, R., AND ACKERMAN, R., Uber die Durchseuchung der landlichen Bevolkerung in der Bun Bundes-

republik Deutschland mit dem Virus der Lymphocytaren Chorio-meningitis, *Zentralbl. Bakteriol. Parasirenkd. Infektionskr. Hyg. Abn I Orig.* **213**:36–48 (1970).

18. BOCHSTAHLER, F. E., CARNEY, P. G., BUSHAR, G., Detection of Junin virus by the polymerase chain reaction, *J. Virol. Methods* 231–235 (1992).

19. BOXAXA, M. C., DE GUERRERO, L. B., AND PARODI, A. S., Viremia en enfermos de fiebre hemorragica, Argentina, *Rev. Assoc. Med Argent.* **79**:230–238 (1965).

20. BUCHMEIER, M. J., LEWICKI, H. A., TOMORI, O., Monoclonal antibodies to lymphocytic choriomeningitis virus react with pathogenic arenaviruses, *Nature* **288**:486–487 (1980).

21. BUCHMEIER, M. J., LEWICKI, H. A., TOMORI, O., Monoclonal antibodies to lymphocytic choriomeningitis and Pichinde viruses: Generation, characterization, and cross-reactivity with other arena-viruses, *Virology* **113**:73–83 (1981).

22. BUCHMEIER, M. J., AND PAREKH, B. S., Protein structure and expression among arenaviruses, in: *Current Topics in Microbiology and Immunology, Arenaviruses, Genes, Proteins, and Expression* (M. B. A. OLDSTONE, ED.), pp. 41–58, Springer-Verlag, New York, 1987.

23. BUCHMEIER, M. J., WELSH, R. M., DUTKO, F. J., AND OLDSTONE, M. B., The virology and immunobiology of lymphocytic chorio-meningitis virus infection, *Adv. Immunol.* **30**:275–331 (1980).

24. BUCKLEY, S. M., AND CASALS, J., Lassa fever, a new virus disease of man from West Africa. II. Isolation and characterization of the virus, *Am. J. Trop. Med. Hyg.* **19**:680–691 (1970).

25. CALISHER, C. H., TZIANABOS, T., LORD, R. D., AND COLEMAN, P. H., Tamiami virus, a new member of the Tacaribe group, *Am. J. Trop. Med. Hyg.* **19**:520–526 (1970).

26. CAREY, D. E., KEMP, G. E., WHITE, H. A., PINNEO, L., ADDY, R. F., FOM, A. L. M. D., STROH, G., CASALS, J., AND HENDERSON, B. E., Lassa fever: Epidemiological aspects of the 1970 epidemic, Jos Nigeria, *Trans. R. Soc. Trop. Med. Hyg.* **66**:402–408 (1972).

27. CASALS, J., Serological reactions with arenaviruses, *Medicina* (Buenos Aires) **37**:59–68 (1977).

28. CENTERS FOR DISEASE CONTROL AND NATIONAL INSTITUTES OF HEALTH, *Biosafety in Microbiological and Biomedical Laboratories*, HHS Publication No. (CDC) 93-8395, Washington, DC: US Government Printing Office (1993).

29. CENTERS FOR DISEASE CONTROL, Management of patients with suspected viral hemorrhagic fever, *Morbid. Mortal. Week. Rep.* **S-3**:27 (1988).

30. CHILD, P. L., MACKENZIE, R. B., VALVERDE, L. P., AND JOHNSON, K. M., Bolivian hemorrhagic fever. A pathologic description, *Arch. Pathol.* **83**:434–445 (1967).

31. CHILDS, J. E., GLASS, G. E., KSIAZEK, T. G., ROSSI, C. A., BERRERA ORO, J. G., AND LEDUC, J. W., Human–rodent contact with lymphocytic choriomeningitis and Seoul viruses in an inner-city population, *Am. J. Trop. Med. Hyg.* **44**:117–121 (1991).

32. CLEGG, J. C. S., Current progress towards vaccines for arena-virus-caused diseases, *Vaccine* **10**:89–95 (1992).

33. CLEGG, J. C. S., Molecular phylogeny of the arenaviruses and guide to published sequence data, in: *The Arenaviridae* (M. S. SALVATO, ED.), pp. 175–187, Plenum Press, New York, 1993.

34. COIMBRA, T. L. M., NASSAR, E. S., BURATTINI, M. N., DE SOUZA, L. T. M., FERREIRA, I. B., ROCCO, I. M., DE ROSA, A., VASCON-CELOS, P., PINHEIRO, F. P., LEDUC, J. W., RICO-HESSE, R., GONZALEZ, J., JAHRLING, P. B., AND TESH, R. B., New arenavirus isolated in Brazil, *Lancet* 391–392 (1994).

35. CONNOLLY, B. M., JENSON, A. B., PETERS, C. J., GEYER, S. J.,

BARTH, J. F., AND MCPHERSON, R. A., Pathogenesis of Pichinde virus infection in strain 13 guinea pigs: An immunocytochemical, virologic, and clinical chemistry study, *Am. J. Trop. Med. Hyg.* **49**:10–23 (1993).

36. CUMMINS, D., FISHER-HOCH, S. P., WALSHE, K. J., MACKIE, I. J., MCCORMICK, J. B., BENNETT, D., PEREZ, G., FARRAR, B., AND MACHIN, S. J., A plasma inhibitor of platelet aggregation in patients with Lassa fever, *Br. J. Hematol.* **72**:543–551 (1989).

37. COMISION NACIONAL COORDINADORA PARA ESTUDIO Y LUCHA CONTRA LA FIEBRE HEMORRAGICA ARGENTINA, Secretaria de Estado de Salvd Publica, Buenos Aires, pp. 1–117 (1966).

38. DOUGLAS, R. G., WIEBENGA, N. H., AND COUCH, P. B., Bolivian hemorrhagic fever probably transmitted by personal contact, *Am. J. Epidemiol.* **82**:85–91 (1965).

39. DOWNS, W. G., ANDERSON, C. R., SPENCE, L., AITKEN, T. H. G., AND GREENHALL, A. H., Tacaribe virus, a new agent isolated from *Artibeus* bats and mosquitoes in Trinidad, West Indies, *Am. J. Trop. Med. Hyg.* **12**:640–646 (1963).

40. DYKEWICIZ, C. A., DATO, V. M., FISHER-HOCH, S. P., HOWARTH, M. V., PEREZ-ORONOZ, G. I., OSTROFF, S. M., GARY JR., H., SCHONBERGER, L. B., AND MCCORMICK, J. B., Lymphocytic choriomeningitis outbreak associated with nude mice in a re-search institute, *J.A.M.A.* **267**:1349–1353 (1992).

41. ELLIOT, L. H., MCCORMICK, J. B., AND JOHNSON, K. M., Inac-tivation of Lassa, Marburg, and Ebola viruses by gamma irradia-tion, *J. Clin. Microbiol.* **16**:704–708 (1982).

42. EMMONS, R. W., CHIN, J., NAYFIELD, C. L., WATERMAN, G. E., DIUMARA, N. J., FLEMING, D. S., ZISKIN, L., GOLGFIELD, M., ALTMAN, R., WOODALL, J., DEIBEL, P., AND HINMAN, A. P., Follow-up on hamster-associated LCM infection, *Morbid. Mor-tal. Week. Rep.* **23**:131–132 (1974).

43. ENRIA, D. A., FERNANDEZ, N. J., BRIGGILER, A. M., Importance of neutralising antibodies in treatment of Argentine haemorrhagic fever with immune plasma, *Lancet* **2**:255 (1984).

44. FISHER-HOCH, S. P., Arenavirus pathophysiology, in: *The Arena-viridae* (M. S. SALVATO, ED.), pp. 299–323, Plenum Press, New York, 1993.

45. FORSTER, U., AND WACHENDORFER, G., Inapparent infection of Syrian hamster with the virus of lymphocytic choriomeningitis, in: *Lymphocytic Choriomeningitis Virus and Other Arenaviruses* (F. LEHMANN-GRUBE, ED.), pp. 113–120, Springer-Verlag, New York, 1973.

46. FRAME, J. D., Clinical features of Lassa fever in Liberia, *Rev. Infect. Dis.* **4**:S783–789 (1989).

47. FRAME, J. D., BALDWIN, J. M., JR., GOCKE, D. J., AND TROUP, J. M., Lassa fever, a new virus disease of man from West Africa. I. Clinical description and pathological findings, *Am. J. Trop. Med. Hyg.* **23**:1131–1139 (1974).

48. FRAME, J. D., JAHRLING, P. B., YALLEY-OGUNRO, J. E., AND MONSON, J. H., Endemic Lassa fever in Liberia. II. Serological and virological findings in hospital patients, *Trans. R. Soc. Trop. Med. Hyg.* **78**:656–660 (1984).

49. FRAME, J. D., YALLEY-OGUNRO, J. E., AND HANSON, A. P., Endemic Lassa fever in Liberia. V. Distribution of Lassa virus activity in Liberia: Hospital staff surveys, *Trans. R. Soc. Trop. Med. Hyg.* **78**:761–763 (1984).

50. FRASER, D. W., CAMPBELL, C. C., MONATH, T. P., GOFF, P. A., AND GREGG, M. B., Lassa fever in the Eastern Province of Sierra Leone, 1970–1972. I. Epidemiologic studies, *Am. J. Trop. Med. Hyg.* **23**:1131–1139 (1974).

51. FREI, K., LEIST, T. P., MEAGER, A., GALLO, P., LEPPERT, D.,

ZINKERNAGEL, R. M., AND FONTANA, A., Production of a β-cell stimulatory factor-2 and interferon gamma in the central nervous system during viral meningitis and encephalitis. Evaluation of a murine model infection and in patients, *J. Exp. Med.* **168**:449–453 (1988).

52. GALLARDO, F., Fiebre hemorragica Argentina: Hallazgos anatomo-patologicos en diez necropsias, *Medicine* **30**(Suppl. 1):77–84 (1970).

53. GARCIA-FRANCO, S., AMBROSIO, A. M., FEUILLADE, M. R., AND MAIZTEGUI, J. I., Evaluation of an enzyme-linked immunosorbent assay for quantitation of antibodies to Junin virus in human sera, *J. Virol. Methods* **19**:299–306 (1988).

54. GONZALEZ, J. P., McCORMICK, J. B., SALUZZO, J. F., HERVE, J. P., JOHNSON, K. M., AND GEORDES, A. J., An arenavirus isolated from wild-caught rodents (*Praomys* species) in the Central African Republic, *Intervirology* **19**:105–112 (1983).

55. GREEN, C. A., KEOGH, H., GORDON, D. H., PINTO, M., AND HARTWIG, E. K., The distribution, identification and naming of the *Mastomys natalensis* species complex in southern Africa (Rodentia: Muridae), *J. Zool. (Lond.)* **192**:17–23 (1980).

56. GRELA, M. E., GARCIAF, C. A., ZANNOLI, V. H., AND BARRERA ORO, J. G., Serologia de la fiebre hemorragica Argentina, II. Comparacion de la prueba indirecta de anticuerpos fluorescentes con la fijacion de complemento, *Acta Bio-Clin. Lationoam.* **9**:141–146 (1975).

57. GSCHWENDER, H. H., AND LEHMANN-GRUBE, F., Antigenic properties of the LCM virus: Virion and complement-fixing antigen, in: *Lymphocytic Choriomeningitis and Other Arenaviruses* (F. LEHMANN-GRUBE, ED.), pp. 26–35, Springer-Verlag, New York, 1973.

58. HANSON, C. V., RIGGS, J. L., AND LENNETTE, E. H., Photochemical inactivation of DNA and RNA viruses by psoralen derivatives. *J. Gen. Virol.* **40**:345–358 (1978).

59. HORTON, J., HOTCHIN, J. E., OLSON, K. B., AND WEBB, H. E., The effects of MP virus infection in lymphoma, *Cancer Res.* **31**:1066–1068 (1971).

60. INSTITUTE OF MEDICINE, *Emerging Infections. Microbial Threats to Health in the United States* (J. LEDERBERG, R. E. SHOPE, AND S. C. OALS, JR., EDS.), National Academy Press, Washington, DC, 1992.

61. INTERNATIONAL COMMITTEE ON TAXONOMY OF VIRUSES, Arenaviruses, *Intervirology* **17**:119–122 (1982).

62. IVANOV, A. P., BASHKIRTSEZ, V. N., AND TKACHENKO, E. A., Enzyme-linked immunosorbent assay for detection of arenaviruses, *Arch. Virol.* **67**:71–74 (1981).

63. JAHRLING, P. B., NICKLASSON, B. S., AND McCORMICK, J. B., Early diagnosis of human Lassa fever by ELISA detection of antigen and antibody, *Lancet* **1**:250–252 (1985).

64. JAHRLING, P. B., AND PETERS, C. J., Lymphocytic choriomeningitis virus: A neglected pathogen of man, *Arch. Pathol. Lab. Med.* **116**:486–488 (1992).

65. JOHNSON, K. M., Arenaviruses, in: *Virology* (B. N. FIELDS, ED.), pp. 1033–1053, Raven Press, New York, 1985.

66. JOHNSON, K. M., Fiebres hemorragicas de America del Sur, *Medicina* **30**(Suppl. 1):99–110 (1970).

67. JOHNSON, K. M., ELLIOTT, L. H., AND HEYMANN, D. L., Preparation of polyvalent viral immunofluorescent intracellular antigens and use in human serosurveys, *J. Clin. Microbiol.* **14**:527–529 (1981).

68. JOHNSON, K. M., HALSTEAD, S. B., AND COHEN, S. N., Hemorrhagic fevers of Southeast Asia and South America: A comparative appraisal, *Prog. Med. Virol.* **9**:105–158 (1967).

69. JOHNSON, K. M., KUNS, M. L., MACKENZIE, R. B., WEBB, P. A., AND YUNKER, C. E., Isolation of Machupo virus from wild rodent *Calomys callosus*, *Am. J. Trop. Med. Hyg.* **15**:103–106 (1966).

70. JOHNSON, K. M., McCORMICK, J. B., WEBB, P. A., SMITH, E., ELLIOTT, L., AND KING, I. J., Lassa fever: Clinical virology of Lassa fever in hospitalized patients, *J. Infect. Dis.* **155**:456–464 (1987).

71. JOHNSON, K. M., TAYLOR, P., ELLIOTT, L. H., AND TOMORI, O., Recovery of a Lassa-related arenavirus in Zimbabwe, *Am. J. Trop. Med. Hyg.* **30**:1291–1293 (1981).

72. JOHNSON, K. M., AND WEBB, P. A., Rodent transmitted hemorrhagic fevers, in: *Diseases Transmitted from Animals to Man,* 6th ed., (W. T. HUBBERT, W. F. McCULLOCH, AND T. R. SCHNURRENBERGER, EDS.), pp. 911–918, Charles C. Thomas, Springfield, IL, 1975.

73. JOHNSON, K. M., WEBB, P. A., AND JUSTINES, G., Biology of Tacaribe-complex viruses, in: *Lymphocytic Choriomeningitis Virus and Other Arenaviruses* (F. LEHMANN-GRUBE, ED.), pp. 241–258, Springer-Verlag, New York, 1973.

74. JOHNSON, K. M., WIEBENGA, N. H., MACKENZIE, R. B., KUNS, M. L., TAURASO, N. M., SHELOKOV, A., WEBB, P. A., JUSTINES, G., AND BEYE, H. K., Virus isolations from human cases of hemorrhagic fever in Bolivia, *Proc. Soc. Exp. Biol. Med.* **118**:113–118 (1965).

75. JUSTIENS, G., AND JOHNSON, K. M., Immune tolerance in *Calomys callosus* infected with Machupo virus, *Nature* **222**:1090–1091 (1969).

76. KEENLYSIDE, R. A., McCORMICK, J. B., WEBB, P. A., SMITH, E., ELLIOTT, L., AND JOHNSON, K. M., Case-control study of *Mastomys natalensis* and *womans* in Lassa virus-infected households in Sierra Leone, *Am. J. Trop. Med. Hyg.* **32**:829–837 (1983).

77. KILEY, M. P., LANGE, J. V., AND JOHNSON, K. M., Protection of rhesus monkeys from Lassa virus by immunisation with closely related arenavirus, *Lancet* **2** (1979).

78. KSIAZEK, T. G., ROLLIN, P. E., JAHRLING, P. B., Enzyme immunosorbent assay for Ebola virus antigens in tissue of infected primates, *J. Clin. Microbiol.* **30**:947–950 (1992).

79. KUNS, M. L., Epidemiology of Machupo virus infection. II. Ecological and control studies of hemorrhagic fever, *Am. J. Trop. Med. Hyg.* **14**:813–816 (1965).

80. LEHMANN-GRUBE, F., *Lymphacytic Choriomeningitis Virus*, Springer-Verlag, New York, 1971.

81. LEHMANN-GRUBE, F., ASSMANN, V., LOLIGUER, C., MOSKOPHIDIS, D., AND LOHLEB, J., Mechanism of recovery from acute virus infection. I. Role of T lymphocytes in the clearance of lymphocytes choriomeningitis virus from spleens of mice, *J. Immunol.* **134**:608–615 (1985).

82. LEIFER, E., GOCKE, D. J., AND BOURNE, H., Lassa fever, a new virus disease of man from West Africa. II. Report of a laboratory-acquired infection treated with plasma from a person recently recovered from the disease, *Am. J. Trop. Med. Hyg.* **19**:677–679 (1970).

83. LEVIS, S. C., SAAVEDRA, M. C., CECCOLI, C., Endogenous interferon in Argentine hemorrhagic fever, *J. Infect. Dis.* **149**:428–433 (1984).

84. LEWIS, A. M., ROWE, W. P., TURNER, H. C., AND HUEBNER, R. J., Lymphocytic-choriomeningitis virus in hamster tumor: Spread to hamsters and humans, *Science* **150**:363–364 (1965).

85. LEWIS, V. J., WALKER, P. D., AND THACKER, W., Comparison of three tests for the serological diagnosis of lymphocytic choriomeningitis virus infection, *J. Clin. Microbiol.* **2**:193–197 (1975).

86. LOZANA, M. E., GHIRINGHELLI, P. D., ROMANOWSKI, V., AND GRAU, O., A simple nucleic acid amplification assay for the rapid

detection of Junin virus in whole blood samples, *Virus Res.* **27**:37–53 (1993).

87. LUCIA, H. L., COPPENHAVER, D. H., HARRISON, R. L., AND BARON, S., The effect of an arenavirus infection on liver morphology and function, *Am. J. Trop. Med. Hyg.* **43**:93–98 (1990).

88. LUKASHEVICH, I. S., Genetic reassortants between African arenaviruses, *Virology* **188**:600–605 (1992).

89. LUNKENHEIMER, K., HUFERT, F. T., AND SCHMITZ, H., Detection of Lassa virus RNA in specimens from patients with Lassa fever by using the polymerase chain reaction, *J. Clin. Microbiol.* **28**:2689–2692 (1990).

90. MACKENZIE, R. B., Epidemiology of Machupo virus infection. I. Pattern of human infection, San Joaquin, Bolivia, 1962–64, *Am. J. Trop. Med. Hyg.* **14**:808–813 (1965).

91. MACKENZIE, R. B., WEBB, P. A., AND JOHNSON, K. M., Detection of complement-fixing antibody after Bolivian hemorrhagic fever, employing Machupo, Junin and Tacaribe virus antigens, *Am. J. Trop. Med. Hyg.* **14**:1079–1084 (1965).

92. MAIZTEGUI, J. I., Argentinian hemorrhagic fever (AHF), in: *Proceedings, Ninth International Congress on Tropical Medicine and Malaria*, Vol. I, p. 31, Athens, 1973.

93. MAIZTRGUI, J. I., AGUIRRE, G. M., SABATTINI, M. S., AND ORO, J. G. B., Actividad de dos "arenavirus" en seres humanos y roedores en un mismo lugar de la zona endemica de fiebre hemorragica Argentina, *Medicina* **1**:509–510 (1971).

94. MAIZTEGUI, J. I., ESTRIBOU, J. P., SABATTINI, M. S., AND ORO, J. G. B., Estudios tendientes a dilucidar el papel del *Mus musculus* en la epidemiologia de la fibre hemorragica Argentina (FHA), *Rev. Soc. Argent. Microbiol.* **2**:186–187 (1970).

95. MAIZTEGUI, J. I., FERNANDLZ, N. J., AND DE DAMILANO, A. J., Efficacy of immune plasma in treatment of Argentine hemorrhagic fever and association between treatment and a late neurological syndrome, *Lancet* **2**:1216–1217 (1979).

96. MAIZTEGUI, J. I., SABITTINI, M. S., AND ORO, J. G. B., Actividad del virus de la coriomeningitis linfocitica (LCM) en el area endemica de fiebre hemorragica Argentina (FHA), *Medicina* **32**:131–137 (1971).

97. MANNWEILER, K., AND LEHMANN-GRUBE, F., Electron microscopy of LCM virus-infected L cells, in: *Lymphocytic Choriomeningitis Virus and Other Arenaviruses* (F. LEHMANN-GRUBE, ED.), pp. 37–48, Springer-Verlag, New York, 1973.

98. MCCORMICK, J. B., AND JOHNSON, K. M., Lassa fever: Historical review and contemporary investigation, in: *Ebola Virus Haemorrhagic Fever* (S. R. PATTYN, ED.), pp. 279–285, Elsevier/North-Holland, Amsterdam, 1978.

99. MCCORMICK, J. B., KING, I. J., WEBB, P. A., JOHNSON, K. M., O'SULLIVAN, R., SMITH, E., AND TRIFPEL, S., A case-control study of the clinical diagnosis and course of Lassa fever, *J. Infect. Dis.* **155**:445–455 (1987).

100. MCCORMICK, J. B., KING, I. J., WEBB, P. A., SCRIBNER, C. L., CRAVEN, R. B., JOHNSON, K. M., ELLIOTT, L. H., AND BELMONT-WILLIAMS, R., Lassa fever. Effective therapy with ribavirin, *N. Engl. J. Med.* **314**:20–26 (1986).

101. MCCORMICK, J. B., WALKER, D. H., KING, I. J., WEBB, P. A., ELLIOTT, L. H., WHITFIELD, S. G., AND JOHNSON, K. M., Lassa virus hepatitis: A study of fatal Lassa fever in humans, *Am. J. Trop. Med. Hyg.* **35**:401–407 (1986).

102. MCCORMICK, J. B., WEBB, P. A., KREBS, J. W., JOHNSON, K. M., AND SMITH, E., A prospective study of epidemiology and ecology of Lassa fever, *J. Infect. Dis.* **155**:437–444 (1987).

103. MERCADO, R., Rodent control programmes in areas affected by Bolivian hemorrhagic fever, *Bull. WHO* **52**:691–695 (1975).

104. METTLER, N. E., CASALS, J., AND SHOFE, R. E., Study of antigenic relationships between Junin virus, the etiological agent of Argentinian hemorrhagic fever, and other arthropodborne viruses, *Am. J. Trop. Med. Hyg.* **12**:647–652 (1963).

105. METTLER, N. E., Estudio realizado con los sueros de la epidemia de fiebre hemorragica Argentina (1963) que no presentaron conversion serologica para virus Junin, *Medicina* **26**:161–169 (1966).

106. METTLER, N. E., *Argentine Hemorrhagic Fever: Current Knowledge*, pp. 1–55, Pan American Health Organization, Scientific Publications No. 183, Washington, DC, 1969.

107. METTLER, N. E., AND CASALS, J., Isolation of St. Louis encephalitis virus from man in Argentina, *Acta Virol.* **15**:148–154 (1971).

108. MEYER, H. M., JOHNSON, R. T., CRAWFORD, I. P., DASCOMB, H. E., AND ROGERS, N. G., Central nervous system syndromes of "viral" etiology: A study of 713 cases, *Am. J. Med.* **29**:334–347 (1960).

109. MONATH, T. P., MAHER, M., CASALS, J., KISSLING, R. E., AND CACCIAPUOTI, A., Lassa fever in the Eastern Province of Sierra Leone, 1970–1972 II. Clinical observations and virological studies on selected hospital cases, *Am. J. Trop. Med. Hyg.* **23**:1140–1149 (1974).

110. MONATH, T. P., MERTENS, P. E., PATTON, R., MOSER, C. R., BAUM, J. J., PINNEO, L., GARY, G. W., AND KISSLING, R. E., A hospital epidemic of Lassa fever in Zorzor, Liberia, March–April 1972, *Am. J. Trop. Med. Hyg.* **22**:773–779 (1973).

111. MONATH, T. P., NEWHOUSE, V. F., KEMP, G. E., SETZER, H. W., AND CACCIAPUTTI, A., Lassa virus isolation from *Mastomys natalensis* rodents during an epidemic in Sierra Leone, *Science* **185**:263–265 (1974).

112. MONSON, M. H., COLE, A. K., AND FRAME, J. D., Pediatric Lassa fever: A review of 33 Liberian cases, *Am. J. Trop. Med. Hyg.* **36**:408–415 (1987).

113. MURPHY, F. A., WEBB, P. A., JOHNSON, K. M., AND WHITFIELD, S. G., Morphological comparison of Machupo with lymphocytic choriomeningitis virus: Basis for a new taxonomic group, *J. Virol.* **4**:535–541 (1969).

114. MURPHY, F. A., WEBB, P. A., JOHNSON, K. M., WHITFIELD, S. G., AND CHAPPELL, W. A., Arenoviruses in Vero cells: Ultrastructural studies, *J. Virol.* **6**:507–518 (1970).

115. MURPHY, E. A., AND WHITFIELD, S. G., Morphology and morphogenesis of arenaviruses, *Bull. WHO* **55**:409–419 (1975).

116. MURPHY, F. A., WHITFIELD, S. G., WEBB, P. A., AND JOHNSON, K. M., Ultrastructural studies of arenaviruses, in: *Lymphocytic Choriomeningitis Virus and Other Arenaviruses* (F. LEHMANN-GRUBE, ED.), pp. 273–285, Springer-Verlag, New York, 1973.

117. NIKLASSON, B. S., JAHRLING, P. B., AND PETERS, C. J., Detection of Lassa virus antigens and Lassa virus-specific immunoglobulins G and M by enzyme-linked immunosorbent assay, *J. Clin. Microbiol.* **20**:239–244 (1984).

118. OLDSTONE, M. B. A., Immunotherapy for virus infection, *Curr. Top. Microbiol. Immunol.* **134**:212–229 (1988).

119. PARODI, A. S., GREENWAY, D. J., RUGIERO, H. R., RIVERO, S., FRIGERIO, M., BARRERA, J. M., METTLER, N., GARZON, F., BOXACA, M., GUERRERO, L., AND NORA, N., Sobre la etiologia del brote epidemico de Junin, *Dia Med.* **30**:2300–2302 (1958).

120. PARODI, A. S., RUGIERO, H. R., GREENWAY, D. L., METTLER, N. E., MARTINEZ, A., BOXACA, M., AND BARRERA, J. M., Aislamiento del virus Junin (FHE) de los acaros de la zona epidemica (*Echinolaelaps echidninus*, Berlese), *Prensa Med. Argent.* **46**:2242–2244 (1959).

121. PETERS, C. J., Arenaviruses, in: *The Textbook of Human Virology* (R. B. BELSHE, ED.), pp. 541–570, Mosby Year Book, St. Louis, MO, 1990.

122. PETERS, C. J., WEBB, P. A., AND JOHNSON, K. M., Measurement of antibodies to Machupo virus by the indirect fluorescent technique, *Proc. Soc. Exp. Biol. Med.* **142:**526–531 (1973).

123. PETTER, F., Les rata a mamelles multiples d'Afrique occidentale et centrale: *Mastomys erythroleucus* (Temminck, 1835) et *M. huberti* (Wroughton, 1908), *Mammalia* **41:**441–443 (1977).

124. PINHEIRO, F. P., SHOPE, R. E., DE ANDRADE, A. H. P., BENSABETH, G., CACIOS, G. V., AND CASALS, J., Amapari, a new virus of the Tacaribe group from rodents and mites of Amapa Territory, Brazil, *Proc. Soc. Exp. Biol. Med.* **122:**532–535 (1966).

125. PINHEIRO, F. P., WOODALL, J. P., DA ROSA A. P. A. T., Studies of arenaviruses in Brazil, *Medicina (Buenos Aires)* **37:**175–181 (1977).

126. PINTOS, I. M., Epidemiologia del "Mal de los Rastrojos," *Sep. An. Com. Invest. Cient. Prov. Buenos Aires* **3:**9–102 (1962).

127. PIROSKY, I. M., ZUCCARINI, J., MOLINELLI, E. A., DEPIETRO, A., ORO, J. G. B., MARTINI, P., AND COPELLO, A. R., *Virosis Hemorragica del Noroeste Bonacrense: Endemo-epidemica, Febril, Enantematica y Leucopenica*, Comision Nacional ad hoc para Estudiar el Brote de 1958, Tallares Graficos del Ministerio de Asistencia Social y Salud Publica, Buenos Aires, 1959.

128. PRICE, M. E., FISHER-HOCH, S. P., CRAVEN, R. B., A prospective study of maternal and fetal outcome in acute Lassa virus infection during pregnancy, *Br. Med. J.* **297:**584–587 (1988).

129. RIVERS, T. M., AND SCOTT, T. F. M., Meningitis in man caused by a filterable virus, *Science* **81:**439–440 (1935).

130. RIVERS, T. M., AND SCOTT, T. F. M., Meningitis in man caused by a filterable virus. II. Identification of the etiological agent, *J. Exp. Med.* **63:**415–432 (1936).

131. RIVIERE, Y., SOUTHERN, P. J., AHMED, R., AND OLDSTONE, M. B. A., Biology of cloned cytotoxic T lymphocytes specific for lymphocytic choriomeningitis virus, V. Recognition is restricted to gene products encoded by the viral sRNA segment, *J. Immunol.* **136:**304–307 (1986).

132. ROWE, W. P., MURPHY, F. A., BERGOLD, G. H., CASALS, J., HOTCHIN, J., JOHNSON, K. M., LEHMANN-GRUBE, F., MIMS, C. A., TRAUB, E., AND WEBB, P. A., Arenoviruses: Proposed name for a newly defined virus group, *J. Virol.* **5:**651–652 (1970).

133. ROWE, W. P., PUGH, W. E., WEBB, P. A., AND PETERS, C. J., Serological relationship of the Tacaribe complex of viruses to lymphocytic choriomeningitis virus, *J. Virol.* **5:**289–292 (1970).

134. RUGIERO, H. R., PARODI, A. S., GOTTA, H., BOXACA, M., OLIVARI, A. J., AND GONZALEZ, E., Fiebre hemorragica epidemica: Infeccion de laboratorio y passage interhummano, *Rev. Asov. Med. Argent.* **76:**413–417 (1962).

135. RUGIERO, H. R., PARODI, A. S., RUGGIERO, H. G., AND MOLTONI, L. R., Fiebre hemorragica Argentina, I. Periodo de encubacion e invasion, *Rev. Assoc. Med. Argent.* **78:**221–226 (1964).

136. SABATTINI, M. S., AND MAIZTEGUI, J. I., Fiebre hemorragicia Argentina, *Medicina* **30**(Suppl. 1):111–128 (1970).

137. SABATTINI, M. S., ORO, J. G. B., MAIZTEGUI, J. I., FERNANDEZ, D., COSTIGIANI, M. S., AND DIAZ, G. E., Aislamiento de un "arenovirus" relacionado con el de coriomeningitis linfocitica (LCM) a partir deun *Mus musculus* capturado en zona endemica de fiebre hemorragica Argentina (FHA), *Rev. Assoc. Argent. Microbiol.* **2:**182–184 (1970).

138. SALAS, R., DE MANZIONE, N., TESH, R. B., RICO-HESSE, R., SHOPE, R. E., BETANCOURT, A., GODOY, O., BRUZUAL, R., PACHECO, M. E., RAMOS, B., TAIBO, M. E., TAMAYO, J. G., JAIMES, E., VASQUEZ, C., ARAOZ, F., AND QUERALES, J., Venezuelan haemorrhagic fever, *Lancet* **2:**1033–1036 (1991).

139. SANCHEZ, A., PIFAT, D. Y., KENYON, R. H., PETERS, C. J., McCORMICK, J. B., AND KILEY, M. P., Junin virus monoclonal antibodies: Characterization and cross-reactivity with other arenavirues, *J. Gen. Virol.* **70:**1125–1132 (1989).

140. SARON, M. E., SHIDANI, B., GUILLON, J. C., AND TRUFFABACHI, P., Beneficial effect of cyclosporin A on the lymphocytic choriomeningitis virus infection in mice, *Eur. J. Immunol.* **14:**1064–1066 (1984).

141. SCHEIDT, W., ACKERMAN, R., AND FELGENHAUER, K., Lymphocytare Choriomeningitis unter dem Bild der Encephalitis lethargica, *Dtsch. Med. Wochenschr.* **93:**940–943 (1968).

142. SCHWARZ, E. R., MANDO, O. G., MAIZTEGUI, J. I., AND VILCHES, A. M., Sintomas y signos iniciales de mayor valor diagnostico en la fiebre hemorragica Argentina, *Medicina* **39**(Suppl. 1):8–14 (1970).

143. SCOTT, T. F. M., AND RIVERS, T. M. Meningitis in man caused by a filterable virus I. Two cases and the method of obtaining a virus from their spinal fluids, *J. Exp. Med.* **63:**397–414 (1936).

144. SHEINBERGAS, M. M., Hydrocephalus due to prenatal infection with the lymphocytic choriomeningitis virus, *Infection* **4:**1–7 (1976).

145. SMADEL, J. E., GREEN, R. H., PALTAUF, R. M., AND GONZALES, T. A., Lymphocytic choriomeningitis: Two human fatalities following an unusual febrile illness, *Proc. Soc. Exp. Biol. Med.* **49:**683–686 (1942).

146. SPEIR, R. W., WOOD, O., LIEBHABER, H., AND BUCKLEY, S. M., Lassa fever, a new virus disease of man from West Africa. IV. Electron microscopy of Vero cell cultures infected with Lassa virus, *Am. J. Trop. Med. Hyg.* **19:**692–694 (1970).

147. STEPHEN, E. L., JONES, D. E., PETERS, C. J., Ribavirin treatment of Toga-, Arena- and Bunyavirus infections in subhuman primates and other laboratory animal species, in: *Ribavirin: A Broad Spectrum Antiviral Agent* (R. A. SMITH AND W. KIRKPATRICK, EDS.), pp. 169–183, Academic Press, New York, 1980.

148. STEPHENSON, C. B., JACOB, J. R., MONTALI, T. J., Isolation of an arenavirus from a marmasoset with Callitrichid hepatitis and its serologic association with disease, *J. Virol.* **65:**3995–4000 (1991).

149. SWANEPOEL, R., LEMAN, P. A., SHEPHERD, A. J., SHEPHERD, S. P., KILEY, M. P., AND McCORMICK, J. B., Identification of Ippy as a Lassa-fever-related virus, *Lancet* **1:**639 (1985).

150. TESH, R. B., JARHLING, P. B., SALAS, R., AND SHOPE, R. E., Description of Guanarito virus (Arenavirid: arenavirus), the etiologic agent of Venezuelan hemorrhagic fever, *Am. J. Trop. Med. Hyg.* **50:**452–459 (1994).

151. TESH, R. B., WILSON, M. L., SALAS, R., DE MANZIONE, N. M. C., TOVAR, D., KSIAZEK, T. G., AND PETERS, C. J., Field studies on the epidemiology of Venezuelan hemorrhagic fever, *Am. J. Trop. Med. Hyg.* **49**(2):227–235 (1993).

152. TRAPIDO, H., AND SANMARTIN, C., Pichinde virus, a new virus of the Tacaribe group from Colombia, *Am. J. Trop. Med. Hyg.* **20:**631–641 (1971).

153. TRAPPIER, S. G., CONATY, A. L., FARRAR, B. B., AUPERIN, D. D., McCORMICK, J. B., AND FISHER-HOCH, S. P., Evaluation of the polymerase chain reaction for diagnosis of Lassa virus infection, *Am. J. Trop. Med. Hyg.* **49**(2):214–221 (1993).

154. TRAUB, E., A filterable virus recovered from white mice, *Science* **81:**298–299 (1935).

155. TRAUB, E., Epidemiology of lymphocytic choriomeningitis in a mouse stock observed for four years, *J. Exp. Med.* **69:**801–817 (1939).

156. VAINRUB, B., AND SALAS, R., Latin American hemorrhagic fever, *Dis. Latin Am.* **8**(1):47–59 (1994).

157. VAN DER GROEN, G., AND ELLIOT, L. H., Use of betapropionolac-

tone inactivated Ebola, Marburg and Lassa intracellular antigens in immunofluorescent antibody assay, *Ann. Soc. Belge Med. Trop.* **62:**49–54 (1982).

159. WALKER, C. M., PAETKAN, V., RAWLS, W. E., AND ROSENTHAL, K. L., Abrogation of anti-Pichinde virus cytotoxic T cell memory by cyclophosphamide and restoration by coinfection or interleukin Z, *J. Immunol.* **135:**1401–1407 (1985).

160. WALKER, D. H., McCORMICK, J. B., JOHNSON, K. M., WEBB, P. A., KOMBA-KONO, G., ELLIOTT, L. H., AND GARDNER, J. J., Pathologic and virologic study of fatal Lassa fever in man, *Am. J. Pathol.* **107:**349–356 (1982).

161. WALKER, D. H., WULFF, H., LANGE, J. V., AND MURPHY, F. A., Comparative pathology of Lassa virus infection in monkeys, guinea-pigs, and *Mastomys natalensis, Bull. WHO* **52:**523–534 (1975).

162. WEBB, P. A., JOHNSON, K. M., HIBBS, J. G., AND KUNS, M. L., Parana, a new Tacaribe complex virus from Paraguay, *Arch. Ges. Virusforsch.* **32:**319–388 (1970).

163. WEBER, E. L., AND BUCHMEIER, M. J., Fine mapping of a peptide sequence containing an antigenic site conserved among arenaviruses, *Virology* **164:**30–38 (1988).

164. WHITE, H. A., Lassa fever: A study of 23 hospital cases, *Trans. R. Soc. Trop. Med. Hyg.* **66:**390–398 (1972).

165. WINN, W. C., JR., MONATH, T. P., MURPHY, F. A., AND WHITFIELD, S. G., Lassa virus hepatitis: Observations on a fatal case from the 1972 Sierra Leone epidemic, *Arch. Pathol.* **99:**599–604 (1975).

166. WINN, W. C., JR., AND WALKER, D. H., The pathology of human Lassa fever, *Bull. WHO* **52:**535–545 (1975).

167. WOOLEY, I. G., ARMSTRONG, C., AND ONSTOTT, R. H., The occurrence in the sera of man and monkeys of protective antibodies against the virus of lymphocytic choriomeningitis as determined by the serum-virus protection test in mice, *Public Health Rep.* **52:**1105–1114 (1937).

168. WORLD HEALTH ORGANIZATION, Vaccination against Argentine haemorrhagic fever, *Weekly Epidemiol. Rec.* **68:**233–236 (1993).

169. WULFF, H., FABIYI, A., AND MONATH, T. P., Recent isolations of Lassa fever from Nigerian rodents, *Bull. WHO* **52:**609–613 (1975).

170. WULFF, H., AND JOHNSON, K. M., Immunoglobulin M and G responses measured by immunofluorescence in patients with Lassa or Marburg virus infections, *Bull WHO* **57:**631–635 (1979).

171. WULFF, H., AND LANGE, J. V., Indirect immunofluorescence for the diagnosis of Lassa fever infection, *Bull. WHO* **52:**429–436 (1975).

172. WULFF, H., LANG, J. V., AND WEBB, P. A., Interrelationships among arenaviruses measured by indirect immunofluorescence, *Intervirology* **9:**344–350 (1978).

173. WULFF, H., McINTOSH, B. M., HAMMER, D. B., AND JOHNSON, K. M., Isolation of an arenavirus closely related to Lassa virus from *Mastomys natalensis* in south-east Africa, *Bull WHO* **55:**441–444 (1977).

174. ZINKERNAGEL, R. M., PFAU, C. J., HENGARTNER, H., AND ALTHAGE, A., Susceptibility to murine lymphocytic choriomeningitis maps to class I MHC genes—a model for MHC/disease associations, *Nature* **316:**814–817 (1985).

13. Suggested Reading

BUCHMEIER, M. J., AND PAREKH, B. S., Protein structure and expression among arenaviruses, in: *Current Topics in Microbiology and Immunology, Arenaviruses, Genes, Proteins, and Expression* (M. B. A. OLDSTONE, ED.), Vol. 133, pp. 41–58, Springer-Verlag, New York, 1987.

International Symposium on Arenaviral Infections of Public Health Importance, *Bull. WHO* **52:**381–766 (1975).

OLDSTONE, M. B. A. (ED.), *Arenaviruses: Biology and Immunotherapy, Current Topics in Microbiology and Immunology*, Vol. 134, Springer-Verlag, New York (1988).

PETERS, C. J., Arenaviruses, in: *The Textbook of Human Virology* (R. B. BELSHE, ED.), pp. 541–570, Mosby Year Book, St. Louis, MO, 1990.

SALVADO, M. S. (ED.), *The Arenaviridae*, Plenum Press, New York, 1993.

CHAPTER 8

Coronaviruses

Arnold S. Monto

1. Introduction

Coronaviridae are a monogeneric family of RNA-containing agents that have been associated etiologically with respiratory illnesses in man and with a number of other diseases in laboratory and domestic animals. They also have been associated with diarrheal disease and other conditions in humans, although for differing reasons these relationships cannot be considered etiologic. The name for the family was adopted to describe the characteristic fringe of crownlike projections seen around the viruses by electron microscopy; these spikes are petal shaped rather than sharp or pointed as is the case with the myxoviruses. Like the myxoviruses, the coronaviruses contain essential lipids and are 80–160 nm in diameter.[28] Unlike them, the coronaviruses are positive-stranded.[66,79] Whereas the animal strains are readily isolated in several different systems, recovery of the human strains has posed major problems, partially related to species specificity.[120] A number of these strains have been isolated only in organ culture of the human respiratory tract. This factor has made it difficult to determine the relationship among isolates and has complicated efforts to understand the role of these viruses in human illness. Therefore, much of the information on the epidemiology of the agents has come from serological studies.

2. Historical Background

The first human coronaviruses were isolated by different techniques in the United States and Britain at approximately the same time. The British Medical Research

Council's Common Cold Research Unit had been studying fluids collected from persons with natural respiratory infections by standard cell culture isolation methods and by inoculating them into human volunteers. Rhinoviruses or other cytopathogenic agents could be recovered from a portion of the fluids.[60] There was an additional substantial portion from which no agents could be isolated but that could still cause colds in the volunteers. Organ cultures of human embryonic trachea or nasal epithelium were then used in a effort to detect the recalcitrant viruses present in the fluids. A specimen, B814, that had been collected in 1960 from a boy with a common cold had not yielded a virus on inoculation into cell culture. After the specimen had been passaged serially three times in human tracheal organ culture, it could still cause colds on inoculation into volunteers, which indicated that replication had taken place.[116]

In Chicago during the winter of 1962, five agents were isolated in primary human kidney cell cultures from specimens collected from medical students with common colds. The viruses were ultimately adapted to WI-38 cultures and exhibited a type of cytopathic effect (CPE) not previously seen. A prototype strain, 229E, was selected for characterization and was found to be RNA-containing, ether-labile, and 89 nm in diameter but distinct serologically from any known myxo- or paramyxoviruses. Sera collected from the five medical students all exhibited a fourfold rise in neutralization antibody titer against 229E.[37]

It became clear that these "novel" viruses were of more than passing significance when organ culture methods were added to standard cell culture techniques in a study of acute respiratory infections of adults conducted at the National Institutes of Health (NIH). Six viruses were found that grew in organ but not cell culture and were ether-labile; on electron microscopy, the agents were shown to resemble avian infectious bronchitis virus (IBV) in structure.[74] The B814 and 229E strains were soon also

Arnold S. Monto • Departments of Epidemiology and International Health, School of Public Health, University of Michigan, Ann Arbor, Michigan 48109.

demonstrated to have a similar structure on electron microscopy and to develop in infected cells by budding into cytoplasmic vesicles.[1,4,36] As a result of the similarity of the human agents to IBV and also to mouse hepatitis virus (MHV), they were collectively considered to represent a group of vertebrate viruses distinct from the myxoviruses antigenically and structurally.[6] The name *coronavirus* was adopted for the group to describe the fringe of projections seen around them on electron microscopy.[28]

Except for 229E, none of the human coronaviruses had been successfully propagated in a system other than organ culture. McIntosh *et al.*[72] reported successful adaptation of two of the NIH isolates, OC (organ culture) 38 and OC43, to the brains of suckling mice. These strains were shown to be essentially identical antigenically but quite distinct from MHV. Only OC38 and OC43 could be so adapted; the other four OC strains resisted such attempts. The IBV was known to exhibit hemagglutination under certain conditions, but no such phenomenon had been demonstrated for the human strains until OC38 and OC43 were adapted to mice. Kaye and Dowdle[53] found that the infected brain preparations would directly and specifically agglutinate red cells obtained from chickens, rats, and mice. This technique greatly expanded the ability to do epidemiologic studies, since it was simple and reproducible.

Subsequent developments included adaptation of OC38 and OC43 to growth in cell monolayers; either mouse brain or organ culture material could be used as sources of virus.[14] Not only was CPE available for reading of neutralization tests, but also the OC38 and OC43 viruses were found to hemadsorb red cells of rats and mice, making available a more precise means of evaluating endpoints in tests involving these organ-culture-derived strains.[51] The other OC strains that could not be adapted to mouse brain resisted adaptation to cell culture. Finally, immune electron microscopy was added to the methods available for identifying the presence of coronaviruses. This highly sensitive technique should improve the ability to detect virus, but it is obviously unsuitable for use in all but the most specialized studies.[52] More recent studies have begun to explain the species specificity of the viruses in terms of cellular receptors.[120]

3. Methodology

3.1. Sources of Mortality Data

Coronaviruses that infect domestic and laboratory animals produce illnesses that are sometimes fatal. In contrast, there is no documented report yet on record of human coronaviruses being involved in a lethal respiratory infection. This situation may be a reflection of the limited number of investigations carried out and the difficulty in isolating the virus. It is known that these agents frequently infect small children and reinfect adults, including persons with chronic respiratory disease.[96] It would be logical to assume that deaths could occasionally occur in these most susceptible segments of the population, but they are probably not very frequent.

3.2. Sources of Morbidity Data

Since coronaviruses usually produce respiratory illnesses indistinguishable from those caused by many other types of viruses, it is not possible to obtain data on morbidity in the absence of laboratory identification of infection. The viruses are difficult to isolate, so most workers have relied on serological techniques to increase the numbers that can be studied. The original investigations into coronavirus infection have usually formed part of overall evaluations of the role of viruses in general in respiratory illnesses. As indicated in the partial listing in Table 1, a variety of different open and closed populations were used for these studies. The 229E strain was originally isolated from medical students in Chicago as part of a long-term study of respiratory illnesses in young adults.[35,37] Employee groups were the source of specimens in the NIH[50,77] and in the studies at Charlottesville, Virginia.[41] Infection was also evaluated in children's homes[56] and boarding schools,[60] among military recruits,[119] and among children hospitalized for severe respiratory illnesses in various parts of the world.[50] Serological methods were used to detect occurrence in persons with

Table 1. Longitudinal Studies on the Epidemiology of Coronavirus Infection in Humans

Location	Population	Virus studied
Chicago, IL[35]	Medical students	229E
Washington, DC[50,77]	Hospitalized children	229E, OC43
Bethesda, MD[50,77]	Adult employees	229E, OC viruses
Atlanta, GA[54,56]	Institutionalized children	229E, OC43
Charlottesville, VA[41]	Working adults	229E, OC43
Tecumseh, MI[23,86]	General community	229E, OC43
Brazil[21]	Nonhospitalized children	229E
Denver, CO[75]	Hospitalized asthmatic children	229E, OC43
N. and S. Carolina[119]	Military	229E, OC43

acute exacerbations of asthma[75] or chronic obstructive respiratory disease.[96] Patterns of coronavirus infection were identified among the general population residing in the Tecumseh, Michigan, community as part of a longitudinal study of respiratory illness.[23,86] Volunteers have continued to be employed, especially to determine characteristics of illness not yet well defined in natural infection because of problems associated with isolation of the viruses.[10,11]

3.3. Serological Surveys

Relatively simple serological techniques are available for two coronaviruses (229E and OC38 or OC43), and surveys of antibody prevalence have been carried out in various parts of the world. Many surveys formed a part of studies directed mainly toward determination of incidence of infection. Information on the prevalence of antibody is available for populations in the United States,[23,41,77] Britain,[11] Brazil,[21] and other parts of the world. A special situation is the presence in man of antibody against coronaviruses of animals. The finding of mouse hepatitis antibodies in military recruits and in children and adults from the general population was surprising when first described in 1964.[39] It is now recognized that this does not indicate past experience with MHV but rather with human coronavirus strains that are known to cross-react with it. Similarly, antibodies in human sera against the hemagglutinating encephalomyelitis virus of swine and the coronavirus of calf diarrhea also appear to represent cross-reactions with OC43 or related strains.[58,59] In contrast, in a survey of antibodies to avian IBV, none could be found in a military population. Low-level antibodies were detected only in a portion of subjects who had close contact with poultry.[82] The virus is not known to cross-react with the human strains.

3.4. Laboratory Methods

3.4.1. Viral Isolation.
Only the 229E strain was originally isolated in cell culture. It was eventually adapted to human embryonic lung cells (WI-38), in which it has been maintained.[37] However, this cell line is not a reliable system for primary isolation of 229E-like agents. Human embryonic intestine (MA177) has proven a suitable cell system, but it is available only in limited quantities.[50] Human coronaviruses not related to 229E were originally isolated in organ cultures of human trachea or lung.[38,74,116,117] The presence of virus was usually detected by electron microscopy, or sometimes by fluorescent antibody (FA) staining of impression smears.[115] Two

strains that are essentially identical, OC38 and OC43, have been adapted to suckling mouse brain and to primary monkey kidney and BS-C-1 cell cultures.[14,51,72] Another cell system, L132, a heteroploid human lung line, has been reported to be suitable for primary isolation of 229E, a related virus (LP), and the B814, the first-described organ culture agent.[9,12] This last finding has not been confirmed by other workers.[14] Similarly, MRC-C cells have been used for 229E-like viruses and human rhabdomyosarcoma cells for propagating 229E and OC43.[99,105]

Since it was conceivable that special conditions of cell cultures were required for primary isolation of these agents, various additives to media were examined[87]; this would be similar to the strict requirements for propagation of the rhinoviruses before the availability of WI-38 cells.[95] The situation is in sharp contrast to that found with the cornoaviruses of animals. Although they are species-specific in their *in vitro* growth characteristics, especially on primary isolation, such isolation is easily accomplished.[80,98,102,111]

Recently a technique for direct detection of the 229E virus by nucleic acid hybridization has been described.[91] The method involves creating a cDNA copy of the nucleocapsid gene. It could be predicted from the sequence data that significant cross-hybridization would not occur with OC43 and that the probe would be able to detect 50 pg of viral RNA. When tested on specimens collected from human volunteers artificially infected with 229E, the probe was shown to be as sensitive but no more sensitive than cell culture.[92] Ease of use may be the major advantage at the moment.

3.4.2. Serological Tests.
Neutralization (N) tests of varying degrees of complexity can be performed for all described coronavirus types. The most involved procedure must be used for those viruses that up to now have never been adapted to systems other than organ cultures.[74] This technique involves incubating serum with known virus and inoculating the mixture into cultures of human trachea. Evidence of N manifest by a reduction in viral yield is determined by electron microscopy. For those coronaviruses adapted to cell cultures, tube- or plaque-reduction N tests are available. WI-38 or L132 cells may be used for both methods with 229E virus; a number of cell lines including primary monkey kidney and BS-C-1 have been used for N tests involving the OC38-43 virus.[8,12,14] Hemadsorption rather than CPE can be used for identification of endpoints with the BS-C-1 cell line.[15,16]

Most seroepidemiologic studies have not used N but rather complement-fixation (CF) or hemagglutination-inhibition (HI) tests as sources of their data. The method of preparing a CF antigen for 229E directly from cell culture harvests was reported along with the original description of the viruses by Hamre and Procknow.[37] By this method, the CF test detected antibody in low titer and for only a short time after infection. This observation was subsequently confirmed in a large study, and it was suggested that the presence of CF antibody in a population could be interpreted as evidence for recent activity of the virus.[21] However, it was also learned that if the antigen was highly concentrated, antibody could be detected at a higher titer, and this antibody persisted in the population so that the CF method could be employed in surveys of prevalence.[8] An indirect HI test for 229E virus using tanned sheep erythrocytes has also been described. The procedure appears to be highly sensitive, and no cross-reactions with OC43 virus were observed.[57]

It was found that CF tests can be satisfactorily performed with OC43 virus using infected suckling mouse brain as antigen.[77] The same mouse brain material can also be used in the HI test for OC43 antibody. In this test, the hemagglutination titer was higher for rat than for chicken erythrocytes but was sufficient with the chicken cells so that they could generally be employed; this is of particular importance in view of the spontaneous agglutination that often complicates working with rat erythrocytes. Serum to be tested did not require treatment with receptor-destroying enzyme but rather standard heat inactivation at 56°C. The agglutination took place equally at various temperatures including room temperature.[53] In addition, a single radial hemolysis test has been developed. It can be used not only for OC43 but also for the nonhemagglutinating 229E by using cations to attach virus to glutaraldehyde-treated red cells.[43]

The enzyme-linked immunosorbent assay (ELISA) for antibody has been adapted for use with coronavirus antigens.[62,103] A number of methods have been used to recognize a significant change in antibody titer between specimens. The test has been performed with both 229E and OC43 viruses as antigens. When sera from individuals infected artificially with other coronaviruses, including organ culture viruses, are tested against these antigens, they are found to cross-react with one or the other but not both. Thus, it is possible that the ELISA test with 229E and OC43 antigens may be able to detect infection with most, if not all, human coronaviruses.[68] Other serological tests have been developed that have been used more in antigenic analyses of the different coronaviruses than in epidemiologic studies. With the indirect FA technique, characteristic cytoplasmic inclusions were demonstrated with 229E, OC43, and even the other coronaviruses grown in organ culture.[69,88] The last were prepared for

testing by making smears of fragments of the infected trachea.[76] An example of granular cytoplasmic fluorescence exhibited by OC43 in LLC-MK2 infected cells is shown in Fig. 1. It has also been possible to demonstrate precipitin lines on gel-diffusion tests with coronavirus antigens concentrated 10- to 50-fold. Two or three precipitin lines were observed by Bradburne[8] in tests with hyperimmune animal or human serum, but others have identified only one such line.[55]

4. Biological Characteristics of the Virus

A growing body of information is available on the relationship of coronavirus structure to patterns of antigenicity and infectivity. Much of the work has been done on the animal strains, principally because of their role in producing a number of economically important diseases, although some investigations have involved 229E and OC43. The viruses contain a single continuous strand of RNA that is about 30 kilobases in length.[108] The nonsegmented genome is of positive polarity.[61,66,79,114]

There are three major proteins. The nucleocapsid protein, N, is enclosed within the viral envelope with the RNA in a helical nucleocapsid. The other two, both glycoproteins, are the membrane glycoprotein, M, and the large spike glycoprotein, S. The latter protein, which forms the distinctive projections of the virus, has been associated with N, HA, and CF tests.[106] Antibodies elicited by it are thought to be associated with protection.[113] OC43 along with some animal viruses also contain a third glycoprotein, the hemagglutinin esterase (HE), which forms smaller spikes on the viral envelope.[46] No neuraminidase has ever been detected and there have been reports of phylogenetic relationships of the HE gene of OC43 to influenza type C.[36,55,121] The lipids associated with the envelope have been well defined.[97]

Recent studies on the cellular receptors involved with coronavirus infection help in understanding the phenomenon of species specificity. The receptor for the mouse hepatitis viruses, MHV-A29, has been identified and cloned. It is a member of the carcinoembryonic family. Nonmurine cells, which had been resistant to infection, when transfected with the receptor became susceptible.[31] Antibody to the receptor also prevented infection.[27] More recently, a receptor for 229E has been identified to be human aminopeptidase N, a cell surface metalloprotease present on certain epithelial cells. This receptor is specific for 229E and not for OC43.[120]

The total number of serological types that infect man has not been defined. Here again, the problem revolves

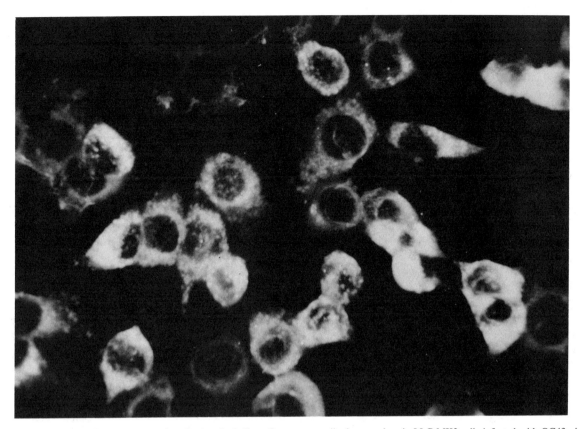

Figure 1. Cytoplasmic fluorescence produced using the indirect fluorescent antibody procedure in LLC-MK2 cells infected with OC43 virus.

around the difficulties encountered is isolating the human coronaviruses. Consequently, there is no way of estimating the proportion of existing types that have already been isolated. It is also difficult to determine the separate antigenic identities of types that grow only in organ culture as compared with those that grow in cell culture. Unfortunately, the situation does not seem to be improving; although there are newer organ culture isolates, the older ones have not been propagated for many years and may actually be lost. The N, CF, HI, gel-diffusion, and immunofluorescent techniques have been used in the antigenic analyses of the older strains by McIntosh et al.,[76] by Bradburne,[8] and by Bradburne and Somerset.[11] As would be expected, results have differed by each of these procedures, with N tests the most specific. However, cross-reactions were commonly demonstrable even by this method using animal antiserum or immune ascitic fluid, indicating that there must be many shared antigens.

An attempt at placing the original groups of human coronavirus isolates in broad groups is shown in Table 2; MHV is included because of its frequent interrelation-

ships with the human strains, and avian IBV is omitted because it is antigenically distinct. The unadapted organ culture strains have been listed separately; it has not been possible to prepare animal antisera against them, and they have been tested only against pairs of sera obtained either

Table 2. Serological Relationships of the Human Coronaviruses

Group	Strains tested with animal antisera		Strains tested with human antisera
I	229E	Closely related	
	229E-like isolates	but not	
	LP, Linder	identical	
	EVS (probably)		
II	OC38	Nearly identical	OC44
	OC43		
	MHV		
Others	B814		OC16
			OC37
			OC48

from individuals naturally infected or from volunteers challenged artificially. Such sera would be expected to be considerably less specific than animal antisera.

There are many additional 229E-related isolates, some originally recovered in cell culture and others in organ culture.[99] The OC43 virus has a low-level cross-reaction with MHV; in some reports, this has been reciprocal and in some one-way. Although B814 virus is quite different from OC43, they both share some antigens in common; again, cross-reactions with 229E are rare. Among the additional viruses, OC44 is closely related antigenically to OC38 and OC43 but has never been successfully adapted to mouse brain or cell cultures. The three other viruses are listed together by exclusion, i.e., not because of any demonstrated relationship to one another but rather because they are less closely related to viruses in the first two groups. Some low-level reactions with the agents in these two groups have been shown to be present, with OC16 virus being the most distinctly different strain.

As indicated above, much of the information on the behavior of 229E and OC43 viruses in populations has come from CF and HI tests. In view of the sharing of antigens among many of the viruses listed in Table 2, the specificity of these procedures must be carefully considered to help in interpreting results. Cross-reactions between 229E and OC43 have been reported only rarely when tested by CF against animal sera. With human serum, heterologous rises in antibody titer have been observed occasionally, but not frequently enough to create problems in studies involving significant numbers of specimens.[13] Of greater practical relevance is the occurrence of cross-reactions between OC43 and the other organ culture viruses. It was postulated that rises in titer detected when using OC43 antigen in seroepidemiologic studies might result either from OC43 infection itself or from infection with one of these related viruses.[86] Indirect evidence that the infecting agent might not be OC43 itself was the dissociation seen between the CF and HI test for OC43 during a particular period of time. Rises in titer by CF should usually be accompanied by rises in titer by HI in the same serum pairs. If this does not ordinarily occur during one time period but does during a second period, it suggests that a related virus but not OC43 was circulating during the first period.[86]

Recent work involving the ELISA test has expanded the above observations. When 229E or OC43 was used as antigen, sera from individuals infected with a wide variety of coronaviruses showed a rise in titer to one or the other virus but not to both.[68] Sera were mainly obtained from studies involving artificial challenge of volunteers, and all

the infecting viruses were found by ELISA to fall into either group I or II, as shown in Table 2. Of particular interest is the fact that new organ culture isolates, which could not be easily adapted to cell culture, all had produced a rise in antibody titer to OC43. This confirms the earlier observation that rises in titer to OC43 by CF and HI might be produced by distinct but related strains. A similar situation exists with 229E-like strains; a number of additional isolates have been identified that are antigenically distinct, and the differences among these related strains may have implications for cross-protection from reinfection.[99] Although the ELISA results suggest that all coronaviruses fall into groups I and II, it is still possible that viruses such as OC16 may prove to be unrelated, since the classification of it and other older viruses cannot now be reevaluated. Work at the molecular level is also helping to determine the relationships among the entire coronavirus family.[45,64]

Data that demonstrate the etiologic role of coronaviruses in respiratory infections are derived from laboratory and field studies. The viruses do interfere with the action of cilia in tracheal organ culture, which suggests that they should have the same effect *in vivo*. In addition, volunteers have been inoculated with essentially all available strains with production of illness.[10,11,99] It has also been possible with 229E to demonstrate that natural infection was statistically related to the production of illness. During the 1967 outbreak of 229E infection in Tecumseh, Michigan, illness was significantly more common among those with infection than among matched subjects without infection.[23] Similarly, 229E infection among Chicago medical students was statistically associated with illness when those with rises in titer were used as their own controls.[35]

5. Descriptive Epidemiology

5.1. Incidence and Prevalence

Coronaviruses are of major importance in common respiratory infections of all age groups. The total impact of coronavirus infections on the general population cannot be calculated at present because not all viral types have been identified. Only 229E and OC43 are amenable to large-scale serological studies; infection rates for other distinct types such as OC16 cannot be determined. The assumption must be made that the former two types are typical of the other viruses. Incidence of infection with these agents exhibits a marked cyclical pattern, so it is to be expected that reported rates will vary based on the

Table 3. Reported Frequency of Infection or Respiratory Illness with 229E and OC43 in Four Locations

Study	Mean incidence of infection with	
	229E	OC43
Chicago medical students[35]	15/100/yr	—
Tecumseh, MI[23,86]	7.7/100/yr	17.1/100/yr
	Proportion of illnesses associated with	
	229E	OC43
Charlottesville, VA, employees[36]	1.7% of illnesses	2.4% of illnesses
Atlanta, GA, children[54,56]	4.3% of illnesses	3.3% of illnesses

number of seasons of high viral activity included in a particular study. Table 3 presents a summary of results obtained in four such studies.

Another approach toward developing a minimal estimate of the total role of coronaviruses in respiratory illnesses comes from a study involving exhaustive laboratory examination, including organ culture, of specimens from 38 common colds. Coronaviruses were isolated from 18.4% of the specimens, but an additional 13%, which were negative in the laboratory, produced colds when given to volunteers.[65] Based on these results, which came from a limited age group, it has been estimated that coronaviruses are responsible for at least 14% of all respiratory illnesses in a general population.[88]

5.1.1. Incidence and Prevalence of 229E Virus. Frequency of 229E illness and infection has been determined in several large-scale investigations. The activity of 229E was found to be of high prevalence in 3 out of 6 years of a study among Chicago medical students. The mean annual incidence of infection during the total period was 15%, based on person-years of observation. The criterion for identification was a reproducible twofold seroconversion determined by CF. There was marked year-to-year variation in infection frequency, ranging from a high of 35% of those tested in 1966–1967 to a low of 1% in 1964–1965. However, nearly 97% of the infections occurred during the months from January to May, often at a time when isolation of rhinoviruses was at a low, and seroconversions for 229E were only rarely accompanied by a rise in titer for another respiratory agent.[35]

The serological study of 229E activity in the community of Tecumseh, Michigan, initially covered 2 years, which included one period of high prevalence. As with the study in Chicago, routine blood specimens were collected

so that infection rates could be determined; however, the study group was composed of individuals of all ages living in their homes. Over the 2 years, infections were detected in 7.7% of individuals tested by CF, as shown in the curve in Fig. 2. However, this appeared to be an underestimate of the actual activity of the virus. Serum specimens had been collected on a regular basis, 6 months apart; rises in titer by CF occurred most frequently in those pairs in which the second specimen was collected in April 1967, clearly indicating the peak period of viral dissemination. Both CF and the more sensitive N test results were combined to give an overall infection rate for the population studied; this rate, 34%, was remarkably similar to the 35% observed in Chicago at the same time. Because of the limited period of viral activity, it was possible to compare illness rates of those infected with persons not infected matched by age and sex; it was estimated that 45% of the infections had produced clinical disease. Thus, the rate of 229E-associated illnesses during the outbreak was 15 per 100 persons studied. Activity in all age groups was apparent, including children under 25 years of age.[23]

In other investigations of 229E activity, attention has been directed mainly toward study of associated illnesses; in such studies, sera have been collected before and after the illness rather than continually on a routine basis as done to determine infection rates. Employees at State Farm Insurance Company, in Charlottesville, Virginia, were studied during a 8-year period for rises in titer for both 229E and OC43. By CF, 229E infection could be related to 3% of the colds that occurred in the winter–spring and to 0.4% of colds that occurred in the summer–fall. There was some year-to-year variation in activity, but

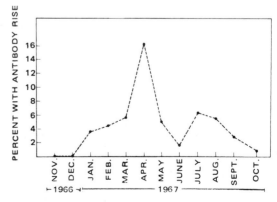

Figure 2. Serological incidence by CF of infection with 229E virus in Tecumseh, Michigan, 1966–1967.

differences in the number of specimens tested from various years did not permit complete identification of cyclical patterns.[41] Employees of the NIH with respiratory illness were studied by both isolation and serology for 229E infection over a 6-year period. Again, attention was specifically directed toward certain segments of the 6 years, and no specimens were tested during other segments. Of particular interest once more is the segment from December 1966 to April 1967. Isolation of rhinoviruses and myxoviruses was uncommon at this time, but respiratory illness continued to occur. During this period, 24% of those persons with colds studied had rises in titer for 229E. As part of the same investigation, paired blood specimens collected from infants and children admitted to the hospital with acute lower respiratory disease during the 1967 period of 229E activity were tested for rise in antibody against the virus, but none was found.[50,77] Healthy children institutionalized in Atlanta, Georgia, were studied from 1960 to 1968; antibody response to 229E was determined by the indirect hemagglutination test. The investigation involved collection of serum specimens related to illness and also routine collection of sera from some nonill individuals. Frequency of infection showed marked variation from year to year. Overall, 4% of colds could be associated with 229E infection, with greatest association in autumn, winter, and spring.[54]

Surveys of prevalence of 229E antibody have also been carried out to document past history of infection, often as parts of longitudinal studies. A general finding is that antibody is present in a significant portion of adults who, despite possessing this antibody, can go on to have reinfection and illness. Reports of antibody prevalence in adults in the United States have varied from 19 to 41%, depending on the type of test used to determine antibody and the time of collection of serum.[25,41,77] Children under 10 years of age exhibited lower mean antibody titers than older children or adults.[23,77] Individual sera from normal healthy adults collected serially in Britain from 1965 through 1970 were tested by Bradburne and Somerset.[11] It is of interest that there was a buildup in sera positive by CF from approximately 17% in specimens collected in October–December 1966 to 62% in those collected in July–September 1967. This suggests that the spring 1967 outbreak that occurred in several parts of the United States may have taken place in Britain as well.

5.1.2. Incidence and Prevalence of OC43 Virus. Populations employed to study infection and illness caused by OC43 virus have generally been the same ones employed to study the occurrence of 229E virus. Kaye *et al.*[56] used the group of institutionalized children in Atlanta, Georgia, to identify infection by means of their HI

test. Infections with the agent were detected in all years of the study, but with definite cyclical variation. Seasons most involved were the winter and spring. Overall, 3% of the illnesses recorded in the 7-year period could be associated with OC43 infection, with a high of 7% in 1960–1961. Interestingly, testing of the sera collected routinely from nonill individuals indicated that an additional equal number of OC43 infections were occurring without the production of symptoms.[56] The Charlottesville study of adult employees was of both OC43 and 229E infections. Here, too, the emphasis was on illness, and it was found in all years studied that OC43 was associated with 5% of colds in the winter–spring and with no illnesses in the summer–fall. Again, there was cyclical variation from year to year in the number of rises in titer detected.[41]

The original isolations of OC38 and OC43 were made in December and January 1965–1966 as part of the study carried out among NIH employees with colds. Testing of sera collected from these employees indicated that during this period, up to 29% of the colds studied were accompanied by rise in titer for OC43. In the children hospitalized with lower respiratory disease, up to 10% of illnesses during this period were associated with such a titer rise. However, it was impossible to show that the relationship to disease was truly etiologic. This finding was in contrast to that seen with 229E, in which no rises in titer were detected in such cases.[73,77]

In the Tecumseh study, occurrence of OC43 infection was determined in the community population over a 4-year period: CF and HI tests were used on all specimens, and N tests were used as an aid in evaluating these results in selected specimens. During the total period, OC43-related infection was detected in 17.1% of the 910 persons studied for 1 year. Most of the infections took place in the winter–spring months of 1965–1966, 1967–1968, and 1968–1969. The only winter–spring period without such activity was in 1966–1967, when the 229E outbreak had taken place. There was good agreement between the CF and HI tests for the 1965–1966 and the 1968–1969 periods but not for 1967–1968. The N test was used to clarify the situation. It was found that most rises in titer for the periods of 1965–1966 or 1968–1969, whether they had occurred by CF or HI or both, were also accompanied by rises in N antibody. In 1967–1968, most CF rises in titer were not accompanied by rises in titer in the HI test, nor was the reverse true; significant change in N antibody in this period was exceedingly rare. It was concluded that the outbreaks of infection in 1965–1966 and 1968–1969 were probably caused by agents closely related to OC43, whereas the 1967–1968 activity was caused by one of the other OC viruses that share some antigens with OC43 but are more distantly related to it. The 1968–1969 outbreak of OC43 infection was nearly as widespread as the prior 229E outbreak, with 25.6% of the population studied showing evidence of infection. Of special note was the fact that children under 5 years of age had the highest infection rates.[86]

Surveys of antibody prevalence have been conducted in several settings using OC43 antigens. McIntosh *et al.*[77] found that children began to acquire antibody to this virus in the first year of life. By the third year of life, more than 50% had antibody present. Among adults, 69% could be demonstrated to have antibody; this indicates, in view of the high incidence of infection with the agents in all age groups, the frequency with which such infections must represent reinfection. The high prevalence of antibody has been confirmed in other studies.[41,47,56] In Britain, Bradburne and Somerset[11] followed prevalence of antibody for OC43 over time, as they also had done with 229E. Each year, the greatest prevalence of antibody was found in the winter–spring period. The single highest point in antibody prevalence was in January–March 1969, at the same time the OC43 outbreak was occurring in some parts of the United States.[11]

5.2. Geographic Distribution

Occurrence of coronavirus infection has now been documented, by either isolation or serology, throughout the world. In the United States, in addition to the studies listed in Table 1, a 229E-like virus has been isolated in California, and OC43 and 229E have been demonstrated to be present in many regions of the country.[93,96] Extensive studies have been carried out by the Common Cold Research Unit, which have demonstrated the presence of the agents in Britain. The activity of 229E virus has been documented in Brazil in an early study of children and adults with and without respiratory illness. Significant rises in antibody titer accompanied nonhospitalized respiratory infection in the children. Prevalence of antibody was determined by CF, and like the situation in some studies in the North Temperate Zone, children had little antibody, whereas 26% of adults were antibody-positive.[21] Later investigations have confirmed the worldwide distribution of these agents.[40,47] These findings suggest that coronaviruses are worldwide in distribution and cause similar types of illness in different localities; such a situation has been noted with many other respiratory viruses.[84] An attempt was actually made to detect rises in antibody titer for 229E in paired sera collected from small children with lower respiratory infection in many tropical parts of the world. No evidence of infection was found, which is hardly surprising, since no rises in titer were found in similar sera collected as part of the same study in Washington, DC.[24,50]

5.3. Temporal Distribution

Because most illnesses caused by coronaviruses are similar to those caused by other respiratory viruses, it is impossible to identify epidemic behavior of the viruses. There is, however, great variation in the frequency of infection on both a seasonal and a cyclical basis. Isolation and rises in antibody titer for all types of coronaviruses have been rare events outside the period from December through May, although this event has been more common in recent studies involving more sensitive techniques.[104] This is the portion of the year in which isolation rates for rhinoviruses and other respiratory viruses often reach their low. An exception to this rule is a study in which frequent rises in titer were detected by ELISA in summer as well.[67] In addition, a cyclical pattern may be discerned when individual virus types are considered. In Fig. 3, data are summarized from five longitudinal studies of coronavirus activity carried out in different parts of the United States. In all studies, some sporadic activity did occur in nearly all years studied, but rises in antibody titer were concentrated in certain years that far exceeded the means for the entire studies. Those periods are indicated as solid black boxes in the figure. The times during which specimens were collected in each investigation are indicated in the figure by the white boxes. Activity of 229E was detected in all four studies at the same time, even though two were in the Midwest and two in the eastern United States. It seems possible, on the basis of these data, to postulate a 2- to 3-year cycle for this agent. The greatest number of infections in Chicago was seen in 1967, after absence of the agent for 3 years, which would suggest a role of herd immunity in determining the time of reappearance of the agent. The findings overall suggest that the serological techniques used detected infection with the various variants of 229E listed in Table 2.

With OC43, the situation is quite different. As with 229E, in no investigation did 2 years with high rates of infection or illness follow one another. A possible exception was in the Tecumseh study. However, the agent that caused the rises in titer in 1967–1968 did not appear as closely related serologically to OC43 as the agent involved in the other two outbreaks. This observation indicates a problem in identifying cycling of OC43 using the serological test employed.

The viruses related to OC43 are apparently more diverse than those related to 229E but still exhibit cross-

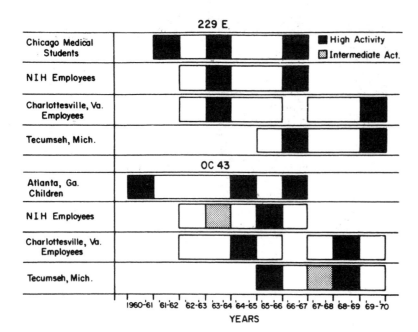

Figure 3. Cyclic behavior of 229E and OC43 viruses observed in five longitudinal studies.

reactions in serological tests; these other viruses may well have cycles of their own that may confuse the situation. In 1964–1965, high activity occurred in Atlanta and Charlottesville. However, in Bethesda, just a short distance away, high activity was not seen in that year but in 1965–1966, the same time as high activity occurred in Michigan, many miles away. In 1968–1969, Charlottesville and Tecumseh data did agree, with very high activity in both area. Thus, cycling of the agents was found in all studies, but the cycles did not agree on specific years. This may be a result of actual differences in patterns of occurrence or a result of differences in the serological techniques used to identify infection, which are of greater importance with OC43 because of the problem of cross-reactions. That cycling of coronaviruses does exist and occurs every 2–4 years with production of many infections suggests that the number of truly different coronaviruses may be relatively small. This situation is unlike that seen with the rhinoviruses, in which cycling has been more difficult to demonstrate, in part because of the large number of serotypes.[18]

5.4. Age

All age groups are involved in infection with OC43 virus. High rates have been noted in children and adults during studies separately examining both groups. In the Tecumseh study, a total population group was followed. During the 1968–1969 outbreak, infection rates were relatively uniform for all age groups, varying from a high of 29.2 per 100 person-years in the 0–4 age group to 22.2 in those over 40 years of age.[86] This finding is quite different from the situation that exists with other respiratory agents, such as respiratory syncytial virus, in which a more distinct decrease in infection rates can be observed with increase in age.[85] The reversal of the pattern of age-specific infection rates customarily associated with the respiratory viruses becomes complete with 229E. Infection with this virus has been more difficult to demonstrate in small children than in adults. In Tecumseh, during the 1966–1967 outbreak, highest age-specific infection rates by CF were found among those 15–29 years of age, following a steady increase in infection frequency from the 0- to 4-year-olds. However, when neutralization tests were used to detect infection, the 15- to 19-year-olds still had high infection rates, but the serial increase to that point among younger age groups was much less steep.[23] This suggests that the apparent sparing of small children with 229E may be an artifact resulting from the relative insensitivity of the young to the serological procedures commonly employed. It would be surprising if two different coronavirus serotypes behaved so differently.

5.5. Other Factors

There is little evidence of a sex differential in infections with the coronaviruses simply because the data have rarely been examined in such a manner. In Tecumseh, adult females experienced higher infection rates with OC43 than adult males, which is in conformity with the usual patterns of all respiratory illnesses.[83] Similarly, female volunteers appeared to be more susceptible to infection with 229E-like strains than males in artificial challenge studies.[19]

In the study by Candeias *et al.*[21] of antibody prevalence, the results were examined by sex, but no significant differences could be observed. There are no data available on occupational or racial susceptibility to infection or on the role of socioeconomic status in influencing rates. Occurrence of infection in closed or special populations, such as military recruits or residents of children's institutions, has been reported.[50,60,119] However, it is at present difficult to determine, based on the relative paucity of information on the behavior of the virus in open populations, whether they exhibit any unique features in other settings. There is a suggestion that OC43 virus might cause acute respiratory disease in military recruits.[119] If this finding is confirmed, it would represent a distinct departure from the types of illness customarily associated with that virus in young civilian adults. The role of the school-age child in dissemination of coronavirus has not yet been clearly defined, but it would be surprising if these infections differed in their transmission pattern so markedly from that documented with the other agents. Because of the high frequency of infection in older children and adults, other sites of dissemination may also be of significance. It has been possible to show that the family unit is of importance in transmission, since clustering of 229E and OC43 infections in families was observed in the Tecumseh study.[23,104]

Although nutritional and genetic factors have not been associated with susceptibility to coronavirus infections, there are clear indications that the viruses are associated with exacerbations of chronic obstructive respiratory disease. such a finding is hardly surprising in view of the high infection rates that have been observed in unselected older adults. It has not yet been demonstrated whether this represents true increased susceptibility to infection or simply a more severe form of expression of the infection when it occurs in an already compromised host. In addition to the situation in older individuals, there is evidence that both OC43 and 229E may trigger acute attacks of wheezing in young asthmatics. In fact, in one study, coronaviruses were the most common agent involved in episodes of wheezy bronchitis.[34,75,81,96]

6. Mechanism and Route of Transmission

The coronaviruses are presumably transmitted by the respiratory route. It has been possible to induce infection experimentally in volunteers by inoculating virus into the nose.[10,117] The virus is most stable at pH 6 0 and low temperature appears to protect it against varied relative humidity.[48,63] No other route of transmission for coronaviruses seems involved in man, although animal coronaviruses are infectious by the fecal–oral route.[111] There is currently no direct evidence to aid in identifying the main mechanisms of transmission. However, it is possible to compare the epidemiologic behavior of the coronaviruses with that of other respiratory agents, the transmission mechanisms of which have been more directly studied. Large-scale outbreaks of coronavirus infections have taken place, as in Tecumseh in 1967.[23] This is much more analogous to the situation seen with influenza than to that with the rhinoviruses. It is likely that the former agent can be transmitted by aerosol as well as by large droplets, which would explain its ability to spread quickly.[29] Rhinoviruses, on the other hand, are thought to be transmitted by large droplet and may at times spread via fomites.[42] It is therefore probable that human coronaviruses can be spread by aerosol as well as by large droplet. Aerosol transmission of avian IBV has actually been documented in poultry.[30]

There is no evidence that any animal reservoir or vector is involved in the maintenance of infection or transmission of the human coronaviruses. Each animal coronavirus appears to be restricted to its own species. The only known exception is the finding of antibody to avian IBV in sera of poultry workers but not of controls.[82]

7. Pathogenesis and Immunity

The incubation period of coronavirus colds is relatively short. In studies involving volunteers, the mean period from inoculation of virus to development of symptoms was from 3.2 to 3.5 days, depending on the strain (range, 2–4 days).[10,117] Following exposure, the virus apparently multiplies superficially in the respiratory tract in a manner similar to that in which multiplication occurs *in vitro*. Nasal airway resistance and temperature of the nasal mucosa increases.[5] Virus excretion usually reaches a detectable level at the time symptoms begin and lasts for 1–4 days. The duration of the illness is from 6 to 7 days on the average, but with some lasting up to 18 days. Serological response either to induced or to naturally acquired infection has been quite variable depending on the infecting strain and the serological test employed. For example, among those experimentally infected with OC38 or OC43 virus who had a cold produced, only 46% had rises in titer by HI and 23% by CF. Fewer than half of those infected with 229E showed a CF rise. It is not clear how the existence of titer or preinfection antibody affects the magnitude of the response detected by these tests. Rises in N antibody titer are easier to detect and have been found with sensitive techniques in all volunteers experimentally infected.[8,11] Use of the ELISA test has given added sensitivity in antibody detection; it is not as yet clear if decreased specificity should be a concern.

An important characteristic of the coronaviruses is their apparent high rate of reinfection, which in volunteers has now been documented to be possible within a year of prior infection.[20] In the Tecumseh study, 81.5% of those infected with OC43 actually possessed prior N antibody.[90] Possession of circulating OC43 HI antibody among the Atlanta children did not appear to play a role in modifying severity of a subsequent illness.[56] With 229E virus, Hamre and Beem[35] demonstrated that the frequency of rise in titer detected by N was inversely proportional to preinfection levels of N antibody, which would indicate that this antibody exerted some protective effect. However, the importance of this N antibody could not be confirmed when infection was detected by CF. Thus, circulating N antibody as measured at present may bear a relationship to modification of infection, but this association is not a very strong one. Since coronavirus infections involve mainly the surface of the respiratory tract, it is likely that secretory IgA antibody plays a more direct role in protection; this had in fact been demonstrated with a swine coronavirus[7] and subsequently with 229E in humans experimentally infected.[19]

8. Patterns of Host Response

The coronaviruses generally produce a coldlike illness that on an individual basis is difficult to distinguish from illness caused by other respiratory viruses. In both induced and natural infections, the most prominent findings have been coryza and nasal discharge, with the discharge being more profuse than that customarily seen with rhinovirus colds.[10] Sore throat has been somewhat less common and in children has been associated with pharyngeal injection.[46] Experimental colds caused by B814 virus were about as severe as those caused by 229E; however, natural OC43 infections caused illnesses with considerably more cough and sore throat than did 229E infections.[30] The mean duration of coronavirus colds, at 6.5 days, is shorter than that seen in rhinovirus colds, at 9.5 days.[10]

There is no clear evidence yet available that coronaviruses cause severe lower respiratory illness in infants and young children. In fact, such infections were more common in one study among the control group than

among the diseased.[77] Mufson *et al.*[90] have associated coronavirus 229E and OC43 infection with acute lower respiratory infections in children at Cook County Hospital. The lack of a comparable control group makes assignment of an etiologic role to these viruses hazardous at present, but the relationship should be sought in the future. The association of OC43 with the acute respiratory disease syndrome in military recruits should also be viewed as tentative.

Clinical disease occurred in no more than 45% of those infected with 229E in Tecumseh during the 1967 outbreak.[23] In Atlanta children, OC43 virus produced illness in about 50% of those infected.[56] It is likely that with increase in age and concomitant experience with these agents, the ratio of clinically apparent to inapparent infection will decrease. As with other respiratory agents, a continuum of severity of symptoms exists among those in whom infection results in disease, and this may also be related to past experience with the viruses.

Coronaviruslike particles have been identified in stools of persons with diarrhea, often in high frequency, and therefore a role in etiology of acute enteric disease has been suggested.[94] This would not be surprising in view of the clear involvement of certain strains in severe diarrheal disease of domestic animals. A general problem in most studies has been the inability to isolate the viruses; typically they have been identified directly by electron microscopy of stool specimens. Only rarely has it been possible to concentrate a specimen sufficiently to characterize the agent contained as a coronavirus.[100] Questions have been raised concerning whether all coronaviruslike particles in stool are truly coronaviruses; in certain situations it is clear that they are not.[107] In addition, although in some studies coronavirus particles are more commonly detected in stools from ill individuals than from healthy controls, in other studies this has not been the case.[22,25,33,109,118] In one study, excretion of these particles were associated more with various exposures than with symptoms of diarrhea.[71]

Multiple sclerosis is another disease entity concerning which evidence has been presented suggesting a role for coronaviruses. Coronaviruses were reported to be cultured from brain tissue of two multiple sclerosis patients.[17] Questions about the species origin of such isolates have been raised.[32] Gene expression of 229E viral RNA in brain tissue of multiple sclerosis patients but not in tissue of controls has also been reported.[112] Some serological results confirm this association, but others do not.[51,70,101] Again, the behavior of certain neurotropic animal coronaviruses is given as support for this association. Finally, coronaviruslike particles have also been observed in renal biopsies from cases of endemic (Balkan) nephropathy. A slow coronavirus infection acquired from pigs has been suggested as being involved.[2]

9. Control and Prevention

It is premature at present to think in terms of control of coronavirus infection by vaccination. Not all viral types have been identified, and some known agents cannot be easily propagated in the laboratory. Thus, preparation of vaccines of the conventional types is impossible. The frequency of reinfection observed with these agents is so high that control by vaccination may not be practical, but it is possible that future studies may allow further characterization of truly protective antibodies. Work on vaccines for the animal viruses is in progress, and these studies may help in understanding issues of protection.[110] Chemoprophylaxis and related measures may be a more practical approach; it has been shown that recombinant α interferon can prevent infections artificially produced in volunteers,[44] and other approaches have been under investigation.[3,26,35,44] There remains environmental control of infection; such efforts have rarely been useful for other respiratory agents, but they may be more efficacious if a practical barrier to transmission can be devised.[78]

10. Unresolved Problems

The major immediate need in coronavirus research lies in the laboratory. If a practical system can be found for isolation and propagation of the viruses, the gaps in understanding the behavior of these agents will quickly be filled. A partial solution would be direct detection of the virus through an ELISA or nucleic acid hybridization; this approach may soon be available.[30,92] Only serological tools are available now for most epidemiologic studies, and even these can be applied to only two different coronavirus types. Therefore, many of the data that have been so laboriously gained give only partial evidence on the total dimensions of the problem—and the problem is almost certainly a very large one. Coronaviruses have been isolated and outbreaks identified in periods of the winter and spring, when rhinoviruses and myxoviruses are uncommon. It appears that during these times, the coronaviruses cause a significant portion of respiratory illnesses. Even discounting suggestions of production of severe disease in young children and in those with chronic respiratory disease, the viruses are important respiratory pathogens simply in terms of numbers of illnesses pro-

duced. They may also be involved in other important diseases. Only through further understanding of the behavior of these agents will it be possible to determine the scope of their activity and the most appropriate means by which control can be attempted.

11. References

1. ALMEIDA, J. D., AND TYRRELL, D. A. J., The morphology of three previously uncharacterized human respiratory viruses that grow in organ culture, *J. Gen. Virol.* **1:**175–178 (1967).
2. APOSTOLOV, K., AND SPASIC, P., Evidence of a viral aetiology in endemic (Balkan) nephropathy, *Lancet* **2:**1271–1273 (1975).
3. BARROW, G. I., HIGGINS, P. G., AL-NAKIB, W., SMITH, A. P., WENHAM, R. B., AND TYRRELL, D. A. The effect of intranasal nedocromil sodium on viral upper respiratory tract infections in human volunteers, *Clin. Exp. Allergy* **20:**45–51 (1990).
4. BECKER, W. B., MCINTOSH, K., DEES, J. H., AND CHANOCK, R. M., Morphogenesis of avian infectious bronchitis virus and a related human virus (strain 229E), *J. Virol.* **1:**1019–1027 (1967).
5. BENDE, M., BARROW, I., HEPTONSTALL, J., HIGGINS, P. G., AL-NAKIB, W., TYRRELL, D. A., AND AKERLUND, A., Changes in human nasal mucosa during experimental coronavirus common colds, *Acta Otalaryngol.* **107:**262–269 (1989).
6. BERRY, D. M., CRUICKSHANK, J. G., CHU, H. P., AND WELLS, R. J. H., The structure of infectious bronchitis virus, *Virology* **23:**403–407 (1964).
7. BOHL, E. H., GUPTA, R. K. P., OLQUIN, M. V. F., AND SAIF, L. J., Antibody responses in serum, colostrum, and milk of swine after infection or vaccination with transmissible gastroenteritis virus, *Infect. Immun.* **6:**289–301 (1972).
8. BRADBURNE, A. F., Antigenic relationships amongst coronaviruses, *Arch. Ges. Virusforsch.* **31:**352–364 (1970).
9. BRADBURNE, A. F., An investigation of the replication of coronaviruses in suspension cultures of L132 cells, *Arch. Ges. Virusforsch.* **37:**297–307 (1972).
10. BRADBURNE, A. F., BYNOE, M. L., AND TYRRELL, D. A. J., Effects of a "new" human respiratory virus in volunteers, *Br. Med. J.* **3:**767–769 (1967).
11. BRADBURNE, A. F., AND SOMERSET, B. A., Coronavirus antibody titres in sera of healthy adults and experimentally infected volunteers, *J. Hyg.* **70:**235–244 (1972).
12. BRADBURNE, A. F., AND TYRRELL, D. A. J., The propagation of "coronaviruses" in tissue culture, *Arch. Ges. Virusforsch.* **28:**133–150 (1969).
13. BRADBURNE, A. F., AND TYRRELL, D. A. J., Coronaviruses of man, *Prog. Med. Virol.* **13:**373–403 (1971).
14. BRUCKOVA, M., MCINTOSH, K., KAPIKIAN, A. Z., AND CHANOCK, R. M., The adaptation of two coronavirus strains (OC38 and OC43) to growth in cell monolayers, *Proc. Soc. Exp. Biol. Med.* **135:**431–435 (1970).
15. BUCKNALL, R. A., KALICA, A. R., AND CHANOCK, R. M., Intracellular development and mechanism of hemadsorption of a human coronavirus, OC43, *Proc. Soc. Exp. Biol. Med.* **139:**811–817 (1972).
16. BUCKNALL, R. A., KING, L. M., KAPIKIAN, A. Z., AND CHANOCK, R. M., Studies with human coronaviruses. II. Some properties of strains 229E and OC43, *Proc. Soc. Exp. Biol. Med.* **139:**722–727 (1972).

17. BURKS, J. S., DEVALD, B. L., JANKOVSKY, L. D., AND GERDES, J. C., Two coronaviruses isolated from central nervous system tissue of two multiple sclerosis patients, *Science* **209**:933–934 (1980).

18. CALHOUN, A. M., JORDAN, W. S., JR., AND GWALTNEY, J. M., JR., Rhinovirus infections in an industrial population. V. Change in distribution of serotypes, *Am. J. Epidemiol.* **99**:58–64 (1974).

19. CALLOW, K. A., Effect of specific humoral immunity and some non-specific factors on resistance of volunteers to respiratory coronavirus infection, *J. Hyg. Camb.* **95**:173–189 (1985).

20. CALLOW, K. A., PARRY, H. F., SERGEANT, M., AND TYRRELL, D. A., The time course of the immune response to experimental coronavirus infection of man, *Epidemiol. Infect.* **105**:435–446 (1990).

21. CANDEIAS, J. A. N., CARVALHO, R. P., DeS., AND ANTONACIO, F., Seroepidemiologic study of coronavirus infection in Brazilian children and civilian adults, *Rev. Inst. Med. Trop.* **14**:121–125 (1972).

22. CAUL, E. O., PAVEL, W. K., AND CLARKE, S. K. R., Coronavirus particles in faece from patients with gastroenteritis, *Lancet* **1**:1192 (1975).

23. CAVALLARO, J. J., AND MONTO, A. S., Community-wide outbreak of infection with a 229E-like coronavirus in Tecumseh, Michigan, *J. Infect. Dis.* **122**:272–279 (1970).

24. CHANOCK, R., CHAMBON, L., CHANG, W., GONCALVES FERREIRA, F., GHARPURE, P., GRANT, L., HATEM, J., IMAN, I., KALRA, S., LIM, K., MADALENGOTIA, J., SPENCE, L., TENG, P., AND FERREIRA, W., WHO respiratory disease survey in children: A serological study, *Bull. WHO* **37**:363–369 (1967).

25. CHANY, C., MOSCOVICI, O., LEBON, P., AND ROUSSET, S., Association of coronavirus infection with neonatal necrotizing enterocolitis, *Pediatrics* **69**:209–214 (1982).

26. COLLINS, A. R., AND GRUBB, A., Inhibitory effects of recombinant human cyctatin C on human coronaviruses, *Antimicrob. Agents Chemother.* **35**:2444–2446 (1991).

27. COMPTON, S. R., STEPHENSEN, C. B., SNYDERS, S. W., WEISMILLER, D. G., AND HOLMES, K. V., Coronavirus species specificity: Murine coronavirus binds to a mouse-specific epitope on its carcinoembryonic antigen-related receptor glycoprotein, *J. Virol.* **66**:7420–7428 (1992).

28. Coronaviruses, *Nature* **220**:650 (1968).

29. COUCH, R. B., DOUGLAS, R. G., JR., LINDGREN, K. M., GERONE, P. J., AND KNIGHT, V., Airborne transmission of respiratory infection with Coxsackievirus A type 21, *Am. J. Epidemiol.* **91**:78–86 (1970).

30. CROUCH, C. F., RAYBOULD, T. J. G., AND ACRES, S. D., Monoclonal antibody capture enzyme-linked immunosorbent assay for detection of bovine enteric coronavirus, *J. Clin. Microbiol.* **19**:388–393 (1984).

31. DVEKSLER, G. S., PENSIERO, M. N., CARDELLICHIO, C. B., WILLIAMS, R. K., JIANG, G. S., HOLMES, K. V., AND DIEFFENBACH, C. W., Cloning of the mouse hepatitis virus (MHV) receptor: Expression in human and hamster cell lines confers susceptibility to MHV, *J. Virol.* **65**:6881–6891 (1991).

32. FLEMING, J. O., EL ZAATARI, F. A., GILMORE, W., BERNE, J. D., BURKS, J. S., STOHLMAN, S. A., TOURTELLOTTE, W. W., AND WEINER, L. P., Antigenic assessment of coronaviruses isolated from patients with multiple sclerosis, *Arch. Neurol.* **45**:629–633 (1988).

33. GERNA, G., PASSARANI, N., BATTAGLIA, M., AND RON-DANELLI, E. G., Human enteric coronaviruses: Antigenic relatedness to human coronavirus OC43 and possible etiologic role in viral gastroenteritis, *J. Infect. Dis.* **151**:796–803 (1985).

34. GUMP, D. W., PHILLIPS, C. A., FORSYTH, B. R., MCINTOSH, K., LAMBORN, K. R., AND STOUCH, W. H., Role of infection in chronic bronchitis, *Am. Rev. Respir. Dis.* **113**:465–474 (1976).

35. HAMRE, D., AND BEEM, M., Virologic studies of acute respiratory disease in young adults. V. Coronavirus 229E infections during six years of surveillance, *Am. J. Epidemiol.* **96**:94–106 (1972).

36. HAMRE, D., KINDIG, D. A., AND MANN, J., Growth and intracellular development of a new respiratory virus, *J. Virol.* **1**:810–816 (1967).

37. HAMRE, D., AND PROCKNOW, J. J., A new virus isolated from the human respiratory tract, *Proc. Soc. Exp. Biol. Med.* **121**:190–193 (1966).

38. HARNETT, G. B., AND HOOPER, W. L., Test-tube organ cultures of ciliated epithelium for the isolation of respiratory viruses, *Lancet* **1**:339–340 (1968).

39. HARTLEY, J. W., ROWE, W. P., BLOOM, H. H., AND TURNER, H. C., Antibodies to mouse hepatitis viruses in human sera, *Proc. Soc. Exp. Biol. Med.* **115**:414–418 (1964).

40. HASONY, H. J., AND MACNAUGHTON, M. R., Prevalence of human coronavirus antibody in the population of southern Iraq, *J. Med. Virol.* **9**:209–216 (1982).

41. HENDLEY, J. O., FISHBURNE, H. B., AND GWALTNEY, J. M., JR., Coronavirus infections in working adults, *Am. Rev. Respir. Dis.* **105**:805–811 (1972).

42. HENDLEY, J. O., WENZEL, R. P., AND GWALTNEY, J. M., JR., Transmission of rhinovirus colds by self-inoculation, *N. Engl. J. Med.* **288**:1361–1364 (1973).

43. HIERHOLZER, J. C., AND TANNOCK, G. A., Quantitation of antibody to non-hemagglutinating viruses by single radial hemolysis: Serological test for human coronaviruses, *J. Clin. Microbiol.* **5**:613–620 (1977).

44. HIGGINS, P. G., PHILLPOTTS, R. J., SCOTT, G. M., WALLACE, J., BERNHARDT, L. L., AND TYRRELL, D. A. J., Intranasal interferon as protection against experimental respiratory coronavirus infections in volunteers, *Antimicrob. Agents Chemother.* **24**:713–715 (1983).

45. HOGUE, B. G., KING, B., AND BRIAN, D. A., Antigenic relationships among proteins of bovine coronavirus, human respiratory coronavirus OC43, and mouse hepatitis coronavirus A59, *J. Virol.* **5**:384–388 (1984).

46. HOLMES, K. V., AND WILLIAMS, R. K. (EDS.), Background paper: Functions of coronavirus glycoproteins, in: *Coronaviruses and Their Diseases* (D. E. CAVANAGH AND T. D. K. BROWN, EDS.), pp. 5–8, Plenum Press, New York, 1990.

47. HOVI, T., KAINULAINEN, H., ZIOLA, B., AND SALMI, A., OC43 strain-related coronavirus antibodies in different age groups, *J. Med. Virol.* **3**:313–320 (1979).

48. IJAZ, M. K., BRUNNER, A. H., SATTAR, S. A., NAIR, R. C., AND JOHNSON-LUSSENBURG, C. M., Survival characteristics of airborne human coronavirus 229E, *J. Gen. Virol.* **66**:2743–2748 (1985).

49. JOHNSON-LUSSENBURG, C. M., AND ZHENG, Q., Coronavirus and multiple sclerosis: Results of a case/control longitudinal serological study, *Adv. Exp. Med. Biol.* **218**:421–429 (1987).

50. KAPIKIAN, A. Z., JAMES, H. D., JR., KELLY, S. J., DEES, J. H., TURNER, H. C., MCINTOSH, K., KIM, H. W., PARROTT, R. H., VINCENT, M. M., AND CHANOCK, R. M., Isolation from man of "avian infectious bronchitis virus-like" viruses (coronaviruses)

similar to 229E virus, with some epidemiological observations, *J. Infect. Dis.* **119:**282–290 (1969).

51. KAPIKIAN, A. Z., JAMES, H. D., JR., KELLY, S. J., KING, L. M., VAUGHN, A. L., AND CHANOCK, R. M., Hemadsorption by coronavirus strain OC43, *Proc. Soc. Exp. Biol. Med.* **139:**179–186 (1972).

52. KAPIKIAN, A. Z., JAMES, H. D., JR., KELLY, S. J., AND VAUGHN, A. L., Detection of coronavirus strain 692 by immune electron microscopy, *Infect. Immun.* **7:**111–116 (1973).

53. KAYE, H. S., AND DOWDLE, W. R., Some characteristics of hemagglutination of certain strains of "IBV-like" viruses, *J. Infect. Dis.* **120:**576–581 (1969).

54. KAYE, H. S., AND DOWDLE, W. R., Seroepidemiologic surgery of coronavirus (strain 229E) infections in a population of children, *Am. J. Epidemiol.* **101:**238–244 (1975).

55. KAYE, H. S., HIERHOLZER, J. C., AND DOWDLE, W. R., Purification and further characterization of an "IBV-like" virus (coronavirus), *Proc. Soc. Exp. Biol. Med.* **135:**457–463 (1970).

56. KAYE, H. S., MARSH, H. B., AND DOWDLE, W. R., Seroepidemiologic survey of coronavirus (strain OC43) related infections in a children's population, *Am. J. Epidemiol.* **94:**43–49 (1971).

57. KAYE, H. S., ONG, S. B., AND DOWDLE, W. R., Detection of coronavirus 229E antibody by indirect hemagglutination, *Appl. Microbiol.* **24:**703–707 (1972).

58. KAYE, H. S., YARBROUGH, W. B., AND REED, C. J., Calf diarrhoea coronavirus, *Lancet* **2:**509 (1975).

59. KAYE, H. S., YARBROUGH, W. B., REED, C. J., AND HARRISON, A. K., Antigenic relationship between human coronavirus strain OC43 and hemagglutinating encephalomyelitis virus strain 67N of swine: Antibody responses in human and animal sera, *J. Infect. Dis.* **135:**201–209 (1977).

60. KENDALL, E. J., BYNOE, M. L., AND TYRRELL, D. A. J., Virus isolation from common colds occurring in a residential school, *Br. Med. J.* **2:**82–86 (1962).

61. KENNEDY, D. A., AND JOHNSON-LUSSENBURG, C. M., Isolation and morphology of the internal component of human coronavirus, strain 229E, *Intervirology* **6:**197–206 (1975–1976).

62. KRAAIJEVELD, C. A., MADGE, M. H., AND MACNAUGHTON, M. R., Enzyme-linked immunosorbent assay for coronaviruses HCV 229E and MHV 3, *J. Gen. Virol.* **49:**83–89 (1980).

63. LAMARRE, A., AND TALBOT, P. J., Effect of pH and temperature on the infectivity of human coronavirus 229E, *Can. J. Microbiol.* **35:**972–974 (1989).

64. LAPPS, W., AND BRIAN, D. A., Oligonucleotide fingerprints of antigenically related bovine coronavirus and human coronavirus OC43, *Arch. Virol.* **86:**101–108 (1985).

65. LARSON, H. E., REED, S. E., AND TYRRELL, D. A., Isolation of rhinoviruses and coronaviruses from 38 colds in adults, *J. Med. Virol.* **5:**221–229 (1980).

66. LOMNICZI, B., AND KENNEDY, I., Genome of infections bronchitis virus, *J. Virol.* **24:**99–107 (1977).

67. MACNAUGHTON, M. R., Occurrence and frequency of coronavirus infections in humans as determined by enzyme-linked immunosorbent assay, *Infect. Immun.* **38:**419–423 (1982).

68. MACNAUGHTON, M. R., MADGE, M. H., AND REED, S. E., Two antigenic groups of human coronaviruses detected by using enzyme-linked immunosorbent assay, *Infect. Immun.* **33:**734–737 (1981).

69. MACNAUGHTON, M. R., THOMAS, B. J., DAVIES, H. A., AND PATTERSON, S., Infectivity of human coronavirus strain 229E, *J. Clin. Microbiol.* **12:**462–468 (1980).

70. MADDEN, D. L., WALLEN, W. C., HOUFF, S. A., LEINIKKI, P. A., SEVER, J. L., HOLMES, K. A., CASTELLANO, G. A., AND SHEKARCHI, I. C., Coronavirus antibodies in sera from patients with multiple sclerosis and matched controls, *Arch. Neurol.* **38:**209–210 (1981).

71. MARSHALL, J. A., THOMPSON, W. L., AND GUST, I. D., Coronavirus-like particles in adults in Melbourne, Australia, *J. Med. Virol.* **29:**238–243 (1989).

72. MCINTOSH, K., BECKER, W. B., AND CHANOCK, R. M., Growth in suckling-mouse brain of "IBV-like" viruses from patients with upper respiratory tract disease, *Proc. Natl. Acad. Sci. USA* **58:**2268–2273 (1967).

73. MCINTOSH, K., BRUCKOVA, M., KAPIKIAN, A. Z., CHANOCK, R. M., AND TURNER, H., Studies of new virus isolates recovered in tracheal organ culture, *Ann. NY Acad. Sci.* **174:**983–989 (1970).

74. MCINTOSH, K., DEES, J. H., BECKER, W. B., KAPIKIAN, A. Z., AND CHANOCK, R. M., Recovery in tracheal organ cultures of novel viruses from patients with respiratory disease, *Proc. Natl. Acad. Sci. USA* **57:**933–940 (1967).

75. MCINTOSH, K., ELLIS, E. F., HOFFMAN, L. S., LYBASS, T. G., ELLER, J. J., AND FULGINITI, V. A., The association of viral and bacterial respiratory infections with exacerbations of wheezing in young asthmatic children, *J. Pediatr.* **82:**578–590 (1973).

76. MCINTOSH, K., KAPIKIAN, A. Z., HARDISON, K. A., HARTLEY, J. W., AND CHANOCK, R. M., Antigenic relationships among the coronaviruses of man and between human and animal coronaviruses, *J. Immunol.* **102:**1109–1118 (1969).

77. MCINTOSH, K., KAPIKIAN, A. Z., TURNER, H. C., HARTLEY, J. W., PARROTT, R. H., AND CHANOCK, R. M., Seroepidemiologic studies of coronavirus infection in adults and children, *Am. J. Epidemiol.* **91:**585–592 (1970).

78. MCLEAN, R. L., General discussion, International Conference on Asian Influenza, *Am. Rev. Respir. Dis.* **83**(Part 2):36–38 (1961).

79. MCNAUGHTON, M. R., AND MADGE, M. H., The genome of human coronavirus strain 229E, *J. Gen. Virol.* **39:**497–504 (1978).

80. MEBUS, C. A., STAIR, E. L., RHODES, M. B., AND TWIEHAUS, M. J., Pathology of neonatal calf diarrhea induced by coronavirus-like agent, *Vet. Pathol.* **10:**45–64 (1973).

81. MERTSOLA, J., ZIEGLER, T., RUUSKANEN, O., VANTO, T., KOIVIKKO, A., AND HALONEN, P., Recurrent wheezy bronchitis and viral respiratory infections, *Arch. Dis. Child.* **66:**124–129 (1991).

82. MILLER, L. T., AND YATES, V. J., Neutralization of infectious bronchitis virus by human sera, *Am. J. Epidemiol.* **88:**406–409 (1968).

83. MONTO, A. S., HIGGINS, M. W., AND ROSS, H. W., The Tecumseh study of respiratory illness. VIII. Acute infection in chronic respiratory disease and comparison groups, *Am. Rev. Respir. Dis.* **111:**27–36 (1975).

84. MONTO, A. S., AND JOHNSON, K. M., Respiratory infections in the American tropics, *Am. J. Trop. Med. Hyg.* **17:**867–874 (1968).

85. MONTO, A. S., AND LIM, S. K., The Tecumseh study of respiratory illness. III. Incidence and periodicity of respiratory syncytial and *Mycoplasma pneumoniae* infections, *Am. J. Epidemiol.* **94:**290–301 (1971).

86. MONTO, A. S., AND LIM, S. K., The Tecumseh study of respiratory illness. VI. Frequency of and relationship between outbreaks of coronavirus infection, *J. Infect. Dis.* **129:**271–276 (1974).

87. MONTO, A. S., AND MAASSAB, H. F., Effect of cyclic nucleotide on coronavirus replication, *Proc. Soc. Exp. Biol. Med.* **161:**181–185 (1979).

88. Monto, A. S., and Rhodes, L. M., Detection of coronavirus infection of man by immunofluorescence, *Proc. Soc. Exp. Biol. Med.* **155:**143–148 (1977).

89. Monto, A. S., and Sullivan, K. M., Acute respiratory illness in the community. Frequency of illness and the agents involved, *Epidemiol. Infect.* **110:**145–160 (1993).

90. Mufson, M. A., McIntosh, K., Chao, R. K., Krause, H. E., Wasil, R., and Mocega, H. E., Epidemiology of coronavirus infections in infants with acute lower respiratory disease, *Clin. Res.* **20:**534 (1972).

91. Myint, S., Harmsen, D., Raabe, R., Characterization of nucleic acid probe for the diagnosis of human coronavirus 229E infections, *J. Med. Virol.* **31:**165–172 (1990).

92. Myint, S., Siddell, S., and Tyrrell, D., Detection of human coronavirus 229E in nasal washings using RNA:RNA hybridisation, *J. Med. Virol.* **29:**70–73 (1989).

93. Oshiro, L. S., Schieble, J. H., and Lennette, E. H., Electron microscopic studies of coronavirus, *J. Gen. Virol.* **12:**161–168 (1971).

94. Payne, C. M., Ray, G., Borduin, V., Minnich, L. L., and Lebowitz, M. D., An eight-year study of the viral agents of acute gastroenteritis in humans: Ultrastructural observations and seasonal distribution with a major emphasis on coronavirus-like particles, *Adv. Exp. Med. Biol.* **218:**579–580 (1987).

95. Pelon, W., Classification of the "2060" viruses ECHO28 and further study of its properties, *Am. J. Hyg.* **73:**36–54 (1961).

96. Phillips, C. A., McIntosh, K., Forsyth, B. R., Gump, D. W., and Stouch, W. H., Coronavirus infections in exacerbations of chronic bronchitis, in: *Twelfth Interscience Conference on Antimicrobial Agents and Chemotherapy*, Atlantic City, New Jersey, p. 3, American Society for Microbiology, Washington, DC, 1972.

97. Pike, B. V., and Garwes, D. J., Lipids of transmissible gastroenteritis virus and their relation to those of two different host cells, *J. Gen. Virol.* **34:**531–535 (1977).

98. Purcell, D. A., and Clarke, J. K., The replication of infectious bronchitis virus in fowl trachea, *Arch. Ges. Virusforsch.* **39:**248–256 (1972).

99. Reed, S. E., The behavior of recent isolates of human respiratory coronavirus *in vitro* and in volunteers: Evidence of heterogeneity among 229E-related strains, *J. Med. Virol.* **13:**179–192 (1984).

100. Resta, S., Luby, J. P., Rosenfeld, C. R., and Siegel, J. D., Isolation and propagation of a human enteric coronavirus, *Science* **229:**978–981 (1985).

101. Riski, H., and Hovi, T., Coronavirus infections of man associated with diseases other than the common cold, *J. Med. Virol.* **6:**259–265 (1980).

102. Saif, L. J., Bohl, E. H., and Gupta, R. K. P., Isolation of porcine immunoglobulins and determination of the immunoglobulin classes of transmissible gastroenteritis viral antibodies, *Infect. Immun.* **6:**600–609 (1972).

103. Schmidt, O. W., Antigenic characterization of human coronaviruses 229E and OC43 by enzyme-linked immunosorbent assay, *J. Clin. Microbiol.* **20:**175–180 (1984).

104. Schmidt, O. W., Allan, I. D., and Cooney, M. K., Rises in titers of antibody to human coronaviruses OC43 and 229E in Seattle families during 1975–1979, *Am. J. Epidemiol.* **123:**862–868 (1986).

105. Schmidt, O. W., Cooney, M. K., and Kenny, G. E., Plaque assay and improved yield of human coronaviruses in a human rhabdomyosarcoma cell line, *J. Clin. Microbiol.* **9:**722–728 (1979).

106. Schmidt, O. W., and Kenny, G. E., Polypeptides and functions of antigens from human coronaviruses 229E and OC43, *Infect. Immun.* **35:**515–522 (1982).

107. Schnagl, R. D., Brookes, S., Medvedec, S., and Morey, F., Characteristics of Australian human enteric coronavirus-like particles: Comparison with human respiratory coronavirus 229E and duodenal brush border vesicles, *Arch. Virol.* **97:**309–323 (1987).

108. Siddell, S., Wege, H., and Ter Meulen, V., The biology of coronaviruses, *J. Gen. Virol.* **64:**761–776 (1983).

109. Sitbon, M., Human-enteric-coronaviruslike particles (CVLP) with different epidemiological characteristics, *J. Med. Virol.* **16:**67–76 (1985).

110. Spaan, W. J. M., Background paper: Progress towards a coronavirus recombinant DNA vaccine, in: *Coronaviruses and Their Diseases* (D. Cavanagh and T. D. K. Brown, eds.), pp. 201–203, Plenum Press, New York, 1990.

111. Stair, E. L., Rhodes, M. B., White, R. G., and Mebus, C. A., Neonatal calf diarrhea: Purification and electron microscopy of a coronavirus-like agent, *Am. J. Vet. Res.* **33:**1147–1156 (1972).

112. Stewart, J. N., Mounir, S., and Talbor, P. J., Human coronavirus gene expression in the brains of multiple sclerosis patients, *Virology* **191:**502–505 (1992).

113. Sturman, L. S., Holmes, K. V., and Behnke, J., Isolation of coronavirus envelope glycoproteins and interaction with the viral nucleocapsid, *J. Virol.* **33:**449–462 (1980).

114. Tannock, G. A., and Hierholzer, J. C., Presence of genomic polyadenylate and absence of detectable virion transcriptase in human coronavirus OC-43, *J. Gen. Virol.* **39:**29–39 (1978).

115. Tyrrell, D. A. J., and Almeida, J. D., Direct electron microscopy of organ cultures for the detection and characterization of viruses, *Arch. Ges. Virusforsch.* **22:**417–421 (1967).

116. Tyrrell, D. A. J., and Bynoe, M. L., Cultivation of a novel type of common-cold virus in organ cultures, *Br. Med. J.* **1:**1467–1470 (1965).

117. Tyrrell, D. A. J., Bynoe, M. L., and Hoorn, B., Cultivation of "difficult" viruses from patients with common colds, *Br. Med. J.* **1:**606–610 (1968).

118. Vaucher, Y. E., Ray, C. G., Minnich, L. L., Payne, C. M., Beck, D., and Lowe, P., Pleomorphic, enveloped, virus-like particles associated with gastrointestinal illness in neonates, *J. Infect. Dis.* **145:**27–36 (1982).

119. Wenzel, R. P., Hendley, J. O., Davies, J. A., and Gwaltney, J. M., Jr., Coronavirus infections in military recruits: Three-year study with coronavirus strains OC43 and 229E, *Am. Rev. Respir. Dis.* **109:**621–624 (1974).

120. Yeager, C. L., Ashmun, R. A., Williams, R. K., Cardellichio, C. B., Shapiro, L. H., Look, A. T., and Holmes, K. V., Human aminopeptidase N is a receptor for human coronavirus 229E, *Nature* **357:**420–422 (1992).

121. Zhang, X., Kousoulas, K. G., and Storz, J., The hemagglutinin/esterase gene of human coronavirus strain OC43: Phylogenetic relationships to bovine and murine coronaviruses and C virus, *Virology* **186:**318–323 (1992).

12. Suggested Reading

Bradburne, A. F., and Tyrrell, D. A. J., Coronaviruses of man, *Prog. Med. Virol.* **13:**373–403 (1971).

Hamre, D., and Beem, M., Virologic studies of acute respiratory disease in young adults. V. Coronavirus 229E infections during six years of surveillance, *Am. J. Epidemiol.* **96:**94–106 (1972).

KAYE, H. S., MARSH, H. B., AND DOWDLE, W. R., Seroepidemiologic survey of coronavirus (strain OC43) related infections in a children's population, *Am. J. Epidemiol.* **94:**43–49 (1971).

McINTOSH, K., KAPIKIAN, A. Z., TURNER, H. C., HARTLEY, J. W., PARROTT, R. H., AND CHANOCK, R. M., Seroepidemiologic studies of coronavirus infection in adults and children, *Am. J. Epidemiol.* **91:**585–592 (1970).

MONTO, A. S., AND LIM, S. K., The Tecumseh study of respiratory illness. VI. Frequency of and relationship between outbreaks of coronavirus infection, *J. Infect. Dis.* **129:**271–276 (1974).

SIDDELL, S., WEGE, H., AND TER MEULEN, V., The biology of coronavirus, *J. Gen. Virol.* **64:**761–776 (1983).

CHAPTER 9

Cytomegalovirus

Anne A. Gershon, Eli Gold, and George A. Nankervis

1. Introduction

Only a small proportion of people throughout the world escape infection with cytomegalovirus (CMV). The age at acquisition, the presence and type of clinical manifestations, and the sites and extent of virus excretion vary, but serological surveys conducted on all continents confirm the ubiquitous distribution of human CMV.

The impression that CMV infection was uniformly associated with severe illness of the newborn and carried a high probability of death or marked damage to the central nervous system has been modified since the availability of laboratory procedures and the performance of prospective studies. It is true that CMV can produce a devastating disease in the newborn, but it has been shown that congenital infection with CMV, although relatively common, is usually clinically inapparent, generally has a benign course, and may be recognized only if laboratory studies are performed.

Most people develop antibody to CMV following unrecognized infection acquired during childhood or the young adult years. A small proportion of normal persons may have a form of infectious mononucleosis or possibly symptoms of a respiratory illness with CMV infection, but clinical disease is much more likely to be evident in those who are immunologically deficient because of neoplastic disease or treatment with immunosuppressants.

Infection with CMV is commonly followed by prolonged, often intermittent periods of virus excretion in the face of high levels of circulating antibody. Dissemination of the agent probably occurs by close contact with an excreter, although spread through the placenta, via blood transfusions, organ transplantation, or in breast milk accounts for a significant proportion of clinically important infections. The pathogenesis of CMV infection has been deduced mainly from clinical observations and epidemiologic studies, but there is reason to believe that primary infection with CMV carries a greater probability of illness with significant residual effect than does recurrent infection. The factors that determine which infected persons are affected and which have no apparent illness remain unclear.

2. Historical Background

Large inclusion-bearing cells originally found in kidney, lung, or liver of infants who died from various causes were considered to be the result of some strange parasite, possibly a protozoan.[22,56,94] Goodpasture and Talbot[67] noted a similarity between the appearance of these strange cells, which they referred to as "cytomegalia," and the intranuclear inclusions seen in the skin lesions of herpesviruses. Shortly thereafter,[206] the viral etiology of these unusual cells was first postulated. Subsequently, Cole and Kuttner[31] transmitted the guinea pig form of cytomegalia to animals, using filtered salivary gland material, indicating a probable viral etiology. There continued to be many descriptions of typical cells in various tissues of infants, especially those who died in the newborn period of a hemorrhagic disease resembling erythroblastosis and referred to as "inclusion body disease," "cytomegalic inclusion disease," or "generalized salivary gland virus infection." A number of reports[58,124,127,213] describing typical inclusion-bearing cells in the urine of symptomatic infants provided a method of diagnosing this disease in neonatal cases. In 1956, three laboratories almost simul-

Anne A. Gershon • Department of Pediatrics, Columbia University College of Physicians and Surgeons, New York, New York 10032. **Eli Gold** • Department of Pediatrics, Harborview Medical Center, University of Washington, Seattle, Washington 98104. **George A. Nankervis** • Department of Pediatrics, Northeastern Ohio University College of Medicine at Children's Hospital Medical Center, Akron, Ohio 44308.

taneously reported the isolation of the etiologic agent employing tissue culture techniques: Smith[178] from salivary gland, Rowe *et al.*[169] from cultures of adenoid tissue (thus AD 169), and Weller *et al.*[209] from a liver biopsy.

3. Methodology

3.1. Mortality

Death from CMV infection is rare, and the cause can usually be recognized only as a result of special virological, serological, or histological examination. Pathological findings consist mainly of nuclear and cytoplasmic inclusions in epithelial cells of many organs.[12] The generalized nature of the infection is impressive. Death rates give no indication of the frequency of this disease.

3.2. Morbidity

Since the syndromes associated with CMV infection are often not clinically distinctive and are usually diagnosed only through laboratory procedures, no estimate can be made of their importance through official health records. Furthermore, most CMV infections are asymptomatic. Such incidence data as are available have come from studies, serological or virological or both, of groups at special risk—pregnant women, neonates, patients undergoing transfusions or organ-transplant surgery, and heterophile-negative patients with a mononucleosislike syndrome.

A national surveillance registry was established in January 1990 for infants with severe symptoms due to congenital CMV.[41] The most common clinical manifestation was petechiae, observed in 50%. Often this symptom was accompanied by hepatomegaly and/or splenomegaly. Intracranial calcifications were present in 43% and thrombocytopenia in 52%. Congenital CMV (usually without symptoms or sequelae) occurs in about 1% of all US births or 40,000 newborns yearly.

3.3. Serological Surveys

The complement-fixation (CF) test was widely used in prevalence studies to determine the experience of different populations with CMV infections[23,51,53,75,92,105,169,175,195] and as an indication of infection in special groups studied prospectively to determine seroconversion (negative to positive) and reinfection rates (fourfold or greater rise in titer) in persons with preexisting antibody.[81,137,161,198,210] The CF test was most commonly used for serological surveys or clinical studies because of the relative ease of

preparing the reagents and performing the test and because CF antibody is of long duration. The AD 169 strain of CMV has usually been employed as the antigen because of its wide range of reactivity.[126] The sensitivity of the test varies depending on the antigen preparation, and antibody levels measured in different laboratories may differ. There are data suggesting the existence of antigenic variants of CMV with which AD 169 does not cross-react, introducing problems with serological surveys based on AD 169 alone.[10,207] Most recently, serological surveys have been based on testing of serum by immunofluorescence and by enzyme-linked immunosorbent assays (EIA).[188,219]

3.4. Laboratory Diagnosis

Increasing numbers of immunocompromised patients, including persons with acquired immunodeficiency syndrome (AIDS) or who have undergone organ transplantation, the need to screen blood donors for past CMV infection, and the availability of antiviral therapy for CMV have led to increasing requirements for rapid and accurate laboratory methods for diagnosis of CMV infection. The methods used include demonstration of the virus or its antigens in putatively infected fluids or tissues and identification of specific antibodies to CMV.

3.4.1. Virus Isolation and Detection. For virus isolation, specimens from the pharynx, buffy coat of peripheral blood, breast milk, urine, stool, tears, cervix, or semen are inoculated onto human fibroblast cultures. Cultures are observed for days to weeks for cytopathic effect (CPE)—the light-microscopic appearance of foci or swollen cells with intranuclear inclusions indicative of viral infection. The time necessary for isolation of virus is inversely proportional to the titer of virus present in the inoculum. For example, heavily infected urine from a newborn with symptomatic CMV infection may yield CPE within a few days. Presumed virus isolates are positively identified by a specific method such as indirect immunofluorescent staining using mouse monoclonal antibodies (MAbs) tagged with fluorescein.[71,201] Utilization of the shell vial technique, employing low-speed centrifugation to enhance inoculation and application of MAbs to immediate early (IE) antigens or CMV before CPE has developed is a rapid diagnostic method utilized by many hospital laboratories.[29,66,146] MAbs to the IE proteins of CMV appear within hours of inoculation of fibroblast cultures.[29]

Molecular methods have been developed for the detection of specific CMV DNA in minute quantities of urine or other secretions. These include polymerase chain reaction (PCR)[37,180,214] and DNA hybridization.[30] These tests take only a few days to complete and are becoming more generally available. Commercial kits exist for *in situ* hybridization procedures that are useful for diagnosis of CMV pneumonia, hepatitis, and gastrointestinal infections.[29]

Strains of CMV can be distinguished by the use of restriction enzyme analysis of viral DNA.[1–3,46,145,181,183,212] This tool is currently limited to the specialized research laboratory but can be utilized, for example, to determine the source of infection and for differentiating reactivation from reinfection.

3.4.2. Serological Tests.

There are several radioimmunoassays (RIA), EIAs, and latex-agglutination (LA) tests that are sensitive and specific for the detection of either immunoglobulin (Ig)M or G CMV antibody and have largely replaced the older serological methods such as complement fixation.[11,26,99,102,103,133,151] Many EIAs are licensed by the FDA for testing of a single dilution of serum; in contrast, LA tests are licensed both for screening for the presence of antibodies and for titration of the amount of antibody present. EIAs for demonstrating CMV IgM are commercially available, but it must be recalled that rheumatoid factors and competition between IgG and IgM may interfere with the results of testing.[29] In addition, reactivation of CMV may also induce low levels of IgM antibody. This problem has caused some laboratories to avoid reliance on the IgM test to distinguish between primary and reactivated infections. It is also important for the physician to determine the specific type of laboratory procedure used for virus detection or the measurement of antibody so that specimens can be collected and transported in an optimum manner and the results properly interpreted.

3.4.3. Diagnostic Pitfalls.

Identification of disease, as opposed to infection, caused by CMV may be difficult. In order to diagnose CMV hepatitis or colitis, for example, a tissue biopsy may be required. For diagnosis of pulmonary infection, however, bronchoalveolar lavage has proven superior to lung biopsy.[29] On the other hand, CMV may be present in a lavage specimen but not necessarily be causative of illness, especially in patients with AIDS in whom the offending organism is likely to be *Pneumocystis carinii*, with which CMV may coexist. Even the demonstration of CMV viremia in an immunocompromised patient is not necessarily indicative of illness due to this virus.[29] In diagnosing CMV infection, the distinction between shedding of virus and causation of symptoms due to CMV must be considered and the interpretation of diagnostic tests individualized for each patient.

4. Biological Characteristics of the Virus

The CMV virion, which is approximately 200 nm in diameter and is one of the largest animal viruses, contains a linear double-stranded DNA approximately 240 kb in size, with a molecular weight of 155,000 kDa.[86,129] The genome is similar in structure to that of herpes simplex but nearly 50% larger. It has been estimated that about 30 of the many proteins coded by CMV become part of the structure of the virion, but the function of the many other proteins remains to be elucidated. Eight glycosylated proteins have been described: one, a 55- to 58-kDa protein known as gB is an important stimulus for both antibody and cellular immunity to CMV.[157] The replicative cycle of CMV is divided into three periods determined by the development of specific genes and proteins, which are important in regulating the synthetic process in a cascading manner. IE or α genes regulate subsequent transcription and translation of the early (β) and late (γ) gene products.[129] As has been mentioned, the detection of some of the early antigens in tissue cultures inoculated with specimens containing CMV has provided a method for rapid diagnosis of viral replication.[29,66,72]

Human CMV produces a permissive infection in cultures of human fibroblasts with the development of a characteristic CPE described in Section 3.4. A low level of virus replication can occur in some lines of human epithelial cells and in cultures of human T or B lymphocytes, especially if the cultures are coinfected with Epstein–Barr virus (EBV). Abortive infection occurs in nonhuman cell cultures, which may show cytopathic change. Virus production, however, is very inefficient in other than human fibroblast cultures, and the number of defective particles far exceeds the yield of complete infectious virus. The so-called dense bodies that make up the cytoplasmic inclusions seen in infected cells are noninfectious defective particles, which have the form of a capsid with a full quota of viral protein but are devoid of viral DNA.

Cell cultures of human or nonhuman origin can be transformed with infectious or UV-treated CMV, and the transformed cells in turn will produce malignant tumors when injected into appropriate animals.[6,64] This sequence of events has provoked considerable concern about the oncogenic potential of CMV in the human, especially if the virus were modified for use as a vaccine or if the host is immune suppressed or immune defective.[174]

Cytomegalovirus induces the development of IgG-Fc membrane receptors in infected human fibroblasts, which presumably have the function of protecting the cell from the effect of cytotoxic antibody and incidentally lead to erroneous results with indirect immunofluorescent studies.[86] Virus remains cell-associated in tissue cultures, although serial propagation may result in preparations with relatively large amounts of infectious virus in the culture supernate. The agent is labile to low pH, fat solvents, and temperature. Preservation is best achieved by rapid freezing and storage at $-60°$ to $-80°C$, preferably in 30–50% sorbitol. Inactivation at $-20°C$ is more rapid than at 4°C, making ordinary refrigerator temperature acceptable for short-term storage of diagnostic specimens such as urine.

5. Descriptive Epidemiology

The earliest information concerning the occurrence of CMV in man came from pathologists in Holland, the United States, France, Germany, Brazil, Hungary, and elsewhere who reported finding typical inclusions in the salivary glands and other organs of infants who died with a syndrome characterized by large liver and spleen, diffuse adenopathy, and jaundice. When a viral etiology of this syndrome was demonstrated and serological methods for diagnosis were developed, it became apparent that infection with CMV was common.

In the United States, 0.5–2.2% of infants are infected *in utero*, and a large number acquire CMV infection from their mothers early in life either during the birth process or via breast milk.[48,165,166,192] Toddlers, particularly those who attend day-care centers, have extremely high rates of acquisition of CMV, reaching in excess of 80%.[1,3,141,143,144] Not only are children infected, but transmission to day-care staff members and parents of children in day care may also occur.[142] By the time the general population reaches puberty, 40–80% have antibody to CMV. Living conditions, especially crowding, sexual practices, use of intravenous drugs, and general hygiene, determine the rate of acquiring CMV infection among young adults. In the United States this averages about 2% per year.[47]

5.1. Prevalence and Incidence

The use of the terms ''prevalence'' and ''incidence'' in CMV infections presents difficulties because the presence of virus in the pharynx or urine or of antibody in the blood might represent (1) a primary infection (i.e., incidence), (2) persistence from an earlier infection (prevalence), or (3) reactivation of a latent infection (prevalence plus incidence). The absence of demonstrable antibody in serum does not necessarily mean that prior infection has not occurred, because antibody titers may have dropped to levels not detected by the test used or there may be antigenic differences between the test virus and the infecting virus. The low frequency of clinical response and the lack of a characteristic syndrome associated with primary infection also contribute to the difficulties of interpreting incidence, reinfection, and prevalence data. In infections occurring *in utero*, in newborns, or in infants, the incidence rate of primary infection can be more reliably calculated.

5.1.1.   Prevalence.   The prevalence of antibodies to CMV in various population groups has indicated that CMV infection is ubiquitous (Table 1). The differences observed have been related to the speed of acquisition of infection in different geographic and socioeconomic settings. In St. Lucia, a survey of antibodies to four herpesviruses revealed that by age 5 years, 83.8% had CMV antibody but only 7.1% had varicella-zoster antibody.[55] The presence of CMV in the urine, cervix, or breast milk has also been used to reflect infection prevalence in special groups. About 1% of newborns in the United States have CMV in the urine.[16,76,117,192,194] Thirteen percent of women undergoing examination because of suspected sexually transmitted disease[95] and 4–28% of pregnant women have CMV in cervical secretions.[132,139,165] Three to twelve percent of the latter have viruria at the time of delivery,[84,137] and a 27% prevalence of CMV in the breast milk of recently delivered CMV-seropositive Australian mothers has been reported.[80]

5.1.2.   Incidence.   The incidence of CMV infection and of clinical disease has been determined in prospective studies of special groups as measured by serological or virological tests or both. Table 2 summarizes several of these reports of CMV infection, and the results are discussed below.

5.1.2a.   Pregnant Women.   Serological studies have shown that approximately 35–80% of women entering the childbearing years have antibody to CMV.[108,137,188,198] The frequency with which CMV can be isolated from the pregnant woman appears to vary with age, presence of antibody to CMV, parity, socioeconomic status, site cultured, and time during gestation at time of study, but in general, viruria has been found in 3–12% of a large number of pregnant women.[57,84,108,132,137,165] In the studies of Stagno *et al.*[188] of over 16,000 young women in Alabama, 35% in an upper socioeconomic group and 77% in a lower socioeconomic group were seropositive with respect to CMV, determined by immunofluorescence.

In a prospective study of 1089 young pregnant women attending the prenatal clinic of a large general hospital, 65% had CF antibody, 124 (11.4%) excreted CMV in their urine on one or more occasions during their

Table 1. Cytomegalovirus Prevalence by Serological Survey: General Populations

Author	Date	Place	Population	CMV seropositive (%)
Hanshaw[75]	1966	Rochester	0–5 mo	35
			5–24 mo	3
			2–6 yr	6
			6–10 yr	9
			10–17 yr	22
			27–40 yr	38
Rowe et al.[169]	1956	Washington	6–24 mo	14
			5–9 yr	33
			>35 yr	81
Stern and Elek[195]	1965	London	6–60 mo	4
			5–10 yr	15
			>35 yr	54
Carlstrom[23]	1965	Stockholm	6–24 mo	36
			5–10 yr	34
			>50 yr	63
Jack and McAuliffe[92]	1968	Melbourne	6–36 mo	22
			10–15 yr	40
			>35 yr	60
Evans et al.[53]	1974	Barbados	1–5 yr	62
			15–25 yr	77
Embil el al.[51]	1969	Nova Scotia	6–12 mo	12
			10–14 yr	14
			>40 yr	52
Krech and Jung[106]	1970	Tanzania	6–8 mo	80
			5–14 yr	100
			>20 yr	99
Shavrina et al.[175]	1973	Leningrad	7–12 mo	67
			11–15 yr	63
			51–60 yr	80
Evans et al.[55]	1979	St. Lucia	0–5 yr	83.5
			6–10 yr	94.9
			11–20 yr	92.5
			21–30 yr	96.2
			31–40 yr	100
			>40 yr	96.9
Yow[219]	1987	Houston	0–1 yr	13.6
			1–2 yr	25.6
			2–3 yr	30.25
			3–4 yr	33.63
			4–5 yr	36.53
			5–6 yr	44.03
			6 yr	50.23
			7–10 yr	75.03

pregnancies, and 16 infants with congenital CMV were born.[13] When these data were further analyzed, it became apparent that 15 of 16 infected infants were delivered by seropositive mothers; only one followed a primary maternal CMV infection. In the Cleveland and other studies it appears that approximately 10% of women with serologi-

cal evidence of previous CMV infection will have a recurrent infection detectable by virus excretion in saliva, urine, or cervical secretions during gestation with transplacental transmission to the fetus about 10% of the time. Thus, a pregnant woman with CMV antibody has approximately a 1% chance of delivering a child with congenital CMV infection.

In a pregnant woman without antibody to CMV, the risk for a primary infection is influenced by living conditions and other factors but not by pregnancy per se. This is well demonstrated by six large prospective studies carried out in Great Britain, Scandinavia, and the United States.[5,69,71,109,187,198] Among a total of 11,394 seronegative pregnant women, 1.15% acquired CMV during the 5- to 7.5-month period of observation with a range in the different studies of 0.7% to 4.1%. The 1.15% average rate of seroconversion in these studies extrapolates to an annual rate of 2.3%, a rate similar to that observed in the nonpregnant population. The 126 women who had documented primary CMV infection during the period of gestation delivered 52 infected infants, a fetal infection rate of 41%. A similar rate of transmission to the fetus was found in the large study of Stagno et al.[188]

Thus, CMV may be transmitted to the infant *in utero* if either maternal primary or recurrent infection occurs during gestation. The overall probability of delivering an infected infant has been discussed and does not differ markedly whether the woman has antibody or not when she becomes pregnant. The risk to the infant, however, in terms of significant disability from sequelae is markedly greater if virus transmission has occurred during a primary maternal infection (this is discussed further in Section 7.1.1). It should be apparent, however, that of all infants born with congenital CMV in a particular population, the proportion resulting from primary maternal infection will depend on the proportion of women in that population who were seronegative at the time they became pregnant. In middle- and upper-income groups in which a large fraction are susceptible to CMV, half or more of the infants born with congenital CMV infection follow primary infections.

In Barbados[53] and Tanzania[106] or similar parts of the world where CMV infection is acquired early in life and nearly 100% have antibody by the end of childhood, one would expect an extremely low incidence of infants to be born with symptomatic congenital CMV infection, assuming that primary infection during pregnancy is a prerequisite for that event. However, in populations in which the opportunity of exposure is high, the small proportion of women without CMV antibody entering childbearing age appear to have a high risk of contracting

Table 2. Incidence of Cytomegalovirus Infection in Prospective Studies

Group	Setting	Time period	Number studied	CMV-Ab[a] negative (%)	Number	Ab[a]	Infected	Rate/100 per period of observation	Reference
					Initial Ab[a] status				
Pregnant women	1. All ages in London	6–9 mo	1040	33.4	270	Neg.	11[b]	4.1	Stern and Tucker[198]
	2. Teenagers, Cleveland	Up to 9 mo	1089	35	379	Neg.	5[b]	1.3	Nankervis et al.[137]
	3. All ages, Alabama								Stagno et al.[188]
	a. Low socioeconomic	6 years	12,140	46.5	5,645	Neg.	77	1.6	
	b. Middle-upper socioeconomic	6 years	4,078	23.4	954	Neg.	19	3.7	Stagno et al.[188]
Young adults	1. English colleges	9 mo	1457	70	713	Neg.	10[b]	1.4	UHP[203]
	2. Marine recruits	14 wk	588	69.7	431	Mixed	4[b]	0.9	Wenzel et al.[210]
Surgical patients	1. Cardiopulmonary bypass	3 mo	212	38.7	82	Neg.	50[b]	61.0	Purcell et al.[161]
					130	Pos.	32[b]	24.6	Henle et al.[81]
	2. Renal transplant	3 mo	44	40.9	18	Neg.	12[c]	66.7	Braun et al.[20]
					26	Pos.	18[c]	69.2	

[a]Antibody.
[b]Fourfold increase in Ab.
[c]Fourfold increase in Ab or positive CMV culture.

primary CMV infection during pregnancy[198] and of delivering an affected infant.

Rates of recovery of CMV from cervical cultures of pregnant women vary from 14% among the Navajo,[132] 18% in Taiwan,[7] 15% in Japan,[139] 12% in Birmingham,[165] 5% in Pittsburgh,[132] to 2% in Seattle.[62] Factors responsible may be racial, seasonal, or more likely socioeconomic. Recovery of CMV from the cervix is more frequent than recovery from urine by a factor of 2 or more, and the simultaneous isolation from both sites is uncommon, although this has not been the experience of all investigators.[137] The frequency of positive cultures of the cervix increases as gestation progresses[132,139,165] and appears to be higher[132] in younger mothers with fewer than four pregnancies. Interestingly, infants born to mothers with proved cervical CMV infections are unlikely to have congenital CMV infection, although such infants have a high risk of acquiring CMV infection during the first few months of life. Reynolds et al.[165] offer the hypothesis that the increasing rate of cervical and urinary excretion as gestation proceeds and the high rate of virus excretion in breast milk are the result of reactivation of latent CMV infection, perhaps provoked by hormonal influences in late pregnancy.

5.1.2b. Young Adults. Prospective studies of young adults have indicated a very low rate of CMV infection that is usually asymptomatic (Table 2). Among

1457 entering freshman college students in the 1960s in five English colleges and universities, 70% lacked CMV antibody; 713 of the 1026 lacking antibody were retested 7 months later and only 10 were found to have acquired antibody, an infection rate of 1.4% in the susceptible group.[203] Only one of these ten developed a mononucleosislike syndrome. A similar study was made of 588 marine recruits undergoing "boot camp" training over 14 weeks at Parris Island.[210] On entry, 69.7% lacked CMV antibody. When a second serum sample was obtained from a group of primarily seronegatives 14 weeks later, four had developed CMV antibody, an infection rate of 0.9%. In contrast to CMV, similar surveys for EBV infection in college and military groups have shown seroconversion rates of 12–13% in those lacking EBV antibody, and 25–75% of these infections have been associated with clinical infectious mononucleosis.[115,171,203] It should be noted, however, that CMV mononucleosis occurs more commonly in the 25- to 35-year age group, so that the group at highest risk has not been included in the prospective studies of CMV infections thus far.

5.1.2c. Children. In the prospective study of sera from 177 infants and children of low- and middle-income families in Houston, it was found that over 25% of children seroconverted to CMV in the first 2 years of life. Thereafter from years 3 to 10, from 3 to 7.5% seroconverted yearly so that 33% had seroconverted by age 10.[219]

5.1.2d. Transfusion and Postsurgical Groups. Following primary infection, CMV is thought to result in latent infection in mononuclear white blood cells, explaining how the virus is at times inadvertently transmitted by blood transfusion.[129,184] CMV can also be transmitted by organ transplantation.[88] The CMV infection rate depends on the initial antibody status of the recipient,[81,161] previous CMV infection in the donor,[20] the number of units of blood transfused,[160] and the immunologic status of the recipient. This last is of special significance because a host compromised by natural or drug-induced immunosuppression appears also to be at high risk of CMV reactivation. Table 2 presents some representative data. A primary infection rate of 66.7% has been recorded following open-heart surgery in those initially lacking CMV antibody, and an antibody rise of 24.6% was observed in renal transplant patients, some of which appeared to be temporally associated with kidney rejection, interstitial pneumonia, or hepatitis.[20,34,35,119,121] Studies in recipients of renal transplants and CMV vaccine have shown that the greatest infection and disease rates occur in CMV seronegative recipients who have been transplanted with a kidney from a CMV seropositive donor.[157]

5.2. Geographic Distribution

Antibody to CMV is prevalent in adults throughout the world, ranging from 37% in Rochester, New York,[75] to 50–60% in Nova Scotia,[51] Berlin,[82] London,[195] and Debrecen, Hungary,[204] and rising to greater than 80% in Barbados[53] and Tanzania.[106] The major differences are related to the speed of acquisition of infection in various geographic and socioeconomic settings. Antibody has been detected in all populations thus far tested, including the remote Tiriyo Indians of Brazil, who essentially lack measles and influenza antibody.[17]

5.3. Age and Sex

Plots of antibody prevalence by age and sex show a similar contour for most populations. There is a loss of transplacentally acquired antibody in the first year of life, a gradual acquisition of antibody throughout childhood, a more rapid increase in proportion with antibody during the young adult years, and then a leveling off in the rate of seroconversion. As shown in Table 1, the most striking difference in data from a range of socioeconomic and racial groups is the age at which most persons have their primary CMV infection (i.e., first develop antibody to CMV). Fewer than half the subjects 15 years of age in

most of the studies conducted in both the Eastern and the Western hemispheres had CMV antibody, but in Barbados, the figure was approximately 80%, and even 100% in Tanzania.

The age at infection may influence the nature of the clinical syndrome produced. Infection during gestation may result in a congenital CMV syndrome, whereas infection delayed to age 20 or over may result in a CMV mononucleosis syndrome; this latter also occurs after blood transfusion in adults. Regardless of age or sex, the majority of infections with CMV in immunologically normal persons are clinically asymptomatic. In contrast, immunocompromised children or adults are at risk to develop severe CMV infections, including pneumonia, colitis, and retinitis (see Sections 8.2.4–8.2.6).

5.4. Temporal Distribution

Seasonal or yearly patterns of CMV infection have not been clearly delineated.

5.5. Occupation

There are several studies of nurses and other health care personnel that attempt to determine if individuals who work in close contact with infected patients acquire CMV more frequently than comparable individuals in other occupations. Many of these studies are flawed by poor compliance, infrequent sampling, selective responses, or absence of proper controls, but in general they show no significant job-related risk (Table 3). A careful prospective study, indicating that nosocomial transmission is unusual, in contrast to the day-care setting, has been published.[39] During a 2-year period, Demmler *et al.*[38,39] prospectively studied possible transmission of CMV in patients and their caretakers in a chronic care pediatric unit and a neonatal nursery. During the study, 2 of 69 (3%) of nurses in the nursery seroconverted. In the chronic care facility, of 188 personnel and 630 patients, there was one instance of possible patient-to-patient spread and one instance of patient-to-caretaker spread. This latter instance was determined by DNA analysis, however, to have been sexually transmitted from husband to wife rather than occurring in the hospital. Careful molecular epidemiology is therefore extremely important in assessing whether nosocomial transmission of CMV has occurred, as demonstrated by other investigations showing that strains of CMV isolated from patients and those excreted by medical personnel or other patients with whom there had been contact were similar[183] or different.[47,212,218]

Table 3. Risk of CMV Infection in Health Care Personnel and Others

Author, date	Observation
Haldane et al., 1969[74a]	Retrospective study based on reponse to questionnaire by 54% of 3670 nurses, which indicated an association between caring for an infant with congenital defects and having such a child.
Yeager, 1975[216]	Seroconversion rate 4.1% among 34 hospital personnel in NICU vs. 7.7% of 31 on General Pediatrics Ward and none of sample without patient contact.
Henneberg et al., 1980[82]	Cytomegalovirus antibody present in 77% of health care workers with patient contract vs. 39% among those not having regular close contact with patients. Sample size 161 individuals.
Ahlfors et al., 1981[5]	Among nurses less than 25 years of age who tended infants for more than 6 months, more had antibody to CMV than a group with less than 6 months of infant nursing service but the same as a group of 163 control women who did not work in the health services.
Dworsky et al., 1983[47]	"Occupational contact (pediatric health care workers) confers no greater risk than faced by young women in the community at large."
Friedman et al., 1984[63a]	Ten of 138 health care workers with patient contact acquired CMV antibody in contrast to 1 of 35 with no patient contact during a 1-year period. "Risk of cytomegalovirus is substantial and appears to be related to the type of patient contact."
Pass et al., 1986[142]	CMV infection occurred in 9 of 20 patients (45%) with a child in day care <18 months at enrollment, who was shedding CMV.
Demmler et al., 1987[39]	Over 2 years, 2 of 69 nurses (3%) in the nursery seroconverted to CMV. In a chronic care facility, 1 of 188 personnel (<1%) seroconverted.

5.6. Race and Socioeconomic Setting

Many variables influence the time at which CMV infections are acquired and the outcome of those infections. People in lower socioeconomic groups, especially if they live in tropical settings, acquire CMV infections earlier in life. In this group, cervical excretion of virus during pregnancy is common and viruria has a high frequency among neonates.

6. Mechanism of Transmission

There are two periods of life when the chances of transmission of CMV are greatest.[87] These are early in life, including gestation and the perinatal period, when there is exposure to maternal CMV; virus in the cervix and urine at the time of delivery in many mothers provides a "sea of CMV" through which the newborn infant must pass, placing the baby at high risk of infection. Later, exposure to breast milk and close contact with other young children results in frequent transmission of CMV. The second period of high transmission is after puberty, when sexual activity begins.

Prolonged excretion of virus in urine, saliva, stool, tears, breast milk, and semen is characteristic following CMV infection whether or not the patient is symptomatic and despite the presence of high levels of circulating antibody and plays an important role in transmission.[112,113] The mechanism responsible for eventually turning off viral synthesis—or, more accurately, turning it off and on, as is frequently observed—is postulated to be related to cell-mediated immune mechanisms, but precise data are not available.

Close or prolonged contact with playmates who are excreting virus is probably the most important method of spread of CMV among children. Studies contrasting antibody prevalence among children living together in large groups and those living at home support this view. Children in English boarding schools had antibody prevalence rates four times those of children attending day school,[195] and a similar magnitude of difference was demonstrable when Swiss nursery school students were compared with a suitable control group.[106] The transmission of CMV from a child to an adult may also occur frequently in the home setting. In a study of 68 Houston families, all of which contained members without antibody to CMV, it was shown that within a period of 3.5 years, members of 37 families acquired CMV antibody.[202] In 10 of the 14 families in which a primary source of infection could be identified, it was a child. This finding is consistent with earlier studies that have shown that infants who acquired CMV by transfusion brought the infection home to their mothers.[216]

Cytomegalovirus infection is spread very efficiently among children, especially those 2–3 years of age in the day-care or play school setting; spread to staff and to parents has also been described.[1,2,90,91,141,143,144] The day-care center has become an important source for the transmission of CMV infection in the United States, where over half the children may be excreting CMV in either urine or saliva with rates as high as 83% among toddlers.[143] Cytomegalovirus has been isolated from toys and objects such as suction bulbs, feeding tubes, diapers, and the Plexiglas surfaces of incubators that have been in contact with infected secretions.[90] Spread probably occurs by direct contact with the secretions of infected individuals, but does not appear to take place by way of aerosol or small droplets among individuals sharing the same room.[141]

The high frequency of antibody among children in underdeveloped countries or in lower-socioeconomic groups is probably the result of crowding and poor sanitation. Interestingly, children of migrant farm workers in upper New York State[118] and 1- to 5-year-olds in Egyptian villages[168] had antibody rates similar to those reported for Barbados[53] and Tanzania.[106]

The low seroconversion rate for CMV in young adults may be due to the low rate of salivary excretion in this age group.[194] In contrast, salivary carrier rates for EBV range from 15–20%.[54]

With the advent of sexual activity, the incidence of transmission of CMV increases. In a study of nonpregnant women attending a clinic for sexually transmitted diseases, primary infection and seropositivity both correlated strongly with sexual activity. Multiple sex partners and a past history of sexually transmitted disease were distinguishing characteristics of a group of women with a yearly acquisition rate of primary CMV more than five times the usual.[25] In male homosexuals with AIDS, nearly 100% have evidence of infection with one or more strains of CMV.[43]

7. Pathogenesis and Immunity

Primary CMV infection has an incubation period of 4 to 6 weeks and is clinically the most important form of the disease. However, there are marked differences in the consequences of the various forms of CMV infection. Congenital infection carries a risk of residual damage, especially to the nervous system, whereas acquired infection, although occasionally associated with illness, is apparently not followed by any disability even when it occurs in the newborn. Maternal primary and reactivation infection during pregnancy is associated with a high probability of infection in the newborn, but sequelae in the infant are more likely to occur after primary maternal infection occurring in the first half of gestation.[188] Not only is recurrent CMV infection during pregnancy unlikely to result in permanent harm to the infected baby, but also recurrent infection at any time is unlikely to be of clinical importance except in the immunodeficient patient. Ho[87] has pointed out that "morbidity of CMV infection may actually be considered a marker for immunosuppression."

The differences in pathogenesis among various types of CMV infections are discussed below.

7.1. Pathogenesis

7.1.1. Neonatal Infection: Congenital. Early in the course of primary infection, CMV is disseminated widely in various organs of the host, as indicated by virus isolation from multiple sites. If the host undergoing primary infection is pregnant, virus may infect the placenta and in some cases penetrate the placenta to infect the fetus. Early studies of infants with what was termed "cytomegalic inclusion disease" showed the association between specific clinical manifestations and evidence of infection with CMV.[208] As laboratory methods for demonstrating infection were more widely utilized, the spectrum of changes occurring in congenital CMV broadened and the "expanded syndrome" was described. Later, as information from prospective studies of groups of unselected newborns accumulated, it became apparent that congenital CMV infection is not invariably associated with severe mental or neurological deficit.[14,32,166,192] Unlike congenital rubella, in which the month of gestation when infection occurs determines the particular organ in which a congenital defect is manifested, this has not been shown of CMV infections during pregnancy.

On the basis of the available data, one can make a number of speculations. Maternal viremia is more likely to occur with primary than with recurrent infection, and if it occurs in the latter, the virus titer may be low. Under certain conditions (e.g., trimester of pregnancy, state of placenta), virus in small quantities may cross the placenta and initiate replication in various fetal tissues. Maternal IgG antibody crosses the placenta as it is formed and interacts with the CMV recently synthesized in the fetus. Whether the fetal infection proceeds and whether clinical manifestations of CMV develop may depend on the relative amounts of virus and immune globulin that interact. Much of the time, even with primary maternal infection, virus replication and dissemination may be inadequate to result in fetal infection; but occasionally, even with recurrent infection, dissemination and fetal infection result, although classic disease is unlikely.

7.1.2. Posttransfusion and Organ Transplant. The pathogenesis of CMV infection following transfusion or organ transplantation is not clear. In many situations several possibilities exist; disease may occur after a primary or a reactivated infection, and if the latter, is dependent on whether the source of virus was the recipient him- or herself, the donor blood, or the transplanted organs of the donor. In transfusion recipients lacking demonstrable CMV antibody and transfused with blood from antibody-positive donors, some of whom have been shown to harbor CMV in their buffy coats,[40] a primary infection seems likely. In the presence of CMV antibody in the recipient, a "mixed-lymphocyte" response between donor and recipient lymphocytes might activate CMV in either.

In renal transplant patients, both reactivation and reinfection are probably important. At least two studies have indicated that the donor kidney may be the source of virus in many posttransplant patients. Ho[88] and Nankervis[136] observed that recipients of kidneys from CMV-seropositive donors had a significantly higher risk of CMV infection than did recipients of kidneys from CMV-negative donors. This was particularly true if the recipients themselves were seronegative.[138] The presence of immunosuppression in these patients probably contributes to the high risk of CMV infection.

7.1.3. Cytomegalovirus Mononucleosis. The epidemiology and pathogenesis of CMV mononucleosis pose questions very similar to those about EBV mononucleosis. In both diseases, the host response appears to be associated with primary infection first acquired in young adult life, most likely through intimate exposure to a pharyngeal or sexual carrier or acute case. Studies indicate that clinical infectious mononucleosis is an immunologic response to EBV infection involving T- and B-cell interactions.[176,205] The atypical lymphocyte may represent in part virus-transformed B-type lymphocytes but mostly T cells reacting to neoantigens induced by EBV on the B-cell membrane. The similarities of the clinical and hematologic picture of CMV mononucleosis to that caused by EBV suggest that a similar immune mechanism may be at play.[167] The major difference is that CMV infection does not produce heterophil antibody. The mean age of patients with CMV mononucleosis is usually older than those with mononucleosis due to EBV, and exudative tonsillitis and cervical adenopathy are less, but enlargement of the liver and spleen are more common.[54,100]

7.1.4. Other Possible Manifestations. Cytomegalovirus has been implicated in the pathogenesis of human atherosclerosis.[74,149,150] Injury to the arterial endothelial lining by CMV and other herpesviruses is postulated to be an initial step in at least some cases of atherosclerosis. Antigens of CMV, viral particles, and CMV DNA have all been demonstrated in surgically removed arterial tissues from patients with this disease.

7.2. Immunity

Antibody to CMV develops regularly following infection and, although levels may fluctuate, appears to persist for life, with the exception of some congenitally infected children who gradually lose their antibody. The primary immune response to CMV is specific, and infection with another member of the herpesvirus family does not cause the appearance of antibody in a previously CMV-antibody-negative subject.

Evidence continues to accumulate, interestingly, that CMV itself is immunosuppressive. The peripheral lymphocytes of patients with primary CMV infection and the mononucleosis syndrome function in a diminished, less-effective manner.[85] There are abnormalities involving cytotoxic, interferon, and proliferative responses of the T lymphocytes.[173] In renal transplant recipients, CMV infection frequently precedes severe infection with opportunistic pathogens, which suggests a CMV-induced immune deficiency. The changes in immune function that occur in patients with AIDS are similar to those associated with CMV, and since almost all patients with AIDS are infected with CMV, a role for CMV in the pathogenesis of AIDS seems possible.[43] In a study of 38 HIV-infected children, mortality and CMV infection were significantly associated.[63] Patients with AIDS, however, have multiple new infections or reinfections with different strains of CMV, each of which results in an extended period of immune suppression. This defect in immunity may provide the opportunity for the AIDS agent to express itself in the form of progressive disease.

Cell-mediated immune mechanisms rather than humoral immunity appear to play the major role in determining the outcome of severe CMV infection. The HLA-restricted cytotoxic lymphocyte (CTL) reaction has emerged as the most likely major factor.[162] The human CTL kills virus-infected cells only if they are HLA-similar and have viral antigen on their surface. Thus CTL probably plays a significant role in recovery from viral infections. Cellular immunity to CMV may also play a role in disease pathogenesis. For example, the CTL response appears in recipients of bone marrow transplants concurrent with the onset of CMV pneumonitis. The most successful form of therapy for these patients is not only administration of the antiviral drug ganciclovir but also CMV immunoglobulin to dampen the cell-mediated immune response to CMV.[73] In normal individuals with CMV infection, function of the CTL is brisk. In patients with AIDS, it is absent.

A desirable goal of vaccination presumably is the induction of the CTL response. Understanding the many cell-mediated immune reactions and their effect on CMV infection and vice versa is the goal of much current research. The hoped-for reward will be effective methods for managing the patient at high risk for severe CMV infection.

8. Patterns of Host Response

Most CMV infections are inapparent and asymptomatic and can be detected only by laboratory study. The frequency and nature of the clinical response appear to depend on the age at which infection is acquired, on the route of infection, and on the immune status of the host. The most common forms of associated clinical syndromes are listed in Table 4.

8.1. Neonatal Infections

8.1.1. Congenital. The risk of infection and the pathogenesis of infection have already been discussed. The association between significant nervous system damage and congenital CMV infection remains unquestioned, and the frequency with which CMV infection acquired *in utero* leads to immediate or late clinical residua as well as the factors related to such an outcome are coming into focus. Early studies suggested a severe outcome after many congenital infections. In a follow-up of 20 children with severe congenital CMV infection, most of whom had overt evidence of disease at the time of birth, only 4 of 15 who survived were considered to have developed normally and 2 others showed possible evidence of retardation.[125] Another group of 12 children with clinical and laboratory evidence of congenital CMV infection were evaluated at the age of 3 to 12 years. Of the 12, 3 were considered of average intelligence, 1 was mildly retarded, 3 were moderately retarded, and 5 were severely retarded.[14]

Several early studies also indicated that infants with no apparent evidence of CMV infection in the newborn period may develop neurological sequelae. Hanshaw,[75] struck by the association between CMV infection and failure of brain growth observed in early studies, tested sera from physically handicapped children who had no history of CMV infection and found a significantly larger proportion with CMV antibody among the group with microcephaly. Baron *et al.*[10] and Nakao and Chiba[135] were unable to confirm these findings. However, in further studies,[196] patients with neurological symptoms compatible with possible congenital infection were found to have CMV antibody significantly more frequently than a selected control group. Longitudinal prospective observations of Reynolds *et al.*[166] eliminated many of the hazards of these retrospective studies. Among 267 neonates with elevated umbilical cord IgM levels, 18 who were clinically well were found to have laboratory evidence of congenital CMV infection. Of 16 who were followed, 9 developed sensorineural hearing loss. Hanshaw *et al.*[78] followed 44 children originally identified by elevated specific CMV IgM levels in cord blood. Of the 44 children, 16, or 36.6%, had developmental abnormalities precluding adequate performance in regular school. In addition, 5 of 40 infected children (12.5%) had sensorineural hearing losses as opposed to only 1 of 44 (2.5%) matched controls. It is currently estimated that of infants with primary congenital CMV infection who are asymptomatic at birth, 15% will go on to develop sequelae during childhood.[61]

Other studies indicate that among infants with congenital CMV infection who appear clinically normal in the neonatal period, the presence of elevated IgM levels (not proved to be specific CMV antibody) may predict an increased incidence of late residual effects. In a sample of nearly 2000 unselected newborn infants, five with both CMV-specific macroglobulin and elevated total IgM were observed for a period of over 2 years.[76] Of these five, two were microcephalic and severely retarded, one showed slow psychomotor development, and two were normal. In a similar survey,[16] three infants found to be excreting CMV in their urine within the first 24 hr of life had elevated IgM levels during the first months of life. None of the three had the classic signs of congenital CMV infection, but one developed mild spasticity in the second year of life, although the other two remained normal. In the most unusual situation, in which two consecutive pregnancies resulted in the birth of CMV-infected infants, one had elevated cord IgM and became markedly retarded, whereas the other, with normal levels of IgM at birth, developed normally.[191]

Table 4. Host Responses to Cytomegalovirus Infections

A. Neontal infections
  1. Congenital
  2. Acquired
B. Infection of children and adults
  1. Hepatitis
  2. Mononucleosis
  3. Pneumonitis
  4. Posttransfusion CMV infection
  5. Posttransplant syndrome
  6. CMV and malignant disease
  7. Other possible syndromes:
    Encephalitis
    (Guillain–Barré syndrome)
    Ulcerative colitis
    Retinitis

The variation in the outcome of congenital CMV in these studies is probably the result of the composition of the groups selected for follow-up. The proportions of the sample resulting from either primary or recurrent infection acquired during pregnancy markedly affects the prognosis in the infants. The vast majority of significant symptoms or disability occur in infants who acquire congenital CMV infection following primary infection during gestation.[109,187]

The risk to the infant following recurrent maternal infection is extremely small. In studies in which cases of congenital CMV are selected at birth because of the presence of symptoms or because of elevated levels of IgM or CMV-IgM in the cord blood, the probability of sequelae will be higher since there is a clear correlation between these particular criteria and an increased likelihood of residual defects. The proportion of infants with congenital CMV infection who develop sensorineural hearing loss ranges from 7 to 25%, depending on the method used in selecting the cases tested.[78,189,190]

In a recently published prospective assessment of 32 symptomatic prospectively followed CMV-infected children whose mean age was 6.6 years, the subjects clearly fell into two groups: those with severe retardation (mean IQ of 29) and those with normal intelligence (mean IQ of 92). There was a strong correlation between microcephaly, neurological abnormalities, and low intelligence, but there was no correlation between low IQ and hearing loss.[32]

Longitudinal studies utilizing sensitive, specific methods for a diagnosis indicate that congenital CMV infection resulting from either primary or recurrent infection during pregnancy occurs in 1% of all newborns.[76,86,109,137,187,188,191] According to a large study in Alabama, about 7% of these infants will be symptomatic, and of those that are symptomatic, 1 in 12 will have fatal infections and 90% of survivors will have sequelae.[61] Follow-up studies of children with subclinical congenital CMV have revealed a variety of late-onset sequelae, the most frequent of which are sensorineural hearing loss, psychomotor retardation, and neuromuscular disorders. The probability of the occurrence of one or more of these abnormalities is 15%. This accounts for some 8000 infants yearly in the United States.[61]

8.1.2. Acquired. It has been observed during the follow-up of infants born to mothers excreting CMV during pregnancy that a significant proportion who were culture negative at birth became infected with CMV during the first few months of life. In a Cleveland study,[137] 21 infants of over 100 CMV-excreting mothers developed viruria, 16 before the age of 14 weeks; others[117,165] have reported similar findings. In one study,[165] infection of the infant during the first few months of life could be correlated with late gestational cervical excretion by the mother, and the conclusion was drawn that the transmission of CMV from the infected cervix during birth is an important route of infection in early life, analogous to that of herpes simplex virus. Transmission of CMV via infected breast milk and from close contacts are also important sources of infection for the young infant.[1,189]

Acquired CMV infection in the young infant is, like the congenital form, chronic, and virus excretion in urine or saliva continues for months. Although the patients usually show no symptoms related to their infection, Nankervis *et al.*[137] found that 7 of the 21 infants in the Cleveland study had symptoms that coincided temporally with the first positive urine culture. Three had interstitial pneumonia, two cases being severe enough to require hospitalization. One patient had cervical and inguinal lymphadenopathy associated with a diffuse maculopapular rash, one had mild hepatosplenomegaly with elevated SGOT [now AST (aspartate aminotransferase)] and SGPT [now ALT (alanine aminotransferase)] and 17% atypical lymphocytes, one had mild hepatosplenomegaly with a diffuse maculopapular rash, and one had only mild hepatosplenomegaly. The research group at Alabama has also described an unusual but severe form of pneumonitis in young infants with CMV infection.[19,186]

Long-term follow-up with pyschometric, neurological, and audiometric testing on infants with acquired disease has indicated that they develop normally.[165]

8.2. Infection of Children and Adults

Most of the large number of children and adults who have antibody to CMV remain well or have a mild, nonspecific illness when they experience their CMV infection. A small proportion, and this figure is probably considerably less than 1%, may have one of a variety of clinical syndromes that have been shown to occur concurrently with CMV infection. These include hepatitis, an infectious mononucleosislike syndrome, various respiratory or gastrointestinal tract symptoms, and even more rarely signs of central nervous system disease. The probability of illness accompanying CMV infection is greater in the immunodeficient person or in those receiving blood transfusions, but illness has also been observed in previously well subjects. The syndromes described often overlap, especially in the transplant recipient or patient with malignant disease. The source of infection in patients with the several clinical forms of CMV appears to be exogenous, but in several instances a number of well-studied cases could have been "reactivated" infections.

8.2.1. Hepatitis. Cytomegalovirus has been associated with liver disease since the original isolation[209] from a biopsy specimen obtained from a child with chorioretinitis, hepatosplenomegaly, and cerebral calcification. Hanshaw *et al.*[77] evaluated 20 asymptomatic CMV-positive children as well as appropriate controls. Abnormal liver function tests were six times more frequent in virus-positive than in control children. Among a group of 22 children and one young adult with an enlarged liver or spleen or both for which an explanation was being sought, nine (39%) had viruria. Cytomegalovirus was also isolated from the liver of one child who died. Further studies[194,197] have indicated that liver involvement, which is part of the congenital CMV syndrome, is not uncommon in patients with acquired CMV, although in the latter group, clinical manifestations may vary from mild transient abnormalities in liver function to severe icteric disease difficult to distinguish from other forms of hepatitis. In general, the prognosis of hepatitis associated with CMV is good; a chronic course appears unlikely, and the few deaths recorded have been in patients with abnormal immune systems.

8.2.2. Mononucleosis. Klemola and Kaariainen[100] reported a series of patients, all but one of whom were adults, who developed an infectious-mononucleosis-like illness characterized by fever and liver involvement but no pharyngitis or significant cervical adenopathy. Laboratory studies revealed a high percentage of atypical lymphocytes in the smears of peripheral blood, a negative Paul–Bunnell test, an increase in cryoimmunoglobulins, and serological evidence of recent CMV infection. This syndrome has been observed in previously healthy individuals, but occurs more frequently in subjects who have received large volumes of transfused blood, as discussed below. Patients generally improve within 6 weeks with a return of normal liver function tests and the disappearance of atypical lymphocytes from the peripheral blood.

Klemola *et al.*[101] have provided rather compelling data to relate heterophil-negative mononucleosis to CMV infection. In a series of studies, they found that among 350 patients with infectious diseases of miscellaneous etiology, only one showed a rise in antibody to CMV (and that patient had Guillain–Barré syndrome); of 90 patients with Paul–Bunnell-positive infectious mononucleosis, none demonstrated a significant titer elevation, but 13 of 18 individuals with febrile, Paul–Bunnell-negative mononucleosis had a fourfold or greater rise in level of CMV CF antibody. Overall, CMV mononucleosis was diagnosed in 9% of the patients referred because of "possible mononucleosis." Spontaneous CMV mononucleosis has subsequently been reported from many laboratories[36,96,104,110,123];

overall, it accounts for 5–7% of the infectious-mononucleosis-type syndrome and shows clinical, hematological, and epidemiologic features similar to those of EBV except that heterophil antibody does not appear.[54]

8.2.3. Posttransfusion. The syndrome of fever with atypical lymphocytes was initially observed in patients who had undergone cardiac surgery using pump oxygenators,[107,211] and the fresh blood used was suspected as the source of virus. Subsequently, CMV has been recovered from the blood of patients with hepatitis,[199] leukemia,[79,93] mononucleosis,[96,111,114] and from immunologically depressed transplant recipients[8,34,59]; but only Diosi *et al.*[40] have been able to isolate CMV from the blood of *healthy* blood donors (2 of 35 tested). A number of laboratories have been unable to repeat the last results, possibly because of the methods of collecting, storing, or testing of the blood samples.[60,98,147]

In one series of 53 patients who underwent open-heart surgery with an extracorporeal pump,[24,60,98,147] 21 had a rise in CMV antibody, but only four developed the typical mononucleosislike syndrome. Other studies[140] have indicated that about a third of such patients show boosts in CMV antibody, with only the occasional patient having an accompanying illness. The volume of blood transfused, the age of the blood, the antibody status of the patient, and mechanical damage to cells by the pump may all contribute to the possibility of the patient showing a CMV antibody response. Prince *et al.*[160] reported a rather direct association between the number of units of blood transfused and CMV antibody conversion; 7% of those who received one unit of blood but 21% of those given multiple transfusions had significant elevations in CMV antibody titer. The rate was over twice as high in immunosuppressed transplant recipients. Preexisting antibody status or use of fresh blood was of little importance.

A boost in CMV antibody followed large-volume transfusions of fresh blood in subjects with no or low titers of preexisting antibody but in none who had high titers. However, among patients who received stored citrated blood, none had antibody rises regardless of the preoperative level.[140] The lack of preoperative antibody and the rather slow rise in antibody titer suggest that the postperfusion syndrome is associated with a primary infection by CMV derived from donor blood, but in some patients, the antibody status and rapid rise make it difficult to rule out the possibility of reactivation of an old infection with antibody boost. Cytomegalovirus infection has occurred in neonates who have undergone single transfusion[215,216] or exchange transfusion.[122]

The probability of primary CMV can be markedly reduced by administering only seronegative blood or blood products to neonates and transplant recipients or by using frozen, deglycerolized, or otherwise treated preparations of blood.[18,217]

8.2.4. Posttransplant. Cytomegalovirus infection following renal transplantation is very common[136] and may be associated with a variety of clinical manifestations or be completely asymptomatic. Fever, leukopenia, hepatitis, arthralgia, retinitis, ulceration of the esophagus, stomach, or bowel, encephalitis, or pneumonia may be present in various combinations in those severely ill. The source of the virus may be endogenous in origin, introduced with blood, or present in the donor kidney. Studies differentiating primary and recurrent infections utilizing DNA fingerprinting have been performed that define the source.[27,28] It is of interest that other herpesviruses, including varicella-zoster virus (VZV), also reactive in patients following renal transplantation.[70,89,120]

The success of a renal transplant depends in large measure on whether or not the recipient develops active CMV infection and whether it is primary or recurrent.[15,85–87,136,148] When an allograft donor is seropositive and the recipient is seronegative, primary infection develops in over 75% of such pairs. If the recipient and donor are both seronegative, primary infection does not occur unless seropositive blood or blood products are administered. The frequency with which CMV shedding occurs with recurrent infection is largely dependent on the immunosuppressant regimen used after transplantation but usually exceeds 50% in seropositive recipients. Illness characterized by fever, chills, joint discomfort, splenomegaly, and lymphopenia is more common in primary infection, although the seropositive recipient who receives antithymocytoglobulin is at high risk for virus shedding and illness. Disseminated CMV accounts for some of the deaths following renal transplantation, but more commonly the associated neutropenia results in secondary bacterial or fungal infections and death. In addition to CMV-associated infection after transplantation, allograft failure is frequent. It is not completely understood whether virus infection leads to graft rejection or vice versa, but the current thinking is that CMV infection influences the retention of the graft in some complex manner. Among the factors that are important are the occurrence of primary CMV infection and the administration of antithymocytoglobulin; this combination leads to the poorest results. The kind and amount of immunosuppressive treatment are the major determinants of the vulnerability of transplanted patients to infections with CMV and other agents. There is a marked decrease in the occurrence of CMV disease in allograft recipients receiving cyclosporin A instead of the previously used immunosuppressive regimens.[44,134] The basis for this difference remains to be defined, but it appears that cyclosporin A is less inhibitory to the immune system than previously used regimens for immunosuppression. Plotkin et al.[157] found improved cadaveric graft survival in 11 renal transplant patients in comparison to 15 unvaccinated renal transplant patients who received CMV plus donor kidneys, suggesting a role for CMV in the pathogenesis of graft-versus-host disease.

Among patients receiving bone marrow transplants, CMV pneumonia with fever and leukopenia carries a poor prognosis, frequently resulting in graft rejection.[131] Again, there appears to be a relationship between the degree of immunosuppression and the severity of CMV infection. Untangling these complex interactions may eventually lead to ways to circumvent the apparent synergistic action of CMV and certain types of immune suppression.

8.2.5. Cytomegalovirus and Malignant Disease. The patient with malignant disease may have a severe or protracted illness with CMV infection[21] but may not have an increased risk of acquiring such an infection. Benyesh-Melnick et al.[13] and Dyment et al.,[49] in separate studies, found low rates of CMV in children with leukemia (2%), and Hanshaw and Weller[75a] isolated CMV from the urine of three patients among 50 with leukemia, lymphoma, or Hodgkin's disease (rate of 6%). Sullivan et al.[200] reported that when seroconversion was used as an indicator of CMV infection, 9 of 16 children had fourfold antibody rises during the course of leukemia. Henson et al.[83] made the observation, which is difficult to interpret, that among children with leukemia, those excreting CMV had more episodes of pneumonitis or fever with rash but not more episodes of hepatitis, fever without rash, or respiratory tract infections than those not shedding virus in the urine.

8.2.6. HIV Infection. Cytomegalovirus has been implicated in a number of infections associated with infection with human immunodeficiency virus (HIV). Although it remains unproven, CMV has been postulated to be a cofactor along with HIV in progression from asymptomatic HIV infection to AIDS, particularly in young infants.[63] In addition, due to immunosuppression, persons with HIV infection are at increased risk to develop severely symptomatic CMV infections, such as retinitis, colitis, and pneumonitis. Moreover, CMV in patients with AIDS has been associated with encephalitis and multiple endocrinopathies.[87]

8.2.7. Other Possible Syndromes. There are several case reports in the literature relating specific illnesses such as encephalitis[42] or ulcerative colitis[116] with CMV infection. In most of these studies, a change in antibody status or the excretion of virus is shown to exist concurrently with a particular disease syndrome; in a few, the evidence for the diagnosis of CMV infection consists of the demonstration of typical intranuclear inclusions in various tissues. The variability in antibody titer and the frequency with which virus can be detected in urine or saliva of chronically infected patients make it very hazardous to interpret two concurrent events as being cause and effect. This is emphasized by the authors of many of the articles cited, but it is only by continuing to collect such data that the final question of etiology can be resolved.

9. Treatment and Prevention

9.1. Treatment

Cytomegalovirus infection is generally benign in the normal individual, but in the congenitally infected infant or immunocompromised host, infection can be severe and may be fatal. Specific therapy, therefore, is rarely necessary or used in immunologically normal hosts. Whether specific therapy could improve the prognosis in congenital infection remains unknown.

Vidarabine, acyclovir, or human interferons, separately or in combination, have been effective in decreasing the infectivity titer of CMV in tissue culture, but in the living patient these antimicrobial agents have little effect on the course of clinical disease.[87,130,131,155,177,182]

Ganciclovir, a nucleoside analogue antiviral drug structurally similar to acyclovir but with greater activity against CMV, has been licensed by the FDA for treatment of severe CMV infections. Patients with retinitis and colitis respond well to ganciclovir, in contrast to those with pneumonitis.[45,68,128,164,172] Unfortunately, only intravenous therapy is available, and since ganciclovir does not cure latent CMV infection, long-term maintenance therapy is usually required after the acute infection has been resolved. Bone marrow transplant patients with pneumonitis have improved responses if they are treated with CMV immunoglobulin as well as ganciclovir.[164]

Despite the availability of effective antiviral therapy, efforts continue, appropriately, to be directed toward prophylaxis. Primary infection can be effectively reduced by administering only seronegative blood or blood products to neonates and transplant recipients or by using frozen, deglycerolized, or otherwise treated preparations.[18,217] Infections may be moderated by judicious use of immunosuppressive agents plus the administration of either plasma or globulin containing high titers of antibody to CMV.[33,159,179] The availability of high-titer monoclonal antibody may lead to improved results with immunoprophylaxis.

Antiviral therapy with ganciclovir has also been successfully employed to prevent CMV infection in bone marrow[172] and heart transplant[128] patients.

9.2. Control and Prevention

Serological monitoring of the pregnant or potentially pregnant woman has been recommended as a method for identifying the woman at risk for primary infection and assisting her in avoiding such a complication.[185] Perhaps more importantly, if the woman has antibody to CMV at the beginning of her pregnancy, she may be reassured that she is highly unlikely to develop a primary CMV infection again and that recurrent CMV infection carries an extremely low risk to the fetus. The seronegative woman would be advised to avoid possible sources of CMV infection, a difficult task since most individuals with infection are completely asymptomatic and therefore not detectable. The high carrier rate of CMV in young children, especially in day-care centers or other aggregations of young children, would suggest prudence in exposing a CMV-seronegative person in such a setting. No methods are available to determine if a particular infant will be infected during maternal primary infection and, if so, whether there will be a serious or asymptomatic infection. Termination of pregnancy is not a reasonable consideration because of the inability to determine with accuracy whether or not the baby will be normal.[188] Thus, the performance of serological testing has limited value, at least until it can be shown that a vaccine will protect the seronegative individual or methods become available to identify the small proportion of infected babies who will become damaged. On the other hand, one can make a case even now for serologic testing due to the reassurance that results for those women who can be told that they are seropositive to CMV.

Since it now appears that primary rather than recurrent CMV infection during pregnancy is vastly more likely to result in fetal infection resulting in residual damage, there is a reasonable basis for considering the eventual use of a vaccine in seronegative girls to prevent severe congenital CMV infections in their eventual offspring. There are only two reports in the entire medical literature documenting congenital CMV infection with sequelae in consecutive pregnancies.[4,170] There are also specific reports of second infected infants being normal.[52,105,191] Moreover, as has been noted, epidemiologic and prospective serological studies performed on large numbers of women have indicated that primary maternal CMV infection is the major culprit leading to sequelae in infants. However, there remains one theoretical area of concern associated with the use of a CMV vaccine.[61,188]

This concern is the malignant potential of the herpesviruses. Members of this group have been clearly implicated as the etiologic agents of malignancy in lower animals, although any strong association between herpes simplex type 2 and cervical carcinoma in the human is no longer accepted except as a possible cofactor in conjunction with papillomaviruses types 16 and 18.[220] Albrecht and Rapp[6] have shown that irradiated CMV is capable of causing transformation in hamster cells *in vitro*. The transformed cells are oncogenic when inoculated into hamsters. The same group of investigators has described *in vitro* growth of human prostate tissue infected *in vivo* with CMV to passage levels higher than those routinely attained by normal cells.[163] They have also demonstrated malignant transformation in one cell line derived from a human prostatic adenocarcinoma that possessed CMV antigenic markers.[64] Nondifferentiated tumors were induced when transformed cells were inoculated subcutaneously into athymic nude mice.

Live attenutated CMV vaccines have been developed and have been tested in clinical trials in humans although they remain unlicensed. Elek and Stern[50] reported results obtained with a live vaccine prepared from AD 169, a standard laboratory strain of CMV. The subcutaneous inoculation of this material into 26 seronegative volunteers was followed by the development of antibody. Mild local signs developed at the injection site in half the subjects and two developed tender adenopathy, one with reactive lymphocytes. None of the subjects demonstrated any disturbance of liver function tests and none excreted virus in throat secretions or urine.

Plotkin *et al.*[152–154,156,157] have developed a live vaccine (Towne strain) by passing an isolate from a congenitally infected infant 125 times in tissue culture. Clinical trials[97,153,193] have indicated that vaccinees develop antibodies and manifest a specific cellular immune response not only to the Towne strain but also to AD 169 and Davis strains of CMV. Local erythema and induration at the site of injection have occurred, but there have been no alterations in blood counts or liver function tests. Attempts to isolate virus postvaccination from throat, urine, buffy coat, semen, or vaginal tampons have been uniformly negative. Antibody has been demonstrated to persist for at least 4 years postvaccination, although the titers have dropped considerably over that time.[65] Glazer *et al.*[65] studied ten Towne vaccine recipients who underwent renal transplantation subsequent to vaccination; CMV was isolated from six patients after transplantation, but the restriction endonuclease patterns of the viral DNA of the four isolates tested differed from those of the vaccine strain. The Towne strain has been shown to be an attenuated CMV strain by inoculation experiments involving healthy human volunteers.[158]

Studies comparing the course of renal transplant patients receiving Towne vaccine or placebo have been reported.[9,65,152,156,157] Patients were vaccinated about 8 weeks prior to transplantation. It has been generally found that the vaccine causes mild local reactions, good antibody responses, and, in the few tested, evidence of cell-mediated immune response to CMV antigens. The rates of CMV excretion in the vaccinated and placebo groups have not differed, but the virus excreted was not the vaccine strain. A scheme based on various clinical and laboratory criteria was used to score the severity of illness; more placebo recipients became ill with CMV infection and the illnesses were more severe. The most striking benefit of the vaccine was in the group of seronegative recipients who received kidneys from CMV-antibody-positive donors. In the renal transplant recipient, the Towne vaccine may be of some benefit in reducing morbidity or mortality, although infection is not prevented. A subunit vaccine that would avoid the complications of virus latency is also under development.[157]

There are no published reports on the effect of the CMV vaccine on primary infection in normal pregnant women, but such studies are in progress. More information is required before it can be decided if there is a role for the currently available vaccines or whether an effort should be made to develop a new product.

10. Unresolved Problems

Over the past 40 years, considerable progress has been made in our understanding of how to diagnose and treat CMV infections as well as the epidemiology of diseases caused by this virus. Unresolved problems continue to be whether there are important biological differences among CMV strains, the relative importance of various routes of transmission, the role of humoral and cell-mediated immunity in protection of CMV, the mechanisms of CMV alteration of host defenses to increase susceptibility to other agents, the oncogenic potential of CMV, usefulness of a vaccine in various clinical settings, efficacy and safety of new antiviral drugs, monoclonal antibody (including human monoclonals) as a treatment modality, and identification of the infected and damaged versus the infected fetus.

11. References

1. ADLER, S. P., Cytomegalovirus transmission among children in day care, their mothers and caretakers, *Pediatr. Infect. Dis.* **7:**279–285 (1988).

2. ADLER, S. P., Molecular epidemiology of cytomegalovirus: Viral transmission among children attending a day care center, their parents, and caretakers, *J. Pediatr.* **112:**366–372 (1988).

3. ADLER, S. P., BAGGETT, J., WILSON, M., LAWRENCE, L., AND McVOY, M., Molecular epidemiology of cytomegalovirus in a nursery: Lack of evidence for nosocomial transmission, *J. Pediatr.* **108:**117–123 (1986).

4. AHLFORS, D., HARRIS, S., IVARSSON, S., AND SVANBERG, L., Secondary maternal cytomegalovirus infection causing symptomatic congenital infection, *N. Engl. J. Med.* **305:**284 (1981).

5. AHLFORS, K., IVARSSON, S. A., JOHNSSON, T., AND SVANBERG, L., Primary and secondary maternal cytomegalovirus infections and their relation to congenital infection, *Acta Paediatr. Scand.* **71:**109–113 (1982).

6. ALBRECHT, P., AND RAPP, F., Malignant transformation of hamster embryo fibroblasts following exposure to ultraviolet irradiated human cytomegalovirus, *Virology* **55:**53–61 (1973).

7. ALEXANDER, E. R., Maternal and neonatal infection with cytomegalovirus in Taiwan (abstract), *Pediatr. Res.* **1:**210 (1967).

8. ARMSTRONG, D., BALAKRISHNAN, S., STEGIR, L., YU, B., AND STENZEL, K. H., Cytomegalovirus infections with viremia following renal transplantation, *Arch. Intern. Med.* **127:**111–115 (1971).

9. BALFOUR, H. H., SACHS, G. W., WELO, P., GEHRZ, R. C., SIMMONS, R. L., AND NAJARIAN, J. S., Cytomegalovirus vaccine in renal transplant candidates: Progress report of a randomized, placebo-controlled, double-blind trial, *Birth Defects* **20:**289–304 (1984).

10. BARON, J., YOUNGBLOOD, L., SIEWERS, C. M. F., AND MEDEARIS, D. N., JR., The incidence of cytomegalovirus, herpes simplex, rubella, and toxoplasma antibodies in microcephalic, mentally retarded, and normocephalic children, *Pediatrics* **44:**932–939 (1969).

11. BECKWITH, D. G., HALSTEAD, D. C., ALPAUGH, K., SCHWEDER, A., BLOUNT-FRONFIELD, D. A., AND TOTH, K., Comparison of a latex agglutination test with five other methods for determining the presence of antibody against cytomegalovirus, *J. Clin. Microbiol.* **21:**328–331 (1985).

12. BECROFT, D. M. O., Prenatal cytomegalovirus infection: Epidemiology, pathology and pathogenesis, in: *Perspectives in Pediatric Pathology* (H. S. ROSENBERG AND J. BERNSTEIN, EDS.), pp. 203–241, Masson, Paris, 1981.

13. BENYESH-MELNICK, M., DESSY, S. L., AND FERNBACH, D., Cytomegaloviruria in children with acute leukemia and in other children, *Proc. Soc. Exp. Biol. Med.* **117:**624–630 (1964).

14. BERENBERG, W., AND NANKERVIS, G. A., Long-term follow up of cytomegalic inclusion disease of infancy, *Pediatrics* **46:**403–409 (1970).

15. BETTS, R. F., The relationship of epidemiology and treatment factors to infection and allograft survival in renal transplantation, *Birth Defects* **20:**87–99 (1984).

16. BIRNBAUM, G., LYNCH, J. L., MARGILETH, A. M., LONERGAN, W. M., AND SEVER, J. L., Cytomegalovirus infections in newborn infants, *J. Pediatr.* **75:**789–795 (1969).

17. BLACK, F. L., WOODALL, J. P., EVANS, A. S., LIEBHABER, H., AND HENLE, G., Prevalence of antibody against viruses in the Tiriyo, an isolated Amazon tribe, *Am. J. Epidemiol.* **91:**430–438 (1970).

18. BRADY, M. T., MILAM, J. D., ANDERSON, D. C., HAWKINS, E. P., SPEER, M. E., SEAVY, D., BLIOU, H., AND YOW, M. D., Use of deglycerolized red blood cells to prevent posttransfusion infection with cytomegalovirus in neonates, *J. Infect. Dis.* **150:**334–339 (1984).

19. BRASFIELD, D. M., STAGNO, S., WHITLEY, R. J., CLOUD, G., CASSELL, G., AND TILLER, R. E., Infant pneumonitis associated with cytomegalovirus, chlamydia, pneumocystis, and ureaplasma: Follow up, *Pediatrics* **79:**76–83 (1987).

20. BRAUN, W. E., NANKERVIS, G. A., BANOWSKY, L. H., PROTIVA, D., BIEKAERT, E., AND McHENRY, M. C., A prospective study of cytomegalovirus infections in 78 renal allograft recipients, *Proc. Dialy. Transpl. Forum.* **6:**8–11 (1976).

21. CANGIR, A., AND SULLIVAN, M., The occurrence of cytomegalovirus infections in childhood leukemia, *J.A.M.A.* **195:**616–622 (1966).

22. CAPPELL, D. F., AND McFARLANE, M. N., Inclusion bodies (protozoan-like cells) in the organs of infants, *J. Pathol. Bacteriol.* **59:**385–398 (1947).

23. CARLSTROM, G., Virologic studies on cytomegalic inclusion disease, *Acta Paediatr. Scand.* **54:**17–23 (1965).

24. CAUL, E. O., CLARKE, S. K. R., MOTT, M. G., PERHAM, T. G. M., AND WILSON, R. S. E., Cytomegalovirus infections after open heart surgery: A prospective study, *Lancet* **1:**771–781 (1971).

25. CHANDLER, S. H., HOLMES, K. K., WENTWORTH, B. B., WIESNER, P. J., ALEXANDER, R., AND HANDSFIELD, H. H., The epidemiology of cytomegaloviral infection in women attending a sexually transmitted disease clinic, *J. Infect. Dis.* **152:**597–605 (1985).

26. CHEUNG, K. S., ROCHE, J. D., CAPEL, W. D., AND LANG, D. J., An evaluation under code of new techniques for the detection of cytomegalovirus antibodies, *J. Clin. Lab. Immunol.* **6:**269–274 (1981).

27. CHOU, S., Acquisition of donor strains of cytomegalovirus by renal transplant recipients, *N. Engl. J. Med.* **314:**1418–1423 (1986).

28. CHOU, S., Reactivation and recombination of multiple cytomegalovirus strains from individual organ donors, *J. Infect. Dis.* **160:**11–15 (1989).

29. CHOU, S., Newer methods for diagnosis of cytomegalovirus infection, *Rev. Infect. Dis.* **12S:**S727–S735 (1990).

30. CHOU, S., AND MERIGAN, T. C., Rapid detection and quantitation of human cytomegalovirus in urine through DNA hybridization, *N. Engl. J. Med.* **308:**921–925 (1983).

31. COLE, R., AND KUTTNER, K. A., A filterable virus present in the submaxillary glands of guinea pigs, *J. Exp. Med.* **44:**855–873 (1926).

32. CONBOY, T. J., PASS, R., STAGNO, S., ALFORD, C. A., MYERS, G. J., BRITT, W. J., AND BOLL, T. J., Early clinical manifestations and intellectual outcome in children with symptomatic congenital cytomegalovirus infection, *J. Pediatr.* **111:**343–348 (1987).

33. CONDIE, R. M., AND O'REILLY, R. J., Prophylaxis of CMV infection in bone marrow transplant recipients by hyperimmune CMV gamma globulin, *Dev. Biol. Scand.* **52:**501–513.

34. COULSON, A. S., LUCAS, Z. J., CONDY, M., AND COHN, R., An epidemic of cytomegalovirus disease in a renal transplant population, *West. J. Med.* **120:**1–7 (1974).

35. CRAIGHEAD, J. E., HANSHAW, J. B., AND CARPENTER, C. B., Cytomegalovirus infection after renal allotransplantation, *J.A.M.A.* **201:**99–102 (1967).

36. DAVIS, L. E., TWEED, G. V., STEWART, J. A., BERNSTEIN, M. T., MILLER, G. L., GRAVELLE, C. R., AND CHIN, D. Y., Cytomegalovirus mononucleosis in a first trimester pregnant female with transmission to the fetus, *Pediatrics* **48:**200–206 (1971).

37. DEMMLER, G., BUFFONE, G. J., SCHIMBOR, C. M., AND MAY, R. A., Detection of cytomegalovirus in urine from newborns by using polymerase chain reaction, *J. Infect. Dis.* **158:**1177–1184 (1988).

38. DEMMLER, G. D., O'NEILL, G. W., AND O'NEILL, G. H., Transmission of cytomegalovirus from husband to wife, *J. Infect. Dis.* **154:**545–546 (1986).

39. DEMMLER, G. D., YOW, M., SPECTOR, S., REIS, S. G., BRADY, M. T., ANDERSON, D. C., AND TABER, L., Nosocomial cytomegalovirus infections within two hospitals caring for infants and children, *J. Infect. Dis.* **156:**9–16 (1987).

40. DIOSI, P., MOLDOVAN, E., AND TOMESCU, N., Latent cytomegalovirus infection in blood donors, *Br. Med. J.* **4:**600–602 (1969).

41. DOBBINS, J. G., AND STEWART, J. A. S., Surveillance of congenital cytomegalovirus disease 1990–1991, *Morbid. Mortal. Week. Rep.* **41:**SS-2 (1992).

42. DORFMAN, L. J., Cytomegalovirus encephalitis in adults, *Neurology* **23:**136–144 (1973).

43. DREW, W. L., AND MINTZ, L., What is the role of cytomegalovirus in AIDS? *Ann. NY Acad. Sci.* **437:**320–324 (1984).

44. DUMMER, J. S., BAHNSON, H. T., GRIFFITH, B. P., HARDESTY, R. L., THOMPSON, M. E., AND HO, M., Early infections in kidney, heart, and liver transplant recipients on cyclosporine, *Transplantation* **36:**259–267 (1983).

45. DUNCAN, S. R., AND COOK, D. J., Survival of ganciclovir-treated heart transplant recipients with cytomegalovirus pneumonitis, *Transplantation* **52:**910–913 (1991).

46. DWORSKY, M., LAKEMAN, F., AND STAGNO, S., Cytomegalovirus transmission within a family, *Pediatr. Infect. Dis.* **3:**236–238 (1984).

47. DWORSKY, M., WELCH, K., CASSADY, G., AND STAGNO, S., Occupational risk for primary cytomegalovirus infection among pediatric health-care workers, *N. Engl. J. Med.* **309:**950–953 (1983).

48. DWORSKY, M., YOW, M., STAGNO, S., PASS, R. F., AND ALFORD, C. A., Cytomegalovorius infection of breast milk and transmission in infancy, *Pediatrics* **72:**295–299 (1983).

49. DYMENT, P. G., ORLANDO, S. J., ISAACS, H. J., AND WRIGHT, H. T., JR., The incidence of cytomegaloviruria and postmortem cytomegalic inclusions in children with acute leukemia, *J. Pediatr.* **72:**533–536 (1968).

50. ELEK, S. D., AND STERN, H., Development of a vaccine against mental retardation caused by cytomegalovirus infection *in utero*, *Lancet* **1:**1–5 (1974).

51. EMBIL, J. A., HALDANE, E. V., MACKENZIE, R. A. E., AND VAN ROOYEN, C. E., Prevalence of cytomegalovirus infection in a normal urban population in Nova Scotia, *Can. Med. Assoc. J.* **101:**730–733 (1969).

52. EMBIL, J. A., OZERE, R. L., AND HALDANE, E. V., Congenital cytomegalovirus infection in two siblings from consecutive pregnancies, *J. Pediatr.* **77:**417–421 (1970).

53. EVANS, A., COX, F., NANKERVIS, G., SHOPE, R., WELLS, A. V., AND WEST, B., A health and seroepidemiological survey of a community in Barbados, *Int. J. Epidemiol.* **3:**167–175 (1974).

54. EVANS, A. S., Infectious mononucleosis and related syndromes, *Am. J. Med. Sci.* **276:**325–339 (1978).

55. EVANS, A. S., COOK, J., KAPIKIAN, A. Z., NANKERVIS, G., SMITH,

A., AND WEST, B., A serological survey of St. Lucia, *Int. J. Epidemiol.* **8:**327–332 (1979).

56. FARBER, S., AND WOLBACK, S. B., Intranuclear and cytoplastic inclusions ("protozoan-like bodies") in salivary glands and other organs of infants, *Am. J. Pathol.* **8:**123–126 (1932).

57. FELDMAN, R., Cytomegalovirus infection during pregnancy, *Am. J. Dis. Child.* **117:**517–521 (1969).

58. FETTERMAN, G. H., A new laboratory aid in the clinical diagnosis of inclusion disease of infancy, *Am. J. Clin. Pathol.* 424–425 (1952).

59. FINE, R. N., GRUSHKIN, C. M., ANAND, S., LIEBERMAN, E., AND WRIGHT, H. T., Cytomegalovirus in children, *Am. J. Dis. Child.* **120:**197–202 (1970).

60. FOSTER, K. M., AND JACK, I., A prospective study of the role of cytomegalovirus in posttransplant mononucleosis, *N. Engl. J. Med.* **280:**1311–1354 (1969).

61. FOWLER, K. B., STAGNO, S., PASS, R. F., BRITT, W. J., BOLL, T. J., AND ALFORD, C. A., The outcome of congenital cytomegalovirus infection in relation to maternal antibody status, *N. Engl. J. Med.* **326:**663–667 (1992).

62. FOY, H. M., KENNY, G. E., WENTWORTH, B. B., JOHNSON, W. L., AND GRAYSTON, J. T., Isolation of mycoplasma hommnis, T-strains, and cytomegalovirus from the cervix of pregnant women, *Am. J. Obstet. Gynecol.* **106:**635–643 (1970).

63. FRENKEL, L., GAUR, S., TSOLIA, M., SCUDDER, R., HOWELL, R., AND KESARWALA, H., Cytomegalovirus infection in children with AIDS, *J. Infect. Dis.* **12S:**S820–S826 (1990).

63a. FRIEDMAN, H. M., LEWIS, M. R., MEMEROFSKY, D. M., AND PLOTKIN, S. A., Acquisition of cytomegalovirus infection among female employees at a pediatric hospital, *Pediatr. Infect. Dis.* **3:**233–235 (1984).

64. GEDER, L., SANFORD, E. J., ROHNER, T. J., AND RAPP, F., Cytomegalovirus and cancer of the prostate: *In vitro* transformation of human cells, *Cancer Treat. Rep.* **61:**139–146 (1977).

65. GLAZER, J. P., FRIEDMAN, H. M., GROSSMAN, R. A., STARR, S. E., BARKER, C. F., PERLOFF, J. L., HUANG, E. S., AND PLOTKIN, S. A., Live cytomegalovirus vaccination of renal transplant candidates, *Ann. Intern. Med.* **91:**676–683 (1979).

66. GLEAVES, C. A., SMITH, T. F., SHUSTER, E. A., AND PEARSON, G. R., Rapid detection of cytomegalovirus in MRC-5 cells inoculated with urine specimens by using low-speed centrifugation and monoclonal antibody to an early antigen, *J. Clin. Microbiol.* **19:**917–919 (1984).

67. GOODPASTURE, E., AND TALBOT, F. B., Concerning the nature of "protozoan-like" cells in certain lesions of infancy, *Am. J. Dis. Child.* **21:**415–421 (1921).

68. GOODRICH, J. M., MORI, M., GLEAVES, C., DU MOND, C., CAYS, M., EBELING, D., BUHLES, W., DEARMOND, B., AND MEYERS, J. D., Early treatment with ganciclovir to prevent cytomegalovirus disease after allogeneic bone marrow transplantation, *N. Engl. J. Med.* **325:**1601–1607 (1991).

69. GRANT, S., EDMOND, E., AND SYME, J., A prospective study of cytomegalovirus infection in pregnancy, *J. Infect. Dis.* **3:**24–31 (1981).

70. GREENBERG, M., FRIEDMAN, H., COHEN, G., OH, S., LASTER, L., AND STARR, S., A comparative study of herpes simplex infections in renal transplant and leukemic patients, *J. Infect. Dis.* **156:**280–287 (1987).

71. GRIFFITHS, P. D., CAMPBELL-BENZIE, A., AND HEATH, R. B., A prospective study of primary cytomegalovirus infection in pregnant women, *Br. J. Obstet. Gynecol.* **87:**308–314 (1980).

72. GRIFFITHS, P. D., PANJWANI, D. D., STIRK, P. R., BALL, M. G., GANCZAKOWSKI, M., BLACKLOCK, H. A., AND PRENTICE, H. G., Rapid diagnosis of cytomegalovirus infection in immunocompromised patients by detection of early antigen fluorescent foci, *Lancet* **2:**1242–1244 (1984).

73. GRUNDY, J. E., Virologic and pathologic aspects of cytomegalovirus infection, *Rev. Infect. Dis.* **12S:**S711–S719 (1990).

74. GYORKEY, F., MELNICK, J. L., GUINN, G. A., GYORKEY, P., AND DEBAKEY, M. E., Herpesviridae in the endothelial and smooth muscle cells of the proximal aorta in arteriosclerotic patients, *Exp. Molec. Pathol.* **40:**328–339 (1984).

74a. HALDANE, E. V., VAN ROOYEN, C. E., EMBIL, J. A., TUPPER, W. C., GORDON, P. C., AND WANKLIN, J. M., A search for transmissible birth defects of virologic origin in members of the nursing profession, *Am. J. Obstet, Gynecol.* **105:**1032–1040 (1969).

75. HANSHAW, J. B., Cytomegalovirus complement-fixing antibody in microcephaly, *N. Engl. J. Med.* **275:**476–479 (1966).

75a. HANSHAW, J. B., AND WELLER, T. H., Urinary excretion of cytomegaloviruses by children with generalized neoplastic disease: Correlation with clinical and histopathologic observations, *J. Pediatr.* **58:**305–311 (1961).

76. HANSHAW, J. B., Congenital cytomegalovirus infection: A fifteen-year perspective, *J. Infect. Dis.* **123:**555–561 (1971).

77. HANSHAW, J. B., BETTS, R. E., SIMON, G., AND BOYNTON, R. C., Acquired cytomegalovirus infection association with hepatomegaly and abnormal liver function tests, *N. Engl. J. Med.* **272:**602–609 (1965).

78. HANSHAW, J. B., SCHEINER, A. P., AND MOXLEY, A. W., School failure and deafness after "silent" congenital cytomegalovirus infection, *N. Engl. J. Med.* **295:**468–470 (1976).

79. HARDEN, D. G., ELSDALE, T. R., YOUNG, D. E., AND ANDROSS, A., The isolation of cytomegalovirus from peripheral blood, *Blood* **30:**120–125 (1967).

80. HAYES, K., DANKS, D. M., AND GIBAS, H., Cytomegalovirus in human milk, *N. Engl. J. Med.* **287:**177–178 (1972).

81. HENLE, W., HENLE, G., SCRIBA, M., JOYNER, C., HARRISON, E., VON ESSEN, R., PALOHEIMO, J., AND KLEMOLA, E., Antibody responses to the Epstein–Barr virus and cytomegalovirus after open-heart and other surgery, *N. Engl. J. Med.* **282:**1068–1074 (1970).

82. HENNEBERG, G., AND ANTONIADIS, G., Serological studies on the incidence of complement-fixing antibodies against cytomegalovirus, in West Berlin, *Zentralbl-Bacteriol.* **213:**416–427 (1970).

83. HENSON, D., SIEGEL, S., FUCILLO, D. A., MATTHEW, E., AND LEVINE, S., Cytomegalovirus infections during acute childhood leukemia, *J. Infect. Dis.* **126:**469–481 (1972).

84. HILDEBRANDT, R. J., SEVER, J. L., MARGILETH, A. M., AND CALLAGHAN, D. A., Cytomegalovirus in the normal pregnant woman, *Am. J. Obstet. Gynecol.* **98:**1125–1128 (1967).

85. HIRSCH, M. S., AND FELSENSTEIN, D., Cytomegalovirus induced immunosuppression, *Ann. N.Y. Acad. Sci.* **437:**8–15 (1984).

86. HO, M., Cytomegalovirus, in: *Cytomegalovius* (M. HO, ED.), Plenum Press, New York, 1982.

87. HO, M., Epidemiology of cytomegalovirus infections, *Rev. Infect. Dis.* **12**(Suppl.):701–710 (1990).

88. HO, M., SUWANSIRKUL, S., DOWLING, J. N., YOUNGBLOOD, L. A., AND ARMSTRONG, J. A., The transplanted kidney as a source of cytomegalovirus infection, *N. Engl. J. Med.* **293:**1109–1112 (1975).

89. HURLEY, J., GREENSLADE, T., LEWY, P., AHMADIAN, Y., AND FIRLIT, C., Varicella zoster infections in pediatric renal transplant patients, *Arch. Surg.* **115:**751–752 (1980).

90. HUTTO, C., LITTLE, A., RICKS, R., LEE, J. D., AND PASS, R. F., Isolation of CMV from toys and hands in a day care center, *J. Infect. Dis.* **154:**527–530 (1986).

91. HUTTO, C., RICKS, R., GARVIE, M., AND PASS, R. F., Epidemiology of cytomegalovirus infections in young children: Day care vs. home care, *Pediatr. Infect. Dis. J.* **4:**149–152 (1985).

92. JACK, I., AND MCAULIFFE, K. C., Sero-epidemiological study of cytomegalovirus infections in Melbourne children and some adults, *Med. J. Aust.* **1:**206–209 (1968).

93. JACK, I., TODD, H., AND TURNER, E. K., Isolation of human cytomegalovirus from the circulating leukocytes of a leukemic patient, *Med. J. Aust.* **1:**210–213 (1968).

94. JESIONEK, A., AND KIOLEMENOGLOU, B., Uber einen Befund von protozoenartigen Gebilden in den Organen eines Feten, *Muench. Med. Wochenschr.* **51:**1905–1907 (1904).

95. JONES, L., DUKE-DUNCAN, P., AND YEAGER, A. S., Cytomegaloviral infections in infant–toddler centers: Centers for the developmentally delayed versus regular day care, *J. Infect. Dis.* **151:**953–955 (1985).

96. JORDON, M. C., ROUSSEAU, W. E., STEWART, J. A., NOBLE, G. R., AND CHIN, T. D. Y., Spontaneous cytomegalovirus mononucleosis, *Ann. Intern. Med.* **79:**153–160 (1973).

97. JUST, M., DUERGIN-WOLFF, A., EMOEDI, G., AND HERNANDEZ, F., Immunization trials with live attenuated cytomegalovirus Towne 125, *Infection* **3:**111–114 (1975).

98. KANE, R. C., ROUSEAU, W. E., NOBLE, G. R., TEGTMEIER, G. E., WULFF, H., HERNDON, H. B., CHIN, T. D. Y., AND BAYER, W. L., A prospective study of cytomegalovirus infection in a volunteer blood donor population, *Infect. Immun.* **11:**719–723 (1975).

99. KIMMEL, N., FRIEDMAN, M. G., AND SAROV, I., Detection of human CMV-specific IgG antibodies by a sensitive solid phase radioimmunoassay and by a rapid-screening test, *J. Med. Virol.* **5:**195–203 (1980).

100. KLEMOLA, E., AND KAARIAINEN, L., Cytomegalovirus as a possible cause of a disease resembling infectious mononucleosis, *Br. Med. J.* **2:**1099–1102 (1965).

101. KLEMOLA, E., KAARIAINEN, R., VON ESSEN, R., HALTIA, K., KOIVUNIEMI, A., AND VON BONSDORFF, C.-H., Further studies on cytomegalovirus mononucleosis in previously healthy individuals, *Acta Med. Scand.* **182:**311–322 (1967).

102. KNEZ, V., STEWART, J. A., AND ZIEGLER, D. W., Cytomegalovirus specific IgM and IgG response in humans studied by radioimmunoassay, *J. Immunol.* **117:**2006–2013 (1976).

103. KRAAT, Y., HENDRIX, R., LANDINI, M., AND BRUGGEMAN, C., Comparison of four techniques for detection of antibodies to cytomegalovirus, *J. Clin. Microbiol.* **30:**522–524 (1992).

104. KRECH, U., JUNG, M., JUNG, F., AND SINGEISEN, C., Virologische und klinische Untersuchungen bei konnatalen und postnatalen Cytomegalien, *Schweiz. Med. Wochenschr.* **98:**1459–1469 (1968).

105. KRECH, U., KONJAJEV, Z., AND JUNG, M., Congenital cytomegalovirus infection in siblings from consecutive pregnancies, *Helv. Paediatr. Acta.* **26:**355–362 (1971).

106. KRECH, U. H., AND JUNG, M., Age distribution of complement-fixing antibodies in Tanzania, 1970, in: *Cytomegalovirus Infections of Man*, pp. 27–28, S. Karger, Basel, 1971.

107. KREEL, I., ZAROFF, L. L., CANTER, J. W., KRASNA, I., AND BARONOFSKY, I. D., Syndrome following total body perfusion, *Surg. Gynecol. Obstet.* **111:**317–321 (1960).

108. KRIEL, R. L., GATES, G. A., WULFF, H., POWELL, N., POLAND,

J. D., AND CHIN, T. D. Y., Cytomegalovirus isolations associated with pregnancy wastage, *Am. Obstet. Gynecol.* **106**:885–892 (1970).

109. KUMAR, M. L., GOLD, E., JACOBS, S. B., EMHART, C. B., AND NANKERVIS, G. A., Primary cytomegalovirus infection in adolescent pregnancy, *Pediatrics* **74**:493–500 (1984).

110. LAMB, S. G., AND STERN, H., Cytomegalovirus mononucleosis with jaundice as presenting sign, *Lancet* **2**:1003–1006 (1966).

111. LANG, D. J., AND HANSHAW, J. B., Cytomegalovirus infection and postperfusion syndrome: Recognition of primary infections in four patients, *N. Engl. J. Med.* **280**:1145–1149 (1969).

112. LANG, D. J., AND KUMMER, J. F., Demonstration of cytomegalovirus in semen, *N. Engl. J. Med.* **287**:756–758 (1972).

113. LANG, D. J., KUMMER, J. F., AND HARTLEY, D. P., Cytomegalovirus in semen: Persistence and demonstration in extracellular fluids, *N. Engl. J. Med.* **291**:121–123 (1974).

114. LANG, D. J., SCOLNICK, E. M., AND WILLERSON, J. T., Association of cytomegalovirus infection with the postperfusion syndrome, *N. Engl. J. Med.* **278**:1147–1149 (1968).

115. LEHANE, D. E., A seroepidemiologic study of infectious mononucleosis: The development of EB virus antibody in a military population, *J.A.M.A.* **212**:2240–2242 (1970).

116. LEVINE, R. S., WARNER, N. E., AND JOHNSON, C. F., Cytomegalic inclusion disease in the gastrointestinal tract of adults, *Ann. Surg.* **159**:37–48 (1964).

117. LEVINSOHN, E. M., FOY, H. M., KENNY, G. E., WENTWORTH, B. B., AND GRAYSTON, J. T., Isolation of cytomegalovirus from a cohort of 100 infants throughout the first year of life, *Proc. Soc. Exp. Biol. Med.* **132**:957–962 (1969).

118. LI, F., AND HANSHAW, B. J., Cytomegalovirus infection among migrant children, *Am. J. Epidemiol.* **86**:137–141 (1967).

119. LOPEZ, C., SIMMONS, R. L., MAUER, S. M., NAJARIAN, J. S., AND GOOD, R. A., Association of renal allograft rejection with viral infection, *Am. J. Med.* **56**:280–289 (1974).

120. LUBY, J., RAMIREZ-RONDA, C., RINNER, S., HULL, A., AND VERGNE-MARINI, P., A longitudinal study of varicella zoster virus infections in renal transplant recipients, *J. Infect. Dis.* **135**:659–663 (1977).

121. LUBY, J. P., BURNETT, W., HULL, A. R., WARE, A. J., SHOREY, J. W., AND PETERS, P. C., Relationship between cytomegalovirus and hepatic function abnormalities in the period after renal transplant, *J. Infect. Dis.* **129**:511–518 (1974).

122. LUTHARDT, T., SIEBERT, H., LOSEL, I., QUEVEDO, M., AND TODT, R., Cytomegalovirus-infektionen bei Kindern mit Blutaustauschtransfusion im Neugeborenenalter, *Klin. Wochenschr.* **49**:81–86 (1971).

123. MANDELL, G. L., Cytomegalovirus mononucleosis, *Del. Med. J.* **43**:155–156 (1971).

124. MARGILETH, A. M., The diagnosis and treatment of generalized cytomegalic inclusion disease of the newborn, *Pediatrics* **15**:270–283 (1955).

125. MCCRACKEN, G. H., JR. SHINEFIELD, H. R., COBB, K., RAUSEN, A. R., DISCHE, M. R., AND EICHENWALD, H. F., Congenital cytomegalic inclusion disease: A longitudinal study of 20 patients, *Am. J. Dis. Child.* **117**:522–539 (1969).

126. MEDEARIS, D. N., JR., Observations concerning human cytomegalovirus infection and disease, *Bull. Johns Hopkins Hosp.* **114**:181–211 (1964).

127. MERCER, R. D., LUSE, S., AND GUYTON, D. H., Clinical diagnosis of generalized cytomegalic inclusion disease, *Pediatrics* **11**:502–514 (1953).

128. MERIGAN, T., RENLUND, D., KEAY, S., BRISTOW, M. R., STARNES, V., O'CONNELL, J. B., RESTA, S., DUNN, D., GAMBERG, P., RATKOVEC, R. M., RICHENBACHER, W. E., MILLAR, R., DUMOND, C., DEAMOND, B., AND SULLIVAN, V., A controlled trial of ganciclovir to prevent CMV disease after heart transplantation, *N. Engl. J. Med.* **326**:1182–1186 (1992).

129. MERIGAN, T., AND RESTA, S., Cytomegalovirus: Where have we been and where are we going? *Rev. Infect. Dis.* **12**(Suppl.):693–700 (1990).

130. MEYERS, J. D., Cytomegalovirus infection following marrow transplantation: Risk, treatment, and prevention, *Birth Defects* **20**:101–117 (1984).

131. MEYERS, J. D., Prevention and treatment of cytomegalovirus infections with interferons and immune globulins, *Infection* **12**:143–150 (1984).

132. MONTGOMERY, R., YOUNGBLOOD, L., AND MEDEARIS, D. N. J., Recovery of cytomegalovirus from the cervix in pregnancy, *Pediatrics* **49**:524–531 (1972).

133. MUSIANI, M., CARPI, C., AND ZERBINI, M., Rapid detection of antibodies against cytomegalovirus induced immediate early and early antigens by an enzyme linked immunosorbent assay, *J. Clin. Pathol.* **37**:122–125 (1984).

134. NAJARIAN, J. L., STRAND, M., FRYD, D. S., FERGUSON, R. M., SIMMONS, R. L., ASCHER, N. L., AND SUTHERLAND, D. E. R., Comparison of cyclosporine versus azathioprine-antilymphocyte globulin in renal transplantation, *Transplant. Proc.* **15**:2463–2468 (1983).

135. NAKAO, T., AND CHIBA, S., Cytomegalovirus and microcephaly, *Pediatrics* **46**:483–484 (1970).

136. NANKERVIS, G. A., Comments on CMV infections in renal transplant patients, *Yale J. Biol. Med.* **49**:27–28 (1976).

137. NANKERVIS, G. A., KUMAR, M. L., COX, E. E., AND GOLD, E., A prospective study of maternal cytomegalovirus infection and its effect on the fetus, *Am. J. Obstet. Gynecol.* **149**:435–440 (1984).

138. NARAQUI, S., JACKSON, G. G., JONASSON, O., AND RUBENIS, M., Search for latent cytomegalovirus in renal allografts, *Infect. Immun.* **19**:699–703 (1978).

139. NUMAZAKI, Y., YANO, N., MORIZUKA, T., TAKAI, S., AND ISHIDA, N., Primary infection with human cytomegalovirus: Virus isolation from healthy infants and pregnant women, *Am. J. Epidemiol.* **91**:410–417 (1970).

140. PALOHEIMO, J. A., VON ESSEN, R., KLEMOLA, E., KAARIAINEN, L., AND SILTANEN, P., Subclinical cytomegalovirus infections and cytomegalovirus mononucleosis after open heart surgery, *Am. J. Cardiol.* **22**:624–630 (1968).

141. PASS, R. E., Epidemiology and transmission of cytomegalovirus, *J. Infect. Dis.* **152**:243–248 (1985).

142. PASS, R. F., HUTTO, C., RICKS, R., AND CLOUD, G. A., Increased rate of cytomegalovirus infection among parents of children attending day-care centers, *N. Engl. J. Med.* **314**:1414–1418 (1986).

143. PASS, R. F., HUTTO, S. C., REYNOLDS, D. W., AND POLHILL, R. B., Increased frequency of cytomegalovirus infection in children in group day care, *Pediatrics* **74**:121–126 (1984).

144. PASS, R. F., AND KINNEY, J. S., Child care workers and children with congenital cytomegalovirus infection, *Pediatrics* **75**:971–973 (1985).

145. PASS, R. F., LITTLE, E. A., STAGNO, S., BRITT, W. J., AND ALFORD, C. A., Young children as a probable source of maternal and congenital cytomegalovirus infection, *N. Engl. J. Med.* **316**:1366–1371 (1987).

146. PAYA, C. V., WOLD, A. D., AND SMITH, T. F., Detection of cytomegalovirus infections in specimens other than urine by the shell vial assay and conventional tube cell cultures, *J. Clin. Microbiol.* **25**:755–757 (1987).

147. PERHAM, T. G. M., CAUL, E. O., CONWAY, P. J., AND MOTT, M. G., Cytomegalovirus infection in blood donors—A prospective study, *Br. J. Haematol.* **20**:307–320 (1971).

148. PETERSON, P. K., BALFOUR, H. H., MARKER, S. C., FRYD, D. S., HOWARD, R. J., AND SIMMONS, R. L., Cytomegalovirus disease in renal allograft recipients: A prospective study of the clinical features, risk factors and impact on renal transplantation, *Medicine* **59**:283–300 (1980).

149. PETRIE, B. L., ADAM, E., AND MELNICK, J. L., Association of herpesvirus/cytomegalovirus infections with human artherosclerosis, *Prog. Med. Virol.* **35**:21–42 (1988).

150. PETRIE, B. L., MELNICK, J. L., ADAM, E., BUREK, J., MCCOLLUM, C. H., AND DEBAKEY, M. E., Nucleic acid sequences of cytomegalovirus in cells cultured from human arterial tissue, *J. Infect. Dis.* **155**:158–159 (1987).

151. PHIPPS, P. H., GREGOIRE, L., ROSIER, E., AND PERRY, E., Comparison of five methods of cytomegalovirus antibody screening of blood donors, *J. Clin. Microbiol.* **18**:1296–1300 (1983).

152. PLOTKIN, S. A., Cytomegalovirus vaccine development: Past and future, *Transpl. Proc.* **23**:85–89 (1991).

153. PLOTKIN, S. A., FARQUHAR, J., AND HORNBERGER, E., Clinical trials of immunization with the Towne 125 strain of human cytomegalovirus, *J. Infect. Dis.* **134**:470–475 (1976).

154. PLOTKIN, S. A., FURUKAWA, T., ZYGRAICH, N., AND HUYGELEN, C., Candidate cytomegalovirus strain for human vaccination, *Infect. Immun.* **12**:521–527 (1975).

155. PLOTKIN, S. A., MICHELSON, S., ALFORD, C. A., STARR, S. E., PARKMAN, P. D., PAGANO, J. S., AND RAPP, F., The pathogenesis and prevention of human cytomegalovirus infection, *Pediatr. Infect. Dis.* **3**:67–74 (1984).

156. PLOTKIN, S. A., STARR, S., FRIEDMAN, H. M., BRAYMAN, K., HARRIS, S., JACKSON, S., TUSTIN, N., GROSSMAN, R., DAFOE, D., AND BARKER, C., Effect of Towne live virus vaccine on cytomegalovirus disease after renal transplant, *Ann. Intern. Med.* **114**:525–531 (1991).

157. PLOTKIN, S. A., STARR, S., FRIEDMAN, H. M., GONCZOL, E., AND BRAYMAN, K., Vaccines for the prevention of human cytomegalovirus infection, *Rev. Infect. Dis.* **12**:S827–838 (1990).

158. PLOTKIN, S. A., WEIBEL, R. E., ALPERT, G., STARR, S. E., FRIEDMAN, H. M., PREBLUD, S. R., AND HOXIE, J., Resistance of seropositive volunteers to subcutaneous challenge with low passage human cytomegalovirus, *J. Infect. Dis.* **151**:737–739 (1985).

159. PRENTICE, H. G., Use of acyclovir for prophylaxis of herpes infections in severely immunocompromised patients, *J. Antimicrob. Chemother.* **12**:153–159 (1983).

160. PRINCE, A. M., SZMUNESS, W., MILLIAN, S. J., AND DAVID, D. S., A serologic study of cytomegalovirus infections associated with blood transfusions, *N. Engl. J. Med.* **284**:1125–1131 (1971).

161. PURCELL, R. H., WALSH, J. H., HOLLAND, P. V., MORROW, A. G., WOOD, S., AND CHANOCK, R. M., Seroepidemiological studies of transfusion-associated hepatitis, *J. Infect. Dis.* **123**:406–413 (1971).

162. QUINNAN, G. V., AND ROCK, A. H., The importance of cytotoxic cellular immunity in the protection from cytomegalovirus infection, *Birth Defects* **20**:245–261 (1984).

163. RAPP, F., GEDER, L., MURASKO, D., LAUSCH, R., HUANG, E., AND WEBBER, M. M., Long-term persistence of cytomegalovirus ge-

164. REED, E. C., BOWDEN, R. A., DANDLIKER, P. S., LILLEBY, K. E., AND MEYERS, J. D., Treatment of cytomegalovirus pneumonia with ganciclovir and intravenous cytomegalovirus immunoglobulin in patients with bone marrow transplants, *Ann. Intern. Med.* **109**:783–788 (1988).

165. REYNOLDS, D. W., STAGNO, S., HOSTY, T. S., TILLER, M., AND ALFORD, C. A. J., Maternal cytomegalovirus excretion and perinatal infection, *N. Engl. J. Med.* **289**:1–5 (1973).

166. REYNOLDS, D. W., STAGNO, S., STUBBS, K. G., DAHLE, A. J., LIVINGSTON, M. M., SAXON, S. S., AND ALFORD, C. A., Inapparent congenital cytomegalovirus infection with elevated cord IgM levels, *N. Engl. J. Med.* **290**:291–296 (1974).

167. RINALDO, C. R., JR., CARNEY, W. P., RICHTER, B. S., BLACK, P. H., AND HIRSCH, J. S., Mechanism of immunosuppression in cytomegaloviral mononucleosis, *J. Infect. Dis.* **141**:488–495 (1980).

168. ROWE, W. A., Adenovirus and salivary gland virus infections in children, in: *Viral Infections of Infancy and Childhood* (H. M. ROSE, ED.), pp. 205–214, Hoeber, New York, 1960.

169. ROWE, W. P., HARTLEY, J. W., WATERMAN, S., TURNER, H. C., AND HUEBNER, R. J., Cytopathogenic agent resembling human salivary gland virus recovered from tissue cultures of human adenoids, *Proc. Soc. Exp. Biol. Med.* **92**:418–424 (1956).

170. RUTTER, D., GRIFFITHS, P., AND TROMPETER, R. S., Cytomegalic inclusion disease after recurrent maternal infection, *Lancet* **2**:1182 (1985).

171. SAWYER, R. N., EVANS, A. S., NIEDERMAN, J. C., AND McCOLLUM, R. W., Prospective studies of a group of Yale University freshmen. 1. Occurrence of infectious mononucleosis, *J. Infect. Dis.* **123**:263–270 (1971).

172. SCHMIDT, G. M., HORAK, D. A., NILAND, J. C., DUNCAN, S. R., FORMAN, S., ZAIA, J., AND CITY OF HOPE STUDY, A randomized, controlled trial of prophylactic ganciclovir for cytomegalovirus pulmonary infection in recipients of allogenic bone marrow transplants, *N. Engl. J. Med.* **324**:1005–1011 (1991).

173. SCHREIER, R. D., RICE, G. P. A., AND OLDSTONE, M. B. A., Suppression of natural killer cell activity and T cell proliferation by fresh isolates of human cytomegalovirus, *J. Infect. Dis.* **153**:1084–1091 (1986).

174. SCHWARTZ, R. S., Immunoregulation, oncogenic viruses, and malignant, lymphomas, *Lancet* **1**:1266–1269 (1972).

175. SHAVRINA, L. V., ASHER, D. M., ILYENKO, V., AND SMORODINTSEV, A. A., Antibodies of cytomegalovirus in the population of Leningrad, *Vopr. Virusol.* **2**:156–159 (1973).

176. SHELDON, P. J., PAPAMICHAIL, M., HEMSTED, E. H., AND HOLBOROW, E. J., Thymic origin of atypical lymphoid cells in infectious mononucleosis, *Lancet* **1**:1153–1155 (1973).

177. SHEPP, D. H., NEWTON, B. A., AND MEYERS, J. D., Intravenous lymphoblastoid interferon and acyclovir for treatment of cytomegaloviral pneumonia, *J. Infect. Dis.* **150**:776–777 (1984).

178. SMITH, M. G., Propagation in tissue cultures of a cytopathogenic virus from human salivary gland virus (SGV) disease, *Proc. Soc. Exp. Biol. Med.* **92**:424–430 (1956).

179. SNYDMAN, D. R., WERNER, B. G., HEINZE-LACEY, B., BERARDI, V. P., TILNEY, N., KIRKMAN, R., MILFORD, E., CHO, S., BUSH, H., LEVEY, A., STROM, T., CARPENTER, C., LEVEY, R., HARMON, W., AND ZIMMERMAN, C., Use of cytomegalovirus immune globulin to prevent cytomegalovirus disease in renal transplant recipients, *N. Engl. J. Med.* **317**:1049–1054 (1987).

nome in cultured cells of prostatic origin, *J. Virol.* **16**:982–990 (1975).

180. SOKOL, D., DEMMLER, G., AND BUFFONE, G., Rapid epidemiologic analysis of cytomegalovirus by using polymerase chain reaction amplification of the L-S junction region, *J. Clin. Microbiol.* **30:**839–844 (1992).

181. SPECTOR, S. A., Transmission of cytomegalovirus among infants in hospital documented by restriction-endonuclease digestion analyses, *Lancet* **1:**378–381 (1983).

182. SPECTOR, S. A., AND KELLEY, E., Inhibition of human cytomegalovirus by combined acylovir and vidarabine, *Antimicrob. Agents Chemother.* **27:**600–604 (1985).

183. SPECTOR, S. A., AND SPECTOR, D. H., Molecular epidemiology of cytomegalovirus infections in premature twin infants and their mother, *Pediatr. Infect. Dis. J.* **1:**405–409 (1982).

184. SPECTOR, S. A., AND SPECTOR, D. H., The use of DNA probes in studies of human cytomegalovirus, *Clin. Chem.* **31:**1514–1520 (1986).

185. STAGNO, S., Isolation precautions for patients with cytomegalovirus infection, *Pediatr. Infect. Dis. J.* **1:**145–147 (1982).

186. STAGNO, S., BRASFIELD, D. M., BROWN, M. B., CASSELL, G., PIFER, L., WHITLEY, R. J., AND TILER, R. E., Infant pneumonitis associated with cytomegalovirus, chlamydia, pneumocystis, and urealoplasma: A prospective study, *Pediatrics* **68:**322–329 (1981).

187. STAGNO, S., PASS, R. F., DWORSKY, M., HENDERSON, R. E., MOORE, E. G., WALTON, P. D., AND ALFORD, C. A., Congenital cytomegalovirus infection: The relative importance of primary and recurrent maternal infection, *N. Engl. J. Med.* **306:**945–949 (1982).

188. STAGNO, S., PASS, R., CLOUD, G., BRITT, G., HENDERSON, R. E., WALTON, P. D., VEREN, D. A., PAGE, F., AND ALFORD, C. A., Primary cytomegalovirus infection in pregnancy. Incidence, transmission to fetus, and clinical outcome, *J.A.M.A.* **256:**1904–1908 (1986).

189. STAGNO, S., PASS, R. F., DWORSKY, M. E., BRITT, W. J., AND ALFORD, C. A., Congenital and perinatal cytomegalovirus infections: Clinical characteristics and pathogenic factors, *Birth Defects* **20:**65–85 (1984).

190. STAGNO, S., REYNOLDS, D. W., AMOS, C. S., DAHLE, A. J., MCCOLLISTER, F. R., MOHINDRA, I., ERMOCILLA, R., AND ALFORD, C. A., Auditory and visual defects resulting from symptomatic and subclinical congenital cytomegaloviral and toxoplasma infections, *Pediatrics* **59:**669–678 (1977).

191. STAGNO, S., REYNOLDS, D. W., LAKEMAN, A., CHARAMELLA, L. J., AND ALFORD, C. A., Congenital cytomegalovirus infection: Consecutive occurrence due to viruses with similar antigenic compositions, *Pediatrics* **52:**788–794 (1973).

192. STARR, J. G., BART, R. D., JR., AND GOLD, E., Inapparent congenital cytomegalovirus infection: Clinical and epidemiologic characteristics in early infancy, *N. Engl. J. Med.* **282:**1075–1077 (1970).

193. STARR, S. E., FRIEDMAN, H. M., GLAZER, J. P., GARRABRANT, T., AND PLOTKIN, S., Immune responses to live cytomegalovirus vaccine (Towne strain) in normal and immunosuppressed volunteers, in: *Interscience Conference on Antimicrobial Agents and Chemotherapy*, American Society for Microbiology, Washington, DC, 1979.

194. STERN, H., Isolation of cytomegalovirus and clinical manifestations of infection at different ages, *Br. Med. J.* **1:**665–669 (1968).

195. STERN, H., AND ELEK, S. D., The incidence of infection with cytomegalovirus in a normal population: A serologic study in greater London, *J. Hyg.* **63:**79–87 (1965).

196. STERN, H., ELEK, S. D., BOOTH, J. C., AND FLECK, D. G., Microbial causes of mental retardation: The role of prenatal infections with cytomegalovirus, rubella virus, and toxoplasma, *Lancet* **2:**443–448 (1969).

197. STERN, H., AND TUCKER, S. M., Cytomegalovirus infection in the newborn and in early childhood: Three atypical cases, *Lancet* **2:**443–448 (1969).

198. STERN, H., AND TUCKER, S. M., Prospective study of cytomegalovirus infection in pregnancy, *Br. Med. J.* **2:**268–279 (1973).

199. STULBERG, C. S., ZUELZER, W. W., PAGE, R. H., TAYLOR, T. E., AND BROUGH, S., Cytomegalovirus infections from lymph node and blood, *Proc. Soc. Exp. Biol. Med.* **123:**976–982 (1966).

200. SULLIVAN, M. P., HANSHAW, J. B., CANGIR, A., AND BUTLER, J. J., Cytomegalovirus complement-fixation antibody levels of leukemic children, *J.A.M.A.* **206:**569–574 (1968).

201. SWENSON, P. D., AND KAPLAN, M. H., Rapid detection of cytomegalovirus in cell culture by indirect immunoperoxidase staining with monoclonal antibody to an early nuclear antigen, *J. Clin. Microbiol.* **21:**669–673 (1985).

202. TABER, L. H., FRANK, A. L., YOW, M. D., AND BAGLEY, A., Acquisition of cytomegaloviral infections in families with young children: A serological study, *J. Infect. Dis.* **151:**948–952 (1985).

203. UNIVERSITY OF HEALTH PHYSICIANS AND PHLS LABORATORIES, Infectious mononucleosis and its relationship to EB virus antibody, *Br. Med. J.* **4:**643–646 (1971).

204. VACZI, L., GONCZOLL, E., LEHEL, F., AND GEDER, L., Isolation of cytomegalovirus and incidence of complement-fixing antibodies against cytomegalovirus in different age groups, *Acta Microbiol. Acad. Sci. Hung.* **12:**115–121 (1965).

205. VERLAINEN, M., ANDERSSON, L. C., AND LALLA, M., T-lymphocyte proliferation in mononucleosis, *Clin. Immunol. Immunopathol.* **2:**114–120 (1973).

206. VON GLAHN, W. C., AND PAPPENHEIMER, A. M., Intranuclear inclusions in visceral disease, *Am. J. Pathol.* **1:**445–465 (1925).

207. WANER, J. L., WELLER, T. H., AND KEVY, S. V., Patterns of cytomegaloviral complement-fixing antibody activity: A longitudinal study of blood donors, *J. Infect. Dis.* **127:**538–543 (1973).

208. WELLER, T. H., AND HANSHAW, J. B., Virologic and clinical observations on cytomegalic inclusion disease, *N. Engl. J. Med.* **266:**1233–1244 (1962).

209. WELLER, T. H., MACAULAY, J. C., CRAIG, J. M., AND WIRTH, P., Isolation of intranuclear inclusion producing agents from infants with illnesses resembling cytomegalic inclusion disease, *Proc. Soc. Exp. Biol. Med.* **94:**4–12 (1957).

210. WENZEL, R. P., MCCORMICK, D. P., DAVIES, J. A., BERLING, C., AND ANDBEAM, W. E., JR., Cytomegalovirus infection: A seroepidemiologic study of a recruit population, *Am. J. Epidemiol.* **97:**410–414 (1973).

211. WHEELER, E. O., TURNER, J. D., AND SCANNELL, J. G., Fever, splenomegaly and atypica lymphocytes: Syndrome observed after cardiac surgery utilizing pump oxygenator, *N. Engl. J. Med.* **266:**454–456 (1962).

212. WILFERT, C. M., HUANG, E. S., AND STAGNO, S., Restriction endonuclease analysis of cytomegalovirus deoxyribonucleic acid as an epidemiologic tool, *Pediatrics* **70:**717–721 (1982).

213. WYATT, J. P., SAXTON, J., LEE, R. S., AND PINKERTON, H., Generalized cytomegalic inclusion disease, *J. Pediatr.* **36:**271–294 (1950).

214. YAMAGUCHI, Y., HIRONAKA, T., KAJIWARA, M., TATENO, E., KITA, H., AND HIRAI, K., Increased sensitivity for detection of

human cytomegalovirus in urine by removal of inhibitors for the polymerase chain reaction, *J. Virol. Methods.* **37:**209–218 (1992).

215. YEAGER, A. S., Transfusion-acquired cytomegalovirus infection in newborn infants, *Am. J. Dis. Child.* **128:**478–483 (1974).

216. YEAGER, A. S., Longitudinal, serological study of cytomegalovirus infections in nurses and in personnel without patient contact, *J. Clin. Microbiol.* **2:**448–452 (1975).

217. YEAGER, A. S., GRUMET, F. C., HAFLEIGH, E. B., ARVIN, A. M., BRADLEY, I. S., AND PROBER, C. G., Prevention of transfusion-acquired cytomegalovirus infections in newborn infants, *J. Pediatr.* **98:**281–287 (1981).

218. YOW, M., LAKEMAN, F., AND STAGNO, S., Use of restriction enzymes to investigate the source of a primary cytomegalovirus infection in a pediatric nurse, *Pediatrics* **70:**713–716 (1982).

219. YOW, M. D., WHITE, N., TABER, L., FRANK, A. L., GRUBER, W. C., MAY, R. A., AND NORTON, H. J., Acquisition of cytomegalovirus infection from birth to 10 years: A longitudinal serologic study, *J. Pediatr.* **110:**37–42 (1987).

220. ZUR HAUSEN, H., Viruses in human cancers, *Science* **254:**1167–1173 (1991).

12. Suggested Reading

HO, M., *Cytomegalovirus: Biology and Infection*, 2nd ed., Plenum Press, New York, 1991.

KRUGMAN, S., KATZ, S., GERSHON, A., AND WILFERT, C., Cytomegalovirus infections, in: *Infections of Infants and Children*, 9th ed. (S. KRUGMAN *ET AL.*, EDS.), pp. 25–45, Mosby, St. Louis, 1992.

STAGNO, S., Cytomegalovirus, in: *Infections of the Fetus and Newborn Infant*, 3rd ed. (J. S. REMINGTON AND J. O. KLEIN, EDS.), pp. 241–281, Saunders, Philadelphia, 1990.

Cytomegalovirus infections: Epidemiology, diagnosis, and treatment strategies, *Rev. Infect. Dis.* **12**(Suppl. 7):S691–S860 (1990).

CHAPTER 10

Epstein–Barr Virus

James C. Niederman and Alfred S. Evans[†]

1. Introduction

Epstein–Barr virus (EBV), a member of the herpes group of viruses, is the cause of heterophil-positive infectious mononucleosis, of most heterophil-negative cases, and of occasional cases of tonsillitis and pharyngitis in childhood. Rarely, it may involve the liver or central nervous system as primary manifestations. This virus is also implicated as having a causal relationship to African Burkitt lymphoma, B- and T-cell malignant lymphomas, including those in patients with the acquired immunodeficiency syndrome (AIDS), and nasopharyngeal cancer. High antibody titers are present in some patients with sarcoidosis, and in systemic lupus erythematosus.

This chapter deals with the epidemiology of EBV infections and the epidemiology of infectious mononucleosis. Infectious mononucleosis can be defined as an acute febrile illness involving children and young adults characterized clinically by sore throat and lymphadenopathy, hematologically by lymphocytosis of 50% or more, of which 10% or more are atypical, and serologically by an elevated absorbed heterophil antibody titer and the development of EBV immunoglobulin M (IgM) and other EBV antibodies.[(14)] This chapter mentions the relationship of high antibody titers to certain chronic and malignant diseases. The relationship of EBV to malignant lymphomas and to nasopharyngeal carcinoma will be discussed in Chapters 30 and 31, respectively.

[†]*Deceased.*

James C. Niederman • Department of Epidemiology and Public Health, Yale University School of Medicine, New Haven, Connecticut 06510.   **Alfred S. Evans** • Department of Epidemiology and Public Health, Yale University School of Medicine, New Haven, Connecticut 06510.

2. Historical Background

In 1889, Emil Pfeiffer of Wiesbaden, Germany described a condition called *Drüsenfieber* (glandular fever), characterized by fever, adenopathy, mild sore throat, and in severe cases enlargement of the liver and spleen.[(112,228)] Since Pfeiffer's original description, as well as Filatov's[(88,89)] in Russia in 1892, antedated by some 30–50 years' recognition of the hematologic changes and the heterophil antibody, it is uncertain whether the patients had definite infectious mononucleosis. However, the clinical details of this febrile syndrome in older children and young adults seem best to fit this diagnosis. There is little doubt about the classic description of the disease made by Sprunt and Evans,[(265)] from Johns Hopkins, in 1920. They defined the disorder in young adults as we now know it, named the disease "infectious mononucleosis," and reported the specific hematologic changes. This description was followed rapidly by similar reports from other workers.[(22,25,36,178,199,267)] A definitive presentation of the hematologic changes was made by Downey and McKinlay[(49)] in 1923.

The next major development was the discovery in 1932 of the heterophil antibody by John R. Paul and William W. Bunnell[(222)] of Yale University. Their report was based on an accidental observation when studying the occurrence of heterophil antibodies in rheumatic fever. The search had been initiated because of the clinical similarity of rheumatic fever and serum sickness and because of the work of Davidsohn[(39)] describing the presence of heterophil antibodies in serum sickness. Among the control subjects for rheumatic fever patients was one who had infectious mononucleosis who was found to have a much higher heterophil antibody titer than was present in any other condition. Paul and Bunnell then continued these observations in three additional cases of infectious mononucleosis and utilized 275 controls for comparison.

Their paper also describes what they believed to be a false-positive heterophil antibody occurring in a patient with aplastic leukemia. A review of the details of this case[71] reveals that the heterophil antibody occurred about 20 days after the administration of several units of blood, and therefore this patient may represent the first case of transfusion infectious mononucleosis. Soon after the discovery of the presence of heterophil antibodies, Davidsohn and Walker[40] reported on the use of guinea pig kidney and of beef cells to absorb serum prior to heterophil testing in order to increase the specificity of the test. Both these procedures have withstood the test of time well and still constitute one of the major criteria of diagnosis. Regular alterations in various liver function tests during acute infectious mononucleosis were recognized in several laboratories in the late 1940s and 1950s,[24,63] even though only 5% of patients had clinical jaundice. This was followed by the discovery of alterations in serum glutamic oxalacetic transaminase (SGOT) and other hepatic enzymes during the course of disease.[245,302]

Search for the etiologic agent of infectious mononucleosis began in the 1920s but met with little success until 1942, when Wising[298] reported the successful transmission of classic infectious mononucleosis to a female medical student volunteer who received 250 ml of blood from a patient ill with the acute disease. This successful experiment was not reproducible by Wising in several other attempts, nor by Bang,[10] who carried out a similar set of volunteer experiments. In 1947 and again in 1950, additional efforts of this sort were carried out at Yale University using whole blood, serum, or throat washings. The results provided suggestive but inconclusive evidence of transmission.[62,64] A third effort without success was reported from Yale University in 1965.[209] Subsequent EBV antibody tests on the sera from these last experiments in 1968 revealed that all volunteers had actually been immune to infection prior to the experiment as indicated by the presence of antibody (J. C. Niederman, unpublished data, 1969).

Repeated attempts in the 1950s to isolate an infectious agent from the throat or blood of patients with infectious mononucleosis using several tissue culture systems, long-term cultures of lymphocytes on a feeder layer, and fluorescent antibody techniques to identify an agent were unsuccessful.[65] Epidemiologically, the key events during this time were the observations of Hoagland, who suggested that the disease might be transmitted by kissing[141] and that the incubation period was of the order of 30–49 days.[143]

Early in 1968, evidence first appeared that EBV was the cause of infectious mononucleosis.[127,206] This virus,

identified by Epstein *et al.*[60] in a culture of African Burkitt's tumor tissue,[230] was found to be a new member of the herpes group of viruses. While working with this agent, a technician in the Henles' laboratory in Philadelphia developed infectious mononucleosis. Her serum, which lacked antibody several months prior to disease, developed EBV antibody during illness, and her lymphocytes, which had previously failed to be cultivated successfully, now grew well in tissue culture and were shown to contain EBV antigen.[127] This serendipitous observation was rapidly confirmed and extended at that time by the Henles in conjunction with Niederman and McCollum[206] of Yale and later in several other prospective studies carried out by the Yale Team[84,204,252] and in one English study.[285] Subsequent investigations established the presence and persistence of EBV in the throat during and after acute infectious mononucleosis,[31,260,299] the occurrence of EBV-specific IgM,[9,116,211,255] and the reproduction of some features of mononucleosis by inoculation of EBV into monkeys.[258,296] Fuller details of the history of infectious mononucleosis have been published elsewhere.[27,71]

3. Methodology

3.1. Mortality

Infectious mononucleosis is rarely a fatal disease; only about 50 fatalities have been reported in normal young adults, usually from central nervous system involvement. About half of primary EBV infections occurring in the rare X-linked lymphoproliferative syndrome have a fatal outcome,[231] as do the rare cases of acute immunoblastic sarcoma.[239] Since these are all rare events, examination of autopsy records or of international indexes of causes of death would therefore give little indication of the occurrence of the disease even though its pathological features are quite characteristic.

3.2. Morbidity

Infectious mononucleosis is not a reportable disease in most states and in most countries. Exceptions are the state of Connecticut, where it has been reportable since 1948,[33] and the United States armed forces, which collect hospitalization data on all diseases.[69] Unless strong emphasis is placed on the need for fulfilling clinical, hematologic, and serological criteria for diagnosis before reporting, the reliability of morbidity data from these sources must be seriously questioned. This requirement is emphasized by the fact that even for the 15–25 age group—in which the disease has its highest incidence, its most characteristic clinical features, and the highest frequency of elevated heterophil antibody tests—only one third of the serum samples sent to a state laboratory for diagnosis of suspected cases were heterophil antibody positive.[66]

To collect morbidity data, special surveys of selected populations for infectious mononucleosis have been carried out in college infirmaries,[84,86,149,204,251] community medical care groups,[124] general practitioners' offices,[145] and by physicians and laboratories serving defined communities.[52,65,121,246] The Centers for Disease Control (CDC) has published a surveillance report on infectious mononucleosis based on data derived from 19 colleges.[29]

The problems of data derived from such surveys are related to the extent to which the numerator or case report reflects the proper diagnosis and to whether adequate surveillance has been carried out with respect to the denominator, the population at risk.

3.3. Serological Surveys

Up to 1968, when the causal association of EBV with infectious mononucleosis was discovered, the heterophil antibody constituted the only serological approach to diagnosis and survey work. Because this is an IgM-type antibody and is transient in nature, it can be used as a serological tool only for incidence data, i.e., during the acute illness, and as an essential diagnostic feature.[206] The specificity of a properly performed quantitative heterophil-antibody test is high provided that the serum has been preabsorbed with guinea pig kidney in the sheep and horse red cell tests, or that the beef hemolysin test has been used. In a test performed in this way, an elevated titer quite accurately reflects the occurrence of infectious mononucleosis even in the absence of clinical and hematologic data and has been utilized as an indicator of infectious mononucleosis in sera sent to hospitals and state laboratories.[66] The major limitation of this approach is the extent to which physicians have sent sera from suspected cases to the diagnostic laboratory for analysis. The increasing use of simple laboratory kits for identifying heterophil-antibody elevations in the physician's office probably results in much less utilization of state and hospital laboratories, so that morbidity data from these sources may greatly underestimate the occurrence of the disease. Thus, utilization of the heterophil antibody as an epidemiologic tool to identify acute illness has high reliability and specificity but low sensitivity. Heterophil-positive cases diagnosed in the Connecticut State Public Health Laboratory alone represented 74.5% of all the reported cases in the state in 1972.

Since the discovery of EBV as the cause of infectious mononucleosis in 1968, many serological surveys in different countries have been made for the presence of antibody to this virus in sera collected from healthy persons, usually employing the indirect immunofluorescence test[125] for viral capsid antigen (VCA). This IgG antibody persists for many years, probably for life.[204,206] Such studies yield prevalence data on prior EBV infection but give no direct indication of the occurrence of clinical infectious mononucleosis. More recently, the EBV-specific IgM antibody has been included in surveys to identify recent infection.[273]

The most accurate information on the incidence of both EBV infections and clinical infectious mononucleosis has come from prospective seroepidemiologic studies of defined populations with close clinical surveillance for the occurrence of suspected cases of infectious mononucleosis and other illnesses. Sera taken at the start of the observation period are tested to define the number of susceptibles, i.e., those lacking EBV antibody; samples showing seroconversion at the end of the observation period will identify the total EBV infection rate; those collected during interim illnesses and tested for EBV and heterophil antibodies will delineate the number and spectrum of clinical illnesses, such as infectious mononucleosis, associated with EBV infection.

3.4. Laboratory Methods

3.4.1. Virus Isolation. The virus cannot be grown in the usual tissue cultures employed for other herpesviruses. The currently available isolation technique is tedious, difficult, and usually confined to research laboratories. It is based on the ability of EBV to transform uninfected human leukocytes into continuous cell lines and the identification of this effect as being caused by EBV. Leukocytes derived from the cords of newborn infants or from persons lacking EBV antibody are employed to ensure absence of EBV antigen in the lymphocytes. Throat washings or other materials to be tested are usually filtered to remove debris and bacteria, then added to the leukocytes, and placed on a placental fibroblast feeder layer. If EBV is present, evidence of transformation is indicated by an abrupt increase in the total number of cells, the production of acid, growth of cells in clumps, and the development of the capacity to be subcultured indefinitely. Usually, transformation occurs 30–90 days after addition of the throat washing.[31,192]

The presence of EBV-associated nuclear antigens

(EBNA) can be demonstrated in acetone-fixed smears of transformed cord cells using an indirect complement-fixation (CF) fluorescence technique,[236] but EBNAs cannot be demonstrated by ordinary immunofluorescence methods, presumably because the virus does not mature sufficiently in such cells. The EBV-transformed leukocytes may also be grown further in culture to prepare a CF antigen as a means of identification; however, this is a laborious method.[292] Robinson and Miller[240] have demonstrated that DNA stimulation occurs early in EBV-infected cord cells and can be detected by increased uptake of radiolabeled thymidine. Several methods of detecting EBV or its genome in blood, lymph nodes, body fluids, and other tissues, including frozen and paraffin sections are now available. These include Southern blot, the polymerase chain reaction (PCR), and the EBV-RNA *in situ* test (EBER), which permits identification of EBV genome in specific cells.[295]

3.4.2. Viral Antibody.   A wide variety of techniques to measure EBV antibody have been developed. Five antibody methods based on immunofluorescence have been used.[132] First, for epidemiologic purposes, the indirect immunofluorescence test of the Henles[125] for VCA has been widely employed in serological surveys as a reliable indicator of susceptibility and immunity to infectious mononucleosis. However, it has not proved very useful as a diagnostic test for infectious mononucleosis because antibody is usually present by the time the patient seeks medical care and rises in titer are detectable in only 15–20% of cases.

Second, antibodies to the D and R components of the "early antigen" (EA) complex are also identified by indirect immunofluorescence. The presence of antibody to the R component and IgG antibody to VCA at titers of ≥1:320 has been considered indicative of an active EBV carrier states.[125,129,134,135] Recent studies suggest that such enhanced EBV activity may persist for 30–104 months after infectious mononucleosis in healthy individuals.[146] Unfortunately, from a diagnostic standpoint, EA antibody is demonstrable in only about 75% of patients with infectious mononucleosis[134]; it also occurs in the sera from patients with Burkitt's and Burkitt's-like lymphoma and nasopharyngeal cancer and reflects replication of the virus in epithelial cells. In nasopharyngeal cancer it is used as both a diagnostic and a screening tool (see Chapter 31).

Third, antibodies to EBNA are detectable by an immunofluorescence technique based on CF; these usually arise only 1 month or more after onset of infectious mononucleosis and after primary infections and probably

persist for life.[128] Their late appearance impairs their usefulness in routine diagnosis. It is now evident that EBNA is an antigen complex consisting of a number of different viral polypeptides, each encoded by a different viral gene.[45,46,138,277] Genes encoding two components of EBNA have been identified: EBNA-1 is encoded by the *Bam* HI-K fragment,[47,136,272] and EBNA-2 maps to *Bam* HI-WY and H.[45,46,138,249] *Bam* HI-K EBNA binds to chromosomes and therefore resembles the EBNA first described by Reedman and Klein.[236] Enzyme-linked immunosorbent assays (ELISA) based on EBNA-1 and EBNA-2 peptides have also been developed for IgG and IgM antibodies.[177] A ratio of <1 between the two tests gave an 98% sensitivity and specificity. Another rapid ELISA test based on EBNA-1 gave positive predictive values of only 87% and negative predictive values of 81%, and its authors suggested it be used only as an adjunct to other tests.[181]

Fourth, there is an indirect immunofluorescence test for EBV-specific IgM antibody, and this is the most useful procedure for the diagnosis of heterophil-negative infectious mononucleosis[51,211,255]; however, in its present form, this test is technically difficult to perform, and it is not currently available in most diagnostic laboratories. Fifth, an immunofluorescence test for EBV-specific IgA antibody, when used with the P3/HR1K cell line as the source of antigen, yields positive results at ≥1:5 at some point in over 85% of cases of infectious mononucleosis.[82] Finally, a membrane fluorescence test has been developed by Klein *et al.*[164-166] Other antibody tests include CF using either the virus[3] or soluble antigens,[96,99] neutralization tests based on contact inhibition,[139–193] and immunodiffusion tests.[214]

3.4.3. Heterophil Antibody Tests.   Four general methods are employed: (1) the classic Paul–Bunnell test,[222] using sheep or horse red cells after absorption of the serum with guinea pig kidney as developed by Davidsohn and Walker[40]; (2) the beef-cell hemolysin test of Bailey and Raffel[6] adopted for diagnostic use by Mason,[180] which does not require absorption of sera with guinea pig kidney because beef-cell hemolysins are absent or at very low titers in normal sera; (3) the enzyme test of Wöllner,[301] in which red-cell receptors for heterophil antibody are specifically removed by treatment with papain or a similar enzyme; and (4) an immune adherence hemagglutination assay (IAHA),[174] which is more sensitive, gives higher titers, and is positive longer than the classic absorbed sheep or horse red cell tests.[82] It has identified heterophil-positive cases in children.[93,217] Recent evidence suggests that heterophil-antibody titers may reach diagnostic levels even after mild or asymptomatic EBV infections, provided serial specimens are tested over a month or so by the horse-cell differential test.[83,142] The test is also useful in childhood EBV infections, which are often mild.

Most state and large diagnostic laboratories employ either the Davidsohn differential absorption test or the beef-cell hemolysin test; the former, using horse cells, is more sensitive, and heterophil antibody to this antigen often persists at diagnostic levels for as long as a year or so.[83,170] The beef hemolysin test is more specific but less sensitive and disappears in 3 months or less; it is perhaps the most reliable test to diagnose infectious mononucleosis during the acute illness. An ELISA has also been developed.[148]

A number of commercial testing kits for the diagnosis of infectious mononucleosis in the physician's office are now available. Most are slide agglutination tests, usually performed at a single dilution and commonly based on the agglutination of formalinized horse cells. Some of these employ guinea pig kidney to remove nonspecific agglutinins from the serum prior to testing. Other tests employ papain-treated red cells in the agglutination test. A spot test using horse red cells and absorption procedures is recommended. Such tests are useful if carried out by trained personnel.

4. Biological Characteristics of the Agent

4.1. The Virus

Epstein–Barr virus is a distinctive member of the herpes group of viruses. It is a double-stranded DNA virus consisting of about 100 genes. The DNA of the virus has been completely sequenced.[7] The virus can infect squamous epithelial cells[237,237a] and B lymphocytes via the C3d receptor, CR2.[90] The EBV genome has been identified also in certain T-cell lymphomas.[35] The infectious particle consists of an a doughnut-shaped central core, an icosohedral-shaped capsid, and an outer envelope. The major antigens expressed in EBV-infected cells include the VCA, EBV-induced membrane antigen (MA), EBV-induced EA, of which there is a diffuse (D) and a restricted form (R), EBNA, of which there are six forms (EBNA 1, 2, 3a, 3b, 3c, and EBNA leader protein EBNA-LP), plus two nonnuclear proteins—lymphocyte membrane protein and the terminal protein.[221,235,237,247] The precise functions of all of these EBNA proteins are not fully known, but EBNA-1 is probably responsible for the maintenance of the virus genome in the episomal state; it is expressed on all virus genome-carrying cells, but is not recognized by cytotoxic T cells.[207] EBNA-2 is involved in the transformation of B cells, which also involves EBNA-3c. All of these antigens are expressed in lymphoblastoid cell lines, whereas EBNA-1 is expressed in Burkitt's lymphoma and nasopharyngeal cells.[91,163] On electron microscopy, EBV appears similar to other herpes-group viruses.[60] Several laboratory strains of EBV have been identified.[194] Earlier evidence suggested the possibility of viral variations in nature,[97] and two viral strains have now been definitely identified based on variations in EB nuclear gene loci and designated as type A (or 1) and type B (or 2).[261] While type A was regarded as worldwide in prevalence and type B restricted primarily to Africa, both are now recognized as widespread. For example, in healthy adults in Memphis, Tennessee, EBV was detected in throat washings of 34 of 157 randomly selected donors and included 50% type A, 41% type B, and 9% with both types.[261] Type A predominates in nasopharygeal cancer in China[30a] and type B in immunodeficient states.[168a]

The virus has been cultivated only in suspension cultures of primate lymphocytes, and most cultures yield only small amounts of extracellular virus. These limitations have made characterization of the physical and chemical properties of EBV very difficult. Epstein–Barr virus is a lymphotropic virus and infects B lymphocytes, which have EBV receptors on their surface[153] that are identical to the C3 receptors[90,303]; there new membrane antigens are induced, to which T cells respond.[229] Cell lymphocytes in continuous cultures established from infectious mononucleosis blood or Burkitt's lymphoma biopsies contain the EBV genome as demonstrated by DNA hybridization or EBNA tests, but only 1–3% have demonstrable VCA.[212,215,236,306] Cell clones grown from such cultures show a similar low percentage of complete virus.[185,305] One line of cells from Burkitt's lymphoma, the P3J line of Pulvertaft,[230] and its cloned derivative, the HR1K, produce more extracellular virus than other lines but fail to induce transformation.[194] Another line, B95-8, derived from EBV-infected marmoset cells by Miller and Lipman,[190] releases about 1000 times more transforming virus but about the same number of viral particles as HR1K. It has been useful in viral characterizations.[122,123]

Some of the biological properties of EBV are important epidemiologically. The capacity for persistence of a lytic infection in the throat provides a source of potential transmission; the low yield of extracellular virus may bear on the need for intimate, oral contact for transmission in young adults.[193,208] The reasons for the higher efficiency of transmission of EBV infection in young children than in adults are unknown but might include the production of more infectious virus in the pharynx of children, more intense exposure, or indirect spread by saliva in settings with poor hygiene.

The capacity for persistence and latency of EBV in a nonproductive form sets the stage for later reactivation under conditions of immunosuppression (e.g., AIDS, malaria, therapeutic immunosuppression in homograft transplantation). Some EBV-related malignant lymphomas such as Hodgkin's disease, some cases of non-Hodgkin's lymphoma, some B-cell and T-cell lymphomas in patients with AIDS, as well as nasopharyngeal carcinoma may be expressions of such reactivation.[79,81,233,234] The long-term persistence of EBV in lymphocytes is of importance epidemiologically in the transmission of infection during blood transfusions to susceptible recipients and via saliva during reactivation. Of great importance is the capacity of EBV to transform uninfected primate lymphocytes, inducing in them the potential for unlimited proliferation; this property was termed "immortalization" by Miller,[184] and the lymphocytes that result are termed "I lymphocytes." The EBV-transformed and -infected cells are B-type lymphocytes.[218,219] Viral induction of new antigens (or unmasking of preexisting ones) such as the membrane antigen of Klein *et al.*[165] may have immunologic consequences in the development of new antibodies, in a graft-versus-host response, and in the induction of cytotoxic T lymphocytes.[153,154,284]

A better understanding is now developing of the dynamics and effects of EBV activity at molecular and cellular levels ("molecular epidemiology") and of the responses of the host under varied conditions of age, concomitant infection, immune status, and genetic constitution. With DNA molecular techniques using cloned fragments of EBV DNA, it is possible to measure serological responses to antigens representing defined regions of the viral genome and to relate these to clinical symptoms.[186,216,221]

4.2. Proof of Causation of Infectious Mononucleosis

The causation of heterophil-positive infectious mononucleosis by EBV has been firmly established. Proof is based on seroepidemiologic and virological evidence and also on partial success in the experimental transmission of infection to monkeys and man.

Seroepidemiologic investigations have repeatedly shown that antibody to EBV of the IgG type has been consistently absent in sera taken prior to the onset of infectious mononucleosis, regularly appears during ill-

ness, and persists for years thereafter.[84,114,206,252,285] The presence of this antibody indicates immunity to clinical infectious mononucleosis, and its absence indicates susceptibility to the disease. Table 1 summarizes 11 prospective studies involving over 5000 children and young adults in support of this relationship. A more recent prospective study in University of Hong Kong students confirms these results in an Asian population.[38] Only 6.8 of entering freshman students lacked EBV antibody, but at the end of the year, 25.8% had asymptomatic seroconversions. No other virus has been found that induces a similar antibody, and no other viral antibody has been demonstrated during heterophil-positive infectious mononucleosis.[82] The occurrence of some heterophil-negative cases of infectious mononucleosis caused by EBV has also been noted in prospective studies.[84,114] Other monolike syndromes are caused by cytomegalovirus and other agents.[73,77,203a]

An EBV-specific antibody of the IgM class has been demonstrated during acute infectious mononucleosis and found to disappear during convalescence, thus indicating that this is a primary response to EBV infection.[9,51,116,255] Both the IgG and IgM EBV-specific antibodies of infectious mononucleosis are distinct from the heterophil antibody.

The virological evidence consists of the appearance of EBV in the oropharynx and in the circulating lymphocytes of patients with acute infectious mononucleosis. The agent has been regularly demonstrated in the pharynx of over 80% of patients during the acute illness.[31,100,173,192,208,227] Recent work has shown that both parotid ductal epithelium and squamous epithelial cells in the oropharynx harbor Epstein–Barr virus DNA and are sites of viral replication and release.[260,299] In addition, EBV antigens

Table 1. Summary of 11 Prospective Studies of Epstein–Barr Virus Infection in Children and Young Adults[a]

EBV antibody status at start	Number	Percentage	Subsequent rate/ 100 per year	
			EBV infection	Clinical infectious mononucleosis[b]
With antibody	3733	70.7	0	
Without antibody	1547	29.3	16.4	7.1
Totals	5280	100	4.6	2.0

[a]From ten studies carried out by Yale investigators[84,114,252] and one by an English team.[285]
[b]Clinical infectious mononucleosis was recognized in 47% of those infected with EBV.

have been found in tonsillar lymphocytes.[292,288] They persist for many months and in several cases have persisted for as long as a year or so. A chronic and intermittent carrier state exists, as suggested by the presence of virus in the oropharynx of 15–20% of healthy seropositive adults.[31,100,269]

Epstein–Barr virus has been regularly demonstrated in lymphocyte cultures from patients with acute infectious mononucleosis, where it may remain in a latent form for years and may be a source of transfusion mononucleosis.[21,44,98,133] The virus has been demonstrated in fresh B lymphocytes from patients with acute infectious mononucleosis[167,243]; up to 0.5% of circulating mononuclear leukocytes are EBV-infected during the acute illness.[243] The appearance and persistence of EBV in the oropharynx following mild or asymptomatic infections provide a large pool of healthy carriers capable of transmitting infection through appropriate exposure.

Efforts to transmit infectious mononucleosis to volunteers using blood, throat washings, or stools from acutely ill patients were made prior to the discovery of EBV in 1968; the results were largely inconclusive or unsuccessful, probably because of the presence of prior immunity in those inoculated. However, there are a few exceptions. Wising[298] successfully transmitted the full-blown disease to a female volunteer by transfusion. Evans[65] and Taylor[278] reported suggestive evidence of successful transmission by inoculation of pooled sera from patients with acute infectious mononucleosis into patients with acute leukemia as a therapeutic effort to induce a remission; the young age of this group probably meant that some were susceptible because they had not been previously exposed. About 50 other experiments in humans were equivocal or unsuccessful.[9,62,64,209] Similarly, earlier efforts to induce infectious mononucleosis in monkeys were not rewarding.[80,264]

Recent studies with this virus in humans have been limited because of concern for the oncogenicity of EBV. Grace et al.[104] repeatedly inoculated partially purified EBV into a terminal cancer patient who lacked prior antibody; both EBV and heterophil antibodies developed. Inoculation of EBV-infected lymphocytes into gibbons has resulted in an exudative tonsillitis and the appearance of EBV antibody.[296] Shope and Miller[258] have induced transient EBV and heterophil antibody in squirrel monkeys inoculated with virus-transformed leukocytes.

In summary, the results of seroepidemiologic, virological, and transmission studies in man and monkeys indicate that EBV is the cause of all cases of heterophil-positive mononucleosis and most heterophil-negative cases.[68,70,73,87]

5. Descriptive Epidemiology

The epidemiologic factors that influence the incidence and distribution of *infection* by Epstein–Barr differ from those that cause *clinical* infectious mononucleosis. The factors influencing infection are related to levels of hygiene and cultural patterns that lead to exposure to saliva, as in prechewing of food for infants in developing countries and salivary exposure in developed countries during intimate oral kissing in young adult life. The factors influencing clinical disease are related to the age at the time of exposure, immune status, genetic factors, and psychosocial variables. These concepts have been summarized.[74] Thus, EBV infection without disease occurs early in life in lower socioeconomic settings throughout the world, whereas clinical infectious mononucleosis is largely limited to developed and higher socioeconomic areas, where exposure and infection are delayed until older childhood and young adult life.

5.1. Prevalence and Incidence

The *prevalence of antibody* to the VCA of EBV has been determined in many countries and in many age groups.[70] Figure 1 indicates the percentage of children in several areas of the world with EBV antibody. In developing and tropical areas, most children have been infected by age 6 years. Because infections with EBV are often mild and asymptomatic in young children, infectious mononucleosis may not be commonly recognized as a clinical entity in such countries. However, more intensive clinical and serological studies, especially those employing newer diagnostic techniques, have permitted identification of both heterophil-positive and heterophil-negative cases of infectious mononucleosis in children in such settings as Singapore[23] and Brazil.[217] The prevalence of EBV antibody in young adults living in different parts of the world is depicted in Fig. 2. A similar socioeconomic pattern exists. It is only when a significant percentage of the population reaches ages 15–25 before exposure to and infection with EBV that infectious mononucleosis emerges as an important clinical entity. This delay in exposure is largely limited to nations with high economic and hygienic levels and to middle and upper socioeconomic classes in any country. The most susceptible college group tested thus far were entering freshman students at Yale University in the period 1958–1963, when nearly 75% were at risk to infectious mononucleosis because they lacked antibody; coincident with programs that broadened admission to include students with widely differing socioeconomic backgrounds, among them many

PERCENT POSITIVE
10
20
30
40
50
60
70
80
90
100
BARBADOS
UGANDA
INDONESIA
MEXICO
HAWAII
BRAZIL
FRANCE
SWEDEN
ENGLAND
CONN.

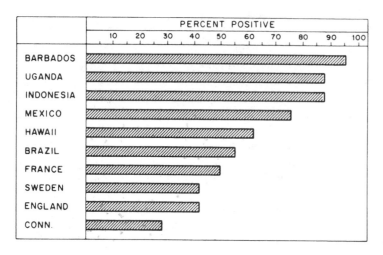

Figure 1. EBV antibody prevalence at age 4–6 years in different populations. Adapted from Evans.[70]

minority groups, the susceptibles decreased to 40–50%. In contrast, fewer than 20% of undergraduate students at the University of Philippines lacked EBV antibody, and all of 145 freshman student nurses in Thailand had antibody.

Hospitalization rates for infectious mononucleosis in the armed forces range from 148 to 250 per 100,000.[69] In the Navy and Marine Corps, for which comparative data are available, it ranks as the fifth most common infectious disease and the fourth most common cause of days lost. Table 2 summarizes data from a recent review on hospitalization in different population groups.[77] On the average, some 15% of cases of infectious mononucleosis were hospitalized for a period of 4.5 days. More recently, the use of short courses of steroids has reduced the percent needing hospitalization to perhaps 5%.

The incidence of clinical infectious mononucleosis is not well documented, since reporting is not obligatory in most states in the United States, and the available data have usually been derived from special surveys such as the community survey in Atlanta, Georgia, where a rate of 45 per 100,000 of heterophil-positive cases was found, and in Olmstead County, Minnesota, which includes the Mayo Clinic, where resident rates were 200 per 100,000.

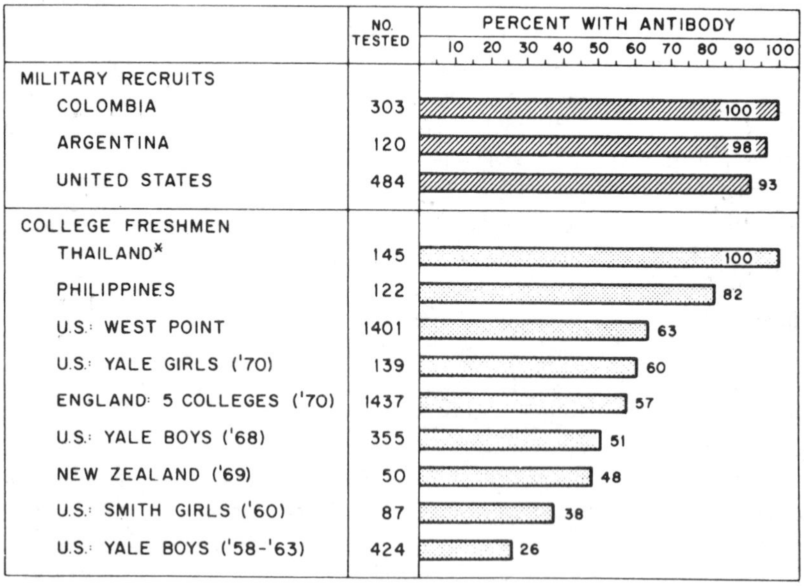

NO. TESTED
PERCENT WITH ANTIBODY
10 20 30 40 50 60 70 80 90 100
MILITARY RECRUITS
COLOMBIA 303 100
ARGENTINA 120 98
UNITED STATES 484 93
COLLEGE FRESHMEN
THAILAND* 145 100
PHILIPPINES 122 82
U.S.: WEST POINT 1401 63
U.S.: YALE GIRLS ('70) 139 60
ENGLAND: 5 COLLEGES ('70) 1437 57
U.S.: YALE BOYS ('68) 355 51
NEW ZEALAND ('69) 50 48
U.S.: SMITH GIRLS ('60) 87 38
U.S.: YALE BOYS ('58-'63) 424 26

Figure 2. EBV antibody prevalence in young adults in different populations. ×, Student nurses. Adapted from Evans.[70]

Table 2. Duration of Hospitalization of Infectious Mononucleosis in Different Populations and Age Groups[a]

Area	Study years	Age of patients	Number	Duration in days
Rochester, MN	1950–69	All	776	5.6
Air Force	1986–91	All		
	Inpatients, active duty		1901	6.3
	Inpatients, other		931	3.9
	Quarters, active duty		708	3.7
			110	4.0
Denmark	1989	0–14	216	4.5
		15–64	527	4.5
			113	

[a]From Evans.[77]

In Denmark, the rate of notified cases is 60 per 100,000.[246] The state of Connecticut requires reporting of infectious mononucleosis, and rates of 48 per 100,000 were recorded in 1972.[33] In 1979, a simplified report form and an aggressive surveillance system were initiated for all reportable diseases, resulting in a rate for infectious mononucleosis of 71.1 per 100,000 in 1983 and 60.7 per 100,000 in 1985 (Connecticut State Health Department, personal communication). This rate of infectious mononucleosis is surpassed only by chickenpox, gonorrhea, and influenza among reportable diseases in Connecticut and greatly exceeds *Salmonella* infections and hepatitis (Table 3). By comparison, the United States rates for these communicable diseases per 100,000 in 1985 were as follows: gonor-

Table 3. Reported Cases per 100,000 Population for Infectious Mononucleosis and Other Reportable Diseases in Connecticut for 1983 and 1985 and in the United States for 1983[a]

Diseases	Connecticut 1983	Connecticut 1985	United States (1983)
Gonorrhoea	318.2	254.5	387.6
Chickenpox	251.7	234.9	99.6
Influenza	217.3	168.9	NR
Infectious mononucleosis	71.1	60.7	NR
Salmonella	34.6	33.9	18.9
Hepatitis			
A	2.9	4.8	9.2
B	14.0	11.9	10.4
Non-A, non-B	1.5	1.1	1.7
Unspecified	1.2	0.8	3.1
Total	18.8	19.3	24.1

[a]Derived from data supplied by the Connecticut Health Department and from Centers for Disease Control.[30] NR, not reportable.

rhea, 387.6; chickenpox, 99.6; hepatitis A, 9.2; hepatitis B, 10.4; non-A, non-B hepatitis, 1.7. Among college students, the rates of infectious mononucleosis are very high, averaging 840 per 100,000 students in 19 colleges in 1971–1972.[29]

The most accurate measure of EBV infection and of disease has been obtained in prospective serological and clinical studies, in which the number of susceptibles, the infection (seroconversion) rate, and the clinical attack rate can be critically defined. Comparative data are available from three prospective investigations of freshman students: at Yale University,[84,204,252] at five English colleges and universities,[285] and at the U.S. Military Academy at West Point, New York.[114] As depicted in Table 4, the incidence rate of EBV *infection* was strikingly similar in all three settings: 12–13% of *susceptible* students were infected with EBV during the freshman year, and of those with known EBV infection, 27.7–74.0% developed clinical infectious mononucleosis. At the U.S. Military Academy, where a prospective investigation was carried on in a single cohort of freshman over 4 years, the EBV infection rate in susceptible cadets was 12.4% in the first year, 24.4% in the second year, 15.1% in the third year, and 30.8% in the fourth college year.[114] Over the 4-year period, 45.9% of susceptible cadets were infected with EBV, and 26.4% of these were known to have clinical infectious mononucleosis; others may have been ill but did not report to the clinic for treatment. The reasons for the varying rates of clinical expression among EBV-infected young adults in similar settings are not known. The variation may be related to the intensity of clinical surveillance, the students' attitude toward the health service, the average time of hospitalization, or various host factors. Some evidence on the influence of psychological factors is emerging[160] (see Section 5.11).

5.2. Epidemic Behavior

True epidemics of infectious mononucleosis that fulfill appropriate diagnostic criteria have not occurred in modern times.[66,144] Earlier, many purported epidemics were described, of which the most impressive are those described by West[297] in the United States in 1896, by Carlson *et al.*[26] in Wisconsin in 1926, and by Moir[195] in the Falkland Islands in 1930. More recent and suggestive outbreaks have been described from an emergency medical hospital[113] and from Oxford, England, reported by Hobson *et al.*[145] A small outbreak involving 9 of 29 staff members in an outpatient clinic was recently reported by Ginsburg *et al.*[103]; however, the source of the outbreak and means of spread were not identified.

Table 4. Epstein–Barr Virus Infection Rates during Freshman Year, and Percentage Clinically Expressed in Different Colleges

Place	Number in study	Susceptible (%)	Infection rate in susceptibles (%)	Clinical infectious mononucleosis (%)
U.S. Military Academy[114]	1401	36	12.3	27.7
Five English schools[290]	1487	43	12.0	59.1
Yale University[251]	355	49	13.1	74.0

The high incidence in military camps during World War II probably reflected the rapid turnover of large numbers of men.[85,286,294] Some reported hospital "outbreaks" are suggestive of a true outbreak[113,145] but in general do not fully meet diagnostic criteria. On a hypothetical basis, the early acquisition of immunity to infectious mononucleosis by mild and inapparent infections with EBV in childhood and the major route of transmission via intimate oral contact in young adults weigh heavily against the occurrence of "epidemic infectious mononucleosis."

The high prevalence rates of EBV antibody in children in developing countries[70] (A. S. Evans, R. P. S. Carvalho, and L. Grossman, unpublished data, 1974), in nurseries,[226] and in orphanages[281] suggest that EBV spreads effectively in young children under circumstances of crowding and poor hygiene to reach almost all susceptibles. However, the contagiousness of infectious mononucleosis has been notoriously low in young adult populations; secondary cases have been rare in roommates of index cases,[65,114,141] in college dormitories,[65,84,252] aboard ship,[223] and on Polaris submarines.[268] The low contagiousness in college populations has been confirmed in recent studies employing the status of EBV antibody as a marker of susceptibility and of infection. Among susceptible and exposed roommates of Yale freshmen with infectious mononucleosis, no evidence of increased risk was found over the susceptible population as a whole[252]; however, there tended to be some aggregation of cases in social clusters in dormitories. In a more critical analysis of this issue at the U.S. Military Academy over a period of 4 years, no evidence of increased spread of EBV infection to susceptible roommates exposed to an index case was detected as compared with susceptible roommates not so exposed.[114] In a family setting, about 10% of exposed and susceptible members will develop EBV infection.[126,152,273,291] With new EBNA detection techniques it has been shown that only one EBV strain is usually present in a given family and that the incidence of secondary infection may be higher than previously thought.[105]

This low level of contagiousness of EBV infection in older children and in young adults of the same sex may be related to a high level of existing immunity and to the need for intimate oral contact. However, in 16 freshman Chinese university students who became EBV infected without clinical disease in their first year, 14 gave no history of intimate kissing.[38] The rate of infection among susceptible persons who are known to have had intimate oral contact with patients who have infectious mononucleosis or with established pharyngeal carriers to EBV has not yet been defined; it may well be high.

5.3. Geographic Distribution

Infection with EBV is worldwide. Antibody to EBV has been demonstrated in every population thus far tested, including very isolated tribes in Brazil,[20] Alaska,[281] and other remote areas[99] where measles and influenza antibody are often lacking. Infection occurs earlier in life in developing countries. A recent study of 94 children in the Republic of China revealed that 78.6% had EBV-VCA IgG antibody by the end of the first year of life and 80.7% were positive by the age of 3.[293]

Clinical infectious mononucleosis occurs most commonly in those hygienic and socioeconomic areas where exposure to and infection with EBV are delayed until older childhood and young adult life. These include Australia, Canada, England, many European countries, New Zealand, Scandinavian countries, and the United States.[70] In contrast, at the University of the Philippines, not a single case was recorded among 5000 admissions to the college infirmary, where laboratory facilities existed[78]; EBV antibody determinations in this college population revealed a very high level of prior immunity. The disease is now being recognized with more careful clinical and diagnostic scrutiny in developing countries.[23,217] The prevalence of EBV antibody had been found to vary in young adults entering the U.S. Military Academy from different areas of the United States.[114] The highest rate of 81.5% was found in cadets resident for 6 years or more in the East South Central states, and the lowest prevalence rate of 51.9% in the West North Central states. Since admission to the academy is based on competitive academic, athletic, and achievement values rather than on any social or economic considerations, a broad range of backgrounds would be expected.

5.4. Temporal Distribution

There is no clear-cut evidence of yearly fluctuations in the incidence of infectious mononucleosis, although appropriate morbidity data are not available to determine this accurately. Since 1948 in Connecticut, a yearly increase in incidence has been noted from 3.9 cases per 100,000 initially to 46.7 in 1967[33] and more recently 60.7 in 1985; this probably reflects increased reporting rather than actual changes in incidence. In one Swedish hospital where the same population was served and the same diagnostic criteria were presumably applied over a period from 1940 to 1957, the hospitalization rate increased from 12 cases per year in 1940–1942 to 110 per year in 1955–1957.[272] Caution must be observed in interpreting hospital data in which there is no defined denominator. No changes in the incidence rates of infectious mononucleosis were noted at Yale University over a 5-year period[84] or in a careful study with a defined population base in Rochester, Minnesota, over the period 1950–1969.[124]

Earlier studies of college students at the U.S. Military Academy[144] and at the University of Wisconsin[65] showed a peak in February, some 4–6 weeks after Christmas vacation, presumably because of increased exposure at that time. However, no clear-cut seasonal pattern has been seen in the CDC Surveillance Reports from 19 colleges and universities.[29] In a community study in Atlanta, Georgia,[121] two peaks were found, one in early fall and a larger one in late winter and early spring, but in Rochester, Minnesota,[124] no seasonal peak was observed.

5.5. Age

The acquisition of EBV antibody by age is shown in Fig. 3 for three different geographic areas. Antibody occurs early in life in economically underdeveloped countries, often reaching close to 100% immunity by age 10. In a prospective study of EBV infections in newborns living in Accra, Ghana, 81% had acquired antibody by age 21 months,[18] but none showed evidence of clinical infectious mononucleosis.[19] In contrast, clinical infectious mononucleosis is clearly a disease of older children and young adults in economically developed countries, with its highest incidence in the 15- to 25-year-old age group. This has been true of data based on hospitalized cases in the United States,[95,121,124,202] France,[263] and Denmark[279];

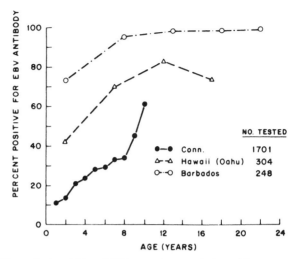

Figure 3. Acquisition of EBV antibody by age in three different areas. Data derived in part from Evans[70] and Jennings.[151]

on heterophil-positive cases identified in state public health laboratories[41,66,200,201,246]; and on recent community surveys in which the population at risk can be defined.[121,124] In results from the Atlanta community survey based on 575 heterophil-positive cases, the highest rate, 345.2 per 100,000, occurred in the 15–19 age group, and the next highest, 122.8, in the 20–24 age group; 27 heterophil-positive cases occurred in the 5–9 age group, and four in the 0–4 age group. Figure 4 shows the distribution of cases in this study. A similar age distribution was observed in the Wisconsin State Laboratory data based on elevated antibody titers in sera from suspected cases sent in for heterophil testing.[66] The peak frequency was in the 20–24 age group, in which 29.6% of the sera were posi-

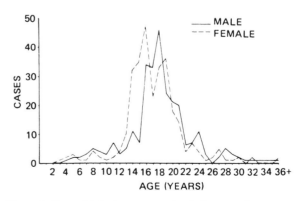

Figure 4. Cases of infectious mononucleosis, by age and sex, metropolitan Atlanta, Georgia, 1968.

tive. Some cases occurred at the extremes of age: 11.9% of the sera from suspected cases were positive in the 5–9 age group and 5.8% in the 65–69 age group. Use of more sensitive heterophil tests and of the EBV IgM tests now permits an increasing number of cases to be identified in childhood.[93,102,217] The mean age of mononucleosis cases was 18.4 years in Wisconsin[66] and 19.3 years in Norway.[200]

In developing countries, the age distribution is shifted downward because of the small number of children who escape infection until an age when the host response is that of recognizable infectious mononucleosis. In a study in São Paulo, Brazil, the average age of 15 heterophil-negative patients was 4.7 years and that of 32 heterophil-positive cases was 13.2 years.[217] The oldest case was age 15; five cases occurred in the 0–2 age group. In this setting, over 90% of the population has EBV antibody by age 10.[28]

Infectious mononucleosis rarely occurs in the elderly.[5] Only 29 reported cases over age 60 were found in a recent review.[254] Pharyngitis, lymphadenopathy, splenomegaly, and atypical lymophocytosis are unusual in this age group but liver involvement is more common.[5a,254]

5.6. Sex

No difference in EBV-antibody prevalence rates by sex has been noted in population surveys.

Infectious mononucleosis occurs equally in both sexes, although girls appear to develop the disease earlier than boys,[121,124,147] with a peak occurring in girls at age 16 and in boys at age 18 (see Fig. 4).

5.7. Race

Epstein–Barr virus *infection* occurs in all ethnic groups, and no evidence of differential susceptibility has been found.

Infectious mononucleosis in developed countries has been rare in blacks, but this probably reflects sociohygienic factors and earlier acquisition of infection rather than any difference in susceptibility. The incidence of the disease in whites in Atlanta, Georgia, was 30 times higher than in blacks.[121] Antibody prevalence to EBV among entering black cadets at the U.S. Military Academy was 85% as compared with 65% among whites.[114] In an analysis of prevalence rates among different ethnic groups in Hawaii,[151] higher rates were observed in Hawaiians and Filipinos than in Caucasians of the same age. However, socioeconomic levels, hygienic habits, and varying

cultural practices in the home cannot be separated from the ethnic backgrounds.

5.8. Occupation

Infectious mononucleosis is most often a disease of the college student and of the white-collar worker.[84,202] It is these persons who are likely to escape infection until young adult life because of higher socioeconomic level or hygienic standards or both. It is also common in the military.[53,68,69]

5.9. Occurrence in Different Settings

The original cases of Pfeiffer[228] suggested that the disease had a familial pattern, and this has been partially borne out by subsequent studies. Analyses of sera from the Cleveland family study[48,126] produced evidence of several cases in three of seven families. There appeared to be a paucity of EBV infections in the 6- to 12-year age group in this setting, with higher rates observed in children under 6 and over 12. Among 75 Canadian families, Joncas and Mitnyan[159] identified 67 persons lacking EBV antibody; in follow-up over approximately 2 years, only 10.5% of these susceptible persons developed EBV antibody. In Sweden, Wahren *et al.*[291] found EBV antibody increases in 7 of 21 members exposed to an index case; 6 of the 21 contacts lacked EBV antibody initially, and three of these seroconverted.

The high rates of infectious mononucleosis in college and military settings have already been noted. However, during the early recruit-training period in the armed forces, infectious mononucleosis is not a common disease, unlike adenovirus, *Mycoplasma pneumoniae*, and other respiratory infections. This is probably because of both a high level of preexisting immunity among recruits and the long incubation period of about 4–7 weeks, so that exposed and susceptible recruits usually develop illness after the end of the 6-week basic training period and after dispersal of recruits to other military assignments. The former point is supported by the finding of an antibody prevalence rate of 85% in entering Marine recruits at Parris Island[171] and of 93% in Army recruits at Fort Jackson, South Carolina (A. S. Evans, R. Jensen, J. C. Niederman, and D. K. Wallace, unpublished data, 1974). Among the Marine recruits at Parris Island whose sera lacked antibody, the infection rate was 18.5 per 1000 over the 16-week training period; in those returning from a 13-month overseas assignment, the EBV infection rate was estimated at 23.8 per 1000. Among the 34 Fort Jackson recruits who lacked EBV antibody, three recruits developed EBV antibody during the 16 weeks of basic and advanced training, a rate of 88 per 1000 recruits.

5.10. Socioeconomic Factors

Socioeconomic settings influence the incidence of both EBV infection and infectious mononucleosis, but in opposite directions. Low socioeconomic groups have high rates of EBV infection early in life but little clinical infectious mononucleosis; high socioeconomic groups have low levels of EBV infection early in life but a high rate of clinical disease that occurs in the 15- to 25-year-old group. Two examples illustrate this effect on infection rates. At the U.S. Military Academy at West Point, the EBV-antibody prevalence rate was 77.1% in cadets coming from families earning under $6000 and only 58.6% among those from families with incomes over $30,000.[114] In New Haven, the antibody prevalence among first-graders in three schools serving a low socioeconomic group was 84.8%, and in three schools serving a high socioeconomic group, it was 37.8% (T. Shope, A. S. Evans, and D. M. Horstmann, unpublished data, 1973). A second serum sample collected from these same children 4–5 years later revealed an EBV seroconversion rate of 50% among susceptible children from lower socioeconomic areas and only 2.4% in susceptible children in the higher socioeconomic group.

5.11. Other Factors

Little is known of the role of nutritional and genetic factors in relation to either EBV infections or infectious mononucleosis. However, it is recognized that ABO blood groups are not correlated with susceptibility to infection or to clinical disease.[114,252] The relationship to HLA antigens has not been clearly established,[253] but HLA-restricted, specific cytotoxic T lymphocytes appear to play a key role in surveillance of the viral carrier state after primary infection.[59,237] It seems likely that genetic control of the immune response plays a role in the severity of clinical illness, in the persistence of virus, and in possible oncogenic sequelae as suggested in recent studies of EBV infections in the X-linked lymphoproliferative syndrome.[231,232,251]

Psychological and behavioral factors may influence the frequency with which clinical infectious mononucleosis develops after EBV infection. When psychological scores taken on entry into the school and subsequent academic achievement were correlated with the prospective serological and clinical data collected from cadets at the United States Military Academy at West Point,[114] the following factors were found to be significantly associ-ated with an increased risk of development of clinical infectious mononucleosis in cadets infected with EBV[160]: (1) having fathers who were "overachievers"; (2) having a strong commitment to a military career; (3) ascribing strong values to various aspects of the training and military career; (4) scoring poorly on indices of relative academic performance; and (5) having strong motivation and doing relatively poorly academically. The same factors seemed to influence the development of heterophil antibody during mild or inapparent EBV infections as well as the duration of clinical illness.

6. Mechanism and Route of Transmission

The major route of transmission of infectious mononucleosis in young adults is probably through intimate oral contact in kissing with the exchange of saliva, as first suggested by Hoagland[141] in 1955 and later confirmed in a controlled study.[65] This concept is supported by three types of circumstantial evidence. First, close personal contact without kissing, as in roommates of infected patients[252] or in such confining environments as a destroyer[223] or a Polaris submarine,[268] rarely leads to secondary cases; this has been true even when the exposed roommate is known to lack EBV antibody and is followed closely over 2 months for the appearance of antibody or of clinical symptoms.[114] Second, a history of intimate oral contact within the appropriate incubation period is common in young adults who develop infectious mononucleosis[144] and occurs statistically more frequently than in healthy controls or patients with acute respiratory infections.[65] Third, the presence of EBV has been demonstrated in the pharynx during acute illness and during convalescence for periods of many months (Table 5).[31,100,192,197,208] In addition, cross-sectional studies of presumably healthy adults have also shown EBV pharyngeal excretion in approximately 20% of young adults.[100,269] One investigation found a leukocyte-transforming factor, presumably EBV, in the throats of 18% of 368 patients attending an outpatient clinic.[32] Transmission of EBV infection may also occur via blood transfusions, usually without illness.[21,98,133,289] Epstein–Barr virus in lymphocyte cultures from patients with acute infectious mononucleosis may remain in a latent form for years and may be a source of transfusion-associated mononucleosis.[44,98,133] This virus has also been demonstrated in cervical secretions[149,259] and in breast milk.[158] The importance of these routes of excretion in transmission of the virus is not known.

This asymptomatic prolonged carrier state after clinical infectious mononucleosis and following inapparent

Table 5. Recovery of Epstein–Barr Virus from Oropharyngeal Excretions of 32 Patients with Infectious Mononucleosis

Days	Throat washings		
	Number tested	Positive	
		Number	%
0–6	5	1	20.0
7–14	20	15	75.0
15–21	8	5	62.5
22–28	10	6	60.0
29–60	13	8	61.5
61–150	12	11	91.7
>150	19	6	31.6
Totals	87	52	59.7

EBV infection serves as the principal source of exposure in young adults. The long duration of virus excretion explains the difficulties in tracing transmission of disease from case to case. Virus excretion occurs in the presence of circulating antibody, which suggests that humoral antibody does not have a major role in the regulation of oropharyngeal shedding. The routes and mechanism of EBV transmission from person to person have been reviewed.[304] Recent studies indicate that both parotid ductal epithelium and squamous epithelial cells in the oropharynx harbor EBV DNA and are sites of viral replication and release,[107,173,197,260,299] and EBV receptors have been demonstrated on human pharyngeal epithelia.[304] Epstein–Barr virus was not found in the urine of 10 acute cases or in the cervices of 175 pregnant or postpartum women.[290]

The mechanism of transmission accounting for the rapid and high rate of acquisition of EBV antibody in nurseries and in young children in low socioeconomic circumstances[97,226,281] is not definitely known. Presumably, transfer of infected saliva on fingers, toys, and other inanimate objects in settings of poor hygiene can account for much of the spread of infection. Perhaps more cell-free virus is released in childhood infections. A recent report of hepatitis caused by EBV in a hemodialysis unit raises the possibility of airborne spread.[34]

From a practical standpoint, the low contagiousness of the disease in young adults eliminates the need for strict isolation procedures.

7. Pathogenesis and Immunity

The incubation period of infectious mononucleosis is 4–7 weeks[65,143] in college students. This estimate is based on well-defined, often single, contacts between an index case and a susceptible contact with intimate oral contact. The virus probably first infects epithelial cells of the oropharynx where it produces infectious virus. The B lymphocyte is then infected.

Studies of the recovery of EBV from oropharyngeal secretions of patients with infectious mononucleosis have revealed that virus shedding occurs during acute illness and from several weeks to months after onset. Transformation of umbilical cord leukocytes into continuous cell lines has been the assay system used for demonstration of the virus, and this transformation has been neutralized by sera containing EBV antibody but not affected by sera lacking this antibody.[192] In addition, transformed leukocytes acquire the EBV genome, demonstrated by nucleic acid hybridization, and express EBV-associated antigens. The virus has been detected in throat washings for prolonged periods and, as indicated in Table 5, is regularly recovered months after clinical illness. In 6 of 19 specimens, the agent was still present over 5 months after disease had occurred, and in one case it was detected 24 months after onset. No special clinical characteristics have as yet been identified in those cases associated with prolonged oropharyngeal virus shedding.

Prior to the onset of definite symptoms in young adults, there is frequently a history of ill-defined complaints, such as malaise and easy fatigue. It has been suggested that an early, abortive infection of this type may occur in children without subsequent development of classic infectious mononucleosis.[65] In an analysis of 100 presumed heterophil-negative cases of infectious mononucleosis involving the 0- to 9-year age group studied in England,[145] the incubation period was shorter than for adults and is estimated at 4–10 days.

The pathogenesis of infectious mononucleosis is intriguing not only because the self-limited clinical disease is manifested primarily in young adults but also because of its relationship to malignant lymphomas and to nasopharyngeal carcinoma.[56,57] An understanding of which immunologic mechanisms turn infectious mononucleosis on and which mechanisms turn it off is important in this context.[250]

Transient depression of delayed hypersensitivity has been described during acute infectious mononucleosis[16,111] and depressed T-cell stimulation[257] by phytohemagglutinin has been recorded. Profound alterations in cell-mediated immunity have been demonstrated by intradermal skin tests, *in vitro* lymphocyte stimulation, and enumeration of absolute numbers of peripheral-blood T and B cells.[179] Lymphocyte responsiveness to a variety of mitogens and antigens was found to be depressed during the first weeks of illness. Serial studies of the interaction of T- and B-cell populations during acute disease indicated that peripheral-blood B cells increase during the first week of illness and return to normal levels several weeks later. In contrast, T cells reach peak values during the second week of disease and remain elevated for approximately 5 weeks.[179]

Recent investigations indicate that both T and B cells may be transformed into atypical lymphocytes characteristic of this disease,[54,103a,179,220] but the predominant atypical lymphocyte later in illness is a reactive T cell. These observations suggest that B cells may be transformed by infection with EBV and T cells transformed as an immunologic response to the viral antigen itself or to altered antigens on the surface of the B cells.

A schematic diagram of the probable pathogenesis of infectious mononucleosis (Fig. 5) summarizes a hypothetical concept. The EBV enters the oropharynx through salivary fluid transfer, probably by kissing in young adults or saliva-contaminated objects in young children. It multiplies locally in parotid ductal epithelium and oropharyngeal squamous epithelial cells and presumably in salivary gland tissue and lingual epithelium.[107,173,197,260] A sore throat accompanies this, often with exudative pharyngotonsillitis. Persistent, intermittent oropharyngeal excre-

tion over many years occurs in approximately 20% following either apparent or inapparent EBV infection,[32,208] which may be greatly increased during immunosuppression.[130,269] The local lymphatics are probably invaded, resulting in cervical lymphadenopathy. The EBV enters the bloodstream from one of these sources, involves B lymphocytes, and spreads hematologically to the liver, producing hepatitis, and to the spleen, causing splenomegaly; involvement of the brain, lung, or other organs occurs rarely.

Epstein–Barr virus has at least three functional effects on B lymphocytes, although it is not known whether separate B-cell populations are involved. One is through the action of virus on a subset of specific antibody-producing cells, the second results from infection and transformation, and the third is a consequence of polyclonal B-cell activation. The first leads to the production of EBV antibody. The second is a complex event related to the existence of EBV receptors on the surface of B cells that are identical to C_{3d} receptors[153,154]; the virus enters and multiplies in the B cells carrying the receptor, producing EBNA-positive cells and a variety of antigens such as early antigen, VCA, and EBNA, against which specific antibodies are produced via other B cells. The EBNA-

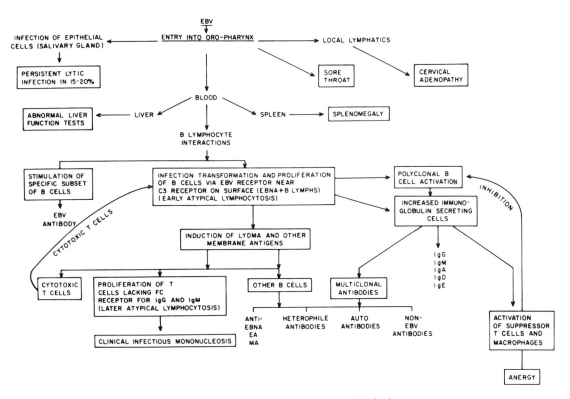

Figure 5. Hypothetical pathogenesis of EBV infection.

positive B cell transforms and proliferates, and this constitutes most of the early atypical lymphocytosis of infectious mononucleosis.[119,179] *In vitro*, the transformed B cell is "immortalized," i.e., made capable of continual multiplication.[185] The EBV induces neoantigens on the surface of some infected B cells, more from adults than children[185,234]; these include lymphocyte-determined membrane antigen (LYDMA) and possibly heterophil antigen of the Paul–Bunnell type, antibodies to which are formed by other B cells. The neoantigens stimulate a B–T cell interaction similar to a graft-versus-host response, which may account for some of the clinical symptoms of the disease.

Based on interactions in cell culture systems, three forms of interaction between EBV and B cells have also been recognized.[237] These are (1) passive latent infection, displayed by tumor biopsy cells and early-passage cell lines derived from the endemic form of Burkitt's lymphoma, in which just EBNA-1 is expressed and which is encoded by the BKRF 1 reading frame of the EBV genome. (It is called passive because there are no detectable virus-induced changes in the phenotype of the infected cell); (2) active latent infection, characteristic of EBV-infected resting B cells *in vitro*, which drives the cells into permanent proliferation and expression of eight EBV proteins (6 nuclear antigens—EBNA 1, 2, 3a, 3b, 3c, and EBNA-LP); and (3) lytic infection, or "virus-producer" lines, with the BZLF 1 gene, an obligatory but insufficient step in initiating the lytic cycle. These lines are either conventional lymphoid cell lines (LCLs) or EBV-positive cell lines that have switched from a passive to an active latent form on long-term serial passage.[237]

The mechanism of production of the clinical features of infectious mononucleosis is not well understood, but they might result from B–T cell interactions, from immune complex deposition or other immunopathologic events, from direct effects of EBV on cells, or from some combination of these mechanisms. The rarity of classic infectious mononucleosis in early childhood might be related to the failure of EBV to evoke some of these immunologic responses, such as induction of neoantigens on the B-cell surface. The T cells that respond to such neoantigens in older children and young adults represent the atypical lymphocytosis that characterizes the later stages of the acute clinical disease and are composed largely of proliferating CD8 T cells that lack the Fc receptor for IgG and IgM.[119] As in other viral infections, some T cells are cytotoxic and lyse infected cells,[12,118,149,157,266] limiting their proliferation and accounting for the low percentage of EBNA-positive B cells demonstrable in the peripheral blood during the acute disease.[167,243]

Very early, 5–18% of the B-cell fraction of blood lymphocytes express six EBNA proteins demonstrated in circulating mitotic B lymphocytes.[241,242] Both VCA and EBNA antibodies persist throughout life, suggesting that small amounts of these antigens regularly become available over long periods for maintenance of detectable antibody. It is now evident that EBNA is an antigen complex consisting of different viral polypeptides encoded by separate viral genes.[136,138,256,275] Recently, genes encoding components of EBNA have been identified: EBNA-1 is encoded in the *Bam* HI-K fragment, and EBNA-2 maps to *Bam* HI-WY and H.[37,45,46,108,138] Monoclonal antigens representing products of such small defined regions of the viral genome are now being used as reagents in seroepidemiologic studies.[75,186,207]

The third functional effect of EBV on B cells is similar to that of pokeweed mitogen but is probably caused by infection of B cells rather than a surface effect and results in a polyclonal B-cell proliferation; an increase in immunoglobulin-secreting cells then occurs, synthesizing IgA early, with IgD, then later IgM and IgG. Multiclonal antibodies are stimulated via other B cells such as sheep-cell antibodies of the Forssman and possibly the Paul–Bunnell type, plus other antibodies unrelated to EBV.[159,284] These antibodies are probably not produced directly by EBNA-positive B cells. In the wake of such B–T cell interactions, there appear T cells and perhaps macrophages, which inhibit the proliferation of B cells[120,238,284] and probably underlie the severe anergy that occurs in infectious mononucleosis.

Proliferation of B cells and the acute disease itself are limited in part by nonspecific killer T cells followed by antigen-specific cytotoxic T cells, along with the lymphokines they produce, and humoral antibody.[238,280] The development of specific cell-mediated immunity to EBV during infectious mononucleosis has been demonstrated.[210,211] Knowledge of the nature of this cell-mediated response is now substantial.[282] Primary infection is associated with a HLA-restricted specific T-cell response against all allogeneic target cells. Nonspecific T cells appear to account for only 35% of this response.[282] The HLA-restricted specific cytotoxic T cells also appear to play a role in maintaining the life-long surveillance of the viral carrier state after primary infection has been cleared for some years.[59] Failure of these immune mechanisms, either on a genetic basis, as in the X-linked lymphoproliferative syndrome of males,[231,232,251] or because of acquired immunodeficiencies, as possible in a fatal case in a 4-year-old girl,[239] can permit uncontrolled B-cell proliferation, leading to an immunoblastic B-cell sarcoma or to other lymphoproliferative responses (Burkitt's lymphoma, primary lymphomas of the central nervous system, lymphomas occurring in donor cells in recipients of bone marrow grafts, and lymphocytic interstitial pneumonitis in children with AIDS).[2,140] Such severe, often fatal, oncogenic and sometimes lytic consequences of EBV infection are of great importance in terms of both human life and an understanding of their pathogenesis. A registry has been set up to record and study them.[115]

The mechanism of heterophil-antibody production is still unexplained, as is the source of the antigen that produces it, but knowledge that its appearance is most common in EBV infections of young adults[70] and that the degree of expression and release of EBV *in vitro* varies in lymphocytes from donors of different ages suggests avenues of investigation. For example, EBV-infected fetal lymphocytes do not have demonstrable VCA but do contain EBNA[189,236] and CF antigens.[188,191] In lymphocytes from adults and from marmosets, EBV may mature more fully, resulting in release of EBV antigens to other lymphocytes. Heterophil antibody may occur in response to membrane-induced antigens of EBV expressing themselves more fully in lymphocytes of young adults than in those of younger children or in fetal lymphocytes. There is preliminary evidence of heterophil-antibody production by lymphocytes cultured from acute cases of infectious mononucleosis,[198] and the various possible mechanisms of its appearance have been reviewed by Kano and Milgrom.[159]

The presence of antibody to the VCA of EBV has been shown to indicate protection against infectious mononucleosis, and its absence indicates susceptibility.[72,84,204,252] The actual antibody that provides immunity is probably the neutralizing antibody,[244] for which tests have recently been developed.[43,139,193,224] One attack of infectious mononucleosis confers a high degree of durable immunity to subsequent attacks of clinical infectious mononucleosis.[27,84,193,206,208] Presumably, subclinical or inapparent EBV infections also confer lasting immunity. Endogenous reactivation of EBV occurs when the immune system is depressed, as in renal transplant patients[269] or in persons with the acquired immune deficiency syndrome.[233]

8. Patterns of Host Response

8.1. Clinical Features

Epstein–Barr virus produces a variety of acute, chronic, and malignant syndromes.[75a] Infectious mononucleosis is the classical manifestation of the acute disease in older children and young adults, but other acute syndromes may occur without any of the typical features of infectious mononucleosis, such as hepatitis, involvement of the central nervous system, disseminated EBV infections, and pneumonitis in infancy.

8.2. Acute Infection

When infection with EBV occurs in childhood, a mild, nonspecific illness or an inapparent infection may develop, both of which are associated with the appearance and persistence of antibody to EBV. If exposure and primary infection are delayed until adolescence or young adulthood, the characteristic clinical picture usually occurs. This consists of fever, pharyngitis, and cervical lymphadenopathy accompanied by splenomegaly in 50% and hepatomegaly in 10%. The pharyngitis is often associated with a whitish or a gray–green exudate having an offensive odor. The eyelids may be swollen, and petechiae occur on the hard palate in 25% of cases.

Abnormalities of liver function tests are a regular feature of infectious mononucleosis, and clinically recognizable jaundice occurs in 5% of cases. Rare manifestations include a variety of central nervous system syndromes (encephalitis, meningoencephalitis, Guillain–Barré syndrome), pneumonitis and pneumonia, thrombocytopenic purpura, myocarditis, ataxia telangiectasia, and nephritis.[17,67,144] Hepatitis and central nervous system involvement may occur in the absence of other features of infectious mononucleosis.[52,109] The major complications include splenic rupture and airway obstruction from exudative pharyngotonsillitis. Over 50 deaths have been reported, mostly from central respiratory failure; recently, cell-associated EBV was detected in the cerebrospinal fluid of a case with complicating meningoencephalitis (J. Schiff, J. Schaeffer, and J. Robinson, personal communication, 1978). An immunologic deficit, especially in cell-mediated immunity, may influence the severity of the clinical response, as suggested by the events of two deaths in a family[11] and a recent severe illness in a woman with a T-cell defect.[262] Acute progressive and often fatal EBV infection rarely occurs, but when it does, it is usually associated with immunodeficiency.[239,270] An X-linked recessive lymphoproliferative syndrome in which EBV has been implicated has been described by Purtilo *et al.*[231,232,251] Both lytic and proliferative manifestations have appeared in one kindred; these include fatal infectious mononucleosis, malignant lymphoma, and agammaglobulinemia.[233,233a]

The frequency with which EBV infections are expressed as clinical illness in young adults has varied in different populations. In a study of a cohort of U.S. Military Academy cadets over a 4-year period, only 26.4% of 201 infected with this agent developed heterophil-positive clinical infectious mononucleosis.[114] The apparent-to-inapparent EBV infection ratio in different years ranged from 1:1 to 1:2.6 in this population. Comparison of the frequency of clinically expressed infectious mononucleosis in freshman students in three different settings is presented in Table 4 (Section 5.1). As mentioned, the reasons for the differences are not known but may relate to the motivation to seek medical care, physical fitness, psychological factors, or concern about the effect of hospitalization on academic and school activities.

The relationships between clinical features and antibody levels in a typical heterophil-positive case in an 18-year-old student are shown in Fig. 6. Following a prodromal period associated with fatigue, fever, and headache over several days, the onset of sore throat, cervical adenopathy, and recurrent fever developed during the second week. Characteristic blood changes were present on the third day after onset, and on culture the patient's lymphocytes contained EBNA CF antigens in a nuclear location. The heterophil antibody titer was negative on the first day of symptoms, rose to 1:14 after guinea pig absorption 2 days later, and then increased to 1:896 on the 15th day. In contrast, Epstein–Barr VCA antibodies of IgG type, undetectable on the third day of illness, were present on the eighth day and rose to a level of 1:320 by the second week; EBV-specific IgM antibodies were demonstrable on the eighth day at a titer of 1:2.5, which then increased to 1:10 by the 15th day.

No direct correlation has been found between the levels of Epstein–Barr VCA and heterophil antibody levels or between VCA, early antigen, and EBNA antibody levels and the severity of clinical symptoms and hematologic changes.[128,134,206] Neither does their persistence correlate with the duration of clinical illness.

Pathologically, the lymph node from acute cases of infectious mononucleosis show paracortical expansion and immunoblastic proliferation accompanied by polymorphous inflammatory cells and Reed–Sternberg-like cells.[1] The histological picture may resemble that of Hodgkin's disease or non-Hodgkin's lymphoma.[79,81]

8.3. "Chronic Mononucleosis Syndrome"

Recognition of the life-long relationship between latent Epstein–Barr virus and the human host suggests that periodic reactivation of the virus may be associated with episodic or chronic illness in some patients.[106] For many years a syndrome called chronic mononucleosis has been sporadically reported as a clinical entity, one accompanied by a high degree of morbidity.[8,15,196,150,161] Several recent series have described patients, usually 20 to 50 years of age, with recurrent illnesses characterized by extreme fatigue, weakness, headache, myalgia, adenopathy, pharyngitis, and low-grade fever; profound constitutional symptoms have had little relation to objective physical findings.[50,155,271,283] A subset of patients have a more severe form of disease with a fulminating clinical course associated with fever, hematologic abnormalities, abnormal immunoglobulin production, and pneumonitis.[187]

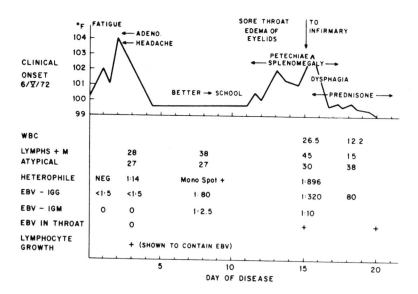

Figure 6. Relationship between clinical and laboratory features of infectious mononucleosis in an 18-year-old male.

Diagnostic findings have revealed that most, but not all, so-called chronic mononucleosis cases have elevated antibody titers to the EBV replicative antigens, VCA and EA. However, except for relative lymphocytosis in some patients, routine laboratory tests are within normal limits; determinations of lymphocyte subpopulations indicate no abnormalities.[50,155,162,271] Development of more specific serological and immunologic markers of active EBV infection is essential for understanding the pathophysiology of this chronic syndrome.[203,277] Two of these are genes encoding different components of EBNA. In a recent report, three patients with symptoms of chronic active infection including recurrent fever, pneumonitis, neutropenia, and hypergammaglobulinemia were found to have high antibody titers to capsid and early antigen, but specifically to lack antibody to K antigen.[187]

At present, 12 patients with the chronic mononucleosis syndrome have been identified who lack EBNA-1 (*Bam* HI-K) antibody.[187,189] These represent 8 of 25 (32%) with the severe form of the syndrome, and there were five who had low antibody titers to this antigen. This occurred in patients despite EBV–VCA IgG antibody titers of 1 : 1000 or over. Four of 39 (10%) patients with the less severe, common "fatigue" syndrome also lacked this EBNA antibody. Overall, 12 of 64 (18%) lacked this antibody, even though VCA and EA antibodies were usually high and antibodies to other EBNA antibodies were present.[187,189] The absence of this antibody was rarely detected in other conditions. Of 33 healthy adults, 15 healthy convalescent infectious mononucleosis adults, 30 persons with nasopharyngeal carcinoma, and 44 with neuropsychiatric disease, none lacked this antibody. It was absent from only 4 of 43 patients with Burkitt's lymphoma. One exception was the case of 12 of 15 children with AIDS in whom this EBNA antibody was not detected. The frequency of this phenomenon in children with AIDS suggests that it is associated with an immunodeficient state but is not dependent on genetic susceptibility. The relation of these findings to the chronic mononucleosis syndrome is unknown at this time, as is the reason antibody is not made. Since some evidence exists that cells can make it, the failure may lie in inability to recognize it or improper presentation to antibody-forming cells. At the very least, it represents an important and diagnostic marker of a subset of persons alleged to have the chronic mononucleosis syndrome.

Henle et al.[131] have also confirmed these findings by investigating antibodies to both EBNA-1 and EBNA-2 antigens. Unlike the usual inversion of the ratio of anti-EBNA-1 to anti-EBNA-2 titers, which is well below 1.0 during the first 6–12 months after infectious mono-

nucleosis and turns to well above subsequently, in patients with the clear-cut chronic syndrome there is a delay or absence of this inversion. Thus, ratios of EBNA-1 to EBNA-2 antibodies under 1.0, 2 years or more after onset of infectious mononucleosis, are suggestive of a defect in the normal recovery process from a primary EBV infection. Such evidence suggests that the syndrome of chronic mononucleosis may consist of several subsets of patients characterized by selective abnormalities. The availability of highly discriminating serological markers could ultimately permit precise diagnostic classification of such clinical categories.

In more recent studies, the putative role of EBV in what is now designated as the "chronic fatigue syndrome" has been greatly reduced.[156,182,274] A variety of different agents may play a role in the pathogenesis of this syndrome, and stress and depression may alter the immune system and reactivate EBV, leading to increased antibody titers.[76] A review of four clusters of cases in different geographic areas showed sufficient differences in clinical presentations that different etiologic agents may have been involved.[175] Almost all patients had returned to preillness status at the end of 3 years. Current evidence thus suggests that, with the possible exception of some severe cases that lack EBNA-1 antibody, the altered titers of EBV antibodies are a result rather than a cause of the syndrome. The terms "chronic infectious mononucleosis" and "chronic EBV infection" should be abandoned or at least confined to a subset of cases characterized by a severe course and the absence or very low titer of EBNA-1 antibody.[187]

9. Diagnosis

The diagnosis of infectious mononucleosis is based on a typical clinical picture with the triad of fever, sore throat, and cervical lymphadenopathy, the occurrence of at least 50% lymphocytosis with at least 10% atypical lymphocytes, and the appearance of heterophil antibodies. The presence of EBV IgM or IgD component early antigen antibodies or both is an absolute requirement in doubtful or heterophil-negative cases. Antibody to VCA is usually present at the time the physician first sees the patient, and only 15–20% of patients will show a subsequent rise in titer; antibodies to early antigen appear later but are present in only 75% of typical cases.[134,137] Antibodies to EBNA arise 1 month or more after illness and usually persist for life.[128] Also, EBV-specific IgM antibodies are demonstrable in 85% of cases during acute illness. Figure 7 depicts the course of EBV-specific IgG

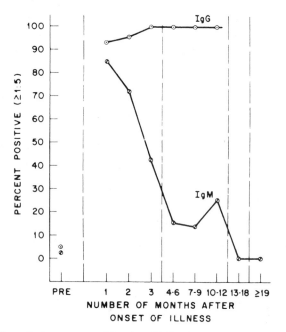

Figure 7. Appearance and duration of IgG and IgM antibodies specific for EBV during infectious mononucleosis. From Evans *et al.*[83]

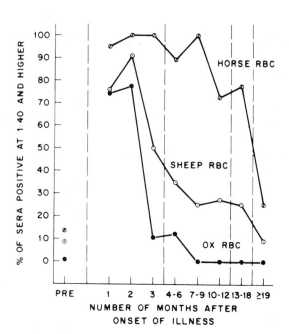

Figure 8. Persistence of heterophil antibodies during infectious mononucleosis. RBC, Red blood cells. From Evans *et al.*[83]

and IgM antibodies during the course of illness. The IgG antibodies persist for years, probably for life; IgM antibodies usually disappear in 3–6 months. At present, the EBV-VCA IgM is the best antibody test for diagnosis in cases that are heterophil-antibody negative[51,83,102,211]; the presence of false positives because of rheumatoid factor must be excluded. ELISA tests are also available for IgG and IgM antibody based on the EBNA-1 and/or EBNA-2 antigen, but other tests seem simpler and more accurate at this time[177,181] (see Section 3.4.2).

Major reliance in diagnosis is placed on heterophil antibodies, which are of the IgM type. Methods in most common use are the sheep- and horse-cell agglutination tests after absorption of the serum with guinea pig kidney to remove Forssman antibody and the beef-cell hemolysin test, which does not require absorption. The appearance and persistence of these test results during acute and convalescent infectious mononucleosis are shown in Fig. 8. The beef hemolysin test is the most specific but has a short duration; the horse-cell test is the most sensitive and most persistent, with positive tests present for a year or more in 75%.[83] The appearance and persistence of antibody to horse cells have been found to follow mild and subclinical episodes of infectious mononucleosis provided that sera are collected over a long time. This test is useful in childhood infections with EBV, which are often hetero-

phil-negative by other tests or when inadequately followed. A new immune adherence heterophil test is highly sensitive and useful in childhood infections,[82,93,174] and in those negative by other heterophil tests.[82]

Development of EBV antibody has also been shown in cases in which the clinical and hematologic characteristics are those of infectious mononucleosis but the heterophil antibody remains persistently negative.[84,204,211] These heterophil-negative, EBV-antibody-positive cases appear to be common in infants and children but are rare in adults.[18,19,97] A characteristic of EBV infection early in life, especially in developing countries when the majority of infections occur, is that clinical infectious mononucleosis is uncommon and the heterophil antibody is usually absent. In developed countries and in the middle and upper socioeconomic groups where infection is common in the age group 15–25, about 15% of patients who fulfill the clinical, hepatic, and hematologic criteria of infectious mononucleosis are said to be heterophil-antibody negative. In these cases the diagnosis is established by the EBV-specific VCA-IgM antibody test.[73,83] This absence may be due partly to failure to employ a sensitive heterophil antibody test, such as the horse-cell test or the immune adherence hemagglutination assay.[82,93,174] The latter test can often demonstrate heterophil antibody in childhood EBV infections and in mild or even inapparent infections in adults.[93,217] Infection with cytomegalovirus (CMV) may produce a clinical picture of heterophil-negative mononucleosis that is hard to distinguish from classic infectious mononucleosis; however, it usually occurs at a later age, and adenopathy and exudative pharyngitis are rare.[73,117,168,169] Epstein–Barr virus may possibly be reactivated during the course of other herpesvirus infections, especially CMV,[172] but EBV does not appear to reactivate other viruses.[92] Infection with EBV can result in development of a false-positive CMV IgM-antibody response; however, the reverse does not occur.[117] *Toxoplasma gondii*, adenovirus, rubella, and hepatitis A infections may also resemble heterophil-negative infectious mononucleosis, but the lymphocytosis is relative and not absolute. The infectious-mononucleosis-like syndromes have been reviewed.[73,82]

Rarely a mononucleosislike, heterophil-antibody-negative syndrome may occur due to herpesvirus type 6 (HHV-6),[203a] as well a similar syndrome early in the course of human immunodeficiency virus (HIV) infection.[94]

10. Control and Prevention

Attempts to control infectious mononucleosis and EBV infections by interrupting the presumed chain of transmission in young adults seem neither realistic nor perhaps desirable in light of our current knowledge. Salivary exchange represents the major route of spread, and there seems little likelihood of interdicting this practice. If poor hygienic conditions promote the spread of EBV infections in young children, then improvement in hygienic and socioeconomic circumstances might reduce their incidence. Unfortunately, control of spread at this time when infection is largely mild and asymptomatic might simply delay exposure to later childhood and young adult life, when the majority of EBV infections are expressed as clinical infectious mononucleosis.

The high degree of protection against infectious mononucleosis provided by natural infection with EBV has suggested that a vaccine capable of evoking similar humoral, cell-mediated, and local immunity might be highly effective.[56] Because of the association between EBV and cancer in African Burkitt's lymphoma and in nasopharyngeal carcinoma (NPC) patients of Chinese descent, search for a vaccine against EBV has been undertaken since 1976 to prevent infection and reduce tumor incidence among high-risk populations in Africa, China, and southeast Asia. The EBV-determined membrane antigen (MA) was proposed as a candidate immunogen[56]

because it was known to elicit neutralizing antibodies in man.[42,42,101,224,225] Progress has been made in defining and preparing MA molecules, and recently a candidate immunogen, an EBV glycoprotein termed GP 340, has been identified; a sensitive assay for measurement of GP 340 antibodies has now been developed.[213] Utilizing cotton-top tamarins (*Saguinus oediupus oediupus*) in which experimental EBV infection regularly causes malignant lymphoma, challenge experiments have been conducted in unvaccinated and vaccinated animals. Those receiving isolated cell membranes (MA) or purified MA glycoprotein (GP 340) intraperitoneally were found to develop protective antibodies measured by ELISA, by indirect immunofluorescence, and by standard neutralization tests of EBV transformation of fetal cord lymphocytes. Tamarins receiving either preparation were protected against 100% tumor-inducing challenge doses of EBV.[56,61] Although problems of procurement and testing of such a preparation remain, the groundwork for developing an effective vaccine against EBV and the technological production of sufficient immunogen appear to be solid recent advances.[13]

A variety of expression vectors for a recombinant vaccine are under study, including adenovirus, baculovirus, canarypoxvirus, and vaccinia virus.[196] The applicability of the evidence of protection in tamarin monkeys against EBV infection and/or tumor development to humans is not known. To this end, a small, encouraging trial of a truncated EBV glycoprotein 340 inserted into a Chinese vaccinia vector has been carried out in China.[110,170a,300] In addition, the development of a purified form of gp 320/220 has led to the initiation of a human phase 1 trial in the United Kingdom.[5]

The cost-effectiveness of an EBV vaccine for infectious mononucleosis has been evaluated but recent data are not available because infectious mononucleosis is not now a reportable disease in the United States and few recent epidemiological studies have been carried out. However, Evans[77] has recently carried out a review using available, but older and incomplete data, on the incidence, severity, and cost of hospitalization. Based on a conservative estimate of incidence of 65 cases per 100,000 population in the United States, a 15% rate of hospitalization for 4.5 days, and $600/day hospitalization charge, a vaccine that cost $40/dose for three doses would not be cost-effective based on hospitalization costs alone. If morbidity and lost time from school and work are included, it would be worth consideration. It would not require a large number of college students to establish the efficacy of an EBV vaccine. If the incidence of EBV infection is 12–13% and 50% of the infected develop the disease, and if the vaccine is 90% effective and the follow-up is 1 year, it would require 184 vaccinated susceptibles and 368 susceptible, unvaccinated (or placebo) control subjects to prove effectiveness in preventing clinical infectious mononucleosis.[176]

Given the possibility that prevention of primary EBV infection might reduce the incidence of EBV-related lymphomas, such as 95% of African Burkitt's lymphoma, 50% or more of Hodgkin's disease, 20% of non-Hodgkin's lymphoma, and 30% of the lymphomas occurring in patients with AIDS, as well as almost all cases of NPC, further vaccine research is certainly justified (see Chapters 30 and 31 on malignant lymphomas are nasopharyngeal carcinoma for evidence of the causal associations of these conditions with EBV). The age at vaccination and perhaps the preparation of the vaccine and its method of administration might have to vary with the EBV-associated disease to be prevented. An oral, live attenuated product that could induce mucosal, humoral, and cell-mediated immunity would seem to be the ideal objective.

11. Unresolved Problems

We do not know why some young adults infected with EBV develop clinical infectious mononucleosis and others do not, although psychosocial factors and genetic susceptibility may play a role. We have not identified the site or mechanism of heterophil antibody production. We lack complete understanding of events that control EBV infection and those that lead to cell transformation, chronicity, reactivation, and malignancy, but molecular techniques are contributing to rapid progress in our understanding of these complex events.[163,237]

Information is meager on the series of events following primary EBV infection in terms of the sites of virus latency and factors exerting a role in asymptomatic chronic virus production and release in antibody-positive persons. There is urgent need for more specific serological and immunologic markers to differentiate chronic latent infections from chronic active infection. Fortunately, opportunities for clarifying the association of EBV with a variety of clinical disorders now exist through the availability of molecular methods utilizing recombinant DNA techniques. Using cloned fragments of Epstein–Barr virus DNA, it is now possible to measure serological responses to antigens representing products of sharply defined regions of the viral genome.[307] Such investigations offer the potential of providing precise information on the function of specific DNA fragments in the process of B-lymphocyte immortalization. Monoclonal antigens representing products of distinct regions of the EBV genome can now be used as reagents in molecular seroepidemiologic studies. It is conceivable that chronic active EBV infection may be the result of altered host immune responses with reduction in levels of antibody to a specific functional protein.[248] In such situations, low levels of essential antibody could result from inadequate antigen production or inadequate recognition by the host immune system. In future investigations it will be important to examine the role of specific monoclonal antigens in cell-mediated immunity to EB virus or to immortalized B cells. Finally, it is important to determine what are the precise immunologic and molecular events that make most EBV infections limited and benign and what events and cofactors lead, albeit rarely, to a variety of malignant lymphomas and NPC.

12. References

1. ABBONDAZO, S. L., SATO, N., STRAUS, S. E., AND JAFFE, E. S., Acute infectious mononucleosis. CD30 (Ki-1) antigen expression and histologic correlations, *Am. J. Clin. Pathol.* **93**:698–702 (1990).

2. ANDIMAN, W., GRADOVILLE, L., HESTON, L., NEYDORFF, R., SAVAGE, M. E., KITCHINGMAN, G., SHEDD, D., AND MILLER, G., Use of cloned probes to detect Epstein–Barr viral DNA in tissues of patients with neoplastic and lymphoproliferative diseases, *J. Infect. Dis.* **148**:967–977 (1983).

3. ARMSTRONG, D., HENLE, G., AND HENLE, W., Complement-fixation tests with cell lines derived from Burkitt's lymphoma and acute leukemias, *J. Bacteriol.* **91**:1257–1262 (1966).

4. ARMSTRONG, J. A., EVANS, A. S., RAO, N., AND HO, M., Viral infections in renal transplant recipients, *Infect. Immun.* **14**:970–975 (1976).

5. ARRAND, J. R., MADEJ, M., CONWAY, M. J., MORGAN, A., TARELTON, J., WALLACE, L., QUALTIERE, L. F., AND MACKETT, M., Purification and characterization of EBV gp 340/220 from a papillomavirus expression system, in: *Vth International Congress on EBV and Associated Diseases.*

5a. AXELROD, P., AND FINESTONE, A. J., Infectious mononucleosis in older patients, *Am. Fam. Physician* **42**:1588–1606 (1990).

6. BAILEY, G. H., AND RAFFEL, S., Hemolytic antibodies for sheep and ox erythrocytes in infectious mononucleosis *J. Clin. Invest.* **14**:228–244 (1935).

7. BAER, R., BANKIER, A. T., BIGGIN, M. D., DEINIGER, P. L., FARRELL, P. J., GIBSON, T. J., HATFULL, F., HUDSON, G. S., SATCHWELL, S. C., SEQUIN, C., TUFFNELL, P. S., AND BARRELO, B. G., DNA sequences and expression of the B95-8 Epstein–Barr virus genome, *Nature* **310**:207–211 (1984).

8. BALLOW, M., SEELEY, J., PURTILO, D. T., ONGE, S. S., SAKAMOTO, K., AND RICKLES, F. R., Familial chronic mononucleosis, *Ann. Intern. Med.* **97**:821–825 (1982).

9. BANATVALA, J. E., BEST, J. M., AND WALLER, D. K., Epstein–Barr virus-specific IgM in infectious mononucleosis, Burkitt lymphoma, and nasopharyngeal carcinoma, *Lancet* **1**:1205–1208 (1972).

10. BANG, J., Forsoeg paa at overfoere mononucleosis infectiosa til mennesket, *Ugeshkr. Laeg.* **105:**499–504 (1943).

11. BAR, R. S., DELOR, J., CLAUSEN, K. P., HURTUBISE, P., HENLE, W., AND HEWETSON, J. F., Fatal infectious mononucleosis in a family, *N. Engl. J. Med.* **290:**363–367 (1974).

12. BAUSCHER, J. C., AND SMITH, R. T., Studies of Epstein–Barr virus–host-relationship: Autochthonous and allogeneic lymphocyte stimulation by lymphoblast cell lines in mixed cell culture, *Clin. Immunol. Immunopathol.* **1:**270–281 (1973).

13. BEALE, A. J., Epstein–Barr virus. Dream or reality of a vaccine, *Nature* **318:**230 (1985).

14. BENDER, C. E., Interpretation of hematologic and serologic findings in the diagnosis of infectious mononucleosis, *Ann. Intern. Med.* **49:**852–865 (1958).

15. BENDER, C. E., Recurrent mononucleosis, *J.A.M.A.* **182:**954–956 (1962).

16. BENTZON, J. W., The effect of certain infectious diseases on tuberculin allergy, *Tubercle* **34:**34–41 (1953).

17. BERKEL, A. I., HENLE, W., HENLE, G., KLEIN, G., ERSOY, F., AND SANAL, O., Epstein–Barr virus related antibody patterns in ataxia–telangiectasia, *Clin. Exp. Immunol.* **35:**196–201 (1979).

18. BIGGAR, R. J., HENLE, W., FLEISCHER, G., BÖCKER, J., LENNETTE, E. T., AND HENLE, G., Primary Epstein–Barr virus infections in African infants. I. Decline of maternal antibodies and time of infection, *Int. J. Cancer* **22:**239–243 (1978).

19. BIGGAR, R. J., HENLE, G., BÖCKER, J., LENNETTE, E. T., FLEISCHER, G., AND HENLE, W., Primary Epstein–Barr virus infections in African infants. II. Clinical and serological observations during seroconversion, *Int. J. Cancer* **22:**244–250 (1978).

20. BLACK, F. L., HIERHOLZER, W. J., PINHEIRO, DEP., EVANS, A. S., WOODALL, J. P., OPTON, E. M., EMMONS, J. E., WEST, B. S., EDSALL, G., DOWNS, W. G., AND WALLACE, G. D., Evidence for persistence of infectious agents in isolated human populations, *Am. J. Epidemiol.* **100:**230–250 (1974).

21. BLACKLOW, N. R., WATSON, B. K., MILLER, G., AND JAKOBSON, B. M., Mononucleosis with heterophile antibodies and EBV virus infection. Acquistion by an elderly patient in hospital, *Am. J. Med.* **51:**549–552 (1971).

22. BLOEDORN, W. A., AND HOUGHTON, J. E. The occurrence of abnormal leukocytes in the blood in acute infections, *Arch. Intern. Med.* **27:**315–325 (1921).

23. BOON, W. H., MUI C. L. Y., AND TOO, M., Infectious mononucleosis in Singapore, *J. Singapore Pediatr. Soc.* **19:**153–161 (1977).

24. BROWN, J. W., CLIFFORD, J. E., SIMS, J. L., AND WHITE, E., Liver function during infectious mononucleosis, *Am. J. Med.* **6:**321–328 (1949).

25. CABOT, R. C., The lymphocytosis of infection, *Am. J. Med. Sci.* **145:**335–339 (1913).

26. CARLSON, G. W., BROOKS, E. H., AND MARSHALL, V. G., Acute glandular fever: Recent epidemic, report of cases, *Wis. Med. J.* **25:**176–178 (1926).

27. CARTER, R. L., AND PENMAN, H. G., The early history of infectious mononucleosis and its relation to "glandular fever," in: *Infectious Mononucleosis* (R. L. CARTER AND H. G. PENMAN, EDS.), pp. 1–18, Blackwell, Oxford, 1969.

28. CARVALHO, R. P. S., EVANS, A. S., FROST, P., DALLDORF, G., CAMARGO, M. E., AND JARMA, M., EBV infection in Brazil. I. Occurrence in normal persons, in lymphomas and in leukemias, *Int. J. Cancer* **11:**191–201 (1973).

29. CENTERS FOR DISEASE CONTROL, Infectious Mononucleosis Surveillance, November 1972.

30. CENTERS FOR DISEASE CONTROL, Annual Summary 1983, *Morbid. Mortal. Week. Rep.* **32:**1–125 (1984).

30a. CHAN, X. Y., PEPPER, S. V. D., AND ARRAND, J. R., Prevalance of the A and B types of Epstein–Barr virus DNA in nasopharyngeal carcinoma biopsies from southern China, *J. Gen. Virol.* **82:**463–466 (1992).

31. CHANG, R. S., AND GOLDEN, H. D., Transformation of human leukocytes by throat washings from infectious mononucleosis patients, *Nature* **234:**359–360 (1971).

32. CHANG, R. S., LEWIS, J. P., AND ABILDGAARD, C. F., Excretors of leucocyte-transforming agents among a human population, *N. Engl. J. Med.* **289:**1328–1329 (1973).

33. CHRISTINE, B. W., Infectious mononucleosis, *Conn. Health Bull.* **82:**115–119 (1968).

34. COREV, L., STAMM, W. E., FEORINO, P. M., BRYAN, J. A., WESELEY, S., GREGG, M. B., AND SOLANGI, K., HB,AG-negative hepatitis in a hemodialysis unit: Relation to Epstein–Barr virus, *N. Engl. J. Med.* **293:**1273–1278 (1975).

35. CRAWFORD, D. H. Epstein–Barr virus related lymphomas, in: *Vth International Symposium on Epstein–Barr Virus and Associated Diseases* (1992).

36. CROSS, J. G., Conditions simulating an acute leukemia (acute benign leukemia), *Minn. Med.* **5:**579–581 (1922).

37. DAMBAUGH, T., HENNESSY, K., CHAMNANKIT, L., AND KIEFF, E., U2 region of Epstein–Barr virus DNA may encode Epstein–Barr nuclear antigen 2, *Proc. Natl. Acad. Sci. USA* **81:**7632–7636 (1984).

38. DAN, R., AND CHANG, R. S., A prospective study of primary Epstein–Barr virus infections among university students in Hong Kong, *Am. J. Trop. Med. Hyg.* **42:**380–385 (1990).

39. DAVIDSOHN, I., Heterophile antibodies in serum sickness, *J. Immunol.* **16:**259–273 (1929).

40. DAVIDSOHN, I., AND WALKER, P. H., The nature of the heterophilic antibodies in infectious mononucleosis, *Am. J. Clin. Pathol.* **5:**455–465 (1935).

41. DAVIDSOHN, R. J. L., A survey of infectious mononucleosis in the North-East Regional Hospital Board area of Scotland, 1960–9, *J. Hyg.* **68:**393–400 (1970).

42. DESCHRYVER, A., KLEIN, G., HEWETSON, J., ROCCHI, A., HENLE, G., HENLE, W., AND POPE, J., Comparison of neutralization tests based on abortive infection or transformation of lymphoid cells and their relation to membrane reactive antibodies (anti-MA), *Int. J. Cancer* **13:**353–362 (1974).

43. DESCHRYVER, A., ROSEN, A., GUNVEN, P., AND KLEIN, G., Comparison between two antibody populations in the EBV system: Anti-MA versus neutralizing antibody activity, *Int. J. Cancer* **17:**8–13 (1976).

44. DIEHL, V., HENLE, G., HENLE, W., AND KOHN, G., Demonstration of a herpes group virus in cultures of peripheral leukocytes from patients with infectious mononucleosis, *J. Virol.* **2:**663–669 (1968).

45. DILLNER, J., KALLIN, B., EHLIN-HENRIKSSON, B., TIMAR, L., AND KLEIN, G., Characterization of a second Epstein–Barr virus-determined nuclear antigen associated with the *Bam* HI-WYH region of EBV-DNA, *Int. J. Cancer* **35:**359–366 (1985).

46. DILLNER, J., KALLIN, B., KLEIN, G., JÖRNVALL, H., ALEXANDER, H., AND LERNER, R., Antibodies against synthetic peptides react with the second Epstein–Barr virus-associated nuclear antigen, *EMBO J.* **4**(7):1813–1818 (1985).

47. DILLNER, J., STERNÅS, K., KALLIN, B., ALEXANDER, H., EHLIN-HENRIKSSON, B., JÖRNVALL, H., KLEIN, G., AND LERNER, R.,

Antibodies against a synthetic peptide identify the Epstein–Barr virus-determined nuclear antigen, *Proc. Natl. Acad. Sci. USA* **81:**4652–4656 (1984).

48. DINGLE, J. H., BADGER, G. F., AND JORDAN, W. S., JR., *Illness in the Home: A Study of 25,000 Illnesses in a Group of Cleveland Families*, The Press of Western Reserve, Cleveland, 1964.

49. DOWNEY, H., AND MCKINLAY, C. A., Acute lymphadenosis compared with acute lymphatic leukemia, *Arch. Intern. Med.* **32:**82–112 (1923).

50. DUBOIS, R. E., SEELEY, J. K., BRUS, I., SAKAMOTO, K., BALLOW, M., HARADA, S., BECHTOLD, T. A., PEARSON, G., AND PURTILO, D. T., Chronic mononucleosis syndrome, *South Med. J.* **77:**1376–1382 (1984).

51. EDWARDS, J. M. B., AND MCSWIGGAN, D. A., Studies on the diagnostic value of an immunofluorescence test for EB virus specific IgM, *Clin. Pathol.* **27:**647–651 (1974).

52. EDWARDS, J. M. B., VANDERVELDE, E. M., COHEN, B. J., AND MCSWIGGAN, D. A., Laboratory diagnosis of EB virus infection in some cases presenting as hepatitis, *J. Clin. Pathol.* **31:**179–182 (1978).

53. ELLENBOGEN, C., AND REINARZ, J. A., The Epstein–Barr virus and its relationship to infectious mononucleosis in Air Force recruits, *Milit. Med.* **140:**371–373 (1974).

54. ENBERG, R. N., EBERLE, B. J., AND WILLIAMS, R. C., Peripheral blood T and B cells in infectious mononucleosis, *J. Infect. Dis.* **130:**104–111 (1974).

55. EPSTEIN, M. A., Epstein–Barr virus—Is it time to develop a vaccine program? *J. Natl. Cancer Inst.* **56:**697–700 (1976).

56. EPSTEIN, M. A., Vaccination against Epstein–Barr virus: Current progress and future strategies, *Lancet* **1:**1425–1427 (1986).

57. EPSTEIN, M. A., AND ACHONG, B. G., Various forms of Epstein–Barr virus infection in man: Established facts and a general concept, *Lancet* **2:**836–839 (1973).

58. EPSTEIN, M. A., AND ACHONG, B. G., Pathogenesis of infectious mononucleosis, *Lancet* **2:**1270–1278 (1977).

59. EPSTEIN, M. A., AND ACHONG, G. G., Introductory considerations, in: *The Epstein–Barr Virus: Recent Advances* (M. A. EPSTEIN AND B. G. ACHONG, EDS.) pp. 2–10, Wiley Medical Publication, New York, 1986.

60. EPSTEIN, M. A., ACHONG, B. G., AND BARR, Y. M., Virus particles in cultured lymphoblasts from Burkitt's lymphoma, *Lancet* **1:**702–703 (1964).

61. EPSTEIN, M. A., MORGAN, A. J., FINERTY, S., RANDLE, B. J., AND KIRKWOOD, J. K., Protection of cottontop tamarins against Epstein–Barr virus-induced malignant lymphoma by a prototype subunit vaccine, *Nature* **318:**287–289 (1985).

62. EVANS, A. S., Experimental attempts to transmit infectious mononucleosis to man, *Yale J. Biol. Med.* **20:**19–26 (1947).

63. EVANS, A. S., Liver function tests in infectious mononucleosis, *J. Clin. Invest.* **27:**106–110 (1948).

64. EVANS, A. S., Further experimental attempts to transmit infectious mononucleosis to man, *J. Clin. Invest.* **29:**508–512 (1950).

65. EVANS, A. S., Infectious mononucleosis in University of Wisconsin students: Report of a five-year investigation, *Am. J. Hyg.* **71:**342–362 (1960).

66. EVANS, A. S. Infectious mononucleosis: Observations from a public health laboratory, *Yale J. Biol. Med.* **34:**261–276 (1961/1962).

67. EVANS, A. S., Complications of infectious mononucleosis: Recognition and management, *Hosp. Med.* **3:**24–25, 28–33 (1967).

68. EVANS, A. S., Infectious mononucleosis: Recent developments, *Gen. Pract.* **60:**127–134 (1969).

69. EVANS, A. S., Infectious mononucleosis in the armed forces, *Milit. Med.* **135:**300–304 (1970).

70. EVANS, A. S., New discoveries in infectious mononucleosis, *Mod. Med.* **1:**18–24 (1974).

71. EVANS, A. S., The history of infectious mononucleosis, *Am. J. Med. Sci.* **267:**189–195 (1974).

72. EVANS, A. S., Commentary—EB virus, infectious mononucleosis and cancer: The closing of the web, *Yale J. Biol. Med.* **47:**113–122 (1974).

73. EVANS, A. S., Infectious mononucleosis and related syndromes, *Am. J. Med. Sci.* **276:**325–339 (1978).

74. EVANS, A. S., Epidemiology of Epstein–Barr virus infection and disease, in: *The Human Herpesvirus* (A. J. NAHMIAS, W. DOUDLE, AND R. SCHINAZI, EDS.), pp. 172–183, Elsevier North-Holland, 1982.

75. EVANS, A. S., Discussion of some recent developments in the molecular epidemiology of Epstein–Barr virus infections, *Yale J. Biol. Med.* **60:**317–319 (1987).

75a. EVANS, A. S., Epstein–Barr virus: An organism for all seasons, in: *Medical Virology VII* (L. M. DE LA MAZA AND E. M. PETERSON, EDS.), pp. 57–98, Elsevier, Amsterdam, 1988.

76. EVANS, A. S., Chronic fatigue syndrome: Thoughts on pathogenesis, *Rev. Infect. Dis.* **13**(Suppl. 1):S56–59 (1991).

77. EVANS, A. S. EBV vaccine: Use in infectious mononucleosis, in: *Vth International Symposium on Epstein–Barr Virus and Associated Diseases* (1992).

78. EVANS, A. S., AND CAMPOS, L. E., Acute respiratory diseases in students at the University of the Philippines, *Bull. WHO* **45:**103–112 (1971).

79. EVANS, A. S., AND COMSTOCK, G. W., Presence of elevated antibody titers to Epstein–Barr virus before Hodgkin's disease, *Lancet* **1:**1183–1186 (1981).

80. EVANS, A. S., EVANS, B. K., AND STURTZ, V., Standards for hepatic and hematologic tests in monkeys: Observations during experiments with hepatitis and mononucleosis, *Proc. Soc. Exp. Biol. Med.* **82:**437–440 (1953).

81. EVANS, A. S., AND GUTENSOHN, N. M., A population-based case-control study of EBV and other viral antibodies among persons with Hodgkin's disease and their siblings, *Int. J. Cancer* **34:**149–157 (1984).

82. EVANS, A. S., AND NIEDERMAN, J. C., EBV-IgA and new heterophile antibody tests in diagnosis of infectious mononucleosis, *Am. J. Clin. Pathol.* **77:**555–560 (1982).

83. EVANS, A. S., NIEDERMAN, J. C., CENABRE, L. C., WEST, B., AND RICHARDS, V. A., A prospective evaluation of heterophile and Epstein–Barr virus-specific IgM antibody tests in clinical and subclinical infectious mononucleosis: Specificity and sensitivity of the tests and persistence of antibody, *J. Infect. Dis.* **132:**546–554 (1975).

84. EVANS, A. S., NIEDERMAN, J. C., AND MCCOLLUM, R. W., Seroepidemiological studies of infectious mononucleosis with EB virus, *N. Engl. J. Med.* **279:**1121–1127 (1968).

85. EVANS, A. S., AND PAUL, J. R., Infectious mononucleosis, in: *Preventive Medicine in World War II*, Vol. V: *Communicable Diseases* (J. B. COATES, JR., ED.), pp. 355–361, Office of the Surgeon General, Department of the Army, Washington, 1960.

86. EVANS, A. S., AND ROBINTON, E. D., An epidemiologic study of infectious mononucleosis in a New England college, *N. Engl. J. Med.* **242:**492–496 (1950).

87. EVANS, A. S., WANAT, J., AND NIEDERMAN, J. C., Failure to demonstrate concomitant antibody changes to viral antigens other

than Epstein–Barr virus (EBV) during or after infectious mononucleosis, *Yale J. Biol. Med.* **56:**203–209 (1983).

88. FILATOV, N. F., *Lektuse ob Ostrikh Infektsion nikh Lolienzynak [Lecture on Acute Infectious Diseases of Children]* U. Deitel, Moscow, 1885; cited by P. J. WISING, *Acta Med. Scand.* [Suppl.] **133:**1–102 (1942).

89. FILATOV, N. F., *Semiotik and Diagnostik de Kinderkrankheiten*, Verlag von Ferdinand Enke, Stuttgart, 1892.

90. FINGERHOTH, J. D., WEIS, J. J., TEDDER, STROMINGER, J. L., BIRO, P. A., AND FEARRON, D. T., Epstein–Barr virus receptor in human B lymphocytes is the C3d receptor CR2, *Proc. Natl. Acad. Sci. USA* **81:**4510–4514 (1984).

91. FISCHER, D. K., ROBERT, M. F., SHEDD, D., SUMMERS, W. P., ROBINSON, J. E., WALAK, J., STEFANO, J., AND MILLER, G. Identification of Epstein–Barr nuclear antigen polypeptide in mouse and monkey cells after gene transfer with a cloned 2.9 kilobase pair subfragment of the genome, *Proc. Natl. Acad. Sci. USA* **81:**43–47 (1984).

92. FLEISCHER, G. R., AND BOLOGNESE, R., Persistent Epstein–Barr virus infection and pregnancy, *J. Infect. Dis.* **147:**982–986 (1983).

93. FLEISCHER, G., LENNETTE, E. T., HENLE, G., AND HENLE, W., Incidence of heterophile antibody responses in children with infectious mononucleosis, *J. Pediatr.* **94:**723–728 (1979).

94. GAINES H., VON SYDOW M., PETERSON P. O., AND LUNDBEGH, P., Clinical picture of primary HIV infection presenting as a glandular-fever-like illness, *Br. Med. J.* **297:**1163–1168 (1988).

95. GARDNER, H. T., AND PAUL, J. R., Infectious mononucleosis at the New Haven Hospital, 1921–46, *Yale J. Biol. Med.* **19:**839–853 (1947).

96. GERBER, P., AND DEAL, D. R., Epstein–Barr virus-induced viral and soluble complement-fixing antigens in Burkitt lymphoma cell cultures. *Proc. Soc. Exp. Biol. Med.* **134:**748–751 (1970).

97. GERBER, P., NKRUMAH, F. K., PRITCHETT, R., AND KIEFF, E., Comparative studies of Epstein–Barr virus strains from Ghana and the United States, *Int. J. Cancer* **17:**71–81 (1976).

98. GERBER, P., PURCELL, R. H., ROSENBLUM, E. N., AND WALSH, J. H., Association of EB-virus infection with the post-perfusion syndrome, *Lancet* **1:**593–596 (1969).

99. GERBER, R., AND ROSENBLUM, E. N., The incidence of complement-fixing antibodies to herpes simplex and herpes-like viruses in man and rhesus monkeys, *Proc. Soc. Exp. Biol. Med.* **128:**541–546 (1968).

100. GERBER, R., GOLDSTEIN, L. I., LUCAS, S., NONOYAMA, M., AND PERLIN, E., Oral excretion of Epstein–Barr viruses by healthy subjects and patients with infectious mononucleosis, *Lancet* **2:**988–989 (1972).

101. GERGELY, L., KLEIN, G., AND ERNBERG, I., Appearance of Epstein–Barr virus-associated antigens in infected Raji cells, *Virology* **45:**10–21 (1971).

102. GINSBURG, C. M., HENLE, W., HENLE, G., AND HORWITZ, C. A., Infectious mononucleosis in children: Evaluation of Epstein–Barr virus specific serological data, *J.A.M.A.* **237:**781–785 (1977).

103. GINSBURG, C. M., HENLE, G., AND HENLE, W., An outbreak of infectious mononucleosis among the personnel of an outpatient clinic, *Am. J. Epidemiol.* **104:**571–575 (1976).

103a. GIULANO, V. J., JASIN, H. E., AND ZIFF, M., The nature of the atypical lymphocyte in infectious mononucleosis. *Clin. Immunol. Immunopathol.* **3:**90–98 (1974).

104. GRACE, J. T., BLAKESLEE, J., AND JONES, R., Induction of infectious mononucleosis in man by the herpes-type virus (HTV) in Burkitt lymphoma cells in tissue culture. *Proc. Am. Assoc. Cancer Res.* **10:**31 (1969).

105. GRATAMA, J. W., OOSTERVEER, M. A., KLEIN, G., AND ERNBERG, I., EBNA size polymorphism can be used to trace Epstein–Barr virus spread within familes, *J. Virol.* **64:**4703–4708 (1990).

106. GRAVES, S., JR., Recurrent infectious mononucleosis, *J. Ky. Med. Assoc.* **1:**790–793 (1970).

107. GREENSPAN, J. S., GREENSPAN, D., LENNETTE, E. T., ABRAMS, D. I., CONANT, M. A., PETERSON, V., AND FREESE, V. K., Replication of Epstein–Barr virus within the epithelial cells of oral hairy leukoplakia, an AIDS-associated lesion, *N. Engl. J. Med.* **313:**1564–1571 (1985).

108. GROGAN, E. A., SUMMERS, W. P., DOWLING, S., SHEDD, D., GRADOVILLE, L., AND MILLER, G., Two Epstein–Barr viral nuclear neoantigens distinguished by gene transfer, serology and chromosome binding, *Proc. Natl. Acad. Sci. USA* **80:**7650–7653 (1983).

109. GROSE, C., HENLE, W., HENLE, G., AND FEORINO, P. M., Primary Epstein–Barr infections in acute neurologic disease, *N. Engl. J. Med.* **292:**392–393 (1975).

110. GU, S., HUANG, T., MIAO, Y., RUAN, L., ZHAO, Y., HAN, C., XIAO, Y., ZHU, J., AND WOLF, H. A preliminary study of the immunogenicity in rabbits and human volunteers of a recombinant vaccinia virus expressing Epstein–Barr membrane antigen, *Chin. Med. Sci. J.* **6:**241–243 (1991).

111. HAIDER, S., COUTINHO, M. D., AND EMOND, R. T. D., Tuberculin anergy and infectious mononucleosis, *Lancet* **2:**74 (1973).

112. HAINEBACH, J., II, Beitrag zur Aetiologie des Pfeiffer' schen Drüsenfiebers, *Dtsch. Med. Wochenschr.* **26:**419–420 (1899).

113. HALCHROW, J. P. A., OWEN, L. M., AND ROGER, N. O., Infectious mononucleosis with an account of an epidemic in E.M.S. Hospital, *Br. Med. J.* **2:**443–447 (1943).

114. HALLEE, T. J., EVANS, A. S., NIEDERMAN, J. C., BROOKS, C. M., AND VOEGTLY, J. H., Infectious mononucleosis at the US Military Academy: A prospective study of a single class over four years, *Yale J. Biol. Med.* **47:**182–195 (1974).

115. HAMILTON, J. K., SULLIVAN, J. L., MAURER, H. S., CRUZI, F. G., PROIRSON, A. J., STENKE, K., FINKELSTEIN, G. Z., LANDING, B., GRUNNET, M., AND PURTILO, D. T., X-linked lymphoproliferative syndrome registry report, *J. Pediatr.* **4:**669–673 (1980).

116. HAMPAR, B., HSU, K. C., MARTOS, L. M., AND WALKER, J. L., Serologic evidence that a herpes-type virus is the etiologic agent of heterophile-positive infectious mononucleosis, *Proc. Natl. Acad. Sci. USA* **68:**1407–1411 (1971).

117. HANSHAW, J. B., NIEDERMAN, J. C., AND CHESSIN, L. N., Cytomegalovirus macroglobulin in cell associated herpes virus infections, *J. Infect. Dis.* **125:**304–306 (1972).

118. HARDY, D. A., AND STEEL, C. M., Cytotoxic potential of lymphocytes stimulated with autochthonous lymphoid cell lines, *Experientia* **27:**1336–1338 (1971).

119. HAYNES, B. F., SCHOOLEY, R. T., GROUSE, J. E., PLAYING-WRIGHT, C. R., DOLIN, R., AND FAUCI, A. S., Characterization of thymus-derived lymphocyte subsets in acute Epstein–Barr virus induced infectious mononucleosis, *J. Immunol.* **122:**699–702 (1979).

120. HAYNES, B. F., SCHOOLEY, R. T., PLAYING-WRIGHT, C. R., GROUSE, J. E., DOLIN, R., AND FAUCI, A. S., Emergence of suppressor cells of immunoglobulin synthesis during acute Epstein–Barr-virus-induced infectious mononucleosis, *J. Immunol.* **123:**2095–2101 (1979).

121. HEATH, C. W., BRODSKY, A. L., AND POTOLSKY, A. I., Infectious

mononucleosis in a general population, *Am. J. Epidemiol.* **95**:46–52 (1972).

122. HENDERSON, E., HESTON, L., GROGAN, E., AND MILLER, G., Radiobiological inactivation of Epstein–Barr virus, *J. Virol.* **25**:51–59 (1978).

123. HENDERSON, E., MILLER, G., ROBINSON, J., AND HESTON, L., Efficiency of transformation of lymphocytes by Epstein–Barr virus, *Virology* **76**:152–163 (1977).

124. HENKE, C. E., KURLAND, L. T., AND ELVEBACK, L. R., Infectious mononucleosis in Rochester, Minn., 1950 through 1969, *Am. J. Epidemiol.* **98**:483–490 (1973).

125. HENLE, G., AND HENLE, W., Immunofluorescence in cells derived from Burkitt's lymphoma, *J. Bacteriol.* **91**:1248–1256 (1966).

126. HENLE, G., AND HENLE, W., Observations on childhood infections with Epstein–Barr virus, *J. Infect. Dis.* **121**:303–310 (1970).

127. HENLE, G., HENLE, W., AND DIEHL, V., Relation of Burkitt's tumor-associated herpes-type virus to infectious mononucleosis, *Proc. Natl. Acad. Sci. USA* **59**:94–101 (1968).

128. HENLE, G., HENLE, W., AND HORWITZ, C. A., Antibodies to Epstein–Barr virus-associated nuclear antigen in infectious mononucleosis, *J. Infect. Dis.* **130**:231–239 (1974).

129. HENLE, W., AND KLEIN, G., Demonstration of two distinct components in the early antigen complex of Epstein–Barr virus-infected cells, *Int. J. Cancer* **8**:272–282 (1971).

130. HENLE, W., AND HENLE, G., Epstein–Barr virus-specific serology in immunologically compromised individuals, *Cancer Res.* **41**:4222–4225 (1981).

131. HENLE, W., HENLE, G., ANDERSSON, J., ERNBERN, I., KLEIN, G., HORWITZ, C. A., MARKLUND, G., RYMO, L., WELLINDER, C., AND STRAUS, S. E., Antibody responses to Epstein–Barr virus-determined nuclear antigen (EBNA)-1 and EBNA-2 in acute and chronic Epstein–Barr virus infection, *Proc. Natl. Acad. Sci. USA* **84**:570–574 (1987).

132. HENLE, W., HENLE, G., AND HORWITZ, C. A., Epstein–Barr virus specific diagnostic tests in infectious mononucleosis, *Hum. Pathol.* **5**:551–565 (1974).

133. HENLE, W., HENLE, G., HARRISON, F. S., JOYNER, C. R., KLEMOLA, E., PALOHEIMO, J., SCRIBA, M., AND VON ESSEN, F., Antibody responses to the Epstein–Barr virus and cytomegaloviruses after open-heart and other surgery. *N. Engl. J. Med.* **282**:1068–1074 (1968).

134. HENLE, W., HENLE, G., NIEDERMAN, J. C., HALTIA, K., AND KLEMOLA, E., Antibodies to early antigens induced by Epstein–Barr virus in infectious mononucleosis, *J. Infect. Dis.* **124**:58–67 (1971).

135. HENLE, W., HENLE, G., PEARSON, G., SCRIBA, M., WAUBKE, R., AND ZAJAC, B. A., Differential reactivity of human serums with early antigens induced by Epstein–Barr virus, *Science* **169**:188–190 (1970).

136. HENNESSY, K., FENNEWALD, S., AND KIEFF, E., A third viral nuclear protein in lymphoblasts immortalized by Epstein–Barr virus, *Proc. Natl. Acad. Sci. USA* **82**:6300–6304 (1985).

137. HENNESSY, K., HELLER, M., VAN SAULTERI, V., AND KIEFF, E., Simple repeat array in Epstein–Barr virus DNA encodes that of the Epstein–Barr nuclear antigen, *Science* **220**:1396–1398 (1983).

138. HENNESSY, K., AND KIEFF, E., A second nuclear protein is encoded by Epstein–Barr virus in latent infection, *Science* **227**:1238–1240 (1985).

139. HEWETSON, J. F., ROCHI, G., HENLE, W., AND HENLE, G., Neutralizing antibodies to Epstein–Barr virus in healthy populations and patients with Infectious mononucleosis, *J. Infect. Dis.* **128**:283–389 (1973).

140. HO, M., MILLER, G., ATCHISON, R. W., BREINIG, M. K., DUMMER, J. S., ANDIMAN, W., STARZL, T. E., EASTMAN, R., GRIFFITH, P. B., HARDESTY, R. L., BAHNSON, H. T., AND ROSENTHAL, J. T., Epstein–Barr virus infections and DNA hybridization studies in post-transplantation lymphoma and lymphoproliferative lesions. The role of primary infection, *J. Infect. Dis.* **152**:876–885 (1985).

141. HOAGLAND, R. J., The transmission of infectious mononucleosis, *Am. J. Med. Sci.* **229**:262–272 (1955).

142. HOAGLAND, R. J., Resurgent heterophil-antibody reaction after infectious mononucleosis, *N. Engl. J. Med.* **269**:1307–1308 (1963).

143. HOAGLAND, R. J., The incubation period of infectious mononucleosis, *Am. J. Public Health* **54**:1699–1705 (1964).

144. HOAGLAND, R., *Infectious Mononucleosis*, Grune & Stratton, New York, 1967.

145. HOBSON, F. G., LAWSON, B., AND WIGFIELD, M., Glandular fever, a field study, *Br. Med. J.* **1**:845–852 (1958).

145a. HOLMES, G. P., KAPLAN, J. E., GANTZ, N. M., KOMAROFF, A. L., SCHONBERGER, L. B., STRAUS, S. E., JONES, J. E., DUBOIS, R. E., CUNNINGHAM-RUNDLES, C., PAHWA, S., TOSATO, S., GANS, L. S., PURTILO, D. T., BROWN, N., CHALLEY, R. J., AND BRUS, I., Chronic fatigue syndome: A working case definition, *Ann. Intern. Med.* **108**:387–389 (1988).

146. HORWITZ, C. A., HENLE, W., HENLE, G., RUDNICK, H., AND LATTS, E., Long-term serological follow-up of patients for Epstein–Barr virus after recovery from infectious mononucleosis, *J. Infect. Dis.* **151**:1150–1153 (1985).

147. HOSKINS, T. W., FLETCHER, W. B., BLAKE, J. M., PEREIRA, M. S., AND EDWARDS, J. M. B., EB virus antibody and infectious mononucleosis in a boarding school for boys, *J. Clin. Pathol.* **29**:42–45 (1976).

148. HSU, J. F., EVANS, A. S., NIEDERMAN, J. C., AND CENABRE, L. C., An enzyme-linked immunosorbant assay (ELISA) for measurement of heterophile antibody, *Yale J. Biol. Med.* **55**:429–436 (1982).

149. HUTT, L. M., HUANG, Y. T., DASCOMB, H. E., AND PAGANO, J. S., Enhanced destruction of lymphoid cell lines by peripheral blood leucocytes taken from patients with acute infectious mononucleosis, *J. Immunol.* **115**:243–248 (1975).

150. ISAACS, R., Chronic infectious mononucleosis, *Blood* **3**:858–861 (1948).

151. JENNINGS, E., *Prevalence of EB Virus Antibody in Hawaii*, MD Thesis, Yale University School of Medicine, New Haven, 1973.

152. JONCAS, J., AND MITNYAN, C., Serological response of the EBV antibodies in pediatric cases of infectious mononucleosis and in their contacts, *Can. Med. Assoc. J.* **102**:1260–1263 (1970).

153. JONDAL, M., AND KLEIN, G., Surface markers on human B and T lymphocytes. II. Presence of Epstein–Barr virus receptors on B lymphocytes, *J. Exp. Med.* **138**:1365–1378 (1973).

154. JONDAL, M., KLEIN, G., OLDSTONE, M. B. S., BOKISH, V., AND YEFENOF, E., Surface markers on human B and T lymphocytes. VII. Association between complement and Epstein–Barr virus receptors on human lymphoid lines, *Scand. J. Immunol.* **5**:401–410 (1976).

155. JONES, J. F., RAY, C. G., MINNICH, L. L., HICKS, M. J., KIBLER, R., AND LUCAS, D. O., Evidence for active Epstein–Barr virus infection in patients with persistent unexplained illnesses: Elevated anti-early antigen antibodies, *Ann. Intern. Med.* **102**:1–7 (1985).

156. JONES, J. F., Serologic and immunologic responses in chronic

fatigue syndrome with emphasis on the Epstein–Barr virus, *Rev. Infect. Dis.* **13**(Suppl. 1):S26–31 (1991).

157. JUNGE, U., DEINHARDT, F., AND HOEKSTRA, J., Stimulation of peripheral lymphocytes by allogeneic and autochthonous mononucleosis lymphocyte cell lines, *J. Immunol.* **106**:1306–1315 (1971).

158. JUNKER, A. K., THOMAS, E. E., RADCLIFFE, A., FORSYTH, R. B., DAVIDSON, A. G., AND RYMO, L., Epstein–Barr virus shedding in breast milk, *Am. J. Med. Sci.* **302**:220–223 (1991).

159. KANO, K., AND MILGROM, F., Heterophile antigens and antibodies in medicine, *Curr. Top. Microbiol. Immunol.* **77**:43–69 (1977).

160. KASL, S. V., EVANS, A. S., AND NIEDERMAN, J. C., Psychosocial risk factors in the development of infectious mononucleosis, *Psychosom. Med.* **41**:445–466 (1979).

161. KAUFMAN, R. E., Recurrences in infectious mononucleosis, *Am. Pract.* **1**:673–676 (1950).

162. KIBLER, R., LUCAS, D. O., HICKS, M. J., POULOS, B. T., AND JONES, J. F., Immune function in chronic active Epstein–Barr virus infection, *J. Clin. Immunol.* **5**:46–54 (1985).

163. KLEIN, G., Viral latency and transformation: The strategy of Epstein–Barr virus, *Cell* **58**:5–8 (1989).

164. KLEIN, G., DIEHL, V., HENLE, G., HENLE, W., PEARSON, G., AND NIEDERMAN, J. C., Relations between Epstein–Barr viral and cell membrane immunofluorescence in Burkitt tumor cells. II. Comparison of cells and sera from patients with Burkitt's lymphoma and infectious mononucleosis, *J. Exp. Med.* **128**:1021–1030 (1968).

165. KLEIN, G., KLEIN, E., CLIFFORD, P., AND STERNSWARD, G., Search for tumor specific immune reactions in Burkitt lymphoma patients by the membrane immunofluorescence reaction, *Proc. Natl. Acad. Sci. USA* **55**:1628–1635 (1966).

166. KLEIN, G., LINDAHL, T., JONDAL, M., LEIBOLB, W., MENÉZES, J., NILSSON, K., AND SUNDSTRÖM, C., Continuous lymphoid cell lines with characteristics of B cells (bone-marrow-derived) lacking the epstein–Barr virus genome and derived from three human lymphomas, *Proc. Natl. Acad. Sci. USA* **71**:3283–3286 (1974).

167. KLEIN, G., SVEDMYR, E., JONDAL, M., AND PERSSON, P. O., EBV-determined nuclear antigen (EBNA)-positive cells in the peripheral blood of infectious mononucleosis patients, *Int. J. Cancer* **17**:21–26 (1976).

168. KLEMOLA, E., HENLE, G., HENLE, W., AND VON ESSEN, R., Infectious mononucleosis-like disease with negative heterophile agglutination test: Clinical features in relation to Epstein–Barr virus and cytomegalovirus antibodies, *J. Infect. Dis.* **121**:608–614 (1970).

168a. KYAW, M. T., HURREN, L., EVANS, L., MOSS, D. J., COOPER, D. A., BENSON, E., ESMORE, D., AND SCULLEY, T. B., Expression of B-type Epstein–Barr virus in HIV-infected patients and cardiac transplant recipients, in: *Vth International Symposium on Epstein–Barr Virus and Associated Diseases* (1992).

169. LANG, D. J., AND HANSHAW, J. B., Cytomegalovirus infection and the postperfusion syndrome: Recognition of primary infection in four patients, *N. Engl. J. Med.* **280**:1145–1149 (1969).

170. LEE, C. L., DAVIDSOHN, I., AND SLABY, R., Horse agglutinins in infectious mononucleosis, *Am. J. Clin. Pathol.* **49**:3–11 (1968).

170a. LEES, J. F., ARRAND, J. E., PEPPER, S. D. V., STEWART, J. P., MACKET, M., AND ARRAND, J. R., The Epstein–Barr virus candidate antigen CP340-220 is highly conserved between virus Type A and B, *Virology* **195**(2):578–586 (1993).

171. LEHANE, D. E., A seroepidemiologic study of infectious mono-

nucleosis: The development of EB virus antibody in a military population, *J.A.M.A.* **212**:2240–2242 (1970).

172. LEMON, S. M., HUTT, L. M., HUANG, Y., BLUM, J., AND PAGANO, J. H., Simultaneous infection with multiple herpesviruses, *J.A.M.A.* **66**:270–276 (1979).

173. LEMON, S. M., HUTT, L. M., SHAW, J. C., LI, J. H., AND PAGANO, J. H., Replication of EBV in epithelial cells during infectious mononucleosis, *Nature* **268**:268–270 (1977).

174. LENNETTE, E. T., HENLE, G., HENLE, W., AND HOROWITZ, C. A., Heterophil antigen in bovine sera detectable by immune adherence hemagglutination with infectious mononucleosis sera, *Infect. Immun.* **19**:923–927 (1978).

175. LEVINE, P. H., JACOBSON, S., POCINKI, A. G., CHENEY, P., PETERSON, D., CONNELLY, R. R., WEIL, R., ROBINSON, S. M., ABLASHI, S. V., AND SALAHUDDIN, S. V. Clinical epidemiologic, and virologic studies in four clusters of the chronic fatigue syndrome, *Arch. Intern. Med.* **152**:1611–1616 (1992).

176. LEVINE, P. H., LUBIN, L. J., AND EVANS, A. S., Evaluating the efficacy of an EBV vaccine: Clinical and epidemiologic considerations, in: *Vth International Symposium on Epstein–Barr Virus and Associated Diseases* (1992).

177. LINDE, A., KALLIN, B., DILLNER, J., ANDERSON, J., JAGDAHL, L., AND WAHREN, B., Evaluation of enzyme-linked immunosorbent assays with two synthetic peptides of Epstein–Barr virus for diagnosis of infectious mononucleosis, *J. Infect. Dis.* **161**:903–909 (1990).

178. LONGCOPE, W. T., Infectious mononucleosis (glandular fever), with a report of ten cases, *Am. J. Med. Sci.* **164**:781–807 (1922).

179. MANGI, R., NIEDERMAN, J. C., KELLEHER, J. E., DWYER, J. M., EVANS, A. S., AND KANTOR, F. S., Depression of cell-mediated immunity during acute infectious mononucleosis, *N. Engl. J. Med.* **291**:1149–1153 (1974).

180. MASON, K. L., An ox cell hemolysin test for the diagnosis of infectious mononucleosis, *J. Hyg.* **49**:471–481 (1951).

181. MATHESON, B. A., CHISHOLM, S. M., AND HO-YEN, D. O., Assessment of rapid ELISA test for detection of Epstein–Barr virus infection, *J. Clin. Pathol.* **43**:691–693 (1990).

182. MATTHEWS, D. A., LANE, T. J., AND MANU, P., Antibodies to Epstein–Barr virus in patients with chronic fatigue syndrome, *South. Med. J.* **84**:832–840 (1991).

183. MAURER, B. A., IMAMURA, T., AND WILBERT, S. M., Incidence of EB virus containing cells in primary and secondary clones of several Burkitt lymphoma cell lines, *Cancer Res.* **30**:2870–2875 (1970).

184. MILLER, G., The oncogenicity of Epstein–Barr virus, *J. Infect. Dis.* **130**:187–205 (1974).

185. MILLER, G., Biology of Epstein–Barr virus, in: *Viral Oncology* (G. KLEIN, ED.), pp. 713–738, Raven Press, New York, 1980.

186. MILLER, G., GROGAN, E. A., FISCHER, D. K., NIEDERMAN, J. C., SCHOOLEY, R. T., HENLE, W., LENOIR, G., AND LIU, C., Antibody responses to two Epstein–Barr virus nuclear antigens defined by gene transfer, *N. Engl. J. Med.* **312**:750–756 (1985).

187. MILLER, G., GROGAN, E., ROWE, D., ROONEY, C., HESTON, L., EASTMAN, R., ANDIMAN, W., NIEDERMAN, J., LENOIR, G., HENLE, W., SULLIVAN, J., SCHOOLEY, R., VOSSEN, J., STAUSS, S., AND ISSEKUTZ, T., Selective lack of antibody to a component of EB nuclear antigen in patients with chronic active Epstein–Barr virus infection, *J. Infect. Dis.* **156**:26–35 (1987).

188. MILLER, G., AND HESTON, L., Expression of Epstein–Barr viral capsid, complement fixing and nuclear antigens in stationary and exponential phase cultures, *Yale J. Biol. Med.* **47**:123–135 (1974).

189. MILLER, G., KATZ, B. Z., AND NIEDERMAN, J. C., Some recent developments in the molecular epidemiology of Epstein–Barr virus infections, *Yale J. Biol. Med.* **60:**307–316 (1986).

190. MILLER, G., AND LIPMAN, M., Release of infectious Epstein–Barr virus by transformed marmoset leucocytes, *Proc. Natl. Acad. Sci. USA* **70:**190–194 (1973).

191. MILLER, G., MILLER, M. H., AND STITT, D., Epstein–Barr viral antigen in single cell clones of two human leucocytic lines, *J. Virol.* **6:**699–701 (1970).

192. MILLER, G., NIEDERMAN, J. C., AND ANDREWS, L. L., Prolonged oropharyngeal excretion of Epstein–Barr virus after infectious mononucleosis, *N. Engl. J. Med.* **288:**229–232 (1973).

193. MILLER, G., NIEDERMAN, J. C., AND STITT, D. A., Infectious mononucleosis: Appearance of neutralizing antibody to Epstein–Barr virus measured by inhibition of formation of lymphoblastoid cell lines, *J. Infect. Dis.* **125:**403–406 (1972).

194. MILLER, G., ROBINSON, J., HESTON, L., AND LIPMAN, M., Differences between laboratory strains of Epstein–Barr virus based on immortalization, abortive infection and interference, *Proc. Natl. Acad. Sci. USA* **71:**4006–4010 (1974).

195. MOIR, J. I., Glandular fever in the Falkland Islands, *Br. Med. J.* **2:**822–823 (1930).

196. MORGAN, A. J., Recent progress in EB vaccine development, in: *Vth International Symposium on Epstein–Barr Virus and Associated Diseases* (1992).

197. MORGAN, D. G., NIEDERMAN, J. C., MILLER, G., SMITH, H. W., AND DOWALIBY, J. M., Site of Epstein–Barr virus replication in the oropharynx, *Lancet* **2:**1154–1157 (1979).

198. MORI, T., KANO, K., AND MILGROM, F., Formation of Paul Bunnell antibodies by cultures of lymphocytes from Infectious mononucleosis, *Cell Immunol.* **34:**289–298 (1977).

199. MORSE, P. F., Glandular fever, *J.A.M.A.* **77:**1403–1404 (1921).

200. MUÑOZ, N., DAVIDSON, R. J. K., WITTHOFF, B., ERICSSON, J. E., AND DE-THE, G., Infectious mononucleosis and Hodgkin's disease, *Int. J. Cancer* **22:**10–13 (1978).

201. NEWALL, K. W., The reported incidence of glandular fever, and analysis of a report of the Public Health Laboratory Service, *J. Clin. Pathol.* **10:**20–22 (1957).

202. NIEDERMAN, J. C., Infectious mononucleosis at the Yale–New Haven Medical Center, 1946–1955, *Yale J. Biol. Med.* **28:**629–643 (1956).

203. NIEDERMAN, J. C., Chronicity of Epstein–Barr virus infection, *Ann. Intern. Med.* **102:**119–121 (1985).

204. NIEDERMAN, J. C., EVANS, A. S., McCOLLUM, R. W., AND SUBRAHMANYAN, L., Prevalence, incidence and persistence of EB virus antibody in young adults, *N. Engl. J. Med.* **282:**361–365 (1970).

205. NIEDERMAN, J. C., LIU, C.-R., KAPLAN, M. H., AND BROWN, N. A., Clinical and serologic features associated with evidence of human herpes-6 infection in three adults, *Lancet* **2:**817–819 (1988).

206. NIEDERMAN, J. C., McCOLLUM, R. W., HENLE, G., AND HENLE, W., Infectious mononucleosis: Clinical manifestations in relation to EB virus antibodies, *J.A.M.A.* **203:**20–209 (1968).

207. NIEDERMAN, J. C., AND MILLER, G., Kinetics of the antibody response to *Bam* HI-K nuclear antigen in uncomplicated infectious mononucleosis, *J. Infect. Dis.* **154:**346–349 (1986).

208. NIEDERMAN, J. C., MILLER, G., PEARSON, H. A., PAGANO, J. S., AND DOWALIBY, J. M., Infectious mononucleosis: EB virus shedding in saliva and the oropharynx, *N. Engl. J. Med.* **294:**1355–1359 (1976).

209. NIEDERMAN, J. C., AND SCOTT, R. B., Studies on infectious mononucleosis: Attempts to transmit the disease to human volunteers, *Yale J. Biol. Med.* **38:**1–10 (1965).

210. NIKOSKELAINEN, J. J., ABLASHI, D. V., ISENBERG, R. A., NEEL, E. U., MILLER, R. G., AND STEVENS, D. A., Cellular immunity in infectious mononucleosis. II. Specific reactivity to Epstein–Barr virus antigens and correlation with clinical and hematologic parameters, *J. Immunol.* **121:**1239–1244 (1978).

211. NIKOSKELAINEN, J., LEIKOLA, J., AND KLEMOLA, E., IgM antibodies specific for Epstein–Barr virus in infectious mononucleosis without heterophile antibodies, *Br. Med. J.* **4:**72–75 (1974).

212. NONOYAMA, M., AND PAGANO, J. S., Homology between Epstein–Barr virus DNA and viral DNA from Burkitt's lymphoma and nasopharyngeal carcinoma determined by DNA–DNA reassociation kinetics, *Nature* **242:**44–47 (1973).

213. NORTH, J. R., MORGAN, A. J., THOMPSON, J. L., AND EPSTEIN, M. A., Purified EB virus M_R 340,000 glycoprotein induces potent virus-neutralizing antibodies when incorporated in liposomes, *Proc. Natl. Acad. Sci. USA* **79:**7504–7508 (1982).

214. OLD, L. J., CLIFFORD, P., BOYSE, E. A., DeHARVEN, E., GEERING, G., OETTGEN, H. F., AND WILLIAMSON, B., Precipitating antibody in human serum to an antigen present in cultured Burkitt lymphoma cells, *Proc. Natl. Acad. Sci. USA* **56:**1699–1704 (1966).

215. PAGANO, J. S., The Epstein–Barr viral genome and its interactions with human lymphoblastoid cells and chromosomes, in: *Viruses, Evolution and Cancer* (K. MARAMOROSCH AND E. KURSTAK, EDS.), pp. 79–116, Academic Press, New York, 1974.

216. PAGANO, J. S., Detection of Epstein–Barr virus with molecular hybridization techniques, *J. Infect. Dis.* **13**(Suppl.):S123–128 (1991).

217. PANNUTI, C. S., CARVALHO, R. P. S., EVANS, A. S., CENABRE, L. C., NETO, A., CAMARGO, M., ANGELO, J. J. P., AND TAKIMOTO, S., A prospective clinical study of the mononucleosis syndrome in a developing country, *Int. J. Epidemiol.* **9:**349–353 (1980).

218. PATTENGALE, P. K., GERBER, P., AND SMITH, R. W., Selective transformation of B lymphocytes by EB virus, *Lancet* **2:**1153–1155 (1973).

219. PATTENGALE, P. K., GERBER, P., AND SMITH, R. W., B-cell characteristics of human peripheral and cord blood lymphocytes transformed by Epstein–Barr virus, *J. Natl. Cancer Inst.* **52:**1081–1086 (1974).

220. PATTENGALE, P. K., SMITH, R. W., AND PERLIN, E., Atypical lymphocytes in acute infectious mononucleosis, *N. Engl. J. Med.* **291:**1145–1148.

221. PEARSON, G. R., AND LUKA, J., Characteristics of the virus-determined antigens, in: *The Epstein–Barr Virus: Recent Advances* (M. A. EPSTEIN AND B. G. ACHONG, EDS.), pp. 48–73, Wiley Medical Publication, New York, 1986.

222. PAUL, J. R., AND BUNNELL, W. W., The presence of heterophile antibodies in infectious mononucleosis, *Am. J. Med. Sci.* **183:**91–104 (1932).

223. PAUL, O., Mononucleosis on board a destroyer, *US Naval Med. Bull.* **44:**614–617 (1945).

224. PEARSON, G., DEWEY, F., KLEIN, G., HENLE, G., AND HENLE, W., Relation between neutralization of Epstein–Barr virus and antibodies to cell membrane antigens induced by the virus, *J. Natl. Cancer Inst.* **45:**989–995 (1970).

225. PEARSON, G., HENLE, G., AND HENLE, W., Production of antigens associated with Epstein–Barr virus in experimentally infected lymphoblastoid cell lines, *J. Natl. Cancer Inst.* **46:**1243–1250 (1971).

226. Pereira, M. S., Blake, J. M., and MacCrae, A. D., EB virus antibody at different ages, *Br. Med. J.* **4**:526–527 (1969).

227. Pereira, M. S., Field, A. M., Blake, J. M., Rodgers, F. G., Bailey, L. A., and Davies, J. R., Evidence for oral excretion of EB virus in infectious mononucleosis, *Lancet* **1**:710–711 (1972).

228. Pfeiffer, E., Drüsenfieber, *Jahrb. Kinderheilkd.* **29**:257–264 (1889).

229. Pope, J. H., Horne, M. K., and Wetters, E. J., Significance of a complement-fixing antigen associated with herpes-like virus and detected in the Raji cell line, *Nature* **228**:186–187 (1969).

230. Pulvertaft, R. J. X., Cytology of Burkitt's tumor (African lymphoma), *Lancet* **1**:238–240 (1964).

231. Purtilo, D. T., Bhawan, J., Hutt, L. M., DeNicola, L., Szymanski, I., Yang, J. P. S., Boto, W., Maier, R., and Thorley-Lawson, D., Epstein–Barr virus infection in the X-linked recessive lymphoproliferative syndrome, *Lancet* **1**: 798–801 (1978).

232. Purtilo, D. T., Hutt, L., Bhawan, J., Yang, J. P. S., Cassel, C., Allegro, S., and Rosen, F. S., Immunodeficiency to the Epstein–Barr virus in the X-linked recessive lymphoproliferative syndrome, *Clin. Immunol. Immunopathol.* **9**:147–156 (1978).

233. Purtilo, D. T., and Lai, P. K., Clinical and immunopathological manifestations and detection of Epstein–Barr virus infection in immune deficient patients, in: *Medical Virology VI* (L. M. de la Maza and E. M. Peterson, eds.), pp. 121–167, Elsevier Publications, Amsterdam, 1987.

233a. Purtilo, D. T., Okano, M., and Grierson, H. L., Immunodeficiency as a risk factor in Epstein–Barr virus-induced malignant diseases, *Environ. Health Perspect.* **88**:225–230 (1990).

234. Qualtiere, L. F., and Pearson, G. R., Solubilization of Epstein–Barr virus induced membrane antigen by limited papain digestion, *Fed. Proc.* **37**:1817 (1978).

235. Raab-Traub, N., and Pagano, J. S., Epstein–Barr virus and its antigens, *Hum. Immunogenet.* **43**:477–498 (1989).

236. Reedman, B. M., and Klein, G., Cellular localization of an Epstein–Barr virus (EBV) associated complement-fixing antigen in producer and non-producer lymphoblastoid cell lines, *Int. J. Cancer* **11**:499–520 (1973).

237. Rickinson, A. B., On the biology of Epstein–Barr virus persistence: A reappraisal, in: *Immunobiology and Prophylaxsis of Human Herpesvirus Infections* (C. Lopez et al.) pp. 137–146, Plenum Press, New York, 1990.

237a. Rickinson, A. B., Yao, O.-Y., and Wallace, L. E., The Epstein–Barr virus as a model of virus–host interactions, *Br. Med. Bull.* **41**:75–79 (1985).

238. Rickinson, A. B., Crawford, D., and Epstein, M. A., Inhibition of the *in vitro* outgrowth of Epstein–Barr virus transformed lymphocytes by thymus dependent lymphocytes from infectious mononucleosis patients, *Clin. Exp. Immunol.* **28**:72 (1977).

239. Robinson, J. E., Brown, N., Andiman, W., Halliday, K., Francke, U., Robert, M. F., Andersson-Anvret, M., Horstmann, D. M., and Miller, G., Diffuse polyclonal B-cell lymphoma during primary infection with Epstein–Barr virus, *N. Engl. J. Med.* **320**:1293–1296 (1980).

240. Robinson, J., and Miller, G., Assay for Epstein–Barr virus based on stimulation of DNA synthesis in mixed leukocytes from human umbilical cord blood, *J. Virol.* **15**:1065–1072 (1975).

241. Robinson, J., Smith, D., and Niederman, J. C., Plasmacytic differentiation of circulating Epstein–Barr virus-infected B lymphocytes during acute infectious mononucleosis, *J. Exp. Med.* **153**:235–244 (1981).

242. Robinson, J., Smith, D., and Niederman, J. C., Mitotic EBNA-positive lymphocytes in peripheral blood during infectious mononucleosis, *Nature* **287**:334–335 (1981).

243. Rocchi, G., DeFelici, A., Ragona, G., and Heinz, A., Quantitative evaluation of Epstein–Barr-virus-infected mononuclear peripheral blood leukocytes in infectious mononucleosis, *N. Engl. J. Med.* **296**:132–134 (1977).

244. Rocchi, G., Hewetson, J., and Henle, W., Specific neutralizing antibodies in Epstein–Barr virus associated diseases, *Int. J. Cancer* **11**:637–647 (1973).

245. Rosalki, S. B., Lwynn, J. T., and Verney, P. T., Transaminase and liver function studies in infectious mononucleosis, *Br. Med. J.* **1**:929–932 (1960).

246. Rosdahl, N., Larsen, S. O., and Thamdrup, B., Infectious mononucleosis in Denmark: Epidemiological observations based on positive Paul–Bunnell reactions 1940–1969, *Scand. J. Infect. Dis.* **5**:163–170 (1973).

247. Rosen, A., Gergely, P., Jondal, M., and Klein, G., Polyclonal Ig production after Epstein–Barr virus infection of human lymphocytes *in vitro*, *Nature* **267**:52 (1977).

248. Rothfield, N. F., Evans, A. S., and Niederman, J. C., Clinical and laboratory aspects of raised virus antibody titers in systemic lupus erythematosus, *Ann. Rheum. Dis.* **32**:238–246 (1973).

249. Rowe, D., Heston, L., Metlay, J., and Miller, G., Identification and expression of a nuclear antigen from the genomic region of the Jijoye strain of Epstein–Barr virus which is missing in its non-immortalizing deletion mutant, P3 HRJ-1, *Proc. Natl. Acad. Sci. USA* **82**:7429–7433 (1985).

250. Royston, I., Sullivan, J. L., Periman, P. O., and Perlin, E., Cell-mediated immunity to Epstein–Barr virus-transformed lymphoblastoid cells in acute infectious mononucleosis, *N. Engl. J. Med.* **293**:1159–1163 (1975).

251. Sakamoto, K., Freed, H., and Purtilo, D., Antibody responses to Epstein–Barr virus in families with the X-linked lymphoproliferative syndrome, *J. Immunol.* **125**:921–925 (1980).

252. Sawyer, R. N., Evans, A. S., Niederman, J. C., and McCollum, R. W., Prospective studies of a group of Yale University freshmen. I. Occurrence of infectious mononucleosis, *J. Infect. Dis.* **123**:263–269 (1971).

253. Schiller, J., and Davey, F. R., Human leukocyte locus A HL-A antigens and infectious mononucleosis, *Am. J. Clin. Pathol.* **62**:325–328 (1974).

254. Schmader, K. E., Van der Horst, C. M., and Klotman, M. E., Epstein–Barr virus and the elderly host, *Rev. Infect. Dis.* **11**:64–73 (1989).

255. Schmitz, H., and Scherer, M., IgM antibodies to Epstein–Barr virus in infectious mononucleosis, *Arch. Ges. Virusforsch.* **37**: 332–339 1972).

256. Sculley, T. B., Walker, P. J., Moss, D. J., and Pope, J. H., Identification of multiple Epstein–Barr virus-induced nuclear antigens with sera from patients with rheumatoid arthritis, *J. Virol.* **52**:89–93 (1984).

257. Sheldon, P. J., Hemsted, E. H., Holborow, E. J., and Papamichael, M., Thymic origin of atypical lymphocytes in infectious mononucleosis, *Lancet* **2**:1153–1155 (1973).

258. Shope, T., and Miller, G., Epstein–Barr virus, heterophile responses in squirrel monkeys inoculated with virus-transformed autologous leucocytes, *J. Exp. Med.* **137**:140–147 (1973).

259. Sixbey, J. W., Lemon, S. M., and Pagano, J. S., A second site for Epstein–Barr virus shedding: The uterine cervix, *Lancet* **11**:1122–1124 (1986).

260. SIXBEY, J. W., NEDRUD, J. G., RAAB-TRAUB, N., HANES, R. A., AND PAGANO, J. S., Epstein–Barr virus replication in oropharyngeal epithelial cells, *N. Engl. J. Med.* **310:**1225–1230 (1984).

261. SIXBEY, J. W., SHIRLEY, P., CHESNEY, P. J., BUNTIN, D. M., AND RESNICK, L., Detection of a second widespread strain of Epstein–Barr virus, *Lancet* **2:**761–765 (1989).

261a. Sixbey, J. W., Lemon, S. M., and Pagano, J. S., A second site for Epstein–Barr virus shedding. The uterine cervix, *Lancet* **2:**1122–1124 (1986).

262. SMITH, H., AND DENMAN, A. M., A new manifestation of infection with Epstein–Barr virus, *Br. Med. J.* **2:**248–249 (1978).

263. SOHIER, R., *La Mononucleose Infectieuse*, Masson, Paris, 1943.

264. SOHIER, R., LEPINE, P., AND SAUTIER, V., Recherches sur la transmission experimentale de la mononucleose au singe et a l'homme, *Ann. Inst. Pasteur* **65:**50–62 (1940).

265. SPRUNT, T. P., AND EVANS, F. A., Mononuclear leukocytosis in reaction to acute infections (in infectious mononucleosis), *Bull. Johns Hopkins Hosp.* **31:**410–417 (1920).

266. STEEL, C. M., AND LING, N. R., Immunopathology of infectious mononucleosis, *Lancet* **2:**861–862 (1973).

267. STEVENSON, E. M. K., AND BROWN, T. G., Infectious mononucleosis: Preliminary investigation of a series of cases, *Glasgow Med. J.* **140:**139–150 (1943).

268. STORRIE, M. D., SAWYER, R. N., SPHAR, R. L., AND EVANS, A. S., Seroepidemiological studies of Polaris submarine crews. II. Infectious mononucleosis, *Milit. Med.* **141:**30–33 (1976).

269. STRAUGH, B., ANDREWS, L., MILLER, G., AND SIEGEL, N., Oropharyngeal excretion of Epstein–Barr virus by renal transplant recipients and other patients treated with immunosuppressant drugs, *Lancet* **1:**234–237 (1974).

270. STRAUS, S. E., Acute progressive Epstein–Barr virus infections, *Annu. Rev. Med.* **43:**437–449 (1992).

271. STRAUS, S. E., TOSATO, G., ARMSTRONG, G., LAWLEY, T., PREBLE, O. T., HENLE, W., DAVEY, R., PEARSON, G., EPSTEIN, J., BRUS, I., AND BLAESE, R. M., Persisting illness and fatigue in adults with evidence of Epstein–Barr virus infections, *Ann. Intern. Med.* **102:**7–16 (1985).

272. STRÖM, J., Infectious mononucleosis—Is the incidence increasing? *Acta Med. Scand.* **168:**35–39 (1960).

272a. SULLIVAN, J. L., AND WODA, B. A., X-linked immunodeficiency syndrome, *Immunodefic. Rev.* **1:**325–347 (1989).

273. SUMAYA, C. V., Primary Epstein–Barr virus infections in children, *Pediatrics* **59:**16–21 (1977).

274. SUMAYA, C. V., Serologic and virologic epidemiology of Epstein–Barr virus: Relevance to chronic fatigue syndrome, *Rev. Infect. Dis.* **13**(Suppl 1)**:**S19–25 (1991).

275. SUMMERS, W. P., GROGAN, E. A., SHEDD, D., ROBERT, M., LIU, C., AND MILLER, G., Stable expression in mouse cells of nuclear neoantigen following transfer of a 3.4 megadalton cloned fragment of Epstein–Barr virus DNA, *Proc. Natl. Acad. Sci. USA* **79:**5688–5692 (1982).

276. SYNDMAN, D. R., RUDDERS, R. A., DOREST, P., SULLIVAN, J. L., AND EVANS, A. S., Infectious mononucleosis in an adult progressing to fatal immunoblastic lymphoma, *Ann. Intern. Med.* **96:**737–742 (1982).

277. SYMPOSIUM, Chronic Epstein–Barr Virus Disease: A workshop held by the National Institute of Allergy and Infectious Diseases, *Ann. Intern. Med.* **103:**951–953 (1985).

278. TAYLOR, A. W., Effects of glandular fever in acute leukemia, *Br. Med. J.* **1:**589–593 (1953).

279. THOMSEN, S., *Studier over Mononucleosis Infectiosa*, Munksgaard, Copenhagen, 1942.

280. THORLEY-LAWSON, D. A., CHESS, L., AND STROMINGER, J. L., Suppression of *in vitro* Epstein–Barr virus infection: A new role for adult human T lymphocytes, *J. Exp. Med.* **146:**495–507 (1977).

281. TISCHENDORF, P., BALAGTAS, R. C., DEINHARDT, F., KNOSPE, W. H., MAYNARD, J. E., NOBLE, G. R., AND SHRAMEK, G. J., Development and persistence of immunity to Epstein–Barr virus in man, *J. Infect. Dis.* **122:**401–409 (1970).

282. TOMKINSON, B. E., MAZIARZ, R., AND SULLIVAN, J. L., Characterization of the T cell-mediated cellular cytotoxicity during acute infectious mononucleosis, *J. Immunol.* **143:**660–670 (1989).

283. TOBI, M., RAVID, Z., MORAG, A., CHOWERS, I., FELDMAN-WEISS, V., MICHAELI, Y., BEN-CHETRIT, E., SHALIT, M., AND KNOBLER, H., Prolonged atypical illness associated with serological evidence of persistent Epstein–Barr virus infection, *Lancet* **1:**61–64 (1982).

284. TOSATO, G., MAGRATH, I., KOSKI, I., DOOLEY, N., AND BLAESE, M., Activation of suppressor T cells during Epstein–Barr virus induced infectious mononucleosis, *N. Engl. J. Med.* **301:**1133–1137 (1979).

285. UNIVERSITY HEALTH PHYSICIANS AND P.H.L.S. LABORATORIES, A joint investigation: Infectious mononucleosis and its relationship to EB virus antibody, *Br. Med. J.* **4:**643–646 (1971).

286. VANDERMEER, R., LUTTERLOH, C. H., AND PILOT, J., Infectious mononucleosis: An analysis of 26 clinical and 340 subclinical cases, *Am. J. Med. Sci.* **210:**765–774 (1945).

287. VELTRI, R. W., McCLUNG, J. E., AND SPRINKLE, P. M., Epstein–Barr nuclear antigen (EBNA) carrying lymphocytes in human palatine tonsils, *J. Gen. Virol.* **32:**455–460 (1976).

288. VELTRI, R. W., McCLUNG, J. E., AND SPRINKLE, P. M., EBV antigens in lymphocytes of patients with exudative tonsillitis, infectious mononucleosis and Hodgkin's disease, *Int. J. Cancer* **21:**683–687 (1978).

289. VIROLAINEN, M., ANDERSON, L. C., LALLA, M., AND VON ESSEN, R., T-lymphocyte proliferation in mononucleosis, *Clin. Immunol. Immunopathol.* **2:**114–120 (1973).

290. VISINTINE, A. M., GERBER, P., AND NAHMIAS, A. J., Leucocyte transforming agent (Epstein–Barr virus) in newborn infants and older individuals, *J. Pediatr.* **89:**571–575 (1976).

291. WAHREN, B., ESPMARK, A., LANTORP, K., AND STERNER, G., EBV antibodies in family contacts of patients with infectious mononucleosis, *Proc. Soc. Exp. Biol. Med.* **133:**934–939 (1970).

292. WALTERS, M. K., AND POPE, J. H., Studies of the EB virus-related antigens of human leucocyte cell lines, *Int. J. Cancer* **8:**32–40 (1971).

293. WANG, P.-S., AND EVANS, A. S., Prevalence of antibodies to Epstein–Barr virus and cytomegalovirus in sera from a group of children in the People's Republic of China, *J. Infect. Dis.* **153:**150–152 (1986).

294. WECHSLER, H. F., ROSENBLUM, A. H., AND SILLS, C. T., Infectious mononucleosis: Report of an epidemic in an army post, *Ann. Intern. Med.* **25:**113–133, 236–265 (1946).

295. WEISS, L. M., MOVHED, L. A., WAMKE, R. A., AND SKLAR, J., Detection of Epstein–Barr val genomes in Reed–Sternberg cells of Hodgkin's disease, *N. Engl. J. Med.* **320:**502–506 (1989).

296. WERNER, J., HAFF, R. F., HENLE, G., HENLE, W., AND PINTO, C. A., Responses of gibbons to inoculation of Epstein–virus, *J. Infect. Dis.* **126:**678–681 (1972).

297. WEST, J. P., An epidemic of glandular fever, *Arch. Pediatr.* **13:**889–900 (1986).

298. WISING, P. J., A study of infectious mononucleosis (Pfeiffer's disease) from the etiological point of view, *Acta Med. Scand. [Suppl.]* **133:**1–102 (1942).

299. WOLF, H., HANSS, M., AND WILMES, E., Persistence of Epstein–Barr virus in the parotid gland, *J. Virol.* **51:**795–798 (1983).

300. WOLF, H., MOTZ, M., KUHBECK, R., SEIBEL, R., JILG, W., BAYLISS, G. J., BARRELL, B., GOLUB, E., ZENG, Y., AND GU, Y., Strategies for the economic production of Epstein–Barr virus proteins of diagnostic and protective value by genetic engineering: A new approach based on segments of virus-encoded gene products, *IARC Sci. Publ.* **63:**525–539 (1984).

301. WÖLLNER, D., Ueber die serologische Diagnose der infektiosen Mononucleose nach Paul–Bunnell mit nativen und fermentierten Hammel Erythrocyten. 2, *Immunitaets Forsch.* **112:**290–308 (1955).

302. WROBLEWSKI, F., Increasing clinical significance of alterations in enzymes in body fluids, *Ann. Intern. Med.* **50:**62–93 (1959).

303. YEFENOF, E., BAKACS, T., EINHORN, L., ERNBERG, I., AND KLEIN, G., Epstein–Barr virus (EBV) receptors, complement receptors and EBV infectibility of different lymphocyte fractions of human peripheral blood, *Cell Immunol.* **35:**34–42 (1978).

304. YOUNG, L. S., CLARK, D., SIXBEY, J. W., AND RICKINSON, A. B., Epstein–Barr virus receptors on human pharyngeal epithelia, *Lancet* **1:**240–242 (1986).

305. ZAJAC, B. A., AND KOHN, G., Epstein–Barr virus antigens, marker chromosomes, and interferon production in clones derived from cultured Burkitt tumor cells, *J. Natl. Cancer Inst.* **45:**399–406 (1970).

306. ZUR HAUSEN, H. H., CLIFFORD, P., HENLE, G., HENLE, W., KLEIN, G., SANTESSON, L., AND SCHULTE-HOLTHAUSEN, H., EB-virus DNA in biopsies of Burkitt tumors and anaplastic carcinomas of the nasopharynx, *Nature* **228:**1056–1057 (1970).

307. ZUR HAUSEN, H., DORREIER, K., EGGER, H., SCHULTE-HOLTHAUSEN, H., AND WOLF, H., Attempts to detect virus-specific DNA in human tumors, II. Nucleic acid hybridization with complementary RNA of human herpes group viruses, *Int. J. Cancer* **13:**657–664 (1974).

13. Suggested Reading

ABLASHI, D. V., FAGGIONI, A., KRUEGER, G. R. F., PAGANA, J. S., AND PEARSON, G. R. (EDS.), *Epstein–Barr Virus and Human Disease*, Humana Press, Clifton, NJ, 1988.

CARTER, R. L., AND PENMAN, H. G. (EDS.), *Infectious Mononucleosis*, Blackwell, Oxford, 1969.

DE THÉ, G., Epidemiology of Epstein–Barr virus and associated diseases in man, in: *The Herpesviruses*, Vol. 1 (B. ROIZMAN, ED.), pp. 25–103, Plenum Press, New York, 1982.

EPSTEIN, M. A., AND ACHONG, B. G. (EDS.), *The Epstein–Barr Virus*, Springer-Verlag, Berlin, Heidelberg, New York, 1979.

EPSTEIN, M. A., AND ACHONG, G. G. (EDS.), *The Epstein–Barr Virus. Recent Advances*, Wiley Publications, New York, 1986.

EVANS, A. S., Epidemiology of Epstein–Barr virus infection and disease, in: *The Human Herpesvirus* (A. J. NAHMIAS, W. R. DOWDLE, AND R. F. SCHINAZI, EDS.), pp. 172–181, Elsevier, New York, 1980.

HENLE, W., AND HENLE, G., Epstein–Barr virus, *Sci. Am.* **241:**48–59 (1979).

KLEIN, G., The Epstein–Barr virus, in *Herpesviruses* (A. S. KAPLAN, ED.), pp. 521–555, Academic Press, New York, 1973.

MILLER, G., Epstein–Barr virus and infectious mononucleosis, *Prog. Med. Virol.* **20:**84–112 (1975).

SCHLOSSBERG, D. (ED.) *Infectious Mononucleosis*, 2nd ed., Springer-Verlag, New York, 1988.

TURZ, T., *Proceedings of the Vth International Symposium on Epstein–Barr Virus and Associated Diseases*, IARC, Lyon, 1993.

CHAPTER 11

Viral Gastroenteritis

Albert Z. Kapikian

1. Introduction

Viral gastroenteritis, with diarrhea or vomiting or both as its major clinical manifestation, affects a broad segment of the population throughout the world. In the developed countries, it is a major cause of morbidity in infants and young children, whereas in the developing countries, it is a major cause of both morbidity and mortality in this same age group.[33] In the Cleveland Family Study, which included some 25,000 illnesses over an approximate 10-year period, infectious gastroenteritis (considered nonbacterial) was the second most common disease experience, accounting for 16% of all illnesses, averaging 1.5 episodes per person per year.[126] It is remarkable that the frequency of episodes was found to be quite similar (1.2–1.9 per person/year) in surveys carried out 20 or 30 years later.[224,393,394] In addition, a winter survey of a sample of US physicians engaged in pediatric practice revealed that gastrointestinal (GI) disturbance was the second most common disease for which children were brought to the physicians' offices, being responsible for 9.5% of all visits.[7] Although deaths from diarrheal illnesses are not a major public health problem in the United States today, scientists at the Centers for Disease Control (CDC) estimated that between 1979 and 1984, 209,000 infants and young children were hospitalized with a diarrheal illness.[247] This resulted in 877,800 inpatient days (4.2 days per hospitalization). In addition, they estimated that between 1973 and 1983, there were an average of 504 deaths annually from diarrheal illness in children 1 month to 5 years of age and that this comprised 2% of the postneonatal deaths in infants 1 to < 12 months of age. In addition, 79% of these deaths occurred before 1 year of age.

On the global scale, the impact of diarrheal disease is staggering; World Health Organization (WHO) statistics reveal that such illnesses account for a large proportion of the total reported deaths in many countries.[324,615] An estimate of the total number of diarrheal episodes during a single year (1975) in children less than 5 years of age in Asia, Africa, and Latin America revealed that over 450 million episodes of diarrhea would occur, and of these 1–4% would be fatal, resulting in the deaths of 5–18 million infants and young children in this 1-year period.[459] In a recent report on a strategy for disease control in developing countries, it was estimated that in Africa, Asia, and Latin America in a 1-year period (1977–1978), there would be 3–5 billion cases of diarrhea and 5–10 million deaths; diarrheas were ranked number one in frequency in the categories of disease and mortality.[593] In addition, from selected studies it was estimated that in Latin America, Africa, and Asia (excluding China), 744 million to 1 billion diarrhea episodes and 4.6 million deaths from diarrhea occur annually in children under 5 years of age.[516]

Despite the great importance of this problem, studies failed to reveal an etiologic agent for the majority of diarrheal illnesses.[100,652] However, the discovery of the 27-nm Norwalk virus in 1972 and discovery of the 70-nm rotavirus in 1973 paved the way for an abundance of new information about viral gastroenteritis.[43,299] The Norwalk virus group has been associated with gastroenteritis outbreaks occurring in school, community, and family settings affecting school-aged children, adults, family contacts, and some young children as well.[215,287,306] The Norwalk virus is now officially classified as a calicivirus following the recent cloning and characterization of its genome.[111,269,270,363] The 70-nm rotaviruses are associated with 35–52% of acute diarrheal diseases of infants and young children requiring hospitalization in developed countries; they occur with equal frequency as etiologic

Albert Z. Kapikian • Epidemiology Section, Laboratory of Infectious Diseases, National Institute of Allergy and Infectious Diseases, National Institutes of Health, Bethesda, Maryland 20892.

285

agents of severe acute gastroenteritis in this same age group in developing countries as well.[41,286] Rotaviruses have consistently been shown to be the single most important etiologic agents of severe diarrheal illness worldwide. This chapter deals primarily with the rotaviruses and the Norwalk group of viruses. Other viral agents that play a role in these syndromes but to a lesser degree are discussed at the end of the chapter.

2. Historical Background

Diarrhea in humans has been documented since pre-Hippocratic times.[324] Discoveries made in the past century in the fields of bacteriology and parasitology resulted in the elucidation of the etiology of only a portion of diarrheal illnesses. However, it soon became apparent that despite the bacteriological and parasitic discoveries, a significant proportion of epidemic and infantile gastroenteritis could not be ascribed to any etiologic agent.[508] By exclusion, it was assumed that many of these infectious gastroenteritides were caused by viruses. In 1945, Reimann, Price, and Hodges[448] described the transmission of gastroenteric illness to volunteers following administration by the respiratory route of nebulized bacteria-free filtrates of throat washings or fecal suspensions from gastroenteritis patients. Gordon et al.,[195] in 1947, induced an afebrile diarrheal illness in volunteers by the oral administration of bacteria-free fecal filtrates and throat washings from gastroenteritis patients; this infectious inoculum was designated the Marcy strain, since it was derived from pooled diarrheal stools obtained from two patients in a gastroenteritis outbreak at Marcy State Hospital near Utica, New York.

In 1948, Kojima et al.[315] induced gastroenteric illness in volunteers following oral administration of bacteria-free fecal filtrates derived from diarrhea cases in the Niigata Prefecture and other districts in Japan; serial passage was achieved, and short-term immunity was demonstrated on challenge with a single strain. Yamamoto et al.,[632] in 1948, also induced diarrheal illness in volunteers (and cats as well) with bacteria-free fecal filtrates derived from an epidemic of gastroenteritis in the Gumma Prefecture. Later, in 1957, Fukumi et al.[177] reported on the relationship between the Niigata Prefecture strain (derived from a pool of stools of several patients with diarrhea as described above and shown to have been infectious in volunteers) and the Marcy strain. In cross-challenge studies, Niigata and Marcy strains were found to be related.

In 1953, Jordan, Gordon, and Dorrance[276] reported the induction of a febrile gastroenteric illness in volunteers following the oral administration of a bacteria-free fecal filtrate derived from a patient with gastroenteritis who was enrolled in the Cleveland Family Study (FS) cited in Section 1; the agent, which was designated the FS strain, was serially passaged in volunteers. Cross-challenge studies in volunteers revealed that the Marcy and FS strains were not antigenically related; in addition, the incubation period of the illness induced by the two strains as well as the clinical manifestations were somewhat different.[276]

Studies on the etiology of severe infantile gastroenteritis also failed to reveal an etiologic agent in the majority of instances. However, in 1943, Light and Hodes[342] were able to induce diarrhea in calves with a filterable agent derived from diarrheal illness during outbreaks of such illness in premature or full-term nurseries. A calf stool that had been lyophilized and stored for over 30 years was later examined by electron microscopy (EM) and found to contain rotavirus.[248] Whether this represented a true calf rotavirus or the human strain passaged in calves could not be determined conclusively; in further studies, the agent was not infectious when administered to a gnotobiotic calf[248] (R. G. Wyatt, unpublished data).

In 1972, Kapikian et al.,[299] employing immune electron microscopy (IEM), discovered 27-nm particles in stool material derived from a gastroenteritis outbreak in Norwalk, Ohio (Fig. 1A). This technique, which had actually been described in 1939 but not used to its fullest potential, enables the direct observation of antigen–antibody interaction by EM.[9,10,14,293] The 27-nm particles were visualized in a stool filtrate derived from a volunteer who had developed illness following administration of the Norwalk agent[299]; the particle-positive specimen had also induced illness in other volunteers on serial passage.[129] The particles were recognized following reaction of the known infectious stool filtrate with a volunteer's convalescent serum.[299] Serological evidence of infection with this particle was also demonstrated by IEM in certain experimentally and naturally infected individuals, and from these and other data it was postulated that the 27-nm particle was the etiologic agent of the Norwalk outbreak.[299] Particles morphologically similar to Norwalk virus—such as the Hawaii, Montgomery County, Taunton, and Snow Mountain agents—were later detected from patients in outbreaks of gastroenteritis by IEM or conventional EM.[80,81,133,551]

In 1973, Bishop et al.,[43,45] employing thin-section EM, discovered 70-nm particles in duodenal biopsies obtained from infants and young children hospitalized with acute gastroenteritis in Australia. Subsequently,

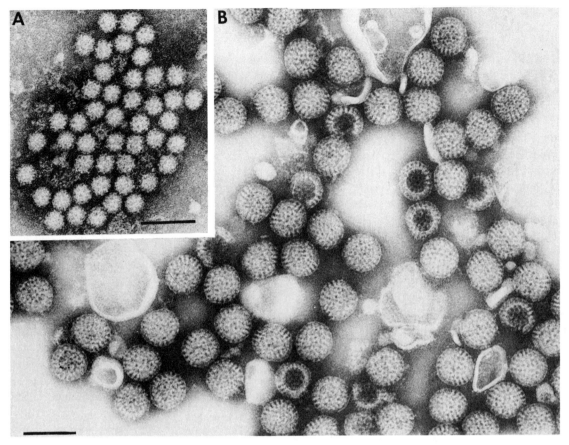

Figure 1. (A) A group of Norwalk virus particles observed after incubation of 0.8 ml of Norwalk stool filtrate (prepared from a stool of a volunteer administered the Norwalk agent) with 0.2 ml of a 1 : 5 dilution of a volunteer's prechallenge serum and further preparation for EM. The quantity of antibody on these particles was rated as 1+. Scale bar, 100 nm. From Kapikian *et al.*[299] (bar added). (B) Human rotavirus particles observed in a stool filtrate (prepared from a stool of an infant with gastroenteritis) after incubation with phosphate-buffered saline and further preparation for EM. The particles appear to have a double-shelled capsid. Occasional "empty" particles are seen. Scale bar, 100 nm. From Kapikian *et al.*[298]

these viruslike particles were found to be readily detectable in stool preparations by EM and became known by various names such as orbivirus, orbiviruslike, reoviruslike agent, duovirus, and infantile gastroenteritis virus, but now are officially designated as rotavirus.[160,163,297,370,436] In a relatively short time, laboratories from all over the world reported in rapid succession the presence of rotavirus in stool specimens from infants and young children with diarrheal illness, and it thus became apparent that this virus was indeed the long-sought major etiologic agent of diarrhea of infants and young children (Fig. 1B).

Two final notes of historical interest: In 1963, Adams and Kraft,[1] using thin-section EM to study intestinal tissue from mice infected with epizootic diarrhea of infant mice virus, described particles very similar to those first observed in 1973 in infants and young children in Austra-

lia. In 1969, Mebus *et al.*,[378] described the presence of reoviruslike particles in stools obtained from calves with a diarrheal illness. Later, both the mouse and calf viruses were found to be antigenically related to human rotavirus.[160,290,291] It is of interest that both the Norwalk virus and the rotavirus could have been discovered much sooner if the concept of "direct virology," using EM, had been applied to appropriate specimens.[304]

3. Methodology Involved in Epidemiologic Analysis

3.1. Sources of Mortality Data

Age-specific mortality data are available in the United States in the Vital Statistics Report prepared by the

National Center for Health Statistics (NCHS) of the Office of Health Research Statistics and Technology, Public Health Service, Hyattsville, Maryland. Specific sources from the NCHS include the National Hospital Discharge Survey (HDS) and the Multiple Causes of Death (MCD) mortality data.[246] The HDS represents a 0.5% sample of all discharges reported in the United States from selected hospitals. The MCD data includes all of the reported deaths in the United States. Since the clinical manifestations of viral diarrheas are not distinctive enough to permit differentiation from many other causes of diarrhea, and since the laboratory diagnosis of infection with viral gastroenteritis agents remains essentially a research tool, it is not yet possible to define the role of specific viruses in overall mortality from diarrhea.

On a worldwide scale, mortality data from diarrheal diseases are available in WHO and Pan American Health Organization publications. Vital statistics from around the world give the overall importance of diarrhea as a cause of death but, for the same reasons noted above, do not specify the role of the newly discovered viruses. Such information is inferred from data from epidemiologic studies. With the emergence of rotaviruses as a major cause of infantile diarrhea, it is generally acknowledged that this group of agents is of major importance as a cause of mortality from diarrheal diseases in the developing countries.

3.2. Sources of Morbidity Data

Since the clinical manifestations of viral gastroenteritis are indistinguishable from many other forms of gastroenteritis, it is not possible to obtain specific morbidity data without the aid of laboratory diagnosis, and as yet such diagnosis remains essentially a research tool. Most viral gastroenteritis morbidity data come from cross-sectional hospital-based studies involving (1) infants and young children admitted for diarrheal illness and (2) outbreaks of gastroenteritis. Such studies undoubtedly provide only a limited view of the total morbidity associated with these viruses, since they include only patients sick enough to come to the hospital or ill persons in selected outbreaks. However, limited morbidity data are available from several pediatric longitudinal studies and from various rotavirus vaccine efficacy field trials.

3.3. Serological Surveys for Rotaviruses and Norwalk Group of Viruses

Serological surveys have been carried out with the rotaviruses and the Norwalk agent to elucidate the prevalence of infection, the pattern of antibody acquisition by age, and the geographic distribution of these agents.[213,291,296,625,651] Rotavirus serology has relied heavily on the complement-fixation (CF) and enzyme-linked immunosorbent assay (ELISA) techniques.[290,291,651] However, with the successful cultivation of human rotaviruses, neutralization assays in tissue culture are now being performed.[490,574]

Until the development of second- and third-generation serological assays, large-scale surveys could not be carried out with Norwalk virus, since the only method available was IEM; this technique was not practical for such studies because it not only was very time-consuming but also required relatively large amounts of antigen, which was in short supply. However, the development of an immune adherence hemagglutination assay (IAHA), a radioimmunoassay-blocking test (RIA), and an ELISA-blocking test made it possible to perform serological surveys with Norwalk virus.[52,180,209,218,241,296] In addition, RIA-blocking and ELISA-blocking tests became available for the Snow Mountain agent as well as an ELISA-blocking test for the Hawaii virus.[349,557,558] A major drawback to serological studies with the Norwalk virus group of agents has been the need to rely on particle-positive stools from humans as the source of antigen. However, a major breakthrough occurred recently when the Norwalk virus genome was cloned, leading to the expression of its capsid protein in the form of viruslike particles in insect cells that were infected with a baculovirus recombinant (Fig. 2).[272] Because of this, a ready source of Norwalk virus antigen is available for the first time.

3.4. Laboratory Methods

3.4.1. Norwalk Group of Viruses

3.4.1a. Antigen Detection. Since this group of viruses has not yet been cultivated in any *in vitro* system, EM remains a mainstay for their recognition from stool specimens. Immune EM entails the reaction of antibody (such as that present in the patient's convalescent serum or in pooled immune serum globulin) with virus particles that may be present in the patient's stool.[282,304] Following centrifugation, the pellet (which contains the antigen–antibody complex) is prepared for examination by negative-stain EM. Antibodies directed against the particle are seen on the surface of the particle, and under appropriate conditions, antibodies induce aggregation of the particles. However, aggregation per se is not indicative of the presence of antibody, since nonspecific aggregation may occur. The presence of antibodies on the particle with or

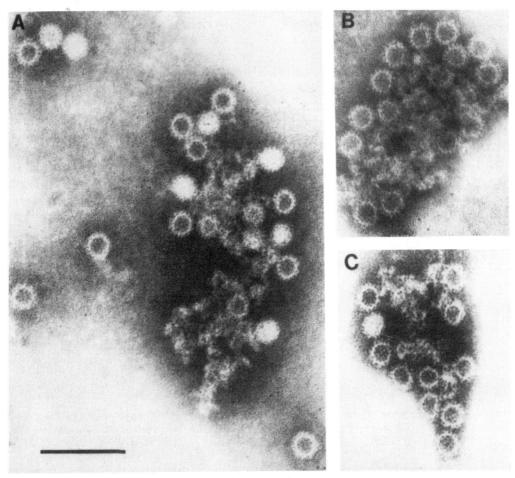

Figure 2. 27 nm viruslike particles observed by direct electron microscopy in 100 microliters of a 1:100 dilution of the baculovirus-expressed recombinant NV capsid protein after staining with 3% phosphotungstic acid. The viruslike particles were predominantly empty, readily seen in most fields, and similar to native NV, except that some had a double-shelled capsid appearance, a characteristic not reported with the native NV. Bar, 100 nm (for panels A, B, and C). From Green et al.[205]

without aggregation enables its differentiation from non-specific matter. The specificity of the reaction must be determined in additional IEM experiments, since stools contain a large amount of particulate matter that may cause considerable confusion. Thus, a serological test as outlined in Section 3.4.1b must be carried out with the putative particle as antigen and paired acute (or pre-) and postinfection sera as the source of antibody to determine whether an increase in antibody to the particle occurred. Such paired sera may be from the patient, another individual in the outbreak, or an individual with a response to a known agent. Such a study should routinely be done under code. The direct examination of stool material without addition of serum may also be carried out if sufficient antigen is present; however, identification or determina-

tion of the significance of such a particle should be carried out by IEM as outlined above.[282,304]

An RIA and an ELISA for detection of the Norwalk virus have been developed[52,180,209,218,241] that are more efficient than IEM. The test is essentially a research tool, since suitable reagents are not generally available. Recently, the availability of antibody to the recombinant Norwalk virus has led to the development of an ELISA for detection of the Norwalk virus.[272] Norwalk virus can also now be detected by reverse transcription and polymerase chain reaction methodology.[18,122,271] An IAHA has also been developed for the Norwalk agent,[296] but it is not efficient for its detection in clinical specimens. An RIA and an ELISA are available for the Snow Mountain agent and an ELISA has been developed for the Hawaii virus,[134,349,557] but EM and IEM remain the only methods for the detection and IEM the only method for the identification of the other members or putative members of the group.

The Norwalk group of agents do not produce illness in any experimental animal.[57,129,130,304,619,626] However, the Norwalk virus infects chimpanzees by the alimentary route as indicated by shedding of antigen and a serological response.[218,619]

3.4.1b. Serological Studies. As noted above, IEM remains a mainstay for studying this group of agents. In this technique, the stool material that contains the particle is incubated with a standard dilution of an acute, or pre-, and a postillness serum specimen, and the amount of antibody coating the particle is scored on a 0–4+ scale.[292,293,299,304] An example of a seroresponse to the Norwalk virus by a volunteer who developed illness fol-

lowing oral administration of Norwalk virus is shown in Fig. 3. The difference in the amount of antibody coating the Norwalk virus following its incubation with the volunteer's prechallenge serum or postchallenge serum is clearly evident. Since numerous spherical particles are detected in stool by EM, it is essential to establish the significance of these objects by IEM employing appropriate paired sera. After a viruslike particle has been detected, a seroresponse should be demonstrated as an initial step in associating this particle with infection.

The development of an RIA- and an ELISA-blocking test for measurement of Norwalk virus antibody has greatly facilitated epidemiologic study of this agent.[52,180,213–215] These assays are as efficient as IEM for detecting a seroresponse and are much more practical, since they are much less time-consuming and also require much less antigen and antibody. In addition, an IAHA for detection

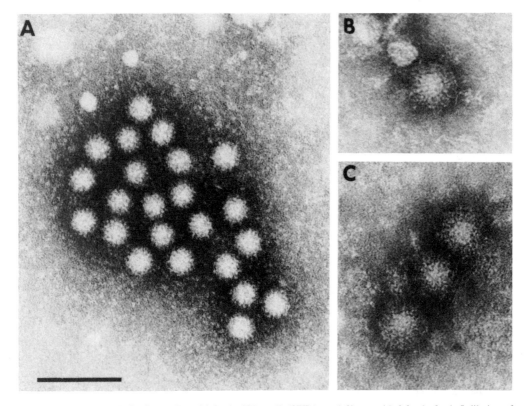

Figure 3. (A) An aggregate observed after incubation of 0.8 ml of Norwalk (8FIIa) stool filtrate with 0.2 ml of a 1:5 dilution of a volunteer's prechallenge serum and further preparation for electron microscopy.[293] This volunteer developed gastroenteritis following challenge with a second-passage Norwalk filtrate that had been heated for 30 min at 60°C.[129] The quantity of antibody on the particles in this aggregate was rated 1–2–2+, and this prechallenge serum was given an overall rating of 1–2+. (B) A single particle and (C) three single particles observed after incubating 0.8 ml of the Norwalk (8FIIa) stool filtrate with 0.2 ml of a dilution of the volunteer's postchallenge convalescent serum and further preparation for EM. These particles are very heavily coated with antibody. The quantity of antibody on these particles was rated 4+, and the serum was given an overall rating of 4+ also.[293] The difference in the quantity of antibody coating the particles in the prechallenge and postchallenge sera of this volunteer is clearly evident. The bar = 100 nm and applies to A, B, and C. From Kapikian *et al.*[293]

of Norwalk virus antibody has been developed.[296] It is not quite as efficient as either IEM or RIA but is, of course, more practical than IEM. The RIA, ELISA, and IAHA are essentially research tools because of the paucity of appropriate Norwalk antigen, which until recently has relied on particle-positive stools as the source of antigen. However, as noted earlier, the availability of baculovirus recombinant-expressed Norwalk capsid protein represents a major breakthrough in the study of these viruses. With this antigen as a precoat, a direct ELISA has been shown to be as specific, sensitive, and efficient as previously described methods for detection of serological evidence of Norwalk virus infection.[205,270,392] RIA- and ELISA-blocking tests have been developed for the Snow Mountain virus and an ELISA-blocking test has been developed for the Hawaii virus.[134,349,557]

3.4.2. Rotavirus

3.4.2a. Antigen Detection. Important progress has finally been made in the cultivation of the fastidious human rotaviruses directly in tissue culture from clinical specimens.[490,574] This was accomplished by pretreatment of the specimen with trypsin (10 μg/ml), incorporation of trypsin into the maintenance medium, use of roller tube cultures of MA104 cells, and incubation at 37°C. Virus can be isolated successfully from over 75% of fecal samples known to be rotavirus positive by other tests. Primary African green or cynomolgus monkey kidney cell cultures are even more efficient than MA104 cells for virus isolation.[236,404,598] However, such methods are not practical for most laboratories since tissue cultures are expensive and, in addition, require a sophisticated laboratory with experienced personnel. In the early studies, EM was the mainstay for detection of human rotavirus in stool specimens since these agents could not be propagated in cell cultures directly from clinical specimens. Although IEM was also employed, the addition of antibody was not essential, since in contrast to the Norwalk group, rotaviruses have a quite distinct morphological appearance, as shown in Fig. 1B.[298] Various simple and more readily available methods for rotavirus detection have been developed as practical alternatives to EM for diagnosis and for research epidemiologic studies. Table 1 shows numerous methods that have been described for rotavirus detection and presents a rating for each on a 1–4+ scale, with 1+ indicating a low degree of efficiency or practicality and 4+ a high degree. Though EM and IEM are very efficient, they are not practical, but nevertheless they remain the "supreme court" of rotavirus detection, since questionable results by any of the assays can usually be resolved by examination of the specimen by EM. In our laboratory, the method of choice at present is the confirm-

atory ELISA, since it is practical and efficient, does not require sophisticated equipment[62,303,304,642] and has a "built-in" control for nonspecific reactions. This is essential because fecal samples may bind nonspecifically to the precoat, which is comprised of rotavirus specific antibody.[94] Without a rotavirus-negative serum as the precoat in a companion well it is not possible to determine if the binding is specific.[441,446,560] Recently, the polymerase chain reaction (PCR) was applied to the detection of rotavirus and shown to be exquisitely sensitive.[631] Commercial kits are now available for the ELISA, latex agglutination (LA), and reverse passive hemagglutination assays.

The selection of technique will depend on the investigator's capabilities and experience and on the availability of appropriate reagents. Regardless of the methods employed, it is striking that most of the rotavirus detection methods do not require *in vitro* cultivation but rather employ "direct virology," a simple concept that as so often occurs in medical research could have been employed years ago but has been applied only relatively recently.[304]

3.4.2b. Serological Studies. Numerous assays have been described for detection of rotavirus antibody. Since in early studies the human rotaviruses could not be grown in cell cultures, the detection of serological responses relied on IEM, utilizing human rotavirus-positive stools as antigen.[298] However, this time-consuming method was soon superseded by the development of a CF test in which particle-rich human stools were used as antigen.[290,291,298] This method was limited by the paucity of human stools containing sufficient particles for the CF assay. It was soon discovered, however, that animal and human rotaviruses shared a common CF antigen, and thus animal strains could be used as substitute CF antigens for detection of infection with human rotavirus.[290,291] This antigenic relationship was of special importance, since certain animal rotaviruses such as a few calf strains, simian strain SA-11, and the "O" agent (from intestinal washings of sheep and cattle) had been propagated efficiently in cell cultures, and thus the quantity of CF antigen was virtually unlimited.[351,352,377,510] With the successful cultivation of human rotaviruses, neutralizing antibodies can now be measured by plaque reduction or by inhibition of cytopathic effect or virus yield in roller tube cultures.[151,236,260,301,302,574,575,624]

A variety of other serological assays have been developed to detect serological evidence of rotavirus infection, such as immunofluorescence (IF), IAHA, hemagglutination inhibition (HI), and ELISA. A summary of the relative efficiency and practicality of serological methods

**Table 1. Efficiency and Practicality of Methods Available
for Detection of Human Rotaviruses from Stool Specimens**[a]

Method	Efficiency[b]	Practicality for large-scale epidemiologic studies (assuming 4+ efficiency)[b]
Electron microscopy (EM)[38,44,62,127,157,298,385,410,417,533]	4+	1+
Immune electron microscopy (IEM)[62,127,297,298,617]	4+	1+
Immunosorbent EM[417]	4+	1+
Complement-fixation (CF) (conventional)[193,317,385,496,520]	1+	4+
Modified CF[656]	3–4+	2–3+
Human fetal intestinal organ culture [with immunofluorescence (IF)][628]	1+	0
Counterimmunoelectro-osmophoresis[202,384,417,520,521,563]	3–4+	4+
Fluorescent virus precipitin test[173,434,647]	4+	1+
Cell culture (cytopathic effect)[398,620]	1+	2–3+
Cell culture (with IF)[6,38,70,398,443,620]	1+	1+
Cell culture (with EM)[443,620]	1+	1+
Centrifugation onto cell culture (with IF)[21,72,396,554,556]	3–4+	1+
Cell culture (with trypsin, roller tubes, 37°C)[8,176,236,489,490,574,575,624]	3–4+	2+
Rescue in cell culture by genetic reassortment[208,217,621]	3+	1+
Gel diffusion[612]	1+	4+
Smears (with IF)[606]	1+	4+
Radioimmunoassay (RIA)[53,109,278,383,633]	4+	3–4+
Enzyme-liked immunosorbent assay (ELISA)[38,62,304,549,638,642,644,645,648]	4+	4+
Immune adherence hemagglutination assay (IAHA)[364,368]	3+	2–3+
Reverse passive hemagglutination assay (RPHA)[480,481]	3–4+	4+
RNA electrophoresis patterns in gels[149,183,252,316]	3+	1+
RNA electrophoresis patterns in silver-stained gels[240,444]	4+	1+
Enzyme-linked fluorescence assay (ELFA)[643]	4+	3+
Ultrasensitive enzymatic radioimmunoassay (USERIA)[235]	4+	3+
Self-contained enzymic membrane immunoassay (SCEMIA)[637]	4+	4+
Solid-phase aggregation of coupled erythrocytes (SPACE)[58]	3–4+	3–4+
Agglutination of coated latex beads (LA)[479,483,580]	3–4+	4+
3′ Terminal labeling of extracted RNA[97]	4+	1+
Dot hybridization[145,164]	4+	1+
Polymerase chain reaction (PCR)[631]	4+	3–4+

[a]From Kapikian et al.,[304] with additions.
[b]On a scale of 1 to 4+, where 1+ indicates a low degree of efficiency or practicality and 4+ indicates a high degree of efficiency or practicality.

for rotavirus is presented in Table 2. The ELISA-blocking and -binding assays are more efficient than CF for detecting rotavirus infection in infants less than 6 months of age and also in adults.[646,649] The CF and ELISA methods are comparable in efficiency for infants and young children 6–24 months of age; IF is almost as efficient as ELISA for rotavirus antibody detection.[646] ELISA appears to be the most efficient of the available methods; however, it is not quite as practical as CF in laboratories where the latter is used routinely for other agents. CF may be employed with confidence if its limits of efficiency are recognized and alternate tests are employed when needed. The ELISA has an additional advantage over CF in that the former permits the measurement of immunoglobulin classes.[373,649,650] In addition, early diagnosis may be made by ELISA, since a specific IgM response may appear as early as 5 or 6 days after onset of illness.[649] Of extreme importance in evaluating serological responses in the under-6-months age group of infants who possess naturally acquired maternal IgG antibodies was the development of an efficient and sensitive ELISA IgA antibody test.[121,344,345] This made it possible to determine if prior rotavirus exposure had occurred because IgA does not cross the placenta. The ELISA IgA has also been used to evaluate the level of coproantibodies as well as local salivary antibodies.[4,268,345] In addition, fecal IgA antibody levels appear to correlate with duodenal IgA antibody levels.[103]

There is no substitute as yet for neutralization assays when serotype-specific antibody is being measured. In this regard, the plaque reduction neutralization assay is

**Table 2.   Efficiency and Practicality of Methods Available
for Detecting Serological Evidence of Human Rotavirus Infection**[a]

Method	Efficiency[b]	Practicality for large-scale epidemiologic studies (assuming 4+ efficiency)[b]
Immune electron microscopy (IEM)[57,159,297,298,317]	4+	<1+
Complement-fixation (CF)[57,138,147,229,290,291,297,298,354]	3–4+	4+
Immunofluorescence (IF)[116,147,160,310,398,424,623,628]	4+	1+
Gel diffusion[612]	Not known	2+
Counterimmunoelectro-osmophoresis[101,384]	Variable	2+
Neutralization of calf rotavirus in cell culture[160,285,301,302,366,612]	2+	1+
Neutralization of human rotavirus or reassortant in cell culture tubes[301,302,621]	4+	1+
Plaque reduction neutralization of human rotavirus or reassortant in cell culture[301,302,621]	4+	1+
Radioimmunoassay (RIA)[53,488]	Not known	3+
Enzyme-linked immunosorbent assay (ELISA)[509,634,646,649–651]	4+	4+
ELISA-based neutralization assay[312a]	4+	2+
Inhibition (neutralization) of fluorescent foci[72,147,553,554,623]	4+	1+
Immune adherence hemagglutination assay (IAHA)[262,365,368]	3–4+	4+
Hemagglutination-inhibition (HI)[154,354,482,506,522]	1+	4+

[a]From Kapikian *et al.*,[304] with additions and modifications.
[b]On a scale of 1 to 4+, where 1+ indicates a low degree of efficiency or practicality and 4+ indicates a high degree of efficiency or practicality.

more sensitive than tube neutralization for detection of antibody, although the latter is somewhat more efficient for detecting a seroresponse (A. Z. Kapikian, unpublished data).

An important addition to dissecting the immune response was the development of a competition-solid-phase immunoassay that measures epitope-specific immune responses to individual rotavirus serotypes.[204,207,361,504] The test serum is used as the blocking reagent, whereas individual monoclonal antibodies serve as the detecting reagent.

4. Biological Characteristics

4.1. Norwalk Group of Viruses

This group of agents is composed of several viruses that: (1) were detected in the stool of patients with gastroenteritis; (2) have not been propagated *in vitro*; (3) are about 27 nm in diameter and do not have a distinctive morphological appearance by EM; (4) have a buoyant density in CsCl of 1.33–1.41 g/cm³; and (5) until recently, had not been characterized regarding the type of nucleic acid in the genome.[295,299,304,551,626] However, in an important breakthrough, two members of the group, first the Norwalk virus and then the Southampton virus (a Norwalk-like virus) were cloned and found to possess a positive-sense single-stranded RNA genome.[269,328,363] The Nor-

walk virus represents the prototype strain of this group of agents.[299]

Some of the biochemical or biophysical characteristics of the Norwalk virus have been elucidated in volunteer studies in which the ability of a treated stool filtrate to induce illness was determined. The Norwalk virus was found to be acid stable (pH 2.7 for 3 hr at room temperature), relative heat stable, and ether stable (20% ether at 4°C for 24 hr).[129]

The Norwalk group of viruses has been characterized antigenically by cross-challenge studies and by IEM or solid-phase (SP) IEM using antisera from infected individuals. Thus, four distinct serotypes are recognized: Norwalk, Hawaii, Snow Mountain, and Taunton viruses, which are numbered 1, 2, 3, or 4 based on historical precedence (Table 3).[328,339,340] By carrying out reciprocal IEM studies on viruslike particles and paired sera obtained from various outbreaks of gastroenteritis in Japan, investigators identified nine antigenic patterns (designated S1–9).[422]

Examination of paired sera from volunteers who had developed illness following challenge with the Norwalk, Hawaii, or Snow Mountain viruses by ELISA demonstrated that this assay could not discriminate serotypically among these strains.[348] This is important, epidemiologically, because it is likely that with shared antigens, the ELISA will not enable a specific etiology to be established. In contrast, a relatively high degree of specificity has been demonstrated by IEM and SPIEM techniques.[328,339,422]

Table 3. Characterisitics of the Norwalk and Norwalk-like Viruses[a,b]

Virus	Size (nm)	Buoyant density in cesium chloride (g/cm³)	Growth in cell culture	Administration[c] of virus induces illness		Particles detected by	Serological studies by	Antigenic relationships [suggested serotype number[(328)]][d]
				Humans	Animal(s)			
Norwalk[(18,122,129,130,180,218,241,271,295,299,349,618)]	27 × 32[e]	1.38–1.41	No	Yes	No	IEM, RIA, IAHA, ELISA, PCR	IEM, RIA, IAHA, ELISA	Distinct [1]
Hawaii[(131,299,551,557,618)]	26 × 29[e]	1.37–1.39	No	Yes	No	IEM, ELISA	IEM, ELISA	Distinct [2]
Montgomery County[(299,551,618)]	27 × 32[e]	1.37–1.41	No	Yes	No	IEM	IEM	Related to Norwalk virus
Snow Mountain[(67,134,371)]	25–26	1.33–1.34	No	NT	NT	IEM, RIA, ELISA	IEM, RIA, ELISA	Distinct [3]
Taunton[(80,81,339)]	32–34	1.36–1.41	No	NT	NT	EM	SPIEM	Distinct [4]

[a]From Kapikian.[(282)]

[b]Also designated small round structured viruses (SRSVs). Reference numbers listed in parentheses. IEM, immune electron microscopy; RIA, radioimmunoassay; IAHA, immune adherence hemagglutination assay; ELISA, enzyme-linked immunosorbent assay; SPIEM, solid-phase immune electron microscopy; PCR, polymerase chain reaction; NT, not tested.

[c]By alimentary route.

[d]By IEM, SPIEM, or virus challenge studies or a combination. There are two other serotyping schemes: one describes four serological groups of these viruses by SPIEM: (1) "SRSV UK1" = Taunton virus; (2) "SRSV UK2" = Norwalk virus; (3) "SRSV UK3" = Hawaii virus; and (4) "SRSV UK4?" = Snow mountain virus = SRSV Japan 9.[(340)] Another describes nine antigenic patterns of SRSV from Japan designated SRSV S1–9.[(422)]

[e]Shortest × longest diameter.

The classification of the Norwalk virus is of special interest because this fastidious agent was initially considered to resemble a parvovirus on the basis of certain properties.[129] However, when it was found later to possess a single structural protein of about 65,000 Da and subsequently, when it was shown that patients with documented calicivirus infections developed serological responses to the Norwalk virus, it seemed almost certain that the Norwalk virus was a calicivirus.[75,107,212,215,282,492,493] This was in spite of the lack of knowledge regarding its nucleic acid content.

Recently, molecular biological studies established that the Norwalk virus was indeed a calicivirus, when it was cloned and sequenced and found to have a genome organization similar to that of various established caliciviruses.[270,272,288,328] It possesses a positive-sense polyadenylated single-stranded RNA of 7642 nucleotides (excluding the poly A tail) that was predicted to encode three open reading frames (ORFs). The second ORF was expressed in insect cells that had been inoculated with a baculovirus recombinant containing ORF2; as noted earlier, empty particles were formed by self-assembly of the expressed protein (Fig. 2).[272] The Southampton virus, another Norwalk-like virus morphologically, which was also cloned, sequenced, and expressed, is also classified as a calicivirus on the basis of its genome organization.[328]

It should also be noted that an earlier attempt at a tentative interim classification of small round fecal viruses based on morphological appearance by EM, two broad groups of viruses were distinguished: featureless or structured, as shown in Table 4.[80] The featureless group included the picornaviruses and the parvoviruslike agents. (The latter included the cockle, Wollan, Ditchling, and Parramatta viruses, each of which had been detected in the feces of individuals with gastroenteritis.) The structured group included the small round structured viruses (SRSVs) composed of the Norwalk, Hawaii, Montgomery County, and Taunton viruses. Other structured viruses included the classical caliciviruses (i.e., those with the characteristic deep surface hollows that are not observed in the Norwalk virus group), astroviruses, minireoviruses, and the Otofuke, Sapporo, and Osaka agents.[386,421,523,545]

Evidence for the etiologic role of the Norwalk group of agents in gastroenteritis differs for each member. In one study, the Norwalk agent was shown to induce illness in 30 (58%) of 52 volunteers[618]; serological evidence of infection has been demonstrated in most volunteers who developed illness as well as in certain individuals who developed illness during the original outbreak.[218,296,299] Further evidence of an etiologic association was the demonstration in volunteers of a close temporal relationship between virus shedding and illness, with maximal shedding at the onset of experimental illness.[550] Only short-term immunity characteristically occurs in volunteers who develop illness after initial challenge[129,426,618]; prechallenge serum antibody titers did not correlate with susceptibility to illness.[426] Volunteers not only developed illness following administration of stool filtrates containing the Hawaii, Montgomery County, or Snow Mountain agents,[397,405] but also demonstrated seroresponses to the challenge virus by IEM.[397,618]

4.2. Rotaviruses

Human rotavirus has been detected in stools of about 35–52% of infants and young children hospitalized with acute gastroenteritis and much less often in older children and adults with this disease.[61,63,114,215,297,304,317,321,341] Rotaviruses have also been found in stools of numerous animals.[473] They have been associated with a diarrheal illness in the bovine calf,[66,377,378,610] infant mouse,[399] piglet,[333,376,453,473,611] foal,[158,569] lamb,[514] young rabbit,[73,435,442,555] monkey,[351,532] newborn deer,[567] newborn antelope,[447] young chimpanzee,[16] young gorilla,[16] young turkey,[32,374,375] chicken,[275,375] young goat,[499] young kitten,[511] young dog, and buffalo calf[401,460,461,473]; pneumoenteric illness has been found in the newborn impala,[153] newborn addax,[153] and newborn gazelle.[153] The offal ("O") agent was derived from mixed intestinal washings from abattoir waste.[352] Until the relatively recent breakthrough that enables the cultivation of most rotaviruses, as described earlier, only human rotavirus Wa, the calf, piglet, and monkey isolates, and the "O" agent had been successfully propagated efficiently in cell culture. With the exception of the SA-11 strain of simian rotavirus, canine rotavirus, and the "O" agent, each of the others has been associated with naturally occurring diarrheal (or pneumoenteric, as already noted) illness in newborns of each respective group.

As shown in Fig. 1B, rotaviruses have a distinctive morphological appearance. Complete particles possess a double-shelled capsid and measure about 70 nm in diameter; single-shelled particles measure about 55 nm in diameter, and within this shell is a third layer, the core, which contains the 11 segments of double-stranded RNA and has a diameter of about 37 nm.[286,371,439a] Cryoelectron microscopic studies demonstrate the presence of 60 spikes about 10–12 nm in length that protrude from the outer capsid.[439a,440] The term rotavirus comes from the Latin word *rota*, meaning wheel, and was suggested because the sharply defined circular outline of the outer capsid gives the appearance of the rim of a wheel placed on short

Table 4. Comparison of Human Caliciviruses with Other Small, Round Particles Shed in Feces by Patients with Acute Gastroenteritis[a]

Virus group	Size (nm)	Characteristics of certain strains in indicated category					Important as cause of:	
		Morphology (previous interim classification)	Nucleic acid and genome organization	Viral proteins	Density in CsCl g/cm³	Growth in cell culture	Severe infantile gastroenteritis	Epidemic gastroenteritis of children and adults
Caliciviruses								
Norwalk group[212,215,295,299,337]	26–32	Spherical, without sharply defined edge; suggestion of surface hollows (structured: SRSV[b])	RNA, calicivirus genome	1 structural protein (58–62kDa)	1.33–1.41	No	No	Yes
UK 1-4 and Japanese virus (also designated as HuCV strains)[77,106,210,327,548]	≈32–35	Prominent surface hollows and may have six-pointed surface star with central hollow (structured: calicivirus)	RNA	1 structural protein (62 kDa)	1.37–1.38	Abortive replication	No	No
Possible caliciviruses								
Otofuke, Sapporo, and Osaka strains[313,421,545]	35–40	Surface projections (structured: SRSV)	Not determined	Not determined	Not determined	No	No	No
Astrovirus[210,326,327]	27–34	Spherical; five or six pointed star without a central hollow (in ~10%); triangular surface hollows; smooth edge (structured: astrovirus)	RNA	Reports vary (90-kDa precursor cleaved to 31-,39-,20-kDa proteins)	1.33–1.42	Yes	No	No
Parvoviruslike viruses: "W,"[12,98,293] Cockle,[11,13,565] Ditchling,[12] Parramatta[92]	23–26	Spherical; smooth outer edge; smooth surface [featureless: probable (parvovirus)]	DNA (cockle virus)	Not determined	1.38–1.40	No	No	No
Small round viruses (SRVs)[11,40,161,314]	23–26	Spherical; smooth outer edge; smooth surface (featureless: SRV)	Not determined	Not determined	Not determined	No	No	No

[a]From Kapikian et al.[287]
[b]SRSV, small round structured virus.

spokes radiating from a wide hub.[155,160,163] Rotaviruses resemble orbiviruses and reoviruses morphologically but differ in their fine structure.[625] Rotaviruses have a density of 1.36 g/cm³ in cesium chloride and are ether stable but acid labile.[286]

The rotavirus genome is comprised of 11 segments of double-stranded RNA.[371] The migration patterns of the RNA segments of rotaviruses as determined by polyacrylamide-gel electrophoresis (PAGE) are of importance not only in the biophysical characterization of these agents but also as epidemiologic probes; they have served as one of the methods of differentiating human and animal rotaviruses as well as various human rotavirus strains.[452] The term "electropherotype" was applied to this method of distinguishing strains.[152] The distinctive RNA migration pattern of rotaviruses has also been used for detection and identification of rotavirus strains from clinical specimens.[149]

Each of the 11 rotavirus genes has now been sequenced.[371] As shown schematically in Fig. 4, the outermost surface of the virion is composed of two proteins, VP7 and VP4. VP4 takes the form of 60 surface projections (spikes) extending about 12 nm from the outer surface (VP1) of the capsid.[439a,440] VP6, which represents over 51% of the virion, makes up the inner layer of the double capsid.[371] Within the inner capsid is the core, which has been described as a third shell, comprised of VP2.[371,439a] The core encloses the genome as well as the viral proteins VP1 and VP3, designated recently as subcore proteins.[439a] Each of the six structural proteins is encoded by a single gene. In addition to the six structural proteins, rotaviruses possess five nonstructural proteins that are found only in infected cells and not in the mature virion.[371] Double-shelled ("complete" or "smooth") particles have a density of 1.36 g/cm³ in CsCl and a sedimentation coefficient of 520–530 S.[371] Single-shelled (rough) particles have a density of 1.38 g/cm³, whereas "empty" particles that have been penetrated by negative stain have a density of approximately 1.29–1.30 g/cm³.[371] Core particles have a density of 1.44 g/cm³ in CsCl.[36,371]

Since rotaviruses share certain properties with the reoviruses and orbiviruses and yet are distinct serologically and in certain biophysical aspects, they have been officially classified as a new genus in the family *Reoviridae*.[370] This family now contains nine genera: orthoreovirus (sigla from respiratory enteric orphan), orbivirus, rotavirus, Coltivirus (sigla from Colorado tick fever), phytoreovirus, Fijivirus, Cypovirus (sigla from cytoplasmic polyhedrosis), aquareovirus, and oryzavirus. Certain members of the first three genera infect humans, whereas the phytoreoviruses and Fijiviruses infect plants, the cytoplasmic polyhedrosis viruses infect insects, and the aquareoviruses infect fish.[370,477,478] Antigenically, rotaviruses are distinct from reoviruses by CF, IEM, and neutralization or RIA, from selected orbiviruses by CF, and from bluetongue virus (an orbivirus) by IEM.[116,278,290,291,298]

Rotaviruses have three major antigenic specificities: group, subgroup, and serotype.[170,289,371] Currently, seven groups (A–G) have been identified.[64] In diagnostic as-

ROTAVIRUS GENE PRODUCTS

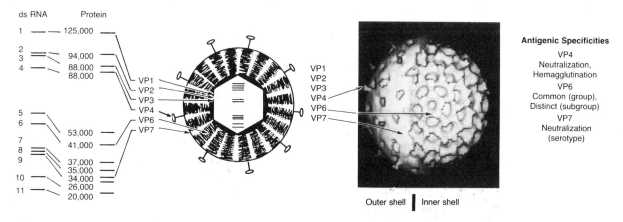

Figure 4. The figure on the left is a schematic representation of the rotavirus double-shelled particle. The figure on the right (from Prasad *et al.*[440]) shows surface representations of the three-dimensional structures of a double-shelled particle (on the left half) and a particle (on the right half) in which most, if not all, of the outer shell and a small portion of the inner shell mass have been removed. The apparent molecular weights derived from sodium dodecyl sulfate PAGE for the SA-11 strain are shown for individual proteins.[371] From Kapikian and Chanock.[286]

says, VP6 (encoded by the sixth gene) is the predominant group antigen. Most epidemiologically important rotaviruses that infect humans and animals share group A specificity. VP6 also mediates subgroup specificity, which continues to be an important epidemiologic marker, especially when serotyping of strains is not feasible. Most group A rotaviruses belong to either subgroup I or II.[19,253] The group A rotaviruses are further divided into serotypes according to the specificity of VP7 (encoded by gene 7, 8, or 9 depending on the strain) and VP4 (encoded by gene 4).[27,165,253,258,277,390] VP7 and VP4 each induce neutralizing antibodies and the separation into serotypes is made by neutralization assay in cell culture.[253,258] At present, 14 VP7 serotypes (also termed G serotypes because VP7 is a glycoprotein) have been described in humans or animals as shown in Table 5.[253] The sharing of VP7 serotype specificity among various human and animal rotaviruses is readily apparent.

A serotyping system for VP4 specificity has recently been suggested. At present, seven VP4 serotypes (termed "P" serotype because VP4 is protease sensitive) and four subtypes have been described on the basis of neutralization as shown in Table 6.[253,500] The genetic homology of the various VP4s has also led to a classification system based on genotype, but the "P" serotype designation is reserved for VP4 neutralization specificity. It should be noted that alphabet letters designate group, Arabic numbers serotype, and Roman numerals subgroup. In addition, a binary system of rotavirus serotype classification is in place to indicate the role of VP4 and VP7 neutralization specificities.[257,258]

As noted above, the rotaviruses that share the group antigen are classified as group A rotaviruses, whereas the others that do not share this antigen are designated as non-group-A rotaviruses.[64] These non-group-A rotaviruses are also known as pararotaviruses, rotaviruslike viruses, novel rotaviruses, antigenically distinct rotaviruses, and adult diarrhea rotaviruses.[64,262] They have been recovered from humans, calves, pigs, lambs, rats, ferrets, and birds.[64,594] The group B rotaviruses have been responsible for large outbreaks of gastroenteritis in China, predominantly in adults.[64,90,262,594] However, the group B rotaviruses have been detected only rarely outside of China.

The human rotaviruses were considered to be rather fastidious agents, defying efficient propagation in any cell- or organ-culture system. A few strains had been cultivated to a limited extent with only a small percentage of cells exhibiting evidence of infection.[153,620,628] However, human rotavirus Wa, a VP7:1, VP4:1A virus, was

Table 5. Serotypic and Genotypic Classification of Group A Rotavirus VP7 and Subgroup Specificities[a]

VP7 (G) serotype[b]	Human rotavirus strains (subgroup)	Animal rotavirus strains (subgroup) [species]
1	Wa, KU, D, M37, RV-4, WI79, K8 (II)	C60, C86 (I) [pig]; T449 (I) [cow]
2	DS-1, S2, KUN, RV-5, 1076 (I)	C134 (I) [pig]
3	P, MO, YO, RV-3, Ito, Nemoto, WI78, McN (II); AU-1, AU228, Ro1845, HCR3 (I); 0264 (I and II)	SA11 (I) [vervet monkey]; MMU18006 (I) [rhesus monkey]; CU-1 K9, RS-15 (I) [dog]; TAKA, Cat2, Cat97, FRV-1 (I) [cat]; H-2 (not I or II), FI-14 (I and II) [horse]; CRW-8, C176 (I) [pig]; R-2 (II), Ala, C11 (I) [rabbit]; Eb (I), EW (not I or II) [mouse][c]
4	ST3, ST4, VA70, Hosokawa, Hochi, 57M (II)	Gottfried, SB-1A (II), BEN-144 (?), SB-2 (I) [pig]
5	IAL28 (II)	OSU, EE, TRF-41 (I) [pig]; H-1 (I) [horse]
6	PA151, PA169 (I)	NCDV, UK, RF, WC3, Q17, OK, ID (I), B641, C486 (?) [cow]
7	None	Ch2 (not I or II) [chicken]; Ty1[d] (not I or II) [turkey]; PO-13 (I) [pigeon]; 993/83 (not I or II) [cow]
8	69M, B37, HAL1271, HAL1166 (I), PA171 (II)	678, J2538, A5 (I) [cow]
9	WI61, F45, 116E (II), Mc323 (I)	ISU-64 (I) [pig]
10	A64 (II), Mc35, I321 (I)	B223, V1005, KK3, 61A, Cr (I) [cow]; Lp14 (?) [sheep]
11	None	YM, A253.1 (I) [pig]
12	L26 (I)	None
13	None	L338 (I) [horse]
14	None	F123 (I) [horse]

[a]Adapted from Hoshino and Kapikian,[253] with additions from Gerna et al.,[189] Gouvea et al.,[199] and Isegawa et al.[267]
[b]VP7 (G) serotype as determined by reciprocal cross-neutralization; in addition, it should be noted that VP7 genotype as determined by comparative amino acid sequence analysis and/or nucleic acid hybridization has been used widely as a means of distinguishing among rotavirus strains and where evaluated, has correlated with serotyping differences by neutralization assay. Thus, genotyping is now frequently used as a proxy method for neutralization for VP7 (G) serotyping.[191,206]
[c]Human rotaviruses with neither subgroup I or II specificity have also been reported but without serotype designation.
[d]Recent amino acid sequence anaysis of the VP7-encoding gene of Ty1 suggests that it may not belong to this serotype.

Table 6. Serotypic and Genotypic Classification of Group A Rotavirus VP4[a]

VP4(P) serotype[b]	VP4 genotype[c]	Human rotavirus strains (G serotype)	Animal rotavirus strains (G serotype) [species]
1A	8	KU, Wa (1); P, YO, MO (3); VA70, Hochi, Hosokawa (4); WI61, F45 (9)	None
1B	4	DS-1, S2, RV-5[d] (2); L26[d] (12)	None
2A	6	M37 (1); 1076 (2); McN, RV-3[d] (3); ST3 (4)	None
2B	6	None	Gottfried (4) [pig]
3A	9	K8 (1); AU-1[d] (3); PA151[d] (6)	FRV-1[d]; Cat2[d] (3) [cat]
3B	14	Mc35 (10)	None
4	10	57M[d] (4); 69M (8)	None
5A	3	HCR3[d], Ro1845 (3)	CU-1, K9 (3) [dog], Cat 97(3) [cat]
5B	3	None	MMU18006 (3) [rhesus monkey]
6	1	None	NCDV, C486, J2538 (6); A5[d] (8) [cow]; SA11-4fM (3) [vervet monkey]
7	5	None	UK, B641, IND[d], OK[d] (6); 61A[d] (10) [cow]
8	11	116E[d] (9); I321[d] (10)	B223, B-11, A44[d], KK3[d], Cr[d] (10) [cow]
9	7	None	CRW-8 Ben-307 (3); BMI-1, SB-1A (4); OSU, TFR-41 (5); YM, A253.1 (11) [pig]; H1[d] (5) [horse]
10	16	None	Eb (3) [mouse]
11	18	PA169 (6); HAL1166 (8)[d]	None
?	2	None	SA11 (3) [vervet monkey]
?	12	None	H-2, FI-14[d], FI-23[d] (3) [horse]
?	13	None	MDR-13 (3/5) [pig]
?	15	None	Lp14 (10) [sheep]
?	17	None	993/83 (7)[d] [cow]
?	19	None	L338 (13) [horse]

[a]Adapted from Hoshino and Kapikian,[253] with additions from Gerna et al.,[189] Gorziglia et al.,[197] Hardy et al.,[234] Isegawa et al.,[267] Nagakomi et al.,[405] Sereno and Gorziglia,[500] and Taniguchi et al.[542]

[b]VP4(P) serotype as determined by reciprocal or one-way cross-neutralization.

[c]VP4 genotype as determined by comparative amino acid sequence analysis and/or nucleic acid hybridization (from Estes and Cohen,[150] with additions).

[d]Not tested by neutralization; relationship suggested by amino acid sequence and/or nucleic acid hybridization analyses.

adapted to grow efficiently in primary African green monkey kidney (AGMK) cells following 11 serial passages in gnotobiotic piglets.[622,623] Pretreatment of this porcine-grown human rotavirus strain with trypsin was required for optimal growth in AGMK cells; low-speed centrifugation of the virus inoculum onto cell cultures was also employed.

Noncultivatable human rotaviruses were successfully rescued following mixed infection of cell cultures with noncultivatable human rotavirus and cultivatable bovine rotavirus, and application of various selective pressures.[208] The cultivatable reassortants had mixed genotypes but also had the neutralization specificity of human rotavirus. Such arduous methods are no longer required to grow human rotaviruses in cell culture directly from clinical specimens since efficient methods are now available, as described in Section 3.4.2.

Experimentally, human rotavirus induces a diarrheal illness in various newborn animals including gnotobiotic calves, gnotobiotic and conventional piglets, rhesus monkeys, gnotobiotic lambs, and suckling mice[31,473]; in addition, subclinical infections were induced in newborn puppy dogs.[472,566] In studies in China, a severe diarrheal illness was induced experimentally in nonhuman primates, the *Tupaia belangeri yunalis*, following administration of a human rotavirus by the alimentary route.[425] Particle-positive stools from calves have been an important source of human rotavirus for biophysical and serological studies following experimental infection.

Firm evidence exists for the association of rotavirus with gastrointestinal illness. The virus has been consistently detected significantly more often in stools from patients 6–24 months old with gastroenteritis than in those without gastroenteritis[63,114] in both hospitalized patients and outpatients. Serological evidence of rotavirus infection has also been observed significantly more often in hospitalized gastroenteritis patients than in hospitalized "controls."[297] The virus is detectable predominantly during the acute phase of illness.[114] In addition, in early studies illness was induced in volunteers with a stool filtrate containing a VP7 serotype 1 strain of human rotavirus.[301,302]

5. Descriptive Epidemiology

5.1. Norwalk Group of Viruses

5.1.1. Incidence and Prevalence Data. Specific incidence data for the Norwalk group in the United States are not available. An estimate of the importance of this group of agents is suggested from various sources. Infectious gastroenteritis was the second most common disease experience in the Cleveland Family Study over an approximate 10-year period,[126] and the Norwalk group was probably associated with some of these illnesses, especially in adults. About one third of all outbreaks of nonbacterial gastroenteritis studied are associated with Norwalk virus infection,[214,215] and in a recent estimate, 65% of nonbacterial gastroenteritis outbreaks were provisionally associated with the Norwalk or a related virus.[306]

It is probable that other members of the group are also responsible for a portion of these illnesses, but appropriate tests are not available for the entire group. In children in developed countries, the Norwalk group is probably not an important cause of severe gastroenteritis. Thus, 27-nm virus particles were present in fewer than 2% of infants and young children hospitalized with diarrhea at the Children's Hospital, Washington, DC, a value not significantly different from that observed in controls,[63] nor was serological evidence of Norwalk virus infection detected in selected diarrhea patients from this study.[296] Similar findings with regard to virus detection were made in Japan in a cross-sectional study of pediatric patients hospitalized with diarrheal illness.[321]

In a prospective study of 28 families over a 2-year period, each of 14 families experienced one outbreak of nonbacterial gastroenteritis in which no enteropathogen could be identified by conventional assay; two were associated with the Norwalk virus.[438] None of 28 infants enrolled at birth developed serological evidence of Norwalk virus infection in the first 2 years of life.

In studies in which the recombinant Norwalk virus was used as antigen, Norwalk virus infections were detected in 49% infants and young children studied retrospectively over a 2-year period in a longitudinal study in Finland.[338] This was an unexpected finding because in other studies in developed countries the Norwalk virus did not appear to readily infect individuals in this age group. It should be noted, however, that illnesses could not be associated with the responses.

Incidence data are available from several longitudinal studies outside the United States. In Bangladesh, the incidence of Norwalk virus infection in children less than 5 years of age as determined by a significant RIA serum antibody rise was 29 per 100 children per year; in addition, it was estimated that 1–2% of the diarrheal episodes in these children (who have 5.6 such episodes per year) were caused by Norwalk virus.[49] Moreover, only 1 of 31 children less than 10 years of age who underwent treatment for dehydrating diarrhea and who did not have rotavirus or bacterial pathogens in their stools had serological evidence of Norwalk virus infection.[49] In a similar study in the San Blas Islands in Panama, 35% of the children under 5 years of age developed a seroresponse to Norwalk virus, and it was suggested that the Norwalk virus infection was associated with mild gastroenteritis in this age group.[466]

Finally, in a longitudinal study of infants and children in three northern communities in Canada, the incidence of Norwalk infection was highest (0.15 infection per child per year) in neonates in the only community with relatively unsafe water supplies.[227]

Prevalence data for Norwalk virus infections are available for various parts of the world. By IAHA, the acquisition of antibody to Norwalk virus and rotavirus was compared in infants and young children in the metropolitan Washington, DC, area, young adults at the University of Maryland, and adults in the metropolitan Washington, DC, area.[296] As shown in Fig. 5, the pattern of antibody acquisition differed markedly for these two viruses. There was a gradual acquisition of Norwalk antibody beginning slowly in childhood and accelerating in the adult period, so that by the fifth decade of life, 50% of the adults possessed Norwalk antibody. In contrast, rotavirus antibody was acquired early in life so that by the 36th month of age, over 90% had such antibody. The gradual acquisition of Norwalk antibody is similar to that observed with hepatitis A virus and certain rhinovirus serotypes in comparable populations.[230,535,536] This pattern of antibody acquisition in a major metropolitan area of a developed country suggests that Norwalk virus is not an important cause of gastroenteritis in infants and young children but rather is associated most often with such illness in older persons. A comparison of the prevalence of IAHA Norwalk and rotavirus antibody in a welfare institution for homeless but otherwise normal children

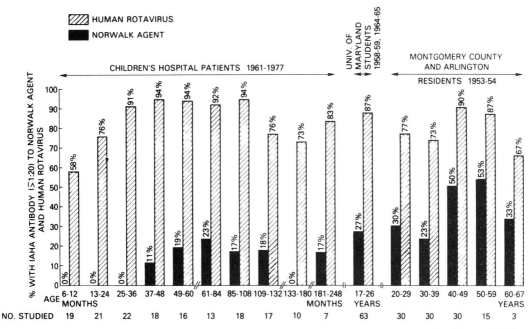

Figure 5. Prevalence of antibody to Norwalk agent and rotavirus by IAHA in three groups. From Kapikian *et al.*[296]

yielded a pattern similar to that just described.[296] In a small study, IAHA antibody to Norwalk agent was also detected in infants, children, and adults in Bangladesh, but the Norwalk antibody prevalence was markedly less than that of rotavirus.[296]

The prevalence of Norwalk antibody was studied in individuals from various parts of the world with the RIA-

blocking assay.[213] As shown in Fig. 6, the prevalence rates in adults in the United States and in certain European and less developed countries were similar, with at least a majority of individuals from each country possessing such antibody. An exception was a highly isolated Ecuadorian Indian tribe in Gabaro in which none of the adults studied had evidence of prior Norwalk infection. This was in

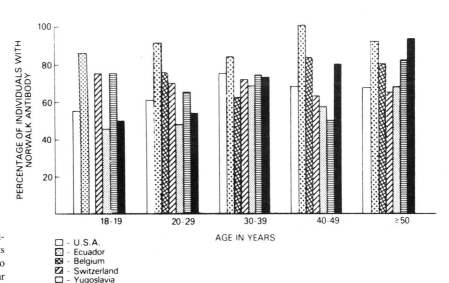

Figure 6. Prevalence by age of serum antibody to Norwalk virus in healthy adults from various parts of the world. Note that no specimens were tested in the 18- to 19-year age group from Belgium. From Greenberg *et al.*[213]

marked contrast to three other less isolated Ecuadorian villages, where approximately 90% had Norwalk antibody. The prevalence of Norwalk antibody in adult male and female homosexuals in the United States was approximately equal (57% and 65%, respectively) and not appreciably different from that in adult blood donors in the United States studied by RIA or in adults studied by IAHA as described above.[213]

Children from the United States, Taiwan, and Yugoslavia acquired antibody more slowly than did children from less well-developed countries such as Ecuador, Bangladesh, Thailand, the Philippines, and Panama.[49,110,137,213,466] This is shown graphically for some of these countries in Fig. 7. The high antibody prevalence in the pediatric age group in Bangladesh and Ecuador (not Gabaro) was unexpected and indicates that the Norwalk or an antigenically related virus infects early in life in at least certain areas of these less-developed countries. Its importance as an etiologic agent of clinical gastroenteritis in this age group remains to be determined; however, as noted above, it does not appear to be an important cause of severe infantile diarrhea. Prevalence data are not available for the

other members of the Norwalk group, since suitable serological assays have yet to be developed or have only recently been developed.

5.1.2. Epidemic Behavior. The Norwalk group of agents is associated with epidemic viral gastroenteritis that occurs in family, school, group, institutional, or community-wide outbreaks affecting adults, school-aged children, family contacts, and some young children as well.[215,305,306,312] Although the term "winter vomiting disease" has been applied to certain outbreaks of epidemic viral gastroenteritis, a clear-cut seasonality does not appear to occur, at least for Norwalk virus-associated outbreaks.[214,215,653]

In the Norwalk outbreak, which occurred in an elementary school, and the Snow Mountain outbreak, which occurred at a resort camp, the primary attack rate was 50 or 55%, respectively, whereas the secondary attack rate was 32 or 11%, respectively.[2,397] The incubation period was short, with an average of 48 hr in secondary cases in the Norwalk, Ohio, outbreak.[2] The Hawaii, Montgomery County, and Southampton viruses were obtained in family outbreaks.[328,551,618] The Taunton agent was derived from a hospital outbreak of gastroenteritis in patients and staff in England; the agent was detected in the stools of 9 (47%) of 19 of the patients.[80,81]

In a systematic study of selected paired sera from 70 outbreaks of nonbacterial gastroenteritis employing the RIA-blocking assay, 24 (34%) appeared related to Norwalk virus.[215] Of the 24 outbreaks, four occurred in each of four settings: recreational camps, cruise ships, contaminated drinking or swimming water, and a community or family; three involved elementary or college students; two, nursing homes; one, shellfish; and two others, adults in other settings. The Norwalk-related outbreaks occurred in both the cooler and the warmer months. The association of such a high percentage of nonbacterial gastroenteritis outbreaks with a single member of this group of viruses was unexpected.

In a review of 74 outbreaks of nonbacterial gastroenteritis studied by the CDC from 1976 to 1980 (including most of the 70 noted above), 42% were associated with Norwalk virus infection; an additional 23% were provisionally associated with the Norwalk virus or a related agent, and the remainder were not related to Norwalk virus infection.[306] The overall importance of the Norwalk virus in unselected outbreaks of gastroenteritis was estimated by a review of 642 gastroenteritis outbreaks reported to the CDC from 1975 to 1981.[305] Fifty-four (9.9%) of 558 such outbreaks for which sufficient data were available to enable a provisional etiologic diagnosis based on clinical and epidemiologic characteristics resembled outbreaks previously linked by laboratory methods to

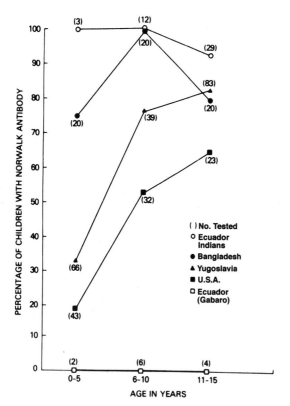

Figure 7. Age-related prevalence of serum antibody to Norwalk virus in children from various countries. From Greenberg *et al.*[213]

Norwalk virus. The provisional role of Norwalk virus in various outbreaks was as follows: 23% of 96 waterborne outbreaks, 4% of 430 food-borne outbreaks, 6 of 9 nursing home outbreaks, 5 of 18 cruise ship outbreaks, and 3 of 5 summer camp outbreaks.

This group of viruses is also frequently implicated as the cause of various food-borne outbreaks of gastroenteritis. They have been cited as the etiologic agent following the ingestion of a myriad of foods such as frosting, salad, celery, melon, potato salad, vermicelli, consomme, fruit salad, coleslaw, tossed salad, cold foods, sandwiches, lettuce, cold cooked ham, oysters, and commercial ice.[141,282] In this regard, in Minnesota, Norwalk and Norwalk-like viruses are associated with food-borne outbreaks of gastroenteritis more frequently (35%) than those caused by any single bacterial agent.[325] The vehicle in these outbreaks was often considered to be a salad item. In Japan, 26 of 38 outbreaks considered nonbacterial were associated with a Norwalk or Norwalk-like particle with the major vehicle of transmission (92%) considered to be oysters.[422]

The Norwalk virus or related viruses have emerged as important causes of gastroenteritis in military personnel in several different parts of the world, as serological evidence of Norwalk virus infection was detected in 10% of U.S. military personnel who developed a diarrheal illness while deployed to South America or West Africa; the most frequently detected pathogens (17%) were the enterotoxigenic *Escherichia coli*.[57a] Norwalk virus infections was demonstrated in 1 of 10 outbreaks of diarrhea among Israeli military personnel.[99]

The impact of diarrheal disease in military personnel was demonstrated more recently in Operation Desert Shield in which 57% of 2002 men who were surveyed by questionnaire and who spent an average of at least 2 months in Saudi Arabia developed at least one episode of diarrhea.[263] Bacterial enteropathogens were identified in the stools of almost one half of 432 individuals with gastroenteritis. Vomiting was reported infrequently as a primary symptom. However, 9 (82%) of 11 military personnel who developed vomiting as a major symptom (two with vomiting and diarrhea, and nine with vomiting alone) demonstrated serological evidence of Norwalk virus infection.

A subset of 404 individuals in a cohort of 883 military personnel who were deployed to Saudi Arabia and Kuwait were evaluated for the incidence of Norwalk virus infection; paired sera obtained prior to or after the 5-month interval of deployment were examined by ELISA using the recombinant Norwalk virus protein as antigen.[264] The impact of gastroenteritis in the military was demonstrated again, since 61% of the entire cohort developed at least one episode of gastroenteritis during the 5-month period. In the subset of 404 troops, serological evidence of Norwalk virus infection was demonstrated in 14.5% of those with vomiting alone or with vomiting and diarrhea, in 6.5% of those with diarrhea alone, and in 3% of the troops without diarrhea or vomiting. Overall, the adjusted incidence of Norwalk virus infection in the entire cohort over the 5-month interval of deployment was 6.2%.

In addition, gastroenteritis disrupted the activities of Navy personnel at sea aboard an aircraft carrier as 25 (19%) of 130 crew members surveyed developed vomiting and/or diarrhea.[502] Norwalk virus infection played a prominent role as 13 (52%) of the 25 individuals developed serological evidence of Norwalk virus infection; a significantly fewer number (24% of 105) without illness had such a response.

The role of Norwalk virus infection in outbreaks of diarrheal illness in families was evaluated during a 1-year prospective study of 28 families that were enrolled at the time of birth of an infant.[438] Fourteen of the families experienced one outbreak of diarrheal disease that could not be associated with a bacterial enteropathogen. Two of those outbreaks were associated serologically with Norwalk virus infection. None of the 28 infants and young children in the study developed serological evidence of Norwalk virus infection.

5.1.3. Geographic Distribution. Norwalk virus appears to have a worldwide distribution because antibody has been detected in populations in the United States, Belgium, Switzerland, Yugoslavia, Bangladesh, Nepal, Japan, Ecuador, Indonesia, Australia, Panama, Thailand, Taiwan, and the Philippines.[5,49,110,137,213–215,218,220,296] The only population studied that lacked detectable antibody was the very isolated Gabaro Ecuadorian Indians, who also lacked antibody to hepatitis B virus (anti-HBc negative).[213] In contrast, they were found to have serum antibody to rotavirus, respiratory syncytial virus, and hepatitis A virus.[213] Gastroenteritis outbreaks associated with Norwalk virus infection have been reported in at least 16 states in the United States.[306]

5.1.4. Temporal Distribution. In developed countries, illness with the Norwalk group of agents was believed to occur predominantly in outbreaks during the cooler months of the year, from the fall through the spring seasons. However, recent studies reveal that Norwalk virus outbreaks occur throughout the year in the United States.[213,214,306] Of 34 outbreaks, 32% occurred in the spring, 29% in the summer, 21% in the fall, and 18% in the winter.[306] The temporal distribution in tropical countries is not known, although in Bangladesh, Norwalk virus infections occurred most often during the cool, dry periods.[49]

5.1.5. Age. During outbreaks, the peak incidence is observed in school-aged children and adults who are in close contact in various group settings.[215] However, close contact is not always essential, since outbreaks have occurred after ingestion of contaminated water or various types of food as noted previously. In the United States, antibody-prevalence data indicate that the Norwalk virus is not an important cause of gastroenteritis in infants and young children, nor has it played an important role in diarrheal illness of early life serious enough to require hospitalization.[63,296,297] Its overall importance in the developing countries is not known. As noted previously, a recent study in which 49% of infants and young children in Finland developed serological evidence of Norwalk virus infection that could not be associated with illness over a 2-year period raises important new questions regarding the natural history of Norwalk virus infection.[338]

5.1.6. Sex, Race, Occupation. There is no evidence of differential susceptibility to this group of agents on the basis of sex, race, or occupation.

5.1.7. Occurrence in Different Settings. Illnesses associated with the Norwalk group of agents tend to occur in sharp outbreaks in families, schools, institutions, or communities and to affect adults, school-aged children, and family contacts as well as some young children. Overall, this epidemic characteristic is one of the main epidemiologic features that differentiates the Norwalk group from the rotaviruses, since the latter are characteristically associated with sporadic gastroenteritis of infants and young children and only infrequently affect older age groups.[286]

5.1.8. Socioeconomic Status. In developed countries, there is no evidence of differential susceptibility to the Norwalk group on the basis of socioeconomic standing. However, the greater prevalence of Norwalk antibody in the pediatric age group in developing countries in comparison to that in developed countries may reflect a role of crowding or other socioeconomic factors in facilitating the spread of this agent.

5.1.8a. In Traveler's Diarrhea. Norwalk virus does not play an important role in the etiology of traveler's diarrhea. Its estimated importance in several studies ranged from 0 to 15%.[47,139]

5.1.8b. In Human Immunodeficiency Virus (HIV) Positive Patients. Norwalk virus infection has been observed in patients who are HIV positive. However, the role of Norwalk virus in the etiology of gastroenteritis in each individual does not appear to be any greater than that observed in non-HIV-infected individuals.[112,279]

5.1.9. Other Factors. The influence of factors such as malnutrition on susceptibility to infection with the Norwalk group is not known. It has been suggested that genetic factors may play a role in determining susceptibility or resistance to infection with Norwalk virus.[426] These factors are discussed in greater detail in Section 7.1.3.

5.2. Rotaviruses

5.2.1. Incidence and Prevalence Data. Rotaviruses have emerged as the major etiologic agents of serious diarrheal disease in infants and children under 2 years of age in practically all areas of the world where this disease has been studied etiologically.[41] The illness rate among family contacts of patients with rotavirus gastroenteritis is low, although subclinical infections in contacts occur frequently.[297,310,537,564] In the metropolitan Washington, DC, area, the pattern of acquisition of rotavirus antibody contrasted sharply to that of the Norwalk agent.[296] By the end of the third year of life, over 90% of infants and young children had acquired rotavirus antibody, a pattern similar to that observed for respiratory syncytial and parainfluenza 3 viruses.[296,309,427] A high prevalence of antibody was maintained into adulthood, probably as a result of frequent reinfection with these agents. In other studies, the acquisition of rotavirus antibody has followed a similar pattern.[69,147,229,291,651]

Cross-sectional epidemiologic studies of hospitalized infants and young children with diarrhea have provided the most striking data regarding the importance of rotaviruses. In a study of patients admitted with a diarrheal illness to Children's Hospital National Medical Center in Washington, DC, from January 1974 to July 1982, 34.5% of 1537 patients shed rotavirus in a stool or rectal swab specimen (Fig. 8).[61] The major role of rotaviruses in diarrheal illnesses requiring hospitalization has also been observed in many other countries of the world, such as Australia, Canada, England, and Japan.[71,114,318,321,385,386] For example, in an Australian study lasting 1 year, 52% of 378 patients admitted with gastroenteritis shed rotavirus.[114] In a Japanese study from December 1974 to June 1981, rotavirus was detected in stools of 45% of 1910 infants and young children hospitalized with diarrhea.[321] A characteristic temporal pattern of rotavirus infections has been observed in temperate climates and is discussed in Sections 5.2.2 and 5.2.4. Sequential infections with human rotavirus have been documented, but it appears that the second infection results in a milder illness.[41]

Although rotaviruses have been implicated in cross-sectional studies as a major cause of gastroenteritis requiring hospitalization of infants and young children, the rates of hospitalization were not known, because of difficulty in

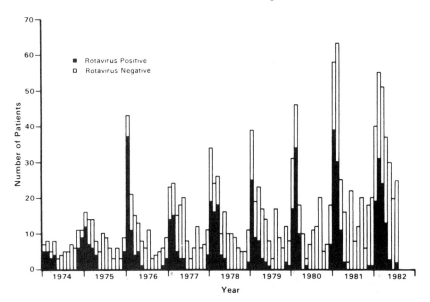

Figure 8. Temporal distribution of rotaviruses detected in stools from infants and young children hospitalized with gastroenteritis at Children's Hospital National Medical Center, Washington, DC, January 1974 (partial)–July 1982. From Brandt *et al.*[61]

estimating the population base of the hospitalized patients. However, the incidence of hospitalization was estimated in the Washington, DC, area because (1) medical care for the population under study was provided by Group Health Association, Inc., and (2) almost all pediatric hospitalizations of the group occurred at the Children's Hospital National Medical Center, which had an ongoing study of the etiology of gastroenteritis.[456] Between January 1977 and March 1979, which included three periods of rotavirus prevalence, about 29,000 patients less than 15 years of age were under surveillance each or part of a year. Thirty-one of the 38 children hospitalized were under 2 years of age. Thirty of the children were studied microbiologically, 19 (63%) of them shed rotavirus, and one developed only serological evidence of infection. From this analysis, it was calculated that 1 in 272 (3.7/1000 per epidemic year) individuals less than 1 year of age and 1 in 451 (2.2/1000 per epidemic year) 12–24 months of age were hospitalized for rotavirus gastroenteritis; this rate dropped precipitously in the 25- to 60-month age group (1 in 5519), and such illness was not observed after 5 years of age. In total, rotavirus infection was associated with 62% of the pediatric hospitalizations for gastroenteritis in this population over a period of 2¼ years. Other agents played a minor role in comparison to the rotaviruses. In the 12- to 24-month age group, 1 of every 3.4 children on the average made an outpatient visit to the clinic for gastroenteritis during each year of the study, and in the less-than-12-month age group, the visits decreased to 1 of every 7.8 infants. The incidence declined sharply in the 25-month to 15-year age group.[456]

The importance of rotaviruses in various other settings in developed countries has also been estimated. In a family study during a 6-year period in Tecumseh, Michigan, the incidence of enteric illnesses was 1.2 per person per year, with the youngest age groups experiencing the highest rates.[393,394] Overall, rotaviruses were detected in stools of 3.8% of those with gastroenteritis symptoms, but the highest rate (10.4%) was in children under 2 years of age. In another study, rotaviruses accounted for a major portion of diarrheal illnesses seen in two private practices and in a hospital in the same locality in Michigan.[323] In both practices, rotavirus was shed by 16% of the children with diarrhea who did not require hospitalization, whereas in the hospital study 32% of the children admitted with diarrhea were rotavirus positive. The frequency of rotavirus diarrhea during two combined rotavirus peak periods in the solo practice and group practice was 30 and 36%, respectively, whereas in the hospital, rotavirus accounted for 48% of the admissions for diarrhea. The incidence of rotavirus diarrhea in the solo practice was highest in infants less than 1 year of age, with an estimate of 0.15 episodes per person per year, whereas in the 1- to 2-year-old age group it was 0.5 such episodes per year. It was of interest that 11% of the patients under 2 years of age with rotavirus diarrhea and only 4% with nonrotavirus diarrhea were hospitalized. Moreover, from the two practices, 38% of the patients with diarrhea who were hospitalized were infected with rotavirus. Extrapolation of this figure to the entire United States indicated that 80,000 infants and young children would be hospitalized annually with rotavirus diarrhea.[265] In Winnipeg, Canada, in a prospective study of diarrheal disease in a cohort of neonates, the most frequently detected enteric pathogens were the rotaviruses, which accounted for 23% of such illness.[228] Over 10% of the rotavirus infections were reinfections, and most were asymptomatic.

In a prospective family study in Virginia, 51% of the families experienced rotavirus infection during the 29-month surveillance period.[458] The 12–23 age group experienced the highest incidence of rotavirus gastroenteritis (40 per 100 person-years) with lower rates in the 6–11 month and 24–35 month age groups (12 per 100 and 13 per 100 person-years, respectively), whereas adults experienced the lowest rate (5 per 100 person-years). The figures from this study were extrapolated nationwide to obtain an indication of the incidence of rotavirus gastroenteritis in the United States annually with the following estimates for the various age groups: 0–11 months, 41,000; 12–23 months, 1,474,000; 24–35 months, 475,000; and 36–60 months, 573,000.[458] It was of interest that: (1) 88% of rotavirus infections were symptomatic in the 36-month and under age group; (2) 1 of 9 ill children who were seen by a physician were hospitalized; and (3) only 0.2% of nonill individuals were rotavirus positive. Demonstration of rotavirus infection in ill individuals at a significantly higher rate than in nonill individuals is the rule in most studies of infants and young children beyond the neonatal period,[63] although exceptions to this pattern have been described.[86]

Information regarding hospital admissions in the United States was also generated by extrapolation from the mean annual number of cases and probable cases of rotavirus gastroenteritis resulting in admission to a children's hospital in a Texas county over a 6-year period. In this analysis, the annual incidence of hospitalization for rotavirus gastroenteritis was estimated to be 11.1 and 5.8 per 1000 children in the first and second years of life, respectively, with a further decline with increasing age.[359] Moreover, it was estimated that 1 of 46 children would require hospitalization for rotavirus gastroenteritis by 18 years of age (the upper age limit for admission) with practically the entire risk accruing during the first 5 years of life. Thus, for the United States, it was estimated that presumptive rotavirus gastroenteritis would result in 110,000 hospitalizations and 583,000 hospital days, annually.

The importance of rotavirus infection as a cause of severe gastroenteritis was demonstrated in a prospective study of acute diarrhea in a cohort of 336 Finnish children from birth to approximately 2½ years of age.[464] Although more than half (55%) of the study group did not develop diarrhea, 26% had one episode and 19% had two or more episodes during the study period (for a total of 248 episodes). With regard to etiology, rotaviruses were the most frequently detected pathogen (26%), followed by adenovirus (4%) and bacterial pathogens (4%); in the remainder, a pathogenic agent could not be identified. The incidence of diarrhea was greatest in the 7–12 month age group (0.24/child per year) and lowest in the 0–6 month (0.09/child per year) and 25–32 month (0.06/child per year) age groups. In addition, rotavirus diarrhea was detected only once in any child with a documented rotavirus-positive episode. The disproportionate influence of rotavirus infections on severe diarrhea was demonstrated when 237 of the 248 diarrheal episodes were tabulated according to a severity score ranging from 0–20: 54% (128 episodes) were considered mild [rated 1–6 (usually not requiring treatment for diarrhea)]; 28% (66 episodes) were classified as moderately severe [rated 7–10 (requiring medical attention but manageable on an outpatient basis)]; and 18% (43 episodes) were considered to be severe [rated $\geq$ 11 (possibly requiring hospitalization)]. With regard to etiology and severity of illness: (1) 75% of the 128 mildest cases failed to yield an etiologic agent (rotavirus was detected in 4%); (2) 41% of the 66 moderately severe cases were associated with rotavirus [3% with adenovirus, 6% with enteropathogenic *E. coli* (EPEC), and 50% were etiologically unresolved]; whereas (3) 75% of the 43 severe cases were associated with rotavirus (5% with adenovirus and 21% were etiologically unresolved). The mean severity scores according to etiology were: rotavirus 11.0 ± 3.7, adenovirus 6.3 ± 4.2, bacteria 6.6 ± 3.5, and unresolved etiology 5.3 ± 3.1. The difference in severity score between the rotavirus-associated illnesses and the other groups was highly significant ($P < 0.0001$). It should be noted that while only about one third of the cases had an etiologic diagnosis (in contrast to almost twice that number in a previously described hospital-based study in Finland), the cases in which a specific etiology was found were generally more severe. Thus, in cases with a score $\geq$ 7, rotavirus was diagnosed in 54% with an overall diagnostic rate of 63%, whereas 75% of the milder cases ($\leq$6) failed to yield a diagnosis.[464]

Acute diarrheal disease occurs frequently in day-care centers, especially in centers caring for children who are not toilet trained,[525] with the rates being highest in the first 4 weeks of attendance. Only 21% of the episodes studied yielded a pathogen, but rotaviruses were detected in 18% of the total 513 episodes.

Rotaviruses are of extreme importance as etiologic agents of diarrheal illness in developing countries. By the end of the third year of life, over 90% of infants and young children acquire rotavirus antibodies, a pattern of infection quite similar to that in developed countries, but the consequences of such infections are markedly different in these settings.[39,49,55,137,147,229,266,291]

The relative role of rotaviruses and bacterial agents in the etiology of gastroenteritis was reported for 6352 patients for a 1-year period, February 1978 to January 1979, at the Matlab Treatment Center in Bangladesh.[51] This study showed that 46% of the patients under 2 years of age shed rotavirus and that 28% shed toxigenic *E. coli*. In the 2-year-and-over age group, bacterial agents were detected more frequently than rotaviruses.

Prospective longitudinal studies in developing countries consistently demonstrate the major role rotaviruses play in the etiology of diarrhea of infants and young children. In a village-based study in northeastern Brazil in which mortality during the first 5 years of life was greater than 14%, diarrhea was listed as either the primary or an associated cause of death in over 52% of the deaths.[225] In selected specimens from diarrheal episodes, enterotoxigenic *E. coli* (ETEC) (21%) and rotavirus (19%) were the most frequently detected pathogens. The highest incidence of diarrhea occurred in the poorest families and reached almost ten per child per year in the 6- to 11-month-old age group.

In Santa Maria Cauque, Guatemala, in 45 infants and young children studied over a 3-year period, the incidence of diarrhea was 7.9 per person per year, with rotavirus associated with 10% of the episodes.[357] The incidence of rotavirus diarrhea was 0.8 per child per year (and of rotavirus infections 1.2 per child per year). One child had only one rotavirus infection, whereas the remainder experienced two to seven rotavirus infections each over the 3-year period.[357] In a sample of 24 of the 45 infants and young children, dehydration occurred 14 times more often in those with rotavirus diarrhea than in those with diarrhea of bacterial, parasitic, or unknown etiology.[630]

In two longitudinal studies in Bangladesh, the disproportionate role of rotavirus as a cause of severe dehydrating diarrhea in comparison to its role in diarrheal illnesses overall was clearly shown. In one study of 197 children 2 to 60 months of age in two contiguous villages, diarrhea was the second most common illness but was the most frequent cause of admission to a treatment center.[48] The incidence of diarrhea was highest in children 2–11 months of age (more than seven per child per year); ETEC was the most frequently detected sole pathogen (27% of episodes), *Shigella* was second (12.8%), and rotaviruses were next (3.8%). However, although rotaviruses were associated with fewer than 4% of diarrheal episodes, they were linked with 39% of episodes with significant dehydration.

In the other village study in Bangladesh, the incidence of diarrhea was highest (>4 per child per year) in the 9- to 11-month age group, decreasing gradually with increasing age to three episodes per child per year for the next 4 years of life, to 0.2 episodes per child per year for children 10 years of age or older.[50] Once again, the ETECs were the most frequently detected pathogens in all age groups (23% of all diarrhea episodes), *Shigella* second (11%), and rotavirus third (5%). Rotaviruses were detected only in children less than 2 years of age and accounted for 11% of the diarrheal episodes in this age group. The peak incidence of rotavirus diarrhea was in the 6- to 12-month age group (0.5 per person per year). It is noteworthy that in children under 2 years of age, the incidence of rotavirus diarrhea was only half that of ETEC diarrhea, but one in six rotavirus diarrheal episodes necessitated a visit to the treatment center, whereas only 1 of 15 ETEC diarrheas required such visits. Furthermore, in children less than 2 years of age who had dehydrating diarrheal episodes, 46% were linked to rotavirus, 24% to ETEC, and 7% to *Shigella*. Moreover, it was estimated that (1) if fluid replacement therapy was not available, (2) if dehydration reached a level of 7.5% or more, (3) if 50% of the children did not survive this level of dehydration, then (4) there would be 6.5 deaths per 1000 children less than 2 years of age per year, of which 44% would be rotavirus associated.

In a study in Cairo, Egypt, of 145 infants and young children under 18 months of age with fatal or potentially fatal diarrhea, the most frequently detected agents that were considered etiologically important were rotavirus (33%), heat-stable ETEC (20%), heat-labile ETEC (11%), EPEC (8%), and *Salmonella* (5%).[507] In contrast, rotaviruses were associated with only 5% of outpatient diarrhea in infants under 1 year of age in a rural Egyptian area.[507,654]

Finally, the influence of age on the severity of symptoms was demonstrated in a study of infants ≤ 12 months of age with acute diarrhea who were brought to an emergency room of a municipal hospital in urban São Paulo.[194] The most frequently detected pathogen was the enteroadherence factor (EAF)-positive classic EPEC (26%), followed by rotavirus (14%), *Salmonella* species (8%), ETEC (7%), and *Shigella* species (5%). The frequency of rotavirus illness was relatively low in the first 3 months of life but increased in the next quartile; however, in the 6- to up to 9-month age group, it almost equaled the frequency of detection of EAF-positive classic EPEC and in the 9- to 12-month age group, it exceeded it. Dehydration occurred most often in the EAF-positive EPEC group (71%) and less frequently in rotavirus (39%) infections.

5.2.2. Rotavirus Subgroups and Serotypes. The epidemiology of rotavirus infection and illness with regard to serotypic diversity is now known primarily because of the development of monoclonal antibodies that recognize the VP7 of most clinical isolates and to a lesser extent because of the widespread application of molecular biological techniques.[253] Prior to this, a rotavirus strain could only be serotyped by techniques not readily adaptable to large-scale epidemiologic surveys, such as neutralization of immunofluorescent foci, plaque reduction neutralization, neutralization of cytopathic effect or virus yield in roller tube tissue culture, or solid-phase immune electron microscopy.[286] Because of this, subgroup specificity became a surrogate for serotyping as among human rotaviruses subgroup I strains almost always belonged to serotype 2, whereas subgroup II strains almost always included strains of serotypes 1, 3, and 4.[253] Overall, in various parts of the world, subgroup II rotaviruses were detected more frequently than subgroup I strains.[15,404,408,527,562,601,651] The molecular epidemiology of rotavirus strains has also been evaluated extensively by determining the electropherotype by gel electrophoresis of the segmented genomic dsRNA of rotavirus.[452] This method does not indicate the serotype but permits study of the variability of strains by PAGE.

However, with the development and availability of VP7 monoclonal antibodies with specificity for individual serotypes, studies of the natural history of human rotavirus serotypes became a reality,[37,104,238,543] because (1) these antibodies could be used to serotype rotavirus strains directly in stool material by ELISA, and (2) there was complete concordance between the serotyping ELISA and neutralization or SPIEM.

The distribution of rotavirus VP7 serotypes was evaluated by ELISA in rotavirus-positive stool specimens obtained from children with acute gastroenteritis over a 6-year period in various countries of Europe, North and South America, Africa, and Asia.[29] This study demonstrated that: (1) 95% of 907 rotavirus-positive stool specimens that were VP7 positive belonged to serotypes 1, 2, 3, or 4; (2) rotavirus type 1 was found most often (54%), followed by serotype 2 (18%), serotype 3 (12%), or serotype 4 (11%); (3) no single serotype was detected exclusively in one location; (4) each of the four serotypes were detected worldwide; (5) the distribution of serotypes differed year by year and in different countries during the same years; and (6) the distribution of serotypes detected in developed and developing countries did not differ.

This general pattern was observed in other studies as well. For example, (1) in a study of hospitalized children in Texas during an 11-year period and in the north central United States during a 9-year period, serotype 1 was predominant overall in Texas, but serotypes 1, 3, and 4 were each predominant in 1 or more years, whereas in the north central United States, serotype 1 was not only predominant overall, but it was the only serotype that pre-

dominated each year[360]; (2) in southeastern New England over nine winter seasons, serotype 1 strains were predominant in seven of nine seasons, and overall accounted for 63% of the typed strains[30]; (3) type 1 strains were predominant overall in various other locations including Australia, Venezuela (short-term), Italy, the Central African Republic, Thailand, Japan, Norway, Finland, West Germany, Switzerland, and Israel, with serotype 2, 3, or 4 predominating intermittently[15,37,46,113,171,182,185,187,201,407,439,505]; (4) in a summary of 28 studies in which over 5419 specimens obtained between 1979 and 1989 from North and South America, Europe, Asia, and Africa were serotyped, the most prevalent serotype was VP7:1 (61%), whereas serotypes 2, 3, and 4 were evenly distributed (10–16%), except that serotype 4 was not detected in Africa[614]; and (5) in a study in London of children admitted to two children's hospitals during a 6-year period, serotype 1 accounted for 60% of the cases, serotype 4 for 24%, serotype 2 for 11%, and serotype 3 for 3%.[416] Moreover, the electrophoretic profiles of 611 strains demonstrated at least 108 different profiles with continuous variation observed throughout the 6-year period. Strikingly, once an electropherotype had disappeared from the population, it never recurred. None of the electrophoretic profiles resembled those of group B or C rotaviruses. No strain appeared to be endemic in these two hospitals.

In a few other locations, the distribution of the four serotypes deviated from this pattern: (1) in an 11-year survey in Venezuela, serotype 3 strains were the most prevalent (25%), followed by serotype 1 (20%), serotype 4 (15%), or serotype 2 (10%)[600]; (2) in Malaysia in specimens obtained discontinuously over a 12-year period, serotype 4 was the predominant strain (71%) with serotype 1 a distant second (15%)[445]; and (3) in Bangladesh (by neutralization assay) over a 2-year period, serotype 3 strains were predominant.[597]

The distribution of serotypes in two day-care centers for infants and young children over three rotavirus seasons was similar to that observed concurrently in hospitalized children: serotype 1 strains accounted for about two thirds and serotype 3 strains for approximately one third of the typed strains.[423] Although it is not possible to determine serotype from the migration pattern of the 11 RNA segments by PAGE or by subgroup analysis, the latter methods are important epidemiologic tools for monitoring the circulation of rotavirus strains.[15,452] Data regarding variability in severity of illness in association with a specific serotype has been inconsistent. For example, in Bangladesh, VP7 serotypes 2 or 3 were associated with more severe dehydrating illness than serotype 1 or 4, but the differences were not of major clinical importance,[34] and in Australia differences in virulence between different rotavirus VP7 serotypes were not observed.[23]

Analysis of the genotype or serotype of clinical specimens according to VP4 specificity is gaining impetus with the use of PCR techniques as well as monoclonal antibodies.[34,102,226,484,500] In studies from various parts of the world, P1A serotypes occur most often.[34,102,226] This was not surprising, because VP7 serotype 1, which is the most frequently detected VP7 serotype, shares VP4 P1A specificity with two other epidemiologically important VP7 serotypes (i.e., 3 and 4). For example, analysis by a PCR-based assay of the VP4 genotype of 89 electrophoretically distinct rotavirus strains that represented each of the electropherotypes detected in 675 rotavirus-positive stool specimens obtained over a 10-year period from Japanese infants with rotavirus diarrhea: (1) the Wa and DS-1 alleles occurred most frequently overall (83.1% and 15.6%, respectively); (2) the M37 and AU-1 alleles were found in none and in 1.3% of the strains, respectively; and (3) the Wa allele was predominant in 7 of the 8 years in which an adequate number of specimens were available, whereas in the most recent year surveyed, 1990–1991, the DS-1 allele was predominant.[226]

5.2.3. Epidemic and Endemic Behavior in Pediatric and Adult Groups. Unlike the situation with the Norwalk group, true outbreaks of rotavirus gastroenteritis occur infrequently. In the temperate climates, rotavirus infections demonstrate a consistent temporal pattern similar to that observed in studies at Children's Hospital, Washington, DC, during 1974–1982 (Fig. 8), which was characterized by a large number of hospitalizations in infants and young children for gastroenteritis during the cooler months of each year.[61,63,71,114,297,318,385] There is also a suggestion of a regional seasonal peak from west to east in the United States.[332] Rotaviruses are also associated with milder bouts of gastroenteritis not requiring hospitalization. For example, in the Children's Hospital study, 22% of 200 outpatients with gastroenteritis studied from November 1975 through June 1978 shed rotavirus.[63] Community outbreaks of rotavirus illness occur uncommonly, since most adults appear to be immune, most likely by virtue of previous rotavirus infection(s). However, subclinical rotavirus infections occur quite commonly in adults.[297,310,537,564] In one study, 22 (55%) of 40 adult contacts of patients hospitalized with gastroenteritis that was associated with rotavirus infection had serological evidence of rotavirus infection at or about the time of their child's admission, whereas only 4 (17%) of 24 control adults whose children also had gastroenteritis but were not infected with rotavirus were found to be infected

with this agent.[310] Only 3 of the 26 adult contacts with rotavirus infection gave a history of an associated gastroenteritis illness. It appears that older siblings or parents might be a source of rotavirus infection for young persons. The frequency of rotavirus infection in contacts also demonstrates the highly contagious nature of rotavirus infection.[142,310,537,564] Intrafamilial spread of rotavirus infection and illness from infected infants and young children to adults has been described at several locations.[219,458]

Although community outbreaks of rotavirus gastroenteritis are uncommon, one unusual outbreak involving not only children and mothers in a play group but also fathers and grandparents was described[455]: all nine children 15 months through 5 years of age, three of five mothers who shared play group activities, four of five fathers, and each of two grandparents developed gastroenteritis. The incubation period was 24–48 hr. The index cases were most likely non-play group siblings who had been cared for (about 48 hr prior to the play group meeting) by the mother in whose home the play group met. The suspected index cases had gastroenteritis, the mother who took care of the index cases developed diarrhea 24 hr after the play group met, and her daughter had onset of diarrhea just before the play group met and vomited during the play group meeting. In all, 18 of 21 persons developed gastroenteritis, and evidence of rotavirus infection was demonstrated in 10 of 11 persons tested for virus in stools, or for a serological response, or both. This unusual outbreak further attests to the contagiousness of rotavirus infection.

Although subclinical infections are the most common outcome of rotavirus infection in adults, rotavirus gastroenteritis has been observed in various groups of adults in various settings in scattered parts of the world.[41,286] Moreover, a high attack rate was observed in several outbreaks in geriatric groups with some fatalities.[108,233,261,353,526]

It should be noted that in China large outbreaks of a severe form of gastroenteritis, in adults predominantly, have been associated with group B rotaviruses.[262,594] The clinical manifestations may be marked by choleralike, watery diarrhea that has been associated with a few deaths in the elderly.[262] As noted earlier, the group B rotaviruses do not share the common group antigen and have been detected almost exclusively only in China.

5.2.4. Geographic Distribution. Rotavirus infection has been detected in virtually all parts of the world.[41,286] In developed countries, rotaviruses have emerged as the major etiologic agents of diarrheal illnesses severe enough to warrant hospitalization. In temperate climates, the pattern has been similar to that described for Washington, DC.[61] In less-developed areas, rotaviruses have been shown to be major etiologic agents of severe diarrheal illness in infants and young children[41,225,286,507,630]; however, it appears that toxigenic *E. coli* also plays an important role in underdeveloped countries.[412] In practically every area of the world studied, rotaviruses have exhibited an important role in acute gastrointestinal disease of the young.

5.2.5.   Temporal Distribution.   In developed countries in the temperate climates, rotavirus infections display a characteristic temporal pattern that peaks in the cooler months of the year.[61,63,71,114,297,385] The pattern for Washington, DC, for an 8-year period, was shown in Fig. 8.[61] A pattern of spread from the western to eastern United States has also been described.[332]

During the months of January and February, 168 (67%) of 250 and 127 (58%) of 219 hospitalized children, respectively, who had diarrhea were rotavirus positive, whereas none of the 256 diarrhea patients studied during the July–October interval shed rotavirus.[61] A similar pattern was observed in outpatients with diarrhea. In a similar study of hospitalized children in Japan over a 6-year period, rotaviruses were shed in the stools by 521 (66%) of 785 diarrhea patients during the cooler months (December, January, February) and by 56% of 549 such patients admitted during the spring (March, April, May).[321] In addition, only 5.6% of 576 children admitted with diarrhea in the summer or autumn months were rotavirus positive. The reason for the seasonal pattern of infection is not known, but it has been suggested that weather-related low relative humidity in the home might facilitate the survival of rotaviruses in the environment,[59] which along with indoor crowding[59] would affect the epidemiology of rotavirus infection.[59,386] This has not been a consistent observation in all settings.[321,391]

The striking seasonal pattern of rotavirus infections described above is not observed in all situations, since a significant number of rotavirus infections has been observed throughout the year in South Africa, during the summer in Taiwan, during the "small rains" in Ethiopia, during most months in the tropical climates but with peak periods during the slightly cooler months, during the summer in a newborn nursery in England, in all seasons in a newborn nursery in Australia, and in the autumn on a United States Indian reservation.[20,138,148,243,350,355,402,495,496,529,556,591] In both nursery studies, most rotavirus-positive infants were symptom-free, a finding that has yet to be explained satisfactorily. Studies of the frequency of rotavirus infections in relation to the amount of rainfall or to relative humidity have led to variable results.[59,321,355,391,591]

5.2.6.   Age.   In studies from various parts of the world, infants and young children (usually 6 months to 2 years of age) experience the highest frequency of rotavirus gastroenteritis that requires hospitalization[63,71,114,318]; infants under 6 months of age have the next highest frequency.[63,105] An unexplained paradox in the epidemiology of rotavirus infection is the low rate of clinical illness in neonates who shed rotavirus.[20,402] In one study, breast-fed infants shed rotavirus significantly less frequently than those who were not breast-fed; however, the effect of breast-feeding on illness could not be determined because most of the rotavirus infections in both the breast-fed and bottle-fed neonates were subclinical.[556] Recent studies indicate that the gene that encodes the outer capsid protein VP4 of neonatal rotavirus strains that persist in newborn nurseries is highly conserved and differs from that of strains recovered from symptomatic infections.[167,196]

Rotavirus gastroenteritis has been reported in older children and adults, who, as noted earlier, may be important in the transmission of infection to infants and young children.[56,57,297,310,564] Rotavirus gastroenteritis has also been observed in geriatric settings as noted earlier.

5.2.7.   Sex.   A somewhat larger number of males than females (M : F = 1.2 : 1.0) were hospitalized for rotavirus gastroenteritis in the Children's Hospital of Washington, DC, study of 1974–1978.[63] A higher frequency of males who were hospitalized with acute rotavirus gastroenteritis was also observed in a Canadian study.[386]

5.2.8.   Race and Socioeconomic Status.   In the Children's Hospital study, 1974–1978, the age distribution of patients admitted to the hospital for gastroenteritis of any etiology was quite different among black and nonblack patients: 59% of all black patients admitted for gastroenteritis were less than 6 months of age.[63] Also, DC residents and Medicaid recipients who were hospitalized with rotavirus infection tended to be younger than non-DC residents and non-Medicaid recipients. In addition, there was a tendency for rotavirus illness to occur earlier in the course of the outbreak in black patients and in DC residents. Transmission of rotavirus might be facilitated by crowding and poor sanitation, and this may explain the earlier appearance of rotavirus infection in DC than in the suburbs, in black patients as compared to nonblack patients, and in Medicaid recipients as compared to non-Medicaid recipients.[63]

Malnutrition appears to be an important factor in increasing the susceptibility of an infant or young child to develop severe clinical manifestations following rotavirus infection.[68] It has been suggested that repeated diarrheal infections may be a prelude to the development of malnutrition by various mechanisms including damage to the intestinal mucosa so that absorptive cells are compromised over an extended period.[356,358] The deleterious effect of malnutrition on the severity of rotavirus infection has been reproduced in the mouse model.[451]

5.2.9. Occurrence in Different Settings. Rotavirus gastroenteritis occurs predominantly in infants and young children with infection occurring by the 36th month of age in almost all children residing in a family setting.[55,147,229,296] Family contacts are also frequently infected with rotavirus, but usually subclinically.[296,310,537,564] Rotavirus infections have also been observed for extended periods in newborn nurseries almost exclusively as subclinical infections.[20,42,93] An exception to this pattern was described in Italy when outbreaks of gastroenteritis were documented in newborn nurseries.[184] In addition, nosocomial rotavirus infections occur commonly.[386,465] In one study, 10 (17%) of the 60 children admitted to the hospital without diarrhea (but during a period of rotavirus prevalence) developed diarrheal illness associated with rotavirus infection while hospitalized.[465] In another hospital study, over a 1-year period, about one of every five rotavirus infections appeared to be hospital acquired.[386] Outbreaks of rotavirus gastroenteritis have been observed in premature infants, in school-aged children, in a home play group setting, in a military group, in nursing homes for the elderly and in geriatric units with a few reported deaths, in adults in South America, and in an isolated South Pacific Island, but characteristically, rotavirus illness occurs sporadically and not in widespread community outbreaks as does the Norwalk group.[41,286] Rotavirus illness is not common beyond the first few years of life. However, as noted earlier, large outbreaks of gastroenteritis have been observed in China, predominantly in adults, associated with group B rotaviruses.[90,262,594]

6. Mechanisms and Route of Transmission

6.1. Norwalk Group of Viruses

Infection with the Norwalk group of agents is most likely spread from person to person by the fecal–oral route. Volunteer studies have established that the Norwalk, Hawaii, and Snow Mountain agents can be transmitted via the oral route, i.e., following the ingestion of stool suspensions containing these infectious agents.[129,130,133,335,397,618] A $10^{-4.7}$ dilution (the highest dilution tested) of a Norwalk virus stool suspension known to be infectious induced illness in volunteers (R. Dolin *et al.*, unpublished data). It is unlikely that this group of agents is transmitted by the respiratory route. Nasopharyngeal washings obtained from a volunteer with experimentally induced Norwalk illness failed to induce illness in three volunteers.[129] However, Norwalk virus has been detected in vomitus[216]; transmission of this group of agents via aerosols generated during projectile vomiting has been suggested as an important mode of spread of these viruses, especially in the hospital setting.[79]

The explosive nature of some of these outbreaks in which large numbers of individuals develop illness in a cluster within 24–48 hr has suggested that a common-source exposure should also be considered in certain outbreaks. Indeed, in a review of 38 outbreaks of gastroenteritis associated with Norwalk virus, 31 (82%) were considered to have originated from a common source of infection.[306] In the Colorado outbreak associated with the Snow Mountain agent, 61% of the 418 cases had onset of illness on a single day.[397]

Epidemiologic analysis revealed that the attack rate increased with consumption of water or ice-containing beverages and that the water supply of the camp was not only inadequately chlorinated but also contaminated by a leaking septic tank; it was therefore suggested that a waterborne agent was responsible for the outbreak. In the Norwalk, Ohio, outbreak, 50% of the students and teachers of an elementary school developed gastroenteritis; it was striking that such illnesses occurred in a 2-day period.[2] Although a common-source exposure was suspected, this could not be established. However, secondary cases among family contacts were observed, and the Norwalk particle was derived from a rectal swab of one such secondary case. Ingestion of contaminated seafood such as oysters was described in early studies as a source of infection with these agents.[220,403] However, outbreaks of Norwalk gastroenteritis have now been associated with the ingestion of many varieties of contaminated food as described earlier, such as cake frosting and salad, as well as contaminated drinking water or swimming in a contaminated lake.[306]

6.2. Rotaviruses

Rotaviruses are also transmitted by the fecal–oral route. Volunteer studies have clearly demonstrated that oral administration of rotavirus-positive stool material can induce a diarrheal illness.[301,302] The rapid acquisition of rotavirus antibody in the first few years of life in all populations studied regardless of hygienic conditions has led to the suggestion that rotaviruses might also be transmitted by the respiratory route.[55,147,229,291,296] Throat gargles obtained from volunteers with an experimentally induced rotavirus diarrheal illness failed to yield rotavirus.[301] However, there are scattered reports of rotavirus antigen being detected in respiratory tract secretions, but most other studies indicate that this is not the usual mode of transmission.[286,655] Although a common-source exposure to rotavirus, such as a contaminated water supply, has been suggested, it is unlikely that such exposure plays a major role in its worldwide occurrence.

The source of infection for the young infant who is not normally in contact with other infants and young children with gastroenteritis is not known with certainty. However, a substantial proportion of parents of rotavirus-infected infants and young children were infected with rotavirus at or about the time of their child's illness; most of these adult infections were subclinical.[297,310,537,564] Thus, an older sibling or family member who is undergoing subclinical rotavirus infection may be the source of infection for the infant or young child with whom he has contact. The highly contagious nature of rotavirus infection may be in part related to the rotavirus' high degree of stability, as demonstrated by the retention of infectivity of calf rotavirus-positive feces that had been kept at room temperature for 7 months.[159] It is likely that human rotavirus is also quite stable and may remain viable in the environment unless destroyed by careful disinfection. The persistence of rotavirus infections in certain newborn nurseries and the frequency of nosocomial rotavirus infection in hospitals provide additional evidence for this possibility.[437] The ability of rotaviruses to persist on different surfaces under various conditions likely contributes to their efficient spread. For example, by PCR, rotavirus was detected in day-care centers on moist surfaces such as water fountains, water-play tables, toilet handles, and telephone receivers, as well as on toy balls, high chair seats, and diaper pail handles.[604] In addition, human volunteers who licked dried preparations of a human rotavirus on a petri dish readily became infected.[595] Infection was prevented if the dried virus was sprayed with a disinfectant. Effective disinfection of contaminated material and care in handwashing may be important measures in containing rotavirus infection, especially in a hospital or day-care setting.[310,322,465,476,512,524,539,540,583,595]

The role, if any, of animals in transmitting rotaviruses to humans is not known. Human rotavirus induces a diarrheal illness in various newborn animals under experimental conditions,[473] and certain naturally occurring animal rotavirus strains may infect humans as whole virions or as reassortants following reassortment with human rotavirus strains.[406,542] However, this type of animal-to-human transmission does not appear to be of clinical or epidemiological relevance.[467]

7. Pathogenesis and Immunity

7.1. Norwalk Group of Viruses

7.1.1. Incubation Period. The incubation period as estimated from various Norwalk outbreaks is 24–48 hr with a range of 4–77 hr.[2,133,306,397] The incubation period in Norwalk virus volunteer studies ranged from 10 to 51 hr, and the illness usually lasted less than 48 hr.[54,129,130,618] The shedding of Norwalk virus by IEM coincided with the onset of illness and usually could not be detected after 72 hr following onset.[550] The incubation period in volunteer studies with the Snow Mountain agent ranged from 19 to 41 hr, with a mean of 27 hr.[133]

7.1.2. Pathogenesis. By light microscopy, biopsies of the proximal small intestine of volunteers with Norwalk or Hawaii virus-induced illness show broadening and blunting of villi, with the mucosa itself being intact histologically; mononuclear cell infiltration and cytoplasmic vacuolization are also observed.[131,497,498] Transmission electron microscopy of the proximal small intestine showed intact epitehlial cells with shortening of microvilli.[3,131,497,498] The extent of the small-intestinal involvement is not known, since studies have included only the proximal small intestine. Histological lesions were not observed in the gastric fundus and antrum or the colonic mucosa following challenge with the Norwalk agent.[605] Brush-border small-intestinal enzyme levels (including alkaline phosphatase, sucrase, and trehalase) were decreased during illness; adenylate cyclase activity was not elevated.[3,54,335] In addition, following Norwalk virus challenge, volunteers experienced marked delays in gastric emptying, which may be the cause of the nausea and vomiting frequently associated with this illness.[380]

7.1.3. Immunity. Volunteer studies with the Norwalk virus have raised rather perplexing questions about the mechanism of immunity. It appears that two forms of clinical immunity exist: one is short term and the other long term.[129,426,618] The former seems to be serotype-specific. For example, volunteers who become ill following administration of Norwalk virus are characteristically resistant to challenge with this virus 6–14 weeks later; in contrast, they are not resistant to challenge with the Hawaii virus nor are Hawaii-virus-infected volunteers resistant to subsequent challenge with Norwalk virus.[129,618]

The situation with regard to long-term immunity was found to be quite different, as indicated when 12 volunteers were challenged with the Norwalk virus on two separate occasions 27–42 months apart, and four were rechallenged 4–8 weeks after the second challenge.[426] Of these 12 volunteers, six developed illness following both the initial challenge and the rechallenge 27–42 months later. In contrast, six volunteers failed to develop illness after the initial challenge or after rechallenge 31–34 months later. Of the six volunteers who developed illness after each of the two sequential challenges, four were challenged a third time 4–8 weeks after the second challenge and only one became ill. Serological studies carried out to clarify this unusual pattern of susceptibility and resistance to Norwalk virus failed to reveal a consistent relationship between the presence or absence of antibody and the subsequent occurrence of illness following challenge. Thus, the presence or absence of serum IEM antibody did not correlate with resistance or susceptibility. It is difficult to explain these findings on the grounds that local intestinal IgA antibody is of prime importance in long-term resistance, since this supposes the existence of two cohorts of individuals, one able and the other unable to sustain the production of local antibody essential for long-term resistance. It has been suggested that other factors that are genetically determined may influence susceptibility to Norwalk infection. For example, there may be a genetically determined specific receptor essential for entry of the Norwalk virus into epithelial cells of the small intestine.[426]

Further evidence for the possible role of nonimmunologic factors in resistance to Norwalk illness was observed when the prechallenge serum and local jejunal antibody levels in 23 volunteers were studied by the RIA-blocking technique.[215] Neither the geometric mean Norwalk antibody titer in serum nor that in jejunal fluid correlated with resistance to illness after challenge. Paradoxically, the prechallenge geometric mean Norwalk antibody titer of jejunal fluid was significantly greater, and such antibody in serum tended to be greater in volunteers who became ill after challenge than in those who did not become ill.[215] A similar paradoxical relationship between prechallenge serum antibody titer and lack of resistance to Norwalk illness in volunteers was reported in another study in which antibody was also measured by RIA.[52]

A more recent study has confirmed the paradoxical effect of serum antibody on protection. In a sequential challenge study in volunteers with the same Norwalk virus inoculum used in the original studies described above, preexisting serum antibody did not confer protection against challenge.[273] Moreover, certain volunteers who had low levels of serum antibody were resistant to challenge. After repeated exposure, however, antibody correlated with protection. In another report, examination of prechallenge sera by EIA failed to show significant correlation between the presence of homotypic serum antibody of $\geq 1:100$ and resistance or susceptibility to challenge with Norwalk, Hawaii, or Snow Mountain virus.[348] Similar observations were made in the natural setting: (1) preexisting serum antibody to Norwalk virus failed to protect medical students from the United States, Puerto Rico, and Mexico against Norwalk virus infection

or illness, and (2) the occurrence of Norwalk virus infection or illness did not correlate with acute phase serum antibody after a common-source exposure of teenagers to a contaminated water supply.[24,274]

7.2. Rotaviruses

7.2.1. Incubation Period. From clinical studies, the incubation period of rotavirus diarrheal illness is estimated to be less than 48 hr.[114] In volunteer studies in which four adults developed a diarrheal illness after oral administration of an untitered stool filtrate containing rotavirus, the incubation period ranged from 1 to 4 days. Virus shedding began the second, third, or fourth day after inoculation and lasted a total of at least 6 days.[301,302]

7.2.2. Pathogenesis.[395] Limited studies of biopsies of the proximal small intestine of a few infants and children hospitalized with rotavirus illness show shortening of the villi, mononuclear cell infiltration in the lamina propria, distended cisternae of the endoplasmic reticulum, mitochondrial swelling, and sparse, irregular microvilli.[251,534] Impaired D-xylose absorption was also observed.[372] In addition, some patients had depressed disaccharidase levels (maltase, sucrase, and lactase).[43]

The pathogenesis of a human rotavirus VP7 serotype 1 strain was studied experimentally in newborn gnotobiotic colostrum-deprived calves that developed illness following intraduodenal administration of this virus.[379,624] Morphological changes in the small intestine proceeded in a cephalocaudad direction: within 2 hr of experimentally induced diarrhea, morphological changes such as denuding of villi and flattening of epithelial cells were observed in the upper small intestine, but rotavirus antigens were not detected by IF; at this time, the lower small intestine was intact, but abundant rotaviral antigens were observed by IF in swollen epithelial cells.[379] Moreover, 7 hr after onset of diarrhea, the lower small intestine demonstrated morphological changes such as denuded villi that were similar to those observed in the upper small intestine earlier; rotaviral antigens also could not be detected by IF. The intestine appeared relatively normal 48 hr after onset of diarrhea. When diarrhea was induced in piglets by human rotavirus, certain functional alterations were observed in the villous epithelial cells of the small intestine: glucose-coupled Na^+ transport was impaired, sucrase activity diminished, and thymidine kinase activity increased, and in contrast, adenylate cyclase and cyclic AMP were not stimulated.[115,178]

More recently, the absorption of the macromolecule human α-lactalbumin during and after acute gastroenteritis in predominantly rotavirus-positive children less than 3 years of age was studied.[249] A significantly greater absorption of proteins was documented 5–8 weeks after the acute phase when compared with the acute phase or controls. Studies in malnourished or normal suckling mice infected with a murine rotavirus showed that malnutrition was associated with more severe symptoms and greater mucosal damage.[570] In addition, although both malnourished and normal mice experienced increased permeability to macromolecules as measured by uptake of ovalbumin, this effect was increased in the malnourished animals.

7.2.3. Immunity.   The correlates of immunity to rotavirus infection and illness have not been clearly elucidated. Results from various studies have yielded inconsistent or conflicting conclusions. However, highlights from various epidemiologic and experimental studies in animals and humans have indicated that antibodies in the intestine or in serum (as a surrogate marker for intestinal antibodies) are important factors in immunity to rotavirus illness).[286,362,559]

The observation was made that newborn calves frequently develop rotavirus diarrhea despite a high level of circulating rotavirus antibody acquired from ingestion of colostrum.[613] This was confirmed experimentally in calves challenged with calf rotavirus and additionally it was shown that antibody in the lumen of the small intestine was of prime importance in protection.[65,613] Similar studies in gnotobiotic lambs examining the relative role of local and systemic rotavirus antibody concluded that antibody in the lumen of the small intestine was the determinant of protection against rotavirus challenge.[513–515] From these and other studies in animals, it appears that antibody in the lumen of the intestine is of prime importance in resistance to rotavirus illness in animals.

The mechanisms of immunity were also studied in 18 volunteers who were administered a human rotavirus VP7 serotype 1 strain orally.[301,302] Five of the 18 individuals shed rotavirus and four of the five developed a diarrheal illness. Examination of the relationship of a moderately high level of prechallenge rotavirus antibody in serum measured by neutralization (to the homotypic or a heterotypic VP7 serotype 2 human rotavirus) to the development of diarrheal illness indicated that such antibody was associated with resistance to the development of illness. The role of local intestinal rotavirus antibody was not clear-cut and needs further evaluation. Two volunteers who developed illness following initial challenge were rechallenged with the same inoculum 19 months later; neither developed a diarrheal illness, although one had mild clinical manifestations.

Recently, the prechallenge sera from these volunteers were reexamined by an epitope-blocking assay that measured antibody to several defined epitopes of VP7 and VP4.[204] A significant protective effect against illness and shedding was associated with the presence of a prechallenge serum antibody titer of $\geq 1:20$ that blocked the binding of a serotype 1 VP7-specific monoclonal antibody that mapped to amino acid 94 in a major antigenic site of serotype 1. A similar protective effect was found for antibodies that blocked binding to a serotype 3 VP7-specific epitope that maps to amino acid 94 on a major antigenic site of serotype 3. Thus, this study not only confirmed the association of protection with serum antibody but also identified a specific epitope associated with such protection. However, in other volunteer studies, the role of serum antibodies has not been consistent.[35,596]

Reinfections with rotavirus occur commonly in adult contacts of patients with rotavirus illness; however, most of these reinfections are subclinical.[297,310,537,564] Sequential rotavirus infections and illnesses have been observed in infants and young children.[172,457,630,651] However, the second infection usually causes a milder illness, indicating that immunity resulted from the initial exposure.[41]

The role of antibodies was also examined in a longitudinal study of healthy 1- to 24-month-old children residents in a "nursery."[91] A serum-neutralizing antibody titer of $1:128$ against VP7 serotype 3 human rotavirus appeared to be protective against gastroenteritis caused by this serotype, whereas levels of $1:64$ or less did not afford significant homotypic protection.

Immunity to a specific serotype appears to be induced after infection with the first homotypic strains. Studies with live attenuated rotavirus vaccines demonstrated that adults develop a significantly greater number of heterotypic serum antibody responses following vaccination with a monovalent strain than do infants less than 6 months of age by epitope-blocking assay.[207] The broadening of the antibody response may explain the epidemiologic observation that a second rotavirus infection regardless of serotype is characteristically less severe than the initial infection.

One of the perplexing areas regarding immunity is the unexplained relative sparing of neonates from rotavirus illness despite frequent infections.[20,93,556] It is suggested that the high levels of transplacental antibodies contribute to this effect. In one study of newborn babies, rotavirus infections occurred significantly less often in breast-fed infants than in bottle-fed infants.[93] The effect of breast-feeding on illness could not be determined, since most of the infections in the breast-fed and bottle-fed infants were subclinical.[20,93,556] The role of breast-feeding on rotavirus infection needs additional study because available data from several studies do not enable a definite conclusion. However, it appears from current data that the effect of breast-feeding on the prevention of rotavirus diarrhea will likely be modest.[192] Whether high levels of circulating rotavirus antibody acquired transplacentally play a role in resistance to disease during early life is not known. However, rotavirus illnesses are observed with moderate frequency in infants less than 6 months of age but beyond the neonatal period, a time when passively acquired circulating antibody is still present but not at as high a level as in neonates.[63,297]

In addition, however, such neonatal subclinical rotavirus infections induced significant protection against severe rotavirus diarrhea for up to 3 years later, but not against rotavirus infection and only limited protection against mild illness.[42] Mechanisms other than neutralizing antibodies such as cytotoxic T lymphocytes may also be important in the immune process.[418] The mechanism of immunity to rotavirus disease in humans needs further study, especially relating to the role of intestinal IgA antibody. Moreover, it was recently suggested that serum IgA antibody to rotavirus reflects the intestinal immune status to rotavirus.[245]

8. Patterns of Host Response

8.1. Norwalk Group of Viruses

8.1.1. Clinical Features. Clinical manifestations observed in the original Norwalk outbreak from which the Norwalk particle was derived demonstrate the key features of this infection. Six hundred and four individuals were considered to be primary or secondary cases: 85% had nausea, 84% vomiting, 62% abdominal cramps, 57%

lethargy, 44% diarrhea, 32% fever, and 5% chills.[2] The duration of signs or symptoms was 12–24 hr; none of the affected individuals was hospitalized. These clinical findings are similar but not identical to those observed in a report describing the findings in 31 of 52 volunteers who developed definite or probable illness following administration of the Norwalk virus.[618] Of the 31 volunteers, 45% had fever (≥ 99.4°F), 81% diarrhea, 65% vomiting, 68% abdominal discomfort, 90% anorexia, 81% headache, and 58% myalgias; clinical manifestations usually lasted 24–48 hr. The diarrheal stools characteristically do not contain gross blood, mucus, or white blood cells.[128] Of 16 volunteers who became ill following Norwalk or Hawaii virus challenge, 14 developed a transient lymphopenia.[132] The illness observed in volunteers was generally mild and self-limited, although one volunteer who vomited about 20 times within a 24-hr period required parenteral fluids.[54,129,130,618] A graphic summary of signs and symptoms of illness observed in two volunteers who developed illness following administration of the Norwalk agent is shown in Fig. 9.[130] The difference in clinical manifestations in these two volunteers who received the same inoculum is striking, since one vomited but did not have diarrhea and the other developed diarrhea but not vomiting. Shedding of Norwalk virus by volunteers as determined by IEM was maximal around the onset of illness and was rarely detected after 3 days following onset.[550] A valid estimate of the ratio of subclinical-to-clinical Norwalk virus infections has not been made. However, serologically proven infection without definite gastroenteric illness has been observed in volunteers challenged with the Norwalk virus[52] (A. Z. Kapikian, unpublished

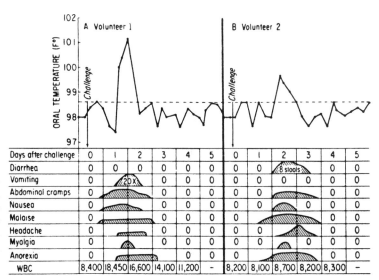

Figure 9. Response of two volunteers to oral administration of stool filtrate derived from a volunteer who received original Norwalk rectal-swab specimen. The height of the curve is directly proportional to the severity of the sign or symptom. Volunteer 1 had severe vomiting without diarrhea, while volunteer 2 had diarrhea without vomiting, although both received the same Norwalk agent inoculum. From Dolin et al.[130]

data). Although the illnesses are usually self-limited, exceptions to this pattern were noted as three middle-aged individuals were hospitalized for severe dehydration during two outbreaks, two elderly debilitated patients died during the course of the illness in a nursing home outbreak, and three patients required intravenous fluids in a nursing home outbreak.[306]

8.1.2. Diagnosis. A specific diagnosis of infection with the Norwalk group is not possible from a patient's clinical presentation. There is no simple available assay for detection of these agents, which have yet to be grown in cell culture or in a suitable animal model. IEM remains the mainstay for detection and identification of the various members of the group as a whole.[292,293,299,304] Various tests have been developed for detection of certain members of this group: an RIA and ELISA (which are even more sensitive than IEM) for detection of Norwalk virus[52,180,218,239,241,349]; an RIA and an ELISA for the Snow Mountain agent[134,349]; and an ELISA for the Hawaii virus.[557] Direct EM examination of negatively stained stool material with or without prior concentration may also be attempted, but this is not usually satisfactory because of the relatively small amounts of virus shed.[62] Although EM examination of stools is a simple and relatively rapid procedure, caution must be used in interpreting the significance of "particles" observed, because stools contain a myriad of small objects that have no relationship to the illness being studied.[292] It is for this reason that carefully controlled IEM studies with appropriate paired sera should be carried out (under code) to determine the significance of the particles. Ideally, a patient's paired sera should show an IEM antibody increase with the observed particle as antigen (see Fig. 3); in addition, if paired sera are not available, careful IEM studies with γ-globulin, paired sera from other individuals in the same outbreak or from other similar outbreaks, or antisera to morphologically similar agents should be studied to determine the significance of the particle in question. Simple aggregation of particles by a serum should not be taken as evidence of a specific response, since certain particles, such as Norwalk virus, aggregate spontaneously without the addition of serum. These nonspecific aggregates may appear to be lightly coated with antibody. Thus, it is essential to quantify the antibody (on a scale of 0–4+) coating the particles even if they are aggregated. If there is any question about the significance of aggregation, the concentration of antigen or antibody should be varied. Such maneuvers should affect both the size of the aggregates and the amount of antibody coating the particles as the reaction proceeds from antigen excess to antibody excess.[292,304] Awareness of the specificity of aggregation is essential, because some stools contain groups of 22-nm "objects" that appear in "aggregate" form with little or no "antibody" on them and that have had no known relationship to the illness being studied.[292,304] These aggregates generally appear similar with regard to size and "antibody" coating when examined with paired acute or convalescent sera.

IEM is the only way to detect serological evidence of infection for a potential new member of the group. Serological diagnosis of Norwalk infection can be made by IEM, but the RIA- and ELISA-blocking tests are more efficient and practical.[52,180,218,241] RIA- and ELISA-blocking tests are available for the Snow Mountain virus and an ELISA-blocking test has been developed for the Hawaii virus.[134,349,557] However, as noted earlier, a major breakthrough in the study of these viruses occurred when the recombinant Norwalk viruslike particle (outer capsid) became available.[272] This recombinant is used as the antigen in a newly developed direct ELISA for detection of a seroresponse.[205,272] It is considered to be as specific, sensitive, and efficient as the native Norwalk virus present in stool that was the source of antigen previously. The availability of an unlimited source of recombinant antigen has already facilitated seroepidemiologic studies of this virus. An immune adherence hemagglutination assay had also proved useful for detecting serological evidence of Norwalk virus infection, but the requirement for a relatively large amount of antigen was a distinct disadvantage for its general use because the source of antigen was a particle-positive stool.[296] The assays for these 27-nm viruses are currently limited to research laboratories because reagents are not yet generally available.

8.2. Rotaviruses

8.2.1. Clinical Features. The three major clinical manifestations observed in infants and young children hospitalized with rotavirus gastroenteritis are vomiting, diarrhea, and dehydration. Signs and symptoms in 72 patients hospitalized with rotavirus diarrhea were compared with those of 78 patients hospitalized with a nonrotavirus diarrheal illness (Table 7).[454] The rotavirus group experienced significantly more vomiting and dehydration. The dehydration was isotonic in 95% of the patients in the rotavirus-infected group and in 77% of the rotavirus-negative group. As determined from history and hospital records, the mean duration of vomiting was also longer in the rotavirus-infected group (2.6 vs. 0.9 days). Diarrhea began later and lasted longer than vomiting in the rotavirus group (mean duration of diarrhea vs. vomiting, 5 days vs. 2.6 days). Once the patient was hospitalized, diarrhea continued for an average of 2.6 days (range 1–9 days) in the rotavirus group and 3.8 days

**Table 7. Clinical Characteristics of 150 Children
Hospitalized with Acute Gastroenteritis**[a]

	Percentage having each clinical finding	
Clinical finding	Rotavirus infection detected (72 patients)	Rotavirus infection not detected (78 patients)
Vomiting	96[b]	58[b]
Fever (°C)		
37.9–39	46	29
> 39	31	33
Total	77	61
Dehydration	83[c]	40[c]
Hypertonic	5	16
Isotonic	95	77
Hypotonic	0	6
Irritability	47	40
Lethargy	36	27
Pharyngeal erythema	49	32
Tonsillar exudate	3	3
Rhinitis	26	22
Red tympanic membrane with loss of landmarks	19	9
Rhonchi or wheezing	8	8
Palpable cervical lymph nodes	18	9

[a]Adapted from Rodriguez et al.[454]
[b]$P < 0.01$.
[c]$P < 0.01$.

(range 1–16 days) in the nonrotavirus group. The duration of hospitalization ranged from 2 to 14 days (mean 4 days) for the rotavirus group. The greatest frequency of rotavirus diarrhea was in the 6- to 24-month age group.

Notable laboratory findings were related to the degree of dehydration.[454] Elevated BUN (>18 mg/dl) and urine specific gravity (>1.025) were observed in 58 and 71%, respectively, of the rotavirus group, frequencies significantly greater than those observed in the nonrotavirus group.

Deaths have been reported in infants and young children with rotavirus illness.[76,114,381,382,385,386] In a study in Canada, 21 deaths were reported between May 1972 and March 1977 in infants and young children with a rotavirus illness. Ten of the 21 children were dead on arrival at the hospital, and ten of the others were moribund and could not be successfully resuscitated on arrival.[76] One child was already in the hospital when he acquired the disease; this patient had congestive cardiomyopathy that contributed to his death. With the exception of this patient and one other, all of the children had been healthy previously. They ranged in age from 4 to 30 months, with a mean of 11 months. Deaths occurred 1 to 3 days after onset of symptoms. The major factor causing death in these children was dehydration and electrolyte imbalance in 16, aspiration of vomitus in 3, and in the remaining 2, seizures

were a contributing factor. The rapid course of the disease is evidenced by the fact that the parents of 16 of the 20 children brought to the hospital had had some contact with a physician during the course of the illness.[76]

Rotaviruses can cause a chronic symptomatic infection in immunodeficient children.[267a,429,491,608] In addition, they may cause serious sequelae that can be life threatening in individuals immunosuppressed for bone marrow suppression.[635] Rotavirus does not appear to have an important role in the etiology of diarrhea in HIV-infected adult patients.[221]

Rotavirus infections have also been associated temporally on rare occasions with other conditions that occurred as isolated cases or single outbreaks such as intussusception,[319,320,400,413] self-limited gastrointestinal bleeding,[120,530,531,547] Henoch–Schoenlein purpura,[120] Reye's syndrome,[475] encephalitis,[475,578] aseptic meningitis,[607] hemolytic uremic syndrome,[616] disseminated intravascular coagulation,[616] elevated serum transaminase levels,[135,537] sudden infant death syndrome,[641] exanthem subitum,[474] Kawasaki syndrome,[144,369] necrotizing enterocolitis and hemorrhagic gastroenteritis in special-care nurseries,[119,231,308,462,463] pneumatosis intestinalis,[308] chronic diarrhea,[156,537] abortion,[74,156] febrile or afebrile convulsions,[341,393,419,581] hyperphosphatasemia,[250] pneumonia,[487] hepatic abscess,[222] acute laryngitis,[414] colitis,[174] benign acute myositis,[237] pancreatitis,[415] and hyperechoic lesions in the basal ganglia.[599] Whether the rare temporal association of these conditions with these ubiquitous viruses is coincidental or significant remains to be determined. However, a primary etiologic relationship appears extremely unlikely, except in the severe illnesses in immunocompromised patients and in some cases of necrotizing enterocolitis and hemorrhagic gastroenteritis in neonates and pneumatosis intestinalis in infancy. Growth of rotavirus has also been described in tissue cultures inoculated with filtrates prepared from intestinal tissue of patients with Crohn's disease, but this observation could not be confirmed.[285,603]

In the volunteer studies in which the VP7 serotype 1 strain was administered orally to volunteers, four developed a diarrheal illness that began 2 to 4 days after inoculation.[301,302] Two of the four volunteers with diarrhea also vomited, one the day after inoculation (2 days before the onset of diarrhea) and the other 3 days after inoculation (the day of onset of diarrhea). The average duration of diarrhea was 2.5 days, with a range of 1 to 4 days. The number of diarrheal stools per illness ranged from 1 to 24, with one volunteer having a maximum of 11 in 1 day. Thus, under experimental as well as natural conditions, adults can develop a rotaviral diarrheal illness. However, subclinical rotavirus infection in adults occurs much more frequently, as demonstrated in one study in which 22 of 50 adult family contacts of pediatric patients hospitalized with rotavirus gastroenteritis themselves developed serological evidence of rotavirus infection at or about the time of their children's hospitalization[310]; however, only three infected parents had a gastroenteric illness at or about the time of their children's illnesses.

8.2.2. Diagnosis. As with the Norwalk group, a specific diagnosis of infection with human rotavirus cannot be made by clinical presentation. Even though rotavirus infections follow a predictable seasonal pattern of high prevalence during the cooler months in temperate climates, a laboratory diagnosis is essential, because other agents may also cause gastroenteritis even during these periods.

Although human rotaviruses can now be grown in cell cultures, this is not a practical method for detection from clinical specimens. Numerous assays have been developed for the detection of rotaviruses, as outlined previously in Table 1. The most widely applied methods aim at detection directly from stool specimens. Electron microscopy is highly specific, because rotaviruses have such a distinct morphological appearance; it is limited, however, by the requirement for an electron microscope as well as a capable operator. It provides the most rapid

diagnosis when dealing with only a limited number of specimens.[62] In addition, it has the advantage of recognizing the non-group-A rotaviruses as well as other viral agents of gastroenteritis in a single specimen.

Other efficient but more practical assays for large numbers of specimens include ELISA, RIA, latex agglutination, and counterimmunoelectro-osmophoresis.[38,62] ELISA is the most practical diagnostic method for large-scale studies and is limited only by the availability of suitable reagents; however, many kits are now available commercially. However, some of these are prone to false-positive reactions. Thus, we recommend that a confirmatory ELISA in which a pre- or postrotavirus immunization serum be used as the solid-phase precoat.[62,304,638,648] In this way, the difference in a specimen's reactivity with the pre- or postserum is determined. Some specimens react equally with both sera, thus preventing a diagnosis. However, without the control serum, such a specimen would be considered positive for rotavirus. A dot hybridization assay as well as the application of PCR technology have further increased the sensitivity of virus detection.[164,631] Thus, there are several efficient and practical methods for detecting rotaviruses; the method of choice will vary according to the resources and experience of individual laboratories. An enzyme immunoassay and a PCR assay have also been developed for detection of infection with group B or group C rotavirus.[146,175,198,409,561]

Rotaviruses can be serotyped (or genotyped) by various methods such as neutralization in cell culture, SPIEM, ELISA, hybridization, sequencing, and PCR.[26,28,37,104,162, 166, 168,185,188,200,201,203,204,206,238,387,501,517,541,544,546,552,576,577, 579] The availability of monoclonal antibodies for each of the four clinically important serotypes has made the ELISA the most practical serotyping method.

Serological evidence of rotavirus infection may be detected by a variety of techniques (see Table 2). Complement fixation is efficient and practical when testing sera from pediatric patients about 6–24 months of age.[304,646] However, it is not as efficient as other techniques when testing sera from patients less than 6 months of age and from adults.[646] Serological evidence of rotavirus infection may be detected in these age groups by ELISA or IF. ELISA has also been employed to measure specific immunoglobulin responses in rotavirus infection.[373,649] An ELISA for serum IgA antibodies (which do not cross the placenta) is particularly useful for measuring antibody responses in infants under 6 months of age because they possess passively acquired maternal IgG antibodies that may confound an IgG ELISA.[121,344,345] In addition, serum IgA antibodies may reflect the immunologic status of the intestine.[117,244,245] Neutralizing antibodies in tissue culture systems are particularly important when serotype-specific antibody rsponses are sought.[12,283,585] A competition-solid-phase immunoassay that detects epitope-specific immune responses to individual serotypes has proven important in determining immune responses to individual epitopes.[204,207,361,504] As long as the limitations of the various methods are recognized, the method of choice will vary according to the resources and experience of individual laboratories.

9. Control and Prevention

9.1. Norwalk Group of Viruses

Methods are not available for the prevention or control of infection or illness with the Norwalk group of agents. Since this group of agents is highly contagious and transmitted by the fecal–oral (or vomitus–oral) route, it is possible that in a family or group setting where one member is ill with this form of gastroenteritis, effective handwashing and disposal or disinfection of contaminated material could decrease the likelihood of transmission. Increased vigilance concerning the purity of drinking water or of water in swimming pools as well as careful attention to hygiene by food handlers and concern regarding the source of uncooked shellfish might also limit the number of outbreaks.

Treatment of gastroenteritis caused by the Norwalk group usually consists of replacement of fluid loss by the administration of liquids orally. Parenteral intravenous fluid therapy is only rarely necessary in this form of self-limited gastroenteritis.[128–130] The impact of this group of agents in debilitated hosts has not been evaluated systematically, although a few deaths in elderly debilitated individuals have been reported.[306] In volunteers, oral administration of bismuth subsalicylate after onset of symptoms significantly reduced the severity and duration of abdominal cramps and the median duration of gastrointestinal symptoms; however, it did not significantly affect the number, weight, or water content of stools.[528]

Because of the paradoxical effect of antibody on the susceptibility or resistance to Norwalk virus infection or illness, it is premature to consider vaccination strategies. However, if the immune mechanisms are elucidated more clearly and antibodies are found to be important in preventing illness, a vaccine may be relevant in preventing epidemic gastroenteritis, especially in special groups such as military personnel and individuals living closely in an institutional setting. The need for such a vaccine in pediatric groups would need further study.

9.2. Rotaviruses

There are no methods for the prevention or control of infection or illness with rotaviruses. Although improved hygiene is of course generally desirable, it does not appear that such measures would markedly affect the transmission of rotavirus infection. As noted earlier, the prevalence of rotavirus infection is similar in both developed and developing countries, regardless of hygienic conditions. Therefore, the development of a rotavirus vaccine is of primary importance.[143]

Treatment of rotavirus gastroenteritis has as its goal the replacement of fluids and electrolytes lost by vomiting and diarrhea. Treatment must be initiated early in the course of illness because rotavirus gastroenteritis can lead to rapid dehydration. Administration of intravenous (IV) fluids and electrolytes are very effective in the treatment of rotavirus dehydration, but this form of therapy is not readily available in many parts of the world. As a result, in seeking alternate approaches, the administration of oral rehydration salts (ORS) solutions was found to be comparable in effectiveness to IV therapy, and thus has provided a major advance in treatment.[281,411,469,471,472,485] In certain instances, when ORS fails to correct the fluid and electrolyte deficit or if the patient is severely dehydrated or in shock, IV therapy must be instituted.

Passive immunization by the oral route with various rotavirus antibody-containing preparations has been effective in selective circumstances in preventing dissemination of rotavirus infections in a nursery or in prevention of nosocomial infections.[22,118,136] Various forms of antibody-containing preparations have been described.[22,118,136,639,640] Passive immunization was effective in the treatment of rotavirus illness in immunodeficient infants and young children and in a recent study in the treatment of normal infants with naturally occurring rotavirus infection.[223,491]

Bismuth subsalicylate (BSS) given orally for 5 days as an adjunct to rehydration therapy in infants and young children was associated with a more rapid recovery from illness.[518,519] The BSS group had shorter times until the last loose stool (57.5 hr vs. 104.5 hr) and the last unformed stool (113 hr vs. 167.7 hr) when compared to the placebo group. Because of the reported correlation between the use of salicylates and Reye's syndrome, the possibility of such an association with BSS or other nonacetylsalicylic acid salicylates was considered and none was found.[518,519]

As noted earlier, rotaviruses represent a major cause of severe diarrhea of infants and young children in both developed and developing countries. Thus, it is clear that a rotavirus vaccine is needed.[283,284,294,300] Although diarrheal illnesses are not a major cause of mortality in the developed countries, they are the leading cause of death in infants and young children in many developing countries, being associated with five to ten million deaths.[593] It has been estimated that in developing countries, rotaviruses are responsible for over 870,000 deaths in children under 5 years of age.[266] It is likely that a rotavirus vaccine would make an important impact on reducing the morbidity and mortality from rotavirus diarrhea in infants and young children.[125]

It has been suggested, however, that enterotoxigenic *E. coli* may be a more important cause of death from diarrhea in developing countries than rotaviruses.[140,334] For example, in the United States in New York City in the early 1900s, there was a staggering infant mortality rate with a large proportion of deaths attributable to outbreaks of summer diarrhea in slum tenements.[334] The infant death rate declined markedly in the next decades, and it has been suggested that this decline was not because of better medical management of summer diarrhea, but rather was the result of development of improved sanitary conditions such as iceboxes, flush toilets, and water supply, which limited bacterial contamination.[334] Indeed, despite the advanced sanitary conditions and high standards of living in the United States today, almost all persons still undergo rotavirus infection by the end of the third year of life, though mortality from diarrheal illnesses is infrequent in the United States. Decline in mortality from diarrheal diseases in developed countries has resulted in part from the availabilty of fluid replacement therapy and possibly better nutrition, but undoubtedly other factors have played a major role, such as the decline in incidence of bacterial diarrheas as sanitation improved.[334] However, from longitudinal and cross-sectional studies in developing countries, it appears that rotaviruses and enterotoxigenic *E. coli* are major causes of clinically significant diarrhea of infants and young children, with rotaviruses being responsible for a disproportionately high percentage of severe illnesses in relation to their overall contribution to diarrhea of any severity.[50,51]

Animal studies cited earlier clearly indicate that antibody in the intestinal lumen plays a major role in resistance to rotavirus disease. In experimentally infected animals, serum rotavirus antibody in the absence of intestinal antibody was not effective in preventing rotavirus illness. Thus, one approach in the control of rotavirus illness may be the encouragement of breast-feeding as a means of providing local antibody to the young infant. Colostrum and milk contain IgA rotavirus antibody, and it may be that such antibody would exert some protective role against rotavirus illness in the infant and young child.[650] If a successful rotavirus vaccine were developed, it might be beneficial to immunize the mother to raise the level of antibodies in her breast milk for transfer to the intestine of the infant. One discouraging aspect of this approach is the high frequency of diarrheal diseases in general, including those associated with rotavirus, in countries where infants and young children are breast-fed for extended periods.[192] However, the nutritional status of the nursing mother may be a critical factor in this observation. This approach to vaccination is not considered to be practical because it would likely entail the parenteral rather than oral administration of a rotavirus vaccine because of the relatively high levels of preexisting maternal antibodies that would neutralize an orally administered vaccine.

The aim of a successful vaccine is to prevent serious illness during the first 2 years of life, when the outcome of such infection may be especially serious or fatal.[87,284,294,300] Thus, the vaccine would be administered orally to induce local IgA antibody within the first 6 months of life or perhaps even neonatally, since it has been estimated that 10 to 40% of infants born in developing countries are in contact with a health care provider only at the time of birth.[232]

Approaches to the development of a rotavirus vaccine range from conventional cell culture cultivation of human or animal rotavirus strains to molecular biological techniques. The most extensively evaluated method involves the "Jennerian" approach in which an antigenically related rotavirus from a nonhuman host such as a calf or monkey is used as the immunizing agent.[283,290,291,294,366,584,588,612] The feasibility of this approach was tested in calves. Calves were inoculated *in utero* with calf rotavirus (serotype 6) or with placebo (or nothing).[260,627,629] Shortly after birth, the calves were challenged with a human rotavirus serotype 1 strain, and it was found that *in utero* infection with calf rotavirus induced resistance to disease caused by challenge with this human rotavirus strain. In contrast, animals that had received placebo (or nothing) developed illness on challenge with the human rotavirus serotype 1 strain soon after birth.[260,627,629] Thus, cross-protection between the calf and human rotavirus was demonstrated, indicating that the bovine virus was sufficiently related antigenically to the human rotavirus strain to induce protection.[629]

Extensive clinical evaluation of a bovine NCDV rotavirus strain has been carried out.[585,587] Efficacy trials with this strain in Finnish children 1 year of age or less demonstrated a protection rate of over 80% against clinically significant diarrhea[586] However, such trials in developing countries yielded less-encouraging results, and therefore this vaccine was withdrawn from further studies.[585] Similarly, a related bovine rotavirus strain (WC3) also demonstrated variable efficacy and was also withdrawn from further study in its current formulation.[585,587]

A rhesus-monkey-derived rotavirus strain has also undergone extensive clinical evaluation as a vaccine candidate.[283,284] This vaccine also demonstrated variable efficacy that ranged from 85% to nil against moderately severe to severe diarrhea. This was attributed to the failure of the vaccine to protect against heterotypic strains in individuals undergoing primary infection.

Because the "Jennerian" concept was not successful in protecting against each of the four epidemiologically important rotavirus serotypes, another approach involving the use of rotavirus reassortants as vaccines was undertaken. Such reassortants were constructed by coinfection of cell cultures with two different rotavirus strains under selective pressure of antibody against one of them.[211,388,389] In this way, the rhesus rotavirus strain was the donor of attenuating genes and a human rotavirus of a specific serotype was the donor of a single gene that codes for the major neutralization protein (VP7). Such single-gene-substitution reassortants have been constructed for human rotavirus serotypes 1, 2, 3, and 4, with the rhesus (MMU18006) and/or bovine (UK) rotavirus strains as the donor of the other ten genes.[211,388,389] Efficacy trials with monovalent or quadrivalent formulations have been completed or are underway with generally encouraging results.[95,124,284,329,330,347,470,589]

In yet another approach, rotavirus strains isolated from neonates with asymptomatic infections may be important for immunoprophylaxis because they appear to be naturally attenuated, and strains belonging to each of the human rotavirus serotypes are now available. A VP7 serotype 1 neonatal strain M37 has been evaluated for efficacy in Finland; it failed to induce protection against VP7 serotype 1 illness in a preliminary study.[167,196,258,259,433,590] Other approaches such as cold-adapted human rotavirus strains or substitution of "virulence" genes with "avirulence" counterparts are under consideration as viable vaccine candidates.[254–256,367] Finally, molecular biological approaches may also yield potential vaccine candidates.[166,169,467]

Thus, it is hoped that an effective immunogen will be developed for rotavirus. However, it should be stressed that since a human rotavirus vaccine has not yet been licensed for general use, effective treatment for rotavirus diarrhea is available in the form of fluid and electrolyte replacement therapy by the oral or parenteral route of administration.[411,471,472,486] Thus, one means of controlling the severe morbidity and mortality from rotavirus diarrhea would be to make available fluids and electrolytes necessary for rehydration. In addition, since this agent is transmitted by the fecal–oral route, careful attention to handwashing, disinfection, and disposal of contaminated material would appear to be one way of limiting the spread of this highly contagious agent, especially in nurseries and hospitals, where nosocomial infections are common.

10. Unresolved Problems and Other Agents or Putative Agents of Viral Gastroenteritis

10.1. Norwalk Group and Other Enteric Viruses

Efforts must be made to determine the number of serotypes responsible for epidemic viral gastroenteritis. Such studies entail careful IEM studies to determine antigenic relationships. The development of radio- and enzyme immunoassays for Norwalk virus has permitted the study of the epidemiologic importance of this agent. However, the cloning, sequencing, and expression of the Norwalk virus and the Southampton virus have opened up a new era of research possibilities that should enable the elucidation of the natural history and the genetic interrelationships of these agents.

A major impediment to the study of the viruses is the inability to propagate them in any cell culture. This is an important problem that must be resolved.

Studies of immunity to Norwalk agent have raised rather perplexing questions because of the paradoxical effect of preexisting antibodies on the susceptibility to illness. It has been suggested that this may be due to a genetic factor, such as a receptor for Norwalk virus, that is lacking in one cohort of individuals and present in the other.[426] The role of local IgA antibody should also be explored further. Resolution of this question is essential before considering vaccine strategies.

Finally, a major unresolved area in the etiology of epidemic viral gastroenteritis is the role of other small round viruses such as astroviruses, the morphologically classical caliciviruses, minireoviruses, the Otofuke-like viruses, and the small round viruses.[286] Some of these agents, such as the astroviruses, have been studied rather intensively.

Astroviruses, which are now classified in the new family *Astroviridae*, are 28 nm in diameter and derive their name from the five- or six-pointed star-shaped configuration observed by negative staining in about 10% of particles.[326] They have a buoyant density of 1.35–1.40 g/cm^3 in cesium chloride[326] and a positive-sense single-stranded RNA genome.[326] Seven distinct serotypes are recognized and each has been grown in cell culture.[326,332a]

These viruses have been linked to epidemic or endemic mild gastroenteritic illness in infants and young children.[242,326,336] They have been associated with outbreaks of gastroenteritis in newborn nurseries and pediatric wards, in community settings, and in nursing homes.[326] Studies of the prevalence of astrovirus antibody have demonstrated a rather rapid acquisition of antibody, so that over 70% of 5-year-old children have acquired antibody.[326] They have only rarely been associated with diarrheal illnesses requiring hospitalization. They were associated with 8.6% of gastroenteritis illnesses in an outpatient pediatric study in Thailand (2% in controls) and in Guatemala in 7.3% of diarrhea episodes (2.4% in diarrhea-free periods) with the highest frequency in children less than 1 year of age in both studies.[242] In day-care center studies in the United States, astroviruses were detected in 4% of the children with diarrhea (less than 1% of the controls).[336]

Astroviruses are transmitted by the fecal–oral route. In volunteer studies, they are of low pathogenicity in adult volunteers, with only 2 of 36 developing vomiting and diarrhea after oral challenge, although over one half developed serological evidence of infection.[327a,388a]

Diagnosis is limited to research laboratories with EM, EIA, and IF used to detect infection.[326] Astroviruses have also been detected in stools of various animals with and without diarrhea.[326]

Caliciviruslike particles, which are about 32–40 nm in diameter and have characteristic cuplike configurations on their surface, have also been studied rather intensively.[106] Gastroenteritis in infants and young children in Japan, England, and Canada has been associated with such particles.[106] In addition, such particles were found in the small intestine of a 22-month-old infant who died of acute gastroenteritis.[161a] As noted earlier, the Norwalk virus is now definitely classified as a calicivirus. Recent molecular biological advances should clarify the relationship of the Norwalk virus with the so-called "classical" (in morphology) caliciviruses.

Another particle, the Otofuke agent, 34–38 nm in diameter with a density of 1.35–1.37 g/cm³, has been associated with an outbreak of gastroenteritis in a work-training facility for mentally deficient persons 15 years of age or older.[573] The antigenically related Sapporo agent was detected in a gastroenteritis outbreak in infants and young children in an orphanage.[313] The 30–32 nm "minireoviruses" have been found in stools of pediatric patients with nosocomial gastroenteritis and with gastroenteritis requiring hospitalization.[386,523] They have recently been found to belong to the calicivirus family by cloning and sequencing techniques.[337] Their role in diarrheal illness needs to be elucidated. Small round viruses

(20–32 nm) have also been found in stools of pediatric patients with diarrhea.[40] Their role in disease is unknown.

Two other groups that have no morphological similarity to the Norwalk virus are the adenoviruses and the coronaviruses. Fastidious adenoviruses, which did not grow in conventional cell cultures known to support the growth of other adenoviruses, have been observed in stools of infants and young children hospitalized with diarrhea.[592] They are referred to as fastidious or enteric adenoviruses. The latter term is somewhat confusing since conventional adenoviruses, which are indistinguishable morphologically by EM, may also be detected in the enteric tract. The fastidious enteric adenoviruses are detected in stools by EM or by an immunoassay.[592]

The fastidious adenoviruses that have been associated with diarrheal illnesses of infants and young children measure 70–80 nm in diameter and are DNA viruses with a buoyant density of 1.34 g/cm³ in cesium chloride.[592] They belong to serotypes 40 and 41 by neutralization and to the F subgroup based on the pattern of DNA fragments produced by the digestion of their nucleic acid by restriction endonucleases as observed on agarose gels by electrophoresis.[592] They can be propagated in adenovirus-5-transformed Graham 293 HEK cells and certain other cells also.[592] They have a worldwide prevalence. About 50% of children acquire antibodies by 4 to 5 years of age.[592] It appears that the fastidious adenoviruses rank second to rotaviruses as viral etiologic agents of diarrhea of infants and young children that requires hospitalization.[571,592] In a Washington, DC, study, 4.4% of children hospitalized with diarrhea and 1.8% of controls shed an adenovirus in stools by EM.[60] In Sweden, fastidious adenoviruses were associated with 7.9% of the acute diarrheal episodes in children who were hospitalized or seen as outpatients.[572] In comparison, rotaviruses were detected in 45% of children with acute diarrheal illnesses. In Korea, 9% of the children hospitalized with diarrhea (mean age 11 months) shed adenoviruses (2% controls), whereas in Italy, 8.3% of hospitalized infants and young children shed adenovirus presumptively.[85,311] Adenoviruses have also been associated with infection and illness in bone marrow transplant patients. They were detected in 12 of 31 bone marrow transplant patients with gastroenteritis, six of whom died.[635]

Adenoviruses have also been detected in stools of 15% of patients hospitalized with gastroenteritis in a Canadian study.[450] In addition, adenoviruses were associated with the deaths of three children with severe gastroenteritis; adenovirus antigen was detected in the jejunal cells of two of the children by IF and in one by thin-section EM.[450,602] Adenoviruses have also been associated with a gastroenteritis outbreak in a long-stay children's ward.[159] They have also been found in small-intestinal fluid of pediatric patients with gastroenteritis[372]; in such patients, D-xylose absorption was impaired. Adenovirus infection has also been associated with intussusception.[96,179]

Coronaviruses are established as etiologic agents of diarrheal disease in many animals, but they have not yet been implicated conclusively as important etiologic agents of infantile gastroenteritis in humans.[78,430] They are difficult to detect conclusively in stools by EM because of the plethora of fringed objects reminiscent of coronaviruslike particles.[78,83,430] There are reports of the detection of coronaviruses in stools by EM in three outbreaks of gastroenteritis in adults and one in neonates, with the particles from one of the adult outbreaks being propagated in organ and cell cultures.[82,84,582] Fringed coronaviruslike particles were also associated with an outbreak of severe hemorrhagic enterocolitis in premature and full-term newborn infants in France, with two deaths.[88] In this same study, similar particles were detected in outbreaks of diarrhea in 3- to 24-month-old children.[88,89] However, these particles have not been described further, and the issue of their nature and origin is still unresolved.[331,428] Human serum has been shown to contain neutralizing antibody to calf coronavirus; however, since the human respiratory coronavirus OC43 and the calf coronavirus share some antigenic relationship, it is not certain whether this antibody is related to OC43 or to another human coronavirus.[307,503] Antibody to human enteric coronaviruslike particles has also been detected in the serum of Australian aborigines by IEM.[494]

In recent studies in Italy and the United States, coronaviruses have been associated with acute infantile gastroenteritis and necrotizing enterocolitis, respectively. In addition, representative particles were characterized and appeared to be true coronaviruses[25,186,449] distinct antigenically from the Breda–Berne group of fringed viruses, which were recently implicated in diarrheal illnesses of humans and calves.[25,609]

Other agents detected in stools of infants and young children include the pleomorphic fringed Breda or Berne-like viruses (toroviruses) with a diameter of 100–140 nm, the 35m picobirnaviruses, and a pestivirus antigen.[181,431,432,609,636] Their contribution to the etiology of viral gastroenteritis needs further study.

10.2. Rotaviruses

Important advances have been made in the development and clinical evaluation of rotavirus vaccines. However, with this progress comes the need to address fundamental questions on the nature of the immune response to rotavirus infections. What is the role of serum or local intestinal antibodies in preventing or modifying rotavirus illness? Must a rotavirus vaccine be administered by the oral route to be effective, or will parenteral immunization also be important? Will the modified "Jennerian" approach, which aims at inducing antibody responses to each of the four epidemiologically relevant serotypes, be effective? Can a rotavirus vaccine be formulated that protects against all episodes of rotavirus diarrhea and not only against all episodes of severe rotavirus diarrhea (with the latter being the current expectation)? Can neonatal rotavirus strains be used as vaccines because they are usually associated with subclinical infections? Can neonates be immunized successfully with rotavirus? Can an oral rotavirus vaccine be given simultaneously with oral poliovirus vaccine without interference?[190] How stable will an attenuated oral rotavirus vaccine strain be after being shed in stools? Will it induce secondary infections or illnesses? Must an oral rotavirus vaccine be administered with buffer because rotaviruses are acid labile at pH 3? Will breast milk interfere with the "take" of an oral rotavirus vaccine? Would vaccinating expectant mothers to boost their breast milk and serum antibody titers to rotavirus have any impact on rotavirus morbidity and mortality in infants? What will be the impact of a rotavirus vaccine on the overall occurrence of severe diarrheal disease and on mortality from diarrheal diseases in a developing country? Why are rotavirus infections predominantly subclinical in neonates? Are there reservoirs for rotaviruses? Practically every animal studied has been found to have a virulent indigenous rotavirus. Although there is evidence of transmission of animal rotaviruses or of human–animal rotavirus reassortants to humans, is this of epidemiologic relevance now or in the future? Will serotypes other than 1, 2, 3, and 4 assume epidemiologic importance? Another area to be resolved concerns the role of rotavirus infection in malnutrition and the effect of malnutrition on rotavirus infection. The possible role of breast milk in prevention or attenuation of rotavirus diarrhea must also be clarified further. Although there is evidence that breast milk can exert an effect on rotavirus shedding, its role in the prevention of rotavirus diarrhea remains to be established. The synergism, if any, between rotaviruses and bacteria needs study. In animals, certain bacteria act synergistically with rotavirus to cause more severe illness than if either were present alone.[568] The worldwide importance of rotaviruses has increased the demand for reagents for study. Suitable reagents for ELISA for detecting human rotaviruses and for typing them are necessary. Finally, the natural history, importance, and distribution of the non-group-A rotaviruses need to be determined. Methods to propagate these strains

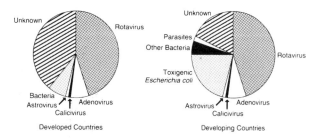

Figure 10. An estimate of the role of various etiologic agents in severe diarrheal illnesses requiring hospitalization of infants and young children in developed and developing countries. From Kapikian.[280]

efficiently in cell culture are needed as well as simple assays to detect and identify them.

11. Summary

Although numerous microbial agents infect the intestinal tract,[412] it is clear that the single most important etiologic agents of severe diarrhea of infants and young children in both developed and developing countries are the group A rotaviruses.[286] The relative role of viral, bacterial, and parasitic agents as a cause of severe diarrhea in infants and young children worldwide is estimated schematically in Fig. 10.[280] Hopefully, the unknown portions of these "pies" will be elucidated in the not-too-distant future.

12. References

1. ADAMS, W. R., AND KRAFT, L. M., Epizootic diarrhea of infant mice: Identification of the etiologic agent, *Science* **141**:359–360 (1963).
2. ADLER, I., AND ZICKL, R., Winter vomiting disease, *J. Infect. Dis.* **119**:668–673 (1969).
3. AGUS, S. G., DOLIN, R., WYATT, R. G., TOUSIMUS, A. J., AND NORTHRUP, R. S., Acute infectious nonbacterial gastroenteritis: Intestinal histopathology. Histologic and enzymatic alterations during illness produced by the Norwalk agent in man, *Ann. Intern. Med.* **79**:18–25 (1973).
4. AIYAR, J., BHAN, K. M., BHANDARI, N., KUMAR, R., RAJ, P., AND SAZAWAL, S., Rotavirus-specific antibody response in saliva of infants with rotavirus diarrhea, *J. Infect. Dis.* **162**:1383–1384 (1990).
5. ALBREY, M. B., AND MURPHY, A. M., Rotavirus and acute gastroenteritis of infants and children, *Med. J. Aust.* **1**:82–85 (1976).
6. ALBREY, M. B., AND MURPHY, A. M., Rotavirus growth in bovine monolayers, *Lancet* **1**:753 (1976).
7. ALDRICH, R. A., Introduction to pediatrics: The change from pediatrics to child health and human development, in: *Brenne-*

mann's *Practice of Pediatrics* (V. C. KELLY, ED.), pp. 1–28, Harper & Row, New York, 1972.
8. ALMEIDA, J. D., HALL, T., BANATVALA, J. E., TOTTERDELL, B. M., AND CHRYSTIE, I. L., The effect of trypsin on the growth of rotavirus, *J. Gen. Virol.* **40**:213–218 (1978).
9. ALMEIDA, J. D., AND WATERSON, A. P., The morphology of virus antibody interaction, *Adv. Viral Res.* **15**:307–338 (1969).
10. ANDERSON, T. F., AND STANLEY, W. M., A study by means of the electron microscope of the reaction between tobacco mosaic virus and its antiserum, *J. Biol. Chem.* **139**:339–344 (1941).
11. APPLETON, H., Small round viruses: Classification and role in food-borne infections, in: *Ciba Foundation Symposium 128. Novel Diarrhoea Viruses*, pp. 108–125, John Wiley & Sons, New York, 1987.
12. APPLETON, H., BUCKLEY, M., THOM, B. T., COTTON, J. L., AND HENDERSON, S., Virus-like particles in winter vomiting disease, *Lancet* **1**:409–411 (1977).
13. APPLETON, H., AND PEREIRA, M. S., A possible virus etiology in outbreaks of food-poisoning from cockles, *Lancet* **1**:780–781 (1977).
14. ARDENNE, M., VON FRIEDRICH-FRESKA, H., AND SCHRAMM, G., Electronenmikrokopischen untersuchung der precipitin Reaktion von tabakmosaik Virus mit kaninchem Anti-serum, *Arch. Ges. Virusforsch.* **2**:80–86 (1941).
15. ARISTA, S., GIOVANNELLI, L., PISTOIA, D., CASCIO, A., PAREA, M., AND GERNA, G., Electropherotypes, subgroups, and serotypes of human rotavirus strains causing gastroenteritis in infants and young children in Palermo, Italy, from 1985 to 1989, *Res. Virol.* **141**:435–448 (1990).
16. ASHLEY, C. R., CAUL, E. O., CLARK, S. K. R., CORNER, B. D., AND DUNN, S., Rotavirus infection of apes, *Lancet* **2**:477 (1978).
17. ASKAA, J., AND BLOCH, B., Infection in piglets with a porcine rotavirus-like virus. Experimental inoculation and ultrastructural examination, *Arch. Virol.* **80**:291–303 (1984).
18. ATMAR, R. L., METCALF, T. G., NEILL, F. H., AND ESTES, M. K., Detection of enteric viruses in oysters by using the polymerase chain reaction, *Appl. Environ. Microbiol.* **59**:631–635 (1993).
19. BACHMANN, P. A., BISHOP, R. F., FLEWETT, T. H., KAPIKIAN, A. Z., MATHAN, M. M., AND ZISSIS, G., Nomenclature of human rotaviruses: Designation of subgroups and serotypes, *Bull. WHO* **62**:501–503 (1984).
20. BANATVALA, J. E., CHRYSTIE, I. L., AND TOTTERDELL, B. M., Rotaviral infections in human neonates, *J. Am. Vet. Med. Assoc.* **173**:527–530 (1978).
21. BANATVALA, J. E., TOTTERDELL, B. M., CHRYSTIE, I. L., AND WOODE, G. N., *In vitro* detection of human rotaviruses, *Lancet* **2**:821 (1975).
22. BARNES, G. L., DOYLE, L. W., HEWSON, P. H., KNOCHES, A. M. L., McLELLAN, J. A., KITCHEN, W. H., AND BISHOP, R. F., A randomized trial of oral gamma globulin in low-birth weight infants infected with rotavirus, *Lancet* **1**:1371–1373 (1982).
23. BARNES, G. L., UNICOMB, L., AND BISHOP, R. F., Severity of rotavirus infection in relation to serotype, monotype, and electropherotype, *J. Paediatr. Child Health* **28**:54–57 (1992).
24. BARON, R. C., GREENBERG, H. B., CUKOR, G., AND BLACKLOW, N. R., Serological responses among teenagers following natural exposure to Norwalk, *J. Infect. Dis.* **150**:531–534 (1984).
25. BATTAGLIA, M., PASSARANI, N., DIMATTEO, A., AND GERNA, G., Human enteric coronaviruses: Further characterization and immunoblotting of viral proteins, *J. Infect. Dis.* **155**:140–143 (1987).
26. BEARDS, G. M., Serotyping of rotavirus by NADP-enhanced enzyme-immunoassay, *J. Virol. Methods* **18**:77–85 (1987).

27. BEARDS, G. M., AND FLEWETT, T. H., Serological characterisation of human rotaviruses propagated in cell cultures, *Arch. Virol.* **80:**231–237 (1984).

28. BEARDS, G. M., PILFOLD, J. N., THOULESS, M. E., AND FLEWETT, T. H., Rotavirus serotypes by neutralization, *J. Med. Virol.* **5:**231–237 (1980).

29. BEARDS, G. M., DESSELBERGER, U., AND FLEWETT, T. H., Temporal and geographical distributions of human rotavirus serotypes, 1983 to 1988, *J. Clin. Microbiol.* **27:**2827–2833 (1989).

30. BEGUE, R. E., DENNEHY, P. H., HUANG, J., AND MARTIN, P., Serotype variation of group A rotaviruses over nine winter epidemics in southeastern New England, *J. Clin. Microbiol.* **30:**1592–1594 (1992).

31. BELL, L. M., CLARK, H. F., O'BRIEN, E. A., KORNSTEIN, M. J., PLOTKIN, S. A., AND OFFIT, P. A., Gastroenteritis caused by human rotavirus (serotype three) in a suckling mouse model, *Proc. Soc. Exp. Biol. Med.* **184:**127–132 (1987).

32. BERGELAND, M. E., MCADARAGH, J. P., AND STOTZ, I., in: *Proceeding of the 26th Western Poultry Disease Conference at University of California, Davis,* pp. 129–130, 1966.

33. BERN, C., AND GLASS, R. I., Impact of diarrheal diseases worldwide, in: *Viral Infections of the Gastrointestinal Tract* (A. Z. KAPIKIAN, ED.), pp. 1–26, Marcel Dekker New York, 1994.

34. BERN, C., UNICOMB, L., GENTSCH, J. R., BANUL, N., YUNUS, M., SACK, R. B., AND GLASS, R. I., Rotavirus diarrhea in Bangladeshi children: Correlation of disease severity with serotype, *J. Clin. Microbiol.* **30:**3234–3238 (1992).

35. BERNSTEIN, D. I., ZIEGLER, J. M., AND WARD, R. L., Rotavirus fecal IgA antibody response in adults challenged with human rotavirus, *J. Med. Virol.* **20:**297–304 (1986).

36. BICAN, P., COHEN, J., CHARPILIENNE, A., AND SCHERRER, R., Purification and characterization of bovine rotavirus cores, *J. Virol.* **43:**1113–1117 (1982).

37. BIRCH, C. J., HEATH, R. L., AND GUST, I. D., Use of serotype-specific monoclonal antibodies to study the epidemiology of rotavirus infection, *J. Med. Virol.* **24:**45–53 (1988).

38. BIRCH, C. J., LEHMANN, N. I., HAWKER, A. J., MARSHALL, J. A., AND GUST, I. D., Comparison of electron microscopy, enzyme-linked immunosorbent assay, solid-phase radioimmunoassay, and indirect immunofluorescence for detection of human rotavirus antigen in faeces, *J. Clin. Pathol.* **32:**700–705 (1979).

39. BIRITWUM, R. B., ISOMURA, S., YAMAGUCHI, H., TOBA, M., AND MINGLE, J. A., Seroepidemiological study of rotavirus infection in rural Ghana, *Ann. Trop. Paediatr.* **4:**237–240 (1984).

40. BISHOP, R. F., Other small virus-like particles in humans, in: *Viral Infections of the Gastrointestinal Tract* (D. A. J. TYRRELL AND A. Z. KAPIKIAN, EDS.), pp. 195–209, Marcel Dekker, New York, 1982.

41. BISHOP, R. F., Natural history of human rotavirus infections, in: *Infections of the Gastrointestinal Tract* (A. Z. KAPIKIAN, ED.), pp. 131–167, Marcel Dekker, New York, 1994.

42. BISHOP, R. F., BARNES, G. L., CIPRIANI, E., AND LUND, J. S., Clinical immunity after neonatal rotavirus infection. A prospective longitudinal study in young children, *N. Engl. J. Med.* **309:**72–76 (1983).

43. BISHOP, R. F., DAVIDSON, G. P., HOLMES, I. H., AND RUCK, B. J., Virus particles in epithelial cells of duodenal mucosa from children with viral gastroenteritis, *Lancet* **2:**1281–1283 (1973).

44. BISHOP, R. F., DAVIDSON, G. P., HOLMES, I. H., AND RUCK, B. J., Detection of a new virus by electron microscopy of fecal extracts from children with acute gastroenteritis, *Lancet* **1:**149–151 (1974).

45. BISHOP, R. F., DAVIDSON, G. P., HOLMES, I. H., AND RUCK, B. J., Evidence for viral gastroenteritis, *N. Engl. J. Med.* **289:**1096–1097 (1973).

46. BISHOP, R. F., UNICOMB, L. E., AND BARNES, G. L., Epidemiology of rotavirus serotypes in Melbourne, Australia, from 1973 to 1989, *J. Clin. Microbiol.* **29:**862–868 (1991).

47. BLACK, R. E., Epidemiology of traveler's diarrhea and relative importance of various pathogens, *Rev. Infect. Dis. (Suppl.)* **1:**S73–S79 (1990).

48. BLACK, R. E., BROWN, K. H., BECKER, S., AND YUNUS, M., Longitudinal studies of infectious diseases and physical growth of children in rural Bangladesh. I. Patterns of morbidity, *Am. J. Epidemiol.* **115:**305–314 (1982).

49. BLACK, R. E., GREENBERG, H. B., KAPIKIAN, A. Z., BROWN, K. H., AND BECKER, S., Acquisition of serum antibody to Norwalk virus and rotavirus in relation to diarrhea in a longitudinal study of young children in rural Bangladesh, *J. Infect. Dis.* **145:**483–489 (1982).

50. BLACK, R. E., MERSON, M. H., HUQ, I., ALIM, A. R. M. A., AND YUNUS, M., Incidence and severity of rotavirus and *Escherichia coli* diarrhoea in rural Bangladesh. Implications for vaccine development, *Lancet* **1:**141–143 (1981).

51. BLACK, R. E., MERSON, M. H., MIZANUR RAHMAN, A. S. M., YUNUS, N., ALIM, A. R. M. A., HUQ, I., YOLKEN, R. H., AND CURLIN, G. T., A 2 year study of bacterial, viral and parasitic agents associated with diarrhea in rural Bangladesh, *J. Infect. Dis.* **142:**660–664 (1980).

52. BLACKLOW, N. R., CUKOR, G., BEDIGIAN, M. K., Immune response and prevalence of antibody to Norwalk enteritis virus as determined by radioimmunoassay, *J. Clin. Microbiol.* **10:**903–909 (1979).

53. BLACKLOW, N. R., CUKOR, G., PANJAVANI, Z., CAPOZZA, F., AND BEDNAREK, F., Simplified radioimmunoassay for detection of rotavirus in pediatric and adult stools, for assessment of duration of antibody to rotavirus in human breast milk, in: *Fourth International Congress for Virology,* p. 464, Centre for Agricultural Publishing and Documentation, Wageningen, 1978.

54. BLACKLOW, N. R., DOLIN, R., FEDSON, D. S., DUPONT, H., NORTHRUP, R. S., HORNICK, R. B., AND CHANOCK, R. M., Acute infectious nonbacterial gastroenteritis: Etiology and pathogenesis. A combined clinical staff conference at the Clinical Center of the National Institutes of Health, *Ann. Intern. Med.* **76:**993–1008 (1972).

55. BLACKLOW, N. R., ECHEVERRIA, P., AND SMITH, D. A., Serological studies with reovirus-like enteritis agent, *Infect. Immun.* **13:**1563–1566 (1976).

56. BOLIVAR, R., CONKLIN, R. H., VOLLETT, J. J., ET AL. Rotavirus in travelers' diarrhea: Study of an adult student population in Mexico, *J. Infect. Dis.* **137:**324–327 (1978).

57. BONSDORFF, C. H. VON, HOVT, T., MAKELA, P., AND MORTTINEN, A., Rotavirus infections in adults in association with acute gastroenteritis, *J. Med. Virol.* **2:**21–28 (1978).

57a. BOURGEOIS A. L., GARDINER, C. H., THORNTON, S. A., BATCHELOR, R. A., BURR, D. H., ESCAMILLA, J., ECHEVERRIA, P., BLACKLOW, N. R., HERRMANN, J. E., AND HYAMS, K. C. Etiology of acute diarrhea among United States military personnel deployed to South America and West Africa. *Am. J. Trop. Med. Hyg.* **48:**243–248 (1993).

58. BRADBURNE, A. F., ALMEIDA, J. D., GARDNER, P. S., MOOSAI, R. B., NASH, A. A., AND COOMBS, R. R. A., A solid phase system (SPACE) for the detection and quantification of rotavirus in faeces, *J. Gen. Virol.* **44:**615–623 (1979).

59. BRANDT, C. D., KIM, H. W., RODRIGUEZ, W. J., ARROBIO, J. O., JEFFRIES, B. C., AND PARROTT, R. H., Rotavirus gastroenteritis and weather, *J. Clin. Microbiol.* **16:**478–482 (1982).

60. BRANDT, C. D., KIM, H. W., RODRIGUEZ, W. J., ARROBIO, J. O, JEFFRIES, B. C., STALLINGS, E. P., LEWIS, C., MILES, A. J., GARDNER, M. K., AND PARROTT, R. H., Adenoviruses and pediatric gastroenteritis, *J. Infect. Dis.* **151:**437–443 (1985).

61. BRANDT, C. D., KIM, H. W., RODRIGUEZ, W. J., ET AL., Pediatric viral gastroenteritis during eight years of study, *J. Clin. Microbiol.* **18:**71–78 (1983).

62. BRANDT, C. D., KIM, H. W., RODRIGUEZ, W. J., ET AL., Comparison of direct electron microscopy, immune electron microscopy, and rotavirus enzyme-linked immunosorbent assay for detection of gastroenteritis viruses in children, *J. Clin. Microbiol.* **13:**976–981 (1981).

63. BRANDT, C. D., KIM, H. W., YOLKEN, R. H., KAPIKIAN, A. Z., ARROBIO, J. O., RODRIGUEZ, W. J., WYATT, R. G., CHANOCK, R. M., AND PARROTT, R. H., Comparative epidemiology of two rotavirus serotypes and other viral agents associated with pediatric gastroenteritis, *Am. J. Epidemiol.* **110:**243–254 (1979).

64. BRIDGER, J. C., Non-group A rotaviruses, in: *Viral Infections of the Gastrointestinal Tract* (A. Z. KAPIKIAN, ED.), pp. 369–407, Marcel Dekker, New York, 1994.

65. BRIDGER, J. C., AND WOODE, G. N., Neonatal calf diarrhoea: Identification of a reovirus-like (rotavirus) agent in faeces by immunofluorescence and immune electron microscopy, *Br. Vet. J.* **131:**528–535 (1975).

66. BRIDGER, J. C., AND WOODE, G. N., Characterization of two particle types of calf rotavirus, *J. Gen. Virol.* **31:**245–250 (1976).

67. BRONDUM, J., SPITALNY, K. C., VOGT, R. L., GODLEWSKI, K., MADORE, H. P., AND DOLIN, R., Snow Mountain agent associated with an outbreak of gastroenteritis in Vermont, *J. Infect. Dis.* **152:**834–837 (1985).

68. BROWN, K. H., GILMAN, R. H., GAFFAR, A., ALAMGIR, S. M., STRIFE, J. L., KAPIKIAN, A. Z., AND SACK, R. B., Infections associated with severe protein-calorie malnutrition in hospitalized infants and children, *Nutr. Res.* **1:**33–46 (1981).

69. BRUSSOW, H., WERCHAU, H., LIEDTKE, W., LERNER, L., MIETENS, C., SIDOTI, J., AND SOTEK, J., Prevalence of antibodies to rotavirus in different age-groups of infants in Bochum, West Germany, *J. Infect. Dis.* **157:**1014–1022 (1988).

70. BRYDEN, A. S., Comparison of electron microscopy and immunofluorescence in cell culture for rotavirus detection, *J. Clin. Pathol.* **33:**413–415 (1980).

71. BRYDEN, A. S., DAVIES, H. A., HADLEY, R. E., FLEWETT, T. H., MORRIS, C. A., AND OLIVER, P., Rotavirus enteritis in the West Midlands during 1974, *Lancet* **2:**241–243 (1975).

72. BRYDEN, A. S., DAVIES, H. A., THOULESS, M. E., AND FLEWETT, T. H., Diagnosis of rotavirus infection by cell culture, *J. Med. Microbiol.* **10:**121–125 (1977).

73. BRYDEN, A. S., THOULESS, M. E., AND FLEWETT, T. H., A rabbit rotavirus, *Vet. Rec.* **99:**323 (1976).

74. BUFFETT-JANVRESSE, C., BERHARD, E., AND MAGARD, H., Responsabilite des rotavirus dans les diarrhees du nourrisson, *Nouv. Presse Med.* **5:**1249–1251 (1976).

75. BURROUGHS, J. N. AND BROWN, F., Presence of a covalently linked protein in calicivirus RNA, *J. Gen. Virol:* **41:**443–446 (1978).

76. CARLSON, J. A. K., MIDDLETON, P. J., SZYMANSKI, M., HUBER, J., AND PETRIC, M., Fatal rotavirus gastroenteritis. An analysis of 21 cases, *Am. J. Dis. Child.* **132:**477–479 (1978).

77. CARTER, M. J., MILTON, I. D., AND MADELEY, C. R., Caliciviruses, *Rev. Med. Virol.* **1:**177–186 (1991).

78. CAUL, E. O., Human coronaviruses, in: *Viral Infections of the Gastrointestinal Tract* (A. Z. KAPIKIAN, ED.), pp. 603–625, Marcel Dekker, New York, 1994.

79. CAUL, E. O., Small round structured viruses: Airborne transmission and hospital control, *Lancet* **343:**1240–1242 (1994).

80. CAUL, E. O., AND APPLETON, H., The electron microscopical and physical characteristics of small round fecal viruses: An interim scheme for classification, *J. Med. Virol.* **9:**257–267 (1982).

81. CAUL, E. O., ASHLEY, C., AND PETHER, J. V. S., "Norwalk"-like particles in epidemic gastroenteritis in the UK, *Lancet* **2:**1292 (1979).

82. CAUL, E. O., AND EGGLESTONE, S. I., Further studies on human enteric coronaviruses, *Arch. Virol.* **54:**107–117 (1977).

83. CAUL, E. O., AND EGGLESTONE, S. I., Coronaviruses in humans, in: *Virus Infections of the Gastrointestinal Tract* (D. A. J. TYRRELL AND A. Z. KAPIKIAN, EDS.), pp. 179–193, Marcel Dekker, New York, 1982.

84. CAUL, E. O., PAVER, W. K., AND CLARKE, S. K. R., Coronavirus particles in faeces from patients with gastroenteritis, *Lancet* **1:**1192 (1975).

85. CEVININI, R., MAZZARACCHIO, R., RUMPIANESI, F., DONATI, M., MORONI, A., SAMBRI, V., AND LAPLACA, M., Prevalence of enteric adenovirus from acute gastroenteritis: A five-year study, *Eur. J. Epidemiol.* **3:**147–150 (1987).

86. CHAMPSAUR, H., HENRY-AMAR, M., GOLDSZMIDT, D., ET AL., Rotavirus carriage, asymptomatic infection, and disease in the first two years of life. II. Serological response, *J. Infect. Dis.* **149:**675–682 (1984).

87. CHANOCK, R. M., WYATT, R. G., AND KAPIKIAN, A. Z., Immunization of infants and young children against rotavirus gastroenteritis—prospects and problems, *J. Am. Vet. Med. Assoc.* **173:**570–572 (1978).

88. CHANY, C., Discussion at Perspectives in Virology, XI Meeting, *Perspect. Virol.* **11:**185–187 (1981).

89. CHANY, C., MUSCOVICI, O., LEBON, P., AND ROUSSET, S., Association of coronavirus infection with neonatal necrotizing enterocolitis, *Pediatrics* **69:**209–214 (1982).

90. CHEN, C. M., HUNG, T., BRIDGER, J. C., AND MCCRAE, M. A., Chinese adult rotavirus is a group B rotavirus, *Lancet* **2:**1123–1124 (1985).

91. CHIBA, S., YOKOYAMA, T., NAKATA, S., MORITA, V., URASAWA, T., TANIGUCHI, K., URASAWA, S., AND NAKAO, T., Protective effect of naturally acquired homotypic and heterotypic rotavirus antibodies, *Lancet* **2:**417–421 (1986).

92. CHRISTOPHER, P. J., GROHMANN, G. S., MILLSOM, R. H., AND MURPHY, A. M., Parvovirus gastroenteritis—A new entity for Australia, *Med. J. Aust.* **1:**121–124 (1978).

93. CHRYSTIE, I. L., TOTTERDELL, B. M., AND BANATVALA, J. E., Asymptomatic endemic rotavirus infections in the newborn, *Lancet* **1:**1176–1178 (1978).

94. CHRYSTIE, I. L., TOTTERDELL, B. M., AND BANATVALA, J. E., False positive rotazyme tests on faecal samples from babies, *Lancet* **2:**1028 (1983).

95. CLARK, H. F., BORIAN, F. E., AND PLOTKIN, S. A., Immune protection of infants against rotavirus gastroenteritis by a serotype 1 reassortant of bovine rotavirus WC3, *J. Infect. Dis.* **161:**1099–1104 (1990).

96. CLARKE, E. J., JR., PHILLIPS, I. A., AND ALEXANDER, E. R., Adenovirus infection in intussusception in children in Taiwan, *J.A.M.A.* **208:**1671–1674 (1969).

97. CLARKE, I. N., AND McCRAE, M. A., A rapid and sensitive method for analysing the genome profiles of field isolates of rotavirus, *J. Virol. Methods* **2:**203–209 (1981).

98. CLARKE, S. K. R., COOK, G. T., EGGLESTONE, S. I., HALL, T. S., MILLER, D. L., REED, S. E., RUBENSTEIN, D., SMITH, A. J., AND TYRRELL, D. A. J., A virus from epidemic vomiting disease, *Br. Med. J.* **3:**86–89 (1972).

99. COHEN, D., AMBAR, R., GREENBERG, Z., BLOCK, C. S., SHIF, I., AND GREEN, M. S., Pilot study of an extended range of potential etiologic agents of diarrhea on the Israel defense forces, *Israel J. Med. Sci.* **28:**49–51 (1992).

100. CONNOR, J. D., AND BARRETT-CONNOR, E., Infectious diarrheas, *Pediatr. Clin. North Am.* **14:**197–221 (1967).

101. COOK, D. A., ZBITNEW, A., DEMPSTER, G., AND GERRARD, J. W., Detection of antibody to rotavirus by counterimmunoelectrophoresis in human serum, colostrum and milk, *J. Pediatr.* **93:** 967–970 (1978).

102. COULSON, B. S., Typing of human rotavirus VP4 by an enzyme immunoassay using monoclonal antibodies, *J. Clin. Microbiol.* **31:**1–8 (1993).

103. COULSON, B. S., AND MENENDYCZ, P. J., Measurement of rotavirus-neutralizing coproantibody in children by fluorescent focus reduction assay, *J. Clin. Microbiol.* **28:**1652–1654 (1990).

104. COULSON, B. S., UNICOMB, L. E., PITSON, G. A., AND BISHOP, R. F., Simple and specific enzyme immunoassay using monoclonal antibodies for serotyping human rotaviruses, *J. Clin. Microbiol.* **25:**509–515 (1987).

105. CRAWLEY, J. M. S., BISHOP, R. F., AND BARNES, G. L., Rotavirus gastroenteritis in infants aged 0–6 months in Melbourne, Australia: Implications for vaccination, *J. Paediatr. Child Health* **29:**219–221 (1993).

106. CUBITT, W. D., Caliciviruses, in: *Viral Infections of the Gastrointestinal Tract* (A. Z. KAPIKIAN, ED.), pp. 549–568, Marcel Dekker, New York, 1994.

107. CUBITT, W. D., BLACKLOW, N. R., HERRMANN, J. E., NOWAK, N. A., NAKATA, S., AND CHIBA, S., Antigenic relationships between human caliciviruses and Norwalk virus, *J. Infect. Dis.* **156:**806–814 (1987).

108. CUBITT, W. D., AND HOLZEL, H., An outbreak of rotavirus infection in a long-stay ward of a geriatric hospital, *J. Clin. Pathol.* **33:**306–308 (1980).

109. CUKOR, G., BERRY, M. K., AND BLACKLOW, N. R., Simplified radioimmunoassay for detection of human rotavirus in stools, *J. Infect. Dis.* **138:**906–910 (1978).

110. CUKOR, G., BLACKLOW, N. R., ECHEVERRIA, P., BEDIGIAN, M. K., PURUGGAN, H., AND BASACA-SEVILLA, V., Comparative study of the acquisition of antibody to Norwalk virus in pediatric populations, *Infect. Immun.* **29:**822–823 (1980).

111. CUBITT, W. D., BRADLEY, D., CARTER, M., CHIBA, S., ESTES, M., SAIF, L., SCHAFFER, F., SMITH, A., STUDDERT, M., AND THIEL, H. J., Caliciviridae, in: *Classification and Nomenclature of Viruses* (Sixth report of the International Committee for Taxonomy of Viruses, in press).

112. CUNNINGHAM, A. L., GROHMAN, G. S., HARKNESS, J., LAW, C., MARRIOTT, D., TINDALL, B., AND COOPER, D. A., Gastrointestinal viral infections in homosexual men who were symptomatic and seropositive for human immunodeficiency virus, *J. Infect. Dis.* **158:**386–391 (1988).

113. DAGAN, R., BAR-DAVID, Y., SAROV, B., KATZ, M., KASSIS, I., GREENBERG, D., GLASS, R. I., MARGOLIS, C.-Z., AND SAROV, I., Rotavirus diarrhea in Jewish and Bedovin children in the Negev region of Israel: Epidemiology, clinical aspects, and possible role of malnutrition in severity of illness, *Pediatr. Infect. Dis. J.* **9:**314–321 (1990).

114. DAVIDSON, G. P., BISHOP, R. F., TOWNLEE, R. R., HOLMES, I. H., AND RUCK, B. J., Importance of a new virus in acute sporadic enteritis in children, *Lancet* **1:**242–245 (1975).

115. DAVIDSON, G. P., BUTLER, D. G., GALL, D. G., PETRIC, M., AND HAMILTON, J. R., Ion transport in enteritis caused by human rotavirus, in: *American Society for Microbiology Annual Meeting*, pp. A-20/1043, American Society for Microbiology, Washington, 1977.

116. DAVIDSON, G. P., GOLLER, I., BISHOP, R. F., TOWNLEE, R. R. W., HOLMES, I. H., AND RUCK, B. J., Immunofluorescence in duodenal mucosa of children with acute enteritis due to a new virus, *J. Clin. Pathol.* **28:**263–266 (1975).

117. DAVIDSON, G. P., HOGG, R. J., AND KIRUBAKARON, C. P., Serum and intestinal immune response to rotavirus enteritis in children, *Infect. Immun.* **40:**447–452 (1983).

118. DAVIDSON, G. P., WHYTE, P. B. D., DANIELS, FRANKLIN, K., NUNAN, H., McCLOUD, P. I., MOORE, A. G., AND MOORE, D. J., Passive immunisation of children with bovine colostrum containing antibodies to human rotavirus, *Lancet* **2:**709–712 (1989).

119. DEARLOVE, J., LATHAM, P., DEARLOVE, B., PEARL, K., THOMSON, A., AND LEWIS, I. G., Clinical range of neonatal rotavirus gastroenteritis, *Br. Med. J.* **286:**1473–1475 (1983).

120. DELAGE, G., McLAUGHLIN, B., AND BERTHIAUME, L., A clinical study of rotavirus gastroenteritis, *J. Pediatr.* **93:**455–457 (1978).

121. DELEM, A., AND VESIKARI, T., Detection of serum antibody responses to RIT 4237 rotavirus vaccine by ELISA and neutralization assays, *J. Med. Virol.* **21:**231–238 (1987).

122. DELEON, R., MATSUI, S. M., BARIC, R. S., HERRMANN, J. E., BLACKLOW, N. R., GREENBERG, H. B., AND SOBSEY, M. D., Detection of Norwalk virus in stool specimens by reverse transcriptase-polymerase chain reaction and non-radioactive oligoprobes, *J. Clin. Microbiol.* **30:**3151–3157 (1992).

123. DE MOL, P., ZISSIS, G., BUTZLER, J. P., MUTWEWINBABO, A., AND ANDRE, F. E., Failure of live, attenuated oral rotavirus vaccine, *Lancet* **2:**108 (1986).

124. DENNEHY, P. H., FOR THE US ROTAVIRUS VACCINE EFFICACY GROUP, Safety and efficacy of an oral tetravalent rhesus rotavirus vaccine (RRV-TV) in healthy infants (Abstract), *Pediatr. Res.* **35:**178A (1994).

125. DE ZOYSA, I., AND FEACHEM, R. G., Interventions for the control of diarrheal diseases among young children: Rotavirus and cholera immunization, *Bull. WHO* **63:**569–583 (1985).

126. DINGLE, J. H., BADGER, G. F., AND JORDAN, W. S., *Illness in the Home: A Study of 25,000 Illnesses in a Group of Cleveland Families*, Case Western Reserve University Press, Cleveland, OH, 1964.

127. DOANE, F. W., Electron microscopy for the detection of gastroenteritis viruses, in: *Viral Infections of the Gastrointestinal Tract* (A. Z. KAPIKIAN, ED.), pp. 101–130, Marcel Dekker, New York, 1994.

128. DOLIN, R., Norwalk-like agents of gastroenteritis, in: *Principles and Practices of Infectious Diseases* (G. L. MANDELL, ED.), pp. 1364–1370, John Wiley & Sons, New York, 1979.

129. DOLIN, R., BLACKLOW, N. R., DuPONT, H., BUSCHO, R. F., WYATT, R. G., KASEL, J. A., HORNICK, R., AND CHANOCK, R. M., Biological properties of Norwalk agent of acute infectious nonbacterial gastroenteritis, *Proc. Soc. Exp. Biol. Med.* **140:**578–583 (1972).

130. DOLIN, R., BLACKLOW, N. R., DuPONT, H., FORMAL, S., BUSCHO,

R. F., KASEL, J. A., CHAMES, R. P., HORNICK, R., AND CHANOCK, R. M., Transmission of acute infectious nonbacterial gastroenteritis to volunteers by oral administration of stool filtrates, *J. Infect. Dis.* **123:**307–312 (1971).

131. DOLIN, R., LEVY, A. G., WYATT, R. G., THORNHILL, T. S., AND GARDNER, J. D., Viral gastroenteritis induced by the Hawaii agent: Jejunal histopathology and seroresponse, *Am. J. Med.* **59:**761–767 (1975).

132. DOLIN, R., REICHMAN, R. C., AND FAUCI, A. S., Lymphocyte populations in acute viral gastroenteritis, *Infect. Immun.* **14:**422–428 (1976).

133. DOLIN, R., REICHMAN, R. C., ROESSNER, K. D., ET AL., Detection by immune electron microscopy of the Snow Mountain agent of acute viral gastroenteritis, *J. Infect. Dis.* **146:**184–189 (1982).

134. DOLIN, R., ROESSNER, D. F., TREANOR, J. J., REICHMAN, R. C., PHILLIPS, M., AND MADORE, H. P., Radioimmunoassay for detection of the Snow Mountain Agent of viral gastroenteritis, *J. Med. Virol.* **19:**11–18 (1986).

135. DOMINICK, H. C., AND MAAS, G., Rotavirus infectionen im kindersalter, *Klin. Padiatr.* **191:**33–39 (1979).

136. EBINA, T., SATO, A., UMEZU, K., ISHIDA, N., OHYAMA, S., OIZUMI, A., AIKAWA, K., KATAGIRI, S., KATSUSHIMA, N., IMAI, A., KETAOKA, S., SUZUKI, H., AND KONNO, T., Prevention of rotavirus infection by oral administration of cow colostrum containing anti-human rotavirus antibody, *Med. Microbiol. Immunol.* **174:**177–185 (1985).

137. ECHEVERRIA, P., BURKE, D. S., BLACKLOW, N. R., CUKOR, G., CHAROENKUL, C., AND YANGGRATOKE, S., Age-specific prevalence of antibody to rotavirus, *Escherichia coli*, heat labile enterotoxin, Norwalk virus, and hepatitis A virus in a rural community in Thailand, *J. Clin. Microbiol.* **17:**923–925 (1983).

138. ECHEVERRIA, P., HO, M. T., BLACKLOW, N. R., ET AL., Relative importance of viruses and bacteria in the etiology of pediatric diarrhea in Taiwan, *J. Infect. Dis.* **136:**383–390 (1977).

139. ECHEVERRIA, P., RAMIREZ, G., BLACKLOW, N. R., KSIAZEK, T., CUKOR, G., AND CROSS, J. H., Travelers' diarrhea among U.S. Army troops in South Korea, *J. Infect. Dis.* **139:**215–219 (1979).

140. EDELMAN, R., AND LEVINE, M. M., Acute diarrheal infections in infants. II. Bacterial and viral causes, *Hosp. Pract.* **15:**96–104 (1980).

141. EDITORIAL, Norwalk agent comes of age, *J. Infect.* **20:**189–192 (1990).

142. EDITORIAL, Viral cross-infections in wards, *Lancet* **1:**1391–1393 (1976).

143. EDITORIAL, Towards a rotavirus vaccine, *Lancet* **2:**619–620 (1981).

144. EDITORIAL, Kawasaki's—Might it be, *Lancet* **2:**1441–1442 (1982).

145. EIDEN, J., SATO, S., AND YOLKEN, R., Specificity of dot hybridization assay in the presence of rRNA for detection of rotaviruses in clinical specimens, *J. Clin. Microbiol.* **25:**1809–1811 (1987).

146. EIDEN, J. J., WILDE, J., FIROOZ-MAND, F., AND YOLKEN, R., Detection of animal and human group B rotaviruses in fecal specimens by polymerase chain reaction, *J. Clin. Microbiol.* **29:**539–543 (1991).

147. ELIAS, M. M., Distribution and titers of rotavirus antibodies in different age groups, *J. Hyg. (Lond.)* **79:**365–372 (1977).

148. ENGLEBERG, N. C., HOLBURT, E. N., BARRETT, T. J., GARY, G. W., JR., TRUJILLO, M. H., FELDMAN, R. A., AND HUGHES, J. M., Epidemiology of diarrhea due to rotavirus on an Indian reservation: Risk factors in the home environment, *J. Infect. Dis.* **145:**894–898 (1982).

149. ESPEJO, R. T., CALDERON, E., AND GONZALEZ, N., Distinct reovirus-like agents associated with acute infantile gastroenteritis, *J. Clin. Microbiol.* **6:**502–506 (1977).

150. ESTES, M. K., AND COHEN, J., Rotavirus gene structure and function, *Microbiol. Rev.* **53:**410–449 (1989).

151. ESTES, M. K., AND GRAHAM, D. Y., Identification of rotavirus of different origins by the plaque reduction test, *Am. J. Vet. Res.* **41:**151–152 (1980).

152. ESTES, M. K., GRAHAM, D. Y., AND DIMITROV, D., The molecular epidemiology of rotavirus gastroenteritis, *Prog. Med. Virol.* **29:**1–24 (1984).

153. EUGSTER, A. K., STROTHER, J., AND HARTFIEL, D. A., Rotavirus (reovirus-like) infection of neonatal ruminants in zoo nursery, *J. Wildl. Dis.* **14:**351–354 (1978).

154. FAUVEL, M., SPENCE, L., BABIUK, L. A., PETRO, R., AND BLOCH, S., Hemagglutination and hemagglutination inhibition studies with a strain of Nebraska calf diarrhea virus (bovine rotavirus), *Intervirology* **9:**95–105 (1978).

155. FLEWETT, T. H., Diagnosis of enteritis virus, *Proc. R. Soc. Med.* **69:**693–696 (1976).

156. FLEWETT, T. H., Clinical features of rotavirus infections, in: *Virus Infections of the Gastrointestinal Tract* (D. A. J. TYRRELL AND A. Z. KAPIKIAN, EDS.), pp. 125–145, Marcel Dekker, New York, 1982.

157. FLEWETT, T. H., BRYDEN, A. S., AND DAVIES, H., Virus particles in gastroenteritis, *Lancet* **2:**1497 (1973).

158. FLEWETT, T. H., BRYDEN, A. S., AND DAVIES, H., Virus diarrhea in foals and other animals, *Vet. Rec.* **97:**477 (1975).

159. FLEWETT, T. H., BRYDEN, A. S., AND DAVIES, H., Epidemic viral enteritis in a long-stay children's ward, *Lancet* **1:**4–5 (1975).

160. FLEWETT, T. H., BRYDEN, A. S., AND DAVIES, H., WOODE, G. N., BRIDGER, J. C., AND DERRICK, J. M., Relationship between virus from acute gastroenteritis of children and newborn calves, *Lancet* **2:**61–63 (1974).

161. FLEWETT, T. H., BRYDEN, A. S., AND DAVIES, H., Diagnostic electron microscopy of faeces. I. The viral flora of the faeces as seen by electron microscopy, *J. Clin. Pathol.* **27:**603–614 (1974).

161a. FLEWETT, T. H., AND DAVIES, H., Caliciviruses in man, *Lancet* **1:**311 (1976).

162. FLEWETT, T. H., THOULESS, M. E., PILFOLD, J. N., BRYDEN, A. S., AND CANDEIAS, J. A. N., More serotypes of human rotaviruses, *Lancet* **2:**632 (1978).

163. FLEWETT, T. H., AND WOODE, G. N., The rotaviruses (brief review), *Arch. Virol.* **57:**1–23 (1978).

164. FLORES, J., BOEGGEMAN, E., PURCELL, R. H., ET AL., A dot hybridization assay for detection of rotavirus, *Lancet* **1:**555–559 (1983).

165. FLORES, J., GREEN, K., GARCIA, D., ET AL., A dot hybridization assay for distinction of human rotavirus serotypes, *J. Clin. Microbiol.* **27:**29–34 (1989).

166. FLORES, J., KAPIKIAN, A. Z., Rotavirus vaccines, in: *Vaccines: New Approaches to Immunological Problems* (R. W. ELLIS, ED.), pp. 255–288, Butterworth-Heinemann, Boston, 1992.

167. FLORES, J., MIDTHUN, K., HOSHINO, Y., ET AL., Conservation of the fourth gene among rotaviruses recovered from asymptomatic newborn infants and its possible role in attenuation, *J. Virol.* **60:**972–979 (1986).

168. FLORES, J., SEARS, J., PEREZ-SCHAEL, I., WHITE, L., GARCIA, D., LANATA, C., AND KAPIKIAN, A. Z., Identification of human rotavirus serotype by hybridization to polymerase chain reaction-generated probes derived from a hyperdivergent region of the

gene encoding outer capsid protein VP7, *J. Virol.* **64:**4021–4024 (1990).

169. FLORES, J., SERENO, M., KALICA, A., KEITH, J., KAPIKIAN, A., AND CHANOCK, R., Molecular cloning of rotavirus genes: Implications for immunoprophylaxis, in: *Modern Approaches to Vaccines: Molecular and Chemical Basis of Virus Virulence and Immunogenicity* (R. M. CHANOCK AND R. A. LERNER, EDS.), pp. 159–164, Cold Spring Harbor Laboratory, Cold Spring Harbor, New York, 1983.

170. FLORES, J., SERENO, M., LAI, C. J., *ET AL.*, Use of single-stranded rotavirus RNA transcripts for the diagnosis of rotavirus infection, the study of genetic diversity among rotaviruses, and the molecular cloning of rotavirus genes, in: *Double-stranded RNA viruses* (R. W. COMPANS AND D. H. L. BISHOP, EDS.), pp. 115–127, Elsevier, Amsterdam, 1983.

171. FLORES, J., TANIGUCHI, K., GREEN, K., PEREZ-SCHAEL, I., GARCIA, D., SEARS, J., URASAWA, S., AND KAPIKIAN, A. Z., Relative frequencies of rotavirus serotypes 1, 2, 3 and 4 in Venezuelan infants with gastroenteritis, *J. Clin. Microbiol.* **26:**2092–2095 (1988).

172. FONTEYNE, J., ZISSIS, G., AND LAMBERT, J. P., Recurrent rotavirus gastroenteritis, *Lancet* **1:**983 (1978).

173. FOSTER, L. G., PETERSON, H., AND SPENDLOVE, R. S., Fluorescent virus precipitin test, *Proc. Soc. Exp. Biol. Med.* **150:**155–160 (1975).

174. FUGITA, Y., HIYOSHI, K., WAKASUGI, N., *ET AL.*, Transient improvement of the West syndrome in two cases following rotavirus, *No To Hattatsu* **20:**59 (1988).

175. FUJII, R., KUZUYA, M., HAMANO, M., YAMADA, M., AND YAMAZAKI, S., Detection of human group C rotaviruses by an enzyme-linked immunosorbent assay using monoclonal antibodies, *J. Clin. Microbiol.* **30:**1307–1311 (1992).

176. FUKOSHO, A., SHIMIZU, Y., AND ITO, Y., Isolation of cytopathic porcine rotavirus in cell roller culture in the presence of trypsin, *Arch. Virol.* **69:**49–60 (1981).

177. FUKUMI, H., NAKAYA, R., HATTA, S., *ET AL.*, An indication as to identity between the infectious diarrhea in Japan and the afebrile infectious nonbacterial gastroenteritis by human volunteer experiments, *Jpn. J. Med. Sci. Biol.* **10:**1–17 (1957).

178. GALL, D. G., Pathophysiology of viral diarrhea, in: *Report of the Seventy-Fourth Ross Conference on Pediatric Research: Etiology, Pathophysiology, and Treatment of Acute Gastroenteritis* (H. J. McCLUNG, ED.), pp. 66–68, Ross Laboratories, Columbus, OH, 1978.

179. GARDNER, P. S., KNOX, E. G., COURT, S. D. M., AND GREEN, C. A., Virus infection and intussusception in childhood, *Br. Med. J.* **2:**697–700 (1962).

180. GARY, G. W., JR., KAPLAN, J. E., STINE, S. E., AND ANDERSON, L. J., Detection of Norwalk virus antibodies and antigen with a biotin–avidin immunoassay, *J. Clin. Microbiol.* **22:**274–278 (1985).

181. GATTI, M. S. V., PESTANO DE CASTRO, A. F., AND FERRAZ, M. M. G., Viruses with bisegmented double-stranded RNA in pig faeces, *Res. Vet. Sci.* **47:**397–398 (1989).

182. GEORGES-COURBOT, M. C., BERAUD, A. M., BEARDS, G. M., CAMPBELL, A. D., GONZALEZ, J. P., GEORGES, A. J., AND FLEWETT, T. H., Subgroups, serotypes and electropherotypes of rotavirus isolated from children in Bangui, Central African Republic, *J. Clin. Microbiol.* **26:**668–671 (1988).

183. GERNA, G., ARISTA, S., PASSARANI, N., SARASINI, A., AND BATTAGLIA, M., Electropherotype heterogeneity within serotypes of

human rotavirus strains circulating in Italy, *Arch. Virol.* **95:**129–135 (1987).

184. GERNA, G., FORSTER, J., PAREA, M., *ET AL.*, Nosocomial outbreak of neonatal gastroenteritis caused by a new serotype 4, subtype 4B human rotavirus, *J. Med. Virol.* **31:**175–182 (1990).

185. GERNA, G., PASSARANI, N., BATTAGLIA, M., AND PERCIVALLE, E., Rapid serotyping of human rotavirus strains by solid-phase immune electron microscopy, *J. Clin. Microbiol.* **19:**273–278 (1984).

186. GERNA, G., PASSARANI, N., BATTAGLIA, M., AND RONDANELLI, E. G., Human enteric coronaviruses: Antigenic relatedness to human coronaviruses OC43 and possible etiologic role in viral gastroenteritis, *J. Infect. Dis.* **151:**796–803 (1985).

187. GERNA, G., PASSARANI, N., SARASINI, A., AND BATTAGLIA, M., Characterization of serotypes of human rotavirus strains by solid-phase immune electron microscopy, *J. Infect. Dis.* **152:**1143–1151 (1985).

188. GERNA, G., SARASINI, A., ARISTA, S., DI MATTEO, A., GIOVANNELLI, L., PAREA, M., AND HALONEN, P., Prevalence of human rotavirus serotypes in some European countries 1981–1988, *Scand. J. Infect. Dis.* **22:**5–10 (1990).

189. GERNA, G., SEARS, J., HOSHINO, Y., STEELE, A. D., NAKAGOMI, O., SARASINI, A., AND FLORES, J., Identification of a new VP4 serotype of human rotavirus, *Virology* **200:**66–71 (1994).

190. GIAMMANCO, G., DE GRANDI, V., LUPO, L., *ET AL.*, Interference of oral polio vaccine on RIT 4237 oral rotavirus vaccine, *Eur. J. Epidemiol.* **4:**121–123 (1988).

191. GLASS, R. I., KEITH, J., NAKAGOMI, O., *ET AL.*, Nucleotide sequence of the structural glycoprotein VP7 gene of Nebraska calf diarrhea virus rotavirus: Comparison with homologous genes from four strains of human and animal rotaviruses, *Virology* **141:**292–298 (1985).

192. GLASS, R. I., STOLL, B. J., WYATT, R. G., HOSHINO, Y., BANU, H., AND KAPIKIAN, A. Z., Observations questioning a protection role for breast-feeding in severe rotavirus diarrhea, *Acta Paediatr. Scand.* **75:**713–718 (1986).

193. GOMEZ-BARRETO, J., PALMER, E. L., NAHMIAS, A. J., AND HATCH, M. H., Acute enteritis associated with reovirus-like agents, *J.A.M.A.* **253:**1857–1860 (1976).

194. GOMES, T. A. T., RASSI, V., MacDONALD, K. L., SILVA RAMOS, S. R. T., TRABULSI, L. R., VIEIRA, M. A. M., GUTH, B. E. C., CANDEIAS, J. A. N., IVEY, C., TOLEDO, M. R. F., AND BLAKE, P. A., Enteropathogens associated with acute diarrheal disease in urban infants in Sao Paulo, Brazil, *J. Infect. Dis.* **164:**331–337 (1991).

195. GORDON, I., INGRAHAM, H. S., AND KORNS, R. F., Transmission of epidemic gastroenteritis to human volunteers by oral administration of fecal filtrates, *J. Exp. Med.* **86:**409–422 (1947).

196. GORZIGLIA, M., HOSHINO, Y., BUCKLER-WHITE, A., BLUMENTALS, I., GLASS, R., FLORES, J., KAPIKIAN, A. Z., AND CHANOCK, R. M., Conservation of amino acid sequence of VP8 and cleavage region of 84-kDa outer capsid protein among rotaviruses recovered from asymptomatic neonatal infection, *Proc. Natl. Acad. Sci. USA* **83:**7039–7043 (1986). [Erratum in *Proc. Natl. Acad. Sci. USA* **84:**2062 (1987).]

197. GORZIGLIA, M., LARRALDE, G., KAPIKIAN, A. Z., AND CHANOCK, R. M., Antigenic relationships among human rotaviruses as determined by outer capsid protein KVP4, *Proc. Natl. Acad. Sci. USA* **87:**7155–7159 (1990).

198. GOUVEA, V., ALLEN, J. R., GLASS, R. I., FANG, Z.-Y., BREMONT, M., COHEN, J., McCRAE, M. A., SAIF, L. J., SINARACHATANANT,

P., AND CAUL, E. O., Detection of group B and C rotaviruses by polymerase chain reaction, *J. Clin. Microbiol.* **29**:519–523 (1991).

199. GOUVEA, V., DE CASTRO, L., TIMENETSKY, M. DO C., GREENBERG, H., AND SANTOS, N., Rotavirus serotype kG5 associated with diarrhea in Brazilian children, *J. Clin. Microbiol.* **32**:1408–1409 (1994).

200. GOUVEA, V., GLASS, R. I., WOODS, P., TANIGUCHI, K., CLARK, H. F., FORRESTER, B., AND FANG, Z. Y., Polymerase chain reaction amplification and typing of rotavirus nucleic acid from stool specimens, *J. Clin. Microbiol.* **28**:276–282 (1990).

201. GOUVEA, V., HO, M. S., GLASS, R., ET AL., Serotypes and electropherotypes of human rotavirus in the USA: 1987–1989, *J. Infect. Dis.* **162**:362–367 (1990).

202. GRAUBELLE, P. C., GENNER, J., MEYLING, A., AND HORNSLETH, A., Rapid diagnosis of rotavirus infections: Comparison of electron microscopy and immunoelectro-osmophoresis for the detection of rotavirus in human infantile gastroenteritis, *J. Gen. Virol.* **35**:203–218 (1977).

203. GREEN, K. Y., JAMES, H. D., JR., AND KAPIKIAN, A. Z., Evaluation of three panels of monoclonal antibodies for the identification of human rotavirus VP7 serotype by ELISA, *Bull. WHO* **68**:601–610 (1990).

204. GREEN, K. Y., AND KAPIKIAN, A. Z., Identification of VP7 epitopes associated with protection against human rotavirus illness or shedding in volunteers, *J. Virol.* **66**:548–553 (1992).

205. GREEN, K. Y., LEW, J. F., JIANG, X., KAPIKIAN, A. Z., AND ESTES, M. K., Comparison of the reactivities of baculovirus-expressed recombinant Norwalk virus capsid antigen with those of the native Norwalk virus antigen in serologic assays and some epidemiologic observations, *J. Clin. Microbiol.* **31**:2185–2191 (1993).

206. GREEN, K. Y., MIDTHUN, K., GORZIGLIA, M., HOSHINO, Y., KAPIKIAN, A. Z., CHANOCK, R. M., AND FLORES, S. J., Comparison of amino acid sequences of the major neutralization protein of four human rotavirus serotypes, *Virology* **161**:153–159 (1987).

207. GREEN, K. Y., TANIGUCHI, K., MACKOW E., AND KAPIKIAN A. Z., Homotypic and heterotypic epitope-specific antibody responses in adult and infant rotavirus vaccinees: Implications for vaccine development, *J. Infect. Dis.* **161**:667–669 (1990).

208. GREENBERG, H. B., KALICA, A. R., WYATT, R. G., JONES, R. W., KAPIKIAN, A. Z., AND CHANOCK, R. M., Rescue of noncultivatable human rotaviruses by gene reassortment during mixed infection with *ts* mutants of a cultivatable bovine rotavirus, *Proc. Natl. Acad. Sci. USA* **78**:420–424 (1981).

209. GREENBERG, H. B., AND KAPIKIAN, A. Z., Detection of Norwalk agent antibody and antigen by solid-phase radioimmunoassay and immune adherence hemagglutination assay, *J. Am. Vet. Med. Assoc.* **173**:620–623 (1978).

210. GREENBERG, H. B., AND MATSUI, S. M., Astroviruses and caliciviruses: Emerging enteric pathogens, *Infect. Agents Dis.* **1**:71–91 (1992).

211. GREENBERG, H. B., MIDTHUN, K., WYATT, R., FLORES, J., HOSHINO, Y., CHANOCK, R. M., AND KAPIKIAN, A. Z., Use of reassortant rotaviruses and monoclonal antibodies to make gene-coding assignments and construct rotavirus vaccine candidates, in: *Modern Approaches to Vaccines: Molecular and Chemical Basis of Virus Virulence and Immunogenicity* (R. M. CHANOCK AND R. A. LERNER, EDS.), pp. 319–327, Cold Spring Harbor Laboratory, Cold Spring Harbor, New York, 1984.

212. GREENBERG, H. B., VALDESUSO, J., KALICA, A. R., WYATT, R. G., MCAULIFFE, V. J., KAPIKIAN, A. Z., AND CHANOCK, R. M., Proteins of Norwalk virus, *J. Virol.* **37**:994–999 (1981).

213. GREENBERG, H. B., VALDESUSO, J., KAPIKIAN, A. Z., CHANOCK, R. M., WYATT, R. G., SZMUNESS, W., LARRICK, J., KAPLAN, J., GILMAN, R. H., AND SACK, D. A., Prevalence of antibody to the Norwalk virus in various countries, *Infect. Immun.* **26**:270–273 (1979).

214. GREENBERG, H. B., VALDESUSO, J., YOLKEN, R. H., GANGAROSA, E., GARY, W., WYATT, R. G., KONNO, T. SUZUKI, H., CHANOCK, M., AND KAPIKIAN, A. Z., Role of Norwalk virus in outbreaks of nonbacterial gastroenteritis, *J. Infect. Dis.* **139**:564–568 (1979).

215. GREENBERG, H. B., WYATT, R. G., KALICA, A. R., YOLKEN, R. H., BLACK, R., KAPIKIAN, A. Z., AND CHANOCK, R. M., New insights in viral gastroenteritis, *Perspect. Virol.* **11**:163–187 (1981).

216. GREENBERG, H. B., WYATT, R. G., AND KAPIKIAN, A. Z., Norwalk virus in vomitus, *Lancet* **1**:55 (1979).

217. GREENBERG, H. B., WYATT, R. G., KAPIKIAN, A. Z., KALICA, A. R., FLORES, J., AND JONES, R., Rescue and serotypic characterization of noncultivatable human rotavirus by gene reassortment, *Infect. Immun.* **37**:104–109 (1982).

218. GREENBERG, H. B., WYATT, R. G., VALDESUSO, J., KALICA, A. R., LONDON, W. T., CHANOCK, R. M., AND KAPIKIAN, A. Z., Solid-phase microtiter radioimmunoassay for detection of the Norwalk strain of acute nonbacterial epidemic gastroenteritis and its antibodies, *J. Med. Virol.* **2**:97–108 (1978).

219. GRIMWOOD, K., ABBOTT, G. P., FERGUSON, D. M., JENNINGS, L. C., AND ALLAN, J. M., Spread of rotavirus within families: A community based study, *Br. Med. J.* **287**:575–577 (1983).

220. GROHMANN, G. S., GREENBERG, H. B., WELCH, B. M., AND MURPHY, A. M., Oyster-associated gastroenteritis in Australia: The detection of Norwalk virus and its antibody by immune electron microscopy and radioimmunoassay *J. Med. Virol.* **6**:11–19 (1980).

221. GROHMANN, G. S., GLASS, R. I., PEREIRA, H. G., MONROE, S. S., HIGHTOWER, A. W., WEBER, R., BRYAN, R. T., Enteric viruses and diarrhea in HIV-infected patients, *N. Engl. J. Med.* **329**:14–20 (1993).

222. GRUNOW, J. E., DUNTON, S. F., AND WANER, J. L., Human rotavirus-like particles in a hepatic abscess, *J. Pediatr.* **106**:73–76 (1985).

223. GUARINO, A., CANANI, R. B., RUSSO, S., ALBANO, F., CANANI, M. B., RUGGERI, F. M., DONELLI, G., AND RUBINO, A., Oral immunoglobulins for treatment of acute rotaviral gastroenteritis, *Pediatrics* **93**:12–16 (1994).

224. GUERRANT, R. L., HUGHES, J. M., LIMA, N. L., AND CRANE, J., Diarrhea in developed and developing countries: Magnitude, special settings, and etiologies, *Rev. Infect. Dis.* **12**:S41–S50 (1990).

225. GUERRANT, R. L., KIRCHOFF, L. V., SCHIELDS, D. S., NATIONS, M. K., LESLIE, J., DE SOUSA, M. A., ARAUJO, J. G., CORREIA, L., SAUER, K. T., MCCLELLAND, K. E., TROWBRIDGE, F. L., AND HUGHES, J. M., Prospective study of diarrheal illnesses in Northeastern Brazil: Patterns of disease, nutritional impact, etiologies, and risk factors, *J. Infect. Dis.* **148**:986–997 (1983).

226. GUNASENA, S., NAKAGOMI, O., ISEGAWA, Y., KAGA, E., NAKAGOMI, T., STEELE, A. D., FLORES, J., AND UEDA, S., Relative frequency of VP4 gene alleles among human rotaviruses recovered over a ten-year period (1982–1991) from Japanese children with diarrhea, *J. Clin. Microbiol.* **31**:2195–2197 (1993).

227. GURWITH, M., WENMAN, W., GURWITH, D., BRUNTON, J., FELTHAM, S., AND GREENBERG, H., Diarrhea among infants and young children in Canada: A longitudinal study in the three northern communities, *J. Infect. Dis.* **147**:685–692 (1983).

228. GURWITH, M., WENMAN, W., HINDE, D., FELTHAM, S., AND GREENBERG, H., A prospective study of rotavirus infection in infants and young children, *J. Infect. Dis.* **144:**218–224 (1981).

229. GUST, I. D., PRINGLE, R. C., BARNES, G. L., DAVIDSON, G. P., AND BISHOP, R. F., Complement-fixing antibody response to rotavirus infection, *J. Clin. Microbiol.* **5:**125–130 (1977).

230. GWALTNEY, J. M., JR., Medical reviews: Rhinoviruses, *Yale J. Biol. Med.* **48:**17–45 (1975).

231. HAFFEJEE, I. E., Neonatal rotavirus infections, *Rev. Infect. Dis.* **13:**957–962 (1991).

232. HALSEY, N., AND GALAZKA, A., The effectiveness of DPT and oral poliomyelitis immunization schedules initiated from birth to 12 weeks of age, *Bull. WHO* **63:**1151–1169 (1985).

233. HALVOSRUD, J., OSTRAVIK, I., An epidemic of rotavirus-associated gastroenteritis in a nursing home for the elderly, *Scand. J. Infect. Dis.* **12:**161–164 (1980).

234. HARDY, M., GORZIGLIA, M., AND WOODE, G. N., The outer capsid protein VP4 of equine rotavirus strain H-2 represents a unique VP4 type by amino acid sequence analysis, *Virology* **193:**492–497 (1993).

235. HARRIS, C. C., YOLKEN, R. H., KROKAN, H., AND HSU, I. C., Ultrasensitive enzymatic radioimmunoassay: Application to detection of cholera toxin and rotavirus, *Proc. Natl. Acad. Sci. USA* **76:**5335–5339 (1979).

236. HASEGAWA, A., MATSUNO, S., INOUYE, S., KONO, R., ISURU-KUBO, Y., MUKOYAMA, A., AND SAITO, Y., Isolation of human rotaviruses in primary cultures of monkey kidney cells, *J. Clin. Microbiol.* **16:**387–390 (1982).

237. HATTORI, H., TORII, S., NAGAFUJI, H., TABATA, Y., AND HATA, A., Benign acute myositis associated with rotavirus gastroenteritis, *J. Pediatr.* **121:**748–749 (1992).

238. HEATH, R., BIRCH, C., AND GUST, I., Antigenic analysis of rotavirus isolates using monoclonal antibodies specific for human serotypes 1, 2, 3 and 4, and SA11, *J. Gen. Virol.* **67:**2455–2466 (1986).

239. HERMANN, J. E., KENT, G. P., NOWAK, N. A., BRONDUM, J., AND BLACKLOW, N. R., Antigen detection in the diagnosis of Norwalk virus gastroenteritis, *J. Infect. Dis.* **154:**547–548 (1986).

240. HERRING, A. J., INGLIS, N. F., OJEH, C. K., SNODGRASS, D. R., AND MENZIES, J. D., Rapid diagnosis of rotavirus infection by direct detection of viral nucleic acid in silver-stained polyacrylamide gels, *J. Clin. Microbiol.* **16:**473–477 (1982).

241. HERRMANN, J. E., NOWAK, N. A., AND BLACKLOW, N. R., Detection of Norwalk virus in stools by enzyme immunoassay, *J. Med. Virol.* **17:**127–133 (1985).

242. HERRMANN, J. E., TAYLOR, D. N., ECHEVERRIA, P., *ET AL.*, Astroviruses as a cause of gastroenteritis in children, *N. Engl. J. Med.* **324:**1757–1760 (1991).

243. HIEBER, J. P., SHELTON, S., NELSON, J. D., LEON, J., AND MOHS, E., Comparison of human rotavirus disease in tropical and temperate settings, *Am. J. Dis. Child.* **132:**853–858 (1978).

244. HJELT, K., GRAUBALLE, P. C., ANDERSEN, L., SCHIOTZ, P. O., HOWITZ, P., AND KRASILNIKOFF, P. A., Antibody response in serum and intestine in children up to six months after a naturally acquired rotavirus gastroenteritis, *J. Pediatr. Gastroenterol. Nutr.* **5:**74–80 (1986).

245. HJELT, K., GRAUBALLE, P. C., PAERREGAARD, A., NIELSEN, O. H., AND KRASILNIKOFF, P. A., Protective effect of preexisting rotavirus-specific immunoglobulin A against naturally acquired rotavirus infection in children, *J. Med. Virol.* **21:**39–47 (1987).

246. HO, M.-S., GLASS, R. I., PINSKY, P. F., AND ANDERSON, L. L., Rotavirus as a cause of diarrheal morbidity and mortality in the United States, *J. Infect. Dis.* **158:**1112–1116 (1988).

247. HO, M.-S., GLASS, R. I., PINSKY, P. F., AND ANDERSON, L. J., Diarrheal deaths in American children. Are they preventable?, *J.A.M.A.* **260:**3281–3285 (1988).

248. HODES, H. L., American Pediatric Society presidential address, *Pediatr. Res.* **10:**201–204 (1976).

249. HOLM, S., ANDERSON, Y., GOTHEFORS, L., AND LINDBERG, T., Increased protein absorption after acute gastroenteritis in children, *Acta Paediatr. Int. J. Paediatr.* **81:**585–588 (1992).

250. HOLT, P. A., STEEL, A. E., AND ARMSTRONG, A. M., Transient hyperphosphatasaemia of infancy following rotavirus infection, *J. Infect.* **9:**283–285 (1985).

251. HOLMES, I. H., RUCK, B. J., BISHOP, R. F., AND DAVIDSON, G. P., Infantile enteritis viruses: Morphogenesis and morphology, *J. Virol.* **16:**937–943 (1975).

252. HOULY, C. A. P., UCHOA, M. M. M., ZAIDAN, A. M. E., DE-OLIVEIRA, F. M., ATHAYDE, M. A. G., ALMEIDA, M. F. L. M., AND PEREIRA, H. G., Electrophoretic study of the genome of human rotavirus from Merceio, Brazil, *Braz. J. Med. Biol. Res.* **19:**33–37 (1986).

253. HOSHINO, Y., AND KAPIKIAN, A. Z., Rotavirus antigens, *Curr. Topics Microbiol. Immunol.* **185:**179–227 (1994).

254. HOSHINO, Y., AND KAPIKIAN, A. Z., Prospects for development of a rotavirus vaccine for the prevention of severe diarrhea in infants and young children, *Trends Microbiol.* **2:**242–249 (1994).

255. HOSHINO, Y., KAPIKIAN, A. Z., AND CHANOCK, R. M., Selection of cold-adapted (ca) mutants of human rotaviruses that exhibit a varying degree of growth restriction *in vitro*, *J. Virol.* **68:**7598–7602 (1994).

256. HOSHINO, Y., SAIF, L. J., KANG, S. Y., SERENO, M., AND KAPIKIAN, A. Z., Genetic determinants of rotavirus virulence studied in gnotobiotic piglets, in: *Vaccine 93. Modern Approaches to New Vaccines Including Prevention of AIDS* (F. BROWN, R. M. CHANOCK, H. S. GINSBERG, AND R. S. LERNER, EDS.), pp. 277–282, Cold Spring Harbor Laboratory Press, Cold Spring Harbor, NY, 1992.

257. HOSHINO, Y., SERENO, M. M., MIDTHUN, K., FLORES, J., CHANOCK, R. M., AND KAPIKIAN, A. Z., Analysis by plaque reduction neutralization assay of intertypic rotaviruses suggests that gene reassortment occurs *in vivo*, *J. Clin. Microbiol.* **25:**290–294 (1987).

258. HOSHINO, Y., SERENO, M. M., MIDTHUN, K., FLORES, J., KAPIKIAN, A. Z., AND CHANOCK, R. M., Independent segregation of two antigenic specificities (VP3 and VP7) involved in neutralization of rotavirus infectivity, *Proc. Natl. Acad. Sci. USA* **82:**8701–8704 (1985).

259. HOSHINO, Y., WYATT, R. G., FLORES, J., MIDTHUN, K., AND KAPIKIAN, A. Z., Serotypic characterization of rotaviruses derived from asymptomatic human neonatal infections, *J. Clin. Microbiol* **21:**425–430 (1985).

260. HOSHINO, Y., WYATT, R. G., GREENBERG, H. B., FLORES, J., AND KAPIKIAN, A. Z., Serotypic similarity and diversity of rotaviruses of mammalian and avian origin as studied by plaque reduction neutralization, *J. Infect. Dis.* **149:**694–702 (1984).

261. HRDY, D. B., Epidemiology of rotaviral infection in adults, *Rev. Infect. Dis.* **9:**461–469 (1987).

262. HUNG, T., CHEN, G., WANG, C., YAO, H., FANG, Z., CHAO, T., CHOU, Z., YE, W., CHANG, X., DEN, S., LIONG, X., AND CHANG, W., Waterborne outbreak of rotavirus diarrhea in adults in China caused by a novel rotavirus, *Lancet* **1:**1139–1142 (1984).

263. HYAMS, K. C., BOURGEOIS, A. L., MERRELL, B. R., ROZMAJZL, P., ESCAMILLA, J., THORNTON, S. A., WASSERMAN, G. M., BURKE, A., ECHEVERRIA, P., GREEN, K. Y., KAPIKIAN, A. Z., WOODY,

J. N., Diarrheal disease during Operation Desert Shield, *N. Engl. J. Med.* **325**:1423–1428 (1991).

264. HYAMS, K. C., MALONE, J. D., KAPIKIAN, A. Z., ESTES, M. K., JIANG, X., BOURGEOIS, A. L., PAPARELLO, S., HAWKINS, R. E., AND GREEN, K. Y., Norwalk virus infection among Desert Storm troops, *J. Infect. Dis.* **167**:986–987 (1993).

265. INSTITUTE OF MEDICINE, Prospects for immunizing against rotavirus, in: *New Vaccine Development. Establishing Priorities. Diseases of Importance in the United States*, Vol. I, pp. 410–423, National Academy Press, Washington, DC, 1985.

266. INSTITUTE OF MEDICINE, The prospects for immunizing against rotavirus, in: *New Vaccine Development. Establishing Priorities. Diseases of Importance in Developing Countries*, Vol. II, pp. 308–316, National Academy Press, Washington, DC, 1986.

267. ISEGAWA, Y., NAKAGOMI, O., BRUSSOW, H., MINAMOTO, N., NAKAGOMI, T., AND VEDA, S., A unique VP4 gene allele carried by an unusual bovine rotavirus strain, 993/83, *Virology* **198**:366–369 (1994).

267a. JARVIS, W. R., MIDDLETON, P. J., AND GELFAND, E. W., Significance of viral infections in severe combined immunodeficiency disease. *Pediatr. Infect. Dis.* **2**:187–192 (1983).

268. JAYASHREE, S., BHAN, M. K., KUMAR, R., RAJ, P., GLASS, R., AND BHANDARI, N., Serum and salivary antibodies as indicators of rotavirus infection in neonates, *J. Infect. Dis.* **158**:1117–1120 (1988).

269. JIANG, X., GRAHAM, D. Y., WANG, K., AND ESTES, M. K., Norwalk virus genome cloning and characterization, *Science* **250**:1580–1583 (1990).

270. JIANG, X., WANG, M., WANG, K., AND ESTES, M. K., Sequence and genome organization of Norwalk virus, *Virology* **195**:51–61 (1993).

271. JIANG, X., WANG, J., GRAHAM, D. Y., AND ESTES, M. K., Detection of Norwalk virus in stool by polymerase chain reaction, *J. Clin. Microbiol.* **30**:2529–2534 (1992).

272. JIANG, X., WANG, M., GRAHAM, D. Y., AND ESTES, M. K., Expression, self-assembly and antigenicity of the Norwalk virus capsid protein, *J. Virol.* **66**:6527–6532 (1992).

273. JOHNSON, P. C., MATHEWSON, J. J., DuPONT, H. L., AND GREENBERG, H. L., Multiple-challenge study of host susceptibility to Norwalk gastroenteritis in US adults, *J. Infect. Dis.* **161**:18–21 (1990).

274. JOHNSON, P. C., HOY, J., MATHEWSON, J. J., ERICSSON, C. D., AND DuPONT, H. L., Occurrence of Norwalk virus infections among adults in Mexico, *J. Infect. Dis.* **162**:389–393 (1990).

275. JONES, R. C., HUGHES, C. S., AND HENRY, R. R., Rotavirus infection in commercial laying hens, *Vet. Rec.* **104**:22 (1979).

276. JORDAN, W. S., GORDON, I., AND DORRANCE, W. R., A study of illness in a group of Cleveland families. VII. Transmission of acute nonbacterial gastroenteritis to volunteers: Evidence for two different etiologic agents, *J. Exp. Med.* **98**:461–475 (1953).

277. KALICA, A. R., GREENBERG, H. B., WYATT, R. G., FLORES, J., SERENO, M. M., KAPIKIAN, A. Z., AND CHANOCK, R. M., Genes of human (strain Wa) and bovine (strain UK) rotaviruses that code for neutralization and subgroup antigens, *Virology* **112**:385–390 (1981).

278. KALICA, A. R., PURCELL, R. H., SERENO, M. M., WYATT, R. G., KIM, H. W., CHANOCK, R. M., AND KAPIKIAN, A. Z., A microtiter solid phase radioimmunoassay for detection of the human reovirus-like agent in stools, *J. Immunol.* **118**:1275–1279 (1977).

279. KALJOT, K. T., LING, J. P., GOLD, J. W. M., LAUGHON, B. E., BARTLETT, J. G., KOTLER, D. P., OSHIRO, L. S., AND GREENBERG, H. B., Prevalence of acute enteric viral pathogens in acquired

immunodeficiency syndrome patients with diarrhea, *Gastroenterology* **97**:1031–1032 (1989).

280. KAPIKIAN, A. Z., Viral gastroenteritis, *J.A.M.A.* **269**:627–630 (1993).

281. KAPIKIAN, A. Z., Response to "Oral Rehydration Therapy for Gastroenteritis," *J.A.M.A.* **270**:578–579 (1993).

282. KAPIKIAN, A. Z., Norwalk and Norwalk-like viruses, in: *Viral Infections of the Gastrointestinal Tract*, 2nd ed. (A. Z. KAPIKIAN, ED.), pp. 471–578, Marcel Dekker, New York, 1994.

283. KAPIKIAN, A. Z., Jennerian and modified Jennerian approach to vaccination against rotavirus diarrhea in infants and young children; an introduction, in: *Viral Infections of the Gastrointestinal Tract*, 2nd ed. (A. Z. KAPIKIAN, ED.), pp. 409–417, Marcel Dekker, New York, 1994.

284. KAPIKIAN, A. Z., Rhesus rotavirus-based human rotavirus vaccines and observations on selected non-Jennerian approaches to rotavirus vaccination, in: *Viral Infections of the Gastrointestinal Tract*, 2nd ed. (A. Z. KAPIKIAN, ED.), pp. 443–470, Marcel Dekker, New York, 1994.

285. KAPIKIAN, A. Z., BARILE, R. G., WYATT, R. H., YOLKEN, R. H., TULLY, J. G., GREENBERG, H. B., KALICA, A. R., AND CHANOCK, R. M., Mycoplasma contamination in cell culture of Crohn's disease material, *Lancet* **2**:466–467 (1979).

286. KAPIKIAN, A. Z., AND CHANOCK, R. M., Rotaviruses, in: *Virology*, 2nd ed. (B. N. FIELDS, D. M. KNIPE, R. M. CHANOCK, J. L. MELNICK, B. ROIZMAN, AND R. E. SHOPE, EDS.), pp. 863–906, Raven Press, New York, 1985.

287. KAPIKIAN, A. Z., ESTES, M. K., AND CHANOCK, R. M., Norwalk group of viruses, in: *Virology*, 3rd ed. (B. N. FIELDS, D. M. KNIPE, P. M. HOWLEY, R. M. CHANOCK, J. L. MELNICK, T. P. MONATH, B. ROIZMAN, AND S. E. STRAUS, EDS.), pp. 783–810, Raven Press, New York, 1996.

288. KAPIKIAN, A. Z., AND ESTES, M. K., Norwalk and related viruses, in: *Encyclopedia of Virology*, Vol. II, (R. G. WEBSTER AND A. GRANOFF, EDS.), pp. 925–930, Academic Press, London, 1994.

289. KAPIKIAN, A. Z., CLINE, W. L., GREENBERG, H. B., WYATT, R. G., KALICA, A. R., BANK, C. E., JAMES, H. D., JR., FLORES, J., AND CHANOCK, R. M., Antigenic characterization of human and animal rotaviruses by immune adherence hemagglutination assay (IAHA): Evidence for distinctness of IAHA and neutralization antigens, *Infect. Immun.* **33**:415–425 (1981).

290. KAPIKIAN, A. Z., CLINE, W. L., KIM, H. W., KALICA, A. R., WYATT, R. G., VanKIRK, D. H., CHANOCK, R. M., JAMES, H. D., JR., AND VAUGHN, A. L., Antigenic relationships among five reovirus-like (RVL) agents by complement fixation (CF) and development of a new substitute CF antigen for the human RVL agent of infantile gastroenteritis, *Proc. Soc. Exp. Biol. Med.* **152**:535–539 (1976).

291. KAPIKIAN, A. Z., CLINE, W. L., MEBUS, C. A., WYATT, R. G., KALICA, A. R., JAMES, H. D., JR., VanKIRK, D., CHANOCK, R. M., AND KIM, H. W., New complement-fixation test for the human reovirus-like agent of infantile gastroenteritis: Nebraska calf diarrhea virus used as antigen, *Lancet* **1**:1056–1069 (1975).

292. KAPIKIAN, A. Z., DIENSTAG, J. L., AND PURCELL, R. H., Immune electron microscopy as a method for the detection, identification, and characterization of agents not cultivable in an *in vitro* system, in: *Manual of Clinical Immunology*, 2nd ed. (N. R. ROSE AND H. FRIEDMAN, EDS.), pp. 70–83, American Society of Microbiology, Washington, DC, 1980.

293. KAPIKIAN, A. Z., FEINSTONE, S. M., PURCELL, R. H., WYATT, R. G., THORNHILL, T. S., KALICA, A. R., AND CHANOCK, R. M.,

Detection and identification by immune electron microscopy of fastidious agents associated with respiratory illness, acute non-bacterial gastroenteritis, and hepatitis A, *Perspect. Virol.* **9**:9–47 (1975).

294. KAPIKIAN, A. Z., FLORES, J., HOSHINO, Y, GLASS, R. I., MID-THUN, K., GORZIGLIA, M., AND CHANOCK, R. M., Rotavirus: The major etiologic agent of severe infantile diarrhea may be controllable by a "Jennerian" approach to vaccination, *J. Infect. Dis.* **153**:815–822 (1986).

295. KAPIKIAN, A. Z., GERIN, J. L., WYATT, R. G., THORNHILL, T. S., AND CHANOCK, R. M., Density in cesium chloride of the 27 nm "8FIIa" particle associated with acute infectious nonbacterial gastroenteritis: Determination by ultracentrifugation and immune electron microscopy, *Proc. Soc. Exp. Biol. Med.* **142**:874–877 (1974).

296. KAPIKIAN, A. Z., GREENBERG, H. B., CLINE, W. L., KALICA, A. R., WYATT, R. G., JAMES, H. D., JR., LLOYD, N. L., CHANOCK, R. M., RYDER, R. W., AND KIM, H. W., Prevalence of antibody to the Norwalk agent by a newly developed immune adherence hemagglutination assay, *J. Med. Virol.* **2**:218–294 (1978).

297. KAPIKIAN, A. Z., KIM, H. W., WYATT, R. G., CLINE, W. L., ARROBIO, J. O., BRANDT, C. D., RODRIGUEZ, W. J., SACK, D. A., CHANOCK, R. M., AND PARROTT, R. H., Human reovirus-like agent as the major pathogen associated with "winter" gastroenteritis in hospitalized infants and young children, *N. Engl. J. Med.* **294**:965–972 (1976).

298. KAPIKIAN, A. Z., KIM, H. W., WYATT, R. G., RODRIGUEZ, W. J., ROSS, S., CLINE, W. L., PARROTT, R. H., AND CHANOCK, R. M., Reovirus-like agent in stools: Association with infantile diarrhea and development of serologic tests, *Science* **185**:1049–1053 (1974).

299. KAPIKIAN, A. Z., WYATT, R. G., DOLIN, R., THORNHILL, T. S., KALICA, A. R., AND CHANOCK, R. M., Visualization by immune electron microscopy of a 27-nm particle associated with acute infectious non-bacterial gastroenteritis, *J. Virol.* **10**:1075–1081 (1972).

300. KAPIKIAN, A. Z., WYATT, R. G., GREENBERG, H. B., KALICA, A. R., KIM, H. W., BRANDT, C. D., RODRIGUEZ, W. J., PARROTT, R. H., AND CHANOCK, R. M., Approaches to immunization of infants and young children against gastroenteritis due to rotaviruses, *Rev. Infect. Dis.* **2**:459–469 (1980).

301. KAPIKIAN, A. Z., WYATT, R. G., LEVINE, M. M., YOLKEN, R. H., VANKIRK, D. H., DOLIN, R., GREENBERG, H. B., AND CHANOCK, R. M., Oral administration of human rotavirus to volunteers: Induction of illness and correlates of resistance, *J. Infect. Dis.* **147**:95–106 (1983).

302. KAPIKIAN, A. Z., WYATT, R. G., LEVINE, M. M., BLACK, R. E., GREENBERG, H. B., FLORES, J., KALICA, A. R., HOSHINO, Y., AND CHANOCK, R. M., Studies in volunteers with human rotaviruses, *Dev. Biol. Stand.* **53**:209–218 (1983).

303. KAPIKIAN, A. Z., AND YOLKEN, R. H., Rotavirus, in: *Principles and Practice of Infectious Diseases*, 2nd ed. (G. L. MANDELL, R. G. DOUGLAS, JR., AND J. E. BENNETT, EDS.), pp. 933–944, John Wiley & Sons, New York, 1985.

304. KAPIKIAN, A. Z., YOLKEN, R. H., GREENBERG, H. B., WYATT, R. G., KALICA, A. R., CHANOCK, R. M., AND KIM, H. W., Gastroenteritis viruses, in: *Diagnostic Procedures for Viral, Rickettsial, and Chlamydial Infections*, 5th ed. (H. LENNETT AND N. J. SCHMIDT, EDS.), pp. 927–995, American Public Health Association, Washington, DC, 1979.

305. KAPLAN, J. E., FELDMAN, R., CAMPBELL, D. S., LOOKABAUGH, C., AND GARY, G. W., The frequency of a Norwalk-like pattern of

illness in outbreaks of acute gastroenteritis, *Am. J. Public Health* **72**:1329–1332 (1982).

306. KAPLAN, J. E., GARY, G. W., BARON, R. C., SINGH, N., SHONBERGER, L. B., FELDMAN, R., AND GREENBERG, H. B., Epidemiology of Norwalk gastroenteritis and the role of Norwalk virus in outbreaks of acute nonbacterial gastroenteritis, *Ann. Intern. Med.* **96**:756–761 (1982).

307. KAYE, H. S., YARBROUGH, W. B., AND REED, C. J., Calf diarrhea coronavirus, *Lancet* **2**:509 (1975).

308. KELLER, K. M., SCHMIDT, H., WIRTH, S., QUEISSER-LUFT, A., AND SCHUMACHER, R., Differences in the clinical and radiologic patterns of rotavirus and non-rotavirus necrotizing enterocolitis, *Pediatr. Infect. Dis. J.* **10**:734–738 (1991).

309. KIM, H. W., ARROBIO, J. O, BRANDT, C. D., JEFFRIES, B. D., PYLES, G., REID, J. L., CHANOCK, R. M., AND PARROTT, R. H., Epidemiology of respiratory syncytial virus infection in Washington, DC. I. Importance of the virus in different respiratory tract disease syndromes and temporal distribution of infection, *Am. J. Epidemiol.* **98**:216–225 (1973).

310. KIM, H. W., BRANDT, C. D., KAPIKIAN, A. Z., WYATT, R. G., ARROBIO, J. O., RODRIGUEZ, W. J., CHANOCK, R. M., AND PARROTT, R. H., Human reovirus-like agent (HRVLA) infection: Occurrence in adult contacts of pediatric patients with gastroenteritis, *J.A.M.A.* **238**:404–407 (1977).

311. KIM, K.-H., YANG, J.-M., JOO, S.-I, CHO, Y.-G., GLASS, R. I., AND CHO, Y.-J., Importance of rotavirus and adenovirus types 40 and 41 in acute gastroenteritis in Korean children, *J. Clin. Microbiol.* **28**:2279–2284 (1990).

312. KJELDSBERG, E., ANESTAD, G., GREENBERG, H., ORSTAVIK, I., PEDERSEN, R., AND SLETTEBO, E., Norwalk virus in Norway: An outbreak of gastroenteritis studied by electron microscopy and radioimmunoassay, *Scand. J. Infect. Dis.* **21**:521–526 (1989).

312a. KNOWLTON, D. R., SPECTOR, D. M., AND WARD, R. L., Development of an improved method for measuring neutralizing antibody to rotavirus. *J. Virol. Methods* **33**:127–134 (1991).

313. KOGASAKA, R., NAKAMURA, S., CHIBA, S., SAKUMA, Y., TERASHIMA, H., YOKOHAMA, T., AND NAKAO, T., The 33- to 39-nm virus-like particles tentatively designated as Sapora agent, associated with an outbreak of acute gastroenteritis, *J. Med. Virol.* **8**:187–193 (1981).

314. KOGASAKA, R., SAKUMA, Y., CHIBA K. S., AKIHARA, M., HORINI, K., AND NAKAS, T., Small round virus-like particles associated with acute gastroenteritis in Japanese children, *J. Med. Virol.* **5**:151–160 (1980).

315. KOJIMA, S., FUKUMI, H., KUSAMA, H., YAMAMOTO, S., SUZUKI, S., UCHIDA, T., ISHIMARAU, T., OKA, T., KURETANI, K., OHMURA, K., NISHIKAWA, F., FUJIMOTO, J., FUJITA, K., NAKANO, A., AND SUNAKAWA, S., Studies on the causative agent of the infectious diarrhea: Records of the experiments on human volunteers, *Jpn. Med. J.* **1**:467–476 (1948).

316. KONNO, T., SATO, T., SUZUKI, H., KITAOKA, S., KATSUSHIMA, N., SAKAMOTO, M., YAZAKI, N., AND ISHIDA, N., Changing RNA patterns in rotaviruses of human origin: Demonstration of a single dominant pattern at the start of an epidemic and various patterns thereafter, *J. Infect. Dis.* **149**:683–687 (1984).

317. KONNO, T., SUZUKI, H., IMAI, A., AND ISHIDA, N., Reovirus-like agent in acute epidemic gastroenteritis in Japanese infants; Fecal shedding and serologic response, *J. Infect. Dis.* **135**:259–266 (1977).

318. KONNO, T., SUZUKI, H., IMAI, A., KUTSUZAWA, T., ISHIDA, N., KATSUSHIMA, N., SAKAMOTO, M., KITAOKA, S., TSUBOI, R., AND

ADACHI, M., A long term survey of rotavirus infection in Japanese children with acute gastroenteritis, *J. Infect. Dis.* **138**:569–576 (1978).

319. KONNO, T., SUZUKI, H., KUTSUZAWA, T., IMAI, A., KATSUSHIMA, N., SAKAMOTO, M., AND KITAOKA, S., Human rotavirus and intussusception, *N. Engl. J. Med.* **297**:945 (1977).

320. KONNO, T., SUZUKI, H., KUTSUZAWA, T., IMAI, A., KATSUSHIMA, N., SAKAMOTO, M., KITAOKI, S., TSUBOI, R., AND ADACHI, M., Human rotavirus infection in infants and young children with intussusception, *J. Med. Virol.* **2**:265–269 (1978).

321. KONNO, T., SUZUKI, H., KATSUSHIMA, N., IMAI, A., TAZAWA, F., KUTSUZAWA, T., KITAOKA, S., SAKAMOTO, M., YAZAKI, N., AND ISHIDA, N., Influence of temperature and relative humidity on human rotavirus infection in Japan, *J. Infect. Dis.* **147**:125–128 (1983).

322. KOOPMAN, J. S., Diarrhea and school toilet hygiene in Cali, Colombia, *Am. J. Epidemiol.* **107**:412–418 (1978).

323. KOOPMAN, J. S., TURKISH, V. J., MONTO, A. S., GOUVEA, V., SRIVASTAVA, S., AND ISAACSON, R. E., Patterns and etiology of diarrhea in three clinical settings, *Am. J. Epidemiol.* **119**:114–123 (1984).

324. KUMATE, J., AND ISIBASH, A., Pediatric diarrheal diseases: A global perspective, *Pediatr. Infect. Dis.* **5**:521–528 (1986).

325. KURITSKY, J. N., OSTERHOLM, M. T., KORLATH, J. A., WHITE, K. E., AND KAPLAN, J. E., A state-wide assessment of the role of Norwalk virus in outbreaks of food-borne gastroenteritis, *J. Infect. Dis.* **151**:568 (1985).

326. KURTZ, J. B., Astroviruses, in: *Viral Infections of the Gastrointestinal Tract* (A. Z. KAPIKIAN, ED.), pp. 569–580, Marcel Dekker, New York, 1994.

327. KURTZ, J., AND CUBITT, K. W. D., Astroviruses and caliciviruses, in: *Enteric Infection: Mechanisms, Manifestations and Management* (M. J. G. FARTHING AND G. T. KEYSCH, EDS.), pp. 205–215, Chapman & Hall, London, 1988.

327a. KURTZ, J. B., LEE, T. W., CRAIG, J. W., AND REED, S. E., Astrovirus infection in volunteers. *J. Med. Virol.* **3**:221–230 (1979).

328. LAMBDEN, P. R., CAUL, E. O., ASHLEY, C. R., AND CLARKE, I. N., Sequence and genome organization of a human small round structured (Norwalk-like) virus, *Science* **259**:516–519 (1993).

329. LANATA, C. F., BLACK, R. E., BURTON, B., MIDTHUN, K., AND DAVIDSON, B., Safety, immunogenicity and efficacy of one or three doses of the rhesus tetravalent rotavirus vaccine in Lima, Peru (Abstract), *Vaccine* **10**:273 (1992).

330. LANATA, C. F., BLACK, R. E., FLORES, J., AND KAPIKIAN, A. Z., Safety immunogenicity, and efficacy of one dose of the rhesus and serotype 1 and 2 human–rhesus reassortant rotavirus vaccines (Abstract), *Vaccine* **10**:272 (1992).

331. LAPORTE, J., AND BOBULESCO, P., Growth of human and canine enteric coronaviruses in a highly susceptible cell line: HRT 18, *Perspect. Virol.* **11**:189–193 (1981).

332. LEBARON, C. W., LEW, J. F., GLASS, R. I., WEBER, J. M., RUIZ-PALAZIOS, G. M., AND THE ROTAVIRUS STUDY GROUP, Annual rotavirus epidemic patterns in North America: Results of a 5-year retrospective survey of 88 centers in Canada, Mexico, and the United States, *J.A.M.A.* **264**:983–988 (1990).

332a. LEE, T. W., AND KURTZ, J. B., Prevalence of human astrovirus serotypes in the Oxford region 1976–1992 with evidence for two new serotypes. *Epid. Infect.* **112**:187–193 (1994).

333. LECCE, J. G., KING, M. W., AND MOCK, R., Reovirus-like agent associated with fatal diarrhea in neonatal piglets, *Infect. Immun.* **14**:816–825 (1976).

334. LEVINE, M. M., AND EDELMAN, R., Acute diarrheal infections in infants. I. Epidemiology, treatment and prospects for immunoprophylaxis, *Hosp. Pract.* **14**:89–100 (1979).

335. LEVY, A. G., WIDERLITE, L., SCHWARTZ, C. J., DOLIN, R., BLACKLOW, N. R., GARDNER, J., KIMBERG, D. V., AND TRIER, J. S., Jejunal adenylate cyclase activity in human subjects during viral gastroenteritis, *Gastroenterology* **70**:321–325 (1976).

336. LEW, J. F., MOE, C. L., MONROE, S. S., ALLEN, J. R., HARRISON, B. M., FORRESTER, B. D., STINE, S. E., WOODS, P. A., HIERHOLZER, J. C., HERRMANN, J. E., BLACKLOW, N. R., BARTLETT, A. V., AND GLASS, R. I., Astrovirus and adenovirus associated with diarrhea in children in day care settings, *J. Infect. Dis.* **164**:673–678 (1991).

337. LEW, J. F., PETRIC, M., KAPIKIAN, A. Z., JIANG, X., ESTES, M. K., AND GREEN, K. Y., Identification of minireovirus as a Norwalk-like virus in pediatric patients with gastroenteritis, *J. Virol.* **68**:3391–3396 (1994).

338. LEW, J. F., VALDESUSO, J., VESIKARI, T., KAPIKIAN, A. Z., JIANG, X., ESTES, M. K., AND GREEN, K. Y., Detection of Norwalk virus or Norwalk-like virus infections in Finnish infants and young children, *J. Infect. Dis.* **169**:1364–1367 (1994).

339. LEWIS, D. C., Three serotypes of Norwalk-like viruses demonstrated by solid-phase immune electron microscopy, *J. Med. Virol.* **30**:77–81 (1990).

340. LEWIS, D. C., Norwalk agent and other small round structural viruses in the UK, *J. Infect.* **23**:220–222 (1991).

341. LEWIS, H. M., PARRY, J. V., DAVIES, H. A., PARRY, R. P., MOTT, A., DOURMASHKIN, R. R., SANDERSON, P. J., TYRRELL, D. A. J., AND VALMAN, H. B., A year's experience of the rotavirus syndrome and its association with respiratory illness, *Arch. Dis. Child.* **54**:339–346 (1979).

342. LIGHT, J. S., AND HODES, H. L., Studies on epidemic diarrhea of the newborn: Isolation of a filtrable agent causing diarrhea in calves, *Am. J. Public Health* **33**:1451–1454 (1943).

343. LOSONSKY, G. A., JOHNSON, G. P., WINKELSTEIN, J. A., AND YOLKEN, R. H., The oral administration of human serum immunoglobulin in immunodeficiency patients with viral gastroenteritis: A pharmacokinetic and functional analysis, *J. Clin. Invest.* **76**:2362–2367 (1985).

344. LOSONSKY, G. A., RENNELS, M. B., LIM, Y., KRALL, G., KAPIKIAN, A. Z., AND LEVINE, M. M., Systemic and mucosal immune responses to rhesus rotavirus vaccine MMU 18006, *Pediatr. Infect. Dis.* **7**:388–393 (1988).

345. LOSONSKY, G. A., AND REYMANN, M., The immune response in primary asymptomatic and symptomatic rotavirus infection in newborn infants, *J. Infect. Dis.* **161**:330–332 (1990).

346. LOSONSKY, G. A., VONDERFECHT, S. L., EIDEN, J., WEE, S. B., AND YOLKEN, R. H., Homotypic and heterotypic antibodies for prevention of experimental rotavirus gastroenteritis, *J. Clin. Microbiol.* **24**:1041–1044 (1986).

347. MADORE, H. P., CHRISTY, C., PICHICHERO, M., LONG, C., PINCUS, P., VOSEFSKY, D., KAPIKIAN, A. Z., DOLIN, R., AND THE ELMWOOD, PANORAMA, AND WESTFALL PEDIATRIC GROUPS, Field trial of rhesus rotavirus or human-rhesus rotavirus reassortant vaccine of VP7 serotype 3 or 1 specificity in infants, *J. Infect. Dis.* **166**:235–243 (1992).

348. MADORE, H. P., TREANOR, J. J., BUJA, R., AND DOLIN, R., Antigenic relatedness among the Norwalk-like agents by serum antibody rises, *J. Med. Virol.* **32**:96–101 (1990).

349. MADORE, H. P., TREANOR, J. J., PRAY, K. A., AND DOLIN, R., Enzyme-linked immunosorbent assays for Snow Mountain and

Norwalk agents of viral gastroenteritis, *J. Clin. Microbiol.* **24:**456–459 (1986).

350. MAIYA, P. P., PEREIRA, S. M., MATHAN, M., BHAT, P., ALBERT, J. M., AND BAKER, J. J., Aetiology of acute gastroenteritis in infancy and early childhood in southern India, *Arch. Dis. Child.* **52:**482–485 (1977).

351. MALHERBE, H. H., HARWIN, R., AND ULRICH, M., The cytopathic effect of vervet monkey viruses, *S. Afr. Med. J.* **37:**407–411 (1963).

352. MALHERBE, H. H., AND STRICKLAND-CHOLMLEY, M., Simian virus SA-11 and the related "O" agent, *Arch. Ges. Virusforsch.* **22:**235–245 (1967).

353. MARRIE, T. J., LEE, S. H. S., FAULKNER, R. S., ETHIER, J., AND YOUNG, C. H., Rotavirus infection in a geriatric population, *Arch. Intern. Med.* **142:**313–316 (1982).

354. MARTIN, M. L., GARY, G. W., JR., AND PALMER, E. L., Comparison of hemagglutination-inhibition, complement-fixation and enzyme-linked immunosorbent assay for quantitation of human rotavirus antibodies, *Arch. Virol.* **62:**131–136 (1979).

355. MATA, L., Rotavirus diarrhea in Costa Rica, in: *4th International Congress for Virology*, p. 469, Centre for Agricultural Publishing and Documentation, Wageningen, 1978.

356. MATA, L., JIMENEZ, P., ALLEN, M. A., VARGAS, W., GARCIA, M. E., URRUTIA, J. J., AND WYATT, R. G., Diarrhea and malnutrition: Breast-feeding intervention in a transitional population, in: *Acute Enteric Infections in Children: New Prospects for Treatment and Prevention* (T. HOLMES, J. HOLMGREN, M. H. MERSON, AND R. MOLLBY, EDS.), pp. 233–251, Elsevier/North Holland, New York, 1981.

357. MATA, L., SIMHON, A., URRUTIA, J. J., KRONMAL, R. A., FERNANDEZ, R., AND GARCIA, B., Epidemiology of rotaviruses in a cohort of 45 Guatemalan Mayan Indian children observed from birth to the age of three years, *J. Infect. Dis.* **148:**452–461 (1983).

358. MATA, L. J., URRUTIA, J. J., AND GORDON, J. E., Diseases and disabilities, in: *The Children of Santa Maria Cauque: A Prospective Field Study of Health and Growth* (L. J. MATA, ED.), pp. 254–292, MIT Press, Cambridge, MA, 1978.

359. MATSON, D. O., AND ESTES, M. K., Impact of rotavirus infection at a large pediatric hospital, *J. Infect. Dis.* **162:**598–604 (1990).

360. MATSON, D. O., ESTES, M. K., BURNS, J. W., GREENBERG, H. B., TANIGUCHI, K., AND URASAWA, S., Serotype variation of human group A rotaviruses in two regions of the USA, *J. Infect. Dis.* **162:**605–614 (1990).

361. MATSON, D. O., O'RYAN, M. L., PICKERING, L. K., CHIBA, S., NAKATA, S., RAJ, P., AND ESTES, M. K., Characterization of serum antibody responses to natural rotavirus infections in children by VP7-specific epitope-blocking assays, *J. Clin. Microbiol.* **30:**1056–1061 (1992).

362. MATSUI, S. M., MACKOW, E. R., AND GREENBERG, H. B., The molecular determinant of rotavirus neutralization and protection, *Adv. Virus Res.* **36:**181–214 (1989).

363. MATSUI, S. M., KIM, J. P., GREENBERG, H. B., SU, W., SUN, Q., JOHNSON, P. C., DUPONT, H. L., OSHIRO, L. S., AND REYES, G. R., The isolation and characterization of Norwalk virus-specific cDNA, *J. Clin. Invest.* **87:**1456–1461 (1991).

364. MATSUNO, S., INOUYE, S., HASEGAWA, A., AND KONO, R., Assay of human rotavirus antibody by immune adherence hemagglutination with a cultivable human rotavirus as antigen, *J. Clin. Microbiol.* **15:**163–165 (1982).

365. MATSUNO, S., INOUYE, S., AND KONO, R., Antigenic relationship between human and bovine rotaviruses as determined by neutral-ization, immune adherence hemagglutination, and complement fixation tests, *Infect. Immun.* **17:**661–662 (1977).

366. MATSUNO, S., INOUYE, S., AND KONO, R., Plaque assay of neonatal calf diarrhea virus and the neutralizing antibody in human sera, *J. Clin. Microbiol.* **5:**1–4 (1977).

367. MATSUNO, S., MURAKAMI, S., TAKAGI, M., HAYASHI, M., INOUYE, S., HASEGAWA, A., AND FUKAI, K., Cold adaptation of human rotavirus, *Virus Res.* **7:**273–280 (1987).

368. MATSUNO, S., AND NAGAYOSHI, S., Quantitative estimation of infantile gastroenteritis virus antigens in stools by immune adherence hemagglutination test, *J. Clin. Microbiol.* **7:**310–311 (1978).

369. MATSUNO, S., UTAGAWA, E., AND SUGIURA, A., Association of rotavirus infection with Kawasaki syndrome, *J. Infect. Dis.* **148:**177 (1983).

370. MATTHEWS, R. E. F., The classification and nomenclature of viruses. Summary of results of meetings of the international committee on taxonomy of viruses in the Hague, September 1978, *Intervirology* **11:**133–135 (1979).

371. MATTION, N. M., COHEN, J., AND ESTES, M. K., The rotavirus proteins, in: *Viral Infections of the Gastrointestinal Tract* (A. Z. KAPIKIAN, ED.), pp. 169–249, Marcel Dekker, New York, 1994.

372. MAVROMICHALIS, J., EVANS, N., MCNEISH, A. S., BRYDEN, A. S., DAVIES, H. A., AND FLEWETT, A., Intestinal damage in rotavirus and adenovirus gastroenteritis assessed by D-xylose malabsorption, *Arch. Dis. Child.* **52:**589–591 (1977).

373. MCLEAN, B., SONZA, S., AND HOLMES, I. H., Measurement of immunoglobulin, A, G and M class rotavirus antibodies in serum and mucosal secretions, *J. Clin. Microbiol.* **12:**314–319 (1980).

374. MCNULTY, M. S., ALLAN, G. M., AND STUART, J. C., Rotavirus infection in avian species, *Vet. Rec.* **103:**319–320 (1978).

375. MCNULTY, M. S., ALLAN, G. M., TODD, D., AND MCFERRAN, J. B., Isolation and cell culture propagation of rotaviruses from turkeys and chickens, *Arch. Virol.* **61:**13–21 (1979).

376. MCNULTY, M. S., PEARSON, G. R., MCFERRAN, J. B., COLLINS, D. S., AND ALLAN, G. M., A reovirus-like agent (rotavirus) associated with diarrhea in neonatal pigs, *Vet. Microbiol.* **1:**55–63 (1976).

377. MEBUS, C. A., KONO, M., UNDERDAHL, N. R., AND TWIEHAUS, M. J., Cell culture propagation of neonatal calf diarrhea (scours) virus, *Can. Vet. J.* **12:**69–72 (1971).

378. MEBUS, C. A., UNDERDAHL, N. R., RHODES, M. B., AND TWIEHAUS, M. J., Calf diarrhea (scours): Reproduced with a virus from a field outbreak, *Univ. Nebraska Res. Bull.* **233:**1–16 (1969).

379. MEBUS, C. A., WYATT, R. G., AND KAPIKIAN, A. Z., Intestinal lesions induced in gnotobiotic calves by the virus of human infantile gastroenteritis, *Vet. Pathol.* **14:**273–282 (1977).

380. MEEROFF, J. C., SCHREIBER, D. S., TRIER, J. S., AND BLACKLOW, N. R., Abnormal gastric motor function in viral gastroenteritis, *Ann. Intern. Med.* **92:**370–373 (1980).

381. MIDDLETON, P. J., Analysis of the pattern of infection, in: *Report of the 74th Ross Conference on Pediatric Research: Etiology, Pathology and Treatment of Acute Gastroenteritis* (H. J. MCCLUNG, ED.), pp. 18–23, Ross Laboratories, Columbus, OH, 1978.

382. MIDDLETON, P. J., Pathogenesis of rotaviral infection, *J. Am. Vet. Med. Assoc.* **173:**544–546 (1978).

383. MIDDLETON, P. J., HOLDAWAY, M. D., PETRIC, M., SZYMANSKI, M. T., AND TAM, J. S., Solid-phase radioimmunoassay for the detection of rotavirus, *Infect. Immun.* **16:**439–444 (1977).

384. MIDDLETON, P. J., PETRIC, M., HEWITT, C. M., SZYMANSKI, M. T., AND TAM, J. S., Counter-immunoelectro-osmophoresis for the

detection of infantile gastroenteritis virus (orbi-group) and antibody, *J. Clin. Pathol.* **29:**191–197 (1976).

385. MIDDLETON, P. J., SZYMANSKI, M. T., ABBOTT, G. D., BORTOLUSSI, R., AND HAMILTON, J. R., Orbivirus acute gastroenteritis of infancy, *Lancet* **1:**1241–1244 (1974).

386. MIDDLETON, P. J., SZYMANSKI, M. T., AND PETRIC, M., Viruses associated with acute gastroenteritis in young children, *Am. J. Dis. Child.* **131:**733–737 (1977).

387. MIDTHUN, K., FLORES, J., TANIGUCHI, K., URASAWA, S., KAPIKIAN, A. Z., AND CHANOCK, R. M., Genetic relatedness among human rotavirus genes coding for VP7, a major neutralization protein, and its application to serotype identification., *J. Clin. Microbiol.* **25:**1269–1274 (1987).

388. MIDTHUN, K., GREENBERG, H. B., HOSHINO, Y., KAPIKIAN, A. Z., WYATT, R. G., AND CHANOCK, R. M., Reassortant rotaviruses as potential live rotaviruses vaccine candidates., *J. Virol.* **53:**949–954 (1985).

388a. MIDTHUN, K., GREENBERG, H. B., KURTZ, J. B., GARY, G. N., LIN, F.-Y. C., AND KAPIKIAN, A. Z., Characterization and seroepidemiology of a type 5 astrovirus associated with an outbreak of gastroenteritis in Marin County, California. *J. Clin. Microbiol.* **31:**955–962 (1993).

389. MIDTHUN, K., HOSHINO, Y., KAPIKIAN, A. Z., CHANOCK, R. M., Single gene substitution rotavirus reassortants containing the major neutralization protein (VP7) of human rotavirus serotype 4., *J. Clin. Microbiol.* **24:**822–826 (1986).

390. MIDTHUN, K., VALDESUSO, J., KAPIKIAN, A. Z., HOSHINO, Y., AND GREEN, K. Y., Identification of serotype 9 human rotavirus by enzyme-linked immunosorbent assay with monoclonal antibodies, *J. Clin. Microbiol.* **27:**2112–2114 (1989).

391. MOE, K., AND SHIRLEY, J. A., The effect of relative humidity and temperature on the survival of human rotavirus in faeces, *Arch. Virol.* **72:**179–186 (1982).

392. MONROE, S. S. STINE, S. E., JIANG, X., ESTES, M. K., AND GLASS, R. I., Detection of antibody of recombinant Norwalk virus antigen (rNV) in specimens from outbreaks of gastroenteritis, *J. Clin. Microbiol.* **31:**2866–2872 (1993).

393. MONTO, A. S., AND KOOPMAN, J. S., The Tecumseh study. XI. Occurrence of acute enteric illness in the community, *Am. J. Epidemiol.* **112:**323–333 (1980).

394. MONTO, A. S., KOOPMAN, J. S., LONGINI, I. M., AND ISSACSON, R. E., The Tecumseh study. XII. Enteric agents in the community, 1976–1981, *J. Infect. Dis.* **148:**284–291 (1983).

395. MOON, H. W., Pathophysiology of viral diarrhea, in: *Viral Infections of the Gastrointestinal Tract* (A. Z. KAPIKIAN, ED.), pp. 27–52, Marcel Dekker, New York, 1994.

396. MOOSAI, R. B., GARDNER, P. S., ALMEIDA, J. D., AND GREENWAY, M. A., A simple immunofluorescence technique for the detection of human rotavirus, *J. Med. Virol.* **3:**189–194 (1979).

397. MORENS, D. M., ZWEIGHAFT, R. M., VERNON, T. M., GARY, G. W., ESLIEN, J. J., WOOD, B. T., HOLMAN, R. C., AND DOLIN, R., A waterborne outbreak of gastroenteritis with secondary person-to-person spread: Association with a viral agent, *Lancet* **1:**964–966 (1979).

398. MORISHIMA, T., NAGAYOSHI, S., OZAKI, T., ISOMURA, S., AND SUZUKI, S., Immunofluorescence of human reovirus-like agent of infantile diarrhoea, *Lancet* **2:**695–696 (1976).

399. MUCH, D., AND ZAJAC, I., Purification and characterization of epizootic diarrhea of infant mice virus, *Infect. Immun.* **6:**1019–1024 (1972).

400. MULCAHY, D. L., KAMATH, K. R., DE SILVA, L. M., HODGES, S., CARTER, I. W., AND CLOONAN, M. J., A two-part study of the aetiological role of rotavirus in intussusception, *J. Med. Virol.* **9:**51–55 (1982).

401. MUNIAPPA, L., GEORGIEV, K. G., DIMITROV, D., MITOUK, B., AND HARALAMBIEV, E. H., Isolation of rotavirus from buffalo calves, *Vet. Rec.* **120:**23 (1987).

402. MURPHY, A. M., ALBREY, M. B., AND CREWE, E. B., Rotavirus infections of neonates, *Lancet* **2:**1149–1150 (1977).

403. MURPHY, A. M., GROHMANN, G. S., CHRISTOPHER, P. J., LOPEZ, W. A., AND MILLSOM, R. H., Oyster food poisoning, *Med. J. Aust.* **2:**439 (1978).

404. NAGUIB, T., WYATT, R. G., MOHIELDIN, M. S., ZAKI, A. M., IMAM, I. Z., AND DUPONT, H. L., Cultivation and subgroup determination of human rotaviruses from Egyptian infants and young children, *J. Clin. Microbiol.* **19:**210–212 (1984).

405. NAGAKOMI, O., ISEGAWA, Y., HOSHINO, Y., ABOUDY, Y., SHIF, I., SILBERSTEIN, I., NAKAGOMI, T., VEDA, K. S., SEARS, J., AND FLORES, J., A new serotype of the outer capsid protein VP4 shared by an unusual human rotavirus strain RO1485 and canine rotaviruses, *J. Gen. Virol.* **74:**2771–2774 (1993).

406. NAKAGOMI, O., AND NAKAGOMI, T., Interspecies transmission of rotaviruses studied from the perspective of genogroup, *Microbiol. Immunol.* **37:**337–348 (1993).

407. NAKAGOMI, O., NAKAGOMI, T., AKATANI, K., IKEGAMI, N., AND KATSUSHIMA, N., Relative frequency of rotavirus serotypes in Yamagata, Japan, over four consecutive rotavirus seasons, *Res. Virol.* **141:**459–463 (1990).

408. NAKAGOMI, O., NAKAGOMI, T., OYAMADA, H., AND SUTO, T., Relative frequency of human rotavirus subgroups 1 and 2 in Japanese children with acute gastroenteritis, *J. Med. Virol.* **17:**29–34 (1985).

409. NAKATA, S., ESTES, M. K., GRAHAM, D. Y., LOOSLE, R., HUNG, T., SHUSHENG, W., SAIF, L. J., AND MELNICK, J. L., Antigenic characterization and ELISA detection of adult diarrhea rotavirus., *J. Infect. Dis.* **154:**448–455 (1986).

410. NAKATA, S., PETRIE, B. L., CALOMENI, E. P., AND ESTES, M. K., Electron microscopy procedure influences detection of rotaviruses, *J. Clin. Microbiol.* **25:**1902–1906 (1987).

411. NALIN, D. R., LEVINE, M. M., MATA, L. *ET AL.*, Comparison of sucrose with glucose in oral therapy of infant diarrhea, *Lancet* **2:**277–279 (1978).

412. NATARO, J. P., AND LEVINE, M. M., Bacterial diarrheas, in: *Viral Infections of the Gastrointestinal Tract* (A. Z. KAPIKIAN, ED.), pp. 697–752, Marcel Dekker, New York, 1994.

413. NICOLAS, J. C., INGRAND, D., FORTIER, B., AND BRICOUT, F., A one-year virological survey of acute intussusception in childhood, *J. Med. Virol.* **9:**267–271 (1982).

414. NIGRO, G., AND MIDULLA, M., Acute laryngitis associated with rotavirus gastroenteritis, *J. Infect.* **7:**81–83 (1983).

415. NIGRO, G., Pancreatitis with hypoglycemia-associated convulsions following rotavirus gastroenteritis, *J. Pediatr. Gastroenterol. Nutr.* **12:**280–282 (1991).

416. NOEL, J. S., BEARDS, G. M., AND CUBITT, W. D., Epidemiological survey of human rotavirus serotypes and electropherotypes in young children admitted to two children's hospitals in northeast London from 1984 to 1990, *J. Clin. Microbiol.* **29:**2213–2219 (1991).

417. OBERT, G., GLOECKLER, R., BURCKARD, J., AND VAN REGENMORTEL, M. H. V., Comparison of immunosorbent electron microscopy, enzyme immunoassay and counterimmunoelectrophoresis for detection of human rotavirus in stools, *J. Virol. Methods* **3:**99–107 (1981).

418. OFFIT, P. H., Virus-specific cellular immune response to intestinal

infection, in: *Viral Infections of the Gastrointestinal Tract* (A. Z. KAPIKIAN, ED.), pp. 89–101, Marcel Dekker, New York, 1994.

419. OHNO, A., TANIGUCHI, K., SUGIMOTO, K., *ET AL.*, Rotavirus gastroenteritis and afebrile infantile convulsion (in Japanese), *No To Hattatsu* **14**:520–521 (1982).

420. OISHI, I., KIMURA, T., MURAKAMI, T., HARUKI, K., YAMAZAKI, K., SETO, Y., MINEKAWA, Y., AND FUNAMOTO, H., Serial observations of chronic rotavirus infection in an immunodeficient child, *Microbiol. Immunol.* **35**:953–961 (1991).

421. OISHI, I., YAMAZAKI, K., MINEKAWA, Y., NISHIMURA, H., AND KITAURA, T., Three-year survey of the epidemiology of rotavirus enteric adenovirus and some small spherical viruses including "Osaka-agent" associated with infantile diarrhea, *Biken J.* **28**:9–19 (1985).

422. OKADA, S., SEKINE, S., ANDO, T., HAYASHI, Y., MURAO, M., YABUUCHI, E., MIKI, T., AND OHASHI, M., Antigenic characterization of small round-structured viruses by immune electron microscopy, *J. Clin. Microbiol.* **28**:1244–1248 (1990).

423. O'RYAN, M. L., MATSON, D. O., ESTES, M. K., BARTLETT, A. V., AND PICKERING, L. K., Molecular epidemiology of rotavirus in children attending day-care centers in Houston, *J. Infect. Dis.* **162**:810–816 (1990).

424. ORSTAVIK, I., FIGENSCHAU, K. J., HAUG, K. W., AND ULSTRUP, J. C., A reovirus-like agent (rotavirus) in gastroenteritis of children, *Scand. J. Infect. Dis.* **8**:1–5 (1976).

425. PANG, Q., LIU, J., WAN, X., QUI, F., AND XU, A., Experimental infection of adult *Tupaia belangeri yunalis* with human rotavirus, *Chin. Med. J.* **96**:85–94 (1983).

426. PARRINO, T. A., SCHREIBER, D. S., TRIER, J. S., KAPIKIAN, A. Z., AND BLACKLOW, N. R., Clinical immunity in acute gastroenteritis caused by the Norwalk agent, *N. Engl. J. Med.* **297**:86–89 (1977).

427. PARROTT, R. H., VARGOSKO, A. J., KIM, H. W., BELLANTI, J. A., AND CHANOCK, R. J., Myxovirus parainfluenza, *Am. J. Public Health* **5**:907–909 (1962).

428. PATEL, J. R., DAVIES, H. A., EDINGTON, N., LAPORTE, J., AND MACNAUGHTON, M. R., Infection of a calf with the enteric coronavirus strain Paris, *Arch. Virol.* **73**:319–327 (1982).

429. PEDLEY, S., HUNDLEY, F., CHRYSTIE, I., McCRAE, M. A., AND DESSELBERGER, U., The genomes of rotaviruses isolated from chronically infected immunodeficient children, *J. Gen. Virol.* **65**:1141–1150 (1984).

430. PENSAERT, M., CALLEBAUT, P., AND COX, E., Enteric coronaviruses of animals, in: *Viral Infections of the Gastrointestinal Tract* (A. Z. KAPIKIAN, ED.), pp. 627–696, Marcel Dekker, New York, 1994.

431. PEREIRA, H. G., FIALHO, A. M., FLEWETT, T. H., TEIXEIRA, J. M. S., AND ANDRADE, Z. P., Novel viruses in human faeces, *Lancet* **2**:103–104 (1988).

432. PEREIRA, H. G., FLEWETT, T. H., CANDEIAS, J. A. N., AND BARTH, O. M., A virus with a bisegmented double-stranded RNA genome in rat (*Oryzomys nigripas*) intestines, *J. Gen. Virol.* **69**:2749–2754 (1988).

433. PEREZ-SCHAEL, I., DAOUD, G., WHITE, L., *ET AL.*, Rotavirus shedding by newborn children, *J. Med. Virol.* **14**:127–136 (1984).

434. PETERSON, M. W., SPENDLOVE, R. S., AND SMART, R. A., Detection of neonatal calf diarrhea virus, infant reovirus-like diarrhea virus and a coronavirus using the fluorescent virus precipitin test, *J. Clin. Microbiol.* **3**:376–377 (1976).

435. PETRIC, M., MIDDLETON, P. J., GRANT, C., TAM, J. S., AND HEWITT, C. M., Lapine rotavirus: Preliminary studies on epizoology and transmission, *Can. J. Comp. Med.* **42**:143–147 (1978).

436. PETRIC, M., TAM, J. S., AND MIDDLETON, P. J., Preliminary characterization of the nucleic acid of infantile gastroenteritis virus (orbivirus group), *Intervirology* **7**:176–180 (1976).

437. PICKERING, L. K., BARTLETT, A. V., 3D, REVES, R. R., AND MORROW, A., Asymptomatic excretion of rotavirus before and after rotavirus diarrhea in children in day care centers, *J. Pediatr.* **112**:361–365 (1988).

438. PICKERING, L. K., DuPONT, H. L., BLACKLOW, N. R., AND CUKOR, G., Diarrhea due to Norwalk virus in families, *J. Infect. Dis.* **146**:116–117 (1982).

439. PONGSUWANNA, Y., TANIGUCHI, K., WAKASUGI, F., SUTIUIJIT, Y., CHIWAKUL, M., WARACHIT, P., JAYAVASU, C., AND URASAWA, S., Distinct yearly change of serotype distribution of human rotavirus in Thailand as determined by ELISA and PCR, *Epidemiol. Infect.* **111**:407–412 (1993).

439a. PRASAD, B. V. V., AND CHIU, W., Structure of rotavirus. In: RAMIG, R F. (ED.), *Rotaviruses*. Springer, Berlin, pp. 9–29 (1994).

440. PRASAD, B. V., WANG, G. J., CLERX, J. P. M., AND CHIU, W., Three-dimensional structure of rotavirus, *J. Mol. Biol.* **199**:269–275 (1988).

441. PREY, M. V., LORELLE, C. A., TAFF, T. A., SONSOUCIE, L., WEBB, M., GARDNER, T. D., AND AQUINO, T. I., Evaluation of three commercially available rotavirus detection methods for neonatal specimens, *Am. J. Clin. Pathol.* **89**:675–678 (1988).

442. PUBLIC HEALTH SERVICE LABORATORY (ENGLAND), Rotavirus in rabbits, *Commun. Dis. Rep.* **32** (1976).

443. PURDHAM, D. R., PURDHAM, P. A., EVANS, N., AND McNEISH, A. S., Isolation of human rotavirus using human embryonic gut monolayers, *Lancet* **2**:977 (1975).

444. PYNDIAH, N., BEGUIN, R., RICHARD, J., CHARLES, M., REY, A., AND BONIFAS, V., Accuracy of rotavirus diagnosis: Modified genome electrophoresis versus electron microscopy, *J. Virol. Methods* **20**:39–44 (1988).

445. RASOOL, N. B., GREEN, K. Y., AND KAPIKIAN, A. Z., Serotype analysis of rotaviruses from different locations in Malaysia, *J. Clin. Microbiol.* **31**:1815–1819 (1993).

446. RATNAM, S., TOBIN, A. M., FLEMMING, J. B., AND BLASKOVIC, P. J., False positive rotazyme results, *Lancet* **1**:345–346 (1984).

447. REED, D. E., DALEY, C. A., AND SHAVE, H. J., Reovirus-like agent associated with neonatal diarrhea in pronghorn antelope, *J. Wildl. Dis.* **12**:488–491 (1976).

448. REIMANN, H. A., PRICE, A. H., AND HODGES, J. H., The cause of epidemic diarrhea, nausea and vomiting (viral dysentery?), *Proc. Soc. Exp. Biol. Med.* **58**:8–9 (1945).

449. RESTA, S., LUBY, J. P., ROSENFELD, C. R., AND SIEGAL, J. D., Isolation and propagation of a human enteric coronavirus, *Science* **229**:978–981 (1985).

450. RETTER, M., MIDDLETON, P. J., TAM, J. S., AND PETRIC, M., Enteric adenoviruses: Detection, replication, and significance, *J. Clin. Microbiol.* **10**:574–578 (1979).

451. RIEPENHOFF-TALTY, M., SUZUKI, H., OFFOR, E., *ET AL.*, Effect of age and malnutrition on rotavirus infection in mice, *Pediatr. Res.* **19**:1250–1257 (1985).

452. RODGER, S. M., BISHOP, R. F., BIRCH, C., McLEAN, B., AND HOLMES, I. H., Molecular epidemiology of human rotaviruses in Melbourne, Australia, from 1973 to 1979, as determined by electrophoresis of genome ribonucleic acid, *J. Clin. Microbiol.* **13**:272–278 (1981).

453. RODGER, S. M., CRAVEN, J. H., WILLIAMS, I., Demonstration of reovirus-like particles in intestinal contents of piglets with diarrhea, *Aust. Vet. J.* **51**:536 (1975).

454. RODRIGUEZ, W. J., KIM, H. W., ARROBIO, J. O., BRANDT, C. D.,

CHANOCK, R. M., KAPIKIAN, A. Z., WYATT, R. G., AND PARROTT, R. H., Clinical features of acute gastroenteritis associated with human reovirus-like agent in infants and young children, *J. Pediatr.* **91**:188–193 (1977).

455. RODRIGUEZ, W. J., KIM, H. W., BRANDT, C. D., YOLKEN, R. H., RICHARD, M., ARROBIO, J. O., SCHWARTZ, R. H., KAPIKIAN, A. Z., CHANOCK, R. M., AND PARROTT, R. H., Common exposure outbreak of type 2 rotavirus gastroenteritis with high secondary attack rate within families, *J. Infect. Dis.* **140**:353–357 (1979).

456. RODRIGUEZ, W. J., KIM, H. W., BRANDT, C. D., BISE, B., KAPIKIAN, A. Z., CHANOCK, R. M., CURLIN, G., AND PARROTT, R. H., Rotavirus gastroenteritis in the Washington, DC area. Incidence of cases resulting in admission to the hospital, *Am. J. Dis. Child.* **134**:777–779 (1980).

457. RODRIGUEZ, W. J., KIM, H. W., BRANDT, C. D., YOLKEN, R. H., ARROBIO, J. O., KAPIKIAN, A. Z., CHANOCK, R. M., AND PARROTT, R. H., Sequential enteric illnesses associated with different rotavirus serotypes, *Lancet* **2**:37 (1978).

458. RODRIGUEZ, W. J., KIM, H. W., BRANDT, C. D., ET AL., Longitudinal study of rotavirus infection and gastroenteritis in families served by a pediatric medical practice: Clinical and epidemiologic observations, *Pediatr. Infect. Dis. J.* **6**:170–176 (1987).

459. ROHDE, J. E., AND NORTHRUP, R. S., Taking science where the diarrhoea is, in: *Acute Diarrhoea in Childhood*. Ciba Foundation Symposium 42 (New Series) pp. 339–366, Elsevier/Excerpta Media/North Holland, 1976.

460. ROSETO, A., LEMA, F., CAVALIERI, F., DIANOUX, K., SITBON, M., FERCHAL, F., LASNERET, J., AND PERIES, J., Electron microscopy detection and characterization of viral particles in dog stools, *Arch. Virol.* **66**:89–93 (1980).

461. ROSETO, A., LEMA, F., SITBON, M., CAVALIERI, F., DIANOUX, L., AND PERIES, J., Detection of rotaviruses in dogs, *Soc. Occup. Med.* **7**:478 (1979).

462. ROTBART, H. A., LEVIN, M. J., YOLKEN, R. H., MANCHESTER, D. K., AND JANTZEN, J., An outbreak of rotavirus-associated neonatal necrotizing enterocolitis, *J. Pediatr.* **103**:454–459 (1983).

463. ROTBART, H. A., NELSON, W. L., GLODE, M. P., TRIFFON, T. C., KOGUT, S. J. H., YOLKEN, R. H., AND HERNANDEZ, J. A., Neonatal rotavirus-associated necrotizing enterocolitis: Case control study and prospective surveillance during an outbreak, *J. Pediatr.* **112**:87–93 (1988).

464. RUUSKA, T., AND VESIKARI, T., A prospective study of acute diarrhoea in Finnish children from birth to 2½ years of age, *Acta Paediatr. Scand.* **80**:500–507 (1991).

465. RYDER, R. W., MCGOWAN, J. E., HATCH, M. H., AND PALMER, E. L., Reovirus-like agent as a cause of nosocomial diarrhea in infants, *J. Pediatr.* **90**:698–702 (1977).

466. RYDER, R. W., SINGH, N., REEVES, W. C., KAPIKIAN, A. Z., GREENBERG, H. B., AND SACK, R. B., Evidence of immunity induced by naturally acquired rotavirus and Norwalk virus infection on two remote Panamanian islands, *J. Infect. Dis.* **151**:91–105 (1985).

467. RYDER, R. W., YOLKEN, R. H., REEVES, W. C., AND SACK, R. B., Enzootic bovine rotavirus is not a source of infection in Panamanian cattle ranchers and their families, *J. Infect. Dis.* **153**:1139–1144 (1986).

468. SABARA, M., BARRINGTON, A., AND BABIUK, L. A., Immunogenicity of a bovine rotavirus glycoprotein fragment, *J. Virol.* **56**:1037–1040 (1985).

469. SACK, D. A., Treatment of acute diarrhoea with oral rehydration solution, *Drugs* **23**:150–157 (1982).

470. SACK, D. A., FOR THE US ROTAVIRUS VACCINE EFFICACY GROUP, Efficacy of rhesus rotavirus monovalent or tetravalent oral vaccines in US children (Abstract), ICAAC, Anaheim, CA, 1992.

471. SACK, D. A., CHOWDHURY, A. M. A. K., EUSOF, A., ET AL., Oral hydration in rotavirus diarrhoea: A double blind comparison of sucrose with glucose electrolyte solution, *Lancet* **2**:280–283 (1978).

472. SACK, R. B., AND RABBANI, G. H., Treatment of diarrheal diseases, in: *Viral Infections of the Gastrointestinal Tract* (A. Z. KAPIKIAN, ED.), pp. 753–775, Marcel Dekker, New York, 1994.

473. SAIF, L. J., ROSEN, B. I., AND PARWANI, A. V., Animal rotaviruses, in: *Viral Infections of the Gastrointestinal Tract* (A. Z. KAPIKIAN, ED.), pp. 279–367, Marcel Dekker, New York, 1994.

474. SAITOH, Y., MATSUNO, S., AND MUKOYAMA, A., Exanthem subitum and rotavirus, *N. Engl. J. Med.* **304**:845 (1983).

475. SALMI, T. T., ARSTILA, P., AND KOIVIKKO, A., Central nervous system involvement in patients with rotavirus gastroenteritis, *Scand. J. Infect. Dis.* **10**:29–31 (1978).

476. SAMADI, A. R., HUQ, M. H., AND AHMED, Q. S., Detection of rotavirus in handwashings of attendants of children with diarrhoea, *Br. Med. J.* **286**:188 (1983).

477. SAMAL, S. K., DOPAZO, C. P., MCPHILLIPS, T. H., BAYA, A., MOHANTY, S. B., AND HETRICK, F. M., Molecular characterization of a rotavirus-like virus isolated from striped bass, *J. Virol.* **64**:5235–5240 (1990).

478. SAMAL, S. K., DOPAZO, C. P., SUBRAMANIAN, K., LUPIANI, B., MOHANTY, S. B., AND HETRICK, F. M., Heterogeneity in the genome RNAs and polypeptides of five members of a novel group of rotavirus-like viruses isolated from aquatic animals, *J. Gen. Virol.* **72**:181–184 (1991).

479. SANDERS, R. C., CAMPBELL, A. D., AND JENKINS, A. F., Routine detection of human rotavirus by latex agglutination: Comparison with latex agglutination, electron microscopy and polyacrylamide gel electrophoresis, *J. Virol. Methods* **13**:285–290 (1986).

480. SANEKATA, T., AND OKADA, H., Human rotavirus detection by agglutination of antibody-coated erythrocytes, *J. Clin. Microbiol.* **17**:1141–1147 (1983).

481. SANEKATA, T., OSHIDA, Y., AND ODA, K., Detection of rotavirus from faeces by reversed passive hemagglutination method, *J. Clin. Pathol.* **32**:963 (1979).

482. SANEKATA, T., OSHIDA, Y., ODA, K., AND OKADA, H., Detection of rotavirus antibody by inhibition of reverse passive hemagglutination, *J. Clin. Microbiol.* **15**:148–155 (1982).

483. SANEKATA, T., OSHIDA, Y., AND OKADA, H., Detection of rotavirus in faeces by latex agglutination, *J. Immunol. Methods* **41**:377–385 (1981).

484. SANTOS, N., RIEPENHOFF-TALTY, M., CLARK, H. F., OFFIT, P., AND GOUVEA, V., VP4 genotyping of human rotavirus in the United States, *J. Clin. Microbiol.* **32**:205–208 (1994).

485. SANTOSHAM, M., BURNS, B., NADKARNI, V., FOSTER, S., GARRETT, S., CROLL, L., O'DONOVAN, J., CROSSON, J., PATHAK, R., AND SACK, R. B., Oral rehydration therapy for acute diarrhea in ambulatory children in the United States: A double-blind comparison of four different solutions, *Pediatrics* **76**:159–166 (1985).

486. SANTOSHAM, M., DAUM, R. S., DILLMAN, L., RODRIGUEZ, J. L., LUQUE, S., RUSSELL, R., KOURANY, M., RYDER, R. W., BARTLETT, A. V., ROSENBERG, A., BENENSON, A. S., AND SOCK, R. R., Oral rehydration therapy of infantile diarrhea: A controlled study of well-nourished children hospitalized in the United States and Panama, *N. Engl. J. Med.* **306**:1070–1076 (1982).

487. SANTOSHAM, M., YOLKEN, R. H., QUIROZ, E., ET AL., Detection

of rotavirus in respiratory secretions of children with pneumonia, *J. Pediatr.* **103**:583–585 (1983).

488. SARKKINEN, H. K., MEURMAN, O. H., AND HALONEN, P. E., Solid-phase radioimmunoassay of IgA, IgG and IgM antibodies to human rotavirus, *J. Med. Virol.* **3**:281–289 (1979).

489. SATO, K., INABA, Y., MIURA, Y., *ET AL.*, Antigenic relationships between rotaviruses from different species as studied by neutralization and immunofluorescence, *Arch. Virol.* **73**:45–50 (1982).

490. SATO, K., INABA, Y., SHINOZAKI, T., FUJII, R., AND MATUMOTO, M., Isolation of human rotavirus in cell cultures, *Arch. Virol.* **69**:155–160 (1981).

491. SAULSBURY, F. T., WINKELSTEIN, J. A., AND YOLKEN, R. H., Chronic rotavirus infection in immunodeficiency, *J. Pediatr.* **97**:61–65 (1980).

492. SCHAFFER, F. L., Caliciviruses, in: *Comprehensive Virology*, Vol. 14 (H. FRAENKEL-CONRAT AND R. R. WAGNER, EDS.), pp. 249–281, Plenum, New York, 1979.

493. SCHAFFER, F. L., AND SOERGEL, M. E., Single major polypeptide of a calicivirus and characterization by polyacrylamide gel electrophoresis and stabilization of virions by cross-linking with dimethyl suberimidate, *J. Virol.* **19**:925–931 (1976).

494. SCHNAGEL, R. D., GRECO, T., AND MOREY, F., Antibody prevalence to human enteric coronavirus-like particles and indications of antigenic differences between particles from different areas, *Arch. Virol.* **87**:331–337 (1986).

495. SCHOUB, B. D., KOORNOF, H. J., LECATSAS, G., PROZESKY, O. W., FREIMAN, I., HARTMAN, E., AND KASSEL, H., Viruses in acute summer gastroenteritis in black infants, *Lancet* **1**:1093–1094 (1975).

496. SCHOUB, B. D., NEL, J. D., LECATSAS, G., LECATSAS, G., GREEF, A., PROZESKY, O. W., HAY, I. T., AND PRINSLOO, J. G., Rotavirus as a cause of gastroenteritis in black South African infants, *S. Afr. Med. J.* **50**:1124 (1976).

497. SCHREIBER, D. S., BLACKLOW, N. R., AND TRIER, J. S., The mucosal lesion of the proximal small intestine in acute infectious nonbacterial gastroenteritis, *N. Engl. J. Med.* **288**:1318–1323 (1973).

498. SCHREIBER, D. S., BLACKLOW, N. R., AND TRIER, J. S., The small intestinal lesion induced by Hawaii agent in acute infectious nonbacterial gastroenteritis, *J. Infect. Dis.* **129**:705–708 (1974).

499. SCOTT, A. C., LUDDINGTON, J., LUCAS, M., AND GILBERT, F. R., Rotavirus in goats, *Vet. Rec.* **103**:145 (1978).

500. SERENO, M. M., AND GORZIGLIA, M., The outer capsid protein VP4 of murine rotavirus strain Eb represents a tentative new P type, *Virology* **199**:500–504 (1994).

501. SETHABUTR, O., HANCHALAY, S., LEXOMBOON, U., BISHOP, R. F., HOLMES, I. H., AND ECHEVERRIA, P., Typing of human group A rotavirus with alkaline phosphatase-labeled oligonucleotide probes, *J. Med. Virol.* **37**:192–196 (1992).

502. SHARP, T. W., HYAMS, K. C., WATTS, D., TROFA, A. F., MARTIN, G. J., KAPIKIAN, A. Z., GREEN, K. Y., JIANG, X., ESTES, M. K., WAACK, M., AND SAVARINA, S. J., Epidemiology of Norwalk virus during an outbreak of acute gastroenteritis aboard a US aircraft carrier, *J. Med. Virol.* **45**:61–67 (1995).

503. SHARPEE, R., AND MEBUS, C. A., Rotaviruses of man and animals, *Lancet* **1**:639 (1975).

504. SHAW, R. D., FONG, K. J., LOSONSKY, G. A., LEVINE, M. M., MALDONADO, Y., YOLKEN, R., FLORES, J., KAPIKIAN, A. Z., VO, P. T., AND GREENBERG, H. B., Epitope-specific immune responses to rotavirus vaccination, *Gastroenterology* **93**:941–950 (1987).

505. SHIF, I., SILBERSTEIN, I., AND BAR-ON, A., Serotypic change in the prevalence of human rotavirus strains in Israel, *Isr. J. Med. Sci.* **28**:98–100 (1991).

506. SHINOZAKI, T., FUJII, R., SATO, K., TAKAHASAKI, E., ITO, Y., AND INABA, Y., Hemagglutinin from human reovirus-like agent, *Lancet* **1**:878 (1978).

507. SHUKRY, S., ZAKI, A. M., SHOUKRY, I., TAGI, M. E., AND HAMED, Z., Detection of enteropathogens in fatal and potentially fatal diarrheas in Cairo, Egypt, *J. Clin. Microbiol.* **24**:959–962 (1986).

508. SIDWELL, R. W., Overview of viral agents in pediatric enteric infections, *Pediatr. Infect. Dis.* **5**:544–545 (1986).

509. SIMHON, A., MATA, L., Anti-rotavirus antibody in human colostrum, *Lancet* **1**:39–40 (1978).

510. SMITH, E. M., ESTES, M. K., GRAHAM, D. Y., AND GERBA, C. P., A plaque assay of the simian rotavirus SA11, *J. Gen. Virol.* **43**:513–519 (1979).

511. SNODGRASS, D. R., ANGUS, K. W., AND GRAY, E. W., A rotavirus from kittens, *Vet. Rec.* **104**:222–223 (1979).

512. SNODGRASS, D. R., AND HERRING, J. A., The activity of disinfectants on lamb rotavirus, *Vet. Rec.* **101**:81 (1977).

513. SNODGRASS, D. R., MADELEY, C. R., WELLS, P. W., AND ANGUS, K. W., Human rotavirus in lambs: Infection and passive protection, *Infect. Immun.* **16**:268–270 (1977).

514. SNODGRASS, D. R., SMITH, W., GRAY, E. W., AND HERRING, J. A., A rotavirus in lambs with diarrhea, *Res. Vet. Sci.* **20**:113–114 (1976).

515. SNODGRASS, D. R., AND WELLS, P. W., Rotavirus infection in lambs: Studies on passive protection, *Arch. Virol.* **52**:201–205 (1976).

516. SNYDER, J. D., AND MERSON, M. H., The magnitude of the global problem of acute diarrhoeal disease: A review of active surveillance data, *Bull. WHO* **60**:605–613 (1982).

517. SONZA, S., BRESCHKIN, A. M., AND HOLMES, I. H., Derivation of neutralizing monoclonal antibodies against rotavirus, *J. Virol.* **45**:1143–1146 (1983).

518. SORIANO-BRUCHER, H., AVENDANO, P., O'RYAN, M., BRAUN, S. D., MANHART, M. D., BALM, T., AND SORIANO, H. A., Bismuth subsalicylate in the treatment of acute diarrhea in children: A clinical study, *Pediatrics* **87**:18–27 (1991).

519. SORIANO-BRUCHER, H. E., AVENDANO, P., O'RYAN, M., AND SORIANO, H., Use of bismuth subsalicylate in acute diarrhea in children, *Rev. Infect. Dis.* **12**:S51–S56 (1990).

520. SPENCE, L., FAUVEL, M., BOUCHARD, S., BABIUK, L., AND SAUNDERS, J. R., Test for reovirus-like agent, *Lancet* **2**:322 (1975).

521. SPENCE, L., FAUVEL, M., PETRO, R., AND BLOCK, S., Comparison of counterimmunoelectrophoresis and electron microscopy for laboratory diagnosis of human reovirus-like agent-associated infantile gastroenteritis, *J. Clin. Microbiol.* **5**:248–249 (1977).

522. SPENCE, L., FAUVEL, M., PETRO, R., AND BLOCK, S., Hemagglutinin from rotavirus, *Lancet* **2**:1023 (1976).

523. SPRATT, H. C., MARKS, M. I., GOMERSALL, M., GILL, P., AND PAI, C. H., Nosocomial infantile gastroenteritis associated with minirotavirus and calicivirus, *J. Pediatr.* **6**:922–926 (1978).

524. SPRINGTHORPE, V. S., GRENIER, J. L., LLOYD-EVANS, N., AND SATTAR, S. A., Chemical disinfection of human rotaviruses: Efficacy of commercially available products in suspension tests, *J. Hyg. (Lond.)* **97**:139–161 (1986).

525. STAAT, M. A., MORROW, A. L., REVES, R. R., BARTLETT, A. V., AND PICKERING, L. K., Diarrhea in children newly enrolled in day-care centers in Houston, *Pediatr. Infect. Dis. J.* **10**:282–286 (1991).

526. STEEL, H. M., GARNHAM, S., BEARDS, G. M., AND BROWN, D. W. G., Investigation of an outbreak of rotavirus infection in geriatric patients by serotyping and polyacrylamide gel electrophoresis (PAGE), *J. Med. Virol.* **37:**132–136 (1992).

527. STEELE, A. D., AND ALEXANDER, J. J., The relative frequency of subgroup I and II rotaviruses in black infants in South Africa, *J. Med. Virol.* **24:**321–327 (1988).

528. STEINHOFF, M. C., DOUGLAS, JR., R. G., GREENBERG, H. B., AND CALLAHAN, D. R., Bismuth subsalicylate therapy of viral gastroenteritis, *Gastroenterology* **78:**1495–1499 (1980).

529. STINTZING, G., TUFVESON, B., HABTE, D., BACK, E., JOHNSSON, T., AND WADSTROM, T., Aetiology of acute diarrhoeal disease in infancy and childhood during the peak season in Addis Ababa 1977: A preliminary report, *Ethiop. Med. J.* **15:**141–146 (1977).

530. STOLL, B. J., GLASS, R. I., BANU, H., HUQ, M. I., KHAN, M. U., AND AHMED, M., Value of stool examination in patients with diarrhoea, *Br. Med. J.* **286:**2037–2040 (1983).

531. STOLL, B. J., GLASS, R. I., HUQ, M. I., KHAN, M. U., HOLT, J. E., AND BANU, H., Surveillance of patients attending a diarrhoeal disease hospital in Bangladesh, *Br. Med. J.* **285:**1185–1188 (1982).

532. STUKER, G., OSHIRO, L., AND SCHMIDT, N. J., Antigenic composition of two new rotaviruses from rhesus monkeys, *J. Clin. Microbiol.* **11:**202–203 (1980).

533. SUZUKI, H., CHEN, G. M., HUNG, T., BEARDS, G. M., BROWN, D. W., AND FLEWETT, T. H., Effects of two staining methods on the Chinese atypical rotavirus, *Arch. Virol.* **94:**305–308 (1987).

534. SUZUKI, H., AND KONNO, T., Reovirus-like particles in jejunal mucosa of a Japanese infant with acute infectious non-bacterial gastroenteritis, *Tohoku J. Exp. Med.* **115:**199–211 (1975).

535. SZMUNESS, W., DIENSTAG, J. L., PURCELL, R. H., HARLEY, E. J., STEVENS, C. E., AND WONG, D. C., Distribution of antibody to hepatitis A antigen in certain adult populations, *N. Engl. J. Med.* **295:**755–759 (1976).

536. SZMUNESS, W., DIENSTAG, J. L., PURCELL, R. H., STEVENS, C. E., WONG, D. C., IKRAM, H., BAR-SHANY, S., BEASLEY, R. P., DESMYTER, J., AND GAON, J. A., The prevalence of antibody to hepatitis A antigen in various parts of the world: A pilot study, *Am. J. Epidemiol.* **106:**392–398 (1977).

537. TALLET, S., MACKENZIE, C., MIDDLETON, P., KERZNER, B., AND HAMILTON, R., Clinical, laboratory and epidemiological features of viral gastroenteritis in infants and children, *Pediatrics* **60:**217–222 (1977).

538. TAM, J. S., KUM, W. W., LAM, B., YEUNG, C. Y., AND NG, M. H., Molecular epidemiology of human rotavirus infection in children in Hong Kong, *J. Clin. Microbiol.* **23:**660–664 (1986).

539. TAN, J. A., AND SCHNAGL, R. G., Inactivation of a rotavirus by disinfectants, *Med. J. Aust.* **1:**19–23 (1981).

540. TAN, J. H., AND SCHNAGL, R. G., Rotavirus inactivated by a hypochlorite-based disinfectant: A reappraisal, *Med. J. Aust.* **1:**550 (1983).

541. TANIGUCHI, K., MORITA, Y., URASAWA, T., AND URASAWA, S., Cross-reactive neutralization epitopes on VP3 of human rotavirus, analysis with monoclonal antibodies and antigenic variants, *J. Virol.* **61:**1726–1730 (1987).

542. TANIGUCHI, K., URASAWA, T., AND URASAWA, S., Species specificity and interspecies relatedness in VP4 genotypes demonstrated by VP4 sequence analysis of equine, feline and canine rotavirus strains, *Virology* **200:**390–400 (1994).

543. TANIGUCHI, K., URASAWA, T., MORITA, Y., GREENBERG, H. B., AND URASAWA, S., Direct serotyping of human rotavirus in stools using serotype 1-, 2-, 3-, and 4-specific monoclonal antibodies to VP7, *J. Infect. Dis.* **155:**1159–1166 (1987).

544. TANIGUCHI, K., URASAWA, S., AND URASAWA, T., Preparation and characterization of neutralizing monoclonal antibodies with different reactivity patterns to human rotaviruses, *J. Gen. Virol.* **66:**1045–1053 (1984).

545. TANIGUCHI, K., URASAWA, S., AND URASAWA, T., Virus-like particles 35 to 40 nm associated with an institutional outbreak of acute gastroenteritis in adults, *J. Clin. Microbiol.* **10:**730–736 (1979).

546. TANIGUCHI, K., WAKASUGI, F., PONGSUWANNA, Y., URASAWA, T., UKAE, S., CHIBA, S., AND URASAWA, S., Identification of human and bovine rotavirus serotypes by polymerase chain reaction, *Epidemiol. Infect.* **109:**303–312 (1992).

547. TAYLOR, P. R., MERSON, M. H., BLACK, R. E., MIZANUR RAHMAN, A. S. M., YUNUS, M. D., ALIM, A. R. M. A., AND YOLKIN, R. H., Oral rehydration therapy for treatment of rotavirus diarrhoea in a rural treatment centre in Bangladesh, *Arch. Dis. Child.* **55:**376–379 (1980).

548. TERASHIMA, H., CHIBA, S. H., SAKUMA, Y., KOGASAKA, R., NAKATA, J. A., MINAMI, R., HORINI, K., AND NAKAO, T., The polypeptides of a human calicivirus, *Arch. Virol.* **78:**1–7 (1983).

549. THOMAS, E. E., PUTERMAN, M. L., KAWANO, E., AND CURRAN, M., Evaluation of seven immunoassays for detection of rotavirus in pediatric stool samples, *J. Clin. Microbiol.* **26:**1189–1193 (1988).

550. THORNHILL, T. S., KALICA, A. R., WYATT, R. G., KAPIKIAN, A. Z., AND CHANOCK, R. M., Pattern of shedding of the Norwalk particle in stools during experimentally induced gastroenteritis in volunteers as determined by immune electron microscopy, *J. Infect. Dis.* **132:**28–34 (1975).

551. THORNHILL, T. S., WYATT, R. G., KALICA, A. R., DOLIN, R., CHANOCK, R. M., AND KAPIKIAN, A. Z., Detection by immune electron microscopy of 26–27 nm virus-like particles associated with two family outbreaks of gastroenteritis, *J. Infect. Dis.* **135:**20–27 (1977).

552. THOULESS, M. E., BEARDS, G. M., AND FLEWETT, T. H., Serotyping and subgrouping of rotavirus strains by the ELISA test, *Arch. Virol.* **73:**219–230 (1982).

553. THOULESS, M. E., BRYDEN, A. S., AND FLEWETT, T. H., Rotavirus neutralization by human milk, *Br. Med. J.* **2:**1390 (1977).

554. THOULESS, M. E., BRYDEN, A. F., FLEWETT, T. H., WOODE, G. N., BRIDGER, J. C., SNODGRASS, G. R., AND HERRING, J. A., Serological relationships between rotaviruses from different species as studied by complement-fixation and neutralization, *Arch. Virol.* **53:**287–294 (1977).

555. THOULESS, M. E., DIGIACOMO, R. F., DEEB, B. J., AND HOWARD, H., Pathogenicity of rotavirus in rabbits, *J. Clin. Microbiol.* **26:**943–947 (1988).

556. TOTTERDELL, B. M., CHRYSTIE, I. L., AND BANATVALA, J. E., Rotavirus infections in a maternity unit, *Arch. Dis. Child.* **51:**924–928 (1976).

557. TREANOR, J. J., MADORE, H. P., AND DOLIN, R., Development of an enzyme immunoassay for the Hawaii agent of viral gastroenteritis, *J. Virol. Methods* **22:**207–214 (1988).

558. TREANOR, J., DOLIN, R., AND MADORE, H. P., Production of monoclonal antibody against the Snow Mountain agent of gastroenteritis by *in vitro* immunization of murine spleen cells, *Proc. Natl. Acad. Sci. USA* **85:**3613–3617 (1988).

559. TRISTRAM, D. A., AND OGRA, P. L., Immunology of the gastro-

intestinal tract, in: *Viral Infections of the Gastrointestinal Tract* (A. Z. KAPIKIAN, ED.), pp. 53–85, Marcel Dekker, New York, 1994.

560. TROONEN, H., False positive rotazyme results, *Lancet* **1**:345 (1984).

561. TSUNEMITSU, H., JIANG, B., AND SAIF, L. J., Detection of group C rotavirus antigens and antibodies in animals and humans by enzyme-linked immunosorbent assays, *J. Clin. Microbiol.* **30**: 2129–2134 (1992).

562. TUFVESSON, B., Detection of a human rotavirus strain different from types 1 and 2—a new subgroup? Epidemiology of subgroups in a Swedish and an Ethiopian community, *J. Med. Virol.* **12**:111–117 (1983).

563. TUFVESSON, B., AND JOHNSSON, T., Immunoelectro-osmophoresis for detection of reo-like virus: Methodology and comparison with electron microscopy, *Acta Pathol. Microbiol. Scand.* **85**: 225–228 (1976).

564. TUFVESSON, B., JOHNSSON, T., AND PETERSON, B., Family infections of reo-like virus, *Scand. J. Infect. Dis.* **9**:257–261 (1977).

565. TURTON, J., APPLETON, H., AND CLOWLEY, J. P., Similarities in nucleotide sequence between serum and faecal human parvovirus DNA, *Epidemiol. Infect.* **105**:197–201 (1990).

566. TZIPORI, S., Human rotavirus in young dogs, *Med. J. Aust.* **2**:922–923 (1977).

567. TZIPORI, S., CAPLE, I. W., AND BUTLER, R., Isolation of a rotavirus from deer, *Vet. Rec.* **99**:398 (1976).

568. TZIPORI, S., SMITH, M., HALPIN, C., MAKIN, T., AND KRAUTH, F., Intestinal changes associated with rotavirus and enterotoxigenic *Escherichia coli* infection in calves, *Vet. Microbiol.* **8**:35–43 (1983).

569. TZIPORI, S., AND WALKER, M., Isolation of rotavirus from foals with diarrhea, *Aust. J. Exp. Biol. Med. Sci.* **56**:453–457 (1978).

570. UHNOO, I. S., FREIHORST, J., RIEPENHOFF-TALTY, M., FISHER, J. E., AND OGRA, P. L., Effect of rotavirus infection and malnutrition on uptake of a dietary antigen in the intestine, *Pediatr. Res.* **27**:153–160 (1990).

571. UHNOO, I., SVENSSON, L., AND WADELL, G., Enteric adenoviruses, *Ballieres Clin. Gastroenterol.* **4**(3):627–642 (1990).

572. UHNOO, I., WADELL, G., SVENSSON, L., AND JOHANSSON, M. E., Importance of enteric adenoviruses 40 and 41 in acute gastroenteritis in infants and young children, *J. Clin. Microbiol.* **20**:365–372 (1984).

573. URASAWA, S., TANIGUCHI, K., URASAWA, T., SAKURADA, N., KAWAMURA, S., AND AKATSUKA, K., Virus particle detected in an institutional outbreak of acute gastroenteritis, *Igaku no Ayumi* **109**(1):25–27 (1979).

574. URASAWA, T., URASAWA, S., AND TANIGUCHI, K., Sequential passages of human rotavirus in MA-104 cells, *Microbiol. Immunol.* **25**:1025–1035 (1981).

575. URUSAWA S., URASAWA T., AND TANIGUCHI, K., Three human rotavirus serotypes demonstrated by plaque neutralization of isolated strains, *Infect. Immun.* **38**:781–784 (1982).

576. URASAWA, S., URASAWA, T., TANIGUCHI, K., CHIBA, S., Serotype determination of human rotavirus isolates and antibody prevalence in pediatric population in Hokkaido, Japan, *Arch. Virol.* **81**:1–12 (1989).

577. URASAWA, S., URASAWA, T., TANIGUCHI, K., *ET AL.*, Validity of an enzyme-linked immunosorbent assay with serotype-specific monoclonal antibodies for serotyping human rotavirus in stool specimens, *Microbiol. Immunol.* **32**:699–708 (1988).

578. USHIJIMA, H., BOSU, K., ABE, T., AND SHINOZAKI, T., Suspected rotavirus encephalitis, *Arch. Dis. Child.* **61**:692–694 (1986).

579. USHIJIMA, H., KOIKE, H., MUKOYAMA, A., HASEGAWA, A., NISHIMURA, S., AND GENTSCH, J., Detection and serotyping of rotaviruses in stool specimens by using reverse transcription and polymerase chain reaction amplification, *J. Med. Virol.* **38**:292–297 (1992).

580. USHIJIMA, H., KONNO, H., KIM, B., SHINOZAKI, T., ARAKI, K., AND FUJII, R., A new latex agglutination test kit for detecting rotavirus in stool from children with gastroenteritis, *Pediatr. Infect. Dis.* **5**:492–493 (1986).

581. USHIJIMA, H., TAJIMA, T., TAGAYA, M., *ET AL.*, Rotavirus and central nervous system (Abstract), *Brain Dev.* **138**:215 (1984).

582. VAUCHER, Y. E., RAY, C. G., MINNICH, L. L., PAYNE, C. M., BECK, D., AND LOWE, P., Pleomorphic, enveloped, virus-like particles associated with gastrointestinal illness in neonates, *J. Infect. Dis.* **145**:27–36 (1982).

583. VAUGHN, J. M., CHEN, Y.-S., AND THOMAS, M. Z., Inactivation of human and simian rotavirus by chlorine, *Appl. Environ. Microbiol.* **51**:391–394 (1986).

584. VESIKARI, T., Progress in rotavirus vaccination, *Pediatr. Infect. Dis.* **4**:612–614 (1985).

585. VESIKARI, T., Bovine rotavirus-based rotavirus vaccines in humans, in: *Viral Infections of the Gastrointestinal Tract* (A. Z. KAPIKIAN, ED.), pp. 419–442, Marcel Dekker, New York, 1994.

586. VESIKARI, T., ISOLAURI, E., D'HONDT, E., DELEM, A., ANDRE, F. E., AND ZISSIS, G., Protection of infants against rotavirus diarrhoea by RIT 4237 attenuated bovine rotavirus strain vaccine, *Lancet* **1**:977–981 (1984).

587. VESIKARI, T., AND KAPIKIAN, A. Z., Rotavirus vaccine development, in: *Modern Vaccinology* (E. KURSTAK, ED.), pp. 213–229, Plenum, New York, 1994.

588. VESIKARI, T., RAUTANEN, T., VARIS, T., BEARDS, G. M., AND KAPIKIAN, A. Z., Rhesus *rotavirus* candidate vaccine: Clinical trial in children vaccinated between 2 and 5 months of age. *Am. J. Dis. Child.* **144**:285–289 (1990).

589. VESIKARI, T., RUUSKA, T., GREEN, K. Y., FLORES, J., AND KAPIKIAN, A. Z., Protective efficacy against serotype 1 rotavirus diarrhea by live oral rhesus–human reassortant rotavirus vaccines with human rotavirus VP7 serotype 1 or 2 specificity, *Pediatr. J. Infect. Dis.* **11**:535–542 (1992).

590. VESIKARI, T., RUUSKA, T., KOUVU, H. P., GREEN, K. Y., FLORES, J., AND KAPIKIAN, A. Z., Evaluation of the M37 human rotavirus vaccine in 2- to 6-month-old infants, *Pediatr. Infect. Dis. J.* **10**: 912–917 (1991).

591. VIERA DE TORRES, B., MAZZALI DE ILJA, R., AND ESPARZA, J., Epidemiological agents of rotavirus infection in hospitalized Venezuelan children with gastroenteritis, *Am. J. Trop. Med. Hyg.* **27**:567–572 (1978).

592. WADDELL, G., ALLARD, A., JOHANSSON, M., SVENSSON, L., AND UHNOO, I., Enteric adenoviruses, in: *Viral Infections of the Gastrointestinal Tract* (A. Z. KAPIKIAN, ED.), pp. 519–547, Marcel Dekker, New York, 1994.

593. WALSH, J. A., AND WARREN, K. S., Selective primary health care. An interim strategy for disease control in developing countries, *N. Engl. J. Med.* **301**:967–974 (1979).

594. WANG, S., CAI, R., CHEN, J., LI, R., AND JIANG, R., Etiologic studies of the 1983 and 1984 outbreaks of epidemic diarrhea in Guangxi, *Intervirology* **24**:140–146 (1985).

595. WARD, R. L., BERNSTEIN, D. I., KNOWLTON, D. R., SHERWOOD,

J. R., YOUNG, E., CUSACK, T. M., RUBINO, J. R., AND SCHIFF, G. M., Prevention of surface-to-human transmission of rotaviruses by treatment with disinfectant spray, *J. Clin. Microbiol.* **29:**1991–1996 (1991).

596. WARD, R. L., BERNSTEIN, D. I., YOUNG, E. C., SHERWOOD, J. R., KNOWLTON, D. R., AND SCHIFF, G. M., Human rotavirus studies in volunteers: Determination of infectious dose and serological response to infection, *J. Infect. Dis.* **154:**871–880 (1986).

597. WARD, R. L., CLEMENS, J. D., SACK, D. A., ET AL., Culture adaptation and characterization of group A rotaviruses causing diarrheal illnesses in Bangladesh from 1985 to 1986, *J. Clin. Microbiol.* **29:**1915–1923 (1991).

598. WARD, R. L., KNOWLTON, D. R., AND PIERCE, M. J., Efficacy of human rotavirus propagation in cell culture, *J. Clin. Microbiol.* **19:**748–753 (1984).

599. WEBER, K., RIEBEL, T. H., AND NASIR, R., Hyperechoic lesions in the basal ganglia: An incidental sonographic finding in neonates and infants, *Pediatr. Radiol.* **22:**182–186 (1992).

600. WHITE, L., GARCIA, D., BOHER, Y., BLANCO, M., PEREZ, M., ROMER, H., FLORES, J., AND PEREZ-SCHAEL, I., Temporal distribution of human rotavirus serotypes 1, 2, 3, and 4 in Venezuelan children with gastroenteritis 1979–1989, *J. Med. Virol.* **34:**79–84 (1991).

601. WHITE, L., PEREZ, I., PEREZ, M., URBINA, G., GREENBERG, H., KAPIKIAN, A., AND FLORES, J., Relative frequency of rotavirus subgroups 1 and 2 in Venezuelan children with gastroenteritis as assayed with monoclonal antibodies, *J. Clin. Microbiol.* **19:**516–520 (1984).

602. WHITELAW, D., DAVIES, H., AND PARRY, J., Electron microscopy of fatal adenovirus gastroenteritis, *Lancet* **1:**361 (1977).

603. WHORWELL, P. J., PHILLIPS, C. A., BEEKEN, W. L., LITTLE, P. K., AND ROESSNER, K. D., Isolation of reovirus-like virus agents from patients with Crohn's disease, *Lancet* **1:**1169–1171 (1977).

604. WILDE, J., VAN, R., PICKERING, L., EIDEN, J., AND YOLKEN, R. H., Detection of rotaviruses in the day care environment by reverse transcription polymerase chain reaction, *J. Infect. Dis.* **166:**507–511 (1992).

605. WILDERLITE, L., TRIER, J. S., BLACKLOW, N. R., AND SCHREIBER, D. S., Structure of the gastric mucosa in acute infections nonbacterial gastroenteritis, *Gastroenterology* **68:**425–430 (1975).

606. WILLIAMS, T., BOURKE, P., AND GURWITH, M., IN: *Program Abstracts, 15th Interscience Conference on Antimicrobial Agents and Chemotherapy*, p. 232, 1975.

607. WONG, C. J., PRICE, Z., AND BRUCKNER, D. A., Aseptic meningitis in an infant with rotavirus gastroenteritis, *Pediatr. Infect. Dis.* **3:**244–246 (1984).

608. WOOD, D. J., DAVID, T. J., CHRYSTIE, I. L., AND TOTTERDELL, B., Chronic enteric virus infection in two T-cell immunodeficient children, *J. Med. Virol.* **24:**435–444 (1988).

609. WOODE, G. N., The Toroviruses: Bovine (Bredavirus) and equine (Bernevirus); the Torovirus-like agents of humans and animals, in: *Viral Infections of the Gastrointestinal Tract* (A. Z. KAPIKIAN, ED.), pp. 581–602, Marcel Dekker, New York, 1994.

610. WOODE, G. N., BRIDGER, J. C., HALL, B., AND DENNIS, M. J., The isolation of a reovirus-like agent associated with diarrhea in colostrum-deprived calves in Great Britain, *Res. Vet. Sci.* **16:**102–105 (1974).

611. WOODE, G. N., BRIDGER, J. C., HALL, G., JONRD, J. M., AND JACKSON, G., The isolation of a reovirus-like agents (rotavirus) from acute gastroenteritis of piglets, *J. Med. Microbiol.* **9:**203–209 (1976).

612. WOODE, G. N., BRIDGER, J. C., JONES, J. M., FLEWETT, T. H., BRYDEN, A. S., DAVIES, H. A., AND WHITE, G. B. B., Morphological and antigenic relationships between viruses (rotaviruses) from acute gastroenteritis of children, calves, piglets, mice and foals, *Infect. Immun.* **14:**804–810 (1976).

613. WOODE, G. N., JONES, J., AND BRIDGER, J., Levels of colostral antibodies against neonatal calf diarrhoea virus, *Vet. Rec.* **97:**148–149 (1975).

614. WOODS, P. A., GENTSCH, J., GOUVEA, V., MATA, M., SIMHON, A., SANTOSHAM, M., BAI, Z.-S., URASAWA, S., AND GLASS, R. I., Distribution of serotypes of human rotavirus in different populations, *J. Clin. Microbiol.* **30:**781–785 (1992).

615. WORLD HEALTH ORGANIZATION, Mortality due to diarrheal diseases in the world, *WHO Week. Epidemiol. Rec.* **48:**409–416 (1973).

616. WORLD HEALTH ORGANIZATION SCIENTIFIC WORKING GROUP, Rotavirus and other viral diarrhoeas, *Bull. WHO* **58:**183–198 (1980).

617. WU, B., MAHONY, J. B., SIMON, G., AND CHERNESKY, M. A., Sensitive solid-phase immune electron microscopy double-antibody technique with gold-immunoglobulin G complexes for detecting rotavirus in cell culture and feces, *J. Clin. Microbiol.* **28:**864–868 (1990).

618. WYATT, R. G., DOLIN, R., BLACKLOW, N. R., DUPONT, H. L., BUSCHO, R. F., THORNHILL, T. S., KAPIKIAN, A. Z., AND CHANOCK, R. M., Comparison of three agents of acute infectious nonbacterial gastroenteritis by cross-challenge in volunteers, *J. Infect. Dis.* **129:**709–714 (1974).

619. WYATT, R. G., GREENBERG, H. B., DALGARD, D. W., ALLEN, W. P., SLY, D. L., THORNHILL, T. S., CHANOCK, R. M., AND KAPIKIAN, A. Z., Experimental infection of chimpanzees with the Norwalk agent of epidemic viral gastroenteritis, *J. Med. Virol.* **2:**89–96 (1978).

620. WYATT, R. G., GILL, V. W., SERENO, M. M., KALICA, A. R., VANKIRK, D. H., CHANOCK, R. M., AND KAPIKIAN, A. Z., Probable *in vitro* cultivation of human reovirus-like agent of infantile diarrhea, *Lancet* **1:**98 (1976).

621. WYATT, R. G., GREENBERG, H. B., JAMES, W. D., PITTMAN, A. L., KALICA, A. R., FLORES, J., AND KAPIKIAN, A. Z., Definition of human rotavirus serotypes by plaque reduction assay, *Infect. Immun.* **37:**110–115 (1984).

622. WYATT, R. G., AND JAMES, W. D., Methods of gastroenteritis virus culture *in vivo* and *in vitro*, in: *Viral Infections of the Gastrointestinal Tract* (A. A. J. TYRRELL AND A. Z. KAPIKIAN, EDS.), pp. 13–35, Marcel Dekker, New York, 1982.

623. WYATT, R. G., JAMES, W. D., BOHL, E. H., THEIL, K. W., SAIF, L. J., KALICA, A. R., GREENBERG, H. B., KAPIKIAN, A. Z., AND CHANOCK, R. M., Human rotavirus type 2: Cultivation *in vitro*, *Science* **207:**189–191 (1980).

624. WYATT, R. G., JAMES, H. D., JR., PITTMAN, A. L., HOSHINO, Y., GREENBERG, H. B., KALICA, A. R., FLORES, J., AND KAPIKIAN, A. Z., Direct isolation in cell culture of human rotaviruses and their characterization into four serotypes, *J. Clin. Microbiol.* **18:**310–317 (1983).

625. WYATT, R. G., KALICA, A. R., MEBUS, C. A., KIM, H. W., LONDON, W. T., CHANOCK, R. M., AND KAPIKIAN, A. Z., Reovirus-like agents (rotaviruses) associated with diarrheal illness in animals and man, *Perspect. Virol.* **10:**121–145 (1978).

626. WYATT, R. G., KAPIKIAN, A. Z., Viral agents associated with acute gastroenteritis in humans, *Am. J. Clin. Nutr.* **30:**1857–1870 (1977).

627. WYATT, R. G., KAPIKIAN, A. Z., AND MEBUS, C. A., Induction of cross-reactive serum neutralizing antibody to human rotavirus in calves after *in utero* administration of bovine rotavirus, *J. Clin. Microbiol.* **18:**505–508 (1983).

628. WYATT, R. G., KAPIKIAN, A. Z., THORNHILL, T. S., SERENO, M. M., KIM, H. W., AND CHANOCK, R. M., *In vitro* cultivation in human fetal intestinal organ culture of a reovirus-like agent associated with nonbacterial gastroenteritis in infants and children, *J. Infect. Dis.* **130:**523–528 (1974).

629. WYATT, R. G., MEBUS, C. A., YOLKEN, R. H., KALICA, A. R., JAMES, H. D., JR., KAPIKIAN, A. Z., AND CHANOCK, R. M., Rotaviral immunity in gnotobiotic calves: Heterologous resistance to human virus induced by bovine virus, *Science* **203:**548–550 (1975).

630. WYATT, R. G., YOLKEN, R. H., URRUTIA, J. J., MATA, L., GREENBERG, H. B., CHANOCK, R. M., AND KAPIKIAN, A. Z., Diarrhea associated with rotavirus in rural Guatemala: A longitudinal study of 24 infants and young children, *Am. J. Trop. Med. Hyg.* **28:**325–328 (1979).

631. XU, L., HARBOUR, D., AND McCRAE, M. A., The application of polymerase chain reaction to the detection of rotaviruses in faeces, *J. Virol. Methods* **27:**29–38 (1990).

632. YAMAMOTO, A., ZENNYOGI, H., YANAGITA, K., AND KATO, S., Research into the causative agent of epidemic gastroenteritis which prevailed in Japan in 1948, *Jpn. Med. J.* **1:**379–384 (1948).

633. YOLKEN, R. H., Avidin–biotin radioimmunoassay for human rotavirus, *J. Infect. Dis.* **148:**942 (1983).

634. YOLKEN, R. H., BARBOUR, B. A., WYATT, R. G., AND KAPIKIAN, A. Z., Immune responses to human rotaviral infection—measurement by enzyme immunoassay, *J. Am. Vet. Med. Assoc.* **173:**552–554 (1978).

635. YOLKEN, R. H., BISHOP, C. A., TOWNSEND, T. R., TIMOTHY, R., BOLYARD, E., BARTLETT, J., SANTOS, G. W., AND SARAL, R., Infectious gastroenteritis in bone-marrow transplant recipients, *N. Engl. J. Med.* **306:**1009–1012 (1982).

636. YOLKEN, R. H., DUBOVI, E., LEISTER, F., REID, R., ALMEIDA-HILL, J., AND SANTOSHAM, M., Infantile gastroenteritis associated with excretion of pestivirus antigens, *Lancet* **1:**517–520 (1989).

637. YOLKEN, R. H., EIDEN, J., AND LEISTER, F., Self-contained enzymic membrane immunoassay for detection of rotavirus antigen in clinical samples, *Lancet* **2:**1305–1307 (1986).

638. YOLKEN, R. H., KIM, H. W., CLEM, T., WYATT, R. G., KALICA, A. R., CHANOCK, R. M., AND KAPIKIAN, A. Z., Enzyme-linked immunosorbent assay (ELISA) for detection of human reovirus-like agent of infantile gastroenteritis, *Lancet* **2:**263–267 (1977).

639. YOLKEN, R. H., LEISTER, F., WEE, S. B., MISKUFF, R., AND VONDERFECHT, S., Antibodies to rotaviruses in chickens' eggs: A potential source of antiviral immunoglobulins suitable for human consumption, *Pediatrics* **81:**291–295 (1988).

640. YOLKEN, R. H., LOSONSKY, G. A. VONDERFECHT, S., LEISTER, F., AND WEE, S. B., Antibody to human rotavirus in cow's milk, *N. Engl. J. Med.* **312:**606–610 (1985).

641. YOLKEN, R. H., AND MURPHY, M., Sudden infant death syndrome associated with rotavirus infection, *J. Med. Virol.* **10:**291–296 (1982).

642. YOLKEN, R. H., AND STOPA, P. J., Analysis of non-specific reactions in enzyme-linked immunosorbent assay testing for human rotavirus, *J. Clin. Microbiol.* **10:**703–707 (1979).

643. YOLKEN, R. H. AND STOPA, P. J., Enzyme-linked fluorescence assay: Ultrasensitive solid-phase assay for detection of human rotavirus, *J. Clin. Microbiol.* **10:**317–321 (1979).

644. YOLKEN, R. H., AND WEE, S. B., Enzyme immunoassays in which biotinylated β-lactamase is used for the detection of microbial antigens, *J. Clin. Microbiol.* **19:**356–360 (1984).

645. YOLKEN, R. H., AND WILDE, J. A., Assays for detecting human rotavirus, in: *Viral Infections of the Gastrointestinal Tract* (A. Z. KAPIKIAN, ED.), pp. 251–278, Marcel Dekker, New York, 1994.

646. YOLKEN, R. H., WYATT, R. G., BARBOUR, B. A., KIM, H. W., KAPIKIAN, A. Z., AND CHANOCK, R. M., Measurement of rotavirus antibody by an enzyme-linked immunosorbent assay blocking assay, *J. Clin. Microbiol.* **8:**283–287 (1978).

647. YOLKEN, R. H., WYATT, R. G., KALICA, A. R., KIM, H. W., BRANDT, C. D., PARROTT, R. H., KAPIKIAN, A. Z., AND CHANOCK, R. M., Use of a free viral immunofluorescence assay to detect human reovirus-like agent in human stools, *Infect. Immun.* **16:**467–470 (1977).

648. YOLKEN, R. H., WYATT, R. G., AND KAPIKIAN, A. Z., ELISA for rotavirus, *Lancet* **2:**818 (1977).

649. YOLKEN, R. H., WYATT, R. G., KIM, H. W., KAPIKIAN, A. Z., AND CHANOCK, R. M., Immunological response to infection with human reovirus-like agent: Measurement of anti-human reovirus-like agent immunoglobulin G and M levels by the method of enzyme-linked immunosorbent assay, *Infect. Immun.* **19:**540–546 (1978).

650. YOLKEN, R. H., WYATT, R. G., MATA, L., URRUTIA, J. J., GARCIA, B., CHANOCK, R. M., AND KAPIKIAN, A. Z., Secretory antibody directed against rotavirus in human milk—Measurement by means of enzyme linked immunosorbent assay, *J. Pediatr.* **93:**916–921 (1978).

651. YOLKEN, R. H., WYATT, R. G., ZISSIS, G. P., BRANDT, C. D., RODRIGUEZ, W. J., KIM, H. W., PARROTT, R., URRUTIA, J. J., MATA, L., GREENBERG, H. B., KAPIKIAN, A. Z., AND CHANOCK, R. M., Epidemiology of human rotavirus types 1 and 2 as studied by enzyme-linked immunosorbent assay, *N. Engl. J. Med.* **299:**1156–1161 (1978).

652. YOW, M. D., MELNICK, J. L., BLATTNER, R. J., STEPHENSON, W. B., ROBINSON, N. M., AND BURKHARDT, M. A., The association of viruses and bacteria with infantile diarrhea, *Am. J. Epidemiol.* **92:**33–39 (1970).

653. ZAHORSKY, J., Hyperemesis hiemis or the winter vomiting disease, *Arch. Pediatr.* **46:**391–395 (1929).

654. ZAKI, A. M., DuPONT, H. D., ALAMY, M. A., *ET AL.*, The detection of enteropathogens in acute diarrhea in a family cohort population in rural Egypt, *Am. J. Trop. Med. Hyg.* **35:**1013–1022 (1986).

655. ZHENG, B. J., CHANG, R. X., MA G. Z., XIE, J. M., LIU, Q., LIANG, X. R., AND NG, M. H., Rotavirus infection of the orapharynx and respiratory tract in young children, *J. Med. Virol.* **34:**29–37 (1991).

656. ZISSIS, G., LAMBERT, J. P., AND DE KEGEL, D., Routine diagnosis of human rotavirus in stools, *J. Clin. Pathol.* **31:**175–178 (1978).

CHAPTER 12

Hantaviruses

James W. LeDuc

1. Introduction

Hantaviruses belong to a recently recognized genus within the family Bunyaviridae. The prototype is Hantaan virus, causative agent of Korean hemorrhagic fever of Korea. This virus was first described in 1978[60]; since then several related viruses have been recovered from humans, rodents, and other small mammals. Human disease caused by these viruses is characterized by acute fever, a tubular renal lesion, and in some instances a life-threatening capillary leak syndrome, often accompanied by hemorrhagic manifestations.

2. History

Acute hemorrhagic and nephropathic clinical syndromes were described and variously named decades before their causative agents were identified. As cited by Casals *et al.*,[12] records of severe, often fatal, hemorrhagic fever have been discovered in Vladivostok, Eastern Siberia, as early as 1912, and it seems likely that similar syndromes, albeit often confused with other causes of fever and hemorrhage, were known in Asia and Europe several centuries prior to this time.

Modern scientific descriptions are credited to Soviet and Japanese scientists working in Siberia and Manchuria in the 1930s.[78] What is now clearly a single disease was called hemorrhagic fever with renal syndrome (HFRS), hemorrhagic nephrosonephritis, Songo fever, and various other names.[27,102] English-speaking physicians first encountered this disease during the Korean conflict in the

This chapter is a revised version of Chapter 12, Karl M. Johnson, *Viral Infections of Humans*, 3rd edition, Plenum Press, 1989.

James W. LeDuc • National Center for Infectious Diseases, Centers for Disease Control and Prevention, Atlanta, Georgia 30333.

early 1950s and named it Korean hemorrhagic fever (KHF). Modern Chinese and Japanese authors prefer the term "epidemic hemorrhagic fever" (EHF). A milder form of illness, with minor hemorrhage and lower mortality, was recognized in Scandinavia, also in the early 1930s, and termed "nephropathia epidemica" (NE).[79]

Hantaviruses are natural parasites of small mammals, and this genus is apparently unique in the Bunyaviridae, which otherwise are uniformly biologically transmitted by arthropods. Infection is transmitted to man by contact with rodent excreta, most likely urine, and occasionally by bite.[116] Routes of transmission among small mammals are not known, although laboratory experiments have clearly documented the susceptibility of laboratory rats to aerosol exposure,[88] and a strong correlation exists among free-living urban rats between increased evidence of wounding and rising antibody prevalence rates, suggesting that transmission by bite may be important.[29] Depending on the virus and host, transmission may occur in rural, urban, or occupational (such as from laboratory rats) settings.

3. Methodology

3.1. Mortality

Clinical recognition of HFRS in the Soviet Union, China, and Korea has proven to be reasonably reliable for estimation of mortality rates. Recent serological confirmation of milder disease has served only to reduce by one third or less previous estimates based on case counting. Mortality in HFRS generally ranges from 5 to 15% and has been reduced in recent years by recourse to dialysis of patients who experience renal shutdown.[12] Case-fatality rates of 10 to 15% were reported from the Russian Federation for the period 1978 to 1992,[127] while mortality in the former Yugoslavia during an outbreak in 1989 was 6.6% (15 deaths of 226 cases).[33a] The case-fatality rate for NE in Sweden was estimated at about 0.2%.[98] A comparison of the clinical courses of NE in Sweden and the Western Russian Federation found the overall frequency of hemorrhagic manifestations was higher among Russian than among Swedish patients (37% vs. 10%), as was the risk of life-threatening complications.[100]

3.2. Morbidity

Estimates of morbidity caused by hantaviruses are still largely derived from reports of clinical cases, although serological confirmation is becoming more frequent as increasing numbers of laboratories acquire diagnostic capabilities. Mainland China has by far the greatest gross morbidity, with nearly 100,000 cases being reported in some years. The Russian Federation recorded nearly 70,000 cases between 1978 and 1992, with peak annual incidence rates of over 11,000 cases (8.0 per 100,000 population),[127] and annual incidence rates of 50 cases per 100,000 in Bashkirtostan.[82] Incidence rates have dropped in recent years in Korea to reports of a few hundred cases annually. Less severe NE-like disease occurs in several hundred patients each year in Scandinavia, and increasing numbers of cases are now recognized in western Europe. Highest incidence rates in Sweden exceeded 20 cases/100,000 population in endemic areas,[84,99] while the prevalence of serum IgG antibodies ranged from about 2% in nonendemic areas to around 8% in endemic regions.[84,86] Incidence of clinical cases and antibody prevalence in an endemic area of Sweden were compared, and the antibody prevalence rate in the oldest age groups was found to be 14 to 20 times higher than the accumulated life-risk of being hospitalized with NE, suggesting that mild or asymptomatic infections are not uncommon.[85] In the Balkan region of Europe, the viruses that cause both severe HFRS and relatively mild NE coexist, although the total disease burden generally does not exceed several hundred cases annually. More than 100 clinical and subclinical infections associated with occupational exposure to laboratory rats have been recognized in Japan, and additional cases have occurred in Korea, the former Soviet Union, Belgium, and the United Kingdom.

3.3. Serological Surveys

Systematic population-based surveys to establish antibody prevalence rates for hantavirus infection have not been widely attempted; however, sufficient examination of selected "at-risk" populations have been conducted to

offer a good indication of the global distribution of the hantaviruses. Methods employed include immunofluorescent antibody (IFA) assays, most often using prototype Hantaan virus-infected cell cultures as antigen, enzyme immunoassays (EIA), and radioimmunoassays (RIA). Populations surveyed are generally rural, frequently with high potential for occupational exposure to rodents (farmers, wood cutters, foresters, and others), patients with clinically compatible or interchangeable disease (leptospirosis, for example), or unselected blood donors. About 12,000 sera from the European parts of the Russian Federation were examined by RIA, with antibodies found in a geographically focal pattern, generally increasing in prevalence with age, and most common in oil production and forestry workers or tractor and truck drivers.[78] Men were more likely to have antibodies than women (1.3 : 1 to 2 : 1), and this ratio increased in the clinically ill to 3 : 1. In a subsequent study in Bashkortostan, antibody prevalence among over 9000 persons reached over 16% in some adult age groups. Seropositive children have been infrequently identified in virtually all serological surveys.

The distribution pattern that has emerged from various surveys suggests that Hantaan virus is most abundant in Asia, including China where the endemic areas are thought to be expanding, and the Korean Peninsula. A closely related or identical strain of Hantaan virus is found in Greece, Bulgaria, Albania, and the former Yugoslavia where it occurs relatively infrequently but produces an especially severe form of HFRS with mortality rates as high as 30% in some areas.[2,32,47]

Puumala virus, cause of nephropathia epidemica, is most abundant across a broad band from Norway, northern Sweden, Finland, and through the Russian Federation to the Ural mountains. Antibody prevalence rates in these endemic areas approach or exceed 10%.[82,84,99] Elsewhere in western Europe, antibody prevalence rates are lower, but clinical cases and positive serological results have now been found in most European countries; and in areas such as France, Belgium, and Germany, where active research programs exist and clinicians recognize the disease, NE is increasingly diagnosed as an important cause of acute disease.[48,121]

Seoul virus, cause of a less severe form of HFRS, is thought to be nearly global in its distribution based on serological surveys of peridomestic rodents,[50] but extensive tests for this specific virus in humans have not been widely attempted. In the United States, however, large surveys were conducted among residents of Baltimore, Maryland, where inner-city rats are known to be heavily infected with a Seoul-like virus.[13,16,17] An antibody prevalence rate of 0.25% was found among 6060 persons with no known risk factors for hantavirus infection except residence in Baltimore.[31] This rate was significantly different from the rate found among patients with proteinuria (1.46%; OR, 3.23; p <0.05), and the rate among dialysis patients with end-stage renal disease (2.76%; OR, 5.03; p < 0.05). Overall, 6.5% of patients with end-stage renal disease due to hypertension were seropositive for a hantavirus, suggesting that hantavirus infection is associated with hypertensive renal disease. Canadian blood donors had 1.4% prevalence of antibodies to hantaviruses (3.5% in the Maritime Provinces), but the exact strain of infecting virus was not determined.[64] In the recent outbreak of hantavirus disease in the southwestern United States, preliminary results indicate that the antibody prevalence rate was approximately 1% (3/270) among residents prior to the outbreak.[9]

3.4. Laboratory Methods

Direct isolation of hantaviruses from acute human specimens or from tissues of suspect small mammalian hosts is difficult and time-consuming. Hantaviruses are slow to adapt to growth in cell culture, and indeed only a few cell lines are known to support hantaviral growth (Vero clone E-6, A549). Infected cell cultures do not exhibit cytopathic effect; consequently, the presence of infectious virus must be established by demonstration of specific viral antigen or nucleic acids, typically by direct IFA assay or polymerase chain reaction (PCR). Primary isolation has been most successful when a combination of techniques were employed. These include inoculation of susceptible seronegative laboratory animals (suckling mice, rats, or adult rats), followed by inoculation of cell cultures and serial blind passage to 50 days or more, with periodic examination by IFA for characteristic cytoplasmic viral antigen or PCR for virus-specific nucleic acids.[43] Suckling mice or rats frequently succumb to infection during the third week following inoculation. Adult rats are asymptomatically infected, but develop specific antibodies by 30 days postinoculation. Lung tissues from seropositive adult rats or tissues from suckling rats or mice can then serve as inoculum for adaptation to growth in cell culture. The benefit of initial passage through laboratory animals is that the starting material for adaptation to cell culture is known to be infectious. A note of caution: Although asymptomatically infected, laboratory animals may shed infectious virus in their excreta and saliva. Consequently, isolation attempts and experimentation that utilizes laboratory animals should only be carried out under biocontainment conditions that minimize the risk of aerosol transmission between animals and to hu-

mans. Strict adherence to biosafety level 3 recommendations is required when handling hantaviruses in the laboratory, and biosafety level 4 precautions, including laminar flow cage covers or bonnets on all animal cages, must be used for animal experimentation.

Examination of tissues for the presence of hantaviral antigen or nucleic acids is often successful, and these procedures continue to improve.[4,81,128,131] Rodent lung tissues are frequently examined by direct IFA in epidemiologic investigations. Antigen-capture EIAs, however, have not been especially useful for detecting hantavirus infections in rodents or human hosts. Recently refined immunohistological procedures can now detect hantaviral antigen with some consistency in fresh or preserved tissues from human or small mammalian hosts.[81] Likewise, recent advances in PCR using genus-reactive primer pairs have been successfully used to detect hantavirus-specific nucleic acid sequences from human autopsy material and from rodent host tissues.[81,131]

A variety of serological procedures have been applied to the hantaviruses. Immunofluorescent antibody assays,[66] EIA,[79] and hemagglutination inhibition tests[6,119] are sensitive assays, but often not specific. The serum dilution–plaque reduction neutralization test[20,51,108] is the accepted method for identification of specific viruses and for determining the infecting virus in situations where antibodies might have derived from infection by different members of the genus.[95,118,121] Most hantaviruses, however, require at least 1 week to form visible plaques, making this test impractical for routine use in many laboratories. Complement fixation[37] and immune adherence hemagglutination,[106] among other procedures, have also been successfully used with the hantaviruses.

4. The Viruses

4.1. Biochemical and Physical Properties

Most work has been done with Hantaan virus, the prototype of the genus, and a strain designated 76-118, recovered from an *Apodemus* rodent, has been used for a majority of the seminal studies. Characterization of this isolate was facilitated by achievement of virus replication in continuously cultivated cells,[26] which permitted clonal purification of the virus by selection and passage of virus plaques under agar.[96]

Hantaan virus was found to be inactivated by several lipid solvents and by exposure to mild acid below pH 5.5.[26] Virion buoyant density is about 1.16 to 1.21, depending on the material used to sediment particles.[73,96]

Analysis of RNA extracted from purified viruses revealed the presence of three distinct single-stranded, negative-sense RNA species having molecular weights of 2.7, 1.2, and 0.6 × 10⁶.[93] This pattern is characteristic of the family Bunyaviridae.

Polypeptides recovered from disrupted virions include a 50,000 molecular weight protein associated with each of the viral RNA species, two glycoproteins of 55,000 and 68,000 molecular weight, and a large protein of about 200,000 Da.[97] These polypeptides represent virion envelope materials and a viral RNA polymerase, respectively. The protein profile of Hantaan virus, although distinct in size, thus conforms in structure and function to that of other members of the family Bunyaviridae.

Detailed analysis of viral RNA also disclosed that the three species separable by size have distinct patterns of oligonucleotides, but that each has a conserved 3′-terminal base sequence of AUC AUC AUC UG, which is distinct from any of those of the previously recognized genera of the Bunyaviridae.[93]

Nucleotide sequence information is now available for many hantaviruses. A comparison of deduced gene products of the large (L), middle-sized (M), and small (S) genome segments of Hantaan, Seoul, Puumala, and Prospect Hill viruses concluded that hantavirus S segments encode the nucleocapsid protein in the virus-complementary sense RNA, with no evidence that they encode nonstructural viral proteins.[1] These same authors determined that the G1 and G2 envelope glycoproteins were encoded in a continuous open reading frame in the virus-complementary sense RNA, without evidence of encoding for nonstructural polypeptides, and that the L segment complementary sense RNA encoded the viral polymerase protein, again with no evidence of nonstructural proteins. The L segment was found to be more highly conserved than the M and S segments, although highly conserved regions were found in both the M and S segments as well.[1]

4.2. Morphology and Morphogenesis

Hantaan and other hantavirus virions were found to be spherical particles 80–160 nm in diameter, averaging about 95 to 122 nm, depending on the particular isolate examined and method of visualization.[40,41,73,126] Particles have a unit membrane and surface projections or spikes, which are hollow cylinders about 12 nm in diameter. Comparative examination of these surface spikes among viruses of different bunyavirus genera revealed differences characteristic of each genus. The Hantaan virus spikes were found to form a unique gridlike pattern of square projections.[71] Thin-section analysis of Hantaan virus particles from virus strains recovered in China disclosed particles averaging 122 nm in diameter.[41] Particles were formed entirely in cell cytoplasm by a process of budding at endoplasmic membranes, often into Golgi cisternae but never at the outer cell membrane. Intracytoplasmic inclusion bodies were occasionally observed.

4.3. Antigenic and Genetic Properties

Antigenic and genetic analyses have clearly established that the genus *Hantavirus* comprises at least seven distinct viruses, with additional ones almost certainly yet to be discovered[18,130] (Figure 1). Early studies found Hantaan virus to be antigenically distinct from all bunyaviruses, arenaviruses, and togaviruses examined, especially those known to cause hemorrhagic disease in humans. Immunofluorescent antigenic relationships, however, were soon reported among a variety of agents recovered from wild and laboratory rodents in many parts of the world. The initial relationship detected was that between Hantaan antigen and antibodies in sera from patients convalescent from nephropathia epidemica in Finland.[61] Antigens of the virus causing this syndrome were then identified in tissues of the vole, *Clethrionomys glareolus*, and it was discovered that they were more virus-specific than those of the prototype Hantaan.[7]

Further examination of many rodent and human isolates using more specific plaque-reduction neutralization methods suggests that the following discernible viruses exist (Tables 1 and 2): Hantaan virus, associated with *Apodemus* rodents and KHF/EHF of Asia and severe HFRS of the Balkans; Seoul virus, associated with *Rattus* rodents virtually worldwide and causing a less severe form of HFRS; Puumala virus, associated with *Clethrionomys glareolus* voles in Europe, especially Scandinavia and western Russian Federation, and causing NE; Prospect Hill virus, associated with *Microtus* voles found in the United States and until recently not thought to be a human pathogen (see Section 9); Thailand virus, associated with rodents (*Bandicota* and *Rattus*) captured in Thailand and yet to be clearly associated with human disease; and Thottopalayam virus, isolated from a shrew captured in India and only recently recognized as a hantavirus.[11,18,130] In addition, Dobrava virus, first reported as an isolate from *Apodemus* rodents captured in the former Yugoslavia,[5,129] and Belgrade virus, isolated from severe HFRS patients, also from the former Yugoslavia,[32] appear to represent two isolates of the same new hantavirus.[109,130] Finally, preliminary indications are that the outbreak of hantavirus disease in the southwestern United

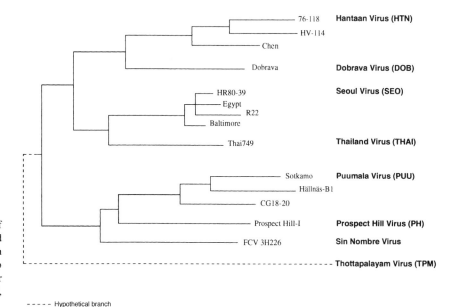

Figure 1. Schematic representation of phylogenetic tree of hantaviruses based on amplification of 333-base pair region of the M segments of various isolates. No data exist for Thottopalayam virus. For more detailed analyses, see refs. 81, 87, and 130.

- - - - - Hypothetical branch

States is due to yet another distinct hantavirus, provisionally named Sin Nombre virus (also referred to as Four Corners virus) and associated with *Peromyscus* rodents.[81] Taken together, the biochemical, morphological, and antigenic properties of Hantaan and related viruses provide the basis for an independent genus within the family Bunyaviridae, for which the name *Hantavirus* has been proposed.[95]

4.4. Biological Properties

Initially, Hantaan virus could only be replicated in tissues of *Apodemus agrarius* mice. Following infection there was a transient viremia followed by the establishment of a persistent silent infection marked by presence of virus and antigen in many tissues, notably the lung, and by continued excretion of virus in urine long after the animals developed virus-specific serum neutralizing antibodies.[59] Subsequently, Hantaan and related agents were successfully propagated in *Clethrionomys glareolus*, other wild rodents, and laboratory rats and mice.[51,68,117,137]

Establishment of the 76-118 virus strain of Hantaan virus in A-549 cells provided the key that led to the rapid characterization of members of the genus.[26] In these cells Hantaan virus produces a chronic, noncytopathic infection detectable by the immunofluorescent techniques. Vi-

Table 1. Names of Various Isolates Appearing in the Scientific Literature and Their Relationship to Recognized Hantaviruses

Hantaan virus isolates	Seoul virus isolates	Puumala virus isolates
76–118 (prototype strain)	HR80-39 (prototype stain)	Sotkamo (prototype strain)
A-9	B-1	CG-1445
Chen	Baltimore Rat	CG-18-20
CG 3880	Brazil	Hallnas-B1
Hojo	Egypt	K-27
HV-114	Girard Point	P360
Jinhae 494, 502	Houston	Poz-Ml
Maaji	Hubei	83-223L
Porogia (Greek)	KI-262	
Fojnica	R-22	
Ussr 452	SR-11	
	Tchoupitoulas	

Table 2. The Known Hantaviruses

Virus	Disease(s) caused[a]	Host	Distribution
Hantaan	Korean hemorrhagic fever	*Apodemus agrarius*	Asia
	Epidemic hemorrhagic fever		
	Hemorrhagic fever with renal syndrome (severe)		
Seoul	Hemorrhagic fever with renal syndrome (mild)	*Rattus norvegicus*	Worldwide
Puumala	Nephropathia epidemica	*Clethrionomys glareolus*	Europe, Russia
Prospect Hill	None recognized	*Microtus pennsylvanicus*	North America
Dobrava/Belgrade	Hemorrhagic fever with renal syndrome (severe)	*Apodemus flavicollis*	Former Yugoslavia
Thai	None recognized	*Bandicota indica*	Asia
Thottopalayam	None recognized	*Suncus murinus* (shrew)	Asia
Sin Nombre	Hantavirus pulmonary syndrome	*Peromyscus maniculatus*	North America

[a]Diseases caused in humans.

rus can be serially propagated by passage of either culture fluids or live cells. Infection of the E-6 clone of Vero cells involves a similar biology, but these cells have superior contact inhibition properties and so are more suitable for detection of infectious centers under agar.

Hantaviruses are readily neutralized by antibodies induced by infection in humans, both wild and laboratory rodents, and other small mammals. Neutralization has been linked to binding on sites of both virionic envelope glycoproteins (G1 and G2).[21] A separate domain in the G2 protein appears to be the site for the viral hemagglutinin.

Hantaviruses, unlike all other genera of the Bunyaviridae, have not been shown to be biologically transmitted by arthropods.[59,116] This fact, together with the so-far-unexplained persistence of infection and virus excretion in immunocompetent rodents and the apparent high host specificity of most of the viruses, provides the framework for current understanding of the biology of this genus.

5. Pathogenesis and Immunity

A central question is whether HFRS and related diseases are the consequence of direct virus-mediated cell damage or of an immunopathologic process. The long incubation period and the pronounced temporal development of renal tubular pathology in the course of illness, together with the inability to easily recover an agent from acutely ill patients, offers support for the second explanation. Work on the pathogenesis of infection, however, has been hampered by the lack of a suitable animal model that faithfully replicates human disease, but emergent data suggest that both fundamental mechanisms may be important. Indeed, patients infected during the outbreak of adult respiratory distress syndrome associated with Sin Nombre

virus showed widespread involvement of endothelial cells and infection of lung, heart, liver, kidney, and spleen tissues as demonstrated by frequent PCR positivity at the time of death. This suggests that direct viral involvement may play a central role in at least this form of hantavirus disease.[81]

Hantaviruses were adapted to replicate in and to kill suckling mice and rats following inoculation.[23,44,80] Death generally occurred between 14 and 18 days after infection, and animals were runted but otherwise grossly unremarkable. Hemorrhage was not manifest. Histological examination of mice, however, revealed changes reminiscent of those described in fatal human disease. These included (1) mononuclear inflammatory lesions in brain, heart (especially the atria), and liver; (2) edema and interstitial pneumonitis in the lungs; (3) focal small hemorrhages in many organs; and (4) marked congestion and hemorrhage in the medulla of the kidneys. Viral antigen was seen in parenchymal cells of brain, heart, lungs, liver, and kidneys. Kupfer cells of the liver were also infected. In addition, viral antigen was detected in capillary endothelial cells of many organs, including lymph nodes, pancreas, and adrenals, where parenchymal cells were generally spared. This finding is of potentially great significance because the early toxic hemorrhagic phase of classic HFRS and in the recently recognized hantaviral pulmonary syndrome of the southwestern United States, both are marked by major loss of plasma proteins from the blood, a phenomenon that leads to hypovolemic shock or acute respiratory distress with large accumulations of interstitial gelatinous protein exudate in patients who die at this stage of illness.[10,24] While the pulmonary involvement is most dramatic in Sin Nombre virus infection, recent studies found that over half the NE patients examined had radiological evidence of pulmonary involvement early in the course of their disease, suggesting that the lung may be a site of viral replication.[69] Indeed, Puumala virus-specific RNA has been detected by PCR in bronchoalveolar lavage samples of NE patients,[35] and Hantaan virus been successfully propagated in human endothelial cells.[90]

Viral antigens have also been detected in trypsin-treated sections of human HFRS tissues stored as fixed specimens for many years.[44] Again, capillary endothelium was a major site of antigen. An immunopathologic component in pathogenesis of disease is possible because of the finding that patients admitted to hospital usually present with IgM and IgG antibodies to viral antigens, both nucleocapsid and glycoprotein.[42,49,60] Further exploration of this possibility is necessary. The classical complement pathway is reportedly activated during HFRS, and this abnormality persists in patients who die.[135] Workers in Finland have detected circulating immune complexes and antiglobulins in patients with NE.[91] Whether those findings signal a significant disease-provoking process or merely represent epiphenomena that accompany a serious virus insult to the infected animal or human remains to be determined.

In any case, it appears that immunity in humans following hantavirus infection is lifelong.[98] Neutralizing antibodies persist for years, and there are no reported cases of clinically manifest reinfection. Whether humans can be persistently infected with these viruses is uncertain. It is a well-accepted fact, however, that medical personnel caring for seriously ill patients do not acquire infection. Thus, virus excretion appears to be minimal at any stage of infection, as supported by the fact that in spite of many attempts, relatively few human isolates have been made.

6. Hantaan Virus and Severe Hemorrhagic Fever with Renal Syndrome

6.1. Descriptive Epidemiology

Infection patterns of distinct members of the hantavirus genus require careful, sometimes inferential, reinterpretation of the epidemiologic literature on this subject. There seems little doubt that Hantaan virus is the etiologic agent of severe HFRS in Asia and that *Apodemus agrarius* is the principal, if not the exclusive, reservoir—vector of infection. In that context, severe HFRS is a disease that occurs in Manchuria, the Soviet Far East, China, and Korea. Other syndromes caused by related viruses having different rodent reservoirs occur in the same geographic areas,[105,139] and the sorting and defining of each of these entities continues. This concept, however, was clearly

recognized by Soviet authors at least 10 years before the isolation of Hantaan virus.[12] They observed that severe or classic HFRS in the Far East occurred in late fall and early winter among persons living along riverine lowlands and that milder disease occurred earlier in the summer–fall among residents of hilly coniferous forests. This differential ecology describes habitats of *Apodemus agrarius* and *Clethrionomys glareolus*, the respective reservoirs of Hantaan and Puumala (cause of NE) viruses.

Over the past decade it has become apparent that viruses closely related or identical to Hantaan virus also exist in the Balkan Region of Europe where they cause severe HFRS, similar to that seen in Asia. *Apodemus agrarius* is not common in this region, and it appears that a related species, *A. flavicollis* is the principal rodent host. The seasonality seen in Asia is distinctive; most human disease is encountered in the warmer months of the year.

The occurrence of severe HFRS in both Asia and the Balkans is strongly occupational. The disease affects mainly adults employed in agriculture and related occupations that require substantial outdoor exposure. In Korea, a small spring cluster of cases is succeeded at harvest time by a larger annual outbreak of disease.[125] Hantaan virus can be recovered most frequently from *Apodemus* rodents at these times, and studies of rodent population dynamics show a biphasic seasonal breeding pattern, which precedes the occurrence of human disease by about 1 month.[55] A similar spring–fall seasonal pattern of disease has been described in China, but interpretation is complicated by the recognition of a mild form of illness in the spring ascribed to Seoul virus harbored by *Rattus norvegicus*.[105] In the Balkans, severe HFRS is most commonly seen in the late spring and summer when shepherds or woodcutters establish outdoor camps, which are subsequently invaded by field mice as discarded food and litter accumulate and serve as attractants for rodents.

6.2. Mechanism and Route of Transmission

Apodemus rodents have been shown to be chronically infected with Hantaan virus. They excrete virus in saliva and feces for a month or more and in urine for many months.[59] Intraspecies virus transmission occurs from one or more of these sources, most likely via aerosolized urine or by bite (see Section 8). Laboratory experiments have clearly shown that less than one plaque-forming unit is required to infect rats with Hantaan, Puumala, or Seoul virus by the aerosol route.[88] Rodent ectoparasites (mites and ticks) were found not to play a role in virus transmission to humans.[59] From these observations, as well as descriptions of circumstances surrounding many focal disease outbreaks, it appears that Hantaan virus infects humans by the respiratory route,[116] and that aerosolized fomites contaminated by infectious *Apodemus* urine or feces are the likely vehicle of transmission. Classic winnowing practices in grain harvesting together with seasonal dry conditions would favor such a mechanism. Transmission of infection to laboratory workers was also documented to occur by the respiratory route.[57,116]

6.3. Patterns of Host Response

6.3.1. Clinical Features.   Clinical HFRS in its classic severe form is a highly distinct entity, difficult to confuse with any other disease. By far the best description of the disease in English is that written by a medical commission during the Korean conflict.[101] Distinct phases of evolution were described, and although not all patients exhibited each of them, the pattern accurately reflected significant physiological changes that characterized Hantaan virus infection (Table 3).

The incubation period of HFRS has been variously estimated at 10–35 days. The longer intervals, in particular, are of significance both in terms of pathogenesis and

for their implications with respect to occurrence of unusual illness among travelers who may be infected on one continent and sick on another.

Abrupt chills and fever most often herald the onset of the initial *febrile phase* of HFRS.[101] Headache, lumbar backache, weakness, anorexia, and general malaise occur individually in about half of all patients, and one or more of these findings accompany fever in nearly all subjects. Conjunctival injection, a dermal flush on the face and upper trunk, and fine petechiae are frequent signs during this phase. Blood leukocytes remain normal or increase (granulocytes), and platelets decrease significantly during the febrile phase of illness. There is frequently evidence of increased vascular permeability and small vessel dysfunction.

At about the fifth to seventh day of illness, fever declines abruptly and other changes occur that define a brief but sometimes fatal *hypotensive phase*. Proteinuria appears and may be severe, urine specific gravity becomes fixed and low, and hematocrit increases in spite of worsening petechiae and gastrointestinal hemorrhage. "Warm" and then classic "cold" shock occur, secondary to loss of plasma protein through capillaries with resultant hypo-

Table 3. Characteristic Course of Clinical Disease for Hemorrhagic Fever with Renal Syndrome[a,b]

Phase	Duration	Predominant signs and symptoms	Laboratory findings
Febrile	3–7 days	Fever, malaise, headache, myalgia, back pain, abdominal pain, nausea, vomiting, facial flush, petechiae (face, neck, trunk), conjunctival hemorrhage	WBC = normal or ↑ Platelets = ↓ Hematocrit = ↑ Urine = proteinuria 1+ → 3+
Hypotensive	2 hr–3 days	Nausea, vomiting, tachycardia, hypotension, shock, visual blurring, hemorrhagic signs, ± oliguria (late)	WBC = ↑ with left shift Platelets = ↓↓ Bleeding time = ↑; PT may be ↑ Hematocrit = ↑↑ Urine = proteinuria 4+ hematuria 1+ hyposthenuria BUN and creatinine = ↑ ing
Oliguric	3–7 days	Oliguria ± anuria BP may ↑; nausea and vomiting may persist; 1/3 with severe hemorrhage (epistaxis, cutaneous, GI, GU, CNS)	WBC = normalizes Platelets = normalize Hematocrit = normalizes, then ↓ Urine = proteinuria 4+ hematuria 1+ → 4+ BUN and creatinine = ↑↑ Na^+↓, K^+↑, Ca^{2+}↓
Diuretic	Days to weeks	Polyuria = 4+ (3–6 liters daily)	BUN and creatinine = normalize electrolytes = possibly abnormal (diuresis) Urine = normalizes
Convalescent	Weeks to months	Strength and function regained slowly	Anemia and hyposthenuria = may persist for months

[a]From McKee *et al.*[73a] Reproduced with permission.
[b]Phases as seen in KHF. All phases may not be present in a given patient.

volemia. At peak hematocrits, if volume and osmotic correction are not achieved rapidly, death ensues.

Patients surviving this 1- to 2-day crisis are not afebrile but become oliguric. This *oliguric phase* generally lasts 4–7 days and may require dialysis to manage metabolic and electrolyte abnormalities. Hypertension and continued bleeding also occur during this period. When renal function returns, there ensues a frank *diuretic phase* of disease, which leads to a prolonged *convalescence phase* that may last for many weeks. Management of fluid and electrolyte balance during diuresis requires great care, and the convalescent period is marked by slow return of renal function to normal in most patients and a gradual return of energy. Clinical residua are thought to be uncommon[92]; however, definitive studies to detect such problems have not been undertaken. Signs and symptoms of disturbed central nervous system function are common in severely ill patients. These include confusion, meningismus, and convulsions, and about 40% of such patients died in one study.[19] Whether Hantaan virus directly affects the brian is not known.

6.3.2. Diagnosis. Isolation of Hantaan virus is difficult, time consuming, and expensive. With current technology it is also insensitive. In contrast, the presence of specific antibodies in sera obtained during the febrile phase of illness renders this method highly reliable, rapid, and relevant to clinical management of patients. Either IgM or IgG antibodies may be measured, and though detection of the former is regarded as definitive, IgG antibodies to Hantaan virus in a patient with classical clinical disease and history of exposure to an "*Apodemus* environment" are highly diagnostic because recurrent disease is unknown. Techniques in routine use include IFA, RIA, EIA (especially IgM capture), and plaque reduction neutralization tests.[3,20,49,51,74,78,95,111] Antibody confirmation of clinically classic cases of HFRS approaches 100%, but less than half of milder suspected infections are specifically diagnosed.[58,60,113] Retrospective examination of patients' sera collected during acute and convalescent stages of disease among United Nations forces serving in Korea found that anti-Hantaan virus-specific IgM antibody was present at or near the time of admission to hospital in most patients, and within 7 days of onset all patients were seropositive by IgM capture immunoassay,[49] confirming both the etiology of the disease described during the Korean Conflict and demonstrating the value of this diagnostic test. More recently, PCR assays have been developed using primer pairs that appear to amplify nearly all known hantaviruses and have been described for both the M and S genomic segments.[4,131] Few clinical applications have been attempted

using PCR as yet (see Section 9); however, this technology holds great promise for future diagnostic applications.

6.4. Treatment and Prevention

Treatment of Hantaan virus-induced HFRS remains primarily supportive, although recent success has been demonstrated in the use of the antiviral drug, ribavirin, to treat HFRS patients.[38] Severe cases test the resources and skills of modern medical management: maintenance of fluid and electrolyte balance, preservation or restoration of circulating blood volume, control of elevated nitrourea and potassium, and detection and treatment of secondary bacterial infection. Advances in patient care have reduced mortality from 15–20% in the 1950s to about 5% or less in recent years.

The broad-spectrum antiviral ribavirin (Virazole), previously found to be dramatically effective in reducing mortality in the arenaviral disease Lassa fever, has been tested in HFRS patients and has proven efficacious. The work originated with demonstration in a lethal mouse model, which found ribavirin treatment offered significant clinical protection.[39] A subsequent double-blind, placebo-controlled clinical trial in Hubei Province, China, found that ribavirin significantly reduced mortality (sevenfold decrease in risk of death) and also significantly reduced the risk of entering the oliguric phase of disease and experiencing hemorrhage. The only ribavirin-related side effect was a well-recognized, fully reversible anemia after completion of therapy.[38]

There is no means for preventing Hantaan virus infection other than human behavior designed to avoid contact with *Apodemus* and fomites contaminated by them. Analysis of a notable outbreak in China underscores this risk.[132] During late 1961, 3.5% of 10,000 residents in eight communities in Anhui Province suffered HFRS. Rodent density was 8.5/100 trap nights; 3 years later this index was only 1.0 and there were 20-fold fewer human illnesses. During the outbreak it was found that persons who did heavy field work had three times the risk of disease compared to those engaged in light field work. Sleeping in the field was a major risk factor and was cumulative: 7% of persons who spent 1–5 nights became ill, whereas 23% of those who slept out for 21 nights or more experienced HFRS. Sleeping on the ground carried a risk double that of using a wooden bed in the fields.

In recent years, progress has been made in development of vaccines to prevent HFRS. Several inactivated vaccines are available or under development in Asia; however, few, if any, have undergone rigorous placebo-controlled efficacy testing.[54,134] Various vaccinia-vectored candidate vaccines have undergone preclinical testing and been shown to elicit neutralizing antibody among laboratory animals, and they hold promise for protection based on limited animal challenge studies.[94] At present, however, these candidates have not entered formal human safety and immunogenicity testing.

7. Puumala Virus and Nephropathia Epidemica

7.1. Descriptive Epidemiology

This disease, originally described in Sweden,[79] is a milder form of HFRS. It is now clear that the causative agent, Puumala virus, is present virtually throughout Europe and Scandinavia west of the Ural mountains.[28,36,75] Thus, HFRS in European Russia is most often NE, clinically, epidemiologically, and virologically.[124] As with HFRS, this disease is fundamentally rural and seasonal, with at least three discernible patterns: (1) sporadic forest-associated cases occurring in summer and early fall related to work or recreation; (2) late fall cases associated with agricultural harvest practices; and (3) winter outbreaks in the former Soviet Union[12] and in Scandinavia associated with invasion of dwellings and food storage structures by infected *Clethrionomys glareolus* rodents.

A virus antigen reactive with antisera to Hantaan virus was first detected in the lungs of *Clethrionomys glareolus* voles in Finland.[7] This antigen, as well as similar preparations from Hantaan virus-infected tissues, was found to be diagnostic for clinical cases of NE.[46] Antigens of NE origin, however, were found to be only weakly reactive with antisera to Hantaan virus, and this pattern has been used to classify infections tentatively in both rodents and humans.[67,112,121,122,140] On this basis, NE infection in humans and/or rodents has been identified from northern Finland to France and Yugoslavia.[107]

The bank vole, *Clethrionomys glareolus*, is clearly the most important reservoir–vector of Puumala virus.[28,36,83,103,115] This rodent lives in forests and is common at forest–field boundaries. These ecological characteristics provide a clear explanation of the observed occupational, seasonal, and sex (males greater than females) distribution of NE. Mechanism and mode of transmission of virus to man are thought to be the same as for Hantaan virus.

Other rodents that have been found to be infected with Puumala or related viruses include *C. rutilus*, *C. rufocanus*, *Apodemus sylvaticus*, *A. flavicollis*, *Microtis arvalis*, *M. agrestis*, and the shrew, *Sorex araneus*, among others.[28,33,36] Caution is required in interpretation of these findings because most reports are based on IFA results, which may not be virus specific. In one study it was found that tissue antigen was more frequent and antibody titers were higher in wild *C. glareolus* than *A. sylvaticus*, suggesting more intense and possibly epidemiologically significant infection in the former species.[103] Elsewhere, *A. flavicollis* has been the source of another hantavirus (Dobrava/Belgrade virus; see Section 10.1), as well as isolates closely related to prototype Hantaan virus.[5,33] Thus, much remains to be learned regarding speciation of rodent hantaviruses, their specific reservoir hosts, and their significance as pathogens for humans.

7.2. Patterns of Host Response

As with Hantaan virus, infection of *C. glareolus* with Puumala virus produces silent, chronic infection. Virus is shed in urine and feces for many days, but experimental intracage transmission occurred only during the period 14 to 28 days after primary infection.[137]

Clinical NE is usually a less severe form of HFRS. The major differences are that NE patients rarely experience clinical shock or complete renal shutdown and have fewer hemorrhagic manifestations, less leukocytosis, and in general a more rapid convalescence. The principal features of HFRS, however, are present in NE, and it is clear that the pathogenesis of the two syndromes differs in degree rather than in mechanism. A study of clinically suspect NE in Finland is instructive. Diagnosis was made by detection of specific antibodies. High fever, proteinuria, increased serum creatinine, polyuria, and thrombocytopenia were recorded in 68–94% of proven cases. The latter three findings were much less common in unconfirmed cases. Hemaconcentration and bradycardia were seen in about one third of NE patients, but not in acute febrile disease of other causes.[46] Comparison of clinical courses of NE between patients infected in Sweden with those infected in Russia found the overall frequency of hemorrhagic manifestations was higher among Russian than among Swedish patients (37% vs. 10%), mortality was higher in Russian patients, and the Russian patients seemed to suffer a higher frequency of life-threatening complications.[100]

Puumala virus RNA has been detected in bronchoalveolar lavage fluids of NE patients by PCR, suggesting that initial viral replication may occur there. Respiratory symptoms have been reported among NE patients,[98] but few detailed studies of pulmonary involvement have been conducted. From 10 to 25% of NE patients have evidence of pulmonary involvement based on conventional chest radiography,[45] but a prospective study of 19 adult patients with NE examined with computed tomography found 10 (53%) with evidence of pulmonary involvement[69] and concluded that pulmonary involvement is a common finding early in the course of NE. These observations are significant in relation to the newly recognized Sin Nombre virus and the adult-onset acute respiratory distress syndrome of the southwestern United States, since preliminary findings suggest genetic similarities between Puumala and Sin Nombre viruses.[81]

8. Seoul Virus and Urban Rat-Borne HFRS

8.1. Descriptive Epidemiology

Although the first indication that HFRS might be present in urban settings in addition to the traditional rural environment came in the early 1960s when 32 cases and one death were reported from Osaka, Japan,[72,110] it was not until nearly 20 years later than Seoul virus was isolated from domestic rats captured in Seoul, Korea.[62,63] This discovery was prompted by the diagnosis of classic HFRS, albeit mild, among urban residents with no rural travel history and no evidence of *Apodemus* rodents near their homes. These patients developed antibodies that were indistinguishable from those to Hantaan virus when measured by IFA assays, and when rats (*Rattus norvegicus* and *R. rattus*) were captured around the patient's homes, they too had similar reacting antibodies. It was only later, using more specific plaque reduction neutralization tests, that it was discovered that these antibodies were to a related yet distinct hantavirus, Seoul virus. Subsequent studies have found urban rat-associated hantaviruses distributed virtually worldwide.[3,17,50,51,53,76,77,104,105,118]

The association between Seoul-related viruses and domestic rats has been most thoroughly studied in Baltimore, Maryland, where infected rats were found widely distributed throughout the city[15,17] and were especially abundant in the lower-income neighborhoods where food and harborage were abundant.[15,16] Seoul virus is clearly enzootic there, with approximately 11% of the rat population becoming infected per month and virtually all of the oldest rats infected.[14] Similar situations undoubtedly occur in other large cities around the world.

Laboratory rats have also been found infected with Seoul-related hantaviruses in Japan, Korea, Belgium, and the United Kingdom, and human infection and diseases among workers in close contact with these animals have been documented.[22,43,70,120,123] Rats of an inbred strain designated LOU were incriminated in Belgium and the United Kingdom. This finding is of great potential significance because this animal is a principal source of cells used to generate antibody-secreting hybridomas, and cell lines derived from tumors of rats were implicated in human infection in the United Kingdom.[70,133] Examination of rat hybridoma cell lines held at the American Type Culture Collection for hantavirus antigens, however, yielded negative results.[52]

8.2. Patterns of Host Response

Seoul virus infection is generally less severe than classic Hantaan-caused HFRS, with the phases of illness characteristically shorter and more difficult to differentiate or skipped completely.[53a] Virtually all patients experience fever, with anorexia, chills, nausea and vomiting, conjunctival injection, and petechiae; abdominal or back pain is also common. Hepatomegaly and laboratory evidence of hepatic dysfunction such as elevated serum transaminase levels are prominent, while renal involvement is less severe than with Hantaan infections. Virtually all patients have proteinuria and most have hematuria. Fatalities are not common.

Until recently, Seoul-like virus disease had not been documented among residents of the United States, perhaps due to a lack of clinical suspicion and an absence of diagnostic tests. Three clinic cases, however, were recently diagnosed in Baltimore following extensive investigations.[30] Perhaps more important was the demonstration that past infection with a rat-borne hantavirus is consistently associated with hypertensive renal disease.[31] This conclusion was based on comparison of the antibody prevalence rates among over 6000 Baltimore residents with no known risk factors for HFRS aside from residence in the city. Their antibody prevalence rate was determined to be 0.25% overall, while that among patients with proteinuria was 1.46% and dialysis patients with end-stage renal disease was 2.76%. Both comparisons yielded significant differences. Hantavirus infection was consistently associated with a diagnosis of hypertensive renal disease, and was unrelated to other chronic renal disease diagnoses. Overall, 6.5% of patients with end-stage renal disease due to hypertension had antibodies to the local isolate of Seoul virus.[31] This association has great relevance in terms of health care costs, since several billion dollars are spent annually in the United States for maintenance of end-stage renal disease patients. If even a small portion of these costs are attributable to hantaviral infections, then opportunities for prevention through rodent control should be cost-effective.

9. Sin Nombre (Four Corners or Muerto Canyon) Virus and Hantavirus Pulmonary Syndrome

9.1. Descriptive Epidemiology

In May 1993, an outbreak of an apparently new disease characterized by unexplained adult respiratory distress syndrome occurred among residents of rural southwestern United States. The outbreak was centered in the Four Corners region, where the states of New Mexico, Arizona, Utah, and Colorado join and primarily affected young, previously healthy adults. By October 1993, 42 cases had been confirmed, including one retrospectively diagnosed from July 1991. Median age of patients was 32 years (range 12 to 69 years) and 52% were males. American Indians accounted for 55% of the cases; of the remainder, 36% were non-Hispanic whites, 7% were Hispanic, and 2% were black. The mortality rate was 62%, and cases were identified from 12 states.[9,10]

Epidemiologic investigations quickly determined that small rodents in the outbreak area were especially abundant in the spring and early summer of 1993, and virological examinations of captured rodents found *Peromyscus maniculatus* to be the species most frequently infected. Actual isolation of Sin Nombre virus has proved to be exceptionally difficult, and PCR techniques, coupled with EIAs, were used to identify infected rodents and humans.[24a,81] Phylogenetic analysis of sequence differences between the Sin Nombre virus segment and other hantaviruses indicated that it represented a distinct, novel hantavirus. Route of transmission from rodents to humans has not been definitely proven, but it is assumed to be by the aerosol route, similar to other hantaviruses. Examination of PCR-amplified products from human cases and from *P. maniculatus* captured in and around patient's homes found identical hantaviral sequences at a given site but surprisingly high levels of genetic diversity among samples taken from different sites.[81] Investigators using the same PCR techniques on preserved tissues of *P. maniculatus* captured in 1983 in California likewise found infected rodents, indicating that this virus is not "new" or recently mutated.[87] Since *P. maniculatus* is one of the most common and widely distributed rodents in North America, it is likely that the additional cases of hantavirus pulmonary syndrome will be recognized in other areas.

9.2. Patterns of Host Response

Hantavirus pulmonary syndrome is characterized by a prodrome of fever, myalgia, and variable respiratory symptoms such as cough, sometimes including headache, abdominal pain, and nausea and vomiting, followed by abrupt onset of acute respiratory distress. Most patients have hemoconcentration and thrombocytopenia on admission, and all cases developed bilateral pulmonary infiltrates within 2 days of hospitalization. Hospitalized patients experienced fever, hypoxia, and hypotension, but unlike other hantavirus diseases, only limited renal involvement has been noted. No sequelae were seen on limited follow-up among the survivors.[10] Serous pleural effusions and heavy edematous lungs are characteristic on postmortem examination, and hantavirus antigens can be detected by immunohistochemistry in preserved tissues of most major organs. Widespread involvement of endothelial cells is evident, especially in the lungs.[10,81] Since Sin Nombre virus was not isolated in a tractable form during the outbreak, serological assays relied initially on the heterologous cross-reactions and later on reactions with a recombinant protein; most patients were initially found with antibodies that reacted at highest titer with Prospect Hill virus, rather than Hantaan, Seoul, and Puumala viruses. Diagnosis has also been confirmed by PCR amplification of specific genomic sequences and by immunohistochemical analysis of postmortem tissues. Intravenous ribavirin was made available as an investigational agent for treatment of hantavirus pulmonary syndrome, but too few patients were treated to determine if it was of benefit.

10. Other Hantaviruses

In addition to the hantaviruses described above, several others have been identified, but often relatively little is known regarding their natural history or ability to cause human disease.

10.1. Dobrava/Belgrade Virus

Dobrava virus was isolated from the yellow-necked mouse, *Apodemus flavicollis*, captured in the northern Slovenian village of Dobrava where severe cases of HFRS had been recognized,[5] and Belgrade virus was isolated from blood and urine of patients suffering from severe HFRS in Serbia.[32] Subsequent molecular characterization of these isolates indicated that they are the same virus.[109,130] Clinical characteristics of the severe HFRS caused by this virus are similar to that seen with classic Hantaan virus; onset of fever, headache, anorexia, diarrhea, nausea and vomiting, and abdominal and back pain, followed by hypotension, hemorrhagic manifestations, and acute renal failure. One of three patients with Belgrade virus died, and the other two had severe diseases requiring dialysis.[32] Dobrava/Belgrade virus cross-reacts to high titers with other hantaviruses when tested by

IFA assays or EIAs, but can be differentiated from other hantaviruses by plaque reduction neutralization tests (Belgrade isolate; Dobrava has not been plaqued) and PCR-restriction fragment length polymorphism methods or direct sequencing.[5,32,129,130]

10.2. Thailand Virus

Thailand virus was originally isolated from lung tissues of *Bandicota indica* and *Rattus norvegicus* captured in Thailand.[25] Antibody prevalence rates were especially high among populations of *B. indica* captured in Kanchanaburi, an inland province in western Thailand. More than 30% of the humans resident in this area also had antihantaviral antibody as measured by IFA, but no attempts were made to associate overt disease with infection.

10.3. Prospect Hill Virus

Prospect Hill virus was first isolated from tissues of the meadow vole, *Microtus pennsylvanicus*, captured in Frederick, Maryland.[65] Subsequently, additional isolations were made from this same species captured in Duluth, Minnesota,[8] and seropositive voles and *Peromyscus* deer mice were found there and in nearby Superior, Wisconsin.[136] Antibodies to Prospect Hill virus were found in 4 of 203 mammalogists examined, but none had a history of HFRS-like illness.[138] Until recently, Prospect Hill virus was considered nonpathogenic for humans; but with the recognition of its close antigenic and genetic relationship to Sin Nombre virus, this assumption has been brought into question.

10.4. Thottopalayam Virus

Thottopalayam virus was first isolated in India from a house shrew, *Suncus murinus*, during investigations of Japanese encephalitis virus.[11] At the time of isolation, and indeed until very recently, almost nothing was known about its taxonomic status or antigenic relatedness to other viruses. Only recently has it been convincingly shown to be a member of the hantaviruses.[18] No attempts have been made to determine the pathogenic potential for Thottopalayam virus for humans.

11. Unresolved Questions

11.1. Hantavirus Immunotypes

It is clear that there are several antigenically and genetically closely related hantaviruses, and it seems likely that others will be identified.[18,130] The recent emergence of hantaviral pulmonary syndrome in the southwestern United States attests not only to the presence of heretofore unrecognized serotypes but also to their pathogenic potential. Candidates targeted for future study include the Thai strain, Latin American observations of human infection, and similar observations from Africa.[25,34,53] The answers will be important both for understanding the epidemiology of infection and for guiding future vaccine development.

11.2. Pathogenesis of Infections in Man

Central questions are whether hantavirus infections have a significant immunopathologic component, whether they invade the central nervous system, and the pathogenesis of the vascular permeability, especially as seen in the pulmonary syndrome due to Sin Nombre virus. Further, the question of permanent renal damage following infection and possible predisposition to subsequent chronic diseases, including hypertensive renal disease, must be addressed, especially in light of the costs for long-term maintenance of end-stage renal disease patients. Finally, an animal model that faithfully replicates the human disease is needed to allow evaluation of vaccines and clinical interventions.

11.3. Diagnostic Tests

Full recognition of the disease burden due to hantavirus infections will not be appreciated until appropriate diagnostic tests are widely available and the disease is routinely considered in attending physicians' differential diagnosis. Measurement of IgM-specific antibodies has been shown clearly to be the method of choice for laboratory confirmation of hantaviral infection,[49] but diagnostic testing is now limited to specialized laboratories. What is needed is a sensitive, specific, inexpensive assay that is widely available for routine use.

11.4. Vaccines

Vaccines for Asian and European forms of HFRS are clearly justified, and at present several candidate vaccines are in various stages of development. At least one is commercially available outside the United States, but definitive efficacy has not been demonstrated.[54] Inactivated candidate vaccines have been produced in North and South Korea and in China, while engineered recombinant candidate vaccines are being developed in the United States and elsewhere.[94,134] Development of live attenuated virus strains would lead to the most economical and effective vaccines, but this approach is blocked by the absence of a suitable animal for study of virus genetics associated with virulence. Nevertheless, there is reason for optimism because of the great progress made in virus characterization over the past decade.

12. References

1. ANTIC, D., KANG, C. Y., SPIK, K., SCHMALJOHN, C., VAPALAHTI, O., AND VAHERI, A., Comparison of the deduced gene products of the L, M and S genome segments of hantaviruses, *Virus Res.* **24:** 35–46 (1992).

2. ANTONIADES, A., GREKAS, D., ROSSI, C. A., AND LEDUC, J. W., Isolation of a hantavirus from a severely ill patient with hemorrhagic fever with renal syndrome in Greece, *J. Infect. Dis.* **156:**1010–1013 (1987).

3. ARIKAWA, J., TAKASHIMA, I., HASHIMOTO, N., MORITA, C., SUGIYAMA, K., MATSUURA, Y., SHIGA, S., AND KITAMURA, T., Epidemiological studies of hemorrhagic fever among urban rats in two islands in Tokyo Bay, Japan, *Acta Virol.* **29:**66–72 (1985).

4. ARTHUR, R. R., LOFTS, R. S., GOMEZ, J., GLASS, G. E., LEDUC, J. W., AND CHILDS, J. E., Grouping of hantaviruses by small (S) genome segment polymerase chain reaction and amplification of viral RNA from wild-caught rats, *Am. J. Trop Med. Hyg.* **47:**210–224 (1992).

5. AVSIC-ZUPANC, T., XIAO, S.-Y., STOJANOVIC, R., STOJANOVIC, R., GLIGIC, A., VAN DER GROEN, G., AND LEDUC, J. W., Characterization of Dobrava virus: A hantavirus from Slovenia, Yugoslavia, *J. Med. Virol.* **38:**132–137 (1992).

6. BRUMMER-KORVENKONTIO, M., MANNI, T., UKKONEN, S., AND VAHERI, A., Detection of hemagglutination-inhibiting antibodies in patients with nephropathia epidemica and Korean Hemorrhagic fever by using Puumala virus cell culture antigen, *J. Infect. Dis.* **153:**997–998 (1986).

7. BRUMMER-KORVENKONTIO, M., VAHERI, A., VON BONSDORFF, C. H., VUORIMIES, J., MANNI, T., PENTTINEN, K., OKER-BLOM, N., AND LAEHDVIRTA, J., Nephropathia epidemica: Detection of antigen in bank voles and serologic diagnosis of human infection, *J. Infect. Dis.* **141:**131–141 (1980).

8. BUREK, K. A., ROSSI, C. A., LEDUC, J. W., AND YUILL, T. M., Serologic and virologic evidence of a Prospect Hill-like hantavirus in Wisconsin and Minnesota, *Am. J. Trop. Med. Hyg.* **51:**286–294 (1994).

9. CENTERS FOR DISEASE CONTROL, Update: Outbreak of hantavirus infection, *Morbid. Mortal. Week. Rep.* **42:**441–443 (1993).

10. CENTERS FOR DISEASE CONTROL, Update: Hantavirus pulmonary syndrome—United States, 1993, *Morbid. Mortal. Week. Rep.* **42:**816–820 (1993).

11. CAREY, D. E., REUBEN, R., PANICKER, K. N., SHOPE, R. E., AND MYERS, R. M., Thottapalayam virus: A presumptive arbovirus isolated from a shrew in India, *Indian J. Med. Res.* **59:**1758–1760 (1971).

12. CASALS, J., HENDERSON, B. E., HOOGSTRAAL, H., JOHNSON, K. M., AND SHELOKOV, A., A review of Soviet viral hemorrhagic fevers, 1969, *J. Infect. Dis.* **122:**437–453 (1970).

13. CHILDS, J. E., GLASS, G. E., KORCH, G. W., ARTHUR, R. R., SHAH, K. V., GLASSER, D., ROSSI, C., AND LEDUC, J. W., Evidence of human infection with a rat-associated *Hantavirus* in Baltimore, Maryland, *Am. J. Epidemiol.* **127:**875–878 (1988).

14. CHILDS, J. E., GLASS, G. E., KORCH, G. W., AND LEDUC, J. W., Prospective seroepidemiology of hantaviruses and population dynamics of small mammal communities of Baltimore, Maryland, *Am. J. Trop. Med. Hyg.* **37:**648–662 (1987).

15. CHILDS, J. E., GLASS, G. E., KORCH, G. W., AND LEDUC, J. W., The ecology and epizootiology of hantaviral infections in small mammal communities of Baltimore: A review and synthesis, *Bull. Soc. Vector Ecol.* **13:**113–122 (1988).

16. CHILDS, J. E., KORCH, G. W., GLASS, G. E., LEDUC, J. W., AND SHAH, K. V., Epizootiology of *Hantavirus* infections in Baltimore: Isolation of a virus from Norway rats, and characteristics of infected rat populations, *Am. J. Epidemiol.* **125:**55–68 (1987).

17. CHILDS, J. E., KORCH, G. W., SMITH, G. A., TERRY, A. D., AND LEDUC, J. W., Geographical distribution and age-related prevalence of antibody to Hantaan-like virus rat population of Baltimore, Maryland, USA, *Am. J. Trop Med. Hyg.* **34:**385–387 (1985).

18. CHU, Y. K., ROSSI, C., LEDUC, J. W., LEE, H. W., SCHMALJOHN, C. S., AND DALRYMPLE, J. M., Serological relationships among viruses in the *Hantavirus* genus, family Bunyaviridae, *Virology* **198:**196–204 (1994).

19. COHEN, M. S., KWEI, H.-E., CHIN, C.-C., AND GE, H.-C., CNS manifestations of epidemic hemorrhagic fever, *Arch. Intern. Med.* **143:**2070–2072 (1983).

20. DALRYMPLE, J. M., Plaque assay and plaque-reduction neutralization tests, in: *Manual of Hemorrhagic Fever with Renal Syndrome* (H. W. LEE AND J. M. DALRYMPLE, EDS.), pp. 102–106, World Health Organization, Seoul, Korea, 1989.

21. DANTAS, J. R., JR., OKUNO, Y., ASADA, H., TAMURA, M., TAKAHASHI, M., TANISHITE, O., TAKAHASI, Y., KURATA, T., AND YAMANISHI, K., Characterization of glycoproteins of viruses causing hemorrhagic fever with renal syndrome (HFRS) using monoclonal antibodies, *Virology* **151:**379–384 (1986).

22. DESMYTER, J., LEDUC, J. W., JOHNSON, K. M., DECKERS, C., BRASSEUR, F., AND VAN YPERSELE DE STRIHOU, C., Laboratory rat associated outbreak of haemorrhagic fever with renal syndrome due to Hantaan-like virus in Belgium, *Lancet* **2:**1445–1448 (1983).

23. DIGLISTIC, G., XIAO, S.-Y., GLIGIC, A., OBRADOVIC, M., STOJANOVIC, R., VELIMIROVIC, D., LUKAC, V., ROSSI, C. A., AND LEDUC, J. W., Isolation of a Puumala-like virus from *Mus musculus* captured in Yugoslavia and its association with severe hemorrhagic fever with renal syndrome, *J. Infect. Dis.* **169:**204–207 (1994).

24. EARLE, D. P., Analysis of sequential physiologic derangements in epidemic hemorrhagic fever, *Am. J. Med.* **16:**690–709 (1954).

24a. ELLIOTT, L. H., KSIAZEK, T. G., ROLLIN, P. E., SPIROPOULOU, C. F., MORZUNOV, S., MONROE, M., GOLDSMITH, C. S., HUMPHREY, C. D., ZAKI, S. R., KREBS, J. W., MAUPIN, G., GAGE, K., CHILDS, J. E., NICHOL, S. T., AND PETERS, C. J., Isolation of the causative agent of *Hantavirus* pulmonary syndrome, *Am. J. Trop. Med. Hyg.* **51:**102–108 (1994).

25. ELWELL, M. R., WARD, G. S., TINGPALAPONG, M., AND LEDUC, J. W., Serologic evidence of Hantaan-like virus in rodents and man in Thailand, *Southeast Asian J. Trop. Med. Public Health* **16:**349–354 (1985).

26. FRENCH, G. R., FOULKE, R. S., BRAND, O. A., EDDY, G. A., LEE, H. W., AND LEE, P.-W., Propagation of etiologic agent of Korean hemorrhagic fever in a cultured continuous cell line of human origin, *Science* **211:**1046–1048 (1981).

27. GAJDUSEK, D. C., Virus hemorrhagic fevers, *J. Pediatr.* **60**:841–857 (1962).

28. GAVRILOVSKAYA, I. N., APEKINA, N. S., MYASNIKOV, Y. A., BERNSHTEIN, A. D., RYLTSEVA, E. V., GORBACHKOVA, E. A., AND CHUMAKOV, M. P., Features of circulation of hemorrhagic fever with renal syndrome (HFRS) virus among small mammals in the European U.S.S.R., *Arch. Virol.* **75**:313–316 (1983).

29. GLASS, G. E., CHILDS, J. E., KORCH, G. W., AND LEDUC, J. W., Association of intraspecific wounding with hantaviral infection in wild rats (*Rattus norvegicus*), *Epidemiol. Infect.* **101**:459–472 (1988).

30. GLASS, G. E., WATSON, A. J., LEDUC, J. W., AND CHILDS, J. E., Domestic cases of hemorrhagic fever with renal syndrome in the United States, *Nephron* **68**:48–51 (1994).

31. GLASS, G. E., WATSON, A. J., LEDUC, J. W., KELEN, G. D., QUINN, T. C., AND CHILDS, J. E., Infection with a ratborne hantavirus in US residents is consistently associated with hypertensive renal disease, *J. Infect. Dis.* **167**:614–620 (1993).

32. GLIGIC, A., DIMKOVIC, N., XIAO, S.-Y., BUCKLE, G. J., JOVANOVIC, D., VELIMIROVIC, D., STOJANOVIC, R., OBRADOVIC, M., DIGLISTIC, G., MICIC, J., ASHER, D. M., LEDUC, J. W., YANAGIHARA, R., AND GAJDUSEK, D. C., Belgrade virus: A new hantavirus causing severe hemorrhagic fever with renal syndrome in Yugoslavia, *J. Infect. Dis.* **166**:113–120 (1992).

33. GLIGIC, A., FRUSIC, M., OBRADOVIC, M., STOJANOVIC, R., HLACA, D., GIBBS, C. J., JR., YANAGIHARA, R., CALISHER, C. H., AND GAJDUSEK, D. C., Hemorrhagic fever with renal syndrome in Yugoslavia: Antigenic characterization of hantaviruses isolated from *Apodemus flavicollis* and *Clethrionomys glareolus*, *Am. J. Trop. Med. Hyg.* **41**:109–115 (1989).

33a. Gligic, A., Stojanovic, R., Obradovic, M., Hlaca, D., Dimkovic, N., Diglisic, G., Lukac, V., Ler, Z., Bogdanovic, R., Antonijevic, B., Ropac, D., Avsic-Zupanc, T., LeDuc, J. W., Ksiazek, T., Yanagihara, R., and Gajdusek, D. C., Hemorrhagic fever with renal syndrome in Yugoslavia: Epidemiologic and epizootiologic features of a nationwide outbreak in 1989, *Eur. J. Epidemiol.* **8**:816–825 (1992).

34. GONZALES, J P., McCORMICK, J. B., BANIN, D., GRUTUN, J. P., MEUNIER, D. Y., DOURNON, E., AND GEORGES, A. J., Serological evidence for Hantaan-related virus in Africa, *Lancet* **2**:1036–1037 (1984).

35. GRANKVIST, O., JUTO, P., SETTERGREN, B., AHLM, C., BJERMER, L., LINDERHOLM, M., TARNVIK, A., AND WADEL, G., Detection of nephropathia epidemica virus RNA in patient samples using a nested primer-based polymerase chain reaction, *J. Infect. Dis.* **165**:934–937 (1992).

36. GRESIKOVA, M., RAJCANI, J., SEKEYOVA, M., BRUMMER-KORVENKONTIO, M., KOZUCH, O., LABUDA, M., TUREK, R., WEISMANN, P., NOSEK, J., AND LYSY, J., Haemorrhagic fever virus with renal syndrome in small rodents in Czechoslovakia, *Acta Virol.* **28**:416–421 (1984).

37. GRESIKOVA, M., AND SEKEYOVA, M., A simple method of preparing a complement-fixing antigen from the virus of haemorrhagic fever with renal syndrome (western type), *Acta Virol.* **22**:272–275 (1988).

38. HUGGINS, J. W., HSIANG, C. M., COSGRIFF, T. M., GUANG, M. Y., SMITH, J. I., WU, Z. O., LEDUC, J. W., ZHENG, Z. M., MEEGAN, J. M., WANG, Q. N., OLAND, D. D., GUI, X. E., GIBBS, P. H., YUAN, G. H., AND ZHANG, T. M., Prospective, double-blind, concurrent, placebo-controlled clinical trial of intravenous riba-virin therapy of hemorrhagic fever with renal syndrome, *J. Infect. Dis.* **164**:1119–1127 (1991).

39. HUGGINS, J. W., KIM, G. R., BRAND, O. M., AND MCKEE, K. T., JR., Ribavirin therapy for Hantaan virus infection in suckling mice, *J. Infect. Dis.* **153**:489–497 (1986).

40. HUNG, T., XIA, S.-M., SONG, G., LIAO, H.-X., CHAO, T.-X., CHOU, Z.-Y., AND HANG, C.-S., Viruses of classical and mild forms of haemorrhagic fever with renal syndrome isolated in China have similar Bunyavirus-like morphology, *Lancet* **1**:589–591 (1983).

41. HUNG, T., XIA, S. M., ZHAO, T-X., ZHAO, J. Y., SONG, G., LIAO, G. X. H., YE, W. W., CHU, Y. L., AND HANG, C. S., Morphological evidence for identifying the viruses of hemorrhagic fever with renal syndrome as candidate members of the bunyaviridae family, *Arch. Virol.* **78**:137–144 (1983).

42. JIANG, Y., LI, Z., AND SONG, G., Seroepidemiologic study of epidemic hemorrhagic fever with renal syndrome in China, *Chin. Med. J.* **94**:221–228 (1981).

43. KITAMURA, T., MORITA, C., KOMATSU, T., SUGIYAMA, K., ARIKAWA, J., SHIGA, S., TAKEDA, H., AKAO, Y., IMAIZUMI, K., OYA, A., HASHIMOTO, N., AND URASAWA, S., Isolation of virus causing hemorrhagic fever with renal syndrome (HFRS) through a cell culture system, *Jpn. J. Med. Sci. Biol.* **36**:17–25 (1983).

44. KURATA, T., TSAI, T. F., BAUER, S. P., McCormick, J. B., Immunofluoresence studies of disseminated Hantaan virus infection of sucking mice, *Infect. Immun.* **41**:391–398 (1983).

45. LAHDEVIRTA, J., Nephropathia epidemica in Finland, a clinical, histological and epidemiological study, *Ann. Clin. Res.* **3**(Suppl. 8):1–154 (1971).

46. LAHDEVIRTA, L., SAVOLA, J., BRUMMER-KORVENKONTIO, M., BERNDT, R., ILLIKAINEN, R., AND VAHERI, A., Clinical and serological diagnosis of nephropathia epidemica, the mild type of haemorrhagic fever with renal syndrome, *J. Infect.* **9**:230–238 (1984).

47. LEDUC, J. W., ANTONIADES, A., AND SIAMPOULOS, K., Epidemiological investigations following an outbreak of hemorrhagic fever with renal syndrome in Greece, *Am. J. Trop. Med. Hyg.* **35**:654–659 (1986).

48. LEDUC, J. W., CHILDS, J. E., AND GLASS, G. E., The Hantaviruses, etiologic agents of hemorrhagic fever with renal syndrome: A possible cause of hypertension and chronic renal disease in the United States, *Annu. Rev. Public Health* **13**:79–98 (1992).

49. LEDUC, J. W., KSIAZEK, T. G., ROSSI, C. A., AND DALRYMPLE, J. M., A retrospective analysis of sera collected by the hemorrhagic fever commission during the Korean Conflict, *J. Infect. Dis.* **162**:1182–1184 (1990).

50. LEDUC, J. W., SMITH, G. A., CHILDS, J. E., PINHEIRO, F. P., MAIZTEGUI, J. I., NIKLASSON, B., ANTONIADES, A., ROBINSON, D. M., KHIN, M., SHORTRIDGE, K. F., WOOSTER, M. T., ELWELL, M. R., ILBERY, P. L. T., KOECH, D., ROSA, E. S. T., AND ROSEN, L., Global survey of antibody to Hantaan-related viruses among peridomestic rodents, *Bull. WHO* **64**:139–144 (1986).

51. LEDUC, J. W., SMITH, G. A., AND JOHNSON, K. M., Hantaan-like viruses from domestic rats captured in the United States, *Am. J. Trop. Med. Hyg.* **33**:992–998 (1984).

52. LEDUC, J. W., SMITH, G. A., MACY, M., AND HAY, R. J., Certified cell lines of rat origin appear free of infection with Hantavirus, *J. Infect. Dis.* **152**:1082–1083 (1985).

53. LEDUC, J. W., SMITH, G. A., PINHEIRO, F. P., VASCONCELSO, P. F.

C., Rosa, E. S. T., and Maiztegui, J. I., Isolation of a Hantaan-related virus from Brazilian rats and serologic evidence of its widespread distribution in South America, *Am. J. Trop. Med. Hyg.* **34:**810–815 (1985).

53a. Lee, H. W., Hemorrhagic fever with renal syndrome in Korea, *Rev. Infect. Dis.* **11**(Suppl. 4.):S864–876 (1989).

54. Lee, H. W., Ahn, C. N., Song, J. W., Baek, L. J., Seo, T. J., and Park, S. C., Field trail of an inactivated vaccine against hemorrhagic fever with renal syndrome in humans, *Arch. Virol.* (Suppl. 1):35–47 (1990).

55. Lee, H. W., Baek, L. J., and Doo, C. D., The study on breeding season of *Apodemus agrarius*, the natural host of Korean hemorrhagic fever, *Korean J. Virol.* **11:**1–5 (1981).

56. Lee, H. W., Baek, L. J., and Johnson, K. M., Isolation of Hantaan virus, the etiologic agent of Korean hemorrhagic fever, from wild urban rats, *J. Infect. Dis.* **146:**638–644 (1982).

57. Lee, H. W., and Johnson, K. M., Laboratory-acquired infections with Hantaan virus, the etiologic agent of Korean hemorrhagic fever, *J. Infect. Dis* **146:**645–651 (1982).

58. Lee, H. W., Lee, P. W., and Baek, L. J., Korean hemorrhagic fever IV. Serologic diagnosis, *Korean J. Virol.* **10:**7–13 (1980).

59. Lee, H. W., Lee, P. W., Baek, L. J., Song, C. K., and Seong, I. W., Intraspecific transmission of Hantaan virus, etiologic agent of Korean hemorrhagic fever, in the rodent *Apodemus agrarius*, *Am. J. Trop. Med. Hyg.* **30:**1106–1112 (1981).

60. Lee, H. W., Lee, P. W., and Johnson, K. M., Isolation of the etiologic agent of Korean hemorrhagic fever, *J. Infect. Dis.* **137:**298–308 (1978).

61. Lee, H. W., Lee, P. W., Lahdevirta, J., and Brummer-Korvenkontio, M., Aetiological relation between Korean haemorrhagic fever and nephropathia epidemica, *Lancet* **1:**186–187 (1979).

62. Lee, H. W., Lee, P.-W., Tamura, M., Tamura, T., and Okuno, Y., Etiological relation between Korean hemorrhagic fever and epidemic hemorrhagic fever in Japan, *Biken J.* **22:**41–45 (1979).

63. Lee, H. W., Park, D. H., Baek, L. J., Choi, K. S., Whang, Y. N., and Woo, M. S., Korean hemorrhagic fever patients in urban areas of Seoul, *Korean J. Virol.* **10:**1–6 (1980).

64. Lee, H. W., Seong, I. W., Baek, L. J., McLeod, D. A., Seo, J. S., and Kang, C. Y., Positive serological evidence that Hantaan virus, the etiologic agent of hemorrhagic fever with renal syndrome, is endemic in Canada, *Can. J. Microbiol.* **30:**1137–1140 (1984).

65. Lee, P. W., Amyx, H. L., Yanagihara, R., Gajdusek, D. C., Goldgarber, D., and Gibbs, C. J., Jr., Partial characterization of Prospect Hill virus isolated from meadow voles in the United States, *J. Infect. Dis.* **152:**826–829 (1985).

66. Lee, P.-W., Gibbs, C. J., Jr., Gajdusek, D. C., and Yanagihara, R., Serotype classification of hantaviruses by indirect immunofluorescent antibody and plaque reduction neutralization tests, *J. Clin. Microbiol.* **22:**940–944 (1985).

67. Lee, P.-W., Svedmyr, A., Gajdusek, D. C., and Yanagihara, R., Antigenic difference between European and East Asian viruses causing haemorrhagic fever with renal syndrome, *Lancet* **2:**256–257 (1981).

68. Lee, P. W., Yanagihara, R., Gibbs, C. J., Jr., and Gajdusek, D. C., Pathogenesis of experimental Hantaan virus infection in laboratory rats, *Arch. Virol.* **88:**57–66 (1986).

69. Linderholm, M., Billstrom, A., Settergren, B., and Tarnvik, A., Pulmonary involvement in nephropathia epidemica as demonstrated by computed tomography, *Infection* **20:**263–266 (1992).

70. Lloyd, G., and Jones, N., Infection of laboratory workers with hantavirus acquired from immunocytomas propagated in laboratory rats, *J. Infect.* **12:**117–125 (1986).

71. Martin, M. L., Lindsey-Regnery, H., Sasso, D.R., McCormick, J. B., and Palmer, E., Distinction between Bunyaviridae genera by surface structure and comparison with Hantaan virus using negative strain electron microscopy, *Arch. Virol.* **86:**17–28 (1985).

72. Matsumoto, Y., Nishino, K., and Yagura, T., A case of suspicious epidemic hemorrhagic fever, *Biken J.* **7:**95–106 (1964).

73. McCormick, J. B., Sasso, D. R., Palmer, E. L., and Kiley, M. P., Morphological identification of the agent of Korean haemorrhagic fever (Hantaan virus) as a member of the Bunyaviridae, *Lancet* **1:**765–768 (1982).

73a. McKee, K. T., Jr., LeDuc, J. W., MacDonald, C., and Peters, C. J., Hemorrhagic fever with renal syndrome—A clinical perspective. *Mil. Med.* **150:**640–647 (1985).

74. Meegan, J. M., and LeDuc, J. W., Enzyme immunoassays, in: *Manual of haemorrhagic fever with renal syndrome* (H. W. Lee and J. M. Dalrymple, eds.), pp. 83–88, World Health Organization, Seoul, Korea, 1989.

75. Montagnac, R., Schillinger, F., Croix, J. C., and Dournon, E., Fievre hemorrhagique avec syndrome renal du au virus Puumala: Deux cas avec hypertriglyceridemia dans le departement de l'Aube, *Presse Med.* **14:**2016–2017 (1985).

76. Morita, C., Matsuura, Y., Morikawa, S., and Kitamura, T., Age-dependent transmission of hemorrhagic fever with renal syndrome (HFRS) virus in rats, *Arch. Virol.* **85:**145–149 (1985).

77. Morita, C., Sugiyama, K., Matsuura, Y., Kitamura, T., Komatsu, T., Akao, Y., Jitsukawa, W., and Sakkakibara, H., Detection of antibody against hemorrhagic fever with renal syndrome (HFRS) virus in sera of house rats captured in port areas of Japan, *Jpn. J. Med. Sci. Biol.* **36:**55–57 (1983).

78. Myasnikov, Y. A., Rezapkin, G. V., Shikova, Z. V., Tkachenko, E. A., Ivanova, A. A., Nurgaleeva, R. G., Stepanenko, A. G., Vereschagin, N. N., Loginov, A. I., Bagan, R. N., Zaitseva, A. A., Levacheva, Z. A., Bobylkova, T.V., Ishcheryakova, A. M., and Boruta, V. V., Antibodies to the HFRS virus in the human population of European RSFSR as detected by radioimmunoassay, *Arch. Virol.* **79:**109–115 (1984).

79. Myhrman, G., Nephropathia epidemica, a new infectious disease in Northern Scandinavia, *Acta Med. Scand.* **140:**52–56 (1951).

80. Nakamura, T., Yanagihara, R., Gibbs, C. J., Jr., and Gajdusek, D. C., Immune spleen cell-mediated protection against fatal Hantaan virus infant mice, *J. Infect. Dis.* **151:**691–697 (1985).

81. Nichol, S. T., Spiropoulou, C. F., Morzunov, S., Rollin, P. E., Ksiazek, T. G., Feldman, H., Sanchez, A., Childs, J., Zaki, S., and Peters, C. J., Genetic identification of hanatavirus associated with an outbreak of acute respiratory illness, *Science* **262:**914–917.

82. Niklasson, B., Hornfeldt, B., Mullaart, M., Settergren, B., Tkachenko, E., Myasnikov, Y. A., Ryltceva, E. V., Leschinskaya, E., Malkin, A., and Dzagurova, T., An epidemiologic study of hemorrhagic fever with renal syndrome in Bashkirtostan (Russia) and Sweden, *Am. J. Trop. Med. Hyg.* **48:**670–675 (1993).

83. Niklasson, B., and LeDuc, J. W., Isolation of nephropathia epidemica agent in Sweden, *Lancet* **1:**1012–1013 (1984).

84. Niklasson, B., and LeDuc, J. W., Epidemiology of nephro-

pathia epidemica in Sweden, *J. Infect. Dis.* **155**:269–276 (1987).

85. NIKLASSON, B., LEDUC, J. W., NYSTROM, K., AND NYMAN, L., Nephropathia epidemica: Incidence of clinical cases and antibody prevalence in an endemic area of Sweden, *Epidemiol. Infect.* **99**:559–562 (1987).

86. NIKLASSON, B., NYMAN, L., LINDE, A., GRANDIEN, M., AND DALRYMPLE, J., An epidemiological survey of nephropathia epidemica in Sweden, *Scand. J Infect. Dis.* **15**:239–245 (1983).

87. NERURKAR, V. R., SONG, K.-J., GAJDUSEK, D. C., AND YANAGI-HARA, R., Genetically distinct hantavirus in deer mice, *Lancet* **342**:1058–1059 (1993).

88. NUZUM, E. O., ROSSI, C. A., STEPHENSON, E. H., AND LEDUC, J. W., Aerosol transmission of Hantaan and related viruses to laboratory rats, *Am. J. Trop. Med. Hyg.* **38**:636–640 (1988).

89. OKUNO, Y., YAMANISKI, K., TAKAHASHI, Y., TANISHITA, O., NAGAI, T., DANTAS, J. R., JR., OKAMOTO, Y., TADANO, M., AND TAKAHASHI, M., Haemagglutination-inhibition test for haemorrhagic fever with renal syndrome using virus antigen prepared from infected tissue culture fluid, *J. Gen. Virol.* **67**:149–156 (1986).

90. PENSIERO, M. N., SHAREFKIN, J. B., DIEFFENBACH, C. W., AND HAY, J., Hantaan virus infection of human endothelial cells, *J. Virol.* **66**:5929–5936 (1992).

91. PENTTINEN, K., LAHDEVIRTA, J., KEKOMAKI, R., ZIOLA, B., SALMI, A., HAUTANEN, A., LINSTROM, P., VAHERI, A., BRUMMER-KORVENKONTIO, M., AND WAGER, W., Circulating immune complexes, immunoconglutinins, and rheumatoid factors in nephropathia epidemica, *J. Infect. Dis.* **143**:15–21 (1983).

92. RUBINI, M. E., JABLON, S., AND MCDOWELL, M. E., Renal residuals of acute epidemic hemorrhagic fever, *Arch. Intern. Med.* **106**:378–387 (1960).

93. SCHMALJOHN, C. S., AND DALRYMPLE, J. M., Analysis of Hantaan virus RNA: Evidence for new genus of Bunyaviridae, *Virology* **131**:482–491 (1983).

94. SCHMALJOHN, C. S., HASTY, S. E., AND DALRYMPLE, J. M., Preparation of candidate vaccinia-vectored vaccines for haemorrhagic fever with renal syndrome, *Vaccine* **10**:10–13 (1992).

95. SCHMALJOHN, C. S., HASTY, S. E., DALRYMPLE, J. M., LEDUC, J. W., LEE, H. W., VON BONSDORF, C.-H., BRUMMER-KORVENKONTIO, M., VAHERI, A., TSAI, T. F., REGNERY, H. L., GOLDGABER, D., AND LEE, P.-W., Antigenic and genetic properties of viruses linked to hemorrhagic fever with renal syndrome, *Science* **227**:1041–1044 (1985).

96. SCHMALJOHN, C. S., HASTY, S. E., HARRISON, S. A., AND DALRYMPLE, J. M., Characterization of Hantaan virions, the prototype virus of hemorrhagic fever with renal syndrome, *J. Infect. Dis.* **148**:1005–1012 (1983).

97. SCHMALJOHN, C. S., HASTY, E., RASMUSSEN, L., AND DALRYMPLE, J. M., Hantaan virus replication: Effects of monensin, tunicamycin and endoglycosidase on the structural glycoprotein, *J. Gen. Virol.* **67**:707–717 (1986).

98. SETTERGREN, B., Nephropathia epidemica (hemorrhagic fever with renal syndrome) in Scandinavia, *Rev. Infect. Dis.* **13**:736–744 (1991).

99. SETTERGREN, B., JUTO, P., WADEL, G., TROLLFORS, B., AND NORRBY, S. R., Incidence and geographic distribution of serologically verified cases of nephropathia epidemica in Sweden, *Am. J. Epidemiol.* **127**:801–807 (1988).

100. SETTERGREN, B., LESCHINSKAYA, E., ZAGIDULLIN, I., FAZLYEVA, R., KHUNAFINA, D., AND NIKLASSON, B., Hemorrhagic fever with renal syndrome: Comparison of clinical course in Sweden and in the Western Soviet Union, *Scand. J. Infect. Dis.* **23**:549–552 (1991).

101. SHEEDY, J. A., FROEB, H. F., BATSON, H. A., CONLEY, C. C., MURPHY, J. P., HUNTER, R. B., CUGELL, D. W., GILES, R. B., BERSHADSKY, S. C., VESTER, J. W., AND YOE, R. H., The clinical course of epidemic hemorrhagic fever, *Am. J. Med.* **16**:619–628 (1954).

102. SMORODINSTSEV, A. A., CHUDADOV, V. G., AND CHURILOV, A. V., *Haemorrhagic Nephroso-nephritis*, Pergamon Press London, 1959.

103. SOMMER, A. I., TRAAVIK, T., MEHL, R., BERDAHL, B. P., AND DALRYMPLE, J. M., Reservoir animals for nephropathia epidemica in Norway: Indications of a major role for the bank vole (*C. glareolus*) in comparison with the woodmouse (*A. sylvaticus*), *J. Hyg.* **94**:123–127 (1985).

104. SONG, G., HANG, C.-S., LIAO, H.-X., AND FU, J.-L., Antigenic comparison of virus strains of mild and classical types of epidemic haemorrhagic fever isolated in China and adaptation of these to cultures of normal cells, *Lancet* **1**:677–678 (1984).

105. SONG, G., HANG, C., LIAO, H., FU, J., GAO, G., QIU, H. AND ZHANG, Q., Antigenic differences between viral strains causing classical and mild types of epidemic hemorrhagic fever with renal syndrome in China, *J. Infect. Dis.* **150**:889–894 (1984).

106. SUGIYAMA, K., MATSUURA, Y., MORITA, C., SHIGA, S., AKAO, Y., KOMATSU, T., AND KITAMURA, T., An immune adherence assay for discrimination between etiologic agents of hemorrhagic fever with renal syndrome, *J. Infect. Dis.* **149**:67–73 (1984).

107. SVEDMYR, A., LEE, P.-W., GOLDGABER, D., YANAGIHARA, R., GAJDUSEK, D. C., GIBBS, C. J., JR., AND NYSTROM, K., Antigenic differences between European and East Asian strains of HFRS virus, *Scand. J. Infect. Dis. (Suppl.)* **36**:86–87 (1982).

108. TAKENAKA, A., GIBBS, C. J., JR., AND GAJDUSEK, D. C., Antiviral neutralizing antibody to Hantaan virus as determined by plaque reduction technique, *Arch. Virol.* **84**:197–206 (1985).

109. TALLER, A. M., XIAO, S.-Y., GODEC, M. S., GLIGIC, A., AVSIC-ZUPANC, T., GOLDFARB, L. G., YANAGIHARA, R., AND ASHER, D. M., Belgrade virus, a cause of hemorrhagic fever with renal syndrome in the Balkans, is closely related to Dobrava virus of field mice, *J. Infect. Dis.* **168**:750–753 (1993).

110. TAMURA, M., Occurrence of epidemic hemorrhagic fever in Osaka city: First cases found in Japan with characteristic feature of marked proteinuria, *Biken J.* **7**:79–94 (1964).

111. TANISHITA, O., TAKAHASHI, Y., OKUNO, Y., YAMANISHI, K., AND TAKAHASHI, M., Evaluation of focus reduction neutralization test with peroxidase–antiperoxidase staining technique for hemorrhagic fever with renal syndrome virus, *J. Clin. Microbiol.* **20**:1213–1215 (1984).

112. TKACHENKO, E. A., DONETS, M. A., REZAPKIN, G. V., DZAGUROVA, T. K., IVANOV, A. P., LESHCHINSKAYA, E. V., RESHETNIKOV, I., DROZKOV, S. G., SLONOVA, R. A., AND SOMOV, G. P., Serotypes of HFRS (haemorrhagic fever with renal syndrome) virus in East European and Far Eastern U.S.S.R., *Lancet* **1**:863 (1982).

113. TKACHENKO, E. A., DZAGUROVA, T. K., LESHCHINSKAYA, E. V., ZAGIDULIN, I. M., USTJUGOVA, I. M., GASANOVA, T. A., REZAPIN, G. V., AND MIASMIKOV, J. A., Serological diagnosis of haemorrhagic fever with renal syndrome in European region of U.S.S.R., *Lancet* **2**:1407 (1982).

114. TKACHENKO, E. A., IVANOV, A. P., DZAGUROVA, T. K., DONETS, M. A., REZAPKIN, G. V., LESHCHINSKAYA, E. V., ZATGIDULIN, U. M., IVANOVA, A. A., STRADUKHIN, O. V., MUSTAFIN, N. M., AND

SAVINOVA, T. I., Immunosorbent assays for diagnosis of haemorrhagic fever with renal syndrome, *Lancet* **2:**257–258 (1981).

115. TRAAVIK, T., SOMMER, A. I., MEHL, R., BERDAHL, B. P., STAVEM, K., HENDERI, O. H., AND DALRYMPLE, J., Nephropathia epidemica in Norway: Antigen and antibodies in rodent reservoirs and selected human populations, *J. Hyg.* **93:**139–146 (1984).

116. TSAI, T. F., Hemorrhagic fever with renal syndrome: Mode of transmission, *Lab. Anim. Sci.* **37:**428–430 (1987).

117. TSAI, T. F., BAUER, S. P., MCCORMICK, J. B., AND KURATA, T., Intracerebral inoculation of suckling mice with Hantaan virus, *Lancet* **1:**503–504 (1982).

118. TSAI, T. F., BAUER, S. P., SASSO, D. R., WHITFIELD, S. G., MCCORMICK, J. B., CARAWAY, T. C., MCFARLAND, L., BRADFORD, H., AND KURATA, T., Serological and virological evidence of a Hantaan virus-related enzootic in the United States, *J. Infect. Dis.* **152:**126–144 (1985).

119. TSAI, T. F., TANG, J. W., HU, S. L., YE, K. L., CHOU, G. L., AND XU, Z. Y., Hemagglutination-inhibiting antibody in hemorrhagic fever with renal syndrome, *J. Infect. Dis.* **150:**895–898 (1984).

120. UMENAI, T., LEE, P. W., TOYODA, T., YOSHINAGA, K., HORIUCHI, T., LEE, H. W., SAITO, T., HONGO, M., NOBUNAGA, T., AND ISHIDA, N., Korean haemorrhagic fever in staff in an animal laboratory, *Lancet* **1:**1314–1316 (1979).

121. VAN DER GROEN, G., PIOT, P., DESMYTER, J., PIOT, P., COLAERT, J., MUYLLE, L., TKACHENKO, E. A., IVANOV, A. P., VERHAGEN, R., AND VAN YPERSELE DE STRIHOU, C., Seroepidemiology of Hantaan-related virus infections in Belgian populations, *Lancet* **2:**1493–1494 (1983).

122. VAN DER GROEN, G., TKACHENKO, E. A. IVANOV, A. P., AND VERHAGEN, R., Haemorrhagic fever with renal syndrome related virus in indigenous wild rodents in Belgium, *Lancet* **2:**110–111 (1983).

123. VAN YPERSELE DE STRIHOU, C., VANDENBROUCK, J. M., LEVY, M., DOYEN, C., COSYNS, J. P., VAN DER GROEN, G., AND DESMYTER, J., Diagnosis of epidemic and sporadic interstitial nephritis due to Hantaan-like virus in Belgium, *Lancet* **2:**8365–8366 (1983).

124. VASYUTA, Y. S., The epidemiology of haemorrhagic fever with renal syndrome in the R.S.F.S.R., *Zh. Mikrobiol. Epidemiol. Immun.* **32:**49–56 (1961).

125. WHI, W. Y., AND WANG, L. H., Epidemiologic study of Korean hemorrhagic fever, *Korean U. Med. J.* **17:**137–144 (1980).

126. WHITE, J. D., SHIREY, F. G., FRENCH, G. R., HUGGINS, J. W., BRAND, O. M., AND LEE, H. W., Hantaan virus, aetiological agent of Korean haemorrhagic fever, has bunyaviridae-like morphology, *Lancet* **1:**768–771 (1982).

127. WORLD HEALTH ORGANIZATION, Haemorrhagic fever with renal syndrome, Russian Federation, *Wkly. Epidemiol. Rec.* **68:**189–191 (1993).

128. XIAO, S.-Y., CHU, Y.-K., KNAUERT, F. K., LOFTS, R., DALRYMPLE, J. M., AND LEDUC, J. W., Comparison of hantavirus isolates using a genus-reactive primer pair polymerase chain reaction, *J. Gen. Virol.* **73:**567–573 (1992).

129. XIAO, S.-Y., DIGLISIC, G., AVSIC-ZUPANC, T., AND LEDUC, J. W., Dobrava virus as a new hantavirus: Evidence by comparative sequence analysis, *J. Med. Virol.* **39:**152–155 (1993).

130. XIAO, S.-Y., LEDUC, J. W., CHU, Y. K., AND SCHMALJOHN, C. S.,

Phylogenetic analyses of virus isolates in the genus *Hantavirus*, family Bunyaviridae, *Virology* **198:**205–217 (1994).

131. XIAO, S.-Y., YANAGIHARA, R., GODEC, M. S., ELDADAH, Z. A., JOHNSON, B. K., GAJDUSEK, D. C., AND ASHER, D. M., Detection of hantavirus RNA in tissues of experimentally infected mice using reverse transcriptase-directed polymerase chain reaction, *J. Med. Virol.* **33:**277–282 (1991).

132. XU, Z.-Y., GUO, C.-S., WU, Y.-L., ZHANG, X.-W., AND LIU, K., Epidemiological studies of haemorrhagic fever with renal syndrome: Analysis of risk factors and mode of transmission, *J. Infect. Dis.* **152:**137–144 (1985).

133. YAMANISHI, K., DANTAS, J. R., TAKAHASHI, M., YAMANOUCHI, T., DOMAE, K., KAWAMATA, J., AND KURATA, T., Isolation of hemorrhagic fever with renal syndrome (HFRS) virus from a tumor specimen in a rat, *Biken J.* **26:**155–160 (1983).

134. YAMANISHI, K., TANISHITA, O., TAMURA, M., ASADA, H., KONDO, K., TAKAGI, M., YOSHIDA, I., KONOBE, T., AND FUKAI, K., Development of inactivated vaccine against virus causing haemorrhagic fever with renal syndrome, *Vaccine* **6:**278–282 (1988).

135. YAN, D., GU, X., WANG, D., AND YANG, S., Studies on immunopathogenesis in epidemic hemorrhagic fever. Sequential observations on activation of the first complement component in sera from patients with epidemic hemorrhagic fever, *J. Immunol.* **127:**1604–1067 (1981).

136. YANAGIHARA, R., Hantavirus infection in the United States: Epizootiology and epidemiology, *Rev. Infect. Dis.* **12:**449–457 (1990).

137. YANAGIHARA, R., AMYX, H. L., AND GAJDUSEK, D. C., Experimental infection with Puumala virus, the etiological agent of nephropathia epidemica, in bank voles (*Clethrionomys glareolus*), *J. Virol.* **55:**34–38 (1985).

138. YANAGIHARA, R., GAJDUSEK, D. C., GIBBS, C. J., JR., AND TRAUB, R., Prospect Hill virus: Serological evidence for infection in mammalogists, *N. Engl. J. Med.* **310:**1325–1326 (1984).

139. YANG, Y., XY, Z. Y., ZHU, Z. Y., AND TSAI, T. F., Isolation of haemorrhagic fever with renal syndrome virus from *Suncus murinus*, an insectivore, *Lancet* **1:**513–514 (1985).

140. ZHANG, C. A., XIE, Y. J., ZHANG, C. A., MCCORMICK, J. B., SANCHEZ, A., ENGELMAN, H. M., CHEN, S. Z., GU, X. S., YANG, W. T., AND ZHANG, J., New isolates of HFRS virus in Sichuan, China and characterization of antigenic differences by monoclonal antibodies, *Lancet* **1:**8493 (1986).

13. Suggested Reading

COSGRIFF, T. M. (ED.), Hemostatic impairment associated with hemorrhagic fever viruses, *Rev. Infect. Dis.* **11**(Suppl. 4):S669–896 (1989).

LAHDEVIRTA, J., Nephropathia epidemica in Finland, *Am. J. Clin. Res.* **3**(Suppl. 8):1–154 (1971).

Symposium on Epidemic Hemorrhagic Fever, *Am. J. Med.* **16:**619–709 (1954).

TRINCSENI, T., AND KELETI, B., *Clinical Aspects and Epidemiology of Haemorrhagic Fever with Renal Syndrome*, Akademiai Kiado, Budapest, 1971.

CHAPTER 13

Viral Hepatitis

Harold S. Margolis, Miriam J. Alter, and Stephen C. Hadler

1. Introduction

Viral hepatitis represents a disease entity caused by at least five unrelated viruses whose primary tissue tropism is the hepatocyte. Hepatitis A is an acute self-limited disease resulting from infection with hepatitis A virus (HAV). Infection with hepatitis B virus (HBV) can produce a chronic infection that places the individual at risk of death from chronic liver disease or primary hepatocellular carcinoma. Delta hepatitis is caused by hepatitis delta virus (HDV), an "incomplete" virus and the only viroid known to infect man. It requires an active HBV infection to replicate. The recent identification of two new hepatitis viruses has almost completely characterized the disease entity previously designated non-A, non-B hepatitis, which was defined by the serological exclusion of known causes of viral hepatitis and their epidemiologic characteristics. Hepatitis C virus (HCV) infection is responsible for the majority of cases of parenterally transmitted non-A, non-B hepatitis and produces a persistent infection that is often associated with chronic liver disease. Hepatitis E virus (HEV) is the cause of almost all cases of enterically transmitted non-A, non-B hepatitis and is not known to produce a persistent infection. However, the serological exclusion of the known hepatitis viruses continues to leave cases of hepatitis that clinically appear to be viral in origin. The majority of these cases of "non-ABCDE" hepatitis appear to be parenterally transmitted, although outbreaks of enterically transmitted disease have been reported.

The overwhelming majority of cases of viral hepa-

titis are due to the viruses described in this chapter. However, a number of other viruses are associated with infections whose manifestations occasionally include hepatitis, especially in certain age groups or in certain parts of the world. These include cytomegalovirus, Epstein–Barr virus, herpes simplex virus, yellow fever virus, rubella, parvovirus, and viruses such as Lassa, Ebola, and Marburg.[149] Because the symptoms produced by all of these viral infections are quite similar, the specific diagnosis of any case of viral hepatitis can only be made by serological testing. Specific serological tests are now commercially available for the diagnosis of HAV, HBV, HCV, HDV, and HEV infection. The diagnosis of "non-ABCDE" hepatitis is based on the serological exclusion of the specific agents responsible for the most common causes of viral hepatitis. However, other causes of hepatitis should be entertained based on associated symptoms and epidemiologic characteristics of the illness.

The currently accepted nomenclature and abbreviations for the viruses, antigens, and antibodies associated with HAV and HBV infection have been reported by an Expert Committee of the World Health Organization. These and the currently used designations for HCV, HDV, and HEV infections are the basis for the following outline:

Hepatitis A

HAV	Hepatitis A virus; etiologic agent of hepatitis A; a picornavirus with a single serotype
anti-HAV	Total antibody to HAV; indicates acute or resolved infection; indicates a protective immune response to infection, passively acquired antibody, or response to vaccination
IgM anti-HAV	IgM-class antibody indicating recent infection with HAV

Harold S. Margolis and Miriam J. Alter • Hepatitis Branch, Division of Viral and Rickettsial Diseases, National Center for Infectious Diseases, Centers for Disease Control and Prevention, Atlanta, Georgia 30333. **Stephen C. Hadler** • Epidemiology and Surveillance Division, National Immunization Program, Centers for Disease Control and Prevention, Atlanta, Georgia 30333.

HAV-RNA	RNA of HAV; detected by nucleic acid amplification and/or hybridization

Hepatitis B

HBV	Hepatitis B virus; etiologic agent of hepatitis B; a hepadnavirus with a single major serotype
HBsAg	Hepatitis B surface antigen; surface glycoprotein coat of HBV; produced in excess of the whole virion; antigen used in hepatitis B vaccine
HBeAg	Hepatitis B e antigen; conformational soluble antigen of HBcAg that correlates with HBV replication and infectivity
HBcAg	Hepatitis B core antigen; nucleoprotein of virus; no commercial test available
anti-HBs	Antibody to HBsAg; indicates a protective immune response to HBV infection or vaccination, or passively acquired antibody
anti-HBe	Antibody to HBeAg; presence in serum of HBV-infected person suggests low titer of virus and lower infectivity
anti-HBc	Total antibody to HBcAg; indicates acute, chronic, or resolved HBV infection; not elicited by vaccination
IgM anti-HBc	IgM-class antibody to HBcAg; indicates recent infection with HBV
HBV-DNA	DNA of HBV; detected by nucleic acid amplification and/or hybridization

Hepatitis C

HCV	Hepatitis C virus; etiologic agent of hepatitis C and majority of cases of what were previously termed parenterally transmitted non-A, non-B hepatitis
anti-HCV	Antibody to HCV; does not differentiate acute infection from resolved or chronic infection
HCV-RNA	RNA of HCV; defines viremia; detected by nucleic acid amplification

Delta hepatitis

HDV	Hepatitis delta virus; etiologic agent of delta hepatitis; only infectious in presence of acute or chronic HBV infection
HDV-Ag	Delta antigen; detectable during early acute HDV infection
anti-HDV	Antibody to delta antigen; indicates acute, resolved, or chronic infection
IgM anti-HDV	IgM-class antibody to HDV; indicates either acute infection or chronic infection with active viral replication
HDV-RNA	RNA of HDV, detected by nucleic acid amplification or hybridization

Hepatitis E

HEV	Hepatitis E virus; etiologic agent of hepatitis E, which was previously designated enterically transmitted non-A, non-B hepatitis
anti-HEV	Antibody to hepatitis E virus; indicates acute or resolved infection
IgM anti-HEV	IgM-class antibody to hepatitis E virus; indicates acute infection

Non-ABCDE hepatitis

Parenterally transmitted	Diagnosis of exclusion; epidemiologic evidence of parenteral or sexual transmission
Enterically transmitted	Diagnosis of exclusion; epidemiologic evidence of fecal–oral transmission

Worldwide, viral hepatitis is one of the most frequently reported diseases, and in almost any country, acute and chronic liver disease due to viral hepatitis accounts for substantial morbidity and mortality. With the characterization of the agents responsible for hepatitis C and hepatitis E, the prevention and control of all forms of viral hepatitis may soon become a reality (Table 1). Transmission of HAV and HBV infection can be prevented by active or passive immunization. Preexposure immunization with hepatitis A vaccine prevents HAV infection and affords the potential for substantially reducing disease incidence, while passive immunization with immune globulin (IG) remains the mainstay of postexposure prophylaxis. HBV infection can be effectively prevented by preexposure or postexposure immunization, and the elimination of HBV transmission is possible with the widespread use of hepatitis B vaccine. In addition, hepatitis B immunization will prevent HDV infection, although there is no effective means to prevent HDV infection in persons with chronic HBV infection.

The characterization of HCV has resulted in the development of diagnostic assays that are being used to routinely screen blood donors and prevent posttransfusion HCV infection in most countries. Whether HCV infection

Table 1. Etiologic Agents of Viral Hepatitis, Outcome of Infection, and Approaches to Disease Prevention

Virus	Classification	Outcome of infection	Major modes of transmission	Prevention methods
HAV	Picornavirus[a]	Acute Resolved	Fecal–oral	Preexposure vaccination Hygiene
HBV	Hepadnavirus	Acute Chronic	Parenteral Sexual	Preexposure/postexposure vaccination
HCV	Flavivirus	Acute Chronic	Parenteral Sexual	Blood donor screening Reducing high-risk behavior (drug use)
HDV	Satellite virus	Acute Chronic	Parenteral	Same as HBV
HEV	Calicivirus[b]	Acute Resolved	Fecal–oral	Hygiene Vaccine (?)

[a]Genus hepatovirus.
[b]Final classification not determined.

can be prevented by immunization remains to be determined, although recent studies suggest that immunization of chimpanzees with recombinant HCV antigens may provide short-term protection and prevent chronic HCV infection. In the meantime, reducing HCV transmission will only occur through public health efforts directed at the prevention of blood-borne infections transmitted by injection drug use.

The primary source of HEV infection appears to be fecally contaminated water, and assurance of a safe water supply is the primary means of prevention. While experiments in nonhuman primates suggest that immunization may prevent infection, questions must be answered concerning the natural history of the immune response to infection in order to develop comprehensive prevention strategies.

Recent reports suggest that other hepatotropic viruses have been identified in patients with non-ABCDE hepatitis, including those with parenterally or enterically transmitted infections. Until these agents are characterized further and their association with disease established, their public health importance and means of prevention remain unclear.

2. Hepatitis A

2.1. Historical Background

Jaundice and other diseases of the liver were described in the writings of a number of ancient societies. Jaundice was generally considered obstructive in origin, although large epidemics associated with military campaigns were described as early as the 17th century and continued through World War II.[346] From the 1940s through the 1950s, extensive efforts to identify the agent(s) associated with viral hepatitis were unsuccessful. Epidemiologic evidence supported two types of hepatitis—infectious hepatitis (i.e., hepatitis A) and serum hepatitis (i.e., hepatitis B)—although a clear distinction could not be made between the two agents until transmission studies were conducted in humans.[184,346] These studies, conducted between 1940 and 1965, clarified clinical and epidemiologic distinctions between the two forms of viral hepatitis, showed the lack of cross-immunity between the two types, and established the efficacy of IG in preventing infectious hepatitis.[184,185,346]

In the early 1970s, the discovery of HBV and development of assays to identify HBV infection further differentiated these two forms of viral hepatitis. However, the critical advance occurred in 1973 when virus particles were identified in stool samples of patients with hepatitis A.[107] Over the next decade, the immunologic and virological aspects of the natural history of HAV infection were defined in nonhuman primates, and the epidemiology of HAV infection was defined using newly developed assays that differentiated acute and resolved infections. HAV was first grown in cell culture in 1979, and vaccines developed from cell culture-derived virus have been shown to effectively prevent hepatitis A. Our ability to eliminate the morbidity associated with hepatitis A awaits the development, implementation, and evaluation of effective vaccination strategies.

2.2. Methodology Involved in Epidemiologic Analysis

2.2.1. Mortality. HAV infection has a low case fatality rate, and mortality from hepatitis A is a poor indicator of disease incidence. Although viral hepatitis has been included in the International Classification of Diseases (ICD) for many decades, neither serological testing nor appropriate coding to distinguish different viral etiologies were available until 1980. The ICD-coded data on hepatitis A mortality should become more reliable in developed countries where serological testing is widely available. Because accurate serological testing is not widely available in developing countries, mortality statistics are unreliable and remain poor indicators of disease incidence.

2.2.2.   Morbidity.   Viral hepatitis is a reportable disease in almost all countries. However, several factors have limited the usefulness of morbidity data in assessing the impact of HAV infection. In developed countries, the reporting of hepatitis A and hepatitis B as different diseases only occurred in the 1970s, and it was not until 1980 that a diagnostic test for hepatitis A became available commercially. In less-developed countries, serological testing is not widely available and each type of viral hepatitis is not reported separately. The completeness of reporting is not known in essentially all countries and probably varies most widely in less-developed areas. Hence, it is not possible to make strict comparisons of reported rates of hepatitis A from different geographic areas.

In spite of these well-recognized limitations, cyclical patterns of increased disease incidence and trends of disease over time are reasonable indicators of the epidemiology of hepatitis A. Data derived from studies using serological testing to define the age-specific etiology of acute cases of viral hepatitis in specific locations can be combined with reported age-specific incidence data to estimate the overall morbidity resulting from hepatitis A.[127]

2.2.3.   Survey and Other Epidemiologic Studies. Epidemiologic studies that used serological assays to differentiate acute from resolved HAV infections have greatly increased the understanding of this disease. Recently, nucleic acid variation within the region of the HAV genome that codes for the capsid (surface) proteins has been used as a marker in epidemiologic studies.[242,273]

Cross-sectional studies of the prevalence of HAV infection in population samples stratified by age, and in some cases by socioeconomic status, have been completed in many parts of the world and provide the most accurate picture of HAV infection.[114,127] Serological testing of representative samples of acute hepatitis cases in children (less than 15 years old) and in adults have been used to define the proportion caused by HAV. In less-developed countries, these data can be combined with

reported age-specific rates of acute hepatitis to estimate disease incidence. Studies to define the risk of HAV infection in certain groups or populations have primarily been confined to persons traveling to countries with high rates of infection. Few data are available to define the current risk of infection in homosexual men, drug users, the developmentally disabled, children attending day care, persons working in day-care settings, or health care workers.

Mechanisms or pathways of disease transmission have been defined by serological testing of persons ill or exposed in outbreaks, and in some instances by analysis of genetic variation of HAV isolates. Serological testing, or HAV-RNA detection and analysis, have improved our knowledge of the epidemiology of common-source outbreaks, community-wide outbreaks, outbreaks associated with day-care centers and institutions for the developmentally disabled, and outbreaks occurring in hospitals.

2.2.4.   Laboratory Methods

2.2.4a.   Isolation and Identification of the Organism.   Primary virus isolation from clinical or environmental specimens (i.e., feces, water, shellfish) has become possible using various primary or continuous cell lines. However, because of the slow growth characteristics of HAV, very long (up to 120 days) adaptation periods are required before either infectious foci or HAV antigen can be detected.[85,194,263] HAV generally does not produce a cytopathic effect in cell culture,[85] and detection of virus requires special techniques such as the radioimmunofocus assay (RIFA), which uses radiolabeled antibody to detect infectious foci of HAV in fixed cells.[85,194] The inability to rapidly culture HAV has precluded the use of this method for the diagnosis of infection or for detection of HAV in environmental samples.

Immune electron microscopy (IEM) was used extensively in early studies to identify HAV,[107] but has been supplanted by more sensitive techniques. These include immunoassays to detect HAV antigen [331] and detection of HAV-RNA following amplification by the polymerase chain reaction (PCR), the latter being the method of choice for detection of low levels of HAV in clinical and environmental samples.[85,158,273] However, the presence of HAV-RNA may not always equate with the presence of infectious virus.

2.2.4b.   Serological and Immunologic Diagnostic Methods.   In the 1980s, immunoassays to detect antibodies to HAV structural antigens (anti-HAV) became available commercially, including those that detected both IgG and IgM antibody (i.e., total anti-HAV), or IgM anti-HAV, and allowed the differentiation of past and current (within prior 6 months) infections.[163,202] To better evaluate the effectiveness of hepatitis A immunization, quantitative assays for anti-HAV with increased sensitivity have been developed, as well as cell culture assays that detect low levels of neutralizing antibody.[193] In addition, nonstructural proteins produced during HAV replication elicit a host-immune response that may be useful in detecting breakthrough infections in persons previously immunized with inactivated hepatitis A vaccines.[274]

2.3. Biological Characteristics of the Organism

HAV is a small (27 nm) noneveloped RNA virus belonging to the family Picornaviridae. The virus has icosahedral symmetry, a buoyant density of 1.33 g/cc, and contains a single-stranded positive-sense RNA of approximately 7500 nucleotides. The virion is composed of at least three major structural polypeptides: VP1, VP2, VP3 (33,000–22,000 Da). Based on the nucleotide sequence, VP4, a protein of approximately 2500 Da should be encoded but has not been identified in mature virions. The genomic organization and replication of HAV appear similar to that of polio and other picornaviruses; viral RNA encodes a single large polyprotein from which structural and nonstructural proteins are subsequently cleaved.[117,192,331]

When compared to other enteroviruses, HAV has essentially no nucleotide or amino acid homology, has not been shown to have an intestinal replication phase, replicates mores slowly in cell culture, and is more resistant to heat inactivation. Although HAV was initially classified in the genus *Enterovirus* (enterovirus 72), it has been reclassified in a separate genus designated *Hepatovirus*.[233]

HAV is stable in the environment, retaining infectivity in feces for at least 2 weeks and having only a 100-fold decline in infectivity over 4 weeks at room temperature.[85,192,222] The virus resists extraction by nonionic detergents, chloroform, or ether, and retains infectivity in pH 1.0 at 38°C for 90 min. HAV is more resistant than poliovirus to heat, being only partially inactivated at 60°C for 1 hr.[85] When suspended in milk at 62.8°C for 30 min, 0.1% infectivity remains, suggesting that pasteurization may not completely inactivate HAV and temperatures of 85–95°C for 1 min are required to completely inactivate HAV in shellfish.[85,231,248] HAV is completely inactivated by formalin (0.02% at 37°C for 72 hr), but appears to be relatively resistant to free chlorine, especially when the virus is associated with organic matter.[85] An outbreak of hepatitis A among swimmers suggested that if a prescribed free chlorine level of 0.3–0.5 ppm existed in the swimming pool, it failed to inactivate HAV in fecally contaminated water.[204] Only sodium hypochlorite, 2% glutaraldehyde, and quaternary ammonia compound (QAC)

with 23% HCl have been effective in reducing the titer of HAV by more than 10^4 on contaminated surfaces.[220]

HAV exists as a single serotype, and monoclonal antibodies that identify overlapping neutralization epitopes react with virus isolates from all parts of the world.[192] The single immunodominant neutralization epitope appears to be highly conformational and is composed of several sites located on VP1 and VP3.[85,192,238,254] Although a high degree of nucleic acid conservation exists among HAV isolates, nucleotide variation in selected regions of VP1 and VP3 has been used to define four human HAV genotypes and three genotypes comprised of HAV isolated from Old World monkeys.[273]

Besides man, the host range of HAV includes nonhuman primates (marmosets, tamarins, owl monkeys, chimpanzees) that have been infected experimentally and have served as animal models of human HAV infection.[263] In addition, several species of Old World monkeys have been shown to be infected with an HAV that is serologically similar to the human virus but genetically distinct and that does not appear to infect humans.[85,239,312]

In 1979, cell culture of HAV in fetal rhesus monkey kidney cells was achieved only after virus was passaged multiple times in marmosets.[85,262] Since then, HAV has been cultivated directly from clinical or environmental samples, but long (4–10 weeks) adaptation periods have been required for detection of significant amounts of HAV antigen in infected cells. The lack of a cytopathic effect allowed serial passage of virus that ultimately produced strains that replicate more rapidly and produce higher virus yields.[85] HAV that has been adapted to grow efficiently in cell culture appears to have mutations in the 2B and 2C areas of the genome that codes for nonstructural proteins.[85] In addition, cytopathic variants of HAV have been isolated in several laboratories, and similar to polio, these variants produce plaques in cell culture and have proven useful in laboratory studies.[85,86]

2.4. Descriptive Epidemiology

Prior to identification of its specific viral etiology and the development of diagnostic tests, hepatitis A was presumed to be the cause of most sporadic and epidemic hepatitis worldwide. While large epidemics of hepatitis A occur in developing countries, it appears that hepatitis E is the cause of epidemics of hepatitis that involved large numbers of people, especially in India and central Asia. Nevertheless, hepatitis A remains an important cause of hepatitis in most countries.

The endemicity of HAV infection is determined by

its predominantly fecal–oral route of transmission. As with all enterically transmitted organisms, endemicity of infection and groups at risk of infection are related inversely to sanitation and hygienic standards and socioeconomic conditions.[127] The clinical expression of HAV infection (i.e., hepatitis A) is highly age dependent. Serological studies have confirmed that less than 10% of infected children under age 6 will become jaundiced, whereas in adults and children above age 5, infection usually causes jaundice in 50–90% of cases.[133,189,192]

2.4.1. Prevalence and Incidence. The age-specific prevalence of anti-HAV worldwide can be used to define several patterns of infection (Fig. 1). Areas with a very high endemicity of infection primarily consist of less developed and developing nations of Asia, Africa, South and Central America, the Pacific Islands, and certain populations within the United States. In these countries and within ethnically defined populations in the United States, the prevalence of HAV infection in adults reaches 90% or higher, almost all older children have serological evidence of prior infection, and most children become infected by age 10 years.[97,106,127,300] However, persons of upper socioeconomic status may not become infected until they are adolescents or young adults.

In more developed countries in Europe and Asia, the endemicity of HAV infection is intermediate to low, and the prevalence of anti-HAV varies widely.[114,127] In countries such as Greece, Italy, and Taiwan, prevalence of past HAV infection in adults reaches 80–90%; but in children under 10 the prevalence is only 20–30% and the major

increase in prevalence of infection occurs between 10 and 19 years.[151,181,298] In Europe and the United States, anti-HAV prevalence in young adults varies from 30 to 70% but is less than 10% in children under age 10. However, low socioeconomic status is associated with high rates (endemicity) of infection.[282]

In some northern European countries and in Japan, HAV infection appears to be disappearing, and the endemicity of infection is considered very low. Thirty to sixty percent of adults older than age 40 have been infected, but the prevalence of infection is less than 10% in young adults and almost nil in children. In both very low and low endemicity areas, there is evidence that a cohort effect accounts for the high prevalence of infection among adults, indicating that HAV infection was much more common in the past, especially among persons who were born prior to World War II.[127]

Data on disease incidence, although often limited by lack of a serological diagnosis, corroborate the patterns of HAV infection. In those countries with a very high or high endemicity of infection, rates of reported hepatitis A vary from <10 to >300 cases per 10^5 per year.[127] In these areas, serological studies of the etiology of acute hepatitis usually show HAV infection to be the predominant cause (up to 90% of cases) of disease in children younger than age 15 years, and to cause little or no disease in adults. In areas of intermediate endemicity, studies in adults show that relatively high proportions (50–80%) of cases of acute hepatitis are due to HAV infection.[127] In low endemicity areas such as western Europe and the United

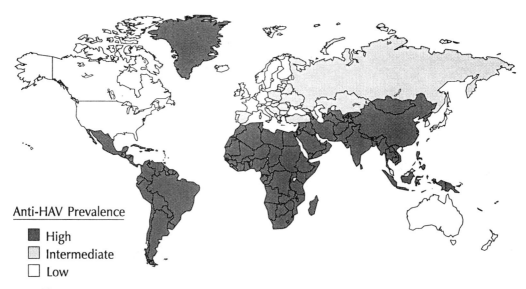

Figure 1. Worldwide distribution and endemicity of hepatitis A virus infection. Adapted from Hadler.[127]

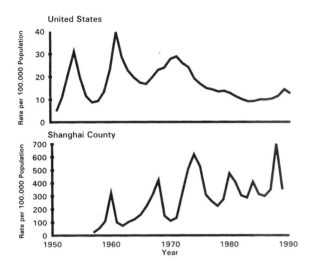

Figure 2. Incidence patterns of reported cases of hepatitis A, United States and Shanghai County, People's Republic of China.

States, the majority of acute hepatitis cases in children are caused by HAV, but in adults this proportion varies from 10 to 50%.[59,127] Rates of hepatitis A are not inherently stable, and periodic widespread epidemics appear to occur in high, intermediate, and low endemicity areas and produce cyclical patterns of disease incidence (Fig. 2).

The United States has a low endemicity of HAV infection based on prevalence of infection and disease incidence. Serological studies have shown low prevalence of infection (less than 10%) in children, increasing up to 60% in older adults, with a cohort effect contributing to the high prevalence in adults.[281] However, infection and disease rates vary markedly in various populations defined by socioeconomic status and ethnicity.[281,282] Some of these populations have high endemic patterns with high infection rates in young children and predictable periodic community-wide epidemics that affect older children and adolescents.[283] Since 1980, serological testing for HAV infection has been employed to define the etiology of acute hepatitis and has greatly increased the reliability of surveillance-based data. In 1992, 23,112 cases of hepatitis A were reported in the United States (incidence rate of $9.05/10^5$ per year) and represents a decline from the most recent epidemic cycle which peaked in 1989 (Fig. 2).[59] Nationally, disease rates may vary over 60-fold within counties in a state, and are highest in those counties with epidemiologic characteristics that include large Hispanic and American Indian populations.[281,282] The most often identified sources of disease transmission include contact with a recognized case of hepatitis A (24%), enrollment or contact with children in day care (14%), and travel to

countries with high rates of HAV infection (6%).[281,282] Fewer than 5% of cases are associated with food- or waterborne outbreaks, and a source of infection cannot be identified in about 50% of cases (Fig. 3).

2.4.2. Epidemic Behavior and Contagiousness. Over the past 20–40 years, nationwide epidemics of hepatitis A have largely disappeared in most countries with a low endemicity of infection, while epidemic cycles continue to occur in countries with either a high or intermediate endemicity. Determinants of the cyclical recurrence of epidemics include the age-specifc proportion of the population susceptible to infection, the interepidemic rate of HAV infection and the likelihood that HAV will be introduced into the susceptible population. Several patterns appear to produce these epidemic cycles, and include: (1) high rates of asymptomatic infection in children under age 5, who serve as the reservoir for infection of susceptible older children; a pattern observed in American Indian populations[283]; (2) periodic introductions of HAV into largely susceptible isolated populations where epidemics occur due to living conditions that facilitate virus transmission and where the high attack rate results in the disappearance of hepatitis A during interepidemic periods; a pattern observed in Alaskan Native villages and some Pacific Island populations[151]; and (3) the periodic introduction of HAV through vehicles such as food in populations where the standard of living has improved and a large proportion of the young adult population remains susceptible, a pattern observed in Shanghai, China, where over 300,000 cases of hepatitis A occurred in young adults following the consumption of shellfish contaminated with HAV.[137]

Community-wide epidemics, with high rates of infection among children and affecting persons of low socioeconomic status, remain common in the United States

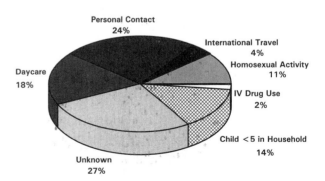

Figure 3. Sources of infection for reported cases of hepatitis A in four sentinel counties, United States, 1991. Source: Sentinel Counties Study of Viral Hepatitis, Centers for Disease Control and Prevention.

and usually occur in 5- to 10-year cycles.[282,285] Smaller outbreaks continue to play a role in disease transmission in the United States and elsewhere. Common-source outbreaks related to contaminated water, shellfish harvested from sewage-contaminated areas, or food contaminated by an infected food handler have been recognized for many years.[55,59,85,92,242,275] Outbreaks among homosexual men and drug users have been recognized in recent years, with infections in the latter probably occurring from person-to-person spread or occasionally from common-source contamination of drugs, although HAV can be transmitted parenterally.[60,82,85,279] In addition, outbreaks continue to occur among adult contacts (parents, caretakers) of diapered children in group day-care settings.[59,133,282]

2.4.3. Geographic Distribution. As described previously, HAV infection occurs worldwide with endemicity patterns defined by the socioeconomic development of the nation (Fig. 1). As the socioeconomic status of the population increases, endemicity declines and the median age of infection increases. In the least-developed countries, HAV transmission occurs exclusively during early childhood, almost the entire population becomes infected before age 10 years, and clinical disease is recognized only in older children. As hygiene/sanitation standards improve, the age of infection shifts progressively to include older children and young adults, particularly in upper socioeconomic groups. This may result in higher rates of clinical disease, and common-source outbreaks and/or nationwide epidemics appear if a substantial proportion of adults remain susceptible to infection. As sanitation and socioeconomic conditions improve further, the infection frequency in all age groups and overall rates of disease decrease; clinical disease is recognized in older children and young adults, and localized outbreaks may occur. Finally, in some developed countries infection rates decline in all age groups and cases of disease occur primarily as importations from areas having a high endemicity of infection. This model, developed to describe polio virus epidemiology in the 1950s, appears to be applicable to HAV infection.

2.4.4. Temporal Distribution. Long-term data from Europe and the United States have shown a marked change in patterns of reported incidence of disease in the past 50 years. Prior to and immediately following World War II, hepatitis rates were high, and nationwide epidemics, presumably of hepatitis A, occurred at 6- to 10-year intervals. In addition, seasonal fluctuations of disease occurred, with peaks in the late summer and early fall. Following World War II, disease rates began to decrease in northern Europe and epidemic cycles disappeared, as did seasonal variation in disease rates. In the United States, epidemic cycles continue to occur but with a lower peak incidence and longer interepidemic periods (Fig. 2). It is believed that improvement in overall sanitation and hygienic standards has led to these changes.

2.4.5. Age. In the United States, the highest incidence of hepatitis A has been in the 5- to 14-year-old age group, but there has been an increase in incidence in older age groups since the 1971 epidemic. Currently, about 60% of cases occur in adolescents and young adults aged 15 to 39 years, and 25% occur in children under age 15 years.[59,281,282] In outbreaks, characteristic age-specific disease patterns are evident. In community-wide outbreaks affecting specific ethnic or low socioeconomic groups, children aged 5 to 14 years have the highest attack rates.[281,283,285] A bimodal distribution may be seen in day-care outbreaks, with children aged 5 to 9 years and adults 25 to 35 years most commonly affected.[133]

Similar age-specific disease patterns occur in countries with a low endemicity of infection in Europe and Asia, although in Scandinavian countries disease is largely limited to adults, since the majority of infections are imported. In populations with a high endemicity of infection, cases occur primarily in children, as all adults are immune as a result of infection in childhood.

2.4.6. Sex. In the United States, the sex ratio among reported hepatitis A cases has been 1.3–1.5 : 1 (male : female) in recent years.[59] This modest male predominance probably represents unequal distribution in persons at risk of infection from drug use, homosexual activity, and international travel.

2.4.7. Race. Hepatitis A occurs in all racial groups, and race per se is not believed to predispose to infection except where related to socioeconomic status or country of origin. In the United States, the prevalence of HAV infection is higher in blacks, hispanics, and American Indians/Alaskan Natives than in whites, but differences largely disappear when controlled for socioeconomic status.[281,282] Prevalence of infection is high in Asian/Pacific Islanders because of birth in areas with a high endemicity of infection.

2.4.8. Occupation. There does not appear to be an increased risk of HAV infection associated with a particular occupational group, although outbreaks of hepatitis A have occurred in persons working with certain nonhuman primates. Employment in a day-care center is a reported source of infection, but no studies have shown that this occupational group has an increased risk of infection compared to the general population. Susceptible persons from low endemic countries who work in countries with a high endemicity of HAV infection are at risk of infection, and this risk increases the longer they reside in the country.[188,292] Studies of health care workers show that they are not at increased risk of infection.[122,217]

2.4.9. Military and Other Settings. Hepatitis A among military personnel stationed in low endemicity countries does not appear to differ from that in the surrounding community. Outbreaks caused by contaminated food, exposure to children in day-care centers, and sporadic disease from person-to-person contact have been reported. However, hepatitis A poses a risk among troops deployed to parts of the world where HAV infection is of high or intermediate endemicity.[24]

2.5. Mechanisms and Routes of Transmission

Studies in humans and experimentally infected non-human primates have shown that HAV is excreted in large quantities (up to 10^8 infectious units per ml) in feces during the late incubation period and first week of clinical illness. Some studies suggest that children may shed virus longer than adults, but there is no evidence of chronic virus excretion.[56,263,304] Viremia occurs from the middle of the incubation period into the early clinical illness; infectivity titers of serum are 10^3- to 10^6-fold lower than those of feces. Virus is also present in saliva with an infectivity that may be 10^4 lower than serum.[81,263] These data are firmly supported by epidemiologic data that have implicated the fecal–oral pathway as the predominant route of HAV transmission and occasionally demonstrated blood-borne transmission.

Ingestion of infectious feces or fomites transferred from person-to-person, or ingestion of contaminated food or water are the major routes of HAV transmission. In addition, blood-borne transmission occurs rarely and can lead to secondary outbreaks among health care workers due to inapparent fecal–oral spread of the virus.[243,276] Transmission among homosexual men has been well documented; whether this occurs through sexual contact or simply by nonsexual intimate contact is not certain.[60,82]

Common-source outbreaks have occurred from contamination of food and various sources of water, including streams, individual wells, or community supplies.[38,85,192] Food-borne outbreaks usually result from contamination by a food handler during the disease incubation period, and poor hygiene and diarrhea enhance the risk of transmission.[55] Implicated foods are usually handled extensively after cooking and/or are eaten uncooked. Such foods have included salads, sandwiches, glazed or iced pastries, and in the past some dairy products.[55,85,196] In addition, foods may become contaminated during growth, harvest, or processing. Outbreaks have been reported throughout the world when filter-feeding bivalve mollusks (clams, oysters, mussels) harvested from sewerage-contaminated waters have been eaten with little or no cooking.[85,92,100,137] In addition, outbreaks have occurred from foods contaminated at the time of harvest or processing that are subsequently served raw, and have included lettuce, strawberries, and raspberries.[85,242,275] Nevertheless, in the United States, disease traceable to contaminated food or water constitutes less than 5% of the overall disease burden.[59,281]

Direct person-to-person transmission by the fecal–oral route accounts for most infections in all parts of the world. Susceptible household contacts have a 10–20% risk of acquiring infection from a family member with acute illness.[192,321] Person-to-person transmission is the predominant pathway in the day-care setting, in community-wide outbreaks, and among homosexual men. The common denominator of close contact in settings with less than optimal hygiene accounts for the ease of HAV transmission.

Transmission in the day-care setting has received close study and provides a model for person-to-person disease transmission.[133] In centers enrolling children in diapers, hepatitis A outbreaks may be common and involve transmission not only among young children but also among adult contacts in the center, at home, or in the community. Infected children are rarely jaundiced, but they transmit HAV to adult contacts who are more likely to become symptomatic and comprise 70–80% of recognized cases. Adult contacts of 1- to 2-year-old children are at highest risk of infection; risk of transmission decreases with increasing age of the child, and contacts of older children (age 5 to 6 years) are not commonly affected. In centers not enrolling children in diapers, outbreaks are uncommon, and spread following the introduction of infection is limited. HAV spreads rapidly but silently among mobile, fecally incontinent children in day care and subsequently to their contacts.

The day-care model indicates that asymptomatic HAV-infected children are highly infectious and efficiently spread disease. Among persons without an apparent source for their HAV infection, almost 50% have a child under 5 years of age residing in the household. This suggests that much community-acquired hepatitis A may be acquired from inapparently infected young children.

Transmission of HAV by blood occurs infrequently and represents a rare cause of posttransfusion hepatitis.[243,276] Outbreaks of hepatitis A in hospitals have been traced to an index case (often an infant) who acquired infection through transfusion. Transmission to hospital staff exposed to feces through breaks in infection control practices have occurred in neonatal intensive care units, with silent transmission among infants and high attack rates among hospital staff.[243,276]

2.6. Pathogenesis and Immunity

The immunologic, virological, and clinical events that occur during HAV infection have been characterized in experimentally infected nonhuman primates and naturally infected humans. The incubation period may range from 14 to 45 days, with a median of 28 days observed in common-source exposures. Studies in experimentally infected nonhuman primates suggest that the incubation period is lengthened with the inoculation of a lower dose of virus or shortened when inoculated parenterally, but that the clinical severity of disease is not dependent on dose and route of inoculation.[210,263] Hepatocytes are the primary site of HAV replication, and intestinal replication has not been identified using highly sensitive detection methods.[180,305] HAV antigen is found primarily in the cytoplasm of hepatocytes, but can also be found in liver macrophages (Kupffer cells). HAV appears in hepatocytes prior to detection in feces and prior to the onset of liver enzyme elevations. HAV is excreted via the biliary system into the feces where it appears in high concentrations from 1 to 2 weeks prior to onset of clinical illness. Virus excretion begins to decline at the onset of clinical illness, and in most persons it has decreased substantially 1 week later. A modest proportion (less than 33%) of those infected may continue to excrete virus during the second and third week of illness, and children may excrete virus for longer periods than adults.[56,276,304] Approximately 15–20% of cases may experience a relapse of symptoms 2 to 8 weeks after the initial illness, and virus excretion may occur during this time.[123,287] Viremia is present from the middle of the incubation period until early in the clinical illness.

In tissue culture, HAV can replicate and be released without cell damage, which suggests that hepatocellular injury is immune mediated. No cellular injury occurs during the high rate of viral replication that occurs early in the infection. Several studies suggest that cytotoxic T cells are involved in the resolution of HAV infection and may contribute to the observed hepatocellular injury.[317] Peripheral-blood lymphocytes have been shown to produce interferon when exposed to HAV-infected cells, and γ interferon production by T lymphocytes has been found during acute infection. Although serum complement levels drop during infection, and complement has been shown to bind to HAV capsid proteins, it is not clear whether complement-mediated cellular injury occurs.[210]

Coincident with the onset of illness, there is a humoral immune response to HAV structural (capsid) proteins. This initially involves the production of IgM antibodies, although IgG- and IgA-class antibodies may also be detected.[192,202] Neutralizing antibodies may be present at or before the onset of illness. However, HAV in feces continues to remain infectious.[193] In addition to the antibody response to structural proteins, antibodies to virus-specific nonstructural proteins are also produced.[274] The virus-specific antibody response is also accompanied by a polyclonal rise in serum immunoglobulins, especially IgM, which is characteristic of HAV infection.

The IgM anti-HAV response is relatively short-lived; by the most sensitive assays such antibodies disappear in 2 to 12 months. Commercial assays for IgM anti-HAV usually are positive within 1 week of onset of illness and become negative after 6 months. After infection, IgG anti-HAV probably persists for life. Nevertheless, serological surveys in countries with a high endemicity of infection show decreasing prevalence of antibody in the oldest age groups.[300,321] However, seronegative individuals in these age groups appear to be protected from clinical disease, and after exposure to HAV they develop IgG but not IgM antibodies.[321] Studies with hepatitis A vaccines indicate that the level of anti-HAV that protects against infection is approximately 20 milli-international units per milliliter (mIU/ml), a level that is not detected by most commercially available antibody tests, unless they have been modified. Currently available immunoassays do not detect antibody in persons receiving passive immunization with immune globulin, although cell culture neutralization assays and epidemiologic studies indicate protective antibody levels. During the early phase of active immunization a substantial proportion of vaccinees are antibody negative as determined with commercially available immunoassays; yet, more sensitive assays detect neutralizing antibody. These findings probably explain the observed lack of infections among older persons in high endemicity areas who have no detectable anti-HAV.

2.7. Patterns of Host Response

2.7.1. Clinical Features.
The spectrum of acute disease due to HAV infection is comparable to that of other types of viral hepatitis and ranges from few or no symptoms to fulminant hepatitis. Important features include the relationship of age to clinical expression of infection and the absence of chronic disease.

The age spectrum of clinical disease has been clarified through serological studies of HAV infection. Children under age 6 generally have mild, often nonspecific, symptoms that include nausea and/or vomiting, malaise, diarrhea in 50–70%, and fever or dark urine in 30–50%.[133] Among those under age 3, fewer than 5% become icteric, compared to 10% of those ages 4 to 6 years. Among infected adults, the majority develop the classic symptoms of malaise, nausea and/or vomiting, loss of appetite, and >75% develop the specific signs of dark urine and/or jaundice.[133,189,192] Data are less complete for children aged 6 to 14 years, but suggest that symptom patterns are comparable to those of adults.

Clinical symptoms of hepatitis A are indistinguishable from those of other types of viral hepatitis. Rapid symptom onset and diarrhea are more characteristic of hepatitis A than hepatitis B, but symptoms cannot be used to distinguish types of hepatitis. Extrahepatic manifestations are uncommon, although arthritis and rashes have been reported prior to onset of illness, and arteritis, meningoencephalitis, and aplastic anemia have been reported following HAV infection.[192]

Hepatitis A accounts for 1.5–27% of cases of fulminant hepatitis in developed areas.[2,214] Hepatitis surveillance in the United States has shown the fatality rate among reported cases to be age dependent. The highest rates are among children under age five (1.5 per 1000 cases) and in persons over age 40 (27.0 per 1000 cases), especially among those with underlying chronic liver disease.[127]

There is no evidence that HAV either causes chronic disease or produces a chronic infection. However, about 15–20% of persons with hepatitis A may have a relapse of illness, and virus shedding during the relapse suggests that these patients may be infectious.[123,287]

2.7.2. Diagnosis. Diagnosis of hepatitis A requires demonstrating liver inflammation or dysfunction by standard biochemical tests (e.g., SGOT, SGPT, bilirubin) and the presence of IgM anti-HAV. Because of the higher frequency of nonspecific symptoms or anicteric infection in young children, the index of suspicion should be high in particular circumstances (i.e., those exposed in day-care centers), and testing for liver enzymes or IgM anti-HAV should be considered when an adult contact of a child has hepatitis A. Clinical symptoms are inadequate to distinguish hepatitis type, and virus-specific serological testing must be done for all cases of presumed viral hepatitis. Commercially available IgM anti-HAV tests remain positive in virtually all cases tested within 2 months of disease onset and allow for the diagnosis based on a single acute or early convalescent serum specimen. Testing for anti-HAV (total) will not differentiate acute from past infection, but is useful for identifying immunity as a result of prior infection or immunization.

2.8. Control and Prevention

Personal hygiene and environmental sanitation to prevent transmission through fecal contamination of food and water or by personal contact have been the primary means to control hepatitis A in any setting. Improved sanitation is presumed to have lowered the incidence of infection and disease in developed countries. In less-developed countries, improvements in sanitation may be expected to shift the average age of infection from young children to older age groups and may paradoxically result in an increase in clinical disease rates and the appearance of epidemics.

Until recently, passive immunization with pooled human immune globulin (IG) containing high titers of anti-HAV has been the only means to provide preexposure or postexposure immunoprophylaxis against hepatitis A. The development and recent licensure of hepatitis A vaccine in a number of countries offer long-term protection against HAV infection and should change prevention strategies. IG is highly effective in preventing symptomatic HAV infection if given before or within 2 weeks after exposure to the virus.[185,332] If given prior to exposure to HAV, IG prevents almost 100% of infections. However, when given after exposure, IG is only 75–85% effective in preventing hepatitis A; infection may occur, but symptoms are attenuated and fecal shedding of virus may be limited.[185,332]

In the United States, IG is primarily recommended, at a dose of 0.02 cc/kg, for postexposure prophylaxis of household and sexual contacts of acute cases of hepatitis A, and for children and staff exposed to hepatitis A in day-care settings or closed institutions. IG is rarely recommended for common-source outbreaks because detection of the outbreak invariably occurs 3 or more weeks after the exposure occurred.[55,64] Use of IG in developing countries is often patterned after recommendations in the United States; however, since most adults in these countries are immune, such use is often inappropriate, and these areas should develop recommendations for use of IG specific to their own disease transmission patterns.

The propagation of HAV in cell culture has allowed the development of inactivated and live attenuated hepatitis A vaccines. Inactivated hepatitis A vaccines have been produced using cell culture-derived HAV that has been formalin-inactivated, as was done for inactivated polio vaccine.[286] These vaccines have proven to be highly immunogenic in adults and children when given in a one-, two-, or three-dose schedule, achieving seroconversion rates approaching 100%.[19,69,155,329] However, the immunogenicity of inactivated hepatitis A vaccine is blunted somewhat by preexisting antibody, such as occurs when vaccine is given with IG or when given to infants of previously infected mothers.[69,190] The clinical consequences of these observations are not known, especially with regard to routine vaccination of infants against hepatitis A.

The high efficacy of these vaccines in preventing hepatitis A in children has been shown in placebo-controlled clinical trials, and in one study a single dose of vaccine was shown to provide complete protection against disease.[156,329] Antibody following immunization has been shown to persist for at least 5 years, and models of antibody decay suggest that protective levels of antibody may last for up to 20 years.[69] However, the number of doses of vaccine required to provide lasting immunity has not been determined, and follow-up studies will be required to define the duration of vaccine efficacy. It is not known whether hepatitis A immunization will provide effective postexposure protection. Limited studies in the chimpanzee model of HAV infectivity suggest that although hepatitis A vaccination 1 to 3 days following inoculation did not prevent infection in all instances, it may attenuate disease expression and prevent fecal excretion of virus.[272] However, appropriately designed clinical trials are required before hepatitis A vaccine could replace IG for postexposure prophylaxis.

Candidate live, attenuated vaccines have been developed using a number of cell culture-adapted strains of HAV.[286] When appropriately attenuated, these vaccines have been shown to have good immunogenicity when administered intramuscularly.[206,230] However, their efficacy in preventing disease has not been evaluated in controlled clinical trials, and the degree to which virus shedding and reversion to wild-type HAV occur has not been evaluated in humans. Ultimately, a number of factors will determine which of the two types of vaccines will be used routinely, including the rate of antibody induction after vaccination, the number of doses required for immunization, the duration of vaccine efficacy or antibody persistence, and the ease with which a particular vaccine can be combined with other vaccine antigens.

Inactivated hepatitis A vaccine recently has been licensed in Europe, Asia, Canada, and the United States. In all of these areas it is recommended for persons (adults, children, the military) who travel or work in countries with a very high, high, or intermediate endemicity of HAV infection, and it is expected that hepatitis A vaccine will replace IG for preexposure prophylaxis in this setting.[69] Most persons would be expected to be protected from HAV infection within 1 month after vaccination. However, IG (0.02 ml/kg) can be given along with hepatitis A vaccine in those persons not vaccinated in sufficient time

to achieve an expected protective antibody response. In addition, for those travelers who choose not to receive hepatitis A vaccine, IG should be given (0.02 ml/kg) for short-term (<3 months) or longer-term exposure (0.06 ml/kg repeated every 4 months).[69]

Widespread hepatitis A immunization has the potential to eliminate this ancient disease when combined with improved socioeconomic status and hygiene. Routine immunization of travelers (adults, children, the military) would be expected to lower infection rates only within this risk group. In some European countries, vaccination of travelers may lower the national incidence of hepatitis A because of the high rate of international travel among adults, the fact that a high proportion of disease is currently attributed to importation by travelers, and the fact that immunized travelers will be protected from community-acquired HAV infection. However, such an immunization strategy would not be expected to lower the overall incidence of HAV infection in a country such as the United States where less than 10% of cases are due to international travel, at least 25% of cases occur in children who serve as a reservoir for infection of adults, and where most infections occur as part of community-wide outbreaks.

In the United States, the public health objective of hepatitis A vaccination is the reduction of disease incidence and possibly eradication of HAV infection through routine infant immunization.[69] However, current studies have not defined an optimum infant vaccination schedule that minimizes the negative effect on immunogenicity of passively acquired maternal antibody. However, for populations that experience periodic epidemics of hepatitis A in children, adolescents, and young adults (i.e., where HAV infection is highly endemic), routine vaccination of children beginning late in the second year of life, combined with "catch-up" vaccination of preschool and school-aged children, is recommended to lower the susceptible population to a level that would not sustain HAV transmission.[69] The ultimate success of routine infant immunization to eliminate HAV transmission is dependent on the availability of an affordable vaccine that can be combined with other antigens and provides long-term protection. Such a strategy would also be appropriate for countries with an intermediate and high endemicity of infection where most persons become infected during childhood.[85,211]

2.9. Unresolved Problems

The demonstrated effectiveness of hepatitis A immunization holds the potential for the eventual eradication of HAV infection. Elimination of HAV transmission occurs when a critical level of herd immunity is present in the population. Population-based studies to define the threshold levels of immunity required to prevent HAV transmission are needed for the development of recommendations for controlling community-wide outbreaks of hepatitis A through immunization. In addition, studies will be needed to determine the long-term efficacy of hepatitis A immunization and the effect of childhood immunization on the incidence of HAV infection in populations with high endemic rates of infection. The production of adequate quantities of vaccine may be limited by the poor growth characteristics of HAV. Development of rapidly replicating strains of HAV, either through clonal selection or site-directed mutagenesis, would facilitate vaccine production. In addition, further studies to define the conformationally correct neutralization site of HAV might allow the production of vaccines by recombinant DNA technology.

3. Hepatitis B

3.1. Historical Background

The existence of parenterally transmitted hepatitis was documented in the late 1800s, but the modern history of hepatitis B begins in the 1930s when the reuse of syringes and needles in venereal disease and diabetic clinics and the injection of measles and mumps convalescent serum were shown to cause jaundice.[346] The transmission of hepatitis from blood products was confirmed in outbreaks of jaundice associated with receipt of yellow fever vaccines containing human plasma and from the transfusion of whole blood.[346] Ultimately the viral etiology of these outbreaks was established through human volunteer studies using filtered plasma from implicated lots of yellow fever vaccine.[99,110,346] About the same time, the existence of two distinct forms of viral hepatitis was suggested, based on the epidemiology of the different modes of disease transmission. However, all attempts to cultivate hepatitis viruses in cell culture or to establish small animal models of disease were unsuccessful. Transmission and cross-challenge studies in human volunteers were required to ultimately differentiate hepatitis B from hepatitis A. The studies conducted by Krugman and associates[184] at the Willowbrook State School showed that the MS-2 strain of hepatitis (i.e., HBV) was almost exclusively transmitted parenterally and differentiated it from the MS-1 strain of hepatitis (i.e., HAV). These studies established the basic tenets of the epidemiology and natural history of HBV infection and laid the framework for its prevention.[346]

In 1965, Blumberg and colleagues described an isoprecipitin in the serum of Australian aborigines,[30] termed Australia antigen, that was later shown to be related to HBV infection.[261] Within a decade following the discovery of Australia antigen, which was ultimately shown to be the surface antigen of HBV, the virus was fully characterized, various antigen and antibody systems were described, and immunoassays for their detection were developed. The availability of serological tests for hepatitis B surface antigen (HBsAg) expanded the understanding of the epidemiology and consequences of HBV infection and rapidly eliminated a major cause of posttransfusion hepatitis through screening of the blood supply.

A number of epidemiologic studies showed an association between primary hepatocellular carcinoma (PHC) and chronic HBV infection, and PHC was shown to be among the leading causes of death in adults worldwide. The realization that HBsAg could serve as an immunogen for the production of anti-HBs, and that this antibody was protective against HBV infection led to the development of prototype hepatitis B vaccines.[142,183,264] Beginning in the 1980s, hepatitis B vaccines were produced and licensed in a number of countries, and by 1990, there was a worldwide effort to eliminate the transmission of HBV and HBV-related chronic liver disease and PHC through routine vaccination of all infants.[71,215,219]

3.2. Methodology Involved in Epidemiologic Analysis

3.2.1. Sources of Mortality Data. Death rates from acute HBV infection are low and cases of fulminant viral hepatitis are usually not distinguished by etiology, especially in less-developed countries. However, PHC mortality data most accurately reflects the incidence of chronic HBV infection, although some portion of cases may be attributable to chronic HCV infection or PHC from other causes. Although testing for HBV markers is not always included as part of the diagnostic workup of PHC, the incidence of this cancer generally serves as an indicator of the relative prevalence of chronic HBV infection. Even in developing countries, the diagnosis of PHC is made with sufficient accuracy to provide a good index of the endemicity of chronic HBV infection, and PHC is included as an ICD diagnosis.

Cirrhosis is a leading cause of death among adults in most developed and developing countries, non-alcohol-related liver disease can be determined from ICD codes, and alcoholic liver disease has been shown to account for <50% of all deaths from cirrhosis in developed countries. In countries with high rates of chronic HBV infection, a large proportion of chronic liver deaths can be presumed to be due to HBV infection, although few studies of etiology-specific causes of death are available.

3.2.2. Sources of Morbidity Data. In countries where most HBV infections occur among adults, acute, symptomatic HBV infection (i.e., hepatitis B) may provide a reliable indicator of the incidence of infection. While the diagnostic reliability of reported cases of hepatitis has increased with widespread use of serological testing, general underreporting and selective underreporting for certain groups at risk of HBV infection may limit the interpretation of this data source.[15] In parts of the world where HBV infection occurs primarily in early childhood, morbidity reporting significantly underestimates the true rate of infection because the majority of infections are asymptomatic. Thus, estimates of the overall incidence of HBV infection cannot be based solely on morbidity (case) data, but must include serological data to define those ages with the highest rates of infection.

3.2.3. Surveys. Extensive serological surveys have been conducted in most parts of the world to define the age-specific prevalence of chronic HBV infection (i.e., HBsAg-positive individuals) and overall prevalence of markers of HBV infection. These data have been used to define three levels HBV infection endemicity (Table 2). The source of data and whether it is representative of the entire population must be ascertained when extrapolating rates of HBV infection for the general population. This is especially true in areas of low endemicity of infection where certain populations may be at higher risk of infection than the general population, and the development of a country-wide profile of HBV infection must take these differences into account.

3.2.4. Laboratory Diagnosis

3.2.4a. Isolation and Identification of the Organism. Although numerous cell lines have been used, human HBV has only been cultured with limited success in lymphoblasts and hepatocytes, but hepatoma cell lines and cultured primary hepatocytes have produced infectious virus following transfection with cloned hepadnaviral DNA. HBV only produces infection in humans and chimpanzees and the host range of other hepadnaviruses appears to be species specific. HBV has been extensively characterized, in spite of the lack of a cell culture system and limited animal models of infection. Its relatively high titers in infected persons or susceptible animals led to the purification and identification of a number of antigens and their corresponding antibodies, and allowed the characterization of the various stages in the natural course of infection.[32,74,143,145,236] HBV has been readily visualized by immune electron microscopy, and viral antigens can be identified in tissue using both immunofluorescent and enzyme-labeled probes. Characterization of the genome led to the identification of HBV-DNA in serum and tissue, and the ability to amplify and sequence HBV-DNA using polymerase chain reaction (PCR)-based methods has led to the identification of a number of genotypic and phenotypic variants.[35,49,119,316,322,341]

3.2.4b. Serological and Immunologic Diagnostic Methods. A variety of immunodiagnostic tests are available that can be used to make the diagnosis of HBV infection.[145] The glycoprotein coat of HBV is designated HBsAg, an antigen that is produced in excess and circulates independent of the virus. The presence of HBsAg in the serum of an individual is indicative of either an acute or chronic infection. Detection of HBsAg has evolved from first-generation immunodiffusion methods to third-generation tests that include the less-sensitive reversed passive hemagglutination assays (RPHA) or the more sensitive radio- (RIA) or enzyme immunoassays (EIA) in a double-antibody sandwich format.[145]

Table 2. Characteristics of Patterns of Endemicity of Hepatitis B Virus Infection[a]

Endemicity of infection	Low	Intermediate	High
Prevalence of chronic infection	0.1–1%	2–7%	8–15%
Prevalence of HBV infection	4–15%	16–55%	40–90%
Perinatal infection[b]	Rare (<10%)	Uncommon (10–20%)	Very common/uncommon[d] (>20%)
Early childhood infection[c]	Rare (<10%)	Common (10–60%)	Very common (>60%)
Adolescent/adult infections	Very common (70–90%)	Common (20–50%	Uncommon (10–20%)

[a]Adapted from Margolis *et al.*,[208] Maynard,[216] and Maynard *et al.*[219]
[b]Infections up to 1 year of age, as estimated percent of total infections.
[c]Infections at ages 1–5 years, as estimated percent of total infections.
[d]In areas where rate of maternal HBeAg positivity is low, perinatal transmission may be uncommon.

While the presence of HBsAg indicates HBV infection, determination of the stage of infection requires additional diagnostic testing or information concerning the duration of test positivity.[145] Chronic HBV infection (i.e., HBV chronic carrier) is identified by the presence of HBsAg for 6 months or the presence of HBsAg without IgM anti-HBc. All HBsAg-positive persons should be considered potentially infectious; however, the relative degree of infectivity can be assessed by other markers such as HBeAg or HBV-DNA.

Antibody to HBsAg (anti-HBs) appears after exposure to HBsAg, either following infection or through vaccination. The test is not routinely used in the evaluation of a person with hepatitis B, but is used to identify persons previously infected with HBV or to determine the immune status of an individual following vaccination.[64,143,145] Anti-HBs concentration is expressed in milli-international units per milliliter (mIU/ml) of antibody.[143] Anti-HBs levels less than 10 mIU/ml represent either biological false positives or nonprotective levels of antibody. Previously unimmunized persons with levels of anti-HBs <10 mIU/ml do not have an anamnestic antibody response following vaccination and are not protected from HBV infection.[111,134]

Hepatitis B core antigen (HBcAg) is a distinct protein that is part of the virus nucleocapsid. HBcAg is not soluble, does not circulate, and a serological test for the antigen is not available.[120] HBcAg is localized in the nuclei of infected hepatocytes and can be identified by immunohistological techniques on liver biopsy specimens.[37,119] Competitive inhibition immunoassays for antibody to HBcAg (anti-HBc) detect both IgM- and IgG-class antibody. Anti-HBc develops soon after the establishment of HBV infection, persists in persons convalescent from infection, and is present in persons with chronic HBV infection.[134,145] This antibody is initially of the IgM class, then switches to predominantly IgG-class antibody that persists along with anti-HBs for the lifetime of persons with a resolved infection. In persons with chronic HBV infection, HBsAg and anti-HBc are both present, the latter being a mixture of IgG- and very low levels of IgM-class antibody. During the early convalescent phase of infection, anti-HBc may be the only serological marker of HBV infection that is detected. This "window phase" of infection occurs as the concentration of HBsAg declines and anti-HBs increases, and may result from the formation of antigen–antibody complexes whose components are not detected by serological testing.[226] Infrequently, this pattern may persist for several years and may be accompanied by ongoing HBV replication but without detectable HBsAg. The presence of anti-HBc alone may

represent decay of anti-HBs that occurs many years after a resolved infection or it may represent a false-positive test.[120,134,177]

Immunoassays for the detection of anti-HBc of the IgM class have become central for the differential diagnosis of acute versus chronic HBV infection.[74] During acute HBV infection IgM anti-HBc becomes detectable almost coincidently with the appearance of HBsAg and persists for 2–6 months. IgM anti-HBc is always present during the "window phase" of resolving acute infection. In the diagnostic workup of acute viral hepatitis, IgM anti-HBc and IgM anti-HAV can differentiate these infections. In addition, IgM anti-HBc serves to identify those persons with acute HBV infection in which HBsAg has become negative by the time of presentation, differentiates the patient with acute HBV infection from those with chronic HBV infection, and identifies patients with chronic HBV infection who may have another etiology for their acute hepatitis. In chronic HBV infection, IgM anti-HBc may persist but at relatively low titers; most of the commercially available tests are configured in such a way as to minimize the likelihood of a positive result in this situation.[74]

A third antigen–antibody system of HBV is hepatitis B e antigen or HBeAg.[120,145,236] HBeAg is a soluble, conformational antigen of HBcAg and its presence is associated with higher titers and likelihood of HBV transmission.[120] HBeAg is present from the onset of acute infection until its resolution. The likelihood that HBeAg is present in persons with chronic HBV infection may vary in different populations because of the age at which chronic infection occurs and the age-dependent rate of HBeAg clearance.[17,212,225] The association between HBeAg and infectivity is related to its association with the presence of HBV-DNA. However, as many as 20% of HBeAg-negative persons have circulating HBV-DNA, including a small proportion with high levels.[191] Determination of HBeAg or anti-HBe should not be considered a part of the routine workup of either acute or chronic HBV infection. It is only useful in selected studies of HBV infectivity or in the clinical management of persons with chronic HBV infection being considered for treatment with interferon.

3.2.4c Nucleic Acid Detection Methods. HBV-DNA in serum can be detected by nucleic acid hybridization or PCR amplification following extraction or immunocapture of HBV.[316,322] These diagnostic methods are primarily used in research; however, monitoring of HBV-DNA has proven useful in the evaluation and management of patients on antiviral therapy for chronic HBV infection. HBV-DNA can be detected in tissue by *in situ* hybridization or following PCR amplification, methods primarily relegated to research situations. In addition, nucleic acid sequence analysis has been used to identify genetic variants of HBV. While only available in research laboratories, nucleic acid sequencing and pattern analysis should be considered in unusual situations of disease transmission or outcome.

3.3. Biological Characteristics of the Organism

Hepatitis B virus is a member of the family Hepadnaviridae, which includes woodchuck hepatitis virus, duck hepatitis virus, and ground squirrel hepatitis virus.[213,299,309] HBV has a very small, circular DNA genome (3200 nucleotides) that is partially double-stranded and replicates through an RNA intermediate that is transcribed by a gene product with reverse transcriptase activity. A notable feature of the virus is its efficient use of genetic information to encode four groups of proteins and their regulatory elements. This is achieved by shifting the reading frames over the same genetic material in order to encode distinct proteins within the same regions.[299,309] The four proteins encoded by the HBV genome are the envelope, the nucleocapsid, the "X," and a DNA polymerase.

Besides limited tissue tropism, narrow host ranges, and the ability to produce a persistent infection, these viruses integrate DNA into host cells and they are oncogenic. During acute and chronic HBV infection, both complete virions (42-nm Dane particles) and free HBsAg (22-nm spheres and tubules) can be found in the serum. The concentration of non-virion-associated HBsAg exceeds the concentration of complete virions by 10^4 or greater.

The HBV genome encodes three envelope proteins: pre-S_1, pre-S_2, and HBsAg. HBsAg is a glycosylated lipoprotein that contains the major site for binding of neutralizing antibody, designated the *a* determinant. In addition, four subtype epitopes have been identified, giving rise to four major serotypes of HBsAg designated *adw*, *ayw*, *adr*, and *ayr*, and additional complex serotypes.[33,83,120] HBsAg subtypes have been shown to have distinct geographic distribution and they have been used in epidemiologic studies to identify patterns of virus transmission.[33,83] Additional clusters of epitopes within HBsAg subtypes have been identified using monoclonal antibodies and have been used to further differentiate HBV transmission patterns in epidemiologic studies.[33] No apparent differences in infectivity or virulence have been attributed to subtypes. Recently, variants of HBV with mutations in the amino acids comprising the *a* determinant have been identified from immunized persons who

subsequently became infected with HBV.[49,57] While the existence of such variants is not unexpected, their true frequency and clinical significance have not been determined, and no epidemiologic studies indicate that these variants have changed vaccine efficacy.

HBV encodes nucleocapsid proteins, which include HBeAg and HBcAg. A precursor protein containing both precore and core sequences is processed in the endoplasmic reticulum to form HBeAg, which is soluble, does not become part of the virion, and is secreted from infected hepatocytes. HBcAg is translated from the core sequence, is the major nucleocapsid protein of HBV, and is not found in serum.[120] A number of mutations have been found in the core or precore regions of the HBV genome. Mutations in the precore gene that stop HBeAg synthesis have been found in chronically infected persons who are HBV-DNA positive but HBeAg negative.[49,322] In some instances, HBV precore variants have been associated with HBV transmission from an HBeAg-negative persons that resulted in fulminant hepatitis.[197,307] However, the mechanism of this type of disease expression and its relation to the observed mutation are unknown. In addition, precore mutations have been identified over time in chronically infected persons, suggesting that they may occur due to immune pressure and might be a means of "escape" from immune surveillance.

Humans appear to be the only natural host for HBV; however, several of the higher nonhuman primates have been shown to be susceptible to infection. The chimpanzee is the primary experimental model for infection, but the gibbon and gorilla have also been shown to be susceptible. In all these species, HBV can produce a chronic infection, but the characteristics and determinants of virus persistence are not clearly defined except on an epidemiologic basis. Age at infection is the most important factor for determining the risk of chronic HBV infection. The risk of chronic HBV infection is 90% in newborns, 30% in children less than 5 years of age, and 2 to 10% in older children, adolescents, and adults.[29,31,116,146,225,294] Male sex, Down's syndrome or depressed cell-mediated immunity from drugs or infection have been shown to increase the risk of chronic infection.

HBV has been shown to retain infectivity for at least 1 month when kept at room temperature, and much longer when frozen. Infectivity is destroyed at 90°C after 1 hr.[174]

3.4. Descriptive Epidemiology

The endemicity of HBV infection varies greatly worldwide (Fig. 4).[106,207,208,216] Endemicity of infection is considered high in those parts of the world where at

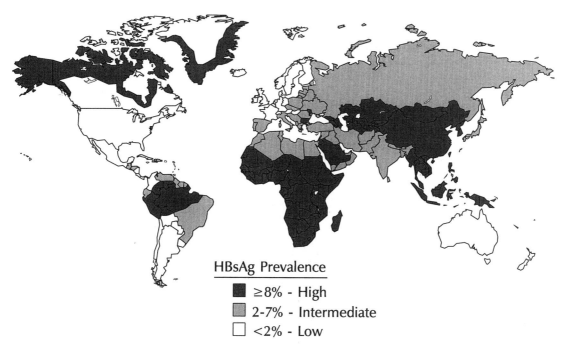

Figure 4. Worldwide distribution and endemicity of chronic hepatitis B virus infection. Adapted from Margolis *et al.*,[208] Maynard,[216] and Maynard *et al.*[219]

least 8% of the population is HBsAg positive. In these areas, 70–90% of the population generally have serological evidence of previous HBV infection. Almost all infections occur either during the perinatal period or early in childhood and account for the high rates of chronic HBV infection in these populations (Fig. 4). Risk of HBV infection continues after the first 5 years of life, but its eventual contribution to the high rate of chronic infection is less significant. Chronic infection with HBV is strongly associated with PHC, and areas with a high endemicity of chronic HBV infection have the highest death rates from this neoplasm.

In most developed parts of the world (Fig. 4) the prevalence of chronic HBV infection is less than 1% and the overall infection rate is 5–7%. Within these areas most infections occur among high-risk adult populations that include parenteral drug users, persons with multiple heterosexual partners, homosexual men, and health care workers.[207,208] The proportion of infant and early childhood infections is low; however, they account for a disproportionately high number of chronic HBV infections.[208,330]

The remaining parts of the world have an intermediate pattern of HBV infection (Fig. 4). The prevalence of HBsAg positivity ranges from 1 to 7%, and serological evidence of past infection is found in 10 to 60% of the

population. In these areas there are mixed patterns of infant, early childhood, and adult transmission.

3.4.1. Prevalence and Incidence. A prevalence of chronic HBV infection ≥8% signifies an area or population with a high endemicity of infection; the overall prevalence of infection usually ranging from 70–90%. In such populations, the majority of acute and chronic infections occur during the first 5 years of life and the annual risk of infection during this period ranges from 5 to 10% (Table 2). Because most early childhood infections are asymptomatic, estimates of infection rates have been obtained from serological studies and rates of reported hepatitis B do not give an accurate picture of the risk of infection. Rates of acute and chronic infection in these areas have remained stable over long periods of time, and any changes would now be attributed to hepatitis B vaccination programs.

In areas with an intermediate endemicity of infection, the prevalence of chronic infection is 1–7% and the overall prevalence of infection ranges from 10 to 60%. The annual risk of infection varies by age and differs by country or population. Secular declines in rate of HBV infection have occurred over the past several decades in countries that have experienced significant socioeconomic changes.[88]

In an area of low endemicity of infection, such as the

United States, the incidence of hepatitis B has generally increased over the past 25 years, reflecting changes in infection rates among various risk groups.[208] In low endemicity areas, most HBV infections occur in adults (Table 2), and changes in the reported incidence of hepatitis B should reflect changes in infection rates. However, the reported incidence of hepatitis B underestimates significantly the rate of HBV infection in infants and children. Large serological studies indicate that at least 5% of the population in the United States has been infected with HBV, and the prevalence of infection varies by racial or ethnic groups, with blacks having a fourfold greater risk of infection than whites.[228]

3.4.2. Epidemic Behavior. Epidemics of hepatitis B are unusual and signal transmission of the virus through direct parenteral exposure. Outbreaks have occurred in the hospital setting when breaks in technique have allowed parenteral exposure through use of shared intravenous fluids or fingerstick devices for blood collections.[256] Transmission of infection from health care workers, either acutely or chronically infected with HBV, to their patients has occurred infrequently and only during the course of surgical or dental procedures.[54,284] Risk factors associated with transmission by the health care worker during invasive procedures have included being HBeAg positive, not using gloves when performing dental procedures, and performing blind intraoperative palpation of suture needles that resulted in fingersticks.[63]

Outbreaks of hepatitis B were common among hemodialysis patients and staff during the 1960s and 1970s because of nonovert percutaneous transmission by blood. Such outbreaks have been largely eliminated through routine HBsAg testing, isolation of HBsAg-positive patients, the use of "universal precautions," and hepatitis B vaccination of staff and patients.[10,65,68]

Outbreaks of hepatitis B have been reported among parenteral drug users. Occasionally these have been associated with a high mortality rates primarily because of coinfection with hepatitis delta virus.[195] Outbreaks related to contaminated blood products, including IG, have been reported from countries where lapses in donor screening and product processing have occurred.

3.4.3. Geographic Distribution. The worldwide distribution of the various patterns of HBV infection has been well defined over the past two decades. As shown in Fig. 4, patterns of HBV endemicity tend to be grouped geographically. Within areas of high endemicity there is usually little variation in disease prevalence, although lower rates of infection may be found among those of the highest socioeconomic status.[88,310] However, there is wide variation in the prevalence of HBeAg positivity;

this variation segregates geographically and determines the age (infancy or early childhood) at which chronic HBV infection is most likely to occur. In most of eastern and southeast Asia, the prevalence of HBeAg among pregnant women is 20–30% and results in high rates of perinatal HBV transmission. However, in Africa and the Middle East, the prevalence of HBeAg in pregnant women is only 5–10% and perinatal HBV transmission is rare.

Within areas of intermediate and low endemicity, the prevalence of infection can vary widely. In a country with an intermediate endemicity of infection such as Israel, widely varying infection and transmission rates reflect the HBV endemicity of the country of origin of immigrant populations.[40] In countries with a low endemicity of infection such as the United States and Canada, Eskimo populations have highly endemic rates of HBV infection,[227] and immigrant populations from high endemic counties continue to transmit HBV infection to their children at a higher rate than the general population.[112,152] In the United States, infection rates also vary by geographic region and probably reflect differences in the contribution of various high-risk populations to the overall infection rates.[208] In South America the endemicity of infection varies from high endemicity in the Amazon basin to low endemicity in the southern part of the continent.[106]

3.4.4. Age. Infants born to HBsAg-positive mothers who are also HBeAg positive have a 70% chance of becoming infected during the perinatal period, and 90% of these infants are at risk of chronic HBV infection.[31,294] The risk of infection in infants born to HBsAg-positive mothers who are HBeAg negative is lower, with 0–10% becoming infected at birth.[294,340] Infants of HBsAg-positive mothers not infected at birth continue to be at high risk of infection (average 30%) over the first 5 years of life, even if their mother is HBeAg negative.[29]

Children are at high risk of becoming infected by 5 years of age if they are born into a family with a chronic HBV carrier.[112,152,258] Although the mother is the most likely source of infection, chronically infected siblings who are usually HBeAg-positive are also a significant source of HBV transmission.[29,112,310] HBV infections acquired during early childhood have a high likelihood of becoming chronic and perpetuate the high rates of chronic infection seen in certain populations. Infections among infants and young children are rarely symptomatic; only 10% of infected children under 10 years of age would be expected to have symptoms consistent with hepatitis B.[225]

The next period for risk of HBV infection occurs among adolescents and young adulthood, with the onset of sexual activity. HBV infections acquired at this age are more likely to be symptomatic, and only 2–10% of infected persons will remain chronically infected.[111,146,208,225]

3.4.5. Sex. Susceptibility to HBV infection has not been related to the sex of infants or children, the likelihood of chronic infection being equal for both sexes. However, a male preponderance of HBV infection is present as these populations become older,[17] and the rates of PHC are highest among men. In areas of low endemicity of infection, such as the United States, the male-to-female ratio among reported adult cases of hepatitis B is almost 2:1 as a result of overrepresentation of males in groups at high risk of acquiring HBV infection (i.e., parenteral drug users, homosexual men).[208] However, among reported cases in teenagers, where the primary mode of transmission is among heterosexual partners, the male-to-female ratio is 1.4:1.[208] In these areas, the higher rates of chronic infection among men probably reflects both their higher risk of HBV infection as well as their biological propensity for HBV persistence.

In areas with a high endemicity of HBV infection, chronically infected women have been shown to have a higher rate of HBsAg clearance with subsequent pregnancies and may partially account for the lower rates of chronic HBV infection observed among adult women.[73]

3.4.6. Race. Although marked differences in rates of HBV infection occur among various racial and ethnic groups, these differences are attributable to epidemiologically defined factors that promote transmission of HBV rather than genetic differences. Infant and early childhood transmission account for the high rates of acute and chronic infection observed in Asian, Pacific Islander, African, and Eskimo populations. In addition, a decline in HBV infection rates has been observed over several generations among Asians who moved from their homeland to a country with a low endemicity of infection. Ethnic Asian women have higher rates of HBeAg positivity, and rates of perinatal transmission are markedly higher than in other racial groups with a high endemicity of HBV infection but with low rates of HBeAg positivity. It is unclear whether these differences are due to genetic factors, the age at HBV infection, or age and frequency of childbearing, since the latter have been shown to effect the clearance of HBeAg.

3.4.7. Occupational Risk. Viral hepatitis has been recognized as a frequent occupational hazard among persons working in laboratories or exposed to blood while caring for patients. Increased risk of HBV infection has been found among health care workers, especially those having frequent contact with blood and/or exposure to needles or sharp instruments.[129] Prior to the availability of hepatitis B vaccine, serological surveys among physi-

cians and dentists have shown a high rate of infection that increases with number of years in practice. Direct parenteral exposures can only be identified as the source of a small proportion of these infections; the majority appear to be a result of inapparent parenteral contacts through breaks in the skin or contamination of mucous membranes with blood or blood-contaminated bodily fluids.

Hepatitis B vaccination and the implementation of barrier techniques to prevent blood exposures ("universal precautions") have markedly reduced the rate of HBV infection among health care workers over the past decade.

Other occupational groups have been identified that have frequent exposures to blood or bodily fluids (i.e., policemen, firefighters, other public safety workers). However, studies in these populations have not shown them to be at increased risk of HBV infection and that their risk of infection comes from nonoccupational sources such as sexual transmission.[61,70,334]

3.4.8. Other Factors. Persons with hepatitis B acquired in selected communities in the United States have been studied to determine the source of their infection.[9,14,208] The groups at highest risk of infection are injection drug users (10–16% of cases), persons with multiple heterosexual partners (12–41% of cases), and homosexual men (14–20% of cases). Sexual transmission of HBV through contact with a known case, multiple heterosexual partners, or homosexual activity is the major identified source of infection in the United States. The remaining groups at risk of infection are health care workers, individuals receiving blood transfusions or blood products, and persons on hemodialysis, although the risk in these groups has diminished markedly due to immunization, donor screening, and intensified infection control practices. The remainder of cases of hepatitis B have no known source for their infection, although they tend to be of low socioeconomic status.[208]

Other groups known to have high rates of HBV infection but who are not represented in overall community-based rates include clients of institutions for the developmentally disabled and prisoners who most often acquire their infection outside of prison due to high-risk behaviors.

3.5. Mechanisms and Routes of Transmission

3.5.1. Direct Parenteral Exposure. Prior to the availability of diagnostic tests, epidemiologic studies attributed the primary source of HBV infection to blood transfusions or accidental contamination of needles and other injected products; thus, the designation of serum hepatitis. Direct transmission via blood transfusion has been eliminated in most countries through donor screening for HBsAg. In some developing countries where HBV is highly endemic, routine donor screening has not been implemented because of the mistaken belief that most recipients of whole blood transfusions would already be immune to infection. However, in these areas children are the primary recipients of blood transfusions and are both more likely to become infected with HBV and more likely to develop chronic infection.[121] In addition, despite donor screening, clotting factor concentrates have been shown to transmit infection unless the manufacturing process includes an inactivation step such as heating or solvent-detergent treatment.

Transmission of HBV has been shown to occur with inadequately sterilized needles and medical instruments or the reuse of needles and syringes without sterilization. However, this has not been a significant mode of transmission in developed countries over the past 25 years, although occasional nosocomial outbreaks of HBV infection have occurred from contamination of multidose vials or automatic fingerstick devices.[256] However, injecting drug users sharing needles still account for a large proportion of cases seen in the United States and other countries with a low endemicity of HBV infection.[195,208] In developing countries, parenteral transmission through contaminated injection equipment remains a problem because of the difficulty in obtaining disposable needles and syringes and the lack of means to adequately sterilize reusable equipment. Immunization programs in most developing countries have taken measures to ensure the use of disposable needles and syringes or the proper sterilization of injection equipment. However, injections given outside of these programs often continue to transmit infection.

One of the most effective modes of transmission occurs at the time of birth in infants of chronically infected mothers. Of infants born to HBsAg-positive/HBeAg-positive mothers, approximately 70% have serological evidence of HBV infection (HBsAg positive) by 6 months of age. The risk of *in utero* infection is only 5–10% and probably represents the low frequency of maternal–fetal bleeding that occurs during gestation rather than true transplacental transmission of virus. The remainder of infections are presumed to occur at the time of delivery via penetration of mucous membranes by HBV. Risk of HBV transmission during the perinatal period correlates directly with the concentration of HBV-DNA in the serum of the mother.

3.5.2. Inapparent Parenteral Exposure. HBV circulates in the blood of a person with acute infection or early in chronic infection at concentrations as high as 10^{10} virions/ml. The primary vehicles for transmission are blood or serous fluids from HBV-infected individuals. Both HBsAg and HBV are present in various body secretions, including saliva and semen, at concentrations approximately 1/1000 lower than that of blood.[25,159,253] The detection of HBV in semen and vaginal secretions provides a mechanism for HBV transmission through both heterosexual and homosexual activity.

Person-to-person spread of HBV can occur in settings involving interpersonal contact over a long period of time, such as when a chronically infected person resides in a household.[34,112,152] In this setting, nonsexual transmission occurs from adults (primarily mothers) to infants, children, and adults, or from children to other siblings or adults. The precise mechanisms of transmission are unknown; however, frequent interpersonal contact of nonintact skin or mucous membranes with blood-containing secretions or perhaps saliva are the most likely means of transmission. Occurrences such as premastication of food; sharing toothbrushes, washcloths, or razors; and the presence of pyoderma or eczematous lesions could facilitate HBV transmission in households. Because of the extremely high concentration of virus in the blood, the number of virions in even very small amounts of blood or body fluids can be quite high. In addition, HBsAg contamination of surfaces is widespread in homes of chronically infected persons,[253] and HBV remains infectious for long periods of time under ambient conditions.

3.6. Pathogenesis and Immunity

Hepatitis B virus must gain access to the circulation either through direct inoculation, by passage through mucous membranes, or through breaks in the integument and must be carried to the liver where primary replication occurs in hepatocytes after an incubation period of 6 weeks to 6 months. Precise definition of the events involved in HBV binding and replication has been hampered due to the lack of a permissive cell culture system. Experimental evidence indicates that the pre-S1 polypeptide of HBsAg contains a hepatocyte attachment site(s), but the corresponding cell receptor has not been identified.[120] In addition, the pre-S2 polypeptide may contain sites involved in hepatocyte binding or insertion either via modified human serum albumin, a glycan-linked peptide, or proteolytic cleavage to expose a fusion peptide.[120] Although HBsAg has been found in other organ tissues, there is little evidence to indicate primary replication in other than hepatocytes. Immunofluorescent studies of biopsies obtained during the course of infection in chimpanzees indicate that early replication occurs in the hepatocyte nuclei with expression of HBcAg limited solely to the nucleus, while HBsAg is detected in hepatocyte cytoplasm and at the cell surface.

Essentially all data indicate that virus clearance and hepatocellular injury that occurs during acute and chronic HBV are immune mediated, and that HBV is not directly cytopathic to hepatocytes. Defining the immunopathological events in acute and chronic HBV infection also has been hampered by the lack of cell culture systems and small animal models for HBV infection.[213] Studies in humans have shown that the HBV nucleocapsid (HBcAg–HBeAg) and envelope contain epitopes that signal cytotoxic T lymphocytes in association with human leukocyte antigen (HLA) class I molecules on the hepatocyte during acute infection, but this response has not been detected during chronic HBV-related hepatitis.[250] In addition, studies in transgenic mice that express HBsAg have shown a role of an HBV-specific cytotoxic lymphocyte response and cytokines in the pathogenesis of fulminant hepatitis B.[18] The pathogenesis of extrahepatic manifestations of HBV infection that appear during the prodromal phase of acute infection (i.e., arthritis, urticaria) or chronic infection (i.e., vasculitis, glomerulonephritis) is thought to be attributable to immune complex formation.[95]

The mechanisms associated with the establishment of virus persistence in HBV infection are not known.[116,299,309] This phenomenon is strongly age related, and it has been suggested that it may relate to the ontogeny of cytokine expression and the possible induction of various classes of helper T cells.

Chronic HBV infection is strongly associated with PHC. Geographically, high rates of chronic infection correlate with high rates of PHC, and prospective longitudinal studies have demonstrated a high relative risk for death from PHC among individuals chronically infected with HBV. In addition, woodchucks chronically infected with woodchuck hepatitis B virus also develop PHC.[27,28]

3.7. Patterns of Host Response

HBV infection can cause a broad spectrum of disease ranging from asymptomatic acute infection to fulminant liver failure and death, and individuals with persistent infection can remain asymptomatic or progress to chronic liver disease, liver failure, or develop PHC. Acute hepatitis B cannot be differentiated from other viral causes of hepatitis on the basis of signs or symptoms. A prodromal phase of disease usually occurs with the most common findings being malaise, weakness, and anorexia. Myalgia and arthralgia without other clinical signs of hepatitis have been described in 10–30% of patients with acute

hepatitis B; in one third of these patients a urticarial or maculopapular rash appears with joint symptoms.[95] Jaundice develops in at least 30% of infected adults and may persist for several weeks to months. Liver enzyme elevations usually occur prior to onset of jaundice and may persist for several weeks after its resolution.

3.7.1. Serological Patterns of Infection. The patterns of expression of the various antigens and antibodies of HBV have been described in a previous section. HBsAg becomes detectable 1 to 2 months prior to onset of clinical illness and is soon followed by the appearance of IgM anti-HBc, such that both are present at the onset of illness. Hepatitis B e antigen and HBV-DNA are present during the course of the acute infection and disappear as the infection resolves. In those infections that resolve, HBsAg begins to decline and eventually disappears, and anti-HBs becomes detectable. Although a "window phase" may occur when neither HBsAg nor anti-HBs are detectable, IgM anti-HBc remains positive during this period and for an additional 6–10 weeks. Ultimately, anti-HBs and anti-HBc persist as markers of a resolved infection.[145]

Among individuals in whom HBV infection persists, both HBsAg and anti-HBc usually remain detectable indefinitely. Hepatitis B e antigen is initially present, but approximately 10% of chronically infected individuals lose HBeAg each year, beginning at a variable time period after the initial infection and related to factors such as age at time of infection.[17] Approximately 1% of chronically infected individuals lose HBsAg each year and appear to resolve their chronic infection.[165] In a small proportion (2–3%) of chronically infected persons, HBsAg and anti-HBs are present and represent incomplete resolution of chronic infection or reinfection with a second subtype (genotype) of HBV.[341]

3.7.2. Clinical Outcomes of Infection. A direct relationship exists between age at infection and the likelihood of symptomatic HBV infection. Infections in infants and children under 5 years of age rarely produce signs or symptoms consistent with viral hepatitis, while 10% of older children and 30% of adults may be become jaundiced, and a higher proportion may have additional signs or symptoms consistent with viral hepatitis.[145,225] The risk of chronic HBV infection has been shown to be inversely related to age and averages 90% for infants infected at birth and 30% for children infected before age 5 years.[29,294,340] Among adults infected with HBV, 2 to 10% remain chronically infected, although these estimates are based on small studies.[146,268,303]

In the vast majority of cases, the severity of the illness associated with acute hepatitis B does not require hospitalization. Only 10–12% are hospitalized in the United States, but rates of hospitalization may be higher in other countries because of established patterns of medical practice and not disease severity.[59] Fulminant hepatitis occurs rarely in infected infants and may be more frequently associated with infections due to HBeAg-negative variants.[307] It is estimated that fulminant liver failure occurs in up to 1% of adults with acute hepatitis B,[59] although this rate may be higher in populations where coinfection with HDV occurs commonly.

Clinical studies indicate that the majority of persons chronically infected with HBV remain healthy and do not suffer the consequences of HBV-related chronic liver disease or PHC, especially if they became infected as adults.[147,224] However, chronic persistent hepatitis, chronic active hepatitis, cirrhosis, or PHC may occur in persons chronically infected with HBV,[224,327] especially in those infected as infants or young children.[150] Prospective studies in Taiwan have estimated that approximately 25% of chronically infected infants or children will die as adults from HBV-associated PHC.[27,28,30] The 5-year survival among persons with HBsAg-positive chronic liver disease is approximately 40% for persons who most likely acquired their infections as adults, and this survival rate appears to be similar to that for persons who most likely acquired their infections as infants or young children.[201,327]

3.8. Control and Prevention

3.8.1. Hepatitis B Immunization. By the early 1970s, several groups showed that HBsAg obtained from the serum of persons chronically infected with HBV could be formulated into a vaccine that induced protective antibodies against HBV infection.[142,183,264] Plasma-derived hepatitis B vaccine was shown to be safe, highly immunogenic, and effective in preventing acute and chronic HBV infections in infants, children, and adults. Subsequently, new-generation hepatitis B vaccines were produced using recombinant DNA technology to express HBsAg in yeast or mammalian cell substrates, and the new vaccines were shown to be comparable to plasma-derived vaccines in preventing acute and chronic infections.[111,260,301,340] Hepatitis B immunization using either type of vaccine has been shown to eliminate HBV transmission and should prevent HBV-related chronic liver disease.

Preexposure vaccination generally requires three doses to induce an immune response that provides long-term protection. Hepatitis B vaccine can be given along with other commonly used vaccines in a variety of schedules that result in excellent immunogenicity and do not interfere with the immunogenicity of the other vaccines.

In addition, infection can be effectively prevented after exposure to HBV (postexposure prophylaxis) through the passive administration of anti-HBs [hepatitis B immune globulin (HBIG)] and hepatitis B vaccine[295] or with vaccine alone.[260,340]

The primary objective of hepatitis B immunization is the prevention of chronic HBV infection and its sequelae. During the past decade, vaccination strategies have changed somewhat, especially in some countries with a low endemicity of infection. In areas of high HBV endemicity, routine vaccination of infants is the only means of interrupting HBV transmission because of the high rates of infant and early childhood infection. In most of these areas, infant vaccines are delivered through the World Health Organization's Expanded Program on Immunization (EPI), and it has been recommended that all infants in these countries be routinely immunized against HBV infection.[223,335] A number of large-scale demonstration projects have shown the feasibility of adding hepatitis B vaccine to the routine infant immunization schedule, with subsequent lowering of the age-specific rate of acute and chronic HBV infection.[79,205]

In areas of high HBV endemicity, the timing of the first dose of hepatitis B vaccine is determined by the need for postexposure immunoprophylaxis to prevent perinatal HBV transmission. Where perinatal transmission accounts for a substantial amount of chronic infections (Table 2), immunization should begin within 12 hr after birth. However, in most of these countries it is not feasible to test pregnant women to identify infants that require postexposure immunoprophylaxis. Thus, all infants should receive a dose of vaccine shown to have postexposure efficacy.[70,339,340] Where the prevalence of HBeAg among pregnant women is low and perinatal transmission accounts for a small proportion of chronic HBV infection (Table 2), the first dose of vaccine can be given soon after birth or it can be given when the infant receives the first of diphtheria–tetanus–pertussis (DTP) vaccine.

Ultimately, the design of hepatitis B vaccination programs, including the timing of vaccine doses, is influenced by the epidemiology of HBV infection and patterns of health care (immunization) delivery. Thus, prevention of perinatal HBV infection is limited to those countries where most infants are born in the hospital or with a birth attendant trained to vaccinate infants.

The success of a hepatitis B vaccination program is determined by the presence of a well-established infrastructure for vaccine delivery. This is illustrated by the recent experience in the United States, a country with a low endemicity of HBV infection where the majority of infections occur among adults (Table 2, Fig. 4). Early recommendations called for immunization of selected groups of adults at high risk of HBV infection, immunization of infants in ethnically defined populations with a high endemicity of infection, and postexposure immunization of infants born to HBsAg-positive women. However, immunization of adults was largely unsuccessful because of the lack of established immunization programs, the poor performance of selective immunization, and the inability to vaccinate prior to the establishment of behaviors that placed the individual at high risk of infection. The failure of selective adult immunization to lower the incidence of HBV infection led to routine infant hepatitis B immunization as the primary implementation strategy.[61]

Although immunization of successive birth cohorts may delay control of HBV transmission for several decades in a country of low HBV endemicity (i.e., the United States), it appears to be the most cost-effective immunization strategy for a number of reasons.[178,209] Routine infant immunization ensures the prevention of HBV infections in populations residing within low endemicity countries that have high rates of early childhood infection (i.e., Eskimos, Asian/Pacific Islanders, infants of immigrant women from high endemicity areas). In addition, high immunization coverage rates are ensured because of the proven vaccine delivery system, and the proven long-term efficacy of hepatitis B immunization should prevent infections in adolescents and young adults. However, a more rapid reduction in HBV transmission could be achieved in low endemicity countries with the addition of routine adolescent vaccination until such time that cohorts of children vaccinated as infants become teenagers.[70] Routine hepatitis B vaccination of adolescents, primarily in school-based programs, has been shown to be feasible in a number of countries (e.g., Canada, Spain, United States) and could provide the foundation for the delivery of other prevention services to this age group.

3.8.2. Long-Term Efficacy of Hepatitis B Immunization. Routine infant vaccination has become the strategy of choice in preventing HBV infection because it provides long-term protection from chronic and acute infection. Cohort- and population-based studies indicate that immunized persons retain a protective immune response for at least 10 years (even if anti-HBs is no longer detectable). Although serologically resolved HBV infections have been found in immunized cohorts of infants, children, or adults at continued risk of HBV infection, no symptomatic infections and almost no chronic HBV infections have been identified (Table 3). In addition, follow-up studies of routine infant hepatitis B vaccination programs in populations with high rates of HBV infection have shown a dramatic reduction in age-specific infection rates and few late infections among immunized infants (Table 4).

All available data indicate that routine infant hepatitis B immunization will provide protection during the

Table 3. Long-Term Protection from Hepatitis B Virus Infection among Cohorts of Children and Adults Known to Have Responded to Hepatitis B Vaccination

Study group (reference)	Follow-up No.[a]	(yr)	Anti-HBs loss[b] (%)	HBV infections Anti-HBc(+)	HBsAg(+)
Postexposure immunoprophylaxis: infants of HBeAg-positive mothers					
Passive–active					
Taiwan[200]	199	5	3	0	0
Taiwan[153]	654	5	9	46	4
United States[296,297]	315	4–11	12	30	0
Active					
China[339]	55	5	17	6	1
Routine preexposure immunization of infants/children					
Senegal[84]	100	6	22	8	4
Alaska[323,324]	600	10	17	4	0
Venezuela[132]	280	6	29	6	0
Preexposure immunization of adults					
Homosexual men[130,131]	634	9	54	48	4
Homosexual men[297]	127	11	61	26	0
Alaskan Eskimos[323,324]	272	10	38	6	0

[a]Number of cases; some studies used person-years of follow-up.
[b]Less than 10 milli-international units/ml of anti-HBs.

Table 4. Effect of Routine Infant Hepatitis B Immunization on Prevalence of Hepatitis B Virus Infection in Populations with a High or Intermediate Endemicity of Infection

| | | | HBV infection (%) | | | |
| | | | Before program | | After program | |
Study (reference)	Immunization strategy	Program duration (yrs)	Chronic	All	Chronic	All
Taiwan[76]	Perinatal	5	10	ND*	2.2	ND*
	Routine infant					
Gambia[136]	Routine infant	5	12	53	0.5	9
Alaska[139]	Perinatal	10	15	37	0	0.8
	Routine infant					
	Catch-up of susceptibles					
American Samoa[205]	Perinatal	4	7	23	1	12
	Routine infant					
	Catch-up of susceptibles					

*ND, not determined.

periods of highest risk of chronic infection in populations with high rates of early childhood infection. Although chronic HBV infection will persist in the previously unimmunized population and eradication of infection will not be attainable for many generations, the elimination of HBV transmission can only be attained through continued immunization of successive birth cohorts.

It is currently not known whether infant hepatitis B immunization will confer lifelong immunity against chronic infection, which is of particular importance where vaccination of infants is expected to prevent adult-acquired infection. Current data indicate that boosters of hepatitis B vaccine are not required for at least the first decade following infant, childhood, or adult immunization.[70] It is not known whether a booster dose of vaccine is required to provide lasting immunity for adolescents entering the period of highest risk for HBV infection, although it appears unlikely. The very long incubation period of HBV infection (40–120 days), coupled with the excellent anamnestic antibody response to HBsAg in previously immunized persons, would appear to limit breakthrough infections to ones that produce limited viremia and do not become persistent.

3.8.3. Treatment of HBV-Related Chronic Liver Disease. Prevention of HBV infection through immunization is the only effective means of eliminating the consequences of chronic infection. However, the moderate success achieved in eradicating or suppressing HBV replication through antiviral therapy could potentially reduce the mortality and morbidity of HBV-related chronic liver disease and shrink the reservoir of potentially infectious individuals. Unfortunately, current antiviral treatment is expensive and its efficacy appears to be limited to a relatively small segment of the population of chronically infected individuals.

Treatment with α interferon has been shown to effectively eliminate viral replication (i.e., loss of HBeAg and HBV-DNA) in 25 to 40% of patients; among patients who lose HBV-DNA, approximately 50% become HBsAg negative and develop anti-HBs.[175,252] However, only a small proportion of persons chronically infected with HBV meet the serological selection criteria for interferon therapy; i.e., HBeAg and HBV-DNA positive and with elevated liver enzymes. In particular, persons infected as infants or children have had a poor response to currently used regimens of interferon therapy. A number of new antiviral agents are being evaluated for their effect in eliminating HBV replication, including β interferon, thymosin, and nucleoside analogues.

Orthotopic liver transplantation has been used as a lifesaving procedure for persons with HBV-related end-stage liver disease. However, in all instances, the transplanted liver becomes reinfected with HBV and the long-term survival for these patients is poorer than for other patients undergoing orthotopic liver transplantation for other causes of end-stage liver disease. For this reason, most centers have limited transplantation for chronic HBV infection unless the patient is placed on long-term, high-dose immunoprophylaxis with HBIG. Liver transplantation does little to reduce the pool of persons infectious for HBV because of the high rate of reinfection.

3.9. Unresolved Problems

The greatest challenge facing the biomedical and public health community is routine infant hepatitis B vaccination worldwide. In almost every country, the appropriate use of hepatitis B vaccine has the potential to eliminate a major cause of liver disease and cancer, yet a mechanism to fund the introduction of a new and effective vaccine into national immunization programs has not been developed. Toward this end, combining hepatitis B vaccine with other antigens (i.e., DTP) would facilitate its delivery, although how this new vaccine (new product) would be funded is unclear. To ensure successful hepatitis B immunization programs, a number of questions will have to be addressed and should include the number of vaccine antigens that can be combined with HBsAg, the extent of the long-term efficacy of hepatitis B immunization, and the role HBV variants might play in late breakthrough infections in immunized populations.

The major challenge of future research remains in understanding the mechanisms that control the establishment of chronic HBV infection. Understanding these mechanisms would aid in the development of effective antiviral therapy or therapeutic vaccines to eliminate the chronic infectious state. Reducing the size of the reservoir of persons chronically infected with HBV would greatly aid the effectiveness of hepatitis B immunization efforts and potentially reduce the burden of chronic liver disease morbidity and mortality related to HBV.

4. Hepatitis C

4.1. Historical Background

Hepatitis C virus was discovered in 1988 and was shown to be the primary etiologic agent of parenterally transmitted non-A, non-B (PT-NANB) hepatitis worldwide.[7,78,186] PT-NANB hepatitis was first recognized in the 1970s, following development of specific serological tests to identify HBV and HAV infection.[108] The term "NANB hepatitis" is applied to those cases of acute hepatitis for which other specific etiologies (HAV, HBV, Epstein–Barr virus, cytomegalovirus, and a variety of other infectious and noninfectious agents that can cause liver inflammation) can be reliably excluded. Although PT-NANB hepatitis was first recognized to be commonly associated with blood transfusion, it is now known to be an important cause of community-acquired viral hepatitis.

4.2. Methodology Involved in Epidemiologic Analysis

4.2.1. Sources of Morbidity and Mortality Data.
PT-NANB hepatitis remains a diagnosis of exclusion. Reported data on disease incidence greatly underestimate actual disease occurrence because of underreporting as well as underutilization of specific tests to rule out hepatitis A and hepatitis B. In the United States, hepatitis C is not a separate reportable disease, but is included among all cases of NANB hepatitis. Although a serological assay for the detection of IgG antibody to HCV (anti-HCV) has been commercially available since May 1990, no assay to reliably distinguish acute infection, such as IgM anti-HCV, has been developed. At least 20% of patients with hepatitis C may not become anti-HCV positive for weeks to months after onset of their clinical illness and an additional 10% remain persistently anti-HCV negative; infection in these patients can only be diagnosed by the detection of HCV-RNA with research-based methods. These limitations prevent using the available serological assays for reporting of acute disease.

4.2.2. Surveys.
Useful estimates of disease morbidity and mortality due to NANB hepatitis are obtained by studies that ensure that all cases of acute hepatitis are systematically tested for markers of acute HAV and HBV infection. Retrospective anti-HCV testing of many of these study populations has provided specific estimates of the disease burden associated with HCV infection. Prospective studies that use serial tests for liver inflammation [alanine aminotransferase (ALT)] to identify cases of viral hepatitis, along with specific serological testing for HAV, HBV, and HCV infection, have been essential to estimate the frequency of clinical and subclinical HCV infection following blood transfusion and to estimate the frequency of chronic hepatitis following HCV infection. Serological studies of special populations, as well as studies of outbreaks, have provided information on pathways of disease transmission.

4.2.3. Laboratory Diagnosis.
The first-generation EIAs (EIA-1) for anti-HCV detected antibody to a nonstructural protein encoded by the HCV genome (c100-3 antigen) that was produced by recombinant DNA technology. The second-generation EIAs (EIA-2) for anti-HCV use three recombinant proteins, two (c100-3 and c33c) coded by the nonstructural region of the genome and one (c22-3) coded by the nucleocapsid region of the genome.[4] Using these assays, anti-HCV has been detected in an average of 70 to 90% of patients with PT-NANB hepatitis.[5,16,101] The greater sensitivity of the EIA-2 compared with the EIA-1 has resulted in a 6–32% increase in the antibody detection rate among patients with PT-NANB hepatitis[1,13,16,176] and in a 59% increase among volunteer blood donors.[172] In this latter group, the EIA-2 has detected an additional 1.4 HCV-positive donors per 1000 tested.[172]

Among patients with HCV infection, the EIA-2 detects anti-HCV in approximately 90%.[16] In addition to its greater sensitivity, the EIA-2 detects antibody earlier in the course of HCV infection; 5 to 6 weeks after onset of hepatitis in 80% of patients compared with 40 to 60% with EIA-1.[16] Anti-HCV remains detectable by EIA-2 long after the primary infection; in contrast, in many patients anti-HCV became undetectable by EIA-1 within a few years after disease onset.[16,26,103]

Interpretation of the results of EIAs that screen for anti-HCV is limited by several factors: (1) these assays will not detect anti-HCV in approximately 10% of persons infected with HCV; (2) these assays do not distinguish between acute or chronic or past infection; (3) in the acute phase of hepatitis C, there may be a prolonged interval between onset of illness and seroconversion; and (4) in populations with a low prevalence of infection, the false-positivity rate is high and the positive predictive value of the tests is low.

As with any screening test, the proportion of repeatedly reactive EIA results that are falsely positive varies depending on the prevalence of infection in the population screened. No true confirmatory test has been developed because authentic HCV proteins are not available. However, supplemental tests to evaluate the specificity of repeatedly reactive results obtained from the screening assays are available. They use testing formats that are different from the screening tests and different methods to produce the synthetic antigens.[172,232] On the basis of these supplemental assays, 70 to 100% of repeatedly reactive anti-HCV results in persons with clinically diagnosed hepatitis C, injection drug use, and a history of blood transfusion are judged true positives.[16,62] In contrast, less than 50% of repeatedly reactive anti-HCV results in persons at low risk of HCV infection, such as volunteer blood donors, are judged true positives.[62,172]

The diagnosis of HCV infection is also possible by detecting HCV-RNA following reverse transcription to cDNA and amplification using the PCR.[148] HCV-RNA can be detected many weeks before the appearance of viral antibodies, and in some persons may be the only evidence of infection. Another research-based tool for the diagnosis of HCV infection is a fluorescent-antibody blocking assay, which detects HCV antigen(s) in liver and antibody to this antigen (anti-HCVAg) in serum of infected patients. Anti-HCVAg can be detected in almost one quarter of PT-NANB hepatitis patients who are persistently anti-HCV (by EIA-2) negative.[16]

4.3. Biological Characteristics

HCV is a single-strand, positive-sense RNA virus with a genome of approximately 9400 nucleotides that has a single open reading frame.[46,148] The 5′-end encodes the nucleocapsid and envelope proteins, followed by the nonstructural proteins that extend to the 3′ end of the genome. Because of similarities to the flavi- and pestivirus ge-

nomes, HCV has been classified as a separate genus in the family *Flaviviridae*.

In vitro translation of the portion of the genome that encodes the structural proteins indicates that it is processed into a minimum of four polypeptides. The amino-terminal end is cleaved to produce an nonglycosylated nucleocapsid protein (p22), followed by two envelope glycoproteins, designated E1 and E2. The amino-terminal end of E2 exhibits up to 58% nucleotide variation among geographically distinct virus isolates, and has been termed the hypervariable region.[46] This high degree of variation suggests that HCV is capable of rapid mutation in order to evade immune detection by the host and may complicate efforts toward the development of a vaccine.

HCV encodes a number of nonstructural proteins, including the NS3 region that contains consensus sequences for two distinct enzymatic functions, which may be involved in polyprotein processing (protease) and in unwinding the RNA genome for replication (helicase). The NS5 region encodes a protein that contains an RNA-dependent RNA polymerase involved in genome replication. Like its pestivirus and flavivirus relatives, HCV does not appear to produce DNA replication intermediates, and integration of the viral genome into the host genome has not been detected.

Comparative nucleotide sequence analysis of different HCV isolates indicates the existence of multiple HCV genotypes.[50,72] At present, there are no consistently applied criteria by which new genotypes are defined, and at least four different systems of nomenclature have been described. The clinical significance of this sequence variability is not well defined. Animal transmission experiments suggest that antibody elicited by infection with one genotype fails to cross-neutralize heterologous virus genotypes. Other studies suggest that phenotypic differences affect disease progression and response to treatment with α interferon.

4.4. Descriptive Epidemiology

The discovery of HCV and the development of a serological test for anti-HCV have shown that this virus is the primary etiologic agent of PT-NANB hepatitis and an important cause of acute and chronic hepatitis worldwide. HCV is a blood-borne virus, and most epidemiologic studies have focused on groups at risk of infection as a result of direct percutaneous exposures. However, many patients infected with HCV do not report such exposures, and the role of inapparent parenteral exposures in HCV transmission, such as through sexual activity and household contact, has not been well defined.

4.4.1. Prevalence and Incidence. The prevalence of anti-HCV is highly variable in the general population.[8] The highest rates are found among persons with repeated direct percutaneous exposures, such as injection drug users and hemophilia patients (60–90%); moderate rates are found among those with smaller but repeated direct or inapparent percutaneous exposures, such as hemodialysis patients (20%); and lower rates are found among those with inapparent parenteral or mucosal exposures, such as persons with high-risk sexual behaviors and sexual and household contacts of infected persons (1–10%), as well as among those with sporadic percutaneous exposures, such as health care workers (1–4%). The lowest rates of anti-HCV are found among those with no high-risk characteristics, such as volunteer blood donors (0.3–1%). These patterns are fairly similar throughout the world, although the range of rates in different segments of the population may vary between countries.

In the United States, hepatitis C is not a separate reportable disease as discussed above, but is included among all cases of NANB hepatitis. Studies that assessed reporting have shown that NANB hepatitis is the most poorly reported of the major hepatitis types[15]; thus, national surveillance data greatly underestimate the acute hepatitis morbidity from NANB hepatitis. More accurate estimates of both NANB hepatitis and hepatitis C are derived from a sentinel surveillance system in four counties in the United States that stimulates reporting and includes complete serological testing.[13,16] Studies in these counties have shown that an average of 25% of cases of acute hepatitis can be classified as NANB hepatitis agents; 21% are estimated to be due to HCV, and up to 4% may be due to an agent or agents not yet identified. In these counties, the estimated incidence of acute hepatitis C remained relatively stable through much of the 1980s, with an average rate of 15/100,000 (corrected for underreporting), but declined by more than 50% between 1989 and 1991.[8,13] The number of cases of transfusion-associated hepatitis C declined significantly after 1985, but this change had little impact on overall disease incidence. The dramatic decline observed since 1989 correlates with a decrease in cases associated with injection drug use, possibly related to safer needle-using practices. After correcting for underreporting and asymptomatic infections, it is estimated that an average of 150,000 HCV infections occurred annually in the United States during the past decade.

4.4.2. Epidemic Behavior. Outbreaks of hepatitis C are limited in scope because of the predominantly blood-borne route of transmission. Historically, outbreaks of PT-NANB hepatitis have been documented among hemodialysis patients, presumably through blood-borne transmission in a highly contaminated environment, among groups of patients having common exposure to blood components, and among injection drug users.[118,126,241] In Japan, a community outbreak of PT-NANB hepatitis was described for which no specific source or mode of transmission was identified.[115] Although secondary transmission to household contacts occurred in some instances, it was usually limited. Retrospective testing of serum samples from some of these outbreaks have shown them to be caused by HCV.[115,241]

4.4.3. Geographic. Worldwide, the reanalyses of prospective studies of transfusion recipients who contracted NANB hepatitis have shown that 46 to 90% of these patients seroconverted to anti-HCV.[1,5,101] The proportion of community-acquired NANB hepatitis reported to be attributable to HCV has ranged from 50 to 75%.[13,16,182,302] Anti-HCV has been detected in 61 to 93% of NANB hepatitis patients with a probable history of percutaneous exposures, and in 31 to 53% of such patients with no history of percutaneous exposures. The wide variation in the proportion of both posttransfusion and community-acquired NANB hepatitis attributed to HCV reflects differences in the populations studied, their sample sizes, the methods by which patients were identified, the case definitions used, the ascertainment of exposure histories, the durations of follow-up, and the testing methods used.

Geographic clustering of high prevalence rates of anti-HCV have been reported from Japan and Egypt in populations lacking commonly recognized risk factors for acquiring hepatitis C.[141,157] The reasons for such clustering have yet to be ascertained, but hypotheses include traditional and nontraditional medical treatment with contaminated equipment and familial aggregation as a result of vertical and horizontal transmission.

4.4.4. Temporal. Studies of community-acquired hepatitis C in Western countries generally show no evidence of seasonality or of epidemic cycles.

4.4.5. Age. Acute hepatitis C may occur in all age groups; however, most cases occur among young adults.[13,16,59] Clinical illness is uncommon among children, but among older adults above age 40 years hepatitis C is often the most common cause of acute hepatitis.[59] This age distribution of disease is likely related to patterns of exposure (injection drug use in young adults; transfusions in older adults) and possibly to age-specific variations in clinical expression of disease.

4.4.6. Sex. There are no consistent sex predilections for hepatitis C other than a higher frequency of cases related to injection drug use among men. Among community-acquired hepatitis C in the United States, the male-to-female ratio is 1:1.[13,16,59]

4.4.7. Race.   Hepatitis C occurs worldwide in all racial/ethnic groups studied. In the United States, the highest proportion of cases is among whites, but the incidence of disease is highest in nonwhite racial/ethnic groups, particularly Hispanic.[13]

4.4.8. Occupation.   A case-control study of patients with acute PT-NANB hepatitis conducted prior to the discovery of HCV found a significant association between acquiring disease and health care employment, specifically patient care or laboratory work.[11] Seroprevalence studies using EIA-1 and supplemental testing have reported anti-HCV rates of 1 to 4% among hospital-based health care workers.[257] In the studies that assessed risk factors for infection, a history of accidental needlesticks among hospital workers and the practice of oral surgery among dentists were associated with anti-HCV positivity.[171,257]

Several case reports have documented the transmission of HCV infection from anti-HCV-positive patients to health care workers as a result of accidental needlesticks or cuts with sharp instruments.[315] In studies that reported on the follow-up of health care workers who sustained percutaneous exposures to blood from anti-HCV-positive patients, the incidence of anti-HCV seroconversion was 3–10%.[170,234]

4.4.9. Other Settings.   No unusual risk of hepatitis C has been observed in the military. Transmission of HCV in hemodialysis units to patients has been documented, as has disease in transplant recipients.[241,251] Standard medical care procedures have not been associated with the transmission of HCV in the United States,[9,11] nor has health care worker-to-patient transmission of HCV been reported. However, the prevalence of anti-HCV among hemodialysis patients averages 20%, and studies have demonstrated an association between anti-HCV positivity and increasing years on dialysis that was independent of blood transfusion.[235] These studies, as well as investigations of dialysis-associated outbreaks of hepatitis C, suggest that HCV may be transmitted between patients in the dialysis center, possibly because of poor infection control practices, including improper cleaning procedures. Other than injection drug use, however, the role of community exposures in the acquisition of HCV infection among these patients has not been explored. Theoretically, nosocomial transmission of HCV is possible if breaks in technique occur or disinfection procedures are inadequate and contaminated equipment is shared between patients.

4.4.10. Other Factors.   The risk of HCV infection is well documented in persons exposed to blood or blood products through transfusion. Prospective studies in the 1970s showed that an average of 10% (range, 7–21%) of persons receiving transfusions screened for HBV in the United States developed acute hepatitis.[6] Risk was lower among recipients of volunteer rather than commercial blood.[6] In the 1980s, it was also shown that the risk of posttransfusion hepatitis could be reduced by screening blood donors with surrogate markers for PT-NANB hepatitis (ALT abnormalities or prior HBV infection).[177,293] By 1991, the combination of surrogate and first-generation anti-HCV testing of donors had reduced the risk of posttransfusion hepatitis C to <1%.[98]

Preparations of pooled clotting factors historically have posed markedly higher risk of infection. Hemophilia patients who have been heavily transfused with non-treated factor concentrates have prevalence rates of anti-HCV exceeding 90%, higher than for any other group studied with the exception of injection drug users.[311]

Persons with no specific source for their acute hepatitis C account for a substantial amount of disease, and low socioeconomic level is associated with a large proportion of these patients.[7,13,16] More than half report histories of some type of high-risk behavior or contact, including imprisonment, use of noninjection illegal drugs, contact with a sexual partner or household member who used injection drugs, one or more sexually transmitted diseases, or injection drug use but not in the 6 months preceding illness.[8] Low socioeconomic level and the other high-risk attributes identified among these patients have been associated with the transmission of a number of infectious diseases, and probably serve as surrogate indicators for routes of transmission; however, the nonspecific nature of these indicators make disease prevention difficult.

4.5. Mechanisms and Routes of Transmission

The most efficient transmission of HCV is associated with direct percutaneous exposure to blood, such as through transfusion of blood or blood products, transplantation of organs from infectious donors, and sharing of contaminated needles among injection drug users. Health care workers experiencing needlestick injuries also are at risk of acquiring HCV infection. However, less than half of reported cases of hepatitis C report a history of possible percutaneous exposures to blood.[13,16] Surveillance data from the CDC Sentinel Counties study show that 4% of reported cases of acute hepatitis C are associated with blood transfusion, 38% with injection drug use, 3% with occupational exposure to blood, and 14% with exposure to a sexual partner or household contact who had hepatitis or to multiple heterosexual partners. In 41%, no specific exposure occurring in the prior 6 months was reported, but

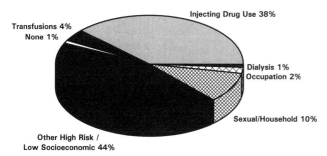

Figure 5. Risk factors associated with reported cases of acute hepatitis C in four sentinel counties, United States, 1991–1993. Preliminary data, Sentinel Counties Study of Viral Hepatitis, Centers for Disease Control and Prevention.

98% of these patients had some lifetime history of high-risk drug or sexual behaviors or had characteristics associated with low socioeconomic level populations (Fig. 5).

HCV circulates at low titers in infected serum,[45] and no methods are available to quantitatively measure infectivity. The inconsistent reports of detection of HCV-RNA in body fluids other than serum and plasma may reflect different titers of virus in the infected persons sampled.[113,325] Although studies that address survival and transmission of HCV in the environment have not been done, rapid degradation of the virus seems to occur when serum containing virus is left at room temperature.[87]

The risk of HCV transmission by sexual or household contact has been evaluated in a limited number of studies. In case-control studies, patients with NANB hepatitis who had no history of percutaneous exposures were more likely than control patients with no liver disease to have an exposure to a sexual partner or household contact with a history of hepatitis or to multiple heterosexual partners.[9,11] These studies found no association between acquiring NANB hepatitis and male homosexual activity. Recently, a seroprevalence study of more than 700 homosexual men in San Francisco found an anti-HCV prevalence of only 1.5% among those who denied injection drug use.[245] In spite of this relatively low prevalence, anti-HCV positivity was directly associated with increasing numbers of both recent and lifetime partners. Anti-HCV seroprevalence rates of 9–12% have been found among non-drug-using female prostitutes.[338] Anti-HCV positivity was significantly associated with increasing numbers of paid partners, sexual activities involving trauma, failure to use a condom, positive serology for syphilis, and increasing years working as a prostitute. Many other studies have examined the prevalence of anti-HCV among groups with different sexual behaviors, but evaluation of their results is difficult because most involved inadequate

sample sizes, did not obtain a history of injection drug use, and/or failed to exclude false-positive anti-HCV results through the use of supplemental testing.

Serological studies also have examined the prevalence of anti-HCV among the sexual and household contacts of patients with chronic hepatitis C, although generally the sample sizes have been inadequate to draw firm conclusions. These studies found anti-HCV rates of 0–18% among sexual partners or spouses who had no other risk factors for hepatitis C[3,102,164,246]; in two of these studies the index patients were coinfected with HCV and human immunodeficiency virus (HIV).[102,246] In one study, anti-HCV also was found in 1% of children and in none of the other family members.[164] Using PCR for the detection of HCV infection, investigators from Japan reported finding HCV-RNA in 12 (32%) of 37 spouses and in 14 (21%) of 67 children of patients with chronic hepatitis C.[240]

An evaluation of the risk of transmission of HCV from infected women to their newborn infants on the basis of published studies is difficult because the small sample sizes and the duration of follow-up and types of serological testing are highly variable. Among studies of infants born to anti-HCV-positive, anti-HIV-negative women, an average of 6% of infants seroconverted or were persistently positive for anti-HCV based on second-generation enzyme immunoassay and supplemental testing.[80,187,199,244,277,328] Studies that used PCR to detect HCV-RNA did not result in the identification of a higher average proportion (7.5%) of infected infants.[187,199,244,277,328] Two studies from Asia suggested that the risk of perinatal transmission is related to the titer of HCV-RNA.[199,244] In one study, transmission occurred at a rate of 36% only from women with titers $\geq 10^{6}$[244]; in the other, transmission occurred only from one woman with a titer of 10^{10}.[199] In the one study that followed 58 infants born to women coinfected with HCV and HIV, the risk of HCV infection was 5%, no higher than for infants born to women with HCV infection alone.[187] In this study, of twins born to a coinfected mother, one developed HCV infection in the absence of HIV and the other developed HIV but not HCV infection.

4.6. Pathogenesis and Immunity

The incubation period for acute hepatitis C following transfusion or accidental needlestick has been reported to average 6–7 weeks,[176,315] but may range from 2 to 26 weeks. Virtually all persons with acute HCV infection may become chronically infected,[16] and chronic liver disease with persistently elevated liver enzymes develops in an average of 67%, independent of the source for infection.[1,5,16,176,302] Of those with chronic liver disease, chronic active hepatitis or cirrhosis has been found in 29 to 76% within several years after onset of their hepatitis,[16,302] and in 88% in the one study that followed patients for an average of 18 years. Histopathologic features commonly observed in biopsy specimens from patients with chronic hepatitis C include bile duct damage, bile duct loss, steatosis, and lymphoid cell aggregation (follicles).[22] Severe lobular necrosis and inflammation and piecemeal necrosis are seen less often. HCV also may be a major contributing cause of hepatocellular carcinoma.[169] In seroprevalence studies, antibody to HCV (anti-HCV) has been detected in 13–74% of patients with hepatocellular carcinoma, and strong association between the presence of anti-HCV and hepatocellular carcinoma has been demonstrated in several case-control studies.[94,161,344]

Hepatitis C may result in persistent infection even in the absence of active liver disease. A recently published study of patients with community-acquired hepatitis C found that of 15 patients in whom hepatitis had resolved, all were persistently HCV-RNA positive by PCR up to 4 years after onset of their acute illness, although in four patients detection of HCV-RNA was intermittent.[16] HCV-RNA also has been detected in anti-HCV-positive blood donors with normal ALT levels.[48,172] One study reported that of four such patients who underwent liver biopsy, all had normal findings.[48]

The extraordinarily high rates of chronic disease and persistent viremia observed in humans suggest that HCV fails to induce an effective neutralizing antibody response. Experimental studies in animals have shown that after a primary HCV infection, rechallenge of convalescent (i.e., HCV-RNA negative) chimpanzees with the same or different strain of HCV resulted in the reappearance of viremia, which was due to infection with the subsequent challenge virus.[103] This reappearance of viremia was associated with mild ALT elevations and histopathologic signs of acute hepatitis. In chimpanzees that were HCV-RNA positive at the time of rechallenge, the virus recovered was identical to that present before challenge; however, serial liver biopsies showed recurrent necroinflammatory changes after rechallenge. All animals were persistently positive for anti-HCV on the basis of second-generation testing.

Although superinfection with a different strain of HCV occurred only in HCV-RNA-negative animals, the failure to detect viral RNA by PCR in animals or humans is not necessarily indicative of recovery or noninfectivity.[26] Inoculation of PCR-negative material from an HCV-infected chimpanzee with chronic ALT elevations caused disease in a naive animal.[26] Reinfection or superinfection with HCV may be a possible explanation for the apparent episodes of posttransfusion or posttransplant hepatitis observed among recipients who were anti-HCV-positive before they received blood or organs from an HCV infectious donor.[251]

4.7. Patterns of Host Response

The spectrum of illness caused by HCV ranges from asymptomatic infection to acute fulminant hepatitis. In prospective studies of transfusion recipients who contracted PT-NANB hepatitis, clinical evidence of acute hepatitis developed in an average of 25%.[6] Of patients with either posttransfusion or community-acquired hepatitis C with acute clinical disease, 75% have peak ALT elevations >15 times the upper limit of normal, and 70% have bilirubin levels ≥3 mg/dl.[13,16] Case-fatality rates are available only for PT-NANB hepatitis as a whole; in the United States, the rate is 1–2%.[13,59] The etiology of fulminant PT-NANB hepatitis remains controversial.[336] In several published case series of fulminant PT-NANB hepatitis from Western countries, few had evidence of HCV infection (or other known viruses), suggesting that a PT-NANB hepatitis agent other than HCV may be responsible for most deaths from acute disease.[198,337] Alternatively, other etiologies, such as hepatotoxins or mutant hepatotropic viruses, have not been definitively excluded.[336] In Japan, however, one study found evidence of HCV infection in most patients with acute liver failure.[342]

The patterns of expression of anti-HCV and HCV-RNA have been described in previous sections. Anti-HCV (by EIA-2) becomes detectable in patients with posttransfusion hepatitis C as early as 5 weeks after exposure and is present in 80% 5 to 6 weeks after onset of hepatitis; in rare instances, seroconversion may not occur for 6–9 months.[16] Among most patients, anti-HCV remains detectable for years after their primary infection, regardless of the outcome of disease.[16] HCV-RNA becomes detectable within several days after experimental inoculation of animals with HCV[26,103]; among patients infected with HCV after transfusion, HCV-RNA is detected by the second week after transfusion. HCV-RNA may be detected for years after the primary infection in most persons.[16]

4.8. Control and Prevention

Routine anti-HCV screening of blood donors (with or without surrogate marker testing) has substantially reduced the incidence of posttransfusion hepatitis C[62,98]; routine anti-HCV screening of donors of organs, tissues, and semen is also recommended.[62] As more sensitive assays are developed and introduced, we can look forward to the virtual elimination of transfusions and transplants as a source of HCV infection, similar to what has occurred for hepatitis B. High-risk behavior modification, such as safer needle-using practices, seems to have reduced the number of hepatitis C cases among injection drug users. However, many persons who acquire hepatitis C cannot be readily identified as belonging to this or other high-risk groups.

Several studies have attempted to assess the value of prophylaxis with immune globulins against posttransfusion NANB hepatitis, but the results are difficult to interpret and compare because of the lack of uniformity in diagnostic criteria, mixed sources of donors (volunteer and commercial), and different study designs (some lack blinding and placebo controls). In some of these studies, immune globulins seemed to reduce the rate of clinical disease but not overall infection rates[173,278,280]; in one, patients receiving immune globulin were less likely to develop chronic hepatitis.[278]

None of these data have been reanalyzed since anti-HCV testing became available, and in only one study was the first dose of immune globulin given after, rather than before, the exposure (i.e., transfusion), making it difficult to assess its value for postexposure prophylaxis. In addition, recent HCV transmission experiments with animals demonstrated that chimpanzees convalescent from HCV infection were not protected against rechallenge with homologous HCV strains,[103] and postexposure prophylaxis with HCV immune globulin failed to protect from infection.[180a]

Studies have shown that α interferon therapy may have a beneficial effect among some patients.[89,93] In these studies, such therapy resulted in marked improvement of serum aminotransferase activity among approximately half of the patients treated, with most of these patients also having improvement in liver histology and a decrease in serum levels of HCV-RNA. However, improvement of aminotransferase activity was sustained among only 10–51% of the patients, although in most of the patients with a sustained response, HCV-RNA remained undetectable.[93] In addition, no clinical, demographic, serum biochemical, serological, or histological features have been identified that reliably predict which patients will respond to treatment and which will have a subsequent long-term remission.[93] Among patients who respond to therapy but later suffer a relapse, a second course of therapy may result in a beneficial response, but whether the second response will be sustained remains unclear.[93]

4.9. Unanswered Questions

4.9.1. Hepatitis C. A number of problems remain to be solved in the area of diagnostic testing for HCV infection. No true confirmatory test for the screening anti-HCV EIA is available because HCV has not been grown in cell culture in sufficient quantities to obtain authentic viral proteins. In addition, there is a need for diagnostic tests that will differentiate acute infection from chronic or resolved infection.

Occupational, sexual, household, and perinatal transmission of HCV clearly occurs. Yet the relative inefficiency of HCV transmission in the absence of large or repeated direct percutaneous exposures to infectious blood remains unexplained. Our understanding of the risks of transmission in these settings is limited by inadequate studies, the need for more sensitive tests for the detection of infection, and our inability to quantitatively measure infectivity.

No neutralizing antibody to HCV has been identified, and available data suggest that few persons recover from HCV infection and that most remain chronically infected with HCV even in the absence of active liver disease. The failure to demonstrate recovery from HCV infection or protective immunity after single or multiple episodes of infection raises major concerns for the development of effective vaccines against HCV, although preliminary data suggest that recombinant vaccine constructs may provide some protection in the chimpanzee model of infection.[77]

4.9.2. Other PT-NANB Hepatitis Agents. Prior to the discovery of HCV, there were both laboratory and epidemiologic data suggesting the existence of more than one PT-NANB hepatitis agent. Such data included the occurrence of multiple episodes of acute NANB hepatitis in the same person, primarily injection drug users; the observation of long and short incubation periods among persons with posttransfusion NANB hepatitis; differences in liver ultrastructural pathology; and differences in physicochemical properties between different PT-NANB hepatitis inocula.[96] In addition, cross-challenge studies in chimpanzees with different inocula seemed to demonstrate no cross-protection between a chloroform-sensitive, ultrastructural tubule-forming NANB agent and a chloroform-resistant, non-tubule-forming NANB agent.[96]

Since the discovery of HCV, most of these observations can be explained by what we now know to be characteristics of this virus. The failure of HCV to induce an effective neutralizing antibody response and to provide protection from reinfection or superinfection with the same or different strains of HCV would explain multiple episodes of disease in the same person as well as the results of animal cross-challenge studies. Furthermore, the fluctuating nature of hepatic inflammation along with the high rate of persistent infection associated with HCV[16] make it difficult to interpret recurrent episodes of disease in humans or animals. However, HCV serological evaluation of chimpanzee inoculation experiments in one laboratory has demonstrated that infection with the chloroform-resistant non–tubule-forming agent failed to elicit any serological evidence of HCV infection,[96] supporting the existence of distinct PT-NANB hepatitis agents.

HCV serological evaluation of patients with PT-NANB hepatitis using both commercial and research-based assays have shown that 10–15% have no evidence of HCV infection and could be classified as non-ABCDE hepatitis.[16] Compared with patients with acute hepatitis C, patients with acute non-ABCDE are less likely to be jaundiced, have lower peak ALT levels, and a lower frequency of acquiring chronic hepatitis (0–29%).[1,16,176] In spite of this propensity for milder illness, some studies have indicated that a non-ABCDE agent is responsible for most PT-NANB fulminant hepatitis.[336] An epidemiologic study of community-acquired disease has shown that 75% of non-ABCDE hepatitis patients had risk factors or characteristics associated with acquiring viral hepatitis.[16] The cause of disease in patients with these profiles is unclear, but the presence of chronic disease, as well as epidemiologic factors consistent with the transmission of infectious disease, lends further support to the existence of at least one other PT-NANB hepatitis agent.

Recently, a new flavivirus has been characterized from patients with PT-NANB hepatitis and designated hepatitis G virus.[199a] However, it is not clear whether this infection is associated with acute or chronic hepatitis.[16a]

5. Delta Hepatitis

5.1. Historical Background

Hepatitis delta virus was discovered by Rizzetto[269] in 1977 and was initially described as a new antigen detectable in patients with HBV-associated chronic liver disease. Studies in chimpanzees in 1979–1980 established that HDV was a transmissible agent, but one dependent on the presence of active HBV infection to cause infection.[270] Serological tests for HDV and antibodies to HDV developed by the early 1980s have been used to characterize the clinical and epidemiologic aspects of infection.[43,109]

The dependence of HDV on HBV replication permits HDV infection to occur in two primary circumstances—

as HDV-HBV coinfection of a person susceptible to HBV infection and as HDV superinfection of a person with chronic HBV infection.[270] A third form of HDV infection—latent infection—may occur in persons who receive liver transplants for fulminant or chronic HBV-HDV liver disease, in which HDV infection of the new liver may occur without the recurrence of HBV infection.[247,318] The nature and consequences of these infections differ. HDV superinfection of an HBV carrier usually leads to establishment of chronic HDV infection and chronic hepatitis, whereas coinfection usually causes self-limited hepatitis. Latent HDV infection does not cause liver disease unless HBV infection recurs, in which case both acute and chronic hepatitis usually ensue.

5.2. Methodology Involved in Epidemiologic Analysis

5.2.1. Morbidity and Mortality Data. Delta hepatitis infection is not a reportable disease in the United States, and precise data are not available regarding numbers of cases or deaths that occur annually. Disease burden has been inferred from studies measuring the frequency of concomitant HDV infection in cases of acute and fulminant hepatitis B and in persons with chronic liver disease. In 1992, in the United States, it was estimated that 7500 cases of acute HDV infection occurred, that 70,000 persons were chronically infected with HDV, and that as many as 1000 persons died annually due to HDV-related chronic liver disease or fulminant HDV hepatitis.[12] However, these data do not offer information on trends of disease over time.

5.2.2. Surveys. Surveys measuring prevalence of HDV infection in various populations with acute or chronic HBV infection provide the most important information on the epidemiology of HDV. Because HDV infection is dependent on active HBV infection, testing of persons positive for HBsAg provides the most useful information. Surveys of the frequency of HDV among persons with asymptomatic chronic HBV infection and in persons with HBV-associated chronic liver disease are used to assess relative disease importance in a general population.[109,259] Surveys of HDV prevalence among HBsAg-positive persons in various groups at high risk of hepatitis B (e.g., parenteral drug users, homosexual men, hemophiliacs, family contacts of HBV carriers) provide useful information on relative frequencies in these groups and on epidemiologic patterns of disease transmission.[218]

5.2.3. Laboratory Diagnosis. Laboratory techniques for cell culture cultivation of HDV are not available, primarily because no *in vitro* system is available to isolate HBV, its host virus. Therefore, laboratory diagnosis is based on tests to detect virus proteins, antibodies to viral proteins, and HDV-RNA in serum and/or liver. Serological tests to detect antibodies to the delta antigen provide the most widely used diagnostic and epidemiologic tools. Both IgG and IgM antibodies develop early in the course of natural HDV infection.[20,43] Tests for total anti-HDV are useful for epidemiologic studies of chronic delta infection and to diagnose acute disease. IgM tests appear to have higher sensitivity in detecting acute disease and can also be used to define the presence of active liver disease in those with chronic infection.[20,53,104] Sensitive RIAs and EIAs for total anti-HDV, IgM anti-HDV, and IgG anti-HDV have been developed, and a test for total anti-HDV is commercially available in the United States.

Tests for delta antigen in both serum and liver specimens are available primarily in research settings. The RIA and EIA tests for delta antigen in serum require prior treatment with detergent to disrupt the viral particles.[109,125] These methods have modest sensitivity and are only positive during the acute phase of infection. Immunoblot assays are more sensitive but only available in research laboratories.[255] Immunoperoxidase and immunofluorescence assays for delta antigen in liver are sensitive and able to detect intrahepatic antigen in cases of chronic HDV hepatitis.[269] Such assays require liver biopsy specimens, but have proven useful in retrospective epidemiologic studies of chronic HDV infection. Molecular cloning of the HDV genome has permitted development of cDNA and ssRNA assays for the delta genome in serum and liver (*in situ* hybridization).[291] These assays have proven sensitive for detection of circulating delta virus and appear useful to differentiate active chronic HDV infection with infectivity to others from inactive HDV infection in HBV carriers.

5.3. Biological Characteristics

Animal studies, primarily in chimpanzees, have provided much of the basis for the biological characterization of HDV and its infection.[265,270] Early studies showed that HDV infection was dependent on HBV infection, only able to cause disease from a coprimary infection with HBV or from the superinfection of an HBV carrier. The natural host range of HDV appears to be restricted to humans. However, experimental coinfections have been produced in chimpanzees, and HDV infection can be established in species that become chronically infected with their respective hepadnavirus (i.e., chimpanzees, woodchucks, ducks).[43,213] Natural HDV infection has not been identified in any of these species to date.

The HDV is a 35- to 38-nm enveloped particle that contains a small circular single-stranded RNA of 500,000

D (1.7 kilobases), a unique internal protein (the delta antigen), and an outer coat of the hepatitis B surface antigen (HBsAg).[41] The virus appears most closely related to certain disease-causing organisms of plants (viroids, satellite RNAs, satellite viruses) based on nucleic acid structure and replication mechanisms.[326] The HDV-RNA forms a rod-shaped tertiary structure and encodes the delta antigen, which has two subspecies of slightly different length, the shorter of which is required for HDV replication and the latter for HDV packaging into viral particles.[306] The HDV-RNA has no homology with HBV-DNA, and the nature of dependence on HBV is not known beyond the need for the HBsAg as the surface coating of the virus particle. The resistance of HDV to heat appears similar to that of HBV; however, its resistance to commonly used disinfectants is not known. Only a single serotype of HDV has been recognized, but sequencing of the HDV genomes reveals 86–90% homology of different isolates, which have been divided into three genotypes: one from North America, Europe and Asia; one from Japan; and one from tropical South America.[58] HDV disease severity varies substantially; disease in northern South America appears to be more severe than that in Europe and North America and to have a different histopathology during acute infection.[52] Disease in some populations in Greece and the Western Pacific appears minimal. Whether such differences are related to strain variations or to other environmental or genetic factors is currently unknown.

5.4. Descriptive Epidemiology

The transmission patterns and epidemiology of HDV infection closely parallel those of HBV infection. In general, HDV infection occurs with highest frequency in those parts of the world with highest HBV endemicity and in those persons who are at highest risk of infection.[43,109] Nevertheless, several departures from this pattern, including low frequency in Asian countries and in homosexual men, suggest important differences in the epidemiology of these viruses.

5.4.1. Prevalence and Incidence. Few data on the incidence of HDV infection are available, and the epidemiology of HDV must be inferred from studies of HDV prevalence in persons with active HBV infection. Studies that evaluate prevalence of anti-HDV in asymptomatic HBV carriers, persons with acute and fulminant hepatitis, and with HBV-chronic liver disease are most widely available. Because HDV augments the severity of HBV infection, the prevalence of HDV tends to be four- to fivefold higher in persons with chronic hepatitis than in asymptomatic HBV carriers, and in persons with fulminant hepatitis B cases than in acute, self-limited hepatitis B.[125,289]

The prevalence of HDV infection worldwide tends to correlate with endemicity of HBV infection. Prevalence in asymptomatic HBV carriers varies from 0 to 25% worldwide, and prevalence among persons with HBV chronic liver disease tends to vary more widely, from 5 to 100%. In low-HBV endemicity areas such as the United States, prevalence of HDV infection is low (0–5%) in HBsAg-positive individuals and 10–25% in persons with HBV-chronic liver disease, but is higher in certain groups at high risk of HBV infection, particularly injection drug users and hemophiliacs.[218] In moderate-HBV endemicity areas, HDV prevalence tends to be higher, reaching 15% in persons with chronic HBV infection and 30–50% in HBV-chronic liver disease cases in areas such as southern Italy.[290] In high-HBV-endemicity areas, HDV prevalence reaches the highest levels (20% in HBsAg-positive persons and up to 90% in HBV-chronic liver disease) in parts of the Amazon Basin.[218] Nevertheless, HDV prevalence in high-HBV-endemicity areas may be highly variable, differing markedly in neighboring tribal groups in Kenya and among persons living in different villages in a single Indian group in Venezuela.[109,135] HDV prevalence is generally low in east and southeast Asia despite high HBV endemicity throughout this region.[259]

In the United States and western Europe, prevalence studies have shown a high risk of HDV infection in HBsAg-positive injection drug users (30–50%) and hemophiliacs but much lower prevalence in other high-HBV-risk groups, including homosexual men, persons with multiple heterosexual partners, the developmentally disabled, hemodialysis patients, and household contacts of HBV carriers.[138,218,255] The low prevalence in homosexual men (usually less than 5% among those with asymptomatic chronic HBV infection) contrasts markedly with that in injection drug users (usually >20%) despite comparable risk of HBV infection. This suggests that HDV is transmitted less efficiently by sexual contact than by blood exposure. In the United States, HDV infection is found in about 5% of cases of acute hepatitis B and in up to 25% of cases of fulminant hepatitis.[218].

5.4.2. Epidemic Behavior. Outbreaks of HDV infection have been recognized primarily in two settings: among injection drug users in the United States and western Europe and in high-HBV-endemicity populations in South America and Africa. Outbreaks among drug abusers usually involve coprimary transmission of HBV and HDV, may cause high mortality (up to 10%) from fulminant hepatitis, and may result in secondary transmission to sexual contacts.[53,195] Outbreaks of HDV superinfection

in persons with chronic HBV infection have been recognized in indigent local populations in Brazil, Venezuela, Colombia, the Central African Republic, and southern Kashmir.[52,128] These outbreaks characteristically affect children and young adults and cause high acute mortality (up to 20%) and chronic morbidity (up to 50%). Transmission is believed to occur primarily via open skin sores and sexual contact. In the latter, whole families chronically infected with HBV may succumb to fulminant hepatitis during the outbreaks. Unique outbreaks have been reported in institutions for the developmentally disabled and in contacts of an infected butcher.

5.4.3. Geographical Distribution. HDV infection has been observed in all parts of the world, with frequency generally corresponding to HBV endemicity (Fig. 6).[43,109,259] Highest HDV prevalence has been observed in widely dispersed high-HBV-endemicity areas including the Amazon Basin and other parts of northern South America, parts of Africa, and Romania. Moderate prevalence of HDV infection has been documented in southern Italy, parts of eastern Europe, the Middle East, Africa, and in some Pacific Islands groups. As noted previously, HDV prevalence is low in the low-HBV-endemic areas of western Europe, North America, and Australia, although prevalence is high in injection drug users in all these areas. The prevalence of HDV infection is very low in east and southeast Asia despite high HBV endemicity throughout this region. The reasons for this are unclear; HDV infection is present in neighboring areas of Asia and in some Pacific Islands, and a high frequency of HDV infection has been observed among injection drug users and prostitutes in Taiwan.[75,255] Possibly this virus has only been recently introduced, via drug abusers, into ethnic Oriental populations.

5.4.4. Temporal Distribution. Few data are available regarding long-term temporal trends of delta infection. HDV infection has now been identified as the cause of fulminant hepatitis in northern Colombia since the 1930s, through detection of antigen in samples of liver from historic studies of yellow fever, and is presumed present in the Amazon Basin of Brazil for a similar duration as the cause of Labrea hepatitis.[52] Delta antibodies have been found in IG lots produced in 1945, and HDV has been implicated as the cause of hepatitis outbreaks in an institution for the developmentally disabled in the 1960s.[218] It has been shown that HDV first appeared in injection drug users in Scandinavia and Australia during the 1970s, suggesting more recent introduction into those regions.[138]

5.4.5. Age. The age-specific frequency of delta infection depends on the predominant age of HBV infection in the region. In the United States and western Eu-

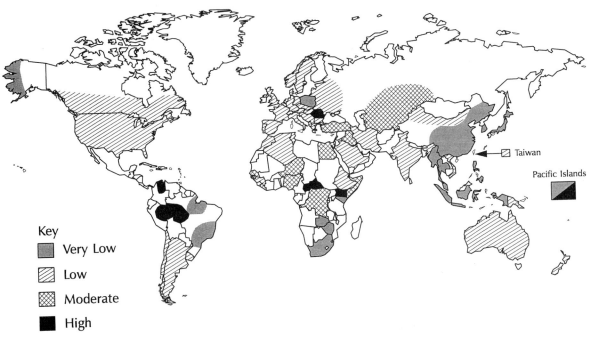

Figure 6. Worldwide distribution of hepatitis D virus infection.

rope, HDV infection, like HBV infection, occurs primarily in young adults.[43,218] In contrast, in South America and other areas in which HBV infection occurs in childhood, delta infection occurs in both children and adults; in these areas, delta antibody prevalence in HBV carriers is usually low in young children (<10 years) but higher in older children, adolescents, and young adults, possibly due to sexual activity.[135]

5.4.6. Sex. Delta infection is observed more commonly in men than in women. The highest HBV risk groups in the United States and Europe are predominantly male (drug users, hemophiliacs, and homosexual men). In high-HBV-endemicity areas, men are more likely to become HBV carriers, and thus to become susceptible to HDV superinfection. Although HDV infection is more frequent in men for these reasons, it has been observed among the female sexual contacts of parenteral drug users and in prostitutes, and in endemic areas infection occurs with comparable frequency among male and female HBV carriers.[135,195]

5.4.7. Race. There is no general racial predilection for HDV infection except as correlates with HBV endemicity; cases have been observed in all racial groups in all parts of the world. Nevertheless, the low frequency among ethnic Oriental populations in Asia and in Alaskan Eskimo populations remains unexplained.

5.4.8. Occupation. HDV infection poses a modest occupational hazard for health care workers who may be exposed to blood from HBV-HDV carriers; however, few actual cases have been reported. Wide use of hepatitis B vaccine should eliminate the risk of HDV infection in this population.

5.4.9. Occurrence in Different Settings. Delta virus infection has not been reported in the military and is uncommon in institutional settings. Nevertheless, an outbreak of HBV infection in a facility for the mentally retarded has been reported in the United States during the 1960s, a period of severe crowding and poor conditions in such facilities.[218] Transmission in this setting appears negligible following deinstitutionalization, reduction in crowding, and improvement in staffing ratios.

Delta virus transmission within the family setting is important in areas of moderate-to-high HBV endemicity. Thus, in southern Italy, high rates of HDV infection are observed among HBV carriers living in the same household as persons with chronic HBV-HDV infection.[43] In addition, during outbreaks of HDV superinfection in South America, risk of acquiring HDV was highest for persons with chronic HBV infection who lived with an index case, accounting for the familial clustering of cases of fulminant hepatitis observed in these regions.[135]

5.4.10. Socioeconomic Factors. Socioeconomic

factors play a role in HDV transmission primarily through their impact on HBV endemicity. HBV infection is highly endemic in most of the poorest populations worldwide, and, with the exception of southeast Asian countries, the HBV carriers in these populations are at risk of HDV infection. Crowding, skin disease, sexual activity, and use of nonsterilized needles all may play a role in HDV transmission in these areas. Medical use of nonsterilized needles recently has been implicated as a cause of high rates of HBV and human immunodeficiency virus infection in Romania and may account for high risk of HDV infection in this population as well.[140]

5.5. Mechanisms and Routes of Transmission

HDV, like HBV, is found in highest concentration in the blood of persons with acute or chronic infection. Virus titers may be extremely high (up to 10^{12} infectious doses/ml) and exceed those of HBV.[265] Because delta antigen is only detectable during acute HDV infection, virus titers may be higher in acute than in chronic infection. There is no consensus as to whether HBeAg status is related to the outcome or relative infectivity in HDV infection. HDV may be presumed present in serum-derived fluids such as wound exudates; however, its presence in other body fluids (semen, saliva, feces) has not been studied.

Transmission occurs by percutaneous exposure to blood containing HDV, either directly or indirectly, and by sexual contact. Direct blood-borne transmission appears highly efficient as shown by high risk in parenteral drug abusers and hemophiliacs. However, indirect exposure to blood through open skin wounds, either directly or indirectly via contaminated environmental surfaces, has presumably been the major transmission route in the outbreaks in South America and in the outbreak in a butcher shop in Australia.[128] Sexual contact can transmit disease but is probably less efficient, as evidenced by the low frequency of HDV infection in homosexual men. HDV infection risk in homosexual men increases with number of sexual partners and HDV infection occurs in prostitutes who deny parenteral drug abuse.[75,255] Perinatal transmission from mother to infant can occur, but only when the mother is HBeAg positive.[345] This combination of markers is uncommon among groups studied to date, and so perinatal transmission is of minimal importance. Casual contact does not result in virus transmission, but HBsAg-positive household contacts of HBV-HDV carriers are at significant risk of infection over long periods of time.[43] HDV, like HBV, is not spread by the fecal–oral or airborne routes.

5.6. Pathogenesis and Immunity

The natural course of HDV infection differs depending on whether it is a coinfection with HBV or a super-infection of a person with chronic HBV infection. Coinfection of an individual susceptible to HBV infection follows exposure to an inoculum containing both HBV and HDV, and the incubation period is 6 to 12 or more weeks, as occurs in natural HBV infection. In chimpanzees, the initial phase of HBV replication in the liver is followed by HDV replication after a variable interval.[265,270] Onset of liver inflammation usually coincides with the appearance of delta antigen in the liver, but it may be biphasic. In coinfection, HDV replication is almost always limited by natural resolution of the HBV infection. Fewer than 5% of such infections lead to the chronic HBV carrier state and persistent HDV infection, although chronic hepatitis may progress rapidly in those who develop chronic infection.[53] The late phase of the infection is marked by disappearance of both HBsAg and delta antigen in liver and serum and development of specific antibodies to both viruses. The development of anti-HBs provides protection against reinfection with both HBV and HDV.

HDV superinfection follows exposure of an HBV carrier to an HDV-positive inoculum, and evidence of HDV infection may appear within 2 to 6 weeks.[270] Replication of HDV may occur immediately because of the established presence of HBV in liver cells, and liver inflammation and acute hepatitis appear rapidly. Acute HDV infection may inhibit HBV replication and occasionally terminate the HBV carrier state.[43] However, in the majority of cases, persistent HBV infection allows HDV infection to persist indefinitely.[288] Although both IgM and IgG anti-HDV appear during acute HDV infection, these antibodies do not signify resolution of infection or elimination of the virus. In the majority of HDV superinfections, delta antigen persists in the liver, HDV-RNA can be detected in blood, and chronic hepatitis develops.[288,291] In some cases, delta antibody persists, although evidence of HDV replication in the liver or HDV-RNA is absent in serum.[291] It is uncertain whether this represents resolved or merely quiescent infection.

The pathogenesis of HDV infection is not known. In HDV infection in western Europe and the United States, the pathology of acute and chronic HDV infections does not differ substantially from that of other types of hepatitis.[320] In South America, however, the acute lesion in HDV superinfection involves a characteristic picture of eosinophilic necrosis and microvesicular fat infiltration, the latter similar to lesions of posttransfusion non-A, non-B hepatitis, tetracycline toxicity, and fatty liver of pregnancy.[52] Evidence from tissue culture and absence of liver disease in latent HDV infection following liver transplantation suggests that HDV is not cytopathic; thus the mechanism in liver injury is currently unknown.[247,306,318]

5.7. Patterns of Host Response

The spectrum of clinical disease in acute HDV infection, whether HDV-HBV coinfection or HDV superinfection, may vary from no illness to acute and fulminant hepatitis. In general, HDV augments the severity of both acute and chronic HBV infection. Current data suggest that 50–70% of cases of HDV co- or superinfection cause an episode of clinical acute hepatitis with jaundice, in contrast to about 20–30% of HBV infections that cause icterus. The risk of fulminant disease may reach 10% for clinical HDV-HBV coinfections and 20% in HDV superinfections.[128,195] In studies in both the United States and Europe, HDV infection has been found in 30–50% of fulminant cases of HBsAg-positive hepatitis, a frequency five times higher than found in acute, uncomplicated HBV infection.[218,289]

The spectrum of illness caused by chronic HDV infection also ranges from no symptoms to chronic active hepatitis, which may progress rapidly to cirrhosis and death from liver failure.[21,271] Delta superinfection in an HBV carrier markedly increases the risk of chronic liver disease; markers of HDV infection are two to ten times more common in persons with chronic hepatitis than in HBV carriers without liver disease.[109] Based on studies in Italy and South America, it is estimated that 50–90% of persons with HDV superinfection develop chronic hepatitis.[42,135,271] Among chronic liver disease patients, those with HDV usually are more likely to have chronic active than chronic persistent disease.[291] Persons with HDV-related chronic active hepatitis often progress steadily to cirrhosis within 5 or fewer years; these patients do not respond to immunosuppressive therapy, although some respond to high doses of α interferon.[42,140,271]

Persons with chronic HBV infection and HDV infection may have no symptoms and lack evidence of liver inflammation or detectable HDV in liver or blood.[21] Nevertheless, a significant proportion (38%) of such persons have been found to have occult chronic liver disease on careful study.

Available evidence suggests that HDV infection may be associated with an increased risk of developing primary hepatocellular carcinoma (PHC). PHC appears to develop at a younger age in HDV-infected patients than in HBV carriers without HDV infection.[43,218] However, some studies of HBsAg-positive cases of PHC have failed to identify HDV in more than a small minority of such cases.

Recent work in recipients of liver transplants has demonstrated the existence of latent HDV infection, in which HDV may persistently infect the new liver of recipients in the absence of HBV infection. This infection appears nonpathogenic; however, when HBV infection also occurs, both acute and progressive chronic hepatitis usually develop rapidly.[247,318]

5.7.1. Clinical Features. The clinical features of acute and chronic HDV infection are similar to those of other types of acute viral hepatitis and chronic hepatitis, and there is no way to distinguish either disease without appropriate serological tests. Acute HDV-HBV may cause, in about 15% of cases, a biphasic illness that is not common in any other type of hepatitis.[53] Several studies have suggested that splenomegaly may be a common feature in chronic HBV infection.[271]

5.7.2. Diagnosis. Delta hepatitis infection is diagnosed by serological testing for HBV and HDV markers in persons with clinical symptoms of acute or chronic hepatitis. Acute HBV-HDV coinfection is verified by positive test for IgM anti-HBc (specific for acute HBV infection) accompanied by the presence of delta antigen or total and/or IgM anti-HDV. Delta antigen is usually present during early acute illness, whereas both IgM and IgG antibodies appear within several days to weeks after onset.[53] The antibody response in such cases is not strong, and accurate diagnosis is best accomplished during acute illness or early convalescence.[53,270] Only a total anti-HDV test is commercially available in the United States and may result in underdiagnosis of this disease.

Acute HDV superinfection is diagnosed by finding HDV markers in an HBsAg-positive person with acute hepatitis but IgM anti-HBc negative (indicative of the chronic HBV infection). Delta antigen may be present in early acute illness, but both the IgM and IgG specific antibodies usually appear rapidly, and the test for anti-HDV usually becomes strongly positive during acute illness.[288] In cases of fulminant hepatitis, HDV markers in serum may be negative, although delta antigen may be demonstrable in liver.[195]

The diagnosis of chronic HDV hepatitis is usually made in HBsAg-positive persons with chronic hepatitis who are also positive for anti-HDV. The most active cases can be distinguished by the presence of greater degrees of liver inflammation, the presence of IgM anti-HDV and HDV-RNA in the serum, and delta antigen in the liver.[42,104]

5.8. Control and Prevention

Measures for control and prevention of HDV infection are identical to those for HBV. General measures to prevent HBV transmission, which include sterilization of needles and instruments that penetrate the skin in a medical care setting and screening of blood for HBsAg, are effective in preventing both HBV and HDV infection. With routine screening of blood donors for HBsAg, posttransfusion HDV infection, like HBV infection, has been uncommon in developed countries; nevertheless, risk for recipients of pooled plasma products (e.g., hemophiliacs) remains.[43]

Hepatitis B vaccine has 85–95% efficacy in preventing HBV infection and is presumed equally effective in preventing HBV-HDV coinfection. Vaccination of persons at risk of HBV infection is the best protective measure against HDV coinfection as well.

For HBV carriers at risk of HDV superinfection, there is no specific prevention that can be offered. Several drug regimens may eliminate the HBV carrier state, but effectiveness is usually less than 50% and is highest in those with most recent infection. In these persons, neither hepatitis B vaccine nor HBIG is effective in preventing HDV infections. The major preventative that can be offered is counseling to avoid exposure to contaminated needles or sexual exposure to HBV-HDV carriers. Nevertheless, such measures are likely to have limited impact among drug abusers or in HBV-HDV endemic areas in which poverty, crowding, insect bites, and open skin lesions facilitate disease transmission.

Treatment of persons with HDV infection with high doses of α interferon has been shown to induce a remission in liver disease in between 25 to 70% during therapy, but many persons relapse when treatment is discontinued.[144] Up to 33% of treated persons may lose HDV-RNA and a smaller proportion lose HBsAg, with a cessation of liver disease. Liver transplant is increasingly done in persons with end-stage HDV-HBV liver disease; HDV and HBV infection recurs in between 50 to 90% of persons, usually with recurrence of liver disease.[247,318]

5.9. Unresolved Problems

Major unanswered questions include defining the biological origin of HDV, its mechanisms for dependency on HBV and for causing disease, and whether more than a single strain of virus exists. Limiting the risk of HDV transmission, particularly among HBV carriers in less-developed countries, is an important public health challenge of the future. Measures to modify the risk of delta carriage or decrease the risk of disease transmission are not available for the highest risk populations. Initial trials testing an anti-HDV vaccine (recombinant delta antigen) have been completed in animal models, but show at best only partial protection; since the delta antigen is sequestered within the virus particle, such a vaccine may have a low likelihood of being effective.

6. Hepatitis E

6.1. Historical Background

Large epidemics of viral hepatitis that epidemiologically resembled hepatitis A have been reported primarily from Asia since the mid-1950s. After the development of serological tests for the diagnosis of HAV infection in the 1970s, retrospective examination of acute and convalescent serum specimens from these early outbreaks indicated that the vast majority of cases were not attributable to HAV infection.[333] These observations defined a new form of non-A, non-B hepatitis that was transmitted via the fecal–oral route. Subsequently, numerous outbreaks of this enterically transmitted non-A, non-B hepatitis were identified in many parts of the world because of the widespread use of IgM testing for the diagnosis of acute hepatitis A. In addition, this new disease had several distinct epidemiologic features that differentiated it from hepatitis A, including a high attack rate among adults and an unusually high case-fatality rate among pregnant women.[160,168] This form of viral hepatitis was initially designated epidemic non-A, non-B hepatitis and then enterically transmitted non-A, non-B (ET-NANB) hepatitis until an etiologic agent was fully characterized in the late 1980s, at which time it was designated hepatitis E.

Clinical material from a number of outbreaks led to the identification, by IEM, of viruslike particles in the feces of patients and the establishment of an animal model of disease in cynomolgus macaques.[47,162] The establishment of the experimental model provided material for the morphological and biophysical characterization of what is now known as hepatitis E virus (HEV). In 1990, HEV was cloned and sequenced, and immunodiagnostic tests were soon developed from recombinant expressed antigens or synthetic peptides.[44,267] Although many questions remain concerning the epidemiology of HEV infection, recent studies suggest that vaccination may become an approach to preventing this disease in those parts of the world where hepatitis E is endemic.

6.2. Methodology Involved in Epidemiologic Analysis

6.2.1. Sources of Mortality Data.

Hepatitis E has primarily been reported from countries of the developing world where the diagnostic capabilities to differentiate this disease from hepatitis A are limited. The primary sources of data are investigations of outbreaks that used diagnostic testing to exclude HBV and HAV infection and more recently have used newly developed diagnostic tests for antibody to HEV (anti-HEV). Serologically defined mortality is not reported and hepatitis E is not included in the ICD-9. However, the unusual characteristic of this infection that results in high maternal mortality during pregnancy has served as an indicator of epidemics of hepatitis E.

6.2.2. Sources of Morbidity Data.

The primary sources of data concerning morbidity from hepatitis E comes from investigations of epidemics and of convenience samples of sporadic cases of acute hepatitis. The limited availability of serological testing for IgM anti-HAV and IgM anti-HBc and more recently anti-HEV, in those parts of the world where hepatitis E is endemic has resulted in a lack of etiology-specific reporting of most types of acute viral hepatitis. In addition, hepatitis E or ET-NANB hepatitis has not been made a reportable disease category. Currently, the only reliable information concerning disease morbidity comes from outbreak investigations and limited studies undertaken to define the relative contribution of hepatitis E and/or ET-NANB hepatitis to the disease burden caused by acute viral hepatitis.

6.2.3. Surveys.

Definition of the prevalence of HEV infection has been hampered until very recently by the lack of reliable serological tests. Well-designed, age-specific, population-based seroprevalence studies are needed to define whether there are significant differences in the relative endemicity of HEV worldwide, as might be inferred from reported outbreaks of the disease. Most currently reported data represent samples of conveniently available populations, including those involved in outbreaks or living in areas where outbreaks have occurred. Thus, these populations may not provide an accurate representation of the prevalence of infection, and these data should be interpreted cautiously.

6.2.4. Laboratory Diagnosis.

Initial studies to detect antibody to HEV (anti-HEV) was accomplished by IEM using virus isolates from fecal specimens or by a fluorescent antibody-blocking assay using HEV antigen in tissue.[23,47,179] Isotype-specific antibodies to HEV have been detected by Western blot analysis using recombinant expressed antigens, and EIAs have been developed using recombinant expressed proteins or synthetic peptides representing immunodominant epitopes of the putative structural proteins of HEV[85,105,343]; most have included a number of distinct antigenic domains from at least two geographically different HEV strains. Supplemental neutralization tests utilizing synthetic peptides have been developed to provide confirmation of anti-HEV activity.

Several studies of antibody persistence following acute hepatitis E suggest that IgG anti-HEV may disappear in some patients after periods as short as 1 year.[203,343] However, detection of antibody persistence may depend on the epitopes included in the immunoassay. Until the natural history of the antibody response to HEV infection is defined, it is difficult to determine the true frequency of HEV infection in any population. At the present time, it is not known whether the IgG anti-HEV detected by these assays represents neutralizing antibody(ies).

Reverse transcription–polymerase chain reaction (RT-PCR) has been used to detect HEV-RNA in experimentally infected animals, as well as in humans with naturally occurring infection.[221,313] HEV-RNA has been obtained by RNA extraction or by immunoprecipitation or immunocapture of the virus prior to amplification by RT-PCR.[36,221] No assays have been developed that will detect HEV antigen in stool specimens, although HEV-Ag has been detected in liver biopsies of experimentally infected animals using fluorescent antibody microscopy.[179]

6.3. Biological Characteristics of the Organism

HEV is a 32- to 34-nm virus that appears to be unstable in harsh environmental conditions such as exposure to high concentrations of salt, repeated freeze-thawing, or pelleting from stool suspensions.[44] The computed sedimentation coefficient is approximately $183S$, and the buoyant density as determined in a potassium tartrate/glycerol gradient is 1.29 g/ml. HEV that sediments at $165S$ has been presumed to be defective.[44]

The absence of a cell culture system for virus propagation has required the use of clinical specimens from experimentally infected nonhuman primates or from naturally infected humans for the characterization of HEV. Molecular cloning has shown that HEV contains a single-stranded positive-sense RNA genome with three open reading frames (ORF). The nonstructural genes appear to be located at the 5′ end and the structural genes at the 3′ end of the genome. ORF1 is approximately 5 kilobases (kb) in length and contains sequences associated with proteins involved in virus replication and processing.[44,267]

ORF2 is approximately 2 kb in size and contains the major structural proteins. ORF3 has 328 nucleotides that appear to code for structural proteins. It overlaps ORF1 by one nucleotide and is completely overlapped by ORF2. The genomic organization of HEV is substantially different from HAV and HCV, since structural and nonstructural proteins are coded by discontinuous, partially overlapping ORFs. In addition, the region coding for nonstructural proteins is located at the opposite end (5' vs. 3') of the genome from that for HAV and HCV.[44]. Recent studies suggest that HEV may belong to a family that includes caliciviruses.[44] Geographically distinct isolates of HEV have shown little genomic variability in those portions of the genome coding for structural proteins, whereas a high degree of variability has been found in regions coding for nonstructural proteins.[44,249,313]

6.4. Descriptive Epidemiology

Prior to the development of serological tests for acute HAV infection, the existence of ET-NANB hepatitis (i.e., hepatitis E) was largely unrecognized. When very large epidemics (10,000–30,000 persons) of enterically transmitted acute hepatitis among adults were recognized in India in the mid-1950s, one theory was that this was hepatitis A but that the inoculum was so great as to overwhelm existing humoral immunity.[229] However, retrospective IgM anti-HAV testing identified these early outbreaks as a new type of viral hepatitis.[333]

Limited studies indicate that a substantial proportion of acute viral hepatitis that occurs in young adults in Asia, the Indian subcontinent, the Middle East, and possibly Africa is caused by HEV.[105,154,167,203,237,308,314,333] The disease has been shown to occur in both epidemic and sporadic forms and is primarily associated with the ingestion of fecally contaminated drinking water. Recurrent epidemics of hepatitis E have been observed in several parts of the world including India, the central Asian republics of the former Soviet Union, and Xinjian province of northwest China. While it appears that HEV infection is endemic in these areas, it has not been determined whether there is a high prevalence of infection in the entire population or whether infection occurs only in persons with certain risk factors. All epidemics reported to date have been associated with drinking water contaminated with feces, usually during or following periods of extensive rain.

Several epidemiologic characteristics tend to distinguish hepatitis E from hepatitis A, and these differences have not been explained even with the availability of diagnostic tests. The majority of symptomatic infections occur among young adults, which is unexpected in areas where enteric infections are highly endemic among children and adults. Seroprevalence studies in endemic areas show surprisingly low rates of infection in children. Another unusual characteristic of hepatitis E is the high case-fatality rate among pregnant women. Mortality has reached as high as 30% and death is the result of liver failure.

6.4.1. Prevalence and Incidence. Prior to the availability of diagnostic tests, studies of outbreaks indicated that hepatitis E only occurred in young adults. However, recent studies have shown that hepatitis E and HEV infection occur in children.[105,124,203] The age-specific prevalence of anti-HEV in countries where the disease is endemic suggests that low rates of infection occur among children, with the highest rate of antibody found in adults 20–30 years of age. Whether these data reflect the true prevalence of infection in these age groups or whether there is a poor humoral response to infection that occurs at an early age is not known.

At this time, the majority of data on the epidemiology of hepatitis E come from surveys of clinical disease conducted during outbreaks. The incidence of disease during large epidemics in India and Nepal was 1400–1650/10^5,[166,229,243] whereas attack rates in smaller outbreaks in two Mexican villages were 5–6,000/10^5,[67] similar to the rates of clinical disease observed in outbreaks in refugee camps.[66] In Rangoon, Myanmar (Burma), the incidence of clinical ET-NANB hepatitis among other persons in households having an index case was 770/10$^{5[237]}$ Population-based studies have not been conducted to determine the incidence of hepatitis E or HEV infection using currently available immunoassays.

In developed countries, including the United States, cases of hepatitis E imported from countries with known epidemic or endemic disease have been detected, but there has been no evidence of transmission to other persons.[91] Small studies have shown a 1–5% prevalence of anti-HEV among blood donors.[90] However, since epidemiologic data were not collected in these studies, it is not known whether these represent true infections or false-positive results.

6.4.2. Epidemic Behavior. The dramatic nature of the epidemics produced by HEV gave rise to its initial designation of epidemic NANB hepatitis. These large epidemics have occurred in India, Nepal, Myanmar (Burma), and China.[36,229,237,243,333] In addition, smaller epidemics have been reported in Kyrghyzia and Tajikistan, Pakistan, Hong Kong, Algeria, Ivory Coast, Somalia, Kenya, and Mexico.[66,67,85,319]

Epidemics have occurred in a wide variety of settings

including urban areas, small towns and villages, and refugee camps. In all instances there has been evidence of poor sanitary conditions with inadequate disposal of feces and contamination of water supplies with fecal material. Population-based studies during epidemic periods have demonstrated clinical attack rates ranging from 0.7% to as high as 10%.[66,166,229,237,243] Secondary attack rates in households have ranged from 0.7 to 2.2%. Case-control studies during epidemics have shown that contact with an ill person outside of the household is also a risk factor for illness.[237] Cases have occurred among health care workers caring for patients during epidemic periods.

The high endemicity of infection in some areas is suggested by the occurrence of 7- to 10-year epidemic cycles that have been documented in central Asia,[85] similar to those observed for hepatitis A. In areas where long-term surveillance has been conducted, outbreaks of hepatitis E have shown a pronounced seasonal distribution most often associated with the local rainy season,[85] and the rapid onset of these epidemics has suggested a common source such as contaminated water.

6.4.3. Geographic Distribution. The precise worldwide distribution of HEV infection has not been determined due to the lack of well-standardized and readily available serodiagnostic tests. However, the geographic distribution of countries where outbreaks and sporadic cases have been reported indicates that this disease may be endemic in developing countries. Epidemics and sporadic cases of hepatitis E have been reported from most of central Asia, southeast Asia and Indonesia, the Middle East, northern Africa, sub-Saharan Africa, and Mexico (Fig. 7). Outbreaks have not been recognized in Europe, the United States, Australia, or South America. However, sporadic cases of hepatitis E have been identified in Mediterranean countries and cases imported from endemic areas have been recognized in the United States.

The recent availability of diagnostic tests for HEV infection should better define the true worldwide distribution of infection. In addition, it will be important to determine the uniformity of the distribution of infection within areas where HEV infection is considered endemic.

6.4.4. Temporal. In all epidemic situations the apparent mode of disease transmission has been by the fecal–oral route, usually through contamination of water supplies. This has usually occurred in those parts of the world where unusually heavy rains occur or monsoon conditions are present; disease incidence is highest during or just after the local rainy season. In many instances the high rates of disease have persisted through two rainy seasons followed by a significant decrease in the number

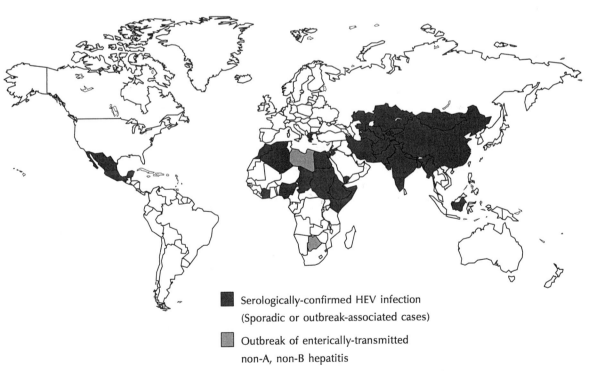

Figure 7. Countries with reported epidemic or endemic cases of hepatitis E virus infection. Adapted from Cromeans *et al.*[85]

of cases and an ending of the epidemic. Epidemics at 5- to 10-year cycles have been documented in the Kathmandu Valley of Nepal and in the central Asian republics of the former Soviet Union.[85] However, no long-term population-based studies have defined whether a temporal pattern of infection occurs in endemic and nonendemic areas.

6.4.5. Age. The mean age for cases of hepatitis E has been 29 years and the highest age-specific prevalence of disease has been between 20 and 30 years of age.[67,167,229,237,243] Clinical attack rates among adults have ranged from 3 to 10%, whereas among children less than 15 years of age the attack rates have ranged from 0.2 to 3%.[66,166,243,319] In studies of secondary attack rates, children appear to be a reservoir for the virus during epidemics, with as many as 60% of the secondary cases being attributable to contact with children.[237] In both epidemic and nonepidemic situations, symptomatic cases are seen among children and usually contribute 5–10% of the total clinical cases observed.[167,237] The true ratio of symptomatic to asymptomatic cases is not known.

6.4.6. Sex. Among persons with symptomatic infection, the male-to-female ratio has ranged from 1:1 to 3:1.[85,166,243] However, epidemiologic studies have shown that various sociological factors may have contributed to the predominance of men in some outbreaks. In the Kathmandu epidemic, in which the male-to-female ratio was 3:1, the highest disease rates were in urban areas where men from rural areas went to obtain work, while women remaining in the rural villages had lower attack rates.[243] In outbreaks in Central Asia, Africa, and Mexico there has not been a predominance of males.[85] On the other hand, an increased awareness of the severe consequences of viral hepatitis among pregnant women may contribute to an overrepresentation of women among clinical cases presenting to hospitals during epidemics of this disease.

6.4.7. Race. There is no evidence that there is any racial selection for this clinical entity.

6.4.8. Other Factors. The most significant factor leading to the spread of hepatitis E is that of poor sanitation. The most severe epidemics have occurred in areas of low socioeconomic conditions where there is an absence of disposal of feces, resulting in contamination of water supplies during periods of rain. The lack of outbreaks or apparent disease in countries where water supplies undergo rigorous purification and decontamination suggests that HEV transmission can be controlled through appropriately protected water supplies. Recent observations suggest that HEV may infect domestic animals such as pigs and cattle. If these observations are confirmed and

an animal reservoir exists for HEV infection, control of this infection may be even more problematic.

6.5. Mechanisms and Routes of Transmission

Investigations of all outbreaks to date have indicated that the mode of transmission is via the fecal–oral route. Case-control studies among persons with hepatitis E, B, and A indicate that percutaneous exposures, either by the direct parenteral route or through sexual exposure, are not implicated as a means of disease transmission.[167,237] Transmission through person-to-person contact has been demonstrated in case-control studies and by secondary attack rates that range from 0.7 to 8.0% in households of cases.[237,319] While several reports have suggested that HEV may be transmitted by food, these studies have not had adequate control groups or serological testing.[85]

6.6. Pathogenesis and Immunity

Experimental HEV infection in nonhuman primates has been induced successfully by parenteral inoculation,[47] and a human volunteer study has shown that HEV is transmitted by the enteral route.[23] Experimental infection in several nonhuman primate models has defined the temporal relationship of virological, immunologic, clinical, and pathomorphological events. Following intravenous inoculation, liver enzyme elevations occur after 24 to 38 days, HEV is detected in feces approximately 2 weeks before the onset of biochemical hepatitis, and virus shedding appears to end when liver enzyme levels return to normal.[47,85] This pattern of virus excretion was also observed in the human volunteer study.[23] Viremia, as indicated by the detection of HEV-RNA in serum, has been detected in experimentally infected animals prior to the elevation of liver enzymes and remains present until the appearance of anti-HEV.[313]

The primary site of replication of HEV has not been determined. An antigen associated with HEV infection (HEV-Ag) appears in the cytoplasm of hepatocytes as early as 10 days postinoculation, most often precedes the period of liver enzyme elevations, and may persist for as long as 21 days.[179] However, the degree of necroinflammatory response during the course of experimental infection is minimal. The primary histopathologic findings are degenerative changes in hepatocytes, acute focal necrosis, and a minimal portal tract inflammatory infiltrate.[91] Pathomorphological changes observed in patients with hepatitis E include cholestasis and glandlike transformation of bile ducts with preservation of the lobular structure, portal inflammation, ballooning degeneration, Kupffer cell hyperplasia, and liver cell necrosis that has varied from single-cell degeneration to bridging necrosis that resolves without chronic sequelae.[166,167]

During the early phase of infection, IgM anti-HEV is present and can be detected for 5–6 months.[105,203] IgG anti-HEV has generally been detected almost simultaneously with IgM antibody.[105,203] Some have suggested that IgG anti-HEV is relatively short-lived[124]; however, IgG anti-HEV activity has been found for at least 1 year after acute infection in most primates and for up to 10 years in experimentally infected primates.[105] The long-term persistence of IgG anti-HEV is further suggested by its detection in asymptomatic persons living in areas where HEV infection has been shown to be endemic.

6.7. Patterns of Host Response

6.7.1. Clinical Features. The incubation period is longer than for hepatitis A, with a range of 22 to 60 days and a mode of 40 days.[85,229] In symptomatic (jaundiced) cases, a 1- to 10-day prodromal phase has been described prior to the onset of symptoms that include nausea (46–85%), dark urine (92–100%), abdominal pain (41–87%), vomiting (50%), pruritus (13–55%), joint pain (28–81%), rash (3%), and diarrhea (3%). Fever and hepatomegaly have been present in over 50% of patients. Peak mean bilirubin and liver enzyme values do not differ from those observed with hepatitis A and hepatitis B.[85,166,167]

A high case-fatality rate among pregnant women has been a consistent feature of hepatitis E. Mortality has ranged from 17 to 33% in pregnant women, while case-fatality rates among nonpregnant women do not differ from those observed for men.[85,243] The highest rates of fulminant hepatitis and death have occurred during the 20th to 32nd week of gestation, as well as during labor. Hemorrhage is the most common cause of death, and high rates of perinatal death are associated with fulminant hepatitis E during pregnancy. However, termination of pregnancy does not improve the clinical status of these women.[85]

6.7.2. Diagnosis. Differentiation of hepatitis E from other forms of acute viral hepatitis can be made only by using specific serological testing, the availability of which is somewhat limited at this time. Testing for IgM anti-HAV and IgM anti-HBc can exclude those causes of viral hepatitis. However, since a well-standardized test for IgM anti-HEV is not available, the diagnosis of hepatitis E in the acutely ill patient remains somewhat problematic. In parts of the world where hepatitis E has been identified and/or is considered endemic, the presence of anti-HEV (total or IgG) can be considered diagnostic of acute infection in a patient who is anti-HCV negative. However, in the patient who is positive for both anti-HCV and anti-HEV, the final diagnosis may depend on a history consistent with the epidemiologic features of the particular disease, including the likely source of infection, incubation period, and eventual clinical outcome. Obtaining testing for IgM anti-HEV may be required for some cases.

Epidemiologic features of cases that should suggest the diagnosis of hepatitis E include a hepatitis A-like illness in an adult living in those parts of the world where hepatitis A is highly endemic and fulminant hepatitis in a pregnant woman. Cases presenting in countries where hepatitis E has not been identified should raise the possibility of imported illness. A complete travel history should be obtained to determine if the individual traveled in an area where HEV infection is known to be endemic. In all such cases, serological testing should be obtained to confirm the diagnosis.

6.8. Control and Prevention

Currently the most important means of preventing hepatitis E is protection of water systems from contamination with fecal material. Epidemiologic evidence indicates that boiling water will interrupt disease transmission. Similiar data concerning chlorination of water are not available, although this strategy should also be used to interrupt disease transmission in epidemic situations.

The prophylactic effect of IG has not been demonstrated. One study using IG produced in India did not show a statistically different rate of disease between exposed individuals who received IG and those who did not receive IG.[160] In addition, it is likely that IGs prepared from parts of the world where hepatitis E is not an endemic disease would not have appropriate levels of antibody. Recently, pooled IG was used during an abating epidemic in Mexico, and the rate of infection among recipients was not lowered from the expected rate of infection.[67]

Recent experiments have shown that an antigen from the capsid region of HEV produced by recombinant DNA technology induced antibodies in cynomolgus macaques, and that some of the animals were protected from infection upon challenge with the wild-type virus.[108a,266,313a] Whether an effective vaccine can be produced that will prevent HEV infection and hepatitis E remains to be determined. The complete picture of the immunologic parameters required for protection from HEV infection are still unknown at this time.

Currently, the best approach to the control of HEV infection is for health care providers to be alert to the existence of possible epidemics. Identification of possible epidemic situations is determined by increasing numbers of cases of acute hepatitis among adults and increasing numbers of hospitalizations for severe hepatitis among pregnant women. If an epidemic is suspected, the diagnosis of hepatitis E should be confirmed with available diagnostic testing and the involved population should be advised against drinking water from implicated sources without boiling or hand chlorination. Since pregnant women are particularly vulnerable to fulminant hepatitis from HEV infection, they should be advised of these sanitary requirements. Foreign travelers to epidemic areas should be particularly cautious with regard to these recommendations and should not be presumed to be protected by IG.

6.9. Unresolved Problems

The recent characterization of HEV and development of diagnostic reagents has gone a long way to define the spectrum and outcome of this infection. The most pressing issues remain the definition of the worldwide and local distribution of infection, vehicles of transmission, possible animal reservoirs for the virus, and the natural history of protective immunity to infection. The availability of this information will greatly assist in developing logical strategies to prevent HEV infection.

7. References

1. AACH, R. D., STEVENS, C. E., HOLLINGER, F. B., MOSLEY, J. W., PETERSON, D. A., TAYLOR, P. E., JOHNSON, R. G., BARBOSA, L. H., AND NEMO, G. J., Hepatitis C virus infection in posttransfusion hepatitis. An analysis with first- and second-generation assays, *N. Engl. J. Med.* **325:**1325–1329 (1991).
2. ACUTE HEPATIC FAILURE STUDY GROUP, Etiology and prognosis in fulminant hepatitis, *Gastroenterology* **77:**A33 (1979).
3. AKAHANE, Y., AIKAWA, T., SUGAI, Y., TSUDA, F., OKAMOTO, H., AND MISHIRO, S., Transmission of HCV between spouses, *Lancet* **339:**1059–1060 (1992).
4. ALBERTI, A., CHEMELLO, L., CAVALLETTO, D., TAGGER, A., CANTON, A. D., BIZZARO, N., TAGARIELLO, G., AND RUOL, A., Antibody to hepatitis C virus and liver disease in volunteer blood donors, *Ann. Intern. Med.* **114:**1010–1012 (1991).
5. ALTER, H., JETT, B. W., POLITO, A. J., FARCI, P., MELPOLDER, J. C., SHIH, J. W., SHIMIZU, Y., AND PURCELL, R. H., Analysis of the role of hepatitis C in transfusion associated hepatitis C, in: *Viral Hepatitis and Liver Disease* (F. B. HOLLINGER, S. M. LEMON, AND H. S. MARGOLIS, EDS.), pp. 396–402, Williams and Wilkins, Baltimore, 1991.

6. ALTER, H. J., PURCELL, R. H., HOLLAND, P. V., FEINSTONE, S. M., MORROW, A. G., AND MORITSUGU, Y., Clinical and serological analysis of transfusion-associated hepatitis, *Lancet* **2:**838–841 (1975).
7. ALTER, H. J., PURCELL, R. H., SHIH, J. W., MELPOLDER, J. C., HOUGHTON, M., CHOO, Q. L., AND KUO, G., Detection of antibody to hepatitis C virus in prospectively followed transfusion recipients with acute and chronic non-A, non-B hepatitis, *N. Engl. J. Med* **321:**1494–1500 (1989).
8. ALTER, M. J., The detection, transmission, and outcome of hepatitis C virus infection, *Infect. Agents Dis.* **2:**155–166 (1993).
9. ALTER, M. J., COLEMAN, P. J., ALEXANDER, W. J., KRAMER, E., MILLER, J. K., MANDEL, E., HADLER, S. C., AND MARGOLIS, H. S., The importance of heterosexual activity in the transmission of hepatitis B and non-A, non-B hepatitis in the United States, *J.A.M.A.* **262:**1201–1205 (1989).
10. ALTER, M. J., FAVERO, M. S., AND MAYNARD, J. E., Impact of infection control strategies on the incidence of dialysis-associated hepatitis in the United States, *J. Infect. Dis.* **153:**1149–1151 (1986).
11. ALTER, M. J., GERETY, R. J., SMALLWOOD, L., SAMPLINER, R. S., TABOR, E., DEINHARDT, F., FROSNER, G., AND MATANOSKI, G. M., Sporadic non-A, non-B hepatitis: Frequency and epidemiology in an urban United States population, *J. Infect. Dis.* **145:**886–893 (1982).
12. ALTER, M. J., AND HADLER, S. C., Delta hepatitis and infection in North America, in: *Hepatitis Delta Virus. Molecular Biology, Pathogenesis and Clinical Aspects* (S. J. HADZIYANNIS, J. M., TAYLOR, AND F. BONINO, EDS.), pp. 243–250, Wiley-Liss, New York, 1993.
13. ALTER, M. J., HADLER, S. C., JUDSON, F. N., MARES, A., ALEXANDER, W. J., HU, P. Y., MILLER, J. K., MOYER, L. A., FIELDS, H. A., BRADLEY, D. W., AND MARGOLIS, H. S., Risk factors for acute non-A, non-B hepatitis in the United States and association with hepatitis C virus infection, *J.A.M.A.* **264:**2231–2235 (1990).
14. ALTER, M. J., HADLER, S. C., MARGOLIS, H. S., ALEXANDER, W. J., HU, P. Y., JUDSON, F. N., MARES, A., MILLER, J. K., AND MOYER, L. A., The changing epidemiology of hepatitis B in the United States; need for alternative vaccination strategies, *J.A.M.A.* **263:**1218–1222 (1990).
15. ALTER, M. J., MARES, A., HADLER, S. C., AND MAYNARD, J. E., The effect of underreporting on the apparent incidence and epidemiology of acute viral hepatitis, *Am. J. Epidemiol.* **125:**133–139 (1987).
16. ALTER, M. J., MARGOLIS, H. S., KRAWCZYNSKI, K., JUDSON, F. N., HU, P. Y., MILLER, J. K., GERBER, M. A., SAMPLINER, R. E., MEEKS, E. L., BEACH, M. J., AND SENTINEL COUNTY CHRONIC NON-A, NON-B HEPATITIS STUDY TEAM, The natural history of community-acquired hepatitis C in the United States, *N. Engl. J. Med.* **327:**1899–1905 (1992).
16a. ALTER, M. J., GALLAGHER, M., MORRIS, T. T., MOYER, L. A., MECKS, E. L., KRAWCZYNSKI, K., KIM, J. P., AND MARGOLIS, H. S., FOR THE SENTINEL COUNTIES VIRAL HEPATITIS STUDY TEAM, Sentinel surveillance for acute non A-E hepatitis in the United States and the role of hepatitis G virus infection, *N. Engl. J. Med.* (in press).
17. ALWARD, W. L. M., MCMAHON, B. J., HALL, D. B., HEYWARD, W. L., FRANCIS, D. P., AND BENDER, T. R., The long-term serological course of asymptomatic hepatitis B virus carriers and the development of primary hepatocellular carcinoma, *J. Infect. Dis.* **151:**604–609 (1985).
18. ANDO, K., MORIYAMA, T., GUIDOTTI, L. G., WIRTH, S., SCHREIBER, R. D., SCHLICHT, H. J., HUANG, S.-N., AND CHISARI, F. V.,

Mechanisms of class I restricted immunopathology. A transgenic mouse model of fulminant hepatitis, *J. Exp. Med.* **178:**1541–1554 (1993).

19. ANDRE, F. E., D'HONDT, E., DELEM, A., AND SAFARY, A., Clinical assessment of the safety and efficacy of an inactivated hepatitis A vaccine: Rationale and summary of findings, *Vaccine* **10**(Suppl. 1):s160–s168 (1992).

20. ARAGONA, M., MACAGUO, S., CAREDDA, F., CRIVELLI, O., LAVARINI, C., BERTOLUSSO, L., FARCI, P., AND RIZZETTO, M., IgM anti-HD: Diagnostic and prognostic significance, *Prog. Clin. Biol. Res.* **234:**243–248 (1986).

21. ARICO, S., ARAGONA, M., RIZZETTO, M., CAREDDA, F., ZANETTI, A., MARINUCCI, G., DIANA, S., FARCI, P., ARNONE, M., CAPORASO, N., ASCIONE, A., DENTICO, P., PASTORE, G., RAIMONDO, G., AND CRAXI, A., Clinical significance of antibody to the hepatitis delta virus in symptomless HBsAg carriers, *Lancet* **2:**356–357 (1985).

22. BACH, N., THUNG, S. N., AND SCHAFFNER, F., The histological features of chronic hepatitis C and autoimmune chronic hepatitis: A comparative analysis, *Hepatology* **15:**572–577 (1992).

23. BALAYAN, M. S., ANDJAPARIDZE, A. G., SAVINSKAYA, S. S., KETILADZE, E. S., BRAGINSKY, D. M., SAVINOV, A. P., AND POLESCHUK, V. F., Evidence for a virus in non-A, non-B hepatitis transmitted via the fecal–oral route, *Intervirology* **20:**23–31 (1983).

24. BANCROFT, W. H., AND LEMON, S. M., Hepatitis A from the military perspective, in: *Hepatitis A* (R. J. GERETY, ED.), pp. 81–100, Academic Press, New York, 1984.

25. BANCROFT, W. H., SNITBHAN, R., SCOTT, R. M., TINGPALAPNG, W. T., WATSON, W. T., TAUTICHAROENYOS, P., KARWACKI, J. J., AND SRIMARUT, S., Transmission of hepatitis B virus to gibbons by exposure to human saliva containing hepatitis B surface antigen, *J. Infect. Dis.* **135:**79–85 (1977).

26. BEACH, M. J., MEEKS, E. L., MIMMS, L. T., VALLARI, D., DUCHARME, L., SPELBRING, J., TASKAR, S., SCHLEICHER, J. B., KRAWCZYNSKI, K., AND BRADLEY, D. W., Temporal relationships of hepatitis C virus RNA and antibody responses following experimental infection of chimpanzees, *J. Med. Virol.* **36:**226–237 (1992).

27. BEASLEY, R. P., Hepatitis B virus. The major etiology of hepatocellular carcinoma, *Cancer* **61:**1942–1956 (1988).

28. BEASLEY, R. P., AND HWANG, L.-Y., Overview on the epidemiology of hepatocellular carcinoma, in: *Viral Hepatitis and Liver Disease* (F. B. HOLLINGER, S. M. LEMON, AND H. S. MARGOLIS, EDS.), pp. 532–535, Williams and Wilkins, Baltimore, 1991.

29. BEASLEY, R. P., AND HWANG, L.-Y., Postnatal infectivity of hepatitis B surface antigen-carrier mothers, *J. Infect. Dis.* **147:**185–190 (1983).

30. BEASLEY, R. P., HWANG, L. Y., LIN, C. C., AND CHIEN, C. S., Hepatocellular carcinoma and hepatitis B virus. A prospective study of 22,707 men in Taiwan, *Lancet* **2:**1129–1133 (1981).

31. BEASLEY, R. P., HWANG, L. Y., STEVENS, C. E., LIN, C. C., HSIEH, F. J., WANG, K. Y., SUN, T. S., AND SZMUNESS, W., Efficacy of hepatitis B immunoglobulin for prevention of perinatal transmission of the hepatitis B virus: Final report of a randomized double-blind, placebo-controlled trial, *Hepatology* **3:**135–141 (1983).

32. BEN-PORATH, E., WANDS, J., GRUIAM, M., AND ISSELBACHER, K., Clinical significance of an enhanced detection of HBsAg by a monoclonal radioimmunoassay, *Hepatology* **4:**803–807 (1984).

33. BEN-PORATH, E., WANDS, J., MARCINIAK, R. A., WONG, A. M., HORNSTEIN, O., RYDER, R., CANLAS, M., LINGAO, A., AND ISSEL-

BACHER, K. J., Structural analysis of hepatitis B surface antigen by monoclonal antibodies, *J. Clin. Invest.* **76:**1338–1347 (1985).

34. BERNIER, R. H., SAMPLINER, R., GERETY, R., TABOR, E., HAMILTON, F., AND NATHANSON, N., Hepatitis B infection in households of chronic carriers of hepatitis B surface antigen: Factors associated with prevalence of infection, *Am. J. Epidemiol.* **116:**199–211 (1982).

35. BERNINGER, M., HAMMER, M., HOYER, B., AND GERIN, J. L., An assay for the detection of DNA genome of hepatitis B virus in serum, *J. Med. Virol.* **9:**57–68 (1982).

36. BI, S.-L., PURDY, M. A., MCCAUSTLAND, K. A., MARGOLIS, H. S., AND BRADLEY, D. W., The sequence of hepatitis E virus isolated directly from a single source during an outbreak in China, *Virus Res.* **28:**223–247 (1993).

37. BIANCHI, L., The immunopathology of acute type B hepatitis, *Springer Semin. Immunopathol.* **3:**421–438 (1981).

38. BLOCH, A. B., STRAMER, S. L., SMITH, D., MARGOLIS, H. S., FIELDS, H. A., MCKINLEY, T. W., GERBA, C. P., MAYNARD, J. E., AND SIKES, K., Recovery of hepatitis A virus from a water supply responsible for a common source outbreak of hepatitis A, *Am. J. Public Health* **80:**428–430 (1990).

39. BLUMBERG, B. S., ALTER, H. J., AND VISNICH, S., A "new" antigen in leukemia sera, *J.A.M.A.* **191:**541–546 (1965).

40. BOGOMOLSKI-YAHALOM, V., GRANOT, E., LINDER, N., ADLER, R., KORMAN, S., MANNY, N., TUR-KASPA, R., AND SHOUVAL, D., Prevalence of HBsAg carriers in native and immigrant pregnant female populations in Israel and passive/active vaccination against HBV of newborns at risk, *J. Med. Virol.* **34:**217–222 (1991).

41. BONINO, F., HOYER, B. H., SHIH, J. W., RIZZETTO, M., PURCELL, R. H., AND GERIN, J. L., Delta hepatitis agent: Structural and antigenic properties of the delta associated particle, *Infect. Immun.* **43:**1000–1005 (1984).

42. BONINO, F., NEGRO, F., BALDI, M., BRUNETTO, M. F., CHIABERGE, E., ROCCA, N., AND ROCCA, G., The natural history of delta hepatitis, *Prog. Clin. Biol. Res.* **234:**145–152 (1986).

43. BONINO, F., AND SMEDILE, A., Delta agent (type D) hepatitis, *Semin. Liver Dis.* **6:**28–33 (1986).

44. BRADLEY, D. W., BEACH, M. J., AND PURDY, M. A., Recent developments in the molecular cloning and characterization of hepatitis C and E viruses, *Microb. Pathog.* **12:**391–398 (1992).

45. BRADLEY, D. W., KRAWCZYNSKI, K., BEACH, M. J., AND PURDY, M. A., Non-A, non-B hepatitis: Toward the discovery of hepatitis C and E viruses, *Semin. Liver Dis.* **11:**128–146 (1991).

47. BRADLEY, D. W., KRAWCZYNSKI, K., COOK, JR., E. H., MCCAUSTLAND, K. A., HUMPHREY, C. D., SPELBRING, J. E., MYINT, H., AND MAYNARD, J. E., Enterically transmitted non-A, non-B hepatitis: Serial passage of disease in cynomolgus macaques and tamarins, and recovery of disease-associated 27 to 34 nm virus-like particles, *Proc. Natl. Acad. Sci. USA* **84:**6277–6281 (1987).

48. BRILLANTI, S., GAIANI, S., MIGLIOLI, M., FOLI, M., MASCI, C., AND BARBARA, L., Persistent hepatitis C viraemia without liver disease, *Lancet* **341:**464–465 (1993).

49. BROWN, J. L., CARMAN, W. F., AND THOMAS, H. C., The clinical significance of molecular variation within the hepatitis B virus genome, *Hepatology* **15:**144–148 (1992).

50. BUKH, J., PURCELL, R. H., AND MILLER, R. H., At least 12 genotypes of hepatitis C virus predicted by sequence analysis of the putative E1 gene of isolates collected worldwide, *Proc. Natl. Acad. Sci. USA* **90:**8234–8238 (1993).

51. BULKOW, L. R., WAINWRIGHT, R. B., MCMAHON, B. J., MIDDAUGH, J. P., JENKERSON, S. A., AND MARGOLIS, H. S., Secular

trends in hepatitis A virus infection among Alaska Natives, *J. Infect. Dis.* **168:**1017–1020 (1993).

52. BUTTRAGO, B., POPPER, H., HADLER, S. C., THUNG, S. N., GERBER, M. A., PURCELL, R. H., AND MAYNARD, J. E., Specific histologic features of Santa Marta hepatitis: A severe form of hepatitis delta virus infection in northern Columbia, *Hepatology* **6:**1285–1291 (1986).

53. CAREDDA, F., ANTINORI, S., RE, T., PASTECCHIA, C., AND MORONI, M., Course and prognosis of acute HDV hepatitis, *Prog. Clin. Biol. Res.* **234:**267–276 (1987).

54. CARL, M., BLAKEY, D. L., FRANCIS, D. P., AND MAYNARD, J. E., Interruption of hepatitis B transmission by modification of a gynecologist's surgical technique, *Lancet* **1:**731–733 (1982).

55. CARL, M., FRANCIS, D. P., AND MAYNARD, J. P., Food-borne hepatitis A: Recommendation for control, *J. Infect. Dis.* **148:**1133–1135 (1983).

56. CARL, M., KANTOR, P. J., WEBSTER, H. M., FIELDS, H. A., AND MAYNARD, J. E., Excretion of hepatitis A virus in the stools of hospital patients, *J. Med. Virol.* **9:**125–129 (1982).

57. CARMAN, W. F., ZANETTI, A. R., KARAYIANNIS, P., WATERS, J., MANZILLO, G., TANZI, E., ZUCKERMAN, A. J., AND THOMAS, H. C., Vaccine-induced escape mutant of hepatitis B virus, *Lancet* **336:**325–329 (1990).

58. CASEY, J. L., BROWN, T. L., COLAN, E. J., WIGNALL, F. S., AND GERIN, J. L., A genotype of hepatitis D virus that occurs in Northern South America, *Proc. Natl. Acad. Sci. USA* **90:**9016–9020 (1993).

59. CENTERS FOR DISEASE CONTROL, Hepatitis Surveillance Report, *55:*1–35 (1994).

60. CENTERS FOR DISEASE CONTROL, Hepatitis A among homosexual men—United States, Canada, and Australia, *Morb. Mortal. Wkly. Rep.* **41:**155, 161–164 (1992).

61. CENTERS FOR DISEASE CONTROL, Hepatitis B virus: A comprehensive strategy for eliminating transmission in the United States through universal childhood vaccination. Recommendations of the immunization practices advisory committee (ACIP), *Morb. Mortal. Wkly. Rep.* **40:**(RR-131)1–20 (1991).

62. CENTERS FOR DISEASE CONTROL, Public Health Service Interagency guidelines for screening donors of blood, plasma, organs, tissues and semen for evidence of hepatitis B and hepatitis C, *Morb. Mortal. Wkly. Rep.* **40**(RR-4):1–17 (1991).

63. CENTERS FOR DISEASE CONTROL, Recommendations for preventing transmission of human immunodeficiency virus and hepatitis B virus to patients during exposure-prone invasive procedures, *Morb. Mortal. Wkly. Rep.* **40:**1–9 (1991).

64. CENTERS FOR DISEASE CONTROL, Protection against viral hepatitis: Recommendation of the immunization practices advisory committee, *Morb. Mortal. Wkly. Rep.* **39:**5–22 (1990).

65. CENTERS FOR DISEASE CONTROL, Update: Universal precautions for prevention of transmission of human immunodeficiency virus, hepatitis B virus, and other bloodborne pathogens in health care settings, *Morb. Mortal. Wkly. Rep.* **37:**377–382, 387–388 (1988).

66. CENTERS FOR DISEASE CONTROL, Enterically transmitted non-A, non-B hepatitis—East Africa, *Morb. Mortal. Wkly. Rep.* **36:**241–244 (1987).

67. CENTERS FOR DISEASE CONTROL, Enterically transmitted non-A, non-B hepatitis—Mexico, *Morb. Mortal. Wkly. Rep.* **36:**597–602 (1987).

68. CENTERS FOR DISEASE CONTROL, Recommendations for prevention of HIV transmission in health-care settings, *Morb. Mortal. Wkly. Rep.* **36**(Suppl. 2S):1–18s (1987).

69. CENTERS FOR DISEASE CONTROL AND PREVENTION, Prevention of hepatitis A through active or passive immunization, *Morb. Mortal. Wkly. Rep.* **45:**1–30 (1996).

70. CENTERS FOR DISEASE CONTROL AND PREVENTION, Update: Recommendations to prevent hepatitis B virus transmission–United States, *Morb. Mortal. Wkly. Rep.* **44:**574–575 (1995).

71. CENTERS FOR DISEASE CONTROL AND PREVENTION, Recommendations of the International Task Force for Disease Eradication, *Morb. Mortal. Wkly. Rep.* **42**(No. RR-16):12 (1993).

72. CHA, T. A., BEALL, E., IRVINE, B., KOLBERG, J., CHIEN, D., KUO, G., AND URDEA, M. S., At least five related, but distinct, hepatitis C viral genotypes exist, *Proc. Natl. Acad. Sci. USA* **89:**7144–7148 (1992).

73. CHAN, G. C. B., YEOH, E. K., YOUNG, B., CHANG, W. K., LIM, W. L., AND LIN, H. J., Effect of pregnancy on the hepatitis B carrier state, in: *Viral Hepatitis and Liver Disease* (F. B. HOLLINGER, S. M. LEMON, AND H. S. MARGOLIS, EDS.), pp. 678–680, Williams and Wilkins, Baltimore, 1991.

74. CHAU, K. H., HARGIE, M. P., DECKER, R. H., MUSHAHWAR, I. K., AND OVERBY, L. R., Serodiagnosis of recent hepatitis B virus infection by IgM class anti-HBc, *Hepatology* **3:**142–149 (1983).

75. CHEN, C. J., HWANG, S. J., FAN, K. Y., CHANG, S. A., CHANG, Y. H., WANA, S. R., HU, W. T., LIAW, Y. F., CHAI, C. Y., CHANG, R., AND HO, M., Seroepidemiology of human T-lymphotropic viruses and hepatitis viruses among prostitutes in Taiwan, *J. Infect. Dis.* **158:**633–635 (1988).

76. CHEN, D.-S., Control of hepatitis B in Asia: Mass immunization program in Taiwan, in: *Viral Hepatitis and Liver Disease* (F. B. HOLLINGER, S. M. LEMON, AND H. S. MARGOLIS, EDS.), pp. 716–719, Williams and Wilkins, Baltimore, 1991.

77. CHOO, Q. L., KUO, G., RALSTON, R., WEINER, A., CHIEN, D., VAN NEST, G., HAN, J., BERGER, K., THUDIUM, K., KUO, C., KANSOPON, J., MCFARLAND, J., TABRIZI, A., CHING, K., MOSS, B., CUMMINS, L. B., HOUGHTON, M., AND MUCHMORE, E., Vaccination of chimpanzees against infection by the hepatitis C virus, *Proc. Natl. Acad. Sci. USA* **91:**1294–1298 (1994).

78. CHOO, Q. L., KUO, G., WEINER, A. J., OVERBY, L. R., BRADLEY, D. W., AND HOUGHTON, M., Isolation of a cDNA clone derived from a bloodborne non-A, non-B viral hepatitis genome, *Science* **244:**359–362 (1989).

79. CHOTARD, J., INSKIP, H. M., HALL, A. J., LOIK, F., MENDY, M., WHITTLE, H., GEORGE, M. O., AND LOWE, Y., The Gambia hepatitis intervention study: Follow-up of a cohort of children vaccinated against hepatitis B, *J. Infect. Dis.* **166:**764–768 (1992).

80. CILLA, G., PEREZ-TRALLERO, E., ITURRIZA, M., CARCEDO, A., AND ECHEVERRIA, J., Maternal–infant transmission of hepatitis C virus infection, *Pediatr. Infect. Dis. J.* **11:**417 (1992).

81. COHEN, J. I., FEINSTONE, S., AND PURCELL, R. H., Hepatitis A virus infection in a chimpanzee: Duration of viremia and detection of virus in saliva and throat swabs, *J. Infect. Dis.* **160:**887–890 (1989).

82. COREY, L., AND HOLMES, K. K., Sexual transmission of hepatitis A in homosexual men., *N. Engl. J. Med.* **302:**435–438 (1980).

83. COUROUCE-PAUTY, A. M., PLANCON, A., AND SOULIER, J. P., Distribution of HBsAg subtypes in the world, *Vox. Sang.* **44:**197–211 (1983).

84. COURSAGET, P., YVONNET, B., CHOTARD, J., SARR, M., VINCELOT, P., N'DOYE, R., DIOP-MAR, I., AND CHIRON, J. P., Seven-year study of hepatitis B vaccine efficacy in infants from an endemic area (Senegal), *Lancet* **2:**1143–1145 (1986).

85. CROMEANS, T., NAINAN, O., V, FIELDS, H. A., FAVOROV, M. O., AND MARGOLIS, H. S., Hepatitis A and E viruses, in: *Foodborne Disease Handbook* (Y. H. HUI, J. R. GORHAM, K. D. MUCELL, AND D. O. CLIVER, EDS.), pp. 1–56, Marcel Dekker, New York, 1994.

86. CROMEANS, T., SOBSEY, M. D., AND FIELDS, H. A., Development of a plaque assay for a cytopathic, rapidly replicating isolate of hepatitis A virus, *J. Med. Virol.* **22**:45–56 (1987).

87. CUYPERS, H. T. M., BRESTERS, D., WINKEL, I. N., REESINK, H. W., WEINER, A. J., HOUGHTON, M., VAN DER POEL, C. L., AND LELIE, P. N., Storage conditions of blood samples and primer selection affect yield of cDNA polymerase chain reaction products of hepatitis C virus, *J. Clin. Microbiol.* **30**:3320–3324 (1992).

88. D'AMELIO, R., MATRICARDI, P. M., BISELLI, R., STROFFOLINI, T., MELE, A., SPADA, E., CHIONNE, P., RAPICETTA, M., FERRIGNO, L., AND PASQUINI, P., Changing epidemiology of hepatitis B in Italy: Public health implications, *Am. J. Epidemiol.* **136**:1012–1018 (1992).

89. DAVIS, G. L., BALART, L. A., SCHIFF, E. R., LINDSAY, K., BODENHEIMER, H. C., JR., PERRILLO, R. P., CAREY, W., JACOBSON, I. M., PAYNE, J., DIENSTAG, J. L., VAN THIEL, D. H., TAMBURRO, C., LEFKOWITCH, J., ALBRECHT, J., MESCHIEVITZ, C., ORTEGO, T. J., GIVAS, A., AND HEPATITIS INTERVENTIONAL THERAPY GROUP, Treatment of chronic hepatitis C with recombinant interferon alpha: A multicenter randomized controlled trial, *N. Engl. J. Med.* **321**:1501–1506 (1989).

90. DAWSON, G. J., CHAU, K. H., CABAL, C. M., YARBOUGH, P. O., REYES, G. R., AND MUSHAHWAR, I. K., Solid-phase enzyme-linked immunosorbent assay for hepatitis E virus IgG and IgM antibodies utilizing recombinant antigens and synthetic peptides, *J. Virol. Methods* **38**:175–186 (1992).

91. DE COCK, K. M., BRADLEY, D. W., SANDFORD, N. L., GOVINDARAJAN, S., MAYNARD, J. E., AND REDEKER, A. G., Epidemic non-A, non-B hepatitis in patients from Pakistan, *Ann. Intern. Med.* **106**:227–230 (1987).

92. DESENCLOS, J. A., KLONTZ, K. C., WILDER, M. H., NAINAN, O., V, MARGOLIS, H. S., AND GUNN, R. A., A multistate outbreak of hepatitis A caused by the consumption of raw oysters, *Am. J. Public Health* **81**:1268–1272 (1991).

93. DI BISCEGLIE, A. M., AND HOOFNAGLE, J. H., Therapy of chronic hepatitis C with alpha-interferon: The answer? or more questions? *Hepatology* **13**:601–603 (1991).

94. DI BISCEGLIE, A. M., ORDER, S. E., KLEIN, J. L., WAGGONER, J. G., SJOGREN, M. H., KUO, G., HOUGHTON, M., CHOO, Q. L., AND HOOFNAGLE, J. H., The role of chronic viral hepatitis in hepatocellular carcinoma in the United States, *Am. J. Gastroenterol.* **86**:335–338 (1991).

95. DIENSTAG, J. L., Immunogenesis of the extrahepatic manifestations of hepatitis B virus infection, *Springer Semin. Immunopathol.* **3**:461–472 (1981).

96. DIENSTAG, J. L., KATKOV, W. N., AND CODY, H., Evidence for non-A, non-B hepatitis agents besides hepatitis C virus, in: *Viral Hepatitis and Liver Disease* (F. B. HOLLINGER, S. M. LEMON, AND H. S. MARGOLIS, EDS.), pp. 349–356, Williams and Wilkins, Baltimore, 1991.

97. DIENSTAG, J. L., SZMUNESS, W., STEVENS, C. E., AND PURCELL, R. H., Hepatitis A virus infection: New insights from seroepidemiologic studies, *J. Infect. Dis.* **137**:328–340 (1978).

98. DONAHUE, J. G., MUNOZ, A., NESS, P. M., BROWN, D. E., JR., YAWN, D. H., MCALLISTER, H. A., JR., REITZ, B. A., AND NELSON, K. E., The declining risk of post-transfusion hepatitis C virus infection, *N. Engl. J. Med.* **327**:369–373 (1992).

99. EDITORIAL, Jaundice following yellow fever vaccine, *J.A.M.A.* **119**:110 (1942).

100. ENRIQUEZ, R., FROSNER, G. G., HOCHSTEIN-MINTZEL, V., RIEDEMANN, S., AND REINHARDT, G., Accumulation and persistence of hepatitis A virus in mussels, *J. Med. Virol.* **37**:174–179 (1992).

101. ESTEBAN, J. I., GONZALEZ, A., HERNANDEZ, J. M., VILADOMIU, L., SANCHEZ, C., LOPEZ-TALAVERA, J. C., LUCEA, D., MARTIN-VEGA, C., VIDAL, X., ESTEBAN, R., AND GUARDIA, J., Evaluation of antibodies to hepatitis C virus in a study of transfusion-associated hepatitis, *N. Engl. J. Med.* **323**:1107–1112 (1990).

102. EYSTER, M. E., ALTER, H. J., ALEDORT, L. M., QUAN, S., HATZAKIS, A., AND GOEDERT, J. J., Heterosexual cotransmission of hepatitis C virus (HCV) and human immunodeficiency virus (HIV), *Ann. Intern. Med.* **115**:764–768 (1991).

103. FARCI, P., ALTER, H. J., GOVINDARAJAN, S., WONG, D. C., ENGLE, R., LESNIEWSKI, R. R., MUSHAHWAR, I. K., DESAI, S. M., MILLER, R. H., OGATA, N., AND PURCELL, R. H., Lack of protective immunity against reinfection with hepatitis C virus, *Science* **258**:135–140 (1992).

104. FARCI, P., GERIN, J. L., ARAGONA, M., LINDSEY, I., CRIVELLI, O., BALESTRIERI, A., SMEDILE, A., THOMAS, H. C., AND RIZZETTO, M., Diagnostic and prognostic significance of the IgM antibody to the hepatitis delta virus, **255**:1443–1446 (1986).

105. FAVOROV, M. O., FIELDS, H. A., PURDY, M. A., YASHINA, T. L., ALEKSANDROV, A. G., ALTER, M. J., YARASHEVA, D. M., BRADLEY, D. W., AND MARGOLIS, H. S., Serologic identification of hepatitis E virus infections in epidemic and endemic settings, *J. Med. Virol.* **36**:246–250 (1992).

106. FAY, O. H., HADLER, S. C., MAYNARD, J. E., AND PINHEIRO, F., Hepatitis in the Americas, *Bull. Pan. Am. Health Org.* **19**:401–408 (1985).

107. FEINSTONE, S. M., KAPIKIAN, A. Z., AND PURCELL, R. H., Hepatitis A: Detection by immune electron microscopy of a virus-like antigen association with acute illness, *Science* **182**:1026–1028 (1973).

108. FEINSTONE, S. M., KAPIKIAN, A. Z., PURCELL, R. H., ALTER, H. J., AND HOLLAND, P. V., Transfusion-associated hepatitis not due to viral hepatitis type A or B, *N. Engl. J. Med.* **292**:767–770 (1975).

108a. FUERST, T. R., YARBOUGH, P. O., ZHANG, Y., MCATEE, P., TAM, A., LIFSON, J., MCCAUSTLAND, K., SPELBRING, J., BRADLEY, D., MARGOLIS, H. S., FRANCOTTE, M., GARCON, N., SLAOUI, M., PRIEELS, J. P., AND KRAWCZYNSKI, K. Prevention of hepatitis E using a novel ORF-2 subunit vaccine: in *Enterically-Transmitted Hepatitis Viruses* (Y. BUISSON, P. COURSAGET, AND M. KANE, EDS.), pp. 384–392, LaSimarre, Tours, France, 1996.

109. FIELDS, H. A., AND HADLER, S. C., Delta hepatitis: A review, *J. Clin. Immunoassay* **9**:128–142 (1986).

110. FINDLAY, G. M., AND MACCALLUM, F. O., Note on acute hepatitis and yellow fever immunization, *Trans. Roy. Soc. Trop. Med. Hyg.* **31**:297–308 (1937).

111. FRANCIS, D. P., HADLER, S. C., THOMPSON, S. E., MAYNARD, J. E., OSTROW, D. G., ALTMAN, N., BRAFF, E. H., O'MALLEY, P., HAWKINS, D., JUDSON, F. N., PENLEY, K., NYLUND, T., CHRISTIE, G., MEYERS, F., MOORE, J. N., JR., GARDNER, A., DOTO, I. L., MILLER, J. H., REYNOLDS, G. H., MURPHY, B. L., SCHABLE, C. A., CLARK, B. T., CURRAN, J. W., AND REDEKER, A. G., Prevention of hepatitis B with vaccine. Report from the Centers for

Disease Control multi-center efficacy trial among homosexual men, *Ann. Intern. Med.* **97:**362–366 (1982).

112. FRANKS, A. L., BERG, C. J., KANE, M. A., BROWNE, B. B., SIKES, R. K., ELSEA, W. R., AND BURTON, A. H., Hepatitis B infection among children born in the United States to southeast Asian refugees, *N. Engl. J. Med.* **321:**1301–1305 (1989).

113. FRIED, M. W., SHINDO, M., FONG, T. L., FOX, P. C., HOOFNAGLE, J. H., AND DI BISCEGLIE, A. M., Absence of hepatitis C viral RNA from saliva and semen of patients with chronic hepatitis C, *Gastroenterology* **102:**1306–1308 (1992).

114. FROSNER, G. G., PAPAEVANGELOU, G., BUTLER, R., IWARSON, S., LINDHOLM, A., COUROUCE-PAUTY, A., HAAS, H., AND DIENHARDT, F., Antibody against hepatitis A in seven European countries, *Am. J. Epidemiol.* **110:**63–69 (1979).

115. FUKUDA, Y., NAGURA, H., TAKAYAMA, T., IMOTO, M., SHIBATA, M., KUDO, T., MORISHIMA, T., MIYAMURA, T., AND NAGAI, Y., Correlation between detection of anti-viral antibody and histopathological disease activity in an epidemic of hepatitis C, *Arch. Virol.* **126:**171–178 (1992).

116. GANEM, D., Persistent infection of humans with hepatitis B virus: Mechanisms and consequences, *Rev. Infect. Dis.* **4:**1026–1047 (1982).

117. GAUSS-MULLER, V., VON DER HELM, K., AND DEINHARDT, F., Translation *in vitro* of hepatitis A virus RNA, *Virology* **137:**182–184 (1984).

118. GERBER, A. R., ENGLENDER, S. J., SELVEY, D., CARLSON, J. F., MATTHEWS, D. L., WEBSTER, H. M., AND CALDWELL, G. G., An outbreak of non-A, non-B hepatitis associated with the infusion of a commercial factor IX complex during cardiovascular surgery, *Vox. Sang.* **58:**270–275 (1990).

119. GERBER, M. A., AND THUNG, S. N., Pathobiology of hepatitis B virus, *Pathobiol. Annu.* **11:**197–228 (1981).

120. GERLICH, W. H., AND HEERMANN, K.-H., Functions of hepatitis B virus proteins and virus assembly, in: *Viral Hepatitis and Liver Disease* (F. B. HOLLINGER, S. M. LEMON, AND H. S. MARGOLIS, EDS.), pp. 121–134, Williams and Wilkins, Baltimore, 1991.

121. GHENDON, Y., AND KANE, M. A., Assuring a hepatitis B-free blood supply in developing countries, in: *Viral Hepatitis and Liver Disease* (F. B. HOLLINGER, S. M. LEMON, AND H. S. MARGOLIS, EDS.), pp. 722–724, Williams and Wilkins, Baltimore, 1991.

122. GIBAS, A., BIEWETT, D. R., SCHOENFELD, D. A., AND DIENSTAG, J. L., Prevalence and incidence of viral hepatitis in health workers in the prehepatitis B vaccination era, *Am. J. Epidemiol.* **136:**603–610 (1992).

123. GLIKSON, M., GALUN, E., OREN, R., TUR-KASPA, R., AND SHOUVAL, D., Relapsing hepatitis A, *Medicine* **71:**14–23 (1992).

124. GOLDSMITH, R., YARBOUGH, P. O., REYES, G. R., FRY, K. E., GABOR, K. A., KAMEL, M., ZAKARIA, S., AMER, S., AND GAFFAR, Y., Enzyme-linked immunosorbent assay for diagnosis of acute sporadic hepatitis E in Egyptian children, *Lancet* **339:**328–331 (1992).

125. GOVINDARAJAN, S., KANEL, G. C., AND PETERS, R. C., Prevalence of delta antibody among chronic hepatitis B virus infected patients in the Los Angeles area: Its correlation with liver biopsy diagnosis, *Gastroenterology* **85:**160–166 (1983).

126. GUYER, B., BRADLEY, D. W., BRYAN, J. A., AND MAYNARD, J. E., Non-A, non-B hepatitis among participants in a plasmapheresis stimulation program, *J. Infect. Dis.* **139:**634–640 (1979).

127. HADLER, S. C., Global impact of hepatitis A virus infection changing patterns, in: *Viral Hepatitis and Liver Disease* (F. B.

HOLLINGER, S. M. LEMON, AND H. S. MARGOLIS, EDS.), pp. 14–20, Williams and Wilkins, Baltimore, 1991.

128. HADLER, S. C., DE MONSON, M., PONZETTO, A., ANZOLA, E., RIVERO, D., MONDOLFI, A., BRACHO, A., FRANCIS, D. P., GERBER, M. A., THUNG, S. N., GERIN, J. L., MAYNARD, J. E., POPPER, H., AND PURCELL, R. H., Delta virus infection and severe hepatitis. An epidemic in the Yucpa Indians of Venezuela, *Ann. Intern. Med.* **100:**339–344 (1984).

129. HADLER, S. C., DOTO, I. L., MAYNARD, J. E., SMITH, J., CLARK, B., MOSLEY, J., EICKHOFF, T., HIMMELSBACH, C. K., AND COLE, W. R., Occupational risk of hepatitis B infection in hospital workers, *Infect. Control* **6:**24–31 (1985).

130. HADLER, S. C., FRANCIS, D. P., MAYNARD, J. E., THOMPSON, S. E., JUDSON, F. N., ECHENBERG, D. F., OSTROW, D. G., O'MALLEY, P. M., ALTMAN, N. L., BRAFF, E. H., SHIPMAN, G. F., COLEMAN, P. J., AND MANDEL, E. J., Long-term immunogenicity and efficacy of hepatitis B vaccine in homosexual men, *N. Engl. J. Med.* **215:**209–214 (1986).

131. HADLER, S. C., JUDSON, F. N., O'MALLEY, P. M., *ET AL.*, Studies of hepatitis B vaccine in homosexual men, in: *Progress in Hepatitis B Immunization* (P. COURSAGET AND M. J. TONG, EDS.), pp. 165–175, John Libby Eurotext, London, 1990.

132. HADLER, S. C., AND MARGOLIS, H. S., Hepatitis B immunization: Vaccine types, efficacy, and indications for immunization, in: *Current Clinical Topics in Infectious Diseases* (J. S. REMINGTON AND M. N. SWARTZ, EDS.), pp. 282–308, Blackwell Scientific Publications, Boston, 1992.

133. HADLER, S. C., AND McFARLAND, L., Hepatitis in day care centers: Epidemiology and prevention, *Rev. Infect. Dis.* **8:**548–557 (1986).

134. HADLER, S. C., MURPHY, B. L., SCHABLE, C. A., HEYWARD, W. L., FRANCIS, D. P., AND KANE, M. A., Epidemiological analysis of the significance of low positive test results on antibody to hepatitis B core antigen, *J. Clin. Microbiol.* **19:**521–525 (1984).

135. HADLER, S. C., RIVERO, D., MONZON, M. A., AND PEREZ, M., Ongoing studies of delta infection in the Yucpa Indians in Venezuela, *J. Med. Virol.* **21:**33A (1987).

136. HALL, A. J., INSKIP, H. M., LOIK, F., *Lancet* **337:**747–750 (1991).

137. HALLIDAY, M. L., KANG, LAI-Y., ZHOU, T., HU, MENG-D., PAN, QI-C., FU, TING-Y., HUANG, YU-S., AND HU, SHAN-L., An epidemic of hepatitis A attributable to the ingestion of raw clams in Shanghai, China, *J. Infect. Dis.* **164:**852–859 (1991).

138. HANSSON, B. G., MOESTRUP, T., WIDELL, A., AND NORDENFELT, E., Infection with delta agent in Sweden: Introduction of a new hepatitis agent, *J. Infect. Dis.* **142:**472–478 (1982).

139. HARPAZ, R., Effectiveness of a comprehensive hepatitis B immunization strategy in preventing hepatitis B virus infections in a population with a high endemicity of infection, unpublished data.

140. HERSH, B. S., POPOVICHI, F., JEZEK, Z., SATTEN, G. A., APETREI, R. C., BELDESCU, N., GEORGE, J. R., SHAPIRO, C. N., GAYLE, H. D., AND HEYMANN, D. L., Risk factors for HIV infection among abandoned Romanian children, *AIDS* **7:**1617–1624 (1993).

141. HIBBS, R. G., CORWIN, A. L., HASSAN, N. F., KAMEL, M., DARWISH, M., EDELMAN, R., CONSTANTINE, N. T., RAO, M. R., KHALIFA, A. S., MOKHTAR, S., FAM, N. S., EKLADIOUS, E. M., AND BASSILY, S. B., The epidemiology of antibody to hepatitis C in Egypt, *J. Infect. Dis.* **168:**789–790 (1993).

142. HILLEMAN, M. R., Plasma-derived hepatitis B vaccine: A breakthrough in preventive medicine, in: *Hepatitis B Vaccines in Clinical Practice* (R. W. ELLIS, ED.), pp. 17–39, Marcel Dekker, New York, 1993.

143. HOLLINGER, F. B., AND DIENSTAG, J. L., Hepatitis viruses, in: *Manual of Clinical Microbiology* (E. H. LANNETE, A. BALOWS, W. H. HAUSLER, JR., AND H. J. SHADOMY, EDS.), pp. 813–835, American Society for Microbiology, Washington, DC, 1985.

144. HOOFNAGLE, J. H., AND DI BISCEGLIE, A. M., Therapy of chronic delta hepatitis: Overview, in: *Hepatitis Delta Virus. Molecular Biology, Pathogenesis, and Clinical Aspects* (S. J. HADZIYANNIS, J. M. TAYLOR, AND F. BONINO, EDS.), pp. 337–344, Wiley-Liss, New York, 1993.

145. HOOFNAGLE, J. H., AND DI BISCEGLIE, A. M., Serologic diagnosis of acute and chronic viral hepatitis, *Semin. Liver Dis.* **11:**73–83 (1991).

146. HOOFNAGLE, J. H., SEEF, L. B., BALES, Z. B., GERETY, R. J., AND TABOR, E., Serologic responses in hepatitis B, in: *Viral Hepatitis* (G. N. VYAS, S. N. COHEN, AND R. SCHMID, EDS.), pp. 219–242, Franklin Institute Press, Philadelphia, 1978.

147. HOOFNAGLE, J. H., SHAFRITZ, D. A., AND POPPER, H., Chronic type B hepatitis and the "healthy" HBsAg carrier state, *Hepatology* **7:**758–763 (1987).

148. HOUGHTON, M., WEINER, A., HAN, J., KUO, G., AND CHOO, Q. L., Molecular biology of the hepatitis C viruses: Implications for diagnosis, development and control of viral disease, *Hepatology* **14:**381–388 (1991).

149. HOWARD, C. R., AND SIMPSON, D. I. H., Viruses other than the hepatitis viruses that cause hepatitides in man, in: *Viral Hepatitis: Laboratory and Clinical Science* (F. DEINHARDT AND J. DEINHARDT, EDS.), pp. 139–156, Marcel Dekker, New York, 1983.

150. HSIEH, C. C., TZONOU, A., ZAVITSANOS, X., KALAMANI, E., LAN, S. J., AND TRICHOPOULOS, D., Age at first establishment of chronic hepatitis B virus infection and hepatocellular carcinoma risk. A birth order study, *Am. J. Epidemiol.* **136:**1115–1121 (1992).

151. HSU, H. Y., CHANG, M. H., CHEN, D. S., LEE, C. Y., AND SUNG, J. L., Changing seroepidemiology of hepatitis A virus infection in Taiwan, *J. Med. Virol.* **17:**297–301 (1985).

152. HURIE, M. B., MAST, E. E., AND DAVIS, J. P., Horizontal transmission of hepatitis B virus infection to United States-born children of Hmong refugees, *Pediatrics* **89:**269–273 (1992).

153. HWANG, L.-Y., LEE, C.-Y., AND BEASLEY, R. P., Five year follow-up of HBV vaccination with plasma-derived vaccine in neonates. Evaluation of immunogenicity and efficacy against perinatal transmission, in: *Viral Hepatitis and Liver Disease* (F. B. HOLLINGER, S. M. LEMON, AND H. S. MARGOLIS, EDS.), pp. 759–761, Williams and Wilkins, Baltimore, 1991.

154. IARASHEVA, D. M., FAVOROV, M. O., IASHINA, T. L., SHAKGIL'DIAN, I. V., UMAROVA, A. A., SOROKINA, S. A., KAMARDINOV, K. K., AND MAVASHEV, V. I., The etiological structure of acute viral hepatitis in Tadzhikistan in a period of decreased morbidity, *Vopr. Virusol.* **36:**454–456 (1991).

155. IINO, S., FUJIYAMA, S., HORIUCHI, K., JYO, K., KUWABARA, Y., SATO, S., SAIKA, S., MORITA, M., ODOH, K., KUZUHARA, S., WATANABE, H., TANAKA, M., AND MIZUNO, K., Clinical trial of a lyophilized inactivated hepatitis A candidate vaccine in healthy adult volunteers, *Vaccine* **10:**323–328 (1992).

156. INNIS, B. L., SNITBHAN, R., KUNASOL, P., LAORAKPONGSE, T., POOPATANAKOOL, W., SUNTAYAKORN, S., SUBNANTAPONG, T., SAFARY, A., AND BOSLEGO, J. W., Protection against hepatitis A by an inactivated vaccine, *J.A.M.A.* **271:**1328–1334 (1994).

157. ITO, S.-I., ITO, M., CHO, M.-J., SHIMOTOHNO, K., AND TAJIMA, K., Massive sero-epidemiological survey of hepatitis C virus: Clustering of carriers on the southwest coast of Tsushima, Japan, *Jpn. J. Cancer Res.* **82:**1–3 (1991).

158. JANSEN, R. W., NEWBOLD, J. E., AND LEMON, S. M., Combined immunoaffinity cDNA-RNA hybridization assay for detection of hepatitis A virus in clinical specimens, *J. Clin. Microbiol.* **22:**984–989 (1985).

159. JENISON, S. A., LEMON, S. M., BAKER, L. N., AND NEWBOLD, J. E., Quantitative analysis of hepatitis B virus DNA in saliva and semen of chronically infected homosexual men, *J. Infect. Dis.* **156:**299–307 (1987).

160. JOSHI, Y. K., BAKU, S., SARIN, S., TANDON, B. N., GANDHI, B. M., AND CHATURVEDI, V. C., Immunoprophylaxis of epidemic non-A, non-B hepatitis, *Indian J. Med. Res.* **81:**18–19 (1985).

161. KAKLAMANI, E., TRICHOPOULOS, D., TZONOU, A., ZAVITSANOS, X., KOUMANTAKI, Y., HATZAKIS, A., HSIEH, C. C., AND HATZIYANNIS, S., Hepatitis B and C viruses and their interaction in the origin of hepatocellular carcinoma, *J.A.M.A.* **265:**1974–1976 (1991).

162. KANE, M. A., BRADLEY, D. W., SHRESTHA, S. M., MAYNARD, J. E., COOK, E. H., MISHRA, R. P., AND JOSHI, D. D., Epidemic non-A, non-B hepatitis in Nepal: Recovery of a possible etiologic agent and transmission studies to marmosets, *J.A.M.A.* **252:**3140–3145 (1984).

163. KAO, H. W., ASCHAVAI, M., AND REDEKER, A. G., The persistence of hepatitis A IgM antibody after acute clinical hepatitis A, *Hepatology* **4:**933–936 (1984).

164. KAO, J. H., CHEN, P. J., YANG, P. M., LAI, M. Y., SHEU, J. C., WANG, T. H., AND CHEN, D. S., Intrafamilial transmission of hepatitis C virus: The important role of infections between spouses, *J. Infect. Dis.* **166:**900–903 (1992).

165. KARAYIANNIS, P., O'ROURKE, S., MCGARVEY, M. J., LUTHER, S., WATERS, J., GOLDIN, R., AND THOMAS, H. C., A recombinant vaccinia virus expressing hepatitis A virus structural polypeptides: Characterization and demonstration of protective immunogenicity, *J. Gen. Virol.* **72:**2167–2172 (1991).

166. KHUROO, M. S., Study of an epidemic of non-A, non-B hepatitis: Possibility of another human hepatitis virus distinct from posttransfusion non-A, non-B type, *Am. J. Med.* **68:**818–824 (1980).

167. KHUROO, M. S., DUERMEYER, W., ZARGAR, S. A., AHANGER, M. A., AND SHAH, M. A., Acute sporadic non-A, non-B hepatitis in India, *Am. J. Epidemiol.* **118:**360–364 (1983).

168. KHUROO, M. S., TELI, M. R., SKIDMORE, S., SOFI, M. A., AND KHUROO, M. I., Incidence and severity of viral hepatitis in pregnancy, *Am. J. Med.* **70:**252–255 (1981).

169. KIYOSAWA, K., SODEYAMA, T., TANAKA, E., GIBO, Y., YOSHIZAWA, K., NAKANO, Y., FURUTA, S., AKAHANE, Y., NISHIOKA, K., PURCELL, R. H., AND ALTER, H. J., Interrelationship of blood transfusion, non-A, non-B hepatitis and hepatocellular carcinoma: Analysis by detection of antibody to hepatitis C virus, *Hepatology* **12:**671–675 (1990).

170. KIYOSAWA, K., SODEYAMA, T., TANAKA, E., NAKANO, Y., FURUTA, S., NISHIOKA, K., PURCELL, R. H., AND ALTER, H. J., Hepatitis C in hospital employees with needlestick injuries, *Ann. Intern. Med.* **115:**367–369 (1991).

171. KLEIN, R. S., FREEMAN, K., TAYLOR, P. E., AND STEVENS, C. E., Occupational risk for hepatitis C virus infection among New York City dentists, *Lancet* **338:**1539–1542 (1991).

172. KLEINMAN, S., ALTER, H., BUSCH, M., HOLLAND, P., TEGTMEIER, G., NELLES, M., LEE, S., PAGE, E., WILBER, J., AND POLITO, A., Increased detection of hepatitis C virus (HCV)-infected blood donors by a multiple-antigen HCV enzyme immunoassay, *Transfusion* **32:**805–813 (1992).

173. KNODELL, R. G., CONRAD, M. E., GINSBURG, A. L., BELL, C. J.,

AND FLANNERY, E. P., Efficacy of prophylactic gammaglobulin in preventing non-A, non-B post-transfusion hepatitis, *Lancet* **1**: 557–561 (1976).

174. KOBAYASHI, H., TSUZUKI, M., KOSHIMUZU, K., TOYAMA, H., YOSHIHARA, N., SHIKATA, T., ABE, K., MIZUNO, K., OTOMO, N., AND ODA, T., Susceptibility of hepatitis B virus to disinfection or heat, *J. Clin. Microbiol.* **20**:214–216 (1984).

175. KORENMANN, J., BAKER, B., WAGGONER, J., EVERHART, J. E., DIBISCEGLIE, A. M., AND HOOFNAGLE, J. H., Long-term remission in chronic hepatitis B after alfa-interferon therapy, *Ann. Intern. Med.* **114**:629–634 (1991).

176. KORETZ, R. L., BREZINA, M., POLITO, A. J., QUAN, S., WILBER, J., DINELLO, R., AND GITNICK, G., Non-A, non-B posttransfusion hepatitis: Comparing C and non-C hepatitis, *Hepatology* **17**:361–365 (1993).

177. KOZIOL, D. E., HOLLAND, P. V., ALLING, D. W., MELPOLDER, J. C., SOLOMON, R. E., PURCELL, R. H., HUDSON, L. M., SHOUP, F. J., KRAKAUER, H., AND ALTER, H. J., Antibody to hepatitis B core antigen as a paradoxical marker for non-A, non-B hepatitis agents in donated blood, *Ann. Intern. Med.* **104**:488–495 (1986).

178. KRAHN, M., AND DETSKY, A. S., Should Canada and the United States universally vaccinate infants against hepatitis B, *Med. Decis. Making* **13**:4–20 (1993).

179. KRAWCZYNSKI, K., AND BRADLEY, D. W., Enterically transmitted non-A, non-B hepatitis: Identification of virus-associated antigen in experimentally infected cynomolgus macaques, *J. Infect. Dis.* **159**:1042–1049 (1989).

180. KRAWCZYNSKI, K. Z., BRADLEY, D. W., MURPHY, B. L., EBERT, J. W., ANDERSON, T. A., DOTO, I. L., NOWOSLAWSKI, A., DUERMEYER, W., AND MAYNARD, J. E., Pathogenetic aspects of hepatitis A virus infection in enterally inoculated marmosets, *Am. J. Clin. Pathol.* **76**:698–706 (1981).

180a. KRAWCZYNSKI, K., ALTER, M. J., TANKERSLEY, D. L., BEACH, M., ROBERTSON, B. H., LAMBERT, S., KUO, G., SPELBRING, J. E., MEEKS, E., SINHA, S., AND CARSON, D. A., Effect of immune globulin on the prevention of experimental hepatitis C virus infection, *J. Infect. Dis.* **173**:822–828 (1996).

181. KREMASTINOU, J., KALAPOTHAKI, V., AND TRICHOPOULOS, D., The changing epidemiologic pattern of hepatitis A infection in urban Greece, *Am. J. Epidemiol.* **120**:703–706 (1984).

182. KROGSGAARD, K., WANTZIN, P., MATHIESEN, L. R., SONNE, J., RING-LARSEN, H., AND THE COPENHAGEN HEPATITIS ACUTA PROGRAMME, Early appearance of antibodies to hepatitis C virus in community acquired acute non-A, non-B hepatitis is associated with progression to chronic liver disease, *Scand. J. Infect. Dis.* **22**:399–402 (1990).

183. KRUGMAN, S., GILES, J. P., AND HAMMOND, J., Hepatitis virus effect of heat on the infectivity and antigenicity of the MS-1 and MS-2 strain, *J. Infect. Dis.* **122**:423–436 (1970).

184. KRUGMAN, S., GILES, J. P., AND HAMMOND, J., Infectious hepatitis: Evidence for two distinctive clinical, epidemiological and immunological types of infection, *J.A.M.A.* **200**:365–373 (1967).

185. KRUGMAN, S., WARD, R., GILES, J. P., AND JACOBS, A. M., Infectious hepatitis, study on effect of gamma globulin and on the incidence of apparent infection, *J.A.M.A.* **174**:823–830 (1960).

186. KUO, G., CHOO, Q. L., ALTER, H. J., GITNICK, G. L., REDEKER, A. G., PURCELL, R. H., MIYAMURA, T., DIENSTAG, J. L., ALTER, M. J., STEVENS, C. E., TEGTMEIER, G. E., BONINO, F., COLOMBO, M., LEE, W. S., KUO, C., BERGER, K., SHUSTER, J. R., OVERBY, L. R., BRADLEY, D. W., AND HOUGHTON, M., An assay for circulat-

ing antibodies to a major etiologic virus of human non-A, non-B hepatitis, *Science* **244**:362–364 (1989).

187. LAM, J. P. H., MCOMISH, F., BURNS, S. M., YAP, P. L., MOK, J. Y. Q., AND SIMMONDS, P., Infrequent vertical transmission of hepatitis C virus, *J. Infect. Dis.* **167**:572–576 (1993).

188. LANGE, W. R., AND FRAME, J. D., High incidence of viral hepatitis among American missionaries in Africa, *Am. J. Trop. Med. Hyg.* **43**:527–533 (1990).

189. LEDNAR, W. M., LEMON, S. M., KIRKPATRICK, J. W., REDFIELD, R. R., FIELDS, M. L., AND KELLY, P. W., Frequency of illness associated with epidemic hepatitis A virus infection in adults, *Am. J. Epidemiol.* **122**:226–233 (1985).

190. LEENTVAAR-KUIJPERS, A., COUTINHO, R. A., BRULEIN, V., AND SAFARY, A., Simultaneous passive and active immunization against hepatitis A, *Vaccine* **10**:s138–s141 (1992).

191. LELIE, P. N., IP, H. M. H., REESINK, H. W., WONG, V. C. W., AND KUHNS, M. C., Prevention of the hepatitis B virus carrier state in infants of mothers with high and low levels of HBV-DNA, in: *Viral Hepatitis and Liver Disease* (F. B. HOLLINGER, S. M. LEMON, AND H. S. MARGOLIS, EDS.), pp. 753–756, Williams and Wilkins, Baltimore, 1991.

192. LEMON, S. M., Type A viral hepatitis: New developments in an old disease, *N. Engl. J. Med.* **313**:1059–1067 (1985).

193. LEMON, S. M., AND BINN, L. N., Serum neutralizing antibody response to hepatitis A virus, *J. Infect. Dis.* **148**:1033–1039 (1983).

194. LEMON, S. M., BINN, L. N., AND MARCHWICKI, R. H., Radio-immunofocus assay for quantitation of hepatitis A virus in cell culture, *J. Clin. Microbiol.* **17**:834–839 (1983).

195. LETTAU, L., MCCARTHY, J. G., SMITH, M. H., HADLER, S. C., MORSE, L. J., UKENA, T., BESSETTE, R., GURWITZ, A., IRVINE, W. G., FIELDS, H. A., GRADY, G. F., AND MAYNARD, J. E., An outbreak of severe hepatitis due to delta and hepatitis B viruses in parenteral drug abusers and their contacts, *N. Engl. J. Med.* **317**:1256–1261 (1987).

196. LEVY, B. S., FONTAINE, R. E., SMITH, C. A., BRINDA, J., HIRMAN, G., NELSON, D. B., JOHNSON, P. M., AND LARSON, O., A large food-borne outbreak of hepatitis A, *J.A.M.A.* **234**:289–294 (1975).

197. LIANG, T. J., HASEGAWA, K., RIMON, N., WANDS, J. R., AND BEN-PORATH, E., A hepatitis B virus mutant associated with an epidemic of fulminant hepatitis, *N. Engl. J. Med.* **324**:1705–1709 (1991).

198. LIANG, T. J., JEFFERS, L., REDDY, R. K., SILVA, M. O., CHEINQUER, H., FINDOR, A., DEMEDINA, M., YARBOUGH, P. O., REYES, G. R., AND SCHIFF, E. R., Fulminant or subfulminant non-A, non-B hepatitis: Hepatitis C and E viruses, *Gastroenterology* **103**:556–562 (1992).

199. LIN, H. H., KAO, J. H., HSU, H. Y., NI, Y. H., YEH, S. H., HWANG, L. H., CHANG, M. H., HWANG, S. C., CHEN, P. J., AND CHEN, D. S., Possible role of high-titer maternal viremia in perinatal transmission of hepatitis C virus, *J. Infect. Dis.* **169**:638–641 (1994).

199a. LINNEN, J., WAGES, J., ZHANG-KECK, Z.-Y., FRY, K. E., KRAWCZYNSKI, K. K., ALTER, H., KOONIN, E., GALLAGHER, M., ALTER, M., HADZIYANNIS, S., KARAYIANNIS, P., FUNG, K., NAKATSUJI, Y., SHIH, J. W-K., YOUNG, L., PIATAK, M., HOOVER, C., FERNANDEZ, J., CHEN, S., ZOU, J-C., MORRIS, T., HYAMS, K. C., ISMAY, S., LIFSON, J. D., HESS, G., FOUNG, S. K. J., THOMAS, H., BRADLEY, D., MARGOLIS, H., KIM, J. P., Molecular cloning and disease association of hepatitis G virus: A transfusion-transmissible agent, *Science* **271**:505–508 (1996).

200. LO, K.-J., LEE, S.-D., TSAI, Y.-T., WU, T.-C., CHAN, C.-Y., CHEN, G.-H., AND YEH, C.-L., Long-term immunogenicity and efficacy of hepatitis B vaccine in infants born to HBeAg-positive HBsAg-carrier mothers, *Hepatology* **8**:1647–1650 (1988).

201. LO, K.-J., TONG, M. J., CHIEN, M.-C., TSAI, Y.-T., LIAW, Y.-F., YANG, K.-C., CHIAN, H., LIU, H.-C., AND LEE, S.-D., The natural course of hepatitis B surface antigen-positive chronic active hepatitis in Taiwan, *J. Infect. Dis.* **146**:205–210 (1982).

202. LOCARNINI, S. A., FERRIS, A. A., LEHMANN, N. I., AND GUST, I., The antibody response following hepatitis A infection, *Intervirology* **8**:309–318 (1977).

203. LOK, A. S. F., KWAN, WAI-K., MOECKLI, R., YARBOUGH, P. O., CHAN, R. T., REYES, G. R., LAI, CHING-L., CHUNG, HAU-T., AND LAI, T. S. T., Seroepidemiological survey of hepatitis E in Hong Kong by recombinant-based enzyme immunoassays, *Lancet* **340**:1205–1208 (1992).

204. MAHONEY, F. J., FARLEY, T. A., KELSO, K. Y., WILSON, S. A., HORNA, J. M., AND MCFARLAND, L. M., An outbreak of hepatitis A associated with swimming in a public pool, *J. Infect. Dis.* **165**:613–618 (1992).

205. MAHONEY, F. J., WOODRUFF, B. A., ERBEN, J. J., COLEMAN, P. J., REID, E. C., SCHATZ, G. C., AND KANE, M. A., Effect of hepatitis B vaccination program on the prevalence of hepatitis B virus infection, *J. Infect. Dis.* **167**:203–207 (1993).

206. MAO, J. S., DONG, D. X., ZHANG, S. Y., ZHANG, H. Y., CHEN, N. L., HUANG, H. Y., XIE, R. Y., CHAI, C. A., ZHOU, T. J., WU, D. M., AND ZHANG, H. C., Further studies of attenuated live hepatitis A vaccine (strain H2) in humans, in: *Viral Hepatitis and Liver Disease* (F. B. HOLLINGER, S. M. LEMON, AND H. S. MARGOLIS, EDS.), pp. 110–111, Williams and Wilkins, Baltimore, 1991.

207. MARGOLIS, H. S., Prevention of acute and chronic liver disease through immunization: Hepatitis B and beyond, *J. Infect. Dis.* **168**:9–14 (1993).

208. MARGOLIS, H. S., ALTER, M. J., AND HADLER, S. C., Hepatitis B: Evolving epidemiology and implications for control, *Semin. Liver Dis.* **11**:84–92 (1991).

209. MARGOLIS, H. S., COLEMAN, P. J., BROWN, R. E., MAST, E. E., SHIENGOLD, S. H., AND AREVALO, J. A., Prevention of hepatitis B virus transmission by immunization: An economic analysis of current recommendations, *J.A.M.A.* **274**:1242–1243 (1995).

210. MARGOLIS, H. S., NAINAN, O., V KRAWCZYNSKI, K., BRADLEY, D. W., EBERT, J. W., SPELBRING, J., FIELDS, H. A., AND MAYNARD, J. E., Appearance of immune complexes during experimental hepatitis A infection in chimpanzees, *J. Med. Virol.* **26**:315–326 (1988).

211. MARGOLIS, H. S., AND SHAPIRO, C. N., Considerations for the development of recommendations for the use of hepatitis A vaccine, *J. Hepatol.* **18**(Suppl. 2):s56–s60 (1993).

212. MARINIER, E., BARROIS, V., LAROUZE, B., LONDON, W. T., COFER, A., DIAKHATE, L., AND BLUMBERG, B. S., Lack of perinatal transmission of hepatitis B virus infection in Senegal, West Africa, *J. Pediatr.* **106**:843–849 (1985).

213. MARION, P. L., TREPO, C., MATSUBARA, K., AND PRICE, P. M., Experimental models in hepadnavirus research: Report of a workshop, in: *Viral Hepatitis and Liver Disease* (F. B. HOLLINGER, S. M. LEMON, AND H. S. MARGOLIS, EDS.), pp. 866–874, Williams and Wilkins, Baltimore, 1991.

214. MATHIESEN, L. R., SKINHOLJ, P., NIELSEN, J. O., PURCELL, R. H.,

WONG, D. C., AND RANEK, L., Hepatitis A, B, and non-A, non-B in fulminant hepatitis, *Gut* **21**:72–77 (1980).

215. MAYNARD, J. E., Hepatitis B: Global importance and need for control, *Vaccine* **8**:18s–20s (1990).

216. MAYNARD, J. E., Hepatitis B vaccine: Strategies for utilization, in: *Hepatitis B Vaccine* (P. MAUPAS AND P. GUESRY, EDS.), pp. 13–19, Elsevier/North Holland Biomedical Press, Amsterdam, 1981.

217. MAYNARD, J. E., Viral hepatitis as an occupational hazard in the health care profession, in: *Viral Hepatitis: A Contemporary Assessment of Etiology, Epidemiology, Pathogenesis and Prevention* (G. N. VYAS, S. N. COHEN, AND R. SCHMID, EDS.), pp. 321–331, The Franklin Institute Press, Philadelphia, 1978.

218. MAYNARD, J. E., HADLER, S. C., AND FIELDS, H. A., Delta hepatitis in the Americas: An overview, *Prog. Clin. Biol. Res.* **234**:493–505 (1986).

219. MAYNARD, J. E., KANE, M. A., AND HADLER, S. C., Global control of hepatitis B through vaccination: Role of hepatitis B vaccine in the expanded programme on immunization, *Rev. Infect. Dis.* **11**:s574–s578 (1989).

220. MBITHI, J. N., SPRINGTHORPE, S., BOULET, J. R., AND SATTAR, S. A., Survival of hepatitis A virus on human hands and its transfer on contact with animate and inanimate surfaces, *J. Clin. Microbiol.* **30**:757–763 (1992).

221. MCCAUSTLAND, K. A., BI, S., PURDY, M. A., AND BRADLEY, D. W., Application of two RNA extraction methods prior to amplification of hepatitis E virus nucleic acid by the polymerase chain reaction, *J. Virol. Methods* **35**:331–342 (1991).

222. MCCAUSTLAND, K. A., BOND, W. W., BRADLEY, D. W., EBERT, J. W., AND MAYNARD, J. E., Survival of hepatitis A virus in feces after drying and storage for 1 month, *J. Clin. Microbiol.* **16**:957–958 (1982).

223. MCGREGOR, A., Round the world. WHO: World Health Assembly, *Lancet* **339**:1287 (1992).

224. MCMAHON, B. J., ALBERTS, S. R., WAINWRIGHT, R. B., BULKOW, L., AND LANIER, A. P., Hepatitis B-related sequelae. Prospective study in 1400 hepatitis B surface antigen-positive Alaska Native carriers, *Arch. Intern. Med.* **150**:1051–1054 (1990).

225. MCMAHON, B. J., ALWARD, W. L. M., HALL, D. B., HEYWARD, W. L., BENDER, T. R., FRANCIS, D. P., AND MAYNARD, J. E., Acute hepatitis B virus infection: Relation of age to the clinical expression of disease and subsequent development of the carrier state, *J. Infect. Dis.* **151**:599–603 (1985).

226. MCMAHON, B. J., BENDER, T. R., BERQUIST, K. R., SCHREEDER, M. T., AND HARPSTER, A. P., Delayed development of antibody to hepatitis B surface antigen for symptomatic infection with hepatitis B virus, *J. Clin. Microbiol.* **14**:130–134 (1981).

227. MCMAHON, B. J., SCHOENBERG, S., BULKOW, L., WAINWRIGHT, R. B., FITZGERALD, M. A., PARKINSON, A. J., COKER, E., AND RITTER, D., Seroprevalence of hepatitis B viral markers in 52,000 Alaskan Natives, *Am. J. Epidemiol.* **138**:544–549 (1993).

228. MCQUILLAN, G. M., TOWNSEND, T. R., FIELDS, H. A., CARROLL, M., LEAHY, M., AND POLK, B. F., The seroepidemiology of hepatitis B virus in the United States, 1976–80, *Am. J. Med.* **87**:5–10 (1989).

229. MELNICK, J. L., A water-borne urban epidemic of hepatitis, in: *Hepatitis Frontiers* (F. W. HARTMAN, G. A. LOGRIPPO, J. G. MATFFER, AND J. BARRON, EDS.), pp. 211–225, Little, Brown and Company, Boston, 1957.

230. MIDTHUN, K., ELLERBECK, E., GERSHMAN, K., CALANDRA, G., KRAH, D., MCCAUGHTRY, M., NALIN, D., AND PROVOST, P.,

Safety and immunogenicity of a live attenuated hepatitis A virus vaccine in seronegative volunteers, *J. Infect. Dis.* **163:**735–739 (1991).

231. MILLARD, J., APPLETON, H., AND PARRY, J. V., Studies on heat inactivation of hepatitis A virus with special reference to shellfish Part 1. Procedures for infection and recovery of virus from laboratory-maintained cockles, *Epidemiol. Infect.* **98:**397–414 (1987).

232. MIMMS, L., VALLARI, D., DUCHARME, L., HOLLAND, P., KURA-MOTO, I. K., AND ZELDIS, J., Specificity of anti-HCV ELISA assessed by reactivity to three immunodominant HCV regions, *Lancet* **336:**1590–1591 (1990).

233. MINOR, P. D., Picornaviridae, in: *Classification and Nomenclature of Viruses: The Fifth Report of the International Committee on Taxonomy of Viruses* (R. I. B. FRANKI, C. M. FAUQUET, D. L. KNUDSON, AND F. BROWN, EDS.), pp. 320–326, Springer Verlag, Wien, 1991.

234. MITSUI, T., IWANO, K., MASUKO, K., YAMAZAKI, C., OKAMOTO, H., TSUDA, F., TANAKA, T., AND MISHIRO, S., Hepatitis C virus infection in medical personnel after needlestick accident, *Hepatology* **16:**1109–1114 (1992).

235. MOYER, L. A., AND ALTER, M. J., Hepatitis C virus in the hemodialysis setting: A review with recommendations for control, *Semin. Dialysis* **7:**124–127 (1994).

236. MUSHAHWAR, I. K., McGRATH, L. C., DRNEC, J., AND OVERBY, L. R., Radioimmunoassay for the detection of hepatitis B e antigen and its antibody. Results of a clinical evaluation, *Am. J. Clin. Pathol.* **76:**692–697 (1981).

237. MYINT, H., SOE, M. M., KHIN, T., MYINT, T. M., AND TIN, K. M., A clinical and epidemiological study of an epidemic of non-A, non-B hepatitis in Rangoon, *Am. J. Trop. Med. Hyg.* **34:**1183–1189 (1985).

238. NAINAN, O. V., BRINTON, M. A., AND MARGOLIS, H. S., Identification of amino acids located in the antibody binding sites of human hepatitis A virus, *Virology* **191:**984–987 (1992).

239. NAINAN, O. V., MARGOLIS, H. S., ROBERTSON, B. H., BALAYAN, M., AND BRINTON, M. A., Sequence analysis of a new hepatitis A virus naturally infecting cynomolgus macaques (Macaca fascicularis), *J. Gen. Virol.* **72:**1685–1689 (1991).

240. NISHIGUCHI, S., FUKUDA, K., SHIOMI, S., ISSHIKI, G., KUROKI, T., NAKAJIMA, S., MURATA, R., AND KOBAYSHI, K., Familial clustering of HCV, *Lancet* **339:**1486 (1992).

241. NIU, M. T., ALTER, M. J., KRISTENSEN, C., AND MARGOLIS, H. S., Outbreak of hemodialysis-associated non-A, non-B hepatitis and correlation with antibody to hepatitis C virus, *Am. J. Kidney Dis.* **4:**345–352 (1992).

242. NIU, M. T., POLISH, L. B., ROBERTSON, B. H., KHANNA, B., WOODRUFF, B. A., SHAPIRO, C. N., MILLER, M. A., SMITH, J. D., GEDROSE, J. K., ALTER, M. J., AND MARGOLIS, H. S., A multistate outbreak of hepatitis A associated with frozen strawberries, *J. Infect. Dis.* **166:**518–524 (1992).

243. NOBLE, R. C., KANE, M. A., REEVES, S. A., AND ROECKEL, I., Posttransfusion hepatitis A in a neonatal intensive care unit, *J.A.M.A.* **252:**2711–2715 (1984).

244. OHTO, H., TERAZAWA, S., SASAKI, N., HINO, K., ISHIWATA, C., KAKO, M., UJIIE, N., ENDO, C., MATSUI, A., OKAMOTO, H., MISHIRO, S., AND THE VERTICAL TRANSMISSION OF HEPATITIS C VIRUS COLLABORATIVE STUDY GROUP, Transmission of hepatitis C virus from mothers to infants, *N. Engl. J. Med.* **330:**744–750 (1994).

245. OSMOND, D. H., CHARLEBOIS, E., SHEPPARD, H. W., PAGE, K., WINKELSTEIN, W., MOSS, A. R., AND REINGOLD, A., Comparison of risk factors for hepatitis C and hepatitis B virus infection in homosexual men, *J. Infect. Dis.* **167:**66–71 (1993).

246. OSMOND, D. H., PADIAN, N. S., SHEPPARD, H. W., GLASS, S., SHIBOSKI, S. C., AND REINGOLD, A., Risk factors for hepatitis C virus seropositivity in heterosexual couples, *J.A.M.A.* **269:**361–365 (1993).

247. OTTOBRELLI, A., MARZANO, A., SMEDILE, A., RECCHIA, S., SA-LIZZONI, M., CORNU, C., LARNY, M. E., OTTE, J. B., DE HEMP-TINNE, B., GEUBEL, A., GRENDELE, M., COLLEDAN, M., GAL-MARINI, D., MARINUCCI, G., DI GIACOMO, C., AGNES, S., BONINO, F., AND RIZZETTO, M., Patterns of hepatitis delta virus reinfection and disease in liver transplantation, *Gastroenterology* **101:**1649–1655 (1991).

248. PARRY, J. V., AND MORTIMER, P. P., The heat sensitivity of hepatitis A virus determined by a simple tissue culture method, *J. Med. Virol.* **14:**277–283 (1984).

249. PARRY, J. V, PERRY, K. R., PANDAY, S., AND MORTIMER, P. P., Diagnosis of hepatitis A and B by testing saliva, *J. Med. Virol.* **28:**255–260 (1989).

250. PENNA, A., CHISARI, F. V., BERTOLETTI, A., MISSALE, G., FOWLER, P., GIUBERTI, T., FIACCADORI, F., AND FERRARI, C., Cytotoxic T lymphocytes recognize an HLA-A2-restricted epitope within the hepatitis B virus nucleocapsid antigen, *J. Exp. Med.* **174:**1565–1570 (1991).

251. PEREIRA, B. J. G., MILFORD, E. L., KIRKMAN, R. L., QUAN, S., SAYRE, K. R., JOHNSON, P. J., WILBER, J. C., AND LEVEY, A. S., Prevalence of hepatitis C virus RNA in organ donors positive for hepatitis C antibody and in the recipients of their organs, *N. Engl. J. Med.* **327:**910–915 (1992).

252. PERRILLO, R. P., SCHIFF, E. R., AND DAVIS, G. L., A randomized controlled trial of interferon alfa-2b alone and after prednisone withdrawal for the treatment of chronic hepatitis B., *N. Engl. J. Med.* **323:**295–301 (1990).

253. PETERSEN, N. J., BARRETT, D. H., BOND, W. H., BERQUIST, K. R., FAVERO, M. S., BENDER, T. R., AND MAYNARD, J. E., Hepatitis B surface antigen in saliva, impetiginous lesions, and the environment in two remote Alaskan villages, *App. Environ. Microbiol.* **32:**572–574 (1976).

254. PING, L.-H., AND LEMON, S. M., Antigenic structure of human hepatitis A virus defined by analysis of escape mutants selected against murine monoclonal antibodies, *J. Virol.* **66:**2208–2216 (1992).

255. POLISH, L. B., GALLAGHER, M., FIELDS, H. A., AND HADLER, S. C., Delta hepatitis. Molecular biology and clinical and epidemiological features, *Clin. Microbiol. Rev.* **6:**211–229 (1993).

256. POLISH, L. B., SHAPIRO, C. N., BAUER, F., KLOTZ, P., GINIER, P., ROBERTO, R. R., MARGOLIS, H. S., AND ALTER, M. J., Nosocomial transmission of hepatitis B virus associated with a spring-loaded fingerstick device, *N. Engl. J. Med.* **326:**721–725 (1992).

257. POLISH, L. B., TONG, M. J., CO, R. L., COLEMAN, P. J., AND ALTER, M. J., Risk factors for hepatitis C virus infection among health care personnel in a community hospital, *Am. J. Infect. Control* **21:**196–200 (1993).

258. PON, E. W., REN, H. X., MARGOLIS, H. S., SCHATZ, G. C., AND DIWAN, A., Hepatitis B virus infection in Honolulu students, *Pediatrics* **92:**574–578 (1993).

259. PONZETTO, A., FORZANI, B., PARRAVICINI, P. P., HELE, C., ZANETTI, A., AND RIZZETTO, M., Epidemiology of delta virus infection, *Eur. J. Epidemiol.* **1:**257–263 (1986).

260. POOVORAWAN, Y., SANPAVAT, S., PONGPUNIERT, W., CHUM-DERMPADETSUK, S., SENTRAKUL, P., AND SAFARY, A., Protective efficacy of a recombinant DNA hepatitis B vaccine in neonates of HBe antigen-positive mother, *J.A.M.A.* **261:**3278–3281 (1989).

261. PRINCE, A. M., An antigen detected in the blood during the incubation period of serum hepatitis, *Proc. Natl. Acad. Sci. USA* **60:**814–821 (1968).

262. PROVOST, P. J., AND HILLEMAN, M. R., Propagation of human hepatitis A virus in cell culture *in vitro, Proc. Soc. Exp. Biol. Med.* **160:**213–221 (1979).

263. PURCELL, R. H., FEINSTONE, S. M., TICEHURST, J. R., DAEMER, R. J., AND BAROUDY, B. M., Hepatitis A virus, in: *Viral Hepatitis and Liver Disease* (G. N. VYAS, J. L. DIENSTAG, AND J. H., HOOFNAGLE, EDS.), pp. 9–22, Grune & Stratton, Orlando, FL, 1984.

264. PURCELL, R. H., AND GERIN, J. L., Hepatitis B subunit vaccine: A preliminary report of safety and efficacy in chimpanzees, *Am. J. Med. Sci.* **270:**395–399 (1975).

265. PURCELL, R. H., SATTERFIELD, W. C., BERGMANN, K. F., SMEDILE, A., PONZETTO, A., AND GERIN, J. L., Experimental hepatitis delta virus infection in the chimpanzee, *Prog. Clin. Biol. Res.* **234:**27–36 (1986).

266. PURDY, M., MCCAUSTLAND, K., KRAWCZYNSKI, K., BEACH, M., SPELBRING, J., REYES, G., AND BRADLEY, D., An expressed recombinant HEV protein that protects cynomologus macaques against challenge with wild-type hepatitis E virus, in: *Immunobiology and Pathogenesis of Persistent Virus Infections.* Savannah, GA, October 24–28, 1992, pp. 41, Elsevier Science Publ., Amsterdam (1993).

267. REYES, G. R., PURDY, M. A., KIM, J. P., LUK, D.-C., YOUNG, L. M., TAM, A. W., AND BRADLEY, D. W., Isolation of a cDNA from the virus responsible for enterically transmitted non-A, non-B hepatitis, *Science* **247:**1335–1339 (1990).

268. RINKER, J., AND GALAMBOS, J. T., Prospective study of hepatitis B in thirty-two inadvertently infected people, *Gastroenterology* **81:**686–691 (1981).

269. RIZZETTO, M., CANESE, M. C., ARICO, S., CRIVELLI, O., TREPO, C., BONINO, R., AND VERME, G., Immunofluorescence detection of a new antigen-antibody system (/anti-) associated to the hepatitis B virus in the liver and in the serum of HBsAg carriers, *Gut* **18:**997–1003 (1977).

270. RIZZETTO, M., CANESE, M. C., GERIN, J. L., LONDON, W. T., SLY, I. D., AND PURCELL, R. H., Transmission of the hepatitis B virus associated delta antigen to chimpanzees, *J. Infect. Dis.* **141:**590–601 (1980).

271. RIZZETTO, M., VERME, G., RECCHIA, S., BONINO, F., FARCI, P., ARICO, S., CALAZIA, R., PICCIOTTO, A., COLOMBO, M., AND POPPER, H., Chronic hepatitis in carriers of hepatitis B surface antigen with intrahepatic expression of the delta antigen. An active and progressive disease unresponsive to immunosuppressive treatment, *Ann. Intern. Med.* **8:**437–441 (1981).

272. ROBERTSON, B. H., D'HONDT, E. H., SPELBRING, J., TIAN, H. W., KRAWCZYNSKI, K. Z., AND MARGOLIS, H. S., Effect of postexposure vaccination in a chimpanzee model of hepatitis A virus infection, *J. Med. Virol.* **43:**249–251 (1994).

273. ROBERTSON, B. H., JANSEN, R. W., KHANNA, B., TOTSUKA, A., NAINAN, O., V, SIEGL, G., WIDELL, A., MARGOLIS, H. S., ISOMURA, S., ITO, K., ISHIZU, T., MORITSUGU, Y., AND LEMON, S. M., Genetic relatedness of hepatitis A virus strains recovered from different geographical regions, *J. Gen. Virol.* **73:**1365–1377 (1992).

274. ROBERTSON, B. J., JIA, X.-Y., TIAN, H., MARGOLIS, H. S., SUM-

275. ROSENBLUM, L. S., MIRKIN, I. R., ALLEN, D. T., SAFFORD, S., AND HADLER, S. C., A multifocal outbreak of hepatitis A traced to commercially distributed lettuce, *Am. J. Public Health* **80:** 1075–1079 (1990).

276. ROSENBLUM, L. S., VILLARINO, M. E., NAINAN, O., V, MELISH, M. E., HADLER, S. C., PINSKY, P. P., JARVIS, W. R., OTT, C. E., AND MARGOLIS, H. S., Hepatitis A outbreak in a neonatal intensive care unit: Risk factors for transmission and evidence of prolonged viral excretion among preterm infants, *J. Infect. Dis.* **164:**476–482 (1991).

277. ROUDOT-THORAVAL, F., PAWLOTSKY, J.-M., THIERS, V., DEFORGES, L., GIROLLET, P.-P., GUILLOT, F., HURAUX, C., AUMONT, P., BRECHOT, C., AND DHUMEAUX, D., Lack of mother-to-infant transmission of hepatitis C virus in human immunodeficiency virus-seronegative women: A prospective study with hepatitis C virus RNA testing, *Hepatology* **17:**722–777 (1993).

278. SANCHEZ-QUIJANO, A., PINEDA, J. A., LISSEN, E., LEAL, M., DIAZ-TORRES, M. A., GARCIA DE PESQUERA, F., RIVERA, F., CASTRO, R., AND MUNOZ, J., Prevention of post-transfusion non-A, non-B hepatitis by non-specific immunoglobulin in heart surgery patients, *Lancet* **1:**1245–1249 (1988).

279. SCHADE, C. P., AND KOMORWSKA, D., Continuing outbreak of hepatitis A linked with intravenous drug abuse in Multnomah County, *Public Health Rep.* **103:**452–459 (1988).

280. SEEFF, L. B., ZIMMERMAN, J. H., WRIGHT, E. L., FINKELSTEIN, J. D., GARCIA-PONT, P., GREENLEE, H. B., DIETZ, A. A., LEEVY, C. M., TAMBURRO, C. H., SCHIFF, E. R., SCHIMMEL, E. M., ZEMEL, R., ZIMMON, D. S., AND MCCOLLUM, R. W., A randomized double-blind controlled trial of the efficacy of immune serum globulin for the prevention of post-transfusion hepatitis. A Veterans Administration cooperative study, *Gastroenterology* **72:**111–121 (1977).

281. SHAPIRO, C. N., COLEMAN, P. J., MCQUILLAN, G. M., ALTER, M. J., AND MARGOLIS, H. S., Epidemiology of hepatitis A: Seroepidemiology and risk groups in the USA, *Vaccine* **10:**s59–s62 (1992).

282. SHAPIRO, C. N., SHAW, F. E., MENDEL, E. J., AND HADLER, S. C., Epidemiology of hepatitis A in the United States, in: *Viral Hepatitis and Liver Disease* (F. B. HOLLINGER, S. M. LEMON, AND H. S. MARGOLIS, EDS.), pp. 214–220, Williams and Wilkins, Baltimore, 1991.

283. SHAW, F. E. J., SHAPIRO, C. N., WELTY, T. K., DILL, W., REDDINGTON, J., AND HADLER, S. C., Hepatitis transmission among the Sioux Indians of South Dakota, *Am. J. Public Health* **80:** 1091–1094 (1990).

284. SHAW, F. E., JR., BARRETT, C. L., HAMM, R., PEARE, R. B., COLEMAN, P. J., HADLER, S. C., FIELDS, H. A., AND MAYNARD, J. E., Lethal outbreak of hepatitis B in a dental practice, *J.A.M.A.* **255:**3260–3264 (1986).

285. SHAW, F. E., JR., SUDMAN, J. H., SMITH, S. M., WILLIAMS, D. L., KAPELL, L. A., HADLER, S. C., HALPIN, T. J., AND MAYNARD, J. E., A community-wide epidemic of hepatitis A in Ohio, *Am. J. Epidemiol.* **123:**1057–1065 (1986).

286. SIEGL, G., AND LEMON, S. M., Recent advances in hepatitis A vaccine development, *Virus Res.* **17:**75–92 (1990).

287. SJOGREN, M. H., TANNO, H., FAY, O., SILEONI, S., COHEN, B. D., BURKE, D. S., AND FEIGHNY, R. J., Hepatitis A virus in stool during clinical relapse, *Ann. Intern. Med.* **106:**221–226 (1987).

MERS, D. F., AND EHRENFELD, E., Antibody response to nonstructural proteins of hepatitis A virus following infection, *J. Med. Virol.* **40:**76–82 (1993).

288. SMEDILE, A., DENTICO, P., ZANETTI, A., SAGNELLI, E., NORDEN-FELT, E., ACTIS, G. C., AND RIZZETTO, M., Infection with the delta agent in chronic HBsAg carriers, *Gastroenterology* **81**: 992–997 (1981).

289. SMEDILE, A., FARCI, P., VERME, G., CAREDDA, F., CARGNEL, A., CAPORASO, N., DENTICO, P., TREPO, C., OPOLON, P., GIMSON, A., VERGANI, D., WILLIAMS, R., AND RIZZETTO, M., Influence of delta infection on the severity of hepatitis B, *Lancet* **2**:9–15, (1982).

290. SMEDILE, A., LAVARINI, C., FARCI, P., ARICO, S., MARINUCCI, G., DENTICO, P., GIULIANI, G., CARGNEL, A., BLANDO, C. V., AND RIZZETTO, M., Epidemiologic patterns of infection with the hepatitis B virus associated delta agent in Italy, *Am. J. Epidemiol.* **117**:223–229 (1983).

291. SMEDILE, A., RIZZETTO, M., DENNISTON, K., BONINO, F., WELLS, F., VERME, G., CONSOLO, F., HOYER, B., PURCELL, R. H., AND GERIN, J. L., Type D hepatitis: The clinical significance of hepatitis D virus RNA in serum as detected by a hybridization-based assay, *Hepatology* **6**:1297–1302 (1986).

292. STEFFEN, R., Risk of hepatitis A in travellers, *Vaccine* **10**:s69–s72 (1992).

293. STEVENS, C. E., AACH, R. D., HOLLINGER, F. B., MOSLEY, J. W., SZMUNESS, W., KAHN, R., WERCH, J., AND EDWARDS, V., Hepatitis B virus antibody in blood donors and the occurrence of non-A, non-B hepatitis in transfusion recipients: An analysis of the transfusion-transmitted viruses study, *Ann. Intern. Med.* **101**: 733–738 (1984).

294. STEVENS, C. E., NEURATH, R. A., BEASLEY, R. P., AND SZMUNESS, W., HBeAg and anti-HBe detection by radioimmunoassay: Correlation of vertical transmission of hepatitis B virus in Taiwan, *J. Med. Virol.* **3**:237–241 (1979).

295. STEVENS, C. E., TAYLOR, P. E., TONG, M. J., TOY, P. T., VYAS, G. N., NAIR, P. V., WEISSMAN J. Y., AND KRUGMAN, S., Yeast-recombinant hepatitis B vaccine. Efficacy with hepatitis B immune globulin in prevention of perinatal hepatitis B virus transmission, *J.A.M.A.* **257**:2612–2616 (1987).

296. STEVENS, C. E., TAYLOR, P. E., TOY, P. E., AND MARGOLIS, H. S., Long-term follow-up of infants receiving postexposure immunoprophylaxis to prevent perinatal hepatitis B virus infection, unpublished data.

297. STEVENS, C. E., TOY, P. T., TAYLOR, P. E., LEE, T., AND YIP, H.-Y., Prospects for control of hepatitis B virus infection: Implications of childhood vaccination and long-term protection, *Pediatrics* **90**:170–173 (1992).

298. STROFFOLINI, T., DECRESCENZO, L., GIAMMANCO, A., INTONAZZO, V., LAROSA, G., CASCIO, A., SARZANA, A., CHIARINI, A., AND DARDANONI, L., Changing patterns of hepatitis A virus infection in children in Palermo, Italy, *Eur. J. Epidemiol.* **6**:84–87 (1990).

299. SUMMERS, J., AND MASON, W. S., Properties of the hepatitis B-like virus as related to their taxonomic classification, *Hepatology* **2**(Suppl.):61–66 (1982).

300. SZMUNESS, W., DIENSTAG, J. L., PURCELL, R. H., STEVENS, C. E., WONG, D. C., IKRAM, H., BAR-SHANY, S., BEASLEY, R. P., DESMYTER, J., AND GAON, J. A., The prevalence of antibody to hepatitis A antigen in various parts of the world: A pilot study, *Am. J. Epidemiol.* **106**:392–398 (1977).

301. SZMUNESS, W., STEVENS, C. E., HARLEY, E. J., ZANG, E. A., OLESZKO, W. R., WILLIAM, D. C., SADOVSKY, R., MORRISON, J. M., AND KELLNER, A., Hepatitis B vaccine: Demonstration of efficacy in a controlled clinical trial in a high risk population in the United States, *N. Engl. J. Med.* **303**:833–841 (1980).

302. TASSOPOULOS, N. C., HATZAKIS, A., DELLADETSIMA, I., KOUTELOU, M. G., TODOULOS, A., AND MIRIAGOU, V., Role of hepatitis C virus in acute non-A, non-B hepatitis in Greece: A 5-year prospective study, *Gastroenterology* **102**:969–972 (1992).

303. TASSOPOULOS, N. C., PAPAEVANGELOU, G. J., SJOGREN, M. H., ROUMELIOTOU-KARAYANNIS, A., GERIN, J. L., AND PURCELL, R. H., Natural history of acute hepatitis B surface antigen-positive hepatitis in Greek adults, *Gastroenterology* **92**:1844–1850 (1987).

304. TASSOPOULOS, N. C., PAPAEVANGELOU, G. J., TICEHURST, J. R., AND PURCELL, R. H., Fecal excretion of Greek strains of hepatitis A virus in patients with hepatitis A and in experimentally infected chimpanzees, *J. Infect. Dis.* **154**:231–237 (1986).

305. TAYLOR, G. M., GOLDIN, R. D., KARAYIANNIS, P., AND THOMAS, H. C., *In situ* hybridization studies in hepatitis A infection, *Hepatology* **16**:642–648 (1992).

306. TAYLOR, J., Introduction to HDV genome replication, in: *Hepatitis Delta Virus. Molecular Biology, Pathogenesis, and Clinical Aspects* (S. J. HADZIYANNIS, J. M. TAYLOR, AND F. BONINO, EDS.), pp. 1–5, Wiley-Liss, New York, 1993.

307. TERAZAWA, S., KOJIMA, M., YAMANAKA, T., YOTSUMOTO, S., OKAMOTO, H., TSUDA, F., MIYAKAWA, Y., AND MAYUMI, M., Hepatitis B virus mutants with precore-region defects in two babies with fulminant hepatitis and their mothers positive for antibody to hepatitis B e antigen, *Pediatr. Res.* **29**:5–9 (1991).

308. TICEHURST, J., POPKIN, T. J., BRYAN, J. P., INNIS, B. L., DUNCAN, J. F., AHMED, A., IQBAL, M., MALIK, I., KAPIKIAN, A. Z., LEGTERS, L. J., AND PURCELL, R. H., Association of hepatitis E virus with an outbreak of hepatitis in Pakistan: Serologic responses and pattern of virus excretion, *J. Med. Virol.* **36**:84–92 (1992).

309. TIOLLAIS, P., CHARNAY, P., AND VYAS, G. N., Biology of hepatitis B virus, *Science* **213**:406–411 (1981).

310. TOUKAN, A. U., SHARAIHA, Z. K., ABU-EL-RUB, O. A., HANOUD, M. K., DAHBOUR, S. S., ABU-HASSAN, H., YACOUB, S. M., HADLER, S. C. MARGOLIS, H. S., COLEMAN, P. J., AND MAYNARD, J. E., The epidemiology of hepatitis B virus among family members in the Middle East, *Am. J. Epidemiol.* **132**:220–232 (1990).

311. TROISI, C. L., HOLLINGER, F. B., HOOTS, W. K., CONTANT, C., GILL, J., RAGNI, M., PARMLEY, R., SEXAUER, C., GOMPERTS, E., BUCHANAN, G., SCHWARTZ, B., ADAIR, S., AND FIELDS, H., A multicenter study of viral hepatitis in a United States hemophilic population, *Blood* **81**:412–418 (1993).

312. TSAREV, S. A., EMERSON, S. U., BALAYAN, M. S., TICEHURST, J., AND PURCELL, R. H., Simian hepatitis A virus (HAV) strain AGM-27: Comparison of genome structure and growth in cell culture with other HAV strains, *J. Gen. Virol.* **72**:1677–1683 (1991).

313. TSAREV, S., EMERSON, S. U., REYES, G. R., TSAREVA, T. S., LEGTERS, L. J., MALIK, I. A., IQBAL, M., AND PURCELL, R. H., Characterization of a prototype strain of hepatitis E virus, *Proc. Natl. Acad. Sci. USA* **89**:559–563 (1992).

313a. TSAREV, S. A., TSAREVA, T. S., EMERSON, S. U., GOVINDARAJAN, S., SHAPIRO, M., GERIN, J. L., AND PURCELL, R. H., Successful passive and active immunization of cynomolgus monkeys against hepatitis E. *Proc. Natl. Acad. Sci. USA* **91**:10198–10202 (1994).

314. TSEGA, E., KRAWCZYNSKI, K., HANSSON, B.-G., AND NORDENFELT, E., Acute sporadic viral hepatitis in Ethiopia: Causes, risk factors, and effects on pregnancy, *Clin. Infect. Dis.* **14**:961–965 (1992).

315. TSUDE, K., FUJIYAMA, S., SATO, S., KAWANO, S., TAURA, Y., YOSHIDA, K., AND SATO, T., Two cases of accidental transmission of hepatitis C to medical staff, *Hepato-Gastroenterology* **39:**73–75 (1992).

316. ULRICH, P. P., BHAT, R. A., SETO, B., MACK, D., SNINSKY, J., AND VYAS, G. N., Enzymatic amplification of hepatitis B virus DNA in serum compared with infectivity testing in chimpanzees, *J. Infect. Dis.* **160:**37–43 (1989).

317. VALLBRACHT, A., MAIER, K., STIERHOF, Y.-D., WIEDMANN, K. H., FLEHMIG, B., AND FLEISCHER, B., Liver-derived cytotoxic T cells in hepatitis A virus infection, *J. Infect. Dis.* **160:**209–217 (1989).

318. VAN THIEL, D. H., FAGINOLI, S., AND WRIGHT, H. I., Liver transplantation and viral hepatitis: The current situation, in: *Hepatitis Delta Virus. Molecular Biology, Pathogenesis and Clinical Aspects* (S. J. HADZIYANNIS, J. M. TAYLOR, AND F. BONINO, EDS.), pp. 377–388, Wiley-Liss, New York, 1993.

319. VELAZQUEZ, O., STETLER, H. C., AVILA, C., ORNELAS, G., ALVAREZ, C., HADLER, S. C., BRADLEY, D. W., AND SEPULVEDA, J., Epidemic transmission of enterically transmitted non-A, non-B hepatitis in Mexico, 1986–1987, *J.A.M.A.* **263:**3261–3285 (1990).

320. VERME, G., AMOROSO, P., LETTIERI, G., PIERRI, P., DAVID, E., SESS, F., RIZZI, R., BONINO, F., RECCHIA, S., AND RIZZETTO, M., A histological study of hepatitis delta virus liver disease, *Hepatology* **6:**1303–1307 (1986).

321. VILLAREJOS, V. M., SERRA, J., ANDERSON-VISONA, K., AND MOSLEY, J. W., Hepatitis A virus infections in households, *Am. J. Epidemiol.* **116:**577–586 (1982).

322. VYAS, G. N., AND ULRICH, P. O., Molecular characterization of genetic variants of hepatitis B virus, in: *Viral Hepatitis and Liver Disease* (F. B. HOLLINGER, S. M. LEMON, AND H. S. MARGOLIS, EDS.), pp. 135–148, Williams and Wilkins, Baltimore, 1991.

323. WAINWRIGHT, R. B., MCMAHON, B. J., BULKOW, L. R., HALL, D. B., FITZGERALD, M. A., HARPSTER, A. P., HADLER, S. C., LANIER, A. P., AND HEYWARD, W. L., Duration of immunogenicity and efficacy of hepatitis B vaccine in a Yupik Eskimo populations, *J.A.M.A.* **261:**2362–2366 (1989).

324. WAINWRIGHT, R. B., MCMAHON, B. J., BULKOW, L. R., PARKINSON, A. J., HARPSTER, A. P., AND HADLER, S. C., Duration of immunogenicity and efficacy of hepatitis B vaccine in a Yupik Eskimo population: Preliminary results of an 8-year study, in: *Viral Hepatitis and Liver disease* (F. B. HOLLINGER, S. M. LEMON, AND H. S. MARGOLIS, EDS.), pp. 762–766, Williams and Wilkins, Baltimore, 1991.

325. WANG, J. T., WANG, T. H., SHEU, J. C., LIN, J. T., AND CHEN, D. S., Hepatitis C virus RNA in saliva of patients with posttransfusion hepatitis and low efficiency of transmission among spouses, *J. Med. Virol.* **36:**28–31 (1992).

326. WANG, K.-S., Structure, sequence and expression of the hepatitis delta viral genome, *Nature* **323:**508–513 (1986).

327. WEISSBERG, J. I., ANDRES, L. L., SMITH, C. I., WEICK, S., NICHOLS, J. E., GARCIA, G., ROBINSON, W. S., MERRIGAN, T. C., AND GREGORY, P. B., Survival in chronic hepatitis B. Analysis of 379 patients, *Ann. Intern. Med.* **101:**613–616 (1984).

328. WEJSTAL, R., ANDERS, W., MANSSON, A. S., HERMODSSON, S., AND NORKRANS, G., Mother-to-infant transmission of hepatitis C virus, *Ann. Intern. Med.* **117:**887–890 (1992).

329. WERZBERGER, A., MENSCH, B., KUTER, B., BROWN, L., LEWIS, J., SITRIN, R., MILLER, W., SHOUVAL, D., WIENS, B., CALANDRA, G., RYAN, J., PROVOST, P., AND NALIN, D., A controlled trial of formalin-inactivated hepatitis A vaccine in healthy children, *N. Engl. J. Med.* **327:**453–457 (1992).

330. WEST, D. J., AND MARGOLIS, H. S., Prevention of hepatitis B virus infection in the United States: A pediatric perspective, *Pediatr. Infect. Dis. J.* **11:**866–874 (1992).

331. WHEELER, C. M., ROBERTSON, B. H., VAN NEST, G., DINA, D., BRADLEY, D. W., AND FIELDS, H. A., Structure of hepatitis A virion: Peptide mapping of the capsid region, *J. Virol.* **58:**307–313 (1986).

332. WINOKUR, P. L., AND STAPLETON, J. T., Immunoglobulin prophylaxis for hepatitis A, *Clin. Infect. Dis.* **14:**580–586 (1992).

333. WONG, D. C., PURCELL, R. H., SREENIVASAN, M. A., PRASAD, S. R., AND PAVRI, K. M., Epidemic and endemic hepatitis in India: Evidence for non-A/non-B hepatitis virus etiology, *Lancet* **2:** 876–878 (1980).

334. WOODRUFF, B. A., MOYER, L. A., O'ROURKE, K. M., AND MARGOLIS, H. S., Blood exposure and risk of hepatitis B virus infection in firefighters, *J. Occup. Med.* **35:**1048–1054 (1993).

335. WORLD HEALTH ORGANIZATION, Progress in the control of viral hepatitis: Memorandum from a WHO meeting, *Bull. WHO* **66:**443–455 (1988).

336. WRIGHT, T. L., Etiology of fulminant hepatic failure: Is another virus involved, *Gastroenterology* **104:**640–653 (1993).

337. WRIGHT, T. L., HSU, H., DONEGAN, E., FEINSTONE, S., GREENBERG, H., READ, A., ASCHER, N. L., ROBERTS, J. P., AND LAKE, J. R., Hepatitis C virus not found in fulminant non-A, non-B hepatitis, *Ann. Intern. Med.* **115:**111–112 (1991).

338. WU, J. C., LIN, H. C., JENG, F. S., MA, G. Y., LEE, S. D., AND SHENG, W. Y., Prevalence, infectivity, and risk factor analysis of hepatitis C virus infection in prostitutes, *J. Med. Virol.* **39:**312–317 (1993).

339. XU, Z. Y., DUAN, S. C., MARGOLIS, H. S., PURCELL, R. H., OUYANG, P. Y., COLEMEN, P. J., ZHUANG, Y. L., XU, H. F., QIAN, S. G., ZHU, Q. R., WAN, C. Y., LIU, C. B., GUN, Z. L., AND THE UNITED STATES–PEOPLE'S REPUBLIC OF CHINA STUDY GROUP ON HEPATITIS B, Long-term efficacy of active postexposure immunization of infants for prevention of hepatitis B virus infection, *J. Infect. Dis.* **171:**54–60 (1995).

340. XU, Z. Y., LIU, C. B., FRANCIS, D. P., PURCELL, R. H., GUN, Z. L., DUAN, S. C., CHEN, R. J., MARGOLIS, H. S., HUANG, C. H., MAYNARD, J. E., AND THE UNITED STATES–CHINA COOPERATIVE STUDY GROUP ON HEPATITIS B, Prevention of perinatal acquisition of hepatitis B virus carriage using vaccine: Preliminary report of a randomized double-blind placebo-controlled and comparative trial, *Pediatrics* **76:**713–718 (1985).

341. YAMAMOTO, K., HORIKITA, M., TSUDA, F., ITOH, K., AKAHANE, Y., YOTSUMOTO, S., OKAMOTO, H., MIYAKAWA, Y., AND MAYUMI, M., Naturally occurring escape mutants of hepatitis B virus with various mutations in the S gene in carriers seropositive for antibody to hepatitis B surface antigen, *J. Virol.* **68:**2671–2676 (1994).

342. YANAGI, M., KANEKO, S., UNOURA, M., MORAKAMI, S., KOBAYASHI, K., SUGIHARA, J., CHUISHI, H., AND MUTO, Y., Hepatitis C virus in fulminant liver failure, *N. Engl. J. Med.* **324:**1895–1896 (1992).

343. YARBOUGH, P. O., TAM, A. W., FRY, K. E., KRAWCZYNSKI, K., MCCAUSTLAND, K. A., BRADLEY, D. W., AND REYES, G. R., Hepatitis E virus: Identification of type-common epitopes, *J. Virol.* **65:**5790–5797 (1991).

344. YU, M. C., TONG, M. J., COURSAGET, P., ROSS, R. K., GOVIN-

DARAJAN, S., AND HENDERSON, B. E., Prevalence of hepatitis B and C viral markers in black and white patients with hepatocellular carcinoma in the United States, *J. Natl. Cancer Inst.* **82:**1038–1041 (1990).

345.   ZANETTI, A., FERRONI, P., MAGLIANO, E. M., PIROVANO, P., LAVARINI, C., MASSORO, A. L., GAVINELLI, R., FABRIS, C., AND RIZZETTO, M., Perinatal transmission of hepatitis B virus and the HBV-associated delta agent from mothers to offspring in northern Italy, *J. Med. Virol.* **9:**139–144 (1982).

346.   ZUCKERMAN, A. J., The history of viral hepatitis from antiquity to the present, in: *Viral Hepatitis: Laboratory and Clinical Science* (F. DEINHARDT AND J. DIENHARDT, EDS.), pp. 3–32, Marcel Dekker, New York, 1983.

CHAPTER 14

Herpes Simplex Viruses 1 and 2

Lawrence R. Stanberry, Daniel M. Jorgensen, and André J. Nahmias

1. Introduction and Social Significance

Herpes simplex virus (HSV) infections are among the most common communicable diseases of humans. The natural history of HSV infection is influenced by two features of the virus: (1) There are two distinct HSV serotypes: HSV-1, which is transmitted chiefly via a nongenital route, and HSV-2, which is most often transmitted sexually or from a mother's genital infection to the newborn; and (2) during initial (primary) infection, HSV establishes a persistent state that is maintained for the life of the host. Periodically, the latent virus can be reactivated to cause symptomatic or subclinical recurrent infections. Hence, HSV infections may range from subclinical to life threatening, and the specific clinical illness will be determined by the portal of virus entry, the competence of the host immune system, and whether the infection is primary or recurrent.[248] Unless resulting from autoinoculation or sexual spread from the mouth to genital sites, HSV-1 infections occur most frequently during childhood and affect most often the mouth, lips, and skin sites above the waist; HSV-2 infections, on the other hand, occur most often during adolescence and young adulthood and involve skin sites below the waist, most often the genitalia. It should be emphasized, however, that both HSV-1 and HSV-2 can cause clinically indistinguishable infections above or below the waist. Most infections in newborns result from the infant passing through an infected birth canal or as an ascending infection; thus, most are due to HSV-2, although HSV-1 maternal genital infection is well documented. Postpartum infection of the newborn is uncommon but may be a particular problem with premature infants and babies born to women with primary, nongenital infection.

Although the majority of infections in individuals without prior exposure to either virus type (primary infections) are subclinical, they tend to be more severe than infections occurring in individuals previously exposed to HSV-1 or HSV-2 or both. The clinical manifestations of either virus may also be more severe in certain types of hosts, e.g., the newborn or immunocompromised patient, and with involvement of certain sites, e.g., the central nervous system.

Although not ubiquitous in all populations studied, infection with these viruses represents a socially significant problem for which no effective vaccine is yet available. HSV-2 infection is now appreciated as a sexually transmissible infection of varying prevalence in different countries of the world and appears to increase the risk of acquiring the human immunodeficiency virus (HIV).[28,59,94,99,100,116] Herpes simplex virus infections of the central nervous system in older children and adults are often fatal or debilitating, and ocular infections at any age may endanger normal vision. Recurrent HSV infections may be asymptomatic or clinically manifest; depending on site, duration, and pain, they can be physically and psychologically distressing. With the greater use of immunosuppressive and cytotoxic drugs and the increasing frequency of acquired immunodeficiency syndrome (AIDS), HSV infections of varying clinical severity are more commonly recognized. In addition, with the advent of new laboratory techniques, including amplification of nucleic acids by polymerase chain reaction and specific serological tests, the total impact of the relationship of herpes simplex viruses to human cancers, abortions, birth defects, and chronic neurological diseases may soon be appreciated.[4,105,124]

Lawrence R. Stanberry • Division of Infectious Diseases, Department of Pediatrics, University of Cincinnati College of Medicine, and Children's Hospital Medical Center, Cincinnati, Ohio 45229. **Daniel M. Jorgensen and André J. Nahmias** • Division of Infectious Diseases, Epidemiology and Immunology, Department of Pediatrics, Emory University School of Medicine, Atlanta, Georgia 30303.

2. Evolutionary and Historical Background

Herpes simplex viruses belong to the Herpesviridae family, a collection of more than 110 DNA viruses that infect a wide range of species from fungi to man.[208,209] The other human herpesviruses are cytomegalovirus (CMV), varicella-zoster virus (VZV), Epstein–Barr virus (EBV), and the newly recognized human herpesvirus 6 (HHV-6) and human herpesvirus 7 (HHV-7).[208,209,217] All of these viruses have the capacity to persist in their natural host, either in neural cells, e.g., HSV and VZV, or in nonneural cells, e.g., CMV, EBV, HHV-6, and HHV-7. The high prevalence in primitive societies of antibodies to the nonsexually transmitted human herpesviruses, in contrast to the low prevalence of antibodies to other nonpersistent viruses,[25] emphasizes the survival advantage conveyed by viral persistence and suggests an early origin for viruses in the herpes family. Moreover, the range of genital HSV-2 infection in adult members of several Amazon Indian tribes (seroprevalence, 2–88%) bespeaks differences in the time at which the virus was introduced into the tribes, as well as variability in their sexual behavior (A. J. Nahmias, A. DeSouse, F. L. Black, H. K. Keesling, and F. K. Lee, unpublished data). Many of the herpesviruses affecting vertebrates have very similar clinicoepidemiologic patterns, such as sexual transmission and the ability to cause encephalitis, keratitis, skin or genital lesions, and disseminated neonatal disease.[158]

HSV-1 and HSV-2 are closely related alpha herpesviruses that share about 50% nucleic acid homology, suggesting that these two viruses evolved from a common ancestral herpesvirus.[112] Based on amino acid sequence analysis, it has been estimated that the two viruses diverged between 8 and 10 million years ago.[81] It has been hypothesized that divergence occurred as a consequence of the acquisition of continual sexual attractiveness by the ancestral human female and the adoption of face-to-face mating practices. It is also possible that HSV-1 is more closely related to the ancestral herpesvirus and that HSV-2 represents recent evolution of the family. This hypothesis is supported in part by the recognition that HSV-1 is almost universally acquired during childhood in lower socioeconomic populations, together with observations indicating that a prior HSV-1 infection does not completely protect the host from acquisition of an HSV-2 infection, whereas the reverse is generally the case. The older observation[163] that a particular HSV type is more likely to be associated with a genital or nongenital site can now be explained by at least two reasons: Not only is there a greater likelihood for HSV-2, in contrast to HSV-1, to recur in genital sites after a primary infection,[48,249] but there is also a much greater ability of HSV-2 to infect individuals with prior HSV-1 infections primarily because of childhood oral infection. In general, such childhood HSV-1 infections protect adolescents and adults from acquiring a genital HSV-1 infection.

Beswick[23] has determined that the terms *herpes* has been used in medicine for at least 25 centuries. The word ἑρπηζ, from the verb ἑρπειυ (to creep), was used by the ancient Greeks to describe spreading cutaneous lesions of varied etiology.[163] The "herpetic eruptions which appear about the mouth at the crisis of simple fever" were first described around 100 AD by a Roman physician, Herodotus. About 1600 years later, herpes of the genital tract was first reported by a French physician, Astruc. By the 19th century, the generally accepted use of the term *herpes* was restricted to certain diseases associated with vesicular eruptions. Early in the 19th century, Willan and Bateman first suggested that herpes labialis, herpes genitalis, and herpes zoster could be differentiated from each other based on clinical features.[23] By the latter part of that century, a recognition of cytopathological differences permitted discrimination between infections of the pox and the herpes groups. In the early part of the 20th century, epidemiologic and experimental studies further supported the distinction between herpes zoster and oral and genital herpes. The studies of Gruter and other European workers showed that specimens obtained from zoster lesions could not be transmitted to the rabbit cornea, in contrast to those obtained from the other two herpetic conditions. Around 1920, a German physician, Lipschutz, maintained that, although biologically related, orolabial herpes and genital herpes were etiologically different.

Over the next 40 years, the experimental host range of herpes simplex viruses was widened to include other laboratory animals, chick embryos, and ultimately cell cultures. The clinical spectrum of HSV infections was expanded to include gingivostomatitis, encephalitis, meningitis, Kaposi's varicelliform eruption, and neonatal disease. It also became appreciated that HSV infections could recur in the presence of demonstrable levels of serum antibodies.

In the early 1960s, Schneweis[222] in Germany and Plummer[191] in England found antigenic differences among HSV strains. By 1967, Nahmias and Dowdle[163] in the United States had demonstrated that the large majority of genital and newborn infections are caused by HSV-2 and that most nongenital infections are caused by HSV-1, relating these clinical findings to the usual mode of transmission of the two virus types. In more recent years, strain differences within each of the two HSV types have been demonstrated in their polypeptides and by restriction endonuclease analysis of their viral DNAs.[36–38,189] The application of modern biochemical and immunologic technology, together with the broadening clinicopathologic and epidemiologic observations in more recent times, has thus provided new approaches to laboratory diagnosis, prevention, and therapy.[46,91,115,166,184,207,214,266]

3. Methodology Involved in Epidemiologic Analysis

3.1. Mortality

Herpes simplex virus infections are not reported nationwide in the United States other than the few fatal cases of HSV encephalitis, which are reported to the Centers for Disease Control (CDC). Mortality from HSV infection occurs primarily in three types of hosts: newborns, older individuals with encephalitis, and those who are compromised by immunologic or skin defects or by severe malnutrition. In the preantiviral era, the overall case fatality rate for neonatal herpes was around 50%.[170] With antiviral treatment, outcome is improved; mortality is rare in infants with localized cutaneous or ocular infection, about 15% mortality with neonatal encephalitis and approximately 60% mortality in newborns with disseminated infection.[274,275] Untreated, children and adults with herpes encephalitis have a 70% case fatality rate, while antiviral treatment with vidarabine or acyclovir has significantly reduced mortality.[273,28]

3.2. Morbidity

The data on morbidity of HSV infections are largely limited to severe, life-threatening HSV disease.[273–275,280] Because there is no requirement for reporting HSV infections in the United States, there exists only unofficial estimates of the incidence and prevalence of HSV disease. Estimates of the frequency of genital or neonatal herpes have been obtained from review of physician, clinic, or hospital records.[14,40,44,260] Use of banked sera demographically representative of the United States population obtained during the National Health and Nutritional Examination Survey II (NHANES II) in the late 1970s and 1980s has provided a more reliable estimate of the prevalence of HSV-1 and HSV-2 infection in the United States.[106,162] It is important to note that any estimate of the true extent of HSV-1 and HSV-2 infections must take into consideration (1) the specificity and sensitivity of the method of diagnosis, (2) the occurrence of inapparent infections, particularly of the mouth and urogenital areas, and (3) the fact that HSV-1 and HSV-2 can infect the same individual at different times and even concomitantly, and

that HSV-1 and/or HSV-2 infection can be recurrent in the same individual. It is therefore important to keep in mind the following possible circumstances in any one individual within a study population: (1) no evidence of infection with either HSV-1 or HSV-2; (2) a primary HSV-1 or HSV-2 infection but without recurrences with either virus; (3) a recurrent infection with one HSV type only; (4) a recurrent infection with one HSV type and a first infection with the other type; or (5) recurrent infections with both HSV-1 and HSV-2. The observation that an individual may acquire an exogenous reinfection with a different strain of the same HSV type[37,38] has provided further complexity to such distinctions.

Information on prevalence and, less often, incidence rates has been derived from clinical, virological, and cytohistopathologic observations and by serological surveys in various populations in different parts of the world.

3.2.1. Clinical. Clinical surveys without laboratory support will be accurate to different degrees, particularly when the information is obtained retrospectively and based primarily on the patient's recollection. Nevertheless, surveys on the prevalence of cold sores or herpetic keratitis are likely to be more reliable than those on genital or intraoral infections. This is because inapparent infections in the two latter sites are common. Although clinically manifest infections are less easily misdiagnosed, microbiological tools are still helpful in distinguishing HSV from nonherpetic causes of genital ulcer disease (e.g., syphilis, chancroid, etc.).[186] Furthermore, it is often difficult to differentiate among primary, nonprimary first (initial), and recurrent infections in order to develop estimates of incidence. In addition, differences in clinical rates between men and women may simply reflect greater ease in recognizing herpetic lesions in males. Skin infections may also be confused clinically with other entities. Infections of other sites such as the central nervous system, cervix, or urethra, as well as cases of eczema herpeticum, require laboratory aids for diagnosis.

3.2.2. Virological. Viral culture (recovery of infectious virus) has proven useful for detecting HSV in both clinically suspect herpetic conditions and clinically inapparent cases.[49,124,258] Type-specific monoclonal antibodies have been used to distinguish between HSV-1 and HSV-2 clinical isolates.[6] The newer methods of restriction enzyme analysis of viral DNA[36–38] have provided an exquisite tool for the fingerprinting of strains within each HSV type, since it appears that all isolates of HSV-1 or HSV-2 are different unless epidemiologically related. Recently, the technique of enzymatic amplification of viral DNA by polymerase chain reaction (PCR) has been used to differentiate strains of HSV-1 from HSV-2, with sensitivity approaching 100%.[91] Demonstration of viral antigens or viral nucleic acids in either symptomatic or asymptomatic patients has also proven useful in establishing the diagnosis of HSV infection.[47,115,283] Another virological approach has been to detect latent virus or viral nucleic acids in the sensory or autonomic nervous system ganglia of human cadavers.[11,12,54,55,269]

3.2.3. Cytohistopathologic. Papanicolaou screening of cervicovaginal smears for cervical cancer has offered another approach to determining the relative frequency of genital herpes in different female populations, since cellular changes associated with herpesviruses can be demonstrated in such smears.[180] This method is about one half to two thirds as sensitive as viral isolation and has been found to be highly specific for genital HSV infection.

Tzanck smears have been used to examine cellular material collected by scraping the base of suspected herpetic lesions. Tzanck or Papanicolaou staining of such material can demonstrate herpesvirus-induced changes in cells but cannot distinguish between HSV-1, HSV-2, and VZV.[143] An improvement on the traditional Tzanck smear has been the use of colloidal gold immunoelectron microscopy to examine suspect herpetic lesions.[75]

Histological examination of biopsy or autopsy specimens may be helpful in cases of neonatal infection, encephalitis, or disseminated HSV infections.[15,92,198] Because other viral infections and some noninfectious conditions may produce similar histopathologic changes, pathologists now utilize immunohistochemical techniques or *in situ* nucleic acid hybridization methods to identify HSV antigens or nucleic acids in biopsy or autopsy specimens.[75,78,115,262] These new methods can be used to distinguish between HSV-1 and HSV-2 infection.

3.3. Serological Surveys

HSV antibody can be detected by several commercially available methods including complement fixation, virus neutralization, and enzyme immunoassays. However, these serological methods cannot differentiate antibodies to HSV-1 and HSV-2 because of the presence of common antigens in the two viruses.[223] Even the recently developed enzyme immunoassays currently offered by commercial reference laboratories cannot accurately distinguish between HSV-1 and HSV-2 infection with both viruses.[7] Several serological methods (which are not commercially available) such as Western Blot and immunodot and enzyme immunoassays using type-specific glycoprotein G monoclonal antibodies can reliably distinguish between HSV-1 and HSV-2.[22,132,133,159,160,267] Studies using these accurate assay methods have revealed that the

majority of individuals infected with either HSV-1 or HSV-2 have no history of symptomatic HSV infection.[24,31,83,123,124,126,169,173,235] A direct correlation between the frequency of HSV-2 antibodies and the number of lifetime sexual partners has also been found in several populations studied in addition to other aspects of sexual behavior; this expands the use of these assays from simple seroepidemiology to the new realm of serosociology.[172,173]

3.4. Laboratory Diagnosis

3.4.1. Virological Methods. The gold standard for the diagnosis of HSV infection remains virus isolation. Clinical specimens for virus isolation should be transported to the virology laboratory and processed as rapidly as possible in order to maximize virus recovery. In situations when prompt transport of the specimen cannot be assured, the sample should be maintained on wet ice ($0°C$) or in the refrigerator until it can be processed. When lengthy delays in processing are anticipated, the specimen should be stored frozen at $-70°C$. Freezing and thawing of the specimen will affect its infectivity, and hence, virus recovery, and no virus will be recovered if the specimen is placed in the $-15°C$ freezing compartment. The development of various transport media, e.g., Leibowitz-Emory,[179] has facilitated the shipment of clinical specimens, since the swab with which the specimen is obtained, when placed in this medium, can be stored and shipped at ambient temperature.

HSV can be readily isolated in a number of tissue culture systems often within 1 to 3 days (about as rapidly as bacterial culture). HSV can be identified in traditional cell culture by its characteristic cytopathic effect. Recently it has been shown that shell vial cultures examined at 16 hr postinoculation by an immunofluorescence assay or DNA probe assay were more rapid, sensitive, and specific than traditional culture procedures.[70] Specific identification and typing of HSV recovered by cell culture can be obtained by neutralization, immunofluorescence, immunoperoxidase, or nucleic acid techniques.[29,175,258]

The choice of specimens for culture will be influenced by the clinical presentation but may include vesicle fluid, swab specimens of mucosal surfaces, urine, cerebrospinal fluid, and biopsy and autopsy material. It has been reported that viral blood cultures can be helpful in identifying HSV infection in neonates and immunocompromised patients.[170,252]

Methods for detection of viral antigens can be used for rapid identification of HSV in clinical specimens. A recently developed enzyme immunoassay has been shown to be twice as sensitive as culture for detecting HSV in late-stage genital lesions and equivalent to culture for the detection of HSV in early-stage genital lesions.[47] For patients with HSV encephalitis, immunofluorescence tests are useful for examining brain biopsies.[178]

New methods for detecting viral nucleic acids using PCR amplification have been shown to be useful in examining cerebrospinal fluid,[115,117,214,266] genital secretions,[46,91] and even fixed and paraffin-embedded human brain specimens.[184]

Herpes simplex virus can be demonstrated in vesicle fluid and biopsy or autopsy specimens by electron microscopic techniques. Traditional electron microscopy can identify enveloped or nonenveloped virus particles; however, this technique cannot distinguish between viruses with similar morphological characteristics such as HSV, CMV, and VZV. The newer method of colloidal gold immunoelectron microscopy does not have this limitation and can distinguish between HSV-1, HSV-2, and VZV.[75]

It has also been possible to demonstrate latent HSV in sensory or autonomic nervous system ganglia obtained from human cadavers. Two methods usually have been employed. The first uses cell culture methods to isolate reactivated HSV from ganglion explants.[11,12,269] With this approach, infectious virus is usually detectable no earlier than 8 days and as late as 40 days after incubation of the ganglion cultures. The second approach is to use molecular biological methods to demonstrate the presence of HSV nucleic acids in the ganglia.[54,55,66,125,254]

3.4.2. Morphological Aids. The intranuclear inclusions and multinucleated giant cells that are seen in Papanicolaou-stained smears of fixed cells, obtained by scraping the margin of herpetic vesicles or ulcers of the skin, mouth, conjunctiva, or cornea, are characteristic of herpesviruses, including HSV and VZV.[180] Only the multinucleated giant cells are usually apparent in Wright- or Giemsa-stained air-dried smears (Tzanck smears). In most instances, there is little difficulty in differentiating clinically between HSV and VZV infection. The inclusions in exfoliated urinary cells may make it difficult, however, to distinguish HSV from CMV infection. Papanicolaou smears of the cervix are particularly helpful in detecting subclinical herpetic cervicitis in women.[160,180] Biopsy or autopsy material should preferably be fixed in Bouin's fixative rather than formalin in order to enhance the demonstration of the characteristic inclusions and giant cells.

3.4.3. Serological Tests and Assays of Cell-Mediated Immunity. Many serological assays can be used to demonstrate HSV antibodies, including neutralization, complement fixation, passive hemagglutination, complement-mediated cytolysis, indirect immunofluores-

cence, and enzyme-linked immunosorbent assays (ELISA). Attempts to determine the HSV antibody types (types 1 and/or 2) have previously involved procedures such as microneutralization, kinetic neutralization, multiplicity analysis, inhibition, passive hemagglutination tests, or radioimmunoassays.[192,204,253] For almost two decades, data obtained by any of these serological methods had to be interpreted with suspicion. This is because of the lack of specificity, in differing degrees, of all of these assays in discriminating between antibodies to the two HSV types, particularly in individuals infected with both viruses. Recently, the Western blot (immunoblot) assay and the use of type-specific glycoproteins, gG1 and gG2, as antigens in immunologic assays has permitted differentiation of prior infection with HSV-1, HSV-2, or both virus types.[8,9,22,132,133] Occasionally, particularly in cases of primary infections, it may not be possible to detect type-specific antibodies in early convalescent sera positive for total HSV antibodies by a standard ELISA.[45]

Convincing evidence of primary infection is obtained when the acute serum shows no antibodies to either HSV type but the convalescent serum is positive for one or the other of the viruses. A fourfold or greater rise in titer between the acute and the convalescent serum cannot be reliably used as evidence of primary infection because such increases may occur as a consequence of recurrent infections. However, if the acute serum is obtained late after onset, an initially low HSV antibody titer that rises rapidly in convalescence is most likely the result of a primary infection. Even a first infection with one type of HSV in an individual previously infected with the other type may not be detected by a rise in total HSV antibodies, but such an initial (nonprimary first) infection can now be detected readily with the type-specific antibody assays.

Indirect immunofluorescence, enzymatic, or radio-immunologic methods have been developed to detect HSV antibodies in the IgM, IgG, and IgA classes of immunoglobulins.[154,175,258] Because maternal IgM antibodies do not cross the placenta, IgM antibodies to HSV can be used to diagnose neonatal herpes in patients without characteristic findings in the eye, throat, or skin. Such IgM antibodies usually appear within 1–4 weeks after birth and persist for at least 6 months. For older individuals the detection of HSV-specific IgM or IgA antibodies when no HSV-specific IgG antibody is present generally indicates primary infection. When IgG antibodies to HSV are present in serum of older patients, the detection of IgM or IgA antibodies cannot be used to differentiate a primary from a recurrent HSV infection, since recurrent infections can induce amnestic IgM and IgA responses.[52,131,154,220,276]

While most cytokine and cellular immune assays are

research tools, studies using these methods have explored the immunologic responses important in control of genital, orolabial, and neonatal herpes.[121,264,265] Skin testing, a simple measure of cell-mediated delayed hypersensitivity, can be elicited with inactivated HSV preparations and correlates well with the presence of serum neutralization antibodies[282]; however, such HSV skin test materials are not currently available for routine use.

4. Biological Characteristics of HSV-1 and HSV-2

Herpes simplex viruses consist of four major morphological components: a centrally located core surrounded by three concentric structures—the capsid, the tegument, and the envelope. The core contains DNA coiled around proteins arranged in the form of a barbell. The icosahedral capsid contains 162 capsomeres and measures around 100 nm. Between the capsid and the envelope is the tegument, composed of fibrillar material. The envelope, derived from nuclear and occasionally other cell membranes, confers a diameter of 150–200 nm on the complete virus particle. The lipid composition of the envelope makes the virus particularly susceptible to ether and other lipid solvents.

The HSV virion contains a linear double-stranded DNA molecule of approximately 150 kilobase pairs. About half of the HSV-1 and HSV-2 DNA sequences are homologous. It is possible to infect cells *in vitro* with "naked" DNA free of virus proteins and to demonstrate defective DNA in some strains of HSV, of possible relevance to the oncogenic potential of HSV. Studies using temperature-sensitive mutants, HSV-1 × HSV-2 intertypic mutants and DNA sequencing have permitted the development of genetic maps. Ongoing work with deletion mutants is defining the biological roles of the 70+ viral genes encoded by the viruses.

HSV-1 encodes for at least 11 glycoproteins, of which gB, gD, gH, gK, and gL are essential for productive infection in cell culture.[52,58,139,241] HSV gG is type-specific, and gE and gI can act as a receptor for the Fc portion of IgG. The glycoproteins on the envelope of the virus appear to be similar to those found on the membranes of infected cells, so immune mechanisms could operate not only on the virus itself but also on cells infected by the virus. Antibodies alone, however, are most often incapable of inhibiting cell-to-cell spread of the virus. Together with complement, or with T lymphocytes, monocytes, or polymorphonuclear leukocytes of seronegative or seropositive individuals, antibodies can lyse HSV-infected cells *in vitro*.[120,220,232,281] Such mecha-

nisms could be operative *in vivo*, since early lysis of infected cells can occur with some of these systems before new viral progeny are produced.

The different lability of HSV-1 and HSV-2 to high-temperature exposure (HSV-2 is more labile at 39°C than HSV-1) and their different susceptibility to certain antiviral agents may potentially affect pathogenesis and therapy. For instance, fever is known to reactivate HSV-1 recurrences but may not be as effective in reactivating HSV-2. The viruses are also labile to low pH, so they are infrequently recovered from the gastrointestinal tract. However, recent studies using nucleic acid methods have shown that HSV-DNA can be detected in peptic ulcers, suggesting that HSV may play a role in that pathogenesis of some cases of peptic ulcer disease.[131] Even though HSV can be cultured from various contaminated surfaces for many hours,[182] the clinicoepidemiologic relevance of this finding remains very doubtful. Like gonorrhea, which can be cultured from a toilet seat for several hours, these laboratory experiments may merely provide a good alibi for how genital herpes might have been acquired.

HSV-1 and HSV-2 have a wide *in vitro* and *in vivo* host range, being able to infect a large variety of cell types of human or animal origin as well as numerous types of experimental animals ranging from mice to monkeys.[247] Variability in the *in vitro* and *in vivo* behavior of the two virus types, as well as of different strains within each type, can be demonstrated in some of these systems. In humans, either HSV type appears to be capable of infecting similar body sites and both can spread from the portal of entry by either intraneuronal or hematogenous dissemination. However, an individual with HSV-2 genital infection is twice as likely to experience recurrent disease as a person with HSV-1 genital infection, and the recurrences are eight to ten times more frequent. In contrast, HSV-1 is more likely to cause recurrent orolabial infections than HSV-2.[50,52,247] HSV-2 also appears to be a more neurovirulent virus than HSV-1, causing more neurological injury in infants surviving neonatal infection and more likely to cause meningitis in patients with primary genital herpes. Alternatively, the increased chance of meningitis with HSV-2 and of encephalitis with HSV-1 may be related to differences in the mode of viral spread, i.e., HSV-2 via the blood to the meninges and HSV-1 via intraneuronal spread to the brain.[53]

As noted earlier, all herpesviruses have the capacity to persist throughout the host's lifetime, providing this group of viruses with a great survival advantage. In the case of HSV, it has been appreciated that recrudescence occurs in individuals with circulating HSV antibodies. Although the possibility of low-level virus multiplication around the site of involvement cannot be completely ruled out as a possible mechanism for viral persistence, the best evidence at present favors latency of the virus in a non-infectious form in sensory and autonomic nervous system ganglia. This conclusion is supported by experimental data in mice, rabbits, guinea pigs, and humans.[11,12,97,247,249] Although infectious virus cannot be demonstrated in the latently infected ganglia, it is possible to detect viral nucleic acids, and with the use of special cultivation methods, infectious virus can be reactivated within several days or weeks (see Section 3.4.1). It is also worth pointing out that HSV-1 and HSV-2 have been found to be oncogenic in various systems.[174,201]

5. Descriptive Epidemiology

The epidemiologic characteristics of herpes simplex viruses can be divided into those reflecting the overall infection rate, as determined by various clinical or laboratory surveys, and those related to particular clinical entities.

5.1. General Epidemiology

The major epidemiologic determinants of HSV infections are age, race, sex, socioeconomic level, and geographic area.[18,68,105,240] In cases of HSV-2 infection, as well as of genital HSV-1 infection, the degree of sexual exposure is particularly important. The interdependency of these variable causes some overlap in the following discussion.

5.1.1. Incidence and Prevalence. A high prevalence rate of HSV (untyped) infection has been found in virological or serological studies of healthy individuals. In 1953, Buddingh and colleagues[39] reported on the recovery of HSV from the mouth of 20% of asymptomatic children 7 months to 2 years of age, 9% of children 3–14 years of age, and 2.4% of adolescents and adults. This high isolation rate has been questioned, because the technique used at that time for identifying HSV was chorioallantoic membrane inoculation, which might yield nonspecific lesions.[228] By use of tissue culture methods, the isolation rate from the mouth in asymptomatic children has been found by later workers to be around 1%[118,132] and that in adults to be between 0.75 and 5%.[85,134,228] In a children's home, 32% of the children with serum HSV antibodies were found to shed oral virus periodically and most were asymptomatic.[42] In studies of normal adults with serum HSV antibodies, serial sampling indicated that HSV could be recovered from oral secretions in up to 50%

in the absence of clinical lesions.[62,86] About 50–80% of seropositive individuals who have various types of organ transplants (e.g., renal, bone marrow, heart) will shed virus asymptomatically from the oral cavity or demonstrate symptomatic herpes infection in various parts of the body within the first month of transplant.[142] Asymptomatic shedding from the genital tract has not been systematically examined in the immunocompromised host; however, some transplant patients have genital recurrences suggesting the likelihood of such an occurrence. The frequency of recurrent labial herpes, as determined by retrospective epidemiologic studies, is summarized in Table 1, and the variability noted is discussed in appropriate sections below.

A study on 80 trigeminal ganglia obtained from cadavers of adults has revealed that 55% contained latent HSV-1.[12] Close to 20% of 21 sacral ganglia were found to contain latent HSV-2.[11] These rates are only slightly less than those expected from serological studies in similar population groups.

The prevalence of clinically recognized genital herpes reported from six sexually transmitted disease clinics in the United States between 1976 and 1977 varied from 0 to 3.2% in women and 0.1 to 7.3% in men.[40] Virological studies of HSV-2 in pregnant women of lower socioeconomic status have revealed genital viral excretion rates of 1:75 to 1:250.[88,263] Rates of genital isolation of HSV-2 as high as 12% have been recorded in prostitutes.[63] Intermittent asymptomatic genital excretion of HSV has been observed at rates of 4 to 14% in women followed prospectively.[1,48,49,88,200] Virus can often be isolated from swabbing the vulva or cervix without apparent lesions. The duration of virus shedding and the frequency of virus isolation from the cervix vary from individual to individual. The prevalence of asymptomatic cervical herpetic infection detected by cytological methods[180] has varied from 0.03 to 6.9% depending on the population studied (see Section 5.2). In asymptomatic males attending a VA hospital, HSV was isolated from the urethra, prostate, or epididymis of 15% of the study group.[41] These high rates of asymptomatic urogenital HSV infections in males have not been found by other workers.[104,155]

Early antibody surveys, which did not differentiate between HSV-1 and HSV-2 antibodies, found antibody prevalence rates in adult populations of 50 to close to 100%, depending on socioeconomic status.[202,240] Later surveys with serological assays, again lacking complete specificity in measuring HSV-1 and HSV-2 antibodies, suggested wide differences in the prevalence rates of HSV-2 (10–70%) in different subpopulations in the world.[167,202,203] More recently, the specific HSV-type

**Table 1. Occurrence of Recurrent Herpes Labialis
as Determined by Retrospective Epidemiology Studies**[a]

Population	Recurrent herpes labialis		
	Number studied	Percentage positive	Reference
Students of health care professions,[b] Philadelphia, PA	1788	38.2	231
Philadephia, PA			
Medical students	343	44.6	229,230
Hospital patients	242	31.5	
Students of health case professions[b]			
South America	1713	16.0	68
Asia	950	17.6	
Africa	404	30.2	
Europe	2085	30.9	
Australia	552	33.0	
North America	4155	37.9	
Patients in a general practice North Wales, England	1855	45.8	87
Patients attending a cancer prevention center, Chicago, IL	423	41.1	c
School of dentistry in Michigan			
Students	731	16.3	285
Faculty	300	30.7	
Students and scientific workers			
Philadelphia, PA	146	35.6	269
Kyoto, Japan	245	24.1	

[a]Adapted from Rawls and Campion-Piccardo.[202]
[b]Students of medicine, dentistry, veterinary medicine, dental hygiene, and nursing.
[c]I.D. Rotkin and co-workers (unpublished observation, 1975).

antibody assays have provided finite information on the prevalence of HSV-1 and HSV-2 infection according to various demographic variables and in several countries (Figs. 1 and 2).[31,43,82,83,105,113,114,123,137,162,169,185,206,235,267]

The incidence of HSV (presumably HSV-1) infections has been measured in children's institutions through repeated sampling over time. In one 6-year study in a children's home, Cesario et al.[42] found that of the 70 initially seronegative children, eight (11.4%) experienced a primary infection while in the home, six of whom had an associated illness. In an earlier Australian study in a home for children, all under 3 years of age, Anderson and Hamilton[3] found that 29 of 43 seronegative children (67.4%) developed HSV antibodies over a 1-year period, 20 of whom had an associated illness. Finally, the incidence of oral HSV infection has been studied in immunocompromised populations, with rates ranging from 40 to 90% in seropositive patients and rates as low as 1.6% in seronegative patients. The discrepancy in incidence between seropositive and seronegative patients indicates that recurrent, and not primary, HSV infections are responsible for most oral herpetic lesions in the immunocompromised population, presumably due to a depressed cell-mediated response.[152] Few data are available on the

reactivation of HSV-2 infection in immunocompromised hosts. However, a recent study on immunosuppressed individuals showed selective clinical reactivation of either HSV-1 or HSV-2 infection in individuals seropositive for both HSV-1 and HSV-2 antibodies. This suggests that other characteristics, such as establishment and duration of latency, rather than immunosuppression alone, may account for the varying rates of reactivated infection (A. J. Nahmias and M. A. Jackson, unpublished data).

Increasing trends in clinically manifest genital herpes are suggested by two retrospective surveys. In a review of computerized records of a middle-class predominantly white community in Minnesota of cases of clinically recognized first-episode genital herpes, the rate was estimated to be 12.5 per 100,000 in 1965 and to rise to 82.3 by 1979.[44] Another retrospective study, conducted by the CDC, extrapolated the United States data from a sample of private physician–patient consultations obtained during the National Disease and Therapeutic Index Survey between 1966 and 1981. The number of patient visits for newly diagnosed infections was estimated to have increased 7.5-fold. However, over a similar period (1964–1984) in a municipal hospital primarily for indigent blacks, we found little variation over the years for the rate

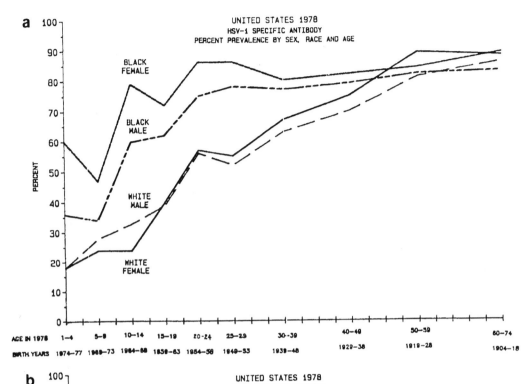

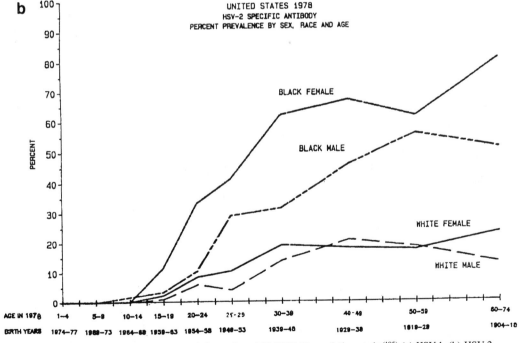

Figure 1. HSV prevalence in the United States from NHANES II population study.[105] (a) HSV-1. (b) HSV-2.

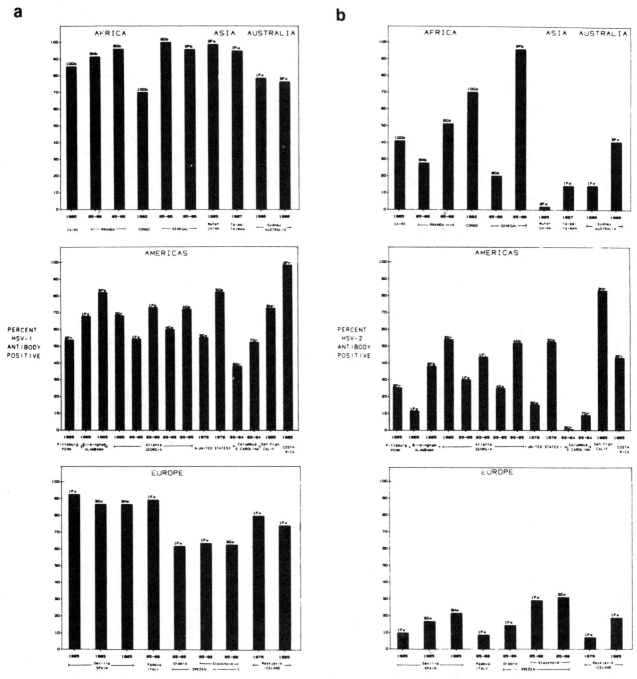

Figure 2. Prevalence of herpes simplex virus antibodies in different populations of the world. (a) HSV-1. (b) HSV-2. Serum source: 1, prenatal clinics; 2, family planning or gynecology clinic; 3, STD clinic; 4, health maintenance organization (HMO); 5, NHANES II (U.S. population); 6, college students (freshmen); 7, college students (seniors); 8, hospital workers or patients; 9, army recruits; 10, general population. Sex: M, males; F, females; G, males and females; H, homosexual males. Race: w, white; b, black; 4, white and black; o, oriental. These surveys were performed in collaboration with the following investigators: S. Beckman, N. Clumeck, A. Cunningham, D. Danielsson, M. Essex, D. Freeman, M. Gudnadottir, M. Guinan, J. Gibson, A. Gustafsson, S. Homberg, H. Jaffe, R. Johnson, W. Louv, M. Schanfield, S. Stagno, S. Thompson, E. Torfason, W. Whittington, C. Wickliffe, W. Wu, and X. Yao.

of cytologically detected genital herpes in asymptomatic women who had a routine Papanicolaou smear, around 1–2 per 1000.[160]

The discrepancy between these findings has been explained by our serological surveys, which indicate little change in the prevalence of HSV-2 antibodies among blacks in the past decade but a comparatively much greater increase among whites. We now believe that two factors were likely to have brought white middle-class individuals to physicians' attention since the early 1970s. Because that group has a much lower prevalence of HSV-1 antibodies by adolescence than blacks (Fig. 1a), their first contact with HSV-2 would have been more likely to result in a primary infection that would tend to cause clinically manifest genital herpes with lesions and more severe systemic disease. Another factor responsible for bringing middle-class whites to the attention of physicians, and for the clinicians diagnosing genital herpes more frequently, was the popularization, both in the medical and lay press, of genital herpes in the 1970s. This in turn led to the misapprehension that genital herpes was much more common in whites than blacks, a confusion engendered by the fact that only clinically manifest genital herpes was measured.

Similar population differences might explain the discrepancies noted in the incidence of neonatal herpes between other workers and ourselves. Corey's group[260] reported in a primarily white middle-class population that there had been a marked increase in the number of diagnosed cases of neonatal herpes in King County, Washington, over a 15-year period. In an unpublished review of neonatal deaths in the United States, CDC workers also found an increase in recent years.[17] However, we have not noted any significant change since 1964 in the rate of neonatal herpes at our municipal hospital, which remains around 1 per 3500 deliveries.[170]

Another way to estimate the incidence rate of herpes in a population is by measuring the antibody seroconversion rate of HSV-1 or HSV-2 in longitudinal studies. We have obtained preliminary results with such studies in four groups, using the type-specific immunodot assay for HSV antibody.[83,101,173] Initial analyses in the groups have revealed the following:

1. In college students (South Carolina), the incidence rate was about 4% per year for HSV-1 and 2% for HSV-2 between freshmen and senior years.
2. In a health maintenance organization (Atlanta), the incidence rate for HSV-2 was about 2% per year.
3. Multipartnered heterosexuals [sexually transmitted diseases (STD) clinic, Atlanta] had an incidence rate of around 5% for HSV-2 over a 6- month follow-up period.

4. Homosexual men (San Francisco) had an incidence rate of around 5% per year in a 7-year follow-up study.

5.1.2. Epidemic Behavior.

Epidemics of diseases associated with HSV have not been described. This is in large part because HSV infections, unlike varicella, are very often asymptomatic. However, clusters of cases in various environments have been reported. Scott[224] and Juretic[108] have reviewed such clusters in families, hospitals, and various closed populations. At least 17 families have been described in which two or more family members were found to be infected over a short period of time. The source of the infection was ascribed to adults with recurrent labial infections or children with primary oral infections.

Outbreaks within hospital or clinics of eczema herpeticum or of gingivostomatitis have been described in England, Yugoslavia, and the United States.[108,140,197] In the contact outbreak of eczema herpeticum, four patients developed the infection in succession within 8 days. In addition, some of the nurses caring for the patients developed herpetic lesions on their hands. In a crowded pediatric ward housing 37 children below 2 years of age, 17 of the infants developed HSV infections over a 1-month period. All of the affected infants had gingivostomatitis, but in three cases this was also associated with herpetic keratitis or skin involvement. An outbreak of gingivostomatitis occurred in 20 of 46 patients handled by a dental hygienist with a herpetic whitlow.[140] The reverse, however, is usually the case in that dental or medical personnel who have contact with patients' oral secretions are at risk for developing whitlow.[36,72,212] Indeed, in a pediatric intensive care unit, two independent outbreaks of HSV-1 infection occurred whose sources were patients who transmitted their virus to health care workers, most often causing herpetic whitlow.[36]

Outbreaks of herpetic stomatitis have also been described in orphanage nurseries, children's homes, and day-care centers. Within a 1-month period, the attack rate of clinically apparent primary infections was around 80%.[89,108] The attack rate in another institution for young infants over an 11-month period was found to be 56%.[3] Although these studies would imply that HSV spreads very readily in closed populations, another report demonstrated that only 10% of susceptible children developed HSV infections over a 6-year period.[42] Of 115 children followed longitudinally in a research day-care center over 12 years, four small clusters of primary infection accounted for 55% of all primary infections, with peak incidence among children 1–2 years of age.[221] Similarly, in a day-care setting in Japan where four outbreaks occurred, the rates of infection in susceptible children ranged from 75 to 100%.[128,129] Several reports have also documented transmission of neonatal herpes in nurseries, most often by inapparent spread of virus from an infected asymptomatic or initially asymptomatic baby to another neonate by a nursery attendant.[77,90,135,216] Several clusters of an unusual form of cutaneous HSV ("herpes gladiatorum") have also been reported.[98,193,225,279] Person-to-person spread from the close body contact associated with wrestling or other sports, such as rugby, has been observed repeatedly. Underlying skin conditions or mat burns increase the susceptibility of the wrestler to this type of skin infection.

5.1.3. Geographic Distribution. Herpes simplex virus infections occur worldwide, even in isolated primitive populations. For example, sera from nearly all of the tribes studies have high rates (>90%) of HSV-1 antibodies. This is in contrast to the low incidence or even absence of antibodies to respiratory-transmitted infections such as influenza and measles. In addition, HSV-2 antibody prevalence varied according to when the virus was introduced and the sexual behavior patterns of the different tribes, as previously noted Nahmias, De Souse, Black, Keesling, and Lee (unpublished data).[25]

In an international study of health care professional students,[68] variability in responses to prior history of herpes labialis was noted (Table 1). Students from South America and Asia appeared to be about half as likely to have experienced (or remembered) this recurrent herpetic infection in comparison to those from Europe, Australia, or North America. A lower rate of recurrent labial herpes and of recovery of latent virus from trigeminal ganglia was found in Japanese than in Americans.[269]

Results of the testing of sera from different sources and parts of the world with an immunodot enzyme assay for HSV-1 and HSV-2 antibodies are presented in Fig. 2. The results with HSV-1 antibody assays (Fig. 2a) fit mostly the expected correlation with socioeconomic status of the populations. The lowest prevalence was found to be among freshmen (40%) and seniors (54%) in a state university in South Carolina.[82] Other low rates (55–65%) were found in a family planning clinic in Pittsburgh (Pennsylvania) and in white pregnant women in Birmingham (Alabama), Atlanta (Georgia), and Orebro and Stockholm (Sweden). The overall rate for United States whites in 1978 was 60%, and the rate in STD clinics in Stockholm was 62%. Rates of 90% and over of HSV-1 antibodies were found in Wuhan (China), Taipei (Taiwan), Seville (Spain), Costa Rica, Senegal, and Rwanda.

Two interesting findings were noted regarding the relative prevalence of HSV-1 and HSV-2 in the different populations (Fig. 2). Whereas all populations had a higher rate of HSV-1 than HSV-2 antibodies, two groups provided exceptions. The homosexual male cohort from San Francisco demonstrated a higher prevalence of HSV-2 than HSV-1 antibodies, reflecting possibly a middle-class socioeconomic group with high sexual activity. In case of the Congo population studies, the rates for HSV-1 and HSV-2 antibodies were essentially the same (70%). Of those individuals who were HSV-2 antibody positive, 43% had antibodies to the HSV-2-specific protein only and the other 57% had antibodies to both HSV-1 and HSV-2 proteins. The finding suggests that HSV-2 protects from the acquisition of HSV-1 infection and that sexual relations must have started at a very early age before the HSV-1 infection, which occurs most commonly before 15 years of age. The different prevalence of HSV-2 antibodies observed in prenatal clinics in various populations of women within a similar age range provides an objective measure of populations at risk, because of their own past sexual behavior or that of their male consorts, of acquiring the HIV.[28,94,99,100,116,172,173]

5.1.4. Temporal Distribution. One report indicates that herpetic skin infections are more common in the summertime, perhaps as a result of sunlight-induced reactivation of the virus.[67] A significant increase in the occurrence in the summer months of genital herpetic infections was also observed in a university student population.[261] This was explained in part by the recent detachment of students from parental restraints. Another report suggests a higher frequency of primary ocular infections during the winter.[56] Otherwise, no clear-cut yearly or seasonal pattern of infection has been demonstrated for either HSV-1 or HSV-2. It is still possible that such variations in infection rate might occur and go unrecognized because of the need for laboratory-based studies to identify the infection.

5.1.5. Age. Newborn infections, most often caused by HSV-2, occur usually as a consequence of maternal genital infection.[170] Thereafter, during childhood, primary HSV-1 infections are more common and are reflected clinically as herpetic gingivostomatitis and less often as infections of the eyes, skin, and central nervous system.

A first infection with HSV-1 can occur in adulthood, primarily as oral or genital infection and infrequently as an infection of the eyes, skin, or central nervous system.[48,53,85,87,88,178,227] Such observations may cause one to conclude that the rise in the prevalence of HSV-1 antibodies in the white men and women in the United States in 1978, from about 40% at age 19 years to 80–90% at age 50 (Fig. 1a), was a result of acquisition of HSV-1 during adulthood. Even though some of this doubling of the rate during adulthood is at least partly a result of some new HSV-1 acquisition (Section 5.1.1), other clinical and epidemiologic information strongly suggests that there is also a cohort effect, i.e., many of the individuals 50 years of age or over actually acquired their HSV-1 infection when they were living under lower socioeconomic conditions 40 or more years earlier.

By the use of virological, cytological, and clinical observations, the peak age for detecting genital HSV infections has been found to be between 20 and 29 years of age.[48,52,165,180,183] The data presented in Fig. 1b on HSV-2 antibody prevalence also show the sharpest change in prevalence rate, whether in blacks or whites, to be in that age group. Blacks, particularly females, show an early rise of HSV-2 antibodies. There is a continued increase in the prevalence of these antibodies in blacks after the age of 30 years, in contrast to whites.

Besides indicating the degree of sexual activity and how early it may have begun as well as reflecting on socioeconomic status, the seroepidemiologic studies of HSV-1 and HSV-2 antibodies, particularly when applied in a prospective manner, can provide important sociological information of current, past, and future predictive value. Indeed, by a simple formula we developed[162] (see also Section 5.2.1), one can also obtain some appreciation of the amount of HSV-1 antibody resulting from genital HSV-1 infection, a reflection of oral–genital contact. Such analyses permit us to make the transition from seroepidemiology to "serosociology" and provide useful objective guides for comparing socioeconomic status and various patterns of sexual behavior in different populations.

5.1.6. Occupation. An occupational risk has been noted in three groups. The first group includes medical, nursing, and dental personnel, who are at higher risk of developing herpetic paronychia.[36,72,212] A second group is comprised of college wrestlers, who have been noted in several studies to have a propensity to acquire "herpes gladiatorum," a special form of skin herpes.[193,225,271] A third group includes prostitutes, who are at high risk of developing genital herpes. The most striking observations are the high prevalence rates of HSV-2 antibodies of close to 70–95% in prostitutes[63] (Fig. 2b) and the absence of HSV-2 antibodies found in nuns. Although 3% of nuns had been found earlier to be HSV-2 antibody positive by an older microneutralization test,[167] the same sera were later found to be HSV-2 antibody negative by the new specific enzyme immunoassay.[132]

5.1.7. Occurrence in Different Settings. HSV-1 is transmitted by personal contact, so that settings with close and prolonged exposure such as families, nurseries, and orphanages can result in a higher rate of infection. Reactivation of HSV-1 infections is seen on hospital wards among patients having a variety of febrile illnesses or cancers, HIV infections, or those being treated with immunosuppressive drugs; it is also commonly reported among high-altitude skiers, presumably because of solar ultraviolet radiation exposure. Prophylactic acyclovir has been reported to be useful in controlling recurrent orolabial HSV infections in immunocompromised patients and skiers.[242,276] At least one report has documented that HSV can be transmitted through transplantation of an infected organ.[65]

Outside the newborn period, HSV-2 infections occur most commonly in settings associated with sexual activity, sexual promiscuity, or sexual abuse.[35,80,147] It is not uncommon to find HSV-2 infections in association with other venereal diseases.[48,88,177] Indeed, the prevalence of HSV antibodies was found to correlate well with a history of some other STD.[169]

5.1.8. Socioeconomic. The higher frequency of HSV infections in lower socioeconomic groups was well established in earlier antibody surveys.[39,240] Although these did not differentiate type 1 and type 2 antibodies, the conclusions for the age groups below 18 years of age seem valid even today, since the antibodies measured were mostly HSV-1 antibodies. Essentially similar findings from measuring HSV-1 antibody rates have been observed in more recent work (Figs. 1, 2a). In the case of HSV-2 antibodies, with both older or newer serological tests, the association found between socioeconomic status and sexual activity has made interpretation more difficult; cultural patterns and number of sexual partners are more likely related to HSV-2 infection than is socioeconomic status per se. One confounding variable is that measures of socioeconomic status, in the United States at least, obtained at the time of serological testing do not necessarily reflect the status when the genital herpes was acquired. Although educational levels are probably better guidelines of current socioeconomic status, neither was found to be related to the rate of genital herpes in the United States population survey.[162]

5.1.9. Other Factors. Trigeminal neurectomy and fever are known to reactivate HSV oral infections.[175] Recurrent herpes labialis has also been reported following dental extraction.[188] Certain forms of immunosuppression, such as are administered to cancer patients or to renal, bone marrow, or heart transplant recipients, are also associated with more frequent recurrent orolabial infections,[10,142,157,199] which tend to be more severe or chronic. What is still unclear is why reactivation of only one of the two HSV types occurs when the individual had serological evidence of infection with both HSV types prior to the immunosuppression.

Severe malnutrition does not appear to alter susceptibility but instead increases markedly the severity of a herpetic infection.[19] Early immunogenetic studies suggested a higher frequency of HLA-A1 antigens in individuals with frequent recurrences of herpes labialis than in the general population.[215] Similar early studies in patients with recurrent ocular infections, however, failed to demonstrate a correlation with any histocompatibility type.[150,286]

Whether strains of herpes simplex virus may differ in their ability to spread, establish latency, or cause diseases of varying types in humans is still unclear. The greater frequency of genital recurrences after a primary infection with HSV-2 than with HSV-1[48,130] has been noted earlier. Our preliminary data indicating that recurrences are more likely *not* to be clinically manifest if the first infection was itself clinically inapparent and that the frequency of recurrences may also be comparatively lower are intriguing.

5.2. Epidemiologic Aspects of Specific Clinical Entities

5.2.1. Genital Infections. For decades, surveys of patients attending STD clinics have provided estimates of the prevalence of genital herpes. As far back as 1880, the rates were found to be 7% in women and 1% in men attending a STD clinic in Hamburg; approximately similar rates have been observed recently in venereal disease (VD) clinics in the United States, England, and Sweden.[40,48,88,177] Recent studies in Western nations have reported the isolation of HSV from 0.3 to 5.4% of males and 1.6 to 10% of females attending STD clinics.[50] Studies in college students[82,109,261] have also indicated that genital herpes is common in this sexually active population, often more frequently diagnosed than syphilis or gonorrhea.

Using cytological methods for the detection of genital HSV infections, rates have varied from 0.03 to 6.9%, depending on the population studied.[180] The highest rates were in women attending VD clinics and in prostitutes. Comparative studies by the same group of workers revealed the rate in indigent women to be 0.31% and in private patients to be 0.02%.[183] HSV has been isolated from the genital tract in 0.25 to 5.0% of patients attending general (non-STD) clinics; many of these individuals were asymptomatic and some had no history of genital herpes.[50] Recent seroepidemiologic studies[105,114,137,169] confirm earlier findings that the majority of genitally infected individuals are asymptomatic and would be missed by surveys solely directed at identifying patients with

genital disease.[88] The use of more sensitive detection methods, such as PCR, is likely to establish that many HSV-2-seropositive subjects without history of genital herpes asymptomatically shed HSV from their genital tract.[46,91,119] With both HSV-1 and HSV-2, at most one third, and often much less, of the individuals with HSV antibodies will give a positive history of any form of herpes. The HSV-2 antibody prevalence was high (70%) in homosexual men attending an STD clinic in Seattle, Washington,[138] but only 22% reported a history of genital or rectal herpes infection. Two other studies purport to indicate that the rates of clinically manifest genital herpes in homosexual men are lower than in heterosexual men.[107,233] Yet, the rates of HSV-2 antibodies in homosexual men in San Francisco are higher than in any other group of men studied (Fig. 2b).

From cross-sectional or longitudinal studies, use of the immunodot enzyme HSV type-specific antibody assay can provide estimates of the prevalence and incidence of genital and nongenital herpes rather than of only HSV-1 and HSV-2 antibodies. The rates of genital herpes would be expected to include individuals with genital HSV-1 as well as those with genital HSV-2 infections. Since in a population at large, HSV-2 infection is very infrequently a result of neonatal or nosocomial infection, the presence of HSV-2 antibodies can be considered to be almost synonymous with past experience with genital herpes (herein epidemiologically defined as being sexually transmitted infections, rather than defined by actual anatomic site of involvement, since lesions on the anus, skin below the waist, or the mouth can be consequences of sexually transmitted herpes). The proportion of genital HSV-1 infection among those with HSV-1 antibodies can be estimated if one accepts the premise that genital HSV-1 infection is almost always a primary infection with the type 1 herpesvirus. The serological assay we employ[132,133] can determine the proportion of HSV-2 antibodies that were a result of primary HSV-2 infections. Since the ratio of primary HSV-1 to primary HSV-2 infection is of the order of 1:1 at most and 1:10 at least,[88] one can estimate the frequency of genital HSV-1 infection within these limits. The point is that, even in populations with 1:1 ratios of primary genital HSV-1 and HSV-2, the impact of genital HSV-1 can be estimated to be as little as 1% or less but no more than 20% of the total prevalence of genital herpes. An estimate of nongenital herpes, i.e., non-sexually transmitted herpes, can also be obtained by subtracting the component estimated, as noted above, to represent genital HSV-1 infection from the total frequency of HSV-1 antibody.

Even though several estimates have been made of

the prevalence of genital and nongenital herpes in the past,[14,44,88] the data presented in Fig. 1 provide a better basis for estimating numbers in the total United States population, at least in the late 1970s. Thus, using population breakdowns for 1978, the NHANES data[162] provide the estimate that approximately 120 million (60%) U.S. citizens of all ages would have been infected with nongenital herpes and around 30 million (15%) with genital herpes. The rates and corresponding numbers vary according to age, sex, race, and region of the country. The NHANES III data point to a continued high risk of HSV-2 among African-American and Hispanic males, with an increased risk among young white males over the previous 10-year period.[106] The HSV-2 seroprevalence data presented in Table 2 provide an estimate of the prevalence of HSV-2 genital infection in selected U.S. populations.

Genital herpes during pregnancy poses a special problem because of the risk of transmitting virus from the mother to the fetus or newborn. Symptomatic genital HSV infection is estimated to occur in about 1% of pregnant women sometime during gestation.[170] Recurrent infections are much more common than primary infection. A recent study showed that for women with a history of recurrent genital herpes prior to pregnancy, the mean number of clinically recognized recurrences increases from 0.97 to 1.26 to 1.63 in the first, second, and third trimesters, respectively.[33] Asymptomatic infection, manifest by shedding of virus from the genital tract, has been noted in 0.65 to 3.0% of viral cultures collected during pregnancy.[5] In the general population, shedding of virus at the time of delivery occurs in 0.01 to 0.39% of pregnant women.[272] For women with a history of genital herpes,

the frequency of HSV shedding during the week prior to delivery is 1.3%.[5] Because the duration of shedding due to recurrent infections is brief, about 1.5 days, antepartum culturing of the maternal genital tract does not accurately predict the infant's risk of exposure at the time of delivery. Such exposure can only be reliably determined if the cultures are collected at the time of delivery. Both symptomatic and asymptomatic primary infection can also occur during pregnancy. Seroprevalence studies in California and Thailand reported the annualized rate for acquiring HSV-2 infection in pregnant women was 0.58% and 0.125%, respectively.[27,205] A recent prospective study among pregnant women and their spouses revealed that 32% of the women had serological evidence of prior HSV-2 infection.[126] Of the couples, 27% were serologically discordant with 9.5% of seronegative women having HSV-2-seropositive husbands. Further, only 56% of the seropositive men reported a history of genital herpes. This suggests that about 5% of women are at risk of acquiring genital herpes during pregnancy. With the advent of specific and sensitive assays to detect serological evidence of HSV-2 infection, it has become possible to screen women to determine who is at risk for primary or recurrent HSV-2 infections during pregnancy.

With increasing sexual promiscuity, genital HSV infection has become a significant problem among sexually active adolescents.[34] Using a glycoprotein G-based assay, a study in Cincinnati found that 7% of teenage males and 16% of teenage females attending an adolescent clinic in Cincinnati were HSV-2 seropositive.[24] For males, age and number of previous STDs correlated with HSV-2 serological status, while for females, only race was

Table 2. HSV-2 Seroprevalence Rates from Population Surveys

Population	Site	Percent HSV-2+	Reference
Pregnant women	California	32	126
Adult women attending a family planning clinic	Pennsylvania	9	31
Sexually active adolescent women	Ohio	16	24
Sexually active adolescent men	Ohio	7	24
College students	South Carolina	4.3	83
Adult women	United States	19.4	105
Adult men	United States	13.2	105
Adult Caucasians	United States	13.3	105
Adult African-Americans	United States	41.0	105
Single people	United States	13.8	105
Divorced/widowed people	United States	35.3	105
Married people	United States	16.1	105
Homo- or bisexual men (HIV+) attending an STD clinic	Maryland	81.3	100
Homo- or bisexual men (HIV−) attending an STD clinic	Maryland	54.5	100
Homosexual men (HIV+)	Washington	66.0	244
Homosexual men (HIV−)	Washington	35.0	244

a positive predictor, with African-American females having a higher HSV-2 seroprevalence than white females (17 vs. 4%).

5.2.2. Oral Infections. The most thorough study of primary herpetic oral infection was performed by Juretic in Yugoslavia.[108] This worker found that over a 10-year period, about 13% of 18,730 children attending outpatient clinics had clinical evidence of oral herpes. No cases were recorded in the first 6 months of life. The frequency distribution of the total number of cases over the next age periods were as follows: 6–12 months, 12%; 1–2 years, 35%; 2–3 years, 23%; 4–5 years, 11%; 5–6 years, 8%. Thereafter, cases continued to be observed at low frequency. No significant sex differences or seasonal variations were observed. This age distribution corresponds well with those noted in South Africa in the severely malnourished children with fatal disseminated HSV infections whose initial site of infection was intraoral herpes.[19] A high rate of herpetic stomatitis (25% of those with HSV-1 antibodies) has also been observed in Navajo Indian children.[16] As previously noted, outbreaks of herpetic stomatitis have occurred in day-care settings in Japan, with infection rates ranging from 73 to 100% in susceptible children.[128] Since several studies have indicated a much lower frequency of symptomatic disease in HSV-1-infected children, it remains unclear whether genetic, nutritional, or socioeconomic factors may be responsible for the clinical manifestations of HSV-1 infection in children.

As noted, symptomatic primary oral herpetic infections can also occur in adolescence and adulthood, and the virus has been associated with about 10% of cases of pharyngitis in student populations.[71,85] Estimates of the incidence of oral herpes in different populations will be eventually obtained from ongoing prospective studies. For the present, a crude extrapolation of available data suggests that about half a million cases of primary oral herpetic infections occur yearly. The large majority of these oral infections would be expected to be caused by HSV-1; however, primary oral infections with HSV-2 are well substantiated (Table 3).

Finally, the incidence of oral herpes has been studied in groups of immunocompromised individuals. Recent data indicate that the rate of infection may be higher than previously recognized.[152]

5.2.3. Labial Herpes. A history of recurrent herpes labialis was recorded in 38% of 1800 students attending professional schools at the University of Pennsylvania.[231] Among the students susceptible to recurrences, new lesions occurred once a month in 5%, at intervals of 2–11 months in 34%, and once a year or less in

61%. Recurrent herpes labialis was also noted to occur three times as frequently in a group of febrile patients as in a group of nonfebrile controls.[86,181] This study corroborated the older findings that up to one half of individuals treated with fever therapy experienced reactivation of a herpetic infection, mostly on the lips. The Perinatal Study of the National Institutes of Health found 1% of pregnant women to have labial herpes at some time during pregnancy. Within any one week, as many as 1% of hospital personnel were found to have recurrent labial herpes and 2% to have HSV asymptomatically in their saliva.[77,170] Estimates based on various studies suggest that about 100 million episodes of recurrent labial herpes occur yearly in the United States.[88]

5.2.4. Herpes of the Eye. About 5% of individuals attending an ophthalmology clinic were noted to have herpetic ocular disease. Primary herpetic infections, commonly associated with conjunctivitis,[56] were found most often in younger age groups. If herpetic involvement of the cornea is the first evidence of herpetic infection, close to half of the patients would be expected to have at least one recurrence of herpetic keratitis within a 2-year period.[101] A review of 141 patients with herpes keratitis revealed that most were men over 40 years of age and 65 (46%) experienced recurrent infections.[20] Of those patients who developed recurrent herpes keratitis, 34% experienced one or more recurrences per year and 68% reported at least one recurrence every two years. These rates were higher than in an earlier study reporting that 24% of patients experienced a recurrence within 1 year and 33% within 2 years.[234] Recurrences were reported most frequently during the winter months and correlated with an increased frequency of respiratory virus infections.[244]

5.2.5. Herpes of the Skin. The prevalence rate of herpetic skin infections was found to be about 1% of 7495 persons over 7 years of age in the county of Skaraborg in Sweden.[93] This investigation, performed during a mass X-ray survey of the lungs, showed little variation in rates according to age or sex. Another Swedish study found that 2% of men and 2.5% of women attending a dermatology clinic in Gothenburg over a 6-year period had clinical evidence of HSV skin infection.[67] In a Canadian study of 79 culture-confirmed cases of HSV infection of the hand, 20 cases of HSV-2 were found, all in adults (mostly women with recurrent genital herpes) and 13 cases of HSV-1 were found, all among children (mostly primary HSV-1 with associated gingivostomatitis); the remaining isolates were not typed. They estimated an incidence of 2.4 cases per 100,000 per year.[84]

Table 3. Clinical Spectrum of Infections Caused by Herpes Simplex Viruses 1 and 2 in Newborns and Older Persons and the Type Isolated from Different Sites and Clinical Conditions[a]

	Number of individuals with HSV		
	Type 1	Type 2	Total
I. Usually mild to moderately severe (persons over 1 month of age)			
A. Urogenital infection			
1. Females (cervix, vulva, vagina, urethra)	51(8[b])	514(1,[d]2[c])	565
2. Males (penis, urethra)	11(2[b])	333(1[d],2[c])	344
B. Nongenital infections			
1. Gingivostomatitis or asymptomatic (mouth)	197	5(3[e])	202
2. Herpes labialis (lips)	125	1	126
3. Keratitis and/or conjunctivitis (cornea and/or conjunctiva)	50	1	51
4. Dermatitis	6	0	6
a. Skin above waist	110(2[e])	8	118
b. Skin below waist	11(3[b],2[e])	118(6[e],2[f])	129
c. Hands or arms	25(2[g])	21(2[f],3[e])	46
C. Latent infections (trigeminal or thoracic ganglia) (sacral ganglia)	0	5	5
II. Usually severe to fatal (persons over 1 month of age)			
A. Meningoencephalitis (brain, spinal cord, CSF)	176	5	181
B. Multiple sclerosis (brain)	0	1	1
C. Eczema herpeticum (skin, lungs)	14	1	15
D. Generalized disease (visceral organs)	3	1(1[h])	4
III. Newborns—localized or generalized infection (skin, eyes, brain, CSF, Visceral organs)	57(7[i])	142(31[i])	199
Totals	862	1156	2018

[a]Typing is done by microneutralization or direct immunofluorescence tests.
[b]Simultaneous isolation of similar HSV type from mouth.
[c]Simultaneous isolation of type 2 HSV from cervix or vulva and type 1 HSV from lip or mouth.
[d]Simultaneous isolation of type 2 HSV from penile lesion and type 1 HSV from eye.
[e]Simultaneous isolation of same HSV type from genitals.
[f]Laboratory or hospital-acquired infection.
[g]Simultaneous isolation of type 2 HSV from sacral ganglia.
[h]Isolated also from brain.
[i]Same HSV type isolated from mother's genital tract.

5.2.6. Respiratory Infections. In a 6-year study involving 293 students attending a health clinic, 142 (11.5%) with respiratory illnesses were found with oral HSV, whereas only 11 (1.1%) asymptomatic students were found to be shedding virus.[85] Most of the respiratory illnesses, principally pharyngotonsillitis, were associated with primary HSV infections.[62,71,134,156,228] A Swedish study found that HSV was the most common pathogen isolated from patients with severe or complicated respiratory tract infection. The investigators reported virus recovery from 37 of 308 consecutive patients.[194]

5.2.7. Neurological Infections. According to limited reports from the CDC, HSV encephalitis is associated with a higher case-fatality rate than encephalitis caused by other viruses, except for rabies. Prospective studies in the United States and England indicate a 70% fatality rate.[136,273] Mortality can be significantly reduced by treatment with vidarabine or acyclovir.[237,273] About one third of cases occur in persons less than 20 years of age and approximately one half in patients older than 50 years. The long-term prognosis is better in patients less than 30 years of age.[273]

The incidence in the United States has been estimated to be 1–2 cases per million population; hence, around 250–500 cases are likely to occur every year in the United States. In Sweden, the incidence for all ages is estimated at 2–4 cases per million persons per year.[236] In addition to encephalitis, HSV may also cause meningitis, transverse myelitis, and sacral radiculopathy syndrome. HSV has been isolated from 0.5–3.0% of patients evaluated for meningitis.[148,238] In patients with viral meningitis, HSV-2 is recovered more frequently than HSV-1.[187,238] Meningitis is a common complication of genital herpes. In a report by Corey et al.,[51] 36% of women and 13% of men with primary genital herpes exhibited symptoms of meningitis; symptoms were severe enough in 6.4% of women and 1.6% of men to necessitate hospitalization. Long-term sequelae of HSV-2 meningitis include periodic headaches associated with genital recurrences and recurrent meningitis.[21]

5.2.8. Neonatal Herpes. HSV infection of the fetus or newborn is estimated to occur in 0.02–0.07% of live births, or about 700–2300 newborns annually in the United States.[272] In Stockholm, 10 of approximately 200,000 Swedish newborns over a 10-year period were diagnosed with neonatal herpes type 2 infection.[76] Data from Seattle, Washington, indicated that cases of neonatal herpes increased from 2.6 per 100,000 births in 1969 to 11.9 per 100,000 births in 1981.[266] During a comparable period, the incidence of neonatal herpes remained relatively constant in Atlanta, Georgia.[17] This apparent difference in the frequency of neonatal infection in different centers is most likely due to the types of maternal populations studied, especially with regard to the relative frequency of primary versus recurrent genital herpes infection during pregnancy. As noted below, the risk of transmission to the newborn is far greater than with maternal primary herpes. It is estimated that in the Atlanta population, 50% of the predominantly African-American pregnant women studied had recurrent genital herpes, versus only 15–20% for the predominantly white Seattle population, thus leaving a greater proportion of the latter group susceptible to acquiring primary herpes, or a nonprimary first episode of clinical herpes, during pregnancy. Moreover, the Atlanta women were less likely to have steady sexual partners and thereby were unlikely to be exposed to genital HSV-2 or oral–genital HSV-1 during the latter stages of pregnancy, around the time of delivery.

Intrauterine infection by placental transfer of virus is rare, and when it occurs it is likely to cause abortions or stillbirths. Cases where intrauterine infection has occurred with concomitant transfer of transplacental antibodies have resulted in neurological, ocular, and/or skin lesions at birth.[102] Both HSV-1 and HSV-2 can produce infection in the newborn. About 15–30% of cases of neonatal herpes are due to HSV-1.[52,272] For both HSV-1 and HSV-2, transmission usually occurs at the time of delivery, when the infant passes through an infected birth canal, although ascending infections have been documented in cases with or without membrane rupture and with or without membrane resealing.[255] Mothers with primary oral herpes or lesions on the skin (particularly the hands or breasts) can transmit HSV-1 to their newborns postnatally; other HSV-1 infections can result from nosocomial spread in nurseries, particularly those nurseries with premature infants.[170]

The risk of HSV-2 transmission from mother to newborn is influenced by whether the mother is experiencing primary or recurrent genital herpes at the time of delivery.[195] For primary infection, the risk to the newborn is high, around 40%, compared to 2–5% for women with recurrent genital herpes. Given the high incidence of recurrent genital herpes in pregnant women,[17,33,35] it is surprising that so few infants develop neonatal herpes. At least two factors probably contribute to the lower risk of neonatal infection in infants born to women with recurrent genital herpes. First, women who acquire primary genital HSV-2 infection prior to pregnancy develop anti-HSV antibodies that are transferred to the fetus.[4] Passively acquired antibodies appear to confer some protection to the infant, preventing or modifying infection in most cases.[4,196] Second, the amount of virus present during recurrent infections is much less than that present during primary infection,[32] suggesting that newborns delivered to women with recurrent genital herpes may be exposed to significantly less virus than infants born to women with primary infection.

While recognition of genital herpes in the pregnant woman is important in predicting the risk to the newborn, most cases of neonatal herpes occur in infants whose mothers have no history of genital infection. For instance, during an 18-month hospital-based surveillance study, the CDC identified 184 cases of neonatal herpes, but only 22% of the mothers had a history of genital HSV infection and only 9% had lesions at the time of delivery.[255]

6. Mechanisms and Routes of Transmission

There are no known animal vectors for the transmission of herpes simplex viruses. Although a few cases of HSV-1 or HSV-2 laboratory-acquired infections have occurred, the major mode of spread appears to be by close personal contact . The source of virus is an individual with a subclinical or clinically apparent primary or recurrent infection. Although it is possible for individuals to autoinfect themselves, in cases of genital herpes, this occurs most usually in children with primary oral infections who fondle their genitals.[164] This fact should caution against the automatic labeling of a case of genital herpes in a child as being a result of sexual abuse.

Exogenous reinfection with the same virus type has been documented[37]; however, the rate of reinfection with the same HSV type appears to be very low in patients with symptomatic recurrent disease.[219] These observations suggest that most symptomatic recurrent HSV infections result from reactivation of latent (endogenous) virus rather than exposure to exogenous virus. The incubation period for primary HSV-1 or HSV-2 infection ranges from 2 to 20 days, with an average of around 6 days. The incubation period for HSV encephalitis has been more difficult to define but may be longer. The duration and magnitude of virus excretion is greater in patients with primary, vis-à-vis recurrent genital herpes (14 days compared to 7 and $10^{4.5}$ pfu compared to $10^{2.5}$ pfu).[32,48]

The major source of virus to the newborn is the mothers' HSV-2 (and occasionally HSV-1) genital infection around the time of delivery.[170] The virus is acquired by an ascending infection if membranes are ruptured or on passage of the infant through the infected birth canal. Postnatal acquisition of the virus from a maternal genital infection, a maternal nongenital infection, and infection in other family members has also been observed.[61,170,284] Acquisition of virus by the newborn from a doctor with a herpetic orolabial infection, who had used oral suction on the baby, has also been reported.[268] Infant-to-infant transmission in a nursery, presumably via personnel hands, has been noted repeatedly.[77,90,135,216] Again, despite the long period of medical knowledge about various forms of nongenital herpes, such as herpes of the eye, information is still lacking regarding pathways of transmission of the virus: e.g., is the first episode of herpetic keratitis acquired as a new infection, reinfection, reactivation, or autoinoculation? How does the virus get to the eye in a person with conjunctivitis? In the case of herpetic skin lesions, virus spread via kissing or from body contact associated with wrestling has been corroborated. Spread by saliva is likely to have been involved in the outbreaks of herpetic somatitis in orphanage nurseries and children's homes noted earlier (Section 5.1.2). Such spread is more directly evidenced by the cases of herpetic paronychia (herpes whitlow) in medical, nursing, or dental personnel who handle the infected oral cavity of patients or contaminated tracheal catheters. Spread via air droplets or infected skin squares has not been well documented. HSV-1 appears to have varying degrees of communicability, as suggested by surveys in closed populations.[3,42,90,108] HSV-1 has been demonstrated to be transmitted exogenously by oral–genital or oral–anal contact as well as by autoinoculation of oral virus to the genitals or other body sites.[60,164]

The sexual transmission of genital HSV infection was postulated during the 19th and early 20th centuries.[23,163] Thereafter, despite a few case reports substantiating such transmission, the venereal route of spread was not accepted generally, most likely as a result of the lack of appreciation of subclinical genital infections. The advent of HSV type differentiation has provided conclusive evidence of sexual transmission.[145,146] Two case-contact studies[137,147] have indicated that the asymptomatic individual, whether having ever had clinically apparent le-

sions or not, is commonly the source of HSV infection to sexual partners; therefore, there is need to caution infected individuals not to assume that they can transmit the virus only when they have lesions. The recently appreciated increased risk of individuals acquiring the HIV if infected with HSV-2[28,94,99,100,116,173] is relevant to this important issue.

In a small series, the risk of developing genital herpes in the female contacts of males with penile herpes was found to be around 60–80%,[114,203] although the risk has not been defined in terms of a single exposure. In studies of larger numbers of couples who had been together for an average of 4 years, about three quarters of the female members were HSV-2 antibody positive when their partner was also HSV-2 antibody positive.[113] In contrast, fewer than half of the male partners of HSV-2-antibody-positive females were found to be positive. In conjunction with a recent vaccine study using the sexual partners of patients with genital herpes it was determined that the annual rate of acquisition of HSV-2 infection was 8%.[144] The attack rate was higher for patients who lacked HSV-1 antibody at entry compared to those who were initially HSV-1 seropositive (15.5% and 5.9%, respectively). Acquisition of primary genital herpes during pregnancy, especially near term, places the newborn at high risk for neonatal herpes. Seroprevalence studies in the United States and in the Far East have estimated the annualized rate for acquiring HSV-2 infection during pregnancy to be about 0.1–0.5%.[27,205] Homosexual transmission between males has been reported, resulting in HSV-1 or HSV-2 perianal lesions.[60,107,138,233] Similar transmission in female homosexuals, although likely to occur, has not yet been described. The transmission of HSV-2 by oral–genital contact, causing intraoral HSV-2 infections, has also been observed.[48,60]

HSV-2 infections of genital sites may occasionally involve neighboring areas such as the perineum, thighs, or buttocks, either by autoinoculation or by contact with infected partners. Recurrences in these sites can occur independent of, or together with, recurrences at genital sites. In addition, HSV-2 infections of the hands have been acquired by medical or nursing personnel from contact with infected patients.

7. Pathogenesis and Immunity

Our understanding of the pathophysiology of mucocutaneous HSV infection is based on clinical observations and experimental animal studies. The pathophysiology of primary mucocutaneous HSV infection is schematically

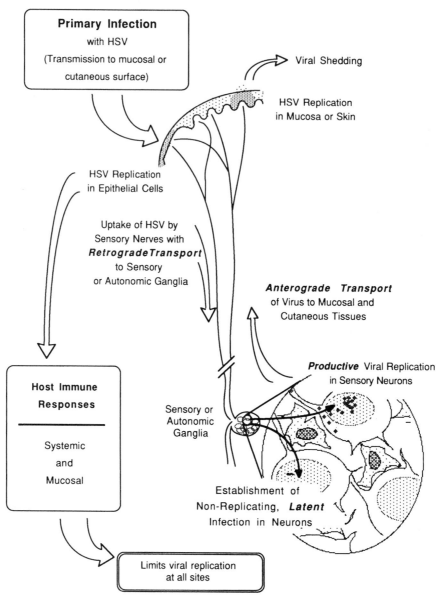

Figure 3. Schematic representation of the pathogenesis of primary mucocutaneous herpes simplex virus infection. Adapted from Stanberry.[248]

represented in Fig. 3. In the susceptible host, primary infection begins when HSV is transmitted to a mucosal or cutaneous surface, generally the oral cavity or genital tract. Two viral glycoproteins, gB and gC, are responsible for attachment of the virion to heparin sulfate, the putative epithelial cell surface receptor.[139] Polyionic compounds such as heparin can interfere with viral infectivity, presumably by preventing virus from binding to the cellular receptor.[171] Other viral glycoproteins, gB, gD, and the hetero-oligomer gH/gI, facilitate virus penetration through fusion of the viral envelope with the cellular membrane.[58] Penetration of HSV may be inhibited by drugs that disrupt the cell membrane cytoskeletal structure, suggesting that virus attachment triggers a microfilament activity required for internalization of the virus. During the fusion process the viral envelope is lost and the nucleocapsid enters the cytoplasm and migrates to the nucleus where replication of the viral DNA occurs.[210] The glycoproteins that are destined to be incorporated into progeny virus are transported to the inner nuclear membrane. The progeny virus are assembled in the nucleus and acquire their envelope by budding through the glycoprotein-rich inner lamellae of the nuclear membrane. The new virions move via the cisternae of the rough endoplasmic reticulum to the Golgi apparatus where posttranslational modification of the viral glycoproteins occurs. The mature virions are packaged in transport vesicles that move from the Golgi apparatus to the cell surface. Transport vesicles fuse with plasmid membrane permitting release of newly replicated virus into the extracellular milieu.

Viral replication occurs in epithelial cells at the portal of entry, but HSV also rapidly enters nerve endings and moves via axoplasmic transport to the nucleus of neurons located within sensory ganglia.[47a,247] Animal studies have shown that, following replication in ganglion cells, virus is transported through unmyelinated sensory nerve fibers back to mucosal and cutaneous sites where further replication produces the characteristic vesicles of primary mucocutaneous HSV infection.[245,250] The interneuronal spread and amplification of virus at the level of the ganglia appear necessary for the development of clinically apparent mucocutaneous HSV infection.[248] The distinction between the individual who experiences subclinical primary infection and the patient who suffers severe symptomatic disease probably relates to the extent to which virus replicates in sensory ganglia and spreads to multiple mucocutaneous locations.

HSV has evolved a novel survival strategy: It is able to persist within sensory neurons in a nonreplicating state throughout the life of the host. Recently, it has been shown that one region of the HSV genome is transcriptionally active during latency. This "latency-associated transcript" (LAT) has been shown to be nonessential for replication in cell culture, but it is required for the production of spontaneous and induced recurrent genital and ocular infections in animals.[249] In humans or animals the nonreplicating latent virus can periodically reactivate to a replication-competent state, resulting in the production of infectious virions. Stimuli such as ultraviolet radiation, trauma, or stress may trigger the reaction process. Following reactivation, virus is transported via selected sensory neurons back to the periphery where further replication in epithelial cells results in subclinical or symptomatic recurrent HSV infections. The pathogenesis of recurrent mucocutaneous HSV infection is schematically presented in Fig. 4.

Host immune responses are important in limiting HSV infection. Patients with immature or compromised immune function generally have more severe primary HSV infections. These patient populations include some

newborns, severely malnourished children, those with measles or HIV infection, severe burns, those receiving immunosuppressive therapy, patients with cancers, and persons with selected immune deficiencies, such as the Wiskott-Aldrich syndrome.[142,157,173,173a] While in the normal host, viremia does not appear to play a role in the pathogenesis of HSV infection; in the immune compromised patient, including the newborn, virus may disseminate to internal organs via hematogenous spread.[19,53,127,252]

HSV infection of the central nervous system (encephalitis) and of the eye have special pathogenic mechanisms by which virus causes disease.[213,259] Brain and mouth isolates taken concurrently in patients with herpes encephalitis have been found to be both identical or different by restriction enzyme analysis. These data suggest that persons previously infected with HSV-1 can develop a central nervous system herpetic infection from their old strain or from a newly acquired HSV-1 infection. Additionally, serological data demonstrating concomitant primary oral and central nervous system infection support the concept that herpes encephalitis outside the newborn period can be a primary infection.[277] In the case of HSV infection of the posterior segment of the eye, it is recognized that, in large part, an immunopathologic process produces the retinal injury.[213]

Unlike HSV infections in adults, neonatal herpes is rarely asymptomatic. The clinical presentation, however, varies greatly. Based on physical findings and diagnostic criteria, three categories of neonatal disease have been developed: (1) disease localized to skin, eye, or mouth; (2) encephalitis, with or without cutaneous lesions; and (3) infection disseminated to multiple organs with or without central nervous system involvement.[161] The outcome of neonatal infection is different for each of these three categories. Recent animal model studies suggest that the pathogenesis of infection may also be different for each of the three categories.[30,79] Factors that probably influence the extent and severity of neonatal disease include the portal of virus entry, viral inoculum, neurovirulence of the virus strain, immunocompetence of the neonate, and perhaps genetic determinants of susceptibility to HSV infection. For neonatal infection the portals of entry may include the umbilical cord, the eye, the oral and nasal orifices, and skin if damaged by trauma or scalp electrode placement. If host and viral factors restrict HSV replication to the site of inoculation, the infant may develop infection limited to the skin, eye, or mouth. Encephalitis without evidence of disseminated disease probably results from intraneuronal spread of virus from mucosal sites in the eye, nose, or mouth to sites within the central nervous system. Disseminated infection with mul-

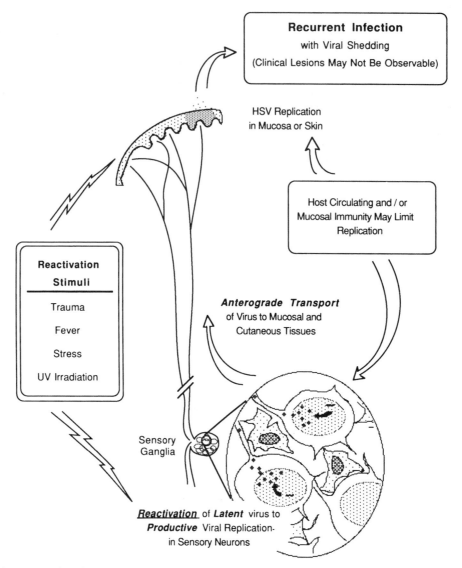

Recurrent Infection
with Viral Shedding
(Clinical Lesions May Not Be Observable)

HSV Replication
in Mucosa or Skin

Host Circulating and / or
Mucosal Immunity May Limit
Replication

**Reactivation
Stimuli**

Trauma

Fever

Stress

UV Irradiation

Anterograde Transport
of Virus to Mucosal and
Cutaneous Tissues

Sensory
Ganglia

Reactivation of ***Latent*** virus to
Productive Viral Replication-
in Sensory Neurons

Figure 4. Schematic representation of the pathogenesis of recurrent mucocutaneous herpes simplex virus infection. Adapted from Stanberry.[248]

tiple organ involvement including the brain is probably due to hematogenous spread, viremia having been documented to occur in some cases of neonatal infection.[170,252] as with other HSV infections, neonatal herpes results in the establishment of a latent infection that may be reactivated to commonly cause recurrent cutaneous infections and infrequently to cause recurrent ocular or central nervous system infections.

The immunology of HSV infection in humans and experimental animals has been reviewed.[120,220,232] In individuals with a primary HSV infection, humoral antibodies can be detected within 1–3 weeks by a variety of neutral-

ization, complement-fixation, complement-mediated cytolysis, antibody-mediated mononuclear cell cytolysis, or passive hemagglutination tests.[232,239] With special serological methods, it is possible to demonstrate an early rise of IgM antibodies to HSV, followed by IgA and IgG antibodies. In the newborn, IgM antibodies to HSV can be detected within 1–4 weeks after birth and are present for 6 months or longer. Recently, more sensitive immunoassays have indicated that IgA antibodies to HSV may be present for very long periods.[154]

In addition to the humoral responses, various assays of cellular immunity *in vitro* have demonstrated a cell-mediated response within 1–2 weeks after onset of infection in both man and experimental animals. Individuals with serum neutralization antibodies usually also demonstrate a delayed hypersensitivity skin test response to HSV antigens.

It is unclear which humoral or cellular factors operate in limiting virus replication in a primary infection or in a newborn without transplacental antibodies. Clinical studies have shown that patients with impaired humoral immunity can effectively control HSV infection, while persons with impaired cell-mediated immunity may have severe or life-threatening HSV disease.[149] Interestingly, persons with AIDS and concurrent HSV infection are not particularly likely to have disseminated HSV disease. Such observations suggest that cell-mediated immunity does not play a major role in preventing generalized dissemination to internal organs or neurogenic dissemination to the brain.

Although HSV-1 or HSV-2 infections in individuals with prior HSV-1 and/or HSV-2 infections tend to be less severe clinically than primary infections, in compromised hosts such infections tend to be more extensive and chronic. Herpes simplex encephalitis, which was once believed to occur only in association with a primary infection, has now been documented in individuals with prior HSV infection.[178] Similarly, although newborns with transplacental HSV antibodies were originally believed to be protected from acquiring HSV infections, many cases of neonatal herpetic disease, particularly of the central nervous system, have been recorded in such infants.[170]

In recurrent HSV infections, all individuals appear to possess neutralizing antibodies in varying titers, usually higher than in those persons without recurrences. The antibodies most often do not rise after a recurrence, and it appears that clinically apparent or inapparent recurrences might be necessary to maintain the neutralizing titer at constant levels in the serum. There are also individuals with frequent herpetic recurrences who possess high levels of IgG, IgA, and IgM antibodies to HSV in their serum and who may demonstrate a significant boost in titers of antibodies in the various classes after a recurrence. The apparent failure of high titer neutralizing antibody to prevent recurrent disease suggests that other (i.e., cellular) immune mechanisms may be more important in controlling the reactivation of disease.

Preliminary studies of subclass immunoglobulin antibodies to HSV suggest that all IgG subclasses are more likely to be found in individuals with recurrences than after a primary infection.[45] It is widely stated that antibodies offer no protection to HSV, because all individuals with recurrent herpes have serum HSV antibodies. It is

often forgotten, however, that about two thirds of those with HSV-1 antibodies do not have any clinically manifest herpes, and two thirds or more of those with HSV-2 antibodies do not manifest clinical herpes.[169,173a] Furthermore, maternal antibodies appear to protect most infants from disseminated herpetic infection and may play a role in preventing neonatal infection altogether.[170,196]

Individuals with recurrences also demonstrate positive responses in a variety of *in vitro* cell-mediated assays. There are conflicting data regarding the correlation between depression of cell-mediated immunity, including such lymphokines as interferon or leukocyte inhibitory factor, and an increased predilection to HSV recurrences. The importance of cell-mediated mechanisms, however, is suggested by studies in immunocompromised individuals who tend to develop more frequent and more severe herpetic recurrences.

In vitro studies indicate that both nonspecific and specific mechanisms are involved in stopping the cell-to-cell spread of HSV, since neutralizing antibodies alone cannot usually prevent this type of viral spread. In addition, *in vitro* observations indicate that neutralizing antibodies are unable to inactivate large amounts of extracellular virus and may operate together with complement or with nonimmune mononuclear or phagocytic cells in lysing HSV infected cells.

8. Patterns of Host Response

Table 3 presents our findings on close to 2000 individuals in whom HSV-1 and/or HSV-2 has been isolated from a variety of body sites. The occurrence in some individuals of simultaneous infections of different sites by HSV-1 and HSV-2 can be noted from the table. It is important to emphasize that although one HSV type is most likely to occur in any one body site, the other HSV type can occasionally infect similar sites. Also worth reemphasizing is that the majority of HSV-1 or HSV-2 infections in the normal host outside the newborn age cause no apparent disease whatsoever. In summary, the immune mechanisms involved in preventing the transmission, spread, and end-organ pathology of HSV infection in humans are complex. It is helpful to view host protection as a five-step process:

1. Preventing establishment of viral infection (humoral immunity).
2. Preventing cell-to-cell transmission of virus locally (cellular immunity).
3. Preventing contiguous mucocutaneous spread or spread to other organs, e.g., the esophagus (cellular immunity).
4. Preventing hematogenous spread to the liver (humoral and cellular immunity).

5. Localization in the liver without a tendency to spread to, or replication within, other distant organs (cellular and humoral immunity).

8.1. Mouth and Respiratory and Gastrointestinal Tracts

The mouth is the most common site of primary HSV-1 infection. The average duration of symptoms for herpes gingivostomatitis is 12 days.[26] Children are most often affected, but oral infections may also occur in older individuals. The spectrum of oral infections in the normal host ranges from inapparent infection to severe gingivostomatitis with extensive ulcerations of the mouth, tongue, and gums, cervical adenopathy, and fever, sometimes necessitating hospitalization for fluid and electrolyte restoration. The patients may occasionally autoinoculate other body sites, such as the face, fingers, or genitals. In the newborn or compromised host, and infrequently in the apparently normal individual, oral infection may extend to involve the larynx, esophagus, or lung and may disseminate to involve the liver and other visceral organs or the brain. Recent studies show that even healthy adults with genital herpes can demonstrate a rise in liver enzymes; fortunately, extensive liver disease has not been documented.[151]

Because of the common presence of HSV in the mouth of asymptomatic individuals, it has been difficult to ascertain how often HSV is the etiology of upper respiratory infections such as rhinitis and pharyngotonsillitis or the cause of recurrent lesions inside the mouth. Studies have suggested that HSV may be etiologically related to these clinical manifestations.[85,270] The large majority of cases of oral herpetic infections is caused by HSV-1, but HSV-2 has also been found to cause such infections as a consequence of fellatio or cunnilingus (Table 1).

8.2. Lips

Labial herpes is the most common form of recrudescence with HSV-1. It is of interest that even though HSV-1 is the definite cause of some primary oral infections, HSV-2 has only rarely been isolated from labial recurrent lesions. This is similar in many respects to the less common recurrence observed with HSV-1 when the virus type infects the genitals. The labial lesions, which are single or multiple, may involve the mucocutaneous junction at or around the same site with every recurrence. Typically the lesions last about 8 days.[243] Occasionally, the herpetic lesions may affect both sides and may be accompanied by involvement of other body sites. The lips

may also be infrequently involved during a primary infection. Extensive local spread or persistence of the lesions for long periods may occur in compromised individuals. The natural history of labial herpes has been well detailed.[243]

8.3. Eyes

The spectrum of ocular involvement includes conjunctivitis, keratitis (superficial and stomal), cataracts, iridocyclitis, panuveitis, and retinitis. Although the more superficial ocular lesions in children and adults are usually associated with HSV-1, a few cases of HSV-2 infection have been recognized. More recently, HSV has been implicated in a newly recognized ocular disease: acute retinal necrosis.[64,110,111] In the newborn, the ocular lesions may be caused by either HSV-1 or HSV-2. However the deeper ocular lesions are almost always caused by HSV-2, whereas cases of superficial keratoconjunctivitis can be caused by either HSV-1 or HSV-2.[79] The well-appreciated ill effects of corticosteroid in ocular herpes are worth reemphasizing here.

8.4. Skin

Herpes simplex vesicles on the skin are usually localized, may assume a zosteriform distribution in some instances, and may result from either primary or recrudescent infection. In the compromised individual, the skin lesions may be more extensive and chronic. In individuals with atopic eczema or with other dermatoses (e.g., Darier's disease), HSV infection can be more generalized and involve both affected and nonaffected areas of the underlying skin condition, resulting in Kaposi's varicelliform eruption (KVE) or eczema herpeticum; even with extensive KVE on the face and near the eyes, little ocular pathology has been reported.[73,197] Occasional recurrences of the entity have been described, and mortality may ensue from viral dissemination to vital organs or from bacterial superinfection.

Traumatic herpes occurs in areas of skin abrasions, burns, or puncture wounds. Laboratory-acquired HSV-1 or HSV-2 infections have occasionally resulted from such skin breaks in the hand (Table 3). Herpetic paronychia (herpetic whitlow) occurring in medical, nursing, or dental personnel and the skin infections occurring in wrestlers (herpes gladiatorum) have been cited. HSV infection of severe burns may be fatal, as it may lead to pneumonia and viral involvement of other vital organs. Herpes simplex infections may also be associated on occasion with the allergic manifestations of erythema multiforme.[227]

In general, skin lesions below the waist (Table 3) are caused by HSV-2 often in association with genital infection, whereas skin lesions above the waist are caused by HSV-1, reflecting the different modes of acquisition of the two virus types. Herpetic lesions of the hand or fingers can result from infection with either virus type.

8.5. Urogenital Tract

The term *herpes progenitalis* should be discarded, since in women genital infections occur commonly in the cervix, and in either symptomatic or asymptomatic males the virus may be cultured not only from the penis but also from the urethra, prostate, and seminal vesicles.[41,104,155,257] The mean duration of illness for clinically apparent primary genital HSV-2 infection is about 17 days for men and 20 days for women. The clinical manifestations of recurrent genital herpes are milder, lasting about 11 days in men and 9 days in women.[50] It should be emphasized that even though males are thought to have a higher rate of clinically manifest genital infections than females, allegedly because of the easier visibility of male genital organs, the frequency of asymptomatic HSV-2 infection appears to be similar in both sexes.[169] It could be that, as in females with nonvisible lesions but with infectious virus recoverable from the vulva or cervix, the male may have nonvisible lesions, with infectious virus on the penile surface. Although HSV has been implicated as an occasional cause of urethritis, prostatitis, and cystitis, its exact role in these diseases remains to be elucidated. Cerebrospinal fluid pleocytosis in the presence or absence of meningeal signs, radiculitis, and myelitis has been occasionally observed in association with genital herpetic infections.[48,53]

8.6. Nervous System

The neurological manifestations associated with HSV infection include encephalitis, meningitis, radiculitis, and myelitis.[53,136] The association of a particular neurological manifestation with HSV-1 or with HSV-2 appears to depend on (1) the host, i.e., whether newborn or older and whether compromised or immunologically normal, and (2) on the prevalence of a particular HSV type at various ages and in different socioeconomic groups. Since HSV-1 is the viral type that has been recovered in all but a few cases of herpetic encephalitis (Table 3), the possibility exists that the two HSV types also differ in their ability to spread to the brain via neuronal or hematogenous routes. Studies using restriction enzyme analysis to characterize orolabial and brain isolates from the same individuals

suggest that encephalitis can result from a primary infection with a unique HSV strain or may be due to reactivation of the latent virus responsible for labial herpes.[277] Further study is needed to substantiate the role of HSV in chronic CNS disease, psychiatric disorders, and Bell's palsy.

8.7. Fetus and Newborn

A wide spectrum of clinical manifestations has been observed with HSV infections in the newborn, which appears to be similar whether HSV-1 or HSV-2 is the causal agent.[170,272] The most severe form is the disseminated disease with involvement of the liver and other visceral organs, frequently including the brain. Herpes simplex encephalitis, which can occur without evidence of viral blood dissemination to visceral organs, is also associated with a poor prognosis. Localized involvement of the eyes, skin, and/or mouth may occur. Although fatality is rare in such cases, some of the infants have been found with neurological and/or ocular sequelae. In contrast to CMV infections, subclinical HSV infections of the newborn are very infrequent.

The association of maternal genital herpetic infection with abortions, stillbirths, and premature delivery has been reported.[170,277] Even though the virus has been recovered from abortus material, the possibility exists that the fetus was contaminated on passage through the mother's infected genital tract, as exemplified by two of our recent cases. The role of HSV in congenital malformations is suggested by the finding of a few cases of microcephaly or chorioretinitis soon after birth.[74,170]

8.8. Cervical Neoplasia

The evidence regarding the oncogenic potential of HSV *in vitro* and *in vivo* has been reviewed in an earlier edition of this book.[160,168] The epidemiologic evidence relating specifically HSV-2 to cervical neoplasia was tarnished by the use of generally poor type-specific antibody assays, inadequate controls, and unavailable statistical methodologies (e.g., multivariate analysis). So-called prospective studies[168] did not actually follow the development of cervical neoplasia in women identified with or without HSV-2 infection. Rather, they involved a retrospective evaluation of women for whom sera had been stored and who later developed cervical neoplasia. A true prospective study had demonstrated that a small number of women[256] identified with a primary HSV-2 infection had a high rate (33%) of developing severe cervical neoplasia. In several hundred women with recurrent genital herpes, the risk was twice that of similar numbers of control women without HSV-2 antibodies.[260]

An important role of papillomaviruses in cervical neoplasia (see Chapter 33, this volume) was known several years after the possible role of HSV was noted. However, as early as 1982, zur Hausen[287] suggested a possible combined role of the two viruses in cervical carcinogenesis. Recently, two well-controlled epidemiologic studies have indicated that HSV-2 may be a co-factor, adding significantly to the risk connected with papillomaviruses.[96,256] Studies in mice also indicate a combined effect of HSV-2 and human papillomavirus types 16 and 18 DNA in producing oncogenic transformation.[57,103]

9. Control and Prevention

The potential methods that might be used in the control of HSV infections can be grouped into (1) prevention of the initial infection in an individual and (2) reduction of the source of virus transmissible to others.

9.1. Prevention of the Initial Infection

9.1.1. Active Immunization. Over the past 50 years there have been numerous unsuccessful attempts to develop HSV vaccines. Most of the early work suffered from two major problems: (1) flaws in the design of clinical trial, and (2) vaccines that were either nonimmunogenic or, at best, poorly immunogenic.[69,153,211,246,258] With the development of new molecular and biochemical methodologies, several major vaccine manufacturers and bioengineering companies are reexamining strategies for developing effective HSV vaccines.[2] Two new approaches currently under development are the use of genetically engineered viral glycoproteins and the use of genetically engineered HSV mutants. Recent studies have shown that vaccines containing HSV glycoprotein D appear both safe and immunogenic even when combined with potent new adjuvants. Clinical trials to evaluate the efficacy of these vaccines in preventing primary HSV infection are currently underway in the United States, Canada, Australia, and Europe. Potential problems that will need to be examined during ongoing and future studies include whether vaccines can induce durable protective immunity and whether they will be well tolerated, and it will be important to demonstrate they are nononcogenic.

9.1.2. Passive Immunization. It has been found in experimental animals that HSV antibodies will prevent infection if administered prior to virus inoculation.[232]

Recent animal studies have shown that administration of high-titer anti-HSV-2 antibodies even 24 hr after viral challenge not only afforded significant protection against primary genital herpes but also significantly reduced the magnitude of latent infection and the frequency of subsequent recurrent disease.[249] Hyperimmune HSV globulin has been prepared for potential use in humans.[13] The protection afforded by immunoglobulin preparations that contain relatively high titers of HSV antibodies has been inconsistent in a very small number of newborns exposed to maternal genital HSV infections at delivery. Although newborns can become infected despite the presence in their serum of transplacentally transmitted antibodies, it has not been determined whether these failures represent the rule rather than the exception.

9.1.3. Interruption of Transmission. One strategy to prevent neonatal herpes is to deliver women with active genital HSV infection by cesarean section. There are compelling data to support this approach if the membranes are intact or have been ruptured for less than 4 hr.[170] Even this approach is not 100% effective, since it is well documented that infants delivered by cesarean section even with intact membranes can still acquire HSV-2 infection.[255] Optimal methods to identify women who are shedding virus asymptomatically at the time of delivery are under study.[46,47] Current rapid diagnostic methods do not provide the sensitivity and specificity to detect low titers of virus present in the genital tract of women during labor.[258] Weekly surveillance cultures of women with a history of genital herpes have been shown to be of no value in predicting the infant's risk of acquiring neonatal infection.[5] This approach is no longer recommended in the management of pregnancies complicated by genital HSV infections.[195] Another approach would be to identify couples in which the mother is HSV-2 negative and the father HSV-2 positive.[113,195]

The use of condoms or vaginal spermicide to prevent transmission of genital infections has not yet been evaluated in humans, although in artificial models the virus does not appear to leak through most condoms and is inactivated by some of these chemicals. Individuals might abstain from sexual contact when partners have obvious lesions. Similarly, close contact with individuals with nongenital lesions on the lips or skin might be avoided. As noted earlier, the problem is that many times the genital or nongenital infections are asymptomatic.

The use of gloves by medical or dental personnel when handling oral catheters or the mouth of patients might prevent acquisition of herpetic paronychia (herpetic whitlow). Neonates with suspect or proven HSV infection should be isolated in view of the many outbreaks reported.

Similarly, isolation of patients with severe herpetic involvement, such as patients with eczema herpeticum, might prevent transmission to other hospitalized patients or to attending personnel. Personnel with herpetic whitlow should abstain from patient care in view of potential transmission from this site to the patient. Similarly close contact, such as oral suctioning by personnel with a cold sore,[282] should be avoided. This unusual circumstance, however, does not justify the routine recommendation of having vital hospital personnel with herpetic cold sores abstain from patient care. Avoidance of close contact, use of face masks, and washing of hands appear sufficient.

9.2. Reducing the Transmissibility of Persons Infected with HSV

An obvious approach to controlling HSV infections is to minimize the contagiousness of persons infected with the virus. One strategy is to reduce the duration of lesions and viral excretion by use of antiviral chemotherapy. The first effective antiviral agent was iododexyuridine for the treatment of active herpetic keratitis, which did not, however, reduce the recurrence rate. Other drugs, including adenine arabinoside (vidarabine), trifluorothymidine, and acyclovir, have also been found to be effective for the treatment of ocular herpes but these do not apparently affect the frequency of recurrences either. Systematically administered vidarabine and acyclovir have been shown in controlled trials to reduce the mortality associated with herpes simplex encephalitis and with neonatal herpes and to be useful in the treatment of various forms of mucocutaneous HSV infections.[48,52,176,218,258,273,278] Acyclovir, topically administered, is also useful primarily for the treatment of primary genital herpetic infections. Continuous use of the oral form of therapy has also been effective in decreasing significantly the frequency of lesions in individuals with very frequent episodes of recurrent genital herpes.[28,258] Although antiviral therapy may reduce the duration of viral excretion, it is unknown whether it interferes with asymptomatic spread of virus. Unfortunately, acyclovir-resistant genital herpes has been recognized even in immunocompetent patients; however, the magnitude of this problem is unknown.[122]

An alternative approach to antiviral treatment is the use of therapeutic vaccines.[226] Throughout the history of HSV vaccine development it has been suggested that enhancement of the immune system of the latently infected host might result in better control of recurrent herpetic infections. The basic premise has been that host response to natural infection is incomplete or inadequate to control herpetic recurrences. Early efforts to enhance immune response to herpes included use of nonspecific stimulants such as the smallpox vaccine as well as autoinoculation with live nonattenuated HSV and a variety of inactivated HSV preparations. Most studies designed to evaluate the efficacy of vaccines and control of recurrent disease lacked appropriate control groups, making interpretation of the results impossible. In one well-controlled trial using an inactivated whole HSV vaccine, investigators found that 70% of the vaccine recipients reported improvement (noticeably fewer recurrences), while 76% of the placebo control group also reported improvement.[246] More recently, animal model studies have provided convincing evidence that HSV subunit vaccines can enhance host immunity, resulting in reduced symptomatic and subclinical recurrent infections.[246,251] The conclusion of these animal studies is supported by a preliminary report of a double-blind, placebo-controlled trial examining the efficacy of an immunogenic HSV-2 glycoprotein D vaccine for the treatment of recurrent genital herpes.[141] In the clinical trial patients with frequent recurrent genital herpes who received the vaccine experienced significantly fewer and less frequent virologically confirmed recurrences over a 12-month period. With these promising results it is likely that further trials of vaccine immunotherapy will be undertaken.

A more effective approach would be to find some means to prevent recurrences completely by restricting the virus in its latent state or in its early state of reactivation. One approach would be to maintain the virus in its latent state in the ganglia. The current availability of several animal models and of human ganglia cultures for latency studies, as well as newer molecular genetic techniques, would permit experimentation with a variety of approaches toward this goal.

10. Unresolved Problems

10.1. Reporting

A system for national reporting of certain forms of HSV infections, such as some forms of ocular infections, should be established, and the reporting of neonatal HSV infections and infections of the genital tract and nervous system should be improved.

10.2. Virological Aspects

These include (1) improving our knowledge of the molecular and genetic aspects of these viruses and (2) further understanding of the basic mechanisms involved with latency and transformation.

10.3. Host Factors

Information needs to be expanded on the genetic and immune factors that might be associated with the severity or the frequency of recurrences of HSV infection in certain individuals.

10.4. Control and Prevention

Further evaluation in the laboratory and in humans of the various possible approaches listed in Section 9 is required.

10.5. Relation to Infection with the Human Immunodeficiency Virus

More epidemiologic studies and further detailed biological investigations are needed to define the increased risk of acquisition of HIV infection in individuals with genital herpes, as well as the potential for employing HSV-2 antibodies as predictors of populations at risk for acquiring HIV infection on the basis of past sexual behavior and as markers for evaluating the effectiveness of sexual behavior modification attempts.

ACKNOWLEDGMENTS

We appreciate the assistance of Drs. William S. Josey, Harry L. Keyserling, and Francis K. Lee in preparing the earlier versions of this chapter. We thank M. S. Orlando, M. J. Kauffman, Ms. A. Mrcozynski and Ms. C. Polliotti for their secretarial assistance. We would also like to thank Dr. Beverly Connelly for creating the schematic representations of primary and recurrent herpes simplex virus infections (Figs. 3 and 4).

11. References

1. ADAM, E., KAUFMAN, R H., MIRKOVIC, R. R., AND MELNICK, J. L., Persistence of virus shedding in asymptomatic women after recovery from herpes genitalis, *Obstet. Gynecol.* **54:**171–173 (1979).

2. ALLEN, W. P., AND HITCHCOCK, P. J. (EDS.), Herpes simplex virus vaccine workshop, *Rev. Infect. Dis.* **13**(Suppl. 11)**:**S891–S979 (1991).

3. ANDERSON, S. G., AND HAMILTON, J., The epidemiology of human herpes simplex infection, *Med. J. Aust.* **1:**308–311 (1949).

4. ARVIN, A. M., Relationship between maternal immunity to herpes simplex virus and the risk of neonatal herpesvirus infection, *Rev. Infect. Dis.* **13**(Suppl. 11)**:**S953–956 (1991).

5. ARVIN, A. M., HENSLEIGH, P. A., PROBER, C. G., AU, D. S., YASUKAWA, L. L., WITTER, A. E., PALUMBO, P. E., PARYANI, S. G., AND YEAGER, A. S., Failure of antepartum maternal cultures to predict the infant's risk of exposure to herpes simplex virus at delivery, *N. Engl. J. Med.* **314:**1561–1564 (1986).

6. ASHLEY, R., Laboratory techniques in the diagnosis of herpes simplex infection, *Genitourin. Med.* **69:**174–183 (1993).

7. ASHLEY, R., CENT, A., MAGGS, V., NAHMIAS, A., AND COREY, L., Inability of enzyme immunoassays to discriminate between infection with herpes simplex virus types 1 or 2, *Ann. Intern. Med.* **115:**520–526 (1991).

8. ASHLEY, R., AND MILITONI, J., Use of densitometric analysis for interpreting HSV serologies based on Western blot, *J. Virol. Methods* **18:**159–168 (1987).

9. ASHLEY, R. L., MILITONI, J., LEE, F., NAHMIAS, A., AND COREY, L., Comparison of Western blot (immunoblot) and glycoprotein G-specific immunodot enzyme assay for detecting antibodies to herpes simplex virus type 1 and 2 in human sera, *J. Clin. Microbiol.* **26:**662–667 (1988).

10. ASTON, D. L., CON, A., AND SPINDLER, M., Herpesvirus hominis infection in patients with certain myeloproliferative and lymphoproliferative disorders, *Br. Med. J.* **4:**46–465 (1972).

11. BARINGER, J. R., Recovery of herpes simplex virus from human sacral ganglions, *N. Engl. J. Med.* **291:**828–830 (1974).

12. BARINGER, J. R., AND SWORELAND, P., Recovery of herpes simplex virus from human trigeminal ganglions, *N. Engl. J. Med.* **288:**648–650 (1973).

13. BARON, S., GEORGIADES, J., AND WORTHINGTON, M., Potential for postexposure prophylaxis of neonatal herpes using passive antibody, in: *The Human Herpesviruses: An Interdisciplinary Perspective* (A. NAHMIAS, X. V. DOWDLE, AND R. SCHINAZI, EDS.), pp. 491–495, Elsevier-North Holland, New York, 1981.

14. BECKER, T. M., BLOUNT, J. H., AND GUINAN, M. E., Genital herpes infections in private practice in the United States, 1966–1981, *J.A.M.A.* **253:**1601–1603 (1985).

15. BECKER, W. B., KIPPS, A., AND MCKENZIE, D., Disseminated herpes simplex virus infection: Its pathogenesis based on virological and pathological studies in 33 cases, *Am. J. Dis. Child.* **115:**1–8 (1968).

16. BECKER, T. M., MAGDER, L., HARRISON, H. R., STEWARD, J. A., HUMPHREY, D. D., HAULER, J., AND NAHMIAS, A. J., The epidemiology of infection with the human herpesvirus in Navajo children, *Am. J. Epidemiol.* **127:**1071–1078 (1988).

17. BECKER, T. M., AND NAHMIAS, A. J., Genital herpes yesterday, today, and tomorrow, *Annu. Rev. Med.* **35:**185–193 (1985).

18. BECKER, W. B., The epidemiology of herpesvirus infection in three racial communities in Cape Town, *S. Afr. Med. J.* **40:**109–111 (1966).

20. BELL, D. M., HOLMAN, R. C., AND PAVAN-LANGSTON, D., Herpes simplex keratitis: Epidemiologic aspects, *Ann. Ophthalmol.* **14:**421–424 (1982).

21. BERGSTROM, T., VAHLNE, A., ALESTIG, K., JEANSSON, S., FORSGREN, M., AND LYKE, E., Primary and recurrent herpes simplex virus type 2-induced meningitis, *J. Infect. Dis.* **162:**322–330 (1990).

22. BERNSTEIN, D. I., BRYSON, Y. J., AND LOVETT, M. A., Antibody response to type-common and type-unique epitopes of herpes simplex virus polypeptides, *J. Med. Virol.* **15:**251–263 (1985).

23. BESWICK, T. S. L., The origin and use of the word herpes, *Med. Hist.* **6:**214–232 (1962).

24. BIRO, F. M., ROSENTHAL, S. L., BERNSTEIN, D. I., AND STANBERRY, L. R., Seroprevalence of HSV types 1 and 2 in sexually active adolescents, *J. Adoles. Health* **15:**60 (1994).

25. BLACK, F. L., Infectious diseases in primitive societies, *Science* **187:**515–518 (1975).

26. BLACK, W. C., The etiology of acute infectious gingivostomatitis (Vincent's stomatitis), *J. Pediatr.* **20:**145–160 (1942).

27. BOUCHER, F. D., YASUKAWA, L. L., BRONZAN, R. N., HENSLEIGH, P. A., ARVIN, A. M., AND PROBER, C. G., A prospective evaluation of primary genital herpes simplex virus type 2 infections acquired during pregnancy, *Pediatr. Infect. Dis. J.* **9**:499–504 (1990).

28. BOULOS, R., RUFF, A. J., NAHMIAS, A., HOLT, E., HARRISON, L., MAGDER, L., WIKTOR, S. Z., QUINN, T. C., MARGOLIS, H., HALSEY, N. A., AND THE JOHNS HOPKINS UNIVERSITY (JHU)/ CENTRE POUR LE DEVELOPPEMENT ET LA SANTE (CDS) HIV STUDY GROUP, Herpes simplex virus type 2 infection, syphilis, hepatitis B virus infection in Haitian women with human immunodeficiency virus type 1 and human T lymphotropic virus type I infections, *J. Infect. Dis.* **166**:418–420 (1992).

29. BRAUTIGAM, A. R., RICHMAN, D. D., AND OXMAN, M. N., Rapid typing of herpes simplex virus isolates by deoxyribonucleic acid: Deoxyribonucleic acid hybridization, *J. Clin. Microbiol.* **12**:226–234 (1980).

30. BRAVO, F. J., MYERS, M. G., AND STANBERRY, L. R., Neonatal herpes simplex virus infection: Pathogenesis and treatment in the guinea pig, *J. Infect. Dis.* **169**:947–955 (1994).

31. BREINIG, M. K., KINGSLEY, L. A., ARMSTRONG, J. A., FREEMAN, D. J., AND HO, M., Epidemiology of genital herpes in Pittsburgh: Serologic, sexual, and racial correlates of apparent and inapparent herpes simplex infections, *J. Infect. Dis.* **162**:299–305 (1990).

32. BROWN, S. A., KERN, E. R., SPRUANCE, S. L., AND OVERALL, JR., J. C., Clinical and virologic course of herpes simplex genitalis, *West. J. Med.* **130**:414–421 (1979).

33. BROWN, S. A., VONTVER, L. A., BENEDETTI, J., CRITCHLOW, C. W., HICKOK, D. E., SELLS, C. J., BERRY, S., AND COREY, L., Genital herpes in pregnancy: Risk factors associated with recurrences and asymptomatic viral shedding, *Am. J. Obstet. Gynecol.* **153**:24–30 (1985).

34. BRYSON, Y. J., Genital herpes in adolescents and young adults, *Adolesc. Med. State Art Rev.* **1**:471–482 (1990).

35. BRYSON, Y., DILLION, M., BERNSTEIN, D. I., RADOLF, J., ZAKOWSKI, P., AND GARRATTY, E., Risk of acquisition of genital herpes simplex virus type 2 in sex partners of persons with genital herpes: A prospective couple study, *J. Infect. Dis.* **167**:942–946 (1993).

36. BUCHMAN, T. G., ROIZMAN, B., ADAMS, G., AND STOVER, G. H., Restriction endonuclease fingerprinting of herpes simplex virus DNA: A novel epidemiological tool applied to a nosocomial outbreak, *J. Infect. Dis.* **138**:488–498 (1978).

37. BUCHMAN, T., ROIZMAN, B., AND NAHMIAS, A., Demonstration of exogenous genital reinfection with herpes simplex virus type 2 by restriction endonuclease fingerprinting of viral DNA *J. Infect. Dis.* **140**:195–304 (1979).

38. BUCHMAN, T. G., ROIZMAN, B., AND NAHMIAS, A. J., Structure of herpes simplex virus DNA and application to molecular epidemiology, *Ann. N.Y. Acad. Sci.* **354**:279–290 (1980).

39. BUDDINGH, G. J., SCHRUM, D. I., LANIER, J. C., AND GUIDAY, D. J., Studies of the natural history of herpes simplex infections, *Pediatrics* **311**:595–610 (1953).

40. CENTERS FOR DISEASE CONTROL, Non-reported sexually transmitted diseases in the United States, *Morbid. Mortal. Week. Rep.* **28**:61–63 (1979).

41. CENTIFANTO, Y. M., DRYLIE, D. M., DEARDOURFF, S. L., AND KAUFMAN, H., Herpesvirus type 2 in the male genitourinary tract, *Science* **178**:318–319 (1972).

42. CESARIO, T. C., POLAND, J. D., AND WULFF, H., Six years experience with herpes simplex virus in a children's home, *Am. J. Epidemiol.* **90**:416–422 (1969).

43. CHRISTENSON, B., BOTTIGER, M., SVERSON, A., AND JEANSSON, S., A 15-year surveillance study of antibodies to herpes to herpes simplex virus types 1 and 2 in a cohort of young girls, *J. Infect.* **25**:147–154 (1992).

44. CHUANG, T., SU, W. P., PERRY, H. O., ILSTRUP, D. M., AND KURLAND, L. T., Incidence and trend of herpes progenitalis: A 15-year population study, *Mayo Clin. Proc.* **58**:426–441 (1983).

45. COLEMAN, R. M., PERETRA, L., BAILEY, P. D., DONDERO, D., WYCLIFFE, C., AND NAHMIAS, A. J., Demonstration of herpes simplex virus type specific antibodies by ELISA, *J. Clin. Microbiol.* **18**:287–291 (1983).

46. CONE, R. W., HOBSON, A. C., PALMER, J., REMINGTON, M., AND COREY, L., Extended duration of herpes simplex virus DNA in genital lesions detected by polymerase chain reaction, *J. Infect. Dis.* **164**:757–760 (1991).

47. CONE, R. W., SWENSON, P., HOBSON, A. C., REMINGTON, M., AND COREY, L., Herpes simplex virus detection from genital lesions: A comparative study using antigen detection (Herpcheck) and culture, *J. Clin. Microbiol.* **31**:1774–1776 (1993).

47a. COOK, M. L., AND STEVEN, J. G., Pathogenesis of herpetic neuritis and ganglionitis in mice: Evidence for intra-axonal transport of infection, *Infect. Immunol.* **7**:272–288 (1973).

48. COREY, L., Genital herpes, in: *The Herpesviruses*, Vol. 4 (B. ROIZMAN AND C. LOPEZ, EDS.), pp. 1–36, Plenum Press, New York, 1986.

49. COREY, L., First-episode, recurrent, and asymptomatic herpes simplex infections, *J. Am. Acad. Dermatol.* **18**:169–172 (1988).

50. COREY, L., Genital herpes, in: *Sexually Transmitted Diseases* (K. K. HOLMES, P-A. MARDH, P. F. SPARLING, AND P. J. WIESNER, EDS.), pp. 391–414, McGraw-Hill, New York, 1990.

51. COREY, L., ADAMS, H. G., BROWN, Z. A., AND HOLMES, K. K., Genital herpes simplex virus infection: Clinical manifestations, course and complications, *Ann. Intern. Med* **98**:958–972 (1983).

52. COREY, L., AND SPEAR, P. G., Infections with herpes simplex viruses, *N. Engl. J. Med.* **314**:686–691, 749–757 (1986).

53. CRAIG, C. P., AND NAHMIAS, A., Different patterns of neurologic involvement with herpes simplex virus types 1 and 2: Isolation of herpes simplex virus type 2 from the buffy coat of two adults with meningitis, *J. Infect Dis.* **127**:365–372 (1973).

54. CROEN, K. D., OSTROVE, J. M., DRAGOVIC, L. J., SMIALED, J. I., AND STRAUS, S. E., Latent herpes simplex virus in human trigeminal ganglia: Detection of an immediate early gene "antisense" transcript by *in situ* hybridization, *N. Engl. J. Med.* **317**:1427–1432 (1987).

55. CROEN, K. D., OSTROVE, J. M., DRAGOVIC, L., AND STRAUS, S. E., Characterization of herpes simplex virus type 2 latency associated transcription in the human sacral ganglia and in cell culture, *J. Infect. Dis.* **163**:23–28 (1991).

56. DAROUGAR, S., WISHART, M. S., AND VISWALINGAM, N. D., Epidemiological and clinical features of primary herpes simplex virus ocular infection, *Br. J. Ophthalmol.* **69**:2–6 (1985).

57. DILUCA, D., PILOTTI, S., AND ROTOLA, A., Simultaneous presence of herpes simplex and human papilloma virus sequences in human genital tumors, *Int. J. Cancer* **40**:763–768 (1987).

58. DINGWELL, K. S., BRUNETTI, C. R., HENDRICKS, R. L., TANG, Q., TANG, M., RAINBOW, A. J., AND JOHNSON, D. C., Herpes simplex virus glycoproteins E and I facilitate cell-to-cell spread *in vivo* and across junctions of cultured cells, *J. Virol.* **68**:834–845 (1994).

59. DIVISION OF STD HIV PREVENTION, SEXUALLY TRANSMITTED DISEASE SURVEILLANCE, 1992, US DEPARTMENT OF HEALTH AND HUMAN SERVICES, PUBLIC HEALTH SERVICE, ATLANTA, CENTERS FOR DISEASE CONTROL AND PREVENTION, JULY 1993.

60. DOLIN, R., GILL, F., AND NAHMIAS, A., Genital herpes simplex virus type I infection-variability in modes of spread, *J. Am. Vener. Dis. Assoc.* **2:**13–16 (1975).

61. DOUGLAS, J., SCHMIDT, O., AND COREY, L., Acquisition of neonatal HSV-1 infection from a paternal source contact, *J. Pediatr.* **103:**908–910 (1983).

62. DOUGLAS, R. G., JR., AND COUCH, R. B., A prospective study of chronic herpes simplex virus infection and recurrent herpes labialis in humans, *J. Immunol.* **104:**289–295 (1970).

63. DUENAS, A., ADAM, E., MELNICK, J. L., AND RAWLS, W. E., Herpesvirus type 2 in a prostitute population, *Am. J. Epidemiol.* **95:**483–489 (1972).

64. DUKER, J. S., AND BLUMENKRANZ, M. S., Diagnosis and management of acute retinal necrosis (ARN) syndrome, *Surv. Ophthalmol.* **35:**327–343 (1991).

65. DUMMER, J. S., ARMSTRONG, J., SOMERS, J., KUSNE, S., CARPENTER, B. J., ROSENTHAL, J. T., AND HO, M., Transmission of infection with herpes simplex virus by renal transplantation, *J. Infect. Dis.* **155:**202–206 (1987).

66. EFSTATHIOU, A. M., SPIVACK, J. G., LAVI, E., AND FRASER, N. W., Detection of herpes simplex virus-specific DNA sequences in latently infected mice and humans, *J. Virol.* **57:**446–455 (1986).

67. EILARD, U., AND HELLGREN, L., Herpes simplex: A statistical and clinical investigation based on 669 patients, *Dermatologica* **130:**101–106 (1965).

68. EMBIL, J. A., STEPHENS, R. G., AND MANUEL, F. R., Prevalence of recurrent herpes labialis and aphthous ulcers among young adults on six continents, *Can. Med. Assoc. J.* **113:**627–630 (1975).

69. ENNIS, F. A., AND MORESCHI, G. I. Prevention of herpes simplex virus infection, in: *The Human Herpesviruses: An Interdisciplinary Perspective* (A. NAHMIAS, W. DOWDLE, AND R. SCHINAZI, EDS.), pp. 441–446, Elsevier-North Holland, New York, 1981.

70. ESPY, M. J., AND SMITH, T. F., Detection of herpes simplex virus in conventional tube cell cultures and in shell vials with a DNA probe kit and monoclonal antibodies, *J. Clin. Microbiol.* **26:**22–24 (1988).

71. EVANS, A. S., AND DICK, E. C., Acute pharyngitis and tonsillitis in University of Wisconsin students, *J.A.M.A.* **190:**699–708 (1964).

72. FEDER, H. M., AND LONG, S. S., Herpetic whitlow, *Am. J. Dis. Child.* **137:**861–863 (1983).

73. FIVENSON, D. P., BRENEMAN, D. L., AND WANDER, A. H., Kaposi's varicelliform eruption, *Arch. Dermatol.* **126:**1037–1039 (1990).

74. FLORMAN, A. L., GENSHON, A. A., BLACKETT, P. R., AND NAHMIAS, A. J., Intrauterine infection with herpes simplex virus: Resultant congenital malformation, *J.A.M.A.* **225:**129–132 (1973).

75. FOLKERS, E., VREESWIJK, J., ORANJE, A. P., WAGENAAR, F., AND DUIVENVOORDEN, J. N., Improved detection of HSV by electron microscopy in clinical specimens using ultracentrifugation and colloidal gold immunoelectron microscopy: Comparison with viral culture and cytodiagnosis, *J. Virol. Meth.* **34:**273–289 (1991).

76. FORSGREN, M., Genital herpes simplex virus infection and incidence of neonatal disease in Sweden, *Scand. J. Infect. Dis.* **69:**37–41 (1990).

77. FRANCIS, D. P., HERMANN, K. L., AND MACMAHON, J. R., Nosocomial and maternally acquired herpesvirus hominis infections: A report of four fatal cases in neonates, *Am. J. Dis. Child.* **129**(8):889–893 (1975).

78. GAFFEY, M. J., BEN-EZRA, J. M., AND WEISS, L. M., Herpes simplex lymphadenitis, *Am. J. Clin. Pathol.* **95:**709–714 (1991).

79. GAMMON, J. A., AND NAHMIAS, A. J., Herpes simplex ocular infections in the newborn, in: *Viral Diseases of the Eye* (R. W. DARRELL, ED.), pp. 46–58, Lea & Febiger, Philadelphia, 1985.

80. GARDNER, M., AND JONES, J. G., Genital herpes acquired by sexual abuse of children, *J. Pediatr.* **104:**243–244 (1984).

81. GENTRY, G. A., LOWE, M., ALFORD, G., AND NEVINS, R., Sequence analysis of herpesviral enzymes suggest an ancient origin for human sexual behavior, *Proc. Natl. Acad. Sci. USA* **85:**2658–2661 (1988).

83. GIBSON, J. J., HORNUNG, C. A., ALEXANDER, G. R., LEE, F. K., POTTS, W. A., AND NAHMIAS, A. J., A cross-sectional study of herpes simplex virus types 1 and 2 in college students: Occurrence and determinants of infection, *J. Infect. Dis.* **162:**306–312 (1990).

84. GILL, M. J., ARLETTE, J., AND BUCHAN, K., Herpes simplex virus infection of the hand: A profile of 79 cases, *Am. J. Med.* **84:**89–93 (1988).

85. GLEZEN, W. P., FERNALD, G. W., AND LOHR, J. A., Acute respiratory disease of university students with special reference to the etiologic role of herpesvirus hominis, *Am. J. Epidemiol.* **101:**111–121 (1975).

86. GREENBERG, M. S., BRIGHTMAN, V. J., AND SHIP, I. I., Clinical and laboratory differentiation of recurrent intraoral herpes simplex virus infections following fever, *J. Dent. Res.* **48:**385–391 (1969).

87. GROUT, P., AND BARBER, V. E., Cold sores—an epidemiological survey, *J. R. Coll. Gen. Pract.* **26:**428–434 (1976).

88. GUINAN, M. E., WOLINSKY, S. M., AND REICHMAN, R. C., Epidemiology of genital herpes simplex virus infection, *Epidemiol. Rev.* **7:**127–146 (1985).

89. HALE, B. D., RENDTORFF, R. C., WALKER, L. C., AND ROBERTS, A. N., Epidemic herpetic stomatitis in an orphanage nursery, *J.A.M.A.* **183:**1068–1072 (1963).

90. HAMMERBERG, O., WATTS, J., CHERNESKY, M., LUCHSINGER, I., AND RAWLS, W., An outbreak of herpes simplex virus type I in an intensive care nursery, *Pediatr. Infect. Dis.* **2:**290–294 (1983).

91. HARDY, D. A., ARVIN, A. A., YASUKAWA, L. L., BRONZAN, R. N., LEWINSOHN, D. M., HENSLEIGH, P. A., AND PROBER, C. G., Use of polymerase chain reaction for successful identification of asymptomatic genital infection with herpes simplex virus in pregnant women at delivery, *J. Infect. Dis.* **162:**1031–1035 (1990).

92. HAYNES, R. E., AZIMI, P. H., AND CRAMBLETT, H. G., Fatal herpesvirus hominis (herpes simplex virus) infections in children—clinical, pathologic and virologic characteristics, *J.A.M.A.* **206:**312–319 (1968).

93. HELLGREN, L., The prevalence of some skin diseases and joint diseases in total populations in different areas of Sweden, *Proc. North. Dermatol. Soc.* **6:**155–162 (1962).

94. HENG, M. C. Y., HENG, S. Y., AND ALLEN, S. G., Co-infection and synergy of human immunodeficiency virus-1 and herpes simplex virus-1, *Lancet* **1:**255–258 (1994).

95. HERRMANN, K. L., AND STEWART, J. A., Diagnosis of herpes simplex virus types 1 and 2, in: *The Human Herpesviruses: An Interdisciplinary Perspective* (A. NAHMIAS, W. DOWDLE, AND R. SCHINAZI, EDS.), pp. 343–350, Elsevier-North Holland, New York, 1981.

96. HILDESHEIM, A., MANN, V., BRINTON, L. A., SZKLO, M., REEVES, W. C., AND RAWLS, W. E., Herpes simplex virus type 2: A possible interaction with human papillomavirus types 16/18 in the development of invasive cervical cancer, *Int. J. Cancer* **49:**335–340 (1991).

97. HILL, T., Herpes simplex virus latency, in: *The Herpesviruses*, Vol. 3 (B. ROIZMAN, ED.), pp. 175–240, Plenum Press, New York, 1985.

98. HOLLAND, E. J., MAHANTI, R. L., BELONGIA, E. A., MIZENER, M. W., GOODMAN, J. L., ANDRES, C. W., AND OSTERHOLM, M. T., Ocular involvement in an outbreak of herpes gladiatorum, *Am. J. Ophthalmol.* **114:**680–684 (1992).

99. HOLMBERG, S. D., STEWART, J. A., GERBER, R., BYERS, R. H., LEE, F. K., O'MALLEY, P. M., AND NAHMIAS, A. J., Prior herpes simplex virus type 2 infection as a risk factor for HIV infection, *J.A.M.A.* **259:**1048–1050 (1988).

100. HOOK, E. W., III, CANNON, R. O., NAHMIAS, A. J., LEE, F. K., CAMPBELL, C. H., JR., GLASSER, D., AND QUINN, T. C., Herpes simplex virus infection as a risk factor for human immunodeficiency virus infection in heterosexuals, *J. Infect. Dis.* **165:**251–255 (1992).

101. HOWARD, G. M., AND KAUFMAN, H. E., Herpes simplex keratitis, *Arch. Ophthalmol.* **67:**373–387 (1962).

102. HUTTO, C., ARVIN, A., JACOBS, R., STEELE, R., STANGO, S., LYRENE, R., WILLETT, L., POWELL, D., ANDERSON, R., WERTHAMMER, J., RATCLIFF, G., NAHMIAS, A., CHRISTY, C., AND WHITLEY, R., Intrauterine herpes simplex virus infection, *J. Pediatr.* **110:**97–101 (1987).

103. IWASAKA, T., YOKOYAMA, M., HAYASHI, Y., AND SUGIMORE, H., Combined herpes simplex virus type 2 and human papillomavirustype 16 or 18 deoxyribonucleic acid leads to oncogenic transformation, *J. Obstet. Gynecol.* **159:**1251–1255 (1988).

104. JEANSSON, S., AND MOLIN, M. L., On the recurrence of genital herpes simplex virus infection: Clinical and virological findings and relation to gonorrhea, *Acta Dermattel. Venereol.* **54:**479–485 (1974).

105. JOHNSON, R. E., NAHMIAS, A. J., MAGDER, L. S., LEE, F. K., BROOKS, C. A., AND SNOWDEN, C. B., A seroepidemiologic survey for the prevalence of herpes simplex virus type 2 infection in the United States, *N. Engl. J. Med.* **321:**7–12 (1989).

106. JOHNSON, R., LEE, F., HAGDU, A., McQUILLAN, G., ARAL, S., KEESLING, S., AND NAHMIAS, A., US genital herpes trends during the first decade of AIDs-prevalences increased in young white and elevated in blacks (submitted).

107. JUDSON, F. N., PENLEY, K. A., ROBINSON, M. E., AND SMITH, J. K., Comparative prevalence rates of sexually transmitted diseases in heterosexual and homosexual men, *Am. J. Epidemiol.* **112:**836–843 (1980).

108. JURETIC, M., Natural history of herpetic infection, *Heir. Paediatr. Acta* **21:**356–368 (1966).

109. KALINYAK, J. E., FLEAGLE, G., AND DOCHERTY, J. J., Incidence and distribution of herpes simplex virus types I and 2 from genital lesions in college women, *J. Med. Virol.* **1:**173–181 (1977).

110. KAPLAN, H. J., LEE, F. K., WILLIG, J. L., REESE, L., HALL, E. C., AND NAHMIAS, A. J., Intraocular antibody production (submitted).

111. KAPLAN, H. J., WILLIG, J. L., CULBERTSON, W. W., HOLLAND, G. N., LEWIS, M. L., BLUEMENKRANZ, M. S., COWAN, G., KREIGER, A. E., LEE, F. K., REESE, L., AND NAHMIAS, A. J., Intraocular antibody production. II. Herpes simplex virus-type 2 causes acute retinal necrosis in the young (submitted).

112. KARLIN, S., MOCARSKI, E. S., AND SCHACHTEL, G. A., Molecular evolution of herpesviruses: Genomic and protein sequence comparisons, *J. Virol.* **68:**1886–1902 (1994).

113. KEYSERLING, H. K., ROBINOWITZ, M., RATCHFORD, R., BAIN, R., AND NAHMIAS, A. J., Herpes simplex virus type antibodies and history of genital herpes among steady couples, in: *Second World Congress on Sexually Transmitted Diseases.*

114. KEYSERLING, H. K., THOMPSON, S., ROBINOWITZ, M., LEE, F. K., PEREIRA, L., COLEMAN, R., BAIN, R., AND NAHMIAS, A. J., Prevalence of genital herpes and/or herpes simplex virus type 2 (HSV-2) antibodies in two obstetric populations (submitted).

115. KIMURA, H., FUTAMURA, M., KITO, H., ANDO, T., GOTO, M., KUZUSHIMA, K., SHIBATA, M., AND MORISHIMA, T., Detection of viral DNA in neonatal herpes simplex virus infections: Frequent and prolonged presence in serum and cerebrospinal fluid, *J. Infect. Dis.* **164:**289–293 (1991).

116. KINGSLEY, L. A., ARMSTRONG, J., RAHMAN, A., HO, M., AND RINALDO, C. R., No association between herpes simplex virus type-2 seropositivity or anogenital lesions and HIV seroconversion among homosexual men, *J. Acquir. Immune Defic. Syndr.* **3:**773–779 (1990).

117. KLAPPER, P. E., CLEATOR, G. M., DENNETT, C., AND LEWIS, A. G., Diagnosis of herpes encephalitis via southern blotting of cerebrospinal fluid DNA amplified by polymerase chain reaction, *J. Med. Virol.* **32:**261–264 (1990).

118. KLOENE, W., BANG, F. B., CHAKRABORTY, S. M., COOPER, M. R., KULEMANN, H., OTA, M., AND SHAH, K. V., A two year respiratory virus survey in four villages in West Bengal, India, *Am. J. Epidemiol.* **92:**307–320 (1970).

119. KOELLE, D. M., BENEDETTI, J., LANGENBERG, A., AND COREY, L., Asymptomatic reactivation of herpes simplex virus in women after the first episode of genital herpes, *Ann. Intern. Med.* **116:**433–437 (1992).

120. KOHL, S., The role of antibody in herpes simplex virus infection in humans, *Curr. Top. Microbiol. Immunol.* **179:**75–88 (1992).

121. KOHL, S., WEST, M. S., PROBER, C. G., SULLENDER, W. M., LOO, L. S., AND ARVIN, A. M., Neonatal antibody-dependent cellular cytotoxic antibody levels are associated with the clinical presentation of neonatal herpes simplex virus infection, *J. Infect. Dis.* **160:**770–776 (1989).

122. KOST, R. G., HILL, E. L., TIGGES, M., AND STRAUS, S. E., Brief report: Recurrent acyclovir-resistant genital herpes in an immunocompetent patient, *N. Engl. J. Med.* **329:**1777–1784 (1993).

123. KOUTSKY, L., ASHLEY, R., HOLMES, K., STEVENS, C., CRITCHLOW, C., KIVIAT, N., LIPINSKI, C., WOLNER-HANSSEN, P., AND COREY, L., The frequency of unrecognized type 2 herpes simplex virus infection among women: Implications for the control of genital herpes, *Sex. Trans. Dis.* **17:**90–94 (1990).

124. KOUTSKY, L. A., STEVENS, C. E., HOLMES, K. K., ASHLEY, R. L., KIVIAT, N. B., CRITCHLOW, C. W., AND COREY, L., Underdiagnosis of genital herpes by current clinical and viral-isolation procedures, *N. Engl. J. Med.* **326:**1533–1539 (1992).

125. KRAUSE, P. R., CROEN, K. D., STRAUS, S. E., AND OSTROVE, J. M., Detection and preliminary characterization of herpes simplex virus type 1 transcripts in latently infected human trigeminal ganglia, *J. Virol.* **62:**4819–4823 (1988).

126. KULHANJIAN, J. A., SOROUSH, V., AU, D. S., BRONZAN, R. N., YASUKAWA, L. L., WEYLMAN, L. E., ARVIN, A. M., AND PROBER, C. G., Identification of women at unsuspected risk of primary infection with herpes simplex virus type 2 during pregnancy, *N. Engl. J. Med.* **326:**916–920 (1992).

127. KUSNE, S., SCHWARTZ, M., BREINIG, M. K., DUMMER, J. S., LEE, R. E., SELBY, R., STARZL, T. E., SIMMONS, R. L., AND HO, M., Herpes simplex virus hepatitis after solid organ transplantation in adults, *J. Infect. Dis.* **163:**1001–1007 (1991).

128. KUZUSHIMA, K., KIMURA, H., KINO, Y., KIDO, S., HANADA, N.,

SHIBATA, M., AND MORISHIMA, T., Clinical manifestations of primary herpes simplex virus type 1 infection in a closed community, *Pediatrics* **87:**152–157 (1991).

129. KUZUSHIMA, K., KUDO, T., KIMURA, H., KIDO, S., HANADA, N., SHIBATA, M., NISHIKAWA, K., AND MORISHIMA, T., Prophylactic oral acyclovir in outbreaks of primary herpes simplex virus type 1 infection in a closed community, *Pediatrics* **89:**379–383 (1992).

130. LAFFERTY, W. E., COOMBS, R. W., BENEDETTI, J., CRITCHLOW, C., AND COREY, L., Recurrences after oral and genital herpes simplex virus infection, *N. Engl. J. Med.* **316:**1444–1449 (1987).

131. LANDERS, D. V., SMITH, J. P., WALKER, C. K., MILAM, T., SANCHEZ-PESCADOR, L., AND KOHL, S., Human fetal antibody-dependent cellular cytotoxicity to herpes simplex virus-infected cells, *Pediatr. Res.* **35:**289–292 (1994).

132. LEE, F. K., COLEMAN, R. M., PERETRA, L., BAILEY, P., TATSUNO, M., AND NAHMIAS, A. J., Detection of herpes simplex virus type-2 specific antibody with glycoprotein G, *J. Clin. Microbiol.* **4:**641–646 (1985).

133. LEE, F. K., PEREIRA, L., GRIFFIN, C., REID, E., AND NAHMIAS, A. J., A novel glycoprotein (gG-1) for detection of herpes simplex virus type I specific antibodies, *J. Virol. Methods* **14:**111–118 (1986).

134. LINDGREN, K. M., DOUGLAS, R. G., AND COUCH, R. B., Significance of herpesvirus hominis in respiratory secretions of man, *N. Engl. J. Med.* **278:**517–523 (1968).

135. LINNEMANN, JR., C. C., BUCHMAN, T. G., LIGHT, I. J., BALLARD, J. L., AND ROIZMAN, B., Transmission of herpes simplex virus type I in a nursery for the newborn: Identification of viral isolates by DNA fingerprinting, *Lancet* **1:**964–966 (1978).

136. LONGSON, M., Herpes encephalitis, in: *Clinical Virology* (E. HEALTH, ED.), pp. 73–86, Pittman Medical, London, 1979.

137. LOSSICK, J. G., WHITTINGTON, W., NIGIDA, S., MILLER, R., MAGDE, L., LEE, F. K., AND NAHMIAS, A. J., Sexual transmission patterns of primary genital herpes infection (submitted).

138. MANN, S. L., MEYERS, J. D., HOLMES, K. L., AND COREY, L., Prevalence and incidence of herpesvirus infections among homosexually active men, *J. Infect. Dis.* **149:**1026–1027 (1984).

139. MANSERVIGI, R., AND CASSAI, E., The glycoproteins of the human herpesviruses, *Comp. Immunol. Microbiol. Infect. Dis.* **14:**81–95 (1991).

140. MANZELLA, J. P., McCONVILLE, J. H., VALENTI, E., MENEGUS, M. A., SWIERKOSZ, E. M., AND ARENS, M., An outbreak of herpes simplex virus type 1 gingivostomatitis in a dental hygiene practice, *J.A.M.A.* **252:**2019–2022 (1984).

141. MEIER, J., COREY, L., BURKE, R. L., SAVARESE, B., BARNUM, G., KOST, R., ADAIR, S., DEKKER, C., AND STRAUS, S. E., Immunotherapy of genital herpes with a recombinant herpes simplex virus type 2 glycoprotein D (gD2) vaccine, a placebo-control trial, *Clin. Res.* **41:**199a (1993).

142. MERIGAN, T., Immunosuppression and herpesviruses, in: *The Human Herpesviruses: An Interdisciplinary Perspective* (A. NAHMIAS, W. DOWDLE, AND R. SCHINAZI, EDS.), pp. 309–316, Elsevier-North Holland, New York, 1981.

143. MERTZ, G. J., Herpes simplex virus, in: *Practical Diagnosis of Viral Infections* (G. J. GALASSO, R. J. WHITLEY, AND T. C. MERIGAN, EDS.), pp. 121–130, Raven Press, New York, 1993.

144. MERTZ, G.J., ASHLEY, R., BURKE, R. L., BENEDETTI, J., CRITCHLOW, C., JONES, C. C., AND COREY, L., Double-blind, placebo-controlled trial of a herpes simplex virus type 2 glycoprotein vaccine in persons at high risk for genital herpes infection, *J. Infect. Dis.* **161:**653–660 (1990).

145. MERTZ, G. J., BENEDETTI, J., ASHLEY, R., SELKE, S. A., AND COREY, L., Risk factors for the sexual transmission of genital herpes, *Ann. Intern. Med.* **116:**197–202 (1992).

146. MERTZ, G. J., COOMBS, R. W., ASHLEY, R., JOURDEN, J., REMINGTON, M., WINTER, C., FAHNLANDER A., GUINAN, M., DUCEY, H., AND COREY, L., Transmission of genital herpes in couples with one symptomatic and one asymptomatic partner: A prospective study, *J. Infect. Dis.* **157:**1169–1177 (1988).

147. MERTZ, G. J., SCHMIDT, O., JOURDEN, J. L., GUINAN, M. E., REMINGTON, M. L., FAHLANDER, A., WINTER, C., HOLMES, K. K., AND COREY, L., Frequency of acquisition of first-episode genital infection with simplex virus from symptomatic and asymptomatic source contacts, *Sex. Transm. Dis.* **12:**133–139 (1985).

148. MEYER, H. M., JOHNSON, R. T., AND CRAWFORD, I. P., Central nervous system syndromes of "viral" etiology, *Am. J. Med.* **29:**334–350 (1960).

149. MEYERS, J. D., FLUORNOY, N., AND THOMAS, E. D., Infection with herpes simplex virus and cell mediated immunity after marrow transplant, *J. Infect. Dis.* **142:**338–345 (1980).

150. MEYERS-ELLIOT, J. H., MAXWELL, W. A., PETIT, T. H., O'DAY, D. M., TERASAKI, P. I., AND BERNOCO, D., HLA antigens in herpes stromal keratitis, *Am. J. Ophthalmol.* **89:**54–57 (1980).

151. MINUK, G. Y., AND NICOLLE, L. E., Genital herpes and hepatitis in healthy young adults, *J. Med. Virol.* **19:**269–275 (1986).

152. MONTGOMERY, M. T., REDDING, S. W., AND LEMAISTRE, C. F., The incidence of oral herpes simplex virus infection in patients undergoing cancer chemotherapy, *Oral Surg. Oral Med. Oral Pathol.* **61:**238–242 (1986).

153. MOREIN, B., AND MERZA, M., Vaccination against herpesvirus, fiction or reality? *Scand. J. Infect.* **78:**110–118 (1991).

154. MORRIS, G., COLEMAN, R. M., BESTS, J. M., BENETATO, B. B., AND NAHMIAS, A. J., Persistence of serum IgG and IgA herpes simplex, varicella-zoster, cytomegalo- and rubella viruses detected by enzyme-linked immunosorbent assays, *J. Med. Virol.* **16:**343–349 (1985).

155. MORRISSEAU, P. M., PHILLIPS, C. A., AND LEADBETTER, G. W., JR., Viral prostatitis, *J. Urol.* **103:**767–769 (1970).

156. MUFSON, M. A., WEBB, P. A., AND KENNEDY, H., Etiology of upper respiratory tract illnesses among civilian adults, *J.A.M.A.* **195:**1–7 (1966).

157. MULLER, S. A., HERRMANN, E. C., JR., AND WINKELMANN, R. K., Herpes simplex infections in hematologic malignancies, *Am. J. Med.* **52:**102–114 (1972).

158. NAHMIAS, A. J., Herpesviruses from fish to man a search for pathobiologic unity, *Pathobiol. Annu.* **2:**153–182 (1972).

159. NAHMIAS, A. J., The evolution (evovirology) of herpesviruses, in: *Viruses: Evolution and Cancer* (E. KURSTAK AND K. MARAMOROSCH, EDS.), pp. 605–624, Academic Press, New York, 1974.

160. NAHMIAS, A. J., ADELUSI, B., NAIB, Z., AND MUTHER, J., Changing concepts on the relation of genital herpes and cervical cancer, in: *Cancer of the Uterine Cervix*, Vol. 8 (E. GRUNDMANN, ED.), pp. 141–149, Gustav Fischer Verlag, Stuttgart, New York, 1985.

161. NAHMIAS, A. J., ALFORD, C. A., AND KORONES, S. B., Infection of the newborn with herpesvirus hominis, *Adv. Pediatr.* **17:**185–226 (1970).

162. NAHMIAS, A. J., BROOKS, C., JOHNSON, R., LEE, F. K., PEREORA, L., GRIFFIN, C., REID, E., AND FOREST, E., Distribution of antibodies to herpes simplex viruses (1 and 2) in the United States as measured by a new antibody type-specific assay (submitted).

163. NAHMIAS, A. J., AND DOWDLE, W. R., Antigenic and biologic

differences in herpesvirus hominis, *Prog. Med. Virol.* **10:**110–159 (1968).

164. NAHMIAS, A. J., DOWDLE, W. R., NAIB, Z. M., JOSEY, W. E., MCCLONE, D., AND DOMESCIK, G., Genital infections with *Herpesvirus hominus* type 1 and 2 in children, *Pediatrics* **42:**659–666 (1968).

165. NAHMIAS, A. J., DOWDLE, W. R., NAIB, Z. M., JOSEY, W. E., MCCLONE, D., AND DOMESCIK, G., Genital infection with type 2 herpesvirus hominis—a commonly occurring venereal disease, *Br. J. Vener. Dis.* **45:**294–298 (1969).

166. NAHMIAS, A. J., DOWDLE, W. R., AND SCHINAZI, R. (EDS.), *The Human Herpesviruses. An Interdisciplinary Perspective*, Elsevier-North Holland, New York, 1981.

167. NAHMIAS, A. J., JOSEY, W. E., NAIB, Z. M., LUCE, C., AND DUFFEY, C., Antibodies to herpesvirus hominis types 1 and 2 in humans. I. Patients with genital herpetic infections, *Am. J. Epidemiol.* **91:**539–546 (1970).

168. NAHMIAS, A., JOSEY, W. E., AND OLESKE, J. M., Epidemiology of cervical cancer, in: *Viral Infections of Humans*, 2nd ed. (A. EVANS, ED.), pp. 653–673, Plenum Press, New York, 1982.

169. NAHMIAS, A. J., KEYSERLING, H. K., BAIN, R., BECKER, T., LEE, F., COLEMAN, M., DRAGALIN, D., PEREIRA, L., WICKCLIFFE, C., WELLS, E., PERRY, L., AND MUTHER, J., Prevalence of herpes simplex virus (HSV) type specific antibodies in a USA prepaid group medical practice population (submitted).

170. NAHMIAS, A. J., KEYSERLING, H. K., AND KERRICK, G. M., Herpes simplex, in: *Infectious Diseases of the Fetus and Newborn Infant* (J. REMINGTON AND J. KLEIN, EDS.), pp. 636–678, W. B. Saunders, Philadelphia, 1983.

171. NAHMIAS, A. J., AND KIBRICK, S., Inhibitory effect of heparin on herpes simplex virus, *J. Bacteriol.* **87:**1060–1066 (1964).

172. NAHMIAS, A. J., LEE, F. K., AND BECKMAN-NAHMIAS, S., Sero-epidemiological and -sociological patterns of herpes simplex virus infection in the world, *Scand. J. Infect. Dis.* **69:**19–36 (1990).

173. NAHMIAS, A., MUTHER, J., AND LEE, F., Herpes simplex type 2-A marker of behavioral and biological risk for the acquisition of HIV infection, 4th International Conference on AIDS, Stockholm, 1988.

173a. NAHMIAS, A., AND NORRILD, B., Herpes simplex viruses I and 2-basic and clinical aspects, *Disease-a-Month* **25**(10):5–49 (1979).

174. NAHMIAS, A. J., AND NORRILD, B., The oncogenic potential of herpes simplex viruses and their association with cervical neoplasia, in: *Oncogenic Herpesviruses*, Vol. 2 (F. RAPP, ED.), pp. 25–46, CRC Press, Boca Raton, FL, 1980.

175. NAHMIAS, A. J., AND ROIZMAN, B., Infection with herpes simplex viruses I and 2, *N. Engl. J. Med.* **289:**667–674 719–725 (1973).

176. NAHMIAS, A. J., AND SCHINAZI, R. F., Pediatric herpes simplex virus infection, in: *Current Therapy in Pediatric Infectious Diseases* (J. NELSON, ED.), pp. 153–157, B. C. Decker, Trenton, NJ, 1986.

177. NAHMIAS, A. J., VON REYN, C. F., JOSEY, W. E., NAIB, Z. M., AND HUTTON, R., Genital herpes simplex virus infection and gonorrhoea—association and analogies, *Br. J. Vener. Dis.* **49:** 306–309 (1973).

178. NAHMIAS, A. J., WHITLEY, R., VISINTINE, A., TAKEI, Y., ALFORD, C. A., AND THE COLLABORATIVE ANTIVIRAL STUDY GROUP, Herpes simplex virus encephalitis: Laboratory evaluations and their diagnostic significance, *J. Infect. Dis.* **145:**829–836 (1982).

179. NAHMIAS, A. J., WICKLIFFE, C., PIPKIN, J., LEIBOVITZ, A., AND

HUTTON, R., Transport media for herpes simplex virus types 1 and 2, *Appl. Microbiol.* **22:**451–454 (1971).

180. NAIB, Z. M., NAHMIAS, A. J., JOSEY, W. E., AND ZAKI, S. A., Relation of cytohistopathology of genital herpesvirus infection to cervical anaplasia, *Cancer Res.* **33:**1452–1463 (1973).

181. NELSON, H. G., Epidemic cold sore, *Irish Med. J.* **68:**527–534 (1975).

182. NERURKAR, L. S., WEST, F., MAY, M., MADDEN, D., AND SEVER, J., Survival of herpes simplex virus in water specimens collected from hot tubs in spa facilities and on plastic surfaces, *J.A.M.A.* **22:**3081–3083 (1983).

183. NG, A. B., REAGAN, J. W., AND YEN, S. S., Herpes genitalis—clinical and cytopathologic experience with 256 patients, *Obstet, Gynecol.* **36:**645–651 (1970).

184. NICOLL, J. A., MAITLAND, N. L., AND LOVE, S., Use of the polymerase chain reaction to detect herpes simplex virus DNA in paraffin sections of human brain at necropsy, *J. Neurol. Neurosurg. Psychiatr.* **54:**167–168 (1991).

185. NICOLLE, L. E., MINUK, G. Y., POSTL, B., LING, N., MADDEN, D. L., AND HOOFNAGLE, J. H., Cross-sectional seroepidemiologic study of the prevalence of cytomegalovirus and herpes simplex virus infection in a Canadian Inuit (Eskimo) community, *Scand. J. Infect. Dis.* **18:**19–23 (1986).

186. O'FARRELL, N., HOOSEN, A. A., COETZEE, K. D., AND VAN DEN ENDE, J., Genital ulcer disease in women in Durban, South Africa, *Genitourin. Med.* **67:**322–326 (1991).

187. OLSON, L. C., BUESCHER, E. L., AND ARTENSTEIN, M. S., Herpesvirus infections of the human central nervous system, *N. Engl. J. Med.* **277:**1271–1277 (1967).

188. OPENSHAW, H., AND BENNETT, H. E., Recurrence of herpes simplex virus after dental extraction, *J. Infect Dis.* **146:**707 (1982).

189. PEREIRA, L., CASSAI, E., HONESS, R., ROIZMAN, B., TRERNI, M., AND NAHMIAS, A., Variability in the structural polypeptides of herpes simplex virus strains: Potential application in molecular epidemiology, *Infect. Immun.* **13:**211–220 (1976).

191. PLUMMER, G., Serological comparison of the herpesviruses, *Br. J. Exp. Pathol.* **45:**135–141 (1964).

192. PLUMMER, G., A review of the identification and titration of antibodies to herpes simplex viruses type 1 and type 2 in human sera, *Cancer Res.* **33:**1469–1476 (1973).

193. PORTER, P. S., AND BAUGMAN, R. D., Epidemiology of herpes simplex among wrestlers, *J.A.M.A.* **194:**998–1000 (1965).

194. PRELLNER, T., FLAMHOLC, L., HAIDL, S., LINDHOLM, K., AND WIDELL, A., Herpes simplex virus—the most frequently isolated pathogen in the lungs of patients with severe respiratory distress, *Scand. J Infect. Dis.* **24:**283–292 (1992).

195. PROBER, C. G., COREY, L., BROWN, Z. A., HENSLEIGH, P. A., FRENKEL, L. M., BRYSON, Y. J., WHITLEY, R. J., AND ARVIN, A. M., The management of pregnancies complicated by genital infections with herpes simplex virus, *Clin. Infect. Dis.* **15:**1031–1038 (1992).

196. PROBER, C. G., SULLENDER, W. M., YASUKAWA, L. L., AU, D. S., YEAGER, A. S., AND ARVIN, A. M., Low risk of herpes simplex virus infections in neonates exposed to the virus at the time of vaginal delivery to mothers with recurrent genital herpes simplex virus infections, *N. Engl. J. Med.* **316:**240–244 (1987).

197. PUGH, R. C. B., DUDGEON, J. A., AND BODIAN, M., Kaposi's varicelliform eruption (eczema herpeticum) with typical and atypical visceral necrosis, *J. Pathol. Bacteriol.* **69:**67–80 (1955).

198. RAGA, J., CHRYSTAL, V., AND COOVADIA, H. M., Usefulness of

clinical features and liver biopsy in diagnosis of disseminated herpes simplex infection, *Arch. Dis. Child.* **59:**820–824 (1984).

199. RAND, K. H., POLLARD, R. B., AND MERIGAN, T. C., Increased pulmonary superinfections in cardiac-transplant patients undergoing primary cytomegalovirus infection, *N. Engl. J. Med.* **298:**951–953 (1978).

200. RATTRAY, M. C., COREY, L., REEVES, W. C., VONTVER, L. A., AND HOLMES, K. K., Recurrent genital herpes among women: Symptomatic vs. asymptomatic viral shedding, *Br. J. Vener. Dis.* **54:**262–265 (1978).

201. RAWLS, W. E., Herpes simplex viruses and their role in human cancer, in: *The Herpesviruses*, Vol. 3 (B. ROISMAN, ED.), pp. 241–255, Plenum Press, New York, 1985.

202. RAWLS, W. E., AND CAMPIONE-PICCARDO, J., Epidemiology of herpes simplex virus type I and type 2, in: *The Human Herpesviruses: An Interdisciplinary Perspective* (A. NAHMIAS, W. DOWDLE, AND R. SCHINAZI, EDS.), pp. 137–152, Elsevier-North Holland, New York, 1981.

203. RAWLS, W. E., GARDNER, H. L., FLANDERS, R. W., LOWRY, S. P., KAUFMAN, R. H., AND MELNICK, J. L., Genital herpes in two social groups, *Am. J. Obstet, Gynecol.* **110:**682–689 (1971).

204. RAWLS, W. E., IWAMOTO, K., ADAM, E., AND MELNICK, J. L., Measurement of antibodies to herpesvirus types 1 and 2 in human sera, *J. Immunol.* **112:**728–736 (1974).

205. RIMDUSIT, P., YOOSOOK, C., SRIVANBOON, S., SIRIMONGKOLKASEM, R., AND PUMEECHOCKCHAI, W., Prevalence of genital herpes simplex infection and abnormal vaginal cytology in late pregnancy in asymptomatic patients, *Int. J. Gynecol. Obstet.* **30:**231–236 (1989).

206. RODU, B., TATE, A. L., LAKEMAN, A. F., MATTINGLY, G., RUSSELL, C. M., AND WHITLEY, R. J., Prevalence of herpes simplex virus antibodies in dental students, *J. Dental Educ.* **56:**206–208 (1992).

207. ROIZMAN, B. (ED.), *The Herpesviruses*, Vols. 1–4, Plenum Press, New York, 1982–1986.

208. ROIZMAN, B., AND BAINES, J., The diversity and unity of herpesvirudae, *Comp. Immunol. Microbiol. Infect. Dis.* **14:**63–79 (1991).

209. ROIZMAN, B., DESROSIERS, R. C., FLECKENSTEIN, B., LOPEZ, C., MINSON, A. C., AND STUDDERT, M. J., The family Herpesviridae: An update, *Arch. Virol.* **123:**425–449 (1992).

210. ROIZMAN, B., AND SEARS, A. E., Herpes simplex viruses and their replication, in: *Virology* (B. N. FIELD AND D. M. KNIPE, EDS.), pp. 1795–1841, Raven, New York, 1990.

211. ROIZMAN, B., WARREN, J., THUNING, C. A., FANSHAW, M. S., NORRILD, B., AND MEIGNER, B., Application of molecular genetics to the design of herpes simplex virus vaccines, *Dev. Biol. Stand.* **52:**287–304 (1982).

212. ROSATO, F. E., ROSATO, E. F., AND PLOTKIN, S. A., Herpetic paronychia—an occupational hazard of medical personnel, *N. Engl. J. Med.* **283:**804–805 (1970).

213. ROUSE, B. T., Immunopathology of herpesvirus infections, in: *The Herpesviruses: Immunopathology and Prophylaxis of Human Herpesvirus Infection* (B. ROIZMAN AND C. LOPEZ, EDS.) pp. 103–119, Plenum Press, New York, 1985.

214. ROWLEY, A., WHITLEY, R., LAKEMAN, F., AND WOLINSKY, S., Rapid detection of herpes simplex virus DNA in cerebrospinal fluid of patients with herpes simplex encephalitis, *Lancet* **1:**440–441 (1990).

215. RUSSELL, A. S., AND SCHLANTR, J., HLA transplantation antigens in subjects susceptible to recrudescent herpes labialis, *Tissue Antigens* **6:**257–261 (1975).

216. SAKAOKA, J., SAHEKI, Y., UZUKI, K., NAKAITA, T., SAITO, H.,

217. SEKINE, K., AND FUJINAGA, K., Two outbreaks of herpes simplex virus type I nosocomial infection among newborns, *J. Clin. Microbiol.* **24:**36–40 (1986).

217. SALAHUDDIN, S. Z., ABLASHI, D. V., MARKHAM, P. E., JOSEPHS, S. F., STURZENEGGER, S., KAPLAN, M., HALLIGAN, G., BIBERFELD, P., WONG-STAAL, F., KRAMARSKY, B., AND GALLO, R. C., Isolation of a new virus, HBLV, in patients with lymphoproliferative disorders, *Science* **234:**596–601 (1986).

218. SCHINAZI, R. F., AND NAHMIAS, A. J., Herpes simplex virus infections, in: *Current Therapy in Internal Medicine* (T. M. BAYLESS, M. C. BRAIN, AND R. M. CHERNIACK, EDS.), pp. 126–132, B. C. Decker, Trenton, NJ, 1984.

219. SCHMIDT, O. W., FIFE, K. H., AND COREY, L., Reinfection is an uncommon occurrence in patients with symptomatic recurrent genital herpes, *J. Infect. Dis.* **149:**645–646 (1984).

220. SCHMID, D. S., AND ROUSE, B. T., The role of T cell immunity in control of herpes simplex virus, in: *Herpes Simplex Virus: Pathogenesis, Immunobiology and Control* (B. T. ROUSE, ED.), pp. 57–74, Springer-Verlag, Berlin, 1992.

221. SCHMITT, D. L., JOHNSON, D. W., AND HENDERSON, F. W., Herpes simplex type 1 infections in group day care, *Pediatr. Infect. Dis. J.* **10:**729–734 (1991).

222. SCHNEWEIS, K. E., Serologische untersuchungen zur Typendifferenzierung des Herpesvirus hominis, *Z. Immunitaetsforsch. Exp. Ther.* **124:**24–48 (1962).

223. SCHNEWEIS, K. E., AND NAHMIAS, A. I., Antigens of herpes simplex virus types 1 and 2-immunodiffusion and inhibition passive hemagglutination studies, *Z. Itnmunitaetsforsch. Exp. Klin.* **141:**471–487, 1971.

224. SCOTT, T. F., Epidemiology of herpetic infections, *Am. J. Ophthalmol.* **43:**134–146 (1957).

225. SELLING, B., AND KIBRICK, S., An outbreak of herpes simplex among wrestlers (herpes gladiatorum), *N. Engl. J. Med.* **270:**979–982 (1964).

226. SHANLEY, J. D., AND STANBERRY, L. R., Immunotherapy of persistent viral infections, *Rev Med. Virol.* **4:**105–118 (1994).

227. SHELLEY, W. B., Herpes simplex virus as a cause of erythema multiforme, *J.A.M.A.* **201:**153–156 (1967).

228. SHERIDAN, P. J., AND HERRMAN, E. C., JR., Intraoral lesions of adults associated with herpes simplex virus, *Oral Surg.* **32:**391–397 (1971).

229. SHIP, I. I., BRIGHTMAN, V. J., AND LASTER, L. L., The patient with recurrent aphthous ulcers and the patient with recurrent herpes labialis: A study of two population samples, *Am. Dent. Assoc.* **75:**645–654 (1967).

230. SHIP, I. I., MILLER, M. F., AND RAM, C., A retrospective study of recurrent herpes labialis (RHL) in a professional population 1958–1971, *Oral Surg.* **44:**723–730 (1977).

231. SHIP, I. I., MORRIS, A. L., DUROCHER, R. T., AND BURKET, L. W., Recurrent aplhhous ulcerations and recurrent herpes labialis in a professional school student population. I. Experience, *Oral Surg.* **13:**1191–1202 (1960).

232. SHORE, S., AND NAHMIAS, A., Immunology of herpes simplex virus infection, in: *Immunology of Human Infection* (A. NAHMIAS AND R. O. REILLY, EDS.), pp. 21–72, Plenum Press, New York, 1982.

233. SHORT, S. L., STOCKMAN, D. L., WOLINSKY, S. M., TRUPEI, M. A., MOORE, J., AND REICHMAN, R. C., Comparative rates of sexually transmitted diseases among heterosexual men, homosexual men, and heterosexual women, *Sex. Transm. Dis.* **11:**271–274 (1984).

234. SHUSTER, J. J., KAUFMAN, H. E., AND NESBURN, A. B., Statistical

analysis of the rate of recurrence of herpesvirus ocular epithelial disease, *Am. J. Ophthalmol.* **91:**328–331 (1981).

235. SIEGEL, D., GOLDEN, E., WASHINGTON, E., MORSE, S. A., FULLILOVE, M. T., CATANIA, J. A., MARIN, B., AND HULLEY, S. B., Prevalence and correlates of herpes simplex infections, *J.A.M.A.* **268:**1702–1708 (1992).

236. SKOLDENBERG, B., Herpes simplex encephalitis, *Scand. J. Infect.* **78:**40–46 (1991).

237. SKOLDENBERG, B., JEANSSON, S., AND WOLONTIS, S., Herpes simplex virus type 2 and acute aseptic meningitis: Clinical features of cases with isolation of herpes simplex virus from cerebrospinal fluids, *Scand. J. Infect. Dis.* **7:**227–232 (1975).

238. SKOLDENBERG, B., ALESTIG, K., BURMAN, L., FORKMAN, A., LOVGREN, K., NORRBY, R., STIERNSTEDT, G., FORSGREN, M., BERGSTROM, T., DAHLQVIST, E., FRYDEN, A., NORLIN, K., OLDING-STENKVIST, E., AND UHNOO, I., Acyclovir versus vidarabine in herpes simplex encephalitis, *Lancet* **1:**707–711 (1984).

239. SMITH, I. W., ADAM, E., MELNICK, J. L., AND RAWLS, W. E., Use of the ^{51}Cr release test to demonstrate patterns of antibody response in humans to herpesvirus types 1 and 2, *J. Immunol.* **109:**554–564 (1972).

240. SMITH, I. W., PEUTHERER, J. F., AND MACCALLIUM, F. O., The incidence of herpesvirus hominis antibody in the population, *J. Hyg.* **65:**395–408 (1967).

241. SPEAR, P., Glycoproteins specified by herpes simplex viruses, in: *The Herpesviruses,* Vol. 3 (B. ROIZMAN, ED.), pp. 315–356, Plenum Press, New York, 1985.

242. SPRUANCE, S. L., HAMILL, M. L., HOGE, W. S., DAVIS, L. G., AND MILLS, J., Acyclovir prevents reactivation of herpes labialis in skiers, *J.A.M.A.* **260:**1597–1599 (1988).

243. SPRUANCE, S. L., OVERALL, J. C., JR., KERN, E. R., KRUEGER, G. C., PLIAM, V., AND MILLER, W., The natural history of recurrent herpes simplex labialis: Implications for antiviral therapy, *N. Engl. J. Med.* **297:**69–75 (1977).

244. STAMM, W. E., HANDSFIELD, H. H., RAMPALO, A. M., ASHLEY, R. L., ROBRETS, P. L., AND COREY, L., The association between genital ulcer disease and acquisition of HIV infection in homosexual men, *J.A.M.A.* **260:**1429–1433 (1988).

245. STANBERRY, L. R., Capsaicin interferes with the centrifugal spread of virus in primary and recurrent genital herpes simplex virus infections, *J. Infect. Dis.* **162:**29–34 (1990).

246. STANBERRY, L. R., Herpes simplex virus vaccines, *Semin. Pediatr. Infect. Dis.* **2:**178–185 (1991).

247. STANBERRY, L. R., Pathogenesis of herpes simplex virus infection and animal models for its study, *Curr. Top. Microbiol. Immunol.* **179:**15–30 (1992).

248. STANBERRY, L. R., Genital and neonatal herpes simplex virus infections: Epidemiology, pathogenesis and prospects for control, *Rev. Med. Virol.* **3:**37–46 (1993).

249. STANBERRY, L. R., Animals models and HSV latency, *Semin. Virol.* **5:**213–219 (1994).

250. STANBERRY, L. R., BOURNE, N., BRAVO, F. J., AND BERNSTEIN, D. K., Capsaicin sensitive peptidergic neurons are involved in the zosteriform spread of herpes virus infection, *J. Med. Virol.* **38:**142–146 (1992).

251. STANBERRY, L. R., BURKE, R. L., AND MYERS, M. G., Herpes simplex virus glycoprotein treatment of recurrent genital herpes, *J. Infect. Dis.* **157:**156–163 (1988).

252. STANBERRY, L. R., FLOYD-REISING, S. A., CONNELLY, B. L., ALTER, S. J., GILCHRIST, M. J. R., RUBIO, C., AND MYERS, M. G., Herpes simplex viremia: Report of eight pediatric cases and review of the literature, *Clin. Infect. Dis.* **18:**401–407 (1994).

253. STAVRAKY, K. M., RAWLS, W. E., CHIAVETTA, J., DONNER, A. P., AND WANKLIN, J. M., Sexual and socioeconomic factors affecting the risk of past infections with herpes simplex virus type 2, *Ann. J. Epidemiol.* **118:**109–121 (1983).

254. STEVENS, J. G., HAARR, L., PORTER, D. D., COOK, M. L., AND WAGNER, E. K., Prominence of the herpes simplex virus latency associated transcript in trigeminal ganglia from seropositive humans, *J. Infect. Dis.* **158:**117–122 (1988).

255. STONE, K. M., BROOKS, C. A., GUINAN, M. E., AND ALEXANDER, E. R., National surveillance for neonatal herpes simplex virus infections, *Sex. Transm. Dis.* **16:**152–156 (1989).

256. STONE, K. M., ZAIDI, A., ROSERO-BIXBY, L., OBERLE, M. W., REYNOLDS, G., LARSEN, S., NAHMIAS, A. J., LEE, F. K., SCHACHTER, J., AND GUINAN, M. E., Sexual behavior, STD, and risk of cervical cancer, *Epidemiology* **6:**409–414 (1995).

257. STRAND, A., VAHLNE, A., SVENNERHOLM, B., WALLIN, J., AND LYCKE, E., Asymptomatic virus shedding in men with genital herpes infection, *Scand. J. Infect. Dis.* **18:**195–197 (1986).

258. STRAUS, S. E., ROONEY, J. F., SEVER, J. L., SEIDLIN, M., NUSINOFF LEHRMAN, S., AND CREMER, K., Herpes simplex virus infection: Biology, treatment, and prevention, *Ann. Intern. Med.* **103:**404–419 (1985).

259. STROOP, W. G., AND SCHAEFER, D. C., Production of encephalitis restricted to the temporal lobes by experimental reactivation of herpes simplex virus, *J. Infect. Dis.* **153:**721–731 (1986).

260. SULLIVAN-BOLYAI, J., HULL, H. F., WILSON, C., AND COREY, L., Neonatal herpes simplex virus infection in King County, Washington, *J.A.M.A.* **250:**3059–3062 (1983).

261. SUMAYA, C. V., MARX, J., AND ULLIS, F., Genital infection with herpes simplex virus in a university student population, *Sex. Transm. Dis.* **7:**16–20 (1980).

262. TAMARU, J., MIKATA, A., HORIE, H., ITOH, K., ASAI, K., HONDO, R., AND MORI, S., Herpes simplex lymphadenitis. Report of two cases with review of the literature, *Am. J. Surg. Pathol.* **14:**571–577 (1990).

263. TEJANI, N., KLEIN, S. W., AND KAPLAN, M., Subclinical herpes simplex genitalis infections in the perinatal period, *Ann. J. Obstet. Gynecol.* **135:**547–556 (1979).

264. TIGGES, M. A., KOELLE, D., HARTOG, K., SEKULOVICH, R. E., COREY, L., AND BURKE, R. L., Human CD8$^+$ herpes simplex virus-specific cytotoxic T-lymphocyte clones recognize diverse virion protein antigens, *J. Virol.* **66:**1622–1634 (1992).

265. TSUTSUMI, H., BERNSTEIN, J. M., RIEPENHOFF-TALTY, M., AND OGRA, P. L., Immune responses to herpes simplex virus in patients with recurrent herpes labialis. II. Relationship between interferon production and cytotoxic responses, *Pediatr. Res.* **20:**905–908 (1986).

266. UREN, E. C., JOHNSON, P. D. R., MONTANARO, J., AND GILBERT, G. L., Herpes simplex virus encephalitis in pediatrics: Diagnosis by detection of antibodies and DNA in cerebrospinal fluid, *Pediatr. Infect. Dis.* **12:**1001–1006 (1993).

267. VANDERHOOFT, S., AND KIRBY, P., Genital herpes simplex virus infection: Natural history, *Semin. Dermatol.* **11:**190–199 (1992).

268. VAN DYKE, R. B., AND SPECTOR, S. A., Transmission of herpes simplex virus type 1 to a newborn infant during endotracheal suctioning for meconium aspiration, *Pediatr. Infect. Dis.* **3:**153–156 (1984).

269. WARREN, K. G., WROBLEWSKA, Z., OKABE, H., BROWN, S. M., GILDEN, D. H., KOPROWSKI, H., RORKE, L. B., SUBAK-SHARPE, J., AND YONEZAWA, T., Virology and histopathology of the trigeminal ganglia of Americans and Japanese, *Can. J. Neurol. Sci.* **5:**425–430 (1978).

270. WEATHERS, D. R., AND GRIFFIN, J. W., Intraoral ulcerations, recurrent herpes simplex, and recurrent aphthae: Two distinct clinical entities, *J. Am. Dent. Assoc.* **81:**81–87 (1970).

271. WHEELER, JR., C. E., AND CABANISS, JR., W. H., Epidemic cutaneous herpes simplex in wrestlers (herpes gladiatorum), *J.A.M.A.* **194:**993–997 (1965).

272. WHITLEY, R. J., Herpes simplex virus infection, in: *Infectious Diseases of the Fetus and Newborn Infant* (S. J. REMINGTON AND J. O. KLEIN, EDS.), pp. 282–305, Saunders, Philadelphia, 1990.

273. WHITLEY, R. J., ALFORD, C. A., HIRSCH, M. S., SCHOOLEY, R. T., LUBY, J. P., AOKI, F. T., HANLEY, D., NAHMIAS, A. J., SOONG, S. J., AND THE NIH-NIAID COLLABORATIVE ANTIVIRAL STUDY GROUP, Herpes simplex encephalitis—vidarabine versus acyclovir therapy, *N. Engl. J. Med.* **324:**144–149 (1986).

274. WHITLEY, R., ARVIN, A., PROBER, C., BURCHETT, S., COREY, L., POWELL, D., PLOTKIN, S., STARR, S., ALFORD, C., CONNOR, J., JACOBS, R., NAHMIAS, A., SOONG, S-J., AND THE NATIONAL INSTITUTE OF ALLERGY AND INFECTIOUS DISEASES COLLABORATIVE ANTIVIRAL STUDY GROUP, A controlled trial comparing vidarabine with acyclovir in neonatal herpes simplex virus infection, *N. Engl. J. Med.* **324:**444–449 (1991).

275. WHITLEY, R., ARVIN, A., PROBER, C., COREY, L., BURCHETT, S., PLOTKIN, S., STARR, S., JACOBS, R., POWELL, D., NAHMIAS, A., SUMAYA, C., EDWARDS, K., ALFORD, C., CADDELL, G., SOONG, S-J., AND THE NATIONAL INSTITUTE OF ALLERGY AND INFECTIOUS DISEASES COLLABORATIVE ANTIVIRAL STUDY GROUP, Predictors of morbidity and mortality in neonates with herpes simplex virus infections, *N. Engl. J. Med.* **324:**450–454 (1991).

276. WHITLEY, R. J., AND GNANN, J. W., Acyclovir: A decade later, *N. Engl. J. Med.* **327:**782–789 (1992).

277. WHITLEY, R. J., LAKEMAN, F., NAHMIAS, A. J., AND ROIZMAN, B., DNA restriction enzyme analysis of herpes simplex virus isolates from patients with encephalitis, *N. Engl. J. Med.* **307:**1060–1062 (1982).

278. WHITLEY, R. J., NAHMIAS, A. J., AND THE NIAID COLLABORATIVE ANTIVIRAL STUDY GROUP, Therapeutic challenges of neo-natal herpes simplex virus infection, *Scand. J. Infect. Dis.* (Suppl.) **47:**97–106 (1985).

279. WHITLEY, R. J., NAHMIAS, A. J., VISINTINE, A. M., FLEMING, C. L., AND ALFORD, C. A., The natural history of herpes simplex virus infection of mother and newborn, *Pediatrics* **66:**489–494 (1980).

280. WHITLEY, R., SOONG, S-J., DOLIN, R., GALASSO, G. J., CH'IEN, L. T., ALFORD, C. A., AND THE NATIONAL INSTITUTE OF ALLERGY AND INFECTIOUS DISEASES COLLABORATIVE ANTIVIRAL STUDY GROUP, Adenine arabinoside therapy of biopsy-proved herpes encephalitis, *N. Engl. J. Med.* **297:**289–294 (1977).

281. WU, L., AND MORAHAN, P. S., Macrophages and other non-specific defenses: Role in modulating resistance against herpes simplex virus, in: *Herpes Simplex Virus: Pathogenesis, Immunobiology and Control* (B. T. ROUSE, ED.), pp. 89–110, Springer-Verlag, Berlin, 1992.

282. YAMAMOTO, Y., A re-evaluation of the skin test of herpes simplex virus, *Jpn. J. Microbiol.* **10:**67–77 (1966).

283. YAMAMOTO, L. J., TEDDER, D. G., ASHLEY, R., AND LEVIN, J. M., Herpes simplex virus type 1 DNA in cerebrospinal fluid of a patient with Mollaret's meningitis, *N. Engl. J. Med.* **23:**1082–1085 (1991).

284. YEAGER, A. S., ASHLEY, R. L., AND COREY, L., Transmission of herpes simplex virus from father to neonate, *J. Pediatr.* **103:**905–907 (1983).

285. YOUNG, S. K., ROWE, N. H., AND BUCHANAN, R. A., A clinical study for the control of facial mucocutaneous herpes virus infections. I. Characterization of natural history in a professional school population, *Oral Surg.* **41:**498–507 (1976).

286. ZIMMERMAN, T. J., MCNEIL, J. T., RICHMAN, A., KAUFMAN, H. E., AND WALTMAN, S., HLA types and recurrent corneal herpes simplex infection, *Invest. Ophthalmol.* **16:**756–757 (1977).

287. ZUR HAUSEN, H., Human genital cancer: Synergism between two virus infections or synergism between a virus infection and initiating events? *Lancet* **2:**1370–1372 (1982).

CHAPTER 15

Human Herpesvirus-6 and Human Herpesvirus-7

Paul H. Levine

1. Introduction

Human herpesvirus type 6 (HHV-6) and human herpesvirus type 7 (HHV-7) are two of the most recently characterized viruses in the group that includes herpes simplex virus 1 (HHV-1), herpes simplex virus 2 (HHV-2), varicella-zoster (HHV-3), Epstein–Barr virus (EBV) (HHV-4), cytomegalovirus (CMV) (HHV-5), and the Kaposi's sarcoma-associated herpesvirus HHV-8.[33a,97a] The impact of HHV-6 and HHV-7 on the human population is largely unknown, although HHV-6 is generally accepted as the primary cause of roseola or exanthem subitum,[158,172] may be the most common pathogenic virus infecting infants and young children,[123,126] and also is etiologically linked to a number of illnesses (see Section 8.1).

This chapter deals primarily with the patterns of infection currently associated with these two viruses, the evidence linking them to specific diseases, and the attempts to characterize the potential importance of strain differences as well as possible population differences that could affect the interpretation of new studies involving these agents.

2. Historical Background

The detection of HHV-6 was first reported in 1986 after its isolation from the peripheral blood cells of six patients with immunosuppressive disorders and/or lymphoproliferative disease.[135] As part of an effort focused

on the pathogenesis of human immunodeficiency virus-1 (HIV-1), evidence of this new herpesvirus was first noted in late 1984 with the observation that the leukocytes from certain immunosuppressed patients underwent "ballooning" and the formation of syncytia, properties that could be transmitted by cell-free culture supernate and other biological methods to cord blood leukocytes.[134] The cytotoxicity of this agent prevented the maintenance of a continuous cell line, but within 2 years it was apparent that this new agent was a previously unidentified herpesvirus. Initially designated human B-lymphotropic virus (HBLV) because of its apparent effect on B lymphocytes, the identification of T cells as the primary virus target and the subsequent infection of a variety of target cells, including megakaryocytes, macrophages, epithelial cells, fibroblast cells, and glioblastoma cells,[2,3,42,111,145,167] led to its subsequent designation as HHV-6.

Shortly thereafter, HHV-7 was isolated from CD4$^+$ T cells obtained from the blood and saliva of a healthy individual[62,168] and from the blood of a patient with chronic fatigue syndrome (CFS).[16,17] The exposure of cells to conditions leading to T-cell activation was apparently important to the process of isolating HHV-7.[63] Comparative analysis of DNA[16,65] and serological studies[169] have shown that HHV-6 and HHV-7 are distinct viruses.

3. Methodology

3.1. Mortality

None of the illnesses related to HHV-6 or HHV-7 are reportable diseases, and therefore the information de-

Paul H. Levine • Viral Epidemiology Branch, National Cancer Institute, Bethesda, Maryland 20892.

scribed below is derived solely from the literature. HHV-6 has been documented to cause fatal disease,[10,11,124,146] but in view of the ubiquity of this virus a fatal outcome of infection is rare.

3.2. Morbidity

Quantitative data on morbidity for HHV-6 and HHV-7 are not available at the present time, although descriptions of case series indicate that HHV-6 is a frequent cause of febrile illness in the pediatric population,[123,126] and HHV-7 is also increasingly associated with clinical disease.[85,,97a,160,162a] Thus far, HHV-7 has only rarely been associated with clinical illness.[85,160]

3.3. Surveys

Numerous serological surveys have been reported for HHV-6[22,23,36,53,91,102,104,105,119,121,138] but few for HHV-7.[37,97a,169,173] Although methodological differences have affected the interpretation of prevalence in different study groups, it is apparent that in Western countries infections with HHV-6 and HHV-7 usually occur in the first 2 years of life. Outbreaks of exanthem subitum, the most common illness etiologically associated with HHV-6, have been reported in hospitals and other group settings,[21,38,83] but these were prior to the discovery of HHV-6 and some may be due to other organisms such as coxsackieviruses. To date, systematic household- or population-based surveys of HHV-6 or HHV-7 have not been conducted and information on the epidemiology remains incomplete.

3.4. Laboratory Diagnosis

3.4.1. Isolation and Identification of the Viruses. Isolation of HHV-6 and HHV-7 from biological samples is most readily accomplished by exposing them to cord blood lymphocytes stimulated with phytohemagglutin or interleukin 2 (IL-2).[63] Identification of the virus is usually performed by polymerase chain reaction (PCR), which can distinguish between HHV-6, HHV-7, and the closely related CMV.[16,96] Monoclonal antibodies can also be used to detect these viruses in infected cells.[57]

3.4.2. Serological Tests. Virtually all data describing patterns of HHV-6 and HHV-7 infection have been derived from studies using antigens prepared from cell lines containing replicating virus. The most readily detected antigen is comparable to the viral capsid antigen of EBV, which was first detected by the Henles[68] using immunofluorescence. The antigen for this assay is derived

from replicating virus, which also produces early antigen, but because antibody titers to early antigen are generally lower than viral capsid antigen antibody titers, results of the test are reported for viral capsid antigen antibodies alone. In addition to the immunofluorescence assay (IFA), an enzyme-linked immunosorbent assay (ELISA), a circle immunoassay (CIA), and a competitive radio immunoassay (RIA) have been utilized in measuring antibodies to HHV-6.[39,138] In a comparison of the IFA, CIA, and RIA assays applied to sera from blood donors, Coyle et al.[39] found a good correlation among the three, although the IFA was less sensitive than the other two. The antigens for all three assays were prepared from J-JHAN cells and presumably are comparable. Concordance between IFA and the ELISA assays has not been so complete, at least in part because the studies have included both patients with abnormal HHV-6 reactivity and healthy donors. In the U.S. population, the general pattern of reactivity based on the ELISA assay has shown similar patterns on a population basis,[138] with strong reactivity in young children and a decline in older age groups. Marked differences, however, were observed between ELISA and immunofluorescence test results in various patients groups,[98,99] probably because different antigens were used. In one study, for example, the ELISA assay using purified virus as antigen correlated well with the IFA in untreated patients with Hodgkin's disease, patients accomplishing long-term remissions, and healthy controls. When disease recurred, however, antibody levels as detected by IFA rose, in contrast to the antiviral antibody detected by ELISA, which remained stable.

4. Biological Characteristics Affecting the Epidemiologic Pattern

HHV-6 and HHV-7, unlike EBV, have a predilection for T lymphocytes rather than B lymphocytes. HHV-6 has a tropism for both $CD4^+$ and $CD8^+$ T cells[111] and has recently been demonstrated to infect natural killer cells.[110] Although studies on the spectrum of cells readily infected by these viruses are limited, as noted previously in vitro studies have shown that HHV-6 can infect megakaryocytes, macrophages, epithelial cells, fibroblast cells, and glioblastoma cells.[2,42,111,145,167] These in vitro findings are supported by clinical observations documenting infection of a wide variety of tissues including liver, lung, and retina (see Section 8.1). Two strains of HHV-6 (A and B) have been identified on the basis of their tropism for different cell lines, their reactivity with various monoclonal antibodies, and their restriction enzyme patterns.[4,33,139,167]

The molecular characterization of HHV-6 and HHV-7, which has been described in detail elsewhere,[16,96] indicates that there is less than 40% homology between these two viruses.[16] The molecular divergence between HHV-6 and HHV-7 is less than that between CMV and these two viruses,[16] and it is apparent that the two strains of HHV-6 (A and B) are more closely related to each other than to HHV-7.[16] Biological differences between the A and B strains of HHV-6 is an area of active research, but biological properties of the two protypes are significantly different.[4,63,139] The prototype isolates primarily used to study group A characteristics have been GS[135] and U1102[45] and the prototype for group B is Z-29.[107] Although the original isolates from group A were from patients in the United States[135] and the Z-29 prototype was isolated from an AIDS patient in Zaire,[107] distinctive geographic patterns have not been identified for the two strains. At the present time, the evidence suggests that variant B is the major etiologic agent of exanthem subitum. It was the only variant identified in 97% of 76 infants with symptomatic primary HHV-6 infections.[43] No consistent association with disease has been found for HHV-6 A. It has been noted that HHV-6 A infection is rare in infants but is the most common strain isolated from adults,[2,13,81,135] with the reason for this age discordancy being unexplained.

Latency, a common feature of herpesvirus, may be important in the pattern of diseases associated with HHV-6. For example, activation of latent HHV-6 is believed by some to be involved in chronic fatigue syndome (CFS)[25] and posttransplantation pneumonitis,[31] among other illnesses (see Section 8.1).

5. Descriptive Epidemiology

5.1. Prevalence and Incidence

HHV-6 and HHV-7 are ubiquitous viruses that usually infect infants and children in the United States at an earlier age than EBV. These agents frequently infect children in the first year of life, shortly after the decline of maternal antibody; the peak incidence of exanthem subitum (ES), the sentinel disease for HHV-6 infection, is from 7 to 13 months of age. Early studies from more than 13 countries[94] suggested more than two thirds of the population was infected with HHV-6 worldwide and that the incidence of infection was highest in the first 2 years of life (Fig. 1). More recent studies support the ubiquity of HHV-6 in most populations, but racial–ethnic–geographic variation has been observed (see Section 5.3), suggesting

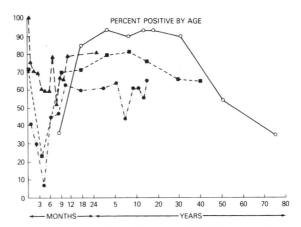

Figure 1. Prevalence of IgG antibodies to HHV-6 by immunofluorescence assay. (From ref. 102 with permission.)

substantial deficiency in our knowledge of the prevalence and incidence of infection outside the developed countries.

Infection with HHV-7 is believed to occur at a slightly later age than with HHV-6,[169] a finding supported by studies in Japan where the highest prevalence (60%) was noted at ages 11–13[173] and where clinical infection with HHV-7 was observed in children older than those with symptomatic HHV-6 infection.[162a] A recent hospital-based series from the United Kingdom showed a parallel age-specific prevalence for HHV-6 and HHV-7,[37] however, and larger representative studies of the general population in a number of countries is needed to allow definitive observations on the epidemiology of HHV-7 other than its high prevalence in the general population.

Both HHV-6 and HHV-7, as with other herpesviruses, remain latent after primary infection and reactivation of HHV-6 during a variety of illnesses has been well documented.[25,31]

5.2. Epidemic Behavior and Contagiousness

The epidemiology of ES reflects that of HHV-6 infection, since this childhood disease is the most recognizable and consistent outcome of HHV-6 infection. Although reported outbreaks of ES, especially those occurring in adults, are most probably due to agents other than HHV-6, certain patterns of infection are suggested by early studies. Exanthem subitum is the most common exanthem seen in infants under 2 years of age. The young age of primary infection documented by the early acquisition of antibody supports the epidemiologic inference that virus transmission is related to the number of close contacts.

For example, a year-round occurrence has been observed,[177] but literature reviews suggest the lowest incidence has been in December[21,38,83] when children have been more isolated. Close contact appears to favor onset of disease, since outbreaks occur where infants share close quarters,[15,40,80] and the advent of day-care centers may be increasing virus transmission in developed countries. The disease is rarely seen after the age of 4 years,[35] when the prevalence of protective antibody is high. The striking similarity between the age-specific incidence of ES and the age at which HHV-6 seroconversion occurs has provided supportive evidence for the role of the virus in this disease. The disease has been reported in adults during an outbreak,[80] but other agents such as coxsackie and echoviruses can make a similar presentation.[35] Because not all cases of pediatric illness due to HHV-6 are associated with rash,[126,152] studies of undefined viral illnesses in children, as well as outbreaks of "ES" affecting adults (indeed, all studies of ES) should now rely on specific serological markers. The incubation period in epidemics has been reported to be 5–15 days,[15,21,40] and there appears to be no predilection by sex.[35]

In the absence of well-documented associated clinical disease, data regarding transmission, incubation periods, and other epidemiologic features are not available for HHV-7.

5.3. Geographic Distribution

Information regarding patterns of infection outside of Japan and the Western world is limited, but geographic and racial–ethnic differences in response to HHV-6 infection have been documented.[26,102,104,129] Not only are there variations in the seroprevalence of HHV-6 by age in different countries and different racial–ethnic groups, but there are also marked variations in the immune response to HHV-6 infection, as measured by antibody titers in seropositive people, in different populations of both healthy and ill individuals. An early study showed lower HHV-6 titers in pregnant Moroccan women compared with those in Sub-Saharan African countries such as Ivory Coast and Congo.[129] Subsequently, a controlled study of healthy young adults living in three diverse geographic locations (Ghana, the United States, and Malaysia)[104] demonstrated that levels of antibody detected by immunofluorescence in Ghanaians were significantly higher than among seropositive donors living in Malaysia; seropositive U.S. donors were intermediate in antibody titers. Significant titer differences among African countries[104,129] provide opportunities to investigate possible environmental and genetic factors contributing to HHV-6 antibody titers.

Information on these geographic and racial–ethnic differences may also be extended through investigation of the possible role of strain differences.

6. Mechanism and Routes of Transmission

HHV-6 and HHV-7 have been identified in salivary gland tissue and saliva,[20,61,69,105] as well as the oropharynx of healthy individuals,[64] and salivary transmission of HHV-6[105] generally has been accepted as the primary route of infection. HHV-6 has also been found in vaginal secretions,[97] which may explain the congenital infection that has been noted as a rare occurrence by Dunne and Demmler.[50] These investigators noted IgM antibody to HHV-6 in 2 (0.28%) of 799 cord bloods randomly selected from a cohort of 3899 samples in which a somewhat larger (0.4%) prevalence of congenital infection with CMV had previously been documented. Kempe et al.[87] were able to produce febrile illnesses in controlled experiments by injecting serum or inoculating throat washings from patients with ES intranasally into rhesus monkeys.

7. Pathogenesis and Immunity

The incubation period for HHV-6-related disease has been best studied during epidemics of ES, where incubation periods of 5–15 days have been noted.[15,21,40] Kempe et al.[87] have reported a 9-day incubation period in a direct transmission experiment using serum from an 18-month-old infant for intravenous inoculation into a 6-month-old susceptible host, but since HHV-7 also causes ES,[160,162a] the specificity of the study for HHV-6 is uncertain.

Immunity appears to be lifelong, with antibody persisting through adulthood; however, HHV-6 can be reactivated by other viral infections or clinical illness,[153] and reactivation has been associated with severe or even fatal illnesses.[31,47] While less well studied, HHV-7 shares many of the pathogenic features of HHV-6 with similar patterns of immunity and reactivation.[97a]

8. Patterns of Host Response

There is a wide spectrum of clinical manifestations associated with primary HHV-6 infection and considerable evidence for other severe clinical disorders, such as fatal pneumonitis or cancer, resulting from reactivation or cell transformation years after primary infection. The classic approach to defining an infectious etiology for a disease has been the Henle–Koch postulates, but modifications have been suggested by Evans,[54] and with newer techniques consistently improving viral detection, the criteria for establishing causation need to be constantly reevaluated. In the absence of an observed disease pattern following purposeful infection with an agent, the most direct evidence for causation is seroconversion to a single agent in association with disease onset. Using this criterion, it is now apparent that HHV-6 is the primary cause of ES, but it is also an occasional cause of infectious mononucleosis and infectious hepatitis. Reactivation of HHV-6, which is much more difficult to document, has been suggested as precipitating CFS,[25,103] Rosai-Dorfman disease,[101] Kikuchi's disease,[88] posttransplantation pneumonitis,[31] and other illnesses (Section 8.1), with the production of IgM antibody being used as the primary evidence for viral pathogenesis. More recently, etiologic studies have emphasized the detection of virus in disease tissue. The PCR assay has been useful because of its sensitivity, and Southern blot has been widely used for improved specificity; but in contrast to *in situ* hybridization neither of these assays indicate which cells are infected, important information for distinguishing an etiologic from a bystander role. In considering the disease associations, therefore, it is important to realize that while HHV-6 produces a wide spectrum of clinical manifestations, ES remains the only disease where HHV-6 is the principal etiologic agent.

8.1. Clinical Features

8.1.1. Illnesses Associated with Seroconversion.
Exanthem subitum is an illness characterized by transient high fever, often to 103°F, and a discrete maculopapular rash. Initially linked to HHV-6 infection by Yamanishi *et al.*,[172] ES is now widely accepted as the most readily identifiable HHV-6-induced illness. Because of its ubiquity, the virus frequently infects infants and apparently causes a variety of acute self-limiting illnesses without the classic features of ES.[67,126,152] Neurological manifestations of HHV-6 have been noted,[47,93,113,136,171,174] and clinical features of meningoencephalitis have presumably been due to direct infection of brain tissue by the virus.[174] In a fatal encephalitis illness following primary HHV-6 infection,[10] however, neither viral antigen nor inflammatory cells could be detected. The investigators in this case surmised that the neurological findings were due to an immunologic phenomenon.

Although ES is the most frequently recognized acute illness associated with primary HHV-6 infection, unusual complications have also been reported. For example, a 6-month-old girl with fever and exanthema associated with seroconversion to HHV-6 developed transient anemia, granulocytopenia, and liver dysfunction with "intralobular spotty necrosis" noted on a liver biopsy. She made a complete recovery after receiving subcutaneous interferon-α.[159] Two other illnesses that are occasionally associated with HHV-6 seroconversion are heterophile-negative mononucleosis[148] and hepatitis.[11,49,78,146,157] Initial reports documented IgM antibody to HHV-6 in EBV-negative cases of infectious mononucleosis,[18,78,148] and current estimates are that 5–10% of EBV-seronegative infectious mononucleosis may result from primary infection with HHV-6 at an older age. Hepatitis appears to be a more frequent outcome of HHV-6 infection and can be observed at all ages with varying degrees of severity, ranging from self-limiting disease in adults[78] to a fatal complication of childhood ES.[11]

Several important case reports have documented unusual illnesses during primary infection with HHV-6, which may, on occasion, have been the etiologic agent. A case of Kawasaki disease[66] was described in a 4-month-old girl with an initial clinical illness suggestive of ES. The patient was initially seronegative to HHV-6, the virus was cultured from an admission blood sample, and HHV-6 antibody appeared in convalescent serum. These findings could be explained at least partially by gamma globulin therapy (which probably included HHV-6 antibody-positive sera). Since the authors were unable to isolate HHV-6 from five other patients with Kawasaki disease and noted other cases in which measles virus[165] and parainfluenza virus[86] were associated with Kawasaki disease, they considered the HHV-6 infection incidental. Until a single Kawasaki agent is identified, however, the possibility remains that Kawasaki disease is an unusual response that can be triggered by any of several infectious agents, including HHV-6. A similar mechanism is postulated for CFS, another HHV-6-associated disease (Section 8.12).

Another unusual outcome of HHV-6 infection is hemophagocytic syndrome,[77] a frequently fatal illness triggered by a number of infectious agents, particularly herpesviruses, and characterized by an aggressive proliferation of activated macrophages throughout the lymphoreticular system.[48] Often associated with immunosuppression,[48] this illness was reported in an 8-month-old female who had a history of fever and a 3-day rash 3 weeks prior to admission for hepatomegaly, pancytopenia, and a bleeding diathesis. Rising antibodies to HHV-6 and virus isolation from cultured mononuclear cells in the absence of EBV, CMV, and adenovirus antibodies strongly suggested HHV-6 as the etiologic agent.

This disorder was initially described as a benign histiocytic proliferation.[132] Nonviral infections (bacterial, fungal, mycobacterial, rickettsial, parasitic) as well as herpes simplex, VZV, EBV, CMV, and adenovirus have been implicated as etiologic agents.[48,125,130]

The diverse possible outcomes of primary infection with HHV-6 may best be illustrated by the fatal outcome in an apparently immunocompetent 13-month-old Vietnamese girl who died of multisystem disease, the terminal event being cardiac failure.[124] Intranuclear viral inclusions were reportedly found in atypical lymphoid cells infiltrating multiple organs, including the heart, lungs, liver, spleen, lymph nodes, bone marrow, thymus, kidney, bladder, gastrointestinal tract, salivary glands, middle ear, peripheral nerve, and skeletal muscle.

HIV-7-associated diseases have been documented more frequently as reagents for identification become increasingly available. Recent reports include a possible etiologic relationship to a variety of diseases, including infectious mononucleosis, ES, and hepatitis.[85,97a,160,162a] Reports of new HHV-6 and HHV-7-associated illnesses can be expected in the near future.

8.1.2. Illnesses Associated with Elevated Antibody Levels and Virus Detection.

Elevated herpesvirus IgM antibody levels, formerly considered to be documentation of a primary viral infection, have now been well documented in virus reactivation.[149] Since few people (outside of China and other countries where HHV-6 infection may be delayed) are uninfected by age 5 years, most adults with IgM antibodies to HHV-6 can be presumed to have reactivation rather than primary infection. IgM antibodies have also been taken to support the diagnosis of a prolonged infectious disease with lymphadenopathy due to HHV-6 in EBV-negative adults.[116] Because IgM antibodies suggest active viral replication,[154] medical complications associated with IgM antibodies to HHV-6 are included in this section. Reactivation of HHV-6 in other viral infections, such as measles,[153] may occur even more easily than reactivation of EBV, thus allowing HHV-6 to play a secondary role in pathogenesis.

A possible role for HIV-6 in abortion was noted by Ando et al.[9] and Kakimoto et al.,[84] who investigated 100 primagravida women and 30 patients with spontaneous abortion in weeks 6–12 of pregnancy. Three of 30 patients had IgM antibody to HHV-6, and viral antigen was detected in aborted villous tissue by an HHV-6 monoclonal antibody, OHV-I, in two of these three patients, suggesting to the investigators that infection or reactivation of HHV-6 had a causative role in spontaneous abortion.

In addition to having a possible adverse effect on pregnancy, which can be considered a state of partial immunosuppression, HHV-6 appears to be an important pathogen in patients with acquired immunodeficiency disease (AIDS) and in patients undergoing bone marrow transplantation. In one series of nine unselected AIDS patients, HHV-6-infected cells were detected in multiple organs associated with lymphocyte infiltrates and probably caused fatal pneumonitis in one patient.[92] Fatal bone marrow suppression has also been observed,[31] and an *in vitro* model for suppression of bone marrow progenitor cells has been evaluated.[91a] Interstitial pneumonitis associated with HHV-6 isolation was observed in two patients, one receiving bone marrow transplantation for testicular carcinoma and the other with myelodysplastic–preleukemia syndrome. In both patients, immunohistochemical staining of involved lung tissue from both patients stained positively for HHV-6, and attempts to identify other possible etiologic agents, including CMV, were negative. Both patients were seropositive for HHV-6 prior to transplantation, and therefore reactivation was presumed to be responsible for the pneumonitis in these two patients. HHV-6 has also been reported to cause fatal encephalitis in a bone marrow recipient with Hodgkin's disease who was seropositive prior to treatment.[47]

Although HHV-6 may cause symptoms in renal transplant recipients, the consequences appear to be less severe. Yoshikawa et al.[175,176] prospectively studied 65 kidney transplant recipients and their donors and noted viremia in nine (14%) of the recipients 2–4 weeks posttransplantation with a significant rise in antibody titer in an additional 27 (42%). All of the recipients were immune to HHV-6 prior to transplantation and virus could not be recovered more than 3 weeks after transplantation, documenting the importance of reactivation as a posttransplantation phenomenon. Skin rashes and fever were noted in approximately one third of the patients at the time of virus isolation.[175] Merlino et al.[114] also noted HHV-6 reactivation during renal transplantation but no apparent impact of reactivation on the success of the kidney transplant.

Other clinical observations associated with the appearance of IgM antibody to HHV-6 include the transformation of an indolent large granular lymphocytic leukemia to an aggressive and ultimately fatal phase.[156] A 75-year-old female had been followed for more than 1 year and her transformation was associated with a conversion of IgM anti-HHV-6 from <1 : 4 to 1 : 120, a rise in IgG anti-HHV-6 from 1 : 1,640 to 1 : 20,000, and detection of HHV-6-specific sequences in the DNA of peripheral blood cells by PCR. The specific role of HHV-6 in this transformation remains speculative, however, since reactivation could have been a secondary effect of the transformation.

Some seroepidemiologic studies initially have linked a number of illnesses to HHV-6 primarily on the basis of elevated IgG antibody titers to the viral capsid antigen. These include sarcoidosis,[19] leukemia,[1] Hodgkin's disease (HD),[36] and Kawasaki disease.[118] As with EBV, however, elevated HHV-6 antibody levels can result from immune dysregulation. Therefore, high antibody levels by themselves, as seen in healthy Ghanaians,[104] for example, are insufficient to support an etiologic relationship between HHV-6 and a specific disease. Following early reports of elevated antibody titers in HD[36] and leukemia,[1] more detailed studies were initiated to determine the consistency and the clinical applications of these apparent associations. In HD, a longitudinal study indicated that the elevated antibody titers developed following immunosuppressive therapy, with untreated patients having similar titers to healthy age- and sex-matched controls.[99] Antibody titers are of potential value clinically in this malignancy, however, since viral capsid antigen levels appeared to fall in patients with long-term remissions but not those with relapses. A case-control study of children with acute leukemia found no antibody elevation in the leukemia patients,[98] suggesting that the earlier reports of increased titers reflected the young age of the leukemia patient population compared to other patient groups in the study.[22,23]

The association of high antibody titers and Kawasaki disease, as discussed above, is more likely to be a manifestation of immune dysregulation than etiology because of the infrequent isolation of HHV-6 from active cases.[66] A similar situation may be responsible for the HHV-6 pattern in sarcoidosis. Biberfeld et al.[19] reported detection of HHV-6 genome in one of five involved lymph nodes from sarcoidosis patients in addition to increased antibody titers, and suggested that sarcoidosis might be an unusual response to more than one infectious agent, including HHV-6. In a study of 138 patients with hematologic malignancies, 127 with oral cancer, and 75 healthy controls in India, Shanavas et al.[141] found an antibody pattern generally similar to that reported elsewhere for all groups except the oral cancer patients. More than half of the oral cancer patients had antibody titers >1 : 640, and the entire group had a geometric mean titer of 1 : 2,042 [in comparison to 1 : 1,208 in acute lymphocytic leukemia (ALL) patients, 1 : 1,109 in HD, 1 : 88 in non-Hodgkin's lymphoma (NHL), and 1 : 47 in normal controls]. No information was provided on clinical status, and since abnormalities of cellular immunity had previously been noted in the patient group,[137] these findings may reflect incidental reactivation such as that reported in HD.[99]

Identification of HHV-6 in involved tissues has been useful in pursuing the leads generated by serological studies. Detection of virus has been reported in tissues obtained for patients with posttransplantation pneumonitis,[29,31] sinus histiocytosis with massive lymphadenopathy (SHML, or Rosai-Dorfman disease),[101] Kikuchi's disease,[88,155] sarcoidosis,[19] cervical cancer,[34] oral cancer,[170] and occasional cases of lymphoproliferative disorders including lymphoma[65,81,82,94,108,112,142,161,162] (see Table 1). The significance of these findings should be interpreted in relationship to the location of the virus.

As noted above, a pathogenetic role in posttransplantation pneumonia has been strongly supported by immunohistochemical staining of lung tissue samples obtained at autopsy of patients with terminal interstitial pneumonia following bone marrow transplantation.[29,31] In Rosai-Dorfman disease, or SHML, a relatively uncommon illness marked by lymphadenopathy, usually self-limited but occasionally fatal,[58,59] HHV-6 was detected in the involved tissues in seven of nine biopsies examined; EBV was detected in the one HHV-6 negative as well as one HHV-6-positive case.[101] Clinically, SHML appears to be a marked immunologic response to an infectious agent, and HHV-6 may well be one important precipitating factor.

Detection of HHV-6 in lymphoma tissues have been infrequent but may signify its etiologic role in rare cases. Jarrett et al.[81] identified HHV-6 in two NHLs (one B cell and one T cell) in a series of 117 lymphoma biopsies (including 29 with HD), and Josephs et al.[82] detected HHV-6 sequences in tissue obtained from 3 of 82 patients with B-cell lymphomas (a follicular large-cell lymphoma, a diffuse small-cell lymphoma in a patient with Sjögren's syndrome, and an African Burkitt's lymphoma). In this series, HHV-6 was not detected in 97 other neoplasms and lymphoid abnormalities, including 15 cases of HD and 22 T-cell lymphomas. Torelli et al.[161,162] subsequently found HHV-6 sequences in 3 of 25 Hodgkin's lymphomas and Gompels et al.[65] identified HHV-6 in 1 of 12 cases of HD, but their use of PCR did not allow localization of the virus-infected cells, an approach that has proven to be extremely useful in EBV-associated malignancies.[7,120,143] Recognizing the potential value of in situ hybridization, Shen et al.[142] identified HHV-6 sequences in 8 of 45 formaldehyde-fixed paraffin-embedded lymphoma samples obtained from Chinese patients with NHL. Some cases of lymphoma had extratumoral and others intratumoral HHV-6, possibly differentiating an etiologic role from a passenger status. A recent report by Maeda et al.[112] confirmed the presence of HHV-6 in HD by immunohistochemistry and in situ hybridization, but, as in the study by Shen et al.,[142] the apparent localization of HHV-6 to macrophages, particularly in lymphoid folli-

Table 1. Illnesses Reported to Have an Association with HHV-6

Disease	Strength, type, and frequency of association	Supporting data	Contrasting data
Exanthem subitum	Very strong[a], consistent	Seroconversion[12,172]	—
Other acute self-limiting pediatric illness	Strong,[a] common	Seroconversion[126,152]	—
Heterophile negative infectious mononucleosis	Strong,[a] occasional	Seroconversion[148]	—
Hepatitis	Strong,[a] rare	Seroconversion,[78] association with exanthem subitum[11]	—
Posttransplantation pneumonia	Moderate,[b] common	Virus detection in involved tissues[31]	—
Rosai–Dorfman disease (sinus histiocytes with massive lymphadenopathy)	Moderate,[c] unknown	Virus detection in tissues by ISH[101]	—
Abortion	Weak,[a] rare	IgM antibodies[9,84]	—
Hodgkin's disease	Weak,[a] rare	Virus detection by PCR[161,162] or serology[36]	Case control/longitudinal study[99]
Cervical cancer	Weak,[a] rare	Detection of DNA in squamous cervical carcinoma and cervical intraepithelial neoplasia[34]	—
Kawasaki disease	Weak,[a] rare	Virus isolation, seroconversion[66]	Numerous non-HHV-6-associated cases
Kikuchi's disease	Moderate,[a] occasional	Antibody pattern, virus detection[51,70,88,95,155]	Absence of DNA in case series[71] EBV detection by *in situ* hybridization[8]
Sarcoidosis	Moderate,[a] occasional	High antibody titers, virus detection[19]	Inconsistent virus detection,[56] normal antibody titers[56]
Acute lymphocytic leukemic	None	High antibody titers (case series)[1]	No difference in a case control series[98]
Other leukemias	Moderate,[b] rare	Malignant transformation during HHV-6 infection[156]	
Thrombotic thrombycytopenic purpora	Moderate,[a] rare	Association with primary HHV-6 infection[117]	
Hemophagocytic syndrome	Moderate, rare	Association with primary HHV-6 infection[77]	
Atypical polyclonal lymphoproliferation	Weak	Virus detection[94]	
Sjögren's syndrome	Weak	Occasional virus detection[82]	Unremarkable serology[14,56]
Ophthalmologic disease	Moderate,[a] rare	Virus detection[127,128,131]	
Neurological disease	Moderate,[a] rare	Association with primary HHV-6 infection[136,174]	—
Chronic fatigue syndrome	Moderate,[b] common	HHV-6-activated lymphocytes,[25] elevated antibody titers[25,100]	—
Oral cancer	Moderate,[a] common	Elevated geometric mean titer vs. controls,[141] detection in involved tissues[170]	—
AIDS	Weak,[b] unknown	*In vitro* activation of HIV[74,109]; elevated antibody titers in HIV-infected people[32,140]	*In vitro* suppression of HIV[30,106] No elevation of antibody titers in HIV-infected people[147] Associated with high CD4 counts[55]
Angioimmunoblastic lymphadenopathy	Weak,[c] unknown	Detection by PCR in involved tissues[108]	

[a]E, etiologic.
[b]R, reactivation.
[c]N, no data.

cles, rather than in the Reed-Sternberg cells, suggests a passenger role for HHV-6 in these cases. Krueger *et al.*[94] also reported HHV-6 genomes in scattered small lymphoid cells of occasional malignant lymphomas and a lymphoproliferative disorder they termed "atypical polyclonal lymphoproliferation," but since the virus was not detected in the tumor immunoblasts, its role in the illness is uncertain. It should be noted that inability to detect a virus by *in situ* hybridization may not mean it is not present, because the probes may be insensitive or directed against the wrong viral markers. For example, EBV remained undetected in paraffin blocks and frozen tissues being evaluated for EBV nuclear antigen (EBNA) and viral capsid antigen until probes against Epstein–Barr early RNAs (EBERs) were eventually successful in detecting the virus.[7,164]

Several techniques have been applied to the study of HHV-6 in Kikuchi's disease (histiocytic necrotizing lymphadenitis), but the role of HHV-6 in triggering this disease remains uncertain.[51,71,75,95,155] *In situ* hybridization studies were used by Sumiyoshi *et al.*[155] to summarize attempts to detect HHV-6 by Southern blot analysis (all cases were negative) and PCR (26 of 27 cases were positive) with the more frequent detection of HHV-6 in lymph nodes with Kikuchi's disease than other disorders (reactive paracortical hyperplasia, nonspecific lymphadenitis, and tuberculous lymphadenitis). In a series of 20 cases studied in three laboratories, HHV-6 was not linked to this disorder.[71] Since HHV-6 was identified in involved lymph nodes in varying proportions (17 of 18 by Eizuru *et al.*[51] and 5 of 17 by Kurata *et al.*[95]), HHV-6 could occasionally precipitate Kikuchi's disease, but its presence as a nonpathogenic bystander is quite likely.

As with EBV, the role of HHV-6 in disorders of immune dysregulation is of great interest since the virus may play a secondary role in disease pathogenesis after being activated by other viruses or by the disease process itself. This possibility was raised early in studies of patients with AIDS,[109] but the clinical and laboratory data regarding the interaction of HHV-6 and HIV are conflicting.[30,122,140,147] Lusso *et al.*[109] and Horvat *et al.*[74] have reported that HHV-6 can transactivate the HIV promoter *in vitro*. On the other hand, Pietroboni *et al.*,[122] Levy *et al.*,[106] and Carrigan *et al.*[30] observed suppression of HIV-1 replication by HHV-6 *in vitro*. In addition, Fox *et al.*[60] found that HIV infection did not affect HHV-6 antibody titers, Spira *et al.*[147] noted no correlation between HHV-6 antibody titers (an apparent surrogate of viral activation) and the course of HIV-related disease, and Fairfax *et al.*[55] showed a correlation of high levels of HHV-6 DNA and high CD4 counts in HIV-infected

men.[55] While some investigators noted higher HHV-6 titers in their HIV-infected patients,[32,140] and HHV-6 is a potentially important and even lethal agent in AIDS patients,[92] its role in the setting of HIV infection requires further study.

Evidence for a pathogenetic role of HHV-6 in CFS has been accumulating. CFS is a constellation of signs and symptoms for which a working research case definition has been developed.[63a,72] This syndrome is frequently associated with elevated HHV-6 antibodies[25,100] and appearance of "ballooning" lymphocytes characteristic of the virus in tissue culture.[25] However, as a diagnosis of exclusion without a specific diagnostic laboratory test, CFS appears to be a heterogeneous entity that can be triggered by a variety of agents.[103] The immune dysregulation found in many patients with CFS[28,90,150] provides an opportunity for reactivation of a variety of viruses,[25,73,100] and therefore it is not surprising that the second isolation of HHV-7 was from a well-documented CFS patient with abnormal circulating lymphocytes suggestive of the "balloon" cells noted in active HHV-6 infection. Not all patients with CFS have clinical or laboratory evidence of immune dysfunction,[24,103] which is a major reason for retaining the term CFS instead of adopting the term "chronic fatigue immune dysfunction syndrome" preferred by many patient support groups. Even in those patients with immune dysfunction, HHV-6 activation is not a consistent finding, but sufficient concern as to its possible role in exacerbating the disease has led to an interest in antiviral agents with anti-HHV-6 activity as a potential tool in CFS therapy (see Section 9).

The ability of HHV-6 to infect human retinas was demonstrated by Reux *et al.*[131] who used monoclonal antibodies (OHV-I for HHV-6 and E13 for CMV) to examine postmortem tissue from two patients developing retinitis in the course of AIDS. Identification of HHV-6- and CMV-infected cells was clearly differentiated in these studies. Qavi *et al.*[127,128] have identified HHV-6 DNA sequences in retinal lesions from AIDs patients, thereby implicating HHV-6 in the development of retinitis in patients with AIDS, a finding also associated with CMV, herpes simplex, and VZV. HHV-6 was also isolated from 1 of 70 corneas from asymptomatic HIV-1-infected individuals and HHV-6 DNA sequences were demonstrated in three HIV-1 culture-positive corneas; however, because no herpesviruses were identified in corneal sections,[128] the significance of the virus detection remains undetermined.

As noted above, neurologic disease can be found in association with primary HHV-6 infection[171] and a variety of manifestations such as meningoencephalitis[174] have been reported. Neurological illness due to HHV-6 has not been well documented in adults, although HHV-6 has joined CMV and EBV as one of the herpesviruses associated with Guillain–Barré syndrome.[113]

8.2. Diagnosis

Diagnosis of active HHV-6 infection can be made by isolation of the virus from peripheral blood containing "balloon cells," IgM antibody, seroconversion to IgG anti-HHV-6, and detection in diseased tissue by *in situ* hybridization or immunohistochemistry. It has been suggested that in older individuals, who may have a significant decline in HHV-6 antibody, "seroconversion" to IgG anti-HHV-6 viral capsid antigen without detection of IgM antibody to HHV-6 is suspect.[5]

As noted, HHV-6 antibody responses have been shown to vary in patients according to disease status, with the pattern of change being partly dependent on the assay being used.[99] Marked differences also can be observed between disease groups. For example, high levels of an early antigen antibody (p41/38) identified through the preparation of monoclonal antibodies were detected in sera from patients with Burkitt's lymphoma and HD disease more frequently than in sera from healthy donors or patients with other illnesses (including CFS) presenting the equally high levels of viral capsid antigen.[79]

9. Prevention and Control of HHV-6 and HHV-7

Uncertainty about the spectrum of disease associated with HHV-6 and HHV-7 has limited the interest in vaccine development. The ubiquity of these two viruses, which are apparently harbored in the salivary glands[102,105,121,168] and readily transmitted from person to person, make other control measures impractical. Primary intervention efforts to date have been directed to the control of HHV-6 replication in patients with CFS, where virus reactivation is considered a possible contributor to the disease process. A variety of antiviral agents have been reported to be effective against HHV-6 *in vitro*[6,27,89,115,133,144,151,163] (for more details, see ref. 166), but no controlled clinical trial demonstrating an effect on HHV-6 *in vivo* has yet been reported.

10. Unresolved Problems

The relatively recent discoveries of HHV-6 and HHV-7 have led to many interesting and unresolved ques-

tions. At the epidemiologic level, the significance of population differences in antibody response and the possible relationship to disease remains to be explored. Clinically, the list of new HHV-6- and HHV-7-related diseases is expected to grow.

The recent evidence suggesting that HHV-6 may have a deleterious effect on the immune system[110] and bone marrow function[46] can be expected to spur research on this virus. In the laboratory, new assays should elucidate the variability of viral strains, specify host responses more accurately, clarify pathogenesis, and accelerate development of clinical markers. Antigen-capture assays currently being developed are intended to measure the amount of circulating viral antigen and may provide a better estimate of the impact of the virus on the patient than antibody assays. Also, the use of PCR to detect virus in serum or plasma[76] may be increasingly important.

A major problem with these increasingly specific and sensitive assays, however, is their interpretation in evaluating viral pathogenesis. Virus detection in involved tissues is becoming increasingly important in studies of etiology and has been cited in the etiologic relationship between HHV-6 and many diseases, including Sjögren's syndrome.[41] Many questions need to be answered about the data produced by this new technology. Is the assay specific for HHV-6? Is the virus detected in cells of a patient with a particular disease important in the pathogenesis of that disease? What percentage of cases have this virus in the relevant cells? If it is 100% of cases, the likelihood of an etiologic role is considerable; but even if the cases are rare, experience with infectious mononucleosis and infectious hepatitis mandates that the occasional detection of HHV-6 in a disease of unknown etiology should receive serious consideration.

10.1. Viral Strain Differences

A major issue in pathogenesis is the disease spectrum associated with different strains of HHV-6. As noted in Section 4, two groups have been proposed based on growth in various cell lines, reactivity to monoclonal antibodies, and restriction enzyme cleavage patterns.[24,63,139] The GS strain, originally isolated from patients with AIDS and lymphoproliferative disease,[135] and others in group A, such as U1102[45] have been associated primarily with those illnesses, whereas Z-29 (isolated from a Zairian donor[107] and those in group B have been associated primarily with ES.[4,13,43,44,139] Dewhurst *et al.*[42] originally had noted that 2 of 15 isolates from children with roseolalike illnesses associated with primary HHV-6 infection predominantly resembled group A isolates as characterized by reactivity with monoclonal antibodies and host cell range *in vitro* (infection of HSB-2 as well as J-JAHN and Molt-3 T-cell lines), but some characteristics of group B were also observed in these isolates. Subsequent studies from this group[43] noted that virtually all illnesses in their infant population were due to group B isolates, and therefore the group A isolates may not be pathogenic in this age group; this finding is of interest in view of the association of group A isolates with disease in adults.[45,135] Presently, the association of one or another HHV-6 group with a specific disease is not totally predictable.

10.2. Laboratory Assay Differences

As noted above, important differences in serology occasionally can be observed in the detection of presumed viral capsid antigen antibody according to the assay used. As with EBV, the production of early antigen by HHV-6[52,79] has complicated the interpretation of serological data. Assays for this antibody are beginning to be applied in seroepidemiologic studies and may improve interpretation of discordant findings in comparisons using ELISA and IFA assays.

NOTE ADDED IN PROOF. Since this article was completed, there have been several reports linking HHV-6 to multiple sclerosis (MS). Immunocytochemical demonstration of HHV-6 early antigen was observed in MS oligodendrocytes compared to a more neuronal staining in controls (*PNAS* 1995; **92**:7440). Additional studies investigating the presence of antibodies to HHV-6 antigens in serum and CSF of well-defined populations at the National Institutes of Health noted that IgM antibodies to an HHV-6 early antigen were significantly elevated in MS patients compared to control groups (S. Jacobson, personal communication). It is possible that at least a subset of MS patients have their illness triggered by HHV-6, although other agents may be involved in MS as well.

11. References

1. ABLASHI, D. V., JOSEPHS, S. F., BUCHBINDER, A., HELLMAN, K., NAKAMURA, S., LLANA, T., LUSSO, P., KAPLAN, M., DAHLBERG, J., MEMON, S., IMAN, F., ABALSHI, K. L., MARKHAM, P. D., KRAMARSKY, B., KRUEGER, G. R. F., BIBERFELD, P., WONG-STAAL, F., SALAHUDDIN, S.Z., AND GALLO, R. C., Human B-lymphotropic virus (human herpes virus-6), *J. Virol. Methods* **21**:29–48 (1988).

2. ABLASHI, D. V., BALACHANDRAN, N., JOSEPHS, S. F., HUNG, C. L., KRUEGER, G. R. F., KRAMARSKY, B., SALAHUDDIN, S. Z., AND GALLO, R. C., Genomic polymorphism, growth properties, and immunologic variations in human herpesvirus-6 isolates, *Virology* **184**:545–552 (1991).

3. ABLASHI, D. V., SALAHUDDIN, S. Z., JOSEPHS, S. F., IMAM, F., LUSSO, P., AND GALLO, R. C., HBLV (or HHV-6) in human cell lines (letter), *Nature* **329**:207 (1987).

4. ABLASHI, D. V., AGUT, H., BERNEMANN, Z., CAMPADELLI-FIUME, G., CARRIGAN, D., CECCERINI-NELLI, L., CHANDRAN, B., CHOU, S., COLLANDRE, H., CONE, R., DAMBAUGH, T., DEWHURST, S., DILUCA, D., FOA-TOMASI, L., FLECKENSTEIN, B., FRENKEL, N., GALLO, R., GOMPELS, U., HALL, C., JONES, M., LAWRENCE, G., MARTIN, M., MONTAGNIER, L., NEIPEL, F., NICHOLAS, J., PELLETT, P. (CORRESPONDING AUTHOR), RAZZAQUE, A., TORRELLI, G., THOMPSON, B., SALAHUDDIN, S., WYATT, L., AND YAMANISHI, K., Human herpesvirus-6 strain groups: A nomenclature, *Arch. Virol.* **129**:363–366 (1993).

5. AGUT, H., Puzzles concerning the pathogenicity of human herpesvirus 6, *N. Engl. J. Med.* **329**:203–204 (1993).

6. AGUT, H., HURAUX, J-M., COLLANDRE, H., AND MONTAGNIER, L., Susceptibility of human herpes-virus 6 to acyclovir and ganciclovir, *Lancet* **2**:626 (1989).

7. AMBINDER, R. F., BROWNING, P. J., LORENZANA, I., LEVENTHAL, B. G., COSENZA, H., MANN, R. B., MACMAHON, E. M. E., MEDINA, R., CARDONA, V., GRUFFERMAN, S., OLSHAN, A., LEVIN, A., PETERSEN, E. A., BLATTNER, W., AND LEVINE, P. H., Epstein–Barr virus and childhood Hodgkin's disease in Honduras and the United States, *Blood* **81**:462–467 (1993).

8. ANAGNOSTOPOULOS, I., HUMMEL, M., KORBJUHN, P., PAPADAKI, T., ANAGNOSTOU, D., AND STEIN, H., Epstein–Barr virus in Kikuchi–Fujimoto disease (letter), *Lancet* **341**:893 (1993).

9. ANDO, Y., KAKIMOTO, K., EKUNI, Y., AND ICHIJO, M., HHV-6 infection during pregnancy and spontaneous abortion (letter), *Lancet* 1289 (1992).

10. ASANO, Y., YOSHIKAWA, T., KAJITA, Y., OGURA, R., SUGA, S., YAZAKI, T., NAKASHIMA, T., YAMADA, A., AND KURATA, T., Fatal encephalitis/encephalopathy in primary human herpesvirus-6 infection, *Arch. Dis. Child* **67**:1484–1485 (1992).

11. ASANO, Y., YOSHIKAWA, T., SUGA, S., YAZAKI, T., KONDO, K., AND YAMANISHI, K., Fatal fulminant hepatitis in an infant with human herpesvirus-6 infection (letter), *Lancet* **335**:862–863 (1990).

12. ASANO, Y., YOSHIKAWA, T., SUGA, S., YAZAKI, T., HATA, T., NAGAI, T., KAJITA, Y., OZAKI, T., AND YOSHIDA, S., Viremia and neutralizing antibody response in infants with exanthem subitum, *J. Pediatr.* **114**:535–539 (1989).

13. AUBIN, J.-T., COLLANDRE, H., CANDOTTI, D., INGRAND, D., ROUZIOUX, C., BURGARD, M., RICHARD, S., HURAUX, J.-M., AND AGUT, H., Several groups among human herpesvirus 6 strains can be distinguished by Southern blotting and polymerase chain reaction, *J. Clin. Microbiol.* **29**:367–372 (1991).

14. BABOONIAN, C., VENABLES, P. J. W., MAINI, R. N., KANGRO, H. O., AND OSMAN, H. K., Antibodies to human herpesvirus-6 in Sjogren's syndrome, *Arthritis Rheum.* **33**:1749–1750 (1990).

15. BARENBERG, L. H., AND GREENSPAN, L., Exanthema subitum (roseola infantum), *Am. J. Dis. Child.* **58**:983–993 (1939).

16. BERNEMAN, Z. N., ABLASHI, D. V., LI, G., EGER-FLETCHER, M., REITZ, M. S., JR., HUNG, C.-L., BRUS, I., KOMAROFF, A. L., AND GALLO, R. C., Human herpesvirus 7 is a T-lymphotropic virus and is related to but significantly different from herpesvirus 6 and human cytomegalovirus, *Proc. Natl. Acad. Sci. USA* **89**:10552–10556 (1992).

17. BERNEMAN, Z. N., GALLO, R. C., ABLASHI, D. V., FRENKEL, N., KATSAFANAS, G., KRAMARSKY, B., AND BRUS, I., Human herpes-

virus 7 (HHV-7) strain JI independent confirmation of HHV-7, *J. Infect. Dis.* **166**:690–691 (1992).

18. BERTRAM, G., DREINER, N., KRUEGER, G. R. F., ABLASHI, D. V., AND SALAHUDDIN, S. Z., Serological correlation of HHV-6 and EBV infections in infectious mononucleosis, in: *Epstein–Barr Virus and Human Disease* (D. V. ABLASHI, A. FAGGIONI, AND G. R. F. KRUEGER, EDS.), pp. 361–367, Humana Press, Clifton, NJ, 1989.

19. BIBERFELD, P., PETRÉN, A.-L., EKLUND, A., LINDEMALM, C., BARKHEM, T., EKMAN, M., ABLASHI, D., AND SALAHUDDIN, Z., Human herpesvirus-6 (HHV-6, HBLV) in sarcoidosis and lymphoproliferative disorders, *J. Virol. Methods* **21**:49–59 (1988).

20. BLACK, J. B., INOUE, N., KITE-POWELL, K., SALAHUDDIN, Z., AND PELLETT, P. E., Frequent isolation of human herpesvirus 7 from saliva, *Virus Res.* **29**(1):91–98 (1993).

21. BREESE, B. B., JR., Roseola infantum (exanthem subitum), *NY State J. Med.* **41**:1854–1859 (1941).

22. BRIGGS, M., FOX, J., AND TEDDER, R. S., Age prevalence of antibody to human herpes virus-6, *Lancet* **1**:1058–1059 (1988).

23. BROWN, N. A., SUMAYA, C. V., LIU, C.-R., ENCH, Y., KOVACS, A., CORONESI, M., AND KAPLAN, M. H., Fall in human herpesvirus-6 seropositivity with age, *Lancet* **2**:396 (1988).

24. BUCHWALD, D., AND KOMAROFF, A. L., Review of laboratory findings for patients with chronic fatigue syndrome, *Rev. Infect. Dis.* **13**:S12–S18 (1991).

25. BUCHWALD, D., CHENEY, P. R., PETERSON, D. L., HENRY, B., WORMSLEY, S. B., GEIGER, A., ABLASHI, D. V., SALAHUDDIN, S. Z., SAXINGER, C., BIDDLE, R., KIKINIS, R., JOLESZ, F. A., FOLKS, T., BALACHANDRAN, N., PETER, J. B., GALLO, R. C., AND KOMAROFF, A. L., A chronic illness characterized by fatigue, neurologic and immunologic disorders, and active human herpesvirus type 6 infection, *Ann. Inter. Med.* **116**:103–113 (1992).

26. BUCHWALD, D., HOOTON, T. M., AND ASHLEY, R. L., Prevalence of herpesvirus, human T-lymphotropic virus type I, and treponemal infection in Southeast Asian refugees, *J. Med. Virol.* **38**:195–199 (1992).

27. BURNS, W. H., AND SANFORD, G. R., Susceptibility of human herpesvirus 6 to antivirals *in vitro*, *J. Infect. Dis.* **162**:634–637 (1990).

28. CALIGIURI, M., MURRAY, C., BUCHWALD, D., LEVINE, H., CHENEY, P., PETERSON, D., KOMAROFF, A. L., AND RITZ, J., Phenotypic and functional deficiency of natural killer cells in patients with chronic fatigue syndrome, *J. Immunol.* **139**:3306–3313 (1987).

29. CARRIGAN, D. R., Human herpesvirus-6 and bone marrow transplantation, in: *Human Herpesvirus-6: Epidemiology, Molecular Biology and Clinical Pathology. Perspectives in Medical Virology*, Vol. 4 (D. V. ABLASHI, G. R. F. KRUEGER, AND S. Z. SALAHUDDIN, EDS.), pp. 281–301, Elsevier, Amsterdam, 1992.

30. CARRIGAN, D. R., KNOX, K. K., AND TAPPER, M. C., Suppression of human immunodeficiency virus type I replication by human herpesvirus-6, *J. Infect. Dis.* **162**:844–851 (1990).

31. CARRIGAN, D. R., DROBYSKI, W. R., RUSSLER, S. K., TAPPER, M. A., KNOX, K. K., AND ASH, R. C., Interstitial pneumonitis associated with human herpesvirus-6 infection after marrow transplantation, *Lancet* **338**:147–149 (1991).

32. CERMELLI, C., MORONI, A., PIETROSEMOLI, P., PECORARI, M., AND PORTOLANI, M., IgG antibodies to human herpesvirus-6 (HHV-6) in Italian people, *Microbiologica* **15**:57–63 (1992).

33. CHANDRAN, B., TIRAWATNAPONG, S., PFEIFFER, B., AND ABLASHI, D. V., Antigenic relationships among human herpesvirus-6 isolates, *J. Med. Virol.* **37**:247–254 (1992).

33a. CHANG, Y., CESARMAN, E., PESSIN, M. S., LEE, F., CULPEPPER, J., KNOWLES, D. M., AND MOORE, P. S., Identification of herpesvirus-like DNA sequences in AIDS-associated Kaposi's sarcoma, *Science* **266**:1865 (1994).

34. CHEN, M., WANG, H., WOODWORTH, C. D., LUSSO, P., BERNEMAN, Z., KINGMAN, D., DELGADO, G., AND DIPAOLA, J. A., Detection of human herpesvirus 6 and human papillomavirus 16 in cervical carcinoma, *Am. J. Pathol.* **145**:1509–1516 (1994).

35. CHERRY, D., Roseola infantum (exanthem subitum), *Textbook of Pediatric Infectious Diseases*, 2nd ed. (R. D. FEIGN AND J. D. CHERRY, EDS.), pp. 1842–1845, Saunders, Philadelphia, 1987.

36. CLARK, D. A., ALEXANDER, F. E., MCKINNEY, P. A., ROBERTS, B. E., O'BRIEN, C., JARRETT, R. F., CARTWRIGHT, R. A., AND ONIONS, D. E., The seroepidemiology of human herpesvirus-6 (HHV-6) from a case-control study of leukaemia and lymphoma, *Int. J. Cancer* **45**:829–833 (1990).

37. CLARK, D. A., FREELAND, J. M. L., MACKIE, P. L. K., JARRETT, R. F., AND ONIONS, D. E., Prevalence of antibody to human herpesvirus 7 by age, *J. Infect. Dis.* **168**:251–252 (1993).

38. CLEMENS, H. H., Exanthem subitum (roseola infantum); Report of eighty cases, *J. Pediatr.* **26**:66–67 (1945).

39. COYLE, P. V., BRIGGS, M., TEDDER, R. S., AND FOX, J. D., Comparison of three immunoassays for the detection of anti-HHV-6, *J. Virol. Methods* **38**:283–295 (1992).

40. CUSHING, H. B., An epidemic of roseola infantum, *Can. Med. Assoc. J.* **17**:905–906 (1927).

41. DECLERCK, L. S., BOURGEOIS, N., KRUEGER, G. R. F., AND STEVENS, W. J., Human herpesvirus-6 in Sjögren's syndrome, in: *Human Herpesvirus-6: Epidemiology, Molecular Biology and Clinical Pathology. Perspectives in Medical Virology*, Vol. 4. (D. V. ABLASHI, G. R. F. KRUEGER, AND S. Z. SALAHUDDIN, EDS.), pp. 303–315, Elsevier, Amsterdam, 1992.

42. DEWHURST, S., CHANDRAN, B., MCINTYRE, K., SCHNABEL, K., AND HALL, C. B., Phenotypic and genetic polymorphisms among human herpesvirus 6 isolates from North American infants, *Virology* **190**:490–493 (1992).

43. DEWHURST, S., MCINTYRE, K., SCHNABEL, K., AND HALL, C. B., Human herpesvirus 6 (HHV-6) variant B accounts for the majority of symptomatic primary HHV-6 infections in a population of U.S. infants, *J. Clin. Microbiol.* **31**(2):416–418 (1993).

44. DILUCA, D., MIRANDOLA, P., SECCHIERO, P., CERMELLI, C., ALEOTTI, A., BOVENZI, P., PORTOLANI, M., AND CASSAI, E., Characterization of human herpesvirus 6 strains isolated from patients with exanthem subitum with or without cutaneous rash (letter), *J. Infect. Dis.* **166**:689 (1992).

45. DOWNING, R. G., SEWANKAMBO, N., SERWADDA, D., HONESS, R., CRAWFORD, D., JARRETT, R., AND GRIFFIN, B. E., Isolation of human lymphotropic herpesvirus from Uganda (letter), *Lancet* **2**:390 (1987).

46. DROBYSKI, W. R., DUNNE, W. M., BURD, E. M., KNOX, K. K., ASH, R. C., HOROWITZ, M. M., FLOMENBERG, N., AND CARRIGAN, D. R., Human herpesvirus-6 (HHV-6) infection in allogeneic bone marrow transplant recipients: Evidence of a marrow-suppressive role for HHV-6 *in vivo*, *J. Infect. Dis.* **167**:735–739 (1993).

47. DROBYSKI, W. R., KNOX, K. K., MAJEWSKI, D., AND CARRIGAN, D. R., Brief report: Fatal encephalitis due to variant B human herpesvirus-6 infection in a bone marrow transplant recipient, *N. Engl. J. Med.* **19**:1356–1360 (1994).

48. DREYER, Z. A. E., DOWELL, B. L., CHEN, H., HAWKINS, E., AND MCCLAIN, K. L., Infection-associated hemophagocytic syndrome, *Am. J. Pediatr. Hematol. Oncol.* **13**:476–481 (1991).

49. DUBEDAT, S., AND KAPPAGODA, N., Hepatitis due to human herpesvirus-6, *Lancet* **2:**1463–1464 (1989).

50. DUNNE, W. M., JR., AND DEMMLER, G. J., Serological evidence for congenital transmission of human herpesvirus 6, *Lancet* **340:**121–122 (1992).

51. EIZURU, Y., MINEMATU, T., MINAMISHIMA, Y., KIKUCHI, M., YAMANISHI, K., TAKAHASHI, M., AND KURATA, T., Human herpesvirus-6 in lymph nodes, *Lancet* **1:**40 (1989).

52. EIZURU, Y., AND MINAMISHIMA, Y., Evidence for putative immediate early antigens in human herpesvirus 6-infected cells, *J. Gen. Virol.* **73:**2161–2165 (1992).

53. ENDERS, G., BIBER, M., MEYER, G., AND HELFTENBEIN, E., Prevalence of antibodies to human herpesvirus 6 in different age groups, in children with exanthema subitum, other acute exanthematous childhood diseases, Kawasaki syndrome, and acute infections with other herpesviruses and HIV, *Infection* **18:**12–15 (1990).

54. EVANS, A. S., Causation and disease: The Henle–Koch postulates revisited, *Yale, J. Biol. Med.* **49:**175–195 (1976).

55. FAIRFAX, M. R., SCHACKER, T., CONE, R. W., COLLIER, A.C., AND COREY, L., Human herpesvirus 6 DNA in blood cells of human immunodeficiency virus-infected men: Correlation of high levels with high CD4 cell counts, *J. Infect. Dis.* **169:**1342–1345 (1994).

56. FILLET, A. M., RAGUIN, G., AGUT, H., BOISNIC, S., AGBO-GODEAU, S., AND ROBERT, C., Evidence of human herpesvirus 6 in Sjögren syndrome and sarcoidosis, *Eur. J. Clin. Microbiol. Infect. Dis.* **11:**564–566 (1992).

57. FO'A-TOMASI, L., AVITABILE, E., KE, L., AND CAMPADELLI-FIUME, G., Polyvalent and monoclonal antibodies identify major immunogenic proteins specific for human herpesvirus 7-infected cells and have weak cross-reactivity with human herpesvirus 6, *J. Gen. Virol.* **75:**2719–2727 (1994).

58. FOUCAR, E., ROSAI, J., AND DORFMAN, R. F., Sinus histiocytosis with massive lymphadenopathy (Rosai-Dorfman disease): Review of the entity, *Semin. Diagn. Pathol.* **7:**19–73 (1990).

59. FOUCAR, E., ROSAI, J., AND DORFMAN, R. F., Sinus histiocytosis with massive lymphoadenopathy. An analysis of 14 deaths occurring in a patient registry, *Cancer* **54:**1834–1840 (1984).

60. FOX, J., BRIGGS, M., AND TEDDER, R. S., Antibody to human herpesvirus 6 in HIV-1 positive and negative homosexual men, *Lancet* **2:**396–397 (1988).

61. FOX, J. D., BRIGGS, M., WARD, P. A., AND TEDDER, R., Human herpesvirus 6 in salivary glands, *Lancet* **336:**590–593 (1990).

62. FRENKEL, N., SCHIRMER, E. C., WYATT, L. S., KATSAFANAS, G., ROFFMAN, E., DANOVICH, R. M., AND JUNE, C. H., Isolation of a new herpesvirus from human CD4$^+$ T cells, *Proc. Natl. Acad. Sci. USA* **87:**748–752 (1990).

63. FRENKEL, N., AND WYATT, L. S., HHV-6 and HHV-7 as exogenous agents in human lymphocytes, in: *Developments in Biological Standardization*, Vol. 76 (F. BROWN, E. E. ESBER, AND M. WILLIAMS, EDS.), pp. 259–265, Karger, Basal, 1992.

63a. FUKUDA, K., STRAUS, S. E., HICKIE, I., SHARPE, M. C., DOBBINS, J. G., KOMAROFF, A., AND THE INTERNATIONAL CHRONIC FATIGUE SYNDROME STUDY GROUP, The chronic fatigue syndrome: A comprehensive approach to its definition and study, *Ann. Intern. Med.* **121:**953–959 (1994).

64. GOPAL, M. R., THOMSON, B. J., FOX, J., TEDDER, R. S., AND HONESS, R. W., Detection by PCR of HHV-6 and EBV DNA in blood and oropharynx of healthy adults and HIV seropositives, *Lancet* **1:**1598–1599 (1990).

65. GOMPELS, U. A., CARRIGAN, D. R., CARSS, A. L., AND ARNO, J., Two groups of human herpesvirus 6 identified by sequence analyses of laboratory strains and variants from Hodgkin's lymphoma and bone marrow transplant patients, *J. Gen. Virol.* **74:**613–622 (1993).

66. HAGIWARA, K., KOMURA, H., KISHI, F., KAJI, T., AND YOSHIDA, T., Isolation of human herpesvirus-6 from an infant with Kawasaki disease (letter), *Eur. J. Pediatr.* **151:**867–868 (1992).

67. HALL, C. B., LONG, C. E., SCHNABEL, K. C., CASERTA, M. T., MCINTYRE, K. M., COSTANZO, M. A., KNOTT, A., DEWHURST, S., INSEL, R. A., AND EPSTEIN, L. G., Human herpesvirus-6 infection in children. A prospective study of complications and reactivation, *N. Engl. J. Med.* **331:**432–438 (1994).

68. HENLE, G., AND HENLE, W., Immunofluorescence in cells derived from Burkitt's lymphoma, *J. Bacteriol.* **91:**1248–1256 (1966).

69. HIDAKA, Y., LIU, Y., YAMAMOTO, M., MORI, R., MIYAZAKI, C., KUSUHARA, K., OKADA, K., AND UEDA, K., Frequent isolation of human herpesvirus 7 from saliva samples, *J. Med. Virol.* **40:**343–346 (1993).

70. HOFFMANN, A., KIRN, E., KUERTEN, A., SANDER, C. H., KRUEGER, G. R. F., AND ABLASHI, D. V., Active human herpesvirus-6 (HHV-6) infection associated with Kikuchi–Fujimoto disease and systemic lupus erythematosus (SLE), *In Vivo* **5:**265–270 (1991).

71. HOLLINGSWORTH, H. C., PEIPER, S. C., WEISS, L. M., RAFFELD, M., AND JAFFE, E. S., An investigation of the viral pathogenesis of Kikuchi–Fujimoto disease. Lack of evidence for Epstein–Barr virus or human herpesvirus type 6 as the causative agents, *Arch. Pathol. Lab. Med.* **118:**134–140 (1994).

72. HOLMES, G. P., KAPLAN, J. E., GANTZ, N. M., KOMAROFF, A. L., SCHONBERGER, L. B., STRAUS, S. E., JONES, J. F., DUBOIS, R. E., CUNNINGHAM-RUNDLES, C., PAHWA, S., TOSATO, G., ZEGANS, L. S., PURTILO, D. T., BROWN, N., SCHOOLEY, R. T., AND BRUS, I., Chronic fatigue syndrome: A working case definition, *Ann. Intern. Med.* **108:**387–439 (1988).

73. HOLMES, G. P., KAPLAN, J. E., STEWART, J. A., HUNT, B., PINSKY, P. F., AND SCHONBERGER, L. B., A cluster of patients with a chronic mononucleosis-like syndrome: Is Epstein–Barr virus the cause? *J.A.M.A.* **257:**2297–2302 (1987).

74. HORVAT, R. T., WOOD, C., AND BALACHANDRAN, N., Transactivation of human immunodeficiency virus promoter by human herpesvirus 6, *J. Virol.* **63:**970–973 (1988).

75. HORWITZ, C. A., AND BENEKE, J., Human herpesvirus-6 revisited, *Am. J. Clin. Pathol.* **99:**533–534 (1993).

76. HUANG, L. M., KUO, P. F., LEE, C. Y., CHEN, J. Y., LIU, M. Y., AND YANG, C. S., Detection of human herpesvirus-6 DNA by polymerase chain reaction in serum or plasma, *J. Med. Virol.* **38:**7–10 (1992).

77. HUANG, L. M., LEE, C. Y., LIN, K. H., CHUU, W. M., LEE, P. I., CHEN, R. L., CHEN, J. M., AND LIN, D. T., Human herpesvirus-6 associated with fatal haemophagocytic syndrome (letter), *Lancet* 60–61 (1990).

78. IRVING, W. L., AND CUNNINGHAM, A. L., Serological diagnosis of infection with human herpesvirus-6, *Br. Med. J.* **300:**156–159 (1990).

79. IYENGAR, S., LEVINE, P. H., ABLASHI, D., NEEQUAYE, J., AND PEARSON, G. R., Seroepidemiological investigations on human herpesvirus 6 (HHV-6) infections using a newly developed early antigen assay, *Int. J. Cancer* **49:**551–557 (1991).

80. JAMES, V., AND FREIER, A., Roseola infantum: Outbreak in a maternity hospital, *Arch. Dis. Child.* **23–24:**54–58 (1948–1949).

81. JARRETT, R. F., GLEDHILL, S., QURESHI, F., CRAE, S. H., MADHOK, R., BROWN, I., EVANS, I., KRAJEWSKI, A., O'BRIEN, C. J., CARTWRIGHT, R. A., VENABLES, P., AND ONIONS, D. E., Identi-

fication of human herpesvirus 6-specific DNA sequences in two patients with non-Hodgkin's lymphoma, *Leukemia* **2:**496–502 (1988).

82. JOSEPHS, S. F., BUCHBINDER, A., STREICHER, H. Z., ABLASHI, D. V., SALAHUDDIN, S. Z., GUO, H. G., WONG-STAAL, F., COSSMAN, J., RAFFELD, M., SUNDEN, J., LEVINE, P., BIGGAR, R., KRUEGER, G. R. F., FOX, R. I., AND GALLO, R. C., Detection of human B-lymphotropic virus (human herpesvirus 6) sequences in B cell lymphoma tissues of three patients, *Leukemia* **2:**132–135 (1988).

83. JURETIĆ, M., Exanthema subitum: A review of 243 cases, *Helv. Paediatr. Acta* **18:**80–95 (1963).

84. KAKIMOTO, K., ANDO, Y., MORIYAMA, I., AND ICHIJO, M., The significance of determination of human herpesvirus-6 (HHV-6) antibody during pregnancy, *Nippon Sanka Fujinka Gakkai Zasshi* **44:**1571–1577 (1992).

85. KAWA-HA, K., TANAKA, K., INOUE, M., SAKATA, N., OKADA, S., KURATA, T., MUKAI, T., AND YAMANISHI, K., Isolation of human herpesvirus 7 from a child with symptoms mimicking chronic Epstein–Barr virus infection, *Br. J. Haematol.* **84:**545–548 (1993).

86. KEIM, D. E., KELLER, E. W., AND HIRSCH, M. S., Mucocutaneous lymph-node syndrome and parainfluenza 2 virus infection, *Lancet* **2:**303 (1977).

87. KEMPE, C. H., SHAW, E. B., JACKSON, J. R., AND SILVER, H. K., Studies on the etiology of exanthema subitum (roseola infantum), *J. Pediatr.* **37:**561–568 (1950).

88. KIKUCHI, M., SUMIYOSHI, Y., AND MINAMISHIMA, Y., Kukuchi's disease (histiocytic necrotizing lymphadenitis), in: *Human Herpesvirus-6: Epidemiology, Molecular Biology and Clinical Pathology. Perspectives in Medical Virology*, Vol. 4 (D. V. ABLASHI, G. R. F. KRUEGER, AND S. Z. SALAHUDDIN, EDS.), pp. 175–183, Elsevier, Amsterdam, 1992.

89. KUKUTA, H., LU, H., AND MATSUMOTO, S., Susceptibility of human herpesvirus 6 to acyclovir (letter), *Lancet* **2:**861 (1989).

90. KLIMAS, N. G., SALVATO, F. R., MORGAN, R., AND FLETCHER, M. A., Immunologic abnormalities in chronic fatigue syndrome, *J. Clin. Microbiol.* **28:**1403–1410 (1990).

91. KNOWLES, W., AND GARDNER, S., High prevalence of antibody to human herpesvirus 6 and seroconversion with rash in two infants, *Lancet* **2:**912–913 (1988).

91a. KNOX, K. K., AND CARRIGAN, D. R., *In vitro* suppression of bone marrow progenitor cell differentiation by human herpesvirus 6 infection, *J. Infect. Dis.* **165:**925–929 (1992).

92. KNOX, K. K., AND CARRIGAN, D. R., Disseminated active HHV-6 infections in patients with AIDS, *Lancet* **343:**577–578 (1994).

93. KONDO, K., NAGAFUJI, H., HATA, A., TOMOMORI, C., AND YAMANISHI, K., Association of human herpesvirus 6 infection of the central nervous system with recurrence of febrile convulsions, *J. Infect. Dis.* **167:**1197–1200 (1993).

94. KRUEGER, G. R. F., MANAK, M., BOURGEOIS, N., ABLASHI, D. V., SALAHUDDIN, S. Z., JOSEPHS, S. F., BUCHBINDER, A., GALLO, R. C., BERTHOLD, F., AND TESCH, H., Persistent active herpes virus infection associated with atypical polyclonal lymphoproliferation (APL) and malignant lymphoma, *Anticancer Res.* **9:**1457–1476 (1989).

95. KURATA, T., IWASAKI, T., SATA, T., WAKABAYASHI, T., YAMAGUCHI, K., OKUNO, T., YAMANISHI, K., AND TAKEI, I., Viral pathology of human herpesvirus 6 infection, in: *Immunobiology and Prophylaxis of Human Herpesvirus Infections* (C. LOPEZ, R. MORI, B. ROIZMAN, AND R. J. WHITLEY, EDS.), pp. 39–47, Plenum Press, New York.

96. LAWRENCE, G. L., CHEE, M., CRAXTON, M. A., GOMPELS, U. A., HONESS, R. W., AND BARRELL, B. G., Human herpesvirus 6 is closely related to human cytomegalovirus, *J. Virol.* **64:**287–299 (1990).

97. LEACH, C. T., NEWTON, E. R., MCPARLIN, S., AND JENSON, H. B., Human herpesvirus 6 infection of the female genital tract, *J. Infect. Dis.* **169:**1281–1283 (1994).

97a. LEVINE, P. H., AND ABLASHI, D. V., The new herpesviruses: HHV-7 and HHV-8 (KS-virus), *Infections in Medicine* (in press).

98. LEVINE, P. H., ABLASHI, D. V., SAXINGER, W. C., AND CONNELLY, R. R., Antibodies to human herpesvirus-6 in patients with acute lymphocytic leukemia, *Leukemia* **6:**1229–1231 (1992).

99. LEVINE, P. H., EBBESEN, P., ABLASHI, D. V., SAXINGER, W. C., NORDENTOFT, A., AND CONNELLY, R. R., Antibodies to human herpes virus type 6 and clinical course in patients with Hodgkin's disease, *Int. J. Cancer* **51:**53–57 (1992).

100. LEVINE, P. H., JACOBSON, S., POCINKI, A. G., CHENEY, P., PETERSON, D., CONNELLY, R. R., WEIL, R., ROBINSON, S. M., ABLASHI, D. V., SALAHUDDIN, S. Z., PEARSON, G. R., AND HOOVER, R., Clinical, epidemiologic, and virologic studies in four clusters of the chronic fatigue syndrome, *Arch. Intern. Med.* **152:**1611–1616 (1992).

101. LEVINE, P. H., JAHAN, N., MURARI, P., MANK, M., AND JAFFE, E. S., Detection of human herpesvirus 6 in tissues involved by sinus histiocytosis with massive lymphadenopathy Rosai-Dorfman disease, *J. Infect. Dis.* **166:**291–295 (1992).

102. LEVINE, P. H., JARRETT, R., AND CLARK, D. A., The epidemiology of human herpesvirus-6, in: *Human Herpesvirus-6: Epidemiology, Molecular Biology and Clinical Pathology* (D. V. ABLASHI, G. R. F. KRUEGER, AND S. Z. SALAHUDDIN, EDS.), pp. 9–23, Elsevier, Amsterdam, 1992.

103. LEVINE, P. H., KRUEGER, G. R. F., KAPLAN, M., BELL, D., DUBOIS, R.E., HUANG, A., QUINLAN, A., BUCHWALD, D., ARCHARD, L., GUPTA, S., JONES, J., STRAUS, S., AND TOSATO, G., The postinfectious chronic fatigue syndrome, in: *Epstein–Barr Virus and Human Disease* (D. V. ABLASHI, A. T. HUANG, J. S. PAGANO, ET AL., EDS.), pp. 405–438, Humana Press, Clifton, NJ, 1989.

104. LEVINE, P. H., NEEQUAYE, J., YADAV, M., AND CONNELLY, R., Geographic/ethnic differences in human herpesvirus-6 antibody patterns, *Microbiol. Immunol.* **36:**169–172 (1992).

105. LEVY, J. A., FERRO, F., GREENSPAN, D., AND LENNETTE, E. T., Frequent isolation of HHV-6 from saliva and high seroprevalence of the virus in the population, *Lancet* **335:**1047–1050 (1990).

106. LEVY, J., LANDAY, A., AND LENNETTE, E., Human herpesvirus 6 inhibits human immunodeficiency virus type 1 replication in cell culture, *J. Clin. Microbiol.* **28:**2362–2364 (1990).

107. LOPEZ, C., PELLETT, P., STEWART, J., GOLDSMITH, C., SANDERLIN, K., BLACK, J., WARFIELD, D., AND FEORINO, P., Characteristics of human herpesvirus-6 (letter), *J. Infect. Dis.* **157:**1271–1273 (1988).

108. LUPPI, M., MARASCA, R., BAROZZI, P., ARTUSI, T., AND TORELLI, G., Frequent detection of human herpesvirus-6 sequences by polymerase chain reaction in paraffin-embedded lymph nodes from patients with angioblastic lymphadenopathy and angioimmunoblastic lymphadenopathy-like lymphoma, *Leuk. Res.* **17:**1003–1111 (1993).

109. LUSSO, P., ABLASHI, D. V., AND LUKA, J., Interactions between HHV-6 and other viruses, in: *Human Herpesvirus-6: Epidemiology, Molecular Biology and Clinical Pathology. Perspectives in Medical Virology*, Vol. 4 (D. ABLASHI, G. R. F. KRUEGER, AND S. Z. SALAHUDDIN, EDS.), pp. 121–133, Elsevier, Amsterdam, 1992.

110. LUSSO, P., MALNATI, M. S., GARZINO-DEMO, A., CROWLEY, R. W., LONG, E. O., AND GALLO, R. C., Infection of natural killer cells by human herpesvirus-6, *Nature* **362**:458–462 (1993).

111. LUSSO, P., MARKHAM, P. D., TSCHACHLER, E., VERONESE, F. D. M., SALAHUDDIN, S. Z., ABLASHI, D. V., PAHWA, S., KROHN, K., AND GALLO, R. C., *In vitro* cellular tropism of human B-lymphotropic virus (human herpesvirus-6), *J. Exp. Med.* **167**:1659–1670 (1988).

112. MAEDA, A., SATA, T., ENZAN, H., TANAKA, K., WAKIGUCHI, H., KURASHIGE, T., YAMANISHI, K., AND KURATA, T., The evidence of human herpesvirus 6 infection in the lymph nodes of Hodgkin's disease, *Virchows. Arch. A Pathol. Anat. Histopathol.* **423**(1):71–75 (1993).

113. MERELLI, E., SOLA, P., FAGLIONI, P., POGGI, M., MONTORSI, M., AND TORELLI, G., Newest human herpesvirus (HHV-6) in the Guillain-Barré syndrome and other neurological diseases, *Acta Neurol. Scand.* **85**:334–336 (1992).

114. MERLINO, C., SINESI, F., MESSINA, M., GIACCHINO, F., AND NEGRO PONZI, A., Infection by human herpesvirus type 6 (HHV-6) and renal transplantation, *Minerva Urol. Nefrol.* **44**:157–153 (1992).

115. NAKAGAMI, T., TAJI, S., TAKAHASHI, M., AND YAMANISHI, K., Antiviral activity of a bile pigment, biliverdin, against human herpesvirus 6 (HHV-6) *in vitro*, *Microbiol. Immunol.* **36**:381–390 (1992).

116. NIEDERMAN, J. C., LIN, C.-R., KAPLAN, M. H., AND BROWN, N. A., Clinical and serological features of human herpesvirus-6 infection in three adults, *Lancet* **2**:817–819 (1988).

117. HISHIMURA, K., AND IGARASHI, M., Thrombotic thrombocytopenic purpura associated with exanthem subitum, *Pediatrics* **60**:260 (1972).

118. OKANO, M., LUKA, J., THIELE, G. M., SAKIYAMA, Y., MATSUMOTO, S., AND PURTILO, D. T., Human herpesvirus-6 infection and Kawasaki disease, *J. Clin. Microbiol.* **27**:2379–2380 (1989).

119. OKUNO, T., TAKAHASHI, K., BALACHANDRA, K., SHIRAKI, K., YAMANISHI, K., TAKAHASHI, M., AND BABA, K., Seroepidemiology of human herpesvirus 6 in normal children and adults, *J. Clin. Microbiol.* **27**:651–653 (1989).

120. PALLESEN, G., HAMILTON-DUTOIT, S. J., ROWE, M., AND YOUNG, L. S., Expression of Epstein–Barr virus latent gene products in tumour cells of Hodgkin's disease, *Lancet* **337**:320–322 (1991).

121. PIETROBONI, G. R., HARNETT, G. B., BUCENS, M. R., AND HONESS, R. W., Isolation of human herpes-virus-6 from saliva, *Lancet* **1**:1059 (1988).

122. PIETROBONI, G. R., HARNETT, G. B., FARR, T. J., AND BUCENS, M. R., Human herpesvirus type 6 (HHV-6) and its *in vitro* effect on human immunodeficiency virus (HIV), *J. Clin. Pathol.* **41**:1310–1312 (1988).

123. PORLOTANI, M., CERMELLI, C., MORONI, A., BERTOLANI, M. F., DI LUCA, D., CASSAI, E., AND SABBATINI, A. M., Human herpesvirus-6 infections in infants admitted to hospital, *J. Med. Virol.* **39**:146–151 (1993).

124. PREZIOSO, P. J., CANGIARELLA, J., LEE, M., NUOVO, G. J., BORKOWSKY, W., ORLOW, S. J., AND GRECO, M. A., Fatal disseminated infection with human herpesvirus-6, *J. Pediatr.* **120**:921–923 (1992).

125. PRUSSIA, P. R., MANSOOR, G. A., EDWARDS, C., AND JORDAN, O., Haemophagocytic syndrome, *West Indian Med. J.* **40**:188–192 (1991).

126. PRUKSANANONDA, P., HALL, C. B., INSEL, R. A., MCINTYRE, K., PELLETT, P. E., LONG, C. E., SCHNABEL, K. C., PINCUS, P. H., STAMEY, F. R., DAMBAUGH, T. R., AND STEWART, J. A., Primary human herpesvirus 6 infection in young children, *N. Engl. J. Med.* **326**:1445–1450 (1992).

127. QAVI, H. B., GREEN, M. T., SEGALL, G. K., LEWIS, D. E., AND HOLLINGER, F. B., Frequency of dual infections of corneas with HIV-1 and HHV-6, *Curr. Eye Res.* **11**:315–323 (1992).

128. QAVI, H. B., GREEN, M. T., AND SEGALL, G. K., HIV-1 and HHV-6 infections of human retina and cornea, in: *Human Herpesvirus-6: Epidemiology, Molecular Biology and Clinical Pathology. Perspectives in Medical Virology*, Vol. 4 (D. V. ABLASHI, G. R. F. KRUEGER, AND S. Z. SALAHUDDIN, EDS.),, pp. 263–280, Elsevier, Amsterdam, 1992.

129. RANGER, S., PATILLAUD, S., DENIS, F., HIMMICH, A., SANGARE, A., M'BOUP, S., ZTOUA-N'GAPORO, A., PRINCE-DAVID, M., CHOUT, R., CEVALLOS, R., AND AGUT, H., Seroepidemiology of human herpesvirus-6 in pregnant women from different parts of the world, *J. Med. Virol.* **34**:194–198 (1991).

130. REINER, A. P., AND SPIVAK, J. L., Hematophagic histiocytosis. A report of 23 new patients and a review of the literature, *Medicine* **67**:369–388 (1988).

131. REUX, I., FILLET, A. M., AGUT, H., KATLAMA, C., HAUW, J. J., AND LEHOANG, P., *In situ* detection of human herpesvirus 6 in retinitis associated with acquired immunodeficiency syndrome (letter), *Am. J. Ophthalmol.* **114**:375–377 (1992).

132. RISDALL, R. J., MCKENNA, R. W., NESBIT, M. E., KRIVIT, W., BALFOUR, H. H., SIMMONS, R. L., AND BRUNNING, R. D., Virus associated hemophagocytic syndrome. A benign histiocytic proliferation distinct from malignant histiocytosis, *Cancer* **44**:993–1002 (1979).

133. RUSSLER, S. K., TAPPER, M. A., AND CARRIGAN, D. R., Susceptibility of human herpesvirus 6 to acyclovir and ganciclovir, *Lancet* **2**:382 (1989).

134. SALAHUDDIN, S. Z., The discovery of human herpesvirus type 6, in: *Human Herpesvirus-6: Epidemiology, Molecular Biology and Clinical Pathology. Perspectives in Medical Virology*, Vol. 4 (D. V. ABLASHI, G. R. F. KRUEGER, AND S. Z. SALAHUDDIN, EDS.), pp. 3–8, Elsevier, Amsterdam, 1992.

135. SALAHUDDIN, S. Z., ABLASHI, D. V., MARKHAM, P. D., JOSEPHS, S. F., STURZENEGGER, S., KAPLAN, M., HALLIGAN, G., BIBERFELD, P., WONG-STAAL, F., KRAMARSKY, B., AND GALLO, R. C., Isolation of a new virus, HBLV, in patients with lymphoproliferative disorders, *Science* **234**:596–601 (1986).

136. SATO, T., INOUE, T., KAJIWARA, M., MIYAZAKI, C., KUSUNOKI, K., AND UEDO, K., Acute encephalopathy following exanthem subitum caused by human herpesvirus-6, *Kansenshogaku Zasshi* **66**:551–554 (1992).

137. SARANATH, D., MUKHOPADHYAYA, R., RAO, R. S., FAKIH, A. R., NAIK, S. J., AND GANGAL, S. G., Cell-mediated immune status in patients with squamous cell carcinoma of the oral cavity, *Cancer* **56**:1062–1070 (1985).

138. SAXINGER, C., POLESKY, H., EBY, N., GRUFFERMAN, S., MURPHY, R., TEGTMEIR, G., PAREKH, V., MEMON, S., AND HUNG, C., Antibody reactivity with HBLV (HHV-6) in U.S. populations, *J. Virol. Methods* **21**:199–208 (1988).

139. SCHIRMER, E. C., WYATT, L. S., YAMANISHI, K., RODRIGUEZ, W. J., AND FRENKEL, N., Differentiation between two distinct classes of viruses now classified as human herpesvirus 6, *Proc. Natl. Acad. Sci. USA* **88**:5922–5926 (1991).

140. SCOTT, D. A., AND CONSTANTINE, N. T., Human herpesvirus type 6 and HIV infection in Africa, *AIDS* **11**:1161–1162 (1990).

141. SHANAVAS, K. R., KALA, V., VASUDEVAN, D. M., VIJAYAKUMAR,

T., AND YADAV, M., Anti-HHV-6 antibodies in normal population and in cancer patients in India, *J. Exp. Pathol.* **6**:95–105 (1992).

142. SHEN, Y. Y., HUANG, A. M., JAHAN, N., MANAK, M., JAFFE, E. S., AND LEVINE, P. H., *In situ* hybridization detection of human herpesvirus-6 in biopsy specimens from Chinese patients with non-Hodgkin's lymphoma, *Arch. Pathol. Lab. Med.* **117**:502–506 (1993).

143. SHIBATA, D., AND WEISS, L. M., Epstein–Barr virus-associated gastric adenocarcinoma, *Am. J. Pathol.* **140**:769–774 (1992).

144. SHIRAKI, K., OKUNO, T., YAMANISHI, K., AND TAKAHASHI, M., Phosphonoacetic acid inhibits replication of human herpesvirus-6, *Antiviral Res.* **12**:311–318 (1989).

145. SIMMONS, A., DEMMRICH, Y., LA VISTA, A., AND SMITH, K., Replication of human herpesvirus 6 in epithelial cells *in vitro*, *J. Infect. Dis.* **166**:202–205 (1992).

146. SOBUE, R., MIYAZAKI, H., OKAMOTO, M., HIRANO, M., YOSHIKAWA, T., SUGA, S., AND ASANO, Y., Fulminant hepatitis in primary human herpesvirus-6 infection, *N. Engl. J. Med.* **324**:1290 (1991).

147. SPIRA, T. J., BOZEMAN, L. H., SANDERLIN, K. C., WARFIELD, D. T., FEORINO, P. M., HOLMAN, R. C., KAPLAN, J. E., FISHBEIN, D. B., AND LOPEZ, C., Lack of correlation between human herpesvirus-6 infection and the course of human immunodeficiency virus infection, *J. Infect. Dis.* **161**:567–570 (1990).

148. STEEPER, T. A., HORWITZ, C. A., ABLASHI, D. V., SALAHUDDIN, S. Z., SAXINGER, C., SALTZMAN, R., AND SCHWARTZ, B., The spectrum of clinical and laboratory findings resulting from human herpesvirus-6 (HHV-6) in patients with mononucleosis-like illnesses not resulting from Epstein–Barr virus or cytomegalovirus, *Am. J. Clin. Pathol.* **93**:776–783 (1990).

149. STEVENS, J. G., Human herpesviruses: A consideration of the latent state, *Microbiol. Rev.* **53**:318–332 (1989).

150. STRAUS, S. E., FRITZ, S., DALE, J. K., GOULD, B., AND STROBER, W., Lymphocyte phenotype and function in the chronic fatigue syndrome, *J. Clin. Immunol.* **13**:30–40 (1993).

151. STREICHER, H. Z. HUNG, C. L., ABLASHI, D. V., HELLMAN, K., SAXINGER, C., FULLEN, J., AND SALAHUDDIN, S. Z., *In vitro* inhibition of human herpesvirus-6 by phosphonoformate, *J. Virol. Methods* **21**:301–304 (1988).

152. SUGA, S., YOSHIKAWA, T., ASANO, Y., NAKASHIMA, T., KOBAYASHI, I., AND YAZAKI, T., Human herpesvirus-6 infection (exanthem subitum) without rash, *Pediatrics* **83**:1003–1006 (1989).

153. SUGA, S., YOSHIKAWA, T., ASANO, Y., NAKASHIMA, T., KOBAYASHI, I., AND YAZAKI, T., Activation of human herpesvirus-6 in children with acute measles, *J. Med. Virol.* **38**:278–282 (1992).

154. SUGA, S., YOSHIKAWA, T., ASANO, Y., NAKASHIMA, T., YAZAKI, T., FUKUDA, M., KOJIMA, S., MATSUYAMA, T., ONO, Y., AND OSHIMA, S., IgM neutralizing antibody responses to human herpesvirus-6 in patients with exanthem subitum or organ transplantation, *Microbiol. Immunol.* **36**:495–506 (1992).

155. SUMIYOSHI, Y., KIKUCHI, M., OHSHIMA, K., YONEDA, S., KOBARI, S., TAKESHITA, M., EIZURA, Y., AND MINAMISHIMA, Y., Human herpesvirus-6 genomes in histiocytic necrotizing lymphadenitis (Kikuchi's disease) and other forms of lymphadenitis, *Am. J. Clin. Pathol.* **99**:609–614 (1993).

156. TAGAWA, S., MIZUKI, M., ONOI, U., NAKAMURA, Y., NOZIMA, J., YOSHIDA, H., KONDO, K., MUKAI, T., YAMANISHI, K., AND KITANI, T., Transformation of large granular lymphocytic leukemia during the course of a reactivated human herpesvirus-6 infection, *Leukemia* **6**:465–469 (1992).

157. TAJIRI, H., NOSE, O., BABA, K., AND OKADA, S., Human herpes-virus-6 infection with liver injury in neonatal hepatitis, *Lancet* **335**:863 (1990).

158. TAKAHASHI, K., SONODA, S., KAWAKAMI, K., MIYATA, K., OKI, T., NAGATA, T., OKUNO, T., AND YAMANISHI, K., Human herpesvirus-6 and exanthem subitum, *Lancet* **1**:1463 (1988).

159. TAKIKAWA, T., HAYASHIBARA, H., HARADA, Y., AND SHIRAKI, K., Liver dysfunction anaemia and granulocytopenia after exanthema subitum (letter), *Lancet* **340**:1288–1289 (1992).

160. TANAKA, K., KONDO, T., TORIGOE, S., OKADA, S., MUKAI, T., AND YAMANISHI, K., Human herpesvirus 7: Another casual agent for roseola (exanthem subitum), *J. Pediatr.* **125**:1–5 (1994).

161. TORELLI, G., MARASCA, R., LUPPI, M., SELLERI, L., FERRARI, S., NARNI, F., MARIANO, M. T., FEDERICO, M., CECCHERINI-NEILLI, L., BONDINELLI, M., MONTAGNANI, G., MONTORSI, M., AND ARTUSI, T., Human herpesvirus-6 in human lymphomas: Identification of specific sequences in Hodgkin's lymphomas by polymerase chain reaction, *Blood* **77**:2251–2258 (1991).

162. TORELLI, G., MARASCA, R., MONTORSI, M., LUPPI, M., BAROZZI, P., CECCHERINI, L., BATONI, G., BENDINELLI, M., AND MUYOMBANO, A., Human herpesvirus 6 in non-AIDS related Hodgkin's and non-Hodgkin's lymphomas, *Leukemia* **6**(Suppl. 3):46S–48S (1992).

162a. TORIGOE, S., KUMAMOTO, T., KOIDE, W., TAYA, K., AND YAMANISHI, K., Clinical manifestations associated with human herpesvirus 7 infection, *Arch. Dis. Child. (England)* **72**(6):518–519 (1995).

163. VIZA, D., ARANDA-ANZALDO, A., ABLASHI, D., AND KRAMARSKY, B., HHV-6 inhibition by two polar compounds, *Antiviral Res.* **18**:27–38 (1992). (Erratum **19**:179, 1992)

164. WEISS, L. M., MOVAHED, L. A., WARNKE, R. A., AND SKLAR, J., Detection of Epstein–Barr viral genomes in Reed–Sternberg cells of Hodgkin's disease, *N. Engl. J. Med.* **320**:502–506 (1989).

165. WHITBY, D., HOAD, J. G., TIZARD, E. J., DILLON, M. J., WEBER, J. N., WEISS, R. A., AND SCHULZ, T. F., Isolation of measles virus from a child with Kawasaki disease, *Lancet* **338**:1215 (1991).

166. WILLIAMS, M. V., HHV-6: Response to antiviral agents, in: *Human Herpesvirus-6: Epidemiology, Molecular Biology and Clinical Pathology. Perspectives in Medical Virology*, Vol. 4 (D. V. ABLASHI, G. R. F. KRUEGER, AND S. Z. SALAHUDDIN, EDS.), pp. 317–335, Elsevier, Amsterdam, 1992.

167. WYATT, C. S., BALACHANDRAN, N., AND FRENKEL, N., Variations in the replication and antigenic properties of human herpesvirus-6 strains, *J. Infect. Dis.* **162**:852–857 (1990).

168. WYATT, L. S., AND FRENKEL, N., Human herpesvirus 7 is a constitutive inhabitant of adult human saliva, *J. Virol.* **66**:3206–3209 (1992).

169. WYATT, L. S., RODRIGUEZ, W. J., BALACHANDRAN, N., AND FRENKEL, N., Human herpesvirus 7: Antigenic properties and prevalence in children and adults, *J. Virol.* **65**:6260–6265 (1991).

170. YADAV, M., CHANDRASHEKRAN, A., VASUDEVAN, D. M., AND ABLASHI, D. V., Frequent detection of human herpesvirus 6 in oral carcinoma, *J. Natl. Cancer Inst.* **86**:1792–1794 (1994).

171. YAMANISHI, K., KONDO, K., MUKAI, T., KONDO, T., NAGAFUJI, H., KATO, T., OKUNO, T., AND KURATA, T., Human herpesvirus 6 (HHV-6) infection in the central nervous system, *Acta Paediatr. Jpn.* **34**:337–343 (1992).

172. YAMANISHI, K., OKUNO, T., SHIRAKI, K., TAKAHASHI, M., KONDO, T., ASANO, Y., AND KURATA, T., Identification of human herpesvirus-6 as a causal agent for exanthem subitum, *Lancet* **1**:1065–1067 (1988).

173. YOSHIKAWA, T., ASANO, Y., KOBAYASHI, I., NAKASHIMA, T.,

YAZAKI, T., SUGA, S., OZAKI, T., WYATT, L. S., AND FRENKEL, N., Seroepidemiology of human herpesvirus 7 in healthy children and adults in Japan, *J. Med. Virol.* **41:**319–323 (1993).

174. YOSHIKAWA, T., NAKASHIMA, T., SUGA, S., ASANO, Y., YAZAKI, T., KIMURA, H., MORISHIMA, T., KONDO, K., AND YAMANISHI, K., Human herpesvirus-6 DNA in cerebrospinal fluid of a child with exanthem subitum and meningoencephalitis, *Pediatrics* **89:**888–890 (1992).

175. YOSHIKAWA, T., SUGA, S., ASANO, Y., NAKASHIMA, T., YAZAKI, T., SOBUE, R., HIRANO, M., FUKUDA, M., KOJIMA, S., AND MATSUYAMA, T., Human herpesvirus-6 infection in bone marrow transplantation, *Blood* **78:**1381–1384 (1991).

176. YOSHIKAWA, T., SUGA, S., ASANO, Y., NAKASHIMA, T., YAZAKI, T., ONO, Y., FUJITA, T., TSUZUKI, K., SUGIYAMA, S., AND OSHIMA, S., A prospective study of human herpesvirus-6 infection in renal transplantation, *Transplantation* **54:**879–883 (1992).

177. ZAHORSKY, J., Roseola infantum, *J.A.M.A.* **61:**1445–1450 (1913).

12. Suggested Reading

ABLASHI, D. V., KRUEGER, G. R. F., AND SALAHUDDIN, S. Z. (EDS.), *Human Herpesvirus-6: Epidemiology, Molecular Biology and Clinical Pathology. Perspectives in Medical Virology,* Vol. 4. Elsevier, Amsterdam, 1992.

LEACH, C. T., SUMAYA, C. V., AND BROWN, N. A., Human herpesvirus-6 clinical implications of a recently discovered ubiquitous agent, *J. Pediatr.* **121:**173–181 (1992).

OREN, I., AND SOBEL, J. D., Human herpesvirus type 6: Review, *Clin. Infect. Dis.* **14:**741–746 (1992).

CHAPTER 16

Influenza Viruses

W. Paul Glezen and Robert B. Couch

1. Introduction

Influenza virus infections are the most important cause of medically attended acute respiratory illness.[143,149,150] Their impact is universal, affecting persons of all ages in all parts of the world,[6,7,268] including both temperate and tropical climates. Acute respiratory diseases, to which influenza is a major contributor, rival gastroenteritis as the principal cause of morbidity and mortality throughout the world. Epidemics occur annually, and, although they vary considerably in severity and intensity, the peak of acute respiratory illness causing persons to seek medical care always coincides with the peak of influenza virus activity. Traditionally, the impact of influenza epidemics has been measured by estimating excess mortality, a finding specific for influenza and primarily occurring among aged and chronically ill persons. However, hospitalization with acute respiratory disease (ARD) may also be considered to be a serious consequence of influenza virus infection worthy of prevention. Recent surveys have shown that only about one quarter of patients hospitalized with ARD during influenza epidemics are 65 years of age or older and that only 31% have underlying chronic conditions for which vaccine is currently recommended.[155] Thus, influenza is a serious threat for the entire population including young and apparently healthy persons.

The major factor responsible for the recurring nature of influenza epidemics is antigenic variation of the surface glycoproteins of influenza viruses. These changes lead to renewed susceptibility of persons infected previously so that reinfection and illness may occur. The current classification of influenza viruses recognizes this antigenic variation, and an appreciation of this is important to understanding the epidemiology of influenza virus infections. Table 1 shows the current nomenclature as well as earlier designations.[341] Three types of influenza viruses infect man, types A, B, and C. Types B and C appear to be unique pathogens of man and demonstrate less antigenic diversity than does type A. Three subtypes of type A have produced human infections; these are classified by the antigenic uniqueness of the surface glycoproteins, the hemagglutinin designated as H, and the neuroaminidase designated as N. Antigenic variants of the type A and B viruses are classified by the geographic site of the isolate and the culture number and the year of isolation. Examples of recent type A and B variants are shown in Table 2.

The features of influenza viruses that allow them to assault communities repeatedly have also confounded efforts to produce effective methods of prophylaxis. It is difficult to construct a vaccine for a constantly changing virus. Currently available vaccines do, however, significantly reduce the risk of influenza virus infection. The greatest handicap of these vaccines is the lack of confidence held by most clinicians and their patients in the ability of the vaccines to reduce morbidity; in the United States, they are regularly delivered to about 30% of those for whom they are recommended.[42] Current research must be aimed toward development of new implements and strategies for the control of influenza.

Recent studies of the epidemiology of influenza have led to changes and clarifications of some established concepts. In this chapter, we emphasize the epidemiology of influenza as it has occurred during the past 20 years. Reference is made to the historic ravages of pandemic influenza and to earlier data and concepts where appropriate.

W. Paul Glezen and Robert B. Couch • Influenza Research Center, Department of Microbiology and Immunology, Baylor College of Medicine, Houston, Texas 77030.

Table 1. Classification of Human Influenza Viruses

Types	Subtypes	Previous designation	Prototype
A	H1N1	Asw, Hsw1N1	A/New Jersey/8/76(H1N1)
		H0N1	A/PR/8/34(H1N1)
		H1N1	A/FM/1/47(H1N1)
	H2N2	A2	A/Japan/305/57(H2N2)
	H3N2	A2	A/Hong Kong/1/168(H3N2)
B	None	—	B/Lee/40
C	None	—	C/Taylor/1233/47

2. History

Epidemics of disease similar to influenza have been described almost from the beginning of recorded history.[329] Although complete descriptions began in the 16th century, accounts from the 10th and 11th centuries probably represent influenza epidemics. In fact, a recent reassessment of the plague of Athens of 430 BC attributes the high mortality to a combination of infection with influenza and the *Staphylococcus* of "toxic shock" syndrome.[212]

The first quantitation of mortality attributed to influenza was recorded by Robert Graves in Dublin in 1837.[156] This was accomplished by counting the new graves in the church cemetery following a severe epidemic and comparing the number with the count for the preceding year. William Farr introduced the concept of "excess mortality" in his vivid description of the London epidemic of 1847.[112] Pearl[266] and Frost[137] first used this concept in the United States to describe the 1918 influenza epidemic, but it was Selwyn Collins who systematically used excess

Table 2. Influenza Virus Variants, 1974–1993

Type A		Type B
H1N1	H3N2	
A/New Jersey/8/76	A/Port Chalmers/1/73	B/Hong Kong/5/72
A/USSR/90/77	A/Victoria/3/75	B/Singapore/222/79
A/Brazil/11/78	A/Texas/1/77	B/USSR/100/82
A/England/333/80	A/Bangkok/1/79	B/Ann Arbor/1/86
A/Chile/1/83	A/Philippines/2/82	B/Victoria/2/87
A/Taiwan/1/86	A/Mississippi/1/85	B/Yamagata/16/88
A/Texas/36/91	A/Sichuan/2/87	B/Panama/45/90
	A/Shanghai/16/89	
	A/Beijing/353/89	
	A/Beijing/32/92	

mortality as an index for recognition of influenza epidemics.[59,60] Collins estimated baseline mortality by calculating weekly arithmetic means, taking into account the seasonal fluctuation, and adjusting for long-term trends. The summary of excess mortality attributed to influenza from 1887 to 1956 is shown in Fig. 1. As a result of his work, excess mortality was regularly reported in the *Morbidity and Mortality Weekly Report* of the Public Health Service. Serfling refined the baseline estimate by deriving a regression function to describe seasonal variation.[300] Choi and Thacker improved the accuracy of the estimate by utilizing time-series forecasts of the expected pneumonia and influenza (P–I) deaths.[51] They found that excess deaths correlated well with the proportion, P–I deaths/total deaths; this proportion is currently reported in the *Morbidity and Mortality Weekly Report*.

The first influenza virus was isolated from chickens with fowl plague in 1901.[28] but it was not recognized that this was an influenza A virus until 1955.[289] This virus is now classified as H7N7. Shope isolated the swine influenza virus in 1931.[301,302] This was a finding of great significance because veterinarians of that era believed that influenza was transmitted to swine during the 1918 pandemic. This virus, which still infects swine, is considered to be closely related to the Great Pandemic strain. Smith, Andrewes, and Laidlaw successfully transmitted a human influenza virus to ferrets in 1933[311]; the observation was confirmed by Francis in 1934.[122] During the next few years, virus was shown to grow in embryonated eggs,[24,125] and virus particles were shown to agglutinate red blood cells.[171] This technology enhanced the ease of virus isolation and permitted the development of both complement-fixation[310] and hemagglutination-inhibition[288] procedures for serological studies. The first influenza B virus was isolated by Francis in 1940,[123] and the first influenza C virus by Taylor in 1947.[326] These descriptions of the causative agents and the practical methods for their study provided the basis for new and expanded studies of influenza. These have included etiologic documentation of the epidemiology, development, and testing of vaccines and antiviral drugs, and modern observations on the molecular virology of influenza virus and specific components of the immune response.

3. Methodology Involved in Epidemiologic Analysis

3.1. Sources of Mortality Data

Data from death certificates are compiled in many parts of the world for estimation of excess deaths that

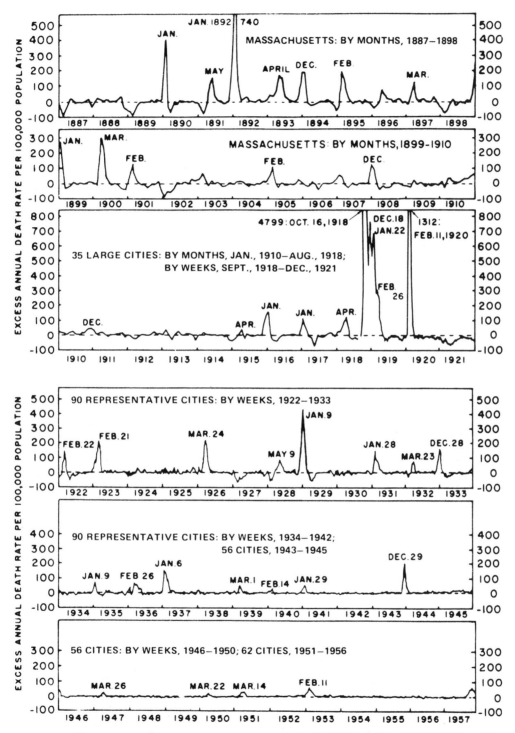

Figure 1. Excess annual death rate in Massachusetts and in representative United States cities, 1887–1956. From Collins.[59]

might be attributed to influenza. Since 1970, the World Health Organization (WHO) has requested mortality data from selected representative nations to provide some estimate of the worldwide impact of influenza.[6,340] In addition, several countries have reported national data, including the United States,[3,51] Great Britain,[55,327] Australia,[25,299] the Netherlands,[316] and Taiwan.[282] Different methods are employed to estimate the expected number of deaths in the absence of influenza so that excess mortality can be derived from the observed number during epidemics. Some countries use the regression model of Serfling[300] to project a smoothed sine curve of expected seasonal mortality. This method was developed in order to produce an advance estimation of the expected number of deaths in the absence of influenza, which then allows the use of the number of deaths rather than rates during an epidemic in order to generate timely reports of the excess of deaths.

Currently in the United States, Choi and Thacker have applied a time-series analysis method that utilizes a seasonal autoregressive integrated moving average model for calculating the expected number of deaths in the absence of influenza.[51] Pneumonia and influenza deaths are presented as the percentage of total deaths. Total deaths and the number of P–I deaths are reported weekly from 121 cities in the United States to provide timely information. In Great Britain, Clifford and her associates estimated excess mortality by a stepwise regression analysis[55]; this method takes into account the effects of other variables such as the report of influenzalike illnesses by practitioners, but the determination of the baseline expected number of deaths is retrospective, so this method cannot be used prospectively to describe the progress of an epidemic.

When national mortality data are available, the total excess mortality is estimated for each epidemic; this includes not only P–I mortality but excess deaths attributed to other causes. Usually deaths attributed to cardiac disorders and other pulmonary causes also occur in excess during influenza epidemics[154,175]; their numbers may be two to three times those attributed to pneumonia and influenza.

3.2. Sources of Morbidity Data

The sources of morbidity data are many and varied. These may be divided into two general categories—those associated with virological studies and those based on clinical diagnosis only. Influenza is not officially reported in most countries, but in the United States,[36] Belgium,[313] France,[79,163] and Great Britain[328] sentinel physicians report numbers of influenzalike illnesses on a weekly basis. In the former USSR, information about all visits to health clinics were reported directly to a central computerized system.[13] Since all absences from school or work for illness were verified by the clinic, the reporting was complete. This provided a unique set of morbidity statistics from which a statistical model was constructed to forecast the sequence of influenza epidemics in major cities and the severity of epidemics. An important source of illness data in the United States has been the analyses of serious morbidity occurring in the Portland health maintenance organization.[10–12] Although commonly used as an index of influenza, absentee data from schools and places of employment have not been consistently helpful; this is also true for tabulations of emergency clinic visits.

The most reliable morbidity data in the United States have come from longitudinal studies of defined populations that have included virological studies. The first of these was the Cleveland Family Study, which covered the decade from 1947 to 1957[185] and was reactivated for the first wave of the H2N2 epidemic of the autumn of 1957.[186] Other long-term morbidity studies with virological confirmation include those of the Influenza Research Center in Houston,[129–136,139,325] The Seattle Virus Watch[161] family[116] and group health care studies,[117–119] the Tecumseh Study in Michigan,[242–245] and the Chapel Hill, North Carolina, pediatric practice[49,86] and day-care center.[169,202] Similar studies have been conducted in New Zealand and Great Britain.[174,181,182] Studies lacking a defined population have also contributed much to morbidity assessments; these include the studies of hospitalized children in Washington, DC,[199] acute respiratory disease at military posts,[105,237] and cross-sectional studies from many other sites.[50,83,99,106,121,153,194,200,215,218,227,248,283,315,342,343]

3.3. Serological Surveys

Serological surveys have contributed significantly to epidemiologic assessments of influenza.[22,105,108,226,259,277,308] Although a variety of procedures have been used,[46,47,291] the hemagglutination-inhibition (HI) test[216] has been used most often for this purpose because the antibody measured is durable and is directed toward a surface glycoprotein, the hemagglutinin, that is most relevant for assessing susceptibility and immunity.

New antigenic variants of influenza identified in the laboratory by monospecific antisera raised in animals must be tested against a representative sample of recently obtained human serum specimens to determine clinical susceptibility. Such data are useful for predicting the likelihood of occurrence and potential severity of subsequent epidemics. These same new variants may be tested against sera from persons who were recently immunized with the currently available vaccine to determine if that vaccine will provide protection against the new variant. If the vaccine-induced antibody is not adequately cross-reactive with the new variant, it may be necessary to alter the vaccine formula if the new variant has shown epidemic potential in early outbreaks. Surveys such as these are regularly conducted by the Influenza Branch of the Centers for Disease for Disease Control and Prevention (CDC).

Serological surveys of persons in different birth cohorts have contributed to knowledge about immune responses to influenza viruses and have provided information about possible recycling of the influenza A subtypes.[81,124,228,229,250] Continuing studies of the specificity of HI antibody responses were stimulated first by the observations of Francis and his associates, who found that the HI antibody responses to infection may be greatest to the antigen of the first variant of a subtype to which the individual was exposed. Subsequent infections with related variants tend to reinforce the response to the first variant. These observations not only have elucidated the pattern of responses that might be expected from individuals after exposure to a series of type A variants but also have provided evidence for recycling of the major subtypes during the past century.

3.4. Laboratory Methods

Laboratory surveillance of influenza is maintained by a network of laboratories coordinated by WHO.[27,267,268] Member laboratories submit recent influenza isolates to the three Collaborating Centers for Influenza: one located at CDC in Atlanta that is designated as the WHO Collaborating Center for Surveillance, Epidemiology and Control of Influenza, and the other two in London and Melbourne, designated as WHO Collaborating Centres for Reference and Research on Influenza. There the important work of systematically testing the isolates for antigenic novelty is conducted. In the United States, most state health department laboratories and many virus diagnostic laboratories contribute to this effort.

Serum antibodies are usually measured by the HI test.[262] For this test, the serum must be pretreated to remove nonspecific inhibitors. Receptor-destroying enzyme (RDE) from *Vibrio cholerae* is usually employed for this purpose, but at times, other procedures are needed. The complement-fixation (CF) test is almost as sensitive

as the HI test, but it will not distinguish between infections with different influenza A subtypes since, as usually performed, it employs the internal nucleoprotein (NP) antigen, which is type specific. The use of the CF test with NP antigen may help distinguish between the antibody response to inactivated vaccine and natural infection, because a rise in CF antibody level above 1 : 8 is uncommon following use of whole-virus vaccines. Neutralizing antibodies are more specific than HI and are the best measure of protective immunity.[128]

The HI test is not sensitive enough to measure antibody in respiratory secretions; therefore, neutralizing antibodies or more sensitive tests such as the radio-immunoassay[305] or enzyme-linked immunosorbent assays (ELISA)[253] must be used. Some laboratories also use serum ELISA for detecting infection.[254,255] ELISA tests are particularly valuable for measuring antibodies specific for immunoglobulin class and subclass.

4. Biological Characteristics of Influenza

4.1. General Properties

Virions of influenza A and B virus are pleomorphic, spherical, and filamentous particles with a diameter of 80–120 nm.[201] The nucleic acid is single-stranded RNA of negative polarity that occurs in eight separate segments. The nucleocapsid exhibits helical symmetry and is surrounded by a lipid-containing envelope. Projecting from the envelope are two species of glycoprotein, the hemagglutinin (HA) and the neuraminadase (NA). Influenza C viruses are less well characterized, but they have only seven gene segments[236] and contain a single surface glycoprotein (HE protein) that exhibits hemagglutination and esterase activity.[201]

Embryonated hens' eggs were the first practical culture system for isolation of influenza viruses and are still used widely. Primary monkey kidney cultures are sensitive for most influenza viruses[15] but have become costly in recent years. Some continuous cell tissue culture lines have been employed recently and have been found to be satisfactory and much less expensive.[27,127] The Madin–Darby line of canine kidney (MDCK) and the rhesus monkey kidney-derived LLC-MK2 line have been used extensively. In general, tissue cultures have been as sensitive as eggs for isolation of influenza A viruses and more sensitive than eggs for influenza B viruses. The main exceptions to these generalizations are the influenza A (H2N2) and influenza C viruses that grow poorly in tissue culture. Influenza virus replication in eggs usually is detected by hemagglutination of avian, human O, or guinea pig red blood cells. Isolates usually are identified in HI tests using monospecific antisera, although CF tests can provide a type designation. A number of animal species may be used for developing antisera for determining type or subtype designation, but ferret antisera are the standard for differentiating between viruses within a subtype. Isolation of influenza virus in tissue culture is usually detected by hemadsorption (HAd) and identified in HAd-inhibition, immunofluorescence, or HI tests with antisera. More recently, rapid diagnostic methods have been used for recognition of influenza virus infection—both immunofluorescence and enzyme-linked immunoassays have been used with or without tissue culture amplification.[92,109,280,319,333]

The eight gene segments of influenza A viruses are known to code for ten proteins (Table 3).[210,256] The PB2, PB1, PA, and NP proteins are involved in transcription and replication of the nucleic acid. The NP, HA, and NA proteins are used for classification. The NP is antigenically stable and provides for a designation of type; there are no cross-reactions between the NP antigens of types A, B, and C influenza virus. The HA and NA consist of trimeric and tetrameric glycoprotein molecules, respectively; each exhibits antigenic variation, the characteristic of types A and B virus that accounts for the epidemiologic pattern of recurring infection and disease.

The M1 matrix protein exhibits antigenic stability similar to that of the NP. It binds to the nucleocapsid and participates in regulation of transcription and nucleocytoplasmic transport.[162,230] It accumulates at the cell surface during the assembly process and may contribute to an ordered assembly and budding process. The M2 protein is a smaller protein and is the product of a second reading frame of gene 7.[210,346] The M2 protein is a

tetrameric structure that spans the virion envelope and creates an ion channel.[173,321] This ion channel serves as a conduit for protons in endocytic vacuoles to penetrate into the virion particle and promote dissociation of the M1 protein and RNP so that transcription can ensue.[168] A second protein (NB protein) of the influenza B virus NA gene may serve a similar role for that virus type.[338] As with gene 7, gene 8 codes for two proteins, NS1 and NS2; they appear to be involved in nucleocytoplasmic transport of RNP.[4,209]

Cellular infection is initiated with attachment of the HA subunit of virions to cell surface glycoproteins containing N-acetylneuraminic acid (NANA). Antibody to the HA will prevent cellular infection. Penetration and uncoating occur via endosomes, with fusion of the viral envelope and plasma membrane resulting from an acid-dependent partial unfolding of the HA stalk.[307] Nucleocapsids are assembled in the nucleus; virion assembly takes place at the plasma membrane, and mature virus is released by budding. The lipid bilayer is acquired during the budding process.

The surface glycoproteins of newly synthesized virions contain NANA as a part of their carbohydrate structure. A major function of the neuraminidase is the enzymatic removal of the NANA residues to disrupt or prevent the occurrence of aggregates and thereby increase the number of free infectious particles. Antibody to the NA glycoprotein will reduce the number of detectable infectious units, presumably through enzyme inhibition and/or antibody-mediated aggregation.[197] Antibody to the HA protein may also lead to reduced numbers of new infectious units, presumably through promoting aggregation.[96] Since newly synthesized viral antigens appear on the cell surface before virion assembly, immune-mediated cell cytoxicity via specifically sensitized lymphocytes or antibody and complement may play a role in control of the infectious process.

4.2. Antigenic Variation and Genetics of Influenza Viruses

As a consequence of their segmented genome, influenza viruses exhibit high frequencies of genetic recombination or, more properly, genetic reassortment *in vitro*.[201,256] In cells simultaneously infected with two viruses of the same type, but with different genetic properties, the assembly process may incorporate RNA segments from either parent into progeny virions. As a consequence, progeny virions may exhibit properties of either or both parent viruses. The occurrence of reassortment between two influenza A viruses is considered to be the

Table 3. The Proteins of Influenza A Virus

Gene segment	Protein	Designation
1	Basic protein 2	PB2
2	Basic protein 1	PB1
3	Acidic protein	PA
4	Hemagglutinin	HA
5	Neuraminidase	NA
6	Nucleoprotein	NP
7	Matrix proteins	M1, M2
8	Nonstructural proteins	NS1, NS2

most likely basis for emergence of new subtypes of type A viruses, the so-called antigenic shift.[214,295] This is based on sequence analysis of the genes of viruses isolated from man and various animal species. These analyses have led to the belief that new human pandemic influenza A viruses derive their HA and NA surface glycoproteins from an avian lineage.[335] A large avian gene pool exists for influenza A viruses. "Humanizing" of these avian viruses could occur through reassortment between human and avian viruses in humans or avian species; however, available evidence suggests that the domestic pig may serve as a "mixing vessel" for the human and avian viruses as the pig appears to represent a permissive host for both viral sources.[335] In this regard, it has been proposed that southern China may represent an epicenter for new pandemic viruses, since intimate contact exists between avians via ducks and duck ponds, domestic pigs, and humans. Nevertheless, it is clear that the appearance of a new influenza A subtype remains an unpredictable event in location and time. The absence of an avian pool of influenza B viruses could explain the absence of subtypes of type B.

Minor antigenic changes in the HA and NA of type A and B viruses occur sequentially with time (antigenic drift). These progressive changes alter antigenic sites for reactivity with human antibodies so that the antibody becomes progressively less reactive with circulating viruses. The resulting renewed susceptibility may then lead to a repeat infection. These drift (variant) viruses are presumed to be selected from naturally occurring mutants by population immunity.[256] In addition to a degree of antigenic uniqueness, these viruses must also possess the virulence properties required for illness production and transmissibility among humans.

Considerable effort has been expended in characterizing the amino acid changes accompanying antigenic drift. These studies have described the degree of change required for a virus to be a new variant in conventional characterizations with polyclonal ferret sera. In terms of the three-dimensional structure, this consists of a change in at least four amino acids in at least two of the four antigenic regions on the globular tip of the HA molecule.[336] Sequence analysis of the genes of a large number of human viruses indicated the existence of considerable heterogeneity in the HA gene among strains.[77] Cocirculation of H1, H3, or B viruses of different lineages may occur and the patterns of HA evolution do not permit a prediction of the strain that will become dominant in a future epidemic. This complexity of strains also complicates decisions for vaccine strain selection. Antigenic drift in the NA has also been shown to involve specific amino acid changes.

5. Descriptive Epidemiology

5.1. Incidence and Prevalence Data

5.1.1. Mortality Data. A statement made by Robert Graves after the influenza epidemic of 1836–1837 is still an accurate description of mortality from influenza.[156] In his lecture, he compared influenza to cholera and stated, "Influenza is not by any means so severe or so rapidly fatal a disease as cholera, but the mortality which it has produced is greater, as *it affects almost every person in society*, while the ravages of cholera were comparatively limited." Although the current mortality does not approach the 5 per 1000 rate seen in the 1918 epidemic, the CDC has estimated that an average of 20,000 persons die each year in the United States.[31] This estimate is based on excess mortality from all causes from data provided by the National Center for Health Statistics.

Figure 2 illustrates the occurrence of excess mortality in the United States and seven other countries reporting to the WHO for the period from 1973 to 1983.[340] The heavy oscillating line represents the expected mortality, and the lighter solid line showed the observed mortality, with high peaks indicating influenza epidemics. In the United States (right-hand graph, Fig. 2, page 481), the most severe epidemics occurred with A/Victoria (H3N2) in early 1976 and with A/Bangkok (H3N2) and A/England (H1N1) in December 1980–January 1981. Excess deaths were recognized in six respiratory disease seasons in the United States but were absent in four. Considerable variability is noted for the other nations.

An analysis of P–I deaths in Houston during this period revealed a noticeable increase in deaths recorded at the city health department yearly coinciding with each influenza epidemic.[154] This was true for all epidemics regardless of severity or intensity or whether the peak occurred as early as mid-December or as late as mid-March. No midwinter increase in number of deaths was noted in the absence of influenza activity. Cross correlations by time-series analysis of P–I deaths with the epidemic curve for virus culture-proven morbidity showed the strongest positive correlation at the 2-week lag ($r = 0.72$, $P < 0.01$). This is illustrated in Fig. 3, which shows the composite curve for P–I mortality peaking 2 weeks after the peak of influenza virus activity. During this time, 62.5% of persons dying with pneumonia or influenza were 65 years of age or older. A survey of clinical status prior to the terminal illness showed that from 10 to 24% were carrying on normal activities for their ages prior to death and that 15 to 31% could care for their own personal needs; the remaining 52 to 67% required nursing care. The low

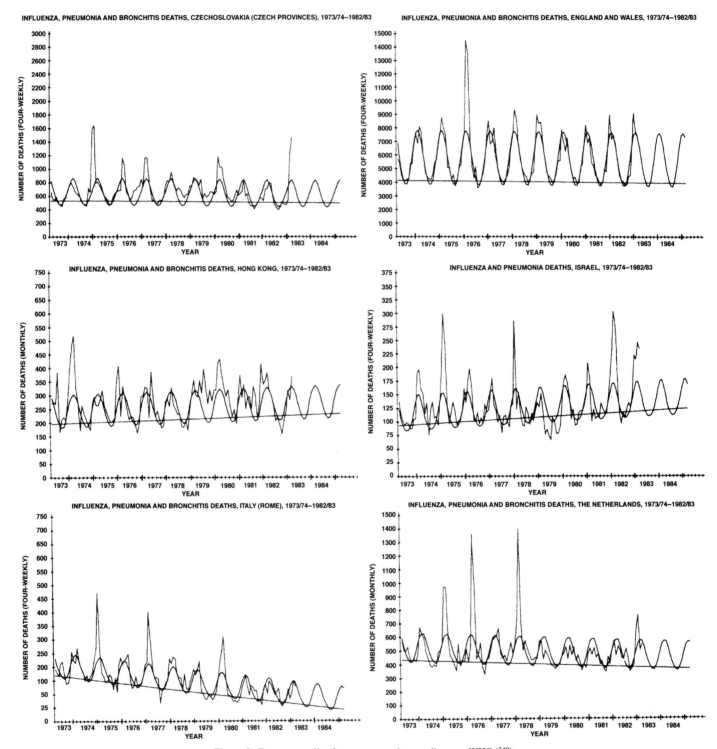

Figure 2. Excess mortality from acute respiratory diseases, WHO.[340]

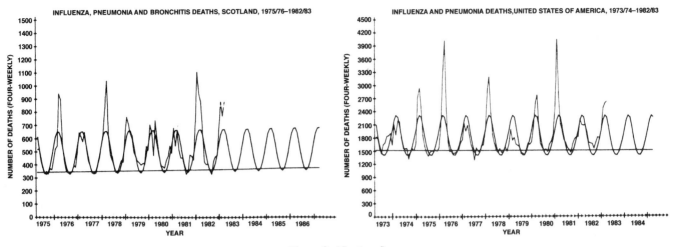

Figure 2. (*Continued*).

est proportion with underlying conditions or receiving nursing care were seen with the severe A/Victoria (H3N2) epidemic of 1976, indicating that this virus caused deaths in many previously healthy persons.

At the national level, excess mortality is a dependable and useful tool for influenza surveillance. Since epi-

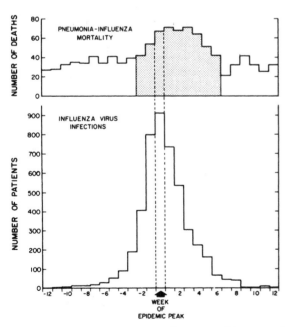

Figure 3. The composite epidemic curve for persons with acute respiratory illness infected with influenza viruses and for deaths attributed to pneumonia and influenza. The number of deaths was plotted in relation to the week of peak influenza activity for each season in Houston, 1974–1981. From Glezen *et al.*[154]

demic influenza is virtually the only condition that will cause noticeable increases in P–I mortality, it carries great specificity. The necessary information is readily available on death certificates, and timely and useful reports can be disseminated during the course of epidemics. Excess mortality is, however, a relatively insensitive measure of the impact of influenza epidemics. During the period from 1966 through 1993, excess mortality was detected in only 19 or 27 epidemic periods despite virological evidence that influenza activity occurred during each respiratory disease season.[29,32–39,41,44] Six of the seasons without excess mortality were years when influenza B was the predominant virus; one was predominantly A (H2N2), and one was A (H1N1).

Several factors may be involved in the variability of detection of these epidemics by excess mortality. An obvious factor is the accumulated immunity of the adult population to influenza B viruses, which decreases the susceptibility of the older and chronically ill population and, therefore, the risk of death. Also, the baseline expected number of deaths may be artifactually high so that epidemics with low mortality go undetected.[143,154] In this regard, none of the statistical methods for estimating the baseline have been validated by observations through a respiratory disease season in which the absence of influenza was confirmed virologically. There is a minimal overlap in the occurrence of outbreaks of influenza and the other major viruses causing respiratory disease, so that influenza is not superimposed on disease caused by the other agents but rather displaces them, making a high midwinter threshold for detection of excess mortality unreasonable. For influenza virus infections to cause a sufficient number of deaths to exceed the epidemic threshold under these circumstances, the epidemics must be intense and must occur simultaneously in most parts of the country.

Another fact that has contributed to the insensitivity of the excess mortality method is the declining rate of mortality from pneumonia and other acute respiratory diseases,[34,35] despite the fact that the overall occurrence of those acute conditions has not decreased. Furthermore, P–I deaths represent a relatively small proportion of deaths that can be correlated with influenza epidemics; about twice as many deaths attributed to heart disease occur in peaks that can be correlated with P–I deaths.[154,175] This may result from the fact that the rules for assigning the cause of death under the International Classification of Diseases, Adapted (ICDA) usually give precedence to cardiac conditions. One study found that only 61% of P–I deaths were reported by the criteria established by the CDC for their weekly reports and less than one quarter would be so classified under the ICDA coding rules.[11]

In Houston, during three epidemics that occurred between 1978 and 1981, the number of deaths among persons hospitalized with ARD was considerably higher than the number coded to P–I from death certificates.[150,271] Surveys of hospital discharge diagnoses during epidemic periods may give a more accurate picture of mortality during influenza epidemics. The average annual death rate for persons hospitalized with ARD during recent influenza epidemics in Houston, if age-adjusted to the United States population, would yield about 20,000 deaths per year, or about twice the current estimate for excess mortality from death certificate data.[155]

5.1.2. Morbidity. Longitudinal studies of influenza virus infections in defined populations have provided the best infection and illness rates. Over the past two decades, three centers have carried out studies in different geographic areas of the United States, Seattle,[116,161] Tecumseh,[243–245] and Houston,[129–136] which have yielded remarkably consistent data despite the fact that the studies differed considerably in their structures and methods of study. Table 4 shows the composite infection and illness rate for the Houston Family Study for the period from 1976 through 1985. The overall infection rate was 33 per 100 person-years in these families with young children. Influenza A (H3N2) viruses produced the highest infection rates for the group as a whole and for infants and parents. Influenza B and influenza A (H1N1) gave consistently high rates in school children but lower rates in children under 2 years of age. The infection rates observed in the other centers were similar for the years when the studies coincided except that influenza B was not epidemic in Seattle in 1976–1977; both influenza A/Victoria (H3N2) and B occurred during the previous season. Somewhat lower infection rates in Tecumseh for the 1976–1981 period reflect the fact that sampling there included all households, not just those with children. These studies amply demonstrate the consistently high infection rates for influenza viruses over the past 20 years.

Although less apparent, influenza C has also yielded consistently high infection rates.[98,180,189,330] Antibody surveys have demonstrated that a majority of children acquire infection by 5 years of age, and virtually all have by 15 years of age. Illnesses associated with influenza C are usually uncomplicated, involving only the upper respiratory tract.

5.1.3. Medically Attended Illness. The impact of influenza on medically attended illness has been illustrated by the community surveillance program of the Influenza Research Center in Houston, where physicians in sentinel clinics regularly obtained cultures from all patients presenting with febrile ARD.[71,146] These sentinel clinics served patients who were representative of 2.5 million persons living in Harris County, Texas. The number of visits per week and the number of cultures that were positive for influenza viruses for the 11-year period from 1974 to 1985 are shown in Fig. 4. Each epidemic is labeled with the predominant virus for that season. Influenza viruses were epidemic each year, and the peak of visits for ARD coincided with the peak of influenza virus activity. All ARD visits to a large health maintenance organization were tabulated for 2 years from 1981 to 1983,[149,155] and the age-specific rates for ARD visits during the most intense periods of influenza virus activity (10 weeks during each season) are shown in Table 5. The risk of a visit for ARD was about 12 per 100 persons among these employed persons and their families. Both viruses that

Table 4. Influenza Virus Infection and Illness Rates[a] for Persons Followed Longitudinally during an 8-Year Period, Houston Family Study, 1976–1984

Age (years)	Number of person-years	Number (rate) of infections	Number (rate) of illnesses
<2	332	118 (35.5)	112 (33.7)
2–5	474	211 (44.5)	178 (37.6)
6–10	300	143 (47.7)	118 (39.3)
11–17	149	60 (40.3)	45 (30.2)
18–24	178	41 (23.0)	35 (19.7)
25–34	651	140 (21.5)	101 (15.5)
≥35	257	54 (21.0)	38 (14.8)
Totals	2341	767 (32.8)	525 (26.3)

[a]Per 100 person-years.

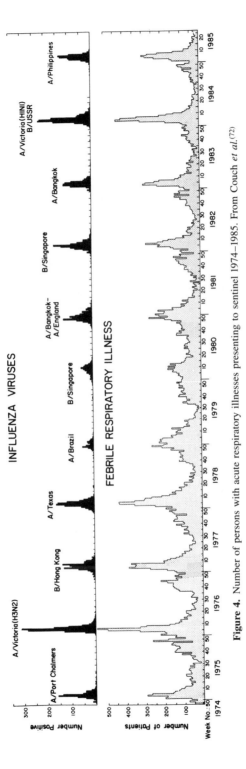

Figure 4. Number of persons with acute respiratory illnesses presenting to sentinel 1974–1985. From Couch *et al.*[72]

Table 5. Rate of Visits by Age for Acute Respiratory Disease for Members of a Health Maintenance Organization during Influenza Epidemics in Houston, Weeks 2–11, 1982 and 1983

Age (years)	Rate per 100 persons for epidemic period	
	B/Singapore, 1982	A/Bangkok (H3N2), 1983
<5	26.6	29.3
5–14	18.5	12.4
15–24	7.7	7.0
25–44	11.9	8.9
≥45	10.8	8.4
All ages	13.7	11.3

were responsible for these outbreaks were returning for the second time—B/Singapore was the predominant virus in 1978–1979, and A/Bangkok (H3N2) was a major virus in the 1980–1981 epidemic.

Long-term studies have documented virologically the role of influenza virus infections as causes of serious morbidity. In Chapel Hill, studies of children presenting to a pediatric practice with acute lower respiratory tract illness over a 10-year period associated influenza viruses with tracheobronchitis[49] particularly, and longitudinal observations of children in day care indicated that influenza virus infections were frequently accompanied by otitis media.[169,170] In Washington, DC,[199] influenza virus infections were associated with over 5% of hospital admissions of children with ARD in 1957–1976. Foy and associates documented influenza virus infection annually among patients of the Group Health Cooperative of Puget Sound with pneumonia at rates ranging from 0.1 to 2.5 per 1000 per year for the period from 1964 to 1975.[117] Over a 9-year period, Evans studied the etiology of acute respiratory infections in college students; influenza was associated with 33.9% of ARD illnesses, 14.7% of acute bronchitis, and 3.6% of pneumonias.[106] Numerous additional cross-sectional studies have added virological evidence of the association of influenza virus infection with pneumonia and other lower respiratory illnesses.[143,144,286,320]

To quantitate the risk of hospitalization with ARD during influenza epidemics, surveys have been conducted in Houston.[143,149,155,271] These surveys have utilized computer-generated lists of discharge diagnoses of patients from a sample of hospitals in Harris County. Each year the peak of ARD hospitalizations of adults and school children has coincided with the peak of influenza virus activity. The occurrence of ARD hospitalizations of infants and preschool children is affected by all of the major respiratory viruses, including respiratory syncytial,

parainfluenza, and influenza viruses. The rate of ARD hospitalizations by 2-week periods for 1978–1982 is shown in Fig. 5. The parainfluenza and respiratory syncytial virus outbreaks that preceded the influenza B epidemics of 1979–1980 and 1981–1982 are evident by the double peaks for children less than 5 years of age. The age-specific rates of persons hospitalized with ARD during the three epidemics of 1978 to 1981 are shown in Table 6. Harris County has a relatively young population, with only 6.1% of the population over 65 years of age. If the rate for 1980–1981 is age-adjusted to the U.S. census, the rate would be 16.9 per 10,000. Barker and Mullooly have found similar rates of hospitalization in a study of patients enrolled in a Portland, Oregon, health maintenance organization,[12] and they found that persons immunized with an antigenically appropriate influenza vaccine had a significantly decreased risk of hospitalization and death.[10]

Priorities for administration of influenza vaccine have been recommended by the Immunization Practices Advisory Committee of the U.S. Public Health Service.[40,43] These recommendations are reviewed each year and altered under appropriate circumstances. The high-risk categories generally include groups of persons who have underlying conditions that put them at special risk for influenza. Highest priority for vaccination is given to adults and children who have chronic disorders of the cardiovascular or pulmonary systems and persons confined to nursing homes and other chronic-care facilities. Vaccine is also recommended for adults and children with chronic metabolic diseases (including diabetes), renal dysfunction, anemia, immunosuppression, or asthma. Children receiving long-term aspirin therapy are recommended to receive vaccine because of the risk of developing Reye's syndrome following influenza infection.

When discharge diagnoses of persons hospitalized with ARD were examined, it was found that only 40.6% of them had one of the underlying illnesses that would call for influenza vaccine by the present standards.[150,155] Thus, current recommendations would have a modest effect on the risk of hospitalization even if fully implemented.

5.2. Epidemic Behavior

Influenza epidemics occur yearly and coincide with the peak occurrence of acute respiratory illnesses that cause persons to seek medical care.[71,146] Infections with both influenza A subtypes, H1N1 and H3N2, and influenza B are detected each year, but usually variants of only one or two are epidemic in a given season.

Some previously accepted concepts of the epidemic

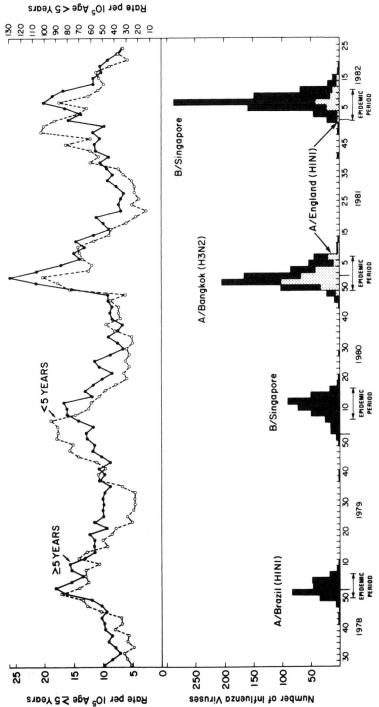

Figure 5. Estimated rates of hospitalizations with acute respiratory disease for children less than 5 years of age and for persons 5 years and older by 2-week periods, and the temporal relationship to influenza epidemics defined by virological surveillance at the Influenza Research Center, Houston, 1978–1982. Adapted from Glezen *et al.*[(149)] and Perrotta *et al.*[(271)]

Table 6. Estimated Rate of Hospitalization of Persons with Acute Respiratory Disease, by Age during Three Influenza Epidemics, Harris County, Texas, 1978–1981

Age (years)	Rate[a] by epidemic period		
	A/Brazil, 1978–1979	B/Singapore, 1979–1980	A/Bangkok–A/England, 1980–1981
<1	50.5	61.4	73.4
1–4	26.7	26.0	35.4
5–9	6.2	7.4	7.4
10–24	5.0	3.4	5.4
25–44	6.4	7.8	11.2
45–54	6.8	7.9	13.2
55–64	8.9	15.9	18.0
≥65	30.4	37.8	58.9
All ages	9.0	10.2	14.2

[a]Per 10,000 persons.

behavior of influenza are no longer considered true. The expectation that a new influenza A subtype or "pandemic strain" would appear at 10- to 14-year intervals is no longer tenable. This was the interval observed between the emergence of the old H0N1, H1N1, and the H2N2 and H3N2 viruses, but with reclassification of the A viruses, the interval since emergence of the previous subtype, H1N1, was almost 40 years, and it has now been almost 20 years since H3N2 viruses appeared. The belief that two A subtypes could not circulate simultaneously has been nullified by the concurrent circulation of both H1N1 and H3N2 viruses since 1978.[193]

A growing body of evidence is against the likelihood that influenza viruses may continue to spread within a population between epidemics. The short generation time and sharp seasonality of influenza make it unlikely that infection can be perpetuated even in a large population (T1 million)[345]; current data support the reintroduction of virus each season.[147] Therefore, the viruses must migrate between the Northern and Southern hemispheres to become epidemic in the respective temperate zones during the cold-weather season.[177]

5.3. Geographic Distribution

Influenza virus infections occur regularly throughout the world. With the availability of rapid air transportation to most areas, it is unlikely that many places escape exposure to most variants. The converse is also true; when a new variant with epidemic potential arises anywhere in the world, it will spread rapidly.[211,267] A recent example of this was the influenza A/Victoria/3/75 (H3N2) strain, which was first isolated in Australia in April 1975 and produced epidemics throughout the Northern hemisphere

in the winter of 1975–1976.[146,269] Other evidence of rapid dissemination is the fact that the same strains have been epidemic each year in the United States[76,120,121,258,304,306] and Great Britain[268–270] with only a few exceptions, and for those exceptions, occurrence was different by only one year. These same patterns have been confirmed in many sites throughout the world by national laboratories[8,20,93,110,111,195,208,279,337] and by WHO.[6] These studies and others[48,107] have also confirmed that infections occur within the tropics with at least the same frequency and severity as temperate zones.

5.4. Temporal Distribution

In the temperate zones influenza occurs during the cold-weather season. Epidemics can occur at any time but usually do not spread extensively unless schools are in session. Viruses that are novel for the population, such as the 1918 pandemic strain and the first wave of influenza A (H2N2) when it appeared in 1957,[211] may peak in autumn, but most peak between December and March. In Houston for the period from 1974 to 1993, the peak of activity occurred in February for 9 of the 19 seasons, in December for three, in January for two, and in November and March for one.[71] During three seasons, 1988–1989, 1990–1991, and 1992–1993, multiple peaks have occurred with activity of influenza B and influenza A (H1N1). Influenza B has peaked first in December and January followed by subsequent smaller peaks of influenza A (H1N1) in February and March. The seasonal occurrence in the tropics is not well defined. Some studies suggest that outbreaks can occur during any time of the year and that multiple outbreaks are common.[93,279] This is the pattern that would be expected if influenza does not persist through the warm-weather seasons in temperate zones, as noted in Section 5.2.

For about one half of the epidemics in Houston, the predominant virus was present during the preceding season, usually as influenza activity was waning.[147,155] This herald wave activity was also noted in Seattle during the period from 1976 to 1980.[116,121] If observations of this type can be systematized by improved virological surveillance, they carry the potential for earlier recognition of new variants with epidemic potential.

5.5. Serological Surveys

Serological surveys have been important for determining the age-specific susceptibility for infection with influenza viruses.[105,108,292] Many important observations have come from such surveys. Shope found in 1936[303] that persons over 12 years of age had antibodies to the influenza virus that he had isolated from swine in 1931. This observation has been confirmed by many investigators[232,292] and has led to the assumption that a virus antigenically similar to the swine influenza virus caused the 1918 pandemic. Davenport and his associates extended these observations and suggested that antibody-forming mechanisms appear to be oriented toward the antigens of the first influenza infection in childhood so that subsequent exposures to related strains result in progressive reinforcement of the primary antibody.[81] For example, in 1976, persons who were born before 1924 had HI titers to the influenza A/New Jersey/76 (H1N1), or swinelike virus, that had persisted for more than 50 years.[264] When the A/USSR/77 (H1N1) virus emerged in 1977, it was found to be antigenically identical to the A/FW/50 (H1N1) virus and similar to other viruses that circulated between 1946 and 1957. Persons who were born between 1941 and 1950, whose first infections were with these viruses, had the highest HI titers to the A/USSR virus when they were tested in 1978.[278]

The testing of serum specimens of older persons that were collected before either the H2N2 or H3N2 viruses had circulated in 1957 and 1968, respectively, has suggested that influenza viruses with antigenically similar hemagglutinins circulated before the turn of the century.[229,250] The data are most persuasive for activity of a virus similar to A/Hong Kong/68 (H3N2) before 1890, including the severe epidemic of 1889–1890.[293] More than 90% of persons born before 1890 had antibodies to this virus in sera collected before the appearance of H3N2 viruses in 1968. Furthermore, persons in that birth cohort ($\geq$75 years of age) had lower morbidity[293] and mortality[176] rates than slightly younger persons (65–74 years of age) during the first epidemic. About 26% of persons born between 1857 and 1877 had antibodies to A/Japan/305/57 (H2N2), suggesting that a similar virus may have been present in the middle of the 19th century.[233]

5.6. Age

All age groups are at risk for infection with influenza viruses. Infants are usually spared infection during the first weeks of life, and this protection has been correlated with maternal antibody levels.[276] In general, influenza A (H3N2) viruses have caused more serious infections in infants than have influenza A (H1N1) or B viruses.[131,136,242,243,245] Preschool children in day care or living with older siblings and school children have highest infection rates.[131,325] Susceptibility decreases with age in relation to previous experience with antigenically similar viruses. Currently, older persons have least resistance to infection with influenza A (H3N2) viruses, which did not appear until 1968, and most resistance to influenza A (H1N1) viruses because antigenically related viruses circulated between 1918 and 1957, giving them prior experience with these antigens. Their susceptibility to influenza B viruses is intermediate; these viruses have circulated throughout their lives, but with gradual antigenic alterations that have resulted in at least partial immunity for many.

When age distributions of infected persons are examined through the course of an epidemic, a consistent shift is observed.[143,146,148] Early in the course of an epidemic, from 35 to 50% of affected persons will be school aged, and as the epidemic progresses, this proportion will decrease, and the proportion of affected preschool children and adults will increase. This age shift suggests that the initial horizontal spread in the community occurs among school children and is followed by vertical spread to their older and younger contacts.

Another type of age shift has occurred regarding persons presenting for medical care with influenza A (H1N1) infections.[152] At the time that H1N1 viruses reappeared in 1977–1978, most of the associated influenzalike illnesses occurred in older school children and young adults. Few medically attended illnesses of persons born before 1951 were documented by virus culture. This distribution probably resulted from immunity generated by H1N1 viruses that were prevalent prior to 1951. The influenza A/USSR/77 (H1N1) virus was identical to a virus isolated in the United States in 1950. As new variants of H1N1 virus appeared (Table 2), the age distribution of culture-positive cases broadened to include increasing proportions of preschool children and adults born before 1951. These observations provide clinical evidence of the divergent evolutionary paths taken by subsequent variants of the identical viruses labeled A/Fort Warren/50 (H1N1) and A/USSR/77(H1N1).

Many other studies have demonstrated the importance of school children as disseminators of influenza in the community and introducers of infection into families. Studies in Cleveland,[184] Houston,[325] Seattle,[116] and Tecumseh[242] have solidified this concept and also recognized the importance of preschool children as well, particularly those in regular day care outside of the home. Longini has introduced a useful mathematical model to estimate the frequency of community-acquired infection and the secondary attack rate within the household.[221,222] Using infection data from both the Seattle and Tecumseh studies, he has shown that the model closely simulates the best available infection data.

A pronounced difference in the age distribution of affected persons is also seen when socioeconomic status is considered.[151] The risk for an infant from a low-income family to experience an illness requiring medical attention is about five times greater than for an infant in a middle-income family. Conversely, adults in low-income families are less likely to seek medical care for influenza virus infections. The same pattern of risk has been observed for children of the two income groups in studies of influenza-related Reye's syndrome.[323,324]

5.7. Race, Sex, and Occupation

Any situation that allows daily mixing of large numbers of susceptible persons will increase the risk of infection. Such high-risk groups include children in day care[202] or boarding schools,[82] school children,[50,184] college students,[215] military recruits,[206,216,237,238] and residents of nursing homes.[265] No special susceptibility has been noted for any racial groups or for either gender.

5.8. Nosocomial Infections

Any infectious agent that spreads rapidly in the community and affects a sizable proportion of the population poses a serious risk for introduction into the hospital or extended-care facility. Numerous reports have documented the occurrence of nosocomial influenza infections and the frequency of serious consequences.[95,144,200] Current recommendations give priority to use of both vaccine and amantadine to decrease the risk of spread in these facilities.[40,43]

5.9. Socioeconomic Status

Studies in Houston have demonstrated that young children in low-income families are at a greater risk for infection than are preschool children of middle-income families.[151,153] Conversely, adults in middle-income settings seem to have a greater risk. This suggests that intense exposure to influenza early in life may provide some measure of protection that persists into adulthood. It also suggests that the pattern of infection in developing countries may resemble the experience of low-income children in the United States and that this early exposure may contribute to the high pneumonia death rates for young children in these areas.[18,286]

6. Mechanism and Route of Transmission

Influenza is transmitted from person to person by transfer of virus-containing respiratory secretions from an infected person to a susceptible recipient. This could be accomplished by a means requiring direct or indirect contact or by inhalation of airborne droplets of respiratory secretions. Although not definitely demonstrated, most infections are probably acquired by inhalation of small (1- to 5-μm-diameter) particles. Epidemiologic evidence supporting significance of airborne spread includes the usual rapid increase to a peak of occurrence of cases in most population groups and the high attack rates when most persons are susceptible.[146,157,158,203] It is not reasonable to suggest that infection could disseminate so rapidly by direct contact in such a brief period of time. More tangible evidence for airborne spread is provided by analysis of a factory outbreak in Britain and an airline-related outbreak in Alaska.[1,248] Risk of infection in the factory shared features with tuberculosis in that risk was not related to proximity of workers within rooms and was increased as numbers of workers in the room increased. In addition, in 1977, a commercial passenger flight was detained for 4½ hr without provision for auxiliary ventilation most of the time. An acutely ill, coughing passenger appears to have been the source of virus that induced influenza in 72% of exposed persons in a pattern indicating a common source. Since most passengers had no direct or indirect contact with the passenger, transmission by the airborne route is the only reasonable possibility.

Additional features of influenza also support airborne transmission. Cough and clinical acute tracheobronchitis are typical of natural influenza, and induction of such a syndrome in volunteers requires inhalation of virus in small particles that will initially deposit in the lower respiratory tract.[65] Studies of naturally occurring cases have indicated the early occurrence of infection of the lower respiratory tract through detection of virus and abnormal pulmonary function tests.[220,334] Finally, studies in volunteers have indicated that the amount of virus required to infect the lower respiratory tract is very small (<5 infectious units); about 100-fold more virus is required to infect the nasopharynx.[205] The existence of a propensity for influenza to spread by the airborne route is supported by animal studies, since influenza had also been shown to be transmitted by the airborne route between infected and susceptible mice and ferrets, the most commonly used experimental animals.[5,297]

Although it seems most likely that the airborne transmission route represents the major route of spread, it is probable that influenza is also transmitted by direct contact. Sometimes, spread of influenza among population groups is slow, and the overall attack rates are low.[158] In addition, the period before and after the epidemic period may be prolonged and involve few persons. Moreover, in the experimental situation, influenza can be induced by intranasal drops, a method simulating direct contact.[63,65] Finally, the ability of influenza to spread to some extent in all human populations where circumstances of exposure must vary greatly suggests that influenza virus can spread by more than a single means.

7. Pathogenesis and Immunity

7.1. Pathogenesis

The incubation period is 1 to 5 days. Inhalation of influenza virus in a small-particle aerosol will result in an equal chance for deposition of virus on the nasopharyngeal or the lowest bronchial passages, whereas direct contact transmission will only lead to nasopharyngeal deposition.[65,205] Although no direct comparative data for man are available, the range of 120 to 300 $TCID_{50}$ required to initiate infection of the nasopharynx of man with H3N2 viruses and the reported induction of infection with as little as 3 $TCID_{50}$ of an H2N2 virus in a small-particle aerosol suggest a greater susceptibility of the lower respiratory tract to infection with influenza virus.[62,205] The existence of a differential in susceptibility of the two parts of the respiratory tract was shown in comparative studies of the two routes of inoculation with adenovirus type 4 and a rhinovirus: the former exhibited a greater infectivity for the lower respiratory tract, whereas the latter was more infectious for the nasopharynx.[65,74] The apparent differential susceptibility of the respiratory tract in combination with data on the probable transmission method suggests that most infections are initiated in the lower respiratory passages.

The mucous blanket covering the respiratory mucosa contains mucoproteins with NANA as part of their carbohydrate structure. The HA of the virus should attach to these molecules, which contain the specific receptor for influenza virus. This binding of HA to mucoproteins in various body fluids was recognized in early studies with influenza viruses.[58] The ability of the neuraminidase to reverse this binding led to adoption of pretreatment of sera and other fluids with neuraminidase (receptor-destroying enzyme) as a basis for removing the inhibitor before testing for HI antibody. In any event, it seems likely that the action of neuroaminidase in a series of binding and elution steps plays a role in liquefying secretions and promoting access of virus to the mucosal epithelium.

The incubation period to onset of virus shedding and illness is 1 to 5 days.[62] Studies in susceptible volunteers inoculated by nasal drops showed that virus shedding into nasal secretions began 1 day before or on the day of onset of illness.[62] Concentrations of virus in secretions then increased over the next 1 to 2 days to a peak of 10^4 to 10^7 $TCID_{50}$/ml during the period of maximal symptoms.[66] Virus concentrations then progressively decreased, and the decrease was accompanied by an improvement in symptomatology. The period of virus detection varied between 3 and 8 days. When virus concentrations in secretions during the peak shedding period and illness severity were compared, a direct correlation between the two was noted. An evaluation of rhinovirus infections of the nasopharynx indicated a direct correlation between the concentration of virus in secretions and the extent of mucosal involvement with infection.[94] Although similar data are not available for influenza infection, it seems likely that severity of illness is a direct consequence of the extent of mucosal involvement with infection.

Less extensive studies of naturally occurring infections have shown findings similar to those in volunteers. Duration of virus shedding after onset of illness is usually 3 to 5 days in adults, with longer periods seen in persons with severe illnesses and in children.[134,157] In a retrospective analysis of virus isolations from children in a family study, Frank saw an average duration of shedding of 9 days with onset as much as 3 days preceding the onsets of illness. Peak virus titers from cases of natural disease are similar to those from artificially infected volunteers, but very high titers (up to 10^9/ml) have been seen in cases of influenza pneumonia.[217]

The quantitative data referred to earlier indicate that virus can replicate in the nasopharynx. Although sputum specimens usually contain virus, there is no certainty that such virus was derived from the lower respiratory tract. Nevertheless, histological study of the lower respiratory tract of naturally occurring disease revealed mucosal pathology, and both histological and immunologic study of terminal bronchi, bronchioles, and alveoli of persons dying with acute influenza indicated virus growth and cellular destruction at the lowermost levels of the respiratory system.[249,334] Moreover, complete histological and virological studies of animal models indicate a capacity of influenza virus to replicate in all parts of the respiratory tract.[213]

Efforts to detect viremia or virus replication in extrapulmonary sites have been largely unsuccessful. Blood cultures from volunteers were reported as frequently positive and invariably negative in two separate studies.[204,317] Attempts to demonstrate viremia in natural cases during the 1957 Asian pandemic were unsuccessful, although virus was isolated from extrapulmonary sites with fatal influenza.[263] Another report suggested that virus might be circulating on peripheral blood lymphocytes.[339] Sporadic reports of isolation of influenza virus from muscle, liver, urine, and cerebrospinal fluid (CSF) in combination with the viremia studies suggest that influenza virus may, on occasion, disseminate, but virus replication as a basis for disease at extrapulmonary sites has not been demonstrated.[196]

The pathology accompanying influenza virus infection results from cellular destruction caused by the infection. Virus infects columnar epithelial cells, which shed virus from their surface for about 8 hr before ceasing metabolic activity and being sloughed into the lumen of the respiratory passage. In organ culture, infection has been shown to disrupt significantly the normal ciliary activity; when this is extensive, the epithelium may be denuded down to the basement membrane.[290] The acute injury is initially accompanied by a polymorphonuclear infiltrate; sputum, although produced in small quantities, may be purulent. Histological studies of influenza in the animal model indicate that a localized edema and a peribronchial and peribronchiolar lymphocytic infiltrate develop.[223] The edema may produce significant narrowing of terminal bronchi and bronchioles. This pathology is presumed to be that accompanying typical cases of acute influenza with tracheobronchitis and the occasional case that exhibits a segmental lung infiltrate on chest roentgenograms. Fatal cases with pure virus pneumonia will exhibit varying degrees of interstitial cellular infiltrates, alveolar edema, and hyalin membrane deposition.[249] Cases occurring in patients with underlying congestive heart failure usually show extensive accumulations of alveolar fluid, whereas pure virus pneumonia in a person otherwise presumed to be healthy is usually characterized by extensive cellular infiltrates and fibrosis. The reparative process for the mucosal surfaces and the lung may require weeks to completely clear the lung pathology.

A major complication of influenza is the development of a secondary bacterial pneumonia. In these cases, viral infection has altered mucociliary clearance and may have impaired pulmonary macrophage function (as demonstrated in the mouse model) so that normal lung defense is inadequate for prevention of secondary invasion by pathogenic bacteria colonizing the nasopharynx.[26,188] Moreover, once a focus is established, impaired polymorphonuclear leukocyte function that accompanies influenza may fail to contain it, and the infection progresses to pneumonia.[231]

The pathogenesis of sinusitis and otitis media that may complicate acute influenza is presumed to be similar to that for pneumonia, although an ability of influenza virus alone to produce these diseases has been suggested. The pathogenesis of other complications of influenza,

including acute encephalitis, Reye's syndrome, and myositis, is uncertain.

7.2. Immunity

Immunity to influenza has been clearly demonstrated in studies involving challenge of human volunteers and in observations of naturally occurring infections. In combination with historical accounts of epidemics, early observations of influenza viruses in man and animal models indicated that immunity could be potent but was apparently limited in duration. Both the degree and duration of immunity to type A and B viruses have been clarified in recent studies. Couch demonstrated a potent immunity to reinfection and illness with the Hong Kong variant of A (H3N2) virus for 4 years.[63] Observations of resistance to reinfection with homologous virus under natural conditions also indicated development of a potent immunity to reinfection and illness.[170] Reinfection can occur, however, as demonstrated by occurrence of serological responses, but these infections do not often lead to clinically significant illness.[314] That this homologous immunity may last for many years was demonstrated by the reappearance in 1977 of the influenza A (H1N1) subtype in human populations. Persons born before 1951 carried a high probability of a prior infection with a closely related virus; although surveillance activities have indicated that such persons were infected with H1N1 viruses during post-1977 epidemics, they rarely developed febrile influenza.[73,152] Thus, it seems clear that a substantial degree of immunity to reinfection and illness with homotypic virus develops following a type A virus infection and persists for decades.

Immunity to heterologous but serologically related type A viruses also develops, and this may persist for years after an infection. In a series of challenges with sequentially appearing type A (H3N2) variant viruses, Couch *et al.* showed that volunteers infected with the A/Hong Kong virus in 1968 were immune to rechallenge with antigenic variants of this virus appearing in the subsequent 6 years.[73] This potent immunity spanned the degree of variation from A/Hong Kong/68 to A/Victoria/75. In a series of studies of heterotypic immunity to natural infections, Gill and Murphy also showed a potent heterotypic immunity that spanned the same interval. Immunity progressively waned, however, as epidemic variants were progressively less cross-reactive in serological tests with the A/Hong Kong virus.[141,142]

Longitudinal studies in families have been somewhat less consistent in demonstrations of heterotypic immunity. Frank *et al.* reported near-complete resistance to reinfection with the A/Bangkok/1/82 (H3N2) viruses among children with a prior documented A/Victoria or A/Texas (H3N2) infection, whereas they had not found resistance among another group of young children from low-income families at an earlier time.[129,132] Fox *et al.* also reported resistance to reinfection with A (H3N2) viruses, but only among family members who had persisting high titers of HI antibody.[116] Frank *et al.* reported a potent resistance to reinfection with A (H1N1) viruses and, more recently, to influenza B that lasted for several years.[129,133] Frank's data with influenza B, however, suggested that immunity is less potent for this virus. Fox *et al.* were unable to detect immunity to either A (H1N1) or B viruses.[116] Thus, although some studies have not confirmed the existence of immunity to reinfection and illness, most indicate that protection develops to related antigenic variants of influenza virus after an infection and that this protection spans a considerable degree of antigenic variation and lasts for several years.

The basis of the immunity to influenza is uncertain; however, considerable information on immune responses and their role in influenza virus infection in humans and animal models has been developed that indicates the probable contribution of a variety of immune mechanisms. The more recently described immune mechanisms are those mediated by cells or mediators of cellular effects. These are cytotoxic T-lymphocyte (CTL) effects and delayed-type hypersensitivity (DTH), natural killer (NK) cell cytotoxicity, macrophage antiviral effects, and interferon. The CTL response develops in humans and mice after infection with influenza virus, reaches a peak before antibody is detectable, and returns to an undetectable baseline.[104,347] Adoptive transfer of active CTLs to infected mice produces a reduction in viral content of lungs and severity of the pneumonia.[344] The CTLs from both mice and men primarily exhibit type specificity, although a population of subtype–variant-specific cells is also present.[21,101,234] The major antigen recognized by cells with type specificity appears to be the NP, whereas those with subtype–variant specificity recognize HA determinants.[89,207,235] DTH cells do not exhibit antiviral effects; their preferential sensitization by inactivated vaccines in unprimed mice leads to enhanced pneumonia and mortality when adoptively transferred to infected mice.[2] This demonstration provides a basis for concern about immunization approaches in unprimed infants. The NK cells have not been shown to play a role in immunity to influenza in mouse systems, although both macrophage and interferon appear capable of contributing to recovery.[219,284] Thus, our present understanding of the various cell-mediated immunity (CMI) functions in influenza suggests a major role for these reactions in recovery from influenza. This understanding does not support a role for CMI in resistance to infection, although, on occasion, rapid onset of a response might abort an infection so that it would be scored at the clinical level as no infection.

Antibody to a number of virus proteins develops after infection. The role of antibody to the HA, NA, M, and NP proteins is best understood. Anti-M and anti-NP exhibit type specificity, and neither antibody is thought to play a role in immunity.[73] Antibody to the NA reduces the number of newly synthesized infectious units released from infected cells; it develops in both serum and respiratory secretions after infection.[252] Presence of anti-NA antibody in serum before infection is associated with a reduced occurrence and severity of illnesses accompanying infection, and it may prevent infection at the clinical level if present in high titer.[70]

Anti-HA antibody is the primary mediator of immunity to infection. It may function in a variety of ways, including neutralization of infectivity, complement-mediated lysis of infected cells, aggregation of virions, and mediation of antibody-dependent cellular cytotoxicity. Anti-HA antibody develops in serum and respiratory secretions after infection; it persists for decades in serum and in nasal secretions.[54,72] Antibody persistence patterns may reflect the predominant antibody class in each of the body fluids. Serum antibody is primarily in the IgG class, whereas antibody in secretions is primarily IgA.[69,73] Use of sensitive assays for IgA and IgG antibody in nasal secretions has suggested that both classes of antibody play a role in resistance of volunteers to intranasal challenge with infectious virus.[69] Natural infection, however, is considered to be acquired primarily by low doses of virus inhaled into the lower respiratory tract remotely after a prior infection (see Section 6). A comparison of IgA and IgG antibody in bronchial lavage fluids after an earlier infection indicated a predominance of IgG antibody.[73] This finding, in combination with the demonstrated prophylactic effect of passive antibody and the data indicating that most or all of IgG in the lower respiratory passages is derived from serum, suggests that serum IgG antibody is the primary mediator for prevention of naturally occurring influenza.[69,72,73]

8. Patterns of Host Response

8.1. Clinical Features

The spectrum of illness associated with influenza virus infection ranges from inapparent infection to fulminant, fatal pneumonia. The severity of illness may depend on the previous experience with antigenically related variants. When a large proportion of the population has at least partial immunity, about 20% of infections will be inapparent, and about 30% will be manifested only by signs and symptoms of upper respiratory tract involvement without fever.[129–136] Febrile upper respiratory illness (URI) or influenzalike illness (sudden onset of fever, chills, sore throat, myalgia, malaise, headache, and hacking, nonproductive cough) will occur in about 50% of infected patients. These illnesses are usually classified as upper respiratory tract, but many persons will have at least tracheitis and pulmonary function test findings that suggest peripheral airway dysfunction.[220] About 5% will progress to frank involvement of the lower respiratory tract disease (LRD) with development of tracheobronchitis or pneumonia. The risk for LRD is much greater for young children and for persons with underlying pulmonary disorders.[150,155,220,331] For younger persons, asthma or reactive airway disorders are more frequent,[145] whereas older adults are likely to have chronic bronchitis or emphysema. Pneumonitis in young children is usually a primary viral pneumonia, whereas for older persons, the pneumonia is often caused by bacterial superinfection.[19,144,224,261,263,272,298] When there is a predominant bacterial pathogen, it is usually *Streptococcus pneumoniae*, *Staphylococcus aureus*, or *Haemophilus influenzae*.

The high rate of excess deaths attributed to cardiac conditions has raised speculation about the ability of an influenza virus infection to precipitate a myocardial infarction[9,275]; this may be possible under certain circumstances, but most cardiac deaths probably result among persons who are marginally compensated and who develop pulmonary edema or pneumonia.

Important complications of influenza virus infections involve the central nervous system. Encephalopathy has been frequent in the past and may be fatal.[84,85,100,114,172,322] Encephalopathy with fatty microvesiculation of the liver (Reye's syndrome) was particularly frequent for children given aspirin for influenza infection[61,179,246] until this association was recognized[178] and the administration of aspirin was discouraged.[281] Acute myositis is another complication that occurs more frequently in children with influenza B infection[88,113,239]; however, both encephalopathy and myositis may occur in adults and may be associated with influenza A infections.[14,17,87,138,240,247]

8.2. Diagnosis

A clinical diagnosis of a sporadic infection with influenza virus is not possible because of the similarity of the illness to those caused by other respiratory viruses. However, during epidemic periods, an influenza virus infection will be the likely cause of illness. When increasing numbers of school children and young adults present to physicians and clinics in the winter with febrile respiratory illnesses, an influenza epidemic is probably beginning. This impression can usually be confirmed by bulletins from regional, state, and national laboratories. So long as this increase in cases continues, influenza infection as a cause is likely. Occurrences of increased school and industrial absences for respiratory disease are supportive of a continuing epidemic, but they may not be noticeably increased until virus activity has peaked.

Virologic confirmation of infection requires demonstration of virus in respiratory secretions or rise in a specific antibody between paired sera. Optimal specimens for demonstration of virus are nasal washes and throat swab specimens; highest isolation rates involve use of both. Virus may be isolated and identified as described in Section 3.4. A number of rapid methods for viral detection have been described and some are now available commercially. These include tests for viral antigens in respiratory secretion specimens by enzyme immunoassay, immunofluorescence of sloughed respiratory epithelial cells, and tests for viral nucleic acid after amplification with the polymerase chain reaction.[52,53,92,280,319,333] A rise in antibody titer may be detected with any of several serological techniques. The CF test is most commonly available and will detect a rise in about 70% of infections; HI is more sensitive and will detect over 80% of infections. Immunodiffusion also is reportedly useful.[46,47,291] Neutralization tests and enzyme immunoassay detect nearly all infections, but these tests are available in research laboratories only.[128,131,253,254]

9. Control

A multitude of different remedies have been prescribed over the centuries of recurring epidemics of influenza to prevent or ameliorate their consequences. None of the numerous measures proposed have survived as effective medical or public health procedures for control of influenza today. Although avoidance of ill persons, crowds, and other circumstances in which influenza might be acquired might theoretically be of value, such methods have not been shown to be effective and are largely impractical in modern urban societies.

Recovery of the influenza virus provided the first means for developing specific immunoprophylaxis. Inactivated whole-virus vaccines prepared in chicken embryos were first tested in the 1940s and have been shown to be effective for prevention of influenza.[126] They remain the principal method for providing a measure of control of influenza today. Early preparations were crude and highly reactogenic; however, improved preparations are now minimally reactogenic and satisfactorily immunogenic for most population groups. Better vaccines have resulted from improved production by use of high-yield recombinant viruses developed by Kilbourne *et al.*, use of various purification processes for removal of nonviral protein, and standardization of preparations of HA content.[71,198] Improved worldwide surveillance through the network of WHO laboratories has allowed earlier recognition of new variants with epidemic potential. Particularly, collaboration between CDC and the Institute of Virology in Beijing, China, has resulted in early recognition of new drift variants on multiple occasions. As a consequence, a better antigenic match between the available influenza vaccine and the prevalent epidemic virus has occurred for several seasons.

Over a number of years, the Commission on Influenza of the Armed Forces Epidemiological Board, Department of Defense, monitored effectiveness of influenza vaccine administered to military personnel and reported protection levels ranging from 70 to 90%.[80] Since that time, influenza virus vaccine has been administered annually to all military personnel, and the reported record of effectiveness in control of epidemic influenza is impressive.[237] Annual immunization of older adults in the United States also has been shown to be effective.[190]

The principal groups for whom vaccine is recommended in most countries are those at risk of developing a complication of influenza and dying as a consequence and those healthy persons in close contact with these high-risk persons.[40,43] High-risk persons are those with a chronic condition that compromises heart or lung function, nursing home residents, those over 65 years of age, and those with chronic metabolic disorders, anemia, renal disease, and immunosuppression. The degree of reported protection has been variable, particularly among the elderly, in whom protection has ranged from 0 to 100%.[64,265] Reasons proposed for this variability include reduced antibody responsiveness, reduction in effectiveness of other specific and nonspecific immune mechanisms, and difficulties in documenting the occurrence and effects of influenza virus infection. Nevertheless, an overall assessment indicates a beneficial effect of vaccine for these groups. Significant reduction of pneumonia hospitalizations of vaccinated elderly persons has been demonstrated,[115] as well as cost saving by vaccine programs directed toward the elderly patients enrolled in managed care programs.[257]

Current policy for use of vaccine in most developed countries is for control of the serious consequences of influenza and not for prevention of epidemic influenza.[30,43] Unfortunately, this policy has never been effectively implemented. In the United States, only 30% of the population for whom vaccination is recommended actually receive it each year.[42] A new policy enacted in May 1993 to reimburse providers for administration of influenza vaccine as well as pneumococcal vaccine for Medicare recipients has improved compliance with vaccine recommendations for persons ≥65 years of age.[45]

Only one attempt has been made to prevent epidemic influenza, and that was for prevention of a threatened swine influenza epidemic in 1976. Almost 25% of the United States population received a vaccine that year, but the swine influenza virus did not spread. The policy in Japan for many years has been to vaccinate all school children to prevent spread of influenza.[97] Evaluations of effectiveness have not been conducted, but influenza epidemics in Japan have not been prevented.

The duration of protection after vaccination is not clearly established. Because early trials suggested the duration of effectiveness was limited, vaccine has been recommended annually in the autumn preceding the respiratory disease season. In a number of studies, effectiveness for 2 years has been documented, and in one for 3 years.[118] Nevertheless, the recommendation for annual immunization remains, even when the vaccine formula is unchanged from one year to the next. Vaccine effectiveness is improved with increasing dose, which stimulates a higher antibody response and is more effective against a homotypic virus than against a heterotypic virus.[238,294,318] Vaccine must be given parenterally; one dose is required for persons with prior exposure to related viruses, and two doses a month apart for persons with no prior exposure to vaccine or related viruses.

Persons who are allergic to eggs and chicken products may experience serious allergic reactions, but, otherwise, side effects are limited to mild local reactions in 10 to 15% and systemic reactions in 1 to 2% of persons during the 2 or 3 days after vaccination.[40,43] Split-product vaccines prepared from disrupted virus are recommended for children because they are less reactogenic than whole-virus vaccines.[43] Influenza vaccines are safe for pregnant women who may be at increased risk for complications of influenza during the third trimester of pregnancy.[103] Young infants also may benefit from the higher antibody levels available for transplacental transmission after maternal immunization.

During the 1976 swine influenza vaccine campaign, an increase in the risk of developing Guillain–Barré syndrome was noted among vaccinees.[296] This reaction occurred within 6 weeks following vaccination at a frequency of almost 1 per 100,000 vaccinees who were over

18 years of age. The risk for older adult vaccinees was seven times greater than that for unvaccinated adults.[287] In three subsequent years when other vaccines were administered, no increased risk for Guillain–Barré syndrome was observed.[187] Furthermore, no increased risk was ever observed for children < 18 years of age or for military personnel.[16,183,285,296] The cause of the 1976 episode remains uncertain, but development of the Guillain–Barré syndrome does not appear to be an inherent risk of the use of influenza virus vaccine.

The drug amantadine hydrochloride and its analogue, rimantadine, are alternatives to vaccine for prevention of influenza; they also are effective for treatment.[40,43] They are effective only for type A virus infections and not against type B infections. Their action stems from the ability to block the ion channel created by the M2 protein that spans the envelope (see Section 4.1). This inhibits dissociation of internal proteins required for initiation of transcription.[23,230] If taken before exposure to influenza A, either drug will prevent 50 to 70% of infections and 70 to 90% of illnesses.[68] Once the drug is stopped, however, susceptibility is restored; thus, it must be taken for the duration of presumed exposure. Amantadine is an acceptable alternative to vaccine for persons who cannot take vaccine or when available vaccine is a poor antigenic match with epidemic viruses. It is effective for aborting outbreaks in nursing homes; in this circumstance, amantadine is recommended for all residents after segregating patients with influenzalike illnesses.[43] Rimantadine has the same efficacy and has been shown to be safe and effective for preventing both influenza A (H1N1) and A (H3N2) in children.[56,78]

A problem with the use of amantadine is the occurrence of side reactions. Central nervous system side effects, such as nausea, drowsiness, dizziness, nervousness, and insomnia, will occur in about 5 to 7% of otherwise healthy adults.[68] Severe side effects including hallucinations and convulsions may occur with the high blood levels sometimes seen in elderly patients. These side effects are less of a problem with rimantadine. In several controlled studies, the frequencies of reported side effects were the same for persons receiving placebo as for those receiving rimantadine.[56,78,91] The difference in the risk of side effects may stem from the fact that amantadine is excreted unchanged in the urine while rimantadine is metabolized in the liver.[167,241]

When amantadine or rimantadine are used as therapy, signs and symptoms of influenza clear more rapidly in treated persons than in persons given placebo if the drug is started within the first 48 hr after onset of illness.[68] Drug-resistant influenza A (H3N2) viruses have emerged during rimantadine or amantadine treatment of uncomplicated influenza in children, adults, and residents of nursing homes.[165,166] Illness associated with drug-resistant virus has occurred in these circumstances. Rimantadine therapy was accompanied by clinical benefits, despite the isolation of resistant virus from some treated patients.[160,165] Emergence of resistant virus as a dominant strain during an epidemic would require that the new variant develop from a virus containing a resistant M2 protein, a highly unlikely circumstance with the present limited use of these drugs.[165] Thus, at present, the risks of emergence of resistant virus do not outweigh the documented benefits of drug prophylaxis and therapy in most circumstances.[241]

Despite the availability of two modalities with proven effectiveness, a vaccine and antiviral therapy, no noticeable progress in control of influenza has been achieved, largely because of inadequate use of both. Wider use of existing inactivated vaccines among groups for whom they are currently recommended is indicated. In addition to healthy persons in close contact with persons with high-risk conditions, we recommend vaccine for all individuals in hospitals and clinics who are involved in patient care and for other groups of persons whose disruption of work would impair essential community services. It is also reasonable to administer vaccine to any person or group of persons for whom there is a desire to prevent illness and work absences caused by influenza.

Newer vaccines, including live influenza virus vaccines, other antivirals, and different strategies for their use offer encouragement for improved control in the future. Subunit vaccines for influenza, including purified NA vaccines, have been proposed for development.[71] The former may permit an increase in dosage so as to obtain an improved immune response that should lead to greater protection. Neuraminidase vaccines have been proposed that may allow an attenuated infection on exposure to natural virus and the resulting benefits of infection-induced immunity.[197] Inactivated vaccines given by nasal drops, sprays, or small-particle aerosol induce local antibody and protection against challenge.[67,332]

Live virus vaccines are also currently receiving emphasis. If a live virus vaccine could be developed that is safe, effective, and induces protection for as long as natural infection, a major new modality for prevention of influenza would be available. Live influenza virus vaccines prepared by passage in chicken embryos have been used extensively in the USSR and China for prevention of influenza in healthy adults.[312] They are reported to be effective for this purpose. The major candidates currently under development are the cold-adapted influenza viruses developed by Maassab.[225] A master cold-adapted strain was obtained by stepwise adaptation of a 1960 A(H2N2) strain in primary chick cells until viral replication was optimal at a low temperature (25°C). This cold-adapted virus is used for preparation of cold-reassortant (CR) vaccines containing the six genes for internal proteins from the modified H2N2 virus and the genes for the HA and NA from putative epidemic viruses. Thus far, these vaccines have been shown to be safe and effective for prevention of influenza in adults and children.[71] They are probably nontransmissible, and their level of protection is at least as good as inactivated vaccines.[75] Direct comparisons in children of CR vaccines with the standard inactivated vaccines have demonstrated that CR vaccines are better, particularly for CR recipients 3 to 9 years of age, who experienced broader and longer-lasting protection against a new variant of influenza A (H1N1).[57,159,273,274] Trivalent preparations of CR strains are undergoing evaluation.[191,251]

Alternative strategies for the use of antiviral preparations, rimantadine and ribavirin, have been proposed (ribavirin is a nucleoside analogue effective against both influenza A and B influenza viruses). Dolin showed that inactivated vaccine prior to and rimantadine prophylaxis during the epidemic period significantly improved protection among nursing home residents.[90] Ribavirin has been given by small-particle aerosol for treatment of acute influenza.[140] It promotes more rapid clearance of virus and symptoms and has not been accompanied by any detectable side effects.

This developing capability for application of a number of new modalities for prevention and treatment makes it possible to consider alternative approaches for control of influenza in the future. In this regard, it seems unlikely that a single modality, such as a live virus vaccine, will be sufficient; simultaneous application of several modalities will probably be required. This approach would not only reduce serious morbidity and mortality, the current goal, but effectively control epidemic influenza and reduce total morbidity. In order to do this, an emphasis must be placed on prevention of spread, and this must include prevention of influenza in school children and children in day care. This population experiences earlier and higher attack rates of influenza than other populations; they are major disseminators of virus throughout the community. In this regard, Monto provided data supporting a degree of effectiveness of inactivated influenza virus vaccine given to school children for this purpose during the 1968 A/Hong Kong epidemic.[242] Closing of schools might also dampen an epidemic; a computer model of this approach suggested that a beneficial effect would ensue.[102] This represents a modality for consideration. A number of other options such as early treatment to reduce spread, quarantine, and disinfection of air have potential, but a more effective use of vaccines and antivirals for prevention of infection seems most likely to achieve the desired effect. A live virus vaccine for use in school children, children in day care, and other populations of healthy persons such as employed persons, prisoners, and military personnel at 3- to 5-year intervals would be a major addition to options for control. Improved inactivated vaccines with greater effectiveness might encourage greater use among high-risk persons, and, finally, more widespread use of antivirals could provide protection for still other populations. Adequate information is now available for designing a new control strategy; availability of these new modalities should make is possible to explore new approaches to control of epidemic influenza.

10. Unresolved Problems

One vexing problem is the current inability to recognize among viruses with novel antigens those variants with epidemic potential. Epidemic potential includes properties of the virus related to the transmissibility and virulence. Some new technologies are being brought to bear on this problem[192]; perhaps in the future it will be possible to predict epidemic potential by examining the amino acid sequence of the hemagglutinin moiety, for example. This could lead to greater lead time in the selection of antigens for vaccines. Implicit in the recognition of new variants with epidemic potential is the necessity for maintaining and improving viral surveillance throughout the world.

Improvement of immunogenicity of vaccines is another important goal for the future. Vaccines are needed that are more immunogenic than natural infection in that they should stimulate a response that is both longer lasting and broader so as to provide protection against a spectrum of antigenic variants of the circulating viruses. Better understanding of the protective immune response, as well as new methods for production of antigens, such as recombinant DNA techniques,[260,309] is needed.

To achieve control of epidemic influenza, the control measures would have to be administered to a large number of people within a short period of time unless strategies can be developed to allow for more efficient use of such measures. If accessible segments of the population, such as school children, children in day care, college students, military personnel, and employed persons, can be rapidly immunized, it would reduce the risk of exposure for high-risk groups and allow more time to provide protection for them. Large field trials will be necessary to establish the efficacy of these strategies.

Some methods for rapid diagnosis of influenza virus infection have been developed,[109] but they have not been widely used. These techniques need to be improved and made generally available. This would improve surveillance as well as allow for effective use of antiviral drugs.

11. References

1. ACHESON, F., AND HEWITT, D., Spread of influenza in a factory, *Br. J. Soc. Med.* **6**:68–75 (1952).
2. ADA, G. L., LIEUNG, K.-N., AND ERTL, H., An analysis of effector T cell generation and function in mice exposed to influenza A or Sendai viruses, *Immunol. Rev.* **58**:5–24 (1981).
3. ALLING, D. W., BLACKWELDER, W. C., AND STUART-HARRIS, C. H., A study of excess mortality during influenza epidemics in the United States, 1968–1976, *Am. J. Epidemiol.* **113**:30–42 (1981).
4. ALONSO-CAPLER, F. V., NEMEROFF, M. E., QUI, V., AND KRUG, R., Nucleocytoplasmic transport: The influenza virus NS1 protein regulates the transport of spliced NS2 virus and its precursor NS1 in RNA, *Genes Dev.* **6**:255–267 (1992).
5. ANDREWES, C. H., AND GLOVER, G. E., Spread of infection from the respiratory tract of the ferret. I. Transmission of influenza A virus, *Br. J. Exp. Pathol.* **22**:91–97 (1941).
6. ASSAAD, F. BEKTIMIROV, T., AND LJUNGARS-ESTEVES, K., Influenza/world experience, in: *The Molecular Virology and Epidemiology of Influenza* (C. STUART-HARRIS AND C. W. POTTER, EDS.), pp. 5–13, Academic Press, Orlando, FL, 1984.
7. ASSAAD, F., COCKBURN, W. C., AND SUNDARESAN, T. K., Use of excess mortality from respiratory diseases in the study of influenza, *Bull. WHO* **49**:219–233 (1973).
8. AUSTIN, F. J., AND MATAIKA, J. V., Influenza virus A (H1N1) in Fiji, *N. Z. Med. J.* **90**:242–244 (1979).
9. BAINTON, D., JONES, G. R., AND HOLE, D., Influenza and ischaemic heart disease—a possible trigger for acute myocardial infarction, *Int. J. Epidemiol.* **7**:231–239 (1978).
10. BARKER, W. H., AND MULLOOLY, J. P., Influenza vaccination of elderly persons. Reduction in pneumonia and influenza hospitalizations and deaths, *J.A.M.A.* **244**:2547–2549 (1980).
11. BARKER, W. H., AND MULLOOLY, J. P., Underestimation of the role of pneumonia and influenza in causing excess mortality, *Am. J. Public Health* **71**:643–645 (1981).
12. BARKER, W. H., AND MULLOOLY, J. P., Pneumonia and influenza deaths during epidemics, *Arch. Intern. Med.* **142**:85–89 (1982).
13. BAROYAN, O. V., RVACHEV, L. A., AND IVANNIKOV, Y. G., *Modeling and Forecasting of Influenza Epidemics for the Territory of the USSR*, Gamaleya Institute of Epidemiology and Microbiology, Moscow, 1977.
14. BARTON, L. L., AND CHALHUB, E. G., Myositis associated with influenza A infection, *J. Pediatr.* **87**:1003–1004 (1975).
15. BAXTER, B. D., COUCH, R. B., GREENBERG, S. B., AND KASEL, J. A., Maintenance of viability and comparison of identification methods for influenza and other respiratory viruses of humans, *J. Clin. Microbiol.* **6**:19–22 (1977).
16. BEGHI, E., KURLAND, L. T., MULDER, D. W., AND WIEDERHOLD, W. C., Guillain–Barré syndrome. Clinicoepidemiologic features

17. BERLIN, B. S., SIMON, N. M., AND BOVNER, R. N., Myoglobinuria precipitated by viral infection, *J.A.M.A.* **227**:1414–1415 (1974).
18. BERMAN, S., AND McINTOSH, K., Selective primary health care: Strategies for control of disease in the developing world. XXI. Acute respiratory infections, *Rev. Infect. Dis.* **7**:674–691 (1985).
19. BISNO, A. L., GRIFFIN, J. P., VAN EPPS, K. A., NIELL, H. B., AND RYTEL, M. W., Pneumonia and Hong Kong influenza: A prospective study of the 1968–1969 epidemic, *Am. J. Biol. Sci.* **261**:251–263 (1971).
20. BOLLEGRAAF, E., Laboratory data on influenza in Canada, 1983–84, *Can. Med. Assoc. J.* **131**:303–304 (1984).
21. BRACIALE, T. J., BRACIALE, V. L., HENKEL, T. J., SAMBROOK, J., AND GETHING, M. J., Cytotoxic T lymphocyte recognition of the influenza hemagglutinin gene product expressed by DNA mediated gene transfer, *J. Exp. Med.* **159**:341–354 (1984).
22. BROWN, P. K., AND TAYLOR-ROBINSON, D., Respiratory virus antibodies in sera of persons living in isolated communities, *Bull. WHO* **34**:895–900 (1966).
23. BUKRINSKAYA, A. G., VORKUNOVA, N. D., KORNILAYEVA, G. V., NARMANBETOVA, R. A., AND VORKUNOVA, G. K., Influenza virus uncoating in infected cells and effect of rimantadine, *J. Gen. Virol.* **60**:49–59 (1982).
24. BURNET, F. M., Influenza virus infections of the chick embryo lung, *Br. J. Exp. Path.* **21**:147–153 (1940).
25. CAMERON, A. S., RODER, D. M., ESTERMAN, A. J., AND MOORE, B. W., Mortality from influenza and allied infections in South Australia during 1968–1981, *Med. J. Aust.* **142**:14–17 (1985).
26. CAMNER, P., JARSTRAND, C., AND PHILIPSON, K., Tracheobronchial clearance in patients with influenza, *Am. Rev. Respir. Dis.* **108**:131–135 (1973).
27. CANIL, K. A., PRATT, D., SUNGU, M. S., AND PHILLIPS, P. A., Influenza surveillance: Alternative laboratory techniques for a developing country, *Bull. WHO* **63**:79–82 (1985).
28. CENTANNI, E., AND SAVONUZZI, O., cited by Stubbs, E. L., Fowl plague, in: *Diseases of Poultry*, 4th ed. (H. E. BIESTER AND L. H. SCHWARTZ, EDS.), Iowa State University Press, Ames, Iowa, 1965.
29. CENTERS FOR DISEASE CONTROL, Influenza-respiratory disease surveillance, Report 83, pp. 1–8 (June, 1967).
30. CENTERS FOR DISEASE CONTROL, United States immunization survey: 1978, HEW Publication No. (CDC), 79–8221 (1979).
31. CENTERS FOR DISEASE CONTROL, Influenza vaccine 1981–1982, *Ann. Int. Med.* **95**:461–463 (1981).
32. CENTERS FOR DISEASE CONTROL, Annual summary, 1978, *Morbid. Mortal. Week. Rep.* **27**:79 (1979).
33. CENTERS FOR DISEASE CONTROL, Influenza mortality surveillance—United States, *Morbid. Mortal. Week. Rep.* **29**:578–584 (1980).
34. CENTERS FOR DISEASE CONTROL, Annual summary, 1980, *Morbid. Mortal. Week. Rep.* **29**:112–115 (1981).
35. CENTERS FOR DISEASE CONTROL, Annual summary 1982, *Morbid. Mortal. Week. Rep.* **31**:120–121 (1983).
36. CENTERS FOR DISEASE CONTROL, Influenza surveillance summary—United States, 1982–1983 season, *Morbid. Mortal. Week. Rep.* **32**:373–377 (1983).
37. CENTERS FOR DISEASE CONTROL, Annual summary 1983, *Morbid. Mortal. Week. Rep.* **32**:121 (1984).
38. CENTERS FOR DISEASE CONTROL, Influenza—United States, 1983–1984 season, *Morbid. Mortal. Week. Rep.* **33**:417–421 (1984).

39. CENTERS FOR DISEASE CONTROL, Influenza—United States, 1984–1985 season, *Morbid. Mortal. Week. Rep.* **34**:440–443 (1985).

40. CENTERS FOR DISEASE CONTROL, ACIP: Prevention and control of influenza, *Morbid. Mortal. Week. Rep.* **35**:317–325 (1986).

41. CENTERS FOR DISEASE CONTROL, Influenza—United States, 1987–88 season, *Morbid. Mortal. Week. Rep.* **37**:497–503 (1988).

42. CENTERS FOR DISEASE CONTROL, Influenza vaccination levels in selected states, behavioral risk factor surveillance system, 1987, *Morbid. Mortal. Week. Rep.* **38**:124–133 (1989).

43. CENTERS FOR DISEASE CONTROL, ACIP: Prevention and control of influenza, *Morbid. Mortal. Week. Rep.* **41**:1–14 (1992).

44. CENTERS FOR DISEASE CONTROL, Update: Influenza activity—United States, 1992–93 season, *Morbid. Mortal. Week. Rep.* **42**:385–387 (1993).

45. CENTERS FOR DISEASE CONTROL, Final results: Medicare influenza vaccine demonstration—selected states, 1988–1992, *Morbid. Mortal. Week. Rep.* **42**:601–604 (1993).

46. CHAKRAVERTY, P., Comparison of haemagglutination-inhibition and single-radial-haemolysis techniques for detection of antibodies to influenza B virus, *Arch. Virol.* **63**:285–289 (1980).

47. CHAKRAVERTY, P., PEREIRA, M. S., AND SCHILD, G. C., Use of single radial diffusion technique for influenza antibody surveys, *Bull. WHO* **49**:327–332 (1973).

48. CHANOCK, R., CHANBON, L., CHANG, W., GONCALVES FERREIRA, F., GHARPURE, P., GRANT, L., HATEM, J., IMAM, I., KALVA, S., LIM, K., MADALENGOITIA, J., SPENCE, L., TENG, P., AND FERREIRA, W., WHO, respiratory disease survey in children, *Bull. WHO* **37**:363–369 (1967).

49. CHAPMAN, R. S., HENDERSON, F. W., CLYDE, W. A., JR., COLLIER, A. M., AND DENNY, F. W., The epidemiology of tracheobronchitis in pediatric practice, *Am. J. Epidemiol.* **114**:786–797 (1981).

50. CHIN, T. D. Y., FOLEY, J. F., DOTO, I. L., GRAVELLE, C. R., AND WESTON, J., Morbidity and mortality characteristics of Asian strain influenza, *Public Health Rep.* **75**:149–158 (1960).

51. CHOI, K., AND THACKER, S. B., An evaluation of influenza mortality surveillance, 1962–1979. 1. Time series forecasts of expected pneumonia and influenza deaths, *Am. J. Epidemiol.* **113**:215–222 (1981).

52. CHOMEL, J. J., REMILLEUX, M. F., MARCHAND, P., AND AYMARD, M., Rapid diagnosis of influenza A. Comparison with ELISA immunocapture and culture, *J. Virol. Methods* **37**:337–344 (1992).

53. CLAAS, E. C. J., VAN MILAAN, A. J., AND SPRENGER, M. J. W., Prospective application of reverse transcriptase polymerase chain reaction for diagnosing influenza infections in respiratory samples from a children's hospital, *J. Clin. Microbiol.* **31**:2218–2221 (1993).

54. CLEMENTS, M. L., AND MURPHY, B. R., Development and persistence of local and systemic antibody responses in adults given live attenuated or inactivate influenza A vaccine, *J. Clin. Microbiol.* **23**:66–72 (1986).

55. CLIFFORD, R. E., SMITH, J. W. G., TILLETT, H. E., AND WHERRY, P. J., Excess mortality associated with influenza in England and Wales, *Int. J. Epidemiol.* **6**:115–128 (1977).

56. CLOVER, R. D., CRAWFORD, S. A., ABELL, T. D., RAMSEY, C. N., GLEZEN, W. P., AND COUCH, R. B., Effectiveness of rimantadine prophylaxis of children within families, *Am. J. Dis. Child.* **140**:706–709 (1986).

57. CLOVER, R., CRAWFORD, S., GLEZEN, W. P., TABER, L. H., MATSON, C. C., AND COUCH, R. B., Comparison of heterotypic protection against influenza A/Taiwan/86 (H1N1) by attenuated and inactivated vaccines to A/Chile/83-like viruses, *J. Infect. Dis.* **163**:300–304 (1991).

58. COLEMAN, M., Immunologic methodology in influenza diagnosis and research—summary of influenza workshop II, *J. Infect. Dis.* **126**:219–230 (1972).

59. COLLINS, S. D., Influenza in the United States, 1887–1956, Public Health Monograph No. 48, Government Printing Office, Washington, DC, 1957.

60. COLLINS, S. D., AND LEHMANN, J., Trends and epidemics of influenza and pneumonia, 1918–1951, *Public Health Rep.* **66**:1487–1505 (1951).

61. COREY, L., RUBIN, R. J., HATTIWICK, M. A., NOBLE, G. R., AND CASSIDY, E., A nationwide outbreak at Reye's syndrome, *Am. J. Med.* **61**:615–625 (1976).

62. COUCH, R. B., Epidemiology of influenza—summary of Influenza Workshop IV, *J. Infect. Dis.* **128**:361–386 (1973).

63. COUCH, R. B., Assessment of immunity to influenza using artificial challenge of normal volunteers with influenza virus, *Dev. Biol. Stand.* **28**:295–306 (1975).

64. COUCH, R. B., AND CATE, T. R., Managing influenza in older patients, *Geriatrics* **38**:61–74 (1983).

65. COUCH, R. B., CATE, T. R., DOUGLAS, R. G., JR., GERONE, P. J., AND KNIGHT, V., Effect of route of inoculation on experimental respiratory viral disease in volunteers and evidence of airborne transmission, *Bacteriol. Rev.* **30**:517–529 (1966).

66. COUCH, R. B., DOUGLAS, R. G., JR., FEDSON, D. S., AND KASEL, J. A., Correlated studies of a recombinant influenza virus vaccine. III. Protection against experimental influenza in man, *J. Infect. Dis.* **124**:473–480 (1971).

67. COUCH, R. B., DOUGLAS, R. G., JR., ROSSEN, R. D., AND KASEL, J. A., Role of secretory antibody in influenza, in: *The Secretory Immunologic System* (D. H. DAYTON, JR., P. A. SMALL, JR., R. M. CHANOCK, H. E. KAUFMAN, AND T. B. TOMASI, JR., EDS.), pp. 93–112, Government Printing Office, Washington, DC, 1971.

68. COUCH, R. B., AND JACKSON, G. G., Antiviral agents in influenza—summary of Influenza Workshop VIII, *J. Infect. Dis.* **134**:516–527 (1976).

69. COUCH, R. B., AND KASEL, J. A., Immunity to influenza in man, *Annu. Rev. Microbiol.* **37**:529–549 (1983).

70. COUCH, R. B., KASEL, J. A., GERIN, J. L., SCHULMAN, J. L., AND KILBOURNE, E. D., Induction of partial immunity to influenza by a neuraminidase-specific vaccine, *J. Infect. Dis.* **129**:411–420 (1974).

71. COUCH, R. B., KASEL, J. A., GLEZEN, W. P., CATE, T. R., SIX, H. R., TABER, L. H., FRANK, A. L., GREENBERG, S. B., ZAHRADNIK, J. M., AND KEITEL, W. A., Influenza: Its control in persons and populations, *J. Infect. Dis.* **153**:431–440 (1986).

72. COUCH, R. B., KASEL, J. A., SIX, H. R., AND CATE, T. R., The basis for immunity to influenza in man, in: *Genetic Variations among Influenza Viruses*, Vol. XXI (D. P. NAYAK, ED.), pp. 535–546, Academic Press, New York, London, 1981.

73. COUCH, R. B., KASEL, J. A., SIX, H. R., CATE, T. R., AND ZAHRADNIK, J. M., Immunological reactions and resistance to infection with influenza virus, in: *Molecular Virology and Epidemiology of Influenza* (C. STUART-HARRIS AND C. PORTER, EDS.), pp. 119–153, Academic Press, Orlando, FL, 1984.

74. COUCH, R. B., KNIGHT, V., HAMORY, B. H., BLACK, S., AND DOUGLAS, R. G., JR., The minimal infectious dose of adenovirus type 4: The case for natural transmission by viral aerosol, *Trans. Am. Clin. Climat. Assoc.* **80**:205–211 (1968).

75. COUCH, R. B., QUARLES, J. M., CATE, T. R., AND ZAHRADNIK, J.

M., Clinical trials with live cold-reassortant influenza virus vaccines, in: *Options for Control of Influenza* (A. P. KENDAL AND P. A. PATRIARCHA, EDS.), pp. 23–241, Alan R. Liss, New York, 1986.

76. COX, N. J., BAI, Z. S., AND KENDAL, A. P., Laboratory-based surveillance of influenza A (H1N1) and A (H3N2) viruses in 1980–81: Antigenic and genomic analyses, *Bull. WHO* **61**:143–152 (1983).

77. COX, N., XU, X., BENDER, C., KENDAL, A., REGNERY, H., HEMPHILL, M., AND ROTA, P., Evolution of hemagglutinin in epidemic variants and section of vaccine viruses, in: *Options for the Control of Influenza* II (C. HANNOUN, A. P. KENDAL, H. D. KLENK, AND F. L. RUBEN, EDS.), pp. 223–230, Excerpta Medica, Amsterdam, 1993.

78. CRAWFORD, S. A., CLOVER, R. D., ABELL, T. D., RAMSEY, C. N., GLEZEN, W. P., AND COUCH, R. B., Rimantadine prophylaxis in children: A follow-up study, *Pediatr. Infect. Dis. J.* **7**:379–383 (1988).

79. DAB, W., QUENEL, P., COHEN, J. M., AND HANNOUN, C., A new influenza surveillance system in France: The ILE-De-France "GROG." 2. Validity of indicators (1984–1989), *Eur. J. Epidemiol.* **7**:579–587 (1991).

80. DAVENPORT, F. M., Control of influenza, symposium on influenza, *Med. J. Aust. [Spec. Suppl.]* **1**:33–38 (1973).

81. DAVENPORT, F. M., HENNESSY, A. V., AND FRANCIS, T., JR., Epidemiologic and immunologic significance of age distribution of antibody to antigenic variants of influenza virus, *J. Exp. Med.* **98**:641–656 (1953).

82. DAVIES, J. R., GRILLI, E. A., AND SMITH, A. J., Influenza A: Infection and reinfection, *J. Hyg. (Camb.)* **92**:125–127 (1984).

83. DAVIS, L. E., CALDWELL, G. G., LYNCH, R. E., BAILEY, R. E., AND CHIN, T. D. Y., Hong Kong influenza: The epidemiologic features of a high school family study analyzed and compared with a similar study during the 1957 Asian influenza epidemic, *Am. J. Epidemiol.* **92**:240–247 (1970).

84. DEIBEL, R., FLANAGAN, T. D., AND SMITH, V., Central nervous system infections. Etiologic and epidemiologic observations in New York State, 1975, *NY State J. Med.* **77**:1398–1404 (1977).

85. DELORME, L., AND MIDDLETON, P. J., Influenza A virus associated with acute encephalopathy, *Am. J. Dis. Child.* **133**:822–824 (1979).

86. DENNY, F. W., AND CLYDE, W. A., JR., Acute lower respiratory tract infections in nonhospitalized children, *J. Pediatr.* **108**:635–646 (1986).

87. DIBONA, F. J., AND MORENS, D. M., Rhabdomyolysis associated with influenza A. Report of a case with unusual fluid and electrolyte abnormalities, *J. Pediatr.* **91**:943–945 (1977).

88. DIETZMAN, D. E., SCHALLER, J. G., RAY, C. G., AND REED, M. E., Acute myositis associated with influenza B infection, *Pediatrics* **57**:255–258 (1976).

89. DILLON, S. B., DEMUTH, S. G., SCHNEIDER, M. A., WESTON, C. B., JONES, C. S., YOUNG, J. F., SCOTT, M., BHATNAGHAR, P. K., LOCASTRO, S., AND HANNA, N., Induction of protective class I MHC-restricted CTL in mice by a recombinant influenza vaccine in aluminum hydroxide adjuvant, *Vaccine* **10**:309–318 (1992).

90. DOLIN, R., BETTS, R. F., TREANOR, S. S., ERG, S. M., O'BRIEN, D. H., ROTH, F.-K., MILLER, P., AND DUFFY, P., Rimantadine prophylaxis of influenza in the elderly, in: *Proceedings of the 23rd Interscience Conference on Antimicrobial Agents and Chemotherapy*, p. 210, American Society for Microbiology, Washington, DC, 1983.

91. DOLIN, R., REICHMAN, R. C., MADORE, H. P., MAYNARD, R., LINTON, P. N., AND WEBBER-JONES, J., A controlled trial of amantadine and rimantadine in the prophylaxis of influenza A infection, *N. Engl. J. Med.* **307**:580–584 (1982).

92. DOMINGUEZ, E. A., TABER, L. H., AND COUCH, R. B., Comparison of rapid diagnostic techniques for respiratory syncytial and influenza A virus respiratory infections in young children, *J. Clin. Microbiol.* **31**:2286–2290 (1993).

93. DORAISINGHAM, S., AND LING, A. E., Acute nonbacterial infections of the respiratory tract in Singapore children: An analysis of three years' laboratory findings, *Ann. Med. Singapore* **10**:69–78 (1981).

94. DOUGLAS, R. G., JR., ALFORD, B. R., AND COUCH, R. B., Atraumatic nasal biopsy for studies of respiratory virus infection in volunteers, *Antimicrob. Agents Chemother.* **8**:340–343 (1968).

95. DOUGLAS, R. G., BETTS, R. F., HRUSKA, J. F., AND HALL, C. B., Epidemiology of nosocomial viral infections, in: *Seminars in Infectious Diseases* (L. WEINSTEIN AND B. N. FIELDS, EDS.), pp. 98–144, Stratton Intercontinental Medical Books, New York, 1979.

96. DOWDLE, W. R., DOWNIE, J. C., AND LAVER, W. G., Inhibition of virus release by antibodies by surface antigens of influenza virus, *J. Virol.* **13**:269–275 (1974).

97. DOWDLE, W. R., MILLAR, J. O., SCHONBERGER, L. B., ENNIS, F. A., AND LA MONTAGNE, J. R., Influenza immunization policies and practices in Japan, *J. Infect. Dis.* **141**:258–264 (1980).

98. DYKES, A. C., CHERRY, J. D., AND NOLAN, C. E., A clinical, epidemiologic, serologic, and virologic study of influenza C virus infection, *Arch. Intern. Med.* **140**:1295–1298 (1980).

99. EASON, R. J., Deaths from influenza A, subtype H1N1 during the 1979 Auckland epidemic, *N.Z. Med. J.* **91**:129–131 (1980).

100. EDELEN, J. S., BENDER, T. R., AND CHIN, T. D. Y., Encephalopathy and pericarditis during an outbreak of influenza, *Am. J. Epidemiol.* **100**:79–84 (1974).

101. EFFORS, R. B., DOHERTY, P. C., GERHARD, W., AND BENNINK, J., Generation of both cross-reactive and virus specific T cell populations after immunization with serologically distinct influenza A viruses, *J. Exp. Med.* **145**:557–568 (1977).

102. ELVEBACK, L. R., FOX, J. P., ACKERMAN, E., LANGWORTHY, A., BOYD, M., AND GATEWOOD, L., An influenza simulation model for immunization studies, *Am. J. Epidemiol.* **103**:152–165 (1976).

103. ENGLUND, J. A., MBAWUIKE, I. N., HAMMILL, H., HOLLEMAN, M. C., BAXTER, B. D., AND GLEZEN, W. P., Maternal immunization with influenza or tetanus toxoid vaccine for passive antibody protection in young infants, *J. Infect. Dis.* **168**:647–656 (1993).

104. ENNIS, F. A., ROOK, A. H., HUA, Q. Y., SCHILD, G. C., RILEY, D., PRATT, R., AND POTTER, C. W., HLA-restricted virus-specific cytotoxic T-lymphocyte responses to live and inactivated influenza viruses, *Lancet* **2**:887–891 (1981).

105. EVANS, A. S., Serologic studies of acute respiratory infections in military personnel, *Yale J. Biol. Med.* **48**:201–209 (1975).

106. EVANS, A. S., DICK, E. C., AND NYSTUEN, K., Influenza in University of Wisconsin students, *Arch. Environ. Health* **6**:62–69 (1963).

107. EVANS, A. S., AND ESPIRITU-CAMPOS, L., Acute respiratory diseases in students at the University of the Philippines, 1964–69, *Bull. WHO* **45**:103–112 (1971).

108. EVANS, A. S., NIEDERMAN, J. C., AND SAWYER, R. N., WITH THE TECHNICAL ASSISTANCE OF WANAT, J., CENABRE, L., SHEPARD, K., AND RICHARDS, V., Prospective studies of a group of Yale University freshman. II. Occurrence of acute respiratory infections and rubella, *J. Infect. Dis.* **123**:271–278 (1971).

109. EVANS, A. S., AND OLSON, B., Rapid diagnostic methods for influenza virus in clinical specimens: A comparative study, *Yale J. Biol. Med.* **55:**391–403 (1982).

110. FANG-ZHENG, S., GUI-FANG, Z., AND XIAO-LIAN, X., Influenza surveillance, *Clin. Med. J.* **96:**349–354 (1983).

111. FANG-ZHENG, S., PEI-JUN, Z., GUI-FANG, Z., MEI-HUA, W., AND JI-MING, Z., Influenza surveillance in Shanghai, *Chin. Med. J.* **97:**339–344 (1984).

112. FARR, W., *Tenth Annual Report of the Registrar General*, H.M.S.O., London, 1847.

113. FARRELL, M. K., PARTIN, J. C., AND BOVE, K. E., WITH THE TECHNICAL ASSISTANCE OF JACOBS, R., AND HILTON, P. K., Epidemic influenza myopathy in Cincinnati in 1977, *J. Pediatr.* **96:**545–551 (1980).

114. FLEWETT, T. H., AND HOULT, J. G., Influenzal encephalopathy and postinfluenzal encephalitis, *Lancet* **2:**11–15 (1958).

115. FOSTER, D. A., TALSMA, A., FURUMOTO-DAWSON, A., OHMIT, S., MARGUILES, J. R., ARDEN, N. H., AND MONTO, A. S., Influenza vaccine effectiveness in preventing hospitalization for pneumonia in the elderly, *Am. J. Epidemiol.* **136:**296–307 (1992).

116. FOX, J. P., HALL, C. E., COONEY, M. K., AND FOY, H. M., Influenza virus: Infections in Seattle families, 1975–1979, *Am. J. Epidemiol.* **116:**212–227 (1982).

117. FOY, H. M., COONEY, M. K., ALLAN, I., AND KENNY, G. E., Rates of pneumonia during influenza epidemics in Seattle, 1964 to 1975, *J.A.M.A.* **241:**253–258 (1979).

118. FOY, H. M., COONEY, M. K., AND MCMAHAN, R., A/Hong Kong influenza immunity three years after immunization, *J.A.M.A.* **226:**758–761 (1973).

119. FOY, H. M., COONEY, M. K., MCMAHAN, R., AND GRAYSTON, J. T., Viral and mycoplasmal pneumonia in a prepaid medical care group during an eight-year period, *Am. J. Epidemiol.* **97:**93–102 (1973).

120. FOY, H. M., HALL, C. E., COONEY, M. K., ALLAN, I., AND FOX, J. P., Influenza surveillance by age and target group, *Am. J. Epidemiol.* **109:**582–586 (1979).

121. FOY, H. M., HALL, C. E., COONEY, M. K., ALLAN, I. D., AND FOX, J. P., Influenza surveillance in the Pacific Northwest 1976–1980, *Int. J. Epidemiol.* **12:**353–356 (1983).

122. FRANCIS, T., JR., Transmission of influenza by a filterable virus, *Science* **80:**457–459 (1934).

123. FRANCIS, T., JR., A new type of virus from epidemic influenza, *Science* **92:**405–408 (1940).

124. FRANCIS, T., JR., DAVENPORT, F. M., AND HENNESSY, A. V., A serological recapitulation of human infection with different strains of influenza virus, *Trans. Assoc. Am. Physicians* **66:**231–239 (1953).

125. FRANCIS, T., JR., AND MAGILL, T. P., Direct isolation of human influenza virus in tissue culture medium and on egg membrane, *Proc. Soc. Exp. Biol. Med.* **36:**134–135 (1937).

126. FRANCIS, T., JR., SALK, J. E., PEARSON, H. E., AND BROWN, P. N., Protective effect of vaccination against induced influenza A, *J. Clin. Invest.* **24:**536–546 (1945).

127. FRANK, A. L., COUCH, R. B., GRIFFIS, C. A., AND BAXTER, B. D., Comparison of different tissue cultures for isolation and quantitation of influenza and parainfluenza viruses, *J. Clin. Microbiol.* **10:**32–36 (1979).

128. FRANK, A. L., PUCK, J., HUGHES, B. J., AND CATE, T. R., Microneutralization test for influenza A and B and parainfluenza 1 and 2 viruses that uses continuous cell lines and fresh serum enhancement, *J. Clin. Microbiol.* **12:**426–432 (1980).

129. FRANK, A. L., AND TABER, L. H., Variation in frequency in natural reinfection with influenza A viruses, *J. Med. Virol.* **12:**17–23 (1983).

130. FRANK, A. L., TABER, L. H., GLEZEN, W. P., GEYER, E. A., MCILWAIN, S., AND PAREDES, A., Influenza B virus infections in the community and the family. The epidemics of 1976–1977 and 1979–1980 in Houston, Texas, *Am. J. Epidemiol.* **118:**313–325 (1983).

131. FRANK, A. L., TABER, L. H., GLEZEN, W. P., PAREDES, A., AND COUCH, R. B., Reinfection with influenza A (H3N2) virus in young children and their families, *J. Infect. Dis.* **140:**829–835 (1979).

132. FRANK, A. L., TABER, L. H., AND PORTER, C. M., Influenza B virus reinfection, *Am. J. Epidemiol.* **125:**576–586 (1987).

133. FRANK, A. L., TABER, L. H., WELLS, C. R., WELLS, J. M., GLEZEN, W. P., AND PAREDES, A., Patterns of shedding of myxoviruses and paramyxoviruses in children, *J. Infect. Dis.* **144:**433–441 (1981).

134. FRANK, A. L., TABER, L. H., AND WELLS, J. M., Individuals infected with two subtypes of influenza A virus in the same season, *J. Infect. Dis.* **147:**120–124 (1983).

135. FRANK, A. L., TABER, L. H., AND WELLS, J. M., Comparison of infection rates and severity of illness for influenza A subtypes H1N1 and H3N2, *J. Infect. Dis.* **151:**73–80 (1985).

136. FRANK, A. L., WEBSTER, R. G., GLEZEN, W. P., AND CATE, T. R., Trial of A/USSR influenza virus hemagglutinin and neuroaminidase subunit vaccine in young adults, in: *Current Chemotherapy and Infectious Diseases* (J. D. NELSON AND C. GRASSI, EDS.), pp. 1337–1338, American Society of Microbiology, Washington, DC, 1980.

137. FROST, W. H., The epidemiology of influenza, *Public Health Rep.* **34(33):**1823–1826 (1919).

138. GAMBOA, E. T., EASTWOOD, A. B., HAYS, A. P., MAXWELL, J., AND PENN, A. S., Isolation of influenza virus from muscle in myoglobinuric polymyositis, *Neurology (Minneap.)* **29:**1323–1335 (1979).

139. GARDNER, G., FRANK, A. L., AND TABER, L. H., Effects of social and family factors on viral respiratory infection and illness in the first year of life, *J. Epidemiol. Commun. Health* **38:**42–48 (1983).

140. GILBERT, B. E., WILSON, S. Z., KNIGHT, V., COUCH, R. B., MELHOFF, T. L., MCCLUNG, H. W., DIVINE, G. W., BARTLET, D. D., COHAN, L. C., GALLION, T. L., AND QUARLES, J. M., Ribavirin small-particle aerosol treatment of influenza in college students, 1981–1983, in: *Clinical Applications of Ribavirin* (R. A. SMITH, V. KNIGHT, AND J. D. SMITH, EDS.), pp. 125–143, Academic Press, New York, 1984.

141. GILL, P. W., AND MURPHY, A. M., Naturally acquired immunity to influenza type A, *Med. J. Aust.* **2:**329–333 (1976).

142. GILL, P. W., AND MURPHY, A. M., Naturally acquired immunity to influenza type A. A further prospective study, *Med. J. Aust.* **2:**761–765 (1977).

143. GLEZEN, W. P., Serious morbidity and mortality associated with influenza epidemics, *Epidemiol. Rev.* **4:**25–44 (1982).

144. GLEZEN, W. P., Viral pneumonia as a cause and result of hospitalization, *J. Infect. Dis.* **147:**765–770 (1983).

145. GLEZEN, W. P., Reactive airway disorders in children. Relationship to respiratory virus infections, *Clin. Chest Med.* **5:**635–643 (1984).

146. GLEZEN, W. P., AND COUCH, R. B., Interpandemic influenza in the Houston area, 1974–76, *N. Engl. J. Med.* **298:**587–592 (1978).

147. GLEZEN, W. P., COUCH, R. B., AND SIX, H. R., The influenza herald wave, *Am. J. Epidemiol.* **116:**589–598 (1982).

148. GLEZEN, W. P., COUCH, R. B., TABER, L. H., PAREDES, A., ALLISON, J. E., FRANK, A., AND ALDRIDGE, C., Epidemiologic observations of influenza B virus infections in Houston, 1976–77, *Am. J. Epidemiol.* **111:**13–22 (1980).

149. GLEZEN, W. P., DECKER, M., JOSEPH, S. W., AND MECREADY, R. G., Acute respiratory disease associated with influenza epidemics in Houston, 1981–1983, *J. Infect. Dis.* **155:**1119–1126 (1987a).

150. GLEZEN, W. P., DECKER, M., AND PERROTTA, D. M., Survey of underlying conditions of persons hospitalized with acute respiratory disease during influenza epidemics in Houston, 1978–1981, *Am. Rev. Respir. Dis.* **136:**550–555 (1987b).

151. GLEZEN, W. P., FRANK, A. L., TABER, L. H., TRISTAN, M. P., VALBONA, C., PAREDES, A., AND ALLISON, J. E., Influenza in childhood, *Pediatr. Res.* **17:**1029–1032 (1983).

152. GLEZEN, W. P., KEITEL, W. A., TABER, L. H., PIEDRA, P. A., COVER, R. D., AND COUCH, R. B., Age distribution of patients with medically attended illness caused by sequential variants of influenza A/H1N1: Comparison to age-specific infection rates, *Am. J. Epidemiol.* **133:**296–304 (1991).

153. GLEZEN, W. P., PAREDES, A., AND TABER, L. H., Influenza in children: Relationship to other respiratory agents, *J.A.M.A.* **243:**1345–1349 (1980).

154. GLEZEN, W. P., PAYNE, A. A., SNYDER, D. N., AND DOWNS, T. D., Mortality and influenza, *J. Infect. Dis.* **146:**313–321 (1982).

155. GLEZEN, W. P., SIX, H. R., FRANK, A. L., TABER, L. H., PERROTTA, D. M., AND DECKER, M., Impact of epidemics on communities and families, in: *Options for the Control of Influenza* (A. P. KENDAL AND P. A. PATRIARCHA, EDS.), pp. 63–73, Alan R. Liss, New York, 1986.

156. GRAVES, R. J., Influenza, in: *System of Clinical Medicine* (R. J. GRAVES AND W. W. GERHARD, EDS.), pp. 462–480, E. Barrington and G. D. Haswell, Philadelphia, 1848.

157. GREENBERG, S. B., COUCH, R. B., AND KASEL, J. A., An outbreak of an influenza type A variant in a closed population: The effect of homologous and heterologous antibody on illness and infection, *Am. J. Epidemiol.* **100:**209–215 (1974).

158. GREGG, M. B., The epidemiology of influenza in humans, *Ann. N.Y. Acad. Sci.* **353:**45–53 (1980).

159. GRUBER, W. C., TABER, L. H., GLEZEN, W. P., CLOVER, R. D., ABELL, T. D., DEMMLER, R. W., AND COUCH, R. B., Live attenuated and inactivated influenza vaccine in school-age children, *Am. J. Dis. Child.* **133:**595–600 (1990).

160. HALL, C. B., DOLIN, R., GALA, C. L., MARKOVITZ, D. M., ZHANG, Y. Q., MADORE, P. H., DISNEY, F. A., TALPEY, W. B., GREEN, J. L., FRANCIS, A. B., AND PICHICHERO, M. E., Children with influenza A infection: Treatment with rimantidine, *Pediatrics* **80:**275–282 (1987).

161. HALL, C. E., COONEY, M. K., AND FOX, J. P., The Seattle virus watch, *Am. J. Epidemiol.* **98:**365–380 (1973).

162. HANKINS, R. W., NAGATA, K., BUTCHER, D. J., POPPLE, S., AND ISHIHAMA, A., Monoclonal antibody analysis of influenza virus matrix protein epitopes involved in transcription inhibition, *Virus Genes* **3:**111–126 (1989).

163. HANNOUN, C., DAB, W., AND COHEN, J. M., A new influenza surveillance system in France: The ILE-DE-France "GROG". 1. Principles and methodology, *Eur. J. Epidemiol.* **5:**285–293 (1989).

164. HAY, A. J., WOLSTENHOLME, A. J., SKEHEL, J. J., AND SMITH, M. H., The molecular basis of the specific anti-influenza action of amantadine, *EMBO J.* **4:**3021–3024 (1985).

165. HAYDEN, F. G., BELSHE, R. B., CLOVER, R. D., HAY, A. J., OAKES, M. G., AND SOO, W., Emergency and apparent transmission of rimantadine-resistant influenza A virus in families, *N. Engl. J. Med.* **321:**1696–1702 (1989).

166. HAYDEN, F. G., AND COUCH, R. B., Clinical and epidemiological importance of influenza A viruses resistant to amantadine and rimantadine, *Rev. Med. Virol.* **2:**89–96 (1982).

167. HAYDEN, F. G., MINOCHA, A., SPYKER, D. A., AND HOFFMAN, H. E., Comparative single-dose pharmacokinetics of amantadine hydrochloride and rimantadine hydrochloride in young and elderly adults, *Antimicrob. Agents Chemother.* **28:**216–221 (1985).

168. HELENIUS, A., Unpacking the incoming influenza virus, *Cell* **69:**577–578 (1992).

169. HENDERSON, F. W., COLLIER, A. M., SANYAL, M. A., WATKINS, J. M., FAIRCLOUGH, D. L., CLYDE, W. A., JR., AND DENNY, F. W., A longitudinal study of respiratory viruses and bacteria in the etiology of acute otitis media with effusion, *N. Engl. J. Med.* **306:**1377–1383 (1982).

170. HENNESSY, A. V., DAVENPORT, F. M., HORTON, R. J. M., NAPIER, J. A., AND FRANCIS, T., JR., Asian influenza: Occurrence and recurrence, a community and family study, *Milit. Med.* **129:**38–50 (1964).

171. HIRST, G. K., The agglutination of red cells by allantoic fluid of chick embryos infected with influenza virus, *Science* **94:**22–23 (1941).

172. HOCHBERG, I. H., NELSON, K., AND JANZEN, W., Influenza type B-related encephalopathy. The 1971 outbreak of Reye syndrome in Chicago, *J.A.M.A.* **231:**817–821 (1975).

173. HOLSINGER, L. J., AND LAMB, R. A., Influenza virus M2 integral membrane protein is a homotetramer stabilized by formation of disulfide bonds, *Virology* **183:**32–43 (1991).

174. HOPE-SIMPSON, R. E., Age and secular distributions of virus-proven influenza in successive epidemics 1961–1976 in Cirencester: Epidemiological significance discussed, *J. Hyg. (Camb.)* **92:**303–336 (1984).

175. HOUSWORTH, W. J., AND LANGMUIR, A. D., Excess mortality from epidemic influenza, 1957–1966, *Am. J. Epidemiol.* **100:**40–48 (1974).

176. HOUSWORTH, W. J., AND SPOON, M. M., The age distribution of excess mortality during A2 Hong Kong influenza epidemics compared with earlier A2 outbreaks, *Am. J. Epidemiol.* **94:**348–350 (1974).

177. HOYLE, L., *The Influenza Viruses*, pp. 262–264, Springer-Verlag, New York, 1968.

178. HURWITZ, E. S., BARRETT, M. J., BRAGMAN, D., GUNN, W. J., SCHONBERGER, L. B., FAIRWEATHER, W. R., DRAGE, J. S. LA-MONTAGNE, J. R., KASLOW, R. A., BURLINGTON, D. B., QUINNAN, G. V., PARKER, R. A., PHILLIPS, K., PINSKY, P., DAYTON, D., AND DOWDLE, N. R., Public Health Service study on Reye's syndrome and medications, *N. Engl. J. Med.* **313:**849–857 (1985).

179. HURWITZ, E. S., NELSON, D. B., DAVIS, C., MORENS, D., AND SCHONBERGER, L. B., National surveillance for Reye syndrome: A five-year review, *Pediatrics* **6:**895–900 (1982).

180. JENNINGS, L. C., MACDIARMID, R. D., AND MILES, J. A. R., A study of acute respiratory disease in the community of Port Chalmers. I. Illnesses within a group of selected families and the relative incidence of respiratory pathogens in the whole community, *J. Hyg. (Camb.)* **81:**49–66 (1978).

181. JENNINGS, L. C., AND MILES, J. A. R., A study of acute respiratory disease in the community of Port Chalmers. II. Influenza A/Port

Chalmers/1/73 intrafamilial spread and the effect of antibodies to the surface antigens, *J. Hyg. (Camb.)* **81**:67–75 (1978).

182. JENNINGS, R., Respiratory viruses in Jamaica: A virologic and serologic study. 3. Hemagglutination-inhibiting antibodies to type B and C influenza viruses in the sera of Jamaicans, *Am. J. Epidemiol.* **87**:440–446 (1968).

183. JOHNSON, D. E., Guillain–Barré syndrome in the US Army, *Arch. Neurol.* **39**:21–24 (1982).

184. JORDAN, W. S., JR., The mechanism of spread of Asian influenza, *Am. Rev. Respir. Dis.* **83**:29–35 (1961).

185. JORDAN, W. S., JR., BADGER, G. F., AND DINGLE, J. A., A study of illness in a group of Cleveland families. XVI. The epidemiology of influenza, 1948–1953, *Am. J. Hyg.* **68**:169–189 (1958).

186. JORDAN, W. S., JR., DENNY, F. W., BADGER, G. F., CURTISS, C., DINGLE, J. H., OSEASOHN, R., AND STEVENS, D. A., A study of illness in a group of Cleveland families. XVII. The occurrence of Asian influenza, *Am. J. Hyg.* **68**:190–212 (1958).

187. KAPLAN, J. E., KATONA, P., HURWITZ, E. S., AND SCHONBERGER, L. B., Guillain–Barré syndrome in the United States, 1979–1980 and 1980–1981, *J.A.M.A.* **248**:698–700 (1982).

188. KASS, E. H., GREEN, G. M., AND GOLDSTEIN, E., Mechanisms of antibacterial action in the respiratory system, *Bacteriol. Rev.* **30**:488–496 (1966).

189. KATAGIRI, S., OHIZUMI, A., AND HOMMA, M., An outbreak of type C influenza in a children's home, *J. Infect. Dis.* **148**:51–56 (1983).

190. KEITEL, W. A., CATE, T. R., AND COUCH, R. B., Efficacy of sequential annual vaccination with inactivated influenza virus vaccine, *Am. J. Epidemiol.* **127**:353–364 (1988).

191. KEITEL, W. A., COUCH, R. B., QUARLES, J. M., CATE, T. R., BAXTER, B., AND MAASSAB, H. F., Trivalent attenuated cold-adapted influenza virus vaccine: Reduced viral shedding and serum antibody responses in susceptible adults, *J. Infect. Dis.* **167**:305–311 (1993).

192. KENDAL, A. P., AND COX, N. J., Forecasting the epidemic potential of influenza virus variants based on their molecular properties, *Vaccine* **3**:263–266 (1985).

193. KENDAL, A. P., JOSEPH, J. R., KOBAYASHI, G., NELSON, D., REYES, C. R., ROSS, M. R., SARANDRIA, J. L., WHITE, R., WOODALL, D. F., NOBLE, G. R., AND DOWDLE, W. R., Laboratory-based surveillance of influenza virus in the United States of 1977–1978. I. Periods of prevalence of H1N1 and H3N2 influenza A strains, their relative rates of isolation, *Am. J. Epidemiol.* **110**:449–461 (1977).

194. KENDAL, A. P., SCHIEBLE, J., COONEY, M. K., CHIN, J., FOY, H. M., AND NOBLE, G. R., Co-circulation of two influenza A (H3N2) antigenic variants detected by virus surveillance in individual communities, *Am. J. Epidemiol.* **108**:308–311 (1978).

195. KENNETT, M. L., DOWNIE, J., WHITE, J., WARD, B. K., MUTTON, J., IRVING, L. G., BIRCH, C. J., AND RODGER, S. M., Influenza in Melbourne, 1982, epidemiology and virology, *Med. J. Aust.* **141**:89–92 (1984).

196. KESSLER, H. A., TRENHOLME, G. M., HARRIS, A. A., AND LEVIN, S., Acute myopathy associated with influenza A/Texas/1/77 infection. Isolation of virus from a muscle biopsy specimen, *J.A.M.A.* **243**:461–462 (1980).

197. KILBOURNE, E. D., LEIEF, F. S., SCHULMAN, J. L., JAHIEL, R. I., AND LAVER, W. G., Antigenic hybrids of influenza virus and their implications, in: *Perspectives in Virology*, Vol. V (M. POLLARD, ED.), pp. 87–106, Academic Press, New York, 1967.

198. KILBOURNE, E. D., SCHULMAN, J. L., SCHOOLER, G., SWANSON, J., AND BUCHER, D., Correlated studies of a recombinant influenza virus vaccine. I. Derivation and characterization of the virus and vaccine, *J. Infect. Dis.* **124**:449–462 (1971).

199. KIM, H. K., BRANDT, C. D., ARROBIO, J. O., MURPHY, B., CHANOCK, R. M., AND PARROTT, R. H., Influenza A and B virus infection in infants and young children during the years 1957–1976, *Am. J. Epidemiol.* **109**:464–479 (1979).

200. KIMBALL, A. M., FOY, H. M., COONEY, M. K., ALLAN, I. D., MATLOCK, M., AND PLORDE, J. J., Isolation of respiratory syncytial and influenza viruses from the sputum of patients hospitalized with pneumonia, *J. Infect. Dis.* **147**:181–184 (1983).

201. KINGSBURY, D. W., Orthomyxo- and paramyxoviruses and their replication, in: *Virology* (B. FIELDS, ED.), pp. 1157–1178, Raven Press, New York, 1985.

202. KLEIN, J. D., COLLIER, A. M., AND GLEZEN, W. P., An influenza B epidemic among children in day-care, *Pediatrics* **58**:340–345 (1976).

203. KLONTZ, K. C., HYNES, N. A., GUNN, R. A., WILDER, M. H., HARMON, M. W., AND KENDAL, A. P., An outbreak of influenza A/Taiwan/1/86 (H1N1) infections at a naval base and its association with airplane travel, *Am. J. Epidemiol.* **129**:341–348 (1980).

204. KNIGHT, V., Discussion of "Viremia in Asian Influenza," *Trans. Assoc. Am. Physicians* **79**:384–386 (1966).

205. KNIGHT, V., COUCH, R. B., AND LANDAHL, H. D., The effect of the lack of gravity on airborne infection during space flight, *J.A.M.A.* **214**:513–518 (1970).

206. KSIAZEK, T. G., OLSON, J. G., IRVING, G. S., SETTLE, C. S., WHITE, R., AND PETRUSSO, R., An influenza outbreak due to A/USSR/77-like (H1N1) virus aborad a U.S. Navy ship, *Am. J. Epidemiol.* **112**:487–494 (1980).

207. KUWANO, K., SCOTT, M., YOUNG, J. F., AND ENNIS, F. A., HA2 subunit of influenza HA1 and H2 subtype induces a protective cross-reactive cytotoxic T lymphocyte response, *J. Immunol.* **140**:1264–1268 (1988).

208. KYRIAZOPOULOU-DALAINA, V., Distribution of influenza viruses in Northern Greece during 1972–1983, *J. Hyg. (Camb).* **93**:263–267 (1984).

209. LAMB, R. A., AND CHOPPIN, P. W., Segment 8 of the influenza virus genome is unique in coding for two polypeptides, *Proc. Natl. Acad. Sci. USA* **76**:4908–4912 (1979).

210. LAMB, R. A., AND CHOPPIN, P. W., Identification of a second protein (M_2) encoded by RNA segment 7 of influenza virus, *Virology* **112**:729–737 (1981).

211. LANGMUIR, A. D., PIZZI, M., TROTTER, W. Y., AND DUNN, F. L., Asian influenza surveillance, *Public Health Rep.* **73**:114–120 (1958).

212. LANGMUIR, A. D., WORTHEN, T. D., SOLOMON, J., RAY, C. G., AND PETERSON, E., The Thucydides syndrome: A new hypothesis for the cause of the plague of Athens, *N. Engl. J. Med.* **313**:1027–1030 (1985).

213. LARSON, E. W., DOMINIK, J. W., ROWBERG, A. H., AND HIGBEE, G. A., Influenza virus population dynamics in the respiratory tract of experimentally infected mice, *Infect. Immun.* **13**:438–447 (1976).

214. LAVER, W. G., AND WEBSTER, R. G., Studies on the origin of pandemic influenza. III. Evidence implicating duck and equine influenza viruses as possible progenitors of the Hong Kong strain of human influenza, *Virology* **51**:383–391 (1973).

215. LAYDE, P. M., ENGELBERG, A. L., DOBBS, H. I., CURTIS, A. C., CRAVEN, R. B., GRAITCER, P. L., SEDMAK, G. V., ERICKSON, J. D., AND NOBLE, G. R., Outbreak of influenza A/USSR/77 at Marquette University, *J. Infect. Dis.* **142**:347–352 (1980).

216. LEBIUSH, M., RANNON, L., AND KARK, J. D., An outbreak of A/USSR/90/77 (H1N1) influenza in army recruits: Clinical and laboratory observations, *Milit. Med.* **147**:43–48 (1982).

217. LEFRAK, E. A., STEVENS, P. M., PITHA, J., BALSINGER, E., NOON, G. P., AND MAYOR, H. D., Extracorporal membrane oxygenation for fulminant influenza pneumonia, *Chest* **66**:385–388 (1974).

218. LENNON, D. R., CHERRY, J. D., MORGENSTEIN, A., CHAMPION, J. C., AND BRYSON, Y. J., Longitudinal study of influenza B symptomatology and interferon production in children and college students, *Pediatr. Infect. Dis.* **2**:212–215 (1983).

219. LEUNA, K. N., AND ADA, G. L., Induction of natural killer cells during murine influenza virus infection, *Immunobiology* **160**:352–366 (1981).

220. LITTLE, J. W., HALL, W. J., DOUGLAS, R. G., JR., MUDHOLKAR, G. S., SPEERS, D. M., AND PATEL, K., Airway hyperreactivity and peripheral airway dysfunction in influenza A infection, *Am. Rev. Respir. Dis.* **118**:295–303 (1978).

221. LONGINI, I. M., JR., AND KOOPMAN, J. S., Household and community transmission parameters from final distributions of infections in households, *Biometrics* **38**:115–126 (1982).

222. LONGINI, I. M., JR., KOOPMAN, J. S., MONTO, A. S., AND FOX, J. P., Estimating household and community transmission parameters for influenza, *Am. J. Epidemiol.* **115**:736–748 (1982).

223. LOOSLI, C. G., The pathogenesis and pathology of experimental airborne influenza virus A infections in mice, *J. Infect. Dis.* **84**:153–168 (1949).

224. LOURIA, D. B., BLUMENFELD, H. L., ELLIS, J. T., KILBOURNE, E. D., AND ROGERS, D. E., Studies on influenza in the pandemic of 1957–1958. II. Pulmonary complications of influenza, *J. Clin. Invest.* **38**:213–265 (1959).

225. MAASSAB, H. F., AND DE BORDE, C., Development and characterization of cold-adapted viruses as live virus vaccines, *Vaccine* **3**:355–369 (1985).

226. MACHIN, S. J., POTTER, C. W., AND OXFORD, J. S., Changes in the antibody status of a population following epidemic infection by influenza virus A/2/Hong Kong/1/68, *J. Hyg. (Camb.)* **68**:497–504 (1970).

227. MARINE, W. M., MCGOWAN, J. E., JR., AND THOMAS, J. E., Influenza detection: A prospective comparison of surveillance methods and analysis of isolates, *Am. J. Epidemiol.* **104**:248–255 (1976).

228. MARINE, W. M., AND THOMAS, J. E., Antigenic memory to influenza A viruses in man determined by monovalent vaccines, *Postgrad. Med. J.* **55**:98–108 (1979).

229. MARINE, W. M., AND WORKMAN, W. W., Hong Kong influenza recapitulation, *Am. J. Epidemiol.* **90**:406–415 (1969).

230. MARTIN, K., AND HELENIUS, A., Nuclear transport of influenza virus ribonucleoproteins: The viral matrix protein (M1) promotes export and inhibits import, *Cell* **67**:117–130 (1991).

231. MARTIN, R. R., COUCH, R. B., GREENBERG, S. B., CATE, T. R., AND WARR, G. A., Effects of infection with influenza virus on the function of polymorphonuclear leukocytes, *J. Infect. Dis.* **144**:279 (1981).

232. MASUREL, N., Swine influenza virus and the recycling of influenza A viruses in man, *Lancet* **2**:244–247 (1976).

233. MASUREL, N., AND MARINE, W. M., Recycling of Asian and Hong Kong influenza A virus hemagglutinins in man, *Am. J. Epidemiol.* **97**:44–49 (1973).

234. MCMICHAEL, A. J., AND ASKONAS, B. A., Influenza virus specific cytotoxic T cells in man: Induction and properties, *Eur. J. Immunol.* **8**:705–711 (1984).

235. MCMICHAEL, A. J., MICHIE, C. A., GATCH, F. M., SMITH, G. L.,

AND MOSS, B., Recognition of influenza A virus nucleoprotein by human cytotoxic T lymphocyte, *J. Gen. Virol.* **67**:719–726 (1986).

236. MEIER-EWERT, H., NAGOLI, A., HERRLER, G., BASAK, S., AND COMPAN, R. W., Analysis of influenza C structural proteins and identification of a virion RNA polymerase in the replication of negative strand viruses, in: *Developments in Cell Biology*, Vol. 7 (D. H. L. BISHOP AND R. W. COMPANS, EDS.), pp. 173–180, Elsevier North-Holland, New York, 1980.

237. MEIKLEJOHN, G., Viral respiratory disease at Lowry Air Force Base in Denver, 1952–1982, *J. Infect. Dis.* **148**:775–784 (1983).

238. MEIKELJOHN, G., KEMPE, C. H., THALMAN, W. G., AND LENETTE, E. H., Evaluation of monovalent influenza vaccines, II. Observations during an influenza A-prime epidemic, *Am. J. Hyg.* **55**:12–21 (1952).

239. MIDDLETON, P. J., ALEXANDER, R. M., AND SZYMANSKI, M. T., Severe myositis during recovery from influenza, *Lancet* **2**:533–535 (1970).

240. MINOW, R. A., GORBACH, S., JOHNSON, B. L., JR., AND DORNFELD, L., Myoglobinuria associated with influenza A infection, *Ann. Intern. Med.* **80**:359–361 (1974).

241. MONTO, A. S., AND ARDEN, N. H., Implications of viral resistance to amantadine in control of influenza A, *Clin. Infect. Dis.* **15**:362–367 (1992).

242. MONTO, A. S., DAVENPORT, F. M., NAPIER, J. A., AND FRANCIS, T., JR., Effect of vaccination of a school-age population upon the course of an A2/Hong Kong influenza epidemic, *Bull. WHO* **41**:537–542 (1969).

243. MONTO, A. S., AND KIOUMEHR, F., The Tecumseh study of respiratory illness. IX. Occurrence of influenza in the community, 1966–1971, *Am. J. Epidemiol.* **102**:553–563 (1975).

244. MONTO, A. S., KOOPMAN, J. S., AND LONGINI, I. M., JR., Tecumseh study of illness. XIII. Influenza infection and disease, 1976–1981, *Am. J. Epidemiol.* **121**:811–822 (1985).

245. MONTO, A. S., AND SULLIVAN, K. M., Acute respiratory illness in the community. Frequency of illness and the agents involved, *Epidemiol. Infect.* **110**:145–160 (1993).

246. MORENS, D. M., SULLIVAN-BOLYAI, J. Z., SLATER, J. E., SCHONBERGER, L. B., AND NELSON, D. B., Surveillance of Reye's syndrome in the United States, 1977, *Am. J. Epidemiol.* **114**:406–416 (1981).

247. MORGENSEN, J. L., Myoglobinuria and renal failure associated with influenza, *Ann. Intern. Med.* **80**:362–363 (1974).

248. MOSER, M. R., BENDER, T. R., MARELOLIS, N. S., NOBLE, G. R., KENDAL, A. P., AND RITTER, D. G., An outbreak of influenza aboard a commercial airliner, *Am. J. Epidemiol* **110**:1–7 (1979).

249. MULDER, J., AND HERS, J. F. P., *Influenza*, Wolters-Noordhoff, Groningen, The Netherlands, 1972.

250. MULDER, J., MASUREL, N., AND WEBBERS, P. J., Pre-epidemic antibody against 1957 strain of asiatic influenza in serum of older people living in the Netherlands, *Lancet* **1**:810–814 (1958).

251. MURPHY, B. R., Use of live attenuated cold-adapted influenza A reassortant virus vaccines in infants, children, young adults, and elderly adults, *Infect. Dis. Clin. Prac.* **2**:174–181 (1993).

252. MURPHY, B. R., KASEL, J. A., AND CHANOCK, R. M., Association with serum anti-neuraminidase antibody with resistance to influenza in man, *N. Engl. J. Med.* **25**:1329–1332 (1972).

253. MURPHY, B. R., NELSON, D. L., AND WRIGHT, P. F., Secretory and systemic immunological response in children infected with live attenuated influenza A virus vaccines, *Infect. Immun.* **36**:1102–1108 (1982).

254. MURPHY, B. R., PHELAN, M. A., AND NELSON, D. L., Hemagglu-

tinin-specific enzyme linked immunosorbent assay for antibodies to influenza A and B viruses, *J. Clin. Microbiol.* **13**:554–560 (1981).

255. MURPHY, B. R., TIERNEY, E. L., BARBOUR, B. A., YOLKEN, R. H., ALLING, R. E., AND CHANOCK, R. M., Use of the enzyme-linked immunosorbent assay to detect serum antibody response of volunteers who received attenuated influenza A virus vaccines, *Infect. Immun.* **29**:342–347 (1980).

256. MURPHY, B. R., AND WEBSTER, R. G., Influenza viruses, in: *Virology* (B. FIELDS, ED.), pp. 1179–1239, Raven Press, New York, 1985.

257. MULLOOLY, J., BENNETT, M., HORNBROOK, M., BARKER, W., WILLIAMS, W., PATRIARCA, P., AND RHODES, P., Cost-effectiveness of influenza vaccination programs in an HMO: The experience of Kaiser Permanente, Northwest Region, in: *Options for the Control of Influenza II* (C. HAMMOUN, A. P. KENDAL, H. D. KLENK, AND F. L. RUBEN, EDS.), pp. 53–62, Excerpta Medica, Amsterdam, 1993.

258. NAKAJIMA, S., COX, N. J., AND KENDAL, A. P., Antigenic and genomic analyses of influenza A (H1N1) viruses from different regions of the world, February 1978 to March 1980, *Infect. Immun.* **32**:287–294 (1981).

259. NASCIMENTO, J. P., KRAWCZUK, M. M., MARCOPITO, L. F., AND BARUZZI, R. G., Prevalence of antibody against influenza A viruses in the Kren-Akorere, an Indian tribe of central Brazil, first contacted in 1973, *J. Hyg. (Camb.)* **95**:159–164 (1985).

260. NAYAK, D. P., AND JABBAR, M. A., Expression of influenza viral hemagglutinin (HA) in the yeast *Saccharomyces cerevisiae:* Biological and antigenic properties of genetically engineered wild type and mutant hemagglutinins, in: *Options for the Control of Influenza* (A. P. KENDAL AND P. A. PATRIARCA, EDS.), pp. 357–373, Alan R. Liss, New York, 1986.

261. NEWTON-JOHN, H. F., YUNG, A. P., BENNETT, N. M., AND FORBES, J. A., Influenza virus pneumonitis: A report of ten cases, *Med. J. Aust.* **2**:1160–1166 (1971).

262. NOBLE, G. R., KAYE, H. S., YARBROUGH, W. B., FIEDLER, B. K., REED, C. J., FELKER, M. B., KENDAL, A. P., AND DOWDLE, W. R., Measurement of hemagglutination-inhibiting antibody to influenza virus in the 1976 influenza vaccine program: Methods and reproducibility, *J. Infect. Dis.* **136**(Suppl.):S429–S434 (1977).

263. OSEASOHN, R., ADELSON, L., AND KAJI, M., Clinicopathologic study of thirty-three fatal class of Asian influenza, *N. Engl. J. Med.* **260**:509–518 (1959).

264. PARKMAN, P. D., HOPPS, H. E., RASTOGI, S. C., AND MEYER, H. M., JR., Summary of clinical trials of influenza virus vaccines in adults, *J. Infect. Dis.* **136**:S722–S730 (1977).

265. PATRIARCA, P. A., WEBER, J. A., PARKER, R. A., HALL, W. N., KENDAL, A. P., BREGMAN, D. J., AND SCHONBERGER, L. B., Efficacy of influenza vaccine in nursing homes. Reduction in illness and complications during an influenza A (H3N2) epidemic, *J.A.M.A.* **253**:1136–1139 (1985).

266. PEARL, R., Influenza studies: On certain general statistical aspects of the 1918 epidemic in American cities, *Public Health Rep.* **34**:1743–1783 (1919).

267. PEREIRA, M. S., Global surveillance of influenza, *Br. Med. J.* **35**:9–14 (1979).

268. PEREIRA, M., ASSAAD, F. A., AND DELON, P. J., Influenza surveillance, *Bull. WHO* **56**:193–203 (1978).

269. PEREIRA, M. S., AND CHAKRAVERTY, P., The laboratory surveillance of influenza epidemics in the United Kingdom 1968–1976, *J. Hyg. (Camb.)* **79**:77–87 (1977).

270. PEREIRA, M. S., AND CHAKRAVERTY, P., Influenza in the United Kingdom 1977–1981, *J. Hyg. (Camb.)* **88**:501–512 (1982).

271. PERROTTA, D. M., DECKER, M., AND GLEZEN, W. P., Acute respiratory disease hospitalizations as a measure of impact of epidemic influenza, *Am. J. Epidemiol.* **122**:468–476 (1985).

272. PETERSDORF, R. G., FUSCO, J. J., HARTER, D. H., AND ALBRINK, W. S., Pulmonary infections complicating Asian influenza, *Arch. Intern. Med.* **103**:262–272 (1959).

273. PIEDRA, P. A., AND GLEZEN, W. P., Influenza in children: Epidemiology, immunity, and vaccines, *Semin. Pediatr. Infect. Dis.* **2**:140–146 (1991).

274. PIEDRA, P. A., GLEZEN, W. P., MBAWUIKE, I., GRUBER, W. C., BAXTER, B. D., BOLAND, F. J., BYRD, R. W., FAN, L. L., LEWIS, J. K., RHODES, L. J., WHITNEY, S. E., AND TABER, L. H., Studies on reactogenicity and immunogenicity of attenuated bivalent cold recombinant influenza type A (CRA) and inactivated trivalent influenza virus (TI) vaccines in infants and young children, *Vaccine* **11**:718–724 (1993).

275. PONKA, A., JALANKO, H., PONKA, T., AND SENVIK, M., Viral and mycoplasmal antibodies in patients with myocardial infarction, *Ann. Clin. Res.* **13**:429–432 (1981).

276. PUCK, J. M., GLEZEN, W. P., FRANK, A. L., AND SIX, H. R., Protection of infants from infection with influenza A virus by transplacentally acquired antibody, *J. Infect. Dis.* **142**:844–847 (1980).

277. PYHALA, R., AND AHO, K., Seroepidemiology of H1N1 infection: The infection and re-infection rate in winter 1978–79, *J. Hyg. (Camb.)* **86**:27–33 (1981).

278. QUINNAN, G. V., SCHOOLEY, R., DOLIN, R., ENNIS, F. A., GROSS, P., AND GWALTNEY, J. M., Serologic responses and systemic reactions in adults after vaccination with monovalent A/USSR/77 and trivalent A/USSR/77, A/Texas/77, B/Hong Kong/72 influenza vaccines, *Rev. Infect. Dis.* **5**:748–757 (1983).

279. RAO, B. L., KADAM, S. S., PAVRI, K. M., AND KOTHAVALE, V. S., Epidemiological, clinical, and virological features of influenza outbreaks in Pune, India, 1980, *Bull. WHO* **60**:639–642 (1982).

280. RAY, C. G., AND MINNICH, L. L., Efficiency of immunofluorescence for rapid detection of common respiratory viruses, *J. Clin. Microbiol.* **25**:355–357 (1987).

281. REMINGTON, P. L., ROWLEY, D., MCGEE, H., HALL, W. N., AND MONTO, A. S., Decreasing trends in Reye's syndrome and aspirin use in Michigan 1979–1984, *Pediatrics* **77**:93–98 (1986).

282. RETALLIAU, H. F., GALE, J. L., BEASLEY, R. P., AND HATTWICK, M. A. W., Excess mortality and influenza surveillance in Taiwan, *Int. J. Epidemiol.* **7**:223–229 (1978).

283. RETAILLIAU, H. F., STORCH, G. A., CURTIS, A. C., HORNE, T. J., SCALLY, M. J., AND HATTWICK, M. A., The epidemiology of influenza B in a rural setting in 1977, *Am. J. Epidemiol.* **109**:639–649 (1979).

284. RODGERS, B. C., AND MIMS, C. A., Interaction of influenza virus with mouse macrophages, *Infect. Immun.* **31**:751–757 (1981).

285. ROSCELLI, J. D., BASS, J. W., AND PANG, L., Guillain–Barré syndrome and influenza vaccination in the US Army, 1980–1988, *Am. J. Epidemiol.* **133**:952–955 (1991).

286. RUUTU, P., HALONEN, P., MEURMAN, O., TORRES, C., PALADIN, F., YAMAOKA, K., AND TUPASI, T. E., Viral lower respiratory tract infections in Filipino children, *J. Infect. Dis.* **161**:175–179 (1990).

287. SAFRANEK, T. J., LAWRENCE, D. N., KURLAND, L. T., CULVER, D. H., WIEDERHOLT, W. C., HAYNER, S. S., OSTERHOLM, M. T., O'BRIEN, P., HUGHES, J. M., AND EXPERT NEUROLOGY GROUP, Reassessment of the association between Guillain–Barré syndrome and receipt of swine influenza vaccine in 1976–1977:

Results of a two-state study, *Am. J. Epidemiol.* **133**:940–951 (1991).

288. SALK, J. E., A simplified procedure of titrating hemagglutinating capacity of influenza virus and the corresponding antibody, *J. Immunol.* **49**:87–98 (1944).

289. SCHAFER, W., Vergleichende sero-immunoglische Untersuchungen ueber die Viren der Influenza und klassichen oefluegel Pest, *Z. Naturforsch* **106**:81–91 (1955).

290. SCHIFF, L. J., Studies on the mechanisms of influenza virus infection in hamster trachea organ culture, *Arch. Ges. Virusforsch.* **44**:195–204 (1974).

291. SCHILD, G. C., PEREIRA, M. S., AND CHAKRAVERTY, P., Single-radial haemolysis: A new method for the assay of antibody to influenza haemagglutinin, *Bull. WHO* **52**:43–50 (1975).

292. SCHILD, G. C., AND STUART-HARRIS, C. H., Serological epidemiological studies with influenza A viruses, *J. Hyg. (Camb.)* **63**:479–490 (1965).

293. SCHOENBAUM, S. C., COLEMAN, M. T., DOWDLE, W. R., AND MOSTOW, S. R., Epidemiology of influenza in the elderly: Evidence of virus recycling, *Am. J. Epidemiol.* **103**:166–173 (1976).

294. SCHOENBAUM, S. C., MOSTOW, S. R., DOWDLE, W. R., COLEMAN, M. T., AND KAYE, H. S., Studies with inactivated influenza vaccines purified by zonal centrifugation. II. Efficacy, *Bull. WHO* **41**:531–535 (1969).

295. SCHOLTISSEK, C., Influenza virus genetics, *Adv. Genet.* **20**:1–36 (1979).

296. SCHONBERGER, L. B., BREGMAN, D. J., SULLIVAN-BOLYAI, J. Z., KEENLYSIDE, R. A., ZIEGLER, D. W., RETAILLIAU, H. F., EDDIS, D. L., AND BRYAN, J. A., Guillain–Barré syndrome following vaccination in the national influenza immunization program, United States, 1966–1967, *Am. J. Epidemiol.* **110**:105–123 (1979).

297. SCHULMAN, J. L., AND KILBOURNE, E. D., Airborne transmission of influenza virus infection in mice, *Nature* **195**:119 (1962).

298. SCHWARZMANN, S. W., ADLER, J. L., SULLIVAN, R. J., JR., AND MARINE, W. M., Bacterial pneumonia during the Hong Kong influenza epidemic of 1968–1969, *Arch. Intern. Med.* **127**:1037–1041 (1971).

299. SCRAGG, R., Effect of influenza epidemics on Australian mortality, *Med. J. Aust.* **142**:98–102 (1985).

300. SERFLING, R. E., Methods for current statistical analysis of excess pneumonia–influenza deaths, *Public Health Rep.* **78**:494–506 (1963).

301. SHOPE, R. E., Swine influenza, I. Experimental transmission and pathology, *J. Exp. Med.* **54**:349–359 (1931).

302. SHOPE, R. E., Swine influenza. III. Filtration experiments and etiology, *J. Exp. Med.* **54**:373–385 (1931).

303. SHOPE, R. E., The incidence of neutralizing antibodies for swine influenza virus in the sera of human beings of different ages, *J. Exp. Med.* **63**:669–684 (1936).

304. SIX, H. R., GLEZEN, W. P., KASEL, J. A., COUCH, R. B., AND GRIFFIS, C., Heterogenicity of influenza viruses isolated from the Houston community during defined epidemic periods, in: *ICN-UCLA Symposia on Molecular and Cellular Biology*, Vol. XXI, *Genetic Variation among Influenza Viruses* (D. NAYAK, ED.), pp. 505–513, Academic Press, New York, 1981.

305. SIX, H. R., AND KASEL, J. A., Radioimmunoprecipitation assay for quantitation of serum antibody to the hemagglutinin of type A influenza virus, *J. Clin. Microbiol.* **7**:165–171 (1978).

306. SIX, H. R., WEBSTER, R. G., KENDAL, A. P., GLEZEN, W. P., GRIFFIS, C., AND COUCH, R. B., Antigenic analysis of H1N1

viruses isolated in the Houston metropolitan area during four successive seasons, *Infect. Immun.* **42**:453–458 (1983).

307. SKEHEL, J. J., BAYLEY, P. M., BROWN, E. B., MARTIN, S. R., WATERFIELD, M. D., WHITE, J. M., WILSON, I. A., AND WILEY, D. C., Changes in the conformation of influenza virus hemagglutinin at the pH optimum of virus-mediated membrane fusion, *Proc. Natl. Acad. Sci. USA* **79**:968–972 (1982).

308. SLEPUSHKIN, A. N., RITOVA, V. V., FEKLISOVA, L. V., FEDOROVA, G. I., OBROSOVA-SEROVA, N. P., SAFONOVA, E. S., KUPRYASHINA, L. M., MOLIBOG, E. V., SLEPUSHKIN, V. A., AND ZHDANOV, V. M., Results of a two-year study of humoral immunity to influenza A and B viruses in children under the age of 14 years in Moscow and its suburbs, *Bull. WHO* **62**:75–82 (1984).

309. SMITH, G. L., BENNINK, J. R., YEWDELL, J. W., SMALL, P. A., JR., MURPHY, B. R., AND MOSS, B., Vaccine virus recombinants expressing influenza virus genes, in: *Options for the Control of Influenza* (A. P. KENDAL AND P. A. PATRIARCA, EDS.), pp. 375–389, Alan R. Liss, New York, 1986.

310. SMITH, W., The complement-fixation reaction in influenza, *Lancet* **2**:1256–1259 (1936).

311. SMITH, W., ANDREWES, C. H., AND LAIDLAW, P. P., A virus obtained from influenza patients, *Lancet* **2**:66–68 (1933).

312. SMORODINCEV, A. A., The efficacy of live influenza vaccines, *Bull. WHO* **41**:585–588 (1969).

313. SNACKEN, R., LION, J., VAN CASTEREN, V., CORNELIS, R., YANE, F., MOMBAERTS, M., AELVOET, W., AND STROOBANT, A. A., Five years of sentinel surveillance of acute respiratory infections (1985–1990): The benefits of an influenza early warning system, *Eur. J. Epidemiol.* **8**:485–490 (1992).

314. SONOGUCHI, T., SAKOH, M., KUMTAS, N., SATSUTA, K., NORIKI, H., AND FUKUMI, H., Reinfection with influenza A (H2N2, H3N2, and H1N1) viruses in soldiers and students in Japan, *J. Infect. Dis.* **153**:33–40 (1986).

315. SPELMAN, D. W., AND MCHARDY, C. J., Concurrent outbreaks of influenza A and influenza B, *J. Hyg. (Camb.)* **94**:331–339 (1985).

316. SPRENGER, M. J. W., MULDER, P. G. H., AND BEYER, W. E. P., Influenza: Relation of mortality to morbidity parameters—Netherlands, 1970–1989, *Int. J. Epidemiol.* **20**:1118–1124 (1991).

317. STANLEY, E. D., AND JACKSON, G. G., Viremia in Asian influenza, *Trans. Assoc. Am. Physicians* **79**:376–387 (1966).

318. STIVER, H. G., GRAVES, P., EICKHOFF, T. C., AND MEIKLEJOHN, G., Efficacy of "Hong Kong" vaccine in preventing "England" variant influenza A in 1972, *N. Engl. J. Med.* **289**:1267–1271 (1973).

319. STOUT, C., LAWRENCE, S., AND JULIAN, S., Evaluation of a monoclonal antibody pool for rapid diagnosis of respiratory viral infections, *J. Clin. Microbiol.* **27**:448–452 (1989).

320. SUGAYA, N., NEROME, K., ISHIDA, M., NEROME, R., NAGAE, M., TAKEUCHI, Y., AND OSANO, M., Impact of influenza virus infection as a cause of pediatric hospitalization, *J. Infect. Dis.* **165**:373–375 (1992).

321. SUGRUE, R. J., AND HAY, A. J., Structural characteristics of the M2 protein of influenza A viruses: Evidence that it forms a tetrameric channel, *Virology* **180**:617–624 (1991).

322. SULKAVA, R., RISSANAN, A., AND PYHALA, R., Post-influenzal encephalitis during the influenza A outbreak in 1979/1980, *J. Neurol. Neurosurg. Psychiatry* **44**:161–163 (1981).

323. SULLIVAN-BOLYAI, J. Z., AND COREY, L., Epidemiology of Reye syndrome, *Epidemiol. Rev.* **3**:1–26 (1981).

324. SULLIVAN-BOLYAI, J. Z., NELSON, D. B., MORENS, D. M., AND SCHONBERGER, L. B., Reye syndrome in children less than one

year old: Some epidemiologic observations, *Pediatrics* **65**:627–629 (1980).

325. TABER, L. H., PAREDES, A., GLEZEN, W. P., AND COUCH, R. B., Infection with influenza A/Victoria virus in Houston families, 1976, *J. Hyg. (Camb.)* **86**:303–313 (1981).

326. TAYLOR, R. M., Studies on survival of influenza virus between epidemics and antigenic variants of the virus, *Am. J. Public Health* **39**:171–178 (1949).

327. TILLETT, H. E., SMITH, J. W. G., AND CLIFFORD, R. E., Excess morbidity and mortality associated with influenza in England and Wales, *Lancet* **1**:793–795 (1980).

328. TILLETT, H. E., AND SPENCER, I. L., Influenza surveillance in England and Wales using routine statistics, *J. Hyg. (Camb.)* **88**:83–94 (1982).

329. TOWNSEND, J. F., History of influenza epidemics, *Ann. Med. History* **5**:533–547 (1933).

330. TUMOVA, B., SCHARFENORTH, H., AND ADAMCZYK, G., Incidence of influenza C virus in Czechoslovakia and German Democratic Republic, *Acta Virol.* **27**:502–510 (1983).

331. UTELL, M. J., AQUILINA, A. T., HALL, W. J., SPEERS, D. M., DOUGLAS, R. G., JR., GIBB, F. R., MORROW, P. E., AND HYDE, R. W., Development of airway reactivity to nitrates in subjects with influenza, *Am. Rev. Respir. Dis.* **121**:233–241 (1980).

332. WALDMAN, R. H., BOND, J. O., LEVITT, L. P., HARTWIG, E. C., PRATHER, E. C., BURATTA, R. L., NEILL, J. S., AND SMALL, P. A., JR., An evaluation of influenza immunization, *Bull. WHO* **41**:543–548 (1969).

333. WANER, J. L., TODD, S. J., SHALABY, H., MURPHY, P., AND WALL, L. V., Comparison of directigen FLU-A with viral isolation and direct immunofluorescence for the rapid detection and identification of influenza A virus, *J. Clin. Microbiol.* **29**:479–482 (1991).

334. WALSH, J. J., DIETLEIN, L. F., LOW, F. N., BURCH, G. E., AND MOGABGAB, W. J., Bronchotracheal response in human influenza, *Arch. Intern. Med.* **108**:376–388 (1961).

335. WEBSTER, R. G., BEAN, W. J., GORMAN, O. T., CHAMBERS, T. M., AND KAWASKA, Y., Evaluation and ecology of influenza A viruses, *Microbiol. Rev.* **56**:152–179 (1992).

336. WILEY, D. C., WILSON, I. A., AND SKEHEL, J. J., Structural identification of the antibody-binding sites of Hong Kong influenza hemagglutinin and their involvement in antigenic variation, *Nature* **289**:373–378 (1981).

337. WILLERS, H., AND HOPKEN, W., Epidemiology of influenza in lower Saxony during the period 1968–1978 with particular emphasis on subtypes A (H3N2) and A (H1N1) in winter 1977–78, *Med. Microbiol. Immunol.* **167**:21–27 (1979).

338. WILLIAMS, M. A., AND LAMB, R. A., Determination of the orientation of an integral membrane protein and sites of glycoprotein by oligonucleotide-directed mutagenesis: Influenza B virus NB glycoprotein lacks a cleavable signal sequence and has an extracellular NH2 terminal region, *Mol. Cell. Biol.* **8**:1186–1196 (1988).

339. WILSON, A. B., PLANTEROSE, D. N., NAGINGTON, J., PARK, J. R., BARRY, R. D., AND COOMBS, R. R. A., Influenza A antigens on human lymphocytes *in vitro* and probably *in vivo*, *Nature* **259**:582–584 (1976).

340. WORLD HEALTH ORGANIZATION, Influenza in the world, *Week. Epidemiol. Rec.* **59**:5–10 (1984).

341. WORLD HEALTH ORGANIZATION, A revision of the system of nomenclature for influenza viruses, *Bull. WHO* **58**:585–591 (1980).

342. WRIGHT, P. F., BRYANT, J. D., AND KARZON, D. T., Comparison of influenza B/Hong Kong virus infections among infants, children, and young adults, *J. Infect. Dis.* **141**:430–435 (1980).

343. WRIGHT, P. F., THOMPSON, J., AND KARZON, D. T., Differing virulence of H1N1 and H2N2 influenza strains, *Am. J. Epidemiol.* **112**:814–819 (1980).

344. YAP, K. L., AND ADA, G. L., The recovery of mice from influenza virus infection. Adoptive transfer of immunity with immune T lymphocytes, *Scand. J. Immunol.* **7**:389–397 (1978).

345. YORKE, J. A., NATHANSON, N., PIANIGIANI, G., AND MARTIN, J., Seasonality and the requirement for perpetuation and eradication of viruses in populations, *Am. J. Epidemiol.* **109**:103–123 (1979).

346. ZEBEDEE, S. L., AND LAMB, R. A., Influenza A virus M2 protein: Monoclonal antibody restriction of virus growth and detection of M2 in virions, *J. Virol.* **62**:2762–2772 (1988).

347. ZINKERNAGEL, R. M., AND DOHERTY, P. C., MHC-restricted cytotoxic T cells: Studies on the biological role of polymorphic major transplantation antigens determining T cell restriction—specificity, function, and responsiveness, *Adv. Immunol.* **27**:51–177 (1979).

12. Suggested Reading

COUCH, R. B., KASEL, J. A., GLEZEN, W. P., CATE, T. R., SIX, H. R., TABER, L. H., FRANK, A. L., GREENBERG, S. B., ZAHRADNIK, J. M., AND KEITEL, W. A., Influenza: Its control in persons and populations, *J. Infect. Dis.* **153**:431–447 (1986).

GLEZEN, W. P., Serious morbidity and mortality associated with influenza, *Epidemiol. Rev.* **4**:25–44 (1982).

HANNOUN, C., KENDAL, A. P., KLENK, H. D., AND RUBEN, F. L. (EDS.), *Options for the Control of Influenza II*, Excerpta Medica, Amsterdam, 1993.

KENDAL, A. P., AND PATRIARCA, P. A. (EDS.), *Options for the Control of Influenza*, Alan R. Liss, New York, 1986.

KILBOURNE, E. D., *Influenza*, Plenum Press, New York, 1987.

SHAW, M. W., ARDEN, N. H., AND MAASSAB, H. F., New aspects of influenza viruses, *Clin. Microbiol. Rev.* **5**:74–92 (1992).

STUART-HARRIS, C. H., SCHILD, G. C., AND OXFORD, J. S., *Influenza—The Viruses and the Disease*, Edward Arnold, Baltimore, 1985.

CHAPTER 17

Measles

Francis L. Black

1. Introduction

"The simplest of all infectious diseases is measles," said Kenneth Maxcy 50 years ago.[110] That was before the virus had been isolated or any serological test had become available. The picture is more complicated than Maxcy knew. The virus can persist in humans, at least in defective form, and the distribution of the disease has been made more complex by an artificially determined pattern of vaccine distribution. However, the distinctive clinical response, the durability of immunity, and the absence of both nonhuman hosts and serologically cross-reactive viruses of humans continue to make measles the ideal basic model of infectious disease epidemiology. Essentially every unvaccinated person still contracts measles,[31,43] and most infections are clinically identifiable by an experienced layperson.

Vaccination programs have reduced the incidence to low levels in developed countries and eliminated it in some areas, such as the British Caribbean.[8b] Statistics from less developed countries are inadequate, but intensified vaccination programs have had substantial impact there, too. The idea that measles can be eradicated has been revived.[8a] This idea depends on the premise that the virus has no effective reservoir. Persistence of virus in subacute sclerosing panencephalitis is unimportant if the virus does not reactivate from this source. Other possible reservoirs are discussed below. Residual pockets of susceptibility often coincide with groups at risk of severe disease, and mortality during the 1989–1991 resurgence of measles in the United States was ten times that of the prevaccine era.[38]

2. Historical Background

The writings of Abu Becr,[5] known by his hometown name of Rhazes, provide the earliest clear description of measles. Rhazes lived in the 10th century, but he quoted other authors on measles from as far back as Al Yehudi, "The Jew," who lived in the 7th century. A Chinese reference to a new rash disease that occurred in 653 AD may also have been measles.[112] Rhazes looked on measles as a relatively severe disease and considered it "more to be dreaded than smallpox." It is curious, then, that measles had not been described earlier; smallpox had been accurately described by Galen in the 2nd century AD.

From the diversity of names for measles, it is evident that identification of this disease did not depend on medical science. Rhazes called measles *Hasbah*, which means "casting out" and carries much the same connotation as "eruption" in English. That name has come to specifically mean measles in Arabic. This stands in contrast to classical Greek, which had no name for measles, and Latin, which came to use the terms *rubeola* and *morbilli* only in the Middle Ages. The fact that the Iberian name, *sarampion*, has a different root supports a post-Roman date for the Latin names. On the other hand, Teutonic languages have a common root name in the Old German *mazer*, which became *Masern* (German), *mislingar* (Icelandic), or measles (English). Linguistic evidence thus suggests that the disease was recognized before the Germanic migrations but after fragmentation of the Roman Empire[137] and concurs with written accounts in pointing to the period between 450 and 650 AD as the time when measles was identified.

Measles requires a human population of a few hundred thousand persons to provide a sufficient supply of new susceptibles for its persistence.[25] Population groups of this size did not exist prior to development of the Middle Eastern river valley civilizations, and measles as

Francis L. Black • Department of Epidemiology and Public Health, Yale University School of Medicine, New Haven, Connecticut 06519.

we know it must have arisen since that time, i.e., since 2500 BC. Rinderpest, the Morbillivirus of cattle, on the basis of molecular relationships,[147] is the most likely parent virus. However, only canine distemper routinely causes persistent infection with reactivation of infectious virus,[6] and it is believed to have been the source of the viruses that appeared recently in seals and porpoises.[32a]

Rhazes seems not to have considered measles infectious but a necessary part of growing up. Sydenham,[154] who described measles in northern Europe in 1670, considered it "atmospheric." However, by 1758, Home[90] must have been thinking in terms of an infection when he attempted to immunize against measles by a procedure analogous to variolation. Even so, it was not clear that every case was contracted from an preceding case. Demonstration of this had to wait another century for Panum and his studies in the Faroe Islands.[132] Panum also defined the 14-day incubation period and showed that infection conferred lifetime immunity. Hirsch[88] built on Panum's work to reach the conclusion that a measles epidemic persists "so long as there are found susceptible individuals affording the poison a soil adapted to its reproduction, whilst it perishes if there be no ground to reproduce itself" (p. 156). The concept of an epidemic cycle involving input of new births and output of immunes to maintain a fluctuating equilibrium was formulated by Hamer[82] in 1906.

The years 1906 to 1954 contributed as much confusion as clarification to our understanding of measles, but in the latter year Enders and Peebles[54] isolated the virus and developed serological tests of immunity. Katz *et al.*[98] soon followed this with an effective vaccine, which was licensed in 1963. Initially, there was reluctance to interfere with the natural immunization process, but this abated by 1967 when evidence of long-term effectiveness of the immunization had accumulated. That was quite a year, as the smallpox eradication program began its triumphant march and eradication of measles also seemed within reach.[144] This expectation, then premature, has been revived 30 years later.[8a]

3. Methodology Involved in Epidemiologic Analysis

3.1. Sources of Mortality Data

Measles mortality rates are tabulated by various national vital statistics publications such as *Vital and Health Statistics* published by the U.S. Department of Health and Human Services and by *World Health Statistics Annual* (Causes of Death 042) published by the World Health Organization. Where case reporting is very incomplete,

mortality data may be a useful surrogate for incidence, but the relationship between the two statistics is variable. Mortality rates are influenced by nutrition, the age at which measles is contracted, and exposure to secondary infections (see Section 5).

3.2. Sources of Morbidity Data

Reporting of measles cases to public health authorities is mandatory in most economically advanced countries. Systematic reporting was first achieved in the 19th century for a few European cities,[143] but only became general in the 20th century. Most of these data, however, are so incomplete that they can only indicate when measles occurred and major changes in frequency over short time periods. Before measles vaccine was widely used, it was possible to estimate the efficiency of reporting on the assumption that the actual average number of cases was close to the number of births. On this basis, reporting to the U.S. Communicable Disease Commission prior to 1967 accounted for 10% of all cases. In Mexico, 3% were reported, and in most other developing countries fewer were reported.

After vaccine reduced the total number of cases, no continuing case was routine. Well-publicized campaigns enhanced interest and more frequent follow-up uncovered antecedent and secondary cases, but the simple method of evaluating efficiency of the system was lost. Systematic bias in reporting persists; for example, cases in school-age children are more likely to come to medical attention than in preschoolers. With fewer valid cases among an unchanged or increasing number of other syndromes that might be confused and modified cases in unsuccessfully vaccinated persons, diagnoses that are not confirmed by laboratory tests have become less reliable (see Section 8.2). These caveats notwithstanding, measles case reporting in many developed and some developing countries has improved greatly. Current statistics generally provide a valid picture of annual fluctuations and good estimates of age distribution. These reports are published regularly in the health bulletins of many developed countries, as in the *Morbidity and Mortality Weekly Reports* in the United States.

3.3. Serological Surveys

Serological methods are essential to confirming and sharpening the precision of reported incidence of measles and in delineating patterns of incidence where reporting is unreliable, as well as to determining adequacy of vaccine coverage. The technical methods are described in standard manuals.[26,123] The serological response to measles is specific, being unmodified by any other human infection. Cross-reactions occur with canine distemper and rinderpest, but humans are not naturally infected by either of these viruses. Wherever measles antibody has been found, there has been a history of measles activity or vaccination, and persons with antibody to the virus hemagglutinin at a specifically defined titer are resistant to infection with wild or attenuated virus.[41] Measles antibody remains detectable for life in nearly everybody who has had the disease and in the majority of vaccine recipients for as long as the vaccine has been in use (Fig. 1). Natural infection produced 99% positivity rates in adult populations of many countries.[27] To permit integration of the results of a serum survey with earlier studies, calibration of the assay against the WHO standard serum[57] should be included.

Enzyme-linked immunosorbent assay (ELISA) kits offer the most convenient test for measles antibody in that they give a rapid result and do not require maintenance of tissue culture or a source of monkey erythrocytes. They can be adapted to test for immunoglobulin M (IgM), and thus for recent infection. Because of cost, however, they are ill suited to the quantitative determinations that are needed to estimate duration of maternal protection and optimal age for vaccination. Also, because they measure the sum of antibodies to all viral proteins and antibody to internal proteins is quantitatively dominant but immunologically secondary, they are less reliable in predicting individual susceptibility than tests based on virus surface antigens.[41] The use of these tests to determine attack rates in relation to prior immune status is inappropriate, because many who have titers below the manufacturer's cutoff have adequate protection and some who have antibody to internal proteins of the virus are not protected.

Antibody to the dominant epitopes of either of the two viral glycoproteins, H and F, provides protection from infection.[155,157] Of these two, naturally acquired anti-H titers are higher, and it usually determines both hemagglutination inhibition (HI) and neutralization (NT) test results. The HI test is based on the fact that red blood cells of Old World monkeys, unlike human erythrocytes and nonprimate cells, carry the receptor for measles virus. The HI test is efficient to perform and yields reliable results.[26,31] However, stringent regulations on the capture, importation, and care of monkeys have made cells for this test difficult to obtain and expensive when available. The NT offers an alternative that correlates more directly with immunity.[7] If the tests are conducted on a plaque reduction or focus count basis, they can be more sensitive than other tests. Such tests require maintenance of a measles-susceptible tissue culture cell line, such as Vero, but modern multiwell plates permit efficiency that, although less than that of the HI test or qualitative ELISA, permit surveys to be carried out on a practical basis. These tests can detect titers that are considerably below those that are protective and they are of value in identifying persons who have experienced primary stimulation but lack protection.[41]

To characterize abnormal reactions to measles virus

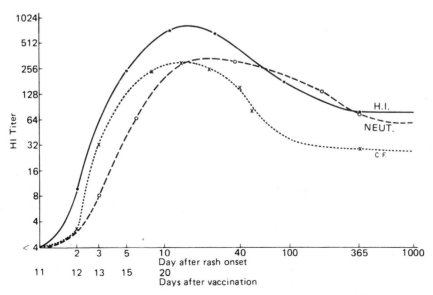

Figure 1. Measles antibody titers in serum relative to time after infection with vaccine (HI) or wild virus (neutralization and CF).

it is necessary to determine the relative strength of reaction to the several viral proteins (see Sections 8.1.2 and 8.1.3) and the several components of cellular response. An antihemolysin assay gave very good predictive correlation with susceptibility in one study,[121] but it has been difficult for others and is little used. Immunoblot tests using dispersed measles virus proteins as antigens are more generally useful.[83] Measurements of the cellular response have been used less often than they might be. A full range of cellular responses can be observed by standard methods. These more elaborate techniques are not needed to diagnose ordinary cases, but they have unexploited potential in deciphering the mechanisms underlying other measles-related syndromes.[74a]

3.4. Other Laboratory Methods

Virus isolation is slow and seldom necessary in diagnosis of ordinary measles. Appropriate material for isolation is difficult to obtain because the virus is cleared rapidly after appearance of the rash and measles is seldom anticipated at earlier stages. No one culture system has been determined to be optimal, but a variety of human or simian untransformed cell lines may serve.[58] Blood, nasal discharge, and urine serve as specimens for isolation or polymerase chain reaction (PCR).

Immunofluorescent techniques applied to cells in the nasal discharge provide a means of rapid diagnosis that, unlike any serological method, is useful during the prodromal period.[149] To localize measles activity in persistently infected tissues electron microscopy,[116] tagged antibody techniques[15] and *in situ*) hybridization[77] have been used. PCR methods have largely replaced isolation.[70] They permit tracing the epidemiologic lineage of a virus, but have not revealed clinically important strain differences. Blood, nasal discharge, or urine serve as specimens for isolation or PCR (see Section 10.2).

4. Biological Characteristics of the Virus

The causative agent of measles is a member of the *Morbillivirus* genus and *Paramyxoviridae* family and is morphologically typical of the group. As in other members of the family, the hemagglutinating (H) and fusion (F) functions are on different glycoproteins coded by separate cistrons.[12] The F protein must be cleaved to become active. Measles H protein recognizes a distinctive cell receptor, the "membrane cofactor protein" CD46.[52,119] When the gene for human CD46 is transferred to measles-resistant mouse cells, the latter become capable of being infected and producing measles virus. The receptor comes in several forms with different proportions in different tissues, and this may determine the susceptibility of specific tissues to the virus.[159a] The fact that our measles vaccines represent strains of virus that have been adapted to chick cells, which lack human CD46, suggests that their hemagglutinin is modified. If so, these vaccines may not elicit maximally effective antibody.

The genome of the prototype Edmonston strain of measles virus has been completely sequenced.[62] The virus envelope carries, as well as H and F proteins, a smaller nonglycosylated M protein that is essential for virion formation. Three internal proteins, N, P, and L, are present in decreasing amounts. The gene that codes for P also codes two other proteins in different reading frames, but these, together with two other small open reading frames, have unknown functions. Sequences of other virus strains, especially strains derived from persistent subacute sclerosing panencephalitis (SSPE), indicate that the virus is highly mutable like other RNA viruses.[11,47,69] The immunodominant epitopes of the H and F proteins are held constant by their need to react with host proteins,[146] but differences in other sequences define virus lineages.[18a,90a,138,158]

The most important biological characteristic of measles virus is infectiousness. The average child was infected with measles at 5 years of age in England and Wales in the prevaccine era.[8] This is later than with parainfluenza 3 virus or rotavirus, but immunity to those viruses is not solid. They can reinfect and thus expand their pool of carriers. Measles virus moves from the relatively few persons who carry it to susceptible persons with greater efficiency than any other clinically significant human disease (see Section 5.2). This infectiousness means that herd protection only occurs when a very high proportion of a population is immune.

The reason for this efficiency can only be deduced by elimination. It is not virus stability; the virus is easily inactivated in the environment. Even in favorable liquid medium half of the infectivity is lost at 37°C in 2 hr.[24] It is inactivated by pH below 5 and by proteolytic enzymes, and thus prevented from infecting via the gut. It does not survive when dried on a surface, and hence is not spread by contaminated articles. However, it does survive drying in microdroplets where it is not subject to distortion by surface tension and airborne transmission is characteristic of its spread.[33] There is no good estimate of the amount of virus excreted by an infected person, but difficulty encountered in isolating virus from cases suggests that the amount is not high. Thus, measles virus efficiency must be associated with processes that occur after the virus reaches a new host. These are likely to reflect the unique measles virus–cell attachment mechanism.

Persistence of measles virus in the human host has been demonstrated by virus isolation only in the case of SSPE (Section 8.1.2). Less definitive evidence suggests that it or a related virus also persists in the lesions of several other chronic diseases (Section 10.2). If so, this virus is usually defective like that isolated from SSPE. If not, measles elimination could not have proceeded as far as it has. However, continued, apparently sporadic, cases of measles reported in the United States raise the possibility that infectious virus may, very occasionally, reactivate from latent infection.

5. Descriptive Epidemiology

Susceptibility to measles is universal among humans and spread of the virus is independent of any vector. The chief variable in the natural distribution of measles incidence is the age at which infection occurs. Widespread vaccine use has protected many, but residual pockets of susceptibility determine the continuing distribution of the disease.

5.1. Incidence

Prior to the introduction of measles vaccine, about 400,000 cases of measles were reported in the United States every year, but 4 million children were born and essentially all of them ultimately developed measles antibody that could only have been acquired as the result of infection. Thus, the mean true number of cases per year was about 4 million. In 1967, extensive use of vaccine reduced the number of cases precipitously and, with some fluctuation, it has remained low (Fig. 2). A substantial increase in 1989–1991 prompted intensified efforts to improve vaccine coverage, which in turn reduced the number of cases reported in 1993 to the new low of about 300. In the latter part of each year, several weeks passed without any reported case.[39] These constitute the first "fade-outs" in the United States, interruptions of continuous transmission.

Achieving this reduction required maintenance of a very high level of vaccine coverage in teenagers, because they gather into large groups at high school and college functions.[75,145] At its beginning, the 1989–1991 epidemic had its highest attack rate here. Immunity in this age group was raised by requiring a second dose of vaccine at school age. It then became clear that our estimate of vaccine coverage in infants was higher than was the case,[166] and

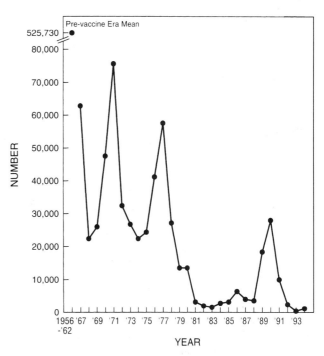

Figure 2. Reported cases of measles in the United States 1956–1994. Data from annual summaries by the Centers for Disease Control as published in *Morbidity and Mortality Weekly Reports* and from U.S. Vital Statistic Annual.

as the epidemic matured, it became most intense in the very young (Fig. 3). The end of the epidemic cannot be attributed to exhaustion of the susceptible pool by experience with the disease; the 40,000 cases that occurred in the whole epidemic represent about 0.05% of the under-20-

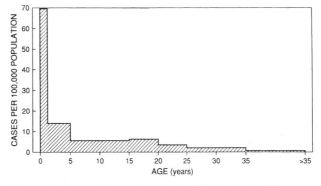

Figure 3. Age-specific attack rate for the United States in 1991. Case incidence is adapted from Atkinson *et al.*[10a] and the 1990 census figures used for population at risk. This represents the last year of an epidemic when it had moved maximally into the youngest age strata.

year-old population. Increased vaccination must be given credit for the success.[40]

This episode provides glowing evidence of the potential of vaccination and represents an important milestone on the road to eradication. Nevertheless, 20% of the 1993 cases were characterized as "imported," and until the infection rate in other countries can be greatly reduced, the United States will not be free of measles. Indeed, the 1994 tally increased to about 1000 cases.

The experience of other developed countries has been similar to that of the United States, but achieving a comparable effect in less developed areas of the world has been much more difficult.[49,108,159] Part of the problem is that poor reporting prevents accurate assessment of what is done, but there is more than that. The window of opportunity for successful vaccination—that is, the time between decline of maternal titer and natural infection—is very narrow in these countries. Even at the optimal age for vaccination, a substantial proportion of the population may be refractory, due to residual passive antibody, while a like number have already been infected.[27] Strenuous efforts have achieved substantial nominal vaccine coverage, but only recently have they greatly reduced measles mortality in the world as a whole.[165]

5.2. Epidemic Behavior

Measles is the model with which much of our knowledge of the epidemic behavior of infectious disease has been developed. In the prevaccine era, in populous temperate parts of the world measles incidence peaked every 2–5 years, persisting at high levels for 3–4 months. However, close scrutiny revealed that a few cases occurred every week throughout the interepidemic period in any

large population. The timing of the peaks was closely associated with the school year, but even before schooling was general the clustering of persons indoors in winter led to increased incidence in the early spring.[143]

Because of its relative simplicity and the challenge of predicting the effects of vaccination campaigns, measles has attracted the interest of mathematical modelers. The first efforts, initiated by Hamer[82] in 1906 and pursued by Bartlett,[13] Griffiths,[74] and others, focused on the periodicity of the epidemics. This work developed the basic premises that over a short period of time the number of people infected will be dependent on the number susceptible as well as on the number who are infectious. The number susceptible is continuously modified by new births and by removal of susceptibles through infection. Since the number infectious in one round depends on the number who were infected in the preceding round, it is relatively easy to develop formulas that will predict oscillating endemic periods as in Fig. 4A. The epidemic waves of these simple models, however, are subject to damping and gradually approach a steady state. To get a more accurate picture of the way epidemics come and go, it is necessary to recognize the roles played by chance and seasonal fluctuation. When an epidemic is in the ebb phase, the number of cases will be small and the number of new contacts highly variable. Bartlett[13] selected values for the number of new cases from pools of random numbers and produced a model that shows strong resemblance to actual measles epidemic curves (Fig. 3B,C). Fine and Clarkson,[55] however, showed that season influences could give the cycle a sufficient kick to maintain amplitude. Bolker and Grenfell[34] and Keeling and Grenfell[98a] have produced general syntheses of these elements.

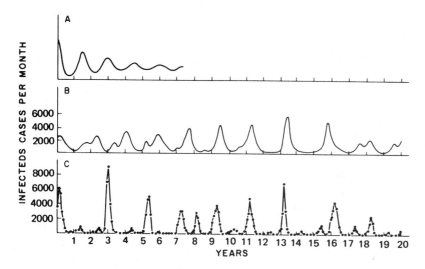

Figure 4. Actual and model measles epidemic curves. (A) Curve derived from the deterministic model; (B) curve derived from the stochastic model; (C) actual curve for Baltimore, Maryland, 1900–1920. Curves are redrawn to a common scale from the work of Bartlett,[13] Griffiths,[74] and Hedrick.[84]

These studies led to the estimate that measles transmission would fail during the ebb period in any community with an average of fewer than 100 cases per week. Actual observations[14] are in agreement with this; an average of 80 cases in North America and at least 106 in England was required for continuity. However, in island communities where breaks in continuity are more obvious and introductions fewer, the critical value was between 260 and 320 cases per week.[25] An implication of this is that measles cannot persist in any population with fewer than 5,000–10,000 births per year. The precise value is dependent on population density, a compact population sustains the infection less well than a more dispersed one (Fig. 5). Vaccination at the age when susceptibility begins effectively removes recipients from the number of susceptible persons entering a community, and 5–10,000 is the number of unvaccinated children that may be permitted while still achieving virus elimination. It is not profitable, however, to pursue this line of reasoning too far as a way to determine the vaccine coverage needed to eliminate measles from a community because the effective community size increases as the susceptible population ages.

Another way to approach the problem is to consider the proportion of a population that must be susceptible if an introduced case is to set off an epidemic. This is measured by the basic reproductive rate, R_0 which is defined as the number of secondary cases to be expected when one infectious person is introduced to a totally susceptible community. It depends both on characteristics of the virus that determine its efficiency and on characteristics of the host population that determine the number of contacts between persons. It is rare to find a totally susceptible population, but R_0 was measured directly in Greenland in 1951 when measles was first imported into that country. A value of 200 was observed.[43] A single measurement must be suspect, however; in this instance the index case went to a community dance on his arrival. Anderson and May[8] give a general approximation for R_0 that can be applied to partially immune populations:

$$R_0 = L - M/A - M$$

where L is the life expectancy, A is the average age at infection, and M is the duration of maternal protection. They used this formula to estimate a value of 16 for R_0 of measles in England and Wales in the prevaccine era. This would imply that if more than 15/16th (94%) of the population could be immunized, measles would die out. However, when the young children were removed from the susceptible pool by vaccination, the mean age at infection went up, the number of person to person contacts went up, R_0 went up, and it became more difficult than predicted to stop the virus. Measles can spread in high schools and colleges where nearly 99% are immune,[41] suggesting that 100 may be a better estimate of R_0 in such a population. This value is more applicable than that of Anderson and May to the current situation in the United States and other countries where the vaccine has been widely used, and it is closer to the value observed in Greenland. On this basis, failure to immunize more than 1% of the population may be expected to lead to failure to eliminate the virus.

A characteristic of the epidemic curve is momentum. The more intense the peak of the epidemic, the farther it will carry the residual number of susceptibles below the equilibrium state before subsiding. In isolated areas where epidemics seldom occur, large numbers of susceptibles accumulate during interepidemic periods and nearly every susceptible person becomes infected in an epidemic. In the 1951 Greenland epidemic, 99.9% of a susceptible population of more than 4000 were infected within six weeks.[43]

5.3. Geographic Distribution

Measles occurs wherever humans live, except in the most remote and isolated populations. In the prevaccine era the average age at infection varied widely from one country to another, but as evidenced by antibody acquisition nearly every person ultimately became infected almost everywhere (Fig. 6). The disease is remarkably uniform and immunity is completely portable over space and time. This does not imply that the virus is invariant, but that key structures are held constant by the need to interact

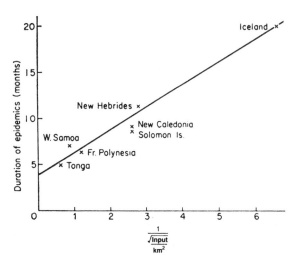

Figure 5. Relationship between average duration of measles epidemics and dispersion of population in areas with about 2000–4000 new susceptible children per year. The abscissa plot of the inverse root of the number of new susceptibles introduced annually per square kilometer represents the mean distance between new susceptible persons.

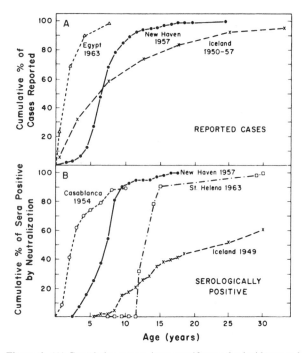

Figure 6. (A) Cumulative reported age-specific measles incidence and (B) age-specific percentage positive by serological test in developed, underdeveloped, thinly populated, and isolated areas. Egypt and Casablanca are typical of the pattern observed in populous underdeveloped areas. New Haven is characteristic of developed countries. In Iceland, the population is relatively dispersed. St. Helena is representative of an isolated population, which last experienced a measles epidemic 12–13 years prior to the time the sera were collected. The two methods of determining the age-specific attack rate give similar results, except that in Iceland a significant part of the population escaped measles altogether, and the proportion immune did not reach more than 60% even in the older age brackets.

with human cells (Section 4). Regional differences are recognized in measles severity and in the frequency of diarrhea and severe ocular involvement. However, these differences are not carried with the virus when it is transported from one area to another, as in outbreaks in U.S. border cities, and it is probable that they reflect differences in the condition of the host not the virus (discussed in Sections 5.5 and 5.11).

5.4. Temporal Distribution

DeJong and Winkler[94] found that measles virus was most stable when the relative humidity was below 40%. There was a second optimum above 80% humidity. These differences may determine whether and how quickly the virus dries in aerosols and, assuming it is relatively stable

once dry, may explain the propensity for measles to spread in the northern winter when homes are heated and very dry and during the hot dry season of Rhazes' Persia[5] or modern Sahelian Africa.[64] However, it may simply be the congregation of humans in inclement weather that explains that fact that these are the times when epidemics most often peak.

5.5. Age

The age at which measles is contracted is modified by three factors, the first of which has three component elements[27]:

1. the duration of maternally derived protection
   a. the mother's antibody titer
   b. the efficiency of transfer across the placenta
   c. the rate of catabolism in the child
2. the propensity of the host to associate in large crowds
3. the proportion of the community that is vaccinated

There may be some ethnically determined variation in maternal titer (Section 5.7), but more generally titers are low in mothers who acquired immunity by vaccination.[104] Most U.S. and Canadian mothers now fall into this group and possess about half as much antibody as was usual when it was first recommended that vaccine only be given at 15 months of age; most children now become susceptible about 7 weeks sooner than previously. Although many children are protected by passive antibody for much of their first 16 months, in 1991 in the United States, 29% of all cases occurred in this narrow age span (Fig. 3). To meet this challenge, vaccination at 12 months is now commonly recommended. Determination of the best age is complex.

Low efficiency of transfer across the placenta may be a factor in South Asia. In other parts of the world the rate with which antibody is degraded in the infant is a more important variable. The half-life of measles antibody is 48 days in the United States[27] and Finland,[140] but it is shorter in less-developed countries. This period is inversely associated with the level of all IgG in the baby's serum, which in turn parallels the frequency of all infections. It seems probable that a child exposed to many infections makes a large variety of IgGs; in order to keep the total blood IgG level in the normal range, catabolism is accelerated and passively acquired antibodies are swept out at an accelerated pace and not replaced until active stimulation occurs. In this way, early susceptibility to measles is strongly correlated with low economic status (Fig. 7). Although this is the overall pattern, individuals vary widely, and a standardized vaccination program,

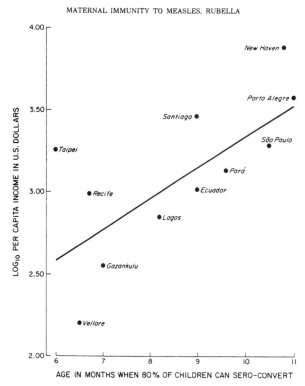

MATERNAL IMMUNITY TO MEASLES, RUBELLA

Figure 7. Relationship between gross per capita product of a country or area and the age at which 80% of vaccinated children seroconverted. $P <$ 0.01. For sources of data, see Black.[27]

by reducing the frequency of contact with infectious persons will raise the average age of continuing cases. Other things being equal, one would expect that the 99% reduction cases seen in the United States would extend the average time that elapses before infection occurs 100-fold. Where the mean age at infection was previously about 5 years after becoming susceptible, it now nominally would be 500, and most susceptible persons would escape throughout a normal life span. This phenomenon has resulted in a marked increase in the mean age at measles, but other conditions are not equal and the effect is less than predicted. Persons who have not been vaccinated on schedule are concentrated in certain parts of the society, and most continuing cases occur in communities that do not have the level of herd protection possessed by the whole.[67]

The age at infection affects measles mortality. Comparison between areas with different age-specific attack rates is complicated by differences in nutrition and frequency of secondary infections, but where epidemics have affected all ages, the mortality rate declines for the first 5 years (Fig. 8). There is then a gradual increase in the mortality to the point where persons over 60 may experience rates comparable to those in babies.[133]

using currently available vaccines, demands that some children be left susceptible for a time if others are not to be vaccinated prematurely.

The aforementioned tendency to associate in progressively larger groups with increasing age means that susceptible persons have progressively greater chance of coming in contact with an infectious individual. In a childcare center a baby may typically be exposed to 20 other children, any of whom may have measles. When the child enters elementary school this number increases to a few hundred; in high school there may be 1000 or 2000 classmates, and regional sports events expose adolescents to several thousand more contacts; and in college there are often 10,000 fellow students, and sports and social activities create exposures to persons from other cities. Even when the absolute number of cases does not increase with age, the number proportionate to the number susceptible usually does.

An effective program of measles vaccination will reduce the number of cases in the targeted age groups and

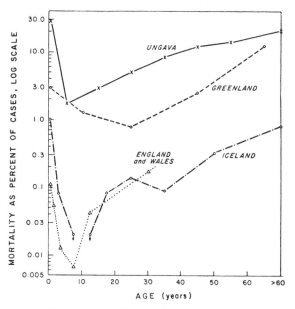

Figure 8. Measles age-specific mortality rates in endemic and virgin-soil areas. (×——×) Ungava[133]; (o- - - -o) Greenland[43]; (o-•-•-o) Iceland[103a]; (△ ···· △) England and Wales.[37a]

5.6. Sex

For all its uniformity across other demographic distinctions, measles is more severe in females than males. Measles mortality is 5% higher in females than males across a broad range of ages and reporting units.[63] It used to be assumed that this was due to parental preference for sons, but Garenne found it in Europe and North America where sex-based differences in care are not supposed to occur. Girls are also more affected than boys by the delayed mortality associated with high titer vaccine.[89] Measles antibody titers are higher in females than males,[35] a characteristic that often mirrors the extent of virus growth. It is not clear at this time whether the sex difference is due to a difference in ability of the virus to grow or a difference in immune systems, but the net effect, though small, is significant.

5.7. Race

It is commonly believed that measles is more severe in races that have had limited experience with it. Classic reports of epidemics in these populations gave at best percentage mortality rates without specific figures, but these high rates were confirmed by more specific data from virgin soil epidemics along the Davis Strait in the Canadian Arctic[133] (4.5% mortality) and on the Xingu river in the Brazilian Amazon[125] (17.4%). Several explanations can be offered, but enhanced susceptibility of individuals within these populations does not seem to be important.[28] Breakdown of social support is a factor.[120] In these particular outbreaks this was manifest by higher mortality before than after arrival of outside support. A general explanation lies in genetic homogeneity of groups subject to high measles mortality. If the virus adapts to individuals, it would find itself adapted to the whole population. This may make it more virulent, when people are homogeneous in their immune capabilities.[28] Elevated mortality rates were not seen in virgin soil epidemics that occurred in Greenland[17] where the people had already mixed with the European gene pool.

5.8. Occupation

Occupation does not affect the ultimate probability of being infected. As noted in Section 5.5, however, activities that increase the number of persons to whom an individual is exposed increase the chance that a susceptible individual will become infected at that time. The military is one such occupation and measles used to be a major problem in this setting (Table 1). This problem

Table 1. Measles Rates in United States Armed Forces

	Cases per 1000 person-years	Deaths
Civil War	32.2	2.0
Spanish–American War	26.1	0.32
World War I	23.8	0.57
World War II	4.7	0.004
Vietnam War, 1966	0.9	—

gradually declined as mixing increased within the national population, and it disappeared when vaccination against measles on induction was made routine.

5.9. Social Setting

Two quite different studies show an association of social factors with measles (Table 2). Family size, father's education, and residential neighborhood influence the probability of early infection.[30] All of these factors may act by increasing the number of contacts. Low income and maternal education are associated with severity of disease.[49] Other evidence of an effect of social setting on measles mortality comes from Africa and the work of Aaby et al.,[3] who found increased mortality rates in large polygamous versus smaller monogamous households in West Africa. Garenne and Aaby[64] related this to proximity, genetic as well as physical, of primary and secondary cases. Infection contracted from a sibling is more dangerous than infection contracted on the street.

The significance of these biological effects pales in relation to the effect of social factors on the probability that a child will be vaccinated at an appropriate time. In the 1989–1991 U.S. epidemic, attack rates in minority groups were seven to ten times higher than those in the majority and most cases occurred in unvaccinated children.[68] For children who are dependent on public immunization clinics, and especially for those whose mothers have regular jobs, the disincentives to vaccination are great.

5.10. Socioecomomic Status

Within a developed country, socioeconomic status is related to many of the factors described above, and hence shows a correlation with the age at which measles infection occurs. However, more dramatic correlations become apparent in international comparisons. There is a strong association between a low level of development and high

**Table 2. Relationship between Social Characteristics
and Measles History and Severity of Disease**

New Haven[a]	Number	Number of siblings	Residential neighborhood[b]	Father's education
Positive history[c]	388	2.75	3.54	11.0 yr
Negtive history	168	1.98	2.31	12.5
P		< 0.001	< 0.001	< 0.001

| | | | Mother | |
Puerto Rico[d]	Number	Case complicated by underlying illness	< 18 yr old	< High school education
Severe measles	16	50%	31%	62%
Not hospitalized measles	39	16%	3%	37%
P		0.009	0.03	0.007

[a]Data from Black and Davis.[30]
[b]Rated on a six point scale with 1 the best and 6 the poorest.
[c]Serologically confirmed histories in children in the first school grade.
[d]Data from Diaz et al.[49]

measles mortality. Part of this can be ascribed to nutrition (Section 5.11) and part to differences in age at infection and duration of maternal protection.[27,44] Per capita income is directly correlated with the duration of maternal protection (Fig. 7) (Section 5.5). Early loss of maternal protection means more cases in children less than 1 year old, who are more likely to die.[163]

5.11. Nutrition

Measles mortality is highest in malnourished populations,[118] but the multifactorial nature of nutrition and confounding with social factors make it impossible to define its importance until specific deficiencies are sorted out. The role of adequate protein and calorie intake remains controversial. Low weight for age is commonly associated with an adverse outcome,[100,109] but Aaby[1] did not find this association in Guinea-Bissau and no nutritionally related difference in response to vaccine was seen in a large Latin American study.[130] This inconsistency suggests that much of the observed effect is indirect or caused by other deficits. Inadequate weight not only makes it difficult to survive the stress of measles, but measles, through loss of appetite, diarrhea, and tissue destruction makes it more difficult for a child to maintain the weight he or she has.[109]

There is no doubt about the importance of vitamin A nutrition in determining the severity and consequences of measles. A deficit of vitamin A explains the association of measles with diarrhea[100] and blindness[59,135] in countries where nutrition is poor. Vitamin A supplementation re-

duces the proportion of severe cases[134] and is effective in treatment of severe disease in countries where vitamin A deficiency is likely[91] Again, a preexisting deficiency is not necessary for vitamin A to be important; measles causes a reduction in serum retinol and may precipitate the deficiency in borderline cases.[37]

6. Transmission

Aerosol is the only important vehicle of transmission. The mucosal cells of the respiratory tract are the source of most disseminated virus. Infected cells do not lyse or die quickly but continue to produce virus for several days. Adjacent cells fuse into syncytia and these slough into the discharge. The sneezing and coughing, so prominent in the prodromal stage, ensure efficient dissemination. Infection in the urinary tract also results in virus discharge via the urine, which may be aerosolized on release. Whereas excretion from the respiratory tract terminates promptly with development of antibodies and the associated rash, excretion in the urine may persist a day or two longer.[73]

Experiments using attenuated virus show the most effective route of infection is via droplets small enough to be carried into the lungs.[102] However, other exposed mucous surfaces can also be infected and most naturally generated droplets may not penetrate as far as the lungs. When the virus is injected, as when vaccine virus is given subcutaneously, the incubation period is shortened by about 2 days. This permits immunization of recently exposed persons.

There is no known reservoir or vector of measles virus, animate or inanimate. Old World primates can become infected, but in the natural course of events this is confined to animals living in proximity to humans.[21] Monkeys caught in the forest or savannah are usually free of measles antibody.

7. Pathogenesis and Immunity

Although commonly classified as an exanthem, measles infection involves many tissues and occurs in lymphatic and mucosal–epithelial cells throughout the body. Infection is followed by an inapparent incubation period that lasts 10–12 days in children, and this together with a 3-day prodromal period gives a total interval until the disease is usually diagnosed of 14 days. It may be slightly shorter in infants or extended to 21 days in adults. Failure to anticipate the extended period in older persons has sometimes invalidated quarantine precautions.[43] As noted in Section 6, administration by injection shortens the inapparent period. Differences in size of the infecting dose also affect the length of this period,[102] but it is doubtful if natural inocula often vary on the scale required to cause significant difference.

The only evidence of infection during the incubation period is a decreasing leukocyte count, especially lower eosinophil and lymphocyte counts.[19,32] The virus replicates in diverse types of leukocyte, and it is plausible that this decline is due to direct virus action.[20] The magnitude of lymphopenia demands that both B and T cells are involved. Both CD4$^+$ and CD8$^+$ T-cell counts may be below normal in the period of rash, a stage when the total lymphocyte count begins to recover.[10]

With the onset of the prodromal period, virus appears in tears, nasal secretions, sputum, and urine. The relative titers of virus in these diverse fluids have not been defined. At this stage, typical measles giant cells can be found in lymph nodes, including tonsils, and in the appendix. Nucleocapsid can be revealed in the buccal mucosa by electron microscopy.[153] Virus occurs freely in the blood at this stage. Neutrophil counts decline, lymphocytes remain depressed, and eosinophils may disappear. Hypersensitivity reactions are suppressed and remain low for about a month. Specifically, the tuberculin reaction may become negative for 30 days.[85] A similar, but lesser, loss of tuberculin sensitivity occurs after vaccination against measles.[56] Electroencephalograms regularly show brain involvement, although it is not certain that this is not an effect of the fever.[129] Koplik spots appear on the oral membranes and sometimes on other mucosal surfaces toward the end of the prodrome and provide the first specific clinical sign of measles.

When the rash appears, both IgG and IgM antibodies become detectable by any of a variety of tests. Peak levels of measles IgM are reached about 10 days after appearance of the rash[142] and IgG peaks at about 30 days. Little is known of the timing of the IgA response, but this antibody is also produced.[18] Responsiveness of lymphocytes to measles antigens is relatively muted, even in the acute phase, and it is not usually demonstrable in the convalescent phase unless sensitized cells are amplified by prior stimulation.[23,101] Free virus is quickly cleared from the blood with the appearance of antibody, but it remains detectable in circulating leukocytes and the urine for another day or two. The total lymphocyte count returns to normal during this period, but neutrophils reach an ebb and recover more gradually.

Like the Koplik spots, cells in the rash macules contain nucleocapsidlike structures.[153] The two lesions show similar pathology except that there are more inflammatory cells in the rash. It is probable these two prototypic signs of measles have a common mechanism, but that lesions are not visible through the skin until an immunologic reaction to virus in the capillary epithelium causes dilation and extravasation. The synchrony between the appearance of rash and circulating antibody suggests an etiologic relation (Fig. 9), and the fact that a rash often fails to appear in immunodeficient persons, including those who have acquired immunodeficiency syndrome (AIDS)[96] supports this assumption.

Once infectious measles virus is cleared from the body, within a few days of the rash, it does not reappear. No instance has been reported in which new cases of measles have been traced to contact with persons who had been infected in the past. Although the active infection ends, excess mortality from a variety of immediate causes may persist for months[100] or even years.[2] Some of this continuing mortality may be due to the nutritional debilitation that accompanies the disease,[109] but chronic immunologic damage is implied.[105]

Immunoglobulin M disappears in about 60 days and IgG falls two- to fourfold in the ensuing months but then remains nearly stationary for life. The stability of IgG titers is reflected by an equally solid and durable immunity. Secondary cases are very rare and usually follow primary cases modified by passive antibody.[141] They seem to entail the same mechanism as much of the measles observed in unsuccessfully vaccinated persons; i.e., reinfection can occur when the initial infection is so muted that the immune response is very weak.[107,150] Passive antibody in adequate titer, whether acquired from the

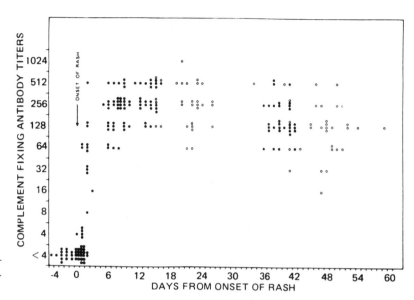

Figure 9. Measles CF antibody titers in Greenlanders relative to the time of onset of measles rash. Data from Bech.[16]

mother or by injection,[97] protects effectively. The maternal antibody is mostly IgG1, and in the normal course of events immunity to measles is closely correlated with the IgG anti-H titer. This relationship breaks down in recipients of the killed virus vaccine, who have adequate anti-H titers but lack anti-F[123] and most cellular immunity. Agammaglobulinemic children usually develop immunity to measles, suggesting that the cellular system can be independently protective.[139] In view of the ability of measles virus to spread *in vitro* in the presence of specific IgG, it seems probable that complement- or T-cell-mediated lysis of infected cells would be needed to eliminate an established infection.

8. Patterns of Host Response

The usual pattern of host response to measles has been described in Section 7, but there are other rarer and generally more serious forms of the disease.

8.1. Clinical and Epidemiologic Features of Unusual Forms

8.1.1. Postinfectious Encephalitis. It has been noted that abnormal encephalographic patterns are usual in acute measles.[66] There is some evidence that this neurological involvement, even without frank encephalitis, may leave minor residua in many children.[60,129] Nevertheless, the virus only occasionally causes acute primary encephalitis, and fully competent measles virus

seldom has been isolated from the brain.[113] Allergic encephalitis occurs more often, about once in 2000 cases of measles in developed countries, but more frequently in older than younger children.[72] Onset of encephalitic signs may occur 1 to 15 days after appearance of rash, with a sharp peak frequency on the sixth day. Its pathology is characterized by numerous small demyelinated plaques and perivascular cuffing in both white and gray matter.[99] Mortality rates are not high, but sequelae are frequent. This encephalitis follows measles less often than mumps, but the consequences are more serious.

Geographic variation in prevalence of measles encephalitis is not sufficient to be visible against the background of different age-specific measles attack rates and different reporting efficiencies. That it is more common in children over 10 years old suggests that it may occur less often in the less-developed countries where most cases are younger.

8.1.2. Subacute Sclerosing Panencephalitis. SSPE is caused by the persistence of defective measles virus in the brain. The syndrome usually becomes evident about 7 years after infection, and its peak incidence occurs at about 9 years of age.[81] It starts with a slow onset of mental deterioration and myoclonic jerks and progresses through convulsions and coma to death in 6 months to a few years. The electroencephalogram shows a characteristic pattern of brief paroxysms. It is regularly accompanied by unusually high titers of serum antibody to all measles virus proteins except M.[162] Antibodies can also be demonstrated in the cerebrospinal fluid (CSF) in titers that are elevated relative to the usual CSF : serum ratio of

1 : 320. Boys are affected more often than girls. In the United States, SSPE was associated with a rural environment and exposure to birds,[80] although no explanation of these associations has been found.

Measles virus has been isolated from SSPE on several occasions but never from cell-free material. All SSPE viruses that have been studied have mutations in the M gene, which blocks production of this protein in normal form. The specific defect varies from one strain to another. Although some virus lineages may be more often associated with SSPE than others, the disease is not caused by a particular strain.[11] It is probable that if the virus gets into the brain and loses its ability to produce M protein, it fails to be assembled on the cell surface and fails, in this "immunologically privileged" site, to induce the local immune reaction needed for its elimination. Hematogenous spread is blocked by the defect, but virus nevertheless spreads from cell to adjacent cell to cause sclerotic plaques. Measles virus antigens and the measles virus genome are demonstrable in the lesions,[76] but lacking mature virions, the virus does not serve as a source of secondary contagion.

In contrast to postinfectious encephalitis, SSPE most frequently follows measles contracted at an early age and is most common in less-developed countries where measles commonly attacks early.[71,95] United States[79] and British[115] data indicate that SSPE most often follows measles infections in children under 2 years old, but the paucity of cases prevents further subdivision of age categories. In these developed countries, SSPE follows 1 in 100,000 to 300,000 measles cases,[81] and with reduction of the annual number of measles cases in these countries below 100,000, the disease has disappeared. In less-developed countries, where most continuing measles cases occur, the incidence per case of measles may be 100 times that which pertained in the United States, and SSPE may occur more frequently than postinfectious encephalitis.[95] If the incidence increases progressively with decreasing age at measles infection, the greatly increased proportion of measles cases in children less than 1 year old in these countries may offer a sufficient explanation for the high incidence.

8.1.3. Atypical Measles after Killed Vaccine. During the 1960s, several killed measles vaccines were widely used with satisfactory short-term results. However, the protection elicited by these vaccines waned after about a year, and recipients became susceptible to a form of measles characterized by centripetal, often hemorrhagic, rash accompanied by pneumonia. It later became apparent that these vaccines induced antibody to the H protein but failed to induce antibody to the virus F protein[124] and failed sensitize key sets of lymphocytes.[61,74a,101] When the killed vaccine recipients were given live vaccine, a strong local reaction ensued and sometimes, in reduced intensity, the symptoms of the atypical disease. Although alarming, this was deemed preferable to the atypical reaction to infection with wild virus. Most such persons have been revaccinated now and their immunity seems solid, but this episode remains an example of the hazard of immunization that does not give the normal balance of immune responses.

8.1.4. Measles in Immunocompromised Persons. Measles in immunocompromised persons is a growing problem because medically induced immunosuppression for cancer therapy and transplantation and an increased number of pediatric AIDS cases has increased the population at risk. Typically this syndrome is associated with a giant cell pneumonia and severe respiratory distress. Often no rash appears. Mortality in one U.S. series was 40%.[10] Children most likely to be immunosuppressed by AIDS are concentrated in that part of our society that is least likely to get vaccine and in certain African countries where measles vaccine coverage is poor. Vaccination of human immunodeficiency virus (HIV)-infected but not severely immunocompromised persons at the usual age is the standard recommendation,[127] but this is not without risk[117] and the resulting immunity may be lost due to subsequent immunosuppression.[128] Furthermore, children with immunodeficiency may have been maintained on exogenous IgG, which blocks the vaccine response. The situation is thus a complicated one in which no one regimen is best for all. When vaccination fails, ribovirin treatment offers possible benefit.[46]

One reason for the very high mortality rate in the 1989–1991 epidemic is that those who were infected included a disproportionate number of immunosuppressed children.

8.2. Diagnosis

8.2.1. Clinical. Unmodified measles infections follow a typical course that is almost always apparent. Koplik spots, whitish lesions on the buccal mucosa, are a particularly useful criterion because they appear early and are not seen in other exanthems. They do not appear in all cases, however. The prodromal period is longer and more intense than that of rubella, and although the lymph nodes are major targets of the measles virus, they are not enlarged as in rubella. Catarrhal symptoms are more prominent than in roseola or scarlatina. The rash is more macular than that of scarlatina, and unlike roseola, fever persists after the rash appears. The rash is centrifugal rather than centripetal as in erythema infectiosum.

Whereas the vast majority of cases used to follow this typical pattern, the disease is less distinctive where the vaccine has been widely used. Continuing cases in inadequately vaccinated persons retain the typical characteristics but are milder.[114,150] Cases in the immunocompromised may fail to develop the rash.[96] Rubella cases, which might cause confusion, have been reduced together with the reduction in measles, but the unchanged prevalence of other infections and of allergic rashes increases the chance that they will be mistaken for measles. Dengue, which may be confused with measles, has increased in prevalence and geographic distribution.[51] All this makes an erroneous diagnosis of measles more likely at a time when valid diagnoses have become more important. In countries where laboratory facilities are available, clinically suspected cases should now be confirmed.

8.2.2. Laboratory. Diagnosis by virus isolation is slow and impractical,[58] but measles antibody appears coincident with the rash with sufficient regularity to be an effective basis for diagnosis.[26] An increase in total antibody can be demonstrated in paired sera collected as little as 2 days apart if the first is taken shortly after onset of rash or separated by a week if the first is collected within 2 weeks. Other infections do not cross-react to cause nonspecific boosts in titer, and reexposure of immune persons to measles virus has no effect on titer. A single test for IgM may be adequate for diagnosis of recent infection, because this antibody is not found more than 60 days after the rash.[106] However, IgM may fail to appear in persons in whom a very low response to earlier vaccination has muted but not blocked the infection. Commercial ELISA tests using IgM-specific tagged label are the most efficient method of measuring IgM, but HI or NT tests also serve if the IgG is first removed with protein A and the IgM titer is taken as the amount of antibody that is then lost on treatment of the serum with mercaptoethanol.[29]

9. Control and Prevention

9.1. Quarantine

Quarantine has been used to control measles for more than a century. The infectiousness of measles virus, however, is such that the best result to be expected is delay of infection until a later date. Only if the quarantine is promptly followed by vaccination is it likely to have a net effect in reducing the incidence of measles.

9.2. Vaccines

9.2.1. Spectrum of Available Vaccines.
Several attenuated strains have been used since 1963 and have left their marks on the immune systems of people of the world.[18a] The Edmonston B vaccine, marketed in the United States from 1963 to 1967, elicited substantial reaction and was commonly administered together with a separate dose of immune serum globulin. The chance for confusion of this vaccine with the contemporary killed vaccines, the chance that the vaccine might have been incorrectly combined with the IgG before injection, and the fact that this product was less well-stabilized than later preparations combine to make suspect a record of vaccination carried out in this period. In 1967 the Edmonston B was displaced by the Swartz and Moraten strains, which have since jointly been viewed as "standard."[103] These vaccines produce a significant reaction in many recipients. The fever is occasionally alarming but almost always resolves without harm. Japanese workers found this reaction unacceptable to their populace and developed still further attenuated strains, which have since been produced in other countries.[126] The "Leningrad" strain, used extensively in the former Communist block, probably retains somewhat greater virulence.[151] The Edmonston–Zagreb and the Japanese Aichi strains have greater potency than the others in the presence of low levels of passive antibody.[92] All current strains were attenuated, at least in part, by adaptation to chick embryo cells.

A consistent difference between wild-virus- and vaccine-induced immunity is that vaccines induce lower antibody titers. Titers elicited by standard vaccines are about twofold lower than naturally induced and those induced by other strains fit on this scale relative to their virulence.[130] Properly administered vaccine seldom gives so low a titer that it fails to protect the recipient, but the period of passive protection in children of vaccinated mothers may be as much as 3 months shorter.[105] As well as opening the potential for measles at relatively early ages, until all mothers are vaccinated, this increases the variance in age at which children become responsive to vaccine. (Prospective vaccines are discussed in Section 10.1.)

9.2.2. Protection of Recipients by Vaccination.
Control of measles by vaccination can be evaluated at two levels: protection of the recipient and establishment of herd immunity. In developed countries, standard vaccine

has had substantial success at the first level, reducing the number of cases at least a 100-fold. Success at the second level has been, at best, tantalizing (see Section 5.1). Current recommendations for the use of this vaccine in the United States call for the first dose of vaccine to be given between 12 and 15 months of age.[40] Earlier, delivery has sometimes been used in high-risk areas, such as inner cities faced with an immediate epidemic, but this is often motivated more by a desire to do something than by good epidemiology. Whenever vaccine is given before the recommended age, the child should not be considered immunized. A second dose is called for in school-age children both by the CDC Advisory Committee on Immunization Practices (ACIP)[40] and the American Academy of Pediatrics (AAPCID).[45] The ACIP advises that this be given at school entry, whereas the AAPCID recommends it be given in middle or high school. The ACIP recognizes no evidence that the second dose acts as a booster, but advocates it to immunize those who, as a result of primary vaccine failure or faulty records, were not immunized in infancy.[87] The implication of the AAPCID recommendation is that they believe titers may wane and that boosting near the age of maximal risk of exposure has enhanced value.

In less developed countries the vaccination program has been much less successful, but a vigorously applied program can prevent a majority of cases. The World Health Organization (WHO)[165] generally recommends that the vaccine be given at 9 months of age in these areas, but many countries follow individually tailored schedules.

While emphasis is appropriately placed on better vaccine coverage, overdoctoring contributes to the problem. Ardent health care workers may administer vaccine before the recommended age, lest there not be a chance at the appropriate time. This makes some children refractory to vaccination when appropriate (Section 9.2.3). Also, immune globulin, used to protect exposed children, invalidates the effectiveness of vaccine administered up to 6 months later,[148] much longer than generally recognized.

In order to improve on the record of standard vaccine, for a while the WHO recommended that a more potent high-titer vaccine be used at an earlier age. The strain most extensively used was developed in Croatia and is referred to as Edmonston–Zagreb.[92] Especially when used in very large doses, this vaccine can override low levels of passive antibody and immunize children earlier in life than the other strains.[163] It seemed to be a solution to the problem of immunization where there is no clear gap between the age when children lose maternal protection and the age when they become infected. The procedure was effective in controlling measles, but the children

were made unusually susceptible to a wide variety of other infections and elevated mortality rates persisted at least 2 years.[4,65,89,165] This seems to reflect the same process as the high continuing mortality seen in children who experience wild measles virus infection at an unusually early age.[2] It may reflect persistence of the immunologic damage caused by measles virus when that damage occurs early.[105] This procedure is now deemed unacceptable.[165]

9.2.3. Building Herd Immunity. Circulation of measles virus has been less affected by vaccination programs than was predicted by simple models. It was hoped that vaccines that give 95–97% postvaccinal seroconversion rates[103,111] would permit establishment of sufficient herd immunity to eliminate the disease quickly.[86,144] Two characteristics frustrated this hope: the extraordinary infectiousness of the virus and the need to delay vaccination until late infancy. The first means that a very high proportion of a population must be immunized to stop virus spread and the second makes it difficult to achieve these high coverage rates. The age at which a child becomes responsive to vaccine and susceptible to the disease is variable from person to person as well as from country to country.[27,159] Withholding vaccine until an age when good immunization rates can be achieved means leaving a substantial number of children unprotected for several months. Individualizing the time of vaccination is not generally practical, and an optimum must be determined for each country. At this age the number of children infected before they are due for vaccination is balanced against the number left susceptible afterward.[36,42,81]

If was often assumed that the vaccine could be given twice in infancy to immunize both those who become susceptible early and late at appropriate times. However, aside from the cost, prohibitive for many less-developed countries, this procedure has both sociological and biological drawbacks. When vaccine is given while too early, the parents may be made falsely secure and the vaccine discredited when measles follows. When children are vaccinated while they will have passive antibody, they can be made refractory to reimmunization. In the latter circumstance, revaccination may produce only a transient secondary response and even twice- or thrice-vaccinated children may remain unprotected.[29,107]

Achieving good vaccine coverage in less-developed countries has been difficult.[68,108,159] Where measles is strongly seasonal, better coverage can sometimes be obtained by means of well-publicized mass campaigns than by trying to immunize each infant at the most appropriate age.[93,131] Such "pulse" vaccination programs abandon the idea of protecting very young children directly but, by reducing the total number of cases, raise the average age when a child is exposed. Although a single properly administered dose of vaccine should be adequate, even a few errors and low failure rates leave enough susceptible individuals to permit continued outbreaks. In the United States, compulsory vaccination at school entry[38] resulted in elimination of the classic peak measles incidence in elementary school children, and revaccination is now proposed for the less-developed countries as well.[159]

10. Unresolved Problems

10.1. Development of a Better Vaccine

The standard measles vaccine is so good that the challenge of developing a better vaccine is daunting. Nevertheless, in 30 years of trying, we have failed to eliminate measles using the standard vaccine. Hope for success has not been abandoned,[8a,65a] but as we lose the solidly naturally immunized adult population through the passage of years, the job becomes more difficult and it is less probable that standard vaccine can do the job. The greatest impediment is the requirement for delay until maternally derived antibody has been lost. This requirement means that measles vaccination cannot be integrated with the other most widely used vaccines, that some children get measles before they are vaccinated, and that some children are vaccinated without being immunized. Other problems may lie in sequence changes in the H protein coincident with adaptation of attenuated strains to chick cells.

A solution would be a vaccine that does not depend on measles virus replication. Such a vaccine could use the H and F genes or gene products of virus that has not been adapted to chick embryo, thus retaining the natural attachment site fitted to human CD46 receptor.[52] Various variants of two distinct methods for achieving these goals are at the stage of laboratory-based tests: (1) cloned vaccines in which viral proteins are grown in baculovirus[155] or other cells[160] and administered in lipid micelles (IS-COMS) that enhance stimulation of CD8$^+$ cells,[159a] or (2) vectored vaccines in which measles virus genes are inserted into a live carrier such as vaccinia,[157,164] canary pox,[156] or BCG.[152] Caution, predicated on the history of untoward effects precipitated by the early killed vaccines and the recent high-titer vaccinations, as well as the need to demonstrate durability of resulting immunity, means that development of all these vaccines must proceed slowly. Even if and when trials are successfully completed, it will be a challenge to keep the cost of such "high-tech" vaccines down to a level where they can be used where the greatest need lies in the less-developed parts of the world.

10.2. Persistent Measles Genomes in Chronic Disease

Measles virus has been associated with several chronic diseases other than SSPE. The strength of association varies, but the evidence from several sources and methods cumulatively suggests that measles virus may frequently be persistent. In each of the diseases described below, a statistically significant association has been found using one or more tests. However, some cases of each disease are negative and some controls are usually positive. This may indicate that none of the methods is specific or it may mean that measles virus genomes persist in normal tissues only less frequently than in diseased counterparts.

10.2.1. Multiple Sclerosis. Multiple sclerosis (MS) is an autoimmune disease in which immune suppression by T cells is relaxed and the myelin attacked.[9] It is quite clear that the precipitating cause is autoimmunity and that there are predisposing genetic factors, but it is not at all clear what triggers the reaction. The thesis that it is triggered by a commonly occurring virus infecting at an unusually late age, i.e., the "polio model," became tenable when Dean and Kurtzke[48] showed that the chance of developing MS is lower in persons of common genetic background who grow up in less-developed communities. That measles or canine distemper, specifically, were involved was suggested by finding slightly elevated mean measles antibody titers in MS sera[6] and measles virus genome sequences in the lesions.[77] However, measles antibodies are not uniquely elevated,[78,122] and positive tests for measles RNA *in situ* were found in some normal brain tissue.[77] The factors that determine the distribution of MS are still moot, but the possibility that measles virus plays a special etiologic role has diminished.

10.2.2. Osteitis Deformans. Osteitis deformans, or Paget's disease of bone, is a syndrome in which osteoclasts become hyperactive in remodeling bone into malfunctional shapes. The disease does not spread from one bone to another, except where they are fused, suggesting that the causative agent spreads by cell-to-cell contact. Osteoclasts, which are responsible for bone resorption, are normally multinucleated, but the degree of multinucleation is excessive in Paget's disease and reminiscent of the giant cells of paramyxovirus infection. Electron microscopy revealed intranuclear tubules in these cells that resemble the nucleocapsids of the *Paramyxoviridae*.[116] Immunofluorescence studies showed reactions with measles monoclonal antibodies but also with antibodies to SV5 and parainfluenza type 3.[15] *In situ* hybridization has given positive reactions with canine distemper probes and the PCR has amplified 187 bases from a canine distemper virus sequence.[70] These sequences differed from prototype canine distemper strains, indicating that the result was not due to laboratory contamination. Set against these positive findings is the fact that serum titers to measles and the other paramyxoviruses in pagetic persons are not abnormal like those in SSPE and MS. The net effect of these several lines of investigation lends strong support to the thesis that some paramyxovirus is involved in Paget's disease, but it is not exclusively measles and not definite that the relationship is causal.[22]

10.2.3. Crohn's Disease.

Chronic inflammatory bowel, Crohn's disease, is a sharply demarcated granulomatous vasculitis occurring in any part of the intestine but most often the ileum and upper large bowel. Capillary endothelial cells in the lesions are often positive for measles virus by immunohistology, *in situ* hybridization and PCR, and they contain nucleocapsidlike structures evident by electron microscopy.[161] One study found that Crohn's disease was likely to follow exposure to measles during the third trimester *in utero*.[53] No evidence contrary to this association or evidence that it does not pertain specifically to measles has appeared, but this may be because the relationship was only recognized recently.

10.2.4. Autoimmune Chronic Active Hepatitis.

Autoimmune chronic active hepatitis is recognized as a hepatitis of distinct etiology. Measles virus genome, identified by *in situ* hybridization with a 50-base measles-virus-specific probe, was found in affected cells in 12 of 18 cases and one of 45 controls.[136]

11. References

1. AABY, P., Malnutrition and overcrowding-exposure in severe measles infection: A review of community studies, *Rev. Infect. Dis.* **10**:478–491 (1988).
2. AABY, P., ANDERSEN, M., AND KNUDSEN, K., Excess mortality after early exposure to measles, *Int. J. Epidemiol.* **22**:156–162 (1993).
3. AABY, P., BUKH, J., LISSE, I. M., AND SMITS, A. J., Measles mortality, state of nutrition, and family structure: A community study in Guinea-Bissau, *J. Infect. Dis.* **147**:693–701 (1983).
4. AABY, P., SAMB, B., SIMONDEN, F., WHITTLE, H., SECK, A. M. C., KNUDSEN, K., BENNETT, J., MARKOWITZ, L., AND RHODES, P., Child mortality after high-titre measles vaccines in Senegal: The complete data set, *Lancet* **338**:1518–1519 (1991).
5. ABU BECR, M., *A Discourse on the Smallpox and Measles* (R. MEAD, TRANS.), J. Brindley, London, 1748.
6. ADAMS, J. M., AND IMAGAWA, D. T., Measles antibodies in

7. ALBRECHT, P., HERRMANN, K., AND BURNS, G. R., Role of virus strain in conventional and enhanced measles plaque neutralization test, *J. Virol. Methods* **3**:251–260 (1981).
8. ANDERSON, R. M., AND MAY, R. M., *Infectious Diseases of Humans. Dynamics and Control*, Oxford Science Publications, Oxford, 1991.
8a. ANONYMOUS, Global measles eradication: target 2010?, *EPI Newsletter* **18**(4):1–3 (1996).
8b. ANONYMOUS, Record five years measles-free, *EPI Newsletter* **18**(6):1–3 (1996).
9. ANTEL, J. P., RICHMAN, D. P., MEDOF, M. E., AND ARNESON, B. G. W., Lymphocyte function and the role of regulator cells in multiple sclerosis, *Neurology* **28**:106–110 (1978).
10. ARNEBORN, P., AND BIBERFELD, G., T-lymphocyte subpopulations in relation to immunosuppression in measles and varicella, *Infect. Immun.* **39**:29–37 (1983).
10a. ATKINSON, W. L., HADLER, S. C., REDD, S. B., AND ORENSTEIN, W. A., Measles Surveillance–United States, 1991, *Morb. Mort. Week. Rpts.* **41**(SS6):1–12 (1992).
11. BACZKO, K., BRINCKMANN, U., PARDOWITZ, I., RIMA, P. K., AND TER MEULEN, V., Nucleotide sequences of the genes encoding the matrix protein of two wild-type measles virus strains, *J. Gen. Virol.* **72**:2279–2282 (1991).
12. BARRETT, T., SUBBARAO, S. M., BELSHAM, G. J., AND MAHY, B. M. J., The molecular biology of the morbilliviruses, in: *The Paramyxoviruses* (D. W. KINGSBURY, ED.), pp. 83–102, Plenum Press, New York, 1991.
13. BARTLETT, M. S., Deterministic and stochastic models for recurrent epidemics, *Third Berkeley Symp. Math. Stat. Prob.* **4**:81–109 (1956).
14. BARTLETT, M. S., The critical community size for measles in the United States, *J. R. Stat. Soc. Ser. A* **123**:37–44 (1960).
15. BASLÉ, M. F., RUSSELL, W. C., GOSWAMI, K. K. A., REBEL, A., GIRAUDON, P., WILD, F., AND FILMON, R., Paramyxovirus antigens in osteoclasts from Paget's bone tissue detected by monoclonal antibodies, *J. Gen. Virol.* **66**:2103–2110 (1985).
16. BECH, V., Studies on the development of complement fixing antibodies in measles patients, *J. Immunol.* **103**:252–253 (1962).
17. BECH, V., Measles epidemics in Greenland, *Arch. Ges. Virusforsh.* **17**:53–56 (1965).
18. BELLANTI, J. A., SANGA, R. I., KLUTINIS, B., BRANDT, B., AND ARTENSTEIN, M. S., Antibody response in serum and nasal secretion in children immunized with inactivated and attenuated measles virus vaccines, *N. Engl. J. Med.* **280**:628–633 (1969).
18a. BELLINI, W. J., ROTA, J. S., AND ROTA, P. A., Virology of measles virus, *J. Infect. Dis.* **170**:S15–S23 (1994).
19. BENJAMIN, B., AND WARD, S. M., Leukocyte response to measles, *Am. J. Dis. Child.* **44**:921–963 (1932).
20. BERG, R. B., AND ROSENTHAL, M. S., Propagation of measles virus in suspensions of human and monkey leukocytes, *Proc. Soc. Exp. Biol. Med.* **106**:581–585 (1961).
21. BHATT, P. N., BRANDT, C. D., WEISS, R. A., FOX, J. P., AND SCHAFFER, M. F., Viral infections of monkeys in their natural habitat in southern India, *Am. J. Trop. Med. Hyg.* **15**:561–566 (1966).
22. BIANCO, P., SILVESTRINI, G., BALLANTI, P., AND BONUCCI, E., Paramyxovirus-like nuclear inclusions identical to those of Pagets disease of bone detected in giant cells of primary oxalosis, *Virchows Arch.* **421**:427–433 (1992).

multiple sclerosis, *Proc. Soc. Exp. Biol. Med.* **111**:562–566 (1962).

23. VAN BINNENDIJK, R. S., POELEN, M. C. M., DE VRIES, P., VOORMA, H. O., OSTERHAUS, A. D. M. E., AND UYTDEHAAG, F. G. C. M., Measles virus-specific human T cell clones. Characterization of specificity and function of CD4$^+$ helper/cytotoxic and CD8$^+$ cytotoxic T cell clones, *J. Immunol.* **142**:2847–2854 (1989).

24. BLACK, F. L., Growth and stability of measles virus, *Virology* **7**:184–192 (1959).

25. BLACK, F. L., Measles endemicity in insular populations: Critical community size and its evolutionary implication, *Theor. Biol.* **11**:207–211 (1965).

26. BLACK, F. L., Measles and mumps, in: *Manual of Clinical Laboratory Immunology*, 4th ed. (N. R. ROSE, E. C. DEMACARIO, J. L. FAHEY, H. FRIEDMAN, AND G. M. PENN, EDS.), pp. 596–599, American Society for Microbiology, Washington, DC, 1992.

27. BLACK, F. L., Measles active and passive immunity in worldwide perspective, *Prog. Med. Virol.* **36**:1–33 (1989).

28. BLACK, F. L., An explanation of high death rates among New World peoples when in contact with Old World diseases, *Perspect. Biol. Med.* **37**:292–307 (1993).

29. BLACK, F. L., BERMAN, L. L., LIBEL, M., REICHELT, C. A., PINHIERO, F. P., TRAVASSOS DA ROSA, A., FIGUEIRA, F., AND SIQUIERA-CAMPOS, E., Inadequate immunity to measles in children vaccinated at an early age: Effect of revaccination, *Bull. WHO* **62**:315–319 (1984).

30. BLACK, F. L., AND DAVIS, D. E. M., Measles and readiness for reading and learning. II. New Haven study, *Am. J. Epidemiol.* **88**:337–344 (1968).

31. BLACK, F. L., AND ROSEN, L., Patterns of measles antibodies in residents of Tahiti and their stability in the absence of re-exposure, *J. Immunol.* **88**:725–731 (1962).

32. BLACK, F. L., AND SHERIDAN, S. R., Blood leukocyte response to live measles vaccines, *Am. J. Dis. Child.* **113**:301–304 (1967).

32a. BLIXENKRONE-MØLLER, M., SHARMA, B., VARSANYI, T. M., HU, A., NORRBY, E., AND KÖVAMEES, J., Sequence analysis of the genes encoding the nucleocapsid protein and phosphoprotein (P) of phocid distemper virus, and editing of the P gene transcript, *J. Gen. Virol.* **73**:885–893 (1992).

33. BLOCK, A. B., ORENSTEIN, W. A., EWING, W. M., SPIN, W. H., MALLISON, G. F., HERRMANN, K. L., AND HINMAN, A. R., Measles outbreak in a pediatric practice: Airborne transmission in an office setting, *Pediatrics* **75**:676–683 (1985).

34. BOLKER, B. M., AND GRENFELL, B. T., Chaos and biological complexity in measles dynamics, *Proc. R. Soc. London Ser. B.* **251**:75–81 (1993)

35. BRODY, J. A., SEVER, J. L., EDGAR, A., AND McNEW, J., Measles antibody titer in multiple sclerosis patients and their siblings, *Neurology* **22**:492–499 (1972).

36. BURROWS, J., AND CRUICKSHANK, J. G., At what age should measles vaccine be given? Report of a small trial in Bulawayo, *Cent. Afr. J. Med.* **22**:45–47 (1976).

37. CABALLERO, B., AND RICE, A., Low serum retinol is associated with increased severity of measles in New York City children, *Nutr. Rev.* **50**:291–292 (1992).

37a. CELERS, J., Problémes de santé publique posés par la rougeole dans les pays favorisés, *Arch. Ges. Virusforsch.* **16**:5–18 (1965).

38. CENTERS FOR DISEASE CONTROL, Measles—United States, 1992, *Morbid. Mortal. Week. Rep.* **42**:378–381 (1993).

39. CENTERS FOR DISEASE CONTROL, Absence of reported measles—United States, November 1993, *Morbid. Mortal. Week. Rep.* **42**:925–926 (1993).

40. CENTERS FOR DISEASES CONTROL, Recommended childhood immunization schedule—United States, *Morbid. Mortal. Week. Rep.* **44**:RR–5 (1995).

41. CHEN, R. T., MARKOWITZ, L. E., ALBRECHT, P., STEWART, J. A., MOFENSON, L. M., PREBLUD, S. R., AND ORENSTEIN, W. A., Measles antibody: Reevaluation of protective titers, *J. Infect. Dis.* **162**:1036–1042 (1990).

42. CHEN, S. L., AND LAM, S. K., Optimum age for measles immunization in Malaysia, *Med. J. Malaysia* **40**:281–288 (1985).

43. CHRISTENSEN, P. E., HENNING, S., BANG, H. O., ANDERSEN, V., JORDAL, B., AND JENSEN, O., An epidemic of measles in southern Greenland, 1951: Measles in virgin soil. II. The epidemic proper, *Acta Med. Scand.* **144**:430–449 (1952).

44. CHRISTIE, C. D., HIRSCH, J. L., ROGALL, B., MERRILL, S., RAMLAL, A. A., KARIAN, V., AND BLACK, F. L., Durability of passive antibody in Jamaican children, *Int. J. Epidemiol.* **19**:698–702 (1990).

45. COMMITTEE ON INFECTIOUS DISEASES, AMERICAN ACADEMY OF PEDIATRICS, Measles: Reassessment of the current immunization policy, *Pediatrics* **84**:1110–1113 (1989).

46. CONNOR, E., MORRISON, S., LANE, J., OLESKE, J., SONKE, R. L., AND CONNOR, J., Safety, tolerance, and pharmacodynamics of systemic ribavirin in children with human immunodeficiency virus infection, *Antimicrob. Agents Chemother.* **37**:532–539 (1993).

47. CROWLEY, J. C., DOWLING, P. C., MENONNA, J., SILVERMAN, J. I., SCHUBACH, D., COOK, S. D., AND BLUMBERG, B. M., Sequence variability and function of measles virus 3′ and 5′ ends and intracistronic regions, *Virology* **164**:498–506 (1988).

48. DEAN, G., AND KURTZKE, J. F., On the risk of multiple sclerosis according to age at immigration to South Africa, *Br. Med. J.* **3**:725–729 (1971).

49. DIAZ, T., NUÑEZ, J. C., RULLAN, J. V., MARKOWITZ, L. E., BARKER, N. D., AND HORAN, J., Risk factors associated with severe measles in Puerto Rico, *Pediatr. Infect. Dis. J.* **11**:836–840 (1992).

50. DIETZ, K., Overall population patterns in the transmission cycle of infectious disease agents, in: *Population Biology of Infectious Diseases* (R. M. ANDERSON AND R. M. MAY, EDS.), pp. 87–102, Springer-Verlag, New York, 1984.

51. DIETZ, V. J., NIEBERG, P., GUBLER, D. J., AND GOMEZ, I., Diagnosis of measles by clinical case definition in dengue-endemic areas—Implications for measles surveillance and control, *Bull. WHO* **70**:745–750 (1992).

52. DÖRIG, R. E., MARCIL, A., CHOPRA, A., AND RICHARDSON, C. D., The human CD46 molecule is a receptor for measles virus (Edmonston strain), *Cell* **75**:295–305 (1993).

53. EKBOM, A., ZACK, M., ADAMI, H-O., AND HEIMICK, C., Is there clustering of inflammatory bowel disease at birth? *Am. J. Epidemiol.* **134**:876–886 (1991).

54. ENDERS, J. F., AND PEEBLES, T. C., Propagation in tissue cultures of cytopathogenic agents from patients with measles, *Proc. Soc. Exp. Biol. Med.* **86**:277–286 (1954).

55. FINE, P. E. M., AND CLARKSON, J. A., Measles in England and Wales—I: An analysis of factors underlying seasonal patterns, *Int. J. Epidemiol.* **11**:5–14 (1982).

56. FIREMAN, P., FRIDAY, G., AND KUMATE, J., Effect of measles vaccine on immunological responsiveness, *Pediatrics* **43**:264–272 (1969).

57. FORSEY, T., HEATH, A. B., AND MINOR, P. D., The 1st international standard for anti-measles serum, *Biologicals* **19**:237–241 (1991).

58. FORTHAL, D. N., BLANDING, J., AARNAES, S., PETERSON, E. M., DE LA MAZA, L. M., AND TILLES, J. G., Comparison of different methods and cell lines for isolating measles virus, *J. Clin. Microbiol.* **31:**695–697 (1993).

59. FOSTER, A., AND SOMMER, A., Childhood blindness from corneal ulceration in Africa: Causes, prevention and treatment, *Bull. WHO* **64:**619–623 (1986).

60. FOX, J. P., BLACK, F. L., AND KOGON, A., Measles and readiness for reading and learning, *Am. J. Epidemiol.* **88:**168–175 (1969).

61. FULGINITI, V. A., AND HELFER, R. E., Atypical measles in adolescent siblings 16 years after killed virus vaccine, *J.A.M.A.* **244:**804–806 (1980).

62. GALINSKY, M. S., Annotated nucleotide sequence and protein sequences for selected *Paramyxoviridae*, in: *The Paramyxoviruses* (D. W. KINGSBURY, ED.), pp. 537–568, Plenum Press, New York, 1991.

63. GARENNE, M., Sex differences in measles mortality: A world review, Harvard School of Public Health Working Paper Series No. 4 (1992).

64. GARENNE, M., AND AABY, P., Pattern of exposure and measles mortality in Senegal, *J. Infect. Dis.* **161:**1088–1094 (1990).

65. GARRENE, M., LEROY, O., BEAU, J-P., AND SENE, I., Child mortality after high-titre measles vaccines: Prospective study in Senegal, *Lancet* **338:**903–906 (1991).

65a. GELLEN, B. G., AND KATZ, S. L., Putting a stop to a serial killer: Measles, *J. Infect. Dis.* **170:**S1–2 (1994).

66. GIBBS, F. A., GIBBS, E. I., CARPENTER, P. R., AND SPIES, H. W., Electroencephalographic abnormality in "uncomplicated" childhood diseases, *J.A.M.A.* **171:**1030–1055 (1959).

67. GINDLER, J. S., ATKINSON, W. A., MARKOWITZ, L. E., AND HUTCHINS, S. S., Epidemiology of measles in the United States in 1989 and 1990, *Pediatr. Infect. Dis. J.* **11:**841–846 (1992).

68. GINDLER, J. S., CUTTS, F. T., BARNETT-ANTINORI, M. E., SWINT, E. B., HADLER, S. C., AND RULLAN, J. V., Successes and failures in vaccine delivery—Evaluation of the immunization delivery system in Puerto Rico, *Pediatrics* **91:**315–320 (1993).

69. GIRAUDON, P., JACQUIER, M. F., AND WILD, T. F., Antigenic analysis of African measles virus field isolates: Identification and localization of one conserved and two variable epitope sites on the NP protein, *Virus Res.* **18:**137–152 (1988).

70. GORDON, M. T., MEE, A. P., ANDERSON, D. C., AND SHARPE, P. T., Canine distemper transcripts sequenced from pagetic bone, *Bone Miner.* **19:**159–174 (1992).

71. GRAY, R. H., CHAR, G., PRABHAKAR, P., BAINBRIDGE, R., AND JOHNSON, B., Subacute sclerosing panencephalitis in Jamaica, *West Indian Med. J.* **25:**27–34 (1986).

72. GREENBERG, M., PELLITERI, O., AND EISENSTEIN, D. T., Measles encephalitis. I. Prophylactic effect of gamma globulin, *J. Pediatr.* **46:**652–647 (1955).

73. GRESSER, I., AND KATZ, S. L., Isolation of measles virus from urine, *N. Engl. J. Med.* **263:**452–454 (1960).

74. GRIFFITHS, D. A., The effect of measles vaccination on the incidence of measles in a community, *J. R. Stat. Soc. Ser. A.* **136:**441–449 (1973).

74a. GRIFFIN, D. E., WARD, B. J., AND ESOLEN, L. M., Pathogenesis of measles infection: A hypothesis for altered immune responses, *J. Infect. Dis.* **170:**S24–31 (1994).

75. GUSTAFSON, T. L., LEVEN, A. W., BRUNELL, P. A., MUELLER, R. G., BUTTERY, C. M. G., AND SEHULSTER, L. M., Measles outbreak in a fully immunized secondary school population, *N. Engl. J. Med.* **316:**771–774 (1987).

76. HAASE, A. T., GANTZ, D., EBLE, B., WALKER, D., STOWRING, L., VENTURA, P., BLUM, H., WIETGREFE, S., ZUPANIC, M., TOURTELLOTTE, W., GIBBS, C. J., JR., NORRBY, E., AND ROZENBLATT, S., Natural history of restricted synthesis and expression of measles virus genes in subacute sclerosing panencephalitis, *Proc. Natl. Acad. Sci. USA* **82:**3020–3024 (1985).

77. HAASE, A. T., VENTURA, P., GIBBS, C. J., JR., AND TOURTELLOTTE, W. W., Measles virus nucleotide sequences: Detection by hybridization *in situ*, *Science* **212:**672 (1981).

78. HAIRE, M., FRASER, F. B., AND MILLAR, J. H. D., Measles and other virus specific immunoglobulins in multiple sclerosis, *Br. J. Med.* **3:**612–615 (1973).

79. HALSEY, N. A., MODLIN, J. F., JABBOUR, J. T., DUBEY, L., EDDINS, D. L., AND LUDWIG, D. D., Risk factors in subacute sclerosing panencephalitis: A case-control study, *Am. J. Epidemiol.* **111:**415–424 (1980).

80. HALSEY, N. A., The optimal age for administering measles vaccine in developing countries, in: *Recent Advances in Immunization*, Pan American Health Organization Scientific Publication 451, pp. 4–13, PAHD, Washington, DC, 1983.

81. HALSEY, N. A., MODLIN, J. F., AND JABBOUR, J. T., Subacute sclerosing panencephalitis (SSPE). An epidemiologic review, in: *Persistent Viruses* (J. G. STEVENS, G. J. TODARO, AND C. F. FOX, EDS.), pp. 101–113, Academic Press, New York, 1978.

82. HAMER, W. H., The Milroy lectures on epidemic disease in England—The evidence of variability and persistence of type, *Lancet* **1:**733–739 (1906).

83. HANKINS, R. W., AND BLACK, F. L., Western blot analyses of measles virus antibody in normal persons and patients with multiple sclerosis, subacute sclerosing panencephalitis or atypical measles, *J. Clin. Microbiol.* **24:**324–329 (1986).

84. HEDRICK, A. W., Monthly estimates of the child population susceptible to measles 1900–1931, *Am. J. Hyg.* **17:**613–636 (1933).

85. HELMS, S., AND HELMS, P., Tuberculin sensitivity during measles, *Acta Tuberc. Scand.* **35:**166–171 (1956).

86. HINMAN, A. R., BRANLING-BENNETT, A. D., AND NIEBURG, P. I., The opportunity and obligation to eliminate measles from the United States, *J.A.M.A.* **242:**1157–1162 (1979).

87. HINMAN, A. R., ORENSTEIN, W. A., AND MORTIMER, E. A., When, where, and how do immunizations fail? *Ann. Epidemiol.* **2:**805–812 (1992).

88. HIRSCH, A., *Handbook of Geographical and Historical Pathology*, Vol. 1, New Sydenham Society, London, 1883.

89. HOLT, E. A., MOULTON, L. H., SIBERRY, G. K., AND HALSEY, N. A., Differential mortality by measles vaccine titer and sex, *J. Infect. Dis.* **168:**1087–1096 (1993).

90. HOME, F., *Medical Facts and Experiments*, A, Millar, London, 1759.

90a. HU, A., SHESHBERADARAN, H., NORRBY, E., AND KÖVAMEES, J., Molecular characterization of epitopes on the measles virus hemagglutinin protein, *Virology* **192:**351–354 (1993).

91. HUSSY, C. D., AND KLEIN, A., A randomized controlled trial of vitamin A in children with severe measles, *N. Engl. J. Med.* **323:**160–164 (1990).

92. IKIC, D. M., Edmonston–Zagreb strain of measles vaccine: Epidemiologic evaluation of Yugoslavia, *Rev. Infect. Dis.* **5:**558–563 (1983).

93. JOHN, T. J., AND STEINHOFF, M. C., Control of measles by annual pulse immunization, *Am. J. Dis. Child.* **138:**299–300 (1984).

94. DE JONG, J. G., AND WINKLER, K. O., Survival of measles virus in air, *Nature* **201:**1054–1055 (1964).

95. KAPIL, A., BROOR, S., AND SETH, P., Prevalence of SSPE: A serological study, *Indian Pediatr.* **29:**731–734 (1992).

96. KAPLAN, L. J., DAUM, R. S., SMARON, M., AND MCCARTHY, C. A., Severe measles in immunocompromised patients, *J.A.M.A.* **267:**1237–1241 (1992).

97. KARELITZ, S., Globulin extract of immune adult serum in prophylaxis of measles, *Proc. Soc. Exp. Biol. Med.* **31:**793–796 (1933).

98. KATZ, S. L., KEMPE, C. H., BLACK, F. L., LEPOW, M. L., KRUGMAN, S. L., HAGGERTY, R. J., AND ENDERS, J. F., Studies on an attenuated measles-virus vaccine. VIII. General summary and evaluation of the results of vaccination, *N. Engl. J. Med.* **273:**180–184 (1960).

98a. KEELING, M. J., AND GRENFELL, B. T., Disease extinction and community size: Modeling the persistence of measles, *Science* **275:**65–67 (1997).

99. KOPROWSKI, H., The role of hypergy in measles encephalitis, *Am. J. Dis. Child.* **103:**273–278 (1962).

100. KOSTER, F. T., CURLIN, G. C., AZIZ, K. M. A., AND HAQUE, A., Synergistic impact of measles and diarrhea on nutrition and mortality in Bangladesh, *Bull. WHO* **59:**901–908 (1981).

101. KRAUSE, P. J., CHERRY, J. D., CARNEY, J. M., NAIDITCH, H. J., AND O'CONNOR, K., Measles-specific lymphocyte reactivity and serum antibody in subjects with different measles histories, *Am. J. Dis. Child.* **134:**565–571 (1980).

102. KRESS, S., SCHLEUDERBERG, A. E., HORNICK, R. B., MORSE, L. J., COLE, J. L., SLATER, E. A., AND McCRUMB, F. R., Studies with live attenuated measles-virus vaccine. II. Clinical and immunological response of children in an open community, *Am. J. Dis. Child.* **101:**701–707 (1961).

103. KRUGMAN, S., GILES, J. P., JACOBS, A. M., AND FRIEDMAN, H., Studies with a further attenuated measles-virus vaccine, *Pediatrics* **66:**471–488 (1965).

103a. LANDLAEKNI, *Heilbrigdisskyslur (Public Health in Iceland)*, Rikisprentsmidjan Gutenberg, Reykjavik, 1941–1950.

104. LENNON, J. L., AND BLACK, F. L., Maternally derived measles immunity in the era of vaccine-protected mothers, *J. Pediatr.* **108:**671–676 (1986).

105. LEON, M. E., WARD, B., KANASHIRO, R., HERNANDEZ, H., BERRY, S., VAISBERG, A., ESCAMILLA, J., CAMPOS, M., BELLOMO, S., AZABACHE, V., AND HALSEY, N. A., Immunologic parameters 2 years after high-titer measles immunization in Peruvian children, *J. Infect. Dis.* **168:**1097–1104 (1993).

106. LIEVENS, A. W., AND BRUNELL, P. A., Specific immunoglobulin M enzyme-linked immunosorbent assay for confirming the diagnosis of measles, *J. Clin. Microbiol.* **24:**391–394 (1986).

107. MARKOWITZ, L. E., ALBRECHT, P., ORENSTEIN, W. A., LETT, S. M., PUGLIESE, T. J., AND FARRELL, D., Persistence of measles antibody after revaccination, *J. Infect. Dis.* **166:**205–208 (1992).

108. MASHAKO, L. M. N., KAPONGO, C. N., NSIBU, C. N., MALAMBA, M., DAVACHI, F., AND OTHEPA, M. O., Evaluation de la couverture vaccinale des enfants moins de 2 ans a Kinshasa (Zaire), *Arch. Fr. Pediatr.* **49:**717–720 (1992).

109. MATA, L., *The Children of Santa Maria Cauque*, M.I.T. Press, Cambridge, MA, 1978.

110. MAXCY, K. F., Principles and methods of epidemiology, in: *Viral and Rickettsial Infections of Man* (T. M. RIVERS, ED.), pp. 128–146, Lippincott, Philadelphia, 1948.

111. MCCORMICK, J. B., HALSEY, N. A., AND ROSENBERG, R., Measles vaccine efficacy from secondary attack rates during a severe epidemic, *J. Pediatr.* **90:**13–16 (1977).

112. MCNEILL, W. H., *Plagues and Peoples: A Natural History of Infectious Diseases*, Doubleday, New York, 1976.

113. TER MEULEN, V., MÜLLER, D., KÄCKELL, Y., KATZ, M., AND MAYERMANN, R., Isolation of infectious measles virus in measles encephalitis, *Lancet* **2:**1172–1175 (1972).

114. MILLER, C. L., Current impact of measles in the United Kingdom, *Rev. Infect. Dis.* **5:**427–432 (1983).

115. MILLER, C., FARRINGTON, C. P., AND HARBERT, K., The epidemiology of subacute sclerosing panencephalitis in England and Wales 1970–1989, *Int. J. Epidemiol.* **21:**998–1006 (1992).

116. MILLS, B. G., AND SINGER, F. R., Nuclear inclusions in Paget's disease of bone, *Science* **194:**201–202 (1976).

117. MITIS, A., HOLLOWAY, A., EVANS, A. E., AND ENDERS, J. F., Attenuated measles vaccine in children with acute leukemia, *Am. J. Dis. Child.* **103:**413–418 (1962).

118. MORLEY, D., The severe measles of West Africa, *Proc. R. Soc. Med.* **57:**846–849 (1969).

119. NANICHE, D., VARIOR-KRISHNAN, G., CERVONI, F., WILD, T. F., ROSSI, B., RABOURDIN-COMBE, C., AND GERLIER, D., Human membrane cofactor protein (CD46) acts as a cellular receptor for measles virus, *J. Virol.* **67:**6025–6032 (1993).

120. NEEL, J. V., CENTERWALL, W. R., CHAGNON, N. A., AND CASEY, H. L., Notes on the effect of measles in a virgin-soil population of South American Indians, *Am. J. Epidemiol.* **91:**418–429 (1970).

121. NEUMANN, P. W., WEBER, J. M., JESSAMINE, A. G., AND O'SHAUGHNESSY, M. V., Comparison of measles antihemolysis test, enzyme-linked immunosorbent assay, and hemagglutination inhibition test with neutralization test for determination of immune status, *J. Clin. Microbiol.* **22:**296–298 (1985).

122. NORRBY, E., Viral antibodies in multiple sclerosis, *Prog. Med. Virol.* **24:**1–39 (1978).

123. NORRBY, E., *Paramyxoviridae*: Measles virus, in: *Laboratory Diagnosis of Infectious Diseases. Principles and Practice* (E. H. LENNETTE, P. HALONEN, AND E. A. MURPHY, EDS.), pp. 525–539, Springer-Verlag, New York, 1988.

124. NORRBY, E., ENDERS-RUCKLE, G., AND TER MEULEN, V., Differences in the appearance of antibodies to structural components of measles virus after immunization with inactivated and live virus, *J. Infect. Dis.* **132:**262–269 (1975).

125. NUTELS, N., Medical problems of newly contacted Indian groups, *Pan. Am. Health Org. Sci. Publ.* **165:**68–76 (1964).

126. OKUNO, Y., UEDA, S., KURIMURA, T., SUZUKI, N., YAMANISHI, K., BABA, K., TAKAHASHI, M., KONOBE, T., SASADA, T., ONISHI, K., AND TAKAKU, K., Studies on further attenuated live measles vaccine. VII. Development and evaluation of CAM-70 measles virus vaccine, *Biken J.* **14:**253–258 (1971).

127. ONORATO, I. M., MARKOWITZ, L. E., AND OXTOBY, M. T., Childhood immunization, vaccine preventable diseases, and infection with HIV, *Pediatr. Infect. Dis. J.* **7:**588–595 (1988).

128. PALOMBO, P., HOYT, L., DeMASIO, K., OLESKE, J., AND CONNOR, E., Population-based study of measles and measles immunization in human immunodeficiency virus-infected children, *Pediatr. Infect. Dis. J.* **11:**1008–1014 (1992).

129. PAMPIGLIONE, G., Prodromal phase of measles: Some neurophysiological studies, *Br. Med. J.* **2:**1296–1300 (1964).

130. PAN AMERICAN HEALTH ORGANIZATION AND THE MINISTRIES OF HEALTH OF BRAZIL, CHILE AND ECUADOR, Seroconversion rates and antibody titer induced by measles vaccine in Latin American children 6–12 months of age, *Bull. Pan. Am. Health Org.* **16:**272–285 (1983).

131. PANNUTI, C. S., MORAES, J. C., SOUZA, V. A. U. F., CAMARGO, M. C. C., AND HIDALGO, N. R. T., Vaccine effectiveness in Saõ Paulo, *Bull. WHO* **69**:557–560 (1991).

132. PANUM, P. L., *Observation Made during the Epidemic of Measles on the Faroe Islands in the Year 1846*, American Publishing Association, New York, 1940.

133. PEART, A. F. W., AND NAGLER, F. P., Measles in the Canadian Arctic, 1952, *Can. J. Public Health* **45**:146–157 (1954).

134. RAHMATHULLAH, L., UNDERWOOD, B. A., THULASIRAJ, R. D., MILTON, R. C., RAMASWAMY, K., RAHMATHULLAH, R., AND BABU, G., Reduced mortality among children in southern India receiving small weekly doses of Vitamin A, *N. Engl. J. Med.* **323**:929–935 (1990).

135. REDDY, V., BHASKARAM, B. P., RAGHURAMULA, N., MILTON, R. C., RAO, V., MADHUSUDAN, J., AND RADHA KRISHNA, K. V., Relationship between measles, malnutrition and blindness: A prospective study in Indian children, *Am. J. Clin. Nutr.* **44**:924–930 (1986).

136. ROBERTSON, D. A. F., ZHANG, S. L., GUY, E. C., AND WRIGHT, R., Persistent measles virus genome in autoimmune chronic active hepatitis, *Lancet* **2**:9–11 (1987).

137. ROLLESTON, J. D., *History of Acute Exanthemata*, Heinemann, London, 1937.

138. ROTA, J. S., HUMMEL, K. B., ROTA, P. A., AND BELLINI, W. J., Genetic variability of the glycoprotein genes of current wild-type measles virus isolates, *Virology* **188**:135–142 (1992).

139. RUCHDESCHEL, J. C., GRAZIANO, K. D., AND MARDINEY, M. R., JR., Additional evidence that the cell-associated immune system is the primary host defense against measles, *Cell. Immunol.* **17**:11–18 (1975).

140. SARVAS, H., SEPPALA, I., KURIKKA, S., SIEGBERG, R., AND MAKELA, O., Half-life of the maternal IgG1 allotype in infants, *J. Clin. Immunol.* **13**:145–151 (1993).

141. SCHAFFNER, W., SCHLEUDERBERG, A. E. S., AND BYRNE, E. B., Clinical epidemiology of sporadic measles in a highly immunized population, *N. Engl. J. Med.* **279**:783–789 (1968).

142. SCHLEUDERBERG, A. E., Immune globulins in human viral infections, *Nature* **205**:1232–1233 (1965).

143. SCHULTZ, F., *Die Epideiologie der Mazern*, Gustaf Fischer, Jena, Germany, 1925.

144. SENCER, D., DULL, H. B., AND LANGMUIR, A. D., Epidemiologic basis for eradication of measles in 1967, *Public Health Rep.* **82**:253–256 (1967).

145. SHASBY, D. M., SHOPE, T. C., DOWNS, H., HERMANN, K. L., AND POLKOWSKI, J., Epidemic measles in a highly vaccinated population, *N. Engl. J. Med.* **296**:585–589 (1977).

146. SHESHBERADARAN, H., AND NORRBY, E., Characterization of epitopes of the measles virus hemagglutinin, *Virology* **152**:58–69 (1986).

147. SHESHBERADARAN, H., NORRBY, E., McCULLOUGH, K. C., CARPENTER, W. C., AND ÖRVELL, C., The antigenic relationship between measles, canine distemper and rinderpest studied with monoclonal antibodies, *J. Gen. Virol.* **67**:1381–1393 (1986).

148. SIBER, G. R., WERNER, B. G., HALSEY, N. A., REID, R., ALMEIDO-HILL, J., GARRETT, S. C., THOMPSON, C., AND SANTOSHAM, M., Interference of immune globulin with measles and rubella immunization, *J. Pediatr.* **122**:204–211 (1993).

149. SMARON, M. F., SARON, E., WOOD, L., McCARTHY, C., AND MORELLO, J. A., Diagnosis of measles by fluorescent antibody and culture of nasopharyngeal secretions, *J. Virol. Methods* **33**:223–229 (1991).

150. SMITH, F. R., CURRAN, A. S., RACITI, A., AND BLACK, F. L.,

151. Reported measles in persons immunologically primed by prior vaccination, *J. Pediatr.* **101**:391–393 (1982).

151. SMORODINTSEV, A. A., BOICHUK, L. M., SHIKINA, E. S., BATANOVA, T. B., BYSTRYAKOVA, L. V., AND PERADZE, T. V., Clinical and immunological response to live tissue culture vaccine against measles, *Acta Virol.* **4**:201–214 (1960).

152. STOVER, C. K., DE LA CRUZ, V. F., FUERST, T. R., BURLEIN, J. E., BENSON, L. A., BENNETT, L. T., BANSAL, G. P., YOUNG, J. F., LEE, M. H., HATFULL, G. F., SNAPPER, S. B., BARLETTA, R. G., JACOBS, W. R., JR., AND BLOOM, B. R., New use of BCG for recombinant vaccines, *Nature* **351**:456–460 (1991).

153. SURINGA, D. W. R., BANK, L. J., AND ACKERMAN, A. B., Role of measles in skin lesions and Koplik spots, *N. Engl. J. Med.* **283**:1139–1142 (1970).

154. SYDENHAM, T., *The Works of Thomas Sydenham*, Vol. 2, Sydenham Society, London, 1922.

155. TAKAHARA, K., HASHIMOTO, H., RI, T., MORI, T., AND YOSHIMURA, M., Characterization of baculovirus-expressed hemagglutinin and fusion glycoproteins of the attenuated measles virus strain AIK-C, *Virus Res.* **26**:167–175 (1992).

156. TARTARGLIA, J., COX, W. I., TAYLOR, J., PINKUS, M., RIVIERE, M., MEIGNIER, B., AND PAOLETTI, E., Highly attenuated poxvirus vectors, *AIDS Res. Hum. Retroviruses* **8**:1445–1447 (1992).

157. TAYLOR, J., PINCUS, S., TARTAGLIA, J., RICHARDSON, C., AL-KHATIB, G., BRIEDIS, D., APPEL, M., NORTON, E., AND PAOLETTI, E., Vaccinia virus recombinants expressing either the measles virus fusion or hemagglutinin glycoprotein protect dogs against canine distemper virus challenge, *J. Virol.* **65**:4273–4274 (1991).

158. TAYLOR, M. J., GODFREY, E., BACZKO, K., TER MEULEN, V., AND RIMA, B. K., Identification of several lineages of measles virus, *J. Gen. Virol.* **72**:83–88 (1992).

159. TULCHINSKI, T. H., GINSBERG, G. M., ABED, Y., ANGELES, M. T., AKUKWE, C., AND BONN, J., Measles control in developing and developed countries—The case for a 2-dose policy, *Bull. WHO* **71**:93–103 (1993).

159a. VARIOR-KRISHNAN, G., TRESCOL-BIÉMONT, M-C., NANICHE, D., RABOURDIN-COMBE, C., AND GERLIER, D., Glycosyl-phosphatidylinositol-anchored and transmembrane forms of CD46 display similar measles receptor binding properties: Virus binding, fusion and replication; down-regulation by hemagglutinin; and virus uptake and endocytosis for antigen presentation by major histocompatibility complex class II molecules, *J. Virol.* **68**:7891–7899 (1994).

160. VARSANYI, T. M., MOREIN, B., LÖVE, A., AND NORRBY, E., Protection against lethal measles infection in mice by immune-stimulating complexes containing the hemagglutinin of fusion proteins, *J. Virol.* **61**:3896–3901 (1987).

161. WAKEFIELD, A. J., PITTILO, R. M., COSBY, S. L., STEPHENSON, J. R., DHILLON, A. P., AND POUNDER, R. E., Evidence of persistent measles virus infection in Crohn's disease, *J. Med. Virol.* **39**:345–353 (1993).

162. WECHSLER, S. L., WEINER, H. L., AND FIELDS, B. N., Immune response in subacute sclerosing panencephalitis: Reduced antibody response to the matrix protein of measles virus, *J. Immunol.* **123**:884–889 (1979).

163. WHITTLE, H. C., ROWLAND, M. G. M., MANN, G. F., LAMB, W. H., AND LEWIS, R. A., Immunization of 4–6 month old Gambian infants with Edmonston–Zagreb measles vaccine, *Lancet* **2**:834–837 (1984).

164. WILD, T. F., BERNARD, A., SPEHNER, D., VILLEVAL, D., AND DRILLIEN, R., Vaccination of mice against canine distemper virus-induced encephalitis with vaccinia virus recombinants en-

coding measles or canine distemper virus antigens, *Vaccine* **11:**438–444 (1993).

165. WORLD HEALTH ORGANIZATION, Expanded program on immunization. Safety of high titer measles vaccine, *Wkly. Epidemiol. Rec.* **67:**357–361 (1992).

166. ZELL, E., Low vaccination levels in U.S. preschool and school-aged children: Retrospective assessment of vaccination coverage, 1991–1992, *J.A.M.A.* **271:**833–839 (1994).

12. Suggested Reading

BARRETT, T., SUBBARAO, S. M., BELSHAM, G. J., AND MAHY, B. W. J., The molecular biology of the morbilliviruses, in: *The Paramyxo-viruses* (D. W. KINGSBURY, ED.), pp. 83–102, Academic Press, New York, 1991.

ENDERS, J. F., Measles virus: Historical review, isolation and behavior in various systems, *Am. J. Dis. Child.* **103:**112–117 (1962).

HALSEY, N. A., Increased mortality after high titer measles vaccines: Too much of a good thing, *Pediatr. Infect. Dis. J.* **12:**462–465 (1993).

MARKOWITZ, L. E., AND ORENSTEIN, W. A., Measles vaccines, *Pediatr. Clin. North Am.* **37:**603–625 (1990).

NORRBY, E., Immunobiology of the paramyxoviruses, in: *The Paramyxoviruses* (D. W. KINGSBURY, ED.), pp. 83–102, Academic Press, New York, 1991.

PANUM, P. L., *Observation Made during the Epidemic of Measles on the Faroe Islands in the Year 1846*, American Publishing Association, New York, 1940.

CHAPTER 18

Mumps

Sandra J. Holmes

1. Introduction

Mumps is an acute contagious disease caused by a paramyxovirus. Infection with the mumps virus typically results in unilateral or bilateral parotitis. However, parotitis may be absent with only nonspecific or respiratory symptoms present, or the infection may be asymptomatic. Serious complications are rare.

2. Historical Background

Mumps was first described by Hippocrates in the 5th century BC as an illness accompanied by swelling around one or both ears and, in some cases, painful swelling of one or both testes. Central nervous system (CNS) involvement was first described in the literature by Hamilton in 1790.[100] There are several speculations as to the etymological origin of the term *mumps*, including the English noun, *mump* (meaning lump); the English verb, *mump* (to be sulky) (may define the facial grimace); or that it is named after the mumbling speech of patients with parotitis.

Although in modern times mumps is generally thought of as a disease of childhood, historically it was considered to be an illness that affected armies during times of mobilization. Following the description from ancient times, our understanding of mumps began with epidemics in the 18th and 19th century, which demonstrated that the disease was contagious and occurred worldwide, primarily among persons living together in crowded settings such as military quarters, ships, prisons, orphanages, and boarding schools. Mumps was the leading cause of days lost from active duty by American troops in France during World War I.[89,90] The annual hospital admission rate for mumps among soldiers in the American army at that time was 55.8 in 1000 average strength, exceeded only by influenza and gonorrhea.[230] The risk among recruits was 2.7 times higher in blacks than in whites. The highest reported rate of mumps in the Army (75.5 in 1000) occurred in 1918 and declined markedly to 6.9 in 1000 in 1943. Mumps outbreaks continued to occur in military settings in the postvaccine era. A high incidence among Soviet military recruits[203] and in U.S. forces in Korea were reported.[4] Recent serological surveys showed that an average of 12–16% of U.S. military recruits entering training during 1989 and 1990 were susceptible to mumps.[140,225,232]

In 1934, Johnson and Goodpasture[131,132] demonstrated that mumps was caused by a virus present in saliva of infected patients. They injected filtered, bacteria-free material from the saliva of patients with epidemic parotitis into the Stenson's ducts of rhesus monkeys and produced nonsuppurative parotitis. A diluted emulsion from the parotids of these monkeys caused parotitis in susceptible children but not in immune controls. Ten years following this landmark discovery, Habel[96] and Enders[63,64] cultivated the virus in developing chick embryo. Their techniques led to the development of an inactivated vaccine that was tested in humans a few years later.[98] This vaccine was the precursor of live attenuated mumps virus vaccines developed in the 1960s in the Soviet Union[226] and in the United States.[31]

This chapter is a revised version of Chapter 17, Harry A. Feldman, *Viral Infections in Humans*, 3rd edition, Plenum Press, 1989.

Sandra J. Holmes • Center for Pediatric Research, Eastern Virginia Medical School, Children's Hospital of The King's Daughters, Norfolk, Virginia 23510-1001.

3. The Agent

Mumps virus is a member of the genus *Paramyxovirus* in the family *Paramyxoviridae*, which also includes Newcastle disease virus and the parainfluenza viruses (Sendai and simian virus 5). The virus is inactivated by heat, 0.2% formalin, ultraviolet light and other agents.

The enveloped mumps virion is pleomorphic and contains a negative-sense RNA genome. The virus contains five major structural proteins that generally parallel those of the other paramyxoviruses. These include the nucleocapsid protein (NP), a putative RNA polymerase protein (P), a matrix protein (M), and two glycoproteins, a hemagglutinin-neuraminidase (HN) and a fusion (F) protein.[273] A sixth protein, a high-molecular-weight polypeptide (L) associated with the nucleocapsid, has been identified by several investigators.[115,171,208] The HN and F glycoproteins, exposed on the surface, are important in the production of an immune response. The HN protein is essential for adsorption of the virus to host cells, and the F protein mediates the fusion of lipid membranes and allows penetration of the viral nucleocapsid into the host cell. Hemagglutination, neuraminidase activity, and infectivity are inhibited by antibodies to HN but not by antibodies to F, which inhibit hemolysis.[197] Antigenic differences among the various strains of mumps virus have been demonstrated using monoclonal antibodies to the major structural proteins.[198,212,221] These antigenic differences, however, do not appear to prevent immune responses that protect against all strains.

Sequence analysis has been used to characterize the structural proteins and has demonstrated similarities with related paramyxoviruses.[59–62,150,255,256] The F and HN proteins of mumps show considerable homology in their nucleotide sequences with the corresponding proteins of Newcastle disease virus and simian virus 5. The serological cross-reactions observed among the paramyxoviruses may be explained by the similarities in the structural proteins involved in the immune response.[179,199,242] Geographical differences in strains also have been demonstrated by the sequence analysis of structural proteins.[275]

4. Methodology Involved in Epidemiologic Analysis

4.1. Sources of Data

The epidemiologic study of mumps is limited by the frequency of asymptomatic infection, the lack of serological testing in cases of parotitis with other etiologies misdiagnosed as mumps, and the failure to report cases. Large outbreaks, as well as specific studies, have provided information on incidence and complications. The experience during World Wars I and II have contributed greatly to our understanding of the epidemiology of mumps.[89,90,146,175] Incidence data are derived from the National Notifiable Diseases Surveillance System (NNDSS). Mumps was first made a nationally reportable disease in 1922, but was removed from the list of reportable diseases in 1950. Mumps was reinstated as a notifiable disease in 1968 following the licensure of a live attenuated mumps virus vaccine. The number of cases of mumps reported each week is published in the *Morbidity and Mortality Weekly Reports*, by the Centers for Disease Control and Prevention (CDC) and incidence data are summarized annually. Surveillance reports covering several years provide a discussion of the data and are published periodically.[39,42,247]

4.2. Serological Surveys

Over the years, the seroprevalence of mumps immunity has been assessed in a number of serological surveys using various methods available at the times the studies were conducted.[18,20,65,111,118,137,140,160,162,191,225,232]

4.3. Laboratory Methods

The diagnosis of mumps can be confirmed by isolation of virus or by serological testing. Serological tests are used to identify persons susceptible to mumps and to measure responses to vaccination.

4.3.1. Virus Isolation. Virus can be recovered from saliva, swabs obtained from the area proximal to the Stensen's duct, blood, urine, and cerebrospinal fluid (CSF) of patients with meningitis.[112,113,200,244,274] Mumps virus can be cultured *in vitro* in primary and continuous mammalian cell lines[110] and in embryonated hen eggs.[96,113,159] Primary rhesus monkey kidney cells are used most often.[189] Not all mumps virus isolates show the characteristic cytopathic effects (CPE), and primary monkey kidney cells may contain other viruses that produce CPE. Therefore, hemadsorption tests that detect viral hemagglutinin and are independent of cytopathology are used to screen for the virus in cell culture.[273] Several techniques including immunofluorescence (IF), complement fixation (CF), hemagglutination inhibition (HI), hemadsorption inhibition, and neutralization can be used to identify isolates as mumps virus.[236]

In most cases, diagnosis is made solely on a clinical basis; however, when laboratory confirmation is sought, serological techniques are used most commonly.

4.3.2. Serological Tests. Serological confirmation for clinical diagnoses of mumps can be obtained by testing paired serum samples using CF, neutralization, HI, hemolysis in gel, or enzyme-linked immunosorbent assays (ELISAs).[78,94,179] The diagnosis is confirmed by a fourfold or greater rise in IgG antibody titer between acute serum obtained shortly after onset of symptoms and convalescent serum obtained 2–4 weeks later. Several problems are inherent in tests comparing acute and convalescent sera including the long period of time in obtaining diagnostic confirmation and the inability to observe a rise in titer when there is a delay in obtaining the acute serum sample. Furthermore, cross-reactions with other paramyxoviruses may occur.[156] Neutralization assays are arduous, and CF and HI are relatively insensitive. False-positive and false-negative results may occur with the hemolysis-in-gel test.[47] The newer ELISA is preferable to the assays developed earlier in that the ELISA is highly sensitive, specific, and rapid, as well as being cost-effective.[165] However, cross-reactions also occur with the ELISA, and a rise in titer may be more difficult to detect because of the earlier appearance of antibody compared to that observed with CF.[179] Assays measuring IgM are advantageous because cross-reactions between mumps and parainfluenza viruses have not been observed for IgM antibody.[95,179,242] Also, these assays are performed on acute sera only, and thus results are obtained rapidly and are not subject to the problems inherent in the comparison of acute and convalescent titers. Mumps-specific IgM antibody is present within the first 5 days of illness, peaks at about 1 week, and persists for at least 6 weeks.[15] The IgM-capture ELISA is a highly sensitive and specific method of detecting mumps infection.[95,215]

In patients with mumps meningitis, antibody in CSF is present within a few days of onset of illness and peaks within 2 weeks. The ELISA or imprint immunofixation assays are used to detect CSF antibody.[242,249] Elevated CSF–serum IgG antibody ratios from samples collected on the same day indicate mumps infection because mumps antibodies are uncommon in CSF in the presence of other infections.[55,81,187,241]

A solid-phase radioimmunoassay to detect mumps-specific IgA in saliva has been used as a rapid diagnostic test.[79] A recent study in which oral fluid samples were used to assess immunity to measles, mumps, and rubella showed a high level of concordance between mumps-specific IgG antibodies in oral fluid samples measured by a modified ELISA and serum antibodies measured by the conventional ELISA.[237]

A commercially available ELISA and a solid-phase IF assay are the most commonly used tests to measure mumps IgG in single serum samples to determine immune status.[236] The serum level of neutralizing antibody is considered a reliable index of immunity, but the traditional methods used to measure neutralizing antibody are labor intensive and time consuming, and thus are unsuitable for screening large numbers of samples. Recently, an enzyme immunoassay to measure mumps-virus-neutralizing antibodies using peroxidase-labeled virus-specific monoclonal antibodies (N_{50}-EIA) was described.[248] The N_{50}-EIA and the ELISA have been shown to be comparable in assessing seroconversion after vaccination.[101] Although the ELISA was more sensitive, it was less specific than the N_{50}-EIA because of cross-reactivity of the mumps virus with parainfluenza viruses.

Skin tests are unreliable and should not be used to assess immunity to mumps.[23,29]

5. Descriptive Epidemiology

5.1. Incidence and Prevalence

In 1993 and 1994, the number of cases of mumps reported to the CDC reached an all-time low (1692 and 1537, respectively). This represents a 99% decrease in the number of cases reported in the United States since 1968 when mumps was reinstituted as a reportable disease following licensure of mumps vaccine in 1967 and 152,209 cases were reported (Fig. 1). Mumps previously had been a reportable disease between 1922 and 1950. Following a period of a record low incidence in the early 1980s, a resurgence of mumps occurred in the United States between 1986 and 1987, which resulted in an almost fivefold increase in the annual incidence rate per 100,000 population (1985, 1.1; 1987, 5.2).[42] This resurgence was attributed primarily to the low vaccination coverage of adolescents and young adults who were born between 1967 and 1977.[48,227] Although vaccine was licensed in 1967, its use was limited until 1977 when it was recommended for routine use in young children.

The number of cases of mumps reported each year is generally considered to be far less than what actually occurs. Reporting efficiency was estimated to be 10% in a recent cost–benefit analysis (unpublished data). Underreporting results in part from the mild nature of the disease and the associated lack of need to seek medical care. In addition, health care providers do not necessarily report cases, and mumps is still not a reportable disease in two states. Subclinical cases are common. In a serological survey conducted in Florida,[160] 25% of household contacts of cases had had inapparent infections. This study also demonstrated an infrequent number of physician con-

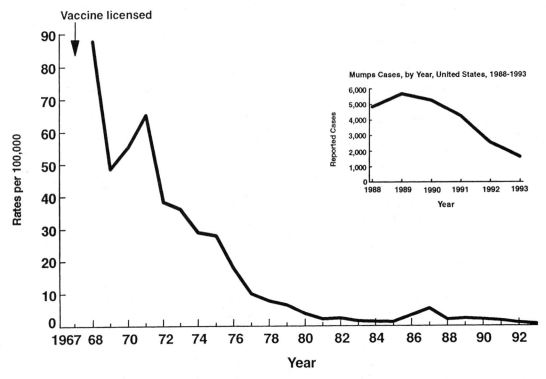

Figure 1. Reported cases per 100,000 population, by year, United States, 1968–1993. (From Centers for Disease Control and Prevention.)

tacts and the failure of physicians to report cases; only 27% of the 126 cases identified were seen by a physician and only 6% were reported. The estimated annual incidence rate based on this study was ten times the rate based on national surveillance figures. On the other hand, cases reported as mumps based on the presence of parotitis without laboratory confirmation may not be mumps, since there are other causes of parotitis. In a Canadian study, mumps could not be confirmed serologically in one third of sporadic cases of parotitis seen by family practitioners.[69] Clinical diagnoses of mumps in cases of parotitis that are epidemiologically linked, however, are considered reliable since other viral infections do not result in outbreaks of parotitis.

5.2. Survey Data

Numerous population surveys have been carried out using questionnaires and interviews, skin tests, and serological tests to assess patterns of susceptibility and immunity in communities.

Harris and co-workers[102] used questionnaires to study the incidence of mumps among health professionals and university faculty members. Cases of mumps were most likely among pediatricians and least likely among the general university faculty. Harris et al.[103] also used questionnaires in household surveys and found that the peak incidence of mumps occurred at 7 years of age, and 74% of the study population reported mumps before 10 years of age.

An early study by Henle et al.[111] in 1951 showed that 70–80% of persons with a history of mumps had positive findings on the CF test and skin tests. Although these tests are now known to be insensitive or unreliable, less than 2% of those who tested positive subsequently acquired mumps.

Early serological surveys were conducted in military populations. Serosurveys in the prevaccine era from 1951 to 1962 using the CF or HI tests showed that 46–76% of army recruits in the United States were immune to mumps.[18,162,191] Recent serosurveys[140,225,232] conducted more than two decades after the licensure of mumps vaccine and using the more sensitive ELISA test found that 84–88% of military recruits were immune. In contrast to World War I data that showed increased susceptibility among men from rural areas,[25] no significant urban–rural differences were found in the later serosurveys conducted in 1962 and in 1989.[18,140] Kelley and co-workers[140] found that black, non-Hispanic recruits were more likely to be seropositive than those from other racial or ethnic groups.

Serosurveys also have been conducted in civilian populations of all ages. Retrospective tests on sera in the World Health Organization (WHO) Serum Bank at Yale University showed wide variation in the proportion of seropositive findings according to geographic area.[20] Among children 5–9 years of age, mumps antibody was not found in any of the children from Alaska and in only 15% of those from Tahiti. In contrast, antibody was detected in more than 50% of 5- to 9-year-olds from New Haven, Connecticut (52%), the Cape Verde Islands (55%), the Bahamas (74%), and Iceland (79%). Among persons of all ages residing in remote islands off Alaska, only 7% of residents of St. Paul Island[170] and 12% of those from St. George Island[206] were seropositive. Similarly, mumps antibody was found in an average of only 33% of members of isolated Indian tribes in South America.[19] In comparison, mumps antibody was found in 90% of the residents of New Haven and 88% of medical students tested in another seroprevalence study using the neutralization test.[23]

Kenny *et al.*[141] measured mumps neutralizing antibody in sera collected between 1970 and 1973 in children 1 to 15 years old in the Dominican Republic, Honduras, the Republic of Panama, and the United States. The proportion of seropositive 1- to 3-year-old children in the Middle American countries and in the United States was similar (20–25%). In the older age groups, significantly higher rates of seropositivity were found among U.S. children compared to children in the Middle American countries. Among children 4 to 6 years old, 70% of children in the United States had mumps antibody compared to approximately 55% of children in the Middle American countries ($p < 0.05$).

Variation in the average age of infection has been found in other serosurveys from other parts of the world. In St. Lucia, 70% of children were found to be seropositive by 4 years of age[51] compared to a majority of children from the Netherlands[253] and from Scotland[190] who remain susceptible at this age. In the Netherlands, most children develop mumps between 4 and 6 years of age, and by 14 years of age, 90% are seropositive. Before vaccination campaigns were initiated in Spain, 53% of 3- to 5-year-old and 61% of 6- to 7-year-old children were immune.[6] The lack of immunity was associated with rural residency, low socioeconomic status, and lack of school attendance or siblings. In the United Kingdom before routine mumps vaccination was initiated in 1988, most cases were in children 6 to 7 years of age.[3] Subse-

quent to the routine vaccination program, the average age of infection has decreased.[192] Serological data from England and Wales from 1990 showed that 70% of 1- to 2-year-olds and 73% of 3- to 4-year-olds were immune.[134]

5.3. Epidemic Behavior and Contagiousness

The term "epidemic parotitis" reflects the fact that the transmission of mumps virus causes outbreaks of mumps infection characterized by parotitis. However, the infection is always generalized rather than localized to the parotid gland and outbreaks may include many cases without parotitis. Outbreaks commonly occur wherever children and young adults aggregate, as in schools, military barracks, and other institutions.

Mumps is less contagious than measles or chickenpox. Hope Simpson[123] demonstrated the infectiousness of mumps by assessing the incidence of cases resulting from every exposure of a susceptible person in the home and concluded that the susceptible-exposure attack rate for mumps was 31.1% compared to 75.6% for measles and 61.0% for chickenpox.

Outbreaks in virgin populations have been used to estimate the contagiousness of mumps. In three isolated island groups off Alaska no prior outbreak of mumps was known, exposure had not occurred for more than 50 years, and approximately 90% of persons tested were seronegative.[170,202,206] Clinical mumps occurred in 35–65% of this population and an additional 20–24% had subclinical infections.

5.4. Geographic Distribution

Mumps occurs worldwide with the exception of isolated island groups and remote, sparsely populated areas. In the United States, mumps outbreaks occur whenever a sufficient number of susceptible persons accumulate to support an outbreak. It follows that the incidence is higher in more densely populated areas. During the military outbreak of 1917–1918, 85% of cases were among soldiers from rural areas,[25,265] although later serosurveys among military recruits found little or no differences between men from urban and rural areas.[18,140]

5.5. Temporal Distribution

In the North Temperate Zone, mumps occurs more frequently in winter and spring than at other times of year. This pattern has not changed since the wide use of vaccine and the associated dramatic reduction in incidence. Seasonal difference are not apparent in tropical areas.

5.6. Age Distribution

In the past, the largest proportion of cases of mumps occurred in school-aged children, 5 to 9 years of age. A shift to older age groups was first observed in 1982.[39] A more marked shift occurred during a resurgence of mumps in 1986–1987 when a disproportionately high number of cases were reported among older children and adolescents aged 10 to 19 years and among young adults.[42] Outbreaks in colleges and universities were reported by three states in 1986–1987.[41] Further evidence of the shift in incidence to older age groups was seen in an unprecedented outbreak in a workplace in 1987 involving persons employed at three Chicago futures exchanges and their household contacts.[138] Although patients ranged in age from 17 to 70 years, 77% of cases occurred in those less than 30 years. The age distribution of cases reported in 1988–1993 (Fig. 2) is similar to the pattern observed in 1986–1987.[247] Since 1987, persons 15 years of age and older have accounted for 36–38% of reported cases with known age compared to an estimated 8% in this age group during the 5 years following licensure (based on data from California, Massachusetts, and New York City).

5.7. Gender Distribution

Males and females are affected with equal frequency. Because mumps in postpubertal males is frequently complicated by orchitis, cases among males are more likely to

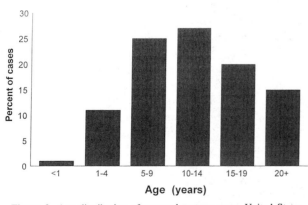

Figure 2. Age distribution of reported mumps cases, United States, 1988–1993. Extrapolated from age distribution of cases with known age. (From Centers for Disease Control and Prevention.)

be seen by physicians and to be reported. The frequency of encephalitis in males has been reported to be three- to fourfold higher than in females; however, the risk of death is equal for male and female patients.[17]

5.8. Race and Occupation

Incidence rates among blacks tend to be higher than among whites. This was first shown in military populations and has been demonstrated subsequently in studies of civilian outbreaks.[267] Based on data from 28 states in which race and ethnicity were reported in at least one half of the cases, incidence rates among black persons ranged from 1.2 to 8.2 times those of other racial groups during the 4-year-period of 1990–1993.[247] Recent seroprevalence studies in military recruits also have shown higher rates of seropositivity among blacks than among other racial groups.[140,232]

Persons engaged in occupations that involve contact with persons in age groups or settings associated with the highest incidence of mumps are likely to have frequent exposures to the mumps virus. A large mail survey conducted in the late 1960s showed that the risk for acquiring mumps from an occupational exposure was highest among practicing pediatricians and lowest among university staff.[102] Among teachers, the risk of mumps was related inversely to the age of their students.

5.9. Occurrence in Different Settings

Any setting in which a pool of susceptible persons are in close contact facilitates the transmission of the mumps virus.

5.9.1. Families and Schools. Families and schools have had an important role in the transmission of mumps.[181] Children exposed to mumps at school introduce the virus into the household where it may spread to susceptible family members. Prior to routine vaccination of young children, the younger children in a family usually acquired mumps at an earlier age than their oldest sibling. Most outbreaks have occurred in school settings, and only until recently the peak incidence has occurred in children during the first 5 years of school attendance.

Outbreaks have occurred in highly vaccinated populations. The majority of cases (77.3%) in an outbreak in Kansas in 1988–1989 occurred among primary and secondary school students of whom nearly 98% had documentation of vaccination.[116] The attack rate was higher among persons who had received only one dose of vaccine compared to those who had received two doses (relative risk, 5.2; 95% CI = 1.0, 206.2). Of the school-age patients who were serologically tested, 47% lacked detectable antibodies to measles, rubella, or both, despite having received measles, mumps, rubella (MMR) vaccine. This proportion is higher than what would be expected from waning immunity and suggests that these vaccinees had failed to seroconvert following vaccination. Thus, primary vaccination failure rather than waning immunity was considered the key factor in this outbreak. Primary vaccination failure also accounted for most cases in a high school in Tennessee in which 98% of students had been vaccinated.[24] However, among 13 vaccinated students for whom paired serum samples were tested, three failed to show IgM antibody responses, suggesting waning immunity.

5.9.2. Military. Historically, epidemiologic data were derived largely from studies of military populations (see Section 2). The seasonality of mumps was recognized by the high proportion (70%) of cases occurring during winter and spring. The incidence was higher among blacks than among whites. Mumps was associated with high rates of hospitalization and days lost from duty. Mortality was low, although deaths resulting from secondary infections such as pneumonia were more common during World War I than in later times when effective treatments for these complications became available.

Seroprevalence studies carried out among military recruits in the prevaccine era found larger proportions of susceptible individuals[18,162,191] than were found in recent serosurveys[140,225,232] (see Section 5.2).

Mumps outbreaks continue to occur in military settings in the postvaccine era. High incidence rates were reported among Russian military recruits[203] and among U.S. forces in Korea.[4]

5.9.3. Hospitals. Mumps is a relatively uncommon cause of nosocomial disease. Sporadic cases resulting from transmission in hospitals have been documented, including transmission from asymptomatic staff.[29,70,87,228] Data from a study in an area with widespread outbreaks of mumps suggested that the introduction of mumps into hospitals by employees or by patients is likely when local incidence is high.[268]

5.10. Other Factors

Socioeconomic status has not been an important risk factor for mumps, except among impoverished populations where overcrowded living conditions facilitate the transmission of mumps virus as well as other infective agents.

6. Mechanisms and Routes of Transmission

Humans are the only known reservoir for mumps virus. The virus can be transmitted following inoculation in nasal or buccal mucosa, suggesting that infection is spread through large droplets containing mumps virus infecting the upper respiratory tract.[112] The spread of mumps among persons in close contact also suggests this mode of transmission. Mumps virus may be present in saliva of infected individuals for up to 7 days before onset of clinical illness[66] and 8–9 days following onset; however, it is usually present from 2 to 3 days before and 4 to 5 days after onset of symptoms.[45] The virus also is present in saliva of persons with inapparent infections.[112] Data from animal studies and from case reports of humans provide evidence that mumps virus can be transferred transplacentally.[84,135,143,153,186]

7. Pathogenesis and Immunity

Mumps is acquired through the respiratory tract. During the incubation period of 16 to 18 days (range, 12–25 days) following exposure, the mumps virus multiplies in the upper respiratory mucosa and spreads to the regional lymph nodes. Virus is infrequently isolated from the blood because viremia develops toward the end of the incubation period and ends with the development of humoral antibody in 3 to 5 days.[200] Viral excretion in saliva clears with the appearance of mumps-specific secretory IgA antibody about 5 days after onset.[45,66,112] In addition to saliva, virus has been isolated from swabs from the Stensen's duct, blood, urine, stool, milk, and CSF. Circulating lymphocytes are probably infected, since they have been shown to be infected in vitro,[56] especially activated T lymphocytes.[72] Viremia leads to parotitis or other salivary gland involvement in approximately two thirds of infections. Parotitis is usually one of the first clinical symptoms; complications may follow by a week or more. The continued dissemination of the virus to distant organs in the presence of increasing humoral antibody may be attributed to infected circulating lymphocytes.[273]

Viral replication in the parotid gland produces periductal interstitial edema and local inflammation, primarily involving lymphocytes and macrophages.[262] Serum and urine amylase levels may be elevated as a result of inflammation and tissue damage in the parotid gland.[219]

Involvement of the CNS is common and may occur whether or not parotitis is present.[142] More than one half of mumps infections involve the CNS as measured by pleocytosis of the CSF.[10,27] Virus can be recovered from CSF early in the course of meningitis.[7,142] Experimental infection in hamsters suggests that mumps virus enters the CSF through the choroid plexus.[272] Virus is distributed through the ventricular pathways and the subarachnoid space by CSF. Also, infected choroidal and ependymal epithelia are desquamated into the CSF.[271] Involvement of the CNS is usually limited to meningeal infection, but encephalitis may occur, probably as a result of viral penetration of brain parenchyma by spread from contiguous ependymal cells that line the ventricular cavities of the brain.[273] Viral entry into neurons allows the virus to spread widely along neuronal pathways. The neuropathic potential of the virus is demonstrated by sequelae such as sensorineural deafness.[67,145] and obstructive hydrocephalus and aqueductal stenosis.[22,229,239] In addition to evidence for viral persistence from laboratory studies of cell cultures[109,172,194,254] and from animal models,[105,272] cellular responses and oligoclonal humoral responses have been shown to persist within CNS of patients, implying continued antigenic stimulation and suggesting the persistence of virus.[80,167,250,273] The association of such persistence with late occurring progressive CNS disease has been suggested.[246]

Renal involvement occurs frequently and is mild. One group of investigators detected viruria in 80% of specimens collected within the first five days,[244] and in another study of young servicemen, viruria persisted in some patients for as long as 25 days following onset.[243] All of these men had some abnormal renal function tests; however, none had generalized edema or hypertension and all had negative cultures and normal renal function at the end of the study.

Involvement of the pancreas is characteristic of experimental mumps, and human pancreatic beta cell cultures have been infected.[205] Pancreatic involvement in patients with mumps is generally limited to epigastric pain; however, extensive damage can occur, resulting in hemorrhagic pancreatitis.[71]

Humoral immunity is most likely the primary mode of protection; however, cellular immunity has been demonstrated. *In vitro* lymphocyte proliferative responses to mumps antigen occur in seropositive individuals[128] and are largely dependent on T lymphocytes with IgG receptors.[174]

Reinfection with mumps virus in persons with naturally acquired immunity is difficult to substantiate by patient history because parotitis can be caused by other agents, and mumps is not usually serologically confirmed. The frequency of asymptomatic infections also is a complicating factor. Outbreaks of mumps in highly vaccinated populations, however, suggest that waning immunity may occur.[24]

Antibodies to mumps are transmitted across the placenta and protect infants of immune mothers for about the first 6 months of life.[73]

8. Patterns of Host Response

8.1. Common Clinical Features

Mumps virus infections may be asymptomatic or associated only with nonspecific or respiratory symptoms in more than one half of persons infected.[50,69] Mumps virus has been isolated from young children with lower respiratory disease.[75] Inapparent infection may be more common in adults than in children, and parotitis may be more common in children. Among persons with clinically apparent disease, host responses vary, depending on which organs may be affected. Swelling and tenderness of the salivary glands are common, primarily the parotid glands; however, sublingual and submandibular glands also may be involved. Parotitis may be unilateral or, more commonly, bilateral, peaking on about the third day of illness and resolving within 10 days. A prodromal period may precede the parotitis by several days with nonspecific symptoms, including fever, malaise, myalgia, headache, and anorexia. In some patients other glandular tissues may be infected, causing epididymo-orchitis,[11] oophoritis and mastitis,[188] pancreatitis,[71] or thyroiditis.[68] Pleocytosis of the CSF is common,[10,27] and aseptic meningitis often is clinically apparent; encephalitis is uncommon. Transient renal abnormalities also are common.[243] Infrequent manifestations include arthritis,[91] myocarditis,[28] and thrombocytopenia.[92] Any of these manifestations may occur before or in the absence of parotitis.

Case reports of complications are important sources of information but do not allow the calculation of rates. Large studies in military populations during World Wars I and II consisted mostly of clinical findings not confirmed by serological testing or virus isolation. An investigation of mumps in a virgin population of Eskimos living on St. Lawrence Island in the Bering Sea in the 1950s provided rates of clinical findings in serologically confirmed cases (Table 1).[202] Mumps infection was confirmed in 82% of the 561 residents. Orchitis and mastitis were age related with sharp increases in frequency at puberty. Orchitis was bilateral in 37% of cases. There were a few women who had lower abdominal pain suggesting oophoritis. In a few cases there were symptoms of thyroiditis. Delirium, vomiting, and high fever were associated with stiff neck in

Table 1. Incidence of Infection and Clinical Disease in a Virgin Soil Outbreak of Mumps in 561 Residents of St. Lawrence Island, Bering Sea[a]

Feature	Number	Percent
1. Mumps infections[b]	460[c]	82
a. Males	300[c]	53
Clinical mumps	205	68
Asymptomatic	95	32
b. Females	261[c]	47
Clinical mumps	158	61
Asymptomatic	103	40
c. Both sexes	561	100
Clinical mumps	363[c]	65
Asymptomatic	198[c]	35
Total infection rate[d]	—	85
2. Clinical mumps	363[c]	65
a. Salivary gland swelling	344	95
b. Stiffness or neck	40	11
c. Scrotal swelling (males)	52	25
d. Swelling of breasts (females)	24	15

[a]Data derived from Philip et al.[202]
[b]Based on serological data.
[c]Of total population of 561.
[d]An additional 3% had disease without antibodies.

some of the patients, but all recovered. One of four deaths reported during the outbreak occurred 2 days following the onset of parotitis in an infant; however, the cause of death was not determined. There were three spontaneous abortions among eight women with clinical mumps in the first trimester of pregnancy and one abortion among 12 women with inapparent infections. No stillbirths or miscarriages occurred among women who acquired infection after the first trimester, and there were no congenital malformations in infants born to mothers with mumps during pregnancy.

8.2. Involvement of the Central Nervous System

Mumps has been implicated as an important cause of aseptic meningitis and encephalitis.[1,13,88,148,149,157,158,176] The high frequency of involvement of the CNS in mumps infection demonstrated by numerous studies conducted over the past 50 years has led to CNS involvement frequently being considered part of the natural history of the disease.[88]

Many studies of CNS involvement have categorized aseptic meningitis and encephalitis together under the term "meningoencephalitis." Thus, it is not surprising that a wide range (<1% to >70%) of rates of CNS manifestations in mumps infections have been re-

ported.[10,16,17,27,30,76,122,177,180,217] Studies that combine cases of aseptic meningitis and encephalitis as meningoencephalitis render it difficult to differentiate between these two CNS complications that have markedly different clinical courses and prognoses. Variability in incidence rates also reflect the case definition used, the skill of the observers, the age distribution of the population, whether cases are hospitalized or are diagnosed as outpatients, and the frequency of the use of lumbar puncture.

CSF pleocytosis most likely occurs in 40–65% of patients with mumps, but only 10–30% have symptoms of meningeal irritation.[10,21,30] The CSF profile consists of normal opening pressure, a predominance of lymphocytes with cell counts commonly between 200 and 600/mm^3 and not infrequently 1000/mm^3, elevated protein in up to 70% of patients, and moderately low glucose concentrations in up to 30% of patients.[7,30,133,147,151,161,270] Aseptic meningitis associated with mumps is a benign disease that resolves spontaneously and is not associated with long-term morbidity. There is no correlation between the level of CSF pleocytosis and the severity of illness or outcome following encephalitis.[147,161,209]

Encephalitis complicating mumps is much less frequent than aseptic meningitis, probably occurring in 0.1% of cases.[147,211] Mumps accounted for a decreasing number of cases of viral encephalitis reported to the CDC between 1967 and 1972 (36% to 13%).[39] Mumps had been considered the most common cause of viral encephalitis until the early 1980s when the incidence of mumps was greatly reduced and other viruses such as herpes simplex, enteroviruses, arboviruses, and varicella-zoster virus became the leading causes of encephalitis. Although mumps encephalitis may be a severe disease, most patients completely recover.[7,30,106,177] Postencephalitis ataxia, behavioral changes, and electroencephalographic abnormalities have been shown to persist for a few weeks.[7] Rarely, permanent neurological sequelae result from mumps encephalitis, including behavioral disorders, seizure disorders, cranial nerve palsies, muscle weakness, ataxia, chronic headaches, aqueductal stenosis, and hydrocephalus.[12,22,133,142,147,151,154,209,238] Myelitis or polyneuritis also may follow mumps encephalitis.[86,163,184,222] Overall, the mortality rate associated with mumps encephalitis is less than 2% and is higher in adults than in children. Neuropathologic findings at autopsy are variable, and many reports come from unconfirmed cases. In most carefully examined cases, there is evidence of both cellular destruction, suggesting a direct effect of the virus, and demyelinization, suggesting an autoimmune process.[88]

Among children with mumps, CNS disease is three-to fourfold more common in males than in females.[7,21,30,133,147,151,161,270]

Sensorineural deafness is an important sequelae of mumps and may occur in the absence of meningitis or encephalitis.[204,252] Deafness usually has a sudden onset, is unilateral in about 80% of cases, and most often is permanent.[67,99,145,196,252] The incidence of deafness associated with mumps has been estimated to be 0.05 per 1000 cases,[67] which may be an underestimate since inapparent mumps can cause deafness.[74] The virus may cause direct damage to the cochlea and to cochlear neurons.[54,166,224] Mumps virus has been isolated from the perilymph of a patient with mumps and sudden unilateral deafness.[266]

8.3. Involvement of the Heart

Myocarditis may result from various viral infections, including mumps.[5,14,57,155,210,240] Among persons hospitalized for mumps, 3% of children and 7% of adults show transient electrocardiographic abnormalities involving the ST segment or atrioventricular conduction defects.[5,14] Myocarditis is rarely symptomatic and is usually self-limited; however, it may be catastrophic.[5,14] Some investigators have suggested that intrauterine infection may result in endocardial fibroelastosis.[58,85,193,213,214,251]

8.4. Orchitis and Sterility

Mumps may be complicated by orchitis in up to 38% of postpubertal males.[11,16,69,175,202,203,265] In most cases, testicular swelling and pain is unilateral, but may be bilateral in up to one third of cases. Follow-up of men who had orchitis during the large outbreaks of mumps that occurred during World Wars I and II did not show an association between orchitis and impotence or sterility. Although testicular atrophy occurred in one third of cases, it did not affect the quality or quantity of sperm produced.[263,264] An increased risk of testicular cancer following mumps orchitis has been reported,[11,139,235] but the significance of this observation is unknown.

8.5. Mumps and Diabetes

An association between the pancreatic involvement that may be associated with mumps infection and diabetes mellitus has been suggested, but a causal relationship remains to be proved.[127,234] Mumps virus can infect and destroy human and rhesus monkey beta cells *in vitro*,[205] and mumps has been shown to cause pancreatitis in 4% of cases.[69] Pancreatic damage has not been documented in case reports of diabetes in children following mumps

infection.[52,77,121,136,144,173,178,195] It has been speculated that mumps and diabetes have a related periodicity based on the occurrence of outbreaks of diabetes months or years following outbreaks of mumps.[234] These studies were uncontrolled and relied on the clinical diagnosis of mumps without supporting laboratory tests. In one study, mumps accounted for 8–22% of the 30% of cases of insulin-dependent diabetes mellitus in which a viral infection may have been the precipitating event.[130] Another study[83] found a seasonal relationship of diabetes with coxsackievirus but not with other viruses. The association of mumps and coxsackievirus with diabetes was not confirmed in a subsequent study.[117] Less than 1% of cases of juvenile onset diabetes followed recent infections with mumps virus in a sample of 1663 diabetic patients in England.[82] A study in pregnant women found no association between the presence of mumps, coxsackie B or respiratory syncytial virus antibodies and diabetes.[168] Other investigators also have failed to find a relationship between antibodies to mumps or other viruses and diabetes mellitus.[9,216,218]

8.6. Other Complications

A variety of other acute and delayed complications have been reported following mumps infection. Arthropathy is a rare complication that occurs most frequently in young men. Gordon and Lauter[91] reviewed the literature on mumps arthritis from the first description in 1850 through the early 1980s. They noted that a total of 32 well-documented cases of mumps arthritis had been reported in the literature since 1924 when 6 cases (0.44%) were reported during an outbreak of 1334 cases of mumps in Paris. A retrospective survey of 2482 patients with mumps in England and Wales failed to identify any cases with arthritis. Arthropathy following mumps may be manifest as arthralgia without signs of inflammation; as polyarticular arthritis, which is often migratory; or as monoarticular arthritis of the knee, hip, or ankle. The arthropathy may last from 2 days to 6 months and is self-limited without sequelae. Mumps virus can replicate in human joint tissue *in vitro*[125,126]; however, the virus has not been isolated from affected joints of patients.

Viruria is a common finding in patients with mumps.[245] In one study all 20 patients evaluated with laboratory-confirmed mumps had abnormal renal function at some time during the acute infection, but all had normal renal function and negative urine cultures within 24 days.[243] Clinically apparent nephritis is uncommon and generally benign[2,164,169,185]; however, fatal cases have occurred.[2,124] Nephritis may result from direct infection of the kidney or immune complex glomerulonephritis.[124]

The most common ocular manifestations of mumps include involvement of the lacrimal gland, which affected 20% of soldiers in a 1903 epidemic, and optic neuritis, which may be associated with involvement of other sites in the CNS.[207] Keratitis, iritis, conjunctivitis, scleritis, tenonitis, and central retinal vein occlusion have been reported rarely.

Although the persistence of mumps virus has been suggested as a factor in neurological and muscular disorders of unknown cause, there is no compelling evidence for a causal association.[46,182,220]

9. Control and Prevention

Although the mumps virus was identified in 1934,[131] little enthusiasm was generated for the development of a vaccine since mumps was considered primarily a mild inconsequential disease of childhood. A decade later, the work of Habel[96] and Enders[64] led to the development of inactivated and live mumps virus vaccines. These investigators cultivated mumps virus in chick embryos and demonstrated that virus that had undergone continuous passage in chick embryo produced inapparent infections resulting in immune responses in monkeys.[64,96,97] An experimental inactivated vaccine was tested in young men at the beginning of an outbreak in 1946, and resulted in fewer cases among vaccinated men compared to unvaccinated controls.[98] During this same period of time, Henle and associates[107,114] also were developing mumps vaccines. They reported using multiple doses of an inactivated vaccine that produced neutralizing antibodies in children.[108] A killed mumps virus vaccine that produced transient immunity was licensed in the United States from 1950 to 1978. In Finland, the incidence of mumps among men in the military was reduced by 94% following the routine use of two doses of an inactivated mumps vaccine.[201] Following the use of this vaccine, cases of mumps occurred most frequently among new recruits in the first month after induction before immunity was established. The observation that vaccines using killed antigens did not produce long-lasting protection led to the further development of live attenuated mumps virus vaccines in the 1960s in the Soviet Union[226] and in the United States.[31] At least ten different mumps vaccine strains have been used in various countries around the world.[49] A live attenuated mumps virus vaccine, using the Jeryl Lynn strain of mumps virus named after the child from whom the strain was first isolated, was licensed in the United States in 1967. It was developed by Buynak and Hilleman.[31] The virus is attenuated by passage in embryonated hens' eggs and in chick embryo cell culture.

Extensive prelicensure field trials were carried out by Stokes and Weibel[231] in Philadelphia and by investigators in other areas.[32,33,53,119,120,233,258–260,278] Overall the seroconversion rate was 97% among children and 93% among adults. Efficacy was approximately 95% over a period of 20 months. Follow-up studies showed that the vaccine-induced antibody persisted for at least 9.5 years.[257] The vaccine was safe and effective when given alone or when coadministered with live measles vaccine. Subsequently, the trivalent MMR vaccine containing the Jeryl Lynn strain of mumps virus and attenuated measles and rubella viruses was tested and found to be safe and effective in infants and young children.[8,152,223,261]

In 1967, when the live attenuated mumps virus vaccine was licensed, mumps control was not a high priority in the public health community. The Advisory Committee on Immunization Practices (ACIP) recommended that the newly licensed mumps vaccine be considered for use in children approaching puberty, for adolescents, and for adults.[35] A series of progressively stronger ACIP statements on mumps control followed,[36–38,43] with routine childhood immunization recommended in 1977.

The most recent recommendations for the prevention of mumps issued in 1989[43] state that all susceptible children 12 months of age and older, adolescents, and adults should be vaccinated unless vaccination is contraindicated. Persons should be considered susceptible to mumps unless they have documentation of (1) physician-diagnosed mumps, (2) adequate immunization with live mumps vaccine on or after their first birthday, or (3) laboratory evidence of immunity. There is no increased risk associated with vaccinating immune persons; thus, those who are unsure of their histories of mumps disease or vaccination should be vaccinated. Adverse effects are infrequent and minor, with parotitis and low-grade fever being the most common. Mild allergic reactions of short duration such as rash, pruritis, and purpura occur infrequently. Vaccine containing the Urabe strain of mumps virus has been associated with postvaccination aseptic meningitis,[26,183,276] and its use has been discontinued in several countries. Vaccination with the Jeryl Lynn vaccine, the only mumps vaccine that has ever been licensed for use in the United States, has not been similarly associated with aseptic meningitis. The evidence for other nervous system disorders such as encephalitis, encephalopathy, and sensorineural deafness was reviewed by the Institute of Medicine[129] and found to be inadequate to accept or reject a causal relationship with mumps vaccine.

Mumps vaccine is contraindicated for pregnant women because of a theoretical risk to the fetus. Mumps vaccine virus has been shown to infect the placenta, but there is no evidence that it causes congenital malformations in humans.[277] Inadvertent vaccination during pregnancy should not be considered an indication for termination of the pregnancy. Because live mumps vaccine is produced in chick embryo cell culture, persons with a history of anaphylactic reactions to eggs should be vaccinated with caution according to established protocols.[93] The vaccine also contains trace amounts of neomycin, and persons with anaphylactic reactions to neomycin should not receive the vaccine. Mumps vaccine should not be administered to immunodeficient persons, but their risk of exposure to mumps virus may be reduced by vaccinating their close susceptible contacts. However, children who are infected with human immunodeficiency virus (HIV) and are asymptomatic should be vaccinated. Also, MMR vaccine should be used when vaccination against measles is indicated for symptomatic HIV-infected children.

A cost–benefit study conducted in 1993–1994 by the CDC and Battelle, Centers for Public Health Research and Evaluation (unpublished data) confirmed the cost effectiveness of MMR vaccine demonstrated by earlier studies.[269]

School immunization laws beginning during the late 1970s have had a dramatic effect on outbreaks in schools and on the overall incidence of mumps. Since 1977 when the ACIP recommended that children routinely be vaccinated against mumps[37] and at which time only five states had immunization laws, the number of states with immunization laws of some kind have increased steadily to 43 states including the District of Columbia in 1993. In 1985, states without laws had twice the incidence rates as states with comprehensive laws requiring proof of immunity against mumps for all students in kindergarten through grade 12.[48] During the resurgence of mumps in 1986, Tennessee, which at the time did not have school immunization laws, reported the highest number of cases (1437) in the United States. Between August 1986 and January 1987, 1060 cases were reported by schools in one county; most cases were among students in high schools and middle schools.[267] Mumps vaccination was not included in immunizations given in public health clinics in Tennessee before 1974. An outbreak in New Jersey in 1983, in which students in the sixth grade were nearly seven times more likely to develop mumps than those in lower grades, reflected the partial school law that covered students from kindergarten through grade 5.[40,44] The number of states with comprehensive school immunization laws doubled between 1988 and 1993, and during the same period incidence declined by 65%.[247] The impact of school laws on

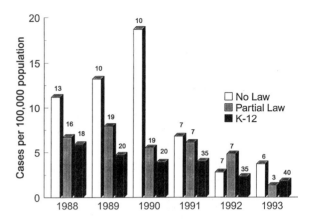

Figure 3. Reported incidence of mumps among children aged 5 to 19 years, by state immunization requirements, United States, 1988–1993. Cases per 100,000 population adjusted for states not reporting mumps (Rhode Island and Mississippi, 1988; New Mexico and Oregon, 1988–1993). Includes District of Columbia. California excluded for 1988 since no age breakdown was available. No law: no requirement for mumps vaccination before entering school. Partial law: includes some but not all children in kindergarten through grade 12. K–12: comprehensive (i.e., day care, Head Start, kindergarten through grade 12) laws requiring proof of immunity before entering school.

incidence rates in the school-aged population is shown in Fig. 3. The incidence of mumps also has decreased in states that have not enacted school immunization laws. The decreased incidence in states with school laws may help to reduce transmission of virus to persons in neighboring states. The two-dose schedule for MMR vaccine beginning in 1989 is also likely to have had an impact on the decreasing incidence of mumps in recent years.

An outbreak in Arizona in 1994 highlights some of the problems that continue to contribute to the transmission of mumps.[34] Cases appeared primarily among children in grades 4 to 8, although about 25% of cases were in older persons. School laws in Arizona do not require a second dose of MMR, and documentation of mumps vaccination has been required only since 1992. Six percent of the children in the school attended by the index case were unvaccinated because of personal or religious exemptions. The index case was believed to be an unvaccinated child, born in Mexico, who had recently returned from a visit to Mexico where he had been exposed to several children with mumps. Mexico and many other countries do not routinely vaccinate against mumps.

An estimated 137 million doses of mumps-containing vaccine were distributed between 1967 and 1993. Since licensure of the vaccine and concomitant with its wide use, the incidence of mumps has been reduced by 99% and record low numbers of cases were reported in 1993 and 1994. A goal to reduce the annual number of cases to 1600 by 1996 as part of the Childhood Immunization Initiative was met; 906 cases were reported in 1995. National surveillance for all vaccine-preventable diseases including mumps has been enhanced. All cases are reported to the NNDSS via the National Electronic Telecommunication System for Surveillance (NETSS) and additional information including vaccination status and clinical data are collected. In the past, efficiency of the reporting of mumps through the NNDSS has been estimated to be only 10%. Factors contributing to underreporting include (1) mild cases of mumps may not be diagnosed, (2) cases of parotitis diagnosed as mumps may not be reported, (3) mumps is not a reportable disease in two states, and (4) approximately one third of cases are subclinical. On the other hand, some cases of parotitis reported as mumps may be caused by other infectious agents.

10. Unresolved Problems

The issue of whether vaccine-induced immunity is lifelong or whether immunity may wane over time remains unresolved. An outbreak in a highly vaccinated population suggested that although most cases appeared to be primary vaccination failures, waning immunity may be a contributing factor.

More knowledge is needed on the role of cellular immunity in the prevention or pathogenesis of mumps virus infection. The timing of the development of cellular immunity following natural or vaccine-induced mumps infection has not been clearly established.

Mumps surveillance needs to be improved. The number of cases of parotitis that do not occur in outbreak settings and are reported as mumps but are not caused by the mumps virus need to be established. Laboratory assessment of such cases is needed to obtain this information.

In the United States, cases of mumps are imported from countries without mumps vaccination programs or occur following exposure to imported cases. At the present time the reduction or elimination of mumps is not a high priority in countries with high incidences of more serious diseases such as measles, diphtheria, and polio. When the control of these more serious vaccine-preventable diseases is improved, the inclusion of mumps vaccination into routine vaccination schedules may be considered.

11. References

1. ADAIR, C. V., GAULD, R. L., AND SMADEL, J. E., Aseptic meningitis, a disease of diverse etiology: Clinical and etiologic studies on 854 cases, *Ann. Intern. Med.* **39:**675–704 (1953).

2. ANDERSON, D. M., AND HUTCHINSON, D. N., Renal damage and virus infection, *Br. Med. J.* **3:**680–681 (1968).

3. ANDERSON, R. M., CROMBIE, J. A., AND GRENFELL, B. T., The epidemiology of mumps in the UK: A preliminary study of virus transmission, herd immunity and the potential impact of immunization, *Epidemiol. Infect.* **99:**65–84 (1987).

4. ARDAY, D. R., KANJARPANE, D. D., AND KELLEY, P. W., Mumps in the US Army 1980–1986: Should recruits be immunized? *Am. J. Public Health* **79:**471–474 (1989).

5. ARITA, M., UENO, Y., AND MASUYAMA, Y., Complete heart block in mumps myocarditis, *Br. Heart J.* **46:**342–344 (1981).

6. ARROYO, M., ALIA, J. M., MATEOS, M. L., CARRASCO, J. L., BALLESTEROS, F., LARDINOIS, R., AND THE PAEDIATRIC COLLABORATIVE GROUP, Natural immunity to measles, rubella and mumps among Spanish children in the pre-vaccination era, *Int. J. Epidemiol.* **15:**95–100 (1986).

7. AZIMI, P. H., CRAMBLETT, H. G., AND HAYNES, R. E., Mumps meningoencephalitis in children, *J.A.M.A.* **207:**509–512 (1969).

8. BALFOUR, H. H., JR., AND AMREN, D. P., Rubella, measles and mumps antibodies following vaccination in children: A potential rubella problem, *Am. J. Dis. Child.* **132:**573–577 (1978).

9. BANATVALA, J. E., BRYANT, J., SCHERNTHANER, G., BORKENSTEIN, M., SCHUBER, E., BROWN, D., DESILVA, L. M., MENSER, M. A., AND SILINK, M., Coxsackie B, mumps, rubella and cytomegalovirus specific IgM responses in patients with juvenile-onset insulin-dependent diabetes mellitus in Britain, Austria, and Australia, *Lancet* **1:**1409–1412 (1985).

10. BANG, H. O., AND BANG, J., Involvement of the central nervous system in mumps, *Acta Med. Scand.* **113:**487–505 (1943).

11. BEARD, C. M., BENSON, R. C., JR., KELALIS, P. P., ELVEBACK, L. R., AND KURLAND, L. T., The incidence and outcome of mumps orchitis in Rochester, Minnesota, 1934 to 1974, *Mayo Clin. Proc.* **52:**3–7 (1977).

12. BEARDWELL, A., Facial palsy due to the mumps virus, *Br. J. Clin. Pract.* **23:**37–38 (1969).

13. BEGHI, E., NICOLOSI, A., KURLAND, L. T., MULDER, D. W., HAUSER, W. A., AND SHUSTER, L., Encephalitis and aseptic meningitis. Olmstead County, Minnesota, 1950–1981: 1. Epidemiology, *Ann. Neurol.* **16:**283–294 (1984).

14. BENGTSSON, E., AND ORNDAHL, G., Complications of mumps with special reference to the incidence of myocarditis, *Acta Med. Scand.* **149:**381–388 (1954).

15. BENITO, R. J., LARRARD, L., LASIERRA, M. P., BENITO, J. F., AND ERDOCIAIN, F., Persistence of specific IgM antibodies after natural mumps infection, *J. Infect. Dis.* **155:**156–157 (1987).

16. BJORVATN, B., AND SKOLDENBERG, B., Mumps, meningitis, and orchitis in Stockholm during 1955–1976: An epidemiological background for a vaccination policy, *Lakartidningen* **75:**2295–2298 (1978).

17. BJORVATN, B., AND WOLONTIS, S., Mumps meningoencephalitis in Stockholm November 1964–July 1971. 1. Analysis of a hospitalized study group. Questions of selection and representivity, *Scand. J. Infect. Dis.* **5:**253–260 (1973).

18. BLACK, F. L., A nationwide serum survey of United States military recruits. 1962. III. Measles and mumps antibodies, *Am. J. Hyg.* **80:**304–307 (1964).

19. BLACK, F. L., HIERHOLZER, W. J. DEP., PINHEIRO, F., EVANS, A. S., WOODALL, J. P., OPTON, E. M., EMMONS, J. E., WEST, B. E., EDSALL, G., DOWNS, W. G., AND WALLACE, G. D., Evidence for persistence of infectious agents in isolated human populations, *Am. J. Epidemiol.* **100:**230–250 (1974).

20. BLACK, F. L., AND HOUGHTON, W. J., The significance of mumps

hemagglutinin inhibition titers in normal populations, *Am. J. Epidemiol.* **85:**101–197 (1967).

21. BOWERS, D., AND WEATHERHEAD, D. S. P., Mumps encephalitis, *Can. Med. Assoc. J.* **69:**49–55 (1953).

22. BRAY, P. F., Mumps—A cause of hydrocephalus? *Pediatrics* **49:**446–449 (1972).

23. BRICKMAN, A., AND BRUNELL, P. A., Susceptibility of medical students to mumps: Comparison of serum neutralizing antibody and skin test, *Pediatrics* **48:**447–450 (1972).

24. BRISS, P. A., FEHRS, L. J., PARKER, R. A., WRIGHT, P. F., SANNELLA, E. C., HUTCHESON, R. H., AND SCHAFFNER, W., Sustained transmission of mumps in a highly vaccinated population: Assessment of primary vaccine failure and waning vaccine-induced immunity, *J. Infect. Dis.* **169:**77–82 (1994).

25. BROOKS, H., Epidemic parotitis as a military disease, *Med. Clin. North Am.* **2:**493–503 (1918).

26. BROWN, E. G., FURESZ, J., DIMOCK, K., YAROSH, W., AND CONTRERAS, G., Nucleotide sequence analysis of Urabe mumps vaccine strain that caused meningitis in vaccine recipients, *Vaccine* **9:**840–841 (1991).

27. BROWN, J. W., KIRKLAND, H. B., AND HEIN, G. E., Central nervous system involvement during mumps, *Am. J. Med. Sci.* **215:**434–441 (1948).

28. BROWN, N. J., AND RICHMAN, S. J., Fatal mumps myocarditis in an 8-month-old child, *Br. Med. J.* **281:**356–357 (1980).

29. BRUNELL, P. A., BRICKMAN, A., O'HARE, D., AND STEINBERG, S., Ineffectiveness of isolation of patients as a method of preventing the spread of mumps: Failure of the mumps skin antigen to predict immune status, *N. Engl. J. Med.* **279:**1357–1361 (1968).

30. BRUYN, H. B., SEXTON, H. M., AND BRAINERD, H. D., Mumps meningoencephalitis: A clinical review of 119 cases with one death, *Calif. Med.* **86:**153–160 (1957).

31. BUYNAK, E. B., AND HILLEMAN, M. R., Live attenuated mumps virus vaccine. 1. Vaccine development, *Proc. Soc. Exp. Biol. Med.* **123:**768–775 (1966).

32. BUYNAK, E. B., HILLEMAN, M. R., LEAGUS, M. B., WHITMAN, J. E., JR., WEIBEL, R. E., AND STOKES, J., JR., Jeryl Lynn strain live attenuated mumps virus vaccine. Influence of age, virus dose, lot, and γ-globulin administration on response, *J.A.M.A.* **203:**63–67 (1968).

33. BUYNAK, E. B., HILLEMAN, M. R., WHITMAN, J. E., JR., WEIBEL, R. E., AND STOKES, J., JR., Measurement and durability of antibody following Jeryl Lynn strain viral mumps vaccine, in: *Vaccines against Viral and Rickettsial Diseases of Man: Proceedings of The First International Conference on Vaccines Against Viral and Rickettsial Diseases in Man* (Washington, DC, November 7–11, 1966) (Scientific Publ. 147), pp. 440–442, Washington, DC, Pan American Health Organization (1967).

34. CARMODY, L., VAN LOON, F., AND ZEITZ, P. S., Mumps outbreak on Yavapai County, Arizona, *Immunization Action News* **1**(2):2 (1994).

35. CENTERS FOR DISEASE CONTROL, Immunization Practices Advisory Committee, Mumps vaccine, *Morbid. Mortal. Week. Rep.* **16:**430–431 (1967).

36. CENTERS FOR DISEASE CONTROL, Immunization Practices Advisory Committee, Mumps vaccine, *Morbid. Mortal. Week. Rep.* **21**(Suppl.):17–18 (1972).

37. CENTERS FOR DISEASE CONTROL Mumps prevention: Recommendations of the Immunization Practices Advisory Committee (ACIP), *Morbid. Mortal. Week. Rep.* **26:**393–394 (1977).

38. CENTERS FOR DISEASE CONTROL, Mumps prevention: Recom-

mendations of the Immunization Practices Advisory Committee (ACIP), *Morbid. Mortal. Week. Rep.* **29:**87–88, 93–94 (1980).

39. CENTERS FOR DISEASE CONTROL, Mumps—United States, 1980–1983, *Morbid. Mortal. Week. Rep.* **32:**545–547 (1983).

40. CENTERS FOR DISEASE CONTROL, Mumps outbreak—New Jersey, *Morbid. Mortal. Week. Resp.* **33:**421–430 (1984).

41. CENTERS FOR DISEASE CONTROL, Mumps outbreaks in university campuses—Illinois, Wisconsin, South Dakota, *Morbid. Mortal. Week. Rep.* **36:**496–498 (1987).

42. CENTERS FOR DISEASE CONTROL, Mumps—United States, 1985–1988, *Morbid. Mortal. Week. Rep.* **38:**101–105 (1989).

43. CENTERS FOR DISEASE CONTROL, Mumps prevention: Recommendations of the Immunization Practices Advisory Committee (ACIP), *Morbid. Mortal. Week. Rep.* **38:**388–392, 397–400 (1989).

44. CHAIKEN, B. P., WILLIAMS, N. M., PREBLUD, S. T., PARKIN, W., AND ALTMAN, R., The effect of a school entry law on mumps activity in a school district, *J.A.M.A.* **257:**2455–2458 (1987).

45. CHIBA, Y., HORINO, K., UMETSU, M., WATAYA, Y., CHIBA, S., AND NAKAO, T., Virus excretion and antibody responses in saliva in natural mumps, *Tohoku J. Exp. Med.* **111:**229–238 (1973).

46. CHOU, S. M., Inclusion body myositis: A chronic persistent mumps myositis? *Human Pathol.* **17:**765–777 (1986).

47. CHRISTENSON, B., AND BÖTTIGER, M., Methods for screening the naturally acquired and vaccine-induced immunity to the mumps virus, *Biologicals* **18:**213–219 (1990).

48. COCHI, S. L., PREBLUD, S. R., AND ORENSTEIN, W. A., Perspectives on the relative resurgence of mumps in the United States, *Am. J. Dis. Child.* **142:**499–507 (1988).

49. COCHI, S. L., WHARTON, M., AND PLOTKIN, S. A., Mumps vaccine, in: *Vaccines* (S. A. PLOTKIN AND E. A. MORTIMER, JR., EDS.), pp. 277–301, Saunders, Philadelphia, 1994.

50. COONEY, M. K., FOX, J. P., AND HALL, C. E., The Seattle virus watch. VI. Observations of infections with and illness due to parainfluenza, mumps and respiratory syncytial viruses and mycoplasma pneumonia, *Am. J. Epidemiol.* **202:**532–551 (1975).

51. COX, M. J., ANDERSON, R. M., BUNDY, D. A. P., NOKES, D. J., AND DIDIER, J. M., Seroepidemiological study of the transmission of the mumps virus in St. Lucia, West Indies, *Epidemiol. Infect.* **102:**147–160 (1989).

52. DACON-VOUTETAKIS, C., CONSTANTINIDIS, M., MOSCHOS, A., VLACHON, C., AND MATSANIOTIS, N., Diabetes mellitus following mumps: Insulin reserve, *Am. J. Dis. Child.* **127:**890–891 (1974).

53. DAVIDSON, W. L., BUYNAK, E. B., LEAGUS, M. B., WHITMAN, J. E., JR., AND HILLMAN, M. R., Vaccination of adults with live attenuated mumps virus vaccine, *J.A.M.A.* **201:**995–998 (1967).

54. DAVIS, L. E., AND JOHNSON, R. T., Experimental viral infection of the inner ear. I. Acute infections of the newborn hamster labrynth, *Lab. Invest.* **34:**349–356 (1976).

55. DEIBEL, R., AND SCHRYVER, G. D., Viral antibody in the cerebrospinal fluid of patients with acute central nervous system infections, *J. Clin. Microbiol.* **3:**397–401 (1976).

56. DUC-NGUYEN, H., AND HENLE, W., Replication of mumps virus in human leukocyte cultures, *J. Bacteriol.* **92:**258–265 (1966).

57. EDITORIAL, Mumps of the heart, *Br. Med. J.* **5481:**187–188 (1966).

58. EDITORIAL, Mumps and the endocardium, *N. Engl. J. Med.* **275:**393 (1966).

59. ELANGO, N., The mumps virus nucleocapsid mRNA sequence and homology among the paramyxoviridae proteins, *Virus Res.* **12:**77–86 (1989).

60. ELANGO, N., Complete nucleotide sequence of the matrix protein mRNA of mumps virus, *Virology* **168:**426–428 (1989).

61. ELANGO, N., KÖVAMEES, J., AND NORRBY, E., Sequence analysis of the mumps virus mRNA encoding the P protein, *Virology* **169:**62–67 (1989).

62. ELANGO, N., VARSANYI, T. M., KÖVAMEES, J., AND NORRBY, E., The mumps virus fusion protein mRNA sequence and homology among the *Paramyxoviridae* proteins, *J. Gen. Virol.* **70:**801–807 (1989).

63. ENDERS, J. F., Techniques of laboratory diagnosis, tests for susceptibility, and experiments on specific prophylaxis, *J. Pediatr.* **29:**129–142 (1946).

64. ENDERS, J. F., LEVENS, J. H., STOKES, J., JR., MARIS, E. P., AND BERENBERG, W., Attenuation of virulence with retention of antigenicity of mumps virus after passage in the embryonated egg, *J. Immunol.* **54:**283–291 (1946).

65. ENNIS, F. A., Immunity to mumps in an institutional epidemic: Correlation of unsusceptibility to mumps with serum plaque neutralizing and hemagglutination-inhibiting antibodies, *J. Infect. Dis.* **119:**654–657 (1969).

66. ENNIS, F. A., AND JACKSON, D., Isolation of virus during the incubation period of mumps infection, *J. Pediatr.* **72:**536–537 (1968).

67. EVERBERG, G., Deafness following mumps, *Acta Otolaryngol.* **48:**397–403 (1957).

68. EYLAND, E., ZMUCKY, R., AND SHEBA, C., Mumps virus and subacute thyroiditis: Evidence of a causal association, *Lancet* **1:**1062–1065 (1957).

69. FALK, W. A., BUCHAN, K., DOW, M., GARSON, J. Z., HILL, E., NOSAL, M., TARRANT, M., WESTBURY, R. C., AND WHITE, F. M. M., The epidemiology of mumps in Southern Alberta, 1980–1982, *Am. J. Epidemiol.* **130:**736–749 (1989).

70. FAOAGALI, J. L., An assessment of the need for vaccination amongst junior medical staff, *N. Z. Med. J.* **84:**147–150 (1976).

71. FELDSTEIN, J. D., JOHNSON, F. R., KALLICK, C. A., AND DOOLAS, A., Acute hemorrhagic pancreatitis and pseudocyst due to mumps, *Ann. Surg.* **180:**85–88 (1974).

72. FLEISHER, B., AND KRETH, H. W., Mumps virus replication in human lymphoid cell lines and in peripheral blood lymphocytes: Preference for T cells, *Infect. Immun.* **35:**25–31 (1982).

73. FLORMAN, A. L., SCHICK, B., AND SCALETTAR, H. E., Placental transmission of mumps and streptococcus MG antibodies, *Proc. Soc. Exp. Biol. Med.* **78:**126–128 (1951).

74. FOWLER, E. R., Deafness in mumps, *Pediatrics* **77:**243–245 (1960).

75. FOY, H. M., COONEY, K. M., HALL, C. E., BOR, E., AND MALETZKY, A. J., Isolation of mumps virus from children with acute lower respiratory tract disease, *Am. J. Epidemiol.* **94:**467–472 (1971).

76. FRANKLAND, A. W., Mumps meningo-encephalitis, *Br. Med. J.* **2:**48–49 (1941).

77. FREEMAN, A. G., Mumps followed by diabetes [letter to the editor], *Lancet* **2:**96 (1962).

78. FREEMAN, R., AND HAMBLING, M. H., Serological studies on 40 cases of mumps virus infection, *J. Clin. Pathol.* **33:**28–32 (1980).

79. FRIEDMAN, M., HADARI, I., GOLDSTEIN, V., AND SAROV, L., Virus-specific secretory IgA antibodies as means of rapid diagnosis of measles and mumps infection, *Israel J. Med. Sci.* **19:**881–884 (1983).

80. FRYDEN, A., LINK H., AND MOLLER, E., Demonstration of CSF lymphocytes sensitized against mumps virus antigens in mumps meningitis, *Acta Neurol. Scand.* **57:**396–404 (1978).

81. FRYDEN, A., LINK, H., AND NORRBY, E., Cerebrospinal fluid and serum immunoglobulins and antibody titers in mumps meningitis and in aseptic meningitis of other etiology, *Infect. Immun.* **21:**852–861 (1978).

82. GAMBLE, D. R., Relationship of antecedent illness to the development of diabetes in children, *Br. Med. J.* **281:**99–101 (1980).

83. GAMBLE, D. R., AND TAYLOR, K. W., Seasonal incidence of diabetes mellitus, *Br. Med. J.* **3:**631–633 (1969).

84. GARCIA, A. G., PEREIRA, J. M., VIDIGAL, N., LOKATO, Y. Y., PEGADO, C. S., AND BRANCO, J. P., Intrauterine infection with mumps virus, *Obstet. Gynecol.* **56:**756–759 (1980).

85. GERSONY, W. M., KATZ, S. L, AND NADAS, A. S., Endocardial fibroelastosis and the mumps virus, *Pediatrics* **37:**430–434 (1966).

86. GHOSH, S., Guillain–Barre syndrome complicating mumps, *Lancet* **1:**895–896 (1967).

87. GLICK, D., An isolated case of mumps in a geriatric population, *J. Am. Geriatr. Soc.* **18:**642–644 (1970).

88. GNANN, J. W., JR., Meningitis and encephalitis caused by mumps virus, in: *Infections of the Central Nervous System* (W. M. SCHELD, R. J. WHITLEY, AND D. T. DURACK, EDS.), pp. 113–125, Raven Press, New York, 1991.

89. GORDON, J. E., AND HEEREN, R. H., The epidemiology of mumps, *Am. J. Med. Sci.* **200:**412–428 (1940).

90. GORDON, J. E., AND KILHAM, L., Ten years in the epidemiology of mumps, *Am. J. Med. Sci.* **218:**338–359 (1949).

91. GORDON, S. C., AND LAUTER, C. B., Mumps arthritis: A review of the literature, *Rev. Infect. Dis.* **6:**338–344 (1984).

92. GRAHAM, D. Y., BROWN, C. H., BENREY, J., AND BUTEL J. S., Thrombocytopenia: A complication of mumps, *J.A.M.A.* **227:**1161–1164 (1974).

93. GREENBERG, M. A., AND BRIX, D. L., Safe administration of mumps–measles–rubella vaccine in egg-allergic children, *J. Pediatr.* **113:**504–506 (1988).

94. GRILLNER, L., AND BLOMBERG, J., Hemolysis-in-gel and neutralization tests for determination of antibodies to mumps virus, *J. Clin. Microbiol.* **4:**11–15 (1976).

95. GUT, J. P., SPIESS, C., SCHMITT, S., AND KIRN, A., Rapid diagnosis of acute mumps infection by a direct immunoglobulin M antibody capture enzyme immunoassay with labeled antigen, *J. Clin. Microbiol.* **21:**346–352 (1985).

96. HABEL, K., Cultivation of mumps virus in the developing chick embryo and its application to studies of immunity to mumps in man, *Public Health Rep.* **60:**201–212 (1945).

97. HABEL, K., Preparation of mumps vaccine and immunization of monkeys against experimental mumps infection, *Public Health Rep.* **61:**1655–1664 (1946).

98. HABEL, K., Vaccination of human beings against mumps: Vaccine administered at the start of an epidemic. 1. Incidence and severity of mumps in vaccinated and control groups, *Am. J. Hyg.* **54:**295–311 (1951).

99. HALL, R., AND RICHARDS, H., Hearing loss due to mumps, *Arch. Dis. Child.* **62:**189–191 (1987).

100. HAMILTON, R., An account of distemper by the common people of England vulgarly called the mumps, *Lond. Med. J.* **11:**190–211 (1790).

101. HARMSEN, T., JONGERIUS, M. C., VAN DER ZWAN, C. W., PLANTINGA, A. D., KRAAIJEVELD, C. A., AND BERBERS, G. A. M., Comparison of a neutralization enzyme immunoassay and an enzyme-linked immunosorbent assay for evaluation of immune status of children vaccinated for mumps, *J. Clin. Microbiol.* **30:**2139–2144 (1992).

102. HARRIS, R. W., KEHRER, A. F., AND ISAACSON, P., Relationship of occupation to risk of clinical mumps in adults, *Am. J. Epidemiol.* **89:**264–270 (1969).

103. HARRIS, R. W., TURNBEULL, C. D., ISAACSON, P., KARSON, D. T., AND WINKELSTEIN, W., JR., Mumps in a northeast metropolitan community. 1. Epidemiology of clinical mumps, *Am. J. Epidemiol.* **88:**224–233 (1968).

105. HAYASHI, K., ROSS, M. E., AND NOTKINS, A. L., Persistence of mumps viral antigens in mouse brain, *Jpn. J. Exp. Med.* **46:**197–200 (1976).

106. HENDERSON, W., Mumps meningo-encephalitis: An outbreak in a preparatory school, *Lancet* **1:**386–388 (1952).

107. HENLE, G., BASHE, W. J., JR., BURGOON, J. S., BURGOON, C. F., STOKES, J., JR., AND HENLE, W., Studies on the prevention of mumps. III. The effect of subcutaneous injection of inactivated mumps virus vaccinees, *J. Immunol.* **66:**561–577 (1951).

108. HENLE, G., CRAWFORD, M. N., HENLE, G., TABIO, H. F., DEINHARDT, F., CHABAU, A. G., AND OLSHIN, I. J., Studies on the prevention of mumps. VII. Evaluation of dosage schedules for inactivated mumps vaccine, *J. Immunol.* **83:**17–28 (1959).

109. HENLE, G., AND DIENHARDT, F., Propagation and primary isolation of mumps virus in tissue culture, *Proc. Soc. Exp. Biol. Med.* **89:**556–560 (1955).

110. HENLE, G., DIENHARDT, F., BERGS, V. V., AND HENLE, W., Studies on persistent infections in tissue culture. I. General aspects of the system, *J. Exp. Med.* **108:**537–560 (1958).

111. HENLE, G., HENLE, W., BURGOON, J. S., BASHE, W. J., JR., AND STOKES, J., JR., Studies on the prevention of mumps. I. The determination of susceptibility, *J. Immunol.* **66:**535–549 (1951).

112. HENLE, G., HENLE, W., WENDELL, K. K., AND ROSENBERG, P., Isolation of mumps virus from human beings with induced apparent or inapparent infections, *J. Exp. Med.* **88:**223–232 (1948).

113. HENLE, G., AND MCDOUGALL, C. L., Mumps meningoencephalitis: Isolation in chick embryos of virus from spinal fluid of a patient, *Proc. Soc. Exp. Biol. Med.* **66:**209–211 (1957).

114. HENLE, G., STOKES, J., JR., BURGOON, J. S., BASHE, W. J., JR., BURGOON, C. F., AND HENLE, W., Studies on the prevention of mumps. IV. The effect of oral spraying of attenuated active virus, *J. Immunol.* **66:**579–594 (1951).

115. HERRLER, G., AND COMPANS, R. W., Synthesis of mumps polypeptides in infected Vero cells, *Virology* **119:**430–438 (1982).

116. HERSH, B. S., FINE, P. E., KENT, W. K., COCHI, S. L., KAHN, L. H., ZELL, E. R., HAYS, P. L., AND WOOD, C. L., Mumps outbreak in a highly vaccinated population, *J. Pediatr.* **119:**187–193 (1991).

117. HIERHOLZER, J. C., AND FARRIS, W. A., Follow-up of children infected in a coxsackievirus B-3 and B-4 outbreak: No evidence of diabetes mellitus, *J. Infect. Dis.* **129:**741–746 (1974).

118. HILDES, J. A., WILT, J. C., PARKER, W. L., STACKIW, W., AND DELAAT, S., Surveys of respiratory virus antibodies in an Artic Indian population, *Can. Med. Assoc. J.* **93:**1015–1018 (1965).

119. HILLEMAN, M. R., Advances in control of viral infections by non-specific measures and by vaccines, with special reference to live mumps and rubella virus vaccines, *Clin. Pharmacol. Ther.* **7:**752–762 (1966).

120. HILLEMAN, M. R., WEIBEL, R. E., BUYNAK, E. B., STOKES, J., JR., AND WHITMAN, J. E., JR., Live, attenuated mumps-virus vaccine. 4. Protective efficacy as measured in a field evaluation, *N. Engl. J. Med.* **276:**252–258 (1968).

121. HINDEN, E., Mumps followed by diabetes, *Lancet* **1:**1381 (1962).

122. HOLDEN, E. M., EAGLES, A. Y., AND STEVENS, J. E., Mumps

123. HOPE SIMPSON, R. E., Infectiousness of communicable diseases in the household, *Lancet* **2:**549–554 (1952).

124. HUGHES, W. T., STEIGMAN, A. J., AND DELON, H. F., Some implications of fatal nephritis associated with mumps, *Am. J. Dis. Child.* **111:**297–301 (1966).

125. HUPPERTZ, H-I., AND CHANTLER, J. K., Restricted mumps virus infection of cells derived from normal human joint tissue, *J. Gen. Virol.* **72:**339–347 (1991).

126. HUPPERTZ, H-I., NIKO, N. P. H., AND CHANTLER, J. K., Susceptibility of normal human joint tissue to viruses, *J. Rheumatol.* **18:**699–704 (1991).

127. HYÖTY, H., LEINIKKI, P., REUNANEN, A., ILONEN, J., SURCEL, H-M., RILVA, A., KÄÄR, M-L., HUUPPONEN, T., HAKULINEN, A., MÄKELÄ, A-L., AND ÄKERBLOM, H. K., Mumps infections in the etiology of type 1 (insulin-dependent) diabetes, *Diabetes Res.* **9:**111–116 (1988).

128. ILONEN, J., Lymphocyte blast formation response of seropositive and seronegative subjects to herpes simplex, rubella, mumps, and measles virus antigens, *Acta Pathol. Microbiol. Scand.* **87:**151–157 (1979).

129. INSTITUTE OF MEDICINE, *Adverse Events Associated with Childhood Vaccines. Evidence Bearing on Causality* (K. R. STRATTON, C. J. HOWE, AND R. B. JOHNSTON, JR., EDS.), pp. 118–186, National Academy Press, Washington, DC, 1994.

130. JOHN, H. J., Diabetes mellitus in children: Review of 500 cases, *J. Pediatr.* **35:**723–744 (1949).

131. JOHNSON, C. D., AND GOODPASTURE, E. W., An investigation of the etiology of mumps, *J. Exp. Med.* **59:**1–19 (1934).

132. JOHNSON, C. D., AND GOODPASTURE, E. W., The etiology of mumps, *Am. J. Hyg.* **21:**46–57 (1935).

133. JOHNSTONE, J. A., ROSS, C. A. C., AND DUNN, M., Meningitis and encephalitis associated with mumps infection. A 10-year survey, *Arch. Dis. Child.* **47:**647–651 (1972).

134. JONES, A. G. H., WHITE, J. M., AND BEGG, N. T., The impact of MMR vaccine on mumps infection in England and Wales, *Comm. Dis. Rep.* **1:**R93–R96 (1991).

135. JONES, J. F., RAY, C. G., AND FULGINITI, V. A., Perinatal mumps infection, *J. Pediatr.* **96:**912–914 (1980).

136. KAHANA, D., AND BERANT, M., Diabetes in an infant following inapparent mumps, *Clin. Pediatr.* **6:**124–125 (1967).

137. KALTER, S. S., AND PRIER, J. E., Immunity to mumps among physicians, *Am. J. Med. Sci.* **229:**161–164 (1955).

138. KAPLAN, K. M., MARDER, D. C., COCHI, S. L., AND PREBLUD, S. R., Mumps in the workplace. Further evidence of the changing epidemiology of a childhood vaccine-preventable disease, *J.A.M.A.* **250:**1434–1438 (1988).

139. KAUFMAN, J. J., AND BRUCE, P. T., Testicular atrophy following mumps: A cause of testis tumour? *Br. J. Urol.* **35:**65–69 (1963).

140. KELLEY, P. W., PETRUCCELLI, B. P., STEHR-GREEN, P., ERICKSON, R. L., AND MASON, C. J., The susceptibility of young adult Americans to vaccine-preventable infections. A national sero-survey of US Army recruits, *J.A.M.A.* **266:**2724–2729 (1991).

141. KENNY, M. T., JACKSON, J. E., MEDLER, E. M., MILLER, S. A., AND OSBORN, R., Age-related immunity to measles, mumps and rubella in Middle American and United States children, *Am. J. Epidemiol.* **103:**174–180 (1976).

142. KILHAM, L., Mumps meningoencephalitis with and without parotitis, *Am. J. Dis. Child.* **78:**324–333 (1949).

143. KILHAM, L., AND MARGOLIS, G., Intrauterine infections induced by mumps virus in hamsters, *Lab. Invest.* **31:**34–41 (1974).

144. KING, R. C., Mumps followed by diabetes [letter to the editor], *Lancet* **2:**1055 (1962).

145. KIRK, M., Sensorineural hearing loss and mumps, *Br. J. Audiol.* **21:**227–228 (1987).

146. KNEELAND, Y., JR., Mumps, in: *Internal Medicine in World War II*, Vol. II (J. B. COATES, JR., ED. IN CHIEF, W. P. HAVENS, JR., ED. FOR INTERNAL MEDICINE), pp. 35–38, Office of the Surgeon General, Department of the Army, Washington, DC, 1963.

147. KOSKINIEMI, M. L., DONNER, M., AND PETTAY, O., Clinical appearance and outcome of mumps encephalitis in children, *Acta Paediatr. Scand.* **72:**603–609 (1983).

148. KOSKINIEMI, M., MANNINEN, V., VAHERI, A., SAINIO, K., EISTOLA, O., AND KARLI, P., Acute encephalitis: A survey of epidemiological, clinical and microbiological features covering a twelve-year period, *Acta Med. Scand.* **209:**115–120 (1981).

149. KOSKINIEMI, M. L., AND VAHERI, A., Acute encephalitis of viral origin, *Scand. J. Infect. Dis.* **14:**181–187 (1982).

150. KÖVAMEES, J., NORRBY, E., AND ELANGO, N., Complete nucleotide sequence of the hemagglutinin-neuraminidase (HN) mRNA of mumps virus and comparison of paramyxovirus HN proteins, *Virus Res.* **12:**87–96 (1989).

151. KRAVIS, L. P., SIGEL, M. M., AND HENLE, G., Mumps meningoencephalitis with special reference to the use of the complement fixation test in diagnosis, *Pediatrics* **8:**204–214 (1951).

152. KRUGMAN, S., MURIEL, G., AND FONTANA, V. J., Combined live measles, mumps, rubella vaccine: Immunological response, *Am. J. Dis. Child.* **121:**380–381 (1971).

153. KURTZ, J. B., TOMLINSON, A. H., AND PEARSON, J., Mumps virus isolated from a fetus, *Br. Med. J.* **284:**471 (1982).

154. LAURENCE, D., AND McGAVIN, D., The complications of mumps, *Br. Med. J.* **1:**94–97 (1948).

155. LEONIDAS, J. C., ATHANASIADES, T., AND ZOUBOULAKIS, D., Mumps myocarditis: Case reports, *J. Pediatr.* **68:**650–653 (1966).

156. LENNETTE, E. H., JENSEN, F. W., GUENTHER, R. W., AND MAGOFFIN, R. L., Serologic responses to para-influenza viruses in patients with mumps virus infection, *J. Lab. Clin. Med.* **61:**780–788 (1963).

157. LENNETTE, E. H., MAGOFFIN, R. L., AND KNOUP, E. G., Viral central nervous system disease, *J.A.M.A.* **179:**687–695 (1962).

158. LEPOW, M. L., CARVER, D. H., WRIGHT, H. T., JR., WOODS, W. A., AND ROBBINS, F. C., A clinical epidemiologic and laboratory investigation of aseptic meningitis during the four-year period. 1955–1958. I. Observations concerning etiology and epidemiology, *N. Engl. J. Med.* **266:**1181–1187 (1962).

159. LEVENS, J. H., AND ENDERS, J. F., The hemoagglutinative properties of amniotic fluid from embryonated eggs infected with mumps virus, *Science* **102:**117–129 (1945).

160. LEVITT, L. P., MAHONEY, D. J., JR., CASEY, H. L., AND BOND, J. O., Mumps in a general population: A seroepidemiologic study, *Am. J. Dis. Child.* **120:**134–138 (1970).

161. LEVITT, L. P., RICH, T. A., KINDE, S. W., LEWIS, A. L., GATES, E. H., AND BOND, J. O., Central nervous system mumps, *Neurology* **20:**829–834 (1970).

162. LIAO, S. J., AND BENENSON, A. S., Immunity status of military recruits in 1951 in the United States. II. Results of mumps complement-fixation tests, *Am. J. Hyg.* **59:**273–281 (1954).

163. LIGHTWOOD, R., Myelitis from mumps, *Br. Med. J.* **1:**484–485 (1946).

164. LIN, C. Y., CHEN, W. P., AND CHIANG, H., Mumps associated with nephritis, *Child Nephrol. Urol.* **10:**68–71 (1990).

165. LINDE, G. A., GRANSTROM, M., AND ORVELL, C., Immunoglobulin class and immunoglobulin G subclass enzyme-linked immunosorbent assays compared with microneutralization assay for serodiagnosis of mumps infection and determination of immunity, *J. Clin. Microbiol.* **25:**1653–1658 (1987).

166. LINDSAY, J. R., DAVEY, P. R., AND WARD, P. H., Inner ear pathology in deafness due to mumps, *Ann. Otol. Rhinol. Laryngol.* **69:**918–935 (1960).

167. LINK, H., LAURENZI, M. A., AND FRYDEN, A., Viral antibodies in oligoclonal and polyclonal IgG synthesized with the central nervous system over the course of mumps meningitis, *J. Neuroimmunol.* **1:**287–298 (1981).

168. MADDEN, D. L., FUCCILLO, D. A., TRAUB, R. G., LEY, A. C., SEVER, J. L., AND BEADLE, E. L., Juvenile onset diabetes mellitus in pregnant women: Failure to associate with coxsackie B1-6, mumps, or respiratory syncytial virus infections, *J. Pediatr.* **92:**959–960 (1978).

169. MASSON, A. M., AND NICKERSON, G. H., Mumps with nephritis, *Can. Med. Assoc.* **97:**866–867 (1967).

170. MAYNARD, J. E., SHRAMEK, G., NOBLE, G. R., DEINHARDT, F., AND CLARK, P., Use of attenuated live mumps virus vaccine during a "virgin soil" epidemic of mumps on St. Paul Island, Alaska, among US military recruits, *Am. J. Epidemiol.* **92:**301–306 (1970).

171. McCARTHY, M., AND JOHNSON, R. T., A comparison of the structural polypeptides of five strains of mumps virus, *J. Gen. Virol.* **46:**15–27 (1980).

172. McCARTHY, M., WOLINSKY, J. S., AND LAZZARINI, R. A., A persistent infection of Vero cells by egg-adapted mumps virus, *Virology* **114:**343–356 (1981).

173. McCRAE, W. M., Diabetes mellitus following mumps, *Lancet* **1:**1300–1301 (1963).

174. McFARLAND, N. F., PEDONE, C. A., MIAGIALI, E. S., AND McFARLIN, D. E., The response of human lymphocyte subpopulations to measles, mumps, and vaccinia viral antigens, *J. Immunol.* **125:**221–225 (1980).

175. McGUINNESS, A. C., AND GALL, E. A., Mumps at Army camps in 1943, *War Med.* **5:**95–104 (1944).

176. McLEAN, D. M., LARKE, R. P. B., COBB, C., GRIFFIS, E. D., AND HACKETT, S. M. R., Mumps and enteroviral meningitis in Toronto, 1966, *Can. Med. Assoc. J.* **96:**1355–1361 (1967).

177. McLEAN, D. M., WALKER, S. J., AND McNAUGHTON, G. A., Mumps meningoencephalitis: A virologic and clinical study, *Can. Med. Assoc. J.* **83:**148–151 (1960).

178. MESSARITAKIS, J., KARABULA, C., KATTAMIS, C., AND MATSANIOTIS, N., Diabetes following mumps in sibs, *Arch. Dis. Child.* **46:**561–562 (1971).

179. MEURMAN, O., HÄNNINEN, P., KRISHNA, R. V., AND ZIEGLER, T., Determination of IgG- and IgM-class antibodies to mumps virus by solid-phase enzyme immunoassay, *J. Virol. Methods* **4:**249–257 (1982).

180. MEYER, H. M., JOHNSON, R. T., CRAWFORD, I. P., DASCOMB, H. E., AND ROGERS, N. G., Central nervous system syndromes of "viral" etiology, *Am. J. Med.* **29:**334–347 (1960).

181. MEYER, M. B., An epidemiologic study of mumps; its spread in schools and families, *Am. J. Hyg.* **75:**259–281 (1962).

182. MILLAR, J. H. D., FRASER, K. B., HAIRE, M., CONNOLLY, J. H., SHIRODARIA, P. V., AND HADDEN, D. S. M., Immunoglobulin M

specific for measles and mumps in multiple sclerosis, *Br. Med. J.* **2:**378–380 (1971).

183. MILLER, E., GOLDACRE, M., PUGH, S., COLVILLE, A., FARRINGTON, P., FLOWER, A., NASH, J., MACFARLANE, L., AND TETTMAR, R., Risk of aseptic meningitis after measles, mumps and rubella vaccine in UK children, *Lancet* **341:**979–982 (1993).

184. MILLER, H. G., STANTON, J. B., AND GIBBONS, J. L., Parainfectious encephalomyelitis and related syndromes, *Q. J. Med.* **25:**467–505 (1956).

185. MONTEIRO, G. E., AND LILLICRAP, C. A., Case of mumps nephritis, *Br. Med. J.* **4:**721–722 (1967).

186. MORELAND, A. F., GASKIN, J. M., SCHIMPFF, R. D., WOODARD, J. C., AND OLSON, G. A., Effects of influenza, mumps and western equine encephalitis virus on fetal rhesus monkeys, *Teratology* **20:**53–64 (1979).

187. MORISHIMA, T., MIYAZU, M., OGAKI, T., ISOMURA, S., AND SUZUKI, S., Local immunity in mumps meningitis, *Am. J. Dis. Child.* **134:**1060–1064 (1980).

188. MORRISON, J. C., GIVENS, J. R., AND WISER, W. L., Mumps oophoritis: A cause of premature menopause, *Fertil. Steril.* **26:**655–660 (1975).

189. MUFSON, M. A., Parainfluenza viruses, mumps virus, Newcastle disease virus, in: *Viral, Rickettsial and Chlamydial Infections*, 6th Ed. (N. J. SCHMIDT AND R. W. EMMONS, EDS.), pp. 669–691, American Public Association, Washington, DC, 1989.

190. NARAYAN, K. M. V., AND MOFFAT, M. A., Measles, mumps, rubella antibody surveillance: Pilot study in Grampian, Scotland, *Health Bull.* **50:**47–53 (1992).

191. NIEDERMAN, J. C., HENDERSON, J. R., OPTON, E. M., BLACK, F. L., AND SKVRONOVA, K. A., A nationwide serum survey of Brazilian military recruits, 1964. II. Antibody patterns with arboviruses, polioviruses, measles and mumps, *Am. J. Epidemiol.* **86:**319–329 (1967).

192. NOKES, D. J., WRIGHT, J., MORGAN-CAPNER, P., AND ANDERSON, R. M., Serological study of the epidemiology of mumps virus infection in north-west England, *Epidemiol. Infect.* **105:**175–195 (1990).

193. NOREN, G. R., ADAMS, P., JR., AND ANDERSON, R. C., Positive skin reactivity to mumps virus antigen in endocardial fibroelastosis, *J. Pediatr.* **62:**604–606 (1963).

194. NORTHROP, R. L., Effect of puromycin and actinomycin D on a persistent mumps virus infection *in vitro*, *J. Virol.* **4:**133–140 (1969).

195. NOTKINS, A. L., Virus-induced diabetes mellitus. Brief review, *Arch. Virol.* **54:**1–17 (1977).

196. OLDFELT, V., Sequelae of mumps-meningoencephalitis, *Acta Med. Scand.* **134:**405–414 (1949).

197. ÖRVELL, C., Immunological properties of purified mumps virus glycoproteins, *J. Gen. Virol.* **41:**517–526 (1978).

198. ÖRVELL, C., The reactions of monoclonal antibodies with structural proteins of mumps virus, *J. Immunol.* **132:**2622–2699 (1984).

199. ÖRVELL, C., RYDBECK, R., AND LÖVE, A., Immunological relationships between mumps virus and parainfluenza viruses studied with monoclonal antibodies, *J. Gen. Virol.* **67:**1929–1939 (1986).

200. OVERMAN, J. R., Viremia in human mumps virus infections, *Arch. Intern. Med.* **102:**354–356 (1958).

201. PENTTINEN, K., CANTELL, K., SOMER, P., AND POIKOLAINEN, A., Mumps vaccination in the Finnish defense forces, *Am. J. Epidemiol.* **88:**234–244 (1968).

202. PHILIP, R. N., REINHARD, K. R., AND LACKMAN, D. B., Observations on a mumps epidemic in a "virgin" population, *Am. J. Hyg.* **69:**91–111 (1959).

203. POSTOVIT, V. A., Epidemic parotitis in adults, *Voen. Med. Zh.* **3:**38–41 (1983).

204. PRASAD, L. N., Complete bilateral deafness following mumps, *J. Laryngol.* **77:**809–811 (1963).

205. PRINCE, G. A., JENSON, A. B., BILLUPS, L. C., AND NOTKINS, A. L., Infection of human pancreatic beta cell cultures with mumps virus, *Nature* **271:**158–161 (1978).

206. REED, D., BROWN, G., MERRICK, R., SEVER, J., AND FELTZ, E., A mumps epidemic on St. George Island, Alaska, *J.A.M.A.* **199:**967–971 (1967).

207. RIFFENBURGH, R. S., Ocular manifestations of mumps, *Arch. Ophthamol.* **66:**739–743 (1961).

208. RIMA, B. K., ROBERTS, M. W., McADAM, W. D., AND MARTIN, S. J., Polypeptide synthesis in mumps virus-infected cells, *J. Gen. Virol.* **46:**501–505 (1980).

209. RITTER, B. S., Mumps meningoencephalitis in children, *J. Pediatr.* **52:**424–432 (1958).

210. ROBERTS, W. C., AND FOX, S. M., III, Mumps of the heart: Clinical and pathologic features, *Circulation* **32:**342–345 (1965).

211. RUSSELL, R. R., AND DONALD, J. C., The neurological complications of mumps, *Br. Med. J.* **2:**27–30 (1958).

212. RYDBECK, R., LÖVE, A., ÖRVELL, C., AND NORRBY, E., Antigenic variation of envelope and internal proteins of mumps virus strains detected with monoclonal antibodies, *J. Gen. Virol.* **67:**281–287 (1986).

213. ST. GEME, J. W., JR., NOREN, G. R., AND ADAMS, P., JR., Proposed embryopathic relation between mumps virus and primary endocardial fibroelastosis, *N. Engl. J. Med.* **275:**339–347 (1966).

214. ST. GEME, J. W., JR., PERALTA, H., FARIAS, E., DAVIS, C. W. C., AND NOREN, G. R., Experimental gestional mumps virus infection and endocardial fibroelastosis, *Pediatrics* **48:**821–826 (1971).

215. SAKATA, H., TSURUDOME, M., HISHIYAMA, M., ITO, Y., AND SUGIURA, A., Enzyme-linked immunosorbent assay for mumps IgM antibody: Comparison of IgM capture and indirect IgM assay, *J. Virol. Methods* **12:**303–311 (1985).

216. SAMANTRAY, S. K., CHRISTOPHER, S., MUKUNDAN, P., AND JONSON, S. C., Lack of relationship between viruses and human diabetes mellitus, *Aust. N. Z. J. Med.* **7:**139–142 (1977).

217. SCHEID, W., Mumps virus and the central nervous system, *World Neurol.* **2:**117–130 (1961).

218. SCHERBAUM, W. A., HAMPL, W., MUIR, P., GLÜCK, M., SEIBLER, J., EGLE, H., HAUNER, H., BOEHM, B. O., HEINZE, E., BANATVALA, J. E., AND PFEIFFER, E. F., No association between islet cell antibodies and coxsackie B, mumps, rubella and cytomegalovirus antibodies in non-diabetic individuals aged 7–19 years, *Diabetologia* **34:**835–838 (1991).

219. SCULLY, C., ECHERSALL, P. D., EDMOND, R. T. D., BOYLE, P., AND BEELEY, J. A., Serum alpha-amylase isoenzyme in mumps: Estimation of salivary and pancreatic isoenzymes in isoelectric focusing, *Clin. Chim. Acta* **113:**281–291 (1981).

220. SEVER, J. L., AND KURTZKE, J. F., Delayed dermal hypersensitivity to measles and mumps antigens among multiple sclerosis and control patients, *Neurology* **19:**113–115 (1969).

221. SERVER, A. C., MERZ, D. C., WAXHAM, M. N., AND WOLINSKY, J. S., Differentiation of mumps virus strains with monoclonal antibody to the HN glycoprotein, *Infect. Immun.* **35:**179–186 (1982).

222. SILVERMAN, A. C., Mumps complicated by a preceding myelitis, *N. Engl. J. Med.* **241**:262–266 (1949).

223. SINHA, S. K., AND CARLSON, S. D., Immune responses of mentally retarded subjects to measles, mumps and rubella vaccines, *Wisc. Med. J.* **74**:S75–S77 (1975).

224. SMITH, G. A., AND GUESSEN, R., Inner ear pathologic features following mumps infection: Report of a case in an adult, *Arch. Otolaryngol.* **102**:108–111 (1976).

225. SMOAK, B. L., NOVAKOSKI, W. L., MASON, C. J., AND ERICKSON, R. L., Evidence for a recent decrease in measles susceptibility among young American adults, *J. Infect. Dis.* **170**:216–219 (1994).

226. SMORODINTSEV, A. A., LUZIANINA, T. Y., AND MIKUTSKAYA, B. A., Data on the efficiency of live mumps vaccine from chick embryo cell cultures, *Acta Virol.* **9**:240–247 (1965).

227. SOSIN, D. M., COCHI, S. L., GUNN, R. A., JENNINGS, C. E., AND PREBLUD, S. R., Changing epidemiology of mumps and its impact on university campuses, *Pediatrics* **84**:779–784 (1989).

228. SPARLING, D., Transmission of mumps (letter to the editor), *N. Engl. J. Med.* **280**:276 (1969).

229. SPATARO, R. F., LIN, S. R., HORMER, F. A., HALL, C. B., AND MCDONALD, J. V., Aqueductal stenosis and hydrocephalus: Rare sequelae of mumps virus infection, *Neuroradiology* **12**:11–13 (1976).

230. STOKES, J., JR., Mumps, in: *Preventive Medicine in World War II*, Vol. IV, *Communicable Diseases Transmitted Chiefly Through Respiratory and Alimentary Tracts* (prepared by the Historical Unit, U.S. Army Medical Service), pp. 135–140, US Government Printing Office, Washington, DC, 1958.

231. STOKES, J., JR., WEIBEL, R. E., BUYNAK, E. B., AND HILLEMAN, M. R., Live attenuated mumps virus vaccine. II. Early clinical studies, *Pediatrics* **39**:363–371 (1967).

232. STRUEWING, J. P., HYAMS, K. C., TUELLER, J. E., AND GRAY, G. C., The risk of measles, mumps, and varicella among young adults: A serosurvey of US Navy and Marine Corps recruits, *Am. J. Public Health* **83**:1717–1729 (1993).

233. SUGG, W. C., FINGER, J. A., LEVINE, R. H., AND PAGANO, J. S., Field evaluation of live virus mumps vaccine, *J. Pediatr.* **72**:461–466 (1968).

234. SULTZ, H. A., HART, B. A., ZICKZNY, M., SCHLESINGER, E. R., Is mumps virus an etiologic factor in juvenile diabetes mellitus? *J. Pediatr.* **86**:654–656 (1975).

235. SWERDLOW, A. J., HUTTLY, S. R. A., AND SMITH, P. G., Testicular cancer and antecedent diseases, *Br. J. Cancer* **55**:97–103 (1987).

236. SWIERKOSZ, E. M., Mumps virus, in: *Manual of Clinical Microbiology*, 5th ed. (W. J. HAUSLER, JR., K. L. HERRMANN, H. C. ISENBERG, AND H. J. SHADOMY, EDS.), pp. 912–917, American Society for Microbiology, Washington, DC, 1991.

237. THIEME, T., PIACENTINI, S., DAVIDSON, S., AND STEINGART, K., Determination of measles, mumps, and rubella immunization status using oral fluid samples, *J.A.M.A.* **272**:219–221 (1994).

238. TIMMONS, G. D., AND JOHNSON, K. P., Aqueductal stenosis and hydrocephalus after mumps encephalitis, *N. Engl. J. Med.* **283**:1505–1507 (1970).

239. THOMPSON, J. A., Mumps: A case of acquired aqueductal stenosis, *J. Pediatr.* **94**:923–924 (1979).

240. THOMPSON, W. M., JR., AND NOLAN, T. B., Atrioventicular dissociation associated with Adams-Stokes syndrome presumably due to mumps myocarditis, *J. Pediatr.* **68**:601–607 (1966).

241. UKKONEN, P., GRANSTRÖM, M. L., RÄSANEN, J., SALONEN, E.

M., AND PENTTINEN, K., Local production of mumps IgG and IgM antibodies in the cerebrospinal fluid of meningitis patients, *J. Med. Virol.* **8**:257–265 (1981).

242. UKKONEN, P., VÄISÄNEN, O., AND PENTTINEN, K., Enzyme-linked immunosorbent assay for mumps and parainfluenza type 1, immunoglobulin G, and immunoglobulin M antibodies, *J. Clin. Microbiol.* **11**:329–323 (1980).

243. UTZ, J. P., HOUK, V. N., AND ALLING, D. W., Clinical and laboratory studies of mumps. IV. Viruria and abnormal renal function, *N. Engl. J. Med.* **270**:1283–1286 (1964).

244. UTZ, J. P., AND SZWED, C. F., Mumps. III. Comparison of methods for detection of viruria, *Proc. Soc. Exp. Biol. Med.* **110**:841–844 (1962).

245. UTZ, J. P., SZWED, C. F., AND KASEL, J. A., Clinical and laboratory studies of mumps. II. Detection and duration of excretion of virus in urine, *Proc. Soc. Exp. Biol. Med.* **99**:259–261 (1958).

246. VAHERI, A., JULKUNEN, I., AND KOSKINIEMI, M. L., Chronic encephalomyelitis with specific increase in intrathecal mumps antibodies, *Lancet* **2**:685–688 (1982).

247. VAN LOON, F. P. L., HOLMES, S. J., SIROTKIN, B. I., WILLIAMS, W. W., COCHI, S. L., HADLER, S. C., AND LINDEGREN, M. L., Mumps—United States, 1988–1993, *Morbid. Mortal. Week. Rep.* **44**(S5–3):1–14 (1995).

248. VAN TIEL, F. H., KRAAIJEVELD, C. A., BALLER, J., HARMSEN, T., OOSTELAKEN, T. A. M., AND SNIPPE, H., Enzyme immunoassay of mumps virus in cell culture with peroxidase-labelled virus specific monoclonal antibodies and its application for determination of antibodies, *J. Virol. Methods* **22**:99–108 (1988).

249. VANDVIK, B., NILSEN, R. E., VARTDAL, F., AND NORRBY, E., Mumps meningitis: Specific and nonspecific antibody responses in the central nervous system, *Acta Neurol. Scand.* **65**:469–487 (1982).

250. VANDVIK, B., NORRBY, E., STEEN-JOHNSON, J., AND STENSVOLD, K., Mumps meningitis: Prolonged pleocytosis and occurrence of mumps virus-specific oligoclonal IgG in the cerebrospinal fluid, *Eur. Neurol.* **17**:13–22 (1978).

251. VOSBURGH, J. B., DIEHL, A. M., LIU, C., LAUER, R. M., AND FABIYI, A., Relationship of mumps to endocardial fibroelastosis: Complement-fixation, hemagglutination-inhibition and intradermal skin tests for mumps in children with and without endocardial fibroelastosis, *Am. J. Dis. Child.* **109**:69–73 (1965).

252. VUORI, M., LAHIKAINEN, E. A., AND PELTONEN, T., Perceptive deafness in connection with mumps: A study of 298 servicemen suffering from mumps, *Acta Otolaryngol.* **55**:231–236 (1962).

253. WAGENVOORT, J. H. T., HARMSEN, M., BOUTAHAR-TROUW, B. J. K., KRAAIJEVELD, C. A., AND WINKLER, K. C., Epidemiology of mumps in the Netherlands, *J. Hyg.* **85**:313–326 (1980).

254. WALKER, D. L., AND HINZE, H., A carrier state of mumps virus in human conjunctiva cells. II. Observations on intracellular transfer of virus and virus release, *J. Exp. Med.* **116**:751–758 (1962).

255. WAXHAM, M. N., ARONOWSKI, J., SERVER, A. C., WOLINSKY, J. S., SMITH, J. A., AND GOODMAN, H. M., Sequence determination of the mumps virus HN gene, *Virology* **164**:318–325 (1988).

256. WAXHAM, M. N., SERVER, A. C., GOODMAN, H. M., AND WOLINSKY, J. S., Cloning and sequencing of the mumps virus fusion protein gene, *Virology* **159**:381–389 (1987).

257. WEIBEL, R. E., BUYNAK, E. B., MCLEAN, A. A., AND HILLEMAN, M. R., Persistence of antibody after administration of monovalent and combined live attenuated measles, mumps, and rubella virus vaccines, *Pediatrics* **61**:5–11 (1978).

258. WEIBEL, R. E., BUYNAK, E. B., STOKES, J., JR., WHITMAN, J. E., JR., AND HILLEMAN, M. R., Evaluation of live attenuated mumps virus vaccine, strain Jeryl Lynn, in: *Vaccines against Viral and Rickettsial Diseases of Man: Proceedings of the First International Conference on Vaccines against Viral and Rickettsial Diseases in Man* (Washington, DC, November 7–11, 1966) (Scientific Publ. 147), pp. 430–437, Washington, DC, Pan American Health Organization (1967).

259. WEIBEL, R. E., STOKES, J., JR., BUYNAK, E. B., LEAGUS, M. B., AND HILLEMAN, M. R., Durability of immunity following administration of Jeryl Lynn strain live attenuated mumps virus vaccine, *J.A.M.A.* **203**:14–18 (1968).

260. WEIBEL, R. E., STOKES, J., JR., BUYNAK, E. B., WHITMAN, J., AND HILLEMAN, M. R., Live attenuated mumps virus vaccine. 3. Clinical and serologic aspects in a field evaluation, *N. Engl. J. Med.* **276**:245–251 (1967).

261. WEIBEL, R. E., VILLAREJOS, V. M., HERNANDEZ, C. G., STOKES, J., JR., BUYNAK, E. B., AND HILLEMAN, M. R., Combined live measles–mumps virus vaccine, *Arch. Dis. Child.* **48**:532–536 (1973).

262. WELLER, T. H., AND CRAIG, J. R., Isolation of mumps virus at autopsy, *Am. J. Pathol.* **25**:1105–1115 (1949).

263. WERNER, C. A., Mumps orchitis and testicular atrophy. I. Occurrence, *Ann. Intern. Med.* **32**:1066–1074 (1950).

264. WERNER, C. A., Mumps orchitis and testicular atrophy. II. A factor in male sterility, *Ann. Inter. Med.* **32**:1075–1086 (1950).

265. WESSELHOEFT, C., AND WALCOTT, C. F., Mumps as a military disease and its control, *War Med.* **2**:213–222 (1942).

266. WESTMORE, G. A., PICKARD, B. H., AND STERN, H., Isolation of mumps virus from the inner ear after sudden deafness, *Br. Med. J.* **1**:14–15 (1979).

267. WHARTON, M., COCHI, S. L., HUTCHESON, R. H., BISTOWISH, J. M., AND SCHAFFNER, W., A large outbreak of mumps in the postvaccine era, *J. Infect. Dis.* **158**:1253–1260 (1988).

268. WHARTON, M., COCHI, S. L., HUTCHESON, R. W., AND SCHAFFNER, W., Mumps transmission in hospitals, *Arch. Intern. Med.* **150**:47–49 (1990).

269. WHITE, C. C., KOPLAN, J. P., AND ORENSTEIN, W. A., Benefits, risks and costs of immunization for measles, mumps, and rubella, *Am. J. Public Health* **75**:739–744 (1985).

270. WILFERT, C. M., Mumps meningoencephalitis with low cerebrospinal-fluid glucose, prolonged pleocytosis and elevation of protein, *N. Engl. J. Med.* **280**:855–859 (1969).

271. WOLINSKY, J. S., BARINGER, J. R., MARGOLIS, G., AND KILHAM, L., Ultrastructure of mumps virus replication in newborn hamster central nervous system, *Lab. Invest.* **31**:402–412 (1974).

272. WOLINSKY, J. S., KLASSEN, T., AND BARINGER, J. R., Persistence of neuroadapted mumps virus in brains of newborn hamsters after intraperitoneal inoculation, *J. Infect. Dis.* **133**:260–267 (1976).

273. WOLINSKY, J. S., AND WAXHAM, M. N., Mumps virus, in: *Virology* (B. N. FIELDS, D. M. KNIPE, R. M. CHANOCK, M. S. HIRSH, J. L. MELNICK, T. P. MONATH, AND B. ROIZMAN, EDS.), pp. 989–1011, Raven Press, New York, 1990.

274. WOLONTIS, S., AND BHORVATN, B., Mumps meningoencephalitis in Stockholm, November 1964–July 1961. II. Isolation attempts for the cerebrospinal fluid in a hospitalized group, *Scand. J. Infect. Dis.* **5**:261–271 (1973).

275. YAMADA, A., TAKEUCHI, K., TANABAYASHI, K., HISHIYAMA, M., AND SUGIURA, A., Sequence variation of the p gene among mumps virus strains, *Virology* **172**:374–376 (1989).

276. YAMADA, A., TAKEUCHI, K., TANABAYASHI, K., HISHIYAMA, M., TAKAHASHI, Y., AND SUGIURA, A., Differentiation of the mumps vaccine strains from the wild viruses by the nucleotide sequences of the P gene, *Vaccine* **8**:553–557 (1990).

277. YAMAUCHI, T., WILSON, C., AND ST. GEME, J. W., JR., Transmission of live, attenuated mumps virus to the human placenta, *N. Engl. J. Med.* **290**:710–712 (1974).

278. YOUNG, M. L., DICKSTEIN, B., WEIBEL, R. E., STOKES, J., JR., BUYNAK, E. B., AND HILLEMANN, M. R., Experience with Jeryl Lynn strain live attenuated mumps virus vaccine in pediatric outpatient clinic, *Pediatrics* **40**:798–803 (1967).

12. Suggested Reading

BAUM, S. G., AND LITMAN, N., Mumps virus, in: *Principles and Practices of Infectious Diseases*, 4th ed. (G. L. MANDELL, J. E. BENNETT, AND R. DOLIN, EDS.), pp. 1496–1501, Churchill Livingstone, New York, 1995.

BRUNELL, P. A., Mumps, in: *Textbook of Infectious Diseases*, 3rd ed. (J. D. CHERRY AND R. D. FEIGIN, EDS.), pp. 1610–1613, Saunders, Philadelphia, 1992.

COCHI, S. L., WHARTON, M., AND PLOTKIN, S. A., Mumps vaccine, in: *Vaccines* (S. A. PLOTKIN AND E. A. MORTIMER, JR., EDS.), pp. 277–301, Saunders, Philadelphia, 1994.

GNANN, J. W., JR., Meningitis and encephalitis caused by mumps virus, in: *Infections of the Central Nervous System* (W. M. SCHELD, R. J. WHITLEY, AND D. T. DURACK, EDS.), pp. 113–125, Raven Press, New York, 1991.

GUTIERREZ, K., Mumps virus, in: *Pediatric Infectious Diseases* (S. S. LONG, L. K. PICKERING, AND C. G. PROBER, EDS.), pp. 1242–1247, Churchill Livingstone, Inc., New York, 1997.

WOLINSKY, J. S., AND WAXHAM, M. N., Mumps virus, in: *Virology* (B. N. FIELDS, D. M. KNIPE, R. M. CHANOCK, M. S. HIRSH, J. L. MELNICK, T. P. MONATH, AND B. ROIZMAN, EDS.), pp. 989–1011, Raven Press, New York, 1990.

CHAPTER 19

Parainfluenza Viruses

W. Paul Glezen and Floyd W. Denny

1. Introduction

The parainfluenza viruses are species of the genus *Paramyxovirus*, family *Paramyxoviridae*.[101] They are exceeded only by respiratory syncytial virus (RSV) as important causes of lower respiratory disease in young children, and they commonly reinfect older children and adults to produce upper respiratory illnesses. There is considerable diversity in both epidemiologic and clinical manifestations of infections by the parainfluenza viruses. Parainfluenza virus type 1 (Para 1) is the principal cause of croup (laryngotracheobronchitis) in children, and parainfluenza virus type 3 (Para 3) is second only to RSV as a cause of pneumonia and bronchiolitis in infants less than 6 months of age. Parainfluenza virus type 2 (Para 2) resembles Para 1 in clinical manifestations, but serious illnesses occur less frequently; infections with parainfluenza virus type 4 (Para 4) are detected infrequently, and associated illnesses are usually inconsequential.

2. Historical Background

Chanock[18] reported the first isolation of parainfluenza virus from human sources in 1956 in Cincinnati; this virus, recovered from children with croup, was designated originally as the "croup-associated" (CA) virus. Two additional parainfluenza strains were identified in 1958 by their ability to adsorb guinea pig erythrocytes onto infected rhesus monkey kidney cells in culture.[21] Because these viruses, first designated "HA-1" and "HA-2," or hemadsorption 1 and 2, shared many biological properties

with CA virus while being antigenically distinct, they were reclassified as parainfluenza viruses: Para 1 (HA-2), Para 2 (CA), and Para 3 (HA-1).[6] Type 4 was first isolated in 1960.[87] During the same period, these viruses were compared with isolates obtained from animals. The Sendai virus,[106] recovered from rodents, was found to share antigens with Para 1 and is classified as a subtype of this strain.[29,30] In 1959, a hemadsorbing virus[137] antigenically similar to Para 3 virus[1] was recovered from cattle with "shipping fever." A simian virus, SV5,[81] has been shown to be related to Para 2 virus.[79,161]

3. Methodology Involved in Epidemiologic Analysis

3.1. Sources of Mortality Data

Limited mortality data are available and consist of sporadic case reports[2,38,170] and fatal pneumonia in severely immunocompromised patients.[32,48,83,164] The failure to identify more fatal cases may be a result of the lability of these viruses in postmortem specimens. Most deaths caused by parainfluenza virus infections are probably related to Para 3 virus infections in young infants, but even in this age group, the etiology of fatal cases is not documented with sufficient frequency to allow a reasonable estimate of the mortality rate in the general population. This problem is magnified in developing countries, where mortality attributed to acute respiratory disease is many times greater than in Europe or North America.[12] Available data suggest that parainfluenza virus infections contribute to this excessive mortality.[10,85,140,148]

3.2. Sources of Morbidity Data

Official morbidity reports do not include acute respiratory infections; furthermore, laboratory diagnosis is required to establish the diagnosis of a parainfluenza virus

W. Paul Glezen • Influenza Research Center, Department of Microbiology and Immunology, Baylor College of Medicine, Houston, Texas 77030. **Floyd W. Denny** • Department of Pediatrics, School of Medicine, University of North Carolina, Chapel Hill, North Carolina 27514.

infection. Morbidity data, therefore, are based on special research studies of the etiology of respiratory disease in various communities around the world.

Standardized techniques have been employed by reliable investigators to provide some estimate of the impact of parainfluenza viruses as causes of acute respiratory disease in diverse populations. Most studies have been cross-sectional studies of the etiology of either illnesses of hospitalized patients or epidemics in closed populations. These studies provide a limited view of the total morbidity resulting from infections with these viruses. Seven investigations of the etiology of acute respiratory disease conducted over extended periods of time and involving large populations merit special mention because of the broader perspective that they provide. These seven studies in Washington, DC,[15,20] Chapel Hill, North Carolina,[33,55,58] Tecumseh, Michigan,[121,123] Seattle, Washington,[43] Houston, Texas,[46,51,57] Tucson, Arizona,[152,168] and Great Britain,[26,110] have differed in methods of patient selection, types of illnesses surveyed, and ages of the subjects, but taken together, they constitute an extensive survey of respiratory illness over an extended time period in widely diverse geographic environments and socioeconomic groups. Numerous other studies have focused on the occurrence of respiratory illness in specialized population groups. Children seen on the wards and in the clinics of hospitals have been studied extensively both in North American and in Great Britain.[53,72,74,78,114,129,158,171] These patients include those with underlying conditions that put them at high risk for serious consequences of respiratory infections.[69,73,90,164] Studies of uncomplicated respiratory disease in institutionalized children[3,22,68,93] daycare groups,[108] and school children[141] have contributed information about the spectrum of disease caused by these agents. Mild illnesses occurring within families have been described in both the United States[59] and Great Britain.[7,75]

The role of parainfluenza viruses as etiologic agents in respiratory illnesses of adults has also been investigated[41,42,62,66,119,124,149,165] (W. P. Glezen, unpublished data). These studies have reported sampling of patients with acute respiratory disease in hospitals, military services, university student health services, and industrial medical facilities. The ability of parainfluenza viruses to infect volunteers has been demonstrated in studies in both the United States[94,136,145,153] and Great Britain.[151,155,156] There have also been studies of the role of respiratory pathogens including the parainfluenza viruses in asthmatic children[17,113] and in patients with chronic bronchitis.[158] Evidence of infections with paramyxoviruses has been obtained in all areas of the world,[107,120,150,160] includ-

ing tropical climates[13,103,122,127] and isolated communities,[16,111] except for the absence of Para 1 and 2 antibody in children in very remote Indian tribes in South America.[14] In developing countries, parainfluenza viruses are second to respiratory syncytial virus as the most commonly identified respiratory pathogens.[10,13,19,85,89,127,140,148,154] As in Europe and North America, Para 3 virus is the most common parainfluenza virus isolated from severely ill infants, but Para 1 and 2 have also been found.[80,89,154]

3.3. Serological Surveys

In an effort to delineate the importance of the parainfluenza viruses as etiologic agents of respiratory disease, numerous serological studies have been conducted. Such serological surveys have frequently sought to determine whether the pattern of occurrences of parainfluenza virus infection is similar in special population groups to that reported elsewhere. Such studies have been useful in demonstrating the ubiquity of these agents and the relative similarity of age at acquisition of antibody in various areas of the world.[16,77,107,120,122,150] Many community studies have utilized both serological and isolation data but have relied heavily on serological data to determine the frequency of infection within the population.[42,43,64,119,121] Care should be taken in the interpretation of studies that rely solely on the measurement of hemagglutination-inhibition (HI) or complement-fixation (CF) antibodies because cross-reactions are frequent among the paramyxovirus group, which includes mumps virus as well as the parainfluenza viruses.

3.4. Laboratory Methods

Most studies have utilized primary monkey kidney tissue culture for isolation of parainfluenza viruses from clinical specimens. A continuous monkey kidney line, LLC-MK2, with trypsin in the medium, has been demonstrated to be equally sensitive and is less expensive than primary tissue cultures.[44] Infection is usually detected by adding guinea pig erythrocytes, which will adsorb to infected tissue culture cells. This property is useful for identification of isolates, which is accomplished by inhibition of hemadsorption with monospecific antisera raised in laboratory animals.[82] Rapid identification can be accomplished by use of fluorescent-tagged antisera for the major hemadsorbing viruses, including Para 1, 2, and 3.[53,167] Antibody in serum and nasal secretions can be measured by neutralization (N), HI, and CF.[23] All the serological tests have been adapted to microtiter techniques.[45,169] The endpoint of titers of N antibodies to parainfluenza viruses can be detected by hemadsorption; Para 3 virus can be adapted readily to produce a recognizable cytopathic effect in continuous cell lines, and the observation of cytopathology can also be used as an endpoint in N tests. Enzyme immunoassays have also been adapted for detection of parainfluenza antigens and antibodies.[91,142]

Sophisticated methods have been used to examine questions of antigenic diversity among the parainfluenza viruses.[27,28,69,70,102,117,133] Batteries of monoclonal antibodies directed toward epitopes of the surface glycoproteins have been used for antigenic characterization, and gene sequencing has been used to define these epitopes. Polymerase chain reaction-based sequencing assays have been adapted to epidemiologic studies of nosocomial infections.[92]

4. Biological Characteristics of the Virus that Affect the Epidemiologic Pattern

The parainfluenza viruses are RNA-containing viruses with a spike-covered lipoprotein envelope.[82] The spikes are formed by two glycoproteins. The larger glycoprotein (HN) is responsible for the hemagglutination and neuraminidase activity.[132] The smaller glycoprotein (F) is associated with fusion of cell membrane and with hemolysis. The active form of the F glycoprotein is generated by enzyme cleavage, which results in two disulfide-linked subunits, F_1 and F_2.[143] The viruses are ether sensitive and acid labile. They apparently are antigenically stable. Although subgroups of the specific types may have recognizable antigenic differences at certain epitopes, these differences are stable and nonprogressive. Conserved epitopes—particularly on the fusion proteins—generate antibodies that neutralize across these subgroups.[28,71] Para 3 virus remains infectious in an aerosol for periods greater than 1 hr.[118] Studies have shown that 10% of virus particles aerosolized are infectious after 1 hr in 20% relative humidity. No systematic studies of the stability of other parainfluenza viruses in aerosol have been reported, but Para 1 virus has been isolated from air samples obtained in the vicinity of infected children.[115]

The biological characteristic of parainfluenza viruses that seems to be the most important determinant of their success as respiratory pathogens is their ability to replicate in the respiratory epithelium without deeper invasion. The virus is extruded from the cell membrane without immediately destroying the cell, which allows continued release of particles from a single cell.[82] Furthermore, Para 3 virus may infect the mucosa of infants in the presence

of maternally derived circulating antibodies[20,53,58,125] and also frequently may reinfect older children who have circulating antibodies.[22] Excretion of virus may be prolonged up to 1 month or longer, even with the second or third infection[46,60,108] (F. A. Loda, A. M. Collier, and W. P. Glezen, unpublished data). Epidemiologic evidence suggests that reinfected children are infectious for their contacts.[22]

The incubation period is short, probably 3 to 6 days, and the virus spreads rapidly to a high percentage of persons in closed populations, which indicates that the virus possesses a high degree of infectiousness. Adult volunteers with preexisting antibody were infected with as little as 1500 median tissue culture infectious doses ($TCID_{50}$).[156] More than half the volunteers infected with the virus developed signs and symptoms of an upper respiratory illness.

5. Descriptive Epidemiology

5.1. Incidence and Prevalence Data

5.1.1. Serological Surveys. Available evidence strongly supports the view that parainfluenza viruses are ubiquitous viruses that infect most persons during childhood. Serological surveys have uniformly demonstrated that 90–100% of children have antibodies to Para 3 virus by age 5 years.[16,64,122] Antibodies to Para 1 and 2 viruses do not develop as rapidly or as universally as do antibodies to Para 3 virus or RSV; however, 74% of the children over 5 years of age, tested in Washington, DC, possessed antibody against Para 1 virus and 59% possessed antibody against Para 2 virus.[128] Similar data have been obtained in serological surveys in a wide variety of geographical areas[16,64,77,107,111,122,127] but not in remote tribes in South America.[14] Serological studies of natives of Tristan da Cunha who moved to the United Kingdom demonstrated the rapidity with which an isolated population may acquire antibodies to these viruses when they move to a heavily populated area.[157]

5.1.2. Association of Para 1 and 2 Viruses with Illnesses. Parainfluenza viruses were isolated from about 3% of persons with acute respiratory disease presenting for medical care in Houston from 1975 to 1980.[57] Table 1 shows the age distribution of these persons. For this period of time, 30% of the parainfluenza infections were type 1 and 10% were type 2. Persons with Para 1 and 2 infections tended to be older than persons infected with Para 3; over 40% of the Para 1 and 2 infections occurred among persons over 5 years of age. Studies of lower

Table 1. Age Distribution of Persons with Medically Attended Illnesses Associated with Parainfluenza Virus Infections, Houston, 1975–1980

Age	Number (%) of patients infected with		
	Type 1	Type 2	Type 3
Months			
0–6	26 (8.2)	7 (6.3)	141 (22.0)
7–12	40 (12.7)	10 (9.0)	118 (18.4)
Years			
1–4	119 (37.7)	47 (42.3)	226 (35.3)
5–9	54 (17.0)	21 (18.9)	63 (9.8)
10–14	15 (4.8)	3 (2.7)	11 (1.7)
15–19	7 (2.2)	5 (4.5)	13 (2.0)
20–44	41 (13.0)	10 (9.0)	32 (4.0)
≥45	14 (4.4)	8 (7.2)	36 (5.6)
Totals	316	111	640

respiratory disease have shown the highest isolation rates of Para 1 and 2 viruses to be between 4 months and 5 years of age.[20,58] The rate for lower respiratory disease associated with Para 1 virus infection in the Chapel Hill pediatric practice was 17 per 1000 children per year for children under 4 years of age, as shown in Fig. 1.[33,55,58] Infants followed in the Tucson study had lower rates in the first 6 months of life, but thereafter the rates were comparable to those reported from Chapel Hill.[168] Lower respiratory disease after that age was relatively uncommon.[125,149]

Studies of outbreaks of Para 1 and 2 viruses in day-care centers and in families suggest a high attack rate in young antibody-negative children. An outbreak of Para 1 virus infections in a residential home, in Washington, DC, resulted in the development of N antibody in 65% of the children without prior antibody.[22] In studies at the same

home for children in Washington, DC, 21 of 49 residents without antibody were infected during a Para 2 outbreak.[93] In two other outbreaks of Para 2 virus infection, one in a day-care center in Chapel Hill[108] and the other in a children's home in Kansas City, Kansas,[68] the infection rates were 79 and 65%, respectively, in antibody-negative children. Most illnesses in both studies were afebrile and involved the upper respiratory tract. These results suggest that Para 1 and 2 viruses are less effective than Para 3 virus in spreading through a susceptible population.[22]

5.1.3. Association of Para 3 Virus with Illnesses. Para 3 virus infections have been detected in all studies of hospitalized children with acute lower respiratory disease,[15,20,114,126,171] and this virus is now recognized to be second only to RSV as a cause of bronchiolitis and pneumonia in infants.[26,53,55,58,125] Few studies of open populations have had denominator data to allow calculation of attack rates. In Chapel Hill, studies of children with lower respiratory disease presenting to a pediatric practice from 1964 to 1975 showed that 15 children per 1000 per year were infected with Para 3 virus each year for the first 3 years of life (Fig. 1).[33,55,58] The rates for lower respiratory disease in Tucson infants were similar in the first 6 months but fell below those in Chapel Hill during the last half of the first year.[168]

In Houston, 121 children were followed from birth for infection with Para 3 virus.[57] About two thirds were infected during each of the first 2 years of life (Table 2). The risk of illness was at least 30 per 100 children per year. Most lower respiratory tract illnesses accompanied primary infection with a risk of 13 per 100 child-years. Over 90% were infected at least once by 24 months of age, and almost 40% had been infected twice.

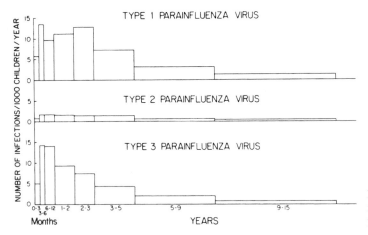

Figure 1. Age-specific attack rates of lower respiratory tract infections caused by parainfluenza viruses in children in a pediatric practice.

**Table 2. Frequency of Infection with Parainfluenza
Virus Type 3 (Para 3) among Children Studied from Birth,
Houston Family Study, 1975–1980**

Age (mo.)	Number of child-years	Number with infection with Para 3 virus			
		Primary	Reinfection	Total (rate)[b]	LRD[a] (rate)
0–12	121	75	6	81 (66.9)	14 (11.6)
13–24	90	30	31	61 (67.8)	9 (10.0)
25–36	63	2	21	23 (36.5)	2 (3.2)
37–48	39	0	13	13 (33.3)	1 (2.6)
49–60	24	0	4	4 (16.7)	0
Totals	337	107	75	182 (54.0)	26 (7.7)

[a]Lower respiratory tract disease.
[b]Per 100 child-years.

A small group of infants in Chapel Hill were followed longitudinally through the first 2 years of life, and 21 of 62 (34%) developed respiratory illnesses associated with Para 3 virus infection; six (9.7 per 100 children) had evidence of involvement of the lower respiratory tract (W. P. Glezen, unpublished data). Of 15 children infected during the first year of life, four were reinfected during the second year; 66% of the total had N antibodies at the end of 2 years, indicating that mild or inapparent infections occurred in a number equal to the number of those presenting with illness. These two studies with intensive follow-up show that lower respiratory tract involvement is common with Para 3 infection occurring during infancy.

Studies in Seattle in a prepaid group health care program have reported isolation rates of Para 3 virus considerably lower than those observed in the Chapel Hill pediatric practice, but in the patients with paired serum specimens, a high percentage showed antibody rises to this virus.[43] In closed populations such as children's homes and nurseries, infections with Para 3 virus were frequent.[3,22] The investigations of Junior Village in Washington, DC, showed that all infants without preexisting antibodies were infected, and 90% of the children with N antibody titers between 1 : 8 and 1 : 32 were reinfected, whereas only one third of those with higher antibody levels were reinfected.[22] In contrast, the studies focused on school-age children[141] and young adults[42,66,119] have revealed a relatively low rate of isolation of parainfluenza viruses, although infections by these agents are well documented in adults.

5.1.4. Para 4 Virus Infections. Para 4 virus has not been isolated frequently, but serological surveys indicate that infection may be common.[54,96,155] Most of these infections are considered to be asymptomatic, but isolation of the virus may be missed in some ill children because of the technical problems in identifying the virus in tissue culture.[54]

5.2. Epidemic Behavior

5.2.1. Para 1 and 2 Viruses. The longest continuous observation of the epidemic behavior of Para 1 virus has been at the Children's Hospital of the District of Columbia. From the initiation of the studies in 1957 until 1961, the virus was endemic.[20] Beginning in 1962, sharp epidemics of Para 1 virus began to occur every 2 years in the autumn of even-numbered years.[15,20] A similar pattern was noted in Great Britain between 1962 and 1977.[26,110] In Chapel Hill, epidemics of Para 1 virus occurred every 2 years from 1964, synchronously with these other epidemics.[55] Epidemics of Para 1 virus occurring in the fall of even-numbered years after 1962 have been described in other reports as well.[72,74,78,114] After 1970, a 3-year lapse occurred before resumption of biannual epidemics.[34] Therefore, since 1973, Para 1 epidemics have occurred in the autumn of odd-numbered years in Houston, accompanied by Para 2 as shown in Fig. 2.[57] In contrast, in other longitudinal studies, including those in Tecumseh,[121] Para 1 virus occurred in the endemic pattern until 1971; however, during the later extension of the Tecumseh studies (1976–1981), the occurrence of Para 1 was confined to autumn and early winter.[123] A more irregular pattern of epidemic occurrence was noted in Seattle, where Para 1 virus epidemics were associated with increased incidence of croup during the winters of 1966–1967, 1967–1968, and 1970–1971.[43] The evidence suggests that Para 1 virus can, at different times and in different places, assume varied patterns of occurrence; however, the predominant pattern in temperate zones is for biannual epidemics.

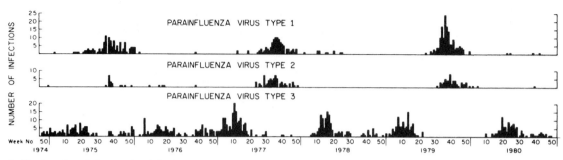

Figure 2. Number of parainfluenza virus infections by week among persons with febrile respiratory illnesses attending sentinel primary care centers, Houston, Texas, 1975–1980.

Para 2 virus appears to be as nearly ubiquitous as Para 1 virus on the basis of serological surveys, but it is not associated as frequently with severe clinical disease as is Para 1 virus. Because many studies depend on the isolation of the agent from patients with lower respiratory disease to establish the presence of the virus in the community, there is less information on the occurrence of Para 2 virus. The available data suggest that it occurs in a sporadic epidemic pattern, often disappearing from the community for fairly long periods of time. In Chapel Hill, Para 2 virus epidemics occurred in the fall of the odd-numbered years[55] and have continued this pattern in Houston from 1975 to 1980.[57] A similar pattern occurred in Washington, DC.[15,99] In Great Britain[26] and in the first study in Tecumseh,[121] Para 2 virus tended to occur in well-defined but somewhat erratic epidemics, and in the Seattle surveillance studies,[43] it was rarely identified. In the second Tecumseh study, Para 2 infections were identified in autumn and early winter, similar to the pattern in Houston and Chapel Hill.[123] In surveillance studies of young children with minor respiratory illness, Para 2 virus has had a distinct epidemic pattern with high attack rates in small, defined populations.[68,93,108]

5.2.2. Para 3 Virus. Infections with Para 3 virus have usually been described as endemic in nature, occurring in all seasons of the year.[15,55] Small outbreaks have been noted, but no predictable periodicity in their occurrence was noted until after a sharp outbreak occurred in Houston in 1977 (Fig. 2).[57] Since that time, Para 3 infections have tended to occur in the spring or during the months when Para 1 and 2 were not prevalent. Although the other major respiratory viruses have caused relatively discrete epidemics, outbreaks of Para 3 virus have occurred concurrently when other viruses were epidemic.[55] This virus was prevalent in Chapel Hill on one occasion each during RSV, influenza virus, or Para 1 virus epidemics.

The high reinfection rate of Para 3 virus allows it to spread in populations containing majorities of people with previous experience with the virus. Observations in Chapel Hill of families with a child under 2 years of age showed that Para 3 virus infected these infants regardless of family constellation, although RSV infections were more likely to occur in infants with an older sibling who might bring the virus into the home (W. P. Glezen, unpublished data). A latent state of infection in adults has been suggested,[60] and this could explain the ability of parents to transmit infection to their infants.

5.3. Geographic Distribution

Available evidence indicates that parainfluenza viruses are found throughout the world and in all areas cause illness in young children. There is remarkable similarity of serological and isolation data obtained from tropical,[13,37,122] temperate,[22,64,150] and arctic[111] climates. However, antibody to Para 1 was found in very few Tiriyo Indians in South America and in none under age 20; in the Xikrin tribe, there was no antibody under age 17. Para 2 antibody was present in Xikrin of all ages but almost entirely absent in other tribes. This suggests that fresh introductions of virus are needed in very remote tribes in which the population base is too small to sustain the infection.[14]

5.4. Temporal Distribution

Parainfluenza viruses can be isolated in any month of the year in both temperate[20,55] and tropical climates.[13,103,127] The most common time for the occurrence of epidemics of Para 1 virus has been during the autumn months.[15,26,43,55,72]

A similar occurrence in the autumn has been noted for Para 2 virus in studies in Chapel Hill[55] and Washington, DC,[15] but studies in Tecumseh[121] first showed a peak occurrence of Para 2 virus in the winter, but later shifted to autumn.[123]

Para 3 virus infections occur endemically throughout the year. During the first 7 years of studies at the Children's Hospital of the District of Columbia, infections with Para 3 virus were associated with respiratory illness in 70 of 81 months.[20] A similar endemicity was noted in studies in the Chapel Hill pediatric practice, where this virus was associated with lower respiratory illness in 63 of 91 months.[55] In Houston, a distinct change in the temporal distribution occurred after the epidemic in the spring of 1977 (Fig. 2).[57] In subsequent years, most of the infections have been detected during spring months and have been essentially absent during the months when Para 1 and 2 were prevalent. Subsequently, a study in England and Wales noted the occurrence of Para 3 infections in the summer from 1978 to 1987.[39] During the same period, investigators in Sydney, Australia, reported a springtime predominance for Para 3 infections.[35]

5.5. Age Distribution

5.5.1. Para 1 and 2 Viruses.
Maternal antibody apparently prevents severe disease in very young infants infected with either Para 1 or 2 virus. Data from studies of lower respiratory disease show few cases of severe disease in infants under 4 months of age.[13,26,43,53,58,171] After 4 months of age, there is a rise in the number of cases of croup and other lower respiratory diseases.[58] This high incidence continues until approximately age 6 years. After a child reaches school age, there is a much lower incidence of lower respiratory symptomatology, and lower respiratory illnesses in persons infected by Para 1 and 2 viruses are distinctly unusual in adolescents[55] and adults,[42,149] although they have been reported.[165]

Studies of milder illness have shown a similar age distribution. Infection and minor clinical illness have been demonstrated more frequently in younger children than in adolescents and adults.[7,59,121] Infections do occur in older persons,[26,42,66,75,119,121,123] most of which presumably represent recurrent infections in persons with antibody. Table 1 shows the frequency of febrile acute respiratory illnesses associated with Para 1 and 2 infections for adults presenting to primary care clinics.[57] The studies conducted in Tecumseh[121,123] showed serological evidence of frequent infections in adults in the age groups that would be likely to include parents of school children. In family studies, infection in adults occurred concurrently with illness in their children, but attack rates in adults were distinctly lower than those in younger family members.[7,59]

5.5.2. Para 3 Virus.
Initial infections with Para 3 virus occur early in life. Among children followed from birth in the Houston Family Study, 62% were infected during the first year of life, and by the end of the second year 92% had been infected at least once and 36% had one or more reinfections.[57] Several investigators have noted that, similar to RSV infections, Para 3 virus infections often occur in the first months of life, when infants still possess circulating antibodies derived from their mothers.[20,26,53,58,125] Infants born with high titers of maternal antibody to Para 3, however, usually are spared during the first months of life, and the risk of lower respiratory disease is greater for primary infection during the second year than for primary infection during the first year.[57] In the studies of lower respiratory disease in the Chapel Hill pediatric group practice, the age-specific attack rate for Para 3 disease paralleled that of RSV.[58] The average annual attack rate ranged from almost 15 to 7 per 1000 children per year for children under 3 years of age.[33] After the first 3 years of life, the incidence of lower respiratory illnesses associated with this virus fell off considerably, but studies in other populations indicate that reinfections are common in older children and adults, but usually are not associated with evidence of lower tract involvement.[22,42,59,62,66]

5.6. Sex

The greater frequency of croup in males has long been recognized.[131] Studies in the Chapel Hill pediatric practice demonstrated infection rates for Para 1 virus of 1.8 lower respiratory illnesses per 100 boys compared to 1.1 per 100 girls.[58] This sex difference disappeared after age 6 years. A similar predominance of serious illness in young males has been noted in other studies of lower respiratory disease.[109] The available information suggests that rates of infection in young boys and girls are the same but that the clinical manifestations of infection are more severe in boys.[55]

There were insufficient numbers of cases of Para 2 virus infections to analyze.

Para 3 virus had identical rates of lower respiratory illness in males and females for children observed in the Chapel Hill pediatric practice.[55] The attack rates for lower respiratory illnesses caused by Para 3 virus infections in the Chapel Hill pediatric practice were 12 and 11 per 1000 children per year for boys and girls under 6 years of age, respectively. In the longitudinal study in Houston, however, boys were more likely to have lower respiratory tract involvement with primary infection.[57]

5.7. Race and Occupation

No differences in infection rates or in the consequences of infection with parainfluenza viruses have been noted in different racial or occupational groups. A South African study suggested that Para 1 and 2 virus infections were predominantly associated with croup in white children, whereas these same viruses were more often associated with pneumonia in black children.[89] This difference may reflect socioeconomic factors rather than racial differences.

5.8. Occurrence in Special Epidemiologic Settings

The frequencies of infections with parainfluenza viruses are greater in studies of young children hospitalized with lower respiratory disease[125] or in day-care facilities.[108] Infections associated with respiratory disease occur but are usually mild or inapparent in school-age children,[141] university students,[42,66] military recruits,[119,165] and adults.[59,62,75,124,149]

Nosocomial infections with respiratory viruses may occur readily if susceptible infants are not kept isolated from children with respiratory disease. Parainfluenza virus infections are detected commonly in studies of hospital-acquired infections in children on pediatric wards[52,126,144,162] and neonatal units.[116] Mufson *et al.*[126] reported infections with Para 3 virus in 18% of infants who were not infected or were infected with some other respiratory pathogen at the time of admission to Cook County Children's Hospital.

Parainfluenza viruses are being reported increasingly in association with severe infections in immunocompromised hosts.[32,48,69,73,83,90,164] Bone marrow transplant units house particularly vulnerable patients. Surveillance with rapid methods of detecting infection is necessary to institute control measures and recognize candidates for antiviral therapy.

5.9. Socioeconomic Status

There is some evidence to suggest that infants from low-income families are more likely to be hospitalized with bronchiolitis or pneumonia in the first months of life than are infants from middle-income families.[55] Both RSV and Para 3 virus are frequent etiologic agents of these illnesses. There may be many factors involved in this situation. The decision to hospitalize an infant may be based on the social milieu in which the child lives and not necessarily on the severity of the presenting signs and symptoms; however, some evidence suggests that young infants in low-income families do have more severe illnesses, and this may result from exposures to relatively large inocula at an earlier age.

Unlike the case with Para 3 virus, there does not appear to be any difference in severity of disease caused by Para 1 and 2 viruses in children of different socio-economic groups,[55] but, as noted in Section 8.1, the clinical manifestations may be different. Croup is the predominant syndrome seen in upper- and middle-income groups, whereas pneumonia is the major manifestations of infection in lower-income groups.

6. Mechanisms and Route of Transmission

Transmission of parainfluenza viruses is by direct person-to-person contact or large-droplet spread. These viruses do not persist long in the environment, but Para 1 virus has been recovered from air samples collected in the vicinity of infected patients,[115] and from 1 to 10% of Para 3 virus particles in aerosol may be viable after 1 hr.[118] Adult volunteers who have had prior natural infection have been reinfected experimentally by inoculation of the upper airway with a coarse aerosol or nasal drops or both.[94,156] The high rate of infection early in life coupled with the frequency of reinfection suggests that the virus spreads readily, that reinfected persons may be infectious, and that a relatively small inoculum is necessary to produce an infection.

There is little evidence of animal reservoirs for human disease and no evidence of vector spread. The Sendai virus is a rodent strain of Para 1 virus, but there is no evidence that it is related to disease in humans.[82] SV5, a simian virus related to Para 2 virus, has been reported to cause human infections on rare occasions[79,163]; reports of human infections must be reviewed critically because this virus often contaminates monkey kidney tissue cultures.[81] A virus that is antigenically similar to human Para 3 virus has been isolated from several animals, particularly cattle[137] and sheep.[76] Bovine Para 3 virus is at least one of the agents commonly associated with an economically important disease of cattle usually called "shipping fever,"[139] but there is no evidence of spread between cattle and man.

7. Pathogenesis and Immunity

7.1. Pathogenesis

Little is known about the pathogenesis of infections with Para 1 and 2 viruses. Spread is by person-to-person contact and by droplet nuclei; the incubation period ranges from 3 to 6 days.[22] Pathological studies have noted the predilection for the marked inflammatory response of glottic tissues[170] that is evident from the serious clinical manifestation (croup) of infections with these viruses. Welliver and his associates have presented evidence for development of virus-specific IgE antibody in nasal secretions during infection with parainfluenza viruses.[163] They suggest that chemical mediators such as histamine may contribute to the inflammatory response, resulting in croup or bronchiolitis. Children who develop croup often are atopic or have a family history of atopy. These children may subsequently develop reactive airway disorders.

Because infection with Para 3 virus may occur early in life in the presence of maternal antibody, the possibility of an immunopathologic process similar to that proposed for RSV disease must be considered. Gardner et al.[53] noted the similarity of the clinical manifestations of infections with these two viruses in early infancy and the need for learning more about the pathogenesis. Although it is well accepted that infection with Para 3 virus may occur in the presence of low levels of maternal antibody, there is no direct evidence that maternal antibody will enhance the pathogenicity of the host–virus interaction. Clinical observations do not support this thesis and, to the contrary, suggest that high levels of maternal antibody are protective.[57] Furthermore, experimental studies in hamsters have not demonstrated enhanced virulence in the presence of passive antibody.[56] In fact, the pulmonary infiltrations were less in animals with passive antibody than in those without; however, animals with primary infection in the presence of passive antibody had significantly greater infiltrate at the time of reinfection. This suggests that passive antibody may dampen some part or parts of the immune response to the primary infection, which allows for development of more disease at the time of reinfection. Studies have demonstrated that the primary serum antibody response to infection is decreased by passive antibody, but it is unknown whether other parameters such as surface antibody production and cellular immune response are affected. Viremia has been reported both with primary infection[138] and with reinfection.[60,86] It is difficult to determine the consequences of this, because clinical and pathological evidence of infection has been noted only in the respiratory tract. Viremia with reinfection was reported to have occurred in the presence of circulating antibodies; more work is necessary to determine the factors inherent in these observations.

7.2. Immunity

Studies with Para 1 virus[145] and Para 2 virus[153] have revealed that immunity is better correlated with the presence of nasal antibody than with the level of serum antibody. The *in vitro* demonstration of intracellular neutralization of Sendai virus with specific IgA antibody provides a mechanism for protection provided by this form of mucosal immunity.[112]

The protective immune response to Para 3 virus infection in humans has not been established. Kasel *et al.*[95] have shown that natural infection stimulates production of antibodies to both surface glycoproteins. Antibody titers to the HN protein are usually higher, and antibodies to the F protein may increase after reinfection. Clinical studies have shown that the risk of reinfection is inversely related to the level of circulating antibodies,[22,57] but the relationship to surface antibody or cellular immunity is unknown. Although reinfection is common, the consequences of subsequent infections are usually minor. Lower respiratory involvement with reinfection is unusual, and mild upper respiratory symptoms are the rule. In experimental infections with Para 3 virus, recovery from primary infection was correlated with interferon production[67] and virus-specific cytotoxicity mediated by T lymphocytes.[100] Antibody and immune interferon probably contribute to defense against reinfection. Interferon activity has been detected in nasal secretions of children with Para 1 infections.[63]

Studies in calves infected with Para 3 virus indicate that maternal antibody derived from colostrum is protective.[47] Calves are usually protected for 10–23 weeks after birth if allowed to suckle. Inactivated vaccines injected parenterally have provided some protection against natural and experimental challenge, but attenuated viruses inoculated intranasally provide better protection.[61] There is some evidence that hyperimmunization of dams with inactivated vaccines to increase antibody titers in colostrum will decrease the morbidity of their offspring.[104]

8. Patterns of Host Response

8.1. Clinical Manifestations

Primary infections with parainfluenza viruses are usually symptomatic, but the clinical manifestations may range from an afebrile upper respiratory illness to severe and life-threatening lower respiratory disease (Table 3). The most characteristic and clinically important syndrome associated with infections with Para 1 and 2 viruses is croup or laryngotracheobronchitis. Para 1 virus was isolated from 20% of patients with croup in studies in the Chapel Hill pediatric practice.[34,58] Para 2 virus is much less frequently associated with croup than is para 1 virus.[20,34,55] Mortality has been reported with both agents but is rare.[13,38,53,129,161] Studies in ambulatory populations

Table 3. Parainfluenza Viruses:
Serotypes and Associated Clinical and Epidemiologic Manifestations

Human serotype	Major clinical syndrome	Peak age (months)	Sex predominance	Periodicity[a]	Antigenically related animal strain
Para 1	Croup	6–24	Male	Epidemic (fall)	Sendai virus (rodents)
Para 2	Croup	6–24	Male	Epidemic (fall)	Simian virus 5 (SV5)
Para 3	Pneumonia and bronchiolitis	0–6	None	Endemic	Bovine Para 3 virus
Para 4	URI	Unknown	Unknown	Endemic	None

[a]Peak season in parentheses.

suggest that most initial infections with Para 1 virus result in febrile upper respiratory illness, whereas initial infections with Para 2 virus result in somewhat less severe upper respiratory illness, with a high proportion being afebrile.[68,108] An exception to this was the epidemic in Finland where most of the croup was caused by Para 2.[105] Reinfection with either of these agents is associated with upper respiratory symptoms indistinguishable from those caused by other viruses.[57,59] The clinical manifestations of infections with Para 1 and 2 viruses appear to be different in developing countries. The association of these viruses with croup, which is well described in developed countries, does not appear to be common in developing countries.[11,13,80,89] Croup is not recognized as frequently, and diphtheria and measles remain the important etiologies when it does occur. Para 1 and 2 virus infections are usually associated with pneumonia and tracheobronchitis.[11,13,80,89] The reasons for this are not understood at this time.

The clinical manifestations of infections with Para 3 virus are varied. In studies in the Chapel Hill pediatric practice of children presenting with lower respiratory disease, the diagnosis varied with age[55]; infants under 1 year of age were likely to present with bronchiolitis or pneumonia, whereas children from 6 to 18 months might often have croup. Older children were usually diagnosed as having tracheobronchitis. In general, though, it should be stated that there was no consistent clinical presentation of Para 3 virus infection. The severe manifestations of Para 3 infection, usually pneumonia, in developing countries are similar to those in developed countries.

Prospective studies of children indicate that primary infection with Para 3 virus is usually symptomatic but often mild.[57,108] In children followed for the first 2 years of life, about one third of primary infections involved the lower respiratory tract, but only 5% of primary infections resulted in lower respiratory illnesses for which families sought medical care (W. P. Glezen, unpublished data).

Frequency of reinfection—both symptomatic and asymptomatic—decreased with age (and experience with the virus). Reinfection was symptomatic among young children at Junior Village 20% of the time,[22] but this frequency is probably higher than that in open populations; the circumstances of living conditions in this nursery could have resulted in a greater inoculum than would occur ordinarily.

Evidence that parainfluenza virus infections may predispose to bacterial infections is accumulating.[25,36] Both otitis media and bacterial tracheitis in children have been associated with these infections. An outbreak of invasive pneumococcal disease in a chimpanzee colony followed infections with Para 3.[88] Animals with Para 3 upper respiratory illness were 5.7 times more likely to develop invasive pneumococcal infection despite the fact that most of the animals had received pneumococcal vaccine.

The incidence of reinfection in adult populations is difficult to determine because of the cross-reactions that may occur by serological testing, particularly if the HI or CF test is used. Studies in Tecumseh have allowed an estimate of infection rates ranging from 11 to 13 per 100 adults per year for types 1, 2, and 3 combined.[123] The viruses have been isolated only rarely from asymptomatic adults (W. P. Glezen, unpublished data).

8.2. Diagnosis

Clinical diagnosis of the etiology of sporadic episodes of acute respiratory disease is difficult; however, infections with parainfluenza viruses may be suspected when characteristic clinical manifestations occur in known epidemiologic patterns.[55] As noted in Section 5.2.1, Para 1 virus infections have produced predictable epidemics of croup in the autumn of even-numbered years in widely scattered geographic areas through 1970. Since 1973, epidemics of both Para 1 and 2 viruses have occurred in the autumn of odd-numbered years.[57] Croup is most common in boys between 6 months and 3 years of age. With knowledge of the patterns of parainfluenza virus infections provided by regional laboratories, it may be possible to predict the occurrence of epidemics of Para 1 or 2 virus infections and thereby establish an estimate of the etiology of respiratory illnesses, particularly when they present with a major clinical manifestation such as croup. In developing countries, Para 1 and 2 infections appear to be less consistently associated with croup. In a Para 1 epidemic in Trinidad, pneumonia was the most common lower respiratory tract diagnosis.[13] In such a setting the clinical disease pattern will be less useful in identifying the occurrence of a Para 1 epidemic. Para 3 virus infections are much less predictable, but this etiology should be considered for infants less than 6 months of age with bronchiolitis or pneumonia occurring at times other than during RSV epidemics.[55] Although medically attended lower respiratory illnesses associated with Para 3 virus have had an even sex distribution,[58] longitudinal studies have found lower tract involvement more common in boys.[57]

Laboratory diagnosis is best accomplished by identification of virus in respiratory secretions, because of cross-reactions that may be observed among parainfluenza viruses when serological tests are used. Cross-reactions appear to be less frequent with tests measuring N antibodies than with those for CF or HI antibodies. Methods for virus isolation and identification are described in Section 3.4. Identification of hemadsorbing viruses can be expedited by the use of immunofluorescent techniques.[53] These viruses can be identified directly in infected epithelial cells shed in respiratory secretions collected from patients by aspiration or washings of the upper respiratory passages. The cells must be washed to separate them from mucus and fixed on slides in six "spots" that are then exposed to specific antisera. Usually, a battery of antisera to parainfluenza viruses and influenza A and B viruses is used along with preimmunization serum from the same animal host. After the slides are washed, fluorescein-tagged antiserum against the animal host globulin is used to counterstain the cells. This method may allow specific diagnosis within a few hours after the specimen is collected. Other antigen detection methods have been used but are not available generally.[142]

9. Control and Prevention Based on Epidemiologic Data

As soon as the clinical importance of the parainfluenza viruses was demonstrated, efforts were begun to develop effective vaccines. Most early efforts involved the use of formalin-inactivated, parenterally administered vaccines prepared from virus pools grown in embryonated hens' eggs.[24,49,84,98] These preparations were given either as a monovalent preparation,[50] as a trivalent parainfluenza vaccine, or as a multivalent vaccine including other respiratory pathogens. Most preparations were aqueous suspensions, but one alum-adsorbed egg-grown Para 3 vaccine was tested.[50] One group employed a monkey-kidney-grown, formalin-killed, alum-precipitated parainfluenza vaccine given both as a trivalent parainfluenza vaccine and as a multivalent respiratory vaccine.[159] A high proportion of subjects developed serum antibody rises with parenterally administered, formalin-inactivated vaccines, particularly antibody-negative subjects. Antibody rises occurred in a smaller proportion of subjects with preexisting antibody, either passively acquired maternal antibody or antibody secondary to natural infection.[43] All studies uniformly failed to show evidence of protection even in subjects with good serum antibody responses. There was no evidence of more severe disease in the subjects given the trivalent aqueous parainfluenza vaccine, an important finding in view of the fact that infants given alum-precipitated RSV vaccine had more severe illness than expected when they were naturally infected.[97,98] Experimental studies in hamsters suggest that vaccines consisting of purified surface glycoproteins may provide better protection than formalin-inactivated whole-virus vaccines.[132]

The demonstration that protective immunity is more closely related to the level of secretory nasal antibody than to serum antibody[145,153] and the actual worsening of disease seen with inactivated RSV vaccine have led to a reevaluation of the use of parenteral respiratory vaccines. Efforts have focused on development of attenuated live vaccines given intranasally. Initial trials of a temperature-sensitive Para 1 mutant in animals were promising,[130] but an attenuated strain suitable for use in humans has not been developed. Cold adaptation of Para 3 virus has resulted in development of strains attenuated for weanling hamsters,[8,31] but passage 18 was not sufficiently attenuated for seronegative young children.[9] Cold passage 45 appears to be attenuated for chimpanzees.[65] It has been shown that aerosol inoculation of an inactivated Para 2 vaccine will safely stimulate nasal secretory antibody in adults.[166] The effectiveness of this method of vaccine administration has not been further tested.

Several new approaches to vaccine development are in progress. Parenteral immunization with purified HN and F subunits of Para 3 protects cotton rats from wild virus challenge.[5] Subunits in lipid vesicles administered intranasally also protect hamsters from challenge.[134] Both microencapsulation of whole virus and recombinant adenoviruses expressing the surface glycoproteins are promising approaches for oral immunization.[40,135]

Epidemiologic data suggest that the most important age group to immunize against Para 1 and 2 viruses is children over 4 months of age. This contrasts with Para 3 virus and RSV, which cause significant illness between 1 and 3 months of age and would require passive or active immunization in early infancy.

For bovine Para 3 virus infections, both inactivated and attenuated vaccines administered parenterally and intranasally have been used in calves.[61] The results are not consistent, but it appears that attenuated strains given intranasally are more effective. Evidence suggests that in calves, vaccine-induced antibody is associated with protection and does not cause more severe disease. Studies in cotton rats have shown that delayed hypersensitivity can be stimulated by both replicating and nonreplicating Para 3 antigens.[4] Until more is known about the pathogenesis of human parainfluenza disease, care must be exercised in the use of any vaccine, particularly in infants.

10. Unresolved Problems

The clinical spectrum of illness, age incidence, and ubiquity of parainfluenza virus infections has been demonstrated. There remains some question about the epidemic patterns associate with parainfluenza viruses that can be clarified by further observations over time. The most important needs at the present time are for more intense study of the pathogenesis of disease and a better understanding of the immune response in order to facilitate the development of effective vaccines. Humoral and cellular responses to natural infection and reinfection must be compared to responses following immunization with attenuated viruses, inactivated whole virus, or purified surface glycoproteins HN and F. The humoral responses, both surface and serum, must be determined for each immunoglobulin class and subclass. The determinants of protective immunity must be defined for natural infection, and efforts made to duplicate the response by active immunization. Immunization of women of childbearing age should be considered as a means of protecting infants by passively acquired antibody during the first months of life.

A major issue is to understand the reasons for the great excess mortality from acute respiratory disease in developing countries. Acute respiratory disease is one of the leading causes of death in children under 5 years of age in the developing world. As many as four million children die each year from pneumonia.[12] Available information suggests that parainfluenza viruses contribute to this problem. Further studies of the factors contributing to the excess mortality are badly needed, and methods for preventing serious respiratory disease must be adaptable to the variety of situations in developing nations.

11. References

1. ABINANTI, R. R., CHANOCK, R. M., COOK, M. K., WONG, D., AND WARFIELD, M., Relationship of human and bovine strains of myxovirus parainfluenza 3, *Proc. Soc. Exp. Biol. Med.* **106**:466–469 (1961).

2. AHERNE, W., BIRD, T., COURT, S. D. M., GARDNER, P. S., AND McQUILLEN, J., Pathological changes in virus infections of the lower respiratory tract in children, *J. Clin. Pathol.* **23**:7–18 (1970).

3. AITKEN, C. J. D., MOFFAT, M. A. J., AND SUTHERLAND, J. A. W., Respiratory illness and viral infection in an Edinburgh nursery, *J. Hyg.* **65**:25–36 (1967).

4. AMBROSE, M. W., AND WYDE, P. R., Parainfluenza virus type 3 (PIV3)-specific delayed type hypersensitivity responses in cotton rats given different PIV3 antigen preparations, *Vaccine* **11**:336–342 (1993).

5. AMBROSE, M. W., WYDE, P. R., EWASYSHYN, M., BONNEAU, A.-M., CAPLAN, B., MEYER, H. L., AND KLEIN, M., Evaluation of the immunogenicity and protective efficacy of a candidate parainfluenza type 3 subunit vaccine in cotton rats, *Vaccine* **9**:505–511 (1991).

6. ANDREWES, C. H., BANG, M. B., CHANOCK, R. M., AND ZHDANOV, B. M., Parainfluenza viruses 1, 2, and 3: Suggested names for recently described myxoviruses, *Virology* **8**:129–130 (1969).

7. BANATVALA, J. E., ANDERSON, T. B., AND REISS, R. B., Parainfluenza infections in the community, *Br. Med. J.* **1**:537–540 (1964).

8. BELSHE, R. B., AND HISSON, F. K., Cold-adaptation of parainfluenza virus type 3: Induction of three phenotype markers, *J. Med. Virol.* **10**:235–242 (1984).

9. BELSHE, R. B., KARRON, R. A., NEWMAN, F. K., ANDERSON, E. L., NUGENT, S. L., STEINHOFF, M., CLEMENTS, M. L., WILSON, M. H., HALL, S. L., TIERNEY, E. L., AND MURPHY, R. R., Evaluation of a live attenuated cold-adapted parainfluenza virus type 3 vaccine in children, *J. Clin. Microbiol.* **30**:2064–2070 (1992).

10. BERMAN, S., Epidemiology of acute respiratory infections in children of developing countries, *Rev. Infect. Dis.* **13**(Suppl.):S454–462 (1991).

11. BERMAN, S., DUENAS, A., BEDOYA, A., CONSTAIN, V., LEON, S., BORRERO, I., AND MURPHY, J., Acute lower respiratory tract illnesses in Cali, Colombia: A two-year preliminary study, *Pediatrics* **71**:210–217 (1983).

12. BERMAN, S., AND MCINTOSH, K., Selective primary health care: Strategies for control of disease in the developing world XXI. Acute respiratory infections, *Rev. Infect. Dis.* **7**:674–691 (1985).

13. BISNO, A. L., BARRATT, N. P., SWANSTON, W. H., AND SPENCE, L. P., An outbreak of acute respiratory disease in Trinidad associated with parainfluenza viruses, *Am. J. Epidemiol.* **91**:68–77 (1970).

14. BLACK, F. L., HIERHOLZER, W. J., DE PINHEIRO, P. F., EVANS, A. S., WOODHALL, J. P., OPTON, E. M., EMMONS, J. E., WEST,

B. S., Edsall, G., Downs, W. G., and Wallace, E. D., Evidence for persistence of infectious agents in isolated human populations, *Am. J. Epidemiol.* **100**:230–250 (1974).

15. Brandt, C. D., Kim, H. W., Chanock, R. M., and Parrott, R. S., Parainfluenza virus epidemiology, *Pediatr. Res.* **8**:422 (1974).

16. Brown, P. K., and Taylor-Robinson, D., Respiratory virus antibodies in sera of persons living in isolated communities, *Bull. WHO* **34**:895–900 (1966).

17. Busse, W. W., Lemanske, R. F., Jr., and Dick, E. C., The relationship of viral infections and asthma, *Chest* **101**(Suppl.): 385S–388S (1992).

18. Chanock, R. M., Association of a new type of cytopathogenic myxovirus with infantile croup, *J. Exp. Med.* **104**:555–576 (1956).

19. Chanock, R., Chambon, L., Chang, W., Goncalves Ferreira, F., Gharpure, P., Grant, L., Hatem, J., Iman, I., Kalra, S., Lim, K., Madalengoitia, J., Spence, L., Teng, P., and Ferreira, W., WHO respiratory disease survey in children. A serological study, *Bull. WHO* **37**:363–369 (1967).

20. Chanock, R. M., and Parrott, R. H., Acute respiratory disease in infancy and childhood: Present understanding and prospects for prevention, *Pediatrics* **36**:21–39 (1965).

21. Chanock, R. M., Parrott, R. H., Cook, K., Andrews, B. E., Bell, J. A., Reichelderfer, T., Kapikian, A. Z., Mastrota, F. M., and Huebner, R. J., Newly recognized myxoviruses from children with respiratory disease, *N. Engl. J. Med.* **258**:207–213 (1958).

22. Chanock, R. M., Parrott, R. H., Johnson, K. M., Kapikian, A. Z., and Bell, J. A., Myxoviruses: Parainfluenza, *Am. Rev. Respir. Dis.* **88**:152–166 (1963).

23. Chanock, R. M., Wong, D. C., Huebner, R. J., and Bell, J. A., Serologic response in individuals infected with parainfluenza viruses, *Am. J. Public Health* **50**:1858–1865 (1960).

24. Chin, J., Magoffin, R. L., Shearer, L. A., Schieble, J. H., and Lennette, E. H., Field evaluation of a respiratory syncytial virus vaccine and a trivalent parainfluenza virus vaccine in a pediatric population, *Am. J. Epidemiol.* **89**:449–463 (1969).

25. Chonmaitree, T., Owen, M. J., Patel, J. A., Hedgpeth, D., Horlick, D., and Howie, V. M., Effect of viral respiratory tract infection on outcome of acute otitis media, *J. Pediatr.* **120**:856–862 (1992).

26. Clarke, S. K. R., Parainfluenza virus infections, *Postgrad Med. J.* **49**:792–797 (1973).

27. Coelingh, K. V., and Tierney, E. L., Antigenic and functional organization of human parainfluenza virus type 3 fusion glycoprotein, *J. Virol.* **63**:375–382 (1989).

28. Coelingh, K. V., and Winter, C. C., Naturally occurring human parainfluenza type 3 viruses exhibit divergence in amino acid sequence of their fusion protein neutralization epitopes and cleavage sites, *J. Virol.* **64**:1329–1334 (1990).

29. Cook, M., Andrews, B. E., Fox, H. H., Turner, H. C., James, W. D., and Chanock, R. M., Antigenic relationships among the "newer" myxoviruses (parainfluenza), *Am. J. Hyg.* **69**:250–264 (1959).

30. Cook, M. K., and Chanock, R. M., *In vivo* antigenic studies of parainfluenza viruses, *Am. J. Hyg.* **77**:150–159 (1963).

31. Crookshanks, F. K., and Belshe, R. B., Evaluation of cold-adapted and temperature sensitive mutants of parainfluenza virus type 3 in weanling hamsters, *J. Med. Virol.* **13**:243–249 (1984).

32. Delage, G., Brochu, P., Pelletier, M., Jasmin, G., and Lapointe, N., Giant-cell pneumonia caused by parainfluenza virus, *J. Pediatr.* **94**:426–429 (1979).

33. Denny, F. W., and Clyde, W. A., Jr., Acute lower respiratory infections in nonhospitalized children, *J. Pediatr.* **108**:635–646 (1986).

34. Denny, F. W., Murphy, T. F., Clyde, W. A., Jr., Collier, A. M., and Henderson, F. W., Croup: An 11-year study in a pediatric practice, *Pediatrics* **7**:871–876 (1983).

35. de Silva, L. M., and Cloonan, M. J., Brief report: Parainfluenza virus type 3 infections: Findings in Sydney and some observations on variations in seasonality world wide, *J. Med. Virol.* **35**:19–21 (1991).

36. Donnelly, B. W., McMillan, J. A., and Weiner, L. B., Bacterial tracheitis: Report of eight cases and review, *Rev. Infect. Dis.* **12**:729–735 (1990).

37. Doraisingham, S., and Ling, A. E., Acute non-bacterial infections of the respiratory tract in Singapore: An analysis of three years' laboratory findings, *Ann. Acad. Med. Singapore* **10**:69–78 (1981).

38. Downham, M. A. P. S., McQuillen, J., and Gardner, P. S., Diagnosis and clinical significance of parainfluenza virus infections in children, *Arch. Dis. Child.* **49**:8–15 (1974).

39. Easton, A. J., and Eglin, R. P., Epidemiology of parainfluenza virus type 3 in England and Wales over a ten-year period, *Epidemiol. Infect.* **102**:531–535 (1989).

40. Ebata, S. N., Prevec, L., Graham, F. L., and Dimock, K., Function and immunogenicity of human parainfluenza virus 3 glycoproteins expressed by recombinant adenoviruses, *Virus Res.* **24**:21–33 (1992).

41. Evans, A. S., Infections with hemadsorption virus in University of Wisconsin students, *N. Engl. J. Med.* **263**:233–237 (1960).

42. Evans, A. S., and Dick, E. C., Acute pharyngitis and tonsillitis in University of Wisconsin students, *J.A.M.A.* **190**:699–708 (1964).

43. Foy, H. M., Cooney, M. K., Maletzky, A. J., and Grayston, J. T., Incidence and etiology of pneumonia, croup and bronchiolitis in preschool children belonging to a prepaid medical care group over a four-year period, *Am. J. Epidemiol.* **97**:80–92 (1973).

44. Frank, A. L., Couch, R. B., Griffis, C. A., and Baxter, B. D., Comparison of different tissue cultures for isolation and quantitation of influenza and parainfluenza viruses, *J. Clin. Microbiol.* **10**:32–36 (1979).

45. Frank, A. L., Puck, J., Hughes, B. J., and Cate, T. R., Microneutralization test for influenza A and B and parainfluenza 1 and 2 viruses that uses continuous cell lines and fresh serum enhancement, *J. Clin. Microbiol.* **12**:426–432 (1980).

46. Frank, A. L., Taber, L. H., Wells, C. R., Wells, J. M., Glezen, W. P., and Paredes, A., Patterns of shedding of myxoviruses and paramyxoviruses in children, *J. Infect. Dis.* **144**:433–441 (1981).

47. Frank, G. H., and Marshall, R. G., Parainfluenza 3 virus infection of cattle, *J. Am. Vet. Med. Assoc.* **163**:858–860 (1973).

48. Frank, J. A., Warren, R. W., Tucker, J. A., Zeller, J., and Wilfert, C. M., Disseminated parainfluenza infection in a child with severe combined immunodeficiency, *Am. J. Dis. Child.* **137**:1172–1174 (1983).

49. Fulginiti, V. A., Eller, J. J., Sieber, O. F., Joyner, J. W., Minamitani, M., and Meiklejohn, G., Respiratory virus immunization. I. A field trial of two inactivated respiratory virus vaccines: An aqueous trivalent parainfluenza virus vaccine and an alum-precipitated respiratory syncytial virus vaccine, *Am. J. Epidemiol.* **89**:435–448 (1969).

50. Fulginiti, V. A., Sieber, O. F., John, T. J., Askin, P., and Umlant, H. J., Jr., Parainfluenza virus immunization. II. The

influence of age and maternal antibody upon successful immunization with an alum-adsorbed parainfluenza type 3 vaccine, *Pediatr. Res.* **1:**50–58 (1967).

51. GARDENER, G., FRANK, A. L., AND TABER, L. H., Effects of social and family factors on viral respiratory infection and illness in the first year of life, *J. Epidemiol. Commun. Health* **38:**42–48 (1984).

52. GARDNER, P. S., COURT, S. D. M., BROCKLEBANK, J. T., DOWNHAM, M. A. P. S., AND WEICHTMAN, D., Virus cross-infections in paediatric wards, *Br. Med. J.* **2:**571–575 (1973).

53. GARDNER, P. S., McQUILLAN, J., McGUCKIN, R. M., AND DITCHBURN, R. K., Observations on clinical and immunofluorescent diagnosis of parainfluenza virus infections, *Br. Med. J.* **2:**7–12 (1971).

54. GARDNER, S. D., The isolation of parainfluenza 4 subtypes A and B in England and serological studies of their prevalence, *J. Hyg.* **67:**545–550 (1969).

55. GLEZEN, W. P., AND DENNY, F. W., Epidemiology of acute lower respiratory disease in children, *N. Engl. J. Med.* **288:**498–505 (1973).

56. GLEZEN, W. P., AND FERNALD, G. W., Effect of passive antibody on parainfluenza virus type 3 pneumonia in hamsters, *Infect. Immun.* **14:**212–216 (1976).

57. GLEZEN, W. P., FRANK, A. L., TABER, L. H., AND KASEL, J. A., Parainfluenza virus type 3: Seasonality and risk of infection and reinfection in young children, *J. Infect. Dis.* **150:**851–857 (1984).

58. GLEZEN, W. P., LODA, F. A., CLYDE, W. A., JR., SENIOR, R. J., SHEAFFER, C. I., CONLEY, W. G., AND DENNY, F. W., Epidemiologic patterns of acute lower respiratory disease of children in pediatric group practice, *J. Pediatr.* **78:**397–406 (1971).

59. GLEZEN, W. P., WULFF, H., LAMB, G. A., RAY, C. G., CHIN, T. D. Y., AND WENNER, H. A., Patterns of virus infections in families with acute respiratory illnesses, *Am. J. Epidemiol.* **86:**350–361 (1967).

60. GROSS, P. A., GREEN, R. H., AND CURNEN, M. C. M., Persistent infection with parainfluenza type 3 virus in man, *Am. Rev. Respir. Dis.* **108:**894–898 (1973).

61. GUTENKUNST, D. E., PATON, I. M., AND VOLANCE, F. J., Parainfluenza 3 vaccine in cattle: Comparative efficacy of intranasal and intramuscular routes, *J. Am. Vet. Med. Assoc.* **155:**1879–1885 (1969).

62. GWALTNEY, J. M., JR., HENDLEY, J. O., SIMON, G., AND JORDAN, W. S., Rhinovirus infections in an industrial population. I. The occurrence of illness, *N. Engl. J. Med.* **275:**1261–1268 (1966).

63. HALL, C. B., DOUGLAS, R. G., JR., SIMON, R. L., AND GEIMAN, J. M., Interferon production in children with respiratory syncytial influenza and parainfluenza virus infections, *J. Pediatr.* **93:**28–32 (1978).

64. HALL, C. E., BRANDT, C. D., FROTHINGHAM, T. E., SPIGLAND, I., COONEY, M. K., AND FOX, J. P., The Virus Watch Program: A continuing surveillance of viral infections in metropolitan New York families. IX. A comparison of infections with several respiratory pathogens in New York and New Orleans families, *Am. J. Epidemiol.* **94:**367–385 (1971).

65. HALL, S. L., SARRIS, C. M., TIERNEY, E. L., LONDON, W. T., AND MURPHY, B. R., A cold-adapted mutant of parainfluenza virus type 3 is attenuated and protective in chimpanzees, *J. Infect. Dis.* **167:**958–962 (1992).

66. HAMRE, D., CONNELLY, A. P., AND PROCKNOW, J. J., Virologic studies of acute respiratory disease in young adults. IV. Virus isolations during four years of surveillance, *Am. J. Epidemiol.* **83:**238–249 (1966).

67. HARMON, A. T., HARMON, M. W., AND GLEZEN, W. P., Evidence of interferon production in the hamster lung after primary or secondary exposure to parainfluenza virus type 3, *Am. Rev. Respir. Dis.* **125:**706–711 (1982).

68. HARRIS, D. J., WULFF, H., RAY, C. G., POLAND, J. D., CHIN, T. D. Y., AND WENNER, H. A., Viruses and disease. II. An outbreak of parainfluenza type 2 in a children's home, *Am. J. Epidemiol.* **87:**419–425 (1968).

69. HEIDEMANN, S. M., Clinical characteristics of parainfluenza virus infection in hospitalized children, *Pediatr. Pulmonol.* **13:**86–89 (1992).

70. HENDRICKSON, K. J., Monoclonal antibodies to human parainfluenza virus type 1 detect major antigenic changes in clinical isolates, *J. Infect. Dis.* **164:**1128–1134 (1991).

71. HENDRICKSON, K. J., AND SAVATSKI, L. L., Genetic variation and evolution of human parainfluenza virus type 1 hemagglutinin neuraminidase: Analysis of 12 clinical isolates, *J. Infect. Dis.* **166:**995–1005 (1992).

72. HERRMANN, E. C., AND HABLE, K. A., Experiences in laboratory diagnosis of parainfluenza viruses in routine medical practice, *Mayo Clin. Proc.* **45:**177–188 (1970).

73. HERZOG, K. D., DUNN, S. P., LANGHAM, M. R., AND MARMON, L. M., Association of parainfluenza virus type 3 infection with allograft rejection in a liver transplant recipient, *Pediatr. Infect. Dis. J.* **8:**534–536 (1989).

74. HOLZEL, A., PARKER, L., PATTERSON, W. H., CARTMEL, D., WHITE, L. L. R., PURDY, R., THOMPSON, K. M., AND TOBIN, J., Virus isolations from throats of children admitted to hospital with respiratory and other diseases, *Br. Med. J.* **1:**614–619 (1965).

75. HOPE-SIMPSON, R. E., AND HIGGINS, P. G., A respiratory virus study in Great Britain: Review and evaluation, *Prog. Med. Virol.* **11:**354–407 (1969).

76. HORE, D. E., STEVENSON, R. G., GILMOUR, N. J. L., VANTISIS, J. T., AND THOMPSON, D. A., Isolation of parainfluenza virus from the lungs and nasal passages of sheep showing respiratory disease, *J. Comp. Pathol.* **78:**259–265 (1968).

77. HORNSLETH, A., Respiratory virus disease in infancy and childhood in Copenhagen 1963–65. An estimation of the etiology based on complement fixation tests, *Acta Pathol. Microbiol. Scand.* **69:**287–303 (1967).

78. HORSTMANN, D. M., AND HSIUNG, G. D., Myxovirus infections and respiratory illnesses in children, *Clin. Pediatr.* **2:**378–386 (1963).

79. HSIUNG, G. D., ISACSON, P., AND TUCKER, G., Studies of parainfluenza viruses. II. Serologic interrelationships in humans, *Yale J. Biol. Med.* **35:**534–544 (1963).

80. HUGHES, J., SINHA, D., COOPER, M., SHAH, K., AND BOSE, S., Lung tap in childhood—bacteria, viruses, and mycoplasmas in acute lower respiratory tract infections, *Pediatrics* **44:**477–485 (1969).

81. HULL, R. N., MINNER, J. R., AND SMITH, J. W., New agents recovered from tissue cultures of monkey kidney cells, *Am. J. Hyg.* **63:**204–215 (1956).

82. JACKSON, G. G., AND MULDOON, R. L., Viruses causing common respiratory infections in man. II. Enteroviruses and paramyxoviruses, *J. Infect. Dis.* **128:**387–469 (1973).

83. JARVIS, W. R., MIDDLETON, P. J., AND GELFAND, E. W., Parainfluenza pneumonia in severe combined immunodeficiency disease, *J. Pediatr.* **94:**423–425 (1979).

84. JENSEN, K. E., PEELER, B. E., AND DULWORTH, W. G., Immunizations against parainfluenza infections, *J. Immunol.* **89:**216–226 (1962).

85. JOHN, T. J., CHERIAN, T., STEINHOFF, M. C., SIMOES, E. A. F., AND JOHN, M., Etiology of acute respiratory infections in children in tropical southern India, *Rev. Infect. Dis.* **13**(Suppl.):S463–S469 (1991).

86. JOHNSON, D. P., AND GREEN, R. H., Viremia during parainfluenza type 3 virus infection of hamsters, *Proc. Soc. Exp. Biol. Med.* **144**:745–748 (1973).

87. JOHNSON, K. M., CHANOCK, R. M., COOK, M. K., AND HUEBNER, R. J., Studies of a new human hemadsorption virus. I. Isolation, properties and characterization, *Am. J. Hyg.* **71**:81–92 (1960).

88. JONES, E. E., ALFORD, P. L., REINGOLD, A. L., RUSSELL, H., KEELING, M. E., AND BROOME, C. V., Predisposition to invasive pneumococcal illness following parainfluenza type 3 virus infection in chimpanzees, *J. Am. Vet. Med. Assoc.* **185**:1351–1353 (1984).

89. JOOSTING, A., HARWIN, R., ORCHARD, M., MARTIN, E., AND GEAR, J., Respiratory viruses in hospital patients on the Witerwatersrand, *S. Afr. Med. J.* **55**:403–409 (1979).

90. JOSEPHS, S., KIM, H. W., BRANDT, C. D., AND PARROTT, R. H., Parainfluenza 3 virus and other common respiratory pathogens in children with human immunodeficiency virus infection, *Pediatr. Infect. Dis. J.* **7**:207–209 (1988).

91. JULKUNEN, I., Serological diagnosis of parainfluenza virus infections by enzyme immunoassay with special emphasis on purity of viral antigens, *J. Med. Virol.* **14**:177–187 (1984).

92. KARRON, R. A., O'BRIEN, K. L., FROEHLICH, J. L., AND BROWN, V. A., Molecular epidemiology of a parainfluenza type 3 virus outbreak on a pediatric ward, *J. Infect. Dis.* **167**:1441–1445 (1993).

93. KAPIKIAN, A. Z., BELL, J. A., MASTROTA, F. M., HUEBNER, R. J., WONG, D. C., AND CHANOCK, R. M., An outbreak of parainfluenza 2 (croup-associated) virus infection, *J.A.M.A.* **183**:324–330 (1963).

94. KAPIKIAN, A. Z., CHANOCK, R. M., REICHELDERFER, T. E., WARD, T. G., HUEBNER, R. J., AND BELL, J. A., Inoculation of human volunteers with parainfluenza virus type 3, *J.A.M.A.* **18**:537–541 (1961).

95. KASEL, J. A., FRANK, A. L., KEITEL, W. A., TABER, L. H., AND GLEZEN, W. P., Acquisition of serum antibody to specific viral glycoproteins of parainfluenza virus 3 in children, *J. Virol.* **52**:828–832 (1984).

96. KILLGORE, G. E., AND DOWDLE, W. R., Antigenic characterization of parainfluenza 4A and 4B by the hemagglutination-inhibition test and distribution of HI antibody in human sera, *Am. J. Epidemiol.* **91**:308–316 (1970).

97. KIM, H. W., CANCHOLA, J. G., BRANDT, C. D., PYLES, G., CHANOCK, R. M., JENSEN, K., AND PARROTT, R. H., Respiratory syncytial virus disease in infants despite prior administration of antigenic inactivated vaccine, *Am. J. Epidemiol.* **89**:422–434 (1969).

98. KIM, H. W., CANCHOLA, J. G., VARGOSKO, A. J., ARROBIO, J. O., DE MEJO, J. L., AND PARROTT, R. H., Immunogenicity of inactivated parainfluenza type 1, type 2, and type 3 vaccines in infants, *J.A.M.A.* **196**:819–824 (1966).

99. KIM, H. W., VARGOSKO, E. J., CHANOCK, R. M., AND PARROTT, R. H., Parainfluenza 2 (CA) virus: Etiologic association with croup, *Pediatrics* **28**:614–621 (1961).

100. KIMMELL, K. A., WYDE, P. R., AND GLEZEN, W. P., Evidence of a T-cell-mediated cytotoxic response to parainfluenza virus type 3 pneumonia in hamsters, *J. Reticuloendothel. Soc.* **31**:71–83 (1982).

101. KINGSBURY, D. W., BRATT, M. A., CHOPPIN, P. W., HANSON, R. P., HOSAKA, Y., terMUELEN, V., NORRBY, E., PLOWRIGHT, W., ROTT, R., AND WUNNER, W. H., Paramyxoviridae, *Intervirology* **10**:137–152 (1978).

102. KLIPPMARK, E., RYDBECK, SHIBUTA, H., AND NORRBY, E., Antigenic variation of human and bovine parainfluenza virus type 3 strains, *J. Gen. Virol.* **71**:1577–1580 (1990).

103. KLOENE, W., BANG, F. B., CHAKRABORTY, S. M., COOPER, M. R., KULEMANN, H., OTA, M., AND SHAH, K. V., A two-year respiratory virus survey in four villages in West Bengal, India, *Am. J. Epidemiol.* **2**:307–320 (1970).

104. KOLAR, J. R., JR., SHECHMEISTER, I. L., AND STRACK, L. E., Field experiments with formalin-killed-virus vaccine against infectious bovine rhinotracheitis, bovine viral diarrhea, and parainfluenza-3, *Am. J. Vet. Res.* **34**:1469–1471 (1973).

105. KORPPI, M., HALONEN, P., KLEEMOLA, M., AND LAUNIALA, K., The role of parainfluenza viruses in inspiratory difficulties in children, *Acta Paediatr. Scand.* **77**:105–111 (1988).

106. KUROYA, M., AND ISHIDA, N., Newborn virus pneumonitis (type Sendai). II. Isolation of a new virus possessing hemagglutinin activity, *Yokohoma Med. Bull.* **4**:217–233 (1953).

107. LAPLACA, M., AND MOSCOVICI, C., Distribution of parainfluenza antibodies in different groups of population, *J. Immunol.* **88**:72–77 (1962).

108. LODA, F. A., GLEZEN, W. P., AND CLYDE, W. A., Respiratory disease in group day care, *Pediatrics* **48**:428–437 (1972).

109. MALETZKY, A. J., COONEY, M. K., LUCE, R., KENNY, G. E., AND GRAYSTON, J. T., Epidemiology of viral and mycoplasmal agents associated with childhood lower respiratory illness in a civilian population, *J. Pediatr.* **78**:407–414 (1971).

110. MARTIN, A. J., GARDNER, P. S., AND McQUILLIN, J., Epidemiology of respiratory viral infection among pediatric inpatients over a six-year-period in northeast England, *Lancet* **2**:1035–1038 (1978).

111. MAYNARD, J. E., FELTZ, E. T., WULFF, H., FORTUINE, R., POLAND, J. D., AND CHIN, T. D. Y., Surveillance of respiratory virus infections among Alaskan Eskimo children, *J.A.M.A.* **200**:927–931 (1967).

112. MAZANEC, M. B., KAETZEL, C. S., LAMM, M. E., AND FLETCHER, D., Intracellular neutralization of virus by immunoglobulin A antibodies, *Proc. Natl. Acad. Sci. USA* **89**:6901–6905 (1992).

113. McINTOSH, K., ELLIS, E. F., HOFFMAN, L. S., LYBASS, T. G., ELLER, J. J., AND FULGINITI, V. A., The association of viral and bacterial respiratory infections with exacerbations of wheezing in young asthmatic children, *J Pediatr.* **82**:578–590 (1973).

114. McLEAN, D. M., BACH, R., LAVKE, R. P. B., AND McNAUGHTON, G. A., Myxoviruses associated with acute laryngotracheobronchitis in Toronto, 1962–1963, *Can. Med. Assoc. J.* **89**:1257–1259 (1963).

115. McLEAN, D. M., BANNATYNE, R. M., AND GIBAN, K., Myxovirus dissemination by air, *Can. Med. Assoc. J.* **96**:1449–1453 (1967).

116. MEISSNER, H. C., MURRAY, S. A., KIERNAN, M. A., SNYDMAN, D. R., AND McINTOSH, K., A simultaneous outbreak of respiratory syncytial virus and parainfluenza virus type 3 in a newborn nursery, *J. Pediatr.* **104**:68–684 (1984).

117. MERSON, J. R., HULL, R. A., ESTES, M. K., AND KASEL, J. A., Molecular cloning and sequence determination of the fusion protein gene of human parainfluenza virus type 1, *Virology* **167**:97–105 (1988).

118. MILLER, W. S., AND ARTENSTEIN, M. S., Aerosol stability of three acute respiratory disease viruses, *Proc. Soc. Exp. Biol. Med.* **125**:222–227 (1967).

119. MOGABGAB, W. J., Acute respiratory illnesses in university (1962–1966), military and industrial (1962–1963) populations, *Am. Rev. Respir. Dis.* **98:**359–379 (1968).

120. MONOZENKO, M. A., BANYSHEVA, A. E., TEMOFEYEVA, G. A., BYSTOYAKAVA, L. V., AND KALINNIKOVA, O. N., Diagnostic value of the complement fixation reaction in viral respiratory infections of infants, *Acta Virol.* **7:**534–541 (1963).

121. MONTO, A. A., The Tecumseh study of respiratory illness. V. Patterns of infection with the parainfluenza viruses, *Am. J. Epidemiol.* **97:**338–348 (1973).

122. MONTO, A. S., AND JOHNSON, K. M., Respiratory infections in the American tropics, *Am. J. Trop. Med. Hyg.* **17:**867–874 (1968).

123. MONTO, A. S., AND SULLIVAN, K. M., Acute respiratory illness in the community. Frequency of illness and the agents involved, *Epidemiol. Infect.* **110:**145–160 (1993).

124. MUFSON, M. A., CHANG, V., GILL, V., WOOD, S. C., ROMANSKY, M. J., AND CHANOCK, R. M., The role of viruses, mycoplasmas and bacteria in acute pneumonia in civilian adults, *Am. J. Epidemiol.* **86:**526–544 (1967).

125. MUFSON, M. A., KRAUSE, H. E., MOCEGA, H. E., AND DAWSON, F. W., Viruses, *Mycoplasma pneumoniae* and bacteria associated with lower respiratory tract disease among infants, *Am. J. Epidemiol.* **1:**192–202 (1970).

126. MUFSON, M. A., MOCEGA, H. E., AND KRAUSE, H. E., Acquisition of parainfluenza 3 virus infection by hospitalized children. I. Frequencies, rates, and temporal data, *J. Infect. Dis.* **128:**141–147 (1973).

127. OLSON, L. C., LEXOMBOON, U., SITHISARN, P., AND NOYES, H. E., The etiology of respiratory tract infections in a tropical country, *Am. J. Epidemiol.* **97:**34–43 (1973).

128. PARROTT, R. H., VARGOSKO, A. J., KIM, H. W., BELL, J. A., AND CHANOCK, R. M., III, Myxoviruses: Parainfluenza, *Am. J. Public Health* **52:**907–917 (1962).

129. PEREIRA, M. S., AND FISHER, O. D., An outbreak of acute laryngotracheobronchitis associated with parainfluenza 2 virus, *Lancet* **2:**790–791 (1960).

130. POTASH, L., LEES, R., GREENBERGER, J. L., HOYRUP, A., DENNEY, L. D., AND CHANOCK, R. M., A mutant of parainfluenza type 1 virus with decreased capacity for growth at 38°C and 39°C, *J. Infect. Dis.* **121:**640–647 (1970).

131. RABE, E. F., Infectious croup. II. "Virus" croup, *Pediatrics* **2:**415–427 (1948).

132. RAY, R., BROWN, V. E., AND COMPANS, R. W., Glycoproteins of parainfluenza type 3: Characterization and evaluation of a subunit vaccine, *J. Infect. Dis.* **152:**1219–1230 (1985).

133. RAY, R., DUNCAN, J., QUINN, R., AND MATSUOKA, Y., Distinct hemagglutinin and neuraminidase epitopes involved in antigenic variation of recent human parainfluenza virus type 2 isolates, *Virus Res.* **24:**107–113 (1992).

134. RAY, R., GLAZE, B. J., MOLDOVIANU, Z., AND COMPANS, R. W., Intranasal immunization of hamsters with envelope glycoproteins of human parainfluenza virus type 3, *J. Infect. Dis.* **157:**648–654 (1988).

135. RAY, R., NOVAK, M., DUNCAN, J. D., MATSUOKA, Y., AND COMPANS, R. W., Microencapsulated human parainfluenza virus induces a protective immune response, *J. Infect. Dis.* **167:**752–755 (1993).

136. REICHELDERFER, T. E., CHANOCK, R. M., CRAIGHEAD, J. E., HUEBNER, R. J., WARD, T. J., TURNER, H. C., AND JAMES, W. D., Infection of human volunteers with type 2 hemadsorption virus, *Science* **128:**779–780 (1958).

137. REISINGER, R. C., HEDDLESTON, K. L., AND MANTHEI, C. A., A myxovirus SF-4 associated with shipping fever of cattle, *J. Am. Vet. Med. Assoc.* **135:**147–152 (1959).

138. ROCCHI, G., ARANGRO-RUIZ, G., GIANNINI, V., JEMOLO, A. M., ANDREONI, G., AND ARCHETTI, I., Detection of viremia in acute respiratory disease of man, *Acta Virol.* **14:**405–407 (1970).

139. ROSNER, S. F., Bovine parainfluenza type 3 virus infection and pasteurellosis, *J. Am. Vet. Med. Assoc.* **159:**1375–1381 (1971).

140. RUUTU, P., HALONEN, P., MEURMAN, O., TORRES, C., PALADIN, F., YAMAOLA, K., AND TUPASI, T. E., Viral lower respiratory infections in Filipino children, *J. Infect. Dis.* **161:**175–179 (1990).

141. SALIBA, G. S., GLEZEN, W. P., AND CHIN, T. D. Y., Etiologic studies of acute respiratory illness among children attending public schools, *Am. Rev. Respir. Dis.* **95:**592–602 (1967).

142. SARKKINEN, H. R., HALONEN, P. E., AND SALMI, A. A., Type specific detection of parainfluenza viruses by enzyme immunoassay and radioimmunoassay in nasopharyngeal specimens of patients with acute respiratory disease, *J. Gen. Virol.* **156:**49–57 (1981).

143. SCHEID, A., AND CHOPPIN, P. W., Two disulfide-linked polypeptide chains constitute the active F protein of paramyxoviruses, *Virology* **80:**54–66 (1977).

144. SIMS, D. G., A two-year prospective study of hospital-acquired respiratory infections on paediatric wards, *J. Hyg. (Camb.)* **86:**335–342 (1981).

145. SMITH, C. B., PURCELL, R. H., BELLANTI, J. A., AND CHANOCK, R. M., Protective effect of antibody to parainfluenza type 1 virus, *N. Engl. J. Med.* **275:**1145 (1966).

146. STARK, J. D., HEATH, R. B., AND CURWEN, M. P., Infection with influenza and parainfluenza viruses in chronic bronchitis, *Thorax* **20:**124–127 (1965).

147. STEINHOFF, M., AND JOHN, J., Acute respiratory infections of children in India, *Pediatr. Res.* **17:**1032–1035 (1983).

148. SUWANJUTHA, S., CHANTAROJANASIRI, T., WATTHANA-KASETR, S., SIRINAVIN, S., RUANGKANCHANASETR, S., HOTRAKITYA, S., WASI, C., AND PUTHAVTHANA, P., A study of nonbacterial agents of acute lower respiratory tract infection in Thai children, *Rev. Infect. Dis.* **12**(Suppl.):S923–S928 (1990).

149. SULLIVAN, R. J., DOWDLE, W. R., MARINE, W. M., AND HIERHOLZER, J. C., Adult pneumonia in a general hospital, *Arch. Intern. Med.* **129:**935–942 (1972).

150. TAI, F.-H., AND CHING, C.-M., Antibody patterns of parainfluenza viruses in human populations on Taiwan, *Am. Rev. Respir. Dis.* **97:**941–945 (1968).

151. TAYLOR-ROBINSON, D., AND BYNOE, M. L., Parainfluenza 2 virus infections in adult volunteers, *J. Hyg.* **51:**407–417 (1963).

152. TAUSSIG, L. M., WRIGHT, A. L., MORGAN, W. J., HARRISON, H. R., RAY, C. G., AND THE GROUP HEALTH MEDICAL ASSOCIATES, The Tucson children's respiratory study. I. Design and implementation of a prospective study of acute and chronic respiratory illness in children, *Am. J. Epidemiol.* **129:**1219–1231 (1989).

153. TREMONTI, L. P., LIN, J. S. L., AND JACKSON, G. C., Neutralizing activity in nasal secretions and serum in resistance of volunteers to parainfluenza virus type 2, *J. Immunol.* **101:**572–577 (1968).

154. TUKEI, P., WAFULA, E., NSANZE, H., BELL, T., HAZLEFF, J., JDINYA-ACHOLA, W., OCHIENG, G., AND ADEMBA, A., Acute respiratory diseases among children of the world: Opportunities for study in East Africa, *Pediatr. Res.* **17:**1060–1062 (1983).

155. TYRRELL, D. A. J., AND BYNOE, M., Studies on parainfluenza type 2 and 4 viruses obtained from patients with common colds, *Br. Med. J.* **1:**471–474 (1969).

156. TYRRELL, D. A. J., BYNOE, M. L., BIRKUM, K., PETERSEN, S., AND PEREIRA, M. S., Inoculation of human volunteers with parainfluenza viruses 1 and 3 (HA$_2$ and HA$_1$), *Br. Med. J.* **2:**909–911 (1959).

157. TYRRELL, D., PETO, M., AND KING, N., Serological studies on infections by respiratory viruses of the inhabitants of Tristan da Cunha, *J. Hyg. (Camb.)* **65:**327–341 (1967).

158. VARGOSKO, A. J., CHANOCK, R. M., HUEBNER, R. J., LUCKEY, A. H., KIM, H. W., CUMMINGS, C., AND PARROTT, R. A., Association of type 2 hemadsorption (parainfluenza 1) virus and Asian influenza A virus with infectious croup, *N. Engl. J. Med.* **261:**1–9 (1959).

159. VELLA, P. P., WEIBEL, R. E., WOODHOUR, A. F., MASCOLI, C. C., LEAGUES, M. B., ITTENSOHN, O. L., STOKES, J., JR., AND HILLEMAN, M. R., Respiratory virus vaccines. VIII. Field evaluation of trivalent parainfluenza virus vaccine among preschool children in families, 1967–1968, *Am. Rev. Respir. Dis.* **99:**526–541 (1969).

160. VIHMA, L., Surveillance of acute viral respiratory diseases in children, *Acta Paediatr. Scand. [Suppl.]* **192:**1–52 (1969).

161. VON EULER, L. V., KANTOR, F. S., AND HSIUNG, G. D., Studies of parainfluenza viruses. I. Clinical, pathological and virological observations, *Yale J. Biol. Med.* **35:**523–533 (1963).

162. WELLIVER, R. C., AND MCLAUGHLIN, S., Unique epidemiology of nosocomial infections in a children's hospital, *Am. J. Dis. Child.* **138:**131–135 (1984).

163. WELLIVER, R. C., WONG, D. T., MIDDLETON, E., JR., SUN, M., MCCARTHY, N., AND OGRA, P. L., Role of parainfluenza virus-specific IgE in pathogenesis of croup and wheezing subsequent to infection, *J. Pediatr.* **101:**889–896 (1982).

164. WENDT, C. H., WEISDORF, D. J., JORDAN, M. C., BALFOUR, H. H., JR., AND HERTZ, M. I., Parainfluenza virus respiratory infection after bone marrow transplantation, *N. Engl. J. Med.* **326:**921–926 (1992).

165. WENZEL, R. P., MCCORMICK, D. P., AND BEAM, W. E., JR., Parainfluenza pneumonia in adults, *J.A.M.A.* **221:**294–295 (1972).

166. WIGLEY, F. M., FRUCHTMAN, M. H., AND WALDMAN, R. H.,

167. Aerosol immunization of humans with inactivated parainfluenza type 2 vaccine, *N. Engl. J. Med.* **283:**1250–1253 (1970).

167. WONG, D. T., WELLIVER, R. C., RIDDLESBERGER, K. R., SUN, M. S., AND OGRA, P. L., Rapid diagnosis of parainfluenza virus infection in children, *J. Clin. Microbiol.* **16:**164–167 (1982).

168. WRIGHT, A. L., TAUSSIG, L. M., RAY, C. G., HARRISON, H. R., HOLBERG, C. J., AND THE GROUP HEALTH MEDICAL ASSOCIATES, The Tucson children's respiratory study. II. Lower respiratory tract illness in the first year of life, *Am. J. Epidemiol.* **129:**1232–1246 (1989).

169. WULFF, H., SOEKEN, J., POLAND, J. D., AND CHIN, T. D. Y., A new microneutralization test for antibody determination and typing of parainfluenza and influenza viruses, *Proc. Soc. Exp. Biol. Med.* **125:**1045–1049 (1967).

170. ZINSERLING, A., Peculiarities of lesion in viral and mycoplasma infections of the respiratory tract, *Virchows Arch.* **356:**259–273 (1972).

171. ZOLLAR, L. M., KRAUSE, H. E., AND MUFSON, M. A., Microbiologic studies on young infants with lower respiratory tract disease, *Am. J. Dis. Child.* **126:**56–60 (1973).

12. Suggested Reading

BALE, J. R. (GUEST EDITOR), Etiology and epidemiology of acute respiratory tract infection in children in developing countries, *Rev. Infect. Dis.* **12**(Suppl.):S861–S1083 (1990).

DENNY, F. W., AND CLYDE, W. A., JR., Acute lower respiratory infections in nonhospitalized children, *J. Pediatr.* **108:**635–646 (1986).

GLEZEN, W. P., Reactive airway disorders in children. Role of respiratory virus infections, *Clin. Chest Med.* **5:**635–643 (1984).

GLEZEN, W. P., AND DENNY, F. W., Epidemiology of acute lower respiratory disease in children, *N. Engl. J. Med.* **288:**498–505 (1973).

HALL, C. B., Parainfluenza viruses, in: *Textbook of Pediatric Infectious Diseases* (R. D. FEIGIN AND J. D. CHERRY, EDS.), pp. 1613–1626, Saunders, Philadelphia, 1992.

CHAPTER 20

Parvovirus B19 Infection

Kevin E. Brown and Neal S. Young

1. Introduction

The human parvovirus B19 (B19) was discovered in 1975. Parvovirus B19 is the only known human pathogenic parvovirus. Unlike many virus infections, the disease manifestation of infection with parvovirus B19 varies widely with the immunologic and hematologic status of the host. In individuals with underlying hemolytic disorders, B19 is the primary cause of aplastic crisis. In immunocompromised patients, persistent B19 viremia may develop that manifests as pure red cell aplasia and chronic anemia, and in the fetus, where the immune response is immature infection, it may lead to fetal death *in utero* or hydrops fetalis. The major disease manifestation in normal, immunocompetent individuals is erythema infectiosum (EI), also called fifth disease or "slapped-cheek" disease. Generally this is an innocuous rash disease of childhood, but in adults it may also be associated with an acute symmetrical polyarthropathy, which can mimic acute rheumatoid arthritis.

2. Historical Background

2.1. The Virus

Parvovirus B19 was discovered by Yvonne Cossart and co-workers[36] in England. They were evaluating tests for hepatitis B surface antigen (HBsAg) using panels of serum samples. One serum (coded 19 in panel B) gave anomalous results, positive in counterimmune electrophoresis (CIE) with human antisera but negative in the more specific radioimmunoassay (RIA) and reverse passive hemagglutination (RPHA) tests that used hyper-

immune animal sera. When the precipitin line from the CIE was excised, electron microscopy (EM) showed the presence of 23-nm particles resembling parvovirus. There was no reactivity with antisera to adenoassociated viruses or to rat parvovirus, and the virus was originally labeled "serum parvoviruslike particle" (SPLV). Approximately 30% of adults had antibody to the new virus detectable by CIE.

The first clinically significant illness associated with B19 infection was hypoplastic crisis in patients with sickle-cell anemia.[94] Sera from such patients contained B19 antigen, detectable by CIE or EM at the time of crisis, and convalescent sera lacked virus but showed evidence of antibody seroconversion.

In 1985, the virus was officially recognized as a member of the *Parvoviridae*, and the International Committee on Taxonomy of Viruses (ICTV) recommended the name B19 to prevent confusion with other viruses (i.e., human papillomavirus).[115]

2.2. Transient Aplastic Crisis

Transient aplastic crisis (TAC) was the first clinical illness associated with B19 infection. The term "aplastic crisis" was coined by Owren[91] to describe the abrupt onset of severe anemia with absent reticulocytes in patients with hereditary spherocytosis. (Hemolytic crises in hereditary spherocytosis patients are associated with increased bone marrow turnover and reticulocyte production.) Also, in contrast to hemolytic crises, aplastic crisis occurred as a single episode in the patients life. In TAC cases there was a common history of a preceding prodromal illness and the occurrence of epidemics in large kindreds of hereditary spherocytosis suggested an infectious etiology.

When stored sera from (over 800) children admitted to a London hospital were examined for B19 antigen by

Kevin E. Brown and Neal S. Young • Hematology Branch, National Heart, Lung, and Blood Institute, Bethesda, Maryland 20892-1652.

CIE, a precipitin line was found in a child with sickle-cell disease suffering a hypoplastic crisis. Five other patients presenting with similar symptoms were also investigated and all had evidence of recent infection with B19 (either antigenemia or seroconversion). All were Jamaican immigrants with sickle-cell disease presenting with aplastic crisis. There was a reduced hematocrit and deficient red cell production in their bone marrow.[94] Retrospective studies of sera from Jamaican sickle-cell patients showed that 86% of TACs were associated with recent parvovirus infection.[114]

2.3. Erythema Infectiosum

This exanthematous rash illness of childhood was probably first described by Robert Willan in 1799 and subsequently illustrated in his 1808 textbook. The disease was rediscovered in Germany, and in 1899 Sticker gave it the name erythema infectiosum. Six years later, Cheinisse[25] classified it as the "fifth rash disease" of the six classical exanthemata of childhood. Subsequently, many outbreaks of fifth disease were documented in the medical literature, but the cause remained a mystery. Often the epidemiologic data suggested "a common source exposure to a highly effective transmitter," and an atypical rubella virus or echovirus was thought to be responsible.[77] However, neither virus could be reproducibly isolated from fifth disease patients.

In 1983, following an outbreak of fifth disease in London, England, 31 of 31 children or adolescents who had been affected had anti-B19-specific immunoglobulin M (IgM) antibody in their serum detectable by RIA.[9] Similar results were obtained in other epidemics of fifth disease worldwide, and parvovirus B19 is now known to be the etiologic agent for EI.[11,90,97]

3. Methodology

3.1. Sources of Mortality Data

Parvovirus B19 is a common infection and death due to B19 must be rare. However, life-threatening B19 infection can occur in patients with underlying hemolytic disease. In one study in Jamaica, of 308 children with homozygous sickle-cell disease followed from birth to age 15, 114 (37%) became infected with B19 and there were four deaths attributable to B19 infection.[113] The prevalence of B19 as a cause of chronic anemia and its contribution to excess mortality in immunocompromised patients is still unknown, although deaths have been reported.[62] It has also been suggested that B19 infection contributes to the mortality of malaria in tropical countries.[55,71]

3.2. Sources of Morbidity Data

Official morbidity reports in the United States do not include EI or parvovirus infection. Morbidity data are therefore based on studies of outbreaks of EI (with and without serological confirmation) or on case reports of confirmed infections. In Britain, parvovirus B19 infections are reported to the Communicable Diseases Surveillance Centre (CDSC). The majority of B19 infections are not reported since laboratory investigation is rarely performed for fifth disease.

3.3. Serological Surveys

Serological surveys were originally performed using the relatively insensitive CIE,[36,37] now superseded by more sensitive RIA or enzyme immunoassays. Due to the inability to grow B19 in standard cell culture systems, until recently there has been a shortage of viral antigen for such serological assays. Attempts have been made to develop assays based on the use of synthetic peptides[50] or fusion proteins in *Escherichia coli*.[81,116] The epitopes presented by these products do not accurately reproduce the epitopes of the native capsids, and the practical results generally have been disappointing. The expression of B19 capsid proteins as virionlike particles using transfected B19 genome into CHO cells,[57] COS-7 cells,[16] and the use of the baculovirus expression system[19,56] appears to have overcome these problems. Results based on these antigens show good correlation with assays based on native virus.[56] The antigens are relatively easy to mass produce and are noninfectious and therefore without hazard to laboratory workers. Serological studies using these recombinant capsids are now in progress.

3.4. Laboratory Methods

3.4.1. Virus Isolation and Detection. There is no suitable method for virus isolation from clinical specimens and the presence of virus relies on the detection of viral DNA by hybridization techniques. B19 DNA can be detected in serum at the time of transient aplastic crisis using dot blot hybridization, and *in situ* hybridization has been used to identify B19 DNA within specific cells. Assays for B19 antigen based on monoclonal antibodies are relatively insensitive ($<10^6$ virus particles/ml) and inferior to DNA hybridization. Electron microscopy can be used to detect B19 in serum with the same sensitivity as antigen assays. Within cells, EM cannot always distinguish intracellular virus from ribosomes.

3.4.2. Serological Assays for B19-Specific Seroprevalence. B19 IgG can be detected by capture assay or indirect assay. Antibody to virus is usually present by the seventh day of illness and probably is lifelong thereafter. However the choice of assay may be critical; one recent epidemiologic study using a gamma-capture RIA showed that only 45% of sickle-cell children with documented B19 infection in the previous 5 years had raised B19 IgG levels.[113] Such assays may markedly underestimate the prevalence of B19 infection.

4. Biological Characteristics of the Agent

4.1. Physical Properties

Parvum is Latin for small, and parvoviruses were the smallest known DNA-containing viruses that infect mammalian cells (only the recently described circoviruses are smaller). On EM, the particles are nonenveloped, 15–28 nm in diameter, and show icosahedral symmetry. Often both "empty" and "full" capsids are visible. Mature infectious parvovirus particles have a molecular weight of 5.6×10^6, a buoyant density in cesium chloride gradients of 1.41 g/ml, and a sedimentation coefficient of 110S.

The DNA in infectious particles made up 19–37% of the total mass of the capsid. The genome size is extremely limited, consisting of a single strand of DNA of approximately 5500 nucleotides. As with other parvoviruses, B19 employs overlapping reading frames to encode its nonstructural protein and two capsid proteins (VP1 and VP2). Parvovirus particles do not appear to contain lipids, carbohydrates, or enzymes.

As a consequence of their lack of an envelope and limited DNA content, parvoviruses are extremely stable to physical inactivation. Parvoviruses are stable at 56°C for ≫60 min (in high concentration, 80°C for 72 hr does not destroy infectivity[15]), or in lipid solvents (ether, chloroform), but can be inactivated by formalin, β-propiolactone, and oxidizing agents. Gamma-irradiation will also inactivate B19, with 1.4 mR producing a 10 $\log_{10}$ reduction in infectivity.[30]

4.2. Morphology

The B19 virion is an icosahedron consisting of 60 copies of the capsid proteins. Most of the capsid protein is VP2, with 5% or less of the larger VP1 protein.[57] Using genetic engineering techniques, the capsid proteins can be expressed in a variety of both mammalian[16,57] and insect cell lines.[19,56] Capsid proteins self-assemble in the absence of B19 DNA, and in these systems protein expression leads to formation of recombinant empty capsids. VP1 is not required for capsid formation.[19,56]

The atomic structure of B19 VP2 empty capsids has recently been resolved to 0.8 nm.[2] The virion surface has a major depression encompassing the fivefold axis, similar to the canyon structure found in RNA-containing icosahedral viruses. In B19 capsids there is also a hollow cylindrical structure about the fivefold axes that appears to penetrate to the inside of the virion. The structural distribution of VP1 in the B19 capsid structure is now known. VP1 capsomers alone will not self-assemble to form icosahedral capsids; rather, they form smaller and irregular structures. In native capsids the VP1 unique region may extend through the fivefold axis cylinder to the outside of the virion.[104]

4.3. B19 Strain Variation

No antigenic variation in B19 has been demonstrated. Small changes of nucleotide sequence have been detected by several investigators using restriction enzyme analysis.[80,82,122] Although isolates could be divided into groups (genome types) with particular enzyme digestion patterns, there was no correlation to specific disease presentations (i.e., purpura, aplastic crisis arthralgia). Among all the isolates studied, sequence divergence was less than 1%, with evidence for less variation in regions where antigenic epitopes are coded.[122] The two prototype B19 isolates, Au and Wi, differed by only 41 bases after complete sequencing.[18]

In a study of 12 viruses isolated in Japan at two different times, 1981 and 1986–1987, the genome type differed during each time period, but B19 viruses with similar genome types disseminated widely in Japan during each period.[121] However, there was no evidence of association with any change in antigenicity or disease characteristics.

5. Descriptive Epidemiology

5.1. Prevalence and Incidence

Parvovirus B19 is a common infection in humans. By age 15, approximately 50% of children have detectable IgG. Infection also occurs in adult life, so that more than 90% of the elderly have detectable antibody.[31] Women of child-bearing age show an annual seroconversion rate of 1.5%.[60] Studies in different countries (United States,[4] France,[37] Germany,[128] Japan[89]) show similar patterns, with a slightly higher prevalence in children from countries such as Brazil[85] and Africa[112] compared to the United Kingdom and the United States. Some isolated tribal populations in Brazil and Africa have a much lower prevalence: a general seroprevalence of 2% on Rodriguez Island, Africa,[112] and 4–10% among the tribesman around Belem, Brazil.[38]

Although antibody is prevalent in the general population, viremia is rare. Among blood donors approximately 1 : 20,000–1 : 40,000 units of blood during epidemic seasons will contain high titers of B19.[32] Screening of pooled samples from blood donors showed that 1 : 3000 units contained detectable B19 DNA by the more sensitive polymerase chain reaction (PCR) technique.[76]

5.2. Epidemic Behavior and Contagiousness

Infections in temperate climates are more common in late winter, spring, and early summer months.[6] Rates of infection may also increase every 3 to 4 years (Fig. 1), and this is reflected by corresponding increases in the major clinical manifestations of B19 infection, TACs, and EI.[113]

The virus readily can be transmitted by close contact. The secondary attack (seroconversion) rate has been calculated in various settings (Table 1), and in one study the secondary attack rate from symptomatic TAC or EI patients to susceptible (IgG negative) household contacts was approximately 50%.[26] In school outbreaks, serological studies are generally not available, but 10–60% of students may develop a rash disease consistent with B19 infection.[52,65,130] The highest secondary attack rates and also annual seroconversion rates even in the absence of known outbreaks of infection are observed among workers with close contact with affected children like day-care providers and school personnel (Table 2). Nosocomial transmission in hospital situations has been described[42,61] but is probably infrequent,[61] especially in patients with persistent disease (Tables 1 and 2). Nevertheless, patients with TAC or persistent infection should be considered

infectious and appropriate precautions taken to limit transmission.

6. Mechanism and Routes of Transmission

Parvovirus B19 DNA has been found in the respiratory secretions of patients at the time of viremia,[26,97] suggesting that B19 infection is generally spread by a respiratory route of transmission. In contrast to other respiratory viruses, however, no site of replication for B19 has been found in the nasopharynx. There is little evidence of virus excretion in feces or urine.[8]

The virus can be found in serum, and infection also can be transmitted by blood and blood products. Parvoviruses, including B19, are very heat resistant, and at high virus concentration can withstand the usual heat treatment (80°C for 72 hr) used to destroy infectivity. In addition, solvent–detergent methods, which only inactivate lipid-enveloped viruses, are ineffective. Parvovirus B19 infection has been transmitted by steam- or dry-heated factor VIII or IX[15,74] and by solvent–detergent-treated factor VIII.[12] Hemophiliacs who received heat-treated factor VIII alone had lower prevalence of B19 antibody and lower rates of seroconversion compared to those receiving nonheat-treated factor.[129]

7. Pathogenesis and Immunity

7.1. Pathogenesis

Parvovirus B19, like all the autonomous parvoviruses, is dependent on mitotically active cells for its own replication. Parvovirus B19 also has a very narrow target cell range and can only be propagated in human erythroid cells. For erythroid cells from bone marrow, susceptibility to parvovirus B19 increases with differentiation; the pluripotent stem cell appears to be spared and the main target cells are erythroid colony-forming cells (CFU-E) and erythroblasts.[120] In erythroid progenitors, the virus is cytotoxic, producing a cytopathic effect with characteristic light[92] and electron microscopic[23,133] changes. Infected cultures are characterized by the presence of giant pronormoblasts or "lantern cells," which are early erythroid cells, 25–32 μm in diameter, with cytoplasmic vacuolization, immature chromatin, and large eosinophilic nuclear inclusion bodies. On EM, virus particles are seen in the nucleus and lining cytoplasmic membranes of infected erythroid progenitors, and infected cells show marginated chromatin, pseudopod formation, and cytoplasmic vac-

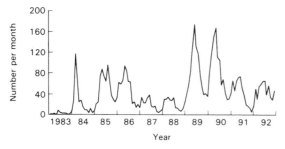

Figure 1. Confirmed cases of parvovirus B19 infection diagnosed at PHLS Virus Reference Division, England, showing a seasonal peak in the first few months of the year, as well as periodic increases every 3–4 years. Data provided by B. J. Cohen, Virus Reference Division, Colindale, London, England.

**Table 1. Secondary Attack Rates for Serologically Confirmed
B19 Infection in Susceptible Individuals**

Setting	Exposure[a]	Contact	Number	Percent	Reference
Schools	EI	School teachers	10/64	16	52
Day care	EI	Day-care workers	2/22	9	52
Pediatric staff	Not known	Staff	10/30	33	96
Hospital staff	TAC	Staff	12/32	37	17
Adult hospital staff	Persistent infection	Staff	0/10	0	61
Household contact	EI/TAC	Family members	59/119	50	26

[a]EI, Erethema infectiosum; TAC transient aplastic crisis.

uolation, typical of cells undergoing apoptosis.[78] The light microscopic findings are also seen in the bone marrow of infected patients.[123]

The basis of the erythroid specificity of parvovirus B19 may be explained by the tissue distribution of the virus cellular receptor, globoside, also known as blood group P antigen.[20] P antigen is found on erythroblasts and megakaryocytes. (It is also present on endothelial cells, which may be targets of viral infection involved in the pathogenesis of transplacental transmission, possibly vasculitis and the rash of fifth disease, and on fetal myocardial cells.[106]) Rare individuals who do not have P antigen on their cells are resistant to B19 infection and their bone marrow cannot be infected with B19 *in vitro*.[22] However, erythroid specificity may also be modulated by specific erythroid transcription factors.[70]

Studies in normal volunteers showed that B19 infection led to an acute but self-limited (4–8 days) cessation of red cell production and a corresponding decline in hemoglobin level.[8,102] In patients with normal erythroid turnover, this short interruption of red cell production does not lead to anemia; but in patients with high red cell turnover, due to hemolysis, blood loss, and so forth, the interruption can precipitate an "aplastic crises." The crisis resolves as virus is "cleared" from the bone marrow by the immune response. In patients who are immunocompromised, the infection may persist and produce chronic pure red cell aplasia.

The infected fetus may suffer severe effects because red blood cell turnover is high and the immune response deficient. During the second trimester there is a great increase in red cell mass. Parvovirus particles can be detected by EM within the hematopoietic tissues of liver and thymus,[44] and B19 DNA and capsid antigen have been detected in the myocardium of infected fetuses.[99,101] In addition, there is evidence that the fetus may develop myocarditis,[79,84] compounding the severe anemia and secondary cardiac failure. By the third trimester, a more effective fetal immune response to the virus may account for the decrease in fetal loss.

The pathogenesis of the rash in EI and polyarthropathy is almost certainly immune complex-mediated. In the volunteer studies, the rash and joint symptoms appeared when viremia was no longer detectable and at the time of development of a detectable immune response.[8] Similar findings have been reported in chronically infected individuals treated with immunoglobulin therapy.[47]

7.2. Immune Response to B19 Infection

Both virus-specific IgM and IgG antibodies are made following experimental[8] and natural[107] B19 parvovirus

**Table 2. Annual Seroconversion Rates
in Different Populations in Nonepidemic Years**

Population	Average seroconversion rate (percent)	Reference
Women of childbearing age	1.5	60
Hospital workers	0.42	1
Contact with children (5–18 years) at work	3.4	1
Contact with children (5–11 years) at home	4.2	1

infection (Fig. 2). Following intranasal inoculation of volunteers, virus can first be detected at days 5–6 and levels peak at days 8–9. IgM antibody to virus appears about 10–12 days after experimental inoculation; IgG antibody appears in normal volunteers about 2 weeks after inoculation.

There is a similar time course in natural infections. In patients with TAC, 10^8–10^{14} genome copies/ml of virus DNA may circulate.[10,27] IgM antibody may be present in patients with TAC at the time of reticulocyte nadir and during the subsequent 10 days; IgG may not be present at the time of reticulocyte depression, but appears rapidly with recovery. Viremia is not detectable in patients with clinical fifth disease (the manifestations are secondary to immune complex formation), and these patients therefore present to medical attention after the period of viremia has passed.

IgM antibody may be found in serum samples for several months after exposure.[5] IgG presumably persists for life and levels rise with reexposure.[8] Transient aplastic crisis due to parvovirus infection does not occur more than once in the life of a sickle-cell patient.[113] Measurable IgA antibodies specific to B19 parvovirus may play a role in protection against infection by the nasopharyngeal route.[41]

In immunocompetent individuals the early antibody response is to the major capsid protein VP2, but as the immune response matures, reactivity to the minor capsid protein, VP1, dominates. Sera from patients with persistent B19 infection typically have antibody to VP2 but not

to VP1.[64] That an immune response to VP1 might be necessary for a protective immunity has been confirmed in animal experiments using recombinant capsid. Rabbits immunized with capsids containing only VP2 produced a strong antibody response, but the sera had low neutralization titers. In contrast, rabbits immunized with capsids containing VP1 produced antibody with neutralizing titers comparable to those produced in humans following acute B19 infection.[105] The role of the cellular immune response in containing parvovirus B19 infection is uncertain. Attempts to detect a cellular response to B19 in lymphocyte proliferative assays have been unsuccessful.[64] Normal recovery from infection correlates with the appearance of circulating specific antivirus antibody, however, and administration of commercial immunoglobulins can cure or ameliorate persistent parvovirus infection in immunodeficient patients.

Persistent B19 parvovirus infection is the result of failure to produce effective neutralizing antibodies by the immunosuppressed host. Perhaps because of the limited number of epitopes presented to the immune system by B19 parvovirus, the congenital immunodeficiency states associated with persistent infection may be clinically subtle, with susceptibility largely restricted to parvovirus, although multiple immune system defects are apparent once directed testing of T- and B-cell function is performed.

8. Patterns of Host Response

8.1. Asymptomatic Infection

Most people with B19 specific antibody have no recollection of any specific symptoms. In one epidemiologic study of a school outbreak, B19 caused asymptomatic infection in approximately 25% of adults.[130] In household contacts of patients with aplastic crises or EI due to B19, 32% (17/52) reported no symptoms.[26] There was more asymptomatic infection in blacks (69%) compared to whites (17%), but the numbers were small and the index infection was different in the two groups (TAC for the black contacts and EI for the white contacts). Parvovirus B19 may go unrecognized in black patients who do not have an underlying hemolytic anemia, since the rash is particularly difficult to see on a dark skin.

8.2. Erythema Infectiosum

As indicated earlier, EI, otherwise known as fifth disease or slapped-cheek disease, is the major manifestation of B19 infection and was well characterized clinically

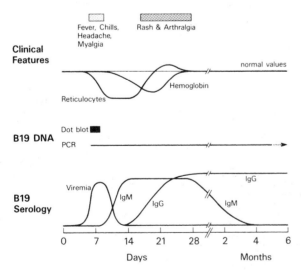

Figure 2. Schematic diagram of the clinical and virological events following B19 parvovirus infection. Data from Patou *et al.*[93] and Potter *et al.*[102]

before the discovery of B19.[3,13] The nonspecific pro-dromal illness often goes unrecognized and may be associated with symptoms of fever, coryza, headache, and mild gastrointestinal symptoms, such as nausea and diarrhea. Two to five days later, the classic slapped-cheek rash appears, a fiery red eruption on the cheek, accompanied by relative circumoral pallor. One to four days after the slapped-cheek rash, the second-stage rash may appear—an erythematous maculopapular exanthemum on the trunk and limbs. As this eruption fades, it takes on a typical lacy appearance. There may be great variation in the dermatological appearance. The classic slapped cheek is more common in children than adults, and the second-stage eruption may vary from a very faint erythema that is easily missed to a florid exanthema and may be transient or recurrent over 1 to 3 weeks. The rash may be accompanied by pruritus, especially on the soles of the feet, which can be the dominant symptom.[130]

8.3. Polyarthropathy Syndrome

In children, B19 infection is usually mild and of short duration. However, in adults and especially in women, there may be arthropathy in approximately 50% of patients.[130] The joints can be painful, often with accompanying swelling and stiffness. The distribution is usually symmetrical; mainly the small joints of hands and feet are involved. Joint symptoms last 1 to 3 weeks, although in 20% of affected women, arthralgia or frank arthritis may persist or recur for more than 2 months, even to 2 years. In the absence of a history of rash, the symptoms may be mistaken for acute rheumatoid arthritis, especially as B19 infection can be associated with transient rheumatoid factor production.[73,83,127] In one study of patients attending an "early synovitis" clinic in England, 19 of 153 (12%) had evidence of recent infection with B19.[127] Parvovirus B19 infection should be considered as part of the differential diagnosis in any patient presenting with acute arthritis, but in contrast to rheumatoid arthritis, B19 infection has not been associated with long-term joint destruction.

8.4. Aplastic Crisis

Transient aplastic crisis is the abrupt cessation of erythropoiesis characterized by reticulocytopenia, absent erythroid precursors in the bone marrow, and precipitous worsening of anemia. TAC was the first clinical illness associated with B19 infection. TAC due to B19 has been described in a wide range of patients with underlying hemolytic disorders, such as hereditary spherocytosis, thalassemia, and red cell enzymopathies such as pyruvate

kinase deficiency and autoimmune hemolytic anemia.[132] TAC can also occur under conditions of erythroid "stress," such as hemorrhage, iron deficiency anemia, and following kidney or bone marrow transplantation.[48,86,88] Acute anemia has been described in hematologically normal persons[54] and a drop in red cell count (and reticulocytes) was seen in healthy volunteers.[8]

Although suffering from an ultimately self-limiting disease, patients with aplastic crisis can be severely ill. Symptoms may include dyspnea, lassitude, and even confusion due to the worsening anemia. Congestive heart failure and severe bone marrow necrosis may develop[34,53] and the illness can be fatal.[113] Aplastic crisis can be the first presentation of an underlying hemolytic disease in a well-compensated patient.[75]

TAC and B19 infection in hematologically normal patients are often associated with changes in the other blood lineages. There may be varying degrees of neutropenia, thrombocytopenia, and transient pancytopenia.[132] Some cases of idiopathic thrombocytopenia purpura[46] and Henoch–Schönlein purpura[66] have been linked to parvovirus B19 infection. Recently, agranulocytosis was also described following B19 infection.[98]

Community-acquired aplastic crisis is almost always due to parvovirus B19[7,113] and should be the presumptive diagnosis in any patient with anemia due to abrupt cessation of erythropoiesis as documented by reduced reticulocytes and bone marrow appearance. In contrast to patients with erythema infectiosum, TAC patients are often viremic at the time of presentation, with concentrations of virus as high as 10^{14} genome copies/ml, and the diagnosis is readily made by detection of B19 DNA in the serum. As B19 DNA levels fall in serum, B19-specific IgM becomes detectable.

TAC is readily treated by blood transfusion. It is a unique event in the patient's life, and following the acute infection immunity is lifelong.

8.5. Infection during Pregnancy

There have been a large number of reports of B19 infection in pregnancy leading to an adverse outcome, either miscarriage or hydrops fetalis.[58] The clinical features have been remarkably similar. In cases where pathological studies were undertaken, the fetuses showed evidence of leukoerythroblastic reaction in the liver and large, pale cells with eosinophilic inclusion bodies and peripheral condensation or margination of the nuclear chromatin. Parvovirus B19 DNA could be detected by DNA dot blot, *in situ* hybridization,[28,110] and parvovirus particles in EM studies.[44] There are, however, many more cases of favorable outcome after confirmed parvovirus B19 infection in pregnancy.[58]

In a prospective British study of 190 women with serologically confirmed B19 during pregnancy, the overall fetal loss rate was 16%, similar to 12–16% fetal loss rates in similar British studies of uninfected women.[103] However, in the second trimester, the fetal loss rate was 12%, compared to 0.6% in unmatched comparison studies. The risk of fetal death due to B19 was estimated at 9%, with a transplacental transmission rate of 33%. Normal healthy infants were born even when there was evidence of intrauterine infection by the presence of B19 IgM in the umbilical cord blood. Infants continue to be observed for possible late sequelae of their infection.

Parvovirus B19 probably causes 10–15% of all cases of nonimmune hydrops. Nonimmune hydrops fetalis is rare (1 in 3000 births), and in approximately 50% of cases the etiology is unknown.[125] In a study of 50 cases, the majority were due to cardiovascular or chromosomal abnormalities, but parvovirus B19 DNA was detected by *in situ* hybridization in the lungs of four fetuses.[100]

8.6. Congenital Infection

No systematic studies have shown evidence for congenital abnormalities following B19 infection. There is one case report of congenital malformation associated with B19 infection[126]; ocular abnormalities similar to those seen in rubella were observed in the 12-week-old fetus and B19 DNA was detected in placental and fetal tissue. (There was no evidence of rubella infection.)

Infants born with chronic anemia following a history of maternal B19 exposure and intrauterine hydrops have recently been described.[21] In these three cases, the virus load was low and localized; B19 virus could be detected in bone marrow samples by PCR but not in concurrent serum samples. All the cases were treated with immunoglobulin therapy. The first case died at 9 months, before completing the immunoglobulin treatment, and B19 DNA was detected in a variety of tissues, including heart, liver, and spleen. The other two cases completed their immunoglobulin therapy, but although B19 DNA could no longer be detected in bone marrow (by PCR), the children remained severely anemia.

8.7. Chronic Bone Marrow Failure

Persistent B19 infection resulting in pure red cell aplasia has been reported in a wide variety of immunosuppressed patients, ranging from patients with congenital immunodeficiency, acquired immunodeficiency syndrome (AIDS), and lymphoproliferative disorders to transplant patients.[49] The stereotypical presentation is with persistent anemia rather than the immune-mediated symptoms of rash or arthropathy. The patients had absent or low levels of B19-specific antibody and persistent or recurrent parvoviremia as detected by B19 DNA in the serum. Bone marrow examination generally revealed the presence of scattered giant pronormoblasts. Administration of immunoglobulin can be beneficial and ameliorative even if not curative.[63] Temporary cessation of maintenance chemotherapy has also led to resolution of the anemia, and in two cases reinstitution did not lead to recurrence,[24,117] suggesting that decreasing the level of immunosuppression may allow the host to produce antibody and resolve the virus infection.

8.8. Vasculitis

The role of parvovirus B19 in vasculitis remains unclear. Several case reports have described positive B19 serology in patients with vasculitis and/or polyarteritis nodosum[35,39,51,67,69,111,118]; however, in each individual report it was uncertain as to whether the association was coincidental or causative. More recently, parvovirus B19 infection has been associated with acute systemic necrotizing vasculitis.[45] Recent infection with parvovirus B19 was indicated in three patients by the presence of both B19 IgM in the serum and B19 DNA in serum and tissues. Treatment with intravenous immunoglobulin led to clearing of the virus and resolution of the patients' symptoms.

A similar report linked parvovirus B19 to Kawasaki disease, a multisystem vasculitis of early childhood. In a recent study from Italy, parvoviral B19 DNA and/or IgM antibodies were found in 10 of 15 patients with Kawasaki disease compared to 0 of 36 control children.[87] The authors do not report on treatment of their cases, but immunoglobulin therapy is known to be beneficial in Kawasaki disease. However, other studies have not shown a relationship between Kawasaki disease and parvoviral infection.[29,131]

8.9. Diagnosis of B19 Infections

The detection of B19 viremia is based on nucleic acid hybridization assays. Parvovirus B19 DNA can be detected in serum at the time of TAC using dot blot hybridization, and *in situ* hybridization has been used to identify B19 DNA within bone marrow and other cells.

In immunocompetent individuals, B19 DNA is only detectable for 2–4 days by dot blot hybridization (Fig. 2), and detection of acute B19 infection is therefore based on IgM assays, ideally performed by the capture technique.[33] (Indirect assays to detect B19 IgM often give false positives due to cross-reacting antibodies or rheumatoid factor.) In a RIA or ELISA, antibody can be detected in over 90% of cases by the third day of TAC or at the time of rash in EI. IgM antibody remains detectable for 2 to 3 months following infection.

Parvovirus B19 IgG can be detected by capture assay or indirect assay. IgG is usually present by the seventh day of illness and probably present for life thereafter. As more than 50% of the population have IgG antibody to B19 infection, detection of B19 IgG is not helpful for the diagnosis of acute infection. Immunocompromised or immunodeficient patients with chronic infection may not mount an immune response to the virus, and testing for B19 antigens or detection of B19 DNA is necessary to document recent infection.

The sensitivity level of detection of B19 has greatly increased by the use of PCR, but at the risk of possible contamination and false-positive results confusing interpretation. Even in immunocompetent persons, B19 DNA may be detectable by PCR for more than 4 months following acute infection.[93] The diagnosis of acute or chronic infection can be made on the basis of standard DNA hybridization in combination with serological assays for B19-specific IgG and/or IgM.

Due to the inability to grow B19 in standard cell culture systems, until recently there has been a shortage of viral antigen for diagnostic assays. However, the expression of B19 capsid proteins as virionlike particles using transfected B19 genome into CHO cells,[57] COS-7 cells,[16] and the use of the baculovirus expression system[19,56] has resolved this difficulty, and commercially available, reliable assays based on these antigens should be available soon.

9. Control and Prevention

The humoral immune response plays a dominant role in the normal immune response to parvovirus, and antibodies are protective in both passive and active immunizations. Human convalescent-phase antisera[64] and commercial immunoglobulin preparations[119] contain neutralizing antibodies to parvovirus, as assessed *in vitro* using erythroid colony systems. In addition, commercial immunoglobulin from normal donors can cure or ameliorate persistent B19 parvovirus infection in immunosuppressed human patients.[47,62]

Prospects for a B19 parvovirus vaccine are good. The immunogen will be a recombinant capsid rather than attenuated or killed virus, because of the difficulty of propagating B19 parvovirus in tissue culture and the potential danger of inadvertently modifying, for the worse, the host range of B19 parvovirus by selection *in vitro*. A recombinant canine parvovirus vaccine has been produced in baculovirus that induces neutralizing antibodies in inoculated animals and has been shown to be protective against virulent canine parvovirus.[72,109] Baculovirus-produced B19 parvovirus capsids induce neutralizing antibodies in experimental animals,[56] even without adjuvant.[14] The presence of VP1 protein in the capsid immunogen appears critical for the production of antibodies that neutralize virus activity *in vitro*, and capsids with supranormal VP1 content are even more efficient in inducing neutralizing activity in immunized animals.[14]

10. Unresolved Problems

10.1. Full Spectrum of Disease

Although B19 is associated with a wide variety of diseases, almost certainly the list is not complete. In addition, the discovery of P antigen as the receptor for B19 suggests a possible role for the virus in diseases previously unsuspected as related to parvoviral infection. P antigen is found on fetal myocardial cells and B19 is known to cause myocarditis in the fetus.[79,84] There have also been two case reports of fatal myocarditis associated with recent B19 infection in young children.[59,108] P antigen is also found on megakaryocytes, and some cases of idiopathic thrombocytopenia have been associated with B19 infection.[46] In one study, 5% (3/16) patients with idiopathic thrombocytopenia had serological evidence of recent B19 infection.[66] These suggestive associations require rigorous confirmation and quantification.

Paroxysmal cold hemoglobinuria, an episodic hemolytic anemia triggered by exposure to cold, is characterized by the development of autoantibodies with P antigen specificity (Donath–Landsteiner antibodies).[68] Paroxysmal cold hemoglobinuria is thought to have a viral etiology, and B19 can now be implicated. It might either induce the development of anti-idiotypic antibodies or stimulate production of antibodies to the B19 receptor complex "neoantigen."

Neurological disease has also been associated with parvovirus infection. Pruritus is not uncommon in fifth disease, and in one study, 50% of patients with serologically confirmed fifth disease experienced neurological symptoms, especially neurasthenia in fingers or toes.[43] One patient developed more significant disease with progressive weakness of one arm. Brachial plexus neuropathy has been described in three other patients with B19 infection.[40,95,124]

In those illnesses where B19 is known to be involved, the full spectrum of disease is still uncertain. Parvovirus B19 is recognized as the cause of acute polyarthropathy, but its role in chronic arthropathy and as a possible trigger for rheumatoid arthritis is still undetermined. Similarly, the role for B19 as a cause of chronic anemia in immunocompromised patients is established, but the role of *in utero* B19 infection inducing constitutional bone marrow failure such as Diamond–Blackfan anemia is still under investigation.

10.2. B19 Vaccine Policy

Although prospects for a vaccine are good and phase 1 trials of one preparation were commenced in 1996, the targets for such a vaccine remain to be determined. Should only patients at high risk of severe or life-threatening disease, such as sickle-cell patients, be protected? Or, in view of the wide variety of disease manifestations affecting all strata of the population, should a universal vaccine policy be pursued? A universal vaccination policy would have the added advantage of possible eradication of B19 from a community, but would have the disadvantages of high cost–benefit ratio.

11. References

1. ADLER, S. P., MANGANELLO, A. M., KOCH, W. C., HEMPFLING, S. H., AND BEST, A. M., Risk of human parvovirus B19 infections among school and hospital employees during endemic periods, *J. Infect. Dis.* **168**:361–368 (1993).

2. AGBANDJE, M., KAJIGAYA, S., MCKENNA, R., YOUNG, N. S., AND ROSSMANN, M. G., The structure of human parvovirus B19 at 8 Å resolution, *Virology* **203**:106–115 (1994).

3. AGER, A. E., CHIN, T. D. J., AND POLAND, J. D., Epidemic erythema infectiosum, *N. Engl. J. Med.* **275**:1326–1331 (1966).

4. ANDERSON, L. J., Role of parvovirus B19 in human disease, *Pediatr. Infect. Dis. J.* **6**:711–178 (1987).

5. ANDERSON, L. J., TSOU, C., PARKER, R. A., CHORBA, T. L., WULFF, H., TATTERSALL, P., AND MORTIMER, P. P., Detection of antibodies and antigens of human parvovirus B19 by enzyme-linked immunosorbent assay, *J. Clin. Microbiol.* **24**:522–526 (1986).

6. ANDERSON, M. J., AND COHEN, B. J., Human parvovirus B19 infections in United Kingdom 1984–86 [letter], *Lancet* **1**:738–739 (1987).

7. ANDERSON, M. J., DAVIS, L. R., HODGSON, J., JONES, S. E., MURTAZA, L., PATTISON, J. R., STROUD, C. E., AND WHITE, J. M., Occurrence of infection with a parvovirus-like agent in children with sickle cell anaemia during a two-year period, *J. Clin. Pathol.* **35**:744–749 (1982).

8. ANDERSON, M. J., HIGGINS, P. G., DAVIS, L. R., WILLMAN, J. S., JONES, S. E., KIDD, I. M., PATTISON, J. R., AND TYRRELL, D. A., Experimental parvoviral infection in humans, *J. Infect. Dis.* **152**:257–265 (1985).

9. ANDERSON, M. J., JONES, S. E., FISHER-HOCH, S. P., LEWIS, E., HALL, S. M., BARTLETT, C. L. R., COHEN, B. J., MORTIMER, P. P., AND PEREIRA, M. S., Human parvovirus, the cause of erythema infectiosum (fifth disease)? [letter], *Lancet* **1**:1378 (1983).

10. ANDERSON, M. J., JONES, S. E., AND MINSON, A. C., Diagnosis of human parvovirus infection by dot-blot hybridization using cloned viral DNA, *J. Med. Virol.* **15**:163–172 (1985).

11. ANDERSON, M. J., LEWIS, E., KIDD, I. M., HALL, S. M, AND COHEN, B. J., An outbreak of erythema infectiosum associated with human parvovirus infection, *J. Hyg. (Lond.)* **93**:85–93 (1984).

12. AZZI, A., CIAPPI, S., ZAKVRZEWSKA, K., MORFINI, M., MARIANI, G., AND MANNUCCI, P. M., Human parvovirus B19 infection in hemophiliacs first infused with two high-purity, virally attenuated factor VIII concentrates, *Am. J. Hematol.* **39**:228–230 (1992).

13. BALFOUR, H. H., Erythema infectiosum (fifth disease). Clinical review and description of 91 cases seen in an epidemic, *Clin. Pediatr.* **8**:721–727 (1969).

14. BANSAL, G. P., HATFIELD, J., DUNN, F. E., WARRENER, P., YOUNG, J. F., TOP, F. H., JR., COLLETT, M. S., ANDERSON, S., ROSENFELD, S., KAJIGAYA, S., AND YOUNG, N. S., Immunogenicity studies of recombinant human parvovirus B19 proteins, in: *Vaccines 92* (F. BROWN, R. M. CHANOCK, H. S. GINSBERG, AND R. A. LERNER, EDS.), PP. 315–319, COLD SPRING HARBOR LABORATORY PRESS, COLD SPRING HARBOR, NY, 1992.

15. BARTOLOMEI CORSI, O., AZZI, A., MORFINI, M., FANCI, R., AND ROSSI FERRINI, P., Human parvovirus infection in haemophiliacs first infused with treated clotting factor concentrates, *J. Med. Virol.* **25**:165–170 (1988).

16. BEARD, C., ST. AMAND, J., AND ASTELL, C. R., Transient expression of B19 parvovirus gene products in COS-7 cells transfected with B19-SV40 hybrid vectors, *Virology* **172**:659–664 (1989).

17. BELL, L. M., NAIDES, S. J., STOFFMAN, P., HODINKA, R. L., AND PLOTKIN, S. A., Human parvovirus B19 infection among hospital staff members after contact with infected patients, *N. Engl. J. Med.* **321**:485–491 (1989).

18. BLUNDELL, M. C., BEARD, C., AND ASTELL, C. R., *In vitro* identification of a B19 parvovirus promoter, *Virology* **157**:534–538 (1987).

19. BROWN, C. S., VAN LENT, J. W., VLAK, J. M., AND SPAAN, W. J., Assembly of empty capsids by using baculovirus recombinants expressing human parvovirus B19 structural proteins, *J. Virol.* **65**:2702–2706 (1991).

20. BROWN, K. E., ANDERSON, S. M., AND YOUNG, N. S., Erythrocyte P antigen: Cellular receptor for B19 parvovirus, *Science* **262**:114–117 (1993).

21. BROWN, K. E., GREEN, S. W., ANTUNEZ DE MAYOLO, J., BELLANTI, J. A., SMITH, S. D., SMITH, T. J., AND YOUNG, N. S., Congenital anaemia after transplacental B19 parvovirus infection, *Lancet* **343**:895–896 (1994).

22. BROWN, K. E., HIBBS, J. R., GALLINELLA, G., ANDERSON, S. M., LEHMAN, E. D., MCCARTHY, P., AND YOUNG, N. S., Resistance to parvovirus B19 infection due to lack of virus receptor (erythrocyte P antigen), *N. Engl. J. Med.* **330**:1192–1196 (1994).

23. BROWN, K. E., MORI, J., COHEN, B. J., AND FIELD, A. M., *In vitro* propagation of parvovirus B19 in primary foetal liver culture, *J. Gen. Virol.* **72**:741–745 (1991).

24. CARSTENSEN, H, ORNVOLD, K., AND COHEN, B. J., Human parvovirus B19 infection associated with prolonged erythroblastopenia in a leukemic child, *Pediatr. Infect. Dis. J.* **8:**56 (1989).

25. CHEINISSE, L., Une cinquième maladie éruptive: Le mégalérythème épidémique, *Sem. Med.* **25:**205–207 (1905).

26. CHORBA, T., COCCIA, P., HOLMAN, R. C., TATTERSALL, P., ANDERSON, L. J., SUDMAN, J., YOUNG, N. S., KURCZYNSKI, E., SAARINEN, U. M., MOIR, R., LAWRENCE, D. N., JASON, J. M., AND EVATT, B., The role of parvovirus B19 in aplastic crisis and erythema infectiosum (fifth disease), *J. Infect. Dis.* **154:**383–393 (1986).

27. CLEWLEY, J. P., Detection of human parvovirus using a molecularly cloned probe, *J. Med. Virol.* **15:**173–181 (1985).

28. CLEWLEY, J. P., COHEN, B. J., AND FIELD, A. M., Detection of parvovirus B19 DNA, antigen, and particles in the human fetus, *J. Med. Virol.* **23:**367–376 (1987).

29. COHEN, B. J., Human parvovirus B19 infection in Kawasaki disease [letter], *Lancet* **344:**59 (1994).

30. COHEN, B. J., AND BROWN, K. E., Laboratory infection with human parvovirus B19 [letter], *J. Infect.* **24:**113–114 (1992).

31. COHEN, B. J., AND BUCKLEY, M. M., The prevalence of antibody to human parvovirus B19 in England and Wales, *J. Med. Microbiol.* **25:**151–153 (1988).

32. COHEN, B. J., FIELD, A. M., GUDNADOTTIR, S., BEARD, S., AND BARBARA, J. A., Blood donor screening for parvovirus B19, *J. Virol. Methods.* **30:**233–238 (1990).

33. COHEN, B. J., MORTIMER, P. P., AND PEREIRA, M. S., Diagnostic assays with monoclonal antibodies for the human serum parvovirus-like virus (SPLV), *J. Hyg. (Lond.)* **91:**113–130 (1983).

34. CONRAD, M. E., STUDDARD, H., AND ANDERSON, L. J., Aplastic crisis in sickle cell disorders: Bone marrow necrosis and human parvovirus infection, *Am. J. Med. Sci.* **295:**212–215 (1988).

35. CORMAN, L. C., AND DOLSON, D. J., Polyarteritis nodosa and parvovirus B19 infection [letter], *Lancet* **339:**491 (1992).

36. COSSART, Y. E., FIELD, A. M., CANT, B., AND WIDDOWS, D., Parvovirus-like particles in human sera, *Lancet* **1:**72–73 (1975).

37. COUROUCE, A. M., FERCHAL, F., MORINET, F., MULLER, A., DROUET, J., SOULIER, J. P., AND PEROL, Y., Human parvovirus infections in France [letter], *Lancet* **1:**160 (1984).

38. DE FREITAS, R. B., WONG, D., BOSWELL, F., DE MIRANDA, M. F., LINHARES, A. C., SHIRLEY, J., AND DESSELBERGER, U., Prevalence of human parvovirus (B19) and rubella virus infections in urban and remote rural areas in northern Brazil, *J. Med. Virol.* **32:**203–208 (1990).

39. DELANNOY, D., BALQUET, M. H., AND SAVINEL, P., [Vasculitis with mixed cryoglobulin in human parvovirus B19 infection (letter)], *Presse Med.* **22:**175 (1993).

40. DENNING, D. W., AMOS, A., RUDGE, P., AND COHEN, B. J., Neuralgic amyotrophy due to parvovirus infection [letter], *J. Neurol. Neurosurg. Psychiatry* **50:**641–642 (1987).

41. ERDMAN, D. D., USHER, M. J., TSOU, C., CAUL, E. O., GARY, G. W., KAJIGAYA, S., YOUNG, N. S., AND ANDERSON, L. J., Human parvovirus B19 specific IgG, IgA, and IgM antibodies and DNA in serum specimens from persons with erythema infectiosum, *J. Med. Virol.* **35:**110–115 (1991).

42. EVANS, J. P., ROSSITER, M. A., KUMARAN, T. O., MARSH, G. W., AND MORTIMER, P. P., Human parvovirus aplasia: Case due to cross-infection in a ward, *Br. Med. J.* **288:**681 (1984).

43. FADEN, H., GARY, G. W., JR., AND KORMAN, M., Numbness and tingling of fingers associated with parvovirus B19 infection [letter], *J. Infect. Dis.* **161:**354–355 (1990).

44. FIELD, A. M., COHEN, B. J., BROWN, K. E., MORI, J., CLEWLEY, J. P., NASCIMENTO, J. P., AND HALLAM, N. F., Detection of B19 parvovirus in human fetal tissues by electron microscopy, *J. Med. Virol.* **35:**85–95 (1991).

45. FINKEL, T. H., TOROK, T. J., FERGUSON, P. J., DURIGON, E. L., ZAKI, S. R., LEUNG, D. Y., HARBECK, R. J., GELFAND, E. W., SAULSBURY, F. T., AND HOLLISTER, J. R., Chronic parvovirus B19 infection and systemic necrotising vasculitis: Opportunistic infection or aetiological agent? *Lancet* **343:**1255–1258 (1994).

46. FOREMAN, N. K., OAKHILL, A., AND CAUL, E. O., Parvovirus-associated thrombocytopenic purpura [letter], *Lancet* **2:**1426–1427 (1988).

47. FRICKHOFEN, N., ABKOWITZ, J. L., SAFFORD, M., BERRY, J. M., ANTUNEZ-DE-MAYOLO, J., ASTROW, A., COHEN, R., HALPERIN, I., KING, L., MINTZER, D., COHEN, B., AND YOUNG, N. S., Persistent B19 parvovirus infection in patients infected with human immunodeficiency virus type 1 (HIV-1): A treatable cause of anemia in AIDS, *Ann. Intern. Med.* **113:**926–933 (1990).

48. FRICKHOFEN, N., ARNOLD, R., HERTENSTEIN, B., WIESNETH, M., AND YOUNG, N. S., Parvovirus B19 infection and bone marrow transplantation, *Ann. Hematol.* **64:**A121–A124 (1992).

49. FRICKHOFEN, N., AND YOUNG, N. S., Persistent parvovirus B19 infections in humans, *Microb. Pathog.* **7:**319–327 (1989).

50. FRIDELL, E., TROJNAR, J., AND WAHREN, B., A new peptide for human parvovirus B19 antibody detection, *Scand. J. Infect. Dis.* **21:**597–603 (1989).

51. GACHES, F., LOUSTAUD, V., VIDAL, E., DELAIRE, L., GUIARD-SCHMID, J. B., LAVOINE, E., NEGRIER, P., AND LIOZON, F., [Periarteritis nodosa and parvovirus B19 infection], *Rev. Med. Interne.* **14:**323–325 (1993).

52. GILLESPIE, S. M., CARTTER, M. L., ASCH, S., ROKOS, J. B., GARY, G. W., TSOU, C. J., HALL, D. B., ANDERSON, L. J., AND HURWITZ, E. S., Occupational risk of human parvovirus B19 infection for school and day-care personnel during an outbreak of erythema infectiosum, *J.A.M.A.* **263:**2061–2065 (1990).

53. GODEAU, B., GALACTEROS, F., SCHAEFFER, A., MORINET, F., BACHIR, D., ROSA, J., AND PORTOS, J. L., Aplastic crisis due to extensive bone marrow necrosis and human parvovirus infection in sickle cell disease [letter], *Am. J. Med.* **91:**557–558 (1991).

54. HAMON, M. D., NEWLAND, A. C., AND ANDERSON, M. J., Severe aplastic anaemia after parvovirus infection in the absence of underlying haemolytic anaemia [letter], *J. Clin. Pathol.* **41:**1242 (1988).

55. JONES, P. H., PICKETT, L. C., ANDERSON, M. J., AND PASVOL, G., Human parvovirus infection in children and severe anaemia seen in an area endemic for malaria, *J. Trop. Med. Hyg.* **93:**67–70 (1990).

56. KAJIGAYA, S., FUJII, H., FIELD, A., ANDERSON, S., ROSENFELD, S., ANDERSON, L. J., SHIMADA, T., AND YOUNG, N. S., Self-assembled B19 parvovirus capsids, produced in a baculovirus system, are antigenically and immunogenically similar to native virions, *Proc. Natl. Acad. Sci. USA* **88:**4646–4650 (1991).

57. KAJIGAYA, S., SHIMADA, T., FUJITA, S., AND YOUNG, N. S., A genetically engineered cell line that produces empty capsids of B19 (human) parvovirus, *Proc. Natl. Acad. Sci. USA* **86:**7601–7605 (1989).

58. KINNEY, J. S., ANDERSON, L. J., FARRAR, J., STRIKAS, R. A., KUMAR, M. L., KLIEGMAN, R. M., SEVER, J. L., HURWITZ, E. S., AND SIKES, R. K., Risk of adverse outcomes of pregnancy after human parvovirus B19 infection, *J. Infect. Dis.* **157:**663–667 (1988).

59. KNISELY, A. S., O'SHEA, P. A., ANDERSON, L. J., AND GARY, G. W., JR., Parvovirus B19 infection, myocarditis, and death in a 3-year-old boy (Abstract), *Pediatr. Pathol.* **8**:665 (1988).

60. KOCH, W. C., AND ADLER, S. P., Human parvovirus B19 infections in women of childbearing age and within families, *Pediatr. Infect. Dis. J.* **8**:83–87 (1989).

61. KOZIOL, D. E., KURTZMAN, G., AYUB, J., YOUNG, N. S., AND HENDERSON, D. K., Nosocomial human parvovirus B19 infection: Lack of transmission from a chronically infected patient to hospital staff, *Infect. Control. Hosp. Epidemiol.* **13**:343–348 (1992).

62. KURTZMAN, G., FRICKHOFEN, N., KIMBALL, J., JENKINS, D. W., NIENHUIS, A. W., AND YOUNG, N. S., Pure red-cell aplasia of 10 years' duration due to persistent parvovirus B19 infection and its cure with immunoglobulin therapy [see comments], *N. Engl. J. Med.* **321**:519–523 (1989).

63. KURTZMAN, G. J., COHEN, B., MEYERS, P., AMUNULLAH, A., AND YOUNG, N. S., Persistent B19 parvovirus infection as a cause of severe chronic anaemia in children with acute lymphocytic leukaemia, *Lancet* **2**:1159–1162 (1988).

64. KURTZMAN, G. J., COHEN, B. J., FIELD, A. M., OSEAS, R., BLAESE, R. M., AND YOUNG, N. S., Immune response to B19 parvovirus and an antibody defect in persistent viral infection, *J. Clin. Invest.* **84**:1114–1123 (1989).

65. LAUER, B. A., MACCORMACK, J. N., AND WILFERT, C., Erythema infectiosum: An elementary school outbreak, *Am. J. Dis. Child.* **130**:252–254 (1976).

66. LEFRÈRE, J. J., COUROUCÉ, A. M., MULLER, J. Y., CLARK, M., AND SOULIER, J. P., Human parvovirus and purpura [letter], *Lancet* **2**:730 (1985).

67. LERUEZ-VILLE, M., LAUGE, A., MORINET, F., GUILLEVIN, L., AND DENY, P., Polyarteritis nodosa and parvovirus B19 [letter], *Lancet* **344**:263–264 (1994).

68. LEVINE, P., CELANO, M. J., AND FALKOWSKI, F., The specificity of the autoantibody in paroxysmal cold haemoglobinuria (PCH), *Transfusion* **3**:278–280 (1963).

69. LI LOONG, T. C., COYLE, P. V., ANDERSON, M. J., ALLEN, G. E., AND CONNOLLY, J. H., Human serum parvovirus associated vasculitis, *Postgrad. Med. J.* **62**:493–494 (1986).

70. LIU, J. M., GREEN, S. W., SHIMADA, T., AND YOUNG, N. S., A block in full-length transcript maturation in cells nonpermissive for B19 parvovirus, *J. Virol.* **66**:4686–4692 (1992).

71. LORTHOLARY, O., ELIASZEWICZ, M., DUPONT, B., AND COUROUCÉ, A. M., Parvovirus B19 infection during acute Plasmodium falciparum malaria [letter], *Eur. J. Haematol.* **49**:219 (1992).

72. LÖPEZ DE TURISO, J. A., CORTÉS, E., MARTINEZ, C., RUIZ DE YBÁNEZ, R., SIMARRO, I., VELA, C., AND CASAL, I., Recombinant vaccine for canine parvovirus in dogs, *J. Virol.* **66**:2748–2753 (1992).

73. LUZZI, G. A., KURTZ, J. B., AND CHAPEL, H., Human parvovirus arthropathy and rheumatoid factor [letter], *Lancet* **1**:1218 (1985).

74. LYON, D. J., CHAPMAN, C. S., MARTIN, C., BROWN, K. E., CLEWLEY, J. P., FLOWER, A. J., AND MITCHELL, V. E., Symptomatic parvovirus B19 infection and heat-treated factor IX concentrate [letter], *Lancet* **1**:1085 (1989).

75. MCLELLAN, N. J., AND RUTTER, N., Hereditary spherocytosis in sisters unmasked by parvovirus infection, *Postgrad. Med. J.* **63**:49–50 (1987).

76. MCOMISH, F., YAP, P. L., JORDAN, A., HART, H., COHEN, B. J., AND SIMMONDS, P., Detection of parvovirus B19 in donated blood: A model system for screening by polymerase chain reaction, *J. Clin. Microbiol.* **31**:323–328 (1993).

77. MORENS, D. M., Fifth disease: Still hazy after all these years, *J.A.M.A.* **248**:553–554 (1982).

78. MOREY, A. L., FERGUSON, D. J., AND FLEMING, K. A., Ultrastructural features of fetal erythroid precursors infected with parvovirus B19 *in vitro*: Evidence of cell death by apoptosis, *J. Pathol.* **169**:213–220 (1993).

79. MOREY, A. L., KEELING, J. W., PORTER, H. J., AND FLEMING, K. A., Clinical and histopathological features of parvovirus B19 infection in the human fetus, *Br. J. Obstet. Gynaecol.* **99**:566–574 (1992).

80. MORI, J., BEATTIE, P., MELTON, D. W., COHEN, B. J., AND CLEWLEY, J. P., Structure and mapping of the DNA of human parvovirus B19, *J. Gen. Virol.* **68**:2797–2806 (1987).

81. MORINET, F., D'AURIOL, L., TRATSCHIN, J. D., AND GALIBERT, F., Expression of the human parvovirus B19 protein fused to protein A in *Escherichia coli*: Recognition by IgM and IgG antibodies in human sera, *J. Gen. Virol.* **70**:3091–3097 (1989).

82. MORINET, F., TRATSCHIN, J. D., PEROL, Y., AND SIEGL, G., Comparison of 17 isolates of the human parvovirus B19 by restriction enzyme analysis. Brief report, *Arch. Virol.* **90**:165–172 (1986).

83. NAIDES, S. J., AND FIELD, E. H., Transient rheumatoid factor positivity in acute human parvovirus infection, *Arch. Intern. Med.* **148**:2587–2589 (1988).

84. NAIDES, S. J., AND WEINER, C. P., Antenatal diagnosis and palliative treatment of non-immune hydrops fetalis secondary to fetal parvovirus B19 infection, *Prenat. Diagn.* **9**:105–114 (1989).

85. NASCIMENTO, J. P., BUCKLEY, M. M., BROWN, K. E., AND COHEN, B. J., The prevalence of antibody to human parvovirus B19 in Rio de Janeiro, Brazil, *Rev. Inst. Med. Trop. São Paulo* **32**:41–45 (1990).

86. NEILD, G., ANDERSON, M., HAWES, S., AND COLVIN, B. T., Parvovirus infection after renal transplant [letter], *Lancet* **2**:1226–1227 (1986).

87. NIGRO, G., ZERBINI, M., KRZYSZTOFIAK, A., GENTILOMI, G., PORCARO, M. A., MANGO, T., AND MUSIANI, M., Active or recent parvovirus B19 infection in children with Kawasaki disease, *Lancet* **343**:1260–1261 (1994).

88. NIITSU, H., TAKATSU, H., MIURA, I., CHUBACHI, A., ITO, T., HIROKAWA, M., ENDO, Y., MIURA, A., FUKUDA, M., AND SASAKI, T., [Pure red cell aplasia induced by B19 parvovirus during allogeneic bone marrow transplantation], *Rinsho Ketsueki.* **31**:1566–1571 (1990).

89. NUNOUE, T., OKOCHI, K., MORTIMER, P. P., AND COHEN, B. J., Human parvovirus (B19) and erythema infectiosum, *J. Pediatr.* **107**:38–40 (1985).

90. OKABE, N., KOBOYASHI, S., TATSUZAWA, O., AND MORTIMER, P. P., Detection of antibodies to human parvovirus in erythema infectiosum (fifth disease), *Arch. Dis. Child.* **59**:1016–1019 (1984).

91. OWREN, P. A., Congenital hemolytic jaundice: The pathogenesis of the "hemolytic crisis," *Blood* **3**:231–248 (1948).

92. OZAWA, K., KURTZMAN, G., AND YOUNG, N., Productive infection by B19 parvovirus of human erythroid bone marrow cells *in vitro*, *Blood* **70**:384–391 (1987).

93. PATOU, G., PILLAY, D., MYINT, S., AND PATTISON, J., Characterization of a nested polymerase chain reaction assay for detection of parvovirus B19, *J. Clin. Microbiol.* **31**:540–546 (1993).

94. PATTISON, J. R., JONES, S. E., HODGSON, J., DAVIS, L. R., WHITE, J. M., STROUD, C. E., AND MURTAZA, L., Parvovirus infections

and hypoplastic crisis in sickle-cell anaemia, *Lancet* **1**:664–665 (1981).

95. PELLAS, F., OLIVARES, J. P., ZANDOTTI, C., AND DELARQUE, A., Neuralgic amyotrophy after parvovirus B19 infection [letter], *Lancet* **342**:503–504 (1993).

96. PILLAY, D., PATOU, G., HURT, S., KIBBLER, C. C., AND GRIFFITHS, P. D., Parvovirus B19 outbreak in a children's ward, *Lancet* **339**:107–109 (1992).

97. PLUMMER, F. A., HAMMOND, G. W., FORWARD, K., SEKLA, L., THOMPSON, L. M., JONES, S. E., KIDD, I. M., AND ANDERSON, M. J., An erythema infectiosum-like illness caused by human parvovirus infection, *N. Engl. J. Med.* **313**:74–79 (1985).

98. PONT, J., PUCHHAMMER-ST:OCKL, E., CHOTT, A., POPOW-KRAUPP, T., KIENZER, H., POSTNER, G., AND HONETZ, N., Recurrent granulocytic aplasia as clinical presentation of a persistent parvovirus B19 infection, *Br. J. Haematol.* **80**:160–165 (1992).

99. PORTER, H. J., HERYET, A., QUANTRILL, A. M., AND FLEMING, K. A., Combined non-isotopic *in situ* hybridisation and immunohistochemistry on routine paraffin wax embedded tissue: Identification of cell type infected by human parvovirus and demonstration of cytomegalovirus DNA and antigen in renal infection, *J. Clin. Pathol.* **43**:129–132 (1990).

100. PORTER, H. J., KHONG, T. Y., EVANS, M. F., CHAN, V. T., AND FLEMING, K. A., Parvovirus as a cause of hydrops fetalis: Detection by *in situ* DNA hybridisation, *J. Clin. Pathol.* **41**:381–383 (1988).

101. PORTER, H. J., QUANTRILL, A. M., AND FLEMING, K. A., B19 parvovirus infection of myocardial cells [letter], *Lancet* **1**:535–536 (1988).

102. POTTER, C. G., POTTER, A. C., HATTON, C. S., CHAPEL, H. M., ANDERSON, M. J., PATTISON, J. R., TYRRELL, D. A., HIGGINS, P. G., WILLMAN, J. S., PARRY, H. F., AND COTES, P. M., Variation of erythroid and myeloid precursors in the marrow and peripheral blood of volunteer subjects infected with human parvovirus (B19), *J. Clin. Invest.* **79**:1486–1492 (1987).

103. PUBLIC HEALTH LABORATORY SERVICE WORKING PARTY ON FIFTH DISEASE, Prospective study of human parvovirus (B19) infection in pregnancy, *Br. Med. J.* **300**:1166–1170 (1990).

104. ROSENFELD, S. J., YOSHIMOTO, K., KAJIGAYA, S., ANDERSON, S., YOUNG, N. S., FIELD, A., WARRENER, P., BANSAL, G., AND COLLETT, M. S., Unique region of the minor capsid protein of human parvovirus B19 is exposed on the virion surface, *J. Clin. Invest.* **89**:2023–2029 (1992).

105. ROSENFELD, S. J., YOUNG, N. S., ALLING, D., AYUB, J., AND SAXINGER, C., Subunit interaction in B19 parvovirus empty capsids, *Arch. Virol.* **136**:9–18 (1994).

106. ROUGER, P., GANE, P., AND SALMON, C., Tissue distribution of H, Lewis and P antigens as shown by a panel of 18 monoclonal antibodies, *Rev. Fr. Transfus. Immunohematol* **30**:699–708 (1987).

107. SAARINEN, U. M., CHORBA, T. L., TATTERSALL, P., YOUNG, N. S., ANDERSON, L. J., PALMER, E., AND COCCIA, P. F., Human parvovirus B19-induced epidemic acute red cell aplasia in patients with hereditary hemolytic anemia, *Blood* **67**:1411–1417 (1986).

108. SAINT-MARTIN, J., CHOULOT, J. J., BONNAUD, E., AND MORINET, F., Myocarditis caused by parvovirus [letter: comment], *J. Pediatr.* **116**:1007–1008 (1990).

109. SALIKI, J. T., MIZAK, B., FLORE, H. P., GETTIG, R. R., BURAND, J. P., CARMICHAEL, L. E., WOOD, H. A., AND PARRISH, C. R., Canine parvovirus empty capsids produced by expression in a baculovirus vector: Use in analysis of viral properties and immunization of dogs, *J. Gen. Virol.* **73**:369–374 (1992).

110. SALIMANS, M. M., VAN DE, R. I., RAAP, A. K., AND VAN ELSACKER-NIELE, A. M., Detection of parvovirus B19 DNA in fetal tissues by *in situ* hybridisation and polymerase chain reaction, *J. Clin. Pathol.* **42**:525–530 (1989).

111. SCHWARZ, T. F., BRUNS, R., SCHRÖDER, C., WIERSBITZKY, S., AND ROGGENDORF, M., Human parvovirus B19 infection associated with vascular purpura and vasculitis [letter], *Infection* **17**:170–171 (1989).

112. SCHWARZ, T. F., GÜRTLER, L. G., ZOULEK, G., DEINHARDT, F., AND ROGGENDORF, M., Seroprevalence of human parvovirus B19 infection in Sao Tomé and Principe, Malawi and Mascarene Islands, *Int. J. Med. Microbiol.* **271**:231–236 (1989).

113. SERJEANT, G. R., SERJEANT, B. E., THOMAS, P. E., ANDERSON, M. J., PATOU, G., AND PATTISON, J. R., Human parvovirus infection in homozygous sickle cell disease, *Lancet* **341**:1237–1240 (1993).

114. SERJEANT, G. R., TOPLEY, J. M., MASON, K., SERJEANT, B. E., PATTISON, J. R., JONES, S. E., AND MOHAMED, R., Outbreak of aplastic crisis in sickle cell anaemia associated with parvovirus-like agent, *Lancet* **2**:595–597 (1981).

115. SIEGL, G., BATES, R. C., BERNS, K. I., CARTER, B. J., KELLY, D. C., KURSTAK, E., AND TATTERSALL, P., Characteristics and taxonomy of *Parvoviridae*, *Intervirology* **23**:61–73 (1985).

116. SISK, W. P., AND BERMAN, M. L., Expression of human parvovirus B19 structural protein in *E. coli* and detection of antiviral antibodies in human serum, *Bio/Technology* **5**:1077–1080 (1987).

117. SMITH, M. A., SHAH, N. R., LOBEL, J. S., CERA, P. J., GARY, G. W., AND ANDERSON, L. J., Severe anemia caused by human parvovirus in a leukemia patient on maintenance chemotherapy, *Clin. Pediatr.* **27**:383–386 (1988).

118. SÖRENSEN, S. F., [Acute vasculitis and arthritis caused by parvovirus B19 infection], *Ugeskr. Laeger* **154**:2032–2033 (1992).

119. TAKAHASHI, M., KOIKE, T., MORIYAMA, Y., AND SHIBATA, A., Neutralizing activity of immunoglobulin preparation against erythropoietic suppression of human parvovirus [letter], *Am. J. Hematol.* **37**:68 (1991).

120. TAKAHASHI, T., OZAWA, K., TAKAHASHI, K., ASANO, S., AND TAKAKU, F., Susceptibility of human erythropoietic cells to B19 parvovirus *in vitro* increases with differentiation, *Blood* **75**:603–610 (1990).

121. UMENE, K., AND NUNOUE, T., The genome type of human parvovirus B19 strains isolated in Japan during 1981 differs from types detected in 1986 to 1987: A correlation between genome type and prevalence, *J. Gen. Virol.* **71**:983–986 (1990).

122. UMENE, K., AND NUNOUE, T., Genetic diversity of human parvovirus B19 determined using a set of restriction endonucleases recognizing four or five base pairs and partial nucleotide sequencing: Use of sequence variability in virus classification, *J. Gen. Virol.* **72**:1997–2001 (1991).

123. VAN HORN, D. K., MORTIMER, P. P., YOUNG, N., AND HANSON, G. R., Human parvovirus-associated red cell aplasia in the absence of underlying hemolytic anemia, *Am. J. Pediatr. Hematol. Oncol.* **8**:235–239 (1986).

124. WALSH, K. J., ARMSTRONG, R. D., AND TURNER, A. M., Brachial plexus neuropathy associated with human parvovirus infection, *Br. Med. J.* **296**:896 (1988).

125. WARSOF, S. L., NICOLAIDES, K. H., AND RODECK, C., Immune and non-immune hydrops, *Clin. Obstet. Gynecol.* **29**:533–542 (1986).

126. WEILAND, H. T., VERMEY-KEERS, C., SALIMANS, M. M., FLEUREN, G. J., VERWEY, R. A., AND ANDERSON, M. J., Parvovirus B19 associated with fetal abnormality [letter], *Lancet* **1**:682–683 (1987).

127. WHITE, D. G., WOOLF, A. D., MORTIMER, P. P., COHEN, B. J., BLAKE, D. R., AND BACON, P. A., Human parvovirus arthropathy, *Lancet* **1**:419–421 (1985).

128. WIERSBITZKY, S., SCHWARZ, T. F., BRUNS, R., JÄGER, G., BITTNER, S., WEIDEMANN, H., DEINHARDT, F., HOTTENTRÄGER, B., ABEL, E., LADSTATTER, L., FRICK, G., AND ROGGENDORF, M., [Seroprevalence of human parvovirus B19 antibodies (Sticker's disease/erythema infectiosum) in the DRG population], *Kinderarztl. Prax.* **58**:185–189 (1990).

129. WILLIAMS, M. D., COHEN, B. J., BEDDALL, A. C., PASI, K. J., MORTIMER, P. P., AND HILL, F. G., Transmission of human parvovirus B19 by coagulation factor concentrates, *Vox. Sang.* **58**:177–181 (1990).

130. WOOLF, A. D., CAMPION, G. V., CHISHICK, A., WISE, S., COHEN, B. J., KLOUDA, P. T., CAUL, O., AND DIEPPE, P. A., Clinical manifestations of human parvovirus B19 in adults, *Arch. Intern. Med.* **149**:1153–1156 (1989).

131. YOTO, Y., KUDOH, T., HASEYAMA, K., SUZUKI, N., CHIBA, S., AND MATSUNAGA, Y., Human parvovirus B19 infection in Kawasaki disease [letter], *Lancet* **344**:58–59 (1994).

132. YOUNG, N., Hematologic and hematopoietic consequences of B19 parvovirus infection, *Semin. Hematol.* **25**:159–172 (1988).

133. YOUNG, N., HARRISON, M., MOORE, J., MORTIMER, P., AND HUMPHRIES, R. K., Direct demonstration of the human parvovirus in erythroid progenitor cells infected *in vitro*, *J. Clin. Invest.* **74**:2024–2032 (1984).

12. Suggested Reading

ANONYMOUS, Risks associated with human parvovirus B19 infection, *Morbid. Mortal. Week. Rep.* **38**:81–88, 93–97 (1989).

ANDERSON, M. J., AND CHERRY, J. D., Parvoviruses, in: *Textbook of Pediatric Infectious Diseases* (R. D. FEIGIN AND J. D. CHERRY, EDS.), pp. 1646–1653, Saunders, Philadelphia, 1987.

BROWN, K. E., HIBBS, J. R., GALLINELLA, G., ANDERSON, S. M., LEHMAN, E. D., MCCARTHY, P., AND YOUNG, N. S., Resistance to parvovirus B19 infection due to lack of virus receptor (erythrocyte P antigen), *N. Engl. J. Med.* **330**:1192–1196 (1994).

BROWN, K. E., YOUNG, N. S., AND LIU, J. M., Molecular, cellular and clinical aspects of parvovirus B19 infection, *Crit. Rev. Oncol. Hematol.* **16**:1–31 (1994).

CHAPTER 21

Poliovirus and Other Enteroviruses

Joseph L. Melnick

1. Introduction

The enterovirus group, named in 1957,[72,285] brought together polioviruses, coxsackieviruses, and echoviruses, all of which inhabit the human alimentary tract. These viruses share a number of clinical, epidemiologic, and ecological characteristics as well as physical and biochemical properties.

Enteroviruses of human origin include the following:

1. Polioviruses: types 1–3.
2. Coxsackieviruses A: 23 types and several variants [coxsackieviruses A1–A24 (coxsackievirus type A23 is the same virus as echovirus 9)].
3. Coxsackieviruses B: types B1–B6.
4. Echoviruses: 31 types [types 1–33 (echovirus 10 has been reclassified as reovirus type 1, and echovirus 28 as rhinovirus type 1A)].
5. Enterovirus types 68–71: These viruses would formerly have been classified as either coxsackievirus or echovirus types (see Section 2).
6. Enterovirus 72: Hepatitis A virus was first provisionally classified as this serotype; it is now considered to be the prototype of a separate picornavirus genus.

In 1963, the name *picornavirus* (*pico* = small; *rna* = ribonucleic acid genome) was introduced as a larger grouping[283] to which not only the enteroviruses but also the rhinoviruses would belong by reason of fundamental similarities in many of their properties. With the advancement of viral classification on the basis of further knowledge of biophysical and biochemical characteristics, the International Committee on Nomenclature of Viruses has officially assigned family status (*Picornaviridae*) to this larger taxon, with *Enterovirus* as one genus, *Rhinovirus* as

another, and two other genera chiefly infecting lower animals—*Aphthovirus* and *Cardiovirus*, including, respectively, the agents of foot-and-mouth disease of cattle and encephalomyocarditis virus of rodents.[74,279]

Hepatitis A virus (HAV) was originally classified as enterovirus type 72,[249] although more recent molecular evidence suggests that it belongs in a separate genus of *Picornaviridae*. In Table 1, the properties that have been determined for HAV are listed alongside the properties of the first known enterovirus, poliovirus type 1. Now that HAV has been classified in a genus close to the enteroviruses, we can make certain predictions about its other properties, the most significant of which are the following: (1) One would not expect there to be chronic carriers of HAV, since carriage of an enterovirus is usually limited to a few weeks, and (2) one would not expect HAV to cause posttransfusion hepatitis, since enterovirus viremia is rarely detected and never lasts longer than 1 or 2 days.

Enteroviral isolates continue to be reported that are not neutralized by any of the known antisera, and undoubtedly some are truly new serotypes. In addition to the enteroviruses of humans, numerous enteroviruses of lower animals are known, and certain plant and insect viruses also have properties similar to those of the picornaviruses, as do some of the RNA-containing bacteriophages.

Many enteroviruses cause disease in man ranging from severe and permanent paralysis to minor undifferentiated febrile illnesses. For all members of the group, however, subclinical infection is far more common than clinically manifest disease. Although certain enteroviruses have been more frequently responsible for epidemics involving a specific syndrome, the same serotypes may at other times and in other places be associated with sporadic infections having different clinical manifestations or producing no symptoms. On the other hand, different viruses may produce the same syndrome. For

Joseph L. Melnick • Division of Molecular Virology, Baylor College of Medicine, Houston, Texas 77030.

Table 1. Comparative Biophysical Properties of Hepatitis Type A Virus and Poliovirus Type 1

Physicochemical characteristics of major virus-particle population	Hepatitis type A virus[a]	Poliovirus type 1
Morphology		
Diameter	28 nm	28 nm
Envelope	None	None
Sedimentation rate	160 S	160 S
Density (CsCl)	1.34 g/ml	1.34 g/ml
Nucleic acid		
Type	Single-stranded RNA	Single-stranded RNA
Length[b]	1.7 μm	2.3 μm
Molecular weight	1.9×10^6	2.6×10^6
Polypeptides (major)— molecular weights	22,000[c] 24,000 29,000	24,000 25,000 34,000

[a]Notwithstanding the overlapping physical properties, hepatitis A has a completely different nucleic acid sequence.
[b]As measured by electron microscopy.
[c]Values of 23,000, 25,500, and 34,000 have also been reported.

these reasons, clinical disease is not a satisfactory basis for classification or as a rule, for diagnosis.

Poliomyelitis is an acute infectious disease that in its serious form affects the CNS. The destruction of motor neurons in the spinal cord results in flaccid paralysis.

The coxsackieviruses produce a variety of illnesses, including aseptic meningitis, herpangina, epidemic myalgia (pleurodynia, Bornholm disease), hand, foot, and mouth disease, myocarditis, pericarditis, pneumonia, rashes, and common colds. They may also have a role in some congenital malformations and perhaps in some forms of diabetes.

Aseptic meningitis, febrile illnesses with or without rash, and common colds are among the diseases caused by echoviruses.

Among the newer enterovirus types, enterovirus 68 has caused lower respiratory illness, enterovirus 70 is the agent of widespread epidemics of acute hemorrhagic conjunctivitis, and enterovirus 71 has caused aseptic meningitis and encephalitis and hand, foot, and mouth disease in a number of countries. (Further details of clinical manifestations of enterovirus infections are given in Section 8.1.)

Because polioviruses can cause the most severe disease of any for which enteroviruses are responsible, these agents have received the most comprehensive study and have served as models in studies of other enteroviruses. In this chapter, therefore, poliovirus studies are used frequently as illustrative of phenomena that also hold true for other enteroviruses.

2. Historical Background

A wealth of detailed information on the development of knowledge of enteroviruses is available in earlier reviews and textbooks.[34,83,179,259,261,262,391] Consequently, only a few of the historic advances in knowledge of the enteroviruses will be mentioned here; for more information, see Paul[368] and Melnick.[268]

Although crippling disease retrospectively recognizable as paralytic poliomyelitis appears in records of early antiquity, it began to be described as a clinical entity only in the late 18th and early 19th centuries and became the subject of intensified study after increasingly severe epidemics began to appear in Europe and North America. Experimental work became possible with the successful transmission of the disease to monkeys in 1908 by Landsteiner and Popper.[226] During the next 40 years, it was shown that the virus was regularly present in stools of patients, that subhuman primates could be infected by the alimentary route, and that strains could be adapted to growth in laboratory rodents, permitting an expansion of laboratory studies. Significant antigenic differences among poliovirus strains were documented, resulting in their separation into three serological types, and it was discovered that polioviruses can be isolated and cultivated in vitro, in cell cultures derived from primate nonneural tissue.

The first strains of what are now known as coxsackievirus subgroup A were isolated by inoculation of infant mice with fecal material from two children suffering from paralysis during an epidemic of poliomyelitis in 1948 in Coxsackie, New York.[84] Additional types, including the first of the coxsackieviruses of subgroup B,[301] were discovered shortly thereafter. Coxsackievirus B agents were associated with a syndrome like nonparalytic poliomyelitis (aseptic meningitis) and also with epidemic myalgia and pleurodynia.[78] Group A and B coxsackieviruses were distinguished by their differing pathological effects in baby mice (see Section 4).

As soon as cultures of human and monkey cells began to be used to search for polioviruses in stool specimens of patients,[102] still more unknown viruses were found that, unlike polioviruses and the newly recognized coxsackieviruses, were not pathogenic for laboratory animals but produced cytopathic effects in cultured cells.[280,392]

It soon became apparent that these agents could be isolated from healthy children[149,167,280,383] as well as from patients with aseptic meningitis[256,280] and that multiple types existed.[256,383] Because the relationship of these newly recognized agents to human disease was unknown, and because they failed to produce illness in laboratory animals, they were called "orphan" viruses or human enteric viruses; later then became known as ECHO (enteric cytopathogenic human orphan) viruses,[71] a name subsequently simplified to "echoviruses."

In addition to their characteristic mouse pathogenicity, certain of the coxsackieviruses were found to grow readily in tissue cultures; other strains, serologically identical with the mouse-pathogenic prototype, failed to produce paralysis in baby mice. Conversely, certain strains of echoviruses were found to be pathogenic for mice. As instances of such overlapping properties accumulated, blurring the initial distinction made between coxsackieviruses and echoviruses, it was recommended[395] that subsequently, as new enterovirus types were discovered, they would simply be assigned sequential numbers, as enterovirus 68, enterovirus 69, and so on. The currently accepted serotypes thus assigned are listed in Table 8 (Section 8).

3. Methodology Involved in Epidemiologic Analysis

3.1. Sources of Mortality Data

In the United States, paralytic poliomyelitis, aseptic meningitis, encephalitis, and any poliovirus that is isolated are "notifiable," i.e., regularly required to be reported to local, state, and national health officers. But among fatalities caused by enteroviruses, only poliomyelitis is generally confirmed by virus isolation. Thus, in deaths from infections by other enteroviruses (such as encephalomyocarditis in infants), the virus responsible may not be recognized unless testing facilities are available. Information reported to the U.S. Centers for Disease Control (CDC) is regularly and promptly circulated in the CDC publication *Morbidity and Mortality Weekly Report* and in special annual CDC surveillance summaries on poliomyelitis, on aseptic meningitis and encephalitis, and on the viruses being isolated and identified.

In worldwide surveillance and reporting programs of the World Health Organization (WHO), also, the most consistently investigated enterovirus-associated fatalities are those involving poliomyelitis. However, increasingly broad and useful information is becoming available on other enterovirus diseases (see Sections 3.2, 5, and 8).

The case-fatality rate for poliovirus infection is not easily determined because of the difficulties in diagnosing nonparalytic infections. In years of high prevalence, the case-fatality rate may appear lower than in years of low prevalence because of the likelihood that nonparalytic poliomyelitis is diagnosed more readily at times of epidemic prevalence. The usual rate varies between 5 and 10% and is highest in the older age groups. In the last major prevaccine epidemics in the United States, in which one third of the cases occurred in patients over 15 years old, two thirds of the deaths were in this age group.

3.2. Sources of Morbidity Data

The sources of morbidity data for enteroviruses are generally far from ideal in that most studies, necessarily conducted in localized or special groups, can only suggest what may occur in the population generally. Results depend on the industriousness of the individual investigator, the specific type of situation (e.g., military or institutional), the means of data collection, and the investigator's knowledge of prior reports in the literature and his ability to relate his findings to them.

Many fruitful enterovirus studies have centered around patients as they came to medical attention or have been occasioned by outbreaks of illness; others have entailed observations of special populations or have consisted of challenge experiments with volunteers. Such investigations can yield valuable knowledge of routes of transmission, types and severity of clinical illness that can be associated with infection, sites and duration of virus excretion, and antibody responses.

But with a group of agents such as the enteroviruses, these studies have certain limitations; large proportions of enterovirus infections are subclinical and do not reach the attention of a physician; a single enterovirus serotype may produce a variety of syndromes, and a similar syndrome may be produced by a number of different enteroviruses as well as by members of other virus groups. In special environments, especially in closed populations—e.g., children's homes or military training centers where close contact, age homogeneity, and unusual stresses are often involved—the patterns of spread of infection may be atypical.

Of considerable significance in filling this gap in the knowledge of viral ecology are studies typified by the Virus Watch Programs,[111,114,212] under which families living in normal circumstances have been observed longitudinally over a period of years, with regular sampling, surveillance, and testing procedures. In an open population, enlisted families with children (or, in later periods of the programs, those with a newborn infant) are maintained under continuing virological surveillance. The principal objectives are to describe the occurrence of enteric and respiratory viral infections and/or possibly related illnesses and, by analysis of such descriptive information, to determine for specific viruses or virus groups the mode and pattern of intrafamilial spread, the relationship of age and immune status to both infection and disease, the nature of associated disease, and the frequency with which it follows infection.

Virus Watch Programs serve as an important guide for experimental design in which the family is taken as the basic epidemiologic unit and in which infection, rather than solely overt illness, is the focus of study. Some of the findings are included in Section 5.

Of continuing value for morbidity data and analyses for the United States are the various publications from the CDC. These include not only the ongoing periodicals such as the *Morbidity and Mortality Weekly Reports* and the various annual surveillance summaries mentioned above but also analyses published by individuals and groups from CDC in the general virological literature.

Another important source of information concerning virus infections on a global scale was initiated in 1963, in the WHO system for collecting and distributing laboratory and epidemiologic information. Virus infections diagnosed by isolation or serology are reported by WHO Reference Centers and national virus laboratories around the world; the data have been consolidated and analyzed by the WHO Virus Unit, Geneva and distributed in *WHO Weekly Epidemiological Records, WHO Quarterly and Annual Reports on Virus Isolations*, and in other special reports. By the end of 1975, 119 laboratories in 47 countries were participating in this scheme. From 1967 through 1975, more than 280,000 reports on viral infections were collected in the WHO virus data file. The volume of data increased from about 20,500 reports received in 1967 to more than 55,000 in 1975. Despite the limitations imposed by a wide diversity of laboratory methods, selective interests and responsibilities of various laboratories (e.g., a necessary sampling bias toward persons with overt illness), and other problems inherent in a program of this scope, these data yield indications of temporal trends in viral infections, marked linkages of certain serotypes to specific clinical syndromes, trends in age incidence, and other new knowledge about enteroviral infections. A report by Assaad and Cockburn[14] on the nonpolio enteroviruses reported during the 4-year period 1967–1970 illustrates the very useful analyses that can be developed on the basis of these WHO records (see Section 5). An update of these analyses, extending the covered period to 8 years (1967–1974), appears in a review by Grist, Bell, and Assaad.[141] More recent reviews have been published in Japan.[318,319,506,507]

Clinical surveys also have a place in providing information about the occurrence of at least one enteroviral illness: paralytic poliomyelitis. Surveys of residual paralysis in young children have provided useful data on the recent history of poliomyelitis in a community, as suggested by Payne.[373] Since the sequelae of paralytic poliomyelitis seem to be distinctive, they have been used as a measure of prior prevalence of the disease[13]

The various types of lameness surveys have been described in WHO publications, and the results of some are discussed in Section 5.3.2. The various merits and disadvantages of different survey methods were reviewed by Evans.[104]

3.3. Serological and Clinical Surveys

The science of serological epidemiology has reached a high degree of sophistication in a relatively short time, thanks to the pioneers who developed this field under exceedingly difficult conditions and with only crude tests available.

Even before the three serotypes of poliovirus were completely delineated, and when tests for antibodies could be conducted only in monkeys, literature reports and serological surveys of a variety of populations around the globe were compared with respect to antibody patterns. Throughout the early studies of poliomyelitis, serological surveys continued to add essential pieces to the puzzle of poliovirus transmission, susceptibility, widely varying geographic and socioeconomic patterns of infection, duration of type-specific immunity, and the shift of populations from endemic to epidemic experience with the polioviruses.[367,369] It was found that the highest percentages of persons possessing poliovirus antibodies were recorded among normal adolescents and adults in tropical areas where some contemporary "authorities" of the period believed that poliomyelitis did not exist.

Today, serological surveys continue to have an important role in maintaining protection against resurgence of epidemic poliomyelitis by making possible the surveillance of immunity levels. Continuing serological surveillance is needed to answer such questions as the following: Are significant proportions of the susceptible age group being reached and protected by vaccination? Do the results of serosurveys parallel the estimates from surveys of immunization history? How well are antibodies induced by vaccination persisting over the years in comparison with their duration after natural infection? Since the answers to these questions may vary from one locality to another[169] and even within different population sectors of the same community,[263] local immunity patterns must be monitored to locate the specific age groups and sectors of the community in which declining antibody levels or failure to obtain vaccination (particularly among disadvantaged inner-city children) is resulting in "protection gaps" (see Sections 5 and 9).

Furthermore, despite the increasing availability of effective poliomyelitis vaccines, there remain a number of developing countries in which vaccination is not yet widely used. Serological surveys can determine whether patterns of naturally acquired immunity are changing and thus can alert the national health authorities to the growing need for vaccination before this need is made disastrously clear by the occurrence of large epidemics.

Serological surveys of varying scope have contributed data from which the history of local exposure to enteroviruses can be read. For example, the tests with serum collected from Eskimos in northern Alaska[19] revealed a population heavily exposed to coxsackieviruses A4 and A10, less experienced with A1, and having no detectable previous experience with coxsackieviruses B1 and B2. In the early years of investigation of the nonpolio enteroviruses, when only a few serotypes were known, the proportion of seroconversions occurring in various age groups during a single year could be obtained by serological surveys conducted in a community before and after the summer–fall season. These serological data, analyzed in conjunction with virus isolations and monitoring of concurrent illness patterns, yielded much of the information now available concerning the age patterns, the ease and routes of dissemination, and the ratio of inapparent infections to clinical illness.[292]

At present, however, general serological screening to detect seroconversions or rises in antibody titer against all possible enteroviruses is a virtual impossibility, for such a general screening would require tests against more than 70 enterovirus antigens, in combined tissue-culture and mouse systems (see Section 3.5). Epidemiologic studies to learn the current pattern of infection with enteroviruses in a community can be conducted much more easily by means of virus isolation and typing, since the availability of reference antiserum pools has vastly reduced the number of tests required to identify an isolate.

In instances when an epidemic caused by a single serotype is in progress, serological surveys can be usefully incorporated into well-planned and specifically targeted prospective or retrospective epidemiologic studies and can provide timely guidance for physicians in making presumptive diagnoses of current illnesses.

Epidemiologic surveillance is an essential part of the control of infectious diseases.[384] However, not all epidemiologic surveillance approaches can be applied to poliomyelitis and other enteroviral diseases because of the large number of infections that are inapparent and the variety of syndromes. Nonetheless, a number of the tools for surveillance have been applied with considerable success in recent years, particularly in conjunction with polio vaccination programs in developing countries with limited laboratory facilities. These tools include more complete and precise reporting of disease, surveys for residual paralysis and studies comparing the effectiveness of various methods by which these disease reports and lameness data are gathered.

3.4. Virus Isolation from Surface Waters as an Indicator of Community Infections

The enteroviruses are excreted so regularly and abundantly in feces that their presence in sewage can provide a great deal of information about the circulation of these viruses in a community. As early as 1940–1945,[254] consecutive tests of sewage samples reflected the seasonal prevalence of polioviruses, showed that these viruses remain infective in flowing sewage for many hours, and demonstrated that the viruses may be continually present in sewage over a period of several months. Taken together with the incidence of clinical polio in the community, the findings also provided a basis for estimating the ratio of inapparent poliovirus infections to clinical poliomyelitis. From the data obtained in New York City, this ratio turned out to be well over $100 : 1$.[254]

Numerous studies[23,32,65,137,201,225,243,287] in developed countries have readily demonstrated the presence of enteroviruses of all subgroups in contaminated streams, in sewage, and in effluents from sewage-treatment plants. In some localities, wild polioviruses have continued to be present long after the introduction of oral poliovirus vaccine. In one metropolitan area, at the same time as virulent polioviruses were revealed in the city streams, there was a small outbreak of paralytic poliomyelitis among unimmunized infants in the area.[201] Of special note is the fact that although methods currently used for sewage treatment may slightly reduce the amount of viruses present, they do not eliminate them from the effluent. In some cases, the concentration of virus in the effluent equals that in the influent sewage or even exceeds it because of disaggregation of virus particles from clumps.

Sampling of sewage reflects not only the presence of enteroviruses but also the changes in predominant serotypes from one period to another[201,287] and has been used recently to assess the impact of oral poliovirus vaccine on the circulation of nonpolio enteroviruses.[169] Isolation of virus from an open lake swimming area served as confirmation of a coxsackievirus B5 epidemic at a summer camp[157] and provided an indication of the extent of the infections.

Portable apparatus and methods amenable to use in the field have been developed for concentrating virus from sewage and from surface water.[271,289,298,309,465] These methods have as their ultimate purpose the monitoring of the viral content of sewage and of water sources, so as to protect communities from viral disease that might be transmitted through drinking and recreational waters. This may become more crucial as water recycling becomes increasingly necessary.[24,500] An important byproduct of this work is providing a more accurate index of the kinds and amounts of enteroviruses present within the community.

3.5. Laboratory Methods

Detailed descriptions of principles and procedures for diagnosis of enterovirus infections have been published.[1,47,59,88,303–309,326,401,425,472,474] Historical information and descriptions of the development of procedures utilizing monkeys as well as cell cultures are also available.[268,305,370]

Epidemiologic knowledge of the enterovirus group—with its numerous members, frequency of silent infection, and variability of clinical manifestations when these do occur—must depend in large part on investigators' ability to identify the viruses isolated and to communicate accurately with others studying the same or similar enteroviruses. This in turn rests on the establishment of a common nomenclature for the viruses and on the development of standard reagents, both reference virus strains and type-specific antisera. Early in studies of enteroviruses, it was recognized that cooperative development of standard reagents was requisite to progress in understanding of enteroviral disease.[71] Under the sponsorship of the WHO, there followed a steadily broadening series of collaborative studies among working virologists around the world to develop such reagents.[264] At present, two sets of lyophilized pools of collaboratively evaluated and standardized typing antisera are available.[297,299,306,494] These are eight pools (A–H) of antisera against 42 enteroviruses that grow readily in cell cultures and seven pools (J–P) of antisera against 19 coxsackievirus A serotypes, including those that have not grown in cell culture and can be cultivated only in newborn mice. Use of these pools greatly facilities the identification of enterovirus isolates[232] (see Section 3.5.1.).

3.5.1. Virus Isolation and Identification. The usual specimens are stools, rectal swabs, and throat swabs.[304] In addition, cerebrospinal fluid (CSF) yields virus often in cases of aseptic meningitis caused by coxsackieviruses or echoviruses but rarely if this syndrome is caused by a poliovirus (i.e., nonparalytic poliomyelitis). Polioviruses have been detected in blood specimens taken very early in infection.[170,298] In children, even after development of symptoms has led to hospitalization, a variety of nonpolio enteroviruses may be detected in the blood, either free in the serum or in mononuclear leukocytes.[380] Virus may also be isolated from vesicle fluids, urine, conjunctival swabs (enterovirus 70), and nasal secretions (which yield coxsackievirus A21 isolates more readily than specimens from the throat). Virus may be recovered from throat swabs taken during the first few days of illness (or of silent infection) and from rectal swabs or stools, often for several weeks. However, some enteroviruses (for example, types 70 and 72) are excreted for short prodromal periods and are difficult to isolate from feces of sick persons. In fatal cases of suspected enteroviral etiology, virus should be sought in the organ system affected as well as in colon contents.

Not all agents recovered from feces or oropharynx are enteroviruses. Others that might be recovered are rotaviruses, reoviruses, adenoviruses, rhinoviruses, and the viruses of measles, mumps, rubella, and herpes simplex. Many of these agents produce distinctive cytopathic effects (CPE), which at once differentiate them from enteroviruses. Some evoke fatal encephalitis in newborn mice.

If the virus laboratory has experienced personnel, a presumptive diagnosis of enteroviral infection can often be made on the basis of the nature of the associated illness (if any), the time of year when the specimen was obtained, the tissue-culture (or mouse) system in which the virus isolate grew, and the characteristic CPE observed in the cultures or the characteristic pathology induced in the mice. There is often value in reporting a presumptive identification without waiting for specific typing of the isolate. Even though specific antiviral therapy is lacking at present, early recognition of probable enteroviral infection can provide information for the management of a patient or of a community outbreak of similar illnesses and may serve to contraindicate administration of unnecessary or undesirable antibiotic therapy and to assist in narrowing the scope of diagnostic tests undertaken during an outbreak of compatible illnesses.

Most of the common enterovirus serotypes that are cytopathogenic can be recovered by inoculation of specimens into primary and passage cultures of monkey kidney cells, but at least one human tissue-culture system must also be included if recovery of all possible types is being attempted. In addition, inoculation of newborn mice must be used if one is to detect those coxsackieviruses of group A that cannot be cultivated in cell cultures. Among the human cell cultures that have been reported as most sensitive for a broad range of enteroviruses are WI-38 (a cell line from human embryonic lung), HEK (primary human embryonic kidney), and various local cell strains. For example, WI-38 cultures more readily support growth of echovirus 30 isolates as well as a number of other echovirus serotypes[73,94,145,196]; however, in initial isolation of polioviruses and coxsackievirus B types, HEK cultures far surpassed either WI-38 or combined HeLa cell and monkey-kidney-cell culture systems.[73] The RD cell line, derived from a human rhabdomyosarcoma,[250] has been shown to support replication of a number of the coxsackievirus group A strains, including types A5 and A6, which previously have been grown only in newborn mice, and several other types that would grow only to low titers in other cell cultures.[77,423] Fortunately, several types that do not grow in RD cells are cultivable in HeLa or other human cell cultures.[423] It should be noted that with RD cells, most strains tested required a second passage to obtain clear-cut CPE. With utilization of RD cell cultures, only coxsackieviruses A1, A19, and A22 remain as types that require newborn mice for their cultivation. The susceptibility of cells to a number of enteroviruses may be increased by treating them with 5-iododeoxyuridine at 50 μg/ml for 3 days prior to exposing the cells to the virus.[21] The BGM (buffalo green monkey) line of African green monkey kidney cells has been reported[80,308] to offer greater sensitivity than primary rhesus or green monkey kidney cells for titration of certain enterovirus types and also for recovery of plaque-forming enteric viruses from sewage and water. Comparative tests with clinical specimens in another laboratory indicated that the line may have limitations in sensitivity for routine isolation of a variety of echovirus types, as compared to primary rhesus monkey kidney and human fetal diploid kidney cells.[424] One potential variable is that the BGM cells obtained commercially and used in testing the clinical specimens may have been less sensitive to virus because of mycoplasma contamination.

With regular use of four types of cell cultures—i.e., adding BGM and RD cultures to the more commonly used human embryonic lung and cynomolgus monkey kidney cells—through two summer–fall seasons, both the frequency and the speed of recovery of enteroviruses from clinical specimens could be increased. Of 2558 specimens (fecal, respiratory, CSF, and blood), 417 yielded an enterovirus. Of the enteroviruses isolated, 18% were detected only in BGM or RD cells. Results were available for the clinician in 42% of the enterovirus-positive specimens by day 2 after inoculation of the cultures, and in 61% by day 4.[79]

Of the higher-numbered enteroviruses, types 70 and 71 were first isolated in human cell cultures and subsequently adapted to monkey kidney cultures. Some strains of enterovirus 71 grow better in suckling mice than in cultured primate cells. Cultivation of enterovirus type 72, hepatitis A virus, now classified as a separate genus of the *Picornaviridae*, is less readily achieved in culture. However, a number of strains have been grown in fetal monkey kidney cells; although they are noncytopathogenic, their growth can be monitored by radioimmunoassay or immunofluorescence.[437] Hepatitis A virus is discussed in detail in Chapter 13.

It is not necessary to attempt to identify the specific serotype for every enterovirus isolate. The decision to proceed with type identification should be made after weighing the specific need that would be served. It is often sufficient for the physician or health officer to know simply that an enterovirus has been isolated. It may be important to determine soon that the isolate is not a poliovirus, because generally the isolation of polioviruses in developed countries now represents only clinically irrelevant, long-term shedding of vaccine virus strains. Excluding poliovirus can be accomplished simply by demonstrating that the isolate fails to be neutralized by a pool of antisera to the three poliovirus serotypes.

For most enteroviruses, specific identification of the serotype rests with serum neutralization testing, although hemagglutination-inhibition (HI), complement-fixation (CF), immunofluorescence, precipitin, and other antigen–antibody reactions may be used in some instances. Although the isolation of an enterovirus is simple and relatively rapid, its specific identification may be slow and expensive if only monospecific hyperimmune sera are used. However, as mentioned above, identification of isolates has been considerably simplified by the development of internationally standardized hyperimmune antisera[150,290] that are now incorporated into combination antiserum pools[297,306,494] constituted in a pattern proposed by Lim and Benyesh-Melnick[239] in such a way that a given antiserum appears in either one, two, or three pools. An unknown enterovirus may be identified by its pattern of neutralization by the pool or pools containing its homotypic antiserum.

Pools A–H were the first sets of these LBM pools to become available. By use of these eight pools in accordance with standardized directions, 42 enteroviruses that grow readily in cell cultures can be correctly identified.[297,306] Standardized equine monovalent antisera against the mouse-grown types of coxsackievirus group A are also available. These have also been combined into additional sets of seven pools (J–P) incorporating 19 coxsackievirus A antisera in a scheme parallel to that described above; they, too, are available in lyophilized form.[299] The LBM pools can be obtained in lyophilized form through the World Health Organization, Geneva. Use of the pools has been described.[297,299,306,494]

The first set of pools served the world's enterovirus laboratories for two decades; a second set of A–H pools was prepared in 1984.[297,306,494] These pools are available in substantial amount, but it is prudent to conserve them; therefore, if an epidemic caused by a single serotype is in progress, use of the single monovalent hyperimmune serum to identify most isolates is indicated.

In use of the LBM pools to identify field isolates, some strains require special attention. For example, antisera for coxsackievirus B3 and echovirus 9 prepared against the prototype strains may not give solid neutralization of all field strains of these serotypes. In such instances, it is important to make early readings to get a clue to the identity of the virus. Neutralization of the virus for a few days may be followed by later breakthrough with some strains. It may be desirable to confirm the identity of the isolate with monovalent antiserum.

With certain strains, efforts at identification by neutralization have been complicated by aggregation of viral particles or other factors, which reduce the access of the virus to specific antibody. For example, as recognized soon after the discovery of the enteroviruses, the prototype echovirus 4 strain (Pesascek) is poorly neutralized by homologous antisera. The Du Toit strain is much more sensitive and is preferred for neutralization tests. The poor neutralization of the Pesascek strain was shown to be related to aggregation of viral particles; virus in nonneutralizable aggregates was found to constitute up to 30% of untreated Pesascek stock preparations but only 0.1% of Du Toit. With monodispersed virus obtained by filtration through millipore membranes of appropriate porosity, efficient neutralization of Pesascek strain can be obtained.[468]

More knowledge continues to come to light regarding neutralization. The Pesascek strain, which has a higher efficiency of plating on human RD cells than on green monkey kidney cells, was more effectively neutralized in tests on RD cells than on monkey cells. After banding of the virus by ultracentrifugation in density gradients, a minor fraction was found that resisted neutralization as tested on either RD or monkey cells. Another fraction containing the main peak of infectivity was reacted with neutralizing antibody and was unable to infect RD cells, but it remained highly infectious for monkey cells.[209] Thus, not only virus aggregation but also a host cell with

different receptors may play a role in neutralization of infectivity.

This phenomenon is still encountered with enteroviruses, as illustrated by the fact that better neutralization of coxsackievirus A isolates has been obtained by treatment with sodium deoxycholate, to increase the access of the virus to antibody, most likely by disaggregation of the virus particles. Isolates of enterovirus 71, also, are often difficult to neutralize, and this has produced problems in recognition of this important pathogen. Successful neutralization of the Swedish strains of enterovirus 71 depended on the use of monodispersed virus,[31] and the etiologic agents of the serious epidemics in Bulgaria and Hungary were identified as strains of this serotype only in special testing of the isolates through cooperative endeavors of WHO Centers for Virus Reference and Research in Moscow, Sofia, Budapest, Berkeley, and Houston.[300] Problems with neutralization of enterovirus 71 strains and also other enterovirus isolates have been met not only by filtration but also by treatment with sodium deoxycholate, ethyl ether, and chloroform. Kapsenberg et al.[199] have found chloroform treatment to be the treatment of choice for routine typing of enterovirus 71 and coxsackievirus types A7 and A16, since much virus is lost by filtration, sodium deoxycholate is cytotoxic and can be used only with virus suspensions of high titer, and removal of the inflammable ethyl ether can be difficult and hazardous. Chloroform treatment was found to be effective with virus suspensions of low titer; chloroform is nonflammable and heavier than water, and it can be removed by simple centrifugation.

A possible mechanism for the development and the elimination of a "nonneutralizable fraction" has been postulated on the basis of investigations with poliovirus. In the course of studies dealing with synthetic lipid vesicles as vehicles for introduction of foreign materials into eukaryotic cells, it has been shown that poliovirus particles can be experimentally encapsulated within synthetic large phospholipid vesicles and that such encapsulated particles are then resistant to type-specific antiserum and are infectious for cells that normally resist infection because of a membrane restriction.[478] In light of these findings, it is possible that the occurrence of such encapsulation during viral replication in nature may be the underlying mechanism causing virus aggregation and the observation of a "nonneutralization fraction" of picornaviruses. Virus particles entirely or partly covered with cell-membrane materials would have an increased tendency to aggregate and would be less available to neutralization by antibody. Such a situation would be compatible with the effects mentioned above in disaggregation of enteroviruses.

In most instances of neutralization problems with cells under fluid medium, plaque reduction tests can be used to detect antibodies even in the presence of non-neutralizable fractions of virus.

Both HI and CF tests have also been used to identify enterovirus serotypes. Only about one third of enteroviruses agglutinate erythrocytes, and this limits the usefulness of this test for routine use. Likewise, all enteroviruses cannot be identified by the CF test. More rapid identification of enterovirus isolates can be made by immunofluorescent staining of viral antigens.[441,445] One technique utilizes combination antiserum pools in conjunction with indirect immunofluorescence to detect type-specific enteroviral antigens in CSF leukocytes of patients with aseptic meningitis[445] and enterovirus type 70 antigen in conjunctival scrapings.[360,361] The results can be available within hours after procurement of the infected cells. In addition to speed, this method has the advantage of associating the enterovirus isolate more directly with the target organ than would isolation from stool or throat. The fluorescent-antibody assay method for polioviruses has been improved by use of tragacanth gum.[200]

If a virus isolate cannot be identified, it may represent a new enterovirus or a mixture of viruses. Therefore, plaque purification of the stock virus or purification by terminal dilution passages may be necessary before typing results are definitive.

Rapid enzyme-linked immunosorbent assay (ELISA) tests have also been developed.[516] A highly specific solid-phase enzyme immunoassay for detection of type 3 poliovirus antigen has been applied in investigations of the type 3 strains involved in an outbreak in Finland (see Section 9). It was hoped that this assay would fill the need for a test permitting the diagnosis of poliovirus infections directly from clinical specimens, but the quantities of poliovirus in fecal specimens were found to be too low. However, the test can be very useful for detection, identification, and quantitation of poliovirus antigen in infected cell cultures.[455]

In tests involving nucleic acid hybridization, virus-specific nucleic acids are bound in a solid phase and then detected by radioactively labeled DNA probes. By use of recombinant DNA procedures, cloned cDNA probes have been developed for detection of enterovirus isolates in cell cultures and are being explored for direct detection of virus in clinical specimens.[178,194,402] The latter method, using the cloned genome of a cardiotropic strain of coxsackievirus B3, is also being explored as a means for clarifying the molecular pathogenesis of enteroviral heart disease and for genetic analysis of the cardiotropic coxsackievirus. For laboratories equipped with an electron microscope, a rapid method dependent on specific clumping of virus particles is available.[336]

The tests for viral genetic material are dependent on detecting viral nucleic acid by dot spot or *in situ* hybridization,[41,45,59,276,376,401,402] but the aforementioned tests are now being supplemented by the polymerase chain reaction (PCR) analysis,[129,330,522] even of paraffin-embedded tissue.[385] The following is an example from one study[129]:

> An assay based on the polymerase chain reaction (PCR) for detection of enteroviral RNA in stool samples was carried out using specimens from 74 patients with aseptic meningitis. The primer pair and probe were derived from the highly conserved 5′ non-coding enterovirus genomic region. Enteroviral RNA was detected in faeces of all 36 patients in whom an enterovirus was isolated from stool. The PCR assay yielded positive results in additionally 3/6 cases where enterovirus diagnoses were obtained by virus isolation from cerebrospinal fluid and/or serological tests. Thus, the positive outcome of the PCR assay was 39 (93%) among the 42 patients with enterovirus diagnoses. Furthermore, 7/19 (37%) cases with an etiology that was not established by other means were positive in the test indicating that the PCR assay may give considerable additional etiological information in patients with aseptic meningitis. The limit of RNA detectability in the PCR assay was about 100 $TCID_{50}$ when highly cytopathogenic enterovirus types (coxsackievirus type B5 and echovirus type 11) were tested. The PCR was negative in all 13 patients with nonenterovirus diagnoses except in one case with a herpes simplex virus type 2 infection. Since enterovirus-specific IgM antibodies could be detected in this case a dual infection seemed probable. All the negative controls, included in the study, were PCR-negative. This study proves the usefulness of the PCR assay for detection of enteroviral RNA in stool samples and suggests that the test may be an alternative to virus isolation for rapid enterovirus diagnosis in patients with aseptic meningitis.

Another area of diagnostic importance is the differentiation of picornavirus isolates from the respiratory tract as rhinoviruses (which are acid-sensitive) or enteroviruses (which are acid-resistant). A reverse transcription PCR test has been devised that clearly classifies these agents.[354] In a recent application,[15] respiratory isolates from persons with acute respiratory disease were analyzed both by the PCR and by acid lability; 91 were found to be rhinoviruses and 44 enteroviruses, by both tests. The only type that was nonclassifiable was echovirus 22.

Reagents have been prepared for incorporating a 5′ noncoding riboprobe to detect a large number of enteroviruses and a VP1 probe to detect the three polioviruses. The methods rapidly detect the group to which the virus belongs.[116,376] Other enterovirus group reagents are being made in rabbits immunized by synthetic peptides from an immunodominant region of the VP1 enteroviral capsid protein.[172] This region exhibits a large homology among the various types that have been sequenced.

3.5.2.   Tests for Antibody.   Testing for presence of type-specific antibody against enteroviruses is feasible only when (1) a known enterovirus isolate from the patient is available and confirmation of the infecting serotype is necessary; (2) a clinical picture such as pleurodynia clearly implicates a small number of antigens (in this case, usually group B coxsackieviruses) against which serum should be tested; (3) an epidemic caused by a single serotype is in progress; or (4) a seroepidemiologic survey is being conducted to determine the community or study-group history of experience with a particular serotype or group (e.g., polioviruses). Otherwise, for initial determination of a current infection in a patient or a locality, virus isolation is far simpler and is recommended.

For any purpose except a serological survey, paired serum specimens are required; the first sample must be taken as early as possible in the course of the illness or infection, the second 3–4 weeks later.

The neutralization test[276] is accurate and type-specific and is at present the test commonly used for most enteroviruses [for polioviruses, the CF test has some advantages (see below)]. Acute and convalescent sera are usually tested simultaneously, using various dilutions of serum against a constant amount of the specific virus. Serum titers are calculated on the basis of the dilution of serum that neutralizes a given amount of virus, and a fourfold or greater rise in antibody titer is considered significant—indicative of an infection during the period covered. However, it should be noted that neutralizing antibody titers may already be high at the time of the onset of clinical symptoms, making interpretation difficult. If neutralizing antibody titers are found to be equally high in both acute and convalescent specimens, the infection might have taken place either recently or many years before, since neutralizing antibody to any of the enteroviruses persists for years if not for life. In addition to the homologous antibody, antibodies against other enterovirus types may appear transiently and at low levels. If only a single serum specimen is available and it is positive, the identification of specific viral antibody in the IgM fraction may be useful in determining whether the infection is recent.[92]

Other serological tests that may be used include CF, HI, immunofluorescence, and passive hemagglutination.

Complement-fixation antibodies appear during the course of an infection but may disappear or drop to a low level within a few months. For most enteroviruses, the CF test has little value for type specificity because of major heterotypic cross-reactions that develop after infection.[220] However, with specimens taken at proper time intervals, the test has been used successfully in the diagnosis of poliovirus infections.[425]

Native (N) and heated (H) CF antigens of poliovirus are described in Section 4.3. In the course of poliomyelitis infection, H antibodies form before N antibodies, and subsequently the level of H antibodies declines first. Early acute-stage sera thus contain H antibodies only; 1–2 weeks later, both N and H antibodies are present; in late convalescent sera, only N antibodies are present. Only a first infection with poliovirus produces strictly type-specific CF responses. Subsequent infections with heterotypic polioviruses recall or induce antibodies, mostly against the heat-stable antigenic components shared by all three types of poliovirus, i.e., against the poliovirus group antigen. Very high CF antibody levels against a poliovirus are strongly suggestive of a recent infection.

After a coxsackievirus infection, patients may develop CF antibodies to a number of both group A and group B agents, and heterotypic echovirus CF antibody responses also are common.[219]

The HI test is relatively easy to perform, and patients who become infected with an enterovirus that hemagglutinates develop homotypic antibody, which may persist for years. They also may develop heterotypic antibody, making the test somewhat nonspecific. As mentioned, the major drawback to the HI test is that only about one third of the known enteroviruses agglutinate erythrocytes; even within a single serotype, some strains hemagglutinate and others do not.[304]

It is possible to detect and titrate serum antibody against enteroviruses by use of an immunofluorescence technique. Infected coverslip cultures for use as a source of antigen may be prepared and kept frozen at −20°C for at least 1 year, making this test readily available for rapid diagnosis.[40,425]

For diagnosis and study of acute hemorrhagic conjunctivitis caused by enterovirus type 70, isolation of the virus has been difficult, and most of the recent outbreaks have been identified solely by serological means. With epidemic spread of the infection and many cases occurring in a short time, new serological methods have been needed to handle large numbers of specimens rapidly.

A microneutralization procedure has been developed that provides convenience and speed in large tests[162] and compares well with other methods. It is more sensitive than the standard tube neutralization tests and the CF test. Although it is less sensitive than the HI test, the latter requires concentrated, partially purified virus as antigen and is difficult to read. Both the prototype virus (J670/71) isolated in Japan in 1971 and the V1250 strain obtained in 1981 from Honduras were used in comparative testing. The recent isolate was a more sensitive antigen than the prototype, as indicated by higher antibody titers in sera of patients from various epidemics. This underscores the need to consider the possibility of enterovirus strain variation in the preparation of test antigen.

Another test, originally developed to screen for rises in antibody to rhinoviruses,[105] has been found to be effective for detecting rises in antibody titer to enteroviruses. This test, a passive hemagglutination test, relies on the use of a coupling reagent, chromic chloride, to attach proteins to indicator erythrocytes. It has made possible the hemagglutination of red blood cells by antigens that otherwise do not demonstrate this property. Although enteroviruses cross-react in this test, and thus specific enterovirus serotypes cannot be distinguished, it can serve for rapid, simple, and useful screening to detect rises in antibody to an enterovirus, i.e., to indicate that an enterovirus infection is present even though the serotype is not known. Such information can be useful, for example, in a community outbreak of illness caused by a single enterovirus serotype that has already been identified. The technique has been utilized as a screening test using paired sera collected during an epidemic of coxsackievirus B5 infections.

Indirect radioimmunoassays (RIAs) of IgM and IgG antibodies have detected significant increases in titer from the acute- to the convalescent-phase specimens from enterovirus-infected patients.[450] The test, however, lacked reliable type specificity. A reverse IgM assay seemed to be type specific, provided the amount of labeled virus was carefully standardized.[450] However, the reverse IgG assay has not yet been made sufficiently sensitive. ELISA tests are also being developed for measuring serum IgA, IgG, and IgM responses.[35]

4. Biological Characteristics of the Virus that Affect the Epidemiologic Pattern

4.1. General Properties

Enteroviruses share the basic properties of *Picornaviridae*[74,264,283,403] including a genome of single-stranded RNA, small size (diameter 22–30 nm), lack of an envelope (i.e., a "naked" nucleocapsid), and insensitivity to ether and other lipid solvents, indicating lack of essential lipids. The molecular weight of the nucleic acid is 2.3–2.8 million, constituting about 30% of the particle mass. It has been estimated that picornaviruses have 12 genes, in contrast to 400 estimated for the large poxviruses. The virus matures in the cytoplasm.

An enterovirus is an icosahedral ribonucleoprotein consisting of five types of molecules: four capsid proteins (VP1–4) and a single molecule of single-stranded RNA. The genome is of "plus-strand" polarity; that is, it has the same polarity as cellular mRNA and, indeed, it serves as mRNA for all viral polypeptides on entry into the cytoplasm. Unlike most DNA-genomic viruses, viruses with an RNA genome do not have the equivalent proofreading and error-editing functions capable of correcting viral polymerase errors, and a large number of faulty nucleotide incorporations accumulate during RNA replication.[479] The impact of error-prone replication is reduced by the relatively small size of the RNA virus genomes; for poliovirus the genome contains only 7442 bases, plus a 3′-terminal poly(A) tail of 60 residues. The molecular biology of enterovirus replication is typical for that of other plus-strand RNA viruses: uptake into the host cell through attachment to a specific cellular receptor, release of genomic RNA, protein synthesis, genome replication, and encapsidation. Each of these steps is under intensive investigation. Infective nucleic acid has been extracted from several enteroviruses and rhinoviruses. Because it is freed of the surface protein antigen, such RNA cannot be neutralized by antiserum against the intact virus. An infectious DNA copy of the poliovirus RNA genome has been studied for the three types of poliovirus,[355,382,432] for coxsackievirus B3,[194] and for echovirus 6.[29]

Although cardioviruses can infect humans, the picornaviruses that infect man are almost always enteroviruses, rhinoviruses, or hepatoviruses (or heparnaviruses). These genera differ in a number of properties, but the most useful and reliable feature in distinguishing them is sensitivity to acid: enteroviruses are stable at acid pH (3–5) for 1–3 hr, whereas rhinoviruses are acid labile. Rhinoviruses multiply chiefly in the nose and throat and can be recovered from these sites but only rarely from fecal specimens. Because of their acid stability, enteroviruses that may have undergone only limited replication in the oropharynx survive transit through the stomach and become implanted in the lower intestinal tract, where they undergo more extensive multiplication. Enteroviruses grow readily in stationary cultures at 36–37°C, but initial growth of rhinoviruses in primary fetal cell cultures is favored when cultures are incubated on roller drums at 33°C. Among the enteroviruses that are cytopathogenic [polioviruses, echoviruses, some coxsackieviruses, and the new enterovirus types (68–71)], growth can usually be obtained readily in primary cultures of human and monkey-kidney cells and in certain cell lines (such as HeLa or, for some serotypes, WI-38); in contrast, most rhinoviruses of man can be recovered initially only in cells of human origin (embryonic human kidney, human diploid cell strains). In cesium chloride, enteroviruses have a density of 1.34 g/ml, whereas rhinoviruses have a density of 1.4 g/ml. Enteroviruses and some rhinoviruses can be stabilized by magnesium chloride against thermal inactivation. Hepatitis A is discussed in Chapter 13 and elsewhere.[166,277]

Although much is known about the enteroviruses, still unresolved is the biochemical basis of virus stability associated with its portal of entry through the enteric tract, the determinants of virus spread in the host, and knowledge of how the virus penetrates its target tissue, thereby causing disease.[327] The key to the early events of infection is determined by a unique cell surface receptor. The receptor plays a key role in the binding, penetration, and uncoating of the virus.[344,387]

The receptors for the polioviruses and rhinoviruses are members of a large group of normal cellular proteins known as the immunoglobulin gene superfamily, which includes antibodies and a number of cell-surface adhesion molecules.[154,213,344,387] The echovirus 1 receptor has recently been identified, and surprisingly it belongs to a different superfamily: the integrin adhesion proteins.[25] These proteins are known to play a role in the interactions between cells and the extracellular matrix.

4.2.   Reactions to Chemical and Physical Agents

Enteroviruses are resistant to all known antibiotics and chemotherapeutic agents, to most laboratory disinfectants, and to lipid solvents (e.g., ether). Treatment with 0.3% formaldehyde, 0.1 N HCl, or free residual chlorine at a level of 0.3–0.5 ppm causes rapid inactivation, but the presence of extraneous organic matter protects the virus from inactivation.[453] Thus, caution must be exercised before carrying over laboratory findings on the chlorination of enteroviruses, often in purified form, to chlorination under natural conditions.

Most enteroviruses are inhibited from propagating in cell cultures by 2-(α-hydroxybenzyl)-benzimidazole (HBB)[448] and by guanidine. However, viral progeny grown in the presence of guanidine become resistant to the drug,[284] and further passage results in selection of strains that are drug dependent.[433]

Exposure of these viruses to a temperature of 50°C destroys them rapidly. However, in the presence of molar magnesium chloride, virtually no detectable inactivation occurs in 1 hr at 50°C.[466] Enteroviruses are stable at freezing temperatures for many years and remain viable for weeks at icebox temperatures (4°C) and for days at room temperature. Their inactivation at all temperatures is inhibited by magnesium chloride; this property has led to the widespread use of $MgCl_2$ as a stabilizer of oral poliovirus vaccine.

Enteroviruses are rapidly inactivated by ultraviolet light and usually by drying, unless special conditions are observed. Vital dyes (neutral red, acridine orange, proflavine), when incorporated into the structure of these viruses, render them readily susceptible to visible light.[421,467]

4.3.   Antigenic Characteristics

All three poliovirus serotypes share some antigens,[101,257] but the polioviruses basically have marked intertypic differences. The epitopes responsible for inducing neutralizing antibodies are located on the three structural proteins (VP1, VP2, and VP3) making up the viral surface, with most of the epitopes clustered on VP1. When the virion is disrupted and the protein purified, the epitope activity is diminished. Five short peptides have been synthesized that contain amino acid sequences for type 1 poliovirus VP1. The peptide containing amino acids 93–103 was capable of eliciting neutralizing antibodies. This same site had previously been identified in type 3 poliovirus as the neutralizing epitope by analyses using viral monoclonal antibodies, although the amino acid sequence of the site is type specific. The other four peptides react specifically with monoclonal antibodies and are capable of priming an animal so that a single inoculation with virus yields antiserum of stable and high neutralizing titer. Some of the type 1 synthetic peptides even primed the host toward type 2 and type 3 viruses.[101]

Even within a single recognized serotype, antigenic differences may occur between different isolates. This is true not only for the polioviruses but also for the coxsackieviruses and the echoviruses.[259,397,400] The considerable differences in antigenic structure among strains within each poliovirus serotype have been clarified in studies in which intratypic serodifferentiation of polioviruses was performed by neutralization or immunodiffusion tests utilizing strain-specific antisera prepared by cross-adsorption with the heterologous strain[459] or by utilizing monoclonal antisera.[357] The techniques are useful in distinguishing vaccine and nonvaccine poliovirus strains and also in identifying relationship among wild strains.

With some enteroviruses, the frequency of antigenic mutation is as high as 1 per 10,000 virions.[379] This may lead to the appearance of prime strains.[258] A prime strain is one that is poorly neutralized by antiserum to the originally characterized (prototype) strain but that induces the production of antibody that neutralizes the prime strain and the prototype strain equally well. The prime strains, however, share complement-fixing (CF) or precipitating antigens that strongly cross-react with prototype strains.

A number of cross relationships exist between several enteroviruses, for example, coxsackieviruses A3 and A8, A11 and A15, A13 and A18; echoviruses 1 and 8, 12 and 29, 6 and 30; and polioviruses 1 and 2 to a minor degree.[152]

For enteroviruses, there has been little documentation of any long-term trends in antigenic alterations of serotypes like those seen with influenza viruses. Such a trend may be suggested, however, by observations that strains of coxsackievirus B5 isolated in 1973 from patients in the United Kingdom are very different antigenically from the prototype B5 virus (Faulkner strain) isolated in 1952. Not only were wide differences observed in immunodiffusion and neutralization tests, but also RNA hybridization procedures showed current human strains to have 100% homology with each other but only 50% homology with the prototype Faulkner strain.[37] This is a difference of the same order as that found between different serotypes of poliovirus. In addition, an enterovirus that causes swine vesicular disease has been shown to be closely related antigenically to human coxsackievirus B5.[37,38,134] In RNA hybridization tests, the porcine virus showed about 50% homology with human strains.

Studies of coxsackievirus B4 isolates identified by use of monoclonal antibodies indicated a high frequency of antigenic variants, estimated to be as high as 1 per 10,000—higher than has been reported for influenza virus but within the range of estimates for other RNA viruses.[379]

Associated with the RNA genomic changes described above are changes in the polypeptide composition of newer coxsackievirus B5 isolates compared to the 1952 prototype virus. Similarly, such differences exist in the swine enteroviruses of 1966–1971 and those of 1972–1973. In the same year, however, a swine virus from France and a human isolate of coxsackievirus B5 have shown virtually identical patterns.[153] Further work is necessary to establish the pathological and epidemiologic significance of these variations.

Complement-fixation antigens are known for each of the three poliovirus serotypes. They may be prepared from tissue culture or infected CNS. Inactivation of the virus by formalin, heat, or ultraviolet light liberates a soluble CF antigen. This antigen is cross-reactive and fixes complement with heterotypic poliomyelitis antibodies. Two type-specific antigens are contained in poliovirus preparations and can be detected by precipitin and CF tests. They are called D (or N) and C (or H). The D antigen occurs as a band in the denser regions of a sucrose density gradient and comprises most of the virus infectivity. The upper band containing the C antigen has little infectivity. The virus in the D zone appears intact in electron micrographs and contains 20–25% RNA, whereas that in the C zone is damaged and contains little or no RNA.[156]

Several coxsackieviruses and echoviruses agglutinate human type O erythrocytes. About one third of the known enteroviruses have this property, and antibodies against the virus can be measured by hemagglutination inhibition.[132,394]

4.4. Host Range *in Vivo* and *in Vitro*

The host range of the enteroviruses varies greatly from one type to the next and even among strains of the same type. They may be readily induced, by laboratory manipulation, to yield variants that have host ranges and tissue tropisms different from those of wild strains; this has led to the development of attenuated poliovaccine strains.

Polioviruses have a very restricted host range among laboratory animals.[33,34] Most strains will infect and cause flaccid paralysis only in monkeys and chimpanzees. Infection is initiated most readily by direct inoculation into the brain or spinal cord. Chimpanzees and cynomolgus monkeys can also be infected by the oral route; in chimpanzees, the infection thus produced is usually asymptomatic. The animals become intestinal carriers of the virus; they also develop a viremia that is quenched by the appearance of antibodies in the circulating blood. Unusual strains have been transmitted to mice or chick embryos.

Most strains can be grown in primary or continuous cell line cultures derived from a variety of human tissues or from monkey kidney, testis, or muscle (see Section 3.5 and below).

The cardinal feature of coxsackieviruses is their infectivity for newborn mice.[83] Certain strains (B1–B6, A7, A9, A16) also grow in monkey-kidney-cell cultures. Some group A strains grow in human amnion cells, HeLa cells, or the RD cell line.[250,423] Coxsackieviruses A1, A19, and A22 have not yet been grown successfully in any cultures and must be cultivated in newborn mice. Chimpanzees and cynomolgus monkeys can be infected sub-clinically; virus appears in the blood and throat for short periods and is excreted in the feces for 2–5 weeks. Type A14 produces poliomyelitislike lesions in adult mice and in monkeys, but in suckling mice, this type produces only myositis. Type A7 strains produce paralysis and severe CNS lesions in monkeys.[82,463]

Group A coxsackieviruses characteristically produce widespread myositis in the skeletal muscles of newborn mice, resulting in flaccid paralysis without other observable lesions.[83,130] In adult mice, surgically denervated muscles can be infected, whereas mature innervated muscles are relatively resistant. Also, when synaptic transmission was chemically inhibited by botulinum toxin, blocking release of acetylcholine, the leg muscles of adult mice became susceptible.[12]

Group B viruses can produce a myositis that is more focal in distribution than that produced by viruses of group A, but they also give rise to a necrotizing steatitis involving principally the maturing fetal fat lobules (e.g., intercapsular pads, cervical and cephalic pads). Encephalitis is found at times; the animals die with paralysis of the spastic type. Some B strains also produce pancreatitis, myocarditis, endocarditis, and hepatitis in both suckling and adult mice. Corticosteroids may enhance the susceptibility of older mice to infection of the pancreas. Normal adult mice tolerate infections with group B coxsackieviruses, but in mice subjected to sustained postweaning undernutrition (marasmus), B3 virus produces severe disease, including persistence of infective virus in the heart, spleen, liver, and pancreas. Lymphoid tissues are markedly atrophic in marasmic animals. Transfer of lymphoid cells from normal mice immunized against the virus provides virus-infected marasmic mice with significant protection against the severe sequelae.[482]

The criteria for classification as a member of the echovirus group included the provision that the prototype strains must fail to produce disease in suckling mice or in monkeys. However, some strains can produce variants that exhibit animal pathogenicity.[182,471] Most of the common echoviruses can be isolated in primary monkey-kidney-cell cultures, but strains of a number of the echovirus serotypes grow more readily in human cell cultures (see Section 3.5).

Parallel to the coxsackieviruslike mouse pathogenicity of some echoviruses, strains of some coxsackievirus types (especially A9) lack mouse pathogenicity, and thus resemble echoviruses. This variability in biological properties is the chief reason that new enteroviruses are no longer being subclassified as echo- or coxsackieviruses.

The five latest enteroviruses all grow in monkey-kidney-cell cultures. Enteroviruses 70 and 71 were isolated in human cell cultures and subsequently adapted to monkey-kidney cultures. Some strains of enterovirus 70 were also adaptable to cultivation in the L strain of murine cells, and some strains of enterovirus 71 grew better in suckling mice than in cultured primate cells. Originally classified as enterovirus 72, HAV has been less readily cultivated. A number of strains of this virus have now been successfully grown in fetal monkey-kidney cells. Although the HAV strains are noncytopathogenic, their growth can be monitored by immunofluorescence or RIA.[437] Evidence suggests that this agent may be the prototype of a new genus of *Picornaviridae*.

The growth of enteroviruses in monolayers of cultured cells is generally associated with a characteristic cytopathic effect.[386] Infected cells round up, show shrinkage and marked nuclear pyknosis, become refractile, and eventually degenerate and fall off the glass surface. In cell cultures covered by fluid nutrient medium, virus may spread via the fluid bathing the cells. Under agar overlay, which confines the spread of virus to a cell-to-adjoining-cell route, plaques of degenerating cells are formed by various members of the enterovirus group in cultures of susceptible cells. Methods that utilize plaque formation have provided very precise quantification of infective virus, required for many laboratory research studies of enteroviruses.[174,304]

The factors underlying cell susceptibility to enteroviruses are basic to understanding host susceptibility to infection. Specific receptor sites, differing for the various enterovirus groups, are located on the surface of the plasma membrane of susceptible cells, where the viruses attach as a prerequisite to penetration and uncoating of the virion.[221,247] Human cells possess a receptor for poliovirus and can therefore be infected and killed by the virus. Rodent cells do not possess the receptor and cannot be infected by polioviruses unless the viruses have been altered by laboratory cultivation. But in human–rodent hybrid cells, possession of a human gene for the poliovirus receptor was found to be sufficient to enable the virus nucleic acid to enter the cell, and once this first step had been taken, the virus then multiplied without the mediation of any further human gene products, the rodent genetic apparatus being sufficient for its needs. The human chromosome carrying the gene concerned with poliovirus reception was identified by making use of human–rodent hybrid cells that differ in their complement of human chromosomes. In these studies, it was found that some of the human chromosomes were shed after hybridization. Only when chromosome 19 was lost did the cells lose their susceptibility to poliovirus.[311] This implies that chromosome 19 carries all the information necessary for acceptance of poliovirus by a human cell and suggests that only a single gene codes for the receptor protein.

4.5. Replication of Enteroviruses

The replication of poliovirus provides useful insights into its transmission, its *in vivo* and *in vitro* host range, and its relatedness to other enteroviruses. The essential theme of viral replication is that specific messenger RNA (mRNA) must be transcribed from the viral nucleic acid for successful expression and duplication of genetic information. Once this is accomplished, viruses use cell components to translate the mRNA. In the replication of poliovirus, all steps are independent of host DNA and occur in the cell cytoplasm.

Normally, soon after a picornavirus attaches to a cell, it is delivered into the cytoplasm as viral RNA freed from its protein shell. The single-stranded genomic RNA then serves as its own mRNA, which is translated to form a single large polypeptide that is subsequently cleaved to produce the various viral capsid polypeptides. Completion of encapsidation produces mature virus particles that are then released when the cell undergoes lysis. The time required from initiation of infection to completion of virus assembly ranges from 5 to 10 hr, depending on pH, temperature, host cell, and number of particles to which the cell is exposed. Yields may be up to 100,000 particles per cell, but only 1 in 1000 particles may be infectious.

Despite the relatively limited information on which the establishment of the enterovirus group was originally based and its subgroups were defined, the validity of these groupings has been borne out by recent studies utilizing sophisticated techniques of modern molecular virology. Comparison of the genomes of representative polioviruses, coxsackieviruses of subgroups A and B, and echoviruses by RNA hybridization has shown at least 5% of the genome to be shared by all enteroviruses tested.[519]

Great strides have been made in recent years in investigating many aspects of the replication and the genetics of enteroviruses and other picornaviruses. The primary structure (sequence) of the genomic RNA has been determined for a number of these viruses.[44,176,208] Complementary DNA has been cloned in a bacterial plasmid,[382,457] and this DNA has been shown to produce infectious RNA.[381] Strategies of viral protein synthesis have been examined and clarified.[16,108,185,443] This work has been reviewed in detail.[211,403]

Double neutralization of virus particles obtained from mixed infections has been observed. Usually—and particularly for antigenically distinct virus types such as influenza viruses A and B or polioviruses types 1 and 2— the doubly antigenic virus proved to be unstable on passage, suggesting that the phenomenon could be explained by phenotypic mixing. Phenotypic mixing has also been found to occur between an echovirus and a coxsackievirus.[183] Virus particles containing antigens of both viruses were obtained that on passage segregated into parental types. The phenotypically mixed particles may be regarded as having an additional surface antigen heterologous to the genetic core or mosaic surface antigens composed of the two parental types. A single particle with two distinct genetic cores (one of each virus) and mixed antigenic coats would also be expected to behave like a phenotypic mixture, i.e., to be doubly neutralizable but on passage breed the pure parental types. The protein of a plant virus (cowpea chlorotic mottle virus) has been shown to be able to encapsidate *in vitro* the genome of poliovirus, to form a particle known as a pseudovirion. When assayed in the presence of DEAE-dextran to facilitate entrance into susceptible human cells in cultures, the polio pseudovirions were about 50 times more infectious than poliovirus RNA itself. This suggests that the encapsidation of the polio genome in a foreign coat protected the RNA from destruction by cellular nucleases.

Polioviruses are included among the agents that during their replication produce deletion mutants called "defective interfering" (DI) particles that interfere with the growth of the normal virus from which they are derived.[68,193] The DI particles have potential clinical significance, since they may play a role in limiting acute viral infections; furthermore, they are present in high concentrations in oral polio vaccines, suggesting the possibility that their presence may be a factor in attenuation.

Because picornaviruses have been so fully studied, and because they are small and relatively simple viruses, they have served and will continue to serve as models for understanding the nature of viruses and their structure, function, replication, and genetics.[9,165,194,403,430]

5. Descriptive Epidemiology

5.1. Key Features of Epidemiology of Enteroviruses

In considering enterovirus epidemiology, it is important to reemphasize that by far the usual infection with almost all of these viruses is a mild or silent episode, and that severe manifestations are relatively uncommon. Paralytic poliomyelitis remained an epidemiologic enigma until this concept was developed.[46,475] Furthermore, it must be kept in mind that clinical features presented by infections with the different serotypes may be similar and also that manifestations of infections with the same serotype may vary widely.

The patterns of incidence and prevalence, the age at the time of infection, and the nature of the host response are the consequences of a number of interdependent variables the common denominator of which is probably the opportunity for exposure, along with the hygienic level under which such exposure occurs. These interdependent variables include geographic area, climate, and socioeconomic setting.

For members of the human enterovirus group, humans are the only known reservoir. Enteroviruses are ubiquitous in tropical and semitropical zones. In temperate climates, they are encountered more commonly during late summer and early fall. Because of their antigenic inexperience, children are the primary targets of enterovirus infections, and thus serve as the main vehicle for their spread, which is chiefly by the fecal–oral route. These viruses are easily spread within family settings and closed institutional populations, where the rate of infection among nonimmune members may reach as high as 80%. In warmer lands where unsanitary and poor socioeconomic conditions prevail, the rate of infection among infants and children may exceed 50%, and two or more enterovirus serotypes may be isolated from a single stool sample. Infants and children living in these circumstances acquire their initial, immunizing experience with most of the locally prevailing enterovirus types earlier in life than those in more "advantaged" environments. Even within a temperate-zone country such as the United States, earlier and more frequent enterovirus infections occur among young children living in warmer areas and poorer socioeconomic conditions. As personal and community hygiene improves in a population, spread of enteroviruses becomes limited, so that increasing numbers of individuals reach later childhood, or even adulthood, without having been infected and immunized by the common serotypes. Island or isolated populations with little contact with the rest of the world also may grow up with little or no experience with some enterovirus types. For example, in some isolated Eskimo communities with little outside contact, the whole population may lack antibodies to certain enteroviruses. The shift in age incidence of poliomyelitis from infants and preschool children to school children and young adults, which has taken place in many parts of the world, has been attributed to improvement in sanitary conditions.

By 2 years of age, regardless of climate, geography, or socioeconomic conditions, most children have already experienced several asymptomatic or mildly symptomatic enterovirus infections. One factor affecting the occurrence of outbreaks of enterovirus infections by various serotypes is the accumulation of a large population of children born since the last widespread outbreak with one serotype, so that there are enough susceptibles to fuel epidemic spread.

Epidemiologic patterns of poliomyelitis are now greatly altered in much of the world by administration of poliovaccines. However, many thousands of cases—estimated at over 100,000 annually—still occur, and even a "polio-free" country that has eliminated polio from its borders cannot afford to be unconcerned. Three phases of poliomyelitis—epidemic, endemic, and vaccine-era—have been observed historically, but they also coexist today in different parts of the world.

5.2. General Epidemiology of Enteroviruses

5.2.1. Incidence and Prevalence. The epidemiologic markers of the *diseases* associated with the enteroviruses are based on the occurrence of a clinical syndrome sufficiently characteristic to be recognized, such as paralysis of poliomyelitis; the lesions of the hand, foot, and mouth syndrome characteristic of infection with echovirus 16 or enterovirus 71; or conjunctivitis characteristic of infection with enterovirus 70 or coxsackievirus A24. Enteroviruses should be sought whenever an outbreak of aseptic meningitis or of an exanthem occurs. The same causative enterovirus will be isolated from many of the patients.

The epidemiologic markers of the current presence in the community of enterovirus *infection* regardless of disease incidence include the prevalence of the virus in the stools of healthy persons or in the sewage from the area. The past circulation of an enterovirus in the area can be determined from the prevalence of antibody as determined by serological surveys.

5.2.2. Epidemic and Endemic Behavior. As documented first by observation of the patterns of poliovirus infections, outbreaks of infection with other enteroviruses occur throughout the world. One set of serotypes may be predominant for a time in some areas, or even worldwide, to be succeeded by other types. In one pattern, the enterovirus circulates endemically within a population, infecting infants or children early in life, and most individuals have acquired antibodies within the first few years of life. Under these conditions, seldom does a population of nonimmune children build to a size sufficient to fuel wide, epidemic spread of the agent. In another type of pattern, the agent goes through cycles of varying length. For a period of a year or more, it circulates widely throughout a population, in a wave of spread that reaches a large proportion of the susceptibles. With a longer period between waves of such an infection, more persons reach a later age before their initial exposure to the agent, a large population of susceptibles again builds up, and the new cycle of infections may reach the magnitude of an epidemic or even a pandemic. Studies described below illustrate the behavior of various enteroviruses in these cycles of spread. The patterns of infection and disease seen in outbreaks of infection by the more recently recognized enteroviruses, types 70 and 71, also are described and discussed in some detail here.

In the prevaccine era, epidemics of poliomyelitis occurred with great frequency in the economically advanced countries of the world. In the United States, the first sizable outbreak was recognized in Vermont in 1894 and involved 132 cases; it was by far the largest number of cases ever reported in any one year anywhere in the world up to that time. In the first half of the 20th century, recurrent outbreaks involved the "developed countries" of the world, such as the United States, Canada, Australia, and European and Scandinavian countries.

Great reductions in epidemic incidence occurred after the introduction of killed-virus poliovaccine in 1955. In the period 1951–1955, an average of almost 38,000 cases of poliomyelitis (approximately 21,000 of them paralytic) were reported annually in the United States; from 1961 to 1965, the annual average was 570 cases (460 paralytic). After live attenuated vaccines were widely administered in the United States during 1962 and 1963, even further reductions were achieved (see Section 5.3.3). Similar reductions from the high numbers of 1951–1955 down to those in 1961–1965 followed vaccination in other countries: in the United Kingdom, from 4381 to 322; in Australia, from 2187 to 154; in Denmark, from 1614 to 77; in Sweden, from 1526 to 28; and in Czechoslovakia, from 1080 to none. In these areas, the numbers of cases have continued to fall. During 1971–1975, an annual average of only 15 cases occurred in the United States, eight cases in the United Kingdom, two in Australia, and none in Denmark or Czechoslovakia.[136,483,486] Thus far into the 1990s, the average annual number is about eight in the United States, all vaccine-associated. There have been no cases caused by wild virus in the Western hemisphere since 1991.[87] Other regions have had similar excellent records of control.

Outbreaks of infection with coxsackieviruses, echoviruses, and certain of the newer enterovirus types have occurred frequently in a wide variety of places and years and continue to do so. Among the enterovirus isolations reported through the WHO virus-reporting system during the 8-year period 1967–1974,[141] the most frequently reported types were: for coxsackieviruses of group A, types 9 and 16; for members of group B, types 3 and 5; and for echoviruses, types 6, 9, 11, and 30.

In 1980, worldwide, coxsackieviruses A9 and A16 again predominated, with a recorded high number of A16 isolations reported; coxsackievirus B4 in 1980 somewhat exceeded B3 and B5 in frequency. Echovirus 30 has had an increasing trend since 1977. In 1980 this type accounted for more than 2400 isolates out of a total of about 5800 isolations of all echovirus types reported. Of the echovirus 30 reports that included clinical information, 76% were from cases involving the central nervous system, especially in the older age groups: of all patients 15–24 years of age yielding echovirus 30 isolates, 90% had CNS disease; this was also true for a large share of those 5–14 and 25–59 years of age. Echoviruses 9 and 11 were also frequently isolated in 1980, but echovirus 7, previously less commonly reported, accounted for more than 500 of the isolates, more than any other type except type 30. Echovirus 22 isolations also have increased in recent years, and more than half of the isolates were from infants less than 6 months old. It is suggested that the prevalence of echovirus 9 may fluctuate in 3- to 4-year cycles, and that of echovirus 6 in even longer cycles. After a peak in 1975, isolations of the latter virus were still decreasing in 1980.

The above-mentioned serotypes generally continue to be commonly reported, although in any single year their predominance may be more marked, or another type may exceed them in frequency for a time. In the United Kingdom during the first 34 weeks of 1978, 449 echovirus type 11 infections were diagnosed—more than the annual totals reported in any of the previous 8 years. Most echovirus 11 infections were reported in infants and children, and slightly more cases in males than females. In 51% of the cases, the nervous system was affected (mainly aseptic meningitis), 33% of the patients had fever or respiratory symptoms, and 10% had gastrointestinal disorders. There were 13 patients with epidemic pleurodynia (Bornholm disease). Ten fatal cases were reported, all in children under 5 years of age, including five neonates.

Enterovirus surveillance data from the United States for 1970–1979 are informative.[322] During the 10-year period, a total of 18,309 enteroviral agents were reported from 18,152 patients. Echoviruses accounted for 58% of the agents reported; group B coxsackieviruses, 24%; polioviruses, 9%; group A coxsackieviruses, 8%; and enterovirus types 68–71, only 0.1%. The types most frequently reported were as follows: echovirus type 9 (more than 2000 isolates); coxsackievirus B5, echovirus 11, echovirus 4, and echovirus 6 (each of the foregoing accounting for more than 1000 isolates); coxsackievirus B2, B4, A9, echovirus 7, poliovirus type 1, coxsackievirus B3, echovirus type 3, and poliovirus type 2 (each of this latter group accounting for more than 500 isolates). These 13 agents accounted for 76% of all isolates reported for the decade.

A retrospective study of the CDC enterovirus surveillance data from 1970 to 1983 has shown that nonpolio enterovirus types that are isolated in March through May in southern and coastal regions of the United States are predictive of types likely to be isolated during the peak enterovirus season. In this study, the six most frequent early isolates accounted for an average of 59%, and always more than 50%, of the isolates in July through December. The study has further indicated that in years when fewer than 1800 enteroviruses are reported, fewer than half of all the enteroviruses for the year are predictable, while in years with 1800 or more enteroviruses isolates reported, more than 69% of the reported isolates might have been predicted.

For 1984, the more common early isolates, identified from March through May in the South Atlantic, West South Central, and Pacific regions of the United States, were echovirus 9 (38 isolates), echovirus 7 (20 isolates), echovirus 30 (15 isolates), and coxsackievirus A9 (13 isolates); coxsackieviruses B4 and B5 were also isolated during the early period. Echovirus 30 was frequently isolated only in January and February. Evaluation of these reports suggested that echoviruses 7 and 9, and coxsackieviruses A9, B4, and B5 would be the common isolates for the year.[51] The final data for 1984 indicated echovirus 9 to be most common, with 17% of all isolates (266/1589), followed by echovirus 11, coxsackievirus B5, echovirus 30, coxsackievirus B2, and coxsackievirus A9.[52]

For 1985, reports indicated that coxsackieviruses A9, B2, and B5, and echoviruses 4, 6, 7, and 11 were the common isolates.[54] For 1991, echovirus 30 was most frequently isolated. Of 710 isolates reported to CDC, 20% (145 isolates) were enterovirus 30. Of the 145 isolates, 62% (90 isolates) were reported from the East Coast. The disease most frequently seen in the echovirus 30 patients was aseptic meningitis.[56,235]

In a recent report,[26] on 274 children less than 2 years of age with aseptic meningitis in Baltimore, enterovirus was recovered in 60% of the patients. The most common isolates were group B coxsackieviruses, echoviruses 4, 7, 11, and 14; coxsackie B2, B4, and B5 were present for several years, but the other viruses were present sporadically, producing substantial numbers of cases in only 1 or

2 years of the 5-year study. Even though infant mice were used in testing for virus (in addition to three cell lines), only three group A coxsackieviruses were isolated.

Perhaps the most detailed investigations of enteroviral aseptic meningitis in recent years have been conducted in Japan, where epidemics have occurred annually during the summer.[507] In 1991, a large-scale epidemic occurred, while in 1992, only a few cases were reported and the frequency of virus isolation from meningitis cases was about one quarter that in the preceding year. Echovirus 30, the type most prevalent for 3 years from 1989 through 1991, decreased markedly in 1992; 24 isolations were made during the period from January to March, another one in June from meningitis cases, and no isolation at all from cases of any other disease after July. Isolations of echovirus 9 and echovirus 6—instead of echovirus 30—increased; the total of these two types accounted for about half of the total isolations from meningitis cases. Echovirus 24 and coxsackievirus B4 were also isolated but at a lower frequency.

There were only a few reports of echovirus 9 isolation during the 5 years after the small epidemic in 1983–1984. Incidence of echovirus 9 infection increased suddenly in 1990, and the epidemics of infections with echovirus 9 as well as echovirus 30 continued for 2 years, until 1992. The echovirus 9 epidemic has further continued even after the echovirus 30 epidemic ceased in 1992. In the districts where echovirus 9 isolation was frequently reported in 1992, reports of isolation were few in 1990–1991. Isolations of echovirus 6 were reported occasionally after the 1985 epidemic but suddenly increased in 1992.

It is noteworthy that in Japan enteroviruses were isolated more frequently from nasopharyngeal samples (41; 68%) than from feces (27; 53%). Even though echoviruses 6 and 9 were found frequently (72%) in the CSF during certain years, the overall enteroviral isolation rate from CSF varied from 7 to 45%.

Also, in Japan, group A coxsackie viruses have been sought each year from 1982 through 1989, from cases of herpangina, with the tests being conducted by inoculation of baby mice in addition to cell cultures.[506] Of 3974 viral isolates, mostly from children under 5 years old, 77% were group A coxsackie viruses; in the order of frequency they were: coxsackievirus A types 4, 10, 5, 6, 2, and 3. Nasopharyngeal specimens were much more apt to be positive than feces (81% vs. 22%). Of the isolates from virus-positive cases, 92% were isolated in mice but only 12% in cell cultures. Of the positive isolates (in addition to the 77% group A coxsackieviruses) 8% belonged to the group B coxsackieviruses and 4% were other enteroviruses, 5% were adenoviruses, and 6% were herpes simplex viruses.

In the course of the Japanese investigations, especially during 1988, over 1000 isolates of echovirus 18 were made.[318] There was a high proportion of young children, 58% being under 3 years of age; 46% suffered from an exanthem and 31% from meningitis. The meningitis patients were slightly older, peaking at 4 to 7 years of age.

For 1991, in the United States, echovirus 30 was most frequently isolated. Of 720 isolates reported to CDC, 20% (145 isolates) were enterovius 30. Of the 145 isolates, 62% (90 isolates) were reported from the East Coast. The disease most frequently seen in the echovirus 30 patients was septic meningitis.[56]

A number of different coxsackievirus and echovirus serotypes have been involved in sizable epidemics, and enteroviruses 70 and 71 have also been responsible for outbreaks in various parts of the world. These outbreaks have been mostly localized to one area, although one epidemic of echovirus 9 infections in the late 1950s had almost worldwide distribution, and enterovirus 70 spread explosively in Asia and Africa during 1969–1973. In 1981, enterovirus 70 epidemics reappeared in Asia and Africa and also spread to the Western hemisphere. Enterovirus 71 has been involved in several outbreaks affecting large numbers of patients in several different nations, but pandemic spread has not been observed.

The 1969–1971 pandemic of acute hemorrhagic conjunctivitis (AHC) involved tens of millions of people. The cause of this widespread disease was recognized as a new enterovirus[217,510] and was designated enterovirus type 70.[302,315]

The disease, generally localized to the eye, is characterized by subconjunctival hemorrhage. Enterovirus 70 is highly contagious, spreading rapidly under crowded and unhygienic conditions: warm, humid, coastal climates seem particularly favorable to its transmission. Intrafamilial spread is common. Some localized outbreaks, especially in developed countries, have centered around eye clinics.

Acute hemorrhagic conjunctivitis seems to have appeared in man first during 1969, in Ghana,[60] and soon thereafter in other African countries. An Asian focus in Java was observed as early as mid-1970. Large epidemics occurred during 1971 in Japan,[217] Singapore,[510] and Morocco.[339] Small outbreaks were seen in Europe beginning in 1971, and later reported European outbreaks were in 1973 in Yugoslavia and in France.[214] Serological surveys in Japan, Ghana, and Indonesia confirmed that the virus was not prevalent before the pandemic and that after the outbreaks antibodies appeared in the populations involved.[218] Multiple epidemics have occurred within a 5-year period in the same regions, particularly in Southeast Asia, suggesting that immunity may be short lived.[509]

Until 1981, virtually no enterovirus 70 infection or disease had been found in Australia and the Americas. Among more than 1000 sera collected in 1971–1974 from residents of the United States, only three had antibodies to the virus.[163] Nevertheless, when the virus was introduced into the United States in 1980, secondary spread did not take place. However, in 1981 this situation changed. Early in 1981, AHC reappeared in some of the countries from which it had been absent for a number of years. The disease spread widely in Africa and Asia, and this time it spread extensively in the Caribbean area, in northern South America, and in Central America during the spring and summer, 1981.[365] In the early autumn, an explosive outbreak occurred in the Miami metropolitan area, involving thousands of cases. In this explosive outbreak, school-aged children in schools of the most heavily affected section of the metropolitan area were more likely to become household index cases than were persons of any other age group. Members of larger households were at greater risk of acquiring AHC. Marked reduction in cases was achieved after exclusion of affected children from schools and educating the public about hygienic precautions (such as separate towels, frequent hand-washing, and avoidance of touching eyes or face of affected family members). This agrees with the earlier reports that enterovirus 70 is transmitted almost exclusively by touching the eye with contaminated fingers or fomites.[509]

Enterovirus 70 isolates obtained from widely separated locales (Asia and the Americas) during the same pandemic period, 1980–1981, were closely related by ribonuclease T_1 oligonucleotide fingerprinting.[204] However, two isolates obtained from Japan and from Morocco during the first AHC pandemic, 1969–1972, although closely related to each other, differed from the current strains by many oligonucleotides. The similarities among contemporaneous strains from distant regions suggest that only one basic genotype of this virus appears to be in circulation, worldwide, at any one time.

A new enterovirus was recovered from the stool of a patient with CNS disease in a 1969–1970 outbreak in California during which an identical strain was recovered from the brain of a patient with fatal encephalitis.[426] The new agent was designated as enterovirus 71.[302] In the years following 1970, this newly recognized virus has moved about the world and has been associated with a variety of clinical manifestations in different regions

Table 2. Some Enterovirus 71 Outbreaks, 1969–1989

Year	Location	Number of patients	Clinical findings	Reference(s)
1969–1973	California	20	Aseptic meningitis Encephalitis	Schmidt et al.[426]
1972	New York State	11	Aseptic meningitis Encephalitis Hand, foot, and mouth disease	Diebel et al.[89]
1972	Australia	49	Aseptic meningitis Rash Acute respiratory infection Polyneuritis	Kennett et al.[202]
1973	Sweden	195	Aseptic meningitis Hand, foot, and mouth disease Aseptic meningitis	Blomberg et al.[31]
1973	Japan	> 3200	Hand, foot, and mouth disease Aseptic meningitis Cases with both syndromes	Hagiwara[144] Miwa[317]
1975	Bulgaria	705	Aseptic meningitis Encephalitis, and some with acute myocarditis Poliolike paralytic disease	Chumakov et al.[64] Shindarov et al.[436]
1977	Rochester, NY	12	Aseptic meningitis Hand, foot, and mouth disease Poliolike transitory paralysis	Chonmaitree et al.[63]
1978	Hungary	1550[a]	Aseptic meningitis Encephalitis Poliolike paralytic disease with persistent flaccid paralysis	Nagy et al.[333]
1979	Lyon, France	5	Acute respiratory infection with CNS involvement	Sohier[440]
1985	Hong Kong		Monoplegia	Samuda et al.[416]
1986	Australia		CNS involvement	Gilbert et al.[128]
1987	USA	45	Poliolike paralytic disease, meningitis, and encephalitis	Alexander et al.[10]
1989	China		Hand, foot, and mouth disease	Zheng et al.[520]

[a] A mixed epidemic of tick-borne encephalitis (chiefly in adults) and enterovirus 71 disease (chiefly in children).

(Table 2). In a 1972 outbreak reported from Australia,[202] aseptic meningitis predominated, and this syndrome also was the predominant illness observed in a 1973 outbreak in Sweden,[31] in which hand, foot, and mouth disease cases also occurred. In Japan, in 1973,[144,317] hand, foot, and mouth disease predominated, but there were some aseptic meningitis cases and some patients with both syndromes. A well-documented outbreak that occurred in Rochester, New York, in 1977, included both aseptic meningitis and hand, foot, and mouth disease, and it also included two cases of poliolike paralytic disease.[63] In view of the variety of host responses and the difficulty of isolating some of the strains of enterovirus 71, it is probable that only a small fraction of the cases that actually occur are diagnosed and reported.[269]

In the severe 1975 epidemic of CNS disease that occurred in Bulgaria, which included numerous polioencephalitis cases as well as cases of meningitis,[300] about 21% developed paralysis, and there were 44 fatalities. Young children were the most severely affected: children under 5 years of age represented 48% of the total cases and constituted 81% of the paralyzed patients and 93% of the patients who died. From specimens obtained from 65 cases, 92 virus isolations were made.[64,436] Of these, 37 came from brain and medulla, 1 from CSF, 10 from mesenteric lymph nodes and tonsils, and 44 from feces. All of the Bulgarian isolates belonged to a single antigenic type, antigenically related to the American and Swedish strains of enterovirus 71.[300] Antibody to enterovirus 71 was found in 72% of 392 patients with a history of poliomyelitislike disease that year, and tests for a rise in antibody titers against poliovirus in convalescent sera collected during the epidemic were uniformly negative. This supports the etiologic association of enterovirus 71 with the epidemic and excludes any possible role of poliovirus. In its very high incidence of paralytic cases, the epidemic in Bulgaria differed considerably from other epidemics. Virus was widely distributed during the epidemic year, infecting a high proportion of the population, but then disappeared. To date, the disease has not reappeared in Bulgaria.

In the large mixed epidemic of CNS disease that occurred in Hungary in 1978,[333] clinical manifestations were predominantly aseptic meningitis, but there were some cases of encephalitis and some fatalities. There were 826 cases of aseptic meningitis and 724 cases of encephalitis with 45 deaths; the encephalitis cases included 13 poliolike cases with flaccid paralysis. All of the poliolike paralytic cases were in children between 8 months and 36 months of age. Children in the first 5 years of life accounted for 52% of the aseptic meningitis cases and 40% of the encephalitis cases; of the 45 fatal encephalitis cases, 27 were in children under 4 years of age. At least two agents were involved: a tick-borne encephalitis virus for the majority of cases in adults and enterovirus 71 for the cases in children regardless of the clinical form of their disease. In the poliolike cases, only enterovirus 71 was isolated, and poliovirus infection was excluded by neutralization tests.

In 1979, enterovirus 71 isolations in Lyon, France, were associated with a small outbreak in children 5–9 years old; their disease was chiefly influenzalike, with CNS involvement only in the most severe cases.[440]

A review of the isolations of enterovirus 71 in the United States has recently been reported from CDC.[10] At least one isolate was made each year, there being 193 culture confirmed infections. The largest number occurred in 1987, with 45 persons in 17 states infected. During 1987, 24 (53%) of 45 patients infected with enterovirus 71 had CNS involvement, with 6 developing paralysis and 18 meningitis. It is noteworthy that the number of cases of paralysis associated with enterovirus 71 in 1987 was about the same as the annual number of oral poliovirus vaccine-associated cases of poliomyelitis reported in the United States since 1980.

With most nonpolio enteroviruses, the pattern of infection for a specific serotype in a particular locality resembles that of poliomyelitis: (1) There may be local serotypes that are endemic, constantly circulating among the few nonimmunes; these are mostly very young children, since virtually all the older children and adults already have protective antibody from previous infections; (2) or specific serotypes may be completely absent from a particular locality or very limited in dissemination for a number of years (see examples below); a population of susceptibles than builds up, and a wave of wide and rapid spread of the virus may occur, reaching a large proportion of all age groups. Dissemination of different serotypes may thus occur in waves, with one predominant type in an area succeeding another from one year to the next or even within the same summer–fall season.

An example of waves of infection with different enterovirus serotypes can be seen in data from the United Kingdom and the Republic of Ireland during the 1960s and early 1970s. A number of echovirus types were isolated sporadically and in relatively small numbers.[483] In 1968, continuing into 1969, there was an epidemic of type 6 infections, and type 6 was by far the one most frequently recovered during 1968 (about 40% of all echovirus isolates reported); in 1969, echovirus type 9 also became epidemic, accounting for about 40% of the echovirus isolates in that year. In the latter epidemic, one fourth to one third of all the cases occurred in Scotland. In 1971, however, about 60% of the echovirus isolates in the United Kingdom were echovirus type 4, which in previous years had been responsible for only about 5% of the echovirus reports. These infections were confined mainly to northern and northwestern regions of England and to Scotland and Ireland. Significantly, in view of the limited previous circulation of echovirus 4, a large share of the isolations were from older children and young adults. It should be noted that the infections selected for laboratory investigation may be biased toward significant illnesses; it seems probable that the younger children were also experiencing widespread infection but that overt illness more frequently brought the older groups to clinical attention. In 1971, echovirus 6 accounted for only 3% of all United Kingdom isolates; echovirus 9 accounted for 2%.

During this period, coxsackievirus A9 was almost always the serotype most frequently reported in the United Kingdom, constituting 35–55% of the group A isolates. Coxsackievirus A16 was frequently reported in most years but tended toward epidemic prevalence every 3 years: peaks of dissemination were recorded in 1964, 1967, 1970, and again in 1973. In 1970, about 57% of the coxsackie A isolates were of the A16 type. Coxsackieviruses B2, B3, B4, and B5 also appeared to be somewhat cyclical in the United Kingdom, in periods of 3–6 years; B1 did not share this periodicity and usually was isolated in lower numbers, although an epidemic of this virus occurred in 1970; type B6 was rarely reported.

Periodicity in enterovirus infections in a particular region can be clearly seen in the data on virus isolations at the Regional Virus Laboratory, Ruchill Hospital, Glasgow, covering a 20-year period, 1957–1976 (Table 3).[141] Although the Glasgow laboratory may isolate 25 or more different enterovirus serotypes in any one year, certain types clearly tend to recur at quite regular intervals. For example, echovirus 9 tended to recur at 4-year intervals, whereas echoviruses 4 and 30 were absent for 7–9 years between outbreaks. As indicated above, the serotypes (echoviruses 4 and 30, in this case) recurring at the longer time intervals and in large epidemics are those for which

Table 3. Infections by Selected Enteroviruses in the Glasgow Area during 20 Years[a,b]

Virus	1957	1958	1959	1960	1961	1962	1963	1964	1965	1966	1967	1968	1969	1970	1971	1972	1973	1974	1975	1976
Coxsackie A7	0	0	37	0	0	0	15	0	0	0	9	1	6	0	0	0	1	0	0	0
Coxsackie B5	2	1	8	5	10	0	0	4	37	0	0	0	2	0	10	2	0	2	1	8
Echo 4	0	0	0	0	0	0	42	12	0	0	0	34	0	0	73	37	0	0	0	0
Echo 6	1	0	3	1	2	22	4	2	9	1	16	9	14	0	1	0	1	8	6	0
Echo 9	3	0	0	119	2	0	1	44	2	0	0	0	38	1	5	0	49	4	0	1
Echo 17	0	0	0	0	0	0	0	0	0	0	1	0	2	23	0	1	0	0	0	0
Echo 19	0	0	0	0	5	0	0	0	0	3	15	20	0	0	4	1	0	9	10	10
Echo 30	0	0	74	6	0	0	0	1	2	92	0	0	0	1	0	1	0	0	19	9

[a]From Grist et al.[141] Used with permission.
[b]Figures represent infections detected by virus isolation at the Regional Virus Laboratory, Ruchill Hospital, Glasgow.

Table 4. Number of Reports of Echovirus Infections by Age (WHO Data 1967–1974)[a]

Age (yr)	Echovirus type 1	3	4	5	6	7	9	11	12	13	14	16	17	18	19	20	21	22	25	30	Total n	%
<1	132	148	156	66	390	300	505	648	112	68	304	58	63	147	169	46	50	353	72	119	3,906	16.9
1–4	132	290	182	87	978	671	1146	937	96	124	348	74	124	193	204	130	89	186	163	391	6,545	28.4
5–14	65	208	660	84	1451	326	1804	727	57	43	203	46	146	251	311	57	78	27	99	1170	7,813	33.9
All children	329	646	998	237	2819	1297	3455	2312	265	235	855	178	333	591	684	233	217	566	334	1680	18,264	79.3
15–24	23	33	277	20	415	68	472	185	12	11	46	11	35	105	110	18	25	3	27	345	2,241	9.7
25–59	30	49	235	28	367	90	580	247	16	22	42	24	30	111	158	16	20	10	28	295	2,398	10.4
≥60	2	3	7	1	30	7	24	21	4	0	3	0	2	3	3	3	1	2	3	6	125	0.5
All adults	55	85	519	49	812	165	1076	453	32	33	91	35	67	219	271	37	46	15	58	646	4,764	20.7
Totals	384	731	1517	286	3631	1462	4531	2765	297	268	946	213	400	810	955	270	263	581	392	2326	23,028	100.0

[a]From Grist et al.[141] Used with permission.

higher proportions of infections have been reported in older children and in adults. That is, in the absence of early childhood exposure, many persons remained susceptible into these older age groups during the longer interval. These parallel age–incidence data can be seen in worldwide virus isolation data for 1967–1974, collected and collated by the WHO (Table 4).[141]

An example of very rapid succession of different enterovirus serotypes in a small community has been well documented.[341,513] In the course of 1968–1971, six outbreaks of febrile viral disease with respiratory and gastrointestinal symptoms occurred in an Israeli kibbutz of about 450 members and were differentiated by correlated clinical, epidemiologic, and laboratory study. In a single summer, there was an extensive outbreak of febrile illnesses, characterized by respiratory and abdominal symptoms, that began in May 1970 and continued without interruption until the beginning of August. During this period, 21 children were ill on two occasions and six were sick three times, with some of the episodes giving the clinical impression of relapses in a disease caused by a single agent. However, these illnesses were found to be caused by successive and overlapping large outbreaks of coxsackievirus B4 and echovirus 9 infections, in which 43 persons were shown to have been infected by both viruses in sequence, at intervals of 3–4 weeks. Echovirus 16 was also found to be involved in a small number of the early illnesses.

5.2.3. Geographic Distribution and Climate.
Enteroviruses are found in all parts of the world. In tropical and semitropical regions, they are widely distributed throughout the year. In temperate climates, they are present at low levels in winter and spring but are encountered far more commonly during summer and fall. Some outbreaks of enteroviral infection have continued from fall into winter months; winter outbreaks have been recorded, but they are rare.

Climate appears to be an important factor in the circulation and prevalence of enteroviruses. Even within the climate range represented in the continental Untied States, healthy children in southern cities harbor a greater abundance of enteroviruses and a wider variety of antigenic types than do those of comparable age in northern cities. Repeated tests on the stools of 136 healthy preschool children in Charleston, West Virginia, over a period of 29 months, indicated that 90% of the cytopathogenic enteroviruses were recovered during the summer and the autumn months and that the incidence was three to six times higher in the lower-socioeconomic than in the mid- to upper-middle class districts; 52% of the viruses recovered belonged to the echovirus group[258] (see Table

8). In areas farther south, such as Phoenix, Arizona,[258] and Louisiana,[125] the prevalence of enteroviruses among healthy children was more evenly distributed throughout the year but still markedly higher during the months of May to October. Subsequent studies on enterovirus excretion rates in young children in Seattle, San Francisco, Minneapolis, Buffalo, Atlanta, and Miami[115,126] have confirmed and extended the earlier results: higher levels of endemicity and longer periods of prevalence were found in the southern cities.

Among the Atlanta and Miami children studied in 1960–1963 and in the earlier investigations in Charleston and Phoenix,[258] average annual virus excretion rates were 7–14%, and far larger numbers of enterovirus serotypes were commonly prevalent than in northern cities. In the normal population under study in the New York Virus Watch Program,[212] only four to six serotypes of cytopathogenic enteroviruses were prevalent at any given period, and the enterovirus isolation rate in fecal specimens from young children (0–5 years of age) was only 2.4%.

Analysis of the seasonal distribution of enterovirus isolations in the United States over the 10-year period 1970–1979[322] presents a classical picture of enterovirus distribution in temperate climates (Fig. 1). The nonpolio enteroviruses were isolated chiefly during the summer and early fall; the average number of enteroviral isolations for each month from June through October was 6.6 times higher than the monthly average for the other 7 months, and 82% of the total isolates were isolated during these peak months. For all three groups of nonpolio enteroviruses so analyzed, August was the peak month of isolation. Polioviruses, however, were isolated throughout the year, as would be expected from a nation in which live poliovaccine is administered year-round.

Among children living in warm climates and poor hygienic conditions, the incidence of infection with one or more enterovirus serotypes may exceed 50%, and mixed infections are common. In a study of infants in Karachi, Pakistan—almost all of whom were less than 2 years of age—approximately 80% of those tested yielded at least one enterovirus.[364] Additional virus serotypes were recovered by mixing a portion of the original swab specimen with type-specific antiserum to block the virus type or types previously isolated from the swab and then inoculating the mixture into tissue cultures. If an isolate was obtained, it was first confirmed as a new type by retesting against antiserum to the previously recovered type or types; if confirmed as new, the serotype was then identified. Of 116 rectal-swab specimens restudied for multiple virus isolation, approximately 45% were positive for at least two different viruses, 14% for three viruses, and each

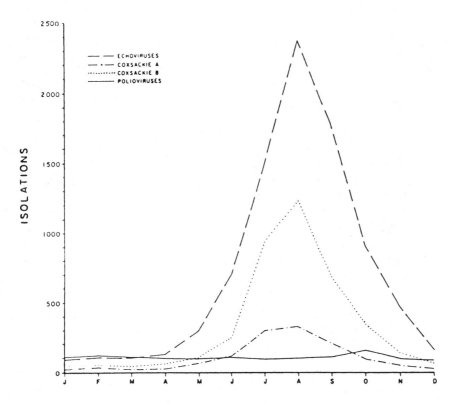

Figure 1. Distribution of infections with enteroviruses in the United States, by the month of onset of illness, from 1970 to 1979. From Moore,[322] with permission.

of two infants had four different viruses in a single swab specimen.[364] Mixed infections may also occur in children living under good socioeconomic circumstances in temperate areas,[73] but they are less frequent.

In some isolated groups, such as certain Eskimo communities,[19] the whole population may lack antibodies to some serotypes (see Fig. 2). In the survey depicted not even the oldest persons, up to 72 years of age, had any serological evidence of past infection with either coxsackievirus B1 or coxsackievirus B2; coxsackievirus A1 apparently had been present some years previously, and coxsackieviruses A4 and A10 were currently present or had been in the very recent past.

On the other hand, a study conducted in the Accra area of Ghana, at a time before polio vaccination was introduced in that country to any significant extent, indicated that infants and young children were experiencing widespread infections with all three types of polioviruses,[359] a pattern typical for a dense population living under poor hygienic conditions. Even among children less than 3 years old, 77% had antibody to poliovirus type 1 and only slightly lower proportions to type 2 and to type 3; 80% were immune to all three types by the age of 6 years. In the same children, coxsackievirus A9 antibody developed at almost the same rate as poliovirus antibodies, and

79% had been infected by 3 years of age and 94% by the age of 6. Coxsackievirus B3, however, was less widespread, infecting only 52% of the children by 4–6 years of age.

The beginnings of a global picture of nonpolio enterovirus prevalence were drawn by Assaad and Cockburn,[14] who analyzed the reports received by WHO for the 4-year period 1967–1970 from laboratories around the world that participate in the WHO reporting system. This analysis was extended to cover 8 years, 1967–1974, by Grist et al.[141] Chiefly temperate regions were represented in these reports. The reports used for the 8-year analysis included about 5200 coxsackievirus A isolates, more than 13,000 coxsackievirus B isolates, and more than 23,000 echoviruses.

These investigators were able to draw comparisons concerning infections by these agents despite the limitations implicit in the nature of the reports—particularly the underrepresentation of coxsackievirus A isolates, since relatively few laboratories now include mice in their isolation systems.

Among the five group B coxsackieviruses under study, B3 and B5 were the most frequently isolated.

No regular yearly pattern, worldwide, was observed in the recurrence of specific types. With each serotype

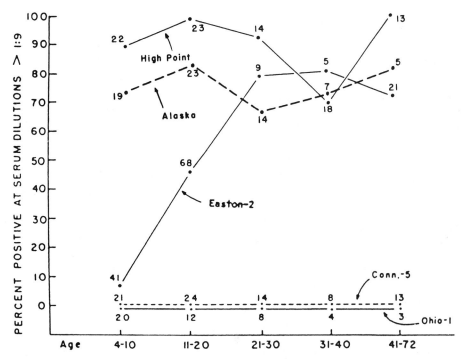

Figure 2. Results of neutralization tests with local Eskimo sera and coxsackieviruses expressed in terms of percentages of sera positive against each of the five coxsackievirus strains. The serotypes represented are High Point (coxsackievirus A4), Alaska (coxsackievirus A10), Easton-2 (coxsackievirus A1), Conn.-5 (coxsackievirus B1), and Ohio-1 (coxsackievirus B2). The numerals at each point show the number of sera tested in the indicated age group. From Banker and Melnick.[19]

analyzed, the number of reported isolations fluctuated greatly from year to year, and even larger fluctuations were observed for individual countries, e.g., of 154 coxsackievirus A9 infections reported by Japan during the period 1967–1970, all but eight were concentrated in 1967.

In this worldwide survey, which tended to focus on virus isolations from sick persons, the clinical manifestations usually included aseptic meningitis, respiratory disease, skin eruptions, undifferentiated febrile illnesses, or gastroenteritis. In the combined reports for the 8-year period 1967–1974, disease of the CNS—commonly aseptic meningitis—was associated with 28% of the coxsackievirus A infections (and with nearly 50% of those identified as A9 infections), with 34% of the coxsackievirus B infections (42% of those identified as infections with B5), and with 56% of the echovirus infections (with much higher percentages of the infections with types 4, 6, 9, and 30: 81, 63, 68, and 83%, respectively). The previously recognized associations of specific syndromes to specific serotypes were also observed, e.g., Bornholm disease (pleurodynia) and myocarditis associated with coxsackieviruses B—particularly in persons over 15 years of age. Although cardiac involvement was more fre-

quently associated with coxsackieviruses B as a group, representing 3.2% of the coxsackievirus B isolations reported, a few cases of cardiac illness were reported with all the coxsackievirus A types tabulated by Grist *et al.*[141] and with most echovirus types, including 3% of the isolates of echovirus 22. Skin eruptions were the most frequent disease associations of coxsackievirus A isolates (especially hand, foot, and mouth disease, seen mainly in children under 5 years of age); 33% of the coxsackievirus A group infections generally—but 82% of A16 infections, 44% of A5, and 41% of A10 infections—were associated with disease of the skin or mucosa or both. Respiratory illnesses accounted for about 20% of the coxsackievirus B isolations but only about 12% of the coxsackievirus A and the echovirus infections; however, with echoviruses 1, 11, 13, and 22, about 25% of the isolations were associated with respiratory disease. There were also small numbers of cases of paralytic CNS disease reported in association with many of the types of each group.

Among the 13 most frequently isolated agents in the United States 10-year study,[322] eight (coxsackievirus B5, poliovirus type 2, and echovirus types 3, 4, 6, 7, 9, and 11) appeared to circulate in an epidemic pattern, with many cases over shorter periods of time. The others showed no particular cyclic recurrence.

Data were available on the clinical syndromes associated with the U.S. isolates for the last 5 years of the period, i.e., 1976–1979. Meningitis was recorded in 35% of the patients; respiratory disease, 21%; encephalitis, 11%; nonspecific febrile illness, 6%; illness with rash, 4%; carditis, 2%; paralytic disease, 1%; and other known syndromes, the remaining 22% of disease reported (percents rounded, making the total more than 100%). Comparing the distribution of isolates from all of the reported cases with the distribution of serotypes associated with each clinical syndrome, specific agents were overrepresented in certain of the syndromes. For example, polioviruses accounted for 53% of the paralytic disease; echoviruses for 78% of the encephalitis and meningitis cases; group B coxsackieviruses for 63% of the carditis cases; group A coxsackieviruses accounted for 32% of the rash illnesses; and coxsackieviruses of groups A and B, and polioviruses were disproportionately represented among cases of respiratory illness (15%, 29%, and 17%, respectively).

5.2.4. Age and Sex. Children are the prime targets of these viruses, and thus serve as the chief vehicle for their spread. In warmer lands and in families living under unsanitary and poor socioeconomic conditions in temperate zones, children are infected very early in life, and more than 90% may have already experienced infections with a number of the locally prevalent enteroviruses before the age of 5 years. In such settings, paralytic poliomyelitis is rarely recognized, and epidemics generally do not occur. When infection is delayed to older childhood and young adult life, the incidence of paralytic poliomyelitis rises, as does the frequency of the more severe manifestations of the other enterovirus infections. The age distribution of echovirus isolations as reported to WHO is illustrated in Table 4 (see Section 5.2.2).

The pattern of age distribution of the nonpolio enteroviruses can be defined by serological studies. For example, an examination of antibodies to coxsackievirus A9 and coxsackieviruses B1–B5 was made in sera obtained from New York Virus Watch families[212] on entry into the program. The results are shown in Table 5. Although comparison by age among virus types can only be suggested on the basis of the relatively small sample, it is clear that coxsackievirus B2, B4, and B5 antibodies were infrequent in the youngest children at the beginning of the Virus Watch observation period. This relative vacuum in immunity was partially filled by subsequent outbreaks of infection with these viruses.

Most of the infections reported to WHO in 1967–1974[141] were in children under 15 years of age: 85% of the coxsackievirus A isolates came from this age group, 79% of the coxsackieviruses B, and 79% of the echoviruses. These overall percentages were virtually identical in both 4-year periods, the total 8-year data changing by only 1 or 2 percentage points from those reported for 1967–1970.

With all the coxsackieviruses (both A and B groups) that were isolated in sufficient numbers for inclusion in the 8-year tabulation, and with most of the echovirus serotypes also, the number of isolates from younger chil-

Table 5. New York Virus Watch: Prevalence of Neutralizing Antibodies to Coxsackieviruses on Entrance to Observation, in Subsequently Invaded and Control Households,[a] by Age and Virus Type[b]

| Coxsackievirus type | Serum neutralizing antibodies by age (yr) | | | | | | | |
| | 0–4 | | 5–19 | | 20+ | | All ages | |
	Number of persons	Percent positive	Number of persons	Percent positive	Number of persons	Percent positive	Number of persons	Percent positive
A9	56	23	57	35	69	59	182	40
B1	10	(20)[c]	2	(0)[c]	14	14	26	15
B2	36	3	37	16	37	30	110	16
B3	14	14	14	36	20	40	48	40
B4	22	5	27	33	24	58	73	33
B5	33	3	53	11	62	24	148	15
All	171	14	190	24	226	41	587	27

[a]Control households were selected because no member had yielded the virus in question and because they were being observed during the period of maximum incidence of infection with the specified virus. The members examined were selected for age (preference given to children) and availability of paired or serial sera bracketing the desired period.
[b]From Kogon et al.[212] Used with permission.
[c]Parentheses are used for percentages based on ten or fewer observations.

dren (0–4 years) was greater than that for the 5- to 14-year-old group and constituted 62% of all the coxsackievirus A isolations reported, 54% of the coxsackievirus B, and 45% of the echovirus isolations. However, with echovirus types 4, 6, 9, and 30, the younger children accounted for a smaller share of the infections, whereas the 5- to 14-year-old group had 44, 40, 40, and 50%, respectively, of the isolations of these types, and the adults, 34, 22, 24, and 28%. Adults and older children also had a significant share of the coxsackievirus B infections reported (types 1–5), with B3 and B5 the most frequently isolated. Irrespective of virus type, the largest number of cases of CNS disease reported in the WHO study[141] occurred in children under 15 years of age, although substantial numbers of adults also had neurological involvement. The largest number of cases of respiratory illness was in children under 5 years of age. Children less than 5 years of age also predominated among the patients with serious CNS disease in the large outbreaks of enterovirus 71 infection in Bulgaria in 1975 and Hungary in 1978.[333,436]

The age distribution of the illnesses in the 10-year surveillance data for the United States[322] was also of interest, confirming and underscoring in these national data the previous views based on small epidemiologic and clinical studies. Most of the reported cases were in young children (56% of the total being under 10 years of age and 26%, under 1 year); this disproportionate predominance of isolates from children under 1 year of age as compared to those 1–9 years old held true for all groups except group A coxsackieviruses, which were found more frequently in the older children. Adults 20 years of age and older accounted for only 16–20% of all patients yielding nonpolio enteroviruses.

When the age distribution of each clinical syndrome was compared with the overall age distribution of the population from whom these isolates were obtained, adults and older children were overrepresented among carditis cases and also among cases of encephalitis and meningitis. The proportion of paralysis was greater than expected among the youngest patients (0 to 4 years) as well as among those in the 20- to 24-year-old and the 35- to 39-year-old groups. Overall, however, the youngest age group (0 to 4 years old) was overrepresented among the milder clinical illnesses. This may reflect both the larger number of specimens collected from this age group and the actual predominance of mild rather than severe illness in young children (except for neonates).

For those enteroviruses on which data have been obtained, the patterns of exposure and spread and of increasing development of antibodies along with increasing age are generally similar, although initial infections occur earlier with some enteroviruses than with others.

Sex sometimes appears to play a role in disease caused by enteroviruses. It has long been known that childhood paralytic poliomyelitis is twice as common in boys as in girls. A similar male predominance has been noted for more severe disease, e.g., CNS disease or myocarditis, than for less severe enteroviral disease, e.g., pleurodynia, hand, foot, and mouth disease, respiratory disease, rash, or undifferentiated febrile illness (see Table 17-2 in Moore and Morens, 1984[324]).

In paralytic poliomyelitis, it has long been recognized that strenuous exercise of an arm or leg tends to localize the paralytic manifestations to that limb. There is also evidence suggesting that heavy physical exertion (in this case, of high-school football players) may predispose infected individuals to more severe enteroviral meningitis.[323]

5.2.5. Occurrence in Families and Closed Ecological Units. Enterovirus infections are highly communicable. Within a community, the viruses generally spread horizontally via preschool children and are found more frequently in families of large size and lower socioeconomic level. However, once an agent has invaded a family, regardless of family size or circumstances, nonimmune family members readily become infected. In the New York Virus Watch Program, spread of coxsackieviruses to susceptible members of the household was high (76%), whereas that of echoviruses was considerably lower (43% of the susceptibles) (Table 6). In addition to

Table 6. New York Virus Watch: Age Distribution of Infections[a]

Virus group	Percentage infected (number observed) by age in years						
	0–1	2–5	6–9	10–19	Mothers	Fathers	All ages
Coxsackie[b]	74 (39)	85 (54)	88 (40)	67 (18)	78 (37)	47 (30)	76 (218)
Echo	43 (7)	68 (25)	58 (24)	22 (18)	24 (17)	24 (17)	43 (108)

[a]From Kogon et al.[212] Used with permission.
[b]The coxsackieviruses sought included only those that could be grown in cell cultures at the time of the studies.

infecting those without prior type-specific antibody, coxsackieviruses also spread to and reinfected one half of the siblings under 10 years of age who already had specific antibody. Only one echovirus reinfection was observed. The frequency of intrafamilial infection may be related in part to the duration of virus excretion by the young index child; in the New York Virus Watch, with a 2-week interval between routine collections, only 16% of those infected with echoviruses yielded virus on more than one occasion, whereas coxsackievirus excretion on more than 1 day was found in 44% of the infections, and excretion for more than a month was not uncommon.

Rapid and extensive spread also occurs in familylike close associations such as those in children's institutions, in cabin groups in summer camps,[157] and in the environment of a small kibbutz.[341,513]

Wide dissemination, in which overt illnesses represent only a "very small tip of a very large iceberg," as first described for poliovirus infections, has been repeatedly documented for a number of enteroviruses. During an epidemic in which 149 inhabitants of a city of 740,000 were hospitalized with echovirus 9 disease, approximately 6%, or 44,000 persons, had an illness compatible with infection with this agent.[409] Among families that had been invaded by the virus, the rate of inapparent or unrecognized infection (based on recovery of the virus) was 18%; however, among 107 persons in families in which no illness was observed, only one person yielded the virus from a stool specimen.

When outbreaks of aseptic meningitis and related illnesses caused by echovirus type 30 began to spread along the Pacific Coast in 1968, the arrival of the virus in the Washington area coincided with the initiation of the Seattle Virus Watch Program, and extraordinary opportunities were available to observe infection and illness among the regularly studied Virus Watch families as related to the community pattern of illness.[145] Sixty-four such families containing 291 members underwent continuing virological surveillance in this period. By virus isolation or serology or both, infection was documented in 70 (79%) of 88 members of 18 families; in the total observed Virus Watch population, the rate was 24%. The invaded families were of slightly larger size, included more children 5–9 years of age, and included only three persons (two adults and one young child) who had antibody prior to the epidemic. (The families escaping infection were by no means totally immune, but prior antibody was present in 13, 14, and 17% of the children in the age groups 2–4, 5–9, and 10–19, respectively). Of the persons observed to shed virus, 47% reported possibly related mild febrile illnesses, few of which were serious enough

to require medical attention; one father in an invaded family did develop aseptic meningitis, almost certainly caused by echovirus 30, although virus could not be isolated from his specimens. Thus, on the basis of the Virus Watch family experiences, there must have been many thousands of echovirus 30 infections in the Seattle area, more than half of which were completely without symptoms, during the period when 44 virologically confirmed cases of echovirus 30 aseptic meningitis occurred in Seattle.[451]

In the normal middle-class families of the New York Virus Watch Program,[212] all the coxsackievirus- and echovirus-associated illnesses observed throughout the study period were mild, and the largest number consisted of upper respiratory disease with or without accompanying enteric illness, rash, or other signs and symptoms. Central nervous system involvement, pleurodynia, pericarditis, and herpangina were not observed. To arrive at an estimate of the number of illnesses attributable to the infection with which they were temporally associated, allowance was made for illnesses "expected" had the observed concurrent infection not been present. For this correction, two types of controls were used: the illness records of matched but virus-free controls and the subject's own illness record before and after the episode of viral infection. For the coxsackievirus infections, these corrected rates of attributable illnesses were 24 or 19% (depending on which type of control is used) and for the echoviruses, 9 or 18%. The observation of respiratory illness in association with coxsackievirus infections has numerous parallels in other studies, particularly in relation to coxsackieviruses of group B. A number of echovirus serotypes have also been incriminated in respiratory or respiratory–enteric diseases.

The mildness of the illnesses associated with infection in the Virus Watch families is noteworthy. Many of the reports on which the more severe disease associations were based have been derived from patient-centered or epidemic-centered investigations. As the authors indicate, the absence of more serious enteroviral disease in this study suggests either that the more severe syndromes are rare—as with poliovirus infections—or that strains infecting the Virus Watch families were of unusually low virulence.

5.2.6. Socioeconomic Setting. The close correlation between low socioeconomic settings and the early acquisition of infection with the enteroviruses has been repeatedly emphasized in both tropical and temperate environments and reflects the general level of hygiene of the group. Fox[110] has predicted that in parallel with the transition of poliomyelitis from endemic or epidemic, the frequency of severe disease associated with coxsackie-virus and echovirus infections may increase as levels of hygiene and sanitation improve. More individuals may escape infection as young children, only to experience more serious clinical manifestations if they become infected in later childhood or adulthood. That this may in fact be occurring is suggested by the increase in reports of epidemics of enterovirus-caused aseptic meningitis, which does not seem to result merely from the increase in reporting because of the growing numbers and improved capabilities of virus-diagnostic laboratories around the world.[141]

5.3. Epidemiologic Patterns of Poliomyelitis

Poliomyelitis can be viewed as having three major epidemiologic phases: endemic, "prevaccine" or epidemic, and "vaccine era." Historically, they represent a temporal sequence, but they all coexist today in different regions of the world. In some crowded, developing areas, chiefly in the tropics, paralytic poliomyelitis continues to be a disease of infancy (truly "infantile paralysis") that is recognized and reported only sporadically. In these populations, virtually all children over 4 years of age are already immune. With the almost universal presence of antibody to all three poliovirus types in women of child-bearing age, passive immunity is transferred from mother to offspring, and many infants subsequently experience their first poliovirus infections while maternal antibodies still provide some protection against paralysis. In addition, because such a large proportion of poliovirus infections takes an inapparent or subclinical form, particularly among infants and young children, the paralytic cases that did occur could go unnoted or unremarked despite the abundance of circulating virus, particularly in populations faced with very high infant and child mortality rates. In the past, the rarity of reporting of clinical paralytic poliomyelitis in the tropics had led many to believe that no poliovirus infections were present there, when in fact the reverse was true: polioviruses were highly endemic, but the infections were largely asymptomatic.

Despite the rarity of reports of poliomyelitis in such areas, paralytic cases do occur and may indeed be increasing in these areas more rapidly than had been believed. In some instances, the reported cases may represent only about 10% of the cases that actually occur.[490] Experience of some 30 years ago foreshadowed this situation, which is being studied today through numerous surveys of residual paralysis in school-age children (see Sections 3.3 and 5.2.2).

5.3.1. Epidemiologic Patterns in Developed Nations in Temperate Zones. The shift from the endemic to the epidemic phase of poliomyelitis was first seen in societies that had advanced systems of hygiene and sanitation and were located in cooler climates. In the latter part of the 19th century and early in the 20th, the urban, industrialized parts of northern Europe and the United States began to suffer from epidemics of paralytic polio that became larger, more frequent, and more severe in an increasing number of localities. The generally accepted explanation, borne out by numerous studies, is that with increased economic development and correspondingly improved resources for community and household hygiene, and with the additional advantage of a temperate-zone climate, the opportunities for immunizing infections among infants and young children were reduced. Therefore, more persons encountered poliovirus for the first time in later childhood or in adult life, at ages when poliovirus infections are more likely to take the paralytic form. Furthermore, the delay in exposure increased the pool of susceptibles, opening the way for rapid and explosive spread of the viruses once they did enter the population, in contrast to the steady endemic transmission of the preceding phase. In some instances this transition from the endemic phase to the epidemic phase took place in an abrupt shift, but in other areas it was characterized by gradual increases in the annual case rates of "sporadic" poliomyelitis.

For example, in the United States, just before inactivated polio vaccine became generally available, the average annual number of cases of paralytic poliomyelitis was approximately 21,000. In epidemics, the peak age incidence was in the 5- to 9-year-olds, and about one third of the cases and two thirds of the deaths occurred in persons over 15 years of age. This was a marked change from the pattern in the great 1916 epidemic, in which approximately 80% of cases were in children under 5 years of age.[231]

In the first half of the 20th century, not only was it the "advantaged" nations that experienced epidemic polio, but it was also the socioeconomically advantaged sectors of the population within these nations that were most at risk. Even within the same city, wider circulation of the wild polioviruses in lower socioeconomic areas with poorer sanitation and hygiene provided more children with immunizing infections at an earlier age and reduced their chances of eventually developing paralytic disease.[291,294] The last outbreaks in the United States before polio vaccine became available included families living in good socioeconomic conditions; the spread of the virus through the community could be traced through young children who might or might not manifest illness; however, a high incidence of paralytic cases occurred among susceptible parents exposed to their virus-carrying children.[252,293,343]

5.3.2. Behavior in Developing Countries in Tropical and Semitropical Regions. The change from an endemic to an epidemic pattern of paralytic poliomyelitis is now being seen in developing countries with rising levels of sanitation, particularly in tropical and semitropical areas. This pattern of poliomyelitis epidemiology can appropriately be called a "prevaccine" phase even where some—but inadequate—vaccination is already being practiced. The "vaccine era" cannot be considered to have arrived in these areas until vaccine administration can be extended to a large proportion of the susceptible population and immunization levels can be sustained in ongoing programs. In a great many areas, even in countries where serious efforts are being made to expand vaccination programs, control of the circulation of wild viruses and of the disease is not yet complete.

Throughout the world in 1983 more than 25,000 cases were reported to the World Health Organization, from 123 countries representing 83% of the population included in the WHO reporting system. A major factor in assessing the worldwide incidence of paralytic polio is the continued gross underreporting of the true numbers of cases in many areas because of deficiencies in facilities and surveillance. Incompleteness of the reported incidence figures can be demonstrated by comparing the reported numbers with the data obtained in special lameness surveys (see below and Section 3.3). On the basis of these comparisons, the numbers of officially reported poliomyelitis cases may represent only about 10% of the actual number of cases that are occurring.

A marked contrast between low numbers of reported cases of poliomyelitis and a relatively high frequency of persons in the population who had residual polio paralysis was first described more than 30 years ago.[371] Among infants in Cairo, Egypt, a striking correlation was discovered between the early age at which clinical polio occurred and the similarly early acquisition of neutralizing antibodies. As part of this study, the apparent rarity of polio among Egyptians was found to represent underreporting of the disease. During the 1940s, only 2 to 11 cases were being reported annually for all of Egypt, but when observations were made of those with late paralytic disease that had never been reported, it turned out that about 1500 cases had been occurring each year. Thus, the annual polio incidence in Egypt during the late 1940s was calculated to be about 7.8 per 100,000 population, not too dissimilar from that in the United States during 1932–1946, when the average annual rate was 7.3 per 100,000.

The findings that were reported in 1952[371] have been confirmed repeatedly in recent studies utilizing surveys of residual paralysis in school-age children as an indication of past occurrences of paralytic poliomyelitis. By gathering information on the prevalence of types of lameness compatible with a history of paralytic poliomyelitis, investigators have estimated the annual incidence of paralytic poliomyelitis during the years just prior to the surveys.[27,161,223,350] For example, in Ghana, although no large epidemics had been reported, the incidence of poliomyelitis based on lameness surveys in 1974 was estimated to have been 116 per 100,000 children 0–5 years of age during the 1960s, that is, about 28 per 100,000 total population.[350] This amounts to about 2100 cases annually, 10 to 20 times the number that had been diagnosed and reported during the same period. This incidence was higher than the rates of paralytic polio seen in the United States and Europe just before the poliovaccines were introduced. Among the children with poliomyelitic lameness in Ghana, about 91% had experienced their acute disease during the first 4 years of life.

Similar surveys conducted in a number of other countries[496] have also indicated that the recent prevalence of paralytic poliomyelitis in children had been much higher than was believed from case reports. These surveys also show that much of the recent paralytic disease occurs at very early ages. In Ghana, 22% of the cases were found to have occurred in the first year of life, and 65% before the age of 2 years; in Nigeria, more than 7% of cases occurred in infants less than 6 months old, and more than 76% took place in the first 2 years of life.

Lameness data from some sections of India, reported in 1979,[489] showed a minimum annual incidence rate of 50 cases per 100,000 children below 6 years of age, or 10 cases per 100,000 population. Extrapolated for the entire country (population close to 700 million), this meant that at least 70,000 children in India were developing paralytic poliomyelitis each year,[488] which was five to seven times the number officially reported in the corresponding period.

Table 7 shows data from various surveys of residual paralysis, listing the reported incidence per 100,000 population during 1976–1980 as compared to the estimated incidence for the same period based on lameness surveys.[490] Representative surveys, with discussions as to the type of information gathered in various localities and the limitations of survey data, have been reviewed.[104,496]

Although the "background" endemic presence of paralytic poliomyelitis in many developing countries, as indicated above, is much larger than hitherto believed, it is also true that a real change in the pattern has taken place in many of these countries,[393,486–488,490,491,493,496] and epi-

Table 7. Mean Annual Reported Incidence of Poliomyelitis per 100,000 Total Population and Incidence Estimated from Lameness Surveys in Selected Countries during 1976–1980[a]

Country	Reported annual incidence	Annual incidence estimated from lameness surveys
Burma	1.1	18
Egypt	1.8	7
Ghana	2.3	31
India	2.1	18
Indonesia	0.1	13
Ivory Coast	1.2	34
Malawi	1.2	28
Thailand	1.7	7
Cameroon	1.2	24
Yemen	3.3	14

[a]From WHO,[490] with permission.

demics are occurring. Of 71 tropical and semitropical countries, 45 reported an incidence of poliomyelitis in 1966 that was three times greater than the average annual incidence for the period 1951–1955.

Ghana again serves as an example: the average annual number of cases reported was only nine during 1951–1955 but increased to 36 in 1961–1965 and to 53 in 1966–1970. Oral polio vaccine was administered from about 1966, but to only a small proportion of urban Ghanian children (estimated at about 20% of the children in Accra). The increase in cases accelerated during 1971–1975 to an annual average of 187; 313 cases were reported in 1976, and the higher rate continued. For 1976–1980, the annual average number of cases reported was 243.[486,491] (On the basis of lameness surveys cited above, it was estimated that the cases actually occurring averaged more than 2000 annually.)[340,350]

In some Central and South American countries where the disease has not yet been satisfactorily controlled, a similar trend has been seen. In Honduras, for example, in the years from 1969 to 1973, the endemic level of poliomyelitis was reported as being 20 to 66 cases each year. Then, in late 1976, after more than a year without a single case being reported, there were 29 cases in 3 months, and during 1977 the number increased to 175 cases with five deaths. Most of the patients were infants. Although some vaccine had been used in Honduras by this time, 72% of the patients had never received any vaccine, and only 9% had received more than a single dose. In 1978 there were 74 cases, but again, in 1979, there was a large increase, to 226 cases. The disease was then brought under

control, as only three cases were reported for 1980, 18 for 1981, and eight each for 1982 and 1983.[496]

Payne[373] and others have shown that there tends to be an inverse relationship between infant mortality rates and the incidence of clinical poliomyelitis. Paul[367] pointed out that when the infant mortality rate in a country drops below 75 per 1000 live births, the incidence of reported polio can be expected to increase. Thus, epidemic polio was a disease of affluent societies in the first half of the 20th century and is now an unwelcome concomitant of improved living standards in developing nations, unless it is controlled by vaccination.

5.3.3. Behavior in the Vaccine Era. The vaccine era for most countries in Europe, North America, Oceania, and some countries in other regions of the world began after 1955, when inactivated poliovirus vaccine was introduced, and was even more firmly established in the 1960s, when live attenuated vaccines became available on a large scale in some countries. Rarely has a serious disease been controlled so quickly and dramatically as was poliomyelitis in these countries. In 1955, the Soviet Union, 23 other European countries, the United States, Canada, Australia, and New Zealand experienced a total of more than 76,000 reported cases of poliomyelitis. Only 12 years later, in 1967, 1013 cases were recorded in these same countries, a reduction of almost 99%.

Even further reductions followed. In 22 industrialized European countries, the total average annual number of poliomyelitis cases in 1966–1970 was 71, and in 1971–1975, this figure was 357; in 1976, there were 151 cases, and in 1977, 145. Adding the data for the same periods from Australia, Canada, Japan, New Zealand, and the United States, these 27 industrialized countries around the world (containing a total of approximately 680 million persons) had total annual case rates of 807 for 1966–1970, 377 in 1971–1975, 165 in 1976, and 169 in 1977.[484,486]

In the United States, from the high prevaccine rates of paralytic polio, between 5 and 10 per 100,000 population (that is, 10,000 to 21,000 paralytic cases per year), cases became far fewer after the killed vaccine came into use, in some years reaching incidence rates as low as 0.5 per 100,000 population. But this still meant that significant numbers of cases were occurring; in 1959, there were 5700 paralytic cases in the nation, and in 1960, there were more than 2500. Concomitant with the use of killed vaccine and the decreasing numbers of cases as compared with the 1955 totals was the finding that some of the cases were in the fully vaccinated. In a study of several thousand paralytic cases, 17% were in children who had received three injections of the killed vaccine. Some of the disappointing results may have been related to problems that have since been corrected in the few small countries (Holland, Sweden, Finland) that have used inactivated vaccines solely through most of the vaccine era.

Following the introduction of live poliovirus vaccine in the United States, the average annual number of paralytic cases decreased precipitously, to 240 during 1961–1965.

In subsequent years, the rates averaged 40 to 50 cases annually, decreasing to 18 cases in 1969. In 1970, however, 32 cases were reported, including 22 from an epidemic in Texas among nonimmunized persons. For the 10 years 1970–1979, the average annual number of cases was 17. In the early to mid-1980s, the average annual number of cases is less than ten per year; for 1986, only five cases were reported. Thus, in recent years case rates have ranged from 0.02 down to less than 0.001 per 100,000 population.

Now, in the well-vaccinated areas of the world, vaccine-era epidemiologic patterns of poliomyelitis are becoming established. These patterns differ from one country to another and to some extent even within the same country.

5.3.3a. Virus Isolation. In a number of areas where routine immunization programs regularly cover almost all infants and children and/or mass vaccination campaigns are conducted regularly and are implemented so as to reach essentially all young children, virtually no cases are being reported, and wild polioviruses are rarely identified. This is true for a number of European, American, Asian, and Western Pacific countries that have extensive and continuing live vaccine programs. Almost all isolates now closely resemble the vaccine strains and are generally presumed to be vaccine progeny. Vaccine viruses are abundantly excreted by the vaccinee and infect unvaccinated

contacts. The rare cases of poliomyelitis that do appear may be caused by imported wild viruses or may in some instances be vaccine-associated (see Section 9). Thus, it seems that within many countries an end to wild poliovirus transmission has indeed been achieved.[87,338,446,447,518]

5.3.3b. Does This Mean that Poliomyelitis Is Being Eradicated? Absence of poliomyelitis cases in many areas has come about, but only within the borders of certain countries. Long-term studies suggest that wild polioviruses may indeed become almost completely eliminated following consistent proper use of vaccine. For example, in Japan since 1964, infants between the ages of 3 months and 18 months have been vaccinated with two doses of trivalent live virus vaccine by routine administration at local health centers over short periods in spring or autumn. Beginning in 1962, a collaborative study conducted under the sponsorship of the Ministry of Health and Welfare has included serological surveys for levels of antibody against polioviruses and virological studies for isolation and identification of polioviruses from the feces of healthy children in periods when the vaccine campaigns were not underway. Although no similar study has been carried out in the United States, a comparison can be made with other U.S. data. Isolation from fecal specimens obtained in 1962–1968 from healthy children in Japan 2 months or more after routine vaccination periods[447] can be compared to American findings of more than a decade earlier (in 1951–1953),[258] made in surveys of healthy preschool children in Charleston, West Virginia, and Phoenix, Arizona, during the period prior to the development of poliovaccines. The results are shown in Tables 8 and 9. The rates of occurrence for nonpolio enteric viruses in Japan were of the same order of magnitude as those found earlier in the U.S. cities, but the effect of systematic

Table 8. Distribution of Enteroviruses Isolated from Healthy Children in Populations of Contrasting Socioeconomic Levels during a Nonepidemic Period (1951–1953)[a]

Population	Number of specimens tested	Percentage yielding viruses			
		Polioviruses	Coxsackieviruses	Echoviruses	All enteroviruses
Charleston, WV					
Lower	597	2.3	2.3	3.7	8.4
Upper	1028	0.5	1.5	0.8	2.7
Phoenix, AZ					
Lower	943	3.0	2.0	8.3	13.3
Upper	399	1.0	1.0	0.3	2.3
Total					
Lower	1540	2.8	2.1	6.6	11.4
Upper	1427	0.6	1.3	0.6	2.6

[a]From Melnick.[258] Used with permission.

Table 9. Poliovirus Isolation from Fecal Specimens from Healthy Children Collected Not Less than 2 Months after the Routine Vaccination[a]

Year	Time of specimen collection	Number of specimens examined	Number of cytopathogenic agents isolated[b]	Poliovirus isolated				Other cytopathogenic agents[b]
				Number[b]	Type			
					1	2	3	
1962	Late summer–early autumn	974	31 (3.2)	1 (0.1)	0	1	0	30 (3.1)
1963	Late summer–early autumn	4954	127 (2.6)	5 (0.1)	0	4	1	122 (2.5)
1964	Late summer–early autumn	2299	81 (3.5)	10 (0.4)	1	2	7	71 (3.1)
	Late autumn–early winter	1803	18 (1.0)	17 (0.9)	4	11	2	1 (0.1)
1965	Late summer–early autumn	2069	174 (5.7)	1 (0.05)	0	1	0	173 (5.6)
	Late autumn–early winter	1770	41 (2.3)	1 (0.06)	0	1	0	40 (2.3)
1966	Late summer–early autumn	3048	107 (5.2)	5 (0.2)	1	1	3	102 (5.0)
	Late autumn–early winter	1831	19 (1.4)	6 (0.3)	1	1	4	13 (0.7)
1967	Late summer–early autumn	1962	131 (6.7)	0	0	0	0	131 (6.7)
	Late summer–early winter	1833	20 (1.1)	2 (0.1)	2	0	0	18 (1.0)
1968	Late summer–early autumn	1504	114 (7.6)	2 (0.1)	2	0	0	112 (7.4)
	Late autumn–early winter	1583	32 (2.0)	3 (0.2)	0	1	2	29 (1.8)

[a]From Takatsu *et al.*[447] Used with permission.
[b]The figures in parentheses indicate the percentage of the total number of specimens examined.

widespread vaccination in Japan is clearly reflected in greatly reduced prevalence of polioviruses.[435] Most of the poliovirus isolates studied in Japan were vaccinelike in their properties and were considered to be vaccine virus progeny.[446]

The United States, like Japan, has for many years relied almost completely on live polio vaccine, and in this country also it appears that there is no longer any endogenous reservoir of wild polioviruses. Wild strains continue to be introduced, particularly from Mexico, but even such imported cases are extremely rare and almost never result in secondary cases. It has been postulated[518] that the use of live poliovaccine has achieved this result by establishing intestinal resistance extensively within the population, reducing the pool of susceptible individuals below the level required for perpetuation of wild polioviruses, so that there has been a true break in the chain of infection.

However, even in well-vaccinated countries that have achieved almost complete elimination of paralytic poliomyelitis by full and proper use of either live attenuated vaccine or inactivated vaccine, there may be important gaps in protection. This vulnerability was shown by the outbreaks of poliomyelitis among persons refusing vaccine on religious grounds in the Netherlands, Canada, and the United States in 1978 and 1979.[49,117,485] A more recent outbreak in a well-vaccinated population occurred in Finland in 1984–1985[118,173,233,245] (see Section 9 for details). These outbreaks were an unfortunate reminder that virulent wild polioviruses do indeed still exist and can readily circulate among susceptibles. Thus, "eradication" of poliomyelitis or of wild polioviruses cannot be considered to be approaching so long as importation of virulent wild polioviruses remains a clear danger against which no nation can be secure except by maintaining high levels of immunity through vaccination.

5.3.3c. Breadth of Immunity. In many countries, the success of live poliovirus vaccination programs has reduced the wild poliovirus circulation so substantially that there are now increasing numbers of people whose immunologic experience with polioviruses is limited to a single vaccine strain of each type. It is believed by many that full and continuing immunization with killed poliovirus vaccine also has a considerable effect in limiting the circulation of wild polioviruses, particularly in populations in developed countries. It is conceivable that the changed ecological situation could act on wild poliovirus populations as a selective mutational pressure toward wide antigenic divergence from the attenuated vaccine strains. If such shifts were to occur, they might permit an increase in the silent circulation of heterologous, but homotypic, wild strains, perhaps reintroducing the risk of paralytic poliomyelitis for individuals lacking sufficiently broad vaccine-induced immunity. In the face of such a development, it would be appropriate to investigate whether successive infections with two different attenuated polioviruses of the same type could provide a broadening of alimentary-tract resistance to wild homotypic viruses.

In some early field studies, children immunized with one strain of type 3 vaccine virus and subsequently given another vaccine strain of the same serotype were susceptible to the second vaccine strain—homotypic but heterologous—but not to the first vaccine strain.[186] These observations raised some concern that the breadth of vaccine-induced immunity may be limited, leaving some vaccines susceptible to successful infection by new wild homotypic strains in areas where little or no wild virus now circulates.

Concern about limitations of the breadth of vaccine-induced immunity was raised again in 1985[233] in light of Finland's experience with an altered strain of type 3 virus that was able to break through to cause an outbreak in a population well vaccinated with inactivated polio vaccine (IPV) (see Section 9.1). The possibility of an alarming degree of variation was underscored by further findings on the divergent properties of the epidemic strain.[173,245] However, the Finnish IPV previously used was recognized to have a low level of immunogenicity for type 3, and other vaccines—both oral polio vaccine (OPV) and the new, more potent IPV—induced high antibody titers against the Finnish epidemic type 3 strain. Furthermore, other similarly divergent type 3 strains have not been found anywhere in the world since the Finnish outbreak.

Thus far, altered strains do not seem to be divergent enough to suggest that the strains in either the live or killed vaccines should be changed in order to provide different and broader antigenic coverage.

5.3.3d. Continuing Cases. At the same time as some nations have been totally free from polio for a number of years, tens of thousands of cases of polio are still reported each year, and many more go unreported; worldwide, there may be more than 250,000 cases annually.[393] The WHO Expanded Programme for Immunization (EPI) was initiated with the objective of reducing morbidity and mortality from six target diseases, including polio, by providing immunization against them for every child in the world. The program activities continue to depend heavily on technical cooperation with and among developing countries, and considerable progress is being made[492,495] (see Section 9.3).

Other aspects of the epidemiology of controlling poliomyelitis in the "vaccine era"—including social failure—are also discussed in Section 9.

6. Mechanisms and Routes of Transmission

Man is the only known reservoir for members of the human enterovirus group, and close human contact appears to be the primary avenue of spread. For almost all these agents, virus can be recovered from the oropharynx and intestine of individuals infected either clinically or subclinically and is generally shed for longer periods (up to a month or more) in stools than in secretions of the upper alimentary tract. Thus, fecal contamination (fingers, table utensils, foodstuffs, milk) is the usual source of infections. However, droplets or aerosols from coughing or sneezing can also be a source of direct or indirect contamination. Coxsackievirus A21 has been shown to be more abundant in nasal secretions than in those from the throat and has been experimentally transmitted from infected volunteers by airborne aerosols produced by natural coughing.[75] Enterovirus 70, the newly recognized agent of acute hemorrhagic conjunctivitis,[315] has thus far been found almost exclusively in conjunctival and throat specimens, but fecal isolations have been reported.

In epidemics of conjunctivitis primarily caused by another enterovirus (e.g., coxsackievirus A24),[512] strains of poliovirus type 1, unlike the Sabin vaccine strains in properties, have also been isolated from conjunctivae. On another occasion, a Sabin-like strain of type 3 was isolated from conjunctivae of a family member who was in contact with an infant vaccinee immediately after the infant received OPV. Polioviruses have not usually been looked for in conjunctivae, and it may be that conjunctival infections with polioviruses are more frequent than has been recognized. It appears that transmission of poliovirus may sometimes occur by this route.

Warm weather favors the spread of virus by increasing human contacts, the susceptibility of the host, or the dissemination of the virus by extrahuman sources. The viruses are most readily spread within the family, and the extent of the intrafamilial infection appears to be closely related to duration of virus shedding, particularly by young children.

During periods of epidemic prevalence, in both rural and urban areas, houseflies (*Musca domestica*) and filth flies (*Phormia regina, Phaenicia sericata, Sarcophagan* species) may be found contaminated with enteroviruses[286,287] and may act as mechanical carriers. The importance of flies in transmission is not easily evaluated, although it is important to note that virus has been found in food naturally contaminated by flies.[469] Virus persists in flies for weeks but does not multiply.[295]

Enteroviruses are present in urban sewage even during periods when no clinically apparent enteroviral infections are being seen in the community. Sewage may serve as a source of contamination not only of flies but also of water supplies used for drinking or bathing, or through its use as fertilizer.[289] Enteroviruses also have been found in aerosols produced by sprinkler irrigation using recycled wastewater.[271]

The presence of human enteric viruses in water has been recognized for several decades, but the full health significance of this pollution has yet to be determined. The problems relate both to the treatment methods themselves and to the criteria for monitoring waters to determine the success of the treatments intended to ensure freedom from health hazards for drinking water, recreational waters, or food-harvesting waters.

The health hazard associated with the discharge of virus-laden sewage into estuaries and oceans became evident with outbreaks of shellfish-associated hepatitis A (caused by enterovirus type 72). Oysters, clams, and mussels can concentrate enteroviruses from water into their tissues. They are filter-feeding organisms; that is, they sieve out suspended food particles from a current of water passing through the shell cavity. Since the entire shellfish is often consumed raw or inadequately cooked, it functions as a passive virus carrier.[289]

A health hazard from contaminated water sources may result from dependence on bacterial standards. The bacterial agents that have been used successfully as sentinels in controlling water carriage of enteric bacterial diseases have been questioned as adequate indicators of viral contaminants. In comparison to the bacterial indicators (total coliform and fecal coliform indices), viruses are far less effectively removed by many of the treatments used for sewage and water and survive much longer, and thus viruses are frequently found in water sources that meet bacterial standards.

Water-borne outbreaks of enteroviruses are not eas-

ily recognized. Many of these viruses cause infections that are usually inapparent, and thus when clinical manifestations do occur they are difficult to trace, and polluted water sources may be missed. The problem of recognizing water transmission is also exacerbated by the fact that viruses can travel long distances from the source of contamination and still be infectious.[272]

Understanding environmental factors controlling enteric virus survival and transport in nature could be a key to understanding the maintenance of infections within a population. Environmental conditions can greatly affect the survival and transport of viruses in nature, and there are numerous routes by which excreted virus may find is way back to the human host (Fig. 3).[289] Many of these routes relate to water as a vehicle in which infectious viruses are carried.[24,288,500]

In most of the world, human wastes are discharged into natural waters with little or no treatment. Thus, there is little reduction in the initial input of viruses. However, in the developed countries a large part of sewage is processed by biological or physicochemical methods before discharge into receiving waters. Although treatment processes can result in large reductions in the concentration of viruses present in raw sewage, substantial numbers usually remain. Disinfection of treated sewage by chlorine is sometimes practiced; although this is effective in reducing bacterial pathogens, viruses are not eliminated by this treatment because of their greater resistance, particularly in the presence of organics.

The average concentration of enteroviruses in sewage appears to be about 100 PFU per liter in the United States, but tenfold more or even higher concentrations may often be found. Because of the widespread use of live poliovaccine, poliovirus is often present in the highest concentration.

In other regions of the world, much higher concentrations of virus have been observed. The average concentration of enteric viruses in sewage in less-developed countries of the world may be 100 times that observed in the United States. However, these concentration differences may reflect to some extent the higher per-capita water consumption in industrialized countries and the consequent greater dilution of virus in wastewater.

Besides being detected in treated sewage and even in drinking water, enteroviruses have also been isolated from marine areas receiving both raw and chlorine-treated sewage. In addition, they have been detected several kilometers from the nearest source of sewage disposal, in bathing beach waters and in waters where sewage sludge is discarded many kilometers from the coast.

Enteroviruses have also been detected in marine sed-

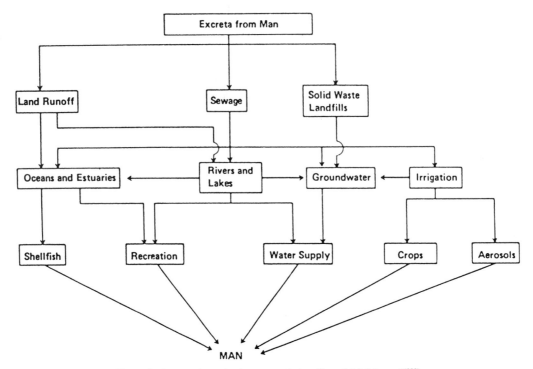

Figure 3. Routes of enteric virus transmission. From Melnick *et al.*[289]

iment in areas of sewage and sludge disposal. The greatest portion of virus pollutants may be associated with solids.[309,310]

With intentional recycling of water already occurring in a number of places in the world, and inadvertent recycling taking place in even more locales, it is clear that public health authorities must be alert to the possibility that water can be a route through which enteroviral diseases are transmitted. During an epidemic of coxsackievirus B5 infections in a boys' summer camp on Lake Champlain, the virus was isolated from water from the lake swimming area.[157] Although in this outbreak the clustering of infection in cabins suggests the principal mode of spread to be person to person, an oral–water–oral route is possible, as well as a rectal–water–oral route.

A strain of coxsackievirus A6 originally obtained from mosquitoes in Fiji has been found to survive in mosquitoes experimentally infected by injection or by feeding on viremic mice, and a small number of transmissions to baby mice by bite were obtained. The function of the mosquito as a true vector was not demonstrated, however, for virus did not multiply within the insect; in both fed and injected mosquitoes, virus titers never exceeded the original level.[246]

7. Pathogenesis and Immunity

7.1. Pathogenesis

The portal of entry of enteroviruses is believed to be the alimentary tract via the mouth. The incubation period (defined as the time from exposure to onset of disease) is usually between 7 and 14 days but may range from 2 to 35 days. After initial and continuing virus multiplication, probably in lymphoid tissue of the pharynx and gut, viremia may occur and in turn lead to further virus proliferation in the cells of the reticuloendothelial system and finally to involvement of the target organs (spinal cord and brain, meninges, myocardium, skin).[34] Usually, the virus is excreted in the stools for several weeks and is present in the pharynx 1–2 weeks postinfection in individuals having either clinical or subclinical infection. Enteroviruses have been isolated from feces, pharyngeal washings, CSF, heart, blood, the CNS, urine, conjunctivae, and lesions of skin or mucous membranes; each enterovirus has its own preferential target organs.

Two or more enteroviruses may propagate simultaneously in the alimentary tract,[364] but under many circumstances, multiplication of one virus may interfere with growth of the heterologous type. Interference with the growth ("take") of live poliovaccine by concurrent infections with other enteroviruses is now well established. If an active enterovirus infection thus prevents the vaccine virus from becoming established in the gut, revaccination will be necessary to confer immunity.

Pathogenesis has been studied most thoroughly for poliomyelitis, the most serious disease caused by any of the enteroviruses.[33,34] The pathogenesis of poliomyelitis may be summarized as follows (see Fig. 4).

Poliovirus may be found in the blood of patients with the abortive form ("minor illness") and can be detected several days before onset of clinical signs of CNS involvement in patients who develop nonparalytic or paralytic poliomyelitis. In orally infected monkeys and chimpanzees, viremia is also regularly present in the preparalytic phase of the disease. Antibodies to the virus appear early in the natural infection and also early in orally infected experimental animals. Antibodies are usually present by the time paralysis appears. In man, viremia has been demonstrated regularly following ingestion of type 2 oral polio vaccine. Free virus is present in the serum between days 2 and 5 after vaccination, and virus bound to antibody can be detected for an additional few days.[296] Bound virus is detected by acid treatment, which inactivates the antibody and liberates active virus.

The virus first multiples in the tonsils, the lymph nodes of the neck, Peyer's patches, and the small intestine. The CNS may then be invaded by way of the circulating blood. In monkeys infected by the oral route, small amounts of antibody prevent the paralytic disease, whereas large amounts are necessary to prevent passage of the virus along nerve fibers. In man also, antibody in low titer in the form of γ-globulin from human donors may prevent paralysis if given before exposure to the virus.[148]

In the experimental infection in monkeys, poliovirus can spread along axons of peripheral nerves to the CNS, and there it continues to progress along the fibers of the lower motor neurons increasingly involving the spinal

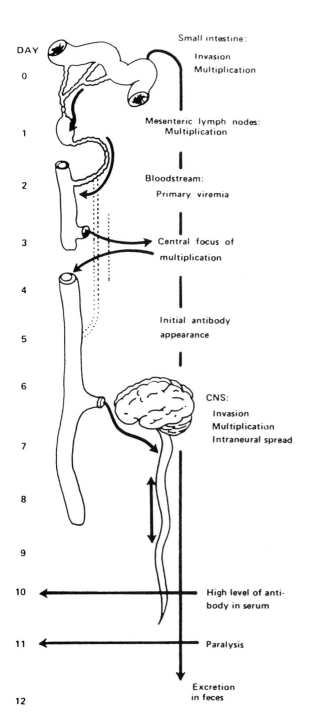

Figure 4. Schematic illustration of the pathogenesis of poliomyelitis (modified from Fenner). Virus enters by way of the alimentary tract and multiplies locally at the initial sites of virus implantation (tonsils, Peyer's patches) or the lymph nodes that drain these tissues, and virus begins to appear in the throat and in the feces. Secondary virus spread occurs by way of the bloodstream to other susceptible tissues, namely, other lymph nodes, brown fat, and the CNS. Within the CNS the virus spreads along nerve fibers. If a high level of multiplication occurs as the virus spreads through the CNS, motor neurons are destroyed, and paralysis occurs. The shedding of virus into the environment does not depend on secondary virus spread to the CNS.

cord or the brain. Neural spread may also occur in children who have inapparent infections at the time of tonsillectomy. In this situation, poliovirus present in the oropharynx may enter nerve fibers exposed during surgery and spread to the brain, resulting in bulbar paralysis. A similar mechanism of virus spread along neural pathways may be responsible for the rare instances of paralysis in a limb recently injected with an irritating material during a period of high poliovirus prevalence. An alternative mechanism for the adverse effects of tonsillectomy has been suggested (see Section 7.2).

Poliovirus invades certain types of nerve cells, and in the process of its intracellular multiplication, it may damage or completely destroy these cells. The anterior horn cells of the spinal cord are most prominently involved, but in severe cases, the intermediate gray ganglia and even the posterior horn and dorsal root ganglia are often affected. Lesions are found as far forward as the hypothalamus and thalamus. In the brain, the reticular formation, the vestibular nuclei, the cerebellar vermis, and the deep cerebellar nuclei are most often affected. The cortex is virtually spared with the exception of the motor cortex along the precentral gyrus.

Poliovirus does not multiply in muscle *in vivo*. Its chief site of action is in the neuron, and the changes that occur in peripheral nerves and voluntary muscles are secondary to destruction of the nerve cell. Changes occur rapidly in nerve cells, from mild chromatolysis to neuronophagia and complete destruction. Cells that are not killed but that lose their function temporarily as a result of edema may recover completely. Inflammation occurs secondary to the attack on the nerve cells; the focal and perivascular infiltrations are chiefly lymphocytes, with some polymorphonuclear cells, plasma cells, and microglia. In addition to pathological changes in the nervous system, hyperplasia and inflammatory lesions of lymph nodes and of Peyer's patches and other lymph follicles in the intestinal tract are frequently observed; interstitial infiltration of the myocardium with leukocytes is common, but necrotizing myocarditis is rare.

The pathogenesis of the nonpolio enteroviruses is similar in the initial stages of infection of the patient, but the target organs vary.[61] For example, some of these viruses may infect the CNS to cause meningitis or paralysis (e.g., echovirus 4, enterovirus 71), whereas others may infect the heart muscle (e.g., coxsackievirus B group) and even the pancreas.

In contrast to other enteroviruses, enterovirus 70 is easily spread by fomites as well as by direct inoculation of conjunctiva from contaminated fingers. Incubation periods are relatively short (12 to 72 hr). The virus replicates preferentially at 33 to 35°C, an adaptation to conjunctival temperatures instead of the higher temperatures of the gut.

7.2. Immunity

Because of the severity of disease caused by poliovirus, immunity to this enterovirus has received most attention. Immunity to the poliovirus type causing the infection is permanent. There may be a low degree of heterotypic resistance induced by infection, especially between type 1 and type 2 polioviruses. This may account for the observation that second attacks of polio have most often involved types 1 and 3.

Passive immunity is transferred from mother to offspring. The maternal antibodies gradually disappear during the first 6 months of life. Passively administered antibody lasts only 3 to 5 weeks.

Virus-neutralizing antibody forms within a few days after exposure to the virus, often before the onset of illness, and persist, apparently, for life.[372] Its formation early in the infection is a result of viral multiplication in the intestinal tract and deep lymphatic structures before invasion of the nervous system. Since antibodies must be present in the blood to prevent the dissemination of virus to the brain and are not effective after this has already occurred, immunization is of value only if it precedes the onset of symptoms referable to the nervous system.

A decrease in resistance to poliovirus accompanies removal of tonsils and adenoids. Preexisting secretory antibody levels in the nasopharynx decrease sharply following operation (particularly in young male children) without any change in antibody levels in serum. Local antibody levels remain low or absent for as long as 7 months. In susceptible seronegative children nasopharyngeal antibody response to polio vaccine develops significantly later and to lower titers in children previously tonsillectomized than in those with intact tonsils. Thus, surgery of this type may eliminate a valuable source of immunocompetent tissue of importance in resistance to poliovirus.

The type of specificity of maternal antibodies has been studied in the laboratory.[282] In cross-protection tests in infant mice born of immunized mothers, coxsackieviruses show the same type specificity as that observed in neutralization (Nt) and complement-fixation (CF) tests. The immunity conferred by mother's milk also proved to be type specific. In humans, also, a passive transfer of neutralizing and CF antibodies from the mother to the offspring occurs.

Circulating serum antibody against enteroviruses is not the only source of protection against infection. The nature of the so-called local or cellular immunity, which is manifested by protection against intestinal reinfection after recovery from a natural infection or after immunization with the live polio vaccine, has not been completely elucidated.

Local or secretory IgA is increasingly recognized as having an important role in defense against enteroviral infections.[351,353,399] The development of both serum and secretory antibody responses to orally administered live polio vaccine and to intramuscular inoculation of killed polio vaccine is shown in Fig. 5.[352]

Enteroviral infections carry an increased risk for persons with deficiencies in either humoral or cell-mediated immunity. In such persons, poliovirus infection may develop in an atypical manner, with an incubation period longer than 28 days, a high mortality rate after a long chronic illness, and unusual lesions in the CNS.[86] A few such persons are included among the vaccine-associated paralytic poliomyelitis cases that have been reported in the United States (see Section 9.2.2). In addition to the devastating consequences to some of these individuals after infection by polioviruses, several studies have documented a number of persistent or fatal infections of immunodeficient persons by echoviruses of several types (echovirus types 30, 19, 9, 33, and 11).[476,521] In most of the patients in these studies, the chief deficit was in the B-cell functions associated with humoral immunity, and the T-cell function was normal (or only secondarily defective). Some patients had been regularly treated from birth into the midteens by administration of human immune serum globulin. A prominent feature of the echo-virus infections was the patients' inability to eradicate the virus from the CSF; some continued to yield virus for up to 3 years.[164,253] The nature and specificity of T-cell-mediated responses in patients to several enteroviruses has been investigated.[133]

The mechanisms involved in viral persistence are not well understood, and *in vitro* systems can provide a useful approach to the identification and study of factors regulating viral persistence. Persistent *in vitro* virus infections have been categorized into two major groups: those in which all of the cells are infected and those in which only a small fraction of cells is infected. A steady-state infection could be established by echovirus 6 (prototype strain D'Amori) in cloned cultured human WISH cells.[127] The infection was maintained for 3 years (more than 125 passages), and essentially all of the cells were infected and expressed viral antigens. Only one other enterovirus, type 72 (hepatitis A), is known to produce a similar prolonged infection.[437,456] Immune serum was not required either to establish or to maintain the echovirus 6 persistence in the cultures, and the infection could not be cured by treatment with excess antibody. Defective virus particles were produced in excess during this persistence, but their role is not yet clear.

8. Patterns of Host Response and Diagnosis

8.1. Clinical Syndromes

Tables 10–14 list the prototype strains of the known enteroviruses together with the illness (if any) in the person yielding the prototype virus. Sections 8.1.1 through 8.1.12 describe the various host responses to enteroviruses including the major clinical manifestations associated with enteroviruses and the serotypes most frequently associated with them. It should be noted that Section 8.1.1—Asymptomatic Infections—is included to emphasize the frequency with which no apparent illness is manifested by the individual infected with an enterovirus. The listing of clinical syndromes in the subsequent sections is not exhaustive, since additional serotypes have been sporadically associated with other syndromes. Further details concerning enteroviral diseases may be found in several reviews on the role of nonpolio enteroviruses in human disease[10,61,62,141,205,304,324,326] and in reviews covering specialized topics such as viral myocardiopathies,[195,227,237,238,481] coxsackievirus infections of newborns,[124,187] virus diseases associated with cutaneous eruptions.[473] respiratory disease viruses,[184] congenital malformations associated with maternal viral infections,[30] and the possible role of viruses in diabetes mellitus.[18,76,347,348]

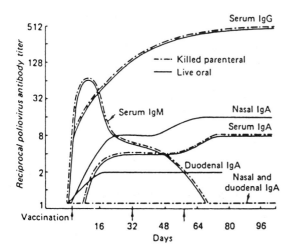

Figure 5. Serum and secretory antibody response to oral administration of live attenuated polio vaccine and to intramuscular inoculation of killed polio vaccine. From Ogra *et al.*,[352] with permission.

Table 10. New Enterovirus Types

Type	Prototype strain	Geographic origin	Illness in person yielding prototype virus	Investigator(s)
68	Fermon	California	Lower respiratory illness (pneumonia and bronchiolitis)[a]	Scheible et al.[422]
69	Toluca-1	Mexico	None[b]	Rosen et al.[396]
70	J670/71	Japan and Singapore	Acute hemorrhagic conjuctivitis (AHC)[c]	Kono et al.,[217] Yin-Murphy and Lim,[510] Mirkovic et al.[315]
71	BrCr	California	Meningitis[d]	Schmidt et al.[426]

[a]Prototype isolated from throat swabs.
[b]Prototype isolated from rectal swabs.
[c]Prototype isolated from conjunctival swabs.
[d]Prototype virus recovered from stool; an identical strain was isolated from the brain of a fatal encephalitis case in the same local outbreak of CNS disease, which occurred in California in 1970. Strains of the same serotype have since been isolated from patients during epidemics of polioencephalitis, meningitis, and hand, foot, and mouth disease.[31,63,64,89,144,202,300,317,333,436,440]

8.1.1. Asymptomatic Infections. Despite the special situations that may hold true for the very young,[61] by far the most common form of infection by any of the enteroviruses is asymptomatic or is manifest by no more than minor malaise. This is true not only for poliovirus infections but also for infections by coxsackieviruses, echoviruses, and the newer enterovirus serotypes. However, careful observation of symptoms may reveal clinical manifestations to occur more frequently than has been generally recognized,[61] and serious clinical syndromes can be associated with many of these agents. The more common of these are discussed in the following sections.

8.1.2. Poliomyelitis. When an individual susceptible to infection is exposed to poliovirus, one of the following responses may occur: (1) inapparent infection without symptoms, (2) mild (minor) illness, (3) aseptic meningitis, or (4) paralytic poliomyelitis. As the disease progresses, one response may merge with a more severe form, often resulting in a biphasic course: a minor illness followed first by a few days free of symptoms and then by the major, severe illness. Only about 1% of infections result in recognized clinical illness.

8.1.2a. Abortive Poliomyelitis. Abortive poliomyelitis is the most common form of the disease. The patient has only the minor illness, characterized by fever, malaise, drowsiness, headache, nausea, vomiting, constipation, or sore throat in various combinations. The patient recovers in a few days. The diagnosis of abortive poliomyelitis cannot be made with assurance, even during an epidemic, except when the virus is isolated or antibody development is measured.

8.1.2b. Nonparalytic Poliomyelitis (Aseptic Meningitis). In addition to the aforementioned symptoms and signs, the patient with the nonparalytic form presents stiffness and pain in the back and neck. The disease lasts 2–10 days, and recovery is rapid and complete. In a small percentage of cases, the disease advances to paralysis. Poliovirus is only one of many viruses that produce aseptic meningitis.

8.1.2c. Paralytic Poliomyelitis. In the absence of virological diagnosis, poliovirus must be suspected if disease occurs in persons associated with paralytic patients, since paralysis is rare in other enterovirus infections. In poliomyelitis, the major illness, when it occurs, may follow the minor illness described above, particularly in young children, but it usually occurs without the antecedent first phase. The predominating sign is flaccid paralysis resulting from lower motor neuron damage. How-

Table 11. Polioviruses[a]

Type	Prototype strain	Geographic origin	Illness in person yielding prototype virus	Investigator(s)
1	Brunhilde	Maryland	Paralytic polio[b]	Howe and Bodian
2	Lansing	Michigan	Fatal paralytic polio[c]	Armstrong
3	Leon	California	Fatal paralytic polio[c]	Kessel

[a]From Melnick et al.[304] Used with permission.
[b]Virus recovered from feces.
[c]Virus recovered from spinal cord.

Table 12. Group A Coxsackieviruses[a,b]

Type	Prototype strain	Geographic origin	Illness in person yielding prototype virus[c]	Investigator
1	Tompkins	Coxsackie, NY	Poliomyelitis[d]	Dalldorf
2	Fleetwood	Delaware	Poliomyelitis[d]	Dalldorf
3	Olson	New York	Meningitis	Dalldorf
4	High Point	North Carolina	Sewage of polio community	Melnick
5	Swartz	New York	Poliomyelitis	Dalldorf
6	Gdula	New York	Meningitis	Dalldorf
7	Parker	New York	Meningitis	Dalldorf
8	Donovan	New York	Poliomyelitis	Dalldorf
9	Bozek	New York	Meningitis	Dalldorf
10	Kowalik	New York	Meningitis	Dalldorf
11	Belgium-1	Belgium	Epidemic myalgia	Curnen
12	Texas-12	Texas	Files in polio community	Melnick
13	Flores	Mexico	None	Sickles
14	G-14	South Africa	None	Gear
15	G-9	South Africa	None	Gear
16	G-10	South Africa	None	Gear
17	G-12	South Africa	None	Gear
18	G-13	South Africa	None	Gear
19	NIH-8663	Japan	Guillain-Barré syndrome	Huebner
20	IH-35	New York	Infectious hepatitis	Sickles
21	Kuykendall; Coe	California	Poliomyelitis,[d] mild respiratory disease[e]	Lennette
22	Chulman	New York	Vomiting and diarrhea	Sickles
24	Joseph	South Afriaca	None	Gear

[a]From Melnick et al.[304] Used with permission.
[b]Cross-reactivity has been observed between A3 and A8, A11 and A15, and A13 and A18.
[c]All isolates were from stools, except for prototypes of A4 and A12, which were isolated from sewage and flies, as indicated. Numerous strains of each of these types were isolated from stools also.
[d]When coxsackieviruses have been isolated from patients with paralytic poliomyelitis, the patient has often been found to have a dual infection, the polioviruses presumably being responsible for the paralytic illness.
[e]The Coe virus was isolated from throat washings.

ever, incoordination secondary to brainstem invasion and painful spasms of nonparalyzed muscles may also occur. The amount of damage and destruction varies from case to case. Muscle involvement is usually maximal within a few days after the paralytic phase begins. The maximal recovery usually occurs within 6 months, but it may take longer.

At times, nonpolio enteroviruses have been associated with cases of poliolike paralytic disease, but this has been uncommon.[123,140] Coxsackievirus A7 has been associated with outbreaks of paralytic disease,[139,463] and enterovirus 71 has been involved in several outbreaks of CNS disease, including poliolike paralysis, with some fatal cases.[269]

Table 13. Group B Coxsackieviruses[a]

Type	Prototype strain	Geographic origin	Illness in person yielding prototype virus[b]	Investigator
1	Conn-5	Connecticut	Meningitis	Melnick
2	Ohio-1	Ohio	Summer grippe	Melnick
3	Nancy	Connecticut	Minor febrile illness	Melnick
4	JVB	New York	Chest and abdominal pain	Sickles
5	Faulkner	Kentucky	Mild paralytic disease with residual atrophy	Steigman
6	Schmidt	Philippine Islands	None	Hammon

[a]From Melnick et al.[304] Used with permission.
[b]All isolates were from stools.

Table 14. Echoviruses^{a,b}

Type	Prototype strain	Geographic origin	Illness in person yielding prototype virus^c	Investigator(s)
1	Farouk	Egypt	None	Melnick
2	Cornelis	Connecticut	Meningitis	Melnick
3	Morrisey	Connecticut	Meningitis	Melnick
4	Pesascek	Connecticut	Meningitis	Melnick
5	Noyce	Maine	Meningitis	Melnick
6	D'Amori	Rhode Island	Meningitis	Melnick
6′	Cox	Ohio	None	Ramos-Alvarez, Sabin
6″	Burgess	Connecticut	Meningitis	Melnick
7	Wallace	Ohio	None	Ramos-Alvarez, Sabin
8	Bryson	Ohio	None	Ramos-Alvarez, Sabin
9	Hill	Ohio	None	Ramos-Alvarez, Sabin
11	Gregory	Ohio	None	Ramos-Alvarez, Sabin
12	Travis	Philippine Islands	None	Hammon, Ludwig
13	Del Carmen	Philippine Islands	None	Hammon, Ludwig
14	Tow	Rhode Island	Meningitis	Melnick
15	CH 96-51	West Virginia	None	Ormsbee, Melnick
16	Harrington	Massachusetts	Meningitis	Kibrick, Enders
17	CHHE-29	Mexico City	None	Ramos-Alvarez, Sabin
18	Metcalf	Ohio	Diarrhea	Ramos-Alvarez, Sabin
19	Burke	Ohio	Diarrhea	Ramos-Alvarez, Sabin
20	JV-1	Washington, DC	Fever	Rosen
21	Farina	Massachusetts	Meningitis	Enders, Kibrick
22	Harris	Ohio	Diarrhea	Sabin
23	Williamson	Ohio	Diarrhea	Sabin
24	DeCamp	Ohio	Diarrhea	Sabin
25	JV-4	Washington, DC	Diarrhea	Rosen
26	Coronel	Philippine Islands	None	Hammon
27	Bacon	Philippine Islands	None	Hammon
29	JV-10	Washington, DC	None	Rosen
30	Bastianni	New York	Meningitis	Plager, Duncan, Lennette
31	Caldwell	Kansas	Meningitis	Wenner, Lennette, von Magnus
32	PR-10	Puerto Rico	Meningitis	Branche
33	Toluca-3	Mexico	None	Rosen, Kern
34	DN-19^d	Texas	Infantile diarrhea	Melnick

^{a}From Melnick *et al.*[304] Used with permission.
^{b}Types 1 and 8 share antigens, type 1 having the broader spectrum.
^{c}All isolates were from stools.
^{d}DN-19 antiserum partially neutralizes coxsackievirus A24, but A24 antiserum does not neutralize DN-19 virus, although it reacts with the virus in CF and gel diffusion tests. Thus, DN-19 should be considered a prime strain of coxsackievirus A24, rather than as a distinct echovirus.

8.1.2d. Progressive Postpoliomyelitis Muscle Atrophy. Finally, a recrudescence of paralysis and muscle wasting has been repeatedly observed in individuals decades after their experience with paralytic poliomyelitis. This phenomenon had been first observed a century ago, but only in recent years have there been sequential studies to follow the progression of neurological deficit. Many of the recently reported cases have been in the United States, where there are approximately 300,000 persons with a history of poliomyelitis. Progressive postpoliomyelitis muscle atrophy (PPMA) is rare, and more frequent causes

of neurological deterioration in these patients arise from the use of crutches or wheelchairs or from spinal deformity.[98] Also, it should be kept in mind that patients with a history of poliomyelitis are no less susceptible to unrelated neuromuscular disorders than the rest of the population.

Nonetheless, PPMA is a specific identifiable syndrome, and there are important questions still to be answered regarding the pathogenesis of these late sequelae. Rather than being a consequence of persistent infection, this later progression of paralysis is thought to result from physiological and aging changes, probably exacerbated by the abnormal burdening of some neuromuscular functions during recovery from the acute disease.[191] The normal decline in anterior horn cells and in muscle strength that occurs with aging is much more pronounced in persons who lost many anterior horn cells and perhaps 50% of their muscle strength when stricken in their acute bout with polio. However, loss of anterior horn cells as a result of aging seldom occurs before age 60, and most of the PPMA cases reported have occurred before this age. Also, there are findings suggesting that rather than loss of anterior horn cells, the syndrome may result from peripheral disintegration of individual nerve terminals in motor units that were reinnervated during recovery following the acute disease. This could occur if the cell bodies lost ability to maintain the metabolic demands of their pathologically increased number of terminal sprouts, an ability that may decline with age.[98]

The intrathecal presence of IgM antibodies to poliovirus has been reported in patients with the postpolio syndrome.[434] This led to the suggestion that the late muscle weakness was caused by a persistent infection of the anterior horn cells of the spinal cord. However, confirmation of this finding is lacking.[254]

8.1.3. Meningitis and Mild Paresis. Fever, malaise, headache, nausea, and abdominal pain are common early symptoms. One to two days later there may be signs of meningeal irritation with stiffness of the neck or back; vomiting may also appear. The disease sometimes progresses to mild muscle weakness that is often confused clinically with paralytic poliomyelitis. Patients almost always recover completely from nonpoliovirus paresis. However, with a number of enteroviruses, there is a risk of serious neurological sequelae among infants infected during their first year of life.[431]

As indicated above, poliovirus infection often is manifested as meningitis or as transient mild paresis. In addition to the polioviruses, almost all coxsackieviruses of both A and B groups, as well as most echoviruses, have been associated to some degree with meningitis and (in very rare instances) with paralytic CNS disease. However, the chief types associated with CNS disease among coxsackieviruses are B1–B6, A7, and A9. A recent report has described a fatal case of meningoencephalitis in which coxsackievirus B5 was clearly implicated not only by serology but also by *in situ* nucleic acid hybridization.[146]

Echoviruses 4, 6, 9, 11, 14, 16, 25, 30, 31, and 33 have been repeatedly associated with meningitis; types 3, 18, and 29 have also been responsible for some outbreaks of this syndrome. Other types have been associated with meningitis only in sporadic cases. With echoviruses 6 and

9, muscle weakness and mild transient paralysis have been observed; echovirus 9 has been recovered in high titer from the medulla of a fatal case.

Among the newer enteroviruses, type 70—the agent of acute hemorrhagic conjunctivitis—in rare instances has been involved in neurological complications including poliomyelitislike illness.[215,216]

An important feature of infections with enterovirus 71 has also been meningitis as well as some cases of more severe CNS disease. As illustrated in Table 2 (Section 5.2.2), this virus has exhibited a variety of clinical manifestations in different regions of the world and at different times.[269] In the California outbreak from which the prototype strain was reported, meningitis predominated, but there were other CNS manifestations including a fatal encephalitis case,[426] and various large outbreaks have been documented that include significant numbers of severe CNS manifestations, some of the cases being fatal (see Table 2). In view of the variety of the illness caused by enterovirus 71 and the difficulty of isolating some of the strains of this virus, it is probable that only a small fraction of the cases that actually occur are ever diagnosed and reported.[269]

The associations of nonpolio enteroviruses with neurological disease have been tabulated by Grist *et al.*[141] and are shown in Table 15. Reviews by Cherry[61,62] and others[10] also provide further details of recent observations.

8.1.4. Pleurodynia (Epidemic Myalgia, Bornholm Disease). Pleurodynia is generally caused by group B coxsackieviruses, rarely by echoviruses (notably types 6 and 9). Fever and chest pain are almost invariably present together; they are usually abrupt in onset, but are sometimes preceded by malaise, headache, and anorexia. The chest pain may be located on either side or substernally, is intensified by movement, and may last from 2 days to 2 weeks. Abdominal pain resulting from involvement of the diaphragm occurs in approximately half the cases; in children, this often takes the place of chest pain and may be the chief complaint. The illness is self-limited, and recovery is complete, although relapses are common.

Coxsackieviruses A have been associated with Bornholm disease less often; the serotypes involved have been A4, A6, and A10. A9 has been associated with this syndrome and also with chronic diseases of muscles and joints.[141] Several reports of pseudocrystalline arrays of picornaviruslike particles in the myocytes of patients with chronic muscle diseases suggest that enteroviruses might have a role in these diseases.[143]

8.1.5. Herpangina. Herpangina is caused chiefly by coxsackievirus group A types 1, 2, 3, 4, 5, 6, 8, 10, and

Table 15. Neurological Disease Associations of Coxsackieviruses, Echoviruses, and "New" Enteroviruses[a]

Syndrome or clinical feature	Virus types[b,c]
Meningitis	Coxsackievirus A**1**, 2, 3, **4**, **5**, **6**, **7***, 8, 9*, 10, 11, **14**, **16**, 17, 18, **22**, 24
	Coxsackieviruses B1*, **2***, **3***, **4***, **5***, 6
	Echoviruses 1, **2***, **3***, **4***, 5, **6**, **7***, **9***, **11***, 12, 13, **14***, 15, **16***, **17***, **18***, **19***, 20, 21, 22, **23 25***, 27, **30***, **31**, 33*
	Enterovirus **71***
Paralytic disease	Coxsackieviruses A4, 6, 7, 9, 11, 14, 21
	Coxsackieviruses B1, **2**, 3, 4, 5, 6
	Echoviruses 1,**2**, 3, **4**, 6, 7, **9**, **11**, 14, **16**, 18, **19**, **30**
	Enterovirus 70, **71**
Encephalitis	Coxsackieviruses A2, 5, 6, 7, **9**
	Coxsackieviruses B1, 2, **3**, 5, **6**
	Echoviruses 2, 3, 4, **6**, 7, **9**, 11, 14, **17**, 18, **19**, 25
	Enterovirus **71**
Ataxia	Coxsackieviruses A4, 7, 9
	Echovirus **9**

[a]From Grist *et al.*[141] Used with permission.
[b]Boldfaced figures indicate virus isolation from cerebrospinal fluid or other parenteral source.
[c]Asterisk indicates outbreak reported with this type.

22.[141] The illness is characterized by an abrupt onset of fever and sore throat. There may be anorexia, dysphagia, vomiting, and abdominal pain. The pharynx is usually hyperemic, and a few (not more than 10–12) characteristic tiny discrete vesicles with a red areola occur on the anterior pillars of the fauces, the posterior pharynx, the palate, uvula, tonsils, or tongue. The illness is self-limited and occurs most frequently in small children.

8.1.6. Hand, Foot, and Mouth Disease. Hand, foot, and mouth disease has been associated particularly with coxsackievirus A16—which continues to predominate as the cause of numerous outbreaks—but A4, A5, A9, and A10 have also been implicated, as well as coxsackievirus B2 and B5. The almost universal sign is enanthem, most commonly on the buccal mucosa. The exanthem soon follows, most frequently on the hands or feet. The intraoral lesions are ulcerative; the lesions on hands and feet are usually vesicular.

As illustrated in Table 2 and discussed in Section 8.1.3, enterovirus 71 has also caused a number of outbreaks of this syndrome since this virus made its appearance in the 1970s.[269] Virus may be recovered not only from the stool and pharyngeal secretions but also from vesicular fluid. A combined syndrome has also been reported in which vesicular lesions and pneumonia are both present.

8.1.7. Respiratory Illness. Numerous enterovirus serotypes have been associated with mild upper respiratory illness; among these are coxsackieviruses A2, A9, A10, A21, A24, and B2–B5.[83,324] Coxsackieviruses have also been implicated in mild lower respiratory infections, particularly in young children. Coxsackievirus A21 has also caused outbreaks of pharyngitis in military recruits and has induced respiratory tract disease in normal adult volunteers. There have been reports, although rare, of fatal pneumonia caused by coxsackievirus A7, in which the virus has been isolated from the lung postmortem.

Echoviruses have also been isolated in association with respiratory illnesses; these include types 1, 9, 11, 19, 20, and 22.[61,261,326] Except in epidemic conditions, a clear-cut causative association of the virus with the illness often cannot be established. The most certain association is with outbreaks in very young children, in whom serious or even fatal lower respiratory tract disease may be involved.

Echovirus 11 frequently caused upper and lower respiratory illness without meningitis in 1982 in Singapore, at a time when echovirus 11 outbreaks in other parts of the world were causing meningitis.[511] It is noteworthy that in the 5-year period prior to the outbreak, the Singapore laboratory had isolated echovirus 11 only once from 1370 clinical specimens. In 1982, the year after the outbreak, 14 of 18 isolates were echovirus 11. The virus was isolated from four fecal samples, four throat swabs, and six nasopharyngeal aspirates, indicating different host responses from that in other parts of the world, probably due to a mutation in the virus.

In a 2-year study of virus isolations from throat swabs of young children with respiratory illnesses who attended a health care unit in a poor section of Rio de Janeiro, enteroviruses were second only to adenoviruses in frequency of isolation. They constituted 25% of the agents found[444] and were apparently most frequent in the younger infants. The serotypes represented among these isolates were echovirus types 6, 7, 9, 11, 17, 18, and 21, coxsackieviruses A7 and B4, and poliovirus (vaccine strain); two unidentified enteroviruses also were among the isolates. Enterovirus 68 has been associated with pneumonia and bronchiolitis in children,[422] and enterovirus 71 with an influenzalike disease.[440]

Respiratory illness induced by enteroviruses cannot be differentiated on clinical grounds from that caused by the viruses more commonly considered to be respiratory tract agents, such as rhinoviruses, parainfluenza viruses, respiratory syncytial virus, or adenoviruses. However, infections with the latter groups of viruses occur most frequently during the winter months, whereas respiratory illness caused by enteroviruses follow the typical enterovirus seasonal pattern of summer and early fall.

8.1.8. Eye Disease. Infections with some coxsackieviruses and echoviruses have been accompanied by conjunctivitis. During an epidemic in Sweden, echovirus type 7 was isolated from conjunctivae[418]; in sporadic cases, other enteroviruses (coxsackievirus B2, echovirus 11, poliovirus 1 and 3) have been recovered from eye specimens. In 1970, however, a large epidemic of acute conjunctivitis occurred in Singapore, with 60,000 cases reported; the agent was identified as a variant of coxsackievirus A24.[240,316] Coxsackievirus A24 outbreaks were also reported from Hong Kong in 1971,[508,509] in 1975 from Singapore,[241] and Hong Kong,[58] and in Taiwan, where outbreaks occurred in 1985, 1988, and 1989.[242] A molecular analysis of coxsackievirus A24 isolates suggested that the variant causing hemorrhagic conjunctivitis was present in Taiwan before the outbreaks. The variant appeared to be derived from a common progenitor prevalent in Taiwan in 1982.[181]

The first outbreak of coxsackievirus A24 hemorrhagic conjunctivitis outside of Southeast Asia occurred in American Samoa in 1986, when 47% of the population were affected. Five years earlier the area had been hit by a similar clinical epidemic but one caused by enterovirus 70[419] (see below).

This coxsackievirus A24 variant characteristically induces mild to severe conjunctivitis, but only in a minority of cases does subconjunctival hemorrhage occur; recovery is usually complete after 1–2 weeks. The virus was isolated from conjunctival swabs or scrapings and also from throat swabs.

In the same period as the first outbreaks of coxsackievirus A24 conjunctivitis were taking place, during 1969–1971, a pandemic of a different form of conjunctivitis was occurring in Africa, Southeast Asia (including Singapore), Japan, and India, involving tens of millions of people. The cause of this widespread disease, termed AHC, was recognized as a new enterovirus[217,510] and was designated enterovirus type 70.[302,315]

Through 1980, virtually no enterovirus 70 infection or disease had been found in Australia and the Americas. However, this changed in 1981, after AHC reappeared in some Asian and African countries from which it had been absent for a number of years. This time the disease spread extensively in the Caribbean, northern South America, and Central America, and three outbreaks were reported in the continental United States as well as one each in Hawaii, American Samoa, and Canada. Extensive study

has been made of an explosive outbreak that occurred in early autumn, 1981, in the Miami metropolitan area.[365]

In 1982, sporadic outbreaks were reported from French Polynesia and other Pacific islands. Later that year, the virus was imported from French Polynesia to Minnesota, where it caused a small outbreak limited to the importer and five other persons who shared a house with her.[222] Even though the importer, a 27-year-old woman, had recovered from the disease before she left Tahiti, she was still able to transmit the virus for at least 8 days after the onset of her symptoms. No other cases were reported that year in the entire United States.

The disease is generally localized to the eye; symptoms include severe eye pain, photophobia, and blurred vision. There is characteristic subconjunctival hemorrhage ranging from discrete petechiae to large blotches of frank hemorrhage covering the bulbar conjunctiva. Corneal involvement, in the form of epithelial keratitis, may occur but is transient. The incubation period is about 24 hr, onset is sudden, and recovery is usually complete within less than 10 days.

Enterovirus 70 has been found almost exclusively in conjunctival and throat specimens, but a few fecal isolations have been reported.[315]

Rarely, neurological complications (acute lumbar radiculomyelopathy, cranial nerve involvement) have been observed.[188,464] Motor paralysis resembling poliomyelitis is the most striking feature, and in some patients (about 1 of 10,000) there is residual paralysis and muscle atrophy. The neurological symptoms develop as much as several weeks after the onset of AHC.[216] This neurological aspect has been reported chiefly in adult males, in contrast to the age pattern of poliomyelitis in most of the populations affected, in which poliomyelitis remains a disease of early childhood. Worldwide, about 100 cases have been reported—a small number among tens of millions of AHC cases. Undoubtedly, cases involving the CNS have gone unreported because they were unrecognized as being related to the AHC episode that occurred some weeks earlier. In any case, the potential for neurovirulence calls for careful attention to any outbreak of enterovirus 70 infection.[215,216] The neurological phase typically begins about 2 weeks after the conjunctivitis, with an acute, hypotonic, areflexic, asymmetric proximal paralysis of the lower limbs, accompanied by fever, malaise, and nerve root pains.[95]

8.1.9. Cardiac Diseases. The etiologic role of the enteroviruses in acute myocarditis and pericarditis is well established, and there are new indications that some chronic cardiovascular diseases may result from coxsackievirus B infections[17,210,377,470] (see also Section 8.1.10).

8.1.9a. *Acute Myocardiopathy and Pericardiopathy.* Coxsackievirus B infections are increasingly recognized as a cause of primary myocardial disease in adults as well as in children.[17,195,227,237] In some series, up to 39% of persons infected with coxsackievirus B5 developed cardiac abnormalities. Coxsackieviruses of group A and echoviruses have also been implicated, but to a lesser degree. In one study of patients with a clinical diagnosis of pericarditis (148 patients), myocarditis (92 patients), or pleurodynia (19 patients), 27% had IgM antibody to one of the group B coxsackieviruses—indicative of current or recent infection—as compared with 8% of the control group.[427]

By means of an ELISA "antibody capture" technique in which a solid phase was coated with antihuman IgM, coxsackievirus-B-specific IgM was demonstrated in 24 of 64 patients with acute pericarditis and 14 of 38 patients with acute myocarditis.[100] On the other hand, another study[151] on patients with myocardial infarction revealed no significant differences between patients and controls in regard to coxsackie B IgM antibodies.

In another study,[230] 910 cardiac patients and 776 noncardiac patients were compared for evidence of coxsackie B virus infections. Of 78 cardiac patients with coxsackie B infection, 30 had pleurodynia, 18 myocarditis, and 20 pericarditis. Of 69 noncardiac patients with coxsackie B infection, 27 presented with pyrexia of unknown origin, 14 with upper respiratory tract infection, and nine with meningitis. Among the cardiac patients, the highest percentage with coxsackie B virus infection was in the 20- to 39-year age group, whereas among the noncardiac patients, it was in the under-19-year age group.

Evidence for a high degree of association of virus with disease has been obtained, usually at autopsy, by demonstration of virus localized in the myocardium, endocardium, and pericardial fluid; the presence of virus at the sites of pathological changes has been demonstrated by immunofluorescence, peroxidase-labeled antibody, or ferritin-labeled antibody. It has been estimated that about 5% of all symptomatic coxsackievirus infections induce heart disease. The virus may affect the endocardium, pericardium, myocardium, or all three. Acute myocardiopathies have been shown to be caused by coxsackieviruses A4, A14, A16, B1–B5, and others, and also by echovirus types 9 and 22.

Monkeys infected with coxsackievirus B4 develop pancarditis with a pathological picture strikingly similar to that of rheumatic heart disease. Myocarditis and cell-mediated immunity also have been induced in primates (*Papio papio*) infected with coxsackievirus B3.[363] In experimental animals, the severity of acute viral myocar-diopathy is greatly increased by vigorous exercise, hydrocortisone, alcohol consumptions, pregnancy, and undernutrition and is greater in males than in females. In human illnesses also, these factors may affect the severity of the disease.

Studies in mice to compare a myocarditic parent strain of coxsackievirus B3 with variants of this strain suggest that the extent of myocarditis induced is affected by some subtle expression of the genome, probably not involving capsid polypeptides. Pathogenicity is influenced also by the haplotype and sex of specific mouse host species.[120]

8.1.9b. *Chronic Cardiovascular Disease.* After acute coxsackievirus carditis, lasting heart damage has been reported with viral antigen, detectable by immunofluorescence, persisting in diseased tissue. These antigens have been reported in diseased cardiac[43] and other tissues,[42] suggesting virus persistence in the cardiovascular system.

Because the direct detection of infectious virus in cardiac disease is difficult, the recent use of newer methods for detection of enteroviruses has brought the disease into clearer focus. These methods include the detection of enteroviral nucleic acid and the use of synthetic and recombinant antigens.[17,175,328] However, in addition to direct destruction of infected cells by enteroviruses, there is evidence that the host response may lead to tissue injury. Thus, some regard the coxsackievirus-induced myocarditis as the result of an autoimmune response.[175]

In serological studies a persisting antigenic stimulus is suggested. Detailed follow-up of two immunocompetent patients with recurrent pericarditis indicated prolonged IgM responses specific for coxsackievirus B. One patient showed an extremely high neutralizing antibody titer to coxsackievirus B5 in 1969 and, starting in 1973, yielded positive specific IgM results on three of four occasions over a period of 6 years. The other patient yielded positive IgM responses to coxsackievirus B4 in six of seven tests over a 2-year period. These findings suggest that consideration should be given to the role of coxsackieviruses and other picornaviruses in inducing persistent infections in humans.[17,328,449]

Animal studies have also been revealing. Findings in weanling mice experimentally infected with coxsackievirus B3 suggest that continuing inflammation observed following infection with this virus is an immunopathologic process[477] resulting from the cytotoxicity of immune spleen cells (T lymphocytes) against virus-infected cardiac cells.[480] A continuing cardiac myonecrosis is the ultimate product of the sensitized lymphocytes.[236,480]

Studies of experimental infection with coxsackievirus B3 have been conducted in outbred mice with defective T-independent B lymphocytes. Some of these mice are "nude mice," which lack a thymus, and others possess a thymus. Athymic mice did not efficiently eliminate infectious virus from their hearts, and the persistent infection was associated with myocardial histopathology and increased mortality rates. Mice with a thymus were able to clear the virus (which in these B-defective animals would be a T-cell-mediated function), and yet many of these animals had continuing myocardial pathology. Thus, there seem to be at least two mechanisms whereby coxsackievirus B3 can cause myocardial histopathology—one dependent on full thymic function, and one dependent on viral persistence.[428]

8.1.10. Neonatal Disease. A number of reviews have dealt with neonatal infections by enteroviruses and also with fetal developmental defects in association with maternal enterovirus infections.[30,61,124,187,205,324] Neonatal infection with group B coxsackieviruses may be acquired transplacentally but more commonly as a contact infection in the newborn nursery. The range of response is from inapparent infection to severe and even fatal disease.[197,224,337] In the symptomatic infant, onset may be marked by lethargy, feeding difficulty, and vomiting with or without fever. In severe cases, myocarditis or pericarditis or both may develop within the first 8 days of life. Cardiac and respiratory embarrassment are indicated by tachycardia, dyspnea, cyanosis, and changes in the electrocardiogram. The clinical course may be rapidly fatal, or the patient may progress to complete recovery. Myocarditis has also been caused by some group A coxsackieviruses and echoviruses.

Coxsackievirus A serotypes have rarely been reported with neonatal illnesses, but echoviruses have been reported frequently in both nursery outbreaks and sporadic infections. Among the echovirus types incriminated are echoviruses 4, 7, 9, 11, 17, 18, 19, 20, 22, and 31. Outbreaks in special-care baby units have included fatal infections with echovirus 7 and, especially, with type 11.[85,332,388,501] Echovirus 11 has also been suspected of association with fetal death.[438]

A recent review of the literature, examining 61 reported cases of neonatal echovirus infections (including 43 caused by echovirus 11) as well as 16 outbreaks in nurseries, has underscored the seriousness of echovirus infections in newborns.[320] Analysis of these cases indicated that if the mother was acutely ill in the 2 weeks before delivery (and particularly in the last week), the illness in the infant was far more likely to be serious or even fatal, in contrast to the milder illnesses if the infant

acquired the infection in the hospital after delivery. Among the vertically infected, severely ill infants, some were judged to have acquired their infection *in utero*, but a larger percentage (63%) gave evidence of having been infected in the immediate perinatal period. Similarly, in the instances in which data were available on this aspect of the nursery outbreaks, the index cases acquired vertically from the mother were more serious than those resulting from secondary spread of the virus in the nursery. The most frequent clinical manifestations in the severe cases reviewed were hepatitis and CNS disease. The frequency of echovirus 11 involvement in these cases may have been influenced in part by the wide circulation of a "prime" strain of echovirus 11 in many parts of the world in the late 1970s.

Results from longitudinal studies in one large New York county[197,224] suggest that the incidence of neonatal disease caused by either echoviruses or coxsackieviruses may be greater than the incidence of perinatally acquired infection caused by herpes simplex virus, an agent believed to affect approximately one infant per 7500 live births or 13 per 100,000.

Enterovirus infection during the summer and fall in the United States is common in neonates. In one prospective study,[187] 75 (13%) of 586 infants, none of whom were found to be excreting nonpolio enteroviruses within 1 day of delivery, acquired infection by a nonpolio enterovirus during the first month of life. Although 79% (59/75) of all enterovirus-positive infants were asymptomatic, 21% (16/75) were ill enough to be readmitted to the hospital with suspected sepsis during their first month. Risk of virus infection among the neonates in this study was associated statistically with lower socioeconomic status and with lack of breast-feeding. Regardless of socioeconomic status, neonates who were not breast-fed were more likely to be infected with an enterovirus but were not more likely to be hospitalized. Among infants who have enterovirus infections in the first month of life, as evidenced by virus isolation, even those who were asymptomatic, at the time of infection may have occult damage to the CNS. In further reports of the prospective study mentioned above, neonatally infected infants were compared with closely matched controls in tests conducted at about 18 months of age. All of the infected infants manifested mild delays in development and in language.

In a study of 602 specimens positive for group B coxsackieviruses during the period 1970–1979,[197] 77 were from infants younger than 3 months of age with nonfatal infections on whom case histories were available. Aseptic meningitis was the most common syndrome (48 of the 77 infants). Eight newborns whose culture specimens were studied during 1970–1981 at the same medical center died of overwhelming group B coxsackievirus disease. These and 33 other fatal cases reported in the literature were reviewed.[197] Three patterns of death were observed: (1) rapid death (12 patients aged 2–27 days); (2) a diphasic illness leading to death at age 8–24 days (11 patients); (3) a progressive illness (18 infants). Myocarditis was present in all the infants who died, and pulmonary hemorrhage and liver necrosis were common (30 and 18 patients, respectively). Among the mothers of these 41 documented fatal cases, 24 had symptomatic illness, mostly mild, in the period from 10 days before to 5 days after their infants' birth. The investigators consider that most fatal cases probably are transmitted transplacentally at term, although infection in the birth canal is also a possibility.

In another recent study, neutralizing antibody to at least one coxsackievirus B serotype was present in ventricular fluids from 4 of 28 newborn infants with severe congenital anatomic defects in the central nervous system. Two of the four did not have detectable levels of the same antibodies in their serum. The mothers of all four infants had serum neutralizing antibody of the same serotype as was found in their infants' ventricular fluid. Of 11 mother–infant pairs who had neutralizing antibody to coxsackievirus B types in both sera, almost half had antibody against more than one serotype. These data suggest a possible association between congenital coxsackievirus B infections and rare severe CNS defects.[121] In newborn and 7-day-old mice, forebrain anomalies (porencephaly or hydranencephaly) resembling those found in human infants can be induced by variants of coxsackievirus B3.[122]

As concerns a possible role of enteroviruses in developmental defects of the fetus, a large prospective study[39] examined enteroviral seroconversions in women during pregnancy in relation to anomalies in their infants. In comparison to matched controls, a higher rate of infection, largely inapparent, was found in women whose offspring were abnormal. There was evidence of an association between maternal infections with coxsackieviruses B2 and B4 and urogenital abnormalities and between coxsackievirus A9 and defects of the digestive system. Cardiovascular anomalies were associated with maternal infections by coxsackievirus B3 and B4; multiple infections with coxsackieviruses during pregnancy increased the likelihood of congenital heart disease in the infant. However, others have found no significant associations of coxsackievirus group B infections with developmental anomalies.[398]

8.1.11. Gastrointestinal Diseases. A number of diseases of the digestive system and related organs have been associated with enterovirus infection.

8.1.11a. Diarrhea. Gastrointestinal upset is commonly reported among associated symptoms in infections by a number of enteroviruses in which other clinical features predominate. In many outbreaks, the enterovirus reported may be merely a passenger virus unrelated to the illness itself. Some enteroviruses (notably echoviruses 1, 2, 4, 6, 7, 11, 14, 18, 19, 20, and 22) have been documented[61,205,324] in relation to outbreaks of diarrhea, but diarrhea was not invariably present in those infected. Echovirus type 20 has been associated with a febrile disease involving both the respiratory and enteric tracts. In an outbreak of gastroenteritis in bone-marrow transplant recipients, however, coxsackievirus A1 was isolated from all seven patients with diarrhea, constituting 50% of the patients in the unit during a 3-week period.[452] A further prospective study of 78 patients in the same unit yielded virus isolations from stools of 22 patients, some of whom were doubly infected. Virus isolates included coxsackieviruses as well as adenoviruses and rotaviruses.[515]

8.1.11b. Hepatitis. Hepatitis A virus, now classified as enterovirus type 72, is one of the most important causes of this disease. Hepatitis is also known to be a part of many severe generalized infections of neonates caused by coxsackieviruses or other enteroviruses.

8.1.11c. Pancreatitis. Pancreatitis may be a part of generalized infections of infants with coxsackievirus B agents. In one study, pancreatic damage was suggested by raised amylase levels in 31% of coxsackievirus B5 infections and 25% of coxsackievirus A9 infections, but not in infections with echovirus types 4 or 6.[335] In experimental infections of mice with coxsackieviruses of the B group, pancreatitis is a long-recognized effect.

8.1.12. Diabetes. Attention has been periodically directed to the possible role of enteroviruses (particularly group B coxsackieviruses) in insulin-dependent type 1 diabetes mellitus on the basis of reports concerning diabetic children and also of studies in experimentally infected animals.[20,69,76,119,155,347] Although the evidence in man is not convincing as yet,[356] coxsackievirus B4 has been isolated from a fatal case of diabetes in a 10-year-old boy, and the disease has been experimentally reproduced in mice with the virus isolated.[517]

In a study of coxsackievirus B antibodies in sera from 166 patients aged 1–17 years with recent-onset insulin-dependent type 1 diabetes mellitus (IDDM), all 166 had a clinical history of recent infectious illness. About 80% had antibodies against at least one coxsackievirus B serotype, but only about 10% of the patients had specific neutralizing IgM antibodies indicating that their infection with a B coxsackievirus had been recent.[99] This study suggests that although coxsackievirus B infection may be one cause for type 1 diabetes, it would explain, at the very most, 10% of the cases.

However, somewhat different conclusions have been drawn by other investigators using the IgM antibody-capture ELISA technique.[18,100,207] Initially, with a small group of patients,[207] they found that 11 of 28 children with type 1 diabetes had evidence of recent infection with coxsackie B viruses, particularly B4, compared with only 16 of 295 matched controls. In a larger study,[18] with 122 patients under age 15 from England, Austria, and Australia, it was found that 37 (30%) of the patients were positive for specific IgM responses to coxsackieviruses B1–5, compared with only 6% of controls. Virus-specific IgM responses were directed against a single serotype, usually coxsackievirus B4 or B5, in 23 of the 37 patients less than 15 years old; among 13 children aged less than 7 years, 10 had monotypic responses. In contrast, there was no evidence of recent infection by mumps, rubella, or cytomegalovirus, since mumps-virus-specific IgM was present in only 1 of 100 children with IDDM and 5 of 139 controls; no rubella- or CMV-specific IgM responses were detected in 60 sera from patients with type 1 diabetes.

8.1.13. Summer Minor Illnesses with or without Exanthems. Enteroviruses are often isolated from patients with acute febrile illnesses of short duration and without distinctive features, occurring during the summer or fall. With some coxsackievirus serotypes, the illness in young children may be accompanied by a rubelliform rash on the face, neck, and chest; it is maculopapular, is not pruritic, and does not desquamate.

Rash is also a common manifestation of infection with echovirus type 9 (less frequently with type 4 and others); the incidence is high in young children and decreases with age. Conjunctivitis may also be present. Echovirus type 16 has been responsible for outbreaks of the maculopapular rash that characterizes "Boston exanthem disease."

8.2. Diagnosis

In view of the wide range of host response to any single enterovirus serotype, from the most common form—silent infection—to the severe diseases that occur, diagnosis must rest on virus isolation and identification and on type-specific antibody response.[61,274,304,324] The significance of enterovirus isolations in association with illness must be critically evaluated, for enteroviruses circulate abundantly, particularly among young children, and coincidental associations are inevitable in some instances. For example, in a series of surveys conducted in Glasgow, enteroviruses were isolated from feces of one fifth up to one half of well day-nursery children less than 5 years of age, even though most specimens were obtained in nonepidemic periods.[142]

Differential diagnosis can present even greater difficulties when a community is invaded simultaneously by several enteroviruses and by other viruses with similar symptomatology, especially when some individuals are infected by both viruses, e.g., St. Louis encephalitis (SLE) virus and an enterovirus. For example, the distinction between an aseptic meningitis or encephalitis caused by SLE virus and that by an enterovirus often cannot be made on clinical grounds; firm diagnosis of singly infected patients, as well as of those with dual enterovirus–SLE infections, can be made only by utilizing the virus laboratory.[378]

To establish etiologic association of an enterovirus with disease, the following criteria can be used: (1) There is a much higher rate of recovery of virus from patients with the disease than from healthy individuals of the same age and socioeconomic level living in the same area at the same time. (2) Antibodies against the virus develop during the course of the disease. If the clinical syndrome can be caused by other known agents, then virological or serological evidence must be negative for concurrent infection with such agents. (3) The virus is isolated in significant concentrations from body fluids or tissues manifesting lesions, e.g., from the CSF in cases of aseptic meningitis.

Diagnosis of an enterovirus infection can be presumptively made in a single serum sample, by ELISA detection of a virus-specific IgM. Coxsackievirus B types have been used as antigen—either as a single antigen (B5)[35] or as a multiple antigen (CB1-5).[329] The IgM test with CB2 as antigen has been put to good use in showing that coxsackie B2 was the cause of an aseptic meningitis outbreak in a high school football team (63% seropositive), but was not responsible for much of the illness seen in other students (12% seropositive).[11] Some diagnostic laboratories are using heat-treated virus or synthetic peptides in ELISA tests, but more work needs to be done before they are introduced into general use.[417]

9. Control and Prevention of Poliomyelitis

Both live and killed poliovirus vaccines have been used widely. Formalin-killed vaccine (Salk),[413] prepared from virus grown in monkey-kidney cultures, is now little used in the United States and is extensively used only in a few European countries (Finland, Sweden, Holland, France) and in some parts of Canada. Trials are being conducted with killed vaccine in some countries, either alone or in combination with live vaccine.[439,442]

The IPV continues to be available in the United States for use under special circumstances (see Sections 9.1.1 and 9.2.2). With vaccines of the potency of those prepared in the 1950s, when intensive studies were made, four inoculations were required for primary immunization, the first three at 4- to 6-week intervals and the fourth 6–12 months later. Booster doses were necessary to maintain immunity. More potent vaccines have been introduced in recent years.[460]

The IPV induces circulating (humoral) antibodies and thus protects the CNS against subsequent invasion by wild virus. Local secretory IgA antibody or cell-mediated immunity is not induced by the killed virus vaccine; hence, wild poliovirus can still multiply in the gut and be a source of infection to others.

After the widespread administration of killed vaccine from 1956 to 1961, paralytic poliomyelitis was greatly reduced, and many thousands of cases of paralytic disease were prevented by its use in various parts of the world. The incidence dropped dramatically in the United States, from 21,000 cases annually seen in prevaccine years (until 1956). However, a few localized epidemics continued to occur, the cases being concentrated in slum areas among unvaccinated preschool children. Some cases continued to occur even in the vaccinated; in a study of several thousand paralytic cases, 17% were in triply vaccinated children. In 1959, there were more than 5700 paralytic cases, and in 1960, despite extensive use of the killed vaccine for almost 5 years, there were still 2545 paralytic cases reported in the United States.

During 1952–1954, some of the first papers were published on attenuation of wild poliovirus for vaccine purposes,[103,255,408] and live virus vaccine candidates were then developed by a number of workers. These efforts came to fruition in 1955–1959, and large-scale field trials were held in many countries under a variety of conditions.[362] Routine use of live oral poliomyelitis vaccines was begun in many countries during the spring of 1960, and vaccines made from the Sabin strains[405] were licensed in the United States in 1961–1962. In the early years of live vaccine immunization, monovalent vaccines incorporating each serotype separately were the most commonly used, but these were soon replaced by trivalent vaccine. As the coverage of the population increased, cases in the United States were dramatically reduced further. From 1961, when almost 1000 cases were reported, the number fell to only 61 by 1965.

The credit for this dramatic reduction in poliomyelitis cases has been claimed for each of the types of polio vaccines separately, but it may have resulted from the combined effects of both vaccines.[429] In the years soon after it was licensed, the killed vaccine was very widely used. Then, in mass immunization efforts during 1962–1964, some 100 million Americans of all ages received the live polio vaccine—about 56% of the total population.[189] The historical pattern of an annual summer increase in cases disappeared in 1964, and the numbers of cases continued to decline despite continued known importations of wild virulent polioviruses.[407]

In 1965, trivalent live vaccine began to be generally used for routine vaccination of children in the United States, and little use was made of the killed vaccine. In fact, its manufacture ceased in the country. With the ongoing reliance on live vaccine, the incidence of paralytic polio has continued to fall. The average annual number of paralytic polio cases during 1966–1970 was 49, and in 1971–1975 it was 15. There were eight cases in 1976, 18 in 1977, and nine in 1978. In 1979 the number increased to 22 (ten of these cases were involved in the outbreak in an unvaccinated group associated with the Netherlands epidemic of 1978). In 1984, only eight cases were reported (one imported; another with onset in 1982).[53] For 1986, there were only five cases of paralytic poliomyelitis in the United States; one of these was an imported case. At this writing, in the 1990s, the average number of cases continued at about six per year, all vaccine-related. The last wild poliovirus was isolated in 1979.

The vast majority of immunizations against poliomyelitis in the United States are in infants and young children as a part of the routine childhood immunization schedules. Schedules as recommended by the Advisory Committee on Immunization Practices (ACIP)[5] for routine administration of polio vaccine to normal infants and children in the United States are shown in Table 16.[5] Primary immunization with OPV for infants should begin at 2 months simultaneously with the first diphtheria–pertussis–tetanus (DPT) inoculation. The second dose follows at 4 months of age, along with the second DPT dose. A third dose of OPV at 6 months, simultaneously with the third DPT dose, is listed as optional by the ACIP for use in situations of higher risk. Previous recommendations had scheduled the combined measles–mumps–rubella (MMR) vaccine for 15 months of age and the DPT and OPV for a separate visit at 18 months.[2] A new feature of the ACIP 1986 recommendation is that OPV and DPT vaccine, together with the MMR vaccine, may be given at age 15 months in a single visit. This simplification is based on a field trial demonstrating that simultaneous adminis-

Table 16. Recommended Poliomyelitis Immunization Schedule for Normal Infants and Children in the United States, 1986[a]

Dose	Live poliovirus vaccine[b]: age	Killed poliovirus vaccine[c]: age/interval[b]
Primary series	2 months	6–12 weeks of age
	4 months	After interval of 4–8 weeks
	6 months[d]	After interval of 4–8 weeks
	15 months	After interval of 6–12 months
Supplementary	4–6 years of age	4–6 years of age

[a]After recommendations of the Advisory Committee on Immunization Practices.[5]
[b]In tropical areas of high endemicity, it is recommended that an initial dose of monovalent type 1 vaccine be given soon after birth.
[c]After recommendations of the Advisory Committee on Immunization Practices.[2] Newer, more potent killed vaccines may require fewer immunization doses.[7]
[d]The ACIP 1986 recommendation[5] includes this dose of OPV as optional.

tration of these vaccines at age 15 months is safe and effective. An OPV booster is recommended for all children entering elementary school. No further boosters are recommended at present.

The primary immunization for children and adolescents consists of two doses of OPV 2 months apart, followed by a third dose 6 months to a year later.[3]

In the right-hand column of Table 16 are shown earlier recommendations of the ACIP[2] for scheduling vaccine doses in situations in which IPV is the vaccine of choice (e.g., immune deficiency in the vaccinee or in the vaccinee's household). It should be noted that newer and more potent killed-vaccine preparations may require fewer immunization doses than indicated in the table.

For tropical countries, particularly those that experience heavy and frequent importations of wild polioviruses, the primary immunization schedules not only should begin early but also, most importantly, should be completed early in infancy.

As regards the household contacts of the OPV recipients, the ACIP has stated:

> When children in the household are given OPV, adults who are not adequately immunized against poliomyelitis are at a very small risk of contracting OPV-associated paralytic poliomyelitis. Because of the overriding importance of ensuring prompt and complete immunization of the child and the extreme rarity of OPV-associated disease in contacts of vaccinees, the ACIP recommends the administration of OPV to a child regardless of the poliovirus-vaccine status of adult household contacts. This is the usual practice in the United States. The responsible adult should be informed of the small risk involved and of precautions to be taken, such as hand washing after changing a diaper. An acceptable alternative, if there is strong assurance that ultimate, full immunization of the child will not be jeopardized or unduly

delayed, is to immunize adults with IPV or OPV, as appropriate to their immunity status, before giving OPV to the child.[4]

Routine immunization for adults residing in the continental United States is not felt to be necessary because of the small risk of exposure.[4] However, adults who are at increased risk because of contact with a patient or who are planning travel to an epidemic or endemic area should be immunized. Pregnancy is neither an indication nor a contraindication to required immunization.

For previously unimmunized or incompletely immunized adult travelers to high-risk areas, IPV is indicated, but if less than 4 weeks is available before protection is needed, a single dose of OPV is recommended.[4] Travelers to high-risk areas who received less than a full primary course should complete the remaining doses of either vaccine, and those who completed a full course of OPV should receive a single booster dose of OPV. Those whose primary course was IPV should receive a dose of either vaccine. If IPV is used exclusively, an additional booster every 5 years may be given if high-risk exposure continues. Similar protection is indicated for health-care and laboratory personnel who might be exposed to wild polioviruses.[4] The ACIP recommends IPV for providers of health care who are at risk of exposure and who have not been fully immunized previously. This recommendation is based not only on the slight risk to adults after OPV administration but also on the possibility that these individuals, in shedding OPV progeny, might expose immunocompromised patients to live virus.[4]

Special precautions (see below and Section 9.2.2) must be observed in considering administration of live vaccine to any person receiving immunosuppressive therapy or to anyone known or suspected to have a defective immune system or to the household contacts of such an individual. Indeed, ACIP[4] has recommended that in a household with any family history of immunodeficiency, no live vaccine should be given to any family member until the immune status of all family members is documented. If protection against poliomyelitis is indicated for such persons, IPV must be used: it can provide protection to the contacts, and it may give some protection for the immune-deficient individual.

Recommendations also are being developed regarding immunizations for children infected with human immunodeficiency virus (HIV), the virus of acquired immunodeficiency syndrome (AIDS).[6] The recommendations for poliovaccines are summarized below:

1. Those with symptomatic HIV infections can be given routine immunizations with IPV and other inactivated vaccines but should not receive any live virus or live bacterial vaccines.

2. For those with previously diagnosed asymptomatic HIV infection, available data suggest that OPV can be given without adverse consequences. However, since their family members may be immunocompromised because of infection with HIV, and since OPV vaccines excrete virus, it may be prudent to use IPV.

3. Uninfected children whose household members are known to be immunosuppressed because of HIV infections should receive IPV for routine immunization, for the reason described above.

Both live and killed poliomyelitis virus vaccines have been used extensively over many years and have been both safe and effective. Nevertheless, a healthy respect should be maintained for the poliovirus strains used as vaccine sources, and great care must be exercised by those undertaking the manufacture or the administration of either vaccine.

In considering initiation and maintenance of vaccine programs, public health officials must pay special attention to the particular setting in which a vaccine is to be used.[180] These matters are discussed in the context of the advantages and disadvantages of killed and live vaccines, which are summarized in Tables 17 and 18.[266]

No medical intervention is absolutely risk-free. Even the most common drugs carry some degree of risk. A public health or medical judgment must be made on the basis of balancing the values and the problems of one procedure against those of another procedure, and also against what may be the greater risks of doing nothing at all.

9.1. Killed Polio Vaccine

9.1.1. Advantages.
When properly prepared and administered, killed vaccine induces good levels of humoral antibodies in a satisfactory proportion of those receiving sufficient dosage and thus protects the vaccinee against paralytic poliomyelitis. It can also provide protection to whole populations and is believed to have limited the circulation of polioviruses in several nations that use it.[410] However, the countries that have used it exclusively with a considerable degree of success are small nations with excellent public health systems, where coverage by the killed vaccine has been wide and frequent, reaching 90% or more of the target population.

Killed vaccine, since it contains no living virus, cannot mutate toward increasing virulence. Because living virus is absent, it is safe to administer killed vaccine to persons with immune deficiency diseases and to their families and to persons undergoing immunosuppressive therapy. Killed vaccine can be combined with DPT vaccines and incorporated into an immunization schedule for infants and young children.

Preparations of IPV used until recently were, in some instances, of low immunogenicity (see below). Newer and more potent killed vaccines have been developed.[460] New production methods, including the new method of growing poliovirus on latex beads, greatly enhance the

---

Table 17. Killed Polio Vaccine: Advantages and Problems[a]

Advantages	Problems
Confers humoral immunity in satisfactory proportion of vaccines if sufficient doses of potent vaccine are given	Early studies indicated a disappointing record in percentage of vaccinees developing antibody after three doses[b]
Can be incorporated into regular pediatric immunization with other injectable vaccines (DPT)	Generally, with the vaccines with have been commercially available, repeated boosters have been required to maintain detectable antibody levels[b]
Absence of living virus excludes potential for mutation and reversion to virulence	Does not induce local (intestinal) immunity in the vaccinee; hence, vaccinees do not serve as a block to transmission of wild polioviruses by the fecal–oral route
Absence of living virus permits its use in immunodeficient or immunosuppressed individuals and their households	More costly than live vaccine in single-dose cost, administration expense, and total amount required, including boosters
Appears to have greatly reduced the spread of polioviruses in small countries where it has been properly used (wide and frequent coverage)	Problems from scarcity of monkey kidneys for growing virus have been overcome by use of continuous-passage Vero monkey cells for vaccine preparation
May prove especially useful in certain tropical areas where live vaccine has failed to "take" in young infants	Use of antigenically potent but virulent polioviruses as vaccine seed creates potential for tragedy if a single failure in virus inactivation were to occur in a batch of released vaccine. This is particularly important since monkey neurovirulence tests are no longer required before release of inactivated vaccine. However, this problem can be overcome by use of attenuated strains for production.

[a]Modified from Melnick.[266] Used with permission.
[b]Some of the disappointing results in the decade after killed vaccine was introduced may have resulted in part from problems that may now have been corrected by development of the more immunopotent van Wezel vaccine.

Table 18. Live Polio Vaccine: Advantages and Problems[a]

Advantages	Problems
Confers both humoral and intestinal immunity, as does natural infection	Being living viruses, the vaccine viruses can mutate and in very rare instances have reverted toward neurovirulence sufficient to cause paralytic polio in recipients or their contacts
Immunity induced may be lifelong	
Induces antibody very quickly in a large proportion of vaccinees	
Oral administration is more acceptable to vaccinees than injection and is easier to accomplish	Vaccine progeny virus spreads to household contacts[b]
Administration does not require use of highly trained personnel	Vaccine progeny virus also spreads to persons in the community who have not agreed to be vaccinated[b]
When stabilized, can retain potency under difficult field conditions with little refrigeration and no freezers	In certain warm-climate countries, induction of antibodies in a satisfactorily high proportion of vaccinees has been difficult to accomplish unless repeated doses are administered; in some areas, where heavy exposure to wild polioviruses often occurs very early in life, even repeated administration has not been effective in providing 100% protection to all infants
Under epidemic conditions, not only induces antibody quickly but also rapidly infects the alimentary tract, blocking spread of the epidemic virus	
Is relatively inexpensive, both to produce the vaccine itself and to administer it, and does not require continued booster doses	
Can be prepared in pretested cell lines, and thus is dependent on monkeys only for neurovirulence testing. Use of pretested continuous cell lines also eliminates theoretical risk of including monkey virus contaminants in the vaccine	Contraindicated in individuals with immunodeficiency diseases, and in those undergoing immunosuppressive therapy, and also in the household associates of persons in either of these groups
	Requires large numbers of monkeys for safety testing

[a]Modified from Melnick.[266] Used with permission.
[b]Some people consider this spread into the community to be an advantage, but the progeny virus excreted and spread by vaccinees is often a mutated virus. Obviously, it cannot be a safety-tested vaccine, licensed for use in the general population.

yield of virus, permitting concentration and standardization of IPV of high potency.[460] Methods for production in greater volumes have also made the vaccine potentially more economical to prepare. A great deal of work has been done[160,411,414,415,460] indicating that only two doses of the newer, more potent IPV may provide immunity in a large proportion of vaccinees.

In a comparative trial of IPV and OPV conducted in the 1980s in the United States,[251] more than 500 children were given one dose of either vaccine at 2, 4, and 18 months of age. The serological status of 439 was assessed at 2, 4, and 6 months, and that of 85 was assessed also at 18 and 20 months of age. At 2 months of age, before vaccine administration, the prevalence and the titers of antibodies against the three poliovirus types were similar in both groups (IPV and OPV recipients), and maternal antibody was shown to be present in a large proportion of both groups: only 12 of the 439 were triple-negative at entry into the study; at 2 months of age, 88–90% of the children had type 1 antibody, 92–96% type 2, and 75–76% type 3. At 6 months of age (2 months after the second dose of vaccine), a minimum of 93% had antibodies to all three types of poliovirus. At 20 months (2 months after the third dose), all but one child had antibodies to all three types.

Geometric mean antibody titers for type 2 and type 3 were significantly higher for the OPV group at 4 months of age. The authors indicate that the lower response to the IPV at this age was probably related to the presence of maternal antibodies, whereas after the first dose of OPV, types 2 and 3 were able to multiply in the intestine in the presence of maternal antibodies and thus induce active antibody by 4 months of age. At 20 months of age, higher geometric mean titers against types 1 and 3 were found in the IPV recipients than in the OPV group. A marked booster effect could be seen after the third dose of IPV.

In another large field trial conducted in Senegal, Africa, two doses of the new IPV induced a good antibody response in more than 90% of the recipients. Those tested were infants who initially had antibody titers ≤4.[442] In this program the new potent IPV was used in a combination vaccine with DPT. During the 1-year study period, approximately 10,000 children received a full set of vaccinations, 6 months apart, and 500 were selected for detailed study. Ages at the first dose were 2 to 8 months, and at the second, 9 to 51 months. Six months after the second dose, type 1, 2, and 3 antibodies were detectable in 97%, 98%, and 90% of the DPT–IPV recipients. In the control group (DPT alone), the corresponding percentages with antibody were 50%, 38%, and 80% presumably as a result of natural infections. However, the results of this field trial in terms of resistance to the development of paralytic poliomyelitis have been somewhat disappointing.[55]

In another study, in Burkina Faso, Africa, two doses of a quadruple DTP–IPV vaccine were found to induce not only bacterial antibodies but also antibodies to the three polioviruses in over 90% of the children.[404] In the majority of the children the polio antibodies persisted for more than 2 years. Even though IPV is more expensive than OPV, the simplified immunization schedule offers advantages, particularly in developing countries.

Although the killed virus vaccine may not confer intestinal resistance to carriage and spread of virulent live virus in the community, the neutralizing antibody it elicits provides protection of the vaccinee against paralysis from wild virulent virus. On a philosophical level, it has the advantage of not introducing into the community any living virus that can spread in an uncontrolled fashion to persons other than those who have sought or agreed to receive the live vaccine.

9.1.2. Disadvantages.

The licensing of killed vaccine was preceded by an immense nationwide trial in the United States, in which vaccine was administered to several hundred thousand children. Yet immediately after the success of the trial had been reported and the vaccine licensed, 61 cases of paralytic poliomyelitis appeared in vaccine recipients and 80 cases in their family contacts. These cases were epidemiologically linked to certain lots of vaccine subsequently found to contain small amounts of live, virulent poliovirus that had been undetected by the manufacturer. The breakdown in safety procedures that allowed the live virus to remain in the final vaccine arose from problems of transferring laboratory procedures to manufacturing production processes, particularly of extrapolating inactivation data. These problems were resolved, and there have been no further reports of any residual live virus problems with any inactivated poliomyelitis vaccines manufactured according to well-standardized procedures.

In the United States, after the early manufacturing defect that sometimes had allowed residual live virus to be present, an extra filtration step was included in the processing. This increased safety but reduced the concentration of viral antigen in the vaccines and consequently diminished their potency during the 1950s—and, indeed, continuing into the 1980s in some instances. In some studies, disappointingly low proportions of vaccinees developed antibodies after three inoculations of the killed vaccine. One such study was conducted in Ontario. A survey in 1969 and 1970 indicated serious gaps in the antibody protection of children who had received only killed vaccine, whereas almost all of those who had received both killed and live vaccines had antibodies to all three poliovirus types.[244]

In Sweden, Finland, the Netherlands, and Nova Scotia, as well as Ontario, the poliomyelitis immunization programs for many years relied on killed vaccine alone. However, they are small countries or provinces, in large part culturally homogeneous and socially advanced. They have excellent governmental health programs covering the entire population, ensuring the administration of full primary vaccination and frequent booster dosage. Particularly in Sweden, Finland, and the Netherlands, killed vaccines have been administered in intensive and regularly maintained immunization programs, achieving vaccine coverage among children that approaches 100%. In Finland and Sweden up to 1977, no paralytic poliomyelitis had been reported for more than a decade, and in the Netherlands, for a number of years only rare sporadic cases had been seen—in unvaccinated persons chiefly within religious communities having low vaccine acceptance rates because of their religious beliefs. It had been believed that wild polioviruses no longer circulate endemically in these countries.

On the other hand, it has also been suggested that the reason so little poliovirus has been found in these nations is that they are adjacent to countries where live vaccine continues to be widely distributed and that opportunities for the importation of wild polioviruses are thus reduced.

In Sweden in 1977, after many years without polio cases, two cases occurred, and a number of poliovirus carriers were detected in the area. The patients had no history of vaccination; this was also true of most of the carriers.

In 1978, the Netherlands experienced an epidemic of paralytic poliomyelis in which a total of 110 cases were reported.[420,487] Of patients who were paralyzed, 67 had spinal paralysis, seven had bulbar paralysis, and six had both bulbar and spinal involvement. All patients were members of a religious group that had refused vaccination. This outbreak occurred in a country with a very high polio vaccine acceptance rate; of persons under 27 years of age, 93% had received between three and six doses of killed-poliovirus vaccine. No cases were reported among vaccinated persons. However, in nursery and primary schools in some of the affected communities, not only were 71% of the nonvaccinated children tested excreting the wild poliovirus, but also about 24% of the vaccinated children were carrying the epidemic wild virus.[420] On the other hand, the 70,000 persons in the religious group that refused vaccine have been viewed—because of their frequent in-group contacts—as a single coherent population. All the cases in the Netherlands in 1978 occurred among this particular religious group of unvaccinated persons, whereas in the far larger number of unvaccinated persons scattered about the country who are not members of these religious groups, no polio cases were reported.

Until 1978, polio cases in the Netherlands had occurred only in the few municipalities with overall vaccination acceptance rates lower than 50–60%. In the 1978 outbreak, however, a particularly surprising finding was that, this time, some of the municipalities involved were those with generally high acceptance of vaccination.[28,485] Thus, some of the 1978 cases occurred in areas where a high degree of herd immunity might have been thought to protect unvaccinated individuals. There also were more cases than in other outbreaks in the "vaccine era," and a higher percentage of patients were older than 4 years.

It is clear that a great deal of population protection has been conferred in the Netherlands by their immunization program. But despite the high vaccination levels of the general population, virulent wild type 1 poliovirus was able to gain a foothold in the country and travel in a narrow stream among susceptibles within these close-knit subpopulations. Schaap et al.[420] in evaluating the epidemic, have noted that the evidence obtained cannot shed light on whether or not the vaccinated contacts who excreted the epidemic virus in their feces played a part in transmission of the virus. They conclude, however, that there were no indications of extensive and persistent circulation of poliovirus among the immunized population or in the unvaccinated persons in the general population who were not part of these religious groups.

In the winter of 1992–1993, the Netherlands suffered another polio outbreak, notwithstanding a vaccination coverage of 97% for the first three doses of IPV.[57] This time the causative virus was a type 3 strain genetically close to a virus isolated in India in 1991. Again, the cases in this new outbreak—68 (mean age 19 years)—were reported among members of a special religious community opposed to vaccination. Because of frequent contact between members of the group who live in the Netherlands and in Canada, the wild virus was again imported into Canada and caused paralysis there also. Of particular note is the isolation of wild poliovirus type 3 from stool specimens of 21 of 45 healthy persons, primarily children of the same religious group, living in a rural community in Canada.[158] The type 3 virus isolated in Canada was identical genetically to that causing the 1992–1993 outbreak in the Netherlands. This importation in 1993 followed a period of several years during which Canada, like the United States, was free of poliomyelitis.

In 1978, the Netherlands epidemic virus also spread to unvaccinated members of associated religious groups in Canada[117,486] and in the United States in 1979.[49] The virulent 1978 strain was an imported one, believed to have come from Turkey,[389] and in none of the countries involved did any cases occur beyond the narrow path of transmission among susceptibles in these interconnected unvaccinated groups.

The polio problem in a well-vaccinated country has been the experience of Finland, where the killed-vaccine

coverage among preschool children was almost 98%.[192] From mid-August 1984 through January 1985, 10 polio cases were documented, nine of them paralytic, and one nonparalytic (meningitis).[173,192,233] All but two of the patients were shown to be excreting type 3 poliovirus. Six of the patients were between the ages of 26 and 48 years; the other four were 4, 6, 12, and 17 years of age. Seven of the patients had received at least three doses of IPV (five had five; two had three), their latest doses having been received from 1977 through 1984. Two had received no vaccine, and one had received only one dose, in 1964. The patients were in widely separated parts of Finland, and subsequent testing of stools of healthy persons and of sewage indicated the epidemic virus to have been evenly spread throughout the country and to have infected at least 100,000 persons.

The inactivated polio vaccine that had been used in Finland's immunization program for over 20 years was known to have relatively weak immunogenicity, particularly in the type 3 vaccine component (Saukett strain), and declining prevalence of antibodies had been noted previously, although no cases had ensued. In a study published as recently as 1982, antibody to type 3 poliovirus was detectable in only 47–60% of Finnish children aged 1–6 years, whereas type 1 antibody was present in 62–67% and type 2 antibody in 83–93% of the children tested in this age group.[228] Because these and other antibody prevalence data indicated low immunogenicity of the IPV in use at that time, a decision was made in September 1984 to use a new and more antigenic IPV preparation beginning in 1986. Following these cases, 1.5 million extra doses of killed polio vaccine were given to persons under 18 years of age, and an oral poliovirus vaccine campaign was mounted that covered about 95% of the entire population; the outbreak was halted in February 1985.

The Finnish epidemic strain of poliovirus type 3 was found to differ in both immunologic and molecular properties from the type 3 Saukett (killed vaccine) strain.[173] Analysis of sera collected before the Finnish outbreak indicated that prevalence of neutralizing antibodies against most of the new type 3 isolates was much lower than that of antibodies to the Saukett strain. However, infants who had previously received two doses of a new, more potent IPV in a comparative vaccine trial had almost equal neutralizing antibodies to the two strains.

Analysis of the amino acid sequence indicated that changes had occurred in the most important 12-amino-acid stretch central to the major antigenic site of poliovirus type 3. The ability of type 3 poliovirus to undergo antigenic drift was also indicated by definite antigenic changes in consecutive isolates from several individuals.

In some studies of this outbreak, antigenic drift was considered to be a major precipitating factor in the outbreak by giving rise to a strain sufficiently divergent to overwhelm the low levels of population immunity to type 3 that were known to exist.[173,233,245]

However, other investigators[117] emphasized the low immunogenicity specifically of the Finnish vaccine[173,233,245] as being an important factor in the outbreak; they question whether the antigenic divergence, albeit wide, had a major role in the origin of the outbreak. They cite studies on sera obtained in the Netherlands after immunization with a potent IPV and in the United States after immunization with OPV showing that serum antibody titers were high against the Finnish epidemic type 3 strain. This parallels the findings by Hovi *et al.*[173] that sera of infants in Finland who had received two doses of a new more potent IPV had almost identical levels of antibodies against the vaccine strain and the epidemic strain. Furthermore, in neighboring Sweden, which has been using a different form of IPV, no type 3 cases were reported at the time of the outbreak in Finland.

The source of Finland's epidemic strain is not known; to date, attempts to identify similarly altered strains in other parts of the world have failed. Thus, it is not known whether this episode foreshadows further variation or more widespread emergence of widely divergent type 3 polioviruses. It is also worth noting that a recent further study of Finland's 1984–1985 type 3 epidemic poliovirus strain, conducted by determination of the nucleotide sequence, included investigation to determine whether the epidemic virus might have been derived from markedly drifted progeny of the type 3 live virus vaccine strain, Leon.[176] The sequence comparisons indicate that the epidemic strain was not derived from the Leon strain.

The Finnish investigative authorities do not consider the outbreak to be a failure of IPV in general, and that country plans to continue with the new and more potent IPV from 1986 on, but without adding OPV to the ongoing vaccine regimen.

Whatever the cause or causes of the outbreak in Finland, if one were to extrapolate the nine cases of paralytic polio in Finland—six of them in fully vaccinated persons—from the Finnish population (less than 5 million persons) to a large population such as that of the United States (more than 233 million persons), the picture would be seen as even more alarming. A parallel outbreak in a population as large as that of the United States would involve 419 cases, 280 in fully vaccinated individuals, and an estimated 4,660,000 healthy carriers excreting the epidemic virus.

For the 80 paralytic cases in the Netherlands 1988 outbreak, extrapolation of the rate from the 14 million population of the Netherlands to the United States population would mean more than 1300 cases of paralytic polio.

It is worth noting that the findings of serological surveys in Finland were predictive of possible vulnerability of the population. This argues for more—and perhaps more frequent—serological surveillance in supposedly secure populations to ensure that protection is elicited and that it persists.

One argument often given for the use of killed virus vaccine is that although serum antibody may be at very low or even undetectable levels, there is an enduring "immunologic memory." This state, not measured by antibody tests, is said to enable the vaccine to make a very quick and high-titer antibody response on further exposure to the homotypic virus.[415] Lack of serum antibody indeed may not indicate complete lack of protection against clinical illness. However, serum antibodies do contribute to prevention of viremia and therefore minimize the possibility that the nervous system will be invaded. It is therefore risky to assume that protection exists when serum antibody cannot be demonstrated.

As regards the needs of developing countries with large health problems and very limited financial resources, it is unfortunate that the cost of the new concentrated IPV vaccine is far greater than that of live OPV vaccine, which continues to be made available to developing countries at a cost of a few cents (U.S.) per dose.[107]

9.2. Live Attenuated Polio Vaccine

9.2.1.   Advantages.   This vaccine is given by the oral route. It infects, multiples, and thus immunizes. In simulating the natural poliovirus infection, it confers long-lasting (possibly even lifelong) immunity. Like the natural infection, it quickly induces the development of circulating antibody.

In addition to inducing humoral immunity, live polio vaccine also induces a state of resistance of the intestinal tract, which subsequently tends to block the spread of circulating poliovirus in the community. Indeed, it is postulated[518] that because of this intestinal resistance, the administration of live polio vaccine has already brought about a break in the chain of transmission of polioviruses within the United States, halting the perpetuation of wild polioviruses in this country. Intestinal resistance after administration of live virus vaccine seems to be dependent on the extent of initial vaccine virus multiplication in the alimentary tract rather than on serum antibody level.

Under epidemic conditions, live vaccine has the advantage of inducing immunity rapidly.[514] Furthermore, by quickly infecting the enteric tract, the vaccine strains tend to preempt this site in many persons in the population, interfering with and halting further spread of the epidemic virus—often within a matter of days, even before the vaccine-induced immunity becomes fully effective.

In a well-vaccinated country, repeated booster doses of live vaccine after the initial childhood series are seldom considered necessary. Since the live vaccine is given orally, it is easy and inexpensive to administer and is more acceptable in many populations. It is also more practical for mass administration and can be readily taken to remote areas and given rapidly without requiring the services of large numbers of skilled personnel. Furthermore, it is much less expensive than the inactivated vaccine, both in terms of single-dose costs and administration and in terms of the total amount needed to establish and maintain adequate immunity.

The live attenuated polioviruses in the vaccine, like most viruses, are unstable except when held at very low temperatures in the frozen state. As a result, present regulations require that for maintenance of potency the vaccine must be stored and shipped frozen, and that after thawing it must be held in the refrigerator at no more than 10°C for a period not to exceed 30 days, after which time it must be discarded. It should be noted in this regard that the infectivity of the enteroviruses can be preserved even when they are heated at 50°C if molar $MgCl_2$ is added, a property that is used in their classification. This finding was quickly applied to the live polio vaccine[281,374] not only in the laboratory but also in the field, where stabilized vaccines were used effectively to halt type 1 and type 3 outbreaks.[514]

In mass poliovirus immunization programs, particularly in developing and tropical countries where it is difficult to maintain or to transport the vaccine under frozen conditions, the stabilized vaccine offers numerous advantages.[314,374] The vaccine strains can be maintained at 4°C for well over a year with no loss in potency, and there is less need to keep the vaccine refrigerated during its administration in the field. Sucrose and sorbitol have also been used as stabilizers, but in comparative studies they have been less effective. In comparative studies, Peetermans et al.[374] concluded that vaccines prepared in molar $MgCl_2$ are still immunogenic after exposure for 1 day at 45°C, 3 days at 37°C, 3 to 6 weeks at +25°C, and more than 18 months at +4°C. In contrast, the vaccine in sucrose retained potency for significantly shorter periods. For developing countries it would be desirable to have an oral vaccine of even greater thermal stability, one that withstands 45°C for 7 days with no significant loss in potency.[497]

Another advantage of live vaccine is that it does not depend on a supply of scarce monkeys, except for neurovirulence testing. Initially, oral poliomyelitis vaccine was prepared in monkey-kidney cultures, but more recently human diploid-cell cultures have been used, and human diploid-cell lines are now licensed for vaccine production. This is desirable because of the possibility that the supply of suitable monkeys from the wild may be greatly curtailed. Also, the use of monkey tissues carries possible hazards (e.g., unknown viral contaminants such as the dangerous Marburg virus) that would not exist with human cells. The most thoroughly studied human diploid-cell lines are WI-38 and MRC-5; these cells have been found free of microbial contamination and can be held frozen until needed for vaccine production. Safety testing of such a cell stock can be far more complete than the testing that is possible within the relatively brief life span of primary cultures such as those from monkey kidney. However, current developments with monkeys bred in captivity, in which the kidneys of newborn animals are used for vaccine production, indicate another economically feasible option.

The safety of oral poliovirus vaccine is frequently questioned in patients infected with HIV. WHO continues to recommend giving infants four doses of oral poliovirus vaccine before the age of 12 months, regardless of the possible presence of HIV infection.[503]

9.2.2. Disadvantages. Problems associated with live polio vaccine relate chiefly to the fact that the vaccine consists of living viruses. Being alive, the vaccine viruses can mutate, and some poliovirus strains excreted by vaccinees, although still attenuated, are less attenuated than the vaccines administered.[260] Vaccine viruses are abundantly excreted and can infect unvaccinated contacts.[22,168,169] Also, serological studies in some populations indicate that vaccine virus spreads beyond the vaccinees. The proportion of individuals found to possess antibodies is considerably greater than would appear to be explainable either by their vaccination histories or by the circulation of wild polioviruses in their communities. Although the spread of live vaccine virus from the vaccinee to household and community contacts is considered by some to be an advantage, in that it may provide "free immunization" to larger numbers of persons, the fact remains that the virus that spreads to the contact is not a licensed vaccine. It is theoretically possible that it could revert sufficiently toward neurovirulence to cause paralytic poliomyelitis in the contacts of the vaccinee.

When inoculated into the monkey spinal cord, all strains of poliovirus can multiply in and destroy motor neurons. The crucial test used to determine whether a strain is sufficiently attenuated and safe for vaccine use is to show that this monkey neurotropism is reduced by one million-fold from that manifested by virulent strains. The techniques used in recognizing and certifying vaccine strains for safety are such that different degrees of neurotropism, even among attenuated strains, can be detected. For example, it has been found that vaccine progeny virus after multiplication in the vaccinees, although still attenuated, may no longer pass the safety tests required of the vaccine itself.[260] As mentioned above, the viruses, particularly type 3, do mutate in the course of their multiplication in vaccinated children, and rare cases of paralytic poliomyelitis have occurred in recipients or in their household contacts.[203] It is for this reason that when a developed nation or region begins administration of oral poliomyelitis vaccines, the initial program should be conducted in intensive campaigns designed for mass immunization of the entire population of an area at the same time. Thus, the polioviruses that would infect virtually the entire population of all ages would be the certified and tested vaccine viruses themselves rather than untested progeny viruses excreted by vaccinees. At the minimum, all susceptible members of a family, including the parents, should be vaccinated simultaneously. Once the bulk of the population in an area has been immunized, subsequent routine immunizations can be limited to babies and young children. This procedure should be sufficient, so long as a large proportion of the children (over 90%) do indeed receive the complete course.

When a developing country begins administration of oral poliomyelitis vaccine, the initial program should be conducted in intensive campaigns designed to immunize the entire child population 3–24 months of age at the same time. If serological surveys or clinical disease indicate risk to younger or older children, the age limits of the populations to be vaccinated may be altered accordingly, from 1 to 36 months of age.

Epidemiologic data indicate that the instances in which the live vaccine virus has reverted to neurovirulence and attacked vaccinees or their close contacts are exceedingly rare. Recently, in the United States, new discussions and arguments have arisen from proposals that killed vaccines be reestablished for general use in this country. The argument for killed vaccines has included reference to the "dangers" of the live vaccine. It is important to reemphasize what has been observed in the United States from the results of administering more than 700 million doses over the past two decades and has repeatedly been stated by many different national and international groups of experts. A vaccine that in the United States during 1973–1984 produced, at most, one case in a recipient or a contact per 2.6 million doses of OPV distributed[54,342] is far from "dangerous" but rather is outstandingly safe, as well as effective. However, it was noted that the associated frequency of paralysis was one case per 520,000 first doses, versus only one case per 12.3 million subsequent doses.[342] Corresponding figures calculated for the United Kingdom since 1967 are similar—about one case in a recipient and two cases in contacts per 10 million doses distributed.[70] These are based on the maximum risk (that is, the risk calculated as though every "vaccine-associated" case in which the vaccine cannot be ruled out were indeed vaccine related). To what extent these time-associated paralytic cases are caused by vaccine virus progeny rather than by wild polioviruses, or by viruses that are not polioviruses, remains a cause of controversy.[406,407,410,411]

Of particular note in this connection are the eventual scientific findings in regard to one of the first vaccine-associated cases in which a huge sum was paid by a manufacturer to a plaintiff in Texas because the court ruled that the epidemiologic evidence supported the view that the patient had been paralyzed by a vaccine virus. It now turns out, from the results obtained by the new laboratory methods, that the virus recovered from the paralyzed child unequivocally possessed the genome of a wild virus and not that of the vaccine virus. Thus, this case clearly illustrates that at least some "vaccine-associated" cases are "associated" merely in time and are not vaccine caused.

Because of the need to monitor the characteristics of polioviruses isolated from persons who have received live vaccine, and particularly from those few who develop paralytic poliomyelitis or whose contacts develop poliomyelitis, there have been many efforts to design laboratory tests capable of indicating whether a poliovirus isolated is vaccine derived or is a wild virus. A number of such "marker" tests have been used, but their results have not given absolute answers, and some investigators consider them of little help, at least in deciding whether a particular case of poliomyelitis is vaccine derived. They have contributed toward designating a particular virus excreted by a vaccinee or a contact as "vaccinelike" or "nonvaccinelike," but the former does not necessarily mean that the virus was derived from the vaccine virus. In some instances wild polioviruses have been found with similar "marker" properties. In any case, results of the marker tests cannot, by themselves, indicate whether vaccine progeny caused the illness. At best, such markers can be used only by highly experienced investigators, together with other information about the history of a particular poliomyelitis case, to make an informed judgment as to whether the case was "probably vaccine caused" or "probably caused by a wild virus."[334]

With the use of the new tools of molecular epidemiology and monoclonal antibodies, advances are being made in resolving some of the controversial cases. The genome of the attenuated type 1 vaccine strain has been found to have only 57 base substitutions of a total genome of 7441 nucleotides.[346] Also, ribonuclease T1-oligonucleotide mapping of poliovirus RNA has been introduced in which wild strains usually can be distinguished from vaccine progeny strains by examination of the two-dimensional electrophoretic migration patterns of fragmented segments of the viral genome. But there are limits, since genomic variation may occur in the process of vaccine virus replication in humans.[203,349] Genomic variation also limits the value of strain identification by polyacrylamide gel electrophoresis of the viral peptides. Wild and vaccine strains are also being identified not only by use of strain-specific absorbed sera[459] but also with monoclonal antibodies.[177,358] However, the degree of virulence of a poliovirus strain (whether vaccine-derived or wild) can only be determined by the neurovirulence test, conducted by inoculating monkeys and evaluating the results according to precise and standardized procedures.

A detailed study was begun in 1969, conducted by a WHO Consultative Group specifically charged "to investigate the possible relationship between acute persisting spinal paralysis and the use of poliomyelitis vaccine."[498] The classification of cases, in general closely following the pattern used in the Neurotropic Viral Diseases Surveillance Program of the U.S. Centers for Disease Control, included "recipient" vaccine-associated case (one in which illness begins 7–30 days after receiving the vaccine), "contact" vaccine-associated case (one in which the patient is known to have been in contact with a vaccinee and becomes ill 7–60 days after the vaccinee received the vaccine), "possible contact" case (no known contact with a vaccinee but occurring in an area and time of mass vaccination), and "no known contact" case (paralysis occurring with no known contact in an area where intensive vaccination is not in progress or where routine vaccination can be given any time during the year).

In an evaluation of the study, a coherent picture could be discerned despite different methods of investigation and reporting among the participating countries. In the first 5 years of the study, during which almost 200 million doses of vaccine were given, 360 cases of paralytic poliomyelitis were reported; 155 were classed as "no known contact" and appeared to have been caused by wild viruses still circulating in these communities. For the

10 years of the investigation conducted in 13 countries, the WHO Consultative Group, reported[499]:

> Altogether 698 cases ... were recorded in a total population of about 509 million over the 10-year study period—an incidence of 0.14 per million per annum. The incidence varied widely between countries.... Consistently over the years, there have been reports of a few cases that were thought to have been causally related to the viruses in the vaccine—especially poliovirus type 3 in vaccine recipients and type 2 in contacts.... The occurrence of a few rare cases among the parents of immunized children suggests that parents without definite evidence of previous effective immunization should be given vaccine at the same time as their children. Overall, the second 5 years of the study confirmed the conclusions reached after the first 5 years, i.e., that live oral poliomyelitis vaccine is one of the safest vaccines in use.

In an extension of the WHO study for a further 5-year period, through 1984, the WHO Consultative Group confirmed "that live poliovaccine is safe and effective in routine use and is indispensable for the control of outbreaks. In the group of six countries with a total population of 403 million, the risk of vaccine-associated paralysis is less than one per million children vaccinated."[66]

Japan is a country that has had large epidemics of poliomyelitis but now has virtually eliminated the disease by OPV immunization. As in many countries, every case is analyzed by laboratory methods.[319] Of 12 cases reported in the 12 years prior to 1991, nine yielded virus; of these, five isolates were vaccine-related. Two wild strains of poliovirus were found to have been introduced into Japan during the 12-year period, 1980–1991, but they did not spread. On the other hand, OPV vaccine strains have been found to spread, which is usual in a country like Japan that has a very high vaccine coverage rate among young children, the prime spreaders. Five percent of the vaccine-related poliovirus isolations were from children over 4 years old, that is, from children older than those receiving OPV. Parallel to the isolation of OPV strains from humans was the recovery of 268 isolates from sewage and river water during the same 12-year period.

In some warm-climate countries, live poliomyelitis vaccines have not induced antibody production in as high a proportion of vaccinees as in areas with more temperate climates.[90] This lower rate of vaccine "take" has been ascribed to various possible factors, including interference from other enteric viruses already present in the intestinal tract. Interference can be an important problem in warm-climate regions where enteroviruses are abundant.[364] Other factors that have been suggested as possibly responsible for this problem include the presence of antibody in breast milk, the presence of cellular resistance in the intestinal tract owing to previous exposure to naturally circulating polioviruses (or perhaps related viruses), or the presence of an inhibitor (saliva) in the alimentary tract of infants in these areas that acts against multiplication of the vaccine virus.[67,90] It also is possible that the persistent diarrhea so common among infants in many tropical areas interferes with implantation of the vaccine viruses. The problem of lower vaccine takes may be overcome by modifying the absolute and relative dosage of the Sabin strains in OPV,[366] or by combining use of IPV and OPV.[275]

The problems of controlling paralytic polio by use of vaccine may be exacerbated by the fact that in some areas the newborn infant has only a short time in which to acquire protective immunity, because exposure comes very early. In some developing and tropical countries poliomyelitis may attack children in their first months of life, and almost all poliomyelitis cases occur within the first 2 years.

For the present, practical experience has shown that a high level of population immunity and significant reductions in incidence can be achieved even in warm-climate countries by use of repeated doses of standard live vaccine alone, if given on a regular basis to infants and young children. After the primary doses of OPV in the first year of life, repeated boostering doses administered years later are important for lasting protection. Neutralizing antibody assays on sera collected from healthy persons under 30 years of age in Israel prior to the 1988 outbreak indicated high antibody levels against the Sabin strains used to produce OPV but very low levels against wild virus from the area. A booster dose of OPV given in 1988 brought the antibodies to high levels against both the Sabin and the wild strains.[135] In their pre-epidemic sera, nonvaccinated adults over 30 years of age—who had lived through the prevaccine period and who, as children, had been exposed to wild viruses as children—were found to have high levels of antibody both to the Sabin strains and to the wild virus. Nevertheless, after receiving a single dose of OPV in 1988, they also responded with an increase in antibody titer.

The above findings indicate that a gap in immunity against wild poliovirus may occur in persons who are vaccinated with OPV in the first year of life and who are not boostered by vaccine or by exposure to wild virus in the early years after vaccination when antibody levels and protection are highest. It seems the better part of wisdom to close this gap in immunity by administering one or more booster doses of OPV.

Special programs were conducted in three areas in tropical Africa that were experiencing incidence rates (as estimated from lameness surveys) ranging from 25 to 62 per 100,000 population in the mid- to late 1970s. Carefully planned, well-implemented programs were instituted[109] in Yaounde, Cameroon, The Gambia, and Abidjan, Ivory Coast. The programs included administration of three doses of oral polio vaccine during the first year of life, 1 month apart starting at 2 to 3 months of age, extensive evaluation of the vaccine coverage achieved,[159] and surveillance of poliomyelitis both before and after the vaccination program.[27,161,223,350] Within a few years the proportion of the children who received three doses of the vaccine, which had been virtually zero except for some (low) coverage in Abidjan, was increased to 50–70%. Even with this less-than-optimal vaccine coverage, the incidence of poliomyelitis decreased significantly in contrast to what has been reported from some other tropical countries, and oral poliovirus vaccine properly administered during the first year of life has been judged to be effective in controlling poliomyelitis in these localities. In Yaounde, for example, by 1980, 50% of the children 12–23 months old had received three doses of vaccine, and from a preimmunization average of 62 cases per 100,000 in 1974 and 1975, the incidence of polio decreased 88% to 7.5 per 100,000 in 1981. As emphasized below, once a vaccine program has been proven to be effective, it must be maintained on a continuing and regular basis.

Live vaccine has been shown to be as effective a means of cutting short an epidemic in a warm-climate country as it is in a temperate-climate country when given, preferably in repeated doses, to a high proportion of the presumed susceptible population in an epidemic area.[67,514]

An approach that has proven very effective in some for those areas where previous programs with live virus vaccines had not achieved full protection is the institution and maintenance of thorough and repeated mass administration of live polio vaccine. This strategy calls for annual national campaigns that include all age groups in which cases are occurring.[406,407] Each child in the age groups chosen is given a dose of live vaccine twice each year, with a 2-month interval between doses, regardless of how many doses of live vaccine he or she may have received before. This mass administration should then reach those missed in previous campaigns and should avoid the problem of record keeping, which often has been a barrier to full coverage in administering the vaccine.

Important questions relative to such a program have been the following: In view of the urgent need for other immunizations as well, would a campaign directed only at poliomyelitis present problems for the overall immunization program?[97] Would the absence of record keeping be detrimental both to the program itself and to the evaluation of its effectiveness?[97] Also, would developing nations be able and willing to make such a commitment for mass administration on two separate occasions each year? Would these nations have the resources to conduct such an annual repetitive mass vaccination program?

It is noteworthy that Brazil initiated such a program in 1980 and markedly reduced the annual incidence of poliomyelitis, from several thousand cases annually in the recent past to only 123 in 1981,[390] 69 in 1982, and only 43 in 1983.[97,407,496] Furthermore, the huge national effort and participation in the polio vaccine program (with some 90,000 vaccination sites set up, 10 times the number of permanent health stations) generated awareness and activity in other immunizations. Cases are no longer reported from Brazil.

Mexico had also been attempting for many years to achieve control of poliomyelitis chiefly through routine vaccination but continued to record considerable numbers of polio cases, an average of almost 700 annually from 1976 through 1980. In 1981, that country undertook mass national immunization campaigns similar to those described above for Brazil. In view of the fact that type 1 poliovirus has been the chief isolate from paralytic cases in Mexico for a number of years, the intensive campaign procedure during 1983 and 1984 was arranged to administer monovalent type 1 vaccine on the first of the two immunization dates, in January, and trivalent vaccine 2 months later, in March.[107] These campaigns have been continued annually and are yielding much greater success both in fuller vaccination coverage of children—to more than 80% in 1982—and in drastically reducing the cases: in 1981, there were only 186 cases, and in 1982, only 57. Again, no cases are now reported from Mexico.

Existing knowledge suggests that retiming of vaccinations also could save lives, especially in those countries where the six diseases targeted by WHO's Expanded Programme on Immunization cause many deaths in the first year of life. A new recommendation is that in these countries trivalent oral polio vaccine be given at birth, along with BCG, and repeated at the same time as DPT, at 6, 10, and 12 weeks of age.[97,147]

However, in certain areas were multiple doses of live polio vaccine have not succeeded in providing 100% protection to young infants, a program for inoculating killed vaccine or for administering killed vaccine together with scheduled feedings of live vaccine should be considered for use until this special problem can be overcome.[229,265,273]

Such a difficulty was encountered in recent years in Gaza and the West Bank, and a program was developed to deal with it, including both killed and live polio vaccines in the immunization schedule. Beginning in 1968, inten-

sive vaccination campaigns have been conducted in these areas, and about 90% of the children received at least three doses of live polio vaccine; some had been given four or five doses. But cases continued to appear. The disease struck particularly at the very young, sometimes with 50% of the cases occurring during the first 12 months of life. Among unvaccinated and incompletely vaccinated children in Gaza, the attack rates per 1000 children up to 3 years of age were 3.90 and 4.71 in epidemic years 1974 and 1976, respectively. But even among fully vaccinated children (some of whom had received up to four or five doses), a number of cases occurred, the attack rates in this group being 0.72 and 0.67 per 1000 children during the respective epidemic years. Thus, the vaccine was still strongly protective, 82% for 1974 and 86% for 1976, but the continued occurrence of some cases among the fully vaccinated was a cause of concern. In early 1978, a program was instituted using both killed and live vaccines.[131]

The rationale for the combined schedule was that under conditions of regular and heavy importation of virus from neighboring countries, there resulted an early and frequent wild virus challenge. Under such conditions, features of both types of vaccine seem to be needed. The killed polio vaccine provides an immediate immunogenic stimulus that is not subject to the interfering or inhibiting factors that prevent live vaccine "takes" in some young infants. Thus, protection can be provided in the critical first weeks or months of life in such areas of high exposure. Live polio vaccine then has time to induce its protective immunity both in the form of circulating humoral antibody and in the form of intestinal immunity. The immunity that ensues is long-lasting, like that which follows the natural infection.

The combined schedule of live and killed polio vaccines has been very effective in these high-risk localities where reimportation of wild virus into the community serves as a repeated challenge to the immunity of very young children. In a relatively short time, this program has transferred Gaza and the West Bank from the list of areas where poliomyelitis has been epidemic to the list of areas where poliomyelitis is now completely controlled.[131,229,267] As of 1993, cases are no longer reported in Gaza, the West Bank, or Israel, where the combination of IPV and OPV is used.[454] This control has been achieved in the face of continuing heavy exposure to polioviruses imported from surrounding countries. As the environment is made more hygienic, and heavy exposure to wild polioviruses no longer occurs in the first months of life, live polio vaccine alone should be able to provide both of the essential elements—sufficient early protection plus enduring immunity—as it has in the developed countries.

A record of combined use of live and killed polio vaccines in a very different epidemiologic setting can be seen in the experience of Denmark. From the beginning of that country's use of polio vaccines, serological surveys covering all age groups have been conducted frequently to monitor the prevalence of antibody in the population. After several years of experience with use of killed polio vaccine alone, Denmark undertook huge nationwide mass vaccination programs with oral polio vaccines in 1963 and in 1966; and in 1968 it was felt necessary to add live polio vaccine to the routine vaccination program.[375,461] The complete series of polio vaccinations in Denmark now consists of three killed vaccine injections at 5, 6, and 15 months of age, given in a triple vaccine in combination with diphtheria and tetanus vaccines, followed by three feedings of live polio vaccine at approximately 2, 3, and 4 years of age.[462] Vaccine coverage is very high: by the time they enter school at 6–7 years of age, 96% of children have had three injections of killed polio vaccine, and 75% have also have three doses of live polio vaccine. An increasing percentage of children (20% in 1975) also have one or more live vaccine doses at school age. By the ninth year of school (14–15 years of age), 98% have had three or four killed vaccine doses and 93% three doses of live vaccine also[462] (I. Petersen and H. von Magnus, personal communication).

The results of the Danish program are excellent. In 1977–1978, serological surveys showed that 76% of infants 6–11 months of age and 95% of those over 3 years of age possessed antibodies to all three poliovirus types. Geometric means of the measurable antibody titers were greater than 1:64 against all three poliovirus types at all ages beyond 1 year[462] (I. Petersen and H. von Magnus, personal communication). Since 1968, when the combined killed and live polio vaccine program began, there have been only two cases of paralytic poliomyelitis in Denmark. Poliomyelitis is no longer a disease in that nation. But the country, by virtue of its location in northern Europe and its freedom from daily massive importations of virulent wild polioviruses, is spared the repeated virus assaults that take place in similarly well-vaccinated sections of the Middle East. The slow schedule of vaccination that provides ample protection for Danish children would not be adequate in the Middle East today.

In making recommendations for vaccination, and also in evaluating vaccine-associated cases, it is important to emphasize the hazards of administering live vaccines to persons with immune-system problems. These include not only patients with immunodeficiency disease but also persons with altered immune states resulting from other diseases or from immunosuppressive therapeutic proce-

dures. In the United States during 1961–1971, there were 73 poliomyelitis cases among vaccine recipients and 37 cases among contacts. Nearly 10% of these cases were in persons with immune-system disorders, an incidence almost 10,000 times greater than in normal persons.[504] Cases associated with vaccination in persons with immune-system problems have continued to be reported, with ten additional patients identified in the nation through 1976.[48,502]

Without discounting the tragedy involved for the individual child and family who suffer in the exceedingly rare instance when an immunodeficient child encounters and succumbs to infection by vaccine or vaccine progeny virus, it must be recognized that some of these children are notoriously subject to infection—which is frequently fatal—by a wide variety of normally benign or avirulent agents.[476,521] Clearly, if compromised immune-system function is known or even suspected in potential vaccinees or their siblings, live poliomyelitis vaccine should not be given. However, unless other circumstances have brought the condition to light, such an immune-system problem is usually not known by the time the first routine vaccine doses are given at about 2 months of age.

9.3. Conclusions on Polio Vaccines and Their Future Use

Progress in polio control, aimed at total eradication, has continued to be made globally.[116,171,234,275,497] The increasing use of vaccine has been accompanied by progress in surveillance and timeliness in reporting. The global goal of eradication was formally set by WHO in 1988, and by the end of 1991, reported cases had decreased by 60%, the result primarily of improved vaccine coverage around the world. In infants, routine coverage, chiefly by OPV, had reached 84% by 1991. For eradication to be achieved, supplemental immunization activities include national vaccination days, increased use of vaccine as a response to local outbreaks, and "mopping-up" vaccination in areas where even a single case occurs.[87]

The advantages and disadvantages of killed and live polio vaccines need to be weighed with respect to the particular setting in which a vaccine is used. Since killed vaccine has proved to be effective in preventing poliomyelitis outbreaks in small countries with very competent and thorough vaccination programs, there may be little reason for them to change. But by the same token, since live virus vaccines have been working superbly for more than 25 years in the United States, the USSR, most other European countries, the United Kingdom, Australia, and Japan,[435] and more recently in China[91] and many other countries around the world, it would be unwise to interfere with these programs, which involve hundreds of millions of persons.

If a nation elected to recommend a change from live to killed vaccine for general use, a new and untried situation would exist in that country's defenses against poliomyelitis. Urgent new questions would have to be raised that have never been fully answered in the particular national settings concerned, especially in large countries with voluntary health systems: Would individuals maintain their immunity adequately by returning for repeated booster injections? Would immunity induced by killed vaccine, even if sufficient to protect the individual from paralytic poliomyelitis, be able to block the circulation of wild viruses in large and mobile populations as effectively as the live vaccine? Would the extra costs for vaccine, for its administration, and for repeated booster inoculations be justified? In nations in which live vaccine has already proved to be effective and safe, are there not other health programs far more urgently in need of the extra funds that would be required for implementation of a killed polio vaccine program?

There is considerable question as to whether the United States could ever achieve the high levels of over 90% polio vaccine acceptance that have been characteristic of those countries that depend on killed virus vaccine. In the United States, immunization ultimately depends on the initiative of the family in seeking care either from a private physician or from a public health clinic. Furthermore, it is difficult to hope that so large a proportion of families would maintain the needed schedule for repeated killed vaccine doses and boosters in a mobile population in which a young family is quite likely to move several times during a child's preschool years, changing physicians at each move. These reservations and concerns have been reemphasized in *The Lancet*, in which several letters were submitted[81,138,325] in response to the many publications arguing for extension of the use of killed poliovirus vaccine, with implications that developed nations currently using live vaccine should change over to the killed poliovirus.[96,411,412]

In some tropical areas where it is known that live poliovirus vaccine has been repeatedly and fully administered to a large proportion of the childhood population and nonetheless has not provided 100% protection, killed virus vaccine can be incorporated into the initial schedule of the poliovirus vaccination programs.[273] Alternatively, consideration could be given to a two-dose schedule with the new, potent IPV administered 6 months apart along with other childhood vaccines by mobile field health units going regularly into remote rural areas, as in a trial con-

ducted in Senegal.[442] Further experience is necessary to determine whether such an immunization program can be regularly pursued and whether adequate protection can be achieved and maintained so as to bring poliomyelitis under complete control. Indeed, with the current availability of potent killed virus polio vaccines, a single dose at 2 months of age, followed by the regular course of live polio vaccine, not only in tropical areas but also in developed countries, should be sufficient to protect children against wild polioviruses and also from the minimal risk of vaccine-associated paralytic poliovirus as a result of live vaccine virus reversion.

The comparisons and choices between killed virus and live virus polio vaccines may become unnecessary in the future, as new types of vaccines are developed. The available vaccines are made from either killed virus or attenuated live virus. The killed virus vaccine has been improved in potency by using purified and concentrated virus as a source of antigen, but it has not yet proven to be able to eliminate poliomyelitis in large regions of the world, and it seems to be too expensive for the developing nations. Attenuated vaccine contains a virus with the undesirable property of mutability, which needs constant meticulous monitoring during its manufacture and specialized laboratory-assisted epidemiologic surveillance in its field use.

Since poliovirus nucleic acid has been analyzed and the amino acid sequence is known, one approach has been to prepare a vaccine that contains no viral nucleic acid but only selected polypeptides that are immunologically active. The major neutralization epitopes of poliovirus types 1 and 3 have been located in the amino acid domain of viral capsid polypeptide 1 (VP1).[313,458]

Other aspects of the molecular biology of live poliovirus vaccines have been reviewed.[36,312,345] Type 3 epitopes have been thoroughly studied.[106,313] An octadecapeptide, containing amino acids 89–100 from the N terminus and with added cysteine residues, induced high levels of type-specific neutralizing antibodies in rabbits. As with a previously synthesized immunogenic hepatitis B polypeptide,[93,270] confirmation through the use of cysteine residues as well as specific amino acid sequence played a role in creating the antiviral antigenic properties of the peptides.[106] For type 1 poliovirus, a major neutralization epitope has been identified in a similar region, namely, amino acid residues 93–104 of VP1.[101,505]

Recombinant DNA techniques have also opened the way to prepare, in *Escherichia coli*, polypeptide vaccines containing only the essential immunogenic components. For a DNA virus like hepatitis B virus, the essential portion of the viral genetic material can be inserted into an

E. coli plasmid or phage. For an RNA virus like poliovirus, the complementary DNA (cDNA) of the virus is first prepared from viral RNA by use of the reverse transcriptase enzyme.[381,382] The entire cDNA of poliovirus, or a part of it, is then inserted into the phage for replication in *E. coli*. Subsequent transcription and translation of the viral genetic material under controlled conditions should yield the desired polypeptide vaccine free of any poliovirus nucleic acid.

In addition to yielding antigenic proteins, recombinant DNA technology can produce infectious virus with tailored, desired properties. As an example, complementary double-stranded DNA copies of three fragments of poliovirus RNA were synthesized and inserted into a bacterial plasmid. Three clones were derived that together provided DNA copies the length of the entire viral genome. The three cDNA clones were joined to produce a single cDNA copy of the entire poliovirus genome within the bacterial plasmid pBR322. When monkey or human cells were transfected with this hybrid plasmid, complete infectious virus was produced.[382]

We can now look forward to experiments in which the cloned cDNA is mutagenized to generate defined poliovirus mutants with specific alterations in the genome. Recombinants have been constructed from parental viruses belonging to different poliovirus serotypes.[8] Recombinants also have been made from virulent and attenuated strains of the same serotype.[9] and the part of the viral gene responsible for the attenuated phenotype has been identified. The availability of such mutants and recombinants will not only expand our knowledge of poliovirus genetics but also should make it possible to plan and manufacture totally attenuated vaccines that can no longer revert to virulence because the virulence genes would have been deleted.[35] When such vaccines become available for widespread use, they may be able not only to block infections with wild polioviruses but also to replace them in nature. If this can be accomplished, then it could herald the end of paralytic poliomyelitis.

9.4. Social Aspects of Polio Vaccine Administration

In most countries in which polio vaccine is administered widely and properly, circulation of wild polioviruses has been suppressed.[206] But these viruses have not been eliminated from the world. The existence of virulent wild polioviruses that can be imported all too easily into even the most fully vaccinated countries has been demonstrated in the 1978 outbreak in the Netherlands,[420] from which the epidemic virus spread to Canada[117] and subsequently, in 1979, to the United States.[49] All three outbreaks in 1978–1979 were in subpopulations inadequately vaccinated or unvaccinated because of religious beliefs. In 1984–1985, a well-vaccinated country, Finland, experienced an outbreak in which some fully vaccinated persons developed paralytic polio.[173,192,233] Even if the future may hold hope for the worldwide eradication of poliomyelitis, at present the most that any region or nation can achieve is elimination of the disease, and perhaps elimination also of endemic virus circulation, within its borders. Virulent wild polioviruses, endemic in many parts of the world and still producing thousands of cases each year, remain a real and present danger.

Although local secretory antibody or an "immunologic memory" may indeed provide protection even when circulating antibody cannot be detected,[353,411] the threat from imported wild virus seems to call for firmer and more assured lines of defense. Furthermore, there remains some question as to whether, if antibodies are undetectable, an "immunologic memory" would function rapidly enough, and with a sufficiently broad response, to produce a solidly protective antibody defense in the face of an infection by a virulent wild virus.[112,113,407] This would be particularly of concern if the wild virus were one like the Finnish type 3 strains of 1984–1985, which displayed wide antigenic differences from the type 3 vaccine strain.[173] It is thus misleading to assume that adequate protection is present when serum antibody cannot be demonstrated.

At least two factors contribute to the development of deficiencies in the immune status of populations: (1) failure to obtain proper vaccination, which can lead to the development of pockets of susceptibles even within a "well-vaccinated" community; and (2) decline of antibody levels in individuals with passage of time after vaccination. Declining antibody titers may become even more common as wild virus circulation is reduced or eliminated. If the trend continues or increases for less solid immunity among children, the way could become open for spread of imported viruses. Thus, if wild polioviruses are introduced, they could circulate widely in the general population, with outbreaks in which paralytic poliomyelitis could once again become prevalent, including nonimmune adolescents and adults. Immunity levels in all age groups can and should be monitored by periodic serological surveys to detect conditions of risk before poliomyelitis epidemics reappear. Because a large proportion of poliovirus infections are silent and subclinical, imported viruses could become widely disseminated before the first sign of their presence appeared in the form of paralyzed children and adults. Such epidemics would be even more tragic than those of the 1940–1950 era, for now—with the means of prevention readily at hand—a resurgence of paralytic poliomyelitis would reflect not "vaccine failures" but social failures. The central-city poverty areas depend chiefly on the public sector for health care delivery. More effective efforts by both public and private practice health professionals are needed to contribute toward community awareness of what care is available, when and how it can be obtained, and why it is needed.

Once considered a disease of the temperate zones, poliomyelitis is now far more prevalent in tropical countries. But excellent vaccines are available, and sufficient knowledge now exists to allow the vaccines, usually alone but sometimes in combination, to be used with a high degree of effectiveness in all environments.[180] In areas where the disease still flourishes, those responsible for public health should intensify their efforts to mount well-targeted immunization campaigns. In a number of developing nations with populations that are entering the epidemic phase of poliomyelitis, determined efforts in immunization programs[492] are yielding good results in reducing the incidence of paralytic poliomyelitis.

A 1985 status report of the Expanded Programme on Immunization of the World Health Organization[495] indicates that the immunization coverage with three doses of oral polio vaccine is decidedly improved, reaching a worldwide average of 44% of the children in the developing countries. The figures for different WHO regions vary widely, from 12% in some regions to 70% in others. Although the level of vaccine coverage is far from ideal, it represents a marked advance over the situation of only a few years ago, when scarcely any vaccine was reaching children in many of these regions. Nonetheless, the report also emphasizes that paralytic polio is still occurring in 1985 in these countries at an annual rate of 265,000 cases.

9.5. Nonspecific Control Measures for Poliomyelitis

It is not possible to list rules for the prevention of poliomyelitis other than vaccination. Quarantine either of patients or of exposed family or intimate contacts is ineffective in controlling the spread of the disease. This is understandable in view of the large number of inapparent and therefore unrecognized infections that occur during an epidemic.

During epidemic periods, children with fever should be given bed rest. Undue exercise or fatigue should be avoided, especially if there is any suspicion of involvement of the nervous system. Elective nose and throat operations and dental extractions should be avoided. Children should not travel unnecessarily to or from epidemic areas. Food and human excrement should be protected from flies. Once the poliovirus type responsible for the epidemic is determined, type-specific monovalent oral polio vaccine should be administered to susceptible persons in the population.

Patients with poliomyelitis can be admitted to general hospitals provided that the hospital regulations are followed. All pharyngeal and bowel discharges are considered infectious.

10. Control of Other Enterovirus Infections

For the nonpolio enteroviruses, no specific control measures are known. Avoidance of contact with patients exhibiting acute febrile illness, especially those with a rash, is advisable for very young children. Members of institutional staffs responsible for caring for infants should be tested to determine whether they are carriers of enteroviruses. This is particularly important during outbreaks among infants.

There have been numerous reports of serious and even fatal enterovirus infections of newborns, both in nursery outbreaks and as sporadic infections believed to have been acquired at birth from a contaminated cervix. A number of different echovirus serotypes (notably echovirus 11) and several coxsackieviruses of the B group have been associated with severe generalized infections of neonates. Hospital personnel need to be particularly alert to even "minor illnesses" compatible with enterovirus infections in mothers delivering babies who enter newborn nurseries or special units, and the staff members of these units also need to be constantly aware of the possible hazards introduced by their own "minor illnesses."[321]

An avenue of infection not routinely considered in avoiding nosocomial infection is oral–oral transmission from the hands of health care personnel during mouth care and gavage feeding procedures. Such a route has been strongly suggested in a study of outbreaks of coxsackievirus B4 infections in wards for children with severe neurological handicaps.[190] The same route should also be considered with neonates and others in special-care situations.

For most enteroviruses, quarantine of patients is not effective. This is understandable in view of the large proportion of enterovirus infections that are inapparent, and the fact that these infections usually spread very rapidly to nonimmune close contacts. However, in the case of enterovirus 70, for example, a degree of control can be achieved. If virus-induced acute hemorrhagic conjunctivitis is present in the community, its ability to be transmitted by fomites must be considered. Extra precautions in eye clinics are indicated, as well as excluding children from school if they are infected or known to have been exposed. Such hygienic precautions as avoidance of touching a virus-infected eye or sharing towels have been judged to be helpful in limiting explosive outbreaks.[365]

Another means of control in special circumstances is to provide passive protection with commercially available γ-globulin. It may be considered for infants threatened by severe nursery outbreaks of infection by group B coxsackieviruses, and it has been protective against other serious enteroviral diseases, e.g., in halting an outbreak of echovirus 11 infections in a special-care unit for hospitalization of especially vulnerable infants such as premature infants or infants undergoing surgery.[331]

γ-Globulin has also proven useful in treating immunodeficient children with a chronic enterovirus infection of the CNS.[253]

11. Unresolved Problems

Although many of the unresolved problems concerning the enteroviruses are described or implied in the foregoing presentation, this section is included to emphasize their importance to persons concerned with public health and epidemiology, who may be ideally situated to contribute uniquely toward solutions for the numerous problems that still remain.

Answers are needed to the "social failures" of vaccine delivery; to the epidemiologic consequences of a highly successful polio vaccine program for one segment of the population while another segment is placed at higher risk by reduction of natural infection at an early age; to the present and potential problems of viruses in drinking and recreational water and the need for public awareness and support of new water-treatment methods that would permit safe recycling of this limited and dwindling resource; and to the possible role of enteroviruses in diseases such as diabetes and cardiovascular disorders, both acute and chronic. We must be alert to the possibility that infections with nonpolio enteroviruses might undergo a shift in age incidence, involving a consequent increase in severity of the illnesses that they cause, in parallel with that which took place with the polioviruses in the late 19th and early 20th centuries. For example, enteroviruses that appear to be increasingly involved with CNS disease, such as enterovirus 71 and certain echoviruses, will need to be studied more fully and watched epidemiologically.

The relative freedom that the Western hemisphere had thus far enjoyed from acute hemorrhagic conjunctivitis (an enterovirus 70 disease) cannot be taken for granted—as was demonstrated in 1981 (see Section 5.2.2). Health care personnel—most particularly those dealing with eye examinations—must be alert for the onset of any outbreak. In any large pandemic of this infection, besides the temporarily incapacitating eye disease, the likelihood of a certain small proportion of more serious CNS complications must also be recognized.

Despite the availability of effective vaccines, poliomyelitis can be expected to be a continuing problem in developing countries in some tropical areas as well as in low socioeconomic groups of some developed countries. Even in countries where vaccine is being more widely used, there remains the question of why live vaccine has been more successful in some tropical areas than in others. Is the problem chiefly interference by other enteroviruses, persistent diarrhea, nonspecific inhibitors, or other unknown host factors that have limited vaccine effectiveness? Or is it the existence of massive exposures of infants in the very first months of life to virulent polioviruses, either indigenous or regularly imported? The major determinants of whether poliomyelitis infection and disease will be controlled are no longer the local presence or absence of wild virus. Instead, control depends on the immunization practices of the country and the degree to which effective and long-lasting immunity has been induced in all segments of the population, without leaving unimmunized or incompletely immunized pockets of susceptibles who remain at risk. Total eradication of poliomyelitis is beginning to be viewed as a possible future goal, which should lead to increasing of immunization efforts. At present, wild virulent polioviruses can all too easily be imported into a supposedly polio-free country, and vigilant maintenance and expansion of vaccination programs will be required for a long time yet. Monitoring to ensure that levels of immunity remain sufficient should not consist merely of being alert for the occurrence of "sentinel" clinical cases but instead should be determined by periodic serum surveys to determine the distribution and level of antibody in the population.

Development of vaccines for selected nonpolio enteroviruses is technically possible. It may be wise to look toward having such vaccines available for certain of these agents, particularly coxsackieviruses of the B group that have a recognized role not only in diseases of the CNS but also in myocarditis of infants and even in adult cardiovascular disorders. Such vaccines may not appear to be required for widespread use in any entire population but could be important for specifically targeted groups at high risk.

ACKNOWLEDGMENT

The able assistance of Miss Verle Rennick in the preparation of this chapter is gratefully acknowledged.

12. References

1. ABRAHAM, R., CHONMAITREE, T., MCCOMBS, J., PRABHAKAR, B., LO VERDE, P. T., AND OGRA, P. L., Rapid detection of poliovirus by reverse transcription and polymerase chain amplification: Application for differentiation between poliovirus and non-poliovirus enteroviruses, *J. Clin. Microbiol.* **31:**395–399 (1993).

2. ADVISORY COMMITTEE ON IMMUNIZATION PRACTICES, Poliomyelitis prevention, *Morbid. Mortal. Week. Rep.* **31:**22–26, 31–33 (1982).

3. ADVISORY COMMITTEE ON IMMUNIZATION PRACTICES (ACIP), General recommendations on immunization, *Morbid. Mortal. Week. Rep.* **32:**1–18 (1983).

4. ADVISORY COMMITTEE ON IMMUNIZATION PRACTICES, Adult immunization, *Morbid. Mortal. Week. Rep.* **33**(Suppl.):1S–68S (1984).

5. ADVISORY COMMITTEE ON IMMUNIZATION PRACTICES (ACIP), New recommended schedule for active immunization of normal infants and children, *Morbid. Mortal. Week. Rep.* **34:**577–579 (1986).

6. ADVISORY COMMITTEE ON IMMUNIZATION PRACTICES (ACIP), Immunization of children infected with human T-lymphotropic virus type III/lymphadenopathy-associated virus, *Morbid. Mortal. Week. Rep.* **35:**595–598, 603–606 (1986).

7. ADVISORY COMMITTEE ON IMMUNIZATION PRACTICES (ACIP), Poliomyelitis prevention: Enhanced-potency inactivated poliomyelitis vaccine—Supplementary statement, *Morbid. Mortal. Week. Rep.* **36:**795–798 (1987).

8. AGOL, V. I., DROZDOV, S. G., FROLOVA, M. P., GRACHEV, V. P., KOLESNIKOVA, M. S., KOZLOV, V. G., RALPH, N. M., ROMANOVA, L. I., TOLSKAYA, E. A., AND VIKTOROVA, E. G., Neurovirulence of the intertypic poliovirus recombinant v3/a1–25: Characterization of strains isolated from the spinal cord of diseased monkeys and evaluation of the contribution of the 3' half of the genome, *J. Gen. Virol.* **66:**309–316 (1985).

9. AGOL, V. I., DROZDOV, S. G., GRACHEV, V. P., KOLESNIKOVA, M. S., KOZLOV, V. G., RALPH, N. M., ROMANOVA, L. I., TOLSKAYA, E. A., TYUFANOV, A. V., AND VIKTOROVA, E. G., Recombinants between attenuated and virulent strains of poliovirus type 1: Derivation and characterization of recombinants with centrally located crossover points, *Virology* **143:**467–477 (1985).

10. ALEXANDER, J. P., BADEN, L., PALLANSCH, M. A., AND ANDERSON, L. J., Enterovirus 71 infections and neurologic disease, United States, 1977–1991, *J. Infect. Dis.* **169:**905–908 (1994).

11. ALEXANDER, J. P., JR., CHAPMAN, L. E., PALLANSCH, M. A., STEPHENSON, W. T., TOROK, T. J., AND ANDERSON, L. J., Coxsackievirus B2 infection and aseptic meningitis: A focal outbreak among members of a high school football team, *J. Infect. Diseases* **167:**1201–1205 (1993).

12. ANDREW, C. G., DRACHMAN, D. B., PESTRONK, A., AND NARAYAN, O., Susceptibility of skeletal muscle to Coxsackie A$_2$ virus infection: Effects of botulinum toxin and denervation, *Science* **223:**714–716 (1984).

13. ANDRUS, J. K., DE QUADROS, C., OLIVE, J.-M., AND HULL, H. F., Screening of cases of acute flaccid paralysis for poliomyelitis eradication: Ways to improve specificity, *Bull. WHO* **70:**591–596 (1992).

14. ASSAAD, F., AND COCKBURN, W. C., Four-year study of WHO virus reports on enteroviruses other than poliovirus, *Bull. WHO* **46:**329–336 (1972).

15. ATMAR, R. L., AND GEORGHIOU, P. R., Classification of respiratory tract picornavirus isolates as enteroviruses or rhinoviruses by using reverse transcription-polymerase chain reaction, *J. Clin. Microbiol.* **31:**2455–2546 (1993).

16. BALTIMORE, D., Picornaviruses are no longer black boxes, *Science* **229:**1366–1367 (1985).

17. BANATVALA, J. E. (ED.), *Viral Infections of the Heart*, Edward Arnold, London, 1993.

18. BANTVALA, J. E., SCHERNTHANER, G., SCHOBER, E., DE SILVA, L. M., BRYANT, J., BORKENSTEIN, M., BROWN, D., MENSER, M. A., AND SILINK, M., Coxsackie B, mumps, rubella, and cytomegalovirus specific IgM responses in patients with juvenile-onset insulin-dependent diabetes mellitus in Britain, Austria, and Australia, *Lancet* **1:**1409–1412 (1985).

19. BANKER, D. D., AND MELNICK, J. L., Isolation of Coxsackie virus (C virus) from North Alaskan Eskimos, *Am. J. Hyg.* **54:**383–390 (1951).

20. BARRETT-CONNOR, E., Is insulin-dependent diabetes mellitus caused by coxsackievirus B infection? A review of the epidemiologic evidence, *Rev. Infect. Dis.* **7:**207–215 (1985).

21. BENTON, W. H., AND WARD, R. L., Induction of cytopathogenicity in mammalian cell lines challenged with culturable enteric viruses and its enhancement by 5-iododeoxyuridine, *Appl. Environ. Microbiol.* **43:**861–868 (1982).

22. BENYESH-MELNICK, M., MELNICK, J. L., RAWLS, W. E., WIMBERLY, I., BARRERA-ORO, J., BEN-PORATH, E., AND RENNICK, V., Studies of the immunogenicity, communicability, and genetic stability of oral poliovaccine administered during the winter, *Am. J. Epidemiol.* **86:**112–136 (1967).

23. BERG, G. (ED.), *Transmission of Viruses by the Water Route*, John Wiley & Sons, New York, 1967.

24. BERG, G., BODILY, H., LENNETTE, E. H., MELNICK, J. L., AND METCALF, T. G. (EDS.), *Viruses in Water*, American Public Health Association, Washington, 1976.

25. BERGELSON, J. M., SHEPLEY, M. P., CHAN, B. M. C., HEMLER, M. E., AND FINBERG, R. W., Identification of the integrin VLA-2 as a receptor for echovirus 1, *Science* **255:**1718–1720 (1992).

26. BERLIN, L. E., RORABAUGH, M. L., HELDRICH, F., ROBERTS, K., DORAN, T., AND MODLIN, J. F., Aseptic meningitis in infants < 2 years of age: Diagnosis and etiology, *J. Infect. Dis.* **168:**888–892 (1993).

27. BERNIER, R. H., Prevalence survey techniques for paralytic polio: An update. Expanded Programme on Immunization Working Paper EPI/GAG 83/10, World Health Organization, Geneva, 1983.

28. BUKERK, H., Poliomyelitis in the Netherlands, in: *International Symposium on Reassessment of Inactivated Poliomyelitis Vaccine, Developments in Biological Standardization*, Vol. 47, pp. 233–240, S. Karger, Basel, 1981.

29. BLACKBURN, R. V., RACANIELLO, V. R., AND RICHTHAND, V. F., Construction of an infectious cDNA clone of echovirus 6, *Virus Res.* **22:**71–78 (1991).

30. BLATTNER, R. J., WILLIAMSON, A. P., AND HEYS, F. M., Role of viruses in the etiology of congenital malformations, *Prog. Med. Virol.* **15:**1–41 (1973).

31. BLOMBERG, J., LYCKE, E., AHLFORS, K., JOHNSSON, T., WOLON-

tis, S., and von Zeipel, G., New enterovirus type associated with epidemic of aseptic meningitis and/or hand, foot, and mouth disease, *Lancet* **2:**112 (1974).

32. Bloom, H. H., Mack, W. N., Krueger, B. J., and Mallmann, W. L., Identification of enteroviruses in sewage, *J. Infect. Dis.* **105:**61–68 (1959).

33. Bodian, D., Poliomyelitis: Pathogenesis and histopathology, in: *Viral and Rickettsial Infections of Man*, 3rd ed. (T. M. Rivers and F. L. Horsfall, Jr., eds.), pp. 479–518, J. B. Lippincott, Philadelphia, 1959.

34. Bodian, D., and Horstmann, D. M., Polioviruses, in: *Viral and Rickettsial Infections of Man*, 4th ed. (F. L. Horsfall, Jr., and I. Tamm, eds.), pp. 430–473, J. B. Lippincott, Philadelphia, 1965.

35. Boman, J., Nilsson, B., and Juto, P., Serum IgA, IgG, and IgM responses to different enteroviruses as measured by a coxsackie B5-based indirect ELISA, *J. Med. Virol.* **38:**32–35 (1992).

36. Brown, F., and Lewis, B. P., Poliovirus attenuation: Molecular mechanisms and practical aspects, *Dev. Biol. Stand.* **78:**1–187 (1993).

37. Brown, F., Talbot, P., and Burrows, R., Antigenic differences between isolates of swine vesicular disease virus and their relationship to Coxsackie B5 virus, *Nature* **245:**315–316 (1973).

38. Brown, F., and Wild, F., Variation in the coxsackievirus type B5 and its possible role in the etiology of swine vesicular disease, *Intervirology* **3:**125–128 (1974).

39. Brown, G. C., and Karunas, R. S., Relationship of congenital anomalies and maternal infection with selected enteroviruses, *Am. J. Epidemiol.* **95:**207–217 (1972).

40. Brown, G. C., and O'Leary, T. P., Fluorescent antibody responses of cases and contacts of hand, foot, and mouth disease, *Infect. Immun.* **9:**1098–1101 (1974).

41. Bruce, D., Al-Nakib, W., Forsyth, M., Stanway, G., and Almond, J. W., Detection of enterovirus using cDNA and synthetic oligonucleotide probes, *J. Virol. Methods* **25:**223–240 (1989).

42. Burch, G. E., Shewey, L. L., and Harb, J. M., Coxsackie B4 viruses and atrial myxoma, *Am. Heart J.* **88:**634–639 (1974).

43. Burch, G. E., Sun, S. C., Chu, K. C., Sohal, R. S., and Colclough, H. L., Interstitial and coxsackievirus B myocarditis in infants and children: A comparative histologic and immunofluorescent study of 50 autopsied hearts, *J.A.M.A.* **203:**1–8 (1968).

44. Butterworth, B. E., A comparison of the virus-specific polypeptides of encephalomyocarditis virus, human rhinovirus-1A, and poliovirus, *Virology* **56:**439–453 (1973).

45. Carstens, J. M., Tracy, S., Chapman, N. M., and Gauntt, C. J., Detection of enteroviruses in cell cultures by using *in situ* transcription, *J. Clin. Microbiol.* **30:**25–35 (1992).

46. Caverly, C. S., Notes of an epidemic of acute anterior poliomyelitis, *J.A.M.A.* **26:**1–5 (1896).

47. Cello, J., Samuelson, A., Stalhandske, P., Svennerholm, T., Jeansson, S., and Forsgren, M., Identification of group-common linear epitopes in structural and nonstructural proteins of enteroviruses by using synthetic peptides, *J. Clin. Microbiol.* **31:**911–916 (1992).

48. Center for Disease Control, *Poliomyelitis Surveillance Summary 1974–76* (1977).

49. Center for Disease Control, Poliomyelitis—Pennsylvania, Maryland, *Morbid. Mortal. Week. Rep.* **28:**49–50 (1979).

50. Centers for Disease Control, *Poliomyelitis Surveillance Summary 1977–78* (1980).

51. Centers for Disease Control, Enterovirus surveillance—United States, 1984, *Morbid. Mortal. Week. Rep.* **33:**388 (1984).

52. Centers for Disease Control, Enterovirus surveillance—United States, 1985, *Morbid Mortal. Week. Rep.* **34:**494–495 (1985).

53. Centers for Disease Control, Annual Summary 1984, *Morbid. Mortal. Week. Rep.* **33:**46–47 (1986).

54. Centers for Disease Control, Poliomyelitis, *Morbid. Mortal. Week. Rep.* **35:**180–182 (1986).

55. Centers for Disease Control, Preliminary report: Paralytic poliomyelitis—Senegal, 1986, *Morbid. Mortal. Week. Rep.* **36:**387–390 (1987).

56. Centers for Disease Control, Aseptic meningitis—New York State and United States, Weeks 1–36, 1991, *Morbid. Mortal. Week. Rep.* **40:**773–775 (1991).

57. Centers for Disease Control, Update: Poliomyelitis outbreak—Netherlands, *Morbid. Mortal. Week. Rep.* **41:**917–919 (1992).

58. Chang, W. K., Liu, K. C., Foo, T. C., Lam, M. W., and Chan, C. F., Acute haemorrhagic conjunctivitis in Hong Kong 1971–1975, *Southeast Asian J. Trop. Med. Public Health* **8:**1–6 (1977).

59. Chapman, N. M., Tracy, S., Gauntt, C. J., and Fortmueller, U., Molecular detection and identification of enteroviruses using enzymatic amplification and nucleic acid hybridization, *J. Clin. Microbiol.* **28:**843–850 (1990).

60. Chatterjee, S., Quarcoopome, C. O., and Apenteng, A., Unusual type of epidemic of conjunctivitis in Ghana, *Br. J. Ophthalmol.* **54:**628–630 (1970).

61. Cherry, J. D., Nonpolio enteroviruses: Coxsackieviruses, echoviruses and enteroviruses, in: *Textbook of Pediatric Infectious Diseases* (R. D. Feigin and J. D. Cherry, eds.), pp. 1316–1365, W. B. Saunders, Philadelphia, 1981.

62. Cherry, J. D., Enteroviruses, in: *Textbook of Pediatric Infectious Diseases*, 3rd ed. (R. D. Feigin and J. D. Cherry, eds.), pp. 1705–1753, W. B. Saunders, Philadelphia, 1992.

63. Chonmaitree, T., Menegus, M. A., Schervish-Swierkosz, E. M., and Schwalenstocker, E., Enterovirus 71 infection: Report of an outbreak with two cases of paralysis and a review of the literature, *Pediatrics* **67:**489–493 (1981).

64. Chumakov, M. P., Voroshilova, M. K., Shindarov, L., Lavrova, I., Gracheva, L., Koroleva, G., Vasilenko, S., Bradvarova, I., Nikolova, M., Gyurova, S., Gacheva, M., Mitov, G., Ninov, M., Tsylka, E., Robinson, I., Frolova, M., Bashkirtsev, V., Martjanova, L., and Rodin, V. I., Enterovirus 71 isolated from cases of poliomyelitis-like disease in Bulgaria, *Arch. Virol.* **60:**329–340 (1979).

65. Clarke, N. A., and Kabler, P. W., Human enteric viruses in sewage, *Health Lab. Sci.* **1:**44–49 (1964).

66. Cockburn, W. C., The work of the WHO Consultative Group in Poliomyelitis Vaccines, *Bull. WHO* **66:**143–154 (1988).

67. Cockburn, W. C., and Drozdov, S. G., Poliomyelitis in the world, *Bull. WHO* **42:**405–417 (1970).

68. Cole, C. N., Defective interfering (DI) particles of poliovirus, *Prog. Med. Virol.* **20:**180–207 (1975).

69. Coleman, T. J., Gamble, D. R., and Taylor, K. W., Diabetes in mice after Coxsackie B$_4$ infection, *Br. Med. J.* **3:**25–27 (1973).

70. Collingham, K. E., Pollock, T. M., and Roebuck, M. O., Paralytic poliomyelitis in England and Wales, 1976–77, *Lancet* **1:**976–977 (1978).

71. Committee on the ECHO Viruses, Enteric cytopathogenic human orphan (ECHO) viruses, *Science* **122:**1187–1188 (1955).

72. COMMITTEE ON THE ENTEROVIRUSES, National Foundation for Infantile Paralysis: The enteroviruses, *Am. J. Public Health* **47:**1556–1566 (1957).

73. COONEY, M. K., HALL, C. E., AND FOX, J. P., The Seattle Virus Watch. III. Evaluation of isolation methods and summary of infections detected by virus isolations, *Am. J. Epidemiol.* **96:**286–305 (1972).

74. COOPER, P. D., AGOL, V. I., BACHRACH, H. L., BROWN, F., GHENDON, Y., GIBBS, A. J., GILLESPIE, J. H., LONBERG-HOLM, K., MANDEL, B., MELNICK, J. L., MOHANTY, S. B., POVEY, R. C., RUECKERT, R. R., SCHAFFER, F. L., AND TYRRELL, D. A. J., Picornaviridae: Second report, *Intervirology* **10:**165–180 (1978).

75. COUCH, R. B., DOUGLAS, R. G., JR., LINDGREN, K. M., GERONE, P. J., AND KNIGHT, V., Airborne transmission of respiratory infection with coxsackievirus A type 21, *Am. J. Epidemiol.* **91:**78–86 (1978).

76. CRAIGHEAD, J. E., The role of viruses in the pathogenesis of pancreatic disease and diabetes mellitus, *Prog. Med. Virol.* **19:**161–214 (1975).

77. CROWELL, R. L., AND GOLDBERG, B., Propagation and assay of group A coxsackieviruses in RD cells, *Abstr. Ann. Meeting Am. Soc. Microbiol.* **44:**208 (1974).

78. CURNEN, E. C., SHAW, E. W., AND MELNICK, J. L., Disease resembling nonparalytic poliomyelitis associated with a virus pathogenic for infant mice, *J.A.M.A.* **141:**894–901 (1949).

79. DAGAN, R., AND MENEGUS, M. A., A combination of four cell types for rapid detection of enteroviruses in clinical specimens, *J. Med. Virol.* **19:**219–228 (1986).

80. DAHLING, D. R., BERG, G., AND BERMAN, D., BGM, a continuous cell line more sensitive than primary rhesus and African green monkey kidney cells for the recovery of viruses from water, *Health Lab. Sci.* **11:**275–282 (1974).

81. DALES, L., Poliovaccine strategy, *Lancet* **1:**286 (1985).

82. DALLDORF, G., Neuropathogenicity of group A Coxsackie viruses, *J. Exp. Med.* **106:**69–76 (1957).

83. DALLDORF, G., AND MELNICK, J. L., Coxsackieviruses, in: *Viral and Rickettsial Infections of Man*, 4th ed. (F. L. HORSFALL, JR., AND I. TAMM, EDS.), pp. 474–512, J. B. Lippincott, Philadelphia, 1965.

84. DALLDORF, G., SICKLES, G. M., PLAGER, H., AND GIFFORD, R., A virus recovered from the feces of "poliomyelitis" patients pathogenic for suckling mice, *J. Exp. Med.* **89:**567–582 (1949).

85. DAVIES, D. P., HUGHES, C. A., MACVICAR, J., HAWKES, P., AND MAIR, H. J., Echovirus-11 infection in a special-care baby unit, *Lancet* **1:**96 (1979).

86. DAVIS, L. E., BODIAN, D., PRICE, D., BUTLER, I. J., AND VICKERS, J. H., Chronic progressive poliomyelitis secondary to vaccination of an immunodeficient child, *N. Engl. J. Med.* **297:**241–245 (1977).

87. DE QUADROS, C. A., ANDRUS, J. K., OLIVE, J.-M., AND DE MACEDO, C. G., Polio eradication from the Western Hemisphere, *Annu. Rev. Publ. Health* **13:**239–252 (1992).

88. DESSELBERGER, U., AND FLEWETT, T. H., Clinical and public health virology: A continuous task of changing pattern, *Prog. Med. Virol.* **40:**48–81 (1993).

89. DIEBEL, R., GROSS, L. L., AND COLLINS, D. N., Isolation of a new enterovirus, *Proc. Soc. Exp. Biol. Med.* **148:**203–207 (1975).

90. DOMOK, I., BALAYAN, M. S., FAYINKA, O. A., SKRTIC, N., SONEJI, A. D., AND HARLAND, P. S. E. G., Factors affecting the efficacy of live poliovirus vaccines in warm climates, *Bull. WHO* **51:**333–347 (1974).

91. DONG, D.-X., Immunization with oral poliovirus vaccine in China, *Prog. Med. Virol.* **31:**212–221 (1985).

92. DORRIES, R., AND TER MEULEN, V., Specificity of IgM antibodies in acute human coxsackievirus B infections, analysed by indirect solid phase enzyme immunoassay and immunoblot technique, *J. Gen. Virol.* **64:**159–167 (1983).

93. DREESMAN, G. R., SANCHEZ, Y., IONESCU-MATIU, I., SPARROW, J. T., SIX, H. R., PETERSON, D. L., HOLLINGER, F. B., AND MELNICK, J. L., Antibody to hepatitis B surface antigen after a single inoculation of uncoupled synthetic HBsAg peptides, *Nature* **295:**158–160 (1982).

94. DUNCAN, I. B. R., A comparative study of 63 strains of echovirus type 30, *Arch. Virusforsch.* **25:**93–104 (1968).

95. EDITORIAL, Neurovirulence of enterovirus 70, *Lancet* **1:**373–374 (1982).

96. EDITORIAL, Polio reconsidered, *Lancet* **2:**1309–1310 (1984).

97. EDITORIAL, Expanded immunisation, *Lancet* **1:**438–439 (1985).

98. EDITORIAL, Late sequelae of polio, *Lancet* **2:**1195–1196 (1986).

99. EGGERS, H. J., MERTENS, T., AND GRUNEKLEE, D., Coxsackie infection and diabetes, *Lancet* **2:**631 (1983).

100. EL-HAGRASSY, M. M. O., BANATVALA, J. E., AND COLTART, D. J., Coxsackie-B-virus-specific IgM responses in patients with cardiac and other diseases, *Lancet* **2:**1160 (1980).

101. EMINI, E. A., JAMESON, B. A., AND WIMMER, E., Priming for and induction of anti-poliovirus neutralizing antibodies by synthetic peptides, *Nature* **304:**699–703 (1983).

102. ENDERS, J. F., WELLER, T. H., AND ROBBINS, F. C., Cultivation of the Lansing strain of poliomyelitis virus in cultures of various human embryonic tissues, *Science* **109:**85–87 (1949).

103. ENDERS, J. F., WELLER, T. H., AND ROBBINS, F. C., Alteration in pathogenicity for monkeys of Brunhilde strain of poliomyelitis virus following cultivation in human tissues, *Fed. Proc.* **11:**467 (1952).

104. EVANS, A. S., Criteria for control of infectious diseases with poliomyelitis as an example, *Prog. Med. Virol.* **29:**141–165 (1984).

105. FAULK, W. P., VYAS, G. N., PHILLIPS, C. A., FUDENBERG, H. H., AND CHISM, K., Passive hemagglutination test for antirhinovirus antibodies, *Nature (New Biol.)* **231:**101–104 (1971).

106. FERGUSON, M., EVANS, D. M. A., MAGRATH, D. I., MINOR, P. D., ALMOND, J. W., AND SCHILD, G. C., Induction by synthetic peptides of broadly reactive, type-specific neutralizing antibody to poliovirus type 3, *Virology* **142:**505–515 (1985).

107. FERNANDEZ DE CASTRO, P. J., [Comments on the geographic distribution of poliomyelitis in the Americas: Notes for a program on continental eradication of poliomyelitis], *Rev. Salud Pub. Mexico* **26:**Issue #3 (1984).

108. FERNANDEZ-TOMAS, C. B., AND BALTIMORE, D., Morphogenesis of poliovirus. II. Demonstration of a new intermediate, the provirion, *J. Virol.* **12:**1122–1130 (1973).

109. FOSTER, S. O., KESSENG-MABEN, G., N'JIE, H., AND COFFI, E., Control of poliomyelitis in Africa, *Rev. Infect. Dis.* **6:**(Suppl. 2):S433–S437 (1984).

110. FOX, J. P., Epidemiological aspects of coxsackie and echo virus infections in tropical areas, in: *Proceedings 7th International Congresses on Tropical Medicine and Malaria*, Vol. 3, pp. 212–213, International Congress on Tropical Medicine and Malaria, Rio de Janeiro, 1964.

111. FOX, J. P., Family-based epidemiologic studies: The second Wade Hampton Frost Lecture, *Am. J. Epidemiol.* **99:**165–179 (1974).

112. FOX, J., Eradication of poliomyelitis in the United States: A

commentary on the Salk reviews, *Rev. Infect. Dis.* **2**:277–281 (1980).

113. FOX, J. P., Modes of action of poliovirus vaccines and relation to resulting immunity, *Rev. Infect. Dis.* **6**(Suppl. 2):S352–S355 (1984).

114. FOX, J. P., HALL, C. E., COONEY, M. K., LUCE, R. E., AND KRONMAL, R. A., The Seattle Virus Watch. II. Objectives, study population and its observation, data processing and summary of illnesses, *Am. J. Epidemiol.* **96**:270–285 (1972).

115. FOERSCHLE, J. E., FEORINO, P. M., AND GELFAND, H. J. M., A continuing surveillance of enterovirus infection in healthy children in six United States cities. II. Surveillance enterovirus isolates 1960–1963 and comparison with enterovirus isolates from cases of acute central nervous system disease, *Am. J. Epidemiol.* **83**:455–469 (1966).

116. FUCHS, F., LEPARC, I., KOPECKA, H., GARIN, D., AND AYMARD, M., Use of cRNA digoxigenin-labelled probes for detection of enteroviruses in humans and in the environment, *J. Virol. Methods* **42**:217–226 (1993).

117. FURESZ, J., ARMSTRONG, R. E., AND CONTRERAS, G., Viral and epidemiological links between poliomyelitis outbreaks in unprotected communities in Canada and the Netherlands, *Lancet* **2**:1248 (1978).

118. FURESZ, J., AND CONTRERAS, G., Poliomyelitis in Finland, *Lancet* **2**:693 (1986).

119. GAMBLE, D. R., TAYLOR, K. W., AND CUMMING, H., Coxsackie viruses and diabetes mellitus, *Br. Med. J.* **4**:260–262 (1973).

120. GAUNTT, C. J., GOMEZ, P. T., DUFFEY, P. S., GRANT, J. A., TRENT, D. W., WITHERSPOON, S. M., AND PAQUE, R. E., Characterization and myocarditis capabilities of coxsackievirus B3 variants in selected mouse strains, *J. Virol.* **52**:598–605 (1984).

121. GAUNTT, C. J., GUDVANGEN, R. J., BRANS, Y. W., AND MARLIN, A. E., Coxsackievirus group B antibodies in the ventricular fluid of infants with severe anatomic defects of the central nervous system, *Pediatrics* **76**:64–68 (1985).

122. GAUNTT, C. J., JONES, D. C., HUNTINGTON, H. W., ARIZPE, H. M., GUDVANGEN, R. J., AND DESHAMBO, R. M., Murine forebrain anomalies induced by coxsackievirus B3 variants, *J. Med. Virol.* **14**:341–355 (1984).

123. GEAR, J. H. S., Nonpolio causes of polio-like paralytic syndromes, *Rev. Infect. Dis.* **6**(Suppl. 2):S379–S385 (1984).

124. GEAR, J. H. S., AND MEASROCH, V., Coxsackievirus infections of the newborn, *Prog. Med. Virol.* **15**:42–62 (1973).

125. GELFAND, H. M., FOX, J. P., AND LEBLANC, D. R., The enteric viral flora of a population of normal children in southern Louisiana, *Am. J. Trop. Med.* **6**:521–531 (1957).

126. GELFAND, H. M., HOLGUIN, A. H., MARCHETTI, G. E., AND FEORINO, P. M., A continuing surveillance of enterovirus infections in healthy children in six United States cities. I. Viruses isolated during 1960 and 1961, *Am. J. Hyg.* **78**:358–375 (1963).

127. GIBSON, J. P., AND RIGHTHAND, V. F., Persistence of echovirus 6 in cloned human cells, *J. Virol.* **54**:219–223 (1985).

128. GILBERT, G. L., DICKSON, K. E., WATERS, M. J., KENNETT, M. L., LAND, S. A., AND SNEDDON, M., Outbreak of enterovirus 71 infection in Victoria, Australia, with a high incidence of neurologic involvement, *Pediatr. Infect. Dis.* **7**:484–488 (1988).

129. GLIMAKER, M., ABEBE, A., JOHANSSON, B., EHRNST, A., OLCEN, P., AND STRANNEGARD, O., Detection of enteroviral RNA by polymerase chain reaction in faecal samples from patients with aseptic meningitis, *J. Med. Virol.* **38**:54–61 (1992).

130. GODMAN, G. C., BUNTING, H., AND MELNICK, J. L., The histo-

pathology of Coxsackie virus infection in mice. I. Morphologic observations with four different viral types, *Am. J. Pathol.* **28**:223–257 (1952).

131. GOLDBLUM, N., SWARTZ, T., GERICHTER, C. B., HANDSHER, R., LASCH, E. E., AND MELNICK, J. L., The natural history of poliomyelitis in Israel, 1949–1982, *Prog. Med. Virol.* **29**:115–123 (1984).

132. GOLDFIELD, M. S., SRIHONGSE, S., AND FOX, J. P., Hemagglutinins associated with certain human enteric viruses, *Proc. Soc. Exp. Biol. Med.* **96**:788–791 (1957).

133. GRAHAM, S., WANG, E. C. Y., JENKINS, O., AND BORYSIEWICZ, L. K., Analysis of the human T-cell response to picornaviruses: Identification of T-cell epitopes close to B-cell epitopes in poliovirus, *J. Virol.* **67**:1627–1637 (1993).

134. GRAVES, J. H., Serological relationship of swine vesicular disease virus and coxsackie B5 virus, *Nature* **245**:314–315 (1973).

135. GREEN, M. S., HANDSHER, R., COHEN, D., MELNICK, J. L., SLEPON, R. S., MENDELSOHN, E., AND DANON, Y. L., Age differences in immunity against wild and vaccine strains of poliovirus prior to the 1988 outbreak in Israel: Evidence supporting the need for a booster immunization in adolescents, *Vaccine* **11**:75–81 (1993).

136. GREGG, M. B., Paralytic poliomyelitis can be eliminated, *Rev. Infect. Dis.* **6**(Suppl. 2):S577–S580 (1984).

137. GRINSTEIN, S., MELNICK, J. L., AND WALLIS, C., Virus isolations from sewage and from a stream receiving effluents of sewage treatment plants, *Bull. WHO* **42**:291–296 (1970).

138. GRIST, N. R., Poliovaccine strategy, *Lancet* **1**:286 (1985).

139. GRIST, N. R., AND BELL, E. J., Enteroviral etiology of the paralytic poliomyelitis syndrome. Studies before and after vaccination, *Arch. Environ. Health* **21**:382–387 (1970).

140. GRIST, N. R., AND BELL, E. J., Paralytic poliomyelitis and nonpolio enteroviruses: Studies in Scotland, *Rev. Infect. Dis.* **6**(Suppl. 2):S385–S386 (1984).

141. GRIST, N. R., BELL, E. J., AND ASSAAD, F., Enteroviruses in human disease, *Prog. Med. Virol.* **24**:114–157 (1978).

142. GRIST, N. R., BELL, E. J., AND REID, E., The epidemiology of enteroviruses, *Scottish Med. J.* **20**:27–31 (1975).

143. GYORKEY, F., CABRAL, G. A., GYORKEY, P. K., URIBE-BOTERO, G., DREESMAN, G. R., AND MELNICK, J. L., Coxsackievirus aggregates in muscle cells of a polymyositis patient, *Intervirology* **10**:69–77 (1978).

144. HAGIWARA, A., TAGAYA, I., AND YONEYAMA, T., Epidemic of hand, foot and mouth disease associated with enterovirus 71 infection, *Intervirology* **9**:60–63 (1978).

145. HALL, C. E., COONEY, M. K., AND FOX, J. P., The Seattle Virus Watch Program. I. Infection and illness experience of Virus Watch families during a community-wide epidemic of echovirus type 30 aseptic meningitis, *Am. J. Public Health* **60**:1456–1465 (1970).

146. HALLAM, N. F., EGLIN, R. P., HOLLAND, P., BELL, E. J., AND SQUIER, M. V., Fatal coxsackie B meningoencephalitis diagnosed by serology and in-situ nucleic acid hybridisation, *Lancet* **2**:1213–1214 (1986).

147. HALSEY, N., AND GALAZKA, A., The efficacy of DPT and oral poliomyelitis immunization schedules initiated from birth to 12 weeks of age, *Bull. WHO* **63**:1151–1169 (1985).

148. HAMMON, W. McD., CORIELL, L. L., AND WEHRLE, P. F., Evaluation of Red Cross gamma globulin as a prophylactic agent for poliomyelitis. IV. Final report of results based on clinical diagnosis, *J.A.M.A.* **51**:1272–1285 (1953).

149. HAMMON, W., McD., LUDWIG, E. H., SATHER, G., AND YOHN, D. S., Comparative studies on patterns of family infections with polioviruses and ECHO virus type 1 on an American military base in the Philippines, *Am. J. Public Health* **47**:802–811 (1957).

150. HAMPIL, B., MELNICK, J. L., WALLIS, C., BROWN, R. W., BRAYE, E. T., AND ADAMS, R. R., JR., Preparation of antiserum to enteroviruses in large animals, *J. Immunol.* **95**:895–908 (1965).

151. HANNINGTON, G., BOOTH, J. C., BOWES, R. J., AND STERN, H., Coxsackie B virus-specific IgM antibody and myocardial infarction, *J. Med. Microbiol.* **21**:287–291 (1986).

152. HARRIS, L. F., HAYNES, R. E., CRAMBLETT, H. G., CONANT, R. M., AND JENKINS, G. R., Antigenic analysis of echoviruses 1 and 8, *J. Infect. Dis.* **127**:63–68 (1973).

153. HARRIS, T. J. R., AND BROWN, F., Correlation of polypeptide composition with antigenic variation in the swine vesicular disease and coxsackie B₅ viruses, *Nature* **258**:758–760 (1975).

154. HARRISON, S. C., Common cold virus and its receptor, *Proc. Natl. Acad. Sci. USA* **90**:783 (1993).

155. HARTIG, P. C., MADGE, G. E., AND WEBB, S. R., Diversity within a human isolate of Coxsackie B4: Relationship to viral-induced diabetes, *J. Med. Virol.* **12**:23–30 (1983).

156. HASEGAWA, A., AND INOUYE, S., Production in guinea pigs of antibodies to cross-reactive antigenic determinants of human enteroviruses, *J. Clin. Microbiol.* **17**:458–462 (1983).

157. HAWLEY, H. B., MORIN, D. P., GERAGHTY, M. E., TOMKOW, J., AND PHILLIPS, C. A., Coxsackievirus B epidemic at a boy's summer camp: Isolation of virus from swimming water, *J.A.M.A.* **226**:33–36 (1973).

158. HEALTH AND WELFARE CANADA, Wild poliovirus isolated in Alberta, 1993, *Can. Commun. Dis. Rep.* **19**:57–58 (1993).

159. HENDERSON, R. H., AND SUNDARESAN, T., Cluster sampling to assess immunization coverage: A review of experience with a simplified sampling methodology, *Bull. WHO* **60**:253–260 (1982).

160. HENNESSEN, W., AND VAN WEZEL, A. L. (EDS.), *Reassessment of Inactivated Poliomyelitis Vaccine: Developments in Biological Standardization*, Vol. 47, S. Karger, Basel, 1981.

161. HEYMANN, D. L., FLOYD, V. D., LUCHNEVSKI, M., KESSENG-MABEN, G., AND MVONGO, F., Estimation of incidence of poliomyelitis by three survey methods in different regions of the United Republic of Cameroon, *Bull. WHO* **61**:501–507 (1983).

162. HIERHOLZER, J. C., BINGHAM, P. G., COOMBS, R. A., STONE, Y. O., AND HATCH, M. H., Quantitation of enterovirus-70 antibody by microneutralization test and comparison with standard neutralization, hemagglutination-inhibition, and complement-fixation tests with different virus strains, *J. Clin. Microbiol.* **19**:826–830 (1984).

163. HIERHOLZER, J. C., HILLIARD, K. A., AND ESPOSITO, J. J., Serosurvey for "acute hemorrhagic conjunctivitis" virus (enterovirus 70) antibodies in the southeastern Untied States, with review of the literature and some epidemiologic implications, *Am. J. Epidemiol.* **102**:533–544 (1975).

164. HODES, D. S., AND ESPINOZA, D. V., Temperature sensitivity of isolates of echovirus type 11 causing chronic meningoencephalitis in an agammaglobulinemic patient, *J. Infect. Dis.* **144**:377 (1981).

165. HOGLE, J. M., CHOW, M., AND FILMAN, D. J., Three-dimensional structure of poliovirus at 2.9 Å resolution, *Science* **229**:1358–1365 (1985).

166. HOLLINGER, F. B., ANDRE, F., AND MELNICK, J. L. (EDS.), International Symposium on Active Immunization against Hepatitis A, *Vaccine* **10**(Suppl. 1):1–174 (1992).

167. HONIG, E. I., MELNICK, J. L., ISACSON, P., PARR, R., MYERS, I. L., AND WALTON, M., An epidemiological study of enteric virus infections: Poliomyelitis, Coxsackie, and orphan (ECHO) viruses isolated from normal children in two socioeconomic groups, *J. Exp. Med.* **103**:247–262 (1956).

168. HORSTMANN, D. M., Control of poliomyelitis: A continuing paradox, *J. Infect. Dis.* **146**:540–551 (1982).

169. HORSTMANN, D. M., EMMONS, J., GIMPEL, L., SUBRAHMANYAN, T., AND RIORDAN, J. T., Enteroviruses surveillance following a community-wide oral poliovirus vaccination program: A seven-year study, *Am. J. Epidemiol.* **97**:173–186 (1973).

170. HORSTMANN, D. M., McCOLLUM, R. W., AND MASCOLA, A. D., Viremia in human poliomyelitis, *J. Exp. Med.* **99**:355–369 (1954).

171. HOVI, T., Remaining problems before eradication of poliomyelitis can be accomplished, *Prog. Med. Virol.* **38**:42–55 (1991).

172. HOVI, T., AND ROIVAINEN, M., Peptide antisera targeted to a conserved sequence in poliovirus capsid protein VP1 cross-react widely with members of the genus *Enterovirus*, *J. Clin. Microbiol.* **31**:1083–1087 (1993).

173. HOVI, T., HUOVILAINEN, A., KURONEN, T., POYRY, T., SALAMA, N., CANTELL, K., KINNUNEN, E., LAPINLEIMU, K., ROIVAINEN, M., STENVIK, M. SILANDER, A., THODEN, C.-J., SALMINEN, S., AND WECKSTROM, P., Outbreak of paralytic poliomyelitis in Finland: Widespread circulation of antigenically altered poliovirus type 3 in a vaccinated population, *Lancet* **1**:1427–1432 (1986).

174. HSIUNG, G. D., AND MELNICK, J. L., Morphologic characteristics of plaques produced on monkey kidney monolayer cultures by enteric viruses (poliomyelitis, Coxsackie and ECHO groups), *J. Immunol.* **78**:128–136 (1957).

175. HUBER, S. A., HAISCH, C., AND LODGE, P. A., Coxsackievirus-induced myocarditis: Model for picornavirus-induced autoimmunity, *Semin. Virol.* **1**:289–295 (1990).

176. HUGHES, P. J., EVANS, D. M. A., MINOR, P. D., SCHILD, G. C., ALMOND, J. W., AND STANWAY, G., The nucleotide sequence of a type 3 poliovirus isolated during a recent outbreak of poliomyelitis in Finland, *J. Gen. Virol.* **67**:2093–2102 (1986).

177. HUMPHREY, D. D., KEW, O. M., AND FEORINO, P. M., Monoclonal antibodies of four different specificities for neutralization of type 1 polioviruses, *Infect. Immun.* **36**:841–843 (1982).

178. HYYPIA, T., STALHANDSKE, P., VAINIONPAA, R., AND PETTERSSON, U., Detection of enteroviruses by spot hybridization, *J. Clin. Microbiol.* **19**:436–438 (1984).

179. INTERNATIONAL POLIOMYELITIS CONFERENCES, *Poliomyelitis: Papers and Discussions Presented at the First through Fifth International Poliomyelitis Conferences* (M. FISHBEIN, ED.), J. B. Lippincott, Philadelphia, 1949, 1952, 1955, 1958, 1961.

180. INTERNATIONAL SYMPOSIUM ON POLIOMYELITIS CONTROL, Pan American Health Organization, Washington, *Rev. Infect. Dis.* **6**(Suppl. 2):S301–S601 (1984).

181. ISHIKO, H., TAKEDA, N., MIYAMURA, K., KATO, N., TANIMURA, M., LIN, K. H., YIN-MURPHY, M., TAM, J. S., MU, G. F., AND YAMAZAKI, S., Phylogenetic analysis of a coxsackievirus A24 variant: The most recent worldwide pandemic was caused by progenies of a virus prevalent around 1981, *Virology* **187**:748–759 (1992).

182. ITOH, H., AND MELNICK, J. L., The infection of chimpanzees with ECHO viruses, *J. Exp. Med.* **106**:677–688 (1957).

183. ITOH, H., AND MELNICK, J. L., Double infections of single cells

with ECHO 7 and Coxsackie A9 viruses, *J. Exp. Med.* **109**:393–406 (1959).

184. JACKSON, G. G., AND MULDOON, R. L., Viruses causing common respiratory infections in man, *J. Infect. Dis.* **127**:328–408 (1973).

185. JACOBSON, M. F., ASSO, J., AND BALTIMORE, D., Further evidence on the formation of poliovirus proteins, *J. Mol. Biol.* **49**:657–669 (1970).

186. JANDA, Z., ADAM, E., AND VONKA, V., Properties of a new type 3 attenuated poliovirus. VI. Alimentary tract resistance in children fed previously with type 3 Sabin vaccine to reinfection with homologous and heterologous type 3 attenuated poliovirus, *Arch. Virusforsch.* **20**:87–98 (1967).

187. JENISTA, J. A., POWELL, K. R., AND MENEGUS, M. A., The epidemiology of neonatal enterovirus infection, *J. Pediatrics* **104**:685–690 (1984).

188. JOHN, T. J., CHRISTOPHER, S., AND ABRAHAM, J., Neurological manifestation of acute haemorrhagic conjunctivitis due to enterovirus 70, *Lancet* **2**:1283–1284 (1981).

189. JOHNS, R. B., FARNSWORTH, S., THOMPSON, H., AND BRADY, F., Two voluntary mass immunization programs using Sabin oral vaccine, *J.A.M.A.* **183**:171–175 (1963).

190. JOHNSON, I., HAMMOND, G. W., AND VERMA, M. R, Nosocomial coxsackie B4 virus infections in two chronic-care pediatric neurological wards, *J. Infect. Dis.* **151**:1153–1156 (1985).

191. JOHNSTON, R. T., Late progression of poliomyelitis paralysis: Discussion of pathogenesis, *Ref. Infect. Dis.* **6**(Suppl. 2):S568–S570 (1984).

192. JOHNSTONE, T., Poliovaccine strategy, *Lancet* **2**:695 (1985).

193. KAJIGAYA, S., ARAKAWA, H., KUGE, S., KOI, T., IMURA, N., AND NOMOTO, A., Isolation and characterization of defective-interfering particles of poliovirus Sabin 1 strain, *Virology* **142**:307–316 (1985).

194. KANDOLF, R., AND HOFSCHNEIDER, P. H., Molecular cloning of the genome of a cardiotropic Coxsackie B3 virus: Full-length reverse-transcribed recombinant cDNA generates infectious virus in mammalian cells, *Proc. Natl. Acad. Sci. USA* **82**:4818–4822 (1985).

195. KANDOLF, R., KLINGEL, K., ZELL, R., SELINKA, H.-C., RAAB, U., SCHNEIDER-BRACHERT, W., AND BULTMANN, B., Molecular pathogenesis of enterovirus-induced myocarditis: Virus persistence and chronic inflammation, *Intervirology* **35**:140–151 (1993).

196. KAPLAN, G. J., CLARK, P. S., BENDER, T. R., FELTZ, E. T., LISTYOUNG, B., NEVIUS, S. E., AND CHIN, T. D. Y., Echovirus type 30 meningitis and related febrile illness: Epidemiologic study of an outbreak in an Eskimo community, *Am. J. Epidemiol.* **92**:257–265 (1970).

197. KAPLAN, M. H., KLEIN, S. W., MCPHEE, J., AND HARPER, R. G., Group B coxsackievirus infections in infants younger than three months of age: A serious childhood illness, *Rev. Infect. Dis.* **5**:1019–1032 (1983).

198. KAPSENBERG, J., Picornaviridae: The Enteroviruses, in: *Laboratory Diagnosis of Infectious Diseases*, Vol. 2 (E. H. LENNETTE, P. HALONEN, AND F. A. MURPHY, EDS.), pp. 692–722, Springer-Verlag, New York, 1988.

199. KAPSENBERG, J. G., RAS, A., AND KORTE, J., Improvement of enterovirus neutralization by treatment with sodium deoxycholate or chloroform, *Intervirology* **12**:329–334 (1979).

200. KEDMI, S., AND KATZENELSON, E., A rapid quantitative fluorescent antibody assay of polioviruses using tragacanth gum, *Arch. Virol.* **56**:337–340 (1978).

201. KELLY, S., WINSSER, J., AND WINKELSTEIN, W., JR., Poliomyelitis and other enteric viruses in sewage, *Am. J. Public Health* **47**:72–77 (1957).

202. KENNETT, M. L., BIRCH, C. J., LEWIS, F. A., YUNG, A. P., LOCARNINI, S. A., AND GUST, I. D., Enterovirus type 71 infection in Melbourne, *Bull. WHO* **51**:609–615 (1970).

203. KEW, O. M., AND NOTTAY, B. K., Molecular epidemiology of polioviruses, *Rev. Infect. Dis.* **6**(Suppl. 2):S499–S504 (1984).

204. KEW, O. M., NOTTAY, B. K., HATCH, M. H., HIERHOLZER, J. C., AND OBIJESKI, J. F., Oligonucleotide fingerprint analysis of enterovirus 70 isolates from the 1980 to 1981 pandemic of acute hemorrhagic conjunctivitis: Evidence for a close genetic relationship among Asian and American strains, *Infect. Immun.* **41**:631–635 (1983).

205. KIBRICK, S., Current status of Coxsackie and ECHO viruses in human disease, *Prog. Med. Virol.* **6**:27–70 (1964).

206. KIM-FARLEY, R. J., SCHONBERGER, L. B., NKOWANE, B. M., KEW, O. M., BART, K. J., ORENSTEIN, W. A., HINMAN, A. R., HATCH, M. H., AND KAPLAN, J. E., Poliomyelitis in the USA: Virtual elimination of disease caused by wild virus, *Lancet* **2**:1315–1317 (1984).

207. KING, M. L., BIDWELL, D., SHAIKH, A., VOLLER, A., AND BANATVALA, J. E., Coxsackie B virus specific IgM responses in children with insulin-dependent (juvenile-onset; type 1) diabetes mellitus, *Lancet* **1**:1397–1399 (1983).

208. KITAMURA, N., SEMLER, B. L., ROTHBERG, P. G., LARSEN, G. R., ADLER, C. J., DORNER, A. J., EMINI, E. A., HANECAK, R., LEE, J. J., VAN DER WERF, S., ANDERSON, C. W., AND WIMMER, E., Primary structure, gene organization, and polypeptide expression of poliovirus RNA, *Nature* **291**:547–553 (1981).

209. KJELLEN, L., AND VON ZEIPEL, G., Influence of host cells on the infectivity and neutralizability of echovirus type 4 (Pesascek), *Intervirology* **22**:32–40 (1984).

210. KLINGEL, K., HOHENADL, C., CANU, A., ALBRECHT, M., SEEMANN, M., MALL, G., AND KANDOLF, R., Ongoing enterovirus-induced myocarditis is associated with persistent heart muscle infection: Quantitative analysis of virus replication, tissue damage, and inflammation, *Proc. Natl. Acad. Sci. USA* **89**:314–318 (1992).

211. KOCH, F., AND KOCH, G., *The Molecular Biology of Poliovirus*, Springer-Verlag, New York, 1985.

212. KOGON, A., SPIGLAND, I., FROTHINGHAM, T. C., ELVEBACK, L., WILLIAMS, C., HALL, C. E., AND FOX, J. P., The Virus Watch Program: A continuing surveillance of viral infections in metropolitan New York families. VII. Observations on viral excretion, seroimmunity, intrafamilial spread and illness association in coxsackievirus and echovirus infections, *Am. J. Epidemiol.* **89**:51–61 (1969).

213. KOIKE, S., ISE, I., SATO, Y., MITSUI, K., HORIE, H., UMEYAMA, H., AND NOMOTO, A., Early events of poliovirus infection, *Semin. Virol.* **3**:109–115 (1992).

214. KONO, R., Apollo 11 disease or acute hemorrhagic conjunctivitis: A pandemic of a new enterovirus infection of the eyes, *Am. J. Epidemiol.* **101**:383–390 (1975).

215. KONO, R., MIYAMURA, K., TAJIRI, E., SASAGAWA, A., PHUAPRADIT, P., ROONGWITHU, N., VEJJAJIVA, A., JAYAVASU, C., THONGCHAROEN, P., WASI, C., AND RODPRASSERT, P., Virological and serological studies of neurological complications of acute hemorrhagic conjunctivitis in Thailand, *J. Infect. Dis.* **135**:706–713 (1977).

216. KONO, R., MIYAMURA, K., TAJIRI, E., SHIGA, S., SASAGAWA, A.,

IRANI, P. F., KATRAK, S. M., AND WADIA, N. H., Neurologic complications associated with acute hemorrhagic conjunctivitis virus infection and its serological confirmation, *J. Infect. Dis.* **129:**590–593 (1974).

217. KONO, R., SASAGAWA, A., ISHII, K., SUGIURA, S., OCHI, M., MATSUMIYA, H., UCHIDA, Y., KAMEYAMA, K., KANEKO, M., AND SAKURAI, N., Pandemic of new type of conjunctivitis, *Lancet* **1:**1191–1194 (1972).

218. KONO, R., SASAGAWA, A., MIYAMURA, K., AND TAJIRI, E., Serologic characterization and sero-epidemiologic studies on acute hemorrhagic conjunctivitis (AHC) virus, *Am. J. Epidemiol.* **101:**444–457 (1975).

219. KRAFT, L. M., AND MELNICK, J. L., Complement fixation tests with homologous and heterologous types of Coxsackie virus in man, *J. Immunol.* **68:**297–310 (1952).

220. KRAFT, L. M., AND MELNICK, J. L., Quantitative studies of the virus-host relationship in chimpanzees after inapparent infection with Coxsackie viruses. II. The development of complement-fixing antibodies, *J. Exp. Med.* **97:**401–414 (1953).

221. KRAH, D. L., AND CROWELL, R. L., A solid-phase assay of solubilized HeLa cell membrane receptors for binding group B coxsackieviruses and polioviruses, *Virology* **118:**148–156 (1982).

222. KURITSKY, J. N., WEAVER, J. H., BERNARD, K. W., MOKHBAT, J. E., HATCH, M. H., OSTERHOLM, M. T., AND PATRIARCA, P. A., An outbreak of acute hemorrhagic conjunctivitis in central Minnesota, *Am. J. Ophthalmol.* **96:**449–452 (1983).

223. LAFORCE, F. M., LICHNEVSKI, M. S., KEJA, J., AND HENDERSON, R. H., Clinical survey techniques to estimate prevalence and annual incidence of poliomyelitis in developing countries, *Bull. WHO* **58:**609–620 (1980).

224. LAKE, A. M., LAUER, B. A., CLARK, J. C., WESENBERG, R. L., AND MCINTOSH, K., Enterovirus infections in neonates, *J. Pediatr.* **89:**787–791 (1976).

225. LAMB, G. A., CHIN, T. D. Y., AND SCARCE, L. E., Isolations of enteric viruses from sewage and river water in a metropolitan area, *Am. J. Hyg.* **80:**320–327 (1964).

226. LANDSTEINER, K., AND POPPER, E., Ubertragung der Poliomyelitis acute auf Affen, *Z. Immunitaetsforsch. Orig.* **2:**377–390 (1909).

227. LANSDOWN, A. B. G., Viral infections and diseases of the heart, *Prog. Med. Virol.* **24:**70–113 (1978).

228. LAPINLEIMU, K., Killed polio virus vaccine in the control of poliomyelitis in Finland, *Ann. Clin. Res.* **144:**199–203 (1982).

229. LASCH, E. E., ABED, Y., ABDULLA, K., EL TIBBI, A. G., MARCUS, O., EL MASSRI, M., HANDSCHER, R., GERICHTER, C. B., AND MELNICK, J. L., Successful results of a program combining live and inactivated poliovirus vaccines to control poliomyelitis in Gaza, *Rev. Infect. Dis.* **6**(Suppl. 2):S467–S470 (1984).

230. LAU, R. C. H., Coxsackie B virus infections in New Zealand patients with cardiac and noncardiac diseases, *J. Med. Virol.* **11:**131–137 (1983).

231. LAVINDER, C. H., FREEMAN, A. W., AND FROST, W. H., Epidemiologic studies of poliomyelitis in New York City and Northeastern United States during the year 1916, *Public Health Bull.* **91:**1–309 (1918).

232. LEE H. S., BOULILIER, J. E., MACDONALD, M. A., AND FORWARD, K. E., A novel approach to enterovirus typing, *J. Med. Virol.* **35:**128–132 (1991).

233. LEINIKKI, P. O., PASTERNACK, A., MUSTONEN, J., TANUIANPAA, P., AND HYOTY, H., Paralytic poliomyelitis in Finland, *Lancet* **2:**507 (1985).

234. LEMON, S. M., AND ROBERTSON, S. E., Global eradication of poliomyelitis: Recent progress, future prospects, and new research priorities, *Prog. Med. Virol.* **38:**42–55 (1991).

235. LEONARDI, G. P., GREENBERG, A. J., COSTELLO, P., AND SZABO, K., Echovirus type 30 infection associated with aseptic meningitis in Nassau County, New York, USA, *Intervirology* **36:**53–56 (1993).

236. LERNER, A. M., Coxsackievirus myocardiopathy, *J. Infect. Dis.* **120:**496–499 (1969).

237. LERNER, A. M., Myocarditis and pericarditis, in: *International Textbook of Medicine* (A. I. BRAUDE, ED.), pp. 1520–1530, W. B. Saunders, Philadelphia, 1981.

238. LERNER, A. M., AND WILSON, F. M., Virus myocardiopathy, *Prog. Med. Virol.* **15:**63–91 (1973).

239. LIM, K. A., AND BENYESH-MELNICK, M., Typing of viruses by combinations of antiserum pools: Application to typing of enteroviruses (Coxsackie and echo), *J. Immunol.* **84:**309–317 (1960).

240. LIM, K. H., AND YIN-MURPHY, M., An epidemic of conjunctivitis in Singapore in 1970, *Singapore Med. J.* **12:**247–249 (1971).

241. LIM, K. H., AND YIN-MURPHY, M., The aetiologic agents of epidemic conjunctivitis, *Singapore Med. J.* **18:**41–43 (1977).

242. LIN, K.-H. WANG, H.-L., SHEU, M.-M., HUANG, W.-L., CHEN, C.-W., YANG, C.-S., TAKEDA, N., KATO, N., MIYAMURA, K., AND YAMAZAKI, S., Molecular epidemiology of a variant of coxsackievirus A24 in Taiwan: Two epidemics caused by phylogenetically distinct viruses from 1985 to 1989, *J. Clin. Microbiol.* **31:**1160–1166 (1993).

243. LUND, E., HEDSTROM, C.-E., AND STRANNEGARD, O., A comparison between virus isolations from sewage and from fecal specimens from patients, *Am. J. Epidemiol.* **84:**282–286 (1966).

244. MACLEOD, D. R. E., ING, W. K., BELCOURT, R. J.-P., PEARSON, E. W., AND BELL, J. R., Antibody status to poliomyelitis, measles, rubella, diphtheria and tetanus, Ontario, 1969–70: Deficiencies discovered and remedies required, *Can. Med. Assoc. J.* **113:**619–623 (1975).

245. MAGRATH, D. I., EVANS, D. M. A., FERGUSON, M., SCHILD, G. C., MINOR, P. D., HORAUD, F., CRAINIC, R., STENVIK, M., AND HOVI, T., Antigenic and molecular properties of type 3 poliovirus responsible for an outbreak of poliomyelitis in a vaccinated population, *J. Gen. Virol.* **67:**899–905 (1986).

246. MAGUIRE, T., The laboratory transmission of Coxsackie A6 virus by mosquitoes, *J. Hyg.* **68:**625–630 (1970).

247. MAPOLES, J. E., KRAH, D. L., AND CROWELL, R. L., Purification of a HeLa cell receptor protein for group B coxsackieviruses, *J. Virol.* **55:**560–566 (1985).

248. MARTINO, T. A., SOLE, M. J., PENN, L. Z., LIEW, C.-C., AND LIU, P., Quantitation of enteroviral RNA by competitive polymerase chain reaction, *J. Clin. Microbiol.* **31:**2634–2640 (1993).

249. MATTHEWS, R. E. F., Classification and nomenclature of viruses. Fourth Report of the International Committee on Taxonomy of Viruses, *Intervirology* **17:**1–199 (1982).

250. MCALLISTER, R. M., MELNYK, J., FINKELSTEIN, J. Z., ADAMS, E. C., JR., AND GARDNER, M. B., Cultivation *in vitro* of cells derived from a human rhabdomyosarcoma, *Cancer* **24:**520–526 (1969).

251. MCBEAN, A. M., THOMAS, M. L., JOHNSON, R. H., GADLESS, B. R., MACDONALD, B., NERHOOD, L., CUMMINS, P., HUGHES, J., KINNEAR, J., WATTS, C., KRAFT, M., ALBRECHT, P., BOONE, E. J., MOORE, M., FRANK, J. A., JR., AND BERNIER, R., A comparison of the serologic responses to oral and injectable trivalent poliovirus vaccines, *Rev. Infect. Dis.* **6**(Suppl. 2):S552–S555 (1984).

252. MCCARROLL, J. R., MELNICK, J. L., AND HORSTMANN, D. M.,

Spread of poliomyelitis infection in nursery schools, *Am. J. Publ. Health* **45:**1541–1550 (1955).

253. MEASE, P. J., OCHS, H. D., COREY, L., DRAGAVON, J., AND WEDGEWOOD, R. J., Echovirus encephalitis/myositis in X-linked agammaglobulinemia, *N. Engl. J. Med.* **313:**758 (1985).

254. MELCHERS, W., DE VISSER, M., JONGEN, P., VAN LOON, A., NIBBELING, R., OOSTVOGEL, P., WILLEMSE, D., AND GALAMA, J., The postpolio syndrome: No evidence for poliovirus persistence, *Ann. Neurol.* **32:**728–732 (1992).

254a. MELNICK, J. L., Poliomyelitis virus in urban sewage in epidemic and in non-epidemic times, *Am. J. Hyg.* **45:**240–253 (1947).

255. MELNICK, J. L., Variation in poliomyelitis virus on serial passage through tissue culture, *Cold Spring Harbor Symp. Quant. Biol.* **18:**278–279 (1953).

256. MELNICK, J. L., Application of tissue culture methods to epidemiological studies of poliomyelitis, *Am. J. Public Health* **44:**571–580 (1954).

257. MELNICK, J. L., Antigenic crossings within poliovirus types, *Proc. Soc. Exp. Biol. Med.* **89:**131–133 (1955).

258. MELNICK, J. L., ECHO viruses, in: *Cellular Biology, Nucleic Acids, and Viruses*, Special Publications of the New York Academy of Sciences, Vol. 5 (THOMAS M. RIVERS, CONSULTING ED.), pp. 365–381, New York Academy of Sciences, New York, 1957.

259. MELNICK, J. L., Advances in the study of the enteroviruses, *Prog. Med. Virol.* **1:**59–105 (1958).

260. MELNICK, J. L., Population genetics applied to live poliovirus vaccine, *Am. J. Publ. Health* **52:**472–483 (1962).

261. MELNICK, J. L., Echoviruses, in: *Viral and Rickettsial Infections of Man*, 4th ed. (F. L. HORSFALL, JR., AND I. TAMM, EDS.), pp. 513–545, J. B. Lippincott, Philadelphia, 1965.

262. MELNICK, J. L., Enteroviruses: Vaccines, epidemiology, diagnosis, and classification, *CRC Crit. Rev. Clin. Lab. Sci.* **1:**87–118 (1970).

263. MELNICK, J. L., Periodic serological surveillance, in: *Serological Epidemiology* (J. R. PAUL AND C. WHITE, EDS.), pp. 142–154, Academic Press, New York, 1973.

264. MELNICK, J. L., Reference materials in virology: The enterovirus example, in: *Proceedings International Conference on Standardization of Diagnostic Materials*, pp. 213–235, Centers for Disease Control, Atlanta, 1974.

265. MELNICK, J. L., *Report to the World Health Organization: Recommendations for the Control of Poliomyelitis in Israel, West Bank and Gaza Strip*, Report No. EM/VIR/7, EPID/54, World Health Organization, Geneva, 1977.

266. MELNICK, J. L., Advantages and disadvantages of killed and live poliomyelitis vaccines, *Bull. WHO* **56:**21–38 (1978).

267. MELNICK, J. L., Towards the eradication of poliomyelitis, in: *Medical Virology: Proceedings International Symposium on Medical Virology* (L. M. DE LA MAZA AND E. M. PETERSON, EDS.), pp. 261–299, Elsevier Biomedical, New York, 1982.

268. MELNICK, J. L., Portraits of viruses: The picornaviruses, *Intervirology* **20:**61–100 (1983).

269. MELNICK, J. L., Enterovirus type 71 infections: A varied clinical pattern sometimes mimicking paralytic poliomyelitis, *Rev. Infect. Dis.* **6**(Suppl. 2):S387–S390 (1984).

270. MELNICK, J. L., New approaches to hepatitis B vaccines, in: *New Approaches to Vaccine Development* (R. BELL AND G. TORRIGIANI, EDS.), pp. 218–236, Schwabe & Co., Basel, 1984.

271. MELNICK, J. L. (ED.), *Enteric Viruses in Water, Monographs in Virology*, Vol. 15, S. Karger, Basel, 1984.

272. MELNICK, J. L., Etiologic agents and their potential for causing waterborne virus diseases, in: *Enteric Viruses in Water: Monographs in Virology*, Vol. 15 (J. L. MELNICK, ED.), pp. 1–16, S. Karger, Basel, 1984.

273. MELNICK, J. L., Recent developments in the worldwide control of poliomyelitis, in: *Control of Virus Diseases* (E. KURSTAK AND R. G. MARUSYK, EDS.), pp. 3–31, Marcel Dekker, New York, 1984.

274. MELNICK, J. L., Enteroviruses, in: *Laboratory of Diagnosis of Viral Infections* (E. H. LENNETTE, ED.), pp. 241–256, Marcel Dekker, New York, 1985.

275. MELNICK, J. L., Poliomyelitis: Eradication in sight, *Epidemiol. Infect.* **108:**1–18 (1992).

276. MELNICK, J. L., Enteroviruses, in: *Manual of Clinical Laboratory Immunology*, 4th ed. (N. R. ROSE, E. C. MACARIO, J. L. FAHEY, H. FRIEDMAN, AND G. M. PENN, EDS.), pp. 631–633, Washington, DC, American Society for Microbiology, 1992.

277. MELNICK, J. L., Properties and classification of hepatitis A virus, *Vaccine* **10**(Suppl. 1):24–26 (1992).

278. MELNICK, J. L., Live attenuated poliovaccines in: *Vaccines*, 2nd ed. (S. A. PLOTKIN AND E. A. MORTIMER, JR., EDS.), pp. 155–204, W. B. Saunders, Philadelphia, 1993.

279. MELNICK, J. L., AGOL, V. I., BACHRACH, H. L., BROWN, F., COOPER, P. D., FIERS, W., GARD, S., GEAR, J. H. S., GHENDON, Y., KASZA, L., LAPLACA, M., MANDEL, B., MCGREGOR, S., MOHANTY, S. B., PLUMMER, G., RUECKERT, R. R., SCHAFFER, F. L., TAGAYA, I., TYRRELL, D. A. J., VOROSHILOVA, M., AND WENNER, H., Picornaviridae, *Intervirology* **4:**303–316 (1974).

280. MELNICK, J. L., AND ÅGREN, K., Poliomyelitis and Coxsackie viruses isolated from normal infants in Egypt, *Proc. Soc. Exp. Biol. Med.* **81:**621–624 (1952).

281. MELNICK, J. L., ASHKENAZI, A., MIDULLA, V. C., WALLIS, C., AND BERNSTEIN, A., Immunogenic potency of MgCl$_2$-stabilized oral poliovaccine, *J.A.M.A.* **185:**406–408 (1963).

282. MELNICK, J. L., CLARKE, N. A., AND KRAFT, L. M., Immunological reactions of the Coxsackie viruses. III. Cross-protection tests in infant mice born of vaccinated mothers. Transfer of immunity through the milk, *J. Exp. Med.* **92:**499–505 (1950).

283. MELNICK, J. L., COCKBURN, W. C., DALLDORF, G., GARD, S., GEAR, J. H. S., HAMMON, W., McD., KAPLAN, M. M., NAGLER, F. P., OKER-BLOM, N., RHODES, A. J., SABIN, A. B., VERLINDE, J. D., AND VON MAGNUS, H., Picornavirus group, *Virology* **19:**114–116 (1963).

284. MELNICK, J. L., CROWTHER, D., AND BARRERA-ORO, J., Rapid development of drug-resistant mutants of poliovirus, *Science* **134:**557 (1961).

285. MELNICK, J. L., DALLDORF, G., ENDERS, J. F., HAMMON, W. McD., SABIN, A. B., SYVERTON, J. T., AND WENNER, H. A., The enteroviruses, *Am. J. Publ. Health* **47:**1556–1566 (1957).

286. MELNICK, J. L., AND DOW, R. P., Poliomyelitis in Hidalgo County, Texas, 1948. Poliomyelitis and Coxsackie viruses from flies, *Am. J. Hyg.* **58:**288–309 (1953).

287. MELNICK, J. L., EMMONS, J., COFFEY, J. H., AND SCHOOF, H., Seasonal distribution of Coxsackie viruses in urban sewage and flies, *Am. J. Hyg.* **59:**164–184 (1954).

288. MELNICK, J. L., AND GERBA, C. P., The ecology of enteroviruses in natural waters, *CRC Crit. Rev. Environ. Control* **10:**65–93 (1980).

289. MELNICK, J. L., GERBA, C. P., AND WALLIS, C., Viruses in water, *Bull. WHO* **56:**499–508 (1978).

290. MELNICK, J. L., AND HAMPIL, B., WHO collaborative studies on enterovirus reference antisera: Fourth report, *Bull. WHO* **48:**381–396 (1973).

291. MELNICK, J. L., AND LEDINKO, N., Social serology: Antibody levels in a normal young population during an epidemic of poliomyelitis, *Am. J. Hyg.* **54**:354–382 (1951).

292. MELNICK, J. L., AND LEDINKO, N., Development of neutralizing antibodies against the three types of poliomyelitis virus during an epidemic period: The ratio of inapparent infection to clinical poliomyelitis, *Am. J. Hyg.* **58**:207–222 (1953).

293. MELNICK, J. L., McCARROLL, J. R., AND HORSTMANN, D. M., A winter outbreak of poliomyelitis in New York City: The complement-fixation test as an aid in rapid diagnosis, *Am. J. Hyg.* **63**:95–114 (1956).

294. MELNICK, J. L., PAUL, J. R., AND WALTON, M., Serologic epidemiology of poliomyelitis, *Am. J. Publ. Health* **45**:429–437 (1955).

295. MELNICK, J. L., AND PENNER, L. R., The survival of poliomyelitis and Coxsackie viruses following their ingestion by flies, *J. Exp. Med.* **96**:255–271 (1951).

296. MELNICK, J. L., PROCTOR, R. O., OCAMPO, A. R., DIWAN, A. R., AND BEN-PORATH, E., Free and bound virus in serum after administration of oral poliovirus vaccine, *Am. J. Epidemiol.* **84**:329–342 (1966).

297. MELNICK, J. L., RENNICK, V., HAMPIL, B., SCHMIDT, N. J., AND HO, H. H., Lyophilized combination pools of enterovirus equine antisera: Preparation and test procedures for the identification of field strains of 42 enteroviruses, *Bull. WHO* **48**:263–268 (1973).

298. MELNICK, J. L., SAFFERMAN, R., RAO, V. C., GOYAL, S., BERG, G., DAHLING, D. R., WRIGHT, B. A., AKIN, E., STETLER, R., SORBER, C., MOORE, B., SOBSEY, M. D., MOORE, R., LEWIS, A. L., AND WELLINGS, F. M., Round robin investigation of methods for the recovery of poliovirus from drinking water, *Appl. Environmental Microbiol.* **47**:144–150 (1984).

299. MELNICK, J. L., SCHMIDT, N. J., HAMPIL, B., AND HO, H. H., Lyophilized combination pools of enterovirus equine antisera: Preparation and test procedures for the identification of field strains of 19 group A coxsackievirus serotypes, *Intervirology* **8**:172–181 (1977).

300. MELNICK, J. L., SCHMIDT, N. J., MIRKOVIC, R. R., CHUMAKOV, M. P., LAVROVA, I. K., AND VOROSHILOVA, M. K., Identification of Bulgarian strain 258 of enterovirus 71, *Intervirology* **12**:297–302 (1979).

301. MELNICK, J. L., SHAW, E. W., AND CURNEN, E. C., A virus from patients diagnosed as non-paralytic poliomyelitis or aseptic meningitis, *Proc. Soc. Exp. Biol. Med.* **71**:344–349 (1949).

302. MELNICK, J. L., TAGAYA, I., AND VON MAGNUS, H., Enteroviruses 60, 70, and 71, *Intervirology* **4**:369–370 (1974).

303. MELNICK, J. L., AND WENNER, H. A., Enteroviruses, in: *Diagnostic Procedures for Viral and Rickettsial Diseases*, 4th ed. (E. H. LENNETTE AND N. J. SCHMIDT, EDS.), pp. 529–602, American Public Health Association, Washington, DC, 1969.

304. MELNICK, J. L., WENNER, H. A., AND PHILLIPS, C. A., Enteroviruses, in: *Diagnostic Procedures for Viral, Rickettsial, and Chlamydial Infections*, 5th ed. (E. H. LENNETTE AND N. J. SCHMIDT, EDS.), pp. 471–534, American Public Health Association, Washington, DC, 1979.

305. MELNICK, J. L., WENNER, H. A., AND ROSEN, L., The enteroviruses, in: *Diagnostic Procedures for Viral and Rickettsial Diseases*, 3rd ed. (E. H. LENNETTE AND N. J. SCHMIDT, EDS.), pp. 194–242, American Public Health Association, Washington, DC, 1964.

306. MELNICK, J. L., AND WIMBERLY, I. L., Lyophilized combination pools of enterovirus equine antisera: New LBM pools prepared from reserves of antisera stored frozen for two decades, *Bull. WHO* **63**:543–550 (1985).

307. MENEGUS, M. A., Enteroviruses, in: *Manual of Clinical Microbiology*, 4th ed. (E. H. LENNETTE, ED.), pp. 743–746, American Society for Microbiology, Washington, DC, 1985.

308. MENEGUS, M. A., AND HOLLICK, G. E., Increased efficiency of group B coxsackievirus isolation from clinical specimens by use of BGM cells, *J. Clin. Microbiol.* **15**:945–948 (1982).

309. METCALF, T. G., AND MELNICK, J. L., Simple apparatus for collecting estuarine sediments and suspended solids to detect solids-associated virus, *Appl. Environ. Microbiol.* **45**:323–327 (1983).

310. METCALF, T. G., RAO, V. C., AND MELNICK, J. L., Solid-associated viruses in a polluted estuary, in: *Enteric Viruses in Water: Monographs in Virology*, Vol. 15 (J. L. MELNICK, ED.), pp. 97–110, S. Karger, Basel, 1984.

311. MILLER, D. A., MILLER, O. J., VAITHILINGHAM, G. D., HASHMI, S., TANTRAVAHI, R., MEDRANO, L., AND GREEN, H., Human chromosome 19 carries a poliovirus receptor gene, *Cell* **1**:167–173 (1974).

312. MINOR, P. D., Review article: The molecular biology of poliovaccines, *J. Gen. Virol.* **74**:3065–3077 (1992).

313. MINOR, P. D., EVANS, D. M. A., FERGUSON, M., SCHILD, G. C., WESTROP, G., AND ALMOND, J. W., Principal and subsidiary antigenic sites of VP1 involved in the neutralization of poliovirus type 3, *J. Gen. Virol.* **65**:1159–1165 (1985).

314. MIRCHAMSY, H., SHAFYI, A., MAHINPOUR, M., AND NAZARI, P., Stabilizing effect of magnesium chloride and sucrose on Sabin live polio vaccine, in: *Vaccinations in the Developing Countries, Developments in Biological Standardization* (R. H. REGANEY, ACTING ED.), Vol. 41, pp. 255–257, S. Karger, Basel, 1978.

315. MIRKOVIC, R. R., KONO, R., YIN-MURPHY, M., SOHIER, R., SCHMIDT, N. J., AND MELNICK, J. L., Enterovirus type 70: The etiologic agent of pandemic acute haemorrhagic conjunctivitis, *Bull. WHO* **49**:341–346 (1973).

316. MIRKOVIC, R. R., SCHMIDT, N. J., YIN-MURPHY, M., AND MELNICK, J. L., Enterovirus etiology of the 1970 Singapore epidemic of acute conjunctivitis, *Intervirology* **4**:119–127 (1974).

317. MIWA, C., YAMADA, F., MATSUURA, A., AND YOSHIZAWA, K., Epidemic of hand, foot and mouth disease in Gifu prefecture in 1973, *Virus* **28**:78–86 (1978).

318. MIYAMURA, K., YAMASHITA, K., YAMADERA, S., KATO, N., AKATSUKA, M., AND YAMAZAKI, S., An epidemic of echovirus 18 in Japan—High association with clinical manifestation of exanthem, *Jpn. J. Med. Sci. Biol.* **43**:51–58 (1990).

319. MIYAMURA, K., YAMASHITA, K., YAMADERA, S., KATO, N., AKATSUKA, M., HARA, M., INOUYE, S., AND YAMAZAKI, S., Poliovirus surveillance: Isolation of polioviruses in Japan, 1980–1991, *Jpn. J. Med. Sci. Biol.* **45**:203–214 (1992).

320. MODLIN, J. F., Perinatal echovirus infection: Insights from a literature review of 61 cases of serious infection and 16 outbreaks in nurseries, *Rev. Infect. Dis.* **8**:918–926 (1986).

321. MODLIN, J. F., POLK, B. F., HORTON, P., ETKIND, P., CRANE, E., AND SPILIOTES, A., Perinatal echovirus infection: Risk of transmission during a community outbreak, *N. Engl. J. Med.* **305**:368–371 (1981).

322. MOORE, M., From the Centers for Disease Control: Enteroviral disease in the United States, 1970–1979, *J. Infect. Dis.* **146**:103–108 (1982).

323. MOORE, M., BARON, R. C., FILSTEIN, M. R., LOFGREN, J. P., ROWLEY, D. L., SCHONBERGER, L. B., AND HATCH, M. H., Asep-

tic meningitis and high school football players, 1978–1980, *J.A.M.A.* **249:**2039–2042 (1983).

324. MOORE, M., AND MORENS, D. M., Enteroviruses, including polioviruses, in: *Textbook of Human Virology*, 2nd ed. (R. B. BELSHE, ED.), pp. 407–483, PSG Publishing Co., Littleton, MA, 1984.

325. MORENS, D. M., Poliovaccine strategy, *Lancet* **1:**285–286 (1985).

326. MORENS, D. M., PALLANSCH, M. A., AND MOORE, M., Polioviruses and other enteroviruses, in: *Textbook of Human Virology*, 2nd ed. (R. B. BELSHE, ED.), pp. 427–497, Mosby Year Book, St. Louis, MO, 1990.

327. MORRISON, L. A., AND FIELDS, B. N., Minireview: Parallel mechanisms in neuropathogenesis of enteric virus infections, *J. Virol.* **65:**2767–2772 (1991).

328. MUIR, P., AND KANDOLF, R., The laboratory diagnosis of enterovirus-induced heart disease, in: *Viral Infections of the Heart* (J. E. BANATVALA, ED.), London, Edward Arnold, pp. 211–230, 1993.

329. MUIR, P., SINGH, N. B., AND BANATVALA, J. E., Enterovirus-specific serum IgA antibody responses in patients with acute infections, chronic cardiac disease, and recently diagnosed insulin-dependent diabetes mellitus, *J. Med. Virol.* **32:**236–242 (1990).

330. MUIR, P., NICHOLSON, F., JHETAM, M., NEOGI, S., AND BANATVALA, J. E., Rapid diagnosis of enterovirus infection by magnetic bead extraction and polymerase chain reaction detection of enterovirus RNA in clinical specimens, *J. Clin. Microbiol.* **31:**31–38 (1993).

331. NAGINGTON, J., GANDY, G., WALKER, J., AND GRAY, J. J., Use of normal immunoglobulin in an echovirus outbreak in a special-care baby unit, *Lancet* **2:**443–446 (1983).

332. NAGINGTON, J., WREGHITT, T. G., GANDY, G., ROBERTON, N. R. C., AND BERRY, P. J., Fatal echovirus 11 infections in outbreak in special-care baby unit, *Lancet* **2:**725 (1978).

333. NAGY, G., TAKATSY, S., KUKAN, E., MIHALY, I., AND DOMOK, I., Virological diagnosis of enterovirus type 71 infections: Experiences gained during an epidemic of acute CNS diseases in Hungary in 1978, *Arch. Virol.* **71:**217–227 (1982).

334. NAKANO, J. H., HATCH, M. H., THIEME, M. L., AND NOTTAY, B., Parameters for differentiating vaccine-derived and wild poliovirus strains, *Prog. Med. Virol.* **24:**178–206 (1978).

335. NAKAO, T., Coxsackie viruses and diabetes, *Lancet* **2:**1423 (1971).

336. NARANG, H. K., AND CODD, A. A., Enterovirus typing by immune electron microscopy using low-speed centrifugation, *J. Clin. Pathol.* **33:**191–194 (1980).

337. NARDI, G., BARONI, M., TANZI, M. L., GRANDI, D., BEVILACQUA, G., AND TEDESCHI, F., Epidemic due to group B Coxsackie viruses in a newborn infants' department, *Ann. Sclavo* **18:**793–808 (1976).

338. NATHANSON, N., Epidemiologic aspects of poliomyelitis eradication, *Rev. Infect. Dis.* **6**(Suppl. 2):308–312 (1984).

339. NEJMI, S., GAUDIN, O. G., CHOMEL, J. J., BAAJ, A., SOHIER, R., AND BOSSHARD, S., Isolation of a virus responsible for an outbreak of acute haemorrhagic conjunctivitis in Morocco, *J. Hyg.* **72:**181–183 (1974).

340. NICHOLAS, D. D., KRATZER, J. H., OFOSU-AMAAH, S., AND BELCHER, D. W., Is poliomyelitis a serious problem in developing countries?—the Danfa experience, *Br. Med. J.* **1:**1009–1112 (1977).

341. NISHMI, M., AND YODFAT, Y., Successive overlapping outbreaks of a febrile illness associated with coxsackie virus type B4 and echo virus type 9 in a kibbutz, *Israel J. Med. Sci.* **9:**895–899 (1973).

342. NKOWANE, B. M., WASSILAK, S. G. F., ORENSTEIN, W. A., BART, K. J., SCHONBERGER, L. B., HINMAN, A. R., AND KEW, O. M., Vaccine-associated paralytic poliomyelitis. United States: 1973 through 1984, *J.A.M.A.* **257:**1335–1340 (1987).

343. NOLAN, J. P., WILMER, B. J., AND MELNICK, J. L., Poliomyelitis: Its highly invasive nature and narrow stream of infection in a community of high socioeconomic level, *N. Engl. J. Med.* **253:**945–954 (1955).

344. NOMOTO, A., Cellular receptors for virus infection, *Semin. Virol.* **3:**77–133 (1992).

345. NOMOTO, A., Recombinant polioviruses as candidates for oral live poliovaccines, *Microbiol. Immunol.* **37:**169–174 (1993).

346. NOMOTO, A., OMATA, T., TOYODA, H., KUGE, S., HORIE, H., KATAOKA, Y., GENBA, Y., NAKANO, Y., AND IMURA, N., Complete nucleotide sequence of the attenuated poliovirus Sabin 1 strain genome, *Proc. Natl. Acad. Sci. USA* **79:**5793–5797 (1982).

347. NOTKINS, A. L., Virus-induced diabetes mellitus, *Arch. Virol.* **54:**1–17 (1977).

348. NOTKINS, A. L., AND YOON, J. W., Virus-induced diabetes mellitus, in: *Concepts in Viral Pathogenesis* (A. L. NOTKINS AND M. B. A. OLDSTONE, EDS.), pp. 241–247, Springer-Verlag, New York, 1984.

349. NOTTAY, B. K., KEW, O. M., HATCH, M. H., HEYWARD, J. T., AND OBIJESKI, J. F., Molecular variation of type 1 vaccine-related and wild polioviruses during replication in humans, *Virology* **108:**405–423 (1981).

350. OFOSU-AMAAH, S., KRATZER, J. H., AND NICHOLAS, D. D., Is poliomyelitis a serious problem in developing countries? Lameness in Ghanaian schools, *Br. Med. J.* **1:**1012–1014 (1977).

351. OGRA, P. L., Mucosal immune response to poliovirus vaccines in childhood, *Rev. Infect. Dis.* **6**(Suppl. 2):S361–S368 (1984).

352. OGRA, P. F., FISHAUT, M., AND GALLAGHER, M. R., Viral vaccination via the mucosal routes, *Rev. Infect. Dis.* **2:**352–369 (1980).

353. OGRA, P. L., AND KARZON, D. T., Formation and function of poliovirus antibody in different tissues, *Prog. Med. Virol.* **13:**156–193 (1971).

354. OLIVE, D. M., AL-MUFTI, S., AL-MULLA, W., KHAN, M. A., PASCA, A., STANEAY, G., AND AL-NAKIB, W., Detection and differentiation of picornaviruses in clinical samples following genomic amplification, *J. Gen. Virol.* **71:**2141–2147 (1990).

355. OMATA, T., KOHARA, M., SAKAI, Y., KAMEDA, A., IMURA, N., AND NOMOTO, A., Cloned infectious complementary DNA of the poliovirus Sabin I genome: Biochemical and biological properties of the recovered virus, *Gene* **32:**1–10 (1984).

356. ORCHARD, T. J., ATCHISON, R. W., BECKER, D., RABIN, B., EBERHARDT, M., KULLER, L. H., LAPORTE, R. E., AND CAVENDER, D., Coxsackie infection and diabetes, *Lancet* **2:**631 (1983).

357. OSTERHAUS, A. D. M. E., VAN WEZEL, A. L., HAZENDONK, T. G., UYTEDEHAAG, F. G. C. M., VAN ASTE, J. A. A. M., AND VAN STEENIS, B., Monoclonal antibodies to polioviruses. Comparison of intratypic strain differentiation of poliovirus type 1 using monoclonal antibodies versus cross-absorbed antisera, *Intervirology* **20:**129–136 (1983).

358. OSTERHAUS, A. D. M. E., VAN WEZEL, A. L., VAN STEENIS, B., DROST, G. A., AND HAZENDONK, A. G., Monoclonal antibodies to polioviruses. Production of specific monoclonal antibodies to the Sabin vaccine strains, *Intervirology* **16:**218–224 (1981).

359. PACSA, S., AND WERBLINSKA, J., Natural immunity of Ghanaian children to polio and coxsackieviruses: Brief report, *Arch. Virusforsch.* **33:**192–193 (1971).

360. PAL, S. R., SZUCS, G., AND MELNICK, J. L., Rapid immuno-

fluorescence diagnosis of acute hemorrhagic conjunctivitis caused by enterovirus 70, *Intervirology* **20:**19–22 (1983).

361. PAL, S. R., SZUCS, G., MELNICK, J. L., KAIWAR, R., BHARDWAJ, G., SINGH, R., GANGWAR, D. N., CHOUDHURY, S., AND JAIN, I. S., Immunofluorescence test for the epidemiological monitoring of acute haemorrhagic conjunctivitis cases, *Bull. WHO* **61:**485–490 (1983).

362. PAN AMERICAN HEALTH ORGANIZATION, *Live Poliovirus Vaccines*, Special Publications of the Pan American Health Organization, Nos. 44 and 50, 1959, 1960.

363. PAQUE, R. E., GAUNTT, C. J., AND NEALON, T. J., Assessment of cell-mediated immunity against coxsackievirus B3-induced myocarditis in a primate model (*Papio papio*), *Infect. Immun.* **31:**470–479 (1981).

364. PARKS, W. P., QUEIROGA, L. T., AND MELNICK, J. L., Studies of infantile diarrhea in Karachi, Pakistan. II. Multiple virus isolations from rectal swabs, *Am. J. Epidemiol.* **84:**469–478 (1967).

365. PATRIARCA, P. A., ONORATO, I. M., SKLAR, V. E. F., SCHONBERGER, L. B., KAMINSKI, R. M., HATCH, M. H., MORENS, D. M., AND FORSTER, R. K., Acute hemorrhagic conjunctivitis. Investigation of a large-scale community outbreak in Dade County, Florida, *J.A.M.A.* **249:**1283–1289 (1983).

366. PATRIARCA, P. A., WRIGHT, P. F., AND JOHN, T. J., Factors affecting the immunogenicity of oral poliovirus vaccine in developing countries: Review, *Rev. Infect. Dis.* **13:**926–939 (1991).

367. PAUL, J. R., Endemic and epidemic trends of poliomyelitis in Central and South America, *Bull. WHO* **19:**737–758 (1958).

368. PAUL, J. R., *A History of Poliomyelitis*, Yale University Press, New Haven, 1971.

369. PAUL, J. R., Development and use of serum surveys in epidemiology, in: *Serological Epidemiology* (J. R. PAUL AND C. WHITE, EDS.), pp. 1–13, Academic Press, New York, 1973.

370. PAUL, J. R., AND MELNICK, J. L., Poliomyelitis, in: *Diagnostic Procedures for Virus and Rickettsial Diseases*, 2nd ed. (E. H. LENNETTE AND N. J. SCHMIDT, EDS.), pp. 53–90, American Public Health Association, New York, 1956.

371. PAUL, J. R., MELNICK, J. L., BARNETT, V. H., AND GOLDBLUM, N., A survey of neutralizing antibodies to poliomyelitis in Cairo, Egypt, *Am. J. Hyg.* **55:**402–413 (1952).

372. PAUL, J. R., RIORDAN, J. T., AND MELNICK, J. L., Antibodies to three different antigenic types of poliomyelitis virus in sera from North Alaskan Eskimos, *Am. J. Hyg.* **54:**275–285 (1951).

373. PAYNE, A. M.-M., Poliomyelitis as a world problem, in: *Papers and Discussions Presented at the Third International Poliomyelitis Conference*, pp. 393–400, J. B. Lippincott, Philadelphia, 1955.

374. PEETERMANS, J., COLINET, G., AND STEPHENNE, J., Activity of attenuated poliomyelitis and measles vaccines exposed at different temperatures, in: *Proceedings Symposium on Stability and Effectiveness of Measles, Poliomyelitis and Pertussis Vaccines*, pp. 61–66, Yugoslav Academy of Sciences and Arts, Zagreb, 1976.

375. PETERSEN, I., Polio antibody patterns in the Danish population since 1954, in: *Proceedings of the 75th Anniversary Meeting*, Statens Serum Instituts, Copenhagen, 1977.

376. PETITJEAN, J., QUIBRIAC, M., FREYMUTH, F., FUCHS, F., LACONCHE, N., AYMARD, M., AND KOPECKA, H., Specific detection of enteroviruses in clinical samples by molecular hybridization using poliovirus subgenomic riboprobes, *J. Clin. Microbiol.* **28:**307–311 (1990).

377. PETITJEAN, J., KOPECKA, H., FREYMUTH, F., LANGLARD, J. M.,

378. SCANU, P., GALATEAU, F., BOUHOUR, J. B., FERRIERE, M., CHARBONNEAU, P., AND KOMAJDA, M., Detection of enteroviruses in endomyocardial biopsy by molecular approach, *J. Med. Virol.* **37:**76–82 (1992).

378. PHILLIPS, C. A., MELNICK, J. L., BARRETT, F. F., BEHBEHANI, A. M., AND RIGGS, S., Dual virus infections: Simultaneous enteroviral disease and St. Louis encephalitis, *J.A.M.A.* **197:**169–172 (1966).

379. PRABHAKAR, B. S., HASPEL, M. V., MCCLINTOCK, P. R., AND NOTKINS, A. L., High frequency of antigenic variants among naturally occurring human coxsackie B4 virus isolates identified by monoclonal antibodies, *Nature* **300:**374–376 (1982).

380. PRATHER, S. L., DAGAN, R., JENISTA, J. A., AND MENEGUS, M. A., The isolation of enteroviruses from blood: A comparison of four processing methods, *J. Med. Virol.* **14:**221–227 (1984).

381. RACANIELLO, V. R., AND BALTIMORE, D., Molecular cloning of poliovirus cDNA and determination of the complete nucleotide sequence of the viral genome, *Proc. Natl. Acad. Sci. USA* **78:**4887–4891 (1981).

382. RACANIELLO, V. R., AND BALTIMORE, D., Cloned poliovirus complementary DNA is infectious in mammalian cells, *Science* **214:**916–918 (1981).

383. RAMOS-ALVAREZ, M., AND SABIN, A. B., Characteristics of poliomyelitis and other enteric viruses recovered in tissue culture from healthy American children, *Proc. Soc. Exp. Biol. Med.* **87:**655–661 (1954).

384. RASKA, K., Editorial: Epidemiologic surveillance in the control of infectious disease, *Rev. Infect. Dis.* **5:**1112–1117 (1983).

385. REDLINE, R. W., GENEST, D. R., AND TYCKO, B., Detection of enteroviral infection in paraffin-embedded tissue by the RNA polymerase chain reaction technique, *Am. J. Clin. Pathol.* **96:**569–571 (1991).

386. REISSIG, M., HOWES, D. W., AND MELNICK, J. L., Sequence of morphological changes in epithelial cell cultures infected with poliovirus, *J. Exp. Med.* **104:**289–304 (1956).

387. REN, R., AND RACANIELLO, V. R., Human poliovirus receptor gene expression and poliovirus tissue tropism in transgenic mice, *J. Virol.* **66:**296–304 (1992).

388. REYES, M. P., OSTREA, E. M., JR., ROSKAMP, J., AND LERNER, A. M., Disseminated neonatal echovirus 11 disease following antenatal maternal infection with a virus-positive cervix and virus-negative gastrointestinal tract, *J. Med. Virol.* **12:**155–159 (1982).

389. RICO-HESSE, R., PALLANSCH, M. A., NOTTAY, B. K., AND KEW, O. M., Geographic distribution of wild poliovirus type 1 genotypes, *Virology* **160:**311–322 (1987).

390. RISI, J. B., JR., The control of poliomyelitis in Brazil, *Rev. Infect. Dis.* **6**(Suppl. 2)**:**S400–S403 (1984).

391. RIVERS, T. M., AND HORSFALL, F. L., JR. (EDS.), *Viral and Rickettsial Infections of Man*, 3rd ed., J. B. Lippincott, Philadelphia, 1959.

392. ROBBINS, F. C., ENDERS, J. F., WELLER, T. H., AND FLORENTINO, G. L., Studies on the cultivation of poliomyelitis viruses in tissue culture. V. The direct isolation and serologic identification of virus strains in tissue culture from patients with nonparalytic and paralytic poliomyelitis, *Am. J. Hyg.* **54:**286–293 (1951).

393. ROBBINS, F. C., AND NIGHTINGALE, E. O., Selective primary health care: Strategies for control of disease in the developing world. IX. Poliomyelitis, *Rev. Infect. Dis.* **5:**957–968 (1983).

394. ROSEN, L., AND KERN, J. K., Hemagglutination and hemagglutination-inhibition with Coxsackie B viruses, *Proc. Soc. Exp. Biol. Med.* **107:**626–628 (1981).

395. ROSEN, L., MELNICK, J. L., SCHMIDT, N. J., AND WENNER, H. A., Subclassification of enteroviruses and echovirus type 34, *Arch. Virusforsch.* **30**:89–92 (1970).

396. ROSEN, L., SCHMIDT, N. J., AND KERN, J., Toluca-1, a newly recognized enterovirus, *Arch. Ges. Virusforsch.* **40**:132–136 (1973).

397. ROSENWIRTH, B., AND EGGERS, H. J., Biochemistry and pathogenicity of echovirus 9: 1. Characterization of the virus particles of strains Barty and Hill, *Virology* **123**:102–112 (1982).

398. ROSS, C. A. C., BELL, E. J., KERR, M. M., AND WILLIAMS, K. A. B., Infective agents and embryopathy in the West of Scotland 1966–70, *Scott. Med. J.* **17**:252–258 (1972).

399. ROSSEN, R. D., KASEL, J. A., AND COUCH, R. B., The secretory immune system: Its relation to respiratory viral infection, *Prog. Med. Virol.* **13**:194–238 (1971).

400. ROSSOUW, E., TSILIMIGRAS, C. S. A., AND SCHOUB, B. D., Molecular epidemiology of a coxsackievirus B3 outbreak, *J. Med. Virol.* **34**:165–171 (1991).

401. ROTBART, H. A., New methods of rapid enteroviral diagnosis, *Prog. Med. Virol.* **40**:96–108 (1991).

402. ROTBART, H. A., LEVIN, M. J., AND VILLARREAL, L. P., Use of subgenomic poliovirus DNA hybridization probes to detect the major subgroups of enteroviruses, *J. Clin. Microbiol.* **20**:1105–1108 (1984).

403. RUECKERT, R. R., Picornaviruses and their replication, in: *Virology* (B. N. FIELDS, D. M. KNIPE, R. M. CHANOCK, J. L. MELNICK, B. ROIZMAN, AND R. E. SHOPE, EDS.), pp. 705–738, Raven Press, New York, 1985.

404. RUMKE, H. C., SCHLUMBERGER, M., FLOURY, B., NAGEL, J., AND VAN STEENIS, B., Serological evaluation of a simplified immunization schedule using quadruple DPT–polio vaccine in Burkina Faso, *Vaccine* **11**:1113–1118 (1993).

405. SABIN, A. B., Oral poliovirus vaccine, *J.A.M.A.* **194**:872–876 (1965).

406. SABIN, A. B., Paralytic poliomyelitis: Old dogmas and new perspectives, *Rev. Infect. Dis.* **3**:543–564 (1981).

407. SABIN, A. B., Oral poliovirus vaccine: History of its development and use and current challenge to eliminate poliomyelitis from the world, *J. Infect. Dis.* **151**:420–436 (1985).

408. SABIN, A. B., HENNESSEN, W. A., AND WINSSER, J., Studies on variants of poliomyelitis virus. I. Experimental segregation and properties of avirulent variants of three immunologic types, *J. Exp. Med.* **99**:551–576 (1954).

409. SABIN, A. B., KRUMBIEGEL, E. R., AND WIGAND, R., ECHO type 9 virus disease: Virologically controlled clinical and epidemiologic observations during 1957 epidemic in Milwaukee with notes on concurrent similar diseases associated with Coxsackie and other ECHO viruses, *Am. J. Dis. Child.* **96**:197–219 (1958).

410. SALK, D., Eradication of poliomyelitis in the United States. III. Poliovaccines—practical considerations, *Rev. Infect. Dis.* **2**:258–273 (1980).

411. SALK, D., AND SALK, J., Vaccinology of poliomyelitis: A review, *Vaccine* **2**:59–74 (1984).

412. SALK, D., VON WEZEL, A. L., AND SALK, J., Induction of long-term immunity to paralytic poliomyelitis by use of noninfectious vaccine, *Lancet* **2**:1317–1321 (1984).

413. SALK, J. E., Poliomyelitis: Control, in: *Viral and Rickettsial Infections of Man*, 3rd ed. (T. M. RIVERS AND F. L. HORSFALL, JR., EDS.), pp. 499–518, J. B. Lippincott, Philadelphia, 1959.

414. SALK, J., One-dose immunization against paralytic poliomyelitis using a noninfectious vaccine, *Rev. Infect. Dis.* **6**(Suppl. 2):S444–S450 (1984).

415. SALK, J., AND SALK, D., Control of influenza and poliomyelitis with killed virus vaccines, *Science* **195**:834–837 (1977).

416. SAMUDA, G. M., CHANG, W. K., YEUNG, C. Y., AND TANG, P. S., Monoplegia caused by enterovirus 71: An outbreak in Hong Kong, *Pediatr. Infect. Dis.* **6**:206–208 (1987).

417. SAMUELSON, A., GLIMAKER, M., SKOOG, E., CELLO, J., AND FORSGREN, M., Diagnosis of enteroviral meningitis with IgG-EIA using heat-treated virions and synthetic peptides as antigens, *J. Med. Virol.* **40**:271–277 (1993).

418. SANDELIN, K., TUOMIOJA, M., AND ERKKILA, H., Echovirus type 7 isolated from conjunctival scrapings, *Scand J. Infect. Dis.* **9**:71–73 (1977).

419. SAWYER, L. A., HERSHOW, R. C., PALLANSCH, M. A., FISHBEIN, D. B., PINSKY, P. F., BROERMAN, S. F., GRIMM, B. B., ANDERSON, L. J., HALL, D. B., AND SCHONBERGER, L. B., An epidemic of acute hemorrhagic conjunctivitis in American Samoa caused by coxsackievirus A24 variant, *Am. J. Epidemiol.* **130**:1187–1198 (1989).

420. SCHAAP, G. J. P., BUKERK, H., COUTINHO, R. A., KAPSENBERG, J. G., AND VAN WEZEL, A. L., The spread of poliovirus in the well-vaccinated Netherlands in connection with the 1978 epidemic, *Prog. Med. Virol.* **29**:124–140 (1984).

421. SCHAFFER, F. L., Binding of proflavine by and photoinactivation of poliovirus propagated in the presence of the dye, *Virology* **18**:412–425 (1962).

422. SCHEIBLE, J. H., FOX, V. L., AND LENNETTE, E. H., A probable new human picornavirus associated with respiratory disease, *Am. J. Epidemiol.* **85**:297–310 (1967).

423. SCHMIDT, N. J., HO, H. H., AND LENNETTE, E. H., Propagation and isolation of group A coxsackieviruses in RD cells, *J. Clin. Microbiol.* **2**:183–185 (1975).

424. SCHMIDT, N. J., HO, H. H., AND LENNETTE, E. H., Comparative sensitivity of the BGM cell line for isolation of enteric viruses, *Health Lab. Sci.* **13**:115–117 (1976).

425. SCHMIDT, N. J., AND LENNETTE, E. H., Advances in the serodiagnosis of viral infections, *Prog. Med. Virol.* **15**:244–308 (1973).

426. SCHMIDT, N. J., LENNETTE, E. H., AND HO, H. H., An apparently new enterovirus isolated from patients with disease of the central nervous system, *J. Infect. Dis.* **129**:304–309 (1974).

427. SCHMIDT, N. J., MAGOFFIN, R. L, AND LENNETTE, E. H., Association of group B coxsackieviruses with cases of pericarditis, myocarditis, or pleurodynia by demonstration of immunoglobulin M antibody, *Infect. Immun.* **8**:341–348 (1973).

428. SCHNURR, D. P., CAO, Y., AND SCHMIDT, N. J., Coxsackievirus B3 persistence and myocarditis in N:NIH(S) II *nu/nu* and +/*nu* mice, *J. Gen. Virol.* **65**:1197–1201 (1984).

429. SCHONBERGER, L. B., KAPLAN, J., KIM-FARLEY, R., MOORE, M., EDDINS, D. L., AND HATCH, M., Control of paralytic poliomyelitis in the United States, *Rev. Infect. Dis.* **6**(Suppl. 2):S424–S426 (1984).

430. SEAL, L. A., AND JAMISON, R. M., Evidence for secondary structure within the virion RNA of echovirus 22, *J. Virol.* **50**:641–644 (1984).

431. SELLS, C. J., CARPENTER, R. L., AND RAY, C. G., Sequelae of central-nervous-system enterovirus infections, *J. Engl. J. Med.* **293**:1–4 (1975).

432. SEMLER, B. L., DORNER, A. J., AND WIMMER, E., Production of an infectious poliovirus from cloned cDNA is dramatically increased by SV40 transcription and replication signals, *Nucleic Acids Res.* **12**:5123–5141 (1984).

433. SERGIESCU, D., HORODNICEANU, F., AND AUBERT-COMBIESCU, A., The use of inhibitors in the study of picornavirus genetics, *Prog. Med. Virol.* **14**:123–199 (1972).

434. SHARIEF, M. K., HENTGES, R., AND CIARDI, M., Intrathecal immune response in patients with the post-polio syndrome, *New Engl. J. Med.* **325**:749–755 (1991).

435. SHIMOJO, H., Poliomyelitis control in Japan, *Rev. Infect. Dis.* **6**(Suppl. 2):S427–S430 (1984).

436. SHINDAROV, L. M., CHUMAKOV, M. P., VOROSHILOVA, M. K., BOJINOV, S., VASILENKO, S. M., IORDANOV, I., KIROV, I. D., KAMENOV, E., LESHINSKAYA, E. V., MITOV, G., ROBINSON, I. A., SIVCHEV, S., AND STAIKOV, S., Epidemiological, clinical, and pathomorphological characteristics of epidemic poliomyelitis-like disease caused by enterovirus 71, *J. Hyg. Epidemiol. Microbiol. Immunol.* **23**:284–295 (1979).

437. SIMMONDS, R. S., SZUCS, G., METCALF, T. G., AND MELNICK, J. L., Persistently infected cultures as a source of hepatitis A virus, *Appl. Environ. Microbiol.* **49**:749–755 (1985).

438. SKEELS, M. R., WILLIAMS, J. J., AND RICKER, F. M., Perinatal echovirus infection, *N. Engl. J. Med.* **305**:1529 (1981).

439. SLATER, P. E., ORENSTEIN, W. A., MORAG, A., AVNI, A., HANDSHER, R., GREEN, M. S., COSTIN, C., YARROW, A., RISHPON, S., HAVKIN, O., BEN-ZVI, T., KEW, O. M., REY, M., EPSTEIN, I., SWARTZ, T. A., AND MELNICK, J. L., Poliomyelitis outbreak in Israel in 1988: A report with two commentaries, *Lancet* **335**:1192–1198 (1990).

440. SOHIER, R., Enterovirus type 71 surveillance: France, *WHO Week Epidemiol. Rec.* **54**:219 (1979).

441. SOMMERVILLE, R. G., Rapid diagnosis of viral infections by immunofluorescent staining of viral antigens in leukocytes and macrophages, *Prog. Med. Virol.* **10**:398–414 (1968).

442. STOECKEL, P., SCHLUMBERGER, M., PARENT, G., MAIRE, B., VAN WEZEL, A., VAN STEENIS, G., EVANS, A., AND SALK, D., Use of killed poliovirus vaccine in a routine immunization program in West Africa, *Rev Infect. Dis.* **6**(Suppl. 2):S463–S466 (1984).

443. SUMMERS, D. F., AND MAIZEL, J. V., JR., Evidence for large precursor proteins in poliovirus synthesis, *Proc. Natl. Acad. Sci. USA* **59**:966–971 (1968).

444. SUTMOLLER, F., NASCIMENTO, J. P., CHAVES, J. R. S., FERREIRA, V., AND PEREIRA, M. S., Viral etiology of acute respiratory diseases in Rio de Janeiro: First two years of a longitudinal study, *Bull. WHO* **49**:129–137 (1983).

445. TABER, L. H., MIRKOVIC, R. R., ADAM, V., ELLIS, S., YOW, M. D., AND MELNICK, J. L., Rapid diagnosis of enterovirus meningitis by immunofluorescent staining of CSF leukocytes, *Intervirology* **1**:127–134 (1973).

446. TAGAYA, I., NAKAO, C., HARA, M., AND YAMADERA, S., Characterization of poliovirus isolates in Japan after the mass vaccination with live oral poliomyelitis vaccine (Sabin), *Bull. WHO* **48**:547–554 (1973).

447. TAKATSU, T., TAGAYA, I., AND HIRAYAMA, M., Poliomyelitis in Japan during the period 1961–1968 after the introduction of mass vaccination with Sabin vaccine, *Bull. WHO* **49**:129–137 (1973).

448. TAMM, I., AND EGGERS, H. J., Differences in the selective virus-inhibitory action of 2-(α-hydroxybenzyl)-benzimidazole and guanidine · HCl, *Virology* **18**:439–447 (1962).

449. TILZEY, A. J., SIGNEY, M., AND BANATVALA, J. E., Persistent coxsackie B virus specific IgM response in patients with recurrent pericarditis, *Lancet* **1**:1491–1492 (1986).

450. TORFASON, E. G., FRISK, G., AND DIDERHOLM, H., Indirect and reverse radioimmunoassays and their apparent specificities in the detection of antibodies to enteroviruses in human sera, *J. Med. Virol.* **13**:13–31 (1984).

451. TORPHY, D. E., RAY, C. G., THOMPSON, R. S., AND FOX, J. P., An epidemic of aseptic meningitis due to echovirus type 30: Epidemiological features and clinical laboratory findings, *Am. J. Public Health* **60**:1447–1455 (1970).

452. TOWNSEND, T. R., BOLYARD, E. A., YOLKEN, R. H., BESCHORNER, W. E., BISHOP, C. A., BURNS, W. H., SANTOS, G. W., AND SARAL, R., Outbreak of coxsackie A1 gastroenteritis: A complication of bone-marrow transplantation, *Lancet* **1**:820–823 (1982).

453. TRASK, J. D., MELNICK, J. L., AND WENNER, H. A., Chlorination of human, monkey-adapted and mouse strains of poliomyelitis virus, *Am. J. Hyg.* **41**:30–40 (1945).

454. TULCHINSKY, T. H., HANDSHER, R., MELNICK, J. L., SHABAAN, D. A., NEUMANN, M., ABED, Y., AND BUDNITZ, D., Immune status to various strains of wild poliovirus among children in Gaza immunized with live attenuated oral vaccine alone compared with combination of live and inactivated vaccines, *J. Viral Dis.* **1**:1–9 (1993).

455. UKKONEN, P., HUOVILAINEN, A., AND HOVI, T., Detection of poliovirus antigen by enzyme immunoassay, *J. Clin. Microbiol.* **24**:954–958 (1986).

456. VALLBRACHT, A., HOFMANN, L., WURSTER, K. G., AND FLEHMIG, B., Persistent infection of human fibroblasts by hepatitis A virus, *J. Gen. Virol.* **65**:609–615 (1984).

457. VAN DER WERF, S., BREGEGERE, F., KOPECKA, H., KITAMURA, N., ROTHBERG, P. G., KOURILSKY, P., WIMMER, E., AND GIRARD, M., Molecular cloning of the genome of poliovirus type 1, *Proc. Natl. Acad. Sci. USA* **78**:5983–5987 (1981).

458. VAN DER WERF, S., WYCHOWSKI, C., BRUNEAU, P., BLONDEL, B., CRAINIC, R., HORODNICEANU, F., AND GIRARD, M., Localization of a poliovirus type 1 neutralization epitope in viral capsid polypeptide VP1, *Proc. Natl. Acad. Sci. USA* **80**:5080–5084 (1983).

459. VAN WEZEL, A. L., AND HAZENDONK, A. G., Intratypic serodifferentiation of poliomyelitis virus strains by strain-specific antisera, *Intervirology* **11**:2–8 (1979).

460. VAN WEZEL, A. L., VAN STEENIS, G., VAN DER MAREL, P., AND OSTERHAUS, A. D. M. E., Inactivated poliovirus vaccine: Current production methods and new developments, *Rev. Infect. Dis.* **6**(Suppl. 2):S335–S340 (1984).

461. VON MAGNUS, H., Polio vaccination in Denmark, in: *Report of the Committee for the Study of Poliomyelitis Vaccines*, Institute of Medicine, National Academy of Sciences USA, 1977.

462. VON MAGNUS, H., AND PETERSEN, I., Vaccination with inactivated poliovirus vaccine and oral poliovirus vaccine in Denmark, *Rev. Infect. Dis.* **6**(Suppl. 2):S471–S474 (1984).

463. VOROSHILOVA, M. K., AND CHUMAKOV, M. P., Poliomyelitis-like properties of AB-IV-Coxsackie A7 group of viruses, *Prog. Med. Virol.* **2**:106–170 (1959).

464. WADIA, N. H., IRANI, P. F., AND KATRAK, S. M., Lumbosacral radiculomyelitis associated with pandemic acute haemorrhagic conjunctivitis, *Lancet* **1**:350–352 (1973).

465. WALLIS, C., HOMMA, A., AND MELNICK, J. L., A portable virus concentrator for testing water in the field, *Water Res.* **6**:1249–1256 (1972).

466. WALLIS, C., AND MELNICK, J. L., Cationic stabilization—A new property of enteroviruses, *Virology* **16**:683–700 (1961).

467. WALLIS, C., AND MELNICK, J. L., Photodynamic inactivation of enteroviruses, *J. Bacteriol.* **89**:41–46 (1965).

468. WALLIS, C., AND MELNICK, J. L., Virus aggregation as the cause of the non-neutralizable persistent fraction, *J. Virol.* **1**:478–488 (1967).

469. WARD, R., MELNICK, J. L., AND HORSTMANN, D. M., Poliomyelitis virus in fly-contaminated food collected at an epidemic, *Science* **101**:491–493 (1945).

470. WEISS, L. M., LIU, X.-F., CHANG, K. L., AND BILLINGHAM, M. E.,

Detection of enteroviral RNA in idiopathic dilated cardiomyopathy and other human cardiac tissues, *J. Clin. Invest.* **90:**156–159 (1992).

471. WENNER, H. A., The ECHO viruses, *Ann. N.Y. Acad. Sci.* **101:**398–412 (1962).

472. WENNER, H. A., Outline of laboratory procedures for the diagnosis of enterovirus infections, in: *Diagnostic Procedures for Viral and Rickettsial Diseases*, 3rd ed. (E. H. LENETTE AND N. J. SCHMIDT, EDS.), pp. 243–258, American Public Health Association, Washington, 1964.

473. WENNER, H. A., Virus diseases associated with cutaneous eruptions, *Prog. Med. Virol.* **16:**269–336 (1973).

474. WHITTLE, H., HAZLETT, D., WOOD, D., AND BELL, C., Immunofluorescence technique for the identification of polioviruses, *Lancet* **39:**429–430 (1992).

475. WICKMAN, I., Studien uber Poliomyelitis acuta: Zugleich ein Beitrag zur Kenntnis der Myelitis acuta (1905). [English translation. *Nervous and Mental Disease Monograph Series*, No. 16, S. Karger, Berlin, 1913].

476. WILFERT, C. M., BUCKLEY, R. H., MOHANAKUMAR, T., GRIFFITH, J. F., KATZ, S. L., WHISNANT, J. K., EGGLESTON, P. A., MOORE, M., TREADWELL, E., OXMAN, M. N., AND ROSEN, F. S., Persistent and fatal central-nervous-system echovirus infections in patients with agammaglobulinemia, *N. Engl. J. Med.* **296:**1485–1489 (1977).

477. WILSON, F. M., MIRANDA, Q. R., CHASON, J. L., AND LERNER, A. M., Residual pathologic changes following murine coxsackie A and B myocarditis, *Am. J. Pathol.* **55:**253–265 (1979).

478. WILSON, T., PAPAHADJOPOULOS, D., AND TABER, R., Biological properties of poliovirus encapsulated in lipid vesicles: Antibody resistance and infectivity in virus-resistant cells, *Proc. Natl. Acad. Sci. USA* **74:**3471–3475 (1977).

479. WIMMER, E., HELLEN, C. U. T., AND CAO, X., Genetics of poliovirus, *Annu. Rev. Genet.* **27:**353–435 (1993).

480. WONG, C. Y., WOODRUFF, J. J., AND WOODRUFF, J. F., Generation of cytotoxic T lymphocytes during coxsackievirus B-3 infection. II. Characterization of effector cells and demonstration of cytotoxicity against viral-infected myofibers, *J. Immunol.* **118:**165–169 (1977).

481. WOODRUFF, J. F., Viral myocarditis. A review, *Am. J. Pathol.* **101:**427–478 (1980).

482. WOODRUFF, J. F., AND WOODRUFF, J. J., Modification of severe coxsackievirus B₃ infection in marasmic mice by transfer of immune lymphoid cells, *Proc. Natl. Acad. Sci. USA* **68:**2108–2111 (1971).

483. WORLD HEALTH ORGANIZATION, *WHO Weekly Epidemiological Record*, 1960–1979.

484. WORLD HEALTH ORGANIZATION, *WHO Weekly Epidemiological Record* and *Quarterly and Annual Reports on Virus Isolations*, 1963–1985.

485. WORLD HEALTH ORGANIZATION, Poliomyelitis surveillance, *Week. Epidemiol. Rec.* **53:**304 (1978).

486. WORLD HEALTH ORGANIZATION, Poliomyelitis in 1977, *Week. Epidemiol. Rec.* **53:**321–327 (1978)

487. WORLD HEALTH ORGANIZATION, Poliomyelitis in 1978, *Week. Epidemiol. Rec.* **54:**361–368 (1979).

488. WORLD HEALTH ORGANIZATION, Poliomyelitis in 1979, Parts 1 and 2, *Week. Epidemiol. Rec.* **55:**361–366, 369–376 (1980).

489. WORLD HEALTH ORGANIZATION, Expanded Programme on Immunization (EPI): Poliomyelitis prevalence surveys, Tamil Nadu, India, 1979, *Week. Epidemiol. Rec.* **56:**131–132 (1981).

490. WORLD HEALTH ORGANIZATION, Poliomyelitis in 1980, Parts 1 and 2, *Week. Epidemiol. Rec.* **56:**329–332, 337–341 (1981).

491. WORLD HEALTH ORGANIZATION, Poliomyelitis in 1981, *Week. Epidemiol. Rec.* **57:**305–310 (1982).

492. WORLD HEALTH ORGANIZATION, Expanded Programme on Immunization, *Bull. WHO* **61:**611–615 (1983).

493. WORLD HEALTH ORGANIZATION, Poliomyelitis in 1982, *Week. Epidemiol. Rec.* **58:**385–390 (1983).

494. WORLD HEALTH ORGANIZATION (MELNICK, J. L., MORDHORST, C. H., AND BEKTIMIROV, T.), Virus Disease Surveillance: New LBM antiserum pools available for typing enteroviruses, *Week. Epidemiol. Rec.* **59:**345 (1984).

495. WORLD HEALTH ORGANIZATION, Expanded Programme on Immunization: Global status report, *Week. Epidemiol. Rec.* **60:**261–263 (1985).

496. WORLD HEALTH ORGANIZATION, Poliomyelitis in 1983, *Week. Epidemiol. Rec.* **60:**173–179 (1985).

497. WORLD HEALTH ORGANIZATION, *Status Report: The Global Eradication of Poliomyelitis*, May 1993, pp. 1–11, World Health Organization, Geneva, Switzerland, 1993.

498. WORLD HEALTH ORGANIZATION COMMITTEE ON POLIOMYELITIS, The relation between acute persisting spinal paralysis and poliomyelitis vaccine (oral): Results of a WHO enquiry, *Bull. WHO* **53:**319–331 (1976).

499. WORLD HEALTH ORGANIZATION CONSULTATIVE GROUP, The relation between acute persisting spinal paralysis and poliomyelitis vaccine—results of a ten-year enquiry, *Bull. WHO* **60:**231–242 (1982).

500. WORLD HEALTH ORGANIZATION SCIENTIFIC GROUP, Human viruses in water, wastewater, and soil, *WHO Tech. Rep. Ser.* **639** (1979).

501. WREGHITT, T. G., GANDY, G. M., KING, A., AND SUTEHALL, G., Fatal neonatal echo 7 virus infection, *Lancet* **2:**465 (1984).

502. WRIGHT, P. F., HATCH, M. H., KASSELBERG, A. G., LOWRY, S. P., WADLINGTON, W. B., AND KARZON, D. T., Vaccine-associated poliomyelitis in a child with sex-linked agammaglobulinemia, *J. Pediatr.* **91:**408–412 (1977).

503. WRIGHT, P. F., KIM-FARLEY, R. J., DE QUADROS, C. A., ROBERTSON, S. E., SCOTT, R. M., WARD, N. A., AND HENDERSON, R. H., Strategies for the global eradication of poliomyelitis by the year 2000, *N. Engl. J. Med.* **325:**1774–1779 (1991).

504. WYATT, H. B., Hypothesis: Poliomyelitis in hypogammaglobulinemics, *J. Infect. Dis.* **128:**802–806 (1973).

505. WYCHOWSKI, C., VAN DER WERF, S., SIFFERT, O., CRAINIC, R., BRUNEAU, P., AND GIRARD, M., A poliovirus type 1 neutralization epitope is located within amino acid residues 94 to 104 of viral capsid polypeptide VP1, *EMBO J.* **2:**2019–2024 (1983).

506. YAMADERA, S., YAMASHITA, K., KATO, N., AKATSUKA, M., MIYAMURA, K., AND YAMAZAKI, S., Herpangina surveillance in Japan, 1982–1989, *Jpn. J. Med. Sci. Biol.* **44:**29–39 (1991).

507. YAMASHITA, K., MIYAMURA, K., YAMADERA, S., KATO, N., AKATSUKA, M., INOUYE, S., AND YAMAZAKI, S., Enteroviral aseptic meningitis in Japan, 1981–1991, *Jpn. J. Med. Sci. Biol.* **45:**151–161 (1992).

508. YIN-MURPHY, M., Viruses of acute haemorrhagic conjunctivitis, *Lancet* **1:**545–546 (1973).

509. YIN-MURPHY, M., Acute hemorrhagic conjunctivitis, *Prog. Med. Virol.* **29:**23–44 (1984).

510. YIN-MURPHY, M., AND LIM, K. H., Picornavirus epidemic conjunctivitis in Singapore, *Lancet* **2:**857–858 (1972).

511. YIN-MURPHY, M., PHOON, M. C., AND BAHARUDI-ISHAK, Echo-

virus type 11 infection in Singapore, *Singapore Med. J.* **25:**38–42 (1984).

512. Yin-Murphy, M., Wong, H. B., and Goh, K. T., Review of the poliomyelitis immunization campaign and syndromes associated with poliovirus in Singapore, *J. Singapore Paed. Soc.* **29(3–4):**99–107 (1987).

513. Yodfat, Y., and Nishmi, M., Epidemiologic and clinical observations in six outbreaks of viral disease in a kibbutz, 1968–1971, *Am. J. Epidemiol.* **97:**415–423 (1973).

514. Yofe, J., Goldblum, N., Eylan, E., and Melnick, J. L., An outbreak of poliomyelitis in Israel in 1961 and the use of attenuated type 1 vaccine in its control, *Am. J. Hyg.* **76:**225–238 (1962).

515. Yolken, R. H., Bishop, C. A., Townsend, T. R., Bolyard, E. A., Bartlett, J., Santos, G. W., and Saral, R., Infectious gastroenteritis in bone-marrow-transplant recipients, *N. Engl. J. Med.* **306:**1009–1012 (1982).

516. Yolken, R. H, and Torsch, V. M., Enzyme-linked immunosorbent assay for detection and identification of coxsackieviruses A, *Infect. Immun.* **31:**742–750 (1981).

517. Yoo, J.-W., Austin, M., Onodera, T., and Notkins, A. L., Virus-induced diabetes mellitus. Isolation of a virus from the pancreas of a child with diabetic ketoacidosis, *N. Engl. J. Med.* **300:**1173–1179 (1979).

518. Yorke, J. A., Nathanson, N., Pianigiani, G., and Martin, J., Seasonality and the requirements for perpetuation and eradication of viruses in populations, *Am. J. Epidemiol.* **109:**102–123 (1979).

519. Young, N. A., Polioviruses, coxsackieviruses, and echoviruses: Comparison of the genomes by RNA hybridization, *J. Virol.* **11:**832–839 (1973).

520. Zheng, Z. M., Zhang, J. H., Zhu, W. P., and He, P. J., Isolation of enterovirus type 71 from the vesicle fluid of an adult patient with hand-foot-mouth disease in China, *Virol. Sinica* **4:**375–382 (1989).

521. Ziegler, J. B., and Penny, R., Fatal echo 30 virus infection and amyloidosis in X-linked hypogammaglobulinemia, *Clin. Immunol. Immunopath.* **3:**347–352 (1975).

522. Zoll, G. J., Melchers, W. J. G., Kopecka, H., Jambroes, G., van der Poel, H. J. A., and Galama, J. M. D., General primer-mediated polymerase chain reaction for detection of enteroviruses: Application for diagnostic routine and persistent infections, *J. Clin. Microbiol.* **30:**160–165 (1992).

13. Suggested Reading

Banatvala, J. E. (ed.), *Viral Infections of the Heart*, Edward Arnold, London, Boston, 1993.

Brown, F., and Lewis, B. P. (eds.), Poliovirus attenuation, *Dev. Biol. Stand.* **78:**1–187 (1993).

Cherry, J. D., Enteroviruses, in: *Textbook of Pediatric Infectious Diseases*, 3rd ed. (R. D Feigin and J. D. Cherry, eds.), pp. 1705–1753, W. B. Saunders, Philadelphia, 1992.

International Symposium on Poliomyelitis Control, Pan American Health Organization, Washington, DC, *Rev. Infect. Dis.* **6**(Suppl. 2) (1984).

Kapsenberg, J., *Picornaviridae*: The enteroviruses, in: *Laboratory Diagnosis of Infectious Diseases*, Vol. 2 (E. H. Lennette, P. Halonen, and F. A. Murphy, eds.), pp. 692–722, Springer-Verlag, New York, 1988.

Melnick, J. L., The picornaviruses, in: *Portraits of Viruses. A History of Virology* (F. Fenner and A. Gibbs, eds.), pp. 147–188, S. Karger, Basel, 1988.

Melnick, J. L., Live attenuated poliovaccines, in: *Vaccines*, 2nd ed. (S. A. Plotkin and E. A. Mortimer, Jr., eds.), pp. 155–204, W. B. Saunders, Philadelphia, 1993.

Melnick, J. L., Poliomyelitis: Eradication in sight, *Epidemiol. Infect.* **108:**1–18 (1992).

Morens, D. M., Pallansch, M. A., and Moore, M., Polioviruses and other enteroviruses, in: *Textbook of Human Virology*, 2nd ed. (R. B. Belshe, ed.), pp. 317–497, Mosby Year Book, St. Louis, MO, 1990.

Paul, J. R., *A History of Poliomyelitis*, Yale University Press, New Haven, 1971.

CHAPTER 22

Rabies

Charles E. Rupprecht and Cathleen A. Hanlon

1. Introduction

Rabies is an acute central nervous system (CNS) disease of mammals that almost invariably results in death. Disease develops following productive infection with a bullet-shaped, enveloped virus that contains single-stranded, nonsegmented, negative-sense RNA.[169] Rabies virus is the type species of the *Lyssavirus* genus, family *Rhabdoviridae*. In the usual scheme of pathogenesis, virus gains entry to the host body via the bite of a rabid animal. Historically in the United States, the primary source of human exposure was the domestic dog, which still predominates as the major reservoir in developing countries. Now, wildlife are primarily affected in developed countries: current rabies reservoirs in the United States include raccoons, skunks, foxes, coyotes, and bats.[88] The incubation period in humans is variable, from < 10 days to > 6 years, but usually on the order of 4–6 weeks following an animal bite.[136] During the incubation period, virus is nearly undetectable and may replicate locally in muscle tissue at the initial site of entry.[34,35] Alternatively, virus may proceed directly and centripetally through the axoplasm of peripheral nerves to the CNS. Once in the nervous system, virus replicates and spreads quickly,[127] with associated dysfunctional clinical signs partially dependent on the affected area and relative severity of infection. Overt disease is initially nonspecific, consisting of signs and symptoms in humans compatible with a "flulike illness," such as fever, headache, and general malaise. Following the prodromal stage, an acute neurological phase may include intermittent insomnia, anxiety, confu-

sion, paresis, percussion myoedema, excitation, agitation, hallucinations, cranial nerve deficits, chorea, dysphagia, hypersalivation, piloerection, priapism, paralysis, and sometimes maniacal behavior.[75] Clinical presentation may also include classic symptoms of paresthesia at the site of bite exposure and hydrophobia (a synonymous term for the human disease)[57] or aerophobia manifesting as phobic pharyngeal spasms following provocative stimuli. The clinical course is acute, with death usually ensuing within days. A form of the disease termed "dumb rabies" may also present as part of the clinical spectrum, with the general sparing of consciousness together with ascending paralysis, progressive unresponsiveness, coma, and death. Once clinical signs are present, there is no cure. Intensive medical support may prolong life, but ultimately death ensues. Exceptions to this are exceedingly rare with only four well-documented cases of human survival from clinical rabies (all with a history of either pre- or postexposure therapy).[1,23,73,109] Two of the four survivors have significant residual neurological impairment. Acquired immunity in the rabies vector species, presumably following subclinical exposure, abortive infection, or survival of overt clinical rabies, is apparently rare, but has been supported to an extent by serological surveys of wildlife and documented occurrences under laboratory conditions.[50,54]

2. Historical Background

Rabies is an ancient malady of uncertain historic origin; it is one of the oldest recognized infectious diseases.[116] As such, it is difficult to provide more than highlights, given its fascination and dread by prior historians; for a more thorough treatise, the interested reader is referred to Steele,[139] Baer,[7] and Wilkinson.[162] Most global vocabularies singularly denote the terms damage,

This chapter is a revised version of Chapter 19, Robert E. Schope, in *Viral Infections of Humans*, 3rd edition; Plenum Press, 1989.

Charles E. Rupprecht • Centers for Disease Control and Prevention, Viral and Rickettsial Zoonoses Branch, Atlanta, Georgia 30333. **Cathleen A. Hanlon** • New York State Department of Health, Zoonoses Program, Albany, New York 12237.

violence, fury, madness, or rage as literally synonymous with this affliction. Many great civilizations refer to a disease akin to rabies, weaving a tapestry of overstated fact, legend, and nightmare, exceedingly out of proportion to its actual pathos. In the not-too-distant past, extreme ostracism, torture, or execution was inflicted on those even remotely suspected of hydrophobia. Cited as far back as the third millennium BC, the pre-Mosaic Eshnunna Code refers to the death of people following a dog bite. Moreover, Democritus in 500 BC recognized rabies in dogs and other domestic animals, as did many other Greek and Roman scholars. Thus, it can be assumed that this viral disease, or a similar contagion, was well-recognized and occurred in parts of Asia, Europe, and, perhaps, Africa, throughout centuries of recorded history.

Early historical accounts focused primarily on individual human or animal case reports. Wildlife rabies was recognized throughout the Middle East since biblical times,[116] but few epizootics were well documented until the Middle Ages. There were multiple accounts of rabid wolf attacks in Franconia during 1271,[139] and in Europe outbreaks were recorded in wolves, foxes, and dogs, especially during the 18th and 19th centuries. These continued sporadically until World War II, when another major epizootic in foxes swept through Europe, from east to west at a rate of about 30–50 km per year, the repercussions of which have been felt up to the present.

Rabies and its related lyssaviruses (Mokola, Lagos Bat, Duvenhage) probably evolved in the Old World. For most of the New World, the disease was largely unknown, at least among dogs, until one of the first descriptions in 1703 from Mexico. Rabies was subsequently identified from the Caribbean by the mid- and late-1700s. By 1753, dogs in Virginia had been affected, and most of the middle American and New England colonies shortly afterward. The American West followed suit with the pioneer movements of the early 1800s, replete with tales of "madstones" and "phobey cats." Rabies was first reported in Peruvian dogs in South America during 1803, although Baer[7] cites Spanish reports from the 1500s onward that associated bat bites and human mortality in Latin America. Thus, the virus was likely imported multiple times into the New World, from the initial Old World circumpolar invasions before recorded history and more extensively during a period of active colonization in the 18th century, concomitant with major domestic animal outbreaks in Europe. Later documentation of antigenic and genetic similarities between both Old and New World terrestrial rabies viruses[121,137] supports the contention of trans-Atlantic introduction by infected animals whose long incubation periods could exceed the length of the

voyage.[136,138] The continued isolation of distinct viruses in Africa, serologically and genetically related to rabies,[17] supports the hypothesis of an African genesis,[86] with adaptive evolution of several lyssaviruses.

Beyond mere geographic documentation, the history of rabies is also punctuated by a long series of observations relating to its treatment, cause, prevention, and diagnosis. For example, about 100 AD, Celsus treated animal bite wounds by cauterization and in 200 AD, Galen recommended amputation. Both had limited success (later supported by laboratory animal research).[139] In 1804, Zinke demonstrated transmission of rabies to a normal dog by inoculation of infected saliva, considering it a toxin.[139] This observation led to institution of muzzle laws and stray dog control, which resulted in rabies elimination from Denmark, Norway, and Sweden by 1826.

In 1879, Galtier utilized the domestic rabbit as a suitable laboratory host for rabies.[139] His observations enabled the classic 1881 experiments in which Pasteur and co-workers (not to overlook the primary contributions of his collaborators Roux, Chamberland, and Thuillier) reported the characterization of a "microbe of infinite smallness" attenuated for the dog but having a uniform (fixed) incubation period in the rabbit.[157] Pasteur distinguished this laboratory form from that of the agent in nature, or street virus. He subsequently used this fixed virus, grown in rabbit spinal cords and dried for varying periods, to give graded doses of noninfectious to fully infectious virus for the immunization of animals. The first human rabies vaccination was administered in 1885 to 9-year-old, Joseph Meister, severely bitten by a presumably rabid dog (based on clinical derangement and pica).[157] This single historic event ushered in the era of therapy, which was reasonably successful by accepted standards of the day. Occasional failures were attributed either to prolonged delays before treatment initiation or to the particular severity of an exposure.

From Pasteur's time, numerous improvements in safety and efficacy of the early nerve tissue origin (NTO) biologicals have been attempted. They were partially frustrated, until the 1940s, by the lack of standardized potency evaluation.[22] In 1908, Fermi developed the first chemically treated vaccine, unfortunately with residual live virus. By 1919, Semple showed that phenol completely inactivated rabies virus without destroying its antigenicity. Semple's vaccine was used extensively for at least 65 years, but was gradually replaced in Latin America and parts of Africa by the suckling mouse NTO vaccine of Fuenzalida in 1955. By circumventing the sensitization to myelin basic proteins found in adult animal brains, the suckling mouse brain-origin vaccine had a lower rate of neuroparalytic reactions. A number of studies from the turn of the century, with renewed investigation in the 1940s, suggested that, besides vaccine, rabies immune serum was also effective as a postexposure treatment (PET) in preventing disease. The essential utility of serum plus vaccine, in combination with local wound treatment, has been demonstrated most aptly in those few instances where human rabies occurred, despite intervention, when one of these facets of accepted protocol was altered or delayed.[145] Major vaccine initiatives during the 1950s and 1960s proceeded away from NTO biologicals, with the development of avian embryo vaccines, initially of dubious potency. Such alternative strategies eventually culminated with the first cell culture vaccine in the 1970s, the human diploid cell vaccine (HDCV), premier in safety and efficacy.[19,22,158] Although the original Pasteurian treatment with attenuated virus was discontinued in the 1950s, inactivated NTO vaccines, from rodents, rabbits, sheep, or goats, are still widely used in many parts of the developing world, despite the availability of safer, potent, albeit more costly alternatives such as the HDCV. The further development of purified duck and chick embryo vaccines and Vero cell vaccines may ultimately compete economically and displace NTO vaccines altogether.

Before the institution of routine laboratory tests for rabies, the clinical presentation of the biting animal provided the primary motivation for human PET. At the beginning of the 20th century, scientists such as Babes and Van Gehuchten had described lesions in the CNS, presumptively related to rabies,[139] but which were nonspecific for a set of other etiologies as well. However, it was in 1903 that Negri described intracytoplasmic acidophilic inclusion bodies (which he believed to be protozoa) in neurons of rabid humans and other mammals, facilitating the early diagnosis of rabies when brain tissue was appropriately collected, fixed, and stained. Histopathologic identification was a key procedure despite the relatively high rate of false negative observations until Goldwasser and Kissling introduced the immunofluorescence diagnostic technique in 1958.[64] That technique is still in routine use today as the global gold standard for rapid, sensitive, and specific rabies diagnosis.

The above historical accomplishments reveal that rabies research efforts focused on the human as victim rather than on control in the animal reservoir. While animals did serve as critical experimental subjects, human vaccination preceded the consideration of dog immunization, which was originally deemed impractical. It was not until 1919 that an attempt at mass vaccination of dogs occurred with Fermi vaccine in Japan, with the focus shifting to the source of exposure for many thousands of persons treated with Pasteurian PET during the three preceding decades. However, problems associated with potency of early inactivated vaccines plagued these first trials. By 1948, a live virus vaccine using an attenuated Flury strain of egg-adapted virus was successfully applied to dogs. A variety of both inactivated and live virus vaccines was subsequently shown to be effective in preventing rabies in all of the major domestic animals. Since the 1950s, stray dog removal and mandatory dog vaccination has resulted in virtual elimination of canine rabies in the United States and other developed nations.[153] This, coupled with improvements to the present day in epidemiologic surveillance, diagnostics, and pre- and postexposure biologics, has consequently resulted in a marked decrease in the numbers of human rabies cases. Nonetheless, an extensive public health infrastructure and allocation of resources are required to minimize domestic animal rabies and maintain the vigilance necessary to recognize potential human rabies exposure for the prompt initiation of prophylaxis.

3. Methodology Involved in Epidemiologic Analysis

3.1. Sources of Mortality Data

Human rabies is a specific notifiable disease in the United States, as in most other countries. Statistics for humans and animals are compiled from local and state health departments by the Centers for Disease Control and Prevention (CDC) and are published in the CDC's *Morbidity and Mortality Weekly Report*, as well as an annual summary. In Europe, numbers and distribution of rabid animals are published quarterly in the *Rabies Bulletin of Europe*. The World Health Organization (WHO), through the Veterinary Public Health Unit in Geneva and the associated Pan American Health Organization in Washington, DC, collects data from countries or political units on deaths in animals, people, number of people exposed, and number of doses of rabies vaccine administered. These data are distributed annually to national governments as a source of data on global rabies occurrence and the relative risk of rabies infection, which, provided that proper epidemiologic surveillance is ongoing, should be nil in countries with no reported cases (Table 1). Notably, the number of human PETs administered in the United States is lacking in these WHO summaries, since reporting is required neither by most states nor on a national level. The emergence of raccoon rabies epizootic in the mid-Atlantic and northeastern United States and a coyote rabies epizootic in Texas[89] has significantly affected the

Table 1. Global Disease Surveillance: Countries/Political Units Reporting No Cases of Rabies[a,b]

Region	Countries
Africa	Mauritius[c]; Libya[c]; Djibouti[c]; Lesotho[c]; Seychelles[c]
Americas	
North	Bermuda; St. Pierre and Miquelon
Caribbean	Anguilla; Antigua and Barbuda; Bahamas; Barbados; Cayman Islands; Dominica; Guadeloupe; Jamaica; Martinique; Montserrat; Netherlands Antilles (Aruba, Bonaire, Curacao, Saba, St. Maarten, and St. Eustatius); St. Christopher (St. Kitts) and Nevis; St. Lucia; St. Martin; St. Vincent and Grenadines; Trinidad; Tobago; Turks and Caicos Islands; Virgin Islands (UK and US)
South	Uruguay[c]
Asia	Bahrain; Hong Kong; Japan; Kuwait; Malaysia (Malaysia-Sabah[c]); Maldives[c]; Singapore; Taiwan
Europe	Cyprus; Denmark; Faroe Islands; Finland; Gibraltar; Greece; Iceland; Ireland; Malta; Norway (mainland); Portugal; Spain (except Ceuta/Melilla); Sweden; United Kingdom (Britain and Northern Ireland)
Oceania	American Samoa; Australia; Cook Islands; Fiji; French Polynesia; Guam; Hawaii; Indonesia (with exception of Java, Kalimantan, Sumatra and Sulawesi); Kiribati; New Caledonia; New Zealand; Niue; Papua New Guinea; Solomon Islands; Tonga; Vanuatu

[a]Based on data from the following publications and other information provided to the Centers for Disease Control and Prevention: World Health Organization,[168] WHO Collaborating Centre for Rabies Surveillance and Research (*Rabies Bulletin Europe*, 1994, Vols. 1 & 2), and Pan American Health Organization. Epidemiological surveillance of rabies in the Americas, 1993, 1994; Vol. 25 (No. 1–12).
[b]Bat rabies may exist in some areas that are reportedly free of terrestrial rabies.
[c]Countries whose classifications may be considered provisional.

incidence and epidemiology of human PET, diverting critical health resources as animal rabies cases rise. To properly assess national needs and to implement rationale reduction of the improper usage of PET, efforts toward creating at least regional surveillance in the United States by the turn of the century are under discussion.

When rabies surveillance is adequate, human death rates may accurately represent incidence rates of human infection. The disease for statistical purposes can be regarded as 100% fatal. (Due to the acute clinical course incidence rates are also virtually equivalent to prevalence rates.) However, in most developing countries, it is widely accepted that human rabies is underreported, and when it is reported, may outnumber documented cases in reservoir species. It is remotely possible that human rabies may be underreported even in the United States, considering that within the past several years, at least six human cases were diagnosed retrospectively from necropsy material.[26-29] The diagnosis of rabies was confirmed in these latter cases using fixed material rather than fresh tissue, which is the preferred diagnostic material. In these cases, it was fortunate that even fixed brain tissue was available, because in a significant proportion of human encephalitis cases of unknown etiology resulting in death, an autopsy may not be performed.

Where canine rabies is present, human exposure and rabies incidence rates appear to be related to the local epizootiology of the disease. However, the relationships between human rabies mortality, rabies exposure, and wildlife rabies cases are less well established. Since the majority of rabies vector species in the United States are free-ranging wildlife, the number of animal rabies cases is variable and biased by surveillance effort. Surveillance, typically based on the intensity and occurrence of animals with suspicious clinical presentations and encountering humans or domestic animals, would tend to be more reliable in detecting cases during an outbreak but less so during normal enzootic periods. This bias may be greatest in localities where the responsibility for rabies-suspect animals, now predominantly sick or nuisance wildlife rather than domestic dogs, does not fall within the purview of an official local response. Historically, a dog control officer responded to domestic animal problems in the community. Without a mandated and funded local response to rabies-suspect wildlife, a "shoot-and-bury" practice may evolve to cope with rabies among wildlife, with the result that few cases are recorded. Disease-specific incidence (or death) rates for animals are generally unavailable, since free-ranging wildlife populations are rarely properly enumerated. Thus, in the United States, representation of the public health magnitude or burden of the disease solely by absolute case numbers of animal rabies is inherently biased and should be qualified in view of its limitations.

Since the incursion of raccoon rabies into New York State in 1990, the annual totals of rabid animals have surpassed historical records, demonstrating the magnitude of the wildlife rabies problem in this geographic region.[89] The total areas affected and the proportion diagnosed rabid among raccoons tested are still increasing, although absolute numbers submitted for testing have declined. Concurrently, human PETs have increased dramatically, from 89 in 1989, prior to the invasion of raccoon rabies, to annual totals of 197, 965, 1123, and 3247, in 1990 through 1993, respectively. based on provisional reports in New York.

3.2. Sources of Morbidity Data

Human rabies case morbidity is largely reflected in mortality data. As cited, there are at least four documented cases of humans surviving clinical rabies, although two resulted in serious neurological sequelae. Morbidity that has otherwise been averted could be assessed through the number of human PETs; yet, due to the lack of required reporting of human PET in the United States, information on potential morbidity prevented by PET is not readily available. It may be approximated, as in developing countries, by the number of vaccine doses and human rabies immune globulin (HRIG) sold or used. Based on these imperfect data (because of the vaccine's utilization in both pre- or postexposure protocols and administration of HRIG on a per-weight basis), current national estimates range between 20,000 to 40,000 human PETs per year. In some states human PET data are already reportable, whereas in others vaccine and HRIG are available only through the state, from which the number of treatments may be approximated. Representative reporting of human PETs would be advantageous for proper assessment of relevant epidemiologic variables, national and regional needs, and risk factors for exposure, as well as for further definition of the overall economic impact of wildlife rabies.

3.3. Serological Surveys

Historically, rabies serological evaluation[134] has been performed by antigen-function assays, such as the mouse neutralization test, by antigen-binding assays, such as the indirect immunofluorescence test, or by antibody-function assays, such as the hemagglutination test. Currently, one of the most widely used serological tests is the rapid fluorescent focus inhibition test (RFFIT). Results are typically reported in relation to a known standard, as international units (IU) per milliliter of sera. Alternatively, an enzyme-linked immunosorbant assay (ELISA) may be applied for simple, rapid screening of large numbers of sera; however, its specificity may be less reliable, especially at low antibody levels, and its utility is determined by the source and quality of crude antigen. The fundamental difference between these tests is that the ELISA is based on recognition of antigen, whereas the RFFIT is a functional test.

The absolute interpretation of rabies serological survey data is not straightforward. There is no known absolute "protective antibody" level for all humans or animals.[134] For example, among a group of animals vaccinated against rabies, antibody level at the time of laboratory challenge is not always a reliable predictor of survival.[119] Moreover, minimum standards for humans are based empirically on presumed protective activity of rabies-specific virus-neutralizing antibody (VNA) for a given exposure scenario and on repeatable values for paired sera as could be readily detected by reference laboratories. It is currently recommended that a RFFIT titer of ≥ 1/5 (an international antibody level of 0.5 IU/ml) or higher be verified in persons at either constant or frequent risk of exposure, at 6-month or 2-year intervals, respectively, as a measure, however flawed, of baseline immunity.[25] A single booster dose is administered if the level is lower, based on a determination of apparent risk (Table 2). Preexposure immunization greatly simplifies human PET due to a priming of the immune response; it is thought to provide protection against unrecognized exposure, which by definition should be negligible. Nevertheless, when preimmunized persons are knowingly exposed to rabies, two booster doses of vaccine, administered on days 0 and 3, are required to induce a sufficient anamnestic response to preclude development of disease; HRIG is contraindicated in that it may interfere with an anamnestic response.[130,145] Preimmunized individuals should remain vigilant in recognizing potential exposures and seek appropriate PET. If actual rabies exposures occur but are unrecognized, the preimmunized individual may still succumb to rabies, albeit rarely, as did a Peace Corps volunteers who was bitten by a puppy that died of a disease compatible with rabies.[24]

As in other viral infections, there is increasing evidence that sublethal rabies virus exposures may occur. The likelihood of this phenomenon may depend on inoculum and strain of virus, the route of inoculation, and host factors. Some authors suggested that vocational pursuits, such as trapping of rabies reservoir furbearers, could lead to low levels of rabies antibodies.[12,59,104] At least one study has documented a titer of 2.3 IU/ml in a Native Alaskan trapper, who reported handling a lifetime total of some 3000 arctic foxes.[59] Heavy outer clothing and gloves may decrease the exposure potential to outright bites, but multiple nonbite exposures per se may still be involved over time during the skinning process.

In contrast to potential acquisition of immunity through nonbite exposures as above, it is conceivable in certain rare instances that low doses of street virus with repeated bite exposures may result in sublethal rabies exposures and act in concert to "vaccinate" a certain small segment of a population at risk, almost in traditional Pasteurian fashion, with live virus. For example, in remote

Table 2. Rabies Immunization—Preexposure and Postexposure[a,b]

Exposure category	Nature of risk	Typical populations	Preexposure regimen
Continuous	Virus present continuously, often in high concentrations. Aerosol, mucous membrane, bite, or nonbite exposure possible. Specific exposures may go unrecognized.	Rabies research lab workers.[c] Rabies biologics production workers.	Primary preexposure immunization course. Serology every 6 months. Booster immunization when antibody titer falls below acceptable level.[c,d]
Frequent	Exposure usually episodic with source recognized, but exposure may also be unrecognized. Aerosol, mucous membrane, bite, or nonbite exposure.	Rabies diagnostic lab workers,[c] spelunkers, veterinarians, and animal control and wildlife workers in rabies epizootic areas. Certain travelers to foreign rabies epizootic areas.	Primary preexposure immunization course. Serology or booster immunization every 2 years[d]
Infrequent (greater then population-at-large)	Exposure nearly always episodic with source recognized. Mucous membrane, bite, or nonbite exposure.	Veterinarians and animal control and wildlife workers in area of low rabies endemicity. Veterinary students.	Primary preexposure immunization course. No routine booster immunizaton or serology.
Rare (population-at-large)	Exposure always episodic, mucous membrane, or bite with source recognized.	US population-at-large, including individuals in rabies-epizootic areas.	No preexposure immunization.

Criteria for preexposure immunication

[a]From Centers for Disease Control.[(25)]

[b]**Preexposure immunization.**   Preexposure immunization consists of three doses of HDCV or RVA vaccine, 1.0 ml, IM (i.e., deltoid area), one each on days 0, 7, and 21 or 28. *Only* HDCV may be administered by the intradermal (ID) route (0.1 ml ID on days 0, 7, and 21 or 28). If an individual will be taking chloroquine or mefloquine for malaria chemoprophylaxsis, a 3-dose rabies vaccination series must be completed before initiation of antimalarials. If this is not possible, the IM dose/route should be used. Administration of routine booster doses of vaccine depends on exposure risk category as noted below.
Postexposure immunization.   All postexposure treatment should begin with immediate thorough cleansing of all wounds with soap and water. **Persons not previously immunized:** HRIG, 20 I.U./kg body weight, one half infiltrated at bite site (if possible), remainder IM; 5 doses of HDCV or RVA, 1.0 ml IM (i.e., deltoid area), one each on days 0, 3, 7, 14, and 28. **Persons previously immunized:** Two doses of HDCV or RVA, 1.0 ml, IM (i.e., deltoid area), one each on days 0 and 3. HRIG should not be administered. Preexposure immunization with HDCV or RVA; prior postexposure prophylaxis with HDCV or RVA; or persons previously immunized with any other type of rabies vaccine and a doumented history of positive antibody response to the prior vaccination.
[c]Judgment of relative risk and extra monitoring of immunization status of laboratory workers is the responsibility of the laboratory supervisor (see U.S. Department of Health and Human Services' *Biosafety in Microbiological and Biomedical Laboratories*, 1993).
[d]Preexposure booster immunization consists of one dose of HDCV or RVA, 1.0 ml/dose, IM (deltoid area), or HDCV, 0.1 ml ID (deltoid). Acceptable antibody level is 1:5 titer (complete inhibition in RFFIT at 1:5 dilution). Boost if titer falls below 1:5.

communities in Amazonia, humans may become a frequent alternative to the preferred bovine prey of vampire bats, a proportion of which may be rabid.[(114)] Depending in part on viral burden, exposure site, particular variants, and so on, many human victims will succumb, but some small fraction of patients may develop immunity from repeated exposures, with or without clinical illness.[(97)]

With regard to rabies in the vector species, there are experimental and field data to support the contention that a proportion of animals may be exposed to rabies, develop measurable antibodies, and are then immune to subsequent challenges.[(50,54,79)] Thus, serological surveys of wildlife in areas with enzootic terrestrial rabies may reveal a low prevalence of antibodies.[(76)] These may occur among animals that are infected and will ultimately succumb to rabies or among individuals that have developed immunity to rabies, presumably following natural sublethal exposure. These naturally occurring antibodies are distinct from those due to purposeful immunization of

free-ranging wildlife via limited trap–vaccinate–release programs,[(115)] regional oral rabies vaccination programs,[(123)] or parenteral vaccination of wildlife with domestic animal vaccines by well-intentioned wildlife rehabilitators.

While there is little evidence among most mammals supporting a significant role for naturally induced immunity, viverrid species provide a classic example of the dynamic, complex, host–parasite relationship that exists in rabies.[(48)] Mongooses introduced for rodent control onto most of the larger Caribbean Islands, including Puerto Rico, Dominican Republic (Haiti), Cuba, and Grenada brought rabies as well. The mongooses proliferated and are now deemed undesirable due in part to predation on native fauna and the risk from rabies in this reservoir species. For example, 79% of the 43 rabies cases diagnosed in Puerto Rico during 1993 were due to mongoose. Although poisoning has been practiced, it is undesirable and ineffective in the long term. Many mongooses (as many as 55% at the end of an epizootic cycle) have evidence of rabies VNA, possibly as a result of sublethal rabies exposure; they may be immune for life.[48] It may be detrimental to remove them through poisoning, which may simple enhance turnover, dispersal, and increase the number of naive, susceptible animals. In view of such levels of naturally occurring, probable herd immunity, these closed populations of mammals should be ideal candidates for oral vaccination as one future means of disease control,[96] especially in a limited island environment.

3.4. Laboratory Methods

Many other encephalitides may be confused with rabies. It is essential that epidemiologic studies be based on carefully documented laboratory confirmation of suspected disease or death from rabies virus infection. As summarized in the historical description, laboratory techniques have evolved gradually over the last century, progressing from reliance on animal inoculation and histopathologic examination to immunofluorescent antibody (FA) diagnostic tests and others. Although variations are in use in many laboratories, the standard techniques approved by the eighth WHO Expert Committee on Rabies[164] and the most current edition of the WHO *Laboratory Techniques in Rabies* are recommended and widely used, to which the interested reader is further directed. Included in these texts are methods for the collection, preparation, and shipping of rabies specimens; tissue and organ removal; examination of brain tissue for Negri bodies; laboratory-animal inoculation; FA procedures for detection of rabies virus antigens; serological determinations; cell culture propagation of virus; electron microscopy; vaccine production; potency and safety determinations; and specific virus identification by antigenic and genetic techniques. For regulations governing specific international or national shipment of potentially infectious specimens, such as rabies, the national or local public health laboratory should be consulted.

In the brain tissue from the rabid patient, Negri bodies are specific intracytoplasmic inclusion bodies seen under the light microscope. The inclusions contain rabies virus nucleoprotein, which accumulates in quantities large enough to stain by the Seller's, Giemsa, or Mann methods and can be microscopically visualized. Within the hippocampus, Ammon's horn is one particular portion of the CNS usually examined, as are the Purkinje cells of the cerebellum and pyramidal cells of the cerebral cortex. Limitations are that Negri bodies may be absent in one quarter to one third of rabies cases, while artifact and inclusions produced by other viruses can be confusing and

differentiated only after considerable experience. Nevertheless, histopathologic examination offers a rapid and inexpensive diagnosis, particularly for developing countries, if the FA test is unavailable.

Animal inoculation with suspect diagnostic material was once widely used to support rabies diagnosis, because of potential ramifications of false-negative results in human rabies exposure situations. Suckling and weanling mice are highly susceptible to rabies infection by intracerebral inoculation. The time from inoculation to illness varies from 7 days to 4 weeks, but rabies antigen in mouse brain may be confirmed as early as 4 days following inoculation. Confirmation of rabies, as the source of illness and mortality, should always be performed because laboratory animals may die of intercurrent colony infections or as a result of infections by other agents in the original clinical specimen.

Murine neuroblastoma cell cultures are as sensitive as laboratory animals for isolation of field strains from saliva and brain of rabid animals and are the method of choice in laboratories where cell culture facilities are available.[117] The presence of antigen in infected cells is demonstrated by the FA test.

The direct FA test uses a fluorescein dye conjugated to rabies immune serum, which in turn is reacted with a fixed impression from brain tissue of the presumed rabid host, including the brainstem, cerebellum, and hippocampus.[64] The antigen–antibody reaction is detected by microscopic observation of fluorescence under light of the appropriate wavelength. The test is rapid and reliable when used by an experienced person in the appropriate manner. The FA method is the test of choice for fresh or frozen tissue, because of its sensitivity, specificity, and economy; alternative methods on fixed tissue,[9,15,49,51,67,80,106,113] while encouraging in their continued development and superiority to Negri body detection alone, are otherwise inadequate in these regards. In Africa and Europe, other lyssaviruses related to rabies may cross-react significantly in the FA test, depending on the potency of the diagnostic conjugate. Specific monoclonal antibodies (MAbs) can be used to distinguish rabies from these related lyssaviruses, but the public health consequences are the same, regardless of exposure to any one of these etiologic agents.

Human antemortem diagnosis can sometimes be made by virus isolation from saliva, antigen detection in biopsy tissue, or demonstration of rabies-specific antibody (without prior vaccination). Selected FA tests are performed on human brain or skin biopsies,[135] the latter usually taken from highly innervated areas such as the nape of the neck, where nerve plexi are prominent. Also, in some human rabies cases, corneal cells obtained from touch impressions may be FA-positive during life, care being taken not to abrade the delicate corneal tissue (a rationale to caution against this method vs. other stated procedures). Antemortem diagnosis may also be made by showing a fourfold or greater rise in serum antibody titer during the illness in the absence of vaccination or administration of rabies immune serum. Demonstration of rabies antibody in the cerebrospinal fluid (CSF) is also a reliable indicator of infection even after vaccination. Vaccination alone does not elicit CSF antibody. During rabies infection, other pathophysiologic changes in the CSF, such as an increase in specific density, pleocytosis, particularly a monocytosis, which are compatible with a viral encephalitis, may also occur.[68,135] Repeated samples should be taken for antibody and antigen detection, because negative results in early samples do not rule out rabies. Moreover, because of the risk of false-negative results with a potential human exposure, intravitam testing of the rabies-suspect animal is inappropriate.

Recent progress has been made in the application of molecular techniques, particularly nucleic acid detection by reverse transcription and amplification of cDNA by the polymerase chain reaction (RT-PCR),[82,99,126] with subsequent generation of viral nucleotide sequences leading to a greater understanding of lyssavirus epidemiology and phylogeny.[16,17,135] However, it is the extreme sensitivity of the RT-PCR technique that also greatly increases the probability of a false-positive diagnosis from laboratory contamination. Moreover, because of lyssavirus heterogeneity, false-negative results can occur if primer selection is inadequate to compensate for heterogeneity; universal primers for all known lyssavirus variants have not been clearly defined nor standardized. Considering these factors, related costs and the considerable expertise required for proper analysis and interpretation, such molecular techniques are not recommended for routine rabies diagnosis at the present time.

The routine sequence characterization of lyssavirus strains from formalin-fixed tissues is possible via the RT-PCR test, but may be difficult, partially as a function of the fixation time and resultant short fragments of nucleic acid.[137] Optimization of these methods, beyond mere qualitative diagnostic methods for viral antigen analysis preserved in such tissue, may be feasible by the adaptation of ELISA techniques that would employ MAbs for retrospective strain identification in archival material.[49] Besides determination of optimal specific reagents, corroborative data would have to be generated on the sensitivity and specificity of this technique in comparison to fresh tissue analysis.

4. Biological Characteristics of the Virus that Affect the Epidemiologic Pattern

Rabies is essentially a dead-end disease in humans. Human-to-human transmission via bite has not been reliably established.[74] A marked exception consists of eight human rabies cases resulting from the surgical implantation of infected corneas from donors that had succumbed to undiagnosed rabies, the most recent reported from Iran.[165]

Biological characteristics of rabies viruses have a measurable, although incompletely understood, effect on epizootiologic patterns among their host species.[154] Initially through classic serology and later by MAb and genetic analysis, it was possible to differentiate serotypes, variants, and genotypes of rabies and related lyssaviruses.[121,137,169] Epizootiologically, such variant viruses are largely perpetuated by different host species. Complex virus–host interactions lead to the emergence of viral characteristics that are beneficial to self-perpetuation among these particular vector species that manifest as regional epizootiological "compartments."[2,28,124] Thus, there are discrete geographic zones of raccoon rabies, skunk rabies, and so forth that are fully transmissible to other mammals within a region but genetically and antigenically distinguishable as a single strain in a particular geographic area, regardless of affected host, be it the reservoir species (e.g., raccoon) or mere "spillover" into an essentially dead-end, but susceptible host (e.g., domestic cat).

As one specific example of rabies virus vagility, coyotes possess many host qualities ideal for the initiation of a rabies epizootic. During 1915–1917, an extensive outbreak of coyote rabies extended over large portions of southeastern Oregon, northeastern California, western Utah, and Nevada.[98] Thereafter, rabies among coyotes was only sporadically reported in the western United States, despite the coyotes' widespread distribution and abundance. Another major focus was detected in Texas in 1988,[40] and virus is currently spreading throughout southern Texas. Antigenic and genetic analysis of isolates obtained from this ongoing outbreak suggests the involvement of a rabies variant historically associated with canine rabies along the U.S.–Mexican border, but now apparently capable of sustained coyote-to-coyote perpetuation.

Certain other critical biological characteristics may relate to sustainability that can be maximized in vector species. One of these may be a predilection for specific neuronal populations responsible for behavioral changes enhancing transmission.[34,35,127,132] Vector species may have a higher probability of retaining the capacity to actively seek other susceptible conspecifics in the early infectious period while still engaging in normal socio-biological behavior.[2,132] An overwhelming rabies encephalitis may occur more rapidly in other host species, resulting in incapacitation prior to opportunities for transmission. By the same token, it may not necessarily be the often-cited fury that assists in effective viral transmission; if paresis and paralysis manifest, a normal individual may initiate an encounter with a clinically abnormal host because of inappropriate behavior in the affected animal. The behavioral changes associated with rabies infection by a particular strain in its respective reservoir species may be unique to the biochemical milieu of that species and consequent alterations on neuronal infection.[149] Additionally, there may be species-dependent differential capacities of rabies virus variants[132] for attaining titers of sufficient magnitude in the salivary gland and saliva for reliable transmission to a host of choice versus no virus or lower concentrations at a portal of exit in hosts other than the reservoir species for that particular variant.[10,34,77,78,154]

Another unique characteristic of rabies virus infection is a prolonged incubation period during which the infection is virtually undetectable.[34] In humans, although the incubation period is generally several weeks to several months, unusually long incubation periods of 6 or more years have been described.[136] This potential for long incubation periods directly influences the management of exposed domestic animals. A rabies-exposed, unvaccinated domestic animal may be euthanized or held in quarantine for 6 months,[32] which extends beyond the majority of documented incubation periods for dogs and cats.

Neurotropism appears to be a critical facet of a successful strategy to partially evade immune detection[34,127,133] and allow for direct influence on host behavior,[132] with the ultimate end to ensure future viral progeny. Under the constraints of mammalian anatomy, there are limited routes available to a virus for transit from an initial portal of entry (i.e., a bite in a peripheral muscle) to a primary portal of exit (i.e., the saliva). One of the most common mechanisms for viral transit to such distant sites may otherwise be through viremia or spread via lymphatics; passage along both of these latter pathways may also initiate a vigorous immune response. An effective blood–brain barrier would limit the utility of viremia in enhancing the spread and perpetuation of a neurotropic virus. All known lyssaviruses utilize neurotropism; direct evasion, as opposed to suppression, mimicry, and so forth, of host immune surveillance appears to be a critical evolutionary strategy.[65] Long incubation periods may be one consequence of reliance on passive flow within neural circuitry.[127] It is also a mechanism to lower the risk of extinction in a host that may pursue a solitary existence for short periods of its life history. Since the reservoir host must eventually seek a breeding partner, an opportunity for transmission is readily available, considering the common use of teeth and oral mucosal contact preparatory to typical mammalian copulation, especially among the carnivores and bats. Following bite inoculation, covert access to the CNS via neuron-to-neuron passage would be one unique mechanism for minimizing immune detection while reaching distant sites important for replication and transmission.[34,131] Dependence on fluid-sucking arthropod vectors is unnecessary, as are additional structure proteins to ensure extramammalian environmental survival, because *in vivo* transmission is largely a direct host-to-host event, for example, polar fox-to-polar fox, red bat-to-red bat, and so on, irrespective of firm constraints imposed by either latitude or season.

5. Descriptive Epidemiology

Rabies is a zoonosis with diverse natural reservoirs.[154] Predominant hosts are bats and mammalian carnivores. These include canids, such as the domestic dog as the principal global reservoir (particularly in equatorial regions of Asia, Latin America, and Africa[100,153,168]; foxes in the circumpolar Arctic, eastern Canada, and portions of New England, Central and Western Europe, the Middle East, and scattered foci throughout the western United States[168]; raccoon dogs in eastern Europe[37,103] and coyotes in the western United States[40] and Latin America[168]; mustelids, such as the skunks, primarily in the central United States and Canada[89]; procyonids, such as the raccoon, in the southeastern, mid-Atlantic, and northeastern United States[89]; and viverrids, such as the mongoose in Asia, Africa, and several Caribbean islands.[48,168] Other conspicuous carnivores, such as wolves, jackals, and all of the felids, can serve as effective short-term transmitters intra- and interspecifically, but insufficient documentation is available to suggest that these groups are capable of sustained perpetuation or serve as reservoir hosts for unique viral variants, as opposed to primary infection by the same. "Bat" rabies per se is a New World phenomenon, described primarily among the insectivorous species of North America and the hematophagous vampires from Latin America.[85,97] Related lyssaviruses have been diagnosed in African and European bat species.[16,86] While the disease is naturally maintained by relatively few taxa, rabies may affect any mammal, such as ungulates, and result in a largely "dead-end" infection. Contrary to popular belief, rodent (mice, rats, etc.) and lagomorph (rabbits, hares, etc.) rabies is uniformly rare.[55] Moreover, rodents have not arisen as "missing link" reservoirs in any region in which domestic dog or wildlife rabies has been controlled.

5.1. Incidence

Rather than a global rabies pandemic, single or multispecies groupings are apparent when disease surveillance is systematically practiced. Combined with historical temporal and geographical data, both antigenic characterization and nucleotide sequence analysis can be used to "compartmentalize" rabies isolates with different reservoirs responsible for their perpetuation and to estimate relative risks of human disease from differential animal exposure (Table 3). For example, during 1993, only three human fatalities occurred, however, 9495 animal rabies cases (an increase of 10% over 1992) were reported in the United States (including the District of Columbia and Puerto Rico), the highest combined total in over 45 years[89] (Fig. 1). In contrast to fulminant canine rabies pre-World War II, > 90% of current animal rabies cases are from wildlife (Fig. 2a, top). Most of this increase resulted from continued spread of a predominant raccoon rabies virus variant, following unrestricted progression of an outbreak initiated by animal translocation during the late 1970s to the Virginias from a nidus in the southeastern United States.[28,89,124] Cases rise as infected individuals

within raccoon populations encounter naive conspecifics, with dramatic results. In 1993, New York state alone reported a total of 2747 rabies cases (as compared to a prior level of 54 in 1989), the most ever recorded for any single state; 86% of the diagnosed animals were raccoons. Epizootics of raccoon rabies in the northeastern and southeastern United States have now approached convergence.[28,124]

Rabies was also reported in other important wildlife, primarily skunks (1640 cases) and foxes (361 cases), in the Midwest and Alaska/northeastern United States, respectively, and bat rabies (759 cases) widely distributed over 46 states and the District of Columbia. Alarmingly, new wildlife rabies outbreaks continue to emerge: 74 cases of coyote rabies were diagnosed in 1993 (71 from southern Texas), with a similar increase for reported dog rabies cases in affected areas.[40,89] Historically, Hawaii is unique to the United States in never having reported a case of indigenously acquired rabies, due in part to its strict isolation and 120-day quarantine policy.

Domestic animals most at risk include those with a lower likelihood of parenteral vaccination but higher potential for rabies exposure (especially if poorly supervised and free-ranging). Consequently, the 291 "spillover" cases of rabies in cats in 1993 (due almost entirely from states also reporting raccoon rabies) represented a 54% increase from 1991 (Fig. 2, bottom); feline rabies often outnumbers its canine counterpart, owing in part to the increased popularity of cats as pets but also lower likelihood of routine vaccination. Difficulties in clinical recognition also increase the public health significance of feline rabies, as early nonspecific signs may easily be compat-

Table 3. Rabies Postexposure Treatment (PET) Guide, United States

Animal type	Evaluation and disposition of animal	Postexposure treatment recommendations
Dogs and cats	Healthy and available for 10 days observation Rabid or suspected rabid Unknown (escaped)	Should not begin PET unless animal develops signs of rabies[a] Initiate PET immediately[b] Consult public health officials
Raccoons, skunks, bats, foxes, other carnivores, woodchucks	Regard as rabid unless geographic area is known to be free of rabies or until animal is proven negative by laboratory tests	Initiate PET.[c] Consider factors such as provocation, suggestive clinical signs, severity of wounds, type of exposure, and timeliness of test results (24–48 hr) for decisions regarding immediate initiation or to delay pending test results
Livestock, rodents, and lagomorphs (rabbits and hares)	Consider individually	Consult public health officials; bites of squirrels, hamsters, guinea pigs, gerbils, chipmunks, rats, mice, other rodents, rabbits, and hares, almost never require PET

[a]If clinical signs compatible with rabies develop during the 10-day confinement and observation period, the animal should be euthanized and tested. Depending on circumstances, initiation of treatment may be delayed pending a positive report, if results may be obtained in 24–48 hr.

[b]If the bite was unprovoked or resulted in severe wounds, treatment of the bitten person should begin immediately with human rabies immune globulin (HRIG) and human diploid cell vaccine (HDCV) or rabies vaccine adsorbed (RVA). Treatment may be discontinued if the test is negative.

[c]If available, the animal should be humanely euthanized and tested as soon as possible. Holding for an observation period is not recommended since the potential shedding period prior to clinical signs has only been determined for dogs and cats.

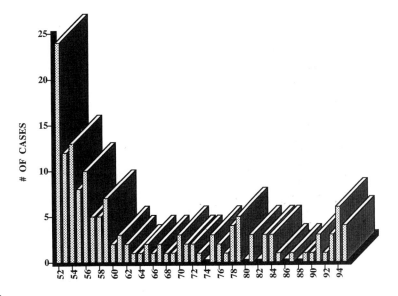

Figure 1. Reported human rabies deaths in the United States, 1952 to 1995.

ible with unrelated differential diagnoses, even if compounded by an abscess compatible with a bite wound in an outdoor cat.[58] By temporal comparison, 130 cases of dog rabies were reported in the United States during 1993, a decrease of 16% from 1991.[89]

In contrast to areas enzootic for dog rabies with a consequent high case load (e.g., estimated at > 25,000 annual human fatalities in India alone), human rabies is rare in developed countries.[168] Between 1980 and 1994, only 24 human deaths were diagnosed in the United States, nine of which were likely acquired abroad.[24,27,30] Of the cases suspected of infection while in the United States, 11 were due to bat rabies virus variants.[26,27,31,33] Prior to 1960, most human rabies cases in the United States originated from the bite of a rabid domestic animal,[105] while a definitive history of animal exposure is often difficult to elicit in the majority of current human fatalities.

5.2. Epidemic Behavior

Application of MAb technology to the study of the *Lyssavirus* genus (previously felt to consist of a relatively homogeneous group of viruses) first provided substantive evidence for considerable antigenic variation within the group, based on the N and G protein reactivities among both fixed and street rabies viruses, and between rabies and related viruses.[43,121] The use of MAbs has been particularly useful in determination of the extent of natural antigenic variation among lyssaviruses, either isolated from a variety of wildlife reservoirs within a fairly restricted geographical area or between continents. In particular, distinctions became much more obvious between those viruses isolated from bats and terrestrial carnivores.[16,121]

5.3. Geographic Distribution

Currently in the United States the most prominent wildlife rabies vectors include raccoons, skunks, foxes, and bats.[89,124] Additionally, there is a foci of gray fox rabies in north-central Texas and a newly emerging epizootic among coyotes in southern Texas.[89,124]

Through MAb application to recent U.S. surveillance data, two fox rabies outbreaks in New York/Vermont and Alaska appear related, as do the two raccoon epizootics in the northeastern and southeastern states.[124] Skunk rabies isolates group as distinct variants defining separate outbreaks in the north central/south central states and California. Additionally, smaller independent foci involve foxes, dogs, and coyotes in Texas and foxes in Arizona.[124] Such investigations, when extended over large geographic areas, can lead to more refined notions of lyssavirus dynamics and improved animal rabies control programs, to distinguish recent viral introductions, potential vaccine failures due to disparity of seed strain and wild variants and so on.

Antigenic or molecular characterization may also be employed to investigate unusual or unexpected lyssavirus

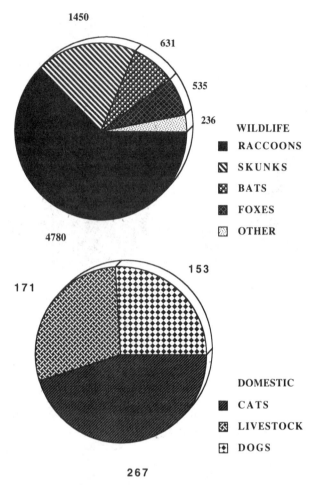

Figure 2. (Top) Reported rabies cases in wild animals in the United States, 1994. (Bottom) Reported rabies cases in domestic animals in the United States, 1994.

mortality, especially in domestic animals or humans that lack an obvious exposure history. The antigenic patterns or nucleotide sequences obtained can be compared with rabies variants from known animal reservoirs. For example, analysis of recent human rabies cases from New York[26] and Texas[33] implicate a strain of virus associated with insectivorous bats, indirectly or directly, in the etiology of infection. Unfortunately, bat rabies cases are not routinely identified to species; thus, case surveillance data alone do not provide adequate clues about unique variants affecting different insectivorous bat species or species groups. More rigorous reporting standards coupled with modern virological analysis will be necessary to provide more than a casual glimpse into epidemiologic cycles of chiropteran lyssaviruses.

In contrast to the United States, Canada, and western Europe where wildlife rabies predominates, in southeast Asia, Latin America, and many developing countries, canids continue to be the principal causative vectors of rabies to humans.[153,168] For example, during 1990 to 1992, of the 531 human cases in Latin America for which the animal was identified, at least 446 (84%) were due to dogs.[168] Additionally, rabies transmission by hematophagous bats, unique to the Americas, is an emerging public health problem, in addition to being a historically important disease of livestock with widespread economic implications.[114,139]

5.4. Temporal Distribution

Due to the exceedingly low occurrence of human rabies mortality in the United States, there is no meaningful trend in the temporal distribution of cases. Moreover, the limitations of a passive surveillance system for animal rabies may obscure temporal trends in the vector species. Given the above limitations, it would be reasonable to investigate a potential association between increased animal movement, human outdoor activities, and a seasonal increase in exposures resulting in PET.

5.5. Age, Sex, Race, Occupation: Socioeconomic, Nutritional, and Genetic Factors

There do not appear to be reliable trends associated with age, sex, race, occupational, or activity risk patterns leading to rabies exposures. Racial, genetic, nutritional, or socioeconomic differences in risk of exposure have not been fully investigated, but there does not appear to be a trend in occurrence or susceptibility with regard to these factors, as well. If reporting of PET were practiced, these epidemiologic characteristics would be analyzed to further delineate sources of exposure and target preventive measures.

6. Mechanisms and Route of Transmission

Rabies is transmitted by virus-laden saliva from the bite of a rabid animal, which may be infectious for days before obvious illness, but not prior to brain involvement.[3,6,34,50,119] If the converse existed, thousands of human mortalities would be obvious following exposure to animals proven nonrabid by routine postmortem diagnostic examination of brain material. Moreover, if a "carrier" state readily existed, in which most infected animals remained healthy while excreting large quantities of infectious material,[54] human rabies would not be rare; the current practice of confinement and observation of the rabies-suspect biting dog or cat would fail as an effective public health measure. Beyond the domestic dog and cat,[32] reliable observation periods for other carnivores have not been routinely established because of the possibility of prolonged shedding of virus and the subjective nature of defining obvious illness in nondomestic animals.[3]

Virus does not penetrate intact skin; hence, touching a rabid animal does not constitute exposure.[25] Other nonbite routes of natural transmission have rarely been documented,[163] and largely remain unimportant epidemiologically. For example, contamination of mucous membranes with infectious material is considered the major nonbite route of exposure. At least eight human corneal transplant recipients have succumbed, in which the donors had died of undiagnosed rabies.[165] Four cases of apparent aerosol transmission have been implicated,[23,163] two individuals working with extremely dense populations of bats in the confines of caves (vs. millions of spelunkers who may do the same), and in two laboratory workers who were exposed to high concentrations of aerosolized rabies virus. These considerations aside, there is a theoretical risk of human-to-human transmission resulting from contact with rabid patients, although none have been documented in recent memory.[74] In some animals, oral transmission has been observed under experimental conditions (as has immunization, more importantly) and has been implicated a few times in the field in Europe and Canada following distribution of modified live rabies virus vaccines intended for oral wildlife rabies vaccination.[152] Although suspected on occasion in both wild and domestic animals, neither the oral route nor other unusual means, such as transplacental transmission, have been definitively implicated as important in human rabies cases to date, partially because of the difficulty in ruling out the traditional bite route as the primary means of exposure, especially in regions with enzootic dog rabies. In the United States, where canine rabies has been controlled, only one of the ten last human rabies cases from 1993 to date contained a history of animal bite exposure. This single individual had been bitten by a dog while living in Mexico; except for one additional human fatality, identified at postmortem, as related to a canine rabies variant from the Caribbean (which concurred with the patient's history of travel to the area), the other eight cases were attributed as unrecognized exposures to viruses associated with bats.[26,27,29,30,31,33]

7. Pathogenesis and Immunity

Following local introduction at the site of a bite, rabies virus may remain localized and undergo limited replication at extraneuronal sites, such as in muscle tissue.[34] Yet, it remains unclear if this "eclipse" period, during which the virus is virtually undetectable, represents a phase of extraneuronal infection[10,34] or coincides with direct infection of the nervous system.[131] During these early phases of rabies virus infection, the neuronal synaptic transfer may occur in the form of ribonucleo-protein–transcriptase complexes,[65] rather than the assembly and passage of complete virions per se, as regularly occurs in later phases of replication. After a relatively quiescent early phase, viral antigen can be found distributed in most parts of the CNS.[35] Can rabies virus modify its microenvironment to create a niche more conducive to viral replication dynamics? A temporal relationship between viral RNA synthesis and immediate-early gene mRNA expression suggests an activation and upregulation mechanism may be responsible for the burst of viral activity after a variable incubation period.[62]

Often considered as the prototype neurotropic virus, multiple nonneural sites have also been associated with viral antigen following centrifugal transport from the CNS.[34] Recent studies suggest that the relative proportions of infectious virions will vary depending on the end organ examined.[10] Typical exit portal tissues important in transmission, such as salivary glands, frequently support high viral titers and contain only a minimal accumulation of matrix and deviant viral production products, while multiplication in other nonneural sites, such as adrenal glands and mucous glands of the nasal mucosa, produce only moderate viral concentrations but large amounts of viral nucleocapsid and anomalous structures.[10]

A relatively unique characteristic of rabies virus infection is consequent behavioral changes that may favor transmission, most notably irritability and the predilection to bite. Other more subtle behavioral changes may occur as well; for example, in the case of raccoon rabies, induction of juvenile vocalizations in an infected adult host has been observed, which may enhance investigation by conspecifics, bringing into close proximity another susceptible host.

Mechanistically, behavioral manifestations of "dumb" or "furious" rabies in a given host may be dependent on the specific accumulation and differential effect of virus in selective brain areas but not necessarily related to the limbic system; a specific immunopathologic component should not be discounted, while recognizing that even severe-combined-immunodeficient mice (devoid of functional B- and T-cell components) will eventually succumb to rabies, albeit after a prolonged incubation period.[125] Host incapacitation during clinical disease may be obvious and dramatic; a recent study demonstrated the role of T lymphocytes in paralysis through experimental infection and selective reconstitution of immunodeficient athymic nude mice.[140] As such, rabies virus has often been characterized as a poorly adapted or imperfect parasite because it typically results in host death. But in an ecological sense, the type of behavioral changes associated with classic rabies may ensure viral perpetuation regardless of ultimate host outcome, provided transmission of progeny virus has occurred before host demise.

While the phenomenon of differential susceptibility of mammalian hosts to various rabies virus variants is clear, the mechanisms underlying such observations are not. For example, approximately 5% of free-ranging raccoons had serum VNA in two Iowa counties, without raccoon rabies.[76] Moreover, while all skunks inoculated with a midwestern skunk isolate succumbed within 4 or 5 weeks, raccoons infected with several greater concentrations of virus did not succumb or shed detectable amounts of rabies virus; 2 of 12 raccoons developed VNA and were the only two that survived subsequent rabies virus challenge with a NYC/GA strain.[77,78] Such observations may partially explain the occurrence of rabies in midwestern skunks, without apparent perpetuation in raccoons. In the eastern United States, raccoons are competent hosts for a different variant, which does not appear to be independently supported by skunks. Additionally, the relative refractoriness of the only North American marsupial, the opossum (*Didelphis virginianus*), to rabies and its paucity of acetylcholine receptors provide experimental data suggesting that host susceptibility may be influenced by receptor occurrence and abundance, to which the rabies virus may bind and be internalized.[8]

Rabies VNA, induced solely by the viral glycoprotein, are believed to play a major role in the pre- and postexposure protection against rabies.[22] However, comparison of absolute VNA titers and mortality in laboratory rodents immunized with whole virus vaccine or viral glycoprotein did not delineate a clear role for VNA in vaccine-induced resistance to disease.[119] Specific discrepancies between VNA titer and rabies protection was also shown in wild carnivores where no firm correlation existed between any known titer level of VNA resulting from oral vaccination and protection per se.[120] Immunization experiments suggest that protective activity may ultimately correlate with a vaccine's ability to induce immunologic memory. Lack of firm correlation between absolute VNA titers and survivorship in immunized animals suggests that, in addition to VNA, other immune effector mechanisms may be involved in the protection against lethal rabies virus infection in either pre- or postexposure situations.[22,134]

Recent studies have demonstrated that in addition to glycoprotein the internal viral ribonucleoprotein (RNP)[169] may also be important to the induction of protective immunity.[41,42] In related studies following two RNP immunizations, primates developed a strong anti-RNP antibody response, were protected against lethal rabies, and were primed for the induction of a VNA response following exposure to antigen.[146] These experiments and others suggest that the RNP plays an important role in inducing both primary protective immunity and cross-protective immunity against infection to heterologous strains. Since RNP is a strong inducer of nonneutralizing antibody, one of the functions of RNP-specific antibodies may be the promotion of viral RNP attachment via Fc receptors to phagocytotic cells, stimulated by virus to produce cytokines such as interferon, which inhibits viral replication. Moreover, it recently has been suggested that the internal nucleoprotein of rabies virus has a superantigenlike effect on the human immune response.[93]

8. Patterns of Host Response

8.1. Clinical Features

The bite of a rabid animal may be too superficial to produce infection. But even if it is effectively transdermal, the bite must be concomitant with the presence of virus, which may be excreted only intermittently. A host is only considered infected once virus has breached the skin or mucous membrane barriers. This is not synonymous with disease but is partially dependent on viral load and variant characteristics within a host environment ranging from fully susceptible to immune. Host immune response following viral exposure may or may not result in the production of detectable amounts of VNA. While in theory only a single virion is necessary, it usually requires numbers of fully infectious particles several orders of magnitude greater to produce disease.

Undisputedly, inapparent infection of long duration may occur (rarely), but has not been demonstrated to be of epidemiologic importance, regardless of widespread speculation to the contrary. Distinct from inapparent or latent infection, the incubation period associated with overt disease can be highly variable and prolonged, during which the host is not infectious. Productive infection, during which input virus replicates, may result in a viral encephalitis that manifests with neurological signs and symptoms. Perhaps the only consistent feature of the resulting clinical presentation is that it is rarely typical from one host to another. The prodromal stage is uniformly acute, lasting some 2 to 4 days, including moderate fever, malaise, anorexia, headache, and nausea. Presenting symptoms in humans are often nonspecific; however, paresthesia at the site of the bite and aerophagia or hydrophobia (spasms of pharyngeal or inspirational musculature induced by air currents or anticipation of swallowing) are highly suggestive of rabies. Physical and mental status deteriorates rapidly (see Section 1). Not discounting classical Pasteurian observations of spontaneous recovery in animals or the remarkable contemporary accounts of the few human survivors,[1,54,73,79,109] the case-fatality ratio in practice still approaches unity, once clinical signs are present. Death is usually attributable to respiratory or cardiac failure.

8.2. Diagnosis

The incidence of more common infectious agents should be compared to the relative rarity of rabies in the establishment of a diagnosis. Pairing of a known animal exposure followed in several weeks time with the appearance of some of the classic signs and symptoms should alert the astute infectious disease professional. However, rabies should also be considered in any suspected viral encephalitis of unknown origin, regardless of a definitive history of animal bite, especially once a compatible clinical syndrome is manifest; an exposure may have been unrecognized, forgotten, or discounted. In humans, the differential diagnosis includes herpesvirus and arboviral encephalitis and poliomyelitis, among others. Although rabies virus infection and pathogenesis can be described experimentally, there remains no practical or reliable method for detecting rabies exposure or infection in humans until the CNS has been extensively infected and there has been centrifugal distribution of virus via peripheral nerves. At this point, rabies may be diagnosed in humans through positive FA results on brain biopsy, full-thickness skin biopsy from the nape of the neck or on a corneal impression, virus isolation from saliva, or detectable antibody in the CSF or serum (the latter from an unvaccinated individual).[135] Negative results do not rule out rabies. Optimal diagnostic material is fresh brain tissue.

9. Control and Prevention

9.1. Epidemiologic Methods

A modern preemptive public health strategy against rabies encompasses the prompt and proper postexposure treatment of humans bitten by infected animals, prevention of disease among companion animal species by preexposure vaccination, and the ecologically sound management and control of rabies among free-ranging reservoirs. Since the conception of oral rabies vaccination (ORV) of wildlife during the 1960s at the Centers for Disease Control and the initial field experiment during 1978 in Switzerland with modified live rabies virus, the distribution of more than 70 million vaccine doses in Europe and North America, primarily targeting the red fox, attests to the utility of this technique, with local disease elimination as one intended result.[5,20,152,167]

Although modified live rabies viruses result in successful oral immunization of foxes, they are comparatively less effective by this route in other carnivores, such as raccoons or skunks, which are the important reservoirs in North America. Moreover, safety questions over the possibility of vaccine-induced disease in nontarget species were an added concern in the evolution of candidate vaccines.[11] In one attempt to solve these dilemmas, attenuated rabies viruses were selected under MAb pressure, which were greatly reduced in virulence but retained the capacity for oral efficacy.[90] In an alternative strategy, a vaccinia-rabies glycoprotein (V-RG) recombinant virus vaccine[84,155,156] was developed that has proved to be an effective oral immunogen in raccoons[118,120] and a variety of other important species[13,107,123,147,148] and provides long-term protection against rabies.[123] Advantages of this recombinant orthopoxvirus vaccine included greater thermostability than attenuated rabies viruses and the inability to cause rabies, because only the cDNA of the surface glycoprotein of the rabies virus was inserted within the recombinant virus.[84] The period of the 1980s was marked by intensive, collaborative, safety evaluations of the V-RG virus under laboratory conditions in North America and Europe[123,144] that culminated in the initial limited field experiments by the end of that decade.[69,70]

The first North American field trials of the V-RG recombinant virus vaccine began during August 1990, on Parramore Island, Virginia, and June 1991, in a 10-km^2 mountainous watershed of Pennsylvania.[123] The major objective of these two consecutive 12-month studies was an environmental assessment of the V-RG virus within rather limited ecosystems. Vaccine was contained in fish-meal polymer baits (containing serological and calciphilic biomarkers),[69,70,81,95] which were distributed by hand in a variety of habitats at densities of 5–10 baits/hectare, in a manner conducive to consumption by a variety of vertebrates. Depending on location and local animal abundance, >50% of baits were disturbed within relatively short intervals (2 days to 2 weeks). Based on tracking pit data, mammalian carnivores (primarily raccoons) were the group largely implicated in bait contact; no avian disturbance of baits was observed. Corroborative evidence of bait consumption by raccoons was provided by detection of biomarkers in >70% of animals sampled[69]; other biomarker-positive mammals included opossums, foxes, black bear and, only rarely, small mammals.[123] No pox-related mortality, morbidity, or other adverse effects were detected in any species, despite intensive field surveillance. Rather, laboratory rabies challenge of recaptured raccoons from field sites demonstrated a duration of immunity between a minimum of 7 months to in excess of 1 year.[123] Additionally, despite a high bait density enhancing availability, there was no history of human or domestic animal interference of vaccine-laden baits in these rather remote sites. Field researchers at greatest potential risk of vaccine contamination with this novel genetically engineered biological demonstrated neither illness, lesions, nor a specific anamnestic response related to presumed viral exposure. Regarding viral stability (ambient temperatures ranged from 11 to >37°C) at sites receiving full direct sunlight, viral inactivation within baits was minimal, decreasing from approximately $10^{8.2}$ to $10^{7.5}$ TCID$_{50}$/ml. These studies were the first attempts at understanding the dynamics of field-released recombinant virus prior to larger-scale efficacy testing in less remote areas. Later field trials in the United States concentrated on actual rabies control among free-ranging raccoons in New Jersey and Massachusetts during 1992 and 1993, respectively, with attempts at the creation of an immune barrier against the introduction of raccoon rabies, and most recently initiated in New York during 1994, evaluating potential control of enzootic raccoon rabies.

In spite of the successes observed over the last decade with vulpine rabies control in Europe[20] and Canada, it is as yet premature to tacitly extend the concept of oral vaccine efficacy to other important rabies reservoirs (e.g., skunks, coyotes, etc.) from these initial efforts in the United States. While preliminary field data have corroborated laboratory observations of vaccine purity, potency, and stability and importantly confirmed its safety (i.e., to date no vaccine-related mortality, morbidity, or other adverse effects have been detected), several critical aspects

of furbearer behavioral ecology require investigation to further address both basic and applied information gaps that prevent extension to actual disease control. Reliable census estimates of population density in a variety of habitats and the validation of these values over large areas should be applied before, during, and after an outbreak to assess overall rabies impact; additionally, host density estimates would allow evaluation of critical bait densities required per unit area. Quantitative knowledge of factors affecting carnivore dispersal, including relative and absolute barriers to movement, will complement disease modeling and potential vaccination strategies. Since viral vaccines capable of replication in the animal will not offer 100% safety (especially in immunosuppressed hosts), disease surveys in the target species should extend beyond the routine etiologic agents (e.g., canine distemper, heavy metal toxicities, etc.), especially those of a vesicular or neurological nature to include the possibility of vaccine-related, orthopoxvirus lesions. Related data concerning population turnover are implicit to relevant concepts of herd immunity. Surveillance methods for detecting and responding to a disease outbreak (e.g., public health submissions, road-kill sampling, etc.) should be compared for their sensitivity and cost-effectiveness. To test the merits of a central point versus selective distribution of biologicals, combinations of a greater array of olfactory, visual, auditory, and tactile cues could be tested as specific lures for mammalian carnivores.

The advantages of greater safety, potency, efficacy, and versatility of newer oral biologicals intended for free-ranging animals (foxes, raccoons, coyotes, domestic dogs in developing countries, African wild dogs,[63] etc.) may gradually supplant less ideal first-generation modified live rabies viruses of three decades ago.[66] While orthopox-based viruses were the first recombinant rabies vaccines exploited in the laboratory and field for wildlife,[47,52,118] other viral and prokaryotic vectors (adenovirus,[36,111] baculovirus,[61,72,110,150] yeast,[87] etc.) or, as production and expression techniques progress, inactivated or synthetic products[53,94,122,128] may also ultimately contribute to disease control as epidemiologic conditions necessitate. Risk analyses and related health economics should demonstrate that the benefits of future genetically engineered vaccines outweigh perceived environmental or public health hazards (genetic transfer between wild-type and recombinant viruses, establishment of new animal reservoirs, etc.) that are often misconstrued or disproportionately weighted when associated with biotechnology. Attention to parental vector characteristics, such as origin, passage history, pathogenesis, and putative unique qualities of the resulting recombinant species, are more objective criteria than concentration on the specific mechanism of vaccine creation per se. Additionally, bait matrices and distribution systems may be tailored to maximize specific attractiveness to the target species, as well as efficacious delivery of the vaccine.[60,83] Regardless of the type of vaccine virus, conventional or recombinant, continued development of relevant experimental animal models is critically needed to properly resolve overt issues related to safety in the immunosuppressed host,[125] prior to field release, especially as the emphasis turns from wildlife in developed countries to free-ranging dogs in the developing world.[166]

Some opponents to wildlife ORV espouse depopulation, rather than vaccination, as a control method, citing the effectiveness of dog rabies control via compulsory muzzling, movement restrictions, and stray animal removal in Europe and North America during the early 1900s, before the utilization of efficacious vaccines en masse. Two components of this technique—animal movement restrictions and muzzling—are inapplicable to wildlife, leaving only depopulation. As a sole mechanism of control, depopulation may decrease the local intensity of rabies or postpone the invasion of a new geographic area, but is unlikely to completely eliminate rabies or permanently protect a region from an impending epizootic.

Some professional groups view a number of wildlife species, including some rabies vector species, as nuisance or pest species, and hence oppose ORV, citing the undesirable potential for increased numbers following elimination of rabies. However, proportionate mortality due to rabies among the most prominent wildlife vectors in the United States, such as raccoons, skunks, and so on, has not been investigated. In Europe, collective data suggest that rabies may not be a significant regulator of population size; there has been no substantial increase in red fox numbers following the elimination of rabies over large geographic areas.[4]

Economic analyses need to be developed to determine the overall economic utility of ORV. In one such study, benefit–cost ratios varied ranging from 2.2–6.8, payback periods varied from 0.7 to 2.1 years, and net profit values ranged between 2.1 and 4.1 million dollars, depending on the projected cost of a suggested intervention in a raccoon rabies epizootic.[151] Despite the relatively few human fatalities associated with rabies in developed countries, huge expenditures directed at prevention activities require extensive public health monies, often competing with other basic programs (e.g., childhood vaccination, etc.) for scarce resources.

Addressing bat rabies control, the importance of bat exclusion from human habitations and better public edu-

cation to facilitate recognition of potential exposures cannot be over emphasized. Overall, bat conservation is not incompatible with basic public health tenets.

9.2. Immunization Concepts and Practice

For practical purposes, rabies is universally fatal once clinical symptoms are present. In the same context, extremely effective human PET consists of vaccine and immune globulin if administered promptly and properly. The extension of PET to companion animals and livestock has only recently been investigated in dogs[39] an sheep.[14]

While current rabies biologicals for humans are much improved and associated with fewer side effects, there has been at least one unexpected report of a cross-reactive human immunodeficiency virus (HIV) serological response in a female, with no HIV transmission risk factors, following preexposure rabies immunization.[108] The HIV ELISA tests were repeatedly, but transiently, positive. A Western blot was indeterminate; the patient was negative for proviral HIV RNA by PCR. Other observations of such false-positives have not been reported to date.

There is increasing documentation of higher than previously reported allergic reactions among persons receiving preimmunization and PET in the United States.[56] Until these albeit predominantly mild, adverse reactions may be controlled through further vaccine refinement, it would be prudent to restrict vaccine administration to circumstances where it is clearly indicated. Further investigations may clarify an observed trend toward fewer complications when the routes, either intradermal or intramuscular for preexposure primary and booster immunizations, are concordant.[56] Additionally, prospective investigation of risk factors for systemic hypersensitivity may lead to recommendations that would minimize the occurrence of adverse reactions.

Continuing safety concerns over the potential utilization of recombinant vaccinia viruses in humans may be partially alleviated by concentration on the deletion of virulence genes.[141] Alternatively, avian pox viruses result in an abortive infection in mammalian cells, but enable the presentation of early gene products to the immune system, resulting in the induction of VNA and an appropriate anamnestic response.[21,142,143]

Newer approaches to immunization may also include genetic therapy. Recently, mice immunized intramuscularly with a plasmid vector expressing the rabies glycoprotein under a SV40 early promoter developed specific cytotoxic lymphocytes, thymic lymphocytes (subset TH1), and rabies VNA and were protected against rabies virus challenge.[170] Although many questions remain regarding the consequences of persistent ''infection'' with a plasmid, the effects of long-term stimulation of the immune system, the genetic stability and safety of foreign promoters and nucleic acids, and the potential benefits are clear, which include simplicity and economy. Obviating the need for booster immunizations would be particularly useful for livestock in extensive range situations, for example, where protection against bovine paralytic rabies via vampire bats in Latin America with current vaccines is prohibitive economically and logistically.

What part may viral heterogeneity play in rabies vaccination failure? Compared to the conserved fixed virus strains that form the basis for rabies vaccine production, considerable diversity is obvious among rabies antigens by MAb analysis. Nevertheless, even recombinant rabies vaccines that only express the viral glycoprotein are able to fully protect animals from severe experimental and field exposure against all of the major rabies variants. After considerable investigation, there is no firm evidence to suggest that recent human rabies deaths are attributable to antigenic variation of street rabies viruses, but rather they appear due to serious omissions in PET protocols.[145]

Besides epidemiologic utility (for which they were originally generated), antirabies MAbs are also useful in the refinement of host immunization and PET concepts. Rabies VNA are induced solely by the glycoprotein and play a major role in the immunoprotection. Hybridomas that secrete rabies virus antigen-specific MAb have been generated and can effectively neutralize fixed and street lyssavirus isolates. These murine MAbs were selected on the basis of isotype, antigen and epitope specificity, virus strain specificity, affinity, and neutralizing activity. Administration of such MAbs protected animals when challenged with a lethal dose of rabies virus in experimental PET models.[129] These data clearly suggest that MAbs, alone or in combination with vaccine, are an effective method of protection against clinically relevant lyssaviruses.[125] Such an approach has several theoretical advantages over presently used hyperimmune sera: (1) in contrast to hyperimmune serum, comparatively small volumes of MAbs would have to be inoculated for equivalent active protein content because specific neutralizing activity per mass of protein is higher, so MAbs may be optimal for lessening the trauma and pain of local wound infiltration with a source of passive antibodies; and (2) safety issues arising from the possibility of adventitious agents associated with human or animal blood products would be alleviated by bulk production in cell culture. Despite the experimental progress shown, the rationale and acceptability of murine MAbs for future human rabies PET

have been questioned on the grounds that a human anti-mouse response to the MAbs would present a significant drawback by the consequent effects on kinetics and antigenic targeting. If murine or other heterologous MAbs are used, human anti-species responses may be expected, but these are not necessarily deleterious because MAbs would only be used once in rabies PET. As is the case for HRIG, MAbs would not be readministred should the person be reexposed in the future; rabies PET in these situations consists of vaccine only on days 0 and 3. Moreover, if such MAbs are recognized as foreign antigens, with an expected shortened serum half-life, this potentiated clearance may be advantageous by minimizing the opportunity for interference with an active immune response on behalf of the vaccinated host, while effectively neutralizing virus, prior to induction of host VNA.

Mere *in vitro* neutralizing capacity alone may not be fully indicative of protective function *in vivo*. Some MAbs that do not neutralize viruses in cell culture are effective when administered postexposure to animals.[91] Besides extracellular neutralization, other useful capabilities of MAbs may also be operative, such as the inhibition of intracellular viral spread and interference in transcription of rabies viral RNA.

10. Unresolved Problems

10.1. Rabies Control in the Stray Dog: Oral Vaccination as a Solution?

During the past 25 years, considerable progress has been made in the laboratory development and field testing of ORV for free-ranging terrestrial wildlife control in Europe and North America. Major species have included the red fox, raccoon, and most recently the domestic dog. Oral vaccines tested in the field to date have included attenuated rabies (e.g., SAD/ERA/SAG) and recombinant glycoprotein (e.g., pox) viruses. Especially in regard to domestic canine vaccination, intended biologicals must at minimum be pure, potent, efficacious (against a virulent, epidemiologically relevant rabies virus challenge), and safe to target and nontarget species, especially humans, at risk for exposure to distributed, edible baits, or through direct, intimate contact with a vaccinated host.[101]

Self-replicating attenuated or recombinant vaccines offer the greatest versatility for oral rabies vaccination of dogs. While a few candidate oral rabies vaccine preparations have demonstrated sound protective immunity in captive dogs, no self-replicating viral system, conventional or otherwise, can be considered completely apathogenic, especially in the context of the immunocompromised host.[101] As an alternative, inactivated oral preparations will maximize safety considerations, yet the significant quantities required for minimal efficacy may be cost-prohibitive unless novel adjuvants, delivery techniques, or production methods are found.[122] Another major drawback of oral vaccination with a nonreplicating agent is the relatively short retention time in the oropharyngeal and gastrointestinal tracts. To retain relative antigen concentrations high enough to induce a protective (non-tolerant) immune response, it may be necessary to arrest the antigen at the cell surface. The tonsillar area, consisting of a single epithelial barrier from the lymphocyte-rich environment of the crypts, would be one attractive area to target for binding, as would the much greater area of the gut-associated lymphoid tissue. To facilitate attachment of rabies immunogens at cellular surfaces, it may be necessary to target specific molecules, via covalent linkage of viral antigens to ligands, that specifically bind particular receptors. For quantity, chimeric genes could be constructed, consisting of a rabies viral gene and a gene coding for the binding domains of a particular ligand, produced in suitable expression systems. Beyond these conceptual points, the lack of background data concerning the fate of subunit vaccines administered orally prevents any likely prediction as to whether lyssavirus antigens should be targeted to antigen-presenting cells or to T or B lymphocytes.

Applied research on the oral or enteric administration of affordable inactivated rabies antigens may diminish the public health concerns related to the current biologicals under consideration for application to free-ranging dogs. However, regardless of the approach, limited laboratory trials of oral canine vaccination may not directly equate to rabies control under field conditions, where a combination of ORV and traditional parenteral vaccination may be necessary. Critical to the success of any proposed campaign is baseline data on the local behavioral ecology of the poorly supervised dog, historical spatiotemporal occurrence of canine rabies cases, and the availability of ample political infrastructural support within the socioeconomic and cultural confines of the human community[101,153] These attributes are often insufficient in many developing countries where ORV may be the most promising new adjunct to successful dog rabies control.

10.2. Alternatives to Current Biologicals for Disease Prevention

The vigilance to minimize human rabies mortality necessitates an extensive public health infrastructure and requires the annual treatment of millions in the developing world and tens of thousands of potential exposure cases throughout the United States; in New York state alone (currently experiencing a raccoon rabies epizootic), more than 3000 people were treated during 1993, at a cost in excess of one million dollars (just for the purchase of anti-rabies biologicals) to prevent the disease. Following a bite from an infected mammal, PET of rabies in humans includes proper wound care and the simultaneous administration of multiple doses of an efficacious rabies vaccine, together with a preformed antibody source, typically HRIG. However, this treatment regimen presents certain obstacles. Modern inactivated cell culture vaccines and HRIG are vastly improved over historical biologicals, but such products are relatively expensive (especially in the developing world where they are most needed), are often in scarce supply, and carry a theoretical risk of adventitious agents. Major concerns in rabies control have concentrated on the need for potent, inexpensive postexposure treatments for humans, especially replacement for costly HRIG, while retaining activity against a variety of lyssaviruses.

In contrast to the relatively impotent equine antirabies serum used in the past that resulted in high adverse reactions, such as serum sickness in up to 40% of human recipients, modern purified equine rabies immunoglobulin (ERIG) products are safe, potent, and affordable, at a fraction of the cost of HRIG.[159,161] They have been used effectively in conjunction with vaccine in human rabies PETs, particularly in developing countries under the continued threat of enzootic dog rabies. Less than 1% of humans have reported adverse events, coupled with complete efficacy.[160] No ERIG has been available in the United States market for several years, although it was licensed into the late 1980s. Due to its potency and lack of apparent significant local or systemic effects, purified ERIG is the only immediate alternative, should the supply of available HRIG be threatened by shortage or contamination. Its use would be considered a temporary antecedent to more novel replacements, such as MAbs.

To be most effective, conventional PET should be applied prior to viral invasion of the nervous system. In experimental infection, analysis by RT-PCR has revealed the presence of rabies virus-specific RNA in the olfactory bulb and cerebral cortex within 6 hr of direct intranasal inoculation. Yet, when animals were treated with a neutralizing MAb up to 24 hr after inoculation, 80% survived a challenge in which all controls died.[125,129] Neither virus nor virus-specific RNA was detected when these survivors were euthanized a month later. These data clearly demonstrate that virus "neutralization" and clearance from the CNS are complex processes.[45] Thus, if produced in a cost-effective manner, antirabies MAbs may be useful in the future PET of humans, as well as economically important, unimmunized domestic animals. Such biologicals, whether murine,[91,102,129] human,[44,46,92,112] chimeric, primatized, or recombinant in origin,[38] that abrogate infection even after virus enters the CNS present exciting new possibilities for therapy with significant advantages over polyclonal rabies immune globulin.

10.3. Treatment of Clinical Human Rabies

While prevention of disease among reservoir hosts and victim alike is still the optimal approach, when this fails, further alternatives regarding treatment of clinical rabies are warranted. The experimental application of rabies immunoglobulin and rabies immunoglobulin fragments, cytosine or adenine arabinoside, interferon, acyclovir, antithymocyte globulin, steroids, vidarabine, tribavirin, inosine pranobex, or ribavirin has not demonstrated any utility to date.[75] However, case histories of recovery after clinical signs suggest that host defenses may be exploited to alter the end stage of this otherwise fatal malady if the mechanism of organic disease can be understood. For example, recent studies indicate that a nitric oxide produced in the CNS via inducible nitric oxide synthase may be one of the critical toxic factors involved in neuronal cell damage during rabies infection.[88] If so, such experimental pharmacological approaches to alter the underlying process may be useful in the design of future treatments of this viral encephalomyelitis, but only if properly coupled with prompt antemortem diagnosis.

10.4. Animal Translocation and Disease Introduction

Given the availability and rapidity of modern transportation, alarming trends are evident in animal translocation on local, regional, and intercontinental levels. Such deliberate translocation, but often with unintended results, carries the significant risk of the emergence of infectious agents into new niches previously unavailable because of zoogeographic barriers. There are limited legitimate reasons for purposeful translocation, such as for species reintroduction, research, education, or zoological captive breeding. Unrestricted importation of nonnative species or translocation of native mammalian species during the incubation stage of the disease poses the threat of rabies virus introduction into rabies-free regions. Within rabies enzootic countries, there is a danger that other bat or terrestrial rabies variants will be introduced into new

areas, such as occurred in the late 1970s with translocation of infected raccoons from the southeastern United States, resulting in the initiation of the mid-Atlantic raccoon rabies epizootic. The same potential looms over the transportation of infected coyotes from Texas to unaffected states for hunting purposes, with the added threat of infection of other canids such as foxes or, more importantly, domestic dogs, as occurred in Florida during 1994. Also confirmed in 1994, there were at least two incidents of international translocation of infected wildlife (i.e., bats and canids), from enzootic rabies areas in the United States to western Europe, which could hamper public health efforts at wildlife rabies control had they escaped. More recently, as part of a novel, exotic animal industry, thousands of Old World bats were brought into the United States for the pet trade. Little is known about the clinical signs, incubation periods, or potentials for transmission of related agents, such as Lagos Bat or Duvenhage viruses to New World hosts via importation of these potential reservoir species; such establishment would be problematic, owing to the questionable vaccine efficacy against nonrabies lyssaviruses. Enlightened and targeted education, enforced legislation, improved regional surveillance, and a rapid, appropriate public health response will minimize the overall dangers associated with animal translocation and will hopefully prevent an infectious nidus from becoming an epizootic.

10.5. Reservoir Population Management: Immunocontraception of Free-Ranging Carnivores?

Rabies is but one of a multitude of factors that arguably may influence the density of free-ranging carnivores over time.[4] One criticism of wildlife rabies vaccination is a concern that by decreasing or eliminating rabies mortality, a program may increase local carnivore populations over time, which in many instances may be undesirable. Historically, lethal means to control carnivore populations have been extensively employed, primarily to decrease economic loss from predation on livestock. Additionally, limiting or reducing carnivore populations may be beneficial for the protection of other, more "valued" animals, including important game, endangered, and threatened species, and to decrease human-nuisance carnivore interactions. Regardless of the potential benefits, sustained reduction of carnivore populations exclusively by lethal means over large regions may be unacceptable due to economic factors, questionable efficaciousness, and ethical considerations. Besides the carnivore issue, there has been a significant renewed impetus to enact population management for other species as well, most notably humans. One of the major strategies has been fertility control by immunocontraception.[18,71] Contraceptive investigations have pursued several avenues: (1) eliciting an immune response to ova, sperm, or zona pellucida antigens; (2) disrupting spermatogenesis; and (3) interfering with reproductive hormones. The ideal would entail a targeted approach, involving either a species-specific antigen or vector, although none has yet been identified. Although the need appears obvious and preliminary efforts are promising, there are serious considerations that must be addressed, including: (1) the levels of herd immunity needed to preclude the persistence of either a subpopulation that evades exposure to these agents (e.g., bait avoidance) and prevent "nonresponders" from gaining a significant reproductive advantage; (2) the nontarget species issue, because of bait competition; and (3) the escape of a transmissible agent and propagated spread beyond an intended "pest" target population (i.e., introduced red foxes in Australia) to other distant geographic regions (i.e., red foxes in North America) by deliberate or unintentional animal or agent translocation. If such concerns could be minimized, a recombinant vaccine could in theory not only immunize against an infectious agent of choice but also simultaneously decrease host abundance as well.

11. References

1. ALVAREZ, L., FAJARDO, R., LOPEZ, E., PEDROZA, R., HEMACHUDA, T., KAMOLVARIN, N., CORTES, G., AND BAER, G., Partial recovery from rabies in a nine-year-old boy, *Pediatr. Infect. Dis. J.* **13:**1154–1155 (1994).
2. ARTOIS, M., AUBERT, M., AND BLANCOU, J., Behavioral ecology of the transmission of rabies, *Ann. Rech. Vet.* **22:**163–172 (1991).
3. AUBERT, M., BLANCOU, J., BARRAT, J., ARTOIS, M., AND BARRAT, M. J., Transmission and pathogenesis of two rabies isolates from the red fox at a 10 year interval, *Ann. Rech. Vet.* **22:**77–93 (1991).
4. AUBERT, M., Control of rabies in foxes: What are the appropriate measures? *Vet. Rec.* **134:**55–59 (1994).
5. AUBERT, M., MASSON, E., ARTOIS, M., AND BARRAT, J., Oral wildlife rabies vaccination field trials in Europe, with recent emphasis on France, *Curr. Topics Microbiol. Immun.* **187:**219–243 (1994).
6. BAER, G. M., Animal models in the pathogenesis and treatment of rabies, *Rev. Infect. Dis.* **10:**S739–750 (1988).
7. BAER, G. M., Rabies—an historical perspective, *Infect. Agents Dis.* **3:**168–180 (1994).
8. BAER, G. M., SHADDOCK, J. H., QUIRION, R., DAM, T. V., AND LENTZ, T. L., Rabies susceptibility and the acetylcholine receptor, *Lancet* **335:**664–665 (1990).
9. BARNARD, B. J. H., AND VOGES, S. I., A simple technique for the rapid diagnosis of rabies in formalin-fixed brain, *Onderstepoort J. Vet. Res.* **49:**193–194 (1982).
10. BALACHANDRAN, A., AND CHARLTON, K., Experimental rabies infection of non-nervous tissues in skunks (*Mephitis mephitis*) and foxes (*Vulpes vulpes*), *Vet. Pathol.* **31:**93–102 (1994).
11. BINGHAM, J., FOGGIN, C. M., GERBER, H., HILL, F. W., KAPPELER, A., KING, A. A., PERRY, B. D., AND WANDELER, A. I., Pathogenicity of SAD rabies vaccine given orally in chacma baboons (*Papio ursinus*), *Vet. Rec.* **131:**55–56 (1992).
12. BLACK, D., AND WIKTOR, T. J., Survey of raccoon hunters for rabies antibody titers: Pilot study, *J. Fla. Med. Assoc.* **73:**517–520 (1986).
13. BLANCOU, J., KIENY, M. P., LATHE, R., LECOCQ, J. P., PASTORET, P. P., SOULEBOT, J. P., AND DESMETTRE, P., Oral vaccination of the fox against rabies using a live recombinant vaccinia virus, *Nature* **332:**373–375 (1986).
14. BLANCOU, J., BALTAZAR, R. S., MOLLI, I., AND STOLTZ, J. F., Effective postexposure treatment of rabies-infected sheep with rabies immune globulin and vaccine, *Vaccine* **9:**432–437 (1991).
15. BOURGON, A. R., AND CHARLTON, K. M., The demonstration of rabies antigen in paraffin-embedded tissues using the peroxidase-antiperoxidase method: a comparative study, *Can. J. Vet. Res.* **51:**117–120 (1986).
16. BOURHY, H., KISSI, B., LAFON, M., SACRAMENTO, D., AND TORDO, N., Antigenic and molecular characterization of bat rabies virus in Europe, *J. Clin. Microbiol.* **30:**2419–2426 (1992).
17. BOURHY, H., KISSI, B., AND TORDO, N., Molecular diversity of the lyssavirus genus, *Virology* **194:**70–78 (1993).
18. BOYLE, D. B., Disease and fertility control in wildlife and feral animal populations: Options for vaccine delivery using vectors, *Reprod. Fertil. Dev.* **6:**393–400 (1994).
19. BRIGGS, D. J., AND SCHWENKE, J. R., Longevity of rabies antibody titre in recipients of human diploid cell rabies vaccine, *Vaccine* **10:**125–129 (1992).
20. BROCHIER, B., KIENY, M. P., COSTY, F., COPPENS, P., BAUDUIN, B., LECOCQ, J. P., LANGUET, B., CHAPPUIS, G., DESMETTRE, P., AFIADEMANYO, K., LIBOIS, R., AND PASTORET, D.-P., Large-scale eradication of rabies using recombinant vaccinia-rabies vaccine, *Nature* **354:**520–522 (1991).
21. CADOZ, M., STRADY, A., MEIGNIER, B., TAYLOR, J., TARTAGLIA, J., PAOLETTI, E., AND PLOTKIN, S., Immunisation with canarypox virus expressing rabies glycoprotein, *Lancet* **339:**1429–1432 (1992).
22. CELIS, E., RUPPRECHT, C. E., AND PLOTKIN, S. A., New and improved vaccines against rabies, in: *New Generation Vaccines* (G. C. WOODROW AND M. M. LEVINE, EDS.), pp. 419–438, Marcel Dekker, New York, 1990.
23. CENTERS FOR DISEASE CONTROL, Rabies in a laboratory worker—New York, *Morbid. Mortal. Week. Rep.* **26:**183–184 (1977).
24. CENTERS FOR DISEASE CONTROL, Human rabies—Kenya, *Morbid. Mortal. Week. Rep.* **32:**494–495 (1983).
25. CENTERS FOR DISEASE CONTROL, Rabies prevention—United States, 1991, *Morbid. Mortal. Week. Rep.* **40**(RR-3):1–19 (1991).
26. CENTERS FOR DISEASE CONTROL, Human rabies—New York, 1993, *Morbid. Mortal. Week. Rep.* **42:**799–806 (1993).
27. CENTERS FOR DISEASE CONTROL, Human rabies—Texas and California, 1993, *Morbid. Mortal. Week. Rep.* **43:**93–96, 1994.
28. CENTERS FOR DISEASE CONTROL, Raccoon rabies epizootic—United States, 1993, *Morbid. Mortal. Week. Rep.* **43:**269–273 (1994).
29. CENTERS FOR DISEASE CONTROL, Human rabies—California, 1994, *Morbid. Mortal. Week. Rep.* **43:**455–458 (1994).

30. CENTERS FOR DISEASE CONTROL, Human rabies—Miami, 1994, *Morbid. Mortal. Week. Rep.* **43:**773–775 (1994).

31. CENTERS FOR DISEASE CONTROL, Human rabies—West Virginia, 1994, *Morbid. Mortal. Week. Rep.* **44:**86–87 (1994).

32. CENTERS FOR DISEASE CONTROL, Compendium of animal rabies control, 1995, *Morbid. Mortal. Week. Rep.* **44**(RR-2)**:**1–9 (1995).

33. CENTERS FOR DISEASE CONTROL, Human rabies—Alabama, Tennessee and Texas, 1994, *Morbid. Mortal. Week. Rep.* **44:**269–272 (1995).

34. CHARLTON, K. M., The pathogenesis of rabies and other lyssa-viral infections: Recent studies, *Curr. Topics Microbiol. Immun.* **187:**95–119 (1994).

35. CHARLTON, K. M., WEBSTER, W. A., AND CASEY, G. A., Skunk rabies, in: *Natural History of Rabies*, 2nd ed. (G. M. BAER, ED.), pp. 307–324, CRC Press, Boca Raton, FL, 1991.

36. CHARLTON, K. M., ARTOIS, M., PREVEC, L., CAMPBELL, J. B., CASEY, G. A., WANDELER, A. I., AND ARMSTRONG, J., Oral rabies vaccination of skunks and foxes with a recombinant human adenovirus vaccine, *Arch. Virol.* **123:**169–179 (1992).

37. CHERKASSKIY, B. L., Roles of the wolf and the raccoon dog in the ecology and epidemiology of rabies in the USSR, *Rev. Infect. Dis.* **10**(Suppl. 4)**:**S634–636 (1988).

38. CHEUNG, S. C., DIETZSCHOLD, B., KOPROWSKI, H., NOTKINS, A. L., AND RANDO, R. F., A recombinant human Fab expressed in *Escherichia coli* neutralizes rabies virus, *J. Virol.* **66:**6714–6720 (1992).

39. CHO, H. C., AND LAWSON, K. F., Protection of dogs against death from experimental rabies by postexposure administration of rabies vaccine and hyperimmune globulin (human), *Can. J. Vet. Res.* **53:**434–437 (1989).

40. CLARK, K. A., NEILL, S. U., SMITH, J. S., WILSON, P. J., WHADFORD, V. W., AND MCKIRAHAN, G. W., Epizootic canine rabies transmitted by coyotes in south Texas, *J. Am. Vet. Med. Assoc.* **204:**536–540 (1994).

41. DIETZSCHOLD, B., WANG, H., RUPPRECHT, C. E., CELIS, E., TOLLIS, M., ERTL, H., HEBER-KATZ, E., AND KOPROWSKI, H., Induction of protective immunity against rabies by immunization with rabies virus nucleoprotein, *Proc. Natl. Acad. Sci. USA* **84:**9165–9169 (1987).

42. DIETZSCHOLD, B., TOLLIS, M., RUPPRECHT, C. E., CELIS, E., AND KOPROWSKI, H., Antigenic variation in rabies and rabies-related viruses: Cross-protection independent of glycoprotein-mediated virus-neutralizing antibody, *J. Infect. Dis.* **156:**815–822 (1987).

43. DIETZSCHOLD, B., RUPPRECHT, C. E., TOLLIS, M., LAFON, M., MATTEI, J., WIKTOR, T. J., AND KOPROWSKI, H., Antigenic diversity of the glycoprotein and nucleocapsid proteins of rabies and rabies-related viruses: Implications for epidemiology and control of rabies, *Rev. Infect. Dis.* **10**(Suppl. 4)**:**S785–798 (1988).

44. DIETZSCHOLD, B., GORE, M., CASALI, P., UEKI, Y., RUPPRECHT, C. E., NOTKINS, A. L., AND KOPROWSKI, H., Biological characterization of human monoclonal antibodies to rabies virus, *J. Virol.* **64:**3087–3090 (1990).

45. DIETZSCHOLD, B., KAO, M., ZHENG, Y. M., CHEN, Z. Y., MAUL, G., FU, Z. F., RUPPRECHT, C. E., AND KOPROWSKI, H., Delineation of putative mechanisms involved in antibody-mediated clearance of rabies virus from the central nervous system, *Proc. Natl. Acad. Sci. USA* **89:**7252–7256 (1992).

46. ENSSLE, K., KURRLE, R., KOHLER, R., MULLER, H., KANZY, E. J., HILFENHAUS, J., AND SEILER, F. R., A rabies-specific human monoclonal antibody that protects mice against lethal rabies, *Hybridoma* **10:**547–556 (1991).

47. ESPOSITO, J. J., KNIGHT, J. C., SHADDOCK, J. H., NOVEMBRE, F. J., AND BAER, G. M., Successful oral vaccination of raccoons with raccoon poxvirus recombinants expressing rabies virus glycoprotein, *Virology* **165:**313–316 (1988).

48. EVERARD, C. O., AND EVERARD, J. D., Mongoose rabies in the Caribbean, *Ann. N.Y. Acad. Sci.* **653:**356–366 (1992).

49. FEIDON, W., KALSER, F., GERHARD, L., DAHME, E., GYLSTROFF, B., WANDELER, A., AND EHRENSBERGER, F., Immunohistochemical staining of rabies virus antigen with monoclonal and polyclonal antibodies in paraffin tissue section, *Zentralbl. Vet. Med.* **35:**247–258 (1988).

50. FEKADU, M., Latency and aborted rabies, in: *The Natural History of Rabies*, 2nd ed. (G. M. BAER, ED.), pp. 191–200, CRC Press, Boca Raton, FL, 1991.

51. FEKADU, M., GREER, P. W., CHANDLER, F. W., AND SANDERLIN, D. W., Use of the avidin–biotin peroxidase system to detect rabies antigen in formalin-fixed paraffin-embedded tissues, *J. Virol. Methods* **19:**91–96 (1988).

52. FEKADU, M., SHADDOCK, J. H., SUMNER, J. W., SANDERLIN, D. W., KNIGHT, J. C., ESPOSITO, J. J., AND BAER, G. M., Oral vaccination of skunks with raccoon poxvirus recombinants expressing the rabies glycoprotein or the nucleoprotein, *J. Wildl. Dis.* **27:**681–684 (1991).

53. FEKADU, M., SHADDOCK, J. H., EKSTROM, J., OSTERHAUS, A., SANDERLIN, D. W., SUNDQUIST, B., AND MOREIN, B., An immune stimulating complex (ISCOM) subunit rabies vaccine protects dogs and mice against street rabies challenge, *Vaccine* **10:**192–197 (1992).

54. FEKADU, M., SUMMER, J. W., SHADDOCK, J. H., SANDERLIN, D. W., AND BAER, G. M., Sickness and recovery of dogs challenged with a street rabies virus after vaccination with a vaccinia virus recombinant expressing rabies virus N protein, *J. Virol.* **66:**2601–2604 (1992).

55. FISHBEIN, D., BELOTTO, A. J., PACER, R. E., SMITH, J. W., WINKLER, W. G., JENKINS, S. R., AND PORTER, K. M., Rabies in rodents and lagomorphs in the United States, 1971–1984: Increased cases in the woodchuck (*Marmota monax*) in mid-Atlantic States, *J. Wildl. Dis.* **22:**151–155 (1986).

56. FISHBEIN, D. B., YENNE, K. M., DREESEN, D. W., TEPLIS, C. F., MEHTA, N., AND BRIGGS, D. J., Risk factors for systemic hypersensitivity reactions after booster vaccinations with human diploid cell rabies vaccine: A nationwide prospective study, *Vaccine* **11:**1390–1394 (1993).

57. FLEMING, G., *Rabies and Hydrophobia*, Chapman and Hall, London, 1872.

58. FOGELMAN, V., FISCHMAN, H. R., HORMAN, J. T., AND GRIGOR, J. K., Epidemiologic and clinical characteristics of rabies in cats, *J. Am. Vet. Med. Assoc.* **202:**1829–1833 (1993).

59. FOLLMANN, E. H., RITTER, D. G., AND BELLER, M., Survey of fox trappers in northern Alaska for rabies antibody, *Epidemiol. Infect.* **113:**137–140 (1994).

60. FRONTINI, M. G., FISHBEIN, D. B., GARZA, R. J., FLORES COLLINS, E., BALDARET TORRES, J. M., QUIRET HUERTA, G., GAMEZ RORIGUEZ, J. J., BELOTTO, A. J., DOBBINS, J. G., LINHART, S. B., AND BAER, G. M., A field evaluation in Mexico of four baits for oral rabies vaccination of dogs, *Am. J. Trop. Med. Hyg.* **47:**310–316 (1992).

61. FU, Z. F., RUPPRECHT, C. E., DIETZSCHOLD, B. D., SAIKUMAR, P., NIU, H. S., BABKA, I., WUNNER, W. H., AND KOPROWSKI, H., Oral vaccination of raccoons (*Procyon lotor*) with baculo-virus-expressed rabies virus glycoprotein, *Vaccine* **11:**925–928 (1993).

62. Fu, Z. F., Weihe, E., Zheng, Y. M., Schafer, M. K., Sheng, H., Corisdeo, S., Rauscher, F. J., Koprowski, H. K., and Dietzschold, B., Differential effects of rabies and borna disease viruses on immediate-early- and late-response gene expression in brain tissues, *J. Virol.* **67**:6674–6681 (1993).

63. Gascoyne, S. C., Laurenson, M. K., Lelo, S., and Borner, M., Rabies in African wild dogs (*Lycaon pictus*) in the Serengeti region, Tanzania, *J. Wildl. Dis.* **29**:396–402 (1993).

64. Goldwasser, R. A., and Kissling, R. E., Fluorescent antibody staining of street and fixed rabies virus antigens, *Proc. Soc. Exp. Biol. Med.* **98**:219–223 (1958).

65. Gosztonyi, G., Dietzschold, B., Kao, M., Rupprecht, C. E., and Koprowski, H., Rabies and borna disease: A comparative pathogenetic study of two neurovirulent agents, *Lab. Invest.* **68**:285–295 (1993).

66. Haddad, N., Ben Khelifa, R., Matter, H., Kharmachi, H., Aubert, M. F., Wandeler, A., and Blancou, J., Assay of oral vaccination of dogs against rabies in Tunisia with the vaccinal strain SADBern, *Vaccine* **12**:307–309 (1994).

67. Hamir, A. N., and Moser, G., Immunoperoxidase test for rabies: Utility as a diagnostic test, *J. Vet. Diagn. Invest.* **6**:148–152 (1994).

68. Hanlon, C. A., Ziemer, E. L., Hamir, A. N., and Rupprecht, C. E., Cerebrospinal fluid analysis of rabid and vaccinia-rabies glycoprotein recombinant, orally vaccinated raccoons (*Procyon lotor*), *Am. J. Vet. Res.* **50**:364–367 (1989).

69. Hanlon, C. A., Hayes, D. E., Hamir, A. N., Snyder, D. E., Jenkins, S., Hable, C. P., and Rupprecht, C. E., Proposed field evaluation of a rabies recombinant vaccine for raccoons (*Procyon lotor*): Site selection, target species characteristics, and placebo baiting trials, *J. Wildl. Dis.* **25**:555–567 (1989).

70. Hanlon, C. A., Buchanan, J. R., Nelson, E., Niu, H. S., Diehl, D., and Rupprecht, C. E., A vaccinia-vectored rabies vaccine field trial: Ante- and post-mortem biomarkers, *Rev. Sci. Tech.* **12**:99–107 (1993).

71. Hanlon, C. A., and Rupprecht, C. E., Considerations for immunocontraception among free-ranging carnivores: The rabies paradigm, in: *Contraception in Wildlife Management* (T. J. Kreeger, ed.), US Government Printing Office, Washington, DC, in press.

72. Hasemann, C. A., and Capra, J. C., High level production of a functional immunoglobulin in a baculovirus expression system, *Proc. Natl. Acad. Sci. USA* **87**:3942–3946 (1990).

73. Hattwick, M. A. W., Weis, T. T., Stechschulte, J., Baer, G. M., and Gregg, M. B., Recovery from rabies: A case report, *Ann. Intern. Med.* **76**:931–942 (1972).

74. Helmick, C. G., Tauxe, R. V., and Vernon, A. A., Is there a risk to contacts of patients with rabies? *Rev. Infect. Dis.* **9**:511–518 (1987).

75. Hemachuda, T., Human rabies: Clinical aspects, pathogenesis, and potential therapy, in: *Lyssaviruses* (C. E. Rupprecht, B. Dietzschold, and H. Koprowski, eds.), pp. 121–143, Springer-Verlag, New York, 1994.

76. Hill, R. E., Beran, G. W., and Clark, W. R., Demonstration of rabies virus-specific antibody in the sera of free-ranging Iowa raccoons (*Procyon lotor*), *J. Wildl. Dis.* **28**:377–385 (1992).

77. Hill, R. E., and Beran, G. W., Experimental inoculation of raccoons (*Procyon lotor*) with rabies virus of skunk origin, *J. Wildl. Dis.* **28**:51–56 (1992).

78. Hill, R. E., Smith, K. E., Beran, G. W., and Beard, P. D., Further studies on the susceptibility of raccoons (*Procyon lotor*) to a rabies virus of skunk origin and comparative susceptibility of striped skunks (*Mephitis mephitis*), *J. Wildl. Dis.* **29**:475–477 (1993).

79. Jackson, A. C., Reimer, D. L., and Ludwin, S. K., Spontaneous recovery from the encephalomyelitis in mice caused by street rabies virus, *Neuropathol. Appl. Neurobiol.* **15**:459–475 (1989).

80. Johnson, K. P., Swoveland, P. T., and Emmons, R. W., Diagnosis of rabies by immunofluorescence in trypsin-treated histologic sections, *J.A.M.A.* **244**:41–43 (1980).

81. Johnston, D. H., Voight, D. R., MacInnes, C. D., Bachmann, P., Lawson, K. F., and Rupprecht, C. E., An aerial baiting system for the distribution of attenuated or recombinant rabies vaccines for foxes, raccoons and skunks, *Rev. Infect. Dis.* **10**:S660–664 (1988).

82. Kamolvarin, N., Tirawatnpong, T., Rattanasiwamoke, R., Tirawatnpong, S., Panpanich, T., and Hemachuda, T., Diagnosis of rabies by polymerase chain reaction with nested primers, *J. Infect. Dis.* **167**:207–210 (1993).

83. Kharmachi, H., Haddad, N., and Matter, H., Tests of four baits for oral vaccination of dogs against rabies in Tunisia, *Vet. Rec.* **130**:494 (1992).

84. Kieny, M. P., Lathe, R., Drillien, R., Spehner, D., Skory, S., Schmitt, D., Wiktor, T. J., Koprowski, H., and Lecocq, J. P., Expression of the rabies virus glycoprotein from a recombinant vaccinia virus, *Nature* **312**:163–166 (1984).

85. King, A., Davies, P., and Lawrie, A., The rabies viruses of bats, *Vet. Microbiol.* **23**:165–174 (1990).

86. King, A. A., Meredith, C. D., and Thomson, G. R., The biology of Southern African lyssavirus variants, *Curr. Topics Microbiol. Immun.* **187**:267–295 (1994).

87. Klepfer, S. R., Debouck, C., Uffelman, J., Jacobs, P., Bollen, A., and Jores, E. V., Characterization of rabies glycoprotein expressed in yeast, *Arch. Virol.* **128**:269–286 (1993).

88. Koprowski, H., Zheng, Y. M., Heber-Katz, E., Fraser, N., Rorke, L., Fu, Z. F., Hanlon, C., and Dietzschold, B., *In vivo* expression of inducible nitric oxide synthase in experimentally induced neurologic diseases, *Proc. Natl. Acad. Sci. USA* **90**:3024–3027 (1993).

89. Krebs, J. W., Strine, T. W., Smith, J. S., Rupprecht, C. E., and Childs, J. E., Rabies surveillance in the United States during 1994, *J. Am. Vet. Med. Assoc.* **207**:1562–1575. (1995).

90. Lafay, F., Benejean, J., Tuffereau, C., Flamand, A., and Coulon, P., Vaccination against rabies: Construction and characterization of SAG2, a double avirulent derivative of SADBern, *Vaccine* **12**:317–320 (1994).

91. Lafon, M., and Lafage, M., Antiviral activity of monoclonal antibodies specific for the internal proteins of N and NS of rabies virus, *J. Gen. Virol.* **68**:3113–3123 (1987).

92. Lafon, M., Edelman, L., Bouvet, J. P., Lafage, M., and Montchatre, E., Human monoclonal antibodies specific for the rabies virus glycoprotein and N protein, *J. Gen. Virol.* **71**:1689–1696 (1990).

93. Lafon, M., Lafage, M., Martinez-Arends, A., Ramirez, R., Vuillier, F., Charron, D., Lotteau, V., and Scott-Algara, D., Evidence for a viral superantigen in humans, *Nature* **358**:607–610 (1992).

94. Lawson, K. F., Johnston, D. H., Patterson, J. M., Hertler, R., Campbell, J. B., and Rhodes, A. J., Immunization of foxes by the intestinal route using an inactivated rabies vaccine, *Can. J. Vet. Res.* **53**:56–61 (1989).

95. Linhart, S. B., Blom, F. S., Dasch, G. J., Roberts, J. D.,

ENGEMAN, R. M., ESPOSITO, J. J., SHADDOCK, J. H., AND BAER, G. M., Formulation and evaluation of baits for oral rabies vaccination of raccoons (*Procyon lotor*), *J. Wildl. Dis.* **27**:21–33 (1991).

96. LINHART, S. B., CREEKMORE, T. E., CORN, J. L., WHITNEY, M. D., SNYDER, B. D., AND NETTLES, V. F., Evaluation of baits for oral rabies vaccination of mongooses: Pilot field trials in Antigua, West Indies, *J. Wildl. Dis.* **29**:290–294 (1993).

97. LOPEZ, A., MIRANDA, P., TEJADA, E., AND FISHBEIN, D. B., Outbreak of human rabies in the Peruvian jungle, *Lancet* **339**:408–411 (1992).

98. MALLORY, L. B., Campaign against rabies in Modoc and Lassen Counties, *Calif. State Board Health Monthly Bull.* **11**:273–277 (1915).

99. MCCOLL, K. A., GOULD, A. R., SELLECK, P. W., HOOPER, P. T., WESTBURY, H. A., AND SMITH, J. S., Polymerase chain reaction and other laboratory techniques in the diagnosis of long incubation periods in Australia, *Aust. Vet. J.* **70**:84–89 (1993).

100. MEBATSION, T., COX, J. H., AND FROST, J. W., Isolation and characterization of 115 street rabies virus isolates from Ethiopia by using monoclonal antibodies: Identification of 2 isolates as Mokola and Lagos bat viruses, *J. Infect. Dis.* **166**:972–977 (1992).

101. MESLIN, F. X., FISHBEIN, D. B., AND MATTER, H. C., Rationale and prospects for rabies elimination in developing countries, in: *Lyssaviruses* (C. E. RUPPRECHT, B. DIETZSCHOLD, AND H. KOPROWSKI, EDS.), pp. 1–26, Springer-Verlag, New York, 1994.

102. MONTANO-HIROSE, J. A., LAFAGE, M., WEBER, P., BADRANE, H., TORDO, N., AND LAFON, M., Protective activity of a murine monoclonal antibody against European bat lyssavirus 1 (EBV1) infection in mice, *Vaccine* **11**:1259–1266 (1993).

103. NYBERG, M., KULONEN, K., NEUVONEN, E., EK, K. C., NUORGAM, M., AND WESTERLING, B., An epidemic of sylvatic rabies in Finland—Descriptive epidemiology and results of oral vaccination, *Acta Vet. Scand.* **33**:43–57 (1992).

104. ORR, P. H., RUBIN, M. R., AND AOKI, F. Y., Naturally acquired serum rabies neutralizing antibody in a Canadian Inuit population, *Arctic Med. Res.* **47**(Suppl. 1):699–700 (1988).

105. PACER, R. E., FISHBEIN, D. B., BAER, G. M., JENKINS, S. R., AND SMITH, J. S., Rabies in the United States and Canada, 1983, *Morbid. Mortal. Week. Rep.* **34**:11SS–27SS (1985).

106. PALMER, D. G., OSSENT, P., SUTER, M. M., AND FERRARI, E., Demonstration of rabies virus antigen in paraffin tissue sections: Comparison of the immunofluorescence technique with the unlabeled antibody enzyme method, *Am. J. Vet. Res.* **46**:283–286 (1985).

107. PASTORET, P. P., BROCHIER, B., LANGUET, B., THOMAS, I., PAQUOT, A., BAULDAIN, B., KIENY, M. P., LECOCQ, J. P., DEBRUYN, J., COSTY, F., ANTOINE, H., AND DESMETTRE, P., First field trial of fox vaccination against rabies using a vaccinia-rabies recombinant virus, *Vet. Rec.* **123**:481–483 (1988).

108. PEARLMAN, E. S., AND BALLAS, S. K., False-positive human immunodeficiency virus screening test related to rabies vaccination, *Arch. Pathol. Lab. Med.* **118**:805–806 (1994).

109. PORRAS, C., BARBOZA, J. J., FUENZALIDA, E., ADAROS, H. L., OVIEDO, A. M., AND FURST, J., Recovery from rabies in man, *Ann. Intern. Med.* **85**:44–48 (1976).

110. PREHAUD, C., TAKEHANAK, K., FLAMAND, A., AND BISHOP, D. H. L., Immunogenic and protective properties of rabies virus glycoprotein expressed by baculovirus vectors, *Virology* **173**:390–399 (1989).

111. PREVEC, L., CAMPBELL, J. B., CHRISTIE, B. S., BELBECK, L., AND GRAHAM, F. L., A recombinant human adenovirus vaccine against rabies, *J. Infect. Dis.* **161**:27–30 (1990).

112. RANDO, R. F., AND NOTKINS, A. L., Production of human monoclonal antibodies against rabies virus, *Curr. Topics Microbiol. Immun.* **187**:195–205 (1994).

113. REID, F. L., HALL, H. N., SMITH, J. S., AND BAER, G. M., Increased immunofluorescence staining of rabies-infected, formalin-fixed brain tissue after pepsin and trypsin digestion, *J. Clin. Microbiol.* **18**:968–971 (1983).

114. REVILIA, L., PHSSARA, F., ORMAECHO, M., VOLANDE, J., RODRIGUEZ, A., RUPPRECHT, C., CHOY, M., AND SNIADACK, D., Survival from rabies infection in the Peruvian jungle, 1994, Field Epidemiology Training Program Scientific Conference Abstract, Centers for Disease Control and Prevention, Atlanta, GA, 24 March 1995, p. 38.

115. ROSATTE, R. C., HOWARD, D. R., CAMPBELL, J. B., AND MACINNES, C. D., Intramuscular vaccination of skunks and raccoons against rabies, *J. Wildl. Dis.* **26**:225–230 (1990).

116. ROSNER, F., Rabies in the Talmud, *Med. Hist.* **18**:198–200 (1974).

117. RUDD, R. J., AND TRIMARCHI, C. V., Development and evaluation of an *in vitro* virus isolation procedure as a replacement for the mouse inoculation test in rabies diagnosis, *J. Clin. Microbiol.* **27**:2522–2528 (1989).

118. RUPPRECHT, C. E., WIKTOR, T. J., JOHNSTON, D. H., HAMIR, A. N., DIETZSCHOLD, B., WUNNER, W. H., GLICKMAN, L. T., AND KOPROWSKI, H., Oral immunization and protection of raccoons (*Procyon lotor*) with a vaccinia–rabies glycoprotein recombinant virus vaccine, *Proc. Natl. Acad. Sci. USA* **83**:7947–7950 (1986).

119. RUPPRECHT, C. E., AND DIETZSCHOLD, B., Perspectives on rabies virus pathogenesis, *Lab. Invest.* **57**:603–606 (1987).

120. RUPPRECHT, C. E., HAMIR, A. N., JOHNSTON, D. H., AND KOPROWSKI, H., Efficacy of vaccinia–rabies glycoprotein recombinant virus vaccine in raccoons (*Procyon lotor*), *Rev. Infect. Dis.* **10**:S803–809 (1988).

121. RUPPRECHT, C. E., DIETZSCHOLD, B., WUNNER, W. H., AND KOPROWSKI, H., Antigenic relationships of lyssaviruses, in: *The Natural History of Rabies* (G. M. BAER, ED.), pp. 69–100, CRC Press, Boca Raton, FL, 1991.

122. RUPPRECHT, C. E., DIETZSCHOLD, B., CAMPBELL, J. B., CHARLTON, K. M., AND KOPROWSKI, H., Consideration of inactivated rabies vaccines as oral immunogens of wild carnivores, *J. Wildl. Dis.* **28**:629–635 (1992).

123. RUPPRECHT, C. E., HANLON, C. A., HAMIR, A. N., AND KOPROWSKI, H., Oral wildlife rabies vaccination: Development of a recombinant virus vaccine, in: *Transactions of the 57th North American Wildlife and Natural Resources Conference—1992* (R. E. MCCABE, ED.), pp. 439–452, Washington, DC, Wildlife Management Institute, 1992.

124. RUPPRECHT, C. E., AND SMITH, J. S., Raccoon rabies: The emergence of an epizootic in a densely populated area, *Semin. Virol.* **5**:155–164 (1994).

125. RUPPRECHT, C. E., SHANKAR, V., HANLON, C. A., HAMIR, A., AND KOPROWSKI, H., Beyond Pasteur to 2001: Future trends in lyssavirus research? in: *Lyssaviruses* (C. E. RUPPRECHT, B. DIETZSCHOLD, AND H. KOPROWSKI, EDS.), pp. 325–340, Springer-Verlag, New York, 1994.

126. SACRAMENTO, D., BOURHY, H., AND TORDO, N., PCR technique as an alternative method for diagnosis and molecular epidemiology of rabies virus, *Mol. Cell. Probes* **5**:229–240 (1991).

127. SCHNEIDER, L. G., Spread of virus within the central nervous

system, in: *The Natural History of Rabies* (G. M. BAER, ED.), pp. 199–216, Academic Press, New York, 1975.

128. SCHNELL, M. J., MEBATSION, T., AND CONZELMANN, K. K., Infectious rabies viruses from cloned cDNA, *EMBO J.* **13**:4195–4203 (1994).

129. SCHUMACHER, C. L., DIETZSCHOLD, B., ERTL, H. C., NIU, H. S., RUPPRECHT, C. E., AND KOPROWSKI, H., Use of mouse anti-rabies monoclonal antibodies in postexposure treatment of rabies, *J. Clin. Invest.* **84**:971–975 (1989).

130. SCHUMACHER, C., ERTL, H., KOPROWSKI, H., AND DIETZSCHOLD, B., Inhibition of immune responses against rabies virus by monoclonal antibodies directed against rabies virus antigens, *Vaccine* **10**:754–760 (1992).

131. SHANKAR, V., DIETZSCHOLD, B., AND KOPROWSKI, H., Direct entry of rabies virus into the central nervous system without prior local replication, *J. Virol.* **65**:2736–2738 (1991).

132. SMART, N. L., AND CHARLTON, K. M., The distribution of challenge virus standard rabies virus versus skunk street rabies virus in the brains of experimentally infected rabid skunks, *Acta Neuropathol.* **84**:501–508 (1992).

133. SMITH, G. L., Virus strategies for evasion of the host response to infection, *Trends Microbiol.* **2**:81–88 (1994).

134. SMITH, J. S., Rabies serology, in: *The Natural History of Rabies*, 2nd ed. (G. M. BAER, ED.), pp. 235–254, CRC Press, Boca Raton, FL, 1991.

135. SMITH, J. S., Rabies virus, in: *Manual of Clinical Microbiology*, 6th ed., (P. R. MURRAY *ET AL.*, EDS.), pp. 997–1003, ASM Press, Washington, DC, 1995.

136. SMITH, J. S., FISHBEIN, D. B., RUPPRECHT, C. E., AND CLARK, K., Unexplained rabies in three immigrants in the United States, *N. Engl. J. Med.* **324**:205–211 (1991).

137. SMITH, J. S., ORCIARI, L. A., YAGER, P. A., SEIDEL, H. D., AND WARNER, C. K., Epidemiologic and historical relationships among 87 rabies virus isolates as determined by limited sequence analysis, *J. Infect. Dis.* **166**:296–307 (1992).

138. SMITH, J. S., AND SEIDEL, H. D., Rabies: A new look at an old disease, in: *Progress in Medical Virology* (J. L. MELNICK, ED.), pp. 82–106, Karger, Basel, 1993.

139. STEELE, J. H., AND FERNANDEZ, P. J., History of rabies and global aspects, in: *The Natural History of Rabies* (G. M. BAER, ED.), pp. 1–24, CRC Press, Boca Raton, FL, 1991.

140. SUGAMATA, M., MIYAZAWA, M., MORI, S., SPRANGRUDE, G. J., EWALT, L. C., AND LODMELL, D. L., Paralysis of street rabies virus-infected mice is dependent on T lymphocytes, *J. Virol.* **66**:1252–1260 (1992).

141. TARTAGLIA, J., PERKUS, M. E., TAYLOR, J., NORTON, E. K., AUDONNET, J. C., COX, W. I., DAVIS, S. W., VANDERHOEVEN, J., MEIGNIER, B., RIVIERE, M., LANGUET, B., AND PAOLETTI, E., NYVAC: A highly attenuated strain of vaccinia virus, *Virology* **188**:217–232 (1992).

142. TAYLOR, J., WEINBERG, R., LANGUET, B., DESMETTRE, P., AND PAOLETTI, E., A recombinant fowlpox virus induces protective immunity in non-avian species, *Vaccine* **6**:497–503 (1988).

143. TAYLOR, J., TRIMARCHI, C., WEINBERG, R., LANGUET, B., GUILLEMIN, F., DESMETTRE, P., AND PAOLETTI, E., Efficacy studies on a canarypox-rabies recombinant virus, *Vaccine* **9**:190–193 (1991).

144. THOMAS, I., BROCHIER, B., LANGUET, B., BLANCOU, J., PEHARPRE, D., KIENY, M. P., DESMETTRE, P., CHAPPUIS, G., AND PASTORET, P. P., Primary multiplication site of the vaccinia–rabies glycoprotein recombinant virus administered to foxes by the oral route, *J. Gen. Virol.* **71**:37–42 (1990).

145. THRAENHART, O., MARCUS, I., AND KREUZFELDER, E., Current and future immunoprophylaxis against human rabies: Reduction of treatment failures and errors, *Curr. Topics Microbiol. Immun.* **187**:173–194 (1994).

146. TOLLIS, M., DIETZSCHOLD, B., VOLIA, C. B., AND KOPROWSKI, H., Immunization of monkeys with rabies ribonucleoprotein (RNP) confers protective immunity against rabies, *Vaccine* **9**:134–136 (1991).

147. TOLSON, N. D., CHARLTON, K. M., STEWART, R. B., CAMPBELL, J. B., AND WIKTOR, T. J., Immune response in skunks to a vaccinia virus recombinant expressing the rabies virus glycoprotein, *Can. J. Vet. Res.* **52**:363–366 (1987).

148. TOLSON, N. D., CHARLTON, K. M., CASEY, G. A., KNOWLES, M. K., RUPPRECHT, C. E., LAWSON, K. F., AND CAMPBELL, J. B., Immunization of foxes against rabies with a vaccinia recombinant virus expressing the rabies glycoprotein, *Arch. Virol.* **102**:297–301 (1990).

149. TSIANG, H., Pathophysiology of rabies virus infection of the nervous system, *Adv. Virus Res.* **42**:375–412 (1993).

150. TUCHIYA, K., MATSUURA, Y., KAWAI, A., ISHIHAMA, A., AND UEDA, S., Characterization of rabies virus glycoprotein expressed by recombinant baculovirus, *Virus. Res.* **25**:1–13 (1992).

151. UHAA, I. J., DATO, V. M., SORHAGE, F. E., BECKLEY, J. W., ROSCOE, D. E., GORSKY, R. D., AND FISHBEIN, D. B., Benefits and costs of using an orally absorbed vaccine to control rabies in raccoons, *J. Am. Vet. Med. Assoc.* **201**:1873–1882 (1992).

152. WANDELER, A. I., Oral immunization of wildlife, in: *The Natural History of Rabies* (G. M. BAER, ED.), pp. 485–503, CRC Press, Boca Raton, FL, 1991.

153. WANDELER, A. I., MATTER, H. C., KAPPELER, A., AND BUDDE, A., The ecology of dogs and canine rabies: A selective review, *Rev. Sci. Tech.* **12**:51–71 (1993).

154. WANDELER, A. I., NADIN-DAVIS, S. A., TINLINE, R. R., AND RUPPRECHT, C. E., Rabies epidemiology: Some ecological and evolutionary perspectives, in: *Lyssaviruses* (C. E. RUPPRECHT, B. DIETZSCHOLD, AND H. KOPROWSKI, EDS.), pp. 297–324, Springer-Verlag, New York, 1994.

155. WIKTOR, T. J., MACFARLAN, R. I., REAGAN, K. J., DIETZSCHOLD, B. D., CURTIS, P. J., WUNNER, W. H., KIENY, M.-P., LATHE, R., LECOCQ, J., MACKETT, M., MOSS, B., AND KOPROWSKI, H., Protection from rabies by a vaccinia virus recombinant containing the rabies virus glycoprotein gene, *Proc. Natl. Acad. Sci. USA* **81**:7194–7198 (1984).

156. WIKTOR, T. J., MACFARLAN, R. I., DIETZSCHOLD, B., RUPPRECHT, C. E., AND WUNNER, W. H., Immunogenic properties of vaccinia recombinant virus expressing the rabies glycoprotein, *Ann. Inst. Past./Virol.* **136E**:405–411 (1985).

157. WIKTOR, T. J., Historical aspects of rabies treatment, in: *World's Debt to Pasteur* (H. KOPROWSKI AND S. PLOTKIN, EDS.), pp. 141–151, Liss, New York, 1985.

158. WIKTOR, T. J., SOKOL, F., KUWERT, E., AND KOPROWSKI, H., Immunogenicity of concentrated and purified rabies vaccine of tissue culture origin, *Proc. Soc. Exp. Biol. Med.* **131**:799–805 (1975).

159. WILDE, H., CHOMCHEY, P., PUNYARATABANDHU, P., PHANUPAK, P., AND CHUTIVONGSE, S., Purified equine rabies immune globulin: A safe and affordable alternative to human rabies immune globulin, *Bull. WHO* **67**:731–736 (1989).

160. WILDE, H., CHOMCHEY P., PRAKONGSRI, S., PUYARATABANDHU, P., AND CHUTIVONGSE, S., Adverse effects of equine rabies immune globulin, *Vaccine* **7**:10–11 (1989).

161. WILDE, H., AND CHUTIVONGSE, S., Equine rabies immune globulin: A product with an undeserved poor reputation, *Am. J. Trop. Med. Hyg.* **42:**175–178 (1990).

162. WILKINSON, L., The development of the virus concept as reflected in corpora of studies on individual pathogens: Rabies—Two millennia of ideas and conjecture of the aetiology of a virus disease, *Med. Hist.* **21:**15–31 (1977).

163. WINKLER, W. G., FASHINELL, T. R., LEFFINGWELL, L., HOWARD, P., AND CONOMY, P., Airborne rabies transmission in a laboratory worker, *J.A.M.A.* **226:**1219–1221 (1973).

164. WORLD HEALTH ORGANIZATION, WHO Expert Committee On Rabies—8th report, *World Health Organ. Tech Rep. Ser.* **824:**1–84 (1992).

165. WORLD HEALTH ORGANIZATION, Two rabies cases following corneal transplantation, *Week. Epidemiol. Rec.* **69:**330 (1994).

166. WORLD HEALTH ORGANIZATION, Report of the 5th consultation on oral immunization of dogs against rabies, Geneva, 20–22 June, 1994, pp. 1–24.

167. WORLD HEALTH ORGANIZATION, Oral immunization of foxes in Europe in 1994, *Week. Epidemiol. Rec.* **70:**89–91 (1995).

168. WORLD HEALTH ORGANIZATION, World Survey of Rabies 28 for Year 1992, Geneva, 1994, pp. 1–24.

169. WUNNER, W. H., The chemical composition and molecular structure of rabies viruses, in: *The Natural History of Rabies*, 2nd ed. (G. M. BAER, ED.), pp. 31–67, CRC Press, Boca Raton, FL, 1991.

170. XIANG, Z. W., SPITALNIK, S., TRAN, M., WUNNER, W. H., CHENG, J., AND ERTL, H. C., Vaccination with a plasmid vector carrying the rabies virus glycoprotein gene induces protective immunity against rabies virus, *Virology* **199:**132–140 (1994).

CHAPTER 23

Respiratory Syncytial Virus

Kenneth McIntosh

1. Introduction and Historical Background

Respiratory syncytial virus (RSV) was first isolated from a chimpanzee with common-cold-like illness.[151] Shortly thereafter, the virus was recovered from young children with severe lower respiratory tract disease in Baltimore.[22,28] Since its initial isolation from infants with respiratory disease 30 years ago, RSV has emerged as the major lower respiratory tract viral pathogen of infancy and early childhood throughout the world.[10,24,25,55,64,121,129,132,201,206] In all geographic areas it is now clear that RSV is the major cause of bronchiolitis and pneumonia in infants and young children. Respiratory syncytial virus presents a special challenge to the epidemiologist, since this virus exhibits a pattern of infection and disease unlike that of any of the other known respiratory tract viral pathogens. Unanswered are many pressing questions concerning the pathogenesis of serious life-threatening disease of the lower respiratory tract produced by this virus during early infancy. A safe, effective vaccine for prevention of serious pediatric RSV illness is not available at this time. Recent studies and

those now in progress, however, offer hope that it should be possible to develop effective and widely available immunoprophylaxis for RSV bronchiolitis and pneumonia.

The RSV is a medium-sized (120–200 nm) enveloped virus that contains a lipoprotein coat and a linear, minus-sense RNA genome.[13,31,32,116,160,161] The virus matures at the limiting membrane of the infected cell.[116,161] The nonsegmented, minus-sense RNA genome contains 10 separate genes, each coding for a separate protein, and the gene order has been established unambiguously as $3'$-NS_1-NS_2-N-P-M-SH-G-F-M_2-L-$5'$.[31,33,35] Its internal helical nucleocapsid contains the linear viral genome to which are complexed many molecules of the nucleocapsid (N) protein and considerably fewer molecules of the phosphoprotein (P) and putative viral polymerase (L). Its outer envelope is lined internally with matrix (M) protein and is studded externally with fusion (F) and attachment (G) glycoprotein projections.[48,114,210] Another protein (M_2) is also present in the viral envelope.[36] Finally, three other proteins are either nonstructural (NS_1 and NS_2) or present in virions (SH) in a position that remains to be determined. The SH (small hydrophobic protein) was recently shown to be a third integral membrane glycoprotein that is inserted into the outer membrane of infected cells.

It is now agreed that RSV and the pneumonia virus of mice (PVM) comprise a group of enveloped RNA viruses distinct from the orthomyxoviruses (the influenza viruses). Both RSV and PVM are more closely related to the paramyxoviruses (parainfluenza viruses, mumps virus, Newcastle disease virus, and measles, rinderpest, and distemper viruses); the former two viruses, however, exhibit important differences and have been assigned to a separate genus of Paramyxoviridae, which is designated the genus *Pneumovirus*.

Kenneth McIntosh • Division of Infectious Diseases, The Children's Hospital, and Department of Pediatrics, Harvard Medical School, Boston, Massachusetts 02115.

2. Methodology Involved in Epidemiologic Analysis

2.1. Sources of Mortality Data

From the serious nature of RSV bronchiolitis and pneumonia of infancy, there is reason to suspect that the virus is a major cause of fatal respiratory tract disease during the first year of life. This view is supported by two reports from the United Kingdom concerning 46 infants and children who died with lower respiratory tract disease; 36 of the patients were less than 1 year of age, and RSV was isolated from 13 of the patients postmortem.[40,59] In another study in the United Kingdom, RSV was recovered from 3 of 12 infants who died with lower respiratory tract disease.[110] One of these infants had bilateral hydronephrosis and another Down's syndrome, and these conditions were thought to contribute to the fatal outcome of their disease. Respiratory syncytial virus or viral antigens have been detected postmortem in the lungs of a varying proportion of infants with the sudden infant death syndrome (SIDS).[15,40,179] Several recent surveys of hospitalized infants and children with RSV infections in the United States and Canada have indicated that mortality rates in modern hospitals are low, between 0.3 and 1.0%.[148,158] Risk factors for mortality that have been identified for nosocomial infection as well as for infection acquired in the home include prematurity, congenital heart disease, bronchopulmonary dysplasia, immunodeficiency, and immunosuppression.[12,87,90,93,125,130,159,166,169,199]

Anderson and colleagues[5] analyzed respiratory deaths in infants and young children reported to the Centers for Disease Control in Atlanta during the years 1975 to 1984 and correlated them with isolation of various respiratory viruses in sentinel laboratories around the country. RSV and influenza were the viruses more frequently associated with "residual" (or excess) respiratory deaths: RSV in infants 1–5 months old (correlation coefficient 0.499, $P < 0.0001$), and influenza in children 24–59 months old (correlation coefficient 0.350, $P < 0.001$). The total number of deaths associated with winter peaks during this interval ranged from 186 to 454 per year for children 1–11 months old.

2.2. Sources of Morbidity Data

Although the importance of RSV in fatal respiratory tract disease in early life, or in any age group for that matter, has not been defined clearly, there is abundant evidence that suggests that this virus is a major cause of serious life-threatening disease of the lower respiratory tract during early life. These data come from cross-sectional studies of pediatric patients admitted to the hospital with a diagnosis of bronchiolitis, pneumonia, bronchitis, or croup, as well as prospectively studied cohorts.

Estimates of the contribution of RSV to pediatric respiratory disease probably represent an underevaluation of the role of this agent. For example, human heteroploid-cell cultures commonly used for the recovery of RSV exhibit marked variation of sensitivity to the agent; at times such cell cultures may be completely resistant to RSV.[206] Another factor that should be considered in assessing the impact of RSV in pediatric respiratory disease is the inefficiency of serological techniques for the detection of infection in young infants.[24,168,180,182,183]

The behavior of RSV in pediatric populations throughout the world has been discerned primarily from the pattern of serious respiratory disease produced by this virus. Although surveillance of seriously ill infants and children can serve as a barometer of the virus in the community, this type of study cannot, in itself, yield an estimate of the incidence of infection, the type and severity of clinical syndromes produced, or the risk of serious illness during infection. Such information, however, has been obtained from prospective studies of families, children in a day-care center, or residents of a semiclosed nursery.[84,97,117,149] Also, risk factors for additional morbidity and extension of hospital stay have been documented during study of nosocomial infection in premature infants and young children in whom the rate of cross-infection has been estimated to vary from 20 to 47% during periods of RSV prevalence.[12,87,90,125,130,159,166]

2.3. Serological Surveys

Serological surveys have been performed using the complement-fixation and neutralization techniques. The former method is relatively less sensitive than the latter. In the first seroepidemiologic survey performed, the proportion of individuals with serum N antibody to RSV increased rapidly with age and reached 80% by 4 years of age in the Baltimore area.[22] In later studies using this technique, it was found that all adults tested possessed serum N antibody, and the level of antibody was significantly higher than that detected in the serum of seropositive children.[117] Finally, newborn infants were shown to possess the same level of serum N antibody as their mothers.[9] In one later study, each of a group of infants 1–5 months of age was found to possess serum N antibody when measured by the plaque-reduction technique; in addition, a small proportion of infants who were 6–7 months of age also possessed such neutralizing activity in their serum.[167] These infants were born after the last epidemic of RSV and presumably had not been infected with the virus. The mean titer of antibody decreased with age, dropping approximately twofold per 3-week interval. This suggested that the antibody measured was passively acquired from the mother.

In summary, serological surveys using techniques of varying sensitivity have shown that all adults possess serum N antibody, as do infants at the time of birth. Antibody found in the serum of newborn infants represents passively acquired antibody; this antibody has a half-life of approximately 3 weeks and usually cannot be detected after 6–7 months of age unless the infant has been infected with RSV.

2.4. Laboratory Methods

Respiratory syncytial virus is very labile, and for this reason, and because during the height of the infection, large amounts of virus and viral antigen are produced in the respiratory tract, most clinical laboratories now rely on antigen detection methods for diagnosis of infection. Respiratory secretions used for culture should be put immediately at 4°C for transport and then inoculated as quickly as possible (within 4 hr if possible) into roller tubes containing HEp-2 cells of known sensitivity to RSV. Freezing and thawing of respiratory tract secretions or nasopharyngeal washings often result in a significant decrease in titer of virus as well as failure to recover virus from such specimens.[10,18]

Both immunofluorescence and enzyme-linked immunosorbent assay (ELISA) have been used with success to detect RSV antigen in respiratory secretions. Both methods have advantages over culture of convenience and speed, and both are of superior sensitivity when applied to specimens subject to delays in delivery. The sensitivity, in relation to culture, of immunofluorescence is 90–100%, and that of ELISA is 80–90%.[138,141]

3. Biological Characteristics of the Viruses that Affect the Epidemiologic Pattern

There is no information that would link various biological properties of RSV with its unusual epidemiologic behavior. Its lability does not seem to limit its spread in a susceptible population.

Antigenic polymorphism among RSV strains was recognized in cross-neutralization tests performed over 30 years ago.[30] The original two RSV antigenic subgroups have now been defined in greater detail using monoclonal antibodies, and it was shown that these subgroups have continued to circulate simultaneously in the community in varying proportion,[2,4,99,154] a pattern that has been observed in urban communities throughout the world.[137] The major share of the antigenic dimorphism initially recognized by the neutralization assay appears to be linked to variation in the viral G surface glycoprotein.[54,200] In tests performed on convalescent sera of infants undergoing primary infection with RSV, the extent of reciprocal reduction in heterologous serum neutralizing antibody titer observed for the antigenic subgroups was approximately threefold.[98] This difference is considerably less than that which occurs with viruses belonging to different influenza A virus subtypes. It appears that the differences among the most divergent RSV strains are equivalent to those that exist between successive antigenic variants within an influenza A virus subtype. This degree of antigen dimorphism may contribute to the occurrence of initial reinfection,[152] but its effect can hardly explain the major aspects of RSV epidemics and patterns of disease. In any case, it would be prudent to include both antigenic variants in a RSV vaccine.

Genetic variation among RSV strains has been studied by the technique of RNA fingerprinting by ribonuclease protection.[197] Analysis of multiple strains demonstrated wide variation in nucleic acid sequences in the G protein within a single urban outbreak, although strains clearly linked by epidemiologic data, such as those in a family or hospital outbreak, were genetically identical.

Persistent infection in tissue culture can be initiated with certain temperature-sensitive mutants of RSV.[176] Nevertheless, the occurrence of a chronic carrier state has not been demonstrated in immunocompetent individuals. The capacity of the virus to reinfect and reinstitute pharyngeal carriage is probably a major mechanism of survival and spread.

4. Descriptive Epidemiology

4.1. Incidence and Prevalence Data

4.1.1. Risk of Infection and Reinfection. Based on serological surveys, infection with RSV appears to be a common occurrence during the first few years of life. In a study performed in Baltimore shortly after isolation of the first human strains of RSV, it was found that 48% of children possessed serum N antibody for the virus by 2 years of age and 77% by 3 years of age.[22] In a study employing the more sensitive plaque-reduction method, it was found that approximately half of the infants who lived through one RSV epidemic in the Washington, DC, area were infected during this period.[167] Furthermore, almost all children who had lived through two successive RSV epidemics were infected, as indicated by the presence of neutralizing antibody in their serum. Thus, the risk of infection for previously uninfected infants and young children is extremely high. In a prospective study in Houston, the infection rate for RSV was 68.8/100 during the first year of life and 82.6/100 during the second year.[66] Virtually all children were infected by 24 months and half experienced two infections. The risk of reinfection decreased to 33.3/100 by 48 months of age.[66] When one considers the overall impact of RSV on pediatric respiratory disease throughout the world, it is likely that initial RSV infection also occurs early in life in most geographic areas.

A high frequency of infection and reinfection was also observed for young children in a day-care center.[96] In this population, 98% of seronegative infants and young children were infected with RSV during the first outbreak in which they were at risk, whereas during two succeeding exposures to the virus the frequency of reinfection was 74% and 65%, respectively. In another longitudinal study, RSV infection was detected in 45% of 36 families studied during one epidemic period.[84] Of interest, 9 of 21 individuals 17–45 years of age became infected when virus was introduced into the family. This observation supports the view that reinfection is a common event even in adults.

In surveys of serum from adults, it was shown that 33–99% of individuals possessed complement-fixing antibody for RSV.[38,92,111] The presence of this antibody in serum is most likely a reflection of the frequent occurrence of reinfection rather than long-term persistence of antibody following a limited, superficial infection of the respiratory tract.

In summary, it appears that most individuals become infected with RSV early in life and few escape infection by this virus during infancy or early childhood. Further, reinfection occurs with appreciable frequency in older children and young adults and probably plays a major role in the spread of virus to the young infant, who is the target host for serious disease. It is highly unlikely that infants with serious RSV disease are the usual source of infection for other young infants.

4.1.2.   Risk of Serious Bronchiolitis or Pneumonia during Infancy.   The risk of RSV illness serious enough to require admission of the affected individual to the hospital has been estimated by several groups of investigators. First, in a study spanning 11 consecutive yearly RSV epidemics in the Washington, DC, area, it was estimated that 1 in 200 infants (0–12 months) developed RSV bronchiolitis and/or pneumonia requiring hospitalization.[121] Second, during a collaborative study in the

United Kingdom involving 10 centers, the estimate for all infants was 1 in 120, whereas in highly industrialized areas the estimate was 1 in 70.[55] For all infants 1 to 3 months of age, the estimate was 1 in 55, and in industrial areas it was 1 in 40. Third, during a prospective study in Houston, 2 of 125 infants were hospitalized during the first 3 months of life for RSV lower respiratory tract disease.[66]

Of course, the toll of RSV in infancy is considerably higher, since for every hospital admission for RSV bronchiolitis or pneumonia there are many infants who develop respiratory disease that is almost as serious as that seen in the individuals admitted to the hospital. In a prospective study of 1179 healthy infants enrolled at birth, 143, or 12.1%, developed lower respiratory tract illnesses during the first year of life in which RSV was identified. There were five additional repeat RSV infections in this cohort, and the 148 episodes amounted in all to 39.5% of all such illnesses studied virologically. Of the 143 first infections, 123, or 86%, were classified as bronchiolitis.[103]

4.1.3.   Risk of Pneumonia and Febrile Disease. Of 90 institutionalized infants and young children, 40% developed pneumonia over a 4-week period during an outbreak of RSV infection that occurred in a welfare nursery in Washington, DC.[117] An additional 53% of the nursery residents developed a febrile illness. The relationship of the virus to illness was determined by cross-sectional analysis in which the illness attack rate during a 5-day period bracketing initial virus isolation (test period) among infants and children who were RSV positive was compared with the attack rate during the same 5-day period among children who were RSV negative. This type of analysis indicated a significant association of virus with febrile illness and with pneumonia. The cross-sectional analysis, however, did not adjust for attributes such as age, sex, race, and duration of residence in the nursery, some of which might influence the occurrence of illness. Therefore, a horizontal analysis was made comparing the febrile illness experiences of those children from whom RSV was isolated with the febrile experiences of the same children 2 weeks before and 2 weeks after a 5-day period (test period) bracketing initial virus isolation. This type of horizontal analysis indicated that the children who were RSV positive experienced an onset of febrile respiratory disease three times more often than the expected number of febrile episodes ($P = 0.001$). This finding indicated a striking association between the recovery of RSV and the onset of a febrile illness in the children involved in the nursery outbreak. In a second outbreak of RSV infection that occurred in the same nursery 6 years later, only 10% of infants and young children developed pneumonia.[118] There is no obvious reason (or reasons) for this difference in response at this time. It should be emphasized that in both outbreaks, the population consisted of a mixture of infants and children, of whom some were probably completely susceptible to both infection and disease, whereas others had experienced several RSV infections previously and undoubtedly had developed some resistance to this virus.

An outbreak of RSV that occurred in a nursery in Taiwan also provided significant information about the pathogenic potential of RSV.[128] The outbreak involved 15 normal infants ranging between 7½ and 12 months of age. The clinical attack rate was 100%; each of the infants developed symptoms of respiratory disease, and 13 of the 15 developed a febrile response that exceeded 38°C. A similar experience was recorded during an outbreak of RSV infection in a home for infants in Stockholm.[195] Thus, it appears that infection with RSV has a high degree of clinical penetrance and that most infections that occur in infancy and early childhood lead to the development of signs and symptoms of respiratory tract disease, and in most instances an associated febrile response occurs. Most adults undergoing reinfection also develop acute respiratory tract disease, but fever is less common that in infants and children.[84]

4.1.4. Role of RSV in Different Clinical Syndromes. In a 13-year surveillance of infants and young children admitted to the Children's Hospital National Medical Center (DC) with lower respiratory tract disease, RSV infection was detected in 43% admitted with the diagnosis of bronchiolitis and in 25% of patients with pneumonia.[121] Respiratory syncytial virus infection was detected less often in the syndromes of bronchitis (10.6%) and croup (9.8%). In contrast, RSV infection was detected in only 5.4% of infants and children seen in the clinic or admitted to the hospital for nonrespiratory illness during this interval. The findings that emerged from this study of approximately 5000 infants and children with lower respiratory tract disease are representative of observations made by others during the past 30 years. Thus, it is clear that RSV is the major cause of bronchiolitis of early infancy. In addition, the virus is a major cause of pneumonia during the first few years of life.

In a study of respiratory tract agents in a private pediatric practice in North Carolina from July 1964 through June 1975, wheezing was identified as a common manifestation of acute respiratory illness in children.[96] The most common cause of wheezing associated with respiratory illness of varying severity in children under 5 years of age was RSV, whereas *Mycoplasma pneumoniae* was the agent most frequently isolated from school-age children. Other bacterial and viral agents were also associated with wheezing not severe enough to require hospitalization. The findings in this study confirmed and further defined the relative importance of RSV (and other viral pathogens and *M. pneumoniae*) in lower respiratory tract disease of varying severity including severe wheezing disease traditionally defined as bronchiolitis. The relative role of the various etiologic agents in this study was essentially the same as that previously established in cross-sectional studies of children coming to a hospital.

4.1.5. Role of RSV in Nosocomial Infection. Respiratory syncytial virus is a major cause of nosocomial infection and a particular hazard for premature infants, infants with congenital heart or lung disease, and infants, children, or adults who are immunodeficient.[12,44,56,65,77,81,87,89,93,125,130,144,146,166,170,216] These individuals are also at high risk of serious illness when infection is acquired in the home. The rate of hospital-acquired infection for infants and children during an RSV season has been reported to range from 26 to 47% in newborn units and from 20 to 40% for older children.[81,87,144,146] In one report of 100 nosocomial infections among infants and children in a children's hospital, 45 were caused by rotavirus and 20 by RSV.[216]

Hospital staff appear to play a major role in nosocomial spread of RSV infection, and this occurs whether or not infected physicians and nurses develop respiratory symptoms.[83,87,157] The likelihood that an individual infant or child will acquire nosocomial RSV infection increases with the duration of stay in the hospital and the number of individuals housed in his or her room.[56,83,85,87,157]

4.1.6. Role of RSV in Infections with Underlying Cardiopulmonary Disease. Numerous studies have pointed out the increased risk of RSV infection in infants and children with compromised cardiac and pulmonary function. An early study of children with congenital heart disease pointed out the particular susceptibility of those with pulmonary hypertension.[130] In this study, the mortality in this group was 44%, nosocomial infection being responsible for a large proportion of the total. A more recent retrospective survey in children with heart disease has described a lower mortality.[148] A total of 740 RSV infections were surveyed over a 7-year period at one hospital, of which 79 occurred in children with congenital heart disease with only two deaths. Nosocomial infections accounted for 25 of these, however, and both deaths. In comparison to all children without congenital heart disease, hospital stays were longer (29.2 days vs. 11.8 days, $P < .001$), there were more intensive care days (9.6 days vs. 3.0 days, $P = 0.001$), and more days on mechanical ventilation (0.9 days vs. 4.1 days, $P < 0.001$). In a large Canadian survey of RSV infections from 1988 to 1991,

260 children with congenital heart disease were hospitalized with RSV infection.[158] The mortality in the group with pulmonary hypertension (53 total) was 9.4% compared to 2% in those without.

Children with underlying lung disease also run the risk of greater morbidity and mortality with RSV infection. This has been found particularly in children with bronchopulmonary dysplasia,[71,142] but is seen also in those with cystic fibrosis,[1] chronic aspiration, reactive airways disease,[135] and other conditions.

4.1.7. Role of RSV Infections in Patients with Immunodeficiency.

It has been recognized for some time that children with defects in lymphocyte immunity shed virus for longer than normal children. Such children can be congenitally immunodeficient, immunosuppressed by various types of chemotherapy, or suffering from the acquired immunodeficiency syndrome (AIDS).[21,199] In one study, 3 of 20 children undergoing chemotherapy and two of five children with congenital immunodeficiency died in the course of RSV infection.[90] Another more recent series of 17 children experiencing RSV infection following liver transplantation included two deaths associated with progressive pulmonary disease. Both children were infected in the immediate postoperative period.[170]

RSV infection has proven to be a particular hazard for patients undergoing bone marrow transplantation, where the mortality in several small series has been 45–50%.[44]

4.2. Epidemic Behavior

One of the most remarkable features of the epidemiology of RSV is the consistent pattern of infection and disease. Other respiratory viruses cause epidemics at irregular intervals or exhibit a mixed endemic–epidemic pattern, but RSV is the only respiratory viral pathogen that produces a sizable epidemic every year in large urban centers.[26,167] This pattern has been observed wherever RSV has been studied. In the temperate areas of the world, RSV epidemics have occurred primarily in the late fall, winter, or spring but never during the summer. In Washington, DC, over an interval of 13 years, RSV was most active during the period from January to April and virtually absent from the community during August and September.[121] Large annual variations in the impact of RSV on the pediatric population of the Washington, DC area were not observed over this 13-year period. During 11 consecutive epidemics studied intensively, the number and proportion of infants and children admitted to the Children's Hospital National Medical Center of DC for RSV lower respiratory tract disease did not vary more

than 2.7-fold.[121] Also remarkable was the consistency with which RSV produced the same clinical pattern of respiratory tract disease, especially bronchiolitis and pneumonia, year after year. These illnesses were barometers of RSV infection in the community; when the virus was at its epidemic peak, hospitalization for bronchiolitis and pneumonia in infants and young children soared.[121] A similar temporal association of RSV infection and serious lower respiratory illness in infants was also documented in Scotland.[70]

More than 15,000 infants and young children were studied for RSV infection during 11 consecutive outbreaks in the Washington, DC, area. Data obtained at monthly intervals during the outbreaks were combined to plot a composite epidemic curve, which showed a normal distribution.[14] Of more than 1000 respiratory disease patients who yielded an RSV isolate during the composite outbreak, 40% shed virus during the peak epidemic month and 82% shed virus during the period encompassing the peak month and the months preceding and following it; this indicated the sharpness of the yearly epidemics of the virus in the Washington, DC, area. During the peak month of the composite epidemic, RSV was recovered from 46% of all inpatients hospitalized with bronchiolitis, from 34% of all patients hospitalized with any type of respiratory disease, and from 32% of patients with respiratory disease who were seen as outpatients in the hospital. Control subjects who were free of respiratory disease rarely yielded RSV (less than 1%). As indicated by virus recovery and/or the development of serum complement-fixing antibody, 70% of bronchiolitis patients and 56% of all respiratory disease inpatients exhibited evidence of RSV infection during the peak epidemic month. A similar epidemic wave was seen in males compared to females and in black compared to nonblack children.

4.3. Geographic Distribution

Respiratory syncytial virus has emerged as the major pediatric respiratory tract pathogen wherever appropriate studies have been performed to detect infection. Furthermore, the epidemic pattern of disease just described for the Washington, DC, area appears to be characteristic of the behavior of this virus in large urban centers throughout the world. In a series of hospital-based surveys of microbial agents associated with acute lower respiratory illnesses in hospitalized infants and children in cities in Argentina, Uruguay, Bangladesh, Thailand, and the Philippines, RSV was consistently the most common agent found, 45–82% of all viruses identified.[6,104,107,178,204] Serological surveys of individuals from remote isolated populations have in each instance revealed evidence of prior RSV infection (R. M. Chanock, unpublished studies).

4.4. Temporal Distribution

In the United States and other temperate areas, RSV epidemics occur in the late fall, winter, or spring but not during the summer.[121,154] The virus is rarely isolated during August or September. In Washington, DC, outbreak peaks were observed to occur in six different calendar months over a period of 13 years.[14,121] The outbreak peak occurred as late as June and as early as December. Each RSV epidemic lasted approximately 5 months. In Chicago, the peaks of three successive outbreaks were separated by an interval of 55–58 weeks.[153] A fourth outbreak occurred after a shorter interval of 39 weeks, and this was followed by a fifth outbreak in which the interval between peaks was prolonged to 62 weeks. Analysis of 13 successive outbreaks that occurred in Washington, DC, suggested that the interval between successive peaks was alternately long (13–16 months) and then short (7–12 months).[121]

In one semitropical area (Trinidad), RSV was found to exhibit a temporal pattern different from that characteristic of temperate regions. Over a 3-year period of surveillance, RSV epidemics occurred in Trinidad during the rainy season, which extended from June through December.[193] The Trinidad epidemics started in June 1964, September 1965, and August 1966.

4.5. Age

In a large cross-sectional study of RSV infection in pediatric patients admitted to the hospital, the peak incidence of RSV bronchiolitis and pneumonia was observed at 2 months of age.[167] Thereafter, the incidence of these diseases decreased with increasing age, more rapidly for bronchiolitis than for pneumonia. Except for the first month of life, the incidence curve of RSV bronchiolitis exhibited a marked downward slope so that by 10–12 months of age few infants were admitted to the hospital with this diagnosis. During the first month, the incidence of RSV bronchiolitis was approximately one third that observed during the second month, the period of peak occurrence of the disease. Although RSV pneumonia occurred most often at 2 months of age, the incidence of this disease decreased rather slowly with increasing age. It was not unusual for the virus to cause pneumonia in older children. It should be emphasized that the age distribution of the most serious RSV disease, i.e., bronchiolitis, is unique among viral infections of man: RSV is the only virus that preferentially produces severe disease and has its maximum impact during the first few months of life.[23,167]

A prospective cohort study in Tucson, Arizona followed 1179 infants from birth through the first year of life and examined rates of lower respiratory tract illnesses.[103] In this group the incidence of RSV infections reached a peak at 3 months of age and remained at about the same level (four lower respiratory tract infections per 100 infants) until the end of the 7th month of life, declining slowly thereafter.

In a prospective study that extended over one RSV epidemic season in Rochester, it was observed that the rate of infection did not vary appreciably with age except during infancy.[84] Infection was monitored by virus isolation, which is generally a more sensitive indicator of reinfection than serology. Infection was detected in 16 of the 36 families under surveillance. In families into which the virus was introduced, approximately 40% of individuals 1 to 45 years of age became infected, and the rate was higher for infants—62%. In these families it appeared that children were primarily responsible for introducing virus into the home.

Age and prematurity also had an influence on the nature of illness produced by RSV. Very young infants, and particularly infants born premature, have been found to have an increased incidence of apneic episodes with RSV infection, as well as a greater tendency to require intensive care, assisted ventilation, and prolonged hospitalization.[19,69,143]

4.6. Sex

Severe RSV disease that requires hospitalization occurs approximately 30% more often among male infants than among female infants.[167]

4.7. Race

In a study of 13 consecutive epidemics in the Washington, DC, area, the proportion of nonblack patients with various forms of lower respiratory disease associated with RSV infection was somewhat higher than that of black patients.[167] A plot of all hospitalized respiratory disease patients indicated that in every age category black patients yielded RSV slightly less often than nonblack patients; however, the age pattern of RSV bronchiolitis was the same for both groups of patients. The nonblack patients were in most instances from the suburbs of the Washington, DC, area, whereas the black patients tended to come from the poorer areas of the District of Columbia immediately surrounding the Children's Hospital.

4.8. Occupation

We have no information about any relationship of RSV infection to occupation.

4.9. Occurrence in Different Settings

Differences in the spread of RSV within different families were noted in a seroepidemiologic study performed in Tecumseh, Michigan.[149] As the number of members in the family increased from three to six persons, there was an increase in the proportion of families in which one or more individuals were infected with RSV. This apparently was a reflection of the additional opportunity for introduction of infection into a larger family. The increased number of persons at risk in larger families also resulted in more multiple infections within the family once the virus was introduced. In the Tecumseh study, only 5.5% of persons in families with three members were infected with RSV, whereas the proportion increased to 16.3% in families that had six members. Data from this study support the view that the virus is introduced into the family unit most often by a school-aged child.

The prospective cohort study in Tucson cited above examined risk factors for RSV-associated lower respiratory tract illness and found that early disease (in the first 2 months of life) was associated with increasing numbers of others sharing the child's bedroom.[103] In the 7- to 9-month group, day care was a significant risk factor.

The possible role of breast-feeding in protecting against RSV-induced lower respiratory tract illness has been the subject of a number of studies. Perhaps the most convincing evidence of a beneficial effect was found in the Tucson study, where breast-feeding was found to lower the risk only during the first 3 months of life and only in infants of mothers with a lower educational level.[103]

4.10. Socioeconomic Status

Data from a study in North Carolina suggested that serious RSV disease occurred more frequently in infants of low socioeconomic status from a rural area than in infants of higher socioeconomic status living in an urban center.[64] In this situation, socioeconomic factors may have influenced the risk of infection during early infancy, and this in turn influenced the pattern of disease. It is likely that delay of primary RSV infection past the age of

greatest vulnerability (i.e., the first 6 months of life) increases the probability that disease will be mild rather than severe. Observations made during the multicenter study in the United Kingdom also indicated that the incidence of serious RSV illness during infancy varied significantly in relation to social class. The incidence was 12-fold higher among infants from the lowest social class compared to those of the highest class.[55]

4.11. Other Factors

We have no information concerning other factors.

5. Mechanisms and Routes of Transmission

Respiratory syncytial virus is clearly spread by infected respiratory secretions. Evidence is accumulating that the major mode of spread is not through droplet nuclei or small-particle aerosols but, rather, by large droplets or through fomite contamination.[75,77,80,82] Spread requires either close contact with infected infants or contamination of the hands with fomites from surfaces to which infected respiratory secretions have spread by mechanical means, and subsequent contact between fingers and nasal or conjunctival mucosa.

In spite of these limitations, however, RSV has proved to be a very common cause of hospital-acquired respiratory infection in nurseries and on pediatric wards during community outbreaks. In one study, 45% of those infants admitted for other causes but remaining in a hospital for 1 week or more and 100% of those remaining for 4 weeks or more, became infected.[78] Many of these patients developed clinically significant illness. In addition, 25–45% of hospital staff regularly become infected during such outbreaks. Attempts to control such outbreaks by use of conventional methods such as gown and mask have been unsuccessful.[76] Nosocomial spread in nurseries and pediatrics wards, however, can be reduced by a combination of precautions that includes limitation of visitors, active surveillance of RSV infection among patients and new admissions, and cohorting of RSV infected patients, combined with a series of measures designed to reduce passive transfer of RSV by the hospital staff to patients, as well as acquisition of RSV infection by the staff.[16,53,190] Precautions that appear to be effective in achieving the latter objectives include strict observance of hand washing and use of gloves and gowns.[127] The separation of nurses into cohorts for the care of infected infants may also assist in the control of RSV infections when these standard methods fail.[131]

A virus that is similar antigenically and biologically to RSV has been recovered from cattle with respiratory disease in Europe, Asia, and the United States.[108,109,163] Related strains have also been recovered from sheep and goats.[194] The animal strains appear to be related antigenically to, but distinct from, the human strains. To date natural infection of humans with animal RSV or vice versa has not been documented.

6. Pathogenesis and Immunity

From information obtained during outbreaks in closed populations, it was estimated that the incubation period from exposure to virus to development of fever and signs of RSV lower respiratory disease was approximately 4.5 days.[117,195] When adults were experimentally infected with RSV, the incubation period averaged 5 days and illness lasted an average of 5½ days.[117,126]

The virus replicates in the nasopharynx, commonly reaching a titer of 10^4 to 10^6 TCID$_{50}$ per milliliter of nasal secretion in young infants[78,79] and a somewhat lower titer in adult volunteers.[145] The titer of virus, as well as the quantity of detectable RSV antigen, in sequential nasal secretions has been measured in infants hospitalized with lower respiratory tract illness caused by RSV.[79,88,139,140,212] In most infants, virus titers begin at a level of $10^{3.5}$ to 10^5 TCID$_{50}$ per milliliter and fall gradually throughout the hospitalization. Some infants continued to have positive cultures as long as 3 weeks after initial hospitalization. In one study there was some correlation between severity of disease (but not age) and duration of virus shedding and, inversely, between age and maximum titer of virus shed.

The titer of virus in the lower respiratory tract is not known, but autopsy studies have shown that viral antigens are found in abundance in fatal RSV pneumonia but only in small amounts in fatal RSV bronchiolitis.[58] In these same studies viral antigens were not found any deeper than the superficial layers of the respiratory epithelium. Several cases of fatal RSV infection have been described in infants or adults who lack cell-mediated immunity.[49,113] In these instances virus spread outside the respiratory tract and into other organs, including kidney, liver, and myocardium. In the presence of normal immunity, productive infection is probably limited to the respiratory epithelium. Respiratory syncytial virus antigens have been detected in circulating mononuclear leukocytes of RSV-infected individuals, but these antigens appear to be the result of abortive rather than productive infection, and this finding lacks confirmation.[39]

In the studies of Hall et al.[78,79] clinical recovery often occurred in spite of the persistence of viral shedding from the upper respiratory tract. In the experience of other investigators, however, viral shedding ended as secretory antibodies began to appear, usually coincident with clinical recovery.[139,140] An intact immune system appears to be necessary for the termination of RSV infection. Patients who lack cell-mediated immunity may become persistently infected with RSV. This has been found in both congenital and acquired immunodeficiency syndromes and in infants and children treated with immunosuppressive drugs.[21,90]

For many years there has been speculation that part of the pathogenesis of respiratory disease during RSV infection involved an immunologic response to the virus.[27,136] Disease is most common during the first 6 months of life, when maternally transmitted serum antibodies are almost universal. It was this coincidence of serum antibody and severe infection that first prompted the suggestion that immunopathologic mechanisms might be involved in RSV bronchiolitis.[27] Since the time that this theory was put forward, however, several studies have uncovered aspects of RSV infection with which it is not consistent. First bronchiolitis is most common between 6 weeks and 6 months of age, with relative sparing during the first 6 weeks of life.[103,167] Second, some babies develop bronchiolitis in the absence of measurable acute-phase serum antibodies.[167] Third, prospective investigations in Houston, Texas, have revealed an association between low umbilical cord antibody titer and the subsequent early development of serious lower tract disease caused by RSV infection, a finding that argues for a protective, rather than a pathogenic, role for serum antibodies.[65] Finally, a recent study of high-titered RSV intravenous immune globulin in high-risk infants demonstrated a clear protective effect. Infants receiving large doses of the product at 4-week intervals developed fewer lower respiratory infections and had fewer hospitalizations and fewer days in the hospital than those receiving a control preparation.[72]

Proliferative lymphocyte responses *in vitro* have been measured by Welliver *et al.*[214] after natural RSV infection of infants. These workers found increased stimulation indices early in convalescence among those with bronchiolitis and asthma, in contrast to little or no increase in activity at the same stage of illness among patients with upper respiratory infection or pneumonia. The suggestion was made that wheezing illness might therefore be related to lymphocyte-mediated delayed hypersensitivity in response to the viral infection. Cranage and Gardner[37] observed a tendency to higher lymphocyte stimulation indices during convalescence in infants with more severe bronchiolitis. In addition, extensive studies in the mouse model of RSV infection have indicated that CD4- and CD8-bearing T lymphocytes contribute both to the histopathology of pneumonia and also to clearance of virus.[3,68] In this mouse model, antibody appears to have little role in the normal recovery process.[67]

A possible role for the alveolar macrophage in both release of cytokines, which might enhance the inflammatory response, and induction of beneficial lymphocyte activity has been postulated.[164,165] It appears likely that RSV reaches as far as the alveoli during the course of lower respiratory illness and that macrophages are infected, with possible consequences to both the local histopathology and the immune response.

The idea that antiviral immunoglobulin E (IgE) might be involved in the pathogenesis of RSV infection has received support from the findings of Welliver *et al.*[215,217,218] These workers measured IgE attached to RSV-infected cells or free in nasal secretions during convalescence of infants from RSV infection. The levels of free IgE found were higher in those infants with asthma and wheezing and correlated with the severity of hypoxia during acute disease. Moreover, histamine, presumably released during the interaction of IgE, RSV, and mast cells in the respiratory epithelium, was detected in secretions of patients with bronchiolitis or RSV-induced asthma but not in secretions of patients with either pneumonia or upper respiratory infections. Furthermore, the incidence of subsequent recurrent wheezing episodes was greatest in individuals who had the highest RSV–IgE response at the time of their RSV bronchiolitis. More recent studies by this same group have shown a highly significant correlation between wheezing during RSV infection and the concentrations of eosinophil cationic protein in the airway.[60] Eosinophil cationic protein is a cytotoxic protein found in eosinophil granules and may have a role in the pathogenesis of asthma. The postulate that this pathogenic mechanism is important in bronchiolitis has received further support from the observation that leukotriene C_4 is found in high concentration in the respiratory secretions during RSV infection, particularly when the infection is accompanied by wheezing.[207]

Finally, it has been proposed that local antigen–antibody complexes might form and induce injury during RSV illness.[140] Complement components have been found on the surface of RSV-infected epithelial cells shed from the respiratory tract during infection.[120] Moreover, *in vitro* studies have shown that RSV–antibody complexes efficiently activated oxidative and arachidonic acid metabolism in neutrophils.[45]

It is important to emphasize that although it appears likely that there is an immunologic component to the pathogenesis of RSV bronchiolitis and pneumonia, a significant share of the pathology probably can be attributed to direct cytopathology. Moreover, the particular clinical presentation of bronchiolitis with its resemblance to asthma in older infants is probably a combination of many factors: immunologic mechanisms outlined above, the peculiar tropism of RSV for bronchiolar epithelium, and the particular anatomy of the infant bronchioles, which, because of their small diameter, obstruct easily in the face of necrosis and edema.[102]

During bronchiolitis there is necrosis and occasionally proliferation of the bronchiolar epithelium and destruction of ciliated epithelial cells. A peribronchiolar infiltrate of lymphocytes, plasma cells, and macrophages appears, with migration of the lymphocytes among the mucosal epithelial cells. Submucosal and adventitial tissues become edematous, and secretion of mucus is excessive. These processes cause obstruction of the small bronchioles with either collapse or emphysema of distal portions of the airway. In those instances in which pneumonia occurs, the interalveolar walls thicken as a result of mononuclear cell infiltration, and the alveolar spaces may fill with fluid. There is usually a patchy appearance to these pathological changes even though disease may be widespread.

Recovery from RSV infection is probably mediated largely by the immune system. In normal children, shedding of virus ceases after 1 to 3 weeks. In contrast, children lacking cell-mediated immunity are unable to terminate their infection and may continue to shed virus for many months.[50] Disappearance of virus in immunologically normal infants and children coincides with the appearance of secretory antibody, largely of the IgA class.[139,140,212] In children with leukemia, who frequently excrete virus for prolonged periods of time, the secretory IgA response is weak or absent.[202] Nevertheless, the responsible immune mechanism in recovery is not known.

Both serum and secretory antibodies are made in response to infection, even by very young children,[24,139,155,180,203] although in the latter group titers achieved are low. For example, infected infants 1–8 months of age develop one fifth to one sixth the concentration of serum antibodies to the F or G glycoprotein of RSV as older infants or young children. Also, the development of serum neutralizing antibodies is reduced approximately 75% compared to older infants and young children. This diminution of the immune response in young infants may play a role in the greater severity of disease in this age group.

Respiratory syncytial virus infection frequently fails

to induce detectable levels of interferon in the nasal secretions.[83,134] This is in contrast to both parainfluenza and influenza virus infections, in which nasal interferon levels are found to be high during the peak of infection and correlate with disappearance of virus. It appears that either interferon plays a minor role in recovery from RSV infection or, possibly, RSV is very sensitive to the small amounts of interferon that are induced.

The mechanisms by which the immune system protects against RSV infection and reinfection are not well understood. It is apparent that immunity is only partially effective. Infection of small infants occurs even in the presence of a moderate level of maternally transmitted serum antibodies, and reinfection at all ages is common, occurring in infants sometimes within a few weeks of recovery from primary infection.[8,10]

Protection against reinfection by respiratory viruses has traditionally been attributed to secretory antibodies and other function of the secretory immune system. Young infants usually develop a lower level of RSV secretory antibodies than older individuals, and the antibodies that are produced are insufficient or functionally unable to neutralize the virus *in vitro*.[139,155] This deficiency may contribute to the severity of disease in early life and may be at least partially responsible for the incompleteness of natural immunity following initial infection. As individuals grow older, however, multiple reinfections induce more substantial temporary immunity.[96] Studies in adult volunteers have shown that immunity to induced infection correlates better with the level of nasal neutralizing IgA antibodies than with serum neutralizing antibodies.[145,213] Nevertheless, such antibodies are only partially protective since volunteers could still be infected and become symptomatic, even in their presence in high concentration.

Evidence that circulating IgG antibody has some protective effect in infants and children has been reviewed above. There are extensive corroborative data from the cotton rat model to show that a sufficient level of circulating antibody protects the lower, but not the upper, respiratory tract against experimental RSV infection.[173,174] On the other hand, prior RSV infection protects both the upper and lower tracts for 6–12 months.

In summary, the precise role of immunity in RSV infection still requires additional clarification. It is likely that it is involved in recovery and that the cellular immune system is most important in this regard. It is also likely that both serum and secretory antibodies play a role in protection against infection and reinfection, but this protection is incomplete and is often overcome under natural conditions. Local immunity appears to be the most impor-

tant in protecting the upper respiratory tract, whereas high levels of serum antibodies play a major role in resistance to RSV infection in the lower respiratory tract.

Older children and adults usually develop less serious illness during reinfection with RSV than do young infants undergoing primary infection. In the Chapel Hill longitudinal day-care study, there was a steady reduction in the occurrence of lower respiratory tract illness, middle ear disease, and fever during second and third infections with RSV.[99]

In the early studies of RSV infection in young infants, as well as in more recent studies, it was observed that the serum antibody response of young infants to this virus was impaired.[24,168,183,203] Also, the local secretory IgA antibodies produced in response to RSV infection during early infancy are reduced in amount and unable to neutralize this virus effectively.[139] A delayed, decreased, or inefficient immunologic response to infection could contribute to the increased severity of RSV disease during infancy, as well as the inadequate defense against reinfection. Factors that might contribute to such an impaired response are immunologic immaturity and immunologic suppression produced by maternally derived serum antibodies. If infection of monocytes and macrophages is important in the pathogenesis of disease,[165] it may be that the capacity of RSV to impair production of various monokines (ICAM-1 and LFA-1),[185] and to stimulate interleukin 1 inhibitor production[184] also predisposes to a diminished curative and subsequently protective immune response.

7. Patterns of Host Response

7.1. Symptoms

The effect of RSV on the host ranges from inapparent infection to severe respiratory tract disease such as bronchiolitis or pneumonia. Response to infection is influenced by age, underlying cardiopulmonary abnormalities, and immunologic status. Bronchiolitis and pneumonia caused by RSV occur in most instances during the first year of life, with the most severe disease requiring prolonged or intensive hospitalization in children under 6 weeks of age.[69] Initially in an RSV infection in infancy and early childhood there are involvement and inflammation of the mucous membranes of the nose and nasopharynx. Paranasal and Eustachian tube obstruction and otitis media occur commonly. In a significant proportion of infections in early infancy, and in a minority of instances in later life, there is an extension of the inflammatory process into the trachea, bronchioles, and the parenchyma of the lung. In young infants, there is a tendency for bronchiolitis and pneumonia to develop, and apnea may occur in a significant proportion of such illness.[19,143,199] Bronchiolar obstruction results in focal areas of atelectasis or emphysema. The clinical picture may be dominated by obstructive bronchiolitis or pneumonitis or a combination of the two patterns. Inflammation and edema of the larynx may occur, but this condition is less frequently observed than bronchiolitis or pneumonia. Older children or adults who undergo reinfection with RSV often develop mild upper respiratory tract illness indistinguishable from that produced by the multitude of other viruses that cause the common cold. However, more serious disease involving the lower respiratory tract may occur and require bed rest.[51,86] This type of febrile response, resembling influenza in severity, occurs more commonly than previously appreciated and is associated with an increase in total airway resistance and altered reactivity to carbachol challenge that lasts at least 8 weeks.[86]

There is some evidence that RSV infection in the elderly is a relatively common cause of febrile bronchitis and severe or even fatal pneumonia.[51,61] A number of RSV outbreaks in nursing homes have been studied. In most instances the evidence for infection has been serological. The importance of such infections in comparison to those with influenza virus has not been completely clarified, but in one study of a concurrent RSV–influenza outbreak, the illness were indistinguishable on clinical grounds.[46,133] In other studies RSV disease in the elderly was also judged to be equivalent in severity to illness caused by influenza A virus.[20,150,192]

It has been suggested that exacerbation of chronic bronchitis and emphysema in adults is frequently associated with reinfection with RSV.[191] Recent studies have failed to yield evidence to support this view. However, exacerbations of wheezing in asthmatic children are often associated with RSV infection.[135]

The long-term prognosis of RSV (or other viral) bronchiolitis and pneumonia during infancy has been the subject of numerous studies. Although there is some disagreement about details, it seems quite clear that a child who has apparently recovered completely from such an illness may retain measurable and symptomatic respiratory abnormalities for many years. One investigation of 18 infants who were hospitalized for bronchiolitis (one half of which was caused by RSV) found only one with normal pulmonary function tests 1 year later.[196] Ten had increased airway resistance, two had increased thoracic gas volumes, and five had both abnormalities. Another study of 23 children 10 years after bronchiolitis found that

although all were symptom-free (a criterion for admission to the study), 20 had some measurable physiological abnormality of lung function or arterial blood gases.[119] Recurrent wheezing has also been observed to be common, occurring in 30 to 50% of infants who were previously hospitalized for RSV bronchiolitis,[177,181,188,220] but wheezing was rarely of great severity. A large case-control study of 200 children and 200 controls 7 years after bronchiolitis or pneumonia (100 of these were caused by RSV) showed overall a significant tendency to recurrent cough, wheezing, school absenteeism, asthma, and bronchitis in the previously hospitalized groups.[147] Post-exercise or pharmacologically induced bronchial lability was also increased despite an absence of symptoms.[74,188] There have been no studies that have examined a possible link between such illnesses during infancy and chronic pulmonary disease in adults. It is important to note that cause and effect have not been demonstrated between RSV infections and subsequent long-term abnormalities. It is possible that infants with clinically severe RSV infections have some underlying diathesis that predisposes them to both acute bronchiolitis and chronic lower-tract abnormalities.

7.2. Diagnosis

Diagnosis of RSV infection can be made by virus isolation, detection of viral antigens in nasopharyngeal washings by ELISA, detection of antigen in exfoliated cells by immunofluorescence, demonstration of a rise in serum antibodies during convalescence, or some combination of these techniques. Examination of exfoliated cells from the respiratory tract by immunofluorescence provides a highly efficient method for the diagnosis of RSV infection.[57] Virus is present in the nasal and pharyngeal secretions of infected individuals and can be isolated with greatest efficiency when specimens are inoculated directly into cell culture without prior freezing.[18,78,79] Human heteroploid cell cultures (HeLa, HEp-2, etc.) represent the most sensitive host systems for recovery of naturally occurring virus. Characteristic syncytial cytopathic changes usually develop 3–10 days after inoculation of clinical specimens containing RSV. Since sublines of HeLa and HEp-2 cells differ considerably in sensitivity to RSV, it is essential that cultures of a sensitive cell line be used in virus isolation attempts and that this sensitivity be periodically verified.[206] Human embryonic kidney and human diploid fibroblast cultures can also be used for virus isolation, but these are generally less sensitive than the heteroploid cell cultures described above. Except in early infancy, the complement-fixation, ELISA, and tissue culture neutralization techniques are relatively efficient serological procedures for diagnosis of RSV infection. Infants less than 7 months of age develop neutralizing, complement-fixing, and ELISA antibodies less often following RSV infection than do older individuals.[24,168,182,183] In young infants, infection is most accurately determined by virus isolation, ELISA, or immunofluorescence.

8. Control and Prevention

8.1. Treatment

Good supportive care is fundamental to the treatment of RSV infections and has clearly made the major impact on mortality from severe bronchiolitis or pneumonia over the past decade. While this is difficult to document precisely, it is illustrated by the difference between early and recent mortality rates in infants and children with congenital heart disease and pulmonary hypertension, namely a fall from 44%[130] to 9.4%.[158] Advances in general pediatric intensive care may well be similarly responsible for the overall low mortality rates in hospitalized children with RSV infections.[71,142,148,158,199] On the other hand, it has been difficult to show an effect of antiviral chemotherapy in the form of aerosolized ribavirin on mortality.

Only one antiviral compound has shown promise as a suitable drug for treatment of RSV disease in infants. Ribavirin (1-α-D-ribofuranosyl-1,2,4-triazole-3-carboxyamine) has activity in tissue culture[105] and in the cotton rat model.[106] When administered by mouth, this drug has mild hepatic and bone marrow toxicity. For RSV infections, it is delivered almost continuously, i.e., 18–20 hr a day, by small-particle aerosol, by which route it appears to have little or no systemic toxicity. Trials of aerosolized ribavirin for experimental RSV infection of adult volunteers[91] and naturally occurring RSV lower respiratory tract disease of infants[73,88,89,189] have indicated a modest but significant beneficial effect on both virus shedding and the clinical illness. Enthusiasm for ribavirin has been dampened by fears of risk to family members and caretakers who are exposed to aersolized drug,[63] and alterations in the schedule have been introduced to minimize this possibility.[43] The evidence of the efficacy of these regimens, however, is limited to data from animal models.[62] Although ribavirin treatment decreases the amount of virus shed by patients and increases the patients' level of oxygenation, there is no evidence that therapy decreases mortality or duration of hospitalization or diminishes the need for supportive therapies.[208]

Recent studies in experimental animals suggest that immunotherapy may also prove useful in the treatment of serious RSV lower respiratory tract disease in infants and young children. Initially, parenteral inoculation of cotton rat RSV convalescent serum or purified human intravenous Ig (IVIG) containing a high titer of RSV neutralizing antibodies and suitable for intravenous administration was shown to be effective in prophylaxis of RSV infection of the respiratory tract in cotton rats.[172,174] Subsequently a therapeutic effect of parenterally administered RSV antibodies was demonstrated in cotton rats and owl monkeys.[94,172] IVIG inoculated parenterally at the height of RSV infection in these animals effected a $10^{-1.7}$ to $10^{-2.7}$ reduction in level of pulmonary virus. Approximately twice as high a concentration of serum neutralizing antibodies was required to achieve a therapeutic effect than a prophylactic effect. Based on these observations, a controlled clinical trial of the therapeutic efficacy of intravenously administered IVIG was performed in infants and young children with RSV pneumonia or bronchiolitis.[95] Treatment with 2 g/kg as a single dose effected a significant reduction in amount of RSV shed within 48 hr and a significant increase in oxygenation within 24 hr. The mean duration of hospitalization, however, was not reduced for the treated patients.

Human intravenous Ig is also effective therapeutically when inoculated directly or delivered by aerosol into the lungs of cotton rats at the height of an RSV infection.[171] Topical administration of IVIG offers an advantage over parenteral inoculation because the therapeutic dose is significantly lower.

8.2. Immunoprophylaxis

Two of the ten proteins of RSV have been shown to play the major role in protective immunity to RSV infection in experimental animals.[41,47,162,210] Both of these antigens are glycoproteins that are displayed on the surface of the lipid outer membrane of the virus.[47,210] The G glycoprotein is thought to mediate attachment of virus to the host cell, whereas the fusion or F glycoprotein is probably responsible for virus penetration and fusion of infected cells with contiguous cells to produce syncytia, a process that gives the virus its name. Certain monoclonal antibodies to the F or G glycoprotein neutralize infectivity of RSV *in vitro* and provide passive protection to RSV infection in cotton rats or mice.[211]

The unusual age distribution of RSV bronchiolitis and pneumonia makes it important to immunize as soon after birth as possible. If a vaccine is to have a major impact on RSV bronchiolitis and pneumonia, it must be capable of stimulating effective host resistance in the first 2–3 months of life, since the peak incidence of disease occurs early in the first year.

Previous attempts to develop a safe and effective

RSV vaccine had met with failure. A formalin-inactivated virus vaccine tested 20 years ago failed to protect against RSV infection or disease.[29,52,118,123] Instead, disease was enhanced during subsequent infection by RSV. The basis for this potentiation of disease by formalin-inactivated virus has never been fully understood, although efforts to mimic the effect in rodent and primate models have shed some light on the problem.[115,175] Rodent models have been the best studied. When cotton rats previously inoculated with formalin-inactivated RSV were challenged intranasally with live RSV, an enhancement of pulmonary histopathology was observed beginning 24 hr after infection and reaching a maximum by the fourth day.[139] Histologically, the lesions resembled an experimental pulmonary Arthus reaction. It is possible that formalin-treated RSV stimulates an unbalanced immune response in which an unusually large proportion of the induced antibodies are directed against nonprotective epitopes on the viral surface glycoproteins. It is also possible that an enhanced, unbalanced cell-mediated immune response plays a role in formalin-inactivated RSV potentiation of disease. Shortly after the vaccine trials 20 years ago, a study of recipients of the inactivated preparation who had not yet been infected with RSV revealed that the vaccine induced a heightened cell-mediated immune response as assayed *in vitro* by lymphocyte transformation.[124] This response was observed to be significantly greater than that which occurred after RSV infection.

After the failure of the formalin-inactivated RSV vaccine, efforts to develop immunoprophylaxis were directed toward the construction of attenuated RSV mutants that could be used in a live vaccine administered topically. Temperature sensitive (*ts*) mutants of RSV were developed, but these viruses could not be used as a live vaccine because of overattenuation or genetic instability, which allowed the mutants to lose their *ts* phenotype.[101,122,219] Immunization by parenteral inoculation of live RSV has also been evaluated, but a field trial indicated this approach to be ineffective.[11] Recent efforts at development of a subunit vaccine have been directed at an immunoaffinity-purified F-glycoprotein vaccine, as well as antigens produced by bioengineering. Immunization of cotton rats with the F-glycoprotein induced complete resistance to RSV infection in the lungs.[209] While some enhanced pulmonary pathology was seen following challenge with RSV,[156] it appeared to be less than and different from the findings following formalin-inactivated vaccines.[100] Similarly, a fusion protein that included the antigenically important parts of both F and G has shown promise as a vaccine in animal models.[17] Clinical trials with the purified F-glycoprotein vaccine have been conducted in children 18–36 months old who have experienced prior RSV infection, and protection in this low-risk group of older children appears to have been induced.[204]

Recombinant technology has been used to develop several live vaccine candidates containing the F and G glycoproteins. Initially, complete cDNA copies of the genes for these glycoproteins were cloned and sequenced.[7,32,42,186] This information was used to construct vaccinia recombinant viruses containing RSV glycoprotein gene sequences, and these recombinants were then evaluated for their potential usefulness in immunoprophylaxis.[41,162,198] Cotton rats infected intradermally with vaccinia–F or vaccinia–G developed a high level of glycoprotein-specific and neutralizing antibodies. Early trials of vaccinia recombinants expressing F and G glycoproteins in chimpanzees, however, have been disappointing. In this species, antibody responses were modest and protection on challenge was marginal.[34]

In view of the real and theoretical difficulties of active immunization against RSV infections in infancy, and in an effort to mimic the natural protection afforded to infants by placentally transferred maternal IgG antibody, passively administered antibody has also received considerable attention as a preventive measure. Successful prophylaxis was achieved recently with a preparation of IVIG that had been prepared from plasma selected for its high mouse-protective antibody titers.[187] The incidence of lower respiratory infection by RSV and both the number and duration of hospital admissions were significantly reduced in the children who received high doses of this preparation.[72]

Research on vaccines and other improved immunoprophylactic agents is continuing in several areas. Improved preparations of antibody, either monoclonal or manufactured from combinatorial libraries, are likely to appear and undergo trials. In addition, both component and live attenuated vaccines are under development. RSV infections remain an important target for prevention because of their importance in the health of children throughout the world.

9. Unresolved Problems

The major unresolved problem in our understanding of RSV concerns the mechanism whereby the virus produces its most frequent and most severe disease manifestations during the first few months of life. The role and relative importance of age, immunologic maturity, immunosuppression, and immunopathology in serious RSV lower respiratory disease of early infancy must be assessed in a more definitive manner than in past studies. Greater insight in these areas is required before the final strategy for prevention or therapy of RSV disease can be planned.

In addition, we need a more complete understanding of the manner in which RSV evades the host's defense mechanisms and reinfects the same person with appreciable frequency over a period of years. Greater comprehension in this area may allow us to manipulate the host's defenses so that the reinfection can be diminished and the quantity of virus available for infection of susceptible infants can be decreased significantly.

10. References

1. ABMAN, S. H., OGLE, J. W., BUTLER-SIMON, N., RUMACK, C. M., AND ACCURSO, F. J., Role of respiratory syncytial virus in early hospitalizations for respiratory distress of young infants with cystic fibrosis, *J. Pediatr.* **113**:826–830 (1988).
2. ÅKERLIND, B., AND NORRBY, E., Occurrence of respiratory syncytial virus subtypes A and B strains in Sweden, *J. Med. Virol.* **19**:241–247 (1986).
3. ALWAN, W. H., RECORD, F. M., AND OPENSHAW, P. J., CD4+ T cells clear virus but augment disease in mice infected with respiratory syncytial virus. Comparison with the effects of CD8+ T cells, *Clin. Exp. Immunol.* **88**(3):527–536 (1992).
4. ANDERSON, L. J., HIERHOLZER, J. C., TSOU, C., HENDRY, R. M., FERNIE, B. F., STONE, Y., AND MCINTOSH, K., Antigenic characterization of respiratory syncytial virus strains with monoclonal antibodies, *J. Infect. Dis.* **151**:627–633 (1985).
5. ANDERSON, L. J., PARKER, R. A., AND STRIKAS, R. L., Association between respiratory syncytial virus outbreaks and lower respiratory tract deaths of infants and young children, *J. Infect. Dis.* **161**(4):640–664 (1990).
6. AVILA, M., SALOMON, H., CARBALLAL, G., EBEKIAN, B., WOYSKOVSKY, N., CERQUEIRO, M. C., AND WEISSENBACHER, M., Isolation and identification of viral agents in Argentinian children with acute lower respiratory tract infection, *Rev. Infect. Dis.* **12**(Suppl. 8):S974–981 (1990).
7. BALL, L. A., YOUNG, K. K. Y., ANDERSON, K., COLLINS, P. L., AND WERTZ, G. W., Expression of the major glycoprotein G of human respiratory syncytial virus from recombinant vaccinia virus vectors, *Proc. Natl. Acad. Sci. USA* **83**:246–250 (1986).
8. BEEM, M., Repeated infections with respiratory syncytial virus, *J. Immunol.* **98**:1115–1122 (1967).
9. BEEM, M., EGERER, R., AND ANDERSON, J., Respiratory syncytial virus neutralizing antibodies in persons residing in Chicago, Illinois, *Pediatrics* **34**:761–770 (1964).
10. BEEM, M., WRIGHT, F. H., HAMRE, D., EGERER, R., AND OEHME, M., Association of the chimpanzee coryza agent with acute respiratory disease in children, *N. Engl. J. Med.* **263**:523–530 (1960).
11. BELSHE, R. B., VAN VORIS, L. P., AND MUFSON, M. A., Parenteral administration of live respiratory syncytial virus vaccine: Results of a field trial, *J. Infect. Dis.* **145**:311–319 (1982).
12. BERKOVICH, S., AND TARANKO, L., Acute respiratory illness in the premature nursery associated with respiratory syncytial virus infections, *Pediatrics* **34**:753–760 (1964).
13. BERTHIAUME, L., JONCAS, J., AND PAVILANIS, V., Comparative structure, morphogenesis and biological characteristics of the respiratory syncytial (RS) virus and the pneumonia virus of mice, *Arch. Ges. Virusforsch.* **45**:39–51 (1974).
14. BRANDT, C. D., KIM, H. W., ARROBIO, J. O., JEFFRIES, B. C., WOOD, S. C., CHANOCK, R. M., AND PARROTT, R. H., Epidemiology of respiratory syncytial virus infection in Washington, D.C., *Am. J. Epidemiol.* **98**:355–364 (1973).
15. BRANDT, C. D., PARROTT, R. H., PATRICK, J. R., KIM, H. W., ARROBIO, J. O., CHANDRA, R., JEFFRIES, B. C., AND CHANOCK, R. M., Epidemiology—SIDS and viral respiratory disease in metropolitan Washington D.C., in: *S.I.D.S. Proceedings of the Francis E. Camps International Symposium on Sudden and Unexpected Death in Infancy* (R. R. ROBINSON, ED.), pp. 117–129, Canadian Foundation for the Study of Infant Death, 1974.
16. BRAWLEY, R. L., Infection control practices for preventing respiratory syncytial virus infections, *Infect. Control Hosp. Epidemiol.* **9**:105–108 (1988).
17. BRIDEAU, R. J., WALTERS, R. R., STIER, M. A., AND WATHEN, M. W., Protection of cotton rats against human respiratory syncytial virus by vaccination with a novel chimeric FG glycoprotein, *J. Gen. Virol.* **70**:2637–2644 (1989).
18. BROMBERG, K., DAIDONE, B., CLARKE, L., AND SIERRA, M. F., Comparison of immediate and delayed inoculation of HEp-2 cells for isolation of respiratory syncytial virus, *J. Clin. Microbiol.* **20**:123–124 (1984).
19. BRUHN, F. W., MOKROHISKY, S. T., AND MCINTOSH, K., Apnea associated with respiratory syncytial virus infection in young infants, *J. Pediatr.* **90**:382–386 (1977).
20. CAPEWELL, A., INGLIS, J. M., AND WILLIAMSON, J., Respiratory syncytial virus infection in the elderly, *Br. Med. J.* **288**:235–236 (1984).
21. CHANDWANI, S., BORKOWSKY, W., KRASINSKI, K., LAWRENCE, R., AND WELLIVER, R., Respiratory syncytial virus infection in human immunodeficiency virus-infected children, *J. Pediatr.* **117**:251–254 (1990).
22. CHANOCK, R., AND FINBERG, L., Recovery from infants with respiratory illness of a virus related to chimpanzee coryza agent (CCA), *Am. J. Hyg.* **66**:291–300 (1957).
23. CHANOCK, R. M., KAPIKIAN, A. Z., MILLS, J., KIM, H. W., AND PARROTT, R. H., Influence of immunological factors in respiratory syncytial virus disease of the lower respiratory tract, *Arch. Environ. Health* **21**:347–355 (1970).
24. CHANOCK, R. M., KIM, H. W., VARGOSKO, A. J., DELEVA, A., JOHNSON, K. M., CUMMING, C., AND PARROTT, R. H., Respiratory syncytial virus: I. Virus recovery and other observations during 1960 outbreak of bronchiolitis pneumonia, and minor respiratory diseases in children, *J.A.M.A.* **176**:647–653 (1961).
25. CHANOCK, R. M., AND PARROTT, R. H., Acute respiratory disease in infancy and childhood: Present understanding and prospects for prevention, *Pediatrics* **36**:21–39 (1965).
26. CHANOCK, R. M., PARROTT, R. H., JOHNSON, K. M., MUFSON, M. A., AND KNIGHT, V., Biology and ecology of two major lower respiratory tract pathogens—RS virus and Eaton PPLO, in: *Perspectives in Virology III* (M. POLLARD, ED.), pp. 257–281, Burgess, Minneapolis, 1963.
27. CHANOCK, R. M., PARROTT, R. H., KAPIKIAN, A. Z., KIM, H. W., AND BRANDT, C. D., Possible role of immunological factors in pathogenesis of RS virus lower respiratory tract disease, in: *Perspectives in Virology VI* (M. POLLARD, ED.), pp. 125–139, Academic Press, New York, 1968.
28. CHANOCK, R., ROIZMAN, B., AND MYERS, R., Recovery from

infants with respiratory illness of a virus related to chimpanzee coryza agent (CCA), *Am. J. Hyg.* **66:**281–290 (1957).

29. CHIN, J., MAGOFFIN, R. L., SHEARER, L. A., SCHIEBLE, J. H., AND LENNETTE, E. H., Field evaluation of a respiratory syncytial virus vaccine and a trivalent parainfluenza virus vaccine in a pediatric population, *Am. J. Epidemiol.* **89:**449–463 (1969).

30. COATES, H. V., ALLING, D. W., AND CHANOCK, R. M., An antigenic analysis of respiratory syncytial virus isolates by a plaque reduction neutralization test, *Am. J. Epidemiol.* **83:**299–313 (1966).

31. COLLINS, P. L., DICKENS, L. E., BUCKLER-WHITE, A., OLMSTED, R. A., SPRIGGS, M. K., CAMARGO, E., AND COELINGH, K. V. W., Nucleotide sequences for the gene junctions of human respiratory syncytial virus reveal distinctive features of intergenic structure and gene order, *Proc. Natl. Acad. Sci. USA* **83:**4594–4598 (1986).

32. COLLINS, P. L., HUANG, Y. T., AND WERTZ, G. W., Identification of a tenth mRNA of respiratory syncytial virus and assignment of polypeptides to the 10 viral genes, *J. Virol.* **49:**572–578 (1984).

33. COLLINS, P. L., HUANG, Y. T., AND WERTZ, G. W., Nucleotide sequence of the gene encoding the fusion (F) glycoprotein of human respiratory syncytial virus, *Proc. Natl. Acad. Sci. USA* **81:**7683–7687 (1984).

34. COLLINS, P. L., PURCELL, R. H., LONDON, W. T., LAWRENCE, L. A., CHANOCK, R. M., AND MURPHY, B. R., Evaluation in chimpanzees of vaccinia virus recombinants that express the surface glycoproteins of human respiratory syncytial virus, *Vaccine* **8**(2):164–168 (1990).

35. COLLINS, P. L., AND WERTZ, G. W., cDNA cloning and transcriptional mapping of nine polyadenylated RNAs encoded by the genome of human respiratory syncytial virus, *Proc. Natl. Acad. Sci. USA* **80:**3208–3212 (1983).

36. COLLINS, P. L., AND WERTZ, G. W., Gene products and genome organization of human respiratory syncytial (RS) virus, in: *Modern Approaches to Vaccines: Molecular and Chemical Basis of Resistance to Viral, Bacterial, and Parasitic Diseases* (R. A. LERNER AND R. M. CHANOCK, EDS.), pp. 297–301, Cold Spring Harbor Laboratory, New York, 1985.

37. CRANGE, M. P., AND GARDNER, P. S., Systemic cell-mediated and antibody responses infants with respiratory syncytial virus infections, *J. Med. Virol.* **5:**161–170 (1980).

38. DOGGETT, J. E., AND TAYLOR-ROBINSON, D., Serological studies with respiratory syncytial virus, *Arch. Ges. Virusforsch.* **15:**601–608 (1965).

39. DOMURAT, F., ROBERTS, N. J., JR., WALSH, E. E., AND DAGAN, R., Respiratory syncytial virus infection of human mononuclear leukocytes *in vitro* and *in vivo*, *J. Infect. Dis.* **152:**895–902 (1985).

40. DOWNHAM, M. A. P. S., GARDNER, P. S., MCQUILLIN, J., AND FERRIS, J. A. J., Role of respiratory viruses in childhood mortality, *Br. Med. J.* **1:**235–239 (1975).

41. ELANGO, N., PRINCE, G. A., MURPHY, B. R., VENKATESAN, S., CHANOCK, R. M., AND MOSS, B., Resistance to human respiratory syncytial virus (RSV) infection induced by immunization of cotton rats with a recombinant vaccinia virus expressing the RSV G glycoprotein, *Proc. Natl. Acad. Sci. USA* **83:**1906–1910 (1986).

42. ELANGO, N., SATAKE, M., COLIGAN, J. E., NORRBY, E., CAMARGO, E., AND VENKATESAN, S., Respiratory syncytial virus fusion glycoprotein: Nucleotide sequence of mRNA, identification of cleavage activation site and amino acid sequence of N-terminus of F_1 subunit, *Nucleic Acids Res.* **13:**1559–1574 (1985).

43. ENGLUND, J. A., PIEDRA, P. A. JEFFERSON, L. S., WILSON, S. Z., TABER, L. H., AND GILBERT, B. E., High-dose, short-duration ribavirin aerosol therapy in children with suspected respiratory syncytial virus infection, *J. Pediatr.* **117:**313–320 (1990).

44. ENGLUND, J. A., SULLIVAN, C. J., JORDAN, M. C., DEHNER, L. P., VERCELLOTTI, G. M., AND BALFOUR, H. H., JR., Respiratory syncytial virus infection in immunocompromised adults, *Ann. Intern. Med.* **109:**203–208 (1988).

45. FADEN, H., KAUL, T. M., AND OGRA, P. L., Activation of oxidative and arachidonic acid metabolism in neutrophils by respiratory syncytial virus antibody complexes: Possible role in disease, *J. Infect. Dis.* **148:**110–116 (1983).

46. FALSEY, A. R., WALSH, E. E., AND BETTS, R. F., Serologic evidence of respiratory syncytial virus infection in nursing home patients, *J. Infect. Dis.* **162**(2):568–569 (1990).

47. FERNIE, B. F., COTE, P. J., AND GERIN, J. L., Classification of hybridomas to respiratory syncytial virus glycoproteins, *Proc. Soc. Exp. Biol. Med.* **171:**266–271 (1982).

48. FERNIE, B. F., AND GERIN, J. L., Immunochemical identification of viral and nonviral proteins of the respiratory syncytial virus virion, *Infect. Immun.* **37:**243–249 (1982).

49. FISHAUT, M., SCHWARTZMAN, J. D., MCINTOSH, K., AND MASTOW, S. R., Behavior of respiratory syncytial virus in infant porcine tracheal organ culture, *J. Infect. Dis.* **138:**644–649 (1978).

50. FISHAUT, M., TUBERGEN, D., AND MCINTOSH, K., Cellular response to respiratory viruses with particular reference to children with disorders of cell-mediated immunity, *J. Pediatr.* **96:**179–186 (1980).

51. FRANSEN, H., GORAN, S., FORSGREN, M., HEIGL, Z., WOLONTIS, S., SVEDMYR, A., AND TUNEVALL, G., Acute lower respiratory illness in elderly patients with respiratory syncytial virus infection, *Acta Med. Scand.* **182:**323–330 (1967).

52. FULGINITI, V. A., ELLER, J. J., SIEBER, O. F., JOYNER, J. W., MINAMITANI, M., AND MEIKLEJOHN, G., Respiratory virus immunization: I. A field trial of two inactivated respiratory virus vaccines; an aqueous trivalent parainfluenza virus vaccine and an alum-precipitated respiratory syncytial virus vaccine, *Am. J. Epidemiol.* **89:**435–448 (1969).

53. GALA, C. L., HALL, C. B., SCHNABEL, K. C., PINCUS, P. H., BLOSSOM, P., HILDRETH, S. W., BETTS, R. F., AND DOUGLAS, R. G., JR., The use of eye–nose goggles to control nosocomial respiratory syncytial virus infection, *J.A.M.A.* **256:**2706–2708 (1986).

54. GARCIA-BARRENO, B., PALOMO, C., PENAS, C., DELGADO, T., PEREZ-BRENA, P., AND MELERO, J. A., Marked differences in the antigenic structure of human respiratory syncytial virus F and G glycoproteins, *J. Virol.* **63**(2):925–932 (1989).

55. GARDNER, P. S., Respiratory syncytial virus infection: Admissions to hospital in industrial, urban, and rural areas: Report to the Medical Research Council Subcommittee on Respiratory Syncytial Virus Vaccines, *Br. Med. J.* **2:**796–798 (1978).

56. GARDNER, P. S., COURT, S. D. M., BROCKLEBANK, J. T., DOWNHAM, M. A. P. S., AND WEIGHTMAN, D., Virus cross-infection in paediatric wards, *Br. Med. J.* **2:**571–575 (1973).

57. GARDNER, P. S., AND MCQUILLIN, J., Application of immunofluorescent antibody technique in rapid diagnosis of respiratory syncytial virus infection, *Br. Med. J.* **3:**340–343 (1968).

58. GARDNER, P. S., MCQUILLIN, J., AND COURT, S. D. M., Speculation on pathogenesis in death from respiratory syncytial virus infection, *Br. Med. J.* **1:**327–330 (1970).

59. GARDNER, P. S., TURK, D. C., AHERNE, W. A., BIRD, T., HOLDAWAY, M. D., AND COURT, S. D. M., Deaths associated with respiratory tract infection in childhood, *Br. Med. J.* **4:**316–320 (1967).

60. GAROFALO, R., KIMPEN, J. L., WELLIVER, R. C., AND OGRA, P. L., Eosinophil degranulation in the respiratory tract during naturally acquired respiratory syncytial virus infection, *J. Pediatr.* **120**(1): 28–32 (1992).

61. GARVIE, D. G., AND GRAY, J., Outbreak of respiratory syncytial virus infection in the elderly, *Br. Med. J.* **281**:1253–1254 (1980).

62. GILBERT, B. E., WYDE, P. R., AMBROSE, M. W., WILSON, S. Z., AND KNIGHT, V., Further studies with short duration ribavirin aerosol for the treatment of influenza virus infection in mice and respiratory syncytial virus infection in cotton rats, *Antiviral Res.* **17**(1):33–42 (1992).

63. GLADU, J. M., AND ECOBICHON, D. J., Evaluation of exposure of health care personnel to ribavirin, *J. Toxicol. Environ. Health* **38**:1–12 (1989).

64. GLEZEN, W. P., AND DENNY, F. W., Epidemiology of acute lower respiratory disease in children, *N. Engl. J. Med.* **288**:498–505 (1973).

65. GLEZEN, W. P., PAREDES, A., ALLISON, J. E., TABER, L. H., AND FRANK, A. L., Risk of respiratory syncytial virus infection for infants from low-income families in relationship to age, sex, ethnic group, and maternal antibody level, *J. Pediatr.* **98**:708–715 (1981).

66. GLEZEN, W. P., TABER, L. H., FRANK, A. L., AND KASEL, J. A., Risk of primary infection and reinfection with respiratory syncytial virus, *Am. J. Dis. Child.* **140**:543–546 (1986).

67. GRAHAM, B. S., BUNTON, L. A., ROWLAND, J., WRIGHT, P. F., AND KARZON, D. T., Respiratory syncytial virus infection in anti-mu-treated mice, *J. Virol.* **65**(9):4936–4942 (1991).

68. GRAHAM, B. S., BUNTON, L. A., WRIGHT, P. F., AND KARZON, D. T., Role of T lymphocyte subsets in the pathogenesis of primary infection and rechallenge with respiratory syncytial virus in mice, *J. Clin. Invest.* **88**(3):1026–1033 (1991).

69. GREEN, M., BRAYER, A. F., SCHENKMAN, K. A., AND WALD, E. R., Duration of hospitalization in previously well infants with respiratory syncytial virus infection, *Pediatr. Infect. Dis. J.* **8**:601–605 (1989).

70. GRIST, N. R., ROSS, C. A. C., AND STOTT, E. J., Influenza respiratory syncytial virus, and pneumonia in Glasgow, 1962–5, *Br. Med. J.* **1**:456–457 (1967).

71. GROOTHUIS, J. R., GUTIERREZ, K. M., AND LAUER, B. A., Respiratory syncytial virus infection in children with bronchopulmonary dysplasia, *Pediatrics* **82**:199–203 (1988).

72. GROOTHUIS, J. R., SIMOES, E. A. F., LEVIN, M. J., HALL, C. B., LONG, C. E., RODRIGUEZ, W. J., ARROBIO, J., MEISSNER, H. C., FULTON, D. R., WELLIVER, R. C., TRISTRAM, D. A., SIBER, G. R., PRINCE, G. A., VAN RADEN, M., HEMMING, V. G., AND THE RESPIRATORY SYNCYTIAL VIRUS IMMUNE GLOBULIN STUDY GROUP, Prophylactic administration of respiratory syncytial virus immune globulin to high-risk infants and young children, *N. Engl. J. Med.* **329**:1524–1530 (1993).

73. GROOTHUIS, J. R., WOODIN, K. A., KATZ, R., ROBERTSON, A. D., McBRIDE, J. T., HALL, C. B., McWILLIAMS, B. C., AND LAUER, B. A., Early ribavirin treatment of respiratory syncytial viral infection in high-risk children, *J. Pediatr.* **117**(5):792–798 (1990).

74. GURWITZ, D., MINDORFF, G., AND LEVISON, H., Increased incidence of bronchial reactivity in children with a history of bronchiolitis, *J. Pediatr.* **98**:551–555 (1981).

75. HALL, C. B., Nosocomial viral respiratory infections: Perennial weeds on pediatric wards, *Am. J. Med.* **70**:670–676 (1981).

76. HALL, C. B., AND DOUGLAS, R. G., Nosocomial respiratory syncytial virus infections: Should gowns and masks be used? *Am. J. Dis. Child.* **135**:512 (1981).

77. HALL, C. B., AND DOUGLAS, R. G., JR., Modes of transmission of respiratory syncytial virus, *J. Pediatr.* **99**:100–103 (1981).

78. HALL, C. B., DOUGLAS, R. G., JR., AND GEIMAN, J. M., Quantitative shedding patterns of respiratory syncytial virus in infants, *J. Infect. Dis.* **132**:151–156 (1975).

79. HALL, C. B., DOUGLAS, R. G., JR., AND GEIMAN, J. M., Respiratory syncytial virus infectious in infants: Quantitation and duration of shedding, *J. Pediatr.* **89**:11–15 (1976).

80. HALL, C. B., DOUGLAS, R. G., JR., AND GEIMAN, J. M., Possible transmission by fomites of respiratory syncytial virus, *J. Infect. Dis.* **141**:98–102 (1980).

81. HALL, C. B., DOUGLAS, R. G., JR., GEIMAN, J. M., AND MESSNER, M. K., Nosocomial respiratory syncytial virus infections, *N. Engl. J. Med.* **293**:1343–1346 (1975).

82. HALL, C. B., DOUGLAS, R. G., JR., SCHNABEL, K. C., AND GEIMAN, J. M., Infectivity of respiratory syncytial virus by various routes of inoculation, *Infect. Immun.* **33**:779–783 (1981).

83. HALL, C. B., DOUGLAS, R. G., JR., AND SIMONS, R. L., Interferon production in adults with respiratory syncytial viral infection, *Ann. Intern. Med.* **94**:53–55 (1981).

84. HALL, C. B., GEIMAN, J. M., BIGGAR, R., KOTOK, D. I., HOGAN, P. M., AND DOUGLAS, R. G., JR., Respiratory syncytial virus infections within families, *N. Engl. J. Med.* **294**:414–419 (1976).

85. HALL, C. B., GEIMAN, J. M., BIGGAR, R., KOTOK, D. I., HOGAN, P. M., AND DOUGLAS, R. G., JR., Control of nosocomial respiratory syncytial viral infections, *Pediatrics* **62**:728–732 (1978).

86. HALL, C. B., HALL, W. J., AND SPEERS, D. M., Respiratory syncytial virus infection in adults: Clinical, virologic, and serial pulmonary function studies, *Ann. Intern. Med.* **88**:203–205 (1978).

87. HALL, C. B., KOPELMAN, A. E., DOUGLAS, R. G., JR., GEIMAN, J. M., AND MEAGHER, M. P., Neonatal respiratory syncytial virus infection, *N. Engl. J. Med.* **300**:393–396 (1979).

88. HALL, C. B., McBRIDE, J. T., GALA, C. L., HILDRETH, S. W., AND SCHNABEL, K. C., Ribavirin treatment of respiratory syncytial viral infection in infants with underlying cardiopulmonary disease, *J.A.M.A.* **254**:3047–3051 (1985).

89. HALL, C. B., McBRIDE, J. T., WALSH, E. E., BELL, D. M., GALA, C. L., HILDRETH, S., TEN EYCK, L. G., AND HALL, W. J., Aerosolized ribavirin treatment of infants with respiratory syncytial viral infection: A randomized double-blind study, *N. Engl. J. Med.* **308**:1443–1447 (1983).

90. HALL, C. B., POWELL, K. R., MacDONALD, N. E., GALA, C. L., MENEGUS, M. E., SUFFIN, S. C., AND COHEN, H. J., Respiratory syncytial viral infection in children with compromised immune function, *N. Engl. J. Med.* **315**:77–81 (1986).

91. HALL, C. B., WALSH, E. E., HRUSKA, J. F., BETTS, R. F., AND HALL, W. J., Ribavirin treatment of experimental respiratory syncytial virus infection, *J.A.M.A.* **249**:2666–2670 (1983).

92. HAMBLING, M. H., A survey of antibodies to respiratory syncytial virus in the population, *Br. Med. J.* **1**:1223–1225 (1964).

93. HARRINGTON, R. D., HOOTON, T. M., HACKMAN, R. C., STORCH, G. A., OSBORNE, B., GLEAVES, C. A., BENSON, A., AND MEYERS, J. D., An outbreak of respiratory syncytial virus in a bone marrow transplant center, *J. Infect. Dis.* **165**(6):987–939 (1992).

94. HEMMING, V. G., PRINCE, G. A., HORSWOOD, R. L., LONDON, W. T., MURPHY, B. R., WALSH, E. E., FISCHER, G. W., WEISMAN, L. E., BARON, P. A., AND CHANOCK, R. M., Studies of passive immunotherapy for infections of respiratory syncytial virus in the respiratory tract of a primate model, *J. Infect. Dis.* **152**:1083–1087 (1985).

95. HEMMING, V. G., RODRIGUEZ, W., KIM, H. W., BRANDT, C. D., PARROTT, R. H., BURCH, B., PRINCE, G. A., BARON, P. A., FINK, R. J., AND REMAN, G., Intravenous immunoglobulin treatment of

respiratory syncytial virus infections in infants and young children, *Antimicrob. Agents Chemother.* **31**:1882–1886 (1987).

96. HENDERSON, F. W., CLYDE, W. A., JR., COLLIER, A. M., AND DENNY, F. W., The etiologic and epidemiologic spectrum of bronchiolitis in pediatric practice, *J. Pediatr.* **95**:183–190 (1979).

97. HENDERSON, F. W., COLLIER, A. M., CLYDE, W. A., JR., AND DENNY, F. W., Respiratory-syncytial-virus infections, reinfections and immunity: A prospective, longitudinal study in young children, *N. Engl. J. Med.* **300**:530–534 (1979).

98. HENDRY, R. M., BURNS, J. C., WALSH, E. E., GRAHAM, B. S., WRIGHT, P. F., HEMMING, B. G., RODRIGUEZ, W. J., KIM, H. W., PRINCE, G. A., MCINTOSH, K., CHANOCK, R. M., AND MURPHY, B. R., Strain-specific serum antibody responses in infants undergoing primary infection with respiratory syncytial virus, *J. Infect. Dis.* **157**:640–647 (1988).

99. HENDRY, R. M., TALIS, A. L., GODFREY, E., ANDERSON, L. J., FERNIE, B. F., AND MCINTOSH, K., Concurrent circulation of antigenically distinct strains of respiratory syncytial virus during community outbreaks, *J. Infect. Dis.* **153**:291–297 (1986).

100. HILDRETH, S. W., BAGGS, R. R., BROWNSTEIN, D. G., CASTLEMAN, W. L., AND PARADISO, P. R., Lack of detectable enhanced pulmonary histopathology in cotton rats immunized with purified F glycoprotein of respiratory syncytial virus (RSV) when challenged at 3–6 months after immunization, *Vaccine* **11**:615–618 (1993).

101. HODES, D. S., KIM, H. W., PARROTT, R. H., CAMARGO, E., AND CHANOCK, R. M., Genetic alteration in a temperature-sensitive mutant of respiratory syncytial virus after replication *in vivo*, *Proc. Soc. Exp. Biol. Med.* **145**:1158–1164 (1974).

102. HOGG, J. C., WILLIAMS, J., RICHARDSON, J. B., MACKLEM, P. J., AND THURLBECK, M. B., Age as a factor in the distribution of lower-airway conductance and in the pathologic anatomy of obstructive lung disease, *N. Engl. J. Med.* **282**:1283–1287 (1970).

103. HOLBERG, C. J., WRIGHT, A. L., MARTINEZ, F. D., RAY, C. G., TAUSSIG, L. M., AND LEBOWITZ, M. D., Risk factors for respiratory syncytial virus-associated lower respiratory illnesses in the first year of life, *Am. J. Epidemiol.* **133**(11):1135–1151 (1991).

104. HORTAL, M., MOGDASYY, C., RUSSI, J. C., DELEON, C., AND SUAREZ, A., Microbial agents associated with pneumonia in children from Uruguay, *Rev. Infect. Dis.* **12**:S915–922 (1990).

105. HRUSKA, J. F., BERNSTEIN, J. M., DOUGLAS, R. G., AND HALL, C. B., Effects of ribavirin on RSV *in vitro*, *Antimicrob. Agents Chemother.* **17**:770–775 (1980).

106. HRUSKA, J. F., MORROW, P. E., SUFFIN, S. C., AND DOUGLAS, R. G., JR., *In vivo* inhibition of respiratory syncytial virus by ribavirin, *Antimicrob. Agents Chemother.* **21**:125–130 (1982).

107. HUQ, F., RAHMAN, M., NAHAR, N., ALAM, A., HAQUE, M., SACK, D. A., BUTLER, T., AND HAIDER, R., Acute lower respiratory tract infection due to virus among hospitalized children in Dhaka, Bangladesh, *Rev. Infect. Dis.* **12**(Suppl. 8):S982–987 (1990).

108. INABA, Y., TANAKA, Y., SATO, K., OMORI, T., AND MATUMOTO, M., Bovine respiratory syncytial virus: Studies on an outbreak in Japan, 1968–1969, *Jpn. J. Microbiol.* **16**:373–383 (1972).

109. JACOBS, J. W., AND EDGINGTON, N., Isolation of respiratory syncytial virus from cattle in Britain, *Vet. Rec.* **87**:694 (1971).

110. JACOBS, J. W., PEACOCK, D. B., CORNER, B. D., CAUL, E. O., AND CLARKE, S. K. R., Respiratory syncytial and other viruses associated with respiratory disease in infants, *Lancet* **1**:871–876 (1971).

111. JENNINGS, R., Adenovirus, parainfluenza virus and respiratory syncytial virus antibodies in the sera of Jamaicans, *J. Hyg.* **70**:523–529 (1972).

112. JOHNSON, K. M., CHANOCK, R. M., RIFKIND, D., KRAVETZ, H. M., AND KNIGHT, V., Respiratory syncytial virus. IV. Correlation of virus shedding, serologic response, and illness in adult volunteers, *J.A.M.A.* **176**:663–667 (1961).

113. JOHNSON, R. A., PRINCE, G. A., SUFFIN, S. C., HORSWOOD, R. L., AND CHANOCK, R. M., Respiratory syncytial virus infection in cyclophosphamide-treated cotton rats, *Infect. Immun.* **37**:369–373 (1982).

114. JONCAS, J. H., BERTHIAUME, L., WILLIAMS, R., BEAUDRY, P., AND PAVILANIS, V., Diagnosis of viral respiratory infections by electron microscopy, *Lancet* **1**:956–959 (1969).

115. KAKUK, T. J., SOIKE, K., BRIDEAU, R. J., ZAYA, R. M., COLE, S. L., ZHANG, J. Y., ROBERTS, E. D., WELLS, P. A., AND WATHEN, M. W., A human respiratory syncytial virus (RSV) primate model of enhanced pulmonary pathology induced with a formalin-inactivated RSV vaccine but not a recombinant FG subunit vaccine, *J. Infect. Dis.* **167**(3):553–561 (1993).

116. KALICA, A. R., WRIGHT, P. F., HETRICK, F. M., AND CHANOCK, R. M., Electron microscopic studies of respiratory syncytial temperature-sensitive mutants, *Arch. Ges. Virusforsch.* **41**:248–258 (1973).

117. KAPIKIAN, A. Z., BELL, J. A., MASTROTA, F. M., JOHNSON, K. M., HUEBNER, R. J., AND CHANOCK, R. M., An outbreak of febrile illness and pneumonia associated with respiratory syncytial virus infection, *Am. J. Hyg.* **74**:234–248 (1961).

118. KAPIKIAN, A. Z., MITCHELL, R. H., CHANOCK, R. M., SHVEDOFF, R. A., AND STEWART, C. E., An epidemiologic study of altered clinical reactivity to respiratory syncytial (RS) virus infection in children previously vaccinated with an inactivated RS virus vaccine, *Am. J. Epidemiol.* **89**:405–421 (1969).

119. KATTAN, M., KEENS, T. G., LAPIERRE, J.-G., LEVISON, H., BRYAN, A. C., AND REILLY, B. J., Pulmonary function abnormalities in symptom-free children after bronchiolitis, *Pediatrics* **59**:683–688 (1977).

120. KAUL, T. N., WELLIVER, R. C., AND OGRA, P. L., Appearance of complement components and immunoglobulins on nasopharyngeal epithelial cells following naturally acquired infection with respiratory syncytial virus, *J. Med. Virol.* **9**:149–158 (1982).

121. KIM, H. W., ARROBIO, J. O., BRANDT, C. D., JEFFRIES, B. C., PYLES, G., REID, J. L., CHANOCK, R. M., AND PARROTT, R. H., Epidemiology of respiratory syncytial virus infection in Washington, D.C.: I. Importance of the virus in different respiratory tract disease syndromes and temporal distribution of infection, *Am. J. Epidemiol.* **98**:216–225 (1973).

122. KIM, H. W., ARROBIO, J. O., BRANDT, C. D., WRIGHT, P., HODES, D., CHANOCK, R. M., AND PARROTT, R. H., Safety and antigenicity of temperature sensitive (TS) mutant respiratory syncytial virus (RSV) in infants and children, *Pediatrics* **52**:56–63 (1973).

123. KIM, H. W., CANCHOLA, J. G., BRANDT, C. D., PYLES, G., CHANOCK, R. M., JENSEN, K., AND PARROTT, R. H., Respiratory syncytial virus disease in infants despite prior administration of antigenic inactivated vaccine, *Am. J. Epidemiol.* **89**:422–434 (1969).

124. KIM, H. W., LEIKIN, S. L., ARROBIO, J., BRANDT, C. D., CHANOCK, R. M., AND PARROTT, R. H., Cell-mediated immunity to respiratory syncytial virus induced by inactivated vaccine or by infection, *Pediatr. Res.* **10**:75–78 (1976).

125. KRASINSKI, K., Severe respiratory syncytial virus infection: Clinical features, nosocomial acquisition and outcome, *Pediatr. Infect. Dis.* **4**:250–257 (1985).

126. KRAVETZ, H. M., KNIGHT, V., CHANOCK, R. M., MORRIS, J. A., JOHNSON, K. M., RIFKIND, D., AND UTZ, J. P., Respiratory syncytial virus. III. Production of illness and clinical observations in adult volunteers, *J.A.M.A.* **176**:657–667 (1961).

127. LECLAIR, J. M., FREEMAN, J., SULLIVAN, B. F., CROWLEY, C. M., AND GOLDMANN, D. A., Prevention of nosocomial respiratory syncytial virus infections through compliance with glove and gown isolation precautions, *N. Engl. J. Med.* **317**:329–334 (1987).

128. LEE, G. C.-Y., FUNK, G. A., CHEN, S.-T., HUANG, Y.-T., AND WEI, H.-Y., An outbreak of respiratory syncytial virus infection in an infant nursery, *J. Formosan Med. Assoc.* **72**:39–46 (1973).

129. LODA, F. A., CLYDE, W. A., GLEZEN, W. P., SENIOR, R. J., SHEAFFER, C. I., AND DENNY, F. W., Studies on the role of viruses, bacteria and *M. pneumoniae* as causes of lower respiratory tract infections in children, *J. Pediatr.* **72**:161–176 (1968).

130. MACDONALD, N. E., HALL, C. B., SUFFIN, S. C., ALEXSON, D., HARRIS, P. J., AND MANNING, J. A., Respiratory syncytial viral infection in infants with congenital heart disease, *N. Engl. J. Med.* **307**:397–400 (1982).

131. MADGE, P., PATON, J. Y., McCOLL, J. H., AND MACKIE, P. L., Prospective controlled study of four infection-control procedures to prevent nosocomial infection with respiratory syncytial virus, *Lancet* **340**:1079–1083 (1992).

132. MARTIN, A. J., GARDNER, P. S., AND McQUILLIN, J., Epidemiology of respiratory viral infection among pediatric inpatients over a six-year period in North-East England, *Lancet* **2**:1035–1038 (1978).

133. MATHUR, U., BENTLEY, D. W., AND HALL, C. B., Concurrent respiratory syncytial virus and influenza A infections in the institutionalized elderly and chronically ill, *Ann. Intern. Med.* **93**:49–52 (1980).

134. McINTOSH, K., Interferon in nasal secretions from infants with viral respiratory tract infections, *J. Pediatr.* **93**:33–36 (1978).

135. McINTOSH, K., ELLIS, E. F., HOFFMAN, L. S., LYBASS, T. G., ELLER, J. J., AND FULGINITI, V. A., The association of viral and bacterial respiratory infections with exacerbations of wheezing in young asthmatic children, *J. Pediatr.* **82**:578–590 (1973).

136. McINTOSH, K., AND FISHAUT, J. M., Immunopathologic mechanisms in lower respiratory tract disease of infants due to respiratory syncytial virus, *Prog. Med. Virol.* **26**:94–118 (1980).

137. McINTOSH, K., HALONEN, P., AND RUUSKANEN, O., Report of a workshop on respiratory viral infections: Epidemiology, diagnosis, treatment, and prevention, *Clin. Infect. Dis.* **16**(1):151–164 (1993).

138. McINTOSH, K., HENDRY, R. M., FAHNESTOCK, M. L., AND PIERIK, L. T., Enzyme-linked immunosorbent assay for detection of respiratory syncytial virus infection: Application to clinical samples, *J. Clin. Microbiol.* **16**:329 (1982).

139. McINTOSH, K., MASTERS, H. B., ORR, I., CHAO, R. K., AND BARKIN, R. M., The immunologic response to infection with respiratory syncytial virus in infants, *J. Infect. Dis.* **158**:24–32 (1978).

140. McINTOSH, K., McQUILLIN, J., AND GARDNER, P. S., Cell-free and cell-bound antibody in nasal secretions from infants with respiratory syncytial virus infection, *Infect. Immun.* **32**:276–281 (1979).

141. McQUILLIN, J., AND GARDNER, P. S., Rapid diagnosis of respiratory syncytial virus infection by immunofluorescent antibody techniques, *Br. Med. J.* **1**:602–605 (1968).

142. MEERT, K., HEIDEMANN, S., LIEH-LAI, M., AND SARNAIK, A. P., Clinical characteristics of respiratory syncytial virus infections in

healthy versus previously compromised host, *Pediatr. Pulmonol.* **7**(3):167–170 (1989).

143. MEERT, K., HEIDEMANN, S., ABELLA, B., AND SARNAIK, A., Does prematurity alter the course of respiratory syncytial virus infection? *Crit. Care Med.* **18**(12):1357–1395 (1990).

144. MEISSNER, H. C., MURRAY, S. A., KIERNAN, M. A., SNYDMAN, D. R., AND McINTOSH, K., A simultaneous outbreak of respiratory syncytial virus and parainfluenza virus type 3 in a newborn nursery, *J. Pediatr.* **104**:680–684 (1984).

145. MILLS, J., VAN KIRK, J. E., WRIGHT, P. F., AND CHANOCK, R. M., Experimental respiratory syncytial virus infection of adults, *J. Immunol.* **107**:123–130 (1971).

146. MINTZ, L., BALLARD, R. A., SNIDERMAN, S. H., ROTH, R. S., AND DREW, W. L., Nosocomial respiratory syncytial virus infections in an intensive care nursery: Rapid diagnosis by direct immunofluorescence, *Pediatrics* **64**:149–153 (1979).

147. MOK, J. Y. Q., AND SIMPSON, H., Outcome of acute lower respiratory tract infection in infants: Preliminary report of seven-year follow-up study, *Br. Med. J.* **285**:333–337 (1982).

148. MOLER, F. W., KAHN, A. S., MELIONES, J. N., CUSTER, J. R., PALMISANO, J., AND SHOPE, T. C., Respiratory syncytial virus morbidity and mortality estimates in congenital heart disease patients: A recent experience, *Crit. Care Med.* **20**(10):1406–1413 (1992).

149. MONTO, A. S., AND LIM, S. K., The Tecumseh study of respiratory illness. III. Incidence and periodicity of respiratory syncytial virus and mycoplasma pneumonia infections, *Am. J. Epidemiol.* **94**:290–301 (1971).

150. MORALES, F., CALDER, M. A., INGLIS, J. M., MURDOCH, P. S., AND WILLIAMSON, J., A study of respiratory infections in the elderly to assess the role of respiratory syncytial virus, *J. Infect.* **7**:236–247 (1983).

151. MORRIS, J. A., BLOUNT, R. E., JR., AND SAVAGE, R. E., Recovery of cytopathogenic agent from chimpanzees with coryza, *Proc. Soc. Exp. Biol. Med.* **92**:544 (1956).

152. MUFSON, M. A., BELSHE, R. B., ORVELL, C., AND NORRBY, E., Subgroup characteristics of respiratory syncytial virus strains recovered from children with two consecutive infections, *J. Clin. Microbiol.* **25**(8):1535–1539 (1987).

153. MUFSON, M. A., LEVINE, H. D., WASIL, R. E., MOCEGA-GONZALEZ, H. E., AND KRAUSE, H. E., Epidemiology of respiratory syncytial virus infection among infants and children in Chicago, *Am. J. Epidemiol.* **98**:88–95 (1973).

154. MUFSON, M. A., ORVELL, C., RAFNAR, B., AND NORRBY, E., Two distinct subtypes of human respiratory syncytial virus, *J. Gen. Virol.* **66**:2111–2124 (1985).

155. MURPHY, B. R., GRAHAN, B. S., PRINCE, G. A., WALSH, E. E., CHANOCK, R. M., KARZON, D. T., AND WRIGHT, P. F., Serum and nasal-wash immunoglobulin G and A antibody response of infants and children to respiratory syncytial virus F and G glycoproteins following primary infection, *J. Clin. Microbiol.* **23**:1009–1014 (1986).

156. MURPHY, B. R., SOTNIKOV, A. V., LAWRENCE, L. A., BANKS, S. M., AND PRINCE, G. A., Enhanced pulmonary histopathology is observed in cotton rats immunized with formalin-inactivated respiratory syncytial virus (RSV) or purified F glycoprotein and challenged with RSV 3–6 months after immunization, *Vaccine* **8**(5):497–502 (1990).

157. MURPHY, D., TODD, J. K., CHAO, R. K., ORR, I., AND McINTOSH, K., The use of gowns and masks to control respiratory illness in pediatric hospital personnel, *J. Pediatr.* **99**:746–750 (1981).

158. NAVAS, L., WANG, E., DE CARVALHO, V., AND ROBINSON, J., Improved outcome of respiratory syncytial virus infection in a high-risk hospitalized population of Canadian children. Pediatric Investigators Collaborative Network on Infections in Canada, *J. Pediatr.* **121**(3):348–354 (1992).

159. NELSON, D. B., NELSON, R. M., AND SCHLOSSER, W., Family acquired respiratory disease in high-risk infants, *Clin. Pediatr.* **19**:325–328 (1980).

160. NORRBY, E., Myxoviridae: Pseudomyxovirus—respiratory syncytial (RS) virus, in: *Handbook Series in Clinical Laboratory Science* (D. SELIGSON, ED.), pp. 401–409, CRC Press, West Palm Beach, FL, 1979.

161. NORRBY, E., MARUSYK, H., AND ORVELL, C., Ultrastructural studies of the multiplication of RS (respiratory syncytial) virus, *Acta Pathol. Microbiol. Scand. [B]* **78**:268 (1970).

162. OLMSTED, R. A., ELANGO, N., PRINCE, G. A., MURPHY, B. R., JOHNSON, P. R., MOSS, B., CHANOCK, R. M., AND COLLINS, P. L., Expression of the F glycoprotein of respiratory syncytial virus by a recombinant vaccinia virus: Comparison of the individual contributions of the F and G glycoproteins of host immunity, *Proc. Natl. Acad. Sci. USA* **83**:7462–7466 (1986).

163. PACCAUD, M. R., AND JACQUIER, C., A respiratory syncytial virus of bovine origin, *Arch. Ges. Virusforsch.* **30**:327–342 (1970).

164. PANUSKA, J. R., CIRINO, N. M., MIDULLA, F., DESPOT, J. E., MCFADDEN, E. R., JR., AND HUANG, Y. T., Productive infection of isolated human alveolar macrophages by respiratory syncytial virus, *J. Clin. Invest.* **86**(1):113–119 (1990).

165. PANUSKA, J. R., HERTZ, M. I., TARAF, H., VILLANI, A., AND CIRINO, N. M., Respiratory syncytial virus infection of alveolar macrophages in adult transplant patients, *Am. Rev. Respir. Dis.* **145**:934–939 (1992).

166. PARROTT, R. H., Respiratory syncytial virus in nurseries, *N. Engl. J. Med.* **300**:430–431 (1979).

167. PARROTT, R. H., KIM, H. W., ARROBIO, J. O., HODES, D. S., MURPHY, B. R., BRANDT, C. D., CAMARGO, E., AND CHANOCK, R. M., Epidemiology of respiratory syncytial virus infection in Washington, D.C.: II. Infection and disease with respect to age, immunologic status, race and sex, *Am. J. Epidemiol.* **98**:289–300 (1973).

168. PARROTT, R. H., VARGOSKO, A. J., KIM, H. W., CUMMING, C., TURNER, H., HUEBNER, R. J., AND CHANOCK, R. M., Respiratory syncytial virus: II. Serological studies over a 34-month period of children with bronchiolitis, pneumonia, and minor respiratory diseases, *J.A.M.A.* **176**:653–657 (1961).

169. PIAZZA, F. M., JOHNSON, S. A., OTTOLINI, M. G., SCHMIDT, H. J., DARNELL, M. E., HEMMING, V. G., AND PRINCE, G. A., Immunotherapy of respiratory syncytial virus infection in cotton rats (Sigmodon fulviventer) using IgG in a small-particle aerosol, *J. Infect. Dis.* **166**(6):1422–1424 (1992).

170. POHL, C., GREEN, M., WALD, E. R., AND LEDESMA-MEDINA, J., Respiratory syncytial virus infections in pediatric liver transplant recipients, *J. Infect. Dis.* **165**(1):166–169 (1992).

171. PRINCE, G. A., HEMMING, V. G., HORSWOOD, R. L., BARON, P. A., AND CHANOCK, R. M., Effectiveness of topically administered neutralizing antibodies in experimental immunotherapy of respiratory syncytial virus infection in cotton rats, *J. Virol.* **61**:1851–1854 (1987).

172. PRINCE, G. A., HEMMING, V. G., HORSWOOD, R. L., AND CHANOCK, R. M., Immunoprophylaxis and immunotherapy of respiratory syncytial virus infection in the cotton rat, *Virus Res.* **3**:193–206 (1985).

173. PRINCE, G. A., HORSWOOD, R. L., CAMARGO, E., KOENIG, D., AND CHANOCK, R. M., Mechanisms of immunity to respiratory syncytial virus in cotton rats, *Infect. Immun.* **42**:81–87 (1983).

174. PRINCE, G. A., HORSWOOD, R. L., AND CHANOCK, R. M., Quantitative aspects of passive immunity to respiratory syncytial virus infection to infant cotton rats, *J. Virol.* **55**:517–520 (1985).

175. PRINCE, G. A., JENSON, A. B., HEMMING, V. G., MURPHY, B. R., WALSH, E. E., HORSWOOD, R. L., AND CHANOCK, R. M., Enhancement of respiratory syncytial virus pulmonary pathology in cotton rats by intramuscular inoculation of formalin inactivated virus, *J. Virol.* **57**:721–728 (1986).

176. PRINGLE, C. R., SHIRODARIA, P. V., CASH, P., CHISWELL, D. J., AND MALLOY, P., Initiation and maintenance of persistent infection by respiratory syncytial virus, *J. Virol.* **28**:199–211 (1978).

177. PULLAN, C. R., AND HEY, E. N., Wheezing, asthma, and pulmonary dysfunction 10 years after infection with respiratory syncytial virus in infancy, *Br. Med. J.* **284**:1665–1669 (1982).

178. PUTHAVATHANA, P., WASI, C., KOSITANONT, U., SUWANJUTHA, S., AND CHANTAROJANASIRI, T., KANTAKAMALAKUL, W., KANTAWATEERA, P., AND THONGCHAROEN, P., A hospital-based study of acute viral infections of the respiratory tract in Thai children, with emphasis on laboratory diagnosis, *Rev. Infect. Dis.* **12**(Suppl. 8):S988–S994 (1990).

179. RAVEN, C., MAVERAKIS, N. H., EVELAND, W. C., AND ACKERMANN, W. W., The sudden infant death syndrome: A possible hypersensitivity reaction determined by distribution of IgG in lungs, *J. Forensic Sci.* **23**:116–138 (1978).

180. RICHARDSON, L. S., YOLKEN, R. H., BELSHE, R. B., CAMARGO, E., KIM, H. W., AND CHANOCK, R. M., Enzyme-linked immunosorbent assay for measurement of serological response to respiratory syncytial virus infection, *Infect. Immun.* **20**:660–664 (1978).

181. ROONEY, J. C., AND WILLIAMS, H. E., The relationship between proved viral bronchiolitis and subsequent wheezing, *J. Pediatr.* **79**:744–747 (1971).

182. ROSS, C. A. C., PINKERTON, I. W., AND ASSAAD, F. A., Pathogenesis of respiratory syncytial virus diseases in infancy, *Arch. Dis. Child.* **46**:702–704 (1971).

183. ROSS, C. A. C., STOTT, E. J., AND CROWTHER, I., Respiratory syncytial virus infection, *Br. Med. J.* **1**:56 (1964).

184. SALKIND, A. R., MCCARTHY, D. O., NICHOLS, J. E., DOMURAT, F. M., WALSH, E. E., AND ROBERTS, N. J., JR., Interleukin-1-inhibitor activity induced by respiratory syncytial virus: Abrogation of virus-specific and alternate human lymphocyte proliferative responses, *J. Infect. Dis.* **163**(1):71–77 (1991).

185. SALKIND, A. R., NICHOLS, J. E., AND ROBERTS, N. J., JR., Suppresed expression of ICAM-1 and LFA-1 and abrogation of leukocyte collaboration after exposure of human mononuclear leukocytes to respiratory syncytial virus *in vitro*. Comparison with exposure to influenza virus, *J. Clin. Invest.* **88**(2):505–511 (1991).

186. SATAKE, M., COLIGAN, J. E., ELANGO, N., NORRBY, E., AND VENKATESAN, S., Respiratory syncytial virus envelope glycoprotein (G) has a novel structure, *Nucleic Acids Res.* **13**:7795–7812 (1985).

187. SIBER, G. R., LESZCYNSKI, J., PEÑA-CRUZ, V., FERREN-GARDNER, C., ANDERSON, R., HEMMING, V. G., WALSH, E. E., BURNS, J., MCINTOSH, K., GONIN, R., AND ANDERSON, L. J., Protective activity of a human respiratory syncytial virus immune globulin prepared from donors screened by microneutralization assay, *J. Infect. Dis.* **165**(3):456–463 (1992).

188. SIMS, D. G., DOWNHAM, M. A. P. S., GARDNER, P. S., WEBB, J. K.

G., AND WEIGHTMAN, D., Study of 8-year-old children with a history of respiratory syncytial virus bronchiolitis in infancy, *Br. Med. J.* **1:**11–14 (1978).

189. SMITH, D. W., FRANKEL, L. R., MATHERS, L. H., TANG, A. T., ARIAGNO, R. L., AND PROBER, C. G., A controlled trial of aerosolized ribavirin in infants receiving mechanical ventilation for severe respiratory syncytial virus infection, *N. Engl. J. Med.* **325**(1):24–29 (1991).

190. SNYDMAN, D. R., GREER, C., MEISSNER, H. C., AND MCINTOSH, K., Prevention of nosocomial transmission of respiratory syncytial virus in a newborn nursery, *Infect. Control Hosp. Epidemiol.* **9:**105–108 (1988).

191. SOMMERVILLE, R. G., Respiratory syncytial virus in acute exacerbations of chronic bronchitis, *Lancet* **2:**1247–1248 (1963).

192. SORVILLO, F. J., HUIE, S. F., STRASSBURG, M. A., HUTSUMYO, A., SHANDERA, W. X., AND FANNIN, S. L., An outbreak of respiratory syncytial virus pneumonia in a nursing home for the elderly, *J. Infect.* **9:**252–256 (1984).

193. SPENCE, L., AND BARRATT, N., Respiratory syncytial virus associated with acute respiratory infections in Trinidadian patients, *Am. J. Epidemiol.* **88:**257–266 (1968).

194. SPRAKER, T. R., AND COLLINS, J. K., Isolation and serologic evidence of a respiratory syncytial virus in bighorn sheep from Colorado, *J. Wildlife Dis.* **22:**416–418 (1986).

195. STERNER, G., WOLONTIS, S., BLOTH, B., AND DE HEVESY, G., Respiratory syncytial virus: An outbreak of acute respiratory illness in a home for infants, *Acta Paediatr. Scand.* **55:**273–279 (1966).

196. STOKES, G. M., MILNER, A. D., HODGES, I. G. C., AND GROGGINS, R. C., Lung function abnormalities after acute bronchiolitis, *J. Pediatr.* **98:**871–874 (1981).

197. STORCH, G. A., PARK, C. S., AND DOHNER, D. E., RNA fingerprinting of respiratory syncytial virus using ribonuclease protection. Application to molecular epidemiology, *J. Clin. Invest.* **83**(6):1894–1902 (1989).

198. STOTT, E. J., BALL, L. A., YOUNG, K. K., FURZE, J., AND WERTZ, G. W., Human respiratory syncytial virus glycoprotein G expressed from a recombinant vaccinia virus vector protects mice against live-virus challenge, *J. Virol.* **60:**607–613 (1986).

199. STRETTON, M., AJIZIAN, S. J., MITCHELL, I., AND NEWTH, C. J., Intensive care course and outcome of patients infected with respiratory syncytial virus, *Pediatr. Pulmonol.* **13**(3):143–150 (1992).

200. SULLENDER, W. M., MUFSON, M. A., ANDERSON, L. J., AND WERTZ, G. W., Genetic diversity of the attachment protein of subgroup B respiratory syncytial viruses, *J. Virol.* **65**(10):5425–5434 (1991).

201. SUTO, T., YANO, N., IKEDA, M., MIYAMOTO, M., TAKAI, S., SHIGETA, S., HINUMA, Y., AND ISHIDA, N., Respiratory syncytial virus infection and its erologic epidemiology, *Am. J. Epidemiol.* **82:**211–224 (1965).

202. TAYLOR, C. E., CRAFT, A. W., KERNAHAN, J., MILLMAN, R., REID, M. M., SCOTT, R., AND TOMS, G. L., Local antibody production and respiratory syncytial virus infection in children with leukaemia, *J. Med. Virol.* **30**(4):277–281 (1990).

203. TOMS, G. L., WEBB, M. S., MILNER, P. D., MINER, A. D., ROUTLEDGE, E. G., SCOTT, R., STOKES, G. M., SWARBRICK, A., AND TAYLOR, C. E., IgG and IgM antibodies to viral glycoproteins in respiratory syncytial virus infections of graded severity, *Arch. Dis. Child.* **64**(12):1661–1665 (1989).

204. TRISTRAM, D. A., WELLIVER, R. C., MOHAR, C. K., HOGERMAN, D. A., HILDRETH, S. W., AND PARADISO, P., Immunogenicity and safety of respiratory syncytial virus subunit vaccine in seropositive children 18–36 months old, *J. Infect. Dis.* **167**(1):191–195 (1993).

205. TUPASI, T. E., LUCERO, M. G., MAGDANGAL, D. N., MANGUBAT, N. W., SUNICO, E. S., TORRES, C. U., DE LEON, L. E., PALADIN, J. F., BAES, L., AND JAVATO, M. C., Etiology of acute lower respiratory tract infection in children from Alabang, Metro Manila, *Rev. Infect. Dis.* **12:**S929–939 (1990).

206. TYRRELL, D. A. J., Discovering and defining the etiology of acute respiratory viral disease, *Am. Rev. Respir. Dis.* **88:**77–81 (1963).

207. VOLOVITZ, B., FADEN, H., AND OGRA, P. L., Release of leukotriene C_4 in the respiratory tract during acute viral infection, *J. Pediatr.* **112:**218–222 (1988).

208. WALD, E. R., DASHEFSKY, B., AND GREEN, M., In re ribavirin: A case of premature adjudication? *J. Pediatr.* **112:**154–158 (1988).

209. WALSH, E. E., HALL, C. B., BRISELLI, M., BRANDRISS, M. W., AND SCHLESINGER, J. J., Immunization with glycoprotein subunits of respiratory syncytial virus to protect cotton rats against viral infection, *J. Infect. Dis.* **155:**1198–1204 (1987).

210. WALSH, E. E., AND HRUSKA, J., Monoclonal antibodies to respiratory syncytial virus proteins: Identification of the fusion protein, *J. Virol.* **47:**171–177 (1983).

211. WALSH, E. E., SCHLESINGER, J. J., AND BRANDRISS, M. W., Protection from respiratory syncytial virus infection in cotton rats by passive transfer of monoclonal antibodies, *Infect. Immun.* **43:**756–758 (1984).

212. WARIS, M., MEURMAN, O., MUFSON, M. A., RUUSKANEN, O., AND HALONEN, P., Shedding of infectious virus and virus antigen during acute infection with respiratory syncytial virus, *J. Med. Virol.* **38**(2):111–1116 (1992).

213. WATT, P. J., ROBINSON, B. S., PRINGLE, C. R., AND TYRRELL, D. A., Determinants of susceptibility to challenge and the antibody response of adult volunteers given experimental respiratory syncytial virus vaccines, *Vaccine* **8**(3):231–236 (1990).

214. WELLIVER, R. C., KAUL, T. N., AND OGRA, P. L., Cell-mediated immune response to respiratory syncytial virus infection: Relationship to the development of reactive airway disease, *J. Pediatr.* **94:**370–375 (1979).

215. WELLIVER, R. C., KAUL, T. N., AND OGRA, P. L., The appearance of cell-bound IgE in respiratory-tract epithelium after respiratory-syncytial-virus infection, *N. Engl. J. Med.* **303:**1198–1202 (1980).

216. WELLIVER, R. C., AND MCLAUGHLIN, S., Unique epidemiology of nosocomial infection in a children's hospital, *Am. J. Dis. Child.* **138:**131–135 (1984).

217. WELLIVER, R. C., SUN, M., RINALDO, D., AND OGRA, P. L., Predictive value of respiratory syncytial virus-specific IgE responses for recurrent wheezing following bronchiolitis, *J. Pediatr.* **109:**776–780 (1986).

218. WELLIVER, R. C., WONG, D. T., SUN, M., MIDDLETON, E., JR., VAUGHAN, R. S., AND OGRA, P. L., The development of respiratory syncytial virus-specific IgE and the release of histamine in nasopharyngeal secretions after infection, *N. Engl. J. Med.* **305:**841–846 (1981).

219. WRIGHT, P. F., BELSHE, R. B., KIM, H. W., VAN VORIS, L. P., AND CHANOCK, R. M., Administration of a highly attenuated, live respiratory syncytial virus vaccine to adults and children, *Infect. Immun.* **37:**397–400 (1982).

220. ZWEIMAN, B., SCHOENWETTER, W. F., AND HILDRETH, E. A., The relationship between bronchiolitis and allergic asthma, *J. Allergy* **37:**48–52 (1966).

CHAPTER 24

Retroviruses—Human Immunodeficiency Virus

William A. Blattner, Thomas R. O'Brien, and Nancy E. Mueller

1. Introduction

The harbinger of the worldwide pandemic of acquired immunodeficiency syndrome (AIDS) was a handful of cases of *Pneumocystis carinii* pneumonia clustered among homosexual men in Los Angeles, California[79]; a short time later, a parallel epidemic of Kaposi's sarcoma was reported. The thread that tied these disparate conditions together was profound immunodeficiency. In a brief time AIDS cases were recognized among hemophiliacs, transfusion recipients, and injection drug users from diverse geographic locales around the world. Subclinical perturbations of lymphocyte subsets were found in high-risk populations; a variety of theories was developed to account for the confusing emerging pattern: lifestyle, immune overload, and an infectious agent. Through the concerted efforts of scientists from different disciplines and countries, the cause was ultimately shown to be a new class of lenti-retrovirus now called human immunodeficiency virus type 1 (HIV-1).

Following the identification of HIV-1, a related but less pathogenic virus was identified in Africa through serological cross-reactivity in patient and population screening. This virus, HIV-2, was subsequently isolated and sequenced with some of its epidemiologic features determined. Investigation of HIV-2, however, has been largely overshadowed by HIV-1 research.

In the decade and a half since AIDS was first recognized, the international scientific research community has undertaken a massive research effort that has yielded many biological insights; epidemiologists have provided vital information about the extent of the epidemic and the modes of transmission of its causative agent. But fundamental gaps in knowledge are barriers to a cure and have blocked vaccine development as well as effective prevention efforts. Thus, despite the best efforts of epidemiologists and other scientists, HIV infection continues to spread worldwide; AIDS threatens to wreak havoc in some developing countries and severely stress health care infrastructures in economically advantaged countries. This chapter will focus on the global nature of HIV/AIDS, the current state-of-knowledge, strategies in progress, and future directions.

2. Historical Background

The discovery of HIV-1 has its scientific roots in research of oncogenic viruses in chickens. Ellerman and Bang[175] in 1908 and Rous[437] in 1911 found that filterable agents were capable of inducing leukemias and sarcomas in chickens. Retroviruses (initially called RNA tumor viruses) are subdivided into the onco- and lentivirus families. In the 1950s and 1960s, the onco-retroviruses were found to be RNA viruses that cause a variety of mammalian cancers and leukemias; these RNA viruses and related lenti-retroviruses caused a variety of pulmonary and neurological conditions primarily of ungulates. Re-

William A. Blattner • Institute of Human Virology, University of Maryland, Medical Biotechnology Center, Baltimore, Maryland 21201-1192. **Thomas R. O'Brien** • Viral Epidemiology Branch, Epidemiology and Biostatistics Program, National Cancer Institute, Bethesda, Maryland 20892. **Nancy E. Mueller** • Department of Epidemiology, Harvard School of Public Health, Boston, Massachusetts 02115.

verse transcriptase, a virally encoded enzyme that promotes the synthesis of a proviral DNA from a viral RNA template was discovered in the early 1970s by Temin[470] and Baltimore[16] and provided critical insights into the life cycle of these viruses. Specialized cell culture techniques that permit the long-term growth of target T lymphocytes through the use of a growth factor called interleukin-2 (IL-2) and the development of highly sensitive assays for reverse transcriptase were essential tools for the discovery of the first human retrovirus, human T-lymphotrophic virus type 1 (HTLV-I).[348] The discovery of HTLV-I in 1979[408] paved the way for the discovery of HIV-1.[198]

With the advent of AIDS in 1981 and the subsequent elucidation of its underlying T-cell deficiency, a retrovirus was proposed as a possible etiologic agent because of the similarities to HTLV in T-cell trophism, pathogenicity, and modes of transmission.[198] The tools and techniques developed in the search for HTLV-I were adapted to successfully detect and grow HIV-1.[18,411] HIV-1, formerly known as HTLV-III or lymphadenopathy-associated virus (LAV) was discovered and reported by Montagnier in 1983.[198] In 1984, Gallo and his co-workers proved HIV-1 to be the etiologic agent of AIDS.[18,201,411,426] HIV-1 was soon linked to a broad spectrum of AIDS-defining disease conditions by epidemiologic studies showing strong risk ratios when AIDS cases were compared to controls from high-risk homosexual populations, hemophiliacs, and transfusion recipients.[208,212] Cohort analyses documented that HIV-1 infection antedated AIDS onset among seroconvertors; they provided population estimates of attributable risk[350,409] as well as insights into modes of HIV-1 transmission.[211]

HIV-2 (formerly HTLV-IV and LAV-2) was reported by Essex, Kanki, and Montagnier[17,118,249] in 1985; it has been linked to some patients (largely in West Africa) with AIDS-like illnesses.[118] This virus is much more closely related to a nonhuman primate virus, simian immunodeficiency virus (SIV), than is HIV-1[113,465]; SIV causes AIDS-like illnesses in some primate species.[293]

The geographic origin of the AIDS epidemic is unknown. The first documented case of AIDS occurred in a British seaman who died of *P. carinii* pneumonia in Manchester, England, in 1959; HIV-1 infection was subsequently demonstrated with the polymerase chain reaction (PCR) assay.[132] Sera obtained from Zaire during the same time period demonstrated (in rare instances) antibodies suggestive of HIV-1 infection.[360] Africans with AIDS or pre-AIDs who were serologically positive for HIV-2 have also been described, including a person who died in 1978.[29,59] A married Portuguese couple was described in whom the husband likely acquired HIV-2 infec-

tion while on military service in Guinea-Bissau between 1966 and 1969. In 1980, the wife, who was also infected with HIV-2, developed a non-Hodgkin's lymphoma, and in 1985, the husband developed opportunistic infections. None of their seven children were seropositive. This report captured several of the epidemiologic features of HIV-2 infection including the propensity for heterosexual transmission, the low perinatal transmission rate, and the apparent long latency for disease.[6]

Studies of the phylogenetic relationships between human and simian immunodeficiency viruses have raised the possibility of a nonhuman primate origin of HIV-1.[465] Isolates from chimpanzees, the only other species susceptible to HIV-1 infection, bear a close phylogenetic relationship to the human virus.[195] Recent data suggest the multiple introduction of HIV-1 viruses into man, presumably from enzootic exposure.[218,246] Sequencing of HIV-2 and an SIV isolate from a sooty mangabey (SIV_{SM}) indicated that these viruses are closely related.[242] An endemic infection of African green monkeys (SIV_{AGM}) with distinguishable SIV subtypes corresponding to species in different parts of Africa has been identified.[3,194,358]

In 1992, Gao and his colleagues[203] isolated and genotyped a new strain of HIV-2 from a West African that was more closely related to SIV_{SM} than to prototype human isolates. This observation argued that HIV-2 and SIV form a single, but highly diverse species of virus that cannot be distinguished by species of origin. Their work further suggested that HIV-2 has been acquired by infection from monkey to human. Although SIV_{SM} is nonpathogenic in sooty mangabeys, an apparent natural host,[284] it has been linked with AIDS-like illnesses in other primate species. African green monkeys also appear to be natural hosts for SIV_{AGM}, but with no related disease.[164] The "epidemic" of AIDS-like disease among SIV_{SM}-infected captive macaques (a nonnatural host) in primate research centers in the United States in the 1970s likely represents a cross-species infection.[359] These observations illustrate that although adaptation may occur between a virus and its host over time with little pathology evident, when transmission occurs in new species the virus is generally much more pathogenic.[65] The homology among growing numbers of HIV isolates from nonhuman primates (predominantly in Africa) show simian viruses (SIVs) that are virtually indistinguishable from human viruses (HIV-2); this has prompted the proposal that these viruses be reclassified as "primate" immunodeficiency viruses.[22,316,343,457]

The nonhuman primate-to-human hypothesis is supported by recent reports of laboratory workers becoming infected with SIV.[261,316] A report from Guinea-Bissau that women involved in the preparation of primate meat for human consumption (with potential blood exposure) were at higher risk for HIV-2 infection[416] offers a mechanism by which this class of virus was introduced into humans (who subsequently spread it through sexual and parenteral routes). The dramatic worldwide spread of the virus could also have a biological basis if a more transmissible strain emerged from a primate source.[370,395]

Whatever the biological basis of the introduction of HIV to humans, global social upheavals such as large population migrations from rural to urban areas and the dissemination of injection drug and crack-cocaine use have fueled the spread. Changes in social mores and sexual behavior have provided an avenue for extended transmission through homo- and heterosexual contact and subsequently via injection drug users and their sexual partners.[266]

The chronology of scientific progress in response to AIDS has been uneven: early scientific breakthroughs have been followed by a disappointing flattening of the knowledge curve. Between 1981 and 1983 the clinical syndrome was described and its multiple manifestations attributed to an underlying immunodeficiency; these findings led to the development of an initial surveillance definition. Risk groups were defined for surveillance purposes with the expectation that an infectious agent was involved and transmitted by sexual routes, blood transfusion, injection drug use, and from mother to infant. Between 1983 and 1984, the etiologic agent was isolated and proven to be the cause of AIDS. Over the next 5 years, rapid progress was made in defining the broader spectrum of HIV-associated diseases, developing an accurate blood screening test, characterizing the virus molecularly, defining certain aspects of pathogenesis, and determining the positive benefits of nucleotide analogue antivirals and prophylactic therapies for opportunistic infections in prolonging the life of those with severe immunodeficiency. Despite these advances, the epidemic has continued to grow; new regional epidemics are emerging in parts of the world (particularly southeast Asia), which initially had been spared. In the United States the epidemic peaked in homosexual men; a second wave involving injection drug users and, more recently, heterosexually active populations has surfaced. Early optimism that a preventive vaccine would be developed has been dampened by the failure of initial animal vaccine experiments, the recognition that the virus is highly mutable, and the discovery of antigenically distinct families of virus in different geographic locales. Current research initiatives are focusing on the arduous task of defining the pathogenetic mechanisms of virus infection and disease causation; fundamental insights are needed to advance the treatment and prevention agenda.

The sophistication of current science would seem to have precluded a worldwide pandemic; the emergence of a totally new class of virus in modern times was unexpected. The AIDS epidemic is a reminder of the complexity of viruses of long latency and their potential to spread widely before an epidemic is detected.

3. Methods of Epidemiologic Analysis

3.1. Sources of Mortality Data

Median survival after an HIV-1-infected person develops AIDS is about 2 years,[55] but survival time varies by the specific AIDS-defining condition. The two systems for recording AIDS mortality in the United States are the Vital Statistics of All Cause Mortality recorded by National Center for Health Statistics (NCHS) and the National AIDS Surveillance maintained by the Centers for Disease Control and Prevention (CDC) through reports from AIDS registries in the states.

In the setting of traditional death certification, HIV-related death reporting is complicated by the broad range of fatal conditions associated with extreme immunodeficiency. There are also inherent difficulties in disentangling the underlying cause of death in groups such as injection drug users where the poor availability of medical services may act as a barrier to quality diagnostic information.[61] Furthermore, because of the social stigma associated with AIDS, physicians may use nonspecific codes such as pneumonia to protect their patients from being publicly reported as AIDS deaths.[61]

The rubrics for AIDS-defining deaths have changed over time and closely parallel changes in the AIDS definition. Beginning in 1983, an existing code for "deficiency of cell-mediated immunity" was adapted to define an AIDS-associated death.[61] In 1986, procedures and codes for death certification were updated to comply with the 1987 revision of the CDC AIDS definition; they included some conditions defined by coincidence of HIV infection with the widespread introduction of a licensed screening assay.[90]

Recent analyses of AIDS mortality in the United States indicate that the combined reports of NCHS death records and CDC AIDS surveillance account for 70 to 90% of HIV-related deaths[61]; the register of AIDS-associated deaths in women may be less efficient, recording as few as 55%,[33] and similar patterns have been reported in England.[325] Many unattributed deaths result from conditions such as sepsis, recurrent pneumonia, tuberculosis, and some AIDS-associated conditions that are not classified under AIDs-specific rubrics. A growing number of deaths are classified as due to illicit drug use; often the underlying AIDS diagnosis is missed.

In most areas of the world, AIDS mortality statistics are incomplete but may be estimated on the basis of underlying HIV prevalence.[1] In some cases special investigations have documented the magnitude of HIV-related mortality; an intensive autopsy study of adults in Abidjan, Côte d'Ivoire, identified AIDS as the leading cause of death.[146] In parts of the developing world where HIV-1 has been recently introduced, such as Asia, mortality statistics may be an especially a poor indication of the magnitude of the underlying HIV epidemic.

3.2. Sources of Morbidity Data

3.2.1. HIV-1.
In 1981, CDC established nationwide surveillance for patients diagnosed with AIDS in the United States. This activity led to the early recognition of the emerging magnitude and slope of the epidemic curve

for AIDS. Case reporting from the states and territories also led to the recognition that persons with different lifestyles and backgrounds were at risk for AIDS. Insights from this coordinated AIDS surveillance effort were of particular value in defining the likelihood that an infectious etiology best explained the emerging epidemic.

A surveillance definition for AIDS was initially proposed in 1983. It was based on the complex constellation of medical conditions, especially opportunistic infections such as *P. carinii* pneumonia, that are indicative of severe underlying immunodeficiency not attributable to known causes.[498] With the growing recognition that AIDS represented a wider range of conditions and with the introduction of the HIV-1 antibody test, the AIDS definition was broadened in 1987 to encompass a larger spectrum of immunodeficiency-related outcomes.[89] The 1993 revision of the CDC AIDS definition (Table 1) includes some additional medical conditions, but more importantly, patients with HIV-1 infection and a CD4 count of 200 or less.[101] The new definition is structured within a staging scheme based on clinical parameters and CD4 level.

Each major revision of the AIDS case definition has

Table 1. 1993 Revised Classification System for HIV Infection and Expanded AIDS Surveillance Case Definition for Adolescents and Adults

CD4+ T-cell categories	Clinical categories		
	(A) Asymptomatic, acute (primary) HIV or PGL	(B) Symptomatic, not (A) or (C) conditions	(C) AIDS-indicator conditions[a]
(1) ≥500/μL (2) 200–499/μL (3) <200/μL AIDS-	Asymptomatic HIV infection; persistent generalized lymphadenopathy (PGL); acute (primary) HIV infection with accompanying illness or history of acute HIV infection	Bacillary angiomatosis; candidiasis, oropharyngeal (thrush); candidiasis (thrush); candidiasis, vulvovaginal; persistent, frequent, or poorly responsive to therapy; cervical dysplasia (moderate or severe)/ cervical carcinoma *in situ*; constitutional symptoms, such as fever (38.5°C) or diarrhea lasting >1 month; hairy leukoplakia, oral herpes zoster (shingles), involving at least two distinct episodes or more than one dermatome; idiopathic thrombocytopenic purpura; listeriosis; pelvic inflammatory disease, particularly if complicated by tubo-ovarian abscess; peripheral neuropathy	Candidiasis of bronchi, trachea, or lungs; candidiasis, esophageal; cervical cancer, invasive[a]; coccidioidomycosis, disseminated or extrapulmonary; cryptococcosis, extrapulmonary; cryptosporidiosis, chronic intestinal (>1 month's duration); cytomegalovirus disease (other than liver, spleen, or nodes); cytomegalovirus retinitis (with loss of vision); encephalopathy, HIV-related; herpes simplex: chronic ulcer(s) (>1 month's duration); or bronchitis, pneumonitis, or esophagitis; histoplasmosis, disseminated or extrapulmonary; isosporiasis, chronic intestinal (>1 month's duration); Kaposi's sarcoma; lymphoma, Burkitt's (or equivalent term); lymphoma, immunoblastic (or equivalent term); lymphoma, primary, of brain; *Mycobacterium avium* complex or *M. kansasii*, disseminated or extrapulmonary); *Mycobacterium*, other species or unidentified species, disseminated or extrapulmonary; *Pneumocystis carinii* pneumonia; pneumonia, recurrent[a]; progressive multifocal leukoencephalopathy; salmonella septicemia, recurrent; toxoplasmosis of brain; wasting syndrome due to HIV

[a]Persons with AIDS-indicator conditions (category C) as well as those with CD4+ T-lymphocyte counts <200/μL (categories A3 or B3) will be reportable as AIDS cases in the United States and Territories, effective January 1, 1993.

Table 2. Characteristics of Persons with Reported AIDS Cases and Percentage Increase in the Number of Cases, by Year of Report—United States, 1992–1993

Characteristic	1993 reported cases		1992 reported cases	Percent increase 1992 to 1993
	No.	(%)		
Sex				
Male	86,986	(84.0)	42,445	105
Female	16,514	(16.0)	6,571	151
Age group (yr)				
13–19	555	(0.5)	177	214
20–24	3,722	(3.6)	1,600	133
25–29	14,680	(14.2)	7,021	109
30–39	47,415	(45.8)	22,358	112
40–49	26,956	(26.0)	12,609	114
50–59	7,514	(7.3)	3,700	103
≥60	2,658	(2.6)	1,551	71
Race/ethnicity[a]				
White, non-Hispanic	47,003	(45.4)	23,305	102
Black, non-Hispanic	36,951	(35.7)	16,582	123
Hispanic	18,318	(17.7)	8,541	114
Asian/Pacific Islander	741	(0.7)	332	123
American Indian/Alaskan Native	320	(0.3)	114	181
HIV-exposure category				
Male homosexual/bisexual contact	48,266	(46.6)	25,864	87
History of injecting-drug users	28,687	(27.7)	12,163	136
Women and heterosexual men	5,745	(5.6)	3,028	90
Person with hemophilia	1,041	(1.0)	360	189
Heterosexual contact[b]	9,288	(9.0)	4,045	130
Transfusion recipients	1,219	(1.2)	710	72
No risk reported	9,254	(8.9)	2,846	—
Region[c]				
Northeast	30,876	(29.8)	13,243	133
Northcentral	10,755	(10.4)	5,656	90
South	34,264	(33.1)	16,588	107
West	24,372	(23.5)	11,914	105
U.S. territories	3,233	(3.1)	1,615	100
Total	103,500	(100.0)	49,016	111

[a]Excluded persons with unspecified race/ethnicity (167 in 1993; 142 in 1992).

[b]Persons whose origin is or who had sex with a person whose origin is a country where heterosexual transmission was presumed to be the predominant mode of HIV transmission (i.e., formerly classified by the World Health Organization as pattern II countries) are no longer automatically classified as having heterosexually acquired AIDS. These persons are classified as "no risk reported."

[c]Northeast = New England and Middle Atlantic regions; Northcentral = East North Central and West North Central regions; South = South Atlantic, East South Central, and West South Central regions; West = Mountain and Pacific regions.

led to significant increases in case numbers.[108] The most recent definitional change (CD4 count of 200 or less) has had dramatic effects on AIDS surveillance because now cases are determined in large part by a laboratory marker of an intermediate outcome rather than by an endstage marker of profound clinical immunodeficiency, the previous benchmark. The 1993 definition has resulted in a 111% increase in reported AIDS cases, with the most remarkable increases occurring among women, minorities, and those with infection attributed to heterosexual contact (Table 2).[101] Since AIDS is generally only reported once to CDC on any individual, the new case definition creates problems for statistical modeling of the epidemic. Methods such as back calculation (a statistical tool for estimating an underlying infection rate) are based on the number of AIDS cases and the incubation period from infection to disease.[54] With the new AIDS definition, persons with a CD4 count of less than 200 will be registered only once and late-stage AIDS-defining illnesses previously reported to CDC will no longer be recorded except as they may appear on the death certificate when an AIDS-related death is reported to the AIDS registry. In the United States, patients are reported using a hierarchical scheme for risk group classification. In some instances this classification may underreport the contribution of an AIDS risk factor. For example, in high drug abuse areas the role of heterosexual transmission may be underestimated in females because infection resulting from sexual transmission may be ascribed to drug abuse.

The emergence of the AIDS pandemic has led to the establishment of World Health Organization (WHO) sponsored surveillance efforts modeled and adapted from the CDC approach.[499] Currently, over 100 nations contribute statistics to this effort. However, the complexity of the disease and its diagnosis combined with the fact that some already common infectious diseases (e.g., tuberculosis) may be manifestations of the syndrome has made accurate statistical data gathering challenging, especially in developing countries.[116] In addition, the stigma associated with AIDS has limited the reporting of cases from some countries that are unwilling to be officially linked to the problem. Despite these limitations, the statistics present a grim picture (albeit an underestimate) of the magnitude of the problem.

Other sources of surveillance information are sometimes available. Because the mortality associated with the diagnosis of AIDS is extremely high, AIDS-related mortality provides a reliable index of morbidity. In some parts of the United States, surveillance data on HIV-1 infection are collected, but there has been resistance to implementing surveillance for HIV-1 infection because of concerns about confidentiality and discrimination.[382] Furthermore, prejudice associated with participation in HIV-1 screening programs makes some high-risk individuals willing to be tested only if it is done anonymously.

3.2.2. HIV-2. In the United States, HIV-2-associated AIDS is reported and noted as such by the CDC in a footnote to aggregated AIDS cases. The prevalence of HIV-2 seropositivity has been estimated in a variety of population surveys.

3.3. Serological Surveys

AIDS case surveillance is supplemented with HIV-1 seroprevalence surveys conducted among various populations, especially those considered at risk for acquiring HIV-1 infection because of circumstances or behaviors: hemophiliacs, blood transfusion recipients, prostitutes, attenders at sexually transmitted disease (STD) clinics, tuberculosis clinics, drug treatment centers, Latino and black women's primary care clinics, and mental health clinics. While these surveys provide useful data for seroprevalence in specific populations, the results cannot be extrapolated to the broader population. Blinded seroprevalence surveys are therefore also conducted of blood donors, hospital patients (excluding those with HIV-associated conditions), childbearing women, applicants for military service and the Job Corps, and other special populations (e.g., prison inmates). Back calculation is applied to AIDS statistics to generate independent measures of seroprevalence and seroincidence. Prospective cohort studies of high-risk populations including homosexual men, hemophiliacs, and STD attenders also provide information about trends in seroincidence. Many of the same population groups have been surveyed worldwide, and because of deficiencies in AIDS case surveillance, they provide a more precise measure of the magnitude of the HIV epidemic in these locales.

3.4. Laboratory Diagnosis

3.4.1. Virus Isolation. Historically, a major challenge in the detection and isolation of HIV-1 was its propensity to lyse the target CD4 cell. Unlike HTLV-I, which promotes the growth of CD4 cells, HIV cell cultures result in the death of these cells. Initial laboratory techniques developed to isolate the virus involved serial propagation in tissue culture by continually coculturing with normal peripheral blood lymphocytes or cord blood lymphocytes.[18,442] Detection of virus relied on assays for virally associated reverse transcriptase in culture media, detection of viral antigen, and electron microscopy, as shown in Fig. 1.[201] A major breakthrough occurred when Popovic and colleagues[411] grew the virus in continuous mature T-cell lines derived from leukemic patients. Virus grown in these lysis-resistant cell lines could be purified and its antigens studied as the basis for subsequent development of serological assays. It is noteworthy that in this first paper describing the growth characteristics of the virus, there was marked variation from patient to patient and from cell line subclone to subclone in virus growth and cytopathic effects.[411] Since these early breakthroughs, refinements in culture techniques have provided insights about the protean variables that define virus growth characteristics and cell trophism. While cellular trophism for HIV-1 has been classified on the basis of growth in CD4/T cells versus macrophages, these characteristics are not absolute; most strains of HIV-1 will grow in either cell type but manifest a strong predilection for one or the other. The methods used for HIV-2 viral isolation are essentially the same as for HIV-1.

Two basic virus types called rapid–high and slow–

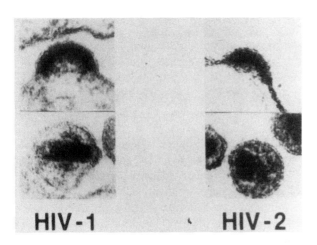

Figure 1. Thin section electrographies of HIV-1 and HIV-2. Mature virion are shown in the lower panel and budding particles in the upper panel for each virus. Photograph courtesy of Dr. Robert C. Gallo.

low (characteristics that describe the *in vitro* growth characteristics of the virus) have been described.[294] Another growth characteristic of HIV-1 relates to whether the viruses are syncytia-inducing (SI) or nonsyncytia-inducing (NSI) variants. NSI types often have slow–low characteristics and account for most initial infections; they may have trophism for either CD4 cells or monocyte macrophages. SI variants are most often detected in the later stages of infection (sometimes in association with accelerated progression). Success rates of over 95% have been achieved in cultures of AIDS patients, and, with improved techniques, similar rates can be obtained in most HIV-1 positives. Virus culture, however, is not a suitable approach for most epidemiologic studies because of cost, technical complexity, and the difficulties of obtaining viable cells in field settings.

A variety of advanced molecular techniques involving amplification of HIV viral RNA or proviral DNA have been developed. The most widely used is PCR, but approaches such as ligase chain reaction[262] and branch chain amplification are also available.[48,173,216,307,453] Current versions of these assays not only detect virus but also quantify the amount present.[69,509] Because of their increasing reliability and formatting for large-scale screening, quantitative amplification tests will likely become important in clinical evaluation of patients and in epidemiologic studies of natural history and disease progression. *In situ* detection and amplification of HIV, which allows evaluation of the particular infected cell and the number of cells involved,[176,220] is valuable for pathogenesis research; newer approaches combining intracellu-

lar amplification and flow cytometry may find applications in clinical and epidemiologic settings.

3.4.2. Serological Diagnostic Methods. During primary infection with HIV-1 the initial buildup of virus in the plasma is followed by the appearance of HIV-1 antibodies. Concomitant with this early phase of infection there is a decline in CD4 cells that can be modest or quite dramatic; immune activation markers such as neopterin and β_2-microglobulin also increase with HIV-1 infection. Measurements of all of these parameters [viral antigen (p24 antigen), viral antibody, lymphocyte subset count (especially CD4 cells and markers of immune activation)] have been used in epidemiologic studies to detect infection and define predictors of endpoints such as AIDS and death.[183,184,205,271,272]

The mainstay of HIV-1 detection is testing for antibody to the virus. HIV-1 antibody testing is a highly predictive of infection[238,514] except when antibody is passively transferred at birth[26,129,152,438] or lost during late-stage infection.[12,67,338] In view of the short time from initial propagation of the virus in continuous T-cell lines to assay licensure, first-generation HIV-1 whole virus lysate tests achieved remarkable sensitivity and specificity.[68,489,494] These assays, however, were insensitive for detecting antibody response during early infection, and false-positive tests occurred because reactions to various viral antigens or contaminating cellular proteins were higher than optimal. The algorithm developed for testing required the presence of a repeatedly reactive whole virus enzyme-linked immunosorbent assay (ELISA) and a positive Western blot test for confirmation (Fig. 2). Strict criteria were established to define a positive Western blot based on the presence of multiple bands representing antibodies to viral envelope and structural antigens.

Newer whole virus HIV-1 screening tests have substantially improved viral purity and often include recombinantly produced viral peptides to enhance sensitivity.[347] Consequently, these assays detect antibody in earlier stages of seroconversion, substantially reducing the number of false negatives. In fact, current screening tests are often more sensitive than Western blot. Nonetheless, there continues to be a period after infection when HIV tests are negative because antibody has not yet developed.[324,441] This window can be especially critical in areas with a high incidence of new infections. Additionally, fears that the variability in HIV-1 might result in false-negative reactions were realized in recent reports of a variant HIV-1 virus isolate, termed "HIV 0" (first isolated in the Cameroons), which is missed by some currently licensed screening assays.[218] A number of manufacturers have developed rapid assays as possible home

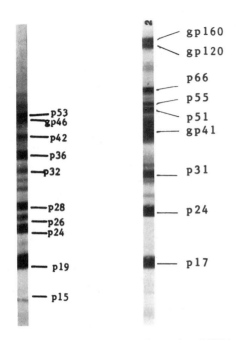

Figure 2. Western blot of major antigenic proteins of HIV-1. Photograph courtesy of Dr. Steven Alexander.

tests kits for HIV infection. Such assays may be useful for epidemiologic field studies and for screening in developing world settings where more sophisticated laboratory equipment may be lacking. However, performance characteristics of some of these tests have not been fully evaluated. Other serological techniques for HIV-1 antibody detection include radioimmune assays, immunofluorescence assays, and hemagglutination approaches.[159]

The methods for serological diagnosis for HIV-2 infection are complicated because HIV-1 and HIV-2 have a nucleotide sequence homology of about 60% for the *gag* and *pol* glycoproteins and about 40% for *env*.[378] Because HIV-1 and HIV-2 differ antigenically, HIV-1 screening tests are unreliable for detecting HIV-2. The amount of cross-reactivity varies between assays and between areas. It also decreases concurrently with the level of immunosuppression, so that HIV-2-associated AIDS cases are less likely to be detected by an HIV-1 assay than are healthy carriers. HIV-2 screening tests have been developed, including combination tests with antigens representing both virus types. Confirmatory testing for HIV-2 also requires HIV-2-specific assays to distinguish the two infections, since HIV-2-infected persons may test positive, indeterminant, or negative by HIV-1 Western blot. Although many early reports of apparent dual infection with HIV-1 and HIV-2 simply represent cross-reactivity, true dual

infection has been confirmed by PCR[406] and viral culture.

Since June 1992, blood donors in the United States have been screened for HIV-2 antibodies using licensed combination enzyme immunoassays (EIAs); persons from HIV-2-endemic areas have also been asked to refrain from donating blood. These actions have reduced the probability of transfusion-acquired HIV-2 infection to essentially zero.[378] Routine screening outside of blood centers is not recommended. CDC guidelines indicate that HIV-2 testing should be used in relation to HIV testing for individuals who are specifically at risk for HIV-2 infection, either by nationality, by sexual exposure to a person from an endemic area, or by birth from an infected mother.[377]

The reported performance of the Food and Drug Administration (FDA)-licensed HIV-1/HIV-2 combination assays for detection of HIV-2 infection is quite high, >99% sensitivity. There are currently no FDA-licensed HIV-2 supplemental tests, but a variety of research tests are available. The CDC-recommended algorithm for screening for both viruses involves use of supplementary testing for those specimens found to be repeatedly reactive on the combination assay, but negative or indeterminate on HIV-1 Western blot or those that are positive but have risk factors for HIV-2 infection (Fig. 3). An alterna-

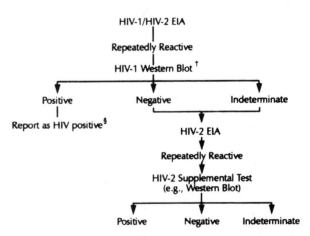

Figure 3. Testing schema for HIV-1 and -2 in the United States. †An immunofluorescence assay (IFA) for HIV-1 antibodies has recently been licensed by the Food and Drug Administration and can be used instead of Western blot. Positive and negative IFA results should be interpreted in the same manner as similar results from Western blot tests. An indeterminate IFA should first be tested by HIV-1 Western blot and then as indicated by the Western blot results. §Perform HIV-2 EIA only if there is an identified risk factor for HIV-2 infection. From Centers for Disease Control.

tive algorithm for an HIV-1 high-risk population has been proposed.[231]

HIV-1 viral antigenemia may occur early in infection or at the end stage of disease, and serological assays can detect the 24,000 molecular weight, p24, viral nucleocapsid antigen. The antigen-capture assay is relatively insensitive compared to techniques such as RNA PCR assays; sensitivity can be enhanced by acid dissociation, which frees the p24 antigen from its complexing antibody.[225]

3.4.3. Markers of Immunologic Status.

Both HIV-1 and HIV-2 are trophic for the CD4$^+$ cells and a sine qua non of infection is perturbations of the CD4$^+$ lymphocyte count. Early in the AIDS epidemic, the CD4 count was used as a surrogate for persons infected with the then unknown etiologic agent.[209] Since the discovery of HIV-1, the CD4 count has been widely used clinically as an intermediate marker of disease and epidemiologically as a staging tool.[145,233] In epidemiologic studies, CD4 has been used to estimate duration of disease following seroconversion[126,127,196,269,356] and to predict AIDS onset.[207] CD4 has now become a tool for defining AIDS in the 1993 CDC AIDS definition.[101]

The CD4$^+$ lymphocyte count is measured most reliably on a fluorescence-activated cell counter (FACS) that uses a laser to activate chemiluminescent-tagged antibodies specific for cell surface markers such as CD4 (T helper) or CD8 (T suppressor). Increased clinical and epidemiologic applications of this technology have led to adaption of instrumentation to this setting and have focused considerable attention on issues of quality control.[283,515] Other immunologic markers include soluble CD4 measurements and markers of immune activation, such as neopterin, β_2-microglobulin, and interferon, that predict disease outcome.[52,62,193,272,515] Other immunologic tools, including delayed hypersensitivity skin testing and complex markers of cell-mediated immunity, have been applied less frequently to epidemiologically defined populations.[283]

4. Biological Characteristics of HIV

The three classes of human retroviruses—oncornaviruses, lentiviruses, and spumaviruses—are enveloped RNA viruses. HIV-1 and HIV-2 are lentiviruses with a single-stranded RNA genome that replicates through a double-stranded DNA proviral intermediate (cDNA) (see Fig. 4).[200,477] Reverse transcriptase, a viral enzyme, catalyzes the unique life cycle of these viruses by converting genomic RNA into cDNA; viral integrase incorporates cDNA into the host genome.[477] Genomic integration

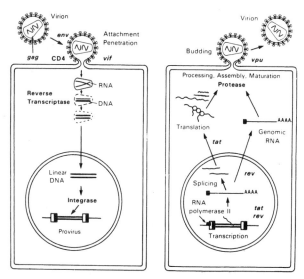

Figure 4. Life cycle of HIV. From archival photos.

allows retroviruses to cause lifelong infection and may play a role in the propensity of human retroviruses to cause diseases of long latency.[200]

Morphologically, HIV-1 is approximately 110 nm in diameter with a thin electron-dense outer envelope and an electron-dense, roughly cone-shaped cylindrical core (Fig. 1). In the production of new HIV particles, retroviral genes are expressed as large overlapping polyproteins that are later processed into functional peptide products by the viral-encoded and cellular proteases.[477] HIV long terminal repeats (LTR) at the 5′ and 3′ ends of the genome contain regulatory elements that promote virion production.[202] The genomic structure of HIV-1 and -2 (Fig. 5) contains the following genes: *gag* (group-specific antigen), whose products form the skeleton for the virion; *pol* (polymerase/integrase), whose products are involved in enzymatic functions of the virus; *env* (envelope), which codes for the external and transmembrane outer envelope elements involved in virus binding; and a series of accessory genes that regulate virus expression.[198–200,221] The *gag* gene products are synthesized from a single message, and the functional proteins are formed through enzymatic cleavage using the viral-encoded protease. These structural elements are the matrix protein of 17,000 Da (p17), the capsid of 24,000 Da (p24), the nucleocapsid of 9000 Da (p9), and the nucleic acid binding protein of 6000 Da (p6). On Western blot these proteins (except for p6) are strongly antigenic and appear as distinct bands with additional bands representing precursor intermediates. The *pol* gene codes for several enzymes: reverse transcriptase

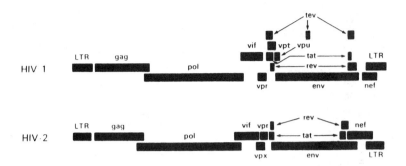

Figure 5. Genomic structure of HIV-1 and -2. HIV viruses have a complex regulatory array. From Gallo and Nerurkar,[202a] with permission.

(involved in RNA to DNA transcription), endonuclease (ribonuclease-H), and integrase, which functions for viral integration. The *env* gene encodes the major components of the viral coat: the surface glycoprotein of 120,000 Da (gp120), the transmembrane component of 41,000 Da (gp41), and the precursor called gp160.

HIV uses a complex process for controlling virion production and propagation through production of a series of early and late regulatory proteins produced through a series of multiply spliced mRNAs. The early regulatory proteins are *tat*, *rev*, and *nef*, and the late regulatory proteins are *vif* and (for HIV-1 only) *vpu*; additional accessory proteins, *vpr* and (for HIV-2 only) *vpx*, influence replication efficiency.[221] HIV *tat* is responsible for enhanced transcription of viral gene products and also has a transactivating repressor function for some genes in the major histocompatibility complex. *Rev* (regulator for HIV) modulates the complex transport of virion components and the translation of viral messages containing the *rev* responsive element in the production of virus particles. *Nef* modulates CD4 receptor expression and facilitates *in vivo* viral replication and pathogenesis. *Vif* modulates infectivity, *vpu* (for HIV-1) and *vpx* (for HIV-2) influence virion maturation and release, and *vpr* facilitates replication.

The life cycle of HIV involves an infection phase (including viral attachment, entry, reverse transcription, and proviral integration) and an expression phase (including transcription, translation, assembly, and budding of the virion)[477] (Fig. 4). The initial stage of infection involves the binding of HIV-1 and -2 to cell surface structures that determine viral trophism. Many of the pathogenic effects of the virus result from immune dysfunction, which compromises the ability of the immune system to perform its myriad functions because of HIV-induced perturbations of cell function and immune regulatory processes. The HIV-1 envelope has a high-affinity binding site specific for the CD4 molecule. The CD4 molecule occurs on mature T-helper lymphocytes and on other cells

of the immune system, particularly circulating and fixed-tissue (dendritic) monocytes, macrophages, and microglia.[143,381,410,413] Other cells where the virus attaches are follicular dendritic cells in lymph nodes, M cells on Peyers patches, and galactosylcerebroside-positive cells in the brain and gut.[488] The CD4 molecule alone is not sufficient for virus binding and determination of the second binding site has been a focus of ongoing research.[77,117,290]

Following attachment, a fusogenic domain on the gp41 appears to be involved in the uncoating and release of virion genetic material into the cell. The viral reverse transcriptase synthesizes a cDNA copy of the virus that is transported to the nucleus: viral integrase is involved in proviral integration. Integration of HIV-1 appears to be random but may be influenced by host chromatic factors.[221] A recent study reported nonrandom integration of HIV-1 in juxtaposition to the c-*fos* oncogene in certain cases of non-B-cell lymphoma in HIV-1-infected persons, a finding awaiting confirmation.[459] Once integrated into the host chromosome, the virus is replicated like a cellular gene into the progeny cells. The proviral genome is the template for new virus production and the expression of new virion production is controlled by viral and host cellular regulatory elements. The first elements produced are the regulatory elements, which then modulate the production of genomic RNA and mRNA necessary for production of virion structural proteins. The process is regulated by the regulatory elements discussed above, with the *rev* protein playing a key role in the switch to virion production.

The morphology, life cycle, and genomic structure of HIV-2 is quite similar to that of the HIV-1 (Figs. 1, 3, and 4); the regulatory genes of HIV-2 differ from those of HIV-1 by coding for the *vpx* protein and lacking the *vpu* protein as noted above. The *env* gene encodes a gp120 membrane and a gp32-40 transmembrane glycoprotein. The *gag* gene encodes a p55 precursor to p24-26 and p15. The *pol* gene products include the p64 and the p53 of reverse transcriptase, p34 of integrase, and p11 of protease.[315] Of note is the report by Simon and colleagues[460] that the circulating viral load in 40 HIV-2-infected adults is significantly lower than that in HIV-1-infected individuals in the same levels of CD4 count.

5. Descriptive Epidemiology

HIV has spread around the world in about 20 years, with the vast majority of infections recorded since the 1980s.[8] While there are few precise data on the prevalence of HIV-1 in the 1970s, archival blood collections from some high-risk groups document the presence of HIV-1 in a small percentage of people during that decade (Fig. 6A). A pattern has been repeated around the world: An initial low level prevalence of the virus in the population is followed by a sometimes explosive, but more often insidious, increase in infections (Fig. 6B). Because of the long latency between infection and clinical AIDS (an estimated median of 8 to 10 years for adults in developed countries), the spread of virus can be quite extensive before the magnitude of the problem is appreciated (as recently evidenced in Thailand and India). Often the introduction of the virus takes place through a core transmission group with subsequent centripetal spread to larger segments of the population.

The spread of the virus has been documented through both standard and molecular epidemiologic techniques. The descriptive epidemiology of HIV and AIDS is a multifaceted tableau involving a complex dynamic that defies simple description. The intercontinental spread among homosexual men was documented from the United States to Europe and the Caribbean early in the epidemic in studies in which sexual contact with a US homosexual was identified as a major risk factor for AIDS at these international sites.[19–21,170,331] Intracontinental spread in Africa has involved long-distance truck drivers and female prostitutes along their routes.[462] Once a toehold is established in a new geographic area, the virus spreads through sexual, parenteral, or mother-to-child routes, often enhanced by factors such as coincident sexually transmitted diseases and socioeconomic upheaval (e.g., worker migrations in Africa).

When the AIDS epidemic was first recognized in the United States in 1981, the strong male predominance reflected cases diagnosed among sexually active male homosexuals and injection drug users. During the 1980s, the pattern in the United States broadened as women and minorities increasingly became affected. The European pattern parallels that seen in the United States except for a

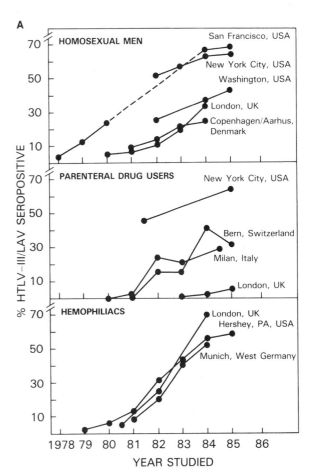

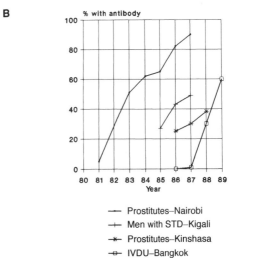

Figure 6. (A) Historic seroprevalence in IDUs, homosexuals, and hemophiliacs. Distribution of HIV seropositivity by year and geographic location in three groups at high risk of AIDS. From Melbye *et al.*,[333a] with permission. (B) HIV-1 antibody prevalence rated over time in high-risk populations. From Piot *et al.*,[406a] with permission.

predominance of injection-drug-use-associated cases in a number of countries and an early wave of cases among migrants from Africa. Worldwide, the AIDS epidemic among sexually active heterosexuals was first recognized in Haiti and in some African countries.[37,138] The number of AIDS cases in Asia belies a burgeoning HIV-1 epidemic driven by injection drug use and heterosexual spread, especially involving prostitutes. In most geographic areas, sexually active persons from the late teens through the 40s are the group most conspicuously affected; the pediatric population experiences infection early in life largely from *in utero* and/or perinatal exposure and in developing countries through breast-feeding.

5.1. Descriptive Epidemiology of HIV and AIDS

5.1.1. Prevalence and Incidence

5.1.1a. Infection. The prevalence of HIV infection varies by age, sex, race, geography, and risk group. The number of persons infected in the world or even in individual countries is difficult to estimate. Prevalence rates are derived either from convenience samples or population-based sampling, but myriad biases compromise the accuracy of these estimates.

HIV-1 in the United States. Table 3 summarizes the results of surveys conducted among various populations in the United States, including blood donors, military and Job Corps recruits, injection drug users in treatment, attenders at clinics for STDs, and newborn infants. Blood

donor screening in the United States has been conducted since 1985.[85] Retrospective analyses of banked samples from blood donors documented the efficacy of procedures to encourage self-deferral of high-risk groups; rates of seropositivity in one San Francisco blood bank fell from 1 in 100 transfused unit before donor deferral to 1 in 700 after donor counseling was implemented.[70] Figure 7 shows the time trends in HIV seroprevalence in first-time blood donors between July 1985 and December 1990. Male rates have consistently been higher than female rates; a decline for males from 0.08% in the early time to 0.025% in 1990 was ascribed to donor deferral of HIV-1 positives as well as more effective donor education.[96,97] Recent estimates indicate approximately 1 in 20,000 donors test positive for HIV.[96,97] The HIV prevalence is lower in blood donor populations than in the general population due to self-deferral of high-risk donors, higher socioeconomic status of donors, and underrepresentation of minority populations where HIV rates are higher.

Applicants for military service in the United States represent another large population with ongoing surveillance. Geography influences HIV-1 seroprevalence rates in this group (Fig. 8); seroprevalence rates are highest in the mid-Atlantic region corresponding to high AIDS incidence areas. The overall prevalence between 1985 and 1990 was 1.2 per 1000 among all recruits.[56,57,430] For military recruits, crude seroprevalence is 1.42 per 1000 for men and 0.66 per 1000 for women with peak prevalence in those aged 25 to 30 years. Among minority populations,

Table 3. Sentinel Populations in the Family of HIV Seroprevalence Surveys[a]

Study population	Number of metropolitan areas	Number of clinics	Seroprevalence (%) Median	Seroprevalence (%) Range[b]
1. Injection drug user	40	61	3.9	0.0–44
2. Persons in tuberculosis treatment	18	35	5.9	0.4–36
3. Women seeking family planning, prenatal care and abortion services	38	146	0.2	0.0–1.7
4. Hospital patients admitted for non HIV-related conditions	21	26	0.7	0.1–7.8
5. Childbearing women	39 states/territories		0.07	0.0–0.58
6. Job Corps entrants	52 states/territories		0.23	0.0–1.11
7. University students	9 universities		0.0	0.0–0.9
8. Prisoners	16 populations with compulsory tests		0.8	0.0–17
9. Civilian applicants for military service	52 states/territories		Black males 0.37 Hispanic males 0.18 White males 0.05 Black females 0.14 Hispanic females 0.09	
10. Blood donors	50 Red Cross centers		Males 0.04 Females 0.01	

[a]Adapted from Brookmeyer and Gail.[54]
[b]The range refers to the range of the median values over metropolitan areas unless otherwise indicated.

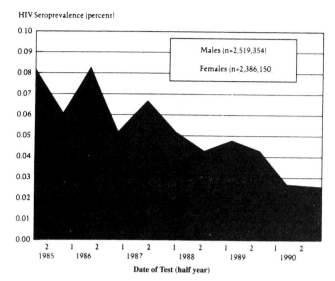

Figure 7. First time blood donor time trends. From Mann *et al.*,[314a] with permission.

rates for black males are 4 per 1000, for Hispanic males 2.2 per 1000; rates among whites are 0.5 per 1000. The changing rates observed in this population reflect the impact of policies that exclude those with a history of drug abuse and homosexual orientation from military service and the effects of changing demographics in recruitment.[56–58,430] These data provide a biased but useful window to understand broad patterns of seroprevalence in the United States.

Among Job Corps applicants, geographic and racial trends are similar to those observed for military recruits.[96,97] Rates range from 5.5 per 1000 for black males to 1.5 per 1000 for white males. Male predominance among Job Corps applicants is lower than in military recruits: 3.4 per 1000 for men and 3.1 per 1000 for women. Because Job Corps applicants are drawn from lower-socioeconomic youths from inner cities where illicit drug use and other high-risk behaviors are more common, the prevalence rates obtained from this group tend to be higher than those in the general population.

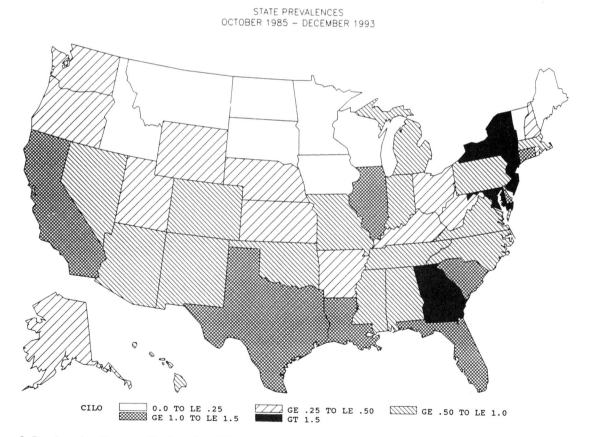

Figure 8. Prevalence in military recruits. Sex-adjusted HIV-1 antibody prevalence (percentage positive) in military applicants. From Centers for Disease Control.

Another source of prevalence data is blinded testing of newborn blood collected routinely to screen for inborn metabolic errors. These tests provide a reasonable estimate of HIV-1 prevalence in women in the reproductive years but are biased by the demographics of those women who achieve term delivery.[219] Prevalence rates by state range between 0.0 and 0.66% with highest rates observed in urban areas of the northeast United States. Racial and ethnic patterns also influence rates; the highest rates of seropositivity are observed among infants born in New York City (1.6% among African American, 1.0% Hispanic children, compared to 0.1% of the Caucasian children).[219]

Other sources of seroprevalence data are surveys of high-risk populations. In the United States, high rates of seropositivity are observed in cohorts of homosexual men, parenteral drug abusers, and hemophiliacs, groups defined by CDC as being at high risk for AIDS[89]; the estimates are highly biased by the nature of recruitment. A 1987 population-based probability estimate in San Francisco of single men 25 to 54 years of age from 19 census tracts at the epicenter of AIDS found no positives among 204 heterosexual men, while 306 of 796 (48.5%) of homosexual and bisexual men were HIV positive.[502,503,505] Other surveys of healthy homosexual men in high and intermediate AIDS incidence areas demonstrated rates of 20–80% in the middle 1980s,[214,240,263,331] but these surveys may be biased by self-selection of volunteers from the more sexually active subpopulation of homosexual men.

Hemophiliacs who used commercial factor VIII concentrate preparations between 1978 and 1985 have rates of HIV-1 antibodies ranging from 30 to over 70%, depending on severity of disease and amount of concentrate used.[277] Figure 9 demonstrates the emergence of the first wave of HIV-1 infections in the late 1970s among those with severe hemophilia and high factor concentrate usage. The pattern of seroconversion between 1981 and 1984 reveals the extremely high hazard associated with continual exposure to the factor concentrate of that day.[277]

Parenteral drug abusers have a median HIV-1 seroprevalence of 3.9%; their rates of positivity vary much more widely than those for homosexual men (Fig. 10). The highest rates (50% or more) are seen in the New York City metropolitan area; lower rates (1–30%) are observed in drug-abusing populations in other areas.[96,97,429]

Prevalence data are also obtained from STD clinics in the United States where the median value was 3.9%, but with values as high as 49.3% among persons entering treatment.[152,412] Rates in this population are influenced by many variables including coincidence of other risk factors such as parenteral drug abuse and high-risk sexual lifestyle.

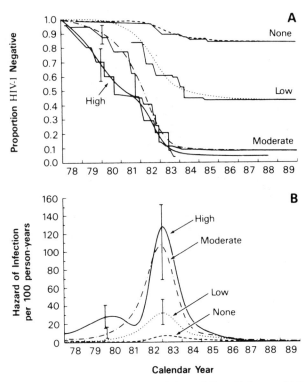

Figure 9. Hazard of HIV prevalence in hemophiliacs. Estimates of (A) the HIV-1-free-survival curve and (B) the corresponding hazard rate for members of the five-center cohort with type A hemophilia, grouped by mean annual dose of factor VIII concentrate used between 1978 and 1984. From Kroner et al.,[277] with permission.

Estimates of the total U.S. prevalence are derived from several sources. The initial estimates used calculations based on HIV prevalence surveys from high-risk populations such as those described above and then extrapolated to the total numbers of persons in the general U.S. population with that lifestyle.[54] Initially 1.0 to 1.5 million persons were estimated to be infected with HIV-1 as of January 1986, but this estimate was disputed because of concerns about the validity of the behavior frequency rates (e.g., a 10% population frequency of homosexuality). Subsequent efforts, utilizing similar strategies with an expanded family of surveys, concluded that 650,000 to 900,000 persons were infected as of January 1986 and 800,000 to 1.2 million as of June 1989.[97] The precision of such estimates has been hampered by the inability to quantify the various biases within each of the survey groups.

A national probability sample has not been feasible because biases discovered in pilot surveys challenged the prudence of doing large-scale surveys of uncertain validity. These biases include underrepresentation of certain high-risk groups such as intravenous drug users whose

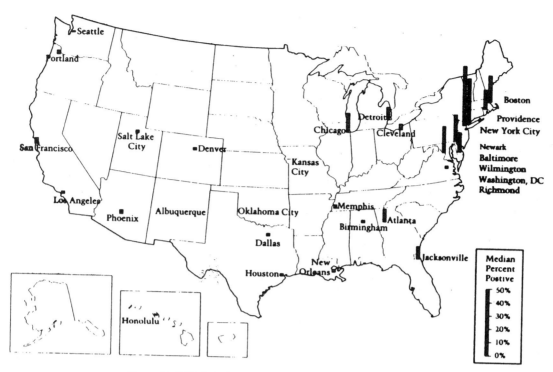

Figure 10. IDUs in United States. From Centers for Disease Control.

residential status is marginal, nonparticipation of persons with high-risk sexual lifestyles, and misclassification of risk behavior.[54] The largest such pilot probability sample in the United States was conducted in Dallas, Texas, and found 15 positives among 1374 sampled individuals, referable to 950,000 persons. The adjusted prevalence was 0.40%, but 20.3% of eligibles did not participate,[54,319] and further studies of the nonresponding population suggested they might have higher risk for HIV. Adjusting for this bias, the final estimated population prevalence was 0.42%.[54,235] Extrapolating from this estimate to the U.S. population, 234,700 infections were predicted, considerably less than that based on other estimation approaches.[54]

Since AIDS statistics are fairly reliable, they offer an opportunity to apply statistical techniques such as back calculation to estimate the number of infected persons required to account for the observed cases of AIDS. A critical measure for developing these estimates is the underlying incubation distribution of time from HIV-1 infection to AIDS. Numerous factors influence this time-to-AIDS estimate including intrinsic factors such as age and risk group (e.g., the high rate of Kaposi's sarcoma among homosexual men) and extrinsic factors such as underdiagnosis, disease-reporting delays, and therapeutic effects that can slow the time to AIDS. Based on back calculation, Brookmeyer and colleagues[53] estimated that

the plausible range of HIV-1 infections in the United States as of January 1991 was 628,000 to 988,000 (Fig. 11). Others have reported similar estimates with modest variation in rate based on differences in the assumptions employed in the backcalculation model.

HIV-1 in the World. The international dimension of the HIV-1 epidemic was first recognized through AIDS cases among migrants from Haiti to the United States and from certain African countries to Europe. The magnitude

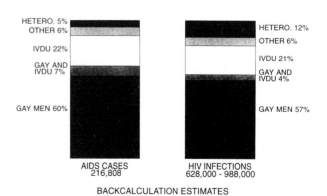

Figure 11. HIV/AIDS cases by risk group. From Brookmeyer *et al.*,[53] with permission.

of the epidemic is of catastrophic proportion in some locales, particularly in Africa, and is projected to reverse gains in longevity and economic growth because of high attack rates in the economically most productive age groups.[8] However, estimating the true prevalence of HIV-1 in developing countries is even more complex than in the United States because of limitations in health care resources and the difficulty of determining prevalence in representative populations. Nonetheless, surveys of sentinel populations such as injection drug users, women of childbearing age, STD clinic attenders, blood donors, and female prostitutes provide useful insights into the pattern of spread, if not a precise estimate of HIV-1 prevalence in the overall population.

Table 4 summarizes data from high-risk groups in different regions and representative locales. Such surveys share many of the same biases as those in the United States. In many locales where heterosexual spread is the major route, female prostitutes are an important core transmission group. Thus the high rates listed in Table 4 for this risk group serve as a harbinger of HIV-1 introduction and spread.

Table 4. Seroprevalence Rates per 100 Persons of Selected Populations in Urban Areas of Designated Nations: 1988–1992[a]

Region/locale	Prostitutes	IDUs	STD clinics	Pregnant women
Africa				
Zambia	NA	NA	60–65	9–30.0
Malawi	0–56.0	NA	62.0	2–18.0
Ethiopia	24–69.0	NA	18.4	2.8
Kenya	36–88.0	NA	23.0	5–15.0
Côte d'Ivoire	7–69.0	NA	22.0	1–12.3
Caribbean/South America				
Argentina	1–8.3	29–36.8	6–18.0	NA
Colombia	0–.3	NA	12–14.0	NA
Dominican Rep.	2–4.3	0.00	3–5.0	0.7–1.5
Jamaica	14.6	NA	0.1–3.0	NA
Martinique	40.0	NA	0.3–4.0	NA
Trinidad	13.0	NA	10–14.0	0.00
Asia				
Thailand, Bangkok	9–33.3	34–45.0	8.8	0–1.2
Thailand, Central	3.3–14.0	43–50.0	36.0	0–2.0
Bombay, India	41.2	3.4	0.8–15.0	1.0
Tamil, India	7.6	NA	0.4–8.5	0.5
Philippines	0.2	NA	0.1	NA
Cambodia	9.2	NA	4.2	NA

[a]Estimates derived from U.S. Bureau of the Census, Center for International Research, HIV/AIDS Surveillance Database, June 1994.

A second core transmission group, injection drug users, also appear with elevated rates early in the course of new epidemics. Because of the high frequency of STDs in these core transmission groups, STD clinic attenders are a useful sentinel for further spread into the population, particularly as a means of tracking the secondary spread of the virus among sexually active networks. Furthermore, coincident STDs appear to augment HIV spread.

Surveys of childbearing women provide additional insights into the spread of the epidemic; they represent a more general population sample, and thus high rates are an indicator of substantial spread into the general sexually active population. High rates in blood donor populations also are an indicator of broader spread into the community. Finally, high rates in medical clinics, particularly tuberculosis clinics in developing countries, indicate a maturing of the epidemic as clinical manifestations of immunodeficiency emerge.[1]

In a few places, carefully crafted surveys have provided reasonable population-based estimates of HIV-1 prevalence,[148,266,366,486] but nonparticipation, unreliable census data, and marked in-country geographic variation (between-urban, periurban, and rural areas) limit the generalization of these data to the total population. Nonparticipation is a particularly vexing problem. In many developing countries, outmigration, particularly of young men seeking employment, has been a major catalyst for spreading HIV-1 from locale to locale. Itinerant workers return to their home districts infected with HIV from exposure to female prostitutes in the urban centers. They represent a high-risk group who are frequently absent from home and therefore not surveyed; consequently their absence contributes to underestimates of seroprevalence.[265] Population mobility further confounds proper census estimates and may result in undercounting, thereby increasing HIV-1 prevalence estimates.

To compensate for these limitations, statistical modeling has been adapted to provide HIV-1 prevalence estimates in developed and developing countries. Back calculation is often used in the United States and Europe where AIDS statistics are largely reliable and estimates of HIV-1 incubation distribution are understood.[54] In areas of the world where the epidemic is the greatest, these parameter estimates are unreliable, so alternate modeling approaches are needed. Compartment models use estimates of HIV-1 prevalence in different risk groups and the size of those risk groups, but these parameters are difficult to measure.[54] Epidemic transmission models are often complex, relying on a series of parameter estimates concerning efficiency of transmission between different groups.[54,267] Other approaches have focused on the probability of

Table 5. Estimated Cumulative Adult HIV Infections by Geographic Area as of January 1, 1992, and 1995[a]

Geographic area	January 1, 1992, estimate	1995 estimate	No. of new infections 1992–1995	Percent increase 1992–1995
North America	1,167,000	1,495,000	328,000	28
Western Europe	718,000	1,186,000	468,000	65
Oceania	28,000	40,000	12,000	43
Latin America	995,000	1,407,000	412,000	41
Sub-Saharan Africa	7,803,000	11,449,000	3,646,000	47
Caribbean	310,000	474,000	164,000	53
Eastern Europe	27,000	44,000	17,000	63
Southeast Mediterranean	35,000	59,000	24,000	69
Northeast Asia	41,000	80,000	39,000	95
Southeast Asia	675,000	1,220,000	545,000	66
Total	11,799,000	17,454,000	5,655,000	48

[a]From Mann *et al.*[(314a)]

ly few AIDS cases (evidence of the long latency from infection to disease). In the United States and , a steady state has been achieved between new ons and AIDS deaths; this results in a deceptive ing of cases and infections. Projections for South a and the Caribbean call for rather severe increases S cases.

.1c. AIDS Mortality Data. HIV-1 has an un- high mortality rate for a viral infection so there is a arallel between AIDS and death. The median time e-1993 AIDS and death is 24 months. In the United Fig. 16 top), AIDS has recently become the leading cause of death among men between the ages of 25 and 44 years. For women (Fig. 16 bottom), AIDS is the fourth leading cause of death in the same age group. Worldwide to date, over 2.5 million have died from AIDS (Table 8).[(1,312)] By 1995, it is estimated that over 6 million will have died with AIDS; by 2000, up to 20 million adults will have died from the disease.[(1,312)] Three quarters of adult deaths and 90% of pediatric deaths will occur on the African continent. While AIDS is known to be among the leading causes of death in the younger age groups in many developed countries (Fig. 17), for reasons already discussed, mortality recording in developing countries is less

"mixing" among compartments of the sexually active population as a means of accessing spread and as a means of projecting prevalence.[(42,322)] All of these modeling approaches are limited by the quality of underlying data, and thus may be more useful for estimating the impact of various interventions than as a means of providing precise estimates of prevalence.[(450)]

Chin and Lwanga[(115)] use seroprevalence surveys as a means of back calculating infection rates and for projecting prevalence; WHO uses this technique for making projections. The Delphi method attempts to reach consensus among experts to derive a range of prevalence estimates.[(116,313)] Figure 12 displays the worldwide geographic distribution of HIV-1 infections in 1991; the totals are estimated to have escalated to approximately 11.7 million in 1992 and approximately 13 million in 1994.

HIV-1 Incidence. Measuring HIV-1 prevalence is fraught with complexity; estimating the incidence of new infections in populations is an even more daunting task. The simplest approach has been to perform serial cross-sectional surveys of the various populations who serve as the source of HIV-1 prevalence estimation. Figure 13

presents an example of serial seroprevalence among pregnant African women and Fig. 7 for the first-time blood donors in the United States. The major challenge in this approach is the need to know the frequency of repeated measurements on the same individual over time and the dynamics of loss of individuals from the cohort. Nonetheless, serial cross-sectional surveys can be used to estimate HIV-1 seroincidence. For example, Brundage and colleagues estimated incidence of HIV infections among miliary recruits based on serial cross-sectional studies. While inexact, such measurements provide insights about relative contributions of subpopulations to the trend. In the Brundage report, African-American populations exhibited the highest estimated rates of new infections; these rates varied geographically in concordance with community prevalence rates.[(56,57)]

Direct measurement of seroincidence in cohorts from different places in the United States and the world have provided insights into the patterns and dynamics of virus dissemination. Many of the populations shown in Fig. 6A represent serial measurements on high-risk cohorts collected (for other purposes) during the early years

Table 6. Cumulative Adult and Pediatric AIDS Cases by Geographic Area as of January 1, 1992[a]

Geographic area	Adult cases reported No.	Adult cases reported %	Adult cases No.	Adult cases %	Pediatric cases No.	Pediatric cases %	Total cases No.	Total cases %
North America	218,989	45.2	257,500	12.8	9,000	1.6	266,500	10.3
Western Europe	63,659	13.1	99,000	4.9	4,000	0.7	103,000	4.0
Oceania	3,509	0.7	4,500	0.2	200	—	4,700	0.2
Latin America	41,603	8.6	173,000	8.6	21,500	3.7	194,500	7.5
Sub-Saharan Africa	144,522	29.9	1,367,000	67.7	520,500	90.7	1,887,500	72.8
Caribbean	7,885	1.6	43,000	2.1	8,000	1.4	51,000	2.0
Eastern Europe	2,236	0.5	2,500	0.1	100	—	2,600	0.1
Southeast Mediterranean	813	0.2	3,500	0.2	400	0.1	3,900	0.1
Northeast Asia	516	0.1	3,500	0.2	300	0.1	3,800	0.1
Southeast Asia	431	0.1	65,000	3.2	9,500	1.7	74,500	2.9
Total	484,163	100.0	2,018,500	100.0	573,500	100.0	2,592,000	100.0

Adult and pediatric cases estimated spans the Adult cases, Pediatric cases, and Total cases columns.

[a]From Mann *et al.*[(314a)]

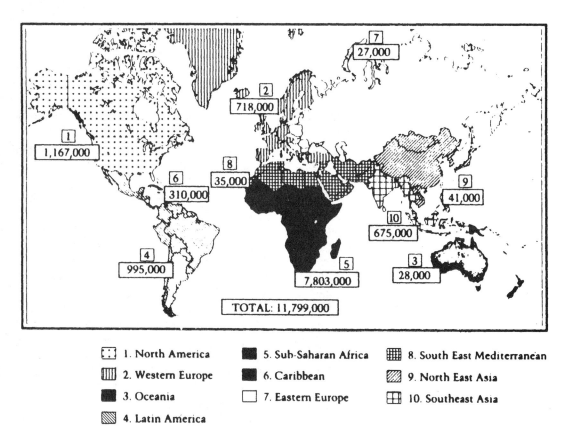

Figure 12. World map of infections. From Mann *et al.*,[(314)] with permission.

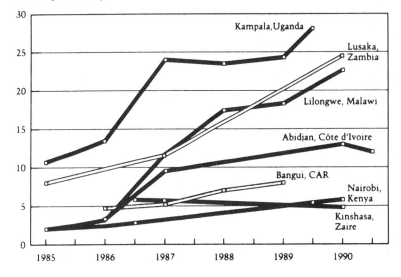

HIV Seroprevalence (percent)

Figure 13. HIV seroprevalence for pregnant women in selected urban areas of Africa, 1985–1990. It includes infection from HIV-11 and/or HIV-2. From Center for International Research, U.S. Bureau of the Census.

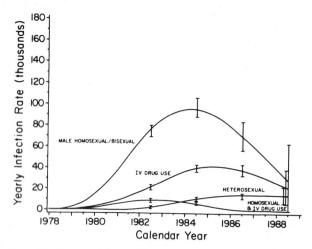

Figure 14. Reconstruction of HIV infection curves in the United States based on back calculation. From Gail and Brookmeyer,[54] with permission.

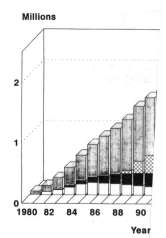

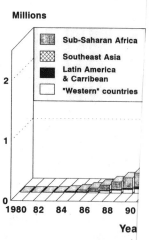

Sub-Saharan Africa
Southeast Asia
Latin America & Carribean
"Western" countries

Figure 15. (Top) Estimated/proj
(Bottom) Estimated/projected ann
114, 114a, 114b, 116a, 116b, 314a.

of the epidemic. The Multicenter Hemophilia Cohort Study (MHCS) summarized in the lower panel shows the very high rates of infections during the early 1980s when most factor VIII concentrates were contaminated with infectious virus; since 1985 (the introduction of effective HIV-1 screening of the blood supply), virtually no new infections have been recorded.[277] The Multicenter AIDS Cohort Study (MACS) provides a measure of HIV-1 incidence infections in homosexual men, albeit among a closed cohort; new infections have tapered off since the mid-1980s (inception of the MACS cohort). Such closed cohorts are biased by the fact that the most susceptible individuals were already infected by the time of cohort initiation. The remaining cohort is also at lower risk due to preventive educational interventions that accompany cohort participation.[354]

More recent studies, undertaken in preparation for possible HIV-1 vaccine trials in the United States and elsewhere, have also measured seroincidence in targeted high-risk populations. In San Francisco, young homosexual men engaged in a high-risk lifestyle (e.g., casual and/or anonymous sex in bars or bathhouses) have incidence rates of 1 to 5% per annum.[291,327,328] In studies of injection drug users in the United States, Thailand, and India, similar rates have been observed.[156,157,445,480,511] High-risk groups such as STD clinic attenders in Trinidad,[119] prostitutes in India and Thailand,[110,280,371,374,463,464] and community cohorts and partners of seropositives in Uganda and Rwanda[5,27] have demonstrated seroincidence rates as high as 30% per year (Thai prostitutes), but more usually in the 3 to 10% range.[110]

Statistical modeling techniques are extensively used for estimating seroincidence and providing projections of future infections; the data obtained are made available to and used by government policymakers and other health planners. Back calculation was used in Fig. 14 to provide an estimate (over time) of the density of HIV-1 infections in different risk groups in the United States. The peak incidence of HIV infections among male homosexuals occurred from 1982 to 1986; the pattern in injection drug users and heterosexual populations show later peaks and ongoing incidence[53]; similar calculations are shown in Fig. 9 for hemophiliacs in the United States.[277] Such broad-based approaches may overlook the importance of geographically clustered epidemics such as the rising trend in HIV-1 incidence in heterosexuals and injection drug users in Washington, DC, reported by Rosenberg and colleagues.[309]

Projections of HIV-1 incidence by region summarized in Fig. 15 provide WHO estimates through the year 2000 using the "forward" calculation method of Chin and Lwanga.[115] This method projects an emerging epidemic in the major population centers of Asia. Table 5 shows estimates of current HIV infections in 10 geographic areas of the world and percentage increases projected in the short term. Because of the complexity of population characteristics in the regions, age-adjusted HIV-1 incidence data have been difficult to estimate. What is clear is that the areas of the world experiencing the greatest increases are the major population centers in countries and regions with the least resources to cope with the projected onslaught.

5.1.1b. Clinical Disease. Surveillance for AIDS provides a useful index to the scope of the emerging pandemic, but the long latency between infection and disease and the wide variation in reporting by countries worldwide limit the accuracy of available data.[92,138,312,349] As of July 1993, 190 countries have reported AIDS cases to the WHO. The numbers presented in Table 6 confirm that the largest proportion of AIDS cases in both adults (67%) and children (90%) have occurred in sub-Saharan Africa. The United States, which accounts for almost 50% of the reported cases to WHO, contributes 10% of the estimated total AIDS burden worldwide. The assumption is that in Europe and the Western hemisphere, 80 to 90% or more of the actual number of cases are reported.[312] For the rest of the world, WHO estimates that these reported cases represent only 20% of the actual total. The January 1992 world estimate of the Global AIDS Policy Coalition (GAPC) places the total AIDS burden at 2.5 million worldwide (see Table 6).[1]

The discrepancy between reported and estimated numbers reflect not only the difficulties with diagnosing AIDS in developing countries that have an inadequate health care infrastructure, but also political circumstances that result in some countries not reporting cases at all.[312] Furthermore, the definition of AIDS in developing countries differs from that in countries with advanced technology that can diagnose exotic fungal and viral pathogens included in the AIDS definition[124,125,147,474] (see Table 1).

Geographic variations in AIDS incidence rates provide a useful context for understanding the maturity and

re
pe
Eu
in
pla
An
in

us
clo
fro
Sta

direction of the AIDS epid
from countries in different
rized in Table 7. Among
total caseload rate per 100,0
Uganda (22.3), Malawi (51.
(28.3), Tanzania (15.5), and
of the Caribbean region
105.7.[312] The low rates i
relatively recent introducti

Figure 15 (bottom)
future AIDS projections b
HIV-1 patterns in Fig. 15
tions in Fig. 15 (bottom) il
of HIV infections in Afri
mortality skyrocket. By c
areas of Asia show massi

Table 7. Cumulative AIDS Cases Reported to the World Health Organization as of July 1, 1993

Region/locale	Case rate[a]	Region/locale	Case rate[a]	Region/locale	Case rate[a]
Africa		America		Europe	
Uganda	22.3	United States of America	19.6	France	7.3
Kenya	24.7	Trinidad and Tobago	19.4	Spain	8.7
Malawi	51.6	Bahamas	105.7	Italy	6.6
Côte d'Ivoire	28.3	Guyana	14.9	Germany	2.1
Zimbabwe	33.5	Barbados	29.4	Switzerland	6.8
Rwanda	37.5	Bermuda	29.8	Monaco	25.9
Burundi	27.4	Saint Lucia	14.6	Southeast Asia	
Congo	84.7	Turks and Caicos Islands	40.0	Thailand	1.2
Togo	18.3	Cayman Islands	23.5	India	0.0
Swaziland	18.4	British Virgin Islands	20.0	Western Pacific	
		Eastern Mediterranean		Australia	2.5
		Sudan	0.6	New Zealand	1.4
		Djibouti	33.3	New Caledonia	2.6

[a]Rate, reported cases/100,000.

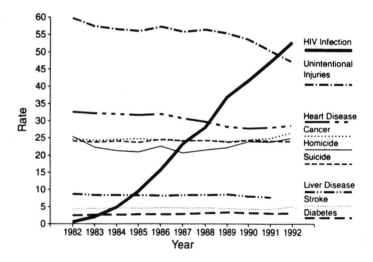

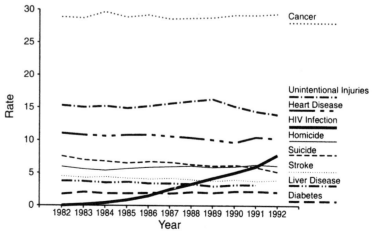

Figure 16. (Top) AIDS as a leading cause of death of males in the United States, 1982–1992. (Bottom) AIDS as a leading cause of death of females in the United States, 1982–1992. Rates per 100,000 population. National vital statistics based on underlying cause of death, using final data for 1982–1991 and provisional data for 1992. Data for liver disease in 1992 are unavailable. From Centers for Disease Control.

Table 8. Adult and Pediatric Deaths from AIDS by Geographic Area as of January 1, 1992, and 1995[a]

Geographic area	January 1, 1992			1995			Proportional increase of total
	Adult	Pediatric	Total	Adult	Pediatric	Total	
North America	214,500	9,000	223,500	475,500	20,500	496,000	2.2
Western Europe	78,500	4,000	82,500	238,000	12,000	250,000	3.0
Oceania	3,500	200	3,700	10,000	500	10,500	2.8
Latin America	166,000	21,000	187,000	406,000	55,000	461,000	2.5
Sub-Saharan Africa	1,312,000	511,000	1,823,000	3,184,000	1,320,500	4,504,500	2.5
Caribbean	41,000	7,500	48,500	117,000	23,000	140,000	2.9
Eastern Europe	2,000	100	2,100	7,500	300	7,800	3.7
Southeast Mediterranean	2,500	400	2,900	10,500	1,500	12,000	4.1
Northeast Asia	3,500	300	3,800	14,000	1,100	15,100	4.0
Southeast Asia	61,500	9,000	70,500	231,000	40,000	271,000	3.8
Total	1,885,000	562,500	2,447,500	4,693,500	1,474,400	6,167,900	2.5

[a]From Mann *et al.*[314a]

reliable. However, a recent survey in the capital of Côte d'Ivoire confirmed that HIV-1 was associated with over a third of all deaths (based on postmortem blood samples) and was similar to the United States as a leading cause of death for persons in the sexually active age group.[146] The Côte d'Ivoire survey illustrates how the lack of adequate medical services and the complexity of diagnosing AIDS can result in significant underestimation of the numbers of HIV-infected persons and attendant mortality.

5.1.2. Epidemic Behavior and Contagiousness. The origins of the HIV epidemic are uncertain, but the virus itself is apparently a newcomer to human populations. And, while the earliest case was documented in 1959, epidemic cases were not detected in the United States until the mid- to late-1970s. The relatively long latent period between infection and disease makes studies focusing on HIV-1 prevalence and incidence the most informative basis for monitoring the epidemic behavior of the virus and its contagiousness. A number of viral and host factors seem to impact on infectiousness and the likelihood of transmission. For example, some strains of virus such as those with nonsyncytia-inducing characteristics and macrophage–monocyte trophism may be more readily transmitted sexually.[516] HIV infectivity may vary widely between persons and infectivity may be greater at the beginning of infection or late in the course of disease, in association with severe immunodeficiency. Thus, developing accurate estimates of virus infectivity on a per-exposure or per-partner basis is difficult.

A number of partner and cohort studies have pro-

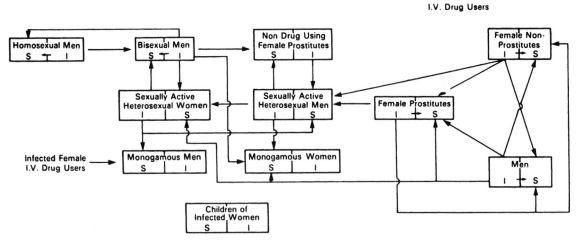

Figure 17. Compartmental model for spread of AIDS. From Gail and Brookmeyer,[54] with permission.

Table 9. Estimates of Infectivity due to Sexual and Parenteral Exposure[a]

Transmission probability from a single contact with an infected partner	
Heterosexual transmission	
Male to female (California Partner Study)	0.001–0.03
Male to female (Transfusion Partner Study)	0.001–0.189
Female to male (STD patients, Kenya)	0.082
Female to male (Thailand military recruits)	0.025–0.075
Homosexual transmission	0.008–0.032
Per partnership transmission probability from a relationship with an infected partner	
Heterosexual transmission	
Male to female	0.10, 0.18, 0.20, 0.22, 0.27, 0.28
Female to male	0.01–0.12
Homosexual transmission	
San Francisco Men's Health Study	.10
Transmission probability from single contact in nonsexual setting	
Blood transfusion	0.60–0.72
Needle stick, health care worker	0.0046–0.009
Laboratory worker	0.0048–0.023
Perinatal transmission	0.12–0.30

[a]Adapted from Brookmeyer and Gail.[54]

vided estimates of the risk of HIV-1 transmission; these are summarized in Table 9. Based on these data, the per-partnership transmission probability is greater for male-to-female transmission than for female-to-male. The range of estimates for transmission after a single sexual contact reflects, in part, the impact of additional cofactors such as stage of infection, level of viremia, coincident ulcerative genital lesions, and so forth. In general, these data confirm that HIV-1 is *not* a highly transmissible agent, with maximal rates of approximately 3 per 100 contacts for male-to-male contacts but with a lower bound of 1 per 10,000 contacts for male-to-female transmission. The highest risk was noted for blood transfusion, where two thirds of exposed individuals in follow-back studies were found to be seropositive if they received a positive unit.[70,483] Another group at high risk for transmission are infants of HIV-1-positive mothers where the transmission probability varies from 1 in 3 to 1 in 10.[178,212] Health care workers who experience a needle stick or laboratory workers working with highly infectious virus have a per-episode rate that resembles that of exposure to an infected heterosexual partner.[386,495] A number of studies have been unable to document any evidence for household transmission (e.g., via casual contact, shared tooth brushes or razors, etc.).[81,84,104,107] Anecdotal cases of health care workers[244] and household transmission associated with cutaneous or mucous membrane exposure have been reported,[30,82,95] though the risk of such transmission is undoubtedly very low.[222]

Despite the fact that HIV-1 is not a highly contagious virus, the epidemic continues to progress at a remarkable rate, embracing new areas and populations of the world. The HIV-1 epidemic is exceedingly complex; it in fact represents a series of interrelated epidemics that have spread within and between communities, with core transmission groups playing a critical role in this process. In the United States, the virus was rapidly disseminated within a large segment of an at-risk population (sexually active homosexual men, particularly those residing in the original epicenters of New York, Los Angeles, and San Francisco), through circumstances of homosexual lifestyle, i.e., anonymous, multiple partner, and bathhouse sex, often involving receptive anal intercourse.[214,240] Rates of infection in these high-risk groups were estimated to be as high as 5–10% per annum at the peak of the early epidemic, but rates declined as the pool of susceptibles was depleted and safer sex campaigns were implemented. New epidemic foci developed, however, as secondary spread of the virus took place between persons from high-risk areas and those from initially low-risk areas. Studies documented early in the epidemic that risk for infection in a "low-risk" area could be traced to sexual contact with someone from a high-risk area.[330,334] Once introduced, the virus is rapidly transmitted among a highly sexually active core transmission group and then secondarily to those with fewer partners. In this initial burst of transmission, there may be a core of viremic persons who are especially infectious; the presence in this core of virus strains that favor sexual transmission may help to explain the initial explosive spread of the virus.

This pattern (ripple effect) is manifested in each of the at-risk groups; any variation is triggered by how the core risk group acquires the virus and the attendant behavior that determines the mode and range of transmission. For example, in Africa urban female prostitutes represent a core transmission group who become infected with the virus and pass it on to their customers; they would include, among others, itinerant laborers and/or long-distance truck drivers from rural areas of the country. Rates of infection can be quite high, with a secondary spread to the rural areas when infected men return to their rural homes and infect their spouses and other sexual partners. This scenario has been repeated in southeast Asia and India where drug trafficking provided a vector for virus introduction into the female prostitute population, with secondary spread into the community. In Thailand, two district epidemics with distinct virus types have been tracked. A clade B-related virus largely spread via injection drug abuse in Bangkok,[383,511] while a clade E virus spread via heterosexual contact among a network of female prostitutes, their military conscript clients, and the sexual partners and spouses of these men.

A variety of mathematical models, especially multicompartmental models, have been developed to account for the myriad parameters that influence epidemic spread of the virus. One such model, depicted in Fig. 17, documents the interrelationships of various compartmentalized risk groups and their potential for sexual and parenteral exposure and virus spread. While these models provide a framework for considering the interrelatedness of various risk factors and cofactors, they are dependent on parameter estimates of transmission that are difficult to measure accurately in the face of myriad cofactors involved in transmission efficiency.

5.1.3. Geographic Distribution. Of the estimated 13–15 million infections in the world as of 1994, over one half are estimated to cluster in sub-Saharan Africa (Table 5 and Fig. 12). Projected new infections will occur in great numbers in sub-Saharan Africa, accounting for half of all new infections by 1995 (Table 5). However, as seen in Fig. 15 (top), by the year 2000, infections in southeast Asia will exceed those in any other part of the world. AIDS case numbers show similar patterns, with 70% of cumulative cases occurring in Africa (1.8 million plus, as of January 1992) (Table 6), while the United States accounts for 10% of the total.

As shown in Fig. 18, the risk group associated with AIDS cases and HIV-1 infection varies geographically. In the United States, homosexual men account for over half of all AIDS cases and injection drug users a quarter, while heterosexual transmission accounts for approximately

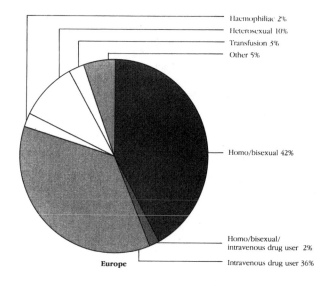

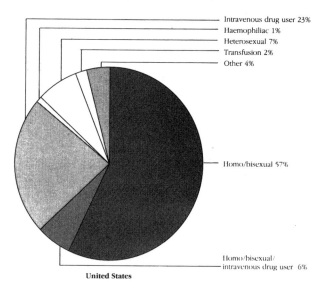

Figure 18. Adult AIDS cases by risk exposure source: Europe, United States, Latin America, and Africa. From U.S. Department of Health and Human Services, 1993, and WHO-EC Collaborating Centre on AIDS, 1992.[111]

7%. The distribution among injection drug abusers varies widely by geographic region, and as seen in Fig. 9, the majority of positives cluster in the mid-Atlantic region. In Europe, the proportion ascribed to homosexual contact and drug abuse are virtually equal but with marked disparity among the countries of Europe. For example, countries of southern Europe, especially Italy and Spain, have a high proportion of cases due to injection drug use; heterosexual contact accounts for 10% of cases. In contrast, in

Africa, heterosexual contact accounts for 94% of all infections with other sources being negligible (Fig. 18). But even within Africa, rates of HIV-1 infection vary widely. As seen in Fig. 19, seroprevalence among low-risk urban populations vary markedly. Countries in the equatorial region, including Zaire, Zambia, Uganda, Kenya, Tanzania, Rwanda, Burundi, and Malawi have rates of 5 to over 10%, as shown in Fig. 19.

The geographic distribution of clinical AIDS incidence in the United States by state is shown in Fig. 20. Among states, New York, New Jersey, Florida, and California have the highest case rates for males, while New York, New Jersey, Connecticut, and Florida have the highest rates for females, representing the impact of injection drug abuse as a major source of infection in this group. The highest rates for the country are observed in the District of Columbia for both males and females, and detailed analysis of this site documents a typical urban minority injection drug use and emerging heterosexual epidemic superimposed on an earlier epidemic among homosexual men. Recent trends in the District suggest an emerging heterosexual epidemic that may be indicative of

similar urban epidemics embedded in larger population bases.[309] For cases of AIDS in the pediatric age group, New York, New Jersey, and Florida are disproportionately represented, reflecting the high incidence among offspring of female parenteral drug abusers in the mid-Atlantic region and the substantial contribution of Haitian immigrants in the Miami, Florida, area.

The geographic distribution of seropositivity for HIV-1 closely parallels the occurrence of AIDS as seen in Fig. 8 for military recruits and in other sentinel populations. Incident infections documented among active duty military and reserve components of the military also tend to cluster geographically in areas of high prevalence.[135] One group with unusual geographic clustering in the United States are parenteral drug abusers. Seroprevalence rates in drug abusers appear to be much more geographically restricted than those observed in other groups (Fig. 11). Seroprevalence in West Coast drug abusers is much lower than rates in the mid-Atlantic region, reinforcing the concept that parenteral drug abusers are a much less mobile population for spreading the virus. Studies of seroprevalence among drug abusers in New York and New

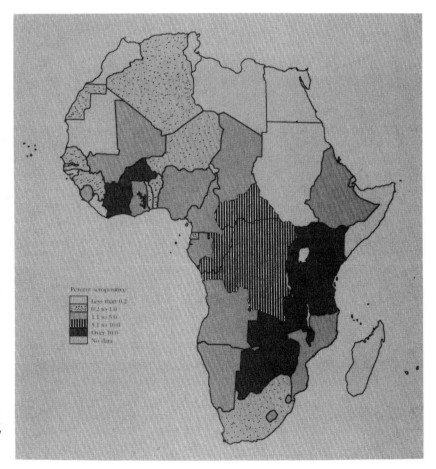

Figure 19. HIV-1 seroprevalence among low-risk urban populations in Africa, circa 1992. From US Bureau of the Census. HIV/AIDS Surveillance Data Base.[111]

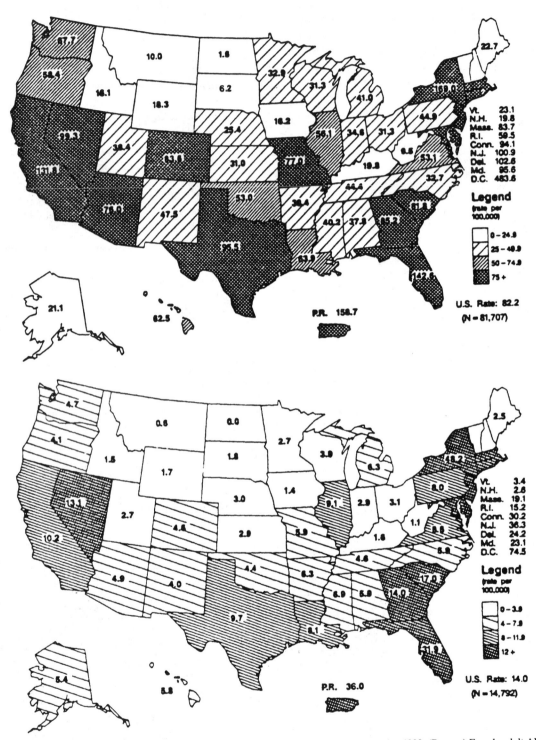

Figure 20. (Top) Male adult AIDS annual rates reported in the United States, October 1992–September 1993. (Bottom) Female adult AIDS annual rates reported in the United States, October 1992–September 1993. From Centers for Disease Control.[103]

Jersey document geographic clustering and a marked gradient in New Jersey from north to south: rates range from 60 to 70% in areas adjacent to lower Manhattan to less than 1% in the southernmost area. High rates are also observed in some subpopulations, particularly in black drug abusers in central New Jersey locations. Historic patterns of drug use may have created a circumstance where travel to high-rate areas of New York City led to the early introduction of the virus into this population.[155,157,490–492] There are also marked geographic variations in HIV seroprevalence among drug users in Europe. In particular, there is a substantial clustering of drug-use-related HIV-1 positivity in southern Europe, especially in Italy; microgeographic surveys there document a pattern of uneven distribution of virus within cities and districts.

These geographic variations are also mirrored in the patterns of seroprevalence among female prostitutes in the United States where the highest rates are observed in the New Jersey–New York metropolitan area and in Miami, with much lower rates in Atlanta and several West Coast centers. The major risk factor among this population was a history of drug use. An additional likely source of exposure is via unprotected heterosexual contact with the male boyfriend or husband of the female prostitute, who often are HIV-1 seropositive due to drug use.

5.1.4. Temporal Distribution. Since 1981, as shown in Fig. 21, cases of AIDS in the United States have dramatically increased, with incidence showing marked rises in the first 5 years and reaching a plateau beginning in 1987. This leveling of AIDS incidence in part represents the impact of improved survival, especially among those with best access to medical care, but it also resulted from slowing of rates of infections from the early days of the epidemic.[197,433] Compelling evidence for a treatment

effect due to antiretroviral nucleotide analogues as well as prophylactic therapies for the complicating infections associated with immunodeficiency was based on careful modeling of case projections and the abrupt downward trends compared to predicted patterns of occurrence among homosexual men.[197,433] As seen in Fig. 21, rates in homosexual men have a slightly downward incidence projection, while among heterosexual cases rates are upward, and in injection drug users the pattern is flat. In the United States alone, over a quarter of a million cases had been reported by 1992, and numbers of cases were projected to double in the next few years (Table 8). In fact, by 1994, over 400,000 cases had been reported.

Sex-specific patterns are also changing dramatically, as shown in Fig. 22. Among women the major shift is an upward trend in cases ascribed to heterosexual transmission (Fig. 22 top), while among males the trend is toward a reduction in the proportion of cases ascribed to homosexual exposure and a substantial rise in cases due to injection drug use (Fig. 22 bottom).

Worldwide this temporal distribution of AIDS cases is mirrored in many geographic locales, with cases of Haiti, Africa, and the United States appearing almost synchronously in the late 1970s and early 1980s, with subsequent appearance of cases in new locales in the world as the epidemic spread (Fig. 14). The clinical manifestations of AIDS differ geographically: *P. carinii* pneumonia and *Mycobacterium avium* are more prominent in the United States and Europe, while bacterial sepsis and *Mycobacterium tuberculosis* are more prominent in developing countries.

Mortality trends parallel AIDS and HIV incidence trends. In the United States, as shown in Fig. 16 (top), AIDS has become the leading cause of death among men be-

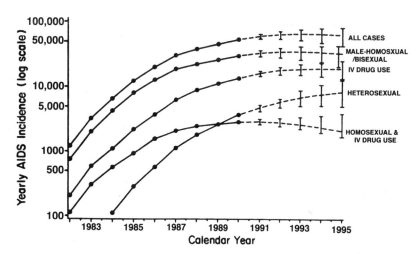

Figure 21. Projections of annual AIDS incidence: 1991–1995 for the entire United States and four transmission groups. From Gail and Brookmeyer,[54] with permission.

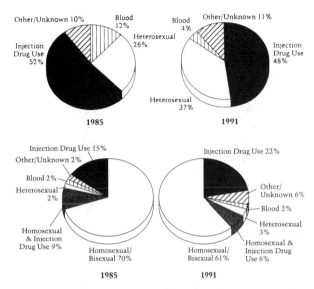

Figure 22. (Top) AIDS cases in the United States in women by mode of transmission, 1985 and 1991. (Bottom) AIDS cases in the United States in men by mode of transmission, 1985 and 1991. From 1990 AIDS/HIV/ STD surveillance report. Pan American Health Organization.[111]

tween the ages of 25 and 44 and ranks eighth in all cause mortality for males, accounting for 1.7% of all deaths based on 1989 statistics.[46] For women (Fig. 16 bottom), AIDS has assumed third position in the 25- to 44-year age group, with a rising trend. Worldwide (Table 8), over 6 million deaths due to HIV/AIDS are projected by the end of 1995, with sub-Saharan Africa accounting for two thirds of the total. As shown in Fig. 23, life expectancy will be substantially reduced in many developing countries of Africa and Asia, and population growth rates will be dampened by the year 2010, possibly leading to a negative population growth rate in Thailand.[111]

HIV seroprevalence from various geographic locales shows marked upward trends. Historically, as shown in Figs. 6A and 11, the appearance of HIV-1 seropositivity in high-risk populations antedated the occurrence of AIDS by a number of years, consistent with the long latency from infection to disease. In some instances infection was apparent in the mid to late 1970s. It would appear that HIV-1 in the United States was first introduced into the population of homosexual men residing in New York and San Francisco. The earliest seropositives in Fig. 6a were observed in 1978 in San Francisco, but extrapolating the rate among New York homosexual men it is likely that as early as 1975 or 1976 as many as 3–5% of the male homosexual population in New York may have been seropositive.[37,468] Modeled estimates of infection among ho-

mosexual men in the United States show similar patterns (Fig. 14), with infections occurring in the mid to late 1970s accounting for the subsequent burst of AIDS cases in the mid 1980s.

The HIV-1 seroprevalence patterns among parenteral drug abusers shows the parallel introduction of the virus into injection drug users in the United States and overseas. Within the New York City metropolitan area, rates of seropositivity among drug abusers closely parallel those in the male homosexual population, and the slope of increased prevalence suggests rapid introduction of the virus. The incidence of HIV-1 in a particular locale reflects the time of introduction of the virus into a population. Thus in 1980 (Fig. 6A), the rate among homosexuals in London, England, and Copenhagen, Denmark was similar to that in San Francisco in 1978, and the lag period in the incidence of AIDS reflects this relatively more recent introduction of the virus into these populations (Fig. 6A). For hemophiliacs, the rising prevalence is dramatic, but as displayed in Fig. 9 a number of factors related to blood product usage and severity of hemophilia contributed to the likelihood of infection.[277] The exponential rise in seroprevalence also points to the fact that during the early 1980s a substantial proportion of factor VIII concentrate lots appear to have been infectious,[185] resulting in large numbers of individuals becoming infected in a relatively short time period.

HIV-1 has spread within the United States among various groups in a centrifugal fashion, with rates of risk group mixing accounting for the timing and extent of further spread (Fig. 15). This pattern has been repeated internationally and has been documented in analytic studies that show that a major risk factor initially for the introduction of the virus into a new population is sexual or drug abuse contact between a person from a high-rate area with someone from a low-risk area. Examples include studies of homosexuals in Denmark and Trinidad where sexual contact with an American was a risk factor for seropositivity[21,331] and recent analyses in Africa of female prostitutes and high-risk groups such as long-distance truck drivers where contacts between high- and lower-risk areas result in viral transmission[462] (see Section 5.2.2). More recently, molecular epidemiologic studies of virus strain in India and Thailand have helped to trace the origin of virus from Africa, and epidemiologic studies have linked some of this transmission to the international drug trafficking.

In Africa the emergence of the AIDS epidemic was almost synchronous with that in the United States, and the pattern of spread of the virus within Africa appears to involve transmission through heterosexual contacts,[142]

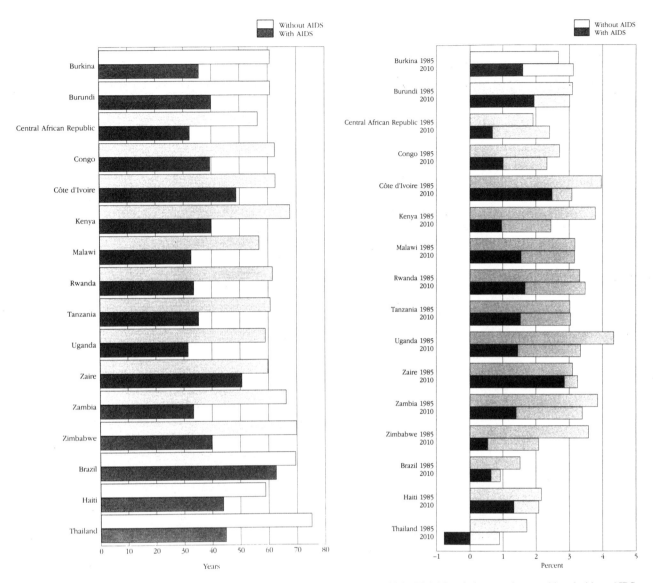

Figure 23. (Left) Life expectancy at birth with and without AIDS for selected countries, 2010. (Right) Population growth rates with and without AIDS for selected countries, 1985 and 2010. For 2010, the portion of the bar "without AIDS" represents additional growth that would take place without AIDS. From Center for International Research, U.S. Bureau of the Census.[111]

particularly in the major urban areas.[275,420] The usual pattern involves the initial introduction of the virus into a major metropolitan area of a country, frequently in association with prostitute contact. Spread continues through secondary contacts of persons within the sexually active community. This pattern was well described in a study in Nairobi, Kenya. In this study, 4% of a cohort of 446 female prostitutes were HIV positive in 1981, and the rate rose to 59% by 1985. Among 450 men attending an STD clinic with genital ulcers, 3% were seropositive in 1981

and 18% were seropositive by 1985.[275] No pregnant women enrolled in the study were positive in 1981, but by 1985, 2% were positive. Rising rates of infection in Nairobi and in a number of other locales (Fig. 13) suggest the continued spread of the virus into the broader population group as males infected via prostitute contact infected their childbearing spouses. A major risk factor for seropositivity in the prostitutes was contact with male customers from an AIDS epidemic area.[5,27]

Although Haitians were among the first to be diagnosed with AIDS in the Americas, no data distinguish whether HIV-1 was brought to Haiti from the United States or vice versa. Molecular analysis of viruses in Haiti place them in the same clade as those in the United States and Europe, but with some variation. The pattern within Haiti suggests a progressive spread via sexual contact similar to the pattern observed in Africa.[393] An unusual feature of the epidemic in Haiti is the fact that initially the majority cases were confined to homosexual or bisexual men, and the predominance of AIDS cases, as in the United States, was in men. Subsequently, the incidence of AIDS has shifted so that the male-to-female ratio is rapidly approaching unity, recapitulating a pattern observed in a number of Caribbean locales where contact between North American homosexual men and bisexual men from the island resulted in initial spread into the homosexual community followed by secondary spread to the sexually active heterosexual community.[21]

Taken together, the data on temporal trends from AIDS surveillance and prospective serosurveys provide strong evidence for the fact that the epidemic of HIV and AIDS is a new phenomenon, and the rate of spread and the level of seroprevalence in some populations result from a synergism of virus transmission through core high-risk groups and coincident sexually transmitted diseases that amplify spread.

5.1.5. Age. The major peak for AIDS incidence, as seen in Fig. 24, occurs in the sexually active age group between 20 and 40 years of age. This pattern is observed among diverse populations including homosexual men in the United States and heterosexually active populations in Africa, Asia, or the Caribbean.[92,314] High rates are also observed in this same age span for parenteral drug

abusers, with a tendency for highest rates to be in the mid- to late-20s.

The second mode of HIV infection occurs in the pediatric age group (Fig. 25). About 20% of children infected through maternal transmission develop symptomatic AIDS in the first year of life; the remainder develop AIDS at a nearly constant rate of 8% per year (Fig. 25 bottom).[14] The children who develop AIDS rapidly may have acquired infection *in utero*, while the second group, with a time to AIDS more similar to adults, may have acquired infection at birth or through nursing.[72] Perinatal exposure will be an important component of the future of the epidemic as increasing numbers of infected women transmit the virus to their offspring.[219]

Seropositivity for HIV shows an age-dependent curve that closely parallels that for AIDS disease incidence. For example, seroprevalence among military and Job Corps recruits and blood donors is highest in the sexually active age group between 20 and 30 years of age, albeit such surveys are highly biased in the representativeness of recruitment.[63,86,136,204] In the population-based pilot survey in Dallas, Texas, the sexually active age group was most affected.[99] Internationally, similar trends are observed in heterosexually active populations in Africa and Southeast Asia. For example, the pattern of HIV-1 prevalence in Uganda shown in Fig. 26 is typical of age-specific patterns in many developing countries where there is a characteristic peak prevalence in the 20- to 30-year age group among men and a shift among females toward a 5- to 10-years younger age peak.[1,118] This younger age peak among females may represent the impact of more efficient male-to-female transmission and the tendency of older males to have sexual contact with younger females.

5.1.6. Sex. In the United States in 1993, 84% of AIDS cases occurred in males (Table 2). As shown in Fig. 16 (top) and Table 2, homosexual and bisexual men constitute the largest proportion of AIDS cases[92] and parenteral drug users account for approximately 27% of all cases. Although male parenteral drug users outnumber female parenteral drug users with AIDS by a margin of 4 to 1, female drug users account for the largest proportion of AIDS cases among females. However, as seen in Fig. 22 (top), the proportion of women with heterosexual exposure as their source of infection has increased from 26% in 1985 to 37%in 1991. Virtually all persons receiving clotting factor concentrate are males because hemophilia is an X-chromosome-linked disorder. Thus, the high proportion of male AIDS cases in the United States results from three factors: male homosexuals are the major U.S. risk group, drug abuse tends to be more prominent in

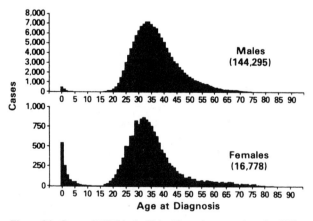

Figure 24. Cases of AIDS in the United States by age and gender, 1981–1990. From Centers for Disease Control.

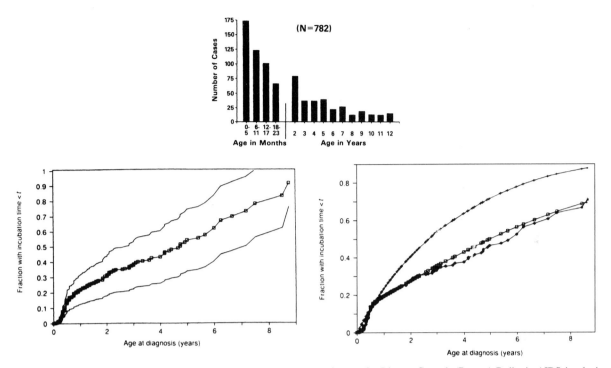

Figure 25. (Top) Pediatric cases of AIDS by age at diagnosis, 1990. From Centers for Disease Control. (Bottom) Pediatric AIDS incubation. Nonparametric maximum likelihood estimates of the conditional distribution of incubation intervals (squares) From Auger *et al.*,[14] with permission.

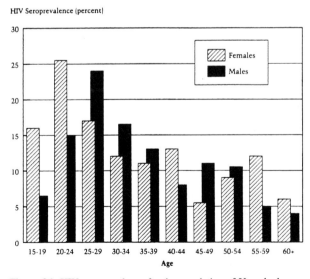

Figure 26. HIV seroprevalence for the population of Uganda, by age and gender, 1987–1988. From Center for International Research, U.S. Bureau of the Census.[111]

young males, and hemophilia is a genetic disorder of men. In contrast, the proportion of male and female transfusion recipients with AIDS is 1.7 to 1.[397] The marked excess of women with heterosexual contact as the major risk factor for infection compared to males is consistent with the growing epidemic of heterosexually transmitted virus from male drug users to their female partners.[157]

The male predominance is also observed in Europe, again reflecting exposure via male homosexual and parenteral drug contact. In Europe, among the subset of patients who are migrants from sub-Saharan Africa diagnosed with AIDS, the ratio of male-to-female cases is almost unity.[314] This reflects the fact that in Africa, AIDS occurs equally in males and females due to heterosexual transmission. In Thailand, males predominate among those with HIV-1 infection due to injection drug use, while in the largely heterosexual epidemic, particularly in northern Thailand, the sex ratio is almost unity. In locales where heterosexual spread predominates, the equal male-to-female ratio appears to result from the more efficient transmission of HIV-1 from males to females and the counterbalancing higher rate of partner change among males.[265,266] Among children in the United States and abroad, the proportion of male-to-female cases also approaches unity. Thus, the disproportionate occurrence of AIDS among males in the United States is an artifact of the population where the virus was first introduced rather than the result of some special propensity of the virus to infect males as opposed to females.

5.1.7. Race and Ethnic Group. One striking aspect of AIDS occurrence in the United States is reflected in the marked difference in disease among various ethnic and racial groups. Recent surveillance data reveal that U.S. blacks account for 35.7% of all AIDS cases and Hispanics for 17%, while these groups compose only 12% and 6%, respectively, of the U.S. population (Table 2). Women represent a higher proportion of infected blacks and Hispanics (52% and 20%, respectively) than do white women, and the cumulative incidence rates for these minorities are 13.3 and 11.1, respectively, times the incidence for white women. This results because heterosexual transmission of HIV is occurring in blacks and Hispanic disproportionate to whites. Recent data for the District of Columbia confirm a pattern of emerging HIV-1 incidence among young minority male and females largely due to heterosexual transmission.[435]

Serosurveys in the United States document similar patterns to those observed in AIDS cases. The largest surveys have been undertaken by the military; data from recruits confirms a disproportionate occurrence of seropositivity among minority populations, a finding confounded by recruitment practices and patterns. However, among active duty and reserve components, which represent prospectively followed cohorts, incidence of new infections is 3.7 times higher in African-American compared to Caucasian troops, and Hispanic rates are also 3.0 higher than in Caucasians.[56] The finding that patterns of prevalence and incidence correlate with geographic centers of high prevalence suggest that these troops may represent a microcosm for new infections from their home communities.

5.1.8. Occupation. Female prostitutes are an important high-risk core group for HIV-1 in many locales.[142] In the United States, where rates vary considerably by geography, parenteral drug use and proximity to an AIDS endemic area are the major risk factors. Among African and Asian prostitutes, proximity or contact with a high-risk area for HIV and history of another sexually transmitted disease are the major risk factors. In many developing countries, female prostitutes are a major core transmission group, and their rates of infection (Tables 3 and 4, Fig. 6B) are a barometer of the introduction and potential for spread of the virus in some populations. Some of the highest incidence rates for HIV-1 are reported among female prostitutes in Africa, Thailand, and India. It is of some interest that in one cohort of female prostitutes in Kenya, some long-term exposed individuals have not

seroconverted; it is possible that they have developed effective immune responses that have prevented infection.[501]

Health care workers have been carefully studied for an excess risk for seropositivity or AIDS, since health care workers are at increased risk for other blood-borne agents such as hepatitis B.[223,495] As outlined in Table 9, the risk for infection among medical and laboratory personnel is low but resembles rates for single contact heterosexual transmission.[386,389,458] Several large cohort studies have been undertaken of intensively followed health care populations involved in the close care of AIDS patients. Compared with the relatively large number of health care workers potentially exposed and the high frequency of documented needle-stick and surgical exposure to the virus, there has been a relative paucity of seropositives or AIDS cases among this occupational group.[74,105]

To date, three laboratory workers involved in producing large quantities of concentrated virus have become infected with HIV-1. One worker became infected after accidental inoculation, a second experienced extensive cutaneous and mucous membrane exposure following a pump malfunction, while the third had no apparent source of exposure.[39] The estimated risk summarized in Table 9 is similar to that for health care workers, but numerous instances of putative parenteral exposure without seroconversion point out that risk for infection is low. Nonetheless, the finding of these infected workers emphasizes the need for carefully following established biosafety guidelines.[493]

5.1.9. Occurrence in Different Settings. Groups at high risk for AIDS include homosexual or bisexual men, injection drug users, and heterosexually active adults (Figs, 16, 19, and 21). While these categories are useful for surveillance purposes, it is the behavior associated with these groups that result in HIV-1 infection. For example, those who are mutually monogamous in their sexual relations or never share needles or "works" in their drug abuse are not at risk for HIV-1 infection. While AIDS surveillance includes a category for "no identified risk," in almost all cases these represent instances of missing information on known risk factors. There have been some reports of cutaneous or mucous membrane exposure in home care medical settings that may have resulted in infection, but these are rare.[191,264] Some instances of "household transmission" have been reported: in one instance, exposure was due to casual handling of injection equipment by a parent of a hemophiliac[102]; in another, one hemophiliac brother may have transmitted virus to a sibling via a blood-contaminated razor.[104] Studies of military populations, both recruits and active duty, have not identified major additional risk factors or evidence for casual spread such as is seen with some infectious agents among military recruits.[86] A compilation of household studies totaling 1167 persons for more than 1700 person-years have identified no instances of infection, with a 95% confidence interval of 0–0.2 infections per 100 person years of follow-up estimated.[192,461] One anecdotal report suggested that the possibility of infection in an older sibling who was seropositive at the time of the survey may have resulted from a bite wound inflicted by a younger infected sibling.[192] These data reinforce the fact that HIV appears to have a low risk for transmission (Table 9).

5.1.10. Socioeconomic Factors. Socioeconomic factors associated with HIV and AIDS are linked to behaviors that put one at high risk for AIDS. For example, parenteral drug users are a socially and economically disadvantaged group, but their risk for transmission results from their parenteral drug exposure. Persons of lower socioeconomic status in the United States have increased rates of HIV infection through heterosexual transmission, but the same group has elevated rates of teenage pregnancy, another indication of high-risk, unprotected sexual behavior. Similarly, rates of sexually transmitted disease, a cofactor for HIV-1 transmission, are higher in economically disadvantaged groups.[100,140,373] Recently, tuberculosis has emerged as a significant complication of immunosuppression induced by HIV-1; tuberculosis rates are highest in disadvantaged populations. Socioeconomic factors have an impact on HIV-1 in Europe, Africa, Asia, and the Western hemisphere, especially among developing nations. In Russia and Romania, a major outbreak of infections is attributable to the reuse of unsterilized needles.[226,270] In Africa and Asia, female prostitutes account for a significant proportion of HIV-1 transmission. Throughout both continents, women from lower socioeconomic status are forced into prostitution to support themselves before marriage.[110,142,374] Thus, they not only become infected themselves but they serve as a core transmission group, often infecting their husbands and children. Paradoxically, in Africa, rates of seropositivity may be higher in men of higher socioeconomic status because they can afford more frequent contact with female prostitutes.[335] The economics of the drug traffic also affect the spread of HIV-1 both within communities and internationally.

5.2. Descriptive Epidemiology of HIV-2

5.2.1. Prevalence and Incidence of HIV-2. Following their initial discovery of serological evidence of HIV-2 among Senegalese prostitutes, the Essex group, in collaboration with others, began to sketch out HIV-2 seroprevalence in Africa. In the Ivory Coast, they screened serum samples from 1508 persons from a variety of groups in different parts of the country. These included healthy female prostitutes and prisoners; diabetics, those with tuberculosis, and psychiatric patients; hotel, prison, and hospital staff; and pregnant women. They found seropositive people in all subgroups, with prevalence ranging from less than 1% in pregnant women to as high as 21% in prostitutes.[153]

However, in the same number of subjects sampled from six countries in Central Africa, no seropositives were identified.[245] A large survey based on cluster sampling in Rwanda in Central Africa conducted by a national group found no seropositives among 2612 samples.[427] Similarly, a sample of 150 aboriginal Sun people were seronegative.[427] None of 20,000 serum samples from Uganda were found to be positive.[165] In Gabon, only 3 of 2042 samples were positive.[506]

Additional areas of significant seropositivity were identified in West Africa including Guinea-Bissau where rates of 4–10% were found among the general population.[365,414] In a population sampling in Senegal, rates were highest in the southern area adjacent to Guinea-Bissau.[289] Two areas of seroprevalence were identified in the southern parts of the continent: Angola and Mozambique.[248,315,444]

In terms of incidence rates, a population-based study using a randomly selected sample in Guinea-Bissau in 1987 found the initial HIV-2 prevalence rate to be 8.9% in adults. Then 330 adults of the initially 594 who were seronegative at first screen were prospectively followed for 2 years for seroconversion. Seven seroconversions occurred for an incidence rate of 1 per 100 person-years.[415] A high-risk population of 1452 Senegalese female prostitutes were followed for the period from 1985 through 1992; in this dynamic cohort individuals were followed 3.2 years on average. The overall incidence rate for both HIV-2 and HIV-1 infection was the same: on average, 1.1 per 100 person-years. However, the incidence rate increased dramatically for HIV-1 during the study period, while that for HIV-2 remained stable.[250] This implies that the HIV-2 infection has been in the "exposing" population for a longer time and is closer to reaching a steady-state condition than is HIV-1. It further implies that the level of infectiousness of HIV-1 is higher, since HIV-1 was much less prevalent in the community.

5.2.2. Epidemic Behavior and Contagiousness of HIV-2. The descriptive epidemiologic data suggest that the spread of HIV-2 is much slower than that of HIV-1. This is most evident in Africa where the epidemic of

HIV-1 has moved rapidly throughout the continent. In contrast, that of HIV-2 has remained much more curtailed. Analytic studies also support the assertion that HIV-2 is less infectious (see Section 5.2.1).

5.2.3.   Geographic Distribution of HIV-2.   The general distribution of HIV-2 endemic populations in Africa as summarized by Markovitz[315] is depicted in Fig. 27. HIV-2 infection is rarely outside of Africa except among persons from or with sexual contact with persons from West Africa.[134,301,315] An apparent exception is Bombay, India, where 10 HIV-2 seropositives were reported from 243 HIV-positive blood samples.[401]

5.2.4.   Temporal Distribution of HIV-2.   A total of 3177 serum samples were collected from 11 West African countries between 1966 and 1977; 2 of 207 samples from the Ivory Coast in 1966 were HIV-2 positive; among the samples collected in 1967, 2 of 197 from Nigeria and 2 from 80 from Gaban were also positive.[258] These observations suggest that the virus has been in the human population for at least several decades. None of 339 samples drawn in 1963 from natives of Cape Verde were positive; in 1987, 1.2% of 335 samples from that population were seropositive.[75]

In the Ivory Coast, surveys of both general population and hospitalized patients were done in 1985 and 1987. There was no evidence of an increase in infection in the former population; there was some suggestion of an increase among those hospitalized, from 1.6% of 120 patients in 1985 to 8.4% of 176 patients in 1987.[384]

5.2.5.   Demographic Characteristics of HIV-2. Few data are available concerning the sex and age distribution of infection of the population level. As noted above, seropositivity is closely linked to exposure to the endemic African populations. The age-prevalence curve of HIV-2 seropositivity in a large population study in the endemic population of Guinea-Bissau shows a substantial increase in positivity with age for both men and women. Rates tend to level after the mid-30s, with slightly higher rates among females (Fig. 28)[415] Among female prostitutes tested in three cities in Senegal, there was an age-dependent increase of seroprevalence through the sixth decade of life in all three cities.[247] This curve is reminiscent of that seen among women in populations with endemic HTLV-I infection (see Chapter 25).

HIV-2 is clearly an occupational hazard for sex workers. In two studies that evaluated the risk of HIV-2 positivity for female prostitutes in endemic areas by social class as indexed by their charge for sexual services and type of business, there was an inverse relationship.[142,153] This increased risk by those women in the lowest economic level likely reflects the seroprevalence of their cliental; it may also reflect less condom use. The risk of

Figure 27. HIV-2 in Africa.

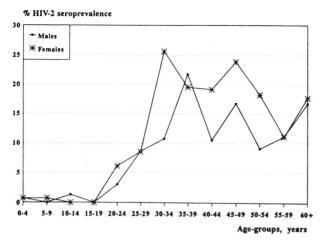

Figure 28. Age prevalence in Guinea-Bissau.

Table 10. Factors that May Influence Heterosexual Transmission of HIV[a]

Routes of transmission
  Mother to child
    In utero
    Perinatal
    Postnatal (breast-feeding)
  Sexual transmission
    Male to male
    Male to female
    Female to male
  Parenteral transmission
    Injection drug use
    Blood and blood product transfusion
  Miscellaneous
    Nosocomial exposure
    Laboratory exposure
Infectivity factors
  Early HIV-1 infection
  Rapid progression to immunodeficiency
  Advanced HIV-1 infection
  Genital ulcer disease
  Pregnancy
  Antiretroviral therapy (may decrease infectivity)
Susceptibility factors
  Genital ulcer disease
  Other sexually transmitted diseases
  Lack of circumcision (men)
  Traumatic sex
  Defloration
  Cervical ectopy

[a]All factors listed are postulated to increase susceptibility or infectivity unless otherwise noted.

occupational exposure to SIV by laboratory workers working with infected monkeys also needs to be considered.[260]

6. Mechanisms and Routes of Transmission

Extensive epidemiologic research has focused on defining the modes of transmission of HIV-1, risk factors for infection, and cofactors that amplify susceptibility. These factors of HIV-1 transmission are listed in Table 10.

6.1. Sexual Transmission of HIV-1

Sexual contact is believed to be the only risk factor for over half of all US adult (both men and women) AIDS cases (Fig. 22 top, bottom) Heterosexual transmission is the major risk factor for transmission in Africa and in parts of Asia and also represents an emerging "second" wave in the United States. While parenteral drug use is a major source of new infections in the United States, sexual contacts of drug users are also at increased risk and represent a potential core transmission group for further spread. The process by which the virus is first introduced into a population and then spread through sexual contact is dynamic. Spread of the virus occurs geographically from primary epicenters to secondary foci as seen in the United States, and Europe in the initial wave of the epidemic among homosexuals and in Africa and Asia by spread from urban epicenters to rural sites. Core transmission groups including injection drug users, female prostitutes, long-distance truck drivers, and persons with a high rate of sexual partner exchange such as those identified in

sexually transmitted disease clinics contribute to ongoing spread of the virus.[263,331]

6.1.1. Homosexual Men. The initial reports of what was later termed AIDS consisted of clusters of *P. carinii* pneumonia and Kaposi's sarcoma in homosexual men[78,80]; suspicion was thus raised from the onset of the AIDS epidemic that these unusual conditions were caused by a sexually transmitted agent. With the isolation of HIV-1 and development of a serological test for antibodies to HIV-1, it was soon apparent that AIDS resulted from HIV-1 infection and that HIV-1 could be transmitted through sexual intercourse. Epidemiologic analyses of AIDS cases in an early case control study conducted by the CDC identified a number of features of the homosexual lifestyle that were associated with being an AIDS case; these findings were the basis for initial preventive recommendations.[239] In this early study, the features associated with these AIDS cases were a history of a large number of sexual partners (often in the hundreds), frequent anonymous sex such in the bathhouse setting, a history of a venereal disease (especially syphilis), and exposure to feces and enteric complications.[239] With the subsequent discovery of HIV-1, additional risk factors were identified that had been missed in this initial study because the control group included a large number of persons who actually were HIV-1 infected. For example, subsequent reports clarified the important role of receptive anal intercourse and nitrite inhalants (a muscle relaxant that facilitate receptive anal intercourse) in HIV-1 infection in homosexual men.[214,263,502]

Receptive anal intercourse, however, appears to be the most efficient and most common mode of acquisition of HIV-1 infection in homosexual men,[263] possibly reflecting trauma-induced tears of the mucosa or the lack of natural barriers, thereby promoting transmission by this route.[263,487] Persons who are the inserting partner in anal sex tend to have a lower rate of seropositivity than those who are primarily receptive in their orientation,[214] and over 90% of the risk for transmission was attributable to receptive anal intercourse.[263] However, rectal sex may also place the inserting partner at risk,[122,263] and receptive oral intercourse with an HIV-1-infected partner may also lead to HIV-1 infection.[296,323] Although less is known about factors that modulate HIV-1 transmission between homosexual men than between men and women, it appears that, similar to reports for heterosexual acquisition of HIV-1, the risk of transmission is associated with advanced immunodeficiency in the transmitting partner[449] and the presence of ulcerative diseases, such as herpes simplex virus type 2,[232] in the susceptible partner. Other practices that cause rectal trauma may also increase the receptive partner's susceptibility to HIV-1 infection.[122,351]

Changes in sexual behaviors (i.e., fewer sex partners, changes in sex practices, use of condoms) played an important role in the decline in incidence of HIV-1 infection in homosexual men (Fig. 14). Unfortunately, recent data suggest that rates of HIV-1 infection and other sexually transmitted diseases are increasing in this population, especially among younger homosexual men.[182,476,504] These infections are associated with high-risk, unprotected sex, often in the "gay bar" and anonymous sex settings with coincident alcohol consumption and/or recreational drug use contributing to risk-taking behavior.[66,255] Behavioral research to understand the basis for continued risk taking and innovative efforts to develop more effective preventive interventions are needed to prevent the transmission of HIV-1 among homosexual men.

6.1.2.   Heterosexuals.   Worldwide, sex between men and women is the most common mode of acquiring HIV. In Africa and Asia, heterosexual transmission ac-

counts for the vast majority of AIDs cases,[1] and in the United States AIDS cases attributed to heterosexual transmission comprise the most rapidly growing category of reported cases.[106] An understanding of the rate at which HIV is transmitted through heterosexual intercourse and the factors that may affect transmission are important to understand the HIV pandemic and to design public health interventions to decrease the incidence of HIV transmission.

6.1.2a.   Heterosexual Transmission of HIV-1 between Steady Sexual Partners.   Studies that assess the prevalence of HIV-1 infection in steady heterosexual partners of HIV-infected persons have produced insights into the efficiency of heterosexual transmission of HIV-1. For example, in retrospective studies of HIV-1-infected hemophiliac men and their female sexual partners, investigators enrolled sexually active couples in whom the female partners generally had no other risks for becoming HIV-1 infected. About 10% of these women were themselves infected with HIV-1.[2,274,339] These results are roughly consistent with studies of heterosexual couples in whom men were infected with HIV-1 during a blood transfusion. One study found that 10 of 55 (18%) female partners of HIV-1-infected male transfusion recipients were themselves infected with the virus[346]; another study found that 7 of 32 (22%) such men had transmitted HIV-1 to their female partners.[376]

Although most male-to-female HIV-1 transmission results from vaginal sex, the per contact risk from rectal sex may be greater than for vaginal intercourse, as heterosexual couples who engage in rectal sex are at increased risk of transmitting HIV-1 infection.[180,387] This may account for the higher prevalence of HIV-1 infection among female partners of bisexual men compared to partners of hemophiliac men or transfusion recipients.[180,387] Higher rates of male-to-female HIV-1 transmission have also been reported in some studies of female partners of injection drug users (IDUs),[469] but it is unclear whether these findings reflect a truly higher risk of heterosexual transmission from male IDUs or additional sources of risk in female partners of IDUs. For example, women who have sex with IDUs may themselves be IDUs or may have other male sexual partners who are HIV-1 infected.

Female-to-male transmission of HIV-1 is less efficient than male-to-female transmission. Heterosexual couples in whom one person has become infected with HIV-1 after receiving a blood transfusion may provide the best comparison of the relative efficiency of transmission since the male and female index cases are of the same risk group. In contrast to the approximately 20% incidence of male-to-female HIV-1 transmission reported among male transfusion recipients and their female sex partners, a study of the male partners of HIV-infected female transfusion recipients found that 2 of 25 (8%) of the males were themselves infected[398]; another study found that 0 of 14 such male partners were infected.[376] A lower rate of female-to-male than male-to-female transmission has also been reported in other risk groups.[181,389] In sum, American and European couples' studies suggest that heterosexual transmission between HIV-1-infected persons and their regular heterosexual partners is relatively inefficient, especially female-to-male transmission.

6.1.2b. Heterosexual Transmission Involving Multiple Partners. In contrast to the relatively low incidence of heterosexual transmission observed in steady sexual partners in the aforementioned studies, researchers in Africa and Asia have demonstrated a high incidence of heterosexual transmission, including female-to-male transmission, in highly sexually active populations. An incidence study in Kenya of female-to-male transmission found that 0.082 of the men developed symptoms of another sexually transmitted disease after a single sexual contact with an infected Kenyan prostitute[73] (Table 9). In a study in Thailand, the estimated probability of HIV-1 transmission from an HIV-1-infected female prostitute to a male sex partner was 0.025 to 0.075.[320] The differences in risk of HIV-1 transmission observed in couples from the developed world compared to highly sexually active persons from the developing world likely reflect biological factors that modulate heterosexual HIV-1 transmission.

6.1.2c. Biological Risk Factors for HIV-1 Heterosexual Transmission. Multiple factors may influence the infectivity of an HIV-1-infected person or the susceptibility of an uninfected person to becoming infected with HIV-1 (Table 10).

6.1.3. Infectivity. Studies suggest a connection between HIV-1 infectivity and the natural history of HIV-1 infection. Infectivity may be high during early infection, before the development of antibodies and/or immunity to HIV-1. Although no direct evidence supports this hypothesis, it is consistent with reports of transient high levels of viremia in patients with acute HIV-1 infection.[141] Infectivity is also related to progression to immunodeficiency. A retrospective and prospective study of HIV-1-infected hemophiliac men and their female sexual partners reported that HIV-1 transmission was more common in partners of men with advanced disease, as measured by low CD4$^+$ cell count, p24 antigenemia, or the presence of AIDS.[211] A study of HIV-1-infected male transfusion recipients found an association between shorter AIDS incubation period and increased infectivity.[376] In a prospective study of heterosexual HIV-1 transmission, European

Study Group[150] investigators observed seroconversions more commonly in partners of persons with symptomatic HIV-1 infection than those with less advanced disease. The association between late-stage infection and infectivity has also been assessed indirectly by use of HIV-1 culture and the PCR. A strong association has been reported between a CD4$^+$ cell count <200 and a positive HIV-1 semen culture.[7] In a study using RNA PCR, cell-free HIV-1 RNA was found more commonly in semen of men with CD4$^+$ cell counts <400.[340]

The association between HIV-1 progression and infectivity likely reflects two phenomena: infectivity may increase as immunodeficiency progresses and rapid progressors may be more infectious than slow progressors throughout the course of HIV-1 infection. Recent data suggest that rapid progressors to AIDS establish and maintain higher levels of infected cells after primary HIV-1 infection than do slow progressors.[287,337] Although the concentration of infected cells may increase over time for both groups, rapid progressors appear to have higher viral loads soon after primary infection than slow progressors have during middle or, possibly even, late infection.

Genital ulcer diseases may increase infectivity of HIV-1-infected persons. A prospective study in Kenya demonstrated a high rate of incident HIV-1 infection in men who develop genital ulcer disease after contact with a prostitute[73]; in the same population of seropositive female prostitutes, HIV-1 was found in the culture from 4 of 36 (11%) ulcers.[45,275] These data demonstrate that HIV-1 is present in some ulcers and suggest that ulcers caused by chancroid, syphilis, or herpes simplex virus type 2 may provide a mechanism to facilitate HIV-1 transmission by altering natural barriers to transmission.

The infectivity of HIV-1-infected women may be increased by pregnancy and related factors. A study of HIV-1 seropositive women attending an STD clinic in Nairobi found that HIV DNA in cervical secretions, detected by PCR, was associated with cervical ectopy, oral contraceptive use, and pregnancy.[120] Another study using HIV-1 culture of cervicovaginal secretion found a higher excretion of HIV-1 in pregnant women compared to nonpregnant women.[224]

Antiretroviral therapy may decrease infectivity, but data are inconsistent. The Anderson study found an inverse association between zidovudine therapy and detection of HIV-1 by culture,[7] but the Krieger study did not.[276]

6.1.4. Susceptibility Factors. Conditions that compromise the barrier function of the vaginal mucosa or penile epithelium may increase susceptibility to HIV-1

infection. In 1986, Kreiss and his colleagues first suggested genital ulceration as an important risk factor in epidemic acquisition of HIV-1 infection by female prostitutes in Kenya[275]; Plummer and his group reported that seroconversion was more common in women from this cohort who had a genital ulcer or *Chlamydia trachomatis* infection.[407] Therefore, STDs that cause inflammation of the genital tract, such as *Chlamydia*, gonorrhea, and *Trichomonas vaginalis* infections, may also lead to increased susceptibility to HIV-1 infection. A 1993 study suggested that because inflammatory STDs are more prevalent than ulcerative STDs, the former may be a more important public health problem in terms of attributable risk of HIV-1 transmission.[279] A cross-sectional study among STD attenders in Jamaica suggested that ulcerative genital lesions were a major risk factor for male susceptibility, while inflammatory lesions such as caused by gonorrhea may amplify female risk more effectively.[189] In areas of high incidence of heterosexually acquired HIV-1 infection, aggressive STD intervention programs should be implemented and evaluated for their impact on HIV-1 transmission.

Other factors that may disrupt genital tract integrity have also been implicated as risk factors for acquiring HIV-1 infection. For example, men who were uncircumcised had an eightfold increase risk of becoming infected and that uncircumcised men who had a genital ulcer had the highest risk of acquiring HIV-1 infection.[73] Some ecological studies have also shown a strong association between the prevalence of HIV-1 infection in regions of Africa and the practice of circumcision.[43] Perhaps the penile foreskin is susceptible to HIV-1 infection because of trauma to the thin preputial epithelium during intercourse and perhaps because of the high density of dendritic fixed tissue macrophages that are thought to be an important target for infection.[425] Traumatic sex also has been postulated as a risk factor for acquisition of HIV-1, as has defloration.[49] Higher rates of male-to-female HIV-1 transmission infection are found among older women; this could be due to increased susceptibility to vaginal trauma[398]; this was not confirmed in a later study.[357] One study reported that among wives of HIV-1-infected men, women with cervical ectopy were more likely to also be infected with HIV-1.[352] In Jamaica, "haircut," a local abrasion of the base of the penis associated with traumatic sex, was associated with increased HIV-1 seropositivity.[189]

In one study, Nairobi prostitutes who had used oral contraceptives were more likely to have become infected with HIV-1 even after number of sexual contacts and condom use were considered[407]; other investigators found a negative association between oral contraceptive use and acquisition of HIV-1 infection.[285]

6.2. Heterosexual Transmission of HIV-2

HIV-2 shares the same routes of transmission as HIV-1. However, based on the available epidemiologic data, HIV-2 has somewhat different transmission patterns. First, HIV-2 appears to be generally less infectious with much less spread than HIV-1, despite the fact that HIV-2 likely has been a human infection for at least as long. Second, HIV-2 is more often transmitted through heterosexual intercourse than homosexually, although this may simply reflect the lack of introduction of the infection into the homosexual population. Risk of seropositivity is associated with history of sexually transmitted diseases as would be expected.[364]

HIV-2, which has a longer incubation period to AIDS than HIV-1, is spreading more slowly than HIV-1 infection in West Africa, even where HIV-2 is the more prevalent virus.[250] HIV-2 is less transmissible through sexual intercourse than HIV-1, possibly reflecting a lower viral load in HIV-2-infected persons during some or all stages of infection.[150] The link between HIV natural history and transmissibility may have an implication for the HIV-1 epidemic.[375] HIV-1 is a rapidly diverging virus with at least six distinct genetic subtypes; preliminary data suggest that HIV-1 subtypes may differ with regard to natural history.[302,496] The HIV-1/HIV-2 model suggests that HIV-1 subtypes with higher viral loads and shorter incubation periods may be transmitted more readily than less virulent varieties, thereby becoming relatively more common with time.

6.3. Transmission via Parenteral Routes

6.3.1. Injection Drug Users. Injection drug use is a major source of HIV-1 infection worldwide (Fig. 6A,B). There is marked geographic variation in occurrence in the United States (Fig. 10), Europe (Fig. 18), and Asia. In the United States (Fig. 14) and in many parts of the world, injection drug use continues to be a major source of new infections and serves as a bridge for secondary transmission via the heterosexual route. Defining the source of transmission in this population is confounded by the difficulties of attributing infection to parenteral versus sexual routes.

A number of cross-sectional studies in the United States, Europe, and more recently in Asia have reported an association between HIV-1 seropositivity and a variety of drug-related activities including frequency of injection, older age, duration of drug use, geographic proximity to a high-rate area, frequency of syringe sharing, and using "shooting galleries," places where drug users buy drugs and rent injection equipment.[112,368,447,475,481] Many of these factors parallel those associated with sexual risk (e.g., bathhouse and anonymous sex) where multiple partners and contact with high-risk individuals is frequent.

Biologically, the means of HIV transmission between IDUs is via injection with a needle that is contaminated by the virus. In many cross-sectional studies, duration of drug use and older age are the predominant risk factors,[112,447,475,481] probably reflecting risk behavior over time. In prospective seroincidence studies, needle sharing is often the predominant risk factor.[368] Younger age also emerges in incidence-based studies, since the highest risk for acquiring infection takes place at the beginning of high-risk behavior when exposure to infected needles may be greatest.[368] Risk for infection from a single injection episode has not been accurately calculated. At a minimum, the per contact risk in settings such as "shooting galleries" where the probability of exposure to contaminated equipment is high may be similar to a needle-stick exposure (Table 9). The average IDU has a high number of such exposures. Furthermore, additional aspects of drug use behavior amplify exposure to blood (e.g., "booting" and other techniques that increase the quantity of blood likely to be exchanged between individuals). Furthermore, bleach treatment has been found to be less effective in disinfecting equipment than originally thought,[385,451] and the fact that the virus can be detected by PCR in needles returned as part of the needle exchange programs supports the concept that such equipment can serve as a vector for transmission, although PCR detection does not directly translate to viable virus.[251,252]

Sexual contact may be an important independent source of HIV infection among drug users. In a recent prospective analysis, subjects with a history of sexual exposure to an HIV-1-infected partner had a fourfold increased risk of HIV infection, independent of needle sharing.[368] Similar to other studies of heterosexual transmission, this risk was higher for women than for men.

Geographic factors also may affect transmission risk since those in close proximity to an HIV-1 endemic area have highest risk.[76] In a seroincidence study, risk of infection was a function of the prevalence of HIV-1 in the community. Thus, rates were higher in areas with high background rates of infection.[368] The incidence of new HIV infections in the community of IDUs can be quite high soon after the initial introduction of the virus. Such "outbreaks" have been observed in Scotland[172,399,511]

and more recently in Thailand and India (Fig. 6B).[361,445] The restricted geographic distribution of seropositivity among drug users in the United States (Fig. 10) probably reflects the fact that members of this risk group are generally not as mobile as homosexual men.

6.3.2. Transfusion Transmission. Blood transfusion was first reported as a possible risk factor for AIDS transmission in 1983,[397] and look-back surveys documented risk in the late 1970s and early 1980s.[71] At the peak of transfusion transmission, before the implementation of donor deferral procedures, it is estimated that as many as 1% of all donations in San Francisco were HIV-1 contaminated.[70] Retrospective studies of transfusion recipients who later developed AIDS have documented the presence of at least one HIV-1 seropositive donor in almost every case. The virus appears to be efficiently transmitted in whole blood, lymphocytes, packed red cells, platelets, and plasma. Other products prepared from blood, such as albumin, gamma globulin, or hepatitis B vaccine, do not appear to be infectious for HIV-1. Antibody against the virus has been documented in immunoglobulin preparations, but, with the exception of one report of two seropositive hypogammaglobulinemic patients who received massive high-dose intravenous exposure to gamma globulin, there is no conclusive evidence for HIV-1 transmission via immunoglobulin.[91,241] In fact, in an experimental study, substantial doses of infectious HIV-1 viral particles added to plasma at the beginning of the immunoglobulin extraction process did not survive and were not detected or cultured in the final product.

Blood and blood products efficiently transmit the HIV-1 virus at a rate estimated to be 66% per infected unit or higher (Table 9). The efficiency by which HIV-1 may be transmitted through transfusion appears to be high. In one large study undertaken prior to HIV-1 blood screening, 200,000 samples were stored and subsequently screened, resulting in 133 trackable persons who had received an infected transfusion. Of 124 recipients without other risk factors, 111 seroconverted, a rate of 89.5% (95% confidence interval—84.1–94.5%).[163] A similar circumstance has been shown experimentally in the inoculation of chimpanzees. These findings have resulted in the CDC recommending that blood recipients be screened for HIV if they were transfused in the late 1970s through 1985, before blood screening was implemented. This recommendation was partly based on the recognition that over 30% of surviving leukemia patients from a New York City center had seroconverted as a result of therapeutic blood product transfusion.[300]

The efficacy of blood donor screening is high and

risk of transmission in the United States is minimal; however, there is continued concern about antibody-negative, virus-positive donors who may not be detected by current screening assays. A handful of seroconverters identified since the implementation of blood bank screening points out that up to 1 in 160,000 blood bank units may still be infectious, due to individuals who are infected but have not yet developed antibodies.[484] Given instances where organ donors have transmitted HIV-1, screening of such donors is also recommended, including antigen screening[372]; this recommendation followed a widely publicized case where multiple recipients of different organs were infected by an antigen-positive, antibody-negative man.[88,98]

6.3.3.   Hemophiliacs.   Clotting factor products, particularly those that are made from pooled plasma of thousands of donors, such as factor VIII concentrate, have been responsible for a worldwide exposure of hemophiliacs to HIV (Fig. 6A). The HIV contamination of these clotting factor products appears to have originated in U.S. plasma collections. Several studies have documented that those users who were the most frequent recipients, because of severe hemophilia, for example (Fig. 9), have the highest rates of seropositivity.[277] Hemophilia A patients who were dependent on factor VIII concentrate products distributed prior to the implementation of heat treatment and routine screening of donors for HIV had a very high likelihood of being HIV seropositive. As summarized in Fig. 9, hazards of HIV-1 infection were enormous in the early 1980s, but were also substantial among severe factor VIII users in the late 1970s.[277] Risk of infection is directly related to dose of non-heat-treated factor concentrate and therefore severity of clotting disorder (Fig. 9).[93,215,277] Hemophilia centers that use commercially available factor concentrate products have uniformly reported seroprevalence figures for hemophilia A patients, ranging between 60 and 90%. In contrast, countries who make their own products from nonrisk donor populations have much lower figures.[333] Seroprevalence is lower for hemophilia B patients who only used factor IX concentrate, probably at least partially due to differences in the procedures by which the products are manufactured. Patients who only use plasma or cryoprecipitate products exclusively made from relatively few donors have a significantly lower risk of being seropositive. (Despite blood donor screening commencing in 1985, there have been rare instances where infection of hemophiliacs has resulted from material contaminated with virus because of missed positive donors in the circumstance of antibody-negative seroconversion.[70,94,367] Factor concentrate is safe from HIV now due to topscreening of donors for HIV and heat treatment of factor concen-

trate.[93] Instances of household transmission of HIV-1 of uninfected hemophiliac brothers have been detected because of the ongoing surveillance of hemophiliacs for seroconversion, which serves as a sentinel for the adequacy of blood donor screening and treatment of blood products.

6.4.   Mother-to-Child Transmission

6.4.1.   HIV-1.   By the year 2000, it is estimated that over a million children (Table 8) will become infected with HIV by maternal transmission[114]; the high HIV-associated mortality in this age group will substantially diminish recent advances in improving childhood survival in some countries. In some urban areas of Africa, the prevalence of HIV in childbearing women is as high as 30% (Fig. 13)[111,292]; 30 to 49% of infants born to infected mothers will themselves become infected due to perinatal transmission or breast-feeding exposure.[169,171,230] In such highly affected areas up to 10% of all infants can be expected to die from HIV-1 infection. In the United States, by early 1994, over 5000 AIDS cases in the pediatric age group had been reported and prospective studies have recorded infant infection rates of 20–30% in the absence of breast-feeding.[9,154] Children from racial minority groups contribute a disproportionate share of total pediatric infections (African-American, 59%; Hispanics, 24%; whites, 17%).[103] Numbers of HIV-infected infants in the United States will rise with the increasing incidence of HIV infections in women due to rising rates of heterosexual spread as exemplified by the reported overall HIV prevalence of about 2% in minority women in urban centers of New York, Baltimore, and Washington.[34,53,432,435] European pediatric cases number over 4500 through 1993, but almost half come from Romania as a result of nosocomial infections in orphanages.[179,226] The rate of mother-to-child transmission in Europe is about 14%; this is lower than the rate in the United States, perhaps due to differences in transmission as a function of disease stage or some other unexplained factor.[177,436] It is noteworthy that in the United States there is a downward trend in HIV-1 incidence in more recent cohort follow-ups.

Numerous prospective cohort and cross-sectional epidemiologic studies have evaluated the efficiency and timing of perinatal HIV transmission. As discussed in Section 6.4.1, descriptive data have documented that the pattern of AIDS in children is biphasic, with a very early peak thought to be due to prenatal infection and a second peak due to perinatal or early life exposure.[500] A number of studies have attempted to document the proportion of infections due to prenatal, perinatal, and postnatal (breast-feeding) exposure. That breast-feeding can result in infection has been documented anecdotally in women infected postpartum due to blood transfusion, in SIV-based animal feeding studies, and by analogy with HTLV-I where breast-feeding is the major route of infection.[10,15,227–229,303,345] The evidence for prenatal infection has been reported in cases where the virus has been detected in culture or by PCR from aborted fetuses or from blood samples taken within 48 hr of birth.[47,133,149,295,394] It has been estimated that 25% of infants of infected mothers acquire HIV infection as a result of perinatal transmission. A number of epidemiologic findings suggest that a substantial risk of infection results from perinatal exposure at the time that the newborn passes through the birth canal.

Several lines of evidence directly support the concept that the majority of HIV transmission occurs during parturition. For example, prolonged rupture of membranes, use of fetal monitors (which abrade the scalp), and episiotomy are associated with increased risk,[178,298,513] while cesarean section reduces risk by approximately 30%.[479] Studies of twins delivered to HIV-infected women[168,210] have documented that if the twins are discordant for HIV infection, the firstborn infant is infected 79% of the time and that the frequency of infection in the firstborn is comparable to that of singleton births. The lower HIV-1 prevalence rate in second-born twins suggests that local factors such as a decrease in potentially infectious blood and mucous in the birth canal may account for this difference. This concept is supported by the finding that only 8% of second-born twins delivered by cesarean section are infected, since such infants would have experienced only minimal exposure to potentially infectious birth canal products. Based on these findings it is estimated that approximately 60% of infections arise due to perinatal exposure and that only 8 to 10% result from prenatal *in utero* infection.

Further support for the concept that the majority of HIV infections occur at parturition come from observations that most HIV-infected infants are normal, without congenital malformations expected with first trimester infections or growth retardation expected with second trimester infections or perturbations in T-cell numbers or proportions.[36,50,362] In summary, these data support the concept that the majority of HIV-1 infections in nonbreast-fed infants occur because of perinatal infection, especially during the birth process where infectious vaginal secretions result in infant exposure and infection.[508] Other factors that may influence the likelihood of pediatric infection are the virus load of the mother and the degree of immune deficiency in the mother as measured by CD4 level. The concept that virus load may be involved in enhanced transmission has been previously raised in the setting of heterosexual transmission where antigenemia

is associated with enhanced transmission. Women with more advanced immunodeficiency and higher virus load are at increased risk for transmitting the virus to their unborn child.[467]

Sophisticated virological analyses of the transmitted virus in some infected infants demonstrated that the virus represented an "escape" mutant, a minor variant strain different from the predominant type in the mother where maternal antibody did not cross-neutralize the child's strain. The role of the mother's immune response in protecting against maternal-to-infant transmission has been controversial, with some studies suggesting that certain immune responses to specific immunologic epitopes contribute to protection against infection, but this concept of a protective antibody reactive to the major neutralizing epitope has not been supported in more recent analyses.[213] Nonetheless, there is continued interest in defining the correlates of immunity for those 70% plus of infants who do not become infected.[243] The degree to which breast-feeding contributes to HIV-1 infection has been difficult to access because breast-feeding is not practiced in the United States and Europe in the setting of HIV-1 infection. Furthermore, the factors that increase perinatal transmission of HIV-1 may well be the same as those contributing to breast-feeding-associated transmission, immune status, presence of escape mutant viruses, and virus load.[439] Further epidemiologic studies coupled with well-conducted laboratory research are needed to clarify the additional factors involved in maternal-to-child transmission.

6.4.2. Perinatal Transmission of HIV-2. HIV-2 infection carries a much lower risk of perinatal transmission than HIV-1. In a population-based study in Guinea-Bissau, of the seven infants born to HIV-2-positive mothers none showed evidence of infection. In comparison to seronegative mothers, no difference was evident in pregnancy outcomes or infant survival for the carriers. On cross-sectional analysis, 18 children of positive mothers were seronegative with one additional child who had died of diarrhea at 4 months of age. If that child were in fact HIV-2 positive, then the rate of perinatal infection would be 3.8%.[417] In a separate prospective study of 82 HIV-2-positive mothers and 102 seronegative mothers, newborns were followed for 20 months. No positive children were identified, although significantly more infants of carrier mothers died within the second year of life and these children had somewhat lower weight than the control children.[11] Similarly, in the Gambia, 931 mother–child pairs were identified in a large population survey as part of a hepatitis B virus vaccination trial. Seven HIV-2-positive mothers were identified; no children in the entire population were positive.[151]

6.5. Other Modes of Transmission

Concern regarding the risk of infection from casual or percutaneous occupational exposure to HIV has grown with the increasing number of such patients in the health care system.[223,495] Current data document that the risk for such transmission is low and that only rarely does occupational exposure result in seropositivity (Table 9). Among 760 health care workers who had direct parenteral or mucous membrane exposure to HIV-infected material, only 2 persons not belonging to any known risk group were found to be seropositive. The risk calculated from this and other similar cohorts is similar to that of a single unprotected heterosexual encounter. While the level of risk is relatively low, the number of health care workers at risk may be very high. In one study, 14% of workers reported possible needle-stick or other percutaneous exposures to HIV, a circumstance that calls for systematic training of medical personnel in the proper techniques for handling instruments for phlebotomy.[495,510]

The risk for transmission may also be linked to the relative infectiousness of the person to whom the health care worker is exposed, but there are no data to support this hypothesis. Several health care workers have seroconverted after extensive mucous membrane or skin exposure to possibly infectious blood or bodily fluids.[83] In one case prolonged contact with a small amount of blood resulted in seroconversion. In another case, a mother was extensively exposed to the secretions of her infant with AIDS. A common denominator of these cases was exposure to blood or bodily fluids of the infected individuals without adequate barrier protection. Despite these cases, the risk of transmission without sexual, parenteral, or perinatal exposure appears to be low.[510] Occasional instances of transmission among household contacts have been reported, including one instance where the brother of a hemophiliac may have been exposed through sharing of a razor, but such instances appear to be very rare.[102,192] Virus has been isolated from saliva and tears. Oral exposure to semen has been linked to possible transmission, but this appears to be very rare.[123] However, no instances of infection associated with kissing has been documented.[217]

7. Pathogenesis and Immunity

7.1. HIV-1

Our understanding of how HIV-1 induces AIDS is incomplete despite substantial insights gained from molecular biology and immunology. The schematic portrayal of the interrelationship between virus expression and the decline in CD4 lymphocytes induced by HIV-1 over time, shown in Fig. 29, inadequately communicates the complexity of the process and the marked between-person variability in this process.[233,392,402,448] *In vitro* HIV-1 mimics, in a crude way, the *in vivo* natural history of the virus in that HIV-1 targets and infects the CD4 cell and causes cell lysis.[160,200] AIDS pathogenesis involves complex interactions of the virus with an immune organ whose very core functions are subverted to promote virus replication and ultimately the death of the host after years of assault.

Prospective and cross-sectional epidemiologic studies have convincingly linked HIV as the etiologic cause of AIDS, but the mechanisms by which the virus causes disease are not yet fully elucidated.[38,448] Sophisticated molecular approaches often coupled to epidemiologically well-defined populations are expanding our understanding of the pathogenic process from infection to end-stage failure of the immune system.[187,404]

7.1.1. Pathogenesis of Acute-Phase Infection. HIV-1 exists as cell-free and cell-associated virus in semen, vaginal fluids, blood, and plasma. Depending on the route of infection, the initial target of infection is either the tissue dendritic cell,[176] the CD4 lymphocyte, or the monocyte–macrophage. For a particular exposure circumstance, sexual versus parenteral, the trophism of the virus for either lymphocytes or macrophage-related cells may be important since the number and availability of different target cells may vary. Furthermore, the virus load involved in a particular exposure may be important to the success of an infection, as suggested by epidemiologic data that indicate that the presence of antigenemia is associated with increased risk for infection.[282,308,507] At least for sexual exposure, the intactness of the mucosa is an important variable in infection, as evidenced by the amplifying effect of coincident, especially ulcerative, genital infections that may disrupt normal protective barriers and enhance inflammatory cell exudates, thus increasing numbers of susceptible target cells. Once infection takes place there is an absence of detailed knowledge about the subsequent trafficking of the virus within the immune system until high-level plasma viremia occurs 2 to 6 weeks following initial exposure.[141,304] In fact, there is some indirect evidence, based on *in vitro* tests of viral specific cell-mediated immunity, that an as yet unknown proportion of exposed individuals may "clear" virus without becoming infected.[35,121] In typical seroconverters, the initial viremia is markedly reduced concomitant with the appearance of a viral-specific cell-mediated immune response; humoral immune responses develop

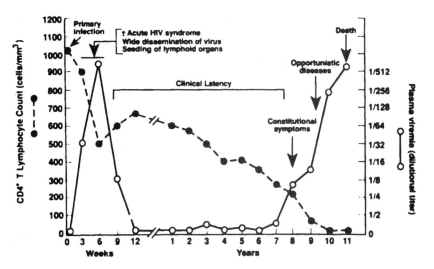

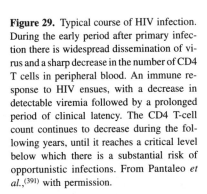

Figure 29. Typical course of HIV infection. During the early period after primary infection there is widespread dissemination of virus and a sharp decrease in the number of CD4 T cells in peripheral blood. An immune response to HIV ensues, with a decrease in detectable viremia followed by a prolonged period of clinical latency. The CD4 T-cell count continues to decrease during the following years, until it reaches a critical level below which there is a substantial risk of opportunistic infections. From Pantaleo *et al.*,[391] with permission.

shortly thereafter.[268] As depicted in Fig. 29, virus load decreases, antibody appears, and CD4 levels decline, with the rate and level of decline markedly varied between individuals.[305,391] It is during this viremic phase that the acute seroconversion syndrome described below occurs.[471]

Sophisticated PCR and/or *in situ* methods of measurement of HIV-1, either in its integrated cDNA form, viral messenger RNA (virus replication), or virion-associated RNA (virus particles), have provided insights into host–virus interactions during the course of HIV-1 infection.[391] Early in infection, there is extensive expression of viral message in the peripheral blood lymphocytes, but once immune mechanisms turn on and virus load diminishes, this activity becomes barely detectable. In contrast, during the period of clinical latency, there is substantial expression in the lymph nodes, and it is only late in the course of disease that virus expression is substantial in both the peripheral blood and lymph node compartments. Virion-associated RNA is substantially elevated during initial viremia and falls substantially to a relatively low but persistently detectable level during clinical latency, only to reemerge at high levels in the later stages of clinical disease.[225,516] Cell-associated viral DNA is detectable throughout the course of infection. Using sophisticated *in situ* techniques, investigators have monitored lymph node architecture throughout the course of disease and have documented an unusual pattern of trapping of virions in the "web" of follicular dendritic cells of the lymph node follicular centers.[176,391] During the progression of HIV disease there is an increasing disruption of lymph node architecture and destruction of the follicles as disease becomes more severe and CD4

cells become increasingly depleted. Thus, the pathogenesis of HIV-1 disease involves concomitant destruction of large numbers of circulating lymphocytes, especially CD4 cells, but also the destruction of the lymphoid organ itself, i.e., stromal tissues and those with specific functional responsibilities for regeneration and immunologic diversity.[391] In the epidemiologic context, this pathogenic process can be measured more crudely in human populations by measuring the between-person variability in time from infection to AIDS in order to understand markers that predict disease progression or long-term survivorship.

The process by which HIV brings about the progressive destruction of the immune system is unknown, but postulated mechanisms include direct killing of cells by the virus,[454] syncytia formation,[380] autoimmune effects, infection and destruction of stem cells and destruction of normal immune regulatory pathways.[455,456] Virus load as measured by a number of quantitative and semiquantitative methods correlate with the progressive disruption of the architecture of the immune organs.[176,391] Virus load as measured by plasma viremia is highest early in infection, with a marked drop in level until the end stages of HIV infection when titers increase progressively. What was not appreciated until recently is that virus replication occurs at every stage of the disease based on *in situ* studies of lymph nodes,[176,342,391] while the level of circulating virus in the peripheral blood is relatively stable at a low level over a considerable period, save for the initial burst of viremia at seroconversion and the subsequent end-stage rise.[225,257] Comparing patients in longitudinal follow-up, the quantitative level of viremia varies between individuals with higher baseline levels in those with rapid pro-

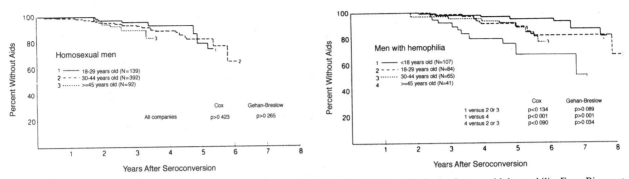

Figure 30. (Left) AIDS incubation distribution in homosexual men. (Right) AIDS incubation distribution in men with hemophilia. From Biggar *et al.*,[32] with permission.

gression to clinical AIDS and low or undetectable levels in those who have longer AIDS-free survival.[287,296,466] The determinants of virus load are also unknown but probably involve host genetic factors, especially immune response genes in the major histocompatibility complex, and cell-mediated immune responses as well as characteristics of the virus phenotype that may reflect growth characteristics of the virus.[13] Cellular immunity, especially CD8 cell-mediated cytotoxicity, has been postulated as a major determinant for immunologic control of virus, and because of the importance of immunogenetic factors in this response, the variation in time to AIDS and even long-term survivorship may reside in this interaction.[158] Other correlates of disease progression are reflected in increasing virus load in lymph nodes and the disruption of lymph node cytoarchitecture, particularly the destruction of the follicular dendritic cell network.[187,392]

The incubation distribution from infection by virus to AIDS occurrence varies from individual to individual but has a media of approximately 8 to 10 years (Fig. 30).[32] In general, when a person is exposed to infectious virus, HIV p24 antigen is detectable within 4 to 6 weeks followed shortly thereafter by the appearance of antibody.[329] In rare instances, HIV virus is detected in the absence of antibodies by virus culture or through the application of the PCR amplification technique.[236,297,369] Recent studies have evaluated highly exposed individuals for evidence of cell-mediated immunity in the absence of detectable virus or antibody.[121] Such cases may result from virus exposure without infection or effective clearing of an abortive infection.[390] Following infection, there is a period of subclinical infection sometimes associated with lymphadenopathy. During this period most persons experience a progressive decline in CD4 cells and other markers of immune function are perturbed.[137,145,391,466] As shown in Fig. 30, the period of silent infection varies

between 8 to 10 years until clinical symptoms or an AIDS-defining opportunistic infection occurs.[32,40] As shown in Fig. 30 (left), progression to clinical AIDS by homosexual men is more rapid than in adult hemophiliacs, a phenomenon explained largely by Kaposi's sarcoma, which almost exclusively occurs in homosexual men. The residual difference may reflect differences in time to AIDS between risk groups.[32,40] The issue of whether progression to AIDS is more rapid among HIV-1-infected patients in developing countries is not resolved, but it is speculated that poor nutrition, coincident parasitic infestations, and other factors could modify the natural history.[341,452] Young age at infection among hemophiliacs is the single most important determinant of time to AIDS (Fig. 30b), as seen among persons infected in preadolescence and young adulthood who progress much less rapidly from infection to AIDS than those who are infected after the age of 45.[32]

The neonatal period offers a more complex pattern for time to AIDS from infection (Fig. 25 bottom). This reflects factors associated with timing of infection, e.g., parturient versus prenatal. Children who progress to clinical AIDS in weeks to months (median of 4.1 months) are thought to have acquired infection *in utero*.[472,473] The remainder of children have a median time to AIDS of approximately 6.1 years.[237] In addition to time of infection, other factors such as viral load and viral strain are postulated as cofactors for time to AIDS.[237,344]

Among children and adolescents infected after birth, the natural history of infection differs from that seen in adults. For example, the time from exposure to depressed CD4 lymphocyte count (e.g., $<200/mm^3$) occurs more rapidly in adults over the age of 30 compared to the rate in children, adolescents, and young adults.[186,400] The time to clinical AIDS, once a CD4 count of under $200/mm^3$ is realized, is longer in younger compared to older HIV-1-

infected individuals.[212,434] The young immune system may have greater "reverse" or more effective response to viral exposure. Race does not appear to be a cofactor for progression to AIDS.[139] The more rapid time to AIDS in Africa might be due to the effect of poor nutrition or other circumstances that weaken immune responses rather than to race per se.[353]

Coincident infections by a number of DNA viruses and the HTLV-I oncornavirus have been postulated to accelerate HIV progression. Cytomegalovirus (CMV) is one such cofactor but age is a confounder, since CMV rates rise with age and time to AIDS is shorter in older persons.[166,423] Human herpes virus 6 (HHV 6) is another candidate, but clear epidemiologic proof is lacking. HTLV-I, but not HTLV-II, coinfection is higher in those with more advanced disease in some cross-sectional studies, and it has been suggested that the immunostimulatory effects of HTLV-I on T lymphocytes causes more rapid CD4 depletion.[21] New subtypes of mycoplasma, based on laboratory experiments, may facilitate the effects of HIV-1 on CD4 destruction. Epidemiologic support for such an association has not been forthcoming and antibiotic prophylaxis has not resulted in benefit.[299,482]

Some immune activation markers such as serum neopterin and β_2-microglobulin levels are predictors of progression to AIDS independent of CD4 level.[128,278,403] Independent quantitative measures of virus load—free p24 antigen, virus culture, and quantitative PCR—correlate with advanced stage of AIDS.[183,212,287404,405] Prospective studies have shown that the level of immunity to virus (e.g., low p24 antibody response to HIV) and elevated virus load early in the course of infection identify individuals at highest risk for subsequent progression to AIDS.[287,326] Host immunogenetic determinants may also impact on rate of disease progression.[456] For example, the major histocompatibility complex appears to influence risk for AIDS. Early studies suggested a heightened risk for Kaposi's sarcoma among individuals carrying the DR-5 locus, but this association has not been subsequently confirmed.[310] Several markers of the major histocompatibility complex have been associated with heightened susceptibility to AIDS or with promotion of long-term survival.[158,311]

7.2. HIV-2

Much less is known about the natural history of HIV-2 than of HIV-1, however, it appears that HIV-2 is less pathogenic. In terms of acute effects, a symptomatic seroconversion has been reported in a 19-year-old woman following sexual transmission. Three weeks following exposure, she experienced nonspecific symptoms. Subsequent medical visits recorded her seroconversion. Her symptoms resolved in a few days and 1 year later she was clinically well with some swollen nodes and two molluscum contagiosum lesions, with no other abnormalities noted.[28]

Early cross-sectional analyses of asymptomatic carriers found some changes in immunologic and hematologic parameters. Marlink et al.[317] evaluated 18 healthy HIV-2 seropositive (mean age, 45 years) and 14 seronegative prostitutes (mean age, 34 years) in Senegal, using a battery of clinical tests. The seropositives had lower CD4 and higher CD8 counts than the seronegatives. In addition, their IgG and β_2-microglobulin levels were increased and their lymphocyte proliferation responses somewhat diminished.[317]

In a similar study in the Gambia, 63 HIV-2-positive prostitutes (mean age, 29.8 years) were compared to 167 seronegatives (mean age, 27.8 years). The same pattern of findings was found. In addition, they found elevated neopterin levels and decreased CD25$^+$ cell counts among the seropositives. The carriers also have a higher prevalence of generalized lymphadenopathy (28.6% vs. 2.4%). Of 5 who were known to have seroconverted within the last 17 months, 4 had generalized lymphadenopathy compared to 6 of 27 who were known to be been infected longer.[396] The same findings held in a study of women at birth in the Ivory Coast.[259]

In a large serosurvey of a rural community in Guinea-Bissau, 21 HIV-2-positives were identified. Of these, 9 had significant clinical findings suggestive of chronic viral infection. In comparison to the 13 asymptomatic carriers, there were a number of differences. These include decreased red and white blood cells, decreased total and CD4 lymphocyte counts, and increased β_2-microglubulin level.[288] These findings were confirmed in a similar comparison of HIV-2-positive subjects in the Gambia.[497] Together, these studies show that there is an increasing abnormal immune and hematologic profile, parallel to that seen in HIV-1 infection, that accompanies the appearance of disease in HIV-2 carriers.

Recently reported data provide stronger evidence that HIV-2 is less pathogenic than HIV-1. In a prospective study of Sengalese prostitutes with incident HIV-1 or HIV-2 infection, investigators found that HIV-1-infected women had a 67% probability of AIDS-free survival 5 years after infection, while HIV-2-infected women had 100% AIDS-free survival to that point.[316] The women infected with HIV-1 were also more likely to progress to low CD4 counts than were women infected with HIV-2.

8. Patterns of Host Response

8.1. HIV-1

8.1.1. Primary Infection.
An acute viral syndrome, characterized as a "flulike" syndrome with fever, rash, myalgias, mouth sores, and occasionally reversible encephalitis (see Section 8.1.3), has been reported among persons undergoing seroconversion. Tindall and Cooper[471] suggest that 50 to 70% of persons with primary HIV-1 infection develop an acute clinical response; others maintain that the proportion is lower.[70] One reason it is difficult to determine the percentage of persons who exhibit clinical evidence of primary HIV-1 infection is that many of the signs and symptoms attributed to primary infection are nonspecific (Table 11). Because symptoms are nonspecific and difficult to distinguish from a variety of acute viral syndromes, such symptoms may or may not bring a patient to medical attention.[131] The CD4+ lymphocyte count may drop during early infection and then rise to levels at or near those present prior to infection. The amount of HIV-1 present in plasma is high during primary HIV-1 infection,[141] but falls to a lower level near the time of development of antibody to HIV-1. The initial immunologic response to HIV-1 infection may be a key to subsequent clinical course, as recent data suggest that the concentration of circulating HIV-1 infected cells may be determined during early infection and that rapid progressors to AIDS establish and maintain higher levels of infected cells than slow progressors.[287]

8.1.2. Chronic Infection.
Although HIV-1-infected persons may have an extended period during which they are free of serious clinical sequelae of HIV infection, the virus does not undergo a period of quiescence but rather continues to replicate in lymph nodes during chronic infection.[176,391] The persistent generalized lymphadenopathy found at some time in a sizable percentage of HIV-infected persons[281] may reflect the lymph node's role as a harbor for HIV-1, but lymphadenopathy in itself is not a reliable prognostic indicator of the development of opportunistic infections or HIV-associated malignancies. The resolution of lymphadenopathy has been found to be an ominous prognostic sign by some investigators[321] but not others.[174]

Depletion of CD4+ lymphocytes is the primary mechanism by which HIV infection leads to other conditions. CD4+ lymphocyte loss occurs throughout the course of HIV-1 infection,[188] but the rate of CD4+ lymphocyte loss varies dramatically between individuals.[60] Because of the CD4+ lymphocyte's role in maintaining cell-mediated immunity, loss of CD4+ lymphocytes leaves an HIV-1-infected person vulnerable to life-threatening opportunistic infections and cancers. HIV infection may also directly cause some conditions.

8.1.3. Opportunistic Infections.
Opportunistic infections are much more likely to cause disease in persons with abnormal resistance than in normal hosts. HIV-infected persons are subject to diseases caused by a large number of commonly encountered parasites, fungi, viruses, and bacteria. In the United States, serious common opportunistic infections include *P. carinii* pneumonia, pulmonary or disseminated *M. tuberculosis* infection, disseminated *M. avium* complex infection, meningitis caused by *Cryptococcus neoformans*, toxoplasmic encephalitis, chorioretinitis due to CMV infection, and many others (Table 1). The HIV epidemic has made common many formerly rare illnesses and fueled a resurgence in tuberculosis incidence and mortality[51] (Fig. 31). Life-threatening infections generally do not develop until an HIV-infected person has suffered considerable loss of CD4+ lymphocytes (<200 total CD4+ lymphocytes per microliter), but less serious conditions, including oral candidiasis (thrush), persistent herpes simplex virus infection, and herpes zoster (shingles), may present earlier in the course of HIV infection.[190]

The incidence of specific HIV-associated opportunistic infections varies by HIV risk group, geographic location, and time. Injection drug users are more likely to suffer *M. tuberculosis* infection than are homosexual

Table 11. Symptoms Reported More Frequently by 39 Homosexual Men Who Seroconverted for HIV-1 Infection than by Homosexual Men Who Remained Seronegative, January 1984 to December 1985[a]

Symptom	Seroconverters (%)	Controls (%)
Fever	77	24
Lethargy	67	24
Malaise	67	20
Sore throat	56	28
Anorexia	56	12
Headache	49	24
Arthralgia	49	20
Weight loss	46	8
Swollen lymph nodes	44	0
Retro-orbital pain	39	4
Dehydration	31	4
Nausea	31	8
Depression	28	0
Irritability	28	0
Truncal rash	23	0

[a]From Tindall and Cooper.[471]

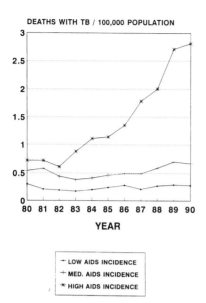

DEATHS WITH TB / 100,000 POPULATION

YEAR

‑ LOW AIDS INCIDENCE
‑ MED. AIDS INCIDENCE
‑ HIGH AIDS INCIDENCE

Figure 31. Deaths with tuberculosis per 100,000 population in persons aged 20 to 49 years in the United States, 1980 through 1990, by state AIDS incidence group. From Braun et al.,[51] with permission.

men,[87] while the incidence of disseminated *M. avium* complex infection does not vary by risk group.[234] Regarding geographic locale, diseases such as tuberculosis and toxoplasmosis are common in Africa, but *P. carinii* pneumonia appears to be rare there.[4,443] The incidence of disease due to endemic mycoses (e.g., histoplasmosis, coccidiomycosis) varies geographically within the United States in association with the geographic distribution of these infections.[446] Concerning secular trends, the incidence of *P. carinii* pneumonia appears to have decreased in response to the introduction of effective chemoprophylaxis (e.g., trimethoprim–sulfamethoxazole) for that infection.[354]

8.1.3a. Neoplasia. Kaposi's sarcoma (KS) and non-Hodgkin's lymphoma (NHL) are strongly associated with infection with HIV-1. In the United States, KS is the most common cancer occurring in HIV-1-infected patients; because of the previous rarity of KS, its appearance in U.S. homosexual men was an early sign of the impending AIDS epidemic. Kaposi's sarcoma tends to appear sooner after HIV infection than serious opportunistic infections or NHL, and the hazard for development of KS is less strongly related to the time since infection or degree of immunosuppression.[422]

Despite the large number of cases of HIV-associated KS observed since 1981, much remains to be learned about this condition. The fact that KS is much more common in HIV-1-infected homosexual men than in other HIV-1-

infected persons suggests that KS may result from infection with a second, perhaps fecally transmitted, agent[23,25]; no agent, however, has been identified. The cell of origin of KS is unknown but has been hypothesized to be the spindle cell.[318] It is not even clear that KS is a true malignancy, as opposed to a hyperplastic proliferation.

The incidence of NHL, especially high-grade B cell lymphoma, is markedly increased in HIV-1-infected persons[24,424] (Fig. 32) and, similar to the general pattern for NHL, increases with age.[421] Compared to KS, NHL is distributed more uniformly among all risk groups and appears later in the course of HIV-1 infection. HIV-1 infection may lead to NHL primarily because of decreased immunosurveillance, but recent evidence suggests that in some instances HIV-1 may cause NHL by insertional activation of an oncogene.[459]

The relationship between HIV-1 infection and development of two other malignancies, cervical cancer and anal cancer, is less clear than that for KS and NHL. The prevalence of cervical cancer precursor states has been shown to be elevated in HIV-1-infected women,[478] and invasive cervical cancer has recently been added to the list of AIDS indicator conditions (Table 1). However, because the risk of cervical cancer is in general strongly linked to infection with human papillomavirus (HPV) and because HPV, like HIV-1, is sexually transmitted, observed associations between HIV-1 infection and cervical cancer development may be confounded. Of 16,824 women reported to CDC as having developed AIDS in 1993, only 40 have invasive cervical cancer,[428] suggesting no strong association between HIV-1 infection and the development

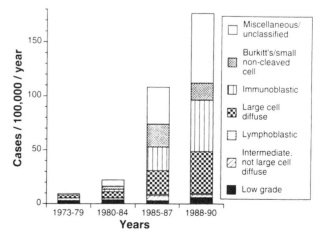

Figure 32. Incidence of non-Hodgkin's lymphoma in unmarried men San Francisco, 1973–1990. With permission from Rabkin (unpublished data).

of cervical cancer. The situation is similar for anal cancer, another malignancy that may be etiologically linked to HPV infection. A high incidence of anal cancer has been reported among AIDS patients.[332,336] However, because homosexual men had an increased incidence of anal cancer before the AIDS epidemic,[144] the relationship between HIV-1 infection and anal cancer is unclear.

8.1.4. Direct Effects.   It appears that some conditions are a direct consequence of HIV-1 infection, rather than resulting from an opportunistic infection or neoplasm. Relatively common direct neurological complications of HIV-1 infection include the AIDS dementia complex, aseptic meningitis, distal sensory polyneuropathy, and autonomic neuropathy.[418] HIV-1 infection of the gastrointestinal tract may cause diarrhea in some persons, although in most cases careful evaluation of the HIV-1-infected patient with diarrhea will result in detection of an opportunistic organism.[161]

8.1.5. Healthy Long-Term HIV-1 Seropositive Men.   The median period from infection to development of clinical conditions that meet the 1987 CDC AIDS definition is 7 to 11 years in the absence of therapeutic intervention.[212,355,440,485] Some HIV-1-infected persons, however, have remained symptom-free and without an appreciable drop in their CD4$^+$ lymphocyte count for an extended period after infection. The San Francisco City Clinic Cohort of homosexual men includes 539 men with a well-documented date of HIV-1 seroconversion prior to 1983, of whom 42 (8%) had been infected for more than 10 years without developing AIDS or a CD4$^+$ lymphocyte count less than 500 cells/mm^3.[60] Buchbinder refers to these men as healthy long-term HIV-positive (HLP). These HLP, however, were not free of evidence of HIV-1 infection; HLP had lower CD4$^+$ lymphocyte counts (mean, 692) than uninfected men and 12% had clinical evidence of HIV-1 infection.

Healthy long-term HIV-positive patients may contain their HIV-1 infection by means of better cell-mediated immunity. Buchbinder *et al.*[60] found that HLP had considerably higher CD8$^+$ lymphocyte counts than uninfected men (or progressors). Koup[268] has reported that the cell-mediated immune response may be the key response for control of primary HIV-1 infection. As discussed above, the concentration of circulating HIV-1-infected cells may be determined during early infection and rapid progressors to AIDS may establish and maintain higher levels of infected cells than slow progressors.[253,254] It is possible, therefore, that long-term survivors of HIV-1 infection mount a more robust cell-mediated immune response to HIV-1 infection that allows them to control the infection for a relatively long period.

8.2. HIV-2

Based on the descriptive and analytic epidemiologic data, it is clear that HIV-2-infected persons can develop AIDS. The clinical manifestations of HIV-2-associated AIDS do not appear to differ in any distinguishable pattern from that of HIV-1 in the same populations.[286,365,379]

It is also clear that HIV-2 carriers have a significantly increased risk of death. A 1-year follow-up study the population in Guinea-Bissau found that 2 of 4 carrier children had died compared to 1.5% of 672 seronegative children; among adults, 6.9% of 58 carriers had died compared to 0.8% of 589 seronegatives.[414] Similarly, in a hospital-based survey using consecutively admitted patients to medical wards in Guinea-Bissau, 113 carriers and 117 age-matched controls were followed for 3 to 15 months following hospitalization. The mortality rate was 72 per 100 person-years for the carriers and 40 per 100 person-years in the noninfected. Of 48 seropositives without AIDS-related symptoms initially, 6 developed symptoms during the follow-up.[363]

9. Control and Prevention

The HIV-1 epidemic has posed a major challenge to disease control and prevention efforts in the United States and worldwide. The challenges are both political and scientific. The political arena has made HIV-1/AIDS a battleground for issues ranging from "family values" as played out in the debate about AIDS and sexual education in public schools, to "gay rights," to issues surrounding health care policy. This politicization has both stimulated action and retarded progress, especially when legitimate scientific issues have not been addressed.

The scientific challenges focus on the difficulties of effectively communicating prevention messages where the motivations for high-risk behavior are inadequately understood. The future of control and prevention efforts will depend not only on improved technologies, such as better drugs, vaccines, or screening tests, but also on a better understanding of the motivations that lead to continued high-risk sexual and drug-using activities.

The three principal modes of HIV-1 transmission are sexual, perinatal, and parenteral (injection drug use and blood transfusion). Techniques for prevention involve educational approaches or technological interventions (e.g., barrier methods, treatments, vaccines, etc.). Given the strong documentation of modes of transmission and insights about natural history, a major future focus in epidemiology will be the assessment of new technologies and behavior interventions for prevention. Interdisciplinary approaches are a key to future progress in prevention of AIDS and HIV-1 infections, especially where rates of infection continue to rise. A technological solution such as an effective preventive vaccine is unlikely to be developed and widely available before the beginning of the 21st century.

9.1. Public Health Approaches

Educational interventions involve generalized public information approaches, targeted information dissemination, and various counseling strategies for groups or individuals. The outcome measures of these interventions usually involve knowledge, attitudes, and behavior in the targeted populations. A major initial goal of most general education campaigns is to provide a basic understanding of how the virus is spread and not spread, since much of the population at large in the United States and worldwide has a substantial concern for HIV-1 as an easily contagious disease.

Many of those infected are from socially marginalized populations: homosexual men, injection drug users, female prostitutes, and so forth. Prejudice encourages a false belief that if an individual is not in the high-risk group, that they are not at risk. At the same time there is a false perception that casual contact with members of these risk groups could result in infection. Public education campaigns have attempted to correct such misconceptions, but high-risk behaviors of individuals such as frequent sexual partner exchange may not be affected by general information campaigns. Remarkably, as the virus has been introduced into new locales, these same false beliefs, born of fear and prejudice, have resurfaced over and over as traditional perceptions clash with the realities of a sexually and parenterally transmitted agent.

Attempts at preventive education are often extremely controversial because they represent an intersection of highly charged biological issues with core moral and ethical standards in society. For this reason, there has been substantial controversy as policy has been developed by government and private sectors to with deal with this problem. During the mid-1980s, the U.S. Surgeon General took a strong but politically controversial stand concerning the need for early and explicit sex education aimed at having a fully informed population as persons enter the sexually active years. The concept of providing education concerning HIV for elementary school students may be logical, but it also raises long-standing social and moral issues of who is responsible for a child's sexual education and the moral and ethical framework of such education. In

particular, curricula that focused on condom use in middle school and high school have raised substantial debate and point to the fact that broad educational campaigns of this sort do not meet the needs of all persons at risk. Ultimately, sexual abstinence is the safest approach and this message is at the core of most school-based programs.

Efforts to target the message are critical. An explicit message to those at high risk such as the sexually promiscuous teenager is appropriate, while for other teenagers a different approach may be more suitable.

A number of advocacy groups, particularly within the homosexual community, have developed explicit educational campaigns that stress a decrease in high-risk behaviors and the use of condoms as a major preventive measure. The best-described efforts have been the campaigns in the homosexual community where explicit educational messages have been communicated to the sexually active members. Such efforts were credited with the substantial reductions in infections in the community in the early to mid-1980s in epicenters such as San Francisco and New York. While there were clear, objective measures of declining seroincidence (Fig. 14), the basis for this success has been argued. Educational campaigns clearly alerted the community to the dangers of the high-risk behaviors summarized above and knowledge, attitudes, and behaviors were reported to be changed. However, the reduction in infections attributable to educational interventions is more difficult to quantify since by the time educational messages were developed and disseminated, a large fraction of those at greatest risk in the East and West Coast epicenters had already acquired infection. Furthermore, while studies have documented evidence that condoms reduce risk, it is also well known that failures take place, often due to a lack of knowledge of the appropriate use of condoms. A series of safe sex standards based on the HIV serostatus of a couple may provide a benchmark for assessing the effectiveness of preventive approaches.[206]

Considerable interest has focused on targeted information campaigns in high-risk communities. It is noteworthy that prevalence rates of infection in sites away from the AIDS epicenters in the United States were lower, possibly reflecting the benefits of such targeted educational efforts, although statistical models have pointed out that infection rates are a function of both behavior and density of infection in the community where exposure might take place.[41]

Individual counseling and partner notification have been used to varying degrees. By law, all persons being tested for HIV-1 in the United States must receive pretest counseling and those who are positive also must receive posttest counseling concerning risk factors and prevention strategies. Such approaches are widely used in clinic settings in the United States and abroad, and prospective studies assessing rates of HIV-1 infection document declining incidence of infections when such interventions are followed, coupled with other strategies such as condom distribution. Those who remain in follow-up experience the greatest reduction in incidence, but their behavioral risk profile may differ from those who abandon follow-up.

Despite wide knowledge of risk factors, new infections continue to occur, reflecting the significant gaps in our understanding of the basis for high-risk sexual and drug use behavior in the setting of HIV-1 risk. Some studies have begun to dissect the correlates of seroconversion and to include alcohol and recreational drug use, which lower the threshold for sexual risk-taking behavior and increase denial of perceived risk, such as occurs among young homosexual men who view HIV/AIDS as a problem of the older generation. For heterosexual couples, substantial risk reduction has been observed through educational programs, especially in the setting of knowledge of HIV status in the partner.[167,388] The high educational level of many subjects in hemophiliac and homosexual cohorts has facilitated information dissemination, while among minority and lower socioeconomic groups in the United States, where the majority of new infections are occurring, educational background and social marginalization adds to the prevention challenge. Studies of HIV prevention in this social context have reported greatest success in the area of drug injection-oriented prevention messages; modification of high-risk sexual behavior has proven more resistant to intervention.

Internationally, the major educational focus has been on prevention of heterosexually acquired infection. Since prostitution and coincident STDs with secondary spread to the spouse account for major sources of infection, the educational message is complex, since condom use in the STD setting has different implications than in the home setting. The Global AIDS Program of the World Health organization has combined its prevention efforts to focus on both HIV and other STDs in recognition of this coincidence.

One focus of HIV-1 prevention research has been defining the frequency and determinants of human sexual behavior that place individuals at high risk of HIV-1 infection. National surveys in the United States and worldwide have chronicled trends in behavior that amplify the potential for HIV-1 transmission. In particular, higher frequency of partner exchange and younger age at first intercourse have been documented in surveys both in the developing and developed world. The reasons for these trends is unclear but has been tied to a breakdown in traditional value systems with increasing population mobility and rising urbanization.

Mathematical modeling has documented an interesting relationship between average frequency of high-risk behavior and variance that argues that a subset of persons in any population account for a major source of variance, i.e., those with unusually high rates of risky behavior or sexual partner exchange. Such observations, now seen in a wide variety of populations, suggest opportunities to target key core transmission groups for HIV-1 and STD transmission. Findings point to the need for further fundamental research on human sexual behavior.

The technological approaches to preventing HIV infection include the development of an accurate and reliable screening test for infection, the dissemination of condoms as a barrier of HIV-1 and other STDs, the development of improved barrier approaches and policies in the health care setting, distribution of sterile needles and/or disinfection approaches to prevent drug abuse spread, the use of antiretrovirals to prevent infection in the mother-to-infant setting and the acute exposure setting, and the search for an effective preventive vaccine.

The development of a reliable blood test for detecting virus infection has had a major impact on HIV-1 prevention efforts. It is well documented that the introduction of the blood test in early 1985, based on the work of those who discovered the virus, saved thousands of lives and created a new tool for HIV-1 prevention. In the blood bank setting (Fig. 9), the introduction of the blood test virtually eliminated new infections in the transfusion and factor concentrate setting.[70] Currently, fewer than 1 in 200,000 units of blood are estimated to be infected in the United States, a substantial decrease from the days when as many as 1 in 100 units were contaminated in some high-rate areas. The blood test has also been shown to be beneficial in counseling discordant couples and sexual contacts of HIV-1 seropositives. Behavior change in the setting of knowledge of HIV-1 status has been substantial, as couples and individuals adopt either safe sex (abstinence) or safer sex (condoms) approaches.[388] It is noteworthy that even in the setting of unprotected sex for the purposes of procreation in discordant couples, the timing and frequency of exposure is substantially reduced.

Condoms are the mainstay of current preventive approaches aimed at reducing HIV-1 transmission. A number of studies have documented that consistent and proper use of condoms can reduce the spread of HIV-1 at least ninefold. Other studies have measured reductions in the occurrence of other STDs, but in some instances the reduction in transmission of other STDs does not predict a similar high level of HIV-1 infection elimination. The failure of condoms to prevent all HIV-1 infections arises from a number of factors. First, condoms are not always used with each sex act. For example, steady partners/spouses may not use condoms because they do not perceive the risk and wives are often overruled by their husbands on the matter. Condom skills are often lacking and improper use results in breakage or other malfunction. Finally, availability in many developing countries is limited by poor distribution infrastructure and local preferences for brands that may or may not be distributed.

Because women are often at the greatest risk for HIV-1 infection in a setting where their male partner is unwilling to use a condom, approaches to allow women greater control in the preventive arena have been a high priority. A female condom has recently been licensed, but its efficacy for preventing HIV-1 infection has not been documented in human trials. Another female-controlled barrier approach involves the use of viricidal spermacides such as sponges containing nonoxynol-9.[273] However, one study suggested that the use of nonoxynol-9 and sponges may enhance HIV-1 transmission, presumably due to vaginal irritation or abrasion.[273] Alternate approaches employing nonoxynol-9 in other mediums are being investigated.

In the clinical and laboratory setting, guidelines for universal infection control precautions have substantially altered clinical practice. Highlights of this preventive strategy include the use of proper barriers (latex gloves, spatter-proof gowns, face shields, etc.), with use of these barriers as dictated by the particulars of exposure risk. Practices such as avoidance of needle recapping and other techniques to avoid "sharps" exposure have also been implemented. Since these guidelines have been adopted as part of Occupational Safety and Health Administration rules, all health workers are required to have annual recertification that insures that these guidelines are widely implemented.

Preventive interventions for parenteral drug abuse are complex and politically charged. Dissemination of sterile needles to drug users has met with resistance in many communities and evaluation of efficacy has varied, but some studies have documented positive benefits, especially when coupled with strong counseling. Approaches that encourage needle disinfection gained substantial community support in the mid to late 1980s, but recent data have cast doubt on the efficacy of this approach unless disinfection times are sufficient to kill all virus. Preventive strategies to increase the availability of drug rehabilitation are based on the observation that duration in treatment was associated with a lower risk of seropositivity. However, high costs of drug rehabilitation have limited this approach.

Prevention of mother-to-child transmission is a ma-

jor challenge as HIV-1 continues to increase in heterosexually active populations. Because of the high incidence of HIV-1 infections, mother-to-child transmission studies offer unique opportunities to assess intervention strategies. Counseling of HIV-1-positive women to avoid pregnancy is reasonable policy, given the high risk for maternal-to-infant transmission and the potential for adverse effects on the mother's immune system in the context of an HIV-1-positive pregnancy. The avoidance of breast-feeding by HIV-1-positive mothers is also important because there is estimated to be an additional 15% risk for infection above that associated with *in utero* and perinatal transmission.[169] However, this recommendation only applies where a safe and reliable alternative feeding source is available, a circumstance lacking in most of the developing world. As outlined above, the majority of mother-to-infant transmissions occur in the perinatal setting. Thus, a number of possible interventions to reduce exposure to potentially infectious materials at the time of birth have been proposed. Cesarean section has been shown to reduce rates of infection in some studies, even when the membranes have already been ruptured,[479] but clinical trials to test this approach have not been reported and operative deliveries could still result in exposure of the infant to maternal blood, resulting in delivery-related infections. Additionally, because of cost, this approach would be impractical as a public health measure in the developing world where the majority of such infections are occurring. Another approach being studied is washing the birth canal during labor and delivery with a viricidal antiseptic.[31] Such an approach employing lavage has been shown to reduce the risk of group B streptococcus infection in Scandinavia.[64]

A recent clinical trial, ACTG 076, involved the administration of the antiretroviral drug zidovudine to women in the second or third trimester, including a bolus during delivery, and postpartum to the infant through the first 6 weeks of life, beginning within 6 hr of delivery. The rate of infection in the placebo group was 25.5% (95% CI, 18.4–32.5) compared to 8.3% (CI, 3.9–12.8) in the treated group. These data indicate that antiretroviral prophylaxis can substantially reduce transmission, but it is unclear which of the three interventions in the trial actually reduced transmission or whether all three interventions affected different aspects of transmission—prenatal or perinatal exposure.[109]

9.2. Vaccines

In the period immediately following the recognition that HIV-1 causes AIDS, there was considerable optimism that an AIDS vaccine would rapidly be forthcoming.

Buoyed by recent successes with synthetic and recombinant vaccines for hepatitis B and driven by concern that a whole virus might be dangerous, emphasis was placed on developing synthetic viral envelope-based subunit vaccines. Initial chimpanzee experiments proved disappointing mainly because of difficulties in titering challenge stocks. Except for occasional instances where animals could be protected against challenge with a strain of the virus identical to that of the vaccine, there was little success. The lack of a small-animal model for these HIV-1 experiments has contributed to the slow progress in the field, although nonhuman primate models employing SIV have provided useful systems.

The barriers to developing a suitable vaccine include the wide variation in virus within and between clades, difficulties in defining a useful correlate of immunity, and difficulties in stimulating appropriate mucosal immunity with parenterally administered vaccines. Plans to proceed with large-scale efficacy trials of first-generation envelope-based subunit vaccines in the United States were halted after several participants in phase I and phase II clinical trials became infected with HIV-1. The decision not to proceed was ultimately based on concerns that a partially effective vaccine might lead vaccinees to resume higher-risk behaviors, resulting in a net negative effect of such a vaccine on rates of new infections. There were also significant concerns that a poorly effective vaccine, while having some benefit from a public health perspective, would preclude the subsequent testing of more effective versions because of sample size considerations. However, efficacy trials are more likely to proceed in developing countries in collaboration with WHO.

Despite the many difficulties, preparation efforts for vaccination campaigns have led to epidemiologic studies to understand the incidence and risk for new infections in high-risk cohorts. Vaccine-related activities have also supported immunologic and virological studies that have, in turn, helped to identify some aspects of the host immune response to viral infection and the virological characteristics of infectious strains (e.g., their monocyte trophism). Studies of the SIV model have also provided insights such as the recent finding that a live attenuated (*nef*-deleted) SIV offers substantial protection against infection. These and other research studies will enhance understanding of the correlates of immunity, which ultimately may lead to an effective vaccine.

9.3. Therapy

Given that approximately one million persons are infected with HIV-1 in the United States and an additional 12 to 15 million are HIV-1 positive worldwide, the need for effective therapy is obvious. Conceptually, therapeutic approaches have focused on agents that attack the virus and others that attempt to modulate the immune response to the virus with the goal of restoring effective immunity or modulating those aspects of immunity that are either harmful to the host or deficient and in need of augmentation.

The chemotherapeutic agents are the more advanced both conceptually and in their documented benefit. The three licensed agents, zidovudine (ZDV), didanosine (ddI), and zalcitabine (ddC) all target the reverse transcriptase of the virus. These nucleotide analogues function by blocking a critical step in the life cycle of HIV-1, reverse transcription. Initial controlled trials with each of the agents have shown short-term benefits, but more recent longer-term follow-up has documented that the benefit of single-agent therapy are transient, as shown in the European Concorde trial.[130,512] The virus enzyme has the capability of mutating to escape the negative effects of the therapy. However, as shown in Fig. 33, the introduction of ZDV in 1987 had an immediate impact on AIDS incidence, an effect that was not seen among those with poor access to care.[197] Current studies are focusing on optimizing the timing in illness, the dose, and combination of agents with the greatest benefit to recipients. The Concorde study, for example, suggested that the benefits of monotherapy with ZDV were transient and that the impetus to use drugs early in the course of illness may not be beneficial in slowing the development of disease and may be associated with adverse side effects.[130] Studies of combinations of the licensed drugs have shown some benefits, but optimal timing has not been well worked out.

Newer agents such as HIV-1 protease inhibitors have shown some preliminary benefit in early phase trials and larger trials are ongoing. Therapies to prevent opportunistic infections have also contributed to the well-being of those with severe immune suppression and at high risk for complicating fungal and protozoan opportunists.

Additional classes of drugs and therapeutic approaches attempt to enhance the immunologic system, but none has proved to be effective. Drugs that attack other functions of the virus, including viral attachment integration and expression, are currently being developed. For the long range, it is likely that combination chemotherapeutic approaches, similar to those used in the treatment of cancer, will be required in the treatment of HIV-1 infection.

An important epidemiologic issue in clinical trials relates to the "frailty bias" introduced in conducting trials in cohorts over time. Given the dynamics of infection

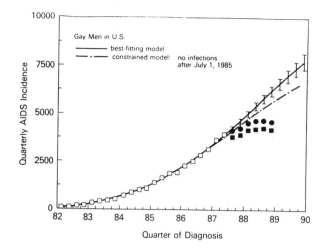

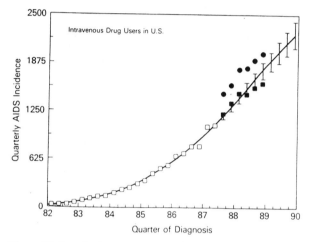

Figure 33. Introduction of ZDV: Impact on AIDS incidence. From Centers for Disease Control.[109]

(Fig. 14), those enrolled recently may represent longer-term survivors than those who have died in the interval since the peak of infection. In this regard, if such biases exist, potential benefits could be missed if participants in trials have a more indolent natural history, which might dampen the ability to monitor therapeutic effects.

9.4. HIV-2

Clearly the measures described above for primary prevention of HIV-1 infection are applicable here. Since the prevalence of HIV-1 infection is so low outside of West Africa, it is important to provide stringent confirmation of positive screening tests in the face of high false-positivity rates.[377] Unlike HIV-1, there is a small-primate model for HIV-2/SIV in macaques.[44,306,419] However, the barriers

facing the full development of a vaccine intervention are enormous, and prevention of infection remains the primary public health goal.

10. Unresolved Problems

In the 10 years since the cause of AIDS was shown to be a lenti-retrovirus, enormous progress has been made in defining the scope of the epidemic and the modes of its transmission. A major unresolved issue is to understand how and why new infections continue to occur despite widespread public understanding of how infection takes place.

Epidemiologic research in the future will expand to focus on evaluating intervention to prevent infections with HIV-1. This research will be multidisciplinary and will involve a strong behavioral component as well as expertise in community-based clinical trials research. Outcomes of such research will include measurements of both biological and behavioral endpoints.

In turn, laboratory measures of infection or resistance to infection following exposure will enhance insights into pathogenesis. Since the epidemic is of relatively recent onset, there are many unanswered questions about the long-term sequelae of HIV-1 infection. In particular, issues of direct oncogenesis or more subtle forms of slowly progressive neurological deterioration need to be studied by assembling and coordinating data collections from various centers in order to assess rarer long-term outcomes. Studies of the factors associated with stable long-term survivorship and definition of host factors and correlates of survivorship may enhance understanding of the mechanisms of pathogenesis.

More precise measures of the virological and host factors associated with HIV-1 spread in communities need to be developed so that issues of primary prevention can be focused with certainty. Issues of infectivity need to be quantified to maximize the protection of those potentially exposed. The issue of variability in strain virulence is also unresolved.

A major focus for epidmeiologic research will come to the fore when prototype vaccines become available. Selecting suitable populations for vaccination and assessing effectiveness of the vaccine will require close coordination between clinicians, epidemiologists, and laboratory scientists. Effective therapies may also be developed and the potential for prevention of spread needs to be fully assessed.

While the current mood in AIDS research is somber as the pace of progress shifts from exponential to incremental, opportunities for progress do exist in the framework of epidemiologically defined populations where interventions, both preventive and therapeutic, can most logically be assessed. The challenge is to bring interdisciplinary approaches built on the framework of defined populations to bear on solving fundamental questions of HIV pathogenesis critical to prevention and treatment in all persons at risk of HIV-1/AIDS.

Many issues are unresolved in the epidemiology of HIV-2 infection. The most important concern the natural history of infection: What is the cumulative incidence of disease? Does this differ between populations? What is the latency distribution? What is the effect of coinfection with other retroviruses? Addressing these questions requires long-term follow-up of infected individuals in diverse areas.[431] HIV-2 also offers a mirror to understanding its pandemic cousin, HIV-1. Elucidating the biological differences between the two viruses and their respective evolution in human hosts may provide clues to potential control of these new human infections.

11. References

1. *AIDS in the World.* MANN, J., TARANTOLA, D. J. M., AND NETTER, T. W. (EDS.), Howard University Press, Cambridge, MA, 1992.

2. ALLAIN, J., LAURIAN, Y., PAUL, D. A., AND SENN, D., Serological markers in early stages of human immunodeficiency virus infection in haemophiliacs, *Lancet* **2:**1233–1236 (1986).

3. ALLAN, J. S., SHORT, M., TAYLOR, M. E., SU, S., HIRSCH, V. M., JOHNSON, P. R., SHAW, G. M., AND HAHN, B. H., Species-specific diversity among simian immunodeficiency viruses from African green monkeys, *J. Virol.* **65:**2816–2828 (1991).

4. ALLEN, S., BATUNGWANAYO, J., KERLIKOWSKE, K., LIFSON, A. R., WOLF, W., GRANICH, R., TAELMAN, H., VAN DE PERRE, P., SERUFILIRA, A., BOGNERTS, J., SLUTKIN, G., AND HOPEWELL, P. C., Two-year incidence of tuberculosis in cohorts of HIV-infected and uninfected urban Rwandan women, *Am. Rev. Respir. Dis.* **146:**1439–1444 (1992).

5. ALLEN, S., LINDAN, C., SERUFILIRA, A., VAN DE PERRE, P., RUNDLE, A. C., NSENGUMUREMYI, F., CARAEL, M., SCHWALBE, J., AND HULLEY, S., Human immunodeficiency virus infection in urban Rwanda. Demographic and behavioral correlates in a representative sample of childbearing women, *J.A.M.A.* **266**(12)**:**1657–1663 (1991).

6. ANCELLE, R., BLETRY O, BAGLIN, A. C., BRUN-VEZINET, F., REY, M. A., AND GODEAU, P., Long incubation period for HIV-2 infection [letter], *Lancet* **1:**688–689 (1987).

7. ANDERSON D. J., O'BRIEN, T. R., POLITCH, J. A., MARTINEZ, A., SEAGE, G. R. III, PADIAN, N., HORSBURGH, C. R., JR., AND MAYER, K. H., Effects of disease stage and zidovudine therapy on the detection of human immunodeficiency virus type 1 in semen, *J.A.M.A.* **267**(20)**:**2769–2774 (1992).

8. ANDERSON, R. M., MAY, R. M., BOILY, M. C., GARNETT, G. P., AND ROWLEY, J. T., The spread of HIV-1 in Africa: Sexual contact patterns and the predicted demographic impact of AIDS, *Nature* **352:**581–589 (1991).

9. ANDIMAN, W. A., SIMPSON, B. J., OLSON, B., DEMBER, L., SILVA, T. J., AND MILLER, G., Rate of transmission of HIV type 1 infection from mother to child and short term outcome of neonatal infection, *Am. J. Dis. Child.* **144:**758–766 (1990).

10. ANDO, Y., NAKANO, S., SAITO, K., SHIMAMOTO, I., ICHIJO, M., TOYAMA, T., AND HINUMA, Y., Transmission of adult T-cell leukemia retrovirus (HTLV-I) from mother to child: Comparison of bottle- with breast-fed babies, *Jpn. J. Cancer Res.* **78:**322–324 (1987).

11. ANDREASSON, P. A., DIAS, F., NAUCLÉR, A., ANDERSSON, S., AND BIBERFELD, G., A prospective study of vertical transmission of HIV-2 in Bissau, Guinea-Bissau, *AIDS* **7:**989–993 (1993).

12. ASCHER, D. P., AND ROBERTS, C., Determination of the etiology of seroreversals in HIV testing by antibody fingerprinting, *J. Acquir. Immune Defic. Syndr.* **6**(3)**:**241–244 (1993).

13. ASCHER, M. S., SHEPPARD, H. W., ARNON, J. M., AND LANG, W., Viral burden in HIV disease, *J. Acquir. Immune Defic. Syndr.* **4**(8)**:**824–825 (1991).

14. AUGER, I., THOMAS, P., DE GRUTTOLA, V., MORSE, D., MOORE, D., WILLIAMS, R., TRUMAN, B., AND LAWRENCE, C. E., Incubation periods for paediatric AIDS patients, *Nature* **336:**575–577 (1988).

15. BABA, T. W., KOCH, J., MITTLER, E. S., GREENE, M., WYAND, M., PENNINCK, D., AND RUPRECHT, R. M., Mucosal infection of rhesus monkeys with cell-free SIV, *AIDS Res. Hum. Retroviruses* **10**(4)**:**351–357 (1994).

16. BALTIMORE, D., RAN-dependent DNA polymerase in virions of RNA tumor viruses, *Nature* **226:**1209–1211 (1970).

17. BARIN, F., DENIS, F., ALLAN, J. S., M'BOOP, S., KANKI, P., LEE, T. H., AND ESSEX, M., Serological evidence for virus related to simian T-lymphotropic retrovirus III in residents of West Africa, *Lancet* **2:**1387–1989 (1985).

18. BARRE-SINOUSSI, F., CHERMANN, J. C., REY, F., NUGEYTRE, M. T., CHAMARET, S., GRUEST, J., DAUGUET, C., AXLER-BLUE, C., VEZINET-BRUN, F., ROUZIOUX, C., ROZENBAUM, W., AND MONTAGNIER, L., Isolation of a T-lymphotropic retrovirus from a patient at risk for acquired immune deficiency syndrome (AIDS), *Science* **220:**868–871 (1983).

19. BARTHOLOMEW, C., RAJU, C., PATRICK, A., PENCO, F., AND JANKEY, N., AIDS on Trinidad, *Lancet* **1:**103 (1984).

20. BARTHOLOMEW, C., RAJU, C. C., AND JANKEY, N., The acquired immune deficiency syndrome in Trinidad: A report on two cases, *West Indian Med. J.* **32:**177–180 (1983).

21. BARTHOLOMEW, C., SAXINGER, W. C., CLARK, J. W., GAIL, M., DUDGEON, A., MAHIBIR, B., HULL-DRYSDALE, B., CLEGHORN, F. R., GALLO, R. C., AND BLATTNER, W. A., Transmission of HTLV-I and HIV among homosexual men in Trinidad, *J.A.M.A.* **257**(19)**:** 2604–2608 (1987).

22. BENVENISTE, R. E., The contributions of retroviruses to the study of mammalian evolution, in: *Molecular Evolutionary Genetics* (R. J. MACINTYRE, ED.), pp. 359–417, Plenum Press, New York, 1985.

23. BERAL, V., BULL, D., DARBY, S., WELLER, I., CARNE, C., BEECHAM, M., AND JAFFE, H., Risk of Kaposi's sarcoma and sexual practices associated with faecal contact in homosexual or bisexual men with AIDS, *Lancet* **339:**632–635 (1992).

24. BERAL, V., PETERMAN, T., BERKELMAN, R., JAFFE, H., AIDS-associated non-Hodgkin lymphoma, *Lancet* **337:**805–809 (1991).

25. BERAL, V., PETERMAN, T. A., BERKELMAN, R. L., AND JAFFE,

H. W., Kaposi's Sarcoma among persons with AIDS: A sexually transmitted infection, *Lancet* **335:**123–128 (1990).

26. BERKELEY, J. S., FOGIEL, P. C., KINDLEY, A. D., AND MOFFAT, M. A., Peripartum HIV seroconversion: A cautionary tale, *Lancet* **340:**58–59 (1992).

27. BERKELEY, S. F., WIDY-WIRSKI, R., OKWARE, S. I., DOWNING, R., LINNAN, M. J., WHITE, K. E., AND SEMPALA, S., Risk factors associated with HIV infection in Uganda, *J. Infect. Dis.* **160** (1):22–29 (1989).

28. BESNIER, J. M., BARIN, F., BAILLOU, A., LIARD, F., CHOUTET, P., AND GOUDEAU, A., Symptomatic HIV-2 primary infection, *Lancet* **335:**798 (1990).

29. BIBERFELD, G., BÖTTIGERM, B., BREDBERG-RADÉN, U., PUTKONEN, P. O., ERICSSON, L., BERGLUND, D., STARUP, C., AND HAKÄUSSON, L., Clinical findings in four HTLV-IV seropositive women from West Africa, *Lancet* **2:**1330 (1986).

30. BIDDLECOM, A. E., LeCLERE, F. B., HARDY, A. M., AND HENDERSHOT, G. E., National study of knowledge of AIDS, testing patterns, and self-assessed risk among health care workers, *J. Acquir. Immune Defic. Syndr.* **5**(11):1131–1136 (1992).

31. BIGGAR, R. J., Preventing mother-to-infant transmission of the human immunodeficiency virus, in press.

32. BIGGAR, R. J., International Registry of Seroconverters. AIDS incubation in 1891 HIV seroconverters from different exposure groups, *AIDS* **4:**1059–1066 (1990).

33. BIGGAR, R. J., AND MELBYE, M., Responses to anonymous questionnaires concerning sexual behavior: A method to examine potential biases, *Am. J. Public Health* **82**(11):1506–1512 (1992).

34. BIGGAR, R. J., AND ROSENBERG, P. S., HIV infection/AIDS in the United States during the 1990s, *Clin. Infect. Dis.* **17s:**S219–S223 (1993).

35. BLACKBURN, R., CLERICI, M., MANN, D., LUCEY, D. R., GOEDERT, J. J., GOLDING, B., SHEARER, G. M., AND GOLDING, H., Common sequence in HIV-1 gp41 and HLA class II beta chains can generate cross-reactive autoantibodies with immunosuppressive potential early in the course of HIV-1 infection, *Adv. Exp. Med. Biol.* **303:**63–69 (1991).

36. BLANCHE, S., ROUZIOUX, C., MOSCATO, M. L., VEBER, F., MAYAUX, M. J., JACOMET, C., TRICOIRE, J., DEVILLE, A., VIAL, M., FIRTION, G., DEGREPY, A., DOUARD, D., ROBIN, H., COURPOTIN, C., CIRARU-VIGNERON, N., LeDEIST, F., AND GRISCELLI, C., A prospective study of infants born to women seropositive for human immunodeficiency virus type 1, *N. Engl. J. Med.* **320:** 1643–1648 (1989).

37. BLATTNER, W. A., BIGGAR, R. J., WEISS, S. H., MELBYE, M., AND GOEDERT, J. J., Epidemiology of human T-lymphotropic virus type III and the risk of the acquired immunodeficiency syndrome, *Ann. Intern. Med.* **103**(5):665–670 (1985).

38. BLATTNER, W. A., GALLO, R. C., AND TEMIN, H. M., Policy forum-HIV causes AIDS, *Science* **241:**515–516 (1988).

39. BLATTNER, W. A., REITZ, M., COLCLOUGH, G., AND WEISS, S., HIV/AIDS in laboratory infected with HIV-III (abstract). Berlin AIDS Meeting 1994; PO-B01-0876.

40. BLAXHULT, A., GRANATH, F., LIDMAN, K., AND GIESECKE, J., The influence of age on the latency period to AIDS in people infected by HIV through blood transfusion, *AIDS* **4**(2):125–129 (1990).

41. BLOWER, S. M., ANDERSON, R. M., AND WALLACE, P., Loglinear models, sexual behavior and HIV: Epidemiological implications of heterosexual transmission, *J. Acquir. Immune Defic. Syndr.* **3**(8):763–772 (1990).

42. BLOWER, S. M., SAMUEL, M. C., AND WILEY, J. A., Sex, power

laws, and HIV transmission, *J. Acquir Immune Defic. Syndr.* **5**(6):633–634 (1992).

43. BONGAARTS, J., REINING, P., WAY, P., AND CONANT, F., The relationship between male circumcision and HIV infection in African populations, *AIDS* **3:**373–377 (1989).

44. BONYHADI, M. L., RABIN, L., SALIMI, S., BROWN, D. A., KOSEK, J., McCURE, J. M., AND KANESHIMA, H., HIV induces thymus depletion *in vivo, Nature* **363:**728–732 (1993).

45. BOOT, J. M., BUITENWERF, J., HUISKAMP, N., AND STOLZ, E., HIV antigen testing, *Lancet* **2:**109 (1988).

46. BORING, C. C., SQUIRES, T. S., AND TONG, T., Cancer statistics, 1992. *CA Cancer J. Clin* **41**(1):19–38 (1994).

47. BORKOWSKY, W., KRASINSKI, K., POLLACK, H., HOOVER, W., KAUL, A., AND ILMET-MOORE, T., Early diagnosis of human immunodeficiency virus infection in children less than 6 months of age: Comparison of polymerase chain reaction, culture, and plasma antigen capture techniques, *J. Infect. Dis.* **166:**616–619 (1992).

48. BOURGOIGNIE, J. J., MENESES, R., ORTIZ, C., JAFFE, D., AND PARDO, V., The clinical spectrum of renal disease associated with human immunodeficiency virus, *Am. J. Kidney Dis.* **12**(2):131–137 (1988).

49. BOUVET, E., DeVINCENZI, I., ANCELLE, R., AND VACHON, F., Defloration as risk factor for heterosexual HIV transmission, *Lancet* **1:**615–616 (1989).

50. BRAMBILLA, D. J., RICH, K. C., LANDAY, A., *ET AL.*, Early differences in lymphocyte subsets between HIV+ and HIV-infants. Preliminary findings (abstract), Ninth Int. Conf. AIDS, June 7–11, 1993.

51. BRAUM, M. M., COTÉ, T. R., AND RABKIN, C. S., Trends in deaths with tuberculosis during the AIDS era, *J.A.M.A.* **269**(22):2865–2868 (1993).

52. BRIGGS, N. C., NATOLI, C., TINARI, N., D'EGIDIO, M., GOEDERT, J. J., AND IACOBELLI, S., A 90-kDa protein serum marker for the prediction of progression to AIDS in a cohort of HIV-1+ homosexual men, *AIDS Res. Hum. Retroviruses* **9**(9):811–816 (1993).

53. BROOKMEYER, R., Reconstruction and future trends of the AIDS epidemic in the United States, *Science* **253:**37–42 (1991).

54. BROOKMEYER, R., AND GAIL M. H., *AIDS Epidemiology: A Quantitative Approach,* Oxford University Press, New York, 1994.

55. BROOKMEYER, R., GAIL, M. H., AND POLK, B. F., The prevalent cohort study and the acquired immunodeficiency syndrome, *Am. J. Epidemiol.* **126:**14–24 (1987).

56. BRUNDAGE, J. F., Epidemiology of HIV infection and AIDS in the United States, *Dermatol. Clin.* **9**(3):443–451 (1991).

57. BRUNDAGE, J. F., BURKE, D. S., GARDNER, L. I., McNEIL, J. G., GOLDENBAUM, M., VISINTINE, R., REDFIELD, R. R., PETERSON, M., AND MILLER, R. N., Tracking the spread of the HIV infection epidemic among young adults in the United States: Results of the first four years of screening among civilian applicants for U.S. military service, *J. Acquir. Immune Defic. Syndr.* **3**(12):1168–1180 (1990).

58. BRUNDAGE, J. F., McNEIL, J. G., MILLER, R. N., GARDNER, L. I., HARRISON, S. M., HAWKES, C., CRAIG, D. B., REDFIELD, R., AND BURKE, D. S., The current distribution of CD4+ T-lymphocyte counts among adults in the United States with human immunodeficiency virus infection: Estimates based on the experience of the U.S. army, *J. Acquir. Immune Defic. Syndr.* **3**(1):92–94 (1990).

59. BRYCESON, A., TOMKINS, A., RIDLEY, D., WARHURST, D., GOLDSTONE, A., BAYLISS, G., TOSWILL, J., AND PARRY, J., HIV-2 associated aids in the 1970s, *Lancet* **2:**221 (1988).

60. BUCHBINDER, S. P., KATZ, M. H., HESSOL, N. A., O'MALLEY, P. M., AND HOLMBERG, S. D., Long-term HIV-1 infection without immunologic progression, *AIDS* **8:**1123–1128 (1994).

61. BUEHLER, J. W., DEVINE, O. J., BERKELMAN, R. L., AND CHEVARLEY, F. M., Impact of the human immunodeficiency virus epidemic on mortality trends in young men, United States, *Am. J. Public Health* **80**(9):1080–1086 (1990).

62. BUGELSKI, P. J., ELLENS, H., HART, T. K., AND KIRSH, R. L., Soluble CD4 and dextran sulfate mediate release of gp120 from HIV-1: Implications for clinical trials, *J. Acquir. Immune Defic. Syndr.* **4**(9):923–924 (1991).

63. BURKE, D. S., BRUNDAGE, J. F., HERBOLD, J. R., BERNER, W., GARDNER, L. I., GUNZENHAUSER, J. D., VOSKOVITCH, J., AND REDFIELD, R. R., Human immunodeficiency virus infections among civilian applicants for United States military service, October 1985 to March 1986 demographic factors associated with seropositivity, *N. Engl. J. Med.* 317:131–136 (1987).

64. BURMAN, L. G., CHRISTENSEN, P., CHRISTENSEN, K., FRYKLUND, B., HELGESSON, A. M., SVENNINGSEN, N. W., AND TULLUS, K., Prevention of excess neonatal morbidity associated with group B streptococci by vaginal chlorhexidine disinfection during labour, *Lancet* **340:**65–69 (1992).

65. BURNET, M., AND WHITE, D. O., *The Natural History of Infectious Disease*, 4th ed., Cambridge University Press, Cambridge, MA, 1990.

66. BURNS, D. N., KRÄMER, A., YELLIN, F., FUCHS, D., WACHTER, H., DIGIOIA, R., SANCHEZ, W. C., GROSSMAN, R. J., GORDIN, F. M., BIGGAR, R. J., AND GOEDERT, J. J., Cigarette smoking: A modifier of human immunodeficiency virus type 1 infection? *J. Acquir. Immune Defic. Syndr.* **4**(1):76–83 (1991).

67. BUSCH, M. P., EBLE, B. E., KHAYAM-BASHI, H., HEILBRON, D., MURPHY, E. L., KWOK, S., SNINSKY, J., PERKINS, H. A., AND VYAS, G. N., Evaluation of screened blood donations for human immunodeficiency virus type 1 by culture and DNA amplification of pooled cells, *N. Engl. J. Med.* **325**(1):1–5 (1991).

68. BUSCH, M. P., EL-AMAD, Z., MCHUGH, T. M., CHIEN, D., AND POLITO, A. J., Reliable confirmation and quantitation of human immunodeficiency virus type I antibody using a recombinant-antigen immunoblot assay, *Transfusion* **13:**129–137 (1991).

69. BUSCH, M. P., HENRARD, D. R., HEWLETT, I. K., MEHAFFEY, W. F., EPSTEIN, J. S., ALLAIN, J. P., LEE, T. H., AND MOSLEY, J. W., Poor sensitivity, specificity, and reproducibility of detection of HIV-1 DNA in serum by polymerase chain reaction, *J. Acquir. Immune Defic. Syndr.* **5**(9):872–877 (1992).

70. BUSCH, M. P., YOUNG M. J., SAMSON, S. M., MOSLEY, J. W., WARD, J. W., AND PERKINS, H. A., Risk of human immunodeficiency virus (HIV) transmission by blood transfusions before the implementation of HIV-1 antibody screening, *Transfusion* **31** (1):4–11 (1991).

71. BYERS, R. H., JR., MORGAN, W. M., DARROW, W. W., DOLL, L., JAFFE, H. W., RUTHERFORD, G., HESSOL, N., AND O'MALLEY, P. M., Estimating AIDS infection rates in the San Francisco cohort, *AIDS* **2:**207–210 (1988).

72. CALDWELL, B., OXTOBY, M., AND ROGERS, M., Proposed CDC pediatric HIV classification system evaluation in an active surveillance system (abstract), Ninth Int. Conf. AIDS, June 6–11, 1993.

73. CAMERON, D. W., SIMONSEN, J. N., D'COSTA, L. J., ET AL., Female to male transmission of human immunodeficiency virus type 1: Risk factors for seroconversion in men, *Lancet* **2:**403–407 (1989).

74. CASTRO, K. G., DOOLEY, S. W., AND CURRAN, J. W., Transmission of HIV-associated tuberculosis to health-care workers, *Lancet* **340:**1043–1044 (1992).

75. CAUSSY, D., TERRINHA, A., EVANS, A. C., REIS, D. D., NUNES, J. F. M., AND BLATTNER, W. A., Changes in HIV-2 seroprevalence in Cape Verde, West Africa, *J. Acquir. Immune Defic. Syndr.* **6:**432–433 (1993).

76. CAUSSY, D., WEISS, S. H., BLATTNER, W. A., FRENCH, J., CANTOR, K. P., GINZBURG, H., ALTMAN, R., AND GOEDERT, J. J., Exposure factors for HIV-1 infection among heterosexual drug abusers in New Jersey treatment centers, *AIDS Res. Hum. Retroviruses* **6**(12):459–467 (1990).

77. CAVACINI, L. A., EMES, C. L., POWER, J., BUCHBINDER, A., ZOLLA-PAZNER, S., AND POSNER, M. R., Human monoclonal antibodies to the V3 loop of HIV-1 gp120 mediate variable and distinct effects on binding and viral neutralization by a human monoclonal antibody to the CD4 binding site, *J. Acquir. Immune Defic. Syndr.* **6**(4):353–358 (1993).

78. CENTERS FOR DISEASE CONTROL, Follow-up on Kaposi's sarcoma and Pneumocystis pneumonia, *MMWR Morbid. Mortal. Week. Rep.* **30**(33):409–420 (1981).

79. CENTERS FOR DISEASE CONTROL, Pneumocystis pneumonia—Los Angeles, *MMWR Morbid. Mortal. Week. Rep.* **30:**250–252 (1981).

80. CENTERS FOR DISEASE CONTROL, Kaposi's sarcoma and Pneumocystis pneumonia among homosexual men—New York City and California, *MMWR Morbid. Mortal. Week. Rep.* **30**(25):305–308 (1981).

81. CENTERS FOR DISEASE CONTROL, Acquired immune deficiency syndrome (AIDS): Precautions for clinical and laboratory staffs, *MMWR Morbid. Mortal. Week. Rep.* **31:**577–580 (1982).

82. CENTERS FOR DISEASE CONTROL, Prospective evaluation of health-care workers exposed via parenteral or mucous-membrane routes to blood and body fluids of patients with acquired immunodeficiency syndrome, *MMWR Morbid. Mortal. Week. Rep.* **33:** 181–182 (1984).

83. CENTERS FOR DISEASE CONTROL, Recommendations for preventing transmission of infection with human T-lymphotropic virus type III/lymphadenopathy-associated virus in the workplace, *MMWR Morbid. Mortal Week. Rep.* **34:**681-6–691-5 (1985).

84. CENTERS FOR DISEASE CONTROL, Update: Evaluation of human T-lymphotropic virus type III/lymphadenopathy-associated virus infection in health-care personnel—United States, *MMWR Morbid. Mortal. Week. Rep.* **34:**575–578 (1985).

85. CENTERS FOR DISEASE CONTROL, Provisional public health service inter-agency recommendations for screening donated blood and plasma for antibody to the virus causing acquired immunodeficiency syndrome, *MMWR Morbid. Mortal. Week. Rep.* **34:**1–5 (1985).

86. CENTERS FOR DISEASE CONTROL, Human T-lymphotropic virus type III/lymphadenopathy-associated virus antibody prevalence in U.S. military recruit applicants, *MMWR Morbid. Mortal. Week. Rep.* **35:**421–424 (1986).

87. CENTERS FOR DISEASE CONTROL, Tuberculosis—United States, 1985—and the possible impact of human T-lymphotropic virus type III/lymphadenopathy-associated virus infection, *MMWR Morbid. Mortal. Week. Rep.* **35:**74–76 (1986).

88. CENTERS FOR DISEASE CONTROL, Human immunodeficiency virus infection transmitted from an organ donor screened for HIV antibody—North Carolina, *MMWR Morbid. Mortal. Week.* **36:** 306–308 (1987).

89. CENTERS FOR DISEASE CONTROL, Revision of the CDC surveillance case definition for acquired immunodeficiency syndrome, *MMWR Morbid. Mortal Week. Rep.* **36**(1s):3s–15s (1987).

90. CENTERS FOR DISEASE CONTROL, Human immunodeficiency virus (HIV) infection codes: Official authorized addendum ICD-9-CM (revision No. 1) effective January 1, 1988, *MMWR CDC Surveill. Summ.* **36**(Nos-7):1–24 (1987).

91. CENTERS FOR DISEASE CONTROL, Lack of transmission of human immunodeficiency virus through Rho(D) immune globulin (human), *MMWR Morbid. Mortal. Week. Rep.* **36**(44):728–729 (1987).

92. CENTERS FOR DISEASE CONTROL, Human immunodeficiency virus infection in the United States: A review of current knowledge, *MMWR Morbid. Mortal. Week. Rep.* **36**(S-6):1–48 (1987).

93. CENTERS FOR DISEASE CONTROL, Survey of non-U.S. hemophilia treatment centers for HIV seroconversions following therapy with heat-treated factor concentrates, *MMWR Morbid. Mortal. Week. Rep.* **36**(9):121–124 (1987).

94. CENTERS FOR DISEASE CONTROL, Perspectives in disease prevention and health promotion. Safety of therapeutic products used for hemophilia patients, *MMWR Morbid. Mortal. Week. Rep.* **37**(29):441–450 (1988).

95. CENTERS FOR DISEASE CONTROL, Guidelines for preventing the transmission of tuberculosis in health-care settings, with special focus on HIV-related issues, *MMWR Morbid. Mortal. Week. Rep.* **39**(RR-17):1–29 (1990).

96. CENTERS FOR DISEASE CONTROL, HIV prevalence estimates and AIDS case projections for the United States: Report based upon a workshop, *MMWR Morbid. Mortal Week. Rep.* **39**(RR16):1–31 (1990).

97. CENTERS FOR DISEASE CONTROL, Estimates of HIV prevalence and projected AIDS cases: Summary of a workshop, October 31–November 1, 1989, *MMWR Morbid. Mortal. Week. Rep.* **39**:110–119 (1990).

98. CENTERS FOR DISEASE CONTROL, Public health service interagency guidelines for screening donors of blood, plasma, organs, tissues, and semen for evidence of hepatitis B and hepatitis C, *MMWR Morbid. Mortal. Week. Rep.* **40**(RR-4):1–17 (1991).

99. CENTERS FOR DISEASE CONTROL, Patterns of sexual behavior change among homosexual/bisexual men—selected U.S. sites, 1987–1990, *MMWR Morbid. Mortal. Week. Rep.* **40**(46):792–794 (1991).

100. CENTERS FOR DISEASE CONTROL, The second 100,000 cases of acquired immunodeficiency syndrome—United States, June 1981–December 1991, *MMWR Morbid. Mortal. Week. Rep.* **41**(2):28–32 (1992).

101. CENTERS FOR DISEASE CONTROL, 1993 revised classification system for HIV infection and expanded surveillance case definition for AIDS among adolescents and adults, *MMWR Morbid. Mortal. Week. Rep.* **41**(17):1–19 (1992).

102. CENTERS FOR DISEASE CONTROL, HIV infection in two brothers receiving intravenous therapy for hemophilia, *MMWR Morbid. Mortal. Week. Rep.* **41**(14):228–231 (1992).

103. CENTERS FOR DISEASE CONTROL, HIV/AIDS surveillance report, *HIV/AIDS Surveill. Rep.* **5**(3) (1993).

104. CENTERS FOR DISEASE CONTROL, HIV transmission between two adolescent brothers with hemophilia, *MMWR Morbid. Mortal. Week. Rep.* **42**(49):948–951 (1993).

105. CENTERS FOR DISEASE CONTROL, Update: Investigation of persons treated by HIV-infected health-care workers—United States, *MMWR Morbid. Mortal. Week. Rep.* **42**(17):329–337 (1993).

106. CENTERS FOR DISEASE CONTROL, Heterosexually acquired AIDS—United States, 1993, *MMWR Morbid. Mortal. Week. Rep.* **43**:155–160 (1994).

107. CENTERS FOR DISEASE CONTROL, Human immunodeficiency virus transmission in household settings—United States, *MMWR Morbid. Mortal. Week. Rep.* **43**(19):347–356 (1994).

108. CENTERS FOR DISEASE CONTROL, Update: Impact of the expanded AIDS surveillance case definition for adolescents and adults on case reporting—United States, 1993, *MMWR Morbid. Mortal. Week. Rep.* **43**:160–170 (1994).

109. CENTERS FOR DISEASE CONTROL, Clinical situations and recommendations for use of ZDV to reduce perinatal transmission, *MMWR Morbid. Mortal. Week. Rep.* **43**(RR-11):7–13 (1994).

110. CELENTANO, D. D., AKARASEWI, P., SUSSMAN, L., SUPRASERT, S., MATANASARAWOOT, A., WRIGHT, N. H., THEETRANONT, C., AND NELSON, K. E., HIV-1 infection among lower class commercial sex workers in Chiang Mai, Thailand, *AIDS* **8**(4):533–537 (1994).

111. CENTER FOR INTERNATIONAL RESEARCH, HIV/AIDS surveillance data base, US Bureau of the Census 1993.

112. CHAISSON, R. E., MOSS, A. R., ONISHI, R., OSMOND, D., AND CARLSON, J. R., Human immunodeficiency virus infection in heterosexual intravenous drug users in San Francisco, *Am. J. Public Health* **77**:169–172 (1987).

113. CHAKRABARTI, L., GUYADER, M., ALIZON, M., DANIEL, M. D., DESROSIERS, R. C., TIOLLAIS, P., AND SONIGO, P., Sequence of simian immunodeficiency virus from macaque and its relationship to other human and simian retroviruses, *Nature* **328**:543–547 (1987).

114. CHIN, J., Current and future dimensions of the HIV/AIDS pandemic in women and children, *Lancet* **336**:2221–2224 (1990).

114a. CHIN, J., Global estimates of HIV infections and AIDS: 1991. *AIDS 1991* **5**(suppl 2):S59–S63 (1992).

114b. CHIN, J., Present and future dimensions of the HIV/AIDS pandemic, in: *Science challenging AIDS* (Plenary lectures of the VII International Conference on AIDS, Florence, 16–21 June 1991), pp. 33–50, Karger, Basel, 1992.

115. CHIN, J., AND LWANGA, S. K., Estimation and projection of adult AIDS cases: A simple epidemiological model, *Bull. World Health Organ.* **69**(4):399–406 (1991).

116. CHIN, J., AND MANN, J., Global surveillance and forecasting of AIDS, *Bull. World Health Organ.* **67**:1–7 (1989).

116a. CHIN, J., REMENYI, M.-A., MORRISON, F., AND BULATAO, R., The global epidemiology of the HIV/AIDS pandemic and its projected demographic impact in Africa. *World Health Stat. Quart.* **45**(2/3):220–227 (1992).

116b. CHIN, J., SATO, P., AND MANN, J. M., Projections of HIV infections and AIDS cases to the year 2000. *Bull. World Health Organ.* **68**(1):1–11 (1990).

117. CHOE, H. R., AND SODROSKI, J., Contribution of charged amino acids in the CDR2 region of CD4 to HIV-1 gp120 binding, *J. Acquir. Immune. Defic. Syndr.* **5**(2):204–210 (1992).

118. CLAVEL, F., GUETARD, D., BRUN-VEZINET, F., CHAMARET, S., REY, M. A., SANTOS-FERREIRA, M. O., LAURENT, A. G., DAUGUET, C., KATLAMA, C., ROUZIOUX, C., KLATZMANN, D., CHAMPALIMAUD, J. L., AND MONTAGNIER, L., Isolation of a new human retrovirus from West African patients with AIDS, *Science* **233**:343–346 (1986).

119. CLEGHORN, F. R., JACK, N., MURPHY, J., EDWARDS, J., MAHABIR, B., PAUL, R., WHITE, F., BARTHOLOMEW, C., AND BLATTNER, W. A., HIV-1 prevalence and risk factors among STD clinic attenders in Trinidad, *AIDS* **9**:389–394 (1995).

120. CLEMETSON, D. B. A., MOSS, G. B., WILLERFORD, D. M., HENSEL, M., EMONYI, W., HOLMES, K. K., PLUMMER, F., NDINYA-ACHOLA, J., ROBERTS, P. L. HILLIER, S., AND KREISS, J. K., Detection of HIV DNA in cervical and vaginal secretions: Prevalence and correlates among women in Nairobi, Kenya, *J.A.M.A.* **269**(22):2860–2864 (1993).

121. CLERICI, M., LEVIN, J. M., KESSLER, H. A., HARRIS, A., BERZOFSKY, J. A., LANDAY, A. L., AND SHEARER, G. M., HIV-specific T-helper activity in seronegative health care workers exposed to contaminated blood, *J.A.M.A.* **271**(1):42–46 (1994).

122. COATES, R. A., CALZAVARA, L. M., READ, S. E., FANNING, M. M., SHEPHERD, F. A., KLEIN, M. H., JOHNSON, J. K., AND SOSKOLNE, C. L., Risk factors for HIV infection in male sexual contacts of men with AIDS or an AIDS-related condition, *Am. J. Epidemiol.* **128**(4):729–739 (1988).

123. COFFIN, J. M., Superantigens and endogenous retroviruses: A confluence of puzzles, *Science* **255**:411–412 (1992).

124. COLEBUNDERS, R., KINTIN, D., FLEERACKERS, Y., DEMEULENAERE, T., VANDENBRUAENE, M., GOEMAN, J., KESTENS, L., FARBER, C. M., AND SOETE, F., Surveillance case definition for AIDS in resource-poor countries, *Lancet* **342**:864–865 (1993).

125. COLEBUNDERS, R., MANN, J. M., FRANCIS, H., BILA, K., IZALEY, L., KAKONDE, N., KABASELE, K., IFOTO, L., NZILAMBI, N., QUINN, T. C., VAN DER GROEN, G., CURRAN, J. W., VERCAUTEREN, G., AND PIOT, P., Evaluation of a clinical case-definition of acquired immunodeficiency syndrome in Africa, *Lancet* **1**:492–494 (1987).

126. COLTON, T., FREEDMAN, L. S., JOHNSON, A. L., AND MACHIN, D., Extending public health surveillance of HIV infection: Information from a five cohort workshop, *Stat. Med.* **12**:2065–2085 (1993).

127. COLTON, T., FREEDMAN, L. S., JOHNSON, A. L., AND MACHIN, D., Markers as time-dependent covariates in relative risk regression, *Stat. Med.* **12**(22):2087–2098 (1993).

128. COLTON, T., FREEDMAN, L. S., JOHNSON, A. L., AND MACHIN, D., Markers paths, in: *Statistics in Medicine* (S. M. GORE AND V. T. FAREWELL, EDS.), pp. 2099–2126, Wiley, New York, 1993.

129. COMEAU, A. M., HSU, H. W., SCHWERZLER, M., MUSHINSKY, G., WALTER, E., HOFMAN, L., AND GRADY, G. F., Identifying human immunodeficiency virus infection at birth: Application of polymerase chain reaction to Guthrie cards, *J. Pediatr.* **123**:252–258 (1993).

130. CONCORDE COORDINATING COMMITTEE, Concorde: MRC/ANRS randomised double-blind controlled trial of immediate and deferred zidovudine in symptom-free HIV infection, *Lancet* **343**:871–881 (1994).

131. COOPER, D. A., GOLD, J., MACLEAN, P., DONOVAN, B., FINLAYSON, R., BARNES, T. G., MICHELMORE, H. M., BROOKE, P., AND PENNY, R., Acute AIDS retrovirus infection. Definition of a clinical illness associated with seroconversion, *Lancet* **1**:537–540 (1985).

132. CORBITT, G., BAILEY, A. S., AND WILLIAMS, G., HIV infection in Manchester, 1959, *Lancet* **336**:51 (1990).

133. COURGNAUD, V., LAURE, F., BROSSARD, A., BIGNOZZI, C., GOUDEAU, A., BARIN, F., AND BRECHOT, C., Frequent and early in utero HIV-1 infection, *AIDS Res. Hum. Retroviruses* **7**(3):337–341 (1991).

134. COUROUCÉ. A. M., Retrovirus study group of the French Society of Blood Transfusion, *AIDS* **2**:261–265 (1988).

135. COWAN, D. N., BRUNDAGE, J. F., AND POMERANTZ, R. S., The incidence of HIV infection among men in the United States army reserve components, 1985–1991, *AIDS* **8**(4):505–511 (1994).

136. COWAN, D. N., POMERANTZ, R. S., WANN, Z. F., GOLDENBAUM, M., BRUNDAGE, J. F., MILLER, R. N., BURKE, D. S., CARROLL, C. A., AND WALTER REED RETROVIRUS RESEARCH GROUP, Human immunodeficiency virus infection among members of the reserve components of the US Army: Prevalence, incidence and demographic characteristics, *J. Infect. Dis.* **162**:827–836 (1990).

137. CROWE, S. M., CARLIN, J. B., STEWART, K. I., LUCAS, C. R., AND HOY, J. F., Predictive value of CD4 lymphocyte numbers for the development of opportunistic infections and malignancies in HIV-infected persons, *J. Acquir. Immun Defic. Syndr.* **4**(8):770–776 (1991).

138. CURRAN, J. W., JAFFE, H. W., HARDY, A. M., MORGAN, W. M., SELIK, R. M., AND DONDERO, T. J., Epidemiology of HIV infection and AIDS in the United States, *Science* **239**:610–616 (1988).

139. CURTIS, J. R., AND PATRICK, D. L., Race and survival time with AIDS: A synthesis of the literature, *Am. J. Public Health* **83**(10):1425–1428 (1993).

140. D'AQUILA, R. T., PETERSON, L. R., WILLIAMS, A. B., AND WILLIAMS, A. E., Race/ethnicity as a risk factor for HIV-1 infection among Connecticut intravenous drug users, *J. Acquir. Immune. Defic. Syndr.* **2**(5):503–513 (1989).

141. DAAR, E. S., MOUDGIL, T., MEYER, R. D., AND HO, D. D., Transient high levels of viremia in patients with primary human immunodeficiency virus type 1 infection, *N. Engl. J. Med.* **324**(14):961–964 (1991).

142. DADA, A. J., OYEWOLE, F., ONOFOWAKAN, R., NASIDI, A., HARRIS, B., LEVIN, A., DIAMONDSTONE, L. S., QUINN, T. C., AND BLATTNER, W. A., Demographic characteristics of retroviral infections (HIV-1, HIV-2 and HTLV-I) among female professional sex workers in Lagos, Nigeria, *J. Acquir. Immune. Defic. Syndr.* **6**:1358–1363 (1993).

143. DALGLEISH, A. G., BEVERLEY, P. C., CLAPHAM, P. R., CRAWFORD, D. H., GREAVES, M. F., AND WEISS, R. A., The CD4 (T4) antigen is an essential component of the receptor for the AIDS retrovirus, *Nature* **312**:763–767 (1984).

144. DALING, J. R., WEISS, N. S., HISLOP, T. G., MADEN, C., COATES, R. J., SHERMAN, K. J., ASHLEY, R. L., BEAGRIE, M., RYAN, J. A., AND COREY, L., Sexual practices, sexually transmitted diseases, and the incidence of anal cancer, *N. Engl. J. Med.* **317**(16):973–977 (1987).

145. DAWSON, J. D., AND LAGAKOS, S. W., Analyzing laboratory marker changes in AIDS clinical trials, *J. Acquir. Immune Defic. Syndr.* **4**(7):667–676 (1991).

146. DE COCK, K. M., BARRERE, B., DIABY, L., LAFONTAINE, M. F., GNAORE, E., PORTER, A., PANTOBE, D., LAFONTANT, G. C., DAGO-AKRIBI, A., ETTE, M., ET AL., AIDS—the leading cause of adult death in the West African city of Abidjan, Ivory Coast, *Science* **249**:793–796 (1990).

147. DE COCK, K. M., LUCAS, S., COULIBALY, D., COULIBALY, I. M., AND SORO, B., Expansion of surveillance case definition for AIDS in resource-poor countries, *Lancet* **342**:437–438 (1993).

148. DE COOK, K. M., ODEHOURI, K., MOREAU, J., KOUADIO, J. C., PORTER, A., BARRERE, B., DIABY, L., AND HEYWARD, W. L., Rapid emergence of AIDS in Abidjan, Ivory Coast, *Lancet* **1**(8660):408–411 (1989).

149. DE ROSSI, A., OMETTO, L., MAMMANO, F., ET AL., Vertical transmission of HIV-1: Lack of detectable virus in peripheral blood cells of infected children at birth, *AIDS* **6**:1117–1120 (1992).

150. DE VINCENZI, I., AND ANCELLE-PARK, R., Heterosexual transmission of HIV: Follow-up of a European cohort of couples (Abstract M. C. 3028), Seventh Int. Conf. AIDS, June 16–21 (1991).

151. DEL MISTRO, A., CHOTARD, J., HALL, A. J., WHITTLE, H., DE ROSSI, A., AND CHIECO-BIANCHI, L., HIV-1 and HIV-2 sero-

prevalence rates in mother-child pairs living in The Gambia (West Africa), *J. Acquir. Immune Defic. Syndr.* **5**:19–24 (1992).

152. DELFRAISSY, J. F., BLANCHE, S., ROUZIOUX, C., AND MAYAUX, M. J., Perinatal HIV transmission facts and controversies, *Immunodefic. Rev.* **3**:305–327 (1992).

153. DENIS, F., GERSHY-DAMET, G., LHUILLIER, M., LEONARD, G., GOUDEAU, A., ESSEX, M., BARIN, F., REY, J. L., MOUNIER, M., SANGARE, A., M'BOUP, S., AND KANKI, P., Prevalence of human T-lymphotropic retroviruses type III (HIV) and type IV in Ivory Coast, *Lancet* **1**:408–411 (1987).

154. DES JARLAIS, D. C., Difficult-to-reuse needles for the prevention of HIV infection among injecting drug users (pamphlet), 1–17 (1992).

155. DES JARLAIS, D. C., AND FRIEDMAN, S. R., HIV infection among persons who inject illicit drugs: Problems and perspectives, *J. Acquir. Immune Defic. Syndr.* **1**(3):267–273 (1988).

156. DES JARLAIS, D. C., AND FRIEDMAN, S. R., AIDS and IV drug use, *Science* **245**:578 (1989).

157. DES JARLAIS, D. C., FRIEDMAN, S. R., AND WARD, T. P., HIV and injecting drug users: Special considerations, in: *Textbook of AIDS Medicine* (S. BRODER, T. C. MERIGAN, JR., AND D. BOLOGNESI, EDS.), pp. 183–191, Williams & Wilkins, Baltimore, 1994.

158. DETELS, R., MANN, D., CARRINGTON, M., HENNESSEY, K., ZUNYOU, W., WILEY, D., VISSCHER, B. R., AND GIORGI, J. V., Persistently seronegative men from whom HIV-1 has been isolated are genetically and immunologically distinct, *J. Acquir. Immune Defic. Syndr.* (in press).

159. DEVEY, R. T., JR., VASUDEVACHARI, M. B., AND LANE, H. C., Serologic tests for human immunodeficiency virus infection, in: *AIDS Etiology, Diagnosis, Treatment and Prevention*, 3rd ed. (V. T. DEVITA, JR., S. HELLMAN, S. A. ROSENBERG, J. CURREN, M. ESSEX, AND A. S. FAUCI, EDS.), pp. 141–155, Lippincott, Philadelphia, 1994.

160. DI RIENZO, A. M., PETRONINI, P. G., GUETARD, D., FAVILLA, R., BORGHETTI, A. F., MONTAGNIER, L., AND PIEDIMONTE, G., Modulation of cell growth and host protein synthesis during HIV infection *in vitro*, *J. Acquir. Immune. Defic. Syndr.* **5**(9):921–929 (1992).

161. DIETERICH, D. T., POLES, M. A., AND LEW, E. A., Gastrointestinal manifestations of HIV disease, in: *Textbook of AIDS Medicine* (S. BRODER, T. C. MERIGAN, JR., AND D. BOLOGNESI, EDS.), pp. 541–554, Williams & Wilkins, Baltimore, 1994.

162. DODD, R. Y., Donor screening for HIV infection—United States model, in: *Blood, Blood Products, and AIDS* (R. MADHOK, C. D. FORBES, AND B. L. EVATT, EDS.), pp. 183–205, Chapman and Hall, London, 1987.

163. DONEGAN, E., STUART, M., NILAND, J. C., *ET AL.*, Infection with human immunodeficiency virus type 1 (HIV-I) among recipients of antibody-positive blood donations, *Ann. Inter. Med.* **113**(10):733–739 (1990).

164. DOOLITTLE, R. F., The simian–human connection, *Nature* **339**:338–339 (1989).

165. DOWNING, R. G., AND BIRYAHWAHO, B., No evidence for HIV-2 in Uganda, *Lancet* **336**:1514–1515 (1990).

166. DREW, W. L., Is cytomegalovirus a cofactor in the pathogenesis of AIDS and Kaposi's sarcoma, *Mt. Sinai J. Med.* **53**:622–626 (1986).

167. DUBLIN, S., ROSENBERG, P. S., AND GOEDERT, J. J., Patterns and predictors of high-risk sexual behavior in female partners of HIV-infected men with hemophilia, *AIDS* **6**(5):475–482 (1992).

168. DULIEGE, A. M., AMOS, C. I., FELTON, S., BIGGAR, R. J., AND GOEDERT, J. J., HIV-1 infection rate and progression of disease in 1st and 2nd born twins born to HIV infected mother; Hypothesis regarding *in utero* and perinatal exposures, Ninth Int. Conf. AIDS, June 7–11, 1993.

169. DUNN, D. T., NEWELL, M. L., ADES, A. E., AND PECKHAM, C. S., Risk of human immunodeficiency virus type 1 transmission through breastfeeding, *Lancet* **340**:585–588 (1992).

170. EBBESEN, P., MELBYE, M., AND BIGGAR, R. J., Sex habits, recent disease and drug use in two groups of Danish male homosexuals, *Arch. Sex. Behav.* **3**:291–300 (1984).

171. EDITORIAL, Risk factors for mother-to-infant transmission on HIV-1, *Lancet* **339**:1007–1012 (1992).

172. EGAN, V. G., CRAWFORD, J. R., BRETTLE, R. P., AND GOODWIN, G. M., The Edinburgh cohort of HIV-positive drug users: Current intellectual function is impaired, but not due to early AIDS dementia complex, *AIDS* **4**:651–656 (1990).

173. EHRLICH, G. D., GREENBERG, S., AND ABBOTT, M. A., Detection of human T-cell lymphomas/leukemia viruses, in: *PCR Protocols: A Guide to Methods and Applications*, pp. 325–336, Academic Press, New York, 1990.

174. EL-SADR, W., MARMOR, M., ZOLLA-PAZNER, S., STAHL, R., LYDEN, M., WILLIAM, D., D'ONOFRIO, S., WEISS, S. H., AND SAXINGER, C. W., Four year prospective study of homosexual men: Correlation of immunologic abnormalities, clinical status and HTLV-III serology, *J. Infect. Dis.* **155**:789–793 (1987).

175. ELLERMAN, V., AND BANG, O., Experimentelle leukämie bei hühnern (Abstract), *Centralbl. f Bakt.* **46**:595–609 (1908).

176. EMBRETSON, J., ZUPANCIC, M., RIBAS, J. L., BURKE, A., RACZ, P., TENNER-RACS, K., AND HAASE, A. T., Massive covert infection of helper T lymphocytes and macrophages by HIV during the incubation period of AIDS, *Nature* **362**:359–362 (1993).

177. EUROPEAN COLLABORATIVE STUDY, Children born to women with HIV-1 infection; Natural history and risk of transmission, *Lancet* **337**:253–260 (1991).

178. EUROPEAN COLLABORATIVE STUDY, Risk factors for mother-to-child transmission of HIV-1, *Lancet* **339**:1007–1012 (1992).

179. EUROPEAN CENTRE FOR THE EPIDEMIOLOGIC MONITORING OF AIDS, *AIDS Surveillance in Europe*. WHO-EC Collaborating Centre on AIDS 40 (1993).

180. EUROPEAN STUDY GROUP, Risk factors for male to female transmission of HIV, *Br. Med. J.* **298**:411–415 (1989).

181. EUROPEAN STUDY GROUP OF HETEROSEXUAL TRANSMISSION OF HIV, Comparison of female to male to female transmission of HIV in 563 stable couples, *Br. Med. J.* **304**:809–813 (1992).

182. EVANS, B. G., CATCHPOLE, M. A., HEPTONSTALL, J., MORTIMER, J. Y., MCCARRIGLE, C. A., NICOLL, A. G., WAIGHT, P., GILL, O. N., AND SWAN, A. V., Sexually transmitted diseases and HIV-1 infection among homosexual men in England and Wales, *Br. Med. J.* **306**:426–428 (1993).

183. EYSTER, M. E., BALLARD, J. O., GAIL, M. H., DRUMMOND, J. E., AND GOEDERT, J. J., Predictive markers for the acquired immunodeficiency syndrome (AIDS) in hemophiliacs: Persistence of P24 antigen and low T4 cell count, *Ann. Intern. Med.* **110**:963–969 (1989).

184. EYSTER, M. E., GAIL, M. H., BALLARD, J. O., AL-MONDHIRY, H., AND GOEDERT, J. J., Natural history of human immunodeficiency virus (HIV) infections in hemophiliacs: Effects of T-cell subsets, platelet counts and age, *Ann. Intern. Med.* **107**:1–6 (1987).

185. EYSTER, M. E., GOEDERT, J. J., SARNGADHARAN, M. G., WEISS, S. H., GALLO, R. C., AND BLATTNER, W. A., Development and early natural history of HTLV-III antibodies in persons with hemophilia, *J.A.M.A.* **253**:2219–2223 (1985).

186. EYSTER, M. E., RABKIN, C. S., HILGARTNER, M. W., ALEDORT, L.

M., RAGNI, M. V., SPRANDIO, J., WHITE, G. C., II, EICHINGER, S., DE MOERLOOSE, P. E., ANDES, W. A., COHEN, A. R., MANCO-JOHNSON, M., BRAY, G. L., SCHRAMM, W., HATZAKIS, A., LEDERMAN, M. M., KESSLER, C. M., AND GOEDERT, J. J., Human immunodeficiency virus-related conditions in children and adults with hemophilia: Rates, relationship to CD4 counts, and predictive value, *Blood* **81**(3):828–834 (1993).

187. FAUCI, A., SCHNITTMAN, S. M., POLI, G., KOENIG, S., AND PANTALEO, G., Immunopathogenetic mechanisms in human immunodeficiency virus (HIV) infection, *Ann. Intern. Med.* **114**:678–693 (1991).

188. FAUCI, A. S., AND ROSENBERG, Z. F., Immunopathogenesis, in: *Textbook of AIDS Medicine* (S. BRODER, T. C. MERIGAN, JR., AND D. BOLOGNESI, EDS.), pp. 55–75, Williams & Wilkins, Baltimore, 1994.

189. FIGUEROA, J. P., BRATHWAITE, A., MORRIS, J., WARD, E., PERUGA, A., BLATTNER, W. A., VERMUND, S. H., AND HAYES, R., Rising HIV-1 prevalence among sexually transmitted disease clinic attenders in Jamaica: Traumatic sex and genital ulcers as risk factors, *J. Acquir. Immune Defic. Syndr.* **7**(3):310–316 (1994).

190. FISCHL, M. A., An introduction to the clinical spectrum of AIDS, in: *Textbook of AIDS Medicine* (S. BRODER, T. C. MERIGAN, AND D. BOLOGNESI, EDS.), pp. 149–159, Williams & Wilkins, Baltimore, 1994.

191. FITZGIBBON, J. E., GAUR, S., FRENKEL, L. D., LARAQUE, F., EDLIN, B. R., AND DUBIN, D. T., Transmission from one child to another of human immunodeficiency virus type I with a zidovudine-resistance mutation, *N. Engl. J. Med.* **329**(25):1835–1841 (1993).

192. FRIEDLAND, G. H., SALTZMAN, B. R., ROGERS, M. F., KAHL, P. A., LESSER, M. L., MAYERS, M. M., AND KLEIN, R. S., Lack of transmission of HTLV-III/LAV infection to household contacts of patients with AIDS or AIDS-related complex with oral candidiasis, *N. Engl. J. Med.* **314**(6):344–349 (1986).

193. FUCHS, D., SPIRA, T. J., HAUSEN, A., REIBNEGGER, G., WERNER, E. R., FELMAYER, G. W., AND WACHTER, H., Neopterin as a predictive marker for disease progression in human immunodeficiency virus type I infection, *Clin. Chem.* **35**(8):1746–1748 (1989).

194. FUKASAWA, M., MIURA, T., HASEGAWA, A., MORIKAWA, S., TSUJIMOTO, H., KITAMURA, T., AND HAYAMI, M., Sequence of simian immunodeficiency virus from African green monkey, a new member of the HIV-SIV group, *Nature* **333**:457–461 (1988).

195. FULTZ, P. N., SIEGEL, R. L., BRODIE, A., MAWLE, A. C., STRICKER, R. B., SWENSON, R. B., ANDERSON, D. C., AND MCCLURE, H. M., Prolonged cd4$^+$ lymphocytopenia and thrombocytopenia in a chimpanzee persistently infected with human immunodeficiency virus type I, *J. Infect. Dis.* **163**:441–447 (1991).

196. FUSARO, R. E., NIELSEN, J. P., AND SCHEIKE, T. H., Marker-dependent hazard estimation: An application to AIDS, *Stat. Med.* **12**:843–865 (1993).

197. GAIL, M. H., ROSENBERG, P. S., AND GOEDERT, J. J., Therapy may explain recent deficits in AIDS incidence, *J. Acquir. Immune Defic. Syndr.* **3**(4):296–306 (1990).

198. GALLO, R., WONG-STAAL, F., MONTAGNIER, L., HASELTINE, W. A., AND YOSHIDA, M., HIV/HTLV gene nomenclature, *Nature* **333**:505 (1988).

199. GALLO, R. C., HIV—the cause of AIDS: An overview on its biology, mechanisms of disease induction, and our attempts to control it, *J. Acquir. Immune Defic. Syndr.* **1**(6):521–535 (1988).

200. GALLO, R. C., Mechanism of disease induction by HIV, *J. Acquir. Immune Defic. Syndr.* **3**(4):380–389 (1990).

201. GALLO, R. C., SALAHUDDIN, S. Z., POPOVIC, M., SHEARER, G. M., KAPLAN, M., HAYNES, B. F., PALKER, T. J., REDFIELD, R., OLESKE, J., SAFAI, B., WHITE, G., FOSTER, P., AND MARICHAM, P. D., Frequent detection and isolation of cytopathic retroviruses (HTLV-III) from patients with AIDS and at risk for AIDS, *Science* **224**:500–503 (1984).

202. GALLO, R. C., WONG-STAAL, F., AND SARIN, P. S., Cellular ONC genes, T-cell leukaemia-lymphoma virus, and leukaemias and lymphomas of man, *Mech. Viral. Leukaemogenesis* 11–37 (1989).

202a. GALLO, R. C., AND NERURKAR, L. S., Human retroviruses: Linkages to leukemia and AIDS, in: *Hematology and Blood Transfusion*, Vol. 35 (R. NETH, ED.), Springer-Verlag, Berlin, 1992.

203. GAO, F., YUE, L., WHITE, A. T., PAPPAS, P. G., BARCHUE, J., HANSON, A. P., GREENE, B. M., SHARP, P. M., SHAW, G. M., AND HAHN, B. H., Human infection by genetically diverse SIV related HIV-2 in West Africa, *Nature* **358**:495–499 (1992).

204. GARDNER, L. I., JR., BRUNDAGE, J. F., MCNEIL, J. G., MILAZZO, M. J., REDFIELD, R. R., ARONSON, N. E., CRAIG, D. B., DAVIS, C., GATES, R. H., LEVIN, L. I., MICHAEL, R. A., OSTER, C. N., RYAN, W. C., BURKE, D. S., AND TRAMONT, E. C., Predictors of HIV-1 disease progression in early- and late-stage patients: The U.S. Army natural history cohort, *J. Acquir. Immune Defic. Syndr.* **5**(8):782–793 (1992).

205. GOEDERT, J. J., Testing for human immunodeficiency virus, *Ann. Intern. Med.* **105**(4):609–610 (1986).

206. GOEDERT, J. J., What is safe sex. Suggested standards linked to testing for HIV, *N. Engl. J. Med.* **316**:1339–1342 (1987).

207. GOEDERT, J. J., BIGGAR, R. J., MELBYE, M., ET AL., Effect of T4 count and cofactors on the incidence of AIDS in homosexual men infected with human immunodeficiency virus, *J.A.M.A.* **257**(3):331–334 (1987).

208. GOEDERT, J. J., BIGGAR, R. J., WEISS, S. H., EYSTER, M. E., MELBYE, M., WILSON, S., GINZBURG, H. M., GROSSMAN, R. J., DIGIOIA, R. A., SANCHEZ, W. C., GIRON, J., EBBESEN, P., GALLO, R. C., AND BLATTNER, W. A., Three-year incidence of AIDS in five cohorts of HTLV-III infected risk group members, *Science* **231**:992–995 (1986).

209. GOEDERT, J. J., BIGGAR, R. J., WINN, D. M., MANN, D. L., STRONG, D. M., DIGIOIA, R. A., GROSSMAN, R. J., SANCHEZ, W. C., KASE, R. G., GREENE, M. H., BYAR, D. P., HOOVER, R. N., AND BLATTNER, W. A., Decreased helper T-lymphocytes in homosexual men: I. Sexual contact in high-incidence areas for the acquired immunodeficiency syndrome, *Am. J. Epidemiol.* **121**:629–636 (1985).

210. GOEDERT, J. J., DULIEGE, A. M., AMOS, C. I., FELTON, S., AND BIGGAR, R. J., High risk of HIV-1 infection for first-born twins, *Lancet* **338**:1471–1475 (1991).

211. GOEDERT, J. J., EYSTER, M. E., BIGGAR, R. J., AND BLATTNER, W. A., Heterosexual transmission of human immunodeficiency virus: Association with severe depletion of T-helper lymphocytes in men with hemophilia, *AIDS Res. Hum. Retroviruses* **3**:355–361 (1987).

212. GOEDERT, J. J., KESSLER, C. M., ALEDORT, L. M., BIGGAR, R. J., ANDES, W. A., WHITE, G. C., II, DRUMMOND, J. E., VAIDYA, K., MANN, D. L., EYSTER, M. E., RAGNI, M. V., LEDERMAN, M. M., COHEN, A. R., BRAY, G. L., ROSENBERG, P. S., FRIEDMAN, R. M., HILGARTNER, M. W., BLATTNER, W. A., KRONER, B. L., AND GAIL, M. H., A prospective study of human immunodeficiency virus type I infection and the development of AIDS in sub-

jects with hemophilia, *N. Engl. J. Med.* **321**(17):1141–1148 (1989).

213. GOEDERT, J. J., MENDEZ, H., DRUMMOND, J. E., ROBERT-GUROFF, M., MINKOFF, H. L., HOLMAN, S., STEVENS, R., RUBINSTEIN, A., BLATTNER, W. A., WILLOUGHBY, A., AND LANDESMAN, S. H., Mother-to-infant transmission of human immunodeficiency virus type 1: Association with prematurity or low anti-gp120, *Lancet* **2**:1351–1355 (1989).

214. GOEDERT, J. J., SARNGADHARAN, M. G., BIGGAR, R. J., WEISS, S. H., WINN, D. M., GROSSMAN, R. J., GREENE, M. H., BODNER, A. J., MANN, D. L., STRONG, D. M., GALLO, R. C., AND BLATTNER, W. A., Determinants of retrovirus (HTLV-III) antibody and immunodeficiency conditions in homosexual men, *Lancet* **2**:711–716 (1984).

215. GOEDERT, J. J., SARNGADHARAN, M. G., EYSTER, M. E., WEISS, S. H., BODNER, A. J., GALLO, R. C., AND BLATTNER, W. A., Antibodies reactive with human T cell leukemia viruses (HTLV-III) in the serum of hemophiliacs receiving factor VIII concentrate, *Blood* **65**(2):492–495 (1985).

216. GREER, C. E., LUND, J. K., AND MANOS, M. M., PCR amplification from paraffin-embedded tissues: Recommendations on fixatives for long-term storage and prospective studies, *PCR Methods Appl.* **1**(1):46–50 (1992).

217. GROOPMAN, J. E., SALAHUDDIN, S. Z., SARNGADHARAN, M. G., MARKHAM, P. D., GONDA, M., SLISKI, A., AND GALLO, R. C., HTLV-III in saliva of people with AIDS-related complex and healthy homosexual men at risk for AIDS, *Science* **226**:447–449 (1984).

218. GÜRTLER, L. G., HAUSER, P. H., EBERLE, J., VON BRUNN, A., KNAPP, S., ZEKENG, L., TSAGUE, J. M., AND KAPTUE, L., A new subtype of human immunodeficiency virus type 1 (MVP-5180) from Cameroon, *J. Virol.* **68**(3):1581–1585 (1994).

219. GWINN, M., PAPPAIOANOU M., GEORGE, J. R., HANNON, W. H., WASSER, S. C., REDUS, M. A., HOFF, R., GRADY, G. F., WILLOUGHBY, A., NOVELLO, A. C., ET AL., Prevalence of HIV infection in childbearing women in the United States. Surveillance using newborn blood samples, *J.A.M.A.* **265**:1704–1708 (1991).

220. HAASE, A. T., Pathogenesis of lentivirus infections, *Nature* **322**:130–136 (1986).

221. HAHN, B. H., Viral genes and their products, in: *Textbook of AIDS Medicine* (S. BRODER, T. C. MERIGAN, JR., AND D. BOLOGNESI, EDS.), pp. 21–43, Williams & Wilkins, Baltimore, 1994.

222. HENDERSON, D. K., HIV transmission in the health care environment. In: *Textbook of AIDS Medicine* (S. BRODER, T. C. MERIGAN, JR., AND D. BOLOGNESI, EDS.), pp. 831–839, Williams & Wilkins, Baltimore, 1994.

223. HENDERSON, D. K., SAAH, A. J., ZAK, B. J., KASLOW, R. A., LANE, H. C., FOLKS, T., BLACKWELDER, W. C., SCHMITT, J., LACAMERA, D. J., MASUR, H., AND FAUCI, A. S., Risk of nosocomial infection with human T-cell lymphotropic virus type III/lymphadenopathy-associated virus in a large cohort of intensively exposed healthy care workers, *Ann. Intern. Med.* **104**:644–647 (1986).

224. HENIN, Y., MANDELBROT, L., HENRION, R., PRADINAUD, R., COULAUD, J. P., AND MONTAGNIER, L., Virus excretion in the cervicovaginal secretions of pregnant and nonpregnant HIV-infected women, *J. Acquir. Immune Defic. Syndr.* **6**(1):72–75 (1993).

225. HENRARD, D. R., PHILLIPS, J., MUENZ, L. R., BLATTNER, W. A., WIESNER, D., EYSTER, M. E., AND GOEDERT, J. J., Natural history of HIV-1 cell-free viremia, *J. Am. Med. Assoc.* **274**(7):554–558 (1995).

226. HERSH, B. S., POPOVICI, F., APETREI, R. C., ZOLOTUSCA, L., BELDESCU, N., CALOMFIRESCU, A., JEZEK, Z., OXTOBY, M. J., GROMYKO, A., AND HEYMANN, D. L., Acquired immunodeficiency syndrome in Romania, *Lancet* **339**:645–649 (1991).

227. HINO, S., Maternal–infant transmission of HTLV-I implication for disease, in: *Human Retrovirology (HTLV-I)* (W. A. BLATTNER, ED.), pp. 363–375, Raven Press, New York, 1990.

228. HINO, S., SUGIYAMA, H., DOI, H., ISHIMARU, T., YAMABE, T., TSUJI, Y., AND MIYAMOTO, T., Breaking the cycle of HTLV-I transmission via carrier mothers' milk (letter), *Lancet* **2**:158–159 (1987).

229. HINO, S., YAMAGUCHI, K., KATAMINE, S., SUGIYAMA, H., AMAGASAKI, T., KINOSHITA, K., YOSHIDA, Y., DOI, H., TSUJI, Y., AND MIYAMOTO, T., Mother-to-child transmission of human T-cell leukemia virus type-I, *Jpn. J. Cancer Res.* **76**:474–480 (1985).

230. HIRA, S. K., KAMANGA, J., BHAT, G. J., MWALE, C., TEMBO, G., LUO, N., AND PERINE, P. L., Perinatal transmission of HIV-I in Zambia, *Br. Med. J.* **299**:1250–1252 (1989).

231. HOLLOMAN, D. L., PAU, C. P., PAREKH, B., SCHABLE, C., ONORATO, I., SCHOCHETMAN, G., AND GEORGE, J. R., Evaluation of testing algorithms following the use of combination HIV-1/HIV-2 EIA for screening purposes, *AIDS Res. Hum. Retroviruses* **9**:147–151 (1993).

232. HOLMBERG, S. D., STEWART, J. A., GERBER, A. R., BYERS, R. H., LEE, F. K., O'MALLEY, P. M., AND NAHMIAS, A. J., Prior herpes simplex virus type 2 infection as a risk factor for HIV infection, *J.A.M.A.* **259**(7):1048–1050 (1988).

233. HOOVER, D. R., GRAHAM, N. M. H., CHEN, B., TAYLOR, J. M. G., PHAIR, J., ZHOU, S. Y. J., AND MUÑOZ, A., Effect of CD4$^+$ cell count measurement variability on staging HIV-1 infection, *J. Acquir. Immune Defic. Syndr.* **5**(8):794–802 (1992).

234. HORSBURGH, C. R., JR., *Mycobacterium avium* complex infection in the acquired immunodeficiency syndrome, *N. Engl. J. Med.* **324**(19):1332–1338 (1991).

235. HORVITZ, D. G., FOLSOM, R. E., EZZATI, T. M., AND MASSEY, J. T., Assessment of non-response bias in the Dallas County HIV survey (in press).

236. IMAGAWA, D. T., LEE, M. H., WOLINSKY, S. M., SANO, K., MORALES, F., KWOK, S., SNINSKY, J. J., NISHANIAN, P. G., GIORGI, J., FAHEY, J. L., DUDLEY, J., VISSCHER, B. R., AND DETELS, R., Human immunodeficiency virus type 1 infection in homosexual men who remain seronegative for prolonged periods, *N. Engl. J. Med.* **320**:1458–1462 (1989).

237. ITALIAN REGISTRY OF HIV INFECTED CHILDREN, Features of children perinatally infected with HIV-1 surviving longer than 5 years, *Lancet* **343**:191–195 (1994).

238. JACKSON, J. B., SANNERUD, K. J., HOPSICKER, J. S., KWOK, S. Y., EDSON, J. R., AND BALFOUR, H. H., JR., Hemophiliacs with HIV antibody are actively infected, *J.A.M.A.* **260**(15):2236–2239 (1988).

239. JAFFE, H. W., CHOI, K., THOMAS, P. A., HAVERKOS, H. W., AUERBACH, D., GUINAN, M. E., ROGERS, M. F., SPIRA, T. J., DARROW, W. W., KRAMER, M. A., FRIEDMAN, S. M., MONROE, J. M., FRIEDMAN-KIEN, A. E., LAUBENSTEIN, L. J., MARMOR, M., SAFAI, B., DRITZ, S. K., CRISPI, S. J., FANNIN, S. L., ORKWIS, J. P., KELTER, A., RUSHING, W. R., THACKER, S. B., AND CURRAN, J. W., National case-control study of Kaposi's sarcoma and *Pneumocystis carinii* pneumonia in homosexual men: Part 1. Epidemiologic results, *Ann. Intern. Med.* **99**(2):145–151 (1983).

240. JAFFE, H. W., DARROW, W. W., ECHENBERG, D. F., O'MALLEY, P. M., GETCHELL, J. P., KALYANARAMAN, V. S., BYERS, R., DREN-

NAN, D. P., BRAFF, E. H., CURRAN, J. W., AND FRANCIS, D. P., The acquired immunodeficiency syndrome in a cohort of homosexual men. A six-year follow-up study, *Ann. Intern. Med.* **103**:210–214 (1985).

241. JANOFF, E. N., JACKSON, S., WAHL, S. M., THOMAS, K., PETERMAN, J. H., AND SMITH, P. D., Intestinal mucosal immunoglobulins during human immunodeficiency virus type I infection, *J. Infect. Dis.* **170**:299–307 (1994).

242. JANVIER, B., BARIN, F., AND MANDRAND, B., A new enzyme immunoassay for titration of antibodies to a cross-reactive epitope of the major core protein of HIV-1 and HIV-2: Diagnosis and prognosis usefulness (abstract), Sixth Int. Conf. AIDS, 1990.

243. JENKINS, M., LANDERS, D., WILLIAMS-HERMAN, D., WARA, D., VISCARELLO, R. R., HAMMILL, H. A., KLINE, M. W., SHEARER, W. T., CHARLEBOIS, E. D., AND KOHL, S., Association between anti-human immunodeficiency virus type I (HIV-1) antibody-dependent cellular cytotoxicity antibody titers at birth and vertical transmission of HIV-1, *J. Infect. Dis.* **170**:308–312 (1994).

244. JONES, D. B., Human immunodeficiency virus exposure among medical students, *Arch. Surg.* **128**:710 (1993).

245. KANKI, P., ALLAN, J., BARIN, F., REDFIELD, R., CLUMECK, N., QUINN, T., MOWOVONDI, F., THIRY, L., BURNY, A., ZAGURY, D., PETAT, E., KOCHELEFF, P., PASCAL, K., LAUSEN, I., FREDRICKSEN, B., CRAIGHEAD, J., M'BOUP, S., FRANCIS, H., ALBAUM, M., TRAVERS, K., MCLANE, M. F., LEE, T. H., AND ESSEX, M., Absence of antibodies to HIV-2/HTLV-4 in six Central African nations, *AIDS Res. Hum. Retroviruses* **3**:317–322 (1987).

246. KANKI, P., AND ESSEX, M., Simian T-lymphotropic viruses and related human viruses, *Vet. Microbiol.* **17**:309–314 (1988).

247. KANKI, P., M'BOUP, S., MARLINK, R., TRAVERS, K., HSIEH, C. C., GUEYE, A., BOYE, C., SANKALÉ, J. L., DONELY, C., LEISENRING, W., SIBY, T., THIOR, I., DIA, M., GUEYE, E. H., N'DOYE, I., AND ESSEX, M., Prevalence and risk determinants of human immunodeficiency virus type 2 (HIV-2) and human immunodeficiency virus type 1 (HIV-1) in West African female prostitutes, *Am. J. Epidemiol.* **136**:895–907 (1992).

248. KANKI, P. J., Biologic features of HIV-2. An update, *AIDS Clin. Rev.*: 17–38 (1991).

249. KANKI, P. J., BARIN, F., M'BOUP, S., ALLAN, J. S., ROMET-LEMONNE, J. L., MARLINK, R., MCLANE, M. F., LEE, T. H., ARBEILLE, B., DENIS, F., AND ESSEX, M., New human T-lymphotropic retrovirus related to simian T-lymphotropic virus type III (STLV-III$_{agm}$), *Science* **232**:238–243 (1986).

250. KANKI, P. J., TRAVERS, K. U., M'BOUP, S., HSIEH, C. C., MARLINK, R. G., GUEYE-NDIAYE, A., SIBY, T., THIOR, I., HERNANDEZ-AVILA, M., SANKALÉ, J. L., NDOYE, I., AND ESSEX, M., Slower heterosexual spread of HIV-2 than HIV-1, *Lancet* **343**:943–946 (1994).

251. KAPLAN, E. H., AND HEIMER, R., A model-based estimate of HIV infectivity via needle sharing, *J. Acquir. Immune Defic. Syndr.* **5**(11):1116–1118 (1992).

252. KAPLAN, E. H., AND HEIMER, R., HIV prevalence among intravenous drug users: Model-based estimates from New Haven's legal needle exchange, *J. Acquir. Immune Defic. Syndr.* **5**(2):163–169 (1992).

253. KAPLAN, J. E., SPIRA, T. J., FISHBEIN, D. B., BOZEMAN, L. H., PINSKY, P. F., AND SCHONBERGER, L. B., A six-year follow-up of HIV-infected homosexual men with lymphadenopathy: Evidence for an increased risk for developing AIDS after the third year of lymphadenopathy, *J.A.M.A.* **260**(18):2694–2697 (1988).

254. KAPLAN, J. E., SPIRA, T. J., FISHBEIN, D. B., PINSKY, P. F., AND SCHONBERGER, L. B., Lymphadenopathy syndrome in homosexual men: Evidence for continuing risk of developing the acquired immunodeficiency syndrome, *J.A.M.A.* **257**(3):335–337 (1987).

255. KASLOW, R. A., BLACKWELDER, W. C., OSTROW, D. G., YERG, D., PALENICEK, J., COULSON, A. H., AND VALDISERRI, R. O., No evidence for a role of alcohol or other psychoactive drugs in accelerating immunodeficiency in HIV-1-positive individuals: A report from the Multicenter AIDS Cohort Study, *J.A.M.A.* **261**(23):424–429 (1989).

256. KASLOW, R. A., CARRINGTON, M., APPLE, R., PARK, L., MUÑOZ, A., SAAH, A. J., *ET AL.*, Influence of combinations of major histocompatibility complex genes on the course of HIV-1 infection (submitted).

257. KATZENSTEIN, D. A., HOLODNIY, M., AND ISRAELSKI, D. M., Plasma viremia in human immunodeficiency virus infection: Relationship to stage disease and antiviral treatment, *J. Acquir. Immune Defic. Syndr.* **5**(2):107–112 (1992).

258. KAWAMURA, M., YAMAZAKI, S., ISHIKAWA, K., KWOFIE, T. B., TSUJIMOTO, H., AND HAYAMI, M., HIV-2 in West Africa in 1966, *Lancet* **1**:385 (1989).

259. KESTENS, L., BRATTEGARD, K., ADJORLOLO, G., EKPINI, E., SIBALLY, T., DIAOLLO, K., GIGASE, H., GAYLE, H., AND DECOOK, K. M., Immunological comparison of HIV-1, HIV-2 and dually-reactive women delivering in Abidjan, Côte d'Ivoire, *AIDS* **6**:803–807 (1992).

260. KHABBAZ, R., ROWE, T., MURPHY-CORB, M., HENEINE, W. M., SCHABLE, C. A., GEORGE, J. R., PAU, C. P., PAREKH, B. S., LAIRMORE, M. D., CURRAN, J. W., KAPLAN, J. E., SCHOCHETMAN, G., AND FOLKS, T. M., Simian immunodeficiency virus needlestick accident in a laboratory worker, *Lancet* **340**:271–273 (1992).

262. KHAN, A. S., HENEINE, W. M., CHAPMAN, L. E., GARY, H. E., JR., WOODS, T. C., FOLKS, T. M., AND SCHONBERGER, L. B., Assessment of a retrovirus sequence and other possible risk factors for the chronic fatigue syndrome in adults, *Ann. Inter. Med.* **118**(4):241–245 (1993).

263. KINGSLEY, L. A., DETELS, R., KASLOW, R., POLK, B. F., RINALDO, C. R., JR., CHMIEL, J., DETRE, K., KELSEY, S. F., ODAKA, N., OSTROW, D., VANRADEN, M., AND VISSCHER, B., Risk factors for seroconversion to human immunodeficiency virus among male homosexuals: Results from the Multicenter Aids Cohort Study, *Lancet* **1**:345–348 (1987).

264. KOENIG, R. E., GAUTIER, T., AND LEVY, J. A., Unusual intrafamilial transmission of human immunodeficiency virus (letter), *Lancet* **2**:627 (1986).

265. KONINGS, E., BLATTNER, W. A., LEVIN, A., BRUBAKER, G., SISO, Z., SHAO, J., *ET AL.*, Sexual behavior survey in a rural area of northwest Tanzania. *AIDS* **8**:987–993 (1994).

266. KONINGS, E., BRUBAKER, G., KIBAURI, A., AND MASSESSA, E., Quantifying sexual behavior in northwest Tanzania, *Tanzania Med. J.* **6**(2):6018–6064 (1991).

267. KOOPMAN, J. S., LONGINI, I. M., JACQUEZ, J. A., *ET AL.*, Assessing risk factors for transmission of infection, *Am. J. Epidemiol.* **1**:486–504 (1991).

268. KOUP, R. A., AND HO, D. D., Shuttling down HIV, *Nature* **370**:416 (1994).

269. KOZIOL, D. E., SAAH, A. J., ODAKA, N., AND MUNOZ, A. A., A comparison of risk factors for human immunodeficiency virus and hepatitis B virus infections in homosexual men, *Ann. Epidemiol.* **3**:434–441 (1993).

270. KOZLOV, A. P., VOLKOVA, G. V., MALYKH, A. G., STEPANOVA, G.

S., AND GLEBOV, A. V., Epidemiology of HIV infection in St. Petersburg, Russia, *J. Acquir. Immune Defic. Syndr.* **6**(2):208–212 (1993).

271. KRÄMER, A., AXMANN, D., AND GOEDERT, J. J., Aalen's linear regression model for detection of time-dependent effects of immunologic AIDS progression markers (in press).

272. KRÄMER, A., BIGGAR, R. J., HAMPL, H., *ET AL.*, Immunologic markers of progression to acquired immunodeficiency syndrome are time-dependent and illness-specific, *Am. J. Epidemiol.* **136** (1):71–80 (1992).

273. KREISS, J., NGUGI, E., HOLMES, K., NDINYA-ACHOLA, J., WAIYAKI, P., ROBERTS, P. L., RUMINJO, I., SAJABI, R., KIMATA, J., FLEMING, T. R., ANZALA, A., HOLTON, D., AND PLUMMER, F., Efficacy of nonoxynol-9 contraceptive sponge use in preventing heterosexual acquisition of HIV in Nairobi prostitutes, *J.A.M.A.* **268**(4):477–482 (1992).

274. KREISS, J. K., KITCHEN, L. W., PRINCE, H. E., KASPER, C. K., AND ESSEX, M., Antibody to human T-lymphotropic virus type III in wives of hemophiliacs. Evidence for heterosexual transmission, *Ann. Intern. Med.* **102**:623–626 (1985).

275. KREISS, J. K., KOECH, D., PLUMMER, F. A., HOLMES, K. K., LIGHTFOOTE, M., PIOT, P., RONALD, A. R., NDINYA-ACHOLA, J. O., D'COSTA, L. J., ROBERTS, P., NGUGI, E. N., AND QUINN, T. C., AIDS virus infection in Nairobi prostitutes. Spread of the epidemic to East Africa, *N. Engl. J. Med.* **314**:414–418.

276. KRIEGER, J. N., COOMBS, R. W., COLLIER, A. C., *ET AL.*, Recovery of human immunodeficiency virus type 1 from semen: Minimal impact of stage of infection and current antiviral chemotherapy, *J. Infect. Dis.* **163**:386–388 (1991).

277. KRONER, B. L., ROSENBERG, P. S., ALEDORT, L. M., ALVORD, W. G., AND GOEDERT, J. J., HIV-1 infection incidence among persons with hemophilia in the United States and Western Europe, 1978–1990, *J. Acquir. Immune Defic. Syndr.* **7**:279–286 (1994).

278. KROWN, S. E., NIEDZWIECKI, D., BHALLA, R. B., FLOMENBERG, N., BUNDOW, D., AND CHAPMAN, D., Relationship and prognostic value of endogenous interferon-α, β₂-microglobulin, and neopterin serum levels in patients with Kaposi sarcoma and AIDS, *J. Acquir. Immune Defic. Syndr.* **4**(9):871–880 (1991).

279. LAGA, M., MANOKA, A., KIVUVU, M., MLELE, B., TULIZA, M., NZILA, N., GOEMAN, J., BEHETS, F., BATTER, V., ALARY, M., HEYWARD, W. L., RYDER, R. W., AND PIOT, P., Non-ulcerative sexually transmitted disease as risk factors for HIV-1 transmission in women: Results from a cohort study, *AIDS* **7**:95–102 (1993).

280. *LANCET*. India: Prostitutes and the spread of AIDS, *Lancet* **335**:1332 (1990).

281. LANG, W., ANDERSON, R. E., PERKINS, H., GRANT, R. M., LYMAN, D., WINKELSTEIN, W., JR., ROYCE, R., AND LEVY, J. A., Clinical, immunologic, and serologic findings in men at risk for acquired immunodeficiency syndrome, *J.A.M.A.* **257**(3):326–330 (1987).

282. LANGE, J. M. A., PAUL, D. A., HUISMAN, H. G., DE WOLF, F., VAN DEN BERG, H., COUTINHO, R. A., DANNER, S. A., VAN DER NOORDAA, J., AND GOUDSMIT, J., Persistent HIV antigenaemia and decline of HIV core antibodies associated with transition to AIDS, *Br. Med. J.* **293**:1459–1462 (1986).

283. LAPOINTE, N., BOUCHER, M., SAMSON, J., AND CHAREST, J., Significant markers in the modulation of immunity during pregnancy and post-partum in a paired HIV positive and HIV negative population (Abstract), Int. Conf. AIDS 1991; **7**(2):195.

284. LAURENT-CRAWFORD, A. G., KRUST, B., RIVIÈRE, Y., DESGRANGES, C., MULLER, S., KIENY, M. P., DAUGUET, C., AND

HOVANESSIAN, A. G., Membrane expression of HIV envelope glycoproteins triggers apoptosis in CD4 cells, *AIDS Res. Hum. Retroviruses* **9**(8):761–773 (1993).

285. LAZZARIN, A., SARACCO, A., MUSICCO, M., NICOLOSI, A., AND THE ITALIAN STUDY GROUP OF HIV HETEROSEXUAL TRANSMISSION, Man-to-woman sexual transmission of the human immunodeficiency virus: Risk factors related to sexual behavior, man's infectiousness, and woman's susceptibility, *Arch. Intern. Med.* **151**:2411–2416 (1991).

286. LE GUENNO, B. M., BARABE, P., GRIFFET, P. A., GUIRAUD, M., MORCILLO, R. J., PEGHINI, M. E., JEAN, P. A., M'BAYE, P. S., DIALLO, A., AND SARTHOU, J. L., HIV-2 and HIV-1 AIDS cases in Senegal: Clinical patterns and immunological perturbations, *J. Acquir. Immune Defic. Syndr.* **4**:421–427 (1991).

287. LEE, T. H., SHEPPARD, H. W., REIS, M., DONDERO, D., OSMOND, D., AND BUSCH, M. P., Circulating HIV-1 infected cell burden from seroconversion to AIDS: Importance of postseroconversion viral load on disease course, *J. Acquir. Immune Defic. Syndr.* **7**(4):381–388 (1994).

288. LEGUENNO, B., PISON, G., ENEL, C., LAGARDE, E., AND SECK, C., HIV-2 infections in a rural Senegalese community, *J. Med. Virol.* **38**:67–70 (1992).

289. LEGUENNO, B., PISON, G., ENEL, C., LEGARDE, E., AND SECK, C., HIV-2 prevalence in three rural regions of Sénégal: Low levels and heterogenous distribution, *Trans. R. Soc. Trop. Med. Hyg.* **86**:301–302 (1992).

290. LEKUTIS, C., OLSHEVY, U., FURMAN, C., THALI, M., AND SODROSKI, J., Contribution of disulfide bonds in the carboxyl terminus of the human immunodeficiency virus type I gp120 glycoprotein to CD4 binding, *J. Acquir. Immune Defic. Syndr.* **5**(1):78–81 (1992).

291. LEMP, G. F., HIROZAWA, A. M., GIVERTZ, D., NIERI, G. N., ANDERSON, L., LINDEGREN, M. L., JANSSEN, R. S., AND KATZ, M., Seroprevalence of HIV and risk behaviors among young homosexual and bisexual men—The San Francisco/Berkley young men's survey, *J.A.M.A.* **272**:449–454 (1994).

292. LEPAGE, P., VAN DE PERRE, P., CAROEL, M., NSENGUMUREMYI, F., NKURUNZIZA, J., BUTZER, J. P., AND SPRECHER, S., Post-natal transmission of HIV from mother to child, *Lancet* **2**:400–401 (1987).

293. LETVIN, N. L., DANIEL, M. D., SEHGAL, P. K., DESROSIERS, R. C., HUNT, R. D., WALDRON, L. M., MACKEY, J. J., SCHMIDT, D. K., CHALIFOUX, L. V., AND KING, N. W., Induction of AIDS-like disease in macaque monkeys with T-cell tropic retrovirus STLV-III, *Science* **230**:71–73 (1985).

294. LEVY, J. A., Pathogenesis of human immunodeficiency virus infection, *Microbiol. Rev.* **57**(1):183–289 (1993).

295. LEWIS, S. H., REYNOLDS-KOHLER, C., FOX, H. E., AND NELSON, J. A., HIV-I in trophoblastic and villous Hofbauer cells, and haematological precursors in eight-week fetuses, *Lancet* **335**: 565–568 (1990).

296. LIFSON, A. R., BUCHBINDER, S. P., SHEPPARD, H. W., SHEPPARD, H. W., MAWLE, A. C., WILBER, J. C., STANLEY, M., HART, C. E., HESSOL, N. A., AND HOLMBERG, S. D., Long-term human immunodeficiency virus infection in asymptomatic homosexual and bisexual men with normal CD4⁺ lymphocyte counts: Immunologic and virologic characteristics, *J. Infect. Dis.* **163**:959–965 (1991).

297. LIFSON, A. R., STANLEY, M., PANE, J., O'MALLEY, P. M., WILBER, J. C., STANLEY, A., JEFFERY, B., RUTHERFORD, G. W., AND SOHMER, P. R., Detection of human immunodeficiency virus DNA using the polymerase chain reaction in a well-characterized

group of homosexual and bisexual men, *J. Infect. Dis.* **161:**436–439 (1990).

298. LINDGREN, S., ANZEN, B., BOHLIN, A. B., AND LIDMAN, K., HIV and child-bearing: Clinical outcome and aspects of mother-to-infant transmission, *AIDS* **5**(9):1111–1116 (1991).

299. LO, S. C., HAYES, M. M., WANG, R. Y. H., PIERCE, P. F., KOTANI, H., AND SHIH, J. W. K., Newly discovered mycoplasma isolated from patients infected with HIV, *Lancet* **33:**1415–1418 (1991).

300. LOH, P. C., MATSUURA, F., AND MIZUMOTO, C., Seroepidemiology of human syncytial virus: Antibody prevalence in the Pacific, *Int. Virol.* **13:**87–90 (1980).

301. LOHSOMBOON, P., YOUNG, N. L., WENINGER, B. G., ATIKIJ, B., LIMPAKARNJANARAT, K., KHABBAZ, R. F., AND KAPLAN, J. E., Nondetection of HTLV-I/II and HIV-2 in Thailand, *J. Acquir. Immune Defic. Syndr.* **7**(2):992–994 (1994).

302. LOUWAGIE, J., McCUTCHAN, F., MASCOLA, J., EDDY, G., FRANSEN, K., PETERS, M., VAN DER GROEN, G., AND BURKE, D., Genetic subtypes of HIV-1, *AIDS Res. Hum. Retroviruses.* **9** (Suppl. 1):S147–S150 (1993).

303. LOVE, R. R., WIEBE, D. A., NEWCOMB, P. A., CAMERON, L., LEVENTHAL, H., JORDAN, V. C., FEYZI, J., AND DeMETS, D. L., Effects of tamoxifen on cardiovascular risk factors in postmenopausal women, *Ann. Intern. Med.* **115:**860–864 (1991).

304. LU, W., EME, D., AND ANDRIEU, J. M., HIV viraemia and seroconversion, *Lancet* **341:**113 (1993).

305. LUZURIAGA, K., McQUILKEN, P., ALIMENTI, A., SOMASUNDARAM, M., HESSELTON, R. A., AND SULLIVAN, J. L., Early viremia and immune responses in vertical human immunodeficiency virus type I infection, *J. Infect. Dis.* **167:**1008–1013 (1993).

306. LÜKE, W., VOSS, G., STAHL-HENNIG, C., COULBALY, C., PUTKONEN, P., PETRY, H., AND HAUSMANN, G., Protection of cynomolgus macaques (*Macaca fascicularis*) against infection with the human immunodeficiency virus type 2 strain ben (HIV-2 ben) by immunization with the virion-derived envelope glycoprotein gp130 (Abstract), *AIDS Res. Hum. Retroviruses* **9:**387–394 (1993).

307. LYNCH, C. E., MADEJ, R., LOUIE, P. H., AND RODGERS, G., Detection of HIV-1 DNA by PCR: Evaluation of primer pair concordance and sensitivity of a single primer pair, *J. Acquir. Immune Defic. Syndr.* **5**(5):433–440 (1992).

308. MacDONELL, K. B., CHMIEL, J. S., POGGENSEE, L., WU, S., AND PHAIR, J. P., Predicting progression to AIDS: Combined usefulness of CD4 lymphocyte counts and p24 antigenemia, *Am. J. Med.* **89:**706–712 (1990).

309. MADELEINE, M. M., WIKTOR, S. Z., GOEDERT, J. J., MANNS, A., LEVINE, P. H., BIGGAR, R. J., AND BLATTNER, W. A., HTLV-I and HTLV-II worldwide distribution: Reanalysis of 4,832 immunoblot results, *Int. J. Cancer* **54**(2):255–260 (1993).

310. MANN, D. L., MURRAY, C., O'DONNELL, M., BLATTNER, W. A., AND GOEDERT, J. J., HLA antigen frequencies in HIV-1 related Kaposi's sarcoma, *J. Acquir. Immune Defic. Syndr.* **3**(S1):S51–S55 (1989).

311. MANN, D. L., MURRAY, C., YARCHOAN, R., BLATTNER, W. A., AND GOEDERT, J. J., HLA antigen frequencies in HIV-1 seropositive disease-free individuals and patients with AIDS, *J. Acquir. Immune Defic. Syndr.* **1:**13–17 (1988).

312. MANN, J. M., Global AIDS into the 1990s, *J. Acquir. Immune Defic. Syndr.* **3**(4):438–442 (1990).

313. MANN, J. M., AIDS—The second decade: A global perspective, *J. Infect. Dis.* **165:**245–250 (1992).

314. MANN, J. M., FRANCIS, H., QUINN, T., ASILA, P. K., BOSENGE, N., NZILAMBI, N., BILA, K., TAMFUM, M., RUTI, K., PIOT, P., McCORMICK, J., AND CURRAN, J. W., Surveillance for AIDS in a Central African city. Kinshasa, Zaire, *J.A.M.A.* **255:**3255–3259 (1986).

314a. MANN, J., TARANTOLA, D. J. M., AND NOTTER, T. W., EDS., *AIDS in the World.* London, Howard University Press, 1992.

315. MARKOVITZ, D. M., Infection with the human immunodeficiency virus type 2, *Ann. Intern. Med.* **118:**211–218 (1993).

316. MARLINK, R., KANKI, P., THIOR, I., TRAVERS, K., EISEN, G., SIBY, T., TRAORE, I., HSIEH, C. C., DIA, M. C., GUEYE, E. H., HELLINGER, J., GUEYE-NDIAYE, A., SANKALÉ, J. L., NDOYE, I., MBOUP, S., AND ESSEX, M., Reduced rate of disease development after HIV-2 infection as compared to HIV-1, *Science* **265:**1587–1590 (1994).

317. MARLINK, R. G., RICARD, D., M'BOUP, S., KANKI, P. J., MONET-LEMONNE, J. L., N'DOYE, I., DIOP, K., SIMPSON, M. A., GRECO, F., CHOU, M., DEGRUTTOLA, V., HSIEH, C., BOYE, C., BARIN, F., DENIS, F., McLANE, M. F., AND ESSEX, M., Clinical, hematologic, and immunologic cross-sectional evaluation of individuals exposed to human immunodeficiency virus type-2 (HIV-2), *AIDS Res. Hum. Retroviruses* **4:**137–148 (1988).

318. MASOOD, R., CAI, J., LAW, R., AND GILL, P., AIDS-associated Kaposi's sarcoma pathogenesis, clinical features, and treatment, *Curr. Opin. Oncol.* **5:**831–834 (1993).

319. MASSEY, J. T., EZZATI, T. M., AND FOLSOM, R., Statistical issues in measuring the prevalence of HIV infection in a household survey. *Proc. Section Survey Res. Meth. Am. Stat. Assoc.* **1:**160–169 (1990).

320. MASTRO, T. D., SATTEN, G. A., NOPKESORN, T., SANGKHAROMYA, S., AND LONGINI, I. M., JR., Probability of female-to-male transmission of HIV-1 in Thailand, *Lancet* **343:**204–207 (1994).

321. MATHUR-WAGH, U., ENLOW, R. W., SPIGLAND, I., WINCHESTER, R. J., SACKS, H. S., RORAT, E., YANCOVITZ, S. R., KLEIN, M. J., WILLIAM, D. C., AND MILDVAN, D., Longitudinal study of persistent generalised lymphadenopathy in homosexual men: Relation to acquired immunodeficiency syndrome, *Lancet* **1:**1033–1038 (1984).

322. MAY, R. M., AND ANDERSON, R. M., Transmission dynamics of HIV infection, *Nature* **326:**137–142 (1987).

323. MAYER, K. H., FALK, L. A., PAUL, D. A., DAWSON, G. J., STODDARD, A. M., McCUSKER, J., SALTZMAN, S. P., MOON, M. W., FERRIANI, R., AND GROOPMAN, J. E., Correlation of enzyme-linked immunosorbent assays for serum human immunodeficiency virus antigen and antibodies to recombinant viral proteins with subsequent clinical outcomes in a cohort of asymptomatic homosexual men, *Am. J. Med.* **83:**208–212 (1987).

324. MAYER, K. H., STODDARD, A.M., McCUSKER, J., AYOTTE, D., FERRIANI, R., AND GROOPMAN, J. E., Human T-lymphotropic virus type III in high-risk, antibody-negative homosexual men, *Ann. Intern. Med.* **104:**194–196 (1986).

325. McCORMICK, A., Trends in mortality statistics in England and Wales with particular reference to AIDS from 1984 to April 1987, *Br. Med. J.* **296:**1289–1292 (1988).

326. McHUGH, T. M., STITES, D. P., BUSCH, M. P., KROWKA, J. F., STRICKLER, R. B., AND HOLLANDER, H., Relation of circulating levels of human immunodeficiency virus (HIV) antigen, antibody to p24, and HIV-containing immune complexes in HIV-infected patients, *J. Infect. Dis.* **158**(5):1088–1091 (1988).

327. McKUSICK, L., COATES, T. J., MORIN, S. F., POLLACK, L., AND HOFF, C., Longitudinal predictors of reductions in unprotected

anal intercourse among gay men in San Francisco: The AIDS behavioral research project, *Am. J. Public Health* **80**(8):978–983 (1990).

328. McKusick, L., Hoff, C. C., Stall, R., and Coates, T. J., Tailoring AIDS prevention: Differences in behavioral strategies among heterosexual and gay bar patrons in San Francisco, *AIDS Educ. Prev.* **3**(1):1–9 (1991).

329. McRae, B., Lange, J. A. M., Ascher, M. S., De Wolf, F., Sheppard, H. W., Goudsmit, J., and Allain, J. P., Immune response to HIV p24 core protein during the early phases of human immunodeficiency virus infection, *AIDS Res. Hum. Retroviruses* **7**(8):637–643 (1991).

330. Melbye, M., Biggar, R. J., Ebbesen, P., Andersen, H. K., and Vestergard, B. F., Lifestyle and antiviral antibody studies among homosexual men in Denmark, *Acta Pathol. Microbiol. Immunol. Scand.* **91**:357–364 (1983).

331. Melbye, M., Biggar, R. J., Ebbesen, P., Sarngadharan, M. G., Weiss, S. H., Gallo, R. C., and Blattner, W. A., Seroepidemiology of HTLV-III antibody in European homosexual men: Prevalence, transmission, and disease outcome, *Br. Med. J.* **289**(6445):573–575 (1984).

332. Melbye, M., Coté, T. R., and Biggar, R. J., High incidence of anal cancer among AIDS patients, *Lancet* **343**:636–639 (1994).

333. Melbye, M., Froebel, K. S., Madhok, R., Biggar, R. J., Sarin, P., Stenbjerg, S., Lowe, G. D., Forbes, C. D., Goedert, J. J., Gallo, R. C., and Ebbesen, P., HTLV-III seropositivity in European hemophiliacs exposed to factor VIII concentrate imported from the U.S.A., *Lancet* **2**:1444–1446 (1984).

333a. Melbye, M., Goedert, J. J., and Blattner, W. A., The natural history of HTLV-III/LAV infection, in: *Current Topics in AIDS, Vol. 1* (M. Gottlieb, D. Jeffries, D. Mildvan, A. Pinching, T. Quinn, and R. Weiss, eds.), pp. 57–93, John Wiley and Sons, London, 1987.

334. Melbye, M., Ingerslev, J., Biggar, R. J., Alexander, S., Sarin, P., Goedert, J. J., Zachariae, E., Ebbesen, P., and Stenbjerg, S., Anal intercourse as a possible factor in heterosexual transmission of HTLV-III to spouse of hemophiliacs, *N. Engl. J. Med.* **312**:857 (1985).

335. Melbye, M., Njelesani, E. K., Bayley, A., Mukelabai, K., Mannuele, J. K., Bowa, F. J., Clayden, S. A., Levine, A., Blattner, W. A., Weiss, R. A., Tedder, R., and Biggar, R. J., Evidence for heterosexual transmission and clinical manifestations of human immunodeficiency virus infection and related conditions in Lusaka, Zambia, *Lancet* **2**:1113–1115 (1986).

336. Melbye, M., Rabkin, C. S., Frisch, M., and Biggar, R. J., Changing patterns of anal cancer incidence in the USA. Period 1940–89, *Am. J. Epidemiol.* **139**(8):772–780 (1994).

337. Mellors, J., Kingsley, L., Gupta, P., *et al.*, Detection of plasma HIV RNA by branched DNA (bDNA) signal amplification predicts early onset of AIDS after seroconversion (Abstract), First Nat. Conf. on Hum. Retroviruses and Related Inf. 1993; 103.

338. Mendez, H., Holman, S., Stevens, R., Wethers, J., Minkoff, H., and Landesman, S., ELISA (E) and Western blot (WB) patterns and a possible marker of seroreversion (SR) in infants born to seropositive (SP) women (Abstract), Int. Conf. AIDS 1989; **5**(T.B.P. 240):327.

339. Merigan, T. C., Amato, D. A., Balsley, J., Power, M., Price, W. A., Benoit, S., Perez-Michael, A., Brownstein, A., Kramer, A. S., Brettler, D., Aledort, L., Ragni, M. V., Andes, W. A., Gill, J. C., Goldsmith, J., Stabler, S., Sanders, N., Gjerset, G., and Lusher, J., Placebo-controlled trial to evaluate zidovudine in treatment of human immunodeficiency virus infection in asymptomatic patients with hemophilia, *Blood* **78**(4):900–906 (1991).

340. Mermin, J. H., Holodniy, M., Katzenstein, D. A., and Merigan, T. C., Detection of human immunodeficiency virus DNA and RNA in semen by the polymerase chain reaction, *J. Infect. Dis.* **164**:769–772 (1991).

341. Mgone, C. S., Mhalu, F. S., Shao, J. F., Britton, S., Sandstrom, A., Bredberg-Raden, U., and Biberfeld, G., Prevalence of HIV-1 infection and symptomatology of AIDS in severely malnourished children in Dar Es Salaam, Tanzania, *J. Acquir. Immune Defic. Syndr.* **4**(9):910–913 (1991).

342. Michael, N. L., Vahey, M., Burke, D. S., and Redfield, R. R., Viral DNA and mRNA expression correlate with the stage of human immunodeficiency virus (HIV) type 1 infection in humans: Evidence for viral replication in all stages of HIV disease, *J. Virol.* **66**(1):310–316 (1992).

343. Miyoshi, I., Yoshimoto, S., Fujishita, M., Taguchi, H., Kubonishi, I., Niiya, K., and Minezawa, M., Natural adult T-cell leukaemia virus infection in Japanese monkeys, *Lancet* **2**:658 (1982).

344. Mofenson, L. M., and Blattner, W. A., Human retroviruses, in: *Textbook of Pediatric Infectious Diseases*, 3rd ed. (R. D. Feigin and J. D. Cherry, eds.), pp. 1757–1788, W. B. Saunders, Philadelphia, 1992.

345. Mok, J., HIV-1 infection: Breast milk and HIV-1 transmission, *Lancet* **341**:930–931 (1993).

346. Montalvo, F. W., Casanova, R., and Clavell, L. A., Treatment outcome in children with malignancies associated with human immunodeficiency virus infection [see comments], *J. Pediatr.* **116**:735–738 (1990).

347. Moore, J. P., The reactivities of HIV-1+ human sera with solid-phase V3 loop peptides can be poor predictors of their reactivities with V3 loops on native gp120 molecules, *AIDS Res. Hum. Retroviruses* **9**(3):209–219 (1993).

348. Morgan, D. A., Ruscetti, F. W., and Gallo, R. C., Selective *in vitro* growth of T-lymphocytes from normal human bone marrows, *Science* **193**:1007–1008 (1976).

349. Morgan, W. M., and Curran, J. W., Acquired immunodeficiency syndrome: Current and future trends, *Public Health Rep.* **101**:459–465 (1986).

350. Moss, A. R., Bacchetti, P., Osmond, D., Krampf, W., Chaisson, R. E., Stites, D., Wilber, J., Allain, J., and Carlson, J., Seropositivity for HIV and the development of AIDS or AIDS related condition: Three year follow up of the San Francisco General Hospital cohort, *Br. Med. J.* **296**:745–750 (1988).

351. Moss, A. R., Osmond, D., Bacchetti, P., Chermann, J. C., Barre-Sinoussi, F., and Carlson, J., Risk factors for AIDS and HIV seropositivity in homosexual men, *Am. J. Epidemiol.* **125**:1035–1047 (1987).

352. Moss, G. B., Clemetson, D., D'Costa, L., Plummer, F. A., Ndinya-Achola, J. O., Reilly, M., Holmes, K. K., Piot, P., Maitha, G. M., Hillier, S. L., Kiviat, N. C., Cameron, C. W., Wamola, I. A., and Kreiss, J. K., Association of cervical ectopy with heterosexual transmission of human immunodeficiency virus: Results of a study of couples in Nairobi, Kenya, *J. Infect. Dis.* **164**:588–591 (1991).

353. Mulder, D. W., Nunn, A. J., Kamali, A., Nakiyingi, J., Wagner, H. U., and Kengeya-Kayondo, J. F., Two-year HIV-1-associated mortality in a Ugandan rural population, *Lancet* **343**:1021–1023 (1994).

355. MUNOZ, A., WANG, M. C., BASS, S., TAYLOR, J. M. G., KINGSLEY, L. A., CHMIEL, J. S., AND POLK, B. F., Acquired immunodeficiency syndrome (AIDS)-free time after human immunodeficiency virus type 1 (HIV-1) seroconversion in homosexual men, *Am. J. Epidemiol.* **130**(3):530–539 (1989).

356. MUNOZ, A., SCHRAGER, L. K., BACELLAR, H., SPEIZER, I., VERMUND, S. H., DETELS, R., SAAH, A. J., KINGSLEY, L. A., SEMINARA, D., AND PHAIR, J. P., Trends in the incidence of outcomes defining acquired immunodeficiency syndrome (AIDS) in the multicenter AIDS cohort study. 1985–1991, *Am. J. Epidemiol.* **137**:423–438 (1993).

357. MURPHY, E. L., WILKS, R., HANCHARD, B., CRANSTON, B., FIGUEROA, J. P., GIBBS, W. N., MURPHY, J., AND BLATTNER, W. A., A case-control study of risk factors for seropositivity to human T-lymphotropic virus type I (HTLV-I) in Jamaica (in press).

358. MYERS, G., AND KORBER, B., The future of human immunodeficiency viruses, in: *The Evolutionary Biology of Viruses* (S. S. MORSE, ED.), pp. 211–232, Raven Press, New York, 1994.

359. MYERS, G., MacINNES, K., AND KORBER, B., The emergence of simian/human immunodeficiency viruses, *AIDS Res. Hum. Retroviruses* **8**(3):373–386 (1992).

360. NAHMIAS, A. J., WEISS, J., YAO, X., LEE, F., KODSI, R., SCHANFIELD, M., MATTHEWS, T., BOLOGNESI, D. P., DURACK, D., MOTULSKY, A., KANKI, P., AND ESSEX, M., Evidence for human infection with an HTLV III/LAV-like virus in Central Africa, 1959, *Lancet* **1**:1279–1280 (1986).

361. NAIK, T. N., SARKAR, S., AND SINGH, N. L., Intravenous drug users—a new high risk group for HIV infection in India, *AIDS* **5**:117–118 (1991).

362. NAIR, P., ALGER, L., HINES, S., SEIDEN, S., HEBEL, R., AND JOHNSON, J. P., Maternal and neonatal characteristics associated with HIV infection of seropositive women, *J. Acquir. Immune Defic. Syndr.* **6**(3):298–302 (1993).

363. NAUCLÉR, A., ALBINO, P., ANERSSON, S., DASILVA, A. P., LINDER, H., ANDREASSON, P. A., AND BIBERFFLD, G., Clinical and immunological follow-up of previously hospitalized HIV-2 seropositive patients in Bissau, Guinea-Bissau, *Scand. J. Infect. Dis.* **24**:725–731 (1992).

364. NAUCLÉR, A., ALBINO, P., DASILVA, A. P., AND BIBERFELD, G., Sexually transmitted diseases and sexual behaviour as risk factors for HIV-2 infection in Bissau, Guinea Bissau, *Int. J. STD AIDS* **4**:217–221 (1993).

365. NAUCLÉR, A., ANDREASSON, P. Å., COSTA, C. M., THORSTENSSON, R., AND BIBERFELD, G., HIV-2 associated AIDS and HIV-2 seroprevalence in Bissau, Guinea-Bissau, *J. Acquir. Immune Defic. Syndr.* **2**:88–93 (1989).

366. NELSON, K. E., CELENTANO, D. D., SUPRASERT, S., WRIGHT, N., EIUMTRAKUL, S., TULYATANA, S., MATANASARAWOOT, A., AKARASEWI, P., KUNTOLBUTRA, S., ROMYEN, S., SIRISOPANA, M., AND THEETRANONT, C., Risk factors for HIV infection among young adult men in northern Thailand, *J.A.M.A.* **270**(8):955–960 (1993).

367. NEUMANN, P., O'SHAUGHNESSY, M., REMIS, R., TSOUKAS, C., LEPINE, D., AND DAVIS, M., Laboratory evidence of active HIV-1 infection in Canadians with hemophilia associated with administration of heat-treated factor VIII, *J. Acquir. Immune Defic. Syndr.* **3**(3):278–281 (1990).

368. NICOLOSI, A., LEITE, M. L. C., MUSICCO, M., MOLINARI, S., AND LAZZARIN, A., Parenteral and sexual transmission of human immunodeficiency virus in intravenous drug users: A study of seroconversion, *Am. J. Epidemiol.* **135**:225–233 (1992).

369. NIELSEN, C., TEGLBJAERG, L. S., PEDERSEN, C., LUNDGREN, J. D., NIELSEN, C. M., AND VESTERGAARD, B. F., Prevalence of HIV infection in seronegative high-risk individuals examined by virus isolation and PCR, *J. Acquir. Immune Defic. Syndr.* **4**(11):1107–1111 (1991).

370. NOWAK, R., Hope or horror? Primate-to-human transplants, *J. NIH Res.* **4**:37–40 (1992).

371. NSUBUGA, P., MUGERWA, R., NSIBAMBI, J., SEWANKAMBO, M., KATABIRA, E., AND BERKLEY, S., The association of genital ulcer disease and HIV infection at a dermatology-STD clinic in Uganda, *J. Acquir. Immune Defic. Syndr.* **3**(10):1002–1005 (1990).

372. NUCHPRAYOON, C., TANPRASERT, S., AND CHUMNIJARAKIJ, T., Is routine p24 HIV antigen screening justified in Thai blood donors? *Lancet* **340**:1041 (1992).

373. NYAMATHI, A., Comparative study of factors relating to HIV risk level of black homeless women, *J. Acquir. Immune Defic. Syndr.* **5**(3):222–228 (1992).

374. NZILA, N., LAGA, M., THIAM, M. A., MAYIMONA, K., EDIDI, B., DYCK, E. V., BEHETS, F., HASSIG, S., NELSON, A., MOKWA, K., ASHLEY, R. I., PIOT, P., AND RYDER, R. W., HIV and other sexually transmitted diseases among female prostitutes in Kinshasa, *AIDS* **5**:715–721 (1991).

375. O'BRIEN, T. R., HIV-2 transmission: Implications for spread of HIV-1, *J.A.M.A.* **271**(12):903–904 (1994).

376. O'BRIEN, T. R., BUSCH, M. P., DONEGAN, E., WARD, J. W., WONG, L., SAMPSON, S. M., PERKINS, H. A., ALTMAN, R., STONEBURNER, R. L., AND HOLMBERG, S. D., Heterosexual transmission of human immunodeficiency virus type 1 from transfusion recipients to their sex partners, *J. Acquir. Immune Defic. Syndr.* **7**:705–710 (1994).

377. O'BRIEN, T. R., GEORGE, J. R., EPSTEIN, J. S., HOLMBERG, S. D., AND SCHOCHETMAN, G., Testing for antibodies to human immunodeficiency virus type 2 in the United States, *MMWR Morbid. Mortal. Week. Rep.* **41**(RR-12):1–9 (1992).

378. O'BRIEN, T. R., GEORGE, J. R., AND HOLMBERG, S. D., Human immunodeficiency virus type 2 infection in the United States: Epidemiology, diagnosis, and public health implications, *J.A.M.A.* **267**:2775–2779 (1992).

379. ODEHOURI, K., DeCOOK, K. M., KREBS, J. W., MOREAU, J., RAYFIELD, M., McCORMICK, J. B., SCHOCHETMAN, G., BRETTON, R., BRETTON, G., OUATTARA, D., HEROIN, P., KANGA, J. M., BEDA, B., NIAMKEY, E., KADIO, A., GARIEPE, E., AND HEYWARD, W. L., HIV-1 and HIV-2 infection associated with AIDS in Abidjan, Côte d'Ivoire, *AIDS* **3**:509–512 (1989).

380. OHKI, K., KISHI, M., NISHINO, Y., SUMIYA, M., KIMURA, T., GOTO, T., NAKAI, M., AND IKUTA, K., Noninfectious doughnut-shaped human immunodeficiency virus type 1 can induce syncytia mediated by fusion of the particles with CD4-positive cells, *J. Acquir. Immune Defic. Syndr.* **4**(12):1233–1240 (1991).

381. OLAFSSON, K., SMITH, M. S., MARSHBURN, P., CARTER, S. G., AND HASKILL, S., Variation of HIV infectibility of macrophages as a function of donor, stage of differentiation, and site of origin, *J. Acquir. Immune Defic. Syndr.* **4**(2):154–164 (1991).

382. OSBORN, J. E., Public health, HIV, and AIDS, in: *Textbook of AIDS Medicine*, (S. BRODER, T. C. MERIGAN, JR., AND D. BOLOGNESI, EDS.), pp. 133–146, Williams & Wilkins, Baltimore, 1994.

383. OU, C. Y., TAKEBE, Y., WENIGER, B. G., LUO, C. C., KALISH, M. L., AUWANIT, W., YAMAZAKI, S., GAYLE, H. D., YOUNG, N. L., AND SCHOCHETMAN, G., Independent introduction of two major HIV-1 genotypes into distinct high-risk populations in Thailand, *Lancet* **341**:1171–1174 (1993).

384. OUATTARA, S. A., MEITE, M., COT, M. C., AND DE-THE, G., Compared prevalence of infections by HIV-1 and HIV-2 during a 2-year period in suburban and rural areas of Ivory Coast, *J. Acquir. Immune Defic. Syndr.* **2**:94–99 (1989).

385. OYAIZU, N., McCLOSKEY, T. W., CORONESI, M., CHIRMULE, N., KALYANARAMAN, V. S., AND PAHWA, S., Accelerated apoptosis in peripheral blood monuclear cell (PBMCs) from human immunodeficiency virus type-1 infected patients and in CD4 cross-linked PBMCs from normal individuals, *Blood* **82**(11):3392–3400 (1993).

386. PADIAN, N., GLASS, S., MARQUIS, I., WILEY, J., AND WINKELSTEIN, W., Heterosexual transmission of HIV in California: Results from a heterosexual partner's study (abstract). *Hepa. Sciet.* 4020 (1988).

387. PADIAN, N., MARQUIS, L., FRANCIS, D. P., ANDERSON, R. E., RUTHERFORD, G. W., O'MALLEY, P. M., AND WINKELSTEIN, W., JR., Male-to-female transmission of human immunodeficiency virus, *J.A.M.A.* **258**:788–790 (1987).

388. PADIAN, N., O'BRIEN, T. R., CHANG, Y. C., GLASS, S., AND FRANCIS, D. P., Prevention of heterosexual transmission of human immunodeficiency virus through couple-counseling, *J. Acquir. Immune Defic. Syndr.* **6**:1043–1048 (1993).

389. PADIAN, N. S., SHIBOSKI, S. C., AND JEWELL, N. P., The effect of number of exposures on the risk of heterosexual HIV transmission, *J. Infect. Dis.* **161**:883–887 (1990).

390. PAN, L. Z., SHEPPARD, H. W., WINKELSTEIN, W., AND LEVY, J. A., Lack of detection of human immunodeficiency virus in persistently seronegative homosexual men with high or medium risks for infection, *J. Infect. Dis.* **164**:962–964 (1991).

391. PANTALEO, G., GRAZIOSI, C., DEMAREST, J. F., BUTINI, L., MONTRONI, M., FOX, C. H., ORENSTEIN, J. M., KOTLER, D. P., AND FAUCI, A. S., HIV infection is active and progressive in lymphoid tissue during the clinically latent stage of disease, *Nature* **362**:355–358 (1993).

392. PANTALEO, G., GRAZIOSI, C., AND FAUCI, A. S., The immunopathogenesis of human immunodeficiency virus infection, *N. Engl. J. Med.* **328**(5):327–335 (1993).

393. PAPE, J. W., LIAUTAUD, B., THOMAS, F., MATHURIN, J. R., STAMAND, M. M., BONCY, M., PEAN, V., PAMPHILE, M., LAROCHE, A. C., DEHOVITZ, J., AND JOHNSON, W. D., The acquired immunodeficiency syndrome in Haiti, *Ann. Intern. Med.* **103**:674–678 (1985).

394. PAPIERNIK, M., BROSSARD, Y., MULLIEZ, N., ROUME, J., BRECHOT, C., BARIN, F., GOUDEAU, A., BACH, J. F., GRISCELLI, C., HENRION, R., AND VAZEUX, R., Thymic abnormalities in fetuses aborted from human immunodeficiency virus type 1 seropositive women, *Pediatrics* **89**(2):297–301 (1992).

395. PEETERS, M., PIOT, P., AND VAN DER GROEN, G., Variability among HIV and SIV strains of African origin, *AIDS* **6**(1):29s–36s (1991).

396. PEPIN, J., MORGAN, G., DUNN, D., GEVAO, S., MENDY, M., GAYE, I., SCOLLEN, N., TEDDERS, R., AND WHITTLE, H., HIV-2 induced immunosuppression among asymptomatic West African prostitutes: Evidence that HIV-2 is pathogenic, but less so than HIV-1, *AIDS* **5**:1165–1172 (1991).

397. PETERMAN, T. A., JAFFE, H. W., FEORINO, P. M., GETCHELL, J. P., WARFIELD, D. T., HAVERKOS, H. W., STONEBURNER, R. L., AND CURRAN, J. W., Transfusion-associated acquired immunodeficiency syndrome in the United States, *J.A.M.A.* **254**:2913–2917 (1985).

398. PETERMAN, T. A., STONEBURNER, R. L., ALLEN, J. R., JAFFE, H. W., AND CURRAN, J. W., Risk of human immunodeficiency virus transmission from heterosexual adults with transfusion-associated infections, *J.A.M.A.* **259**:55–58 (1988).

399. PETERS, D., REID, M. M., AND GRIFFIN, S. G., Edinburgh drug users: Are they injecting and sharing less? *AIDS* **521**:528 (1994).

400. PEZZOTTI, P., REZZA, G., LAZZARIN, A., ANGARANO, G., SINICCO, A., AIUTI, F., ZERBONI, R., SALASSA, B., GAFA, S., PRISTERA, R., COSTIGLIOLA, P., ORTONA, L., BARBANERA, M., TIRELLI, U., CANESSA, A., VIALE, P., CASTELLI, F., AND CAPUTO, S. L., Influence of gender, age, and transmission category on the progression from HIV seroconversion to AIDS, *J. Acquir. Immune Defic. Syndr.* **5**(7):745–747 (1992).

401. PFÜTZNER, A., DIETRICH, U., VON EICHEL, U., VON BRIESEN, H., BREDE, H. D., MANIAR, J. K., AND RÜBSAMEN-WAIGMANN, H., HIV-1 and HIV-2 infections in a high risk population in Bombay, India: Evidence for the spread of HIV-2 and presence of a divergent HIV-1 subtype, *J. Acquir. Immune Defic. Syndr.* **5**:972–977 (1992).

402. PHILLIPS, A. N., LEE, C. A., ELFORD, J., JANOSSY, G., AND KERNOFF, P. B. A., The cumulative risk of AIDS as the CD4 lymphocyte count declines, *J. Acquir. Immune Defic. Syndr.* **5**(2):148–152 (1992).

403. PHILLIPS, A. N., LEE, C. A., ELFORD, J., WEBSTER, A., JANOSSY, G., TIMMS, A., BOFILL, M., AND KERNOR, P. B. A., More rapid progression to AIDS in older HIV-infected people: The role of CD4+ T-cell counts, *J. Acquir. Immune Defic. Syndr.* **4**(10):970–975 (1991).

404. PIATAK, M., JR., LUK, K. C., SAAG, M. S., KAPPES, J. C., YANG, L. C., LIFSON, J. D., CLARK, S. J., HAHN, B. H., AND SHAW, G. M., Viral dynamics in primary HIV-1 infection, *Lancet* **341**:1099 (1993).

405. PIATAK, M., JR., SAAG, M. S., YANG, L. C., CLARK, S. J., KAPPES, J. C., LUK, K. C., HAHN, B. H., SHAW, G. M., AND LIFSON, J. D., High levels of HIV-1 in plasma during all stages of infection determined by competitive PCR, *Science* **259**:1749–1754 (1993).

406. PIENIAZEK, D., PERALTA, J. M., FERREIRA, J. A., KREBS, J. W., OWEN, S. M., SION, F. S., FILHO, C. F., SERENO, A. B., DE SA, C. A., WENINGER, B. G., ET AL., Identification of mixed HIV-1/HIV-2 infections in Brazil by polymerase chain reaction, *AIDS* **5**:1293–1299 (1991).

406a. PIOT, P., LAGA, M., RYDER, R., PERRIENS, J., TEMMERMAN, M., HEYWARD, W., AND CURRAN, J. W., The global epidemiology of HIV infection: Continuity, heterogeneity, and change, *J. Acquir. Immune. Defic. Syndr.* **3**(4):403–412 (1990).

407. PLUMMER, F. A., SIMONSEN, J. N., CAMERON, D. W., NDINYA-ACHOLA, J. O., KREISS, J. K., GAKINYA, M. N., WAIYAKI, P., CHEANG, M., PIOT, P., RONALD, A. R., ET AL., Cofactors in male–female sexual transmission of human immunodeficiency virus type 1, *J. Infect. Dis.* **163**:233–239 (1991).

408. POIESZ, B. J., RUSCETTI, F. W., GAZDAR, A. F., BUNN, P. A., MINNA, J. D., AND GALLO, R. C., Detection and isolation of type-C retrovirus particles from fresh and cultured lymphocytes of a patient with cutaneous T-cell lymphoma, *Proc. Natl. Acad. Sci. USA* **77**:7415–7419 (1980).

409. POLK, B. F., FOX, R., BROOKMEYER, R., KANCHANARAKSA, S., KASLOW, R., VISSCHER, B., RINALDO, C., AND PHAIR, J., Predictors of the acquired immunodeficiency syndrome developing in a cohort of seropositive homosexual men, *N. Engl. J. Med.* **316**:61–66 (1987).

410. POPOVIC, M., AND GARTNER, S., Isolation of HIV-I from monocytes but not T-lymphocytes, *Lancet* **2**:916 (1987).

411. POPOVIC, M., SARNGADHARAN, M. G., READ, E., AND GALLO, R. C., Detection, isolation, and continuous production of cytopathic retroviruses (HTLV-III) from patients with AIDS and pre-AIDS, *Science* **224:**497–500 (1984).

412. POTTERAT, J. J., Lying to military physicians about risk factors for HIV infections, *J.A.M.A.* **257**(13):1727 (1987).

413. POTTS, B. J., MAURY, W., AND MARTIN, M. A., Replication of HIV-1 in primary monocyte cultures, *Virology* **175:**465–476 (1990).

414. POULSEN, A., AABY, P., FREDERIKSEN, K., KVINDESDAL, B., MØLBAK, B., DIAS, F., AND LAURITZEN, E., Prevalence of and mortality from human immunodeficiency virus type 2 in Bissau, West Africa, *Lancet* **1:**817–831 (1989).

415. POULSEN, A. G., AABY, P., GOTTSCHAU, A., KVINESDAL, B. B., DIAS, F., MOLBAK, K., AND LAURITZEN, E., HIV-2 infection in Bissau, West Africa, 1987–1989: Incidence, prevalence, and routes of transmission, *J. Acquir. Immune Defic. Syndr.* **6:**941–948 (1993).

416. POULSEN, A. G., AABY, P., SOARES DA GAMA, M., AND DIAS, F., HIV-2 in people over 50 years in Bissau, prevalence and risk factors, Seventh Int. Conf. AIDS (Abstract), July 19–24, 1992.

417. POULSEN, A. G., KNINESDAL, B. B., AABY, P., LISSE, I. M., GOTTSCHAU, A., MOLBAK, K., DIAS, F., AND LAURITZEN, E., Lack of evidence of vertical transmission of human immunodeficiency virus type 2 in a sample of the general population in Bissau, *J. Acquir. Immune Defic. Syndr.* **5:**25–30 (1992).

418. PRICE, R. W., AND WORLEY, J. M., Neurological complications of HIV-1 infection and AIDS, in: *Textbook of AIDS Medicine* (S. BRODER, T. C. MERIGAN, JR., AND D. BOLOGNESI, EDS.), pp. 489–505, Williams & Wilkins, Baltimore, 1994.

419. PUTKONEN, P., THORSTENSSON, R., GHAVAMZADEH, L., ALBERT, J., HILD, K., BIBERFELD, G., AND NOORBY, E., Prevention of HIV-2 and SIV$_{sm}$ by passive immunization in cynomolgus monkeys, *Nature* **352:**436–438 (1991).

420. QUINN, T. C., MANN, J. M., CURRAN, J. W., AND PIOT, P., AIDS in Africa: An epidemiologic paradigm, *Science* **234:**955–963 (1986).

421. RABKIN, C. S., AND BLATTNER, W. A., HIV infection and cancers other than non-Hodgkin's lymphoma and Kaposi's sarcoma, *Cancer Surv.* **10:**151–160 (1991).

422. RABKIN, C. S., AND GOEDERT, J. J., Risk of non-Hodgkin's Lymphoma and Kaposi's sarcoma in homosexual men, *Lancet* **336:**248–249 (1990).

423. RABKIN, C. S., HATZAKIS, A., GRIFFITHS, P. D., PILLAY, D., RAGNI, M. V., HILGARTNER, M. W., AND GOEDERT, J. J., Cytomegalovirus infection and risk of AIDS in HIV-infected hemophilia patients, *J. Infect. Dis.* **168:**1260–1263 (1993).

424. RABKIN, C. S., HILGARTNER, M., HEDBERG, K. W., ALEDORT, L. M., HATZAKIS, A., EICHINGER, S., EYSTER, M. E., WHITE, G. C., II, KESSLER, C. M., LEDERMAN, M. M., DE MOERLOOSE, P. E., BRAY, G. L., COHEN, A. R., ANDES, W. A., MANCO-JOHNSON, M., SCHRAMM, W., KRONER, B. L., BLATTNER, W. A., AND GOEDERT, J. J., Incidence of lymphomas and other cancers in HIV-infected and HIV-uninfected patients with hemophilia, *J.A.M.A.* **267**(8):1090–1094 (1992).

425. RAPPERSBERGER, K., GARTNER, S., SCHENK, P., STINGL, G., GROH, V., TSCHACHLER, E., MANN, D. L., WOLFF, K., KONRAD, K., AND POPOVIC, M., Langerhans' cells are an actual site of HIV-1 replication, *Intervirology* **29:**185–194 (1988).

426. RATNER, L., GALLO, R. C., AND WONG-STAAL, F., HTLV-III, LAV, ARV are variants of same AIDS virus, *Nature* **313:**636–637 (1985).

427. RWANDAN HIV SEROPREVALENCE STUDY GROUP, Nationwide community-based serological survey of HIV-1 and other human retrovirus infections in a Central African country, *Lancet* **1:**941–943 (1989).

428. REARDON, J., JOHNSON, D., ALDERETE, E., AND GREEN, D. A., Review of invasive cervical cancer cases for AIDS surveillance, *J. Acquir. Immune Defic. Syndr.* **7**(6):631 (1994).

429. ROBERT-GUROFF, M., WEISS, S. H., GIRON, J. A., JENNINGS, A. M., GINZBURG, H. M., MARGOLIS, I. B., BLATTNER, W. A., AND GALLO, R. C., Prevalence of antibodies to HTLV-I, -II, and -III in intravenous drug abusers from an AIDS endemic region, *J.A.M.A.* **255**(22):3133–3137 (1986).

430. ROBERTS, C. R., FIPPS, D. R., BRUNDAGE, J. F., WRIGHT, S. E., GOLDENBAUM, M., ALEXANDER, S. S., AND BURKE, D. S., Prevalence of human T-lymphotropic virus in civilian applicants for the United States armed forces, *Am. J. Public Health* **82**(1):70–73 (1992).

431. ROMIEU, I., MARLINK, R., KANKI, P., M'BOUP, S., AND ESSEX, M., HIV-2 link to AIDS in West Africa, *J. Acquir. Immune Defic. Syndr.* **3:**220–230 (1990).

432. ROSENBERG, P. S., BIGGAR, R. J., AND GOEDERT, J. J., Declining age at HIV infection in the United States, *N. Engl. J. Med.* **330**(11):789–790 (1994).

433. ROSENBERG, P. S., GAIL, M. H., SCHRAGER, L. K., VERMUND, S. H., CREAGH-KIRK, T., ANDREWS, E. B., WINKELSTEIN, W., JR., MARMOR, M., DES JARLAIS, D. C., BIGGAR, R. J., AND GOEDERT, J. J., National AIDS incidence trends and the extent of zidovudine therapy in selected demographic and transmission groups, *J. Acquir. Immune Defic. Syndr.* **4**(4):392–401 (1991).

434. ROSENBERG, P. S., GOEDERT, J. J., AND BIGGAR, R. J., Effect of age at seroconversion on the natural AIDS incubation distribution, *AIDS* **8:**803–810 (1994).

435. ROSENBERG, P. S., LEVY, M. E., BRUNDAGE, J. F., PETERSEN, L. R., KARON, J. M., FEARS, T. R., GARDNER, L. I., GAIL, M. H., GOEDERT, J. J., BLATTNER, W. A., RYAN, C. C., VERMUND, S. H., AND BIGGAR, R. J., Population-based monitoring of an urban HIV/AIDS epidemic: Magnitude and trends in the District of Columbia, *J.A.M.A.* **268**(4):495–503 (1992).

436. ROSSI, P., MOSCHESE, V., WIGZELL, H., BROLIDEN, P. A., AND WAHREN, B., Mother-to-infant transmission of HIV, *Lancet* **335:**359–360 (1990).

437. ROUS, P., A sarcoma of the fowl transmissible by an agent separable from the tumor cells, *J. Exp. Med.* **13:**397 (1911).

438. ROY, M. J., DAMATO, J. J., AND BURKE, D. S., Absence of true seroconversion of HIV-1 antibody in seroreactive individuals, *J.A.M.A.* **269**(22):2876–2879 (1993).

439. RUFF, A. J., COBERLY, J., HALSEY, N. A., BOULOS, R., DESORMEAUX, J., BURNLEY, A., JOSEPH, D. J., MCBRIEN, M., QUINN, T., LOSIKOFF, P., O'BRIEN, K. L., LOUIS, M. A., AND FARZADEGAN, H., Prevalence of HIV-1 DNA and p24 antigen in breast milk and correlation with maternal factors, *J. Acquir. Immune Defic. Syndr.* **7**(1):68–73 (1994).

440. RUTHERFORD, G. W., LIFSON, A. R., HESSOL, N. A., DARROW, W. W., O'MALLEY, P. M., BUCHBINDER, S. P., BARNHART, J. L., BODECKER, T. W., CANNON, L., DOLL, L. S., HOLMBERG, S. D., HARRISON, J. S., ROGERS, M. F., WERDEGAR, D., AND JAFFE, H. W., Course of HIV-1 infection in a cohort of homosexual and bisexual men: An 11 year follow-up study, *Br. Med. J.* **301:**1183–1188 (1990).

441. SALAHUDDIN, S. Z., GROOPMAN, J. E., MARKHAM, P. D., SARNGADHARAN, M. G., REDFIELD, R. R., MCLANE, M. F., ESSEX, M.,

SLISKI, A., GALLO, R. C., POPOVIC, M., ORNDORFF, S., FLADA-GAR, A., PATEL, A., AND GOLD, J., HTLV-III in symptom-free seronegative persons, *Lancet* **2**:1418–1420 (1984).

442. SALAHUDDIN, S. Z., MARKHAM, P. D., POPOVIC, M., ET AL., Isolation of infections human T-cell leukemia/lymphotropic virus type III (HTLV-III) from patients with acquired immunodeficiency syndrome (AIDS) or AIDS-related complex (ARC) and from healthy carriers: A study of risk groups and tissue sources, *Proc. Natl. Acad. Sci. USA* **82**:5530–5534 (1985).

443. SANDE, M. A., AND VOLBERDING, P. A., *Pneumocystis carinii* pneumonia—current concepts, in: *The Medical Management of AIDS*, 3rd ed. (M. A. SANDE AND P. A. VOLBERDING, EDS.), pp. 261–283, W. B. Saunders, Philadelphia, 1992.

444. SANTOS-FERREIRA, M. O., COHEN, T., LOURENCO, M. H., AL-MEIDA, M. J. M., CHAMARET, S., AND MONTAGNIER, L., A study of seroprevalence of HIV-1 and HIV-2 in six provinces of People's Republic of Angola: Clues to the spread of HIV infection, *J. Acquir. Immune Defic. Syndr.* **3**:780–786 (1990).

445. SARKAR, S., MOOKERJEE, P., ROY, A., NAIK, T. N., SINGH, J. K., SHARMA, A. R., SINGH, Y. I., SINGH, P. K., TRIPATHY, S. P., AND PAL, S. C., Descriptive epidemiology of intravenous heroin users—a new risk group for transmission of HIV in India, *J. Infect.* **23**:201–207 (1991).

446. SAROSI, G. A., Endemic mycoses in HIV infection, in: *The Medical Management of AIDS*, 3rd ed. (M. A. SANDE AND P. A. VOLBERDING, EDS.), pp. 311–318, W. B. Saunders, Philadelphia, 1992.

447. SASSE, H., SALMASO, S., AND CONTI, S., Risk behaviors for HIV-1 infection in Italian drug users: Report from a multicenter study, *J. Acquir. Immune Defic. Syndr.* **2**(5):486–496 (1989).

448. SCHNITTMAN, S. M., GREENHOUSE, J. J., PSALLIDOPOULOS, M. C., BASELER, M., SALZMAN, N. P., FAUCI, A. S., AND LANE, H. C., Increasing viral burden in CD4⁺ T cells from patients with human immunodeficiency virus (HIV) infection reflects rapidly progressive immunosuppression and clinical disease, *Ann. Intern. Med.* **113**:438–443 (1990).

449. SEAGE, G. R., III, MAYER, K. H., HORSBURGH, C. R., JR., HOLMBERG, S. D., MOON, M. W., AND LAMB, G. A., The relation between nitrite inhalants, unprotected receptive anal intercourse, and the risk of human immunodeficiency virus infection, *Am. J. Epidemiol.* **135**(1):1–11 (1992).

450. SEITZ, S., BERNSTEIN, R., HOLTGRAVE, D., MOORE, M., STA-NECKI, K., AND WAY, P., AIDS workshop for public health specialists, *AIDS*: 1–52 (1994).

451. SHAPSHAK, P., MCCOY, C. B., RIVERS, J. E., CHITWOOD, D. D., MASH, D. C., WEATHERBY, N. L., INCIARDI, J. A., SHAH, S. M., AND BROWN, B. S., Inactivation of human immunodeficiency virus-1 at short time intervals using undiluted bleach, *J. Acquir. Immune Defic. Syndr.* **6**(2):218–219 (1993).

452. SHARKEY, S. J., SHARKEY, K. A., SUTHERLAND, L. R., AND CHURCH, D. L., Nutritional status and food intake in human immunodeficiency virus infection, *J. Acquir. Immune Defic. Syndr.* **5**(11):1091–1098 (1992).

453. SHEPPARD, H. W., Polymerase chain reaction, *Sixth Ann. Conf. on Human Retrovirus Testing*, 14–15, 1991.

454. SHEPPARD, H. W., AND ARCHER, M. S., AIDS and programmed cell death, *Immunol. Today* **12**(11):423 (1991).

455. SHEPPARD, H. W., AND ASCHER, M. S., The relationship between AIDS and immunologic tolerance, *J. Acquir. Immune Defic. Syndr.* **5**(2):143–147 (1992).

456. SHEPPARD, H. W., ASCHER, M. S., MCRAE, B., ANDERSON, R. E., LANG, W., AND ALLAIN, J., The initial immune response to HIV and immune system activation determine the outcome of HIV disease, *J. Acquir. Immune Defic. Syndr.* **4**(7):704–712 (1991).

457. SHERMAN, M. P., DUBE, S., SPLCER, T. P., KANE, T. D., LOVE, J. L., SAKSENA, N. K., IANNONE, R., GIBBS, C. J., JR., YANAGI-HARA, R., DUBE, D. K., AND POIESZ, B. J., Sequence analysis of an immunogenic and neutralizing domain of the human T-cell lymphoma/leukemia virus type I gp46 surface membrane protein among various primate T-cell lymphoma/leukemia virus isolates including those from a patient with both human T-cell lymphoma/leukemia virus type I-associated myelopathy and adult T-cell leukemia, *Cancer Res.* **53**(24):6067–6073 (1993).

458. SHIBOSKI, S. C., AND JEWELL, N. P., Statistical analysis of the time dependence of HIV infectivity based on partner study data, *J. Am. Stat. Assoc.* **87**(418)360–372 (1992).

459. SHIRAMIZU, B., HERNDIER, B., AND MCGRATH, M. S., Identification of a common clonal human immunodeficiency virus integration site in human immunodeficiency virus-associated lymphomas, *Cancer Res.* **54**:2069–2072 (1994).

460. SIMON, F., MATHERON, S., TAMALET, C., LOUSSERT-AJAKA, I., BARTCZAK, S., PÉPIN, J. M., DHIVER, C., GAMBIA, E., ELBIM, C., GASTAUT, J. A., SAIMOT, A. G., AND BRUN-VÉZINET, F., Cellular and plasma viral load in patients infected with HIV-2, *AIDS* **7**:1411–1417 (1993).

461. SIMONDS, R. J., HOLMBERG, S. D., HURWITZ, R. L., COLEMAN, T. R., BOTTENFIELD, S., CONLEY, L. J., KOHLENBERG, S. H., CAS-TRO, K. G., DAHAN, B. A., SCHABLE, C. A., RAYFIELD, M. A., AND ROGERS, M. F., Transmission of human immunodeficiency virus type 1 from a seronegative organ and tissue donor, *N. Engl. J. Med.* **326**(11):726–732 (1992).

462. SIMONSEN, J. N., PLUMMER, F. A., NGUGI, E. N., BLACK, C., KREISS, J. K., GAKINYA, M. N., WAIYAKI, P., D'COSTA, L. J., NDINYA-ACHOLA, J. O., PIOT, P., AND RONALD, A., HIV infection among lower socioeconomic strata prostitutes in Nairobi, *AIDS* **4**(2):139–144 (1990).

463. SINGH, Y. N., MALAVIYA, A. N., TRIPATHY, S. P., CHAUDHURI, K., BHARGAVA, N. C., AND KHARE, S. D., HIV serosurveillance among prostitutes and patients from a sexually transmitted diseases clinic in Delhi, India, *J. Acquir. Immune Defic. Syndr.* **3**(3):287–289 (1990).

464. SIRAPRAPASIRI, T., THANPRASERTSUK, S., AND RODKLAY, A., Risk factors for HIV among prostitutes in Chiangmai, Thailand, *AIDS* **5**:579–582 (1991).

465. SMITH, T. F., SRINIVASAN, A., SCHOCHETMAN, G., MARCUS, M., AND MYERS, G., The phylogenetic history of immunodeficiency viruses, *Nature* **333**:573–575 (1988).

466. SPIRA, T. J., KAPLAN, J. E., FEORINO, P. M., WARFIELD, D. T., FISHBEIN, D. B., AND BOZEMAN, L. H., Human immunodeficiency virus viremia as a prognostic indicator in homosexual men with lymphadenopathy syndrome, *N. Engl. J. Med.* **317**(17): 1093–1094 (1987).

467. ST. LOUIS, M. E., KAMENGA, M., BROWN, C., NELSON, A. M., MANZILA, T., BATTER, V., BEHETS, F., KABAGABO, U., RYDER, R. W., AND OXTOBY, M., Risk for perinatal HIV-1 transmission according to maternal immunologic, virologic, and placental factors, *J.A.M.A.* **269**(22):2853–2859 (1993).

468. STEVENS, C. E., TAYLOR, P. E., ZANG, E. A., MORRISON, J. M., HARLEY, E. J., DE RODRIQUEZ, C. S., BACINO, C., TING, R. C., BODNER, A. J., SARNGADHARAN, M. G., GALLO, R. C., AND RUBINSTEIN, P., Human T-cell lymphotropic virus type III infection in a cohort of homosexual men in New York City, *J.A.M.A.* **255**:2167–2172 (1986).

469. STIEGBIEGEL, N. H., MAUDE, D. W., FEINER, C. J., HARRIS, C. A., SALTZMAN, B. R., AND KLEIN, R. S., Heterosexual transmission of HIV infection (Abstract 4057), Fourth Int. Conf. AIDS, 1988.

470. TEMIN, H. M., AND MIZUTANI, S., RNA-dependent DNA polymerase in virions of rous sarcoma virus, *Nature* **226**:1211–1213 (1970).

471. TINDALL, B., AND COOPER, D. A., Primary HIV infection: Host responses and intervention strategies, *AIDS* **5**(1):1–14 (1991).

472. TOVO, P. A., DE MARTINO, M., GABIANO, C., CAPPELLO, N., D'ELIA, R., LOY, A., PLEBANI, A., ZUCCOTTI, G. V., DALLACASA, P., AND FERRARIS, G., Prognostic factors and survival in children with perinatal HIV-1 infection, *Lancet* **339**:1249–1253 (1992).

473. TURNER, B. J., DENISON, M., EPPES, S. C., HOUCHENS, R., FANNING, T., AND MARKSON, L. E., Survival experience of 789 children with acquired immunodeficiency syndrome, *Pediatr. Infect. Dis. J.* **12**:310–320 (1993).

474. VAN DE PERRE, P., NZARAMBA, D., NTILIVAMUNDA, A., UWIMANA, A., LEPAGE, P., ROUVROY, D., MAUREL, R., GATSINZI, T., AND BUGINGO, G., AIDS definition for Africa, *Lancet* **2**:99–100 (1987).

475. VAN DEN HOEK, J. A. R., COUTINHO, R. A., VAN HAASTRECHT, H. J. A., VAN ZADELHOFF, A. W., AND GOUDSMIT, J., Prevalence and risk factors of HIV infections among drug users and drug-using prostitutes in Amsterdam, *AIDS* **2**(1):55–60 (1988).

476. VAN DEN HOEK, J. A. R., DE WIT, J. B. F., KEET, I. P. M., FENNEMA, J. S. A., VAN DEN BERGH, H. S. P., SANDFORT, T. G. M., VAN GRIENSVEN, G. J. P., AND COUTINHO, R. A., Increase in unsafe sex in both young and older homosexual men, Ninth Int. Conf. AIDS, June 6–11, 1993.

477. VARMUS, H., Retroviruses, *Science* **240**:1427–1435 (1988).

478. VERMUND, S. H., KELLEY, K. F., KLEIN, R. S., FEINGOLD, A. R., SCHREIBER, K., MUNK, G., AND BURK, R. C., High risk of human papillomavirus infection and cervical squamous intraepithelial lesions among women with symptomatic human immunodeficiency virus infection, *Am. J. Obstet. Gynecol.* **165**(2):392–400 (1991).

479. VILLARI, P., SPINO, C., CHALMERS, T. C., LAU, J., AND SACKS, H. S., Cesarean section to reduce perinatal transmission of human immunodeficiency virus: A meta-analysis, *Online J. Curr. Clin. Trials.* **74** (1993).

480. VLAHOV, D., ANTHONY, J. C., MUÑOZ, A., MAGOLICK, J., NELSON, K. E., CELENTANO, D. D., SOLOMON, L., AND POLK, B. F., The Alive study, a longitudinal study of HIV-1 infection in intravenous drug users, *J. Drug Issues* **21**(4):755–771 (1991).

481. VLAHOV, D., MUÑOZ, A., ANTHONY, J. C., COHN, S., CELENTANO, D. D., AND NELSON, K. E., Association of drug injection patterns with antibody to human immunodeficiency virus type 1 among intravenous drugs users in Baltimore, Maryland, *Am. J. Epidemiol.* **132**(5):847–856 (1990).

482. WANG, R. Y. H., SHIH, J. W. K., GRANDINETTI, T., PIERCE, P. F., HAYES, M. M., WEAR, D. J., ALTER, H. J., AND LO, S. C., High frequency of antibodies to mycoplasma penetrans in HIV-infected patients, *Lancet* **340**:1312–1316 (1992).

483. WARD, J. W., DEPPE, D. A., SAMSON, S., PERKINS, H., HOLLAND, P., FERNANDO, L., FEORINO, P. M., THOMPSON, P., KLEINMAN, S., AND ALLEN, J. R., Risk of human immunodeficiency virus infection from blood donors who later developed the acquired immunodeficiency syndrome, *Ann. Intern. Med.* **106**(1):61–62 (1987).

484. WARD, J. W., HOLMBERG, S. D., ALLEN, J. R., COHN, D. L., CRITCHLEY, S. E., KLEINMAN, S. H., LENES, B. A., RAVENHOLT, O., DAVIS, J. R., QUINN, M. G., AND JAFFE, H. W., Transmission of human immunodeficiency virus (HIV) by blood transfusions screened as negative for HIV antibody, *N. Engl. J. Med.* **318**:473–478 (1988).

485. WARD, J. W., SCHABLE, C., DICKINSON, G. M., MAKOWKA, L., YANAGA, K., CARUANA, R., CHAN, H., SALAZAR, F., SCHOCHETMAN, G., AND HOLMBERG, S., Acute human immunodeficiency virus infection. Antigen detection and seroconversion in immunosuppressed patients, *Transplantation* **47**(4):722–724 (1989).

486. WAWER, M. J., SERWADDA, D., MUSGRAVE, S. D., KONDE-LULE, J. K., MUSAGARA, M., AND SEWANKAMBO, N. K., Dynamics of spread of HIV-1 infection in a rural district of Uganda, *Br. Med. J.* **303**:1303–1306 (1991).

487. WEBER, J. N., CLAPHAM, P. R., WEISS, R. A., PARKER, D., ROBERTS, C., DUNCAN, J., WELLER, I. V., CARNE, C. A., TEDDER, R. S., PINCHING, A. J., AND CHEINGSONG-POPOV, R., Human immunodeficiency virus infection in two cohorts of homosexual men: Neutralising sera and association of anti-gag antibody with prognosis, *Lancet* **1**:119–122 (1987).

488. WEISS, R. A., The virus and its target cells, in: *Textbook of AIDS Medicine* (S. BRODER, T. C. MERIGAN, JR., AND D. BOLOGNESI, EDS.), pp. 15–20, Williams & Wilkins, Philadelphia, 1994.

489. WEISS, S. H., Laboratory detection of human immunodeficiency viruses, in: *Aids and Other Manifestations of HIV Infection* (G. WORMSER, R. STAHL, AND E. BOTTANE, EDS.), pp. 270–293, Noyes, 1987.

490. WEISS, S. H., FRENCH, J., HOLLAND, B., PARKER, M. A., LINGREENBURG, A., AND ALTMAN, R., HTLV-I/II co-infection is significantly associated with risk for progression to AIDS among HIV+ intravenous drug abusers (Abstract), Int. Conf. AIDS 1989.

491. WEISS, S. H., GINZBURG, H. M., GOEDERT, J. J., CANTOR, K., ALTMAN, R., ROBERT-GUROFF, M., GALLO, R. C., AND BLATTNER, W. A., Risk factors for HTLV-III infection among parenteral drug users (PDU), *Proc. ASCO* **5**:3 (1986).

492. WEISS, S. H., GINZBURG, H. M., SAXINGER, W. C., CANTOR, K. P., MUNDON, F. K., ZIMMERMAN, D. H., AND BLATTNER, W. A., Emerging high rates of human T-cell lymphotropic virus type I (HTLV-I) and HIV infection among U.S. drug abusers (DA) (Abstract), *Epidemiology* **F.6.5**:211 (1988).

493. WEISS, S. H., GOEDERT, J. J., GARTNER, S. POPOVIC, M., WATERS, D., MARKHAM, P., VERONESE, F. D. M., GAIL, M. H., BARKLEY, W. E., GIBBONS, J., GILL, F. A., LEUTHER, M., SHAW, G. M., GALLO, R. C., AND BLATTNER, W. A., Risk of human immunodeficiency virus (HIV-1) infection among laboratory workers, *Science* **239**:68–71 (1988).

494. WEISS, S. H., GOEDERT, J. J., SARNGADHARAN, M. G., BODNER, A. J., THE AIDS SEROEPIDEMIOLOGY COLLABORATIVE WORKING GROUP, GALLO, R. C., AND BLATTNER, W. A., Screening test for HTLV-III (AIDS-agent) antibody: Specificity, sensitivity, and applications, *J.A.M.A.* **253**:221–225 (1984).

495. WEISS, S. H., SAXINGER, C. W., RECHTMAN, D., GRIECO, M. H., NADLER, J., HOLMAN, S., GINZBURG, H. M., GROOPMAN, J. E., GOEDERT, J. J., MARKHAM, P. D., GALLO, R. C., BLATTNER, W. A., AND LANDESMAN, S. H., HTLV-III infection among health care workers: Association with needle-stick injuries, *J.A.M.A.* **254**(15):2089–2093 (1985).

496. WENIGER, B. G., TANSUPHASWADIKUL, S., YOUNG, N. L., *ET AL.*, Phenotypical differences in clinical HIV/AIDS disease with Thailand HIV-1 genotypes A and B (Abstract). First Nat. Conf. on Hum. Retroviruses and Related Inf. 1993; 105.

497. WHITTLE, H., EGBOGA, A., TODD, J., MORGAN, G., ROLFE, M.,

SABALLY, S., WILKINS, A., AND CORRAH, T., Immunological responses of Gambians in relation to clinical stage of HIV-2 disease, *Clin. Exp. Immunol.* **93**:45–50 (1993).

498. WORLD HEALTH ORGANIZATION, Acquired immunodeficiency syndrome (AIDS): WHO/CDC case definition for AIDS, *WHO Week. Epidemiol. Rec.* **61**:69–72 (1986).

499. WORLD HEALTH ORGANIZATION, World Health statistics annual, *World Health Stat. Annu.*: 30–32 (1992).

500. WILFERT, C. M., WILSON, C., LUZURIAGA, K., AND EPSTEIN, L., Pathogenesis of pediatric human immunodeficiency virus type I infection, *J. Infect. Dis.* **170**:286–292 (1994).

501. WILLERFORD, D. M., BWAYO, J. J., HENSEL, M., EMONYI, *ET AL.*, Human immunodeficiency virus infection among high-risk seronegative prostitutes in Nairobi, *J. Infect. Dis.* **167**(6):1414–1417 (1994).

502. WINKELSTEIN, W., JR., LYMAN, D. M., PADIAN, N., GRANT, R., SAMUEL, M., WILEY, J. A., ANDERSON, R. E., LANG, W., RIGGS, J., AND LEVY, J. A., Sexual practices and risk of infection by the human immunodeficiency virus: The San Francisco Men's Health Study, *J.A.M.A.* **257**(3):321–325 (1987).

503. WINKELSTEIN, W., JR., SAMUEL, M., PADIAN, N. S., WILEY, J. A., LANG, W., ANDERSON, R. E., AND LEVY, J. A., The San Francisco men's health study: III. Reduction in human immunodeficiency virus transmission among homosexual/bisexual men. 1982–86, *Am. J. Public Health* **77**:685–689 (1987).

504. WINKELSTEIN, W., JR., WILEY, J. A., OSMOND, D., COATES, T., SHEPPARD, H. W., PAGE, K., *ET AL.*, The San Francisco Young Men's Health Study, Ninth Int. Conf. AIDS, June 6–11, 1993.

505. WINKELSTEIN, W., JR., WILEY, J. A., PADIAN, N. S., SAMUEL, M., SHIBOSKI, S., ASCHER, M. S., AND LEVY, J. A., The San Francisco men's health study: Continued decline in HIV seroconversion rates among homosexual/bisexual men, *Am. J. Public Health* **78**(11):1472–1474 (1988).

506. WINKLER, E., HOLTEN, I., MEYER, A., REHLE, T., GARIN, D., MEFANE, C., PARRY, J. W., AND SCHMITZ, H., Seroepidemiology of human retroviruses in Gabon, *AIDS* **3**:106–107 (1989).

507. WITTEK, A. E., PHELAN, M. A., WELLS, M. A., VUJCIC, L. K., EPSTEIN, J. S., LANE, H. C., AND QUINNAN, G. V., JR., Detection of human immunodeficiency virus core protein in plasma by enzyme immunoassay. Association of antigenemia with symptomatic disease and T-helper cell depletion, *Ann. Intern. Med.* **107**:286–292 (1987).

508. WOFSY, C. B., COHEN, J. B., HAUER, L. B., PADIAN, N. S., MICHAELIS, B. A., EVANS, L. A., AND LEVY, J. A., Isolation of AIDS-associated retrovirus from genital secretions of women with antibodies to the virus, *Lancet* **1**:527–529 (1986).

509. WOOD, R., DONG, H., KATZENSTEIN, D. A., AND MERIGAN, T. C., Quantification and comparison of HIV-1 proviral load in peripheral blood mononuclear cells and isolated CD4$^+$ T cells, *J. Acquir. Immune Defic. Syndr.* **6**(3):237–240 (1993).

510. WORMSER, G. P., RABKIN, C. S., AND JOLINE, C., Frequency of nosocomial transmission of HIV infection among health care workers, *N. Engl. J. Med.* **319**(5):307–308 (1988).

511. WRIGHT, N. H., VANICHSENI, S., AKARASEWI, P., WASI, C., AND CHOOPANYA, K., Was the 1988 HIV epidemic among Bangkok's injecting drug users a common source outbreak? *AIDS* **8**(4):529–532 (1994).

512. YARCHOAN, R., KLECKER, R. W., WEINHOLD, K. J., MARKHAM, P. D., LYERLY, H. K., DURACK, D. T., GELMANN, E., LEHRMAN, S. N., BLUM, R. M., BARRY, D. W., SHEARER, G. M., FISCHL, M. A., MITSUYA, H., GALLO, R. C., COLLINS, J. M., BOLOGNESI, D. P., MYERS, C. E., AND BRODER, S., Administration of 3'-azido-3'-deoxythymidine, an inhibitor of HTLV-III/LAV replication, to patients with AIDS or AIDS-related complex, *Lancet* **1**:575–580 (1986).

513. YOUCHAH, J., MINKOFF, H., LANDESMAN, S., *ET AL.*, Longer duration of ruptured membranes is associated with increased risk of vertical transmission of HIV infection (Abstract), *Am. J. Obstet. Gynecol.* **507**:415 (1994).

514. ZAAIJER, H. L., VAN EXEL-OEHLERS, P. J., KRAAIJEVELD, T., ALTENA, E., AND LELIE, P. N., Early detection of antibodies to HIV-1 by third-generation assays, *Lancet* **340**:770–772 (1992).

515. ZANGERLE, R., FUCHS, D., REIBNEGGER, G., FRITSCH, P., AND WACHTER, H., Markers for disease progression in intravenous drug users infected with HIV-1, *AIDS* **5**(8):985–991 (1991).

516. ZHU, T., MO, H., WANG, N., NAM, D. S., CAO, Y., KOUP, R. A., AND HO, D. D., Genotypic and phenotypic characterization of HIV-1 in patients with primary infection, *Science* **261**:1179–1181 (1994).

12. Suggested Reading

BRODER, S., MERIGAN, T. C., JR., AND BOLOGNESI, D., *Textbook of AIDS Medicine*. Baltimore, Williams & Wilkins, 1994.

MYERS, G., AND KORBER, B., The future of human immunodeficiency viruses, in: *The Evolutionary Biology of Viruses* (S. S. MORSE, ED.), pp. 211–232, Raven Press, New York, 1994.

MARKOVITZ, D. M., Infection with the human immunodeficiency virus type 2, *Ann. Intern. Med.* **118**:211–218 (1993).

O'BRIEN, T. R., GEORGE, J. R., AND HOLMBERG, S. D., Human immunodeficiency virus type 2 infection in the United States: Epidemiology, diagnosis, and public health implications, *J.A.M.A.* **267**:2775–2779 (1992).

CHAPTER 25

Retroviruses—Human T-Cell Lymphotropic Virus

Nancy E. Mueller and William A. Blattner

1. Introduction

Retroviruses comprise a class of enveloped RNA viruses that have been long postulated to be a cause of various human diseases based on numerous animal models. These animal retroviruses were described by Rous and colleagues in the first decade of this century as the cause of sarcomatous neoplasms in chickens. Subsequently, mammalian retroviruses were discovered first in mice by Gross and colleagues, and subsequently in many species including cats, ungulates, and nonhuman primates. The range of diseases observed in animals includes malignancies, particularly leukemias and lymphomas, but also sarcomas, brain tumors, and breast cancers, as well as diseases of the immunologic system.[215] Despite the many examples of retrovirus-associated disease in animals, intensive laboratory-based studies, particularly in the late 1950s through mid-1970s, failed to identify a candidate human retrovirus. By then, most investigators had abandoned the search and had focused their attention instead on the transforming genes, termed *oncogenes*, which were essentially transduced host genes incorporated in some acutely transforming retroviruses. However, with the report in 1980 by Poiesz in Gallo's laboratory of the first human retrovirus—human T-cell lymphotropic virus type I (HTLV-I), also known as T-cell leukemia/lymphoma virus type I—a major new chapter in human virology was opened.[212] A second HTLV (HTLV-II) was identified in 1982,[113] and understanding of its epidemiology is slowly evolving. Understanding of the virology and immunology of both of these viruses has been substantially fostered by the explosion of research following the discovery of the human immunodeficiency virus (HIV).

2. Historical Background

In 1970, Temin[261] and Baltimore[7] reported in classic papers published back-to-back in *Nature* the discovery of a novel reverse transcriptase enzyme associated with retroviruses, for which they were jointly awarded the Nobel prize in 1975. This enyzme is the sine qua non of this class of virus. The virally encoded reverse transcriptase enzyme reverses the normal flow of nucleotide transcription, allowing a DNA chain to be copied from the RNA template. Subsequently, double-stranded DNA is formed and integrated into the somatic DNA of the cell. Once integrated, the virus is replicated by subverting the normal cellular machinery to the production of additional RNA copies that are either expressed as viral proteins necessary for virus particle packaging or as the RNA genetic material to be packaged into viral particles. The essential feature of this life cycle with particular epidemiologic relevance is the capacity of this virus to be integrated into the genome of the host cell. This ability to establish latency undoubtedly explains the fact that these viruses are associated with diseases that take years to decades to develop.

The classification of retroviruses has evolved primarily from the study of animal retroviruses and is based on morphology and disease association. The three classes

Nancy E. Mueller • Department of Epidemiology, Harvard School of Public Health, Boston, Massachusetts 02115. **William A. Blattner** • Institute of Human Virology, University of Maryland, Medical Biotechnology Center, Baltimore, Maryland 21201-1192.

of retroviruses are foamy viruses, oncoviruses, and lentiviruses. The foamy viruses are syncytia-forming viruses whose relevance to human disease is unknown. Oncoviruses are associated primarily with hemopoietic malignancies. Lentiviruses have traditionally been associated with chronic degenerative arthritis but also with aplastic anemia.

The HTLV-I is the prototype human retrovirus.[212,292] It is etiologically linked to a novel form of leukemia–lymphoma seen in endemic areas first described as a disease entity by Takatsuki and colleagues as adult T-cell leukemia/lymphoma (ATL).[257] Gessain, working with de-Thé and colleagues, reported an association between HTLV-I and an unusual neurological syndrome sharing some features with multiple sclerosis, known as tropical spastic paraparesis (TSP).[61] Osame *et al.*[206] reported that all patients with a spinal spastic paraparesis, which appears to be endemic to Kagoshima (Japan), were HTLV-I seropositive and termed the condition HTLV-I-associated myelopathy (HAM). Typical cases from other populations in which HTLV-I is endemic were identified, and it was agreed that HTLV-I associated TSP and HAM were identical.[221] Other groups of investigators, notably those led by Miyoshi[134] and Essex,[55] have suggested that HTLV-I has immunosuppressive properties, a suggestion that has been verified by subsequent observations.[193,247] Finally, an indirect leukemogenic role for HTLV-I in B-cell chronic lymphocytic leukemia has been postulated by Mann, Blattner, and colleagues.[171]

The second human retrovirus, HTLV-II, also reported by Gallo and colleagues, has not been clearly linked to a particular human disease.[113] The first isolate came from a man with an extremely rare disease known as T-cell hairy leukemia. Among the oncoviruses, these two HTLVs form a subgroup with the bovine leukemia virus.[30] The two can also be grouped with closely homologous primate viruses such as the simian T-lymphotropic virus type I (STLV-I) into the primate T-lymphotropic virus group (PTLV).[260]

3. Methods for Epidemiologic Analysis

3.1. Sources of Morbidity Data

Morbidity data on HTLV-I infected persons are sporadic and data on mortality are not available. This in part reflects the fact that the diseases associated with HTLV-I are rare (and those associated with HTLV-II undefined), have an apparently long latent period between exposure and outcome, and are difficult to classify. The difficulty in disease classification reflects the overlap between clinical diagnoses associated with HTLV-I and those of related diseases as well as the requirement for evidence of HTLV-I infection. For example, ATL presents as a T-cell malignancy that resembles other T-cell malignancies. The original cases from which the virus was isolated in the United States were clinically classified as mycosis fungoides and Sezary syndrome, respectively. Only after months of futile searching for additional positives among persons with cutaneous T-cell lymphoma was it realized that the virus was not the mycosis fungoides virus, and that the cases, in fact, had distinctive features shared with cases of ATL, first recognized in Japan.[21,23] A series of nationwide surveys conducted in Japan have provided a useful overview of the descriptive epidemiology of the prototypic cases occurring in that population.[254,262–264] However, variants may be less likely to be recognized. Further, since the manifestations and course of ATL may differ between endemic areas, as is seen in the Caribbean,[161] the identification of ATL cases in a given population may require the involvement of experienced clinicians and pathologists from endemic areas.

The neurological syndrome HAM–TSP is similarly protean in its manifestation and may be confused with some other myelopathies found in the tropics, such as those associated with ingestion of plant-related neurotoxins. Less specific outcomes such as HTLV-I uveitis present even greater surveillance problems. The full spectrum of HTLV-I-associated morbidity can only be defined in prospective cohort studies across a range of endemic populations.

3.2. Surveys

Seroprevalence surveys for HTLV-I antibodies have provided insights concerning the distribution of the virus infection in various populations. Large-scale population-based surveys have been undertaken in conjunction with research cohort analyses, blood bank screening, and archived serum banks. In general, there is considerable concordance between disease occurrence and virus seroprevalence.

3.3. Laboratory Diagnosis

3.3.1. Virus Isolation. The isolation of the first human retrovirus culminated a search for such an agent involving many decades of research. Critical to this process was the ability of researchers to grow cells of mature phenotype.[60] In the mid-1970s, while conducting research on colony-stimulating factors of myeloid cells,

Morgan and Russetti in Gallo's laboratory discovered a lymphokine that stimulated the growth of mature T-helper lymphocytes. This T-cell growth factor has subsequently been termed interleukin-2 (IL-2). The availability of this T-cell growth factor was a major breakthrough in the initial isolation of HTLV-I. Isolation of human retroviruses involves the application of highly sophisticated tissue culture and detection techniques.[229] The techniques for isolation of HTLV-I and the properties of this virus make its isolation far from routine. However, the techniques for propagation of the virus in tissue culture lend themselves to effective production of highly purified virus and have supported substantial advances in understanding the molecular biology of these viruses, including the complete genetic sequencing of HTLV-I.[234]

Virus isolation by cell culture for HTLV-I and -II involves the primary culture of separated lymphocytes from infected individuals or the coculture of these cells with either normal adult or cord blood lymphocytes that have been stimulated with mitogens and maintained with IL-2. Subsequent detection of the virus can involve assays for reverse transcriptase, viral antigen detection utilizing monoclonal antibodies, or electron microscopy (see Fig. 1). Because the tissue culture techniques involved are difficult to standardize, time-consuming, and produce isolation rates that vary between laboratories, this approach is generally unsuitable for epidemiologic applications.

HTLV-I targets several types of T cells including immature T cells, but the major target is the CD4+ lymphocyte, while CD8 cells are the primary target for HTLV-II, although CD4 cells are also susceptible.[97] As an alternative to virus isolation, the polymerase chain reac-

tion (PCR) has been applied with substantial implications for HTLV research. While initial approaches with this technique were plagued with the potential for contamination, recent advances have minimized this problem and have greatly facilitated the process for distinguishing the two virus types. By amplifying homologous regions of the viruses with a single primer pair and then employing restriction enzymes to produce distinguishable fragments based on size, a single PCR can distinguish HTLV-I from -II. This ability to certifiably distinguish HTLV-I from -II has allowed the establishment of panels of patient samples for use, in turn, in the development of serological assays for distinguishing HTLV-I and -II in epidemiologic studies.[53,79,156]

3.3.2. Serological and Immunologic Diagnostic Methods. A variety of techniques are utilized in the detection of antibodies to the HTLV-I virus for epidemiologic and diagnostic work. Virtually all HTLV-I-antibody-positive individuals (or carriers) are also virus-positive, reflecting the persistence of these viruses by virtue of their ability to integrate into the genome of target cells. One difficulty in assessing the specificity of and sensitivity of human retrovirus assays is the absence of an absolute gold standard. Despite this limitation, the predictive value of these assays appears to be quite good. The usual procedure is to employ one of the screening assays. Samples scoring "screen positive" are further tested to verify that the positivity is repeatedly present and then to employ one of several confirmatory approaches. The major antigens of the virus are shown in Table 1. These include the group specific core proteins (*gag*), the envelope (*env*), and the transactivating protein (*tax*).

The prototype assays for detection of HTLV-I widely used in the United States involve the enzyme-linked immunosorbent assay (ELISA) technique, utilizing whole disrupted virus.[34,279] These assays have generally performed well.[28,125] Because of the cross-reactivity between HTLV-I and -II, supplementary assays that distinguish between the two have been developed and are in

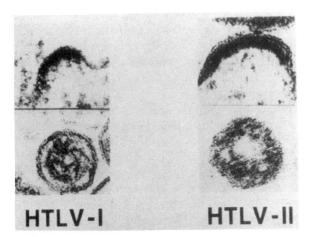

Figure 1. Thin-section electron micrographs of HTLV-I and -II viral particles. Mature virus are shown in lower panel and virus budding particles in the upper panel. Photographs courtesy of Dr. Robert C. Gallo.

Table 1. Major Viral Antigens of HTLV-I and Their Corresponding Gene Regions

Envelope[a]	*gag*	*tax*
gp-61/69 (precursor)	p55 (precursor)	p40
gp46	p24	
p21 (transmembrane)	p19	
	p15	

[a]Numbers refer to molecular weight × 10^3. gp, glycoproteins.

wide use in the United States where both infections are present.

In Japan, the particle agglutination assay (PA) is widely used for detection of HTLV-I antibodies. This assay, using purified viral antigens enriched for *env* proteins and coated on gelatin particles,[59] performs extremely well in that population.[104,127,138] The sensitivity for detection of seroconversion varies between assays.[35a,174]

Confirmation procedures are discussed in detail by Weiss.[279] The technique of Western blot, also known as immunoblot, permits visualization of the component antibody reactivities (see Fig. 2). A major difficulty with Western blot confirmation for HTLV is that antibody reactivity to the heavy *env* proteins is often not detected and a common criterion for positivity requires reactivity to at least one antibody for both core and *env* proteins. The *env* antibodies are selectively picked up by radioimmunoprecipitation assay, a more difficult procedure involving precipitation of labeled purified disrupted virus. For the PA method for HTLV-I, confirmation by immunofluorescence assay is often used[176]; titering is also useful as the initial confirmation step.

For epidemiologic study of HTLV-II, serological assays including a combination of whole virus and recombinantly produced peptides to detect virus infection are employed, some of which have enhanced the detection of HTLV-II.[25,37,42,147,165,220] The Western blot is the standard confirmatory test,[34,162] and recent Western blot technology uses synthetic peptides for confirming positivity and distinguishing virus type.[164,165] Prior to the availability of newer assays for distinguishing virus type, an algorithm based on HTLV-I Western blots was applied for epidemiologic studies. This algorithm was based on the observation that there are differences between HTLV-I and -II in reactivity to core antigens. Specifically, p19 reactivity is weak or absent in HTLV-II-positive individuals (see Fig. 2), while it is strong in HTLV-I-infected persons.[168,282] The specificity of this approach employing PCR-certified samples was approximately 90%.[282]

Competitive binding assays were used in early research. This technique has the potential for false positivity if the competing antigen is contaminated with the same proteins that are detected in the original assay (e.g., cellular membrane antigens) and is very time-consuming. Other assays for virus positivity include *in situ* hybridization where molecular probes are applied to potentially infected cells to search for RNA expression of viral genetic material. Hybridization analysis also has been used to detect integrated DNA of viral genomes in the cells of tumor tissue.

4. Biological Characteristics

Morphologically, HTLV-I and -II have a diameter of about 100 nm with a outer envelope and a roughly spherical core (Fig. 1). The internal core is comprised of the ribonucleoprotein complex of genomic RNA with viral reverse transcriptase and other internal structural proteins. The external lipid envelope has surface projections formed by viral glycoproteins. During budding of the mature virions from the cell surface, a portion of the cell membrane is incorporated in the viral envelope. Structurally, the total provirus genome of HTLV-I is composed of 9032 nucleotides, and HTLV-II has 8932 nucleotides with 65% homology including two identical sequences at each end, the long terminal repeats (LTR), as shown in Fig. 3.[20,234] Retroviral genes generally code for large overlapping polyproteins that are later processed into functional peptide products by the virally encoded protease.[274] The genome contains the three coding areas common to all retroviruses of *gag*, *env*, and the polymerase (*pol*) gene, which encodes reverse transcriptase, protease, endonuclease (ribonuclease), and integrase. HTLV-I has an

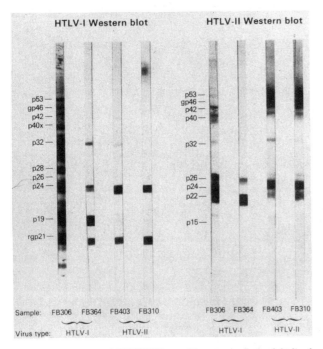

Figure 2. HTLV-I and HTLV-II Western blot patterns. In the right-hand panel, HTLV-I sera react with full complement of bands, while HTLV-II sera react with few bands and absent or markedly diminished p19 reactivity to HTLV-I viral lysate with added recombinant p21e antigen. In the left-hand panel, HTLV-II sera react strongly, while HTLV-I have decreased reactivity to HTLV-II lysate.

Figure 3. Genome structure of HTLV-I and -II.

additional open reading frame sometimes called the "X region," which codes for regulatory proteins including the *tax* (p40), *rex* (p27),[30] and several alternatively sliced, novel mRNAs including p21,[14] "*rof*," and "*tof*."[38] HTLV-II also codes for *tax* (p37) and *rex* (p26/p24) with similar function.[30] The two viruses differ in that HTLV-I preferentially infects CD4 cells and HTLV-II CD8 cells.

The *tax* protein is an astonishingly biologically active factor that is central to the life cycle of the virus. It is responsible for the transactivation of virus transcription and induces the expression of a variety of important host cellular genes including that for IL-2, and of its counterpart, IL-2Rα, which combines with IL-2Rβ to form the high-affinity IL-2 receptor on activated CD4+ T cells, the major target for infection.[72,155,240] It also transactivates the genes for granulocyte–macrophage colony-stimulating factor,[200] the c-*fos* gene,[58] an array of early response genes,[124] the human lymphotoxin gene,[268] and the parathyroid hormone-related protein gene,[278] while it *trans*-represses the β-polymerase gene.[109] Marriott *et al.*[175] have found that extracellular *tax* can induce IL-2Rα expression and lymphocyte proliferation at very low levels. In contrast, the role of the *rex* protein is that of a *trans*-regulator that modulates RNA processing in balance with *tax* activity.[293] The function of the other X region proteins is under investigation.

5. Descriptive Epidemiology

The epidemiologic pattern of HTLV-I infection is that of a chronic endemic infection of very low infectivity. The distribution of the viral infection is highly geographically restricted, showing a characteristic age-dependent seroprevalence curve with an excess among older women and with a high level of microepidemicity.[21,191]

5.1. Epidemiology of HTLV-I Infection

5.1.1. Prevalence and Incidence. Current data concerning the prevalence of HTLV-I infection are limited primarily to cross-sectional serosurveys. The distribution of HTLV-I within populations appears to be largely determined by ethnic and social patterns, since transmission

occurs primarily within family settings outside of subpopulations with a high risk of blood exposure.[190] Thus, estimates based on a limited number of samples can be quite unreliable and an overall population rate cannot necessarily be extrapolated to subpopulations and vice versa. Further, since seroprevalence is strongly age-dependent, the age structure of a population will influence its crude summary rate. This is most evident in comparing populations living under poor conditions with relatively few elderly persons, such as in parts of Africa, with an economically advantaged population with high life expectancy, such as in Japan. Selective sampling on the basis of sex will also bias estimates. Early data were further restricted because of serological cross-reactivity with HTLV-II as well as other problems in the assay systems. Nonetheless, at this time the general geographic boundaries of endemic populations have been drawn and broad estimates of seroprevalence can be made.[168]

The most thoroughly studied population is that of Japan, beginning with a nationwide survey of blood donor populations in 1982.[84] This comprehensive study determined that the highly endemic populations were restricted to the southern islands of Kyushu, Shikoku, and Okinawa (in the Ryukyu chain), corresponding to the areas where ATL occurred. Subsequent studies have confirmed these findings.[4,39,74,111,135,140,169,219,244,252–254] The highest population seropositivity of 18% has been reported for the Ainu people, an aboriginal population living in the northern island of Hokkaido.[102]

The prototypic Japanese strain is termed "cosmopolitan." However, the infection appears to be rare in the rest of Asia,[12,101,142,143,157,159,223,296] with the exception of the finding of 17 seropositives among 746 Aeta, an isolated aboriginal population in the Philippines.[101] Evidence of a variant HTLV has been found in a serosurvey of specimens from the Lake Lindu valley in Indonesia.[149] There have been a few reports of HAM–TSP in India.[6]

A second locus of a divergent strain of HTLV-I infection in the Pacific has been identified among isolated, aboriginal Melanesian populations[26,168,180,230,288,289]; this strain does not appear to be derived from the cosmopolitan strain.[287] The highest rate is reported among the Hagahai of 14%, a remote, recently contacted hunter–horticulturist group from the highland fringe of Papua, New Guinea.[287]

The major endemic area in the New World is in the Caribbean including Jamaica, Trinidad and Tobago, Martinique, Barbados, St. Lucia, Haiti, French Guiana, and the Dominican Republic.[2,24,29,40,44,183,195,196,217] The infection is virtually isolated among people of African descent in these populations.[8] The seroprevalence is relatively low, less than 5%. In Central and South America, some HTLV-I infection is also seen, primarily in areas adjacent to the Caribbean including Colombia, Venezuela, Surinam, Guyana, and Brazil.[41,181,294] Of note, the infection is seen both in people of African origin as well as in isolated Amerindian groups. It has been reported that there is an unusually high prevalence of HAM–TSP in Tumaco, an endemic island population of Colombia.[222]

It is likely that HTLV-I was carried to the Caribbean via the slave trade from West Africa. Seroprevalence surveys in Africa support this assertion, although early studies were plagued with problems in assays.[15] There is evidence that the infection is endemic at very low levels in Gabon, Cameroon, southern Chad, Equatorial Guinea, Guinea Bissau, Ivory Coast, Benin, Ghana, and Zaire.[16,45,46,50,95,159,163,210,275] The infection has also been reported among Pygmies from the Central African Republic[66] and among the population of the Seychelles, off the east coast of Africa.[153]

Surveys in the Middle East are limited. However, a focus of HTLV-I infection has been identified among Jews from Mashadi, Iran—a highly inbred group.[182,239] The origin of the infection in this group is unknown, but the area historically was a crossroads from the Far East to Europe and the infection may have been introduced by intermarriage with a migrant population from an endemic area.

In the United States and Europe, the scattering of infection that is found is essentially limited to migrants and their descendants from endemic areas,[19,23,32,91,154,211,284] without evidence of infection in indigenous populations. An exception is the confirmation of a very low prevalence of infection among Alaskan natives.[118] As would be expected, some persons using intravenous drugs have been found to be seropositive, although HTLV-II infection appears to be much more common in this group.[154]

5.1.2. Epidemic Behavior and Contagiousness. No epidemic of HTLV-I infection has been recorded. All seroconversions that have been documented are due to the established routes of transmission (see Section 6.1). Thus, HTLV-I appears to be a virus maintained in restricted human populations with an endemic pattern of occurrence. Recent analyses of historic serum collections from viral endemic areas support the concept that virus infection is persistent and patterns of occurrence, dating back over 15–20 years, appear unchanged. For example, in a recent analysis of HTLV-I seropositivity in sera collected in Barbados in 1972, rates of seropositivity, patterns of seroprevalence, and risk factors for positivity were indistinguishable from those of current studies.[217] This supports the concept that HTLV-I is an old virus with an endemic pattern of infection and persistent routes of trans-

mission, which have maintained the virus over a long period of time in geographically restricted populations. Other than the potentially new or heightened incidence of HTLV-I among drug abusers and the suggestion that rates of seropositivity among male homosexuals at high risk for HIV-1 in Trinidad may also have elevated rates of HTLV-I seroprevalence,[10] evidence for epidemic spread of the virus is not observed in any of the populations studied. This is consistent with the findings of the transmission studies that suggest that HTLV-I is an extremely difficult infection to transmit, being much less infectious than either HIV-1 or hepatitis-B virus, although sharing their major routes of transmission (see Section 6.1).

5.1.3. Geographic Distribution. Table 2 summarizes the geographic distribution of HTLV-I endemic populations as based on seroepidemiology. In parallel, the history of the evolution of HTLV-I is being reconstructed, based on nucleotide sequence analysis of isolates from these diverse populations. The HTLVs (like its fellow PTLVs) are distinguished by an exceedingly low level of genetic drift *in vivo*. This is reflected in the high level of genetic consistency among isolates maintained in cell culture from the same geographic area or within a related strain. Three major types of HTLV-I have been identified by phylogenetic analysis as summarized by Miura *et al.*[184]: the Melanesian strain from Papua New Guinea, the Solomon Islands, and Australian aboriginals; the Zairian strain seen in parts of Africa; and the cosmopolitan strain. The latter has three major subtypes: A, Ainu and other Japanese, Caribbean, Colombia, Chile, and India; B, Japan and India; and C, Caribbean and Africa. There is less than 0.01 nucleotide substitution per site between the Ainu isolate and several other type A isolates from the Carib-

Table 2. Patterns of HTLV-I Seroprevalence among Population Groups by Apparent Level of Endemicity

Highly endemic (≥ 15%)[a]
  Japan: Kyushu, Shikoku, Okinawa, Ainu Aborigines
  South Pacific: Papua New Guinea, Solomon Islands, Australian Aborigines
  South America: Brazil (Bahia)
Intermediately endemic (5–14%)
  Caribbean: Jamaica, Trinidad, Martinique, French Guiana
  West Africa: Gabon, Cameroon, Equatorial Guinea, Ivory Coast
  South America: Andes Mountain highland Indians
Low seroprevalence (1<5%)
  Caribbean: Barbados
  West Africa: Southern Chad, Nigeria
  America: Alaska Eskimos
  Philippines: Aeta Aborigines

[a]Among adults ≥ 40 years of age in the general population.

bean. In contrast, there is an estimated 8.2% difference between a recent Melanesian strain (Melanesia 1) and the Japanese cosmopolitan type A isolates.[65]

This internal fidelity of the virus in combination with the observation that HTLV-I is an ancient infection, as evidenced by its predilection for endemicity in isolated aboriginal populations,[102] provides the unparalleled opportunity not only to trace the geographic evolution of the virus through its phylogeny, but also to follow it through migrating human populations who brought it along as baggage, so to speak. These elegant ventures in biological archeology are converging in their findings, as represented by the dendrogram derived by Miura *et al.*[184] (Fig. 4). Of note is the interleaf of the STLV-I isolates along with their human counterparts. The fact that the Melanesian isolates come from sequestered populations who had no prior contact with Africans or Japanese suggest that this is an independent, rather than secondary strain of the virus,[287] and a similar isolate from an Australian Aborig-

inal has been reported.[11] Together, these findings suggest that STLV-I and HTLV-I originated in the Indo-Malay region, were disseminated to West Africa, brought to the Caribbean by slave trade,[67] and brought to Japan either independently or via Africa. The identification of isolated endemic Indian populations in South America suggest migration of infected Mongoloid populations to the New World. The geographic distribution of HTLV-I strains are shown in Fig. 5. The recent finding that isolates from India are closely aligned with those from Japan rather than Melanesia suggest alternatively that the point of origin lies in the South Pacific.[184] Of special interest is the fact that the STLV-I and HTLV-I isolates do not segregate independently, suggesting that horizontal transmission between nonhuman primates and humans "... continued to occur over long periods of time and on different continents."[228]

5.1.4. Temporal Distribution. There has been interest in whether the patterns in HTLV-I seroprevalence represent temporal changes in the virus distribution or are the stable result of endemic virus infection. Several authors have speculated that the age-dependent increase in seroprevalence results from a cohort effect. Since the great majority of population-based seroprevalence data available is cross-sectional, it has been difficult to address this question. The question of temporal distribution of the infection consists of two separate issues. The first concerns the stability of the *shape* of the unusual age-specific seroprevalence curve and the second the *level* of seroprevalence.

Concerning the first question, the evidence suggests that the unusual shape of the curve among adults—with a slow and parallel increase for both men and women from about age 20 to age 50 years, followed by a plateau among men but a continuing increase among older women (Fig. 6)—is a consistent feature of the infection. This is true in Japan,[135,243] the Caribbean,[197,217] South America,[267] and Africa.[16] There are not sufficient data from the endemic populations in the Pacific Basin at present to evaluate, although data from a relatively large study are consistent.[287] Further, in the population-based cohort study in Miyazaki Prefecture in Japan, the age and sex distribution of the seroconversions that have occurred during 9 years of follow-up are consistent with the age curve, that is, predominantly among older women.[242] This has also been reported in a population in the Yaeyama district in Okinawa, who were tested in 1980 and retested in 1989–1990. Seven seroconversions were observed, all in persons aged 40 or older, with the rate somewhat higher in women.[188] The fact that a similar curve is seen for monkeys naturally infected with their counterpart retrovirus,

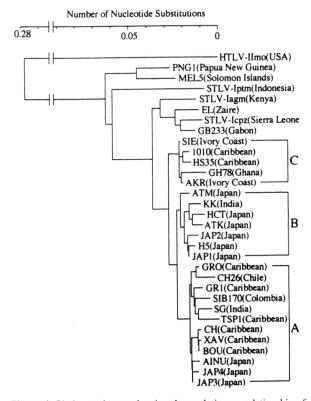

Figure 4. Phylogenetic tree showing the evolutionary relationship of HTLV-I/STLV-I in the world, including recently sequenced isolates from native Indian, Colombian, and Chilean individuals and from the Ainu in Japan. A scale on the tree is the estimated number of nucleotide substitutions per site and the horizontal branch length indicates the genetic distance. From Miura *et al.*,[184] with permission.

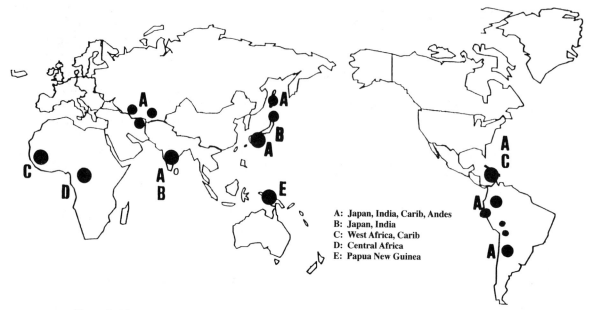

Figure 5. The geographic distribution of strains of HTLV-I. Adapted from S. Sonoda, with permission.

A: Japan, India, Carib, Andes
B: Japan, India
C: West Africa, Carib
D: Central Africa
E: Papua New Guinea

STLV-I, also argues for the validity of the observed curve.[94]

Concerning the second question, there is variation in endemic areas in terms of the level of seroprevalence. In those areas where the level is relatively high, the shape is more defined, a "mature" curve. It is likely that it requires several generations in an endemic population with a high level of intermarriage to achieve this shape.[190] There is evidence that the basal rate of the age curve, representing

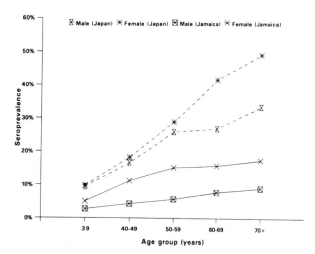

Figure 6. Age-specific seroprevalence of HTLV-I by sex in the Miyazaki cohort (Japan) and Jamaican food handlers.[195]

primarily perinatal infection, has been decreasing in the past few decades in Japan.[122,188,269] Oguma *et al.*[202] have reported that seroprevalence rates decreased between 1986 and 1990 in Kumamoto in population groups other than women aged 50 or more. Among Japanese migrants to Hawaii, there has been a decrease in prevalence with each successive generation.[19,91] These recent changes in Japanese population likely reflect secular changes in the practice and duration of breast-feeding and perhaps in the greater use of condoms. An effect of other factors such as better nutrition has also been proposed.

5.1.5. Age. One of the most consistent features of HTLV-I seroprevalence is the strong age-dependent rise in seropositivity displayed in Fig. 6. Typically, a rate of 3–7% is observed in equal proportion of males and females under age 20, followed by a slow increase for both males and females after age 20. This increase in seroprevalence among women continues throughout the rest of the life span. However, among men the rate plateaus or rises more slowly after age 50, resulting in significantly higher seroprevalence in older women compared with men. This pattern has been documented in virtually all studies in Japan and in most studies from the Caribbean basin, including the southern United States. Based on our knowledge of transmission (see Section 6.1), it appears that the relationship of seroprevalence to age reflects the cumulative effects beginning with mother-to-child transmission and extending into adulthood through sexual and blood exposure.

5.1.6. Sex. Among children and adolescents, the sex ratio is equal, reflecting perinatal transmission. During the reproductive years, the sex ratio can be equal as seen in the Miyazaki cohort or a female predominance as seen in the Jamaican food-handlers cohort (Fig. 6). This difference between the two populations may reflect different sexual behavior or the higher prevalence of transfusion with pregnancy in the latter cohort. The explanation for the diverging seroprevalence between the sexes at about age 50 is unclear. However, a similar curve is seen for seroprevalence of another sexually transmitted latent infection, herpes simplex virus type 2.[110] It may reflect an increased susceptibility of women to sexually acquired infections via changes in the genital tract following menopause. It may also reflect a relatively greater efficiency of transmission of the virus from men to women that is age-related.[112,242]

5.1.7. Race. A major determinant of HTLV-I seropositivity is race or, more precisely, ethnicity. On the cosmopolitan island of Trinidad where persons of Asian and African ancestry are equally represented, HTLV-I seropositivity is virtually confined to the black population.[8,183] In the United States, a similar pattern is observed for southern African-Americans and among Hawaiian-Japanese. The reasons for this clustering do not appear to be related to susceptibility factors but rather reflect patterns of transmission, particularly from mother to child, as well as assortative mating. For example, among migrant populations in Hawaii, risk for seropositivity is not related to being Japanese per se, but rather reflects the tendency of virus to cluster among persons with links to viral endemic areas of Japan. This high level of clustering within specific populations also underlines the very low level of transmission outside of the family.

5.1.8. Occupation. There is no documentation of any occupational association with HTLV-I other than among sex workers. In an early study of ATL cases in Japan, "outdoor occupations" were associated with seropositivity.[263] In Miyazaki Prefecture, seroprevalence was also found to be associated with farming and fishing occupations.[243,244] However, this likely reflects occupational correlates of endemic subpopulations. There is no report of infection acquired via occupational exposure to HTLV-I-infected patients or in the laboratory setting other than one case of accidental self-injection of blood from an ATL patient.

5.1.9. Socioeconomic Factors. Among endemic populations in the Caribbean area, there has been a consistent finding that HTLV-I carriers are of a somewhat lower social class than expected as indexed by measures of housing, hygiene, and educational level.[22,183,195,217,267]

The interpretation of these findings may reflect the ethnicity of infected progenitors, increased sexual exposure, or increased transfusion because of poorer health or higher parity. It does not appear that environmental factors per se are related to risk of infection.

5.2. Epidemiology of HTLV-II Infection

HTLV-II was first isolated in 1982 in the United States from a patient with a rare form of T-cell leukemia called hairy cell leukemia of T cells.[113] The close homology between HTLV-I and -II resulted in problems in detecting and distinguishing HTLV-II, because initially competitive binding assays, which are difficult to apply in epidemiologic studies, were used to document the high prevalence of HTLV-II in injection drug abusers (IDU).[218] Subsequently, the worldwide epidemiology of HTLV-II began to be characterized with the introduction of new Western blot techniques, sensitive recombinant peptide assays, and PCR techniques for detecting and distinguishing virus types.[168]

The first major focus of HTLV-II infection was identified among IDUs in the United States, United Kingdom, and Italy.[156,218,295] The high rates of HTLV-II in IDUs, even among African-American IDUs where HTLV-I might be expected to be elevated, has raised the possibility that HTLV-II is more efficiently transmitted by this route.[31,158] Retrospective surveys of IDU from the late 1960s confirm a high prevalence and argue that HTLV-II infection is preferentially transmitted by the intravenous route.[17] Prospective studies have indicated that IDUs are at increased risk for new infections, and certain practices that increase the potential for blood exposure are linked to these incident infections. Rates as high as 1 to 2% per year have been reported in some cohorts.[277]

The discovery that Amerindians residing in North, Central, and South America had high rates of HTLV-II infection provided the first evidence for a reservoir of infections.[80,90,146,160,216] The pattern of occurrence with marked geographic clustering resembled the pattern observed for HTLV-I.[216] Numerous tribes have been surveyed, but only some are HTLV-II positive. Those with substantial prevalence include the Seminoles in South Florida,[160] the Pueblo and Navajo in New Mexico,[90] the Guyami Indians in Northwestern Panama[146] (but not other Guyami enclaves in southwest Panama or various tribes in other parts of Panama[80,216], and some tribes in Colombia, Brazil, and Argentina.[18,98,170] Included in these surveys are collections of sera obtained years ago as part of anthropologic studies. In one such study out of a dozen tribes in various locales surveyed, HTLV-II infection was identified in only two very remote tribes from the interior of Brazil sharing a common linguistic pattern.[170]

Because of the origin of the Amerindian population through a series of migrations of peoples from Asia, there has been considerable interest in defining the pattern of HTLV-II infection among populations of this region. Surveys in various Chinese populations and among Ainu peoples in Russia have not identified any infected individuals. A recent report of HTLV-II among some hospitalized women in northern Mongolia raises the possibility that this may represent the "missing link" between Native American populations and their Asian ancestors.[73]

Another puzzling observation is the recent report of isolated instances of HTLV-II infection in some areas of West and equatorial Africa. One notable report described a high prevalence among a group of pygmies from Zaire with serological but not molecular evidence of HTLV-II infection.[69] Other investigators have reported HTLV-II infection in West Africa, and molecular characterization suggests a distinct variant.[47,63,64,69,70] The pockets of HTLV-II infection observed in European drug abusers appear to represent infections introduced from the United States via sharing of injection equipment among drug abusers.[295] The main foci of infection are in Italy and Spain.[56,273,295]

5.2.1. Molecular Epidemiology. HTLV-II virus isolates from various populations indicate that there are two basic families of HTLV-II that differ molecularly by approximately 2 to 4%.[73] The strain of virus isolated from Mongolia closely resembles those in one of the two families, a finding that is consistent with the hypothesis that HTLV-II came to the New World as part of human early migration.[73] The two recognized families may represent their evolution in the New World or may be the result of successive migrations of different forerunners of the Amerindians. In view of what is known about the slow rate of evolution of HTLV-I, it is implausible that HTLV-II evolved independently from a common prototype HTLV in the time since humans first migrated to the New World; this raises the possibility of a primate intermediary.[13] To date, HTLV-II has not been isolated from any primate species including their most likely reservoir, New World monkeys.

5.2.2. Demographic Features. The age-dependent rise in HTLV-II seroprevalence (Fig. 7) shows a pattern similar to that of HTLV-I, except that there are no differences between males and females at any age,[276] in contrast with the pattern observed for HTLV-I (Sections 5.1.4–6), where there can be a female excess postadolescence and a continued rise only among females after age 50. Among IDUs, elevated seropositivity rates in older

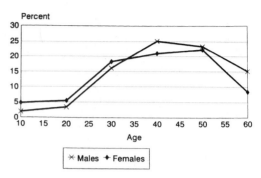

Figure 7. Age-specific seroprevalence of HTLV-II from population survey of Guyami Indians from Panama by sex. In contrast to HTLV-I, where female rates exceed male rates, in this survey, male and female rates overlap, showing a rising prevalence beginning in adolescence. From Vitek *et al.*[276]

age groups have been linked to the sharing of "eyedropper" injection equipment.[214] Immunosilent HTLV-II infection has not been documented.

6. Mechanisms and Routes of Transmission

6.1. HTLV-I

HTLV-I shares many of the same routes of transmission as HIV-1 but is considerably less infectious. Its low efficiency of transmission likely reflects the fact that HTLV-I is a highly cell-associated infection. The major routes of transmission include sexual, particularly male-to-female; perinatal, primarily through breast milk; and parenteral, via transfusion of whole blood (Table 3).

The evidence for sexual transmission initially came from the observation that there was significant clustering of HTLV-I seropositivity among couples in endemic populations.[19,91,112,191,253] There was a case report of apparent

Table 3. HTLV: Modes of Transmission

	HTLV-I	HTLV-II
Mother to infant		
Transplacental or during birth	Yes	Yes
Breast milk	Yes	Possible
Sexual		
Male to female	Yes	Yes
Female to male	Yes	Yes
Male to male	Probable	Unknown
Parenteral		
Blood transfusion of cellular products	Yes	Yes
Intravenous drug use	Yes	Yes

sexual transmission by a white Frenchman who had been diagnosed with HAM–TSP subsequent to transfusion from a HTLV-I-positive donor. Analysis of blood drawn from his wife 6 and 9 months after his diagnosis documented her seroconversion to HTLV-I.[71] Brodine *et al.*[27] has reported female-to-male transmission among a group of U.S. Marines married to seropositive Okinawan women. In a population in Okinawa that was rescreened after an interval of 9 years, 3 seroconversions occurred among 183 married couples. These included 1 of 17 seronegative wives of positive husbands and, from the 100 doubly negative couples, one woman with a history of transfusion and one man with no identified source of infection.[188]

The risk of sexual transmission has been shown to be related to the viral status of the exposing partner. In cross-sectional analyses, Ho *et al.*[91] reported that among the 130 couples included in the Japanese migrant study population in Hawaii, the relative risk of women to be HTLV-I seropositive increased in parallel to their husbands' titer. Kashiwagi *et al.*[121] reported that among 106 couples in Okinawa in which the husband was positive, 83.7% of women whose husbands had antibody to the *tax* protein compared with 77.2% among those whose seropositive husbands were anti-*tax* negative. In the prospective follow-up of 534 married couples in the Miyazaki cohort study, Stuver *et al.*[36,242] documented seven seroconversions over 5 years. All but one occurred within discordant couples; that is, where only one partner was seropositive. The seroconversions occurred primarily among postmenopausal women whose husbands had high antibody titer by PA, had antibody to the *tax* protein, and were older than 60 years. (Two of the seroconversions were among men whose wives were carriers.) On cross-sectional analysis of all couples, greater parity was also independently associated with seropositivity among the wives; for men, only longer length of marriage was also predictive of positivity. Continuing follow-up of this population for three more years has added four additional seroconversions: three women and one man each in a discordant couple; the same pattern of risk factor holds (S. O. Stuver *et al.*, personal communication).

These findings are consistent with earlier reports of high maternal antibody titer and antibody to *tax* as predictors of transmission from HTLV-I-positive mothers to their infants.[81,232] Both of these viral markers are positively associated with proviral load[103,237] and the expression of the *tax* protein is part of viral replication. The finding that seroconversions are occurring at older ages, especially among women, implies that sexual transmission occurs at a very low frequency, after many years of exposure, and that postmenopausal women are at increased risk of infection, or alternatively that older men become more infectious.

A number of studies have evaluated the role of sexual behavior and of sexually transmitted diseases (STD). In a study conducted in a STD clinic in Jamaica, there was a correlation between number of lifetime sexual partners among women and seropositivity, while for men, the presence of genital ulcers was linked to infection. Other measures of sexual activity were also associated with HTLV-I seropositivity.[196] Several studies have documented a higher than expected prevalence of HTLV-I antibodies among female sex workers in endemic populations; as in other populations, the prevalence increases with age. Hyams *et al.*[96] prospectively followed a female prostitute population in Lima, Peru, over a 3-year period with 17 seroconversions observed, or 1.6% per person-year. Seropositivity was associated with both demographic factors (being Mountain Indian or being born in the mountains) and sexual risk factors. In a similar study conducted in a different STD clinic in the same city, risk was associated with length of prostitution, the presence of an STD, and the lack of consistent condom use; risk was higher (but not significantly so) among women born in highland areas.[68] In a recent study of 100 homosexual or bisexual men from Trinidad, HTLV-I positivity was associated with a larger number of partners and duration of homosexuality, even when adjusted for age.[10] However, a similar study in Jamaica did not find a significant increase among homosexual–bisexual men nor association with any risk factors.[194] Overall, these findings provide some evidence that STDs may contribute to the risk of HTLV-I infection; however, in general, the association with sexual behavior is much less strong than is seen for HIV-1 infection.

A second route of transmission is from mother to child, primarily through prolonged breast-feeding. The concordance of antibody status between mother and child was clearly documented in Okinawa in an early study by Kajiyama *et al.*[112] These investigators identified 719 seropositives in a population survey and then traced and obtained blood samples from their children (including adults). In analysis on 434 children for whom both parents' status was known, only those children whose mothers were seropositive were themselves seropositive (86 of 352), whereas of the 82 children whose mothers were negative and fathers positive, none were themselves positive.

Since HTLV-I is a highly cell-associated infection with no evidence of cell-free viremia, the mechanism of mother-to-child transmission appears to be primarily through exposure to infected lymphocytes in breast milk.[172,196] Experimental studies have also documented the capacity of breast milk from antibody-positive mothers to transmit the virus to the offspring of seronegative Japanese macaque birth mothers.[129] The risk of transmission appears to increase with length of nursing. In a prospective study of 277 children who were born to seropositive mothers in Kagoshima and followed for more than 24 months, the seroconversion rate among children breast-fed for more than 6 months (1 of 3) was greater than that in children nursed for shorter periods (1 of 23). Among 151 bottle-fed babies, 9 (6%) seroconverted. Similar proportions were found among 210 exposed children based on retrospective data: 3 of 67 breast-fed for shorter intervals, 19 of 136 breast-fed for longer intervals, and 0 of 7 bottle-fed. In the combined data, the relative risk of seroconversion among breast-fed children nursed for more than 6 months was 3.7, $P = 0.02$.[256] Similar findings were reported by Hirata et al.,[85] where 7 of 61 children nursed for less than 7 months seroconverted compared with 11 of 36 children nursed for longer periods. (Only 2 of 39 children nursed less than 4 months seroconverted.) Wiktor et al.[281] also found in Jamaica that the risk for HTLV-I positivity in breast-fed children of carrier mothers who were nursed for more than 6 months was 3.2 times that in infants nursed for a shorter time.

Taken together, these findings suggest that there is a window of protection afforded from maternal antibody for children exposed via breastmilk. In the rabbit model for milk-borne transmission of HTLV-I, passive immunization has also been shown to prevent infection.[233] Perinatal transmission can occur independently of breast-feeding. The reported rates vary between 0 and 13%. The mechanism of this is unclear but does not appear to reflect intrauterine infection as reflected by cord blood PCR results.[123] Maternal viral load as evidenced by culture assay,[128] high antibody titer,[81,281] and the presence of antibody to the *tax* protein appear to be predictive of mother-to-child transmission.[85,116,232] The relative risk associated with anti-*tax* in these three studies varies from 2.5 to 8.4. Once a mother has a seropositive child, the observed risk to subsequent children is relatively high.[82]

Based on these observations, a prefecture-wide intervention program to interrupt transmission by breast milk is now underway in Nagasaki.[82] Perinatal transmission sets the baseline for seroprevalence in each birth cohort since there is no evidence of other sources of transmission among children.[144] Thus perinatal transmission creates the pool of carriers for subsequent sexual transmission. The observed decreased in seroprevalence among younger Japanese noted above, in parallel to a temporal decrease

in the length of breast-feeding,[254] argues for the gradual diminution of the infection in coming generations in Japan.

Transfusion-mediated transmission of HTLV-I has been well documented. In a classic study by Okochi and colleagues, 60% of recipients of one to two seropositive units seroconverted.[204] Only recipients of whole blood seroconverted. This observation supports the concept that HTLV-I is closely cell-associated and that spread of cell-free virus, if it occurs, is probably rare. Among recipients of positive whole blood, seroconversion occurred within 4–8 weeks of transfusion. That transmission requires exposure to infected lymphocytes is also supported by findings that HTLV-I seropositivity is not elevated among Japanese hemophiliacs who receive plasma product concentrate involving thousands of donors.[76] This finding has been corroborated in the Jamaican population.[24]

Blood donation screening is now underway in several endemic areas as well as in the United States and parts of Europe. The efficacy of this intervention in preventing transfusion-acquired infection has been demonstrated in Japan[115]; however, its effect on the overall population infection rates is likely to be minor.

Infection via other routes appears to be extremely rare. Infection with HTLV-I through needle sharing among drug abusers in the United States occurs, but HTLV-II infection by that route is much more common.[17] There is no evidence of transmission among children.[144] Health care and laboratory workers who are exposed through a needle stick of skin or mucous membrane exposure have never been documented to seroconvert. There is a single report of a Japanese health care worker who seroconverted following a "microtransfusion" when a loaded syringe of HTLV-I-infected blood punctured his foot.

6.2. HTLV-II

The possible modes of HTLV-II transmission are listed in Table 3. Evidence for mother-to-child HTLV-II transmission is circumstantial. While HTLV-II has been detected in breast milk of seropositive mothers,[77] prospective studies of HTLV-II transmission from mother-to-child have not been performed. HTLV-II-seropositive intravenous drug-using mothers who bottle-fed their infants have not transmitted the virus to their offspring.[117] However, among the Guyami Indians of Panama, the 1 to 2% prevalence among preadolescent children is similar to that reported in Jamaica where HLTV-I is prevalent. As seen in Fig. 7, rising rates of HTLV-II infection are seen in the adolescent and postadolescent period.[276] In a cross-sectional analysis of households, children are more likely to be positive when the mother is seropositive than when she is negative.[148,276]

Data supporting sexual transmission of HTLV-II are also circumstantial. If a husband was positive, there was a high rate of concordance for HTLV-II seropositivity for the female spouse in studies performed in Panama and New Mexico, a finding consistent with sexual transmission.[89,276] Studies of female prostitutes in the United States have documented high rates of HTLV-II, but the majority of positives are also IDUs.[126]

The strongest data support a role for parenteral transmission of HTLV-II. Transfusion transmission of HTLV-II is well documented.[88,246] Over half of HTLV infections among blood donors are due to HTLV-II, and this rate might be higher since current HTLV-I-based screening tests are not totally sensitive for detecting HTLV-II infections. Look-back surveys of recipients of known HTLV-II-positive units of blood have documented that approximately half of recipients seroconverted.[49] Longer shelf life of blood is associated with diminished transmission, a finding identical to that for HTLV-I, supporting the concept that the major source of infectious virus is infected lymphocytes.

Intravenous drug use accounts for the majority of infections in the United States. For unexplained reasons, HTLV-II appears to be much more readily transmitted among IDUs than is HTLV-I, as is evidenced by the very high prevalence of HTLV-II compared with HTLV-I in African-American IDUs. This may indicate biological differences between HTLV-I versus HTLV-II in the efficiency of transmission in the drug abuse setting.[156,284] Risk factors for HTLV-II transmission include sharing of drug abuse "works," especially with an eyedropper syringe.[214] Since this injection device was supplanted in the 1960s with the disposable syringe, exceptionally high rates of seropositivity in older drug abusers may have resulted from the substantial blood exposure through this earlier practice. More recently, blood exposure through the practice of booting as a means of distributing drug among needle sharers has been associated with recent seroconversions.

7. Pathogenesis and Immunity of HTLV-I

The natural history of HTLV-I infection is beginning to be understood. As with other latent infections, the risk of virally induced disease appears to be related to age and route of infection, as well as the immune competency of the host. There is no evidence that the various disease outcomes of the infection reflect specific sequence motifs of the various HTLV-I strains.[43,52,131,136] HTLV-I preferentially infects the activated T-helper cell (CD4$^+$/CD25$^+$), which is the phenotype of classic ATL. However, other cells including CD8$^+$ T cells, monocyte–macrophages, microglial cells, and dendritic cells can also be infected.[1,167] These different cell populations likely play a role in the various disease outcomes.

The overall disease risk associated with the infection appears to be generally rather low but not inconsequential. Based on current knowledge, the cumulative lifetime risk of any HTLV-I-related disease appears to be between 5 and 6%. This is considerably less than that estimated for chronic hepatitis B virus carriers in the United States of 37%.[57] In this ancient human infection, a mutual accommodation between the host and virus appears to have been achieved, with little acute pathophysiology evident; this contrasts sharply with the highly pathogenic effects of the HIV-1, a newly introduced human retrovirus.

There is no clinical syndrome recognized in relationship to acute seroconversion to HTLV-I. The length of time between exposure to infection and seroconversion can vary from 1 to 2 months, as seen in transfusion studies,[204] to 18 months, as seen in perinatal studies.[198] An early and serious disease that can follow seroconversion is HAM–TSP. The onset may be as short as 18 weeks,[71] with the median time between transfusion exposure to HTLV-I and onset being 3.3 years.[207] However, the lifetime cumulative risk of HAM–TSP among carriers is low; based on Japanese data, Kaplan et al.[120] estimated it to be 1 in 400. The mean age at onset is 43 years and there is a 3:1 predominance of women to men.[207] Thus risk of HAM–TSP appears to be related to adult exposure, particularly via transfusion and especially among women.

ATL, on the other hand, occurs after a long latency. This is an aggressive, lymphoproliferative malignancy of T lymphocytes with monoclonally integrated HTLV-I genome.[291] In Japan, the mean age of diagnosis of 58 years with a male excess of 1.9:1. The modal age of diagnosis is the sixth decade for men and the seventh decade for women.[255] The cumulative risk of ATL among carriers based on Japanese data is about 5%.[137,265]

Tajima and Hinuma[250] have proposed that perinatal infection is an important risk factor for the development of ATL. This proposal is based on the unimodal distribution of cases (Fig. 8), suggesting an early common age of exposure,[48] the sex ratio being much more like that for seroprevalence among children than for the female predominance seen in adults. Indirect evidence to support the modifying role of age at infection and risk of HAM–TSP and ATL comes from Bartholomew et al.[9] In an ongoing

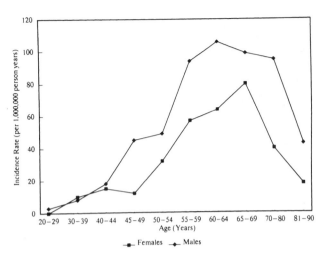

Figure 8. Age-specific incidence rate of ATL in Kyushu Prefecture, 1986–87, by sex. Adapted from Tajimi *et al.*[255]

study in Trinidad/Tobago, they found that all of 20 mothers of ATL patients were HTLV-I seropositive compared with only 3 of 10 mothers of HAM–TSP patients. This finding is consistent with the hypothesis that ATL is related to perinatal infection and HAM–TSP with adult infection.

This follows the analogy provided by the natural animal model of feline leukemia where early age at infection with the feline leukemia virus is associated with an increased risk of malignancy; a similar analogy can be drawn from hepatocellular carcinoma and hepatitis B virus.[54,192] It is likely that oncogenesis in ATL follows a protracted stepwise course beginning with a polyclonal expansion of HTLV-I infected CD4+/CD25+ cells and evolving to a monoclonal expansion and progression. Since the site of integration of monoclonal provirus is essentially random in ATL and there is no evidence of oncogene activation, the role of the *tax* protein is likely key to development of malignancy. Of note, the pX region that codes for *tax* is always included in the integrated genome in ATL tissue.[139,203]

An apparent preleukemic or high-risk state has been identified based on the presence of circulating abnormal lymphocytes on peripheral blood smear. These cells (sometimes called "flower cells") are very similar to the ATL leukemia cells, are CD4+/CD25+, and are found in a small proportion of healthy carriers. These are thought to contain HTLV-I provirus.[178,179] Of interest, very similar cells are seen in STLV-I-infected monkeys.[201] Yamaguchi *et al.*[285] reported on five HTLV-I carriers who presented with skin lesions and a low level of abnormal lymphocytes (0.5–2%). Of these, two developed ATL after intervals of 5 and 13 years and four had concurrent

evidence of immune suppression. The same group[286] later described 15 individuals with polyclonally integrated HTLV-I provirus, immune suppression, and a low level of abnormal lymphocytes (≥0.2%). During 3 years of follow-up, one carrier progressed to ATL. Kinoshita[130] followed 18 carriers with high levels of abnormal lymphocytes (10–40%) and T lymphocytosis; 14 had monoclonally integrated provirus. Of 13 followed from 1 to 7 years, three developed ATL, three had persistent T lymphocytosis, and in seven there was a regression in the level of abnormal lymphocytes and their lymphocyte count normalized. Tomonaga *et al.*[266] reported that of 108 carriers with abnormal lymphocytes identified in a population screening in Nagasaki, monoclonal provirus was present in 12, for an estimated prevalence of 2–3% among all carriers. They then followed a total of 30 carriers with monoclonal provirus and abnormal lymphocytes for a median of 50 months; of these, three developed ATL.

Part of the natural history of this infection is some compromise of cellular immunity, particularly in older carriers. In the Miyazaki cohort, delayed-type hypersensitivity of purified protein derivative (PPD) was measured on 126 consecutively seen subjects of whom 31% were HTLV-I seropositive. The carriers were more likely to be low or nonresponders to the challenge than the seronegatives, with a relative risk of 6.3. The effect was 12-fold among those aged 60 years or more.[247] This finding was confirmed in a subsequent study of a hospitalized elderly population in Miyazaki City using both PPD and passive hemagglutination (PHA) as challenge antigens,[193] and in a much larger sample of the cohort.[280a] Additional evidence of diminished cellular immunity among HTLV-I carriers is seen in studies of antibody response the Epstein–Barr virus (EBV). Imai and Hinuma[99] evaluated the antibody profile of ATL patients, asymptomatic carriers, and seronegatives. They found that the ATL cases had a pattern consistent with immunosuppression[92] and an intermediate pattern among the carriers in comparison to those uninfected with HTLV-I.

The mechanism underlying the immune suppression is unknown but may be mediated by the cytokines secreted by HTLV-I-infected T lymphocytes. It has been reported that infected cells *in vitro* release immunosuppressive factors or lose specific cytotoxic activity.[213,238,259] Further, lymphocytes from asymptomatic HTLV-I carriers spontaneously proliferate, suggesting antigen activation.[141] In addition, these cells secrete significantly elevated levels of IL-6 and tumor necrosis factor-alpha (TNF-α) into culture supernate,[151] as well as leukemia inhibitory factor.[152]

Among asymptomatic carriers, antibody levels in serial samples over a 10-year period were persistent and at stable titer.[23] Serum antibodies are predominantly to the viral *gag* proteins, although antibodies to the *env* and other antigens are detectable as well. Titers of antibody are highest in patients affected with HAM–TSP, somewhat lower in ATL, and generally lowest in asymptomatic carriers.

In HTLV-1 carriers in the Miyazaki cohort, the relationship among the various viral markers including antibody titer by the PA assay, antibody to recombinant *tax* protein, proviral level by semiquantitative PCR, as well as the prevalence of abnormal lymphocytes determined on peripheral blood smear has been evaluated. There is a strong positive association with the prevalence ($\geq 0.6\%$) of abnormal lymphocytes and a high level of proviral DNA; the relative risk was 12 or more in men of all ages and younger women. Among the women ≥ 55 years, the association was twofold. Of interest, men were twice as likely to have high proviral levels and to have abnormal lymphocytes.[248] There is a strong correlation between antibody titer and proviral level[103,181]; there is also a strong correlation between antibody titer and the presence of anti-*tax*.[237] However, neither antibody titer nor the presence of anti-*tax* are independent predictors of the presence of abnormal lymphocytes. In this regard, antibody does not appear to play a directly protective role for a latent viral infection. Rather, elevated antibody levels appear to reflect the amount of viral protein seen by the host in relation to viral reactivation. These higher titers may also reflect the reciprocal relationship between the expression of CD4 Th1 and 2 arms of the immune response,[189] and as such represent a risk state for virally associated diseases, parallel to that seen for the EBV.[92]

There is some evidence that genetic factors may influence the host response to HTLV-I. In Kagoshima there appear to be two ethnic groups from southern Kyushu who differ in their response to HTLV-I, using an *in vitro* culture assay ("high responders and low responders"), and these differences seem to segregate with HLA haplotypes. The high-responder phenotypes are reported to be associated with risk of HAM–TSP and the low with ATL.[272] The latter phenotype is also more common in asymptomatic carriers[271] and, as such, may reflect an originally infected ethnic subgroup in which some immune adaptation has occurred. Further, this genetic association is also seen in endemic populations in other parts of this world.[241]

8. Patterns of Host Response

8.1. HTLV-I

8.1.1. ATL. The discovery that HTLV-I is the etiologic agent of ATL represents the remarkable convergence of scientific discoveries made half a world away from each other.[60] ATL was first reported by Takatsuki and colleagues.[257] The epidemiology of ATL and its clinical features have been carefully described in a series of reports from the T- and B-Cell Malignancy Study Group in Japan, published in the 1980s.[255,262–264]

Extensive molecular studies have documented the universal presence of the monoclonally integrated HTLV-I provirus into the genome of the tumor cells; the restriction enzyme patterns seen are consistent with random integration. No pattern of oncogene overexpression has been identified. As in many malignancies, alteration of the expression of p53 is seen in some cases.[35,227] No consistent karyotype is found; however, almost all of 107 cases reviewed had clonal chromosomal abnormalities, with translocations of 14q32 or 14q11 the most commonly seen.[236]

ATL is characterized as an acute, mature T-cell lymphoproliferative malignancy with a rapidly progressive course. Four clinical subtypes have been characterized in Japan. The diagnostic criteria for each are shown in Table 4.[236] The major clinical features of this syndrome include leukemia, hypercalcemia, bone marrow involvement, generalized lymphadenopathy, cutaneous involvement, hepatosplenomegaly, and occasionally lytic bone lesions. Histopathologically, cells of this tumor are described as having a diffuse pleomorphic morphology of phenotypically mature T lymphocytes, usually of the activated helper (CD4$^+$/CD25$^+$) subtype.[108,263] Patients generally present at an advanced stage of clinical disease. Opportunistic infections are often present at diagnosis and contribute to the rapid course of the disease.

In the Caribbean, ATL cases have an earlier age peak and a more lymphomatous pattern. Smoldering ATL is not seen.[161,173] Of note, ATL diagnosed among migrants from the Caribbean in the Brooklyn, New York, population had an older age peak than that seen in their native community.[280] Thus, environmental factors appear to influence the onset and course of the disease; likely factors include nutrition and the presence of other infections.

8.1.2. Other Malignancies. Two cases of B-cell chronic lymphocytic leukemia have been identified in Jamaica in which the immunoglobulin of the tumor cells react to specific HTLV-I antigens but no viral genome is detected, suggesting an indirect role for the virus.[171]

There are a number of reports of other malignancies occurring in conjunction with HTLV-I infection, but few systematic data are available. Of 394 cancer patients seen in Kumamoto who did not have a history of transfusion, 15.5% were HTLV-I seropositive. This was significantly higher than that expected based on population data. There was no significant correlation with specific types of malig-

Table 4. Diagnostic Criteria for Clinical Subtype of ATL[a]

	Smoldering[b]	Chronic	Lymphoma	Acute
Anti-HTLV-I positive	+	+	+	+
Lymphocyte (× 10⁴/l)	< 4	≥ 4[c]	< 4	•
Abnormal T lymphocytes	≥ 5%	+[d]	≤ 1%	+[d]
LDH	≤ 1.5N	≤ 2N	•	•
Corrected Ca (mmol/liter)	< 2.74	< 2.74	•	•
Histology-proven lymphadenopathy	No	•	+	•
Tumor lesion				
Skin	••	•	•	•
Lung	••	•	•	•
Lymph node	No	•	Yes	•
Liver	No	•	•	•
Spleen	No	•	•	•
CNS	No	No	•	•
Bone	No	No	•	•
Ascites	No	No	•	•
Pleural effusion	No	No	•	•
GI tract	No	No	•	•

[a]Adapted from Shimoyama.[236]
[b]N, normal upper limit. •, No essential qualification except terms required for other subtype(s). ••, No essential qualification if other terms are fulfilled, but histology-proven malignant lesion(s) is required in case abnormal T lymphocytes are less than 5% in peripheral blood.
[c]Accompanied by T lymphocytosis (3–5 × 10⁴/l or more).
[d]In case abnormal T lymphocytes are less than 5% in peripheral blood histology-proven tumor lesion is required.

nancy.[5] Miyazaki et al.[185] compared the HTLV-I seroprevalence among 226 women with gynecologic malignancies in Kumamoto to that among women in the general population. There was a significantly increased risk associated with HTLV-I positivity for cervical cancer in women aged less than 60 years (relative risk, 2.9) and for vaginal cancer (relative risk, 7.4). Given the immune suppression seen in carriers, it is not surprising that opportunistic malignancies related to other oncogenic viruses should occur. Since both cervical and vaginal cancer are associated with human papillomaviruses, they fit this role. There is also a scattering of reports of Kaposi's sarcoma occurring in conjunction with HTLV-I infection.[114,298]

8.1.3. HAM–TSP. The second major disease in which infection with HTLV-I plays a causal role is HAM–TSP, a slowly progressive myelopathy affecting primarily the pyramidal tracts and, to a slight extent, the sensory system. Like multiple sclerosis, this is primarily a demyelinating disease, but unlike multiple sclerosis, there do not appear to be major affective manifestations of the syndrome, the pathological picture is somewhat different, and "flares" of disease activity are not prominent. In general, the natural history of disease is decribed as a slowly progressive degenerative syndrome initially manifested as paraparesis, urinary incontinence, impotence, and difficulty with muscle strength or gait. Over a number

of years, there is a slow progression in symptoms until patients are virtually immobilized. Pathological features show demyelination with often marked lymphocytic infiltration of the meninges and perivascular cuffing. The geographic occurrence of cases coincides with that of HTLV-I infection.

The association of some cases of what was then known as TSP with HTLV-I was first reported by Gessain et al.[61] in 1985, when they noted that 10 of 17 cases in Martinique were HTLV-I seropositive. The next year, Osame et al.[205] reported that similar cases occurring in Kagoshima and throughout Japan[208] were also HTLV-I-positive, and they termed the syndrome HAM. Subsequently, it was agreed that HTLV-I-associated TSP and HAM were identical.

The age of onset in younger (mean age in Japan, 43 years) than those for ATL and there is a strong female predominance (3 : 1). This observation is in accord with the generally higher risk among women who are more prone to autoimmune or immune-mediated diseases. Unlike ATL, risk of HAM–TSP is clearly associated with transfusion of HTLV-I-contaminated blood,[71,206,209] within an interval reportedly as short as 18 weeks. The introduction of blood screening for HTLV-I in Japan in 1986 led to a substantial decrease (16%) in newly reported cases of HAM–TSP in the next 2 years.[207] Yoshida[290] and Imamura[100] have demonstrated that the virus isolated from HAM cases and from ATL cases are identical. Since most cases (85% in Japan) occur after childhood, it appears that HAM–TSP is primarily related to primary HTLV-I infection during adulthood, and resembles a hyperimmune response to the retrovirus.[93] Specifically, patients have high levels of cytotoxic T lymphocytes in the cerebrospinal fluid as well as in the peripheral blood; these cells recognize products of the pX region of the HTLV-I.[107] HAM–TSP patients have high antibody titers against the virus, especially to the *tax* protein, and high proviral load.[62,132,133,150] On autopsy, there is evidence of chronic diffuse inflammation in both the gray and white matter.[105] As noted by Usuku *et al.*,[272] there appears to be an association with HLA haplotypes with risk of HAM–TSP related to "high responders" to the infection.

8.1.4. Other Associated Diseases. HTLV-I is also associated with a variety of autoimmune-type syndromes. The evidence linking these diseases to the virus is generally weaker than for ATL and HAM–TSP and the rate of occurrence among carriers is undefined.

In Jamaica, LaGrenade *et al.*[145] described the infective dermatitis syndrome among "failure-to-thrive" children born to HTLV-I-positive mothers. These children experience refractory generalized eczema and infection with saprophytic staphylococci and streptococci, which must be controlled with continuing antibiotic therapy. Infectious dermatitis is generally first evident in early childhood and can persist to adulthood; cases have also been diagnosed in adolescence. The subsequent development of ATL has been documented.[75] Of note, this syndrome is not seen among HTLV-I-seropositive children in Japan, suggesting that concomitant poor maternal health seen in the Caribbean may contribute to its occurrence.

The virus is also associated with a chronic inflammatory arthropathy in which proviral DNA has been detected in synovial fluid and tissue cells.[231] Uveitis in younger carriers has been described in which infiltrating lymphocytes with HTLV-I provirus have been isolated from the anterior chamber of the eye.[186] HTLV-I carriers appear to be more likely to have chronic renal failure, perhaps reflecting damage from immune complexes.[74,270] Other diseases reported to be associated with HTLV-I infection include polymyositis of skeletal muscle,[187] T-lymphocyte alveolitis,[245] and Sjögren's syndrome.[51] Carriers also appear to be at increased risk of opportunistic infections such as *Pneumocystis carinii*[249] and *Strongyloides stercoralis*.[199] Iwata *et al.*[105a] has reported a significantly higher mortality rate among HTLV-I carriers (relative risk, 1.89 in men and 1.94 in women) in the Nagasaki area over a mean follow-up of 5.3 years.

8.2. HTLV-II

Although epidemiologic proof that HTLV-II causes disease has not been developed, some candidate conditions are listed in Table 5. Since HTLV-I has been linked to T-cell malignancies, the relationship of HTLV-II to such lymphoproliferative conditions has been highlighted by the fact that the original isolation of the virus was from a patient said to have T-cell hairy cell leukemia.[113] Subsequent surveys have found no consistent association of similar cases with HTLV-II.[224–226] Large granulocytic cell leukemia, also a T-cell malignancy with an natural killer cell phenotype, has been reported in a few cases to be HTLV-II positive, but larger surveys of this syndrome have not confirmed an association.[78,166] In New Mexico, with a large Amerindian population, there was no detectable increase in leukemias and lymphomas in a population with high rates of HTLV-II infection.[86] A case of HTLV-II-associated mycosis fungoides has been reported,[297] and dermatopathic lymphadenopathy with eosinophilia was noted in some IDUs coinfected with HTLV-II and HIV-1.[119] *In vitro* spontaneous lymphocyte proliferation, a marker for HTLV-I infection, has also been reported in HTLV-II carriers.[283]

HTLV-II-associated neurological diseases resembling tropical spastic paraparesis/HTLV-associated myelopathy (TSP–HAM), sometimes with ataxic symptoms, have been observed,[87,106] including a few linked to blood transfusion; the oligoclonal bands in the cerebrospinal fluid have specific reactivity for HTLV-II, again mirroring the pattern for HTLV-I and TSP–HAM. A survey of medical conditions in relationship to HTLV-II among Guyami Indians in Panama identified HTLV-II-seropositive cases of adult polyarthritis, eczema of the skin, and asthma, and childhood glomerular nephritis (R. Guisti and F. Gracia, unpublished data). The etiologic relationship of these cases to HTLV-II remains to be confirmed. A possible link of HTLV-II to pulmonary disease in IDUs has been suggested in one study (S. H. Weiss *et al.*, unpublished data). Despite these attempts, no concrete disease association has been detected and HTLV-II remains a true orphan virus without clear disease association.

Table 5. Suspect HTLV-II-Associated Diseases

Glomerular nephritis in children	$+^a$
T-hairy cell/large granulocytic leukemia	++
HAM/TSP	++
Eczema	+
HTLV-associated arthritis	+/−
Asthma	+

^{a}+, Possible association; ++, likely association; +/−, weak association.

9. Control and Prevention

9.1. HTLV-I

The major primary preventive measure currently in place involves screening of blood to prevent transfusion transmission, in some endemic areas, as well as in non-endemic areas, including the United States. Guidelines for counseling of seropositives have been developed by the Centers for Disease Control and Prevention.[33] The cost-effectiveness of screening in low-prevalence areas is under debate.[3] Perinatal screening to prevent postnatal transmission via breast-feeding is being assessed in Nagasaki. There is some evidence that breast-feeding for less than 6 months is not associated with increased transmission, but the data are not conclusive. However, this is important in low-income endemic areas. The control of the common STDs and the use of condoms appear to be the approaches most likely to reduce sexual transmission.

Therapeutic attention to HTLV-I has focused on chemotherapy of ATL, which has a poor prognosis. Experimental therapy using targeted monoclonal antibodies to a tumor-associated cell surface protein has been attempted in a few cases, but without conclusive evidence of benefit. Japanese investigators have reported clinical benefit of corticosteroid treatment for some HAM–TSP patients, but not all centers have documented therapeutic efficacy.

9.2. HTLV-II

Prevention recommendations for HTLV-II are similar to those for HTLV-I: blood should be screened prior to transfusion and positive donor units should be discarded.[34] HTLV-II-positive mothers should be discouraged from breast-feeding where practicable to prevent mother-to-infant transmission. In tropical settings, where breast-feeding is a necessity, the risk of HTLV-II infection is low compared to mortality, for example, from diarrheal diseases in non–breast-fed infants. The use of condoms is recommended for couples discordant for HTLV-II serostatus.[33]

Based on various experimental and animal model data, an HTLV-II vaccine appears feasible, but such a vaccine may be unwarranted in the absence of data suggesting a public health risk for this agent.[20]

10. Unresolved Problems

10.1. HTLV-I

In summary, there has been a consolidation of epidemiologic data concerning geographic distribution, disease patterns, and transmission risk patterns for HTLV-I. Future work should focus on analytic studies in well-defined populations addressing remaining issues in the natural history of this infection, including any sequelae of sexually acquired infections. Such studies should incorporate biomarkers of immune and viral status, and the potential role of coinfection with other pathogenic latent infections should be considered. Development of a more targeted program for perinatal transmission should be considered, since it is only a subset of infected mothers who transmit the infection. The optimum length of breast-feeding should also be considered. The characterization of the subgroup of mothers who transmit the infection independent of breast milk is of high priority.

In terms of treatment, research should focus on the identification of carriers at high risk of developing ATL because their development of malignancy may be aborted by targeted intervention. The lessons learned in the course of HIV-1 research may be applicable in this effort. The application of virus-specific interventions for HAM–TSP and the other autoimmune type diseases may likewise be feasible.

Finally, the possibility for development of an HTLV-I vaccine exists. The existence of the analogue STLV-I model for testing and evaluation of immunologic control is an advantage to any such undertaking. However, in view of the highly reclusive nature of HTLV-I, which spends most of its life cycle hidden deep in the genome of target cells, the prospects for an effective vaccine are uncertain. Again, the lessons learned from the current program to develop a vaccine against the HIV-1 may have direct applicability to that for the HTLV-I. Given the rarity of clinical outcomes known to be associated with HTLV-I, in the event that an effective vaccine were developed, the public health impact of a vaccine program would need to be carefully assessed in each affected population.

10.2. HTLV-II

Much less is known about the epidemiology of HTLV-II than for HTLV-I. There is a need to expand knowledge about the distribution of HTLV-II in different parts of the world, and especially to clarify the nature of the HTLV-II virus in Africa. Analytic studies are needed to identify the various risk factors for transmission and quantify their contribution, and to understand why transmission occurs so efficiently in the intravenous drug use setting. There is a need to better understand the equal male-to-female age-specific seroprevalence of HTLV-II observed in the Panama study and to determine whether this represents more efficient female-to-male transmission than is seen for HTLV-I. A major gap is the lack of information concerning what, if any, diseases are associated with HTLV-II.

11. References

1. AKAGI, T., HOSHIDA, Y., YOSHINO, T., TERAMOTO, N., KONDO, E., HAYASHI, K., TAKAHASHI, K., Infectivity of human T-lymphotropic virus type I to human nervous tissue cells *in vitro*, *Acta Neuropathol.* **84:**147–152 (1992).

2. ALLAIN, J. P., HODGES, W., EINSTEIN, M. H., GEISLER, J., NEILLY, C., DELANEY, S., HODGES, B., LEE, A., Antibody to HIV-I, HTLV-I, and HCV in three populations of rural Haitians, *J. Acquir. Immune Defic. Syndr.* **5:**1230–1236 (1992).

3. ANONYMOUS, HTLV-I—A screen too many? *Lancet* **336:**1161 (1990).

4. AOKI, T., MIYAKOSHI, H., KOIDE, H., YOSHIDA, T., ISHIKAWA, H., SUGISAKI, Y., MIZUKOSHI, M., TAMURA, K., MISAWA, H., HAMADA, C., TING, R. C., ROBERT-AUROFF, M., AND GALLO, R. C., Seroepidemiology of human T-lymphotropic retrovirus type 1 (HTLV-1) in residents of Nigata prefectur, Japan. Comparative studies by indirect immunofluorescence microscopy and enzyme-linked immunosorbent assay, *Int. J. Cancer* **35:**301–306 (1985).

5. ASOU, N., KUMAGAI, T., UEKIHARA, S., ISHII, M., SATO, M., SAKAI, K., NISHIMURA, H., YAMAGUCHI, K., TAKATSUKI, K., HTLV-I seroprevalence in patients with malignancy, *Cancer* **58:**903–907 (1986).

6. BABU, P. G., GNANAMUTHU, C., SARASWATHI, N. K., NERURKAR, V. R., YANAGIHARA, R., JOHN, T. J., HTLV-I-associated myelopathy in South India, *AIDS Res. Hum. Retroviruses* **9:**499–500 (1993).

7. BALTIMORE, D., RNA-dependent DNA polymerase in virions of RNA tumour viruses, *Nature* **226:**1209–1211 (1970).

8. BARTHOLOMEW, C., CHARLES, W., SAXINGER, C., BLATTNER, W., ROBERT-GUROFF, M., RAJU, C., RATAN, P., INCE, W., QUAMINA, D., BASDEO-MAHARAJ, K., AND GALLO, R. C., Racial and other characteristics of human T-cell leukemia/lymphoma (HTLV-I) and AIDS (HTLV-III) in Trinidad, *Br. Med. J.* **290:**1243–1246 (1985).

9. BARTHOLOMEW, C., EDWARDS, J., JACK, N., CORBIN, D., MURPHY, J., CLEGHORN, F., WHITE, F., BLATTNER, W., Studies on maternal transmission of HTLV-I in patients with adult T-cell leukaemia (ATL) in Trinidad and Tobago, *Sixth International Conference on Human Retrovirology: HTLV*, May 14–19, 1994 (Abstract).

10. BARTHOLOMEW, C., SAXINGER, W. C., CLARK, J. W., GAIL, M., DUDGEON, A., MAHABIR, B., HYLL-DRYSDALE, B., CLEGHORN, F., GALLO, R. C., AND BLATTNER, W. A., Transmission of HTLV-I and HIV among homosexuals in Trinidad, *J.A.M.A.* **257:**2604–2608 (1987).

11. BASTIAN, I., GARDNER, J., WEBB, D., AND GARDNER, I., Isolation of a human T-lymphotropic virus type I strain from Australian Aboriginae, *J. Virol.* **67:**843–851 (1993).

12. BATSUURI, J., DASHNYAM, B., MAIDAR, J., BATTULGA, D., DORJSUREN, D., AND ISHIDA, T., Absence of human T-lymphotropic retrovirus type-1 (HTLV-I) in different populations of Mongolia, *Scand. J. Infect. Dis.* **25:**398–399 (1993).

13. BENVENISTE, R. E., The contributors of retroviruses to the study of mammalian evolution, in: *Molecular Evolutionary Genetics* (R. J. MACINTYRE, ED.), pp. 359–417, Plenum Press, New York, 1985.

14. BERNEMAN, Z. N., GARTENHAUS, R. B., REITZ, M. S., JR., BLATTNER, W. A., MANNS, A., HANCHARD, B., IKEHARA, O., GALLO, R. C., AND KLOTMAN, M. E., Expression of alternatively spliced human T-lymphotropic virus type I pX mRNA in infected cell lines and in primary uncultured cells from patients with adult T-cell leukemia/lymphoma and healthy carriers, *Proc. Natl. Acad. Sci. USA* **89:**3005–3009 (1992).

15. BIGGAR, R. J., SAXINGER, C., GARDINER, C., COLLINS, W. E., LEVINE, P. H., CLARK, J. W., NKRUMAH, F. K., GALLO, R. C., AND BLATTNER, W. A., Type I HTLV antibody in urban and rural Ghana, West Africa, *Int. J. Cancer* **34:**215–219 (1984).

16. BIGGAR, R. J., NEEQUAYE, J. E., NEEQUAYE, A. R., ANKRA-BADU, G. A., LEVINE, P. H., MANNS, A., TAYLOR, M., DRUMMOND, J., AND WATERS, D., The prevalence of antibodies to the human T-lymphotropic virus (HTLV) in Ghana, West Africa, *AIDS Res. Hum. Retroviruses* **9:**505–511 (1993).

17. BIGGAR, R. J., BUSKELL-BALES, Z., YAKSHE, P. N., CAUSSY, D., GRIDLEY, G., AND SEEFF, L., Antibody to human retroviruses among drug users in the East Coast American cities, 1972–1976, *J. Infect. Dis.* **163:**57–63 (1991).

18. BIGLIONE, M., GESSAIN, A., QUIRUELAS, S., FAY, O., TABORDA, M. A., AND FERNANDEZ, E., Endemic HTLV-II infection among Tobas and Matacos Amerindians from North Argentina, *J. Acquir. Immune Defic. Syndr.* **6(6):**631–633 (1993).

19. BLATTNER, W. A., NOMURA, A., CLARK, J. W., HO, G. F. Y., NAKAO, Y., GALLO, R., AND ROBERT-GUROFF, M., Modes of transmission and evidence for viral latency from studies of HTLV-I in Japanese migrant populations in Hawaii, *Proc. Natl. Acad. Sci. USA* **83:**4895–4898 (1986).

20. BLATTNER, W. A., Human T-lymphotropic viruses: HTLV-I and HTLV-II, in: *Textbook of AIDS Medicine* (S. BRODER, T. C., MERIGAN, JR., AND D. BOLOGNESI, EDS.), pp. 887–908, Baltimore, Williams and Wilkins, 1994.

21. BLATTNER, W. A., BLAYNEY, D. W., ROBERT-GUROFF, M., SARNGADHARAN, M. G., KALYANARAMAN, V. S., SARIN, P., JAFFE, E. S., AND GALLO, R. C., Epidemiology of human T-cell leukemia/lymphoma virus (HTLV), *J. Infect. Dis.* **147:**406–416 (1983).

22. BLATTNER, W. A., SAXINGER, C., RIEDEL, D., HULL, B., TAYLOR, G., CLEGHORN, F., GALLO, R., BLUMBERG, B., AND BARTHOLOMEW, C., A study of HTLV-I and its associated risk factors in Trinidad and Tobago, *J. Acquir. Immune Defic. Syndr.* **3:**1102–1108 (1990).

23. BLATTNER, W. A., KALYANARAMAN, V. S., ROBERT-GUROFF, M., LISTER, T. A., GALTON, D. A. G., SARIN, P. S., CRAWFORD, D., GREAVES, M., AND GALLO, R. C., The human type-C retrovirus, HTLV, in blacks from the Caribbean region, and relationship to adult T-cell leukemia/lymphoma, *Int. J. Cancer* **30:**257–264 (1982).

24. BLATTNER, W. A., Epidemiology of HTLV-1 and associated diseases, in: *Human Retrovirology: HTLV* (W. A. BLATTNER, ED.), pp. 251–265, Raven Press, New York, 1990.

25. BLOMBERG, J., ROBERT-GUROFF, M., BLATTNER, W. A., AND PIPKORN, R., Type- and group-specific continuous antigenic determinants of synthetic peptides for serotyping of HTLV-I and -II infection, *J. Acquir. Immune Defic. Syndr.* **5(3):**294–302 (1992).

26. BRABIN, L., BRABIN, B. J., DOHERTY, R. R., GUST, I. D., ALPERS, M. P., FUJINO, R., IMAI, J., AND HINUNA, Y., Patterns of migration indicate sexual transmission of HTLV-I infection in non-pregnant women in Papua, New Guinea, *Int. J. Cancer* **44:**59–62 (1989).

27. BRODINE, S. K., OLDFIELD, E. C., III, CORWIN, A. L., THOMAS, R. J., RYAN, A. B., HOLMBERG, J., MALGAARD, C. A., GOLBECK, A. L., RYDEN, L. A., BENENSON, A. S., ROBERTS, C. R., AND BLATT-

NER, W. A., HTLV-I among U.S. marines stationed in a hyperendemic area: Evidence for female-to-male sexual transmission, *J. Acquir. Immune Defic. Syndr.* **5**:158–162 (1992).

28. BUSCH, M. P., LAYCOCK, M., KLEINMAN, S. H., WAGES, JR., J. W., CALABRO, M., KAPLAN, J. E., KHABBAZ, R. F., AND HOLLINGSWORTH, C. G., Accuracy of supplementary serologic testing for human T-lymphotropic virus types I and II in US blood donors. Retrovirus Epidemiology Donor Study, *Blood* **83**:1143–1148 (1994).

29. CALENDAR, A., GESSAIN, A., ESSEX, M., AND DE-THÉ, G., Seroepidemiology of HTLV in the French West Indies: Antibodies in blood donors and lymphoproliferative diseases who do not have AIDS, *Ann. N.Y. Acad. Sci.* **437**:175–176 (1984).

30. CANN, A. J., AND CHEN, I. S. Y., Human T-cell leukemia viruses type I and II, in: *Fields Virology* (3rd ed.) (N. FIELDS, D. M. KNIPE, AND P. M. HOWLEY, EDS.), pp. 1849–1880, Raven Press, New York, 1995.

31. CANTOR, K. P., WEISS, S. H., GOEDERT, J. J., AND BATTJES, R. J., HTLV-I/II seroprevalence and HIV/HTLV coinfection among U.S. intravenous drug users, *J. Acquir. Immune Defic. Syndr.* **4**(5):460–467 (1991).

32. CATOVSKY, D., GREAVES, M. F., ROSE, M., ROSE, M., GALTON, D. A., GOOLDEN, A. W., McCLUSKEY, D. R., WHITE, J. M., LAMPERT, I., BOURIKAS, G., IRELAND, R., BROWNELL, A. I., BRIDGES, J. M., BLATTNER, W. A., AND GALLO, R. C., Adult T-cell lymphoma-leukaemia in Blacks from the West Indies, *Lancet* **1**:639–643 (1982).

33. CENTERS FOR DISEASE CONTROL, Guidelines for counseling persons infected with human T-lymphotropic virus type I (HTLV-I) and type II (HTLV-II), *Ann. Intern. Med.* **118**(6):448–454 (1993).

34. CENTERS FOR DISEASE CONTROL, Licensure of screening tests for antibody to human T-lymphotropic virus type I, *Morbid. Mortal. Week. Rep.* **37**:736, 740, 745, 747 (1988).

35. CESARMAN, E., CHADBURN, A., INGHIRAMI, G., GAIDANO, G., AND KNOWLES, D. M., Structural and functional analysis of oncogenes and tumor suppressor genes in adult T-cell leukemia/lymphoma shows frequent p53 mutations, *Blood* **80**:3205–3216 (1992).

35a. CHEN, Y.-M., GOMEZ-LUCIA, E., OKAYAMA, A., TACHIBANA, N., LEE, T.-H., MUELLER, N., AND ESSEX, M., Antibody profile of early HTLV-I infection, *Lancet* **336**:1214–1216 (1990).

36. CHEN, Y.-M., OKAYAMA, A., LEE, T.-H., TACHIBANA, N., MUELLER, N., AND ESSEX, M., Sexual transmission of HTLV-I associated with the presence of the anti-*tax* antibody, *Proc. Natl. Acad. Sci. USA* **88**:1182–1186 (1991).

37. CHEN, Y. M. A., LEE, T. H., WIKTOR, S. Z., SHAW, G. M., MURPHY, E. L., BLATTNER, W. A., AND ESSEX, M., Type-specific antigens for serological discrimination of HTLV-I and HTLV-II infection, *Lancet* **336**:1153–1155 (1990).

38. CIMINALE, V., PAVLAKIS, G. N., DERSE, D., CUNNINGHAM, C. P., AND FELBER, B. K., Complex splicing in the human T-cell leukemia virus (HTLV) family of retrovirus: Novel mRNAs and proteins produced by HTLV type I, *J. Virol.* **66**:1737–1745 (1992).

39. CLARK, J. W., ROBERT-GUROFF, M., IKEHARA, O., HENZAN, E., AND BLATTNER, W. A., The human T-cell leukemia/lymphoma virus type-I (HTLV-I) in Okinawa, *Cancer Res.* **45**:2849–2852 (1985).

40. CLEGHORN, F. R., CHARLES, W., BLATTNER, W., AND BARTHOLOMEW, C., Adult T-cell leukemia in Trinidad and Tobago, in: *Human Retrovirology: HTLV* (W. A. BLATTNER, ED.), pp. 185–190, Raven Press, New York, 1990.

41. CORTES, E., DETELS, R., ABOULAFAIA, D., LI, X. L., MOUDGIL, T., ALAIN, M., BONECKER, C., GONZAGA, A., OYAFUSO, L., TONDO, M., BOITE, C., HAMMERSHLAK, N., CAPITANI, C., SLAMON, D. J., AND HO, D. D., HIV-I, HIV-2, and HTLV-I infection in high-risk groups in Brazil, *N. Engl. J. Med.* **320**:953–958 (1990).

42. COSSEN, C., HAGENS, S., FUKUCHI, R., FORGHANI, B., GALLO, D., AND ASCHER, M., Comparison of six commercial human T-cell lymphotropic virus type I (HTLV-I) enzyme immunoassay kits for detection of antibody to HTLV-I and -II, *J. Clin. Microbiol.* **30**(3):724–725 (1992).

43. DAENKE, S., NIGHTINGALE, S., CRUICKSHANK, J. K., AND BANGHAM, C. R. M., Sequence variants of human T-cell lymphotropic virus type I from patients with tropical spastic paraparesis and adult T-cell leukemia do not distinguish neurological from leukemic isolates, *J. Virol.* **64**(3):1278–1282 (1990).

44. DE-THÉ, G., GESSAIN, A., ROBERT-GUROFF, M., *ET AL.*, Comparative seroepidemiology of HTLV-I and HTLV-III in French West Indies and some African countries, *Cancer Res.* **45**:4633s–4636s (1985).

45. DELAPORT, E., DUPONT, A., PEETERS, M., JOSSE, R., MERLIN, M., SCHRIJVERS, D., HAMONO, B., BEDJABAGA, L., CHERINGOU, H., BOYER, F., BRUN-VÉZINET, F., AND LAROUZÉ, B., Epidemiology of HTLV-1 in Gabon (western equatorial Africa), *Int. J. Cancer* **42**:687–689 (1988).

46. DELAPORT, E., PEETERS, M., DURAND, J.-P., DUPONT, A., SCHRIJVERS, D., BEDJABAGA, L., HONORÉ, C., OSSARI, S., TREBUCQ, A., JOSSE, R., AND MERLIN, M., Seroepidemiological survey of HTLV-1 infection among randomized populations of western central African countries, *J. Acquir. Immune Defic. Syndr.* **2**:410–413 (1989).

47. DELAPORT, E., MONPLAISIR, N., LOUWAGIE, J., PEETERS, M., MARTIN-PRÉVEL, Y., LOUIS, J.-P., TREBUCQ, A., BEDJABAGA, L., OSSARI, S., HONORÉ, C., LAROUZÉ, B., AURIOL, L., VAN DER GROEN, G., AND PIOT, P., Prevalence of HTLV-I and HTLV-II infection in Gabon, Africa: Comparison of the serological and PCR results, *Int. J. Cancer* **49**:373–376 (1991).

48. DOLL, R., An epidemiologic perspective of the biology of cancer, *Cancer Res.* **38**:3573–3583 (1978).

49. DONEGAN, E., BUSCH, M. P., GALLESHAW, J. A., SHAW, G. M., AND MOSLEY, J. W., Transfusion Safety Study Group: Transfusion of blood components from a donor with human T-lymphotropic virus type II (HTLV-II) infection, *Ann. Intern. Med.* **113**:555–556 (1990).

50. DUMAS, M., HOUINATO, D., VERDIER, M., ZOHOUN, T., JOSSE, R., BONIS, J., ZOHOUN, I., MASSONGBODJI, A., AND DENIS, F., Seroepidemiology of human T-cell lymphotropic virus in Benin (West Africa), *AIDS Res. Hum. Retroviruses* **7**:447–451 (1991).

51. EGUCHI, K., MATSUOKA, N., IDA, H., NAKASHIMA, M., SAKAI, M., SAKITO, S., KAWAKAMI, A., TERADA, K., SHIMADA, H., KAWABE, Y., FUKUDA, T., SAWADA, T., AND NAGATAKI, S., Primary Sjögren's syndrome with antibodies to HTLV-I: Clinical and laboratory features, *Ann. Rheum. Dis.* **51**:769–776 (1992).

52. EHRLICH, G. D., ANDREWS, J., SHERMAN, M. P., GREENBERG, S. J., AND POIESZ, B. J., DNA sequence analysis of the gene encoding the HTLV-I P21E transmembrane protein reveals inter- and intraisolate genetic heterogeneity, *Virology* **186**:619–627 (1992).

53. EHRLICH, G. D., GLASER, J. B., LaVIGNE, K., QUAN, D., MILDVAN, D., SNINSKY, J. J., KWOK, S., PAPSIDERO, L., AND POIESZ, B. J., Prevalence of human T-cell leukemia/lymphoma virus (HTLV) type II infection among high-risk individuals: Type-

specific identification of HTLVs by polymerase chain reaction, *Blood* **74**(5):1658–1664 (1989).

54. ESSEX, M., AND GUTENSOHN (MUELLER) N., A comparison of the pathobiology and epidemiology of cancers associated with viruses in humans and animals, *Prog. Med. Virol.* **27**:114–126 (1981).

55. ESSEX, M., MCLANE, M., TACHIBANA, N., FRANCIS, D. P., AND LEE, T.-H., Seroepidemiology of human T-cell leukemia virus in relation to immunosuppression and the acquired immunodeficiency syndrome, in: *Human T-cell Leukemia/Lymphoma Virus* (R. C. GALLO, M. ESSEX, AND L. GROSS, EDS.), pp. 355–362, Cold Spring Harbor Laboratory, Cold Spring Harbor, NY, 1984.

56. ESTEBANEZ, P., SARASQUETA, C., NAJERA, R., CONTRERAS, G., PÉREZ, L., FITCH, K., AND NICENTE, A., Prevalence of HIV-1, HIV-2, and HTLV-I/II in Spanish seamen, *J. Acquir. Immune Defic. Syndr.* **5**(3):316–317 (1992).

57. FRANCIS, D. P., ESSEX, M., AND MAYNARD, J. E., Feline leukemia virus and hepatitis B virus: A comparison of late manifestations, *Prog. Med. Virol.* **27**:127–132 (1981).

58. FUJII, M., SASSONE-CORSI, P., AND VERMA, I. M., C-*fos* promoter *trans*-activation by the tax_1 protein of human T-cell leukemia virus type I, *Proc. Natl. Acad. Sci. USA* **85**:8526–8530 (1988).

59. FUJINO, R., KAWATO, K., IKEDA, M., MIYAKOSHI, H., MIZUKOSHI, M., AND IMAI, J., Improvement of gelatin particle agglutination test for detection of anti-HTLV-I antibody, *Jpn. J. Cancer Res.* **82**:367–370 (1991).

60. GALLO, R. C., DE-THÉ, G. B., AND ITO, Y., Kyoto workshop on some specific recent advances in human tumor virology, *Cancer Res.* **41**:4738–4739 (1981).

61. GESSAIN, A., BARIN, F., VERNANT, J. C., GOUT, O., MAURS, L., CALENDER, A., AND DE-THÉ, G., Antibodies to human T-lymphotropic virus type-I in patients with tropical spastic paraparesis, *Lancet* **2**:407–410 (1985).

62. GESSAIN, A., SAAL, F., AND GOUT, O., High human T-cell lymphotropic virus type-I proviral DNA load with polyclonal integration in peripheral blood mononuclear cells of French West Indian, Guinean and African patients with tropical spastic paraparesis, *Blood* **75**:428–433 (1990).

63. GESSAIN, A., FRETZ, C., KOULIBALY, M., *ET AL.*, Evidence of HTLV-II infection in Guinea, West Africa, *J. Acquir. Immune Defic. Syndr.* **6**(3):324–325 (1993).

64. GESSAIN, A., TUPPIN, P., KAZANJI, M., MARICLERE, P., MANIEUX, R., GARCIA-CALLEJA, J. M., SALLA, R., TEKARA, F., MILLAN, J., DE-THÉ, G., AND GESSAIN, A., A new HTLV type II subtype A isolate in an HIV type 1-infected prostitute from Cameroon, Central Africa, *AIDS Res. Hum. Retroviruses* **11**:989–993 (1995).

65. GESSAIN, A., YANAGIHARA, R., FRANCHINI, G., GARRUTO, R. M., JENKINS, C. L., AJDUKIEWICZ, A. B., GALLO, R. C., AND GAJDUSEK, D. C., Highly divergent molecular variant of human T-lymphotropic virus type I from isolated populations in Papua New Guinea and the Solomon Islands, *Proc. Natl. Acad. Sci. USA* **88**:7694–7698 (1991).

66. GESSAIN, A., HERVÉ, V., JEANNEL, D., GARIN, B., MATHIOT, C., AND DE-THÉ, G., HTLV-I but not HTLV-II found in pygmies from Central African Republic, *J. Acquir. Immune Defic. Syndr.* **6**:1373–1375 (1993).

67. GESSAIN, A., GALLO, R. C., AND FRANCHINI, G., Low degree of human T-cell leukemia/lymphoma virus type I genetic drift *in vivo* as a means of monitoring viral transmission and movement of ancient human populations, *J. Virol.* **66**:2288–2295 (1992).

68. GOTUZZO, E., SÁNCHEZ, J., ESCAMILLA, J., CARRILLO, C., PHIL-

LIPS, I. A., MOREYRA, L., STAMM, W., ASHLEY, R., ROGGEN, E. L., KREISS, J., PIOT, P., AND HOLMES, K. K., Human T-cell lymphotropic virus type I infection among female sex workers in Peru, *J. Infect. Dis.* **169**:754–759 (1994).

69. GOUBAU, P., DESMYTER, J., GHESQUIERE, J., AND KASEREKA, B., HTLV-II among pygmies, *Nature* **359**:201 (1992).

70. GOUBAU, P., DESMYTER, J., SWANSON, P., REYNDERS, M., SHIH, J., SURMOUNT, I., KAZADI, K., AND LEE, H., Detection of HTLV-I and HTLV-II infection in Africans using type-specific envelope peptides, *J. Med. Virol.* **39**:28–32 (1993).

71. GOUT, O., BAULAC, M., GESSAIN, A., SEMÀH, F., SAAL, F., PÉRIÈS, J., CABRAL, C., FOUCAULT-FRETZ, C., LAPLANE, D., SIGAUX, F., AND DE-THÉ, G., Rapid development of myelopathy after HTLV-1 infection acquired by transfusion during cardiac transplantation, *N. Engl. J. Med.* **322**:383–388 (1990).

72. GREENE, W. C., BOHNLEIN, E., AND BALLARD, D. W., HIV-1, HTLV-1 and normal T-cell growth: Transcriptional strategies and surprises, *Immunol. Today* **10**:272–278 (1989).

73. HALL, W., TAKAHASHI, H., AND LIU, C., Multiple isolates and characteristics of human T-cell leukemia virus type II, *J. Virol.* **66**:2456–2463 (1992).

74. HANADA, S., UEMATSU, T., IWAHASHI, M., NOMURA, K., UTSUNONUYA, A., KODAMA, M., ISHIBASHI, K., TERADA, A., SAITO, T., MAKINO, T., UOZUMI, K., KNWAZURU, Y., OTSUKA, M., HARADA, R., HASHIMOTO, S., AND SAKURANI, T., The prevalence of human T-cell leukemia virus type I infection in patients with hematologic and nonhematologic diseases in an adult T-cell leukemia-endemic area of Japan, *Cancer* **64**:1290–1295 (1989).

75. HANCHARD, B., LAGRENADE, L., CARBERRY, C., FLETCHER, V., WILLIAMS, E., CRANSTON, B., BLATTNER, W. A., AND MANNS, A., Childhood infective dermatitis evolving into adult T-cell leukaemia after 17 years [letter], *Lancet* **338**:1593–1594 (1991).

76. HATTORI, T., IKEMATSU, S., CHOSA, T., YAMAMOTO, S., MATSUOKA, M., FUKUTAKE, K., ROBERT-GUROFF, M., AND TAKATSUKI, K., Anti-HTLV-III and anti-HTLV-I antibodies an T-cell subsets in hemophiliacs living in HTLV-I endemic and nonedemic areas of Japan, *Acta Haematol.* **77**:25–29 (1987).

77. HENEINE, W., WOODS, T., GREEN, D., FUKODA, K., GIUSTI, R., CASTILLO, L., ARMIEN, B., GRACIA, F., AND KAPLAN, J. G., Detection of HTLV-II in breastmilk of HTLV-II infected mothers, *Lancet* **340**:1157–1158 (1992).

78. HENEINE, W., CHAN, W. C., LUST, J. A., SINHA, S. D., ZAKI, S. R., KHABBAZ, R. F., AND KAPLAN, J. E., HTLV-II infection is rare in patients with large granular lymphocyte leukemia, *J. Acquir. Immune Defic. Syndr.* **7**(7):736 (1994).

79. HENEINE, W., KHABBAZ, R. F., LAL, R. B., AND KAPLAN, J. E., Sensitive and specific polymerase chain reaction assays for diagnosis of human T-cell lymphotropic virus type I (HTLV-I) and HTLV-II infections in HTLV-I/II-seropositive individuals, *J. Clin. Microbiol.* **30**(6):1605–1607 (1992).

80. HENEINE, W., KAPLAN, J. E., GRACIA, F., LAL, R., LEVINE, P. H., AND REEVES, W. C., HTLV-II endemicity among Guaymi Indians in Panama, *N. Engl. J. Med.* **324**(8):565 (1991).

81. HINO, S., DOI, H., YOSHIKUNI, H., SUGIYAMA, H., ISHIMAU, T., YAMABE, T., TSUJI, Y., AND MIYANOTO, T., HTLV-I carrier mothers with high-titer antibody are at high risk as a source of infection, *Jpn. J. Cancer Res.* **78**:1156–1158 (1987).

82. HINO, S., KATAMINE, S., KAWASE, K., MIYAMOTO, T., DOI, H., TSUJI, Y., AND YAMABE, S., Intervention of maternal transmission of HTLV-I in Nagasaki, Japan, *Leukemia* **8**(Suppl. 1):68s–70s (1994).

83. HINO, S., SUGIYAMA, H., DOI, H., ISHIMARU, T., YAMABE, T., TSUYI, Y., AND MUJAMOTO, T., Breaking the cycle of HTLV-1 transmission via carrier mothers' milk, *Lancet* **2**:158–159 (1987).

84. HINUMA, Y., KOMODA, H., CHOSA, T., KONDO, T., KOHAKURA, M., TAKENAKA, T., KIKUCHI, M., ICHIMARU, M., YUNOKI, K., SATO, I., MATSUO, R., TAKIUCHI, Y., UCHINO, H., AND HANAOKA, M., Antibodies to adult T-cell leukemia-virus-associated antigen (ATLA) in sera from patients with ATL and controls in Japan: A nation-wide seroepidemiologic study, *Int. J. Cancer* **29**:631–635 (1982).

85. HIRATA, M., HAYASHI, J., NOGUCHI, A., NAKASHIMA, K., KAJIYAMA, W., KASHIWAGI, S., AND SAWADA, T., The effects of breastfeeding and the presence of antibody to p40^tax protein of human T-cell lymphotropic virus type-I on mother-to-child transmission, *Int. J. Epidemiol.* **21**:989–994 (1992).

86. HJELLE, B., MILLS, R., SWENSON, S., MERTZ, G., KEY, C., AND ALLEN, S.,Incidence of hairy cell leukemia, mycosis fungoides, and chronic lymphocytic leukemia in a first known HTLV-II-endemic population, *J. Infect. Dis.* **163**:435–440 (1991).

87. HJELLE, B., APPENZELLER, O., MILLS, R., ALEXANDER, S., TORREZ-MARTINEZ, N., JAHNKE, R., AND ROSS, G., Chronic neurodegenerative disease associated with HTLV-II infection, *Lancet* **339**:645–646 (1992).

88. HJELLE, B., MILLS, R., MERTZ, G., AND SWENSON, S., Transmission of HTLV-II via blood transfusion, *Vox Sang.* **59**:119–122 (1990).

89. HJELLE, B., CYRUS, S., AND SWENSON, S. G., Evidence for sexual transmission of human T lymphotropic virus type II, *Ann. Intern. Med.* **116**(1):90–91 (1992).

90. HJELLE, B., SCALF, R., AND SWENSON, S., High frequency of human T-cell leukemia–lymphoma virus type II infection in New Mexico blood donors: Determination by sequence-specific oligonucleotide hybridization, *Blood* **76**(3):450–454 (1990).

91. HO, G. Y., NOMURA, A., NELSON, K., LEE, H., POLK, B. F., AND BLATTNER, W., Declining seroprevalence and transmission of HTLV-I in Japanese families who immigrated to Hawaii, *Am. J. Epidemiol.* **134**:981–987 (1991).

92. HO, M., MILLER, G., ATCHISON, R. W., BREINIG, M. K., DUMMER, S., ANDIMAN, W., STARZL, T. E., EASTMAN, R., GRIFFITH, B. P., HARDESTY, R. L., BAHNSON, H. T., HAKALA, T. R., AND ROSENTHAL, J. T., Epstein–Barr virus infections and DNA hybridization studies in post-transplantational lymphoma and lymphoproliferative lesions: The role of primary infection, *J. Infect. Dis.* **152**: 876–886 (1985).

93. HÖLLSBERG, P., AND HAFLER, D. A., Pathogenesis of diseases induced by human lymphotropic virus type I infection, *Semin. Med. Beth Israel Hosp. (Boston)* **328**(16):1173–1182 (1993).

94. HUNSMANN, G., SCHNEIDER, J., SCHMITT, J., AND YAMAMOTO, N., Detection of serum antibodies to adult T-cell leukemia virus in non-human primates and in people in Africa, *Int. J. Cancer* **32**:329–332 (1983).

95. HUNSMANN, G., BAYER, H., SCHNEIDER, J., SCHMITZ, H., KERN, P., DIETRICH, M., BÜTTNER, D. W., GOUDEAU, A. M., KULKARNI, G., AND FLEMING, A. F., Antibodies to ATLV/HTLV-1 in Africa, *Med. Microbiol. Immunol.* **173**:167–170 (1984).

96. HYAMS, K. C., PHILLIPS, I. A., TEJADA, A., WIGNALL, F. S., ROBERTS, C. R., AND ESCAMILLA, J., Three-year incidence study of retroviral and viral infection in a Peruvian prostitute population, *J. Acquir. Immune Defic. Syndr.* **6**:1353–1357 (1993).

97. IJICHI, S., RAMUNDO, M. B., TAKAHASHI, H., AND HALL, W. W., *In vivo* cellular trophisms of human T-cell leukemia virus type II (HTLV-II), *J. Exp. Med.* **176**:293–296 (1992).

98. IJICHI, S., ZANINOVIC, V., LEON-S, F. E., KATAHIRA, Y., SONODA, S., MIURA, T., HAYAMI, M., AND HALL, W. W., Identification of human T-cell leukemia virus type IIB infection in the Wayu, an aboriginal population of Colombia, *Jpn. J. Cancer Res.* **84**:1215–1218 (1993).

99. IMAI, J., AND HINUMA, Y., Epstein–Barr virus-specific antibodies in patients with adult T-cell leukemia (ATL) and healthy ATL virus-carriers, *Int. J. Cancer* **31**:197–200 (1983).

100. IMAMURA, J., TSUJIMOTO, A., OHTA, Y., HIROSE, S., SHIMOTOHNO, K., MIWA, M., AND MIYOSHI, J., DNA blotting analysis of human retroviruses in cerebrospinal fluid of spastic paraparesis patients: The viruses are identical to human T-cell leukemia virus type-1 (HTLV-1), *Int. J. Cancer* **42**:221–224 (1988).

101. ISHIDA, T., YAMAMOTO, K., AND OMOTO, K., A seroepidemiological survey of HTLV-1 in the Philippines, *Int. J. Epidemiol.* **17**:625–628 (1988).

102. ISHIDA, T., YAMAMOTO, K., OMOTO, K., IWANAGA, M., OSATO, T., AND HINUMA, Y., Prevalence of a human retrovirus in native Japanese: Evidence for a possible ancient origin, *Infection* **11**:153–157 (1985).

103. ISHIHARA, S., OKAYAMA, A., STUVER, S., HORINOUCHI, A., MURAI, K., SHIOIRI, S., KUBOTA, T., YAMASHITA, R., TACHIBANA, N., AND MUELLER, N., Association of HTLV-I antibody profile of asymptomatic carries with proviral DNA levels of peripheral blood mononuclear cells, *J. Acquir. Immune Defic. Syndr.* **7**:199–203 (1994).

104. ISHIZAKI, J., OKAYAMA, A., TACHIBANA, N., YOKOTA, T., SHISHIME, E., TSUDA, K., AND MUELLER, N., Comparative diagnostic assay results for detecting antibody to HTLV-I in Japanese blood donor samples: Higher positivity rates by particle agglutination assay, *J. Acquir. Immune Defic. Syndr.* **1**:340–345 (1988).

105. IWASAKI, Y., OHARA, Y., KOBAYASHI, I., AND AKIZUKI, S.-I., Infiltration of helper/inducer T-lymphocytes heralds central nervous system damage in human T-cell leukemia virus infection, *Am. J. Pathol.* **140**(5):1003–1008 (1992).

105a. IWATA, K., ITO, S.-I., SAITO, H., ITO, M., NAGATOMO, M., YAMOSAKI, T., YOSHIDA, S., SUTO, H., AND TAJIMA, K., Mortality among inhabitants of an HTLV-I endemic area in Japan, *Jpn. J. Cancer Res.* **25**:231–237 (1994).

106. JACOBSON, S., LEHKY, T., NISHIMURA, M., ROBINSON, S., McFARLIN, D., AND DHIB-JALBUT, S., Isolation of HTLV-II from a patient with chronic, progressive neurological disease clinically indistinguishable from HTLV-I-associated myelopathy/tropical spastic paraparesis, *Am. Neurol. Assoc.* **14**:392–396 (1993).

107. JACOBSON, S., McFARLIN, D. E., ROBINSON, S., VOSKUHL, R., MARTIN, R., BREWAH, A., NEWELL, A. J., AND KOENIG, S., HTLV-I-specific cytotoxic T lymphocytes in the cerebrospinal fluid of patients with HTLV-I-associated neurological disease, *Ann. Neurol.* **32**:651–657 (1992).

108. JAFFE, E. S., BLATTNER, W. A., BLAYNEY, D. W., BUNN, P. A., COSSMAN, J., ROBERT-GUROFF, M., AND GALLO, R. C., The pathologic spectrum of HTLV-associated leukemia/lymphoma in the United States, *Am. J. Surg. Pathol.* **8**:263–275 (1984).

109. JEANG, K-T., WIDDEN, S. G., SEMMES, O. J., IV, AND WILSON, S. H., HTLV-I *trans*-activator protein, tax, is a transrepressor of the human β-polymerase gene, *Science* **247**:1082–1084 (1990).

110. JOHNSON, R. E., NAHMIAS, A. J., MAGDER, L. S., LEE, F. J., BROOKS, C. A., AND SNOWDEN, C. A., A seroepidemiologic survey of the prevalence of herpes simplex virus type I infection in the United States, *N. Engl. J. Med.* **321**:7–12 (1989).

111. KAJIYAMA, W., KASHIWAGA, S., NOMURA, H., IKEMATSU, H.,

HAYASHI, J., AND IKEMATSU, W., Seroepidemiologic study of antibody to adult T-cell leukemia virus in Okinawa, Japan, *Am. J. Epidemiol.* **123**:41–47 (1985).

112. KAJIYAMA, W., KASHIWAGI, S., IKEMATSU, H., HAYASHI, J., NOMURA, H., AND OKOCHI, K., Intrafamilial transmission of adult T-cell leukemia virus, *J. Infect. Dis.* **154**:851–857 (1986).

113. KALYANARAMAN, V. S., SARNGADHARAN, M. G., ROBERT-GUROFF, M., BLAYNEY, D., GOLDE, D., AND GALLO, R. C., A new subtype of human T-cell leukemia virus (HTLV-II) associated with a T-cell variant of hairy cell leukemia, *Science* **218**(5):571–573 (1982).

114. KAMADA, Y., IWAMASA, T., MIYAZATO, M., SUNAGAWA, K., AND KUNISHIMA, N., Kaposi sarcoma in Okinawa, *Cancer* **70**(4):861–868 (1992).

115. KAMIHIRA, S., NAKASIMA, S., OYAKAWA, Y., MORIUTI, Y., ICHIMARU, M., OKUDA, H., KANAMURA, M., AND OOTA, T., Transmission of human T-cell lymphotropic virus type I by blood transfusion before and after mass screening of sera from seropositive donors, *Vox Sang.* **52**:43–44 (1987).

116. KAMIHIRA, S., TORIYA, K., AMAGASAKI, T., MOMITA, S., IKEDA, S., YAMADA, Y., TOMONAGA, M., ICHIMARU, M., KINOSHITA, K.-I., AND SAWADA, T., Antibodies against p40^{tax} gene product of human T-lymphotropic virus type-I (HTLV-I) under various conditions of HTLV-I infection, *Jpn. J. Cancer Res.* **80**:1066–1071 (1989).

117. KAPLAN, J. E., ABRAMS, E., SHAFFER, N., CANNIER, R. O., KAUL, A., KRASINSKY, K., BAMJI, M., HARTLEY, T. M., ROBERTS, B., KILBOURNE, B., THOMAS, P., ROGERS, M., HENEINE, W., AND THE NYC PERINATAL HIV TRANSMISSION COLLABORATIVE STUDY, Low risk of mother-to-child transmission of human T lymphotropic virus type II in non-breast-fed infants, *J. Infect. Dis.* **166**:892–895 (1992).

118. KAPLAN, J. E., LAL, R. B., DAVIDSON, M., LANIER, A. P., AND LAIRMORE, M. D., HTLV-I in Alaska Natives, *J. Acquir. Immune Defic. Syndr.* **6**:327–328 (1993).

119. KAPLAN, M. H., HALL, W. W., SUSIN, M., PAHWA, S., SALAHUDDIN, S. Z., HEILMAN, C., FETTEN, J., CORONESI, M., FARBER, B. F., AND SMITH, S., Syndrome of severe skin disease, eosinophilia and dermatopathic lymphadenopathy in patients with HTLV-II complicating human immunodeficiency virus infection, *Am. J. Med.* **91**:300–309 (1991).

120. KAPLAN, J. E., OSAME, M., KUBOTA, H., IGATA, A., NISHITANI, H., MAEDA, Y., KHABBAZ, R. F., AND JANSSEN, R. S., The risk of development of HTLV-I-associated myelopathy/tropical spastic paraparesis among persons infected with HTLV-I, *J. Acquir. Immune Defic. Syndr.* **3**:1096–1101 (1990).

121. KASHIWAGI, S., KAJIYAMA, W., HAYASHI, J., NOGUCHI, A., NAKASHIMA, K., NOMURA, H., IKEMATSU, H., SAWADA, T., KIDA, S., AND KOIDE, A., Antibody to p40^{tax} protein of human T-cell leukemia virus I and infectivity, *J. Infect. Dis.* **161**:426–429 (1990).

122. KASHIWAGI, S., KAJIYAMA, W., HAYASHI, J., NOMURA, H., AND IKEMATSU, H., No significant changes in adult T-cell leukemia virus infection in Okinawa, Japan, after intervals of 13 and 15 years, *Jpn. J. Cancer Res.* **77**:452–455 (1986).

123. KAWASE, K.-I., KATAMINE, S., MORIUCHI, R., MIYAMOTO, T., KUBOTA, K., IGARASHI, H., DOI, H., TSUJI, Y., YAMABE, T., AND HINO, S., Maternal transmission of HTLV-I other than through breast milk: Discrepancy between the polymerase chain reaction of core blood samples for HTLV-I and the subsequent seropositivity of individuals, *Jpn. J. Cancer Res.* **83**:968–977 (1992).

124. KELLEY, K., DAVIS, P., MITSUYA, H., IRVING, S., WRIGHT, J., GRASSMAN, R., FLECKENSTEIN, B., WANO, Y., GREENE, W., AND SIEBENLIST, U., A high proportion of early response genes are constitutively activated in T-cell by HTLV-I, *Oncogene* **7**:1463–1470 (1992).

125. KHABBAZ, R. F., HARTLEY, T. M., LAIRMORE, M. D., AND KAPLAN, J. E., Epidemiologic assessment of screening test for antibody to human T-lymphotropic virus type I (HTLV-I), *Am. J. Public Health* **30**:190–192 (1990).

126. KHABBAZ, R. F., DARROW, W. W., HARTLEY, T. M., WITTE, J., COHEN, J., FRENCH, J., GILL, P. S., POTTERAT, J., SIKES, R. K., RÉICH, R., KAPLAN, J., AND LAIRMORE, M. D., Seroprevalence and risk factors for HTLV-I/II infection among female prostitutes in the United States, *J.A.M.A.* **263**(1):60–64 (1990).

127. KINOSHITA, T., IMAMURA, J., NAGAI, H., ITO, M., ITO, S.-I., IKEDA, S., NAGATOMO, M., TAJIMA, K., AND SHIMOTOHNO, K., Absence of HTLV-I infection among seronegative subjects in an endemic area of Japan, *Int. J. Cancer* **54**:16–19 (1993).

128. KINOSHITA, K., AMAGASAKI, T., HINO, S., DOI, H., YAMANOUCHI, K., BAN, N., MOMITA, T., IKEDA, S., KAMIHIRA, S., ICHIMARU, M., KATAMINE, S., MIYAMOTO, T., TSUJI, Y., ISHIMARU, T., YAMABE, T., ITO, M., AND KAMURA, S., Milk-borne transmission of HTLV-I from carrier mothers to their children, *Jpn. J. Cancer Res.* **78**:674–680 (1987).

129. KINOSHITA, K., YAMANOUCHI, K., IKEDA, S., MOMITA, S., AMAGASAKI, T., SODA, H., ICHIMARU, M., MORIUSHI, R., KATAMINE, S., MIYAMOTO, T., AND HINO, S., Oral infection of a common marmoset with human T-cell leukemia virus type-I (HTLV-I) by inoculating fresh human milk of HTLV-I carrier mother, *Jpn. J. Cancer Res.* **76**:1147–1153 (1985).

130. KINOSHITA, K., AMAGASAKI, T., IKEDA, S., SUZUYAMA, J., TORIYA, K., NISHINO, K., TAGAWA, M., ICHIMARU, M., KAMIHIRA, S., YAMADA, Y., MOMITA, S., KUSANO, M., MORIKAWA, S., FUJITA, S., UEDA, Y., ITO, N., AND YOSHIDA, M., Preleukemic state of adult T-cell leukemia: Abnormal T-lymphocytosis induced by human adult T-cell leukemia-lymphoma virus, *Blood* **66**:120–127 (1985).

131. KINOSHITA, T., TSUJIMOTA, A., AND SHIMOTOHNO, K., Sequence variations in LTR and *env* regions of HTLV-I do not discriminate between the virus from patients with HTLV-I-associated myelopathy and adult T-cell leukemia, *Int. J. Cancer* **47**:491–495 (1991).

132. KIRA, J.-I., ITOYAMA, Y., KOYANAGI, Y., TATEISHI, J., KISHIKAWA, M., AKIZUKI, S.-I., KOBAYASHI, I., TOKI, N., SUEISHI, K., SATO, H., SAKAKI, Y., YAMAMOTO, N., AND GOTO, I., Presence of HTLV-I proviral DNA in central nervous system of patients with HTLV-I-associated myelopathy, *Ann. Neurol.* **31**:39–45 (1992).

133. KIRA, J.-I., NAKAMURA, M., SAWADA, T., KOYANAGI, Y., OHORI, N., ITOYAMA, Y., YAMAMOTO, N., SAKAKI, Y., AND GOTO, I., Antibody titers to HTLV-I-p40^{tax} protein and gag-env hybrid protein in HTLV-I-associated myelopathy/tropical spastic paraparesis: Correlation with increased HTLV-I proviral DNA load, *J. Neurol. Sci.* **107**:98–104 (1992).

134. KOBAYASHI, M., MIYOSHI, I., SONOBE, H., TAGUCHI, H., AND KUBONISHI, I., Association of *Pneumocystis carinii* pneumonia and scabies, *J.A.M.A.* **248**:1973 (1982).

135. KOHAKURA, M., NAKADA, K., YONAHARA, M., KOMODA, H., IMAI, J., AND HINUMA, Y., Seroepidemiology of the human retrovirus (HTLV/ATLV) in Okinawa where adult T-cell leukemia is highly endemic, *Jpn. J. Cancer Res.* **77**:21–23 (1986).

136. KOMURIAN, F., PELLOQUIN, F., AND DE-THÉ, G., *In vivo* genomic

variability of human T-cell leukemia virus type I depends more upon geography than upon pathologies, *J. Virol.* **65**(7):3770–3778 (1991).

137. KONDO, T., KONO, H., MIYAMOTO, N., YOSHIDA, R., TOKI, H., MATSUMOTO, I., HARA, M., INOUE, H., INATSUKI, A., FUNATSU, T., YAMANO, N., BANDO, F., IWAO, E., MIYOSHI, I., HINUMA, Y., AND HANAOKA, M., Age- and sex-specific cumulative rate and risk of ATLL for HTLV-I carriers, *Int. J. Cancer* **43**:1061–1064 (1989).

138. KONDO, S., MISHIRO, N., UMEKI, K., KOTANI, T., SETOGUCHI, M., SHINGU, T., AND OHTAKI, S., Human T-lymphotropic virus type I (HTLV-I) DNA in an HTLV-I endemic area (Miyazaki, Japan), *Transfusion* **34**:449–450 (1994).

139. KORBER, B., OKAYAMA, A., DONNELLY, R., TACHIBANA, N., AND ESSEX, M., Polymerase chain reaction analysis of defective human T-cell leukemia virus type I proviral genomes in leukemic cells of patients with adult T-cell leukemia, *J. Virol.* **65**(10):5471–5476 (1991).

140. KOSAKA, M., LISHI, Y., HORIUCHI, N., NAKAO, K., OKAGAWA, K., SAITO, S., MINAMI, Y., AND KATOH, K., A cluster of human T-lymphotropic virus type-I carriers found in the southern district of Tokushima prefecture, *Jpn. J. Clin. Oncol.* **19**:30–35 (1989).

141. KRAMER, A., JACOBSON, S., REUBEN, J. F., MURPHY, E. L., WIKTOR, S. Z., CRANSTON, B., FIGUEROA, J. P., HANCHARD, B., MCFARLIN, D., AND BLATTNER, W. A., Spontaneous lymphocyte proliferation in symptom-free HTLV-I positive Jamaicans, *Lancet* **2**:923–924 (1989).

142. KUO, T.-T., CHAN, H.-L., SU, I.-J., EIMOTO, T., MAEDA, Y., KIKUCHI, M., CHEN, M.-J., KUAN, Y.-Z., CHEN, W.-J., SUN, C.-F., SHIH, L.-Y., CHEN, J.-S., AND TAKESHITA, M., Serological survey of antibodies to the adult T-cell leukemia virus-associated antigen (HTLV-A) in Taiwan, *Int. J. Cancer* **36**:345–348 (1985).

143. KURIMURA, T., TSUCHIE, H., KOBAYASHI, S., HINUMA, Y., IMAI, J., LOPEZ, C. B., NITIYANANT, P., PETCHCLAI, B., DOMINGUEZ, C. E., KOIMAN, I., AND WURYDAIE, S., Sporadic cases of carriers of human T-lymphotropic virus type 1 in southeast Asia, *Jpn. J. Med. Sci. Biol.* **39**:25–28 (1986).

144. KUSHUHARA, K., SONODA, S., TAKAHASHI, K., TOKUGAWA, K., FUKUSHIGI, J., AND UEDA, K., Mother-to-child transmission of human T-cell leukemia virus type I (HTLV-I): A fifteen year follow-up study in Okinawa, Japan, *Int. J. Cancer* **40**:755–757 (1987).

145. LAGRENADE, L., HANCHARD, B., FLETCHER, V., CRANSTON, B., AND BLATTNER, W., Infective dermatitis of Jamaican children: A marker for HTLV-I infection, *Lancet* **336**:1345–1347 (1990).

146. LAIRMORE, M. D., JACOBSON, S., GRACIA, F., DE, B. K., CASTILLO, L., LARREATEGUI, M., ROBERTS, B. D., LEVINE, P. H., BLATTNER, W. A., AND KAPLAN, J. E., Isolation of human T-cell lymphotropic virus type 2 from Guaymi indians in Panama, *Proc. Natl. Acad. Sci. USA* **87**:8840–8844 (1990).

147. LAL, R. B., HENEINE, W., RUDOLPH, D. L., PRESENT, W. B., HOFHIENZ, D., HARTLEY, T. M., KHABBAZ, R. F., AND KAPLAN, J. E., Synthetic peptide-based immunoassays for distinguishing between human T-cell lymphotropic virus type I and type II infections in seropositive individuals, *J. Clin. Microbiol.* **29**(10):2253–2258 (1991).

148. LAL, R. B., OWEN, S. M., SEGURADO, A. A. C., AND GONGORA-BIACHI, R. A., Mother-to-child transmission of human T-lymphotropic virus type-II (HTLV-II), *Ann. Intern. Med.* **120**(4):300–301 (1994).

149. LAL, R. B., LIPKA, J. J., FOUNG, S. K. H., HADLOCK, K. G., REYES, G. R., AND CARNEY, W. P., Human T-lymphotropic virus type I/II in Lake Lindu Valley, Central Sulawesi, Indonesia, *J. Acquir. Immune Defic. Syndr.* **6**:1067–1070 (1993).

150. LAL, R. B., GIAM, C.-Z., COLIGAN, J. E., AND RUDOLPH, D. L., Differential immune responsiveness to the immunodominant epitopes of regulatory proteins (tax and rex) in human T-cell lymphotropic virus type I-associated myelopathy, *J. Infect. Dis.* **169**:496–503 (1994).

151. LAL, R. B., AND RUDOLPH, D., Constitutive production of interleukin-6 and tumor necrosis factor-α from spontaneously proliferating T cells in patients with human T-cell lymphotropic virus type-I/II, *Blood* **78**(3):571–574 (1991).

152. LAL, R. B., RUDOLPH, D., BUCKNER, C., PARDI, D., AND HOOPER, W. C., Infection with human T-lymphotropic viruses leads to constitutive expression of leukemia inhibitory factor and interleukin-6, *Blood* **81**(7):1827–1832 (1993).

153. LAVANCHY, D., BOVET, P., HOLLANDA, J., SHAMLAYE, C. F., BURCZAK, J. D., AND LEE, H., High seroprevalence of HTLV-I in the Seychelles, *Lancet* **337**:248–249 (1991).

154. LEE, H. H., SWANSON, P., AND ROSENBLATT, J. D., Relative prevalence and risk factors of HTLV-I and HTLV-II infection in U.S. blood donors, *Lancet* **337**:1435–1439 (1991).

155. LEE, T. H., COLIGAN, J. E., SODROWSKI, J. G., HASELTINE, W. A., SALAHUDDIN, S. Z., WONG-STAAL, F., GALLO, R. C., AND ESSEX, M., Antigens encoded by the 3′ terminal region of human T-cell leukemia virus: Evidence for a functional gene, *Science* **225**:57—61 (1984).

156. LEE, T. H., SWANSON, P., SHORTY, V. S., ZACK, J. A., ROSENBLATT, J. D., AND CHEN, I. S., High rate of HTLV-II infection in seropositive IV-drug abusers in New Orleans, *Science* **244**:471–475 (1989).

157. LEE, S. Y., YAMAGUCHI, K., TAKATSUKI, K., KIM, B. K., PARK, S. L., AND LEE, M., Seroepidemiology of human T-cell leukemia virus type-1 in the Republic of Korea, *Jpn. J. Cancer Res.* **77**:250–254 (1986).

158. LEE, H. H., WEISS, S. H., BROWN, L. S., MILDVAR, D., SHORTY, V., SARAVOLATZ, L., CHU, A., GINZBURG, H. M., MARKOWITZ, N., DESJARLAIS, D. C., BLATTNER, W. A., AND ALLAIN, J.-P., Patterns of HIV-1 and HTLV-I/II in intravenous drug abusers from the middle atlantic and central regions of the USA, *J. Infect. Dis.* **162**:347–352 (1990).

159. LEVINE, P. H., BLATTNER, W. A., CLARK, J., TARONE, R., MALONEY, E. M., MURPHY, E. M., GALLO, R. C., ROBERT-GUROFF, M., AND SAXINGER, W. C., Geographic distribution of HTLV-1 and identification of a new high-risk population, *Int. J. Cancer* **42**:7–12 (1988).

160. LEVINE, P. H., JACOBSON, S., ELLIOTT, R., CAVALLERO, A., COLCLOUGH, G., DORRY, C., STEPHENSON, C., KNIGGIE, R. M., DRUMMOND, J., NISHIMURA, M., TAYLOR, M. E., WIKTOR, S., AND SHAW, G. M., HTLV-II infection in Florida Indians, *AIDS Res. Hum. Retroviruses* **9**(2):123–127 (1993).

161. LEVINE, P. H., MANNS, A., JAFFE, E. S., COLCLOUGH, G., CAVALLARO, A., REDDY, G., AND BLATTNER, W. A., The effect of ethnic differences on the pattern of HTLV-I-associated T-cell leukemia/lymphoma (HATL) in the United States, *Int. J. Cancer* **56**:177–181 (1994).

162. LILLEHOJ, E. P., ALEXANDER, S. S., DUBRULE, C. J., WIKTOR, S., ADAMS, R., TAI, C.-C., MANNS, A., AND BLATTNER, W. A., Development and evaluation of an HTLV-I leukemia virus type-I serologic confirmatory assay incorporating a recombinant envelope polypeptide, *J. Clin. Microbiol.* **28**(12):2653–2658 (1990).

163. LILLO, F., VARNIER, O. E., SABBATANI, S., FERRO, A., AND MENDEZ, P., Detection of HTLV-I and not HTLV-II infection in Guinea Bissau (West Africa), *J. Acquir. Immune Defic. Syndr.* **4**:541–542 (1991).

164. LIPKA, J. J., SANTIAGO, P., CHAN, L., REYES, G. R., SAMUEL, K. P., BLATTNER, W. A., SHAW, G. M., HANSON, C. V., SNINSKY, J. J., AND FOUNG, S. K. H., Modified Western blot assay for confirmation and differentiation of human T-cell lymphotropic virus types I and II, *J. Infect. Dis.* **164**:400–403 (1991).

165. LIPKA, J. J., MIYOSHI, I., HADLOCK, K. G., REYES, G. R., CHOW, T. P., BLATTNER, W. A., SHAW, G. M., HANSON, C. V., GALLO, D., CHAN, L., AND FOUNG, S. K. H., Segregation of human T-cell lymphotropic virus type I and II infections by antibody reactivity to unique viral epitopes, *J. Infect. Dis.* **165**:268–272 (1992).

166. LOUGHRAN, T. P., JR., COYLE, T., SHERMAN, M. P., STARKEBAUM, G., EHRLICH, G. D., RUSCETTI, F. W., AND POIESZ, B. J., Detection of human T-cell leukemia/lymphoma virus, type II, in a patient with large granular lymphocyte leukemia, *Blood* **80**(5):1116–1119 (1992).

167. MACATONIA, S. E., CRUICKSHANK, J. K., RUDGE, P., AND KNIGHT, S. C., Dendritic cells from patients with tropical spastic paraparesis are infected with HTLV-I and stimulate autologous lymphocytes proliferation, *AIDS Hum. Res. Retroviruses* **8**:1699–1706 (1992).

168. MADELEINE, M. M., WIKTOV, S. Z., GOEDERT, J. J., MANNS, A., LEVINE, P. H., BIGGAR, R. J., AND BLATTNER, W. A., HTLV-I and HTLV-II world-wide distribution: Reanalysis of 4,832 immunoblot results, *Int. J. Cancer* **54**:255–260 (1993).

169. MAEDA, Y., FURUKAWA, M., TAKEHARA, Y., YOSHIMURA, K., MIYAMOTO, K., MATSUURA, T., MORISHIMA, Y., TAJIMA, K., OKOCHI, K., AND HINUMA, Y., Prevalence of possible adult T-cell leukemia virus-carriers among volunteer blood donors in Japan: A nation-wide study, *Int. J. Cancer* **33**:717–720 (1984).

170. MALONEY, E. M., BIGGAR, R. J., NEEL, J. V., NEEL, J. V., TAYLOR, M. E., HAHN, B. H., SHAW, G. M., AND BLATTNER, W. A., Endemic human T-cell lymphotropic virus type II infection among isolated Brazilian Amerindians, *J. Infect. Dis.* **166**:100–107 (1992).

171. MANN, D. L., DESANTIS, P., MARK, G., PFEIFER, A., NEWMAN, M., GIBBS, N., POPOVIC, M., SARNGADHARAN, M. G., GALLO, R., CLARK, J., AND BLATTNER, W., HTLV-I associated B-cell chronic lymphocytic leukemia: Indirect role for retrovirus in leukemogenesis, *Science* **236**:1103–1106 (1987).

172. MANNS, A., WILKS, R. J., MURPHY, E. L., HAYNES, G., FIGUEROA, J. P., BARNETT, M., HANCHARD, B., AND BLATTNER, W. A., A prospective study of transmission by transfusion of HTLV-I and risk factors associated with seroconversion, *Int. J. Cancer* **51**:886–891 (1992).

173. MANNS, A., CLEGHORN, F. R., FALK, R. T., HANCHARD, B., JAFFE, E. S., BARTHOLOMEW, C., HARTGE, P., BENICHOU, J., BLATTNER, W. A., AND THE HTLV LYMPHOMA STUDY GROUP, Role of HTLV-I in development of non-Hodgkin lymphoma in Jamaica and Trinidad and Tobago, *Lancet* **342**:1447–1450 (1993).

174. MANNS, A., MURPHY, E. L., WILKS, R. J., HAYNES, G., FIGUEROA, J. P., HANCHARD, B., PALKER, T. J., AND BLATTNER, W. A., Early antibody profile during HTLV-I seroconversion, *Lancet* **337**:181–182 (1991).

175. MARRIOTT, S. J., TRURUH, D., AND BRADY, J. N., Activation of interleukin-2 receptor alpha expression by extracellular HTLV-I Tax_1 protein: A potential role in HTLV-I pathogenesis, *Oncogene* **7**:1749–1755 (1992).

176. MATSUMOTO, C., MITSUNAGA, S., OGUCHI, T., MITOMI, Y., SHIMADA, T., ICHIKAWA, A., WATANBE, J., AND NISHIOKA, K., Detection of human T-cell leukemia virus type I (HTLV-I) provirus in an infected cell line and in peripheral mononuclear cells of blood donors by the nested double polymerase chain reaction method: Comparison with HTLV-I antibody test, *J. Virol.* **64**:5290–5294 (1990).

177. MATSUMURA, M., KUSHIDA, S., AMI, Y., SUGA, T., UCHIDA, K., KAMEYAMA, T., TERANO, A., INOUE, Y., SHIRAKI, H., OKOCHI, K., SATO, H., AND MIWA, M., Quantitation of HTLV-I provirus among seropositive blood donors: Relation with antibody profile using synthetic peptide. *Int. J. Cancer* **55**:220–222 (1993).

178. MATUTES, E., ROBINSON, D., O'BRIEN, M., HAYNES, B. F., ZOLA, H., AND CATOVSKY, D., Candidate counterpart of Sezary cells and adult T-cell lymphoma-leukemia cells in normal peripheral blood: An ultrastructure study with the immunogold method and monoclonal antibodies, *Leuk. Res.* **7**:787–801 (1983).

179. MATUTES, E., DALGLEISH, A. G., WEISS, R. A., JOSEPH, A. P., AND CATOVSKY, D., Studies in healthy human T-cell leukemia/lymphoma virus (HTLV-1) carriers form the Caribbean, *Int. J. Cancer* **38**:41–45 (1986).

180. MAY, J. P., STENT, G., AND SCHNAGL, R. D., Antibody to human T-cell lymphotropic virus type 1 in Australian Aborigines, *Med. J. Aust.* **149**:104 (1988).

181. MERINO, F., ROBERT-GUROFF, M., CLARK, J., BIONDO-BRACHO, M., BLATTNER, W., AND GALLO, R. C., Natural antibodies to human T-cell leukemia/lymphoma virus in healthy Venezuelan populations, *Int. J. Cancer* **34**:501–506 (1984).

182. MEYTES, D., SCHOCHAT, S., LEE, H., NADEL, G., SIDI, Y., CERNEY, M., SWANSON, P., SHAKLAI, M., KILIM, Y., ELGAT, M., CHIN, E., DANON, Y., AND ROSENBLATT, J. D., Serological and molecular survey for HTLV-I infection in a high-risk Middle Eastern group, *Lancet* **336**:1533–1535 (1990).

183. MILLER, G. J., PEGRAM, S. M., KIRKWOOD, B. R., BECKLES, G. L. A., BYAM, N. T. A., CLAYDEN, S. A., KINLEN, L. J., CHAN, L. C., CARSON, D. C., AND GREAVES, M. F., Ethnic composition, age and sex, together with location and standard of housing as determinants of HTLV-I infection in an urban Trinidadian community, *Int. J. Cancer* **38**:801–808 (1986).

184. MIURA, T., FUKUNAGA, T., IGARASHI, T., YAMASHITO, M., IDO, E., FUNAHASHI, S.-I., ISHIDA, T., WASHIO, K., UEDA, S., HASHIMOTO, K.-I., YOSHIDA, M., OSAME, M., SINGHAL, B. S., ZANINOVIC, V., CARTIER, L., SONODA, S., TAJIMA, K., INA, Y., GOJOBORI, T., AND HAYAMI, M., Phylogenetic subtypes of human T-lymphotropic virus type I and their relations to the anthropological background, *Proc. Natl. Acad. Sci. USA* **91**:1124–1127 (1994).

185. MIYAZAKI, K., YAMAGUCHI, K., TOHYA, T., OHBA, T., TAKATSUKI, K., AND OKAMURA, H., Human T-cell leukemia virus type I infection as an oncogenic and prognostic risk factor in cervical and vaginal carcinoma, *Obstet. Gynecol.* **77**(1):107–110 (1991).

186. MOCHIZUKI, M., WATANABE, T., YAMAGUCHI, K., TAKATSUKI, K., YOSHIMURA, K., SHIRAO, M., NAKASHIMA, S., MORI, S., ARAKI, S., AND MIYATA, N., HTLV-I uveitis: A distinct clinical entity cased by HTLV-I, *Jpn. J. Cancer Res.* **83**:236–239 (1992).

187. MORGAN, O. ST. C., RODGERS-JOHNSON, P., MORA, C., AND CHAT, G., HTLV-1 and polymyositis in Jamaica, *Lancet* **2**:1184–1187 (1989).

188. MOROFUJI-HIRATA, M., KAJIYAMA, W., NAKASHIMA, K., NOGUCHI, A., HAYASHI, J., AND KASHIWAGI, S., Prevalence of antibody

to human T-cell lymphotropic virus type I in Okinawa, Japan, after an interval of 9 years, *Am. J. Epidemiol.* **137:**43–48 (1993).

189. MOSMANN, T. R., AND MOORE, K. W., The role of IL-10 in crossregulation of T_H1 and T_H2 responses, *Immunol. Today* **12**(3):A49–53 (1991).

190. MUELLER, N., TACHIBANA, N., STUVER, S., OKAYAMA, A., SHISHIME, E., ISHIZAKI, J., MURAI, K., SHIOIRI, S., AND TSUDA, K., Epidemiologic perspectives of HTLV-I, in: *Human Retrovirology: HTLV* (W. A. BLATTNER, ED.), pp. 281–293, Raven Press, New York, 1990.

191. MUELLER, N., The epidemiology of HTLV-I infection, *Cancer Causes Control* **2:**37–52 (1991).

192. MUELLER, N., EVANS, A. S., AND LONDON, T., Viruses, in: *Cancer Epidemiology and Prevention*, 2nd ed. (D. SCHOTTENFELD AND J. F. FRAUMENI, JR., EDS.), pp. 502–531, Oxford University Press, New York, 1996.

193. MURAI, K., TACHIBANA, N., SHIOIRI, S., SHISHINE, E., OKAYAMA, A., ISHIZAKI, J., TSUDA, K., AND MUELLER, N., Suppression of delayed-type hypersensitivity to PPD and PHA in elderly HTLV-1 carriers, *J. Acquir. Immune Defic. Syndr.* **3:**1006–1009 (1990).

194. MURPHY, E. L., GIBBS, W. N., FIGUEROA, J. G., BAIN, B., LA GRENADE, L., CRANSTON, B., AND BLATTNER, W. A., Human immunodeficiency virus and human T-lymphotropic virus type I infection among homosexual men in Kingston, Jamaica, *J. Acquir. Immune Defic. Syndr.* **1:**143–149 (1988).

195. MURPHY, E. L., FIGUEROA, J. G., GIBBS, W. N., HOLDING-COBHAM, M., CRANSTON, B., MALLEY, K., BRODNER, A. J., ALEXANDER, S. S., AND BLATTNER, W. A., Human T-lymphotropic virus type I (HTLV-I), seroprevalence in Jamaica: I. Demographic determinants, *Am. J. Epidemiol.* **133:**1114–1124 (1991).

196. MURPHY, E. L., FIGUEROA, J. P., GIBBS, W. N., BRATHWAITE, A., HOLDING-COBHAM, M., WATERS, D., CRANSTON, B., HANCHARD, B., AND BLATTNER, W. A., Sexual transmission of human T-lymphotropic virus type I (HTLV-I), *Ann. Intern. Med.* **111:**555–560 (1989).

197. MURPHY, E. L., HANCHARD, B., FIGUEROA, J. P., GIBBS, W. N., LOFTERS, W. S., CAMPBELL, M., GOEDERT, J. J., AND BLATTNER, W. A., Modeling the risk of adult T-cell leukemia/lymphoma in persons infected with human lymphotopic virus type 1, *Int. J. Cancer* **43:**250–253 (1989).

198. NAKANO, S., ANDO, Y., ICHIJO, M., ET AL., Search for possible route of vertical and horizontal transmission of adult T-cell leukemia virus, *Jpn. J. Cancer Res.* **75:**1044–1045 (1984).

199. NEWTON, R. C., LIMPUANGTHIP, P., GREENBERG, S., GAM, A., AND NEVA, F. A., *Strongyloides stercoralis* hyperinfection in a carrier of HTLV-I virus with evidence of selective immunosuppression, *Am. J. Med.* **92:**202–208 (1992).

200. NIMER, S. D., GASSON, J. C., HU, K., SMALBERG, I., WILLIAMS, J. L., CHEN, I. S. Y., AND ROSENBLATT, J. D., Activation of the GM-CSF promoter by HTLV-I and -II tax proteins, *Oncogene* **4:**671–676 (1989).

201. NODA, Y., ISHIKAWA, K.-I., SASAGAWA, A., HONJO, S., MORI, S., TSUJIMOTO, H., AND HAYAMI, M., Hematologic abnormalities similar to the preleukemic state of adult T-cell leukemia in African green monkeys naturally infected with simian T-cell leukemia virus, *Jpn. J. Cancer Res.* **77:**1227–1234 (1986).

202. OGUMA, S., IMAMURA, Y., KUSUMOTO, Y., NISHIMURA, Y., YAMAGUCHI, K., TAKATSUKI, K., TOKUDOME, S., AND OKUMA, M., Accelerated declining tendency of human T-cell leukemia virus type I carrier rates among younger blood donors in Kumamoto, Japan, *Cancer Res.* **52:**2620–2623 (1992).

203. OHSHIMA, K., KIKUCHI, M., MASUDA, Y.-I., KOBARI, S., SUMIYOSHI, Y., EGUCHI, F., MOHTAI, H., YOSHIDA, T., TAKESHITA, M., AND KIMURA, N., Defective provirus form of human T-cell leukemia virus type I in adult T-cell leukemia/lymphoma: Clinicopathological features, *Cancer Res.* **51:**4639–4642 (1991).

204. OKOCHI, K., SATO, H., AND HINUMA, Y., A retrospective study on transmission of adult T-cell leukemia virus by blood transfusion: Seroconversion in recipients, *Vox Sang.* **46:**245–253 (1984).

205. OSAME, M., USUKU, K., IZUMO, S., ET AL., HTLV-1 associated myelopathy, a new clinical entity, *Lancet* **1:**1031–1032 (1986).

206. OSAME, M., IZUMO, S., IGATA, A., MATSUMOTO, M., MATSUMOTO, T., SONODA, S., MITSUTOSHI, T., AND YOSHISADA, S., Blood transfusion and HTLV-1 associated myelopathy, *Lancet* **2:**104–105 (1986).

207. OSAME, M., JANSSEN, R., KUBOTA, H., NISHITANI, H., IGATA, A., NAGATAKI, S., MORI, M., GOTO, I., SHIMABUKURO, H., KHABBAZ, R., AND KAPLAN, J., Nationwide survey of HTLV-1 associated myelopathy/tropical spastic paraparesis in Japan: Association with blood transfusion, *Ann. Neurol.* **28:**50–59 (1990).

208. OSAME, M., AND IGATA, A., The history of discovery and clinicoepidemiology of HTLV-1 associated myelopathy (HAM), *Jpn. J. Med.* **28:**412–414 (1989).

209. OSAME, M., IGATA, A., USUKU, K., ROSALES, R. I., AND MATSUMOTO, M., Mother-to-child transmission in HTLV-1 associated myelopathy, *Lancet* **1:**106 (1987).

210. OUATTARA, S. A., GODY, M., AND DE-THÉ, G., Prevalence of HTLV-1 compared to HIV-1 and HIV-2 antibodies in different groups in the Ivory Coast (west Africa), *J. Acquir. Immune Defic. Syndr.* **2:**481–485 (1989).

211. PARKS, W. P., LENES, B. A., TOMASULO, P. A., SCHIFF, E. R., PARKS, E. S., SHAW, G. M., LEE, H., YAN, H.-Q., LAI, S., HOLLINGSWORTH, C. G., NEMO, G. J., AND MOSLEY, J. W., The Transfusion Safety Study Group, *J. Acquir. Immune Defic. Syndr.* **4:**89–96 (1991).

212. POIESZ, B. J., RUSCETTI, F. W., GAZDAR, A. F., BUNN, P. A., MINNA, J. D., AND GALLO, R. C., Detection and isolation of type-C retrovirus particles from fresh and cultured lymphocytes of patients with cutaneous T-cell lymphoma, *Proc. Natl. Acad. Sci. USA* **77:**7415–7419 (1980).

213. POPOVIC, M., FLOMENBERG, N., VOLKMAN, D., Alteration of T-cell functions by infection with HTLV-1 or HTLV-II, *Science* **226:**459–462 (1984).

214. PROIETTE, F., VIAHOV, D., ALEXANDER, S., TAYLOR, E., KIRBY, A., AND BLATTNER, W. A., Correlates of HTLV-II/HIV-1 seroprevalence and incidence of HTLV-II infection among intravenous drug users, *Int. Conf. AIDS* July 19–24:2 (1992).

215. RAUSCHER, F. J., AND O'CONNOR, T. E., Virology, in: *Cancer Medicine* (J. F. HOLLARD AND E. FREI, III, EDS.), Lea & Febiger, Philadelphia, 1973.

216. REEVES, W. C., LEVINE, P. H., CUEVAS, M., QUIROZ, E., MALONEY, E. M., AND SAXINGER, C. W., Seroepidemiology of human T-cell lymphotropic virus in the Republic of Panama, *Am. J. Trop. Med. Hyg.* **42**(4):374–379 (1990).

217. REIDEL, D. A., EVANS, A. S., SAXINGER, C., AND BLATTNER, W. A., A historical study of human T-lymphotropic virus type I transmission in Barbados, *J. Infect. Dis.* **159:**603–609 (1989).

218. ROBERT-GUROFF, M., WEISS, S. H., GIRON, J. A., JENNINGS, A. M., GINZBURG, H. M., MARGOLIS, I. B., BLATTNER, W. A., AND GALLO, R. C., Prevalence of antibodies to HTLV-I, -II, and -III in intravenous drug abusers from an AIDS endemic region, *J.A.M.A.* **255**(22):3133–3137 (1986).

219. ROBERT-GUROFF, M., KALYANATAMAN, V. S., BLATTNER, W. A., POPOVIC, M., SARNGADHARAN, M. G., MAEDA, M., BLAYNEY, D., CATOVSKY, D., BUNN, P. A., SHIBATA, A., NAKAO, Y., ITO, Y., AOKI, T., AND GALLO, R. C., Evidence for human T-cell lymphoma-leukemia virus infection in family members of human T-cell lymphoma-leukemia positive T-cell leukemia-lymphoma patients, *J. Exp. Med.* **157:**248–258 (1983).

220. ROBERTS, B. D., FOUNG, S. K. H., LIPKA, J. J., KAPLAN, J. E., HADLOCK, K. G., REYES, G. R., CHAN, L., HENEINE, W., AND KHABBAZ, R. F., Evaluation of an immunoblot assay for serological confirmation and differentiation of human T-cell lymphotropic virus types I and II, *J. Clin. Microbiol.* **31**(2)**:**260–264 (1993).

221. ROMAN, G., AND OSAME, M., Identity of HTLV-I-associated tropical spastic paraparesis and HTLV-I-associated myelopathy, *Lancet* **1:**651 (1988).

222. ROMÁN, G. C., ROMÁN, L. N., SPENCER, P. S., AND SCHOENBERG, B. S., Tropical spastic paraparesis: A neuroepidemiological study in Colombia, *Ann. Neurol.* **17:**361–365 (1985).

223. ROONGPISUTHIPONG, A., SUPHANIT, I., YAMAGUCHI, K., NAKAMITSU, M., AND YOSHIKI, K., Very low seroprevalence of HTLV-I in patients with gynecologic disorders in Thailand, *J. Acquir. Immune Defic. Syndr.* **5:**1066–1067 (1992).

224. ROSENBLATT, J. D., GIORGI, J. V., GOLDE, D. W., EZRA, J. B., WU, A., WINBERG, C. D., GLASPY, J., WACHSMAN, W., AND CHEN, I. S. Y., Integrated human T-cell leukemia virus II genome in CD8$^+$ T cells from a patient with "atypical" hairy cell leukemia: Evidence for distinct T and B cell lymphoproliferative disorders, *Blood* **71**(2)**:**363–369 (1988).

225. ROSENBLATT, J. D., GASSON, J. C., GLASPY, J., ET AL., Relationship between human T-cell leukemia virus-II and atypical hairy cell leukemia: Serologic study of hairy cell leukemia patients, *Leukemia* **1**(4)**:**397–401 (1987).

226. ROSENBLATT, J. D., GOLDE, D. W., WACHSMAN, W., GIORGI, J. V., JACOBS, A., SCHMIDT, G. M., QUAN, S., GASSON, J. C., AND CHEN, I. S. Y., A second isolate of HTLV-II associated with atypical hairy cell leukemia, *N. Engl. J. Med.* **315:**372–377 (1986).

227. SAKASHITA, A., HATTORI, T., MILLER, C. W., SUZUSHIMA, H., ASOU, N., TAKATSUKI, K., AND KOEFFLER, H. P., Mutations of the p53 gene in adult T-cell leukemia, *Blood* **79**(2)**:**477–480 (1992).

228. SAKSENA, N. K., SHERMAN, M. P., YANAGIHARA, R., DUBE, D. K., AND POIESZ, B. P., LTR sequence and phylogenetic analyses of a newly discovered variant of HTLV-I isolated from the Hagahai of Papua New Guinea, *Virology* **189:**1–9 (1992).

229. SALAHUDDIN, S. Z., MARKHAM, P. D., POPOVIC, M., SARNGADHARAN, M. G., ORNDORFF, S., FLADAGAR, A., PATEL, A., GOLD, J., AND GALLO, R. C., Isolation of infectious human T-cell leukemia/lymphotropic virus type III (HTLV-III) from patients with acquired immunodeficiency syndrome (AIDS) or AIDS-related complex (ARC) and from healthy carriers: A study of risk groups and tissue sources, *Proc. Natl. Acad. Sci. USA* **82:**5530–5534 (1985).

230. SANDERS, R. C., MAI'IN, P. M., ALEXANDER, S. S., LEVIN, A. G., BLATTNER, W. A., AND ALPERS, M. P., The prevalence of antibodies to human T-lymphotropic virus type I in different population groups in Papua New Guinea, *Arch. Virol.* **130:**327–334 (1993).

231. SATO, K., MARUYAMA, I., MARUYAMA, Y., KITAJIMA, I., NAKAJIMA, Y., HIGAKI, M., YAMAMOTO, K., MIYASAKA, N., OSAME, M., AND NISHIOKA, K., Arthritis in patients infected with human T lymphotropic virus type I: Clinical and immunopathologic features, *Arthritis Rheum.* **34**(6)**:**714–721 (1991).

232. SAWADA, T., TOHMATSU, J., OBARA, T., KOIDE, A., KAMIHIRA, S., ICHIMARU, M., KASHWAGI, S., KAJIYAMA, W., MATSUMURA, N., KINOSHITA, K., YANO, M., YAMAGUCHI, K., KIYOKAWA, T., TAKATSUKI, K., TAGUCHI, H., AND MIYOSHI, I., High-risk of mother-to-child transmission of HTLV-1 in p40^{tax} antibody-negative mothers, *Jpn. J. Cancer Res.* **80:**506–508 (1989).

233. SAWADA, T., IWAHARA, Y., ISHII, K., TAGUCHI, H., HOSHINO, H., AND MIYOSHI, I., Immunoglobulin prophylaxis against milk-borne transmission of human T-cell leukemia virus type I in rabbits, *J. Infect. Dis.* **164:**1193–1196 (1991).

234. SEIKI, M., HATTORI, S., HIRAYAMA, Y., AND YOSHIDA, M., Human adult T-cell leukemia virus: Complete nucleotide sequence of the provirus genome integrated in leukemia cell DNA, *Proc. Natl. Acad. Sci. USA* **80:**3618–3622 (1983).

235. SHIMOTOHNO, K., TAKAHASHI, Y., SHIMIZU, N., TAKANO, M., MIWA, M., SUGIMURA, T., Nucleotide sequence analysis of human T-cell leukemia virus type II, in: *Retroviruses in Human Lymphoma/Leukemia* (M. MIWA ET AL., EDS.), pp. 165–175, Tokyo/VNU Science Press, Ultrecht, 1985.

236. SHIMOYAMA, M., AND THE LYMPHOMA STUDY GROUP (1984–87), Diagnostic criteria and classification of clinical subtypes of adult T-cell leukaemia-lymphoma, *Br. J. Haematol.* **79:**428–437 (1991).

237. SHIOIRI, S., TACHIBANA, N., OKAYAMA, A., ISHIHARA, S., TSUDA, K., ESSEX, M., STUVER, S. O., AND MUELLER, N., Analysis of anti-tax antibody of HTLV-I carriers in an endemic area in Japan, *Int. J. Cancer* **53:**1–4 (1993).

238. SHIRAKAWA, F., TANAKA, Y., ODA, S., CHIBA, S., SUZUKI, H., ETO, S., AND YAMASHITA, U., Immunosuppressive factors from adult T-cell leukemia cells, *Cancer Res.* **46:**4458–4462 (1986).

239. SIDI, Y., MEYTES, D., SHOHAT, B., FENIGI, E., WEISBORT, Y., LEE, H., PINKHAS, J., AND ROSENBLATT, J. D., Adult T-cell lymphoma in Israeli patients of Iranian origin, *Cancer* **65:**590–593 (1990).

240. SODOROSKI, J. G., ROSEN, C. A., AND HASELTINE, W. A., *Trans*-acting transcriptional activation of the long terminal repeat of human T-lymphotropic viruses in infected cells, *Science* **225:**381–385 (1984).

241. SONODA, S., YASHIKI, S., FUJIYOSHI, T., ARIMA, N., TANAKA, H., IZUMO, S., AND OSAME, M., Ethnically defined immunogenetic factors involved in the pathogenesis of HTLV-I associated diseases, *Bio Bull.* **5:**79–88 (1993).

242. STUVER, S. O., TACHIBANA, N., OKAYAMA, A., SHIOIRI, S., TSUNETOSHI, Y., TSUDA, K., AND MUELLER, N., Heterosexual transmission of human T-cell leukemia/lymphoma virus type I among married couples in South-western Japan: An initial report from the Miyazaki Cohort Study, *J. Infect. Dis.* **167:**57–65 (1993).

243. STUVER, S. O., TACHIBANA, N., AND MUELLER, N., A case-control study of factors associated with HTLV-I infection in southern Miyazaki, Japan, *J. Natl. Cancer Inst.* **84:**867–872 (1992).

244. STUVER, S. O., TACHIBANA, N., OKAYAMA, A., ROMANO, F., YOKOTA, T., AND MUELLER, N., Determinants of HTLV-I seroprevalence in Miyazaki Prefecture, Japan: A cross-sectional study, *J. Acquir. Immune Defic. Syndr.* **5:**12–18 (1992).

245. SUGIMOTO, M., NAKASHIMA, H., WATANABE, S., UYAMA, E., TANAKA, F., ANDO, M., AND ARAKI, S., T-lymphocyte aveolitis in HTLV-1 associated myelopathy (letter), *Lancet* **2:**1220 (1987).

246. SULLIVAN, M. T., WILLIAMS, A. E., FANG, C. T., GRANDINETTI,

T., POIESZ, B. J., AND EHRLICH, G. D., Transmission of human T-lymphotropic virus type I and II by blood transfusion: A retrospective study of recipients of blood components (1983 through 1988), *Arch. Intern. Med.* **151:**2043–2048 (1991).

247. TACHIBANA, N., OKAYAMA, A., ISHIZAKI, J., YOKOTA, T., SHISHIME, E., MURAI, K., SHIOIRI, S., TSUDA, K., ESSEX, M., AND MUELLER, N., Suppression of tuberculin skin reaction in healthy HTLV-I carriers from Japan, *Int. J. Cancer* **42:**829–831 (1988).

248. TACHIBANA, N. OKAYAMA, A., ISHIHARA, S., SHIORI, S., MURAI, K., TSUDA, K., GOYA, N., MATSUO, Y., ESSEX, M., STUVER, S., AND MUELLER, N., High HTLV-I proviral DNA level associated with abnormal lymphocytes in peripheral blood from asymptomatic carriers, *Int. J. Cancer* **51:**593–595 (1992).

249. TAGUCHI, H., KOBAYASHI, M., AND MIYOSHI, I., Immunosuppression by HTLV-I infection, *Lancet* **337:**308 (1991).

250. TAJIMA, K., AND HINUMA, Y., Epidemiological features of adult T-cell leukemia virus, in: *Pathophysicological Aspects of Cancer Epidemiology* (G. MATHE AND P. RIZENSTEIN, EDS.), pp. 75–87, Pergamon Press, Oxford, 1984.

251. TAJIMA, K., TOMINAGO, S., SUCHI, T., ET AL., Epidemiological analysis of the distribution of antibody to adult T-cell leukemia-virus-associated antigen: Possible horizontal transmission of adult T-cell leukemia virus, *Jpn. J. Cancer Res.* **73:**893–901 (1982).

252. TAJIMA, K., TOMINAGO, S., SUCHI, T., FUKUI, K., KOMODA, H., AND HINUMA, Y., HTLV-I carriers among migrants from an ATL-endemic area to ATL non-endemic metropolitan areas in Japan, *Int. J. Cancer* **37:**383–387 (1986).

253. TAJIMA, K., KAMURA, S., ITO, S-I., ITO, M., NAGATAMA, M., KINOSHITA, K., AND IKED, S., Epidemiological features of HTLV-I carriers and incidence of ATL in an ATL-endemic island: A report of the community-based cooperative study in Tsushima, Japan, *Int. J. Cancer* **40:**741–746 (1987).

254. TAJIMA, K., ITO, S.-I., AND TSUSHIMA AND THE ATL STUDY GROUP, Prospective studies of HTLV-1 and associated disease in Japan, in: *Human Retrovirology: HTLV* (W. A. BLATTNER, ED.), pp. 267–280, Raven Press, New York, 1990.

255. TAJIMA, K., THE T- AND B-CELL MALIGNANCY STUDY GROUP, AND COAUTHORS OF THE FOURTH NATIONWIDE STUDY OF ADULT T-CELL LEUKEMIA/LYMPHOMA (ATL) IN JAPAN, Estimates of risk of ATL and its geographical and clinical features, *Int. J. Cancer* **45:**237–243 (1990).

256. TAKAHASHI, K., TAKEZAKI, T., OKI, T., KAWAKAMI, K., YASHIKI, S., FUJIYOSHI, T., USUKU, K., AND MUELLER, N., Inhibitory effect of maternal antibody on mother-to-child transmission of human T-lymphotropic virus type I, *Int. J. Cancer* **49:**673–677 (1991).

257. TAKATSUKI, K., UCHIYAMA, T., SAGAWA, K., AND YODOI, J., Adult T-cell leukemia in Japan, in: *Topics in Hematology* (S. SENO, F. TAKAKU, AND S. IRINO, EDS.), pp. 73–77, Excerpta Medica, Amsterdam, 1977.

258. TAKAYANAGUI, O. M., CANTOS, J. L. S., AND JARDIM, E., Tropical spastic paraparesis in Brazil, *Lancet* **337:**309 (1991).

259. TANAKA, Y., SHIRAKAWA, F., ODA, S., CHIBA, S., SUZUKI, H., ETO, S., AND YAMASHITA, U., Effects of immunosuppressive factors produced by adult T cell leukemia cells on B cell responses, *Jpn. J. Cancer Res.* **78:**1390–1399 (1987).

260. TEICH, N., Taxonomy of retroviruses, in: *Molecular Biology of Tumor Viruses: RNA Tumor Viruses* (R. WEISS, N. TEICH, H., VARMUS AND J. COFFIN, EDS.), pp. 25–208, Cold Spring Harbor Laboratory, Cold Spring Harbor, NY, 1984.

261. TEMIN, H. M., AND MIZUTANI, S., RNA-dependent DNA polymerase in virions of Rous sarcoma virus, *Nature* **226:**1211–1213 (1970).

262. THE T- AND B-CELL MALIGNANCY STUDY GROUP, Statistical analysis of immunologic, clinical and histopathologic data on lymphoid malignancies in Japan, *Jpn. J. Clin. Oncol.* **11:**15–38 (1981).

263. THE T- AND B-CELL MALIGNANCY STUDY GROUP, Statistical analyses of clinicopathological, virological and epidemiological data on lymphoid malignancies with special reference to adult T-cell leukemia/lymphoma: A report of the second nationwide study of Japan, *Jpn. J. Clin. Oncol.* **15:**517–535 (1985).

264. THE T- AND B-MALIGNANCY STUDY GROUP, The third natiowide study on adult T-cell leukemia/lymphoma (ATL) in Japan: Characteristic patterns of HLA antigens and HTLV-I infection in ATL patients and their relatives, *Int. J. Cancer* **41:**505–512 (1988).

265. TOKUDOME, S., TOKUNAGA, O., SHIMAMOTO, Y., MIYAMOTO, Y., SUMIDA, I., KIKUCHI, M., TAKESHITA, M., IKEDA, T., FUJIWARA, K., YOSHIHARA, M., YANAGAWA, T., AND NISHIZUMI, M., Incidence of adult T-cell leukemia/lymphoma among human T-lymphotropic virus type I carriers in Saga, Japan, *Cancer Res.* **49:**226–228 (1989).

266. TOMONAGA, M., IKEDA, S., KINOSITA, K., MOMITA, S., KAMIHIRA, S., KANDA, N., ITO, M., ITO, S., NAKATA, K., KINOSHITA, H., SHIMOTOHNO, K., AND OSHIBUCHI, T., The prevalence rate of monoclonal proliferation of HTLV-I infected T-lymphocytes (HCMPT) among HTLV-I carriers [Abstract], Proceedings from the Fifth International Conference on Human Retrovirology; HTLV, Kumamoto, Japan, 1992.

267. TRUJILLO, J. M., CONCHA, M., MUNOZ, A., BERGONZOLI, G., MORA, C., BORRERO, I., GIBBS, C. J., JR., AND ARANGO, C., Seroprevalence and cofactors of HTLV-I infection in Tumaco, Colombia, *AIDS Res. Hum. Retroviruses* **8:**651–657 (1992).

268. TSCHACHLER, E., BÖHNLEIN, E., FELZMANN, S., AND REITZ, JR., M. S., Human T-lymphocyte virus type I *tax* regulates the expression of the human lymphotoxin gene, *Blood* **81:**95–100 (1993).

269. UEDA, K., KUSUHARA, K., TOKUGAWA, K., MIYAZAKI, C., YOSHIDA, C., TOKUMURA, K., SONODA, S., AND TAKAHASHI, K., Cohort effect on HTLV-I seroprevalence in southern Japan, *Lancet* **2:**1337 (1989).

270. UEMATSU, T., HANADA, S., NOMURA, K., ET AL., The incidence of anti-ATLA (antibodies to adult T-cell leukemia-associated antigen) among hemodialysis patients in the Kagoshima district with specific reference to blood transfusions, *J. Jpn. Soc. Dial. Ther.* **19:**757–762 (1986).

271. UNO, H., KAWANO, K., MATSUOKA, H., AND TSUDA K., HLA and adult T-cell leukaemia: HLA-linked genes controlling susceptibility to human T-cell leukaemia virus type 1, *Clin. Exp. Immunol.* **71:**211–215 (1988).

272. USUKU, K., SONADA, S., OSAME, M., YASHIKI, S., TAKAHASHI, K., MATSUMOTO, M., SAWADA, T., TSUJI, K., TARA, M., AND IGATA, A., HLA haplotype-linked high immune responsiveness against HTLV-I in HTLV-I-associated myelopathy: Comparison with adult T-cell leukemia/lymphoma, *Ann. Neurol.* **23**(Suppl.): S143–S150 (1988).

273. VALLEJO, A., AND GARCIA-SAIZ, A., Isolation and nucleotide sequence analysis of human T-cell lymphotropic virus type II in Spain, *J. Acquir. Immune Defic. Syndr.* **7**(5):517–518 (1994).

274. VARMUS, H., Retroviruses, *Science* **244:**1427–1435 (1988).

275. VERDIER, M., DENIS, F., SANGARÉ, A., ET AL., Prevalence of antibody to human T-cell leukemia virus type 1 (HTLV-1) in populations of Ivory Coast, west Africa, *J. Infect. Dis.* **160:**363–370 (1989).

276. VITEK, C. R., GRACIA, F., GIUSTI, R., FUKUDA, K., GROEN, D. B., CASTILLO, L. L., ARMIEN, B., KHABBAZ, R. F., LEVINE, P. H.,

AND KAPLAN, J. E., Evidence for sexual and mother-to-child transmission for HTLV-II among Guaymi Indians, Panama, *J. Infect. Dis.* **171:**1022–1026 (1995).

277. VLAHOV, D., KHABBAZ, R., COHN, S., GALAI, N., TAYLOR, E., AND KAPLAN, J., Risk factors for HTLV-II seroconversion among injecting drug users in Baltimore, *AIDS Res. Hum. Retroviruses* **10**(4)**:**448 (1994).

278. WATANABE, T., YAMAGUCHI, K., TAKATSUKI, K., OSAME, M., AND YOSHIDA, M., Constitutive expression of parathyroid hormone-related protein (PTHrP) gene in HTLV-I carriers and adult T-cell leukemia patients which can be *trans*-activated by HTLV-I *tax* gene, *J. Exp. Med.* **172:**759–765 (1990).

279. WEISS, S. H., Laboratory detection of human retroviral infection, in: *AIDS and Other Manifestations of HIV Infection,* 2nd ed. (G. P. WORMSER, ED.), pp. 95–116, Raven Press, New York, 1992.

280. WELLES, S. L., LEVINE, P. H., JOSEPH, E. M., GOBERDHAN, L. J., LEE, S., MIOTTI, A., CERVANTES, J., BERTONI, M., JAFFE, E., AND DOSIK, H., An enhanced surveillance program for adult T-cell leukemia in central Brooklyn, *Leukemia* **8:**S111–115 (1994).

280a. WELLES, S. L., TACHIBANA, N., OKAYAMA, A., SHIOIRI, S., ISHI-HARA, S., MURAI, K., AND MUELLER, N. E., Decreased reactivity to PPD among HTLV-I carriers in relation to virus and hema-tologic status. *Int. J. Cancer* **56:**337–340 (1994).

281. WIKTOR, S. Z., PATE, E. J., MURPHY, E. L., PALKER, T. J., CHAM-PEGNIE, E., RAMLAL, A., CRANSTON, B., HANCHARD, B., AND BLATTNER, W. A., Mother-child transmission of human T-cell lymphotropic virus type I (HTLV-I) in Jamaica: Association with antibodies to envelope glycoprotein (gp46) epitopes, *J. Acquir. Immune Defic. Syndr.* **6:**1162–1167 (1993).

282. WIKTOR, S. Z., ALEXANDER, S. S., SHAW, G. M., WEISS, S., MURPHY, E. S., WILKS, R. J., SHORTY, V. J., HANCHARD, B., AND BLATTNER, W. A., Distinguishing between HTLV-I and HTLV-II by Western blot, *Lancet* **335:**1533 (1990).

283. WIKTOR, S. Z., JACOBSON, S., WEISS, S. H., MCFARLIN, D. E., JACOBSON, S., SHAW, G. M., SHORTY, V. J., AND BLATTNER, W. A., Spontaneous lymphocyte proliferation in HTLV-II infection, *Lancet* **337:**327–328 (1991).

284. WILLIAMS, A. E., FANG, C. T., SLAMON, D. J., POIESZ, B. J., SANDLER, G., DARR II, W. F., SHULMAN, G., MCGOWAN, E. I., DOUGLAS, D. K., BOWMAN, R. J., PEETOOM, F., KLEINMAN, S. H., LENES, B., AND DODD, R. Y., Seroprevalence and epidemiologi-cal correlates of HTLV-I infection in U.S. blood donors, *Science* **240:**643–646 (1988).

285. YAMAGUCHI, K., NISHIMA, H., KOHROGI, H., JONO, M., MIYA-MOTO, Y., AND TAKATSUKI, K., A proposal for smoldering adult T-cell leukemia: A clinicopathologic study of five cases, *Blood* **62:**758–766 (1983).

286. YAMAGUCHI, K., KIYOKAWA, T., NAKADA, K., YUL, L. S., ASOU, N., ISHII, T., SANADA, I., SUKI, M., YOSHIDA, M., MATUTSI, E., CATOVSKY, D., AND TAKASUKI, K., Polyclonal integration of HTLV-I proviral DNA in lymphocytes from HTLV-I seropositive individuals: An intermediate state between the healthy carrier state and smouldering ATL, *Br. J. Haem.* **68:**169–174 (1988).

287. YANAGIHARA, R., Human T-cell lymphotropic virus type I infec-tion and disease in the Pacific basin, *Hum. Biol.* **64:**843–854 (1992).

288. YANAGIHARA, R., JENKINS, C. L., ALEXANDER, S. S., MORA, C. A., AND GARRUTO, R. M., Human T-lymphotropic virus type I infection in Papua, New Guinea: High prevalence among the Hagahai confirmed by Western analysis, *J. Infect. Dis.* **162:**649–654 (1990).

289. YANAGIHARA, R., AJDUKIEWICZ, A. B., GARRUTO, R. M., SHAR-

LOW, E. R., WU, X. Y., ALEMAENA, O., SALE, H., ALEXANDER, S. S., AND GAJDUSEK, D. C.,Human T-lymphotropic virus type I infection in the Solomon Islands, *Am. J. Trop. Med. Hyg.* **44:**122–130 (1991).

290. YOSHIDA, M., OSAME, M., USUKU, K., MATSUMOTO, M., AND IGATA, A., Viruses detected in HTLV-1 associated myelopathy and adult T-cell leukemia are identical on DNA blotting, *Lancet* **1:**1085–1086 (1987).

291. YOSHIDA, M., SEIKI, M., YAMAGUCHI, K., AND TAKATSUKI, K., Monoclonal integration of human T-cell leukemia provirus in all primary tumors of adult T-cell leukemia suggests causative role of human T-cell leukemia virus in the disease, *Proc. Natl. Acad. Sci. USA* **81:**2534–2537 (1984).

292. YOSHIDA, M., MIYOSHI, I., AND HINUMA, Y., Isolation and char-acterization of retrovirus from all lines of human adult T-cell leukemia and its implication in disease, *Proc. Natl. Acad. Sci. USA* **79:**2031–2035 (1982).

293. YOSHIDA, M., AND FUJISAWA, T.-I., Positive and negative regula-tion of HTLV-I gene expression and their roles in leukemogenesis in ATL, in: *Advances in Adult T-cell Leukemia and HTLV-I Research* (K. TAKATSUKI, Y. HINUMA, AND M. YOSHIDA, EDS.), pp. 217–235, Gann Monograph on Cancer Research No. 39, Japan Scientific Societies Press, Tokyo, and CRC Press, Boca Raton, 1992.

294. ZANINOVIC, V., SANZON, F., LOPEZ, F., VELANDIA, G., BLANK, A., BLANK, M., FUJIYAMA, C., YASHIKI, S., MATSUMOTO, D., KATAHIRA, Y., MIYASHITA, H., FUJIYOSHI, T., CHAN, L., SAWADA, T., MIURA, T., HATAMI, M., TAJIMA, K., AND SONODA, S., Geo-graphic independence of HTLV-I and HTLV-II foci in the Andes Highland, the Atlantic coast, and the Orinoco of Colombia, *AIDS Res. Hum. Retroviruses* **10**(1)**:**97–101 (1994).

295. ZELLA, D., MORI, L., SALA, M., FERRANTE, P., CASOLI, C., MAGNANI, G., ACHILLI, G., CATTANEO, E., LORI, F., AND BER-TAZZONE, U., HTLV-II infection in Italian drug abusers, *Lancet* **336:**575–576 (1990).

296. ZENG, Y., LAN, X. Y., FANG, J., WANG, P. Z., WANG, Y. R., SUI, Y. F., WANG, Z. T., HU, R. J., AND HINUMA, Y., HTLV antibody in China, *Lancet* **1:**799–800 (1984).

297. ZÜCKER-FRANKLIN, D., HOOPER, W. C., AND EVATT, B. L., Hu-man lymphotropic retroviruses associated with mycosis fun-goides: Evidence that human T-cell lymphotropic virus type II (HTLV-II) as well as HTLV-I may play a role in the disease, *Blood* **80**(6)**:**1537–1545 (1992).

298. ZÜCKER-FRANKLIN, D., HUANG, Y. Q., GRUSKY, G. E., AND FRIEDMAN-KIEN, A. E., Kaposi's sarcoma in a human immuno-deficiency virus-negative patient with asymptomatic human T-lymphotropic virus type I infection, *J. Infect. Dis.* **167:**987–989 (1993).

12. Suggested Reading

FIELDS, B. N., KNIPE, D. M., AND HOWLEY, P. M. (EDS.), *Fields Virology,* 3rd ed., Chapter 59, Raven Press, New York, 1995.

GESSAIN, A., AND GOUT, O., Chronic myelopathy association with human T-lymphotropic virus type I (HTLV-I), *Ann. Intern. Med.* **117:**933–946 (1992).

TAKASUKI, K., HINUMA, Y., AND YOSHIDA, M., *Advances in Adult T-cell Leukemia and HTLV-I Research,* Gann Monograph on Cancer Re-search No. 39, Japan Scientific Societies Press, Tokyo, CRC Press, Boca Raton, 1992.

CHAPTER 26

Rhinoviruses

Jack M. Gwaltney, Jr.

1. Introduction

Rhinoviruses are the most important common-cold viruses to be discovered. The name *rhinovirus* reflects the prominent nasal involvement seen in infections with these viruses. The large rhinovirus genus, which is a member of the *Picornavirus* family, contains over 100 different immunotypes. The discovery of the rhinoviruses led to the realization that the common cold is an enormously complex syndrome. The number of antigenically distinct rhinoviruses is so large that one can be infected with a different rhinovirus each year and still not experience all the known types in a lifetime. The antigenic diversity of the rhinovirus group has proved an insurmountable obstacle to rhinovirus vaccine development. It is now known that the cellular receptor site for rhinovirus is shielded from the immune system, eliminating it as a target for vaccines and further discouraging prospects for control of rhinovirus colds by this approach. Recent work on rhinovirus has focused on understanding pathogenesis and on developing control measures such as chemoprophylaxis, chemotherapy, and interruption of transmission.

2. Historical Background

Rhinovirus colds may have affected humans and higher primates for many thousands of years, although natural rhinovirus colds have not been documented in nonhuman primates.[83] A closely related member of the *Picornavirus* family, poliovirus, is known to have caused human disease in ancient times, so it is probable that rhinoviruses were in existence then also. Colds were a nuisance in early civilization; then, as now, many useless remedies were proposed for their treatment. In 400 BC, Hippocrates noted that bleeding was a frequently used, although worthless, treatment for colds. In the first century, Pliny the Younger prescribed "kissing the hairy muzzle of a mouse" for colds. The first sound epidemiologic knowledge about acute respiratory disease came with the observations that sea voyagers and the inhabitants of isolated communities were free of colds while not in contact with the outside world but developed colds when such contact was reestablished. This led to the important conclusion that colds are contagious.

Direct evidence of the infectious nature of colds came in 1914 from the volunteer studies of Kurse,[140] who produced experimental colds in volunteers by intranasal inoculation of cell-free filtrates of nasal secretions from persons with colds. Similar experiments by Dochez *et al.*[50] in 1930 confirmed that colds could be transmitted by bacteria-free filtrates, suggesting that the responsible agents might be viruses. At the same time, epidemiologic studies of acute respiratory disease in populations had been started. Van Loghem[227] measured the incidence of colds and observed their relationship to the seasons. Frost and Gover[72] made the perceptive observation that common respiratory disease appearing during the months of high prevalence, September to March, was composed of a series of short epidemics of irregular sequences and magnitude. This suggested that colds were caused by a variety of different agents occurring in succession. In the 1940s and 1950s, long-term studies of colds in the home by Dingle *et al.*[48] yielded precise information on attack rates by age and the importance of the home as a site for transmission of respiratory infections. During the same period, a group at the Common Cold Research Unit at Salisbury, England, headed by Andrewes and later Tyrrell, was vigorously pursuing questions related to the etiology and epidemiology of colds.[2] Colds were successfully transmitted in volunteers using nasal secretions that were later shown

Jack M. Gwaltney, Jr. • Department of Internal Medicine, University of Virginia School of Medicine, Charlottesville, Virginia 22908.

to contain rhinoviruses. Attempts at the time to establish growth of the virus in artificial culture were unsuccessful.

Specific work on rhinoviruses began in 1956 when Pelon *et al.*[175] and Price,[185] working separately, reported the isolation of a new virus that was subsequently given the designation *rhinovirus 1A*. Within a few years, Ketler *et al.*,[137] using the highly sensitive human embryonic lung cells developed by Hayflick and Moorhead[113] and employing growth methods developed at the Salisbury Common Cold Unit,[226] isolated a number of different serological types, indicating that the rhinovirus group would not be small. Epidemiologic studies conducted by Hamre and Procknow[102] during the same period established that rhinoviruses were responsible for a significant amount of acute respiratory disease. Specific rhinovirus infection rates and the finding of recurrent fall peaks of rhinovirus colds were reported from a longitudinal study by Gwaltney *et al.*[87] In further studies of rhinovirus epidemiology by Monto,[158] Dick *et al.*,[46] and Hendley *et al.*,[116] the importance of the family setting and of schoolchildren in particular in favoring rhinovirus transmission was demonstrated. Couch *et al.*[38] noted the surprisingly small amount of virus necessary to initiate experimental infections in volunteers. This group also provided important information on the pathogenesis[54] and immunology of rhinovirus infections.[193] In 1967, a collaborative program directed by Kapikian *et al.*[132] assigned numbers 1A–55 to the rhinovirus types then known. In 1971, a second phase of this program added types 56–89.[133] Results of a third phase of the numbering program completed in 1987 has extended the numbering system to include 100 rhinovirus types.[97] More recently, work has focused on understanding routes of viral transmission[85,92,117] and mechanisms of pathogenesis.[214,222,223] Also, the structure of the viral shell[194,195] and the composition of the viral genome[22] have determined and a new therapeutic approach consisting of the simultaneous administration of an antiviral agent and of compounds that block the action of selected inflammatory mediators has shown promise.[82]

3. Methodology Involved in Epidemiologic Analysis

3.1. Surveillance and Sampling

Longitudinal studies of rhinovirus epidemiology have provided data on rhinovirus attack rates. Surveillance of a population of young adults at an insurance company in Charlottesville, Virginia, was conducted by collecting illness data on symptom-record cards in conjunction with weekly personal contact by a nurse-epidemiologist.[87] This nurse also collected samples at the time of illness. In addition, samples were obtained weekly from asymptomatic persons in a randomly selected sample of the study population. In another study, families from representative segments of the population in Tecumseh, Michigan, were surveyed by weekly telephone contact with a single household respondent who provided illness information for the family.[160,165–167] In the third investigation, mothers of families with newborn infants in a group health cooperative in Seattle, Washington, recorded illness information on their families and were visited twice weekly for routine sampling.[70] In the latter two studies, specimens were collected during home visits by a nurse-epidemiologist when illness was reported to the study team by telephone. Specimens for viral culture are usually collected from adults by nasal swabs or nasal washes. In young children, nasopharyngeal aspirates have been reported to be superior to nasal swabs for rhinovirus isolation.[41]

3.2. Methods of Virus Isolation, Propagation, and Identification

Cell culture is the standard method for rhinovirus isolation and propagation. Rhinoviruses grow best at temperatures of 33–34°C under conditions of motion[130] and will not grow in embryonated eggs or suckling mice. Most epidemiologic studies have employed human embryonic lung cells, strains W138 and MRC5, or strains of human embryonic lung cells originated by the laboratory conducting the study. Rhinovirus cytopathic effect in W138 and MRC5 cell cultures is readily discernible, making these easy systems with which to work. The sensitivity of these cells to rhinoviruses appears to be similar to that of the nasal mucosa of volunteers. Volunteer challenge experiments comparing rhinovirus median human and tissue culture infectious dose (HID_{50} and $TCID_{50}$) have shown 1 HID_{50} to be equivalent to 0.03–0.75 $TCID_{50}$.[51]

There are problems, however, with the use of human embryonic lung-cell cultures. The sensitivity to rhinovirus of cell strains of different origin may vary 100-fold or more, for poorly understood reasons.[13] Also, different lots of the same strain, such as W138, may have unpredictable variations in rhinovirus sensitivity that are unexplained.[84] Interpretations of rhinovirus morbidity data must take these variations into account, since rates of rhinovirus-associated illness are directly related to the sensitivity of the cell cultures used.

Rhinoviruses will grow in other cell lines and strains derived from human and primate tissues, including rhesus monkey kidney, human embryonic kidney, and KB. The sensitivity of these cells for rhinoviruses tends to be less consistent than that of W138 cells. A strain of HeLa cells with enhanced sensitivity to rhinoviruses has been developed and proven useful for propagation of antigen and for serological procedures.[31] These M-HeLa cells have been used to grow rhinovirus harvests with exceptionally high titers (10^9 PFU/ml)[33] and to prepare large quantities of antigen in suspension cultures.[220] Certain rhinovirus serotypes were recovered from original specimens with M-HeLa cells but not with human diploid-cell cultures.[34,143] All the first 55 numbered rhinovirus types have been plaqued using a method that employs HeLa cells and an agarose overlay containing medium with added magnesium and DEAE-dextran.[66]

The earlier division of rhinoviruses into H and M strains on the basis of growth in cells of human or monkey origin has been of limited epidemiologic importance. M strains tend to grow better in cell culture and thus were more easily recovered with the less sensitive systems used in earlier studies.[90] Consideration should be given to the greater ease of recovery of M strains when epidemiologic data are being evaluated, since this variable could result in overestimation of the importance of M rhinoviruses. Recent work has shown that H strains can be adapted to grow in monkey-kidney cells, suggesting that the division into H and M strains is not based on major differences in the biological properties of rhinoviruses.[52]

Organ cultures of fetal human trachea and other ciliated epithelium have been used to isolate rhinoviruses that did not grow initially in cell culture.[122,225] Comparison of the sensitivity for rhinovirus isolation of standard cell culture and of organ culture has failed to show clear superiority of the organ-culture system[118]; both systems are necessary for optimal recovery of these viruses. Once isolated in organ culture, rhinoviruses can usually be adapted to cell culture. The organ-culture strains have been found to be types that have also been recovered in cell culture. Because of the limited supply of fetal material, it has not been possible to use organ-culture systems in large epidemiologic studies.

The use of the polymerase chain reaction with nucleic acid probes has also been adapted to detection of rhinovirus in clinical specimens.[6,129] The sensitivity of this method compared to sensitive human embryonic lung cell cultures and its practicability for epidemiologic studies have not been well defined. Also, an enzyme-linked immunosorbent assay (ELISA) has been developed for detection of rhinovirus.[44]

Experimental infections with human rhinoviruses have been produced in chimpanzees[45] and gibbons,[183] and a variant of human rhinovirus type 2 has been adapted to replicate in the lungs of Balb/c mice.[232] Rhinoviruses have been isolated from cattle,[157] and respiratory viruses with characteristics similar to those of human rhinoviruses have been recovered from cats[40] and horses.[49]

3.3. Methods Used for Serological Surveys and Antibody Measurements

The multiplicity of rhinovirus types and their relative immunologic specificity have prevented the general use of serological techniques for measuring infection rates. Serological study of infection rates is possible, however, when the types of rhinoviruses circulating in small populations, such as families, are known from viral cultures. Testing for the presence of rhinovirus antibody has been done with the neutralization (N) test. The N test has been used to identify specific antigenic types of viruses and to measure antibody in human serum and nasal secretions. An ELISA has recently been developed for measuring rhinovirus antibody in serum and nasal secretions that was reported to correlate well with the N test.[8]

In experimental rhinovirus infection, virus shedding was found to be more sensitive than antibody response as a means of detecting infection,[114] whereas in studies of natural infections, either procedure alone identified only about two thirds of the diagnosed infections.[9] In family studies, 20–40% of total infections were detected only by serology in persons who had both tests performed.[46,116]

For typing rhinoviruses, hyperimmune rhinovirus antisera have been produced in a number of animal species, including rabbits, guinea pigs, calves, goats, and baboons. Some goat and calf antisera have contained cytotoxic substances that have caused difficulties in the interpretation of N test results.[31] The large number of rhinovirus serotypes has led to the use of antisera pools for serotype identification. An efficient method of antisera pooling is the combinatorial method.[135] Serological identifications of rhinoviruses in large epidemiologic studies can be done with pooled antisera used in microneutralization systems.[78,139]

The accepted standard for serological identity of an unknown rhinovirus is neutralization of virus concentrations ranging from 10 to 300 $TCID_{50}$ by 20 units of antibody.[130] For measuring N antibody in human serum and nasal washings, it is necessary to use small doses of virus (3–30 $TCID_{50}$) for the test to have satisfactory sensitivity.[57]

4. Characteristics of the Virus that Affect the Epidemiologic Pattern

4.1. Physical and Biochemical Characteristics

Rhinoviruses have physical and biochemical properties that put them in the picornavirus family (Table 1).[172,173,206,228] The human rhinovirus virion is a 30-nm-diameter, nonenveloped particle with a shell composed of

Table 1. Characteristics of Rhinovirus

Physical and biochemical

Size: 30 nm

Shape: capsid with icosahedral symmetry with proposed structure of 60 copies each of four polypeptides (VP1–VP4)

Nucleic acid: single-stranded RNA of $2.6 \pm 0.1 \times 10^6$ daltons (30% of total particle mass)

Ether: resistant

Acid: labile (pH 3–5)

Virus: synthesis and maturation in cytoplasm

Biological

Optimal temperature of growth 33–35°C and restriction of growth at 37°C

Inability to survive and replicate in the intestinal tract

Survival on skin and environmental surfaces

Two receptor families for host cells

Antigenic

Native antigenicity: type-specific (D antigenicity)

One hundred or more numbered native antigenic types

Direct and indirect antigenic relationships between some native antigenic types demonstrable with hyperimmune sera

Altered antigenicity (by heat or urea): cross-reactive between types (C antigenicity)

three proteins (VP1, VP2, VP3).[195] The rhinovirus shell is more loosely packed than that of enterovirus, accounting for rhinovirus' greater buoyant density and its susceptibility to inactivation on acid exposure. X-ray defraction studies of the rhinovirus shell have disclosed the presence of a depression on the surface at the junction between the plateau of VP1 and those of VP2 plus VP3.[194] This depression contains the recognition site for the host cell receptor.

The genome of several rhinovirus types has been sequenced, that of rhinovirus type 14 being 7209 nucleotides long.[22] Rhinovirus genomes have been found to share 45 to 62% homology with poliovirus genomes. Similarity in physical nature of the two groups may help explain similarities in epidemiologic behavior, i.e., increased prevalence in late summer and fall and possible spread by direct contact with infectious secretions.

4.2. Biological Characteristics

The biochemical basis for the optimal temperature range for rhinovirus growth is unknown, but this property may be of major epidemiologic importance (Table 1). The mean temperature of nasal mucosa, 33–35°C, corresponds to the optimal temperature for rhinovirus replication. At 37°C, virus yields fall to 10–50% of optimum.[215] In natural infection in man, rhinovirus concentrations are higher in nasal secretions than in pharyngeal secretions,

saliva, or secretions obtained by simulated coughs and sneezes.[117] Attempts to isolate rhinovirus from blood have not been successful,[54,59] nor does rhinovirus survive and replicate in the intestinal tract. Studies of rhinovirus survival in the gut suggest that the temperature of 37°C may be a decisive factor in inhibiting growth, although gastrointestinal secretions and transmit time may also have adverse effects on virus survival.[25] On the basis of these observations, it may be possible that one reason for the different pathogenic and epidemiologic behavior of enteroviruses and rhinoviruses is the difference in the optimal temperature for growth of the two groups of viruses.

Rhinoviruses have been divided into three groups on the basis of their cellular receptors.[28] Ninety-one of the viral immunotypes, the "major group," use the intercellular adhesion molecule-1 (ICAM-1) cellular receptor.[77,212] Another 10 immunotypes (1A, 1B, 2, 29–31, 44, 47, 49, 62) use another unknown receptor, while type 87 requires sialic acid for attachment. ICAM-1 binds into the depression on the viral surface, a site that is inaccessible to antibody.[172] Viral attachment to cellular receptor can be blocked when an immunoglobulin G (IgG) molecule binds to the surface of the virus in a position that spans the canyon.[208] ICAM-1 expression in fibroblasts is induced by some cytokines and inhibited by others.[182]

4.3. Antigenic Characteristics

Rhinoviruses in their native state contain type-specific surface antigens (Table 1). By means of atomic resolution,[194] four previously recognized neutralizing immunogenic regions[202] have been identified as external protrusions on the viral shell. On the basis of collaborative programs, rhinoviruses have been classified as serotypes 1–100 and subtype 1A.[97,131,132] Using antisera for types 1–89, it was possible to identify over 90% of wild rhinovirus strains recovered in three epidemiologic studies.[97,159] This suggests that most rhinovirus immunotypes, at least those currently circulating in the United States, have now been identified and that new types are not continuously emerging.

The criterion for the selection of numbered prototype viruses was the absence of cross-neutralization with other prototype candidates using animal hyperimmune antiserum at dilutions of 1:2–20 in a standard N test. There was a virtual absence of cross-reactions with the antisera that were used in the numbering program. Recent work with high-titered hyperimmune antisera, discussed below, has disclosed antigenic relationships among some of the numbered types that were not discovered in the collaborative program. Despite these findings, which are discussed in the next paragraph, the large number of antigenically different types of rhinoviruses is undoubtedly an important characteristic of the group, influencing epidemiologic behavior and accounting for the frequency of rhinovirus colds.

In an early study, antigenic relationships among different rhinovirus types were reported, using hyperimmune bovine antisera in N tests.[65] The bovine antisera were later recognized to contain anticellular antibody. When this antibody was removed, the antigenic cross-reactions largely disappeared.[30] More recently, potent monotypic animal antisera were used to demonstrate both reciprocal and one-way cross-reactions among numbered rhinovirus types studied.[35,198] The cross-reactions were usually minor. A number of these relationships were indirect and were demonstrable only by primary immunization with one rhinovirus type followed by immunization with a different but related type.

The importance of cross-reactions in immunity in humans is currently unknown, and the results of work in this area are contradictory. Neutralization tests carried out with paired sear from patients have usually not shown significant cross-reactions following natural rhinovirus infections.[102] On the other hand, in a study of experimental infections in volunteers, heterotypic antibody responses were relatively common after infection with some types.[67]

The native antigenicity of rhinoviruses can be altered by experimental means. Treatment at pH 5 at 56°C or in 2 M urea produces virus particles that react in immunodiffusion and CF tests with heterologous types.[146] When the virus is in this C-antigenic state, which results from a configurational change that exposes normally hidden determinants, it is unable to attach to cell receptors. This alteration in antigenicity, which also occurs after virus attachment to host cells, may be an important step in the initiation of infection[145] but probably plays no role in immunity to infection.

5. Descriptive Epidemiology

5.1. Incidence and Prevalence of Infection

5.1.1. Age- and Sex-Specific Infection and Illness Rates.
Rhinovirus infections are the most common of the acute respiratory infections[32,90] and probably the most common of all acute infections of humans. Infection rates based on virus isolations from routine specimens from family members in Seattle with and without symp-

Table 2. Rhinovirus Infection Rates: Calculated from Surveillance and Sampling of All Persons—Well and Ill

Location	Age (yr)	Person-year of observation	Infections per person-year
Seattle, Washington[32]	0–1	144	1.21
	2–5	135	0.54
	6–9	22	0.55
	Mothers	208	0.20
	All ages	510	0.59
Chicago, Illinois[101]	19–32	466	0.74[a]
Charlottesville, Virginia[87]	16–45	500	0.77[b]

[a]Rhinovirus isolation percentages for well and ill persons 1.5 and 25.4%, respectively; sampling interval of well persons 6 weeks; data collected over four periods of 9 months and adjusted to annual rates.
[b]Rhinovirus isolation percentages for well and ill persons 2.1 and 23.3%, respectively; sampling interval of well persons arbitrarily adjusted to 6 weeks; data collected over 1 year.

toms were 0.59 per person-year (Table 2). Rates in this population ranged from 1.21 in the 0 to 1-year age group to 0.20 in mothers; values were intermediate in children 2–9 years of age. Similar data collected from medical students in Chicago[101] and insurance company employees in Charlottesville[87,90] gave rhinovirus infection rates of 0.74 and 0.77 per person-year, respectively.

True rhinovirus infection rates are probably higher than reported, since currently available rhinovirus culture methods lack optimal sensitivity (see Section 3). The overall rhinovirus infection rates of 0.74 and 0.77 per person-year in Chicago and Charlottesville, respectively, are probably minimum values for the true incidence of rhinovirus infections in young adults. Adjustment of the Seattle rates for children to those measured for young adults in Chicago and Charlottesville gives projected rhinovirus infection rates in young children of up to 1.5 per person-year. Of particular interest was the increase in incidence of rhinovirus infections in females 20–39 years in the Michigan population[166] and 16–24 in the Charlottesville population. These findings may relate to the importance of young children in disseminating rhinovirus in the home, particularly to mothers. This is discussed in Section 5.2.1.

Rhinovirus illness rates have been measured in long-term studies of families and insurance company workers. The estimated incidence of rhinovirus respiratory illness in the Tecumseh, Michigan, study for all ages was 0.83 per person per year.[166] The annual incidence in different age groups based on actual viral isolation results ranged from 0.59 in 0- to 4-year-olds to 0.09 in persons over 40 years of age (Table 3). Data collected from the insurance company population of young adults yielded a rhinovirus illness

Table 3. Rhinovirus Illness Rates:
Calculated from Surveillance and Sampling of Persons with Colds

Location	Age (yr)	Person-years of observation	Number of respiratory illnesses per person-year	Number of rhinovirus illnesses per person-year
Tecumseh, Michigan[166]	0–4	539	4.9	0.59
	5–19	1541	2.8	0.13
	20–39	1523	2.2	0.21
	40+	1757	1.6	0.09
Charlottesville, Virginia[87,90]	Males			
	16–24	240	2.2	0.51[a]
	25–34	204	2.1	0.50
	35–44	111	2.3	0.54
	45+	24	2.2	0.51
	All males	579	2.2	0.51
	Females			
	16–24	477	2.6	0.60
	25–34	237	2.1	0.49
	34–44	84	2.1	0.49
	45+	24	1.3	0.31
	All females	822	2.4	0.55
	All persons	1401	2.3	0.53

[a]All rates calculated using rhinovirus isolation percentage of 23.3% (observed: 22.9% in males, 23.6% in females).

rate of 0.53.[87] Rates for males and females derived from this study were 0.51 and 0.55, respectively. The higher rate in females reflected a greater incidence of total colds in females and not an increased incidence of rhinovirus recovery from females, since the rhinovirus isolation percentages from males and females were not different. The reason for the differences in rhinovirus illness rates in these studies is not clear but may relate to variables such as the methods of surveillance, criteria used in counting colds, and varying sensitivities of the cell cultures used for virus recovery.

5.1.2. Prevalence of Antibody and Geographic Distribution. Studies of the prevalence of rhinovirus antibody support the conclusion that rhinovirus infections begin in early childhood and continue into adult life (Fig. 1). Antibody to the various rhinovirus types begins to appear at a early age and increases in prevalence throughout childhood and adolescence.[99,163,218,224] The prevalence of antibody reaches a peak in young adults (mean percentage positive: 50%), probably reflecting the effect of exposure to young children in the home.[98] Antibody prevalence then declines to a slightly lower level that persists throughout adulthood. Studies of antibody in sera collected serially from the same person show persistence of antibody at relatively stable levels for years.[218] The mechanisms by which rhinovirus serum antibody levels persist are unknown and could include inherent stability

of antibody formed initially, recurrent antigenic stimulation from infection with the same or related types, or both. The slight decrease in prevalence of antibody after the early adult peak (Fig. 1) suggests that a decline in antibody occurs when viral exposure is lessened. Limited work has also shown that artificially induced N antibody in nasal secretions may persist for at least 330 days following intranasal vaccination.[17]

Information is also available on the prevalence in adults of serum N antibody to each of the different serotypes, 1A–55. In the groups studied, antibody was present in all the types tested (Fig. 2).[98] The prevalence of antibody ranged from a low of approximately 10% to a high of approximately 80%, and there was no sharp dividing point between types associated with high and low antibody prevalence.

Studies of rhinovirus-antibody prevalence in specimens from many different parts of the world have shown that rhinoviruses have a worldwide distribution.[219] Broadly speaking, there were differences in prevalence of antibody among countries for any particular virus tested. Rhinovirus-antibody prevalence in tropical areas is equal to or greater than that in the temperate zone.

5.1.3. Seasonal Distribution of Infections. In an early epidemiologic study of acute respiratory disease in which virological methods were not available. Frost and Gover[72] noted that "during the season of high preva-

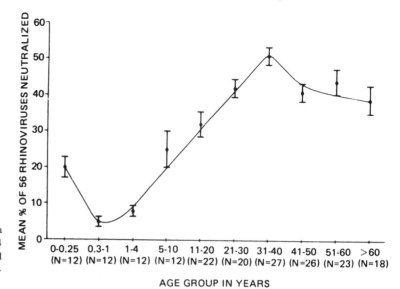

Figure 1. Distribution of N antibody in human sera according to age. A total of 184 sera were tested at 1:4 dilutions versus rhinovirus types 1A–55. The vertical brackets represent the S.E.M. With permission of Hamparian *et al.*[98]

lence, from September to March, inclusive, the incidence curve [for colds] in each locality exhibited a series of oscillations, constituting a succession of epidemics, each of several weeks' duration, rather irregular in sequence and magnitude, but clearly not attributable to mere chance fluctuation." The data from this study showed that one of the recurrent epidemic peaks of colds occurred in the early fall, usually in September. Later, in the Cleveland family study of minor illness, a September peak of colds was a prominent feature of the seasonal pattern of illness, although no respiratory viruses could be associated with this period.[48] Studies using virus cultures have now shown

that rhinoviruses account for a major part of this early fall outbreak of colds that annually initiates the respiratory disease season (Fig. 3),[87] although this has not been observed in all locations.[166] In adults with colds in the eastern United States, rhinovirus infection rates reached their highest annual point (3.5 illnesses/1000 per day, 1.28 per person-year) in September. At this time, rhinoviruses accounted for approximately 40% of all colds and greater than 90% of diagnosed colds. Rhinovirus infection rates fell and remained low (1–1.5/1000 per day) in the winter and early spring. A second peak of rhinovirus illness occurred in April and May. Although total respiratory

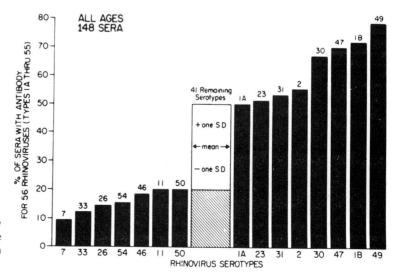

Figure 2. Percentages of human sera with N antibody to rhinovirus types 1A–55. A total of 148 sera were tested at a 1:4 dilution. With permission of Hamparian *et al.*[98]

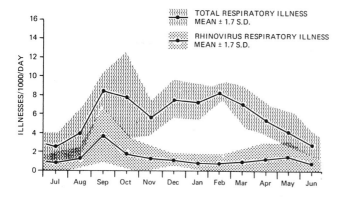

Figure 3. Total and rhinovirus respiratory illness rates (± 1.7 S.D.) in young adults. Data collected over a 7-year period (1963–1969). Adapted from Gwaltney *et al.*[87]

rates were falling in the spring, rhinoviruses were associated with a substantial fraction of all colds during that time. Throughout the summer, rhinovirus infections continued to account for an important part of all colds, although respiratory illness rates reach their lowest point at this time.

In the tropics, the respiratory disease season coincides with the rainy season, beginning in May and June and ending in November and December.[164] Rhinovirus infections were most prevalent during the rainy season in Panama.[163] In the continuously humid climate of Fortaleza, Brazil, rhinoviruses were recovered from 17% of young children with colds, and over a 2-year period, the prevalence of infection appeared to show an inexact correlation with the amount of rainfall.[7] In arctic locations, where the respiratory disease season coincides with cold weather as in temperate climates, rhinovirus outbreaks have been observed, but precise patterns have not been studied.[231]

Although there is a well-established correlation between the lowered temperatures during the fall, winter, and spring months and the increased occurrence of acute respiratory disease during that period,[125] there is no evidence to support a direct causal relationship between thermal cold and increased rates of infection.[81] As to meteorological effects that specifically influence rhinovirus infections, a thorough study of weather and colds showed that none of nine weather variables including temperature had a distribution remotely resembling the autumn (presumed rhinovirus) peak of colds.[144] This is in keeping with the observations from two long-term studies of rhinovirus infections in which prominent September peaks of rhinovirus colds were associated with mild seasonal fall weather and not with the more severe cold of winter that requires heating of homes.[81,162] More direct evidence on this question comes from a volunteer study

with rhinovirus type 15 in which exposure to thermal cold showed no adverse effect on susceptibility to experimental infection or severity of illness.[58]

The reason for seasonal variation in the incidence of colds remains a mystery.[81] Speculations include the idea that cold weather, like rain in the tropics, leads to crowding indoors, thus providing better conditions for virus spread.[126] Also, school openings in the fall bring together into large groups a segment of the population susceptible to rhinoviruses and other respiratory viruses. There has also been speculation on the effect of weather changes on virus survival and infectivity. Changes in humidity have been shown to influence the survival of respiratory viruses.[14] Rhinovirus survives best at relative humidities of over 55%. In temperate areas of the United States, such as Charlottesville, Virginia, indoor relative humidity remains in the favorable range for rhinovirus survival from April through October, which is the period of highest rhinovirus prevalence.[81]

5.1.4. Distribution of Immunotypes. A tally of rhinovirus immunotypes in the United States based on published studies revealed wide dispersal of most types throughout the country.[100] Of the first 55 numbered types, only type 5, a virus first isolated in England, had not been recovered in the United States. The serological survey cited earlier[98] showed antibody to type 5 virus in sera from United States populations. Thus, the conclusion that rhinovirus types are widely distributed throughout the United States and the world is supported by both virus-isolation and serological data.

The current impression, based on longitudinal studies, is that multiple types circulate in a geographic area at any given time with no discernible pattern to their appearance or reappearance.[89,100] Over several years, some types were endemic, whereas others appeared only once or twice. It has been proposed that certain rhinovirus types might possess a higher degree of infectivity than others, increasing their importance as a cause of colds and making them prime candidates for inclusion in vaccines.[161] Analysis of the frequency of isolation of the various rhinovirus types, however, does not show a sharp division between "common" and "uncommon" types. Also, types most commonly encountered in one study have not necessarily been the same as those in other studies. From the analysis of combined data from several studies, it was not possible to designate a particular year as a nationwide epidemic year for a particular type, nor was it possible to detect pathways of rhinovirus transmission by type across the country.[100]

Long-term studies have shown a gradual change with time in the overall distribution of immunotypes in a given geographic location.[21,69] Immunotypes with lower numbers, which in general were discovered earlier, have been replaced by higher-numbered, "newer" types and by strains that could not be typed with available antisera. The reason for the shift in types in a given area over time appears to be the large number of stable immunotypes in existence and not the rapid emergence of new types of rhinovirus.[97]

5.2. Occurrence in Different Settings

5.2.1. Family. A major site for rhinovirus spread in civilian populations is the home.[46,116,158] The characteristic epidemiologic pattern in this setting is for a schoolchild or child in day care to introduce virus into the home, after which transmission occurs to other members of the family (Fig. 4). Secondary infections are most common in young children and mothers, but all members of the household including fathers, other adults working

outside the home, and adolescents are affected. Intervals of 2–5 days are commonly seen between onsets of cases.

In one study, total respiratory illness rates were highest in preschool children.[116] Rates in housewives were similar to those in schoolchildren and were consistently higher than rates in adults working outside the home. During the height of the epidemic, the frequency of rhinovirus infection as determined by culture and serology was similar in all age groups. Later, in October, total illness rates were seen to decline in adults and older children, while young children continued to have frequent colds for which no etiology could be established. The presence of children in the home was associated with total respiratory illness infection rates for adults that were higher than for adults who did not have this exposure. At the height of a September peak of illness, rhinovirus respiratory illness rates for all family members, adults and children, were approximately 8/1000 per day (2.92 per person-year), calculated on the basis of rhinovirus causing 40% of fall colds.

In one family study, the rhinovirus attack rates for two epidemic types were 25 and 50%,[46] and the attack rate for type 16 in an outbreak in families in a small Alaskan community was nearly 70%.[231] In another study, the secondary attack rates for members of families into which a rhinovirus had been introduced were inversely proportional to preexposure serum antibody levels: 71, 50, and 21% of persons with titers of ≤2, 4, and 8–32, respectively, were infected.[114,116] Based on the results of a study of colds in the tropics,[158] the secondary attack rate with type 39 was calculated to be 56% in antibody-free persons.[116]

5.2.2. Schools. A key study has shown that rhinoviruses spread efficiently among children in nursery

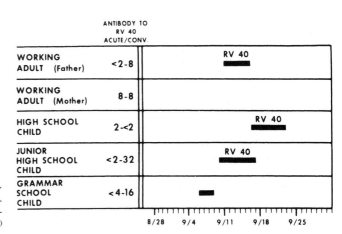

Figure 4. A family outbreak of colds caused by rhinovirus type 40. (■) Periods of symptomatic illness; (RV 40) positive virus culture. The diagnosis of rhinovirus infection in the index case (grammar-school child) was made by serology. Adapted from Hendley *et al.*[116]

school,[9] thus establishing transmission in school as an important step in rhinovirus dissemination in civilian populations. Spread of some types in the schoolroom was extensive, involving up to 77% of children. However, half the serotypes introduced into the groups showed no evidence of spread. The reason for the lack of spread of some types is unknown, but the authors concluded that it was not related to characteristics of the associated illness, patterns of virus shedding, or levels of immunity. Spread was most pronounced during March and April, a recognized time of increased prevalence of rhinovirus colds. The study unfortunately did not extend through the September peak of rhinovirus infections.

Rhinovirus activity has been observed in dayschool groups at various grade levels[176] and in boarding school, university, and medical school populations.[91,101,134,156,181] Rhinoviruses are a prominent cause of morbidity in these groups, although information on their specific epidemiologic behavior is not available. Presumably, spread in older children, adolescents, and young adults who are part of closed populations such as boarding schools occurs among roommates, friends, members of athletic teams, and the like.

5.2.3. Military. Rhinoviruses account for a large amount of the morbidity associated with upper respiratory tract infections in military populations.[68,128,156] In a prospective study of Navy and Marine recruits, 90% of the men developed rhinovirus infection during a 28-day period in basic training, giving an attack rate for this period of 11.7 per person-year![191] Of these infections, 75% occurred within the first 2 weeks of training, and simultaneous or closely spaced infections with two different serotypes in the same man were common. The epidemiologic behavior of the numbered rhinoviruses in military populations is generally similar to that in civilians, showing a constantly changing mosaic of different types.[168]

6. Mechanisms and Routes of Transmission

Although considerable work has been done on the question, the exact mechanism by which rhinoviruses are passed from person to person is unknown.[115] As discussed above, children are the most important reservoir of the virus, and home and school are the places where spread most often occurs. In volunteer experiments, close personal contact appears to be necessary for virus to spread efficiently from an infected to a susceptible subject.[42,121] These facts alone suggest that spread is most often by some type of short-range exposure to infectious secretions. Information on the various steps in the sequence of transmission is best evaluated in relation to the question of spread by direct manual contact with infectious secretions versus spread by contact with virus in contaminated aerosols of large or small particle sizes.[85]

Virus shedding, the first step in the sequence, occurs primarily from the nose. Under experimental conditions, the amount of rhinovirus in the nasopharyngeal washes of volunteers peaked (832 $TCID_{50}$/ml) on the third day after inoculation and then fell to low levels that persisted for up to 2 weeks.[54] Some volunteers showed a different pattern of nasal shedding characterized by delayed onset and slow buildup over 7 days to relatively low maximum virus concentrations (41 $TCID_{50}$/ml). Comparisons of rhinovirus concentrations in respiratory secretions from subjects with natural colds have shown that the quantity of virus in nasal mucus tends to be 10- to 100-fold greater than in pharyngeal secretions.[117] Also, virus was present only 50% of the time in saliva, where it was found in low concentrations. In keeping with the relative scarcity of rhinovirus in saliva was the finding that virus was infrequently recovered from simulated coughs and sneezes.

The relatively poor yield of virus in saliva can be interpreted as evidence against spread through the air, since aerosols produced by coughing and sneezing are mainly of oral origin, coming primarily from the pool of saliva in the anterior part of the mouth.[12,127] On the other hand, the idea of nasal mucus as a direct source of transmissible virus is appealing because of the relatively high titers of virus in mucus and the great potential for people with colds to contaminate the environment, including fingers, with this substance. Rhinovirus has been recovered from the hands of 40–90% of adults with natural[117] and experimental colds[43,92,187] and from 6 and 15% of selected objects in the environment of persons with experimental and natural colds, respectively.[92,187] Information obtained on the second step in transmission, virus survival in the environment, indicates that rhinovirus in concentrations found in nasal mucus survives regularly for up to 3 hr on skin and a variety of surfaces such as wood, plastic, steel, Formica, and hard fabrics.[117]

Evidence in favor of spread through the air comes from experiments in which biological tracers, the spores of *Bacillus mycoides*, were placed in the nose. These experiments showed that blowing the nose and especially sneezing could produce droplets containing the tracer that were small enough to remain airborne and yet in the size range (3–16 μm) that is likely to be trapped in the nose.[15] Rhinovirus survival in aerosol is enhanced by low temperature and high humidity.[120]

Whatever the method of transfer, virus must reach an appropriate portal of entry to complete the sequence of events leading to successful spread. Under experimental conditions, small quantities of rhinovirus (the HID_{50} equivalent to 0.032–0.75 $TCID_{50}$) placed in the nose in coarse drops will efficiently initiate infection.[51] There is indirect evidence that similar small amounts of virus may initiate infection under natural conditions.[114] Experimental infections have been produced by the inhalation of rhinovirus aerosols with particle sizes in the true droplet nuclei range (0.3–2.5 μm) but require approximately 20-fold greater concentrations of virus than intranasal challenge. Thus, it appears that the nasal mucosa is more susceptible to rhinovirus than is the lower respiratory tract.[38] In this experiment, it was not possible to exclude the possibility that infection resulted from the fraction of the viral aerosol that was deposited in the nose rather than that reaching the lower respiratory tract. Experimental rhinovirus colds have also been produced by dropping small amounts of virus on the conjunctiva,[20,117] indicating that the eye may be another portal of entry for rhinovirus. In contrast, rhinovirus placed in the mouth does not readily initiate infection.[117] In related experiments in which infected and susceptible volunteers kissed under controlled conditions, oral contact was an inefficient method of causing spread.[180]

From the results of the work cited above, it appears that rhinovirus must reach the nasal mucosa for efficient initiation of infection. Observations carried out on adults at medical conferences and in Sunday school show that normal behavior includes placing fingers into the nose or onto the conjunctiva with regularity.[117] Episodes in which finger contact with nasal and conjunctival mucosa occurred were measured on the average of two per 3 person-hours of observation. This type of behavior provides sufficient opportunity for accidental self-inoculation if the fingers are contaminated with virus. The alternative method of spread, transmission via airborne particles with deposition in the respiratory tract, is also feasible. The average adult is effectively exposed by inhalation to large amounts (approximately 10 liters/min) or air; thus, small concentrations of virus in the air may be sufficient to transmit infection.

Indirect evidence on the relative importance of these different methods of spread under natural conditions has been obtained in studies of experimental infections. In one study, airborne transmission of rhinovirus did not occur across a wire mesh barrier from infected to susceptible volunteers in closed barracks.[51] In another, infected volunteers who engaged in singing and other activity designed to create infectious aerosols failed to spread rhinovirus to susceptible subjects confined in the same closed room.[42] More recently, transmission models have been

developed for the hand contact/self-inoculation and aerosol routes of rhinovirus transmission. The hand contact model was shown to be quite efficient in one study, with 11 of 15 volunteers infected after brief hand-to-hand contact compared to 1 of 12 infected after exposure by large-particle and none of 10 after small-particle aerosol.[92] The hand contact model has been used to determine the usefulness of viricidal hand treatment,[93] an environmental disinfectant,[86] and viricidal nasal tissues.[112]

Another transmission model, based on an antarctic hut setting has accomplished experimental rhinovirus transmission by aerosol.[151] In this model in which elbow restraints were used to prevent finger-to-nose contact,[47] a linear relation was observed between transmission rates and the number of hours of exposure between donors and recipients. A large pool of coughing donors and a long period of exposure is required for transmission to occur with this model.

While the transmission models allow speculation about what might occur under natural conditions, they cannot provide definitive answers to that question. To discover the natural routes of rhinovirus transmission, the performance of selected intervention methods must be tested in the natural setting (Table 4).[85] Two such intervention studies in a natural setting have addressed the hand route of rhinovirus transmission. In one, contact prophylaxis with a viricidal hand treatment was associate with a 60% reduction in total colds and the elimination of rhinovirus colds in the treated group.[115] In the other study, a programmed reduction in the self-inoculatory behavior of young children was associated with a 45% reduction in the incidence of asthmatic attacks and a 47% reduction in the laboratory-confirmed respiratory virus infection rate.[37] No attempts to interrupt rhinovirus transmission in a natural setting have been reported using methods that would block aerosol spread.

Two studies of contact prophylaxis with natural in-

**Table 4. Postulates to Test
a Hypothesis of Microbial Transmission**

1. Infectious microorganism must be produced in infected host at proposed anatomic source.
2. It must be present in secretions or tissues that are shed from host by proposed route.
3. It must be present and survive in or on the appropriate environmental substance or object.
4. The contaminated environmental substance or object must reach the proposed portal of entry.
5. Interruption of transmission by the hypothesized route must prevent spread of infection under natural conditions.

terferon, while not designed to address transmission routes, nevertheless provide useful insight into the question.[60,104] In these studies, interferon was applied topically into the nose and was associated with marked reduction in the natural rhinovirus infection rate, implicating either finger-to-nose and/or large-particle aerosol as the natural routes of spread. Small-particle aerosols reach the lower airway and lungs, and thus the intranasal instillation of interferon would not be expected to prevent infection at these sites. Thus, in summary, a limited amount of direct evidence suggests that rhinovirus is transmitted by direct hand contact or by a combination of this route and large particle aerosol.

7. Pathogenesis

The incubation period of experimental rhinovirus colds is 16 to 24 hr,[171] but in some cases may extend for up to several days.[51,54] Virus may be recovered from nasal pharyngeal washes in small amounts by 24 hr after inoculation. Virus concentrations then rise rapidly to peak values on days 2 and 3. Maximal virus shedding is followed within 24 hr by the release of large quantities of protein from the mucous membrane.

The virus' ability to evade mucociliary clearance and other nonimmunologic defenses of the nasal passages appears to be important in the initiation of infection. Thus, small inocula of virus placed into the nasal passages of nonimmune persons routinely lead to infection.[51,81,216] In a study employing serial brush biopsies of selected sites in the upper airway, point inoculation of the nasal passage with rhinovirus by way of one tear duct was followed by transport of virus to the posterior nasopharynx and initiation of infection at that site.[230] Infection remained localized in the nasopharynx in some patients but usually spread forward to one or both nasal cavities over several days. Viral shedding persisted for up to 3 weeks. After an experimental challenge, maximum clinical illness usually occurs during the first 4 days of infection.

Infection of the nasal cavity with rhinovirus produces little or no detectable damage to the nasal epithelium as determined by histological examination of nasal biopsies (Fig. 5),[52,229] although occasional ciliated epi-

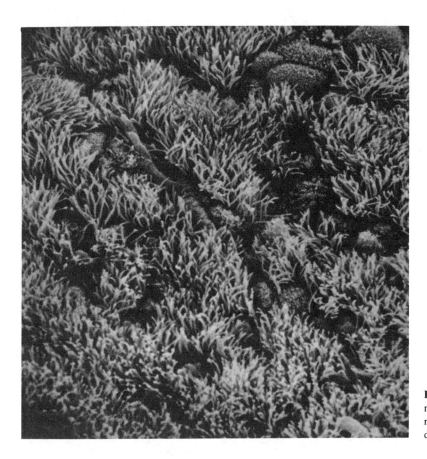

Figure 5. Scanning electron photomicrograph of a nasal biopsy from a volunteer with an experimental rhinovirus cold, showing no evidence of cellular damage.

thelial cells containing rhinovirus antigen have been found in nasal secretions of infected volunteers,[223] and nasal mucociliary flow rates are decreased.[197] This led to the suggestion that the viral infection acts primarily to trigger inflammatory responses by the host, which, in turn, lead to the symptomatic illness.[216,223]

The role of various mediators and neurogenic reflexes in the pathogenesis of rhinovirus colds is now being studied. This is being approached by measuring mediator concentrations in respiratory secretions of persons with colds, by blocking mediator activity with specific compounds, and by challenging volunteers with mediators instilled into the upper airway. Using these approaches, several mediators have been associated with symptom occurrence in rhinovirus colds, including bradykinin and lysylbradykinin,[61,171] prostaglandin,[61,210] histamine,[61,62,73] and interleukin-1 (J. M. Gwaltney and D. Proud, personal communication). Also, parasympathetic[74] and alpha adrenergic[211] pathways have been implicated in rhinovirus pathogenesis.

In addition to the nasal cavity, rhinovirus colds also affect the lower airway, the middle ear, and the paranasal sinuses. In reports on children with wheezing[124] and adults with chronic bronchitis or asthma,[141] rhinovirus was recovered more often from sputum than from the nose or throat, suggesting that viral replication was occurring in the lower respiratory tract. A study using a sampling device designed to minimize upper airway contamination of specimens suggested that rhinovirus replication was occurring in the large airways of volunteers with experimental rhinovirus colds,[95] although it was not entirely possible to exclude the possibility that the specimens had been contaminated by upper airway secretions. There still remains no direct evidence on the question as might be obtained by transtracheal aspiration or lung puncture.

Rhinovirus infections have been implicated as an important precipitant of asthmatic attacks in children.[119,123,124,148,153,154] The mechanism for this is unknown, but a decrease in granulocytic β-adrenergic and H_2 histamine receptor responses has been observed in volunteers with peripheral airway obstruction associated with experimental rhinovirus infection.[18] In another study, 4 of 19 young adults with mild to moderate asthma had decreases in FEV_1 and increases in histamine sensitivity during experimental rhinovirus infection.[94]

Rhinovirus infections have also been associated with periods of acute exacerbation in patients with chronic bronchitis.[63,149,213] A decline in pulmonary function has been observed in patients with chronic obstructive pulmonary disease in association with rhinovirus infection.[207] However, the abnormalities have been mild and transient. Similar changes in pulmonary function have been seen in cigarette smokers[71] and healthy adults[10,26] with rhinovirus infection. The mechanisms by which rhinovirus infection might alter pulmonary function are unknown. Direct invasion of the lower respiratory tract by the virus is a possibility, but reflex mechanisms or secondary bacterial infection might also play a role.

In addition to asthma and chronic bronchitis, there have been multiple reports of rhinovirus infection in patients, especially children, with other diseases of the lower respiratory tract.[29,39,75,179] The possibility cannot be excluded that concurrent infection with other viral or bacterial pathogens may have been present and caused the illness seen in some cases. The opinion of several workers in the field has been that rhinoviruses are not an important cause of viral pneumonia, croup, and bronchiolitis in children.[11,76,169,184,231]

Rhinovirus colds have been associated with the frequent development of abnormalities in eustachian tube function and middle ear pressures in young adults with experimental rhinovirus colds.[147] Also, rhinovirus has been recovered from middle ear aspirates of patients with acute otitis media.[5,80] These findings support the clinical and epidemiologic impression that colds have a major role in the pathogenesis of otitis media. Rhinovirus infections also have recently been shown to cause abnormalities in the paranasal sinuses. In one study, a third of young adults with experimental rhinovirus colds had acute abnormalities of the sinus cavities detected by magnetic resonance imaging.[222]

8. Immunity

Work on the immunology of rhinovirus infections has focused on humoral immunity, particularly the role of antibody in respiratory secretions. Serum N antibody titers rise in up to 75–80% of persons with natural or experimental rhinovirus colds[23,88,90,116]; once present, antibody in serum is well maintained.[218] The level of naturally acquired serum N antibody prior to natural or experimental challenge is inversely proportional to the subsequent infection rate. Under conditions of exposure to rhinovirus in the home, naturally acquired serum antibody at a level of 8 was associated with a sharp reduction in the infection rate, and serum antibody levels of ≥16 were associated with solid immunity.[116] With artificial challenge, it is possible to infect, although at a reduced rate, volunteers who have higher titers of naturally acquired serum antibody. In one study using relatively small challenge doses of virus (0.05–50 $TCID_{50}$), no infections occurred in volunteers with prechallenge titers of 64 or higher.[114] In other studies in which the infecting inocula contained more virus (17–10,000 $TCID_{50}$), infections were observed in volunteers with prechallenge titers of up to 512, presumably as a result of the overwhelming of normal immunity by an artificially large virus challenge.[27,170]

The findings cited above do not necessarily indicate that serum N antibody is the primary immune mechanism responsible for resistance to rhinovirus, since naturally acquired serum antibody is found in close association with antibody in nasal secretions.[171,178] The ratio of nasal secretion to serum N antibody after recent infection (approximately 1 : 2) appears to be higher than that after remote infection (approximately 1 : 16), suggesting a decline in nasal secretion antibody with time.[27] Actual measurements of nasal immunoglobulin concentrations have confirmed that significant falls in titers did occur over a 5-month period after infection.[121]

Attempts have been made to determine the relative importance and specific roles of serum and nasal secretion antibody in protection against rhinovirus infection. Naturally acquired antibody in nasal wash specimens and serum was associated with resistance to "infection" if present in sufficient titer before artificial challenge, but it did not appear to modify "illness" or virus shedding.[27] Because of the close association of naturally acquired antibody in serum and nasal secretions, the findings of this study did not answer the questions posed. Another approach to the problem was to administer inactivated rhinovirus vaccine by either the parenteral or the intranasal route to elicit selectively nasal secretion or serum antibody or both. Vaccine given intranasally in large amounts led to the production of antibody in both serum and nasal secretions, whereas parenteral vaccination resulted primarily in serum-antibody production.[177,178] Intranasal challenge with rhinovirus at a later date resulted in the reduction of *illness* and virus shedding only in volunteers who received the intranasal vaccine.[17,177,178] In these studies, intranasal vaccination was not associated with a clear-cut reduction in *infection* rate determined by antibody response. Therefore, this work suggested that the primary effect of nasal antibody was to modify illness and reduce virus shedding. This conclusion is in conflict with that of the investigation cited above[27] and of other reports that have found that the major effect of humoral (serum) immunity was prevention of infection and not modification of illness.[54,114,116,170] Other studies have reported on finding an association between naturally acquired[87] and vaccine-induced[56] serum antibody and reduction of illness and, in the latter study, diminished virus shedding. Thus, currently available data from studies of the relative importance of nasal and serum antibody associated with immunity are not in complete agreement; further work is necessary to provide a clear understanding in this area.

Naturally acquired neutralizing activity against rhinovirus in serum has been found to sediment primarily in the 5–7 S region and to be associated with fractions containing immunoglobulin A (IgA) and IgG.[27,192] After recent experimentally induced rhinovirus infection or intranasal vaccination with inactivated rhinovirus vaccine, neutralizing activity has also been associated with 19 S IgM.[27,138,192]

Under normal conditions, nasal secretions contain 12 different identifiable proteins found in serum as well as six antigenic components not present in serum.[193] Secretory IgA, the most abundant protein in nasal secretions, is synthesized locally at sites adjacent to the mucosa and accounts for 30% or more of total protein in nasal secretions. Rhinovirus-neutralizing activity in nasal secretions is associated primarily with IgA in 9–11S fractions, although secretory IgA is not entirely homogenous in its sedimentation characteristics, being found also in 7 and 19 S regions.[138] The symptomatic period of rhinovirus illness is associated with considerable transudation of serum proteins, including IgG, into nasal secretions.[19,193] After cessation of illness, the concentration of serum proteins in nasal secretions falls rapidly; at this time, the IgA concentration begins a progressive sustained increase. The IgA that appears during this period is not associated with an increase in specific neutralizing activity for the infecting virus. Specific N antibody first appears in nasal secretions and serum at approximately 2 weeks in volunteers lacking detectable antibody. Antibody concentrations increase most rapidly between the third and fourth weeks, by which time virus shedding is completed. Volunteers with preexisting serum antibody may show rises in nasal antibody titers by as early as day 7. Neutralizing antibody to rhinovirus has also been found in tears and parotid saliva, where it is associated with the IgA fraction.[59]

Because of the sequence of events described above, it is felt that recovery from rhinovirus infection and illness is not dependent on humoral immunity.[27] It has been shown that interferon is released into respiratory secretions during the course of experimental rhinovirus infection. This has led to the suggestion that in rhinovirus colds, as in other viral infections, interferon may have an important role in recovery.[24]

Limited work has been done on the role of cellular immunity in rhinovirus infection. Natural-killer-like cytotoxic cells were induced in peripheral blood mononuclear leukocytes incubated with rhinovirus.[142] Also, crossimmunotype reactivity was elicited in murine lymphocytes from mice immunized with either of two rhinovirus types.[103]

9. Patterns of Host Response

9.1. Clinical Features

Rhinoviruses produce a typical common cold characterized by rhinorrhea, nasal obstruction, sneezing, pharyngeal discomfort, and cough. The medial length of natural illness in young adults is 7 days, with peak symptomatology occurring on the second and third days of illness.[88] Symptoms last up to 2 weeks in one fourth of cases and may be prolonged to 1 month, although secondary bacterial infection may play a role when this occurs. Volunteers with experimental rhinovirus colds have had an average ($\pm$ SD) of 23 g ($\pm$ 22) of nasal secretions over the first 5 days of illness.[174] The profile of rhinovirus illness can be distinguished from that of influenza by the relative severity of systemic complaints and cough that occur with influenza (Fig. 6). Rhinovirus colds differ from group A β-hemolytic streptococcal pharyngitis in having more nasal involvement and cough and less severe and prolonged pharyngeal discomfort. This information is unfortunately of limited value to the clinician. In the individual patient, it is impossible to distinguish, on clinical grounds, rhinovirus colds from those caused by other common respiratory viruses.

In children, rhinoviruses also produce the common cold syndrome.[11,190] Whether rhinoviruses cause more serious disease in children, such as viral pneumonia, croup, and bronchiolitis, is still not clear. As discussed in Section 7, the prevailing opinion is that rhinoviruses, unlike parainfluenza viruses and respiratory syncytial virus, do not commonly cause these diseases.

Cough is a prominent feature of rhinovirus colds in patients of all ages, indicating that involvement of the lower respiratory tract of some type does occur. The frequency and duration of cough are markedly increased in cigarette smokers, particularly females, with rhinovirus colds.[88] Also, it has been reported that up to 40% of exacerbations in patients with chronic bronchitis may be associated with rhinovirus infections.[63,141,149,213]

Rhinoviruses are among the respiratory viruses that precipitate asthmatic attacks in children.[119,148,154] They appear to play an especially important role in causing wheezing in older children.[123,124,154] Multiple serotypes have been implicated.[154] Also, asthmatic children have been found to experience a significantly greater number of viral respiratory infections, primarily caused by rhino-

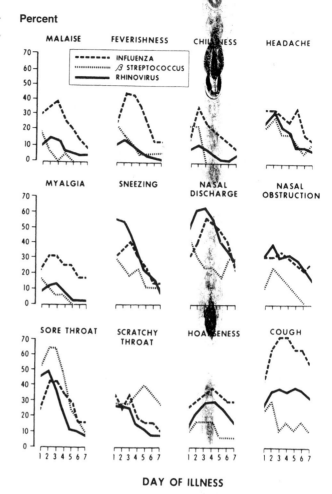

Figure 6. Comparison of symptom profiles of rhinovirus colds (139 cases), type A₂ influenza (33 cases), and group A β-hemolytic streptococcal pharyngitis (17 cases). Adapted from Gwaltney et al.[88]

viruses, than do nonasthmatic controls.[152] These findings are of interest in view of an earlier report that volunteers with a history of allergy have enhanced susceptibility of experimental colds.[126]

During acute rhinovirus illness in volunteers, there is a modest increase in circulating neutrophils.[23] Later in the infection, moderate elevations in the erythrocyte sedimentation rate may occur. The diagnosis of rhinovirus infection is best accomplished by isolation of the virus from nose-and-throat-swab or nasal-wash specimens. There is currently limited availability of facilities for the laboratory diagnosis of rhinovirus infections in routine medical practice.[179]

9.2. Apparent–Inapparent Infection Ratios

Data based on virus isolations are available from several studies for calculating apparent–inapparent infection ratios for rhinoviruses. The results, which are in good agreement, indicate that the majority of rhinovirus infections are associated with symptomatic respiratory illness. The percentages of rhinovirus infections associated with illness were 63% in families,[136] 88% in medical students,[101,102] 69% in insurance company employees,[87] and 70–74% in miliary trainees.[128,168] Thus, the ratio of apparent to inapparent infections is approximately 3:1.

10. Control and Prevention

A vaccine approach to prevention of rhinovirus colds, a long sought after goal,[3,36,55,56,69,96,155,170,186,198,214] has not been successful. The antigenic diversity of the group has proved too much of an obstacle for currently available vaccine technology (Table 5). Other approaches that have been investigated include antiviral agents for prophylaxis and treatment, combined antiviral antimediator treatment, receptor blockade, and interruption of transmission.

Interferon was the first antiviral substance shown to provide effective prophylaxis against rhinovirus in humans.[150] Recombinant interferon-α_2 applied topically in the nasal cavity is a highly efficacious way of preventing both experimental and natural rhinovirus colds. The use of interferon in this way has reduced infection rates by up to 80% and clinical illness rates by 60–100% following experimental virus challenge.[107,196,200] In the natural setting, a strategy of contact prophylaxis for persons in the home exposed to family members with rhinovirus colds

reduced rhinovirus illness rates by 80%.[60,104] This approach may prove to have practical value in clinical practice.

On the other hand, long-term administration of topical intranasal interferon does not appear feasible because of side effects including nasal dryness and stuffiness, blood-tinged nasal mucus, pinpoint bleeding sites, and occasional small ulcerations.[64,110,111] These side effects, which occur after 5 to 7 days of interferon administration, are reversible after discontinuation of the drug. Also, studies of the therapeutic effect of topical intranasal interferon given alone have not been promising. When administered as early as 28 hr after virus challenge, there has been an inconsistent effect on clinical illness and nasal mucus weights, although viral titers in nasal secretions were reduced.[108]

Control of rhinovirus infection by chemoprophylaxis and chemotherapy with compounds other than interferon has also been under investigation.[209] In early work, rhinoviruses were found to be susceptible to 2-(α-hydroxybenzyl)benzimidazole and related compounds that have specific actions on virus replication. Since then, a number of other compounds with activity against rhinoviruses *in vitro* have been discovered.[4,16,79,188,199,201,204,205,217] Most have not been effective in volunteers given experimental virus challenge.[106,188,203,221] Two members of a group of compounds called capsid binders, which bind to the hydrophobic pocket within the rhinovirus shell, have shown prophylactic but not therapeutic activity in volunteers with experimental infection.[1,105]

A new approach to treating rhinovirus colds is based on the idea that effective treatment requires simultaneously suppression of viral replication and blocking of the associated inflammatory events triggered by the infection. In volunteers with experimental rhinovirus colds, early treatment with topical intranasal interferon-α_{2b} and ipratropium combined with oral naproxen reduced the number of full-blown colds that developed in the treated subjects and significantly lowered symptom scores.[82] Another approach that has attracted interest is to block viral attachment with either monoclonal antibody to the cellular (ICAM-1) receptor or to treat with artificial soluble ICAM-1 to bind virus before it reaches natural receptor. One study in which monoclonal antibody was used prophylactically in experimental rhinovirus colds showed promising results.[109] No reports of human trials with soluble receptor have been published.

Interrupting viral spread also remains an area of interest. In one controlled, blinded study conducted under natural conditions in the home, the treatment of the fingers with a virucidal solution containing iodine reduced the

Table 5. Approaches to Control of Rhinovirus Colds

Vaccines
  Not feasible with current technology
Chemoprophylaxis
  Interferon-α_{2b} effective in experimental and natural colds
  Capsid binders (R61837, R77975) effective in experimental colds
Chemotherapy
  Combined antiviral antimediator treatment effective in experimental colds
Receptor blockade
  Monoclonal antibody effective as prophylaxis in experimental cold
  Soluble receptor, no reports of human testing
Interruption of transmission
  Antiviral hand treatment and training to reduce self-inoculatory behavior effective as prophylaxis in natural colds

incidence of all colds by 60% and eliminated laboratory-proven rhinovirus colds.[115] However, the iodine solution is not practical for routine use, and other virucidal treatments for use on the hands that are effective, safe, and cosmetically acceptable have not been found for testing. In another study, training children to avoid self-inoculatory behavior resulted in a reduction in cold-associated asthmatic attacks and the laboratory-proven respiratory virus infection rate.[37]

11. Unresolved Problems

Although the atomic structure of rhinovirus is now known, much remains to be learned about how the virus interacts with its human host. Information is being obtained on the sites of infection, types of disease, and roles of mediators and neurogenic reflexes in pathogenesis but knowledge in these areas is limited. Also, the question of how rhinovirus colds are naturally transmitted has not been fully answered. Since the prospects for developing successful rhinovirus vaccines do not appear good, prevention of viral transmission may offer the best hope for an epidemiologic approach to control. Also, better therapy for rhinovirus colds may be possible through the development of combined antiviral antimediator treatment.

12. References

1. AL-NAKIB-W., HIGGINS, P. G., BARROW, G. I., TYRRELL, D. A. J., ANDRIES, K., VANDEN BUSSCHE, G., TAYLOR, N., AND JANSSEN, P. A. J., Suppression of colds in human volunteers challenged with rhinovirus by a new synthetic drug (R61837), *Antimicrob. Agents. Chemother.* **33**:522 (1989).
2. ANDREWES, C., *The Common Cold*, W. W. Norton, New York, 1965.
3. ANDREWES, C., TYRRELL, D. A. J., STONES, P. B., BEALE, A. J., ANDREWS, R. D., EDWARD, D. G., GOFFE, A. P., DOGGETT, J. E., HOMER, R. F., CRESPI, R. S., AND CLEMENTS, E. M. B., Prevention of colds by vaccination against a rhinovirus: A report by the scientific committee on common cold vaccines, *Br. Med. J.* **1**:1344 (1965).
4. ANDRIES, K., DEWIND, B., SNOEKS, J., AND WILLEBRORDS, R., Lack of quantitative correlation between inhibition of replication of rhinoviruses by an antiviral drug and their stabilization, *Arch. Virol.* **106**:51 (1989).
5. AROLA, M., RUUSKANEN, O., ZIEGLER, T., MERTSOLA, J., NÄNTÖ-SALONEN, K., PUTTO-LAURILA, A., VILJANEN, M. K., AND HALONEN, P., Clinical role of respiratory virus infection in acute otitis media, *Pediatrics* **86**:848 (1990).
6. ARRUDA, E., AND HAYDEN, F. G., Detection of human rhinovirus RNA in nasal washings by PCR, *Mol. Cell. Probes* **7**:373 (1993).
7. ARRUDA, E., HAYDEN, F. G., McAULIFFE, J. F., DE SOUSA, A., MOTA, S. B., McAULIFFE, M. I., GEIST, F. C., CARVALHO, E. P.,

FERNANDES, M. C., GUERRANT, R. L., AND GWALTNEY, J. M., JR., Acute respiratory viral infections in ambulatory children of urban northeast Brazil, *J. Infect. Dis.* **164**:252 (1991).
8. BARCLAY, W. S., CALLOW, K. A., SERGEANT, M., AND AL-NAKIB, W., Evaluation of an enzyme-linked immunosorbent assay that measures rhinovirus-specific antibodies in human sera and nasal secretions, *J. Med. Virol.* **25**:475 (1988).
9. BEEM, M. O., Acute respiratory illness in nursery school children: A longitudinal study of the occurrence of illness and respiratory viruses, *Am. J. Epidemiol.* **90**:30 (1969).
10. BLAIR, H. T., GREENBERG, S. B., STEVENS, P. M., BILUNOS, P. A., AND COUCH, R. B., Effects of rhinovirus infection on pulmonary function of healthy human volunteers, *Am. Rev. Respir. Dis.* **114**:95 (1976).
11. BLOOM, H. H., FORSYTH, B. R., JOHNSON, K. M., AND CHANOCK, R. M., Relationship of rhinovirus infection to mild upper respiratory disease. 1. Results of a survey in young adults and children, *J.A.M.A.* **186**:38 (1963).
12. BOURDILLON, R. B., AND LIDWELL, O. M., Sneezing and the spread of infection, *Lancet* **2**:365 (1941).
13. BROWN, P. K., AND TYRRELL, D. A. J., Experiments on the sensitivity of strains of human fibroblasts to infection with rhinovirus, *Br. J. Exp. Pathol.* **45**:571 (1964).
14. BUCKLAND, F. E., AND TYRRELL, D. A. J., Loss of infectivity on drying various viruses, *Nature* **195**:1063 (1962).
15. BUCKLAND, F. E., AND TYRRELL, D. A. J., Experiments on the spread of colds. 1. Laboratory studies on the dispersal of nasal secretion, *J. Hyg. (Camb.)* **62**:365 (1964).
16. BUCKNALL, R. A., SWALLOW, D. L., MOORES, H., AND HARRAD, J., A novel substituted guanidine with high activity *in vitro* against rhinovirus, *Nature* **246**:144–145 (1973).
17. BUSCHO, R. F., PERKINS, J. C., KNOPF, H. L. S., KAPIKIAN, A. Z., AND CHANOCK, R. M., Further characterization of the local respiratory tract antibody response induced by intranasal instillation of inactivated rhinovirus 13 vaccine, *J. Immunol.* **108**:169 (1972).
18. BUSH, R. K., BUSSE, W., FLAHERTY, D., WARSHAUER, D., DICK, E. C., AND REED, C. E., Effects of experimental rhinovirus 16 infection on airways and leukocyte function in normal subjects, *J. Allergy Clin. Immunol.* **61**:80 (1978).
19. BUTLER, W. T., WALDMANN, T. A., ROSSEN, R. D., DOUGLAS, R. G., JR., AND COUCH, R. B., Changes in IgA and IgG concentrations in nasal secretions prior to the appearance of antibody during viral respiratory infection in man, *J. Immunol.* **105**:584 (1970).
20. BYNOE, M. L., HOBSON, D., HORNER, J., KIPPS, A., SCHILD, G. C., AND TYRRELL, D. A. J., Inoculation of human volunteers with a strain of virus from a common cold, *Lancet* **1**:1194 (1961).
21. CALHOUN, A. M., JORDAN, W. S., JR., AND GWALTNEY, J. M., JR., Rhinovirus infections in an industrial population. V. Change in distribution of serotypes, *Am. J. Epidemiol.* **99**:58 (1974).
22. CALLAHAN, P. L., MIZUTANI, S., AND COLONNO, R. J., Molecular cloning and complete sequence determination of RNA genome of human rhinovirus type 14, *Proc. Natl. Acad. Sci. USA* **82**:732–736 (1985).
23. CATE, T. R., COUCH, R. B., AND JOHNSON, K. M., Studies with rhinoviruses in volunteers: Production of illness, effect of naturally acquired antibody, and demonstration of a protective effect not associated with serum antibody, *J. Clin. Invest.* **43**:56 (1964).
24. CATE, T. R., DOUGLAS, R. G., JR., AND COUCH, R. B., Interferon and resistance to upper respiratory virus illness, *Proc. Soc. Exp. Biol. Med.* **131**:631 (1969).

25. CATE, T. R., DOUGLAS, R. G., JR., JOHNSON, K. M., COUCH, R. B., AND KNIGHT, V., Studies on the inability of rhinovirus to survive and replicate in the intestinal tract of volunteers, *Proc. Soc. Exp. Biol. Med.* **124:**1290 (1967).

26. CATE, T. R., ROBERTS, J. S., RUSS, M. A., AND PIERCE J. A., Effects of common colds on pulmonary function, *Am. Rev. Respir. Dis.* **108:**858 (1973).

27. CATE, T. R., ROSSEN, R. D., DOUGLAS, R. G., JR., BUTLER, W. T., AND COUCH, R. B., The role of nasal secretion and serum antibody in the rhinovirus common cold, *Am. J. Epidemiol.* **84:**352 (1966).

28. COLONNO, R. J., Virus receptors: The Achilles' heel of human rhinoviruses, *Adv. Exp. Med. Biol.* **312:**61 (1992).

29. CHERRY, J. D., DIDDAMS, J. A., AND DICK, E. C., Rhinovirus infections in hospitalized children: Provocative bacterial interrelationships, *Arch. Environ. Health* **14:**390 (1967).

30. CONANT, R. M., AND HAMPARIAN, V. V., Rhinoviruses: Basis for a numbering system. II. Serologic characterization of prototype strains, *J. Immunol.* **100:**114 (1968).

31. CONANT, R. M., SOMERSON, N. L., AND HAMPARIAN, V. V., Rhinovirus: Basis for a numbering system. 1. HeLa cell for propagation and serologic procedures, *J. Immunol.* **100:**107 (1968).

32. COONEY, M. K., HALL, C. E., AND FOX, J. P., The Seattle virus watch. III. Evaluation of isolation methods and summary of infections detected by virus isolations, *Am. J. Epidemiol.* **96:**286 (1972).

33. COONEY, M. K., AND KENNY, G. E., Immunogenicity of rhinoviruses, *Proc. Soc. Exp. Biol. Med.* **133:**645 (1969).

34. COONEY, M. K., AND KENNY, G. E., Demonstration of dual rhinovirus infection in humans by isolation of different serotypes in human heteroploid (HeLa) and human diploid fibroblast cell cultures, *J. Clin. Microbiol.* **5:**202 (1977).

35. COONEY, M. K., KENNY, G. E., TAM, R., AND FOX, J. P., Cross relationships among 37 rhinoviruses demonstrated by virus neutralization with potent monotypic rabbit antisera, *Infect. Immun.* **7:**335 (1973).

36. COONEY, M. K., AND WISE, J. A., Heterotypic stimulation of rhinovirus antibodies in rabbits, in: *Abstracts Annual Meeting American Society for Microbiology*, p. 114, American Society for Microbiology, Washington, 1973.

37. CORLEY, D. L., GEVIRTZ, R., NIDEFFER, R., AND CUMMINS, L., Prevention of postinfectious asthma in children by reducing self-inoculatory behavior, *J. Pediatr. Psychol.* **12:**519 (1987).

38. COUCH, R. B., CATE, T. R., DOUGLAS, R. G., JR., GERONE, P. J., AND KNIGHT, V., Effect of route of inoculation on experimental respiratory viral disease in volunteers and evidence for airborne transmission, *Bacteriol. Rev.* **30:**517 (1966).

39. CRAIGHEAD, J. E., MEIER, M., AND COOLEY, M. H., Pulmonary infection due to rhinovirus type 13, *N. Engl. J. Med.* **281:**1403 (1969).

40. CRANDELL, R. A., A description of eight feline picornaviruses and an attempt to classify them, *Proc. Soc. Exp. Biol. Med.* **126:**240 (1967).

41. CRUZ, R. R., QUINONEZ, E., DE FERNANDEZ, A., AND PERALTA, F., Isolation of viruses from nasopharyngeal secretions: Comparison of aspiration and swabbing as means of sample collection, *J. Infect. Dis.* **156:**415 (1987).

42. D'ALESSIO, D., DICK, C. R., AND DICK, E. C., Transmission of rhinovirus type 55 in human volunteers, in: *International Virology 2* (J. L. MELNICK, ED.), p. 115, S. Karger, Basel, 1972.

43. D'ALESSIO, D. J., PETERSON, J. A., DICK, C. R., AND DICK, E. C., Transmission of experimental rhinovirus colds in volunteer married couples, *J. Infect. Dis.* **133:**28 (1976).

44. DEARDEN, C. J., AND AL-NAKIB, W., Direct detection of rhinoviruses by an enzyme-linked immunosorbent assay, *J. Med. Virol.* **23:**179 (1987).

45. DICK, E. C., Experimental infection of chimpanzees with human rhinovirus type 14 and 43, *Proc. Soc. Exp. Biol. Med.* **127:**1079 (1968).

46. DICK, E. C., BLUMER, C. R., AND EVANS, A. S., Epidemiology of infections with rhinovirus types 43 and 55 in a group of University of Wisconsin student families, *Am. J. Epidemiol.* **86:**386 (1967).

47. DICK, E. C., JENNINGS, L. C., MINK, K. A., WARTGOW, C. D., AND INHORN, S. L., Aerosol transmission of rhinovirus colds, *J. Infect. Dis.* **156:**442 (1987).

48. DINGLE, J. H., BADGER, G. F., AND JORDAN, W. S., JR., Patterns of illness, in: *Illness in the Home*, p. 19, Western Reserve University, Cleveland, 1964.

49. DITCHFIELD, W. J. B., Rhinoviruses and parainfluenza viruses of horses, *J. Am. Vet. Med. Assoc.* **155:**384 (1969).

50. DOCHEZ, A. R., SHIBLEY, G. S., AND MILLS, K. C., Studies in the common cold. IV. Experimental transmission of the common cold to anthropoid apes and human beings by means of a filtrable agent, *J. Exp. Med.* **52:**701 (1930).

51. DOUGLAS, R. G., JR., Pathogenesis of rhinovirus common colds in human volunteers, *Ann. Otol. Rhinol. Laryngol.* **79:**563 (1970).

52. DOUGLAS, R. G., JR., ALFORD, B. R., AND COUCH, R. B., Atraumatic nasal biopsy for studies of respiratory virus infection in volunteers, *Antimicrob. Agents Chemother.* **8:**340 (1968).

53. DOUGLAS, R. G., JR., CATE, T. R., AND COUCH, R. B., Growth and cytopathic effect of H type rhinoviruses in monkey kidney tissue culture, *Proc. Soc. Exp. Biol. Med.* **123:**238 (1966).

54. DOUGLAS, R. G., JR., CATE, T. R., GERONE, P. J., AND COUCH, R. B., Quantitative rhinovirus shedding patterns in volunteers, *Am. Rev. Respir. Dis.* **94:**159 (1966).

55. DOUGLAS, R. G., JR., AND COUCH, R. B., Attenuation of rhinovirus type 15 for humans, *Nature* **223:**213 (1969).

56. DOULGAS, R. G., JR., AND COUCH, R. B., Parenteral inactivated rhinovirus vaccine: Minimal protective effect, *Proc. Soc. Exp. Biol. Med.* **139:**899 (1972).

57. DOUGLAS, R. G., JR., FLEET, W. F., CATE, T. R., AND COUCH, R. B., Antibody to rhinovirus in human sera. I. Standardization of a neutralization test, *Proc. Soc. Exp. Biol. Med.* **127:**497 (1968).

58. DOUGLAS, R. G., JR., LINDGREN, K. M., AND COUCH, R. B., Exposure to cold environment and rhinovirus common cold: Failure to demonstrate effect, *N. Engl. J. Med.* **279:**743 (1968).

59. DOUGLAS, R. G., JR., ROSSEN, R. D., BUTLER, W. T., AND COUCH, R. B., Rhinovirus neutralizing antibody in tears, parotid saliva, nasal secretions and serum, *J. Immunol.* **99:**297 (1967).

60. DOUGLAS, R. M., MOORE, B. W., MILES, H. B., DAVIES, L. M., GRAHAM, N. M. H., RYAN, P., WORSWICK, D. A., AND ALBRECHT, J. K., Prophylactic efficacy of intranasal alpha$_2$-interferon against rhinovirus infections in the family setting, *N. Engl. J. Med.* **314:**65 (1985).

61. DOYLE, W. J., BOEHM, S., AND SKONER, D. P., Physiologic responses to intranasal dose-response challenges with histamine, methacholine, bradykinin, and prostaglandin in adult volunteers with and without nasal allergy, *J. Allergy Clin. Immunol.* **86:**924 (1990).

62. DOYLE, W. J., MCBRIDE, T. P., SKONER, D. P., MADDREN, B. R.,

GWALTNEY, J. M., JR., AND UHRIN, M., A double-blind, placebo-controlled clinical trial of the effect of chlorpheniramine on the response of the nasal airway, middle ear and eustachian tube to provocative rhinovirus challenge, *Pediatr. Infect. Dis. J.* **7:**229 (1988).

63. EADIE, M. B., STOTT, E. J., AND GRIST, N. R., Virological studies in chronic bronchitis, *Br. Med. J.* **2:**671 (1966).

64. FARR, M., GWALTNEY, J. M., JR., ADAMS, K. F., AND HAYDEN, F. G., Intranasal interferon-α_2 for prevention of natural rhinovirus colds, *Antimicrob. Agents Chemother.* **26:**31–34 (1984).

65. FENTERS, J. D., GILLUM, S. S., HOLPER, J. C., AND MARQUIS, G. S., Serotypic relationships among rhinoviruses, *Am. J. Epidemiol.* **84:**10 (1966).

66. FIALA, M., Plaque formation by 55 rhinovirus serotypes, *Appl. Microbiol.* **16:**1445 (1968).

67. FLEET, W. F., DOUGLAS, R. G., JR., CATE, T. R., AND COUCH, R. B., Antibody to rhinovirus in human sera. II. Heterotypic responses, *Proc. Soc. Exp. Biol. Med.* **127:**503 (1968).

68. FORSYTH, B. R., BLOOM, H. H., JOHNSON, K. M., AND CHANOCK, R. M., Patterns of illness in rhinovirus infections of military personnel, *N. Engl. J. Med.* **269:**602 (1963).

69. FOX, J. P., Is a rhinovirus vaccine possible? *Am. J. Epidemiol.* **103:**345 (1976).

70. FOX, J. P., HALL, C. E., COONEY, M. K., LUCE, R. E., AND KRONMAL, R. A., The Seattle virus watch. II. Objectives, study population and its observation, data processing and summary of illnesses, *Am. J. Epidemiol.* **96:**270 (1972).

71. FRIDY, W. W., INGRAM, R. H., HIERHOLZER, J. C., AND COLEMAN, M. T., Airways function during mild viral respiratory illnesses: The effect of rhinovirus infection in cigarette smokers, *Ann. Intern. Med.* **80:**150 (1974).

72. FROST, W. H., AND GLOVER, M., The incidence and time distribution of common colds in several groups kept under continuous observation, in: *Papers of Wade Hampton Frost, M.D.* (K. F. MAXCY, ED.), p. 359, Commonwealth Fund, New York, 1941.

73. GAFFEY, M. J., GWALTNEY, J. M., JR., SASTRE, A., DRESSLER, W. E., SORRENTINO, J. V., AND HAYDEN, F. G., Intranasal and oral antihistamine treatment of experimental rhinovirus colds, *Am. Rev. Respir. Dis.* **136:**556 (1987).

74. GAFFEY, M. J., HAYDEN, F. G., BOYD, J. C., AND GWALTNEY, J. M., JR., Ipratropium bromide treatment of experimental rhinovirus infection, *Antimicrob. Agents Chemother.* **32:**1644 (1988).

75. GEORGE, R. B., AND MOGABGAB, W. J., Atypical pneumonia in young men with rhinovirus infections, *Ann. Intern. Med.* **71:**1073 (1969).

76. GLEZEN, W. P., LODA, F. A., CLYDE, W. A., SENIOR, R. J., SHEAFFER, C. I., CONLEY, W. G., AND DENNY, F. W., Epidemiologic patterns of acute lower respiratory disease of children in a pediatric group practice, *J. Pediatr.* **78:**397 (1971).

77. GREVE, J. M., DAVIS, G., MEYER, A. M., FORTE, C. P., YOST, S. C., MARLOR, C. W., KAMARCK, M. E., AND MCCLELLAND, A., The major human rhinovirus receptor is ICAM-1, *Cell* **56:**839 (1989).

78. GWALTNEY, J. M., JR., Micro-neutralization test for identification of rhinovirus serotypes, *Proc. Soc. Exp. Biol. Med.* **122:**1137 (1966).

79. GWALTNEY, J. M., JR., Rhinovirus inhibition by 3-substituted triazinoindoles, *Proc. Soc. Exp. Biol. Med.* **133:**1148 (1970).

80. GWALTNEY, J. M., JR., Virology of middle ear, *Ann. Otol. Rhinol. Laryngol.* **80:**365 (1971).

81. GWALTNEY, J. M., JR., The Jeremiah Metzger lecture: Climatol-ogy and the common cold, *Trans. Am. Clin. Climatol. Assoc.* **96:**159–175 (1984).

82. GWALTNEY, J. M., JR., Combined antiviral and antimediator treatment of rhinovirus colds, *J. Infect. Dis.* **166:**776 (1992).

83. GWALTNEY, J. M., JR., Historical eras of the common cold, in: *Contemporary Issues in Infectious Diseases*, Vol. 10. *Viral Infections: Diagnosis, Treatment and Prevention* (M. A. SANDE AND R. K. ROOT, EDS.), pp. 1–13, Churchill Livingstone, New York, 1993.

84. GWALTNEY, J. M., JR., AND EDMONDSON, W. P., JR., Etiology and Epidemiology of Acute Respiratory Disease, Annual Progress Report to the Commission on Acute Respiratory Disease of the Armed Forces Epidemiological Board, Contract No. DADA-49-007-MD-1000, September 15, 1968.

85. GWALTNEY, J. M., JR., AND HENDLEY, J. O., Rhinovirus transmission: One if by air, two if by hand, *Am. J. Epidemiol.* **107:**357 (1978).

86. GWALTNEY, J. M., JR., AND HENDLEY, J. O., Transmission of experimental rhinovirus infection by contaminated surfaces, *Am. J. Epidemiol.* **116:**828 (1982).

87. GWALTNEY, J. M., JR., HENDLEY, J. O., SIMON, G., AND JORDAN, W. S., JR., Rhinovirus infections in an industrial population. I. The occurrence of illness, *N. Engl. J. Med.* **275:**1261 (1966).

88. GWALTNEY, J. M., JR., HENDLEY, J. O., SIMON, G., AND JORDAN, W. S., JR., Rhinovirus infections in an industrial population. II. Characteristics of illness and antibody response, *J.A.M.A.* **202:**494 (1967).

89. GWALTNEY, J. M., JR., HENDLEY, J. O., SIMON, G., AND JORDAN, W. S., JR., Rhinovirus infections in an industrial population. III. Number and prevalence of serotypes, *Am. J. Epidemiol.* **87:**158 (1968).

90. GWALTNEY, J. M., JR., AND JORDAN, W. S., JR., Rhinoviruses and respiratory disease, *Bacteriol. Rev.* **28:**409 (1964).

91. GWALTNEY, J. M., JR., AND JORDAN, W. S., JR., Rhinoviruses and respiratory illness in university students, *Am. Rev. Respir. Dis.* **93:**362 (1966).

92. GWALTNEY, J. M., JR., MOSKALSKI, P. B., AND HENDLEY, J. O., Hand-to-hand transmission of rhinovirus colds, *Am. Intern. Med.* **88:**463 (1978).

93. GWALTNEY, J. M., JR., MOSKALSKI, P. B., AND HENDLEY, J. O., Interruption of experimental rhinovirus transmission, *J. Infect. Dis.* **142:**811 (1980).

94. HALPERIN, S. A., EGGLESTON, P. A., BEASLEY, P., SURATT, P., HENDLEY, J. O., GROSCHEL, D. H. M., AND GWALTNEY, J. M., JR., Exacerbations of asthma in adults during experimental rhinovirus infection, *Am. Rev. Respir. Dis.* **132:**976–980 (1985).

95. HALPERIN, S. A., EGGLESTON, P. A., HENDLEY, J. O., SURATT, P. M., GROSCHEL, D. H. M., AND GWALTNEY, J. M., JR., Pathogenesis of lower respiratory tract symptoms in experimental rhinovirus infection, *Am. Rev. Respir. Dis.* **128:**806–810 (1983).

96. HAMORY, B. H., HAMPARIAN, V. V., CONANT, R. M., AND GWALTNEY, J. M., JR., Human responses to two decavalent rhinoviruses vaccines, *J. Infect. Dis.* **132:**623 (1975).

97. HAMPARIAN, V. V., COLONNO, R. J., COONEY, M. K., DICK, E. C., GWALTNEY, J. M., JR., HUGHES, J. H., JORDAN, W. S., JR., KAPIKIAN, A. Z., MOGABGAB, W. J., MONTO, A., PHILLIPS, C. A. (IN ABSENTIA), RUECKERT, R. R., SCHIEBLE, J. H., STOTT, E. J., AND TYRRELL, D. A. J. (IN ABSENTIA). A collaborative report: Rhinoviruses—Extension of the numbering system from 89 to 100, *Virology* **159:**191 (1987).

98. HAMPARIAN, V. V., CONANT, R. M., AND THOMAS, D. C., Rhino-

virus Reference Laboratory, *Annual Contract Progress Report to the National Institute of Allergy and Infectious Diseases*, National Institutes of Health, Bethesda, Maryland, 1970.

99. HAMPARIAN, V. V., LEAGUS, M. B., HILLEMAN, M. R., AND STOKES, J., JR., Epidemiologic investigations of rhinovirus infections, *Proc. Soc. Exp. Biol. Med.* **117**:469 (1964).

100. HAMRE, D., Rhinoviruses in: *Monographs in Virology 1* (J. L. MELNICK, ED.), p. 52, S. Karger, Basel, 1968.

101. HAMRE, D., CONNELLY, A. P., JR., AND PROCKNOW, J. J., Virologic studies of acute respiratory disease in young adults. IV. Virus isolations during four years of surveillance, *Am. J. Epidemiol.* **83**:238 (1966).

102. HAMRE, D., AND PROCKNOW, J. J., Viruses isolated from natural common colds among young adult medical students, *Am. Rev. Respir. Dis.* **88**:277 (1963).

103. HASTINGS, G. Z., ROWLANDS, D. J., AND FRANCIS, M. J., Proliferative responses of T cells primed against human rhinovirus to other rhinovirus serotypes, *J. Gen. Virol.* **72**:2947 (1991).

104. HAYDEN, F. G., ALBRECHT, J. K., KAISER, D. L., AND GWALTNEY, J. M., JR., Prevention of natural colds by contact prophylaxis with intranasal alpha$_2$-interferon, *N. Engl. J. Med.* **314**:71–75 (1986).

105. HAYDEN, F. G., ANDRIES, K., AND JANSSEN, P. A. J., Safety and efficacy of intranasal pirodavir (R77975) in experimental rhinovirus infection, *Antimicrob. Agents Chemother.* **36**:727 (1992).

106. HAYDEN, F. G., AND GWALTNEY, J. M., JR., Prophylactic activity of intranasal enviroxime against experimentally induced rhinovirus type 39 infection, *Antimicrob. Agents Chemother.* **21**:892–897 (1982).

107. HAYDEN, F. G., AND GWALTNEY, J. M., JR., Intranasal interferon-α$_2$ for prevention of rhinovirus infection and illness, *J. Infect. Dis.* **148**:543–550 (1983).

108. HAYDEN, F. G., AND GWALTNEY, J. M., JR., Intranasal interferon-α$_2$ treatment of experimental rhinoviral colds, *J. Infect. Dis.* **150**:174–179 (1984).

109. HAYDEN, F. G., GWALTNEY, J. M., JR., AND COLONNO, R. J., Modification of experimental rhinovirus colds by receptor blockade, *Antiviral Res.* **9**:233 (1988).

110. HAYDEN, F. G., GWALTNEY, J. M., JR., AND JOHNS, M. E., Prophylactic efficacy and tolerance of low dose in interferon-α$_2$ in natural respiratory viral infections, *Antiviral Res.* **5**:111 (1985).

111. HAYDEN, F. G., MILLS, S. E., AND JOHNS, M. E., Human tolerance and histopathologic effects of long-term administration of intranasal interferon-α$_2$, *J. Infect. Dis.* **148**:914–921 (1983).

112. HAYDEN, G. F., HENDLEY, J. O., AND GWALTNEY, J. M., JR., The effect of placebo and virucidal paper handkerchiefs on viral contamination of the hand and transmission of experimental rhinoviral infection, *J. Infect. Dis.* **152**:403 (1985).

113. HAYFLICK, L., AND MOORHEAD, P. S., The serial cultivation of human diploid cell strains, *Exp. Cell Res.* **25**:585 (1961).

114. HENDLEY, J. O., EDMONDSON, W. P., JR., AND GWALTNEY, J. M., JR., Relation between naturally acquired immunity and infectivity of two rhinoviruses in volunteers, *J. Infect. Dis.* **125**:243 (1972).

115. HENDLEY, J. O., AND GWALTNEY, J. M., JR., Mechanisms of transmission of rhinovirus infections, *Epidemiol. Rev.* **10**:242 (1988).

116. HENDLEY, J. O., GWALTNEY, J. M., JR., AND JORDAN, W. S., JR., Rhinovirus infections in an industrial population. IV. Infections within families of employees during two fall peaks of respiratory illness, *Am. J. Epidemiol.* **89**:184 (1969).

117. HENDLEY, J. O., WENZEL, R. P., AND GWALTNEY, J. M., JR., Transmission of rhinovirus colds by self-inoculation, *N. Engl. J. Med.* **288**:1361 (1973).

118. HIGGINS, P. G., ELLIS, E. M., AND WOOLLEY, D. A., A comparative study of standard methods and organ culture for the isolation of respiratory viruses, *J. Med. Microscop.* **2**:109 (1969).

119. HILLEMAN, M. R., REILLY, C. M., STOKES, J., JR., AND HAMPARIAN, V. V., Clinical epidemiologic findings in coryzavirus infections, *Am. Rev. Respir. Dis. Suppl.* **88**:274 (1963).

120. HOLMES, M. J., DEIG, E. F., WILLIAMS, J. A., AND EHRESHMANN, D. W., Stability of airborne rhinovirus type 2 under atmospheric and physiological conditions, *Abstr. Annu. Meet. Am. Soc. Microbiol.* **Q**:18 (1976).

121. HOLMES, M. J., REED, S. E., STOTT, E. J., AND TYRRELL, D. A. J., Studies of experimental rhinovirus type 2 infections in polar isolation and in England, *J. Hyg. (Camb.)* **76**:379 (1976).

122. HOORN, B., AND TYRRELL, D. A. J., On the growth of certain "newer" respiratory viruses in organ cultures, *Br. J. Exp. Pathol.* **46**:109 (1965).

123. HORN, M. E. C., AND GREGG, I., Role of viral infection and host factors in acute episodes of asthma and chronic bronchitis, *Chest* **63**:44S (1973).

124. HORN, M. E. C., REED, S. E., AND TAYLOR, P., The role of viruses and bacteria in acute wheezy bronchitis in childhood: A study of sputum, *Arch. Dis. Child.* **54**:587 (1979).

125. IPSEN, J., *Relationships of Acute Respiratory Disease to Measurements of Atmospheric Pollution and Local Meteorological Conditions, Final Report to the Department of Health, Education, and Welfare*, Public Health Service, Bureau of State Services (October 5, 1960–March 30, 1964), Henry Phipps Institute, University of Pennsylvania, Contract No. PH86-63-25, 1965.

126. JACKSON, G. G., DOWLING, H. F., AND MULDOON, R. L., Acute respiratory diseases of viral etiology. VII. Present concepts of the common cold, *Am. J. Public Health* **52**:940 (1962).

127. JENNISON, M. W., Atomisation of mouth and nose secretions into the air as revealed by high speed photography, *Summ. Proc. Am. Assoc. Adv. Sci.* **17**:106 (1942).

128. JOHNSON, K. M., BLOOM, H. H., FORSYTH, B. R., AND CHANOCK, R. M., Relationship of rhinovirus infection to mild upper respiratory disease. II. Epidemiologic observations in male military trainees, *Am. J. Epidemiol.* **81**:131 (1965).

129. JOHNSTON, S. L., SANDERSON, G., PATTEMORE, P. K., SMITH, S., BARDIN, P. G., BRUCE, C. B., LAMBDEN, P. R., TYRRELL, D. A. J., AND HOLGATE, S. T., Use of polymerase chain reaction for diagnosis of picornavirus infection in subjects with and without respiratory symptoms, *J. Clin. Microbiol.* **31**:111 (1993).

130. KAPIKIAN, A. Z., Rhinoviruses, in: *Diagnostic Procedures for Viral and Rickettsial Infections*, 4th ed. (E. H. LENNETTE AND N. J. SCHMIDT, EDS.), pp. 603–640, American Public Health Association, New York, 1969.

131. KAPIKIAN, A. Z., Rhinoviruses, in: *Strains of Human Viruses* (M. MAJER AND S. A. PLOTKIN, EDS.), pp. 193–228, S. Karger, Basel, 1972.

132. KAPIKIAN, A. Z., CONANT, R. M., HAMPARIAN, V. V., CHANOCK, R. M., CHAPPLE, P. J., DICK, E. C., FENTERS, J. D., GWALTNEY, J. M., JR., HAMRE, D., HOPLER, J. C., JORDAN, W. S., JR., LENNETTE, E. H., MELNICK, J. L., MOGABGAB, W. J., MUFSON, M. A., PHILLIPS, C. A., SCHIEBLE, J. H., AND TYRRELL, D. A. J., Rhinoviruses: A numbering system, *Nature* **213**:761 (1967).

133. KAPIKIAN, A. Z., CONANT, R. M., HAMPARIAN, V. V., CHANOCK, R. M., DICK, E. C., GWALTNEY, J. M., JR., HAMRE, D., JORDAN,

W. S., Jr., Kenny, G. E., Lennette, E. H., Melnick, J. L., Mogabgab, W. J., Phillips, C. A., Schieble, J. H., Stott, E. J., and Tyrrell, D. A. J., A collaborative report: Rhinoviruses—extension of the numbering system, *Virology* **43**:524 (1971).

134. Kendall, E. J. C., Bynoe, M. L., and Tyrrell, D. A. J., Virus isolations from common colds occurring in a residential school, *Br. Med. J.* **2**:82 (1962).

135. Kenny, G. E., Cooney, M. K., and Thompson, D. J., Analysis of serum pooling schemes for identification of large numbers of viruses, *Am. J. Epidemiol.* **91**:439 (1970).

136. Ketler, A., Hall, C. E., Fox, J. P., Elveback, L., and Cooney, M. K., The virus watch program: A continuing surveillance of viral infections in metropolitan New York families. VII. Rhinovirus infections: Observations of virus excretion, intrafamilial spread and clinical response, *Am. J. Epidemiol.* **90**:244 (1969).

137. Ketler, A., Hamparian, V. V., and Hilleman, M. R., Characterization and classification of ECHO 28-rhinovirus-coryzavirus agents, *Proc. Soc. Exp. Biol. Med.* **110**:821 (1962).

138. Knopf, H. L. S., Perkins, J. C., Bertran, D. M., Kapikian, A. Z., and Chanock, R. M., Analysis of the neutralizing activity in nasal wash and serum following intranasal vaccination with inactivated type 13 rhinovirus, *J. Immunol.* **104**:566 (1970).

139. Kriel, R. L., Wulff, H., and Chin, T. D. Y., Microneutralization test for determination of rhinovirus and coxsackievirus A antibody in human diploid cells, *Appl. Microbiol.* **17**:611 (1969).

140. Kruse, W., Die Erregen von Husten und Schnupfen (The etiology of cough and nasal catarrh), *Munch. Med. Wochenschr.* **61**:1574 (1914).

141. Lambert, H. P., and Stern, H., Infective factors in exacerbations of bronchitis and asthma, *Br. Med. J.* **3**:323 (1972).

142. Levandowski, R. A., and Horohov, D. W., Rhinovirus induces natural killer-like cytotoxic cells and interferon alpha in mononuclear leukocytes, *J. Med. Virol.* **35**:116 (1991).

143. Lewis, F. A., and Kennett, M. L., Comparison of rhinovirus-sensitive HeLa cells and human embryo fibroblasts for isolation of rhinoviruses from patients with respiratory disease, *J. Clin. Microbiol.* **3**:528 (1976).

144. Lidwell, O. M., Morgan, R. W., and Williams, R. E. O., The epidemiology of the common cold. IV. The effect of weather, *J. Hyg. (Camb.)* **63**:427 (1965).

145. Lonberg-Holm, K., and Hoble-Harvey, J., Comparison of *in vitro* and cell-mediated alteration of a human rhinovirus and its inhibition by sodium dodecyl sulfate, *J. Virol.* **12**:19 (1973).

146. Lonberg-Holm, K., and Yin, F. H., Antigenic determinants of infective and inactivated human rhinoviruses type 2, *J. Virol.* **12**:114 (1973).

147. McBride, T. P., Doyle, W. J., Hayden, F. G., and Gwaltney, J. M., Jr., Alterations of the eustachian tube, middle ear, and nose in rhinovirus infection, *Arch. Otolaryngol. Head Neck Surg.* **115**:1054 (1989).

148. McIntosh, K., Ellis, E. F., Hoffman, L. S., Lybass, T. G., Eller, J. J., and Fulginiti, V. A., The association of viral and bacterial respiratory infections with exacerbations of wheezing in young asthmatic children, *J. Pediatr.* **82**:578 (1973).

149. McNamara, M. J., Phillips, I. A., and Williams, O. B., Viral and *Mycoplasma pneumoniae* infections in exacerbations of chronic lung disease, *Am. Rev. Respir. Dis.* **100**:19 (1969).

150. Merigan, T. C., Hall, T. S., Reed, S. E., and Tyrrell, D. A. J., Inhibition of respiratory virus infection by locally applied interferon, *Lancet* **1**:563 (1973).

151. Meschievitz, C. K., Schultz, S. B., and Dick, E. C., A model for obtaining predictable natural transmission of rhinoviruses in human volunteers, *J. Infect. Dis.* **150**:195 (1984).

152. Minor, T. E., Baker, J. W., Dick, E. C., DeMeo, A. N., Ouellette, J. J., Cohen, M., and Reed, C. E., Greater frequency of viral respiratory infections in asthmatic children as compared with their nonasthmatic siblings, *J. Pediatr.* **85**:472 (1974).

153. Minor, T. E., Dick, E. C., Baker, J. W., Ouellette, J. J., Cohen, M., and Reed, C. E., Rhinovirus and influenza type A infections as precipitants of asthma, *Am. Rev. Respir. Dis.* **113**:149 (1976).

154. Minor, T. E., Dick, E. C., DeMeo, A. N., Ouellette, J. J., Cohen, M., and Reed, C. E., Viruses as precipitants of asthmatic attacks in children, *J.A.M.A.* **227**:292 (1974).

155. Mogabgab, W. J., Upper respiratory illness vaccines—Perspectives and trials, *Ann. Intern. Med.* **57**:526 (1962).

156. Mogabgab, W. J., Acute respiratory illnesses in university (1962–1966), military and industrial (1962–1963) populations, *Am. Rev. Respir. Dis.* **98**:359 (1968).

157. Mohanty, S. B., Lillie, M. G., and Albert, T. F., Experimental exposure of calves to a bovine rhinovirus, *Am. J. Vet. Res.* **30**:1105 (1969).

158. Monto, A. S., A community study of respiratory infections in the tropics. III. Introduction and transmission of infections within families, *Am. J. Epidemiol.* **88**:69 (1968).

159. Monto, A. S., Bryan, E. R., and Ohmit, S., Rhinnovirus infections in Tecumseh, Michigan: Frequency of illness and number of serotypes, *J. Infect. Dis* **156**:43 (1987).

160. Monto, A. S., and Cavallaro, J. J., The Tecumseh study of respiratory illness II. Patterns of occurrence of infection with respiratory pathogens, 1965–1969, *Am. J. Epidemiol.* **94**:280 (1971).

161. Monto, A. S., and Cavallaro, J. J., The Tecumseh study of respiratory illness. IV. Prevalence of rhinovirus serotypes, 1966–1969, *Am. J. Epidemiol.* **96**:352 (1972).

162. Monto, A. S., Cavallaro, J. J., and Keller, J. B., Seasonal patterns of acute infection in Tecumseh, Mich., *Arch. Environ. Health* **21**:408 (1970).

163. Monto, A. S., and Johnson, K. M., A community study of respiratory infections in the tropics. II. The spread of six rhinovirus isolates within the community, *Am. J. Epidemiol.* **88**:55 (1968).

164. Monto, A. S., and Johnson, K. M., Respiratory infections in the American tropics, *Am. J. Trop. Med. Hyg.* **17**:867 (1968).

165. Monto, A. S., Napier, J. A., and Metzner, H. L., The Tecumseh study of respiratory illness. I. Plan of study and observations on syndromes of acute respiratory disease, *Am. J. Epidemiol.* **94**:269 (1971).

166. Monto, A. S., and Sullivan, K. M., Acute respiratory illness in the community. Frequency of illness and the agents involved, *Epidemiol. Infect.* **110**:145 (1993).

167. Monto, A. S., and Ullman, B. M., Acute respiratory illness in an American community: The Tecumseh study, *J.A.M.A.* **227**:164 (1974).

168. Mufson, M. A., Bloom, H. H., Forsyth, B. R., and Chanock, R. M., Relationship of rhinovirus to mild upper respiratory disease. III. Further epidemiologic observations in military personnel, *Am. J. Epidemiol.* **83**:379 (1966).

169. Mufson, M. A., Krause, H. E., Mocega, H. E., and Dawson, F. W., Viruses, *Mycoplasma pneumoniae* and bacteria associated with lower respiratory tract disease among infants, *Am. J. Epidemiol.* **91**:192 (1970).

170. MUFSON, M. A., LUDWIG, W. M., JAMES, H. D., JR., GAULD, L. W., ROURKE, J. A., HOLPER, J. C., AND CHANOCK, R. M., Effect of neutralizing antibody on experimental rhinovirus infection, *J.A.M.A.* **186:**578 (1963).

171. NACLERIO, R., GWALTNEY, J., HENDLEY, O., BAUMGARTEN, C., KAGEY-SOBOTKA, A., BARTENFELDER, D., BEASLEY, P., LICHTENSTEIN, L., AND PROUD, D., Preliminary observations on mediators in rhinovirus-induced colds, in: *New Dimensions in Otorhinolaryngology-Head and Neck Surgery* (E. MYERS, ED.), p. 341, Elsevier, New York, 1985.

172. OLSON, N. H., KOLATKAR, P. R., OLIVEIRA, M. A., CHENG, R. H., GREVER, J. M., MCCLELLAND, A., BAKER, T. S., AND ROSSMANN, M. G., Structure of a human rhinovirus complexed with its receptor molecule, *Proc. Natl. Acad. Sci. USA* **90:**507 (1993).

173. PALMENBERG, A. C., Sequence alignments of picornaviral capsid proteins, in: *Molecular Aspects of Picornavirus Infection and Detection* (B. L. SEMLER AND E. EHRENFELD, ED.), p. 211, American Society for Microbiology, Washington, 1989.

174. PAREKH, H. H., CRAGUN, K. T., HAYDEN, F. G., HENDLEY, J. O., AND GWALTNEY, J. M., JR., Nasal mucus weights in experimental rhinovirus infection, *Am. J. Rhinol.* **6:**107 (1992).

175. PELON, W., MOGABGAB, W. J., PHILLIPS, I. A., AND PIERCE, W. E., A cytopathogenic agent isolated from naval recruits with mild respiratory illness, *Proc. Soc. Exp. Biol. Med.* **94:**262 (1957).

176. PEREIRA, M. A., ANDREWS, B. E., AND GARDNER, S. D., A study on the virus aetiology of mild respiratory infections in the primary school child, *J. Hyg. (Camb.)* **65:**475 (1967).

177. PERKINS, J. C., TUCKER, D. N., KNOPF, H. L. S., WENZEL, R. P., HORNICK, R. B., KAPIKIAN, A. Z., AND CHANOCK, R. M., Evidence for protective effect of an inactivated rhinovirus vaccine administered by the nasal route, *Am. J. Epidemiol.* **90:**319 (1969).

178. PERKINS, J. C., TUCKER, D. N., KNOPF, H. L. S., WENZEL, R. P., KAPIKIAN, A. Z., AND CHANOCK, R. M., Comparison of protective effect of neutralizing antibody in serum and nasal secretions in experimental rhinovirus type 13 illness, *Am. J. Epidemiol.* **90:**519 (1969).

179. PERSON, D. A., AND HERRMANN, E. C., JR., Experiences in laboratory diagnosis of rhinovirus infections in routine medical practice, *Mayo Clin. Proc.* **45:**517 (1970).

180. PETERSON, J. A., D'ALESSIO, D. J., AND DICK, E. C., Studies on the failure of direct oral contact to transmit rhinovirus infection between humans volunteers, in: *Abstracts, Annual Meeting American Society for Microbiology*, p. 214, American Society for Microbiology, Washington, DC, 1973.

181. PHILLIPS, C. A., MELNICK, J. L., AND GRIM, C. A., Rhinovirus infections in a student population: Isolation of five new serotypes, *Am. J. Epidemiol.* **87:**447 (1968).

182. PIELA-SMITH, T. H., BROKETA, G., HAND, A., AND KORN, J. H., Regulation of ICAM-1 expression and function in human dermal fibroblasts by Il-4, *J. Immunol.* **148:**1375 (1992).

183. PINTO, C. A., AND HAFF, R. F., Experimental infection of gibbons with rhinovirus, *Nature* **224:**1310 (1969).

184. PORTNOY, B., ECKERT, H. L., AND SALVATORE, M. A., Rhinovirus infection in children with acute lower respiratory disease: Evidence against etiological importance, *Pediatrics* **35:**899 (1965).

185. PRICE, W. H., The isolation of a new virus associated with respiratory clinical disease in humans, *Proc. Natl. Acad. Sci. USA* **42:**892 (1956).

186. PRICE, W. H., Vaccine for the prevention in humans of cold-like symptoms associated with the JH virus, *Proc. Natl. Acad. Sci. USA* **43:**790 (1957).

187. REED, S. E., An investigation of the possible transmission of rhinovirus colds through indirect contact, *J. Hyg. (Camb.)* **75:**249 (1975).

188. REED, S. E., AND BYNOE, M. L., The antiviral activity of isoquinoline drugs for rhinoviruses *in vitro* and *in vivo*, *J. Med. Microbiol.* **3:**346 (1970).

190. REILLY, C. M., HOCH, S. M., STOKES, J., MCCLELLAND, L., HAMPARIAN, V. V., KETLER, A., AND HILLEMAN, M. R., Clinical and laboratory findings in cases of respiratory illness caused by coryzaviruses, *Ann. Intern. Med.* **57:**515 (1962).

191. ROSENBAUM, M. J., DE BERRY, P., SULLIVAN, E. J., PIERCE, W. E., MUELLER, R. E., AND PECKINPAUGH, R. O., Epidemiology of the common cold in military recruits with emphasis on infections by rhinovirus types 1A, 2, and two unclassified rhinoviruses, *Am. J. Epidemiol.* **93:**183 (1971).

192. ROSSEN, R. D., DOUGLAS, R. G., JR., CATE, T. R., COUCH, R. B., AND BUTLER, W. T., The sedimentation behavior of rhinovirus neutralizing activity in nasal secretion and serum following the rhinovirus common cold, *J. Immunol.* **97:**532 (1966).

193. ROSSEN, R. D., KASEL, J. A., AND COUCH, R. B., The secretory immune system: Its relation to respiratory viral infection, in: *Progress in Medical Virology* (J. L. MELNICK, ED.), pp. 194–238, S. Karger, Basel, 1971.

194. ROSSMAN, M. G., ARNOLD, E., ERICKSON, J. W., FRANKENBERGER, E. A., GRIFFITH, J. P., HECHT, H.-J., JOHNSON, J. E., KAMER, G., LUO, M., MOSSER, A. G., RUECKERT, R. R., SHERRY, B., AND VRIEND, G., Structure of a human common cold virus and functional relationship to other picornaviruses, *Nature* **317:**145 (1985).

195. RUECKERT, R. R., Picornaviruses and their replication, in: *Virology* (B. N. FIELDS, D. M., KNIPE, R. M. CHANOCK, J. L. MELNICK, B. ROIZMAN, AND R. E. SHOPE, EDS.), p. 705, Raven Press, New York, 1985.

196. SAMO, T. C., GREENBERG, S. B., PALMER, J. M., COUCH, R. B., HARMON, M. W., AND JOHNSON, P. E., Intranasally applied recombinant leukocyte A interferon in normal volunteers. II. Determination of minimal effective and tolerable dose, *J. Infect. Dis.* **150:**181 (1984).

197. SASAKI, Y., TOGO, Y., WAGNER, H. N., JR., HORNICK, R. B., SCHWARTZ, A. R., AND PROCTOR, D. F., Mucociliary function during experimentally induced rhinovirus infection in man, *Ann. Otol.* **82:**203 (1973).

198. SCHIEBLE, J. H., FOX, V. L., LESTER, F., AND LENNETTE, E. H., Rhinoviruses: An antigenic study of the prototype virus strains, *Proc. Soc. Exp. Biol. Med.* **147:**541 (1974).

199. SCHLEICHER, J. B., AQUINO, F., RUETER, A., RODERICK, W. R., AND APPELL, R. N., Antiviral activity in tissue culture systems of *bis*-benzimidazoles, potent inhibitors of rhinoviruses, *Appl. Microbiol.* **23:**113 (1972).

200. SCOTT, G. M., WALLACE, J., GREINER, J., PHILLPOTTS, R. J., GAUCI, C. L., AND TYRRELL, D. A. J., Prevention of rhinovirus colds by human interferon alpha-2 from *Escherichia coli*, *Lancet* **2:**186 (1982).

201. SHANNON, W. M., ARNETT, G., AND SCHABEL, F. M., JR., 3-Deazauridine: Inhibition of ribonucleic acid virus-induced cytopathogenic effect *in vitro*, *Antimicrob. Agents Chemother.* **2:**159 (1972).

202. SHERRY, B., MOSSER, A. G., COLONNO, R. J., AND RUECKERT,

R. C., Use of monoclonal antibodies to identify four neutralization immunogens on a common cold picornavirus, human rhinovirus 14, *J. Virol.* **57**:246–257 (1985).

203. SHIPOWITZ, N. E., BOWER, R. R., SCHLEICHER, J. B., AQUINO, F., APPELL, R. N., AND RODERICK, W. R., Antiviral activity of a *bis*-benzimidazole against experimental rhinovirus infection in chimpanzees, *Appl. Microbiol.* **23**:117 (1972).

204. SIDWELL, R. W., HUFFMAN, J. H., ALLEN, L. B., MEYER, R. B., JR., SHUMAN, D. A., SIMON, L. N., AND ROBINS, R. K., *In vitro* antiviral activity of 6-substituted 9-β-D-ribofuranosylpurine-3′,5′-cyclic phosphates, *Antimicrob. Agents Chemother.* **5**:652 (1974).

205. SIDWELL, R. W., HUFFMAN, J. H., KHARE, G. P., ALLEN, L. B., WITKOWSKI, J. T., AND ROBINS, R. K., Broad-spectrum antiviral activity of virazole: 1-β-D-ribofuranosyl-1,2,4-triazole-3-carboxamide, *Science* **177**:705 (1972).

206. SKERN, T., DUECHLER, M., SOMMERGRUBER, W., BLAAS, D., AND KUECHLER, E., The molecular biology of human rhinoviruses, *Biochem. Soc. Symp.* **53**:63 (1987).

207. SMITH, C. B., KANNER, R. E., GOLDEN, C. A., KLAUBER, M. R., AND RENZETTI, A. D., Effect of viral infections on pulmonary function in patients with chronic obstructive pulmonary diseases, in: *18th Interscience Conference on Antimicrobial Agents and Chemotherapy*, p. 11, Washington, DC, 1978.

208. SMITH, T. J., OLSON, N. H., CHENG, R. H., LIU, H., CHASE, E. S., LEE, W. M., LEIPPE, D. M., MOSSER, A. G., RUECKERT, R. R., AND BAKER, T. S., Structure of human rhinovirus complexed with Fab fragments from a neutralizing antibody, *J. Virol.* **67**:1148 (1993).

209. SPERBER, S. J., AND HAYDEN, F. G., Chemotherapy of rhinovirus colds, *Antimicrob. Agents Chemother.* **32**:409 (1988).

210. SPERBER, S. J., HENDLEY, J. O., HAYDEN, F. G., RIKER, D. K., SORRENTINO, J. V., AND GWALTNEY, J. M., JR., Effects of naproxen on experimental rhinovirus colds, *Ann. Inter. Med.* **117**:37 (1992).

211. SPERBER, S. J., SORRENTINO, J. V., RIKER, D. K., AND HAYDEN, F. G., Evaluation of an alpha agonist alone and in combination with a nonsteroidal anti-inflammatory agent in the treatment of experimental rhinovirus colds, *Bull. NY Acad. Med.* **65**:145 (1989).

212. STAUNTON, D. E., MERIUZZI, V. J., ROTHLEIN, R., BARTON, R., MARLIN, S. D., AND SPRINGER, T. A., A cell adhesion molecule, ICAM-1, is the major surface receptor for rhinoviruses, *Cell* **56**:849 (1989).

213. STENHOUSE, A. C., Rhinovirus infection in acute exacerbations of chronic bronchitis: A controlled prospective study, *Br. Med. J.* **3**:461 (1967).

214. STOTT, E. J., DRAPER, C., STONES, P. B., AND TYRRELL, D. A. J., Absence of heterologous antibody responses in human volunteers after rhinovirus vaccination, *Arch. Ges. Virusforsch.* **28**:90 (1969).

215. STOTT, E. J., AND KILLINGTON, R. A., Rhinoviruses, *Annu. Rev. Microbiol.* **26**:503 (1972).

216. SYMPOSIUM ON RHINOVIRUS PATHOGENESIS, *Acta Otoarlyngol. Suppl. (Stockh.)* **413**:1–45 (1984).

217. TAMM, I., AND CALIGURI, L. A., 2-(α-Hydroxybenzyl)-benzimidazole and related compounds, in: *The International Encyclopedia of Pharmacology and Therapeutics*, Section 61 (D. J. BAUER, ED.), Pergamon Press, New York, 1972.

218. TAYLOR-ROBINSON, D., Studies on some viruses (rhinoviruses) isolated from common colds, *Arch. Ges. Virusforsch.* **13**:281 (1963).

219. TAYLOR-ROBINSON, D., Respiratory virus antibodies in human sera from different regions of the world, *Bull. WHO* **32**:833 (1965).

220. THOMAS, D. C., CONANT, R. M., AND HAMPARIAN, V. V., Rhinovirus replication in suspension cultures of HeLa cells, *Proc. Soc. Exp. Biol. Med.* **133**:62 (1970).

221. TOGO, Y., SCHWARTZ, A. R., AND HORNICK, R. B., Failure of a 3-substituted triazinoindole in the prevention of experimental human rhinovirus infection, *Chemotherapy* **18**:17 (1973).

222. TURNER, B. W., CAIL, W. S., HENDLEY, J. O., HAYDEN, F. G., DOYLE, W. J., SORRENTINO, J. V., AND GWALTNEY, J. M., JR., Physiologic abnormalities in the paranasal sinuses during experimental rhinovirus colds, *J. Allergy Clin. Immunol.* **90**:474 (1992).

223. TURNER, R. B., HENDLEY, J. O., AND GWALTNEY, J. M., JR., Shedding of infected ciliated epithelial cells in rhinovirus colds, *J. Infect. Dis.* **145**:849–853 (1982).

224. TYRRELL, D. A. J., Rhinoviruses, in: *Virology Monographs 2* (S. GARD, C. HALLAUER, AND K. F. MYER, EDS.), p. 67, Springer-Verlag, New York, 1968.

225. TYRRELL, D. A. J., BYNOE, M. L., AND HOORN, B., Cultivation of "difficult" viruses from patients with colds, *Br. Med. J.* **1**:606 (1968).

226. TYRRELL, D. A. J., SHEFF, M. D., AND PARSONS, R., Some virus isolations from common colds. III. Cytopathic effects in tissue cultures, *Lancet* **1**:239 (1960).

227. VAN LOGHEM, J. J., Epidemiologische bijdrage tot de kennis van de ziekten der ademhalingsorganen, *Ned. Tijdschr. Geneeskd.* **72**:666 (1928).

228. WILDY, P., Classification and nomenclature of viruses, in: *Monographs in Virology* (J. L. MELNICK, ED.), p. 56, S. Karger, Basel, 1971.

229. WINTHER, B., FARR, B., TURNER, R. B., HENDLEY, J. O., GWALTNEY, J. M., JR., AND MYGIND, N., Histopathologic examination and enumeration of polymorphonuclear leukocytes in the nasal mucosa during experimental rhinovirus colds, *Acta Otolaryngol. Suppl. (Stockh.)* **413**:19–24 (1984).

230. WINTHER, B., GWALTNEY, J. M., JR., MYGIND, N., TURNER, R. B., AND HENDLEY, J. O., Intranasal spread of rhinovirus (RV) following point-inoculation of the nasal mucosa, in: *Program and Abstracts of the Twenty-fifth Interscience Conference on Antimicrobial Agents and Chemotherapy*, p. 249, American Society for Microbiology, Washington, DC, 1985.

231. WULFF, H., NOBLE, G. R., MAYNARD, J. E., FELTZ, E. T., POLAND, J. D., AND CHIN, T. D. Y., An outbreak of respiratory infection in children associated with rhinovirus types 16 and 29, *Am. J. Epidemiol.* **90**:304 (1969).

232. YIN, F. H., AND LOMAS, N. B., Establishment of a mouse model for human rhinovirus infection, in: *6th International Congress of Virology*, p. 90, ICV, Sendai, Japan, 1984.

13. Suggested Reading

ANDREWES, C., *The Common Cold*, W. W. Norton, New York, 1965.

DINGLE, J. H., BADGER, G. F., AND JORDAN, W. S., JR., Patterns of illness, in: *Illness in the Home* (J. H. DINGLE, G. F. BADGER, AND W. S.

JORDAN, JR., EDS.), p. 19, Western Reserve University, Cleveland, 1964.

FROST, W. H., AND GOVER, M., The incidence and time distribution of common colds in several groups kept under continuous observation, in: *Papers of Wade Hampton Frost, M.D.* (K. F. MAXCY, ED.), p. 359, Commonwealth Fund, New York, 1941.

GWALTNEY, J. M., JR., Historical eras of the common cold, in: *Contemporary Issues in Infectious Diseases*, Vol. 10. *Viral Infections: Diag-nosis, Treatment and Prevention* (M. A. SANDE AND R. K. ROOT, EDS.), pp. 1–13, Churchill Livingstone, New York, 1993.

ROSSMANN, M. G., ARNOLD, E., ERICKSON, J. W., FRANKENBERGER, E. A., GRIFFITH, J. P., HECHT, H.-J., JOHNSON, J. E., KAMER, G., LUO, M., MOSSER, A. G., RUECKERT, R. R., SHERRY, B., AND VRIEND, G., The structure of a human common cold virus (rhinovirus 14) and its functional relations to other picornaviruses, *Nature* **317:**145 (1985).

CHAPTER 27

Rubella

Sandra J. Holmes and Walter A. Orenstein

1. Introduction

Rubella (German measles) generally is a mild viral illness that may be associated with fever, lymphadenopathy, and a generalized rash. An estimated 50% or more of infections are subclinical. The most serious consequences of rubella occur in infants of women infected early in pregnancy. These infants are at high risk for congenital rubella syndrome (CRS), which may include a number of abnormalities.

2. Historical Background

In the mid-18th century, German physicians first described rubella as a specific clinical entity and named it Rotheln.[34,59] It was not until 1815 that it was described in the English literature as a separate disease.[132] However, until the late 19th century, most physicians continued to consider it a mild form of measles or a combination of measles and scarlet fever. In other countries the disease was frequently referred to as German measles because of the German medical community's great interest in the disease and probably because the German name was difficult to pronounce. In 1866, Veale, a Scottish physician, described 30 cases of German measles and recommended a "short and euphonious" name that could be easily pronounced, namely, rubella, a Latin word meaning "little red."[225] Finally, in 1881, at the International Congress of Medicine in London, it was agreed that rubella was not a variation of measles or scarlet fever but a distinctive disease.

In 1914, Hess[90] suggested a viral etiology for rubella on the basis of transmission studies in Rhesus monkeys; however, this was not confirmed until 1938, when Hiro and Tasaka[94] produced rubella in children by inoculating them with nasal washings. In 1942, Habel[80] infected monkeys with nasal washings and blood from infected humans. In volunteer studies in the 1950s, Krugman and associates[117,118] showed that viremia occurs in the preeruptive stage and that the infection can occur without rash.

Rubella was generally thought of as a mild, self-limited, inconsequential disease of childhood and received little attention for the next 60 years. Following an outbreak of rubella in Australia in 1940, Sir Norman Gregg, an Australian ophthalmologist, reported an unusual number of infants born with cataracts and noted that nearly all of the infants were born of mothers who had been infected with rubella during the first trimester of their pregnancies.[77] Gregg also noted other ocular defects and cardiac lesions in the affected infants. The medical community was slow in accepting Gregg's findings and questioned their validity in an editorial in *Lancet* in 1944.[49] However, Gregg had created a heightened awareness, and during the next few years confirmatory reports came from Australia,[212,213] the United States,[52,178,181,232] and the United Kingdom.[134] These early reports further defined the constellation of abnormalities that characterize CRS, including cataracts and other ocular abnormalities, cardiac defects, deafness, microcephaly, and mental retardation.

Early retrospective studies, which were based on hospitalized patients, yielded estimates of stillbirths or congenital abnormalities as high as 90% in infants whose mothers were reported to have had rubella in the first trimester of pregnancy. Lower rates were reported by the first prospective studies in the 1950s,[105,130,131,200] which

This chapter is a revised version of Chapter 23, Dorothy M. Horstmann, in *Viral Infections of Humans*, 3rd edition, Plenum Press, 1989.

Sandra J. Holmes • Center for Pediatric Research, Eastern Virginia Medical School, Children's Hospital of The King's Daughters, Norfolk, Virginia 23510-1001. **Walter A. Orenstein** • Centers for Disease Control and Prevention, The National Immunization Program, Atlanta, Georgia 30340.

included nonepidemic years and relied on clinical diagnoses of rubella, and thus underestimated the incidence of CRS. Subsequent prospective studies in which maternal rubella infections were serologically confirmed showed rates of CRS in the first trimester of up to 90%.[138,147,148]

The recognition of the teratogenic potential of maternal rubella infection established rubella as a serious problem and led to increased efforts to isolate the etiologic agent. A landmark breakthrough occurred in 1962 when two groups of investigators, Weller and Neva[231] at Harvard and Parkman, Buescher, and Artenstein[163] at the Walter Reed Army Institute of Research, independently isolated the rubella virus. The discovery of the rubella virus came at the onset of a pandemic of rubella in Europe in 1962–1963, which spread to the United States in 1964–1965, and allowed for an extensive virological, serological, and epidemiologic investigation. These findings were presented at a Rubella Symposium in 1965.[115] The isolation of the virus together with the tragedy of the pandemic sparked an intensive worldwide effort to develop a vaccine. In February 1969, the International Conference on Rubella Immunization provided a forum to discuss the scientific body of knowledge of rubella and the newly developed vaccines.[116] Later that year, the first live attenuated rubella vaccine was licensed in the United States.[146] Widespread use of rubella vaccines has resulted in a marked decline in the incidence of rubella and CRS.

3. Methodology

3.1. Mortality Data

Since death from rubella is a rare event, mortality data are not important in understanding the epidemiology of the disease.

3.2. Morbidity Data

In the United States, rubella and CRS became nationally reportable to the National Notifiable Disease Surveillance System (NNDSS) in 1966, and the Centers for Disease Control and Prevention (CDC) established the National Congenital Rubella Syndrome Registry (NCRSR) in 1969. The reporting efficiency for clinical cases of rubella is estimated to be only 10 to 20% for several reasons: (1) the disease is usually so mild that medical care may not be sought; (2) the clinical manifestations are not highly specific, and thus, cases frequently are not diagnosed as rubella; and (3) reporting of diagnosed cases by physicians is limited, as is true for other notifiable diseases that are mild and not thought to have great public health importance. Perhaps the most serious omissions are the subclinical cases, which are contagious and in pregnant women may result in damage to the fetus. It has been estimated that there may be as many subclinical as clinical cases, and both are important in terms of the spread of disease and the potential for intrauterine infection and CRS in infants of infected mothers.

3.3. Serological Surveys

Serological surveys of healthy population groups have been of major importance in mapping the epidemiology of rubella and documenting differences in its behavior in various parts of the world. These surveys provide information on age-specific immunity, which identifies target age groups for immunization and helps to monitor the effectiveness of vaccination programs and the persistence of immunity.[99]

In an extensive collaborative survey in 1967 sponsored by the World Health Organization (WHO),[177] 80 to 87% of women 17 to 22 years of age tested in continental Europe, Great Britain, Japan, and Australia had antibodies against rubella. A second collaborative study in 1968 in Caribbean and Middle and South American populations[46] similarly showed that 80% of women of childbearing age had rubella antibodies; however, in certain island groups, smaller proportions were seropositive (discussed in Section 5.3). Serological surveys in the United States conducted during the prevaccination era showed that 80 to 92% of young adults had detectable antibody, and the findings were similar in surveys conducted before and after the 1964 pandemic.[155,195–197,237] Interpretation of postvaccination serological surveys has been complicated by variations in test sensitivity and the generally lower levels of antibody induced by vaccination. Nevertheless, postvaccination era serological surveys have indicated that 75 to 93% of young adults were immune.[41,44,159,180,191,209,211,226] When vaccination status was considered, 87 to 96% of vaccinated persons had detectable antibody compared to 70 to 80% of unvaccinated persons.[121,158,180]

3.4. Laboratory Methods

3.4.1. Virus Isolation. Rubella virus can be grown in cultures of primary cells or cell lines of mammalian origin.[40,87,164] Rubella virus growing in cell cultures produces inconsistent cytopathic effects dependent on a number of variables. Thus, a method using its property of interference with the growth of a variety of other viruses, especially enterovirus, is commonly used to isolate rubella virus. Material from human blood or throat swabs is inoculated into primary African green monkey kidney (AGMK) tissue cultures that are challenged after 9–12 days with a virus, such as echovirus 11, that normally induces cytopathic effects and destroys the cell sheet. If rubella virus is present, it interferes with the growth of the challenge virus and no cytopathic effect develops. Cell lines, such as Vero, a continuous cell line derived from AGMK cells, also are used in the interference method.[169] Rubella virus in tissue culture also can be detected by measuring the production of hemagglutinin or complement-fixing (CF) antigens and by techniques using immuno-fluorescent staining.

3.4.2. Serological Tests.

During the last 25 years, a number of serological tests using different methodologies and variations of the same methodology have been developed for the measurement of antibodies to rubella virus (Fig. 1). These include the neutralization test (NT), hemagglutination inhibition (HAI or HI),[127,128] passive hemagglutination (PHA),[11] radioimmunoassay (RIA),[144] hemolysis in gel (HIG),[119] indirect immunofluorescence (IFA),[39] enzyme-linked immunosorbent assay (ELISA),[184,227] and latex agglutination (LA).[135] The National Committee for Clinical Laboratory Standards (NCCLS) establishes guidelines for serological tests used to detect rubella antibody.[203] The HAI test was adopted by the NCCLS and WHO[240] in 1979 as the reference assay method for rubella antibody. There was a high correlation between results obtained by the HAI and the traditional rubella NT.[125,189] Although no longer used for routine testing, the HAI is used as a reference method to establish a calibration standard for other rubella assays that are more rapid,

less technical, and compare to the HAI in sensitivity and specificity. Many of these tests are available in commercial kits. Determination of which test should be used depends on the purpose for testing, i.e., to determine immunity (detect total or IgG antibody to rubella) or to establish a diagnosis (quantitative comparison of acute and convalescent IgG antibody or detection of IgM antibody). Several tests can be used to measure IgM-specific antibody; the ELISA and the IgM capture ELISA are used frequently.[233] Recently, immunoblot assays have been used to detect antibodies of all immunoglobulin classes as well as to determine the specific epitopic immune response.[241] Numerous studies have compared the various tests used to measure rubella antibody.[31,43,56,103,135–137,204,210,223]

4. Biological Characteristics of the Virus

Rubella is an ether-sensitive virus, readily inactivated by a variety of chemical agents, by a pH below 6.8 or above 8.1, and by UV irradiation. Rubella virus is heat labile; inactivation is rapid at 56°C and slower at 37°C and below. It is best preserved at −60°C and below. In thin sections of infected tissue-culture cells examined by electron microscopy, the virus appears as spherical particles measuring 50–70 nm in diameter with 30-nm electron-dense cores; the virions can be seen budding into intracellular vesicles or directly from the marginal membrane.[153] Released virus is covered with projections 5–6 nm in length that cause the particles to hemagglutinate certain fowl red blood cells and human type O red blood cells.[96]

Rubella, an RNA virus with an envelope, is classified with the togaviruses. Not only does the morphological picture of rubella resemble that of the alpha group of togaviridae (group A arboviruses), but there are other similarities as well. The rubella nucleocapsid core sediments at 150 S and contains a single strand of infectious RNA that sediments at 40 S. The virion contains three major polypeptides, E1, E2, and C.[169,229] Both E1 and E2 are glycosylated, and E2 is made up of several closely related glycopolypeptides. By use of monoclonal antibodies, E1 has been shown to be involved in hemagglutination and lysis of red cells.

There appears to be only one antigenic type of rubella virus; no cross-reactions with alphaviruses or other members of the togavirus group have been found.[143] However, minor differences between strains have been shown,[45] and there is evidence of differences in the biological behavior of different strains. These include the reported low teratogenicity of those causing epidemics in

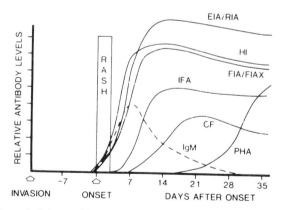

Figure 1. Schema of immune responses in acute rubella infection as measured by various tests. EIA, enzyme immunoassay; RIA, radioimmunoassay; HI, hemagglutination inhibition; FIA/FIAX, IFA, fluorescence assays; CF, complement fixation; PHA, passive hemagglutination.[88]

mainland Japan,[113,221] variations in the capacity to induce interferon *in vitro*,[4] and transmissibility of some strains to rabbit fetuses.[112]

Advances in molecular biology, specifically polymerase chain reaction (PCR) and nucleotide sequencing, have been used to differentiate the genomes of multiple strains of rubella virus. Recently, Frey and Abernathy[66] identified characteristic nucleotide sequences of the RA27/3 strain, the strain used exclusively in rubella-containing vaccine distributed in the United States since 1979; these sequences were maintained in isolates of rubella virus from recent vaccinees. This technology can be used to identify the origin of rubella virus, i.e., wild-type or vaccine-type, recovered from persons with conditions that may be attributable either to natural rubella infection or vaccination.

5. Descriptive Epidemiology

5.1. Incidence and Prevalence

The incidence of rubella infection depends on the number of susceptible persons at risk, i.e., lacking immunity, and the degree of exposure to wild virus circulating within a community. Prospective serological studies of children in institutions have shown that virtually 100% of susceptible children living in cottages in which rubella occurred became infected as secondary and tertiary waves occurred.[101] The situation is analogous to spread in families[68] and in military installations,[19,124] where close contact also results in a similar high degree of communicability. Among susceptible college students, a somewhat lower infection rate (64%) has been recorded.[55]

The introduction of vaccines in 1969 resulted in a dramatic reduction in the incidence of rubella infection. This reduction is demonstrated by the incidence data from 1928 to 1983 in ten selected areas of the United States (Fig. 2). The last large epidemic occurred in 1964–1965 and resulted in an estimated 12.5 million cases of rubella,

20,000 cases of CRS, and more than 11,000 fetal deaths as a result of spontaneous or therapeutic abortions.[157] Nationwide data, available since rubella became a reportable disease in 1966, show that incidence dropped sharply in the early 1970s and held steady from 1974 to 1979 during which an average of 15,000 cases were reported annually. Throughout most of the 1980s, incidence declined steadily to a record low number of cases (225) reported in 1988, representing a 94% decline from 1980.[129] In 1989 to 1991, a resurgence of rubella occurred; in 1989, the number of cases almost doubled (396); in 1990, the number increased by nearly threefold (1125); and in 1991, they increased further (1401). In 1990, 26 outbreaks were reported, occurring mostly in unvaccinated adults and children.[27] In 1991, large outbreaks (60 to >400 cases) occurred among unvaccinated children and young adults in religious communities in at least four states.[13,27,28,106] Following the 3-year period of resurgence of rubella, low numbers of cases were reported in 1992 (160) and 1993 (190). In 1994, there was a 19% increase in incidence over 1993; more than one half of the 227 cases reported in 1994 occurred in an outbreak in Massachusetts, primarily among unvaccinated young adults in prisons, homeless shelters, and the Latino community. An all-time low of 128 cases were reported in 1995.

The incidence of CRS roughly parallels the incidence of rubella in individuals over 15 years of age (Fig. 3). The expected increase in the incidence of CRS was observed during and immediately following the recent resurgence of rubella. The number of CRS cases reported to the NCRSR went from 2 in 1989 to 25 in 1990 and 33 in 1991. Only a few indigenous cases have been reported annually since then, reflecting the concomitant decline in incidence.

The risk for congenital rubella depends on when during the gestational period the infection is acquired. Lundstrom[130] reviewed 15 prospective studies that were conducted prior to the isolation of rubella virus in 1962 and the development of serological tests for diagnosis, and estimated the risk for CRS to be 33% after maternal

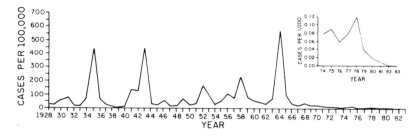

Figure 2. Incidence rates of rubella in ten selected areas of the United States, 1928–1983, which are Maine, Rhode Island, Connecticut, New York City, Ohio, Illinois, Wisconsin, Maryland, Washington, and Massachusetts.[6]

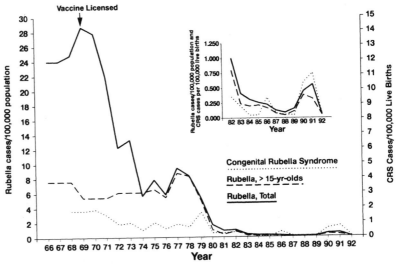

Figure 3. Incidence rates for reported cases of rubella and congenital rubella syndrome in the United States. Average-annual estimate in the United States based on data from Illinois, Massachusetts, and New York City for the 3-year periods 1966–1968, 1969–1971, and 1972–1974. Data for 1975–1992 reflect age-specific rates for the total number of cases reported in the United States.

infection in the first month of pregnancy, 25% in the second month, 9% in the third month, and 4% in the fourth month. These estimates were substantially lower than those that came from the early retrospective studies (Australia) showing that almost all infants born to mothers who had rubella in the first 2 months of pregnancy had CRS. The estimates of CRS derived from the early prospective studies were inaccurate because they were based on clinical diagnoses of rubella, which are unreliable, and thus, undoubtedly included cases that were not rubella. Later prospective studies in the United Kingdom,[147] in which maternal diagnosis was serologically established, confirmed the high rates of CRS associated with maternal rubella early in pregnancy that had been suggested by the early retrospective studies. Among infants born to mothers with rubella during the first 12 weeks of gestation, 86% were infected and 72% of those infected had CRS, yielding an overall risk of 62% for having a child with CRS. Both the infection rate in infants and the proportion of infected infants with CRS decreased as gestational time increased. Maternal infection during the first 10 weeks of gestation resulted in infection in all infants and CRS in 90%, whereas maternal infection during weeks 11 to 12 or weeks 13 to 16 resulted in infection in 73 and 52% of infants, respectively, of whom 50 and 33%, respectively, had CRS. Only infants infected during the first 10 weeks of gestation had defects in addition to sensorineural deafness. The risk for deafness associated with infection between 17 to 18 weeks gestation was 7% among 71% of infants followed for 2 years. Following a rubella outbreak in 1991 among the Amish in Pennsylvania, 70% of infants

infected during the first trimester had defects.[138] Deafness may not be apparent at birth and hearing tests may not be done until later in life. In the United Kingdom, only 14% of 57 deaf children born with CRS between 1978 and 1982 had been tested in the first 6 months of life.[234] In a study [166] in which children were reevaluated at 6 to 8 years of age, the risk for defects among infants infected during the first 16 weeks of pregnancy increased from 34 to 71%. CRS is rare following maternal infection between 17 and 24 weeks gestation[79,148,167]; nevertheless, fetal infection without defects associated with maternal infection in the third trimester can occur.[37]

5.2. Epidemic Behavior

Prior to the introduction of rubella vaccine in 1969, incidence followed a distinctive epidemic cycle, rising and falling every 2 to 4 years with sizable epidemics occurring approximately every 6 to 9 years and major epidemics occurring at intervals up to 21 years (Fig. 2). Rubella remained endemic with yearly occurrence of cases and periodic epidemics in countries without universal childhood vaccination, including Australia,[141] Brazil,[187] Israel,[57,97,214,215] Japan,[219,220] and the United Kingdom.[71]

A puzzling epidemiologic feature has been the sudden eruption of extensive outbreaks after long intervals of time in countries such as the United States. In the intervening periods, the disease followed its usual epidemic cycle. This pattern succeeded in immunizing approximately 85% of the population by 15 to 19 years of age. It is unknown why a particular epidemic suddenly gathered force and resulted in the extraordinary onslaught that occurred during the 1964 epidemic. No significant antigenic differences among viral strains have been documented, yet conceivably viruses with altered biological characteristics and an enhanced capacity to spread may account for such behavior. Host factors also may be involved. There is evidence that infected persons differ in their capacity to transmit rubella, and "spreaders" and "nonspreaders" have been identified in an epidemiologic study of outbreaks of the disease in Hawaii.[84,133] It is likely that there are other unknown factors that may be implicated in the dissemination of rubella virus.

5.3. Geographic Distribution

Although serological surveys have documented the worldwide distribution of rubella infection,[32] there are a number of perplexing features that characterize the behavior of the disease in certain island populations (Fig. 4). Some of these are illustrated by the history of the disease on Taiwan,[68,69,75] a semitropical island with a dense population and extensive communication with the rest of the world. Large epidemics of rubella occurred there in 1944, 1957–1958, and 1968–1969. In the latter two outbreaks, the disease began in the north, where attack rates were highest, and moved toward the south, where the rates were considerably lower and the occurrence of cases

ceased, although significant numbers of susceptible individuals remained. A serological survey that was conducted prior to the 1968–1969 outbreak provided evidence that the virus had not been circulating since the last epidemic in 1957–1958; none of the children born in the 10 years since the previous epidemic had acquired antibodies. Similarly, serological tests in 1971 showed no change in antibody patterns of school children since the 1968–1969 epidemic, confirming surveillance reports that showed an absence of cases since the outbreak. Thus, by 1971, for unexplained reasons, rubella had not succeeded in becoming endemic on the island of Taiwan. However, during the 1970s, a shift to endemicity emerged.

The situation in Taiwan was not unique. Similar results were found in 1972 in Barbados, where no one born there during the previous 10 years had rubella antibodies despite the likelihood of introduction of the virus associated with an influx of some 150,000 tourists yearly.[54] Factors such as population size and density and climate do not seem to provide the answers to this enigmatic epidemiologic behavior. Rubella was found to be endemic on Quemoy, a much smaller and more isolated island near Taiwan, where a serological survey showed that 97% of persons aged 6 to 50 years possessed rubella antibody.[69] Information about the past occurrence of cases of rubella on Quemoy was not available, but no cases occurred during the survey and none had been known to occur in the recent months prior to the survey.

For more than a century, rubella has been endemic in Iceland,[201] a small and sparsely populated island with far less suitable conditions for establishing endemicity than Taiwan or Barbados. Clearly, the epidemiology of rubella on islands is different from its behavior on the mainland, but no common pattern emerges.

5.4. Temporal Distribution

Although sporadic cases of rubella occur throughout the year in the United States, the largest number of cases occur in late winter and spring, both in years of high and low incidence. The reason for this consistent seasonal pattern is not understood.

5.5. Age and Sex

In the prevaccine era, rubella in the United States was primarily a disease of school-aged children, with a peak incidence in the 5- to 9-year age group; the disease was uncommon in preschool children, but cases occurred in adolescents and young adults.[237] As vaccination rates among prepubertal children increased during the 1970s,

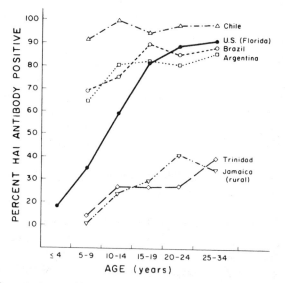

Figure 4. Age-specific rubella HI antibody patterns in different population groups. Reprinted from Horstmann and Liebhaber.[99]

the peak incidence of the disease shifted to 15- to 24-year-olds.[18] Thereafter, when further declines brought the number of reported cases down to <1000, the highest attack rates shifted to the <5- and ≥20-year age groups.[21] During the resurgence of rubella from 1989 to 1991, the incidence rate increased most in persons ≥15 years of age.[27] In 1991, when the resurgence reached its peak, 71% of cases were in children and young adults 5 to 29 years of age, reflecting outbreaks that occurred in religious communities where large numbers of unvaccinated young persons congregated in schools and other settings in which they were in close contact. In 1992–1993, when a new record low number of cases were reported, the incidence shifted to an older age group; an average 30% of cases occurred in persons ≥30 years of age compared to 9% in this age group reported in 1991.[28a] In 1994, when incidence increased by nearly 20%, 47% of cases were among persons ≥30 years of age.

The age at which rubella infection occurs in various parts of the world shows distinct differences as revealed by seroepidemiologic surveys in the late 1960s.[46] Figure 4 compares age-specific rubella antibody patterns in five countries where cases of rubella and CRS were not reported. In the South American countries, infection occurred at an early age, and by 5 to 9 years, 90% of the population may have been immune, as in Chile. In contrast, among island populations such as those of Trinidad and Jamaica[46] and Hawaii (not shown),[81] infection occurred at later ages, resulting in immunity in only 30–40% of the population over 20 years of age. Seroprevalence studies in the late 1970s and 1980s in Africa[73] showed that >80% of children were immune by 10 years of age, a pattern of immunity similar to that seen in South America.

No appreciable differences in attack rates by sex are apparent in children, but more cases are reported in women than in men. This latter finding may relate to the greater emphasis placed on identifying rubella cases in women of childbearing ages because of the risk of CRS in their offspring.

5.6. Other Factors

No racial or ethnic differences in incidence have been shown, although the characteristic rash is more difficult to diagnose in persons with dark skin. A lower incidence of congenital rubella was found in outbreaks in Japan[219,220]; however, it seems more likely that this resulted from relative avirulence of the strain rather than from ethnic differences. There is a trend toward higher incidence in lower socioeconomic groups, probably because of increased exposure in crowded homes.

6. Mechanisms and Routes of Transmission

As with many infections, the exact mode of transmission is not clear, but transmission by close person-to-person contact is apparent and airborne transmission is likely.[133] The virus is present in oropharyngeal secretions and is spread via the respiratory route. The period of communicability is estimated to be from 5 to 7 days before to 3 to 5 days after the appearance of clinical signs, but is greatest just before and on the day of onset. Infants with congenital infection may shed virus from the nose, throat, and urine for a year or more and are effective sources of contagion for susceptible contacts. In women, the virus can be recovered from the genital tract[193]; however, the significance of the presence of virus in this site with respect to transmission is unknown.

7. Pathogenesis and Immunity

Following implantation of the virus on the mucosa of the upper respiratory tract, primary multiplication is thought to occur in the respiratory epithelium, local lymph nodes, or in both sites. This is followed by viremia and shedding of virus from the throat. Rubella virus has been detected in pharyngeal secretions and in the blood for as long as a week before the rash, which develops after a 14- to 21-day incubation period.[76,101] During the few days before onset, the virus also has been recovered from leukocytes, conjunctival swabs, and synovial fluid.[64,86]

Viremia apparently results in wide distribution of the virus, including dissemination to the respiratory tract as suggested by the coryza, cough, and conjunctivitis that sometimes occur. The virus has been recovered from skin lesions within 24 hr of appearance of rash[85]; however, despite this, the rash is probably a manifestation of an antigen–antibody reaction rather than a result of direct viral damage.

Virus is transmitted to the fetus by hematogenous spread during maternal viremia. During the course of maternal viremia, the placenta is seeded with virus, followed by development of inflammatory foci in the chorionic villi, granulomatous changes, and necrosis.[47,53] Infected chorionic cells may break off and act as emboli to target organs.[219] The presence of the virus in the female genital tract raises the possibility of involvement by the ascending route as well.

The mechanisms by which rubella virus induces pathological changes in fetal organs and tissues are not clearly understood. Small size is a striking feature of infants infected *in utero* and has been shown by Naeye and

Blanc[154] to represent an actual diminution in the total number of cells. One hypothesis is that the reduction in the size of infants with CRS results from the dropping out of infected clones of cells, which are known to have a shortened lifespan.[202]

Postnatal infection induces both circulating antibodies and cell-mediated immune responses[16,88] and generally results in lifetime immunity. However, asymptomatic reinfection has been shown to occur in approximately 4% of persons with low levels of naturally acquired antibodies who are exposed during epidemics,[55,100] and frank clinical rubella rarely has been documented in persons with natural or vaccine-induced immunity.[60,236] There is close correlation between antibody levels and resistance: reinfection occurs primarily in persons with low or absent antibody levels. In addition, there appear to be qualitative as well as quantitative differences between the resistance acquired through exposure to the wild virus and that acquired by vaccination, since reinfection rates in vaccinees were found to be ten times higher than in persons with naturally acquired immunity with comparably low antibody titers.[100]

Immune responses in infants infected *in utero* are different from those induced by postnatal infection.[1,205] It is evident that the fetus can make specific IgM after the 16th week. At birth, IgM is present and continues to rise in titer for approximately 6 months, and it may persist through the first year of life. Maternal IgG antibody declines over the first few months after birth as the infant's own IgG antibody response to rubella rises. Some infants have immune system defects that result in the failure to produce antibody.[206] Among others with normal serological responses early in life, approximately 20% may lose antibody detectable by the HAI by the time they are 5 years old.[35] Hardy[83] reported a seronegative child with CRS who developed clinical rubella at 5 years of age after exposure at school. Similarly, apparent and inapparent reinfections in adults who were diagnosed as having CRS in infancy have been noted. Menser and co-workers[142] described a woman with CRS who apparently was reinfected during pregnancy and gave birth to an infant with CRS. The reasons for the more rapid decline in antibody levels and the loss of humoral immunity in persons with CRS are not clear. There also is evidence that cellular immunity is impaired in CRS,[67] and that the degree of impairment is associated with the time of infection; cell-mediated responses were found to be most impaired when infection occurred early in gestation.[17] Defective cell-mediated immunity may explain the long persistence of virus excretion in infected infants.

8. Patterns of Host Response

In postnatally acquired rubella, the ratio of inapparent to apparent infections has been estimated to be from 1 : 1 to as high as 6 : 1.[15,68,100,101] Age probably influences the clinical expression of infection, although the relationship is not as clear as in diseases such as poliomyelitis and hepatitis. Based on observations during an epidemic on the Pribiloff Islands, Brody[14] concluded that there were two groups of adults involved in the outbreak, one whose members had lost detectable antibody and experienced reinfection, largely inapparent, and another group who had their first encounters with the virus and exhibited a high clinical attack rate. In two prospective studies of young adult Hawaiian military recruits at Fort Ord, California, there was a 1.9 : 1 ratio of inapparent infections to apparent cases in one epidemic in which the men were screened intensively for clinical signs, and a 3.7 : 1 ratio in a subsequent outbreak during which clinical surveillance was less assiduous.[100]

8.1. Clinical Manifestations

8.1.1. Acquired Infection. The incubation period ranges from 12 to 23 days, but most frequently is from 14 to 16 days. Rash is the most prominent clinical feature and the first evidence of the disease in 95% of affected children.[86] Adults frequently experience a prodromal syndrome lasting several days, characterized by malaise, low-grade fever, and tender, swollen postauricular and posterior cervical lymph nodes. Mild sore throat, coryza, cough, and conjunctivitis may be present in the more severe cases. The rash begins on the face, spreads rapidly to the chest and abdomen, and within a day or two extends to the extremities; it is pink, maculopapular, and not distinctive in appearance. Initially, the lesions are discrete but later tend to coalesce, particularly on the face. The fact that the rash is always present on the face aids in making the diagnosis. The most characteristic clinical feature of rubella is involvement of specific lymph nodes, i.e., postauricular, suboccipital, and posterior cervical, which are often enlarged and tender, sometimes as long as a week before onset. The disease is usually mild, lasting no more than a few days, but in rare instances complications such as thrombocytopenic purpura, encephalitis, and other central nervous system involvement may occur.[228] A far more common complication is joint involvement, i.e., arthralgia or arthritis, particularly in women. The incidence of joint involvement increases with age and may occur in up to 70% of symptomatic adult females. Rubella virus has been recovered from synovial fluid of patients with the acute disease[64] and in several instances from individuals with chronic arthritis in the absence of evidence of clinical rubella.[30,74] One group of investigators who reported recovering rubella virus from synovial fluid of patients with chronic arthritis[74] later retracted their report because of a methodological problem with their assay.[149a] In a later study they were unable to isolate the virus from patients with rheumatoid arthritis.[149a] Recovery of the virus from peripheral blood lymphocytes of individuals with chronic arthritis also has been reported.[29,30]

8.1.2. Congenital Infection. Like postnatally acquired rubella, the congenital infection may be asymptomatic and without consequence. Infants with CRS may have one or more characteristic features (Table 1). The most common abnormalities are cardiac lesions (most frequently patent ductus arteriosus), ocular lesions (cataracts, retinopathy), sensorineural hearing loss, low birth weight at term, purpuric and petechial skin lesions associated with thrombocytopenia, central nervous system in-

Table 1. Patterns of Host Responses to Rubella

I. Acquired rubella
  A. Inapparent infection
  B. Rubella without rash
  C. Rubella with rash
  D. Complications
    1. Joint involvement (arthralgia, arthritis)
    2. Encephalitis
    3. Thrombocytopenic purpura
II. Congenital rubella
  A. Inapparent infection
  B. Congenital rubella syndrome with single or multiple organ involvement
    1. Cardiovascular lesions (patent ductus arteriosus, ventricular septal defect, coarctation of the aorta, pulmonary stenosis, myocarditis)
    2. Eye defects (cataracts, retinopathy, microphthalmia, glaucoma, severe myopia)
    3. Hearing impairment (deafness)
    4. Growth retardation (low birth weight at term)
    5. Thrombocytopenic purpura
    6. Hepatosplenomegaly
    7. Jaundice (regurgitative)
    8. Hepatitis, hepatosplenomegaly
    9. Central nervous system involvement (psychomotor retardation, meningoencephalitis, microcephaly, aseptic meningitis, central language disorders, behavioral disorders, spastic diplegia)
    10. Bone lesions (long-bone radiolucencies, bone malformations)
    11. Genitourinary tract (undescended testicle, renal lesions)

volvement (microcephaly), hepatosplenomegaly, and lesions of the long bones. Psychomotor retardation may become apparent later. Multiple system involvement is frequent, especially when infection occurs during the second month of fetal life when organs are developing.

Several disorders may appear years later. Diabetes mellitus is relatively common,[70,140] whereas hypothyroidism, hyperthyroidism, and other hormonal abnormalities and progressive panencephalitis occur rarely.[34,198,239] The prognosis for infants who have multiple abnormalities at birth is poor, and mortality is particularly high during the first year of life.[36] During the 1964–1965 epidemic, approximately 10% of infants born with CRS died in the neonatal period.[157]

8.2. Serological Responses

Antibodies to rubella virus develop promptly and can sometimes be detected on the day of onset. The IgM and IgG classes rise rapidly; IgG persists, but IgM, which is present in lower titer than IgG, begins to wane after a week and is no longer detectable by the commonly used tests beyond approximately 8 to 12 weeks.[88] The presence of IgM is a marker of recent or current infection and is useful in confirming acute rubella. Reinfection is usually accompanied by substantial rises in IgG antibody and the absence of IgM antibody or occasionally by minimal IgM responses.

The time course of appearance of antibodies detected by various serological tests is shown in Fig. 1. The patterns of response measured by ELISA, RIA, HI, and fluorescence assays are similar, but ELISA and RIA tend to give somewhat higher titers. Antibodies measured by LA also parallel HI, but at a slightly lower level.[137] CF antibodies do not appear until a week or more after onset.

9. Control

9.1. Vaccine Development

Efforts to develop a vaccine followed the isolation of rubella virus in tissue culture in 1962. At first, a killed-virus vaccine was considered to be the only safe approach, but it soon became apparent that inactivation of the virus by various means inevitably led to loss of antigenicity. Attention was therefore turned to attenuation of the virus by serial passage in tissue culture, and by 1966, Meyer and co-workers[146] had successfully developed an experimental live-virus vaccine using an attenuated strain with 77 passages of the virus in primary GMK cells, "high-passage virus-77" (HPV-77). A further five passages in duck-embryo (DE) tissue culture resulted in HPV-77(DE) vaccine that was licensed in the United States in 1969 after extensive field trials demonstrated that it was safe and immunogenic.[92] A similar vaccine, HPV-77(DK), using dog kidney, also was licensed.[145] Shortly thereafter, another vaccine, developed by Prinzie and co-workers,[176] that used the Cendehill strain was licensed. This strain was isolated in GMK and passaged 51 times in primary rabbit kidney cells. A vaccine developed by Plotkin and Buser using the RA27/3 strain[170,171] was passaged 25 to 30 times in human diploid WI-38 cells. The RA27/3 vaccine was licensed in the United States in 1979 and replaced the previously used vaccines. The vaccines are similar in that they induce seroconversion in approximately 95% of susceptible individuals. However, the RA27/3 vaccine has advantages over the other strains because like the wild virus infection, it has the ability to infect by a natural route, i.e., intranasally.[172]

Only vaccines containing the RA27/3 strain are available in the United States, including a single-antigen rubella vaccine (Meruvax II), a vaccine containing measles and rubella antigens (M-R-Vax II), and the most commonly used vaccine containing measles, mumps, and rubella antigens (MMR II).

9.2. Responses to Rubella Vaccines

9.2.1. Clinical Reactions. Rubella vaccination may cause viremia with fever, lymphadenopathy, sore throat, and headache.[12,58,65,139] These side effects are rarely seen in children, but are common (up to 50% of women) and more severe in adults. As arthralgia or arthritis are frequently part of rubella disease, especially in adults, it is not surprising that joint complaints are the most troublesome side effects of vaccination in adults. Arthralgia (pain, stiffness) is more common than arthritis (swelling, redness, heat, or limitation of movement), following both natural rubella and immunization.[104,122,168,174,216,217,224] Fingers, wrists, and knees are the primary sites affected, occasionally with joint effusions. Although less frequent than after natural rubella, transient arthralgia and arthritis are common following vaccination in adult women, but occur infrequently in children.[183] The previously used duck embryo and dog kidney vaccines resulted in higher proportions of vaccinees with joint symptoms.[5,207,230] The RA27/3 strain has been associated with transient joint symptoms in 10 to 40% of susceptible postpubertal women, but infrequently results in such reactions in children or adult males.[168,174,224]

Chronic, persistent, or recurrent, arthritis or arthral-

gia has been reported following vaccination. Although most of these reports followed vaccination with the duck embryo or dog kidney vaccines, chronic arthritis following the RA27/3 vaccine has been described by a group of investigators in Canada in case reports[216] and in a comparative case series.[217] Rubella virus was isolated from peripheral blood mononuclear cells in two women evaluated for chronic arthropathy following postpartum immunization with RA27/3 vaccine.[216] Using culture techniques similar to those used by the Canadian investigators and PCR assays, Frenkel and associates[65a] did not detect rubella virus in blood components or nasopharyngeal specimens among six persons with persistent symptoms following immunization with RA27/3 vaccine and one person following clinical rubella, or in blood components or synovial fluid of 11 children with juvenile rheumatoid arthritis. In the comparative study, Tingle and associates[217] reported that 2 of 44 (5%) women had continuing or recurrent arthritis or arthralgia 18 to 24 months following vaccination compared with 7 of 23 (30%) women who had arthritis (4) or arthralgia (3) 18 to 24 months after natural infection. These findings have not been replicated by other investigators, although in one double-blind controlled study,[174] the follow-up period of 6 weeks was too brief to identify recurring cases. Preliminary results from a recently completed prospective study in Canada suggest that there were no statistically significant differences in the persistence of arthritis–arthralgia between women given rubella vaccine postpartum and those given placebo.[178a] A recent retrospective cohort study in the United States[177a] based on a review of medical records found no association between rubella vaccination in women and chronic arthropathy. A follow-up clinical evaluation of women with joint complaints identified in this study, including an examination of peripheral blood lymphocytes, is underway. Another retrospective study[204a] in which women with joint complaints following postpartum vaccination were identified and examined found no clinical evidence of chronic arthropathy and no rubella virus was isolated from peripheral blood samples.

The possible association between rubella vaccination and chronic arthritis–arthralgia is complicated because arthritis may develop from a number of causes and the natural history of the disease is not well understood. In a 1991 review of the adverse effects of rubella vaccination, a committee of the Institute of Medicine[104] (IOM) reported that the association between vaccination and chronic arthritis is biologically plausible because of the clear association between natural infection and chronic arthritis and because rubella virus has been recovered from peripheral blood leukocytes and synovial fluid of women with persistent arthritis following vaccination, although whether the virus was of the wild type or vaccine related was not determined. The IOM committee concluded that "the evidence is consistent with a causal relationship between the currently used rubella vaccine strain (RA27/3) and chronic arthritis in adult women although evidence is limited in scope and confined to reports from one institution" (p. 197).

The IOM report was published before results of the recent studies[65a,177a,178a,204a] that failed to find an association between rubella vaccination and chronic arthritis were available.

Although infrequent, polyneuropathy was the most common serious event following vaccination with the HPV-77(DE), HPV-77(DK), or Cendehill strains.[186] A few cases of optic neuritis,[109,110] diffuse myelitis,[9,95] and facial paresis[152] have been reported, and in most cases these events followed vaccination with HPV-77(DE), HPV-77(DK), or Cendehill vaccines.

9.2.2. Shedding of Virus.
In most vaccinees, virus excretion from the throat occurs briefly and in relatively small amounts. However, transmission is so rare as to be a negligible hazard even for susceptible pregnant women exposed to their vaccinated children.[192] In the past, viremia rarely had been documented in vaccinees,[3] but its occurrence was substantiated by transmission of infection to the fetus in susceptible pregnant women who had inadvertently been given rubella vaccine.[150] Virus has been recovered from breast milk of women vaccinated postpartum. Transmission of the virus to the infant has been documented, but the infection was asymptomatic and transient and usually did not evoke a lasting serological response.[114,120]

9.2.3. Serological Responses.
Seroconversion following vaccination occurs in approximately 95% of susceptible vaccinees. The patterns of response are similar to those induced by natural infection (Fig. 1), but titers tend to be somewhat lower and to decline more rapidly. RA27/3 vaccine behaves more like natural infection than the previously used vaccines. It results in higher and more persistent antibody levels[102] and secretory IgA in the nasopharynx,[3,156] and resistance to reinfection is greater.[78] The RA27/3 strain induces secretory IgA in the nasopharynx both after intranasal and after subcutaneous vaccination, although to a lesser extent after the latter.[172]

9.3. Rubella Vaccination Strategies: Their Impact on Rubella and Congenital Rubella

9.3.1. Epidemiologic Approach.
In the United States, universal immunization of children was adopted in

1969 when the first rubella vaccine was licensed. The objective of this vaccination policy was to prevent CRS indirectly by inducing a high degree of herd immunity, and thus blocking circulation of the virus in the segment of the population with the highest incidence of infection, i.e., young children, who commonly served as sources of infection for pregnant women. Immunization of children 1 year of age and older therefore became the main focus of the vaccination program that interrupted the epidemic cycle and greatly reduced the incidence of rubella. Incidence further declined following the widespread use of MMR vaccine.[6,7] From 1969 to 1990, nearly 179 million doses of rubella-containing vaccine were distributed in the United States.

The declining incidence was accompanied by a changing age distribution. Prior to the use of vaccine, the highest attack rate was in children less than 10 years of age (77% of cases), but during the late 1970s it shifted to persons 15 to 19 years of age (Fig. 5).[18,21] By 1976, there was a 65% decline in incidence; however, the proportion of susceptible persons over 15 years of age remained about the same. Control efforts had been directed primarily at preschool- and elementary school-age children, but rubella outbreaks continued to occur among young adults in settings such as the military,[19,38] colleges,[22] prisons,[23] and places of employment, including hospitals.[20,72,159,173] In 1975–1977, more than 60% of cases occurred in young adults compared with only 23% in 1966–1968.[7] The incidence of CRS roughly paralleled the incidence of rubella in individuals over 15 years of age (Fig. 3). Strategies to eliminate rubella and CRS were initiated in the late 1970s, including increased efforts to immunize postpubertal females,[93] efforts to increase coverage in preschool

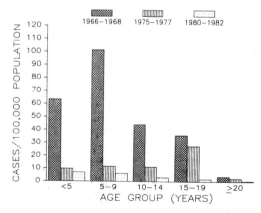

Figure 5. Average annual incidence rates of rubella by age group in selected areas of the United States, 1966–1992, including Illinois, Massachusetts, and New York City. Prevaccine years were 1966–1968.[6]

children, and introduction of state laws in 1979–1980 requiring immunization for school entry. Extension of these laws in many states to include kindergarten through grade 12 resulted in further declines in the incidence of rubella. Following the augmented efforts to vaccinate postpubertal females, the proportion of cases of rubella among persons older than 15 years decreased to <40% during 1981 to 1983, representing a greater than 24% reduction in the proportion of cases in this age group from 1975 to 1977. The decline in rubella in this older age group was accompanied by a decreased incidence of CRS (Fig. 3). Combined, these strategies resulted in a record low number of cases (225) reported in 1988, representing a 94% decline from 1980 and a 99% decline from 1969.[129,157,175] The recommendation in 1989 for a two-dose measles vaccination schedule, preferably given as MMR,[24] provided further protection against rubella.

However, despite these strategies, rubella again increased between 1989 and 1991, although the incidence was approximately 10-fold less than that observed in 1979 and 35-fold less than 1969. In 1990, 26 outbreaks were reported, occurring mostly in unvaccinated adults and children.[27] In 1991, large outbreaks (60 to >400 cases) occurred among unvaccinated children and young adults in religious communities in at least four states.[13,27,28,106] The expected increase in the incidence of CRS was observed during and immediately following this resurgence of rubella. Notably, the largest increases in incidence during 1989 and 1990 occurred in persons older than 15 years, accounting for 55 to 60% of cases. The outbreaks in Amish communities in 1991 involved mostly unvaccinated children and young adults 5 to 29 years of age. The decreased number of cases in the past 3 years has been associated with a shift back to the older age groups.

9.3.2.   The Direct Approach.   Prior to 1988, the United Kingdom, Ireland, and a number of other countries adopted immunization programs for the prevention of rubella and CRS using a direct, selective approach whereby vaccine was administered only to school-aged girls (11–14 years) and to susceptible women of childbearing age.[48,107,218] The rationale for this selective approach was that the duration of vaccine-induced immunity was unknown, and thus immunizing women closer to or of childbearing age would have an immediate effect by providing a higher level of immunity to women who are at risk for pregnancy. Although selective vaccination reduced the proportion of young women who were seronegative, many remained susceptible, and the effect on the incidence of the disease was less striking than in the United States where universal immunization of children has been in effect since licensure of the vaccine. The implementa-

tion of a childhood immunization policy for MMR in 1988 in the United Kingdom has had a major positive impact not only on children under 5 years of age but also on women of childbearing age.[149] As in the United States, the epidemic cycle has been broken and incidence has reached a low endemic level. Similar to the United Kingdom, selective vaccination policy in Ireland failed to provide immunity to a sufficient number of women to control CRS, and in 1988 Ireland also adopted a policy intended to eradicate rubella by vaccinating children at 15 months of age and again at 10 to 14 years.[107] Finland changed to a two-dose vaccination schedule for children in 1982 following 7 years of vaccinating only 13-year-old girls and postpartum women.[222] Peltola and associates reported that indigenous measles, mumps, and rubella were eliminated in Finland after 12 years of a two-dose MMR vaccination program with coverage in excess of 95%.[168a]

In Australia, selective immunization of females at 12 to 13 years of age begun in 1971 had a positive impact on the incidence of CRS.[139,141,208] However, as was the case in other countries where vaccine policies were selective for adolescent females, a proportion of women of childbearing age continued to be susceptible. In 1992, 10% of women aged 15 to 44 years were found to be susceptible to rubella, and in the same year, rubella increased with many cases occurring among women in this age group.[179] Infant immunization was introduced in 1989 to supplement the selective policy of immunizing teenage girls. The effect of this has not been seen because a catch-up program was not included and outbreaks continue to occur among school age children. A two-dose MMR schedule for males and females, beginning at 12 months with a second dose at 10 to 16 years of age, was implemented in 1995 (M. A. Burgess, personal communication).

9.4. Vaccination during Pregnancy

Because rubella vaccine virus can cross the placenta and infect the fetus and possibly be teratogenic, vaccination of pregnant women always has been contraindicated. However, despite this recommendation, many women have been inadvertently vaccinated shortly before conception or soon after, often before they were aware of being pregnant. In order to monitor the risk of maternal vaccination to the fetus, the CDC established a Vaccine in Pregnancy (VIP) registry in 1971.[8,25] There were no cases of CRS among 290 infants born to susceptible or immune women vaccinated during pregnancy with Cendehill or HPV-77 vaccines between 1971 and 1979 or among 562 infants born to susceptible or immune women vaccinated with RA27/3 between 1979 and 1988. In April 1989, the CDC discontinued the VIP registry because of the absence of any cases of CRS among women vaccinated in pregnancy over the 17-year period of surveillance. However, because of the theoretical risk, rubella vaccination remains contraindicated during pregnancy.[26] The results of the U.S. VIP are consistent with the experience in the United Kingdom and Germany.[50,199]

9.5. Persistence of Vaccine-Induced Immunity

The ultimate objective of immunization against rubella is prevention of intrauterine infection and CRS. Thus, the persistence of vaccine-induced antibodies is of prime importance. The Advisory Committee on Immunization Practices (ACIP) recommends that children 12 months of age and older be immunized against rubella. Immunization at this early age means that to be effective throughout the childbearing years, antibodies and immunity must last for approximately 40 years. However, most children are receiving a second dose of rubella antigen, because MMR is the vaccine recommended and used for the required two doses of measles-containing vaccine. In 1993, the recommendation for the first dose of MMR was changed from 15 months of age to 12 to 15 months. A second dose of MMR is recommended at school entry or at entry to middle or junior high school.[24,33] Nevertheless, even with a second dose of rubella vaccine administered at these older ages, immunity must still persist for some 30 to 35 years.

Serological studies of vaccinees followed for as long as 18 years after administration of Cendehill or RA27/3 vaccine have shown persistence of protective levels of antibody in 90 to 98% of individuals.[91,102,108,160,162,171] Although variable, persistence of antibody following HPV-77 has been found to be somewhat less than after the other vaccines.[102] All of the vaccines produce less vigorous and probably less persistent antibody responses than does naturally acquired infection after which immunity has been shown to last for several decades[61] and is generally considered to last a lifetime. However, the RA27/3 strain produces antibody responses that more closely resemble responses to infection with the wild virus, producing stronger and more lasting antibody responses than the other strains.[91,102] Also of note, many individuals who originally seroconverted and subsequently lacked antibody when measured by the HAI may have been immune because antibody was present when more sensitive assays, such as the ELISA, were used for serological testing. Also, when these individuals were revaccinated, they demonstrated a booster response (ab-

sence of IgM antibody and rapid rise and decline in IgG antibody).[89,194] The extent to which asymptomatic boosting from exposure to wild virus contributes to the long-term persistence of vaccine-induced immunity is unknown.

9.6. Reinfection

Reinfection has been documented in persons with low levels of antibody both following natural infection and after vaccination.[2,3,63,82,98,100,161,188,236] The replication of the virus may be limited to the upper respiratory tract because the host's responses act on the virus before it invades the blood, but symptomatic viremia (rash, arthritis) also may be present.[236] Reinfection has been assessed by challenge studies in which persons with vaccine-induced antibodies whose titers had declined to low or undetectable levels were given either RA27/3[3,78,82,161] or a wild, unattenuated strain of rubella virus.[188] Previous investigations[235] have shown that reinfection, determined by marked rises in IgG antibody levels and in some individuals by oropharyngeal excretion of virus, could be readily induced by challenge with wild-type rubella virus. When the RA27/3 strain was used for intranasal challenge, viremia was found in only 1 of 19 persons, in contrast to 8 of 10 seronegative persons similarly challenged.[161] However, when challenge was with a laboratory strain of wild rubella virus (Howell) given intranasally, viremia was detected between days 5 and 12 postinoculation in three of six vaccinees, and one of the three developed classic rubella.[188] All of these individuals had seroconverted following vaccination several years previously but had lost detectable antibodies by the HAI test. The virus challenge dose was large (1000 TCI_{50}). Nevertheless, the results indicate the possibility that exposure to wild virus can induce viremia and clinical infection in vaccinees in whom immunity has waned.[60]

The majority of reinfections are asymptomatic and are of special concern in pregnant women because of the potential threat of viremia to the fetus. Such a possibility has been confirmed by case reports from numerous countries of infants with congenital infection born to mothers who were previously infected by wild virus or who had been vaccinated.[10,37,42,51,62,63,98,126,151,165,182,185,190,242] A prospective study in the United Kingdom[151] investigated the outcome of 47 pregnancies from 1987 to 1991 in which rubella reinfection was established. None of the mothers had a documented history of vaccination. In 42 aborted fetuses or infants tested, laboratory evidence of infection was present in fetal tissue in two cases and in one infant who remained asymptomatic at 1 year of age. Thus, available data suggest that maternal reinfection occurs and can result in CRS in the infant. Although the risk for CRS associated with maternal reinfection is unknown, it appears to be low, certainly much lower than the risk for CRS following primary maternal infection. The frequency of reinfection in vaccinated individuals depends on the durability of vaccine-induced immunity and on the degree of circulation of wild virus. There have been rare case reports of transmission of vaccine virus to immune individuals resulting in subclinical[111] and in symptomatic reinfection.[238] Reinfection can be eliminated by the same methods used to eliminate primary rubella disease and CRS, namely, a comprehensive vaccination policy including high coverage for the first dose of MMR in children 12 to 15 months of age and for the second dose in school-age children or in adolescents, as well as continued efforts to use every opportunity to vaccinate susceptible women of childbearing age and other adults, especially those who are likely to have close contact with women of childbearing age, such as health care workers.

10. Unresolved Problems and Current Approaches

Although our knowledge has increased vastly since the isolation of the rubella virus in 1962, much remains to be learned about the epidemiology of rubella, including its enigmatic behavior in certain island populations. Also, more information is needed on the role of genetic factors in influencing the transmission of infection and the course of outbreaks. For example, a small number of individuals, "spreaders," have been found to have a high potential for transmitting virus to susceptible individuals.[84]

An adequate animal model that imitates clinical rubella infection in humans has not been developed. Thus, the mechanisms by which rubella virus causes a number of syndromes or conditions, such as rubella arthritis, postinfectious encephalopathy, CRS, and late-onset endocrine and neurological disorders, is not fully understood. One such disorder is progressive rubella panencephalitis (PRP).[239] PRP, a rare late complication of rubella infection, is most often seen in persons with CRS, but also can occur after naturally acquired rubella in childhood. Many clinical and laboratory features of PRP are similar to subacute sclerosing panencephalitis, a late sequela of measles.[34] PRP is characterized by a prolonged asymptomatic period followed by neurological deterioration in the second to third decade of life, leading to death. The mechanism by which the virus persists to ultimately cause this lethal effect in the central nervous system is unknown.

Whether PRP occurs after vaccine-induced infection also is unknown.

Perhaps the most vexing unresolved problem that deserves ongoing attention is the association of rubella infection with arthritis after postnatally acquired rubella. Knowledge of the mechanism by which rubella infection affects joints and results in arthritis–arthralgia is key to understanding the association between rubella vaccination and similar joint involvement and to developing an alternate rubella vaccine that is free of the potential to cause this adverse event. There are a number of promising strategies for the development of such a vaccine, including peptide and subunit vaccines and live virus altered by recombinant DNA techniques. However, demonstration of the pathogenesis of rubella-induced arthropathy in an animal model is clearly needed to identify the optimal type of rubella vaccine.

Immunization programs have been increasingly successful in preventing rubella and CRS. Yet, the interpretation of declining incidence must take into account underreporting of both acquired rubella and CRS. In the United States, it is estimated that as few as 10 to 15% of clinical cases are reported. There are a number of explanations for underreporting, including that patients with mild clinical rubella may not seek medical care, physicians may not report mild cases that they do not consider to be important to public health, and cases are likely to be misdiagnosed as other rash illnesses when laboratory confirmation is not sought. Furthermore, there are an estimated two or more inapparent infections for every clinically recognized infection. An enhanced system for reporting cases of rubella to the National Notifiable Disease Surveillance System has been initiated that will improve reporting and provide detailed case level information that will allow a more accurate appraisal of vaccination programs.

Despite the overall marked decrease in rubella incidence since the widespread use of vaccine, outbreaks have continued to occur in unvaccinated populations.[13,28,28a,129] It is evident that in order to prevent the introduction and spread of rubella virus, close to 100% immunity rates are necessary. Presently, this goal is being actively pursued. The Childhood Immunization Initiative, designed to achieve complete immunization in 90% of children under 2 years of age by 2000, was launched in 1993. Extensive resources have been committed to this program. In addition, a greater number of states are requiring a second MMR vaccination for school entry. As increasing numbers of children are immunized and as these children reach adulthood, the spread of rubella virus will be greatly reduced. However, because outbreaks of rubella continue to occur among adults, increased efforts also are needed to vaccinate susceptible persons of all ages, especially women of childbearing age. Whenever possible, contacts with the health care system should be utilized to assess immunity and vaccinate susceptible persons in order to avoid missed opportunities to vaccinate.[123] Institutions that include populations of young adults, such as colleges, the military, and workplaces, especially health care facilities, can help reduce the incidence of rubella and CRS by assessing immunity in their constituents and vaccinating those who are susceptible. The large outbreak in Massachusetts in 1994 in which 23% of cases occurred in prisons demonstrates the need for correctional institutions to review their policies with respect to the prevention of rubella. An additional 24% of cases in the Massachusetts outbreak occurred among unvaccinated young adults in the Latino community. Extensive efforts are needed to vaccinate immigrants from countries without immunization policies for rubella. In addition, greater efforts are needed to reach unvaccinated populations such as the Amish, who do not necessarily object to vaccination but simply avoid modern health care. Together, these strategies for the prevention of rubella and CRS are highly likely to be successful.

In 1996 the CDC embarked on a rubella elimination program. A working group of outside experts met with CDC staff to discuss issues and strategies to eliminate indigenous rubella in the United States. This program also will address the problem of imported cases and work cooperatively with the Pan American Health Organization to encourage use of rubella vaccine in neighboring countries.

11. References

1. ALFORD, C. A., Studies on antibody in congenital rubella infections, *Am. J. Dis. Child.* **110**:455–463 (1965).
2. BALFOUR, H. H., JR., GROTH, K. E., EDELMAN, C. K., AMREN, D. P., BEST, J. M., AND BANATVALA, J. E., Rubella viraemia and antibody responses after rubella vaccination and reimmunization, *Lancet* **1**:1078–1080 (1981).
3. BANATVALA, J. E., BEST, J. M., O'SHEA, S., AND DUDGEON, J. A., Persistence of rubella antibodies after vaccination: Detection after experimental challenge, *Rev. Infect. Dis.* **7**:S86–S90 (1985).
4. BANATVALA, J. E., POTTER, J., AND BEST, J. M., Interferon response to Sendai and rubella viruses in human foetal cultures, leukocytes and placental cultures, *J. Gen. Virol.* **13**:193–201 (1971).
5. BARNES, E. K., ALTMAN, R., AUSTIN, S. M., AND DOUGHERTY, W. J., Joint reactions in children vaccinated against rubella. Study II: Comparison of three vaccines, *Am. J. Epidemiol.* **95**:59–66 (1972).
6. BART, K. J., ORENSTEIN, W. A., PREBLUD, S. R., AND HINMAN, A. R., Universal immunization to interrupt rubella, *Rev. Infect. Dis.* **7**:S177–S184 (1985).
7. BART, K. J., ORENSTEIN, W. A., PREBLUD, S. R., HINMAN, A. R., LEWIS, F. L., JR., AND WILLIAMS, N. M., Elimination of rubella and congenital rubella from the United States, *Pediatr. Infect. Dis.* **4**:14–21 (1985).
8. BART, S. W., STETLER, H. C., PREBLUD, S. R., WILLIAMS, W. A., ORENSTEIN, W. A., BART, K. J., HINMAN, A. R., AND HERRMANN, K. L., Fetal risk associated with rubella vaccine: An update, *Rev. Infect. Dis.* **7**:S95–S102 (1985).
9. BEHAN, P. O., Diffuse myelitis associated with rubella vaccination, *Br. Med. J.* **1**:166 (1977).
10. BEST, J. M., BANATVALA, J. E., MORGAN-CAPNER, P., AND MILLER, E., Fetal infection after maternal reinfection with rubella: Criteria for defining reinfection, *Br. Med. J.* **299**:773–775 (1989).
11. BIRCH, C. J., GLAUN, B. P., HUNT, V., IRVING, L. G., AND GUST, I. D., Comparison of passive haemagglutination and haemagglutination-inhibition techniques for detection of antibodies to rubella virus, *J. Clin. Pathol.* **32**:128–131 (1979).
12. BOTTINGER, M., AND HELLER, L., Experiences from vaccination and revaccination of teenage girls with three different rubella vaccines, *J. Biol. Stand.* **4**:107–114 (1976).
13. BRISS, P. A., FEHRS, L. J., HUCHESON, R. H., AND SCHAFFNER, W., Rubella among the Amish: Resurgent disease in a highly susceptible community, *Pediatr. Infect. Dis. J.* **11**:955–959 (1992).
14. BRODY, J. A., SEVER, J. L., MCALISTER, R., SCHIFF, G. M., AND CUTTING, R., Rubella epidemic on St. Paul Island in the Pribilofs, 1963. I. Epidemiologic, clinical, and serologic findings, *J.A.M.A.* **191**:619–623 (1965).
15. BUESCHER, E. L., Behavior of rubella virus in adult populations, *Arch. Ges. Virusforsch.* **16**:470–476 (1965).
16. BUIMOVICI-KLEIN, E., AND COOPER, L. Z., Cell-mediated immune response in rubella infections, *Rev. Infect. Dis.* **7**:S123–128 (1985).
17. BUIMOVICI-KLEIN, E., LANG, P. B., ZIRING, P. R., AND COOPER, L. Z., Impaired cell-mediated immune response in patients with congenital rubella: Correlation with gestational age at time of infection, *Pediatrics* **64**:620–626 (1979).
18. CENTERS FOR DISEASE CONTROL, Rubella—United States 1977–1979, *Morbid. Mortal. Week. Rep.* **28**:374–375 (1979).
19. CENTERS FOR DISEASE CONTROL, Rubella in Air Force recruits—Texas, 1977–1978, *Morbid. Mortal. Week. Rep.* **29**:73–75 (1980).
20. CENTERS FOR DISEASE CONTROL, Rubella outbreak in an office building—New Jersey, *Morbid. Mortal. Week. Rep.* **29**:517–518 (1980).
21. CENTERS FOR DISEASE CONTROL, Rubella and congenital rubella syndrome—United States, 1983–1984, *Morbid. Mortal. Week. Rep.* **33**:528–531 (1984).
22. CENTERS FOR DISEASE CONTROL, Rubella in colleges—United States, 1983–1984, *Morbid. Mortal. Week. Rep.* **34**:228–231 (1985).
23. CENTERS FOR DISEASE CONTROL, Rubella outbreaks in prisons—New York City, West Virginia, California, *Morbid. Mortal. Week. Rep.* **34**:615–618 (1985).
24. CENTERS FOR DISEASE CONTROL, Measles prevention: Recommendations of the Immunization Practices Advisory Committee (ACIP), *Morbid. Mortal. Week. Rep.* **38**:1–18 (1989).
25. CENTERS FOR DISEASE CONTROL, Rubella vaccination during pregnancy—United States, 1971–1988, *Morbid. Mortal. Week. Rep.* **38**:289–293 (1989).
26. CENTERS FOR DISEASE CONTROL, Rubella prevention: Recommendations of the Immunization Practices Advisory Committee (ACIP), *Morbid. Mortal. Week. Rep.* **39**:1–18 (1990).

27. CENTERS FOR DISEASE CONTROL, Increase in rubella and congenital rubella syndrome—United States, 1988–1990, *Morbid. Mortal. Week. Rep.* **40:**93–99 (1991).

28. CENTERS FOR DISEASE CONTROL, Outbreaks of rubella among the Amish—United States, 1991, *Morbid. Mortal. Week. Rep.* **40:**264–265 (1991).

28a. CENTERS FOR DISEASE CONTROL, Rubella and congenital rubella syndrome—United States, January 1, 1991–May 7, 1994, *Morbid. Mortal. Week. Rep.* **43:**391, 397–401 (1994).

29. CHANTLER, J. K., FORD, D. K., AND TINGLE, A. J., Persistent rubella infection and rubella associated arthritis, *Lancet* **1:**1323–1325 (1982).

30. CHANTLER, J. K., TINGLE, A. J., AND PETTY, R., Persistent rubella virus infection associated with chronic arthritis in children, *N. Engl. J. Med.* **313:**1117–1123 (1985).

31. CHERNESKY, M. A., DeLONG, D. J., MAHONY, J. B., AND CASTRICIANO, S., Differences in antibody responses with rapid agglutination tests for the detection of rubella antibodies, *J. Clin. Microbiol.* **23:**772–776 (1986).

32. COCKBURN, W. D., World aspects of the epidemiology of rubella, *Am. J. Dis. Child.* **118:**112–122 (1969).

33. COMMITTEE ON INFECTIOUS DISEASES, AMERICAN ACADEMY OF PEDIATRICS, Measles: Reassessment of the current immunization policy, *Pediatrics* **84:**110–112 (1989).

34. COOPER, L. Z., The history and medical consequences of rubella, *Rev. Infect. Dis.* **7:**S2–S10 (1985).

35. COOPER, L. Z., FLORMAN, A. L., ZIRING, P. R., AND KRUGMAN, S., Loss of rubella hemagglutination inhibition antibody in congenital rubella, *Am. J. Dis. Child.* **122:**397–403 (1971).

36. COOPER, L. Z., AND KRUGMAN, S., Clinical manifestations of postnatal and congenital rubella, *Arch. Ophthalmol.* **77:**434–439 (1967).

37. CRADOCK-WATSON, J. E., RIDEHALGH, M. K., ANDERSON, M. J., PATTISON, J. R., AND KANGRO, H. O., Outcome of asymptomatic infection with rubella virus during pregnancy, *J. Hyg.* **87:**147–154 (1981).

38. CRAWFORD, G. E., AND GREMILLION, D. H., Epidemic measles and rubella in Air Force recruits: Impact of immunization, *J. Infect. Dis.* **144:**403–410 (1981).

39. CREMER, N. E., HAGENS, S. J., AND COSSEN, C., Comparison of the hemagglutination inhibition test and an indirect fluorescent-antibody test for detection of antibody to rubella virus in human sera, *J. Clin. Microbiol.* **11:**746–747 (1980).

40. CUNNINGHAM, A. L., AND FRASER, J. R. E., Persistent rubella virus infection of human synovial cells cultured *in vitro*, *J. Infect. Dis.* **151:**638–645 (1985).

41. DALES, L. G., AND CHIN, J., Public health implications of rubella antibody levels in California, *Am. J. Public Health* **72:**167–172 (1982).

42. DAS, E. D., LAKHANI, P., KURTZ, J. B., HUNTER, N., WATSON, B. E., CARTWRIGHT, K. A., CAUL, E. O., AND ROOME, A. P., Congenital rubella after previous maternal immunity, *Arch. Dis. Child.* **65:**545–546 (1990).

43. DIMECH, W., BETTOLI, A., ECKERT, D., FRANCIS, B., HAMBLIN, J., KERR, T., RYAN, C., AND SKURRIE, I., Multicenter evaluation of five commercial rubella virus immunoglobulin G kits which report in international units per milliliter, *J. Clin. Microbiol.* **30:**633–641 (1992).

44. DORFMAN, S. F., AND BOWERS, C. H., JR., Rubella susceptibility among prenatal and family planning clinic populations, *Mt. Sinai J. Med.* **52:**248–252 (1985).

45. DORSETT, P. H., MILLER, D. C., GREEN, K. Y., AND BYRD, F. I., Structure and function of the rubella virus proteins, *Rev. Infect. Dis.* **7:**S150–S156 (1985).

46. DOWDLE, W. R., FERREIRA, W., DE SALLES GOMES, L. F., KING, D., KOURANY, M., MADALENGOITIA, J., PEARSON, E., SWANSTON, W. H., TOSI, H. C., AND VILCHES, A. M., WHO collaborative study on the sero-epidemiology of rubella in Caribbean and Middle and South American populations in 1968, *Bull. WHO* **42:**419–422 (1970).

47. DRISCOLL, S. G., Histopathology of gestational rubella, *Am. J. Dis. Child.* **118:**49–53 (1969).

48. DUDGEON, J. A., Selective immunization: Protection of the individual, *Rev. Infect. Dis.* **7:**S185–S190 (1985).

49. EDITORIAL, Rubella and congenital malformations, *Lancet* **1:**316 (1944).

50. ENDERS, G., Rubella antibody titers in vaccinated and nonvaccinated women and results of vaccination during pregnancy, *Rev. Infect. Dis.* **7:**S103–S107 (1985).

51. ENDERS, G., CALM, A., AND SCHAUB, J., Rubella embryopathy after previous maternal rubella vaccination, *Infection* **12:**56–58 (1984).

52. ERICKSON, C. A., Rubella early in pregnancy causing congenital malformations of eyes and heart, *J. Pediatr.* **25:**281–283 (1944).

53. ESTERLY, J. R., AND OPPENHEIMER, E. H., Pathological lesions due to congenital rubella, *Arch. Pathol.* **87:**380–388 (1969).

54. EVANS, A. S., COX, F., NANKERVIS, G., OPTON, E. M., SHOPE, R. E., WELLS, A. V., AND WEST, B., A health and seroepidemiological survey of a community in Barbados, *Int. J. Epidemiol.* **3:**167–175 (1974).

55. EVANS, A. S., NIEDERMAN, J. C., AND SAWYER, R. N., Prospective studies of Yale University freshmen. II. Occurrence of acute respiratory infections and rubella, *J. Infect. Dis.* **123:**271–278 (1971).

56. FERRARO, M. J., KALLAS, W. M., WELCH, K. P., AND LAU, A. Y., Comparison of a new, rapid enzyme immunoassay with a latex agglutination test for qualitative detection of rubella antibodies, *J. Clin. Microbiol.* **25:**1722–1724 (1987).

57. FOGEL, A., GERICHTER, C. B., RANNON, L., BERNHOLTZ, B., AND HANDSHER, R., Serologic studies in 11,460 pregnant women during the 1972 rubella epidemic in Israel, *Am. J. Epidemiol.* **103:**51–59 (1976).

58. FOGEL, A., MOSHKOWITZ, A., RANNON, L., AND GERICHTER, C. B., Comparative trials of RA 27/3 and Cendehill rubella vaccines in adult and adolescent females, *Am. J. Epidemiol.* **93:**392–398 (1971).

59. FORBES, J. A., Rubella, historical aspects, *Am. J. Dis. Child.* **118:**5–11 (1969).

60. FORREST, J. M., MENSER, M. A., AND HONEYMAN, M. C., Clinical rubella eleven months after vaccination, *Lancet* **2:**399–400 (1972).

61. FORREST, J. M., SLINN, R. F., NOWAK, M. J., AND MENSER, M. A., Duration of immunity to rubella, *Lancet* **1:**1013 (1971).

62. FORSGREN, M., CARLSTROM, G., AND STRAMGERT, K., Case of congenital rubella after maternal reinfection, *Scand. J. Infect. Dis.* **11:**81–83 (1979).

63. FORSGREN, M., AND SOREN, L., Subclinical rubella reinfection in vaccinated women with rubella-specific IgM response during pregnancy and transmission of virus to the fetus, *J. Infect. Dis.* **17:**337–341 (1985).

64. FRASER, J. R., CUNNINGHAM, A. L., HAYES, K., LEACH, R., AND LUNT, R., Rubella arthritis in adults. Isolation of virus, cytology

and other aspects of the synovial reaction, *Clin. Exp. Rheumatol.* **1**:287–293 (1983).

65. FREESTONE, D. S., PRYDIE, J., HAMILTON-SMITH, S. G., AND LAURENCE, G., Vaccination of adults with Wistar RA27/3 rubella vaccine, *J. Hyg.* **69**:471–477 (1971).

65a. FRENKEL, L. M., NIELSEN, K., GARAKIAN, A., JIN, R., WOLINSKY, J. S., AND CHERRY, J. D., A search for persistent rubella virus infection in persons with chronic symptoms after rubella and rubella immunization and in patients with juvenile rheumatoid arthritis, *Clin. Infect. Dis.* **22**:287–294 (1996).

66. FREY, T. K., AND ABERNATHY, E. S., Identification of strain-specific nucleotide sequences in the RA 27/3 rubella virus vaccine, *J. Infect. Dis.* **168**:854–864 (1993).

67. FUCCILLO, D. A., STEELE, R. W., HENSEN, S. A., VINCENT, M. M., HARDY, J. B., AND BELLANTI, J. A., Impaired cellular immunity to rubella virus in congenital rubella, *Infect. Immun.* **9**:81–84 (1974).

68. GALE, J. L., DETELS, R., KIM, K. S. W., BEASLEY, R. P., CHEN, K. P., AND GRAYSTON, J. T., The epidemiology of rubella on Taiwan. III. Family studies in cities of high and low attack rates, *Int. J. Epidemiol.* **1**:261–265 (1972).

69. GALE, J. L., GRAYSTON, J. T., BEASLEY, R. P., DETELS, R., AND KIM, K. S. W., The epidemiology of rubella on Taiwan. 1968–1969 epidemic, *Int. J. Epidemiol.* **1**:253–260 (1972).

70. GINSBERG-FELLNER, F., WITT, M. E., FEDUN, B., TAUB, F., DOBERSON, M. J., McEVOY, R. C., COOPER, L. Z., NOTKINS, A. L., AND RUBINSTEIN, P., Diabetes mellitus and autoimmunity in patients with the congenital rubella syndrome, *Rev. Infect. Dis.* **7**:S170–S176 (1985).

71. GOLDWATER, P. N., QUINEY, J. R., AND BANATVALA, J. E., Maternal rubella at St. Thomas' Hospital: Is there a need to change British vaccination policy? *Lancet* **2**:1228–1230 (1978).

72. GOODMAN, A. K., FRIEDMAN, S. M., BEATRICE, S. T., AND BART, S. W., Rubella in the work place: The need for employee immunization, *Am. J. Public Health* **77**:725–726 (1987).

73. GOMWALK, N. E., AND AHMAD, A. A., Prevalence of rubella antibodies on the African continent, *Rev. Infect. Dis.* **11**:116–121 (1989).

74. GRAHAME, R., ARMSTRONG, R., SIMMONS, N., WILTON, J. M., DYSON, M., LAURENT, R., MILLIS, K., AND MIMS, C. A., Chronic arthritis associated with the presence of intrasynovial rubella virus, *Ann. Rheum. Dis.* **42**:2–13 (1983).

75. GRAYSON, J. T., GALE, J. L., AND WATTEM, R. H., The epidemiology of rubella on Taiwan. I. Introduction and description of the 1957–1958 epidemic, *Int. J. Epidemiol.* **1**:245–252 (1972).

76. GREEN, R. H., BALSAMO, M. K., GILES, J. P., KRUGMAN, S., AND MIRICK, G. S., Studies on the natural history and prevention of rubella, *Am. J. Dis. Child.* **110**:348–365 (1965).

77. GREGG, N. M., Congenital cataract following German measles in the mother, *Trans. Ophthalmol. Soc. Aust.* **3**:35–46 (1941).

78. GRILLNER, L., Immunity to intranasal challenge with rubella virus two years after vaccination: Comparison of three vaccines, *J. Infect. Dis.* **133**:637–641 (1976).

79. GRILLNER, L., FORSGREN, M., BARR, B., BOTTIGER, M., DANIELSSON, L., AND DEVERDIER, C., Outcome of rubella during pregnancy with special reference to the 17th–24th weeks of gestation, *Scand. J. Infect. Dis.* **15**:321–325 (1983).

80. HABEL, K., Transmission of rubella to *Macacus mulatta* monkeys, *Public Health Rep.* **57**:1126–1139 (1942).

81. HALSTEAD, S. B., DIWAN, A. R., AND ODA, A. I., Susceptibility to rubella among adolescents and adults in Hawaii, *J.A.M.A.* **210**: 1881–1883 (1969).

82. HARCOURT, G. C., BEST, J. M., AND BANATVALA, J. E., Rubella-specific serum and nasopharyngeal antibodies in volunteers with naturally acquired and vaccine-induced immunity after intranasal challenge, *J. Infect. Dis.* **142**:145–155 (1980).

83. HARDY, J. B., SEVER, J. L., AND GILKESON, M. R., Declining antibody titers in children with congenital rubella, *J. Pediatr.* **75**: 213–220 (1969).

84. HATTIS, R. P., HALSTEAD, S. B., HERRMANN, K. L., AND WITTE, J. J., Rubella in an immunized island population, *J.A.M.A.* **223**: 1019–1021 (1973).

85. HEGGIE, A. D., Pathogenesis of the rubella exanthem: Isolation of rubella virus from the skin, *N. Engl. J. Med.* **285**:664–666 (1971).

86. HEGGIE, A. D., AND ROBBINS, F. C., Natural rubella acquired after birth, clinical features and complications, *Am. J. Dis. Child.* **118**: 12–17 (1969).

87. HERRMANN, K. L., Rubella virus, in: *Diagnostic Procedures for Viral, Rickettsial and Chlamydial Infections* (E. H. LENETTE AND N. J. SCHMIDT, EDS.), pp. 725–766, American Public Health Association, Washington, DC, 1979.

88. HERRMANN, K. L., Available rubella serologic tests, *Rev. Infect. Dis.* **7**:S108–S112 (1985).

89. HERRMANN, K. L., HALSTEAD, S. B., AND WIEBENGA, N. H., Rubella antibody persistence after immunization, *J.A.M.A.* **247**:193–196 (1982).

90. HESS, A. F., German measles (rubella): An experimental study, *Arch. Intern. Med.* **13**:913–916 (1914).

91. HILLARY, I. B., AND GRIFFITH, A. H., Persistence of rubella antibodies 15 years after subcutaneous administration of Wistar 27/3 strain live attenuated rubella virus vaccine, *Vaccine* **2**:274–276 (1984).

92. HILLEMAN, M. R., BUYNAK, E. V., WHITMAN, J. E., WEIBEL, R. E., AND STOKES, J., Live attenuated rubella virus vaccines: Experiences with duck embryo cell preparations, *Am. J. Dis. Child.* **118**:166–171 (1969).

93. HINMAN, A. R., BART, K. J., ORENSTEIN, W. A., AND PREBLUD, S. R., Rational strategy for rubella vaccination, *Lancet* **1**:39–40 (1983).

94. HIRO, Y., AND TASAKA, S., Die Roteln sind eine viruskrankheit, *Monatsschr. Kinderheilkd.* **76**:328–332 (1938).

95. HOLD, S., HUDGINS, D., KRISHNAN, K. R., AND CRITCHLEY, E. M. R., Diffuse myelitis associated with rubella vaccination, *Br. Med. J.* **2**:1037–1038 (1976).

96. HOLMES, I. H., WARK, M. C., AND ANDWARKBURTON, M. F., Is rubella an arbovirus? II. Ultrastructural morphology and development, *Virology* **37**:15–25 (1969).

97. HORNSTEIN, L., AND BEN-PORATH, E., Rubella antibodies in women of childbearing age during an epidemic and two years thereafter, *Is. J. Med. Sci.* **12**:1180–1193 (1976).

98. HORNSTEIN, L., LEVY, U., AND FOGEL, A., Clinical rubella with virus transmission to the fetus in a pregnant woman considered to be immune, *N. Engl. J. Med.* **310**:1415–1416 (1988).

99. HORSTMANN, D. M., AND LIEBHABER, H., Rubella, in: *Serological Epidemiology* (J. R. PAUL AND C. WHITE, EDS.), pp. 83–97, Academic Press, New York, 1973.

100. HORSTMANN, D. M., LIEBHABER, H., LEBOUVIER, G. L., ROSENBERG, D. A., AND HALSTEAD, S. B., Rubella: Reinfection of vaccinated and naturally immune persons exposed in an epidemic, *N. Engl. J. Med.* **283**:771–778 (1970).

101. HORSTMANN, D. M., RIORDAN, J. T., OHTAWARA, M., AND NIEDERMAN, J. C., A natural epidemic of rubella in a closed population—Virologic and epidemiological observations, *Arch. Ges. Virusforsch.* **16**:438–487 (1985).

102. HORSTMANN, D. M., SCHLUEDERBERG, A., EMMONS, E., EVANS, B. K., RANDOLPH, M. F., AND ANDIMAN, W. A., Persistence of vaccine-induced immune responses to rubella: Comparison with natural infection, *Rev. Infect. Dis.* **7:**S80–S85 (1985).

103. INOUYE, S., SATOH, K., AND TAJIMA, T., Single-serum diagnosis of rubella by combined use of the hemagglutination inhibition and passive hemagglutination tests, *J. Clin. Microbiol.* **23:**388–391 (1986).

104. INSTITUTE OF MEDICINE, *Adverse Effects of Pertussis and Rubella Vaccines* (C. P. HOWSON, C. J. HOWE, AND H. V. FINEBERG, EDS.), pp. 187–205, National Academy Press, Washington, DC, 1991.

105. JACKSON, A. D. M., AND FISCH, L., Deafness following maternal rubella. Results of a prospective investigation, *Lancet* **2:**1241–1244 (1958).

106. JACKSON, B. M., PAYTON, T., HORST, G., HALPIN, T. J., AND MORTENSON, B. K., An epidemiological investigation of a rubella outbreak among the Amish of Northeastern Ohio, *Public Health Rep.* **108:**436–439 (1993).

107. JENNINGS, S., AND THORNTON, L., The epidemiology of rubella in the Republic of Ireland, *Communic. Dis. Rep. Rev.* **3:**R115–R117 (1993).

108. JUST, M., JUST, V., BERGER, R., BURKARDT, F., AND SCHILT, U., Duration of immunity after rubella vaccination: A long-term study in Switzerland, *Rev. Infect. Dis.* **7:**S91–S94 (1985).

109. KAZARIAN, E. L., AND GAGER, W. E., Optic neuritis complicating measles, mumps and rubella vaccination, *Am. J. Ophthamol.* **86:**544–547 (1978).

110. KLINE, L. B., MARGULIES, S. L., AND OH, S. J., Optic neuritis and myelitis following rubella vaccination, *Arch. Neurol.* **39:**443–444 (1982).

111. KLOCK, L. E., SPRUANCE, S. L., BAILEY, A., MCQUARRIE, H. G., HEBERSTROM, R. M., SHARP, H. C., AND MUTH, C. B., A clinical and serological study of women exposed to rubella-virus vaccinees, *Am. J. Dis. Child.* **123:**465–468 (1972).

112. KONO, R., HAYAWAKA, Y., AND ISHII, K., Experimental vertical transmission of rubella virus in rabbits, *Lancet* **1:**343–347 (1969).

113. KONO, R., HIRAYAMA, M., SUGISHITA, C., AND MIYAMURA, K., Epidemiology of rubella and congenital rubella infection in Japan, *Rev. Infect. Dis.* **7:**S56–S63 (1985).

114. KROGH, V., DUFFY, L. C., WONG, D., ROSENBRAND, M., RIDDLESBERGER, K. R., AND OGRA, P. L., Postpartum immunization with rubella virus vaccine and antibody response in breast-feeding infants, *J. Lab. Clin. Med.* **113:**695–699 (1989).

115. KRUGMAN, S. (ED.), Rubella symposium, *Am. J. Dis. Child.* **110:**345–476 (1965).

116. KRUGMAN, S. (ED.), International conference on rubella immunization, *Am. J. Dis. Child.* **118:**2–410 (1969).

117. KRUGMAN, S., AND WARD, R., Rubella: Demonstration of neutralizing antibody in gamma globulin and re-evaluation of the rubella problem, *N. Engl. J. Med.* **259:**16–19 (1958).

118. KRUGMAN, S., WARD, R., JACOBS, K. G., AND LAZAR, M., Studies on rubella immunization—I. Demonstration of rubella without rash, *J.A.M.A.* **151:**285–288 (1953).

119. KURTZ, J. B., MORTIMER, P. P., MORGAN-CAPNER, P., SHAFI, M. S., AND WHITE, G. B. B., Rubella antibody measured by radial hemolysis: Characteristics and performance of a simple screening method for use in diagnostic laboratories, *J. Hyg.* **84:**213–222 (1980).

120. LASONSKY, G. A., FISHAUT, J. M., AND OGRA, P. L., Effect of immunization against rubella on lactation products. II. Maternal–neonatal interactions, *J. Infect. Dis.* **145:**661–666 (1982).

121. LAWLESS, M. R., ABRAMSON, J. S., HARLAN, J. E., AND KELSEY, D. S., Rubella susceptibility in sixth graders: Effectiveness of current immunization practice, *Pediatrics* **65:**1086–1089 (1980).

122. LEE, P. R., BARNETT, A. F., SCHOLER, J. F., BRYNER, S., AND CLARK, W. H., Rubella arthritis: A study of twenty cases, *Calif. Med.* **93:**125–128 (1960).

123. LEE, S. H., EWERT, D. P., FREDERICK, P. D., AND MASCOLA, L., Resurgence of congenital rubella syndrome in the 1990s, *J.A.M.A.* **267:**2616–2620 (1992).

124. LEHANE, D. E., NEWBERG, N. R., AND BEAM, W. E., JR., Evaluation of rubella herd immunity during an epidemic, *J.A.M.A.* **213:**2236–2239 (1970).

125. LENNETTE, E. H., SCHMIDT, N. J., AND MAGOFFIN, R. L., The hemagglutination-inhibition test for rubella: A comparison of its sensitivity to that of neutralization, complement fixation, and fluorescent antibody tests for diagnosis of infection and determination of immunity status, *J. Immunol.* **99:**785–793 (1967).

126. LEVINE, J. B., BERKOWITZ, C. D., AND ST. GEME, J. W., Rubella virus reinfection during pregnancy leading to late-onset congenital rubella syndrome, *J. Pediatr.* **100:**589–591 (1982).

127. LIEBHABER, H., Measurement of rubella antibody by hemagglutination inhibition. I. Variables affecting rubella hemagglutination, *J. Immunol.* **104:**818–825 (1970).

128. LIEBHABER, H., Measurement of rubella antibody by hemagglutination inhibition. II. Characteristics of an improved HAI test employing a new method for the removal of non-immunoglobulin HA inhibitors from serum, *J. Immunol.* **104:**826–834 (1970).

129. LINDEGREN, M. L., FEHRS, L. J., HADLER, S. C., AND HINMAN, A. R., Update: Rubella and congenital rubella syndrome, 1980–1990, *Epidemiol. Rev.* **13:**341–348 (1991).

130. LUNDSTROM, R., Rubella during pregnancy—A follow-up study of children born after an epidemic of rubella in Sweden, 1951, with additional investigations on prophylaxis and treatment of maternal rubella, *Acta. Paediatr.* **51**(Suppl. 133):1–110 (1962).

131. MANSON, M. M., LOGAN, W. P. D., AND LOY, R. M., Rubella and other virus infections during pregnancy, in: *Reports on Public Health and Medical Subjects*, No. 101, Ministry of Health, H.M.S.O., London, 1960.

132. MANTON, W. G., Some accounts of rash liable to be mistaken for scarlitina, *Med. Trans. R. Coll. Physicians* **5:**149 (1815).

133. MARKS, J. S., SERDULA, M. K., HALSEY, N. A., GUNARATNE, M. V., CRAVEN, R. B., MURPHY, K. A., KOBAYASHI, G. Y., AND WIEBENGA, N. H., Saturday night fever: A common-source outbreak of rubella among adults in Hawaii, *Am. J. Epidemiol.* **114:**574–583 (1981).

134. MARTIN, S. M., Congenital defects and rubella, *Br. Med. J.* **1:**855 (1945).

135. MAYO, D. R., SIRPENSKI, S. P., AND MARKOWSKI, M. A., Microtiter latex agglutination—quantitative and qualitative equivalent to hemagglutination inhibition for detection of rubella antibody, *Diagn. Microbiol. Infect. Dis.* **5:**55–59 (1986).

136. MEEGAN, J. M., EVANS, B. K., AND HORSTMANN, D. M., Comparison of the latex agglutination test with the hemagglutination test, enzyme-linked immunosorbent assay, and neutralization test for detection of antibodies to rubella virus, *J. Clin. Microbiol.* **16:**644–649 (1982).

137. MEEGAN, J. M., EVANS, B. K., AND HORSTMANN, D. M., Use of enzyme immunoassays and the latex agglutination test to measure the temporal appearance of immunoglobulin G and M antibodies after natural infection or immunization with rubella virus, *J. Clin. Microbiol.* **18:**745–748 (1983).

138. MELLINGER, A. K., CRAGAN, J. D., ATKINSON, W. L., WILLIAMS, W. W., KLEGER, B., KIMBER, R. A., TAVRIS, D., High incidence of congenital rubella syndrome after a rubella outbreak, *Pediatr. Infect. Dis. J.* **14:**573–578 (1995).

139. MENSER, M. A., FORREST, J. M., BRANSBY, R. D., AND COLLINS, E., Rubella vaccination in Australia: Experiences with the RA27/3 rubella vaccine and results of a double-blind trial in schoolgirls, *Med. J. Aust.* **2:**85–88 (1978).

140. MENSER, M. A., FORREST, J. M., HONEYMAN, M. C., AND BURGESS, J. A., Diabetes, HLA-antigens and congenital rubella, *Lancet* **2:**1508–1509 (1974).

141. MENSER, M. A., HUDSON, J. R., MURPHY, A. M., AND UPFOLD, L. J., Epidemiology of congenital rubella and results of rubella vaccination in Australia, *Rev. Infect. Dis.* **1:**S37–S41 (1985).

142. MENSER, M. A., SLINN, R. F., DODS, L., HERTZBERG, R., AND HARLEY, J. D., Congenital rubella in a mother and son, *Aust. Paediatr. J.* **4:**200–202 (1968).

143. METTLER, N. E., PETERELLI, R. I., AND CASALS, J., Absence of cross reactions between rubella virus and arbovirus, *Virology* **36:**503–504 (1968).

144. MEURMAN, O. H., VILJANEN, M. K., AND GRANFORS, K., Solid-phase radioimmunoassay of rubella virus immunoglobulin M antibodies: Comparison with sucrose density gradient centrifugation test., *J. Clin. Microbiol.* **5:**257–262 (1977).

145. MEYER, H. M., PARKMAN, P. D., HOBBINS, T. E., LARSON, H. E., DAVIS, W. S., SIMSARIAN, J. P., AND HOPPS, H. E., Attenuated rubella viruses: Laboratory and clinical characteristics, *Am. J. Dis. Child.* **118:**155–165 (1969).

146. MEYER, H. M., PARKMAN, P. D., AND PANOS, T. C., Attenuated rubella virus. II. Production of an experimental live-virus vaccine and clinical trial, *N. Engl. J. Med.* **275:**575–580 (1966).

147. MILLER, E., Rubella in the United Kingdom, *Epidemiol. Infect.* **107:**21–42 (1991).

148. MILLER, E., CRADOCK-WATSON, J. E., AND POLLACK, T. M., Consequences of confirmed rubella at successive stages of pregnancy, *Lancet* **1:**871–874 (1982).

149. MILLER, E., WAIGHT, P. A., VURDIEN, J. E., JONES, G., TOOKEY, P. A., AND PECKHAM, C. S., Rubella surveillance to December 1992: Second joint report from the PHLS and National Congenital Rubella Surveillance Programme, *Communic. Dis. Rep. Rev.* **3**(3):R35–R40 (1993).

149a. MIMS, C. A., STOKES, A., AND GRAHAME, R., Synthesis of antibodies, including antiviral antibodies, in the knee joints of patients with arthritis, *Ann. Rheum. Dis.* **44:**734–737 (1985).

150. MODLIN, J. E., HERRMANN, K. L., BRANDLING-BENNETT, A. D., EDDINS, D. L., AND HAYDEN, G. F., Risk of congenital abnormality after inadvertent rubella vaccination of pregnant women, *N. Engl. J. Med.* **294:**972–974 (1976).

151. MORGAN-CAPNER, P., MILLER, E., VURDIEN, J. E., AND RAMSAY, M. E. B., Outcome of pregnancy after maternal reinfection with rubella, *Communic. Dis. Rep. Rev.* **1:**R57–R59 (1991).

152. MORTON-KUTE, L., Rubella vaccine and facial paresthesias, *Ann. Intern. Med.* **102:**563 (1985).

153. MURPHY, F. A., HALONEN, P. A., AND HARRISON, A. K., Electron microscopy of the development of rubella virus in BHK-21 cells, *J. Virol.* **2:**1223–1227 (1968).

154. NAEYE, R. L., AND BLANC, W., Pathogenesis of congenital rubella, *J.A.M.A.* **194:**1277–1283 (1965).

155. NATIONAL COMMUNICABLE DISEASE CENTER, Rubella surveillance, Report No. 1, June 1969.

156. OGRA, P. L., KERR-GRANT, D., UMANA, G., DZIERBA, J., AND WEINTRAUB, D., Antibody response in serum and nasopharynx after infections with rubella virus, *N. Engl. J. Med.* **285:**1333–1339 (1971).

157. ORENSTEIN, W. A., BART, K. J., HINMAN, A. R., PREBLUD, S. R., GREAVES, W. L., DOSTER, S. W., STETLER, H. C., AND SIROTKIN, B., The opportunity and obligation to eliminate rubella from the United States, *J.A.M.A.* **251:**1988–1994 (1984).

158. ORENSTEIN, W. A., HERRMANN, K. L., HOLMGREEN, P., BERNIER, R., BART, K. J., EDDINS, D. L., AND FIUMARA, N. J., Prevalence of rubella antibodies in Massachusetts school children, *Am. J. Epidemiol.* **124:**290–298 (1986).

159. ORENSTEIN, W. A., HESELTINE, P. N. R., LeGAGNOUX, S. J., AND PORTNOY, B., Rubella vaccine and susceptible hospital employees. Poor physician participation, *J.A.M.A.* **245:**711–713 (1981).

160. O'SHEA, S., BEST, J. M., BANATVALA, J. E., MARSHALL, W. C., AND DUDGEON, J. A., Rubella vaccination persistence of antibodies for up to 16 years, *Br. Med. J.* **285:**253–255 (1982).

161. O'SHEA, S., BEST, J. M., AND BANATVALA, J. E., Viremia, virus excretion, and antibody responses after challenge in volunteers with low levels of antibody to rubella virus, *J. Infect. Dis.* **148:**639–647 (1983).

162. O'SHEA, S., BEST, J. M., BANATVALA, J. E., MARSHALL, W. C., AND DUDGEON, J. A., Persistence of rubella antibody 8–18 years after vaccination, *Br. Med. J.* **288:**1043 (1984).

163. PARKMAN, P. D., BUESCHER, E. L., AND ARTENSTEIN, M. S., Recovery of rubella virus from army recruits, *Proc. Soc. Exp. Biol. Med.* **111:**225–230 (1962).

164. PARKMAN, P. D., BUESCHER, E. L., ARTENSTEIN, M. S., McCOWN, J. M., MUNDON, F. K., AND DRUZD, A. D., Studies of rubella. I. Properties of the virus, *J. Immunol.* **93:**595–607 (1964).

165. PARTRIDGE, J. W., FLEWETT, T. H., AND WHITEHEAD, J. E., Congenital rubella affecting an infant whose mother had rubella antibodies before conception, *Br. Med. J.* **282:**187–188 (1981).

166. PECKHAM, C. S., Clinical and laboratory study of children exposed *in utero* to maternal rubella at successive stages of pregnancy, *Arch. Dis. Child* **47:**571–577 (1972).

167. PECKHAM, C. S., Congenital rubella in the United Kingdom before 1970: The prevaccine era, *Rev. Infect. Dis.* **7:**S11–S16 (1985).

168. PELTOLA, H., AND HEINONEN, O. P., Frequency of true adverse reactions to measles–mumps–rubella vaccine: A double-blind placebo-controlled trial in twins, *Lancet* **1:**939–942 (1986).

168a. PELTOLA, H., AND HEINONEN, O. P., VALLE, M., PAVNIO, M., VIRTANEN, M., KARANKO, V., AND CANTELL, K., The elimination of indigenous measles, mumps, and rubella from Finland by a 12-year, two-dose vaccination program, *N. Engl. J. Med.* **331:**1397–1402 (1994).

169. PETTERSON, R. F., OKER-BLOM, C., KALKKINEN, N., KALLIO, A., ULMANEN, I., KAARIAINEN, L., PARTANEN, P., AND VAHERI, A., Molecular and antigenic characteristics and synthesis of rubella virus structural proteins, *Rev. Infect. Dis.* **7:**S140–S149 (1985).

170. PLOTKIN, S. A., Attenuation of RA27/3 rubella virus in WI-38 human diploid cells, *Am. J. Dis. Child.* **118:**178–185 (1969).

171. PLOTKIN, S. A., AND BUSER, F., History of RA27/3 rubella vaccine, *Rev. Infect. Dis.* **7:**S77–S88 (1985).

172. PLOTKIN, S. A., FARQUHAR, J. D., AND OGRA, P. L., Immunologic properties of RA27/3 rubella virus vaccine, *J.A.M.A.* **225:**585–590 (1973).

173. POLK, B. F., WHITE, J. A., DeGIROLAMI, P. C., AND MODLIN, J. F., An outbreak of rubella among hospital personnel, *N. Engl. J. Med.* **303:**541–545 (1980).

174. POLK, B. F., MODLIN, J. F., WHITE, J. A., AND DeGIROLAMI, P. C., A controlled comparison of joint reactions among women receiving one of two rubella vaccines, *Am. J. Epidemiol.* **115**:19–25 (1982).

175. PREBLUD, S. R., SERDULA, M. K., FRANK, J. A., JR., BRANDLING-BENNETT, A. D., AND HINMAN, A. R., Rubella vaccination in the United States: A ten-year review, *Epidemiol. Rev.* **2**:171–194 (1980).

176. PRINZIE, A., HUYGELEN, C., GOLD, J., FARQUHAR, J., AND McKEE, J., Experimental live attenuated rubella virus vaccine: Clinical evaluation of Cendehill strain, *Am. J. Dis. Child.* **118**:172–177 (1969).

177. RAWLS, W. E., MELNICK, J. L., BRADSTREET, C. M. P., BAILEY, M., FERRIS, A. A., TIEHMANN, N., NAGLER, F. P., FURESZ, J., KONO, R., OHTAWARA, M., HALONEN, P., STEWART, J., RYAN, J. M., STRAUSS, J., ZDRAZILEK, J., LEERHOY, J., VON MAGNUS, H., SOHIER, R., AND FERREIRA, W., WHO collaborative study on the seroepidemiology of rubella, *Bull. WHO* **37**:79–88 (1967).

177a. RAY, P., BLACK, S., SHINEFELD, H., DILLON, A., SCHWALBE, J., HOLMES, S., CHEN, R., WILLIAMS, W., AND COCHI, S., Retrospective cohort evaluation of chronic rubella vaccine arthropathy, 34th Interscience Conference on Antimicrocial Agents and Chemotherapy, American Society for Microbiology, Orlando, Florida, October 4–7, 1994 (Abstract).

178. REESE, A. B., Congenital cataract and other anomalies following German measles in the mother, *Am. J. Ophthamol.* **27**:483–487 (1944).

178a. Report of an international meeting on rubella vaccines and vaccination, 9 August 1993, Glasgow, United Kingdom, *J. Infect. Dis.* **170**:507–509 (1994).

179. RICHARDSON, D., Rubella immunisation status of year 8 females in the Midwest and Gascoyne, Western Australia, *Communic. Dis. Intell.* **12**:308–310 (1993).

180. ROBINSON, R. G., DUDENHOEFFER, F. E., HOLROYD, H. J., BAKER, L. R., BERNSTEIN, D. I., AND CHERRY, J. D., Rubella immunity in older children, teenagers, and young adults: A comparison of immunity in those previously immunized and those unimmunized, *J. Pediatr.* **101**:188–191 (1982).

181. RONES, B., The relationship of German measles during pregnancy to congenital ocular defects, *Med. Ann. D. C.* **13**:285–287 (1944).

182. ROSS, R., HARVEY, D. R., AND HURLEY, R., Re-infection and congenital rubella syndrome, *Practitioner* **236**:246–251 (1992).

183. ROWLANDS, D. F., AND FREESTONE, D. S., Vaccination against rubella of susceptible schoolgirls in Reading, *J. Hyg.* **69**:579–586 (1971).

184. SANDER, J., AND NIEHAUS, C., Screening for rubella IgG and IgM using an ELISA test applied to dried blood on filter paper, *J. Pediatr.* **106**:457–461 (1985).

185. SAULE, H., ENDERS, G., ZELLER, J., AND BERNSAU, U., Congenital rubella infection after previous immunity of the mother, *Eur. J. Pediatr.* **147**:195–196 (1988).

186. SCHAFFNER, W., FLEET, W. F., KILROY, A. W., LEFKOWITZ, L. B., HERRMANN, K. L., THOMPSON, J., AND KARZON, D. T., Polyneuropathy following rubella immunization. A follow-up study and review of the problem, *Am. J. Dis. Child.* **127**:684–688 (1974).

187. SCHATZMAYR, H. G., Aspects of rubella infection in Brazil, *Rev. Infect Dis.* **7**:S53–S55 (1985).

188. SCHIFF, G. M., YOUNG, B. C., STEFANOVIC, G. M., STAMLER, E. F., KNOWLTON, D. R., GRUNDY, B. J., AND DORSETT, P. H., Challenge with rubella virus after loss of detectable vaccine-induced antibody, *Rev. Infect. Dis.* **7**:S157–S163 (1985).

189. SCHLUEDERBERG, A., HORSTMANN, D. M., ANDIMAN, W. A., AND RANDOLPH, M. F., Neutralizing and hemagglutination-inhibiting antibodies to rubella virus as indicators of protective immunity in vaccinees and naturally immune individuals, *J. Infect. Dis.* **138**:877–883 (1978).

190. SCHOUB, B. D., BLACKBURN, N. K., O'CONNELL, K., KAPLAN, A. B., AND ADNO, J., Symptomatic rubella reinfection in early pregnancy and subsequent delivery of an infected but minimally involved infant, *S. Afr. Med. J.* **78**:484–485 (1990).

191. SCHUM, T. R., NELSON, D. B., DUMA, M. A., AND SEDMAK, G. V., Increasing rubella seronegativity despite a compulsory school law, *Am. J. Public Health* **80**:66–69 (1990).

192. SCOTT, H. D., AND BYRNE, E. B., Exposure of susceptible pregnant women to rubella vaccinees. Serologic findings during the Rhode Island immunization campaign, *J.A.M.A.* **215**:609–612 (1971).

193. SEPPPALA, M., AND VAHERI, A., Natural rubella infection of the female genital tract, *Lancet* **1**:46–47 (1974).

194. SERDULA, M. K., HALSTEAD, S. B., WIEBENGA, N. H., AND HERRMANN, K. L., Serological response to rubella revaccination, *J.A.M.A.* **251**:1974–1977 (1984).

195. SEVER, J. L., The epidemiology of rubella, *Arch. Ophthalmol.* **77**:427–429 (1967).

196. SEVER, J. L., FUCCILLO, D. A., GILKESON, M. R., LEY, A., AND TRAUB, R., Changing susceptibility to rubella, *Obstet. Gynecol.* **32**:365–369 (1968).

197. SEVER, J. L., SCHIFF, G. M., AND HUEBNER, R. J., Frequency of rubella antibody among pregnant women and other human and animal populations, *Obstet. Gynecol.* **23**:153–159 (1964).

198. SEVER, J. L., SOUTH, M. A., AND SHAVER, K. A., Delayed manifestations of congenital rubella, *Rev. Infect. Dis.* **7**:S164–169 (1985).

199. SHEPPARD, S., SMITHELLS, R. W., DICKSON, A., AND HOLZEL, H., Rubella vaccination and pregnancy: Preliminary report of a national survey, *Br. Med. J. (Clin. Res.)* **292**:727 (1986).

200. SIEGEL, M., AND GREENBERG, M., Fetal death, malformation and prematurity after maternal rubella: Results of prospective study, 1949–1958, *N. Engl. J. Med.* **262**:389–393 (1960).

201. SIGURJONSSON, J., Rubella and congenital deafness, *Am. J. Sci.* **240**:712–720 (1961).

202. SIMONS, M. J., Congenital rubella: An immunological paradox? *Lancet* **2**:1275–1278 (1968).

203. SKENDZEL, L. P., CARSKI, T. R., HERRMANN, K. L., KIEFER, D. J., NAKAMURA, R. M., NUTTER, C. D., AND SCHAEFER, L. E., Evaluation and performance criteria for multiple component test products intended for the detection and quantitation of rubella IgG antibody, *Natl. Comm. Clin. Lab. Stand.* **12**:1–24 (1992).

204. SKENDZEL, L. P., WILCOX, K. R., AND EDSON, D., Evaluation of assays for the detection of antibodies to rubella. A report based on data from the College of American Pathologists surveys 1982, *Am. J. Clin. Pathol.* **80**(Suppl.):594–598 (1983).

204a. SLATER, P. E., BEN-ZVI, T., FOGEL, A., EHRENFELD, M., AND EVER-HADANI, S., Absence of an association between rubella vaccination and arthritis in underimmune postpartum women, *Vaccine* **13**:1529–1532 (1995).

205. SOOTHHILL, J. F., HAYES, K., AND DUDGEON, J. D., The immunoglobulins in congenital rubella, *Lancet* **1**:1385–1388 (1966).

206. SOUTH, M. A., MONTGOMERY, J. K., AND RAWLS, W. D., Immune deficiency in rubella and other viral infections, *Birth Defects* **11:**234–238 (1975).

207. SPRUANCE, S. L., AND SMITH, C. B., Joint complications associated with derivatives of HPV-77 rubella vaccine, *Am. J. Dis. Child.* **122:**105–111 (1971).

208. STANLEY, F. J., SIM, M., WILSON, G., AND WORTHINGTON, S., The decline in congenital rubella syndrome in Western Australia: An impact of the the school girl vaccination program?, *Am. J. Public Health* **76:**35–37 (1986).

209. STEHR-GREEN, P. A., COCHI, S. L., PREBLUD, S. R., ORENSTEIN, W. A., Evidence against increasing rubella seronegativity among adolescent girls, *Am. J. Public Health* **80:**88 (1990).

210. STORCH, G. A., AND MYERS, N., Latex-agglutination test for rubella antibody: Validity of positive results assessed by response to immunization and comparison with other tests, *J. Infect Dis.* **149:**459–464 (1984).

211. STRASSBURG, M. A., IMAGAWA, D. T., FANNIN, S. L., TURNER, J. A., CHOW, A. W., MURRAY, R. A., AND CHERRY, J. D., Rubella outbreak among hospital employees, *Obstet. Gynecol.* **57:**283–288 (1981).

212. SWAN, C., TOSTEVIN, A. L., AND BLACK, G. H. B., Final observations on congenital defects in infants following infectious diseases during pregnancy, with special reference to rubella, *Med. J. Aust.* **2:**889–908 (1946).

213. SWAN, C., TOSTEVIN, A. L., MOORE, B., MAYO, H., AND BLACK, G. H. B., Congenital defects in infants following infectious diseases during pregnancy. With special reference to the relationship between German measles and cataract, deaf-mutism, heart disease and microcephaly, and to the period of pregnancy in which the occurrence of rubella is followed by congenital abnormalities, *Med. J. Aust.* **11:**201–210 (1943).

214. SWARTZ, T. A., An extensive rubella epidemic in Israel, 1972: Selected epidemiologic characteristics, *Am. J. Epidemiol.* **103:**60–66 (1976).

215. SWARTZ, T. A., HORNSTEIN, L., AND EPSTEIN, I., Epidemiology of rubella and congenital rubella infection in Israel, a country with a selective immunization program, *Rev. Infect. Dis.* **7:**S42–S46 (1985).

216. TINGLE, A. J., CHANTLER, J. K., POT, K. H., PATY, D. W., AND FORD, D. K., Postpartum rubella immunization: Association with development of prolonged arthritis, neurological sequelae, and chronic rubella viremia, *J. Infect. Dis.* **52:**606–612 (1985).

217. TINGLE, A. J., ALLEN, M., PETTY, R. E., KETTYLS, G. D., AND CHANTLER, J. K., Rubella-associated arthritis. I. Comparative study of joint manifestations associated with natural rubella infection and RA 27/3 rubella immunisation, *Ann. Rheum. Dis.* **45:**110–114 (1986).

218. TOBIN, J. O., SHEPPARD, S., SMITHELLS, R. W., MILTON, A., NOAH, N., AND REID, D., Rubella in the United Kingdom, 1970–1983, *Rev. Infect. Dis.* **7:**S47–S52 (1985).

219. TONDURY, Y. G., AND SMITH, D. W., Fetal rubella pathology, *J. Pediatr.* **68:**867–879 (1966).

220. UEDA, K., NONAKA, S., YOSHIKAWA, H., SASAKI, K., SEGAWA, H., FUJII, H., TASAKI, H., SHIN, H., TOKUGAWA, K., AND SATO, T., Seroepidemiologic studies of rubella in Fukuoka in Southern Japan during 1965–1981: Rubella epidemic pattern, endemicity and immunity gap, *Int. J. Epidemiol.* **12:**450–454 (1983).

221. UEDA, K., SASAKI, F., TOKUGAWA, K., SEGAWA, K., AND FUJII, H., The 1976–1977 rubella epidemic in Fukuoka city in Southern Japan: Epidemiology and incidences of complications among 80,000 persons who were school children at 28 primary schools and their family members, *Biken J.* **27:**161–168 (1984).

222. UKKONEN, P., AND VON BONSDORFF, C.-K., Rubella immunity and morbidity: Effects of vaccination in Finland, *Scand. J. Infect. Dis.* **20:**255–259 (1988).

223. VAANANEN, P., HAIVA, V.-M., KOSKELA, P., AND MEURMAN, O., Comparison of a simple latex agglutination test with hemolysis-in-gel, hemagglutination inhibition, and radioimmunoassay for detection of rubella virus antibodies, *J. Clin. Microbiol.* **21:**793–795 (1985).

224. VALENSIN, P. E., ROSSOLINI, G. M., CUSI, M. G., ZANCHI, A., CELLESI, C., AND ROSSOLINI, A., Specific antibody patterns over a two-year period after rubella immunization with RA 27/3 live attenuated vaccine, *Vaccine* **5:**289–294 (1987).

225. VEALE, H., History of an epidemic of Rotheln, with observations on its pathology, *Edinburgh Med. J.* **12:**404–414 (1866).

226. VOGT, R. L., AND CLARK, S. W., Premarital rubella vaccination program, *Am. J. Public Health* **75:**1088–1089 (1985).

227. VOLLER, A. D., BIDWELL, D., AND BARTLETT, A., Enzyme linked immunosorbent assay, in: *Manual of Clinical Immunity*, 2nd ed. (N. R. ROSE AND H. FRIEDMAN, EDS.), pp. 359–371, American Society for Microbiology, Washington, DC, 1980.

228. WAXHAM, M. N., AND WOLINSKY, J. S., Rubella virus and its effect on the central nervous system, *Neurol. Clin.* **2:**367–385 (1984).

229. WAXHAM, M. N., AND WOLINSKY, J. S., A model of the structural organization of rubella virions, *Rev. Infect. Dis.* **7:**S133–S139 (1985).

230. WEIBEL, R. E., STOKES, J., JR., BUYNAK, E. B., AND HILLEMAN, M. R., Influence of age on clinical responses to HPV77 duck rubella vaccine, *J.A.M.A.* **222:**805–807 (1972).

231. WELLER, T. H., AND NEVA, F. A., Propagation in tissue culture of cytopathic agents from patients with rubella-like illness, *Proc. Soc. Exp. Biol. Med.* **11:**215–225 (1962).

232. WESSELHOEFT, C., Rubella (German measles), *N. Engl. J. Med.* **236:**943–950, 978–988 (1947).

233. WIELAARD, F., DENISSEN, A., VAN ELLESWUK-VAN DER BERG, J., AND VAN GEMERT, G., Clinical validation of an antibody capture anti-rubella IgM ELISA, *J. Virol. Methods* **10:**349–354 (1985).

234. WILD, N. J., SHEPPARD, S., SMITHELLS, R. W., HOLZEL, H., AND JONES, G., Onset and severity of hearing loss due to congenital rubella infection, *Arch. Dis. Child.* **64:**1280–1283 (1989).

235. WILKINS, J., LEEDOM, J. M., PORTNOY, B., AND SALVATORE, M. A., Reinfection with rubella virus despite live vaccine induced immunity. Trials of HPV-77 and HPV-80 live rubella virus vaccines and subsequent artificial and natural challenge studies, *Am. J. Dis. Child.* **118:**275–294 (1969).

236. WILKINS, J., LEEDOM, J. M., SALVATORE, M. A., AND PORTNOY, B., Clinical rubella with arthritis resulting from reinfection, *Ann. Intern. Med.* **77:**930–932 (1972).

237. WITTE, J. J., KARCHMER, A. W., CASE, G., HERRMANN, K. L., KASSANOFF, I., AND NEILL, J. S., Epidemiology of rubella, *Am. J. Dis. Child.* **118:**107–111 (1969).

238. WOLF, J. E., EISEN, J. E., AND FRAIMOW, H. S., Symptomatic rubella reinfection in an immune contact of a rubella vaccine recipient, *So. Med. J.* **86:**91–93 (1993).

239. WOLINSKY, J. S., BERG, B. O., AND MAITLAND, C. J., Progressive rubella panencephalitis, *Arch. Neurol.* **33:**722–723 (1976).

240. WORLD HEALTH ORGANIZATION, LAB/82. 1. Evaluation of candi-

date international reference methods for the rubella hemagglutination-inhibition tests. Report of a collaborative study, 1982.

241. ZHANG, T., MAURACHER, C. A., MITCHELL, L. A, AND TINGLE, A. J., Detection of rubella virus-specific immunoglobulin G (IgG), IgM, and IgA antibodies by immunoblot assays, *J. Clin. Microbiol.* **30:**824–830 (1992).

242. ZOLTI, M., ZION, B. R., BIDER, D., MASHIACH, S., AND FOGEL, A., Rubella-specific IgM in reinfection and risk to the fetus, *Gynecol. Obstet. Invest.* **30:**184–185 (1990).

12. Suggested Reading

CHERRY, J. D., Rubella, in: *Textbook of Infectious Diseases*, 3rd ed. (J. D. CHERRY AND R. D. FEIGIN, EDS.), pp. 1792–1817, W. B. Saunders, Philadelphia, 1992.

GERSHON, A. A., Rubella virus (German measles), in: *Principles and Practices of Infectious Diseases*, 3rd ed. (G. L. MANDELL, R. G. DOUGLAS, JR., AND J. R. BENNETT, EDS.), pp. 1242–1247, Churchill Livingstone, New York, 1990.

International Symposium on the Prevention of Congenital Rubella Infection, *Rev. Infect. Dis.* **7**(Suppl. 1):S1–S215 (1985).

MALDONALDO, Y. A., Rubella virus, in: *Pediatric Infectious Diseases* (S. S. LONG, L. K. PICKERING, AND C. G. PROBER, EDS.), pp. 1228–1237, Churchill Livingstone, New York, 1997.

PLOTKIN, S. A., Rubella vaccine, in: *Vaccines* (S. A. PLOTKIN AND E. A. MORTIMER, EDS.), pp. 303–336, W. B. Saunders, Philadelphia, 1994.

PREBLUD, S. R., AND ALFORD, C. A., Rubella, in: *Infectious Diseases of the Fetus and Newborn Infant*, 3rd ed. (J. S. REMINGTON AND J. O. KLEIN, EDS.), pp. 196–240, W. B. Saunders, Philadelphia, 1983.

WOLINSKY, J. S., Rubella, in: *Virology*, 2nd ed. (B. N. FIELDS, D. M. KNIPE, R. M. CHANOCK, M. S. HIRSCH, J. L. MELNICK, T. P. MONATH, AND B. ROIZMAN, EDS.), pp. 815–838, Raven Press, New York, 1990.

CHAPTER 28

Smallpox

The End of the Story?

Abram S. Benenson

1. Introduction

1.1. Smallpox

From the earliest days of recorded history, humankind has suffered from the depredations of the smallpox virus. In 1796, Edward Jenner discovered a method for countering this implacable enemy. This weapon unfortunately was only used effectively in localized areas of the globe, while smallpox persisted in guerrilla attacks over the rest of the world, with occasional, more aggressive assaults. In 1966, the World Health Organization (WHO) mounted a coordinated global search and destroy operation, hunting and destroying the enemy in its remote hiding places. The last human casualty in this search was found in Somalia when Ali Maow Maalin developed smallpox on October 26, 1977. Over the next 2 years, intensive searches were carried out worldwide to ensure that no traces of any live virus could be found other than that which was imprisoned in deep-freeze storage in approved research laboratories. But, like other miscreants, the disease sought any opportunity to again enjoy the warmth of the human body, and 10 months later, in 1978, succeeded to escape from confinement in a Birmingham, England, research laboratory; it killed a medical photographer whose place of work was close to the smallpox laboratory; her mother was infected but survived; the promising virologist, an expert on the virus, committed suicide.[5]

Since that time, the virus has been successfully incarcerated, held under very tight security, and those countries

that had virus sent the stocks they did not destroy to the WHO Collaborating Centers [the Centers for Disease Control and Prevention (CDC), Atlanta, Georgia, and the Research Institute for Viral Preparations, Moscow, Russia], awaiting the decision of WHO for final action: Should this worst enemy of man now be executed and the global eradication of a vile disease be effected?

In 1983, the Committee on Orthopoxvirus Infections[11] noted that DNA fragments of three strains of variola virus had been cloned into recombinant plasmids and that mapping of cleavage sites of additional strains of variola and for additional endonucleases was under way; thus, the needs for diagnostic purposes and for additional studies could be met without any exposure to infectious virus.

In 1986, the Committee[12] noted that the workers in Atlanta, Georgia, and in the Public Health Laboratory Service Centre for Applied Microbiology and Research, Porton Down, Salisbury, United Kingdom, had prepared plasmids encoding the DNA of two strains of variola major, two of alastrim and one strain of African variola minor. The cross-linked terminal fragments of none had been cloned; the committee considered that it was not essential to clone the cross-linked terminal fragments, but that it would be desirable that material from a West African variola strain be added.

With passage of time, all known strains of variola virus were concentrated in Atlanta and Moscow, and newer scientific techniques became available for dissecting the virus. In December 1990, the Ad Hoc Committee on Orthopoxvirus Infections[13] recommended that at least two cloned representative strains of variola virus be fully sequenced and that then all variola viruses kept by these

Abram S. Benenson • Graduate School of Public Health, San Diego State University, San Diego, California 92182.

two collaborating centers should be destroyed by 31 December, 1993. They also recommended that:

> All recombinant plasmids and other related materials that contain variola virus DNA sequences should be destroyed at the same time as the variola virus stocks, provided that the Technical Committee ... is satisfied that sufficient sequence information is available, and serious scientific objections have not been raised.
>
> In the interim, all recombinant plasmids that contain variola DNA sequences should be registered with WHO. These plasmids may only be provided to requesting scientists after informing WHO, and on the strict understanding that they must not be distributed to third parties or used in laboratories handling other orthopoxviruses.

The designation of a specific date for execution evoked strong feelings, resulting in a reprieve. One camp[9] argued that the disease is gone and the cloned plasmids provide the information that might be necessary for diagnostic purposes in case a pox disease appears. The relatively scarce resources available for study of infectious diseases, in money and in secure facilities, should be expended for the study of new emerging diseases rather than on one that is no longer a threat to the public health. The availability of the cloned plasmids for further research would be controlled by WHO, which would demand a promise that these plasmids must not be used in laboratories handling other orthopoxviruses (lest coinfection result in a virulent virus) and that the plasmids must not be distributed to third parties. This represents the public health attitude.

The other group[8] represents the scientific attitude. These virologists are concerned over the loss of the possibility of determining, at a molecular level, the complex process by which this virus evades the human defense mechanisms, including the coding for proteins that mimic or interfere with host immune and regulatory functions. Comparison of the mechanisms or genes involved during smallpox virus infection with those in infections with monkeypox and vaccinia, viruses with broad host ranges unlike variola with its host range restricted to man, could provide understanding of the genetic basis of host selection. The virologists point out that the variola strains that had been sequenced have all been passaged in eggs, which may have induced some changes, and they are anxious that truly wild strains be sequenced. They are anxious for the opportunity to carry out studies beyond the knowledge of the sequence of the almost 200,000 base pairs in the virus and would like decisions on the destruction of the virus to be deferred for at least 10 years.

These arguments evoked a reprieve, and December 31, 1993, passed with the virus stocks intact. However, the Ad Hoc Committee, in a meeting on September 9, 1994,[1] recognized the arguments for more research on the intact virus but recommended unanimously that the virus be destroyed on June 30, 1995. The committee felt that the risk that the virus might fall into the hands of terrorists outweighs the potential value of continued research. The committee recommended that the cloned fragments be held in two repositories—at CDC and at the Russian State Research Center of Virology and Biotechnology in Koltsovo[2]—and, to prevent any attempt to recreate the virus or incorporate smallpox genes into a related virus, that DNA not be sent to any laboratory working on other pox viruses and that no more than 20% of the fragments be sent to any one laboratory.[1]

These recommendations were presented to the WHO Executive Committee in January 1995, and to the World Health Assembly meeting in May 1995; it is expected that the date for execution will be 30 June, 1995, as recommended. Until the final decision for destruction comes from that supreme court, the viruses continue to be held under tight security in Atlanta and Moscow (where the facility is "guarded by a regiment of ex-army officers").[10]

The situation at the moment can be summed up: The culprit has been neutralized by the intervention of his close relative, vaccinia virus; he has been apprehended and is being held in cold storage; all his troops in the field have been destroyed. Representatives of violent (i.e., virulent) as well as more benign groups are held in two repositories where selected individuals are being thoroughly dissected, their excreta analyzed, and their genetic components categorized, seeking to define that which makes this particular culprit so selective for the human and so destructive to his host! Soon, man will, for the first time, deliberately exterminate a biological species.

David Bishop, director of the Institute of Virology and Environmental Microbiology in Oxford, is quoted as feeling certain that "a cousin of smallpox will emerge from the wild at some stage."[2] In 1802, 4 years after Jenner reported the effectiveness of vaccinia in preventing smallpox, Bryce[3] expressed much the same sentiment when he wrote, "smallpox may again be imported from some remote corner where the influence of cowpox was unknown, or it may originate *de novo* ... and hold a course among mankind nearly as terrific as that described by authors who relate the ravages of this dreadful disease."

1.2. Late Breaker—January 1995

At the meeting of the Executive Board of the WHO in Geneva on January 18, 1995, the recommendation of the Ad Hoc Committee on Orthopoxvirus Infections that variola virus be destroyed on June 30, 1995, was expected to be confirmed. However, because of pressure from "several regions of the world to delay the decision," Dr. Jesus Kumate of Mexico, the board chairman, aborted the discussion and proposed reopening the issue at a future board meeting. The impetus for this action came from defense agencies; since no mechanism exists for verifying that no country has clandestine supplies of variola virus intended for offensive purposes, it is necessary to develop improved defenses against its use. Also, viral destruction would not permit the application of the new research techniques that have been developed. Perhaps most important would be the need to know the efficacy of new virucidal agents against this virus.

To paraphrase Mark Twain, the obituary of variola was premature. It had received a reprieve of at least 1 year, if not more. Man will not yet have succeeded in intentionally eradicating a biological species![1a,9a]

1.3. Late Breaker—January 1996

The culprit is again on death row! On January 24, 1996, the Executive Board of the WHO recommended that the smallpox virus stocks held in the United States and Russia be destroyed by June 30, 1999, after having gained a broader consensus among scientists and political leaders. The board also recommended that WHO stockpile 500,000 doses of smallpox vaccine and that vaccinia seed virus be stored should smallpox reappear. These recommendations were considered by the WHO at its annual meeting in May 1996.[1b]

1.4. Final Sentence—May 1996

On May 25, 1996, the WHO, composed of representatives of 190 countries, voted unanimously to accept the recommendations of the Executive Board of the WHO that the variola virus stocks held in Atlanta, Georgia, and in Koltsovo, Russia, be scheduled for destruction on June 30, 1999. Final approval by the WHO at their May 1999 meeting will be required. A stock of smallpox vaccine for 500,000 people will be kept in case the disease reappears, and the seed virus for the production of more vaccine will be held in the laboratory in Bilthoven, Netherlands. Hopefully, this will rid the globe of the activities of this vicious particle of antihuman material.[1c]

2. Other Poxvirus Infections

2.1. Vaccinia

Routine vaccination of children against smallpox was discontinued in the United States in early 1971–1972,

with the result that the vaccine distribution was restricted to those working with poxviruses and the armed forces. Now, essentially all people under 22 years of age are susceptible to infection. However, vaccinia virus is undergoing a resurrection, since it is used increasingly as the vector for recombinant vaccines[7]; this carries with it the potential for contact spread of vaccinia to nonimmunes, especially those with eczema. The Immunization Practices Advisory Committee of CDC[4] recommends vaccination of all laboratory workers who directly handle cultures or animals contaminated or infected with vaccinia or other orthopoxviruses that infect man; it may be considered for other health care personnel, such as doctors and nurses, whose contact with these viruses is limited to contaminated dressings and are at much lower risk of infection.

In the United States, smallpox vaccine can only be obtained from the CDC for civilians. The instructions that accompany the vaccine should be followed rigorously. Vaccination is repeated if a major reaction does not develop; revaccinations are carried out every 10 years.

2.2. Monkeypox

Monkeypox is a disease occurring among residents in tropical rain forest areas of central and western Africa, with most cases occurring in Zaire. Cases closely resemble smallpox except for a pronounced lymph node enlargement in patients with monkeypox. Intensive studies by WHO[6] in 1982–1984 in Zaire found a case-fatality rate of 11% among those who had not been vaccinated against smallpox, a secondary attack rate of 15.7% among unvaccinated household contacts, and 0.6% among the vaccinated. In only one household there were four probable successive person-to-person transmissions. This disease is not considered a concern as a source of possible resurgence of epidemic poxvirus disease because it has remained restricted to forest areas and successive person-to-person transmissions were so rare.

3. References

1. ALTMAN, L. K., Destruction of smallpox virus backed in WHO committee, *The New York Times*, September 10, 1994, p. 5.
1a. ALTMAN, L. K., Lab samples of smallpox win reprieve, *The New York Times*, January 19, 1995, p. A15.
1b. ALTMAN, L. K., Final stock of the smallpox virus now nearer to extinction in labs, *The New York Times*, January 25, 1996, p. A1.
1c. ALTMAN, L. K., Health group votes to kill last viruses of smallpox, *The New York Times*, May 26, 1996, p. 4, sec. 1 (final edition).
2. ANONYMOUS, WHO panel: Death penalty for smallpox, *Science* **265:**1647 (1994).
3. BRYCE, J., *Practical Observations on the Inoculation of Cow-*

pox, Pointing Out a Test of a Constitutional Affection in Those Cases in Which the Local Inflammation Is Slight and in Which No Fever Is Perceptible, William Couch, Edinburgh, 1802.

4. CENTERS FOR DISEASE CONTROL AND PREVENTION, Vaccinia (smallpox) vaccine: Recommendations of the Immunization Practices Advisory Committee (ACIP), *Morbid. Mortal. Week. Rep.* **40**(RR-14):1–10 (1991).

5. FENNER, F., HENDERSON, D. A., ARITA, I., JEZEK, Z., AND LADNYI, I. D., *Smallpox and Its Eradication*, pp. 1097–1099, World Health Organization, Geneva, 1988.

6. FENNER, F., HENDERSON, D. A., ARITA, I., JEZEK, Z., AND LADNYI, I. D., *Smallpox and Its Eradication*, pp. 1287–1319, World Health Organization, Geneva, 1988.

7. HAMILTON, A., KINCHINGTON, D., GREENAWAY, P. J., DUMBEL, K., Recombinant bacterial plasmids containing inserts of variola DNA, *Lancet* **2**:1356–1357 (1985).

8. JOKLIK, W. K., MOSS, B., FIELDS, B. N., BISHOP, D. H. L., AND SANDAKHCHIEV, L. S., Why the smallpox virus should not be destroyed, *Science* **262**:1225–1226 (1993).

9. MAHY, B. W. J., ALMOND, J. W., BERNS, K. I., CHANOCK, R. M., LVOV, D. K., PETTERSSON, R. F., SCHATZMAYR, H. G., AND FENNER, F., The remaining stocks of smallpox virus should be destroyed, *Science* **262**:1223–1224 (1993).

9a. MAURICE, J., Virus wins stay of execution, *Science* **267**:450 (1995).

10. NEERGAARD, L., Smallpox virus vials prompt moral debate, *San Diego Union-Tribune*, June 20, 1993, page A-10.

11. WORLD HEALTH ORGANIZATION, Orthopoxvirus surveillance: Post-smallpox eradication policy, *Week. Epidemiol. Rec.* **58**:149–154 (1983).

12. WORLD HEALTH ORGANIZATION, Committee on Orthopoxvirus Infections: Report of the fourth meeting, *Week. Epidemiol. Rec.* **61**:289–293 (1986).

13. WORLD HEALTH ORGANIZATION, Post-smallpox eradication policies, *Week. Epidemiol. Rec.* **67**:3–4 (1992).

CHAPTER 29

Varicella–Herpes Zoster Virus

Thomas H. Weller

1. Introduction

1.1. Definition

Varicella–zoster virus (*Herpesvirus varicellae*), commonly abbreviated to "V-Z virus" or "VZV," is the etiologic agent of two diseases of man, varicella and herpes zoster. Varicella (chickenpox) is a ubiquitous, contagious, generalized exanthematous disease of seasonally epidemic propensities that follows primary exposure of a susceptible person, most often a child. Herpes zoster (shingles) is an endemic sporadic disease, most frequent in elderly people, characterized by the appearance of a unilateral, painful, vesicular eruption localized to the dermatome innervated by a specific dorsal root or extramedullary cranial ganglion. In contrast to varicella, which follows primary exogenous contact with the causative virus, zoster reflects endogenous activation of a VZV infection that has survived in latent form following an attack of varicella. The two clinical entities are not as distinct as is customarily assumed. The patient with zoster frequently develops a disseminated varicelliform eruption; rarely, the person with varicella may exhibit a zosteriform concentration of lesions.

The human herpesviruses are grouped into three subfamilies: the *Alphaherpesvirinae*, the *Betaherpesvirinae*, and the *Gammaherpesvirinae*.[201] VZV along with herpesvirus types 1 and 2 (HSV 1 and 2) constitute the *Alphaherpesvirinae*, all characterized by a cytolytic cytopathology and latent neuronal infection. While VZV is considered to be a single type, on restriction endonuclease analysis epidemiologically unrelated strains show genomic differences.[88,122,174] The development of knowledge of VZV has been reviewed.[193]

1.2. Social Significance

Until the recent approval of a varicella vaccine, chickenpox stood alone as the nonvaccine-preventable classic childhood viral disease. In the United States, by adulthood nearly all individuals have experienced an attack of varicella; the annual incidence of varicella therefore approximates the birth cohort. It is estimated that in 1994 there were 4 million live births, indicating that some 3.9 million cases of varicella occurred. The population-based study of zoster in Minnesota[149] revealed an annual incidence rate of 1.3 per thousand; based on an estimated U.S. population in 1994 of 261.7 million, some 340,000 cases of zoster occur annually.

In 1985, the economic burden in the United States of illness caused by VZV was analyzed.[93] The yearly cost of cases of varicella in immunocompetent individuals was estimated to be $77.6 million and the cost of 303,300 cases of zoster was estimated to be $126 million. The analysis also considered immunoincompetent, high-risk individuals, i.e., those with lymphomas or leukemia plus the iatrogenically suppressed recipients of organ transplants; the estimated annual cost of this group was an additional $69.2 million. Since 1985, the use of therapeutic modalities has increased exponentially and biological immunosuppression has further augmented the size of the high-risk group. Prevention and control of nosocomial VZV infections in groups of high-risk patients are now a major economic consideration. For example, varicella prevention in one community hospital for 1986 cost $55,943.[189] The current economic burden imposed by VZV has not been studied; it would not be surprising to find that it now approaches a billion dollars a year.

Thomas H. Weller • Department of Tropical Public Health, Harvard School of Public Health, Boston, Massachusetts 02115. *Present address*: 56 Winding River Road, Needham, Massachusetts 02192.

2. Historical Background

2.1. Clinical Recognition

Herpes zoster was described in premedieval times. Varicella, however, was not differentiated from smallpox (variola) until the end of the 19th century. It is of interest that Osler[136] in 1892 deemed it necessary to emphasize that "there can be no question that varicella is an affection quite distinct from variola and without at present any relation whatever to it." Although the global eradication of smallpox was certified in 1979, differential diagnosis continues to pose problems. Investigation of cases of variola reported since 1979 has revealed that the majority have in fact been varicella.

The infectious nature of varicella was demonstrated in 1875 by Steiner,[169] who induced the disease in volunteers by inoculating them with vesicular fluid from patients with chickenpox. Herpes zoster was experimentally transmitted in similar fashion by Kundratitz[103] in 1925, with the production of varicelliform cutaneous lesions.

2.2. Association of Varicella with Herpes Zoster

Interest in the relationship between varicella and herpes zoster dates from the clinical observations of von Bokay[184] in 1888 that susceptible children acquired varicella after contact with persons with herpes zoster. (Subsequently, this association was repeatedly observed; the report of the School Epidemics Committee of Great Britain[160] in 1938, for example, linked 18 outbreaks of varicella to an exposure to zoster.) Additional evidence in support of the monistic etiologic concept of the two clinical entities slowly accrued. Tyzzer[180] in 1906 described superbly the histopathology of the cutaneous lesion in varicella, with its characteristic intranuclear inclusion-containing multinucleate giant cells. An identical histopathologic picture for the skin lesion of zoster was recorded by Lipschütz[110] in 1921. Technically difficult studies in the 1930s by several workers who employed vesicular-fluid material as antigen suggested an immunologic relationship. In the early 1940s, Zinsser[215] and Sabin[156] alluded to the probable close relationship of the two etiologic agents; emphasis at that period focused on the theoretical existence of strains of virus that possessed differing dermatropic or neurotropic affinities. Garland[56] suggested in 1943 that zoster reflected activation of a latent varicella virus, a mechanism similar to that obtaining with the virus of herpes simplex, and he should be credited with first expressing this now generally accepted thesis.

2.3. Isolation and Propagation of the Etiologic Agent of Varicella–Zoster

Substantiation of the view that varicella and herpes zoster have a common etiology followed the isolation and propagation *in vitro* by us of the etiologic virus.[191,195] By 1958, we[194,196,197] had completed various investigations on agents recovered from patients with varicella and with herpes zoster. These studies revealed no differences in the biological or immunologic attributes of viruses isolated from patients with the two clinical entities. We therefore referred to the agent as *varicella–zoster (V-Z) virus* and concluded that "the accumulation of epidemiologic and laboratory evidence in support of the hypothesis that a single etiologic agent is responsible for varicella and herpes zoster appears so impressive that the burden of proof must now logically shift to those who desire to refute the monistic concept."[196]

The concept that cases of herpes zoster reflect activation of a preexisting latent VZV is now generally accepted.

Parenthetically, confusion with HSV is to be avoided; although VZV and HSV are both members of the herpes group of viruses, the relationship is distant, and the agents are distinct.

3. Methodology Involved in Epidemiologic Analysis

3.1. Sources of Mortality Data

In the United States, deaths attributable to chickenpox and to herpes zoster are coded separately and are recorded annually in *Vital Statistics of the United States*, National Center for Health Statistics. Deaths from varicella, most often resulting from a specific pneumonia or a diffuse hemorrhagic process, have a high probability of accurate attribution; however, those resulting from a postinfectious varicella encephalitis are subject to errors of diagnosis.

The trend, noted in the second edition of this book, of a decrease in deaths attributable to varicella and an increase in deaths attributable to zoster continues. In 1972, there were 122 deaths from varicella and 86 from zoster; for 1981, the figures, respectively, were 84 and 174. For the 10-year period ending in 1981, 1000 varicella deaths and 1308 zoster deaths were recorded.[32] Data are available for the 9 years 1982–1990; there were 667 deaths from varicella and 1064 deaths from zoster. In England and Wales it is estimated that zoster causes over 100 deaths a year.[131]

3.2. Sources of Morbidity Data

3.2.1. Reporting of Cases

3.2.1a. Requirement for Notification in the United States. Varicella prior to 1972 was not reportable nationally in the United States. Although chickenpox has been notifiable in certain states for many years—for example, since 1910 in Massachusetts—in 23 states the disease was not notifiable in 1993.[33] Herpes zoster is generally not reportable.

3.2.1b. Inadequacies of Notification. Only a fraction of the cases of varicella occurring in the United States are reported. In 1936, 200,000 families in 28 cities in the United States were surveyed to establish the occurrence of communicable diseases; the notification rate of chickenpox cases was estimated to be 20–40% of the actual number.[37] In 1947, Feemster, as cited by Gordon,[74] concluded that only 25% of the cases in Massachusetts were reported. This comparatively high rate has not been maintained as interest in infectious diseases has waned.

In 1979, an estimate from the Centers for Disease Control indicated that only 6% of all cases were reported.[148] In 1993, 134,722 cases of varicella were recorded[33]; if the current assumption that the actual number of cases per year is now of the order of 3.9 million is correct, then the reporting system is becoming progressively inadequate. That the current situation is unsatisfactory and extremely variable from state to state is shown by the fact that in 1993, six states reported fewer than 200 cases.[33]

3.2.1c. Occurrence of Inapparent Infections. In contrast to an infection such as mumps, most cases of varicella are clinically apparent. However, in a fraction of infected persons, probably less than 5%, the exanthem is so sparse and transient that its occurrence, especially in the dark-skinned patient, may be missed by the qualified observer. From an analysis of secondary and tertiary attack rates after home exposure, Ross[153] concluded that no more than 4% of cases are unrecognized. The visibility of the disease is indicated by the study by Ross of the number of pocks per child as established by a single home visit on the sixth day to the ninth day of illness. In four groups, mean counts of pocks per child varied from 207 to 510, with a range of 10 to 1968.

Although Ross's conclusions generally obtain, exceptions occur in situations where passive or active immunity plays a role. It has been shown that infants are uniformly susceptible to infection but, if passively protected by maternal antibody, may have few or no pocks.[12] In subjects who have had varicella in the past, exposure to VZV may initiate a second infection; an antibody response without symptoms is common,[5] but rarely a clinically apparent infection may ensue.[66,193]

Herpes zoster, reflecting endogenous viral reactivation, may be covert and be revealed only by demonstration of an antibody response, or may present as pain with or without an overt segmental cutaneous eruption.

3.2.2. Clinical Observation and Retrospective Histories.
Data on the communicability of varicella derive primarily from the careful study of exposed susceptible children in households, in schools, or on hospital wards. The accuracy of a negative history of varicella depends on the informant and on the time span since the period of maximum risk. The parent of the young child is a better informant than is the subject at an older age. However, in a large-scale retrospective study on the occurrence of chickenpox in the United States,[37] it was concluded that the "forgetting factor" became important for subjects as young as 9 years of age, a period when the parent would be the major informant. Inability to remember past events certainly accounts for the low exposure attack rate of 12% observed by Hope-Simpson[90] in intimately exposed "susceptibles" aged 15 years or more. Ross[153] reported an attack rate of 5% in "susceptible" parents in contrast to a rate of 87% in "susceptible" children. Conversely, the attack rate in children with "positive" histories was 7%. In a serological investigation of children with "negative" histories, one third of those tested were reactive with a VZV antigen.[65] Recent studies of three groups of hospital personnel with no history of varicella revealed that 61%,[89] 82%,[128] and 100%[111] were seropositive.

3.3. Serological Surveys

3.3.1. Methodology.
The two classic serological tests for VZV infection, i.e., complement fixation (CF) and neutralization (N) now have been supplanted. Antibody detected by CF is transient and there may be cross-reactions with HSV. The N requires materials usually not available in the diagnostic laboratory. Various hemagglutination tests have been developed. Comment is here limited to two procedures that have proved useful in serological surveys.

One uses infected cells as antigen with demonstration of antibody by immunofluorescence [indirect fluorescent antibody (IFA), FAMA] and the other is an enzyme-linked immunosorbent assay procedure (ELISA) with commercially available antigens.

The FAMA test[62,204] is generally accepted as a specific and sensitive indicator of VZV immune status. In brief, cells from a VZV-infected culture are harvested and frozen for future use. Suitable dilutions of cells and of test serum are reacted for 30 minutes in wells in microtiter U-plates. After suitable washings with intervening centrifugation, goat antihuman γ-globulin fluorescein conjugate is added and reacted for 30 minutes. After additional washings and centrifugation, the cells are suspended in glycerol–saline, drops transferred to a slide, and a coverslip applied. If positive, cells are brightly outlined when visualized by fluorescence microscopy. One modification of the FAMA method avoids the hazards of infectious material by using glutaraldehyde-fixed VZV-infected cells.[213] Another approach eliminates the use of a fluorescence microscope by employing a peroxidase conjugate to visualize VZV antigen in air-dried infected cells.[84]

With the increasing availability of automated equipment for the performance of ELISA and the commercial availability of varicella antigens, it is probable that ELISA will become the standard technique for serological surveys. Studies have indicated that ELISA[50,54,61,72,162] yields results comparable to FAMA. A comparative study of ELISA results with FAMA results showed a sensitivity for ELISA of 86%, a specificity of 99%, a predictive value of a positive test of 99%, and a predictive value for a negative test of 82%.[40]

Radioimmunoassay procedures are equal or superior in sensitivity to other techniques but require special equipment and radioisotopes.[29,208]

3.3.2. Limitations of the Seroepidemiologic Approach. Two problems, sensitivity and specificity, deserve consideration in the interpretation of seroepidemiologic data. Of the commonly used tests, CF is the least sensitive. Within months after a primary attack of VZV, titers of antibody detectable by CF begin to fall and may reach subdetectable levels. The problem of specificity relates to nonspecific rises in titer of VZV antibodies that occur in the person infected with HSV, and vice versa.[99,154] However, heterologous responses, as, for example, a rise in VZV titer in a patient infected with HSV, occur only when the subject has previously been infected with the heterologous agent[158,159]; in each instance, the rise in titer to the current agent greatly exceeds that to the heterologous virus. In general, the more sensitive the serological test, the greater the probability of the detection of low-order heterotypic rises.[50] Demonstration on molecular analysis that a VZV and a HSV envelope glycoprotein share antigenic epitopes provides an explanation for the observed serological cross-reactivity.[44] Although VZV–HSV cross-reactions pose diagnostic problems, the problems are not important in seroepidemiologic surveys.

3.4. Laboratory Methods

3.4.1. Isolation of Virus. The definitive diagnosis of varicella or herpes zoster is based on the isolation of the etiologic agent. Since common laboratory animals are not overtly susceptible, cultures of susceptible cells must be used. Although VZV exhibits less host specificity *in vitro* than *in vivo*, cultures of human cells are the most sensitive indicators of VZV; a further consideration is that confluent, transparent, well-organized sheets of cells provide optimal conditions for microscopic detection of the focal lesions induced by VZV. These criteria are satisfied by actively growing cultures of fibroblasts, kidney cells, or amnion cells, all of human origin. One of these culture systems is usually employed for the isolation of VZV.

Varicella–zoster virus can be recovered with ease by inoculation of cultures of human cells with the aspirated contents of vesicular lesions. It is to be noted that swabbing of vesicles to collect fluid is undesirable; swabs have been shown to reduce viral titers.[109] Selection of the young "dew-drop" lesion containing clear fluid is important. Chances of recovery of virus decrease as the lesion becomes pustular, and VZV, in contrast to variola virus, cannot be isolated from crusts or scabs. Virus can be readily isolated during the first 3 days of the exanthem in varicella; isolation of virus thereafter is rare unless the disease is atypical, as in the immunologically compromised host. Virus can be recovered from the cutaneous lesions of patients with zoster for a longer period of time, i.e., at times to the seventh day or later after the appearance of the segmental vesiculopustular process.

The focal cytopathic process produced by VZV in cultures of susceptible cells has to be differentiated from the transiently focal lesions of the virus of herpes simplex or the more persistent focal lesions induced by the human cytomegaloviruses (CMVs). The VZV focal lesion in a sheet of susceptible cells may be apparent as early as 2 days after inoculation or as late as 3–4 weeks. Microplaques develop, observable under 100× magnification as consisting of a few small, refractile cells. Each microplaque enlarges slowly by involvement of contiguous cells, with progressive degeneration of the older central portion. Very little infectious virus is released from the infected cells, and the number of foci increases slowly. However, peripheral extension of the process continues, with eventual involvement of most of the cell sheet. Stained preparations reveal intranuclear inclusions in involved cells, particularly at the marginal plaque interface.

Although the focal lesions produced on inoculation of cell cultures appear in the original passage, for reasons unknown, to demonstrate viremia by recovery of VZV from circulating white cells, blind subcultures are necessary.[137] Confirmation of identity of the isolate, and in particular differentiation from CMV, can be accomplished immunologically by use of specific VZV antisera in direct fluorescent antibody (FA) tests or by performance of CF or other serological tests in which the antigen derives from the infected culture. For technical details, the reader is referred to a standard reference on diagnostic procedures.[187]

3.4.2. Rapid Diagnostic Approaches in Urgent Clinical Situations. The diagnosis of the patient presenting with a vesiculopustular eruption requires that varicella and varicelliform eruptions caused by HSV be differentiated from those caused by variola–vaccinia viruses. Simple procedures can be applied to differentiate the two groups of etiologic agents. Examination of stained smears from the base of a fresh lesion, or of a biopsy thereof, will, if the patient is infected with VZV or HSV, reveal multinucleate giant cells and other cells with intranuclear inclusions as originally described by Tyzzer[180]; neither giant cells nor intranuclear inclusions are present in the lesions of variola and vaccinia. (This technique is usually referred to as the Tzanck test.)

If viral isolation is not feasible, VZV can be differentiated from HSV immunologically by examination of material from the cutaneous lesions. Antigenically reaction material has been demonstrated in extracts of crusts of involuting lesions as late as 14 days after onset of varicella and 23 days after the onset of zoster.[181] Antibody for this purpose has in the past derived from selected human VZV convalescent sera known to be devoid of HSV reactivity or from sera from immunized animals. Monoclonal antibodies against VZV-specified proteins or glycoproteins are now commercially available; this advance should simplify and expedite the investigation of lesional materials.[51,76,190]

Various techniques can be applied to identify VZV antigens in samples of vesicle fluid or extracts of crusts. With countercurrent immunoelectrophoresis, results may be obtained in 2 hr.[53] Equally rapid results may be obtained with direct or indirect immunofluorescence or immunoperoxidase methods.[214] A different approach to rapid diagnosis useful in the acute phase of illness is the demonstration of the VZV-specified enzyme deoxythymidine kinase in vesicle fluids or in sera.[98]

The introduction of the polymerase chain reaction (PCR) to detect specific VZV DNA in clinical materials provides a rapid approach to diagnosis. The method is applicable to both fresh and stored material, including material from involuting crusted cutaneous lesions.[41,100,132]

In the prevention of nosocomial varicella, rapid determination of the immune status of contacts is essential. Gershon and co-workers[62a] have recently reported that a commercially available latex agglutination test is better than FAMA for this purpose, being sensitive, rapid, simple, and relatively inexpensive. In practice, examination of several dilutions of the serum under examination is essential to avoid a false-negative prozone phenomenon.

4. Biological Characteristics of the Virus that Affect the Epidemiologic Pattern

4.1. Latency in the Human Host: Primary Infection, Latency, and Reactivation

The VZV shares with other herpesviruses the capacity to persist in the body after the primary infection. Decades later, the virus may again become manifest with renewed replication and the production of the clinical syndrome of herpes zoster. Immunologically inexperienced persons in intimate contact with the patient with herpes zoster can contract varicella. Thus, VZV is epidemiologically unique in that it can persist for years in a mobile human incubator, roam globally, and when host defense mechanisms decay again, replicate and initiate an outbreak of the primary disease.

It is now apparent that viral–host interactions may continue following a primary attack of varicella. Reinfection often follows exposure to heterologous strains of VZV, usually without symptoms and demonstrable only on detection of a rise in titer of VZV antibodies.[5] Infrequently, such an exposure may produce overt disease, i.e., a second attack of varicella.[5,43,66,179,193] Of particular interest is the recent report of the occurrence of varicella during pregnancy in women seropositive for VZV.[122a] Genomic differences between varicella isolates have been demonstrated by restriction-endonuclease analysis,[173] providing a useful molecular epidemiologic tool; each epidemiologically distinct isolate proved to be unique. It is probable that antigenic diversity and host immune status together determine whether reinfection is an overt or a covert process.

The demonstration of rising levels of the IgA VZV antibody in the absence of symptoms in a population of older adults suggests that in any case endogenous viral replication often resumes as host defenses wane. The duration of the asymptomatic containment period is variable. I have depicted the dynamic nature of the natural history of infection with VZV[193] (see Fig. 1).

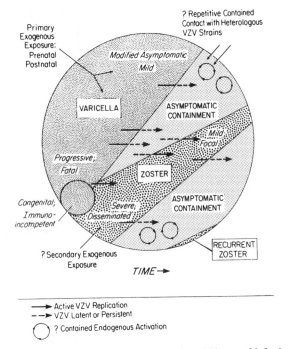

Figure 1. Diagrammatic summary of the natural history of infection with varicella–zoster virus. Two variables are depicted: time and clinical severity. Time is depicted on the horizontal and severity on the vertical plane. In the competent host the containment period is of the order of decades; however, in the immunoincompetent person the two clinical processes may merge without an intervening asymptomatic interval. The containment period after congenital varicella is typically of short duration. During the asymptomatic containment period, episodes of endogenous viral replication probably occur, and contact with heterologous exogenous strains may stimulate host defenses, usually in the absence of overt disease. From Weller,[193] reprinted with permission.

4.2. Failure of Varicella–Zoster Virus to Persist in Scabs or Fomites: Limited Period of Communicability

In contrast to the variola–vaccinia viruses, which survive for long periods in scabs, VZV cannot be recovered from crusts or from the involuting lesions. The period of communicability is therefore discrete and coincident with the duration of vesicular lesions. Room dust is not infectious, and terminal disinfection is not indicated.

5. Descriptive Epidemiology

5.1. Incidence and Prevalence Data

5.1.1. General Comments on Varicella. In modern industrialized societies in the temperate areas of the world, almost all persons contact varicella, usually during childhood. In 1993,[33] 134,722 cases of varicella were reported in the United States. Age was recorded for about one fourth of these cases; 94% occurred under age 15. Figure 2 summarizes the cases reported annually per 100,000 population in Massachusetts from 1910 through 1992; these data reflect an unknown degree of underreporting.

Varicella is a benign disease in the immunologically uncompromised child in the United States; the few deaths that do occur usually reflect illness in the infant or the adult. However, the category of high-risk individuals, wherein varicella may be lethal, continues to increase in size as immunosuppressive measures are employed therapeutically or to enhance acceptance of an organ transplant, as the life expectancy of patients with malignancies increases, and as the prevalence of human-immunodeficiency-virus (HIV)-induced immunodeficiency increases. Yet as indicated in Table 1, which summarizes mortality data by decades for Massachusetts, the downtrend in varicella-associated deaths continues—a tribute to therapeutic and preventive advances.

In tropical regions and the less-developed regions of the world, the situation with respect to varicella is less clear. Morbidity and mortality statistics are essentially nonexistent. However, as noted below, fragmentary evidence suggests that climatic factors combine to inhibit transmission and to extend susceptibility into the older age groups as compared with industrialized societies.

5.1.2. General Comments on Herpes Zoster. Herpes zoster is a sporadic endemic disease reflecting

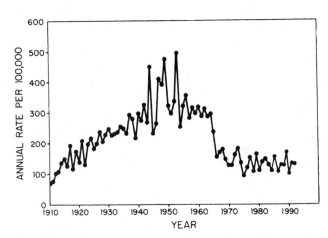

Figure 2. Reported cases of chickenpox per 100,000 population per year in Massachusetts from 1910 through 1992. Source: Dr. George Grady, formerly Assistant Commissioner for Laboratories and Communicable Disease Control, and Dr. Susan Lett, Bureau of Communicable Disease Control, Massachusetts Department of Public Health.

Table 1. Deaths from Chickenpox and Mean Annual Death Rate per 100,000 Population, by Decades, Massachusetts, 1910–1979[a]

Years	Mid-decade population	Deaths	Mean annual death rate
1910–1919	3,701,000	48	0.13
1920–1929	4,158,000	83	0.20
1930–1939	4,361,000	69	0.16
1940–1949	4,511,000	39	0.08
1950–1959	4,853,000	30	0.06
1960–1969	5,343,255	22	0.04
1970–1979	5,789,478	17	0.03

[a]Source: Dr. George Grady, formerly Assistant Commissioner for Laboratories and Communicable Disease Control, Massachusetts Department of Public Health.

reactivation of a latent host-contained infection with VZV, with incidence determined by factors influencing the host–parasite relationship. The primary determinant of prevalence is the age composition of the population, with particular reference to time elapsed since acquisition of the primary VZV infection. In a panel practice of some 3500 persons in England observed over a period of 16 years by Hope-Simpson,[91] herpes zoster occurred at an annual rate of 3.4 per 1000 persons. Under similar circumstances in a panel practice in Scotland, McGregor[123] recorded an annual rate of 4.8 per 1000.

The first population-based study of herpes zoster has been reported from Rochester, Minnesota.[149] Between 1945 and 1959, 590 residents experienced zoster, an average annual incidence rate of 1.3 per 1000 person-years. The rate increased with age; for those over 75 years the rate was 4 per 1000 person-years. The rate observed in Minnesota is similar to that recently reported from England where over a 9-year period the annual rate varied between 1.3 and 1.6 per 1000 population.[70] The age-related activity of VZV in a population in a temperate region is presented graphically in Fig. 3.

5.1.3. Risk of Infection in Susceptible Persons. Although varicella has a justifiable reputation as a highly contagious disease, intimate rather than casual contact is required for a high transmission rate. In an analysis based on reported cases of varicella in New York City, the infectivity of chickenpox in households and in society as expressed as annual exposures divided by average number of susceptibles was 0.61 and 0.12, respectively.[209] In comparison with measles (rubeola), which was arbitrarily assigned an infectivity potential of 100%, chickenpox was 80% as infectious in households and 46% as infectious in society.

The most accurate data on the infectivity of varicella derive from experimental studies on attempts to modify varicella in which carefully observed control populations of susceptible children were at risk. In such a study, Ross[153] observed a secondary attack rate of 87% in susceptible siblings in households following the introduction of the primary case of varicella; a tertiary attack rate of 71% was observed among those siblings (i.e., 13% of the original group) who had escaped infection on exposure to the primary household contact.

Hope-Simpson[90] summarized data on the infectiousness of varicella in a semirural area in Gloucestershire, England; the attack rate among exposed susceptibles, aged 0–15 years, in homes was 61%. In a hospital environment, Gordon[74] cites an experience with a dozen outbreaks of varicella on the wards in which the attack rate among 81 susceptibles was 68%.

The patient with herpes zoster is also infectious. Anecdotal evidence suggests that the patient with zoster poses less of a risk than does the patient with varicella. It has been postulated that the zoster patient is less infectious because the lesions are circumscribed and often covered by clothing. The validity of this thesis is mitigated by the fact that approximately one third of patients with zoster exhibit disseminated cutaneous lesions, indicating the occurrence of a viremia.[135] Perhaps of more import is the fact that patients with zoster often possess significant titers of specific antibody at the time the exanthem is fully developed[196]; therefore, partial neutralization of free infectious virus in the evolving lesion may occur *in vivo* before it is released into the environment.

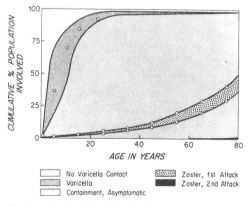

Figure 3. Age-related activity of varicella-zoster virus in a population in a temperate region. Data on varicella are from Preblud in the United States,[146] and data on zoster are from Hope-Simpson[91] in England. From Weller,[193] reprinted with permission.

5.2. Epidemic Behavior

5.2.1. Varicella. In industrialized societies in temperate regions, varicella is endemic in the total population but epidemic in susceptible clustered subgroups and exhibits a characteristic seasonal fluctuation in incidence. Data from Massachusetts reveal a pattern of low levels of endemicity at the beginning of the 80-year period that started in 1910 (Fig. 2). Gordon,[74] who examined the data from Massachusetts in 1962, concluded that the pattern to that time was one of mildly fluctuating increasing endemicity, with superimposed recurring epidemics about equally divided between outbreaks at intervals of 2 years and 3 years; two had an interval of 4 years. The increase in annual rates from 1910 to the 1940s was attributed by Gordon to a gradual improvement in reporting.

With data since 1962 now available, in retrospect it appears that the peak rates of over 400/100,000 in 1944, 1947, 1949, and 1953 in Massachusetts, with intervening extreme fluctuations, reflect the dislocations of the population in the war years followed by postwar spurts in the birth rate. A similar but less marked period of fluctuation in annual incidence is reflected by the data for the period of World War I and immediately thereafter. Thus, except for the war periods, reported rates reflect a relatively smooth increase in annual incidence between 1910 and 1940, little influenced by epidemics, with a tendency to plateau at a level of 300/100,000 between 1940 and 1964. Between 1964 and 1966, rates in Massachusetts fell abruptly to the current level of under 200/100,000 per year. For the 8-year period 1985–1992, the average rate was 131/100,000 per year. Thus, varicella in Massachusetts is at its lowest ebb in half a century, a phenomenon not unique to this state, for data from New York City reveal a parallel decrease in annual incidence.[209]

The apparent rise and fall in the level of endemicity of varicella in Massachusetts and in New York is of interest. Much of the decline in prevalence observed since 1965 is probably artifactual due to deterioration of the reporting process. This explanation is supported by authorities in New York City. However, as noted below, current low levels of prevalence are paralleled by an altered pattern of seasonal incidence and, to a lesser extent, of age-specific attack rates; therefore, the cyclical phenomenon is not solely an artifact of reporting. It is likely that the recent decrease in prevalence also reflects a declining birth rate and a decrease in the relative number of susceptible children. Other hypotheses deserve examination. Has, for example, some subtle shift in the host–parasite relationship occurred? Is the disease now associated with fewer pocks and thus more often overlooked,

perhaps because the average child is better nourished or is less likely to experience another concurrent infectious process? Has the virus become attenuated in nature? These questions are unanswerable at present. Time will reveal whether current low levels of prevalence will persist or whether rates will again gradually rise and then fall, indicating the existence of long-term cyclic fluctuations in the prevalence of varicella heretofore unrecognized.

In contrast to the endemic nature of varicella in the population at large, epidemics follow introduction of varicella into intimately associated subgroups or clusters of susceptibles. Typical is the school situation described by Wells and Holla[198]; of 67 susceptible children in grades 0–4, 61 contracted the disease in an explosive outbreak. Within elements of larger population groups, as in different boroughs in New York City, levels of endemicity vary annually, reflecting the varying summation of epidemic outbreaks in clusters of susceptibles[113] within each borough.

5.2.2. Herpes Zoster. Herpes zoster is an endemic disease that shows no seasonal pattern and appears in "epidemic proportions" only when induced iatrogenically in specialized groups of patients such as those under chemotherapy because of malignancy. In this instance, the situation usually reflects chance concurrent reactivation in each patient and not person-to-person transmission. However, in the event of the appearance of a cluster of cases of "disseminated zoster" in a clinical setting in elderly immunodepleted individuals, and in the light of current knowledge of the heterogeneity of the VZV genome, the possibility of reinfection must be considered.[141]

5.3. Geographic Distribution

5.3.1. Varicella. Varicella and herpes zoster occur throughout the world. There is evidence that in tropical regions, as contrasted to temperate regions, varicella is seen more often in adults. Annual rates for varicella in the United States Army in World War II in the Latin American area, where a large number of soldiers from Puerto Rico were stationed, were 1.41 and 2.27 per 1000, respectively, in 1944 and 1945; the comparable figures for the continental United States were 0.71 and 0.61 for the same years.[171] Brunell[24] notes that a majority of pregnant women with varicella observed in New York are of Caribbean origin; of a group studied by Siegel,[164] 52% were Puerto Ricans. In Sri Lanka, varicella was observed to be at least as common in adults as in children.[121] In an Indian village observed over a period of 5 years, 63% of cases of varicella occurred in persons over the age of 15 years.[166] Investigation of cases of varicella in 1976 in the State of Kerala, India, carried out in the terminal phase of the smallpox eradication program, revealed that 50% of cases occurred in those 15 years of age or older.[199]

Recent observations regarding the tendency for varicella to be an adult disease in the tropics are striking. An outbreak occurred in Denmark in a group of 256 immigrants who were Tamil refugees from Sri Lanka.[101] Within the first few months after arrival, 44% developed varicella, representing 38% of the adults and 68% of the children.

Two outbreaks of varicella occurred among soldiers from Puerto Rico sent to San Antonio, Texas, for training.[114] The attack rate in one outbreak was 30%, and serological testing of 810 adult Puerto Rican recruits for VZV antibody indicated that 42% were seronegative. Investigation of the age distribution of antibodies against VZV in sera from 1810 people from the Caribbean island of St. Lucia revealed that less than 10% of the population had experienced a VZV infection before the age of 15 years.[56a]

Three hypotheses have been proposed to explain the different age-specific incidence of varicella in the tropics. Transmission may be slowed by the relative isolation of familial clusters in rural areas.[199] It has been suggested that there may be "epidemiologic interference" with other prevalent viruses, especially HSV.[166] Probably of equal or greater import is a decreased transmission potential because of ambient temperatures that enhance the lability of VZV outside the human host.

As emphasized by Black *et al.*[18] VZV can persist because of its latent capacity (i.e., zoster) on introduction into a geographically isolated, small human population; a single isolated Indian tribe in the Amazon, when surveyed serologically, was found to have levels of reactivity comparable to those in the United States.

5.3.2. Herpes Zoster. Available data do not suggest geographic differences in the distribution of herpes zoster. If, however, varicella is acquired at a higher mean age in tropical areas, herpes zoster may also occur at a relatively older age. In Kerala, India, herpes zoster is said to be almost unknown.[199]

5.4. Temporal Distribution

5.4.1. Varicella. In the United States, varicella regularly shows a striking seasonal distribution. At the national level the incidence typically peaks between March and May with annual lows in September and October. Reported cases by month in the United States for the period 1986–1993 are depicted in Fig. 4. [In evaluating data on temporal distribution in terms of date of transmission, corrective factors for the duration of the incubation period (10–21 days) and for the lag in reporting must be considered.]

A similar seasonal distribution of cases occurs in Massachusetts, but analysis indicates that peak incidence is modified slightly by the level of prevalence. Gordon[74] analyzed data from Massachusetts for 1956–1961, a period of high prevalence; the monthly peak came in March. Examination of data for 1968–1973, a period of lower prevalence, shows the monthly peak to be in May. These

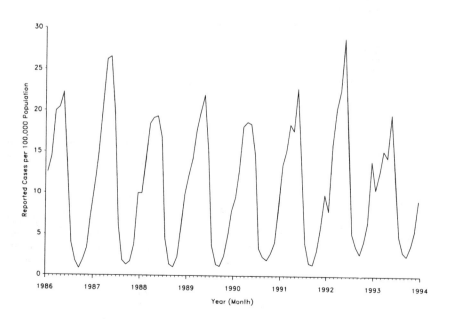

Figure 4. Reported cases of varicella per 100,000 population per month, United States, 1986–1993. Source: Centers for Disease Control.[33]

observations suggest that transmission differs seasonally in years of high incidence as contrasted to periods of low incidence. In years of high incidence, a greater percentage of the pool of susceptibles is infected earlier in the transmission season. During periods of low incidence, there is a temporal lag in progression of varicella through a community with extension of the seasonal epidemic into the late spring. A similar conclusion was reached by Yorke and London,[209] who calculated that in New York City, the mean monthly contact rates for 14 years of high incidence are about 5% higher in September and October ($P < 0.01$) and about 4% lower in May and June ($P < 0.05$) than the mean monthly contact rates for 16 years of low incidence. Using the seasonal data on varicella in New York City, we have summarized the mean monthly cases reported for two 5-year periods, one of high and one of low incidence (see Fig. 5). For the periods of low incidence in New York City, the seasonal pattern of gradually increasing rates that peak in the late spring is even more striking than in Massachusetts.

Varicella in the United States is primarily a disease of winter and spring. The customary explanation of the seasonal pattern relates to aggregation of susceptible children in schools in the fall, the introduction of the agent, and its subsequent dissemination to contacts in the classroom and to susceptible siblings in the home. This epidemiologic pattern is dominant in the United States. It is questionable whether physical clustering of susceptibles should be considered the sole factor influencing transmission. Do similar patterns obtain when exposed susceptibles cluster under climatic conditions that are less favorable to the survival of VZV in the environment, as, for example, in summer schools and camps?

5.4.2. Herpes Zoster. Herpes zoster has no temporal proclivities, appearing sporadically throughout the year and independently of the incidence of varicella,[91,149] although recent evidence suggests that a high intensity of varicella transmission may suppress viral reactivation, i.e., zoster, possibly by enhancing immunity in older adults.[58]

5.5. Age

5.5.1. Varicella. Varicella, in industrialized societies in temperate climates, is primarily a disease of childhood. In Massachusetts, 60–65% of reported cases are in the 5- to 9-year age group, and 20–25% of cases occur in children under 5 years of age (see Fig. 6). In the United States, varicella can be considered as an occupational hazard of nursery and primary school attendance that few children escape. Figure 6 summarizes data for Massachusetts indicating that varicella tends to be acquired at a younger age during periods of high incidence than during periods of low incidence. This trend, although not pro-

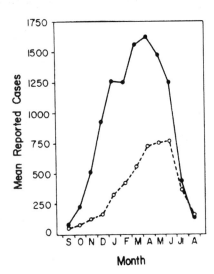

Figure 5. Reported cases of chickenpox, mean numbers by month, New York City, for two 5-year periods, one of high incidence, i.e., September 1937 to August 1942 (●), and one of low incidence, i.e., September 1966 to August 1971 (○). Prepared from tabular data appended by Yorke and London.[209]

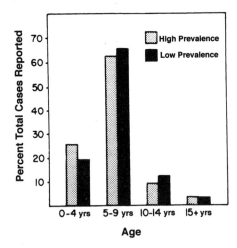

Figure 6. Distribution of reported cases of chickenpox by age in Massachusetts during a 4-year period of relatively high incidence (1962–1965; mean annual rate 281.9/100,000; total cases 59,652) and during a 4-year period of relatively low incidence (1969–1972; mean annual rate 138.9/100,000; total cases 31,164). Source: Massachusetts Department of Public Health, Division of Communicable Diseases, Dr. Nicholas J. Fiumara, former Director.

**Table 2. Age-Specific Rates per 1000 for Reported Cases
of Chickenpox in Massachusetts for 2 Years of High Incidence
and for 2 Years of Low Incidence**[a]

Year	Rate by age group					Total	
	< 5 yr	5–9 yr	10–14 yr	15–19 yr	> 20 yr	Cases	Rate
High prevalence							
1960	7.8	20.9	2.8	0.5	0.1	16,352	3.2
1962	8.0	20.4	2.6	0.4	0.1	16,359	3.1
Low prevalence							
1970	2.5	9.2	1.6	0.2	0.03	7,115	1.3
1971	3.0	8.2	1.5	0.2	0.1	6,996	1.2

[a]Source: Massachusetts Department of Public Health, Division of Communicable Diseases; Dr. Nicholas J. Fiumara, former Director.

nounced, is also apparent on examination of age-specific attack rates for 2 years of high and 2 years of low incidence (Table 2).

Recent studies indicate that in the United States and the United Kingdom varicella is being acquired more frequently at an older age. In England and Wales, cases and deaths more than doubled in adults in the 1971–1990 period.[131] This change probably reflects the immigration of susceptible adults from tropical areas (see Section 5.3.1).

Although data on age-specific risks of complications of varicella are incomplete, analysis of reports on the age-specific distribution of varicella encephalitis and of varicella-associated deaths indicates a markedly increased risk for subjects over 20 years of age and to a lesser degree for infants under 1 year old.[146]

Varicella may occur in the neonatal period as a consequence of infection acquired either *in utero* or at birth if the mother is in the acute phase of the disease shortly prior to or during labor.[129] Brunell[23] demonstrated that VZV antibody crosses the placental barrier. With a sensitive serological indicator (FAMA), the antibody titer of the infant at birth has been shown to parallel that of the mother; the level then falls, and 50% of positive infants have no demonstrable titer by 5½ months of age.[63] The nature of varicella in the near postnatal period mirrors the passively acquired immune state. In the absence of such immunity, infantile varicella may be severe and even fatal. With such protection, infants may experience a modified illness with few pocks or appear to escape completely. However, the presence of passively acquired or actively administered antibody does not prevent infection; immunologic studies of outbreaks in an institutional setting indicated that 100% of infants under 2 months of age became infected. In these infants the modulated disease resulted in reduced humoral and cellular immune re-

sponses as compared to those observed in older children.[12] Such infants rarely may experience a second attack of varicella, or, as noted below, a few may develop zoster during childhood.[193]

The age distribution of varicella in rural, underdeveloped, upland, populated areas may differ from that in industrialized countries and from that in the lowland tropics. In the course of an epidemic in a rural mountain village in Guatemala, 85% of cases were in children under 6 years of age, and varicella was common in children under 2 years of age.[157] Mention has previously been made of the frequency of varicella in adults in the lowland tropics.

5.5.2. Herpes Zoster. Herpes zoster may occur at any age in the subject who has had a prior attack of varicella but is age-related, with risk increasing sharply in the later decades of life. McGregor[123] and Hope-Simpson[91] reviewed data on age distribution based on observations in their panel practices in Scotland and in England. McGregor recorded 81 cases of zoster over a 7-year period in a practice of 2400 patients; 61 (75%) of these were in patients over 45 years of age. Hope-Simpson tabulated data on 192 cases observed over a 16-year period in a population of 3534 persons (Table 3). Incidence rose gradually to age 29, tended to plateau through the ages of 30–49, and thereafter again increased. These conclusions parallel those of the previously cited population based study of zoster in Minnesota that investigated 590 cases. Whereas the average annual incidence rate was 1.3 per 1000 person years, the rate rose to 4 per 1000 person-years in those over 75 years of age. A recent mathematical analysis of data deriving from eight published studies concluded that the observed pattern of zoster cases is explainable on an age-related increasing probability of reactivation.[57]

Table 3. Distribution by Age of 192 Cases of Herpes Zoster Observed over a 16-Year Period in a Population of 3534 People in England[a]

Age (yr)	Population	Number of cases	Rate per 1000	Annual incidence per 1000
0–9	510	6	11.8	0.74
10–19	455	10	22.0	1.38
20–29	412	17	41.3	2.58
30–39	491	18	36.6	2.29
40–49	492	23	46.7	2.92
50–59	454	37	81.5	5.09
60–69	350	38	108.6	6.79
70–79	263	27	102.7	6.42
80–89	99	16	161.6	10.10
90–99	8	0	—	—
Total or mean:	3534	192	54.3	3.39

[a]After Hope-Simpson.[91] By personal communication, Hope-Simpson states that the population in each age group represents the mean obtained from six of the annual censuses taken during the 16-year period.

Sporadic cases of zoster have been reported in the past in children. A population-based study of 173 episodes of zoster in children and adolescents now provides an epidemiologic analysis of this situation. The estimated relative risk of childhood zoster among those who had chickenpox in the first year of life was 2.9 to 20.9 times higher than among those who had chickenpox after the first year of life.[78] Childhood zoster follows primary contact with VZV *in utero* or early in life. A recent study indicated that 7 of the 12 children who developed zoster before 7 years of age had had chickenpox in the first year of life.[179a] The shortened containment period prior to reactivation of the virus may reflect the impaired response of an immature immune system or the diminished antigenic stimulus associated with a modulated mild primary infection.[193]

5.6. Sex

No convincing data suggesting differences in susceptibility by sex have been recorded for either varicella or herpes zoster. The incidence of zoster in women over 70 years of age as observed in a clinical setting is a reflection of the longer life expectancy of the female.

5.7. Race

There is no indication of differing racial susceptibilities to varicella. The previously noted occurrence of varicella pneumonia in adult black or Puerto Rican immigrants to the continental United States is explainable on the basis of diminished opportunities for infection during childhood in their countries of origin. In contrast to the situation with varicella, a recent report indicates that blacks living in North Carolina have a significantly lower risk of developing zoster than do whites.[158a]

5.8. Occupation

Since varicella is predominantly a disease of childhood in this country, no specific occupational associations have been described. For the susceptible adult in the United States, in decreasing order, parenthood, primary school teaching, and patient-associated medical professions could be considered as occupations of high risk. In Sri Lanka, where varicella in adults is common, many cases occur in hospital personnel.[121] Trauma has been described repeatedly as a precipitating factor in herpes zoster. In the Minnesota study trauma conferred a relative risk of 1.9, a risk below that of preexisting malignancy (6.1) and above that of radiation therapy (1.5).[149] Physically hazardous occupations, particularly for the elderly, probably involve an increased risk of precipitating zoster.

5.9. Occurrence of Varicella in Different Settings

5.9.1. In the Home and in Schools. As previously emphasized, varicella is highly infectious on intimate contact. Introduction of VZV into the home or into a primary school with a large percentage of susceptible children characteristically results in an epidemic, usually with exhaustion of the susceptible population by the second or third cycle of transmission.

5.9.2. In the Military. Varicella formerly was of little consequence in the military of the United States with its predominantly immune population. In World War I, there were only 1757 hospital admissions with varicella for the entire United States Army.[171] In terms of mandays lost from duty, 31,534 in all, varicella was a relatively insignificant cause; the noneffectiveness rate for measles was 62 times higher, and for mumps 129 times higher than that recorded for chickenpox. Similarly, in World War II, varicella posed no problem,[73] between 1942 and 1945, 10,664 cases were reported in the U.S. Army, an annual incidence rate of 0.42 per 1000 average strength. No deaths from varicella occurred in the army during the same period.

Currently, varicella is more important in the U.S. military, an indication of the changing composition of the armed forces. Annual hospital admissions for varicella increased more than fourfold in the army between 1980 and 1988, and more than 18-fold among navy enlisted personnel between 1975 and 1988.[75] The 1988 varicella hospitalization incidence rate was four times greater for the army and eight times greater for the navy than the estimated national rates for patients 20 years or older. While personnel from tropical areas had significantly higher risks of hospitalization, the authors raise the question of whether other causal factors are now operative in the military.

5.9.3. In Hospitals. The concentration of patients in centers for the treatment of cancer and in organ transplantation units provides a population at high risk when exposed to VZV.[43] Unavoidable exposure may follow endogenous reactivation of virus, i.e., zoster, or the introduction of a patient in the preeruptive phase of varicella. In such an event, the virus may spread to susceptible patients in an almost mysterious manner. In two such outbreaks, studies of the air-handling system implicated airborne transmission of infectious droplet nuclei to distant points.[80,107]

5.10. Socioeconomic Status

Little information has been published on the epidemiologic role of socioeconomic factors. Ström[175] carried out a retrospective analysis in Sweden of 2000 children born in 1939 and of 2000 born in 1949 in which the children were divided into three social classes. In contrast to pertussis, scarlet fever, and measles, which were less frequent in the 1949 group, varicella had occurred in the 1939 and 1949 groups with equal frequency (67 and 69% by history) by age 12, peaking at 5–6 years of age. No differences in attack rates were noted in the three social classes.

However, on *a priori* grounds in temperate climates the risk of acquisition of varicella early in childhood

would appear directly related to family size and to density of housing and of classroom populations.

5.11. Other Factors

5.11.1. Nutrition. Although there is no evidence that malnutrition influences susceptibility to varicella, the severity of both pathological conditions may be enhanced when infection occurs. In studies on six children in the recovery phase of kwashiorkor, the onset of varicella reduced intake of food and diminished nitrogen retention.[206] In a rural Guatemalan village, varicella was commonly accompanied by diarrhea. In children under 5 years of age with varicella, diarrhea was observed three times more frequently than in an uninfected control group.[157] Although only 12.5% of well-nourished children with varicella developed diarrhea, this symptom developed in 75% of those who were moderately malnourished. Five of 50 children with varicella-associated diarrhea developed kwashiorkor; one subsequently died.

5.11.2. Genetic Factors. Hook *et al.*[89] described an unusual situation wherein five children in a family contracted severe varicella within a 2-week interval; three developed pneumonia and two died. McKusick,[125] referring to this episode, called attention to his observation that varicella may be severe and even fatal in persons with cartilage–hair hypoplasia (CHH), a recessively inherited skeletal dysplasia. Lux *et al.*[117] studied two patients with CHH, both of whom had severe varicella; one had successive crops of vesicles over a 9-day period and the other for 14 days. Immunologic studies revealed normal humoral responses. A cellular immune defect residing in the small lymphocytes was reflected by chronic neutropenia, lymphopenia, diminished delayed skin hypersensitivity, diminished responsiveness of lymphocytes *in vitro* to phytohemagglutinin and to allogeneic cells, and delayed rejection of a skin allograft. These findings have been confirmed in studies on 28 persons with CHH.[183]

As cited by Lux *et al.*,[117] varicella is also severe in patients with another hereditary immunologic disease, Nezelof's syndrome, in which there is an autosomal recessive lymphopenia. Progressive varicella has been observed in three patients with the Wiskott–Aldrich syndrome, a condition with impaired cellular and humoral immunity.[185] Life-threatening infections with VZV, HSV, and CMV were observed in a patient who had a lack of natural killer cells and killer cell functions.[17]

5.11.3. Psychological Factors. Although unconfirmed, the report that a psychic meditator could voluntarily and temporarily suppress cellular immune responses to VZV antigens raises the question of the possible influence of psychological factors on the duration of the containment phase in infections with varicella–zoster virus.[167]

5.11.4.   Iatrogenic Factors

5.11.4a.   Varicella.   Varicella may be life-threatening in children with leukemia and other malignant diseases.[34]; with advances in therapy now prolonging the life of the leukemic child, severe infections are a relatively common complication. Corticosteroid,[34,60] immunosuppressive, and cytotoxic therapy may convert a benign illness into a fulminating disease.[83] Varicella may be fatal in children who have received organ transplants. In a group of 13 liver transplant recipients who developed varicella, two died in spite of intravenous acyclovir therapy.[124] Varicella occurred in 8 of 83 children who had received renal transplants; two died in spite of intravenous acyclovir therapy.[118]

5.11.4b.   Herpes Zoster.   Immunosuppressive or antimetabolite therapy may disturb the delicate balance that maintains VZV in a latent state. Rifkind[151] studied 73 patients for 21–37 months after renal transplantation; six (8%) aged 12–40 years developed herpes zoster. In other series of renal transplant patients, zoster developed in 7%[133] and 13%.[102]

Bone marrow transplantation, following immunologic ablation, carries a high risk of infection with or reactivation of VZV. In one series of 1394 patients, 16.6% had overt infection, with 23 deaths.[112] The labile nature of the VZV–host relationship in transplant recipients is shown by fluctuations in antibody titer in asymptomatic subjects[116]; of a group of 73 individuals receiving bone marrow transplants who had preexisting antibody titers, 21 (28%) manifested rises in titers of antibody in the absence of symptoms.[115] A subclinical viremia is common in such individuals, with virus documented by PCR in the peripheral blood mononuclear cells of 19% of 37 bone marrow transplant recipients.[205] In this study in seropositive patients, 41% developed zoster or had evidence of subclinical VZV replication; zoster developed in 31% of 51 recipients, the majority within the first year after transplantation.

Certain types of malignancy, especially Hodgkin's disease and lymphatic leukemia, are recognized as inducers of attacks of herpes zoster, a relationship enhanced by chemotherapy. Analysis of a group of 1132 hospitalized children with cancer revealed that 9% developed herpes zoster; the attack rate was highest in Hodgkin's disease, occurring in 22% of 97 cases.[47] The advent of antiviral therapy has improved the prognosis in cancer patients as shown by a review of 288 cases at St. Jude's Hospital in Memphis.[48] In untreated patients, pneumonitis developed in 28%, being seen most often in leukemics, and had

a 25% mortality rate. Acyclovir stopped progression of skin lesions in 18 patients; none developed pneumonitis after 2 days of treatment. Zoster occurs more frequently in patients with rheumatoid arthritis receiving methotrexate therapy than in the general population.[2]

5.11.5.   Concurrent Infections

Human Immunodeficiency Virus Infection.   Biological immunosuppression induced by HIV in patients with acquired immunodeficiency syndrome (AIDS) enhances the severity of concurrent VZV infections, occasionally with bizarre presentations.[79] The cutaneous eruption may be more severe and hemorrhagic, and pulmonary and hepatic involvement are more common.[79,144,168] Zoster occurs more frequently. In a cohort study of homosexual men, the incidence of zoster was significantly higher among HIV-seropositive individuals (29.4 cases per 1000 person-years) than in seronegative individuals (2.0 cases per 1000 person-years).[27] In a group of hemophiliacs with AIDS, zoster appeared as a illness of the young, developing in patients from 4 to 27 years of age.[182]

Rather surprisingly in view of the reduced occurrence of childhood varicella in tropical countries, zoster in Kinshasa, Zaire, was found to be a predictor of HIV infection.[36] Of 146 patients with a history of zoster, 91% were HIV seropositive.

Peculiar disseminated hyperkeratotic nodular skin lesions may develop in HIV-positive patients after prolonged acyclovir therapy. From such lesions acyclovir-resistant VZV may be recovered.[79,112a,117a] Analogous to the occurrence of CMV retinitis in AIDS patients, VZV may also cause an acute retinal necrosis.[86]

6.   Mechanisms and Routes of Transmission

6.1.   Varicella

6.1.1.   Congenital Infections.   Following the landmark report of Laforet and Lynch[104] in 1947 of an infant with multiple congenital defects attributed to maternal varicella, similar cases gradually have accumulated in the literature. In 1987, Higa and co-workers[87] summarized data on 52 infants whose mothers had contracted varicella during pregnancy. The mothers of infants with malformations had contracted varicella within the first 20 weeks of gestation, whereas most of the mothers whose offspring developed zoster in the early postnatal period had varicella after 21 weeks of gestation.

Alkalay *et al.*[1] summarized the clinical features of 22 cases confirmed immunologically by prolonged persistence of specific IgG antibodies or the presence of specific IgM antibodies at birth. The lesions are diverse in nature. Cicatricial skin lesions that correspond to a dermatomal distribution are common as are neurological abnormalities of many types. Eye and skeletal anomalies occur in over 50% of cases. Not all infected infants are symptomatic. In one prospective study the syndrome occurred in only 1 of 11 infants born of women with first-trimester varicella, although a second infant had unexplained microcephaly at 1 year of age.[142]

An analysis of case reports produced an estimated 95% confidence interval of 0.5 to 6.5% for the risk of fetal defects after first-trimester varicella infection and 0 to 1.1% after second- or third-trimester infection.[147] More extensive studies establish the risk of embryopathy after maternal varicella infection in the first 20 months of pregnancy as about 2%.[44a,143a] Recently, after ultrasound studies of a potentially infected fetus at 22 weeks' gestation revealed renal abnormalities, VZV was isolated from amniotic fluid samples.[145] In contrast to the well-established congenital varicella syndrome; few if any cases occur after maternal zoster.

Despite these reports, the congenital varicella syndrome is rare. Two large-scale prospective studies involving a total of 433 varicella-complicated pregnancies failed to establish an association.[120,165] A subsequent cohort study in New York City, involving a 5-year follow-up of children born after maternal varicella, revealed no associated malformations.[164]

Onset of varicella in an infant within 10 days of birth is arbitrarily considered as indicating congenital transmission. Perinatal infection occurs less frequently than would be expected. Meyers[129] collected data on 43 cases of congenital varicella; the case-fatality rate was 10%. Reports were available on 46 women with varicella in the last 17 days of pregnancy who were delivered of live infants at term; only 11 (24%) infants had overt varicella. It has been suggested that VZV may not regularly cross the placenta.[52]

6.1.2. Postnatal Transmission. The typical case of varicella is probably infectious for 1–2 days prior to the appearance of the generalized eruption and for 4–5 days thereafter, i.e., until the last crop of vesicles has evolved to the purulent and crusted state. The period of infectivity is prolonged in the immunoincompetent host, who exhibits successive crops of cutaneous lesions over a period of 1 week or more. Airborne droplet infection has been emphasized in the past as an important mode of transmission. Evidence on the role of intimate association suggests that infections are also spread by direct contact and less frequently by indirect contact. The duration of infectivity of droplets containing labile virus must be relatively limited.

The mechanisms whereby virus is shed and spreads to susceptible individuals are ill defined. The diffuse cutaneous lesions and the observation by Tyzzer[180] that the earliest histological changes in such lesions are in the capillary endothelium indicate that a viremia occurs. Confirmation by viral isolation in cell culture now has been obtained in the preeruptive phase of varicella in normal children as early as 5 days before the onset of rash by using hematogenous monocytes as inocula.[9,137] A peculiarity of these isolates is that characteristic VZV cytopathology appears only after subculture.

Virus is present in high titer in the fresh cutaneous lesions. Comparable lesions also occur in the respiratory, digestive, and urinary tracts. Thus, theoretically, various routes exist for dissemination of virus. However, virus has been recovered with regularity only from the cutaneous lesions. Gold[71] could not recover virus from respiratory tract secretions that were obtained daily from three children, beginning in the latter part of the incubation period and continuing to after appearance of the rash. From an additional case, a single isolate was made from pharyngeal secretions collected 1 day after the exanthem developed. However, it appears that classic cultural systems for the isolation of VZV are relatively insensitive. Ozaki *et al.*[138] used cultures of human embryonic lung and isolated VZV from 5 of 117 pharyngeal swabs from varicella patients. In 1991, this group detected VZV DNA by PCR in throat swab material from 26% of patients during the incubation period and from 90% after clinical onset.[100,139]

6.1.3. Vectors and Animal Reservoirs. There is no indication that arthropod vectors play a role in the transmission of VZV. Animal reservoirs are unknown and unlikely to exist; VZV is highly host-specific. Old World monkeys do not develop generalized disease on inoculation with VZV, although a varicellalike disease, produced by a virus closely related to but distinct from VZV, has been described in macaques.[19] Although monkeys have not been shown to be susceptible, anthropoid apes apparently can contract varicella if in contact with a human case.[200] Thus, theoretically, the higher apes could maintain VZV in nature.

6.2. Herpes Zoster

The individual with herpes zoster is infectious and can initiate an outbreak of varicella in susceptibles. The mechanism of transmission is even less well defined than in varicella, although probably it is basically similar. While rare, airborne transmission of virus from a patient with localized zoster may infect susceptible hospital employees.[96] Not sufficiently appreciated is the fact that in herpes zoster, a chickenpoxlike dissemination of cutaneous lesions, apparently reflecting a transient viremia, is a common occurrence. In a prospective study in Sweden of 100 consecutive hospitalized patients with herpes zoster, Oberg and Svedmyr[135] observed the appearance of disseminated cutaneous lesions—usually several days after development of the segmental process—in 33% of their patients.

7. Pathogenesis and Immunity

7.1. Varicella

7.1.1. Pathogenesis. Circumstantial rather than factual evidence and analogies to experimental models of other virus–host interactions provide the basis for hypotheses bearing on the sequence of events following infection of the human host. Entry of the virus is probably via the mucosa of the oropharynx, the upper respiratory tract, the conjunctiva, or, less likely, the skin. Viral replication occurs intracellularly at the site of lodgement, with subsequent cycles of replication involving contiguous cells and leading to dissemination via the blood and lymphatics. Data on growth curves of VZV *in vitro* suggest that multiple cycles of focal replication (perhaps in reticuloendothelial cells) occur during the relatively prolonged incubation period. Early in this stage, nonspecific host immune responses as well as a developing specific immunity partially contain the virus. After 7 days or more, these responses are overwhelmed, the viremia increases quantitatively, and prodromal symptoms develop, followed by cutaneous lesions. Successive crops of vesicles probably reflect a cyclic viremia. In the immunologically competent subject, specific humoral and cellular immune responses thereafter rapidly terminate the viremic phase, although direct cell-to-cell extension with enlargement of established focal lesions continues briefly.

Information on the nature of the immune mechanisms that contain the infectious process is accruing. Rises in titer of immunoglobulin classes IgG, IgM, and IgA are demonstrable by the FAMA method within 5 days of onset.[25] The different subclasses of IgG are variable as to timing of appearance and decay.[8] Suspensions of lymphocytes and monocytes from VZV-immune donors have an enhanced capacity to inactivate VZV when examined *in vitro* by a plaque-reduction technique.[67] Mononuclear cells from immune donors exposed to VZV antigen exhibit a blastogenic response[95] that is specific and not elicited by antigens derived from other members of the

herpesvirus group.[210] A comparison of the blastogenic response of tonsillar and of peripheral blood lymphocytes to VZV antigen revealed the former to be more reactive.[21] Further, some individuals showed evidence of mucosal cell-mediated immunity in the absence of humoral immunity, suggesting that Waldeyer's ring may play a role as a protective barrier.

Varicella–zoster virus induces production of interferon *in vitro* and is inhibited by it[4]; the amount of interferon required for inhibition is directly related to the multiplicity of infection.[150] Interferon can be demonstrated in varicella and zoster vesicular fluids[3]; the appearance and amount correlate directly with the development of the cellular infiltrate in the lesion.[170]

As previously mentioned, the natural history of varicella is influenced by various host factors. Additionally, physical factors that injure the skin, including trauma and ultraviolet light,[68] may produce focal concentrations of lesions. Thus, whereas the well-cared-for baby shows no concentration of lesions in the diaper area, in infants with a constantly wet excoriated anogenital skin surface, as high as 68% of the total-body pocks count may be concentrated in that region.[153]

7.1.2. Incubation Period. The average incubation period of varicella is 14 or 15 days. In the carefully studied series of Ross,[153] the range was 10–23 days, with 99% of cases falling between 11 and 20 days.

7.1.3. Immunity. As indicated above (see Section 4.1), the concept that an attack of varicella confers a lasting immunity is now known to be incorrect. Rarely, virologically confirmable second attacks occur. More frequent are asymptomatic infections indicated by a rise in titers of VZV antibody.

7.2. Herpes Zoster

7.2.1. Pathogenesis

7.2.1a. General. An understanding of the pathogenesis is accruing. Following an attack of varicella, the virus is contained by the host to the extent that symptoms do not occur, usually for many years. Then extensive uncontained replication may again occur with a resulting herpes zoster. Epidemiologically, the clear-cut evidence that cases of zoster occur sporadically throughout the year with no temporal or spatial relationship to epidemics of varicella documents endogenous reactivation as the usual source of virus. This thesis is now supported by the molecular characterization of viral isolates recovered from a patient who had varicella and later developed zoster; the varicella and zoster isolates on endonuclease analysis and Southern blot hybridization of the viral DNA proved to be identical.[174]

Evidence is accruing that Hope-Simpson's thesis that VZV is harbored in sensory ganglia during the containment period is correct.[91] Although efforts to recover infectious virus from dorsal root ganglia from asymptomatic varicella-immune adults have been unsuccessful, the application of molecular hybridization techniques have demonstrated VZV RNA[92] and VZV DNA[69] in trigeminal ganglia harvested randomly from asymptomatic adults. It has been established that this is a common finding. Using PCR techniques, VZV DNA was detected in trigeminal ganglia from 10 of 11 seropositive adults and in the thoracic ganglia from 12 of 14 adults.[119] In another study, VZV DNA was detected in 9 of 13 geniculate ganglia.[55] The type of cell harboring the virus is controversial, with reports that only neuronal or only satellite cells are involved. Recent studies of VZV in ganglion cells indicate that the state of latency is characterized by limited transcription with selected expression of multiple genes.[35,126] During an acute attack of zoster, virus can be recovered from the affected ganglia. Hope-Simpson further suggested that when defense mechanisms deteriorate below the threshold of viral containment, virions replicate in the sensory ganglion and are then transported antidromically down the sensory nerve and released around sensory nerve endings in the skin, producing characteristic clusters of zoster vesicles; virus thereafter can be shed into the environment. That this is not the complete picture is apparent from the previously cited data on the common appearance of disseminated cutaneous lesions within a few days of appearance of the segmental eruption; a viremia must often occur. Indeed, in four of eight patients virus could be isolated from leukocyte-rich plasma immediately before, or coincident with, the onset of cutaneous dissemination.[46]

7.2.1b. Intrinsic Factors. Zoster develops when the host defense mechanisms decay below the containment level. Evidence that the level of cellular immune activity is an important factor is increasing. The person who develops zoster has significant levels of humoral antibody at onset; in some, a further increase in titer beyond an initial high level may not be demonstrable on examination of paired sera. The antibody response, being secondary in nature, might be expected to lack a significant IgM macroglobulin component. However, an IgM response often occurs, with a frequency of demonstration ranging from 20 to 78%.[7,25,39,82] In the past some serological tests have revealed apparent differences in the antibody pattern developing after varicella and after zos-

ter. An explanation now derives from the molecular dissection of the several glycoproteins specified by VZV; antibodies to some are absent or present in low titer in V sera but are prominent in Z sera.[106,190]

A declining level of cellular immunity is a major factor in the termination of the phase of viral containment. The VZV-specific cellular immunity is measurable by studies *in vitro* of lymphocyte stimulation on exposure to VZV and by reactivity on skin testing with VZV antigens. Age-related studies of these responses show a diminution in skin-test reactivity after 40 years of age and of lymphocyte stimulation responses after 60 years of age.[15,28,130] These age-related changes correlate with the increasing risk of zoster in the later decades of life. In a like manner, impaired cellular immunity explains the increased risk of zoster at any age in individuals who have hematopoietic or reticuloendothelial cancer or who are immunosuppressed.[6,7,64,155]

The occurrence of zoster in infancy or early childhood reflects a special situation; either the mother had varicella during pregnancy or the subject had varicella during the first year of life.[78] As previously noted, such an occurrence suggests that the response of an immature immune system was inadequate[193] or that a modulated primary illness provided a reduced antigenic stimulus.

7.2.1c. Extrinsic Factors. Iatrogenic herpes zoster is not a new entity. An association with administration of metallic drugs, especially lead and arsenic, has long been established. Only recently, however, has depression of immunologic activity become a desirable therapeutic objective in medicine; whether achieved by physical, biological, or chemotherapeutic means, the result entails an enhanced risk that zoster will appear if the subject has had a prior attack of varicella.

The role of trauma in the pathogenesis of zoster has been controversial. Hope-Simpson[91] established an association in only 2 of 192 cases and considered that these episodes might have been coincidental. However, the population-based study of Ragozzino *et al.*[149] indicated that trauma increases risk by a factor of 1.9.

7.2.2. Incubation Period. There is no information on the interval between endogenous reactivation and the appearance of symptoms. Exceptionally, pain in the dermatome subsequently involved may develop 10–14 days prior to the appearance of the eruption; usually the interval is 2–4 days.

7.2.3. Immunity. There are numerous reports in the literature of the occurrence of second and rarely of third attacks of herpes zoster. Hope-Simpson[91] summarized data on age incidence and concluded that if a cohort of 1000 people were to live to be 85 years old, half would have experienced one attack of herpes zoster, 10 would have had two attacks, and possibly one would have had a third attack.

8. Patterns of Host Response

8.1. Clinical Patterns

8.1.1. Varicella

8.1.1a. Prodromal Symptoms.   Adults may have 1 or 2 days of fever and malaise prior to appearance of the rash. In children, premonitory symptoms are usually mild, and appearance of the exanthem is often the first evidence of illness.

8.1.1b. Exanthem.   There are three aspects of the vesicular eruption that are of diagnostic significance: the nature and evolution of individual lesions, the concurrent presence of lesions at different stages of development, and the distribution of the process over the body. Individual lesions develop as small, irregular, rose-colored macules in the center of which appears a delicate 1- to 4-mm "dewdroplike" vesicle containing clear fluid. This may rupture; if not, within a few hours the contents become purulent and then dry and crusted. Successive crops of new lesions appear over a period of 2–4 days so that characteristically at the peak of the illness there are in any one area cutaneous lesions at all stages of evolution and of resolution. Lesions appear first on the scalp and then on the trunk. The distribution of the lesions is centripetal. The greatest concentration is on the trunk, and the distal extremities are the least involved. It is misleading to assume that lesions are confined to the skin; similar involvement of the mucosa of the oropharynx and vagina can be seen, and it is probable that lesions comparable to those in the skin develop to a varying degree in each patient in the respiratory and the gastrointestinal tracts.

8.1.1c. Course.   In the immunocompetent child, the illness is typically benign, accompanied by mild malaise, and reflected by pruritus and fever (100–102°F) for 2–3 days. In the adult, systemic symptoms are more marked, and varicella pneumonia that may be life-threatening is a common complication. In one series of 114 consecutive cases in a military population (average age 25 years), lung involvement was demonstrable roentgenographically in 16%.[188] Secondary bacterial infection of the cutaneous lesions occurs frequently. Studies of the complications leading to hospitalization have been reported. One reviewed complications in children; 15 of 96 studied were immunocompromised.[49] Viral encephalitis,

bacterial infection, and Reye's syndrome were common causes. The other, a population-based study, investigated normal individuals of all ages admitted to a local hospital, examined a comparable national sample, and determined age-specific incidence rates.[77] The common causes of hospitalization were bacterial superinfections in children younger than 5 years old, varicella encephalitis (mainly acute cerebellar ataxia) and dehydration in 5- to 9-year-olds, and varicella pneumonia in adults.

Vascular thromboses are a recognized complication of varicella. During the eruptive phase, deep necrotic thromboses may develop that are associated with transient protein C and protein S deficiencies.[70a,133a] Hemiparesis due to a cerebrovascular thrombus may develop a few weeks after a primary infection.[20a] Cases of VZ-associated myelopathy confirmed by the demonstration of VZV DNA in the spinal fluid have been recorded,[68a] as have cases of aseptic meningitis in the absence of cutaneous lesions.[42a]

In the immunologically compromised subject, all clinicopathologic manifestations may be enhanced. Focal pocklike lesions may occur throughout the gastrointestinal and respiratory tracts and in the liver and spleen; there may be widespread vascular damage with prominent hemorrhagic lesions.[34]

Parenthetically, it is to be noted that Reye's syndrome may develop after viral infections, particularly following varicella or influenza. The administration of salicylate compounds as an antipyretic during the viral infection appears responsible in large part for the occurrence of Reye's syndrome. Since the issuance of the Surgeon General's warning in 1982 of the precipitating role of salicylates, the use of aspirin as an antipyretic in varicella has decreased, and the incidence of Reye's syndrome has fallen significantly.[20]

8.1.1d. Inapparent Infections.   Inapparent infections obviously occur, although, as noted earlier, observational studies suggest a frequency of less than 5% in susceptible children.

8.1.2. Herpes Zoster

8.1.2a. Exanthem.   The varicelliform eruption in zoster is characteristically unilateral and initially sharply limited in a band or patchlike distribution to the dermatome (or, rarely, associated dermatomes) supplied by a specific dorsal root or extramedullary cranial nerve ganglion. Within the segmental area of localization, lesions may be scattered and few or may be so numerous as to form an almost confluent large plaque. Hope-Simpson[91] noted that the dermatomes most frequently involved in herpes zoster—i.e., the region supplied by the fifth cranial nerve, and the trunk supplied by ganglia extending from the third dorsal to the second lumbar segments—are those areas wherein the lesions of varicella are also most prominent. The lesions of zoster rarely involve the extremities.

8.1.2b. Course. Prior to the appearance of the eruption, there may be pain and extreme paresthesia in the involved segment. The lesions appear in crops and, although often larger in comparison, evolve and resolve as in varicella, but at a slower pace, so that virus may be recovered from vesicles for as long as a week after appearance of the eruption. Scabs, as in varicella, are noninfectious but may persist for 2 weeks or more. A regional lymphadenopathy is a characteristic feature if the segmental process is extensive. Severe postherpetic neuralgia that is refractory to treatment is a distressing complication. The likelihood of the occurrence of chronic postzoster neuralgia increases markedly with advancing age.[172a]

It is to be noted that evidence of central nervous system involvement with a spinal fluid pleocytosis is common and that motor involvement with weakness or paralysis may occur, but that a fatal outcome attributable to zoster per se is extremely rare. Reference should be made elsewhere for a discussion of other complications.

8.1.2c. Inapparent Infections. The term "zoster sine herpete" has been applied to situations wherein unilateral focal zosterlike premonitory pain is experienced but vesicles fail subsequently to develop over the involved area.[97] Such transitory imbalances in the host–virus relationship would be epidemiologically unimportant unless accompanied by patterns of viral dissemination now unrecognized.

8.2. Diagnosis

With the eradication of variola and the general discontinuance of smallpox vaccination, the practitioner is no longer faced with a sometimes troublesome and always pressing differential diagnosis. Currently the major problem is to differentiate the vesiculopapular lesions produced by VZV from those caused by HSV; the latter may present as a generalized eruption (eczema herpeticum) or as a circumscribed zosterlike process. This differential diagnosis is complicated by the recent demonstration that in the patient with varicella, HSV may also be active in the absence of lesions attributable to it. Landry and Hsiung investigated 12 patients with varicella; both HSV and VZV were isolated from three.[105] The problem is further complicated by the fact that the lesions are similar histologically and that in the interpretation of serological findings it is to be remembered that the viruses have antigens in common. However, as discussed above, these problems

are resolvable if appropriate reagents and procedures are employed.

The eruption of varicella should also be distinguished from that of rickettsial pox,[99a] of infections caused by certain coxsackieviruses, of secondary syphilis in some forms, and from dermatitis herpetiformis. The focal process of zoster, most frequently simulated by herpes simplex, may also be confused with impetigo contagiosa.

9. Control and Prevention

9.1. General Concepts

Specific antiviral chemotherapeutic agents are now available. A vaccine for varicella was approved for use in the United States by the Food and Drug Administration (FDA) in March 1995. Additionally, there is a suggestion that the vaccine may prevent zoster in the elderly by enhancing specific cellular immunity.

9.2. Interruption of Transmission

9.2.1. Isolation and Quarantine in Homes and Schools. Neither isolation nor quarantine is effective in interrupting the spread of the disease in groups of susceptible children; therefore, both practices generally have been abandoned. Since practically all persons will eventually acquire the disease, there has been merit perceived in the acquisition of varicella in childhood, for the illness then will be less severe and less socially disruptive than if experienced in adulthood. The availability of a varicella vaccine may alter this view. For the same reason, the movement of adults with herpes zoster is not usually constrained.

9.2.2. Protection of Groups at Special Risk. Varicella is a potentially fatal disease in susceptible children who are undergoing immunosuppressive therapy or who concurrently have malignant diseases such as leukemia or Hodgkin's disease. Efforts should be made to minimize infectious contacts with such high-risk persons by protective isolation procedures. Contact with adults with herpes zoster and with nonimmune children who might be incubating varicella should be precluded if possible. However, the appearance of a case among patients on a ward poses problems, even with the strictest of precautions. The customary solution is to remove all known susceptibles for a 3-week period. Children at high risk with a known exposure should be given zoster-immune globulin (see Section 9.3.1) and should be observed so that chemotherapy can be initiated immediately if varicella develops.

9.3. Modification or Prevention of Varicella

9.3.1. Administration of Specific Antibody.

Ross[153] demonstrated in 1962 that administration to susceptible children exposed to varicella of γ-globulin obtained from adults at large did not prevent disease but modified the illness as indicated by fewer pocks and a diminished febrile response. Since patients with zoster develop high titers of specific antibody, a logical extension of this observation was to utilize zoster-immune globulin (ZIG) prepared from blood collected from zoster convalescent patients with high VZV CF titers[26]; a 2-ml dose given within 3 days of exposure prevented varicella in exposed susceptible children. Collaborative studies followed to evaluate ZIG as a preventive measure in children at high risk. A review of the results of administration of ZIG to 358 exposed persons at high risk revealed an overall attack rate of 22%[30] Gershon *et al.*[65] found that use of ZIG in 15 exposed immunocompromised children, known to be susceptible by the FAMA test, resulted in subclinical infection in five, attenuated overt disease in nine, and severe illness in one. The period between exposure and administration is critical; a Norwegian study records a 97% protection rate if the period is 3 days or less and a 50% protection rate if the delay is 5 days or more.[207]

The supply of ZIG was limited. A partial solution to this problem was provided by the utilization of outdated plasma for globulin preparation that had significant VZV antibody titers; the product was termed VZIG.[212] The VZIG was shown to be as effective as ZIG[211] and was licensed in 1981. Produced by the Massachusetts Public Health Biological Laboratories, VZIG is distributed nationwide through the American Red Cross Blood Services; a list of distribution centers for VZIG and the recommendations of the Immunization Practices Advisory Committee for its use have been published.[31] In brief, priorities for use are: (1) susceptible, immunocompromised children within 96 hr after significant exposure; (2) newborns of mothers who developed chickenpox within 5 days before and 48 hr after delivery; (3) newborns exposed postnatally; and (4) susceptible immunocompromised adults. According to Dr. Jean Leszczynski of the Biological Laboratories, the number of vials of VZIG distributed annually has increased from 10,524 vials in 1982 to 58,331 in 1994. A modified immune globulin that can be administered intravenously is available.[143]

9.3.2. Use of Interferon and of Transfer Factor.

As reviewed in previous editions, interferon and transfer factor have proved useful in the therapy of VZV infections. However, interest in these entities has waned with the advent of specific antiviral compounds.

9.3.3. Antiviral Chemotherapy.

Since publication of the last edition, acyclovir, 9-(2-hydroxy-ethoxy-methyl) guanine, has supplanted vidarabine as the drug of choice in treating VZV infections, although a double-blind comparative study in patients with disseminated zoster indicates an almost equal effectiveness.[203] Parenteral acyclovir was introduced for the treatment of zoster in 1979.[161] Later, topical and oral preparations became available. The experience with acyclovir, a selective inhibitor of VZV, has been reviewed.[13a,202] In general, acyclovir therapy by the intravenous route is recommended for all immunocompromised patients with varicella, i.e., the high-risk group, with resulting reduction in morbidity and mortality. Administration early in the course of the disease is important to prevent dissemination.[134] Varicella pneumonia in adults and in pregnant women responds to intravenous acyclovir.[22,81] Straus[172] recommends intravenous acyclovir for zoster in immunocompromised subjects including AIDS patients with multidermatomal zoster. In zoster, early treatment shortens the duration of infection; there are discordant opinions whether such treatment affects the probability of development of postzoster neuralgia, although use of oral acyclovir shortens the duration of chronic zoster-associated pain.[91a]

Treatment by the intravenous route carries a risk of renal and of nervous system toxicity.[13] In contrast, acyclovir by the oral route has little toxicity and does not require hospitalization. However, as emphasized by Straus[172] high-dose oral acyclovir (4 g/day in five divided doses) barely achieves the serum concentration (1 to 3 μg/ml) necessary to inhibit VZV replication. Nevertheless, treatment of adults with high-dose oral acyclovir given at the onset of varicella markedly shortens the duration of the illness and reduces the number of lesions.[45,186] In a trial of high-dose oral acyclovir in 25 immunocompromised children, dissemination was prevented in 23; in two it was necessary to initiate intravenous therapy.[127] Varicella in normal children treated with oral acyclovir is reduced in severity and duration.[42] However, it is recommended that oral acyclovir not be used routinely for the treatment of varicella in otherwise healthy children.[38] VZV may become resistant to acyclovir after its prolonged use as in patients with AIDS.[94,140] Resistance has been attributed to a mutation in the VZV thymidine coding sequence that greatly decreases the capacity of the virus to phosphorylate antiviral nucleoside analogues.[152]

Current studies of new drugs, both related and unrelated to acyclovir, suggest that more effective approaches to the therapy of VZV infections are in the offing.[172] Famciclovir, a drug similar in action to acyclovir, is now available in the United States for the treatment of zoster.

9.3.4. Varicella Vaccine. In 1974, Takahashi *et al.*[177,178] developed a live virus vaccine attenuated by serial passage of VZV in cultures of human and guinea pig embryonic cells. His Oka-strain vaccine was first used in high-risk children and was then approved for general use in Japan in 1986 and in Korea in 1988; by 1992, approximately one million children in these two countries had received Oka-strain vaccine.

In the United States, a collaborative study was initiated in 1980 with the objective of determining if a live Oka-strain vaccine was protective as well as safe in high-risk children. Thereafter, the study was expanded with trials in normal children and in adults. A Merck Oka-strain vaccine, Varivax, has now been approved by the FDA. This product has been the subject of extensive studies by the Collaborative Varicella Vaccine Study Group of the National Institute of Allergy and Infectious Diseases, as well as by other groups of workers. Reference may be made to a history of the development of the vaccine[59] and to a summary of the prolonged discussions[122b] that took place prior to its licensure.

Guidelines for the use of varicella vaccine are being formulated. It is expected that one dose of vaccine will be recommended for children 1 through 12 years of age and that two doses given 4 to 8 weeks apart will be recommended for those 13 years or older. With this schedule, 70 to 90% of vaccinated children will be protected from an overt infection on a subsequent exposure; a mild form of varicella is characteristic in those developing an infection. The optimal dosage of the vaccine is under investigation. Two doses 1 month apart provide better protection in children than one dose,[187a] while two doses were required in adults to obtain optimal seroconversion rates.[131a]

As most infections with VZV are relatively benign, there has been a question as to whether a vaccine would be cost-effective. Two recent analyses conclude that the routine immunization of children in the United States would be cost-effective.[91b,109b]

The concern that a live VZV vaccine might increase the severity or risk of zoster now appears invalid. Whether a childhood vaccine might postpone the time of acquisition of infection to adulthood will be determined only by years of observation, although the induced immunity seems persistent.

As zoster is related to decay of cellular immunity with age, the possibility that the vaccination of the elderly might prevent zoster is under investigation. Administra-

tion of a booster dose of Oka-strain vaccine in the elderly has been shown to enhance specific cellular immune responses; 28 of 33 (85%) adults aged 55–65 years who had a negative proliferative cellular immune reaction had developed a positive cell-mediated response when retested 2 months after vaccination.[16]

In another study it was found that after zoster there was an increase in specific T-cell responses and that a similar response occurred in elderly people after immunization with Oka-strain vaccine.[85] Subsequently, 202 seropositive individuals who were 55 to 87 years old received vaccine with a specific T-cell response that persisted during the 2 years of observation.[108] The majority of vaccinees in this cohort still showed enhanced levels of VZV-responding T cells when examined 4 years after immunization, although in 10% there was no response.[109a] Whether this response correlates with the prevention of zoster will require years of observation.

10. Unresolved Problems

10.1. Prevention and Treatment

Varicella is second only to gonorrhea in incidence among reportable diseases in the United States. A varicella vaccine is now available with obvious social and economic benefits if widely used. However, such an intrusion on the relationship between human beings and varicella virus may produce new problems. Prolonged observational studies are needed to assess the relative risks and benefits of this promising development. The possibility that the vaccine may also prevent age-related zoster warrants further investigation. The treatment of zoster neuralgia is unsatisfactory.

10.2. Pathogenesis

Evidence has now accumulated to incriminate declining levels of specific cellular immunity as the primary factor leading to the termination of the containment period and to the renewed replication of VZV with resulting zoster. Yet the routes and mechanisms of viral movement remain obscure. From the practical standpoint the question is whether booster doses or vaccine in the elderly will reverse the decay of cellular immunity and prevent zoster. The pathogenesis of postzoster neuralgia is unclear.

10.3. Epidemiologic Unknowns

Although airborne transmission has now been documented, the relative role of this means of spread as contrasted to direct or indirect contact remain unresolved. Failure to culture VZV from throat washings suggests that cultures are relatively insensitive indicators of virus. How can the technique be improved?

11. References

1. ALKALAY, A. L., POMERANCE, J. J., AND RIMAIN, D. L., Fetal varicella syndrome, *J. Pediatr.* **111:**320–323 (1987).
2. ANTONELLI, M. A. S., MORELAND, L. W., AND BRICK, J. E., Herpes zoster in patients with rheumatoid arthritis treated with weekly low-dose methotrexate, *Am. J. Med.* **90:**295–298 (1991).
3. ARMSTRONG, R. W., GURWITH, M. J., WADDELL, D., AND MERIGAN, T. C., Cutaneous interferon production in patients with Hodgkin's disease and other cancers infected with varicella or vaccinia, *N. Engl. J. Med.* **283:**1182–1187 (1970).
4. ARMSTRONG, R. W., AND MERIGAN, T. C., Varicella–zoster virus: Interferon production and comparative interferon sensitivity in human cell cultures, *J. Gen. Virol.* **12:**53–54 (1971).
5. ARVIN, A. M., KOROPCHAK, C. M., AND WITTEK, A. E., Immunologic evidence of reinfection with varicella–zoster virus, *J. Infect. Dis.* **148:**200–205 (1983).
6. ARVIN, A. M., POLLARD, R. B., RASMUSSEN, L. E., AND MERIGAN, T. C., Selective impairment of lymphocyte reactivity to varicella zoster virus antigen among untreated patients with lymphoma, *J. Infect. Dis* **137:**531–540 (1978).
7. ARVIN, A. M., POLLARD, R. B., RASMUSSEN, L. E., AND MERIGAN, T. C., Cellular and humoral immunity in the pathogenesis of recurrent herpes viral infections in patients with lymphoma, *J. Clin. Invest.* **65:**869–878 (1980).
8. ASANO, Y., HIROISHI, Y., ITAKURA, N., HIROSE, S., KAJITA, Y., NAGAL, T., YAZAKI, T., AND TAKAHASHI, M., Immunoglobulin subclass antibodies to varicella zoster virus, *Pediatrics* **80:**933–936 (1987).
9. ASANO, Y., ITAKURA, N., HIROSHI, Y., HIROSE, S., NAGAI, T., OZAKI, T., YAZAKI, T., YAMANISHI, K., AND TAKAHASHI, M., Viremia is present in incubation period in nonimmunocompromised children with varicella, *J. Pediatr.* **106:**69–71 (1985).
10. ASANO, Y., NAGAI, T., MIYATA, T., YAZAKI, T., ITO, S., YAMANISHI, K., AND TAKAHASHI, M., Long-term protective immunity of recipients of the Oka-strain live varicella vaccine, *Pediatrics* **75:**667–671 (1985).
11. BABA, K., YABUUCHI, H., OKUNI, H., AND TAKAHASHI, M., Studies with live varicella vaccine and inactivated skin test antigen; protective effect of vaccine and clinical application of the skin test, *Pediatrics* **61:**550–555 (1978).
12. BABA, K., YABUUCHI, H., TAKAHASHI, M., AND OGRA, P., Immunologic and epidemiologic aspects of varicella infection acquired during infancy and early childhood, *J. Pediatr.* **100:**881–885 (1982).
13. BALFOUR, H. H., JR., Editorial. Acyclovir therapy for herpes zoster: Advantages and adverse effects, *J.A.M.A.* **255:**387–388 (1986).
13a. BALFOUR, H. H., JR., Current management of varicella zoster infections, *J. Med. Virol. Suppl.* **1:**74–81 (1993).
14. BARRETT, M. J., HURWITZ, E. S., SCHONBERGER, L. B., AND ROGERS, M. F., Changing epidemiology of Reye syndrome in the United States, *Pediatrics* **77:**598–602 (1986).
15. BERGER, R., FLORENT, G., AND JUST, M., Decrease of the lymphoproliferative response to varicella–zoster antigen in the aged, *Infect. Immun.* **32:**24–27 (1981).
16. BERGER, R., LUESCHER, D., AND JUST, M., Enhancement of varicella–zoster specific immune responses in the elderly by boosting with varicella vaccine, *J. Infect. Dis.* **149:**647 (1984).
17. BIRON, C. A., BYRON, K. S., AND SULLIVAN, J. L., Severe herpesvirus infections in an adolescent without natural killer cells, *N. Engl. J. Med.* **320:**1731–1735 (1989).
18. BLACK, F. L., HIERHOLZER, W. J., PINHEIRO, F. deP., EVANS, A. S., WOODALL, J. P., OBTON, E. M., EMMONS, J. E., WEST, B. S., EDSALL, G., DOWNS, W. G., AND WALLACE, G. D., Evidence for persistence of infectious agents in isolated human populations, *Am. J. Epidemiol.* **100:**230–250 (1974).
19. BLAKELY, G. A., LOURIE, B., MORTON, W. G., EVANS, H. H., AND KAUFMANN, A. F., A varicella-like disease in macaque monkeys, *J. Infect. Dis.* **127:**617–623 (1973).
20. BLENDON, R. J., The changing epidemiology of Reye's syndrome in the United States; further evidence for a public health success, *J.A.M.A.* **260:**3178–3180 (1988).
20a. BODENSTEINER, J. B., HILLE, M. R., AND RIGGS, J. E., Clinical features of vascular thrombosis following varicella, *Am. J. Dis. Child.* **146:**100–102 (1992).
21. BOGGER-GOREN, S., BERNSTEIN, J. M., GERSHON, A. A., AND OGRA, P. L., Mucosal cell-mediated immunity to varicella-zoster virus: Role in protection against disease, *J. Pediatr.* **105:**195–199 (1984).
22. BROUSSARD, R. C., PAYNE, D. K., AND GEORGE, R. B., Treatment with acyclovir of varicella pneumonia in pregnancy, *Chest* **99:**1045–1047 (1991).
23. BRUNELL, P. A., Placental transfer of varicella–zoster antibody, *Pediatrics* **38:**1034–1038 (1966).
24. BRUNELL, P. A., Varicella-zoster infections in pregnancy, *J.A.M.A.* **199:**315–317 (1967).
25. BRUNELL, P. A., GERSHON, A. A., UDUMAN, S. A., AND STEINBERG, S., Varicella–zoster immunoglobulins during varicella, latency, and zoster, *J. Infect. Dis.* **132:**49–54 (1975).
26. BRUNELL, P. A., ROSS, A., MILLER, L. H., AND KUO, B., Prevention of varicella by zoster immune globulin, *N. Engl. J. Med.* **280:**1191–1194 (1969).
27. BUCHBINDER, S. P., KATZ, M. H., HESSOL, N. A., LIU, N. A., LIU, J. Y., O'MALLEY, P. M., UNDERWOOD, R., AND HOLMBERG, S. D., Herpes zoster and human immunodeficiency virus infection, *J. Infect. Dis.* **166:**1154–1156 (1992).
28. BURKE, B. L., STEELE, R. W., BEARD, O. W., WOOD, J. S., CAIN, T. D., AND MARMER, D. J., Immune responses to varicella–zoster in the aged, *Arch. Int. Med.* **142:**291–293 (1982).
29. CAMPBELL-BENZIE, A., HEATH, R. B., RIDEHALGH, M. K. S., AND CRADOCK-WATSON, J. E., A comparison of indirect immunofluorescence and radioimmunoassay for detecting antibody to varicella–zoster virus, *J. Virol. Methods* **6:**135–140 (1983).
30. CENTERS FOR DISEASE CONTROL, Zoster immune globulin, *Morbid. Mortal. Week. Rep.* **25:**211–212 (1976).
31. CENTERS FOR DISEASE CONTROL, Varicella–zoster immune globulin distribution, 1981–1983. Recommendations of the Immunization Practices Committee on use of immune globulin for the prevention of chickenpox, *Morbid. Mortal. Week. Rep.* **33:**81–90 (1984).
32. CENTERS FOR DISEASE CONTROL, Annual summary, 1983. Reported morbidity and mortality in the United States, *Morbid. Mortal. Week. Rep.* **32**(54): 1–125 (1984).

33. CENTERS FOR DISEASE CONTROL, Summary of notifiable diseases, United States, 1991, *Morbid. Mortal. Week. Rep.* **40**(53): 1–64 (1992).

34. CHEATHAM, W. J., WELLER, T. H., DOLAN, T. F., JR., AND DOWER, J. C., Varicella; report of two fatal cases with necropsy, virus isolation, and serologic studies, *Am. J. Pathol.* **32**:1015–1035 (1956).

35. COHRS, R., MAHALINGAM, R., DUELAND, A. N., WOLF, W., WELLISH, M., AND GILDEN, D. H., Restricted transcription of varicella–zoster virus in latently infected human trigeminal and thoracic ganglia, *J. Infect. Dis.* **166**(Suppl. 1):S24–29 (1992).

36. COLBUNDERS, R., MANN, J. M., FRANCIS, H., BILA, K., IZALEY, L., KAKONDE, N., QUINN, T. C., CURRAN, J. W., AND PIOT, P., Herpes zoster in African patients; a clinical predictor of human immunodeficiency virus infection, *J. Infect. Dis.* **157**:314–318 (1988).

37. COLLINS, S. D., WHEELER, R. E., AND SHANNON, R. D., *The Occurrence of Whooping Cough, Chickenpox, Mumps, Measles, and German Measles in 200,000 Surveyed Families in 28 Large Cities*, Special Study Series, No. 1, Division of Public Health Methods, NIH, USPHS, Washington, DC, 1942.

38. COMMITTEE ON INFECTIOUS DISEASES, AMERICAN ACADEMY OF PEDIATRICS, The use of oral acyclovir in otherwise healthy children with varicella, *Pediatrics* **91**:674–676 (1993).

39. CRADOCK-WATSON, J. E., RIDEHALGH, M. K. S., AND BOURNE, M. S., Specific immunoglobulin responses after varicella and herpes zoster, *J. Hyg. (Camb.)* **82**:319–336 (1979).

40. DEMMLER, G. J., STEINBERG, S. P., BLUM, G., AND GERSHON, A. A., Rapid enzyme-linked immunosorbent assay for detecting antibody to varicella–zoster virus, *J. Infect. Dis.* **157**:211–212 (1988).

41. DLUGOSCH, D., EIS-HUBINGER, A. M., KLEIM, J. P., KAISER, R., BIERHOFF, E., AND SCHNEWEIS, K. E., Diagnosis of acute and latent varicella–zoster virus infections using the polymerase chain reaction, *J. Med. Virol.* **35**:136–141 (1991).

42. DUNKLE, L. M., ARVIN, A. M., WHITLEY, R. J., ROTBART, H. A., FEDER, H. M., JR., FELDMAN, S., GERSHON, A. A., LEVY, M. L., HAYDEN, G. F., McGUIRT, P. V., HARRIS, J., AND BALFOUR, H. H., JR., A controlled trial of acyclovir for chickenpox in normal children, *N. Engl. J. Med.* **325**:1539–1544 (1991).

42a. ECHEVARRIA, J. M., CASAS, I., TENORIO, A., DE ORY, F., AND MARTINEZ-MARTIN, P., Detection of varicella–zoster virus-specific DNA sequences in cerebrospinal fluid from patients with acute aseptic meningitis and no cutaneous lesions, *J. Med. Virol.* **43**:331–335 (1994).

43. ECKSTEIN, R., JEHN, U., AND LOY, A., Endemic chickenpox on a cancer ward, *J. Infect. Dis.* **149**:829–830 (1984).

44. EDSON, C. M., HOSLER, B. A., RESPESS, R. A., WATERS, D. J., AND THORLEY-LAWSON, D. A., Cross reactivity between herpes simplex virus glycoprotein B and a 63,000 dalton varicella–zoster virus envelope glycoprotein, *J. Virol.* **56**:333–336 (1985).

44a. ENDERS, G., MILLER, M., CRADOCK-WATSON, J., BOLLEY, I., AND RIDEHALGH, M., Consequences of varicella and herpes zoster in pregnancy; prospective study of 1739 cases, *Lancet* **343**:1547–1550 (1994).

45. FEDER, H. M., JR., Treatment of adult chickenpox with oral acyclovir, *Arch. Intern. Med.* **150**:2061–2065 (1990).

46. FELDMAN, S., CHAUDARY, S., OSSI, M., AND EPP, E. A., A viremic phase for herpes zoster in children with cancer, *J. Pediatr.* **91**:597–600 (1977).

47. FELDMAN, S., HUGHES, W. T., AND KIM, H. Y., Herpes zoster in children with cancer, *Am. J. Dis. Child.* **126**:178–184 (1973).

48. FELDMAN, S., AND LOTT, L., Varicella in children with cancer: Impact of antiviral therapy and prophylaxis, *Pediatrics* **80**:465–472 (1987).

49. FLEISHER, G., HENRY, W., McSORLEY, M., ARBETER, A., AND PLOTKIN, S., Life threatening complications of varicella, *Am. J. Dis. Child.* **135**:896–899 (1981).

50. FORGHANI, B., SCHMIDT, N. J., AND DENNIS, J., Antibody assays for varicella–zoster virus: Comparison of enzyme immunoassay with neutralization, immune adherence hemagglutination and complement fixation, *J. Clin. Microbiol.* **8**:545–552 (1978).

51. FOUNG, S. K. H., PERKINS, S., KOROPCHAK, C., FISHWILD, D. M., WITTEK, A. E., ENGLEMAN, E. G., GRUMET, F. C., AND ARVIN, A. M., Human monoclonal antibodies neutralizing varicella–zoster virus, *J. Infect. Dis.* **152**:280–285 (1985).

52. FREY, H. M., BIALKIN, G., AND GERSHON, A. A., Congenital varicella: Case report of a serologically proved long-term survivor, *Pediatrics* **59**:110–112 (1977).

53. FREY, H. M., STEINBERG, S. P., AND GERSHON, A. A., Diagnosis of varicella–zoster infections, in: *The Human Herpesviruses. An Interdisciplinary Perspective* (A. J. NAHMIAS, W. R. DOWDLE, AND R. F. SCHINAZI, EDS.), pp. 351–362, Elsevier, New York, 1981.

54. FRIEDMAN, M. G., HAIKIN, H., LEVENTON-KRISS, S., JOFFE, R., GOLDSTEIN, V., AND SAROV, I., Detection of antibodies to varicella zoster virus by radio-immunoassay and enzyme immunoassay techniques, *Med. Microbiol. Immunol.* **166**:177–186 (1978).

55. FURUTA, Y., TAKASU, Y., FUKUDA, S., SATO-MATSUMURA, K. C., INUYAMA, Y., HONDO, R., AND NAGASHIMA, K., Detection of varicella–zoster DNA in human geniculate ganglia by polymerase chain reaction, *J. Infect. Dis.* **166**:1157–1159 (1992).

56. GARLAND, J., Varicella following exposure to herpes zoster, *N. Engl. J. Med.* **228**:336–337 (1943).

56a. GARNETT, G. P., COX, J. M., BUNDY, D. A. P., DIDIER, J. M., AND ST. CATHERINE, J., The age of infection with varicella–zoster virus in St. Lucia, West Indies, *Epidemiol. Infect.* **110**:361–372 (1993).

57. GARNETT, G. P., AND GRENFELL, B. T., The epidemiology of varicella–zoster virus infections: A mathematical model, *Epidemiol. Infect.* **108**:495–511 (1992).

58. GARNETT, G. P., AND GRENFELL, B. T., The epidemiology of varicella–zoster infections: The influence of varicella on the prevalence of zoster, *Epidemiol. Infect.* **108**:513–528 (1992).

59. GERSHON, A. A., Varicella vaccine; still at the cross roads, *Pediatrics* **90**(Suppl.):144–148 (1992).

60. GERSHON, A. A., BRUNELL, P. A., DOYLE, E. F., AND CLAPPS, A. A., Steroid therapy and varicella, *J. Pediatr.* **81**:1034 (1972).

61. GERSHON, A. A., FREY, H. M., STEINBERG, S. P., SEEMAN, M. D., BIDWELL, D., AND VOLLER, A., Determination of immunity to varicella using an enzyme-linked-immunosorbent assay, *Arch. Virol.* **70**:169–172 (1981).

62. GERSHON, A. A., AND KRUGMAN, S., Seroepidemiologic survey of varicella: Value of specific fluorescent antibody test, *Pediatrics* **56**:1005–1008 (1976).

62a. GERSHON, A. A., LaRUSSA, P., AND STEINBERG, S., Detection of antibodies to varicella–zoster virus using a latex agglutination assay, *Clin. Diagn. Virol.* **2**:271–277 (1994).

63. GERSHON, A. A., RAKER, R., STEINBERG, S., TOPF-OLSTEIN, B., AND DRUSIN, L. M., Antibody to varicella–zoster virus in parturient women and their offspring during the first year of life, *Pediatrics* **58**:692–696 (1976).

64. GERSHON, A. A., AND STEINBERG, S. P., Cellular and humoral immune responses to varicella–zoster virus in immunocompro-

mised patients during and after varicella–zoster infections, *Infect. Immun.* **25:**170–174 (1979).

65. GERSHON, A., STEINBERG, S., AND BRUNELL, P. A., Zoster immune globulin: A further assessment, *N. Engl. J. Med.* **290:**243–245 (1974).

66. GERSHON, A. A., STEINBERG, S. P., GELB, L., AND THE NATIONAL INSTITUTE OF ALLERGY AND INFECTIOUS DISEASES COLLABORATIVE VARICELLA VACCINE STUDY GROUP, Clinical reinfection with varicella–zoster virus, *J. Infect. Dis.* **149:**137–142 (1984).

67. GERSHON, A. A., STEINBERG, S., AND SMITH, M., Cell-mediated immunity to varicella–zoster virus demonstrated by viral inactivation with human leucocytes, *Infect. Immun.* **13:**1549–1553 (1976).

68. GILCHREST, B., AND BADEN, H. P., Photodistribution of viral exanthems, *Pediatrics* **54:**136–138 (1974).

68a. GILDEN, D. H., BEINLICH, B. R., RUBINSTIEN, E. M., STOMMEL, E., SWENSON, R., RUBINSTEIN, D., AND MAHALINGAM, R., Varicella–zoster myelitis, an expanding syndrome, *Neurology* **44:**1818–1823 (1994).

69. GILDEN, D. H., VAFAI, A., SHTRAM, Y., BECKER, Y., DEVLIN, M., AND WELLISH, M., Varicella–zoster virus DNA in human sensory ganglia, *Nature* **306:**478–480 (1983).

70. GLYNN, C., CROCKFORD, G., DAVAGHAN, D., CARDNO, P., PRICE, D., AND MILLER, J., Epidemiology of shingles, *J. R. Soc. Med.* **83:**617–619 (1990).

70a. GOGOS, C. A., APOSTOLIDOU, E., BASSARIS, H. P., AND VAGENAKIS, A. G., Three cases of varicella thrombophlebitis as a complication of varicella zoster infection, *Eur. J. Clin. Microbiol. Infect. Dis.* **12:**43–45 (1993).

71. GOLD, E., Serologic and virus-isolation studies of patients with varicella or herpes–zoster infection, *N. Engl. J. Med.* **274:**181–185 (1966).

72. GOLDBERG, R. D., AND SAROV, I., Enzyme-linked immunosorbent assay for detection of antibodies to varicella–zoster virus, *Isr. J. Med. Sci.* **16:**111–117 (1980).

73. GORDON, J. E., General considerations of modes of transmission, in: *Preventive Medicine in World War II*, Vol. IV, *Communicable Diseases*, p. 27, Department of the Army, Washington, DC, 1958.

74. GORDON, J. E., Chickenpox: An epidemiological review, *Am. J. Med. Sci.* **244:**362–389 (1962).

75. GRAY, G. C., PALINKAS, L. A., AND KELLEY, P. W., Increasing incidence of varicella hospitalizations in United States army and navy personnel; are today's teenagers more susceptible? Should recruits be vaccinated? *Pediatrics* **86:**867–873 (1990).

76. GROSE, C., EDWARDS, D. P., FRIEDRICHS, W. E., WEIGLE, K. A., AND McGUIRE, W. L., Monoclonal antibodies against three major glycoproteins of varicella–zoster virus, *Infect. Immun.* **40:**381–388 (1983).

77. GUESS, H. A., BROUGHTON, D. D., MELTON, L. J., III, AND KURLAND, L. T., Chickenpox hospitalizations among residents of Olmsted County, Minnesota, 1962 through 1981. A population based study, *Am. J. Dis. Child.* **138:**1055–1057 (1984).

78. GUESS, H. A., BROUGHTON, D. D., MELTON, L. J., III, AND KURLAND, L. T., Epidemiology of herpes zoster in children and adolescents: A population based study, *Pediatrics* **76:**512–517 (1985).

79. GULICK, R. M., HEATH-CHIOZZI, M., AND CRUMPACKER, C. S., Varicella–zoster virus disease in patients with human immunodeficiency virus infection, *Arch. Derm.* **126:**1086–1088 (1990).

80. GUSTAFSON, T. L., LAVELY, G. B., BRAWNER, E. R., JR., HUTCHESON, R. H., WRIGHT, P. F., AND SCHAFFNER, W., An

81. HAAKE, D. A., ZAKOWSKI, P. C. HAAKE, D. L., AND BRYSON, Y. J., Early treatment with acyclovir for varicella pneumonia in otherwise healthy adults: Retrospective controlled study and review, *Rev. Infect. Dis.* **12:**788–798 (1990).

82. HACHAM, M., LEVENTON-KRISS, S., AND SAROV, I., Enzyme-linked immunosorbent assay for detection of virus-specific IgM antibodies to varicella–zoster virus, *Intervirology* **13:**214–222 (1980).

83. HAGGERTY, R. J., AND ELEY, R. C., Varicella and cortisone, *Pediatrics* **18:**160–162 (1956).

84. HAIKIN, H., LEVENTON-KRISS, S., AND SAROV, I., Antibody to varicella–zoster virus-induced membrane antigen; immunoperoxidase assay with air-dried target cells, *J. Infect. Dis.* **140:**601–604 (1979).

85. HAYWARD, A., LEVIN, M., WOLF, W., ANGELOVA, G., AND GILDEN, D., Varicella–zoster virus specific immunity after herpes zoster, *J. Infect. Dis.* **163:**873–875 (1991).

86. HELLINGER, W. C., BOLLING, J. P., SMITH, T. F., AND CAMPBELL, R. J., Varicella–zoster virus retinitis in a patient with AIDS-related complex; case report and brief review of the acute retinal necrosis syndrome, *Clinical Infect. Dis.* **16:**208–212 (1993).

87. HIGA, K., DAN, K., AND MANABE, H., Varicella-zoster virus infections during pregnancy: Hypothesis concerning the mechanisms of congenital malformations, *Obstet. Gynecol.* **69:**214–222 (1987).

88. HONDO, R., YOGO, Y., KURATA, T., AND AOYAMA, Y., Genome variation among varicella–zoster isolates derived from different individuals and from the same individuals, *Arch. Virol.* **93:**1–12 (1987).

89. HOOK, E. B., ORANDI, M., TEN BENSEL, R. W., SCHAMBER, W. F., AND ST. GEME, J. W., JR., Familial fatal varicella, *J.A.M.A.* **206:**305–311 (1968).

90. HOPE-SIMPSON, R. E., Infectiousness of communicable diseases in the household (measles, chickenpox, and mumps), *Lancet* **2:**549–554 (1952).

91. HOPE-SIMPSON, R. E., The nature of herpes zoster; a long-term study and a new hypothesis, *Proc. R. Soc. Med.* **58:**9–20 (1965).

91a. HUFF, J. C., DRUCKER, J. L., CLEMMER, A., LASKIN, O. L., CONNOR, J. D., BRYSON, Y. J., AND BALFOUR, H. H., JR., Effect of oral acyclovir on pain resolution in herpes zoster: A reanalysis, *J. Med. Virol. Suppl.* **1:**93–96 (1993).

91b. HUSE, D. M., MEISSNER, H. C., LACEY, M. J., AND OSTER, G., Childhood vaccination against chickenpox: An analysis of benefits and costs, *J. Pediatr.* **124:**869–874 (1994).

92. HYMAN, R. W., ECKER, J. R., AND TENSER, R. B., Varicella–zoster virus RNA in human trigeminal ganglia, *Lancet* **2:**814–816 (1983).

93. INSTITUTE OF MEDICINE COMMITTEE ON ISSUES AND PRIORITIES FOR NEW VACCINE DEVELOPMENT, *Diseases of Importance in the United States*, Appendix J: Prospects for immunizing against *Herpesvirus varicellae*, pp. 313–341, National Academy of Sciences, Washington, DC, 1985.

94. JACOBSON, M. A., BERGER, T. O., FIKRIG, S., BECHERER, P., MOOHR, J. W., STANAT, S. C., AND BIRON, K. K., Acyclovir-resistant varicella zoster virus after chronic oral acyclovir therapy in patients with acquired immunodeficiency syndrome (AIDS), *Ann. Intern. Med.* **112:**1187–1191 (1990).

95. JORDAN, G. W., AND MERIGAN, T. C., Cell-mediated immunity to varicella–zoster virus: *In vitro* lymphocyte responses, *J. Infect. Dis.* **130:**495–501 (1974).

outbreak of airborne nosocomial varicella, *Pediatrics* **70:**550–556 (1982).

96. JOSEPHSON, A., AND GOMBERT, M. E., Airborne transmission of nosocomial varicella from localized zoster, *J. Infect. Dis.* **158:** 238–241 (1988).

97. JUEL-JENSEN, B. E., A new look at infectious diseases: Herpes simplex and zoster, *Br. Med. J.* **1:**406–410 (1973).

98. KALLANDER, C. F. R., GRONOWITZ, J. S., AND OLDING-STEN-KVIST, E., Rapid diagnosis of varicella–zoster virus infection by detection of viral deoxythymidine kinase in serum and vesicle fluid, *J. Clin. Microbiol.* **17:**280–287 (1983).

99. KAPSENBERG, J. G., Possible antigenic relationships between varicella–zoster virus and herpes simplex virus, *Arch. Ges. Virusforsch.* **15:**67–73 (1964).

99a. KASS, E. M., SZANIAWSKI, W. K., LEVY, H., LEACH, J., SRI-NIVASAN, K., AND RIVES, C., Rickettsialpox in a New York city hospital, 1980 to 1989, *N. Engl. J. Med.* **331:**1612–1617 (1994).

100. KIDO, S., OZAKI, T., ASADA, H., HIGASHI, K., KONDO, K., HAYA-KAWA, Y., MORISHIMA, T., TAKAHASHI, M., AND YAMANISHI, K., Detection of varicella–zoster virus (VZV) DNA in clinical samples from patients with VZV by the polymerase chain reaction, *J. Clin. Microbiol.* **29:**76–79 (1991).

101. KJERSEM, H., AND JEPSEN, S., Varicella among immigrants from the tropics, a health problem, *Scand. J. Soc. Med.* **18:**171–174 (1990).

102. KORANDA, F. C., DEHMEL, E. M., KAHN, G., AND PENN, I., Cutaneous complications in immunosuppressed renal homograft recipients, *J.A.M.A.* **229:**419–424 (1974).

103. KUNDRATITZ, K., Experimentelle Ubertragung von Herpes Zoster auf den Menschen und die Beziehungen von Herpes zoster zu Varicellen, *Monatsschr. Kinderheilkd.* **29:**516–522 (1925).

104. LAFORET, E. G., AND LYNCH, C. L., JR., Multiple congenital defects following maternal varicella. Report of a case, *N. Engl. J. Med.* **236:**534–537 (1947).

105. LANDRY, M. L., AND HSIUNG, G. D., Diagnosis of dual herpes virus infection: Varicella and herpes simplex viruses, in: *The Human Herpesviruses* (A. J. NAHMIAS, W. R. DOWDLE, AND R. F. SCHINAZI, EDS.), pp. 652–653, Elsevier, New York, 1981.

106. LARKIN, M., HECKELS, J. E., AND OGILVIE, M. M., Antibody response to varicella–zoster virus surface glycoproteins in chickenpox and shingles, *J. Gen. Virol.* **66:**1785–1793 (1985).

107. LECLAIR, J. M., ZAIA, J. A., LEVIN, M. J., CONGDON, R. G., AND GOLDMANN, D. A., Airborne transmission of chickenpox in a hospital, *N. Engl. J. Med.* **302:**450–453 (1980).

108. LEVIN, M. J., MURRAY, M., ROTBART, H. A., ZERBE, G. O., WHITE, C. J., AND HAYWARD, A. R., Immune response of elderly individuals to live attenuated varicella vaccine, *J. Infect. Dis.* **166:**253–259 (1992).

109. LEVIN, M. J., LEVENTHAL, S., AND MASTERS, H. A., Factors influencing quantitative isolation of varicella virus, *J. Clin. Microbiol.* **19:**880–883 (1984).

109a. LEVIN, M. J., MURRAY, M., ZERBE, G. O., WHITE, C. J., AND HAYWARD, A. R., Immune responses of elderly persons 4 years after receiving a live attenuated varicella vaccine, *J. Infect. Dis.* **170:**522–526 (1994).

109b. LIEU, T. A., COCHI, S. L., HALLORAN, M. E., SHINEFIELD, H. R., HOLMES, S. J., WHARTON, M., AND WASHINGTON, A. E., Cost-effectiveness of a routine varicella vaccination program for US children, *J.A.M.A.* **271:**375–381 (1994).

110. LIPSCHÜTZ, B., Untersuchungen uber die Atiologie der Krankheiten des Herpesgruppe (Herpes zoster, Herpes genitalis, Herpes febrilis), *Arch. Dermatol. Syph. Orig.* **136:**428–482 (1921).

111. LIPTON, S. V., AND BRUNELL, P. A., Management of varicella exposure in a neonatal intensive care unit, *J.A.M.A.* **261:**1782–1784 (1989).

112. LOCKSLEY, R. M., FLOURNOY, N., SULLIVAN, K. M., AND MEYERS, J. D., Infection with varicella–zoster virus after marrow transplantation, *J. Infect. Dis.* **152:**1172–1181 (1985).

112a. LOKKE JENSEN, B., WEISMANN, K., MATHIESEN, L., AND KLEM THOMSEN, H., Atypical varicella–zoster infection in AIDS, *Acta Derm. Venereol. (Stockh.)* **73:**123–125 (1993).

113. LONDON, W. P., AND YORKE, J. A., Recurrent outbreaks of measles, chickenpox, and mumps. I. Seasonal variations in contact rates, *Am. J. Epidemiol.* **98:**453–468 (1972).

114. LONGFIELD, J. N., WINN, R. E., GIBSON, R. L., JUCHAU, S. V., AND HOFFMAN, P. V., Varicella outbreaks in army recruits from Puerto Rico. Varicella susceptibility in a population from the tropics, *Arch. Intern. Med.* **150:**970–973 (1990).

115. LONNQVIST, B., LJUNGMAN, P., BOLME, P., GAHRTON, G., RING-DEN, O., AND WAHREN, B., Frequency of reactivation of varicella–zoster virus (VZV) infection with and without symptoms after bone marrow transplantation, *Exp. Hematol.* **13**(Suppl. 17):72 (1985).

116. LUBY, J. P., RAMIREZ-RONDA, C., RINNER, S., HULL, A., AND VERGNE-MARINI, P., A longitudinal study of varicella–zoster virus infections in renal transplant recipients, *J. Infect. Dis.* **135:**659–663 (1970).

117. LUX, S. E., JOHNSTON, R. B., AUGUST, C. S., SAY, B., PENCHA-SZADEH, V. B., ROSEN, F. S., AND MCKUSICK, V. A., Chronic neutropenia and abnormal cellular immunity in cartilage-hair hypoplasia, *N. Engl. J. Med.* **282:**231–236 (1970).

117a. LYALL, E. G. H., OGILVIE, M. M., SMITH, N. M., AND BURNS, S., Acyclovir-resistant varicella zoster and HIV infection, *Arch. Dis. Child.* **70:**133–135 (1994).

118. LYNFIELD, R., HERRIN, J. T., AND RUBIN, R. H., Varicella in pediatric renal transplant recipients, *Pediatrics* **90:**216–220 (1992).

119. MAHALINGAM, R., WELLISH, M. C., DUELAND, A. N., COHRS, R. J., AND GILDEN, D. H., Localization of herpes simplex virus and varicella zoster virus DNA in human ganglia, *Ann. Neurol.* **31:**444–448 (1992).

120. MANSON, M. M., LOGAN, W. P. D., AND LOY, R. M., *Rubella and Other Virus Infections during Pregnancy*, Report 101, pp. 1–101, Ministry of Health, Her Majesty's Stationery Office, London, 1960.

121. MARETIC, A., AND COORAY, M. P. M., Comparisons between chickenpox in a tropical and a European country, *J. Trop. Med. Hyg.* **66:**311–315 (1963).

122. MARTIN, J. H., DOHNER, D. E., WELLINGHOFF, W. J., AND GELB, L. D., Restriction endonuclease analysis of Varicella–zoster vaccine virus and wild-type DNAs, *J. Med. Virol.* **9:**69–76 (1982).

122a. MARTIN, K. A., JUNKER, A. K., THOMAS, E. E., VAN ALLEN, M. I., AND FRIEDMAN, J. M., Occurrence of chickenpox during pregnancy in women seropositive for varicella–zoster virus, *J. Infect. Dis.* **170:**991–995 (1994).

122b. MARWICK, C., Lengthy tale of varicella vaccine development finally nears a clinically useful conclusion, *J.A.M.A.* **273:**833–836 (1995).

123. MCGREGOR, R. M., Herpes zoster, chickenpox and cancer in general practice, *Br. Med. J.* **1:**84–87 (1957).

124. MCGREGOR, R. S., ZITELLI, B. J., URBACH, A. H., MALATACK, J. J., AND GARTNER, J. C., Varicella in pediatric orthotopic liver transplant recipients, *Pediatrics* **83:**256–261 (1989).

125. MCKUSICK, V. A., Fatal varicella (letter to the editor), *J.A.M.A.* **207:**370 (1969).

126. MEIER, J. L., HOLMAN, R. P., CROEN, K. D., SMIALEK, J. E., AND STRAUS, S. E., Varicella–zoster virus transcription in human trigeminal ganglia, *Virology* **193**:193–200 (1993).

127. MESZNER, Z., NYERGES, G., AND BELL, A. R., Oral acyclovir to prevent dissemination of varicella in immunocompromised children, *J. Infection* **26**:9–15 (1993).

128. MEURISSE, V., MILLER, E., AND KENSIT, J., Varicella in maternity units, *Lancet* **1**:1100–1101 (1990).

129. MEYERS, J. D., Congenital varicella in term infants; risk reconsidered, *J. Infect. Dis.* **129**:215–217 (1974).

130. MILLER, A. E., Selective decline in cellular immune response to varicella–zoster in the elderly, *Neurology (Minneap.)* **30**:582–587 (1980).

131. MILLER, E., VURDIEN, J., AND FARRINGTON, P., Shift in age in chickenpox, *Lancet* **1**:308–309 (1993).

131a. NADER, S., BERGEN, R., SHARP, M., AND ARVIN, A. M., Age-related differences in cell-mediated immunity to varicella–zoster virus among children and adults immunized with live attenuated varicella vaccine, *J. Infect. Dis.* **171**:13–17 (1995).

132. NAHASS, G. T., GOLDSTEIN, B. A., ZHU, W. Y., SERFLING, U., PENNEYS, N. S., AND LEONARDI, C. L., Comparison of Tzanck smear, viral culture, and DNA diagnostic methods in detection of herpes simplex and varicella–zoster infection, *J.A.M.A.* **268**:2541–2544 (1992).

133. NARAQUI, S., JACKSON, G. G., JONASSON, O., AND YAMASHI-ROYA, H. M., Prospective study of prevalence, incidence, and source of herpes virus infections in patients with renal allografts, *J. Infect. Dis.* **136**:531–540 (1977).

133a. NGUYEN, P., REYNAUD, J., POUZOL, P., MUNZER, M., RICHARD, O., AND FRANCIS, P., Varicella and thrombotic complications associated with transient protein C and protein S deficiencies in children, *Eur. J. Pediatr.* **153**:646–649 (1994).

134. NYERGES, G., MESZNER, Z. GYARMATI, E., AND KERPEL-FRONIUS, S., Acyclovir prevents dissemination of varicella in immunocompromised children, *J. Infect. Dis.* **157**:309–311 (1988).

135. OBERG, G., AND SVEDMYR, A., Varicelliform eruptions in herpes zoster; some clinical and serological observations, *Scand. J. Infect. Dis.* **1**:47–49 (1969).

136. OSLER, W., *The Principles and Practice of Medicine*, p. 65, Appleton, New York, 1892.

137. OZAKI, T., ICHIKAWA, T., MATSUI, Y., NAGAI, T., ASANO, Y., YAMINISHI, K., AND TAKAHASHI, M., Viremic phase in nonimmunocompromised children with varicella, *J. Pediatr.* **104**:85–87 (1984).

138. OZAKI, T., MATSUI, Y., ASANO, Y., OKUNO, T., YAMANISHI, K., AND TAKAHASHI, M., Study of virus isolation from pharyngeal swabs in children with varicella, *Am. J. Dis. Child.* **143**:1448–1450 (1989).

139. OZAKI, T., MIWATA, H., MATSUI, Y., KIDO, S., AND YAMANISHI, K., Varicella–zoster virus DNA in throat swabs, *Arch. Dis. Child.* **66**:333–334 (1991).

140. PAHWA, S., BIRON, K., LIM, W., SWENSON, P., KAPLAN, M. H., SADICK, N., AND PAHWA, R., Continuous varicella–zoster infection associated with acyclovir resistance in a child with AIDS, *J.A.M.A.* **260**:2879–2882 (1988).

141. PALLETT, A., AND NICHOLLS, M. W. N., Varicella–zoster. Reactivation or reinfection?, *Lancet* **1**:160 (1986).

142. PARYANI, S. G., AND ARVIN, A. M., Intrauterine infection with varicella–zoster virus after maternal varicella, *N. Engl. J. Med* **314**:1542–1546 (1986).

143. PARYANI, S. G., ARVIN, A. M., KOROPCHAK, C. M., WITTEK, A.

E., AMYLON, M. D., DOBKIN, M. B., AND BUDINGER, M. D., Comparison of varicella–zoster antibody titers in patients given intravenous immune serum globulin or varicella–zoster immune globulin, *J. Pediatr.* **105**:200–205 (1984).

143a. PASTUSZAK, A. L., LEVY, M., SCHICK, B., ZUBER, C., FELDCAMP, M., GLADSTONE, J., BAR-LEVY, F., JACKSON, E., DONNENFELD, A., MESHINO, W., AND KAREN, G., Outcome after maternal varicella in the first 20 weeks of pregnancy, *N. Engl. J. Med.* **330**:901–905 (1994).

144. PERRONNE, C., LAZANAS, M., LEPORT, C., SIMON, F., SALMON, D., DALLOT, A., AND VILDE, J.-L., Varicella in patients infected with the human immunodeficiency virus, *Arch. Dermatol.* **126**:1033–1036 (1990).

145. PONS, J.-C., ROZENBERG, F., IMBERT, M.-C., LEBRON, P., OLIVENNES, F., LELAIDIER, C., STRAUB, N., VIAL, M., AND FRYDMAN, R., Prenatal diagnosis of second-trimester congenital varicella syndrome, *Prenat. Diagn.* **12**:975–976 (1992).

146. PREBLUD, S. R., Age-specific risks of varicella complications, *Pediatrics* **68**:14–17 (1981).

147. PREBLUD, S. R., COCHI, S. L., AND ORENSTEIN, W. A., Letter to the editor, *N. Engl. J. Med.* **315**:1416–1417 (1986).

148. PREBLUD, S. R., AND D'ANGELO, L. J., Chickenpox in the United States, 1972–1977, *J. Infect. Dis.* **140**:257–260 (1979).

149. RAGOZZINO, M. W., MELTON, L. J., III, KURLAND, L. T., CHU, C. P., AND PERRY, H. O., Population-based study of herpes zoster and its sequelae, *Medicine* **61**:310–316 (1982).

150. RASMUSSEN, L., HOLMES, A. R., HOFMEISTER, R., AND MERIGAN, T., Multiplicity-dependent replication of varicella–zoster virus in interferon treated cells, *J. Gen. Virol.* **35**:361–368 (1977).

151. RIFKIND, D., The activation of varicella–zoster virus infections by immunosuppressive therapy, *J. Lab. Clin. Med.* **68**:463–474 (1966).

152. ROBERTS, G. B., FYFE, J. A., GAILLARD, R. K., AND SHORT, S. A., Mutant varicella–zoster thymidine kinase; correlation of clinical resistance and enzyme impairment, *J. Virol.* **65**:6407–6413 (1991).

153. ROSS, A. H., Modification of chickenpox in family contacts by administration of gamma globulin, *N. Engl. J. Med* **267**:369–376 (1962).

154. ROSS, C. A. C., SUBAK SHARPE, J. H., AND FERRY, P., Antigenic relationship of varicella–zoster and herpes simplex, *Lancet* **2**:708–711 (1965).

155. RUCKDESCHEL, J. C., SCHIMPFF, S. C., SMYTH, A. C., AND MARDINEY, M. R., JR., Herpes zoster and impaired cell-associated immunity to the varicella–zoster virus in patients with Hodgkin's disease, *Am. J. Med.* **62**:77–85 (1977).

156. SABIN, A. B., Neurotropic virus diseases of man, *J. Pediatr.* **19**:445–451 (1941).

157. SALOMON, J. B., GORDON, J. E., AND SCRIMSHAW, N. S., Studies of diarrheal disease in Central America. X. Associated chickenpox, diarrhea and kwashiorkor in a highland Guatemalan village, *Am. J. Trop. Med.* **15**:997–1002 (1966).

158. SCHAAP, G. J. P., AND HUISMAN, J., Simultaneous rise in complement fixing antibodies against herpesvirus hominis and varicella–zoster virus, *Arch. Ges. Virusforsch.* **17**:495–503 (1968).

158a. SCHMADER, K., GEORGE, L. K., BURCHETT, B. M., PIEPER, C. F., AND HAMILTON, J. D., Racial differences in the occurence of zoster, *J. Infect. Dis.* **171**:701–704 (1995).

159. SCHMIDT, N. J., LENNETTE, E. H., AND MAGOFFIN, R. L., Immunologic relationship between herpes simplex and varicella–zoster viruses demonstrated by complement-fixation, neutralization, and fluorescent antibody tests, *J. Gen. Virol.* **4**:321–328 (1969).

160. SCHOOL EPIDEMICS COMMITTEE OF GREAT BRITAIN, *Epidemics in Schools*, Medical Research Council, Special Report Series No. 227, London, His Majesty's Stationery Office, 1938.

161. SELBY, P. J., POWLES, R. L., JAMESON, B., KAY, H. E. M., WATSON, J. G., THORNTON, R., MORGENSTERN, G., CLINK, H. M., MCELWAIN, T. J., PRENTICE, H. G., CORRIGHAM, R., ROSS, M. G., HOFFBRAND, A. V., AND BRIGDEN, D., Parenteral acyclovir therapy for herpesvirus infection in man, *Lancet* **2:**1267–1270 (1979).

162. SHANLEY, J., MYERS, M., EDMOND, B., AND STEELE, R., Enzyme-linked immunosorbent assay for detection of antibody to varicella–zoster virus, *J. Clin. Microbiol.* **15:**208–211 (1982).

163. SHEHAB, Z., AND BRUNELL, P. A., Enzyme-linked immunosorbent assay for susceptibility to varicella, *J. Infect. Dis.* **148:**472–476 (1983).

164. SIEGEL, M., Congenital malformations following chickenpox, measles, mumps, and hepatitis: Results of a cohort study, *J.A.M.A.* **226:**1521–1524 (1973).

165. SIEGEL, M., AND FUERST, H. T., Low birth weight and maternal virus disease: A prospective study of rubella, measles, mumps, chickenpox, and hepatitis, *J.A.M.A.* **197:**680–684 (1966).

166. SINHA, D. P., Chickenpox—a disease predominantly affecting adults in West Bengal, India, *Int. J. Epidemiol.* **5:**367–374 (1976).

167. SMITH, G. R., MCKENZIE, J. M., MARMER, D. J., AND STEELE, R. W., Psychologic modulation of the human immune response to varicella zoster, *Arch. Intern. Med.* **145:**2110–2112 (1985).

168. SORIANO, V., BRU, F., AND GONZALEZ-LAHOS, J., Fatal varicella hepatitis in a patient with AIDS, *J. Infect.* **25:**107 (1992).

169. STEINER, Zur Inokulation der Varicellen., *Wien. Med. Wochenschr.* **25:**306 (1875).

170. STEVENS, D. A., FERRINGTON, R. A., JORDAN, G. W., AND MERIGAN, T. C., Cellular events in zoster vesicles; relation to clinical course and immune parameters, *J. Infect. Dis.* **131:**509–515 (1975).

171. STOKES, J., JR., Chickenpox, in: *Preventive Medicine in World War II*, Vol. IV, *Communicable Diseases, Transmitted Chiefly Through Respiratory and Alimentary Tracts*, (E. C. HOFF AND R. M. HOFF, EDS.), pp. 55–56, Department of Army, Washington, DC, 1958.

172. STRAUS, S. E., Shingles; sorrows, salves, solutions, *J.A.M.A.* **269:**1836–1839 (1993).

172a. STRAUS, S. E., Overview: The biology of varicella–zoster virus infection, *Ann. Neurol. Suppl.* **35:**S-4–8 (1994).

173. STRAUS, S. E., HAY, J., SMITH, H., AND OWENS, J., Genome differences among varicella–zoster virus isolates, *J. Gen. Virol.* **64:**1031–1041 (1983).

174. STRAUS, S. E., REINHOLD, W., SMITH, H. A., RUYECHAN, W. T., HENDERSON, D. K., BLAESE, R. M., AND HAY, J., Endonuclease analysis of viral DNA from varicella and subsequent zoster infections in the same patient, *N. Engl. J. Med.* **311:**1362–1364 (1984).

175. STRÖM, J., Social development and declining incidence of some common epidemic diseases in children; a study of the incidence in different age groups in Stockholm, *Acta Paediatr. Scand.* **56:**159–163 (1967).

176. TAKAHASHI, M., Current status and prospects of live varicella vaccine, *Vaccine* **10:**1007–1014 (1992).

177. TAKAHASHI, M., OKUNO, Y., OTSUKA, T., OSAME, J., TAKAMIZAWA, A., SASUDO, T., AND KUBO, T., Development of a live attenuated varicella vaccine, *Biken J.* **18:**25–33 (1975).

178. TAKAHASHI, M., OTSUKA, T., OKUNO, Y., ASANO, Y., YAZAKI, T., AND ISOMURA, S., Live vaccine used to prevent the spread of varicella in children in hospital, *Lancet* **2:**1288–1290 (1974).

179. TALBOT, G. H., SKROS, M., FISHER, M., AND FRIEDMAN, H., Immunologic evidence of reinfection with varicella–zoster virus, *J. Infect. Dis.* **149:**1035–1036 (1984).

179a. TERADA, K., KAWANO, S., YOSHIHIRO, A., MIYASHIMA, H., AND MORITA, T., Characteristics of herpes zoster in otherwise normal children, *Pediatr. Infect. Dis. J.* **12:**960–961 (1993).

180. TYZZER, E. E., The histology of the skin lesions in varicella, *Philipp. J. Sci.* **1:**349–372 (1906).

181. UDUMAN, S. A., GERSHON, A. A., AND BRUNELL, P. A., Rapid diagnosis of varicella–zoster by agar-gel diffusion, *J. Infect. Dis.* **126:**193–195 (1972).

182. VERROUST, F., LEMAY, D., AND LAURIAN, Y., High frequency of herpes zoster in young hemophiliacs, *N. Engl. J. Med.* **316:**166–167 (1987).

183. VIROLAINEN, M., SAVILAHTI, E., KATILA, I., AND PERHENTUPA, J., Cellular and humoral immunity in cartilage-cell hypoplasia, *Pediatr. Res.* **12:**961–966 (1978).

184. VON BOKAY, J., Uber den aetiologischen Zussammenhang der Varicellen mit gewissen Fallen von Herpes zoster, *Wien. Klin. Wochenschr.* **22:**1323–1326 (1909).

185. WADE, N. A., LEPOW, M. L., VEAZEY, J., AND MEUWISSEN, H. J., Progressive varicella in three patients with Wiskott–Aldrich syndrome; treatment with adenine arabinoside, *Pediatrics* **75:**672–675 (1985).

186. WALLACE, M. R., BOWLER, W. A., MURRAY, N. B., BRODINE, S. K., AND OLDFIELD, E. C., III, Treatment of adult varicella with oral acyclovir. A randomized, placebo-controlled trial, *Ann. Intern. Med.* **117:**358–363 (1992).

187. WANER, J. L., AND WELLER, T. H., Varicella–zoster virus, in: *Diagnostic Procedures for Viral, Rickettsial, and Chlamydial Infections*, 6th ed. (N. J. SCHMIDT AND R. W. EMMONS, EDS.), pp. 379–406, American Public Health Association, Washington, DC, 1979.

187a. WATSON, B., BOARDMAN, C., LAUFER, S., PIERCY, S., TUSTIN, N., OLALEYE, D., CNAAN, A., AND STARR, S. E., Humoral and cell-mediated immune responses in healthy children after one or two doses of varicella vaccine, *Clin. Infect. Dis.* **20:**316–319 (1995).

188. WEBER, D. M., AND PELLECCHIA, J. A., Varicella pneumonia; study of prevalence in adult men, *J.A.M.A.* **192:**572–573 (1965).

189. WEBER, D. J., RUTALA, W. A., AND PARHAM, C., Impact and costs of varicella prevention in a university hospital, *Am. J. Public Health* **78:**19–23 (1988).

190. WEIGLE, K. A., AND GROSE, C., Molecular dissection of the humoral immune response to individual varicella–zoster viral proteins during chickenpox, quiescence, reinfection, and reactivation, *J. Infect. Dis.* **149:**741–749 (1948).

191. WELLER, T. H., The propagation *in vitro* of agents producing inclusion bodies derived from varicella and herpes zoster, *Proc. Soc. Exp. Biol. Med.* **83:**340–346 (1953).

192. WELLER, T. H., Varicella and herpes zoster; a perspective and overview, *J. Infect. Dis.* **166**(Suppl. 1)**:**S1–6 (1992).

193. WELLER, T. H., Varicella and herpes zoster. Changing concepts of the natural history, control, and importance of a not so benign virus, *N. Engl. J. Med.* **309:**1362–1368, 1434–1440 (1983).

194. WELLER, T. H., AND COONS, A. H., Fluorescent antibody studies with agents of varicella and herpes zoster propagated *in vitro*, *Proc. Soc. Exp. Biol. Med.* **86:**789–794 (1954).

195. WELLER, T. H., AND STODDARD, M. B., Intranuclear inclusion bodies in cultures of human tissue inoculated with varicella vesicle fluid, *J. Immunol.* **68:**311–319 (1952).

196. WELLER, T. H., AND WITTON, H. M., The etiologic agents of

varicella and herpes zoster; serologic studies with the viruses as propagated *in vitro*, *J. Exp. Med.* **108:**869–890 (1958).

197. WELLER, T. H., WITTON, H. M., AND BELL, E. J., The etiologic agents of varicella and herpes zoster; isolation, propagation, and cultural characteristics *in vitro*, *J. Exp. Med.* **108:**843–868 (1958).

198. WELLS, M. W., AND HOLLA, W. A., Ventilation in the flow of measles and chickenpox through a community, *J.A.M.A.* **142:** 1337–1344 (1950).

199. WHITE, E., Chickenpox in Kerala, India, *Indian J. Public Health* **22:**141–151 (1950).

200. WHITE, R. J., SIMMONS, L., AND WILSON, R. B., Chickenpox in young anthropoid apes; clinical and laboratory findings, *J. Am. Vet. Med. Assoc.* **161:**690–692 (1972).

201. W.H.O. MEETING, Diagnosis of human herpesviruses: Memorandum from a W.H.O. meeting, *Bull. World Health Organ.* **69:**277–283 (1991).

202. WHITLEY, R. J., AND GNANN, J. W., JR., Acyclovir: A decade later, *N. Engl. J. Med.* **327:**782–789 (1992).

203. WHITLEY, R. J., GNANN, J. W., JR., HINTHORN, D., LIU, C., POLLARD, R. B., HAYDEN, F., MERTZ, G. J., OXMAN, M., SOONG, S.-J., AND THE NIAID COLLABORATIVE ANTIVIRAL STUDY GROUP, Disseminated herpes zoster in the immunocompromised host: A comparative trial of acyclovir and vidarabine, *J. Infect. Dis.* **165:**450–455 (1992).

204. WILLIAMS, V., GERSHON, A., AND BRUNELL, P. A., Serologic response to varicella–zoster membrane antigens measured by indirect immunofluorescence, *J. Infect. Dis.* **130:**669–672 (1974).

205. WILSON, A., SHARP, M., KOROPCHAK, C. M., TING, S. F., AND ARVIN, A. M., Subclinical varicella–zoster viremia, herpes zoster, and T lymphocyte immunity to varicella–zoster antigens after bone marrow transplantation, *J. Infect. Dis.* **165:**119–126 (1992).

206. WILSON, D., BRESSANI, R., AND SCRIMSHAW, N. S., Infection and nutritional status. I. Effect of chickenpox on nitrogen metabolism in children, *Am. J. Clin. Nutr.* **9:**154–158 (1961).

207. WINSNES, R., Efficacy of zoster immunoglobulin in prophylaxis of varicella in high-risk patients, *Acta Paediatr. Scand.* **67:**77–82 (1978).

208. WREGHITT, T. G., TEDDER, R. S., NAGINTON, J., AND FERNS, R. B., Antibody assays for varicella virus: Comparison of competitive enzyme-linked immunosorbent assay (ELISA), competitive radioimmunoassay (RIA), complement fixation, and indirect fluorescence assays, *J. Med. Virol.* **13:**361–370 (1984).

209. YORKE, J. A., AND LONDON, W. P., Recurrent outbreaks of measles, chickenpox, and mumps. II. Systematic differences in contact rates and stochastic effects, *Am. J. Epidemiol.* **98:**469–482 (1973).

210. ZAIA, J. A., LEARY, P. L., AND LEVIN, M. J., Specificity of the blastogenic response of human mononuclear cells to herpesvirus antigens, *Infect. Immun.* **20:**646–651 (1978).

211. ZAIA, J. A., LEVIN, M. J., PREBLUD, S. R., LESZCZYNSKI, J., WRIGHT, G. G., ELLIS, R. J., CURTIS, A. C., VALERIO, M. A., AND LE GORE, J., Evaluation of varicella–zoster immune globulin: Protection of immunosuppressed children after household exposure to varicella, *J. Infect. Dis.* **147:**737–743 (1983).

212. ZAIA, J. A., LEVIN, M. J., WRIGHT, G. G., AND GRADY, G. F., A practical method for preparation of varicella–zoster immune globulin, *J. Infect. Dis.* **137:**601–604 (1978).

213. ZAIA, J. A., AND OXMAN, M. N., Antibody to varicella–zoster virus-induced membrane antigen: Immunofluorescence assay using monodisperse glutaraldehyde-fixed target cells, *J. Infect. Dis.* **136:**519–530 (1977).

214. ZIEGLER, T., Detection of varicella–zoster viral antigens in clinical specimens by solid phase enzyme immunoassay, *J. Infect. Dis.* **150:**149–154 (1984).

215. ZINSSER, H., Immunology of infections by filterable virus agents, in: *Virus and Rickettsial Diseases*, pp. 89–117, Harvard University Press, Cambridge, 1940.

12. Suggested Reading

12.1. Comprehensive Review of the Epidemiology of Varicella

GORDON, J. E., Chickenpox: An epidemiological review, *Am. J. Med. Sci.* **244:**362–389 (1962).

12.2. General Summary of Recent Advances

GELB, L. D., Varicella–zoster virus, in: *Virology*, 2nd ed. (B. N. FIELDS AND D. M. KNIPE, EDS.), pp. 2011–2054, Raven Press, New York, 1990.

PART III

Viral Infections and Malignant Diseases

CHAPTER 30

Epstein–Barr Virus and Malignant Lymphomas

Alfred S. Evans[†] and Nancy E. Mueller

1. Introduction

Epstein-Barr virus (EBV) is a versatile DNA herpesvirus associated with a wide variety of clinical syndromes. It has been called a "virus for all seasons" because of this varying role in so many conditions.[66] Primary infection is characterized by infectious mononucleosis (IM) (Chapter 10), and reactivated infections are associated with a wide variety of malignant conditions involving both epithelial and lymphoproliferative disorders. Nasopharyngeal carcinoma (NPC) is the major epithelial tumor and is discussed in Chapter 31. This chapter will deal primarily with Burkitt's lymphoma (BL), the first and most extensively evaluated EBV-related malignant lymphoma. In addition, other EBV-associated lymphomas will be briefly discussed. These include Hodgkin's disease (HD), and non-Hodgkin's lymphoma (NHL), either de novo or occurring in immunosuppressed states including post-organ transplantation, the X-linked lymphoproliferative syndrome, and the acquired immunodeficiency syndrome (AIDS). EBV plays differing pathogenetic roles in these malignancies, and the molecular events responsible for these differences are just being unraveled. The epidemiologic criteria on which the virus–cancer relationships

This chapter is a revised version of Chapter 24, by George Miller, in the second edition, and Chapter 27, by Alfred S. Evans and Guy de-Thé, in the third edition, *Viral Infections of Humans*, Plenum Press, New York, 1989.
[†]*Deceased.*

Alfred S. Evans • Department of Epidemiology and Public Health, Yale University School of Medicine, New Haven, Connecticut 06510. **Nancy E. Mueller** • Department of Epidemiology, Harvard School of Public Health, Boston, Massachusetts 02115.

are based have been recently reviewed,[10] and Zur Hausen[312] has summarized knowledge at the molecular level.

2. Biological Characteristics of the Virus

2.1. Structure of the Virus

Epstein-Barr virus is a very complex DNA virus of the family Herpesviridae. Its biological and chemical features have recently been reviewed by Miller[190] and others.[231,240,257] The virion genome is a linear double-stranded DNA molecule of about 175 kilobases whose complete sequence has been established.[140] The inner capsid of the virus has 162 capsomeres, the major protein being the viral capsid antigen. Mature virions possess a lipoglycoprotein envelope, some glycoprotein structures of which can be recognized by the immune system and can elicit neutralizing antibodies. Comparison of the genomes of viruses originating from different patients provided no evidence of disease-specific substrains, but recent work by Miller *et al.*[194] has shown that many genotypes can be recognized by molecular biological techniques. Different genotypes have been identified in different patients with IM, but the significance in relation to this syndrome or BL is not yet known, nor is the origin of the genotypic variants. In addition, studies of the composition of EB nuclear antigen (EBNA-2) reveal two types of EBV, designated as type A and type B (sometimes called types 1 and 2). Their distribution geographically and in various EBV-related tumors has been explored.[271,275,282] The details of the complex molecular composition of EBV will not be discussed here, but rapid advances are being made constantly.

A book[62] and several reviews[231,261,312] cover some of these advances. In both BL and NPC cells, as well as in lymphocytes immortalized by EBV, the viral genome is present in multiple copies (10–50) as circularly closed episomal DNA.[140] Integration of a single DNA copy into the host genome has also been reported to occur, but not consistently.[4] The significance of integration is still unknown.

2.2. Cell–Virus Relationship

Epstein-Barr virus has a dual tropism for lymphocytes and epithelial cells. B lymphocytes represent the target cells in the blood for EBV, which attaches to the complement receptor (CR2).[76] *In vitro* viral infection is followed by immortalization of B lymphocytes into permanent lymphocytic cell lines (LCL). The proliferating cells are infected latently, and only a very small proportion sustain active viral replication, indicating that the B cells are capable of controlling, at least partly, the expression of the replicative functions of EBV. No *in vitro* cellular system that is fully permissive for viral replication is presently available. Klein[144] has postulated that the reservoir of latent virus is in small, long-lived B cells that only express EBNA-1. There is evidence *in vivo*, however, that the virus is excreted in saliva[219] and is produced in epithelial cells of the oropharynx.[281,301] Indeed, more sensitive methods of viral detection suggest that replication in epithelial cells is probably a continuous process, with various levels of intensity, and that virus from this source may be important in the maintenance of our immunity to EBV.[262] A receptor for the virus has been found on epithelial cells,[280] and evidence for persistence of the virus in the parotid gland has been reported.[301] Multiplication of EBV in epithelial cells infected *in vitro* has been demonstrated.[283] The demonstration in AIDS patients of EBV replication on the epithelial cells of the lateral part of the tongue suggests that this might be a natural site of viral replication.[88] This situation is also reminiscent of Marek's disease virus, a chicken herpesvirus that replicates in epithelial cells of the feather follicles, infects T lymphocytes nonproductively, and causes fatal proliferation of T cells *in vivo*. It should also be noted that EBV-infected lymphocytes *in vitro* become susceptible to infection and multiplication by the human immunodeficiency virus (HIV),[204] and this situation might also occur *in vivo*.

2.3. Epstein–Barr Virus-Determined Antigens

As discussed in Chapter 10, a wide variety of antigens have been identified in EBV-infected cells or transformed cells, mainly by immunofluorescent procedures by which antibodies to them are measured. These antigens have been recently reviewed.[240] These include viral capsid antigen (VCA) and membrane antigens (MA), both of which are detected in cells producing virus particles and are classified as late antigens because their expression is inhibited in the presence of inhibitors of viral DNA synthesis; early antigens (EA), which are synthesized early in cell infection and are divided into diffuse (D) and restricted (R) components based on morphological staining and sensitivity to methanol; and EBNA, which is expressed in every cell containing the viral genome and therefore serves as an immunologic marker for the viral DNA. The EBNA is routinely demonstrated using the anticomplement immunofluorescence test, but recent studies of the EBV genome by Miller *et al.*[189,193] and others[132] suggest it consists of six antigenic components,[105] antibody to one of which is missing in some cases of the chronic EBV syndrome.[109,194,195] The EBNA antigens are variously expressed on various BL cells. All of them, which includes EBNA-1 through -6 and latent membrane protein (LMP-1), are expressed on *in vitro* immortalized LCLs: EBNA-1 on fresh BL cells and EBNA-1 and LMP-1 in about half of NPC biopsy cells.[144] It is important to emphasize that EBNA-1 is the primary, if not only antigen expressed on fresh BL cells and during type I and type II latent infections, and is mediated through a promoter called Fp. It is believed that EBNA-1-bearing cells are not recognized by cytotoxic T cells,[139,261,271] so that the proliferation of such cells is not controlled in BL. EBNA-1 apparently has two functions: to maintain the EBV episome and to regulate gene expression, both by transactivation and repression.[273] The names and functions of the various EBNA antigens, as well as EBV virus–host cell interactions, are discussed in Chapter 10 and in recent reviews.[240,257,261]

2.4. Latency and Cellular Transformation by EBV

When EBV is added *in vitro* to lymphocytes derived from antibody-negative persons or from umbilical cord lymphocytes, the cells acquire the ability to grow continuously in culture and to grow in semisolid medium, a process referred to as immortalization of cells. Only about 1 in 50 viral genomes produced in tissue culture is estimated to be capable of immortalization, and only 20–30% of a cell population enriched in B lymphocytes can be immortalized. The EBNA-2 antigen, which is coded by the *Bam* HI-Y-H region, is probably required for the initiation of the immortalization process, and one cell line, the P3HR-1, which is incapable of immortalization, has been found to have lost this EBNA-2 gene.[22,196]

Rickinson[261] has recently reviewed latent infection of EBV, as also discussed in Chapter 10. Briefly, latency is associated with EBNA-1 expression. The expression of six nuclear antigens (EBNA-1, -2, -3a, -3b, -3c and EBNA leader protein EBNA-LP) plus two nonnuclear proteins, LMP 1 and 2, is variable. Two mechanisms apparently are involved in maintaining herpesviruses in the latent state, one for viral persistence and one to prevent replicative functions leading to latency. For EBV, the former is done by interaction between EBNA 1 and a region of the EBV DNA called *opi* P, permitting the viral genome to remain extrachromosomal and to partition in concert with cell division.[231] The switch from latency to expression of viral replicative function is brought about by the ZEBRA gene product. There must exist a mechanism for inhibiting the action of ZEBRA, probably at the transcriptional level. A fine, short review of viral latency and transformation has also been published by Klein.[144]

The immortalization process appears to be quite complex and probably involves the cooperation of several EBNA and latent membrane proteins in inducing latency and immortalization.[189] Morphological changes consisting of enlargement, clumping, and rapid cell growth are preceded by the appearance of EBNA and stimulation of cellular DNA synthesis.[80,263] The usual target cell for immortalization is a small resting lymphocyte derived from bone marrow.[105,128] In African BL, there is evidence suggesting that the target cells for EBV are B-cell precursers, including cells that have not initiated D–J joining, an early event in B-cell differentiation.[176] There is evidence that an intact EBV genome is required to immortalize the cell, and immortalized cells contain the entire EBV genome.[289]

Epstein–Barr virus can also immortalize nonhuman primate B cells,[73,198,300] and the virus produces a lymphoma in nonhuman primates[64,191,197,277] that has many of the characteristics of human EBV infections including a spectrum of responses in inoculated marmosets varying from inapparent infection to transient lymphoid hyperplasia to malignant lymphoma; animals that live for a long time also develop elevated antibody titers to EBV. Only a fraction of cells in LCLs contain viral particles or EBV VCA; nonetheless, all the cells contain viral genome. If single cells are plated from such a line, all the daughter clones contain virus.[199] This finding demonstrates that the viral genome is present in cells that are not making virus, but its expression is somehow inhibited. Transformation and other immunologic features of EBV infection are further discussed in Chapter 10, and the host–virus

interrelationships have been summarized by Rickinson *et al.*[262] The mystery as to how a common, ubiquitous virus such as EBV could become an oncogenic factor under certain circumstances is being slowly resolved by studies in molecular virology and by monoclonal antibody techniques. This virus is now emerging as a unique model for viral and chemical carcinogenesis, since EBV appears to act in a way similar to that proposed for chemical carcinogens.[155]

Great variability is seen in the time of appearance and in the duration of antibodies to the various EBV antigens. The VCA appears very early; the IgM antibody to it also appears early and lasts only a few months, although it can reappear with viral reactivation, but the IgG antibody persists throughout life at quite stable levels except when reactivation occurs.[122] Early antigen antibodies then appear and are specifically transient. Antibodies to EBNA antigens are usually delayed in appearance, although not always so, and antibodies to different EBNA antigens appear at different times and with varying frequencies.[218] Such antibodies appear to be genetically defined[193] as also evidenced by the failure of the EBNA 1 or *BAM* HI-K antibody to appear in 18% of patients classified as having chronic EBV infection including 32% with the severe form.[194,195]

After the primary infection, there follows a long period during which the EBV infection is latent in B lymphocytes present in circulating blood, lymph nodes, and spleen and during which replication of the virus occurs in epithelial cells of the oropharynx and serves as a source of infection of others during intimate oral contact. Latency is the process by which a virus is present intracellularly in an unexpressed or partially expressed state from which it can be activated *in vivo* in states of immunosuppression, such as that in kidney transplant patients given immunosuppressive drugs or in patients infected with HIV. Increased secretion of virus usually occurs in the oropharynx at these times, and increases in antibody titers occur, especially to EA and VCA antigens.[5,40,288] This latency can be reactivated *in vitro* when such lymphocytes are cultivated, leading to the establishment of long-term lymphyblastoid cell lines. Many compounds have been shown to induce EB viral expression in latently infected cells, such as halogenated pyrimidines[79,103] and tumor promoters such as the phorbal ester TPA,[313] which represents the most active compound of croton oil and is extracted from seeds of the *Euphorbia* plant. Extracts from other *Euphorbia* species, especially that derived from tung oil, which is commonly present in many varnishes, are also powerful activators of EBV in Raji cells.[124]

3. Burkitt's Lymphoma

3.1. Introduction

Burkitt's lymphoma is a malignant lymphoma of children occurring in endemic and sporadic form. The endemic form occurs in the central parts of Africa and in New Guinea in areas associated with a very high malaria infection rate occurring early in life (holoendemic malaria) and with infection and transformation of B lymphocytes by EBV, also in the first few years of life. These are high-incidence areas for the tumor in which the epidemiologic features resemble those of falciparum malaria in their geographic distribution. Such areas are characterized as hot, wet lowlands in which mosquitoes abound. Children over 2 years old and under 20 years old are primarily affected, with a median age of about 8 years, and clinically the tumor most often involves the jaw and the abdominal organs. A small percentage of cases in these endemic areas are not associated with EBV and are similar to the sporadic cases that occur worldwide in low frequency and are unrelated to malaria. Even in developed countries, some 20–30% of the cases fulfilling the pathological criteria of BL show high EBV antibody titers and/or the EBV genome in the tumor cells. The other cases have normal or absent EBV antibody and lack the genome in the tumor cells; such cases must have a different etiologic basis.

A common feature of all BL cases, as well as several other B-cell malignancies, is the presence of one of three chromosomal shifts involving chromosome 8, usually transposing with chromosome 14, less commonly with chromosome 2 or 22.

A viral oncogene, c-*myc*, and possibly other viral oncogenes are intimately involved in the pathogenesis of BL and probably other monoclonal B-lymphocyte malignancies.

B-cell malignancies involving EBV and a chromosomal shift occur most commonly in three settings, each associated with impairment of cell-mediated immunity: (1) in renal and other organ transplant patients given immunosuppressive drugs; (2) in patients infected with HIV, which destroys CD4+ helper T lymphocytes, usually in association with the clinical manifestations of AIDS but sometimes without the typical features, especially in B-cell malignancies of the central nervous system; and (3) less commonly, in other types of acquired or congenital immunodeficiency (the X-linked immunodeficiency syndrome, for example). In addition to these malignant lymphomas, other polyclonal B-cell proliferative lymphomas involving normal B lymphocytes may occur, such as acute immunoblastic lymphomas, and may also be fatal.

This section deals primarily with the epidemiology and pathogenesis of African (endemic) BL in which the causal involvement with EBV is based on high antibody titers, the presence of the EBV genome in tumor cells, the ability of the virus to transform and "immortalize" human B lymphocytes, and the experimental reproduction of a malignant lymphoma in cotton-top marmosets and owl monkeys.

From an epidemiologic viewpoint, this tumor also gives leads for evaluating the origin and causes of any tumor.[155] If one accepts that c-*myc* activation represents the origin of the tumorous transformation of the B lymphocytes in humans, then from an epidemiologic and public health viewpoint, both early infection with EBV and holoendemic malaria in equatorial Africa represent causal or risk factors on which intervention can be attempted. An understanding of the multistep pathogenesis of this and other human tumors is of major importance in efforts at their control.

3.2. Historical Background

When Denis Burkitt described in the *British Journal of Surgery* in 1958[29] a tumor involving the jaw in African children with no histological diagnosis being agreed on, it aroused very little interest among cancer researchers and was passed by, completely ignored. Then O'Conor and Davies[226] described the syndrome as a distinct histological entity, recognizing it as a lymphomatous tumor. Further interest was aroused when Burkitt and O'Conor[33] and O'Conor[225] described the clinical, epidemiologic, and histopathologic features of the tumor and Wright[302] delineated its cytological and histochemical characteristics.

Burkitt's lymphoma is a malignant lymphoma comprising a monomorphic outgrowth of undifferentiated lymphoid cells with little variation in size and shape, an amphophilic cytoplasm with clear vacuoles, and a noncleaved nucleus containing two to five basophilic nucleoli. At low magnification, a "starry sky" pattern is frequently observed, caused by macrophage infiltration of this rapidly growing tumor.[10] In the new working formulation of NHL, BL belongs to the high-grade, malignant lymphoma, small noncleaved cell group.[269] Burkitt's lymphoma consists of a clonal proliferation of lymphocytes from the B-cell subset, all of which synthesize heavy-chain Ig, predominately of the μ subtype.[142] Immunoglobulin light-chain expression is also observed in most but not all cases of BL tumors.[247]

The epidemiologic investigation of BL started with Denis Burkitt in 1959. Over a period of 2 years he mailed illustrated leaflets throughout the African continent and carefully plotted the localization of cases on the map of Africa in his office. This led to the emergence of what is now called the lymphoma belt, which extends from West to East Africa between 10° north and 10° south of the equator plus a tail running down to the eastern and southern coast of Africa (Fig. 1).[29–31] The geographical restriction of the tumor suggested specific environmental factors, and Burkitt thought that a study of the edge of the belt should help to discover the factor operating on one side of the line of demarcation but not on the other. During a 10-week, 10,000-mile safari he visited 60 hospitals in eight African countries, accompanied by Ted Williams. They found the existence of an altitude barrier that dropped progressively as distance from the equator increased. The tumor did not appear to occur when the mean temperature at any time of year fell below 15°C.[31] The observation made in East Africa prompted Burkitt to investigate western and central Africa, relatively flat countries but with climatic differences related to temperature variations that were enormous. The information that emerged was that the tumor was common in the forests of West Africa but rare in the semidesert northern part of West Africa. The determining factor appeared to be rainfall, with a threshold of 50 cm of rain per year. Thus, the two emerging climatic factors, temperature and humidity,[29,31] suggested that an insect-vectored virus could be involved.

Therefore, the insect-borne viruses associated with yellow fever, o'nong-nyong fever, and others were investigated, but no significant association was found.[98] This vectored virus hypothesis turned out to be wrong, and the suggestion that holo- or hyperendemic malaria could be involved in relation to the geographical distribution was first proposed by Kafuko and Burkitt.[130] However, no evidence was forthcoming at that time to substantiate this concept, and it was thought that the immunosuppressive action of intense malaria could enhance the potential oncogenicity of the newly discovered EBV.

After an unsuccessful search for a virus in the tumor cells in the biopsies of BL, Epstein and Barr[60] succeeded in culturing such biopsies, which led to the development of permanent cell lines, known today as permanent lymphoid lines[60]; in these cell lines they observed viruslike particles resembling herpesviruses under the electron microscope.[61] At the same time and in the same issue of the *Lancet*, Pulvertaft[248] also reported the successful growth of tumor cell lines from BL tissues in the laboratory. In 1966, Henle and Henle[106] opened up the possibility of seroepidemiologic and diagnostic studies for BL by developing a simple immunofluorescent test specific for antibody to the newly designated EBV. They showed that sera from BL patients had much higher antibody titers to EBV than different controls in both west and east Africa.[106] The detection of EBV DNA in lymphomatous cells, as demonstrated by Zur Hausen *et al.*[314] and the experimental production in 1973 of lymphomas in nonhuman primates that were in many respects similar to BL by two groups, Shope *et al.*[277] and Epstein *et al.*[64] strongly suggested that EBV had an oncogenic potential in both human and nonhuman primates. A prospective seroepidemiologic study of 42,000 children in Uganda by de-Thé *et al.*[53] provided strong epidemiologic evidence that EBV played a critical causal role in the pathogenesis of African BL; this was the first evidence of this kind that a virus plays a causal role in a human cancer.

From the pioneer work of Manolov and Manlova[182] and subsequent cytogenetic investigations of Zech *et al.*[308] and Yunis *et al.*,[307] it became apparent that certain nonrandom chromosomal changes were characteristic of BL cells and might represent a critical step in the development of the malignant cellular clone.

3.3. Methodology

3.3.1. Mortality and Morbidity Data. The Burkitt safari in 1961[30] was a "premiere" in cancer epidemiology, leading to important discoveries. But more classic epidemiologic methodology, involving the use of tumor

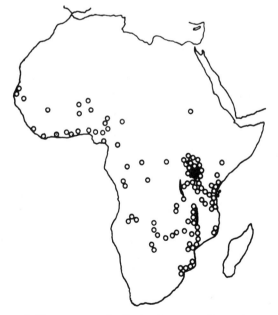

Figure 1. Known tumor distribution in Africa. From Burkitt and Wright.[34]

registries and cancer records in hospitals, has served as the source of case finding and therefore permitted the evaluation of the prevalence of the tumor in different parts of Africa and around the world (Fig. 1). Some areas, such as the West Nile district of Uganda or the North Mara district of Tanzania, were intensively studied for the incidence of BL on a regular basis since the early 1960s. The prospective Ugandan study that took place between 1970 and 1979 represents a unique methodological avenue to establish the causative role of a virus in the development of human tumors.[53,81]

A detailed analysis of the incidence of BL in different areas of the world is difficult to do because of problems related to nomenclature and classification as well as to the completeness and specificity of reporting. In 1969 a standard set of histological criteria were laid down by a WHO Expert Committee,[10] and in 1982 the Working Formulation for Non-Hodgkin's Lymphoma followed the same definitions.[269] Cases reported before 1969 must be regarded with suspicion, as the diagnostic criteria had not been established. Another problem is that tumor registry and death certificate data have usually followed the International Classification of Diseases, and it was not until the ninth edition, which came into use in 1979, that BL received its own rubric (200.2), which allowed it to be separated from other neoplasms classified as "lymphosarcoma and reticulum cell sarcoma." Even so, classification by the fourth digit is rarely done except in special studies.

3.3.2. Serological Surveys.
Seroepidemiologic studies aimed at establishing both the age-specific prevalence and incidence of the EBV antibody[107,131] represented major techniques to evaluate the association between EBV and two tumors, BL and NPC.[50] Measurement of IgG antibody to the VCA of EBV using the immunofluorescent test developed by the Henles[107] has been the most common marker of infection. Tests are often done using a single serum dilution of 1:5 or 1:10, although quantitative titers are needed to identify persons with or at risk of BL. More specialized surveys have added the IgM test to reflect current infections, and some have used measurement of the EA complex for the same purpose.

3.3.3. Laboratory Methods and Diagnosis.
Serological tests include the detection of antibodies to the different structural viral antigens as well as viral-induced but nonstructural antigens. Practically, the serological diagnosis of EBV-associated BL rests on the presence of high titers of antibodies to VCA and, to a lesser extent, EA antigens. In fact, it was found that antibodies to EA were directly related to the titer of antibodies to VCA.[107] Antibodies to the six components of EBNA are also being measured. The detection of EBV DNA fingerprints in the

nucleus of tumorous cells is also an essential part of the assessment of an EBV relationship.[170,224,314] Sophisticated molecular techniques are now being used to identify the various antigens of EBV,[240,257] including the Southern blot and the polymerase chain reaction (PCR) for viral genome fragment and for the transcripts in latently infected cells and EB RNA (EBER).[176,231,235,261] The most sensitive of these is the EBER test. The PCR assays can be carried out on frozen or paraffin sections, making historical excursions into the presence of EBV in various tissues possible.

Other methodological approaches involve cytogenetics and molecular biology, which permit study of both cellular gene rearrangements and the presence of EBV DNA in the tumor cells. Methods for the molecular analysis of chromosomal translocations have yielded new diagnostic and epidemiologic tools for lymphomoid neoplasia.[176,178,179]

3.4. Relationship of Epstein–Barr Virus to Burkitt's Lymphoma

3.4.1. Demonstration of EBV in the Tumor.
Biopsies of BL from African patients regularly contain EBV genome detectable by nucleic acid hybridization with radioactive complementary RNA or DNA probes.[224,314] A rare lymphomatous tumor in Africa that is histologically compatible with BL may fail to show the viral genome (about 1–2% of those diagnosed as BL). In BL tumors occurring outside Africa, EBV DNA is associated with approximately 20% of those tested, and the remainder are EBV-genome-negative.[309]

When EBV DNA is present in BL tumor cells, there are between 6 and 100 copies of the viral genome per cell. Using the Southern transfer technique, Sugden[289] has found that the tumors contain a large portion of the viral DNA, probably all of it, and not just a fragment of the genome. The viral DNA in BL cells is mainly in the form of superhelically twisted circular molecules; integrated DNA may also be present.[134]

Viral-specific RNA has also been found in BL biopsies.[48] These transcripts represent about 3–6% of the genome, and they map at various locations on the viral DNA.

Cell lines containing EBV antigens (EBNA and often VCA and EA) and the viral genome can be regularly established from BL. In the initial biopsy from which the cell lines are derived, only EBNA is detectable, and there is usually no evidence of mature virus or VCA. However, these appear within several days after the cells have been cultivated *in vitro*. A small number of continuous cell lines have been established from genome-negative tumors; these lines lack EBV antigens or DNA. Both EBV-positive and -genome-negative lymphoblastoid cell lines from BL have characteristics of B lymphocytes. Most contain intracellular and surface IgG, and they often secrete Ig. They also contain a receptor for the third component of complement.

3.4.2. Seroepidemiologic Relationships.

Seroepidemiologic surveys for antibodies against the VCA have demonstrated that EBV is worldwide in its distribution. There is no major difference in the prevalence of antibodies at the lowest dilution tested between BL patients and controls, although, as discussed below, there are significant differences in the level of antibody between tumor cases and controls. The rate and age of acquisition of antibodies also vary in controls from different socioeconomic levels and geographic areas. For example, in East Africa, nearly 80% of 2- to 4-year-olds have antibodies to VCA; by contrast, only 40% of sera from 2- to 4-year-olds of American lower socioeconomic classes are positive for antibodies to the same antigen.[107] This result strongly suggests that in the endemic area, infection is acquired at a very early age (see Fig. 2). A review of the causal relationship of EBV and BL in Africa and of seroepidemiologic studies of this relationship in low- and intermediate-incidence areas has recently been published by de-Thé.[51] They previously analyzed international seroepidemiologic studies of EBV antibody in Africans (Uganda), Chinese (Singapore), Indians (Singapore), and Caucasians (Nancy, France) by age and sex.[52]

Studies in Ghana by Biggar *et al.*[14,15] have shown that maternal antibodies against VCA are acquired via the transplacental route by all infants. These disappear by 6 months of age. Between 6 and 18 months of age, 80% of the infants studied developed their own antibodies to EBV

without any clinically discernable illness. It was notable that these silent primary EBV infections were accompanied by the transient appearance of antibody to the R component of early antigen, whereas primary infection of adults, i.e., IM, has classically been associated with an anti-D (EA) response. Anti-R (EA) is also seen in infantile EBV infections in America.[77]

As part of the prospective study in the West Nile district of Uganda, the rate at which susceptibles in East Africa became infected has been investigated: 50% of the susceptibles aged 0–5 years acquired antibodies in the 18 months between a first and a second serum sample.[53] The infection was not as readily transmitted in older subjects, for the rate of seroconversion of susceptibles was 37% in the age group 6–10 and nil in the age group 11–15.[156] There did not seem to be any marked difference in the prevalence of EBV antibody in BL patients and in a wide variety of control groups including subjects matched for age, sex, and tribe. Siblings and neighbors all seem to have approximately the same frequency of antibody.

Thus, the picture that emerged from the initial seroepidemiologic study was that infection with EBV was ubiquitous. However, marked differences were encountered in the levels of antibody titers between BL and control populations. For example, the Henles and their collaborators[107] found that of 139 patients with BL, 81.3% had antibody titers of 1:160 or higher, and the geometric mean titer was 1:275. In contrast, only 14% of the positive titers were at 1:160 or higher in the 489 persons in the control group, and the geometric mean titer was only 1:37. Patients with other cancers, including patients with lymphomas, do not have elevated EBV antibodies as a rule, although some 30–40% of patients with Hodgkin's disease have a higher frequency of elevated titers and a higher geometric mean titer than controls.[56,69,72,158] The association of elevated EB VCA antibody titers with BL is most striking in the endemic area of Africa, where the great majority of BL cases (96%) have EBV DNA in their tumor cells and where elevated VCA antibody titers are found in over 80%.

In contrast to this has been the association of EBV in American and European BL.[65] In the Americas some 421 cases of BL have been reported to the American Burkitt's lymphoma registry over the 8 years since its inception in 1972[160,161] (P. Levine, personal communication, 1983), of which 256 were confirmed by National Cancer Institute (NCI) pathologists as being indistinguishable from African BL; another 143 cases derived from the Surveillance, Epidemiology and End Results (SEER) Program of the NCI have been analyzed in more detail.[160,161] Fewer than 30% of the American BL cases tested have EBV viral

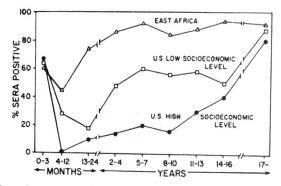

Figure 2. Age distribution of antibodies to EBV in East Africa and the United States. From Henle *et al.*[107]

genome demonstrable in the tumor tissues. In a review of the American experience, Evans[65] estimated that some 20% of the tumors that fulfill the clinical and histopathologic criteria for BL occur in the absence of evidence of EBV infection and of malaria. The serological features of 119 cases tested for EBV-VCA antibody among 257 confirmed cases studied by P. Levine (personal communication, 1983) are shown in Table 1. Overall, 20% lacked demonstrable EBV antibody, 45% had normal levels, and 34% had elevated titers (1 : 320 or higher); the distribution of absent, low, and high EBV antibody levels within the various age groups was quite similar except that 48% of the 29 BL cases over age 19 had elevated titers.

Preliminary evidence from France in 67 Caucasian BL patients studied for EBNA in tumor cells indicated its presence in only 24%, and in 42 children in Lyon with BL, only 15% were positive.[242] In Algeria, serological evidence in 31 childhood cases of BL suggested that 80% were EBV related. De-Thé[51] has published a comparison of the incidence of EBV-related BL in children aged 4–14 in different geographic areas, which is reproduced here as Table 2. A comparison of the incidence of non-EBV-related tumors in different geographic areas suggests that the incidence of EBV-negative BL in equatorial Africa is similar to the overall incidence in Europe and North America, i.e., around 0.2 to 0.5/100,000 per year. As will be discussed in more detail under Section 3.1.3, the point of chromosomal break in chromosome 8 varies by geographic area, by age at the time of infection, and by environmental factors such as malaria,[154,176,179] and this point of translocation is related to the extent to which EBV is involved in tumor pathogenesis.

Elevated antibody titers to EBV are not a feature[56] of malignancy in general but also have been encountered in other conditions such as sarcoidosis[35] and systemic lupus erythematosis[270] often in conjunction with raised antibody levels to certain other viruses. The explanation for this is not clear, but it may be related to the polyclonal proliferation of B cells associated with some of these conditions or to subtle disturbances in immune control. More specific elevations in EBV antibody titers to EBV and not other viruses are found in 30–40% of Hodgkin's disease cases at the time of diagnosis,[69,72] and these elevated titers have been found to precede the diagnosis of the tumor in some cases by several years,[214] as discussed later in this chapter.

Few studies have attempted to compare the titers of antibodies to viruses other than EBV in patients with BL. One study showed that antibody titers to cytomegalovirus (CMV) and varicella–zoster (VZ) were approximately fourfold higher in BL patients than in matched controls.[119] However, this finding has not been confirmed in other laboratories. In the prospective study in Uganda, there were no differences between BL patients and controls in antibody titers to herpes simplex, CMV, or measles virus in sera taken either before or after tumor development.[53]

3.4.3. Prospective Seroepidemiologic Study of BL. Under the auspices of the International Agency for Research on Cancer and under the direction of G. de-Thé, a prospective study was conducted in the West Nile district of Uganda to determine the relationship of EBV infection to BL.[53] From 1972 to the end of 1974, serum samples were taken from 42,000 children aged 0–8 years, with 85% of the eligible children being bled. They were

Table 1. Epstein–Barr VCA Ig Antibody Levels in American Burkitt's Lymphoma[a]

Sex	Number	Percentage distribution of EBV antibody level		
		< 10	10–160	< 320
Male	80	26	44	30
Female	39	8	49	43
Total	119	20	45	34

[a]Based on EBV antibody tests on 119 of 257 cases of confirmed BL cases reported to the American BL Registry[161] (P. Levine, personal communication).

Table 2. Evaluation of the Incidence of Epstein–Barr Virus Associated Lymphoma (BL) in Children Aged 4–14 Years in Different Geographical Areas[a]

Area	NHML[b] incidence/100,000 per year (percentage of childhood tumors)	BL/NHML (%)	EBV-associated BL/all BL (%)	Incidence of EBV-associated BL/100,000 per year
East Africa Uganda	9–15 (about 80)	95	97	About 12
North Africa Algeria	2–4 (30)	47	About 85	About 1–2
Industrialized countries (France, USA)	1–2 (9)	30	About 10–15	About 0.04–0.08

[a]From de-Thé.[51]
[b]NHML, non-Hodgkin malignant lymphoma.

then followed for 5 years in the initial report and then later until 1979, when the political and economic conditions no longer permitted further work. In the initial study group, 14 cases of BL occurred in the study period, a rate of approximately 7 per 100,000 per year, and two additional cases were later identified.[81] This rate of occurrence was lower than expected and was associated with a declining incidence of BL in the study area. All the patients with BL had antibodies to EBV in the initial sample taken between 7 and 54 months in the initial study before the diagnosis of BL was made (see Fig. 3). The main findings were that pre-BL sera showed higher EBV-VCA IgG antibody titers than age/sex/locality-matched controls: 12 of 13 EBV-associated and histologically proven BL cases had pre-BL sera with VCA titers higher than or as high as any control. The increased risk of developing BL was estimated to be approximately 30 times for children who had VCA titers two dilutions or more above the geometric mean titer of the corresponding normal population group standardized for age, sex, and locality.[50,53] Figure 4 gives a schematic presentation of the results.[65]

The presence of EBNA and EBV DNA sequences was established in 9 of 10 confirmed cases from whom frozen biopsies were available, including one with a normal antibody titer in the pre-bleed. Many of these had between 40 and 116 genomes per cell. It was also found that VCA titers did not increase over the initial levels at the time of tumor onset, indicating that the high titers did not result from the disease process itself but reflected a long-standing situation, since the pre-BL blood had been obtained 7 months to 6 years before tumor onset. In contrast to VCA antibodies, those to the EA and EBNA antigens of EBV, as well as those to herpes simplex, CMV, and measles virus, were not elevated in the pre-BL sera as compared to control sera. In seven cases, EBV-EA antibodies developed after tumor onset, but without change in VCA titers. The final results of the study indicate a high degree of significance ($P = 0.002$) between the VCA titers in pre-BL sera and those of control sera. The same pattern was seen in patients bled long before tumor onset and those bled closer to onset. As suggested in Fig. 4, candidates for later development of BL in equatorial Africa are recruited from those 10–25% with the highest EBV-VCA antibody titers among the general population of children. Since the highest incidence among the 5- to 9-year-olds averages about 20/100,000 per year, then only 0.2% of

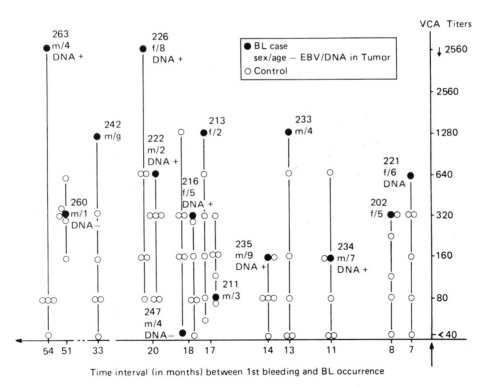

Figure 3. Epstein–Barr VCA titers in sera collected from BL cases prior to tumor manifestation and from controls, by length of interval between bleeding and case detection.

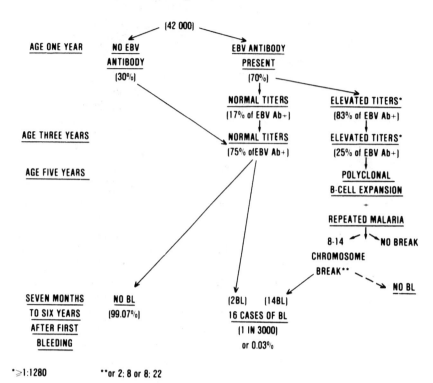

Figure 4. Epstein–Barr virus (EBV) and African Burkitt's lymphoma (BL) in children aged 1–8 years in the West Nile area of Uganda. From Evans.[65]

those in this high-risk group will eventually develop the tumor, thus emphasizing the need for other cofactors in the pathogenesis of BL.

3.5. Descriptive Epidemiology

3.5.1. Incidence.   In the endemic areas of Africa, BL is the most common childhood cancer and may reach

an incidence as high as 4 to 5 cases per 100,000 population (ages 0–15) in certain circumscribed geographic locales. Within Kenya, for example, as illustrated in Table 3, the incidence of childhood lymphoma varies markedly among tribes in different geographic areas. This variation in incidence is not seen in adult lymphoma or in squamous cell carcinoma and suggests the operation of localized promoting factors in the environment. In Kenya, the highest inci-

Table 3. Tribal Distribution of Lymphomas and Epitheliomas in Kenya[a]

		Lymphomas[b]				Squamous-cell carcinoma (adults and children, per 100,000)[b]
		Children		Adults		
Ethnic groups	Tribes	Number	Per 100,000	Number	Per 100,000	
Bantu	Coastal	31	8.2	11	2.9	8.3
Nilotic	Luo	48	6.3	21	2.8	5.7
Bantu	Luhya	30	4.5	13	2.0	6.6
Bantu	Kisii	9	3.5	9	3.5	9.7
Bantu	Kamba	18	3.1	28	4.6	9.2
Bantu	Kikuya	31	2.0	53	3.3	9.5
Nilo-Hamitic	Kalenjin	5	1.2	9	2.2	8.5

[a]From Dalldorf et al.[47]
[b]This is not the annual incidence per 100,000 but the number of tumors identified in the Nairobi Cancer Registry during 1957–1962 inclusive, related to the standardized tribal breakdown of the population in Kenya.

the highest incidence occurs in the Bantu tribes living along the coast, whereas a low incidence (approximately one eighth the incidence among the coastal tribes) occurs in tribes living in central highlands.[49] Similar variations in incidence have been described in the northern districts and low rates in the southwestern districts.[34]

In certain areas of Uganda, for example, the Mengo districts in central Uganda, there appears to have been a decrease in incidence. In the Mengo districts, the rate was 0.77 per 100,000 in 1959–1968.[206,208] By contrast, in other areas of Uganda, for example, the northern region, the incidence remained stable at a rate of 5 to 10 per 100,000 per year.

In many areas of the world, a clinical pathological entity resembling BL has been identified in some countries: Malaysia, Colombia, and Brazil,[32,38,42] for example, in which the incidence is intermediate between the high-incidence areas of Africa and New Guinea, where malaria is common, and the developed, nonmalarial areas of the world. The incidence of childhood BL is shown in Table 2.[51]

There appears to be a gradient in BL incidence from a low in industrialized countries to a high in Africa and other tropical areas where malaria is holoendemic. In evaluating this varying incidence of BL in various parts of the world, the hypothesis was presented that BL occurs worldwide at a low incidence rate and unrelated to EBV or malaria but involving the activation of specific oncogenes, such as the c-*myc* oncogene, and a chromosomal shift.[51] In areas where EBV infection occurs very early in life and malaria is holoendemic, this background incidence is greatly amplified. However, it should be recognized that in the United States EBV appears to play a role in the pathogenesis of some 30% of the cases of BL, as shown by high EBV antibodies and/or EBV DNA in the tumor tissue, and occurs in the absence of malaria.[65,161]

3.5.2. Geographic Factors.
The incidence of BL in various geographic areas has been mentioned above. The clinical features, molecular genetics, chromosomal breakpoint, environmental factors (malaria), age at the time of infection, and the association with EBV vary in different geographic areas.[97,177,181,276] As chromosomal breakpoints appear to play a dominant role in determining these differences in BL patterns, they will be discussed in more detail in Section 3.7.3. Its incidence and various clinical features in Africa, Algeria, Europe, the United States, the Middle East, and Japan are well covered in various chapters of a book entitled *Burkitt's Lymphoma: A Human Cancer Model*,[155] which also sums up many other aspects of the disease. A survey of BL cases reported from different geographic areas up to 1980[156] provides

data on the age and sex distribution of 640 cases in various areas including 269 cases in North America, 47 in Central America, 45 in Asia, 181 in Europe, and 86 in the Middle East plus two in Australia and one in Central America. The geographic features in EBV-related BL in Africa and other high-incidence areas are characterized as being hot, wet, rural lowlands. On the basis of his personal surveys, Burkitt suggested that endemic areas in Africa for the tumor are lower than 3000 feet above sea level and have an annual rainfall of more than 40 inches.[29–31] The distribution in Africa, as mapped by Burkitt, is shown in Fig. 1. Even within highly endemic areas, microepidemics appear to occur that result in a marked clustering of cases in space and time.[206–209,243] Such clustering has also been noted in the United States, for example, in a cluster of BL in four young adults with EBV-associated tumors who lived within 30 miles of one another in rural Pennsylvania, two of whom had shared the same household.[129]

3.5.3. Age and Sex.
The ages of 661 patients of African BL recorded in Uganda by Burkitt are given in Fig. 5.[29] The median age is remarkably constant from one endemic area to another. For example, the median ages ranged from 7.7 to 9.2 years in Uganda, Nigeria, Ghana, and New Guinea.[5] Today, the age distribution is shifting upward in some parts of Africa as the socioeconomic conditions are improved and malaria is brought under better control. The age distribution in other parts of the world is compared with that in Africa in Fig. 6.[239] In North America the 0–8 peak is less sharp, and cases are spread out more into the older childhood ages. In the series of 260 cases reviewed by Lenoir *et al.*,[156] 10% of the 269 cases in North America and 181 cases from Europe were in the 16–19 age bracket, and 24.5% and 6.6%, respectively, were over 20 years old.

Burkitt's lymphoma occurs more often in males than

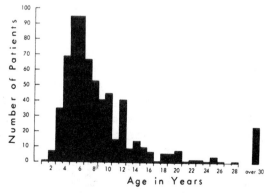

Figure 5. Age distribution of 661 patients recorded in Uganda. From Burkitt and Wright,[34] p. 7.

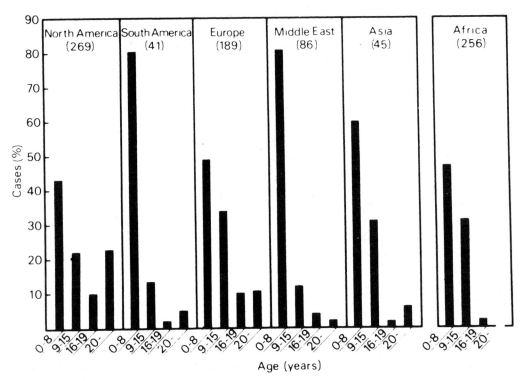

Figure 6. Age distribution of Burkitt's lymphoma from regions throughout the world compared with 256 African cases reported by Olweny *et al.*[229] from Uganda. From Parkin *et al.*[239]

females worldwide. In Africa the male–female ratio is 2.3 : 1. In the 256 confirmed cases reported to the American Burkitt's Lymphoma Registry,[160,161] there were 170 male and 86 females, a male–female ratio of about 2 : 1, but this sex differential disappeared after age 13. In the United States, males with BL tended to be older, with about 18 cases in the 1–3 age group and rising to over 30 cases in the 10- to 12-year-old group, after which it resembled the female distribution. Among females, about ten cases are reported in each age group up to age 20.

3.5.4. Other Risk Factors and Cofactors: Malaria.   The geographic distribution of BL corresponds to that of holoendemic malaria.[29,30,206,208] Although malaria has a much wider distribution throughout the world than EBV-related BL, a closer look at the geographic distribution of BL and of falciparum malaria, particularly that of intense transmission of falciparum characterizing the holoendemic areas, demonstrated a striking association between the two. This was true both on a macroscale in terms of its worldwide distribution and on a microscale, as exemplified by the correlation of malaria parasitemia with the incidence of BL on a district basis in Uganda.[206–209] Nearly all the epidemiologic characteristics of BL of the EBV-related African type can be explained on the basis of

relationships to the intensity of the host response to *Plasmodium falciparum*. These include (1) the geographic distribution of BL in tropical Africa and the lowlands of New Guinea, (2) the age distribution, with most cases in the active childhood years between 5 and 9, virtually no cases under 2, and relatively few after adolescence, (3) older age of onset in those who have migrated from low-incidence areas to high-incidence areas for BL, (4) low incidence in urban areas despite better access to diagnostic medical facilities, (5) steady decline in incidence in areas where death rates from malaria are declining, and (6) time–space clustering.[208] The role of malaria in the pathogenesis of BL is discussed in Section 3.7.1.

3.5.5. Genetic Factors.   One possible role of genetics is related to malaria, and another to immune deficiency, as discussed below. Pike *et al.*[243] studied the frequency of the sickle-cell hemoglobin heterozygous trait in patients with BL as compared with hospitalized controls and normal control subjects. They found AS hemoglobin in approximately 17% of Burkitt's tumor patients compared with 24–29% of controls. Since the sickle-cell trait is associated with protection against malaria, this observation is compatible with the hypothesis that sickle-cell trait may also be somewhat protective against BL. However, there are contradictory reports on the magnitude of the effect of the sickle-cell trait on the incidence of BL, and the most rigorously controlled studies do not achieve statistical significance.[209]

3.5.6. Socioeconomic Factors. In Africa, early infection appears to play a role in the pathogenesis of BL, and this is substantiated by a number of studies such as that of Biggar *et al.*[14,15] who found that 44% of normal infants were EBV-antibody positive by the age of 12 months, 62% by the age of 15 months, and 81% by 21 months. Factors attending early infection were low socioeconomic conditions, poor personal hygiene, prechewing of the baby food by the mother, with about 40% of mothers and family members excreting large amounts of EBV in the saliva, large families, crowding, and poor housing.[65,206–209] In addition, baseline data on socioeconomic characteristics of 715 children aged 3 years or less from the West Nile prospective seroepidemiologic study were analyzed in relation to the presence of elevated VCA titers. With adjustment for all other factors, the relative risk for the presence of high titer (≥1 : 640) versus low titer (1 : 10–80) was 1.7 (95% confidence interval, 0.87, 3.1) for lower social class children.[135]

3.6. Mechanism and Route of Transmission

The geographic and climatic distribution of endemic BL corresponds to the distribution of falciparum malaria. This factor, acting in conjunction with early infection with EBV through close contact with mothers excreting EBV in the oropharynx and who may prechew food to give to the infant, may account for the pattern of transmission of BL in areas holoendemic for malaria and for the geographic distribution of the tumor. As discussed in Chapter 10, the epithelial cells of the oropharynx are probably the site of viral replication[262] and the prevalence of oropharyngeal carriers is high in African women because of both low hygienic conditions and immunosuppression attending repeated malarial infection. Infection with EBV in early life is probably ubiquitous, but malaria varies in intensity from place to place. It is attractive to speculate that the first BL cells may emerge in these oropharyngeal sites of active viral replication and B-cell infection, or is carried into the bone marrow where B-cell precursors are infected and lead to chromosomal translocations.[176]

3.7. Pathogenesis

The main ingredients in the pathogenesis of African BL are holoendemic malaria, EBV infection with B-cell transformation in infancy, chromosomal translocations, and activation of c-*myc* and other ongenes. There may be a

phase of initiation caused by infection with EBV early in life under circumstances of poor hygiene, high rate of salivary excretion, prechewing of food, and other factors discussed in Section 3.5.6; then a phase of promotion occurs characterized by events listed in Table 4[65] and in which chromosomal translocations and c-*myc* activation are the key events.

In nonmalarial settings, EBV plays a role in only some cases, and the process is dependent on other factors that lead to B-cell proliferation, chromosomal changes, and oncogene activation.

3.7.1. Role of Malaria. Malaria may act in three phases as proposed by Klein[143] and modified by Morrow[206,209]: (1) primary EBV infection is generally a necessary factor leading to immortalization of B lymphocytes; (2) chronic severe falciparum malaria infection leads to an intense host response with proliferation of the lymphoreticular system and particularly of B lymphocytes, providing a much higher statistical opportunity for an abnormal cell with specific chromosomal abnormalities to emerge; (3) the emergence of this new cell is no longer controlled by the normal growth-inhibiting mechanisms. However, no significant difference has been observed in EBV reactivity between high- and low-altitude areas in East Africa, which reflect malaria incidence, nor has the level of antibody to malarial antigens been significantly different between BL patients and controls.[209] However, the latter studies have had serious design defects, and the measurement of malaria antibody has not been very accurate. More recently, a number of serological tests have been developed for specific antigenic components of the malaria parasite, which may permit resolution of this issue using the serial sera collected during the prospective study in the West Nile district.[53] In regard to

Table 4. African Burkitt's Lymphoma: Factors Associated with the Promotion Phase[a]

1. Polyclonal expansion of Epstein-Barr virus-infected B lymphocytes
2. High Epstein-Barr virus viral capsid antibody titers
3. Holoendemic malarial infection stimulating further B-cell proliferation
4. Other mitogenic factors enhancing proliferation or depressing the suppressor T-cell response
5. Occurrence of a chromosomal translocation
6. Activation of c-*myc* and perhaps other oncogenes
7. Age and sex (childhood and male sex)
8. Genetic susceptibility to tumor development
9. Nutritional state(?)
10. Other unknown epidemiologic and/or molecular factors

[a]Modified from Evans.[65]

the possible effect of malaria on the immune response to EBV, more recent studies by Moss *et al.*[210] in malaria-endemic and -nonendemic areas of Papua New Guinea suggest that the constant malarial burden specifically impairs EBV-specific T-cell immunity without significantly altering anti-EBV antibody titers. This important observation allows speculation that such impaired T-cell immunity fails to control the *in vivo* proliferation of EBV-infected lymphocytes in children who have high EBV-VCA antibody titers and who are candidates for the development of BL. In addition, the demonstration that antigens from *Plasmodium falciparum* have mitogenic activity on B cells provides a mechanism for further stimulation of EBV-infected B-cell proliferation.[89] Malaria may increase the number of B cells in different stages of maturation, and cells at different stages may be more susceptible to chromosomal translocation, as suggested by Magrath and co-workers[176] and supported by experimental studies of malarial infection of B-cell lymphogenesis in mice.[232]

3.7.2. Epstein–Barr Virus. The role of EBV in endemic and nonendemic BL has already been discussed in relation to early infection, B-cell immortalization, and, in some persons with severe malaria, the emergence of a chromosomal shift and the appearance of a monoclonal B-cell capable of growth to Burkitt's lymphoma. The relationship to EBV has also been reviewed by Purtilo *et al.*[253,255] and more recently by others at the molecular level.[144,176,177,290] EBV appears to act as an initiator in several fashions: it increases the number of target B cells, probably B-cell precursors, increases immunoglobulin enhancer function via functions of the nuclear gene, EBNA-1, and in this way increases the possibility that a translocation may occur in these rapidly dividing cells. It may contribute directly to c-*myc* deregulation. Magrath and co-workers[176,177,179,181] have contributed much to our understanding of the pathogenesis of this tumor and the role of translocations in different environmental settings, as has Lenoir *et al.*[154]

3.7.3. Chromosomal Translocations. B cells derived from patients with BL, as well as the majority of other B-cell neoplasms, regularly show chromosomal changes irrespective of whether EBV is involved or not.[182,308] The most common reciprocal translocation is between the immunoglobulin heavy-chain sequences of chromosome 8q to 14q+, occurring in about 75% of BL.[46] Less common variants are those in which the same fragment of 8q is exchanged either with the kappa light-chain sequence of chromosome 2, t(2:8), seen in some 9%, or with the lambda light-chain sequence of chromosome 22, t(8:22), in 16%.[46,287] One of these three translocations has been regularly found in over 100 cell lines

from BL patients, irrespective of the geographic origin of the tumor or its relationship to EBV.[156,287] The three translocations have been observed by Lenoir *et al.*,[156] in BL cases from France, North Africa, Central Africa, and Japan. One line derived from an EBV-negative African BL designated as BJAB was found to have no anomaly involving chromosome number 8, but the diagnosis in this case remains uncertain on the basis of clinical history, morphology, and surface genotype.

Using molecular techniques, closer examination has demonstrated that the breakpoints of the chromosomal translocation show much variation in different geographic areas. For example, the breakpoint on chromsome 8 can be as much as several hundred kilobases upstream of c-*myc*, within the immediate 5′ flanking sequences, in the 5′ region itself, or as much as several hundred kilobases downstream.[176,177] The chromosome breakpoints on the 8:14 translocation range from far upstream of c-*myc* to within the first intron of the gene itself. Variations in this breakpoint in 92 tumors from different geographic areas has been revealed by Southern blot analysis and show dramatic differences. Breakpoints in the far 5′ region on chromosome 8 occurred in about 75% of tumors from Ghana in equatorial Africa but in only 9% of tumors from the United States.[276] Different breakpoints from these were found in South American countries, such as Brazil, Argentina, and Chile. The relationship of the tumors with far 5′ translocations to EBV also varied in different areas, being higher in Ghana, and then in Brazil, Argentina/Chile, and the United States, in that order. However, no clear relationship was found to the site of translocation. Too few tumors from different geographic areas have thus far been studied to reach firm conclusions. However, Magrath *et al.*[176] state that the data clearly indicate that in each region of the world BL has characteristic clinical features, breakpoint location, and EBV association, but within a given area there is a mixture of subtypes and a different mix of these features. No clear association of EBV to the clinical or molecular pattern has been found.

Because the distal end of the long arm of chromosome 8 is involved in translocation with chromosome 14, 22, or 2, each of which carries a locus for human immunoglobulins, it was speculated that the human immunoglobulin genes might play an important role in the chromosome rearrangements observed in BL and in the pathogenesis of the disease itself.[46] Further studies using probes specific for the variable and constant regions of the human heavy-chain locus on chromosome 14 indicated that this locus is split at various sites by chromosomal translocation and that the genes for the variable regions translocate to the involved chromosome 8 while the genes for the constant regions remain on the involved chromosome 14. The expressed human heavy chain locus in BL was found to remain on the normal chromosome 14, indicating that the translocation directly involves the unexpressed heavy-chain locus.[46]

The specificity of the t(8:14), t(8:22), and t(2:8) translocations for BL appears to be high but not exclusive to BL. Mitelman and Levan[201] reviewed the evidence and found a rare occurrence in non-Burkitt's malignant lymphoproliferative disorders. Sigaux *et al.*[279] noted on the basis of morphometric measurements of malignant cells that rearrangement of 8q24 is found mainly in lymphomas that show a morphological continuum extending from BL to immunoblastic lymphomas. These might all correspond to cell proliferation of the B-cell lineage; EBV-infected and immortalized B cells held in continuous culture also may exhibit some chromosomal changes, but not of the t(8:14) type, as well as cellular changes.[287] Although such immortalized, rapidly growing cells are not regarded as malignant, they can produce a fatal immunoblastic sarcoma, and many lines that have become aneuploid are tumorigenic in immunosuppressed mice, suggesting that B cells can acquire a malignant phenotype in the absence of specific translocations. Even when such t(8:14) translocations occur, it is generally believed that this change is not sufficient itself to make the cell malignant. The probable necessary role of c-*myc* oncogene in this process is discussed below.

3.7.4. Viral Oncogenes. The human homologue, c-*myc*, of the v-*myc* oncogene present in avian myeloblastosis virus is located on chromosome 8 at the band q24 and is translocated to the heavy-chain locus on the 14q⁺ chromosome in BL with the t(8:14) chromosome translocation.[46] Although the major oncogene involved in the pathogenesis of BL is c-*myc*, other oncogenes are probably involved in the process.[151,152] The c-*myc* oncogene becomes juxtaposed close to the immunoglobulin locus. The crossover point occurs 5′ from the c-*myc* gene, sometimes truncating the gene and removing its normal promoter/leader position.[152] Molecular analysis of the c-*myc* oncogene indicates that it is formed by three exons separated by two introns. It has two promoters separated by approximately 160 nucleotides, and its transcripts originate at two initiation points. When the c-*myc* oncogene is decapitated of its first exon because of breakpoints within the first intron, new cryptic promoters are activated within the first intron. The c-*myc* product in BL with or without c-*myc* rearrangements is identical to the normal c-*myc* product.[46] The essential abnormality of c-*myc* in BL may be an inappropriate expression of its function rather than a supranormal level of activity, and immunoglobulin en-

hancer function may play an important role in c-*myc* expression in Burkitt's cells, but other elements are also able to transactivate this oncogene.[176] The product of a tumor suppressor gene, *p53*, also plays a role in BL, as it appears to do in many other malignancies; approximately 40% of BL specimens studied have shown *p53* mutations.[12]

In other translocations, the c-*myc* gene seems to be differently arranged. In each of the fully analyzed cases, some structural change has occured in the promoter/leader region of the c-*myc* gene, and the translocation is accompanied by a subtle deregulation of the c-*myc* transcript, either enhancing its expression or altering the promoter site usually involved in its transcription.[151,152] Croce[46] points out that whereas the normal c-*myc* oncogene on normal chromosome 8 is transcriptionally silent, the c-*myc* gene involved in each of the three different chromosomal translocations is transcribed constitutively at elevated levels. Thus, the consequence of the three different chromosomal translocations is deregulation of transcription of the c-*myc* gene involved in the chromosomal translocation. The expression of the activated c-*myc* oncogene in these translocations appears to be B-cell specific. This observation is consistent with the interpretation that tissue-specific enhancing elements within the immunoglobulin genes are responsible for c-*myc* deregulation.

The malignant cells of most African BL cases appear to express only cytoplasmic or membrane-bound immunoglobulin, whereas those of sporadic cases secrete other immunoglobulins, thus suggesting that the two types of BL might involve cells at different stages of B-cell differentiation.[46] In African BL, the great majority are characterized by rearrangement of the heavy-chain locus in the Jh region or 5′ to it, whereas sporadic BL might have the rearrangements in a heavy-chain switch region—Su, Sr, or Sa 9.[46] These switch regions are normally involved in heavy-chain isotype switching. In most African BL, the break on chromosome 8 is 5′ to the c-*myc* oncogene, often some 30 kilobases 5′ to the translocated c-*myc* gene; in contrast, sporadic BL rearrangements often involve the first intron of the c-*myc* oncogene.

Experiments with fibroblast cells as transformation targets indicate that c-*myc* may not be the only oncogene involved in the transformation process in BL and that B-*lym* or c-*fdgr* may also play a role in the process. The normal c-*myc* gene is thought to be tightly regulated by cell-specific growth factors. The exact biochemical function of the BL oncogenes, their regulation at the genetic and biochemical level, and whether they represent both necessary and sufficient events in the multistep process of BL oncogenesis are the goals of current investigations that hope to provide a complete genetic and biochemical understanding of the molecular pathogenesis of the tumor.

In vitro studies on whether EBV infection and c-*myc* activation are sufficient for the tumorigenic conversion of B cells have been carried out by Lombardi *et al.*[169] They introduced activated c-*myc* genes into human EBV-infected lymphoblastoid lines derived from *in vitro* infection of normal cord blood or directly from infected peripheral blood from AIDS patients. In both cell lines they found that the constitutive expression of exogenous c-*myc* caused negative regulation of endogenous c-*myc* expression that was accompanied by changes in growth properties typical of transformed cells and acquisition of tumorigenicity in immunodeficient mice. In their view, EBV infection and c-*myc* activation were sufficient for the tumorigenic conversion of B cells *in vitro*. They suggest that these two steps of infection and activation may also be involved in the *in vivo* development of BL. Chromosomal translocations typical of BL were not observed, but if the c-*myc* oncogene was already activated at the time of introduction, they were probably not needed at this time, nor was EBV then essential to the process. The changes did not show a clear Burkitt's-like morphology, so there is a question as to whether the cells truly represent BL.

3.8. Patterns of Host Response

3.8.1. Clinical and Pathological Features. The overwhelming majority of African or endemic cases of BL still present years later with the same facial (jaw) and/or abdominal involvement first noted by Burkitt[29] and Burkitt and O'Conor.[33] In 430 cases in Ghana from 1968 to 1978, facial involvement alone occurred on presentation in 39.3% of the cases, abdominal disease alone in 38.6%, and both facial and abdominal involvement in 20.3%.[223] In only 2.3% was the presenting anatomic site neither facial nor abdominal. In those with facial tumors, maxillary bone involvement occurred in 63.8%, and mandibular in 45.5%. In most such patients the tumor starts in the region of the alveolar process of the jaw, leading to expansion of the alveolus, loosening of the teeth, and subsequent displacement of the teeth, particularly of the molars and pre-molars with later displacement and shedding. In maxillary involvement there is often invasion and destruction of the antrum with extension into the nasopharynx and the orbit of the eye. The term "dental anarchy" has been used to describe the clinical and X-ray appearance in such cases. Often they are unable to close the mouth and have dysphagia and signs of airway obstruction. Abdominal involvement may include many organs in the retroperitoneum such as the kidneys, ovaries, liver, and gastrointestinal tract. Children may present with or without ascites, and rapid abdominal swelling may occur. The central nervous system (CNS) is the third most common form of initial presentation and may occur as (1) meningeal involvement, (2) cranial neuropathies, (3) altered consciousness, or (4) paraplegias.[223] Bone marrow involvement is now being recognized in about 13% of the cases.

The age range of the cases from Uganda and Ghana ranged from 2 to 20 years old, with no case under 1 year and with an age peak from 6 to 8 years old. The mean age of patients with facial lesions was 7 years, and with abdominal disease, 9 years. Males predominated over females, especially among cases with facial or jaw involvement. In Ghana there has been an increase in the number of males presenting with abdominal involvement.

Relapses following treatment occurred in about half the cases prior to the routine administration of intrathecal methotrexate, and meningeal involvement was the most significant feature during relapse.

In Europe, BL is the most frequent childhood tumor of the non-Hodgkin's malignant lymphoma type and constitutes 5–10% of this category with a male–female ratio of 3.7 : 1. An association with EBV is found in about 15% of the cases, with an age distribution similar to endemic BL.[242] Abdominal masses occur as the initial presentation in 70% of cases and jaw involvement in only 4%. Appropriate therapy today gives the expectancy that more than 80% will be cured. Of patients alive 8 months after complete remission with no evidence of disease, 90% are expected to be cured.

Despite histological identity and the common chromosomal translocations, cases of BL in the United States show a number of important differences in clinical features, epidemiology, and biology to African BL.[159–161,239] It is a rare disease (1–2 per million children per year) compared with Africa BL (100), its distribution is unrelated to climate and geography as in Africa, and it is uncommonly associated with EBV in the United States. Clinically, jaw tumors are rare and bone marrow involvement is common. Nasopharyngeal tumor occurs in 10% of United States cases but not in African patients. Abdominal involvement is common in both areas. With therapy, about 50% of both African and American cases show prolonged survival. Their differing relationship to EBV suggests the possibility that the tumors have different cellular origin, perhaps arising at different points in the lymphocyte differentiation pathway.

After several years of confusion and variation, a

standard histological definition was established in 1967 by a group of experts under the auspices of the WHO and subsequently published in 1969.[10] The consensus of the expert committee was that BL was a recognizable pathological entity with cellular features distinctive from those of other poorly differentiated lymphomas and leukemias.

Burkitt's lymphoma is a diffuse lymphoma composed of monomorphic, small, noncleaved lymphoid cells of uniform maturity with a round nucleus, three or four nucleoli, and a moderate amount of well-defined, deeply basophilic cytoplasm, which usually contains clear vacuoles.[242] It is classically characterized by a strikingly monotonous proliferation of "primitive cells" with round to oval nuclei and two to five prominent basophilic nucleoli. The nuclei approximate in median diameter those of the benign "starry-sky" macrophages that are often, but not always, present in the tumor.[10] The observation that the neoplastic cells are in some way related to B lymphocytes of normal germinal centers has led to categorization of the tumor in a NCI-sponsored workshop[269] as "small noncleaved cell lymphoma," with provision for noting the presence or absence of follicular involvement.

3.8.2. Diagnostic Laboratory Characteristics.
The demonstration of the DNA of EBV in tumor cells by biopsy or at autopsy by the EBNA test of Reedman and Klein[259] or by hybridization techniques[314] establishes that EBV-related BL tumors are in accord with other clinical and pathological criteria. Careful histological and chromosomal studies are needed to establish the true character of non-EBV-related BL. The serological features of EBV-related BL consist of titers for the EBV-VCA IgG antibody of 1 : 320 or higher, and about 70% will also show elevated titers to the EA of EBV.

3.9. Control and Prevention

3.9.1. General Approach.
Accepting the role of early infection with EBV and holoendemic malaria as the initiators of African BL, one can postulate three lines of control. First, early EBV infection of infants might be modified if socioeconomic, cultural, and hygienic conditions could be improved. Improved standards of living, accompanied by health education, might lead to delay in exposure to the virus in the family setting. Changing the cultural habit of prechewing the baby food might curtail the salivary spread of EBV. Improvement in personal hygiene by introduction of soap, and its frequent usage, especially hand washing and washing of eating and nursing utensils for the baby, may diminish fomite spread of EBV. These may be long-term and impractical projects in the African environment where BL occurs, but the use of soap might be feasible. The benefits of its frequent and proper application may lower transmission of diarrheal (especially rotavirus) and many respiratory diseases (respiratory syncytial virus, rhinovirus). Small, tiltable, simple plastic water dispensers may economize on water so that as little as 50 ml may be adequate for hand washing. Such actions, if effective, may delay the time of exposure to EBV to a time that infection does not entail the risk of BL because of better immunologic response but might eventually lead to the appearance of IM if the delay in exposure should reach older children and young adult age groups.

The second general approach would be to modify or eliminate holoendemic malaria, which both stimulates B cells to greater proliferation, thus increasing the chance of a chromosomal translocation, or enhances EBV replication by suppressing cytotoxic T-cell responses to the virus.[210] Malarial control could be approached through insecticide control of the anopheles mosquito that transmits malaria, screening and other environmental measures, suppressing malaria infection with antimalarial prophylactic drugs in the less than 10-year-old population at risk to BL, or prophylaxis with an immunization against malaria such as those vaccines under development and testing.

An attempt to evaluate one of these approaches, control of malaria by chemoprophylaxis, was carried out by Geser and associates in the 0- to 10-year age group in the North Mara region of Tanzania, and the results have been reviewed.[82] After a surveillance system was established for detecting cases of BL through hospitals and clinics beginning in 1964 and carrying out baseline malarial surveys in 1974, 1975, and 1976 in six sample groups in the lowlands, with repeated examination of the sentinel group of about 80 children four times over the 4-year period prior to the chemosuppression trial, the trial of chemoprophylaxis against malaria was begun. From 1978 to 1982, chloroquine tablets, were administered twice monthly to approximately 100,000 children living in 130 communal villages. The prevalence of EBV antibody was monitored in serological surveys carried out in 1976 and 1983. The results indicated that the incidence of BL in the period from 1964 to 1971 averaged 4.4 per 100,000 per year with wide fluctuations from 2.8 to 6.9 (representing from 4–11 cases per year). Following the initiation of chloroquine prophylaxis, it fell steadily to a minimum of 0.5/100,000 per year in 1980–1981, then rose slightly to 2.8 in 1983. However, a decline in cases had been noted as early as 1971, which was a peak year with an incidence rate of 6.9 cases per 100,000, and the incidence actually fell as low as 1.1 per 100,000 in 1977 just prior to the trial.

No significant difference was seen in the prevalence or geometric mean titer of EBV antibody in sera obtained in the 1976 and 1982 serological surveys. A high level of prevalence of EBV antibody, between 97 and 98%, was found in both males and females at both survey points. Females had a slightly higher geometric mean titer in both surveys, which was significant ($P = 0.001$) after adjustment for age. Surveys for malaria parasitemia showed a marked decrease from 41% in 1976 to 14% in 1978, with a gradual increase after that to 61% by 1983 despite continuation of chloroquine distribution. About 90% of the parasites seen in the slides were *Plasmodium falciparum*. In discussion of the results the authors state: "It remains to be determined to what extent the statistically highly significant decline in the BL incidence rate observed in North Mara from 1972 to 1983 reflects a real change in the epidemiology of BL or a result of deficiency in case reporting."[82] The significance was determined by fitting a regression line to the incidence data using the method of least squares in which the slope was highly significant ($P = 0.001$) during the period from 1964 to 1983, during which the incidence declined an average of 0.14 per year.

However, when one examines the data in the period before and after chemoprophylaxis was begun, the data do not seem impressive or consistent: in the 5 years preceding the trial, the rates per 100,000 per year starting with 1973 were 4.2, 2.9, 3.4, 3.3, and 1.1; in the 5 years after the trial began, the rates were 1.6, 3.0, 0.5, 0.5, 0.9, and 1.8. The 5-year average was 2.5 before and dropped to 1.3 after chemosuppression was begun. Accepting a real decline in BL, the authors conclude that since a declining rate of BL occurred prior to chemosuppression, malaria control could have not been the basic cause of the decline in BL incidence. They also conclude that the study has not provided clear-cut evidence for or against the hypothesis that malaria is a causal factor in BL.

Given the equivocal results of this large trial, one may be hesitant to initiate another large trial to prove that malaria is a causal factor or in the hopes of reducing BL incidence. However, the fall of BL incidence in many areas of Africa in the past few years, sometimes in areas of vigorous malaria campaigns and sometimes not, indicates that several factors must be operative in the general decline of BL. This same issue will be a problem in determining the effect on BL incidence of an effective EBV vaccine, even if one can be successfully developed and administered to a large group of children prior to infection, and with the goal of determining the causal role of

EBV in the pathogenesis of BL. Another problem is the increasing resistance of *Plasmodium falciparum* to chloroquine chemoprophylaxis and therapy. Nonetheless, much attention is being devoted to the development and testing of EBV vaccines, and this progress will now be discussed.

3.9.2. Vaccine

3.9.2a. Background. Following the discovery of EBV in Burkitt's lymphoma by Epstein *et al.* in 1964,[60] progress was rapid in developing various antibody and antigen tests that permitted a better definition of its possible role in causation. By 1976, it seemed essential to consider the possibility of development of an antiviral vaccine designed to prevent infection and thereby to reduce the incidence of the tumor. Such an effective vaccine would also add strong evidence that EBV was indeed an important, and perhaps essential, factor in the pathogenesis of African BL. Although it was recognized at that time that the search for more information on the virus and its biological actions was of unquestionable scientific importance, it was felt that such efforts would considerably enhance work directed toward intervention against the virus. The first proposals for a vaccine were therefore put forward by Epstein in 1976.[59] The proposals drew attention to the precedents for antiviral vaccination in cancer that were provided by vaccination against the lymphoproliferative tumors in Marek's disease of chickens, caused by a herpesvirus of that species, as well as prevention of the experimental lymphomas caused by *Herpes saimiri* in subhuman primates. Epstein has sketched out how such a vaccine might be developed. The urgent goals for an EBV program as conceived then were (1) the study of EBV membrane antigen as the appropriate immunogen, (2) the establishment of breeding colonies for the cotton-top marmoset as the susceptible experimental animal of choice (which has since been put on the endangered-species list), and (3) the investigation of new methods for enhancing antigenicity. In subsequent years developments in molecular virology have opened up these possibilities to a remarkable extent. Much progress has been made in the development of glycoprotein 340 of the EBV envelope as a vaccine, its insertion into various viral carriers, and its effect in preventing tumor development in the tamarin models, and the results of a small trial in China have been published.[91] These developments have been reviewed by several authors[63,67,159,205] and in Chapter 10 of this volume. Further trials with a purified glycoprotein 340 antigen are anticipated in the near future.[3]

3.9.2b. Requirements for Vaccine against EBV. Any vaccine for human use against a putative oncogenic virus must be free of viral DNA and would be best based on an appropriate subunit. In the EBV system, the virus-neutralizing antibodies are those directed against the virus-determined cell surface membrane antigen (MA). Molecular studies of MA have identified two high-molecular-weight glycoprotein components, gp340 and gp270. One cell line derived from marmosets, the B95-8 line of Miller *et al.*,[196] expresses gp340 almost exclusively, thus providing an important source for vaccine studies. Further work yielded methods for immunoassay, isolation, and improvement of the immunogenicity of gp340. Tests of immunogenicity and protection in a susceptible experimental animal were required.

The only susceptible experimental animals that produce lesions after EBV inoculation are the owl monkey (*Aotus*) and the cotton-top tamarin (*Saguinus oedipus*). However, the owl monkey was found to include at least ten different karyotypes and to show variability to various infections. The cotton-top tamarin was thus chosen for vaccine testing, and improved methods of management and breeding provided flourishing colonies; nevertheless, their supply is limited and their use must be directed at essential experiments.

Although earlier experiments with gp340 showed it to have poor antigenicity, even when used with various adjuvants, incorporation into artificial liposomes plus lipid A resulted in improved immunogenicity in the cotton-top tamarin. This antigen was effective in inducing active immunity in tamarin monkeys against subsequent infection with live virus[75]; however, the adjuvant proved too toxic for human use. Large-scale experiments in tamarins that solidly confirm the protection provided by the gp340 vaccine may then open the way for trials in humans. A three-step trial in humans was proposed: (1) determine whether it would protect young college adults against IM, which has a high incidence in the United States; (2) attempt to prevent or reduce the incidence of BL in highly endemic areas; and (3) ultimately, test the vaccine for prevention of NPC carcinoma in high-risk groups.[63] An ideal vaccine would be an attenuated live or recombinant vaccine that would induce humoral, local, and cell-mediated immunity. The prevention of primary infection of epithelial cells in the oropharynx is a key to this. However, the focus thus far has been on using an injectable glycoprotein antigen. This has prevented tumors in tamarin monkeys, but the applicability of this model to humans is not known. No trial has been carried out to prevent BL by a vaccine.

The difficulties in proving that an EBV vaccine would be protective against BL, thus lending strong proof of its essential role in endemic BL, are now compounded by the rapidly decreasing incidence of BL in previously highly endemic areas of Africa, thus requiring a very large test and control group. Recall that only 16 BL tumors were identified in 42,000 children followed in the prospective study carried out by de-Thé *et al.*[53] in the West Nile area of Uganda over a 9-year period of observation (1970–1979).[81] Serological studies would be required in any vaccine trial to show that all the test children lacked EBV antibody at the start; this would mean that the vaccine must be administered prior to exposure in the first 2 years of life. These formidable obstacles, the naturally occurring decrease in tumor incidence, and the 75% current cure rate of BL at modest cost using three-drug therapy and intrathecal prophylaxis would all raise the question as to whether an EBV vaccine trial against African BL would yield significant results and be economically justifiable.

4. Hodgkin's Disease

4.1. Introduction

Hodgkin's disease (HD) is an unusual lymphoma, both in its biological and epidemiologic characteristics. The pathology of HD is distinguished by the presence of the giant Reed–Sternberg cells (RSC) with multiple, hyperlobated nuclei and their mononuclear variants in a characteristically reactive cellular environment. The apparently malignant RSC are usually few in number and have a very low mitotic index.[272] The nonmalignant infiltrating cells surrounding the RSC include rosetting activated T-helper cells, eosinophils, macrophages, histiocytes, interdigitating reticulum cells, plasma cells, and fibroblasts. In contrast, the cardinal feature of the NHLs is the monoclonal proliferation of an abnormal lymphocyte population, accompanied to some degree by a reactive cellular environment. Cases at the interface of these pathological axes can be difficult to diagnose.[295] Based on the appearance of the RSC and of the cellular milieu, HD is subclassified into four major categories: lymphocyte predominance (LP), nodular sclerosis (NS), mixed cellularity (MC), and lymphocyte depletion (LD). The nodular form of LP, which is quite rare, appears to be an independent etiologic entity.[246]

Clinically, the treatment of HD in younger persons is one of the true success stories in cancer therapy with the advent of total nodal radiotherapy and combination chemotherapy.[133] The 5-year relative survival rate for all ages in 1983–1988 was 77.3% compared to 51.3% for the NHL.[188] Essentially all of this difference is seen among patients less than 55 years old.

4.2. Historical Background

Hodgkin's disease was first described by Thomas Hodgkin in his paper, "On some morbid appearance of the absorbent glands and spleen" read before the Medical–Chirurgical Society in London in 1832. As reviewed by Kaplan, the question of whether HD is a true lymphoma or an infectious disease syndrome was a matter of debate for more than a century.[133] As noted by Grufferman,[90] HD was classified as an infectious disease by the International Classification of Disease as late as the 1940s.

Epidemiologic analysis of the disease began with the observation of MacMahon[171] who pointed out the bimodality of the age-specific incidence curve in the United States and Europe with an initial peak among young adults (Fig. 7). Noting the differences in the risk factors for the two peak age groups, MacMahon proposed that these represented at least two separate etiologic entities. He further proposed that the disease in young adults was caused by a biological agent of low infectivity, while among the elderly the causes were probably similar to that of the other lymphomas.[172] This dual hypothesis sparked a substantial amount of research and speculation by epidemiologists, clinicians, and pathologists alike, as noted in a series of commentaries in the *Lancet* on the "Hodgkin's maze."[78] More recently, both epidemiologic and laboratory findings have converged toward a viral etiology for HD, with the EBV clearly a causal agent in a substantial proportion of cases.[213]

4.3. Methodology

4.3.1. Mortality and Morbidity. Data for the United States are available from the SEER Program of the

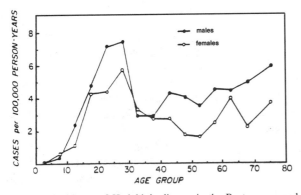

Figure 7. Incidence of Hodgkin's disease in the Boston area and Worcester, MA. SMSAs in 1973–1977. Closed circles, males; open circles, females. From Gutensohn (Mueller) and Cole.[95]

NCI[187] and international data from the International Agency for Research on Cancer.[215]

4.3.2. Laboratory Methods. Both seroepidemiology and the detection of viral genome or gene products in tumor tissue have been used to evaluate the association of the EBV with HD. [While a few studies suggested an association with the human herpesvirus-6, this finding has not been supported by subsequent studies (see Chapter 15).]

4.3.2a. Seroepidemiology. In the seroepidemiologic studies, the pattern of antibody response to the EBV has been compared to that in control populations. These studies have incorporated assays for antibodies to the various antigens of the virus. These include primarily the VCA, the EA, and to a lesser extent, EBNA complex. Some studies have also measured the presence of the immunoglobulin classes of IgM or IgA against specific antigens. Future studies will likely evaluate the presence and level of antibodies against the components of the EBNA complex, especially EBNA-1 and EBNA-2, since the presence of these antigens differ among the EBV-associated malignancies.[145]

This strategy is largely based on the discovery that elevated antibody titers to the VCA were a strong risk factor for incidence of BL in the prospective study in the West Nile district of Uganda described above. Since the EBV is a highly prevalent infection, the simple assay for seropositivity is not, in itself, informative. Rather, the research question has been whether there is evidence that the host response to EBV infection as reflected in the antibody profile is associated with HD. Therefore, the analyses have generally compared cases for the prevalence of elevated titers of antibodies against the VCA (and in some cases EBNA), which are normally present in established EBV infections, and the prevalence of anti-EA, which is indicative of a reactive infection.[190]

Almost all of these studies have been case-control studies using various case series of HD patients and a variety of control groups including hospital controls, population controls, or sibling controls. Since the cases in these studies already had been diagnosed with HD at the time blood was obtained, the question of whether the findings simply reflected the effects of the immune dysfunction seen in HD on latent EBV infection arises. To address this point, we developed a collaboration involving several large population-based serum banks that was established to identify blood specimens drawn *prior* to the diagnoses of HD.[214] Using a nested case-control design,[110] the cohorts were linked to records of diagnoses of HD made subsequent to the initial blood sample. Controls were identified from within the cohorts and matched for age, sex, race, and date of initial specimen. The specimens were retrieved and assayed for EBV antibodies.

4.3.2b. Molecular Assays. More recently, with the advent of highly sensitive molecular probes, assays for the direct detection of viral genome or of viral-encoded proteins or transcripts have been used to evaluate the relationship of EBV with HD. These include the slot and Southern blot,[298] *in situ* hybridization,[2] and PCR[294] assays to detect the presence of viral genome fragments; PCR and *in situ* hybridization to detect the abundant small EB-encoded nuclear RNA transcripts called EBERs,[303] which are actively transcribed in latently infected cells; and *in situ* hybridization with monoclonal antibodies to detect the presence of the latent viral gene products, LMP-1 and EBNA-2.[236] In addition, the clonality of the integrated EBV genome has been assayed using probes for the EBV terminal repeats.[299]

Some of these assays are more sensitive in frozen tissue samples than in those embedded in paraffin. As noted above, the EBER PCR assays are viewed as the most sensitive in detecting EBV genome—even in paraffin-embedded tissues—and the most specific when combined with *in situ* hybridization to the RSC. The detection of LMP-1, which has become a fingerprint of EBV-positive HD, is not universally positive in EBV-positive RSC, and is therefore somewhat less sensitive when used as the only assay.[115]

4.4. Relation of EBV to Hodgkin's Disease

The implication that EBV might be involved in HD was suggested by the consistent finding in six cohort studies involving nearly 42,000 young adults with serologically confirmed IM of an approximate threefold increased risk of HD.[37,43,148,200,216,268] Since IM occurs in about half of primary EBV infections in young adults,[99] this association could simply reflect the increased risk of both HD and IM in young adults of higher social class (see 4.5.3). However, RSC are sometimes seen in IM.[133] Further, in a cohort of 83 IM cases followed for as long as 9 years, a substantial proportion were found to have elevated titers of antibody against the major EBV antigens, similar to the prospective serological study of HD discussed below. In addition, a similar serological pattern was seen in an immunosuppressed 69-year-old man with chronic EBV infection who was followed for 4 months before he was given the diagnosis of HD.[123]

4.4.1. Serological Evidence. Beginning in 1970, a number of investigators evaluated the antibody profile of

HD cases following diagnosis in comparison to controls in a variety of populations. With only two exceptions of studies among children,[149,278] studies have found that the cases had a higher prevalence of high titers of anti-VCA IgG antibody (relative risk generally fourfold or higher) or a higher geometric mean titer as summarized in Table 5. (The titer levels are not necessarily comparable between studies because of differing cell lines used in assays or different methods.) These findings suggest that HD cases had either more severe initial EBV infections or more frequent viral reactivation.[222] In addition, in ten studies that also measured antibodies against the EA (Table 6), all but that of Hesse *et al.*[117] found that cases had elevated titers against the EA, indicative of viral activation. These findings were true for cases from both developing and developed populations and both following and prior to the onset of therapy. Both anti-EA-D and anti-EA-R were elevated, but the associations with the former were stronger.

These observations were extended by our evaluation of specimens obtained from cases *before* the diagnosis of HD.[214] Taken together, the serological analyses point to an enhanced level of EBV replication as part of the pathogenesis in HD, which both precedes and accompanies the development of the tumor. In this study, the EBV serology of specimens obtained on average more than 4 years before the diagnosis of 43 cases was compared to that of 96 matched controls. The relative risk of HD associated with elevated levels of antibodies against the VCA were 2.6, 3.7, and 0.22 for IgG, IgA, and IgM, respectively. That for anti-EBNA was 4.0 and for anti-EA-D, 2.6. The association with IgG and IgA titers against the VCA were not apparent in young adult cases (15–39 years) nor was that for IgG found for the NS cases. The findings for IgM and anti-EBNA held for all age groups and for both NS and MC cases. These associations were stronger in specimens obtained at least 3 years before diagnosis than in those obtained closer to diagnosis. A similar study was conducted in Finland, which included 6 HD cases. Although the data for these cases and their controls were not shown, the authors state that risk of HD was associated with increased antibody response to the EBNA complex.[153]

4.4.2. Molecular Evidence. Recently, more direct evidence of EBV involvement in the etiology of HD has been gained by the application of molecular hybridization assays. Table 7 summarizes the major studies among the plethora of those looking for EBV fingerprints in HD tissue. The reports by Weiss *et al.*[298,299] provided the first concrete evidence that monoclonal EBV was detectable in HD tissue and localized to the RSC in 4 of 21

Table 5. The Number of Subjects and Their EBV-VCA Antibody Prevalence Rate according to Study and the Relative Risk (RR) of Hodgkin's Disease Associated with the Prevalence of High Titers following Diagnosis

Author	Cases				Controls				RR
	No.	Rate	GMT	High titers[a]	No.	Rate	GMT	High titers	
Goldman[85]	57	0.58	—	0.12	54	0.61	—	—	—
Johansson[127]	60	0.95	100	0.47	47	0.89	43	0.17	4.3
Levine[158]	63	—	367	0.35[d]	85	—	91	0.05	10.9
De Schryver[56]	17	—	60	0.24	63	—	23	0.17	1.5
Henle[108]	489	0.89	105	0.40	294	0.84	55	0.14	4.1
Henderson[104]	142	0.94	92	0.47	142	0.93	53	0.25	2.6
Langenhuysen[150]e	25	0.92	1580	0.88[c]	25	0.92	585	0.60	4.9
Hirshaut[121]	51	0.82	141	0.16[c]	45	0.89	—	0.09	1.9
Rocchi[264]	100	0.98	177	0.68	100	0.91	34	0.12	15.6
Gotlieb-Stematsky[86]	67	0.91	67	0.46	186	0.66	9	0.01	79.2
Hilgers[120]	43	0.95	2420	—	43	0.95	401	—	1.9
Hesse[117]e	185	0.95	272	0.79	185	0.95	141	0.61	2.4
Evans[71]	67	1.00	146	0.31[b]	162	—	51	0.04	11.9
Lange[149]e	27	0.63	116	—	71	0.61	101	—	—
Ten Napel[291]e	15	0.94	3180	—	17	1.00	924	—	—
Mochanko[202]	37	1.00	69	0.27	40	0.83	44	0.07	4.6
Evans[72]	70	0.97	110	0.35[b]	70	0.84	30	0.03	17.4
Shope[278]	15	0.73	103	0.20	24	0.58	73	0.25	0.8
Wutzler[304]	57	0.96	166	—	57	0.98	80	—	—
Bogger-Goren[17]	21	0.95	46	0.38	77	0.66	9	0	∞
Evans[69]	304	0.86	176	0.39[b]	276	0.87	58	0.14	4.0

[a] 1:160 unless noted otherwise.
[b] ≥1:320.
[c] ≥1:640.
[d] ≥1:640.
[e] Sera collected before initiation of treatment; all other studies involved treated patients.
Studies listed in sequential order.

Table 6. The Number of Subjects and Their EBV-EA Antibody Prevalence Rate according to Study and the Relative Risk (RR) of Hodgkin's Disease Associated with the Presence of Antibodies following Diagnosis

Author	Cases				Controls				RR
	No.	Rate[a]	GMT	High titers[b]	No.	Rate	GMT	High titers	
De Schryver[56]	15	0.27	—	0.07	0	—	—	—	—
Henle[108]	458	0.30[c]	30	—	1718	0.03	20	—	13.9
Rocchi[264]	100	0.50[c]	34	0.28	100	0.02	5	0	49.0
Gotlieb-Stematsky[86]	63	0.27	—	—	101	0	—	—	—
Hilgers[120]	43	0.91	52	—	43	0.81	9	—	2.4
Hesse[117]g	176	0.84	63	0.26[e]	176	0.82	55	0.27	1.2
Evans[71]	42	0.21	—	—	11	0.09	—	—	2.7
Lange[149]g	17	0.53[c]	—	—	43	0.09	—	—	11.4
Wutzler[304]d	57	0.30	—	0.14	57	0.11	—	—	3.6
Evans[69]f	304	0.46	13	0.30[c]	276	0.23	8	0.11	2.8

[a] 1:5.
[b] ≥1:40.
[c] ≥1:10.
[d] ≥1:20.
[e] ≥1:160.
[f] EA-D only.
[g] Sera collected before initiation of treatment.
Studies listed in sequential order.

Table 7. Summary of Larger Studies on the Detection of Epstein–Barr Viral Genome or Gene Products in Tissue from HIV-1-Negative Hodgkin's Disease Cases[a]

Author	Number of cases	Number positive (%)	Method used	Specific for RSC?	Clonal EBV?	Notes
Herbst[114]	198	114 (58)	PCR, in situ	19/28	—	No variation by age and sex
Jarrett[125]	95	43 (45)	SB	—	25 of 26	48 cases selected for age and histology; positivity highest among children and older adults; 30 of 30 were EBV type 1
Pallesen[237]	96	47 (49)	In situ-LMP-1, ZEBRA, EA, VCA, MA	Yes	—	All cases negative for CMV; 3 LMP-1+ are ZEBRA+; none positive for EA, VCA, MA
Delsol[55]	107	37 (35)	In situ; in situ-LMP-1, EBNA-2	Yes	—	Positivity highest in MC; of 12 cases tested at diagnosis and relapse, all remained concordant for EBV status; 1 LMP-1+ became LMP− and 1 LMP-1+ became +/−; no correlation with serology or short-term prognosis; of 13 positive cases, all were EBNA2 negative
Fellbaum[74]	187	66 (35)	PCR	—	—	Positivity highest in MC; no association with survival
Niedobitek[221]	116	33 (28)	In situ-EBER, p53	Yes	—	Positivity increases with histological grade; 37 positive for p53 in RSC independent of histology, less frequent in EBV+ (21%) then EBV− (36%), $P = 0.12$

[a]Studies listed in sequential order; CMV, cytomegalovirus.

specimens tested. This discovery has been confirmed in a large number of subsequent reports. The detection rate increased with the use of more sensitive methods. Overall, the findings for the almost 2000 HD cases studied suggest that 30–50% of HD cases are EBV-positive. When tested, RSC have contained the EBV genome or gene products and the great majority have clonal EBV. Pallesen et al.[236] further demonstrated that the EBV-positive RSC expressed a restricted latent infection phenotype of LMP-1+/EBNA-2−, both viral gene products being normally coexpressed in LCLs.[141] This important finding has also been confirmed by subsequent investigators. The restricted latent phenotype of EBV found in HD differs from that seen in other EBV-positive lymphomas. However, it is found in EBV-positive NPC; these two malignancies also share similar transcriptional programs[54] and similar serological patterns of elevated anti-EA(D), anti-EBNA, and IgA antibodies against the VCA.[214]

EBV status appears to be stable over time. Delsol et al.[55] reported that EBV status was consistent in subsequent biopsies at relapse (range 14–126 months) in 12 cases of which 7 were initially EBV-positive. Two EBV-genome positive cases showed substantial reduction or loss of LMP-1 staining in their later biopsy. Coates et al.[41] evaluated sequential biopsies in three EBV-positive pa-

tients (range 2–10 years) and found all remained positive at about the same level of the initial biopsy. In terms of EBV strains in HD, almost all those typed were type 1,[1,25,125] although three type 2 and four "hybrid strains" have been reported.[25,165] In general, the number of virus episomes per cell is low.[286]

The question of the specificity of the EBV-positivity association was raised by reports of EBV-positive (but LMP-1−) small lymphocytes in both EBV-positive and negative HD,[1,11,39,114,126,137,183,297] which are also present at a low frequency in normal lymph nodes[219]; further, clonal EBV has been detected in reactive hyperplasia biopsies.[163,182] However, the consistency of the finding of clonal episomal EBV of a unique phenotype expressing the oncogenelike LMP-1 in a substantial proportion of HD cases in many patient populations throughout the world argues strongly for a causal role.

The presence of both LMP-1 and EBER indicate that the episomal viral genome is not silent but that the latent state is actively maintained. The transcript for EBNA-1, the virus genome maintenance protein, is made.[54] In a new technique developed by Grasser and Murray, and reported by Young,[306] a monoclonal antibody against EBNA-1, which can be used on histological material, has detected EBNA-1 in the RSC of EBV-positive HD, as well as in biopsy specimens from other EBV-associated malignancies.

Pallesen *et al.*[237] evaluated whether the BZLF1 gene product ZEBRA was expressed in 47 LMP-1+ HD biopsies. This product induces the switch from latency to the lytic cycle and virus replication. They found that it was rarely expressed—only three were positive—and no structural viral proteins were detected. This observation suggested to these investigators that the latent state of the episomal EBV genome in HD is not severely impaired; rather, that the infrequent activation of replication is impaired, resulting in an abortive viral productive cycle. Thus, the maintenance of the malignant state appears not to be related to EB viral replication per se; rather, it may involve the transforming properties of LMP-1, which inhibits terminal differentiation by up-regulation of *bcl*-2 protein expression.[112] The lack of expression of EBNA-2, a target for cytotoxic lymphocytes, places HD in the family of EBV-associated malignancies with restricted patterns of latent gene expression.

These findings imply that there are two types of HD: EBV-genome positive and EBV-genome negative. However, a number of characteristics of the disease appear to be independent of EBV status. EBV status has not been found to be an independent predictor of prognosis,[55,74,296] nor does it appear to be related to the expression of several cellular antigens, generally present on RSC.[274] In addition, EBV status does not appear to be associated with the presence of *bcl*-2 or c-*myc* overexpression, with p53 mutation, nor with the presence of the t(14,18) translocation.[126,220,221]

The question then arises whether EBV positivity correlates with those risk factors that have pointed to the role of a viral exposure in HD in the past. The answer is that no correlation is evident in the current data available. Few data are available on serological profiles for antibodies to the EBV and the presence of EBV genome or gene products. Ohshima *et al.*[228] reported a single case of a 8-year-old boy with EBV-positive HD. Their detailed serological pattern (anti-VCA-IgG 1:1280, -IgA 1:10, -IgM 1:10; anti-EA 160; anti-EBNA 1:40) was consistent with that reported in the prediagnosis study.[214] Brousset *et al.*[28] and Delsol *et al.*[55] in an overlapping series of 107 cases, concluded there was no association with EBV positivity and a serological pattern of reactivation that they defined as anti-VCA > 1:640, anti-EA > 1:40, anti-EBNA > 1:160. Of the data shown, only 1 of 35 EBV-negative and none of 16 EBV-positive cases had this rather extreme pattern. In a pilot study using both EBER and LMP-1 probes on 19 cases from our early case-control study,[69] five EBV-positive cases did not differ from 17 EBV-negative cases for IgG titers against the VCA, EA-R, or EA-D.[162]

The data concerning histology and EBV status from studies with at least 30 specimens are summarized in Table 8. Since the sensitivity of assays varies between studies, the most valid comparison is within studies. In general, the EBV-positivity rate increases with histological grade, although the data for LD cases are sparse and inconsistent. Thus, of the two most common subtypes, MC cases generally had higher rates than NS cases, with a range of 10 to 100% in studies with at least 10 MC cases tested; however, the rates among NS cases were sometimes high and covered the same range. Murray *et al.*[217] noted that among EBV-positive specimens, the proportion of RSC that is positive for LMP-1 also increases with

Table 8. Summary of Larger Studies on the Prevalence of Epstein–Barr Virus Genome or Gene Products Detected in Tissue from HIV-Negative Hodgkin's Disease Cases by Histological Classification

First author	LP % (No.)	NS % (No.)	MC % (No.)	LD % (No.)
Herbst[116]	33 (3)	15 (26)	0 (10)	—
Uhara[294]	25 (4)	24 (17)	33 (9)	0 (1)
Herbst[114]	50 (12)	61 (109)	58 (64)	25 (4)
Libetta[164]	60 (5)	41 (22)	50 (6)	0 (1)
Pallesen[236]	10 (10)	32 (50)	96 (24)	—
Brousset[28]	0 (5)	18 (22)	46 (26)	—
Herbst[113]	0 (1)	70 (27)	72 (18)	0 (1)
Masih[183]	42 (12)	52 (21)	69 (13)	83 (6)
Jarrett[125]	NS[e]	40 (45)	52 (25)	NS
Gledhill[84]a	100 (2)	27 (26)	17 (6)	100 (1)
Weiss[297]	0 (14)	33 (12)	75 (8)	50 (2)
Coates[41]	0 (7)	29 (24)	13 (16)	0 (8)
Brocksmith[27]	60 (5)	54 (41)	73 (11)	—
Delsol[55]	0 (10)	10 (40)	60 (55)	—
Khan[137]b	0 (2)	15 (26)	67 (3)	0 (2)
Herbst[115]	100 (2)	42 (24)	56 (18)	50 (2)
Fellbaum[74]	31 (13)	28 (98)	50 (68)	12 (8)
Murray[217]	8 (12)	50 (24)	86 (7)	100 (3)
Ambinder[1]c				
Honduras	100 (1)	100 (3)	100 (6)	—
USA	0 (2)	13 (15)	86 (7)	—
Chang[39]d	—	100 (7)	100 (20)	60 (5)
Carbone[36]	(5) (0)	10 (20)	64 (11)	100 (3)
Niedobitek[221]	11 (9)	27 (75)	39 (31)	0 (1)
Khan[138]	0 (10)	24 (38)	68 (22)	14 (7)

[a]Subset of above.
[b]RSC.
[c]Pediatric cases.
[d]Cases from Peru.
[e]NS, not stated.

histological grade. Sandvej et al.[274] have noted that since LMP-1-positive cells should be rapidly removed by an intact T-cell system, their prevalence in RSC occurs in proportion to immune dysfunction. In the two studies involving cases from economically developing countries—Honduras[1] and Peru[39]—where patients generally present with more advanced disease, the positivity rate was notably high. These observations suggest that the factors affecting or reflecting immune competency are associated with EBV-positivity status.

Table 9 summarizes data on EBV-positivity and age. Gledhill et al.[84] concluded that EBV positivity increased with age at diagnosis. However, this conclusion conflicts with the data presented in the paper. Jarrett et al.[125] noted in their series that samples from young adults were rarely positive; however, this is not apparent in other studies. Herbst et al.[114] in a study of 198 patients, reported that there was no significant association between EBV status and age, as did Brocksmith et al.[27] in a study of 57 patients. Further, Coates et al.[41] in a study of 55 patients, reported that the mean age of EBV-positive cases was 36 years (range 18–56), which did not differ from that of negative cases, 34 years (12–63). Thus, in general, there does not appear to be any consistent association with age. The very high rate of positivity in children from economically developing populations is noteworthy.

In summary, based on prior work we would predict that the EBV-positive HD cases also have elevated antibodies to the EBV. The few data reported have not supported this hypothesis, but none have included data on antibodies other than anti-VCA and anti-EA (IgG). The exception is the single case reported by Ohshima et al.[228] of an 8-year-old boy with HD. In that case, the child had a pattern quite similar to what we had found in prediagnosis specimens.[214] In our study of adult cases, the strongest predictors were high titers against the EBNA complex and low or absent IgM titers against the VCA. In the former study, the child had anti-EBNA titers at 1:40, which the authors define as "elevated," high anti-EA, very high IgG anti-VCA, and detectable IgA and IgM anti-VCA.[228]

We would predict that the disease in young children (less than 10 years) is likely to be EBV-related, since risk appears to be related to very early infection[96] (S. Grufferman, personal communication). In Ambinder et al.,[1] the EBV-positivity rate of tumors was higher in young children (4 of 7) than in those aged 10–14 years (5 of 18). Among adults, the risk factor associations are strongest for the young adult age group, which is also distinguished by a high proportion of NS cases. We would predict that EBV positivity would be strongest in that age group and histological category. However, in the prediagnosis serology study, elevated antibodies against EBNA and low levels of IgM antibody to VCA did not appear to correlate with these characteristics, although the data are sparse (Table 10). We would also predict that EBV genome status would correlate with risk factor data related to age at infection. However, no data are available at present to our knowledge to evaluate this prediction. In sum, the EBV–HD story is currently fragmented and inconsistent and will require the integration of epidemiologic, serological, and pathological data with EBV biomarkers for clarification to occur.

Table 9. Summary of Studies on the Prevalence of Epstein–Barr Virus Genomes or Gene Products Detected in Tissue from HIV-Negative Hodgkin's Disease Cases by Broad Age Groups

First author	Population	Children/adolescents (<15 years) % (No.)	Young adults (15–39 years) % (No.)	Middle age (40–49 years) % (No.)	Older adults (≥50 years) % (No.)
Bignon[16]	France		55 (11)	50 (2)	33 (3)
Libetta[164]	UK	38 (16)		50 (10)	
Jarrett[125]	UK	54 (13)	20 (44)		71 (38)
Gledhill[84]a		100 (2)	26 (23)	0 (3)	43 (7)
Ambinder[1]	Honduras	100 (11)			
	USA	56 (16)			
Boyle[25]	Australia		37 (19)	0 (1)	14 (7)
Chang[39]	Peru	100 (19)	83 (6)	80 (5)	100 (2)
Khan[138]	UK	24 (25)	33 (36)	41 (17)	

aSubset of above.

Table 10. Relative Risk^a of Hodgkin's Disease Associated with Elevated Titers of Antibodies against Epstein–Barr Virus, Estimated by Matched Analysis in Patients Grouped according to Age and Histology

Epstein–Barr virus antibody	Age group (yr)			Histology	
	19–39 (N = 14)	40–54 (N = 19)	≥55 (N = 10)	Nodular sclerosis (N = 14)	Mixed cellularity (N = 18)
VCA					
IgG (≥1:320)	0.89 (0.20–3.7)	5.6 (1.2–25.7)	5.3 (0.79–34.8)	0 (NS)^b	3.7 (1.1–12.2)
IgA (≥1:20)	1.4 (0.16–12.6)	3.3 (0.96–11.6)	6.6 (1.0–42.6)	2.6 (0.58–12.0)	4.1 (1.0–16.4)
IgM (≥1:10)	0 (NS)	0.38 (0.06–2.6)	0 (NS)	0 (NS)	0 (NS)
EBNA (≥1:80)	6.5 (0.87–48.3)	2.6 (0.33–20.2)	6.0 (0.90–40.1)	∞ (NS)	3.3 (0.74–14.7)
EA					
Diffuse (≥1:10)	2.1 (0.55–8.3)	6.0 (1.4–26.7)	1.2 (0.24–5.5)	1.4 (0.26–8.0)	2.9 (0.88–9.6)
Restricted (≥1:40)	0.54 (0.14–2.0)	5.5 (1.3–23.0)	4.4 (0.65–29.3)	0.55 (0.08–3.7)	2.5 (0.88–7.4)

^{a}Relative risk (90% confidence interval), adjusted for IgM against VCA.
^{b}NS denotes not significant.

4.5. Descriptive Epidemiology

4.5.1. Incidence. HD is a rare disease and accounts for less than 1% of all malignancies in the United States. Based on SEER data for 1990, the incidence rate in the United States is 2.8 per 100,000 person-years, or about 7900 cases per year.[187] Because of its bimodal age incidence curve, most cases occur in persons between the ages of 15 and 50 years of age.[186] Incidence is generally more common among males, particularly in the second mode, and more common among whites than among blacks, 3.0 versus 2.7 cases per 100,000 person-years, respectively.

Glaser and Swartz[83] recently analyzed incidence data collected by the NCI from 1969 through 1980 to evaluate whether there were secular trends in incidence in the two modes. They corrected these data for diagnostic misclassification using an algorithm derived from data from the Repository Center for Lymphoma Clinical Studies. They concluded that rates of HD in young adults had been increasing, while those for older adults were lower than previously observed and showed no secular changes. This observation is consistent with the hypothesis that the etiology of the disease in young adults is independent from that in older adults, as first proposed by MacMahon.[171]

4.5.2. Age and Sex. Hodgkin's disease in the developed world is a disease of young adults. Typically in such populations there is a bimodal age-specific incidence curve with an initial peak about age 25 and a second among the elderly (Fig. 7). In contrast in economically developing populations, there is an initial peak among boys in early childhood with relatively low rates among young adults. However, in essentially all populations there is an increase in incidence with age after age 50. Intermediate and evolving patterns can also be seen.[44,213]

4.5.3. Socioeconomic and Other Risk Factors. In parallel with the geographic variation in the age-incidence curve with the level of economic development, factors indicative of social class and level of hygiene are associated with risk of HD within age groups.[93,213] Among children, risk is associated with lower social class markers such as lower parental income and higher housing density.[96] This suggests that risk is associated with early primary infection with EBV. Among young adults, the most consistent finding is an inverse association with childhood sibship size. Risk is also associated with higher maternal education and with low density of housing in childhood.[95] These factors are consistent with protection from early infectious exposure and susceptibility to a "late" primary EBV infection during adolescence. A similar pattern of risk factors is seen in the age group between the two peaks—40 to 54 years—suggesting that these cases had experienced primary EBV infection as adults, perhaps by transmission from their children. However, among older persons, there is no clear association with social class factors that would influence age at infection with EBV.[92]

4.6. Mechanism of Transmission

There is no evidence of direct person-to-person transmission of HD itself. Early reports of clusters, such as in high schools, have not been confirmed by subsequent studies.[94] To the extent that EBV is involved, the usual route of transmission would be by oropharyngeal exposure, as described in Chapter 10.

4.7. Pathogenesis and Patterns of Host Response

The disease is accompanied by a complex deficiency in cellular immunity that precedes treatment and persists to some degree following remission. These have been extensively reviewed by Romagnani and co-workers[175,267] and Slivnick et al.[284] and appear to be secondary to chronic production of immune mediators; RSC in culture express mRNA and proteins of various cytokines and cytokine receptors.[57] The origin of the RSC has long been a matter of controversy.[211] Following an international symposium considering this question in 1987, Drexler and Leber[58] concluded that phenotypically these are lymphoid cells "frozen in a state of activation." Herbst and Niedobitek[112] have pointed out that often these cells do not express functional antigen receptors; thus, there is a dissociation of phenotype with genotype. The highly reactive cellular microenvironment in involved nodes is accompanied by a reciprocal reduction of circulating T-helper cells, apparently by displacement from the peripheral blood to the lymphoid organs.[266] All these observations suggest that a fundamental characteristic of HD is that of a sustained immune response to a chronic lymph-node-based antigenic stimulation.

The consistently altered EBV latent phenotype and the elevated EBV antibody profiles in HD patients point to dysfunctional immune control of the EBV. The epidemiologic evidence points to age at infection as being an important modifier of risk of HD.[93] In disadvantaged populations, risk is seen among children, especially boys who are generally more susceptible to complications of infections in comparison to girls. The sex ratio seen is young children is comparable to that seen with BL in Africa in the age group. Conversely, in economically advantaged populations, risk of HD in young adults and middle-aged persons is indexed by factors associated with susceptibility to late infections. Since in general both very early and late infections with common childhood infections tend to be more severe, this suggests that severity (viral load) of primary EBV infection is involved in the pathogenesis of EBV-positive HD in these age groups. Among older persons, an enhanced level of endogenous reactivation of latent EBV, mediated by reduced immune competency with age, may be involved in pathogenesis.

Either scenario could lead to chronic antigenic stimulation by the EBV antigens. In 1983, Smithers[285] postulated a mechanism by which this could act in the pathogenesis of HD. He noted that "... we are bound to look at the evidence for the effect of prolonged pressures on the cell-mediated arm of the immune system for feedback failure of restraint in influencing the development of this disease." It may result in alternation in normal gene expression of both host response[211] and in the EBV itself.

4.8. Control and Prevention

No measures are available at present. A vaccine for EBV may prevent those cases associated with severe primary infection if this hypothesis is true, but would not affect those cases associated with chronically reactivated infection unless the vaccine stimulated sustained cellular immunity to the infection.

5. Lymphoma in Immunodeficiency

The importance of immune surveillance is stressed by the observation that individuals with immune deficiencies may develop uncontrolled proliferation of EBV-immortalized cells involving a polyclonal expansion of the normal B-cell population, leading to an acute and usually fatal lymphoblastic sarcoma; this may occur either during a primary infection, such as IM, or in reactivated infections. Sometimes a Burkitt's-like tumor with typical chromosomal changes results.

The three groups at highest risk to these lymphoproliferative disorders are those with a genetically determined immunodeficiency, such as the X-linked lymphoproliferative syndrome described by Purtilo and associates,[252] patients given immunosuppressive drugs, as in renal transplant patients, and patients with immunodeficiency involving CD4+ helper T cells destroyed by chronic HIV infection.[251,253,254] There is evidence that persons infected with HIV are at increased risk to the development of malignant lymphoma, especially involving the brain, and of the aggressive NHL type,[167] although HD is also seen. These are predominantly high-grade B-cell neoplasms classified as diffuse large-cell lymphomas but also as small, noncleaved, lymphoblastic and true Burkitt's-type lymphomas, some of which carry specific chromosomal translocations.[287] The diffuse large-cell lymphomas are very similar to those occurring in other immunodeficient individuals and involve patients with severely altered immune parameters.[253] They represent polyclonal proliferation and have usually been EBV associated. The Burkitt's-type tumors are more often seen in those patients who have previously had generalized reactive lymphadenopathy and in whom the the immune dysfunction is less severe. This suggests that in the latter patients the T-cell functions are still capable of controlling EBV-induced polyclonal B-cell proliferation. However, multiple factors that trigger B-cell subsets, such as antigenic stimuli associated with concurrent infections, could lead to lymphoid hyperplasia or generalized lymphadenopathy and eventually to the development of a true monoclonal neoplasia.

5.1. X-Linked Lymphoproliferative Disease

The occurrence of EBV-positive lymphoma in the X-linked lymphoproliferative syndrome has been documented by Purtilo and his associates.[252] The syndrome involves a complex of immunodeficiencies resulting in a defect of T-cell immune regulation of B cells, accompanied by a failure to make antibodies against the EBNA complex. Following EBV infection, about two thirds develop a fatal IM and 20% develop a EBV-positive lymphoma.[256]

5.2. Organ Transplant Recipients

In parallel, patients who are immunosuppressed as recipients of organ or bone marrow transplants are at risk for EBV-positive lymphoma. In a multicenter database including 45,141 kidney and 7634 heart transplant recipients, the risk of NHL was 0.2% and 1.2%, respectively, during the first year. Subsequent annual risks dropped but remained substantially above that expected based on general population rates. Risk was increased for those who appeared to receive more aggressive immunosuppressive therapy.[230] As reviewed by List et al.,[168] the EBV appears to be intimately involved in this process. The immunosuppressive treatment to allow tolerance of the transplant results in loss of control of EBV infection. Serologically this is reflected in an enhanced production of antibodies against the VCA (IgG, IgM, IgA) and EA but low or absent antibody against EBNA.[122] These malignancies

typically have integrated monoclonal latent EBV genome,[136] and they express the individual EBNAs and LMP-1.[87,305] Since EBNA2 and LMP-1 are normally targets for EBV-immune cytotoxic T-cell response, these tumors often regress with restoration of immune function.

5.3. AIDS

Non-Hodgkin's lymphoma is the most frequent type of opportunistic malignancy seen in HIV-1-infected persons. These lymphomas are typically of B-cell origin with high-grade histology, primarily immunoblastic and BL, with primary lymphoma of the brain not uncommon. These lymphomas appear to occur independently of risk factors for HIV infection and the risk increased in parallel with decreasing immune function.[8,157,227] Occult NHL is not unusual to find at autopsy.[203] It is estimated that with antiviral therapy the probability of developing NHL is 29% after 36 months of treatment.[244] Hodgkin's disease also occurs about fivefold more frequently than expected in AIDS.[118] This risk appears to be higher among intravenous drug users.[265]

In relation to the EBV, about half of the NHL in AIDS patients are EBV-genome positive. This includes virtually all primary CNS lymphomas, 35–40% of BL, and 50–60% of non-BL lymphomas.[20,100,102,173] There is evidence of frequent spontaneous activation of EBV via detection of replicative antigens, especially in immunoblastic lymphomas.[235] Both type A and B strains are found.[21,24]

Table 11 summarizes the findings for EBV-positivity in HIV-positive HD patients. In contrast to HIV-negative patients, the level of positivity is quite high, 80–100%. This is consistent with the hypothesis that EBV-positivity is directly correlated with the degree of immunosuppression.

Table 11. Summary of Studies on the Detection of Epstein–Barr Viral Genome or Gene Products in Tissue from Hodgkin's Disease Occurring in HIV-I-Infected Patients[a]

Study	No. of cases	No. positive (%)	Method used	Specific for RS cells?	Clonal EBV?	Note
Uccini[292]	12	10 (83)	SB; in situ	Yes	4 of 4	
Audouin[7]	16	16 (100)	In situ—LMP-1	Yes	—	Labeling stronger than for non-HIV cases
Boyle[25]	2	1 (50)	In situ, PCR	Yes	—	EBV-type 2
Boiocchi[19]	7	6 (86)	SB; in situ; in situ-LMP-1, EBNA-2	Yes	4 of 5	All EBNA-2 negative
Hamilton-Dutoit[102]	11	11 (100)	In situ-EBER	Yes	—	—

[a]Studies are listed in sequential order.

6. Unresolved Problems

Although there are great advances in our understanding of the role of EBV, malaria, chromosomal translocations, and oncogenes in endemic (African) BL, we are just beginning to understand how they are regulated at the genetic and biochemical level, their temporal sequence, and the nature of their interaction. Nor do we know all of the factors that are necessary and sufficient for the development of BL.

We know little of what initiates B-cell proliferation, chromosomal changes, and c-*myc* activation in BL in America and in other sporadic settings in the absence of EBV infection early in life or holoendemic malaria. We also know little of the epidemiology and risk factors involved in sporadic BL and of the importance of immunodeficiency in such cases, but it is clear that many EBV-related lymphomas occur in persons with AIDS. We do not fully understand the reasons for the rapid decline of the incidence of African BL in many areas.

Many questions remain to be answered at the molecular level, such as the possible role of strain or genotypic variation of EBV in the production of BL, the influence of genetic factors in the virus and in the host on the patterns of disease, and the factors at the molecular level that lead to B-cell transformation and malignancy. The role of cell-mediated immunity in BL and other B-cell tumors is not fully known, nor are the ways to bolster it.

In terms of the role of EBV in the etiology of HD, the recent molecular findings present a paradox. That is, while those risk factors that reflect age and severity of infection vary with age at diagnosis and to some degree the more favorable histological subtypes, the prevalence of EBV "fingerprints" do not. In fact, the EBV is more commonly found in the more aggressive histological subtypes, and while found among all age groups, it is less evident in young adult disease. Further, EBV status does not appear to correlate with prognosis, with the expression of host oncogenes, or with the presence of T-cell receptor rearrangements, suggesting that EBV itself is not a necessary cause of the disease. The challenge is to reconcile these apparently disparate findings integrating serological, demographic, and molecular information and to determine whether EBV-negative HD represents a separate diseases entity.

The role of EBV in *de novo* NHL needs to be systematically explored. It is likely that it is a causal factor in a subset of patients who are endogenously immunosuppressed. Since EBV positivity tends to be higher in T-cell NHL,[101,233] attention needs to be paid to this subtype. Finally, the prospective serological follow-up of HIV-infected populations in conjunction with the development of EBV-positive NHL offers a laboratory for understanding EBV oncogenesis that cannot be ignored.

7. References

1. AMBINDER, R. A., BROWNING, P. J., LORENZANA, I., LEVENTHAL, B. G., COSENZA, H., MANN, R. B., MACMAHON, E. M. E., MEDINA, R., CARDONA, V., GRUFFERMAN, S., OLSHAN, A., LEVIN, A., PETERSEN, E. A., BLATTNER, W., AND LEVINE, P. H., Epstein–Barr virus and childhood Hodgkin's disease in Honduras and the United States, *Blood* **81**:462–467 (1993).

2. ANAGNOSTOPOULOS, I., HERBST, H., NIEDOBITEK, G., AND STEIN, H., Demonstration of monoclonal EBV genomes in Hodgkin's disease and KI-1-positive anaplastic large cell lymphoma by combined southern blot and *in situ* hybridization, *Blood* **74**:810–816 (1989).

3. ANONYMOUS, Cancer vaccine slated for tests, *Science* **259**:753 (1993).

4. ANVRET, M., KARLSSON, A., AND BJURSELL, G., Evidence for integration of EBV genome in Raji cellular DNA, *Nucleic Acids Res.* **12**:1149–1161 (1984).

5. ARMENIAN, H. K., AND LILIENFIELD, A. M., The distribution of incubation periods of neoplastic disease, *Am. J. Epidemiol.* **99**:92–101 (1974).

6. ARMSTRONG, J. A., EVANS, A. S., RAO, N., AND HO, M., Viral infections in renal transplant recipients, *Infect. Immun.* **14**:970–975 (1976).

7. AUDOUIN, J., DIEBOLD, J., AND PALLESEN, G., Frequent expression of Epstein–Barr virus latent membrane protein-1 in tumour cells of Hodgkin's disease in HIV-positive patients, *J. Pathol.* **167**:381–384 (1992).

8. BERAL, V., PETERMAN, T., BERKELMAN, R., AND JAFFE, H., AIDS-associated non-Hodgkin lymphoma, *Lancet* **337**:805–809 (1991).

9. BERARD, C. W., Morphological definition of Burkitt's tumor: Historical review and present status, in: *Burkitt's Lymphoma: A Human Cancer Model* (G. M. LENOIR, G. T. O'CONOR, AND C. L. M. OLWENY, EDS.), pp. 31–35, IARC, Lyon, 1985.

10. BERARD, C. W., O'CONOR, G. T., THOMAS, L. B., AND TORLONI, H. (EDS.), Histological definition of Burkitt's tumor, *Bull. WHO* **41**:601–607 (1969).

11. BHAGAT, S. K. M., MEDEIROS, L. J., WEISS, L. M., WANG, J., RAFFELD, M., AND STETLER-STEVENSON, M., *bcl-2* expression in Hodgkin's disease: Correlation with the t(14,18) translocation and Epstein–Barr virus, *Am. J. Clin. Pathol.* **99**:604–608 (1993).

12. BHATIA, K. G., GUTIERREZ, M., HUPPI, K., SIWARSKI, D., AND MAGRATH, I. T., The spectrum of p53 mutations in Burkitt's lymphoma differs from that in solid tumors, *Cancer Res.* **52**:4273–4276 (1992).

13. BIGGAR, R. J., Cancer in acquired immunodeficiency syndrome: An epidemiological assessment, *Semin. Oncol.* **17**:251–260 (1990).

14. BIGGAR, R. J., HENLE, W., FLEISHER, G., PROCKER, J., LENNETTE, E. T., AND HENLE, G., Primary Epstein–Barr virus infections in African infants I. Decline of maternal antibodies and time of infection, *Int. J. Cancer* **22**:239–243 (1978).

15. BIGGAR, R. J., HENLE, G., BOCHER, J., LENNETTE, E. T., FLEISHER, G., AND HENLE, W., II, Clinical and serological observations during seroconversion, *Int. J. Cancer* **22**:244–250 (1978).

16. BIGNON, Y.-J., BERNARD, D., CURÉ, H., FONCK, Y., PAUCHARD, J., TRAVADE, P., LEGROS, M., DASTUGUE, B., AND PLAGNE, R., Detection of Epstein–Barr viral genomes in lymph nodes of Hodgkin's disease patients, *Mol. Carcinog.* **3:**9–11 (1990).

17. BOGGER-GOREN, S., ZAIZOV, R., VOGEL, R., LEVENTON-KRISS, S., SAYAR, Y., AND GOTLIEB-STEMATSKY, T., Clinical and virological observations in childhood Hodgkin's disease in Israel, *Isr. J. Med. Sci.* **19:**989–991 (1983).

18. BOIOCCHI, M., CARBONE, A., DE RE, V., AND DOLCETTI, R., Is the Epstein–Barr virus involved in Hodgkin's disease? *Tumori* **75:**345–350 (1989).

19. BOIOCCHI, M., DE RE, V., GLOGHINI, A., VACCHER, E., DOL-CETTI, R., MARZOTTO, A., BERTOLA, G., AND CARBONE, A., High incidence of monoclonal EBV episomes in Hodgkin's disease and anaplastic large-cell KI-1-positive lymphomas in HIV-1-positive patients, *Int. J. Cancer* **54:**53–59 (1993).

20. BORISCH CHAPPUIS, B., MÜLLER, H., STUTTE, J., HEY, M. M., HÜBNER, K., AND MÜLLER-HERMELINK, H. K., Identification of EBV-DNA in lymph nodes from patients with lymphadenopathy and lymphomas associated with AIDS, *Virchows Arch. B* **58:**199–205 (1990).

21. BORISCH, B., FINKE, J., HENNIG, I., DELACRÉTAZ, F., SCHNEIDER, J., HEITZ, PH. U., AND LAISSUE, J. A., Distribution and localization of Epstein–Barr virus subtypes A and B in AIDS-related lymphomas and lymphatic tissue of HIV-positive patients, *J. Pathol.* **168:**229–236 (1992).

22. BORNKAMM, G., HUDEWENTZ, J., FREESE, U. K., AND ZIMBER, U., Deletion of the nontransforming Epstein–Barr virus strain P3HR-1 causes fusion of large internal repeat to the DSL region, *J. Virol.* **43:**952–968 (1982).

23. BORNKAMM, G., STEIN, H., LENNART, K., RUGGELBERG, F., BARTELS, H. F., AND ZUR HAUSEN, H., Attempts to demonstrate virus-specific sequences in human tumors. IV. EB viral DNA in European Burkitt lymphoma and in immunoblastic lymphadenopathy with excessive plasmacytosis, *Int. J. Cancer* **17:**177–181 (1976).

24. BOYLE, M. J., SEWELL, W. A., SCULLEY, T. B., APOLLONI, A., TURNER, J. J., SWANSON, C. E., PENNY, R., AND COOPER, D. A., Subtypes of Epstein–Barr virus in human immunodeficiency virus-associated non-Hodgkin lymphoma, *Blood* **78:**3004–3011 (1991).

25. BOYLE, M. J., VASAK, E., TSCHUCHNIGG, M., TURNER, J. J., SCULLEY, T., PENNY, R., COOPER, D. A., TINDALL, B., AND SEWELL, W. A., Subtypes of Epstein–Barr virus (EBV) in Hodgkin's disease: Association between B-type EBV and immuno-compromise, *Blood* **81:**468–474 (1993).

26. BRICHACEK, B., HIRSCH, I., SIBL, D., VILIKUSOVA, E., AND VONKA, V., Presence of Epstein–Barr virus DNA in carcinomas of the palatine tonsil, *J. Natl. Cancer Inst.* **71:**809–815 (1984).

27. BROCKSMITH, D., ANGEL, C. A., PRINGLE, J. H., AND LAUDER, I., Epstein–Barr viral DNA in Hodgkin's disease: Amplification and detection using the polymerase chain reaction, *J. Pathol.* **165:**11–15 (1991).

28. BROUSSET, P. B., CHITTAL, S., SCHLAIFER, D., ICART, J., PAYEN, C., RIGAL-HUGUET, F., VOIGHT, J.-J., AND DELSOL, G., Detection of Epstein–Barr virus messenger RNA in Reed–Sternberg cells of Hodgkin's disease with biotinylated probes on specially processed modified acetone methyl benzoate xylene (ModAMeX) sections, *Blood* **77:**1781–1786 (1991).

29. BURKITT, D. P., A sarcoma involving the jaws in African children, *Br. J. Surg.* **46:**218–223 (1958).

30. BURKITT, D. P., A children's cancer dependent on climatic factors, *Nature* **194:**232–234 (1962).

31. BURKITT, D. P., Determining the climatic limitations of a childrens tumor common in Africa, *Br. Med. J.* **2:**1019–1023 (1962).

32. BURKITT, D. P., Burkitt's lymphoma outside the known endemic areas of Africa and New Guinea, *Int. J. Cancer* **2:**562–565 (1967).

33. BURKITT, D., AND O'CONOR, G. T., Malignant lymphoma in African children. I. A clinical syndrome, *Cancer* **14:**258–269 (1961).

34. BURKITT, D., AND WRIGHT, D. H., Geographical and tribal distribution of the African lymphoma in Uganda, *Br. Med. J.* **5487:**569–573 (1966).

35. BYRNE, E. B., EVANS, A. S., FOUTS, D. W., AND ISREAL, H. L., A seroepidemiological study of Epstein–Barr virus and other antigens in sarcoidosis, *Am. J. Epidemiol.* **97:**355–363 (1973).

36. CARBONE, A., GLOGHINI, A., ZANETTE, I., CANAL, B., RIZZO, A., AND VOLPE, R., Co-expression of Epstein–Barr virus latent membrane protein and vimentin in "aggressive" histological subtypes of Hodgkin's disease, *Virchows Arch. A Pathol. Anat.* **422:**39–45 (1993).

37. CARTER, C. D., BROWN, T. M., JR., HERBERT, J. T., AND HEATH, C. W., JR., Cancer incidence following infectious mononucleosis, *Am. J. Epidemiol.* **105:**30–36 (1977).

38. CARVALHO, R. P. S., EVANS, A. S., FROST, P., DALLDORF, G., CAMARGO, M. F., AND JAMRA, M., EBV infections in Brazil, I. Occurrence in normal persons, in lymphomas, and in leukemias, *Int. J. Cancer* **11:**191–201 (1973).

39. CHANG, K. L., ALBÚJAR, P. F., CHEN, Y.-Y., JOHNSON, R. M., AND WEISS, L. M., High prevalence of Epstein–Barr virus in the Reed–Sternberg cells of Hodgkin's disease occurring in Peru, *Blood* **81:**496–501 (1993).

40. CHANG, R. S., LEWIS, J. S., REYNOLDS, R. D., SULLIVAN, M. J., AND NEUMAN, J., Oropharyngeal excretion of Epstein–Barr virus by patients with lymphproliferative disorders and by recipients of renal homografts, *Ann. Intern. Med.* **88:**111–123 (1978).

41. COATES, P. J., SLAVIN, G., AND D'ARDENNE, A. J., Persistence of Epstein–Barr virus in Reed–Sternberg cells throughout the course of Hodgkin's disease, *J. Pathol.* **164:**291–297 (1991).

42. Collected reports of cases of Burkitt's lymphoma from countries outside the endemic area, *Int. J. Cancer,* **2:**550–609 (1969).

43. CONNELLY, R. R., AND CHRISTINE, B. W., A cohort study of cancer following infectious mononucleosis, *Cancer Res.* **34:**1172–1178 (1974).

44. CORREA, P., AND O'CONNOR, G. T., Epidemiologic patterns of Hodgkin's disease, *Int. J. Cancer* **8:**192–201 (1971).

45. CRAWFORD, D. H., Epstein–Barr virus: Does it cause cancer? *J. Trop. Med. Hyg.* **95:**229–238 (1992).

46. CROCE, C. M., Chromosome translocations and human cancer, *Cancer Res.* **46:**6019–6023 (1986).

47. DALLDORF, G., LINSELL, C. A., BARNHART, F. C., AND MARTYN, R., An epidemiologic approach to the lymphomas of African children and Burkitt's sarcoma of the jaws, *Perspect. Biol. Med.* **7:**435–449 (1964).

48. DAMBAUGH, T., NKRUMRAH, F. K., BIGGAR, R. J., AND KIEFF, E., Epstein–Barr virus RNA in Burkitt tumor tissue, *Cell* **16:**313–322 (1979).

49. DAVIES, J. N., ELMES, S., HUTT, M. S. R., MTIMAVALZE, L. A. R., OWOR, R., AND SHARPER, L., Cancer in an African community, 1897–1956: An analysis of the records of the Mengo Hospital, Kampala, Uganda, Part 2, *Br. Med. J.* **1:**336–341 (1964).

50. DE-THÉ, G., Epidemiology of Epstein–Barr virus and associated

diseases in man, in: *The Herpesviruses*, Vol. 1 (B. ROIZMAN, ED.), pp. 25–103, Plenum Press, New York, 1978.

51. DE-THÉ, G., Epstein–Barr virus and Burkitt's lymphoma worldwide. The causal relationship revisited, in: *Burkitt's Lymphoma: A Human Cancer Model* (G. M. LENOIR, G. T. O'CONOR, AND C. L. M. OLWENY, EDS.), pp. 165–176, IARC, Lyon, France, 1985.

52. DE-THÉ, G., DAY, N. E., GESER, A., AND LAVOUE, M. F., Seroepidemiological of the Epstein–Barr virus: Preliminary analysis of an international study—A review, in: *Oncogenesis and Herpesviruses*, 11 (G. DE-THÉ, M. A. EPSTEIN, AND H. ZUR HAUSEN, EDS.), pp. 3–16, IARC, Lyon, France, 1975.

53. DE-THÉ, G., GESER, A., DAY, N. E., TUBER, P. M., WILLIAMS, E. H., BEIR, D. P., SMITH, P. G., DEAN, A., BORNKAMM, G. W., FEORINO, P., AND HENLE, W., Epidemiological evidence for causal relationship between Epstein–Barr virus and Burkitt's lymphoma from Ugandan prospective study, *Nature* **274**:756–761 (1978).

54. DEACON, E. M., PALLESEN, G., NIEDOBITEK, G., CROCKER, J., BROOKS, L., RICKINSON, A. B., AND YOUNG, L. S., Epstein–Barr virus and Hodgkin's disease: Transcriptional analysis of virus latency in the malignant cells, *J. Exp. Med.* **177**:339–349 (1993).

55. DELSOL, G., BROUSSET, P., CHITTAL, S., AND RIGAL-HUGUET, F., Correlation of the expression of Epstein-Barr virus latent membrane protein and *in situ* hybridization with biotinylated *Bam*HI-W probes in Hodgkin's disease, *Am. J. Pathol.* **140**:247–253 (1992).

56. DESCHRYVER, A., KLEIN, G., HENLE, G., HENLE, W., CAMERON, H. M., SANTESSA, L., AND CLIFFORD, P., EB-virus associated serology in malignant diseases. Antibody levels to viral capsid antigens (VCA), membrane antigens (MA), and early antigens (EA) in patients with various neoplastic disease, *Int. J. Cancer* **9**:353–364 (1972).

57. DREXLER, H. G., Recent results on the biology of Hodgkin and Reed–Sternberg cells. I. Biopsy material, *Leuk. Lymphoma* **8**:283–313 (1992).

58. DREXLER, H. G. AND LEBER, B. F., The nature of the Hodgkin cell. Report of the First International Symposium on Hodgkin's Lymphoma, Kohl, FRG, October 2–3, 1987, *Blut* **56**:135–137 (1987).

59. EPSTEIN, M. A., Epstein–Barr virus—is it time to develop a vaccine program? *J. Natl. Cancer Inst.* **56**:697–700 (1976).

60. EPSTEIN, M. A., AND BARR, Y. M., Cultivation *in vitro* of human lymphoblasts from Burkitt's malignant lymphoma, *Lancet* **1**:252–253 (1964).

61. EPSTEIN, M. A., ACHONG, B. G., AND BARR, Y. M., Virus particles in cultured lymphoblasts from Burkitt's lymphoma, *Lancet* **1**:702–703 (1964).

62. EPSTEIN, M. A., AND ACHONG, B. G. S., (EDS.), *The Epstein–Barr Virus: Recent Advances*, Wiley Medical Publication, New York, 1986.

63. EPSTEIN, M. A., AND MORGAN, A. J., Prevention of endemic Burkitt's lymphoma, in: *Burkitt's Lymphoma: A Human Cancer Model* (G. M. LENOIR, G. T. O'CONOR, AND C. L. M. OLWENY, EDS.), pp. 293–302, IARC, Lyon, France, 1985.

64. EPSTEIN, M. A., HUNT, R. D., AND RABIN, H., Pilot experiments with EB virus in owl monkeys (*Aotus trivirgatus*). 1. Reticuloproliferative disease in an inoculated animal, *Int. J. Cancer* **12**:309–318 (1973).

65. EVANS, A. S., Epidemiology of Burkitt's lymphoma: Other factors, in: *Burkitt's Lymphoma: A Human Cancer Model* (G. M.

LENOIR, G. T. O'CONOR, AND C. L. M. OLWENY, EDS.), pp. 197–204, IARC, Lyon, France, 1985.

66. EVANS, A. S., Epstein–Barr virus: An organism for all seasons, in: *Medical Virology VII New York* (L. M. D. L. MAZA AND E. M. PETERSON, EDS.), pp. 57–97, Elsevier, Amsterdam, 1988.

67. EVANS, A. S., EBV vaccine: Use in infectious mononucleosis, in: *Vth International Symposium on Epstein–Barr Virus and Associated Diseases*, (T. TURSZ, J. S. PAGANO, D. V. ABLASHI, G. DE THÉ, G. LENOIR, AND G. R. PEARSON, EDS.) pp. 593–598, Annecy, France; John Libby Eurotext, London, 1992.

68. EVANS, A. S., AND DE-THÉ, G., Burkitt's lymphoma, in: *Viral Infections of Humans: Epidemiology and Control*, 3rd ed. (A. S. EVANS, ED.), pp. 713–735, Plenum Press, New York, 1989.

69. EVANS, A. S., AND GUTENSOHN, N. M., A population-based case-control study of EBV and other viral antibodies among persons with Hodgkin's disease and their siblings, *Int. J. Cancer* **34**:149–157 (1984).

70. EVANS, A. S., AND MUELLER, N., Viruses and cancer. Causal associations, *Ann. Epidemiol.* **1**:71–92 (1989).

71. EVANS, A. S., CARVALHO, R. P. S., FROST, P., JAMRA, M., AND POZZI, D. H. B., Epstein–Barr virus infections in Brazil. II. Hodgkin's disease, *J. Natl. Cancer Inst.* **61**:19–26 (1978).

72. EVANS, A. S., KIRCHHOFF, L. V., PANUTTI, C. S., CARVALHO, R. P. S., AND MCCLELLAND, K. E., A case control study of Hodgkin's disease in Brazil. II. Seroepidemiological studies in cases and family members, *Am. J. Epidemiol.* **112**:609–618 (1980).

73. FALK, L., WOLFE, L., DEINHARDT, F., PACIGA, J., DOMBOS, L., KLEIN, G., HENLE, W., AND HENLE, G., Epstein–Barr virus: Transformation of non-human lymphocytes *in vitro*, *Int. J. Cancer* **13**:353–376 (1974).

74. FELLBAUM, C., HANSMANN, M.-L., NIEDERMEYER, H., KRAUS, I., ALAVAIKKO, M. J., BLANCO, G., AINE, R., BUSCH, R., PÜTZ, B., FISCHER, R., AND HÖFLER, H., Influence of Epstein–Barr virus genomes on patient survival in Hodgkin's disease, *Am. J. Clin. Pathol.* **98**:319–323 (1992).

75. FINERTY, S., TARLTON, J., MACHETT, M., CONWAY, M., ARRAND, J. R., WATKINS, P. E., AND MORGAN, A. J., Protective immunization against Epstein–Barr virus induced disease in cotton top marmosets using virus envelope gp340 produced from bovine papilloma virus expression vector, *J. Gen. Virol.* **3**:449–453 (1992).

76. FINGERDROTH, J. D., WEIR, J. J., TEDDER, T. F., STROMINGER, J. L., BIRO, P. A., AND FEARSON, D. T., Epstein–Barr virus receptor of human B lymphocytes is the C3d receptor CR2, *Proc. Natl. Acad. Sci. USA* **81**:4510–4514 (1984).

77. FLEISHER, G., HENLE, W., HENLE, G., LENNETTE, E. T., AND BIGGAR, R. J., Primary infection with Epstein–Barr virus in infants in the United States: Clinical and serological observations, *J. Infect. Dis.* **139**:553–558 (1979).

78. Further in the Hodgkin's maze, *Lancet* **1**:1053–1054 (1971).

79. GERBER, P., Activation of Epstein–Barr virus by 5-bromodeoxyuridine in virus free human cells, *Proc. Natl. Acad. Sci. USA* **69**:83–85 (1972).

80. GERBER, P., AND HOYER, B. H., Induction of cellular DNA synthesis in human leucocytes by Epstein–Barr virus, *Nature* **231**:46–47 (1976).

81. GESER, A., DE-THÉ, G., LENOIR, G., DAY, N. E., AND WILLIAMS, E. H., Final case reporting from the Ugandan prospective study of the relationship between EBV and Burkitt's lymphoma, *Int. J. Cancer* **29**:397–400 (1982).

82. GESER, S., AND BRUBAKER, G., A preliminary report of epidemi-

ological studies of Burkitt lymphoma, Epstein–Barr virus infection, and malaria in North Mara, Tanzania, in: *Burkitt's Lymphoma: A Human Cancer Model* (G. M. Lenoir, G. T. O'Conor, and C. L. M. Olweny, eds.), pp. 205–216, IARC, Lyon, France, 1985.

83. Glaser, S., and Swartz, W. G., Time trends in Hodgkin's disease incidence: The role of diagnostic accuracy, *Cancer* **66:**2196–2204 (1990).

84. Gledhill, S., Gallagher, A., Jones, D. B., Krajewski, A. S., Alexander, F. E., Klee, E., Wright, D. H., O'Brien, C., Onions, D. E., and Jarrett, R. F., Viral involvement in Hodgkin's disease: Detection of clonal type A Epstein–Barr virus genomes in tumour samples, *Br. J. Cancer* **64:**227–232 (1991).

85. Goldman, J. M., and Aisenberg, A. C., Incidence of antibody to EB virus, herpes simplex, and cytomegalovirus in Hodgkin's disease, *Cancer* **26:**327–331 (1970).

86. Gotlieb-Stematsky, T., Vonsover, A., Ramot, B., Zaizov, R., Nordan, U., Aghai, E., Kende, G., and Modan, M., Antibodies to Epstein–Barr virus in patients with Hodgkin's disease and leukemia, *Cancer* **36:**1640–1645 (1975).

87. Gratama, J. W., Zutter, M. M., Minarovits, J., Oosterveer, M. A. P., Thomas, E. D., Klein, G., and Ernberg, I., Expression of Epstein–Barr virus-encoded growth-transformation-associated proteins in lymphoproliferations of bone-marrow transplant recipients, *Int. J. Cancer* **47:**188–192 (1991).

88. Greenspan, J. S., Greenspan, D., Lennette, E. T., Abrams, D. T., Conant, M. A., Peterson, V., and Freese, U. K., Replication of Epstein–Barr virus within the epithelial cells of oral "hairy" leukoplakia, an AIDS-associated lesion, *N. Engl. J. Med.* **313:**1564–1571 (1985).

89. Greenwood, B. M., and Vick, R., Evidence for a malaria antigen in human malaria, *Nature* **257:**592–594 (1975).

90. Grufferman, S., Hodgkin's disease, in *Cancer Etiology and Prevention* (D. Schottenfeld and J. F. Fraumeni, Jr., eds.), pp. 739–753, W. B. Saunders, Philadelphia, 1982.

91. Gu, S., Huang, T., Miao, Y., Ruan, L., Zhao, Y., Han, C., Xiao, Y., Zho, J., and Wolf, H., A preliminary study of the immunogenicity in rabbits and human volunteers of a recombinant vaccinia virus expressing Epstein–Barr membrane antigen, *Chin. Med. Sci. J.* **6:**241–243 (1991).

92. Gutensohn (Mueller), N., Social class and age at diagnosis of Hodgkin's disease: New epidemiologic evidence for the "two-disease" hypothesis, *Cancer Treat. Rep.* **66:**689–695 (1982).

93. Gutensohn (Mueller), N., and Cole, P., Epidemiology of Hodgkin's disease in the young, *Int. J. Cancer* **19:**595–604 (1977).

94. Gutensohn (Mueller), N., and Cole, P., Epidemiology of Hodgkin's disease, *Semin. Oncol.* **7:**92–102 (1980).

95. Gutensohn (Mueller), N., and Cole, P., Childhood social environment and Hodgkin's disease, *N. Engl. J. Med.* **292:**135–140 (1981).

96. Gutensohn (Mueller), N., and Shapiro, D. S., Social class factors among children with Hodgkin's disease, *Int. J. Cancer* **30:**433–435 (1982).

97. Gutierrez, M., Bhatia, K., Barriga, F., Dies, B., Muriel, F. S., de Andreas, M.-L., Epelman, S., Risueno, C. and Magrath, I., Molecular epidemiology of Burkitt's lymphoma from South America: Differences in breakpoint location and Epstein–Barr virus association from tumors in other world regions, *Blood* **79:**3261–3266 (1992).

98. Haddow, A. J., Epidemiological evidence suggesting an infec-

tive element in the etiology, in: *Burkitt's Lymphoma* (D. P. Burkitt and D. H. Wright, eds.), pp. 198–209, E. S. Livingston, Edinburgh, 1970.

99. Hallee, T. J., Evans, A. S., Niederman, J. C., Brooks, C. M., and Voegtly, J. H., Infectious mononucleosis at the US Military Academy: A prospective study of a single class over four years, *Yale J. Biol. Med.* **47:**182–195 (1974).

100. Hamilton-Dutoit, S. J., Pallesen, G., Franzmann, M. B., Karkov, J., Black, F., Skinhøj, P., and Pedersen, C., AIDS-related lymphoma: Histopathology, immunophenotype, and association with Epstein–Barr virus as demonstrated by *in situ* nucleic acid hybridization, *Am. J. Pathol.* **138:**149–163 (1991).

101. Hamilton-Dutoit, S. J., Pallesen, G., A survey of Epstein–Barr virus gene expression in sporadic non-Hodgkin's lymphomas: Detection of Epstein–Barr virus in a subset of peripheral T-cell lymphomas, *Am. J. Pathol.* **140:**1315–1325 (1992).

102. Hamilton-Dutoit, S. J., Raphael, M., Audouin, J., Diebold, J., Lisse, I., Pedersen, C., Oksenhendler, E., Marelle, L., and Pallesen, G., *In situ* demonstration of Epstein–Barr virus small RNAs (EBER 1) in acquired immunodeficiency syndrome-related lymphomas: Correlation with tumor morphology and primary site, *Blood* **82:**619–624 (1993).

103. Hampar, B., Derge, J. G., Martos, L. M., and Walker, J. L., Synthesis of Epstein–Barr virus after the reactivation of the viral genome in a virus-negative human lymphoblastoid cell (Raji) made resistant to five bromodeooxyuridine, *Proc. Natl. Acad. Sci. USA* **60:**78–82 (1972).

104. Henderson, B. E., Dworsky, R., Menck, H., Alena, B., Henle, W., Henle, G., and Terasaki, P., Case-control study of Hodgkin's disease. II. Herpesvirus group antibody titers and HL-A type, *J. Natl. Cancer Inst.* **51:**1443–1447 (1973).

105. Henderson, E., Robinson, J., Frank, A., and Miller, G., Epstein–Barr virus: Transformation of lymphocytes separated by size or exposed to bromodeoxyuridine and light, *Virology* **82:**196–205 (1977).

106. Henle, G., and Henle, W., Immunofluorescence in cells derived from Burkitt's lymphoma, *J. Bacteriol.* **91:**1248–1256 (1966).

107. Henle, G., Henle, W., Cliford, P., Diehl, V., Kafuko, G. W., Kirya, B. G., Klein, G., Morrow, R. H., Minube, G. M., Pike, M. C., Tukei, P. M., and Ziegler, J. L., Antibodies to Epstein–Barr virus in Burkitt's lymphoma and control groups, *J. Natl. Cancer Inst.* **43:**1147–1157 (1969).

108. Henle, W., and Henle, G., Epstein–Barr virus-related serology in Hodgkin's disease, *Natl. Cancer Inst. Monogr.* **36:**79–84 (1973).

109. Henle, W., Henle, G., Andersson, J., Ernberg, I., Klein, G., Horwitz, C. A., Marklund, G., Rymo, L., Wellinder, C., and Strauss, S. E., Antibody responses to Epstein–Barr virus-determined nuclear antigen (EBNA-1 and EBNA-2) in acute and chronic Epstein–Barr virus infection, *Proc. Natl. Acad. Sci. USA* **83:**570–574 (1987).

110. Hennekens, C. H., and Buring, J. E., *Epidemiology in Medicine*, Little Brown and Company, Boston, MA, 1987.

111. Henness, K., Heller, M., Van Santen, V., and Kieff, E., Simple repeat array in Epstein–Barr virus DNA encoded part of the Epstein–Barr nuclear antigen, *Science* **220:**1396–1398 (1983).

112. Herbst, H., and Niedobitek, G., Epstein–Barr virus and Hodgkin's disease, *Int. J. Clin. Lab. Res.* **23:**13–16 (1993).

113. Herbst, H., Dallenbach, F., Hummel, K., Niedobitek, G., Pileri, S., Müller-Lantzsch, N., and Stein, H., Epstein–Barr

virus latent membrane protein expression in Hodgkin and Reed–Sternberg cells, *Proc. Natl. Acad. Sci. USA* **88**:4766–4770 (1991).

114. HERBST, H., NIEDOBITEK, G., KNEBA, M., HUMMEL, M., FINN, T., ANAGNOSTOPOULOS, I., BERGHOLZ, M., KRIEGER, G., AND STEIN, H., High incidence of Epstein–Barr virus genomes in Hodgkin's disease, *Am. J. Pathol.* **137**:13–18 (1990).

115. HERBST, H., STEINBRECHER, E., NIEDOBITEK, G., YOUNG, L. S., BROOKS, L., MÜLLER-LANTZSCH, N., AND STEIN, H., Distribution and phenotype of Epstein–Barr virus-harboring cells in Hodgkin's disease, *Blood* **80**:484–491 (1992).

116. HERBST, H., TIPPELMANN, G., ANAGNOSTOPOULOS, I., GERDES, J., SCHWARTING, R., BOEHM, T., PILERI, S., JONES, D. B., AND STEIN, H., Immunoglobulin and T-cell receptor gene rearrangements in Hodgkin's disease and Ki-1-positive anaplastic large cell lymphoma: Dissociation between phenotype and genotype, *Leuk. Res.* **13**:103–116 (1989).

117. HESSE, J., LEVINE, P. H., EBBESEN, P., CONNELLY, R. R., AND MORDHORST, C. H., A case-control study on immunity to two Epstein–Barr virus-associated antigens and to herpes simplex virus and adenovirus in a population-based group of patients with Hodgkin's disease in Denmark, 1971–1973, *Int. J. Cancer* **19**:45–58 (1977).

118. HESSOL, N. A., KATZ, M. H., LIU, J. Y., BUCHBINDER, S. P., RUBINO, C. J., AND HOLMBERG, S. D., Increased incidence of Hodgkin disease in homosexual men with HIV infection, *Ann. Intern. Med.* **117**:309–311 (1992).

119. HILGERS, J., DEAN, A. G., AND DE-THÉ, G. B., Elevated immunofluorescence titers to several herpes viruses in Burkitt's lymphoma patients. Are high titers unique? *J. Natl. Cancer Inst.* **54**:49–51 (1975).

120. HILGERS, F., AND HILGERS, J., An immunofluorescence technique with counterstain on fixed cells for the detection of antibodies to human herpesviruses; Antibody patterns in patients with Hodgkin's disease and nasopharyngeal carcinoma, *Intervirology* **7**:309–327 (1976).

121. HIRSHAUT, Y., REAGAN, R. I., PERRY, S., DEVITA, V., JR., AND BARILE, M. F., The search for a viral agent in Hodgkin's disease, *Cancer* **34**:1080–1089 (1974).

122. HO, M., MILLER, G., ATCHISON, R. W., BREINIG, M. K., DUMMER, J. S., ANDIMAN, W., STARZL, T. E., EASTMAN, R., GRIFFITH, B. P., HARDESTY, R. L., BAHNSON, H. T., HAKALA, T. R., AND ROSENTHAL, J. T., Epstein–Barr virus infections and DNA hybridization studies in posttransplant lymphoma and lymphoproliferative lesions: The role of primary infection, *J. Infect. Dis.* **152**:876–886 (1985).

123. HORWITZ, C. A., HENLE, W., HENLE, G., RUDNICK, H., AND LATTS, E., Long-term serological follow-up of patients for Epstein–Barr virus after recovery from infectious mononucleosis, *J. Infect. Dis.* **151**:1150–1153 (1985).

124. ITO, Y., KAWANISHI, M., HIIRAYAMA, T., AND TAKABAGASHI, S., Combined effects of the extracts from Croton tiglium, Euphobia lathyres or Euphorbia tizucalle and N-butyrate on Epstein–Barr virus expression in human lymphoblastoid P3HR-1 and Raji cells, *Cancer Lett.* **12**:175–180 (1981).

125. JARRETT, R. F., GALLAGHER, A., JONES, D. B., ALEXANDER, F. E., KRAJEWSKI, A. S., KELSEY, A., ADAMS, J., ANGUS, B., GLEDHILL, S., WRIGHT, D. H., CARTWRIGHT, R. A., AND ONIONS, D. E., Detection of Epstein–Barr virus genomes in Hodgkin's disease: Relation to age, *J. Clin. Pathol.* **44**:844–848 (1991).

126. JIWA, N. M., KANAVAROS, P., VAN DER VALK, P., WALBOOMERS, J. M. M., HORSTMAN, A., VOS, W., MULLINK, H., AND MEIJER, C. J.

L. M., Expression of *c-myc* and *bcl-2* oncogene products in Reed–Sternberg cells independent of presence of Epstein–Barr virus, *J. Clin. Pathol.* **46**:211–217 (1993).

127. JOHANSSON, B., KLEIN, G., HENLE, W., AND HENLE, G., Epstein–Barr virus (EBV)-associated antibody patterns in malignant lymphoma and leukemia, *Int. J. Cancer* **6**:450–462 (1970).

128. JONDAL, M., AND KLEIN, G., Surface markers on human B and T lymphocytes. II. Presence of Epstein–Barr virus receptors on B lymphocytes, *J. Exp. Med.* **138**:1365–1378 (1973).

129. JUDSON, S. C., WERNER, W., AND HENLE, G., A cluster of Epstein–Barr-virus-associated Burkitt's lymphoma, *N. Engl. J. Med.* **297**:464–468 (1977).

130. KAFUKO, G. W., AND BURKITT, D. P., Burkitt lymphoma and malaria, *Int. J. Cancer* **6**:1–9 (1970).

131. KAFUKO, G. W., HENDERSON, B. E., KIRYA, B. G., MANUBE, G. M. R., TUKEI, P. M., DAY, N. E., HENLE, G., HENLE, W., MORROW, R. H., PIKE, M. C., SMITH, P. G., AND WILLIAMS, E. H., Epstein–Barr virus antibody levels in children from the West Nile District of Uganda, *Lancet* **1**:706–709 (1972).

132. KALLIN, B., DILLNER, J., ERNBERG, I., EHLIN-HENRIKSSON, B., ROSEN, A., HENLE, W., HENLE, G., AND KLEIN, G., Four virally determined nuclear antigens are expressed in Epstein–Barr virus-transformed cells, *Proc. Natl. Acad. Sci. USA* **83**:1499–1503 (1986).

133. KAPLAN, H. S., *Hodgkin's Disease*, Harvard University Press, Cambridge, MA, 1980.

134. KASCHUKA-DIERICH, C., ADAMS, A., LINDAHL, T., BORNKAMM, G., BJURSELL, G., KLEIN, G., GIOVANELLA, B., AND SING, S., Intracellular forms of Epstein–Barr virus DNA in human tumor cells *in vivo*, *Nature* **260**:302–306 (1976).

135. KASULE, O., The epidemiology of childhood Epstein–Barr virus infection in Uganda in relation to the risk of Burkitt's lymphoma, doctoral thesis, School of Public Health, Harvard University, 1988.

136. KATZ, B. Z., RAAB-TRAUB, N., AND MILLER, G., Latent and replicating forms of Epstein–Barr virus DNA in lymphoma and lymphoproliferative diseases, *J. Infect. Dis.* **160**:589–598 (1989).

137. KHAN, G., COATES, P. J., GUPTA, R. K., KANGRO, H. O., AND SLAVIN, G., Presence of Epstein–Barr virus in Hodgkin's disease is not exclusive to Reed–Sternberg cells, *Am. J. Pathol.* **140**:757–762 (1992).

138. KHAN, G., NORTON, A. J., AND SLAVIN, G., Epstein–Barr virus in Hodgkin disease: Relation to age and subtype, *Cancer* **71**:3124–3129 (1993).

139. KHANNA, R., BURROWS, S. R., KURILLA, M. G., JACOB, C. A., MISKO, I. S., SCULLEY, T. B., KIEFF, E., AND MOSS, D. J., Localization of Epstein–Barr virus cytotoxic T-cell epitopes using recombinant vaccinia: Implications for vaccine development, *J. Exp. Med.* **176**:169–176 (1992).

140. KIEFF, E., DAMBAUGH, T., HELLER, M., KING, W., CHEUNG, A., VAN SANTEN, V., HUMMEL, M., BEISEL, C., FENNEWALD, S., HENNESSY, K., AND HEINEMAN, T., The biology and chemistry of Epstein–Barr virus, *J. Infect. Dis.* **146**:506–517 (1982).

141. KIEFF, E., AND LIEBOWITZ, D., Epstein–Barr virus and its replication, in: *Virology*, 2nd ed. (B. FIELDS AND D. M. KNIPE, EDS.), pp. 1889–1920, Raven Press, New York, 1990.

142. KLEIN, E., KLEIN, G., NADKARM, J. S., WIGZILL, H., AND CLIFFORD, P., Surface IgM-kappa specificity on cells derived from a Burkitt's lymphoma, *Lancet* **2**:1068–1070 (1967).

143. KLEIN, G., Epstein–Barr virus, malaria, and Burkitt's lymphoma, *Scand. J. Infect. Dis. Suppl.* **36**:15–23 (1982).

144. KLEIN, G., Epstein–Barr virus and its association with human

disease: An overview, in: *Epstein–Barr Virus and Human Disease—1988*, (D. V. ABLASHI, A. FAGGIONI, AND G. R. F. KRUEGERET, EDS.), pp. xvii–xxvii, Humana Press, Clifton, NJ, 1988.

145. KLEIN, G., The paradoxical coexistence of EBV and the human species, *Epstein–Barr Virus Rep.* **1**:1–5 (1994).

146. KLEIN, G., LINDAHL, T., JONDAL, W., MENEZES, J., NILSSON, K., AND SUNDSTROM, C., Continuous lymphoblastoid cell lines with characteristics of B cells (bone marrow derived) lacking the EBV genome and derived from 3 human lymphomas, *Proc. Natl. Acad. Sci. USA* **71**:3283–3286 (1974).

147. KNECHT, H., ODERMATT, B. F., BACHMANN, E., TEIXEIRA, S., SAHLI, R., HAYOZ, D., HEITZ, P., AND BACHMANN, F., Frequent detection of Epstein–Barr virus DNA by the polymerase chain reaction in lymph node biopsies from patients with Hodgkin's disease without genomic evidence of B- or T-cell clonality, *Blood* **78**:760–767 (1991).

148. KVÅLE, G., HØIBY, E. A., AND PEDERSEN, E., Hodgkin's disease in patients with previous infectious mononucleosis, *Int. J. Cancer* **23**:593–597 (1979).

149. LANGE, B., ARBETER, A., HERVETSON, J., AND HENLE, W., Longitudinal study of Epstein–Barr virus antibody titers and excretion in pediatric patients with Hodgkin's disease, *Int. J. Cancer* **22**:521–527 (1978).

150. LANGENHUYSEN, M. M., CAZEMIER, T., HOUWEN, B., BROUWERS, T. M., HALIE, M. R., THE, T. H., AND NIEWEG, H. O., Antibodies to Epstein–Barr virus, cytomegalovirus, and Australia antigen in Hodgkin's disease, *Cancer* **34**:262–267 (1974).

151. LEDER, P., The state and prospect for molecular genetics in Burkitt's lymphoma, in: *Burkitt's Lymphoma: A Human Cancer Model*, (G. M. LENOIR, G. T. O'CONOR, AND C. L. M. OLWENY, EDS.), pp. 465–468, IARC, Lyon, France, 1985.

152. LEDER, P., Translocations among antibody genes in human cancer, in: *Burkitt's Lymphoma: A Human Cancer Model*, (G. M. LENOIR, G. T. O'CONOR, AND C. L. M. OLWENY, EDS.), pp. 341–358, IARC, Lyon (1985).

153. LEHTINEN, T., LUMIO, J., DILLNER, J., HAKAMA, M., KNEKT, P., LEHTINEN, M., TEPPO, L., AND LEINIKKI, P., Increased risk of malignant lymphoma indicated by elevated Epstein–Barr virus antibodies—a prospective study, *Cancer Causes Control* **4**:187–193 (1993).

154. LENOIR, G. M., Role of the virus, chromosomal translocations and cellular oncogenes in the aetiology of Burkitt's lymphoma, in: *The Epstein–Barr Virus: Recent Advances* (M. A. EPSTEIN AND B. G. ACHONG, EDS.), pp. 184–205, Wiley Medical Publication, New York, 1986.

155. LENOIR, G. M., O'CONOR, G., AND OLWENY, C. L. M., (EDS.), *Burkitt's Lymphoma: A Human Cancer Model*, IARC, Lyon, France, 1985.

156. LENOIR, G. M., PHILIP T., AND SOHIER, R., Burkitt-type lymphoma: EBV association and cytogenetic markers in cases from various geographic locations, in: *Pathogenesis of Leukemias and Lymphomas* (I. T. MCGRATH, G. T. O'CONOR, AND B. RAMOT, EDS.), pp. 283–295, Raven Press, New York (1984).

157. LEVINE, A. M., AIDS-related malignancies: The emerging epidemic, *J. Natl. Cancer Inst.* **85**:1382–1397 (1993).

158. LEVINE, P. H., ABLASHI, D. V., AND BERARD, C. W., Elevated antibody titers to Epstein–Barr virus in Hodgkin's disease, *Cancer* **27**:416–421 (1971).

159. LEVINE, P. H., AND BLATTNER, W. A., The epidemiology of human virus-associated hematologic malignancies, *Leukemia* **6**(Suppl. 3):54s–59s (1992).

160. LEVINE, P. H., CONNELLY, R. R., AND MCKAY, F. W., Burkitt's lymphoma cases reported to the American Burkitt's Lymphoma Registry compared with population-based incidence and mortality data, in: *Burkitt's Lymphoma: A Human Cancer Model* (G. M. LENOIR, G. T. O'CONNOR, AND C. L. M. OLWENY, EDS.), pp. 217–224, IARC, Lyon, France, 1985.

161. LEVINE, P., KAMARAJU, L. S., CONNELLY, R. R., BERARD, C. W., DORFMAN, R. F., MACGRATH, I., AND EASTON, J. M., The American Burkitt's Lymphoma Registry. Eight years experience, *Cancer* **49**:1016–1022 (1982).

162. LEVINE, P., PALLESEN, G., EBBESEN, P., HARRIS, N., MUELLER, N., AND EVANS, A., An evaluation of Epstein–Barr virus antibody patterns and the detection of viral markers in the biopsies of patients with Hodgkin's disease, *Int. J. Cancer* **59**:48–50 (1994).

163. LEYVRAZ, S., HENLE, W., CHAHINIAN, A. P., PERLMANN, C., KLEIN, G., GORDON, R. E., ROSENBLUM, M., AND HOLLAND, J. F., Association of Epstein–Barr virus with thymic carcinoma, *N. Engl. J. Med.* **312**:1296–1299 (1985).

164. LIBETTA, C. M., PRINGLE, J. H., ANGEL, C. A., CRAFT, A. W., MALCOLM, A. J., AND LAUDER, I., Demonstration of Epstein–Barr viral DNA in formalin-fixed, paraffin-embedded samples of Hodgkin's disease, *J. Pathol.* **161**:255–260 (1990).

165. LIN, J.-C., LIN, S.-C., DE, B. K., CHAN, W. P., AND EVATT, B. L., Precision of genotyping of Epstein–Barr virus by polymerase chain reaction using three gene loci (EBNA-2, EBNA-3C, and EBER): Predominance of type A virus associated with Hodgkin's disease, *Blood* **81**:3372–3381 (1993).

166. LINDAHL, T., KLEIN, G., REEDMAN, B. M., JOHANSON, B., AND SINGH, S., Relationship between Epstein–Barr virus (EBV) DNA and EBV-determined nuclear antigen (EBNA) in Burkitt lymphoma biopsies and other lymphoproliferative malignancies, *Int. J. Cancer* **13**:764–772 (1974).

167. LIPSOMB, H., TATSUMI, E., HARADA, S., SONNABEND, J., WALLACE, J., YETZ, J., DAVIS, J., MCCLAIN, K., METROKA, C., TUBBS, R., AND PURTILO, D., Epstein–Barr virus, chronic lymphadenopathy, and lymphomas in male homosexuals with acquired immunodeficiency syndrome (AIDS), *AIDS Res.* **1**:59–83 (1983).

168. LIST, A. F., GRECO, A., AND VOGLER, L. B., Lymphoproliferative diseases in immunocompromised hosts: The role of Epstein–Barr virus, *J. Clin. Oncol.* **5**:1673–1689 (1987).

169. LOMBARDI, L., NEWCOMB, E., AND DALLA-FAVEA, R., Pathogenesis of Burkitt's lymphoma: Expression of an activated c-*myc* oncogene cancer tumorgenic conversion of EBV infected human B lymphocytes, *Cell* **49**:161–170 (1987).

170. LUKA, J., LINDAHL, T., AND KLEIN, G., Purification of the Epstein–Barr virus nuclear antigen from transformed human lymphoid cell lines, *J. Virol.* **27**:604–611 (1978).

171. MACMAHON, B., Epidemiologic evidence on the nature of Hodgkin's disease, *Cancer* **10**:1045–1054 (1957).

172. MACMAHON, B., Epidemiology of Hodgkin's disease, *Cancer Res.* **26**:1189–1200 (1966).

173. MACMAHON, E. M. E., GLASS, J. D., HAYWARD, S. D., MANN, R. B., BECKER, P. S., CHARACHE, P., MCARTHUR, J. C., AND AMBINDER, R. F., Epstein–Barr virus in AIDS-related primary central nervous system lymphoma, *Lancet* **338**:969–973 (1991).

174. MADEJ, M., CONWAY, M. J., MORGAN, A. J., SWEET, J., WALLACE, L., QUALTIERE, L. F., ARRAND, J. R., AND MACKETT, M., Purification and characterization of Epstein–Barr virus GP340/

220 produced by a bovine papillomavirus virus expression vector system, *Vaccine* **10:**777–782 (1992).

175. MAGGI, E., PARRONCHI, P., MACCHIA, D., PICCINNI, M.-P., SIMONELLI, C., AND ROMAGNANI, S., Role of T cells in the pathogenesis of Hodgkin's disease, *Int. Rev. Exp. Pathol.* **33:**141–164 (1992).

176. MAGRATH, I., Molecular basis of lymphogenesis, *Cancer Res.* **52**(Suppl. 19):5529s–5540s (1992).

177. MAGRATH, I. T., The pathogenesis of Burkitt's lymphoma, *Adv. Cancer Res.* **55:**133–270 (1990).

178. MAGRATH, I., BARRIGA, F., MCMANAWAY, M., AND SHIRAMIZU, B., The molecular analysis of chromosomal translocations as a diagnostic, epidemiological and potentially prognostic tool in lymphoid neoplasia, *J. Virol. Methods* **21:**275–289 (1988).

179. MAGRATH, I., JAIN, V., AND BHATIA, K., The molecular epidemiology of Burkitt's lymphoma, in: *Epstein–Barr Virus and Associated Diseases*, Vol. 25 (T. TURSZ, J. S. PAGANO, D. V. ABLASHI, G. DE-THÉ, G. LENOIR, AND G. R. PEARSON, EDS.), pp. 337–396, London, Colloque INSERM\John Libbey, 1993.

180. MAGRATH, I. T., AND SARIBAN, E., Clinical features of Burkitt's lymphoma in the USA, in: *Burkitt's Lymphoma: A Human Cancer Model* (G. M. LENOIR, G. T. O'CONOR, AND C. L. M. OLWENY, EDS.), pp. 119–127, IARC, Lyon, France, 1985.

181. MAGRATH, I. T., African Burkitt's lymphoma. History, biology, clinical features, and treatment, *Am. J. Pediatr. Hematol. Oncol.* **13:**222–246 (1991).

182. MANOLOV, G., AND MANLOVA, Y., Experiments with fluorescent chromosome staining in Burkitt tumors, *Hereditas* **68:**235–244 (1971).

183. MASIH, A., WEISENBURGER, D., DUGGAN, M., ARMITAGE, J., BASHIR, R., MITCHELL, D., WICKERT, R., AND PURTILO, D. T., Epstein–Barr viral genome in lymph nodes from patients with Hodgkin's disease may not be specific to Reed–Sternberg cells, *Am. J. Pathol.* **139:**37–43 (1991).

184. MATSUO, T., HELLER, M., PETTI, L., O'SHIRO, E., AND KEIF, E., Persistence of the entire Epstein–Barr virus genome integrated with human lymphocyte DNA, *Science* **226:**1322–1325 (1984).

185. MEEKER, T. C., SHIRAMIZU, B., KAPLAN, L., HERNDIER, B., SANCHEZ, H., GRIMALDI, J. C., BAUMGARTNER, J., RACHLIN, J., FEIGAL, E., ROSENBLUM, M., AND MCGRATH, M.S., Evidence for molecular subtypes of HIV-associated lymphoma: Division into peripheral monoclonal, polyclonal and central nervous system lymphoma, *AIDS* **5:**669–674 (1991).

186. MILLER, B. A., RIES, L. A. G., HANKEY, B. F., KOSARY, C. L., AND EDWARDS, B. K. (EDS.), *Cancer Statistics Review: 1973–1989*, NIH Pub. No. 92-2789, National Cancer Institute, Bethesda, MD, 1992.

187. MILLER, B. A., RIES, L. A. G., HANKEY, B. F., KOSARY, C. L., HAARAS, A., DEVESA, S. S., AND EDWARDS, B. K. (EDS.), *SEER Cancer Statistics Review: 1973–1990*, NIH Pub. No. 93-2789, Bethesda, MD, 1993.

188. MILLER, G., Burkitt lymphoma, in: *Viral Infections of Humans: Epidemiology and Control*, 2nd ed. (A. S. EVANS, ED.), pp. 599–619, Plenum Press, New York, 1982.

189. MILLER, G., Epstein–Barr virus, in: *Field's Virology*, (B. N. FIELD, ED.), pp. 563–589, Raven Press, New York, 1985.

190. MILLER, G., Epstein–Barr virus: Biology, pathogenesis and medical aspects, in: *Virology*, 2nd ed. (B. FIELDS AND D. M. KNIPE, EDS.), pp. 1921–1958, Raven Press, New York (1990).

191. MILLER, G., AND COOP, D., Epstein–Barr viral nuclear antigen (EBNA) in tumor cell imprints of experimental lymphoma of marmosets, *Trans. Assoc. Am. Physicians* **87:**205–218 (1974).

192. MILLER, G., ENDERS, J. F., LISCO, L., AND KOHN, H. I., Establishment of lines from normal blood leukocytes by cocultivation with a leukocyte line derived from a leukemic line, *Proc. Soc. Exp. Biol. Med.* **132:**247–252 (1969).

193. MILLER, G., GROGAN, E., FISCHER, D. K., NIEDERMAN, J. C., SCHOOLEY, R. T., HENLE, W., LENOIR, G., AND LUI, C. R., Antibody responses to two Epstein–Barr nuclear antigens defined by gene transfer, *N. Engl. J. Med.* **312:**750–755 (1985).

194. MILLER, G., GROGAN, E., ROWE, D., ROONEY, C., HESTON, L., EASTMAN, R., ANDIMAN, W., NIEDERMAN, J., LENOIR, G., HENLE, W., SULLIVAN, J., SCHOOLEY, R., VOSSER, J., STRAUSS, S., AND ISSEKUTZ, T., Selective lack of antibody to a component of EB nuclear antigen in patients with chronic Epstein–Barr infection virus, *J. Infect. Dis.* **156:**26–35 (1987).

195. MILLER, G., KATZ, B. Z., AND NIEDERMAN, J. C., Some recent developments in the molecular epidemiology of Epstein–Barr virus infections, *Yale J. Biol. Med.* **60:**307–316 (1987).

196. MILLER, G., ROBINSON, J., HESTON, L., AND LIPMAN, M., Differences between laboratory strains of Epstein–Barr virus based on immortalization, abortive infection, and interference, *Proc. Natl. Acad. Sci. USA* **71:**4006–4010 (1974).

197. MILLER, G., SHOPE, T., COOPE, D., WALTERS, L., PAGANO, J., BORNKAMM, G. W., AND HENLE, W., Lymphoma in cotton-top marmosets after inoculation with Epstein–Barr virus: Tumor incidence, histologic spectrum, antibody responses, demonstration of viral DNA, and characterization of viruses, *J. Exp. Med.* **145:**948–967 (1977).

198. MILLER, G., SHOPE, T., LISCO, H., STITT, D., AND LIPMAN, M., Epstein–Barr virus: Transformation, cytopathogenic changes, and viral antigens in squirrel monkeys and marmoset leukocytes, *Proc. Natl. Acad. Sci. USA* **69:**383–387 (1972).

199. MILLER, M. H., STITT, D., AND MILLER, G., Epstein–Barr viral antigen in single cell clones of two human leukocyte lines, *J. Virol.* **6:**699–701 (1970).

200. MILLER, R. W., AND BEEBE, G. W., Infectious mononucleosis and the empirical risk of cancer, *J. Natl. Cancer Inst.* **50:**315–321 (1973).

201. MITELMAN, F., AND LEVAN, G., Clustering of aberrations to specific chromosomes in human neoplasms. II. A survey of 287 neoplasms, *Hereditas* **82:**167–174 (1976).

202. MOCHANKO, K., FEJES, M., BREAZAVSCEK, D. M., SUAREZ, A., AND BACHMANN, A. E., The relation between Epstein–Barr virus antibodies and clinical symptomatology and immuno-deficiency in patients with Hodgkin's disease, *Cancer* **44:**2067–2070 (1979).

203. MOHAR, A., ROMO, J., SALIDO, F., JESSURUN, J., PONCE DE LEÓN, S., REYES, E., VOLKOW, P., LARRAZA, O., PEREDO, M. A., CANO, C., GÓMEZ, G., SEPÚLVEDA, J., AND MUELLER, N., The spectrum of clinical and pathological manifestations of AIDS in a consecutive series of autopsied patients in Mexico, *AIDS* **6:**467–473 (1992).

204. MONTAGNIER, L., GRUEST, J., CHAMARET, S., DAUGNET, C., AXLER, C., GUETAAD, D., NUGEYRE, M. T., BARRESIMOUSSI, F., CHERMANN, T. C., BRUNET, J. B., KLATZMAN, D., AND GLUCKMAN, J. C., Adaptation of the lymphadenopathy virus (LAV) to replication in EBV-transformed lymphoblastoid cell lines, *Science* **225:**63–66 (1987).

205. MORGAN, A. J., Epstein–Barr virus vaccines, *Vaccine* **10:**563–571 (1992).

206. MORROW, R. H., Burkitt's lymphoma, in: *Cancer Epidemiology and Prevention* (D. SCHOTTENFELD AND J. FRAUMENI, EDS.), pp. 779–794, W. B. Saunders, Philadelphia, 1982.

207. MORROW, R. H., PIKE, M. C., SMITH, P. G., ZIEGLER, J. L., AND KISUULE, A., Burkitt lymphoma: A time–space cluster of cases in Bwamba county of Uganda, *Br. Med. J.* **2**:491–492 (1971).

208. MORROW, R. H., PIKE, M. C., AND SMITH, P. G., Further studies of space–time clustering in Uganda, 1957–1968, *Br. Cancer J.* **35**:668–673 (1977).

209. MORROW, R. H., JR., Epidemiological evidence for a role of falciparum malaria in the pathogenesis of Burkitt's lymphoma, in: *Burkitt's Lymphoma: A Human Cancer Model*, (G. M. LENOIR, G. T. O'CONOR, AND C. L. M. OLWENY, EDS.), pp. 177–186, IARC, Lyon, France, 1985.

210. MOSS, D. J., BURROWS, S. R., CASTELINO, D. J., KASE, G., POPE, J. H., RICKENSON, A. B., ALPERS, M. P., AND HEYWOOD, P. F., A comparison of Epstein–Barr virus-specific T-cell immunity in malaria endemic and non-endemic regions of Papua, New Guinea, *Int. J. Cancer* **31**:727–732 (1983).

211. MUELLER, N., An epidemiologist's view of the new molecular biology findings in Hodgkin's disease, *Ann. Oncol.* **2**(Suppl. 2):23–28 (1991).

212. MUELLER, N., Epidemiologic studies assessing the role of Epstein–Barr virus in Hodgkin's disease, *Yale J. Biol. Med.* **60**:321–328 (1987).

213. MUELLER, N., Hodgkin's disease, in: *Cancer Epidemiology and Prevention*, 2nd ed. (D. SCHOTTENFELD AND J. F. FRAUMENI, JR., EDS.), pp. 893–919, Oxford University Press, New York, 1996.

214. MUELLER, N., EVANS, A., HARRIS, N. L., COMSTOCK, G. W., JELLUM, E., MAGNUS, K., ORENTREICH, N., POLK, B. F., AND VOGELMAN, J., Hodgkin's disease and the EBV: Altered antibody patterns before diagnosis, *N. Engl. J. Med.* **320**:696–701 (1989).

215. MUIR, C., WATERHOUSE, T., POWELL, M. J., AND WHELAN, S. (EDS.), *Cancer Incidence in Five Continents*, Vol. 5, International Agency for Research on Cancer, Lyon, 1987.

216. MUÑOZ, N., DAVIDSON, R. J. L., WITTHOFF, B., ERICSSON, J. E., AND DE-THÉ, G., Infectious mononucleosis and Hodgkin's disease, *Int. J. Cancer* **22**:10–13 (1978).

217. MURRAY, P. G., YOUNG, L. S., ROWE, M., AND CROCKER, J., Immunohistochemical demonstration of the Epstein–Barr virus-encoded latent membrane protein in paraffin sections of Hodgkin's disease, *J. Pathol.* **166**:1–5 (1992).

218. NIEDERMAN, J. C., AND MILLER, G., Kinetics of the antibody responses to Bam HI-K nuclear antigen in uncomplicated infectious mononucleosis, *J. Infect. Dis.* **154**:346–349 (1986).

219. NIEDERMAN, J. C., MILLER, G., PEARSON, H. A., PAGANO, J. S., AND DOWALIBY, J. M., EB virus shedding in saliva and the oropharynx, *N. Engl. J. Med.* **294**:1335–1359 (1976).

220. NIEDOBITEK, G., HERBST, H., YOUNG, L. S., BROOKS, L., MASUCCI, M. G., CROCKER, J., RICKINSON, A. B., AND STEIN, H., Patterns of Epstein–Barr virus infection in non-neoplastic lymphoid tissue, *Blood* **79**:2520–2526 (1992).

221. NIEDOBITEK, G., ROWLANDS, D. C., YOUNG, L. S., HERBST, H., WILLIAMS, A., HALL, P., PADFIELD, J., ROONEY, N., AND JONES, E. L., Overexpression of p53 in Hodgkin's disease: Lack of correlation with Epstein–Barr virus infection, *J. Pathol.* **169**:207–212 (1993).

222. NILSSON, K., AND KLEIN, G., Phenotypic and cytogenetic characteristics of human B lymphoid cell lines and their relevance for the etiology of Burkitt's lymphoma, *Adv. Cancer Res.* **37**:319–380 (1982).

223. NKRUMAH, F. K., AND OLWENY, C. L. M., Clinical features of Burkitt's lymphoma: The African experience, in: *Burkitt's Lymphoma: A Human Cancer Model* (G. M. LENOIR, G. T. O'CONOR, AND C. L. M. OLWENY, EDS.), pp. 87–95, IARC, Lyon, France, 1985.

224. NONOYAMA, M., HUANG, C. H., PAGANO, J. S., KLEIN, G., AND SINGH, S., DNA of Epstein–Barr virus detected in tissue of Burkitt's lymphoma and nasopharyngeal carcinoma, *Proc. Natl. Acad. Sci. USA* **70**:3265–3268 (1973).

225. O'CONOR, G. T., Malignant tumors in African children. II. A pathological entity, *Cancer* **14**:270–283 (1961).

226. O'CONOR, G. T., AND DAVIES, J. N. P., Malignant tumors in African children with special reference to malignant lymphoma, *J. Pediatr.* **56**:526–535 (1960).

227. OBRAMS, G. I., AND GRUFFERMAN, S., Epidemiology of HIV associated non-Hodgkin lymphoma, *Cancer Surv.* **10**:91–102 (1991).

228. OHSHIMA, K., KIKUCHI, M., EGUCHI, F., MASUDA, Y., SUMIYOSHI, Y., MOHTAI, H., TAKESHITA, M., AND KIMURA, N., Analysis of Epstein–Barr viral genomes in lymphoid malignancy using Southern blotting, polymerase chain reaction and *in situ* hybridization, *Virchows Arch. B* **59**:383–390 (1990).

229. OLWENY, C. M., KATUNGOLE-MBIDDE, E., OTIM, D., LWANGA S. K., MAGRATH, I., AND ZIEGLER, J. L., Long term experience with Burkitt's lymphoma in Uganda, *Int. J. Cancer* **26**:261–267 (1980).

230. OPEIZ, G., AND HENDERSON, R., FOR THE COLLABORATIVE TRANSPLANT STUDY, Incidence of non-Hodgkin lymphoma in kidney and heart transplant recipients, *Lancet* **342**:1514–1516 (1993).

231. ORLOWSKI, R., AND MILLER, G., Topological effect of EBNA-1 on oriP, in: *Immunobiology and Prophylaxis of Human Herpesvirus Infections* (C. LOPEZ, R. MORI, B. ROIZMAN, AND R. J. WHITLEY, EDS.), pp. 115–124, Plenum Press, New York, 1990.

232. OSMOND, D. G., PRIDDLE, S., AND RICO-VARGAS, S., Proliferation of B cell precursors in bone marrow of pristane-conditioned and malaria infected mice: Implications for B cell oncogenesis, *Curr. Top. Microbiol. Immunol.* **166**:149–157 (1990).

233. OTT, G., OTT, M. M., AND MÜLLER-HERMELINK, H. K., EBV DNA in nodal and extranodal non-Hodgkin's lymphomas: Impact of cell lineage, morphology, and site of origin, *Toxicol. Lett.* **67**:341–351 (1993).

234. PAGANO, J. S., Detection of Epstein–Barr virus with molecular hybridization techniques, *J. Infect. Dis.* **13**(Suppl.):123s–128s (1991).

235. PALLESEN, G., HAMILTON-DUTOIT, S. J., ROWE, M., LISSE, I., RALFKIAER, E., SANDVEJ, K., AND YOUNG, L. S., Expression of Epstein–Barr virus replicative proteins in AIDS-related non-Hodgkin's lymphoma cells, *J. Pathol.* **165**:289–299 (1991).

236. PALLESEN, G., HAMILTON-DUTOIT, S. J., ROWE, M., AND YOUNG, L. S., Expression of Epstein–Barr virus latent gene products in tumour cells of Hodgkin's disease, *Lancet* **337**:320–322 (1991).

237. PALLESEN, G., SANDVEJ, K., HAMILTON-DUTOIT, S. J., ROWE, M., AND YOUNG, L. S., Activation of Epstein–Barr virus replication in Hodgkin and Reed–Sternberg cells, *Blood* **78**:1162–1165 (1991).

238. PALLESEN, G., HAMILTON-DUTOIT, S. J., AND SANDVEJ, K., EBV-related lymphomas and Hodgkin's disease, in: *Vth International Symposium on Epstein–Barr Virus and Associated Diseases* (T. TURSZ, J. S. PAGANO, D. V. ABLASHI, G. DE-THÉ, G. LENOIR, AND G. R. PEARSON, EDS.), pp. 419–423, Annecy, France, John Libby Eurotext, London, 1992.

239. PARKIN, D. M., SOHIER, R., AND O'CONOR, G. T., Geographic distribution of Burkitt's lymphoma, in: *Burkitt's Lymphoma: A Human Cancer Model* (G. M. LENOIR, G. T. O'CONOR, AND C. L. M. OLWENY, EDS.), pp. 155–164, IARC, Lyon, France, 1985.

240. PEARSON, G. R., AND LUKA, J., Characteristics of the virus-determined antigens, in: *The Epstein–Barr Virus: Recent Advances* (M. A. EPSTEIN, AND B. G. ACHONG, EDS.), pp. 48–73, Wiley Medical, New York (1986).

241. PEARSON, G. R., VROMAN, B., CHASE, B., SCULLEY, T., HUMMEL, M., AND KIEF, E., Identification of polypeptide components of the Epstein–Barr virus early antigen complex with monoclonal antibodies, *J. Virol.* **47:**193–201 (1983).

242. PHILIP, T., Burkitt's lymphoma in Europe, in: *Burkitt's Lymphoma: A Human Cancer Model* (G. M. LENOIR, G. T. O'CONOR, AND C. L. M. OLWENY, EDS.), pp. 107–118, IARC, Lyon, France, 1985.

243. PIKE, M. C., MORROW, R. H., KISUULE, A., AND MAFIGIRI, J., Burkitt's lymphoma and sickle cell trait, *Br. J. Prev. Soc. Med.* **24:**39–41 (1970).

244. PLUDA, J. M., VENSON, D. J., TOSATO, G., LIETZAU, J., WYVILL, K., NELSON, D. L., JAFFE, E. S., KARP, J. E., BRODER, S., AND YARCHOAN, R., Parameters affecting the development of non-Hodgkin's lymphoma in patients with severe human immunodeficiency virus infection receiving antiretroviral therapy, *J. Clin. Oncol.* **11:**1099–1107 (1993).

245. PLUDA, J. M., YARCHOAN, R., JAFFE, E. S., FEURSTEIN, I. M., SOLOMAN, D., STEINBERG, S. M., WYVILL, K. M., RAUBITSCHEK, A., KATZ, D., AND BRODER, S., Development of non-Hodgkin's lymphomas in a cohort of patients with severe immunodeficiency disease (HIV) infection of long term antiretroviral therapy, *Ann. Intern. Med.* **113:**276–282 (1990).

246. POPPEMA, S., KAISERLING, E., AND LENNERT, K., Hodgkin's disease with lymphocyte predominance, nodular type (nodular paragranuloma) and progressively transformed germinal centres: A cytohistological study, *Histopathology* **3:**295–308 (1979).

247. PREUD'HOMME, J. L., FLANDRIN, G., DANIEL, M. T., AND BRONET, J. C., Burkitt's tumor cells in acute leukemia, *Blood* **46:**990–992 (1975).

248. PULVERTAFT, R. J. V., Cytology of Burkitt's tumor (African lymphoma), *Lancet* **1:**238–240 (1964).

249. PURTILO, D. T. (ED.), *Immune Deficiency and Cancer: Epstein–Barr Virus and Lymphoproliferative Malignancies*, Plenum Press, New York, 1984.

250. PURTILO, D. T., AND KLEIN, G., Introduction to Epstein–Barr virus and lymphoproliferative diseases in immunodeficient individuals, *Cancer Res.* **41:**4209 (1981).

251. PURTILO, D. T., AND LAI, P. K., Clinical and immunopathological manifestations and detection of Epstein–Barr virus infection in immune deficient patients, in: *Medical Virology VI* (L. M. D. L. MAZA, AND E. M. PETERSON, EDS.), pp. 121–167, Elsevier, Amsterdam, 1987.

252. PURTILO, D. T., BHAWAN, J., HUTT, L. M., DE NICOEA, L., SZYMANSKI, I., YANG, J. P. S., BOTO, W., MARER, R., AND THORLEY-LAWSON, D., Epstein–Barr virus infection in the X-linked lymphoproliferative syndrome, *Lancet* **1:**798–801 (1978).

253. PURTILO, D. T., MANOLOV, G., MANLOVA, Y., HARADA, S., LIPSCOMB, H., AND TATSULMI, E., Role of Epstein–Barr virus in the etiology of Burkitt's lymphoma, in: *Burkitt's Lymphoma: A Human Cancer Model* (G. M. LENOIR, G. T. O'CONOR, AND C. L. M. OLWENY, EDS.), pp. 231–248, IARC, Lyon, France, 1985.

254. PURTILO, D. T., OKANO, M., AND GRIERSON, H. L., Immune deficiency as a risk factor in Epstein–Barr virus-induced malignant diseases, *Environ. Health Perspect.* **88:**225–230 (1990).

255. PURTILO, D. T., STROBACH, R. S., OKANO, M., AND DAVIS, J. R., Epstein–Barr virus-associated lymphoproliferative disorders, *Lab. Invest.* **67:**5–23 (1992).

256. PURTILO, D. T., YASUDA, N., GRIERSON, H. L., OKANO, M., BRICHACEK, B., AND DAVIS, J., X-linked lymphoproliferative syndrome provides clues to the pathogenesis of Epstein–Barr virus-induced lymphomagenesis, in: *Unusual Occurrences as Clues to Cancer Etiology* (R. W. MILLER, S. WATANABE, J. F. FRAUMENI, JR., T. SUGIMURA, S. TAKAYAMA, AND H. SUGANO, EDS.), pp. 149–158, Japan Scientific Societies Press, Tokyo, 1988.

257. RAAB-TRAUB, N., AND PAGANO, J. S., Epstein–Barr virus and its antigens, *Immunol. Ser.* **43:**477–498 (1989).

258. RABKIN, C. S., BIGGAR, R. J., AND HORM, J. W., Increasing incidence of cancers associated with the human immunodeficiency virus epidemic, *Int. J. Cancer* **47:**692–696 (1991).

259. REEDMAN, B. M., AND KLEIN, G., Cellular localization of an Epstein–Barr virus (EBV) associated complement-fixing antigen in producer and non-producer lymphoblastoid cell lines, *Int. J. Cancer* **11:**499–520 (1973).

260. REYNOLDS, P., SAUNDERS, L. D., LAYEFSKY, M. E., AND LEMP, G. F., The spectrum of acquired immunodeficiency syndrome (AIDS)-associated malignancies in San Francisco, 1980–1987, *Am. J. Epidemiol.* **137:**19–30 (1993).

261. RICKINSON, A. B., On the biology of Epstein–Barr virus persistence: A reappraisal, in: *Immunobiology and Prophylaxis of Human Herpesvirus Infections* (C. LOPEZ, R. MORI, B. ROIZMAN, AND R. J. WHITLEY, EDS.), pp. 137–146, Plenum Press, New York, 1990.

262. RICKINSON, A. B., YAO, Q. Y., AND WALLACE, L. E., The Epstein–Barr virus as a model of virus–host interaction, *Br. Med. J.* **41:**75–79 (1985).

263. ROBINSON, J., AND MILLER, G., Assay for Epstein–Barr virus based on stimulation of DNA synthesis in mixed leukocytes from human umbilical cord blood, *J. Virol.* **15:**1065–1072 (1975).

264. ROCCHI, G., TOSATO, G., PAPA, G., AND RAGONA, G., Antibodies to Epstein–Barr virus-associated nuclear antigen and to other viral and non-viral antigens in Hodgkin's disease, *Int. J. Cancer* **16:**323–328 (1975).

265. ROITHMANN, S., TOURANI, J.-M., AND ANDRIEU, J.-M., Hodgkin's disease in HIV-infected intravenous drug abusers, *N. Engl. J. Med.* **323:**275–276 (1990).

266. ROMAGNANI, S., DEL PRETE, G. F., MAGGI, E., BOSI, A., BERNARDI, F., PONTICELLI, P., DI LOLLO, S., AND RICCI, M., Displacement of T lymphocytes with the "helper/inducer" phenotype from peripheral blood to lymphoid organs in untreated patients with Hodgkin's disease, *Scand. J. Haematol.* **31:**305–314 (1983).

267. ROMAGNANI, S., FERRINI, P. L. S., AND RICCI, M., The immune derangement in Hodgkin's disease, *Semin. Hematol.* **22:**41–55 (1985).

268. ROSDAHL, N., LARSEN, S. O., AND CLEMMESEN, J., Hodgkin's disease in patients with previous infectious mononucleosis: 30 years' experience, *Br. Med. J.* **2:**253–256 (1974).

269. ROSENBERG, S. A. (CHAIRMAN), National Cancer Institute sponsored study of classification of non-Hodgkin's lymphomas. Summary and description of a working formulation for clinical usage, *Cancer* **49:**2112–2135 (1982).

270. ROTHFIELD, N. F., EVANS, A. S., AND NIEDERMAN, J. C., Clinical

and laboratory studies of raised antibody titers in systemic lupus erythematosus, *Ann. Rheum. Dis.* **32**:238–246 (1973).

271. ROWE, M., YOUNG, L. S., CALDWALLADER, K., PETTI, L., KIEFF, E., AND RICKINSON, A. B., Distinction between Epstein–Barr virus type A (EBNA 2A) and type B (EBNA 2B) isolates extends to EBNA 3 family of nuclear proteins, *J. Virol.* **63**:1031–1039 (1989).

272. ROWLEY, J. D., Chromosomes in Hodgkin's disease, *Cancer Treat. Rep.* **66**:639–643 (1982).

273. SAMPLE, J., HENSON, E. B., AND SAMPLE, C., The Epstein–Barr virus nuclear protein 1 promoter in type I latency is autoregulated, *J. Virol.* **66**:4654–4661 (1992).

274. SANDVEJ, K. B., HAMILTON-DUTOIT, S. J., AND PALLESEN, G., Influence of Epstein–Barr virus encoded latent membrane protein 1 on the expression of CD23 antigen, ICAM-1 and LFA-3 in Hodgkin and Reed–Sternberg cells: A morphometric analysis, *Leuk. Lymphoma* **9**:95–101 (1993).

275. SCULLY, T. B., APPPOLOLLONI, A., HURREN, L., MOSS, D. J., AND COOPER, D. A., Coinfection with A- and B-type Epstein–Barr virus in human immunodeficiency virus-positive subjects, *J. Infect. Dis.* **162**:643–648 (1990).

276. SHIRAMIZU, B., BARRIGA, F., NEEQUAYE, J., JAFRI, A., DALLA-FAVERA, R., GUITTIEREZ, M., LEVINE, P., AND MAGRATH, I., Patterns of chromosomal breakpoint locations in Burkitt's lymphoma: Relevance to geography and EBV association, *Blood* **77**:1516–1526 (1991).

277. SHOPE, R., DeCHAIRO, D., AND MILLER, G., Malignant lymphoma in cotton-top marmosets following inoculation of Epstein–Barr virus, *Proc. Natl. Acad. Sci. USA* **70**:2487–2491 (1973).

278. SHOPE, T. C., KHALIFA, S., SMITH, S. T., AND CUSHING, B., Epstein–Barr virus antibody in childhood Hodgkin's disease, *Am. J. Dis. Child.* **136**:701–703 (1982).

279. SIGAUX, F., BERGER, R., BERNHEIM, A., VALENSI, F., DANIEL, M. T., AND FLANDRIN, G., Malignant lymphoma with band 8q14 chromosome abnormality: A morphologic continuum extending from Burkitt's lymphoma to immunoblastic lymphoma, *Br. Med. J.* **57**:393–405 (1984).

280. SIXBEY, J. W., DAVIS, D. S., YOUNG, L. S., HUTT-FLETCHER, L., TEDDAR, T. F., AND RICKINSON, A. B., Human epithelial expression of an Epstein–Barr virus receptor, *J. Virol.* **68**:805–811 (1987).

281. SIXBEY, J. W., NEDRUD, J. G., RAAB-TRAUB, N., HANES, R. A., AND PAGANO, J. S., Epstein–Barr virus replication in oropharyngeal epithelial cells, *N. Engl. J. Med.* **310**:1225–1230 (1984).

282. SIXBEY, J. W., SHIRLEY, P., CHESNEY, P. J., BUNTIN, D. M., AND RESNICK, L., Detection of a second widespread strain of Epstein–Barr virus, *Lancet* **2**:761–765 (1989).

283. SIXBEY, J. W., VESTERINEN, E. H., NEDRUD, J. G., RAAB-TRAUB, N., WALTON, L. A., AND PAGANO, J. S., Replication of Epstein–Barr virus in human epithelial cells infected *in vitro*, *Nature* **306**:480–483 (1983).

284. SLIVNICK, D. J., ELLIS, T. M., NAWROCKI, J. F., AND FISHER, R. I., The impact of Hodgkin's disease on the immune system, *Semin. Oncol.* **17**:673–682 (1990).

285. SMITHERS, D., On some general concepts in oncology with special references to Hodgkin's disease, *Int. J. Radiat. Oncol. Biol. Phys.* **9**:731–738, 1983.

286. STAAL, S. P., AMBINDER, R., BESCHORNER, W. E., HAYWARD, G. S., AND MANN, R., A survey of Epstein–Barr virus DNA in lymphoid tissue, *Am. J. Clin. Pathol.* **91**;1–5 (1989).

287. STEEL, C. M., MORTEN, J. E. N., AND FOSTER, E., The cyto-genetics of human lymphoid malignancy: Studies in Burkitt's lymphoma and Epstein–Barr virus transformed lymphoblastoid cell lines, in: *Burkitt's Lymphoma: A Human Cancer Model* (G. M. LENOIR, G. T. O'CONOR, AND C. L. M. OLWENY, EDS.), pp. 265–292, IARC, Lyon, France, 1985.

288. STRAUCH, B., SIEGEL, N., ANDREWS, L., AND MILLER, G., Oropharyngeal excretion of Epstein–Barr virus by renal transplant recipients and other patients treated with immunosuppressive drugs, *Lancet* **1**:234–237 (1974).

289. SUGDEN, B., Comparison of Epstein–Barr viral DNA's in Burkitt lymphoma biopsy cells and in cells clonally transformed *in vitro*, *Proc. Natl. Acad. Sci. USA* **74**:4651–4655 (1977).

290. SUGDEN, B., The molecular biology of Epstein–Barr virus, in: *Vth International Symposium on Epstein–Barr Virus and Associated Diseases* (T. TURSZ, J. S. PAGANO, D. V. ABLASHI, G. deTHÉ, G. LENOIR, AND G. R. PEARSON, EDS.), p. 57, Annency, France, John Libby Eurotext, 1992.

291. TEN NAPEL, H. H., THE, T. H., VAN EGTEN-BUKER, J., DeGAST, G. C., HALIE, M. R., AND LANGENHUYSEN, M. M., Discordance of the Epstein–Barr virus (EBV) specific humoral and cellular immunity in patients with malignant lymphomas: Elevated antibody titers and lowered *in vitro* lymphocyte reactivity, *Clin. Exp. Immunol.* **34**:338–346 (1978).

292. UCCINI, S., MONARDO, F., RUCO, L. P., BARONI, C. D., FAGGIONI, A., AGLIANO, A. M., GRADILONE, A., MANZARI, V., VAGO, L., COSTANZI, G., CARBONE, A., AND DE RE V., High frequency of Epstein–Barr virus genome in HIV-positive patients with Hodgkin's disease, *Lancet* **333**:1458 (1989).

293. UCCINI, S., MONARDO, F., STOPPACCIARO, A., GRADILONE, A., AGLIANÒ, A. M., FAGGIONI, A., MANZARI, V., VAGO, L., COSTANZI, G., RUCO, L. P., AND BARONI, C. D., High frequency of Epstein–Barr virus genome detection in Hodgkin's disease of HIV-positive patients, *Int. J. Cancer* **46**:581–585 (1990).

294. UHARA, H., SATO, Y., MUKAI, K., AKAO, I., MATSUNO, Y., FURUYA, S., HOSHIKAWA, T., SHIMOSATO, Y., AND SAIDA, T., Detection of Epstein–Barr virus DNA in Reed–Sternberg cells of Hodgkin's disease using the polymerase chain reaction and *in situ* hybridization, *Jpn. J. Cancer Res.* **81**:272–278 (1990).

295. VARIAKOJIS, M. D., AND ANASTASI, J., Unresolved issues concerning Hodgkin's disease and its relationship to non-Hodgkin's lymphoma, *Am. J. Clin. Pathol.* **99**:436–444 (1993).

296. VESTLEV, P. M., PALLESEN, G., SANDVEJ, K., HAMILTON-DUTOIT, S. J., AND BENDTZEN, S. M., Prognosis of Hodgkin's disease is not influenced by Epstein–Barr virus latent membrane protein, *Int. J. Cancer* **50**:670–671 (1992).

297. WEISS, L. M., CHEN, Y.-Y., LIU, X.-F., AND SHIBATA, D., Epstein–Barr virus and Hodgkin's disease: A correlative *in situ* hybridization and polymerase chain reaction study, *Am. J. Pathol.* **139**:1259–1265 (1991).

298. WEISS, L. M., MOVAHED, L. A., WARNKE, R. A., AND SKLAR, J., Detection of Epstein–Barr viral genomes in Reed–Sternberg cells of Hodgkin's disease, *N. Engl. J. Med.* **320**:502–506 (1989).

299. WEISS, L. M., STRICKLER, J. G., WARNKE, R. A., PURTILO, D. T., AND SKLAR, J., Epstein–Barr viral DNA in tissues of Hodgkin's disease, *Am. J. Pathol.* **129**:86–91 (1987).

300. WERNER, G., HENLE, G., PINTO, C. A., HAFF, R. F., AND HENLE, W., Establishment of continuous lymphoblast cultures from leukocytes of gibbons (*Hylobates* lar.), *Int. J. Cancer* **10**:557–567 (1972).

301. WOLF, H., HAUS, M., AND WILMES, E., Persistence of Epstein–Barr virus in the parotid gland, *J. Virol.* **51**:795–798 (1984).

302. WRIGHT, B. H., Cytology and histochemistry of the Burkitt lymphoma, *Br. J. Cancer* **17**:50–55 (1963).

303. WU, T.-C., MANN, R. B., CHARACHE, P., HAYWARD, S. D., STAAL, S., LAMBE, B. C., AND AMBINDER, R. F., Detection of EBV gene expression in Reed–Sternberg cells of Hodgkin's disease, *Int. J. Cancer* **46**:801–804 (1990).

304. WUTZLER, P., FÄRBER, I., SPRÖSSIG, M., SAUERBREI, A., WUTKE, K., HÖCHE, D., RÜDIGER, K.-D., WÖCKEL, W., AND SCHEIBNER, K., Antibodies against herpesviruses in patients with Hodgkin's disease, *Arch. Geschwulstforsch* **53**:417–422 (1983).

305. YOUNG, L., ALFIERI, C., HENNESSY, K., EVANS, H., O'HARA, C., ANDERSON, K. C., RITZ, J., SHAPIRO, R. S., RICKINSON, A., KIEFF, E., AND COHEN, J. I., Expression of Epstein–Barr virus transformation-associated genes in tissues of patients with EBV lymphoproliferative disease, *N. Engl. J. Med.* **321**:1080–1085 (1989).

306. YOUNG, L. S., Ludwig/CRC REV meeting, *Epstein–Barr Virus Rep.* **1**:10–12 (1994).

307. YUNIS, J. J., OKEN, M. M., KAPLAN, M. E., ENSRUD, K. M., HOWE, R. R., AND THEOLOGIDES, A., Distinctive chromosomal abnormalities in histologic subtypes of non-Hodgkin's lymphoma, *N. Engl. J. Med.* **317**:1231–1236 (1982).

308. ZECH, L., HAGLUND, A. N., NILSSON, K., AND KLEIN, G., Characteristic chromosomal abnormalities in biopsies and lymphoid cell lines from patients with Burkitt and non-Burkitt lymphoma, *Int. J. Cancer* **17**:47–56 (1976).

309. ZIEGLER, J. L., ANDERSSON, M., KLEIN, G., AND HENLE, W., Detection of Epstein–Barr virus in American Burkitt's lymphoma, *Int. J. Cancer* **17**:701–706 (1976).

310. ZIEGLER, J. L, MINER, R. C., ROSENBAUM, E., LENETTE, E. T., SHILLETOE, E., CASAVANT, C., DEW, W. L., MINTZ, L., GERSHOW, J., GREENSPAN, J., BECKSTEAD, J., AND YAMAMOTO, K., Outbreak of Burkitt's-like lymphoma in homosexual men, *Lancet* **2**:631–633 (1982).

311. ZUR HAUSEN, H., Papilloma virus/host cell interaction in the pathogenesis of anorectal cancer, in: *Origins of Human Cancer: A Comprehensive Review* (H. H. HIATT, J. D. WATSON, AND J. A. WINSTED, EDS.), pp. 685–688, Cold Spring Harbor Laboratory Press, Cold Spring Harbor, NY, 1991.

312. ZUR HAUSEN, H., Viruses in human cancers, *Science* **254**:1167–1173 (1991).

313. ZUR HAUSEN, H., O'NEILL, F. J., FREESE, U. K., AND HECKER, E., Persisting oncogenic herpesvirus by tumor promoter, TPA, *Nature* **272**:373–375 (1978).

314. ZUR HAUSEN, H., SCHULTE-HOLTHAUSEN, H., KLEIN, G., HENLE, W., HENLE, G., CLIFFORD, P., AND SANTESSON, L., EB-virus DNA in biopsies of Burkitt tumors and anaplastic carcinomas of the nasopharynx, *Nature* **228**:1056–1057 (1970).

CHAPTER 31

Nasopharyngeal Carcinoma

G. de-Thé

1. Introduction

The geographic distribution of the main types of human cancers is such that variations up to 100- and even 1000-fold exist for the incidence of specific tumors, suggesting that environmental factors play a critical role in their development.[54] Furthermore, if chemicals are generally believed to play the major role, microbiological agents should not be overlooked. Because carcinogenesis is accepted by most epidemiologists to represent a multistep and multifactorial process,[37,172] both chemical and biological carcinogens mays interact in such a process.

As seen in Chapter 30, the Epstein–Barr virus (EBV) appears to act as an initiator in the causation of Burkitt's lymphoma (BL) and malaria as a promoter (both are environmental agents).[19,35,44,165] Nasopharyngeal carcinoma (NPC)[115,117] is another example of a close association between the EBV and a human tumor, but the role of the EBV in NPC is not yet fully understood, in spite of the fact that the level of association is stronger the world over in contrast to that observed for Burkitt's-type lymphomas in different geographic environments.[36,137,168]

2. Historical Background

Clifford,[25] on the basis of the work of Derry,[27] Smith and Dawson,[198] Krogman,[132] and Wells,[213,214] stated that the oldest known pathological specimens of NPC were derived from inhabitants of northeast Africa and the Middle East from the period 3500–3000 BC. Reviewing the evidence, Ho[103] concluded that only one of the Romano-Egyptian cases described by Smith and Dawson[198] could have been a nasopharyngeal cancer. In

Europe, Durand-Fardel[55] is generally credited as giving the first clinical description of a case of NPC, and Michaux[149] as reporting the first histologically proven case.

In China, NPC has been recognized since the early part of this century as the "Kwantung (Guangdong) tumor," stressing the high frequency of this neoplasm in Kwantung, the southernmost province. Ho,[103] in his search for a description of the disease in early Chinese medical writings, could find only a fatal disease called *shih ying*, also known as *shih yung*, both meaning literally "loss of nutrition." The description given in the *Encyclopaedia of Chinese Medical Terms*, edited by Wu,[220] of the clinical picture of *shih ying* is consistent with that of the "mainly metastatic type" of NPC[92] (see Section 8.2.1). However, no mention was made in the encyclopedia of when the disease was first described. The apparent lack of a full description of NPC in early Chinese medical writings may be because the disease is largely confined to the south, whereas practically all the early writings were by physicians in northern and central China.[101]

In parts of Southeast Asia, it was some time before it was realized that the malignant deposits in the neck lymph nodes were secondary to a nasopharyngeal primary tumor, and many continued to describe such neoplasms as "reticuloendothelioma lymphoglandulae colli lateralis." It was probably Digby *et al.*,[52] in 1930, who first drew attention to the unusual frequency of NPC among Chinese in Hong Kong and over large parts of China in a remarkably detailed description of the clinical and pathological features of 103 cases in which it was stated categorically that the tumors arose in the epithelium of the nasopharynx.[140]

While pathologists in Europe and North America debated the classification and histogenesis of the neoplasms, notably the so-called lymphoepithelioma and transitional-cell carcinomas, New and Kirch, and later others including Digby *et al.*,[52] concluded that all were

G. de-Thé • The Pasteur Institute, 75015 Paris, France.

variants of squamous-cell carcinoma—a view since upheld by ultrastructural studies.[68,202]

Before 1962, there was no real attempt to study the epidemiology of NPC. The medical profession seemed to be quite satisfied with the hypothesis advanced by Dobson[53] that the high frequency of NPC in Chinese was related to the poorly ventilated houses in which they lived, inhaling the carcinogen-laden domestic smoke. This was later disputed by Ho[103] when he found that the frequency of NPC in Chinese fisherfolk who lived practically all their lives in boats and cooked their food in the open was significantly higher than that in the rest of the Chinese population in Hong Kong, the majority of whom lived in congested dwellings on land.

Because of this he began to suspect traditional ingestants rather than inhalants to be the major risk factor, and among them salted fish was most suspected because it is one item of food traditionally fed to southern Chinese during early childhood. It is this period when the genesis of NPC was suspected to begin in the majority of cases as judged from the age-specific incidence curve for this population.[104]

Serendipity prevailed in the discovery by Old *et al.*,[166] in 1966, in NPC patients; taken as controls for BL in an explanation of EBV-precipitating antibodies, they exhibited much higher titers than the BL patients. This observation opened the highly rewarding field of EBV-related NPC.

3. Methodology Involved in Epidemiologic and Virological Studies

3.1. Sources of Mortality Data

Mortality data are derived from cause-of-death statements on death certificates. The figures available for nasopharyngeal cancer* for a wide variety of countries have been aggregated by the World Health Organization,[217] and age-specific rates, but not age-adjusted rates, provided. In general, the disease is exceedingly rare. Further, national mortality data rarely give figures for racial groups within a country. Although rates for administrative divisions such as provinces are often available on request, these are usually based on small numbers of cases and are subject to considerable statistical fluctuation.

---

*All rates in Section 3 are age-adjusted to the world population distribution[206] and are expressed per 100,000 per annum. In this chapter, *nasopharyngeal cancer* refers not only to the carcinomas (NPC) but also to neoplasms of other cell types, e.g., chordoma, multiple myeloma, and the malignant lymphomas.

3.2. Sources of Morbidity (Incidence) Data

Morbidity data for nasopharyngeal cancer are obtained from cancer registries, being derived from reports on newly diagnosed cases of cancer occurring in a defined population. Regrettably, such information is generally lacking for most of South America, Africa, Asia, and Oceania, and for large parts of Europe. Even though there are large gaps in geographic coverage,[155] the morbidity from cancer is probably better measured than for any other chronic disease and for many infectious and acute processes. Available morbidity data of good quality are published in the *Cancer Incidence in Five Continents* monographs[114,205,206] and are presented in Fig. 1. It will be observed that for the vast majority of cancer registries, not only are the age-adjusted rates below 1 per 100,000 per annum, but also many rates are based on fewer than 10 cases.

3.3. Sources of Relative-Frequency Data

Relative-frequency data, indicating the proportion of nasopharyngeal cancer observed in a series of patients with all types of cancer, are derived from the files of pathology and radiotherapy departments. The sources of bias in measuring relative frequency as well as in mortality and morbidity data have been analyzed.[90,155]

3.4. Serological Surveys

Two types of serological surveys have been carried out to unravel the relationship between the Epstein–Barr herpesvirus (EBV) and NPC. The first type is represented by case-control studies aimed at evaluating the humoral immune response of the NPC patients compared to that of various controls.[46,47,88] The second type is population-based surveys aimed at establishing the epidemiologic characteristics of the virus in populations at different risk for EBV-associated diseases. We have been conducting such a study covering Chinese in Hong Kong, Indians, Chinese and Malays in Singapore, Nilotic tribes in Uganda, and Caucasians in France.[41]

3.4.1. Selection of Groups. In the first type of studies, controls were usually selected among patients of the same age group and sex with tumors other than NPC. Whenever possible, normal subjects such as volunteer blood donors also served as controls. In these seroepidemiologic studies, the choice of groups to be bled was a matter of compromise between representativeness and feasibility and included volunteers in maternity and child clinics, primary and secondary schools, universities, and

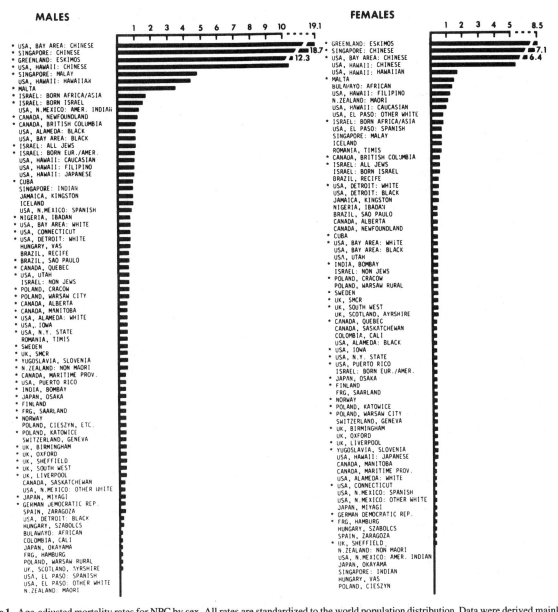

Figure 1. Age-adjusted mortality rates for NPC by sex. All rates are standardized to the world population distribution. Data were derived mainly from *Cancer Incidence in Five Continents*, 1976.[114] *Rates based on more than 10 cases.

army and police groups. Soon it was found that such selected groups were not a good representation of the population at large. Successful efforts were then made to obtain representative samples of the general population, and randomly selected families were visited and interviewed and the eligible members bled.

3.4.2. Serological Tests. Three techniques were mainly utilized in EBV serology: immunofluorescence,

complement fixation (CF), and immunoprecipitation. Two classes of immunoglobulins, IgG and IgA, appear to play an important role in the relationships between EBV and NPC.

3.4.2a. Immunofluorescence Tests. Immunofluorescence (IF) was the main tool of EBV serology used for the detection of four different groups of EBV-specific antibodies directed against viral capsid antigen (VCA), early antigen (EA), membrane antigen (MA), and Epstein–Barr nuclear antigen (EBNA).

The VCA Test.   As described by Henle and Henle,[81] the VCA test is an indirect IF test, detecting intracellular structural antigens (viral capsid antigens) in EBV-producing cell lines. In routine testing, one detects IgG antibodies and a positive test (i.e., $\geq 1:10$) merely reflects that the person concerned has been infected and has reacted to the infection. The titers of VCA antibodies obtained in various laboratories on the same sera can vary from one to three dilutions; the causes of such variations are multiple, the main ones being the lymphoblastoid-cell line and the proportion of VCA-positive cells at the time of testing.[73] All serological tests carried out within the International Agency for Research on Cancer seroepidemiologic survey were done using the Jijoye cell line as the source of antigen. Large antigen batches were prepared to minimize the variation of titers related to the percentage of IF-positive cells. The small variations still observed were evaluated and, if necessary, were corrected during statistical analysis.

The EA Test.   Also described by Henle and co-workers,[80,87] the EA test detects "early antigens" produced within a few hours after superinfection by EBV of non-virus-producing lymphoblastoid lines such as Raji. Early antigen synthesis does not require DNA replication and appears to consist of two different antigens: a diffuse (D) antigen and a restricted (R) one.[83] Antibodies against EA reflect an *active infection* of the organism concerned.

The MA Test.   Developed and extensively used by Klein *et al.*,[127,129] the MA test detects EBV-determined membrane-bound antigens. The test is done on live cells, in contrast to the VCA and EA tests, in which acetone fixation is used. The MAs are comprised of at least three different antigenic components.[124] Antibodies against MA seem to be of clinical and prognostic value in BL and NPC patients.[124,127]

The EBNA Test.   As described by Reedman and Klein,[176] the EBNA test detects the EBV-determined nuclear antigen(s) that appears to be related to the soluble antigen, as detected by the CF test.[46,130,136] This nuclear antigen appears a few hours after EBV infection[9,224] and is present in EBV-transformed lymphoid cells even when they are not virus producers.

EBNA is also detected in BL[177] and NPC[39,110] biopsies as well as in NPC biopsies transplanted into nude mice.[128]

3.4.2b. Complement-Fixation Tests.   Complement fixation was used in EBV serology with either particulate antigen, i.e., semipurified virus,[69] or soluble antigen, extracted from non-virus-producing cell lines.[200,207]

These CF antigens seem to divide into three components: one is sedimentable and heat-resistant, corresponding to a structural antigen, one is heat-labile and sedimentable, and one is heat-stable and soluble, the last two components being nonstructural antigens.[211] Antibodies against particulate antigen(s) parallel those directed against VCA, whereas antibodies directed against soluble antigen from a non-virus-producing line (Raji) appear to parallel those directed against EBNA and seem to reflect a prolonged infection with this virus. The CF soluble (CF/S) antibodies develop only weeks or months after infectious mononucleosis.[199,208]

3.4.2c. Immunoprecipitation Test.   This test for EBV was developed by Old *et al.*[166] using the Ouchterlony immunodiffusion technique. This immunoprecipitation (IP) antigen seems to be closely related, if not identical, to CF/S antigen.[178] This test needed large amounts of antigen and is no longer used.

3.4.2d. Antibodies of the IgA Class.   As discussed in Section 3.6.2b, IgA antibodies are becoming increasingly useful in NPC management. Their detection is possible in the VCA, EA, and MA indirect IF tests, simply by using fluorescein-labeled immunoglobulins directed against purified human IgA.

3.4.2e. Antibody-Dependent Cellular Cytotoxicity Assay.   In this assay, EBV-superinfected Raji cells labeled with ^{51}Cr are used as target cells. Peripheral blood lymphocytes from baboons serve as effector cells. To determine antibody-dependent cellular cytotoxicity (ADCC), the cytotoxicity figure of target cells incubated with control, negative serum was subtracted from the figure obtained with target cells incubated in a given dilution of test serum. Final dilution of test serum mediating a significant ($P < 0.05$ in the Student's *t* test) increase of cytotoxicity of target cells with corresponding dilution of the negative control serum is chosen as the serum titer.[147,170]

3.5. Sociological Surveys

When no clear lead to the etiology of a neoplasm or a plausible hypothesis to test exists, it is often useful to survey the way of life of groups at high and low risk for the disease to determine whether there are habits or exposures that may be pertinent to disease development. Such surveys are usually best conducted by anthropologists or sociologists. Since the sum of lifestyle embraces so many factors, it is essential to describe the current etiologic hypothesis of carcinogenesis to such investigators so that their studies may have specific areas of focus. The very detailed analysis of the way of life in the Caspian region of Iran carried out in connection with esophageal cancer in that area typifies the possible scope of such investigations.[131]

Because of the long induction time for cancer, attempts should be made to determine conditions as they were 20–30 years ago as well as those prevailing at the time of the survey.

3.6. Laboratory Diagnosis

3.6.1. Histopathology. Diagnosis depends on the demonstration of the neoplasm in a biopsy of the nasopharynx.[1] Biopsies from many areas and at several times may be needed, since the tumorous process often infiltrates underneath an apparently normal mucosa. The microscopic appearance of the neoplasms was reviewed by Shanmugaratnam and Muir[186] and by Yeh.[225] The World Health Organization (WHO) published an illustrated classification of upper respiratory tract neoplasms, including those of the nasopharynx, with the appropriate ICD-O codes.[218] Pathologists were urged to use the WHO classification at the NPC conference in Kyoto[115] and are increasingly doing so. This is essential to be able to compare the disease in different geographic areas. The vast majority of NPCs arise from the nasopharyngeal epithelium and should be considered, irrespective of their appearance on light microscopy, as variants of squamous-cell carcinomas. This statement is based on electron-microscopic studies that have revealed the presence of tonofibrils and other epithelial markers in most of the histological variants.[68,201,202] Any of the histological types of NPC may be infiltrated by varying amounts of lymphocytes: this feature is not restricted to the lympho-epitheliomas and may be pertinent to the association with EBV, a lymphotropic virus. For the experienced pathologist working in a high-incidence area, diagnosis rarely poses any problems; however, the Schminke variant composed of single or small loose aggregations of malignant cells set in a dense lymphoid stroma and the rarer spindle-cell and clear-cell variants may cause difficulties for those unfamiliar with the region. Although the diagnosis, particularly of the Regaud variant of the neoplasm, can usually be made with confidence from the biopsied secondary deposits in a neck lymph node, this practice is to be deprecated, since it does not indicate unequivocally the site of the primary tumor, and nasopharyngeal biopsy is simpler to perform.

The pathologist working in a low-incidence area should ask himself, when confronted by a nasopharyngeal biopsy or a secondary deposit in an upper neck lymph node, "Is this a Schminke variant?" and "What about the nasopharynx?"

3.6.2. Serology

3.6.2a. IgG Antibodies. There is a characteristic EBV antibody pattern associated with NPC. Patients with undifferentiated carcinoma of the nasopharynx, regardless of their ethnic or geographic origin, have high EBV VCA, EA, and EBNA serological reactivities (IgG class) when compared to patients with other tumors (OT) or to normal subjects (NS).[47] As can be seen in Table 1, the EBV reactivity that separates best between NPC and either OT or NS relates to antibodies directed against EA. This regularity in humoral response of NPC patients to EBV EA, regardless of origin,[47] contrasts with the situation of BL patients from Africa, who have high EBV reactivities and high EBV DNA content, whereas patients from the United States have inconsistently elevated EBV reactivities and EBV-genome content in tumor cells.[137,168]

Antibodies against EA[87] are of special interest, since they reflect an active EBV infection. In NPC patients, as in patients with infectious mononucleosis, EA antibodies are mostly directed against the diffuse (D) component of the EA complex, whereas in BL, patients' EA antibodies are directed against the restricted (R) component.[83,126] The EA reactivity in NPC patients is also stage-dependent, and Henle *et al.*[88] shared the opinion that reactivity to D component probably reflects the degree of lymph-node involvement.

Antibodies against CF/S antigen are already very high in Chinese NPC patients in stage I[48] (see Table 1), in contrast to BL patients, in whom CF/S antibodies are low at an early stage and increase with clinical deterioration.[199,200] The CF antibodies against CF/S antigen increase slightly from stage I to stage V of NPC and in both diseases appear to have a prognostic value and can be used as a useful clinical guide. Antibodies against MA are also regularly high in NPC patients.[29]

As can be seen in Fig. 2, the serological response to VCA, EA, and EBNA antigens increases with the severity of the disease, from stages I to V of Ho.[46,86,88]

3.6.2b. IgA Antibodies in Serum and Saliva of NPC Patients. Following the observations of Wara *et al.*[212] that NPC patients had very high levels of IgA antibodies in their sera, Henle and Henle[82] found that such IgA antibodies were directed against EBV-determined antigens (VCA and EA-D). These authors found that among healthy donors or patients with tumors of the ear, nose, and throat (ENT) other than NPC, only 5% had antibodies to VCA, and none to EA.[82,116] Ho *et al.*[107] confirmed the importance of IgA antibodies directed to EA in NPC patients, whereas Desgranges *et al.*[30] and Pearson *et al.*[169] extended their observations to non-Chinese NPC patients, namely, Tunisian and American cases. This IgA

Table 1. Epstein–Barr Virus Serological Reactivities in Nasopharyngeal Carcinoma Patients, Patients with Other Tumors, and Normal Subjects Originating from Three Geographic Areas[a]

Group and EBV reactivities		NPC patients GMT	NPC patients Number of sera	OT patients GMT	OT patients Number of sera	NS GMT	NS Number of sera
Hong Kong	VCA	1316		376		119	
Chinese	EA	182		8		9	
	CF	88		18		21	
	EBNA	776		86		ND	
			49		39		45
Tunisian	VCA	1677		190		114	
Arabs	EA	119		8		7	
	CF	75		ND		24	
	EBNA	593		142		ND	
			65		65		65
Paris	VCA	978		166		91	
Caucasians	EA	194		6		7	
	CF	32		10		10	
	EBNA	118		ND		ND	
			18		37		40

[a]From de-Thé et al.[47] Abbreviations here and in Tables 2 and 3: (CF) Complement fixation; (CF/S) CF soluble; (EA) early antigen; (EBNA) Epstein–Barr nuclear antigens; (EBV) Epstein–Barr virus; (GMT) geometric mean titer; (ND) not done; (NPC) nasopharyngeal carcinoma; (NS) normal subjects; (OT) other tumors; (VCA) viral capsid antigen.

response increases with advancing stages of the disease (see Table 2) and represents systemic (7 S) antibodies restricted to EBV, since IgA antibody titers to herpes simplex virus did not differ between NPC patients and controls. Zeng et al.[235] detected Ig A/VCA in NPC patients from eight provinces and cities in China.

Desgranges et al.[30] and Ho et al.[98,100] extended these serological studies to the salivas of NPC patients, which were found to contain high levels of secretory IgA (11 S) specific for VCA and EA-D in 75 and 35% of the salivas of NPC patients, respectively. As can be seen in Fig. 3, IgA antibodies to EBV were found confined to the saliva of 50% of NPC patients but were not detected in saliva from patients with infectious mononucleosis (IM), BL, or other tumors. Further, Desgranges et al.[31] found that plasma cells surrounding the epithelial tumor cells contained IgA antibody molecules that were secondarily found to be specific for EBV VCA, whereas epithelial tumor cells were seen to exhibit the secretory piece at their surface (x, a) (Figs. 4 and 5). These results, confirmed by Ho et al.,[98,99] raise the possibility that the EBV IgA present in NPC saliva has its origin in the tumor itself, possibly representing blocking antibodies. The presence of serum and salivary IgA to VCA and EA could have a practical value not only for the diagnosis of NPC (with a very high sensitivity and specificity) but also as a marker for prognosis in the management of the disease.[51]

3.6.3. Viral Markers in NPC Biopsies. There is a regular presence of viral fingerprints in NPC tumor tissue. Epstein–Barr DNA sequences have been found by DNA–DNA hybridization, DNA–cRNA hybridization, and DNA–DNA reassociation kinetics in most NPC biopsies.[164,167,23a] Because EBV is a lymphotropic virus and NPC is an epithelial tumor (see Section 3.6.1), it was believed that the EBV DNA detected was localized in lymphoid cells, which are regularly present in NPC tumors and not in the epithelial tumor cells. However, the opposite was demonstrated by Wolf et al.,[216] showing that EBV DNA sequences were predominantly in epithelial cells, and confirmed by Desgranges et al.,[32] who detected EBV DNA in separated epithelial cell populations from NPC biopsies originating from different geographic areas. Only undifferentiated types of NPC regularly showed detectable EBV DNA and EBNA.[3,4]

Today, EBV molecular fingerprints can be detected in fresh NPC biopsies and histopathologic sections of fixed NPC tissues, using an in situ hybridization technique. When nested PCR is used, the small nuclear EBV-encoded RNAs (EBERs) are regularly detected, along with EBNA-1 mRNA, LMP-1, and to a lesser extent LMP-2, mRNAs.[15,16,161,221] EBV-specific nuclear antigen (EBNA) was detected by indirect IF in touch smears of NPC biopsies (see Fig. 6).

In parallel to the transcription of viral genes men-

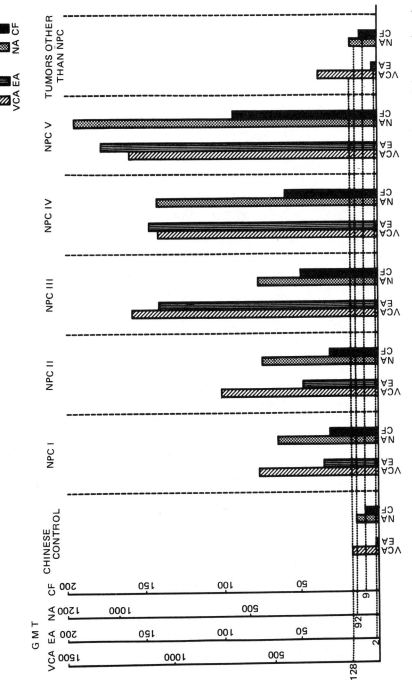

Figure 2. Histogram giving the various EBV antibody reactivities (IgG class) in Chinese NPC patients at different stages of the disease and, for comparison, the reactivities of normal subjects (Chinese controls) and of Chinese patients with tumors other than NPC. (GMT) Geometric mean titer; (VCA) viral capsid antigen; (NA) nuclear antigen; (CF) complement-fixation (soluble) antigens. From de-Thé et al.[46]

Table 2. Epstein–Barr VCA and EA IgG and IgA Geometric Mean Titers in Sera from Chinese NPC Patients in Different Stages of the Disease, in Patients with Other Tumors, and in Controls[a]

Subjects	Number of sera tested	VCA IgA		EA IgA		VCA IgG		EA IgG	
		GMT	S.E.M.	GMT	S.E.M.	GMT	S.E.M.	GMT	S.E.M.
NPC Stage I	38	44.4	0.6	10	0.5	512	0.4	48.5	0.7
NPC Stage II	42	69.6	0.5	12.3	0.6	803.4	0.4	54.9	0.4
NPC Stage III	50	85.7	0.6	23	0.6	1097.5	0.3	97	0.6
NPC Stage IV	50	56.6	0.6	20	0.6	630.4	0.4	84.4	0.7
NPC Stage V	35	52.8	0.9	16.8	0.9	724.1	0.5	101.6	0.5
OT	50	3.3	0.1	2.5	0.0	279.2	0.3	3.3	0.3
NS	50	2.8	0.1	2.5	0.0	92.1	0.3	2.2	0.1

[a]For abbreviations, see Table 1 footnote; (S.E.M.) standard error of the mean.

tioned above, EBNA-1 (but *not* the EBNA-2) antigen is regularly expressed in NPC tumor cells. LMP-1 antigen appears to be expressed at easily detectable levels only in a subset of NPC tumors, even when LMP-1 mRNA can be detected by nested PCR.[15,161] Klein *et al.*[128] successfully grafted NPC tumor biopsies to nude mice and observed that these grafted tumors contained only epithelial cells and regularly exhibited EBNA. These epithelial characteristics were confirmed by electron microscopy.[49] After treatment with 5-iodo-2-deoxyuridine [idoxuridine (IUdR)] or 5-bromodeoxy-uridine (BUdR), or after superinfection with HR1-EBV, epithelial tumor cells can synthesize EA,[74] suggesting that such epithelial cells have surface receptors for EBV and that, under certain circumstances, the viral genome can be derepressed in tumor cells. Full

replication of EBV was observed in epithelial tumor cells after passage in nude mice, and the virus was isolated.[204]

4. Biological Characteristics of Epstein–Barr Virus in Its Relationship with Nasopharyngeal Carcinoma

The herpesvirus discovered by Epstein *et al.*[57,58] in 1964 in a culture derived from a BL biopsy is a widespread and silent parasite of human B-lymphoid cells. The lack of a permissive cell system allowing *in vitro* virus titration and virus cloning makes virological studies on the EBV rather difficult. *In vivo*, the EBV infects B lymphocytes in circulating blood, lymph nodes, spleen, and other sites, where the viral infection becomes latent.

The B lymphocytes appear to be the main target for EBV infection. When B- and T-lymphoid populations are separated from the blood of EBV-seropositive subjects, only B-cell populations give rise to permanent lymphoblastoid cell lines.[224] The EBV receptors in these B lymphocytes (CD21) are undistinguishable from the C3d receptors. Epithelial cells in the nasopharynx are also obvious candidates to replicate the virus. According to Lemon *et al.*[135] EBV replicates in epithelial cells of the ororhinopharynx during IM. Furthermore, epithelial cells of the nasopharynx seem to have C3d receptors,[226] suggesting that they also have EBV receptors.

The lymphoid cells regularly present in the submucosa of the nasopharynx might play a critical role in tumor development, since the EBV is mainly lymphotropic and this tumor arises only in the area of the lympho-epithelium. The B or T nature of the lymphocytes present in the tumor is a matter of controversy: Yata *et al.*[222,223] found that both B and T cells were present in tumor tissue

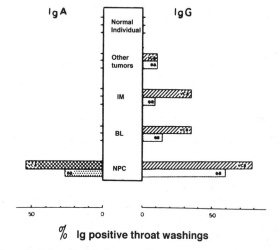

Figure 3. IgA and IgG in throat washings from patients with NPC, BL, IM, other tumors, and normal individuals. From Desgranges *et al.*[30]

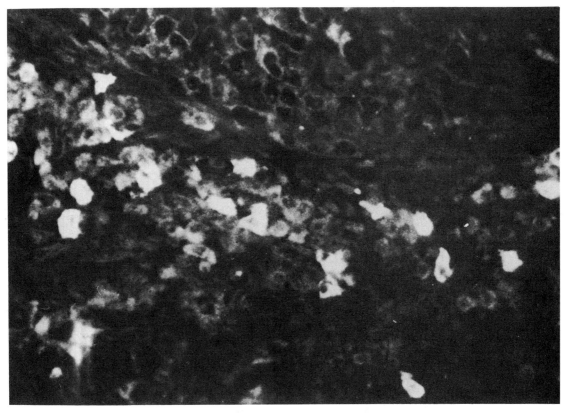

Figure 4. IgA (α) in plasmocytes surrounding epithelial-tumor cells. ×300. From Desgranges *et al.*[30]

but that a relative increase of B cells was observed in the lymphoid-depleted areas as well as in direct contact with epithelial cells. In contrast, Jondal and Klein[120] found that the lymphocytes present in tumor biopsies were mostly of T type. The study of the lymphocytes of the normal nasopharyngeal mucosa and in NPC is therefore of great importance for understanding the pathogenesis of such tumor as well as the cell-mediated immune response of NPC patients to EBV.[21,138]

Latent infection by EBV is present in every population around the world, but age at primary infection varies greatly among ethnic groups and geographic areas. As is true for other herpesviruses, EBV reinfection might occur, as well as reactivation of a latent infection by factors that are unknown at present. One abnormal condition that "turns on" viral replication is the transfer of circulating lymphocytes to *in vitro* culture, leading to the establishment of lymphoblastoid lines. This phenomenon (called "lymphoblastoid transformation"[11]) is of as great interest to oncovirologists as it is to immunologists. Although it is poorly understood (see the review by Klein[125]), there is

agreement that such a phenomenon is caused by the presence of EBV-infected lymphocytes.[40,45,163]

Does "lymphoblastoid transformation" occur *in vivo*, and are such transformed cells eliminated by immunosurveillance mechanisms? Expression of viral-determined antigens occurs in BL tumor cells, where the EBNA and MA (see Section 3.4.2a)[127,177] can be detected.

In vitro (i.e., in lymphoblastoid lines), viral replication is also controlled by unknown factors, synthesis of VCA, EA, and other antigens taking place in a changing proportion of cells. Such variation in "viral expression" is line-dependent, and also time-dependent for most lines.[45,71] Early functions include the EBNA, MA, and EA, and those appearing later include structural VCA and probably the IP antigens.

The mode of transmission of EBV is not fully established, but saliva appears to play an important role in horizontal transmission, since infectious and transforming EBV has regularly been found in the saliva of IM patients and of seropositive normal subjects.[63,70,150,160]

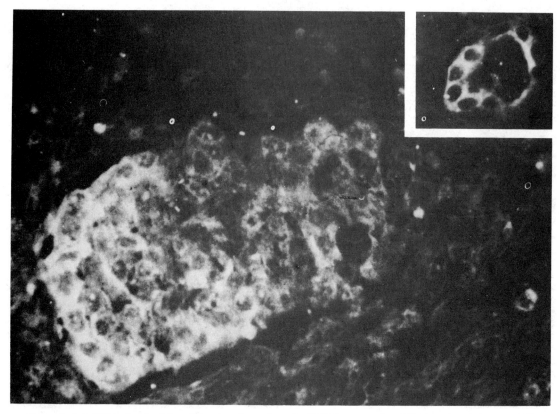

Figure 5. IgA (SP) in the glandular acini and on the surface of epithelial-tumor cells. ×300. From Desgranges *et al.*[30]

Sociocultural customs or habits in which exchange of saliva between adults and young infants takes place could play a role in EBV infection in early infancy. Breast-feeding may play a role in EBV transmission, since milk does contain EBV-infected cells, as demonstrated by the establishment of lymphoblastoid lines from human milk (A. C. Feller, personal communication). All data available indicate that EBV infection is not transmitted through the placenta,[60,84,171] but there is no reason that congenital EBV infection could not occur in the presence of severe malaria infection, which is known to damage the placenta.

The full spectrum of host response to EBV in the human organism is still ill defined. This virus causes Paul–Bunnell-positive IM in young adults.[60,84,85] Most primary infections, however, are asymptomatic, and this virus would have been considered as a mild, inoffensive parasite if it had not been associated with two malignancies: BL and NPC. We saw in Section 3.4 that the association between EBV and NPC is based on serological data and on the evidence for the presence of EBV fingerprints (DNA and EBNA) in tumor cells of both BL and NPC.

This characteristic pattern is found in NPC patients originating from various parts of the world (see de-Thé *et al.*[47] and Table 1), and such consistency is of particular importance, since NPC is a distinctive pathological entity that cannot be readily confused with other conditions. The lack of similar serological reactivities in patients with other carcinomas localized in nearby tissues as the oro- and hypopharynx (Table 3) and, even in patients with tumors of the nasopharynx other than carcinomas,[28] indicates that a peculiar relationship must exist between EBV and NPC.

The nature of the association between EBV and NPC remains to be established. The regular presence of viral fingerprints in tumor cells and the specific immune response to EBV in NPC patients in regions with high, intermediate, or low risk for the disease makes the passenger hypothesis very unlikely.[47] That EBV is the causative agent of most IM is universally accepted (see Chapter 10). Further evidence for the causal nature of the relationship between EBV and BL lies in the results of the prospective seroepidemiologic study in Uganda, indicating that children who are to develop BL have high VCA titers years prior to tumor onset, with an increased relative

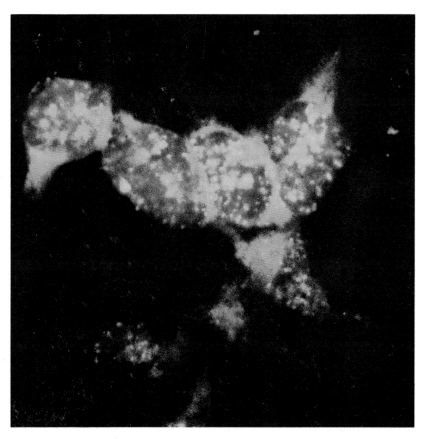

Figure 6. Touch smear from an NPC biopsy originating from Hong Kong (Reg. No. 74/1173), stained for EBNA by the anticomplement immunofluorescence test (ACIF) using an NPC serum (Tu 125) having no detectable antinuclear factor. Note the coarse, granular aspect of the EBNA. The positive cells here are believed to be epithelial.

**Table 3. Geometric Mean Titers of Epstein–Barr Virus Reactivities
in Different Groups of Chinese Sera**[a]

Type of sera	EBV reactivities				
	VCA Jijoye	EA Raji	EBNA (ACIF) Raji	CF/S Raji	Number of sera
NPC patients	937.3	93.3	605.4	49.9	91
Tumor other than NPC	296.2	3.9	115.2	11.6	37
Normal subjects	128	2.2	92.1	9.0	47

[a]For abbreviations, see Table 1.

risk of 30 for children having VCA titers 2 dilutions higher than the mean of the titers of the sex/age/locality-matched normal children[33–35,39,41,42,44,46–49] (see Chapter 30). That early infection by EBV plays the role of an initiating factor in BL development has been hypothesized by us.[35,44] The situation for NPC should be entirely different.

5. Descriptive Epidemiology

5.1. Incidence, Frequency, and Geographic Distribution

For the vast majority of population groups, the incidence rates for nasopharyngeal cancer are below 1 per 100,000 per annum. In persons of Chinese descent, much higher rates have been found, and morbidity rates have recently become available for selected Chinese populations. As can be seen in Fig. 7, the largely similar rates for Chinese in Zhongshan (Chungshan) County of Kwangtung (Guangdong), Hong Kong and the Bay Area of California are almost twice those for Chinese in Hawaii. The risk in Singapore Cantonese is significantly higher than for the remainder of the Singapore Chinese.[189] The rates in Shanghai and in Japan are half those observed in Hawaii.

With respect to the principal Chinese-language groups living in Singapore, the rates for both sexes in the Cantonese are twice those of the Teochew (known as Chiu Chau in Hong Kong) and the Hokkien (who come from the province of Fujian, which is to the north of and in continuity with the Chiu Chau district, northeast of Hong Kong). Since the Cantonese have a much lower risk of esophageal cancer than the Hokkien and Teochew, it is unlikely that these differences could reflect differential reporting among the various groups.[184]

In his studies of Hong Kong Chinese, Ho[91,103,105]

noted that the lowest incidence rate* was found in Chinese originating from the central coastal provinces (4.0) and that a 3–4 times higher incidence was observed in persons originating from Kwangtung (Guangdong) Province. The Chiu Chau area of this province, located to the northeast of Hong Kong, showed a rate of 8.0. On further examination of the Hong Kong population,[104] it was noted that nasopharyngeal cancer incidence among the Cantonese "boat people" was higher (age-adjusted rates of 54.7 for males and 18.8 for females) than that among the land dwellers (18.8 for males and 10.2 for females). The sex-incidence ratio of NPC in the boat group was 2.9 males to 1 female, even higher than for land dwellers. The boat people are so called because they lived on boats up to recent times. They now have resettled on land. They speak their own distinctive dialects of Cantonese and have been present in the Hong Kong region for time unknown.

If incidence data from China are incomplete,[121] the Atlas of Cancers in China[8] represents a unique source of data, showing that the distribution of NPC in mainland China is mainly restricted to the southern provinces of Kwantung and East Guang-Xi. Such a distribution in fact parallels the ethnic distribution of the Ham-Cantonese, at highest risk for NPC, and neighboring groups such as the Zhuang, Hokkien, and others at significantly lower risk. Already, relative-frequency material published in 1959 suggested an increase from north to south.[108] As is apparent in Fig. 7, incidence in the metropolitan area of Shanghai seems to be considerably lower than mortality in the Chungshan (Zhongshan) County of Kwangtung (Guangdong) Province. Although data quality are probably not strictly comparable, nonetheless the data are consistent with lower incidence rates in the north. There are few migrants from northern China in Southeast Asia to form a comparison group. Data from Taiwan show a twofold

---

*These rates are not strictly comparable to those in Fig. 1, being adjusted to the 1961 Hong Kong census population.

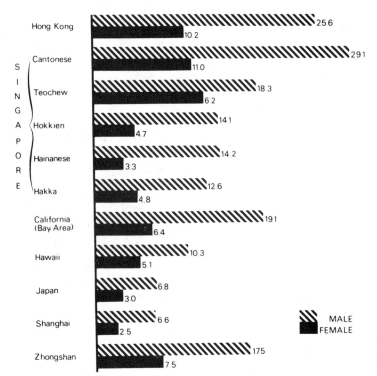

Figure 7. Age-adjusted morbidity for NPC for selected Chinese populations, by sex (see the text). Figures for Zhongshan (Chungshan) County are mortality.

excess of risk for males born in China compared to those born in Taiwan.[144] However, the unusual sex ratio of the incidence rates (M, 11.4; F, 11.7) among the China-born undermines confidence in these findings.

The Malay people of Singapore, who in general use the medical facilities much less than the other groups, have comparatively high rates (see Fig. 1). This group tends to live in the rural districts of the island and has a way of life that varies considerably from that of the Chinese. The incidence rates in the Caucasoid Indian and Pakistani moiety of the Singapore population are very low (see Fig. 1).

Incidence rates for all ethnic groups living on Hawaii are high, but these are based on very small numbers (see Fig. 1).

High frequencies of NPC have been reported in Greenland Eskimos with some admixture of Caucasian blood[162] (Fig. 1) and in Alaskan natives.[14] Unpublished tabulations from the Northwest Territories of Canada show that this disease is considerably more common than elsewhere in the country, probably because of an excess risk in the Eskimo population. Why these Mongoloid groups should have such high rates when northern Chinese and Japanese do not remains to be determined. Are cultural factors rather than racial stock, per se, more

important in determining NPC risk? In Israel, although the cancer is at a much lower incidence level, there are still interesting differences among Jews coming from various parts of the world. The highest rates are noted in the Jews born in Asia or Africa (M, 1.7; F, 0.7), but rates in Jews born in Israel (M, 1.5; F, 0.5) are almost as high. For those born in America or Europe, rates are low. Many of the Jews from Africa migrated from Morocco, Algeria, and Tunisia, where there is evidence (see below) that NPC has a higher frequency. Minimum incidence rates in Kuwait are also higher (M, 2.2; F, 1.0).

Most of the other information on the distribution of NPC comes from relative-frequency series (Table 4). This has been summarized by Muir and Shanmugaratnam.[154,157] Nasopharyngeal carcinoma is common in many populations of Southeast Asia such as Malays, Thais, Vietnamese, and Javanese in which Chinese admixture and racial intermingling cannot be excluded, although it is very rare in Japanese and probably infrequent in the populations of Korea,[225,228] Mongolia, and northern China.[108] Ho[91] has clearly shown that the Macaonese in Hong Kong, the descendants of Portuguese settlers in Macao who intermarried with Chinese from Kwantung (Guangdong), have a much higher frequency of NPC than the rest of the non-Chinese population.

Table 4. Relative Frequency of Nasopharyngeal Carcinoma in African and Asian Populations[a,b]

Population group	Cases M	Cases F	Relative frequency (%) M	Relative frequency (%) F	Population group	Cases M	Cases F	Relative frequency (%) M	Relative frequency (%) F
Malaysia					Vietnamese	105	58		
Chinese	150	77	11.5	6.0		163		3.7	
Malays	54	20	10.6	4.2	Algeria				
Indians	7	2	1.3	0.4	Algerians	364		5.0	
Thailand (Bangkok)					Europeans	83		3.5	
Chinese	27		15.9		Tunisia				
Chinese/Thai	20		10.3		Tunisians			6.7	2.9
Thais	29		4.6		Europeans			2.5	0.8
Northern Thais	34	19	3.7	2.1	Congo (Brazzaville)	1	0	0.4	0
(Chiang Mai)					Senegal (Dakar)	1	0	0.1	0
Sabah					Ivory Coast (Abidjan)	2	0	0.5	0
Chinese	9	0	8.9	0	Iraq (Baghdad)	73	—	1.3	—
Indigenous	11	3	9.2	3.4	Egypt (Alexandria)	26	—	1.3	—
Sarawak					Morocco	95	46	8.5	3.8
Chinese	6	5	4.8	3.8	Sudan	87	14	7.8	1.2
Malays	1	1	3.2	4.2	Zanzibar				
Dayaks	8	4	8.6	4.9	Negroes	1		0.4	
Java					Arabs	1		1.3	
Chinese	31	3	18.2	1.4	Mozambique	3	0	0.7	0
Javanese	108	44	10.3	2.9	(Lorenço Marques)				
Formosa	1260	446	23.2	5.2	Nigeria (Ibadan)	5	3	0.7	0.4
Filipinos (Philippines)	182		2.9		South Africa (Cape	4	1	0.5	0.1
Filipinos (Hawaii)	8	0	2.5	0	Province; coloreds)				

[a]Relative frequency is taken here as the percentage of NPC among malignancies observed in the referred populations.
[b]For references, see Muir.[154]

Clifford[23,24] has claimed that NPC is of high relative frequency among head and neck malignancies treated in Nairobi. However, Linsell[145] noted a more modest relative frequency of 2.3% in Kenya biopsy material in 1957–1961, this being greatest in the Kalenjin tribe (3.2%) and least in the coastal Bantu (0.2%). Computing crude incidence rates from Clifford's material, Muir[154] noted a maximum rate of 1.3 in male Nandi. This rate is not age adjusted, and it is likely that there is considerable underdiagnosis. Nevertheless, it is doubtful that the "true" age-adjusted rate would be greater than 5.

In Uganda, Schmauz and Templeton[180] noted a higher minimum age-adjusted incidence rate in Nilotics and para-Nilotics than in the Bantu–Sudanic groups. In general, the north of the country showed a higher incidence than did the south, despite the better hospital and transport systems in the south.

Relative-frequency studies from Algeria, Morocco, and Tunisia suggest a moderately elevated NPC level. In Tunisia, Zaouche[229] noted that in 23 months, 156 nasopharyngeal cancers were diagnosed at the Institut National de Cancérologie (Institut Salah Azaiz) in Tunis. If these cases are related to the entire Tunisian population, the crude minimum rates for males and females are 2.6 and 0.9, respectively. Since the period covered represents when the institute was begun, it is likely that there was gross underreporting.

In summary, the disease is very common in southern Chinese, Eskimos, and native Greenlanders (an Eskimo population with admixture of Caucasian blood). The Chinese-mixed populations of Southeast Asia have a lower incidence rate. Rare in Japan and infrequent in northern China, NPC is of moderate incidence in the Maghreb (Tunisia, Algeria, and Morocco), in Kuwait (Y. T. Omar, personal communication), in Sudan,[153] in northern parts of Kenya and Uganda, and in Israel, in the Israel-born, and in migrants from North Africa. Elsewhere, it is rare.

5.2. Epidemic Behavior

There have not been exceptional reports of clustering in nasopharyngeal cancer.[215] This does not preclude a part-viral etiology because, as with varying and lengthy latent period, the cases resulting from an infectious process occurring every few years would be unlikely to appear in cyclical fashion.

5.3. Sex and Age

Most series show a male preponderance of around twofold.

The curves of age-specific incidence for nasopharyngeal cancer in Chinese populations in Hong Kong and Singapore are very similar in that they show a fairly steep rise from 20–24 years of age to about 50–54, with a fall thereafter.[103] This could be interpreted as representing a cessation of exposure at around 40 years of age, a cohort effect such as has been noted for breast cancer,[13] or the exhaustion of a pool of susceptibles.

The curve for NPC in Swedes is dissimilar in that the rise occurs two decades later and continues until age 70–74.[103] Bailar[10] noted that NPC tends to appear at an earlier age than most other forms of cancer in Caucasian populations. However, comparison of the age distribution of patients in areas where NPC is common and elsewhere is difficult because it is likely that each series contains some systemic malignant lymphomas that first presented in the nasopharyngeal region, and such neoplasms are proportionately much more common in low-incidence populations.[182,186,225]

Green et al.[76] point out that within the United States, nasopharyngeal cancer has three age peaks with racial and epidemiologic differences held to reflect different etiologies. Under the age of 20, nasopharyngeal cancer incidence was seven times higher in blacks than in whites, although these rates are based on small numbers. This trimodal age curve comprised a small peak under age 10 resulting from rhabdomyosarcoma of the nasopharynx in whites; a second mode at 15–24 of NPC, more pronounced in blacks than in whites; and a third, larger peak in older adults, the rates being higher in blacks than in whites until middle life and lower thereafter. However, these curves were quite overshadowed after the age of 25 by the rates among the United States Chinese.*

It is of interest to note that in Tunisia, Kuwait, and the Sudan, the age curve appears to be bimodal, 20% of cases in the Sudan appearing before the age of 20 years.[20,179] In

Constantine, Algeria,[134] the peak frequency was in the age group 15–24 with a tenfold male excess. Here, 18% of neoplasms were seen below the age of 15.

5.4. Occupation

Analysis of mortality rates in males by occupation in Taiwan[143] revealed excess risks in those engaged in salt production, national defense, public service, and mining. The higher risk in salt workers and miners did not, of course, account for the overall high rate, and the higher rates in the other two groups could be related to the high proportion of mainlanders engaged in these occupations (see Fig. 7). The "boat people" of Hong Kong are largely fishermen, but fishermen elsewhere do not have a high risk of NPC. The bizarre finding of three cases of NPC in Canadian bush pilots could result from chance.[5]

Henderson and co-workers' case-control study in the United States[79] showed that those occupationally exposed to fumes, smoke, and chemicals were at a higher risk. Within the United States, those areas with a high mortality from NPC were also prone to bladder cancer, a correlation that Fraumeni and Mason[66] hold to be consistent with the finding that British workers in leather and shoe factories are also more likely to have both nasal cavity and bladder neoplasms.

5.5. Change of Risk on Migration

The critical role of environmental cofactors is supported by the observation that Caucasian French are at a sixfold higher risk for NPC when born or having sojourned in their youth in high NPC risk areas such as North Africa.[118] Migrants often take critical parts of their environment with them in the form of diet and culture,[181,219] and this appears to be highly relevant for migrant Chinese in the Americas.

Buell[17,18] reviewed California death certificates for 1955–1964 and found that 5 of 273 white male descendants with NPC were born in the Philippines or China, areas where the disease is endemic. The expected number, based on death certificates for other cancer deaths, was much less than one. Analysis of name and birthplace of parents of the descendents showed that all were of Caucasian stock.

Fraumeni and Mason[66] noted that the mortality from NPC was 26-fold higher in Chinese males and 22-fold higher in Chinese females than in the black and white populations of the United States. During the study period, 1950–1969, there was, however, a diminution in risk. This decline could be related to changed food habits or other

---

*Ho observed that whereas the age-specific incidence curves for Hong Kong and U.S. Chinese males showed a sharp rise after the third decade of life and a peak at the sixth, those for both U.S. white and black males showed the rise to occur after the fourth decade and the peak at the seventh. This difference might reflect Chinese being exposed to carcinogenic initiators earlier in life than the other racial groups. It is also likely that the risk factors involved were also different.

environmental factors or to an increase in the proportion of low-risk central and northern Chinese in the United States following 1945. Prior to this period, immigration was largely confined to high-risk southern Chinese. King and Haenszel[122] examined mortality among the foreign and native-born Chinese in the United States. There was an indication of a lower risk for Erdai (the United-States-born Chinese). However, Shanmugaratnam and Tye[187] have suggested that the rate is increased for Singapore-born Chinese, but this question is being reexamined using the most recent Singapore Cancer Registry data. Any rise in incidence among Indians and Pakistanis living in Singapore (see Fig. 1) would be of greatest interest, but this has not been observed.

5.6. Environmental Factors

Polunin[175] reviewed the ways of life of peoples with high risk of nasopharyngeal cancer and was unable to identify any single distinguishing feature. Indeed, Muir and Oakley[156] commented on the large differences in lifestyle among Chinese, Malays, and Dayaks living in Sarawak (Borneo) who had, nevertheless, substantially the same frequency of the disease.

Although there has been no lack of speculation and anecdote since the disease was first characterized, Shanmugaratnam and Higginson[183,185] were the first to conduct a retrospective survey on a group of patients with histologically proven primary NPC. This study was essentially negative. A further case-control study involving 379 Singapore Chinese patients with NPC, together with 595 patients with other ENT diseases and 1044 patients with diseases other than cancer or ENT disease, showed that NPC patients differed significantly from both groups of controls in that they showed stronger associations for both personal and family history of nasal illnesses, the use of traditional Chinese medicines for the nose and throat, and exposure to smoke from antimosquito coils.[188] These latter exposures are not necessarily causal and could be interpreted as reflecting a more traditional household.

Armstrong et al.[7] noted that NPC risk in urban Chinese in Selangor, Malaysia, was elevated in those whose lifestyle was based on a low socioeconomic status, eating few fresh foods, with little variety in meals, and living in old, poor-quality housing.

Ho[92,95,97,102,104,106,111] suggested that salted fish might be an etiologic factor, being a traditional food commonly consumed by rich and poor southern Chinese from early childhood, especially those from Kwangtung (Guangdong), whether resident in China or abroad. An unusual dietary item in central and northern China, this salted fish contains appreciable quantities of nitrosodimethylamine.[64,65,109,203] The fact that NPC has been known to be common among southern Chinese for well over 50 years suggests the influence of a traditional rather than a modern environment.[104] Further, the factor is more likely to be ingested than inhaled, since in Hong Kong NPC is twice as common in "boat people" who live in boats and cook in the open than in land dwellers mostly living in congested apartments.[92,104] The sharp rise in age-specific incidence after the age of 20–24 years suggests the importance in Chinese patients of early childhood.[104]

Since adults will not be able to recall diet in childhood, let alone infancy and weaning, such data have to be obtained by interview of mothers and older relatives of young patients. Thus, Anderson et al.[2] interviewed 24 Hong Kong Chinese NPC patients diagnosed before the age of 25 years. Most interviews took place in the homes where they were born, in the presence of older members of their families. The only foods eaten by all subjects, and worth consideration, were laap cheung (Cantonese pork sausage), salted fish, tau si (a black fermented product made of salted soya beans), and dried squid. Salted fish was the most common item and the only one fed to babies. In childhood, the NPC patients had rarely or never been fed vegetables or fruits. Most, since childhood, were stated to have been sickly, inactive, withdrawn, and choosy about their food. It was suggested that consumption of salted fish and vitamin C deficiency in early childhood could be important environmental factors and that a certain personality type may be associated with an increased risk. Furthermore, the study eliminated household inhalants and aerial contaminants as likely factors.

Geser et al.[72] undertook a case-control study of Chinese NPC patients admitted to the Queen Elizabeth Hospital, Hong Kong, the controls being inpatients with other cancers. Healthy members of the households of cases and controls were also interviewed to obtain information unbiased by the experience of having cancer. Positively associated with NPC were: belonging to the four lowest occupational classes, practicing Buddhism or ancestor worship, having religious altars in the house, and a history of previous illness of the ear or nose after the age of 15 years. Negatively associated factors included eating of bread and tinned food and the use of spices. Weaning habits were compared in the households of NPC patients and those of controls by asking women who had ever breast-fed a child about food supplements given to the baby during and immediately after weaning. Salted fish was given to babies just after weaning more often in households with an NPC case than in control households.

Multivariate analysis showed that traditional lifestyle and the consumption of salted fish during weaning were independent risk factors. This analysis also demonstrated that two or three of the many expressions of a traditional lifestyle included in the study could account for the total increase in NPC risk associated with this way of life, although the authors concluded that it is quite possible that other factors, as yet unidentified, are just as important.

Armstrong et al.[6] studied 100 Chinese histologically confirmed NPC patients (65 males and 35 females) and 100 neighborhood control subjects individually matched to the patients on ethnic subgroup, sex, and age to within 5 years in Malaysia. All interviews were conducted in the homes of the participants by two trained Chinese interviewers using a standard questionnaire. Most of the patients were Chinese born in Malaysia. Analysis of the data showed that daily consumption of salted fish in childhood carried a relative risk of 17.4 (95% confidence interval = 2.7, 111.1) compared with nonconsumption.

Yu et al.[227] carried out a study in 250 Hong Kong Chinese patients aged under 35 years and an equal number of control subjects matched for sex and age to within 5 years. The control subjects were friends of the patients. All interviews were conducted in person by a trained interviewer. The mothers of both patients and controls, when available, were also interviewed concerning the weaning habits and the dietary habits of the participants at age 10 and between 1 and 2 years. Data analysis showed the relative risk for consuming salted fish at least once a week compared with less than once a month at 10 years of age was 37.7 (confidence interval = 14.1, 100.4). The age of 10 years was chosen because recollection of events before that age might not be reliable.

The two above-mentioned case-control studies carried out independently in southern Chinese in Malaysia and Hong Kong have shown a highly significant association between salted fish intake, especially during childhood, and NPC, and the workers estimated that the majority of NPC cases in southern Chinese can be attributed to consumption of this food early in life. Furthermore, the finding suggests that exposure to smoke or dust is unable to account for the majority of the disease in this high-risk population.

Lin et al.,[143] in a case-control study in Taiwan, noted that there was a significant excess of smokers among patients, the relative risk for persons smoking more than 20 cigarettes a day compared with those who had never smoked being greater than twofold. Working under poorly ventilated conditions was found to enhance risk, as did the use of herbal drugs and nasal balms or oils.

Clifford[24] suggested that the smoke and benzpyrene-containing soot so prominent in the huts of Kenyan groups with a raised frequency of the disease could be a causal factor, but the disease was found to be more common in males, who spend much less time in the huts.

Henderson et al.[79] undertook a case-control study on 156 patients and 267 controls in California, finding a highly significant increased relative risk (RR) of 1.8 for a prior history of ENT disease, occupational exposure to fumes (RR 2.0), smoke (RR 3.0), and chemicals (RR 2.4). In this study, over 50% were Chinese, and most of the others whites. The white and Chinese patients were analyzed separately, and the results were the same. This study is particularly valuable in that a group of patients with cancers elsewhere in the pharynx were also interviewed; unlike those with nasopharynx cancer, these clearly showed the effects of tobacco and alcohol. The history of previous ENT disease was common to patients with cancer both of the nasopharynx and elsewhere in the pharynx and is probably nonspecific. Exposure to fumes was most common in cooks among the Chinese, but studies in Singapore and Hong Kong have not shown such an association; among Chinese patients and controls, there was no significant difference in the current use of salted fish, but frequency of use was significantly greater in patients.[78]

Because a number of case-control studies carried out in the high-risk groups concluded that a traditional way of life represented a critical risk factor for NPC,[72,185] it was felt that an anthropological study of the traditional way of life and especially the food habits could be a valuable approach. In southern China where Cantonese Chinese (Han) exhibit a high risk for this tumor, some ethnic minorities, such as the Miao, Yao, and Bouy, living in the same area appear to have a much lower risk. Hubert and de-Thé[112] therefore formulated the hypothesis that food habits might be involved in the NPC genesis. Anthropological community-based research conducted at the family level appeared to be paramount, since family clustering was observed as early as 1972 by Ho.[103] The first step in this approach was to conduct a comparative anthropological study of NPC families in the three high-risk groups for NPC, namely, southern Chinese, Maghrebian Arabs, and Greenland Eskimos, in order to determine common or similar factors or effects that could represent an NPC risk factor among the three groups. The field work was carried out by A. Hubert, a food habit anthropologist, using the classical method of participant observation and key informant interviewing. Twenty families with NPC were selected in each geographical area (southern China, Macao, Tunisia, Greenland) for this pilot anthropological study, thus permitting observations of household life and lengthy conversations with key informants. The following subjects were covered: a genealogy of the household with a history of NPC, a history of the disease as perceived by the patient and his/her family (to try to point out putative cofactors), the life history of the patient and family to note all changes, moves, occupation, data on habitat past and present, observation on hygiene and body techniques, such as physical behavior in the household, notion of cleanliness, purity, and dirt, gestures of affection such as kissing on the mouth of babies and premasticating food for infants, diet and food with description of the cooking area, conversation with members of the family completed with visual observation on table manners, special diet, types of meals, change in diet during life, and weaning, in order to obtain a dietary history as complete as possible, and traditional therapy if any, including traditional medicines. The conclusion of such a pilot anthropological survey was that the three groups used daily preserved foodstuffs that could contain chemical carcinogens. In southern China and Greenland these include salted and unsalted dry fish preparations, pickled vegetables, preserved berries in Greenland, and preserved meat in Tunisia. The mode of preservation in the three areas included the drying, mostly in the sun or in the wind, the length of time of exposure in the open depending on climate conditions, thickness of the food, season, and so forth. In most places no particular attention is paid to cleanliness, and flies are numerous, especially around fish and meat. A slight modification of this drying involves a slight putrefaction of the fish before salting to keep its flesh soft. This latter type of preserved fish is more expensive and therefore less commonly eaten. It must be noted that dry fish prepared by Greenland Eskimos is never salted and that large fish are usually gutted. The dry meat prepared by Tunisians (qaddid) involves a few hours of dipping fresh mutton in a spice mixture also used as stewing base or touklia. Salting has been used since very ancient times for preserving food. Vegetables are preserved in salt or in brine in both China and Tunisia but not in Greenland.[113]

Fermentation is a very ancient way of preserving food, sometimes requiring the use of salt to control the process. In Tunisia, meat preserved in olive oil undergoes fermentation, the chemistry of which still remains unknown. In southern China, small fish and seafood are salted and allowed to ferment for several weeks or months. In Greenland, seal oil is obtained by fermentation of seal blubber in sealskin containers left outside the home. Fermentation occurs, and the blubber becomes fluid and transparent, developing a very strong smell and taste reminiscent of strong cheese. Thus, the anthropological surveys of the three high-risk groups for NPC

showed that the dry fish preparation, salted or unsalted, the pickled vegetables, and the preserved meat could represent risk indicators for NPC.

Representative food samples were collected in the three above-mentioned groups and were analyzed for the presence of volatile nitrosamines. As seen in Table 5,[174] the presence of N-nitrosodimethylamine (NDMA) was detected in salted and dried fish of southern China and in radish, roots, and Chinese cabbage in brine. The presence of NDMA, N-nitrosopyrrolodine (NPYR), and N-nitrosopiperidine (N-PIP) was also found in some preserved food from Tunisia, namely, qaddid, touklia, and harissa. Touklia and harissa are used daily as stewing base and spice mixture in Tunisian food. A later case-control epidemiologic study in Tunisia showed that qaddid, harissa, and a rapid transition from breast milk to an adult diet at weaning represented significant risk factors.[119] Similarly in Greenland, dry but not salted fish preparations contained various amounts of NDMA. Of importance here is that a great variation was observed in different preparations of the same preserved food, indicating that the method of preparation is critical for the presence or absence of volatile nitrosamines. The preserved foods prepared in a more traditional way may contain higher contents of such volatile nitrosamines.

5.7. Genetic Factors

Descriptive epidemiology has established that persons of southern Chinese descent, no matter where they live, or Southeast Asian populations who have genetic similarities to southern Chinese, e.g., the Malays, are at high risk for this cancer. The disease is less common in north China and rare in the Japanese, who are also of north Mongoloid origin.

Familial aggregation of NPC cases takes place both in high-risk areas, as repeatedly stressed by Ho,[91,95,96,103] and in low-risk areas.[158,215]

Walshe,[210] in a review of the physical anthropology of races with a high incidence of nasopharyngeal cancer, suggested that in the absence of a single physical characteristic common to populations of those countries with a high incidence of the disease and not found in others, the most profitable approach to the problem would be to concentrate on patients and controls. One approach was to study blood genetic markers (e.g., red cell antigens, red cell enzymes, HLA) to determine whether NPC patients had a characteristic genetic profile.[195]

Red cell blood group analyses were unrewarding,[77] but genetic studies involving 25 red cell enzymes and five serum protein systems indicated that NPC tends to con-

Table 5. Occurrence of Volatile Nitrosamines in Food Samples Collected in Tunisia, Southern China, and Greenland and Their Intake Frequency[a]

Food sample and intake frequency per week			Number of samples analyzed and nitrosamines detected (level μg/kg)
Tunisia			
"Qaddid" (dried mutton in olive oil)	>3	1	NDMA (23), NPYR (3,4)
"Touklia" (stewing base)	>3	1	NDMA (12), NPIP (43), NPYR (5.8)
Turnips fermented in brine	>3	1	NDMA (3.0)
"Harissa" spice mixture	>3	1	NDMA (trace), NPYR (trace)
Others		6	None
China			
Hard salted and dried fish	>3	14	NDMA (0 to 133)
Soft salted and dried fish	2	1	NDMA (4.0)
"Lap cheung" sausages	2	1	NDMA (1.2)
"Lap yok" salted pork	2	1	NDMA (0.5)
Green mustard leaves in brine	3	1	NDMA (13), NPIP (14), NPYR (18)
Radish roots in brine	>3	2	NDMA (4.3 to 6.0), NPYR (2.4)
Chinese cabbage in brine	>3	3	NDMA (0.6 to 6.1), NPIP (0 to 9.2), NPYR (5.5 to 96)
Fermented soya bean paste	>3	3	NPYR (0 to 5.8)
Others	2	6	None
Greenland			
"Mikialak" dried cod	1	1	NDMA (8.6)
"Uuvaq" dried polar cod	>3	1	NDMA (26)
"Aalissaqaq" Atlantic cod	>3	1	NDMA (29)
"Amassat" dried fish capelin	>3	1	NDMA (38)
Raw fjord seal liver	1	1	NDMA (trace)
Others		7	None

[a]From Poirier et al.[174]

centrate in genetically distinct subpopulations of Chinese in Singapore.[123] More significant differences in human leukocyte antigen (HLA) types were found between Chinese NPC patients and controls. An HLA profile associated with increased risk of NPC consisted of an increase in the frequency of the first-locus antigen (HLA-A2) and a deficit in the frequency of antigens detected at the second locus ("blank").[195] This blank was later demonstrated to represent a new HLA antigen named Sin-2,[192–194] now named B46. The haplotype A2-B-Sin-2 was shown to carry an increased relative risk of 1.96.[196] When attention was directed to newly diagnosed cases, it was found that the haplotype AW19-BW17 was more frequently seen in NPC than in controls.[191] Among long-term survivors, BW17 was found to be significantly decreased in frequency, as though this marker were associated with poor prognosis.[191]

Betuel et al.[12] have obtained evidence suggesting that the second-sublocus deficit is also seen among Tunisian NPC patients, but to a much lesser extent than in Chinese.

Multiple cases of NPC among sibs of large families represent a unique opportunity to investigate genetic markers associated with the disease.

In an international collaborative study[146] involving 30 sibships with two or more NPC cases, and originating from the Guangxi Autonomous Region of the People's Republic of China, but also from Hong Kong, Singapore, and Malaysia, evidence was obtained of the existence of a gene (or genes) closely linked to the HLA region that confers a greatly increased risk for the disease (maximum likelihood estimate 20.9; 95% interval: 5.1 to infinity).

The high-risk HLA pattern is not present in all NPC patients, and conversely, such a pattern is present in some persons with no NPC. The most likely interpretation is that the HLA data reveal the existence of NPC-disease-susceptibility gene(s). A reasonable postulate[148] is that the putative disease-susceptibility genes function as immune-response genes.

5.8. Epidemiologic Behavior of Epstein–Barr Virus

Epstein–Barr virus infection is present in every population around the world, but age at primary infection seems to vary with socioeconomic level. There seems to exist a gradient of prevalence of infection in young children from cold to tropical countries.[61]

From data accumulated in experimental viral oncology, age at infection, dose and route of infection, and genetic susceptibility were known to be critical factors. It was therefore felt essential to have reliable data on the epidemiology of EBV in populations at different risk for EBV-related diseases (IM, BL, NPC). An international collaborative seroepidemiologic study on EBV among Chinese (high risk for NPC, nil for IM and BL), Ugandans (high risk for BL, low for NPC, nil for IM), and Europeans (high risk for IM, low to nil for NPC and BL) took place in Hong Kong, Singapore, Uganda, and France.[41] In a first phase, volunteers from various sources were accepted: maternity and child clinics, army and police, university students, and others. Since such groups did not cover all age groups and did not represent the populations at large, randomly selected families were visited, interviewed, and bled.

The age at primary infection by EBV varies markedly among the groups studied. As seen in Fig. 8, in the 2- to 3-year age group, 97% of Ugandans were EBV-antibody positive, compared with only 20% of Singapore Chinese and 30% of Indians. Such differences tapered off at around 10 years of age, when 100% of Ugandans, 75% of Chinese, 85% of Indians, and 65% of Europeans were found to be EBV-antibody positive. An unexpected finding was that Hong Kong Chinese might have earlier EBV infection than Singapore Chinese (G. de-Thé, N. E. Day, A. Geser, J. H. C. Ho, M. F. Lavove, M. J. Simons, R. Sohier, and P. Tukei, unpublished data).

Ugandan children respond to EBV infection early in life by a strong humoral response. The geometric mean titers of VCA antibodies in the 1- to 5-year age group reached or passed 420 (Fig. 9). After this initial stress, Ugandan children show a dramatic fall of detectable EBV antibodies, possibly because of formation of antigen–antibody complexes or loss of immunity.

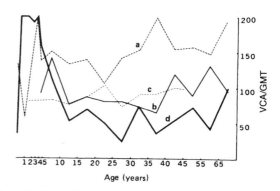

Figure 9. Geometric mean antibody titers against EBV VCA in the different age and ethnic groups (a–d, identified in the caption for Fig. 8). Note that the Ugandans, after a very high peak between 1 and 3 years of age, become the lower responders, whereas the Indians exhibit a relatively stable and high humoral response against the virus.

Indians in Singapore, who are at very low risk for NPC, BL, and IM, have the strongest immune response to EBV VCA.

The aforementioned differences in the infection rate and the immune response to EBV among ethnic groups and geographic areas may reflect cultural differences. They may also reflect a genetically dependent immune response to EBV antigens. The latter is an attractive hypothesis, since NPC has a strong genetic component (see above) and since genetics play an important role in determining the susceptibility of animals to oncogenic viruses in general[67,141,142] and to herpesviruses in particular (Marek's disease).

6. Mechanism of Transmission

Although primary EBV infection is probably transmitted through saliva, current serological and epidemiologic evidence suggests that NPC is a consequence of reactivation rather than a primary infection. The history of a higher frequency of upper respiratory infections in NPC patients than in controls, prior to tumor development, raises the possibility that some respiratory virus occurring in adult life might either reactivate EBV locally or make epithelial cells permissive to EBV infection in genetically susceptible persons, or both. At present, EBV is known to infect both B lymphocytes and epithelial cells of the oropharynx. Of particular relevance here is the recent observation by Sixbey and Yao,[197] that IgA antibodies induce a shift in EBV targeting from lymphoid to epithelial cells.

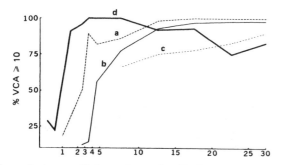

Figure 8. Age-specified prevalence of antibodies to EBV VCA in Singapore Indians (a), Singapore Chinese (b), Caucasians (c), and Ugandans (d). Note the important difference between 1 and 5 years of age in these groups.

7. Pathogenesis

The evidence available thus far strongly suggests that EBV is involved etiologically in the development of NPC but that it does not represent the sole cause of this cancer. Although the various forms of serological reactivity to EBV in NPC patients can be interpreted as either cause or effect, the regular presence of EBV fingerprints in epithelial tumor cells leaves no doubt that there is a strong link between the two. Evidence of close contacts and cytoplasmic bridges between epithelial and lymphoid cells both in normal nasopharyngeal mucosa and in NPC[68,209] might indicate a mode of intercellular interaction through which EBV might operate.

That chemical factors play a part in the carcinogenesis has been suggested in the past by Clifford[26] and more recently by Ho[96,97,104] and Poirier et al.[174]

Figure 10 is a diagram of the possible interplay between the EBV, the environmental carcinogens (which could determine the geographic distribution of NPC, and an NPC-susceptibility gene. These three factors could influence the relationship between epithelial and lymphoid cells in nasopharyngeal mucosa and induce the development of dysplasia, which in turn would induce an IgA-specific response. This would permit not only the diagnosis of NPC precancerous lesions but also a shift in EBV targeting, resulting in epithelial EBV infection, evolving eventually into carcinoma *in situ* and full carcinoma.

8. Patterns of Host Response

8.1. Clinical Course of Nasopharyngeal Carcinoma

The clinical course of NPC varies widely in duration. Without any form of specific therapy or with only palliative or inadequate radiation therapy, the patient may live from a few months to 13 years from the date of diagnosis.[94,101]

The presenting symptoms depend on the location within the nasopharynx of the primary tumor, its tendency to invade neighboring structures, the direction of invasion, and its predilection to metastasize to regional lymph nodes. Thus, the signs and symptoms in order of frequency (see Table 6) are cervical nodal enlargement, nasal symptoms (obstruction, postnasal discharge, epistaxis), aural symptoms secondary to eustachian tubal obstruction (impairment of hearing of the conductive type, with or without tinnitus and serous otitis media), involvement of cranial nerves (V, especially its maxillary branch, VI, IX, X, XI, XII, and upper cervical sympathetics), persistent headache, and stiffness of the jaw as a result of lateral spread of the tumor to the pterygoid muscles. About one fifth of cases have more than one presenting symptom; the frequency distribution is shown in Table 6.

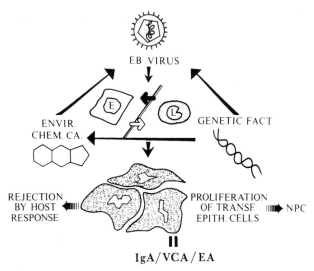

Figure 10. Proposed pathogenesis of NPC resulting from the interplay between EBV, exposure to carcinogens from preserved foods, and host genetic factor (NPC susceptibility gene); such an interplay would have an effect on the physiological interactions of the epithelial cells/lymphoid elements. The resulting dysplasia, or carcinoma *in situ*, would be recognized by the underlying lymphoid tissue, through an IgA response paralleling the risk of developing a clinical tumor. The lower part of the figure shows that only a small proportion of IgA-carrying individuals will develop clinical NPC, with most clones rejected by the host with parallel disappearance of IgA antibodies markers.

Table 6. Order of Frequency of Signs and Symptoms in Nasopharyngeal Carcinoma Patients: All Cases (513)[a]

Presenting symptoms	Frequency[b]	Percentage
Nodal enlargement	288	43.31
Nasal	132	19.85
Aural	133	20.00
Pain	85	12.78
Cranial nerve impairment	20	3.01
Miscellaneous		
Hawking	4	0.60
Trismus	3	0.45
Totals	665	100.00

[a]From Ho.[104]

[b]Some patients had more than one major presenting symptom.

8.2. Clinical Types of Disease

The biological behavior of the tumor determines the clinical types of the disease, which may be classified as (1) metastatic, (2) invasive, and (3) combined.[101]

8.2.1. Metastatic Type. The metastatic type, constituting about 33% of all cases, is characterized by the appearance of metastases, initially in the cervical lymph nodes (with the upper nodes involved before the lower) and later in distant organs, or, much less commonly, the two may appear about the same time, while the primary tumor apparently does not extend locally, remaining confined to the nasopharynx throughout. Only in exceptional cases, probably fewer than 0.1% of patients, have distant metastases been observed after a course of radiation therapy without cervical-nodal metastases being detected before the course or afterward. The spread may be confined to the cervical nodes for 1–3 years and in some for as long as 5 years or more.

Survival is longest with this type, and it is also with this variety that occasional patients are encountered in whom the tumor is apparently controlled for long periods, following what is normally considered as inadequate radiation therapy.

The common sites of distant metastases are the skeleton, especially the spine, liver, lung, and skin. The lymph nodes below the clavicles to as low as the femoral and inguinal nodes may be involved. Epidural and meningeal metastases occur, but metastases in the brain have not been encountered, although the brain is susceptible to direct invasion by the upward spread of the primary tumor or by adjacent meningeal metastases. The brain is believed to be devoid of a lymphatic system but is nevertheless a common site of metastases from carcinomas of bronchus and breast. It is possible that NPC may be peculiar in that blood-borne metastases require the presence of lymphatics for their establishment.

8.2.2. Invasive Type. The invasive type occurs in only about 8% of all cases. There is evidence of direct spread to adjacent muscles, bones, cranial nerves, paranasal sinuses, the orbit, veins, and venous sinuses at the base of the skull. Cervical nodal metastases are either insignificant or not detected, even until the death of the patient. However, in some patients, hematogenous metastases occur, usually after intracranial spread has become evident. Presumably, these metastases are the result of tumor invasion of the basal venous sinuses that communicate with the internal jugular vein and perivertebral plexus of veins.

8.2.3. Combined Type. The combined type, constituting about 59% of all cases, is characterized by a combination of direct spread of the primary tumor and the appearance of cervical nodal metastases. These may occur at about the same time or one after the other, with varying intervals in between. What causes the change in behavior in the latter group is not known.

9. Control and Prevention

The treatment of NPC with localized disease or only regional spread is by megavoltage radiation therapy. Ho[104] has shown that for tumors clinically confined to the nasopharynx (stage I), prophylactic cervical lymph nodal irradiation does not confer improved survival or tumor-control prospects. Cervical irradiation should therefore be withheld until nodal metastasis becomes clinically evident except in patients who are unlikely to attend regularly the follow-up clinic.

There is a place for adjuvant chemotherapy following a course of regional radiation therapy for patients with cervical nodal involvement down to the supraclavicular fossae (stage IV), because over 50% of such patients showed clinical evidence of distant metastases within 18 months of the completion of a course of radiation therapy. Until the immune status (tumor-specific and EBV-specific) of NPC patients is better understood, immunotherapy is not recommended, although research in this direction must be encouraged.

Transfer factor (TF) from donors with EBV antibody activity has been tried by Goldenberg[75] in a prospective randomized double-blind clinical trial in Hong Kong, with no significant antitumor effect.

The cumulative proportional actuarial survival and relapse-free survival after commencement of treatment by clinical stage according to Ho's classification[101] for Chinese patients treated by 4.5 MeV photons at Queen Elizabeth Hospital, Hong Kong, during 1969–1973 are shown in Fig. 11.

Prevention of NPC should concentrate on groups suspected from epidemiologic studies to be at very high risk. These include first-degree relatives of NPC patients, especially those with more than one member of the family affected.[95,96] Apart from educating the general public to discontinue the practice of giving salted fish to babies (which will take a long time), groups at high risk should have their sera tested for EBV reactivities (especially IgA VCA), which has been found to be a useful screening test for NPC.[99,230,231,233,235] Persons found to have elevated EBV serology should have a thorough clinical examination, including the radiological examination recommended by Ho,[91,93] initially and thereafter at yearly inter-

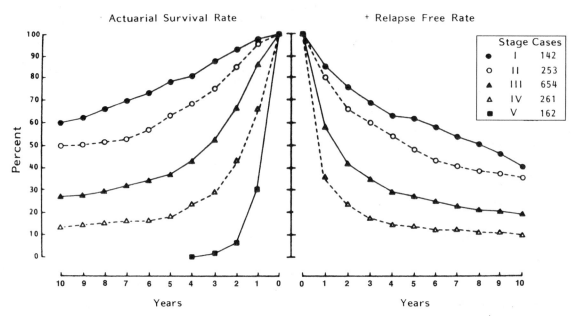

Figure 11. Ten-year treatment results of 1472 NPC patients by stage, 1969–1973, Queen Elizabeth Hospital, Hong Kong. Relapse-free rate denotes cases relapse-free after initial treatment. From Ho.[104]

vals or more frequently when there are suggestive symptoms.[231,235] However, negative clinical and radiological evidence of a tumor does not exclude NPC. Multiple biopsies of the nasopharynx should be done in the presence of a positive serological EBV IgA VCA test and repeated subsequently when the titer rises.

Education of the public, as well as the medical profession, on the early signals of the disease is essential in high-risk populations. A painless lump high up in the neck in an adult southern Chinese should be considered NPC until proved otherwise. This is as much an aphorism for NPC as a lump in the breast in a female over 40 years is for breast carcinoma in the Western world.

If inspection of the lateral pharyngeal recesses (fossa of Rosenmüller) by the use of a Yankauer speculum were routinely employed in examination, fewer tumors would be considered occult. In the mainly submucosal primary tumors, the earliest detectable sign is a hemorrhagic finely granular or velvety patch, which can be missed if only the postnasal mirror is used. Exfoliative cytology has not been found by Ho (unpublished data) to be sufficiently reliable to replace biopsy from multiple sites as a diagnostic method to detect "occult" carcinomatous areas.

A longitudinal multicenter study was carried out between 1979 and 1986 simultaneously in Paris, Tunis, and Hong Kong, with the aim of evaluating EBV serology as markers for prognosis of relapses after successful therapy. Patients who achieved prolonged (> 18 months)

complete remission after radiotherapy exhibited significant differences regarding IgA–EA antibodies between those who relapsed and exhibited increasing IgA–EA antibody titers and those who remained disease-free[43,51] (see Fig. 12).

10. Unresolved Problems and Challenges for the Future

The main and most fascinating problem in NPC research is to unravel the unique interplay between the three factors associated with the development of this tumor, namely, the EBV, environmental chemical carcinogens, and the putative NPC-susceptibility gene. We shall review them briefly, stressing that there are two opposing priorities in the challenge of controlling NPC: the long-term vaccine avenue adopted by the Western world and the pragmatic early detection scheme of cancerous and precancerous lesions as adopted in South China by Zeng et al.[231]

10.1. Role of Epstein–Barr Virus in the Management and Control of Nasopharyngeal Carcinoma

The role of a specific EBV gene(s) in the transformation of epithelial cells of the nasopharynx remains an open

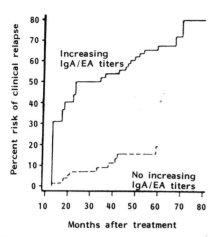

Figure 12. Cumulative risk of relapse after complete remission for a year and then exhibiting or not exhibiting an increase (≥ 1 dilution) of IgA/EA titers. From de-Thé *et al.*[43] and de-Vathaire *et al.*[51]

question.[197] In contrast to BL, the viral intervention in NPC pathogenesis might take place during a late stage on dysplastic epithelial cells, already altered by chemical carcinogens. We proposed such an alternative,[38] following the data of Ng *et al.*,[159] that EBV molecular fingerprints were only occasionally seen in precancerous lesions, while they were regularly detected in tumorous cells. The interesting observation of Sixbey and Yao[197] that EBV target may shift from the usual B lymphocytes to epithelial cells in the presence of IgA–VCA antibodies also supports a late intervention of EBV genes in an ultimate step toward NPC. If this hypothesis were to be confirmed, the presence of IgA antibodies to VCA, and even more so to EA, would reflect an ongoing process from precancerous to cancerous clones in the nasopharynx.

The 10-year prospective study carried out in South China by Zeng *et al.*[231] clearly indicates that the risk of developing NPC in asymptomatic IgA–VCA carriers either fades out if IgA antibodies disappear or increases

significantly in parallel with the increase in IgA–VCA antibodies titers (see Table 7). Thus, precancerous lesions, reflected by the presence of IgA antibodies, can be rejected by the host, recognizing them as foreign. This is no longer the case when fully tumorous clones emerge. Further research in this area could be of both fundamental and clinical interest.

The control of NPC represents a major public health problem in south China and in large parts of Southeast Asia.[42] To achieve early detection, our colleague Zeng Yi and his team from the People's Republic of China have conducted several seroepidemiologic surveys in the adult population of the Guang-xi Autonomous Region. Table 8 gives some results of these serological surveys in both rural and urban populations.[230,231,233,237] The clinical stages of NPC cases, as detected in both the rural Zangwu and in the urban Wuzhou after early detection was implemented, showed that EBV–IgA serological surveys represented a major improvement with a significant increase in the proportions of NPC diagnosed at stage I or II (from 26 to 89%). In turn, early diagnosis permitted better management of NPC by radiotherapy, since early clinical stages lead to much better survival rates 5- to 7-years posttherapy). Of even greater interest for the management and control of NPC were the results obtained in the 10-year follow-up of asymptomatic individuals with IgA–VCA antibodies,[50,231] as seen in Table 7, increase in IgA–VCA antibody titers carried an 18% risk of developing NPC within 3 years, whereas those who lost their IGA–VCA antibodies showed no risk of developing such a tumor. Individuals with stable IgA–VCA antibody titers exhibited a 1.3% risk of developing the disease.

These data indicate that IgA–VCA (see Table 9) and to a greater extent IgA–EA represent immediate-risk markers for NPC, and the question arises whether such serological markers reflect the presence of either precancerous or cancerous cells in the nasopharynx in numbers undetectable by usual means. The individuals who carried

Table 7. Stability and Fluctuations of IgA/VCA Antibodies over 3 Years in Relation to Risk of Developing NPC among IgA/VCA-Positive Individuals in Zang-Wu County (1979–1982)[a]

Number IgA/VCA positive	Stability (no change in IgA/VCA antibody)	Retroversion (loss of IgA/VCA antibody)	Increase[b]	Decline[b]	Up and down
1138 (100%)	455 (40.0%)	398 (35.0%)	81 (7.1%)	162 (14.2%)	42 (3.7%)
Total NPC patients detected	6[c]	0	15[d]	0	0

[a]From Zeng *et al.*[238] and de-Thé and Zeng.[50]
[b]By fourfold in titers.
[c]These six cases represent 1.32% of the 455 individuals with stable titers and 28% of the 21 NPC detected.
[d]These 15 cases represent 18% of the 81 subjects with increasing IgA/VCA titers and 71% of the 21 NPC detected.

Table 8. In Wuzhou City (W) and Zangwu Rural County (Z) NPC Clinical Stages at Diagnosis Either before or during the IgA/VCA Serological Screening and for Comparison during Follow-up Studies[a]

| | | Clinical stages | | | | Early diagnosis |
	N	I	II	III	IV	(stage I + II)
Outpatients without screening	W = 3374	27	846	2043	458	26%
	Z = 59	0	11	24	24	19%
Original screening	W = 18	10	6	2	0	89%
	Z = 15	5	5	3	2	67%
Follow-up studies	W = 29[b]	5	20	4	0	86%
	Z = 21	5	12	4	0	81%

[a]From Zeng *et al.*[231,235]
[b]Among those 29, 23 (80%) were detected within 5 years of the original mass screening

such markers indeed showed some cytological and histopathologic abnormalities including severe metaplasias and dysplasias, which could represent precancerous lesions. However, one could propose that individuals with these markers already have clones of epithelial tumor cells in their nasopharyngeal mucosa. If this were the case, it would indicate that, as in the case of subclinical prostate carcinoma, the frequency of *in situ* carcinomas of the nasopharynx exceeds by a factor of 2 or 3 the number of clinical NPC. This would imply that the organism is capable of recognizing and rejecting the majority of *in situ* nasopharyngeal carcinomas (see Fig. 10). The fact that IgA antibodies against EBV do not protect but, on the contrary, reflect a major risk factor in NPC development is comparable to the situation observed in BL with IgG antibodies to VCA–EBV.[47]

Can an EBV vaccine be contemplated for NPC?[34,89] Live vaccine, using turkey herpesvirus, has been found very efficient in preventing Marek's disease.[22] Laufs[133] has demonstrated that a killed vaccine for *Herpes saimiri* (HVS) can prevent HVS-induced lymphomas in cotton-top marmosets.

A prototype subunit vaccine is being developed in the United Kingdom,[56,59,151] including gp340, obtained

by genetic engineering. The question of the scientific importance and public health value of a vaccine to prevent EBV-associated diseases (infectious mononucleosis, Burkitt's lymphoma, and nasopharyngeal carcinoma) was recently discussed.[38,62,139,152] Clinical trials soon should be aimed at preventing mononucleosis. Vaccination to prevent nasopharyngeal carcinoma remains a far-reaching goal. The major problem with EBV and NPC is that it would be most difficult to vaccinate everyone prior to primary infection, i.e., in the first years of life, and then wait for 40–50 years to evaluate the effect. The only practical possibility would be to investigate the immune status (both humoral and cell-mediated immunity) prior to clinical onset. If a specific cell-mediated immunity were induced by such subunit vaccines, treatment of IgA-seropositive individuals at risk might be considered.

Antiviral chemotherapy, a meager possibility at present, might become a developing field in the coming decade and help in the control of NPC.

10.2. Chemical and Environmental Factors in the Causation of Nasopharyngeal Carcinoma

Our laboratory studies on food samples collected from Cantonese Chinese both in Macao and Wuzhou, but also from Greenland Eskimos and Tunisian Arabs, detected moderate or low levels of nitroso compounds in certain preserved meats and vegetables from the three areas.[174] Such *N*-nitroso compounds can be formed endogenously in humans through nitrosation in the stomach by bacteria containing nitrate to nitrite reductase enzymes. Comparing NPC and control patients from France and North Africa, we observed higher proportions of nitrate reductase bacteria in the flora of the Rosenmüller fossae of Maghrebian patients and controls than in the French counterparts.[173] Furthermore, after such nitrosa-

Table 9. Detection Rate and Incidence of NPC in IgA–VCA-Positive and -Negative Individuals in Zang-Wu, PRC[a]

	N screened (20,726)	IgA/VCA+ (1,136)	IgA/VCA− (19,590)
NPC detected over 10-year follow-up	57	53 (93%)	4 (7%)
Yearly incidence per 100,000	27.5	466	2

[a]From Zeng *et al.*[231]

tion, the volatile nitrosamine contents increased in parallel with genotoxic activities in a number of food specimens from south China and Tunisia.[173]

To further investigate the role of *in vitro* nitrosation in humans, a case-control study has been implemented in south China looking for urine nitrates by the *N*-proline test.[234] The results indicated that healthy individuals in high NPC areas had a significantly higher endogenous nitrosation rate, with higher nitrate concentration in urines, than healthy persons living in low NPC areas.[234] Furthermore, the diet in low NPC areas appeared to contain protective agents (probably vitamin C) missing in the high NPC areas. These studies must be pursued to see if a vitamin C intervention among asymptomatic IgA–VCA carriers could reduce significantly NPC incidence.

Furthermore, EBV-reactivating substances have been found in Chinese traditional medical herb preparations and industrial products, such as boat wood paints in Wuzhou.[238] We similarly searched for EBV-inducing compounds in preserved foods from south China, North Africa, and the Arctic and observed that dried salted fish from China and traditional harissa (a spice mixture from North Africa) induced EBV replication in cultures latently infected by EBV.[174,190] Chemical purification and characterization of the active compound in the traditional Tunisian harissa mixture was undertaken by G. Bouvier who identified a fraction containing a lignin complex, as active as TPA. One could suggest that such compounds could easily be retained in the Rosenmüller recesses and together with nitrate reductase bacteria could induce locally an activation of EBV latency. Thus, a number of new avenues merit exploration involving interactions between EBV and environmental factors. Furthermore, the NPC-susceptibility gene needs to be localized and characterized and its impact on either the immune response to EBV or the metabolism of the environmental carcinogens determined.

10.3. Conclusion

Deciphering of the interplay between environmental and host factors represents a major challenge not only for NPC but for the many human tumors where both environmental (viral and chemical) and genetic factors appear to interact in their pathogenesis. In this context, NPC represents a unique model for unraveling the causation and pathogenesis of human tumors toward a better preventive future.

ACKNOWLEDGMENTS
The typing and editorial help of Mrs. Liliane Lozano is acknowledged.

11. References

1. ALI, M. Y., AND SHANMUGARATNAM, K., Cytodiagnosis of nasopharyngeal carcinoma, *Acta Cytol.* **11**:51–60 (1967).
2. ANDERSON, E. N., JR., ANDERSON, M. L., AND HO, J. H. C., A study of the environmental backgrounds of young Chinese nasopharyngeal carcinoma patients, in: *Nasopharyngeal Carcinoma: Etiology and Control* (G. DE-THÉ AND Y. ITO, EDS.), pp. 231–239, IARC Scientific Publication No. 20, IARC, Lyon, France, 1978.
3. ANDERSSON-ANVRET, M., FORSBY, N., KLEIN, G., AND HENLE, W., Studies on the occurrence of Epstein–Barr virus DNA in nasopharyngeal carcinomas in comparison with tumors of other head and neck regions, *Int. J. Cancer* **20**:702–707 (1977).
4. ANDERSSON-ANVRET, M., FORSBY, N., KLEIN, G., AND HENLE, W., The association between undifferentiated nasopharyngeal carcinoma and Epstein–Barr virus shown by correlated nucleic acid hybridization and histopathological studies, in: *Nasopharyngeal Carcinoma: Etiology and Control* (G. DE-THÉ AND Y. ITO, EDS.), pp. 347–358, IARC Scientific Publication No. 20, IARC, Lyon, France, 1978.
5. ANDREWS, P. A. I., AND MICHAELS, L., Nasopharyngeal carcinoma in Canadian bush pilots, and aviator's cancer, *Lancet* **2**:85, 640 (1968).
6. ARMSTRONG, R. W., ARMSTRONG, M. J. YU, M. C., AND HENDERSON, B. E., Salted fish and inhalants as risk factors for nasopharyngeal carcinoma in Malaysian Chinese, *Cancer Res.* **43**:2967–2970 (1983).
7. ARMSTRONG, R. W., KANNAN-KUTTY, M., AND ARMSTRONG, M. J., Self-specific environments associated with nasopharyngeal carcinoma in Selangor, Malaysia, *Soc. Sci. Med.* **12**:149–156 (1978).
8. *Atlas of Cancer mortality in the People's Republic of China,* edited by the editorial Committee for the Atlas of Cancer mortality in the People's Republic of China, China Map Press, Beijing, 1979.
9. AYA, T., AND OSATO, T., Early events in transformation of human cord leukocytes by Epstein–Barr virus: Induction of DNA synthesis, mitosis and the virus associated nuclear antigen synthesis, *Int. J. Cancer* **14**:341–347 (1974).
10. BAILAR, J. C., III, Race, environment and family in the epidemiology of cancer of the nasopharynx, in: *Cancer of the Nasopharynx,* Vol. I (C. S. MUIR AND K. SHANMUGARATNAM, EDS.), pp. 101–105, UICC Monograph Series, Munksgaard, Copenhagen, 1967.
11. BENYESH-MELNICK, M., FERNBARCH, D. J., AND LEWIS, R. T., Studies on human leukemia. I. Spontaneous lymphoblastoid transformation of fibroblastic bone marrow cultures derived from leukemic and non-leukemic children, *J. Natl. Cancer Inst.* **31**:1311–1331 (1963).
12. BETUEL, H., CAMMOUN, M., COLOMBANI, J., DAY, N. E., EL-LOUZ, R., AND DE-THÉ, G., The relationship between nasopharyngeal carcinoma and the HL-A system among Tunisian, *Int. J. Cancer* **16**:249–254 (1975).
13. BJARNASON, O., DAY, N., SNAEDAL, G., AND TULINIUS, H., The effect of year on the breast cancer age incidence curve in Iceland, *Int. J. Cancer* **13**:689–696 (1974).
14. BLOT, W. J., LANIER, A., AND FARUMENI, J. F., JR., Cancer mortality among Alaskan natives 1960–1969, *J. Natl. Cancer Inst.* **55**:547–554 (1975).
15. BROOKS, L., YAO, P. Y., RICKINSON, A. B., AND YOUNG, L. S., Epstein–Barr virus latent gene transcription in nasopharyngeal carcinoma cells: Co-expression of EBNA-1, LMP.1, and LMP.2 transcripts, *J. Virol.* **66**:2689–2697 (1992).

16. BROUSSET, P., BUTET, V., CHITTAL, S., SELVES, J., AND DELSOL, G., Comparison of *in situ* hybridization using different non-isotopic probes for detection of Epstein–Barr virus in nasopharyngeal carcinoma and immunohistochemical correlation with anti-latent membrane protein antibody, *Lab. Invest.* **67**:457–464 (1992).

17. BUELL, P., Nasopharyngeal cancer in Chinese of California, *Br. J. Cancer* **19**:459–470 (1965).

18. BUELL, P., Race and place in the etiology of nasopharyngeal cancer: A study based on California death certificates, *Int. J. Cancer* **11**:268–272 (1973).

19. BURKITT, D. P., The trail to a virus, in: *Oncogenesis and Herpesviruses* (P. M. BIGGS, G. DE-THÉ, AND L. N. PAYNE, EDS.), pp. 345–348, IARC Scientific Publication No. 2, IARC, Lyon, France, 1972.

20. CAMMOUN, M., VOGT-HOERNER, G., AND MOURALI, N., Les tumeurs du naso-pharynx en Tunisie: Étude anatomo-clinique de 143 observations, *Tunis Med.* **49**(3):131–141 (1971).

21. CHAN, S. H., CHEW, T. S., GOH, E. H., SIMONS, M. J., AND SHANMUGARATNAM, K., Impaired general cell-mediated immune functions *in vivo* and *in vitro* in patients with nasopharyngeal carcinoma, *Int. J. Cancer* **18**(2):139–144 (1976).

22. CHURCHILL, A. E., PAYNE, L. N., AND CHUBB, R. C., Immunization against Marek's disease using a live attenuated virus, *Nature* **221**:744–747 (1969).

23. CLIFFORD, P., Carcinoma of the nasopharynx in Kenya, *East Afr. Med. J.* **42**:373–396 (1965).

24. CLIFFORD, P., Malignant diseases of the nasopharynx and paranasal sinuses in Kenya, in: *Cancer of the Nasopharynx* (C. S. MUIR AND K. SHANMUGARATNAM, EDS.), pp. 82–94, UICC Monograph Series, Vol. 1, Munksgaard, Copenhagen, 1967.

25. CLIFFORD, P., A Review on the epidemiology of nasopharyngeal carcinoma, *Int. J. Cancer* **5**:287–309 (1970).

26. CLIFFORD, P., Carcinogens in the nose and throat: Nasopharyngeal carcinoma in Kenya, *Proc. R. Soc. Med.* **65**:682–686 (1972).

27. DERRY, D. E., Anatomical report (B), in: *Archaeological Survey of Nubia* (Bulletin No. 3), pp. 40–42, Egyptian Ministry of Finance, Cairo, 1909.

28. DESCHRYVER, A., FRIBERG, S., JR., KLEIN, G., HENLE, W., HENLE, G., DE-THÉ, G., CLIFFORD, P., AND HO, H. C., Epstein–Barr associated antibody patterns in carcinoma of the post-nasal space, *Clin. Exp. Immunol.* **5**:443–459 (1969).

29. DESCHRYVER, A., KLEIN, G., HENLE, G., HENLE, W., CAMERON, H., SANTESSON, L., AND CLIFFORD, P., Epstein–Barr virus associated serology in malignant disease: Antibody levels to viral capsid antigens (VCA), membrane antigens (MA) and early antigens (EA) in patients with various neoplastic conditions, *Int. J. Cancer* **9**:353–364 (1972).

30. DESGRANGES, C., DE-THÉ, G., HO, J. H. C., AND ELLOUZ, R., Neutralizing EBV-specific IgA in throat washings of nasopharyngeal carcinoma (NPC) patients, *Int. J. Cancer* **19**:627–633 (1977).

31. DESGRANGES, C., LI, J. Y., AND DE-THÉ, G., EBV specific secretory IgA in saliva of NPC patients: Presence of secretory piece in epithelial malignant cells, *Int. J. Cancer* **20**:881–886 (1977).

32. DESGRANGES, C., WOLF, H., DE-THÉ, G., SHANMUGARATNAM, K., ELLOUZ, R., CAMMOUN, N., KLEIN, G., AND ZUR HAUSEN, H., Nasopharyngeal carcinoma X—Presence of Epstein–Barr genomes in epithelial cells of tumours from high and medium risk areas, *Int. J. Cancer* **16**:7–15 (1975).

33. DE-THÉ, G., Is Burkitt's lymphoma (BL) related to a perinatal infection by Epstein–Barr virus? *Lancet* **1**:335–338 (1977).

34. DE-THÉ, G., Epstein–Barr virus: Is it time to discuss a vaccine? *Biomed. Comm.* **28**:15–17 (1978).

35. DE-THÉ, G., The epidemiology of Burkitt's lymphoma: Evidence for a causal association with Epstein–Barr virus—a review, *Am. J. Epidemiol.* **1**:32–57 (1979).

36. DE-THÉ, G., The role of the Epstein–Barr virus in human diseases: Infectious mononucleosis (IM), Burkitt's lymphoma (BL), nasopharyngeal carcinoma (NPC), in: *Viral Oncology* (G. KLEIN, ED.), pp. 769–797, Raven Press, New York, 1979.

37. DE-THÉ, G., Multistep carcinogenesis, Epstein–Barr virus and human malignancies, in: *Viruses in Naturally Occurring Cancers* (M. ESSEX, G. TODARO, AND H. ZUR HAUSEN, EDS.), pp. 11–21, Cold Spring Harbor Conferences on Cell Proliferation, Vol. 7, Cold Spring Harbor Laboratory, Cold Spring Harbor, NY, 1980.

38. DE-THÉ, G., Epidemiology, pathogenesis and prevention of EBV associated malignancies, in: *The Epstein–Barr virus and associated diseases: Proceedings of the Vth International Symposium* (T. TURSZ, J. S. PAGANO, D. V. ABLASHI, G. DE THÉ, G. LENOIR, AND G. R. PEARSON, EDS.), pp. 15–30, Colloques INSERM/John Libbey Eurotext Ltd., Paris, 1993.

39. DE-THÉ, G., ABLASHI, D. V., LIABEUF, A., AND MOURALI, N., Nasopharyngeal carcinoma (NPC). VI. Presence of an EBV nuclear antigen in fresh tumour biopsies: Preliminary results, *Biomedicine* **19**:349–352 (1973).

40. DE-THÉ, G., AMBROSIONI, J. C., HO, H. C., AND KWAN, H. C., Lymphoblastoid transformation and presence of herpes-types viral particles in a Chinese nasopharyngeal tumour cultured *in vitro*, *Nature* **221**:770–771 (1969).

41. DE-THÉ, G., DAY, N. E., GESER, A., HO, J. H. C., LAVOUE, M. F., SIMONS, M. J., SOHIBA, R., AND TUKBI, P., Epidemiology of the Epstein–Barr virus: Preliminary analysis of an international study, in: *Oncogenisis and Herpesviruses II* (G. DE-THÉ, M. A. EPSTEIN, AND H. ZUR HAUSEN, EDS.), pp. 3–16, IARC Scientific Publication No. 11, Vol. 2, IARC, Lyon, France, 1975.

42. DE-THÉ, G., DESGRANGES, C., ZENG, Y., WANG, P. C., BORN-KAMM, G. W., ZHU, J. S., AND SHANG, M., Search for precancerous lesions and EBV markers in the nasopharynx of IgA positive individuals, in: *Nasopharyngeal Carcinoma*, Vol. 5 (E. GRUNDMANN, G. R. F. XRUEGER, AND D. V. ABLASH, EDS.), pp. 111–117, Gustav Fisher Verlag, Stuttgart, 1981.

43. DE-THÉ, G., DE-VATHAIRE, F., SANCHO-GARNIER, H., DE-THÉ, H., SCHWAAB, G., MICHEAU, C., BSCHWEOU, F., ROCHE, N., ELLOUZ, R., AND HO, J. H. C., IgA/EA: Prognostic marker for relapse among NPC patients with complete remission after radiotherapy, in: *Epstein–Barr Virus and Human Diseases* (P. LEVINE, D. ABLASHI, AND G. PEARSON, EDS.), pp. 43–45, Humana Press, Clifton, NJ, 1987.

44. DE-THÉ, G., GESER, A., DAY, N. E., TUKEI, P. M., WILLIAMS, E. H., BERI, D. P., SMITH, P. G., DEAN, A. G., BORNKAMM, G. W., FEORINO, R., AND HENLE, W., Epidemiological evidence for causal relationship between Epstein–Barr virus and Burkitt's lymphoma from Ugandan prospective study, *Nature* **274**:756–761 (1978).

45. DE-THÉ, G., HO, H. C., KWAN, H. C., DESGRANGES, C., AND FAVRE, M. C., Nasopharyngeal carcinoma (NPC). I. Types of cultures derived from tumour biopsies and non-tumorous tissues of Chinese patients with special reference to lymphoblastoid transformation, *Int. J. Cancer* **6**:189–206 (1970).

46. DE-THÉ, G., HO, J. H. C., ABLASHI, D. V., DAY, N. E., MACARIO, A. J. L., MARTIN-BERTHELON, M. C., PEARSON, G., AND SOHIER, R., Nasopharyngeal carcinoma IX. Antibodies to EBAN and

correlation with response to other EBV antigens in Chinese patients, *Int. J. Cancer* **16:**713–721 (1975).

47. DE-THÉ, G., LAVQUÉ, M. F., AND MUENZ, I., Differences in EBV antibody titres of patients with nasopharyngeal carcinoma originating from high, intermediate and low incidence areas, in: *Nasopharyngeal Carcinoma: Etiology and Control* (G. DE-THÉ AND Y. ITO, EDS.), pp. 471–481, IARC Scientific Publication No. 20, IARC, Lyon, France, 1978.

48. DE-THÉ, G., SOHIER, R., HO, J. H. C., AND FREUND, R., Nasopharyngeal carcinoma IV. Evolution of complement fixing antibodies during the course of the disease, *Int. J. Cancer* **12:**368–377 (1973).

49. DE-THÉ, G., VUILLAUMI, M., GIOVANELLA, A. B. C., AND KLEIN, G., Epithelial characteristics of tumour cells in nasopharyngeal carcinoma (NPC) passaged in nude mice—an ultrastructural study, *J. Natl. Cancer Inst.* **57:**1101–1105 (1976).

50. DE-THÉ, G., AND ZENG, Y., Population screening for EBV markers toward improvement of nasopharyngeal carcinoma, in: *The Epstein–Barr Virus Recent Advances* (M. A. EPSTEIN AND B. G. ACHONG, EDS.), pp. 236–249, William Heinemann Medical Books, London, 1986.

51. DE VATHAIRE, F., SANCHO-GARNIER, H., DE-THÉ, H., PUO-DELOUP, C., SCHWAAB, G., HO, J. H. C., ELLOUZ, R., MICHRAU, C., CAMMOUN, M., CACHIN, Y., AND DE-THÉ, G., Prognostic value of EBV markers in the clinical management of nasopharyngeal carcinoma: A multicenter follow-up study, *Int. J. Cancer* **42:**176–181 (1988).

52. DIGBY, K. H., THOMAS, G. H., AND HSTU, S. T., Notes on carcinoma of the nasopharynx, *Caduceus* **9:**45–64 (1930).

53. DOBSON, W. C., Cervical lymphosarcoma (letter to the editor), *Chin. Med. J.* **38:**786 (1924).

54. DOLL, R., Epidemiology of cancer: Current perspectives, *Am. J. Epidemiol.* **104:**396–408 (1976).

55. DURAND-FARDEL, Cancer du pharynx—Ossification dans is substance musculaire du coeur, *Bull. Soc. Anat.(Paris)* **12:**73–80 (1837).

56. EPSTEIN, M. S., The Epstein–Barr virus: Retrospective reflections and future prospects, in: *The Epstein–Barr virus and associated diseases: Proceedings of the Vth International Symposium* (T. TURSZ, J. S. PAGANO, D. V. ABLASHI, G. DE THÉ, G. LENOIR, AND G. R. PEARSON, EDS.), pp. 3–14, Colloques INSERM/John Libbey Eurotext Ltd., Paris, 1993.

57. EPSTEIN, M. A., AND ACHONG, B. G., Recent progress in BB virus research, *Annu. Rev. Microbiol.* **31:**421–415 (1977).

58. EPSTEIN, M. A., ACHONG, B. G., AND BARR, Y. M., Virus particles in cultured lymphoblasts from Burkitt's lymphoma, *Lancet* **1:**702–703 (1964).

59. EPSTEIN, M. A., MORGAN, A. J., FINERTY, S., RANDLE, B. J., AND KIRKWOOD, J. K., Protection of cottontop tamarins against EB virus-induced malignant lymphoma by a prototype subunit vaccine against MP gp 340, *Nature* **318:**287–289 (1985).

60. EVANS, A. S., The history of infectious mononucleosis, *Am. J. Med. Sci.* **267:**189–195 (1974).

61. EVANS, A. S., New discoveries in infectious mononucleosis, *Mod. Med.* **1:**18–24 (1974).

62. EVANS, A. S. EBV vaccine: Use in infectious mononucleosis, in: *The Epstein–Barr virus and associated diseases: Proceedings of the Vth International Symposium* (T. TURSZ, J. S. PAGANO, D. V. ABLASHI, G. DE THÉ, G. LENOIR, AND G. R. PEARSON, EDS.), pp. 593–598, Colloques INSERM/John Libbey Eurotext Ltd., Paris, 1993.

63. EVANS, A. S., AND NIEDERMAN, J. C., Epidemiology of infectious mononucleosis, in: *Oncogenesis and Herpesviruses* (P. M. BIGGS, G. DE-THÉ, AND L. N. PAYNE, EDS.), pp. 351–356, IARC Scientific Publication No. 2, IARC, Lyon, France, 1972.

64. FONG, Y. Y., AND CHAN, W. C., Bacterial production of dimethyl nitrosamine in salted fish, *Nature* **243:**421–422 (1973).

65. FONG, Y. Y., AND WALSH, E. O., Carcinogenic nitrosamines in Cantonese salt-dried fish, *Lancet* **1:**1032–1033 (1971).

66. FRAUMENI, J. F., JR., AND MASON, T. J., Cancer mortality among Chinese Americans, 1950–69, *J. Natl. Cancer Inst.* **52:**659–665 (1974).

67. GALLO, R. C., SAXINGER, W. C., GALLAGHER, R. E., GILLESPIE, D. A., AULAKH, G. S., AND WO NG-STAAL, F., Some ideas on the origin of leukemia in man and recent evidence for the presence of type-C viral related information, in: *Origins of Human Cancer* (H. HIATT, J. D. WATSON, AND J. A. WINSTEN, EDS.), *Cold Spring Harbor Conferences on Cell Proliferation*, Vol. 4, Book B, pp. 1253–1285, Cold Spring Harbor Laboratory, Cold Spring Harbor, NY, 1977.

68. GAZZOLO, L., DE-THÉ, G., AND VUILLAUME, M., Nasopharyngeal carcinoma, II. Ultrastructure of normal mucosa, tumor biopsies and subsequent epithelial growth *in vitro*, *J. Natl. Cancer Inst.* **48:**73–86 (1972).

69. GERBER, P., AND BIRCH, S., Complement-fixing antibodies in sera of human and non-human primates to viral antigens derived from Burkitt's lymphoma cells, *Proc. Natl. Acad. Sci. USA* **58:**478–484 (1967).

70. GERBER, P., NONOYAMA, M., LUCAS, S., PERLIN, E., AND GOLD-STEIN, L., Oral excretion of EBV by healthy subjects and patients with infectious mononucleosis, *Lancet* **2:**988–989 (1972).

71. GERBER, P., WHANG-PENG, J., AND MONROE, J., Transformation and chromosome changes induced by Epstein–Barr virus in normal human leukocyte cultures, *Proc. Natl. Acad. Sci. USA* **63:**740–747 (1969).

72. GESER, A., CHARNAY, N., DAY, N. E., DE-THÉ, G., AND HO, H. C., Environmental factors in the etiology and nasopharyngeal carcinoma: Report on a case-control study in Hong Kong, in: *Nasopharyngeal Carcinoma: Etiology and Control* (G. DE-THÉ AND Y. ITO, EDS.), pp. 213–229, IARC Scientific Publication No. 20, IARC, Lyon, France, 1978.

73. GESER, A., DAY, N. E., DE-THÉ, G., CHEW, T. S., FREUND, R. J., KWAN, H. C., LAVOUÉ, M. F., SIMKOVIC, D., AND SOHIER, R., The variability in immunofluorescent viral capsid antigen antibody testing in population survey of Epstein–Barr virus infection, *Bull. WHO* **50:**389–400 (1974).

74. GLASER, R., DE-THÉ, G., LENOIR, G., AND HO, J. H. C., Superinfection of nasopharyngeal carcinoma epithelial tumour cells with Epstein–Barr virus, *Proc. Natl. Acad. Sci. USA* **73:**960–963 (1976).

75. GOLDENBERG, G. J., BRANDES, L. J., LAU, W. H., MILLER, A. B., WALL, C., AND HO, J. H. C., Cooperative trial of immunotherapy for nasopharyngeal carcinoma with transfer factor from donors with Epstein–Barr virus antibody activity, *Cancer Treat. Rep.* **69:**761–767 (1985).

76. GREEN, M. H., FRAUMENI, J. F., JR., AND HOOVER, R., Nasopharyngeal cancer among young in United States: Racial variations by cell type, *J. Natl. Cancer Inst.* **58:**1267–1270 (1977).

77. HAWKINS, B. R., SIMONS, M. J., GOH, E. H., CHIA, K. B., AND SHANMUGARATNAM, K., Immunogenetic aspects of nasopharyngeal carcinoma. II. Analysis of ABO, Rhesus and MNS's red cell systems, *Int. J. Cancer* **13:**116–121 (1974).

78. HENDERSON, B. E., AND LOUIE, E., Discussion of risk factors for nasopharyngeal carcinoma, in: *Nasopharyngeal Carcinoma: Eti-*

ology and Control (G. DE-THÉ AND Y. ITO, EDS.), pp. 251–260, IARC Scientific Publication No. 20, IARC, Lyon, France, 1978.

79. HENDERSON, B. E., LOUIE, E., JING, J. S. H., BUELL, P., AND GARDNER, M. B., Risk factors associated with nasopharyngeal carcinoma, *N. Engl. J. Med.* **295**:1101–1106 (1976).

80. HENLE, G., Antibodies to EBV-induced early antigens in infectious mononucleosis, Burkitt's lymphoma and nasopharyngeal carcinoma, in: *Recent Advances in Human Tumor Virology and Immunology* (W. NAKAHARA, K. NISHOKA, T. HIRAYAMA, AND T. ITO, EDS.), pp. 343–359, University of Tokyo, 1971.

81. HENLE, G., AND HENLE, W., Immunofluorescence in cells derived from Burkitt's lymphoma, *J. Bacteriol.* **91**:1248–1256 (1966).

82. HENLE, G., AND HENLE, W., Epstein–Barr virus-specific IgA serum antibodies as an outstanding feature of nasopharyngeal carcinoma, *Int. J. Cancer* **17**:1–7 (1976).

83. HENLE, G., HENLE, W., AND KLEIN, G., Demonstration of two distinct components in the early antigen complex of EBV infected cells, *Int. J. Cancer* **8**:272–282 (1971).

84. HENLE, W., AND HENLE, G., Epstein–Barr virus: The cause of infectious mononucleosis, in: *Oncogenesis and Herpesviruses* (P. M. BIGGS, G. DE-THÉ, AND L. N. PAYNE, EDS.), pp. 269–274, IARC Scientific Publication No. 2, IARC, Lyon, France, 1972.

85. HENLE, W., AND HENLE, G., Epstein–Barr virus and infectious mononucleosis, *N. Engl. J. Med.* **288**:263–264 (1978).

86. HENLE, W., HENLE, G., BURTIN, P., CACHIN, Y., CLIFFORD, P., DE SCHRYVER, A., DE-THÉ, G., DIEHL, V., HO, H. C., AND KLEIN, G., Antibodies to Epstein–Barr virus in nasopharyngeal carcinoma, other head and neck neoplasms and control groups, *J. Natl. Cancer Inst.* **44**:225–231 (1970).

87. HENLE, W., HENLE, G., ZAJAC, B. A., PEARSON, G., WAUBKE, R., AND SCRIBA, M., Differential reactivity of human serums with early antigens induced by Epstein–Barr virus, *Science* **169**:188–190 (1970).

88. HENLE, W., HO, H. C., AND KWAN, H. C., Antibodies to Epstein–Barr virus related antigens in nasopharyngeal carcinoma: Comparison of active cases with long-term survivors, *J. Natl. Cancer Inst.* **51**:361–369 (1973).

89. HIGGINSON, J., DE-THÉ, G., GESER, A., AND DAY, N. E., An epidemiological analysis of cancer vaccines, *Int. J. Cancer* **7**:565–574 (1971).

90. HIGGINSON, J., AND MACLENNAN, R., The world pattern of cancer incidence, in: *Modern Trends in Oncology—Research Progress* (R. W. RAVEN, ED.), pp. 9–27, Butterworths, London, 1973.

91. HO, H. C., Nasopharyngeal carcinoma in Hong Kong, in: *Cancer of the Nasopharynx* (C. S. MUIR AND K. SHANMUGARATNAM, EDS.), pp. 58–63, UICC Monograph Series, Vol. 1, Munksgaard, Copenhagen, 1967.

92. HO, H. C., Incidence of nasopharyngeal cancer in Hong Kong, UICC Bulletin, *Cancer* **9**(2):5–8 (1971).

93. HO, H. C., Radiologic diagnosis of nasopharyngeal carcinoma with special reference to its spread through the base of skull, in: *Cancer of the Nasopharynx* (C. S. MUIR AND K. SHANMUGARATNAM, EDS.), pp. 238–246, UICC Monograph Series, Vol. 1, Medical Examination Publishing, New York, 1972.

94. HO, H. C., Head and neck—Radiologic diagnosis, *J.A.M.A.* **220**:396 (1972).

95. HO, H. C., Current knowledge of the epidemiology of nasopharyngeal carcinoma—a review, in: *Oncogenesis and Herpesviruses* (P. M. BIGGS, G. DE-THÉ, AND L. N. PAYNE, EDS.), pp. 357–366, IARC Scientific Publication No. 2, IARC, Lyon, France, 1972.

96. HO, H. C., Epidemiology of nasopharyngeal carcinoma (NPC), in: *Proceedings of the 1st Asian Cancer Conference* (H. HIRIYAMA, ED.), pp. 357–366, Shima, Tokyo, 1973.

97. HO, H. C., Epidemiology of nasopharyngeal carcinoma, *J. R. Coll Surg. (Edinburgh)* **20**:223–235 (1975).

98. HO, H. C., KWAN, H. C., AND NG, M. H., Immunohisto-chemistry of local immunoglobulin production in nasopharyngeal carcinoma, *Br. J. Cancer* **37**:514–519 (1978).

99. HO, H. C., KWAN, H. C., NG, M. H., AND DE-THÉ, G., Serum IgA antibodies to Epstein–Barr virus capsid antigen preceding symptoms of nasopharyngeal carcinoma, *Lancet* **1**:436–437 (1978).

100. HO, H. C., NG, M. H., AND KWAN, H. C., IgA antibodies to Epstein–Barr virus viral capsid antigens in saliva of nasopharyngeal carcinoma patients, *Br. J. Cancer* **35**:888–890 (1977).

101. HO, J. H. C., The natural history and treatment of nasopharyngeal carcinoma (NPC), in: *Oncology*. Vol. 4, *Proceedings of the 10th International Cancer Congress* (R. LEE-CLARK, R. W. CUMLEY, J. E. MCCAY, AND M. COPELAND, EDS.), pp. 1–14, Year Book Medical Publishers, Chicago, 1970.

102. HO, J. H. C., Genetic and environmental factors in nasopharyngeal carcinoma, in: *Recent Advances in Human Tumor Virology and Immunology, Proceedings of the First International Cancer Symposium of the Princess Takamatsu Cancer Research Fund* (W. NAKAHARA, K. NISHIUKA, T. HIRAYAMA, AND Y. ITO, EDS.), pp. 275–295, University of Tokyo Press, Tokyo, 1971.

103. HO, J. H. C., Nasopharyngeal carcinoma (NPC), in: *Advances in Cancer Research* (G. KLEIN, S. WEINHOUSE, AND A. HADDOW, EDS.), pp. 57–92, Academic Press, New York and London, 1972.

104. HO, J. H. C., An epidemiologic and clinical study of nasopharyngeal carcinoma, *Int. J. Radiat. Oncol. Biol. Phys.* **4**:181–198 (1978).

105. HO, J. H. C., Cancer in Hong Kong: Some epidemiological observations, in: *Third Symposium on Epidemiology and Cancer Registries in the Pacific Basin Held in Maui, Hawaii, January 19–23, 1981*, pp. 37–54, NCI Monograph 62, NIH, Bethesda, MD, 1982.

106. HO, J. H. C., HUANG, D. P., AND FONG, Y. Y., Salted fish and nasopharyngeal carcinoma in southern Chinese, *Lancet* **1**:626–628 (1978).

107. HO, J. H. C., NG, M. H., KWAN, H. C., AND CHAN, J. C. W., Epstein–Barr virus specific IgA and IgG serum antibodies in nasopharyngeal carcinoma, *Br. J. Cancer* **34**:655–660 (1976).

108. HU, C. H., AND YANG, C., A decade of progress in morphologic pathology, *Chin. Med. J.* **79**:409–422 (1959).

109. HUANG, D. P., GOUGH, T., AND HO, J. H. C., Analysis for volatile nitrosamines of salt-preserved foodstuffs traditionally consumed by southern Chinese, in: *Nasopharyngeal Carcinoma: Etiology and Control* (G. DE-THÉ AND Y. ITO, EDS.), pp. 309–314, IARC Scientific Publication No. 20, IARC, Lyon, France, 1978.

110. HUANG, D. P., HO, J. H. C., HENLE, W., AND HENLE, G., Demonstration of Epstein–Barr virus-associated nuclear antigen in nasopharyngeal carcinoma cells from fresh biopsies, *Int. J. Cancer* **14**:580–588 (1974).

111. HUANG, D. P., SAW, D., TEOH, T. B., AND HO, J. H. C., Carcinomas in rats fed with Cantonese salted marine fish, in: *Nasopharyngeal Carcinoma: Etiology and Control* (G. DE-THÉ AND Y. ITO, EDS.), pp. 315–328, IARC Scientific Publication No. 20, IARC, Lyon, France, 1978.

112. HUBERT, A., AND DE-THÉ, G., Comportement alimentaire, modes de vie et cancer du rhinopharynx (NPC), *Bull. Cancer (Paris)* **69**(5):476–482 (1982).

113. HUBERT, A., JEANNEL, D., TUPPIN, P., AND DE-THÉ, G., Anthropology and epidemiology: A pluridisciplinary approach of environmental factors of nasopharyngeal carcinoma, in: *The Epstein–Barr virus and associated diseases: Proceedings of the Vth International Symposium* (T. TURSZ, J. PAGANO, G. DE-THÉ, D. ABLASHI, G. LENOIR, AND G. PEARSON, EDS.), pp. 775–788, INSERM and John Libbey Eurotext, Paris, 1993.

114. INTERNATIONAL AGENCY FOR RESEARCH ON CANCER (IARC), *Cancer Incidence in Five Continents*, Vol. III (J. WATERHOUSE, C. S. MUIR, P. CORREA, AND G. POWELL, EDS.), IARC Scientific Publication No. 15, IARC, Lyon, France, 1976.

115. INTERNATIONAL AGENCY FOR RESEARCH ON CANCER (IARC), *Nasopharyngeal Carcinoma: Etiology and Control*, (G. DE-THÉ AND Y. ITO, EDS.), IARC Scientific Publication No. 20, IARC, Lyon, France, 1978.

116. INTERNATIONAL AGENCY FOR RESEARCH ON CANCER (IARC), *Annual Report*, pp. 75–76, IARC Lyon, France, 1978.

117. INTERNATIONAL UNION AGAINST CANCER [UNION INTERNATIONAL CONTRE LE CANCER (UICC)], *Cancer of the Nasopharynx*, Monograph Series, Vol. 1. (C. S. MUIR AND K. SHANMUGARATNAM, EDS.), Munksgaard, Copenhagen, 1967.

118. JEANNEL, D., GHNASSIA, M., HUBERT, A., SANCHO-GARNIER, H., CROGNIER, E., AND DE-THÉ, G., Increased risk of undifferentiated nasopharyngeal carcinoma among males of French origin born in North Africa (Maghreb), *Int. J. Cancer* **54**:536–539 (1993).

119. JEANNEL, D., HUBERT, A., DE-VATHAIRE, F., ELLOUZ, R., CAMMOUN, M., BEN SALEM, M., SANCHO-GARNIER, H., AND DE-THÉ, G., Diet, living conditions and nasopharyngeal carcinoma in Tunisia. A case control study, *Int. J. Cancer* **46**:421–425 (1990).

120. JONDAL, M., AND KLEIN, G., Classification of lymphocytes in nasopharyngeal carcinoma (NPC) biopsies, *Biomedicine* **23**:163–165 (1975).

121. JUNG, P. F., AND YU, C., Nasopharyngeal cancer in China, *Postgrad. Med.* **33**:A77–A82 (1963).

122. KING, H., AND HAENSZEL, K., Cancer mortality among foreign and native-born Chinese in the United States, *J. Chron. Dis.* **26**:623–646 (1972).

123. KIRK, R. L., BLAKE, N. M., SERJEANTSON, S., SIMONS, M. J., AND CHAN, S. H., Genetic components in susceptibility to nasopharyngeal carcinoma, in: *Nasopharyngeal Carcinoma: Etiology and Control*, (G. DE-THÉ AND Y. ITO, EDS.), pp. 283–297, IARC Scientific Publication No. 20, IARC, Lyon, France, 1978.

124. KLEIN, G., EBV associated membrane antigens, in: *Oncogenesis and Herpesviruses* (P. M. BIGGS, G. DE-THÉ, AND L. N. PAYNE, EDS.), pp. 295–301, IARC Scientific Publication No. 2, IARC, Lyon, France, 1972.

125. KLEIN, G., The Epstein–Barr virus, in: *The Herpesviruses* (A. KAPLAN, ED.), pp. 521–555, Academic Press, New York, 1973.

126. KLEIN, G., *Viral Oncology*, Raven Press, New York, 1979.

127. KLEIN, G., CLIFFORD, P., KLEIN, E., AND STJERNSWÄRD, J., Search for tumor-specific immune reactions in Burkitt lymphoma patients by the membrane immunofluorescence reaction, *Proc. Natl. Acad., Sci. USA* **55**:1628–1635 (1966).

128. KLEIN, G., GIOVANELLA, B. C., LINDAHL, T., FIALKOW, P. J., SINGH, S., AND STEHLIN, J., Direct evidence for the presence of Epstein–Barr virus DNA and nuclear antigen in malignant epithelial cells from patients with anaplastic carcinoma of the nasopharynx, *Proc. Natl. Acad. Sci. USA* **71**:4737–4741 (1974).

129. KLEIN, G., PEARSON, G., HENLE, G., HENLE, W., DIEHL, V., AND NIEDERMAN, J. C., Relation between Epstein–Barr viral and cell membrane immunofluorescence in Burkitt tumor cells. I. Dependence of cell membrane immunfluorescence on presence of EB virus, *J. Exp. Med.* **128**:1011–1020 (1968).

130. KLEIN, G., AND VONKA, V., Relationship between the Epstein–Barr virus (EBV) determined, complement fixing antigen and the nuclear antigen (EBNA) detected by anti-complement fluorescence, *Int. J. Cancer* **53**:1645–1646 (1974).

131. KMET, J., AND MAHBOUBI, E., Esophageal cancer in the Gaspian littoral of Iran: Initial studies, *Science* **175**:846–853 (1972).

132. KROGMAN, W. M., The skeletal and dental pathology of an early Iranian site, *Bull. Hist. Med.* **8**:28–48 (1940).

133. LAUFS, R., Immunisation of marmoset monkeys with a killed oncogenic herpesvirus, *Nature* **249**:571–572 (1974).

134. LEMAIGRE, G., DIEBOLD, J., TEMMIM, L., ARSENIEV, L., LECHARPENTIER, Y., ALLOUACHE, A., DELAITRE, B., AND ABELANET, R., Carcinome du nasopharynx chez les sujets jeunes—Étude clinique, anatomique et ultrastructurale de 50 cas observés dans l'est algérien, *Nouv. Press Med.* **6**:3509–3513 (1977).

135. LEMON, S. M., HUTT, L. M., SHAW, J., LI, J. L., AND PAGANO, J., Replication of EBV in epithelial cells during infectious mononucleosis, *Nature* **268**:268–270 (1977).

136. LENOIR, G., MARTIN-BERTHELON, M. C., FAVRE, M. C., AND DE-THÉ, G., Characterization of EBV antigens. I. Biochemical analysis of the complement-fixing soluble antigen and relationship with EBNA, *J. Virol.* **17**:672–674 (1976).

137. LEVINE, P., CHO, B., CONNELLY, R., DE VITA, C., BERARD, C., O'CONOR, G., AND DORHMAN, R., The American Burkitt's lymphoma registry: A progress report, *Ann. Intern. Med.* **83**:82–83 (1975).

138. LEVINE, P. H., DE-THÉ, G., BRUGERE, J., SCHWAAB, G., MOURALI, N., HEBERMAN, R. B., AMBROSIONI, J. C., AND REVOL, P., Immunity to antigens associated with a cell line derived from nasopharyngeal carcinoma (NPC) in non-Chinese NPC patients, *Int. J. Cancer* **17**:155–160 (1976).

139. LEVINE, P. H., LUBIN, J. H., AND EVANS, A. S., An Epstein–Barr virus (EBV) vaccine: Evaluating the efficacy of an EBV vaccine: Clinical and epidemiologic considerations, in: *The Epstein–Barr virus and associated disease: Proceedings of the Vth International Symposium*, (T. TURSZ, J. S. PAGANO, D. V. ABLASHI, G. DE THÉ, G. LENOIR, AND G. R. PEARSON, EDS.), pp. 585–592, Colloques INSERM/John Libbey Eurotext, Paris, 1993.

140. LIANG, P. C., Studies on nasopharyngeal carcinoma in the Chinese: Statistical and laboratory investigation, *Chin. Med. J.* **83**:373–390 (1964).

141. LILLY, F., The inheritance of susceptibility of the Gross leukemia virus in mice, *Genetics* **53**:529–539 (1966).

142. LILLY, F., Mouse leukemia: A model of a multiple-gene disease, *J. Natl. Cancer Inst.* **49**:927–934 (1972).

143. LIN, T. M., CHEN, K. P., LIN, C. C., HSU, M. M., TU, S. M., CHANG, T. C., JUNG, P. F., AND HIRAYAMA, T., Retrospective study on nasopharyngeal carcinoma, *J. Natl. Cancer Inst.* **51**:1403–1408 (1975).

144. LIN, T. M., HSU, M. M., CHENG, K. P., CHIANG, T. C., JUNG, P. F., AND HIRAYAMA, T., Morbidity and mortality of cancer of the nasopharynx in Taiwan, *Gann Monogr.* **10**:137–144 (1972).

145. LINSELL, C. A., Cancer in Kenya, in: *Cancer in Africa* (P. CLIFFORD, C. A. LINSELL, AND G. L. TIMMS, EDS.), pp. 7–12, East African Publishing House, Nairobi, 1968.

146. LU, S. J., DAY, N. E., DEGOS, I., LEPAGE, V., WANG, P. H., CHAN, S., SIMONS, H., MACKNIGHT, B., EASTON, D., ZENG, Y., DE-THÉ, G., The genetic basis for carcinoma of the nasopharynx (NPC):

Evidence to linkage to the HLA region, *Nature* **346:**470–471 (1990).

147. MATHEW, G. D., QUALTIERE, L. F., NEEL, H. B., III, AND PEARSON, G. R., IgA antibody, antibody-dependent cellular cytotoxicity and prognosis in patients with nasopharyngeal carcinoma, *Int. J. Cancer* **27:**175–180 (1981).

148. MCDEVITT, H. O., AND BODMER, W. J., HL-A, immune response genes, and disease: Occasional survey, *Lancet* **1:**1269–1275 (1974).

149. MICHAUX, L., Carcinome de base du cráne, cited by Godtfredson, E., Ophthalmologic and neurologic symptoms of malignant nasopharyngeal tumours: Clinical study comprising 454 cases, with special reference to histo-pathology and possibility of earlier recognition, *Acta Psychiatr. Scand. Suppl.* **34:**1–323 (1944).

150. MILLER, G., NIEDERMAN, J. C., AND STILL, D. A., Infectious mononucleosis: Appearance of neutralizing antibody to Epstein–Barr virus measured by inhibition of formation of lymphoblastoid cell lines, *J. Infect. Dis.* **125:**403–406 (1972).

151. MORGAN, A. J., Epstein–Barr virus vaccines, in: *Vaccine* **10**(9): 563–571 (1992).

152. MORGAN, A. J., Recent progress in EB vaccine development, in: *Vth International Symposium on Epstein–Barr Virus and Associated Diseases*, (EBV and Human Diseases–1992) (T. TURSZ, J. PAGANO, G. DE-THÉ, M. ABLASHI, G. LENOIR, AND G. PEARSON, EDS.), pp. 231–235, INSERM and John Libbey Eurotext, Paris, 1993.

153. MUIR, C. S., Nasopharyngeal carcinoma in non-Chinese populations with special reference to South East Asia and Africa, *Int. J. Cancer* **8:**351–363 (1971).

154. MUIR, C. S., Nasopharyngeal carcinoma in non-Chinese populations, in: *Oncogenesis and Herpesviruses* (P. M. BIGGS, G. DE-THÉ, AND L. N. PAYNE, EDS.), pp. 367–371, IARC Scientific Publication No. 2, IARC, Lyon, France, 1972.

155. MUIR, C. S., Geographical differences in cancer patterns, in: *Host Environmental Interactions in the Etiology of Cancer in Men* (R. DOLL AND I. VODOPIJA, EDS.), pp. 1–13, IARC Scientific Publication No. 7, IARC, Lyon, France, 1973.

156. MUIR, C. S., AND OAKLEY, W. F., Nasopharyngeal carcinoma in Sarawak (Borneo), *J. Laryngol.* **81:**197–207 (1967).

157. MUIR, C. S., AND SHANMUGARATNAM, K., The incidence of nasopharyngeal cancer in Singapore, in: *Cancer of the Nasopharynx* (C. S. MUIR AND K. SHANMUGARATNAM, EDS.), pp. 47–53, UICC Monograph Series, No. 1, Munksgaard, Copenhagen, 1967.

158. NEVO, S., MEYER, W., AND ALTMAN, M., Carcinoma of nasopharynx in twins, *Cancer* **28:**807–809 (1971).

159. NG, M. H., AND SHAM, J. S. T., Development of nasopharyngeal carcinoma: A working hypothesis. EBV and Human Diseases. Proceeding of the 1992 EBV meeting, in: *The Epstein–Barr virus and associated diseases: Proceedings of the Vth International Symposium* (T. TURSZ, J. PAGANO, G. DE-THÉ, M. ABLASHI, G. LENOIR, AND G. PEARSON, EDS.), pp. 497–506, INSERM and John Libbey Eurotext, Paris, 1993.

160. NIEDERMAN, J. C., AND EVANS, A. S., Infectious mononucleosis, in: *Serological epidemiology*, (A. S. EVANS, ED.), pp. 119–132, Academic Press, New York, 1973.

161. NIEDOBITEK, G., YOUNG, L. S., SAM, C. K., BROOKS, L., PRASAD, U., AND RICKINSON, A. B., Expression of Epstein–Barr virus genes and of lymphocyte activation molecules in undifferentiated nasopharyngeal carcinomas. *Am. J. Pathol.* **140:**879–887 (1992).

162. NIELSEN, N. H., MIKKELSEN, F., AND HART-HANSEN, J. P., Nasopharyngeal cancer in Greenland: Incidence in an arctic Eskimo population, *Acta Pathol. Macrobiol. Scand. A* **85:**850–858 (1977).

163. NILSSON, K., KLEIN, G., HENLE, G., AND HENLE, W., The role of EBV in the establishment of lymphoblastoid cell lines from adult and foetal lymphoid tissue, in: *Oncogenesis and Herpesviruses* (P. M. BIGGS, G. DE-THÉ, AND L. N. PAYNE, EDS.), pp. 285–290, IARC Scientific Publication No. 2, IARC, Lyon, France, 1972.

164. NOMOYAMA, M., AND PAGANO, J. S., Homology between Epstein–Barr viruses DNA and viral DNA from Burkitt's lymphoma and nasopharyngeal carcinoma determined by DNA–DNA reassociation kinetics, *Nature* **242:**44–47 (1973).

165. O'CONOR, G. T., Persistent immunological stimulation as a factor in oncogenesis with special reference to Burkitt's tumor, *Am. J. Med.* **48:**279–285 (1970).

166. OLD, L. J., BOYSE, E. A., OETTIGEN, H. F., OLD, L. J., BOYSE, E. A., OETTGEN, H. F., DE HARVEN, E., GEERING, C., WILLIAMSON, B., AND CLIFFORD, L., Precipitating antibody in human serum to an antigen present in cultured Burkitt's lymphoma cells, *Proc. Natl. Acad. Sci. USA* **56:**1699–1704 (1966).

167. PAGANO, J. S., HUANG, C. H., KLEIN, G., DE-THÉ, G., SHANMUGARATNAM, K., SIMONS, M. J., AND YAN, C. S., Homology of Epstein–Barr viral DNA in nasopharyngeal carcinoma from Kenya, Taiwan, Singapore and Tunis, in: *Oncogenesis and Herpesviruses II* (G. DE-THÉ, M. A. EPSTEIN, AND H. ZUR HAUSEN, EDS.), pp. 191–193, IARC Scientific Publication No. 11, Vol. 2, IARC, Lyon, France, 1975.

168. PAGANO, J. S., HUANG, C. H., AND LEVINE, P., Absence of Epstein–Barr viral DNA in American Burkitt's lymphoma, *N. Engl. J. Med.* **289:**1395–1399 (1973).

169. PEARSON, G. R., COATES, H. L., NEEDL, H. B., LEVINE, P., ABLASHI, D., AND EASTON, J., Clinical evaluation of EBV serology in American patients with nasopharyngeal carcinoma, in: *Nasopharyngeal Carcinoma: Etiology and Control*, (G. DE-THÉ AND Y. ITO, EDS.), pp. 439–448, IARC Scientific Publication No. 20, IARC, Lyon, France, 1978.

170. PEARSON, G. R., AND ORR, T. W., Antibody-dependent lymphocyte cytotoxicity against cells expressing Epstein–Barr virus antigens, *J. Natl. Cancer Inst.* **56:**485–488 (1976).

171. PEREIRA, M. S., FILLD, A. M., BLAKE, J. M., RODGERS, F. G., AND BAILEY, L. A., Evidence for oral excretion of EB virus in infectious mononucleosis, *Lancet* **1:**710–711 (1972).

172. PETO, R., Epidemiology, multistage models, and short-term mutagenicity tests, in: *Origins of Human Cancer* (H. HIATT, J. D. WATSON, AND J. A. WINSTEN, EDS.), *Cold Spring Harbor Conferences on Cell Proliferation*, Vol. 4, Book C, pp. 1403–1428, Cold Spring Harbor Laboratory, Cold Spring Harbor, NY, 1977.

173. POIRIER, S., BOUVIER, G., MALAVEILLE, C., OHSHIMA, H., SHAO, Y. M., HUBERT, A., ZENG, Y., DE-THÉ, G., AND BARTSCH, H., Volatile nitrosamine levels and genotoxicity in food samples from high risk areas for nasopharyngeal carcinoma (NPC) before and after nitrosation, *Int. J. Cancer* **44:**1088–1094 (1989).

174. POIRIER, S., OHSHIMA, H., DE-THÉ, G., HUBERT, A., BOURGADE, M. C., AND BARTSCH, H., Volatile nitrosamine levels in common foods from Tunisia, South China and Greenland. High-risk areas for nasopharyngeal carcinoma (NPC), *Int. J. Cancer* **39:**293–296 (1987).

175. POLUNIN, I., The ways of life of people with high rates of nasopharyngeal carcinoma, in: *Cancer of the Nasopharynx* (C. S. MUIR AND K. SHANMUGARATNAM, EDS.), pp. 106–111, UICC Monograph Series, Vol. 1, Munksgaard, Copenhagen, 1967.

176. REEDMAN, B. M., AND KLEIN, G., Cellular localization of an

Epstein–Barr virus (EBV) associated complement-fixing antigen in producer and nonproducer lymphoblastoid cell lines, *Int. J. Cancer* **11**:499–520 (1973).

177. REEDMAN, B. M., KLEIN, G., POPE, J. H., WALTERS, M. K., HILGERS, J., SMITH, S., AND JOHANSSON, B., Epstein–Barr virus associated complement fixing and nuclear antigen in Burkitt's lymphoma biopsies, *Int. J. Cancer* **13**:755–763 (1974).

178. REEDMAN, B. M., POPE, J. H., AND MOSS, D. J., Identify of the soluble EBV associated antigens of human lymphoid cell lines, *Int. J. Cancer* **9**:172–181 (1972).

179. SAAD, A., Observations on nasopharyngeal carcinoma in the Sudan, in: *Cancer in Africa* (C. A. LINSELL AND G. L. TIMMS, EDS.), pp. 281–285, East African Publishing House, Nairobi, 1968.

180. SCHMAUZ, R., AND TEMPLETON, A. C., Nasopharyngeal cancer in Uganda, *Cancer* **29**:610–621 (1972).

181. SCOTT, G. C., AND ATKINSON, L., Demographic features of the Chinese population in Australia and the relative prevalence of nasopharyngeal cancer among Caucasians and Chinese, in: *Cancer of the Nasopharynx* (C. S. MUIR AND K. SHANMUGARATNAM, EDS.), pp. 64–72, UICC Monograph Series, Vol. 1, Munksgaard, Copenhagen, 1967.

182. SHANMUGARATNAM, K., Nasopharyngeal carcinoma in Asia, in: *Racial and Geographical Factors in Tumour Incidence* (A. A. SHIVAS, ED.), pp. 169–188, Pfizer Medical Monograph, Vol. 2, Edinburgh University Press, Edinburgh, 1967.

183. SHANMUGARATNAM, K., Studies on the etiology of nasopharyngeal carcinoma, *Int. Rev. Exp. Pathol.* **10**:361–413 (1971).

184. SHANMUGARATNAM, K., Cancer in Singapore—Ethnic and dialect group variations in cancer incidence, *Singapore Med. J.* **14**:68–81 (1973).

185. SHANMUGARATNAM, K., AND HIGGINSON, J., Aetiology of nasopharyngeal carcinoma: Report on a retrospective survey in Singapore, in: *Cancer of the Nasopharynx* (C. S. MUIR AND K. SHANMUGARATNAM, EDS.), pp. 130–137, UICC Monograph Series, Vol. 1, Munksgaard, Copenhagen, 1967.

186. SHANMUGARATNAM, K., AND MUIR, C. S., Nasopharyngeal carcinoma: Origin and structure, in: *Cancer of the Nasopharynx* (C. S. MUIR AND K. SHANMUGARATNAM, EDS.), pp. 153–162, UICC Monograph Series, Vol. 1, Munksgaard, Copenhagen, 1967.

187. SHANMUGARATNAM, K., AND TYE, C. Y., A study of nasopharyngeal cancer among Singapore Chinese with special reference to migrant status and specific community (dialect group), *J. Chron. Dis.* **23**:433–441 (1970).

188. SHANMUGARATNAM, K., TYE, C. Y., GOH, E. H., AND CHIA, K. B., Etiological factors in nasopharyngeal carcinoma: A hospital-based, retrospective case-control questionnaire study, in: *Nasopharyngeal Carcinoma: Etiology and Control*, (G. DE-THÉ AND Y. ITO, EDS.), pp. 199–212, IARC Scientific Publication No. 20, IARC, Lyon, France, 1978.

189. SHANMUGARATNAM, K., AND WEE, A., "Dialect group" variations in cancer incidence among Chinese in Singapore, in: *Host Environment Interaction in the Etiology of Cancer in Humans* (R. DOLL AND I. BODOPIJA, EDS.), pp. 67–82, IARC Scientific Publication No. 7, IARC, Lyon, France, 1973.

190. SHAO, Y. M., POIRIER, S., OHSHIMA, H., MALAVEILLE, C., ZENG, Y., DE-THÉ, G., AND BARTSCH, H., Epstein–Barr virus activation in Raji cells by extracts of preserved food from high risk areas for nasopharyngeal carcinoma, *Carcinogenesis* **9**:1455–1457 (1988).

191. SIMONS, M. J., CHAN, S. H., WEE, G. B., SHANMUGARATNAM, K.,

GOH, E. H., HO, J. L. C., CHAU, J. C. W., DARMALINGAM, S., PRASAD, U., BETUEL, H., DAY, N. E., AND DE-THÉ, G., Nasopharyngeal carcinoma and histocompatibility antigens, in: *Nasopharyngeal Carcinoma: Etiology and Control*, (G. DE-THÉ AND Y. ITO, EDS.), pp. 271–282, IARC Scientific Publication No. 20, IARC, Lyon, France, 1978.

192. SIMONS, M. J., DAY, N. E., WEE, G. B., CHAN, S. H., SHANMUGARATNAM, K., AND DE-THÉ, G., Immunogenetic aspects of nasopharyngeal carcinoma (NPC). III. HL-A type as a genetic marker of NPC predisposition to test the hypothesis that EBV is an aetiologic factor in NPC, in: *Oncogenesis and Herpesviruses II* (G. DE-THÉ, M. A. EPSTEIN, AND H. ZUR HAUSEN, EDS.), pp. 249–258, IARC Scientific Publication No. 11, Vol. 2, IARC, Lyon, France, 1975.

193. SIMONS, M. J., DAY, N. E., WEE, G. B., SHANMUGARATNAM, K., HO, H. C., WONG, S. H., TI, T. K., YONG, N. K., DARMALINGAM, S., AND DE-THÉ, G., Nasopharyngeal carcinoma. V. Immunogenetic studies of South East Asian ethnic groups with high and low risk for the tumor, *Cancer Res.* **34**:1192–1195 (1974).

194. SIMONS, M. J., WEE, G. B., DAY, N. E., CHAN, S. H., SHANMUGARATNAM, K., AND DE-THÉ, G., Immunogenetic aspects of nasopharyngeal carcinoma (NPC). IV. Probable identification of an HL-A second antigen associated with a high risk for NPC, *Lancet* **1**:142–143 (1975).

195. SIMONS, M. J., WEE, G. B., DAY, N. E., DE-THÉ, G., MORRIS, P. J., AND SHANMUGARATNAM, K., Immunogenetic aspects of nasopharyngeal carcinoma. I. Differences in HL-A antigen profiles between patients and comparison groups, *Int. J. Cancer* **13**:122–134 (1974).

196. SIMONS, M. J., WEE, G. B., GOH, E. H., CHAN, S. H., SHANMUGARATNAM, K., DAY, N. E., AND DE-THÉ, G., Immunogenetic aspects of nasopharyngeal carcinoma, IV. Increased risk in Chinese of nasopharyngeal carcinoma associated with a Chinese-related HLA profile (A2, Singapore 2), *J. Natl. Cancer Inst.* **57**:977–980 (1976).

197. SIXBEY, J. W., AND YAO, Q. Y., Immunoglobulin A induced shift of Epstein–Barr virus tropism, *Science* **255**:1578–1580 (1991).

198. SMITH, G. E., AND DAWSON, W. R., *Egyptian Mummies*, p. 157, Allen and Unwin, London, 1924.

199. SOHIER, R., AND DE-THÉ, G., Fixation du complément avec un antigène soluble: Differences d'activité importantes entre les sérums de lymphoma de Burkitt, de cancer du rhinopharynx et de mononucléose infecteuse, *C. R. Acad. Sci. [D]* **273**:121–124 (1971).

200. SOHIER, R., AND DE-THÉ, G., Evolution of complement-fixing antibody titers with the development of Burkitt's lymphoma, *Int. J. Cancer* **9**:524–528 (1972).

201. SVOBODA, D. J., KIRCHNER, K. R., AND SHANMUGARATNAM, K., Ultrastructure of nasopharygeal carcinomas in American and Chinese patients: An application of electron microscopy to geographic pathology, *Exp. Mol. Pathol.* **4**:189–204 (1965).

202. SVOBODA, D. J., KIRCHNER, K. R., AND SHANMUGARATNAM, K., The fine structure of nasopharyngeal carcinoma, in: *Cancer of the Nasopharynx* (C. S. MUIR AND K. SHANMUGARATNAM, EDS.), pp. 163–171, UICC Monograph Series, Vol. 1, Munksgaard, Copenhagen, 1967.

203. TERRACINI, B., MAGEE, P. N., AND BARNES, J. M., Hepatic pathology on low dietary levels of dimethylnitrosamine, *Br. J. Cancer* **21**:565–599 (1967).

204. TRUMPER, P. A., EPSTEIN, M. A., GIOVANELLA, G. C., AND FINERTY, S., Isolation of infectious EB virus from epithelial tumor

cells of nasopharyngeal carcinoma, *Int. J. Cancer* **20**:655–662 (1977).

205. Union International Contre Le Cancer, *Cancer Incidence in Five Continents—Technical Report* (R. Doll, P. Payne, and J. Waterhouse, eds.), UICC, Geneva, 1966.

206. Union International Contre Le Cancer, *Cancer Incidence in Five Continents*, Vol. 2 (R. Doll, C. Muir, and J. Waterhouse, eds.), UICC, Geneva, 1970.

207. Vonka, V., Benyesh-Melnick, M., Lewis, R. T., and Wimberly, I., Some properties of the soluble (S) antigen of cultured lymphoblastoid cell lines, *Arch. Ges. Virusforsch.* **31**:113–124 (1970).

208. Vonka, V., Vlchova, I., Zavadova, H., Kouba, K., Lasovska, J., and Duben, J., Antibodies to EB virus capsid antigen and to soluble antigen of lymphoblastoid cell in infectious mononucleosis patients, *Int. J. Cancer* **9**:529–535 (1972).

209. Vuillaume, M., and de-Thé, G., Nasopharyngeal carcinoma. II. Ultrastructure of different growths leading to lymphoblastoid transformation *in vitro*, *J. Natl. Cancer Inst.* **51**:67–80 (1973).

210. Walshe, R. J., The physical anthropology of races with a high incidence of nasopharyngeal cancer, in: *Cancer of the Nasopharynx* (C. S. Muir and K. Shanmugaratnam, eds.), pp. 112–118, UICC Monograph Series, Vol. 1, Munksgaard, Copenhagen, 1967.

211. Walters, M. K., and Pope, J. H., Studies of the EB virus-related antigens of human leukocyte cell lines, *Int. J. Cancer* **8**:32–40 (1971).

212. Wara, W. M., Wara, D. W., Phillips, T. L., and Ammahh, A., Elevated IgA in carcinoma of the nasopharynx, *Cancer* **35**:1313–1315 (1975).

213. Wells, C., Ancient Egyptian pathology, *J. Laryngol.* **77**:261–265 (1963).

214. Wells, C., Two mediaeval cases of malignant disease, *Br. Med. J.* **1**:1611–1612 (1964).

215. Williams, E. H., and de-Thé, G., Familial aggregation in nasopharyngeal carcinoma (letter to the editor), *Lancet* **2**:295 (1974).

216. Wolf, H., zur Hausen, H., and Becker, V., EB viral genomes in epithelial nasopharyngeal carcinoma cells, *Nature (New Biol.)* **244**:245–257 (1973).

217. World Health Organization, *Mortality from Malignant Neoplasms 1955–1965*, WHO, Geneva, 1970.

218. World Health Organization, *Histological Typing of Upper Respiratory Tract Tumours: International Classification of Tumours*, No. 19 (K. Shanmugaratnam and L. H. Sobin, eds.), WHO, Geneva, 1978.

219. Worth, R. M., and Valentine, R., Nasopharyngeal carcinoma in New South Wales, Australia, in: *Cancer of the Nasopharynx* (C. S. Muir and K. Shanmugaratnam, eds.), pp. 73–76, UICC Monograph Series, Vol. 1, Munksgaard, Copenhagen, 1967.

220. Wu, C. H. (ed.), *The Encyclopaedia of Chinese Medical Terms*, Vol. 1, p. 756, Commercial Press, Shanghai, 1921 (in Chinese).

221. Wu, T. C., Mann, R. B., Epstein, J. I., MacMahon, E., Lee, W. A., Charache, P., Hayward, S. D., Kurman, R. J., Hayward, G. S., and Ambinder, R. F., Abundant expression of EBER-1 small nuclear RNA in nasopharyngeal carcinoma. A morphological distinctive target for detection of EBV in formalin fixed paraffin embedded specimen, *Am. J. Pathol.* **138**:1461–1469 (1991).

222. Yata, J., Desgranges, C., de-Thé, G., and Tachibana, T., Lymphocytes in infectious mononucleosis: Properties of the lymphoblastoid lines, *Biomedicine* **19**:479–484 (1973).

223. Yata, J., Desgranges, C., de-Thé, G., and Tachibana, T., Nasopharyngeal carcinoma. VII. Lymphocyte subpopulations in the blood and tumour tissue, *Biomedicine* **21**:244–250 (1974).

224. Yata, J., Desgranges, C., Nakagawa, T., Favri, M. C., and de-Thé, G., Lymphoblastoid transformation and kinetics in the appearance of Epstein–Barr viral nuclear antigen (EBNA) in cord blood B cells by Epstein–Barr virus infection, *Int. J. Cancer* **15**:377–384 (1975).

225. Yeh, S., Histology of nasopharyngeal cancer, in: *Cancer of the Nasopharynx* (C. S. Muir and K. Shanmugaratnam, eds.), pp. 147–152, UICC Monograph Series, Vol. 1, Munksgaard, Copenhagen, 1967.

226. Young, D., Sixbey, J. W., Clarck, D., and Rickinson, A. B., Epstein–Barr virus receptors on human pharyngeal epithelia, *Lancet* **1**:240–242 (1986).

227. Yu, M. C., Ho, J. H. C., Lai, S. H., and Henderson, B. E., Cantonese style salted fish as a cause of nasopharyngeal carcinoma: Report of a case-control study in Hong Kong, *Cancer Res.* **46**:956–961 (1986).

228. Yun, I. S., A statistical study of tumors among Koreans, *Cancer Res.* **9**:370–371 (1949).

229. Zaouche, A., *Les tumeurs malignes de la sphère ORL en Tunisie: A propos des 644 tumeurs des voies aerodigestives superieures à l'Institut National de Carcinologie de Tunis due 1.10.67 au 15.8.69*, Thesis, Paris, 1970.

230. Zeng, Y., Seroepidemiological studies of nasopharyngeal carcinoma in China, *Adv. Cancer Res.* **44**:121–138 (1985).

231. Zeng, Y., Hong, D., Jianming, Z., Naiqin, H., Pingjun, L., Wenjun, P., Yuying, H., Yue, L., Peizhong, W., and de Thé, G., A 10-year prospective study on nasopharyngeal carcinoma in Wuzhou city and Zangwu County, Guangxi, China, in: *The Epstein–Barr virus and associated diseases: Proceedings of the Vth International Symposium* (T. Tursz, J. S. Pagano, D. V. Ablashi, G. de-Thé, G. Lenoir, and G. R. Pearson, eds.), pp. 735–742, Colloques INSERM/John Libbey Eurotext Publishing, Paris, 1993.

232. Zeng, Y., Gong, C. H., Jan, M. G., Fun, Z., Zhang, L. G., and Li, H. Y., Detection of Epstein–Barr virus IgA/EA antibody for diagnosis of nasopharyngeal carcinoma by immunoautoradiography, *Int. J. Cancer* **31**:599–601 (1983).

233. Zeng, Y., Liu, Y. X., Liu, C. R., Chen, S. W., Wei, J. N., Zhu, J. N., and Zai, H. J., Application of immunoenzymatic method and immunoautoradiographic method for the mass survey of nasopharyngeal carcinoma, *Intervirology* **133**:166–168 (1980).

234. Zeng, Y., Oshima, H., Bouvier, G., Roy, P., Jianming, Z., Li, B., Brouet, I., de-Thé, G., and Bartsch, H., Urinary excretion of nitrosamino acids and nitrate by inhabitants of high- and low-risk areas for nasopharyngeal carcinoma in Southern China, *Cancer Epidemiol. Biomarkers Prev.* **2**:195–200 (1993).

235. Zeng, Y., Shang, M., Liu, C. R., Chen, S. W., Wei, J. N., Zhu, J. S., and Zai, H. G., Detection of anti-Epstein–Barr virus IgA in NPC patients in 8 provinces and cities in China, *Chin. Oncol.* **1**:2–7 (1979).

236. Zeng, Y., Shen, S., Gudhua, P., Ma, J. L., Zhang, Q., Zhao, M. G., and Dong, H. J., Application of anticomplement immunoenzymatic method for the detection of EBNA in carcinoma cells and normal epithelial cells from the nasopharynx, in: *Nasopharyngeal Carcinoma Cancer Campaign*, Vol. 5 (E. Grundmann, G. R. F. Krueger, and D. V. Ablashi, eds.), pp. 237–245, Gustav Fischer Verlag, Stuttgart, 1981.

237. Zeng, Y., Zhang, L. G., Li, H. Y., Jan, M. G., Zhang, Q., Wu, Y.

C., WANG, Y. S., AND SU, G. R., Serological mass survey for early detection of nasopharyngeal carcinoma in Wu-Zhou City, China, *Int. J. Cancer* **29:**139–141 (1982).

238. ZENG, Y., ZHONG, J. M., LI, L. Y., WANG, P. Z., TANG, H., MA, Y. R., ZHU, J. S., PAN, W. J., LIO, Y. X., WEI, Z. N., CHEN, J. Y., MO, Y. K., LI, E. J., AND TAN, B. F., Follow-up studies on Epstein–Barr virus IgA/VCA antibody positive persons in Zangwu County, China, *Intervirology* **20:**190–194 (1983).

239. ZUR HAUSEN, H., SCHULTE-HOLTHAUSEN, H., KLEIN, G., HENLE, W., HENLE, G., CLIFFORD, P., AND SANTESSON, L., EBV DNA in biopsies of Burkitt tumours and anaplastic carcinomas of the nasopharynx, *Nature* **228:**1056–1058 (1970).

12. Suggested Reading

DE-THÉ, G., Epidemiology, pathogenesis and prevention of EBV associated malignancies, in: *The Epstein–Barr Virus and Associated Diseases.* pp. 15–30, (T. TURSZ, J. PAGANO, M. ABLASHI, G. DE-THÉ, G. LENOIR, AND G. PEARSON, EDS.), INSERM and John Libbey Eurotext Publishing, Paris, 1993.

DE-THÉ, G., Epidemiology of Epstein–Barr virus and associated diseases in man, in: *Herpesviruses*, Vol. 1 (B. ROIZMAN, ED.), pp. 25–103, Plenum Press, New York, 1982.

DE-THÉ, G., AND YTO, Y. (EDS.), *Nasopharyngeal Carcinoma: Etiology and Control*, IARC Scientific Publication No. 20, IARC, Lyon, France, 1978.

GZUNDMAN, E., KZUEGER, G., ZUNDMAN, E., KZUEGER, G. R. F., AND ABLASHI, D. V. (EDS.), *Nasopharyngeal Carcinoma Cancer Campaign*, Vol. 5, Gustav Fischer Verlag, Stuttgart, 1981.

LEVINE, P., ABLASHI, D., AND PEARSON, G. (EDS.), *Epstein–Barr Virus and Human Diseases*, Humana Press, Clifton, NJ, 1987.

MUIR, C. S., Nasopharyngeal carcinoma in non-Chinese populations with special reference to South East Asia and Africa, *Int. J. Cancer* **8:**351–363 (1971).

MUIR, C. S., AND SHANMUGARATNAM, K. (EDS.), *Cancer of the Nasopharynx*, UICC Monograph Series, No. 1, Munksgaard, Copenhagen, 1967.

CHAPTER 32

Hepatocellular Carcinoma Caused by Hepatitis B Virus

Joseph L. Melnick

1. Introduction

Although hepatocellular carcinoma (HCC) is rare in Europe and North America, it is very common in many areas of the world, especially in Africa and southeast Asia, where it is among the ten most common human cancers. It is estimated that there are more than 500,000 new patients with liver cancer each year. Over 30 new cases of hepatoma occur per 100,000 population annually in the high-prevalence areas, but the yearly incidence is less than 5 per 100,000 in most countries of North America, Europe, and Australia. Hepatocellular carcinoma is more common in males than among females. Its prevalence increases with age, although in high-risk populations, the disease also occurs in younger age groups.

In China, HCC ranks third among cancer deaths, with 112,000 HCC deaths per year, representing 40% of the world's HCC deaths. The peak age is in adults aged 40 to 60 years. The hepatitis B virus (HBV) carrier rate in China is 10%. The enormity of this burden is indicated by the 120 million HBV carriers in the population. Each year, 3 million new virus carriers are added to the Chinese population; of these, 300,000 will die from liver diseases, and HCC is the cause of 112,000 of these deaths.[35,80,87]

In some areas, dietary aflatoxin is a risk factor for HCC in certain areas,[83,91] but the evidence is universal for a consistent and specific causal association between HBV and HCC; up to 80% of such cancers worldwide are believed to be attributable to this virus.[8,24,43,46,55,58,76,87] Hepatitis B is widespread in areas of the world where macronodular cirrhosis and HCC are common and where the development of the carrier state occurs most frequently. A prospective study has clearly shown that HBV carriers have an over 100-fold excess risk of developing HCC relative to noncarriers, and the lifetime risk of HCC in a chronic HBV carrier approaches 50% in males.[3,4]

Further evidence linking HBV and HCC comes from the discovery of similar viruses in animals—the woodchuck (*Marmota monax*), Beechey ground squirrel (*Spermophilus beecheyi*), and Pekin duck (*Anas domesticus*)—which share many biophysical and biochemical properties with HBV. Like humans with HBV, many of these species are carriers of their unique hepadnaviruses, and HCC occurs in many of these chronic carriers.[36,39,54,72]

The association between HBV and HCC is further supported by the detection of the integrated HBV genome in HCC from patients in many parts of the world.[21,59,66,94] Although it is clear that the hepadnaviruses cause HCC, the mechanism of oncogenesis is not clear.

In some parts of the world, notably Japan, hepatitis C virus (HCV), is also associated epidemiologically in patients with liver cancer who have no evidence of prior infection with HBV.[50] However, in Senegal[14] and other areas, including the United States,[50] HCV does not seem to be related to cases of liver cancer. Thus, even though the prevalence of HCV in the populations of Japan and the United States is similar, only in Japan is HCV associated with the liver cancer. In other countries, notably Greece, when recombinant and synthetic peptides derived from structural and nonstructural regions of the HCV genome have been used in the serological diagnosis of HCV infection, the relative risk relating anti-HCV to HCC is elevated, but it is much less than that linking HBV to

Joseph L. Melnick • Division of Molecular Virology, Baylor College of Medicine, Houston, Texas 77030.

HCC.[93] The reasons for the variations in different populations are unclear, but the heterogeneity of HCV has been suggested to be an important factor.[79]

Recent studies have used polymerase chain reaction (PCR) amplification of cDNA (obtained by reverse transcription of HCV RNA) to determine HCV viremia, and thus support the opinion that HCV infection may also play a role in HCC in some countries. In one such study involving Africans with HCC,[10] the results indicated that a small but significant percentage had a current HCV infection (positive viremia). The mechanism leading to HCC is not clear, but the patients positive for HCV RNA were older (mean, 52 years) than the negative patients (mean, 40 years), suggesting that the mechanism for HCC caused by HCV differs from that of HBV. A similar age difference had been noted previously for HCC patients in Taiwan, where patients positive for HCV antibodies but negative for hepatitis B surface antigen (HBsAg) had a higher onset age (mean 60 years) than those negative for anti-HCV but positive for HBsAg (mean, 49 years).[92]

2. Historical Background

The relationship between HBV and HCC was first noted in the early 1970s when it was found that HBV antigenemia corresponded geographically with HCC. The prevalance of antigenemia was found to be higher in cases of HCC than in controls in both high- and low-incidence areas for HCC. These epidemiologic relationships were well summarized by Szmuness[76] in 1978 when he presented evidence for the causal relationship between the two. These findings were further supported in 1981 by the prospective study of Beasley *et al.*[4] in which they showed the risk of HCC over a 5-year period in persons with HBV antigenemia to be 217 times that in the absence of antigenemia. These epidemiologic observations were strengthened by laboratory studies[23] demonstrating in 1979 HBV DNA sequences integrated into the chromosomal DNA of hepatoma cells and by further data in the early 1980s.[9,12,25,66] Finally, although the tumor could not be reproduced in experimental animals with human HBV, natural infections of several animals with their own hepatitis viruses were observed to lead to tumor development. This was first shown in woodchucks in 1978[73] and then in 1980 in ground squirrels in California.[38] Infection of Pekin ducks in the United States without tumor was also reported in 1980.[41] These viruses resemble human HBV and, together with it, make up the hepadnavirus group.

3. Methodology

3.1. Mortality and Morbidity Data

In the United States there are no uniform data on incidence. Data from several state and local registries are included in one publication on *Cancer Incidence in Five Continents*.[85] Uniform data derived from the Third National Cancer Survey are available from seven metropolitan areas and two states for 1969–1971.[15] More recently, cancer data have been collected from cancer registries in a program called Surveillance, Epidemiology, and End Results (SEER); periodic monographs based on these data are published by the National Cancer Institute. Time trends suggesting increasing incidence, as reported in selected cancer registries, were published in 1980.[63] On a global scale, cancer incidence data derived from over 75 registries in five continents have been published by the International Agency for Cancer Research (IARC)[85] and further refined in a publication of the International Union against Cancer.[53] Unfortunately, data are least available in areas of the world where HCC has its highest incidence.

3.2. Serological Surveys

Surveys for HBV antigenemia and for antibodies to HBV have been carried out widely in both the developed and developing world. The methods involved in such surveys are discussed in Chapter 13, Viral Hepatitis.

3.3. Laboratory Methods

The definitive diagnosis of hepatocellular carcinoma is based on histological examination of liver tissue obtained by biopsy or at autopsy. The diagnoses are usually made late in the course of the disease. Earlier diagnoses can be made if α-fetoprotein (AFP) can be demonstrated in the serum in concentrations over 500 ng/ml and if other causes of such elevations can be excluded. This test has been a useful diagnostic and screening tool in China.[68]

4. Biological Characteristics of the Agent

4.1. Molecular Biology of HBV in HCC

Various cellular changes have been found in selected cases of HCC, which include (1) integration of HBV genome fragments near specific regulatory genes of cell growth; (2) amplification of specific growth regulatory genes; (3) specific rearrangement or activation of on-cogenes within the HBV-infected hepatocyte; and (4) loss of heterozygosity at one or more chromosomes. More than one change seems necessary for HCC to develop, and the involvement of each step may not be the same after HBV infection of different populations

That HBV is causally related to the development of HCC is supported by the presence of HBV DNA sequences integrated into the chromosomal DNA of hepatoma cells from HBsAg-positive patients.[9,12,25,66] The structure of those viral sequences has been studied by Southern blotting. The number of integration sites varies among tumors, ranging from very few to as many as 12. Hepatocellular carcinoma tissues usually contain only integrated HBV sequences. Fine-structure analysis of integrated HBV DNA sequences in HCC samples[17,61,89] and in established cell lines, notably PLC/PRF/5 cells,[17,31,96] has been accomplished by the use of cloned DNA of the whole virus or of different regions of its genome. The integrated HBV DNA was found to exist primarily as subgenomic fragments. Marked rearrangements of the viral genomic sequences were frequently observed in primary tumors as well as cell lines. The 11-base-pair direct repeat sequence located at both sides of the cohesive ends of the viral genome appeared to be a preferred region for virus-specific integration.[18]

The mechanism by which HBV is involved in liver oncogenesis remains enigmatic. Both direct and indirect roles for HBV may exist. There is no evidence to suggest that HBV carries a direct transforming gene. The site of integration of HBV DNA in HCC appears to be random, rather than the activation of a specific adjacent cellular oncogene by a promoter insertion mechanism. Chronic infection with woodchuck hepatitis virus (WHV) is also associated with the development of HCC in a high proportion of woodchucks, reaching 100% in one study.[54] The c-*myc* and N-*myc* loci are preferred regions of integration for WHV, resulting in the activation of those *myc* proto-oncogenes.[20,27,86] HBV insertional activation of cellular proto-oncogenes has not been described. However, there are examples of two HCCs in which HBV was integrated directly within an exon of the retinoic acid receptor β gene,[16] whereas in another tumor, HBV was integrated within an intron of the cyclin A gene.[184] Whether the altered expression of those genes contributed to the development of the tumors remains unclear. More important in the etiology of HBV-related cancer may be the gross chromosomal alterations that are associated with HBV infection.[61] The deletion of one allele of the tumor suppressor gene p53, a common genetic alteration in subsets

of HCCs,[69] may in some instances be mediated by an HBV integration event.[95]

The general effects of *cis*-acting HBV-enhancer sequences or *trans*-acting viral proteins on host gene expression are under investigation.[33,87] The intracellular accumulation of HBV large S protein in the livers of transgenic mice leads to the development of "ground-glass" hepatocytes and the development of liver tumors.[13] The finding that both the HBV-encoded X gene product and 3′-truncated S gene product (commonly generated during HBV integration) can transactivate the expression of viral and cellular genes [11,29,62] may suggest a role for these transactivators is HBV-mediated carcinogenesis. Transgenic mice with the X gene controlled by its natural HBV enhancer develop liver tumors,[30] while transgenic mice with the X gene controlled by the α-1-antitrypsin[34] regulatory region remain free of liver damage. The important cellular targets of viral transactivators remain unknown.

It is becoming apparent that all cases of HCC induced by HBV may not have the same underlying mechanism. Both viral and host factors are likely important.[19,48,51,68] Concurrent with virus integration or viral protein expression, a host immune response to virus-infected cells may contribute to cell injury, inflammation, and liver regeneration—processes known to be risk factors for HCC. Because the role of HBV in the etiology of HCC differs among individual tumors, the responsible genetic alterations mediated by HBV would be expected to occur in subsets of hematomas, rather than a single mechanism being common to all HCCs.

In addition to the above, other factors such as HCV or aflatoxin may play a role in some populations. Aflatoxin, which has been implicated epidemiologically but only in some areas, is highly mutagenic and carcinogenic for the liver in experimental animals; in these studies, mutagenicity has correlated with carcinogenicity. The mechanism of action appears to work through cellular oncogenes that are activated not only in experimental liver carcinoma but also in the preneoplastic stages of HCC.[70] Exposure of people to both aflatoxin and HBV seems to increase the risk for developing HCC to a far greater degree than exposure to either factor alone.[88] Experiments to determine activation of oncogenes or mutation of the p53 suppressor gene as mechanisms of action are in progress.[26,65]

4.2. Animal Models of HBV Infection and HCC

Further evidence for a causal relationship between HBV and hepatoma comes from a group of animal viruses that share many of the biophysical and biochemical properties of human HBV, the prototype virus for the hepadnavirus family.[60] The hepadnaviruses are a unique group, all being circular, double-stranded DNA viruses with an incomplete segment in one of the strands. They have in common the ability to replicate via an RNA intermediate. Other members of the family are WHV, ground squirrel hepatitis virus (GSHV), and duck hepatitis B virus (DHBV). These natural animal models provide further strong evidence that hepadnaviruses are oncogenic. The WHV, GSHV, and DHBV are all characterized by transmission from mother to offspring and a chronic carrier state. A very high proportion of carrier woodchucks and ground squirrels develop HCC, which also occurs in carrier ducks but at a lower frequency. Integration of the viral DNA occurs in the tumors of all these animals.

Woodchuck hepatitis virus, the first animal hepadnavirus discovered,[73] has the following characteristics in common with HBV: (1) infection results in persistence of the virus with the accumulation of large amounts of viral coat protein circulating in the blood as 20- to 25-nm spherical or tubular particles; (2) 40- to 50-nm-diameter particles containing small, partially double-stranded circular DNA and DNA polymerase circulate in the blood as complete virions; and (3) both chronic hepatitis and hepatoma are observed. In fact, a much higher incidence of hepatoma formation occurs in woodchucks infected with WHV[39,56,73] than is observed in man with HBV. Approximately one third of infected animals in captivity develop HCC per year, whereas no tumors have been observed in uninfected animals. In animals bearing hepatomas, the surface antigen of WHV is found not in the tumor cells but in the surrounding hepatocytes, analogous to the situation with HBV antigens in human hepatomas.

The surface and core of WHV have antigenic determinants called WHsAg and WHcAg, respectively, to denote their relationship to the corresponding antigens of HBV. Anti-HBc is present in the serum of humans with HBV infections; similarly, anti-WHc is found in WHV-infected woodchucks. The pattern of reaction of these antibodies with WHcAg and HBcAg by immunodiffusion indicates that the core components of the two viruses share major determinants and that each has unique antigenic determinants as well. The presence of similar determinants is also shown by the results of electron microscopic examination of the precipitin lines; viral cores present in crude homogenates of liver obtained from infected woodchucks or humans are aggregated by virus-positive sera from either woodchucks or humans. As found in HBV infections, cores obtained from intact WHV in the serum share determinants with cores purified from infected liver. The core antigenic specificity shared by HBV and WHV reflects a group-specific antigen for this class of virus.[22,36]

The high degree of cross-reactivity seen between the core antigen systems is not seen between the surface antigen systems.[22,72] However, there is some similarity between the surface constituents when the cross tests are carried out by radioimmunoassay.

A small region (100 to 150 base pairs) of strong nucleic acid homology was detected by hybridization. The degree of nucleic acid homology found (3 to 5% of the genomes) is on the order of that detected among some of the tumorigenic papovaviruses (SV40 and polyoma).

A second virus, GSHV, with many of the unique characteristics of human HBV, was found in ground squirrels in northern California.[38] Initial studies of young squirrels had indicated that little hepatitis accompanied GSHV infection and had failed to demonstrate associated hepatomas.[37] Although both WHV and GSHV infect members of the squirrel family and are related genetically, it was believed at first that the two had different roles as pathogens. However, as the investigations unfolded, they indicated that WHV and GSHV have a distinct resemblance to each other and to HBV in association of chronic infection with both hepatitis and HCC in the host species.[36,40]

Duck hepatitis B virus was first detected in duck sera in China, collected because Chinese investigators had noted that hepatoma occurs in these fowl, particularly in areas of high prevalence of human hepatoma. Summers and co-workers [41] then observed a similar virus in about 10% of Pekin ducks (*Anas domesticus*) from flocks in the United States, but hepatoma has not been observed in these flocks. Based on the virus morphology, virion DNA polymerase activity, size and structure of the virus genome, and association between virus in the serum and large amounts of viral DNA in the liver, this agent is now included in the hepadnavirus family.

Pekin ducks in the United States had their origin in ducks brought from China about 100 years ago, suggesting that the virus has been maintained in Pekin ducks in the United States since then. Pekin ducks are generally held in commercial flocks for less than 2 years, whereas the life span of these birds is over 10 years. To determine whether HCC develops, it will be necessary to maintain the animals for many years. HCC has not been detected in virus-infected Pekin ducks after several years of observation in several laboratories around the world,[36] and the virus has not been detected in other ducks.

Long-term study of hepadnavirus animal models has revealed the following common factors of HCC develop

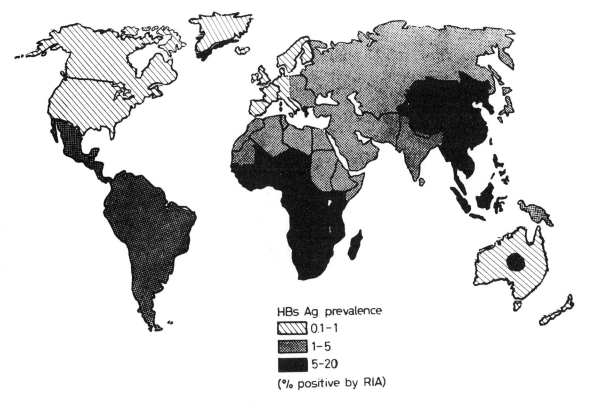

Figure 1. Prevalence of HBsAg in the world in 1980.

ment: (1) present or past infection with the hepadnavirus; (2) a long incubation period, with HCC occurring in the latter half of the life span; (3) hepatitis with degeneration and regeneration of hepatocytes; and (4) presence of integrated viral DNA in some of the carcinoma tissues.[36] Although these animal models provide a means for studying the development of liver cancer, the final and crucial evidence in humans will come from the application of the new hepatitis B vaccines (see Section 9).

5. Descriptive Epidemiology

5.1. Prevalence, Incidence, and Geographic Distribution

In areas of the world where there is a high prevalence of HBV carriage, there is a corresponding high prevalence of hepatoma. In some developing countries the entire population becomes infected with HBV, and as many as 25% die of cirrhosis and liver cancer.[87]. About 300 million virus carriers exist worldwide, but there are large differences in the prevalence of HBs antigenemia throughout the world. Prevalence rates as high as 10% are found in

central and South Africa and in Southeast Asia. The second highest rates, between 2 and 5%, exist in other regions of Asia, in North Africa, and in the Middle East. Rates are 1–2% in South America, southern and eastern Europe, and European Russia. The lowest rates, 0.1–0.5%, occur in western Europe, North America, Australia, and New Zealand. It is striking that the prevalence of hepatoma follows the same geographic pattern of distribution as that of persistent HBV infection. Figures 1 and 2 illustrate the epidemiologic information available in 1980, which established the relationship between HBV infection and development of HCC.[43]

In 18 high-incidence areas of Africa and Asia, hepatoma patients have positive prevalence rates for HBsAg between 37 and 90%, some 10 to 15 times higher than most controls (range 2.6–24.5%).[49,76] In low-incidence areas the discrepancy is even greater, on the order of a 20- to 40-fold higher prevalence of antigenemia in the liver cancer cases than in the controls. However, the ratio of hepatoma cases to carriers of HBsAg has been estimated to be about the same in Africa, Asia, and the United States, approximately 1 case per 250 male carriers per year, for both the high- and the low-incidence areas.[57]

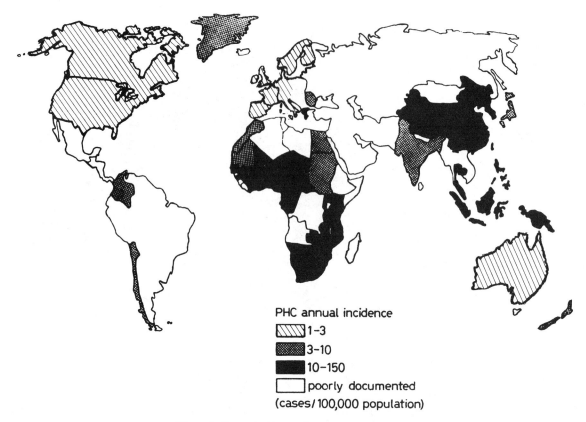

PHC annual incidence

```
[////] 1-3
[::::] 3-10
[████] 10-150
[    ] poorly documented
```
(cases/100,000 population)

Figure 2. Known incidence of HCC in the world in 1980.

A large cooperative study[49] in which 12 investigators from 10 countries in Asia, Africa, and the Pacific region studied approximately 15,000 serum samples lends support to the above concept. Their results indicate the consistently high frequency of antigenemia in hepatoma patients. In this study, the association of detectable HBsAg with hepatoma was even somewhat higher than that found in acute or chronic hepatitis or cirrhosis.

One early study[78] showed that 70% of the African and 41% of the American hepatoma patients were persistently infected with HBV. These percentages were sevenfold higher than in the control population in Africa and over 40 times higher than in the control population in the United States. If those HCC patients with anti-HBs are added, then 97% of the African patients and 74% of the American patients gave evidence of HBV infection, either past or present.

The epidemiologic characteristics of HBV and primary HCC have been compared with those of another virus that has been associated with cancer in man, namely, the Epstein–Barr virus and Burkitt's lymphoma.[44] In each instance, the natural viral infection occurs early in life, preceding by several years the development of the corresponding cancer. A period of 20 to 40 years is usually observed between initiation of infection and appearance of the carcinoma (Fig. 3).

In a prospective study of 22,707 Chinese males in Taiwan followed for an average of 6.2 years, Beasley, Hwang and colleagues[3,4] demonstrated a pronounced excess of deaths from HCC and cirrhosis among the 15.2% who were HBsAg-positive. All but 3 of the 116 men who subsequently died of hepatoma were in the original group of 3454 men who were HBsAg-positive. The annual incidence of hepatoma remained essentially constant over 5 years, implying that new liver cancers were continuously developing. It is noteworthy that 27% of the HBsAg-positive patients who died of hepatoma presumably did not have underlying cirrhosis when they entered the study. Thus, liver cancer can originate in HBsAg carriers even if they do not have underlying cirrhosis. None of 30 HBsAg-negative males known to have cirrhosis when they were recruited developed hepatoma.

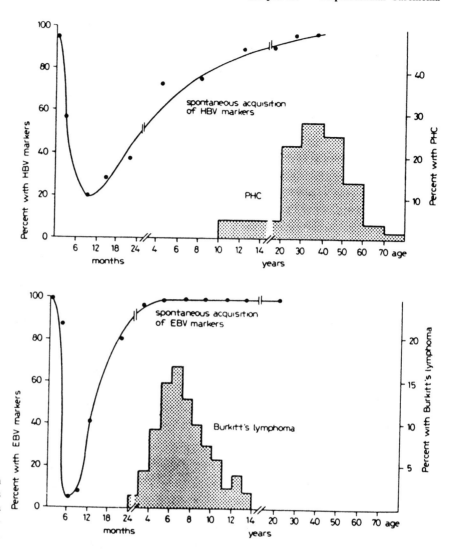

Figure 3. (Top) Cumulative infections with HBV and development of HCC; from Maupas and Melnick.[44] (Bottom) Comparable data for Epstein–Barr virus and Burkitt's lymphoma; adapted from Henle and co-workers.

This study provided further evidence that HBV, rather than cirrhosis, promotes the development of HCC.

The available evidence may be summarized as follows. The increased risk is the result of infection during early life, when the immune system is immature. The probability that HBV infection will lead to the carrier state is inversely related to age of infection. Among infected newborns, the likelihood of becoming a carrier is approximately 95%.[6,52] The risk of an infected child becoming a carrier decreases within the first few years of life.[3] In adults the risk is less than 10%. As HCC is most likely to occur in adults who experienced HBV infection at a very early age and became carriers, vaccination to be maximally effective must be carried out during the first week of life (see Section 9).

5.2. Age and Sex

The incidence of HCC increases with increasing age throughout most of life, with some tailing off at the end.[3] In Africa, the average age of HCC is under 40, whereas in Asia the mean age is over 50 years.[3]

Once infected with HBV, males are more likely to remain persistently infected and to become carriers, whereas females are more likely to be transiently infected and to develop anti-HBs. The sequence of events leading to HCC also shows this trend.[7,67,81] Although both sexes may be equally infected at birth, sons are more apt to develop chronic active hepatitis (by a ratio of approximately 1.5 : 1). More males than females develop cirrhosis of the liver (by a ratio of 3 : 1), and an even greater proportion develop liver cancer (by a ratio of 6:1). Thus, at each progressive stage of liver disease, the ratio of males to females appears to double so that hepatoma becomes predominantly a disease of males.[81]

5.3. Race

A revealing investigation was carried out on HBV markers and excess mortality from HCC in 666 Chinese permanently living in New York City.[77] Approximately 90% of the subjects were born outside the United States, but 51% of the foreign-born subjects had lived in this country for at least 10 years. HBsAg was detected in 9.3% of the subjects tested, anti-HBs in 57%, and anti-HBc in 8.8%; these rates are 10–40 times higher than those found in other ethnic groups living in the area. Nearly all HBV infections were of the subtypes prevalent in Asia. The incidence of anti-HBc alone or anti-HBs alone increased with age, but the incidence of HBsAg and coexisting anti-HBs decreased with age. The age-standardized death rate from hepatoma among Chinese men, but not women, was four times higher than that among black men and ten times higher than that among white men in the city. The findings that Chinese-Americans are at very high risk for both HBV infection and HCC support the hypothesis that these two conditions are causally associated.

5.4. Socioeconomic Factors

Hepatocellular carcinoma occurs most commonly in developing countries where infection with HBV occurs early in life, probably related to poor hygienic settings. In developed countries with higher socioeconomic levels and good hygiene, white HBV-carrier mothers are less apt to transmit HBV to their infants, although an active infection in the latter part of pregnancy may result in intrauterine infection of the fetus.

5.5. Other Factors

Although there are millions of HBV carriers in the world, hepatoma occurs in very few. This situation is not unusual in viral infections; wild poliovirus produces paralysis in only 1% of those infected.[47] Which risk factors other than exposure to HBV are associated with development of hepatoma are not known, but age of infection and perhaps the dose of virus have been incriminated. Factors such as aflatoxins and nitrosamines, which are thought by some to be related to the development of HCC and which are uncommon in low-incidence areas, do not appear to be essential for the development of this disease in human beings.

6. Mechanisms and Routes of Transmission

In Senegal, mothers of hepatoma patients were four times more apt to have a persistent HBV infection than fathers.[32] In contrast, mothers and fathers of matched controls showed no differences and had the same relatively low rates as the fathers of the hepatoma patients. Studies in other areas have demonstrated HBsAg prevalences of 71 to 86% in mothers of HCC patients.[2] When infection occurs at birth from a viremic mother, particularly one positive for hepatitis B e antigen, the exposure of the infant to the mother's infectious blood is high indeed.

Observations such as those above suggested that mothers with persistent HBV infections were more likely to produce offspring at high risk of developing HCC. This observation is in keeping with the vertical transmission of HBV from mothers to their infants that commonly occurs in geographic areas of high hepatoma prevalence.[6,64,71] In Senegal, babies become infected about seven times more frequently if born of an antigenemic mother; the infection does not occur *in utero* or perinatally but is acquired during the first few years of life. In Taiwan, transmission is more rapid; HBV-carrier women immediately transmit the infection to their newborns in the perinatal period. In contrast, white HBV-carrier mothers in the United States or Europe seldom transmit the infection to their infants. But if a white woman develops acute hepatitis B during the last trimester of pregnancy, her baby is usually infected perinatally and becomes a persistent carrier of virus. In the United States and the United Kingdom, vertical transmission among blacks is intermediate in frequency between the high rates among ethnic Asians and the very low rates among whites.

7. Pathogenesis and Pathology

The emerging concept of pathogenesis of primary liver cancer is that HBV infection in early life leads to chronic antigenemia → chronic hepatitis → cirrhosis → hepatoma. In some instances, the chronic hepatitis and cirrhosis stages may be skipped. Morphologically the outcome of acute hepatitis may be complete resolution, recurrence, massive necrosis, chronic persistent or chronic aggressive hepatitis, resolution with scarring, or cirrhosis.

An AFP level >500 ng/ml is diagnostic for HCC if it persists for over a month without other evidence of HCC and where pregnancy, active hepatitis, and embryoplastic tumors of the genital glands can be excluded. The early diagnosis of HCC, when the carcinoma is tiny, hinges mainly on the AFP determination, but AFP may also be present in a few patients with tumors of the alimentary tract or other diseases, especially hepatitis and liver cirrhosis in its active phase. Nevertheless, the detection of AFP has played an important role in the early diagnosis of HCC, particularly in China.[90]

The mechanism of malignant transformation in cirrhosis is not fully understood, but its onset is heralded by liver cell dysplasia (cellular enlargement, nuclear pleomorphism, and multinucleation of liver cells) occurring in foci inside cirrhotic nodules or sometimes occupying an entire nodular area. Dysplastic cells are thought to be precursors of cancer cells, although this has not been proved in man. Of 100 chronic HBsAg carriers, 10 usually develop posthepatitis cirrhosis, of which two to three cases end in fatal HCC.

In one representative study,[23] liver specimens from fatal cases were taken from tumor and nontumor areas of both liver lobes. Histopathology revealed that HCC was strongly associated with cirrhosis. Multilobular cirrhosis of the postnecrotic type occupied the liver parenchyma, and dense fibrosis and voluminous regenerative nodules were observed. Nodules of well-differentiated trabecular carcinoma were scattered over this cirrhotic background. Most of the HBsAg-positive cells were present in the cirrhotic areas and confined to one lobule while adjacent lobules were negative. A diffuse distribution of HBsAg was occasionally found, but more often HBsAg-positive cells were clustered in a segment of the lobule. This irregular distribution of HBsAg-producing cells indicates that replication of HBV may take place in scattered lobules of the cirrhotic parenchyma, a factor to be considered when examining small fragments of liver tissue or samples containing mainly neoplastic tissue.

8. Patterns of Host Response

Hepatitis B infection can lead to an asymptomatic carrier state, acute hepatitis, chronic hepatitis, cirrhosis, and liver cancer. These acute and chronic manifestations are discussed in Chapter 13. Cirrhosis is a frequent but not necessary step in the pathogenesis of HCC. The clinical onset in the majority of patients is characterized by vague abdominal symptoms, but if cirrhosis is present there may be a rapid deterioration in the clinical condition. Abdomi-

nal pain, an abdominal mass, weight loss, and, in two thirds of the patients, enlargement of the liver are other later features of the clinical disease.

9. Control and Immunization

Many investigators have begun to work on interrupting the neonatal transmission, postponing the time of primary infection or eliminating it entirely by passive or active hepatitis B immunization programs. This seems to be the most rational approach to the control of HCC.

The lack of a normal cell-culture system for propagating HBV has meant that the first vaccines had to be prepared from HBsAg isolated and purified from the plasma of chronic carriers. Safety considerations include the possible presence of other viruses that might have infected the donors of the blood from which the vaccine was purified. Thus, the vaccine preparations are treated by methods that inactivate viruses (formalin, heat, urea, proteolytic enzymes), and then they are tested for safety in susceptible chimpanzees.[45]

World Health Organization (WHO) advisory groups have emphasized the tremendous advancement in hepatitis B prophylaxis with the introduction and licensing of hepatitis B vaccine made from purified and inactivated HBsAg. Over 30 million doses of plasma-derived hepatitis B vaccine have been distributed worldwide, and there are at least a dozen manufacturers of vaccine around the world. The vaccines have an impressive record of safety. In addition, several vaccines manufactured by recombinant DNA technology are now on the market, and additional manufacturers are expected to enter the market in the near future. The recombinant DNA vaccines are equivalent to plasma-derived vaccines in terms of safety, immunogenicity, and efficacy. Neither type of vaccine currently offers any advantage over the other in these respects. It appears that plasma-derived vaccines will continue to play an essential role in hepatitis B control programs worldwide for the foreseeable future, particularly because of relative cost and availability. There has been a dramatic decrease in the price of hepatitis B vaccines to less than $1 per dose as of this writing, when purchased for public health use. Thus, many countries in areas hyperendemic for hepatitis B may now begin the development and implementation of large-scale vaccination programs.

The most important means to control hepatitis B on a global scale and to reduce mortality from chronic sequelae of the infection, including cirrhosis and HCC, is the large-scale immunization of infants. The WHO Advisory Group in 1987 recommended that hepatitis B vaccination be integrated into the WHO's Expanded Program in Immunization (EPI), a recommendation that was soon accepted. This is important, particularly for countries with carrier rates of HBsAg greater than 5–10%.

Three doses of hepatitis B vaccine are recommended by the WHO Advisory Group. Each dose should be administered by injection into the thigh of infants. The first dose should be given as soon after birth as possible. Although programs should aim at injection of the first dose of vaccine within the first week of life, it may be given subsequently at any time, but the earlier the better to prevent the high-HCC-risk virus-carrier state from developing. It is also desirable that the first dose of vaccine be given simultaneously with the first EPI immunization, i.e., BCG and oral poliovaccine (OPV). The second dose of hepatitis B vaccine should be given from 4 to 12 weeks after the first dose, as it best fits into the EPI schedule of the particular region [usually with the second dose of OPV and the first dose of diphtheria–pertussis–tetanus (DPT)]. A third dose of vaccine is necessary to achieve high levels of antibody and prolonged protection. There is considerable latitude regarding the timing of this dose. This can be given from 2 to 12 months after the second dose, at a time when it best fits into the region's EPI schedule of administration of DPT, OPV, and measles vaccines. There is no interference of OPV with hepatitis B vaccine or vice versa.

Hepatitis B immune globulin may be of additional value in immunization programs for infants,[5] but the cost of its inclusion into large-scale immunization programs precludes its use in most countries.

The duration of protection from infection after primary immunization is not known, but the early results are very encouraging.[5,42,74,75] Immunity is believed to persist as long as at least 10 international units of anti-HBs per milliliter can be detected in the serum. However, immunity may persist longer because immediate anamnestic responses have been demonstrated after revaccination even when anti-HBs was no longer detectable. Duration of the persistence of anti-HBs after primary immunization can be predicted from the anti-HBs titers achieved by the primary immunization, and this is practiced in high-risk groups in some countries. Further studies are needed to determine if and when booster hepatitis B immunizations should be given.

In the United States, even though HCC is not a common disease, perinatal transmission of HBV does occur. When it does, there follow the same high morbidity and mortality that occur in other parts of the world. It has recently been advocated that routine screening of pregnant women for HBsAg and anti-HBc should be conducted in the United States and that infants born at risk should be immunized, beginning in the perinatal period, with either the plasma-derived or recombinant vaccine.[1,28]

Even after hepatitis B immunization programs have been started, cases of HCC are expected to continue to occur for many decades in those infected early in life, before the advent of the vaccine. However, the continuing development of hepatitis B vaccines and the favorable results being achieved in inducing antibodies and protection in man are good omens for the future. The day may not be too far off when an important cancer of man will be preventable by use of a vaccine for a common virus infection.

10. Unresolved Problems

The major question is the mechanism by which HBV is involved in oncogenesis. Although rapid advances are being made at the molecular level, the pathogenesis still remains enigmatic. In some areas, aflatoxin may be a cofactor.

The possibility that the current trials of HBV in the prevention of HCC will actually be effective must await years of follow-up of immunized and unimmunized infants.

In contrast to the universal relationship of HBV to HCC, the association of HCV infection with HCC varies from low to high in different populations, being particularly high in Japan for patients positive for HCC but negative for HBsAg. While the pathogenesis of HCV-related HCC remains unclear, the strong association in certain populations suggests that the virus plays a role in them.

11. References

1. AREVALO, J. A., AND WASHINGTON, A. E., Cost-effectiveness of prenatal screening and immunization for hepatitis B virus, *J.A.M.A.* **259**:365–369 (1988).
2. BEASLEY, R. P., Hepatitis B virus as the etiologic agent in hepatocellular carcinoma—epidemiologic considerations, *Hepatology* **2**:21S–26S (1982).
3. BEASLEY, R. P., AND HWANG, L.-Y., Epidemiology of hepatocellular carcinoma, in: *Viral Hepatitis and Liver Disease* (G. N. VYAS, J. L. DIENSTAG, AND J. H. HOOFNAGLE, EDS.), pp. 209–224, Grune & Stratton, Orlando, FL, 1984.
4. BEASLEY, R. P., HWANG, L.-Y., LIN, C.-C., AND CHIEN, C.-S., Hepatocellular carcinoma and hepatitis B virus: A prospective study of 22,707 men in Taiwan, *Lancet* **2**:1129–1133 (1981).

5. BEASLEY, R. P., HWANG, L.-Y., STEVENS, C. E., LIN, C.-C., HSIEH, F.-J., WANG, K.-Y., SUN, T. S., AND SZMUNESS, W., Efficacy of hepatitis B immune globulin for prevention of perinatal transmission of the hepatitis B virus carrier state: Final report of a randomized double-blind, placebo-controlled trial, *Hepatology* **3:**135–141 (1983).

6. BEASLEY, R. P., TREPO, C., STEVENS, C. E., AND SZMUNESS, W., The e antigen and vertical transmission of hepatitis B surface antigen, *Am. J. Epidemiol.* **105:**94–98 (1977).

7. BLUMBERG, B. S., AND LONDON, W. T., Hepatitis B virus and primary hepatocellular carcinoma: Relationship of "icrons" to cancer, in: *Viruses in Naturally Occurring Cancers* (M. ESSEX, G. TODARO, AND H. ZUR HAUSEN, EDS.), pp. 401–421, Cold Spring Harbor Laboratory, Cold Spring Harbor, New York, 1980.

8. BLUMBERG, B. S., AND LONDON, W. T., Hepatitis B virus and the prevention of primary cancer of the liver, *J. Natl. Cancer Inst.* **74:**267–273 (1985).

9. BRECHOT, C., POURCEL, C., LOUISE, A., RAIN, B., AND TIOLLAIS, P., Presence of integrated hepatitis B virus DNA sequences in cellular DNA of human hepatocellular carcinoma, *Nature* **285:** 533–535 (1980).

10. BUKH, J., MILLER, R. H., KEW, M. C., AND PURCELL, R. H., Hepatitis C virus RNA in southern African blacks with hepatocellular carcinoma, *Proc. Natl. Acad. Sci. USA* **90:**1848–1851 (1993).

11. CASELMANN, W. H., MEYER, M., KEKULE, A. S., LAUER, U., HOFSCHNEIDER, P. H., AND KOCHY, R., A *trans*-activator function is generated by integration of hepatitis B virus preS/S sequences in human hepatocellular carcinoma DNA, *Proc. Natl. Acad Sci USA* **87:**2970–2974 (1990).

12. CHEN, D. S., HOYER, B. H., NELSON, J., PURCELL, R. H., AND GERIN, J. L., Detection and properties of hepatitis B viral DNA in liver tissues from patients with hepatocellular carcinoma, *Hepatology* **2:**42S–46S (1982).

13. CHISARI, F. V., FILIPPI, P., McLACHLAN, A., MILICH, D. R., RIGGS, M., LEE, S., PALMITER, R. D., PINKERT, C. A., AND BRINSTER, R. L., Expression of hepatitis B virus large envelope polypeptide inhibits hepatitis B surface antigen secretion in transgenic mice, *J. Virol.* **60:**880–887 (1986).

14. COURSAGET, P., LEBOULLEUX, D., CANN, P. LE, BAO, O., AND COLL-SECK, A. M., Hepatitis C virus infection in cirrhosis and primary hepatocellular carcinoma in Senegal, *Trans. R. Soc. Trop. Med. Hyg.* **86:**552–553 (1992).

15. CUTLER, S. J., AND YOUNG, J. L., JR., Third National Cancer Survey. Incidence data, in: *NCI Monograph 41,* DHEW Publication No. (NIH) 75-785, NCI, Bethesda, MD, 1975.

16. DEJEAN, A., BOUGUELERET, L., GRZESCHIK, K.-H., AND TIOLLAIS, P., Hepatitis B virus DNA integration in a sequence homologous to v-*erb*-A and steroid receptor genes in a hepatocellular carcinoma, *Nature* **322:**70–72 (1986).

17. DEJEAN, A., BRECHOT, C., TIOLLAIS, P., AND WAIN-HOBSON, S., Characterization of integrated hepatitis B viral DNA cloned from a human hepatoma and the hepatoma-derived cell line PLC/PRF/5, *Proc. Natl. Acad. Sci. USA* **80:**2505–2509 (1983).

18. DEJEAN, A., SONIGO, P., WAIN-HOBSON, S., AND TIOLLAIS, P., Specific hepatitis B virus integration in hepatocellular carcinoma DNA through a viral 11-base-pair direct repeat, *Proc. Natl. Acad. Sci. USA* **81:**5350–5354 (1984).

19. DE THÉ, H., MARCHIO A., TIOLLAIS, P., AND DEJEAN, A., A novel steroid thyroid hormone receptor-related gene inappropriately expressed in human hepatocellular carcinoma, *Nature* **330:**667–670 (1987).

20. FOUREL, G., TREPO, C., BOUGUELERET, L., HENGLEINS, B., PONZETTO, A., TIOLLAIS, P., AND BUENDIA, M.-A., Frequent activation of N-*myc* genes by hepadnavirus insertion in woodchuck liver tumours, *Nature* **347:**294–298 (1990).

21. GANEM, D., AND VARMUS, H. E., The molecular biology of the hepatitis B viruses, *Annu. Rev. Biochem.* **56:**651–693 (1987).

22. GERIN, J. L., TENNANT, B. C., PONZETTO, A., PURCELL, R. H., AND TYERYAR, F. J., The woodchuck animal model of hepatitis B-like virus infection and disease, *Prog. Clin. Biol. Res.* **143:**23–28 (1983).

23. GOUDEAU, A., MAUPAS, P., COURSAGET, P., DRUCKER, J., CHIRON, J.-P., DENIS, F., MAR, I. D., AND N'DIAYE, P. D., Detection of hepatitis B virus antigens in hepatocellular carcinoma tissues, *Prog. Med. Virol.* **27:**77–87 (1981).

24. HARRISON, T. J., CHEN, J.-Y., AND ZUCKERMAN, A. J., Hepatitis B and primary liver cancer, in: *Cancer Treatment Reviews,* Vol. 13 (K. HELLMANN AND S. K. CARTER, EDS.), pp. 1–16, Academic Press, London, 1986.

25. HINO, O., KITAGAWA, T., KOIKE, K., KOBAYASHI, M., HARA, M., MORI, W., NAKASHIMA, T., HATTORI, N., AND SUGANO, H., Detection of hepatitis B virus DNA in hepatocellular carcinomas in Japan, *Hepatology* **4:**90–95 (1984).

26. HSIA, C. C., KLEINER, D. E., JR., AXIOTIS, C. A., BISCEGLIE, A. D., NOMURA, A. M. Y., STEMMERMANN, G. N., AND TABOR, E., Mutations of p53 gene in hepatocellular carcinoma: Roles of hepatitis virus and aflatoxin contamination in the diet, *J. Natl. Cancer Inst.* **84:**1638–1641 (1992).

27. HSU, T.-Y., MOROY, T., ETIEMBLE, J., LOUISE, A., TREPO, C., TIOLLAIS, P., AND BUENDIA, M.-A., Activation of c-*myc* by woodchuck hepatitis virus insertion in hepatocellular carcinoma, *Cell* **55:**627–635 (1988).

28. KANE, M. A., HADLER, S. C., MARGOLIS, H. S., AND MAYNARD, J. E., Routine prenatal screening for hepatitis B surface antigen, *J.A.M.A.* **259:**408–409 (1988).

29. KEKULE, A. S., LAUER, U., WEISS, L., LUBER, B., AND HOFSCHNEIDER, P. H., Hepatitis B virus transactivator HBx uses a tumour promoter signalling pathway, *Nature* **361:**742–745 (1993).

30. KIM, C. M., KOIKE, K., SAITO, I., MIYAMURA, T., AND JAY, G., HBx gene of hepatitis B virus induces liver cancer in transgenic mice, *Nature* **351:**317–320 (1991).

31. KOSHY, R., KOCH, S., VON LORINGHOVEN, A. F., KAHMANN, R., MURRAY, K., AND HOFSCHNEIDER, P. H., Integration of hepatitis B virus DNA: Evidence for integration in the single-stranded gap, *Cell* **34:**215–223 (1983).

32. LAROUZE, B., LONDON, W. T., SAIMOT, G., WERNER, B. G., LUSTBADER, E. D., PAYET, M., AND BLUMBERG, B. S., Host responses to hepatitis-B infection in patients with primary hepatic carcinoma and their families: A case-control study in Senegal, West Africa, *Lancet* **2:**534–538 (1976).

33. LEE, T.-H., AND BUTEL, J. S., *Trans*-activation of the HBV enhancer by the X gene of HBV, in: *Abstracts of the Conference on Hepatitis B Viruses,* p. 19, Cold Spring Harbor Laboratory, Cold Spring Harbor, NY, 1987.

34. LEE, T.-H., FINEGOLD, M. J., SHEN, R.-F., DEMAYO, J. L., WOO, S. L. C., AND BUTEL, J. S., Hepatitis B virus transactivator X protein is not tumorigenic in transgenic mice, *J. Virol.* **64:**5939–5947 (1990).

35. LIU, C. B., XU, Z. Y., CAO, H. L., SUN, Y. D., JING, Q., CHENG, D. B., CHEN, M. H., AND XIE, Y. F., A field study of the incidence of acute and chronic viral hepatitis and the mortality of hepatitis-related diseases and hepatocellular carcinoma, *Chin. J. Virol.* **1**(Suppl.):1–8 (1991).

36. MARION, P. L., Use of animal models to study hepatitis B virus, *Prog. Med. Virol.* **35**:43–75 (1988).

37. MARION, P. L., KNIGHT, S. S., SALAZAR, F. H., POPPER, H., AND ROBINSON, W. S., Ground squirrel hepatitis virus infection, *Hepatology* **3**:519–527 (1983).

38. MARION, P. L., OSHIRO, L. S., REGNERY, D. C., SCULLARD, G. H., AND ROBINSON, W. S., A virus in Beechey ground squirrels that is related to hepatitis B virus of humans, *Proc. Natl. Acad. Sci. USA* **77**:2941–2945 (1980).

39. MARION, P. L., AND ROBINSON, W. S., Hepadna viruses: Hepatitis B and related viruses, *Curr. Top. Microbiol. Immunol.* **105**:91–121 (1983).

40. MARION, P. L., VAN DAVELAAR, M. J., KNIGHT, S. S., SALAZAR, F. H., GARCIA, G., POPPER, H., AND ROBINSON, W. S., Hepatocellular carcinoma in ground squirrels persistently infected with ground squirrel hepatitis virus, *Proc. Natl. Acad. Sci. USA* **83**:4543–4546 (1986).

41. MASON, W. S., SEAL, G., AND SUMMERS, J., A virus of Pekin ducks with structural and biological relatedness to human hepatitis B virus, *J. Virol.* **36**:829–836 (1980).

42. MAUPAS, P., CHIRON, J.-P., GOUDEAU, A., COURSAGET, P., PERRIN, J., BARIN, F., DENIS, F., AND MAR, B. D., Active immunization against hepatitis B in an area of high endemicity. Part II: Prevention of early infection of the child, *Prog. Med. Virol.* **27**:185–201 (1981).

43. MAUPAS, P., AND MELNICK, J. L. (EDS.), Hepatitis B virus and primary hepatocellular carcinoma, *Prog. Med. Virol.* **27**:1–210 (1981).

44. MAUPAS, P., AND MELNICK, J. L., Hepatitis B infection and primary liver cancer, *Prog. Med. Virol.* **27**:1–5 (1981).

45. MCAULIFFE, V. J., PURCELL, R. H., AND GERIN, J. L., Type B hepatitis: A review of current prospects for a safe and effective vaccine, *Rev. Infect. Dis.* **2**:470–492 (1980).

46. MELNICK, J. L., Hepatitis B virus and liver cancer, in: *Viruses Associated with Human Cancer* (L. A. PHILLIPS, ED.), pp. 337–367, Marcel Dekker, New York, 1983.

47. MELNICK, J. L., AND LEDINKO, N., Development of neutralizing antibodies against the three types of poliomyelitis virus during an epidemic period. The ratio of inapparent infection to clinical poliomyelitis, *Am. J. Hyg.* **58**:207–222 (1953).

48. MIYAKI, M., SATO, C., GOTANDA, T., MATSUI, T., MISHIRO, S., IMAI, M., AND MAYUMI, M., Integration of region X of hepatitis B virus genome in human primary hepatocellular carcinomas propagated in nude mice, *J. Gen. Virol.* **67**:1449–1454 (1986).

49. NISHIOKA, K., LEVIN, A. G., AND SIMONS, M. J., Hepatitis B antigen, antigen subtypes, and hepatitis B antibody in normal subjects and patients with liver disease: Results of a collaborative study, *Bull. WHO* **521**:293–300 (1975).

50. NISHIOKA, K., WATANABE, J., FURUTA, S., TANAKA, E., IINO, S., SUZUKI, H., TSUJI, T., YANO, M., KUO, G., CHOO, Q. L., HOUGHTON, M., AND ODA, T., A high prevalence of antibody to hepatitis C virus in patients with hepatocellular carcinoma in Japan, *Cancer* **67**:429–433 (1991).

51. OCHIYA, T., FUJIYAMA, A., FUKUSHIGE, S., HATADA, I., AND MATSUBARA, K., Molecular cloning of an oncogene from a human hepatocellular carcinoma, *Proc. Natl. Acad. Sci. USA* **83**:4993–4997 (1986).

52. OKADA, K., KAMIYAMA, I., INOMATA, M., IMAI, M., MIYAKAWA, Y., AND MAYUMI, M., E antigen and anti-e in the serum of asymptomatic carrier mothers as indicators of positive and negative transmission of hepatitis B virus to their infants, *N. Engl. J. Med.* **294**:746–749 (1976).

53. OKADA, K., AND BEASLEY, R. P., Epidemiology of hepatocellular carcinoma, *UICC Rep.* **17**:9–30 (1982).

54. POPPER, H., ROTH, L., PURCELL, R. H., TENNANT, B. C., AND GERIN, J. L., Hepatocarcinogenicity of the woodchuck hepatitis virus, *Proc. Natl. Acad. Sci. USA* **84**:866–870 (1987).

55. POPPER, H., SHAFRITZ, D. A., AND HOOFNAGLE, J. H., Relation of the hepatitis B virus carrier state to hepatocellular carcinoma, *Hepatology* **7**:764–772 (1987).

56. POPPER, H., SHIH, J. W. K., GERIN, J. L., WONG, D. C., HOYER, B. H., LONDON, W. T., SLY, D. L., AND PURCELL, R. H., Woodchuck hepatitis and hepatocellular carcinoma: Correlation of histologic with virologic observations, *Hepatology* **1**:91–98 (1981).

57. PRINCE, A. M., Discussion, in: *Viral Hepatitis* (G. N. VYAS, S. N., COHEN, AND R. SCHMID, EDS.), pp. 460–463, Franklin Institute Press, Philadelphia, 1978.

58. PRINCE, A. M., Hepatitis B virus and primary liver cancer, in: *Accomplishments in Cancer Research*, pp. 110–118, J. B. Lippincott, Philadelphia, 1985.

59. ROBINSON, W. S., Hepatitis B virus, in: *Virology* (B. N. FIELDS, D. M. KNIPE, R. M. CHANOCK, J. L. MELNICK, B. ROIZMAN, AND R. E. SHOPE, EDS.), pp. 1384–1406, Raven Press, New York, 1985.

60. ROBINSON, W. S., MARION, P., FEITELSON, M., AND SIDDIQUI, A., The hepadna virus group: Hepatitis B and related viruses, in: *Viral Hepatitis—1981 International Symposium* (W. SZMUNESS, H. J. ALTER, AND J. E. MAYNARD, EDS.), pp. 57–68, Franklin Institute Press, Philadelphia, 1982.

61. ROGLER, C. E., SHERMAN, M., SU, C. Y., SHAFRITZ, D. A., SUMMERS, J., SHOWS, T. B., HENDERSON, A., AND KEW, M., Deletion in chromosome 11p associated with a hepatitis B integration site in hepatocellular carcinoma, *Science* **230**:319–322 (1985).

62. ROSSNER, M. T., Review: Hepatitis B virus x-gene product: A promiscuous transcriptional activator, *J. Med. Virol.* **36**:101–117 (1992).

63. SARACCI, R., AND REPETTO, F., Time-trends of primary liver cancer: Indication of increased incidence in selected cancer registry populations, *J. Natl. Cancer Inst.* **65**:241–247 (1980).

64. SCHWEITZER, I. L., Infection of neonates and infants with the hepatitis B virus, *Prog. Med. Virol.* **20**:27–48 (1975).

65. SCORSONE, K. A., ZHOU, Y.-Z., BUTEL, J. S., AND SLAGLE, B. L., p53 Mutations cluster at codon 249 in hepatitis B virus-positive hepatocellular carcinomas from China, *Cancer Res.* **52**:1635–1638 (1992).

66. SHAFRITZ, D. A., SHOUVAL, D., SHERMAN, H. I., HADZIYANNIS, S. J., AND KEW, M. C., Integration of hepatitis B virus DNA into the genome of liver cells in chronic liver disease and hepatocellular carcinoma. Studies in percutaneous liver biopsies and post-mortem tissue specimens, *N. Engl. J. Med.* **305**:1067–1073 (1981).

67. SHERLOCK, S., *Diseases of the Liver and Biliary System*, 6th ed., Blackwell, Boston, 1981.

68. SLAGLE, B. L., LEE, T.-H., AND BUTEL, J. S., Hepatitis B virus and hepatocellular carcinoma, *Prog. Med. Virol.* **39**:167–203 (1992).

69. SLAGLE, B. L., ZHOU, Y.-Z., AND BUTEL, J. S., Hepatitis B virus integration event in human chromosome 17p near the p53 gene identifies the region of the chromosome commonly deleted in virus-positive hepatocellular carcinomas, *Cancer Res.* **51**:49–54 (1991).

70. SOMAN, N. R., AND WOGAN, G. N., Activation of the c-Kir-*ras* oncogene in aflatoxin B₁-induced hepatocellular carcinoma and adenoma in the rat: Detection by denaturing gradient gel electrophoresis, *Proc. Natl. Acad. Sci. USA* **90**:2045–2049 (1993).

71. STEVENS, C. E., BEASLEY, R. P., TSUI, J., AND LEE, W. C., Vertical

transmission of hepatitis B antigen in Taiwan, *N. Engl. J. Med.* **292:**771–774 (1975).

72. SUMMERS, J., Three recently described animal virus models for human hepatitis B virus, *Hepatology* **1:**179–183 (1981).

73. SUMMERS, J., SMOLEC, J. M., AND SNYDER, R., A virus similar to human hepatitis B virus associated with hepatitis and hepatoma in woodchucks, *Proc. Natl. Acad. Sci. USA* **75:**4533–4537 (1978).

74. SUN, T.-T., CHU, Y.-R., HSIA, C.-C., WEI, Y.-P., AND WU, S.-M., Strategies and current trends of etiologic prevention of liver cancer. *UCLA Symp. Mol. Cell. Biol.* **40:**283–292 (1986).

75. SUN, T.-T., CHU, Y.-R., NI, Z.-Q., LU, J.-H., HUANG, F., NI, Z.-P., PEI, X.-F., YU, Z.-I., AND LIU, G.-T., A pilot study on universal immunization of newborn infants in an area of hepatitis B virus and primary hepatocellular carcinoma prevalence with a low dose of hepatitis B vaccine, *J. Cell. Physiol.* **4**(Suppl.)**:**83–90 (1986).

76. SZMUNESS, W., Hepatocellular carcinoma and the hepatitis B virus: Evidence for a causal association, *Prog. Med. Virol.* **24:**40–69 (1978).

77. SZMUNESS, W., STEVENS, C. E., IKRAM, H., MUCH, M. I., HARLEY, E. J., AND HOLLINGER, F. B., Prevalence of hepatitis B virus infection and hepatocellular carcinoma in Chinese-Americans, *J. Infect. Dis.* **137:**822–829 (1978).

78. TABOR, E., GERETY, R. J., VOGEL, C. L., BAYLEY, A. C., ANTHONY, P. P., CHAN, C. H., AND BARKER, L. F., Hepatitis B virus infection and primary hepatocellular carcinoma, *J. Natl. Cancer Inst.* **58:**1197–1200 (1977).

79. TABOR, E., AND KOBAYASHI, K., Hepatitis C, a causative infectious agent of non-A, non-B hepatitis: Prevalence and structure—summary of a conference on hepatitis C virus as a cause of hepatocellular carcinoma, *J. Natl. Cancer Inst.* **84:**86–90 (1992).

80. TANG, Z. Y., WU, M. C., AND XIA, S. S., *Primary Liver Cancer*, Springer, Berlin, 1989.

81. TONG, M. J., THURSBY, M. W., LIN, J.-H., WEISSMAN, J. V., AND McPEAK, C. M., Studies on the maternal–infant transmission of the hepatitis B virus and HBV infection within families, *Prog. Med. Virol.* **27:**137–147 (1981).

82. TWU, J.-S., AND SCHLOEMER, R. H., Transcriptional *trans*-activating function of hepatitis B virus, *J. Virol.* **61:**3448–3453 (1987).

83. VAN RENSBERG, S. J., COOK-MOZAFFARI, P., VAN SCHALKWYK, D. J., VAN DER WATT, J. J., VINCENT, T. J., AND PURCHASE, I. F., Hepatocellular carcinoma and dietary aflatoxin in Mozambique and Transkei, *Br. J. Cancer* **51:**713–726 (1985).

84. WANG, J., CHENIVESSE, X., HENGLEIN, B., AND BRECHOT, C., Hepatitis B virus integration in a cyclin A gene in a hepatocellular carcinoma, *Nature* **343:**555–557 (1990).

85. WATERHOUSE, J., MUIR, C., AND CORREA, P. (EDS.), *Cancer Incidence in Five Continents*, Vol. 3, IARC Publication No. 15, IARC, Lyon, France, 1976.

86. WEI, Y., FOUREL, G., PONZETTO, A., SILVESTRO, M., TIOLLAIS, P., AND BUENDIA, M.-A., Hepadnavirus integration: Mechanisms of activation of the N-myc2 retrotransposon in woodchuck liver tumors, *J. Virol.* **66:**5265–5278 (1992).

87. WEN, Y.-M., XU, Z.-Y., MELNICK, J. L. (EDS.) Viral hepatitis in China: Problems and control strategies, *Monogr. Virol.* **19:**1–159 (1992).

88. WILD, C. P., SHRESTHA, S. M., ANWAR, W. A., AND MONTESANO, R., Field studies of aflatoxin exposure, metabolism and induction of genetic alterations in relation to HBV infection and hepatocellular carcinoma in the Gambia and Thailand, *Toxicol. Lett.* **64/65:** 455–461 (1992).

89. YAGINUMA, K., KOBAYASHI, H., KOBAYASHI, M., MORISHIMA, T., MATSUYAMA, K., AND KOIKE, K., Multiple integration site of hepatitis B virus DNA in hepatocellular carcinoma and chronic active hepatitis tissues from children, *J. Virol.* **61:**1808–1813 (1987).

90. YEH, F.-S., AND SHEN, K.-N., Epidemiology and early diagnosis of primary liver cancer in China, *Adv. Cancer Res.* **47:**297–329 (1986).

91. YEH, F. S., YU, M. C., MO, C. C., LUO, S., TONG, M. J., AND HENDERSON, B. E., Hepatitis B virus, aflatoxins, and hepatocellular carcinoma in southern Guangxi, China, *Cancer Res.* **49:**2506–2509 (1989).

92. YU, M.-W., YOU, S. L., CHANG, A.-S., LU, S.-N., LIAW, Y.-F., AND CHEN, C.-J., Association between hepatitis C virus antibodies and hepatocellular carcinoma in Taiwan, *Cancer Res.* **51:**5621–5625 (1991).

93. ZAVITSANOS, X., HATZAKIS, A., KAKLAMANI, E., TZONOU, A., TOUPADAKI, N., BROEKSMA, C., CHRISPEELS, J., TROONEN, H., HADZIYANNIS, S., HSIEH, C.-C., ALTER, H., AND TRICHOPOULOS, D., Association between hepatitis C virus and hepatocellular carcinoma using assays based on structural and nonstructural hepatitis C virus peptides, *Cancer Res.* **52:**5364–5367 (1992).

94. ZHOU, Y.-Z., BUTEL, J. S., LI, P.-J., FINEGOLD, M. J., AND MELNICK, J. L., Integrated state of subgenomic fragments of hepatitis B virus DNA in hepatocellular carcinoma from mainland China, *J. Natl. Cancer Inst.* **79:**223–231 (1987).

95. ZHOU, Y.-Z., SLAGLE, B. L., DONEHOWER, L. A., VAN TUINEN, P., LEDBETTER, D. H., AND BUTEL, J. S., Structural analysis of a hepatitis B virus genome integrated into chromosome 17p of a human hepatocellular carcinoma, *J. Virol.* **62:**4224–4231 (1988).

96. ZIEMER, M., GARCIA, P., SHAUL, Y., AND RUTTER, W. J., Sequence of hepatitis B virus DNA incorporated into the genome of a human hepatoma cell line, *J. Virol.* **53:**885–892 (1985).

12. Suggested Reading

BEASLEY, R. P., AND HWANG, L.-Y., Epidemiology of hepatocellular carcinoma, in: *Viral Hepatitis and Liver Disease* (G. N. VYAS, J. L. DIENSTAG, AND J. H. HOOFNAGLE, EDS.), pp. 209–224, Grune & Stratton, Orlando, FL, 1984.

BLUMBERG, B. S., AND LONDON, W. T., Hepatitis B virus and primary hepatocellular carcinoma: Relationship of "icrons" to cancer, in: *Viruses in Naturally Occurring Cancers* (M. ESSEX, G. TODARO, AND H. ZUR HAUSEN, EDS.), pp. 401–421, Cold Spring Harbor Laboratory, Cold Spring Harbor, New York, 1980.

HARRISON, T. J., CHEN, J.-Y., AND ZUCKERMAN, A. J., Hepatitis B and primary liver cancer, in: *Cancer Treatment Reviews*, vol. 13 (K. HELLERMAN AND S. K. CARTER, EDS.), pp. 1–16, Academic Press, London, 1986.

MAUPAS, P., AND MELNICK, J. L. (EDS.), Hepatitis B virus and primary hepatocellular carcinoma, *Prog. Med. Virol.* **27:**1–210 (1981).

MELNICK, J. L., Hepatitis B virus and liver cancer, in: *Viruses Associated with Human Cancer* (L. A. PHILLIPS, ED.), pp. 337–367, Marcel Dekker, New York, 1983.

WEN, Y.-M., XU, Z.-Y., AND MELNICK, J. L., Viral hepatitis in China: Problems and control strategies, *Monogr. Virol.* **19:**1–159 (1992).

YEH, F.-S., AND SHEN, K.-N., Epidemiology and early diagnosis of primary liver cancer in China, *Adv. Cancer Res.* **47:**297–329 (1986).

CHAPTER 33

Human Papillomaviruses

Mark H. Schiffman and Robert D. Burk

1. Introduction

The papillomaviruses are small, double-stranded DNA viruses that infect epithelia, causing a variety of lesions including warts, intraepithelial neoplasia, and carcinoma.[227] Species-specific papillomaviruses infect many kinds of animals, including mammals and birds.[142,246]

Intensive epidemiologic study of the human papillomaviruses (HPVs) began in the mid-1980s, with the cloning and demonstration of HPV genomes in cervical carcinoma tissue.[66,85] It has been the subsequent task of epidemiologists, working in conjunction with molecular biologists, to validate and apply HPV measurement techniques to population-based studies. Epidemiologic evidence has recently been added to the extensive experimental data demonstrating an oncogenic potential for certain HPV types.[108a] This chapter on HPV represents an introduction to a rapidly growing area of cancer virus epidemiology.

The epidemiology of HPV is complicated by three factors: (1) limitations in laboratory techniques to measure HPV infection, (2) the large number of types of HPV, and (3) the highly variable and sometimes prolonged clinical course of infection.

With regard to laboratory techniques, reliable serological assays are not available to define cumulative lifetime HPV incidence.[83] Because HPVs do not grow in culture, HPV infection is most accurately measured by detection of HPV DNA sequences in currently infected tissues. The restriction of measuring only current HPV infection introduces a challenge to epidemiologic studies of the long-term natural history of HPV and the role of HPV infection in causing disease.

The HPVs are a large family of viruses, including more than 70 types that have been cloned and characterized to date.[56,255] Characterized HPV types are assigned sequential numbers, e.g., type 69 or 70, without reference to genetic relatedness. Viral type distinctions, based on DNA sequence differences, are important. Papillomaviruses appear to have evolved quite slowly in concert with their host species.[38,255] Animal papillomaviruses are not known to infect humans and vice versa.[255] Among the HPVs, each type is associated with specific anatomic sites of infection and a distinct natural history.[227] Some HPV types infect mainly nongenital cutaneous sites and other types predominantly infect oral, anal, and/or genital mucosal epithelium. Because HPV types vary in tissue tropism and biological behavior, a comprehensive review of HPV epidemiology would need to discuss each combination of HPV type and body site separately. The available data are too sparse to permit such a comprehensive approach, but similarities between the different types of HPV infections permit some useful generalizations.

The broad clinical spectrum of HPV infection further complicates epidemiologic research. HPV infections may result in a wide variety of outcomes ranging from inapparent lesions to carcinoma. For instance, genital HPV infections are usually asymptomatic and may not be diagnosed clinically.[223] Most diagnosed genital infections are benign, self-limited warty lesions (raised or flat) that disappear within months or a few years. Multiple concurrent or sequential infections are common. HPV-related carcinomas, linked to specific types of HPV, are typically diagnosed many years following initial infection. If the most subtle indications of HPV infection are counted, the cumulative lifetime probability of infection with at least one type of HPV is extremely high for all individuals.[223]

Mark H. Schiffman • Epidemiology and Biostatistics Program, National Cancer Institute, Bethesda, Maryland 20892. Robert D. Burk • Departments of Pediatrics, Microbiology, and Immunology and Obstetrics and Gynecology, Albert Einstein College of Medicine of Yeshiva University, Bronx, New York 10461.

Epidemiologic estimates of HPV incidence or point prevalence grossly underestimate the cumulative incidence.

In summary, the plethora of HPV types, the common occurrence of inapparent infections, the rarity and long latency of more serious HPV-related outcomes, and the difficulty in measuring lifetime exposure to HPV complicate the epidemiologic study of HPV infections. As a result, most epidemiologic research on HPV infections is still motivated by and focused on the priority of understanding and preventing the relatively uncommon outcome of invasive cervical carcinoma. Epidemiologists are attempting to confirm that the genital–mucosal types of HPV cause most cases of cervical carcinoma worldwide and to determine etiologic cofactors acting with HPV in cervical cancer pathogenesis. A role for HPV in other genital and nongenital epithelial carcinomas is also being explored. Epidemiologists are just beginning to study this family of common epithelial pathogens from an infectious disease perspective, where infection is the study "outcome."

2. Historical Background

The papillomaviruses are an ancient group of slowly evolving viruses that produce warts, a long-recognized clinical outcome.[227,255] However, the intensive epidemiologic study of HPV only began in the 1980s, once the link with cervical carcinoma was made. This background section reviews earlier work leading to the recent progress regarding HPV epidemiology.

2.1. Early Research on Animal Papillomaviruses

Animal papillomavirus-related lesions have been recognized for centuries, for example, equine papillomas were described in the 9th century AD. Canine and equine papillomas were transmitted experimentally in 1898 and 1901, respectively.[142,246] In the 20th century, rabbit and bovine papillomaviruses have provided the most important animal models of papillomavirus infections.

In the 1930s, Shope, Rous, and others conducted studies of cottontail rabbit papillomavirus (CRPV) infections. These studies are considered landmarks of DNA tumor virology.[227] Virus recovered from naturally occurring cutaneous horns (papillomas) of the cottontail rabbits produced papillomas in domestic rabbits. The papillomas in domestic rabbits often progressed to carcinoma, particularly following the application of coal tar or other carcinogenic cofactors.[246] Decades later, DNA recovered

from purified cottontail rabbit virus particles was shown to be infectious.[227]

Bovine papillomaviruses (BPV) types 1 and 2 produce large, benign fibroepithelial papillomas of the skin in cattle.[142,246] Despite dermal involvement, only the epithelial components of the papillomas contain virus. Similar fibroblastic reactions are seen in papillomavirus infections of deer, sheep, and elks, but not in HPV infections. It appears that papillomaviruses with dermal involvement tend to have a broader experimental host range than the viruses causing purely epithelial lesions. For example, BPV-1 and -2 can infect horses, causing an experimental equivalent of equine sarcoid.[142]

BPV-4 produces warts of the upper alimentary tract of cattle. The occurrence of alimentary carcinomas in cattle with alimentary BPV-4 infections is linked to the concurrent consumption of bracken fern growing in high-risk areas.[110] Analogous to the example of CRPV infection and coal tar application, BPV-4 infection and bracken fern ingestion serve as a model of viral–chemical interactions in carcinogenesis. However, in contrast to CRPV-associated carcinomas, BPV-4 DNA has not been detected in the bovine alimentary carcinomas.[31,107]

Much of the early work on papillomavirus immunization focused on the protection of cattle herds from BPV-1 and -2 infections.[32,142] Molecular biologists have studied BPV-1 as one of the major model systems of papillomavirus biology, because virions can be obtained in plentiful amounts from bovine warts and the virions or isolated viral DNA can transform and replicate in cultured cells.[31,107]

2.2. Early Clinical Studies of Human Papillomaviruses

Visible warts have been recognized in humans for millennia and documented at least as far back as the first century AD, in the writings of Celsus, a Roman physician.[26] In the 1890s, the transmission of human nongenital cutaneous warts from person to person was experimentally confirmed.[197] Transmission by cell-free filtrate was first demonstrated in 1907.[197] In contrast, the person-to-person infectivity of genital HPV infections was demonstrated most strongly by observational, not experimental, studies. Barrett et al.[9] described the incidence of genital warts in 24 women, 4 to 6 weeks following the return of their husbands from military service. The husbands all reported a history of recent sexual contact with another partner as well as new penile warts.

The relationship of cervical carcinoma, now known

to be an HPV-associated neoplasm, to sexual behavior has been suspected for over a century and well-established by epidemiologic studies since the 1960s.[184] The variable "lifetime number of different sexual partners" was shown by epidemiologists to be the major sexual risk factor, not "lifetime number of sexual acts," strongly suggesting an infectious rather than traumatic etiology. In the 1970s, herpes simplex virus type 2 (HSV-2) was thought to be the likely viral agent based on serological studies. However, molecular biology studies were never able to confirm the presence of HSV DNA in cervical carcinoma tissues. Subsequent work has suggested that HSV-2 is more likely to be a correlate or cofactor of HPV infection in the etiology of cervical carcinoma.[98]

Other early, important clinical studies of HPV focused on epidermodysplasia verruciformis, a rare disease characterized by disseminated polymorphic skin lesions that often progress to skin carcinomas.[185] Epidermodysplasia verruciformis was first described in 1922 and, although only a few hundred cases have been reported, the patients have been extensively studied.[185] Some cases appear to be familial and there has been a long-standing interest in characterizing the immunodeficiency leading to the persistent skin lesions, which contain a multitude of unusual, related types of HPV (see Section 7).

2.3. Early Cytopathology Studies of Human Papillomaviruses

HPV infection produces characteristic morphological changes that can be diagnosed by light microscopic examination of routinely prepared tissue biopsies or cytological preparations. Pathological studies of HPV have focused predominantly on detection of cervical infections in the context of cancer cytology screening programs. The 40-year trend in cytopathologic research has been to ascribe an increasingly large spectrum of cervical pathology to HPV infection. HPV induces pathological changes of the cervix that are usually benign, but some lesions are at increased risk of progression to carcinoma. As will be discussed in Section 8.2, it is debated whether there are any clear-cut microscopically apparent distinctions between the benign lesions and those that will progress to carcinoma.

One of the earliest advances in HPV cytopathology occurred in the mid-1950s, when Koss and Durfee described a morphological abnormality of squamous cells, which they termed *koilocytotic atypia*.[132] *Koilos* is a Greek word meaning *hollow*. Synonyms for koilocytotic atypia include "condylomatous atypia" and "warty atypia."

Koilocytes have hyperchromatic, enlarged, sometimes multiple nuclei surrounded by perinuclear clear zones (halos). Twenty years later, Meisels and Fortin[160] and Purola and Savia[196] were the first groups to propose formally that flat cervical lesions demonstrating koilocytotic atypia were cervical equivalents of condyloma acuminatum (genital warts), already known by then to be caused by HPV.

Pursuing a separate area of research in the 1960s, Richart and colleagues[199] noted the continuity of cervical cancer precursor lesions and introduced the concept of "cervical intraepithelial neoplasia (CIN)." CIN was defined as a continuum (CIN 1, CIN 2, CIN 3) of progressively more severe preinvasive changes. The intraepithelial neoplasia concept has been adapted to several other epithelial tissues relevant to HPV studies; thus, pathologists have defined intraepithelial neoplasia of the penis (PIN), vulva (VIN), vagina (VAIN), and anus (AIN).

Through the 1970s and 1980s, many pathologists began to appreciate that the cytopathologic effects of HPV classified as koilocytotic atypia were difficult to distinguish from "precancerous" changes traditionally called CIN 1. This led to the combination of the two cytological diagnoses for the purposes of cancer screening, in the new Bethesda System of cervical cytology,[235] discussed in Section 8.2. The grouping together of HPV cytopathologic changes and CIN means, in effect, that a continuum of cytopathologic change is recognized ranging from the earliest morphological features of HPV infection to invasive cervical carcinoma. Nevertheless, as emphasized in Section 8, only a very small minority of women with early, low-grade lesions develop cancer.

2.4. Molecular Biological Advances Linking Human Papillomavirus Infection and Cervical Carcinoma

The association of HPV infection and invasive cervical carcinoma was initially suggested by molecular biologists in the late 1970s and was confirmed in the early 1980s. zur Hausen and colleagues[269] found HPV DNA of previously uncharacterized types in cervical carcinomas, by probing DNA from these tissues with HPV-6 and -11 isolated from venereal warts. Immunocytochemical staining of neoplastic cervical lesions confirmed the presence of HPV-related antigens.[139]

The early molecular biological studies of HPV established the plurality of HPV types and the association of specific types with different tissues and degrees of neoplasia. HPV DNA was detected in the majority of cervical neoplastic lesions, ranging from CIN 1 to invasive carcinoma. HPV DNA was also found in cell lines derived from cervical carcinomas, most notably HeLa cells, one of the first established and most widely used cell lines.[269] With somewhat less consistency and lower prevalence, HPV DNA was also detected in other genital tumors (vulvar, vaginal, penile, and anal). A different spectrum of HPV types was found in common nongenital cutaneous warts,[227] and yet another set of HPV types was found in the skin lesions of patients with epidermodysplasia verruciformis.[185]

2.5. Earliest Epidemiologic Studies Using Human Papillomavirus DNA Testing Methods

Once molecular biologists had demonstrated in small studies that HPV DNA was present in cervical cancer tissue using Southern blot analysis, epidemiologists sought to confirm these results in rigorously designed population studies. Early epidemiologic studies were hampered by difficulties in obtaining adequate cervical specimens from nondiseased control women. Over several years, HPV DNA measurement techniques were adapted to enable testing of noninvasively obtained cervical specimens collected at the time of pelvic examination by swab, scrape, brush, or lavage. Many different combinations of specimen collection and HPV testing methods were used. In fact, different epidemiologic studies rarely used similar collection and testing strategies, making interstudy comparisons quite difficult.[18] Interassay comparisons during this developmental period, mostly informal and unpublished, were often quite discouraging, showing poor methodological reliablity. As a corollary, use of these inadequate collection protocols and inaccurate hybridization methods led to attenuated associations between HPV infection and cervical neoplasia, casting doubt for a time on the validity of the early case series that had claimed strong associations with HPV.[178] The effect of HPV exposure misclassification on early studies of HPV and cervical neoplasia was so profound that it merits consideration by all epidemiologists using newly developed assays that are incompletely validated.[80,219]

More standardized DNA test methods, permitting reliable large-scale detection of a full spectrum of HPV types in exfoliated cell specimens collected at the time of anogenital or oral examination, have been available since about 1990.[214] The improved testing techniques now being used by most epidemiologists tend to yield roughly comparable results when performed expertly on adequate specimens. However, in practical terms, the use of varying combinations of specimen collection and DNA testing methods still limits the ease of interstudy comparisons. Thus, the epidemiologist studying HPV must be thoroughly familiar with the major details of viral testing.

3. Methodology Involved in Epidemiologic Analysis

The epidemiologic study of HPV infection has been shaped and limited by available methods of diagnosing and defining infection. Accordingly, this section will briefly summarize the status of HPV DNA detection methods. Cytopathologic diagnosis of HPV infections and serological assays will be discussed in less detail; due to lower sensitivity, they currently serve an ancillary role to HPV DNA testing in epidemiologic studies of HPV. The detection of HPV particles by electron microscopy or HPV proteins by immunoperoxidase techniques will not be reviewed. Although important early studies linking HPV and genital neoplasia relied on these last two methods,[139,143] little subsequent epidemiologic research has been based on the techniques.

3.1. Human Papillomavirus DNA Detection

Currently, the diagnosis of HPV infection relies on DNA detection methods applied to tissue biopsies or exfoliated cell specimens. The detectability of HPV DNA in a single specimen (point prevalence) clearly differs from lifetime exposure to HPV (cumulative incidence). For conceptual precision, the term *HPV infection* must always be referred to specific diagnostic criteria, including the tissue tested, the specimen collection method, and the DNA testing protocol.

3.1.1. Collection of Specimens. Various, unstandardized methods of collecting tissue specimens for HPV testing are still used, even for well-studied body sites such as the cervix. Biopsies of mucosal or cutaneous lesions can be tested for HPV DNA by either *in situ* hybridization or after DNA extraction from the biopsy by any of the techniques described below. However, it is unethical to biopsy healthy women in order to establish the prevalence of infection in normal (control) populations. Thus, formal epidemiologic studies involving normal controls require noninvasive specimen collection techniques. In methodological work focused mainly on the cervix, epidemiologists have used swabs, brushings, scrapes, and lavages to obtain cells noninvasively for HPV DNA testing.[214] Similar techniques have been applied to oral, anal, and urethral mucosal sampling.

Keratinized epithelial surfaces such as the vulva,

penis, or nongenital skin have proven difficult to test reliably. Skin surfaces do not lend themselves to broad sampling by lavage. Focal sampling is less desirable because infections can be missed. HPV tends to infect tissues in a pattern of multiple discrete foci, not diffusely for a given tissue. Also, because of keratinization, skin scrapes yield less testable DNA than mucosal surfaces. Epidemiologic studies of HPV infections in males have been particularly hindered by these technical problems.

At present, the proper choice of a noninvasive cell collection method for HPV DNA testing depends on the body site and the performance of the HPV test method chosen. When HPV testing methods like Southern blot analysis are chosen, requiring a substantial amount of specimen (> 5 μg of cellular DNA), then a cervicovaginal lavage may provide optimal testing.[87] However, when more sensitive methods like consensus primer polymerase chain reaction (PCR) are chosen, the choice of collection method becomes less important, because PCR-based detection and typing of multiple HPV types require very little material.[93] With the current availability of PCR and other amplification strategies, the usefulness of DNA-rich specimens such as lavages may now be greatest when the goal is to sample a broad area like the cervicovaginal or oral cavities, obtain multiple aliquots of specimen for repeated testing, or quantitate amount of viral DNA by a direct (nonamplified) hybridization test.

Currently, epidemiologists are attempting to validate self-collection methods for cervical–vaginal HPV testing, using either home lavage kits, tampons, or self-applied vaginal swabs.[73,169,172] Large published epidemiologic projects, however, have continued to rely on investigator-obtained specimens.

3.1.2. DNA Assay Methods. Essentially there are two categories of HPV DNA detection methods: those that identify nucleic acids directly and those that amplify nucleic acids first and then detect the amplified product. In the first category are Southern blot, filter in situ hybridization (FISH), true in situ hybridization, dot blot (ViraPap/ViraType in particular), and hybrid capture. Though some important early epidemiologic studies used FISH, the technique is very inaccurate and obsolete.[214] The only amplification methods currently used for HPV epidemiology are PCR-based techniques.

Southern blot DNA hybridization was the original reference standard HPV assay because of its well-defined analytic sensitivity (amount of target viral DNA required for detection approximately ≥ 0.5 pg) and because of optimal HPV type identification provided by combined hybridization and restriction fragment analysis.[148] In practice, however, the presence of multiple or variant HPV types, incomplete cutting of the DNA by the restriction enzymes, aberrant migration of the DNA for whatever reason, weak signals, and high background signals can create diagnostic problems and misclassification of both HPV DNA detection and typing.[18] The problems with Southern blot are worsened when the amount and quality of specimen is only marginally adequate, as when a cervical specimen derived from part of a scrape or swab is analyzed.[93] The Southern blot technique, done expertly with an adequate quantity of specimen, remains an important reference standard for HPV typing of clinical specimens. However, the technical expertise and labor required have limited widespread epidemiologic use of this method.

True in situ hybridization methods have the advantage of demonstrating the location of HPV DNA within the tested tissues and the target cells. In situ tests are applied to histopathologic (or less commonly cytological) slides, which can then be viewed under the microscope. Several commercial in situ test kits have been developed.[155] Typically, they require new sections of histopathologic material or specially prepared cytological smears for optimal testing. The analytic sensitivity of in situ hybridization is generally lower than for other techniques.[147] Also, it is difficult to determine a variety of HPV types using in situ methods, because testing of multiple sections with type-specific probes is often impractical.

A reliable in situ (or other) method to test archival Papanicolaou (Pap) smears for HPV DNA would permit historical cohort studies of the natural history of cervical HPV infection. Such a method is not yet widely available, although one group recently reported successful HPV testing of a large group of archival smears in Greenland and Denmark.[233]

An additional application of in situ hybridization is to test for HPV RNA, thereby correlating morphology with transcription of specific HPV genes within the epithelium.[243] For example, researchers have determined that virion assembly, requiring the transcription of "late" regions of the HPV genome, takes place in the more differentiated cells within the superficial layers of the epithelium (explaining the difficulty of growing HPV in cell culture). RNA in situ assays are time consuming and technically difficult and have not been applied to large epidemiologic studies.

One of the simplest and fastest HPV detection methods is the dot blot. Denatured DNA is bound to a filter and probed with DNA or RNA sequences complementary to the desired target. In contrast to Southern blot analysis, DNA is not cleaved with restriction enzymes or electrophoresed. The first HPV assay to be approved by the Food and Drug Administration (FDA) for clinical use in the United States was the ViraPap/ViraType dot blot system.[147] The assay system contained RNA probes to detect HPV DNA presence (ViraPap) and type (ViraType) for the earliest identified genital HPV types: 6, 11, 16, 18, 31, 33, and 35. These types (excluding 6 and 11) are associated with a majority of cases of high-grade CIN and cervical carcinoma, but fewer than half of low-grade lesions (CIN 1 and koilocytotic atypia).[149] Cytologically normal women have an even wider range of HPV types than women with low-grade lesions. Thus, screening of normal populations with ViraPap greatly underestimated the population prevalence of HPV DNA, because of limited type range and analytic sensitivity.[11] Accordingly, an expanded probe set called "HPV Profile" was developed and FDA-approved by the same company (Digene Diagnostics) in 1994, containing the additional types 42, 43, 44, 45, 51, 52, and 56. Few epidemiologic studies employed this improved 14-type dot–blot test, because it was quickly supplanted in 1995 by the approval of the Hybrid Capture test (Digene).

Hybrid Capture is a nonradioactive and rapid liquid RNA–DNA hybridization technique.[23,47] In a liquid medium, denatured target HPV DNA is probed with type-specific RNA. Any resultant RNA–DNA duplexes are captured and detected using monoclonal antibodies that bind only to DNA–RNA hybrids. The technique currently permits semiquantitative estimation of the DNA from up to 16 types of HPV, either singly or in combination. Results from the first masked field studies using the technique have demonstrated good intra- and interlaboratory reliability, with analytic sensitivity comparable to Southern blot analysis.[75,217] Hybrid Capture testing requires less specimen than Southern blot analysis but much more than PCR. A current limitation of this method is the lack of control for the amount of test specimen collected. Therefore, it is impossible to distinguish a true negative from a specimen without sufficient material. Moreover, DNA quantitation can be affected by varying specimen amounts.

PCR-based DNA amplication methods[88] have been successfully applied to the epidemiologic study of HPV infection. The methods use multiple rounds of *in vitro* DNA synthesis with target-specific DNA primer sets, to achieve the logarithmic amplification of specific fragments of DNA. The large amounts of viral DNA generated by PCR can be typed by ethidium bromide staining or by hybridization-based detection and typing strategies such as dot blots.[88] A variety of specimen types have been tested successfully by PCR methods, with the exception of extremely bloody specimens (which inhibit the amplification enzyme) and archival blocks fixed in acidic fixatives such as Bouin's solution (which degrade DNA).[90] Because PCR-based analyses include a DNA amplification step, specimen-to-specimen contamination in the clinic or, more importantly, PCR amplication product contamination in the laboratory are real concerns. PCR analyses require very careful specimen collection, processing, and testing, with extensive negative controls, to monitor and prevent false-positive results.[88]

Not all PCR-based systems are the same, because of variability in primer sets, amplification conditions, and detection/typing methods.[214] There are two distinct kinds of PCR-based systems: those using consensus primers and those using type-specific primers.[88] The most promising PCR methods for screening and epidemiologic studies employ consensus (also known as general) primer sets designed to amplify DNA from most genital HPV types in a single amplification procedure. Two widely used systems are the L1 consensus primer method developed by Manos and colleagues[153] and the general primer system of Walboomers and colleagues.[234] Promising epidemiologic results have also been reported using other systems, for example, an E1 consensus primer method.[203] Once amplified, the DNA product can be typed using either restriction fragment analysis or by hybridization with type-specific oligonucleotides. Most recently, semiquantitative PCR test systems have been described, which permit a very crude classification of viral load.[53,154,171]

3.1.3. Comparisons of Different Viral DNA Assay Methods. Rapid improvement in HPV DNA test methods has been achieved over the past 10 years. The improvement has resulted, in part, from studies using aliquots of the same clinical specimens to compare different techniques or laboratories. Many of these studies were recently reviewed,[214] but a few of the conclusions are worth mentioning.

The first FDA-approved test, ViraPap/ViraType, compared well to a Southern blot reference standard, with sensitivity and specificity exceeding 90% for the seven types included in the assay.[126,214] However, the limitation to seven types is a serious one for epidemiologists interested in the broader spectrum of HPV infections. The reformatted "profile" dot–blot test that includes an expanded 14-type probe set (types 6, 11, 42, 43, and 44 as one probe mixture; types 16, 18, 31, 33, 35, 45, 51, 52, and 56 in another) is more useful and effectively rendered ViraPap obsolete for investigators wishing to use a dot–blot test.

Currently, the three commonly available, nonamplified HPV DNA test methods that detect 14 plus HPV types (Southern blot, dot blot, and hybrid capture) give roughly comparable results when performed expertly on ample specimens. In one multitest comparison of aliquots of ample cervical lavage specimens, HPV DNA positivity and typing were found to be roughly equivalent for the three techniques.[215] However, PCR is clearly more sensitive (but less specific for cervical neoplasia) than any of the nonamplified methods, especially when scant specimens are tested. PCR testing of cytologically normal populations typically leads to HPV prevalence rates two to four times as high as those generated by the nonamplified techniques (M. Schiffman *et al.*, unpublished data). The increased sensitivity of PCR is less evident among women with CIN and cancer because their specimens tend to contain relatively high levels of HPV DNA.

It has been suggested that consensus PCR primers are not as sensitive for specific HPV types as the corresponding type-specific PCR primers might be, perhaps due to differences in concentrations of specific primers or to mutual interference of the different primers in the consensus "cocktail."[251] The proportion of women who harbor HPV DNA at levels detectable only by ultrasensitive type-specific PCR is an unresolved and important issue in assessing HPV prevalence and natural history.[249]

Finally, the relative performance of available assays may vary somewhat in different populations, because of the spectrum and intensities of HPV infections found. Specifically, populations characterized by many very low-level infections, barely detectable at the level of analytic sensitivity of the common "direct detection" (nonamplified) tests, or a high prevalence of unusual HPV types may tend to produce method-dependent results that accentuate the sensitivity advantage of PCR (R. Burk, unpublished data).

3.2. Cytopathologic Diagnosis of Human Papillomavirus Infections

HPV can also be diagnosed based on the histological or cytological recognition of characteristic cytopathic effects induced by the virus. Morphologically, HPV infections of the skin or mucosae can produce epithelial hyperplasia and a cytopathic effect in the superficial epithelial layers referred to as koilocytotic atypia.[132] Koilocytotic atypia, the hallmark of productive HPV infection, is characterized by nuclear hyperchromasia and wrinkling associated with thick-walled perinuclear cavities. Virions in crystalline array are visible by electron microscopy in the nuclei of degenerated cells. Hyperkeratosis (acellular keratin) or parakeratosis (miniature nucleated squamous cells) may be seen at the epithelial surface. The spiny or prickle layers may be particularly thickened (acanthosis),

particularly in exophytic (raised) warts induced by certain types of HPV.

Recognition of exfoliated koilocytes, as part of cervical cancer screening, is considered pathognomonic for HPV infection. Analogous screening of other sites for HPV infection is also possible, for example, anal cytological screening can be performed in populations at high risk for anal carcinoma, such as homosexual men.

Nevertheless, epidemiologic studies of anogenital HPV infection cannot rely solely on cytological diagnosis of HPV, because it is neither sufficiently accurate nor reliable. For example, based on conventional cervical cytological diagnostic practices, only about 10–20% of HPV infections detectable by consensus primer PCR have concurrently abnormal cytological diagnoses of koilocytotic atypia. Cytological diagnosis of HPV is dependent on subjective interpretation of subtle morphological criteria.[228] It is especially precarious to compare the prevalence of HPV cytological effects across populations or over time, as diagnostic criteria tend to change.

3.3. Clinical Diagnosis of Human Papillomavirus Infections

At the clinical level, it is possible to recognize some anogenital HPV infections macroscopically. Skin warts have a characteristic raised appearance. Even when the lesions are flat and uncolored, on mucosal or skin surfaces, HPV-induced lesions may be diagnosed by application of 5% acetic acid (vinegar). HPV-induced lesions are usually "acetowhite," meaning they turn white after vinegar is applied. The mechanism of acetowhitening is not known for sure, although several theories exist.[198] Acetowhitening is frequently used by clinicians examining the cervix, vagina, and vulva under magnification (colposcopy) following an abnormal cytological result, in order to identify and biopsy lesions. Cervicography is a relatively new clinical method of cervical cancer screening that also depends on acetowhitening to diagnose HPV in magnified photographic images of the cervix.[89,248] Because it represents a low-cost version of colposcopy, cervicography is adaptable to epidemiologic studies. Like colposcopy, anoscopy and peniscopy are possible in populations at risk of anal or penile carcinoma.[188]

Directed biopsies have become the clinical reference standard for defining the severity of anogenital HPV-related disease. However, the choice of biopsy site and the pathological diagnosis of resultant biopsies tend to be variable and subjective.[68] In particular, acetowhitening is not specific for HPV infection. Many other types of epithelial conditions are also acetowhite. In fact, there is no true reference standard for defining the severity of HPV-related anogenital pathology, as clinical, microscopic, and molecular diagnostic methods are all prone to some degree of error.

3.4. Serological Assays

HPV serological assay development and validation have lagged behind advances in DNA testing, because of the lack of an abundant source of HPV virions and native antigens and the difficulties in defining the host response to HPV.[83] The mucosal antibody response to epithelial HPV infection is even less understood.[63]

If HPV DNA results are taken as the current reference standard, few reliable epidemiologic insights have been derived from seroepidemiologic studies of HPV published to date. However, this situation is changing as HPV seroepidemiology progresses. Thus, a brief review of serological methods will be presented and selected seroepidemiologic data will be presented in appropriate sections below.

To date, the use of whole or disrupted papillomaviruses as antigens have been limited by the lack of virion supply. Only a few serological studies have employed whole virus, specifically HPV-1 from plantar warts,[193] HPV-11 from human tumors produced in athymic mice,[206] and BPV-1 virions from cattle fibropapillomas.[6]

Most antigens for seroepidemiologic studies of HPV have been produced as bacterial fusion proteins,[112] chemically synthesized peptides,[61] or *in vitro* translated proteins.[175,258] The antigens are derived from particular HPV genomic open reading frames (ORF) or parts of an ORF. The most fully studied ORFs have been E2, E6, E7, and L1.[83] For each ORF, several different epitopes have been tested.

Seroepidemiologic studies have focused on HPV types 16 and 18, two of the most prevalent cancer-associated types, and HPV-6 and -11, the most common HPV types in genital warts. Researchers have attempted to define "HPV type-specific" epitopes as well as "type-shared" or "group-specific" epitopes reflecting exposure to any HPV.[6] The existence of apparently type-specific epitopes has been demonstrated by producing the same proteins (e.g., L1) from a series of HPV types and showing that subjects' antibody responses are restricted to a single HPV type or a few closely related types.[112]

A variety of test formats have been employed, including enzyme-linked immunosorbent assays (ELISA), Western blot, and radioimmunoprecipitation assays (RIPA). Mainly, IgG and IgA antibodies have been assayed in serum, with a few studies of IgA in cervical secretions.[63]

Based on linear epitopes, associations were observed between HPV seropositivity and HPV-associated disease states like CIN[61,244] or carcinoma containing HPV-16 DNA,[62,92,175,258] but not with HPV infection per se. In the last few years, the major emphasis has switched to assay systems presenting conformational epitopes.

Specifically, investigators have proposed that the neutralizing antibody response to HPV is directed predominantly against conformational antigens, created by the folding of the capsid (L1 and L2) proteins during virion assembly. These antigens usually are not completely represented by linear epitopes. A number of investigators have successfully synthesized HPV capsids (pseudovirions or viruslike particles) for viral types 1, 6, 11, 16, 18, 31, 33, and 45.[33,34,123] Data derived using these viruslike particles suggest that the assays provide a useful serological measure of recent type-specific HPV infection, independent of disease status.[123] However, no serological measure of lifetime exposure to HPV is yet on the horizon.

4. Biological Characteristics of the Viruses that Affect the Epidemiologic Pattern

The molecular biology of HPV and other papillomaviruses, as model DNA tumor viruses, is under intensive study. The reader is referred elsewhere for a recent review,[271] since only a few critical details will be mentioned here.

HPVs are nonenveloped, double-stranded DNA viruses of approximately 8000 bases[227] (Fig. 1). The assembled 55-nm virus has icosahedral symmetry with 72 capsomers. The genome is circular and only one strand contains ORFs that are transcribed. There are eight ORFs and an upstream regulatory region (URR), also called the long control region (LCR) or noncoding region (NCR), which contains an origin of replication and *cis*-acting transcriptional regulatory elements. The early region contains six ORFs (E1, E2, E4, E5, E6, and E7) that encode

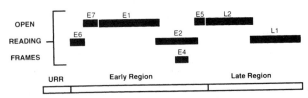

Figure 1. The genome of HPV-16 is composed of approximately 8000 base pairs, including six ORFs in the early region and two ORFs in the late region, as well as an upstream regulatory region (URR). Kindly provided by J. Schiller and Doug Lowy.

proteins necessary for viral replication and cell transformation. E6 and E7 are currently considered the major transforming proteins of the oncogenic types of HPV. In contrast, E6 and E5 serve this role in BPVs. The late region codes for the two structural proteins of the viral capsid, L1 (the major one) and L2.

The maturation of HPV virions is tightly linked to the differentiation program of the epithelial host tissues, such that HPVs cannot be grown readily in cell cultures,[165] nor can HPV be grown conventionally in animal hosts. This limitation has restricted studies of HPV pathogenesis and has limited antigen availability for immunologic studies. As one exception, HPV-11 has been successfully grown in epithelial tissues explanted under the renal capsule of nude mice, but, for unknown reasons, other HPV types have not been successfully grown in this model.[136] Recently, researchers have reported limited success in growing HPV-16 in epithelial raft systems that permit some epithelial differentiation.[165] Thus, *in vitro* studies of HPV pathogenesis may eventually be possible.

According to the current taxonomy of the International Committee on Taxonomy of Viruses (ICTV), the papillomaviruses are members of the genus *Papillomavirus*, which, along with viruses of the genus *Polyomavirus*, comprise the family *Papovaviridae* (an acronym for papillomavirus, polyomavirus, and vacuolating agent SV40). The notion of grouping papillomaviruses and polyomaviruses together has been debated since its conception by Melnick in 1962.[164] The papovavirus concept seems outdated since biochemical and molecular biological studies have revealed fundamental differences in the genomic organization and biology of the two genera.[255]

Currently, then, HPVs form a species within the genus *Papillomavirus*. Within the species, HPV types (not called strains) are defined based on proportion of DNA sequence homology. In the 1970s and 1980s, HPV types were determined by liquid hybridization, reassociation kinetics assays.[45] Viruses with less than 50% cross-hybridization to previously typed HPVs were called new types (and numbered sequentially in order of acceptance by the Papillomavirus Nomenclature Committee centered at the German Cancer Research Center in Heidelberg). Of note, 50% cross-hybridization represents much more than 50% nucleotide sequence homology. A subtype (assigned letters) was defined as a papillomavirus that cross-hybridized under stringent conditions with a given HPV prototype but had a restriction endonuclease pattern distinct from the prototype pattern. A variant was defined as differing from the prototype by an undefined but limited number of nucleotide sequences.[255]

In 1991, it was agreed by the Papillomavirus Nomenclature Committee that for a novel HPV isolate to be recognized as a new type its entire genome must be cloned and the nucleotide sequence of three regions (URR, E6, and L1) should demonstrate less than 90% nucleotide sequence identity with established papillomavirus types. Only a few new HPV types have been assigned since the introduction of this rigorous definition, which demands extensive DNA sequencing work.

Over 70 types of HPV have already been formally recognized in the past 15 years, although the entire DNA sequence is not known for all of them.[60] Many more types exist based on the results of epidemiologic studies employing broad-spectrum HPV detection methods, such as low-stringency Southern blot hybridization or consensus primer PCR. Such studies routinely find a significant proportion of uncharacterized types in study populations, particularly among women with minimal or no cytological abnormalities.[10,114] Uncharacterized (also called "novel" or "related") types test positive only under "lower stringency" test conditions that do not require exact type matches (i.e., the conditions permit some intertype cross-hybridization), but test negative or weakly positive under stringent conditions for characterized types.

When uncharacterized types are found in cervical or other carcinomas, they are usually cloned and sequenced because of their potential clinical importance. However, the pace of finding new types in cervical carcinomas has slowed over the past few years, as the majority of cancer-related HPV types have been cloned. Most remaining uncharacterized types of cervical HPV are probably not associated with cancer but, rather, with low-grade lesions or cytologically normal specimens.

As the plethora of HPV types was recognized, an underlying taxonomy gradually became evident. The data from epidemiology, laboratory assays, and genetic evolutionary analyses are highly congruent in suggesting that HPVs can be divided into two major groups, nongenital cutaneous and mucosal, each with several subgroups.

One proposed phylogenetic tree is shown in Fig. 2.[255] This tree was based on the nucleotide and amino acid sequence alignments of the E6 ORFs of 48 HPV types, using parsimony algorithms that search for the minimum number of genetic events needed to infer the most parsimonious (shortest) tree from a set of sequences. Thus, the phylogenetic tree graphically displays both the degree of similarity between different papillomaviruses and the most probable lineage of genetic change. More definitive trees can be constructed for HPV types for which the complete nucleotide sequence is known, resulting in some minor shifts, for example, the placement of

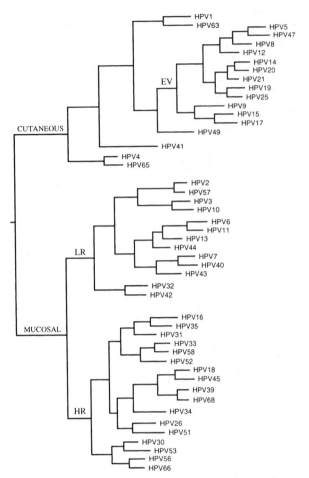

Figure 2. Phylogenetic tree constructed from the alignment of 384 nucleotides in the HPV-E6 genes using maximum parsimony analysis (PAUP 3.0). To root the tree, BPV type 1 was used as an outgroup (not shown). The branch labeled "cutaneous" contains HPV types that group together and are found to infect the cutaneous epithelium. The branch labeled "mucosal" contains HPV types that group together and are found to infect the mucosal (e.g., genital) epithelium. EV, epidermodysplasia veruciformis group; LR, "low risk for cancer" mucosal group; HR, "high risk for cancer" mucosal group. From Van Ranst *et al.*[255]

are ancient viruses that have evolved quite slowly along with their animal hosts.[255] Thus, the branchings of the phylogenetic trees between different HPV types represent many hundreds of thousands of years.

The taxonomy of the HPVs, as shown in the phylogenetic tree (Fig. 2), agrees closely with the epidemiologic data associating specific types of HPV with different tissues and grades of disease (Table 1). Specifically, the HPV types that tend to infect skin appear as a phylogenetically related group in the trees. The rare cutaneous types found mainly in patients with the immunodeficiency disease epidermodyplasia verruciformis are also shown in the tree to be closely related viruses.[185]

Among the mucosal HPV types, low-risk types found mainly in benign warty lesions cluster as a distinct group, separate from the high-risk types associated with invasive carcinomas. This concordance is evident in a comparison of the phylogenetic tree in Fig. 2 to epidemiologic data from large cross-sectional studies demonstrating the association between HPV type and grade of cervical neoplasia.[114,149]

Laboratory studies of HPV biology also support the phylogenetic and epidemiologic data. The cancer-associated

Table 1. HPV Types Grouped by Degree of Genetic Relatedness and Disease Associations[a]

Types	Location and characteristics of lesions
1, 63	Skin; mainly plantar warts
2, 27, 29, 57; 3, 10, 28	Skin; flat and common warts
4, 65	Skin; common warts; pigmented warts
5, 8, 12, 14, 19–23, 25, 36, 46, 47 9, 15, 17; 37, 38; 24; 49	Skin; most isolates are from patients with epidermodysplasia verruciformis
6, 11; 13, 44	Genital tract, esp. warts (6, 11, 44), and oral cavity (13)
7, 40	Skin; common warts; Butcher's warts
16, 31, 35	Genital tract neoplasia
18, 39, 45, 68	Genital tract neoplasia
26, 51	Genital tract neoplasia and skin
30, 53	Larynx
33, 52, 58	Genital tract neoplasia

[a]Modified from Shah and Howley,[227] van Ranst *et al.*,[255] and de Villiers.[57a] Some of these groupings are controversial due to inadequate or apparently discordant data.

HPV-2 and -57 in a more proximal, separate branch of the phylogenetic tree closest among the mucosal group to the nongenital cutaneous group.[255] In addition, more finely branching trees have been established for HPV-16 and -18 by comparing variants of these two important, cancer-associated types.[39,104,183]

By making some assumptions regarding mutation rates, the expanded phylogenetic trees can be used to estimate the evolutionary timeline of the viruses.[39,183,255] From these analyses, it appears that the papillomaviruses types of HPV that have been examined (e.g., 16 and 18) transform cells in culture more efficiently than low-risk types (6 or 11).[95,107] At the biochemical level, the transforming proteins of cancer-associated types bind to cellular tumor suppressor proteins more efficiently than the same proteins from low-risk types.[9,5107]

Epidemiologic studies have made use of the natural groupings of HPV types to permit analyses that would be impractical if each type were to be considered separately. For example, the odds ratios associating HPV and varying grades of CIN are often presented for groupings of HPV types such as "cancer-associated types," "low-risk types," "uncharacterized types," and "negative." These groupings are practical and necessary but may lead to the loss of true distinctions that do exist between the individual viruses in a group. For example, HPV type 18 is more associated with adenocarcinoma of the cervix than HPV-16,[230] although the two types are commonly combined in epidemiologic analyses.[138]

5. Descriptive Epidemiology

5.1. Prevalence and Incidence Data

As demonstrated by the preceding methodological sections, diagnostic variables preclude a simple answer to the frequently asked question: "How common is HPV infection?" The answer must be referred to a particular epithelial tissue, to a particular HPV type or group of types, to a specific specimen collection and testing protocol, and to a defined population. It is safe to state that HPV infections are extremely common in the general population. The most reliable HPV prevalence estimates have been generated for the genital region, particularly the cervicovaginal epithelium, with few data available for other tissues. Incidence rates of cervical HPV infection are currently being estimated from ongoing longitudinal studies.

5.1.1. Prevalence and Incidence of Cervical Human Papillomavirus Infections. Even when discussion of HPV infection is restricted to the most-studied tissue, the cervical epithelium, each method of diagnosing HPV infection yields a different prevalence. Most investigators view the prevalence of HPV infection of the cervix as a pyramid or iceburg, with a wide base of inapparent infections and a narrow tip of severe HPV-induced malignant cases. For example, in one cross-sectional study of young sexually active women, the point prevalence of HPV DNA was 46% when a consensus primer PCR-based technique capable of detecting a wide spectrum of HPV types was used to test both cervical and introital speci-

mens.[11] Less sensitive dot–blot (ViraPap) testing of the same specimens for seven HPV types yielded a prevalence of 11%. Less than 3% of the young women had concurrent cytological evidence of HPV infection and/or CIN. None had cervical carcinoma.

This section will focus on the prevalence of cervical HPV infection defined by DNA testing and cytology. Serological measures of genital-type HPV infection will be discussed in Section 5.1.4. The prevalence of HPV-associated CIN and carcinoma will be discussed in Section 8.

However defined, the HPV prevalence of a population appears to depend on the age and sexual practices of the population. In general, young sexually active women appear to experience the highest HPV prevalences.[79,146,163] Older monogamous women are much less likely to be HPV positive. Thus, it is imprecise to think in terms of a simple summary estimate of cervical HPV prevalence without reference to age and sexual behavior, but some ranges can be given.

Using the cytological diagnosis of koilocytotic atypia as the measure of HPV infection, the point prevalence varies from almost 10% to less than 1%, depending on the population characteristics and the diagnostic criteria employed. Most commonly, point prevalences of 2–3% are reported, with an additional 1% prevalence of CIN 1 [which is combined with koilocytotic atypia as low-grade squamous intraepithelial lesion (SIL) in the Bethesda system].[235] Grossly raised genital warts are infrequently observed on the cervix compared with the flat lesions of koilocytotic atypia/CIN 1.

Moving "down" the pyramid, the prevalence of cervical HPV infection in cytologically normal women as defined by HPV DNA detection appears to be two to four times more common than koilocytotic atypia in the same population, when a nonamplified DNA assay is used. The ratio increases to five to ten times more DNA diagnoses than cases of koilocytotic atypia when consensus primer PCR is used.[213] Correspondingly, HPV DNA prevalences of 5–10% are most commonly reported in cervical screening populations using nonamplified DNA assays, with prevalences of 15–30% typical for PCR-based surveys. However, at the extremes, HPV DNA prevalences can vary from less than 5% to nearly 100%, depending on the analytic sensitivity of the assay and the risk profile of the study group.[24,55]

HPV prevalence tends to decline with age, as will be discussed in Section 5.5. Despite high prevalence rates of HPV DNA among young sexually active women, HPV DNA is not as commonly found using standard techniques (e.g., consensus primer PCR or Southern blot) in the general population of women over 30–40 years of age,[163,171,223,252] at least in the United States and Europe.

The general population prevalence of HPV infections as defined by ultrasensitive, type-specific PCR assays is not known, but probably is higher than the estimates generated by even consensus primer PCR-based methods.[148] Some early prevalence studies using ultrasensitive PCR suggested, for example, that HPV-16 is virtually ubiquitous at very low levels in cervical specimens. This impression may stem from sample contamination[88] or could reflect a biological truth related to HPV latency or low-level viral persistence. Although controversial views exist, few if any laboratory groups are currently working on resolving the true prevalence of HPV infection at the lowest levels of detection (i.e., by directly comparing the results of consensus primer with type-specific PCR).

Most HPV prevalence estimates lump HPV types together. Specific types of cervical HPV can only be reliably distinguished at the molecular, not microscopic or clinical, level. The individual prevalences of the 20 plus HPV types known to infect the cervix are not well defined.

Table 2 presents data from large-scale HPV testing of cytologically normal women screened at Kaiser Permanente health clinics in Portland, Oregon (M. Schiffman, A. Lorincz, and M. Manos, unpublished data). Two different testing methods were used to characterize this population. The first series of women was tested for HPV by consensus primer PCR, with typing of the amplification product by dot–blot hybridization. The second set was tested for 16 specific types of HPV by Hybrid Capture. Using either test method, HPV-16 was observed to be the most common individual type among cytologically normal women. Most HPV types individually accounted for only a small proportion (< 10%) of the total infections. Unknown types as a group still accounted for a sizable fraction of infections among cytologically normal women tested by PCR.

The prevalence of concurrent infections with multiple types of HPV is dependent on test method. The detection of more than one type of HPV in a single specimen is most commonly observed with consensus primer PCR methods capable of detecting a wide range of types. The proportion of multiple infections typically approaches 20–30% of all infected women using PCR[10,99] and 10–15% by low-stringency Southern blot or Hybrid Capture.[114,201] Concurrent, multiple infections appear to be associated with an increased risk of cytologically evident CIN.[171,216] However, preliminary data suggest that multiple HPV types may be less common in cervical carcinoma than in CIN. It is not known yet whether the distribution of HPV types in multiple infections is random or reveals a tendency of certain types to favor or exclude each other's presence.

It is important in natural history studies to distinguish incident (newly acquired) cervical HPV infections from prevalent infections, because prevalence reflects the duration of infection as well as the incidence rate. However, this distinction has proven difficult for HPV. To distinguish HPV incidence from recurrence requires the knowledge at the outset that a woman has not been previously infected. With no marker of lifetime HPV exposure, the possibility of viral reactivation from a "latent" state below the level of current molecular detection must always be considered as an alternative interpretation of apparent incidence. In particular, there are serological data suggest-

Table 2. Prevalence of Individual HPV Types among Cytologically Normal Women, Assayed by L1 Consensus Primer PCR or Hybrid Capture in Cervical Specimens from Two Random Samples of Portland Kaiser Gynecology Patients (Individual Prevalence Estimates Expressed as Percent of Total Positives)

	PCR (n = 453)	Hybrid capture (n = 12,366)
Total HPV positivity	17.6%	4.6%[a]
HPV types (% of positives)		
6/11	9.7%	9.0%
16	14.6%	16.8%
18	1.2%	7.9%
31	4.8%	9.0%
33	1.2%	5.4%
35	1.2%	5.1%
39	2.4%	12.1%
40	2.4%	N/A
42	N/A[b]	15.8%
43	N/A	6.7%
44	N/A	7.0%
45	4.8%	7.9%
51	3.6%	12.3%
52	6.1%	11.9%
53	7.3%	N/A
54	7.3%	N/A
55	3.6%	N/A
56	7.3%	11.4%
58	4.8%	5.8%
Unknown or uncharacterized[c]	23.0%	N/A
Multiple types	26.1%	26.1%

[a]Hybrid Capture was performed for 16 specific types of HPV; thus, the total percentage of positivity cannot be compared to PCR without adjusting for which types were included in the respective tests.
[b]N/A, not assayed.
[c]Types not characterized at the type of original PCR testing. Several of these types have since been assigned numbers, e.g., HPV-66 and -68, or have been found to be subtypes or variants of known types.

ing high prevalence of antibodies to genital HPV types in young children.[111] These data would argue that the study of HPV incidence in young adults is actually the study of reactivation of infection transmitted initially at birth or at least reinfection with the same HPV types. However, elevated seroprevalences reported in children must be viewed with caution, given the absence of seropositivity in sexually inexperienced individuals using more specific, viruslike particle assays.[2]

Despite these theoretical concerns regarding HPV prevalence versus incidence, preliminary evidence from ongoing natural history studies suggests that age-specific incidence rates for cervical HPV infection parallel the age-specific prevalence rates presented above. In particular, the apparent incidence of cervical HPV infection, as measured by new DNA detection or first cytological diagnosis under observation, is highest in sexually active young women, mirroring the prevalence rates of HPV.[100,209] Since prevalence equals incidence multiplied by duration, this implies that the duration of most infections is rather short.

5.1.2. Prevalence and Incidence of Genital Warts.
Systematic HPV DNA testing of normal epithelium of tissues other than the cervix has rarely been performed. This is partly because of specimen collection difficulties discussed in Section 3 and partly because of weaker associations with other cancer outcomes. Discussions of the prevalence of HPV infection of noncervical tissues must rely mainly on clinical outcomes, although it is clear that a clinical spectrum of HPV infections exists in anal, vulvar, and penile infections, as for the cervix.[187,207]

A clinically evident outcome of certain genital HPV infections is exophytic (raised, not flat) genital or venereal warts, also called condyloma acuminatum. The overwhelming majority of exophytic genital warts of all sites (vulva, penis, perineum, anus) contain HPV types 6, 11, and related types.[207] Thus, the prevalence and incidence of clinically diagnosed genital warts can be viewed as a measure of infection with these HPV types, although the true infection rates are likely much higher.

Genital warts are very commonly observed in young men and women, such that they are the most common sexually transmitted viral diseases diagnosed in the United States and the United Kingdom.[14] Based on U.S. national surveillance data, there were over 200,000 first office visits (an insensitive measure of incidence) for genital warts in 1984. In the United Kingdom between 1971 and 1982, the reported incidence of genital warts ranged from about 25–100 per 100,000 population. The prevalence of genital warts in various sexually transmitted disease clinics has ranged from 4 to 13%.

In women, according to one clinical series, genital warts are most commonly found in the following sites (in rough order of decreasing prevalence)[205]: the posterior part of introitus, labia minora and clitoris, labia majora, perineum, anus, vagina, urethra, and cervix. In men, the order of frequency is reported to be the frenum, corona and glans, prepuce (foreskin), meatus, shaft, anus, and scrotum.[205]

Genital warts are commonly multicentric.[238] Moreover, vulvar warts are very commonly found in association with HPV infections of the vagina and cervix, as diagnosed by clinical examination, pathology, or DNA detection.[182,236] Thus, it is valid to view genital HPV infection as a "field phenomenon" or syndrome affecting several adjacent tissues, at least in some cases.[182]

5.1.3. Prevalence and Incidence of Nongenital Cutaneous Warts.
Few if any epidemiologic surveys have investigated the prevalence of HPV infection of the skin by molecular analysis. Thus, rates of nongenital cutaneous HPV infection must be estimated from studies of warts, but skin warts are so common and benign that they have been of little interest to epidemiologists.

One of the largest population surveys of the prevalence of skin diseases in the general population was conducted as part of the U.S. Health and Nutrition Examination Survey (HANES) of 1971–1974.[113] Over 20,000 subjects were examined, representing a stratified random sample of the noninstitutionalized population from 1 to 74 years of age. Skin warts (all types were combined) ranked as the seventh leading type of significant skin pathology diagnosed, with an estimated population prevalence of 850 per 100,000. From other studies it is known that common hand warts are more frequently seen than plantar or flat warts, but precise statements are not trustworthy, because the relative frequencies of wart types and positions vary widely by study.[205]

5.1.4. Prevalence and Incidence of Other Human Papillomavirus Infections.
The vagina, vulva, penis, anus, and larynx are tissues susceptible to infection with the "genital" or "mucosal" types of HPV that infect the cervix. Genital warts were discussed in Section 5.1.2 and respiratory papillomas are described in Section 8.1.3. HPV infection also causes flat intraepithelial lesions of the vagina, vulva, penis, and anus, analogous to CIN, although these tissues are relatively resistant to cancer outcomes.

The prevalence of vaginal HPV infection cannot be easily separated from cervical HPV infection, because it is usually not feasible to sample the vagina without collecting exfoliated cervical cells. A few small studies of HPV DNA prevalence in women with hysterectomies have found vaginal HPV infection to be common.[150] The range of HPV types detected by cervicovaginal lavage is virtually identical to the types found in cervical swabs and scrapes.[257] Flat, koilocytotic lesions of the vaginal epithelium are well recognized. However, surveillance of vaginal lesions is not as rigorous as cervical screening, because vaginal carcinoma is a rare tumor compared with cervical carcinoma.[22] Thus, although the prevalences of HPV infection of the cervix and vagina are likely similar, the risk of developing carcinoma is highly tissue specific.

The prevalence of HPV infection of the vulva must be considered as two separate subsites, the introital mucosa of the labia minorum versus the cornified skin of the labia majorum. As discussed in Section 5.1.2, the viruses that cause exophytic genital warts (6, 11, 42, 44, and related types) commonly infect both subsites of the vulva. The viruses associated with cervical carcinoma (16, 18, 26, 31, 33, 35, 39, 45, 51, 52, 54, 55, 56, 58, 59, 64, 68, and related types) also infect the labia minora with high prevalence, comparable to the cervix,[11] although high-grade intraepithelial and invasive vulvar neoplasia are relatively rare compared to cervical neoplasia. The urethra may also be infected.[263] The labia majora are much more difficult to test for HPV DNA, but HPV-containing flat lesions are commonly observed, and small studies based on vulvar biopsies have suggested that cancer-associated types of HPV may be detected within normal-appearing skin of the vulva.[78]

Similarly in the male, infection of the penis with cancer-associated types of virus appears to be very common, although reliable prevalence estimates are not available. As with the vulvar skin, noninvasive sampling of the penile epithelium is difficult. Few if any community-based surveys of penile HPV infection, equivalent to cervicovaginal screening, have been reported. Nonetheless, prevalence estimates of HPV infection (as measured by PCR-based assays of HPV DNA) in young men attending sexually transmitted disease clinics have ranged up to 84%.[264] HPV-containing, acetowhite lesions of the penis can be found in a high percentage, perhaps even a majority, of men whose sexual partners have CIN.[8,247]

HIV infection of the anus has been studied particularly in the context of human immunodeficiency virus (HIV)–HPV interactions in homosexual men.[124,187] HPV infection of the anal epithelium is very common in homosexual men, as diagnosed by HPV DNA detection or by morphological characteristics in cytological or histological specimens. The types of HPV in anal lesions are genital types, with type 16 the most prevalent. The prevalence of HPV infection of the anus in the general male population is unknown. A few studies of anal carcinoma

in women have suggested correlations with cervical carcinoma in the same populations, implying a common etiology most likely to be anogenital HPV infection.[162]

HPV of anogenital mucosa (as well as nongenital cutaneous) types can also infect the oral cavity and other regions of the upper aerodigestive tract.[145] HPV can be found in biopsies from a variety of lesions of the oral cavity and appears to be the etiologic agent for one type of benign lesion called focal epithelial hyperplasia.[84] However, a causal role for HPV in the etiology of squamous cell carcinoma of the oral cavity has not been established.[81,105,286] The special case of laryngeal papillomas will be mentioned in Section 8.1.

Noninvasive methods to sample the normal oral epithelium for HPV DNA testing are only now being developed. A comparison of three methods indicated that a mouthwash was the superior technique.[144] Although one report suggested a prevalence of 60% for HPV DNA in the oral cavity, such high prevalences have not been confirmed. In fact, in one study using oral lavage and consensus primer PCR, HPV DNA was detected in only 17% of HIV-positive and 7% of HIV-negative subjects. (R. Burk, unpublished data).

HPV infection has been reported in various other human tissues, including the esophagus,[42] lung,[29] ovary (L. Gregoire, unpublished data), prostate,[159] and bladder.[3] These associations should be considered preliminary until confirmed in multiple laboratories.

Because each type of HPV has a predilection for specific anatomic sites and viral infection may be transient, lifetime "whole-body" prevalence rates can only be assessed immunologically. However, as mentioned above, assays reflecting lifetime exposure to HPV are not yet on the horizon.

5.2. Epidemic Behavior

It appears that HPV infections are usually transmitted by person-to-person contact. It is likely (but unproven) that minor epithelial damage is a prerequisite to transmission, permitting the virus to reach the target basal cells. Fomite transmission of HPV appears possible, based on a few studies finding HPV DNA on underclothes and gynecologic equipment[76] and based on the common transmission of plantar warts among school children. Similarly, airborne transmission of viral particles appears plausible in very special settings, such as laser ablation of genital warts in a gynecologist's office, with resultant virus-containing laser plumes.[77] The viability of HPV on environmental surfaces under varying conditions and the absolute risk of transmission are unknown. The evidence for nonsexual transmission of HPV, particularly vertical transmission, has recently been reviewed.[35]

The age-specific prevalences of different types of HPV infection support the view that most infections are passed from infected to uninfected individuals. Common cutaneous warts (HPV-2 and -4, and other types) are most frequently seen in school-age children and adolescents.[197] Plantar warts have a later modal age peak in adolescence and early adulthood.[197] Male and female genital HPV infections (as measured by DNA, cytological diagnosis, or presentation of overt genital warts) peak in early adulthood concurrent with the age of usual onset of sexual intercourse.[79,213] These peaks in age-specific point prevalences appear to represent the combined effects of exposure and immunity.

5.3. Geographic Distribution

5.3.1. Cervical Human Papillomavirus Infections. The prevalence of cervical HPV infection as diagnosed microscopically cannot be compared easily across countries because of differences in terminology and diagnostic criteria, as well as the lack of well-organized registries. In the absence of valid cytological comparisons, HPV researchers have attempted to assess geographic prevalence differences using HPV DNA tests.

The geographic distribution of HPV DNA detection has been studied mainly in correlation with cervical cancer incidence rates to determine whether variation in prevalences of HPV measured by DNA would be reflected in cancer rates. In earlier studies, geographic differences in HPV did not correlate consistently with geographic differences in cervical cancer incidence.[128] However, more recent geographic correlation studies using sensitive PCR DNA testing methods to detect a wide spectrum of HPV types have generally observed HPV prevalence to "fit" the population risk of cervical carcinoma. For example, in a comparison of Colombia and Spain, the Columbian women had higher HPV prevalence when measured by PCR, in accordance with a eightfold higher risk of carcinoma.[176] Similarly, within the United States, a higher-risk urban clinic population had higher HPV prevalence than a suburban middle-class population.[10,99] Of note, many of these geographic correlations have not been able to take into account the different efficacies of regional cervical cancer screening programs, a major determinant of geographic variation in rates of neoplasia.

5.3.2. Other Human Papillomavirus Infections. The reported prevalence of genital warts is roughly the same in the United States and the United Kingdom, although diagnostic differences may confound this comparison.[133] More informative comparisons between widely divergent geographic regions have not been reported.

Although no formal comparisons of the prevalence of nongenital cutaneous HPV infections in different countries have been published, one British survey observed considerable regional variation in wart prevalence, with higher rates in the north.[265]

5.4. Temporal Distribution

5.4.1. Cervical Human Papillomavirus Infections. In many study populations, the prevalence of HPV infection as measured by DNA or cytology decreases strongly with age (see Section 5.5.1), from a peak at 15–25 years.[213] Part of the explanation for a decreasing age trend in HPV prevalence might be a "cohort effect," specifically an increase over time in the amount of HPV in the population. Given that the risk of acquiring cervical HPV infection is highest in young women, if the risk in young women today is higher than it was 10 years ago, and much higher than 20 years ago, and so forth, then the resultant age pattern in women of all ages observed today would be a decrease in HPV prevalence with age. The pattern produced in cross-sectional data by a cohort effect would be difficult to distinguish from the pattern produced by sexual acquisition/immunologic clearance hypothesis discussed in Section 5.5.1. In both cases, a cross-sectional decrease in HPV prevalence with age would be observed.

The possibility of a cohort increase in cervical HPV prevalence may be supported by concurrent increases in the incidence of genital warts and by some scant data on increasing prevalence of cervical cytological diagnoses of koilocytotic atypia.[71] However, cervical cytological data are prone to confounding changes in diagnostic criteria over time.

Time trend data regarding cervical HPV infection as measured by DNA are not available because standardized specimen collection and HPV testing protocols have been only recently developed. The few studies that have attempted to assess cervical HPV DNA prevalence in archival pathology specimens have found a comparable proportion of HPV positivity in all time periods, suggesting that the proportion of cervical neoplasia associated with HPV infection has not changed.[14]

5.4.2. Other Human Papillomavirus Infections. The incidence of genital warts appears to have increased substantially over the past few decades, at least in the United States and the United Kingdom.[14,36,133] The number of cases of condylomata diagnosed by physicians in the United States increased about fivefold between 1966 and 1984, with most of the increase occurring prior to 1976. A similar increase in incidence was observed in the United Kingdom between 1971 and 1982. The increase in the incidence of genital warts over time supports the possibility of a similar temporal increase in rates of cervical HPV infection.

No current time trend data on nongenital cutaneous warts are available, but an increasing prevalence of warts in the middle of the century has been suggested by some early references.[197]

5.5. Age

5.5.1. Cervical Human Papillomavirus Infections.

At the simplest level, it appears that cervical HPV infection rates measured by DNA detection or cytology decline sharply with age, from a peak at 16–25 years of age.[171,213,252] The point prevalence and incidence patterns for HPV DNA and HPV-related cytological diagnoses (low-grade SIL) are shown in Fig. 3. The prevalence and incidence data tend to parallel each other, as one would have predicted from the evidence demonstrating a relatively short duration of most HPV infections (see below). Moreover, the DNA data and the cytological diagnoses yield similar trends, demonstrating the strong association between the two (molecular and microscopic) diagnoses of HPV infection.

The prevalence of HPV infection in women under 16 years of age is difficult to assess, because this is the minimum age of informed consent for most epidemiologic studies requiring gynecologic examinations. Based on limited DNA test data collected on virgins, it appears that cervical HPV infection is extremely uncommon in virginal girls and women.[72,146]

As a corollary, the very high incidence and prevalence rates of HPV in sexually active women aged 16 to 25, shown in Fig. 3, are consistent with an "epidemic curve," a rapid rise in rates of infection following first (sexual) exposure. Under this epidemic curve hypothesis, the profound drop in cervical HPV prevalence in women over 30 might be due to immunologic clearance or suppression of existing infections, combined with less exposure to new HPV types because of fewer new sexual partners. The "sexual acquisition/immunologic clearance" explanation of the HPV age trend is supported by prospective data suggesting rapid, sexual acquisition of HPV[100] and by other prospective data indicating that most HPV infections are only transiently detectable.[100,209] The transience of HPV DNA detection suggests, in turn, that most infections are self-limited. In contrast, the early

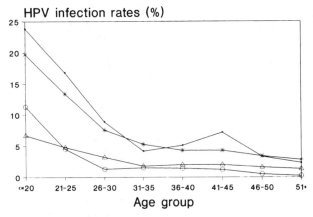

Figure 3. Age-specific HPV infection rates in a population of 23,000 women enrolled during routine cervical cytological screening at Kaiser, Portland clinics (M. H. Schiffman *et al.*, unpublished data). The line marked with triangles indicates the age-specific prevalence of low-grade SIL in the study population. The line marked with circles demonstrates the age-specific incidence rates of low-grade SIL (170 cases total) occurring within a 2-year follow-up of the 17,500 women with normal enrollment smears and no previous history of SIL. The starred line represents the age-specific point prevalences of HPV DNA detection at enrollment among the same 17,500 cytologically normal women. HPV DNA was assayed primarily by Hybrid Capture, with about 1000 women tested instead by L1 consensus primer PCR. The line marked with dots shows the age-specific incidence rates of new HPV DNA detection in 2000 women originally negative for HPV by both DNA tests and cytology, who were tested again after a median of about 1 year. In all four curves, young women have much higher rates of HPV infection than older women. Prevalence and incidence rates are quite similar, suggesting a short duration of detectable infection.

serological data showing high prevalences of genital HPV antibodies in young girls and boys,[111] if confirmed, would imply a more complicated natural history scheme including early exposure, molecularly undetectable viral persistence, and sexual behavior-related viral reactivation.

5.5.2. Other Human Papillomavirus Infections.

Age trends in other genital HPV infections (e.g., condyloma acuminatum) tend to parallel the trends in cervical infections, with the peak rates occurring between ages 15 and 25, corresponding to the usual onset of sexual intercourse. Thus, genital warts in both men and women tend to peak at these ages,[14,133,205] as does the prevalence of penile HPV DNA detection and the occurrence of flat HPV-containing lesions of the penis.[264]

Age-specific prevalence rates of nongenital cutaneous HPV infections have been less well documented, despite (or perhaps because of) the extremely high frequency of these lesions. Most of the reports are quite old, raising the possibility that the situation might have changed. In the HANES study of 1971–1974, the prevalence of nongenital cutaneous warts was observed to peak in the 12- to 17-year age group and to decline thereafter.[113] The age peak in males (18–24 years) was later than in females (12–17 years). Plantar warts were not distinguished from common cutaneous warts; thus, these prevalence estimates are not site or HPV-type specific, A recent population-based British study of nongenital cutaneous warts was consistent with the HANES study, finding a prevalence of 3.9% at age 11 and 4.9% at age 16.[265] Again, the sites and kinds of warts seen were not distinguished. From other, school-based series, it appears that the peak age of plantar warts occurs a few years later than the peak age of common warts.[106]

5.6. Sex

5.6.1. Genital Human Papillomavirus Infections. Reliable prevalence estimates for genital HPV infections in males are more difficult to obtain than for females. It appears from the available data, which are sometimes conflicting, that genital HPV infections are about equally common in both sexes.[14,133,253]

For more subtle aspects of genital HPV infections, it is very difficult to test men. HPV-induced penile lesions can be found in the urethral meatus, on the glans and shaft of the penis, and on the scrotum.[7] Unlike the localized, easily exfoliated mucosal surface of the cervix, the cornified epithelium of the penis and scrotum does not permit a reliable, standardized collection of relevant cells. Thus, although urethral meatal and coronal sulcus swabs have been assayed with positive results by nonamplified DNA test methods, the prevalence estimates of HPV determined by such surveys may not be accurate estimates of true infection rates.

Since the advent of PCR-based HPV test methods, analysis of HPV infection in men is more feasible. A few recent studies using PCR have yielded roughly similar HPV prevalence estimates for men and women attending the same sexually transmitted diseases clinics. Moreover, the few serological studies with relevant data have found little difference in HPV antibodies to genital types between males and females.[112] Seroprevalences in males may be slightly lower.

5.6.2. Nongenital Cutaneous Human Papillomavirus Infections. The HANES data from 1971–1974 suggest that males have slightly higher prevalence of nongenital cutaneous warts (10.3 per 1000) than females (7.2 per 1000).[113] The male excess was strikingly evident in only one age group, namely 18–24 (21.5 per 1000 for males compared to 4.0 per 1000 for females). Other surveys have not generally observed a male excess in wart prevalence. In fact, a female excess for plantar warts among teenagers has been reported.[239]

5.7. Race and Ethnicity

The correlation of HPV prevalence with race, controlling for known risk factors for infection, is not clear. To study the correlation of HPV with race demands random, strictly comparable sampling of different racial populations. Clinic-based studies are prone to confounding by correlates of race, such as differences in socioeconomic status and patterns of self-referral. Diagnostic accuracy of skin lesions may also vary by skin color.

Thus, increased rates of CIN and cervical carcinoma among African-Americans, Hispanics, and Native Americans (compared to non-Hispanic whites) are intertwined with socioeconomic differences in risk (intertwined in turn with a variety of correlated behaviors).[212] In contrast to the cervical cancer rates, the frequency of genital warts[14,133] and nongenital cutaneous warts[211,265] have been reported to be higher in whites than in blacks. Thus, there is no firm pattern of racial differences in susceptibility to HPV infection, at last if all types are considered as a group. Some studies, but not all, suggest that specific, racially correlated HLA haplotypes are associated with altered risk of cervical carcinoma.[4,65,91,259] The results imply that racially associated differences in susceptibility to specific HPV infections could have a biological basis.

5.8. Occupation

The only professions linked to HPV infection are those associated with an increased risk of exposure to virus. Thus, prostitutes are at an increased risk of clinical outcomes of genital HPV infection,[184] although the point prevalence of HPV DNA in surveys of immunocompetent (HIV-uninfected) prostitutes is not necessarily elevated,[137] possibly suggesting immunity in some women following intense and chronic exposure.

Curiously, meat and fish handlers are at increased risk of infection with HPV type 7.[122,210] It is a mystery why an HPV infection should be associated with handling animal tissue, but chronic maceration and trauma may be responsible, permitting viral entry to the germinal epithelium. Still, the source of the virus is unknown. Person-to-person transmission does not appear to be involved.

5.9. Socioeconomic Status

Although the evidence is not strong, cervical HPV prevalence appears to be increased in women of lower socioeconomic status,[10] in accordance with their higher risk of cervical carcinoma.[212] One study of nongenital cutaneous warts also observed increasing prevalence in children of lower socioeconomic status.[265]

5.10. Other Factors Influencing the Prevalence of Genital Human Papillomavirus Infections

Genital HPV infection appears to be sexually transmitted, but additional factors may increase the risk of infection given exposure. These cofactors promoting infection are less certain and more poorly understood than the sexual and demographic factors described above.

Current pregnancy, especially advanced pregnancy, has been found inconsistently to increase the detection of cervical HPV DNA.[99,223,232] Here, the distinction between increased prevalence of DNA detection and increased incidence of infection becomes relevant, as it is difficult to imagine why pregnancy per se would increase the incidence of infection. More likely, an increased prevalence in pregnancy would suggest the possibility of viral reactivation. Similarly, oral contraceptive use may slightly increase HPV DNA detection, as it has been positively correlated with cervical HPV prevalence in some but not all studies.[99,146,150,174] Although the effects may be weak, the influences of endogenous and exogenous hormones on HPV natural history deserve further study, because parity is a probable independent risk factor for cervical neoplasia.[17] In addition, there is some laboratory support for hormonal influences on HPV.[167]

Although smoking may have a role in later stages of cervical and vulvar carcinogenesis,[21,97,216] smoking is not strongly associated with prevalence of cervical HPV infection once correlations with sexual behavior are taken into account.[90] Nutritional and immunologic variables influencing susceptibility to genital types of HPV infection are unknown. The special case of clinically apparent immunosuppression (e.g., due to HIV infection) is an exception,[102,151,256] but one that is again confounded by the distinction between new viral infection and increased detectability of recrudescent infection.

6. Mechanisms and Routes of Transmission

6.1. Evidence for Sexual Transmission of Genital Types of Human Papillomavirus

Decades ago, the sexual transmission of genital warts was demonstrated in the wives of soldiers returning from long tours of duty.[9] It is well accepted now that sexual partners of women or men with HPV-related lesions are more likely themselves to have clinical or cellular evidence of HPV infection.[7,8,13] Some series suggest that the majority of regular sexual partners of infected individuals are themselves infected, although the lesions are often subtle and not noticed.[13,135]

Demonstrating that HPV is transmitted sexually has been more difficult at the molecular level, because the amount of error in early HPV DNA testing methods limited studies of the determinants of infection.[219] However, based on a series of more recent studies using improved HPV tests, it now is clear that cervical HPV DNA detection is sexually associated, as expected.[79,99,146] Specifically, the prevalence of cervical HPV DNA increases with reported numbers of different sexual partners, particularly recent partners. Although HPV type-specific analyses have been limited by small sample sizes, the prevalences of both cancer-associated and non–cancer-associated HPV types are associated with sexual history.

It appears that genital HPV infections as measured by DNA are transmitted rather easily between sexual partners, based on a few indirect lines of reasoning. The scant HPV DNA acquisition data suggest that having a few new male sexual partners leads to high prevalences of cervical HPV DNA within months (M. Schiffman et al., unpublished data). Also, in a few epidemiologic studies, the epidemiologic variables "*lifetime* numbers of *different* sexual partners" or "*recent* numbers of *different* sexual partners" predict risk of prevalent cervical HPV infection better than "*lifetime* numbers of *regular* sexual partners," suggesting that long-term association with a partner is not necessary for transmission.[99] Finally, the age curve of cervical HPV prevalence, with a peak at ages 15 to 25, suggests that the transmission of HPV infection to the cervix occurs soon after the initiation of sexual intercourse.[213]

As a possibly related point, the prevalence of HPV infection does not rise linearly in most studies with increasing lifetime numbers of sexual partners; rather, it reaches a "plateau."[99] In some studies, the plateau is reached with fewer numbers of partners in higher-risk populations.[99] The reasons for the plateau are not clear but may include higher HPV prevalence rates among partners as well as immunity to infection following repeated exposure.

DNA sequence variation analyses, including direct sequence determination and single-strand conformation polymorphism analysis, have only recently been applied to the study of genital specimens from male–female sexual partners.[103,267] These studies confirmed the presence of identical HPV-16 variants in six of ten sexual couples;

however, four couples had mismatching HPV-16 variants. Sequence variation studies will be increasingly important in future HPV transmission research.

6.2. Nonsexual Transmission of Genital Human Papillomavirus

Although the sexual transmissibility of cervical HPV infection is supported by multiple lines of evidence, the contribution of nonsexual routes of transmission remains unresolved and controversial.[35] Early reports of HPV seropositivity in virgins, even among young children, raised the possibility of nonsexual (perhaps vertical) transmission of genital HPV types, although not necessarily to the cervix.[83] Studies using more specific serological assays have so far contradicted the early results.[2] Fomite transmission has been postulated, based on the detection of HPV DNA in gynecologic settings and on underclothes, but transmission by fomites has never been proven.[76] There is an appreciable prevalence of cervical HPV infection as measured by DNA in women reporting only one lifetime sexual partner.[146] HPV infection in such reportedly monogamous women may reflect their partners' sexual experiences with other partners, or misreporting of monogamy by the women, or, conceivably, nonsexual routes of transmission.

Unfortunately, there are only scant data regarding cervical HPV DNA detection in virgins. Some groups have found no cervical HPV DNA in virgins.[72,146] Others have observed some cervical HPV,[261] while noting the difficulty in defining virginity using brief, standardized questionnaires. The transmission of laryngeal HPV infections (often types 6 and 11) supports the notion that vertical transmission of genital HPV infections is possible.[225] The occurrence of laryngeal papillomas in children has been associated with genital warts in the mother, and cesarean section has been associated with some protection against laryngeal papillomas. Moreover, the types of HPV in laryngeal papillomas (types 6 and 11) are the same as in genital warts. However, the reported increase in the incidence of genital warts has not been followed by an appreciable increase in laryngeal papillomatosis.

Although vertical transmission of genital types of HPV is certainly possible, the frequency of vertical transmission is still controversial and unknown.[35,204] HPV DNA is sometimes detected in the oral cavities and on the genital skin of children born to currently infected mothers. However, DNA detection in the neonatal period does not necessarily imply infection. The duration of viral DNA detection in the genital and oral epithelia is unclear. Some investigations have observed high prevalence of HPV in infants weeks to months after birth,[186] while others have not. The possible protection afforded by passive immunization in the context of concurrent vertical viral exposure is unclear.

More studies of cervical HPV infection assessed by DNA and serology in young virgin girls and women could help clarify the frequency of nonsexual transmission of cervical HPV. To date, these studies have been limited to clinical settings where a compelling clinical rationale exists, such as pediatric clinics investigating possible sexual abuse.[94]

6.3. Transmission of Nongenital Cutaneous Types of Human Papillomavirus

Based on earlier experimental work at the turn of the century, direct skin-to-skin transmission is known to be the predominant mechanism of transmission of skin warts.[197] Activities associated with minor skin trauma promote person-to-person transmission and subsequent autoinoculation.[239]

One common source outbreak of warts was reported many years ago, however. Female workers sharing the same pot of "bone glue" in a box factory contracted warts from an infected co-worker.[156]

Given that direct contact is the route of transmission of most nongenital cutaneous HPV infections, one puzzle worth considering is the mode of transmission of the rare HPV types found in epidermodysplasia verruciformis. Only a few hundred cases of the condition have been reported.[185] The reservoir of infection with these rare cutaneous types is unknown. Although data are lacking, the types of HPV commonly found in patients with epidermodysplasia verruciformis may be common commensal organisms in human skin, suppressed immunologically by most individuals.[14]

7. Pathogenesis and Immunity

7.1. Pathogenesis

7.1.1. Definitions Related to Studies of Human Papillomavirus Pathogenesis.
The use of imprecise terminology that implies a knowledge of the stage of HPV infection is common in the medical literature. For instance, the term "latent" has been used variably to refer to (1) detectable HPV DNA with no cytological abnormalities, (2) the period between first HPV infection and disease diagnosis, and (3) a state of persistent viral infection in which the viral genome is present but infectious

particles are not produced and the cell is not destroyed.[170] Other terms, such as "subclinical," also remain ambiguous. Subclinical has been defined as disease without symptoms. However, since HPV infections seldom cause symptoms apart from warts that may not be recognized, the term has little descriptive value.

To facilitate the discussion of HPV pathogenesis, a new classification system has been proposed based on the diagnostic method that establishes the HPV infection.[170] *Molecular* evidence of HPV is established by the demonstration of viral particles, nucleic acids, or proteins by methods such as electron microscopy, PCR or other nucleic acid hybridization techniques, Western blot, or immunocytochemistry. *Cellular* (or microscopic) disease requires morphological evidence of viral cytopathic effect in cytological and/or histological specimens. *Clinical* disease is diagnosed based on symptoms or signs observed on physical examination with or without aids such as magnification and acetowhite staining. A fourth category of *immunologic* diagnosis may be added to this schema.

Each of the diagnostic modalities is prone to some inaccuracy, with false-positive and false-negative results. Error aside, sensitivity increases as the method of diagnosis shifts from the clinical to the cellular to the molecular level. A diagnosis established by clinical criteria should be confirmable by pathological diagnosis and molecular methods. Likewise, cellular evidence of disease should be confirmable by molecular methods.

With these definitions, what little is known about HPV pathogenesis can be discussed. The pathogenesis of genital HPV infection will be emphasized, with reference to nongenital cutaneous warts where appropriate.

7.1.2. Pathogenesis of Genital Human Papillomavirus Infection at a Molecular Level. Most of what is known about HPV molecular biology derives from studies of a few genital HPV types linked to genital warts (HPV 6 and 11) and cervical carcinoma (HPV 16 and 18). It is assumed but not proven that the initial site of HPV infection is germinal cells in the basal layer of the epithelium. Virus probably reaches the germinal cells secondary to minor epithelial injuries. HPV-induced lesions appear from scanty genetic data to be monoclonal, suggesting that each lesion derives from a single infected germinal cell.[27] Early viral transcripts are detectable in the basal and parabasal layers of the epithelium, whereas capsid production and virion assembly occur in the more superficial layers of the differentiated epithelium.[243]

The protein products of the E6 and E7 ORFs of HPV appear to be principally responsible for HPV neoplastic effects. The E6 and E7 proteins have been shown to be cooperative transforming proteins *in vitro*.[221] The E6 protein binds to and promotes the degradation of the p53

tumor suppressor protein by forming a complex requiring the cellular protein E6-AP.[108,260] The E7 protein binds to and inactivates the retinoblastoma tumor suppressor protein.[67] The transforming proteins of cancer-associated HPV types such as types 16 and 18 have greater binding affinities for tumor suppressor proteins and greater *in vitro* transforming abilities than E6 and E7 proteins from low-risk HPV types like HPV-6 and -11.[221]

The HPV genome is maintained in the cell nucleus and is usually episomal.[52] As an exception, in invasive cervical carcinomas, integration of HPV DNA into the host genome is found in the majority of cases, particularly in cancers containing HPV-18.[52] The relative expression of E6 and E7 proteins is maintained in HPV-associated carcinomas, in association with integration of HPV into the cell genome. Integration tends to occur at fragile sites throughout the cell genome.[194] Although one report suggested preferential integration near the c-*myc* locus,[46] the frequency of site-specific integration is unclear. However, with reference to the viral genome, integration in carcinomas is definitely not random. Regulatory elements and the E6 and E7 ORFs are always preserved, with frequent disruption during integration of the E1 and E2 genes that normally inhibit E6 and E7.[69] Thus, continuous production of E6 and E7 proteins appears to have a role in HPV carcinogenicity and, in fact, the E6 and E7 regions of the HPV genome are transcriptionally active in HPV-associated cervical carcinomas and derived cell lines.[69,227]

It is not yet known whether the interactions of HPV-transforming proteins with tumor suppressor proteins explain the spectrum of HPV growth-altering effects, from the production of warts and early intraepithelial neoplasia to the rare development of carcinoma. Other early proteins of HPV, or other biochemical activities of E6 and E7, could also be important. For example, E5 may down-regulate the MHC-1 antigen presentation system, perhaps as a mechanism of immune evasion. (F. V. Cromme *et al.*, personal communication).

At the molecular level, human cancers result from an accumulation of genetic mutations. In the molecular pathogenesis of cervical carcinoma, persistent infection with an oncogenic HPV expressing E6 and E7 can be viewed as the "first hit." Additional somatic genetic changes then occur. In particular, loss of heterozygosity studies suggest that tumor suppressors on chromosome regions 11q[237] and 3p[44,130] may be involved in cervical cancer development.

Although HPV infection occasionally results in carcinoma, the usual molecular pathogenesis of HPV infections is quite benign. Cervical HPV DNA detection usually disappears within months to a few years of detection.[100,209]

Among a group of cytologically normal women measured twice 9 to 30 months apart, most women with cervical HPV infection at the first measurement were either HPV DNA-negative or had lost the original HPV DNA type at the second measurement.[100,209] The longer the interval between first and second measurement, the smaller the probability of a repeat HPV-positive result with the same HPV type. The data appeared to describe a viral "survival curve," characterized by a rather steady, slow resolution of viral infection.

Natural history studies of HPV molecular pathogenesis may be complicated by poorly understood variation in detectability of the virus. In a few repeated-testing studies employing both direct and amplified methods, HPV DNA was observed to be intermittently detectable.[58,173,222] If confirmed, these data raise questions about the molecular state and location of the HPV DNA when undetectable.

Nonetheless, if most HPV infections diagnosed at the molecular level are transient, while few are persistent, then HPV DNA persistence may be an early important step in the development of genital lesions such as warts and neoplasia.[25] Accordingly, women developing CIN under observation are much more likely than controls to have apparently persistent rather than transient DNA detection. For example, one study observed 43% HPV DNA persistence in 114 cases who had developed incident CIN under observation compared with 8% persistence in 400 control women who remained cytologically normal.[100]

The determinants of HPV DNA persistence are largely unknown. Four categories of risk factors seem most plausible: viral factors such as type, transcriptional state, or viral load; target cells (truly germinal versus committed to differentiation); host immunologic response, either genetically (e.g., HLA haplotype) or environmentally determined; and environmental cofactors such as hormonal factors, smoking, and concurrent infection with other sexually transmitted diseases.

With regard to viral characteristics affecting persistence, preliminary data suggest that cancer-associated types of HPV may persist longer than non-cancer-associated types.[100] The role of copy number (viral load) on persistence is currently under study by several groups, with preliminary data suggesting that high levels of virus predict increased rates of persistence (A. Hildesheim et al., unpublished data). Specific immunologic or genetic determinants of HPV DNA persistence are not yet known, and in a recent exploration of possible behavioral influences on HPV persistence (such as smoking and oral contraceptive use) no statistically significant influences were found.[100] In summary, the determinants of HPV persistence may be subtle and difficult to uncover epidemiologically.

7.1.3. Pathogenesis of Genital Human Papillomavirus Infection at a Cellular Level. In cattle, experimental inoculation of BPV by scarification leads to microscopically evident epidermal papillomatosis in about 2–3 months.[142,246] The pathogenesis of human skin warts may have a similar incubation period (see Section 7.1.4). Comparable experimental data are not available for genital HPV infection. However, some preliminary observational data on the pathogenesis of cervical HPV infections at the cellular level are now available from one of the large cohorts under study (M. Schiffman and A. Lorincz, unpublished data). As shown in Fig. 4, among cytologically normal women who are HPV DNA positive at enrollment using the Hybrid Capture assay, the absolute risk of incident abnormal smears rises to a very high level (approximately 25% of smears taken) at 1–2 years following enrollment and declines thereafter, returning to baseline (1–2% of smears taken) at about 4 years. This peak at 1–2 years may approximate the incubation period of cervical HPV infection.

Based on these same data, the cumulative incidence of cytological abnormalities following HPV infections detectable by Hybrid Capture testing may exceed 50% within 4 years compared with about 10% in initially HPV-negative women. The absolute risk of cytological abnormalities following all infections detectable by PCR methods would probably tend to be lower than the risk estimated by the less sensitive (more specific) Hybrid Capture test.

Observational studies that attempt to define the transition from HPV infection defined molecularly by HPV DNA to infection defined microscopically raise the important issue of diagnostic accuracy. The cellular diagnosis of HPV infection is subjective and prone to significant error.[229] The earliest and mildest cellular changes associated with cervical HPV infection are extremely subtle and nonspecific.

Therefore, it is not clear how to interpret data indicating that concurrent cytological abnormalities can be found in only a minority of the women with HPV DNA at the level of detection of Southern blot or consensus primer PCR. One explanation is that very small areas of DNA-shedding cytopathic cells are missed by standard cellular diagnostic methods. Perhaps many HPV DNA-positive women have lesions too small to be diagnosed. In this interpretation, the incidence of CIN following HPV infection shown in Fig. 4 could represent the growth of a focus of HPV-infected cells to cytologically detectable size. Alternatively, HPV DNA might be detectable before and after the briefer appearance of cytopathic changes, i.e., HPV DNA might persist in normal-appearing cells. In fact, in situ tests do sometimes detect HPV DNA but not

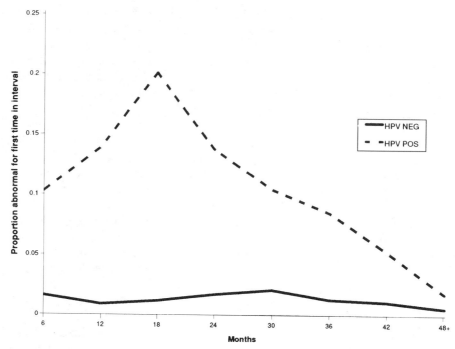

Figure 4. Risk of new diagnosis of SIL by HPV DNA test result at enrollment (M. H. Schiffman *et al.*, unpublished data). A cohort of 17,654 women with normal cytological diagnoses and no known past history of SIL were tested for 16 types of cervical HPV DNA using primarily Hybrid Capture. For each subsequent half-year time interval, the numbers of incident SIL diagnoses were calculated as a proportion of all women in the cohort obtaining smears during the interval. The relative and absolute risks of SIL were greatly increased in the 1279 women testing positive at enrollment for HPV compared with the larger group testing HPV negative. The greatest risk of SIL following HPV DNA detection appeared in the second year of follow-up.

RNA in normal-appearing cells in the midst of or adjacent to abnormal tissue. In summary, it is not clear whether molecularly diagnosed HPV infections, especially high levels of viral DNA, imply the existence of cellular abnormalities somewhere in the infected tissue. Conversely, once cytological misclassification is ruled out, HPV DNA is found in 90% or more of cytologically diagnosed HPV infections, at least when employing the diagnostic criteria common in the United States.[229]

Cancer-associated types of HPV are associated with a higher risk of developing cytologically evident lesions than are non-cancer-associated types of HPV[134,242] (M. Schiffman *et al.*, unpublished data). Apart from HPV type, the determinants of developing cytologic abnormalities following infection are not known. In one case-control study restricted to HPV DNA-positive women, no associations of risk of developing a cytologically evident abnormality were observed with socioeconomic status, lifetime numbers of sexual partners, age at first intercourse, or oral contraceptive use. Smoking was a possible, weak cofactor given infection. Increasing parity was a risk factor for cytological abnormality independent of infection.[216]

A final important point about pathogenesis at the cellular level: It appears that the expansion of the spiny layer of the epithelium characterizing HPV lesions, particularly exophytic ones, results from decreased cell differentiation and reduced squamous cell sloughing, rather than increased rate of cell turnover.[240,241] As a result, the proliferative compartment (the number of cells dividing) is increased in HPV-infected epithelia, but the rate of cell division may not be.

7.1.4. Pathogenesis of Genital Human Papillomavirus Infection at a Clinical Level. Most HPV-induced cellular abnormalities are transient and barely detectable at the clinical level. For example, flat HPV lesions of the penis are usually not recognizable, although they can be seen by peniscopy as acetowhite areas following the application of acetic acid.[264] Similarly, cervical intra-epithelial neoplasia is asymptomatic and requires colposcopy to be seen.

The most commonly diagnosed clinical outcomes of

HPV infection are nongenital cutaneous and genital warts. The incubation period for clinically evident common skin warts after experimental inoculation averages about 4 months (range 1–20 months).[197] The minimum incubation period for the development of exophytic (raised) genital warts appears to be approximately 1–2 months.[9] The incubation period seen in human studies corresponds well to the 2–3 months needed for experimental BPV fibropapillomas to appear.[142,246] The incubation period of flat, genital HPV lesions in humans has not been defined experimentally, but appears to be months to a few years as well, based on preliminary studies (see Section 7.1.3).

Warts tend to grow from an initially inapparent size to an appreciable but delimited lesion that then persists or regresses over the course of a few months to a few years. Clinically evident wart disease has been noted to worsen during pregnancy[191] and in response to immunosuppression, for example, secondary to renal transplantation.[1,74]

A much rarer, delayed clinical outcome of HPV infection is carcinoma, particularly cervical carcinoma. HPV infection (of specific types) appears to be necessary but far from sufficient for the development of most cases (greater than 80%) of cervical carcinoma.[270] Infection with HPV explains a smaller fraction of the other HPV-related malignancies, whose etiologies may be more multifactorial than the etiology of cervical carcinoma. The pathogenesis of cervical carcinoma following HPV infection is discussed in Section 8.

7.2. Immunity

HPV infections are extremely common, while severe cases of warts or HPV-related neoplasia are relatively rare in comparison. Immunity to HPV probably controls most infections, but little is known about the details of immunity to HPV.

HPV infections, with the exception of carcinomas, are limited to mucosal and cutaneous epithelium above the basement membrane. There is no known viremia. It is unknown whether and how systemic or local immune factors affect HPV infections. Also unclear is whether infection with one type of HPV protects against infection with other types.

Much of what is known about HPV immunology derives from animal papillomavirus studies and small investigations of HPV infections in immunodeficient individuals. Unfortunately, early human seroepidemiologic studies using incompletely validated assays yielded confusing results, and not much is known about normal cellular immune responses to HPV infection.

7.2.1. Lessons from Animal Papillomavirus. In

both rabbits and cattle, papillomavirus infections are usually acquired by the young.[142,246] Warts persist for variable periods then regress, leaving the hosts immune to reinfection with the same virus types.

In the rabbit, development of a systemic immune reaction to papillomavirus infection is supported by the following observations[142,246]: (1) All papillomas regress concurrently once regression begins, (2) suppression of the immune system inhibits regression, while vaccination with tumor preparations increases regression frequency, (3) regressing papillomas have a mononuclear infiltrate, and (4) rabbits with regressing papillomas are resistant to reinfection with CRPV DNA, while rabbits with persistent papillomas are not resistant to additional infection.

Multiple bovine fibropapillomas also regress simultaneously, accompanied by infiltrating mononuclear cells, and with resistance to reinfection.[142,246] Early therapeutic vaccination trials in cattle were not successful, but recent results have been promising.[32] Potential animal modes for HPV vaccine development now include (in addition to BPV and CRPV) canine oral papillomavirus and rhesus monkey papillomavirus.[19]

7.2.2. Human Papillomavirus Infections in Immunodeficient Patients. An increased prevalence and severity of wart disease has been observed in patients with a variety of immune-related disorders. For example, an excess in cutaneous and genital warts among subjects with iatrogenic immunosuppression following renal transplant has been demonstrated.[1,14,74,192]

As mentioned above, patients with the rare syndrome epidermodysplasia verruciformis are subject to extensive HPV skin lesions containing characteristic types of HPV.[185] Furthermore, the lesions are prone to develop into squamous cell carcinoma. Although a specific immune dysfunction has been postulated as the cause of epidermodysplasia verruciformis, one has not been identified.[86]

HIV infection provides perhaps the most important example of the effect of immunosuppression on HPV infection. Individuals infected with HIV through sexual contacts are likely to be exposed to genital types of HPV as well, because both viruses are sexually transmissible and prevalent in sexually active populations. Thus, an increase in HPV-related clinical outcomes would be expected in certain HIV-infected cohorts, even if there were no causal connection. However, the increased diagnosis of HPV in HIV-infected individuals (by molecular, cellular, or clinical means) is striking and apparently related to immunosuppression.

Specifically, HIV infection is associated with a very high prevalence of HPV DNA detection that increases further with increasing immunosuppression (as measured by decreasing CD4 counts).[102,161,256] Concurrently, CIN is commonly diagnosed. Parallel situations are seen in studies of anal infection among homosexual men.[161] It is not known whether HIV immunosuppression permits reactivation of previously suppressed HPV infection, as opposed to allowing rapid infection or reinfection from environmental sources.

Although the association between HIV infection, HPV infection, and anogenital intraepithelial neoplasia is established, a causal role for immunosuppression in the risk of progression of intraepithelial lesions to invasive carcinoma is less clear. Anal carcinoma rates are increasing in the homosexual male population, probably related to HIV infection.[187,189] In contrast, cervical carcinoma rates are not greatly elevated in HIV-infected female cohorts, although the length of follow-up of HIV-infected female cohorts is still limited (T. Cote, unpublished data). It is possible that HIV immunosuppression speeds the normally decades-long progression from intraepithelial neoplasia to invasive carcinoma,[213] but HIV immunosuppression could act mainly to increase the prevalence and persistence of CIN precursors.

7.2.3. Serological Studies of Genital Human Papillomavirus Infections.
Epidemiologists and immunologists are attempting to differentiate the serological responses constituting natural immunity to HPV infections from biomarkers of infection or disease progression. The scant data from animal studies do not indicate a major role for antibody responses in wart regression.[142,246] Immunization of animals with papillomavirus capsid proteins blocks reinfection but does not appreciably affect existing warts. The role of mucosal antibody in blocking reinfection in animals has not been well studied.

No convincing evidence of a protective serological response in humans has been reported. Rather, several groups have reported elevated HPV antibody levels in subjects with clinically evident HPV-associated disease compared with controls.[83] For example, elevated titers to HPV 16 E7 are observed in about half of HPV-16-associated cervical carcinoma patients.[112,129,175,258] In this instance, the serological response appears to be a biomarker of recent overexpression of one of the key viral transforming proteins.

The few reports of elevated HPV seroprevalence in women with CIN are not clearly marking immunity. Some of the associations are not HPV type-specific,[61] yet they demonstrate that even early intraepithelial neoplasia can generate a systemic immune response.

Recent epidemiologic studies using viruslike particles as antigens are providing more important data.[34] The majority of women with CIN, cancer, genital warts, or nongenital cutaneous warts have antibodies to the same HPV types found in their lesions. With regard to HPV-16, women with very low probability of exposure, assessed by DNA testing and interview, have a very low HPV-16 seroprevalence, while those infected with other genital types of HPV (not necessarily the types most related genotypically) have an intermediate seroprevalence.[262] Persistent cervical HPV DNA positivity, even in the absence of cytological abnormalities, is highly predictive of seropositivity. Because antibody positivity to HPV-16 viruslike particles is reasonably specific, it is possible to use the assay to address the issue of whether HPV is associated with other neoplasms and with nonsexual modes of transmission. Investigations so far have confirmed the associations of HPV infection with anal cancer[96] and subtypes of vulvar cancer (A. Hildesheim et al., unpublished data). However, seropositivity in men with penile cancer is elevated irregularly depending on study population. Regarding HPV transmission, the reported lack of seropositivity in virginal women weakens somewhat the possibility of common, nonsexual transmission.[2]

7.2.4. Cellular Immune Responses to Human Papillomavirus Infections.
Based on the animal experiments and the immunosuppression data, it is assumed that the key immune response involved in the clearance of HPV infections is cell mediated. As in animals,[142,246] once multiple human warts of a single kind begin to regress, they regress concurrently,[49] although warts of different kinds (e.g., plantar vs. common skin) may not regress concurrently. While genetic defects affecting humoral immunity have little effect on the natural history of HPV infections, deficiencies of cell-mediated immunity profoundly elevate the prevalence of HPV infection assessed clinically, microscopically, or molecularly.[49,115] Histologically, regressing flat skin warts show infiltration with mononuclear leukocytes and other features suggestive of a cell-mediated immune response.[205]

The two classes of cells thought to be involved in the cellular immune response to HPV are antigen-presenting cells and lymphocytes. The usual antigen-presenting cells of the epithelium are the Langerhans cells. Langerhans cells are less numerous near HPV-induced epithelial lesions compared to normal epithelium; thus, Langerhans cell abnormalities may be one aspect of permissive immune reactions to HPV.[168]

Investigators have also attempted to define lymphocyte abnormalities in patients with HPV infections.[51,116,181,202] No specific dysfunctions have been widely accepted, although the general dysfunctions found in HIV infection and transplant-associated immunosuppression were mentioned above. The possible associations of specific HLA locus haplotypes with the risk of invasive cervical carcinoma, if confirmed, might yield indirect evidence of the role of specific aspects of cell-mediated immunity in the natural history of HPV infections.[65] In the study of HPV and cell-mediated immunity, it may be important to study local infiltrating lymphocytes rather than circulating lymphocytes. However, this theoretical requirement, which would make population-based epidemiologic studies nearly impossible by requiring tissue specimens rather than blood from all subjects, is not proven.

8. Patterns of Host Response

Different types of HPV infection may be associated with a wide range of epithelial response, from no apparent lesion to invasive carcinoma. A summary of the main disease associations of selected HPV types is presented in Table 1.

Section 7 focused on the development of HPV-induced epithelial lesions (warts and intraepithelial neoplastic lesions). This section will focus on the subsequent natural history of the different lesions, particularly the best-studied one, cervical intraepithelial neoplasia (CIN).

8.1. Warts

8.1.1. Nongenital Cutaneous Warts. Skin warts are a well-known and common response to infection with HPV types 1, 2, 4, and related types. There are several kinds of warts, with varying classifications whose full description is beyond the scope of this chapter.

The most frequent skin wart is the common variety (verruca vulgaris), which occurs mainly on the back of the hands and fingers, but can occur anywhere on the skin.[205] Common warts are firm papules ranging in size from less than a millimeter to over a centimeter in diameter. They may occur singly or in groups. About two thirds of common warts spontaneously regress within 2 years.[156] Malignant conversion is extremely rare.[205]

Deep plantar warts (verruca plantaris) are most commonly associated with HPV type 1, which is phylogenetically and molecularly quite distinct from the viral types more commonly found in common skin warts.[254] Plantar warts are usually single but may be multiple. These lesions are sharply demarcated from the adjacent normal skin and possess a smooth hyperkeratotic collar. Regression is common but variable, and may decrease with age.[205]

A less common kind of skin wart is the flat wart (verruca plana), which is smooth and flat or mildly elevated. Flat warts occur multiply, mainly on the face, back of the hands, and the shins. They may disappear suddenly after weeks or months or persist for years.[205]

Common skin warts can occur rarely on the genital skin, but no association has yet been reported between the occurrence of nongenital cutaneous warts and genital warts or CIN.[205] Multicentric nongenital cutaneous wart disease is very common, sometimes with mixed kinds of warts.[205] Multicentric wart disease of the same kinds is apparently related to autoinoculation by scratching, handwashing, and nailbiting.[239]

8.1.2. Genital Warts. Almost all exophytic genital warts, also called condylomata acuminata, contain HPV types 6, 11, or related types.[226] Like warts of nongenital skin, they tend to regress spontaneously and only rarely are associated with malignant progression. However, genital warts can be multifocal, can become very large, and may recur after treatment, creating significant discomfort and psychosexual morbidity in some patients.

8.1.3. Respiratory Papillomas. Respiratory (laryngeal) papillomatosis is a rare disease with a bimodal age occurrence.[119] Thirty to 50% of cases become evident before the age of 5 years (juvenile-onset), linked most probably to vertical transmission from the maternal genital tract of HPV-6 or -11.[225] Another group of cases associated with the same HPV types occurs in adulthood (adult-onset), with an unproven mode of infection. From an epidemiologic perspective, the adult-onset respiratory papillomas are distinct from the juvenile-onset variety. Adult-onset respiratory papillomas are more often solitary, recur less frequently, do not tend to spread, and exhibit a male predominance.[119]

Respiratory papillomas most commonly affect the larynx, particularly the true vocal cords, resulting in hoarseness as the major symptom. Interestingly, the true vocal cords contain a junction (transformation zone) between stratified squamous epithelium and columnar epithelium, very similar to that found on the cervix. The papillomas uncommonly undergo malignant transformation, although one report noted cases of malignancy following irradiation therapy.[152] Although they are benign, the papillomas may be life threatening due to airway obstruction, particularly in young children. Laryngeal papillomas tend to recur in many cases despite repeated excision; thus, this rare presentation of warts is among the most severe and debilitating.

8.2. Cervical Intraepithelial Neoplasia

The natural history of HPV-induced lesions has been studied most extensively for cervical infections, because of the association with cervical carcinoma. Therefore, this topic will be more extensively reviewed.

The cervix, though a well-studied epithelial model, is a special histological case (as are the true vocal cords and anus). The transformation zone of the cervix is defined as the anatomic site at which the glandular columnar epithelium of the endocervix is replaced by stratified squamous epithelium, through the process of squamous metaplasia. The transformation zone is particularly susceptible to neoplastic transformation following HPV infection. In vivid contrast, the neighboring mature vaginal epithelium of the ectocervix, though prone to HPV infection, very rarely develops carcinoma.[22] In the laboratory, cultured normal endocervical epithelial cells transformed by HPV-16 develop into lesions resembling carcinoma *in situ*, whereas HPV-16-transformed ectocervical cells develop a lesion with only mild dysplastic change.[245] The reasons for the special susceptibility of the transformation zone are unknown.

8.2.1. Defining CIN. The morphological response of the cervical epithelium to HPV infection is usually not a raised wart but, rather, a flat lesion with warty colposcopic and cytopathic features. As a corollary, the flat lesions of the cervix contain mostly types of HPV not found in exophytic venereal warts (e.g., 16, 18, 26, 31, 33, 35, 39, 45, 51–56, 58, 59, 64, 66, and 68), although flat lesions may also contain the types associated with exophytic genital condylomata (e.g., 6, 11, 42).

The HPV-induced, flat intraepithelial lesions of the cervix form a gradient of risk of malignancy, first formalized by Richart and colleagues when they introduced the term cervical intraepithelial neoplasia.[199] Usually, CIN is classified as CIN 1 (mild dysplasia), CIN 2 (moderate dysplasia), and CIN 3 (encompassing severe dysplasia and carcinoma *in situ*). Several other overlapping nomenclatures have also been applied, as summarized in Fig. 5. As the figure implies, the spectrum of CIN has no perfect "cutpoints" and is, therefore, a subjectively defined gradient that is difficult to categorize reproducibly. Moreover, histopathologic and cytopathologic definitions of CIN vary somewhat because of the different types of specimens.

As mentioned in Section 3.2, lesions designated as CIN 1 are virtually indistinguishable from koilocytotic atypia. The recent combination of CIN 1 and koilocytotic atypia as low-grade squamous intraepithelial lesions (LSIL), in the Bethesda classification of cytopathology, is supported by epidemiologic data demonstrating that the two subsumed diagnoses share the same broad HPV type spectrum, have similar demographic characteristics such as early age peaks, and share a transient and benign

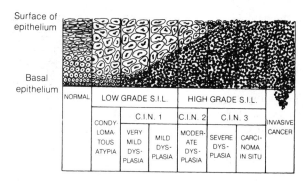

Figure 5. Cervical cancer precursor lesions. Early cervical cancer "precursors" are characterized by superficial cellular abnormalities that represent the koilocytotic changes of HPV infection. More severe, true precursors demonstrate a progressive extension toward the surface of undifferentiated, malignant-appearing cells. This morphological spectrum of changes has been categorized using several nomenclatures, most recently the Bethesda system. Adapted from A. Ferenczy.

natural history.[213] In centers where the distinction is still made, the cytological diagnosis of koilocytotic atypia is usually two to three times more common than CIN 1.[218]

Though oversimplified, the histopathology schema of CIN shown in Fig. 5 is a useful starting point for epidemiologists beginning to conceptualize the natural history of HPV infection and cervical neoplasia. In normal epithelium, proliferation is restricted to the basal layer, the so-called germinal epithelium. The epithelial cells differentiate on a fixed and orderly program that leads to their programmed death and sloughing. In low-grade lesions, squamous differentiation in the more superficial layers of the epithelium becomes abnormal but the cells continue to differentiate, such that there is only a minimal effect on the expansion of the proliferative (non-differentiated or "immortalized") compartment. Thus, in low-grade lesions, the proliferative layer remains less than one third of the full thickness of the epithelium. Fundamentally, low-grade lesions can be thought of as viral infections that are common, minor, and usually transient.

8.2.2. Progression of Low-Grade to High-Grade CIN. Although low-grade lesions caused by HPV are usually transient, women with low-grade cervical lesions are nonetheless at substantially increased risk of developing higher-grade CIN, the accepted precursor to invasive carcinoma.[179] High-grade CIN usually emerges in women with low-grade CIN, but may sometimes arise in women with preceding, repeatedly normal cytology or only equivocal lesions.[125,134] In the United States, high-grade CIN lesions are an order of magnitude less common than low-grade lesions, with an age peak that is 5–10 years later. The low prevalence of high-grade CIN compared with low-grade CIN may not be as pronounced in regions with deficient cervical cancer screening and treatment, where high-grade lesions are permitted to develop and accumulate.

Terminology for high-grade CIN is shown in Fig. 5. A one-third to two-thirds replacement of the epithelium with proliferating cells is termed CIN 2. Virtual full-thickness replacement of the epithelium with proliferating cells, with loss of epithelial differentiation, is termed CIN 3, which includes carcinoma *in situ*. In CIN 2 and CIN 3, in addition to expansion of the proliferative compartment, there is also an increased appearance of progressively more serious nuclear abnormalities. CIN 2 and CIN 3 are combined for cytology in the Bethesda system because (1) the categories are not reliably distinguished in practice and (2) both diagnoses lead to the same practical outcome, i.e., immediate colposcopic referral and biopsy, with ablative treatment if confirmed.

A diagnosis of low-grade CIN predicts an increased risk of a subsequent diagnosis of CIN 2–3, but it is not totally clear just how frequently women with low-grade lesions progress. In one representative study of 555 women newly diagnosed with cytological evidence of HPV infection (CIN 1), 16% of women progressed to CIN 3 or invasive carcinoma ($n = 2$) within the median follow-up of 48 months, 22% had persistent low-grade cytological abnormalities, and 62% regressed.[179] Other studies have found higher rates of progression, but with more severe entry criteria such as multiple consecutive abnormal cytologic smears.[199]

To reiterate an important caution, although prospective studies have demonstrated that women with CIN 1 are at increased risk of developing CIN 2–3 somewhere on the cervix, no one is sure that low-grade *lesions* progress to CIN 2–3 lesions. CIN 2–3 lesions are usually found more proximal to the endocervical canal than low-grade lesions, which are commonly further out on the ectocervix. It has recently been suggested, therefore, that CIN 2–3 might arise from normal-appearing or atypical metaplastic epithelium at the internal margin of low-grade lesions.[125]

Whether or not CIN 2–3 lesions emerge directly from low-grade lesions, the type of HPV found in the cervix appears to predict the risk of the woman progressing to a diagnosis of CIN 2–3. This association has been verified in both cross-sectional[12,114,149] and prospective[30,121] studies. While many HPV types are found in low-grade lesions, fewer types are found in CIN 2 and even fewer in CIN 3. Generally the types of HPV associated with CIN 3 are the same types found in invasive carcinomas. In lower-grade lesions, the cancer-associated HPV types increase steadily in relative prevalence with the severity of the diagnosis.

A crucial current question is, "What else besides HPV type predicts the risk of progression to a diagnosis of high-grade cervical neoplasia?" HPV infection and resultant low-grade lesions are very common, while CIN 2–3 is much less common. This suggests that additional factors must be critical in determining whether HPV infections resolve or progress. The three categories of factors that seem most likely a priori to be responsible for progression include (1) viral factors, (2) host factors, and (3) environmental cofactors.

With regard to viral factors apart from viral type, it appears that high levels of HPV DNA are more closely linked to the prevalence of CIN 2–3 than are low levels.[53,154,171] Time since first infection may also be important.[177] The degree of nuclear atypia may increase with the duration of cervical HPV infections. As a possible mechanism of increased nuclear atypia, it has been suggested that HPV infection increases the mutability of cellular DNA.[270]

The most important host factors related to progression in diagnosis from low-grade HPV lesions to CIN 2–3 are probably immunologic. From animal models, it appears that the regression versus progression of HPV-induced lesions is dictated mainly by the cell-mediated immune response.[142,246] Assays of cell-mediated immunity adaptable to large epidemiologic studies are needed to permit exploration of successful and unsuccessful immune responses to HPV. In one investigation, for example, low levels of interleukin 2 (IL-2) production by lymphocytes exposed *in vivo* to HPV-16 E6 and E7 peptides were found in women with CIN 2–3 or cancer compared with HPV-16-infected women who had not developed high-grade disease (T. Tsukui, personal communication). Investigators are now looking for a simpler assay of this apparently important Th1 response.

Other important host influences on risk of progression to CIN 2–3 could include parity, which might act by influencing immunity or by nutritional or traumatic mechanisms.[177] Age has not been shown to be a strong predictor of risk of progression to CIN 2–3, once the severity of the initial diagnosis is taken into account.[179]

The best a priori candidates for environmental cofactors for the development of CIN 2–3 are the established risk factors for cervical carcinoma that do not appear to be mere proxies for HPV infection. Smoking would seem to be a good candidate as a cofactor because smoking has both mutagenic and immunologic effects. A role for smoking has some epidemiologic support.[216] However, smoking has not been found to be associated with the development of CIN 2–3 in a few recent, well-designed studies, including a prospective study.[134,177] Other possible cofactors for the development of CIN 2–3 might be oral contraceptive use,[216] selected nutrient deficiencies,[28] and concurrent infection with other sexually transmitted agents such as chlamydia.[54,134,177]

8.3. Invasive Cervical Carcinoma

Cervical carcinoma can be considered the most serious known outcome of HPV infection. Cervical carcinoma is the second most common cancer of women worldwide and the fifth most common cancer in humans.[190] The worldwide incidence of cervical carcinoma is about 440,000 cases per year. In the United States, where cervical cytological screening and treatment of CIN has reduced the incidence of invasive carcinoma, only 15,000 invasive cases will occur per year.[15] However, after decades of declining U.S. rates, the incidence and mortality from cervical carcinoma may now be slowly rising again in younger women, particularly white women.[131,200] An increase could relate to increases in HPV prevalence that started decades ago (the cohort effect mentioned in Section 5.4), but this has not been established. A similar recent increase in cervical cancer rates has been observed in other countries, such as Great Britain.

It is known from histological evidence and a few early follow-up studies that invasive cervical carcinoma develops frequently from CIN 3. However, the absolute risk of untreated CIN 3 developing into invasive disease is still argued, with estimates averaging about 30% but ranging from 10 to 90%. Similarly, the risk factors for invasion are poorly known. CIN 3 and invasive carcinoma share nearly all the same risk factors, except for age. The modal age of invasive carcinoma is 10 years later than for CIN 3 and, conceivably, invasion could be a stochastic process requiring an additional, random genotoxic event with few prominent risk factors appreciable by epidemiologic methods.

In any case, the epidemiologic risk factors for invasive cervical carcinoma among HPV-infected women are the same as mentioned above for CIN 3. Smoking is a theoretically attractive cofactor,[97] but smoking was not linked to risk of carcinoma among HPV-positive women in two recent large case-control studies among Latin populations.[17,70]

Hormonal influences appear to be important, with a possible role for oral contraceptive use and high parity.[17] The etiologic supporting roles of nutritional deficiencies of vitamin C, carotenoids, and folate deserve more epidemiologic attention,[195] as do the roles of other concurrently infecting sexually transmitted diseases and cervical–vaginal inflammation in general.[54,98]

The prevalence of HPV DNA in women with invasive cervical carcinoma is extremely high wherever the issue has been intensively studied. The most definitive estimates come from a study of 1050 cervical carcinomas from over 20 countries, tested by L1 consensus primer PCR.[16] Over 85% of cervical carcinomas from each country contained HPV DNA, with the inclusion of "possible" infections raising the proportions even higher.

Accordingly, a "cancer-associated" or "high-risk" group of genital HPV types can be defined as those found with appreciable prevalence, alone, in invasive cervical carcinomas.[82,148] Based on this broadest definition, the current list of cancer-associated HPV types would include at least types 16, 18, 26, 31, 33, 35, 39, 45, 51, 52, 54–56, 58, 59, 64, and 68. Other more restrictive definitions of "cancer-associated" are possible, based on relative risk or attributable proportion calculations.[149] Regardless of definition, HPV-16 is consistently the most common cancer-associated type. As referenced in Section 4, the epidemiologic definition of cancer-associated types correlates well with the relative transforming properties of the viral types as defined *in vitro* and with "phylogenetic" studies grouping HPV types by genetic relatedness.[254,255]

Several epidemiologic studies using serological assays have observed elevated seropositivity in women with invasive cervical carcinoma.[83,175,258] The seropositivity observed in cancer patients appeared to be an insensitive biomarker, in that typically about half of the patients whose tumors contained DNA of the same viral type being assayed serologically are seronegative. A few of the assays, however, have reasonable specificity that would support their use in further epidemiologic surveys of invasive carcinomas.[92,123,258]

The vast majority of cervical carcinomas are HPV-containing, squamous cell tumors, but some important exceptions should be mentioned. Adenocarcinomas are increasing in absolute and relative frequency, especially among younger women.[127] As a result, adenocarcinomas now constitute about 20% of cervical carcinoma in the United States.[127] Adenocarcinomas are disproportionately associated with HPV type 18 compared to squamous carcinomas, but this association is not absolute in that HPV-16 is still the most common type found in adenocarcinomas in many series.[16,230]

HPV-negative cervical carcinoma appears to exist as a separate entity, with a worsened prognosis[59] and possibly distinct molecular features.[117] However, the condition is relatively rare (15% or less of tumors). It is not yet clear whether HPV infection can be lost during pathogenesis. Probably some cervical carcinoma arises independent of HPV. Given the multifactorial etiology of virtually all major cancers, it is rather surprising that HPV infection accounts for as many cases of this carcinoma as it does.

8.4. Other Carcinomas Associated with Human Papillomavirus Infections

8.4.1. Other Genital Carcinomas. Genital HPV infection is associated with carcinomas and precursor lesions of all parts of the female and male lower genital tracts, including vaginal, vulvar, penile, and anal neoplasia. Analogous to the concept of CIN for the cervix, pathologists diagnose a spectrum of HPV-related intraepithelial cytopathologic changes termed vulvar intraepithelial neoplasia (VIN), vaginal intraepithelial neoplasia (VAIN), penile intraepithelial neoplasia (PIN), and anal intraepithelial neoplasia (AIN). Multifocal intraepithelial genital lesions, associated with HPV infection of cancer-associated types, are often observed. Accordingly, some authors have referred to a lower genital neoplasia syndrome.[182]

However, the proportion of other genital carcinomas attributable to HPV infection may be lower than for cervical carcinoma; that is, the other genital carcinomas may be more multifactorial than cervical carcinoma. The point is best illustrated by the example of vulvar carcinoma. It appears that there are different histological subtypes of squamous cell vulvar carcinoma, with distinct epidemiologic profiles.[50,141] While typical squamous cell vulvar carcinoma in older women is not HPV associated, the less common basaloid and warty subtypes found more often in younger women are HPV associated.[141] The HPV-associated vulvar carcinomas are often found adjacent to intraepithelial lesions that may represent vulvar HPV infections.

The clinical appearance of intraepithelial lesions of the skin of the vulva and penis is particularly subtle and variable. Both papular and macular, pigmented and unpigmented lesions are found, associated with a variety of genital HPV types.[5] Acetowhitening is a useful but nonspecific diagnostic aid. The amount of elevation in risk of subsequent carcinoma posed by these lesions on the vulva or penis is unclear. When multiple lesions occur with a histological appearance resembling intraepithelial neoplasia, the term Bowenoid papulosis is applied.[224] Bowenoid papulosis occurs mainly in patients under 40 years of age and is thought to be benign, although HPV-16 is the

predominant HPV type.[224] Bowen's disease, in contrast, occurs at generally later ages and is synonymous with PIN 3 or VIN 3. Bowen's disease occurs most often as a solitary plaque that is slowly progressive.[224] Large epidemiologic studies of intraepithelial, HPV-associated lesions of the penis and vulva have not been reported; thus, prevalences and risks of progression are not well defined. The role of HPV infection in the etiology of penile cancer is especially unclear.[37] In some regions, chronic inflammation appears to be a more important risk factor.[20]

8.4.2. Nongenital Carcinomas Associated with Human Papillomavirus Infections. Although HPV DNA has been found in a variety of nongenital carcinomas, the etiologic role of HPV infection is still speculative. Certain rare verrucous (warty) carcinomas of the larynx and oral cavity are accepted to be HPV-related.[120] HPV appears to have a possible role in some other squamous cell carcinomas of the aerodigestive tract, including some oral carcinomas, nasal carcinomas, tongue carcinomas, and tonsillar carcinomas.[57,145] Animal models of oral HPV infection associated with malignancy exist.[40] A variety of benign lesions of the oral cavity have been linked to HPV infection, including cases of focal epithelial hyperplasia, papillary hyperplasia, fibrous hyperplasia, lichen planus, and leukoplakia.[40] However, the established epidemiologic risk factors for oral carcinoma are alcohol and tobacco; thus, the interaction of HPV with these chemical agents must be resolved.

The role of HPV in carcinoma of the esophagus is unresolved. Several studies conducted mainly in high-risk areas of China have found HPV DNA in a sizable proportion of esophageal carcinomas and precursor lesions.[41,101] Curiously, other investigations have not confirmed this association.[250] The possibility of HPV carcinogenesis of the esophagus is intriguing, given the known model of BPV-4 in bovine alimentary carcinogenesis.[110]

Some of the associations of HPV with other neoplasms, like adenocarcinoma of the prostate,[159] transitional cell carcinoma of bladder,[3] and skin and lung carcinoma,[29] are controversial. Still others, like the reported association of HPV infection with colon carcinoma, are frankly suspect.[180,231] For a very common virus like HPV, false-positive test results and specimen contamination may be difficult to distinguish from true infection. The role of HPV infections in carcinogenesis of these organs is an area of active study, with no firm conclusions as yet. Finally, some rare epithelial tumors have been linked to HPV infection, including conjunctival carcinomas and periungual skin carcinomas.[157]

HPV-16 is the type of HPV most commonly reported in carcinomas other than the cervix, with fewer reports of the other cervical cancer-associated HPV types. In some instances, aberrant HPV variants of usually nononcogenic types (e.g., HPV-6 and -11) have been reported in carcinomas of the vulva, penis, and larynx.[64,118,158]

9. Control and Prevention

9.1. Primary Prevention of Human Papillomavirus Infections

Prevention of transmission of HPV infections appears nearly unachievable, given the ubiquity of HPV infections. The exception is that genital HPV transmission might be reduced by condom use or changes in sexual behavior. However, condom use cannot entirely prevent the spread of genital HPV infections, because genital HPV infections in the male are not limited to the penile skin. Viral spread from scrotal and perineal lesions would not be prevented by condom use, although it appears that these sites of infection are uncommon compared to penile lesions.[207,208]

Ultimately, advances in HPV immunology might make most HPV infections and clinical sequelae vaccine-preventable, including genital warts and even cervical carcinoma.[49] Since HPV is a central cause of most cervical neoplasia, it is reasonable to consider HPV immunization as the ultimate primary preventive strategy for eliminating most cervical carcinoma. The protection of cattle herds from BPV infection by vaccination serves as a successful animal model.[32] Use of the hepatitis B vaccine in Asia to reduce the incidence of hepatocellular carcinoma may serve as a public health model. Currently, the most promising candidate vaccines include the viruslike particles now available in various laboratories for HPV types 6, 16, 18, 31, 33, and 45.

9.2. Treatment of Human Papillomavirus Infections

Until vaccines are available, HPV infections and HPV-induced lesions will continue to be highly prevalent. HPV control efforts will be aimed at preventing carcinoma, particularly cervical carcinoma. Thus, research into the proper treatment of HPV-induced early lesions can be seen as research into the prevention of cervical carcinoma.

In the United States, the standard response to the detection of HPV-associated, precancerous cervical lesions is either expectant management (monitoring to observe whether regression occurs) or destruction of the lesion and the remaining cervical transformation zone by cryosurgery, laser, or loop electrocautery.[266] Some groups have attempted to treat HPV infections of the cervix chemically with topical, destructive chemical applications, or vitamin A analogues.[166] Trials of oral micronutrient supplementation to promote lesion regression are underway, but so far have been unsuccessful.[43]

For genital wart disease, destruction of the lesions remains the major clinical modality, although the indications for therapy and the efficacy of current techniques can be questioned. Eradication of virus is not an achievable goal through destructive means. Destructive methods include chemical applications, such as podophyllin or acids, and surgical means, including laser, cryosurgery, and electrosurgical loop excision. Trials of adjunctive interferon or antimetabolite administration have been conducted for severe recurrent genital warts, with variable success.

Nongenital cutaneous warts, when treated, can be destroyed by a wide variety of destructive methods including keratolytic agents, cytotoxic agents, chemotherapy, immunotherapy, cryosurgery, and various other surgical removal techniques.[109]

9.3. Use of Human Papillomavirus DNA Testing in Cervical Cancer Screening

Since most women with cervical HPV infections on a given day are healthy and will remain so, routine testing of all women simply for the presence of the virus is unlikely to be useful. Nevertheless, more specific uses of HPV testing as a screening tool in selected populations will undoubtedly be important.

It has been demonstrated that HPV testing can be used in several ways to clarify and triage inconclusive or "atypical" Pap smear diagnoses.[47,48,220,229] The borderline between normal and abnormal cervical cytological diagnoses has never been clear, as evidenced by the plethora of uninformative terms such as "class 2 Pap" or "atypia." Vague, inconclusive diagnoses such as atypia may account for 20% or more of diagnoses in some screening centers. Even under the new Bethesda system, which has attempted to sharpen and restrict the number of inconclusive diagnoses,[140] at least 3 to 10% of women who are screened are given a borderline diagnosis of atypical squamous cells of undetermined significance (ASCUS). The triage and treatment of women with borderline smears consumes a disturbingly disproportionate amount of clinical resources.

There are at least two possible settings in which HPV testing can be used to clarify inconclusive cytological diagnoses: (1) quality control of the diagnosis of ASCUS in cytopathology laboratories, and (2) clarification and triage of the individual inconclusive cytological result.

Clinical trials are needed to establish the usefulness of HPV testing in two other important and theoretically appealing applications: general screening of older women and triage of koilocytotic atypia and CIN 1. HPV screening at older ages could be used to supplement Pap smears as a means of defining women at high risk.[163,171] Because HPV DNA prevalence in cytologically normal women declines sharply with age, while HPV prevalence in women with cervical neoplasia remains very high regardless of age, the positive predictive value of finding HPV DNA rises with age.[171] Moreover, the use and accuracy of the Pap smear declines with age, due to inadequate sampling of receding transformation zones (false negatives) and overdiagnosis of atrophy-related changes (false positives). Thus, HPV detection and typing in older women might define a small subset of patients who remain at appreciable risk and could benefit from closer surveillance. Postmenopausal women who are cytologically normal and have a negative HPV DNA test might be at very low risk for the development of cervical carcinoma and could be screened infrequently.

The diagnosis of koilocytotic atypia or CIN 1, by cytology or histology, is a poor marker of risk for the development of cervical carcinoma, because progression to invasive carcinoma is rare, even in untreated women. Many lesions regress spontaneously within months of detection. Colposcopically directed biopsy and ablation of all koilocytotic atypia and CIN 1 represents overtreatment, with high costs and some associated morbidity. Natural history studies indicate that lesions associated with cancer-associated types of HPV are the most likely to progress.[30,121] Consequently, a large, prospective randomized clinical trial is underway in the United States to determine whether testing patients with koilocytotic atypia and CIN 1 (and ASCUS) diagnoses for both HPV type and a measure of viral load could result in safe and cost-effective clinical management, reducing the morbidity that can result from overtreatment.

10. Unresolved Problems

Recent epidemiologic studies of HPV have made firm progress in clarifying (1) the high percentage of cervical neoplasia attributable to HPV infection, (2) the sexual determinants of HPV infection of the cervix, and (3) the benign natural history of most HPV infections and early HPV-induced lesions. Most work will continue to address cervical HPV infections, but it will be important

to verify or exclude the role of HPV in carcinomas of other sites. Comparisons should be made of HPV natural history in different tissues. In particular, why is the transformation zone of the cervix so prone to HPV carcinogenesis?

In cervical studies, an attempt should be made to study cervical HPV incidence rather than HPV prevalence, by enrolling sexually inexperienced women for repeated measurements in follow-up studies. As a related point, epidemiologists must study young couples to fully address issues of sexual transmission of cervical HPV infection. Transmission studies will need to be intensive and relatively small, until reliable measurements of penile HPV infection are validated. Expected advances in the diagnosis of viral type variants and in HPV seroepidemiology should greatly aid transmission and natural history studies.

The seroepidemiology of HPV must be fully explored, using the new assays employing conformationally correct capsid antigens (viruslike particles). In particular, the prevalence of seropositivity to cancer-associated HPV types in young children must be clarified, to understand how common early nonsexual transmission is.

The importance of HPV persistence in the pathogenesis of neoplasia must be verified and the determinants of persistence identified. If carcinomas arise only from persistent infections, not from molecularly inapparent (latent) infections that suddenly reactivate, then prevention of carcinomas should be achievable by screening for virus at ages preceding the usual onset of carcinoma. The study of HPV persistence requires repeated measurements and extremely reliable HPV testing. It will be necessary to test variants of HPV types to permit, for example, the distinction of new HPV-16 infections from persistent ones.[104]

Through continued prospective studies, the decrease in cervical HPV infection rates with increasing age should be better understood. The separate contributions to the age trend of cohort effects and immunologic suppression must be distinguished. Any cohort effect of increasing HPV infection in currently younger women might predict an increase in invasive cervical carcinoma in the future. Immunologic suppression as an explanation for the age trend would be a much more reassuring interpretation of the high HPV rates in younger women today.

A particularly pressing research priority is to understand the natural history of early HPV-induced cervical lesions, in an attempt to explain why a few infected women progress to cervical carcinoma. Viral, host, and environment factors affecting the risk of progression must be addressed.

With regard to viral factors, HPV DNA detection

will continue for at least the next few years to be the major definition of cervical HPV infection. It will be important to understand how subtleties of DNA detection, such as the amounts of viral DNA, reflect and predict natural history. Perhaps high-intensity infection can be used as a cross-sectional predictor of HPV persistence and progression.[53,154,171] The epidemiologic measurement of HPV RNA must be standardized to permit an examination of HPV activity in addition to mere viral presence.

As the highest priority, HPV immunology is likely to occupy epidemiologists studying cervical HPV infection over the next decade. In the immediate future, the interactions of multiple HPV types in mixed cervical infections should be clarified, as one pathway to understanding HPV immunity. Assays of cell-mediated immunity must be developed and applied. The ultimate goal will be to define the successful immune response to HPV infection, in the hope that cancer-preventive immunity can be stimulated by vaccination.[49]

11. References

1. ALLOUB, M. I., BARR, B. B. B., MCLAREN, K. M., SMITH, I. W., BUNNEY, M. H., AND SMART, G. E., Human papillomavirus infection and cervical intraepithelial neoplasia in women with renal allografts, *Br. Med. J.* **298**:153–156 (1989).

2. ANDERSSON-ELLSTROM, A., DILLNER, J., HAGMAR, B., SCHILLER, J., AND FORSSMAN, L., No serological evidence for nonsexual spread of HPV16, *Lancet* **344**:1435 (1994).

3. ANWAR, K., PHIL, M., NAIKI, H., NAKAKUKI, K., AND INUZUKA, M., High frequency of human papillomavirus infection in carcinoma of the urinary bladder, *Cancer* **70**:1967–1973 (1992).

4. APPLE, R. J., ERLICH, H. A., KLITZ, W., MANOS, M. M., BECKER, T. M., AND WHEELER, C. M., HLA DR-DQ associations with cervical carcinoma show papillomavirus-type specificity, *Nat. Genet.* **6**:157–162 (1994).

5. AYNAUD, O., IONESCO, M., AND BARRASSO, R., Penile intraepithelial neoplasia. Specific clinical features correlate with histologic and virologic findings, *Cancer* **74**:1762–1767 (1994).

6. BAIRD, P. J., Serological evidence for the association of papillomavirus and cervical neoplasia, *Lancet* **2**:17–18 (1983).

7. BARRASSO, R., HPV-related genital lesions in men, in: *The Epidemiology of Human Papillomavirus and Cervical Cancer* (N. MUNOZ, F. X. BOSCH, K. V. SHAH, AND A. MEHEUS, EDS.), pp. 85–92, IARC Scientific Publications, Lyon, 1992.

8. BARRASSO, R., DE BRUX, J., CROISSANT, O., AND ORTH, G., High prevalence of papillomavirus-associated penile intraepithelial neoplasia in sexual partners of women with cervical intraepithelial neoplasia, *N. Engl. J. Med.* **317**: 916–923 (1987).

9. BARRETT, T. J., SILBAR, J. D., AND MCGINLEY, J. P., Genital warts—a venereal disease, *J.A.M.A.* **154**:333–334 (1954).

10. BAUER, H. M., HILDESHEIM, A., SCHIFFMAN, M. H., GLASS, A. G., RUSH, B. B., SCOTT, D. R., CADELL, D. M., KURMAN, R. J., AND MANOS, M. M., Determinants of genital human papilloma-

virus infection in low-risk women in Portland, Oregon, *Sex. Transm. Dis.* **20**:274–278 (1993).

11. BAUER, H. M., TING, Y., GREER, C. E., CHAMBERS, J. C., TASHIRO, C. J., CHIMERA, J., REINGOLD, A., AND MANOS, M. M., Genital human papillomavirus infection in female university students as determined by a PCR-based method, *J.A.M.A.* **265**:472–477 (1991).

12. BERGERON, C., BARRASSO, R., BEAUDENON, S., FLAMANT, P., CROISSANT, O., AND ORTH, G., Human papillomaviruses associated with cervical intraepithelial neoplasia. Great diversity and distinct distribution in low- and high-grade lesions, *Am. J. Surg. Pathol.* **16**:641–649 (1992).

13. BERGMAN, A., AND NALICK, R., Prevalence of human papillomavirus infection in men: Comparison of the partners of infected and uninfected women, *J. Reprod. Med.* **37**:710–712 (1992).

13a. BERNARD, H. U., CHAN, S. Y., MANOS, M. M., ONG, C. K., VILLA, L. L., DELIUS, H., PEYTON, C. L., BAUER, H. M., AND WHEELER, C. M., Identification and assessment of known and novel human papillomaviruses by polymerase chain reaction amplification, restriction fragment length polymorphisms, nucleotide sequence, and phylogenetic algorithms, *J. Infect. Dis.* **170**:1077–1085 (1994).

14. BERNSTEIN, S. G., VOET, R. L., GUZICK, D. S., MELANCON, J. T., RONAN-COWEN, L., LIFSHITZ, S., AND BUCHSBAUM, H. J., Prevalence of papillomavirus infection in colposcopically directed cervical biopsy specimens in 1972 and 1982, *Am. J. Obstet. Gynecol.* **151**:577–581 (1985).

15. BORING, C. C., SQUIRES, T. S., TONG, T., AND MONTGOMERY, S., Cancer statistics, 1994, *CA Cancer J. Clin.* **44**:7–26 (1995).

16. BOSCH, F. X., MANOS, M. M., MUNOZ, N., SHERMAN, M. E., JANSEN, A., PETO, J., SCHIFFMAN, M. H., MORENO, V., SHAH, K. V., AND THE IBSCC STUDY GROUP, Prevalence of HPV DNA in cervical cancer in 22 countries, *J. Natl. Cancer Inst.* **87**:796–802 (1995).

17. BOSCH, F. X., MUNOZ, N., DE SANJOSE, S., IZARZUGAZA, I., GILI, M., VILADIU, P., TORMO, M. J., MOREO, P., ASCUNCE, N., GONZALEZ, L. C., TAFUR, L., KALDOR, J. M., GUERRERO, E., ARISTIZABAL, N., SANTAMARIA, M., ALONSO DE RUIZ, P., AND SHAH, K., Risk factors for cervical cancer in Columbia and Spain, *Int. J. Cancer* **52**:750–758 (1992).

18. BRANDSMA, J., BURK, R. D., LANCASTER, W. D., PFISTER, H., AND SCHIFFMAN, M. H., Inter-laboratory variation as an explanation for varying prevalence estimates of human papillomavirus infection, *Int. J. Cancer* **43**:260–262 (1989).

19. BRANDSMA, J. L., Animal models for HPV vaccine development, *Papillomavirus Rep.* **5**:105–111 (1994).

20. BRINTON, L. A., LI, J. Y., SHOU-DE, R., HUANG, S., SHENG, X. B., BAI-GAO, S., ZHE-JUN, Z. S., SCHIFFMAN, M. H., AND DAWSEY, S., Risk factors for penile cancer: Results from a case-control study in China, *Int. J. Cancer* **47**:504–509 (1991).

21. BRINTON, L. A., NASCA, P. C., MALLIN, K., BAPTISTE, M. S., WILBANKS, G. D., AND RICHART, R. M., Case-control study of cancer of the vulva, *Obstet. Gynecol.* **75**:859–866 (1990).

22. BRINTON, L. A., NASCA, P. C., MALLIN, K., SCHAIRER, C., ROSENTHAL, J., ROTHENBERG, R., YORDAN, E., AND RICHART, R., Case-control study of *in situ* and invasive carcinoma of the vagina, *Gynecol. Oncol.* **38**:49–54 (1990).

23. BROWN, D. R., BRYAN, J. T., CRAMER, H., AND FIFE, K. H., Analysis of human papillomavirus types in exophytic con-

dylomata acuminata by hybrid capture and Southern blot techniques, *J. Clin. Microbiol.* **31:**2667–2673 (1993).

24. BURGER, M. P. M., HOLLEMA, H., GOUW, A. S. H., PEITERS, W. J. L. M., AND QUINT, W. G. V., Cigarette smoking and human papillomavirus in patients with reported cervical cytological abnormality, *Br. Med. J.* **306:**749–752 (1993).

25. BURK, R. D., KADISH, A. S., CALDERIN, S., AND ROMNEY, S. L., Human papillomavirus infection of the cervix detected by cervicovaginal lavage and molecular hybridization: Correlation with biopsy results and Papanicolaou smear, *Am. J. Obstet. Gynecol.* **154:**982–989 (1986).

26. BURNS, D. A., Warts and all—the history and folklore of warts: A review, *J. R. Soc. Med.* **85:**37–40 (1992).

27. BUSCEMA, J., SHAH, K., HSU, S., ROSENSHEIN, N., AND WOODRUFF, J. D., Genetic investigation of the cellular origin of vulvovaginal condylomata acuminata, *Cancer Cells* **5:**245–247 (1987).

28. BUTTERWORTH, C. E., HATCH, K. D., MACALUSO, M., COLE, P., SAUBERLICH, H. E., SOONG, S. J., BORST, M., AND BAKER, V. V., Folate deficiency and cervical dysplasia, *J.A.M.A.* **267:**528–533 (1992).

29. BYRNE, J. C., TSAO, M. S., FRASER, R. S., AND HOWLEY, P. M., Human papillomavirus-11 DNA in a patient with chronic laryngotracheobronchial papillomatosis and metastatic squamous-cell carcinoma of the lung, *N. Engl. J. Med.* **317:**873–878 (1987).

30. CAMPION, M. J., McCANCE, D. J., SINGER, A., AND CUZICK, J., Progressive potential of mild cervical atypia: Prospective cytological, colposcopic, and virological study, *Lancet* **2:**237–240 (1986).

31. CAMPO, M. S., Cell transformation by animal papillomaviruses, *J. Gen. Virol.* **73:**217–222 (1992).

32. CAMPO, M. S., GRINDLAY, G. J., O'NEIL, B. W., CHANDRACHUD, L. M., McGARVIE, G. M., AND JARRETT, W. F. H., Prophylactic and therapeutic vaccination against a mucosal papillomavirus, *J. Gen. Virol.* **74:**945–953 (1993).

33. CARTER, J. J., HAGENSEE, M., TAFLIN, M. C., LEE, S. K., KOUTSKY, L. A., AND GALLOWAY, D. A., HPV-1 capsids expressed *in vitro* detect human serum antibodies associated with foot warts, *Virology* **195:**456–462 (1993).

34. CARTER, J. J., HAGENSEE, M. B., LEE, S. K., McKNIGHT, B., KOUTSKY, L. A., AND GALLOWAY, D. A., Use of HPV 1 capsids produced by recombinant vaccinia viruses in an ELISA to detect serum antibodies in people with foot warts, *Virology* **199**(2):284–291 (1994).

35. CASON, J., KAYE, J. N., AND BEST, J. M., Non-sexual acquisition of human genital papillomaviruses, *Papillomavirus Rep.* **6:**1–7 (1995).

36. CENTERS FOR DISEASE CONTROL, Condyloma acuminatum—United States: 1966–1981, *Morbid. Mortal. Week. Rep.* **32:**306–308 (1983).

37. CHAN, K. W., LAM, K. Y., CHAN, A. C. L., LAU, P., AND SRIVASTAVA, G., Prevalence of human papillomavirus types 16 and 18 in penile carcinoma: A study of 41 cases using PCR, *J. Clin. Pathol.* **47**(9):823–826 (1994).

38. CHAN, S. Y., BERNARD, H. U., ONG, C. K., CHAN, S. P., HOFMANN, B., AND DELIUS, H., Phytogenetic analysis of 48 papillomavirus types and 28 subtypes and variants: A showcase for the molecular evolution of DNA viruses, *J. Virol.* **66:**5714–5725 (1992).

39. CHAN, S. Y., HO, L., ONG, C. K., CHOW, V., DRESCHER, B.,

DURST, M., TER MEULEN, J., VILLA, L., LUANDE, J., MGAYA, H. N., AND BERNARD, H. U., Molecular variants of human papillomavirus type 16 from four continents suggest ancient pandemic spread of the virus and its coevolution with humankind, *J. Virol.* **66**(4):2057–2066 (1992).

40. CHANG, F., SYRJANEN, S., KELLOKOSKI, J., AND SYRJANEN, K., Human papillomavirus (HPV) infections and their associations with oral disease, *J. Oral Pathol. Med.* **20:**305–317 (1991).

41. CHANG, F., SYRJANEN, S., SHEN, Q., HONGXIU, J., AND SYRJANEN, K., Human papillomavirus (HPV) DNA in esophageal precancer lesion and squamous cell carcinoma from China, *Int. J. Cancer* **45:**21–25 (1990).

42. CHANG, F., SYRJANEN, S., WANG, L., AND SYRJANEN, K., Infectious agents in the etiology of esophageal cancer, *Gastroenterology* **103:**1336–1348 (1992).

43. CHILDERS, J. M., CHU, J., VOIGT, L. F., FEIGL, P., TAMIMI, H. K., FRANKLIN, E. W., ALBERTS, D. S., AND MEYSKENS, F. L., JR., Chemoprevention of cervical cancer with folic acid: A phase III Southwest Oncology Group intergroup study, *Cancer Epidemiol. Biomarkers Prev.* **4:**155–159 (1995).

44. CHUNG, G. T., HUANG, D. P., LO, K. W., CHAN, M. K., AND WONG, F. W., Genetic lesions in the carcinogenesis of cervical cancer, *Anticancer Res.* **12:**1485–1490 (1992).

45. COGGINS, J. R., AND ZUR HAUSEN, H., Workshop on papillomaviruses and cancer, *Cancer Res.* **39:**545–546 (1979).

46. COUTURIER, J., SASTRE, G. X., SCHNEIDER, M. S., LABIB, A., AND ORTH, G., Integration of papillomavirus DNA near *myc* genes in genital carcinomas and its consequences for proto-oncogene expression, *J. Virol.* **65:**4534–4538 (1991).

47. COX, J. T., LORINCZ, A. T., SCHIFFMAN, M. H., SHERMAN, M. E., KURMAN, R. J., HUSSEIN, M., AND CULLEN, A., Human papillomavirus testing by hybrid capture appears to be useful in triaging women with a cytologic diagnosis of atypical squamous cells of undetermined significance, *Am. J. Obstet. Gynecol.* **172:**946–954 (1995).

48. COX, J. T., SCHIFFMAN, M. H., WINZELBERG, A. J., AND PATTERSON, J. M., An evaluation of HPV testing as a part of referral to colposcopy clinic, *Obstet. Gynecol.* **80:**389–395 (1992).

49. CRAWFORD, L., Prospects for cervical cancer vaccines, *Cancer Surv.* **16:**215–229 (1993).

50. CRUM, C. P., Carcinoma of the vulva: Epidemiology and pathogenesis, *Obstet. Gynecol.* **49:**448–454 (1992).

51. CUBIE, H. A., NORVAL, M., CRAWFORD, L., BANKS, L., AND CROOK, T., Lymphoproliferative response to fusion proteins of human papillomaviruses in patients with cervical intraepithelial neoplasia, *Epidemiol. Infect.* **103:**625–632 (1994).

52. CULLEN, A. P., REID, R., CAMPION, M., AND LORINCZ, A. T., Analysis of the physical state of different human papillomavirus DNAs in intraepithelial and invasive cervical neoplasia, *J. Virol.* **65:**606–612 (1991).

53. CUZICK, J., TERRY, G., HO, L., HOLLINGWORTH, T., AND ANDERSON, M., Human papillomavirus type 16 DNA in cervical smears as a predictor of high-grade cervical intraepithelial neoplasia, *Lancet* **339:**959–960 (1992).

54. DE SANJOSE, S., MUNOZ, N., BOSCH, F. X., REIMANN, K., PEDERSEN, N. S., ORFILA, J., ASCUNE, N., GONZALEZ, L. C., TAFUR, L., GILI, M., LETTE, I., VILADIU, P., TORMO, M. J., MOREO, P., SHAH, K., AND WAHREN, B., Sexually transmitted agents and cervical neoplasia in Colombia and Spain, *Int. J. Cancer* **56:**358–363 (1994).

55. DE SANJOSE, S., SANTAMARIA, M., ALONSO DE RUIZ, P., ARISTIZABAL, N., GUERRERO, E., CASTELLSAGUE, X., AND BOSCH, F. X., HPV types in women with normal cervical cytology, in: *The Epidemiology of Human Papillomavirus and Cervical Cancer* (N. MUNOZ, F. X. BOSCH, K. V. SHAH AND A. MEHEUS, EDS.), pp. 75–84, IARC Scientific Publications, Lyon, 1992.

56. DE VILLIERS, E. M., Minireview: Heterogeneity of the human papillomavirus group, *J. Virol.* **63:**4898–4903 (1989).

57. DE VILLIERS, E. M., Papilloma viruses in cancers and papillomas of the aerodigestive tract, *Biomed. Pharmacother.* **43:**31–36 (1989).

57a. DE VILLIERS, E. M., Human pathogenic papillomavirus types, in: *Current Topics in Microbiology and Immunology, Vol. 186, Human Pathogenic Papillomaviruses* (H. ZUR HAUSEN, ED.), pp. 1–12, Springer-Verlag, Heidelberg, 1994.

58. DE VILLIERS, E. M., WAGNER, D., SCHNEIDER, A., WESCH, H., MUNZ, F., MIKLAW, H., AND ZUR HAUSEN, H., Human papillomavirus DNA in women without and with cytological abnormalities: Results of a 5-year follow-up study, *Gynecol. Oncol.* **44:**33–39 (1992).

59. DEBRITTON, R. C., HILDESHEIM, A., DELAO, S. L., BRINTON, L. A., SATHYA, P., AND REEVES, W. C., Human papillomavirus and other influences on survival from cervical cancer in Panama, *Obstet. Gynecol.* **81:**19–24 (1993).

60. DELIUS, H. J., Evolution of papillomaviruses, in: *Human Pathogenic Papillomaviruses* (H. ZUR HAUSEN, ED.), pp. 13–31, Springer Verlag, Heidelberg, 1994.

61. DILLNER, J., DILLNER, L., ROBB, J., WILLEMS, J., JONES, I., LANCASTER, W., SMITH, R., AND LERNER, R., A synthetic peptide defines a serologic IgA response to a human papillomavirus-encoded nuclear antigen expressed in virus-carrying cervical neoplasia, *Proc. Natl. Acad. Sci. USA* **86:**3838–3841 (1989).

62. DILLNER, J., WIKLUND, F., LENNER, P., EKLUND, C., FREDRIKSSON-SHANAZARIAN, V., SCHILLER, J. T., HIBMA, M., HALLMANS, G., AND STENDAHL, U., Antibodies against linear and conformational epitopes of human papillomavirus type 16 that independently associate with incident cervical cancer, *Int. J. Cancer* **60:**1–6 (1995).

63. DILLNER, L., BEKASSY, Z., JONSSON, N., MORENO-LOPEZ, J., AND BLOMBERG, J., Detection of IgA antibodies against human papillomavirus in cervical secretions from patients with cervical intraepithelial neoplasia, *Int. J. Cancer* **43:**36–40 (1989).

64. DILORENZO, T. P., TAMSEN, A., ABRAMSON, A. L., AND STEINBERG, B. M., Human papillomavirus type 6A DNA in the lung carcinoma of a patient with recurrent laryngeal papillomatosis is characterized by a partial duplication, *J. Gen. Virol.* **73**(2):423–428 (1992).

65. DUGGAN-KEEN, M., KEATING, P. J., CROMME, F. V., WALBOOMERS, J. M. M., AND STERN, P. L., Alterations in major histocompatibility complex expression in cervical cancer: Possible consequences for immunotherapy, *Papillomavirus Rep.* **5:**3–9 (1994).

66. DURST, M., GISSMANN, L., IKENBERG, H., AND ZUR HAUSEN, H., A papillomavirus DNA from a cervical carcinoma and its prevalence in cancer biopsy samples from different geographic regions, *Proc. Natl. Acad. Sci. USA* **80:**3812–3815 (1983).

67. DYSON, N., HOWLEY, P. M., MUNGER, K., AND HARLOW, E., The human papilloma virus-16 E7 oncoprotein is able to bind to the retinoblastoma gene product, *Science* **243:**934–937 (1989).

68. EDITORIAL, Histopathological grading of cervical intraepithelial neoplasia (CIN)—is there a need for change?, *J. Pathol.* **159:**273–275 (1989).

69. EL AWADY, M. K., KAPLAN, J. B., O'BRIEN, S. J., AND BURK, R. D., Molecular analysis of integrated human papillomavirus 16 sequences in the cervical cancer cell line SIHA, *Virology* **159:**389–398 (1987).

70. ELUF-NETO, J., BOOTH, M., MUNOZ, N., BOSCH, F. X., MEIJER, C. J., AND WALBOOMERS, J. M., Human papillomavirus and invasive cervical cancer in Brazil, *Br. J. Cancer* **69:**114–119 (1994).

71. EVANS, S., AND DOWLING, K., The changing prevalence of cervical human papilloma virus infection, *Aust. NZ J. Obstet. Gynecol.* **30**(4):375–377 (1990).

72. FAIRLEY, C. K., CHEN, S., TABRIZI, S. N., LEETON, K., QUINN, M. A., AND GARLAND, S. M., The absence of genital human papillomavirus DNA in virginal women, *Int. J. STD AIDS* **3:**414–417 (1992).

73. FAIRLEY, C. K., CHEN, S., TABRIZI, S. N., QUINN, M. A., MCNEIL, J. J., AND GARLAND, S. M., Tampons: A novel patient-administered method for the assessment of genital human papillomavirus infection, *J. Infect. Dis.* **165:**1103–1106 (1992).

74. FAIRLEY, C. K., SHEIL, A. G. R., MCNEIL, J. J., UGONI, A. M., DISNEY, A. P. S., GILES, G. G., AND AMISS, N., The risk of anogenital malignancies in dialysis and transplant patients, *Clin. Nephrol.* **41**(2):101–105 (1994).

75. FARTHING, A., MASTERSON, P., MASON, W. P., AND VOUSDEN, K. H., Human papillomavirus detection by hybrid capture and its possible clinical use, *J. Clin. Pathol.* **47:**649–652 (1994).

76. FERENCZY, A., BERGERON, C., AND RICHART, R. M., Human papillomavirus DNA in fomites on objects used for the management of patients with genital human papillomavirus infections, *Obstet. Gynecol.* **74:**950–954 (1989).

77. FERENCZY, A., BERGERON, C., AND RICHART, R. M., Human papillomavirus DNA in CO2 laser-generated plume of smoke and its consequences to the surgeon, *Obstet. Gynecol.* **75:**114–118 (1990).

78. FERENCZY, A., MITAO, M., NAGAI, N., SILVERSTEIN, S. J., AND CRUM, C. P., Latent papillomavirus and recurring genital warts, *N. Engl. J. Med.* **313:**784–788 (1985).

79. FISHER, M., ROSENFELD, W. D., AND BURK, R. D., Cervicovaginal human papillomavirus infection in suburban adolescents and young adults, *J. Pediatr.* **119:**821–825 (1991).

80. FRANCO, E. L., The sexually transmitted disease model for cervical cancer: Incoherent epidemiologic findings and the role of misclassification of human papillomavirus infection, *Epidemiology* **2:**98–106 (1991).

81. FRAZER, I. H., LEONARD, J. H., SCHONROCK, J., WRIGHT, R. G., AND KEARSLEY, J. H., HPV DNA in oropharyngeal squamous cell cancers: Comparison of results from four DNA detection methods, *Pathology* **25:**138–143 (1993).

82. FUCHS, P. G., GIRARDI, F., AND PFISTER, H., Human papillomavirus DNA in normal, metaplastic, preneoplastic and neoplastic epithelia of the cervix uteri, *Int. J. Cancer* **41:**41–45 (1988).

83. GALLOWAY, D., Serological assays for the detection of HPV antibodies, in: *The Epidemiology of Human Papillomavirus and Cervical Cancer* (N. MUNOZ, F. X. BOSCH, K. V. SHAH, AND A. MEHEUS, EDS.), pp. 147–161, IARC Scientific Publications, Lyon, 1992.

84. GARLICK, J. A., CALDERON, S., BUCHNER, A., AND MITRANI-ROSENBAUM, S., Detection of human papillomavirus (HPV) DNA in focal epithelial hyperplasia, *J. Oral Pathol. Med.* **18:**172–177 (1989).

85. GISSMANN, L., WOLNIK, L., IKENBERG, H., KOLDOVSKY, U., AND SCHNURCH, H. G., Human papillomavirus types 6 and 11 DNA

sequences in genital and laryngeal papillomas and in some cervical cancers, *Proc. Natl. Acad. Sci. USA* **80:**560–563 (1983).

86. GLINSKI, W., OBALEK, S., JABLONSKA, S., AND ORTH, G., T cell defect in patients with epidermodysplasia verruciformis due to human papillomavirus type 3 and 5, *Dermatologica* **162:**141–147 (1981).

87. GOLDBERG, G. L., VERMUND, S. H., SCHIFFMAN, M. H., RITTER, D. B., SPITZER, C., AND BURK, R. D., Comparison of cytobrush and cervicovaginal lavage sampling methods for the detection of genital human papillomavirus, *Am. J. Obstet. Gynecol.* **161:**1669–1672 (1989).

88. GRAVITT, P. E., AND MANOS, M. M., Polymerase chain reaction-based methods for the detection of human papillomavirus DNA, in: *The Epidemiology of Human Papillomavirus and Cervical Cancer* (N. MUNOZ, F. X. BOSCH, K. V. SHAH, AND A. MEHEUS, EDS.), pp. 121–133, IARC Scientific Publications, Lyon, 1992.

89. GREENBERG, M. D., CAMPION, M. J., AND RUTLEDGE, L. H., Cervicography as an adjunct to cytologic screening, *Obstet. Gynecol. Clin. North Am.* **20**(1)**:**13–29 (1993).

90. GREER, C. E., LUND, J. K., AND MANOS, M. M., PCR amplification from paraffin-embedded tissues: Recommendations on fixatives for long-term storage and prospective studies, *PCR Methods Appl.* **1:**46–50 (1991).

91. GREGOIRE, L., LAWRENCE, W. D., KUKURUGA, D., EISENBREY, A. B., AND LANCASTER, W. D., Association between HLA-DQB1 alleles and risk for cervical cancer in African-American women, *Int. J. Cancer* **57:**504–507 (1994).

92. NINDL, I., BENITEZ-BRIBIESCA, L., BERUMEN, J., FARMANARA, N., FISHER, S., GROSS, A., LOPEZ-CARILLO, L., MULLER, M., TOMMASINO, M., VAZQUEZ-CURIEL, A., ET AL., Antibodies against linear and conformational epitopes of the human papillomavirus (HPV) type 16 E6 and E7 oncoproteins in sera of cervical cancer patients, *Arch. Virol.* **137**(3–4)**:**341–353 (1994).

93. GUERRERO, E., DANIEL, R. W., BOSCH, X., CASTELLSAGUE, X., MUNOZ, N., GILI, M., VILADIU, P., NAVARRO, C., ZUBIRI, M. L., ASCUNCE, N., GONZALEZ, L. C., TAFUR, L., IZARZUGAZA, I., AND SHAH, K. V., Comparison of virapap, Southern hybridization, and polymerase chain reaction methods for human papillomavirus identification in an epidemiological investigation of cervical cancer, *J. Clin. Microbiol.* **30**(11)**:**2951–2959 (1992).

94. GUTMAN, L. T., ST. CLAIRE, K., HERMAN-GIDDENS, M. E., JOHNSTON, W. W., AND PHELPS, W. C., Evaluation of sexually abused and nonabused young girls for intravaginal human papillomavirus infection, *Am. J. Dis. Child.* **146:**694–699 (1992).

95. HECK, D. V., YEE, C. L., HOWLEY, P. M., AND MUNGER, K., Efficiency of binding the retinoblastoma protein correlates with the transforming capacity of the E7 oncoproteins of the human papillomaviruses, *Proc. Natl. Acad. Sci. USA* **89:**4442–4446 (1992).

96. HEINO, P., EKLUND, C., FREDRIKSSON-SHANAZARIAN, V., GOLDMAN, S., SCHILLER, J. T., AND DILLNER, J., Association of serum immunoglobulin G antibodies against human papillomavirus type 16 capsids with anal epidermoid carcinoma, *J. Natl. Cancer Inst.* **87:**437–440 (1995).

97. HERRERO, R., BRINTON, L. A., REEVES, W. C., BRENES, M. M., TENORIO, F., DE BRITTON, R. C., GAITAN, E., GARCIA, M., AND RAWLS, W. E., Invasive cervical cancer and smoking in Latin America, *J. Natl. Cancer Inst.* **81**(3)**:**205–211 (1989).

98. HILDESHEIM, A., BRINTON, L. A., SZKLO, M., REEVES, W. C., AND RAWLS, W. E., Herpes simplex virus 2: A possible interaction with human papillomavirus types 16/18 in the development of invasive cervical cancer, *Int. J. Cancer* **49:**335–340 (1991).

99. HILDESHEIM, A., GRAVITT, P., SCHIFFMAN, M. H., KURMAN, R. J., BARNES, W., JONES, S., TCHABO, J. G., BRINTON, L. A., COPELAND, C., EPP, J., AND MANOS, M. M., Determinants of genital human papillomavirus infection in low-income women in Washington, DC, *Sex. Transm. Dis.* **20:**279–285 (1993).

100. HILDESHEIM, A., SCHIFFMAN, M. H., GRAVITT, P., GLASS, A. G., GREER, C., ZHANG, T., SCOTT, D. R., RUSH, B. B., LAWLER, P., SHERMAN, M. E., KURMAN, R. J., AND MANOS, M. M., Persistence of type-specific human papillomavirus infection among cytologically normal women in Portland, Oregon, *J. Infect. Dis.* **169:**235–240 (1994).

101. HIPPELAINEN, M., ESKELINEN, M., LIPPONEN, P., CHANG, F., AND SYRJANEN, K., Mitotic activity index, volume corrected mitotic index and human papillomavirus suggestive morphology are not prognostic factors in carcinoma of the oesophagus, *Anticancer Res.* **13:**677–682 (1993).

102. HO, G. Y. F., BURK, R. D., FLEMING, I., AND KLEIN, R. S., Risk of genital human papillomavirus infection in women with human immunodeficiency virus-induced immunosuppression, *Int. J. Cancer* **56**(6)**:**788–792 (1994).

103. HO, L., TAY, S. K., CHAN, S. Y., AND BERNARD, H. U., Sequence variants of human papillomavirus type 16 from couples suggests sexual transmission with low infectivity and polyclonality in genital neoplasia, *J. Infect. Dis.* **168:**1–7 (1993).

104. HO, L., CHAN, S. Y., CHOW, V., CONG, T., TAY, S. K., VILLA, L. L., AND BERNARD, H. U., Sequence variants of human papillomavirus type 16 in clinical samples permit verification and extension of epidemiological studies and construction of a phylogenetic tree, *J. Clin. Microbiol.* **29**(9)**:**1765–1772 (1991).

105. HOLLADAY, E. B., AND GERALD, W. L., Viral gene detection in oral neoplasms using the polymerase chain reaction, *Am. J. Clin. Pathol.* **100:**36–40 (1993).

106. HOLLINGSWORTH, T. H., The incidence of warts and plantar warts amongst school children in East Anglia, *Med. Officer* 55–59 (1955).

107. HOWLEY P. M., Papillomaviridae and their replication, in: *Virology* (B. N. FIELDS AND D. M. KNIPE, EDS.), Raven Press, New York, 1990.

108. HUIBREGTSE, J. M., SCHEFFNER, M., AND HOWLEY, P. M., Localization of the E6-AP regions that direct human papillomavirus E6 binding, association with p53, and ubiquitination of associated proteins, *Mol. Cell. Biol.* **13:**4918–4927 (1993).
IARC Monograph on the Evaluation of Carcinogenic Risks to Humans: Vol. 64. Human Papillomaviruses. IARC Scientific Publications, Lyon, France, 1995.

109. JANNIGER, C. K., Childhood warts, *Pediatr. Dermatol.* **50:**15–16 (1992).

110. JARRETT, W. F. H., MCNEIL, P. E., GRIMSHAW, W. T. R., SELMAN, I. E., AND MCINTYRE, W. I. M., High incidence area of cattle cancer with a possible interaction between an environmental carcinogen and a papilloma virus, *Nature* **274:**215–217 (1978).

111. JENISON, S. A., YU, X., VALENTINE, J. M., KOUTSKY, L. A., CHRISTANSEN, A. E., BECKMANN, A. M., AND GALLOWAY, D. A., Evidence of prevalent genital-type human papillomavirus infections in adults and children, *J. Infect. Dis.* **62:**60–69 (1990).

112. JOCHMUS-KUDIELKA, I., SCHNEIDER, A., BRAUN, R., KIMMIG, R., KOLDOVSKY, U., SCHNEWEIS, K. E., SEEDORF, K., AND GISSMANN, L., Antibodies against the human papillomavirus type 16 early proteins in human sera: Correlation of anti-E7 reactivity with cervical cancer, *J. Natl. Cancer Inst.* **81:**1698–1704 (1989).

113. JOHNSON, M. L. T., AND ROBERTS, J., Skin conditions and related

need for medical care among persons 1–74 years: United States, 1971–1974, *Natl. Center Health Stat.* **11**(212):1–72 (1978).

114. KADISH, A. S., HAGAN, R. J., RITTER, D. B., GOLDBERG, G. L., ROMNEY, S. L., KANETSKY, P. A., BEISS, B. K., AND BURK, R. D., Biologic characteristics of specific human papillomavirus types predicted from morphology of cervical lesions, *Hum. Pathol.* **23**(11):1262–1269 (1992).

115. KADISH, A. S., ROMNEY, S. L., LEDWIDGE, R., TINDLE, R., FERNANDO, G. J. P., AND ZEE, S. Y., Cell-mediated immune responses to E7 peptides of human papillomavirus type 16 are dependent on the HPV type infecting the cervix whereas serological reactivity is not type-specific, *J. Gen. Virol.* **75**(9):2277–2284 (1994).

116. KADISH, A. S., ROMNEY, S. L., LEDWIDGE, R., TINDLE, R., FERNANDO, G. J., ZEE, S. Y., VAN RANST, M. A., AND BURK, R. D., Cell-mediated immune responses to E7 peptides of human papillomavirus (HPV) type 16 are dependent on the HPV type infecting the cervix whereas serological reactivity is not type-specific, *J. Gen. Virol.* **75** (Part 9):2277–2284 (1994).

117. KAEBLING, M., BURK, R. D., ATKIN, N. B., AND KLINGER, H. P., Loss of heterozygosity in the region of the TP53 gene on chromosome 17p in HPV-negative human cervical carcinomas, *Lancet* **340**:140–142 (1992).

118. KASHER, M. S., AND ROMAN, A., Characterization of human papillomavirus type 6B DNA isolated from an invasive squamous carcinoma of the vulva, *Virology* **165**:225–233 (1988).

119. KASHIMA, H. K., SHAH, F., LYLES, A., GLACKIN, R., MUHAMMAD, N., TURNER, L., VAN ZANDT, S., WHITT, S., AND SHAH, K., A comparison of risk factors in juvenile-onset and adult-onset recurrent respiratory papillomatosis, *Laryngoscope* **102**:9–13 (1992).

120. KASPERBAUER, J. L., O'HALLORAN, G. L., ESPY, J. J., SMITH, T. F., AND LEWIS, J. E., Polymerase chain reaction (PCR) identification of human papillomavirus (HPV) DNA in verrucous carcinoma of the larynx, *Laryngoscope* **103**:416–420 (1993).

121. KATAJA, V., SYRJANEN, K., SYRJANEN, S., MANTYJARVI, R., YLISKOSKI, M., SAARIKOSKI, S., AND SALONEN, J. T., Prospective follow-up of genital HPV infections: Survival analysis of the HPV typing data, *Eur. J. Epidemiol.* **6**(1):9–14 (1990).

122. KEEFE, M., AL-GHAMDI, A., COGGON, D., MAITLAND, N. J., EGGER, P., AND KEEFE, C. J., Cutaneous warts in butchers, *Br. J. Dermatol.* **130**:9–14 (1994).

123. KIRNBAUER, R., HUBBERT, N. L., WHEELER, C. M., BECKER, T. M., LOWY, D. R., AND SCHILLER, J. T., A virus-like particle ELISA detects serum antibodies in a majority of women infected with human papillomavirus type 16, *J. Natl. Cancer Inst.* **86**:494–499 (1994).

124. KIVIAT, N. B., CRITCHLOW, C. W., HOLMES, K. K., KUYPERS, J., SAYER, J., DUNPHY, C., SURAWICZ, C., KIRBY, P., WOOD, R., AND DALING, J. R., Association of anal dysplasia and human papillomavirus with immunosuppression and HIV infection among homosexual men, *AIDS* **7**:43–49 (1993).

125. KIVIAT, N. B., CRITCHLOW, C. W., AND KURMAN, R. J., Reassessment of the morphological continuum of cervical intraepithelial lesions: Does it reflect different stages in the progression to cervical carcinoma? in: *The Epidemiology of Human Papillomavirus and Cervical Cancer* (N. MUNOZ, F. X. BOSCH, K. V. SHAH, AND A. MEHEUS, EDS.), pp. 59–66, IARC Scientific Publications, Lyon, 1992.

126. KIVIAT, N. B., KOUTSKY, L. A., CRITCHLOW, C. W., GALLOWAY, D. A., VERNON, D. A., PETERSON, M. L., MCELHOSE, P. E., PENDRAS, S. J., STEVENS, C. E., AND HOLMES, K. K., Comparison of Southern transfer hybridization and dot filter hybridization for detection of cervical human papillomavirus infection with types 6, 11, 16, 18, 31, 33, and 35, *Am. J. Clin. Pathol.* **94**(5):561–565 (1990).

127. KJAER, S. K., AND BRINTON, L. A., Adenocarcinomas of the uterine cervix: The epidemiology of an increasing problem, *Epidemiol. Rev.* **15**:486–498 (1993).

128. KJAER, S. K., AND JENSEN, O. M., Comparison studies of HPV detection in areas at different risk for cervical cancer, in: *The Epidemiology of Human Papillomavirus and Cervical Cancer* (N. MUNOZ, F. X. BOSCH, K. V. SHAH, AND A. MEHEUS, EDS.), pp. 243–249, IARC Scientific Publications, Lyon, 1992.

129. KOCHEL, H. G., MONAZAHIAN, M., SIEVERT, K., HOHNE, M., THOMSSEN, C., TEICHMANN, A., ARENDT, P., AND THOMSSEN, R., Occurrence of antibodies to L1, L2, E4 and E7 gene products of human papillomavirus types 6b, 16 and 18 among cervical cancer patients and controls, *Int. J. Cancer* **48**:682–688 (1991).

130. KOHNO, T., TAKAYAMA, H., HAMAGUCHI, M., TAKANO, H., YAMAGUCHI, N., TSUDA, H., HIROHASHI, S., VISSING, H., SHIMIZU, M., OSHIMURA, M., Deletion mapping of chromosome 3p in human uterine cervical cancer, *Oncogene* **8**:1825–1832 (1993).

131. KOSARY, C. L., SCHIFFMAN, M. H., AND TRIMBLE, E. L., Cervix uteri, in: *SEER Cancer Statistics Review: 1973–1990* (B. A. MILLER, L. A. G. RIES, B. F. HANKEY, C. L. KOSARY, A. HARRAS, S. S. DEVESA, AND B. K. EDWARDS, EDS.), Section V, National Cancer Institute NIH Pub. No. 93-2789, Bethesda, 1994.

132. KOSS, L. G., From koilocytosis to molecular biology: The impact of cytology on concepts of early human cancer, *Mod. Pathol.* **2**(5):526–535 (1989).

133. KOUTSKY, L. A., GALLOWAY, D. A., AND HOLMES, K. K., Epidemiology of genital human papillomavirus infection, *Epidemiol. Rev.* **10**:122–163 (1988).

134. KOUTSKY, L. A., HOLMES, K. K., CRITCHLOW, C. W., STEVENS, C. E., PAAVONEN, J., BECKMANN, A. M., DEROUEN, T. A., GALLOWAY, D. A., VERNON, D., AND KIVIAT, N. B., A cohort study of the risk of cervical intraepithelial neoplasia grade 2 or 3 in relation to papillomavirus infection, *New Engl. J. Med.* **327**(18):1272–1278 (1992).

135. KREBS, H. B., AND SCHNEIDER, V., Human papillomavirus-associated lesions of the penis: Colposcopy, cytology, and histology, *Obstet. Gynecol.* **70**(3):299–304 (1987).

136. KREIDER, J. W., HOWETT, M. K., WOLFE, S. A., BARTLETT, G. L., ZAINO, R. J., SEDLACEK, T. V., AND MORTEL, R., Morphological transformation *in vivo* of human uterine cervix with papillomavirus from conylomata acuminata, *Nature* **317**(17):639–641 (1985).

137. KREISS, J. K., KIVIAT, N. B., PLUMMER, F. A., ROBERTS, P. L., WAIYAKI, P., NGUGI, E., AND HOLMES, K. K., Human immunodeficiency virus, human papillomavirus, and cervical intraepithelial neoplasia in Nairobi prostitutes, *Sex. Transm. Dis.* **19**:54–59 (1992).

138. KURMAN, R. J., SCHIFFMAN, M. H., LANCASTER, W. D., REID, R., JENSON, A. B., TEMPLE, G. F., AND LORINCZ, A. T., Analysis of individual human papillomavirus types in cervical neoplasia: A possible role for type 18 in rapid progression, *Am. J. Obstet. Gynecol.* **159**(2):293–296 (1988).

139. KURMAN, R. J., SHAH, K. H., LANCASTER, W. D., AND JENSON, A. B., Immunoperoxidase localization of papillomavirus antigens in cervical dysplasia and vulvar condylomas, *Am. J. Obstet. Gynecol.* **140**:931–935 (1981).

140. KURMAN, R. J., AND SOLOMON, D., *The Bethesda System for*

Reporting Cervical/Vaginal Cytologic Diagnoses, Springer-Verlag, New York, 1994.

141. KURMAN, R. J., TOKI, T., AND SCHIFFMAN, M. H., Basaloid and warty carcinomas of the vulva: Distinctive types of squamous cell carcinoma frequently associated with human papillomavirus, *Am. J. Surg. Pathol.* **17:**133–145 (1993).

142. LANCASTER, W. D., AND OLSON, C., Animal papillomaviruses, *Microbiol. Rev.* **46**(2):191–207 (1982).

143. LAVERTY, C. R., RUSSELL, P., HILLS, E., AND BOOTH, N., The significance of noncondylomatous wart virus infection of the cervical transformation zone. A review with discussion of two illustrative cases, *Acta Cytol.* **22:**195–201 (1978).

144. LAWTON, G., THOMAS, S., SCHONROCK, J., MONSOUR, F., AND FRAZER, I., Human papillomaviruses in normal oral mucosa: A comparison of methods for sample collection, *J. Oral Pathol. Med.* **21:**265–269 (1992).

145. LEE, N. K., RITTER, D. B., GROSS, A. E., MYSSIORECK, D. J., KADISH, A. S., AND BURK, R. D., Head and neck squamous cell carcinomas associated with human papillomaviruses and an increased incidence of cervical pathology, *Otolaryngol. Head Neck Surg.* **99:**296–301 (1988).

146. LEY, C., BAUER, H. M., REINGOLD, A., SCHIFFMAN, M. H., CHAMBERS, J. C., TASHIRO, C. J., AND MANOS, M. M., Determinants of genital papillomavirus infection in young women, *J. Natl. Cancer Inst.* **83:**997–1003 (1991).

147. LORINCZ, A. T., Human papillomavirus detection tests, in: *Sexually Transmitted Diseases*, 2nd ed. (K. K. HOLMES, P. A. MARDH, P. F. SPARLING, P. J. WIESNER, W. CATES, S. M. LEMON, AND W. E. STAMM, EDS.), pp. 953–959, McGraw-Hill, New York, 1990.

148. LORINCZ, A. T., Detection of human papillomavirus DNA without amplications: Prospects for clinical utility, in: *The Epidemiology of Human Papillomavirus and Cervical Cancer* (N. MUNOZ, F. X. BOSCH, K. V. SHAH, AND A. MEHEUS, EDS.), pp. 85–92, IARC Scientific Publications, Lyon, 1992.

149. LORINCZ, A. T., REID, R., JENSON, A. B., GREENBERG, M. D., LANCASTER, W., AND KURMAN, R. J., Human papillomavirus infection of the cervix: Relative risk associations of 15 common anogenital types, *Obstet. Gynecol.* **79:**328–337 (1992).

150. LORINCZ, A. T., SCHIFFMAN, M. H., JAFFURS, W. J., MARLOW, J., QUINN, A. P., AND TEMPLE, G. F., Temporal associations of human papillomavirus infection with cervical cytologic abnormalities, *Am. J. Obstet. Gynecol.* **162:**645–651 (1990).

151. MAIMAN, M., FRUCTHER, R. G., GUY, L., CUTHILL, S., LEVINE, P., AND SERUR, E., Human immunodeficiency virus infection and invasive cervical carcinoma, *Cancer* **71:**402–406 (1993).

152. MAJOROS, M., DEVINE, K. D., AND PARKHILL, E. M., Malignant transformation of benign laryngeal papillomas in children after radiation therapy, *Surg. Clin. North Am.* **43:**1049–1061 (1963).

153. MANOS, M. M., TING, Y., WRIGHT, D. K., LEWIS, A. J., BROKER, T. R., AND WOLINSKY, S. M., Use of polymerase chain reaction amplification for the detection of genital human papillomavirus, *Mol. Diagn. Hum. Cancer* **7:**209–214 (1989).

154. MANSELL, M. E., HO, L., TERRY, G., SINGER, A., AND CUZICK, J., Semi-quantitative human papillomavirus DNA detection in the management of women with minor cytological abnormality, *Br. J. Obstet. Gynaecol.* **101**(9):807–809 (1994).

155. MASIH, A. S., STOLER, M. H., FARROW, G. M., AND JOHANSSON, S. L., Human papillomavirus in penile squamous cell lesions: A comparison of an isotopic RNA and two commercial nonisotopic DNA *in situ* hybridization methods, *Arch. Pathol. Lab. Med.* **117:**302–307 (1993).

156. MASSING, A. M., AND EPSTEIN, W. L., Natural history of warts, *Arch. Dermatol.* **87:**74–78 (1963).

157. McDONNELL, J. M., MAYR, A. J., AND MARTIN, W. J., DNA of human papillomavirus type 16 in dysplastic and malignant lesions of the conjunctiva and cornea, *New Engl. J. Med.* **320**(22):1442–1446 (1989).

158. McGLENNEN, R. C., GHAI, J., OSTROW, R. S., LABRESH, K., SCHNEIDER, J. F., AND FARAS, A. J., Cellular transformation by a unique isolate of human papillomavirus type 11, *Cancer Res.* **52:**5872–5878 (1992).

159. McNICOL, P. J., AND DODD, J. G., High prevalence of human papillomavirus in prostate tissues, *J. Urol.* **145:**850–853 (1991).

160. MEISELS, A., AND FORTIN, R., Condylomatous lesions of the cervix and vagina I. Cytologic patterns, *Acta Cytol.* **20**(6):505–509 (1976).

161. MELBYE, M., PALEFSKY, J., GONZALES, J., RYDER, L. P., NIELSEN, H., BERGMANN, O., PINDEORG, J., AND BIGGAR, R. J., Immune status as a determinant of human papillomavirus detection and its association with anal epithelial abnormalties, *Int. J. Cancer* **46:**203–206 (1990).

162. MELBYE, M., AND SPROGEL, P., Aetiological parallel between anal cancer and cervical cancer, *Lancet* **338:**657–659 (1991).

163. MELKERT, P. W. J., HOPMAN, E., VAN DEN BRULE, A. J. C., RISSE, E. K. J., VAN DIEST, P. J., BLEKER, O. P., HELMERHORST, T., SCHIPPER, M. E. I., MEIJER, C. J. L. M., AND WALBOOMERS, J. M. M., Prevalence of HPV in cytomorphologically normal cervical smears, as determined by the polymerase chain reaction, is age-dependent, *Int. J. Cancer* **53:**919–923 (1993).

164. MELNICK, J. L., Papova virus group, *Science* **135:**1128–1130 (1962).

165. MEYERS, C., FRATTINI, M. G., HUDSON, J. B., AND LAIMINS, L. A., Biosynthesis of human papillomavirus from a continuous cell line upon epithelial differentiation, *Science* **257:**971–973 (1992).

166. MEYSKENS, F. L., JR., SURWIT, E., MOON, T. E., CHILDERS, J. M., DAVIS, J. R., DORR, R. T., JOHNSON, C. S., AND ALBERTS, D. S., Enhancement of regression of cervical intraepithelial neoplasia II (moderate dysplasia) with topically applied all-*trans*-retinoic acid: A randomized trial, *J. Natl. Cancer Inst.* **86:**539–543 (1994).

167. MONSONEGO, J., MAGDELENAT, H., CATALAN, F., COSCAS, Y., ZERAT, L., AND SASTRE, X., Estrogen and progesterone receptors in cervical human papillomavirus related lesions, *Int. J. Cancer* **48:**533–539 (1991).

168. MORELLI, A. E., SANANES, C., DIPAOLA, G., PAREDES, A., AND FAINBOIM, L., Relationship between types of human papillomavirus and Langerhans' cells in cervical condyloma and intraepithelial neoplasia, *Am. J. Clin. Pathol.* **99:**200–206 (1993).

169. MORRISON, E. A., GOLDBERG, G. L., HAGAN, R. J., KADISH, A. S., AND BURK, R. D., Self-administered home cervicovaginal lavage: A novel tool for the clinical–epidemiological investigation of genital human papillomavirus infections, *Am. J. Obstet. Gynecol.* **167**(1):104–107 (1992).

170. MORRISON, E. A. B., Classification of human papillomavirus infection, *J.A.M.A.* **270**(4):453 (1993).

171. MORRISON, E. A. B., HO, G. Y. F., VERMUND, S. H., GOLDBERG, G. L., KADISH, A. S., KELLY, K. F., AND BURK, R. D., Human papillomavirus infection and other risk factors for cervical neoplasia: A case-control study, *Int. J. Cancer* **49:**6–13 (1991).

172. MOSCICKI, A. B., Comparison between methods for human papillomavirus DNA testing: A model for self-testing in young women, *J. Infect. Dis.* **167:**723–725 (1993).

173. MOSCICKI, A. B., PALEFSKY, J., SMITH, G., SIBOSHSKI, S., AND SCHOOLNIK, G., Variability of human papillomavirus DNA testing in a longitudinal cohort of young women, *Obstet. Gynecol.* **82:**578–585 (1993).

174. MOUGIN, C., SCHAAL, J. P., BASSIGNOT, A., MADOZ, L., COAQUETTE, A., LAURENT, R., AND LAB, M., Detection of human papillomavirus and human cytomegalovirus in cervical lesions by *in situ* hybridization using biotinylated probes, *Biomed. Pharmarcother.* **45:**353–357 (1991).

175. MULLER, M., VISCIDI, R., SUN, Y., GUERRERO, E., HILL, P. M., SHAH, F., BOSCH, X., MUNOZ, N., GISSMANN, L., AND SHAH, K. V., Antibodies to HPV-16 E6 and E7 proteins as markers for HPV-16-associated invasive cervical cancer, *Virology* **187:**508–514 (1992).

176. MUNOZ, N., BOSCH, F. X., DESANJOSE, S., TAFUR, L., IZARZUGAZA, I., GILI, M., VILADIU, P., NAVARRO, C., MARTOS, C., ASCUNCE, N., GONZALEZ, L. C., KALDOR, J. M., GUERRERO, E., LORINCZ, A., SANTAMARIA, M., DERUIZ, P. A., ARISTIZABAL, N., AND SHAH, K., The causal link between human papillomavirus and invasive cervical cancer: A population-based case-control study in Columbia and Spain, *Int. J. Cancer* **52:**743–749 (1992).

177. MUNOZ, N., BOSCH, F. X., DESANJOSE, S., VERGARA, A., DEL-MORA, A., MUNOZ, M. T., TAFUR, L., GILI, M., IZARZUGAZA, I., VILADIU, P., NAVARRO, C., ALONSO DE RUIZ, P., ARISTIZABAL, N., SANTAMARIA, M., ORFILA, J., DANIEL, R. W., GUERRERO, E., AND SHAH, K. V., Risk factors for cervical intraepithelial neoplasia grade III/ carcinoma *in situ* in Spain and Columbia, *Cancer Epidemiol. Biomarkers Prev.* **2:**423–431 (1993).

178. MUNOZ, N., BOSCH, X., AND KALDOR, J. M., Does human papillomavirus cause cervical cancer? The state of the epidemiological evidence, *Br. J. Cancer* **57**(1):1–5 (1988).

179. NASIELL, K., ROGER, V., AND NASIELL, M., Behavior of mild cervical dysplasia during long-term follow-up, *Obstet. Gynecol.* **67:**665–669 (1986).

180. NUOVO, G., Evidence against a role for human papillomavirus in colon neoplasms, *Arch. Surg.* **126:**656 (1991).

181. OBALEK, S., GLINSKI, W., HAFTEK, M., ORTH, G., AND JABLONSKA, S., Comparative studies on cell-mediated immunity in patients with different warts, *Dermatologica* **161:**73–83 (1980).

182. OKAGAKI, T., Female genital tumors associated with human papillomavirus infection, and the concept of genital neoplasm–papilloma syndrome (GENPS), *Pathol. Annu.* **19**(2):31–62 (1993).

183. ONG, C. K., CHAN, S. Y., CAMPO, M. S., FUJINAGA, K., MAVROMARA, P., PFISTER, H., TAY, S. K., MEULEN, J. T., VILLA, L. L., AND BERNARD, H. U., Evolution of human papillomavirus type 18: An ancient phylogenetic root in Africa and intratype diversity reflect coevolution with human ethnic groups, *J. Virol.* **67:**6424–6431 (1993).

184. ORIEL, J. D., Sex and cervical cancer, *Genitourin. Med.* **64:**81–89 (1988).

185. ORTH, G., FAVRE, M., BREITBURD, F., CROISSANT, O., JABLONSKA, S., OBALEK, S., JARZABEK-CHORZELSAK, M., AND RZESA, G., Epidermodysplasia verruciformis: A model for the role of papillomaviruses in human cancer, in: *Cold Spring Harbor Conference on Cell Proliferation, Vol. 7*, Cold Spring Harbor, New York, 1980.

186. PAKARIAN, F., KAYE, J., CASON, J., KELL, B., JEWERS, R., DERIAS, N. W., RAJU, K. S., AND BEST, J. M., Cancer associated human papillomaviruses: Perinatal transmission and persistence, *Br. J. Obstet. Gynaecol.* **101:**514–517 (1994).

187. PALEFSKY, J., Human papillomavirus infection among HIV-infected individuals, implications for development of malignant tumors, *Hematol. Oncol. Aspects HIV Dis.* **5**(2):357–370 (1991).

188. PALEFSKY, J. M., GONZALES, J., GREENBLATT, R. M., AHN, D. K., AND HOLLANDER, H., Anal intraepithelial neoplasia and anal papillomavirus infection among homosexual males with group IV HIV disease, *J.A.M.A.* **263**(21):2911–2916 (1990).

189. PALEFSKY, J. M., HOLLY, E. A., GONZALES, J., BERLINE, J., AHN, D. K., AND GREENSPAN, J. S., Detection of human papillomavirus DNA in anal intraepithelial neoplasia and anal cancer, *Cancer Res.* **51:**1014–1019 (1991).

190. PARKIN, D. M., PISANI, P., AND FERLAY, J., Estimates of the worldwide frequency of eighteen major cancers in 1985, *Int. J. Cancer* **54:**594–606 (1993).

191. PATSNER, B., BAKER, D. A., AND ORR, J. W., Human papillomavirus genital tract infections during pregnancy, *Clin. Obstet. Gynecol.* **33:**258–267 (1990).

192. PETRY, K. U., SCHEFFEL, D., BODE, U., GABRYSIAK, T., KOCHEL, H., KUPSCH, E., GLAUBITZ, M., NIESERT, S., KUHNLE, H., AND SCHEDEL, I., Cellular immunodeficiency enhances the progression of human papillomavirus-associated cervical lesions, *Int. J. Cancer* **57:**836–840 (1995).

193. PFISTER, H., AND ZUR HAUSEN, H., Seroepidemiological studies of human papilloma virus (HPV-1) infections, *Cancer* **21:**161–165 (1978).

194. POPESCU, N. C., AND DIPAOLO, J. A., Preferential sites for viral integration on mammalian genome, *Cancer Genet. Cytogenet.* **42:**157–171 (1989).

195. POTISCHMAN, N., Nutritional epidemiology of cervical neoplasia, *J. Nutr.* **123:**424–429 (1993).

196. PUROLA, E., AND SAVIA, E., Cytology of gynecologic condyloma acuminatum, *Acta Cytol.* **21:**26–31 (1977).

197. RASMUSSEN, K. A., *Verrucae Plantares: Symptomatology and Epidemiology*, Copenhagen, 1958.

198. REID, R., HERSCHMAN, B. R., CRUM, C. P., FU, Y. S., BRAUN, L., SHAH, K. V., AGRONOW, S. J., AND STANHOPE, C. R., Genital warts and cervical cancer, *Am. J. Obstet. Gynecol.* **149:**293–303 (1984).

199. RICHART, R. M., AND BARRON, B. A., A follow-up study of patients with cervical dysplasia, *Am. J. Obstet. Gynecol.* **105**(3):386–393 (1969).

200. RIES, L. A. G., MILLER, B. A., AND HANKEY, B. F., *SEER Cancer Statistics Review: 1973–1991—Tables and Graphs*, NIH Pub. No. 94-2789, Bethesda, 1994.

201. RITTER, D. B., KADISH, A. S., VERMUND, S. H., ROMNEY, S. L., VILLARI, D., AND BURK, R. D., Detection of human papillomavirus deoxyribonucleic acid in exfoliated cervicovaginal cells as a predictor of cervical neoplasia in a high-risk population, *Am. J. Obstet. Gynecol.* **159**(6):1517–1525 (1988).

202. ROCHE, J. K., AND CRUM, C. P., Local immunity and the uterine cervix: Implications for cancer-associated viruses, *Cancer Immunol. Immunother.* **33:**203–209 (1991).

203. ROHAN, T., MANN, V., McLAUGHLIN, J., HARNISH, D. G., YU, H., SMITH, D., DAVIS, R., SHIER, R. M., AND RAWLS, W., PCR-detected genital papillomavirus infection: Prevalence and association with risk factors for cervical cancer, *Int. J. Cancer* **49:**856–860 (1991).

204. ROMAN, A., AND FIFE, K., Human papillomavirus DNA associated with foreskins of normal newborns, *J. Infect. Dis.* **153:**855–861 (1986).

205. ROOK, A., Virus and related infections, in: *Textbook of Dermatology*, 4th ed. (A. ROOK, D. S. WILKINSON, F. J. G. EBLING, R. H.

CHAMPION, AND J. L. BURTON, EDS.), pp. 668–679, Blackwell Scientific Publications, London, 1986.

206. ROSE, R. C., BONNEZ, W., REICHMAN, R. C., AND GARCEA, R. L., Expression of human papillomavirus type 11 L1 protein in insect cells: *In vivo* and *in vitro* assembly of viruslike particles, *J. Virol.* **67**:1936–1944 (1993).

207. ROSEMBERG, S. K., Sexually transmitted papillomaviral infection in men, *Dermatol. Clin.* **9**(2):317–331 (1991).

208. ROSEMBERG, S. K., HERMAN, G., AND ELFONT, E., Sexually transmitted papillomaviral infection in the male, *Urology* **37**(5): 437–440 (1991).

209. ROSENFELD, W. D., ROSE, E., VERMUND, S. H., SCHREIBER, K., AND BURK, R. D., Follow-up evaluation of cervicovaginal human papillomavirus infection in adolescents, *J. Pediatr.* **121**:301–311 (1992).

210. RUDLINGER, R., BUNNEY, M. H., GROB, R., AND HUNTER, J. A. A., Warts in fish handlers, *Br. J. Dermatol.* **120**:375–381 (1989).

211. SCHACHNER, L., LING, N. S., AND PRESS, S., A statistical analysis of a pediatric dermatology clinic, *Pediatr. Dermatol.* **1**(2):157–164 (1983).

212. SCHAIRER, C., BRINTON, L. A., DEVESA, S. S., ZIEGLER, R. G., AND FRAUMENI, J. F., Racial differences in the risk of invasive squamous-cell cervical cancer, *Cancer Causes Control* **2**:283–289 (1991).

213. SCHIFFMAN, M. H., Recent progress in defining the epidemiology of human papillomavirus infection and cervical neoplasia, *J. Natl. Cancer Inst.* **84**(6):394–398 (1992).

214. SCHIFFMAN, M. H., Validation of HPV hybridization assays: Correlation of FISH, dot blot, and PCR with Southern blot, in: *The Epidemiology of Human Papillomavirus and Cervical Cancer* (N. MUNOZ, F. X. BOSCH, K. V. SHAH, AND A. MEHEUS, EDS.), pp. 169–179, IARC Scientific Publications, Lyon, 1992.

215. SCHIFFMAN, M. H., Epidemiology of cervical human papillomaviruses, in: *Human Pathogenic Papillomaviruse* (H. ZUR HAUSEN, ED.), pp. 55–81, Springer Verlag, Heidelberg, 1994.

216. SCHIFFMAN, M. H., BAUER, H. M., HOOVER, R. N., GLASS, A. G., CADELL, D. M., RUSH, B. B., SCOTT, D. R., SHERMAN, M. E., KURMAN, R. J., WACHOLDER, S., STANTON, C. K., AND MANOS, M. M., Epidemiologic evidence showing that human papillomavirus infection causes most cervical intraepithelial neoplasia, *J. Natl. Cancer Inst.* **85**:958–964 (1993).

217. SCHIFFMAN, M. H., KIVIAT, N. B., BURK, R. D., SHAH, K. V., DANIEL, R. W., LEWIS, R., KUYPERS, J., MANOS, M. M., SCOTT, D. R., SHERMAN, M. E., KURMAN, R. J., STOLER, M. H., GLASS, A. G., RUSH, B. B., MIELZYNSKA, I., AND LORINCZ, A. T., Accuracy and interlaboratory reliability of HPV DNA testing using hybrid capture, *J. Clin. Microbiol.* **33**:545–550 (1995).

218. SCHIFFMAN, M. H., KURMAN, R. J., BARNES, W., AND LANCASTER, W. D., HPV infection and early cervical cytological abnormalities in 3175 Washington, DC women, in: *Papillomaviruses* (P. HOWLEY AND T. R. BROKER, EDS.), pp. 81–88, Wiley-Liss, New York, 1990.

219. SCHIFFMAN, M. H., AND SCHATZKIN, A., Test reliability is critically important to molecular epidemiology: An example from studies of human papillomavirus infection and cervical neoplasia, *Cancer Res.* **54**:1944s–1947s (1994).

220. SCHIFFMAN, M. H., AND SHERMAN, M. E., HPV testing can be used to improve cervical cancer screening, in: *Molecular Markers of Early Detection of Cancer*, (S. SRIVASTAVA, ED.), pp. 265–277, Futura Publishing, Armonk, NY, 1994.

221. SCHLEGEL, R., PHELPS, W. C., ZHANG, Y. L., AND BARBOSA, M., Quantitative keratinocyte assay detects two biological activities

of human papillomavirus DNA and identifies viral types associated with cervical carcinoma, *EMBO J.* **7**(10):3181–3187 (1988).

222. SCHNEIDER, A., KIRCHHOFF, T., MEINHARDT, G., AND GISSMANN, L., Repeated evaluation of human papillomavirus 16 status in cervical swabs of young women with a history of normal Papanicolaou smears, *Obstet. Gynecol.* **79**(5):683–688 (1992).

223. SCHNEIDER, A., AND KOUTSKY, L. A., Natural history and epidemiological features of genital HPV infection, in: *The Epidemiology of Cervical Cancer and Human Papillomavirus* (N. MUNOZ, F. X. BOSCH, K. V. SHAH, AND A. MEHEUS, EDS.), pp. 25–52, IARC Scientific Publications, Lyon, 1992.

224. SCHWARTZ, R. A., AND JANNIGER, C. K., Bowenoid papulosis, *J. Am. Acad. Dermatol.* **24**(2):261–264 (1991).

225. SHAH, K., KASHIMA, H., POLK, F., SHAH, F., ABBEY, H., AND ABRAMSON, A., Rarity of cesarean delivery in cases of juvenile-onset respiratory papillomatosis, *Obstet. Gynecol.* **68**:795–799 (1986).

226. SHAH, K. V., Genital warts, papillomaviruses, and genital malignancies, *Annu. Rev. Med.* **39**:371–379 (1988).

227. SHAH, K. V., AND HOWLEY, P. M., Papillomaviruses, in: *Virology* (B. H. FIELDS, D. M. KNIPE, AND P. M. HOWLEY, EDS.), pp. 2077–2109, Raven Press, New York, 1995.

228. SHERMAN, M. E., SCHIFFMAN, M. H., EROZAN, Y. S., WACHOLDER, S., AND KURMAN, R. J., The Bethesda system: A proposal for reporting abnormal cervical smears based on the reproducibility of cytopathologic diagnoses, *Arch. Pathol.* **116**:1155–1158 (1992).

229. SHERMAN, M. E., SCHIFFMAN, M. H., LORINCZ, A. T., MANOS, M. M., SCOTT, D. R., KURMAN, R. J., KIVIAT, N. B., STOLER, M., GLASS, A. G., AND RUSH, B. B., Towards objective quality assurance in cervical cytopathology: Correlation of cytopathologic diagnoses with detection of high risk human papillomavirus types, *Am. J. Clin. Pathol.* **101**:182–187 (1994).

230. SHROYER, K. R., Human papillomavirus and endocervical adenocarcinoma, *Hum. Pathol.* **24**:119–120 (1993).

231. SHROYER, K. R., KIM, J. G., MANOS, M. M., GREER, C. E., PEARLMAN, N. W., AND FRANKLIN, W. A., Papillomavirus found in anorectal squamous carcinoma, not in colon adenocarcinoma, *Arch. Surg.* **127**:741–744 (1992).

232. SMITH, E. M., JOHNSON, S. R., JIANG, D., ZALESKI, S., LYNCH, C. F., BRUNDAGE, S., ANDERSON, R. D., AND TUREK, L. P., The association between pregnancy and human papilloma virus prevalence, *Cancer Detect. Prev.* **15**(5):397–402 (1991).

233. SMITS, H. L., TIEBEN, L. M., TJONG-A-HUNG, S. P., JEBBINK, M. F., MINNAAR, R. P., JANSEN, C. L., AND TER SCHEGGET, J., Detection and typing of human papillomaviruses present in fixed and stained archival cervical smears by a consensus polymerase chain reaction and direct sequence analysis allow the identification of a broad spectrum of human papillomavirus types, *J. Gen. Virol.* **73**:3263–3268 (1992).

234. SNIJDERS, P. J. F., VAN DEN BRULE, A. J. C., SCHRIJNEMAKERS, H. F. J., SNOW, G., MEIJER, C. J. L. M., AND WALBOOMERS, J. M. M., The use of general primers in the polymerase chain reaction permits the detection of a broad spectrum of human papillomavirus genotypes, *J. Gen. Virol.* **71**:173–181 (1990).

235. SOLOMON, D., The 1988 Bethesda system for reporting cervical/vaginal cytological diagnoses, *J.A.M.A.* **262**(6):931–934 (1989).

236. SPITZER, M., KRUMHOLZ, B. A., AND SELTZER, V. L., The multicentric nature of disease related to human papillomavirus infection of the female lower genital tract, *Obstet. Gynecol.* **73**(3):303–307 (1989).

237. SRIVATSAN, E. S., MISRA, B. C., VENUGOPALAN, M., AND WILCZYNSKI, S. P., Loss of heterozygosity for alleles on chromosome 11 in cervical carcinoma, *Am. J. Hum. Genet.* **49**:868–877 (1991).

238. STANBRIDGE, C. M., AND BUTLER, E. B., Human papillomavirus infection of the lower female genital tract: Association with multicentric neoplasia, *Int. J. Gynecol. Pathol.* **2**(3):264–274 (1983).

239. STEELE, K., IRWIN, W. G., AND MERRETT, J. D., Warts in general practice, *Irish Med. J.* **82**(3):122–124 (1989).

240. STEINBERG, B. M., Laryngeal papillomatosis is associated with a defect in cellular differentiation, *Ciba Found. Symp.* **120**:208–220 (1986).

241. STEINBERG, B. M., ABRAMSON, A. L., HIRSCHFIELD, L., KAHN, L. B., AND FREIBERGER, I., Vocal cord polyps: Biochemical and histologic evaluation, *Laryngoscope* **95**(11):1327–1331 (1985).

242. STELLATO, G., NIEMINEN, P., AHO, H., VESTERINEN, E., VAHERI, A., AND PAAVONEN, J., Human papillomavirus infection of the female genital tract: Correlation of HPV DNA with cytologic, colposcopic, and natural history findings, *Eur. J. Gynaecol. Oncol.* **13**:262–267 (1994).

243. STOLER, M. H., AND BROKER, T. R., *In situ* hybridization detection of human papillomavirus DNAs and messenger RNAs in genital condylomas and a cervical carcinoma, *Hum. Pathol.* **17**:1250–1258 (1986).

244. STRICKLER, H. D., DILLNER, J., SCHIFFMAN, M. H., MANOS, M. M., GLASS, A. G., GREER, C. E., SCOTT, D. R., SHERMAN, M. E., AND KURMAN, R. J., A seroepidemiologic study of HPV infection and incident cervical squamous intraepithelial lesions, *Virol. Immunol.* **7**:169–177 (1994).

245. SUN, Q., TSUTSUMI, K., KELLEHER, M. B., PATER, A., AND PATER, M. M., Squamous metaplasia of normal and carcinoma in situ of HPV 16-immortalized human endocervical cells, *Cancer Res.* **52**:4254–4260 (1992).

246. SUNDBERG, J. P., Papillomavirus infections in animals, in: *Papillomaviruses and Human Disease*, (K. SYRJANEN, L. GISSMANN AND L. KOSS, EDS.), pp. 40–103, Springer-Verlag, Heidelberg, 1987.

247. TABRIZI, S. N., TAN, J., QUINN, M., BORG, A. J., AND GARLAND, S. M., Detection of genital human papillomavirus (HPV) DNA by PCR and other conventional hybridisation techniques in male partners of women with abnormal Papanicolaou smears, *Genitourin. Med.* **68**:370–373 (1992).

248. TAWA, K., FORSYTHE, A., COVE, K., SALTZ, A., PETERS, H., AND WATRING, W. G., A comparison of the Papanicolaou smear and the cervigram: Sensitivity, specificity, and cost analysis, *Obstet. Gynecol.* **71**:229–235 (1988).

249. TIERNEY, R. J., ELLIS, J. R. M., WINTER, H., KAUR, S., WILSON, S., WOODMAN, C. B. J., AND YOUNG, L. S., PCR for the detection of cervical HPV16 infection: The need for standardization, *Int. J. Cancer* **54**:700–701 (1993).

250. TOH, Y., KUWANO, H., TANAKA, S., BABA, K., MATSUDA, H., SUGIMACHI, K., AND MORI, R., Detection of human papillomavirus DNA in esophageal carcinoma in Japan by polymerase chain reaction, *Cancer* **70**(9):2234–2238 (1992).

251. TUCKER, R. A., JOHNSON, P. R., REEVES, W. C., AND ICENOGLE, J. P., Using the polymerase chain reaction to genotype human papillomavirus DNAs in samples containing multiple HPVs may produce inaccurate results, *J. Virol. Methods* **43**:321–334 (1993).

252. VAN DEN BRULE, A. J. C., WALBOOMERS, J. M. M., DU MAINE, M., KENEMANS, P., AND MEIJER, C. J. L. M., Difference in prevalence of human papillomavirus genotypes in cytomorphologically normal cervical smears is associated with a history of cervical intraepithelial neoplasia, *Int. J. Cancer* **48**:404–408 (1991).

253. VAN DOORNUM, G. J. J., PRINS, M., JUFFERMANS, L. H. J., HOOYKAAS, C., VAN DEN HOEK, J. A. R., COUTINHO, R. A., AND QUINT, W. G. V., Regional distribution and incidence of human papillomavirus infections among heterosexual men and women with multiple sexual partners; A prospective study, *Genitourin. Med.* **70**(4):240–246 (1994).

254. VAN RANST, M., KAPLAN, J. B., AND BURK, R. D., Phylogenetic classification of human papillomaviruses: Correlation with clinical manifestations, *J. Gen. Virol.* **73**:2653–2660 (1992).

255. VAN RANST, M. A., TACHEZY, R., DELIUS, H., AND BURK, R. D., Taxonomy of the human papillomaviruses, *Papillomavirus Rep.* **3**:61–65 (1993).

256. VERMUND, S. H., KELLY, K. F., KLEIN, R. S., FEINGOLD, A. R., SCHREIBER, K., MUNK, G., AND BURK, R. D., High risk of human papillomavirus infection and cervical squamous intraepithelial lesions among women with symptomatic human immunodeficiency virus infection, *Am. J. Obstet. Gynecol.* **165**:392–400 (1991).

257. VERMUND, S. H., SCHIFFMAN, M. H., GOLDBERG, G. L., RITTER, D. B., WELTMAN, A., AND BURK, R. D., Molecular diagnosis of genital human papillomavirus infection: Comparison of two methods used to collect exfoliated cervical cells, *Am. J. Obstet. Gynecol.* **160**:304–308 (1989).

258. VISCIDI, R. P., SUN, Y., TSUZAKI, B., BOSCH, F. X., MUNOZ, N., AND SHAH, K. V., Serologic response in human papillomavirus-associated invasive cervical cancer, *Int. J. Cancer* **55**:780–784 (1993).

259. WANK, R., AND THOMSSEN, C., High risk of squamous cell carcinoma of the cervix for women with HLA-DQw3, *Nature* **352**:723–725 (1991).

260. WERNESS, B. A., MUNGER, K., AND HOWLEY, P. M., Role of the human papillomavirus oncoproteins in transformation and carcinogenic progression, *Important Adv. Oncol.* **1991**:3–18 (1991).

261. WHEELER, C. M., PARMENTER, C. A., HUNT, C., BECKER, T. M., GREER, C. E., HILDESHEIM, A., AND MANOS, M. M., Determinants of genital human papillomavirus infection among cytologically normal women attending the University of New Mexico student health center, *Sex. Transm. Dis.* **20**:286–289 (1993).

262. WIDEROFF, L., SCHIFFMAN, M. H., KIRNBAUER, R., HUBBERT, N., NONNEMACHER, B., LOWY, D., LORINCZ, A. T., MANOS, M. M., GLASS, A. G., SCOTT, D. R., SHERMAN, M. E., KURMAN, R. J., BUCKLAND, J., TARONE, R. E., AND SCHILLER, J., Validation of a serologic marker for persistent HPV 16 infection, *J. Infect. Dis.* **172**:1425–1430 (1995).

263. WIENER, J. S., AND WALTHER, P. J., A high association of oncogenic human papillomaviruses with carcinomas of the female urethra: Polymerase chain reaction-based analyses of multiple histological types, *J. Urol.* **151**:49–53 (1994).

264. WIKSTROM, A., LIDBRINK, P., JOHANSSON, B., AND VON KROGH, G., Penile human papillomavirus carriage among men attending Swedish STD clinics, *Int. J. STD AIDs* **2**:105–109 (1991).

265. WILLIAMS, H. C., POTTIER, A., AND STRACHAN, D., The descriptive epidemiology of warts in British schoolchildren, *Br. J. Dermatol.* **128**:504–511 (1993).

266. WRIGHT, T. C., GAGNON, S., RICHART, R. M., AND FERENCZY, A., Treatment of cervical intraepithelial neoplasia using the loop electrosurgical excision procedure, *Obstet. Gynecol.* **79**:173–178 (1992).

267. XI, L. F., DEMERS, W., KIVIAT, N. B., KUYPERS, J., BECKMANN, A. M., AND GALLOWAY, D. A., Sequence variation in the noncoding region of human papillomavirus type 16 detected by single-strand conformation polymorphism analysis, *J. Infect. Dis.* **168:**610–617 (1993).

268. YEUDALL, W. A., Human papillomaviruses and oral neoplasia, *Oral Oncol. Eur. J. Cancer* **28B:**61–66 (1992).

269. ZUR HAUSEN, H., Papillomaviruses in human cancer, *Cancer* **59:**1692–1696 (1987).

270. ZUR HAUSEN, H., Viruses in human cancers, *Science* **254:**1167–1173 (1991).

271. ZUR HAUSEN, H. (ED.), *Current Topics in Microbiology and Immunology, Vol. 186, Human Pathogenic Papillomaviruses,* Springer Verlag, Heidelberg, 1994.

12. Suggested Reading

IARC monographs on the evaluation of carcinogenic risks to humans, vol. 64: Human papillomaviruses. IARC, Lyon, France, 1995.

MUNOZ, N., BOSCH, F. X., SHAH, K. V., AND MEHEUS, A., *The Epidemiology of Human Papillomavirus and Cervical Cancer*, No. 119, IARC Scientific Publications, Lyon, 1992.

SHAH, K. V., AND HOWLEY, P. M., Papillomaviruses, in: *Virology, 3rd ed.* (B. N. FIELDS, D. M. KNIPE, AND P. M. HOWLEY, EDS.), pp. 2077–2109, Raven Press, New York, 1995.

ZUR HAUSEN, H., *Current Topics in Microbiology and Immunology, Vol. 186, Human Pathogenic Papillomaviruses*, Springer-Verlag, Heidelberg, 1994.

PART IV

Viral Infections and Chronic Diseases

CHAPTER 34

Chronic Neurological Diseases Caused by Slow Infections

David M. Asher and Clarence J. Gibbs, Jr.

1. Introduction

Until recently, most subacute and chronic progressive degenerative diseases of the human nervous system were of unknown etiology. They generally occur sporadically, though some are familial. None can be cured. When kuru was found to be caused by a serially transmissible filterable agent,[80] it established that a chronic degenerative central nervous system (CNS) disease of humans was a "slow" infection.[211] Since that demonstration, several other chronic diseases of the human brain have also been attributed to slow infections, either with unconventional self-replicating agents resembling that causing kuru[79] (the nature of which remains disputed[195,202]) or with common conventional viruses (Table 1). In this chapter, we review both types of slow infection, beginning with those caused by conventional viruses.

2. Subacute Sclerosing Panencephalitis

2.1. Introduction

Subacute sclerosing panencephalitis (SSPE) is a progressive fatal disease affecting the CNS in children and young adults and caused by a persistent infection with measles virus. The strains of measles virus isolated from brains of patients with SSPE differ substantially from

those found in acute measles infections but are clearly derived from them.

2.2. Historical Background

SSPE was described as an epidemic inclusion-body encephalitis in 1933 by Dawson,[59] who attempted unsuccessfully to find an infectious agent, and later by others.[193,225] Structures suggestive of measles viral nucleocapsids were recognized in brains of patients with SSPE by electron microscopy in 1965[26]; measles virus was further implicated by the finding of elevated levels of antiviral antibody in serum and cerebrospinal fluid (CSF) and viral antigens in brain tissue[54] and then confirmed by isolation of the virus from the brains of patients with SSPE.[48,121,122,191,192]

2.3. Methodology

The U.S. Registry of SSPE (Section 2.5) remains a useful source of information. Reports from other countries, where measles is much more common than in the United States, now provide most new information about SSPE. Strains of measles virus associated with SSPE have been isolated by cocultivation of viable brain tissue with susceptible cell cultures and viral genomes have been characterized by molecular amplification and nucleic acid sequencing (Section 2.7).

2.4. Biological Characteristics of the Virus

Complete measles virus particles are not found in the brains of patients with SSPE; however, when infected

David M. Asher and Clarence J. Gibbs, Jr. • Laboratory of Method Development, Center for Biologics Evaluation and Research, Food and Drug Administration, Rockville, Maryland 20852-1448; and Laboratory of Central Nervous System Studies, National Institute of Neurological Disorders and Stroke, National Institutes of Health, Bethesda, Maryland 20892.

Table 1. Degenerative Diseases of the Human Nervous System Caused by Slow Infections with Conventional Viruses[a]

Disease	Virus
Persistent virus in a normal host	
Subacute sclerosing panencephalitis[b]	Measles virus
Progressive rubella panencephalitis[b]	Rubella virus
? Progressive tick-borne encephalitis[b]	TBE virus
Rabies	Rabies virus
Persistent virus in an immune-deficient host	
Progressive multifocal leukoencephalopathy[b]	JC Papovavirus
AIDS dementia/encephalopathy	HIV
Subacute adenoviral encephalitis	Adenovirus 32
Chronic enteroviral meningoencephalomyelitis	Picornaviruses (polio, coxsackie, other)
Demyelination, vasculitis following viral infection (?autoimmune)	
Postinfectious encephalomyelitis	Measles, mumps, influenza viruses
	Varicella–zoster virus
	Vaccinia virus
Guillain–Barré syndrome	Influenza virus (inactivated)
Other delayed effect of conventional virus infection	
Encephalopathy after perinatal infection	Rubella virus
Childhood hydrocephalus	? Mumps virus
Postencephalitic parkinsonism	Japanese encephalitis virus
	Western equine encephalitis virus
Shingles ganglionitis/granulomatous angiitis	Varicella-zoster virus

[a]Some chronic diseases attributed to conventional virus infections are tentative, and the list is not exhaustive.
[b]Discussed in text.

brain cells are cultivated together with permissive cells, complete measles virus often emerges. The matrix (M) protein required for the final assembly and budding of virus from the host cells is usually not detected in brains of patients[13]; however, the full complement of measles viral genetic material needed to code for all proteins, including M, is present and functional. Recent studies suggest that the M proteins encoded by SSPE-associated isolates of measles virus are consistently unable to bind to nucleocapsid,[119] due to a variety of mutations in the M-protein gene,[40,41,118] most of which are U-to-C shifts, possibly caused by a cellular RNA-modifying activity.[249] Mutations are also found in other genes of SSPE-associated measles virus.[106,142,204]

2.5. Descriptive Epidemiology

SSPE occurs throughout the world. Never common, it has become exceedingly rare in the United States. Data for more than 570 cases with onset in 1956 or later have been compiled in the U.S. National SSPE Registry initiated by Jabbour and colleagues[129] and later maintained by Dr. Paul R. Dyken, Department of Neurology, University of South Alabama Medical Center, Mobile.[68] The disease has been diagnosed in patients aged less than a year old[172] and older than 30 years,[37] but it primarily affects children and younger adolescents. More than 85% of cases in the National SSPE Registry had onset between 5 and 15 years of age.[172] The average age of onset of SSPE in U.S. cases reported before 1980 was about 10 years; between 1980 and 1984, the age rose almost to 14 years. In most series of SSPE, at least twice as many boys as girls were affected,[169] the incidence was higher among rural children than city children, and SSPE was more common among children with two or more siblings, of lower socioeconomic status,[61] and mentally retarded children.[169] Acquisition of measles before the age of 18 months increases the risk of SSPE substantially,[150,172] especially during the first year of life,[169,208,216] though this may not be true in all populations.[218] (Exposure to birds and other animals has been reported with abnormal frequency in histories of patients with SSPE; the reason for that is not clear.[112] SSPE was once especially common in the southeastern United States, the Ohio River Valley, and some New England states. Recently, cases have been more frequent in the western United States, especially among Hispanic children[68]; during the past 5 years, most of the cases reported to the U.S. SSPE Registry were in immigrant children living in New York City or California, and only two native-born patients were recognized (P. R. Dyken, personal communication).

Mean annual incidence rates of SSPE in the United States have fallen markedly since 1960, from 0.61 cases per million persons under age 20 years to an estimated 0.06 cases in 1980.[44] In recent years, only two or three new cases were registered each year from the entire United States (P. R. Dyken, personal communication). This drop seems clearly attributable to the progressive decline in measles cases following introduction of live attenuated measles vaccine in the United States in 1963.

Children with SSPE typically have recovered from measles several years before the onset of neurological disease. Measles may have been either mild or severe. Some patients with SSPE had measles pneumonia, but none has had a history of acute measles encephalitis. The mean interval between measles and onset of SSPE in the United States was formerly about 7 years,[44] but more recently it increased to 12 years.[68] The risk was estimated at 8.5 SSPE cases per million cases of acute measles for a 6-year period during which the estimated risk after measles vaccination was only 0.7 cases per million doses of vaccine. Of the first 566 patients with SSPE diagnosed in the United States after 1969 and reported to the National Registry, only 14% had a history of measles vaccination without also having clinically apparent measles, while a roughly equal number had no history of either acute measles or immunization.[44] Lack of measles vaccination itself was a highly significant risk factor in a case-control study of SSPE.[112,113] In a recent survey of SSPE in England and Wales, the protective effect of measles vaccination was especially striking: the overall relative risk of SSPE after measles compared to vaccine was 29, and after measles before the age of one year the risk was more than 100 compared with vaccine.[73,169] It has not been determined whether SSPE in vaccinated children resulted from persistent infection with attenuated vaccine virus (which seems increasingly unlikely), from unrecognized measles infection preceding vaccination, or from vaccine failure and subsequent undiagnosed measles. In any case, the overwhelming protection afforded by measles vaccination in preventing SSPE and other complications of measles is now beyond doubt.

2.6. Clinical Illness, Pathogenesis, Histopathologic Manifestations, and Immunity

The onset of SSPE is usually insidious, marked by subtle changes in behavior and deterioration of school work, followed by more overtly bizarre behavior, and, finally, by frank dementia. Patients have no fever, photophobia, or other findings of acute encephalitis except for occasional complaints of headache. Diffuse neurological disease becomes progressively more severe. Massive repetitive myoclonic jerks, generally symmetrical, especially involving the axial musculature and occurring at 5- to 10-second intervals, typically begin later, though they may be a presenting sign.[136] The myoclonic jerks appear to be abnormal movements rather than epileptic seizures, but true convulsions can also occur at any stage of illness, including onset.[143] In addition to myoclonic jerks, which tend to disappear as disease progresses, a variety of other abnormal movements and dystonias have also been observed. Cerebellar ataxia may be noted as well.[67,68] Chorioretinitis, retinopathy, and optic atrophy may appear, sometimes even before behavioral changes[131,254]; cortical blindness may also appear early.[132] Dementia progresses to stupor and coma, sometimes with autonomic insufficiency. Patients may be rigid or spastic with decorticate postures or may be flaccid.

Although progression of SPE is variable, in at least 60% of patients with SSPE the course is inexorable and rapid. Total duration of illness may be as short as a month,[169,209] but most patients survive for 1 to 3 years after diagnosis, with a mean of about 18 months.[68] Occasional patients have shown spontaneous improvement and have lived for more than 10 years.

2.6.1. Pathology. The histopathology of SSPE consists of inflammation, necrosis, and repair.[67,68,207] Brain biopsy performed in the early stages of SSPE shows mild inflammation of meninges and a panencephalitis involving cortical and subcortical gray matter as well as white matter, with cuffs of plasma cells and lymphocytes around blood vessels and increased numbers of glia throughout. Neuronal loss may not be marked until later in the course of illness. Loss of myelin secondary to neuronal degeneration may be apparent. Intranuclear "Cowdry type A" inclusion bodies surrounded by clear halos noted by Dawson[59] may be seen in sections stained with hematoxylin and eosin, located within the nuclei of neurons, astrocytes, and oligodendrocytes, but inclusions are sometimes difficult to find. By electron microscopy the inclusions are seen to contain tubular structures[26] typical of the nucleocapsids of paramyxoviruses. Measles viral antigens can be demonstrated by labeled-antibody techniques within the inclusions as well as in cells without inclusions. Lesions may be unevenly distributed throughout the brain, and biopsy is not always diagnostic, particularly if only a small sample of tissue is obtained. The same findings of inclusion-body panencephalitis are generally present in the brain at autopsy; however, late in disease it may be difficult to find typical areas of inflammation, and the main histopathologic changes are necrosis and gliosis. The disease is believed to begin in the cortical gray matter, then progressing to white matter and subcortical gray matter, and then to lower structures.[67] Persistent infection of lymphoid tissues with measles virus has also been claimed,[27,74–76] although those tissues show no pathological changes.

2.6.2. Pathogenesis. SSPE results from a persistent infection with measles virus. Two hypotheses have been put forward to explain how a usually self-limited viral infection might occasionally cause a slow infection of the CNS: either a mutant virus or abnormal host. It has been proposed that the mutations detected in the RNA of SSPE-associated measles may render the virus more likely to establish persistent infection; however, it remains unclear whether viral genomic abnormalities in SSPE isolates cause the disease or result from long persistence of the infection. Although a geographic–temporal cluster of SSPE cases has been reported,[16,209] suggesting a common source of infection with an especially virulent strain of measles virus, that is unusual. It has also been theorized that patients with SSPE have some predisposing subtle immune deficiency[17]; the increased risk of SSPE in children who had measles during infancy supports that hypothesis. However, the occurrence of SSPE in only one of identical twins suggests that other environmental factors play a role as well.[14,236]

The recent demonstration of measles nucleocapsid and RNA in vascular endothelial cells of patients with SSPE[128,139] raises the possibility that the virus remained latent outside the CNS, penetrating the parenchyma only later in the incubation period, either from those cells or from lymphoid cells.[74–76]

2.6.3. Immunity. The pathology of SSPE indicates that, in addition to increased amounts of antibodies to measles virus, there is an active cell-mediated immune response to infection. B cells,[176] thought to be secreting antibodies to measles virus,[245] and CD4-bearing T lymphocytes are present in parenchymal infiltrates and perivascular cuffs[176]; and secretion of various cytokines—tumor necrosis factor, several interleukins, gamma interferon, and lymphotoxin—has been demonstrated.[120,177] The modest clinical improvement of some patients with SSPE after treatment with immunomodulators (Section 2.8) suggests that immunopathogenic mechanisms play some role in progression of disease.

2.7. Laboratory and Differential Diagnosis

The blood is normal except for elevated titers of antibodies to measles virus; antibodies are of the IgG and IgM classes and are directed against all the proteins of measles virus except the M protein. Examination of the CSF is most useful for establishing the diagnosis of SSPE. Cell content of the CSF is generally normal, although stained sediments have been reported to show plasma cells. Total protein content of the CSF is normal or only slightly elevated; however, the gamma globulin fraction is greatly elevated (usually comprising at least 20% of total protein in CSF), resulting in a paretic type of colloidal gold curve. When the CSF is examined by electrophoresis or isoelectric focusing, oligoclonal bands of immunoglobulin are often observed. IgG and IgM antibodies to measles virus, not normally found in unconcentrated CSF, make up most of the immunoglobulin, and these may often be detected in dilutions of 1:8 or more. The complement-fixation test, hemagglutination inhibition, immunofluorescence, and other serological tests, including enzyme-linked immunosorbent assay (ELISA),[146] have been satisfactory for demonstrating antibodies in CSF. The normal ratio of titer in serum to titer in CSF

is reduced (below 200) for measles antibodies, while serum–CSF ratios are normal for other viral antibodies and for albumin, indicating that the increased amounts of measles antibodies in CSF of patients with SSPE result from synthesis within the nervous system and that the blood–brain barrier is normal.[223]

The electroencephalogram (EEG) is also useful in supporting the diagnosis of SSPE, although early in disease it may be normal or show only moderate nonspecific slowing.[99] In the myoclonic stage most patients with SSPE have episodes of "suppression-burst" in which high-amplitude slow and sharp waves recur at intervals of 3 to 5 seconds on a slow background; however, that pattern is not unique to SSPE.[100] Later in illness the EEG becomes increasingly disorganized with high-amplitude random dysrhythmic slowing; in terminal disease the amplitude may fall.[67,68]

Computed tomograms or magnetic resonance images of patients with SSPE may show variable cortical atrophy and ventricular enlargement, and there may be focal or multifocal low-density lesions in white matter. However studies may be normal, especially early in disease.[65]

Brain biopsy is not usually needed to diagnose SSPE. When performed, it often shows the typical histopathologic findings described above. Examination of frozen sections by the immunofluorescence technique may demonstrate the presence of measles viral antigens. Measles virus may be isolated from biopsy tissue by culturing fragments of brain or separated cells together with cells susceptible to measles virus and then propagating the mixed cultures for several serial passages. Persistence of measles virus infection in the cultures can be demonstrated by labeled antibody techniques before complete virus appears. Even in highly experienced laboratories many specimens failed to yield complete virus, though measles antigens could be demonstrated in cultures.[135]

Reverse transcriptase–polymerase chain reaction (RT-PCR) detected various regions of the measles virus RNA in brain tissues of patients with SSPE,[101] and that technique may be useful for diagnosis under some circumstances. *In situ* RT-PCR with labeled-probe hybridization[128] has successfully demonstrated the genome of measles virus in paraffin-embedded brain tissues from patients with SSPE (Fig. 1). Nucleic acid hybridization techniques without amplification also detected the measles viral genome in SSPE.[74] Measles virus and viral RNA were purportedly found in lymphoid tissues and circulating blood lymphocytes of SSPE patients,[74–76] although that finding was not confirmed by others.[205]

2.7.1. Differential Diagnosis. In the diagnosis of SSPE, as for other slow infections, it is most important to

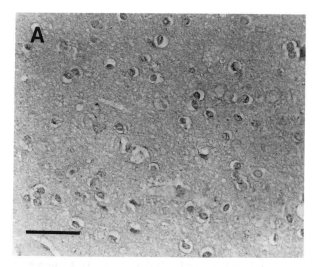

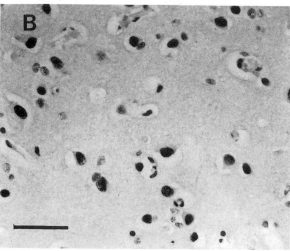

Figure 1. (A) Brain of patient with Creutzfeldt–Jakob disease (as control). (B) Brain of patient with subacute sclerosing panencephalitis. Both sections underwent *in situ* RT-PCR with measles-virus-specific primers followed by *in situ* hybridization with a digoxygenin-labeled measles cDNA probe (also prepared by PCR). Hybridization was detected by immunocytochemistry using enzyme-labeled antibody to digoxygenin. Hematoxylin counterstain. Bars = 50 μm. Preparations kindly provided by Dr. Stuart Isaacson.

rule out treatable illnesses, such as abscesses and tumors. Various cerebral storage diseases and nonstorage poliodystrophies, leukodystrophies, and demyelinating diseases of childhood also produce progressive dementia with seizures and paralysis resembling SSPE.[65] Early in the course of illness, SSPE must be distinguished from atypical acute viral encephalitides.[238] Other slow viral infections, such as Creutzfeldt–Jakob disease and pro-

gressive rubella panencephalitis, must be considered in appropriate age groups. The presence of a typical EEG pattern is suggestive of SSPE, as are unusually high levels of measles antibodies in serum. The diagnosis is practically confirmed if measles antibodies are detected in unconcentrated CSF.

2.8. Prevention, Treatment, and Control

2.8.1. Prevention. Immunization with attenuated measles virus vaccines prevents the overwhelming majority of cases of SSPE and has effectively eliminated the disease from the United States.[44,68]

2.8.2. Therapy. Information on current therapeutic trials should be sought through the U.S. National SSPE Registry. Two therapeutic regimens, alone or combined, are claimed to stabilize the progression of SSPE, though not to cure the disease. It has been reported that treatment with inosiplex increased the number of patients with prolonged survival and possibly some clinical improvement in degree of disability.[68,69] (A study conducted through the SSPE Registry treated with 100 mg of inosiplex per kilogram body weight per day in divided doses; others claimed success with lower doses.[5]) Several groups claimed that progression of SSPE slowed after intrathecal or intraventricular injections with alpha interferon[87,171,214,244] or combined treatment with oral inosiplex and intrathecal or intraventricular interferon.[86,251,252] However, in another study, combined inosiplex and interferon therapy caused no improvement in CSF abnormalities,[165] and no clinical effect of inosiplex alone was apparent.[16] (Cimetidine was also reported to stabilize the progression of SSPE; that was attributed to the drug's immunodulatory effect.[4]) An attempt to improve the immune response of an SSPE-affected identical twin by transfusion with cells from her normal sister failed.[14] No therapy has produced either cure or even marked improvement in SSPE. The judicious use of anticonvulsants for seizures and antibiotics for secondary bacterial infections, careful alimentation, physical therapy, prevention of decubiti, and other elements of supportive care may prolong survival and improve the quality of life.

2.8.3. Precautions. Since no complete virus particles have been detected in patients with SSPE, their tissues, secretions, and excretions should pose no danger of infecting others, and they require no special precautions by family or caregivers. In spite of the recent failure to confirm the finding of measles virus in blood of patients with SSPE,[205] universal precautions should be observed for handling blood and body fluids of these and all other patients.[181]

2.9. Unresolved Problems

Although accumulating evidence favors the hypothesis that the viral mutations observed in SSPE isolates, particularly in the M-protein gene, are somehow involved in the persistence of measles virus in the CNS and the pathogenesis of disease, the causal association remains to be firmly established. It also seems likely that residual maternal antibodies to measles or relative immunologic immaturity predisposes measles-infected infants to develop SSPE later in childhood, but the mechanism for that is completely unknown. The possibility that vaccination with attenuated measles virus itself causes SSPE can now be dismissed.

3. Progressive Rubella Panencephalitis

3.1. Introduction

Progressive rubella panencephalitis[247] is an exceedingly rare chronic encephalitis caused by persistent infection with rubella virus.

3.2. Historical Background

Progressive rubella panencephalitis was first described by Lebon and Lyon in 1974,[147] and fewer than 20 cases have been reported since then. Rubella virus was isolated from the brain of a patient with the disease by Cremer and co-workers.[55]

3.3. Methodology, Characteristics of the Virus, and Descriptive Epidemiology

Because progressive rubella panencephalitis is so exceedingly rare, there is no registry for the disease, and isolates of rubella virus associated with it have been little studied. The disease was reviewed recently.[247] All patients were males and between the ages of 8 and 21 years at onset; most had typical stigmata of the congenital rubella syndrome, including cataracts, deafness, and mental retardation, but several had recovered from childhood rubella.[247] Progressive rubella panencephalitis has apparently not been recognized in the United States at all in recent years, presumably because of the success of childhood immunization in preventing rubella, though cases continue to occur elsewhere[108] where rubella is more common.

3.4. Clinical Illness

The onset of progressive rubella panencephalitis resembles that of SSPE,[224,230,246] with insidious changes in behavior and deterioration in intellectual performance. This is followed by dementia and other signs of multifocal brain disease, including seizures, cerebellar ataxia, and spastic weakness. Myoclonus and other abnormal movements may occur,[1,230] but this is not as common as in SSPE. Retinopathy, similar to that of acute rubella, and optic atrophy may be found.[230,246] The course of progressive rubella panencephalitis is also similar to that in SSPE, progressing to coma, spasticity, bulbar involvement, and death in 2 to 5 years.

The peripheral blood is normal in progressive rubella panencephalitis except for elevated titers of antibodies to rubella virus. The CSF shows normal or slightly elevated cell content; CSF protein is slightly elevated, with marked increase in globulin, which may make up more than 50% of total protein.[224] Oligoclonal electrophoretic bands of globulin are found in CSF of patients with progressive rubella panencephalitis; the bands resemble those in SSPE but consist of antibodies to rubella viral antigens.[246] Antibodies to rubella virus are readily detectable in CSF, often at dilutions of 1:8 or higher. The complement-fixation, hemagglutination-inhibition, and ELISA techniques should be satisfactory for testing spinal fluid. Most of the antibodies to rubella are IgG, although some IgM antibodies have also been detected early in the course of the disease; the serum–CSF ratio of antibody titers to rubella virus is reduced,[246] while ratios of titers to measles and other viruses are normal.

The EEG shows generalized slowing with occasional high-voltage activity, but the suppression-burst pattern of SSPE has not been seen in progressive rubella panencephalitis. Encephalograms show enlargement of all ventricles, especially the fourth, with prominent atrophy of the cerebellum.

Histopathologic changes in brains of patients with progressive rubella panencephalitis are similar to those in SSPE, with cuffs of lymphocytes and plasma cells around blood vessels, glial nodules in the cortex, some loss of neurons, and an increase in astrocytes throughout gray and white matter. The histopathology differs from that of SSPE in two important respects: in progressive rubella panencephalitis no inclusion bodies have been recognized, and deposits of material that stains with the periodic acid-Schiff reaction are found around vessels in subcortical white matter.

Rubella virus was isolated from brain cell cultures of

one patient with progressive rubella panencephalitis by cocultivation with susceptible cells as well as by propagation of explanted cells alone.[55,230] Other attempts at isolation from brain tissues were not successful.[224] Rubella virus was also isolated from separated blood lymphocytes of a patient with the disease by cocultivation with susceptible cells.[248]

Differential diagnosis of progressive rubella panencephalitis is the same as that for SSPE. The presence of stigmata of congenital rubella syndrome or a history of German measles in a young male with progressive neurological disease suggests the diagnosis. Elevated levels of rubella antibodies in serum and the presence of rubella antibodies in the spinal fluid (with reduction of normal serum–CSF ratio for rubella antibodies) should establish the diagnosis of progressive rubella panencephalitis. Isolation of rubella virus from blood lymphocytes may be attempted. Brain biopsy should not be needed to establish the diagnosis.

3.5. Control and Prevention

Patients with progressive rubella panencephalitis pose no substantial risk of infection to others, although it seems reasonable to avoid exposing rubella-susceptible persons to their blood. Rubella virus has not been detected in urine, and secretions have apparently never been studied.

3.6. Unresolved Problems

Because of its extreme rarity and ease of prevention by immunization, even less is known about pathogenetic mechanisms of progressive rubella panencephalitis than SSPE.

4. Progressive Multifocal Leukoencephalopathy

4.1. Introduction and Historical Background

Progressive multifocal leukoencephalopathy (PML)[10] is a progressive demyelinating disease of the nervous system caused by activation of a latent infection with the JC papovavirus[190,255] in immunosuppressed subjects, in whom it is invariably fatal. PML was first recognized as a complication of leukemia and Hodgkin's disease.[10] It is now most often recognized as an opportunistic infection with acquired immunodeficiency syndrome (AIDS) in adults and, rarely, children. In 1964, structures resembling virions of papovaviruses were detected by elec-

tron microscopy in oligodendrocytes of patients with PML,[255,256] and 6 years later the causative agent, the "JC" virus (JCV), was isolated in cultures of fetal spongioblasts (glial precursor cells).[190]

4.2. Methodology

Isolation techniques, requiring human fetal brain cultures, are cumbersome. Negative-stain electron microscopy, sometimes augmented by incubation of extracts with immune serum, has also been useful for detection of virus.[232] Both methods have been largely supplanted by assays for viral proteins[88,232] or, more recently, viral DNA.[3,35,151,227]

4.3. Biological Characteristics of the Virus

JCV is a member of the papovavirus family in the genus *Polyomavirus*—a double-stranded DNA with unenveloped icosahedral nucleocapsid sharing some antigens with other members of the genus. "BK" virus (BKV), another virus of the same genus, originally isolated from the urine of an immunosuppressed renal transplant patient,[85] also infects humans, and has been isolated from brain tumors,[219] though the causal connection is not definite. [Another papovavirus related to JCV, simian virus 40 (SV40) of monkeys, was implicated in two early cases of PML,[231] but apparently not since then.[70]] JCV strains causing PML are heterogeneous and presumably generated from ordinary "archetypal" strains during asymptomatic persistence,[127] rather than from especially neurovirulent progenitors. JCV strains isolated from brain share certain similarities in DNA sequences and differ from strains found in kidney,[11,12] especially in the noncoding regulatory region, prompting the hypothesis that mutations in regulatory genes play an important role in establishing viral latency and reactivation in the nervous system.[151] Recently a synergy was demonstrated between a host protein that activates promoters of JCV, expressed on oligodendrocytes, and the JCV large-tumor antigen,[200] suggesting another potential mechanism by which lytic infection of those cells might be established.

4.4. Descriptive Epidemiology

JCV (as well as BKV) most often causes silent infections of normal children and adolescents.[189,217] PML occurs many years after those initial infections, affecting primarily adults with immune systems compromised for a variety of reasons: congenital or acquired immunodefi-

ciency syndromes, leukemias and lymphomas, immunosuppressive therapy for tumors or transplantation, miliary
tuberculosis, sarcoidosis, and others. A few patients, suspected of having some unrecognized predisposing condition, had PML without other diagnosis (recently reviewed[134,151]). Possibly because of the lower prevalence
of latent infections with JCV in the early years of life, only
a few cases of PML have been recognized in children with
congenital immunodeficiencies[134] and AIDS.[21,226] The
epidemiology of PML reflects that of its underlying conditions; the recent spread of AIDS has resulted in a dramatic
increase in recognized cases of PML as well as a reduction
in mean age of patients (from the sixth to the fifth decade
of life) and relative increase in proportion of males affected.[215] Although PML remains rare overall, it is a
common opportunistic infection among AIDS patients,
affecting more than 4% in some series.[133]

4.5. Clinical Illness, Diagnosis, Pathology, Pathogenesis, and Immunity

PML causes a variety of clinical abnormalities including pareses, sensory deficits, seizures, dementia, dys-
arthria and other bulbar signs, and cerebellar signs, reflecting the multifocal lesions. Progression is relatively
rapid, and patients typically survive for less than a year;
average survival in one series of AIDS patients was less
than 3 months after onset of PML.[133] Occasional prolonged survival has been reported.[115]

The diagnosis of PML is suggested by signs of multifocal neurological disease in an immunodeficient subject.
Cranial imaging typically reveals multiple noncompress-
ing lesions in various areas of the brain,[237] enlarging as
disease progresses.[134] The presence of serum antibodies
to JCV in AIDS patients was of no value in diagnosing
PML.[98] JCV genomic sequences were successfully amplified by PCR from specimens of CSF patients with
PML,[97,116,174,221] although that did not always provide a
sensitive diagnostic test. One study demonstrated papovavirus particles in CSF of patients by electron microscopy.[185]

The histopathology of PML[201] is characterized by
diffusely scattered demyelinated lesions, variable in size
and distribution, generally most numerous in the cerebral
hemispheres, but also present in brain stem and cerebellum, while tending to spare the spinal cord. Myelin
sheaths within the lesions are degenerated and replaced by
lipid-bearing phagocytes, leaving neuronal axis cylinders.
Oligodendroglia at the margins of lesions are abnormal,
with enlarged degenerated nuclei bearing eosinophilic

inclusions. Astrocytes may also be involved, enlarged
with abnormal or multiple nuclei. Perivascular lymphocytic cuffing has often been found. In electron micrographs of AIDS patients with PML, virions of papovavirus were observed not only in oligodendrocytes and
astroglia but also in macrophages and neurons.[24,168] The
histopathologic findings of PML in AIDS have not differed from those in PML with other predisposing conditions.[145]

The pathogenesis of PML is under intense study.
JCV has been found in kidneys and circulating B lymphocytes of asymptomatically infected subjects.[222] An attractive hypothesis is that latently infected B lymphocytes
are activated during immune suppression and carry JCV
from the blood into the brain to infect oligodendrocytes
and astrocytes, causing loss of myelin and functional
impairments of neurons.[151]

The lymphocytes, primarily T cells, found in areas of
inflammation in the brains of patients with PML, are
reported to contain JCV consistently but only rarely are
infected with human immunodeficiency virus (HIV), opportunistic viruses other than JCV, or toxoplasma.[111] In
addition, phagocytosis of papovaviral particles by macrophages has been observed.[168] The role of those cells and
of JCV-infected neurons[24] in the pathogenesis of PML
remains uncertain. Major histocompatibility complex antigens appeared to be increased on endothelial cells, oligodendroglia, and astrocytes of subjects with PML in a
fashion similar to other viral infections, suggesting
that the lack of circulating T cells rather than impaired
response to infection by antigen-presenting cells within
the CNS is most important in the pathogenesis of
PML.[2]

Results of therapy for PML, even after early diagnosis, have been discouraging,[133] although recent reports
suggested that treatment with alpha interferon, with or
without cytarabine, was followed by clinical improvement.[53,213]

4.6. Unresolved Problems

In spite of continued research efforts, the pathogenesis of PML remains only partially understood. The
impairment of the immune system obviously plays an
important role, and changes in the viral genome may be
involved as well. At the moment, prevention and treatment of the underlying predisposing immunodeficiency
states seem the only promising approaches to control of
PML. Specific treatment for PML should also be improved.

5. Other Chronic Conventional Viral Infections of the Nervous System

5.1. Chronic Tick-Borne Encephalitis

The virus of tick-borne encephalitis (TBE), a member of the *Flavivirus* group of small enveloped RNA viruses, usually causes an acute meningoencephalomyelitis (Russian spring–summer encephalitis). Some patients recover from acute encephalitis but develop progressive neurological disease months or years later.[7] Chronic progressive TBE has been reported from Russia and, less often, Japan. Chronic progressive TBE may be associated with movement or seizure disorders, of which epilepsia partialis continua (Kozhevnikov's epilepsy[144]) is the most common; paralytic disorders, often with brain stem involvement[183], or mixed syndromes have also been described. The seizure and movement disorders may stabilize or remit, but the progressive paralytic syndrome is usually fatal. The CSF in one case of chronic TBE contained elevated levels of antibodies to TBE virus. Brain tissue shows panencephalitis without inclusion bodies by light microscopy or viruslike structures by electron microscopy. Isolation of TBE virus from brain tissue or CSF of patients with chronic post-TBE syndromes has been claimed, but most attempts to isolate the virus failed.[7]

5.2. Rasmussen's Encephalitis

Patients with similar types of chronic encephalitis not associated with any known viral infection have been recognized throughout the world. In North America, a syndrome of seizures (especially epilepsia partialis continua), spastic paralysis, and mental retardation associated with chronic encephalitis was described by Rasmussen and colleagues[184,199] in children, adolescents, and young adults. Patients had no history of preceding acute encephalitis. Computed tomography of patients with Rasmussen's encephalitis may show cerebral cortical atrophy or ventricular dilatation. When brain tissue is resected, a panencephalitis without inclusion bodies or viruslike particles is found. Recent studies, using labeled-probe hybridization, implicated the Epstein–Barr virus in two cases of Rasmussen's encephalitis[228] and cytomegalovirus in several others[194]; these findings were not confirmed.[72] A recent study using PCR failed to amplify the genomes of cytomegalovirus, Epstein–Barr virus, and several other viruses known to infect the CNS during childhood from brains of children with Rasmussen's encephalitis (K. L. Pomeroy and D. M. Asher, unpublished). Efforts to isolate viruses from Rasmussen's encephalitis by a variety of conventional techniques were also unsuccessful.[7] Recent studies suggest that Rasmussen's encephalitis may result from an autoimmune process associated with antibodies to a glutamate receptor.[202a]

5.3. Other Conventional Viruses

As summarized in Table 1, a number of other viruses, discussed elsewhere in the text, can also cause CNS disease long after primary infection, including herpesviruses, rabies virus, and the retroviruses HIV and human T-cell lymphotropic virus type I.

6. Slow Infections with Unconventional Agents: The Subacute Spongiform Encephalopathies

6.1. Introduction

The subacute spongiform encephalopathies comprise a group of diseases that include Creutzfeldt–Jacob disease (CJD), the Gerstmann–Sträussler–Scheinker syndrome (GSS) (a similar disease that may be considered a variant of CJD), and kuru of humans, as well as several neurological diseases of animals: scrapie of sheep and goats, transmissible mink encephalopathy, chronic wasting disease of American deer and elk, and bovine spongiform encephalopathy ("mad cow disease") affecting cattle, zoo ungulates, and domestic cats[240] (Table 2). The relationship of a recently described fatal familial insomnia syndrome[84,152,162,164,173] to the spongiform encephalopathies is discussed below. CJD, the spongiform encephalopathy most commonly encountered, occurs more often than formerly thought[78,79]; it is of special importance

Table 2. Subacute Spongiform Encephalopathies

Humans
  Kuru
  Creutzfeldt–Jakob disease[a]
  Gerstmann–Sträussler–Scheinker syndrome[a]
Animals (naturally infected)
  Scrapie (sheep, goats)
  Transmissible mink encephalopathy (mink)
  Bovine spongiform encephalopathy (cattle, zoo ungulates, cats)
  Chronic wasting disease (American elk, deer)

[a]CJD and GSS can be considered variants of a similar infection in subjects of different genetic backgrounds.
(Fatal familial insomnia, proposed to be a similar disease, is associated with a mutation in the same PRNP gene implicated in familial CJD–GSS; the disease has been transmitted to mice but not to nonhuman primates.)

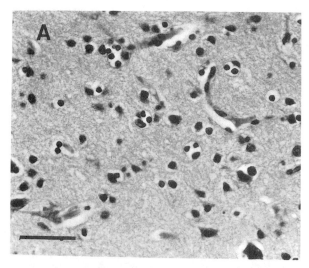

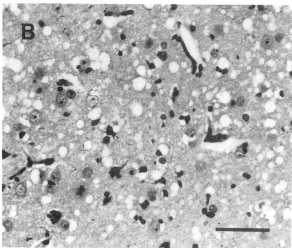

Figure 2. (A) Cerebral cortex of control patient with no neurological disease. (B) Cerebral cortex of patient with spongiform encephalopathy (Gerstmann–Sträussler–Scheinker syndrome). Hematoxylin and eosin stain. Bars = 50 μm.

because it is inevitably fatal and has been iatrogenically transmitted to patients[28,31] and possibly to medical personnel.[20] In brains of humans and animals with spongiform encephalopathies there is usually a variable degree of vacuolation (Fig. 2)—a finding that gave the group of diseases its name.

The spongiform encephalopathies are caused by filterable self-replicating agents that have been called "unconventional viruses,"[77] "prions,"[196] or "infectious amyloids."[79] Since the structure of the infectious particles that transmit the spongiform encephalopathies remains in dispute, we refer to them here simply as "agents."

6.2. Historical Background

Thirty-seven years ago, research among isolated Melanesian peoples living in the mountainous eastern highlands region of Papua New Guinea, where they had a Neolithic culture and practiced ritual endocannibalism (consuming the bodies of dead relatives), led to the recognition that slow infections can cause progressive degenerative diseases of the human nervous system.[80] Gajdusek and Zigas[83] described a previously unknown disease of the nervous system called *kuru*, meaning "trembles," among people of the Fore linguistic group and their neighbors. Kuru had reached hyperendemic proportions, causing over half of all deaths among the Fore in 1957. Kuru was characterized clinically by cerebellar ataxia with less prominent signs of degeneration of basal ganglia and diffuse involvement of other areas of the brain, but without dementia or convulsions. On neuropathologic examination, brains of kuru patients revealed widespread destruction of gray matter without prominent involvement of motor neurons, marked cerebellar degeneration, and frequent amyloid plaques, especially in the cerebellum, without the inflammation or inclusion bodies then thought to be characteristic of CNS infections.

In 1959, Hadlow[110] noted that kuru resembled scrapie, a "slow" infection of sheep,[211] caused by a transmissible agent with unique properties.[110] In 1965, Gajdusek, Gibbs, and Alpers[80] transmitted kuru to chimpanzees by inoculating them with suspensions of human brain tissue after incubation periods lasting almost 2 years. Later, kuru was transmitted to a variety of other animals, and limited characterization demonstrated its etiologic agent to resemble that causing scrapie in animals. Three years later, Gibbs and co-workers[95] showed that sporadic CJD was also caused by a filterable transmissible agent of similar properties, and, within a few more years, brains of subjects with familial forms of CJD,[158] including its GSS variant,[157] were shown to contain transmissible agents as well.

6.3. Methodology

Because kuru has a relatively stereotypical clinical presentation and course and is confined to a single geographic area, case finding is believed to be relatively complete and accurate. Clinical and epidemiologic information concerning CJD and GSS, which are not reportable to public health authorities, although much greater in amount, is probably less reliable. Definitive diagnosis of the spongiform encephalopathies requires brain tissues, and autopsies are frequently not performed on patients.

Although cell cultures can be infected with the agents of the spongiform encephalopathies and some useful information has been obtained from such cultures, particularly concerning the synthesis of the "prion" protein or amyloid,[42,43] they show no recognizable cytopoathic effects and have not replaced animals in assays of infectivity. Because of the time and great expense required to perform complete titrations of infectivity using animals, most investigators now attempt to estimate amounts of infectious agent from the average incubation periods or survival times of small groups of animals inoculated with only the lowest dilutions of material, resulting in a loss of precision. These methodological problems impede progress in understanding the spongiform encephalopathies and their etiologic agents.

6.4. Descriptive Epidemiology (Including Iatrogenic and Other Mechanisms of Transmission)

Kuru has all but disappeared following the elimination of cannibalism from the New Guinea highlands; no documented cases were ever recognized except among the Fore and neighboring peoples.[77–79]

CJD occurs throughout the world.[159] In most populations, at least 90% of all cases of CJD are sporadic, occurring in subjects with no family history of presenile dementia or of probable iatrogenic exposure to the CJD agent. About 10% of patients in typical populations have familial CJD (FCJD).[9] CJD is found in several relatively isolated populations with a very high prevalence of 30 cases per million per year or more; such foci are due to clusters of FCJD. The most distinctive variant of familial spongiform encephalopathy is the GSS syndrome of familial cerebellar ataxia, late-onset dementia, and kurulike amyloid plaques. FCJD and GSS are discussed below.

Our early review of 1435 cases of CJD case histories from the literature and our own experience[159] yielded estimates of annual incidence rates between 1 and 2 cases per million population per year; since the average duration of CJD in our series was less than a year from onset of illness to death, the prevalence rates were about the same. Other series estimated annual mortalities in the United States between 1973 and 1977 of 0.26 per million, varying by state from 0 to 0.60; intensive surveys of the Boston metropolitan area (0.43) with no obvious change over a 20-year period and parts of New York City (0.24) gave similar results. Annual incidences of 0.09 per million in England and Wales, 0.40 in Iceland, 0.32 in France, and 1.2 in Fukuoka, Japan, have been reported. Most population groups in Israel had incidences of CJD of approximately one case per million, while CJD occurred at a much higher rate among Libyan Jews (equivalent to 31 cases per million population per year). Similar high rates were later observed in villages in Slovakia. (Patients with CJD in both foci have the same mutation in the prion protein gene, described below.) Smaller areas of high incidence have been described elsewhere. Although the incidence of CJD across the whole population may be low, age-specific incidence is substantial, and among older adults dying with dementia, CJD is not uncommon; in one series of autopsied patients in California, CJD was one fifth as common as Alzheimer's disease and half as common as Pick's disease.[210]

In our first study, the mean age at death was 51 years for FCJD (patients with FCJD were 15% of the total, presumably overrepresented due to selection bias) and 58 years for sporadic CJD cases.[159] Our more recent summary of cases confirmed by transmission of disease to animals gave similar results.[29] For 232 cases of sporadic CJD from which disease was transmitted, the patients' mean age at onset was 60 years. Familial cases tended to be younger at onset. Three patients with FCJD associated with inserted tandem repeats in the 51/91 region of the open reading frame of the prion protein gene had a mean age at onset of only 33 years, while patients with FCJD associated with a more common point mutation in the human PrP gene (PRNP) at codon 200 (200^{Lys}) had an average age at onset of 57 years, almost the same as that of sporadic cases.

Males and females are affected with CJD in approximately equal numbers, although in several series males predominated, perhaps due to some selection bias. Our recent summary of confirmed cases found a male-to-female ratio of 1.05 : 1.[29]

Mechanisms of natural transmission of human CJD are not understood except for iatrogenic cases. There is no convincing evidence of transplacental (mother-to-child) transmission of CJD. FCJD occurs as frequently in offspring of affected fathers as in those of affected mothers. Kuru was never observed in children of affected mothers unless the children themselves had been exposed to the disease during cannibalism.[141] Experimental spongiform encephalopathies also are not spread transplacentally. Two accounts of conjugal pairs with CJD have been published,[130,161] both more suggestive of common exposure to infection than of spouse-to-spouse transmission.

Iatrogenic transmission of CJD has been demonstrated many times.[28,31] The earliest suggestive report described CJD in three patients who had neurosurgery performed in the same operation suite.[179,241] Later, the accidental transmission of CJD to the recipient of a corneal graft was recognized.[66] Soon afterward, CJD was diagnosed in two young patients with epilepsy more than a year after they had electrocorticography with a probe electrode previously used in the brain of an older subject with CJD[22]; even years after that event, the incriminated electrode, cleaned with benzene and sterilized with alcohol and formaldehyde several times, remained contaminated and experimentally transmitted CJD to a chimpanzee.[92] The most frequently recognized sources of iatrogenically spread CJD are human hormones prepared from pooled pituitary glands of cadavers[28,31]—growth hormone[117] and, less often, gonadotropin[52]; more than 60 cases of CJD have already been reported in recipients of those preparations, with minimum incubation periods longer than 20 years.[31] CJD has been transmitted to monkeys from one growth hormone preparation as well.[91] CJD was also recognized in recipients of lyophilized human dural grafts.[31,45–47,63,155,156,160,180,229] Kuru continues to occur occasionally in subjects exposed to infected tissues by cannibalism in the late 1950s,[141] suggesting that incubation periods in excess of 30 years can be expected in iatrogenic CJD as well.

Recent reports suggest that CJD may have been accidentally transmitted from infected patients or their tissues to medical personnel. More than 20 health care workers have been recognized with CJD, including a neurosurgeon,[82] a pathologist,[107] and two histopathology technicians[170,212] (one known to have processed tissues of a CJD patient), as well as several nurses or nursing assistants, dentists, and an oral surgeon.[20] Although the number of health care workers with CJD remains modest when compared to the thousands of other patients with spongiform encephalopathies in the medical literature,

prudence dictates that medical and laboratory personnel exercise great care to minimize the accidental transmission of infection. Recently more than ten young patients in the U.K. and France were recognized with a new variant CJD (nv-CJD) characterized by a unique clinical picture and the histopathological finding of "florid" amyloid plaques in the brain[240a]; the only new risk factor they shared was a potential exposure to the agent of bovine spongiform encephalopathy. This disturbing new observation forces reconsideration of the danger posed to humans by contact with TSE agents of animals.

6.5. Properties of the Transmissible Agents

As noted above, the structure and replication of the etiologic agents of the spongiform encephalopathies remain a subject of intense controversy. The spongiform encephalopathy agents are highly resistant to inactivation by exposure to heat, ultraviolet and ionizing radiations,[96] or a variety of chemicals that disinfect conventional viruses.[93] They elicit no detectable antibodies or cell-mediated immunity either in susceptible hosts or in resistant animals repeatedly injected with infected tissues with or without adjuvants.[77] This constellation of properties is unique among infectious agents (Tables 3 and 4).

Prusiner was an early proponent of a theory that the spongiform encephalopathy agents must have a unique structure and replicative mechanism,[89,148] and that they probably consist of protein without any nucleic acid genome. He proposed that, since these agents are "proteinaceous" (inactivated by some proteases[51]), they be called "prions."[196] He later concluded that the protease-

Table 3. Atypical Properties of the Spongiform Encephalopathy Agents

Physical and chemical properties	Biological properties

Resistant to inactivation by
  Aldehydes (formaldehyde, glutaraldehyde)
  Beta-propiolactone
  Nucleases (RNase, DNase)
  Heat (incompletely inactivated at 100°C)
  Ionizing radiation
  Ultraviolet radiation
	No eclipse phase
	Doubling time of days (5.2 days for scapie in hamster brain)
	Long incubation period (2 months to >30 years)
	Chronic progressive course (months to years)
	Relentless unremitting course
	Fatal outcome

At least partially sensitive to inactivation by
  Sodium hydroxide
  Sodium hypochlorite
  Formic acid
  Commercial phenolic product
	"Degenerative" histopathology:
	  neuronal vacuolation, loss; glial proliferation, hypertrophy; amyloid
	  accumulation, plaques.
	No inflammation, inclusions, primary demyelination.

Infectivity not rigorously proved to be associated with any
  identified unique physical structure
EM shows tubulovesicular structures in plastic embedded thin
  sections, round particles in negatively stained extracts
Amyloid fibrils ("prion" protein) are host-coded
	No evidence of immune response to infectious agent, but no immune
	  deficiency: no detectable antibodies or cell-mediated response to
	  infectious agent; no change in course of illness with immune suppressive
	  or potentiating treatment.
	No production of or response to interferon or interference with replication of
	  >30 viruses.

**Table 4. Viruslike Properties
of the Spongiform Encephalopathy Agents**

Filtration behavior: Retained by average pore diameter 25 nm (scrapie), 100 nm (kuru, CJD)

Titration of infectivity: All individuals of sensitive host species succumb after inoculation with low dilutions of suspensions

Replication of infectivity: 10 (5) to 10 (11) LD50/g in brain tissue

Pathogenesis: After subcutaneous or intraperitoneal inoculation, scrapie replicates first in lymph nodes, spleen before CNS

Adaptation to new host species: Shortened incubation period on serial passage in species

Host genetic control of incubation period and susceptibility: See text

Strainlike agent behavior: Host range, incubation period, distribution of histopathology, and presence of amyloid plaques typical of different "strains"; selection of multiple strains from "wild" stock by limiting dilution; strains "breed true" on serial passsage in given host species; mutationlike sudden changes in agent properties occur occasionally, then breed true

Interference: Slow-growing strains impede replication of fast-growing strains of scrapie in mice

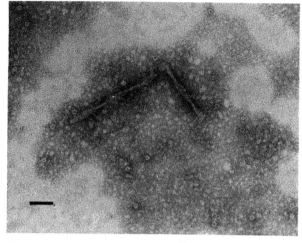

Figure 3. Scrapie-associated fibrils extracted from brain of a hamster with by limited digestion with proteinase. Phosphotungstic acid negative stain. Bar = 80 nm. Preparation kindly provided by Kitty Pomeroy.

resistant amyloid protein that accumulates in the brains of humans and animals with spongiform encephalopathies was likely to be the prion.[25,195]

Gajdusek stressed that the spongiform encephalopathies are not the only degenerative CNS diseases in which amyloid proteins accumulate in the brain, a more common one being Alzheimer's disease (AD); he agreed that the most parsimonious explanation for the unique spectrum of physical and biological behavior of the spongiform encephalopathy agents is that they lack nucleic acid, but suggested that the agents more appropriately be called "infectious amyloids," in contrast to the "noninfectious amyloid" of AD, and even predicted that under some circumstances the amyloid of AD might also serve to catalyze its own synthesis from a normal precursor protein[79] as that of CJD is posited to do. Intriguing hypotheses for possible mechanisms by which abnormal proteins might catalyze their own synthesis by protein-based coding mechanisms[242,243] or by some kind of modulation of an infectious protein by host-encoded nucleic acids[233,234] have been proposed.

The leading contender as the putative prion or infectious amyloid is an abnormal protein that accumulates in the brains of most if not all humans and animals with spongiform encephalopathies. That protein was first demonstrated morphologically by Merz and colleagues[166] in 1981 as abnormal filamentous structures called "scrapie-associated filaments" (SAF) in negatively stained extracts of brains of scrapie-infected rodents (Fig. 3), and later in humans with CJD.[167] Soon afterward, Prusiner and Bolton[25] found an abnormal protein band (called prion pro-

tein 27-30 or PrP27-30) in protease-treated detergent extracts separated by gel electrophoresis (Fig. 4) and stained with silver. Antisera prepared against PrP reacted with SAF as well, establishing that they are essentially the same.[64] The PrP27-30 was found to be the truncated cleavage product of a larger protein, originally called PrP33-35 or later PrP^{Sc} in scrapie and PrP^{CJD} in CJD; that

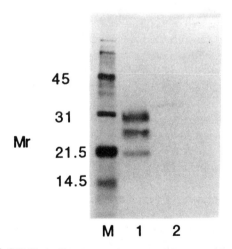

Figure 4. PrP ("prion" protease-resistant protein) extracted from the brain of a patient with Creutzfeldt–Jakob disease (lane 1). Proteins separated by electrophoresis in polyacrylamide gel, transferred to nitrocellulose membrane and Western immunoblotted with labeled antibody to hamster PrP. Lane 2 contains control protein extract from the brain of a subject with no neurological disease. M, molecular weight markers. M$_r$, apparent molecular weights. Preparation kindly provided by Eric Liu and Rachel Sherman.

protein has the same primary structure as a normal "control" protein (PrP^C) expressed in all subjects,[182] although they differ somewhat in physical properties, with PrP^{Sc} and PrP^{CJD} resistant to limited protease digestion while PrP^C is sensitive. PrP^C, secreted at the cell surface and rapidly catabolized by cells, is apparently not essential for life (at least in experimental animals),[197] and its normal function remains unknown.

Prusiner and colleagues[38,235] demonstrated that PrP plays an important role in the pathogenesis of the spongiform encephalopathies and in susceptibility of the host to disease. PrP accumulates in the brains of most if not all subjects with spongiform encephalopathies, forming an important component, probably the major part, of the amyloid plaques. In animals the most important gene controlling incubation period of spongiform encephalopathies is closely linked if not identical to that encoding PrP,[38,235] located on the short arm of chromosome 20 in humans. Mice of genotypes originally resistant to hamster-adapted strains of scrapie into which a portion of the PrP gene from hamsters was engineered then acquired susceptibility to those strains, and the agent recovered from mice had hamster-adapted biological properties.[198,206] Animals not expressing PrP^C (so-called PrP-knockout mice with a disruption engineered into the PrP gene) developed and behaved normally, but they have thus far been resistant to scrapie disease.[36,197] (The role of the human prion protein gene in FCJD is considered below.)

The PrP^{Sc}/PrP27-30 proteins have usually been found in partially purified preparations containing infectivity, and the amounts of PrP detected have often correlated with estimated titers of infectivity after a variety of treatments.[195] Increased expression in mice of an engineered PrP gene with a mutation similar to one frequently found in patients with GSS caused a disease resembling scrapie,[124] although apparently no replicating infectious scrapie agent has been convincingly demonstrated in those mice.

Other authorities remain skeptical about the "all-protein" or prion hypothesis and are unconvinced by supporting evidence marshaled by its proponents.[49,50,154,202] Reinterpretation of irradiation-inactivation kinetic studies suggested that the presence of a small nucleic acid genome was not excluded.[203] No confirmed studies have demonstrated the physical size of the infectivity-bearing agent to be less than that of a small virus,[93] or that a purified protein, uncontaminated with some nucleic acid from an infected host, has replicated the scrapie agent. In several experimental systems, the kinetics of PrP formation failed to correlate with kinetics of infectivity.[56,57,175,250] Tissues of the PrP-knockout mice discussed above, genet-

ically manipulated so that they cannot synthesize the PrP-precursor protein (PrP^C), although not susceptible to overt scrapie disease, nonetheless may have supported replication of the infectious scrapie agent as assayed in wild-type mice, albeit to a much lower titer and possibly in a form more susceptible to inactivation by heat.[36]

"Strains" of the scrapie agent have properties (incubation period, distribution of histopathologic lesions, and amyloid plaque formation) that "breed" true on passage in inbred mice[34] and show sudden changes resembling the genetic mutations of conventional pathogens.[137] (Strains of CJD agent have also been postulated.[60,94]) This biological behavior encourages some authorities to maintain the hypothesis that the agents of spongiform encephalopathies must also contain nucleic acid genomes,[202] however difficult to detect. However, no viruslike particles nor any unique nonhost nucleic acids have been convincingly linked to infectivity in tissues containing the spongiform encephalopathy agents, although intriguing structures have been seen by electron microscopy in thin sections of infected brain[58] and smaller ones in negatively stained extracts of tissue.[188] At the present time it seems reasonable to conclude that no hypothetical structure for the transmissible agents of spongiform encephalopathy has been rigorously proven yet.

6.6. Pathogenesis

The portals of entry for the spongiform encephalopathy agents are not yet clear. Scrapie agent has been found in intestines and mesenteric lymph nodes of asymptomatic naturally infected sheep, presumably reflecting an early site of replication during the incubation period. Monkeys have been experimentally infected by feeding them tissues contaminated with the agents of scrapie, kuru, and CJD. Those findings suggest that the oral route may be an effective portal of infection.[90] Kuru was clearly acquired by exposure to infected tissues during the practice of ritual cannibalism, suggesting that either the gastrointestinal tract or mucocutaneous surfaces were portals of entry for the agent.[79] Failure to infect a chimpanzee after introduction of a very large dose of kuru agent by direct intragastric intubation was interpreted as evidence that lesions in the oral mucosa or integument were the more likely portal of infection rather than the intact gastrointestinal tract. Iatrogenic cases of CJD have been acquired after brain and eye surgery, as well as after repeated hypodermic injections of contaminated hormones, demonstrating that those routes were effective portals of entry. As noted above, transplacental spread from mother to fetus seems unlikely. Portals of entry in most cases of CJD are unknown.

Portals of exit of the agents also remain conjectural. A few reports of CJD agent found in human blood[153] and urine[220] remain unconfirmed. In a small number of experimental studies, other secretions were not infectious. Brain and adjacent tissues (pituitary, dura, optic nerve, and eye) were the apparent portals of exit in iatrogenically transmitted cases. CSF and several internal organs of CJD patients were sometimes found to contain the agent, constituting potential iatrogenic portals of exit.[78,79]

Spread of the agents in the body are also not completely understood. After experimental introduction under the skin or into the peritoneal cavities of animals, the agents appear to replicate first in local lymphoid tissues and later in more distant reticuloendothelial organs. In rodents, a "viremia" has been demonstrated repeatedly, but our attempts to find viremia in primates and humans with kuru and CJD have been negative.[6,29,79] Some evidence suggests that in experimentally infected rodents the scrapie agent ascends from the portal of entry to the CNS via peripheral nerves.[138]

As noted above, no antibodies or cell-mediated immune responses to the infectious agents have ever been convincingly demonstrated. The only evidence to suggest that the host can mount any defense response is that in some animals experimentally infected with scrapie agent, after an initial period of replication, infectivity disappears from lymphoid organs.[109] However, no evidence suggests any effective response to infection within the nervous system.[6]

6.7. Clinical Features and Diagnosis

The clinical diagnosis of CJD is often difficult, the most frequent differential diagnostic possibility being AD. Illness usually begins with subjective awareness of sensory or mental changes, often with anxiety and vaguely abnormal sensorimotor functions, with little obvious impairment in daily activities. Later in the illness, the patient has unequivocal sensorimotor disturbances, often including ataxia and inappropriate mental function, that interfere with day-to-day activities. In the late stage of CJD, there is incapacitating dementia, usually with myoclonus (which may begin in the middle stage). Characteristic changes in EEG are often useful in diagnosis. Other than EEG changes and typical histopathologic lesions in the brain at cerebral biopsy, the only laboratory abnormality useful in premortem diagnosis has been the presence of abnormal peptides in the CSF of CJD patients (not found in CSF of patients with AD) by two-dimensional

electrophoresis. These peptides were recently identified by sequencing and immunoblotting as derived from the "14-3-3" protein.[124a]

The course of CJD is variable but inexorable. Duration of illness is less than 1 year in the great majority of cases. In our most recent series, patients with sporadic CJD survived for an average of 8 months after onset, with a median survival of 4.5 months. About 10% of patients have survived for more than a year, often in a prolonged vegetative state. Patients with FCJD tended to live longer after onset than those with sporadic CJD, varying from a mean of 11 months (for 19 subjects with the PRNP 200^{Lys} mutation) to 102 months for three subjects with inserted tandem repeats in the 52/92 region. Death comes from the complications common to terminal neurological diseases: aspiration with pneumonia, thromboembolic events, and so forth. No treatments have been of any demonstrated benefit.

As noted above, a cerebral biopsy showing typical histopathologic changes and PrP may offer definitive premortem diagnosis; however, biopsy can be recommended only when other potentially treatable diagnoses are also considered. The procedure may be useful in ruling out the presence of fungal or parasitic infections, carcinomatous meningitis, granulomatous cerebral angiitis, or viral encephalitis, especially herpes encephalitis. Even when there is no probable treatment based on results anticipated from a cerebral biopsy, the procedure may occasionally be justified to establish the prognosis of a dementing disease (e.g., CJD vs. AD). When biopsy is indicated, its goal should be to provide sufficient uncrushed tissue for definitive diagnostic studies. Arrangements for microbiological, histological, electron microscopic, and other studies should be made in advance. When the diagnosis of CJD has been confidently established on clinical grounds and no other potentially treatable disease remains in the differential diagnosis, we cannot recommend cerebral biopsy. On occasion the histopathologic lesions in the brain vary in distribution, so that a frontal cortical biopsy may be nondiagnostic. Anatomic diagnosis is made with greater accuracy by autopsy.[19]

6.8. Pathology

Definitive diagnosis of CJD must be made from brain tissue. Typical changes are vacuolation and loss of neurons, proliferation and hypertrophy of astrocytes, accumulation of amyloid protein (PrP), and other more variable changes including secondary degeneration of myelin and microgliosis. Vacuolation varies, increasing from scattered single neuronal vacuoles to spongiform changes resulting from more extensive vacuolation of neurons—less commonly in astrocytes—or severe status spongiosus involving the entire parenchyma; in extreme cases the spongy cortex collapses into an atrophic, sclerotic remnant. Changes are often prominent in neocortex, striatum, thalamus, cerebellar cortex, dentate nucleus, and sometimes in motor pathways. The severity of the changes varies from case to case and also in different areas of the same brain.[15]

Amyloid plaques resembling those seen in kuru can be found in 10 or 15% of brains of CJD patients, especially in the cerebellum, and their presence has been required for the diagnosis of GSS. The PrP–amyloid of which plaques are composed has been detected by detergent–protease extraction and gel electrophoresis with Western immunoblotting, by electron microscopy, or by immunohistochemical staining (enhanced after treatment with formic acid[140]) in more than 80% of brains. The plaques of CJD–GSS are not stained by labeled antibodies to the amyloid of the more common AD and vice versa.

Peculiar tubulovesicular particles, most often in postsynaptic processes, were first noted in brains of animals with experimental scrapie[58,178] and later in the other spongiform encephalopathies including CJD in humans.[149] They have not been reported in AD. Their significance is not known. The smaller particles recently described in extracts of brain tissue[188] have not been identified in embedded sections.

6.9. Familial Spongiform Encephalopathies (Including the GSS Syndrome and Considering the Fatal Familial Insomnia Syndrome)

In many series, 5 to 10% of CJD patients have a family history of presenile dementia. Most pedigrees of FCJD suggest an autosomal dominant mode of inheritance: the disease occurs without skipping generations and affects approximately half the siblings of propositi, both males and females in equal numbers. GSS also shows an autosomal dominant pattern of inheritance.[123]

The genetic basis for FCJD–GSS appears to reside in a series of mutations in the gene coding for the PrP precursor protein, currently designated the PRNP gene, on the short arm of human chromosome 20, PRNP has an open reading frame of 759 nucleotides (253 codons), in which 10 point mutations and a variety of insertions have already been linked to FCJD or GSS. The importance of the analogous gene in spongiform encephalopathies of animals was first noted for experimental scrapie (and later for CJD) in mice,[38,39] in which the "sinc" gene, long known to control incubation periods of scrapie incubation

in mice,[126] was found to be closely linked if not identical to the PrP gene; a similar close linkage was found between the PrP and scrapie incubation period genes of sheep.[125] In mice of some genotypes infected with certain strains of scrapie agent, the incubation periods exceeded the natural life span of the animal, producing a phenotype resistant to disease though not to inapparent infection.[62]

Owen and co-workers discovered an abnormal restriction endonuclease pattern in the PRNP genes of affected members in a family with typical autosomal dominant FCJD[187] that proved to be a large insertion coding for six extra repeats[186] in a region of five normal octapeptide tandem repeats between codons 51 and 91. Goldgaber and co-workers showed that patients from other families with FCJD lacked that abnormality,[105] prompting a search for other associated mutations. Hsaio, Prusiner, and colleagues found that a point mutation (single nucleotide change) in codon 102 of the PRNP gene (102^{Leu}, changing the encoded amino acid from proline to leucine) was linked to GSS.[123] Their finding was confirmed by Goldgaber and co-workers, who also found another point mutation in PRNP at codon 200 (200^{Lys}) that cosegregated with FCJD.[104] These are the two most common point mutations associated with familial spongiform encephalopathies. Several other insertions in the 51/91 octapeptide repeat region, besides the six-octapeptide insertion first discovered, have been associated with FCJD (coding for five to eight extra octapeptides) and one with GSS (coding for nine extra octapeptides).[79,102] Different point mutations have also been demonstrated in association with FCJD and GSS; in addition to the most common 200^{Lys} mutation, FCJD in other kindreds has been linked to each of three other point mutations (178^{Asn}, 180^{Ile}, 210^{Ile}), and GSS has been linked to point mutations at five other codons (105^{Leu}, 117^{Val}, 145^{Stop}, 198^{Ser}, 217^{Arg}), as well as to the much more common 102^{Leu} mutation. Other mutations in the PRNP gene associated with FCJD and GSS probably await discovery.

Recently, fatal familial insomnia (FFI), an inherited syndrome with an autosomal dominant pattern of occurrence characterized by progressive severe insomnia and dysautonomia with selective atrophy of two thalamic nuclei, was described first in a large northern Italian kindred and later in several others.[152,162,164] Patients with FFI have ataxia, myoclonus, and other signs resembling those of CJD–GSS, and a few affected patients had spongiform changes in the cerebral cortex. Those findings prompted the hypothesis that FFI might be a new prion disease; indeed, protease-resistant PrP was detected in brains of patients with FFI, although it apparently differed somewhat from the PrP found in CJD–GSS patients.[164] In patients with FFI, a mutation was found in the PRNP gene at codon 178 (178[Asn]), identical to that found in some kindreds with FCJD. However, the two groups of patients *differed* in PRNP sequences at another codon of the abnormal allele, 129 (a codon encoding methionine in some normal subjects and valine in others); FFI patients had 129[Val] on the 178[Asn] allele, while FCJD patients had 129[Met].[103,163] It remains unknown how the same point mutation in one codon of a gene might interact with different, otherwise normal polymorphic nucleotides in another codon of the same gene to produce different clinical illnesses. FFI may also differ from FCJD in another important way: to date, no agent transmitting encephalopathy to primates has been demonstrated in brain tissues of FFI patients. Transmission of disease to mice from brains of patients with FFI has been claimed.[52a,220a] This is an area of active research.[30,103,163]

In addition to the methionine–valine polymorphism found at codon 129 and a methionine–arginine polymorphism at codon 232 in normal subjects, there appear to be other normal polymorphisms in the PRNP gene at codons 117 and 124 ("silent" differences where both variants encode the same amino acid) and in the 51/91 octapeptide repeat region, where both deletions and short insertions of less than five extra repeats have been detected in healthy subjects.[79] It is not yet known if these polymorphisms play any role in neurological diseases.

Not all subjects with mutations in the PRNP gene have expressed disease, even in affected families. While penetrance appears to be quite high for GSS patients with the 102[Leu] mutation and for FCJD patients with the 178[Asn] mutation, fewer than 60% of subjects with the 200[Lys] mutation get CJD, at least by the usual expected age. It is not known if unaffected family members bearing those mutations in the PRNP gene have "inapparent" infections, perhaps due to exceptionally long incubation periods, or if they have escaped infection. In families with FFI, several members with the 178[Asn] mutation have also survived past the age of 60 without showing signs of illness.

6.10. Prevention

No convincing evidence suggests that excretions or external secretions of patients are hazardous and no empirical or experimental evidence shows the disease to be contagious or transmissible by aerosol; therefore, strict isolation of patients seems unnecessary. Since tissue, especially brain tissue, of patients with CJD and its variants (including FCJD and GSS) may contain large amounts of infectious agent, and considering the established risk of

iatrogenic spread to other patients and possible risk to medical personnel, precautions must be observed in handling such tissues.[8] Special care must be taken to avoid exposing patients to materials contaminated with tissues of other patients and to protect medical personnel against sharp injuries with such materials. Guidelines for safe handling of contaminated tissues and instruments in surgery, autopsy and laboratory have been suggested.[8,18,33,81,239] In addition, adoption of universal precautions for handling human, blood, body fluids (especially CSF, known to contain the infectious agent), and tissues, codified in the OSHA Final Rule on Occupational Exposure to Bloodborne Pathogens,[181] should reduce considerably the risk of accidental transmission of the disease. To the three methods long used for inactivation of the spongiform encephalopathy agents[8]—heat (steam autoclaving at 132°C for at least 1 hr, or incineration where feasible), chlorine bleach (5.25% sodium hypochlorite exposure for at least 1 hr), and NaOH (1 or 2N) exposure for at least 1 hr[32]—two other methods have recently been added: (1) tissues for histology may be treated by exposure to 96% formic acid for 1 hr after formalin fixation,[33] and (2) contaminated materials may be inactivated by exposure to a commercial aqueous acid–phenolic disinfectant at $\geq$10% concentration for at least half an hour.[71] Other methods of disinfection, including exposure to alcohols, formaldehyde, and a variety of other liquid sterilants as well as to ethylene oxide gas, have been ineffective.

6.11. Unresolved Problems

The nature of the infectious agents causing the spongiform encephalopathies remains in doubt and elucidating their structure and mechanism of replication remains the single most important goal of research. That would be facilitated by an assay for infectivity not requiring live animals and long incubation periods. The interaction between the transmissible agents and host genetics and susceptibility must be clarified, particularly considering a recent tendency to include other neurological diseases in the category of prion dementia on the basis of associated mutations in the PRNP, regardless of whether any infectious agent has been implicated.[30] Experience with the successful eradication of kuru and other transmissible infectious diseases dictates that increased efforts be undertaken to determine the sources of infection of the great majority of patients with spongiform encephalopathies who have neither family histories nor iatrogenic exposures, so that transmission can be interrupted, if that is possible. (Particularly in places where scrapie of sheep and bovine spongiform encephalopathy are common, it must be confirmed that, as generally assumed, products contaminated with the agents of those diseases do not constitute a source of human infection.) Elimination of spongiform encephalopathy agents from the body before the onset of CNS disease and effective therapy of established CNS infections are goals for the distant future.

7. Concluding Comment

The discovery that kuru was caused by a slow infection opened an era in the study of degenerative diseases of the human nervous system that has not yet ended. In addition to those diseases discussed here and elsewhere in this volume, other slow infections of humans surely remain to be discovered.

8. References

1. ABE, T., NUKADA, T., HATANAKA, H., TAJIMA, M., MIRAWA, M., AND USHIJIMA, H., Myoclonus in a case of suspected rubella panencephalitis, *Arch. Neurol.* **40:**98–100 (1983).

2. ACHIM, C. L., AND WILEY, C. A., Expression of major histocompatibility complex antigens in the brains of patients with progressive multifocal leukoencephalopathy, *J. Neuropathol. Exp. Neurol.* **51:**257–263 (1992).

3. AKSAMIT, A. J., JR., Nonradioactive *in situ* hybridization in progressive multifocal leukoencephalopathy, *Mayo Clin. Proc.* **68:**899–910 (1993).

4. ANLAR, B., GUCUYENER, K., IMIR, T., YALAZ, K., AND RENDA, Y., Cimetidine as an immunomodulator in subacute sclerosing panencephalitis: A double blind, placebo-controlled study, *Pediatr. Infect. Dis. J.* **12:**578–581 (1993).

5. ANLAR, B., YALAZ, K., IMIR, T., AND TURANLI, G., The effect of inosiplex in subacute sclerosing panencephalitis: A clinical and laboratory study, *Eur. Neurol.* **34:**44–47 (1994).

6. ASHER, D. M., Slow viral infections of the central nervous system, in: *Infections of the Central Nervous System* (W. M. SCHELD, R. J. WHITLEY, AND D. T. DURACK, EDS.), pp. 145–166, Raven Press, New York, 1991.

7. ASHER, D. M., AND GAJDUSEK, D. C., Virologic studies in chronic encephalitis, in: *Chronic Encephalitis and Epilepsy: Rasmussen's Syndrome* (F. ANDERMANN, ED.), pp. 147–158, Butterworth-Heinemann, Boston, 1991.

8. ASHER, D. M., GIBBS, C. J., JR., AND GAJDUSEK, D. C., Slow viral infections: Safe handling of the agents of the subacute spongiform encephalopathies, in: *Laboratory Safety: Principles and Practice* (B. M. MILLER, D. H. M. GRÖSCHEL, J. H. RICHARDSON, D. VESLEY, J. R. SONGER, R. D. HOUSEWRIGHT, AND W. E. BARKLEY, EDS.), pp. 59–71, American Society for Microbiology, Washington, DC, 1986.

9. ASHER, D. M., MASTERS, C. L., GAJDUSEK, D. C., AND GIBBS, C. J., JR., Familial spongiform encephalopathies, in: *Genetics of Neurological and Psychiatric Disorders* (S. KETY, L. ROWLAND, R. SIDMAN, AND S. MATTHYSSE, EDS.), pp. 273–291, Raven Press, New York, 1983.

10. ASTRÖM, K., MANCALL, E. L., AND RICHARDSON, E. P., Progressive multifocal leukoencephalopathy. A hitherto unrecognized complication of chronic lymphatic leukemia and Hodgkin's disease, *Brain* **81:**93–111 (1958).

11. AULT, G. S., AND STONER, G. L., Two major types of JC virus defined in progressive multifocal leukoencephalopathy brain by early and late coding region DNA sequences, *J. Gen. Virol.* **73:**2669–2678 (1992).

12. AULT, G. S., AND STONER, G. L., Human polyomavirus JC promoter/enhancer rearrangement patterns from progressive multifocal leukoencephalopathy brain are unique derivatives of a single archetypal structure, *J. Gen. Virol.* **74:**1499–1507 (1993).

13. BACZKO, K., LIEBERT, U. G., BILLETER, M., CATTANEO, R., BUDKA, H., AND TER MEULEN, V., Expression of defective measles virus genes in brain tissues of patients with subacute sclerosing panencephalitis, *J. Virol.* **59:**472–478 (1986).

14. BAKHEIT, A. M., AND BEHAN, P. O., Unsuccessful treatment of subacute sclerosing panencephalitis treated with transfusion of peripheral blood lymphocytes from an identical twin, *J. Neurol. Neurosurg. Psychiatry* **54:**377–378 (1991).

15. BECK, E., DANIEL, P. M., MATTHEWS, W. B., STEVENS, D. L., ALPERS, M. P., ASHER, D. M., GAJDUSEK, D. C., AND GIBBS, C. J., JR., Creutzfeldt–Jakob disease: The neuropathology of a transmission experiment, *Brain* **92:**699–716 (1969).

16. BEERSMA, M. F., GALAMA, J. M., VAN DRUTEN, H. A., RENIER, W. O., LUCAS, C. J., AND KAPSENBERG, J. G., Subacute sclerosing panencephalitis in The Netherlands—1976–1990, *Int. J. Epidemiol.* **21:**583–588 (1992).

17. BEHAN, P. O., AND BEHAN, W. M. H., Immunological abnormalities and immunotherapeutic attempts in subacute sclerosing panencephalitis, in: *Immunology of Nervous System Infections: Progress in Brain Research,* Vol. 59 (P. O. BEHAN, V. TER MEULEN, AND F. C. ROSE, EDS.), pp. 149–162, Elsevier, Amsterdam, 1983.

18. BELL, J. E., AND IRONSIDE, J. W., How to tackle a possible Creutzfeldt–Jakob disease necropsy, *J. Clin. Pathol.* **46:**193–197 (1993).

19. BELL, J. E., AND IRONSIDE, J. W., Neuropathology of spongiform encephalopathies in humans, *Br. Med. Bull.* **49:**738–777 (1993).

20. BERGER, J. R., AND DAVID, N. J., Creutzfeldt–Jakob disease in a physician—a review of the disorder in health care workers, *Neurology* **43:**205–206 (1993).

21. BERGER, J. R., SCOTT, G., ALBRECHT, J., BELMAN, A. L., TORNATORE, C., AND MAJOR, E. O., Progressive multifocal leukoencephalopathy in HIV-1-infected children, *AIDS* **6:**837–841 (1992).

22. BERNOULLI, C., SIEGFRIED, J., BAUMGARTNER, G., REGLI, F., RABINOWICS, T., GAJDUSEK, D. C., AND GIBBS, C. J., JR., Danger of accidental person-to-person transmission of Creutzfeldt–Jakob disease by surgery, *Lancet* **1:**478–479 (1977).

23. BLISARD, K., DAVIS, L., HARRINGTON, M., LOVELL, J., KORNFELD, M., AND BERGER, M., Pre-mortem diagnosis of Creutzfeldt–Jakob disease by detection of abnormal cerebrospinal fluid proteins, *J. Neurol. Sci.* **99:**75–81 (1990).

24. BOLDORINI, R., CRISTINA, S., VAGO, L., TOSONI, A., GUZZETTI, S., AND COSTANZI, G., Ultrastructural studies in the lytic phase of progressive multifocal leukoencephalopathy in AIDS patients, *Ultrastruct. Pathol.* **17:**599–609 (1993).

25. BOLTON, D. C., MCKINLEY, M. P., AND PRUSINER, S. B., Identification of a protein that purifies with the scrapie prion, *Science* **218:**1309–1311 (1982).

26. BOUTEILLE, M., FONTAINE, C., VERENNE, C., AND DELARUE, J., Sur un cas d'encephalite subaigue a inclusions. Etude anatamo-clinique et ultrastructurale, *Rev. Neurol. (Paris)* **118:**454–458 (1965).

27. BROWN, H. R., GOLLER, N. L., RUDELLI, R. D., DYMECKI, J., AND WISNIEWSKI, H. M., Postmortem detection of measles virus in non-neural tissues in subacute sclerosing panencephalitis, *Ann. Neurol.* **26:**263–268 (1989).

28. BROWN, P., Iatrogenic Creutzfeldt–Jakob disease, *Aust. NZ J. Med.* **20:**633–635 (1990).

29. BROWN, P., GIBBS, C. J., JR., RODGERS-JOHNSON, P., ASHER, D. M., SULIMA, M. P., BACOTE, A., GOLDFARB, L. G., AND GAJDUSEK, D. C., Human spongiform encephalopathy: The NIH series of 300 cases of experimentally transmitted disease, *Ann. Neurol.* **35:**513–529 (1994).

30. BROWN, P., KAUR, P., SULIMA, M.P., GOLDFARB, L. G., GIBBS, C. J., JR., AND GAJDUSEK, D. C., Real and imagined clinicopathological limits of prion dementia, *Lancet* **341:**127–129 (1993).

31. BROWN, P., PREECE, M. A., AND WILL, R. G., Friendly fire in medicine—hormones, homografts, and Creutzfeldt–Jakob disease, *Lancet* **340:**24–27 (1992).

32. BROWN, P., ROHWER, R., AND GAJDUSEK, D. C., Sodium hydroxide disinfection of Creutzfeldt–Jakob disease virus, *N. Engl. J. Med.* **310:**727 (1984).

33. BROWN, P., WOLFF, A., AND GAJDUSEK, D. C., A simple and effective method for inactivating virus infectivity in formalin-fixed tissue samples from patients with Creutzfeldt–Jakob disease, *Neurology* **40:**887–890 (1990).

34. BRUCE, M. E., AND DICKINSON, A. G., Biological evidence that scrapie agent has an independent genome, *J. Gen. Virol.* **68:**79–89 (1987).

35. BUCKLE, G. J., GODEC, M. S., RUBI, J. U., TORNATORE, C., MAJOR, E. O., GIBBS, C. J., JR., GAJDUSEK, D. C., AND ASHER, D. M., Lack of JC viral genomic sequences in multiple sclerosis brain tissue by polymerase chain reaction, *Ann. Neurol.* **32:**829–831 (1992).

36. BÜELER, H., AGUZZI, A., SAILER, A., GREINER, R. A., AUTENRIED, P., AGUET, M., AND WEISSMANN, C., Mice devoid of PrP are resistant to scrapie, *Cell* **73:**1339–1347 (1993).

37. CAPE, C. A., MARTINEZ, A. J., ROBERSTON, J. T., HAMILTON, R., AND JABBOUR, J. T., Adult onset of subacute sclerosing panencephalitis, *Arch. Neurol.* **28:**124–127 (1973).

38. CARLSON, G. A., KINGSBURY, D. T., GOODMAN, P. A., COLEMAN, S., MARSHALL, S. T., DEARMOND, S., WESTAWAY, D., AND PRUSINER, S. B., Linkage of prion protein and scrapie incubation time genes, *Cell* **46:**503–511 (1986).

39. CARLSON, G. A., WESTAWAY, D., GOODMAN, P. A., PETERSON, M., MARSHALL, S. T., AND PRUSINER, S. B., Genetic control of prion incubation period in mice, in: *Novel Infectious Agents and the Central Nervous System, Ciba Foundation Symposium,* Vol. 135 (G. BOCK AND J. MARSH, EDS.), pp. 84–99, John Wiley and Sons, Chichester, 1988.

40. CATTANEO, R., SCHMID, A., BILLETER, M. A., SHEPPARD, R. D., AND UDEM, S. A., Multiple viral mutations rather than host factors cause defective measles virus gene expression in a subacute sclerosing panencephalitis cell line, *J. Virol.* **62:**1388–1397 (1988).

41. CATTANEO, R., SCHMID, A., REBMANN, G., BACZKO, K., TER MEULEN, V., BELLINI, W. J., ROZENBLATT, S., AND BILLETER, M. A., Accumulated measles virus mutations in a case of subacute sclerosing panencephalitis: Interrupted matrix protein reading frame and transcription alteration, *Virology* **154:**97–107 (1986).

42. CAUGHEY, B., RACE, R. E., AND CHESEBRO, B., Detection of prion protein mRNA in normal and scrapie-infected tissues and cell lines, *J. Gen. Virol.* **69:**711–716 (1988).

43. CAUGHEY, B., RACE, R. E., ERNST, D., BUCHMEIER, M. J., AND CHESEBRO, B., Prion protein biosynthesis in scrapie-infected and uninfected neuroblastoma cells, *J. Virol.* **63:**175–181 (1989).

44. CENTERS FOR DISEASE CONTROL, Subacute sclerosing panencephalitis—United States, *Morbid. Mortal. Week. Rep.* **31:**585–588 (1982).

45. CENTERS FOR DISEASE CONTROL, Rapidly progressive dementia in a patient who received a cadaveric dura mater graft, *Morbid. Mortal. Week. Rep.* **36:**49–50, 55 (1987).

46. CENTERS FOR DISEASE CONTROL, Update: Creutzfeldt–Jakob disease in a patient receiving a cadaveric dura mater graft, *Morbid. Mortal. Week. Rep.* **36:**324–325 (1987).

47. CENTERS FOR DISEASE CONTROL, Update: Creutzfeldt–Jakob disease in a second patient who received a cadaveric dura mater graft, *Morbid. Mortal. Week. Rep.* **38:**37–38, 43 (1989).

48. CHEN, T. T., WATANABE, I., ZEMAN, W., AND MEALEY, J., Subacute sclerosing panencephalitis: Propagation of measles virus from brain biopsy in tissue culture, *Science* **163:**1193–1194 (1969).

49. CHESEBRO, B., Spongiform encephalopathies—PrP and the scrapie agent, *Nature* **356:**560 (1992).

50. CHESEBRO, B., AND CAUGHEY, B., Scrapie agent replication without the prior protein, *Curr. Biol.* **3:**696–698 (1993).

51. CHO, H., Requirement of a protein component for scrapie infectivity, *Intervirology* **14:**213–216 (1980).

52. COCHIUS, J. I., HYMAN, N., AND ESIRI, M. M., Creutzfeldt–Jakob disease in a recipient of human pituitary-derived gonadotrophin—a second case, *J. Neurol. Neurosurg. Psychiatry* **55:**1094–1095 (1992).

52a. COLLINGE, J., PALMER, M. S., SIDLE, K. C., GOWLAND, I., MEDORI, R., IRONSIDE, J., AND LANTOS, P., Transmission of fatal familial insomnia to laboratory animals, *Lancet* **346:**569–570 (1995).

53. COLOSIMO, C., LEBON, P., MARTELLI, M., TUMMINELLI, F., AND MANDELLI, F., Alpha-interferon therapy in a case of probable progressive multifocal leukoencephalopathy, *Acta Neurol. Belg.* **92:**24–29 (1992).

54. CONNOLLY, J. H., ALLEN, I. V., AND HURWITZ, L. J., Measles-virus antibody and antigen in subacute sclerosing panencephalitis, *Lancet* **1:**542–544 (1967).

55. CREMER, N. E., OSHIRO, L. S., WEIL, M. L., LENNETTE, E. H., ITABASHI, H. H., AND CARNEY, L., Isolation of rubella virus from brain in chronic progressive panencephalitis, *J. Gen. Virol.* **29:**143–153 (1975).

56. CZUB, M., BRAIG, H. R., AND DIRINGER, H., Pathogenesis of scrapie: Study of the temporal development of clinical symptoms, of infectivity titres and scrapie-associated fibrils in brains of hamsters infected intraperitoneally, *J. Gen. Virol.* **67:**2005–2009 (1986).

57. CZUB, M., BRAIG, H. R., AND DIRINGER, H., Replication of the scrapie agent in hamsters infected intracerebrally confirms the pathogenesis of an amyloid-inducing virosis, *J. Gen. Virol.* **69:**1753–1756 (1988).

58. DAVID-FERREIRA, J. F., DAVID-FERREIRA, K. L., GIBBS, C. J., JR., AND MORRIS, J. A., Scrapie in mice: Ultrastructural observations in the cerebral cortex, *Proc. Soc. Exp. Biol. Med.* **28:**313–320 (1968).

59. DAWSON, J. R., Cellular inclusions in cerebral lesions of lethargic encephalitis, *Am. J. Pathol.* **9**:7–16 (1933).

60. DESLYS, J.-P., LASMÉZAS, C., AND DORMONT, D., Selection of specific strains in iatrogenic Creutzfeldt–Jakob disease, *Lancet* **343**:848–849 (1994).

61. DETELS, R., BRODY, J. A., McNEW, J., AND EDGAR, A. H., Further epidemiologic studies of subacute sclerosing panencephalitis, *Lancet* **2**:11–14 (1973).

62. DICKINSON, A. G., FRASER, H., AND OUTRAM, G. W., Scrapie incubation time can exceed natural lifespan, *Nature* **256**:732–733 (1979).

63. DIRINGER, H., AND BRAIG, H. R., Infectivity of unconventional viruses in dura mater, *Lancet* **1**:439–440 (1989).

64. DIRINGER, H., GELDERBLOM, H., HILMERT, H., ÖZEL, M., AND EDELBLUTH, C., Scrapie infectivity, fibrils and low molecular weight protein, *Nature* **306**:476–478 (1983).

65. DUDA, E., HUTTENLOCHER, P., AND PATRONAS, N., CT of subacute sclerosing panencephalitis, *Am. J. Neurol. Res.* **1**:35–38 (1980).

66. DUFFY, P., COLLINS, G., DEVOE, A. G., STREETEN, B., AND COHEN, D., Possible person-to-person transmission of Creutzfeldt–Jakob disease, *N. Engl. J. Med.* **290**:693 (1974).

67. DYKEN, P. R., Subacute sclerosing panencephalitis. Current status, *Neurol. Clin.* **3**:179–196 (1985).

68. DYKEN, P. R., CUNNINGHAM, S. C., AND WARD, L. C., Changing character of subacute sclerosing panencephalitis in the United States, *Pediatr. Neurol.* **5**:339–341 (1989).

69. DYKEN, P. R., SWIFT, A., AND DURANT, R. H., Long-term follow up of patients with subacute sclerosing panencephalitis treated with inosiplex, *Ann. Neurol.* **11**:359–364 (1982).

70. EIZURU, Y., SAKIHAMA, K., MINAMISHIMA, Y., HAYASHI, T., AND SUMIYOSHI, A., Re-evaluation of a case of progressive multifocal leukoencephalopathy previously diagnosed as simian virus 40 (SV40) etiology, *Acta Pathol. Jpn.* **43**:327–332 (1993).

71. ERNST, D. R., AND RACE, R. E., Comparative analysis of scrapie agent inactivation methods, *J. Virol. Methods* **41**:193–201 (1993).

72. FARRELL, M. A., CHENG, L., CORNFORD, M. E., GRODY, W. W., AND VINTERS, H. V., Cytomegalovirus and Rasmussen's encephalitis, *Lancet* **337**:1551–1552 (1991).

73. FARRINGTON, C. P., Subacute sclerosing panencephalitis in England and Wales: Transient effects and risk estimates, *Stat. Med.* **10**:1733–1744 (1991).

74. FOURNIER, J. G., GERFAUX, J., JORET, A. M., LEBON, P., AND ROZENBLATT, S., Subacute sclerosing panencephalitis: Detection of measles virus sequences in RNA extracted from circulating lymphocytes, *Br. Med. J. [Clin. Res.]* **296**:684 (1988).

75. FOURNIER, J. G., LEBON, P., BOUTEILLE, M., GOUTIERS, F., AND ROZENBLATT, S., Subacute sclerosing panencephalitis: Detection of measles virus RNA in appendix lymphoid tissue before clinical signs, *Br. Med. J. [Clin. Res.]* **293**:523–524 (1986).

76. FOURNIER, J. G., TARDIEU, M., LEBON, P., ROBAIN, O., PONSOT, G., ROZENBLATT, S., AND BOUTEILLE, M., Detection of measles virus RNA in lymphocytes from peripheral-blood and brain perivascular infiltrates of patients with subacute sclerosing panencephalitis, *N. Engl. J. Med.* **313**:910–915 (1985).

77. GAJDUSEK, D., Unconventional viruses and the origin and disappearance of kuru, *Science* **197**:943–960 (1977).

78. GAJDUSEK, D. C., Subacute spongiform encephalopathies: Transmissible cerebral amyloidoses caused by unconventional viruses, in: *Virology*, 2nd ed., Vol. 2 (B. N. FIELDS AND D. M. KNIPE, EDS.), pp. 2289–2324, Raven Press, New York, 1990.

79. GAJDUSEK, D. C., Infectious amyloids. Subacute spongiform encephalopathies as transmissible cerebral amyloidoses, in: *Virology*, 3rd ed., Vol. 2 (B. N. FIELDS, D. M. KNIPE, P. M. HOWLEY, R. M. CHANOCK, J. L. MELNICK, T. P. MONATH, B. ROIZMAN, AND S. E. STRAUS, EDS.), pp. 2851–2900, Raven Press, New York, 1994.

80. GAJDUSEK, D. C., GIBBS, C. J., JR., AND ALPERS, M., Experimental transmission of a kuru-like syndrome in chimpanzees, *Nature* **209**:794–796 (1966).

81. GAJDUSEK, D. C., GIBBS, C. J., JR., ASHER, D. M., BROWN, P., DIWAN, A., HOFFMAN, P., NEMO, G., ROHWER, R., AND WHITE, L., Precautions in medical care of and in handling materials from patients with transmissible virus dementia (Creutzfeldt–Jakob disease), *N. Engl. J. Med.* **297**:1253–1258 (1977).

82. GAJDUSEK, D. C., GIBBS, C. J., JR., EARLE, K., DAMMIN, G. J., SCHOENE, W. C., AND TYLER, H. R., Transmission of subacute spongiform encephalopathy to the chimpanzee and squirrel monkey from a patient with papulosis maligna of Köhlmeyer-Degos, *Excerpta Medica Int. Congr. Ser.* **319**:390–392 (1974).

83. GAJDUSEK, D. C., AND ZIGAS, V., Degenerative disease of the central nervous system in New Guinea: Epidemic occurrence of "kuru" in the native population, *N. Engl. J. Med.* **257**:974–978 (1957).

84. GAMBETTI, P., PETERSEN, R., MONARI, L., TABATON, M., AUTILIO GAMBETTI, L., CORTELLI, P., MONTAGNA, P., AND LUGARESI, E., Fatal familial insomnia and the widening spectrum of prion diseases, *Br. Med. Bull.* **49**:980–994 (1993).

85. GARDNER, S. D., FIELD, A., COLEMAN, D., AND HULME, B., New human papovirus (B.K.) isolated from urine after renal transplantation, *Lancet* **1**:1253–1257 (1971).

86. GASCON, G., YAMANI, S., CROWELL, J., STIGSBY, B., NESTER, M., KANAAN, I., AND JALLU, A., Combined oral isoprinosine–intraventricular alpha-interferon therapy for subacute sclerosing panencephalitis, *Brain Dev.* **15**:346–355 (1993).

87. GASCON, G. G., YAMANI, S., CAFEGE, A., FLOCK, L., AL SEDAIRY, S., PARHAR, R. S., CROWELL, J., NESTER, M., KANAAN, I., AND JALLU, M. A., Treatment of subacute sclerosing panencephalitis with alpha interferon, *Ann. Neurol.* **30**:227–228 (1991).

88. GERBER, M. A., SHAH, K. V., THUNG, S. N., AND ZU RHEIN, G. M., Immunohistochemical demonstration of common antigen of polyomaviruses in routine histologic tissue sections of animals and man, *Am. J. Clin. Pathol.* **73**:795–797 (1980).

89. GIBBONS, R. A., AND HUNTER, G. D., Nature of the scrapie agent, *Nature* **215**:1041–1043 (1967).

90. GIBBS, C. J., JR., AMYX, H. L., BACOTE, A., MASTERS, C. L., AND GAJDUSEK, D. C., Oral transmission of kuru, Creutzfeldt–Jakob disease, and scrapie to nonhuman primates, *J. Infect. Dis.* **142**:205–207 (1980).

91. GIBBS, C. J., JR., ASHER, D. M., BROWN, P. W., FRADKIN, J. E., AND GAJDUSEK, D. C., Creutzfeldt–Jakob disease infectivity of growth hormone derived from human pituitary glands, *N. Engl. J. Med.* **328**:358–359 (1993).

92. GIBBS, C. J., JR., ASHER, D. M., KOBRINE, A., AMYX, H. L., AND GAJDUSEK, D. C., Transmission of Creutzfeldt–Jakob disease to a chimpanzee by electrodes contaminated during neurosurgery, *J. Neurol. Neurosurg. Psychiatry* **57**:757–758 (1994).

93. GIBBS, C. J., JR., AND GAJDUSEK, D. C., Transmission and characterization of the agents of spongiform virus encephalopathies: Kuru, Creutzfeldt–Jakob disease, scrapie and mink encephalopathy, *Res. Publ. Assoc. Res. Nerv. Ment. Dis.* **49**:383–410 (1971).

94. GIBBS, C. J., JR., GAJDUSEK, D. C., AND AMYX, H., Strain

variation in the viruses of Creutzfeldt–Jakob disease and kuru, in: *Slow Transmissible Diseases of the Nervous System*, Vol. 1 (S. B. PRUSINER AND W. J. HADLOW, EDS.), pp.87–110, Academic Press, New York, 1979.

95. GIBBS, C. J., JR., GAJDUSEK, D. C., ASHER, D. M., ALPERS, M. P., BECK, E., DANIEL, P. M., AND MATTHEWS, W. B., Creutzfeldt–Jakob disease (spongiform encephalopathy): Transmission to the chimpanzee, *Science* **161**:388–389 (1968).

96. GIBBS, C. J., JR., GAJDUSEK, D. C., AND LATARJET, R., Unusual resistance to ionizing radiation of the viruses of kuru, Creutzfeldt–Jakob disease and scrapie, *Proc. Natl. Acad. Sci. USA* **75**:6268–6270 (1978).

97. GIBSON, P. E., KNOWLES, W. A., HAND, J. F., AND BROWN, D. W., Detection of JC virus DNA in the cerebrospinal fluid of patients with progressive multifocal leukoencephalopathy, *J. Med. Virol.* **39**:278–281 (1993).

98. GILLESPIE, S. M., CHANG, Y., LEMP, G., ARTHUR, R., BUCHBINDER, S., STEIMLE, A., BAUMGARTNER, J., RANDO, T., NEAL, D., RUTHERFORD, G., ET AL., Progressive multifocal leukoencephalopathy in persons infected with human immunodeficiency virus, San Francisco, 1981–1989, *Ann. Neurol.* **30**:597–604 (1991).

99. GIMENEZ-ROLDAN, S., MARTIN, M., MATEO, D., AND LOPEZ-FRAILE, I. P., Preclinical EEG abnormalities in subacute sclerosing panencephalitis, *Neurology* **31**:763–776 (1981).

100. GLOOR, P., KALABAY, O., AND GIARD, N., The electroencephalogram in diffuse encephalopathies: Electroencephalographic correlates of grey and white matter lesions, *Brain* **91**:779–801 (1968).

101. GODEC, M. S., ASHER, D. M., SWOVELAND, P. T., ELDADAH, Z. A., FEINSTONE, S. M., GOLDFARB, L. G., GIBBS, C. J. J., AND GAJDUSEK, D. C., Detection of measles virus genomic sequences in SSPE brain tissue, *J. Med. Virol.* **30**:237–244 (1990).

102. GOLDFARB, L. G., BROWN, P., AND GAJDUSEK, D. C., The molecular genetics of human transmissible spongiform encephalopathy, in: *Prion Diseases of Humans and Animals* (S. B. PRUSINER, J. COLLINGE, J. POWELL, AND B. ANDERTON, EDS.), pp. 139–153, Ellis Horwood, New York, 1992.

103. GOLDFARB, L. G., PETERSEN, R. B., TABATON, M., BROWN, P., LEBLANC, A. C., MONTAGNA, P., CORTELLI, P., JULIEN, J., VITAL, C., PENDELBURY, W. W., HALTIA, M., WILLS, P. R., HAUW, J. J., MCKEEVER, P. E., MONARI, L., SCHRANK, B., SWERGOLD, G. D., AUTILIO-GAMBETTI, L., GAJDUSEK, D. C., LUGARESI, E., AND GAMBETTI, P., Fatal familial insomnia and familial Creutzfeldt–Jakob disease: Disease phenotype determined by a DNA polymorphism, *Science* **258**:806–808 (1992).

104. GOLDGABER, D., GOLDFARB, L. G., BROWN, P., ASHER, D. M., BROWN, W. T., LINN, W. S., TEENER, J. W., FEINSTONE, S. M., RUBENSTEIN, R., KASCSAK, R. J., BOELLARD, J. W., AND GAJDUSEK, D. C., Mutations in familial Creutzfeldt–Jakob disease and Gerstmann–Sträussler–Scheinker's syndrome, *Exp. Neurol.* **106**:204–206 (1989).

105. GOLDGABER, D., TEENER, J. W., GOLDFARB, L. G., ASHER, D. M., BROWN, P. W., FEINSTONE, S., AND GAJDUSEK, D. C., No Msp 1 polymorphism in the open reading frame of the PrP gene in patients with familial Creutzfeldt–Jakob disease, *Alz. Dis. Assoc. Disord.* **2**:311 (1988).

106. GOMBART, A. F., HIRANO, A., AND WONG, T. C., Expression and properties of the V protein in acute measles virus and subacute sclerosing panencephalitis virus strains, *Virus Res.* **25**:63–78 (1992).

107. GORMAN, D. G., BENSON, D. F., VOGEL, D. G., AND VINTERS, H.

V., Creutzfeldt–Jakob disease in a pathologist, *Neurology* **42**:463 (1992).

108. GUIZZARO, A., VOLPE, E., LUS, G., BRAVACCIO, F., COTRUFO, R., AND PAOLOZZI, C., Progressive rubella panencephalitis. Follow-up EEG study of a case, *Acta Neurol. (Napoli)* **14**:485–492 (1992).

109. HADLOW, W., KENNEDY, R., JACKSON, T., WHITFORD, H., AND BOYLE, C., Course of experimental scrapie virus infection in the goat, *J. Infect. Dis.* **129**:559–567 (1974).

110. HADLOW, W. J., Scrapie and kuru, *Lancet* **2**:289–290 (1959).

111. HAIR, L. S., NUOVO, G., POWERS, J. M., SISTI, M. B., BRITTON, C. B., AND MILLER, J. R., Progressive multifocal leukoencephalopathy in patients with human immunodeficiency virus, *Hum. Pathol.* **23**:663–667 (1992).

112. HALSEY, N., MODLIN, J., JABBOUR, J., DUBEY, L., EDDINS, D., AND LUDWIG, D., Risk factors in subacute sclerosing panencephalitis: A case-control study, *Am. J. Epidemiol.* **111**:415–424 (1980).

113. HALSEY, N. A., AND MODLIN, J. F., Subacute sclerosing panencephalitis, *Pediatr. Neurol.* **7**:151 (1991).

114. HARRINGTON, M. G., MERRIL, C. R., ASHER, D. M., AND GAJDUSEK, D. C., Abnormal proteins in the cerebrospinal fluid of patients with Creutzfeldt–Jakob disease, *N. Engl. J. Med.* **315**:279–283 (1986).

115. HEDLEY-WHITE, E. T., SMITH, B. P., TYLER, H. R., AND PETERSON, W. P., Multifocal leukoencephalopathy with remission and five year survival, *J. Neuropathol. Exp. Neurol.* **25**:107–116 (1966).

116. HENSON, J., ROSENBLUM, M., ARMSTRONG, D., AND FURNEAUX, H., Amplification of JC virus DNA from brain and cerebrospinal fluid of patients with progressive multifocal leukoencephalopathy, *Neurology* **41**:1967–1971 (1991).

117. HINTZ, R., MACGILLIVRAY, M., JOY, A., AND TINTNER, R., Fatal degenerative neurological disease in patients who received pituitary-derived human growth hormone, *Morbid. Mortal. Week. Rep.* **34**:359–366 (1985).

118. HIRANO, A., AYATA, M., WANG, A. H., AND WONG, T. C., Functional analysis of matrix proteins expressed from cloned genes of measles virus variants that cause subacute sclerosing panencephalitis reveals a common defect in nucleocapsid binding, *J. Virol.* **67**:1848–1853 (1993).

119. HIRANO, A., WANG, A. H., GOMBART, A. F., AND WONG, T. C., The matrix proteins of neurovirulent subacute sclerosing panencephalitis virus and its acute measles virus progenitor are functionally different, *Proc. Natl. Acad. Sci. USA* **89**:8745–8749 (1992).

120. HOFMAN, F. M., HINTON, D. R., BAEMAYR, J., WEIL, M., AND MERRILL, J. E., Lymphokines and immunoregulatory molecules in subacute sclerosing panencephalitis, *Clin. Immunol. Immunopathol.* **58**:331–342 (1991).

121. HORTA-BARBOSA, L., FUCCILLO, D. A., LONDON, W. T., JABBOUR, J. T., ZEMAN, W., AND SEVER, J. L., Isolation of measles virus from brain cell cultures of two patients with subacute sclerosing panencephalitis, *Proc. Soc. Exp. Biol. Med.* **132**:272–277 (1969).

122. HORTA-BARBOSA, L., FUCCILLO, D. A., SEVER, J. L., AND ZEMAN, W., Subacute sclerosing panencephalitis: Isolation of measles virus from a brain biopsy, *Nature* **221**:974 (1969).

123. HSIAO, K., BAKER, H. F., CROW, T. J., POULTER, M., OWEN, F., TERWILLIGER, J. D., WESTAWAY, D., OTT, J., AND PRUSINER, S. B., Linkage of a prion protein missense variant to Gerstmann–Sträussler syndrome, *Nature* **338**:342–345 (1989).

124. HSIAO, K. K., SCOTT, M., FOSTER, D., GROTH, D. F., DEARMOND, S. J., AND PRUSINER, S. B., Spontaneous neurodegeneration in transgenic mice with mutant prion protein, *Science* **250:**1587–1590 (1990).

124a. HSICH, G., KENNEY, K., GIBBS, C. J., JR., LEE, K. H., AND HARRINGTON, M. G., The 14-3-3 brain protein in cerebrospinal fluid as a marker for transmissible spongiform encephalopathies, *N. Engl. J. Med.* **335:**924–930 (1996).

125. HUNTER, N., FOSTER, J. D., DICKINSON, A. G., AND HOPE, J.,Linkage of the gene for the scrapie-associated fibril protein (PrP) to the Sip gene in Cheviot sheep, *Vet. Rec.* **124:**364–366 (1989).

126. HUNTER, N., HOPE, J., McCONNELL, I., AND DICKINSON, A. G., Linkage of the scrapie-associated fibril protein (PrP) gene and Sinc using congenic mice and restriction fragment length polymorphism analysis, *J. Gen. Virol.* **68:**2711–2716 (1987).

127. IIDA, T., KITAMURA, T., GUO, J., TAGUCHI, F., ASO, Y., NAGASHIMA, K., AND YOGO, Y., Origin of JC polyomavirus variants associated with progressive multifocal leukoencephalopathy, *Proc. Natl. Acad. Sci. USA* **90:**5062–5065 (1993).

128. ISAACSON, S. H., ASHER, D. M., GAJDUSEK, D. C., AND GIBBS, C. J., JR., Detection of RNA viruses in archival brain tissue by *in situ* RT-PCR amplification and labeled-probe hybridization, *Cell-Vision/J. Anal. Morphol.* **1:**25–28 (1994).

129. JABBOUR, J. T., DUENAS, D. A., SEVER, J. L., KREBS, H. M., AND HORTA-BARBOSA, L., Epidemiology of subacute sclerosing panencephalitis (SSPE), *J.A.M.A.* **220:**959–972 (1972).

130. JELLINGER, K., SEITELBERGER, F., HEISS, W.-D., AND HOLCZABEK, W., Konjugale Form der subakuten spongiöse Enzephalopathie (Jakob–Creutzfeldt Erkrankung), *Wien. Klin. Wochenschr.* **84:**245–249 (1972).

131. JOHNSTON, H. M., WISE, G. A., AND HENRY, J. G., Visual deterioration as presentation of subacute sclerosing panencephalitis, *Arch. Dis. Child.* **55:**899–901 (1980).

132. KABRA, S. K., BAGGA, A., AND SHANKAR, V., Subacute sclerosing panencephalitis presenting as cortical blindness, *Trop. Doct.* **22:**94–95 (1992).

133. KARAHALIOS, D., BREIT, R., DAL CANTO, M. C., AND LEVY, R. M., Progressive multifocal leukoencephalopathy in patients with HIV infection: Lack of impact of early diagnosis by stereotactic brain biopsy, *J. Acquir. Immune Defic. Syndr.* **5:**1030–1038 (1992).

134. KATZ, D. A., BERGER, J. R., HAMILTON, B., MAJOR, E. O., AND POST, M. J., Progressive multifocal leukoencephalopathy complicating Wiskott–Aldrich syndrome. Report of a case and review of the literature of progressive multifocal leukoencephalopathy with other inherited immunodeficiency states, *Arch. Neurol.* **51:**422–426 (1994).

135. KATZ, M., AND KOPROWSKI, H., The significance of failure to isolate infectious viruses in cases of subacute sclerosing panencephalitis, *Arch. Virol.* **41:**390–393 (1973).

136. KHWAJA, G. A., GUPTA, M., AND SHARMA, D. K., Subacute sclerosing panencephalitis, *J. Assoc. Physicians India* **39:**928–933 (1991).

137. KIMBERLIN, R. H., COLE, S., AND WALKER, C. A., Temporary and permanent modifications to a single strain of mouse scrapie on transmission to rats and hamsters, *J. Gen. Virol.* **68:**1875–1881 (1987).

138. KIMBERLIN, R. H., AND WALKER, C. A., Pathogenesis of experimental scrapie, in: *Novel Infectious Agents and the Central Nervous System. Ciba Foundation Symposium*, Vol. 135 (G. BOCK AND J. MARSH, EDS.), pp.37–62, Wiley and Sons, Chichester, 1988.

139. KIRK, J., ZHOU, A. L., McQUAID, S., COSBY, S. L., AND ALLEN, I. V., Cerebral endothelial cell infection by measles virus in subacute sclerosing panencephalitis: Ultrastructural and in situ hybridization evidence, *Neuropathol. Appl. Neurobiol.* **17:**289–297 (1991).

140. KITAMOTO, T., OGOMORI, K., TATEISHI, J., AND PRUSINER, S. B., Formic acid pretreatment enhances immunostaining of cerebral and systemic amyloids, *Lab. Invest.* **57:**230–236 (1987).

141. KLITZMAN, R. L., ALPERS, M. P., AND GAJDUSEK, D. C., The natural incubation period of kuru and the episodes of transmission in three clusters of patients, *Neuroepidemiology* **3:**3–20 (1984).

142. KOMASE, K., HAGA, T., YOSHIKAWA, Y., AND YAMANOUCHI, K., Complete nucleotide sequence of the phosphoprotein of the Yamagata-1 strain of a defective subacute sclerosing panencephalitis (SSPE) virus, *Biochim. Biophys. Acta* **1129:**342–344 (1992).

143. KORNBERG, A. J., HARVEY, A. S., AND SHIELD, L. K., Subacute sclerosing panencephalitis presenting as simple partial seizures, *J. Child Neurol.* **6:**146–149 (1991).

144. KOZHEVNIKOV, A. Y., A particular type of cortical epilepsy (epilepsia corticalis sive partialis continua), in: *Chronic Encephalitis and Epilepsy: Rasmussen's Syndrome*, Vol. 42 .(F. ANDERMANN, ED.), pp. 245–261, Butterworth-Heinemann, Boston, 1991.

145. KUCHELMEISTER, K., GULLOTTA, F., BERGMANN, M., ANGELI, G., AND MASINI, T., Progressive multifocal leukoencephalopathy (PML) in the acquired immunodeficiency syndrome (AIDS). A neuropathological autopsy study of 21 cases, *Pathol. Res. Pract.* **89:**163–173 (1993).

146. LAKSHMI, V., MALATHY, Y., AND RAO, R. R., Serodiagnosis of subacute sclerosing panencephalitis by enzyme linked immunosorbent assay, *Indian J. Pediatr.* **60:**37–41 (1993).

147. LEBON, P., AND LYON, G., Non-congenital rubella encephalitis, *Lancet* **2:**468 (1974).

148. LEWIN, P., Scrapie: An infective peptide?, *Lancet* **1:**748 (1972).

149. LIBERSKI, P. P., BUDKA, H., SLUGA, E., BARCIKOWSKA, M., AND KWIECINSKI, H., Tubulovesicular structures in Creutzfeldt–Jakob disease, *Acta Neuropathol. (Berl.)* **84:**238–243 (1992).

150. LUCAS, K. M., SANDERS, R. C., RONGAP, A., RONGAP, T., PINAI, S., AND ALPERS, M. P., Subacute sclerosing panencephalitis (SSPE) in Papua New Guinea: A high incidence in young children, *Epidemiol. Infect.* **108:**547–553 (1992).

151. MAJOR, E. O., AMEMIYA, K., TORNATORE, C. S., HOUFF, S. A., AND BERGER, J. R., Pathogenesis and molecular biology of progressive multifocal leukoencephalopathy, the JC virus-induced demyelinating disease of the human brain, *Clin. Microbiol. Rev.* **5:**49–73 (1992).

152. MANETTO, V., MEDORI, R., CORTELLI, P., MONTAGNA, P., TINUPER, P., BARUZZI, A., RANCUREL, G., HAUW, J. J., VANDERHAEGHEN, J. J., MAILLEUX, P., BUGIANI, O., TAGLIAVINI, F., BOURAS, C., RIZZUTO, N., LUGARESI, E., AND GAMBETTI, P., Fatal familial insomnia—clinical and pathologic study of five new cases, *Neurology* **42:**312 –319 (1992).

153. MANUELIDIS, E. E., KIM, J. H., MERICANGAS, J. R., AND MANUELIDIS, L., Transmission to animals of Creutzfeldt–Jakob disease from human blood, *Lancet* **2:**896–897 (1985).

154. MANUELIDIS, L., SKLAVIADIS, T., AND MANUELIDIS, E. E., Evidence suggesting that PrP is not the infectious agent in Creutzfeldt–Jakob disease, *EMBO J.* **6:**341–347 (1987).

155. MARTINEZ LAGE, J. F., POZA, M., AND TORTOSA, J. G., Creutz-feldt–Jakob disease in patients who received a cadaveric dura mater graft—Spain, 1985–1992, *Morbid. Mortal. Week. Rep.* **42:**560–563 (1993).

156. MARTINEZ LAGE, J. F., SOLA, J., POZA, M., AND ESTEBAN, J. A., Pediatric Creutzfeldt–Jakob disease—Probable transmission by a dural graft, *Childs Nerv. Syst.* **9:**239–242 (1993).

157. MASTERS, C. L., GAJDUSEK, D. C., AND GIBBS, C. J., JR., Creutzfeldt–Jakob disease virus isolation from the Gerstmann-Sträussler syndrome, with an analysis of the various forms of amyloid deposition in the virus-induced spongiform encepha-lopathies, *Brain* **104:**559–588 (1981).

158. MASTERS, C. L., GAJDUSEK, D. C., GIBBS, C. J., JR., BERNOULLI, C., AND ASHER, D. M., Familial Creutzfeldt–Jakob disease and other familial dementias: An inquiry into possible modes of transmission of virus-induced familial disease, in: *Slow Trans-missible Diseases of the Nervous System*, Vol. I (S. B. PRUSINER AND W. J. HADLOW, EDS.), pp. 143–194, Academic Press, New York, 1979.

159. MASTERS, C. L., HARRIS, J. O., GAJDUSEK, D. C., GIBBS, C. J., JR., BERNOULLI, C., AND ASHER, D. M., Creutzfeldt–Jakob dis-ease: Patterns of worldwide occurrence and the significance of familial and sporadic clustering, *Ann. Neurol.* **5:**177–188 (1979).

160. MASULLO, C., POCCHIARI, M., MACCHE, G., ALEMA, G., PIAZZA, G., AND PANZERA, M. A., Transmission of Creutzfeldt–Jakob disease by dural cadaveric graft, *J. Neurosurg.* **71:**954–955 (1989).

161. MATTHEWS, W. B., Epidemiology of Creutzfeldt–Jakob disease in England and Wales, *J. Neurol. Neurosurg. Psychiatry* **38:**210–213 (1975).

162. MEDORI, R., MONTAGNA, P., TRITSCHLER, H. J., LEBLANC, A., CORTELLI, P., TINUPER, P., LUGARESI, E., AND GAMBETTI, P., Fatal familial insomnia—a second kindred with mutation of prion protein gene at codon-178, *Neurology* **42:**669–670 (1992).

163. MEDORI, R., AND TRITSCHLER, H. J., Prion protein gene analysis in 3 kindreds with fatal familial insomnia (FFI)—codon-178 mutation and codon-129 polymorphism, *Am. J. Hum. Genet.* **53:**822–827 (1993).

164. MEDORI, R., TRITSCHLER, H. J., LEBLANC, A., VILLARE, F., MANETTO, V., CHEN, H. Y., XUE, R., LEAL, S., MONTAGNA, P., CORTELLI, P., TINUPER, P., AVONI, P., MOCHI, M., BARUZZI, A., HAUW, J. J., OTT, J., LUGARESI, E., AUTILIO GAMBETI, L., AND GAMBETTI, P., Fatal familial insomnia, a prion disease with a mutation of codon-178 of the prion protein gene, *N. Engl. J. Med.* **326:**444–449 (1992).

165. MEHTA, P. D., KULCZYCKI, J., PATRICK, B. A., SOBCZYK, W., AND WISNIEWSKI, H. M., Effect of treatment on oligoclonal IgG bands and intrathecal IgG synthesis in sequential cerebrospinal fluid and serum from patients with subacute sclerosing panencephalitis, *J. Neurol. Sci.* **109:**64–68 (1992).

166. MERZ, P. A., SOMERVILLE, R. A., WISNIEWSKI, H. M., AND IQBAL, K., Abnormal fibrils from scrapie-infected brain, *Acta Neuropathol. (Berl.)* **54:**63–74 (1981).

167. MERZ, P. A., SOMERVILLE, R. A., WISNIEWSKI, H. M., MANUE-LIDIS, L., AND MANUELIDIS, E. E., Scrapie-associated fibrils in Creutzfeldt–Jakob disease, *Nature* **306:**474–476 (1983).

168. MESQUITA, R., PARRAVICINI, C., BJORKHOLM, M., EKMAN, M., AND BIBERFELD, P., Macrophage association of polyomavirus in progressive multifocal leukoencephalopathy: An immunohisto-chemical and ultrastructural study. Case report, *APMIS* (Copen-hagen) **100:**993–1000 (1992).

169. MILLER, C., FARRINGTON, C. P., AND HARBERT, K., The epide-miology of subacute sclerosing panencephalitis in England and Wales 1970–1989, *Int. J. Epidemiol.* **21:**998–1006 (1992).

170. MILLER, D., Creutzfeldt–Jakob disease in histopathology techni-cians, *N. Engl. J. Med.* **318:**853–854 (1988).

171. MIYAZAKI, M., HASHIMOTO, T., FUJINO, K., GODA, T., TAYAMA, M., AND KURODA, Y., Apparent response of subacute sclerosing panencephalitis to intrathecal interferon alpha, *Ann. Neurol.* **29:**97–99 (1991).

172. MODLIN, J. F., HALSEY, N. A., EDDINS, D. L., CONRAD, J. L., JABBOUR, J. T., CHIEN, L., AND ROBINSON, H., Epidemiology of subacute sclerosing panencephalitis, *J. Pediatr.* **94:**231–236 (1979).

173. MONARI, L., CHEN, S. G., BROWN, P., PARCHI, P., PETERSEN, R. B., MIKOL, J., GRAY, F., CORTELLI, P., MONTAGNA, P., GHETTI, B., GOLDFARB, L. G., GAJDUSEK, D. C., LUGARESI, E., GAMBETTI, P., AND AUTILIO GAMBETTI, L., Fatal familial insomnia and familial Creutzfeldt–Jakob disease—different prion proteins determined by a DNA polymorphism, *Proc. Natl. Acad. Sci. USA* **91:**2839–2842 (1994).

174. MORET, H., GUICHARD, M., MATHERON, S., KATLAMA, C., SAZDOVITCH, V., HURAUX, J. M., AND INGRAND, D., Virological diagnosis of progressive multifocal leukoencephalopathy: Detec-tion of JC virus DNA in cerebrospinal fluid and brain tissue of AIDS patients, *J. Clin. Microbiol.* **31:**3310–3313 (1993).

175. MURAMOTO, T., KITAMOTO, T., TATEISHI, J., AND GOTO, I., Accu-mulation of abnormal prion protein in mice infected with Creutz-feldt–Jakob disease via intraperitoneal route—a sequential study, *Am. J. Pathol.* **143:**1470–1479 (1993).

176. NAGANO, I., NAKAMURA, S., YOSHIOKA, M., AND KOGURE, K., Immunocytochemical analysis of the cellular infiltrate in brain lesions in subacute sclerosing panencephalitis, *Neurology* **41:** 1639–1642 (1991).

177. NAGANO, I., NAKAMURA, S., YOSHIOKA, M., ONODERA, J., KO-GURE, K., AND ITOYAMA, Y., Expression of cytokines in brain lesions in subacute sclerosing panencephalitis, *Neurology* **44:** 710–715 (1994).

178. NARANG, H. K., ASHER, D. M., POMEROY, K. L., AND GAJDUSEK, D. C., Abnormal tubulovesicular particles in brains of hamsters with scrapie, *Proc. Soc. Exp. Biol. Med.* **184:**504–509 (1987).

179. NEVIN, S., MCMENEMY, W. H., BEHRMAN, D., AND JONES, D. P., Subacute spongiform encephalopathy: A subacute form of en-cephalopathy attributed to vascular dysfunction (spongiform ce-rebral atrophy), *Brain* **83:**519–564 (1960).

180. NISBET, T. J., MACDONALDSON, I., AND BISHARA, S. N., Creutz-feldt–Jakob disease in a second patient who received a cadaveric dura mater graft, *J.A.M.A.* **261:**111 (1989).

181. OCCUPATIONAL SAFETY AND HEALTH ADMINISTRATION, UNITED STATES DEPARTMENT OF LABOR, Occupational exposure to bloodborne pathogens; final rule (29 CFR Part 1910.1030), *Fed-eral Register* **56:**64175–64182 (1991).

182. OESCH, B., WESTAWAY, D., WÄLCHI, M., MCKINLEY, M. P., KENT, S. H. B., AEBERSOLD, R., BARRY, R. A., TEMPST, P., TEPLOW, D. B., HOOD, L. E., PRUSINER, S. B., AND WEISSMAN, C., A cellular gene encodes scrapie PrP 27-30 protein, *Cell* **40:**735–746 (1985).

183. OGAWA, M., OKUBO, H., TSUJI, Y., YASUI, N., AND SOMEDA, K., Chronic progressive encephalitis occurring 13 years after Russian spring-summer encephalitis, *J. Neurol. Sci.* **19:**363–373 (1973).

184. OGUNI, H., ANDERMANN, F., AND RASMUSSEN, T. B., The natural history of the syndrome of chronic encephalitis and epilepsy: A study of the MNI series of forty-eight cases, in: *Chronic Encepha-litis and Epilepsy: Rasmussen's Syndrome* (F. ANDERMANN, ED.), pp. 7–21, Butterworth-Heinemann, Boston, 1991.

185. OREFICE, G., CAMPANELLA, G., CICCIARELLO, S., CHIRIANNI, A., BORGIA, G., RUBINO, S., MAINOLFI, M., COPPOLA, M., AND PIAZZA, M., Presence of papova-like viral particles in cerebrospinal fluid of AIDS patients with progressive multifocal leukoencephalopathy. An additional test for *in vivo* diagnosis, *Acta Neurol. (Napoli)* **15**:328–332 (1993).

186. OWEN, F., POULTER, M., LOFTHOUSE, R., COLLINGE, J., CROW, T. J., RISBY, D., BAKER, H. F., RIDLEY, R. M., HSIAO, K., AND PRUSINER, S. B., Insertion in prion protein gene in familial Creutzfeldt–Jakob disease, *Lancet* **1**:51–52 (1989).

187. OWEN, F., POULTER, M., LOFTHOUSE, R., CROW, T. J., RISBY, D., BAKER, H. F., AND RIDLEY, R. M., A rare MspI polymorphism in the human prion gene in a family with a history of early onset dementia, *Neurosci. Lett. Suppl.* **32**:S53 (1988).

188. ÖZEL, M., AND DIRINGER, H., Small virus-like structure in fractions from scrapie hamster brain, *Lancet* **343**:894–895 (1994).

189. PADGETT, B. L., AND WALKER, D. L., Prevalence of antibodies in human sera against JC virus, an isolate from a case of progressive multifocal leukoencephalopathy, *J. Infect. Dis.* **127**:467–470 (1973).

190. PADGETT, B. L., ZU RHEIN, G. M., WALKER, D. L., ECKROADE, R. J., AND DESSEL, B. H., Cultivation of papova-like virus from human brain with progressive multifocal leukoencephalopathy, *Lancet* **1**:1257–1260 (1971).

191. PAYNE, F. E., AND BAUBLIS, J. V., Measles virus and subacute sclerosing panencephalitis, *Perspect. Virol.* **7**:179–195 (1971).

192. PAYNE, F. E., BAUBLIS, J. V., AND ITABASHI, H. H., Isolation of measles virus from cell cultures of brain from a patient with subacute sclerosing panencephalitis, *N. Engl. J. Med.* **281**:585–589 (1969).

193. PETTE, H., AND DÖRING, G., Ueber ein heimliche Panencephalomyelitis von Character des Encephalitis japonica, *Dtsch. Z. Nervenheilk.* **149**:7–44 (1939).

194. POWER, C., POLAND, S. D., BLUME, W. T., GIRVIN, J. P., AND RICE, G. P., Cytomegalovirus and Rasmussen's encephalitis, *Lancet* **336**:1282–1284 (1990).

195. PRUSINER, S., Scrapie prions, *Annu. Rev. Microbiol.* **43**:345–374 (1989).

196. PRUSINER, S. B., Novel proteinaceous infectious particles cause scrapie, *Science* **216**:136–144 (1982).

197. PRUSINER, S. B., GROTH, D., SERBAN, A., KOEHLER, R., FOSTER, D., TORCHIA, M., BURTON, D., YANG, S. L., AND DEARMOND, S. J., Ablation of the prion protein (PrP) gene in mice prevents scrapie and facilitates production of anti-PrP antibodies, *Proc. Natl. Acad. Sci. USA* **90**:10608–10612 (1993).

198. PRUSINER, S. B., SCOTT, M., FOSTER, D., PAN, K. M., GROTH, D., MIRENDA, C., TORCHIA, M., YANG, S. L., SERBAN, D., CARLSON, G. A., HOPPE, P. C., HESTANAY, D., AND DEARMOND, S. J., Transgenetic studies implicate interactions between homologous PrP isoforms in scrapie prion replication, *Cell* **63**:673–686 (1990).

199. RASMUSSEN, T. B., Chronic encephalitis and seizures: Historical introduction, in: *Chronic Encephalitis and Epilepsy: Rasmussen's Syndrome* (F. ANDERMANN, ED.), pp. 1–4, Butterworth-Heinemann, Boston, 1991.

200. RENNER, K., LEGER, H., AND WEGNER, M., The POU domain protein Tst-1 and papoviral large tumor antigen function synergistically to stimulate glia-specific gene expression of JC virus, *Proc. Natl. Acad. Sci. USA* **91**:6433–6437 (1994).

201. RICHARDSON, E. P., Progressive multifocal leukoencephalopathy, *N. Engl. J. Med.* **265**:815–823 (1961).

202. ROHWER, R., The scrapie agent: "A virus by any other name," *Curr. Top. Microbiol. Immunol.* **172**:195–232 (1991).

203. ROHWER, R. G., Estimation of scrapie nucleic acid MW from standard curves for virus sensitivity to ionizing radiation, *Nature* **320**:381 (1986).

204. SCHMID, A., SPIELHOFER, P., CATTANEO, R., BACZKO, K., TER MEULEN, V., AND BILLETER, M. A., Subacute sclerosing panencephalitis is typically characterized by alterations in the fusion protein cytoplasmic domain of the persisting measles virus, *Virology* **188**:910–915 (1992).

205. SCHNEIDER-SCHAULIES, S., KRETH, H. W., HOFMANN, G., BILLETER, M., AND TER MEULEN, V., Expression of measles virus RNA in peripheral blood mononuclear cells of patients with measles, SSPE, and autoimmune diseases, *Virology* **182**:703–711 (1991).

206. SCOTT, M. R., FOSTER, D., MIRENDA, C., SERBAN, D., COUFAL, F., WÄLCHLI, M., TORCHIA, M., GROTH, D., CARLSON, G., DEARMOND, S. J., HESTANAY, D., AND PRUSINER, S. B., Transgenic mice expressing hamster prion protein produce species-specific scrapie infectivity and amyloid plaques, *Cell* **59**:847–857 (1989).

207. SEVER, J. L., Persistent measles infection of the CNS: Subacute sclerosing panencephalitis, *Rev. Infect. Dis.* **5**:467–473 (1983).

208. SHAIKH, N. J., AND RODRIGUES, J. J., Serological studies on subacute sclerosing panencephalitis, *Indian J. Pediatr.* **58**:833–835 (1991).

209. SIE, T. H., WEBER, W., FRELING, G., GALAMA, J., SPAANS, F., AND VLES, J., Rapidly fatal subacute sclerosing panencephalitis in a 19-year-old man, *Eur. Neurol.* **31**:94–99 (1991).

210. SIEDLER, H., AND MALAMUD, N., Creutzfeldt–Jakob disease: Clinicopathologic report of 15 cases and review of the literature (with special reference to a related disorder designated as subacute spongiform encephalopathy, *J. Neuropathol. Exp. Neurol.* **22**:381–402 (1963).

211. SIGURDSSON, B., Observations on three slow infections of sheep, *Br. Med. J. [Clin. Res.]* **110**:255–270, 307–322, 341–354 (1954).

212. SITWELL, L., LACH, B., ATACK, E., AND ATACK, D., Creutzfeldt–Jakob disease in histopathology technicians, *N. Engl. J. Med.* **318**:854 (1988).

213. STEIGER, M. J., TARNESBY, G., GABE, S., MCLAUGHLIN, J., AND SCHAPIRA, A. H., Successful outcome of progressive multifocal leukoencephalopathy with cytarabine and interferon, *Ann. Neurol.* **33**:407–411 (1993).

214. STEINER, I., WIRGUIN, I., MORAG, A., AND ABRAMSKY, O., Intraventricular interferon treatment for subacute sclerosing panencephalitis, *J. Child Neurol.* **4**:20–23 (1989).

215. STONER, G. L., WALKER, D. L., AND WEBSTER, H. D., Age distribution of progressive multifocal leukoencephalopathy, *Acta Neurol. Scand.* **78**:307–312 (1988).

216. SUSSMAN, J., AND COMPSTON, D. A., Subacute sclerosing panencephalitis in Wales, *Q. J. Med.* **878**:23–34 (1994).

217. TAGUCHI, F., KAJIOKA, J., AND MIYAMURA, T., Prevalence rate and age of acquisition of antibodies against JC virus and BK virus in human sera, *Microbiol. Immunol.* **26**:1057–1064 (1982).

218. TAKASU, T., KONDO, K., AHMED, A., YOSHIKAWA, Y., YAMANOUCHI, K., TSUCHIYA, M., MURAKAMI, N., AND UEDA, S., Elevated ratio of late measles among subacute sclerosing panencephalitis patients in Karachi, Pakistan, *Neuroepidemiology* **11**:282–287 (1992).

219. TAKEMOTO, K. K., RABSON, A. S., MULLARKEY, M. F., BLAESE, R. M., GARON, C. F., AND NELSON, D., Isolation of papovavirus

from brain tumor and urine of a patient with Wiskott–Aldrich syndrome, *J. Natl. Cancer Inst.* **53**:1205–1207 (1974).

220. TATEISHI, J., Transmission of Creutzfeldt–Jakob disease from human blood and urine into mice, *Lancet* **2**:1074 (1985).

220a. TATEISHI, J., BROWN, P., KITAMOTO, T., HOQUE, Z. M., ROOS, R., WOLLMAN, R., CERVENAKOVA, L., AND GAJDUSEK, D. C., First experimental transmission of fatal familial insomnia, *Nature* **376**:434–435 (1995).

221. TELENTI, A., MARSHALL, W. F., AKSAMIT, A. J., SMILACK, J. D., AND SMITH, T. F., Detection of JC virus by polymerase chain reaction in cerebrospinal fluid from two patients with progressive multifocal leukoencephalopathy, *Eur. J. Clin. Microbiol. Infect. Dis.* **11**:253–254 (1992).

222. TORNATORE, C., BERGER, J. R., HOUFF, S. A., CURFMAN, B., MEYERS, K., WINFIELD, D., AND MAJOR, E. O., Detection of JC virus DNA in peripheral lymphocytes from patients with and without progressive multifocal leukoencephalopathy, *Ann. Neurol.* **31**:454–462 (1992).

223. TOURTELLOTTE, W. W., MA, B. I., BRANDES, D. B., WALSH, M. J., AND POTVIN, A. R., Quantification of de novo central nervous system IgG measles antibody synthesis in SSPE, *Ann. Neurol.* **9**:551–556 (1981).

224. TOWNSEND, J. J., BARINGER, J. R., WOLINSKI, J. S., MALAMUD, N., MEDNICK, J. P., PANITCH, H. S., SCOTT, R. A. T., OSHIRO, L. S., AND CREMER, N. E., Progressive rubella panencephalitis: Late onset after congenital rubella, *N. Engl. J. Med.* **292**:990–993 (1975).

225. VAN BOGAERT, L., AND DEBUSSCHER, J., Sur la sclerose de la substance blanche des hemispheres (Spielmeyer), *Rev. Neurol. (Paris)* **71**:679–701 (1939).

226. VANDERSTEENHOVEN, J. J., DBAIBO, G., BOYKO, O. B., HULETTE, C. M., ANTHONY, D. C., KENNY, J. F., AND WILFERT, C. M., Progressive multifocal leukoencephalopathy in pediatric acquired immunodeficiency syndrome, *Pediatr. Infect. Dis. J.* **11**:232–237 (1992).

227. VON EINSIEDEL, R. W., FIFE, T. D., AKSAMIT, A. J., CORNFORD, M. E., SECOR, D. L., TOMIYASU, U., ITABASHI, H. H., AND VINTERS, H. V., Progressive multifocal leukoencephalopathy in AIDS: A clinicopathologic study and review of the literature, *J. Neurol.* **240**:391–406 (1993).

228. WALTER, G. F., AND RENELLA, R. R., Epstein–Barr virus in brain and Rasmussen's encephalitis, *Lancet* **1**:279–280 (1989).

229. WEBER, T., TUMANI, H., HOLDORFF, B., COLLINGE, J., PALMER, M., KRETZSCHMAR, H. A., AND FELGENHAUER, K., Transmission of Creutzfeldt–Jakob disease by handling of dura mater, *Lancet* **341**:123–124 (1993).

230. WEIL, M. L., ITABASHI, H. H., CREMER, N. E., OSHIRO, L. S., LENNETTE, E. H., AND CARNEY, L., Chronic progressive panencephalitis due to rubella virus simulating subacute sclerosing panencephalitis, *N. Engl. J. Med.* **292**:994–998 (1975).

231. WEINER, L. P., HERNDON, R. M., NARAYAN, O., JOHNSON, R. T., SHAH, K., RUBINSTEIN, L. J., PREZIOSI, T. J., AND CONLEY, F. K., Isolation of virus related to SV40 from patients with progressive multifocal leukoencephalopathy, *N. Engl. J. Med.* **286**:385–390 (1972).

232. WEINER, L. P., NARAYAN, O., PENNEY, J. B., JR., HERNDON, R. M., FERINGA, E. R., TOURTELLOTTE, W. W., AND JOHNSON, R. T., Papovavirus of JC type in progressive multifocal leukoencephalopathy. Rapid identification and subsequent isolation, *Arch. Neurol.* **29**:1–3 (1973).

233. WEISSMANN, C., A "unified theory" of prion propagation, *Nature* **352**:679–683 (1991).

234. WEISSMANN, C., BÜELER, H., FISCHER, M., AND AGUET, M., Role of the PrP gene in transmissible spongiform encephalopathies, *Intervirology* **35**:164–175 (1993).

235. WESTAWAY, D., GOODMAN, P. A., MIRENDA, C. A., MCKINLEY, M. P., CARLSON, G. A., AND PRUSINER, S. B., Distinct prion proteins in short and long scrapie incubation period mice, *Cell* **51**:651–662 (1987).

236. WHITAKER, J. N., SEVER, J. L., AND ENGEL, W. K., Subacute sclerosing panencephalitis in only one of identical twins, *N. Engl. J. Med.* **287**:864–866 (1972).

237. WHITEMAN, M. L., POST, M. J., BERGER, J. R., TATE, L. G., BELL, M. D., AND LIMONTE, L. P., Progressive multifocal leukoencephalopathy in 47 HIV-seropositive patients: Neuroimaging with clinical and pathologic correlation, *Radiology* **187**:233–240 (1993).

238. WHITLEY, R. J., Viral encephalitis, *N. Engl. J. Med.* **323**:242–250 (1990).

239. WIGHT, A. L., Prevention of iatrogenic transmission of Creutzfeldt–Jakob disease, *Lancet* **341**:1543 (1993).

240. WILESMITH, J. W., Epidemiology of bovine spongiform encephalopathy and related diseases, *Arch. Virol.* **S7**:245–254 (1993).

240a. WILL, R. G., IRONSIDE, J. W., ZEIDLER, M., COUSENS, S. N., ESTIBEIRO, K., ALPEROVITCH, A., POSER, S., POCCHIARI, M., HOFMAN, A., AND SMITH, P. G., A new variant of Creutzfeldt-Jakob disease in the UK, *Lancet* **347**:921–925 (1996).

241. WILL, R. G., AND MATTHEWS, W. B., Evidence for case-to-case transmission of Creutzfeldt–Jakob disease, *J. Neurol. Neurosurg. Psychiatry* **45**:235–238 (1982).

242. WILLS, P. R., Potential psuedoknots in the PrP-encoding messenger RNA, *J. Theor. Biol.* **159**:523–527 (1992).

243. WILLS, P. R., Self-organization of genetic coding, *J. Theor. Biol.* **162**:267–287 (1993).

244. WIRGUIN, I., STEINER, I., BRENNER, T., AND ABRAMSKY, O., Intraventricular interferon treatment for subacute sclerosing panencephalitis, *Ann. Neurol.* **30**:227 (1991).

245. WIRGUIN, I., STEINITZ, M., SICSIC, C., ABRAMSKY, O., AND BRENNER, T., Synthesis of antibodies against measles virus and myelin by *in vitro* stimulated B-cells derived from patients with subacute sclerosing panencephalitis, *Immunol. Lett.* **38**:55–58 (1993).

246. WOLINSKY, J. S., Progressive rubella panencephalitis, in: *Handbook of Clinical Neurology*, Vol. 34 (P. J. VINKEN AND G. W. BRUYN, EDS.), pp. 331–341, Elsevier/North-Holland, Amsterdam, 1978.

247. WOLINSKY, J. S., Subacute sclerosing panencephalitis, progressive rubella panencephalitis, and multifocal leukoencephalopathy, *Res. Publ. Assoc. Res. Nerv. Ment. Dis.* **68**:259–268 (1990).

248. WOLINSKY, J. S., DAU, P. C., BUIMOVICI-KLEIN, E., MEDNICK, J., BERG, B. O., LANG, P. B., AND COOPER, L. Z., Progressive rubella panencephalitis: Immunovirological studies and results of isoprinosine therapy, *Clin. Exp. Immunol.* **35**:397–404 (1979).

249. WONG, T. C., AYATA, M., UEDA, S., AND HIRANO, A., Role of biased hypermutation in evolution of subacute sclerosing panencephalitis virus from progenitor acute measles virus, *J. Virol.* **65**:2191–2199 (1991).

250. XI, Y. G., INGROSSO, L., LADOGANA, A., MASULLO, C., AND POCCHIARI, M., Amphotericin-B treatment dissociates *in vivo* replication of the scrapie agent from PrP accumulation, *Nature* **356**:598–601 (1992).

251. YAGI, S., MIURA, Y., MIZUTA, S., WAKUNAMI, A., KATAOKA, N., MORITA, T., MORITA, K., ONO, S., AND FUKUNAGA, M., Chrono-

logical SPECT studies of a patient with subacute sclerosing panencephalitis, *Brain Dev.* **15:**141–145 (1993).

252.   YALAZ, K., ANLAR, B., OKTEM, F., AYSUN, S., USTACELEBI, S., GURCAY, O., GUCUYENER, K., AND RENDA, Y., Intraventricular interferon and oral inosiplex in the treatment of subacute sclerosing panencephalitis, *Neurology* **42:**488–491 (1992).

253.   YUN, M., WU, W., HOOD, L., AND HARRINGTON, M., Human cerebrospinal fluid protein database—Edition 1992, *Electrophoresis* **13:**1002–1013 (1992).

254.   ZAGAMI, A. S., AND LETHLEAN, A. K., Chorioretinitis as a possible very early manifestation of subacute sclerosing panencephalitis, *Aust. NZ J. Med.* **21:**350–352 (1991).

255.   ZU RHEIN, G. M., AND CHOU, S.-M., Particles resembling papovaviruses in human cerebral demyelinating disease, *Science* **148:**1477–1479 (1965).

256.   ZU RHEIN, G. M., Association of papova-virions with a human demyelinating disease (progressive multifocal leucoencephalopathy), *Prog. Med. Virol.* **11:**185–247 (1969).

Index